Compound type	*Functional group*	*Simple example*	*Name ending*
Carboxylic acid	$-\overset{\mathrm{O}}{\overset{\Vert}{\mathrm{C}}}-\mathrm{O}-\mathrm{H}$	$\mathrm{CH_3CH_2CH_2}\overset{\mathrm{O}}{\overset{\Vert}{\mathrm{C}}}-\mathrm{OH}$ Butanoic acid	*-oic acid*
Carboxylic acid chloride	$-\overset{\mathrm{O}}{\overset{\Vert}{\mathrm{C}}}-\mathrm{Cl}$	$\mathrm{CH_3CH_2}\overset{\mathrm{O}}{\overset{\Vert}{\mathrm{C}}}-\mathrm{Cl}$ Propanoyl chloride	*-yl chloride*
Amide	$-\overset{\mathrm{O}}{\overset{\Vert}{\mathrm{C}}}-\mathrm{N}\langle$	$\mathrm{CH_3}\overset{\mathrm{O}}{\overset{\Vert}{\mathrm{C}}}-\mathrm{NH_2}$ Ethanamide	*-amide*
Amine	$\rangle\mathrm{C}-\mathrm{N}\langle$	$\mathrm{CH_3CH_2NH_2}$ Ethylamine	*-amine*
Nitrile	$-\mathrm{C}\equiv\mathrm{N}$	$\mathrm{CH_3C}\equiv\mathrm{N}$ Ethanenitrile (acetonitrile)	*-nitrile*
Nitro	$\rangle\mathrm{C}-\overset{+}{\mathrm{N}}(=\mathrm{O})-\mathrm{O}^-$	$\mathrm{CH_3CH_2NO_2}$ Nitroethane	None
Sulfide	$\rangle\mathrm{C}-\mathrm{S}-\mathrm{C}\langle$	$\mathrm{CH_3}-\mathrm{S}-\mathrm{CH_3}$ Dimethyl sulfide	*sulfide*
Sulfoxide	$\rangle\mathrm{C}-\underset{+}{\overset{\mathrm{O}^-}{\overset{\vert}{\mathrm{S}}}}-\mathrm{C}\langle$	$\mathrm{CH_3}-\underset{+}{\overset{\mathrm{O}^-}{\overset{\vert}{\mathrm{S}}}}-\mathrm{CH_3}$ Dimethyl sulfoxide	*sulfoxide*
Sulfone	$\rangle\mathrm{C}-\underset{\underset{\mathrm{O}^-}{\vert}}{\overset{\mathrm{O}^-}{\overset{\vert}{\mathrm{S}^{2+}}}}-\mathrm{C}\langle$	$\mathrm{CH_3}-\underset{\underset{\mathrm{O}^-}{\vert}}{\overset{\mathrm{O}^-}{\overset{\vert}{\mathrm{S}^{2+}}}}-\mathrm{CH_3}$ Dimethyl sulfone	*sulfone*
Organometallic	$\rangle\mathrm{C}-\mathrm{M}$ M = metal	$\mathrm{CH_3}-\mathrm{Li}$ Methyllithium	None

ORGANIC CHEMISTRY

ORGANIC CHEMISTRY

JOHN McMURRY *Cornell University*

Brooks/Cole Publishing Company *Monterey, California*

Brooks/Cole Publishing Company
A Division of Wadsworth, Inc.

Printed in the United States of America

10 9 8 7 6 5 4 3 2

Library of Congress Cataloging in Publication Data

McMurry, John.
Organic chemistry.

(The Brooks/Cole series in chemistry)
Includes index.
1. Chemistry, Organic. I. Title. II. Series.
QD251.2.M43 1984 547 83-7744

ISBN 0-534-01204-3

Sponsoring Editor: Michael Needham
Editorial Assistant: Lorraine McCloud
Production Service: Phyllis Niklas
Production Coordinator: Joan Marsh
Manuscript Editor: Gloria Joyce
Interior and Cover Design: Stan Rice
Interior Illustration: Vantage Art, Inc.
Typesetting: Jonathan Peck Typographers, Ltd.

On the Cover: The cover drawings of a molecule recently synthesized at Cornell University were provided by Molecular Design Limited, Hayward, California. They were generated by the MACCS™ and SPACFIL computer programs for the storage, retrieval, modeling, and display of chemical structures. (Additional airbrush background by Tim Keenan.)

PREFACE

The 1970s and early 1980s were a time of rapid change for the science of organic chemistry; physical organic chemists provided us with a deeper understanding of organic reactivity, and synthetic organic chemists provided us with many selective new reactions and reagents. Organic chemistry texts also underwent changes in this period, but of a largely pedagogical sort. Thus, good teaching texts today make effective use of two-color printing and three-dimensional airbrushed art; they use many stereochemical drawings; and they provide teaching aids such as chapter summaries, reaction summaries, and extensive problem sets of graded difficulty.

My goal in writing this book has been to combine coverage of the many exciting scientific advances of the last decade with the equally important pedagogical advances that have been made. I have attempted to write a lucid, readable, and effective *teaching* text while providing an accurate and up-to-date view of organic chemistry as it is understood and practiced in the mid-1980s.

Organization

This book uses a dual functional-group/reaction-mechanism organization. The primary organization is by functional group, beginning in Chapter 5 with the simple (alkenes) and going on to the more complex. Within this primary organization, however, heavy emphasis is placed on explaining the mechanistic similarities of reactions. Indeed, many chapters, such as Chapter 22, "Aldehydes and Ketones: Nucleophilic Addition Reactions," and Chapter 24, "Carboxylic Acid Derivatives and Nucleophilic Acyl Substitution Reactions," even have dual functional-group/reaction-mechanism titles.

Organic molecules and organic reactions are presented as early as possible. After a brief review of structure, bonding, and molecular properties in Chapters 1 and 2, organic molecules and functional groups are introduced in Chapter 3. An introduction to the nature of organic reactions then follows in Chapter 4.

Insofar as possible, the topics have been arranged in a modular way. Thus, the chapters on spectroscopy are grouped together (Chapters 11–13), the chapters on aromatic chemistry are grouped together (Chapters 14–16), and the chapters on carbonyl compounds are grouped together (Chapters 21–26). This organization not only brings cohesiveness to these subjects, it allows the instructor the flexibility to teach them in an order different from that used in the book. For example, some instructors might want to cover carbonyl-group chemistry before aromatic-ring chemistry.

Classification of Reaction Types

In an effort to demonstrate to students the logic of organic chemistry, reactions are classified into three fundamental types: polar, radical, and pericyclic. Chapter 4, "Organic Reactions," sets the stage for this classification by explaining the three reaction types and giving the characteristics of each type. As each new reaction is introduced in later chapters, its classification is discussed, and analogies are drawn to earlier material. Chapter 17, "Organic Reactions: A Brief Review," is a unique chapter provided as a short breather halfway through the book to reexamine and reinforce the classification scheme.

The Lead-Off Reaction: Addition of HBr to Alkenes

Many students attach great importance to a text's lead-off reaction because it is the first reaction they see and it is discussed in such detail. I have chosen a simple polar reaction—the addition of HBr to an alkene—as the lead-off to illustrate general principles of organic reactions. This choice has the advantage of being relatively simple (prior knowledge of stereochemistry and kinetics is not required, as it is when a nucleophilic substitution reaction is used as lead-off), yet it is also an important polar reaction on a ubiquitous functional group. As such, I believe that this reaction serves to introduce students to functional-group chemistry better than does a lead-off reaction such as free-radical alkane halogenation.

Coverage

The coverage of this book is up-to-date, reflecting many of the important advances of the past decade. For example, the advantages of using lithium diisopropylamide as a base for carbonyl-group alkylations are discussed; selenoxide elimination as a method of introducing double bonds is covered; pyridinium chlorochromate as an oxidant for converting primary alcohols into aldehydes is introduced; and a full chapter is devoted to the consequences of orbital symmetry in organic chemistry, treating this subject not as a special topic but as an integral part of the science.

In addition, many other important topics receive broad coverage in this text. ^{13}C NMR is introduced as a routine spectroscopic tool. The chemistry of synthetic polymers is presented in a unified manner along with a discussion of how polymer structure and chemistry can be related to physical properties. The organic chemistry of nucleic acids is presented, including a section on how "gene machines" work.

Spectroscopy

Spectroscopy is treated in this book as a tool, not as a specialized field of study itself. For this reason, the subject is presented (Chapters 11–13) only after students have gained sufficient knowledge of organic chemistry to benefit from use of the tool. Infrared, ^{13}C and ^{1}H NMR, ultraviolet, and mass spectroscopies are all introduced in preliminary fashion. They are then expanded in sections of later chapters as the need arises in connection with discussion of new functional groups.

Unique Treatment of Carbonyl-Group Chemistry

The chemistry of the various carbonyl-containing functional groups is treated in a unique and innovative manner that emphasizes the mechanistic similarities of many seemingly different reactions. Indeed, I believe that the treatment of carbonyl groups in this book is the best currently available. Chapter 21, "Chemistry of Carbonyl Compounds: An Overview," sets the tone by classifying carbonyl-group reactions into four general types. In this way, the artificial distinction made in many texts between fundamentally similar processes, such as the aldol condensation reaction of a ketone and the Claisen condensation reaction of an ester, is avoided.

Organic Synthesis

Organic synthesis is presented as a teaching device to help students learn to organize and work with the large body of factual information that is organic chemistry. Toward this end, two short sections in Chapters 7 and 16 explain the thought processes involved in working synthesis problems. The value of starting from what is known and logically working backwards one step at a time is emphasized.

Nomenclature

IUPAC nomenclature is used throughout the text. For the most part, this involves the use of systematic names, although IUPAC-approved nonsystematic names such as acetic acid, acetone, and phenol are also employed. Though it may sometimes be grating to the ears of a practicing chemist to hear propionaldehyde referred to as propanal, students find systematic names much easier to learn.

Pedagogy

In addition to the above features, every effort has been made to make this book as effective, clear, readable, and *friendly* as possible:

Chapters are relatively short, and numerous subheadings are used.

Paragraphs start with summary sentences.

Transitions between paragraphs and between topics are smooth.

Concept density is low. New concepts are introduced only as needed and are immediately illustrated with concrete examples.

Extensive cross-referencing to earlier material is used.

Extensive use is made of three-dimensional airbrushed art and stereochemical formulas.

A second color is used to indicate the changes that occur during reactions.

Numerous summaries are included, both within chapters and at the end of every chapter.

More than 1200 problems are included, both within the text of each chapter and at the end of each chapter. These include drill and thought problems of varying levels of difficulty.

An innovative vertical format is used to explain reaction mechanisms. The mechanisms are printed vertically, while an explanation of the changes occurring in each step is given next to the reaction arrow. This format allows the student to see easily what is occurring at each step of a reaction without having to jump back and forth between text and structures.

Much use is made of biological examples to emphasize the close relationship between organic chemistry and biochemistry.

Short biographies of important chemists are given to help humanize the material, a technique first used by Louis Fieser in his classic texts of the 1950s. Discoverers of "name reactions" and Nobel-prize-winning organic chemists are cited.

Solutions Manual and Study Guide

A carefully prepared *Solutions Manual and Study Guide* accompanies this text. Written by Susan McMurry, this companion volume answers all in-text and end-of-chapter problems and explains in detail how the answers are obtained. In addition, the following important supplemental materials are included: a section on the nomenclature of polyfunctional compounds, a glossary, summaries of name reactions, summaries of methods for preparing functional groups, summaries of the uses of important reagents, tables of spectroscopic information, and a list of suggested readings.

ACKNOWLEDGMENTS

It is a great pleasure to thank the many people whose help and suggestions were so valuable. Foremost is my wife Susan who read, criticized, and improved all of the many drafts, and who was my constant companion throughout all stages of this book's development. Among the reviewers providing valuable comments were Robert A. Benkeser (Purdue University), Donald E. Bergstrom (University of North Dakota), Weston T. Borden (University of Washington), Larry Bray (Miami Dade Community College), William D. Closson (State University of New York, Albany), Paul L. Cook (Albion College), Otis Dermer (Oklahoma State University), Linda Domelsmith, David Harpp (McGill University), David Hart (Ohio State University), Norbert Hepfinger (Rensselaer Polytechnic Institute), Werner Herz (Florida State University), Paul R. Jones (North Texas State University), Thomas Katz (Columbia University), Paul E. Klinedinst, Jr. (California State University, Northridge), James G. Macmillan (University of Northern Iowa), Monroe Moosnick (Transylvania University), Harry Morrison (Purdue University), Cary Morrow (University of New Mexico), Wesley A. Pearson (St. Olaf College), Frank P. Robinson (University of Victoria), Neil E. Schore (University of California, Davis), Gerald Selter (California State University, San Jose), Ernest Simpson (California State Polytechnic University, Pomona), Walter Trahanovsky (Iowa State University), Harry Ungar (Cabrillo College), Joseph J. Villafranca (Pennsylvania State University), Daniel P. Weeks (Seton Hall University), Walter Zajac (Villanova University), and Vera Zalkow (Kennesaw College).

In addition, special thanks are due Mildred Newhouse and Virginia Severn for typing and drawing all structures in the manuscript; Phyllis Niklas for her fine work as production editor; and Mike Needham, Jim Leisy, and the entire staff at Brooks/Cole for a thoroughly professional job. It was a pleasure working with them all.

A NOTE FOR STUDENTS

We have the same goals. Yours, as students, is to learn organic chemistry. Mine, as author, is to do everything possible to help you learn. It's going to require some work on your part, but the following hints should prove helpful:

Don't read the text immediately. As you begin each new chapter, look it over first. Find out what topics will be covered. Read the introductory paragraphs and then turn to the end of the chapter and read the summary. You'll be in a much better position to understand new material if you first have a general idea of where you are heading. Many people find it best to read chapters twice. First read the chapter rapidly, making checks or comments in the margin next to important points, and then return for an in-depth study.

Work the problems. There are no shortcuts here; working problems is the only way to master the material (and the only way to prove to yourself that you have mastered it). The in-text problems are placed at the ends of sections to provide immediate practice on material just covered. The end-of-chapter problems provide both additional drill and real challenges to be solved. Answers to selected in-text problems are provided in the Appendix. Full answers and explanations for all problems are given in the accompanying *Solutions Manual and Study Guide.*

Ask questions. Faculty members and teaching assistants are there to help you. Most of them will turn out to be genuinely nice people with a sincere interest in seeing you learn.

Use molecular models. Organic chemistry is a three-dimensional science. Although this book uses many careful drawings to help you visualize molecules, there is no substitute for building a molecular model and turning it in your hands.

Use the study guide. The *Solutions Manual and Study Guide* that accompanies this text gives complete solutions to all problems and provides supplementary material. Included are a large glossary, summaries of reactions, and summaries of reagents. Find out now what is there so that you will know where to go when you need help.

Good luck. I sincerely hope you enjoy learning organic chemistry and come to see the logic and beauty of its structure. I would be glad to receive comments and suggestions from any who have learned from this book.

BRIEF CONTENTS

CONTENTS

1 STRUCTURE AND BONDING 1

2 BONDING AND MOLECULAR PROPERTIES 34

3 THE NATURE OF ORGANIC COMPOUNDS: ALKANES 55

4 ORGANIC REACTIONS 94

5 ALKENES: STRUCTURE AND REACTIVITY 121

ALKENES: REACTIONS AND SYNTHESIS 156

7 ALKYNES 194

8 INTRODUCTION TO STEREOCHEMISTRY 225

9 ALKYL HALIDES 266

10 REACTIONS OF ALKYL HALIDES: NUCLEOPHILIC SUBSTITUTION REACTIONS 289

11 STRUCTURE DETERMINATION: MASS SPECTROSCOPY AND INFRARED SPECTROSCOPY 338

12 STRUCTURE DETERMINATION: NUCLEAR MAGNETIC RESONANCE SPECTROSCOPY 365

13 CONJUGATED DIENES AND ULTRAVIOLET SPECTROSCOPY 408

16 ARENES: SYNTHESIS AND REACTIONS OF ALKYLBENZENES 520

17 ORGANIC REACTIONS: A BRIEF REVIEW 551

18 ALICYCLIC MOLECULES 563

19 ETHERS, EPOXIDES, AND SULFIDES 604

20 ALCOHOLS AND THIOLS 630

21 CHEMISTRY OF CARBONYL COMPOUNDS: AN OVERVIEW 670

22 ALDEHYDES AND KETONES: NUCLEOPHILIC ADDITION REACTIONS 686

23 CARBOXYLIC ACIDS 737

24 CARBOXYLIC ACID DERIVATIVES AND NUCLEOPHILIC ACYL SUBSTITUTION REACTIONS 765

27 CARBOHYDRATES 882

28 ALIPHATIC AMINES 920

29 ARYLAMINES AND PHENOLS 960

30 ORBITALS AND ORGANIC CHEMISTRY: PERICYCLIC REACTIONS 995

31 AMINO ACIDS, PEPTIDES, AND PROTEINS 1031

32 LIPIDS 1069

33 HETEROCYCLES AND NUCLEIC ACIDS 1097

34 SYNTHETIC POLYMERS 1139

CHAPTER 1
STRUCTURE AND BONDING

Organic chemicals are in us and all around us. The proteins that make up our hair, skin, and muscles are organic chemicals; the nucleic acids, RNA and DNA, that control our genetic heritage are organic chemicals; and the foods we eat—fats, proteins, and carbohydrates—are organic chemicals.

Carbon is the element on which life is based, and **organic chemistry,** the study of carbon compounds, has long attracted special attention. In the eighteenth century, however, as chemistry was evolving from an art into a science, the study of carbon compounds advanced more slowly than did the study of metal-containing compounds. There were good reasons for this; the first known organic compounds were derived from animal and vegetable sources and were difficult to crystallize and purify. Even when pure, most organic compounds were difficult to work with, in part because many were more sensitive to decomposition than were metal-containing compounds derived from mineral sources. In 1770, the Swedish chemist Torbern Bergman was the first person to express this difference between "organic" and "inorganic" substances, and the phrase organic chemistry soon came to mean the chemistry of compounds from living organisms. To many chemists at the time, the only explanation for the difference in behavior between organic and inorganic compounds was that organic compounds contained a peculiar and undefinable "vital force" as a result of their derivation from living sources. One consequence of the presence of this vital force, chemists believed, was that organic compounds could not be prepared and manipulated in the laboratory as could inorganic compounds.

Although the vitalistic theory was believed by many influential chemists, its acceptance was by no means universal, and it is doubtful that the development of organic chemistry was much delayed. As early as 1816, the theory received a heavy blow when Michel Chevreul[1] found that soaps, prepared by the reaction of alkali with animal fat, could be separated into several pure organic compounds, which he termed "fatty acids." Thus, for the first time, one organic compound (fat) had been converted into others (fatty acids plus glycerin) without the intervention of an outside vital force.[2]

$$\text{Animal fat} \xrightarrow[H_2O]{NaOH} \text{Soap} + \text{Glycerin}$$

$$\text{Soap} \xrightarrow{H_3O^+} \text{“Fatty acids”}$$

A little more than a decade later, in 1828, the vitalistic theory suffered still further when Friedrich Wöhler[3] discovered that it was possible to convert the "inorganic" salt, ammonium cyanate, into the previously known "organic" substance, urea.

[1]Michel Eugène Chevreul (1786–1889); b. Angers, France; Paris, Muséum d'Histoire Naturelle, professor of physics, Lycée Charlemagne (1813); professor of chemistry (1830).

[2]In the equation that follows, a *single* arrow, indicating an actual reaction, is used. Later in this book you will also see forward and back arrows, $\rightleftarrows$, indicating equilibrium, and a double arrow, $\Rightarrow$, indicating an imagined transformation.

[3]Friedrich Wöhler (1800–1882); b. Escherheim; studied at Heidelberg (Gmelin); professor, Göttingen (1836–1882).

$$NH_4^{+\,-}OCN \xrightarrow{\text{Heat}} H_2N{-}\overset{\overset{\displaystyle O}{\|}}{C}{-}NH_2$$

Ammonium cyanate Urea

By the mid-nineteenth century, the weight of evidence was clearly against the vitalistic theory and, in 1848, William Brande[4] wrote in a paper that "No definite line can be drawn between organic and inorganic chemistry . . . any distinctions . . . must for the present be merely considered as matters of practical convenience calculated to further the progress of students." Chemistry is unified, and the same basic principles that explain the simplest inorganic compounds also explain the most complex organic molecules. However, the division between organic and inorganic chemistry, which began for historical reasons, maintains its "practical convenience . . . to further the progress of students."

Organic chemistry, then, is the study of the compounds of carbon. Carbon, which has atomic number 6, is a second-period element whose position in an abbreviated periodic table is shown in Table 1.1. Although carbon is the principal element, almost all organic compounds also contain hydrogen, and many contain nitrogen, oxygen, phosphorus, sulfur, chlorine, and other elements.

TABLE 1.1 An abbreviated periodic table

	Group								
Period	IA	IIA		IIIA	IVA	VA	VIA	VIIA	0
1	H								He
2	Li	Be		B	C	N	O	F	Ne
3	Na	Mg		Al	Si	P	S	Cl	Ar
4	K	Ca	Sc Ti V Cr Mn Fe Co Ni Cu Zn	Ga	Ge	As	Se	Br	Kr

Why is carbon special? What is it that sets carbon apart from all other elements in the periodic table? The answers to these questions are complex but have much to do with the unique ability of carbon atoms to bond together forming long chains and rings. Carbon, alone of all elements, is able to form an immense diversity of compounds, from the simple to the staggeringly complex: from methane, containing one carbon, to DNA, which can contain hundreds of *billions*. Nor are all carbon compounds derived from living organisms. Chemists in the past 100 years have become extraordinarily sophisticated in their ability to synthesize new organic compounds in the laboratory. Medicines, dyes, polymers, plastics, food additives, pesticides, and a host of other substances—all are prepared in the laboratory, and all are organic chemicals. Organic chemistry is a science that touches the lives of all, and its study can be a fascinating undertaking.

[4]William Thomas Brande (1788–1866); b. London; lecturer in chemistry, London (1808); Royal Institution (1813–1854).

1.1 Quantum mechanics

Throughout the nineteenth century, and into the twentieth, chemists and physicists sought to understand the nature of the atom. A major breakthrough came in 1926 when the theory of **quantum mechanics** was proposed independently by Paul Dirac, Werner Heisenberg, and Erwin Schrödinger.[5] The formulations proposed by all three are mathematical expressions that describe the motion of an electron around a nucleus, but Schrödinger's is the one most commonly used by chemists. The Schrödinger equation offers a detailed description of the electronic structure of atoms and molecules in terms of the energies of electrons, and says that the motion of an electron around the nucleus can be described mathematically by what is known as a **wave equation.** This means that the kind of mathematical expression used for describing the motion of electrons is the same kind of expression that is used to describe the motion of waves in a fluid.

The solution to a wave equation is known as a **wave function,** and, if we were able to determine the wave function for every electron in an atom, we would have a complete electronic description of that atom. In practice, wave equations for electrons in atoms are mathematically so complex that only approximate solutions can be obtained. These approximate solutions agree so well with experimental facts, however, that quantum mechanics is a universally accepted theory for understanding atomic structure.

1.2 Atomic orbitals

How can we interpret wave functions in terms of physical reality? A good way of viewing a wave function is to think of it as an expression that predicts the volume of space around a nucleus where an electron can be found. We can never know the exact position of an electron at a given moment, but the wave function tells us where we would be most likely to find it.

The volume of space in which an electron is most likely to be found is called an **orbital.** It's often helpful to think of an orbital in terms of a time-lapse photograph of an electron's movement around the nucleus. Such a photograph would show the orbital as a kind of blurry cloud indicating where the electron has been. This electron cloud does not have a discrete boundary, but for practical purposes we can set the limits of an orbital by saying that it represents the space where an electron spends *most* (90–95%) of its time.

What shapes do orbitals have? The exact shape and size of an electron's orbital depend on its energy level. The lowest-energy electrons, for example, occupy the 1*s* orbital. The *s* atomic orbitals are spherical and have the nucleus of the atom at their center, as shown in Figure 1.1. Relative energy levels of different kinds of atomic orbitals are shown in Figure 1.2.

[5]Erwin Schrödinger (1887–1961); b. Vienna, Austria; University of Vienna (1910); assistant, University of Vienna (1910); assistant to Max Wein, University of Stuttgart, Germany (1920); professor of physics, University of Zurich, Berlin, Graz, Dublin; Nobel prize in physics (1933).

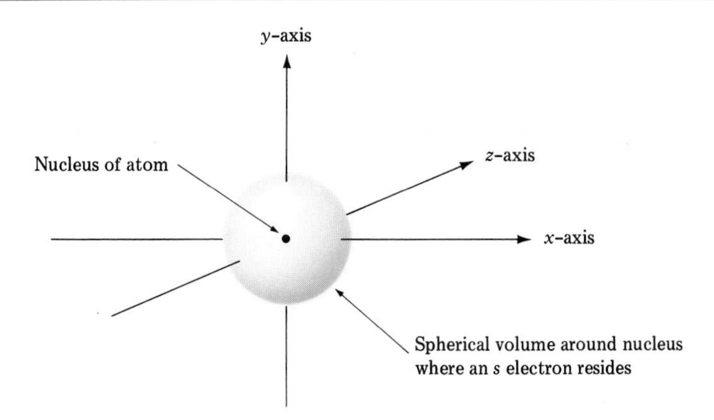

Figure 1.1. An s atomic orbital.

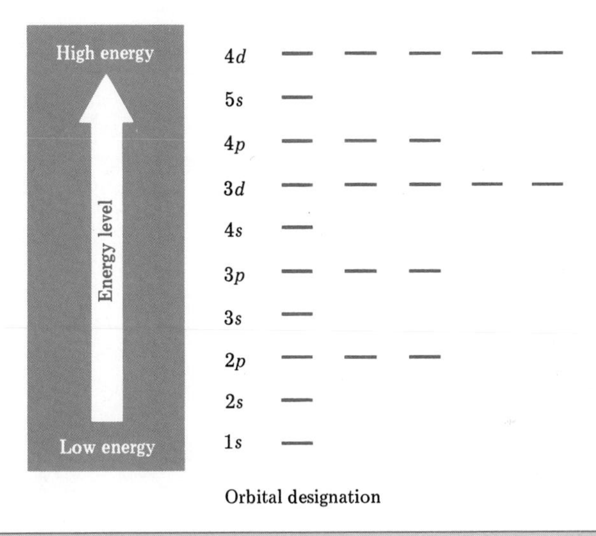

Figure 1.2. Energy levels of atomic orbitals.

Next in energy after the 1s electrons are the 2s electrons. Because they are higher in energy, 2s electrons are farther from the positively charged nucleus on average, and their spherical orbital is somewhat larger than that of 1s electrons.

The 2p electrons are next higher in energy. As Figure 1.3 indicates, there are three 2p orbitals, each of which is roughly dumbbell shaped. The three 2p orbitals are of equal energy and are oriented in space such that each is perpendicular to the other two. They are denoted $2p_x$, $2p_y$, and $2p_z$, depending upon which axis they lie. Note that the plane passing between the two lobes of a p orbital is a region of *zero electron density*. It is called a **nodal plane,** and has certain consequences with respect to chemical bonding that will be taken up in Chapter 30.

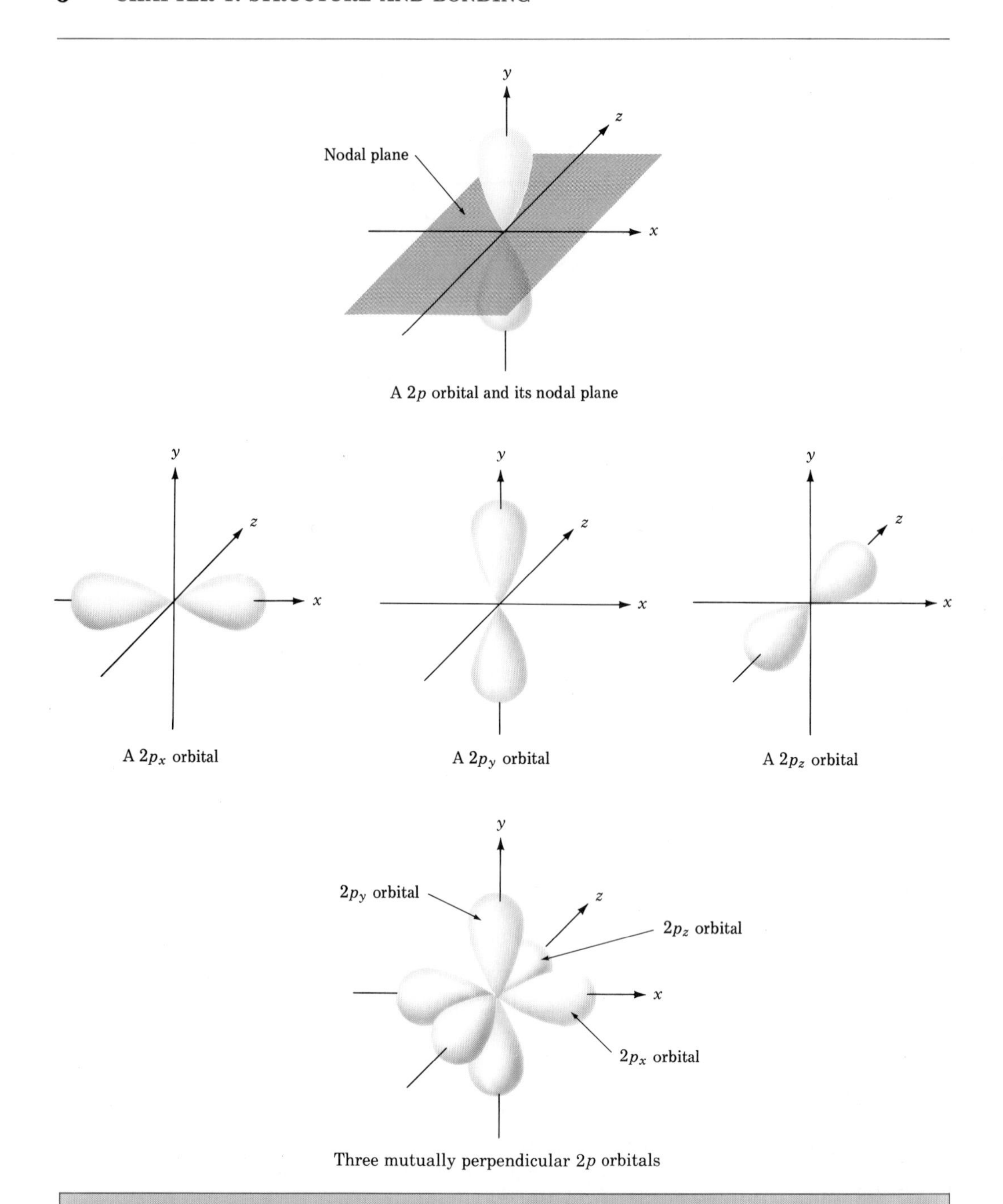

Figure 1.3. Shapes of the 2*p* orbitals.

Still higher in energy are the 3*s* orbital (spherical), 3*p* orbitals (dumbbell shaped), 4*s* orbital (spherical), and 3*d* orbitals. There are five 3*d* orbitals of equal energy, one of which is shown in Figure 1.4. The *d* orbitals do not play as important a role in organic chemistry as do the *s* and *p*

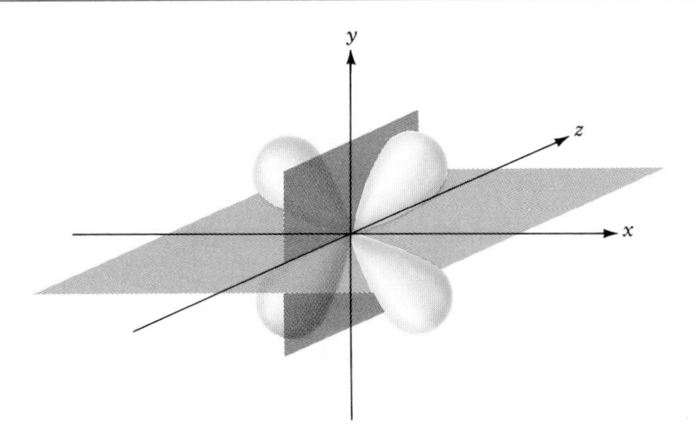

Figure 1.4. A $3d$ orbital.

orbitals, and they will not be discussed in detail. Note that the $3d$ orbital shown has four lobes and two nodal planes.

1.3 Electronic configuration of atoms

The lowest-energy arrangement, or **ground-state electronic configuration,** of any atom can be found by using our knowledge of atomic orbitals and their energy levels. The available electrons are assigned to the proper orbitals by following three rules:

1. The orbitals of lowest energy are filled first (the ***aufbau* principle**).
2. Only two electrons can occupy the same orbital, and they must be of opposite spin[6] (the **Pauli exclusion principle**).
3. If two or more empty orbitals of equal energy are available, one electron is placed in each until all are half-full (**Hund's rule**).

Let's look at some examples to see how these rules are applied. Hydrogen, the lightest element, has only one electron, and we place that electron in the lowest available orbital, giving hydrogen a $1s$ ground-state configuration. Carbon has six electrons, and a ground-state configuration $1s^2 2s^2 2p_x 2p_y$ is arrived at by applying our three rules.[7] These and other examples are shown in Table 1.2 (p. 8).

PROBLEM

1.1 Give the ground-state electronic configuration for the following elements:
(a) Boron (b) Phosphorus (c) Iron (d) Selenium

[6]For the purposes of quantum mechanics, electrons can be considered to spin around an axis much as the earth spins. This spin can have two equal and opposite orientations, which will be denoted as up ↑ and down ↓.

[7]A superscript is used here to represent the number of electrons at a particular energy level; $1s^2$ indicates that there are two electrons at the $1s$ energy level. No superscript is used when there is only one electron.

TABLE 1.2 Ground-state electronic configurations of some elements

Element	*Atomic number*	*Configuration*	*Element*	*Atomic number*	*Configuration*
Hydrogen	1	1*s* ↑	Lithium	3	2*s* ↑ 1*s* ↑↓
Carbon	6	2*p* ↑ ↑ — 2*s* ↑↓ 1*s* ↑↓	Neon	10	2*p* ↑↓ ↑↓ ↑↓ 2*s* ↑↓ 1*s* ↑↓
Sodium	11	3*s* ↑ 2*p* ↑↓ ↑↓ ↑↓ 2*s* ↑↓ 1*s* ↑↓	Argon	18	3*p* ↑↓ ↑↓ ↑↓ 3*s* ↑↓ 2*p* ↑↓ ↑↓ ↑↓ 2*s* ↑↓ 1*s* ↑↓

1.4 Development of chemical bonding theory

By the mid-nineteenth century, with the vitalistic theory of organic chemistry dead and with the distinction between organic and inorganic chemistry nearly gone, chemists began to probe the forces holding molecules together. In 1858, August Kekulé[8] and Archibald Couper[9] independently proposed that, in all organic compounds, carbon always has four "affinity units." That is, carbon is *tetravalent;* it always forms four bonds when it joins other elements to form compounds. Furthermore, said Kekulé, carbon atoms can bond to each other to form extended chains. Shortly after the tetravalent nature of carbon was proposed, extensions to the Kekulé–Couper theory were made when the possibility of multiple bonding between atoms was suggested. Emil Erlenmeyer[10] proposed a carbon-to-carbon *triple* bond for acetylene, and Alexander Crum Brown[11] proposed a carbon-to-carbon *double* bond for ethylene. In 1865, Kekulé provided another major advance in bonding theory when he postulated that carbon chains can double back on themselves to form rings of atoms.

Perhaps the most significant early advance in understanding bonding in organic molecules was the contribution made independently by Jacobus van't Hoff[12] and Joseph Le Bel.[13] Although Kekulé had satisfactorily described the tetravalent nature of carbon, chemistry was viewed in an

[8]Friedrich August Kekulé (1829–1896); b. Darmstadt; University of Giessen (1847); studied under Liebig, Dumas, Gerhardt, and Williamson; assistant to Stenhouse, London; professor, Heidelberg (1855), Ghent (1858), and Bonn (1867).

[9]Archibald Scott Couper (1831–1892); b. Kirkintilloch, Scotland; studied at the universities of Glasgow and Edinburgh (1852) and with Wurtz in Paris; assistant in Edinburgh (1858).

[10]Richard A. C. E. Erlenmeyer (1825–1909); b. Wehen, Germany; studied in Giessen and in Heidelberg; professor, Munich Polytechnicum (1868–1883).

[11]Alexander Crum Brown (1838–1922); b. Edinburgh; studied at Edinburgh, Heidelberg, and Marburg; professor, Edinburgh (1869–1908).

[12]Jacobus Henricus van't Hoff (1852–1911); b. Rotterdam; studied at Polytechnic at Delft, Leyden, Bonn, Paris, and received doctorate at Utrecht (1874); professor, Utrecht, Amsterdam (1878–1896), Berlin; Nobel prize (1901).

[13]Joseph Achille Le Bel (1847–1930); b. Péchelbronn, Alsace; studied in the École Polytechnique, and at the Sorbonne; industrial consultant.

essentially two-dimensional way until 1874. In that year, van't Hoff and Le Bel added a third dimension to our conception of molecules. They proposed that the four bonds of carbon have specific spatial direction. Van't Hoff went even further and correctly proposed that the four atoms to which carbon is bonded sit at the corners of a tetrahedron, with carbon in the center. A representation of a tetrahedral carbon atom is shown in Figure 1.5.

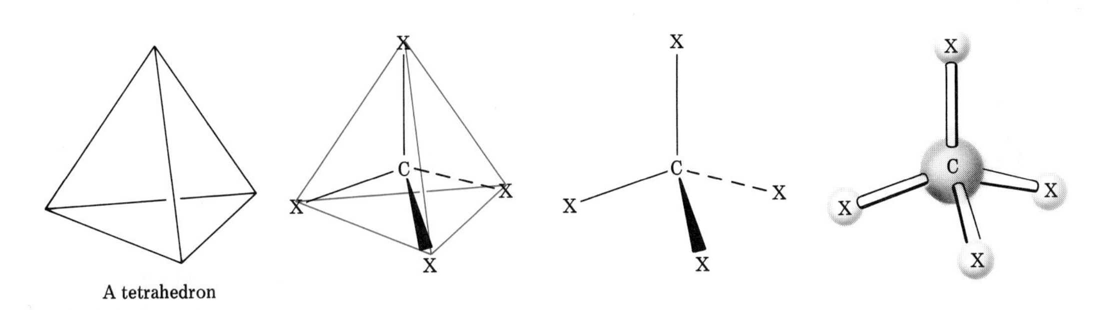

Figure 1.5. Van't Hoff's tetrahedral carbon atom. The heavy wedged line comes out of the plane of the paper; the normal lines are in the plane; and the broken line goes back behind the plane.

1.5 The modern picture of chemical bonding: ionic bonds

What is our modern picture of chemical bonding? Why do atoms bond together, and how does our quantum mechanical view of the atom describe bonding? The *why* question is relatively easy to answer: Atoms form bonds because the compound that results is more stable than the alternative arrangement of isolated atoms. Energy is always *released* when a chemical bond is formed. The *how* question is more difficult. To answer it, we need to know more about the properties of atoms.

We know through empirical observation that a filled octet in the outer electron shell imparts a special stability to the inert-gas elements in Group 0: Ne (2 + *8*); Ar (2 + 8 + *8*); Kr (2 + 18 + *8*). We also know that the chemistry of many elements with *nearly* inert-gas configurations is dominated by attempts to achieve the stable inert-gas electronic makeup. The alkali metals in Group I, for example, have single *s* electrons in their outer shells. By losing this electron, they can achieve the inert-gas configuration. The measure of this tendency to lose an electron is called the **ionization energy, IE,** and it is expressed in kilocalories per mole (kcal/mol). Alkali metals, at the far left of the periodic table, have low ionization energies and are said to be **electropositive.** The elements at the middle and far right of the periodic table have a lesser tendency to lose an electron and therefore have higher IEs. In other words, a low IE corresponds to the ready loss of an electron, and a high IE corresponds to the difficult loss of an electron. Table 1.3 lists some ionization energies.

TABLE 1.3 Ionization energies of some elements

Element (electronic configuration)			*Cation (electronic configuration)*	*Ionization energy (kcal/mol)*
Li	$(1s^2 2s)$	$\xrightarrow{-e^-}$	Li^+ $(1s^2$—same as He)	125
Na	$(1s^2 2s^2 2p^6 3s)$	$\xrightarrow{-e^-}$	Na^+ $(1s^2 2s^2 2p^6$—same as Ne)	118
K	$(\ldots 3s^2 3p^6 4s)$	$\xrightarrow{-e^-}$	K^+ $(\ldots 3s^2 3p^6$—same as Ar)	100
C	$(1s^2 2s^2 2p^2)$	$\xrightarrow{-e^-}$	C^+ $(1s^2 2s^2 2p)$	259
C^+	$(1s^2 2s^2 2p)$	$\xrightarrow{-e^-}$	C^{2+} $(1s^2 2s^2)$	562
F	$(1s^2 2s^2 2p^5)$	$\xrightarrow{-e^-}$	F^+ $(1s^2 2s^2 2p^4)$	401
Ne	$(1s^2 2s^2 2p^6)$	$\xrightarrow{-e^-}$	Ne^+ $(1s^2 2s^2 2p^5)$	497

Just as the electropositive alkali metals at the left of the periodic table have a tendency to form *positive* ions by *losing* an electron, the halogens (Group VIIA elements) at the right of the periodic table have a tendency to form *negative* ions by *gaining* an electron. By so doing, the halogens can achieve an inert-gas configuration. The measure of this tendency to gain an electron is called the **electron affinity, EA** (also expressed in kilocalories per mole). The elements on the far right of the periodic table have much higher electron affinities than those on the far left (i.e., they have a much greater tendency to add an electron), and are called **electronegative.** Table 1.4 lists the electron affinities of a number of elements.

TABLE 1.4 Electron affinities of some elements

Element (electronic configuration)			*Anion (electronic configuration)*	*Electron affinity (kcal/mol)*
Li	$(1s^2 2s)$	$\xrightarrow{+e^-}$	Li^- $(1s^2 2s^2)$	~13.6
Na	$(1s^2 2s^2 2p^6 3s)$	$\xrightarrow{+e^-}$	Na^- $(1s^2 2s^2 2p^6 3s^2)$	~ 5.0
C	$(1s^2 2s^2 2p^2)$	$\xrightarrow{+e^-}$	C^- $(1s^2 2s^2 2p^3)$	28.9
F	$(1s^2 2s^2 2p^5)$	$\xrightarrow{+e^-}$	F^- $(1s^2 2s^2 2p^6$—same as Ne)	79.6
Cl	$(\ldots 3s^2 3p^5)$	$\xrightarrow{+e^-}$	Cl^- $(\ldots 3s^2 3p^6$—same as Ar)	83.2

The simplest kind of chemical bonding occurs between an electropositive element (low IE) and an electronegative element (high EA). For example, when sodium metal (IE = 118 kcal/mol) reacts with chlorine gas (EA = 83.2 kcal/mol), sodium donates an electron to chlorine to form sodium ions and chloride ions. The product, sodium chloride, is said to have **ionic bonding.** That is, the ions are held together purely by electrostatic attraction; the unlike charges on the two ions attract each other. The structure of the sodium chloride crystal lattice shown in Figure 1.6 illustrates how each

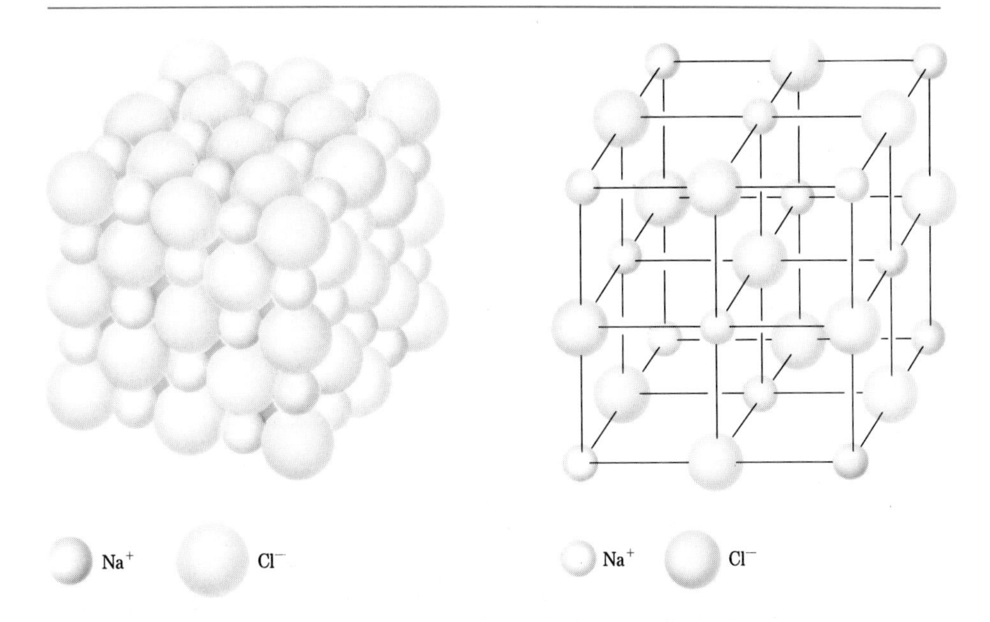

Figure 1.6. Ionic bonding in a sodium chloride crystal. Each sodium ion is surrounded by six chloride ions, and each chloride ion is surrounded by six sodium ions.

positively charged sodium ion is surrounded by negatively charged chloride ions and vice versa. A similar situation exists for many other ionic species such as potassium fluoride (K^+F^-) and lithium bromide (Li^+Br^-). This picture of the ionic bond was first proposed by Walter Kössel[14] in 1916 and satisfactorily accounts for the chemistry of many inorganic salts.

1.6 The covalent bond

Elements that can readily attain an inert-gas configuration by gaining or losing an electron form ionic bonds. How, though, do the elements in the middle of the periodic table form bonds? Let's look at methane, CH_4, as an example. Certainly the bonding in methane is not ionic, since it would be very difficult for carbon ($1s^22s^22p^2$) either to gain or to lose *four* electrons to achieve an inert-gas configuration.[15] In fact, carbon bonds to other atoms, not by donating electrons, but by *sharing* them. Such bonds are called **covalent bonds** and were first proposed in 1916 by G. N. Lewis.[16] The covalent bond is the most important bond in organic chemistry.

[14]Walter Ludwig Julius Paschen Heinrich Kössel (1888–1956); b. Berlin; assistant in physics in Heidelberg (1910) and Munich (1913); professor of physics, Kiel (1921), Danzig (Poland) (1932–1945), and Tübingen (1947).

[15]The electronic configuration of carbon can be written either as $1s^22s^22p^2$ or as $1s^22s^22p_x2p_y$. Both notations are correct, but the latter is more informative since it indicates which of the three equivalent p orbitals are involved.

[16]Gilbert Newton Lewis (1875–1946); b. Weymouth, Mass.; Ph.D. Harvard (1899); professor, Massachusetts Institute of Technology (1905–1912), University of California, Berkeley (1912–1946).

A simple shorthand way of indicating covalent bonds in molecules is to use what are known as **Lewis structures,** in which an atom's outer-shell electrons are represented by dots. Thus, hydrogen has one dot (1*s*), carbon has four dots ($2s^22p^2$), oxygen has six dots ($3s^23p^4$), and so on. A stable molecule results whenever the inert-gas configuration has been achieved for all atoms, as in the following examples:

$$4\,\cdot\dot{\underset{\cdot}{C}}\cdot + 4\,H\cdot \longrightarrow \begin{array}{c} H \\ H{:}\underset{\cdot\cdot}{\overset{\cdot\cdot}{C}}{:}H \\ H \end{array}$$

Methane (CH_4)

$$2\,H\cdot + \cdot\underset{\cdot}{\ddot{O}}{:} \longrightarrow \begin{array}{c} H{:}\underset{\cdot\cdot}{\ddot{O}}{:} \\ H \end{array}$$

Water (H_2O)

$$2\,H\cdot + \cdot\underset{\cdot\cdot}{\ddot{O}}\cdot + H^+ \longrightarrow \begin{array}{c} + \\ H{:}\underset{\cdot\cdot}{\ddot{O}}{:}H \\ H \end{array}$$

Hydronium ion (H_3O^+)

$$3\,H\cdot + \cdot\underset{\cdot\cdot}{\dot{N}}\cdot \longrightarrow \begin{array}{c} H \\ H{:}\underset{\cdot\cdot}{\overset{\cdot\cdot}{N}}{:}H \end{array}$$

Ammonia (NH_3)

$$3H\cdot + \cdot\underset{\cdot}{\dot{C}}\cdot + \cdot\underset{\cdot}{\ddot{O}}{:} + H\cdot \longrightarrow \begin{array}{c} H \\ H{:}\underset{\cdot\cdot}{\overset{\cdot\cdot}{C}}{:}\underset{\cdot\cdot}{\ddot{O}}{:} \\ H\;\;H \end{array}$$

Methanol (CH_3OH)

Lewis structures are valuable because they make electron "book-keeping" possible and constantly remind us of the number of outer-shell electrons (valence electrons) we are dealing with. Simpler still is the use of **"Kekulé" structures,** also called **line-bond structures,** in which a two-electron bond is indicated simply by a line. In Kekulé structures, the pairs of nonbonding outer-shell electrons are often ignored, but we must still be mentally aware of their existence. Some of the molecules already considered are shown in Table 1.5.

PROBLEM

1.2 Write Lewis structures for these molecules:

(a) $CHCl_3$, chloroform
(b) H_2S, hydrogen sulfide
(c) CH_3NH_2, methylamine
(d) BH_3, borane
(e) NaH, sodium hydride
(f) CH_3OCH_3, dimethyl ether

TABLE 1.5 Lewis and Kekulé structures of some simple molecules

Name	*Lewis structure*	*Kekulé structure*
Water (H_2O)	H:Ö: H	H—O | H
Ammonia (NH_3)	H H:N:H	H | H—N—H
Methane (CH_4)	H H:C:H H	H | H—C—H | H
Methanol (CH_3OH)	H H:C:Ö: H H	H | H—C—O | | H H

PROBLEM

1.3 Which of the indicated bonds would you predict to be covalent and which ionic?

(a) CH_3—Cl

(b) Cl—Ca—Cl

(c) CH_3—NH_2

(d) F—B—F (with a third F on B)

(e) K—F

(f) H_2N—NH_2

1.7 Molecular orbital theory

How does quantum mechanics describe this shared or covalent bond? There is more than one answer to this question because there is more than one way to view the problem, but the most generally satisfactory method for dealing with organic compounds is **molecular orbital (MO) theory.**

The major postulate of molecular orbital theory is that, when a covalent bond is formed, the atomic orbitals on different individual atoms *combine* to form *molecular orbitals*. These molecular orbitals hold the bonding (shared) electrons and are so named because they are a property of the whole molecule, not of a single atom within the molecule. We can picture molecular orbital formation as occurring by an *overlapping* of atomic orbitals. For the hydrogen molecule, we can imagine two hydrogen atoms, each with an atomic 1*s* orbital, pressing together. As the two spherical atomic orbitals approach each other and combine, a new, egg-shaped molecular orbital results. This molecular orbital is filled by two electrons, one from each hydrogen:

$$(\text{H}\cdot) + (\text{H}\cdot) \longrightarrow (\text{H}\cdot)(\cdot\text{H}) = (\text{H}:\text{H})$$

1s atomic orbital — 1s atomic orbital — H_2 molecular orbital

This new arrangement of electrons in a hydrogen molecule is considerably more stable than the original arrangement of individual atoms.

During the reaction $2\text{H}\cdot \rightarrow H_2$, 104 kcal/mol of energy is *released.* Since the product H_2 molecule has 104 kcal/mol *less* energy than the starting $2\text{H}\cdot$, we say that the product is more stable than the starting material, and that the new H—H bond formed has a **bond strength** of 104 kcal/mol. We can look at this in another way by saying that we would have to put 104 kcal/mol of energy (heat) *into* the H—H bond in order to break it into two hydrogen atoms. Figure 1.7 shows the relative energy levels of the different orbitals.

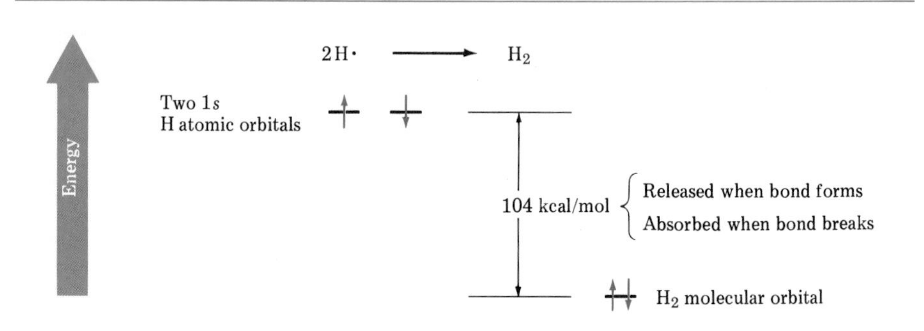

Figure 1.7. Energy levels of H_2 orbitals.

If the two positively charged nuclei in the hydrogen molecule are too close together, they will repel each other electrostatically; if the nuclei are too far apart, they will not be able to share the bonding electrons adequately. Thus, there is an optimum distance between the two nuclei that leads to maximum stability. This optimum distance in the hydrogen molecule is 0.74 angstrom (Å) [1 Å = 10^{-8} centimeter (cm)] and is called the **bond length.** Every covalent bond formed has a characteristic bond strength and bond length.

One further point that should be considered in this description of the hydrogen molecule is that an orbital seems to have disappeared. We began forming the hydrogen molecule by combining *two* atomic orbitals, each of which, if filled, could have held two electrons, for a total of four. We ended up, however, with what seems to be *one* molecular orbital that can hold only two electrons. In fact, an orbital hasn't disappeared; we simply haven't paid it much attention. When we combine a pair of atomic orbitals, a pair of molecular orbitals is produced. One of the molecular orbitals is lower in energy than the starting atomic orbitals, and the other molecular orbital is correspondingly higher in energy. We can represent this as in Figure 1.8.

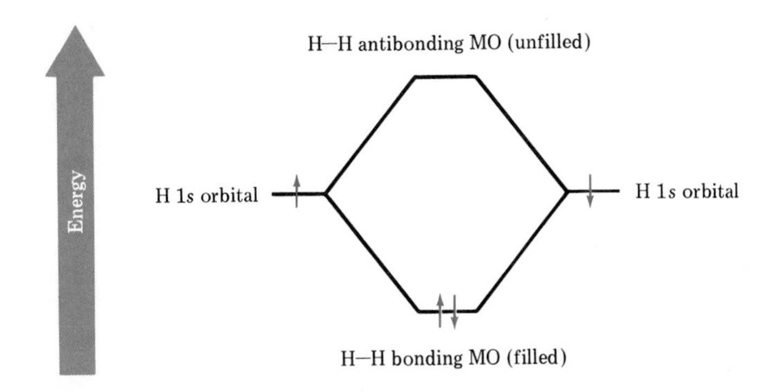

Figure 1.8. Molecular orbitals of H_2. The combination of two hydrogen 1*s* atomic orbitals leads to the formation of two molecular orbitals. The lower-energy (bonding) molecular orbital is filled, and the higher-energy (antibonding) molecular orbital is unfilled.

Two hydrogen 1*s* orbitals overlap to form two molecular orbitals, and both electrons occupy the lower-energy one, which we call the **bonding MO.** A higher-energy **antibonding MO** is also produced, but it remains unoccupied. In many chemical reactions, we do not need to be concerned with the presence of antibonding orbitals. This is not always true, however, and we will see in Chapter 30 that antibonding orbitals can be of critical importance.

The bonding molecular orbital in the hydrogen molecule has the elongated egg shape that we might get by pressing two spheres together. A cross section cut by a plane through the middle of the H—H bond reveals a circle; in other words, the H—H bond is *cylindrically symmetrical,* as shown in Figure 1.9.

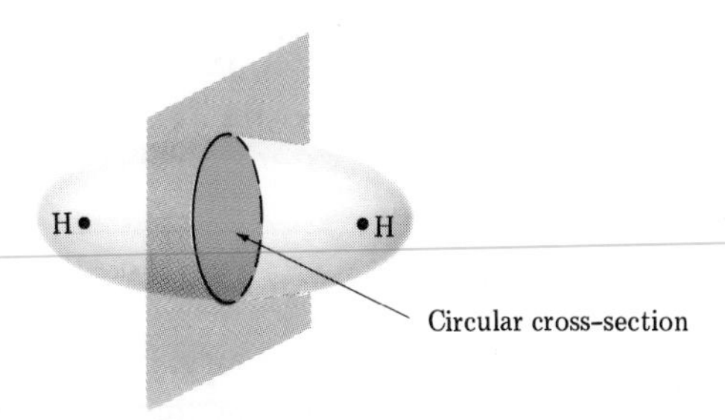

Figure 1.9. The cylindrically symmetrical H—H bond.

Bonds that have circular cross-sections and are formed by head-on overlap of two atomic orbitals are called **sigma (σ) bonds.** There are other types of bonds, however. Let's consider the fluorine molecule, F_2. A fluorine atom has seven outer-shell electrons and the electronic configuration $1s^22s^22p^5$. By bonding together, two fluorine atoms can each achieve stable outer-shell octets:

$$:\ddot{\underset{..}{F}}\cdot + \cdot\ddot{\underset{..}{F}}: \longrightarrow :\ddot{\underset{..}{F}}:\ddot{\underset{..}{F}}:$$

Unlike the hydrogen atom, however, a fluorine atom has an unshared $2p$ electron rather than a $1s$ electron. How can two p orbitals come together to form a bond? The general answer to this question was provided by Linus Pauling[17] in 1931 when he stated the **principle of maximum orbital overlap.** According to this principle, the strongest bond will be formed when the two orbitals achieve maximum overlap. There are two geometric possibilities for p orbital overlap in the fluorine molecule: The p orbitals can be oriented in a head-on fashion to form a sigma bond, or they can overlap in a sideways fashion to form what is called a **pi (π) bond,** shown in Figure 1.10.

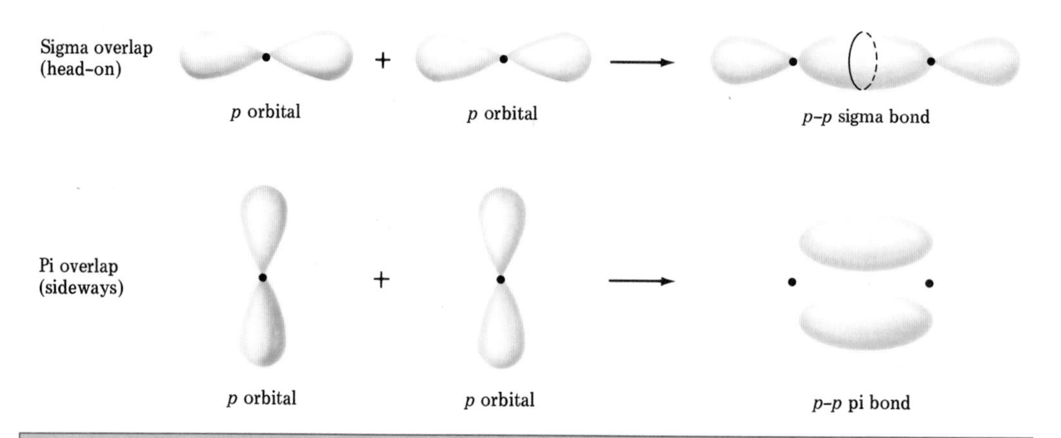

Figure 1.10. The formation of sigma and pi bonds (*p* orbital overlap).

It is difficult to predict which kind of bonding leads to maximum overlap and is favored, but it turns out that sigma bonding is usually more efficient than pi bonding. Fluorine therefore forms a sigma molecular bond between two $2p$ orbitals. The new F—F bond formed has a bond strength of 38 kcal/mol and a bond length of 1.42 Å.

1.8 Hybridization—sp^3 orbitals and the structure of methane

The bonding in both the hydrogen molecule and the fluorine molecule is fairly straightforward, but the situation becomes more complicated when we turn to organic molecules with tetravalent carbon atoms. Let's start with a simple case and consider methane, CH_4. Carbon has the ground-state electronic configuration $1s^2 2s^2 2p_x 2p_y$. The outer shell has four electrons, two of which are paired in the $2s$ orbital, and two of which are unpaired and occupy different $2p$ orbitals:

[17]Linus Pauling (1901–); b. Portland, Oregon; Ph.D. California Institute of Technology (1925); professor, California Institute of Technology (1925–1967); University of California, San Diego; Professor Emeritus, Stanford University (1974–); Nobel prize (1954, 1963).

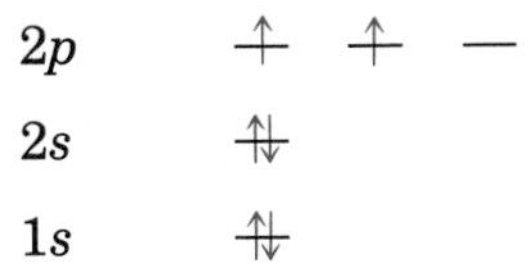

2p	↑ ↑ —
2s	↑↓
1s	↑↓

Ground-state electronic configuration of carbon

The first question we face is immediately apparent. How can carbon form four bonds if it only has *two* unpaired electrons? Why doesn't carbon bond to *two* hydrogen atoms to form CH_2? In fact, CH_2 is a known compound. It is, however, highly unstable and reactive, and has only fleeting existence. We can see why carbon prefers to form four bonds instead of two by looking at the amount of energy released in forming CH_2 versus forming CH_4. By experimental measurement, we know that a typical C—H bond has a strength of approximately 100 kcal/mol. Thus, the reaction of a carbon atom with two hydrogen atoms to form CH_2 should be energetically favored by about 200 kcal/mol.

$$\cdot\ddot{C}\cdot + 2\,H\cdot \longrightarrow H{:}\ddot{C}{:}H + \sim 200 \text{ kcal/mol}$$

Alternatively, however, carbon can adopt an electronic configuration *different* from the ground-state configuration. By promoting one electron from the 2*s* orbital into the vacant $2p_z$ orbital, carbon can achieve the new configuration $1s^2 2s 2p_x 2p_y 2p_z$. This new configuration has one electron placed in a higher-energy orbital than the ground state, and 96 kcal/mol of energy is required to accomplish the electron promotion. The new configuration is called an **excited state.**

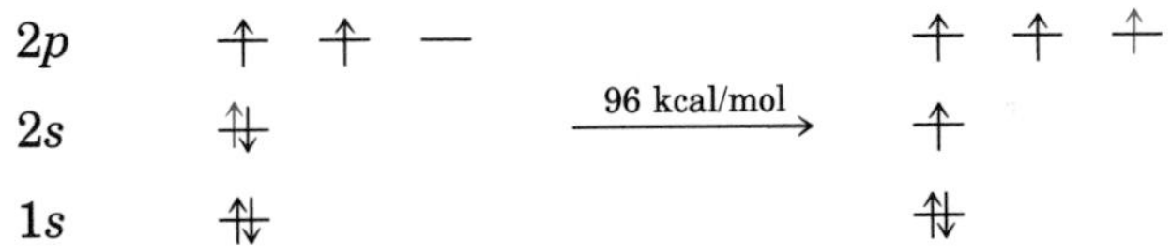

	Ground-state carbon		Excited-state carbon
2p	↑ ↑ —		↑ ↑ ↑
2s	↑↓	$\xrightarrow{\text{96 kcal/mol}}$	↑
1s	↑↓		↑↓

In the excited state, carbon has *four* unpaired electrons and can form *four* bonds with hydrogen. Although 96 kcal/mol is required to promote the 2*s* electron to a 2*p* orbital, this energy loss is more than offset by the formation of four stable C—H bonds rather than two. In Lewis structures,

$$\cdot\ddot{C}\cdot \xrightarrow{\text{96 kcal/mol}} \cdot\dot{\underset{\cdot}{C}}\cdot \xrightarrow{4\,H\cdot} H{:}\underset{\underset{H}{\cdot\cdot}}{\overset{\overset{H}{\cdot\cdot}}{C}}{:}H + \sim 400 \text{ kcal/mol}$$

$$\text{Net energy change} = (400 - 96) \text{ kcal/mol} \approx 300 \text{ kcal/mol}$$

What is the nature of the four C—H bonds in methane? Since excited-state carbon uses *two* kinds of orbitals for bonding purposes, we might expect methane to have *two* kinds of C—H bonds. In fact this is not the case.

A large amount of evidence shows that all four C—H bonds in methane are *identical*. How can we explain this?

The answer was provided by Linus Pauling in 1931. Pauling showed that a combination of an *s* orbital and three *p* orbitals can be mathematically mixed or **hybridized** to form four equivalent new atomic orbitals that are spatially oriented toward the corners of a tetrahedron. We know that electrons tend to repel each other since they are negatively charged, and it's not surprising that they seek to be as far away from each other as possible; this is exactly what the tetrahedral geometry allows for. These new tetrahedral orbitals are called ***sp*³ hybrids,**[18] since they are mathematically constructed from three *p* orbitals and one *s* orbital (see Figure 1.11).

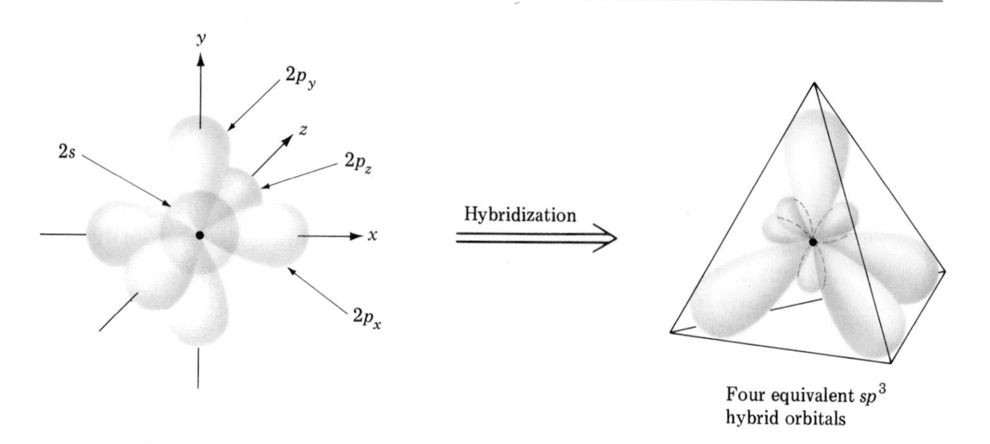

Figure 1.11. The formation of sp^3 hybrid orbitals.

The concept of hybridization explains *how* carbon forms four equivalent tetrahedral bonds but does not answer the question of *why* it does so. Viewing a cross section of an sp^3 hybrid orbital suggests the answer. When an *s* orbital hybridizes with three *p* orbitals, the resultant hybrids are *unsymmetrical* about the nucleus. One of the two lobes of an sp^3 orbital is much larger than the other, and as a result sp^3 hybrid orbitals form much stronger bonds than do unhybridized *s* or *p* orbitals. This asymmetry of sp^3 orbitals arises because of a property of orbitals that we have not yet considered. When the wave equation for a *p* orbital is solved, the two lobes turn out to have opposite algebraic signs, + and −. Thus, when a *p* orbital hybridizes with an *s* orbital, one lobe is *additive* with the *s* orbital, but the other lobe is *subtractive*. The resultant hybrid orbital (Figure 1.12) is therefore strongly oriented in one direction. The chemical consequences of these algebraic signs will be examined in the discussion of orbital symmetry in Chapter 30.

We describe the sp^3 hybrid as a *directed* orbital, and we find that it is capable of forming very strong bonds by overlapping the orbitals of other

[18]Note that the superscript used to identify an sp^3 hybrid orbital tells how many of each type of atomic orbital combine in the hybrid; it does not tell how many electrons occupy that orbital.

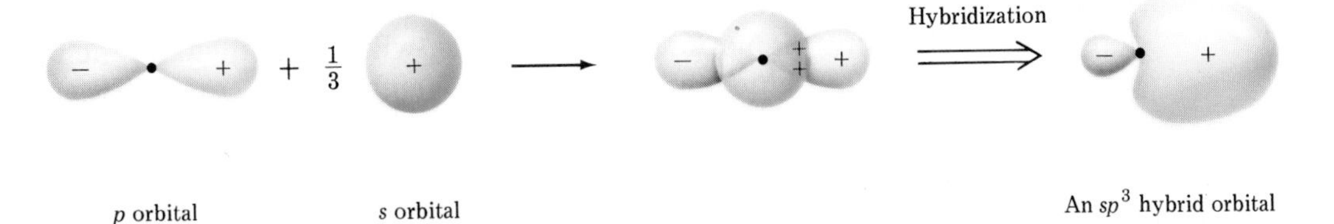

Figure 1.12. The formation of an sp^3 hybrid orbital by overlap of a *p* orbital with part of an *s* orbital. Overlap of the *s* orbital with the positive *p* lobe is additive, but overlap with the negative *p* lobe cancels out.

atoms. For example, the overlap of a carbon sp^3 hybrid orbital with a hydrogen 1*s* orbital gives a strong C—H bond (Figure 1.13).

The C—H bond in methane has a measured bond strength of 104 kcal/mol and a bond length of 1.10 Å. Since the four orbitals have a specific geometry, we can also define a third important physical property of pairs of bonds, called the **bond angle.** The angle formed by each H—C—H is exactly 109.5°, the tetrahedral angle. Methane has the structure shown in Figure 1.14.

Before ending this discussion of the methane structure, it should again be pointed out that no orbitals are "lost" during hybridization. An

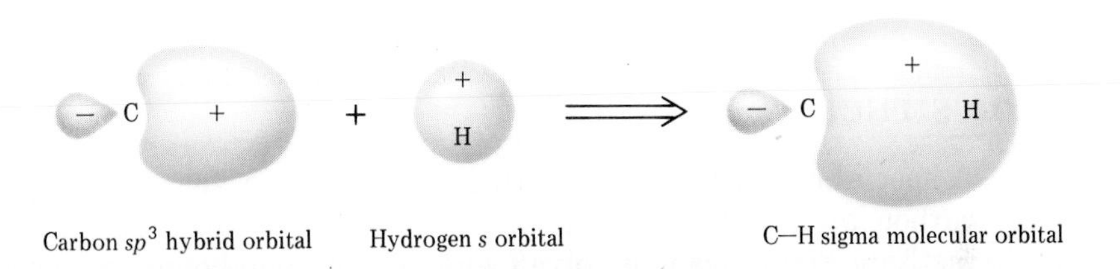

Figure 1.13. The formation of a C—H bond by overlap of a carbon sp^3 hybrid orbital with a hydrogen 1*s* orbital.

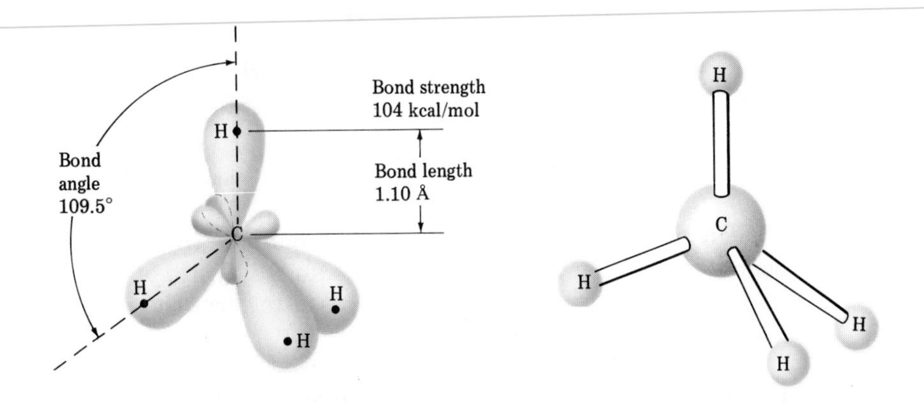

Figure 1.14. The structure of methane.

sp^3-hybridized carbon provides four hybrid orbitals and four electrons. Four hydrogen atoms provide four 1*s* orbitals and four electrons. Methane therefore has four bonding sigma molecular orbitals, which are filled, and four antibonding orbitals, which are unfilled. All eight MOs are shown in Figure 1.15.

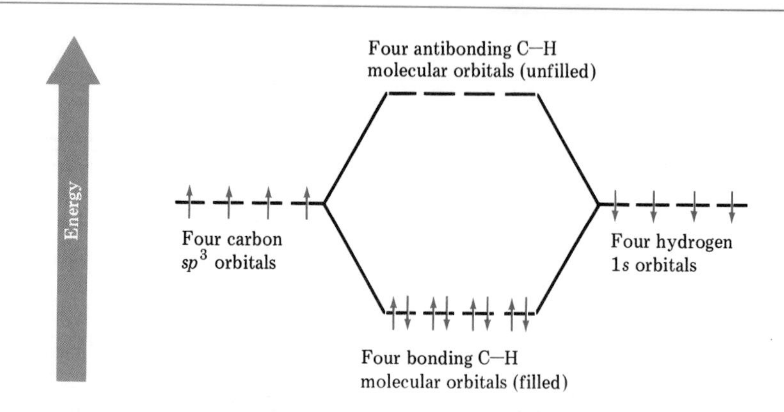

Figure 1.15. Molecular orbitals of methane.

1.9 The structure of ethane

A special characteristic of carbon is that it can form stable bonds to other carbon atoms. Exactly the same kind of hybridization that explains the methane structure is involved when one carbon bonds to another to form a chain. Ethane, C_2H_6, is the simplest molecule containing a carbon–carbon bond:

```
  H H          H  H
H:C:C:H      H—C—C—H        CH3CH3
  H H          H  H
```

Some representations of ethane

We can picture the ethane molecule by assuming that the two carbon atoms bond to each other by sigma overlap of an sp^3 hybrid orbital from each. The remaining six sp^3 orbitals are then used to form the six C—H bonds, as shown in Figure 1.16.

The C—H bonds in ethane are similar to those in methane, though a bit weaker (98 kcal/mol for ethane versus 104 kcal/mol for methane). The C—C bond is 1.54 Å long and has a strength of 88 kcal/mol. All the bond angles of ethane are very near the tetrahedral value, 109.5°.

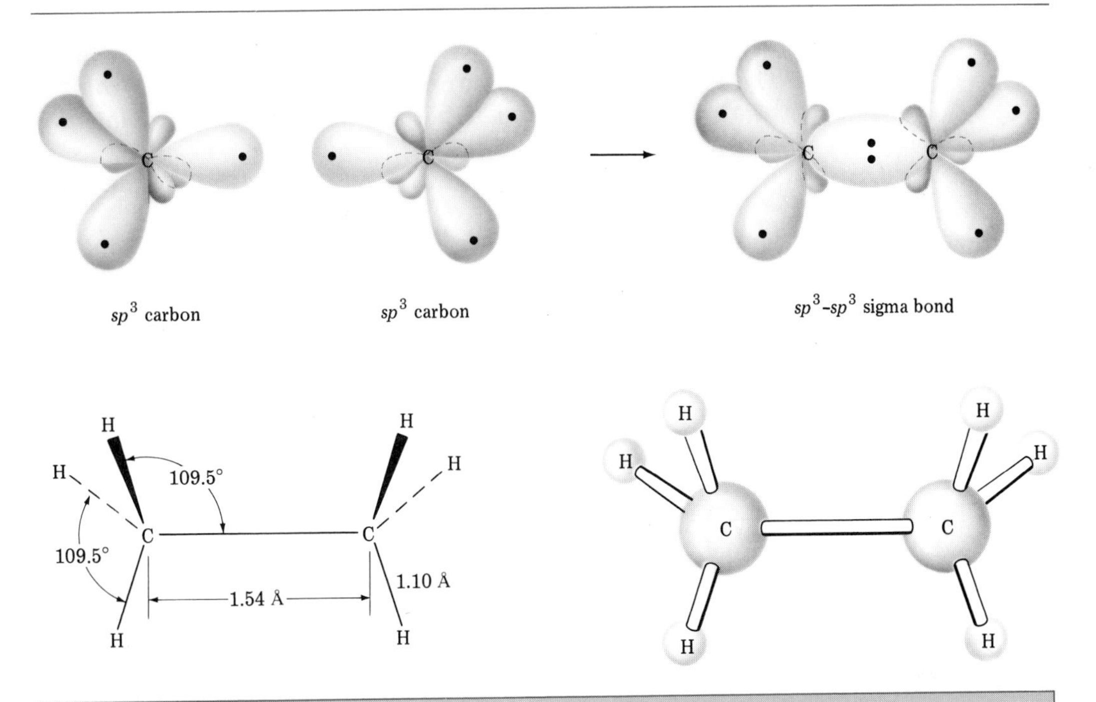

Figure 1.16. Structure of ethane.

1.10 Hybridization—sp^2 orbitals and the structure of ethylene

Although sp^3 hybridization is the most common electronic state of carbon found in organic chemistry, it is not the only possibility. For example, let's look at ethylene, C_2H_4. It was recognized over 100 years ago that ethylene carbons can be tetravalent only if the two carbon atoms are linked by a *double* bond. How can we explain the carbon–carbon double bond in molecular orbital terms?

```
 H H          H       H
 ‥ ‥           \     /          H       H
H:C::C:H        C=C             H>C=C<H
 ‥ ‥           /     \
 H H          H       H

               Top view         Side view
                        Ethylene
```

When we formed sp^3 hybrid orbitals to explain the bonding in methane, we first promoted an electron from the 2*s* orbital of ground-state carbon to form excited-state carbon with four unpaired electrons. We then mathematically mixed the four singly occupied atomic orbitals to construct four sp^3 hybrids. Imagine instead that we mathematically combine the 2*s* orbital with only *two* of the three available $2p$ orbitals. Three hybrid orbitals, which we call **sp^2 hybrids,** result, and one unhybridized $2p$ orbital remains unchanged. The three sp^2 orbitals lie in a plane at angles of 120° to each other, and the remaining p orbital is perpendicular to the sp^2 plane, as shown in Figure 1.17.

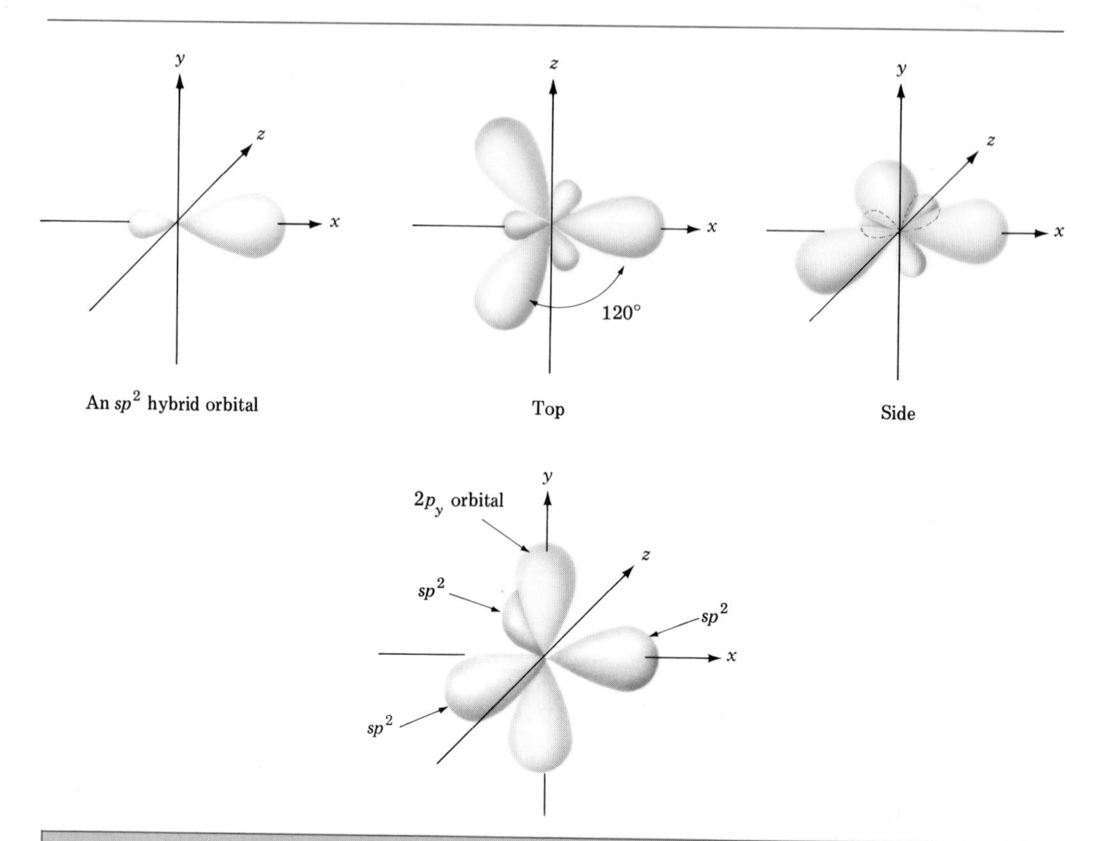

Figure 1.17. An sp^2-hybridized carbon.

As with sp^3 hybrid orbitals, the sp^2 hybrids are strongly oriented in a specific direction and can form strong bonds. If we allow two sp^2-hybridized carbons to approach each other, they can form a strong sigma bond by sp^2–sp^2 overlap. When this occurs, the unhybridized p orbitals on each carbon also approach each other in the correct geometry for sideways overlap to form a pi bond. The combination of sp^2–sp^2 sigma overlap and $2p$–$2p$ pi overlap results in the formation of a carbon–carbon double bond (Figure 1.18).

To complete the structure of ethylene, we need only allow four hydrogen atoms to sigma bond to the remaining sp^2 orbitals:

120° predicted

120° predicted

Ethylene (sigma bonds only)

Ethylene (sigma and pi bonds)

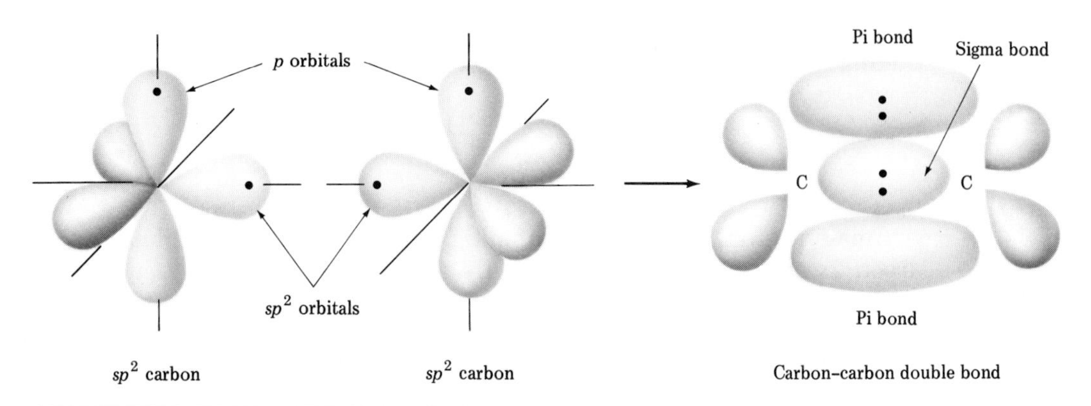

Figure 1.18. Orbital overlap in a carbon–carbon double bond.

From this we can predict that ethylene should be a planar (flat) molecule with H—C—H and H—C—C bond angles of approximately 120°, and this prediction has been verified by experimental observation. Ethylene is indeed flat, and it has H—C—H bond angles of 116.6° and H—C—C bond angles of 121.7°. Each C—H bond has a length of 1.076 Å and a strength of 103 kcal/mol. The central carbon–carbon double bond has a length of 1.33 Å and a strength of 152 kcal/mol. These values for the carbon–carbon double bond are different from those for the carbon–carbon single bond in ethane (1.54 Å and 88 kcal/mol, respectively), and it seems reasonable that the ethylene double bond should be both shorter and stronger than the ethane single bond (see Figure 1.19).

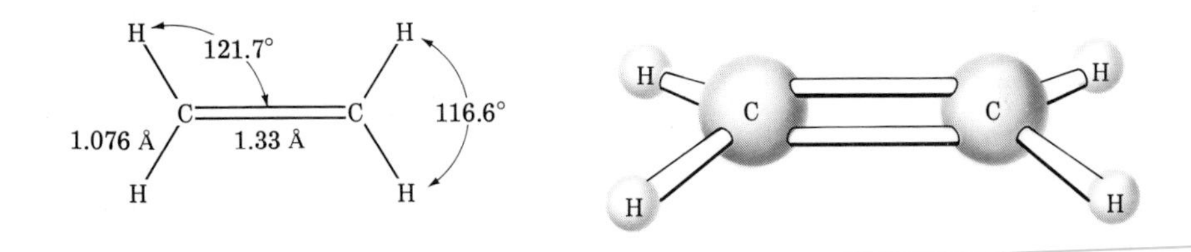

Figure 1.19. Bonding in ethylene.

PROBLEM

1.4 Draw all of the bonds in propene, $CH_3CH{=}CH_2$. Indicate the hybridization of each carbon and describe the orbital overlap of each bond. What geometry do you predict for propene?

1.11 Hybridization—*sp* orbitals and the structure of acetylene

In addition to its ability to form single and double bonds, carbon can form a third kind of bond. Acetylene, C_2H_2, can be satisfactorily pictured only if we assume that it contains a carbon–carbon *triple* bond, and we must construct yet another kind of hybrid orbital, an **sp hybrid,** to explain its manner of bonding.

H:C:::C:H H—C≡C—H

Acetylene

Imagine that, instead of combining with two or three *p* orbitals, the carbon 2*s* orbital hybridizes with only a single *p* orbital. Two *sp* hybrid orbitals result, and two *p* orbitals remain unchanged. These *sp* orbitals are linear; they are 180° apart on the *x*-axis. The remaining two *p* orbitals are perpendicular on the *y*-axis and the *z*-axis, as Figure 1.20 shows.

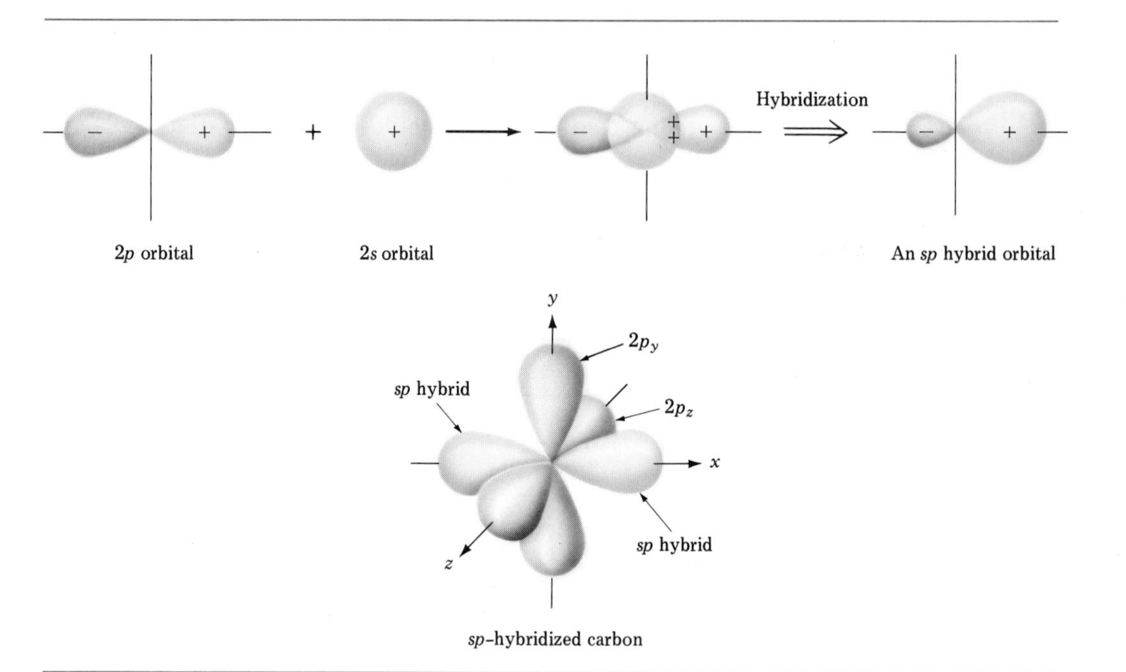

Figure 1.20. An *sp*-hybridized carbon atom.

If we allow two *sp*-hybridized carbon atoms to approach each other, *sp* orbitals from each carbon can overlap head-on to form a strong *sp–sp* sigma bond. The p_z orbitals from each carbon are properly situated to form a p_z–p_z pi bond by sideways overlap, and the p_y orbitals can overlap similarly to form a p_y–p_y pi bond. We have formed one sigma bond and two pi bonds, a net carbon–carbon triple bond. The remaining *sp* hybrid orbitals can sigma bond to hydrogen 1*s* orbitals to complete the acetylene molecule (Figure 1.21).

With *sp* hybridization, we can predict that acetylene should be a linear molecule with an H—C—C bond angle of 180°, and this has been verified experimentally. It has been found that the acetylene carbon–hydrogen bond length is 1.06 Å and the carbon–carbon bond length is 1.20 Å. The C—H bond strength has been measured at 125 kcal/mol and the C≡C bond

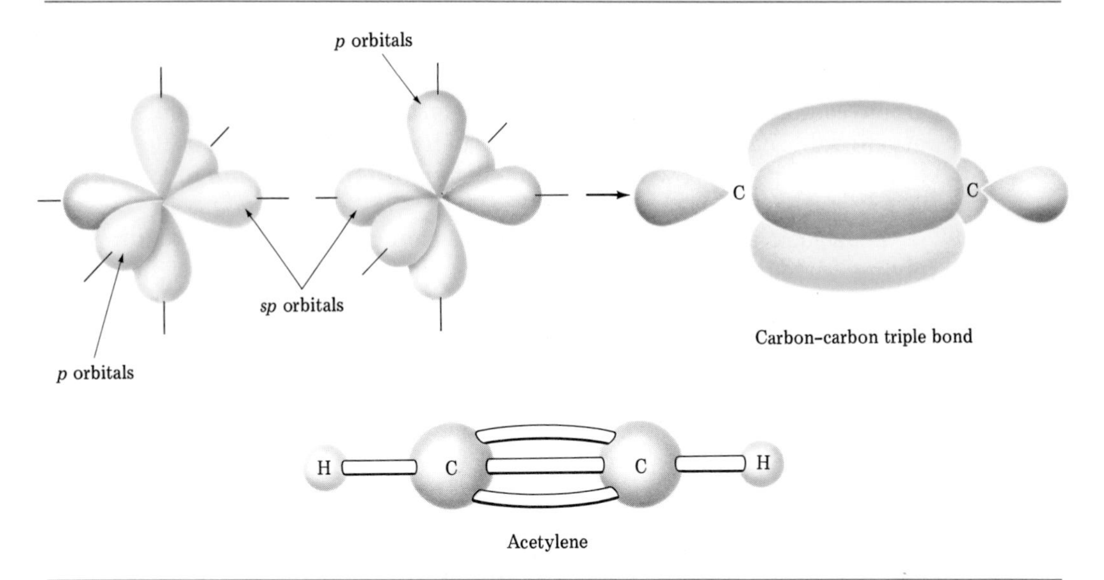

Figure 1.21. Bonding in acetylene.

strength at 200 kcal/mol. It is not surprising to find that the triple bond is so short and so strong; in fact, these values (1.20 Å and 200 kcal/mol) are the shortest and strongest known for any carbon–carbon bond. A comparison of *sp*, sp^2, and sp^3 hybridization is given in Table 1.6.

TABLE 1.6 Comparison of carbon–carbon and carbon–hydrogen bonds in methane, ethane, ethylene, and acetylene

Molecule	*Bond*	*Bond strength (kcal/mol)*	*Bond length (Å)*
Methane, CH_4	$C_{sp^3}—H_{1s}$	104	1.10
Ethane, CH_3CH_3	$C_{sp^3}—C_{sp^3}$	88	1.54
	$C_{sp^3}—H_{1s}$	98	1.10
Ethylene, $H_2C{=}CH_2$	$C_{sp^2}{=}C_{sp^2}$	152	1.33
	$C_{sp^2}—H_{1s}$	103	1.076
Acetylene, $HC{\equiv}CH$	$C_{sp}{\equiv}C_{sp}$	200	1.20
	$C_{sp}—H_{1s}$	125	1.06

PROBLEM

1.5 Draw all the bonds in propyne, $CH_3C{\equiv}CH$. Indicate the hybridization of each carbon and describe the orbital composition of each bond. What is the expected geometry of propyne?

PROBLEM

1.6 Carry out a similar analysis on 1,3-butadiene, $H_2C{=}CH—CH{=}CH_2$, and predict its shape.

1.12 Hybridization of other atoms: nitrogen

The description of covalent bonding developed so far is not restricted to carbon compounds. All covalent bonds formed by other elements in the periodic table can be described in terms of hybrid orbitals. The situation can become more complex when elements heavier than carbon are involved, but the general principles remain the same.

Let's look at ammonia, NH_3, as an example of covalent bonding involving nitrogen. A nitrogen atom has the ground-state electronic configuration $1s^22s^22p_x2p_y2p_z$, and we might expect nitrogen to combine with three hydrogen atoms:

$$\cdot\ddot{\text{N}}\cdot + 3\,\text{H}\cdot \longrightarrow \text{H}:\ddot{\text{N}}:\text{H} \;\; (\text{with H below N}) \quad \text{or} \quad \text{H—}\ddot{\text{N}}\text{—H} \;\; (\text{with H below N})$$

Since the three unpaired electrons of nitrogen occupy half-filled $2p$ orbitals, it is possible that hydrogen $1s$ orbitals might overlap those $2p$ orbitals to form three sigma bonds. Since the $2p$ orbitals are at right angles to each other, ammonia might be expected to have H—N—H bond angles of 90°. In fact, this picture is wrong. The experimentally measured H—N—H bond angle in ammonia is 107.1°, nearly the tetrahedral value (109.5°). Nitrogen hybridizes to form four sp^3 orbitals, *exactly as carbon does.* Since nitrogen has five outer-shell electrons, one sp^3 orbital is occupied by two electrons and the other three sp^3 orbitals each have one electron. Sigma overlap of these three nitrogen sp^3 hybrid orbitals with hydrogen $1s$ orbitals completes the ammonia molecule (Figure 1.22). Thus, ammonia is a tetrahedral molecule with geometry very similar to that of methane. The N—H bond length is 1.01 Å, and the bond strength is 103 kcal/mol.

Ammonia is tetrahedral because of stability—such a shape is the lowest-energy form of all possible alternatives. The energy required to hybridize the nitrogen from the ground-state configuration to sp^3 configuration is more than offset by the added strength of bonding to sp^3

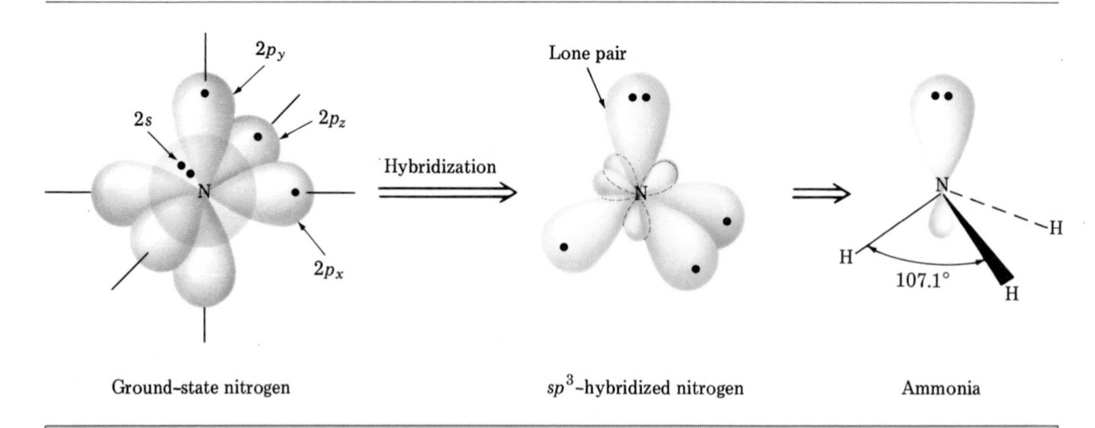

Figure 1.22. Hybridization of nitrogen in ammonia.

orbitals (strongly directed, good overlap), versus bonding to p orbitals (poorly directed, poor overlap). Note also that sp^3 hybridization is preferred over alternatives such as sp^2 or sp hybridization. Only with sp^3 hybridization can the four outer-shell electron pairs on the ammonia nitrogen be as far away from each other as possible. The unshared electron pair on nitrogen, the **lone pair,** occupies nearly as much space as an N—H bond and is very important in the chemistry that ammonia exhibits.

PROBLEM

1.7 Describe the bonding in the nitrogen molecule, N_2. You may assume that both nitrogens are sp hybridized.

1.13 Hybridization of other atoms: oxygen and boron

We saw in ammonia that nonbonding lone-pair electrons can occupy hybrid orbitals just as bonding electron pairs can. The same phenomenon is seen again in the structure of water, H_2O. Ground-state oxygen has the electronic configuration $1s^2 2s^2 2p_x^2 2p_y 2p_z$, and oxygen is divalent:

$$2\,\mathrm{H}\cdot + \cdot\ddot{\underset{\cdot\cdot}{\mathrm{O}}}\cdot \longrightarrow \mathrm{H}:\ddot{\underset{\cdot\cdot}{\mathrm{O}}}:\mathrm{H}$$

We can imagine several modes of bonding in water:

1. Overlap of the two unhybridized oxygen p orbitals with hydrogen $1s$ orbitals. The two oxygen lone pairs would remain in a $2s$ and a $2p_x$ orbital.
2. Hybridization of oxygen into two sp orbitals for bonding. The lone pairs would then both remain in the two unhybridized p orbitals.
3. Hybridization of oxygen into four sp^3 orbitals.

Only the third mode, the hybridization of oxygen into sp^3 orbitals, allows strong bonds and maximum distance between the outer-shell electrons. The oxygen in water is therefore sp^3 hybridized, as illustrated in Figure 1.23.

Measurements on water indicate that the oxygen does not have perfect sp^3 hybrid orbitals. The actual H—O—H bond angle of 104.5° is somewhat less than the predicted tetrahedral angle, but we can explain this by

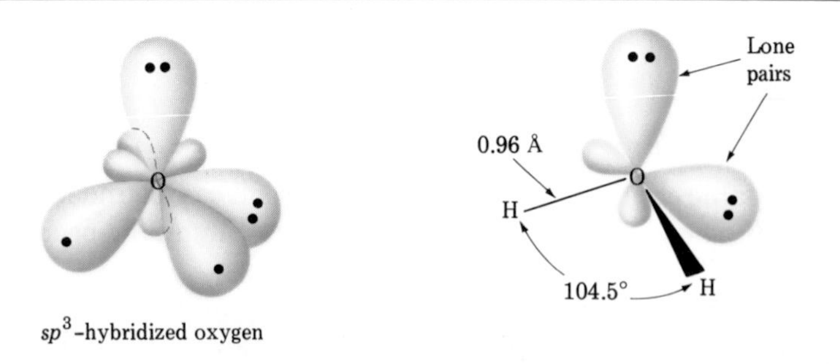

Figure 1.23. The structure of water. The oxygen atom is sp^3 hybridized.

assuming that there is a repulsive interaction between the two lone pairs that forces them apart and thus compresses the H—O—H angle.

One last example of orbital hybridization that we will consider is found in molecules such as boron trifluoride, BF_3. Since boron has only three outer-shell electrons ($1s^22s^22p_x$), it can form a maximum of three bonds. We can promote a $2s$ electron into a $2p_y$ orbital and then hybridize in some manner, but it is not possible to complete a stable octet for boron.

$$3\,:\ddot{\underset{\cdot\cdot}{F}}\cdot \;+\; \cdot\dot{\underset{\cdot}{B}} \longrightarrow \;:\ddot{F}:\!B(:\ddot{F}:)_2 \;=\; BF_3 \text{ (B bonded to three F)}$$

Since there are no lone-pair electrons, we might predict that boron will hybridize in such a way that the three B—F bonds will be as far away from each other as possible. This implies sp^2 hybridization and predicts a planar structure for BF_3. Each fluorine bonds to a boron sp^2 orbital, and *the remaining p orbital on boron is left vacant.* Boron trifluoride has exactly this predicted structure (see Figure 1.24).

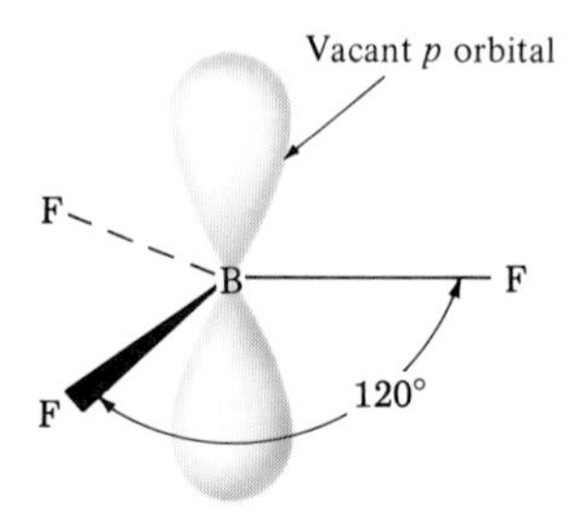

Figure 1.24. The structure of boron trifluoride. Boron is sp^2 hybridized.

PROBLEM

1.8 What shape would you expect each of these species to have?

(a) PH_3 (b) AlH_3 (c) CO_2
(d) CO_3^{2-} (e) CH_3—S—CH_3 (f) CH_3OH

1.14 Summary

Organic chemistry is the study of carbon compounds. For historical reasons, a division into organic and inorganic chemistry occurred, but there is no scientific reason behind the division.

The electronic structure of atoms can be described mathematically by the Schrödinger equation. Electrons can be considered to occupy orbitals centered around the nucleus. Different orbitals have different energy levels and different shapes. The electronic configuration of atoms can be found by assigning electrons to the proper orbitals, beginning with the lowest-energy ones. The *s* orbitals are spherical, and *p* orbitals are dumbbell shaped.

There are two basic kinds of chemical bonds—ionic bonds and covalent bonds. Ionic bonds are based on the electrostatic attraction of unlike charges and are commonly found in inorganic salts. Covalent bonds are formed when an electron pair is shared between two atoms. This electron sharing can be considered to occur by overlap of two atomic orbitals to give a new molecular orbital. Bonds that have a circular cross-section and are formed by head-on overlap are called sigma bonds, whereas bonds formed by sideways overlap of two *p* orbitals are called pi bonds.

In order to form bonds in organic molecules, carbon first hybridizes to an excited-state configuration. In compounds containing carbon–carbon single bonds, carbon is sp^3 hybridized; there are four equivalent sp^3 hybrid orbitals with tetrahedral geometry. In compounds containing double bonds, carbon is sp^2 hybridized; there are three equivalent sp^2 hybrid orbitals with planar trigonal geometry and one unhybridized *p* orbital. The carbon–carbon double bond is formed when two sp^2-hybridized carbon atoms bond together. In compounds containing triple bonds, carbon is *sp* hybridized; there are two equivalent *sp* hybrid orbitals with linear geometry and two unhybridized *p* orbitals. The carbon–carbon triple bond results when two *sp*-hybridized carbon atoms bond together.

Other atoms such as nitrogen, oxygen, and boron also hybridize in order to form stronger bonds. The nitrogen atom in ammonia and the oxygen atom in water are sp^3 hybridized, and the boron atom in boron trifluoride is sp^2 hybridized.

WORKING PROBLEMS

There is no surer way to learn organic chemistry than by working problems.

Learning organic chemistry requires familiarity with a large number of diverse facts. Each page in this book presents new information to be digested and correlated with what has come before. Certainly, learning all this information requires careful reading and rereading, but this alone is not enough; you must *work with* the information and be able to *use* your knowledge in new situations. Problems give you the opportunity to do this.

Each chapter in this book provides many problems of different sorts. The in-chapter problems are placed for immediate reinforcement of new ideas just learned. The end-of-chapter problems provide additional practice and are of two types: drill and thought. The early ones are primarily of the drill type and provide an opportunity for you to practice your command of the fundamentals. Later problems tend to be more thought provoking, and many are real challenges to your depth of understanding.

As you study organic chemistry, take the time to work the problems. Work the ones you can, and ask for help on the ones you can't. If you're stumped by a particular exercise, check the accompanying answer book for an explanation that will help clarify the source of difficulty. Working problems takes effort, but the payoff in knowledge and understanding is immense.

ADDITIONAL PROBLEMS

1.9 Give the ground-state electronic configurations of the following elements. (Carbon, for example, is $1s^2 2s^2 2p^2$.)

(a) Sodium (b) Aluminum (c) Arsenic
(d) Silicon (e) Chromium

1.10 Write Lewis (electron-dot) structures for these molecules:

(a) $H—C\equiv C—H$ (b) AlH_3 (c) $CH_3—S—CH_3$
(d) $H_2C{=}CHCl$ (e) $H_2C{=}CH—CH{=}CH_2$ (f) $CH_3—C({=}O)—O—H$

1.11 Convert the following Kekulé (line-bond) structures into molecular formulas. For example,

$$H_3C—CH_3 \text{ (H—C—C—H with three H on each C)} = C_2H_6$$

(a) Phenol

(b) Aspirin ($O—C(=O)—CH_3$, COOH)

(c) Vitamin C (CH_2OH, H—C—OH, HO, OH)

(d) Nicotine (CH_3)

(e) $C—OCH_2CH_2\overset{+}{N}H(CH_2CH_3)_2\ Cl^-$ (NH_2)
Novocain

(f) Glucose (CH_2OH, HO, OH)

1.12 Convert the following molecular formulas into Kekulé structures that are consistent with valence rules:

(a) C_3H_8 (b) CH_5N
(c) C_2H_6O (2 possibilities) (d) C_3H_7Br (2 possibilities)
(e) C_2H_4O (3 possibilities) (f) C_3H_9N (4 possibilities)

1.13 Indicate the kind of hybridization you might expect for each carbon atom in these molecules:

(a) Propane, $CH_3CH_2CH_3$
(b) 2-Methylpropene, $(CH_3)_2C{=}CH_2$
(c) 1-Buten-3-yne, $H_2C{=}CH{-}C{\equiv}CH$
(d) Cyclobutene,
```
H       H
 \     /
  C = C
  |   |
 H2C—CH2
```
(e) Dimethyl ether, CH_3OCH_3
(f) Toluene,
```
         H
         |
H       C       CH3
 \    /   \\   /
  C          C
  ||         |
  C          C
 /   \     //  \
H      C        H
       |
       H
```

1.14 What kind of hybridization would you expect for the following?
(a) The oxygen in dimethyl ether, $CH_3{-}O{-}CH_3$
(b) The nitrogen in dimethylamine, CH_3NHCH_3
(c) The boron in trimethylborane, $(CH_3)_3B$

1.15 On the basis of your answers to Problem 1.14, what bond angles would you expect for the following?
(a) The C—O—C angle in $CH_3{-}O{-}CH_3$
(b) The C—N—C angle in CH_3NHCH_3
(c) The C—N—H angle in CH_3NHCH_3
(d) The C—B—C angle in $(CH_3)_3B$

1.16 What shape would you expect these species to have?
(a) The ammonium ion, NH_4^+
(b) Trimethylborane, $(CH_3)_3B$
(c) Benzene,
```
         H
         |
H       C       H
 \   //   \   /
  C          C
  |          ||
  C          C
 /  \\     /   \
H      C        H
       |
       H
```
(d) Trimethylphosphine, $(CH_3)_3P$

1.17 Consider the molecules SO_2, SO_3, and the ion SO_4^{2-}.
(a) Write Lewis structures for each.
(b) Predict the shape of each.

1.18 Write Lewis structures for these molecules:
(a) $TiCl_4$
(b) $CH_3{-}Be{-}CH_3$
(c)
```
CH3—P—CH3
    |
    CH3
```

1.19 Indicate the kind of hybridization you might expect for each carbon atom in these molecules:

(a) Acetic acid,

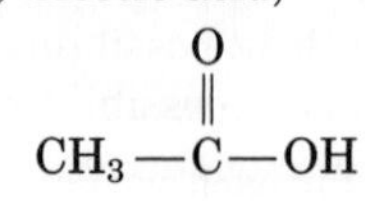

(b) 3-Buten-2-one,

$H_2C{=}CH{-}\overset{\overset{\displaystyle O}{\|}}{C}{-}CH_3$

(c) Acrylonitrile,

$H_2C{=}CH{-}C{\equiv}N$

(d) Benzoic acid,

$C_6H_5{-}\overset{\overset{\displaystyle O}{\|}}{C}{-}OH$

1.20 What kind of hybridization would you expect for the following?

(a) The nitrogen in aniline,

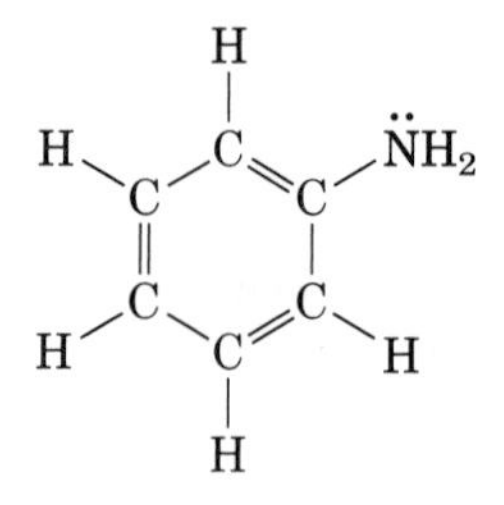

(b) The nitrogen in pyridine,

$C_5H_5\ddot{N}$

(c) The beryllium in dimethylberyllium,

$CH_3{-}Be{-}CH_3$

(d) The phosphorus in trimethylphosphine,

$(CH_3)_3P\colon$

1.21 On the basis of your answers to Problem 1.20, what bond angles do you expect for the following?

(a) The C—N—H angle in aniline
(b) The C—Be—C angle in $(CH_3)_2Be$
(c) The C—P—C angle in $(CH_3)_3P$

1.22 Identify the bonds indicated as either ionic or covalent.

(a) $CH_3O{-}H$ (b) $CH_3O{-}Na$ (c) $H{-}ONO_2$

(d) $CH_3\overset{\overset{\displaystyle O}{\|}}{C}{-}Cl$ (e) $CH_3O{-}Cl$ (f) $Na{-}NH_2$

(g) $CH_3\overset{\overset{\displaystyle O}{\|}}{C}O{-}CH_3$ (h) $F{-}F$ (i) $CH_3\overset{\overset{\displaystyle O}{\|}}{C}{-}O{-}NH_4$

1.23 Order the following sets of bonds according to their increasing ionic character:

(a) $H_3C{-}Cl$, $Cl{-}Cl$, $NH_4{-}Cl$

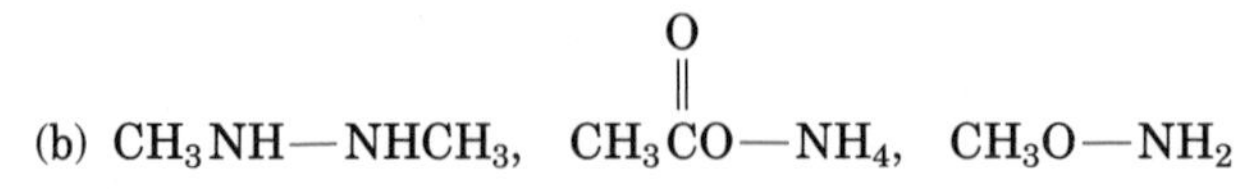

(b) $CH_3NH{-}NHCH_3$, $CH_3\overset{\overset{\displaystyle O}{\|}}{C}O{-}NH_4$, $CH_3O{-}NH_2$

(c) $CH_3\overset{\overset{\displaystyle O}{\|}}{C}—O—H$, $H_3C—H$, $H—Cl$

(d) $Br—CH_3$, $Li—CH_3$, $(CH_3)_3Si—CH_3$

1.24 Although almost all stable organic species have tetravalent carbon atoms, species with trivalent carbon atoms are known to exist. *Carbocations* are one such class of compounds. What hybridization might you expect the carbon atom to have? What geometry would this lead to? What relationship do you see between a carbocation and a trivalent boron compound?

$$\begin{array}{c} H \\ | \\ H—C^+ \\ | \\ H \end{array}$$

A carbocation

1.25 *Carbanions,* $H_3C:^-$, and *radicals,* $H_3C\cdot$, are also known. What hybridization and geometries might you predict for these species?

1.26 Divalent species called *carbenes* are known to be capable of fleeting existence. Methylene, $:CH_2$, is the simplest carbene. Methylene can exist in either of two electronic states. The two unshared electrons can either be spin-paired in a single orbital, or unpaired in different orbitals. Predict the type of hybridization you would expect carbon to adopt in singlet (spin-paired) methylene and triplet (spin-unpaired) methylene. Draw pictures of each, and indicate the types of carbon orbitals present.

CHAPTER 2 BONDING AND MOLECULAR PROPERTIES

2.1 Drawing chemical structures

In the Kekulé structures we have been using, a line between atoms represents the two electrons in a bond. These structures have served chemists well for many years and comprise a universal chemical language. A chemist in China and a chemist in England may not speak each other's language, but a chemical structure means the same to both of them.

Most organic chemists find themselves drawing many structures each day, and it would soon become awkward if every bond and atom had to be indicated. For example, cholesterol, $C_{27}H_{46}O$, has 77 different chemical bonds uniting the 74 atoms. Cholesterol can be drawn showing each bond and atom, but this is a time-consuming process, and the resultant drawing is cluttered.

Chemists have therefore devised a shorthand way of drawing line-bond structures that greatly simplifies matters. The rules for this shorthand are simple:

1. Carbon atoms are not usually shown. Instead, a carbon atom is implicitly assumed to be at each intersection of two lines (bonds) and at the end of each line. Occasionally, carbon atoms are indicated for emphasis.
2. Carbon–hydrogen bonds are not usually shown. Since carbon almost always has a valence of 4, we mentally supply the correct number of hydrogen atoms to fill the valence of each carbon.
3. All elements other than carbon and hydrogen are indicated.

Table 2.1 (pp. 36–37) gives examples.

PROBLEM

2.1 Convert these shorthand structures into molecular formulas:

(a)

Pyridine

(b) O

Cyclohexanone

(c) (d)

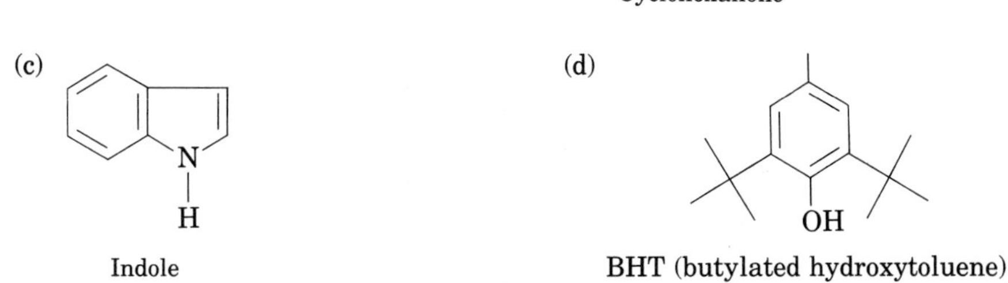

Indole

BHT (butylated hydroxytoluene)

PROBLEM

2.2 Propose Kekulé line-bond structures that satisfy these molecular formulas:
(a) C_5H_{12} (b) C_2H_7N (c) C_3H_6O (d) C_4H_9Cl

TABLE 2.1 Kekulé and shorthand structures for several compounds

Compound	*Kekulé structure*	*Shorthand structure*
Butane, C_4H_{10}	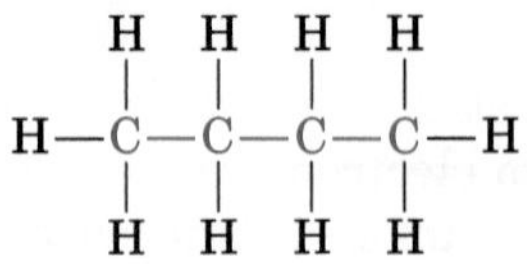 H—C—C—C—C—H (each C bearing two H)	 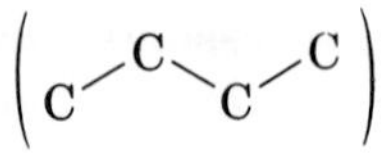 (C–C–C–C)
Chloroethylene (vinyl chloride), C_2H_3Cl	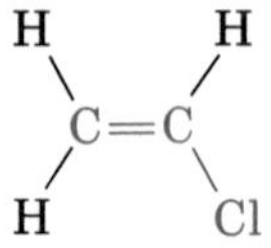 $H_2C{=}CHCl$	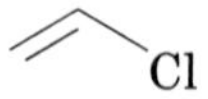 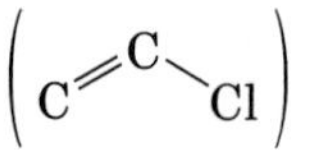 (C=C–Cl)
2-Methyl-1,3-butadiene (isoprene), C_5H_8	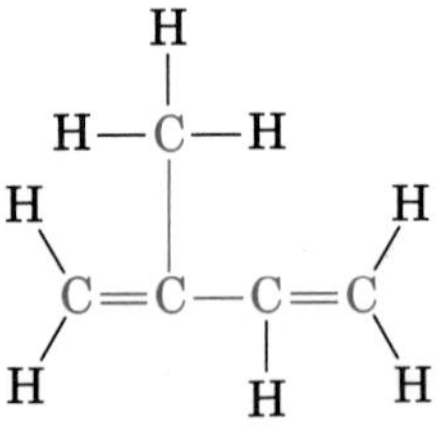 $H_2C{=}C(CH_3){-}CH{=}CH_2$	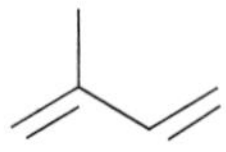
Cyclohexane, C_6H_{12}[a]	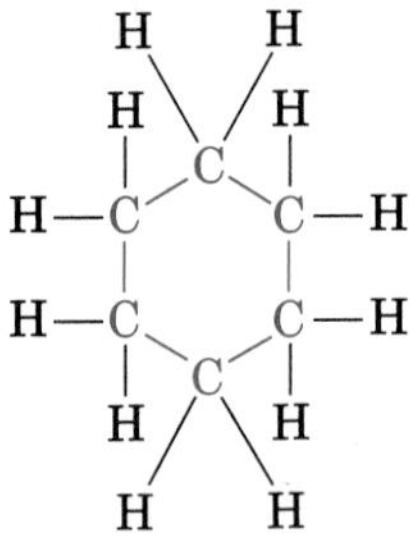	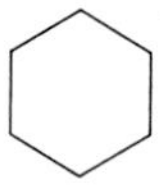
Aspirin, $C_9H_8O_4$	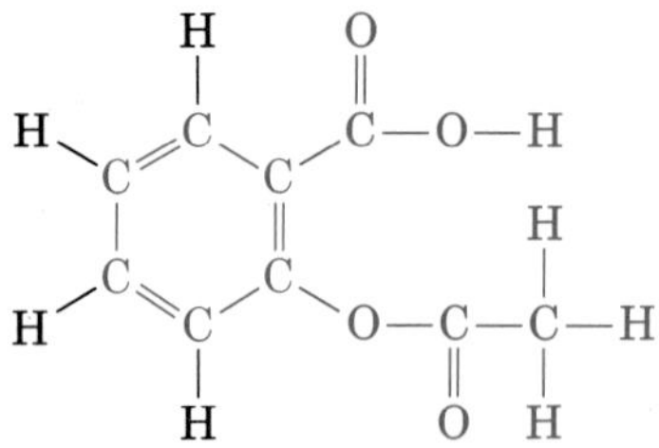	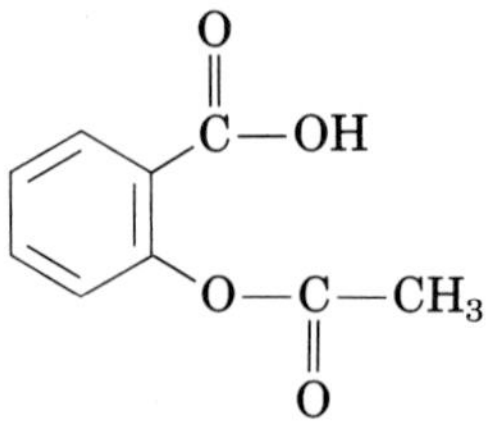 C—OH, O—C—CH_3 (with C=O)

[a]Note that cyclohexane has a *ring* of carbon atoms.

Compound	*Kekulé structure*	*Shorthand structure*
Vitamin A, $C_{20}H_{30}O$	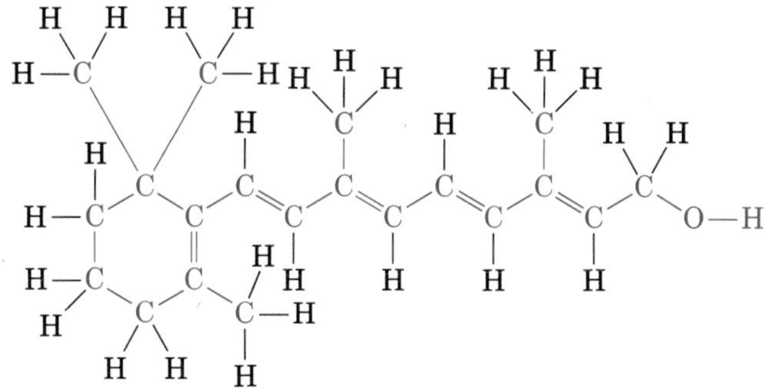	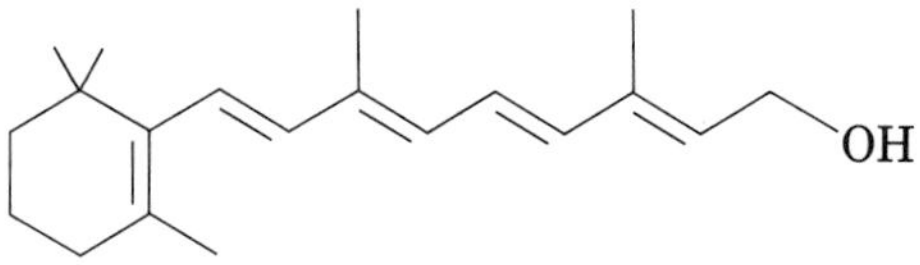
Cholesterol (a steroid), $C_{27}H_{46}O$		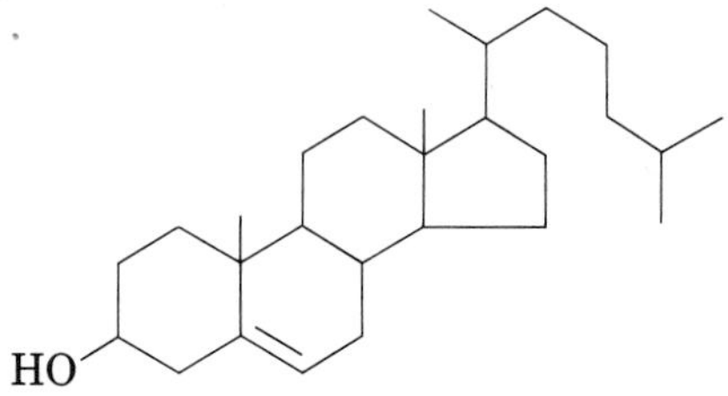

2.2 Molecular models

Another technique that simplifies the chemist's task is the use of molecular models. Organic chemistry is a three-dimensional science, and molecular shape often plays a crucial role in determining the chemistry a compound undergoes. With practice, you can learn to see many spatial relationships even when viewing two-dimensional drawings, but there is no substitute for building a molecular model and turning it in your hands for different perspectives. Many kinds of models are available, some of them at relatively modest cost, and every student should have ready access to a set of models as an adjunct to studying this book.

Research chemists generally prefer to use either space-filling models such as Corey–Pauling–Koltun (CPK™) Molecular Models, or skeletal models such as Dreiding Stereomodels™. Both are quite expensive but are precisely made to reflect accurate bond angles, intramolecular distances, and atomic radii. CPK models are generally preferred for examining the degree of crowding within a molecule. Dreiding models, on the other hand, allow the user to measure bond angles and interatomic distances more readily. Figure 2.1 shows models of acetic acid, CH_3COOH.

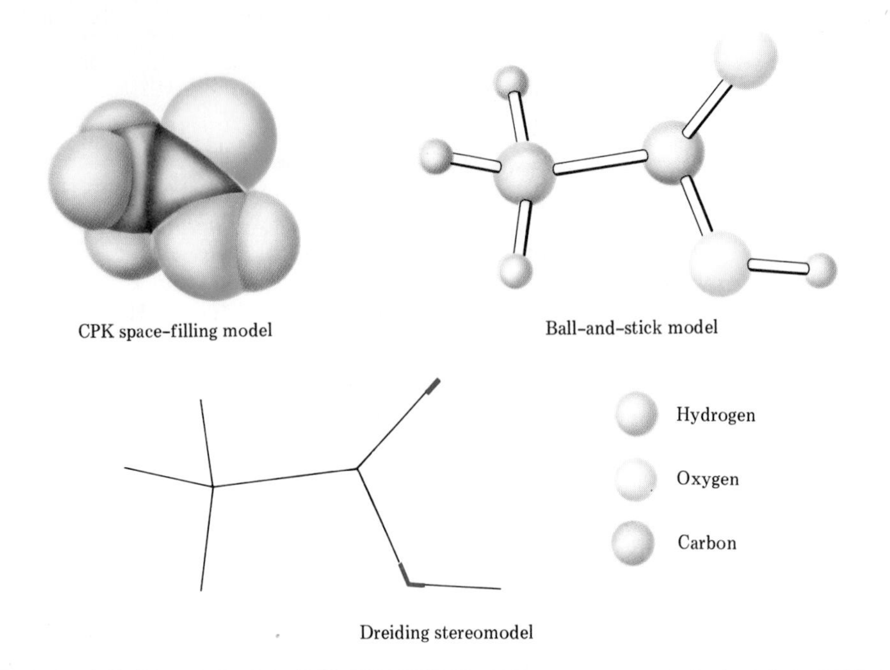

Figure 2.1. Molecular models of acetic acid, CH_3COOH.

2.3 Formal charges

Most organic molecules can be accurately represented by Kekulé line-bond structures. In certain cases, however, electron bookkeeping requires that we attach **formal electrical charges** to specific atoms within a molecule. For example, because of the nature of the electron distribution, acetonitrile oxide (CH_3CNO) must be represented as having a positive charge on nitrogen and a negative charge on oxygen:

$$CH_3{-}C{\equiv}\overset{\oplus}{N}{-}\ddot{\underset{\cdot\cdot}{O}}{:}^{\ominus}$$

Acetonitrile oxide

Let's see why showing these charges is necessary. In the normal covalent bond, each atom donates one electron. Although the bonding electrons are shared by both atoms, each atom may still be considered to "own" one electron for bookkeeping purposes. In methane, for example, there are four C—H bonds totaling eight electrons. Carbon donated half of these bonding electrons and may still be considered to own half. Since an isolated carbon atom has four outer-shell electrons and the methane carbon still owns half of eight, or four, outer-shell electrons, the methane carbon is electrically neutral—it has neither gained nor lost electrons.

The same is true for ammonia, which has three covalent N—H bonds. Atomic nitrogen has five outer-shell electrons. The ammonia nitrogen still owns five (one from each of three shared N—H bonds plus two in the lone pair) and is therefore electrically neutral. To express this in a general way, we can say that each atom in a molecule can be assigned a formal charge, which is equal to the number of outer-shell electrons in an isolated atom less the number of electrons still owned by that atom in the molecule.

$$\text{Formal charge} = \begin{pmatrix}\text{Number of}\\ \text{outer-shell electrons}\\ \text{in a free atom}\end{pmatrix} - \begin{pmatrix}\text{Number of}\\ \text{outer-shell electrons}\\ \text{in bound atom}\end{pmatrix}$$

$$= \begin{pmatrix}\text{Number of}\\ \text{outer-shell}\\ \text{electrons}\end{pmatrix} - \begin{pmatrix}\text{Half the}\\ \text{number of}\\ \text{bonding electrons}\end{pmatrix} - \begin{pmatrix}\text{Number of}\\ \text{nonbonding}\\ \text{electrons}\end{pmatrix}$$

For the methane carbon,

$$\cdot\dot{\underset{\cdot}{C}}\cdot + 4\,H\cdot \longrightarrow H{:}\underset{H}{\overset{H}{\ddot{\underset{\cdot\cdot}{C}}}}{:}H$$

$$\begin{aligned}
&\text{Carbon outer-shell electrons} &&= 4\\
&\text{Methane bonding electrons} &&= 8\\
&\text{Methane nonbonding electrons} &&= 0\\
&\boxed{\text{Formal charge} = 4 - \tfrac{8}{2} - 0 = 0}
\end{aligned}$$

For the ammonia nitrogen,

$$\cdot\ddot{\underset{\cdot}{N}}\cdot + 3\,H\cdot \longrightarrow H{:}\underset{H}{\ddot{\underset{\cdot\cdot}{N}}}{:}H$$

$$\begin{aligned}
&\text{Nitrogen outer-shell electrons} &&= 5\\
&\text{Ammonia bonding electrons} &&= 6\\
&\text{Ammonia nonbonding electrons} &&= 2\\
&\boxed{\text{Formal charge} = 5 - \tfrac{6}{2} - 2 = 0}
\end{aligned}$$

Although most molecules have no formal charges on their atoms, this is not always the case. Let's look again at acetonitrile oxide:

$$CH_3C{\equiv}\overset{\oplus}{N}{-}\overset{\ominus}{O} = H_3C{:}C{:::}\overset{\oplus}{N}{:}\overset{\ominus}{\underset{\cdot\cdot}{\ddot{O}}}{:}$$

For the nitrogen, we have

$$\begin{aligned}
&\text{Nitrogen outer-shell electrons} &&= 5\\
&\text{Bonding electrons} &&= 8\\
&\text{Nonbonding electrons} &&= 0\\
&\boxed{\text{Formal charge} = 5 - \tfrac{8}{2} - 0 = +1}
\end{aligned}$$

For the oxygen, we have

Oxygen outer-shell electrons	=	6
Bonding electrons	=	2
Nonbonding electrons	=	6

$$\text{Formal charge} = 6 - \tfrac{2}{2} - 6 = -1$$

This result tells us that we must write acetonitrile oxide with a formal positive charge on nitrogen and a formal negative charge on oxygen. *Although the acetonitrile oxide molecule as a whole is neutral, specific atoms are charged.* We call such molecules **dipolar molecules,** and this dipolar character often has chemical consequences.

PROBLEM

2.3 Calculate formal charges for the atoms in the following molecules:

(a) Diazomethane, $H_2C{=}N{=}\ddot{\underset{..}{N}}$

(b) Nitromethane, $H_3C{-}N(=\ddot{O}\cdot)(-\ddot{\underset{..}{O}}:)$

(c) Methyl isocyanide, $H_3C{-}N{\equiv}C:$

2.4 Polarization and electronegativity

Relatively few of the molecules we will encounter are dipolar. Most organic molecules, however, are **polar.** This means that, although we cannot assign full formal charges to atoms, the *electron distribution in certain bonds is unsymmetrical.* The electrons are unequally shared by two nuclei.

Thus far, chemical bonding has been portrayed as an either/or situation; a given bond is either covalent or ionic. The concept of polar covalent bonds can be most easily understood, however, by considering bonding to be a *continuum* of possibilities between a perfectly covalent bond with a symmetrical electron distribution on the one hand, and a perfectly ionic bond on the other (Figure 2.2). The carbon–carbon bond in ethane, for example, is symmetrical and therefore perfectly covalent; the two bonding

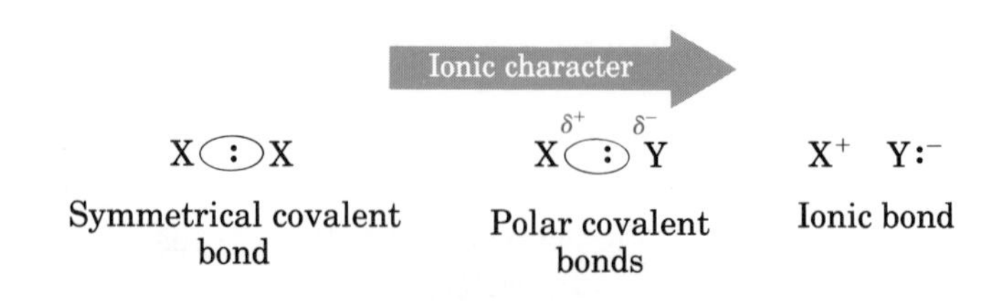

Figure 2.2. The continuum from covalent to ionic bonds. The symbol δ means *partial* charge.

TABLE 2.2 **Relative electronegativities of some common elements**[a]

Element	*Electronegativity*	*Element*	*Electronegativity*
H	2.2	Mg	1.3
Li	1.0	Al	1.6
Be	1.6	Si	1.9
B	2.0	P	2.2
C	2.5	S	2.6
N	3.0	Cl	3.1
O	3.4	Br	3.0
F	4.0	I	2.6
Na	1.0		

[a]The data in this table are Pauling-type electronegativity values as calculated by A. L. Allred, *J. Inorg. Nucl. Chem.*, **17:** 215 (1961). The values are on an arbitrary scale, with H = 2.2 and F ≐ 4.0, and have been rounded off to two significant figures. Carbon has an electronegativity value on this scale of 2.5; any element more electronegative than carbon has a value greater than 2.5, and any element less electronegative than carbon has a value less than 2.5.

electrons are equally shared between the two equivalent carbon atoms. The bond in sodium chloride, by contrast, is purely ionic and is the result of electrostatic attraction between positively charged sodium ions and negatively charged chloride ions. In between these two extremes lie the great majority of chemical bonds, in which the electrons are attracted *somewhat* more strongly by one atom than by the other. Such bonds are polar.

Bond polarity is due to the intrinsic **electronegativity** of the atoms involved. As shown in the electronegativity table (Table 2.2), carbon and hydrogen have similar electronegativities, and C—H bonds are therefore relatively nonpolar. Elements on the *right* side of the periodic table, such as oxygen, fluorine, and chlorine, are more electronegative than carbon; that is, they attract electrons more strongly than carbon. Thus, when carbon bonds to one of these elements, the bond is polarized so that the bonding electrons are drawn more toward the electronegative atom than toward carbon. This leaves carbon with a *partial positive charge* (denoted by δ^+; δ is the Greek letter delta) and the electronegative atom with a *partial negative charge* (δ^-). For example, a C—Cl bond is polar:

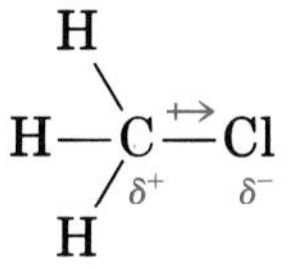

```
     H
      \  +→
  H—C—Cl
      / δ+   δ−
     H
```

Chloromethane

The arrow +→ is used to indicate the direction of polarity. By convention, *electrons move with the arrow;* the tail of the arrow is electron-poor (δ^+) and the head of the arrow is electron-rich (δ^-).

Elements on the *left* side of the periodic table are less electronegative than carbon and attract electrons less strongly. Thus, when carbon bonds to one of these elements, the bond is polarized so that carbon bears a partial negative charge and the other atom bears a partial positive charge. Organometallic compounds such as tetraethyllead, the "lead" in gasoline, provide good examples of this kind of polar bond.

$$\begin{array}{c} \overset{\delta^-}{CH_2}CH_3 \\ | \\ CH_3\overset{\delta^-}{CH_2}-\overset{\delta^+}{Pb}-\overset{\delta^-}{CH_2}CH_3 \\ | \\ \underset{\delta^-}{CH_2}CH_3 \end{array}$$

Tetraethyllead

When we speak of an atom's ability to cause bond polarization, we use the term **inductive effect.** Electropositive elements such as lithium and magnesium inductively *donate* electrons, whereas electronegative elements such as oxygen and chlorine inductively *withdraw* electrons. Inductive effects are of great importance and play a major role in our understanding of chemical reactivity. We will encounter them many times throughout this text in explanations of a wide variety of chemical phenomena.

2.5 Dipole moment

Since individual bonds are often polar, molecules as a whole are often polar also. This overall polarity results from the summation of all individual bond polarities and lone-pair contributions in the molecule. The measure of this net molecular polarity is a quantity called the **dipole moment.** We can view dipole moments in the following way: Assume that there is a center of gravity of all positive charges (nuclei) in a molecule. Assume also that there is a center of gravity of all negative charges (electrons) in the molecule. If these two centers do not coincide, then the molecule is electrically unsymmetrical and has a net polarity. The dipole moment, μ (Greek mu), is defined as the magnitude of a unit charge e times the distance d between the centers, and is expressed in debye units (D):

$$\mu = (e) \times (d) \times (10^{18})$$

where

$$\begin{aligned} e &= \text{Electric charge in electrostatic units (esu)} \\ d &= \text{Distance in centimeters} \end{aligned}$$

For example, if one proton and one electron (charge $e = 4.8 \times 10^{-10}$ esu) are separated from each other by 1 Å (10^{-8} cm), then

TABLE 2.3 **Dipole moments of some compounds**

Compound	*Dipole moment (D)*	*Compound*	*Dipole moment (D)*
NaCl	9.0	NH_3	1.47
$CH_3-N^{\oplus}(=O)-O^{\ominus}$ Nitromethane	3.46	CH_4	0
		CCl_4	0
		CH_3CH_3	0
CH_3Cl	1.87	Benzene	0
H_2O	1.85		
CH_3OH	1.70		
$H_2C=\overset{\oplus}{N}=N^{\ominus}$ Diazomethane	1.50	BF_3	0

$$\mu = (4.8 \times 10^{-10})(10^{-8}\ \text{cm})(10^{18}) = 4.8\ \text{D}$$

Experimentally, it is relatively easy to measure dipole moments, and examples are given in Table 2.3.

Sodium chloride (NaCl) has an extraordinarily large dipole moment because it is ionic. Nitromethane (CH_3NO_2) also has a large dipole moment, because it has formal charges on two atoms (i.e., it is dipolar). Water and ammonia (Figure 2.3) also have strong dipole moments and this too is easily explained. The electronegativity table (Table 2.2) shows that both oxygen and nitrogen are electron-withdrawing relative to hydrogen. In addition, we must not forget the lone-pair electrons, which normally make large contributions to the overall dipole moment since they have no atom attached to them to "neutralize" their negative charge.

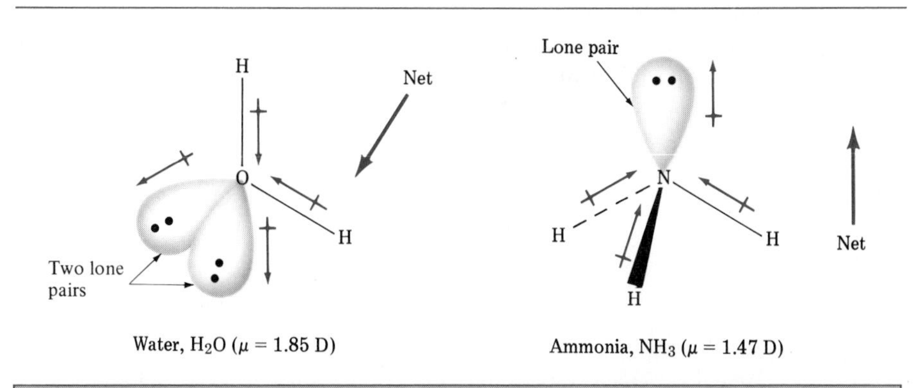

Figure 2.3. Dipole moments of water and ammonia.

By contrast, methane, tetrachloromethane, and ethane have zero dipole moments. Because of the symmetrical structures of these molecules, the individual bond polarities exactly cancel each other.

Methane	Tetrachloromethane	Ethane
(μ = 0 D)	(μ = 0 D)	(μ = 0 D)

PROBLEM

2.4 Account for the observed dipole moments of (a) methanol (CH_3OH, 1.70 D); (b) diazomethane (CH_2N_2, 1.50 D); and (c) benzene (0 D) on the basis of their known structures.

Benzene

2.6 Acids and bases: the Brønsted–Lowry definition

Acidity and **basicity** are related to the concepts of electronegativity and polarity just described, and it is a good idea to review these important topics at this point. We will soon see that the acid–base behavior of organic molecules helps explain much of their chemistry.

According to the **Brønsted–Lowry definition,** an acid is a substance that donates a proton (H^+), and a base is a substance that accepts a proton. For example, when HCl gas dissolves in water, an acid–base reaction occurs. Hydrogen chloride donates a proton, and water acts as a base to accept the proton. The products of the reaction, H_3O^+ and Cl^-, are called the **conjugate acid** and the **conjugate base,** respectively. Other common mineral acids such as sulfuric acid, nitric acid, and hydrogen bromide behave similarly, as do organic carboxylic acids such as acetic acid, CH_3COOH (Section 23.3).

$$\underset{\text{Acid}}{H{-}\ddot{\underset{..}{Cl}}{:}} + \underset{\text{Base}}{H{-}\ddot{O}{:}{-}H} \rightleftharpoons \underset{\text{Conjugate acid}}{H_3O^{+}} + \underset{\text{Conjugate base}}{{:}\ddot{\underset{..}{Cl}}{:}^{-}}$$

Acids differ in proton-donating ability, and we can measure these differences by determining the extent of their reaction with water. Strong acids such as HCl react almost completely with water, whereas weaker acids such as acetic acid react only slightly. Since these reactions are equilibrium processes, we can describe them using **equilibrium constants, K_{eq},**

$$HA + H_2O \rightleftharpoons H_3O^+ + A^-$$

$$K_{eq} = \frac{[H_3O^+][A^-]}{[HA][H_2O]}$$

where HA represents any acid.[1]

In the customary dilute solution used for measuring K_{eq}, the concentration of water, $[H_2O]$, remains nearly unchanged at approximately 55.5*M*. We can therefore rewrite the equilibrium expression using a new term called the **acidity constant, K_a.** The acidity constant for any generalized acid, HA, is simply the equilibrium constant multiplied by the molar concentration of water:

$$HA + H_2O \rightleftharpoons H_3O^+ + A^-$$

$$K_a = K_{eq}[H_2O] = \frac{[H_3O^+][A^-]}{[HA]}$$

Strong acids force the equilibrium to the *right,* and have large acidity constants, whereas weaker acids have their equilibrium toward the *left,* and have smaller acidity constants. We normally express acid strengths by quoting pK_a values, where the pK_a is equal to the negative logarithm of the acidity constant:

$$pK_a = -\log K_a$$

A strong acid (high acidity constant, K_a) has a *low* pK_a; conversely, a weak acid (low K_a) has a *high* pK_a. Table 2.4 lists the pK_a's of some common acids in order of strength.

[1]Recall that brackets, [ ], refer to the concentration of the enclosed species expressed in moles per liter.

TABLE 2.4 Relative strength of some common acids and their conjugate bases

	Acid	*Name*	pK_a	*Conjugate base*	*Name*	
Weak acid	CH_3CH_2OH	Ethanol	16.00	$CH_3CH_2O^-$	Ethoxide ion	Strong base
	H_2O	Water	15.74	HO^-	Hydroxide ion	
	HCN	Hydrocyanic acid	9.2	CN^-	Cyanide ion	
	CH_3COOH	Acetic acid	4.72	CH_3COO^-	Acetate ion	
	HF	Hydrofluoric acid	3.2	F^-	Fluoride ion	
	HNO_3	Nitric acid	−1.3	NO_3^-	Nitrate ion	
Strong acid	HCl	Hydrochloric acid	−7.0	Cl^-	Chloride ion	Weak base

Although we have only considered acids thus far, the same arguments can be used to measure the relative strengths of bases. Thus, the conjugate base of a strong acid must be a weak base, since it has little affinity for protons; similarly, the conjugate base of a weak acid must be a strong base, since it has a high affinity for protons. For example, chloride ion (the conjugate base of the strong acid, HCl) is a weak base. Acetate ion, however (conjugate base of the weaker acid, CH_3COOH), is a stronger base, and hydroxide ion (conjugate base of the weak acid, H_2O) is a still stronger base. Table 2.4 also lists the relative strengths of several common bases.

$$\underset{\text{Stronger acid}}{H{-}Cl} + \underset{\text{Stronger base}}{HO^-} \rightleftharpoons \underset{\text{Weaker acid}}{H_2O} + \underset{\text{Weaker base}}{\underset{\text{Chloride ion}}{Cl^-}}$$

$$\underset{\text{Stronger acid}}{CH_3COOH} + \underset{\text{Stronger base}}{HO^-} \rightleftharpoons \underset{\text{Weaker acid}}{H_2O} + \underset{\text{Weaker base}}{\underset{\text{Acetate ion}}{CH_3COO^-}}$$

2.7 Acids and bases: the Lewis definition

The Brønsted–Lowry concept of acidity is a useful one that can be extended to include organic compounds and, indeed, all compounds containing hydrogen. Of even more use to the organic chemist, however, is the Lewis definition of acids and bases: A **Lewis acid** is a substance that accepts an electron pair; a **Lewis base** is a substance that donates an electron pair. The Lewis definition of acidity is of wide applicability and includes not only proton donors but many other species as well. A proton (hydrogen ion) is a Lewis acid because it has a vacant *s* orbital and needs a pair of electrons to fill its empty valence shell. Such compounds as BF_3 and $AlCl_3$ are also Lewis acids, because they too have vacant orbitals that can accept electron pairs from Lewis bases, as Figure 2.4 shows.

Trivalent boron compounds such as BF_3 have only six electrons in their outer shells and thus can accept electron pairs from donors (Lewis bases) to form stable acid–base complexes. Similarly, $AlCl_3$ has only six electrons in its outer shell and is a powerful Lewis acid.

Perhaps more surprising is the fact that many transition-metal compounds, such as $TiCl_4$, $ZnCl_2$, $FeCl_3$, and $SnCl_4$, are excellent Lewis acids. The bonding in such metal compounds is more complex, since it involves *d*-orbital hybridization, but all Lewis acids have vacant, low-energy orbitals that can accept electron pairs.

The Lewis definition of basicity is self-explanatory and is quite similar to the Brønsted–Lowry definition: A Lewis base has a lone pair of electrons that it can donate to a Lewis acid in forming a new bond. Thus, H_2O, with its two lone pairs of electrons on oxygen, serves as a Lewis base by donating an

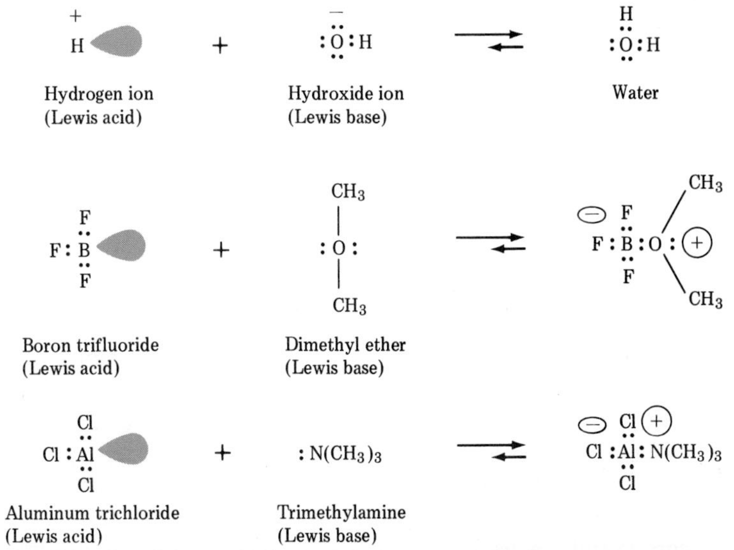

H^+ + $:\ddot{O}:H^-$ ⇌ $H:\ddot{O}:H$

Hydrogen ion (Lewis acid) | Hydroxide ion (Lewis base) | Water

BF_3 + $CH_3-\ddot{O}-CH_3$ ⇌ $F_3\overset{\ominus}{B}:\overset{\oplus}{O}(CH_3)_2$

Boron trifluoride (Lewis acid) | Dimethyl ether (Lewis base)

$AlCl_3$ + $:N(CH_3)_3$ ⇌ $Cl_3\overset{\ominus}{Al}:\overset{\oplus}{N}(CH_3)_3$

Aluminum trichloride (Lewis acid) | Trimethylamine (Lewis base)

Figure 2.4. Lewis acids and Lewis bases.

electron pair to a proton in forming the hydronium ion, H_3O^+:

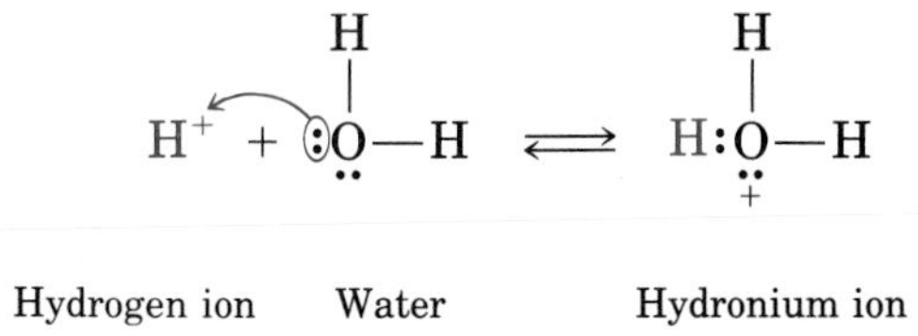

$$H^+ + :\ddot{O}(H)-H \rightleftharpoons H:\overset{+}{\ddot{O}}(H)-H$$

Hydrogen ion | Water | Hydronium ion

In a more general sense, most oxygen- and nitrogen-containing organic compounds are good Lewis bases, since they have lone pairs of available electrons. In the following formulas for organic Lewis bases, R represents an organic group.

$R-\ddot{O}-R$	Ether	$R-\ddot{O}-H$	Alcohol
$R_3N:$	Amine	$R_2C=\ddot{O}$	Ketone

PROBLEM

2.5 Explain by formal-charge calculations why the following molecules have the charges indicated:

(a) $F_3\overset{\ominus}{B}-\overset{\oplus}{O}(CH_3)_2$ (b) $Cl_3\overset{\ominus}{Al}-\overset{\oplus}{N}(CH_3)_3$

PROBLEM

2.6 Which of the structures in (a)–(f) are Lewis acids, and which are Lewis bases?

(a) $CH_3-\ddot{P}(CH_3)-CH_3$ (b) $MgBr_2$ (c) $CH_3-\ddot{S}-CH_3$

(d) $CH_3-\underset{\displaystyle CH_3}{\underset{|}{B}}-CH_3$ (e) $CH_3C\equiv N:$ (f)

```
              H
              |
     H      C      H
       \  //  \   /
        C       C
        |       ||
        C       C
       / \\   /  \
     H     N:      H
```

2.8 Analysis of organic compounds

In the late eighteenth century, only 30 or so elements were known, and the chemist faced with a newly isolated compound had great difficulty even identifying the elements present. A series of experiments on combustion carried out in 1772–1777 by Antoine Lavoisier[2] provided the first real breakthrough in the analysis of organic compounds. Lavoisier's techniques, though suitable for determining the identity of elements present in organic compounds, were by no means accurate enough to determine the elements' relative proportions.

The second breakthrough in organic analysis came in 1831 when Justus von Liebig[3] devised a method that is still used today. The key to Liebig's method was the recognition that organic compounds are efficiently burned on contact with red-hot copper oxide. For example, oxidation of benzene, C_6H_6, proceeds according to the following equation:

$$C_6H_6 + 15\ CuO \xrightarrow{900°C} 6\ CO_2 + 3\ H_2O + 15\ Cu$$

The water produced is swept by a stream of oxygen gas into a tube filled with calcium chloride, which retains the water. By weighing the tube before and after combustion, one can accurately determine the amount of water formed. The CO_2 produced passes through the $CaCl_2$ tube into a separate tube containing potassium hydroxide, KOH, where it is absorbed. Again, the amount of CO_2 present can be determined by weighing the tube before and after combustion, and the percentage composition of carbon and hydrogen present in the original sample can then be determined. The Liebig technique cannot directly determine the amount of oxygen present in a compound. If, however, no other elements are detected and the combined percentages of carbon and hydrogen do not total 100, then the percentage of oxygen present is taken as the difference.

Let's assume, for example, that we have analyzed a 0.55 gram (g) sample of a colorless organic liquid obtained by the distillation of wine. On weighing the $CaCl_2$ and KOH tubes, we find that 0.66 g H_2O and 1.037 g CO_2 have been formed. We can then calculate the percentages of carbon and hydrogen in the unknown sample by using the proper formulas:

[2]Antoine Lavoisier (1743–1794); b. Paris; studied at College Mazarin; considered the founder of modern chemistry; guillotined during French Revolution.

[3]Justus von Liebig (1803–1873); b. Darmstadt; Ph.D. at Erlangen in 1822; professor, Giessen (1824–1852), Munich.

$$\text{Weight of H in sample} = \text{Weight of } H_2O \times \frac{\text{Molecular weight of } H_2 \text{ (2.016)}}{\text{Molecular weight of } H_2O \text{ (18.016)}}$$

$$= (0.66)(0.112) = 0.074 \text{ g H}$$

$$\% \text{ H in sample} = \frac{\text{Weight of H}}{\text{Weight of sample}} = \frac{0.074}{0.55} = 13.44\%$$

$$\text{Weight of C in sample} = \text{Weight of } CO_2 \times \frac{\text{Molecular weight of C (12.01)}}{\text{Molecular weight of } CO_2 \text{ (44.01)}}$$

$$= (1.037)(0.273) = 0.283 \text{ g C}$$

$$\% \text{ C in sample} = \frac{\text{Weight of C}}{\text{Weight of sample}} = \frac{0.283}{0.55} = 51.47\%$$

Since the percentages of carbon and hydrogen add up to only 64.91%, we can assume that our sample also contains 35.09% oxygen. The next step is to determine the *atomic ratios* by dividing the percentage of each element by its atomic weight:

$$\begin{aligned} \text{C:}&\ 51.47\% \div 12.011 = 4.28 \\ \text{H:}&\ 13.44\% \div 1.008 = 13.33 \\ \text{O:}&\ 35.09\% \div 16.00 = 2.193 \end{aligned}$$

The atomic ratio of elements in our sample is C, 4.28 : H, 13.33 : O, 2.193, which reduces to C, 1.95 : H, 6.1 : O, 1. Rounding these numbers off gives us the **empirical formula** C_2H_6O. *Analysis gives us only atomic ratios.* To determine the **molecular formula,** which may be a multiple of the empirical formula, we also need to determine the molecular weight. In the present case, any higher multiple of C_2H_6O would be impossible by the rules of valency. Since carbon has a valence of four, an organic compound with n carbons can have no more than $2n + 2$ hydrogens (2 hydrogen atoms per carbon plus 1 hydrogen at each end of the chain):

$$H - \left(\begin{array}{c} H \\ | \\ C \\ | \\ H \end{array} - \begin{array}{c} H \\ | \\ C \\ | \\ H \end{array} \cdots \begin{array}{c} H \\ | \\ C \\ | \\ H \end{array} \right) - H$$

C_nH_{2n+2}

A formula such as $C_4H_{12}O_2$ for our unknown would be impossible, and we might therefore recognize our liquid to be ethyl alcohol, CH_3CH_2OH.

The Liebig method of analysis for carbon and hydrogen, and a similar method of analysis for nitrogen, introduced by Jean Dumas[4] in 1830, were remarkable achievements at the time, and they testify to the extraordinary experimental skill of their originators. They were, however, limited in their

[4]Jean Baptiste André Dumas (1800–1884); b. Alais, France; professor, École Polytechnique (1835), École de Médecine, Sorbonne.

usefulness by the large sample sizes required. Often it is practically impossible to obtain more than milligram amounts of new compounds, and an analysis that destroys half-gram amounts at a time is unthinkable.

The major limiting factor in the Liebig and Dumas analyses was the accuracy of the analytical balances used to weigh the samples and the collection tubes. As the science of chemistry developed, however, scientific instrumentation became more sophisticated. Under the leadership of Fritz Pregl,[5] a microbalance of great precision was developed, which allowed highly accurate weighing of submilligram amounts. Pregl further refined all aspects of the Liebig method and, in 1911, introduced a method of microanalysis that could be carried out on 5–10 milligram (mg) samples. For his accomplishments, Pregl received the Nobel prize in 1923.

Today, microanalysis of organic compounds is still carried out by the methods pioneered by Liebig, Dumas, and Pregl, although the techniques have become highly automated. Modern chemists do not consider a new compound to be fully characterized until accurate combustion analyses have been carried out, and many chemical journals still require such data before they publish new work.

PROBLEM

2.7 Calculate the percentage of each element in these molecular formulas:
(a) Benzene, C_6H_6
(b) Laetrile, $C_{14}H_{15}NO_7$
(c) Quinine, $C_{20}H_{24}N_2O_2$
(d) Diethylstilbestrol, $C_{18}H_{20}O_2$

PROBLEM

2.8 Citral is the chemical responsible for the odor of lemon. Combustion analysis shows citral to contain 78.9% C and 10.6% H. Assuming that the remainder is due to oxygen, what is the empirical formula of citral? If citral has a molecular weight of 152, what is its molecular formula?

PROBLEM

2.9 Squalene, isolated from shark oil, was submitted for combustion analysis. A sample weighing 8.00 mg gave 25.6 mg CO_2 and 8.75 mg H_2O. Calculate the empirical formula of squalene. If squalene has a molecular weight of 410, what is its molecular formula?

2.9 Summary

Chemists normally draw line-bond structures using a shorthand method in which carbons and most hydrogen atoms are not indicated. A carbon atom is assumed to be at the ends and at the intersections of lines (bonds), and the correct number of hydrogens is mentally supplied. For example,

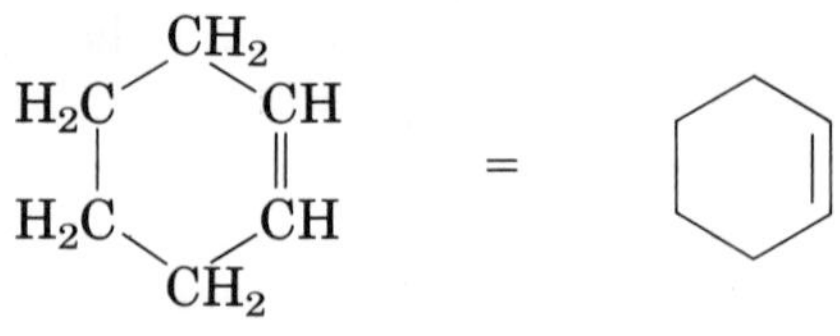

CH_2
H_2C CH
H_2C CH
CH_2
=

Cyclohexene

[5]Fritz Pregl (1869–1930); b. Laibach, Austria; Ph.D. Graz (1893); professor, Innsbruck, Graz; Nobel prize in chemistry (1923).

We use ⊕ and ⊖ signs to indicate the presence of formal charges on atoms. The concept of formal charges on specific atoms in neutral compounds is a bookkeeping matter that allows us to keep track of all outer-shell electrons. The equation for determining formal charge is

$$\text{Formal charge} = \begin{pmatrix}\text{Number of valence electrons}\\ \text{in the free atom}\end{pmatrix} - \begin{pmatrix}\text{Number of electrons}\\ \text{in the bound atom}\end{pmatrix}$$

Most covalent bonds are polar. Bond polarity is a result of unsymmetrical electron sharing and is due to the intrinsic electronegativity of atoms. For example, a carbon–chlorine bond is polar because chlorine attracts the shared electrons more strongly than carbon does. Carbon–metal bonds, however, are usually polarized in the opposite sense, since carbon attracts electrons more strongly than most metals. Carbon–hydrogen bonds are relatively nonpolar.

$$\overset{\delta^+}{\text{C}} \underset{\leftarrow\!\!+}{-} \overset{\delta^-}{\text{Cl}} \qquad \overset{\delta^-}{\text{C}} \underset{\leftarrow\!\!+}{-} \overset{\delta^+}{\text{Li}}$$

Molecules as a whole are also polar, and the amount of polarization of a molecule is called the dipole moment. Dipole moments represent a summation of all individual polar effects in a molecule and can be measured experimentally.

The concepts of acidity and basicity are related to polarity and electronegativity. A Lewis acid is a compound that has a low-energy unfilled orbital and accepts an electron pair; BF_3, $AlCl_3$, and H^+ are examples. A Lewis base is a compound that donates an unshared electron pair; NH_3 and H_2O are examples. Many organic molecules that contain oxygen and nitrogen are weak Lewis bases.

The analysis of organic compounds can be carried out accurately on milligram amounts of sample. The organic material is burned, and the combustion products are weighed to give information that can be used to establish the empirical formula of an unknown.

ADDITIONAL PROBLEMS

2.10 Convert the following structures into shorthand drawings:

(a)

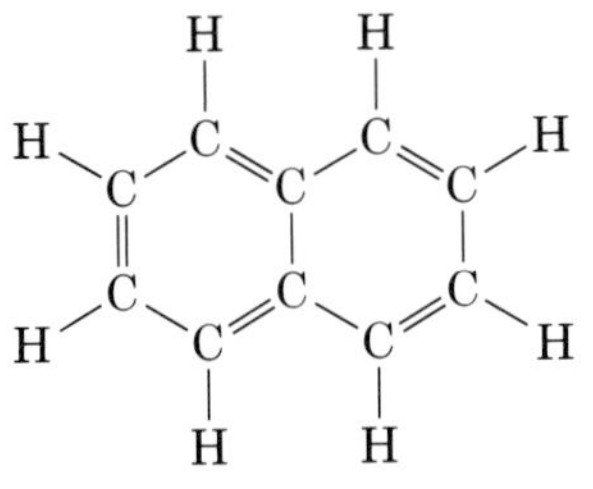

Naphthalene

(b)

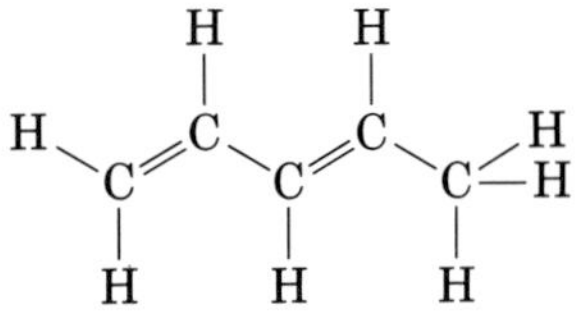

1,3-Pentadiene

(c)

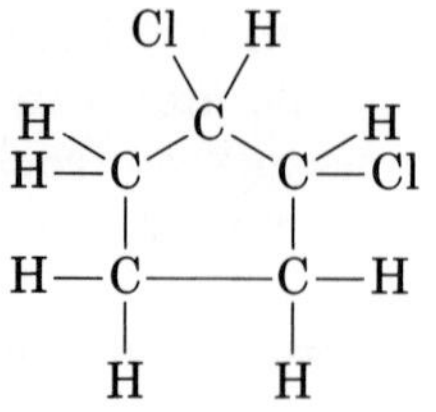

1,2-Dichlorocyclopentane

(d)

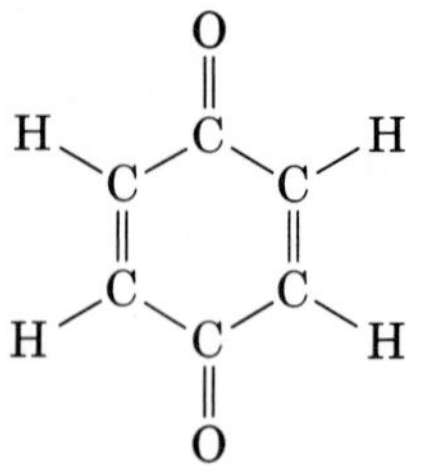

Quinone

(e)

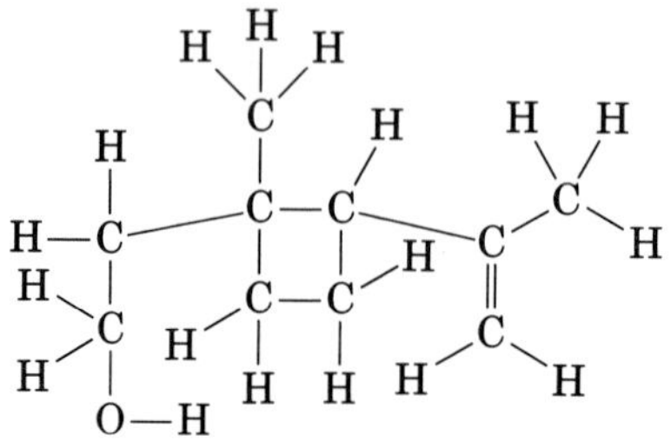

Grandisol
(an insect hormone)

2.11 Convert these shorthand drawings into Kekulé structures:

(a) CN

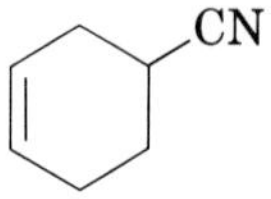

(b) Cl

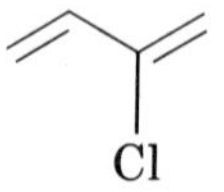

(c) CHO

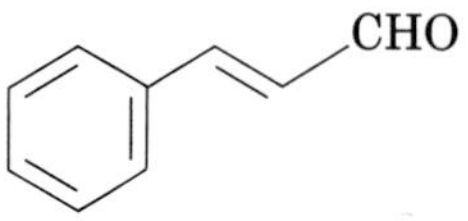

(d) COOH

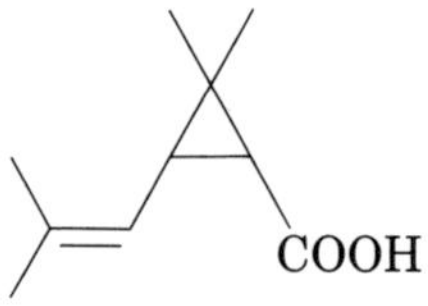

2.12 Calculate the formal charges on the atoms indicated:

(a) $(CH_3)_3O : BF_4$

(b) $H_2\ddot{C}—N{\equiv}N:$

(c) $H_2C{=}N{=}\ddot{N}:$

(d) $:\ddot{O}{=}\ddot{O}—\ddot{O}:$

(e) $H_2\ddot{C}—P(C_6H_5)_3$

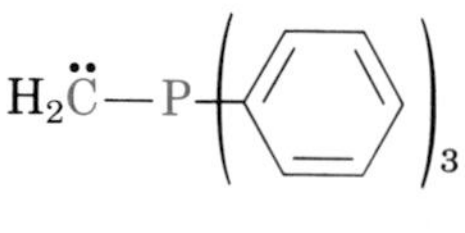

(f) :O:

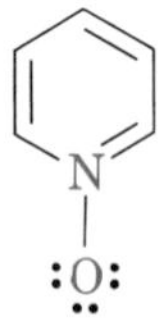

2.13 Indicate the expected direction of the dipole moments of these molecules:

(a) NH_2

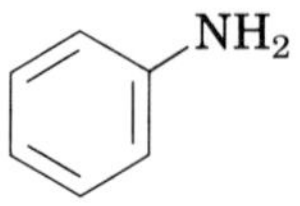

Aniline

(b) $(CH_3)_4Si$

Tetramethylsilane

(c) Cl / Cl C=C H \ H

cis-1,2-Dichloroethylene

(d) Cl / H C=C H \ Cl

trans-1,2-Dichloroethylene

(e) LiH

(f) $F_3B{-}N(CH_3)_3$

(g) C_6H_5OH (phenol)

(h) 1,2-$C_6H_4(OH)_2$

(i) HO–C_6H_4–OH (para)

2.14 Identify the acids and bases in these reactions:

(a) $CH_3OH + H^+ \longrightarrow CH_3\overset{+}{O}H_2$

(b) $CH_3OH + {}^-NH_2 \longrightarrow CH_3O^- + NH_3$

(c) $CH_3\overset{O}{\overset{\|}{C}}CH_3 + TiCl_4 \longrightarrow CH_3{-}\overset{\overset{+}{O}{-}\overset{-}{Ti}Cl_4}{\overset{\|}{C}}{-}CH_3$

(d) Cyclohexanone (H, H on C-2 and H, H on C-6) + NaH ⟶ enolate anion (carbanion at C-6 bearing H, Na^+) + H_2

(e) Morpholine (N–H) + BH_3 ⟶ morpholine–$\overset{-}{B}H_3$ adduct (H, $\overset{+}{N}$)

(f) $(CH_3)_3O^+\ BF_4^-$ + pyridine ⟶ N-methylpyridinium (CH_3, N^+) BF_4^- + CH_3OCH_3

2.15 Calculate the percentage of each element in the following formulas:

(a) Aspirin, $C_9H_8O_4$
(b) Patchouli alcohol (patchouli oil), $C_{15}H_{26}O$
(c) Muscone (musk oil), $C_{16}H_{30}O$
(d) Morphine, $C_{17}H_{19}NO_3$
(e) Strychnine, $C_{21}H_{22}N_2O_2$

2.16 α-Pinene, the main constituent of turpentine, has been shown by mass spectroscopy to have a molecular weight of 136. Combustion analytical data indicate that α-pinene contains 88.3% C and 11.6% H by weight. What is the molecular formula of α-pinene?

2.17 Jasmone, an odoriferous compound isolated from the jasmine flower, is valued for its use in perfumery. A pure sample of jasmone analyzes for 80.7% C and 9.7% H. Its molecular weight is 164. What other element is probably present in jasmone? What is the molecular formula of jasmone?

2.18 Myrcene is a colorless oil isolated from the oil of bay leaves. A sample of myrcene weighing 0.50 g was submitted for combustion analysis according to the classical Liebig method. It was found that 1.60 g CO_2 and 0.54 g H_2O were produced on oxidation. What is the empirical formula of myrcene? If myrcene has a molecular weight of 136, what is its molecular formula?

2.19 Progesterone, the so-called "pregnancy hormone," was isolated by Adolf Butenandt in 1934. By extracting the ovaries of 50,000 sows, Butenandt was able to isolate 20 mg of the pure hormone. Microanalysis of a 0.005 g sample by the Pregl techniques leads to the production of 0.0147 g CO_2 and 0.0041 g H_2O. What is the empirical formula for progesterone? Since we now know the molecular weight of progesterone to be 314, what is the molecular formula?

2.20 The Dumas method of analysis for nitrogen content in a molecule ultimately involves measuring the amount of nitrogen gas that is produced. Application of the gas laws tells us that 28 mg N_2 (1.0 millimole, mmol) has a volume of 22.4 milliliters (mL). By measuring the amount of N_2 derived from a sample of known weight, we can arrive at a value for the percentage of nitrogen in the sample.

Cadaverine, an aptly named amine with a molecular weight of 102, was analyzed for C, H, and N. Analysis of a 0.040 g sample yielded 0.086 g CO_2, 0.051 g H_2O, and 8.6 mL N_2 gas collected at standard temperature and pressure. How many grams of N_2 gas were produced? What is the percentage of N in the 0.040 g sample? Using this information, derive the molecular formula for cadaverine. What olfactory properties would you expect this molecule to possess?

2.21 Dimethyl sulfone has a high dipole moment (μ = 4.4 D). Calculate the formal charges present on oxygen and sulfur, and suggest a geometry that is consistent with the observed dipole moment.

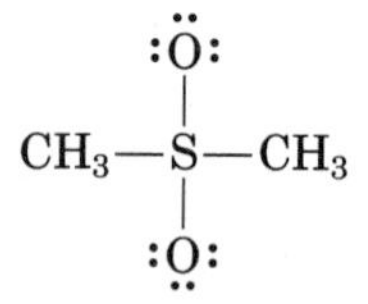

```
       :Ö:
        |
 CH3—S—CH3
        |
       :Ö:
```

Dimethyl sulfone

CHAPTER 3
THE NATURE OF ORGANIC COMPOUNDS: ALKANES

According to *Chemical Abstracts* (the invaluable service that abstracts and indexes the chemical literature), there are more than 7 million known organic compounds. It has been estimated that each organic chemist will, during his or her lifetime, prepare several hundred additional new compounds. Each of these compounds has its own unique physical properties such as melting point and boiling point, and each has its own unique chemical reactivity.

Chemists have learned through many years of experience that organic compounds can be classified into groups according to the kinds of structural features they contain, and that the chemical reactivity of the members of a certain group is often predictable. Instead of 7 million compounds with random reactivity, there are several dozen general classes of organic compounds whose chemistry is, at least to a first approximation, predictable.

3.1 Functional groups

The structural features that allow us to class compounds together by reactivity are called **functional groups.** A functional group is a part of a larger molecule; it is composed of an atom or group of atoms having a characteristic chemical reactivity. Chemically, a given functional group behaves in approximately the same way in all molecules of which it is a part. For example, one of the simplest functional groups is the carbon–carbon double bond. In Chapter 1 we examined the electronic formulation of a carbon–carbon double bond, and we saw that the bond consists of two parts. The carbon atoms are sp^2 hybridized and bond to each other by forming both a sigma bond (head-on overlap of an sp^2 orbital from each carbon) and a pi bond (sideways overlap of a *p* orbital from each carbon), as Figure 3.1 shows.

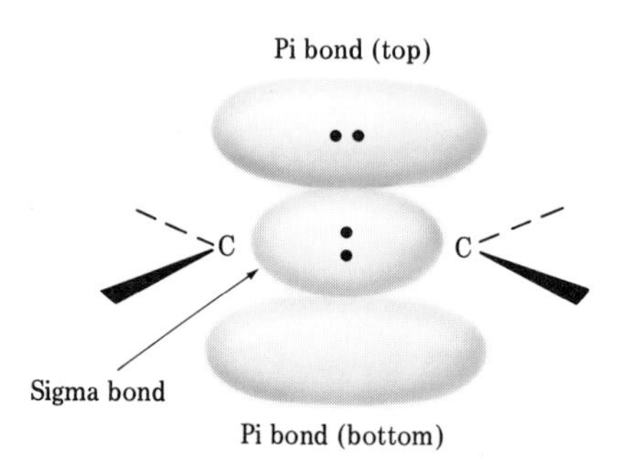

Figure 3.1. A carbon–carbon double bond.

Since the electronic nature of the carbon–carbon double bond changes little from one molecule to another, the bond's chemical reactivity also changes little. Ethylene, the simplest compound with a double bond, undergoes reactions that are remarkably similar to those of cholesterol, a much more complicated molecule. Both, for example, react with bromine to give *addition* products (Figure 3.2).

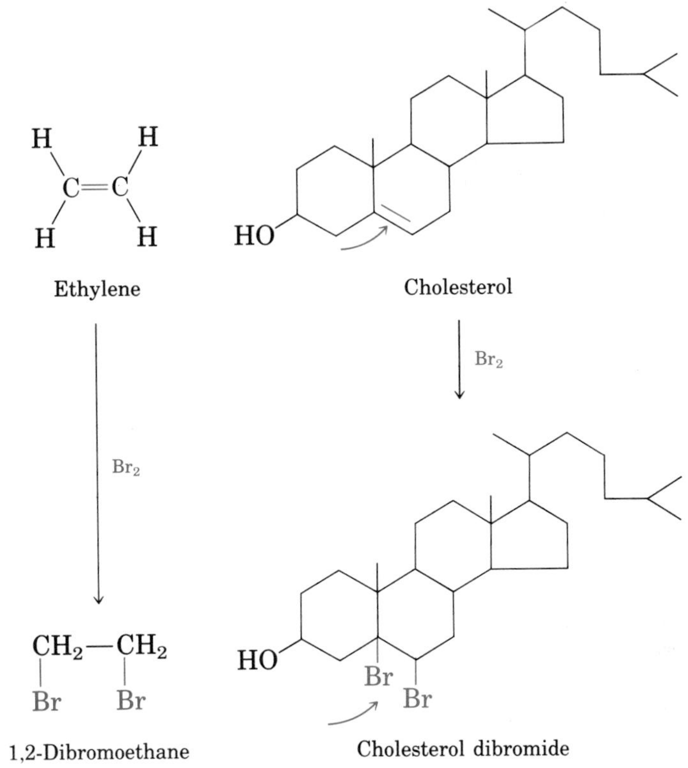

H H
C=C
H H

Ethylene

↓ Br_2

$CH_2—CH_2$
Br Br

1,2-Dibromoethane

HO

Cholesterol

↓ Br_2

HO
Br
Br

Cholesterol dibromide

Figure 3.2. The reaction of ethylene and cholesterol with bromine to give addition products.

This example is typical; the chemistry of a molecule is largely determined by the functional groups it contains. Table 3.1 lists many of the common functional groups found in organic molecules and gives simple examples of their occurrence. Look carefully at this table to see the wide variety of functional-group types that are found in organic compounds. Some functional groups, such as alkenes, alkynes, and aromatic rings, have only carbon and hydrogen and differ only in the number and kind of carbon–carbon multiple bonds they contain. Other functional groups have halogens, and still others have oxygen. The carbonyl functional group, consisting of a carbon–oxygen double bond, is unquestionably the most important functional group in organic chemistry, and many different subcategories exist, depending on what other atoms are bonded to the carbonyl carbon. We will see in Chapters 21–26, however, that all carbonyl-containing compounds behave similarly in many respects, regardless of their exact structure.

Nitrogen-, sulfur-, and metal-containing functional groups are also well known, and each specific group has a characteristic chemical reactivity. As we progress in our discussion of chemical reactions, the functional groups in Table 3.1 and their characteristic chemical reactivities will become second

TABLE 3.1 Some functional groups

Compound type	*Functional group*	*Simple example*	*Name ending*
Alkene (double bond)	$>C=C<$	$CH_3CH=CH_2$ Propene	*-ene*
Alkyne (triple bond)	$-C\equiv C-$	$CH_3C\equiv CH$ Propyne	*-yne*
Arene (aromatic ring)	(benzene ring)	C_6H_6 Benzene	None
Halide	$-\overset{\vert}{\underset{\vert}{C}}-X$ X = F, Cl, Br, I	CH_3CH_2I Iodoethane	None
Alcohol	$-\overset{\vert}{\underset{\vert}{C}}-O-H$	CH_3CH_2OH Ethanol	*-ol*
Ether	$-\overset{\vert}{\underset{\vert}{C}}-O-\overset{\vert}{\underset{\vert}{C}}-$	$CH_3CH_2OCH_2CH_3$ Ethoxyethane (diethyl ether)	*ether*
Carbonyl	$>C=O$		
Aldehyde	$-\overset{\vert}{\underset{\vert}{C}}-C(H)=O$	$CH_3-\overset{O}{\overset{\Vert}{C}}-H$ Ethanal (acetaldehyde)	*-al*
Ketone	$-\overset{\vert}{\underset{\vert}{C}}-C(=O)-\overset{\vert}{\underset{\vert}{C}}-$	$CH_3-\overset{O}{\overset{\Vert}{C}}-CH_3$ 2-Propanone (acetone)	*-one*
Ester	$-\overset{O}{\overset{\Vert}{C}}-O-\overset{\vert}{\underset{\vert}{C}}-$	$CH_3\overset{O}{\overset{\Vert}{C}}-OCH_2CH_3$ Ethyl ethanoate (ethyl acetate)	*-oate*

Compound type	*Functional group*	*Simple example*	*Name ending*
Carboxylic acid	$-\overset{\overset{O}{\|}}{C}-O-H$	$CH_3CH_2CH_2\overset{\overset{O}{\|}}{C}-OH$ Butanoic acid	*-oic acid*
Carboxylic acid chloride	$-\overset{\overset{O}{\|}}{C}-Cl$	$CH_3CH_2\overset{\overset{O}{\|}}{C}-Cl$ Propanoyl chloride	*-yl chloride*
Amide	$-\overset{\overset{O}{\|}}{C}-N\langle$	$CH_3\overset{\overset{O}{\|}}{C}-NH_2$ Ethanamide	*-amide*
Amine	$\rangle C-N\langle$	$CH_3CH_2NH_2$ Ethylamine	*-amine*
Nitrile	$-C\equiv N$	$CH_3C\equiv N$ Ethanenitrile (acetonitrile)	*-nitrile*
Nitro	$\rangle C-\overset{+}{N}(=O)-O^-$	$CH_3CH_2NO_2$ Nitroethane	None
Sulfide	$\rangle C-S-C\langle$	CH_3-S-CH_3 Dimethyl sulfide	*sulfide*
Sulfoxide	$\rangle C-\underset{+}{\overset{\overset{O^-}{\|}}{S}}-C\langle$	$CH_3-\underset{+}{\overset{\overset{O^-}{\|}}{S}}-CH_3$ Dimethyl sulfoxide	*sulfoxide*
Sulfone	$\rangle C-\underset{\underset{O^-}{\|}}{\overset{\overset{O^-}{\|}}{S^{2+}}}-C\langle$	$CH_3-\underset{\underset{O^-}{\|}}{\overset{\overset{O^-}{\|}}{S^{2+}}}-CH_3$ Dimethyl sulfone	*sulfone*
Organometallic	$\rangle C-M$ M = metal	CH_3-Li Methyllithium	None

nature. Much of this book is devoted to the study of these functional groups, and most of the chemistry we will be studying is the chemistry of functional groups.

PROBLEM

3.1 Circle and identify the functional groups present in each of the following molecules:

(a) Cl–C_6H_4–CH(CCl_3)–C_6H_4–Cl

DDT

(b) $C_6H_5CH_2CH(NH_2)COOH$

Phenylalanine

(c) $CH_2{=}CH{-}CHO$

Acrolein

(d) $C_6H_5CH{=}CH_2$

Styrene

3.2 Alkanes and alkyl groups

We have already examined the chemical bonding in ethane. The carbon–hydrogen bonds result from the overlap of hydrogen 1*s* orbitals with carbon sp^3 hybrid orbitals, and the carbon–carbon bond results from overlap of two carbon sp^3 orbitals. It is possible to continue mentally replacing carbon–hydrogen bonds with carbon–carbon bonds and in so doing to generate a large number of molecules of *increasing chain length,* as Figure 3.3 shows.

CH_4 (Methane) ⇒ $H{-}CH_2{-}CH_3$ (Ethane) ⇒ $H{-}CH_2{-}CH_2{-}CH_3$ (Propane) ⇒ $H{-}CH_2{-}CH_2{-}CH_2{-}CH_3$ (Butane) ⇒ $H{-}CH_2{-}CH_2{-}CH_2 \cdots CH_2{-}H$

Methane Ethane Propane Butane

Figure 3.3. The increasing chain length of normal alkanes, C_nH_{2n+2}.

The compounds generated are called **straight-chain alkanes** or **normal alkanes,** and all have the general formula C_nH_{2n+2}, where n is any integer. These compounds are also occasionally referred to as **aliphatic,** derived from the Greek word *aleiphas*, "fat." As we shall see in Chapter 32, animal fats do indeed contain long carbon chains similar to alkanes. A given alkane may be depicted arbitrarily in a great many ways. For example, the straight-chain four-carbon alkane is called butane and may be represented by *any* of the structures in Figure 3.4. These structures do not imply any particular geometry for butane; they indicate only that butane has a chain of four carbons.

$$CH_3—CH_2—CH_2—CH_3 = CH_3—CH_2—CH_2(—CH_3) = CH_3—CH_2(—CH_2—CH_3) = CH_3(CH_2)_2CH_3$$

$$= CH_3—CH_2—CH_2—CH_3 \text{ (drawn with two vertical } CH_3 \text{ groups)} = CH_3\,CH_2\,CH_2\,CH_3 \text{ (zigzag)} = \text{(line structure)}$$

Figure 3.4. Some representations of butane. The molecule is the same regardless of how it is drawn.

In practice we soon tire of drawing bonds at all and refer to butane as $CH_3CH_2CH_2CH_3$ or simply as n-C_4H_{10} (here n signifies *normal,* straight-chain butane). Alkanes are named according to the number of carbon atoms in the chain, as shown in Table 3.2.

The naming system used is a rational one. With the exception of the first four compounds—methane, ethane, propane, and butane—whose names have historical roots, the alkanes are named from Greek numbers according to how many carbons are present. The suffix *-ane* is added to the end of each name to indicate that the molecule identified is an alkane.

If one hydrogen atom is removed from an alkane, the residue is called an **alkyl group.** Alkyl groups are named by replacing the *-ane* ending of the

TABLE 3.2 **Alkane names**

Number of carbons (n)	*Name*	*Formula* (C_nH_{2n+2})	*Number of carbons (n)*	*Name*	*Formula* (C_nH_{2n+2})
1	Methane	CH_4	14	Tetradecane	$C_{14}H_{30}$
2	Ethane	C_2H_6	15	Pentadecane	$C_{15}H_{32}$
3	Propane	C_3H_8	20	Eicosane	$C_{20}H_{42}$
4	Butane	C_4H_{10}	21	Heneicosane	$C_{21}H_{44}$
5	Pentane	C_5H_{12}	22	Docosane	$C_{22}H_{46}$
6	Hexane	C_6H_{14}	23	Tricosane	$C_{23}H_{48}$
7	Heptane	C_7H_{16}	30	Triacontane	$C_{30}H_{62}$
8	Octane	C_8H_{18}	31	Hentriacontane	$C_{31}H_{64}$
9	Nonane	C_9H_{20}	32	Dotriacontane	$C_{32}H_{66}$
10	Decane	$C_{10}H_{22}$	40	Tetracontane	$C_{40}H_{82}$
11	Undecane	$C_{11}H_{24}$	50	Pentacontane	$C_{50}H_{102}$
12	Dodecane	$C_{12}H_{26}$	60	Hexacontane	$C_{60}H_{122}$
13	Tridecane	$C_{13}H_{28}$			

TABLE 3.3 **Some straight-chain alkyl groups**

Alkane	*Alkyl group*	*Example*
CH_4 Methane	CH_3— Methyl	CH_3OH Methyl alcohol
CH_3CH_3 Ethane	CH_3CH_2— Ethyl	$CH_3CH_2NH_2$ Ethylamine
$CH_3CH_2CH_3$ Propane	$CH_3CH_2CH_2$— or n-C_3H_7 Propyl	$CH_3CH_2CH_2Li$ Propyllithium
$CH_3(CH_2)_8CH_3$ Decane	$CH_3(CH_2)_8CH_2$— or n-$C_{10}H_{21}$ Decyl	$CH_3(CH_2)_8CH_2NH_2$ Decylamine

parent alkane by the *-yl* ending. For example, removal of a hydrogen from methane, CH_4, generates the **methyl group,** CH_3—. The combination of alkyl groups with some of the functional groups listed earlier allows us to name many hundreds of thousands of compounds systematically. For example, the removal of a *terminal* hydrogen atom from the *n*-alkanes (again, *normal* alkanes) produces the series of alkyl groups shown in Table 3.3.

CH_4 (H—C—H with H above and below)
Methane

H—C(H)(H)— (wavy bond)
Methyl group[1]

CH_3—Br
Bromomethane (methyl bromide)

CH_3—OH
Methanol (methyl alcohol)

CH_3—NH_2
Methylamine

CH_3—O—C(=O)—CH_3
Methyl acetate

Methane and ethane each have only one kind of hydrogen; that is, the four hydrogens in CH_4 are all equivalent, and the six hydrogens in C_2H_6 are all equivalent. No matter which of the four methane hydrogens or six ethane hydrogens we remove, only one kind of methyl group and one kind of ethyl group result. The situation becomes more complex in higher alkanes, however. Propane and butane have two kinds of hydrogens, types A and B; pentane and hexane have three kinds of hydrogens, types A, B, and C; and so on (Figure 3.5).

We mentally generated the homologous series of straight-chain alkanes (Table 3.2) by successively replacing a terminal hydrogen of a lower alkane with a methyl group. It is equally possible to imagine replacing *internal*

[1]The symbol ⦚ indicates that the partial organic structure shown is bonded to an unspecified partial structure.

$$\underset{A}{CH_3}-\underset{B}{CH_2}-\underset{A}{CH_3}$$

Propane

$$\underset{A}{CH_3}-\underset{B}{CH_2}-\underset{B}{CH_2}-\underset{A}{CH_3}$$

Butane

$$\underset{A}{CH_3}\underset{B}{CH_2}\underset{C}{CH_2}\underset{B}{CH_2}\underset{A}{CH_3}$$

Pentane

$$\underset{A}{CH_3}\underset{B}{CH_2}\underset{C}{CH_2}\underset{C}{CH_2}\underset{B}{CH_2}\underset{A}{CH_3}$$

Hexane

Figure 3.5. Propane and higher alkanes. Letters denote different types of hydrogens.

hydrogen atoms with alkyl groups and to generate thereby a vast number of **branched-chain alkanes.**

Beginning with propane, for example, we can replace a terminal hydrogen by a methyl group to generate butane, or we can replace an internal hydrogen by a methyl group to generate the branched four-carbon alkane, 2-methylpropane or *isobutane*. Both butane and isobutane have the same formula, C_4H_{10}. Compounds with the same molecular formula but different chemical structures are called **isomers.**

$$H_3C-CH_2-CH_3 \longrightarrow H_3C-CH_2-CH_2-CH_3 \text{ (Butane)}$$

$$H_3C-CH_2-CH_3 \longrightarrow H_3C-CH(CH_3)-CH_3 \text{ (2-Methylpropane (isobutane))}$$

Butane

2-Methylpropane
(isobutane)

There are an enormous number of possibilities for branching in the alkane series. Although there is only one methane, one ethane, and one propane, there are *two* butane isomers, *three* pentane isomers, *five* hexane isomers, and so on. As Table 3.4 shows, there are more than 62 trillion possible isomers of $C_{40}H_{82}$! Fortunately, not all have been characterized.

TABLE 3.4 Possible alkane isomers

<table>
<tr><th>Formula</th><th>Isomers</th><th></th><th></th></tr>
<tr><td>C_1H_4</td><td>CH_4</td><td></td><td></td></tr>
<tr><td>C_2H_6</td><td>CH_3CH_3</td><td></td><td></td></tr>
<tr><td>C_3H_8</td><td>$CH_3CH_2CH_3$</td><td></td><td></td></tr>
<tr><td>C_4H_{10}</td><td>$CH_3CH_2CH_2CH_3$</td><td>CH_3
│
CH_3CHCH_3</td><td></td></tr>
<tr><td>C_5H_{12}</td><td>$CH_3CH_2CH_2CH_2CH_3$</td><td>CH_3
│
$CH_3CHCH_2CH_3$</td><td>CH_3
│
$CH_3—C—CH_3$
│
CH_3</td></tr>
<tr><td>C_6H_{14}</td><td>$CH_3CH_2CH_2CH_2CH_2CH_3$</td><td>CH_3
│
$CH_3CHCH_2CH_2CH_3$</td><td>CH_3
│
$CH_3CH_2CHCH_2CH_3$</td></tr>
<tr><td></td><td>CH_3 CH_3
│ │
$CH_3CH—CHCH_3$</td><td>CH_3
│
$CH_3—C—CH_2CH_3$
│
CH_3</td><td></td></tr>
</table>

Formula	*Number of isomers*	*Formula*	*Number of isomers*
C_7H_{16}	9	$C_{12}H_{26}$	355
C_8H_{18}	18	$C_{15}H_{32}$	4,347
C_9H_{20}	35	$C_{20}H_{42}$	366,319
$C_{10}H_{22}$	75	$C_{30}H_{62}$	4,111,846,763
$C_{11}H_{24}$	159	$C_{40}H_{82}$	62,491,178,805,831

PROBLEM

3.2 Propose structures that meet the following descriptions:
(a) Five isomers with formula C_8H_{18}
(b) Two isomeric esters with formula $C_5H_{10}O_2$
(c) Two isomeric nitriles with formula C_4H_7N

PROBLEM

3.3 How many isomers are there that have the following structures?
(a) Alcohols with formula C_3H_8O
(b) Bromoalkanes with formula C_4H_9Br
(c) Dichloroalkanes with formula $C_4H_8Cl_2$

3.3 Nomenclature of alkanes

In earlier times when relatively few pure organic chemicals were known, new compounds were named at the whim of their discoverer. Thus urea (CH_4N_2O) is a pure crystalline substance isolated from urine; morphine

($C_{17}H_{19}NO_3$) is an analgesic (a painkiller) isolated by Sertürner[2] in 1805 from the opium poppy and named after Morpheus, the Greek god of dreams; and citric acid ($C_6H_8O_7$) was isolated by Scheele[3] from the lemon, a citrus fruit. As the science of organic chemistry slowly grew in the nineteenth century, so too did the need for a systematic method for unambiguously naming organic compounds.

Structural drawings provide an unambiguous means for one chemist to describe a specific compound to another chemist, but drawing structures is time-consuming, and structures do not lend themselves readily to information storage and retrieval systems.

If an ideal and universally acceptable system of nomenclature were worked out, it would be possible for a chemist to name uniquely every chemical compound so that every other chemist would understand what is meant. Equally important, it would be possible for a chemist to read a name and attach a unique structure to that name. Great strides have been made toward accomplishing these goals, though perfection has not yet been attained. The system of nomenclature we will use in this book has been devised by the International Union of Pure and Applied Chemistry, IUPAC. IUPAC rules are available for unambiguously naming all but the most complex structures and for dealing with all functional groups. As we cover new functional groups in later chapters, the applicable IUPAC rules of nomenclature will be given. For the present, let's see how we can name branched-chain alkanes.

According to IUPAC rules, most branched-chain alkanes can be named by following four steps. For a few very complex alkanes, a fifth step is needed.

Step 1. Find the parent hydrocarbon:

a. Find the *longest continuous carbon chain* present in the molecule and use the name of that chain as the parent name. The longest chain may not always be apparent from the manner of writing; you may have to "turn corners."

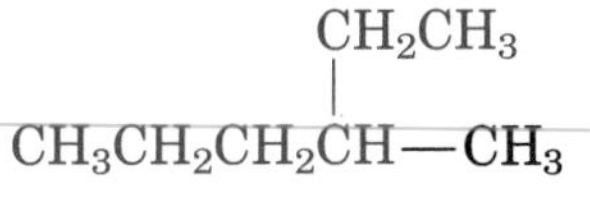

```
                CH2CH3
                |
CH3CH2CH2CH—CH3          Named as a substituted hexane

        CH3
        |
        CH2
        |
CH3—CHCH—CH2CH3          Named as a substituted heptane
          |
          CH2CH2CH3
```

[2]Friedrich Wilhelm Adam Sertürner (1783–1841); b. Neuhaus, Germany; apothecary in Paderborn, Eimbeck, and Hameln.

[3]Karl Wilhelm Scheele (1742–1786); b. Stralsund, Sweden; apothecary in Gothenburg, Malmö, Stockholm, Uppsala, and Köping.

b. If two different chains of equal length are present, select the one with the larger number of branch points as the parent:

```
CH3
|
CH3CHCHCH2CH2CH3
    |
    CH2CH3
```

Named as a hexane with *two* substituents

```
CH3
|
CH3CH—CHCH2CH2CH3
      |
      CH2CH3
```

NOT

as a hexane with *one* substituent

Step 2. Number the atoms:

a. Beginning at the end *nearer the first branch point,* number each carbon atom in the longest chain you have identified:

```
       1 CH3
         |
       2 CH2
         |
CH3—CHCH—CH2CH3
     3  |4
        CH2CH2CH3
        5   6   7
```

NOT

```
       7 CH3
         |
       6 CH2
         |
CH3—CHCH—CH2CH3
     5  |4
        CH2CH2CH3
        3   2   1
```

The first branch occurs at C3 in the proper numbering system. In the improper system, it occurs at C4.

b. If there is branching the same distance away from both ends of the chain, begin numbering at the end nearer the *second* branch point:

```
     9   8
     CH3CH2        CH3  CH2CH3
         |          |    |
CH3—CHCH2CH2CH—CHCH2CH3
     7  6   5   4    3  2   1
```

NOT

```
     1   2
     CH3CH2        CH3  CH2CH3
         |          |    |
CH3—CHCH2CH2CH—CHCH2CH3
     3  4   5   6    7  8   9
```

The longest chain in this example is a nonane, and the first branch from either end is at C3. In the correct numbering system, however, the second branch is at C4; in the incorrect system, the second branch is at C6.

Step 3. Number the substituents:

a. Using the numbering system you have decided is correct, assign a number to each substituent corresponding to its point of attachment to the main chain.

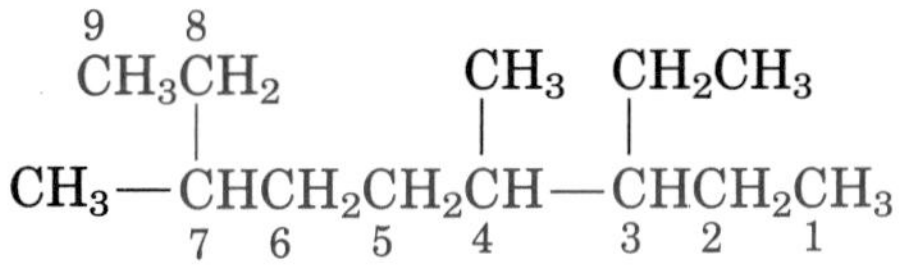

```
       9   8
       CH3CH2          CH3  CH2CH3
           |            |    |
CH3—CHCH2CH2CH—CHCH2CH3
      7  6   5   4     3  2   1
```

Named as a nonane

Substituents: On C3, CH_2CH_3 (3-ethyl)
On C4, CH_3 (4-methyl)
On C7, CH_3 (7-methyl)

b. It may sometimes happen that there are two substituents on the same carbon. This presents no problem; simply assign them both the same number. (There must always be as many numbers in the name as there are substituents.)

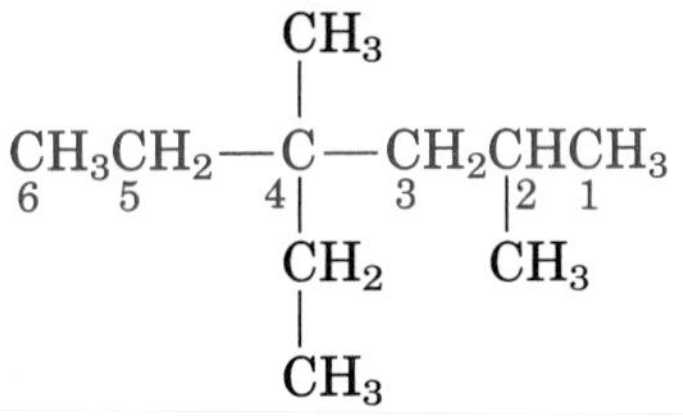

```
              CH3
               |
CH3CH2—C—CH2CHCH3
  6    5     4|    3    |2  1
              CH2     CH3
               |
              CH3
```

Named as a hexane

Substituents: On C2, CH_3 (2-methyl)
On C4, CH_3 (4-methyl)
On C4, CH_2CH_3 (4-ethyl)

Step 4. Write out the name: If two or more *different* side chains are present, cite them in alphabetical order and write out the complete name of the compound. Use the prefixes *di-*, *tri-*, *tetra-*, and so forth if the same substituent occurs more than once, but do not use these prefixes for alphabetizing purposes. Full names for some of the examples we have been using follow.

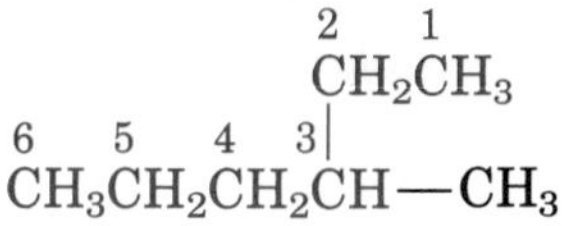

```
                2   1
               CH2CH3
6   5    4   3|
CH3CH2CH2CH—CH3
```

3-Methylhexane

```
             1 CH3
                |
             2 CH2
                |
CH3—CHCH—CH2CH3
          3  |4
             CH2CH2CH3
              5    6    7
```

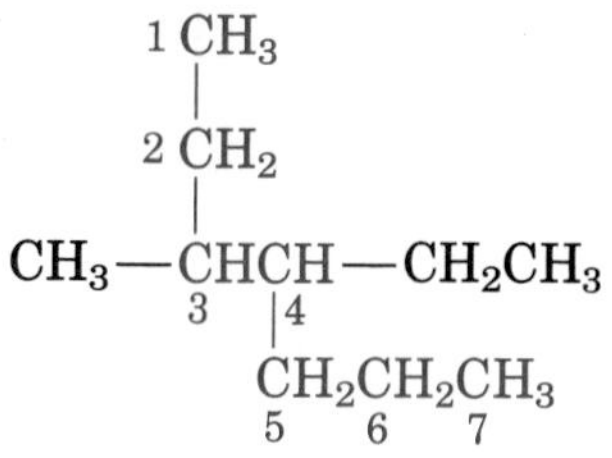

4-Ethyl-3-methylheptane

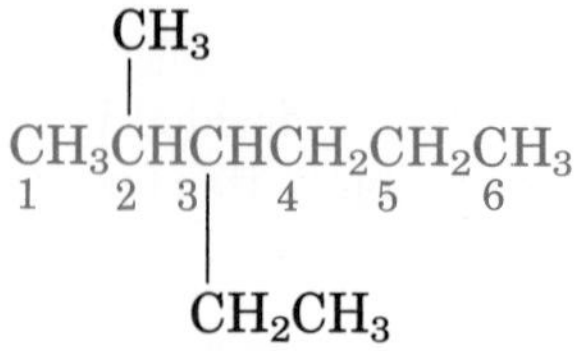

$$\begin{array}{l} \quad\ CH_3 \\ \quad\ \ | \\ CH_3CHCHCH_2CH_2CH_3 \\ 1\ \ \ 2\ \ 3|\ \ \ 4\ \ \ 5\ \ \ 6 \\ \qquad\quad | \\ \qquad CH_2CH_3 \end{array}$$

3-Ethyl-2-methylhexane

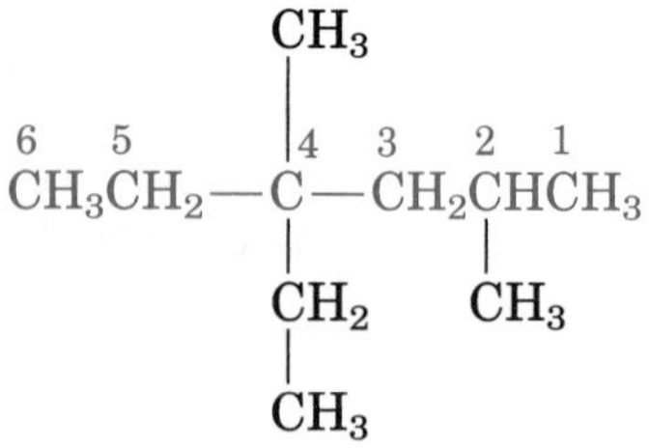

$$\begin{array}{c} CH_3 \\ | \\ \overset{6\ \ \ 5}{CH_3CH_2}-\overset{4}{C}-\overset{3\ \ \ 2\ \ 1}{CH_2CHCH_3} \\ |\qquad\qquad | \\ CH_2 \qquad CH_3 \\ | \\ CH_3 \end{array}$$

4-Ethyl-2,4-dimethylhexane

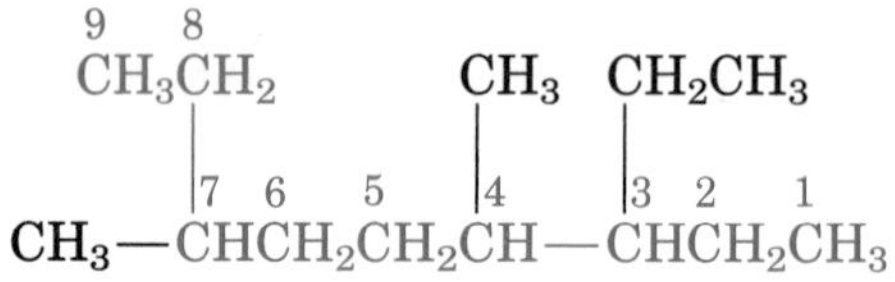

$$\begin{array}{l} \overset{9\ \ 8}{CH_3CH_2} \qquad\quad CH_3\ \ CH_2CH_3 \\ \qquad | \qquad\qquad\quad | \qquad | \\ CH_3-\overset{7}{C}H\overset{6}{C}H_2\overset{5}{C}H_2\overset{4}{C}H-\overset{3}{C}H\overset{2}{C}H_2\overset{1}{C}H_3 \end{array}$$

3-Ethyl-4,7-dimethylnonane

Proper application of these four steps allows us to name many thousands of organic compounds. In some particularly complex cases, however, a further step is necessary. It occasionally happens that a substituent of the main chain has sub-branching:

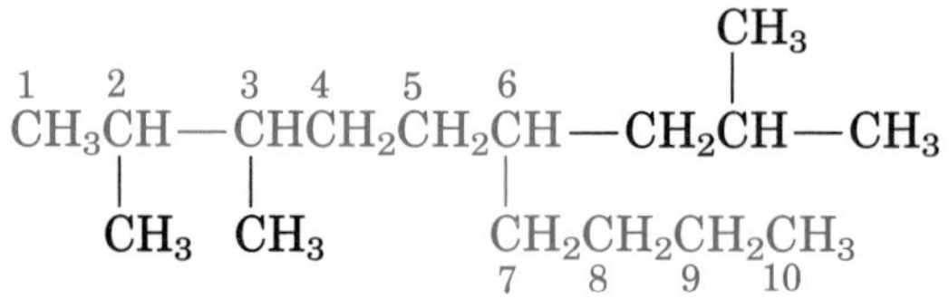

$$\begin{array}{l} \qquad\qquad\qquad\qquad\qquad\qquad\qquad CH_3 \\ \qquad\qquad\qquad\qquad\qquad\qquad\qquad\ | \\ \overset{1}{C}H_3\overset{2}{C}H-\overset{3}{C}H\overset{4}{C}H_2\overset{5}{C}H_2\overset{6}{C}H-CH_2CH-CH_3 \\ \qquad\ |\qquad |\qquad\qquad\quad | \\ \quad CH_3\ \ CH_3 \qquad \underset{7}{C}H_2\underset{8}{C}H_2\underset{9}{C}H_2\underset{10}{C}H_3 \end{array}$$

Named as a 2,3,6-trisubstituted decane

In this case, the substituent at C6 is a four-carbon unit with a sub-branch. To name the compound fully, we must first name the complex substituent.

Step 5. Name the complex substituent: Complex substituents are named by application of the four primary steps exactly as if they were compounds themselves. In the present case, our complex substituent is a substituted propane:

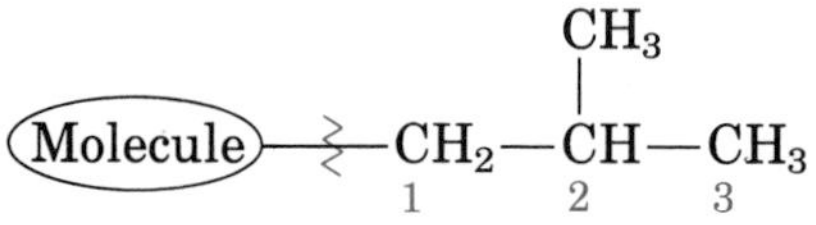

$$\begin{array}{l} \qquad\qquad\qquad\qquad\quad CH_3 \\ \qquad\qquad\qquad\qquad\qquad | \\ \text{(Molecule)}-\!\!\wr\!-\underset{1}{C}H_2-\underset{2}{C}H-\underset{3}{C}H_3 \end{array}$$

We begin numbering at the point of attachment to the main chain and find that we have a 2-methylpropyl substituent. This is set off in parentheses when the name of the complete hydrocarbon is given:

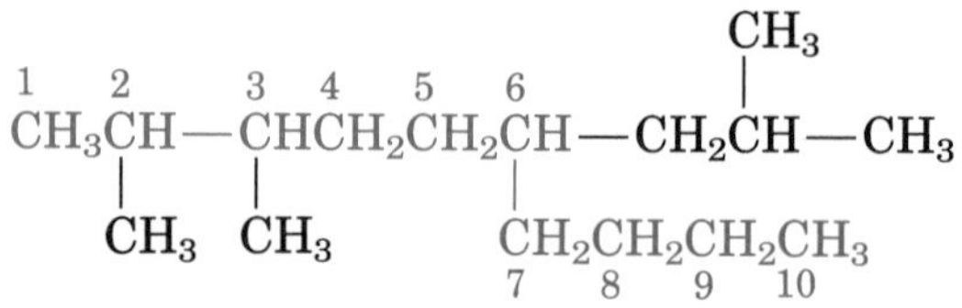

$$\begin{array}{l} \qquad\qquad\qquad\qquad\qquad\qquad\qquad CH_3 \\ \qquad\qquad\qquad\qquad\qquad\qquad\qquad\ | \\ \overset{1}{C}H_3\overset{2}{C}H-\overset{3}{C}H\overset{4}{C}H_2\overset{5}{C}H_2\overset{6}{C}H-CH_2CH-CH_3 \\ \qquad\ |\qquad |\qquad\qquad\quad | \\ \quad CH_3\ \ CH_3 \qquad \underset{7}{C}H_2\underset{8}{C}H_2\underset{9}{C}H_2\underset{10}{C}H_3 \end{array}$$

2,3-Dimethyl-6-(2-methylpropyl)decane

As a further example:

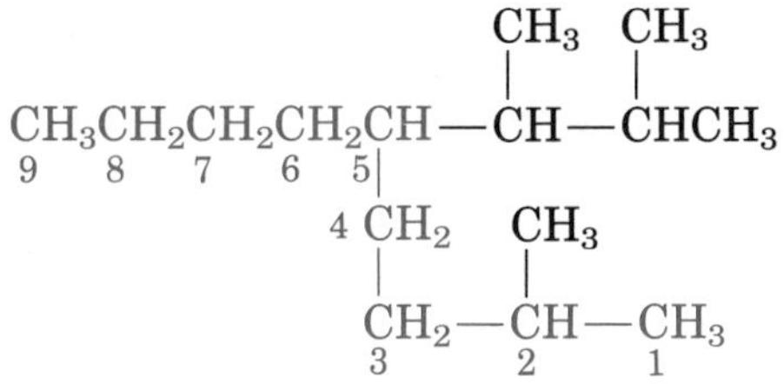

```
                          CH3  CH3                    CH3  CH3
                          |    |                      |    |
CH3CH2CH2CH2CH—CH—CHCH3              —§—CH—CHCH3
 9   8   7   6   5|                                       1    2  3
              4 CH2   CH3
                |     |                       5-(1,2-Dimethylpropyl)-
              CH2—CH—CH3
               3    2    1
```

2-Methyl-5-(1,2-dimethylpropyl)nonane

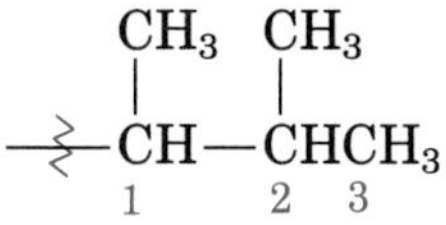

PROBLEM

3.4 Provide proper IUPAC names for the following compounds:

(a) The three isomers of C_5H_{12}

(b) 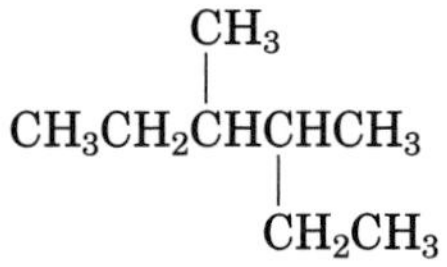
```
            CH3
            |
CH3CH2CHCHCH3
          |
          CH2CH3
```

(c) 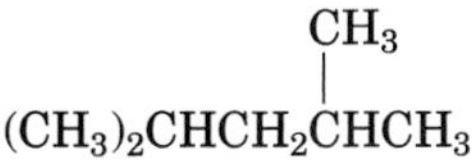
```
                   CH3
                   |
(CH3)2CHCH2CHCH3
```

(d) 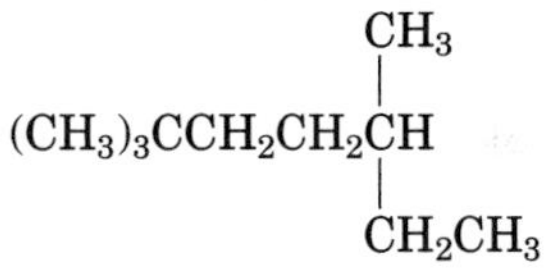
```
                       CH3
                       |
(CH3)3CCH2CH2CH
                       |
                       CH2CH3
```

PROBLEM

3.5 Draw structures corresponding to these IUPAC names:

(a) 3,4-Dimethylnonane
(b) 3-Ethyl-4,4-dimethylheptane
(c) 2,2-Dimethyl-4-propyloctane
(d) 2,2,4-Trimethylpentane

PROBLEM

3.6 The following names are incorrect. Draw the structures they represent, explain why the names are incorrect, and give correct names.

(a) 1,1-Dimethylpentane
(b) 3-Methyl-2-propylhexane
(c) 4,4-Dimethyl-3-ethylpentane
(d) 5-Ethyl-4-methylhexane
(e) 2,3-Methylhexane
(f) 3-Dimethylpentane

3.4 Nomenclature of alkyl groups

Earlier in this chapter, we saw that straight-chain alkyl groups are formed by the removal of a terminal hydrogen atom from straight-chain alkanes. It is equally possible to generate an enormous number of *branched* alkyl groups by removing *internal* hydrogen atoms. For example, there are two possible three-carbon alkyl groups and four possible four-carbon alkyl groups, as Figure 3.6 shows. The possibilities expand at an enormous rate as the number of carbon atoms increases, but all alkyl groups can be rationally and uniquely named by carefully following the steps we have discussed.

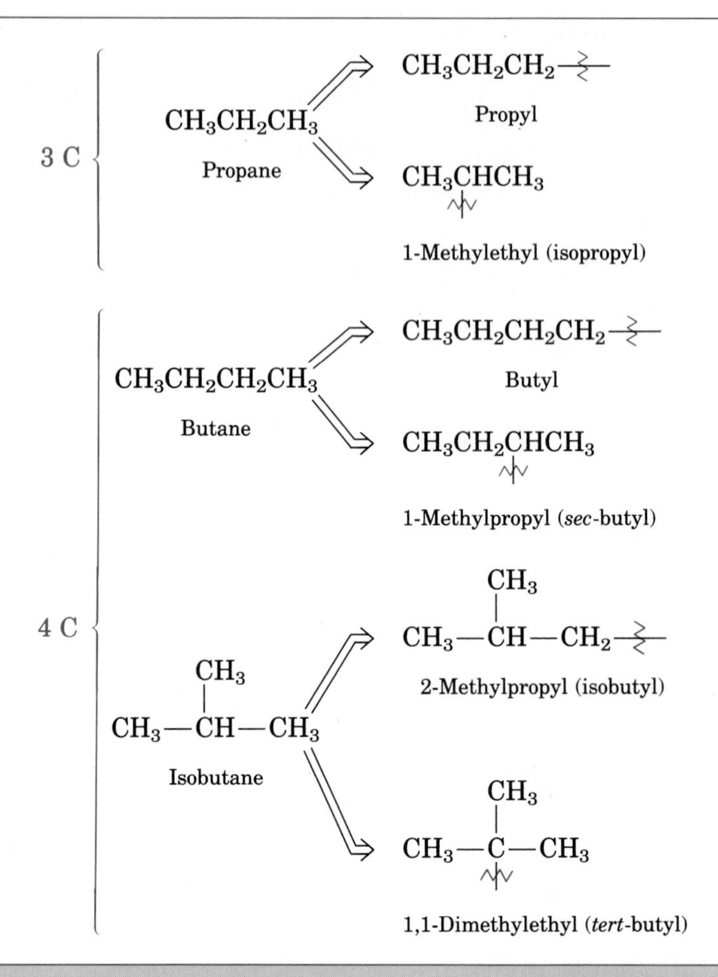

Figure 3.6. Generation of branched-chain alkyl groups from *n*-alkanes.

For historical reasons, some of the simpler alkyl groups have retained nonsystematic or *trivial* names:

1. Three-carbon unit:

$$\begin{array}{l} CH_3 \\ \quad \backslash \\ \quad CH- \\ \quad / \\ CH_3 \end{array}$$

Isopropyl

2. Four-carbon units:

$$\begin{array}{c} CH_3CH_2CH- \\ | \\ CH_3 \end{array}$$

sec-Butyl (for *sec*ondary)

$$\begin{array}{c} CH_3CHCH_2- \\ | \\ CH_3 \end{array}$$

Isobutyl

$$\begin{array}{c} CH_3 \\ | \\ CH_3-C- \\ | \\ CH_3 \end{array}$$

tert-Butyl or *t*-butyl (for *tert*iary)

3. Five-carbon units:

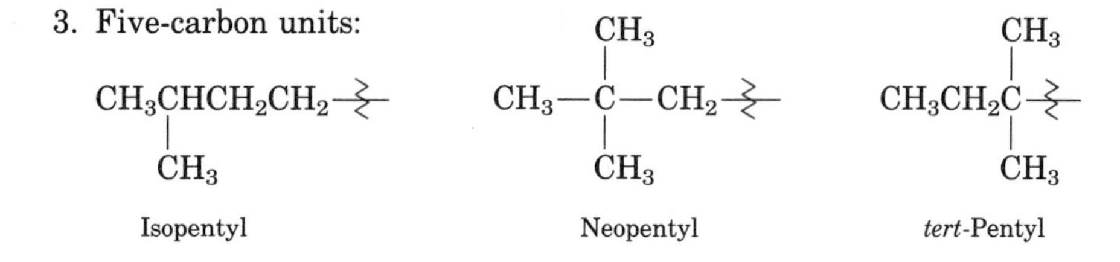

Isopentyl Neopentyl *tert*-Pentyl

These names are so well entrenched in chemical journals, textbooks, chemical catalogs, and common usage that the IUPAC system of nomenclature makes allowance for them. Thus

$$\begin{array}{c} CH_3\ \ CH_3 \\ \diagdown\ \diagup \\ CH \\ | \\ CH_3CH_2CH_2CHCH_2CH_2CH_3 \end{array}$$

4-(1-Methylethyl)heptane

may be properly named 4-isopropylheptane. There is no recourse but to memorize these trivial names; fortunately they are few.

One further word of explanation is necessary: The prefixes *sec-* (for secondary) and *tert-* (for tertiary) refer to the *degree of alkyl substitution* at the carbon atom in question. There are four possible substitution patterns for carbon:

$$\begin{array}{c} H \\ | \\ R-C-H \\ | \\ H \end{array}$$ *Primary* carbon (1°); one substituent on carbon

$$\begin{array}{c} H \\ | \\ R-C-H \\ | \\ R \end{array}$$ *Secondary* carbon (2°); two substituents on carbon

$$\begin{array}{c} R \\ | \\ R-C-H \\ | \\ R \end{array}$$ *Tertiary* carbon (3°); three substituents on carbon

$$\begin{array}{c} R \\ | \\ R-C-R \\ | \\ R \end{array}$$ *Quaternary* carbon (4°); four substituents on carbon

The symbol R is used here and throughout this text to represent a generalized alkyl group. The group R may stand for methyl, ethyl, propyl, or any of an infinite number of other alkyl groups.

We often use these terms when speaking, and their meanings must become second nature. For example, when we say "The product of the reaction is a tertiary alcohol," we are talking about

```
      R
      |
   R—C—OH
      |
      R
```

PROBLEM

3.7 Draw and name the eight possible five-carbon alkyl groups (pentyl isomers).

PROBLEM

3.8 Draw and name alkanes that meet the following descriptions:
(a) An alkane with three tertiary carbons
(b) An alkane with two isopropyl groups
(c) An alkane with one quaternary and one secondary carbon
(d) A secondary chloride

PROBLEM

3.9 Identify the kinds of carbon and hydrogen as primary, secondary, tertiary, or quaternary:
(a) $(CH_3)_2CHCH_2C(CH_3)_3$
(b) $(CH_3)_2CHCH(CH_3)CH_2CH_2CH_3$
(c) $CH_3CH_2C(CH_3)_2C(CH_3)_2CH_2CH_3$

3.5 Occurrence of alkanes: petroleum

Many alkanes occur naturally in the plant and animal world. For example, cabbage leaves contain nonacosane (n-$C_{29}H_{60}$), and the wood oil of the Jeffrey pine common to the Sierra Nevada contains heptane. Beeswax contains, among other things, hentriacontane (n-$C_{31}H_{64}$). By far the major sources of alkanes are the world's natural gas and petroleum deposits. Laid down eons ago, these natural deposits are derived from the decomposition of organic matter of largely marine origin. Natural gas consists chiefly of methane, but ethane, propane, butane, and isobutane are also present. These simple hydrocarbons are used in great quantities to heat our homes, cook our food, and fuel some of our industries. Petroleum is a highly complex mixture of hydrocarbons that must be *refined* into different fractions before it can be used. Refining is a complicated process that includes both distillation of petroleum into cuts of different boiling points (bp), and chemical manipulation of the molecules to form new compounds. Refining begins by **fractional distillation** of crude petroleum into three principal cuts: straight-run gasoline (bp 30–200°C), kerosene (bp 175–300°C), and gas oil (bp 275–400°C). Finally, distillation under reduced pressure gives lubricating oils and waxes, and leaves an undistillable tarry residue of asphalt. (See Figure 3.7.)

A major part of the petroleum consumed is used for automobile fuel, and the simple distillation of petroleum into fractions with different boiling

	Fraction	Composition
Petroleum	Asphalt	
	Lubricating oil Waxes	
	Gas oil (bp 275–400°C)	C_{14}–C_{25} hydrocarbons
	Kerosene (bp 175–300°C)	C_{11}–C_{14} hydrocarbons
	Straight-run gasoline (bp 30–200°C)	C_5–C_{11} hydrocarbons
	Natural gas	C_1–C_4 hydrocarbons

Figure 3.7. The products of petroleum refining.

points is only the first step. Straight-run gasoline is a rather poor fuel, because of the phenomenon of engine knock. In the common internal-combustion engine, the downward intake stroke of the piston causes a precisely metered mixture of air and fuel to be drawn into the cylinder. The piston then compresses this mixture on its upward stroke and, at a certain point just before the end of this compression, the spark plug ignites the mixture, and smooth combustion occurs. Not all fuels burn equally well, however. When poor fuels are used, combustion can be initiated in an uncontrolled manner by a hot surface in the cylinder before the spark plug fires. This *preignition,* detected as an engine knock or ping, can destroy the engine in short order by putting excessive and irregular forces on the crankshaft and by raising engine temperature.

The fuel **octane number** is the measure by which the antiknock properties of fuels are judged. It was recognized early that straight-chain hydrocarbons are far more prone to induce engine knock than are highly branched compounds. Heptane, which is a particularly bad fuel, is assigned a base value of 0 octane number; 2,2,4-trimethylpentane (trivially known as isooctane), which has excellent antiknock characteristics, is given a rating of 100.

$$CH_3CH_2CH_2CH_2CH_2CH_2CH_3$$

Heptane

Octane number = 0

$$\begin{array}{c} CH_3\quad CH_3 \\ |\qquad\quad | \\ CH_3\underset{\underset{CH_3}{|}}{C}CH_2CHCH_3 \end{array}$$

2,2,4-Trimethylpentane (isooctane)

Octane number = 100

Since straight-run gasoline has a high percentage of unbranched alkanes and is therefore a poor fuel, petroleum chemists have devised sophisticated methods for producing higher-quality fuels. Two such methods are known as **catalytic cracking** and **catalytic reforming.** The actual chemistry taking place is extremely complex, but, in essence, catalytic cracking involves taking the high-boiling kerosene cut (C_{11}–C_{14}) and "cracking" it into smaller molecules suitable for use in gasoline. The process takes place on a silica–alumina catalyst at temperatures of 400–500°C, and the major products are light hydrocarbons in the C_3–C_5 range. These small hydrocarbons are then catalytically recombined to yield useful C_7–C_{10} alkanes. Fortunately, the C_7–C_{10} molecules that are produced are highly branched and are perfectly suited for use as high-octane fuels.

Catalytic reforming is a process by which straight-chain alkanes present in straight-run gasoline are converted into aromatic molecules such as toluene and benzene. Aromatics have high octane ratings and are therefore desirable components of gasoline.

3.6 Properties of alkanes

Alkanes are often referred to as **paraffin** hydrocarbons. The word paraffin is derived from the Latin *parum affinis* ("slight affinity"), which aptly describes the behavior of alkanes; alkanes show little chemical affinity for other molecules and are chemically inert to most reagents. As Table 3.5 indicates, the alkanes show regular increases in both boiling point and

TABLE 3.5 **Physical properties of some alkanes**

Number of carbons	*Alkane*	*Melting point (°C)*	*Boiling point (°C)*	*Density (g/mL)*
1	Methane	−182.5	−164.0	0.5547
2	Ethane	−183.3	−88.6	0.509
3	Propane	−189.7	−42.1	0.5005
4	Butane	−138.3	−0.5	0.5788
5	Pentane	−129.7	36.1	0.6262
6	Hexane	−95.0	68.9	0.6603
7	Heptane	−90.6	98.4	0.6837
8	Octane	−56.8	125.7	0.7025
9	Nonane	−51.0	150.8	0.7176
10	Decane	−29.7	174.1	0.7300
11	Undecane	−25.6	195.9	0.7402
12	Dodecane	−9.6	216.3	0.7487
13	Tridecane	−5.5	235.4	0.7564
14	Tetradecane	5.9	253.7	0.7628
20	Eicosane	36.8	343.0	0.7886
30	Triacontane	65.8	450.0	0.8097
4	Isobutane	−159.4	−11.7	0.579
5	Isopentane	−159.9	27.85	0.6201
5	Neopentane	−16.5	9.5	0.6135
8	Isooctane	−107.4	99.3	0.6919

melting point as molecular weight increases. This regularity is reflected in other properties; the average carbon–carbon bond parameters are nearly the same in all alkanes, with bond lengths of 1.54 ± 0.01 Å and bond strengths of 85 ± 3 kcal/mol. The carbon–hydrogen bond parameters are also nearly constant at 1.09 ± 0.01 Å and 95 ± 3 kcal/mol.

Table 3.5 also shows that increased branching has the effect of lowering the boiling point. Thus, pentane boils at 36.1°C, isopentane (2-methylbutane) has one branch and boils at 27.85°C, and neopentane (2,2-dimethylpropane) has two branches and boils at 9.5°C. Similarly, octane boils at 125.7°C, whereas isooctane (2,2,4-trimethylpentane) boils at 99.3°C. This effect can be understood by looking at what occurs during boiling.

Nonpolar molecules such as alkanes are weakly attracted to each other by intermolecular **van der Waals forces.** These forces operate only at very small distances and result from *induced polarization* of the electron clouds in molecules. On the average, the electron distribution in a molecule is uniform over a period of time. At any given instant, however, the electrons are not uniformly distributed. One side of a molecule may, by chance, have a slight excess of electrons relative to the opposite side. When that occurs, the molecule has a *temporary dipole moment.* This temporary dipole in one molecule causes a nearby molecule to adopt a temporarily opposite dipole, and a tiny attraction is induced between the two molecules (Figure 3.8). These temporary dipoles have a fleeting existence and are constantly changing, but the cumulative effect of an enormous number of these interactions produces attractive forces sufficient to cause the molecules to stay in the liquid state rather than the gaseous state. Only when sufficient heat energy is applied to overcome these forces does the liquid boil.

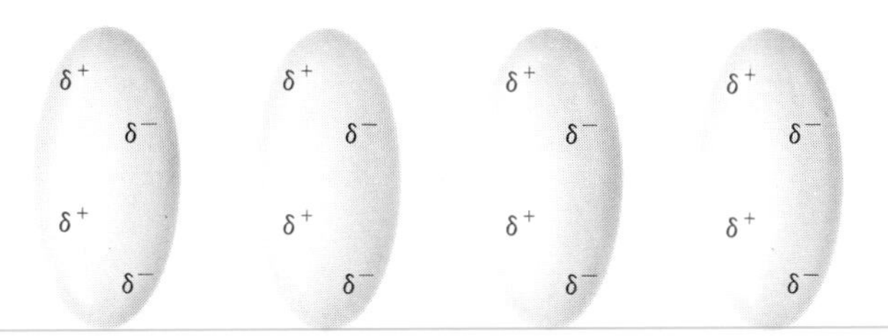

Figure 3.8. Attractive van der Waals forces caused by temporary dipolar attraction between molecules.

We might expect that van der Waals forces would increase as molecule size increases, and this expectation is borne out in the alkane series. Although other factors are also involved, at least part of the increase in boiling point on going up the alkane series is due to increased van der Waals forces. The effect of branching can also be explained by invoking van der Waals forces: Branched alkanes are more nearly spherical than straight-chain alkanes. As a result, they have smaller surface areas, fewer van der Waals forces, and consequently lower boiling points.

3.7 Conformations of ethane

We saw earlier that methane has a tetrahedral structure, and that the carbon–carbon bonds in alkanes result from sigma overlap of two tetrahedral carbon sp^3 orbitals. Let's now look into the three-dimensional consequences of such bonding. What are the spatial relationships between the hydrogens on one carbon and the hydrogens on the other carbon?

We know that sigma bonds result from the head-on overlap of two atomic orbitals, and that they are *cylindrically symmetrical.* In other words, a cross section of a sigma orbital is circular (Figure 3.9). As a consequence of this sigma bond symmetry, there is *free rotation* around the carbon–carbon single bond. Bond overlap is exactly the same for all geometric arrangements of the hydrogens (Figure 3.10).

The different arrangements of atoms caused by rotation about a single bond are called **conformations,** and a specific structure is called a **conformer** (**conform**ational iso**mer**). Unlike other kinds of isomers, however, different conformers cannot usually be isolated, since they interconvert too rapidly.

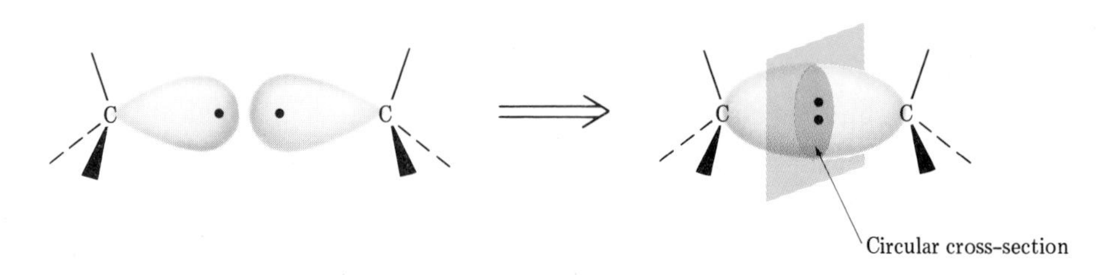

Figure 3.9. Sigma bond symmetry.

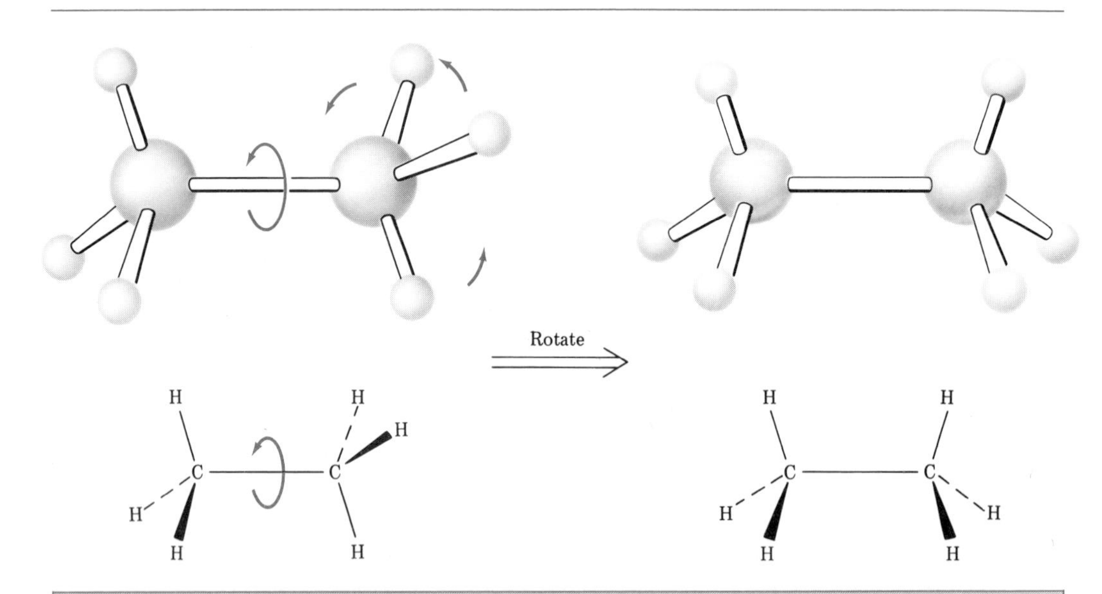

Figure 3.10. Some conformations of ethane.

Chemists have adopted two ways of representing conformational isomers. **Sawhorse representations** view the carbon–carbon bond from an oblique angle and indicate spatial orientation by employing heavy tapered lines for substituents coming out of the page, normal lines for substituents in the plane of the page, and dashed lines for substituents going back behind the plane of the page. **Newman[4] projections** view the carbon–carbon bond directly end-on and represent the two carbon atoms by a circle. Substituents on the front carbon are represented by lines going to the center of the circle, and substituents on the rear carbon are represented by lines going to the edge of the circle. The advantage of Newman projections is that the relationship between substituents on the different carbon atoms is easily seen (Figure 3.11).

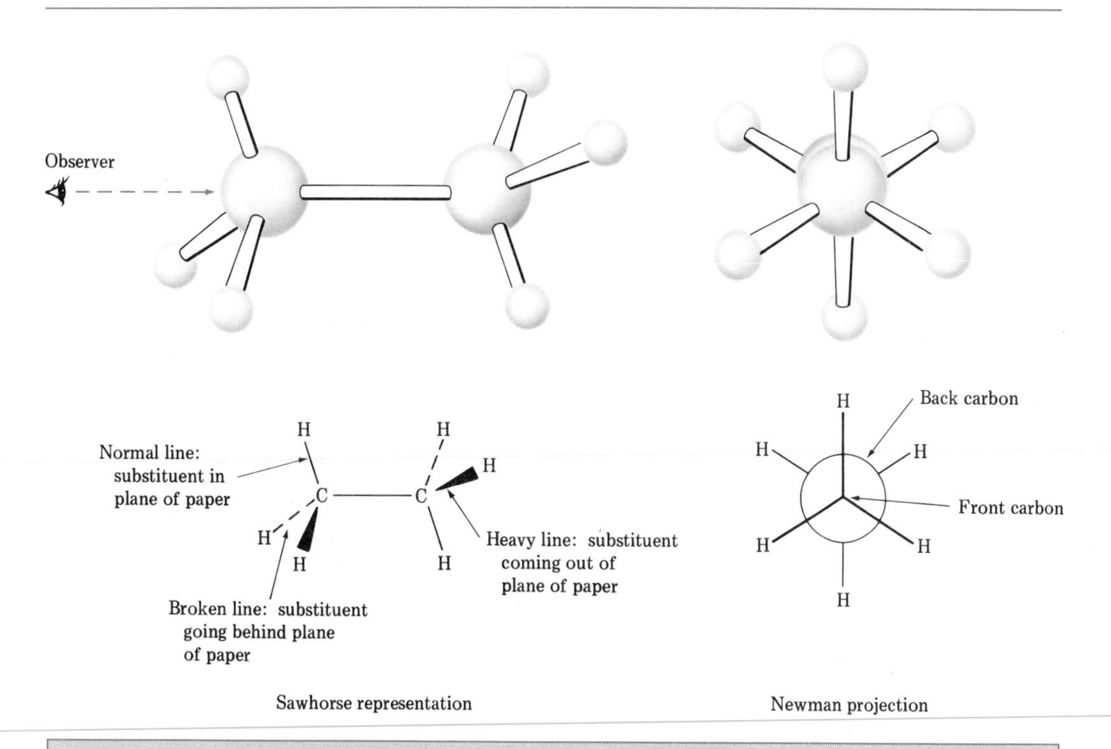

Figure 3.11. Sawhorse and Newman projections of ethane.

In view of the symmetry of the sigma bond, it is perhaps surprising that we do not observe *perfectly* free rotation in ethane. Experiments show that there is a slight (2.9 kcal/mol) barrier to rotation due to the fact that some conformations are more stable than others. The lowest-energy, most stable conformation is called **staggered** and has all six carbon–hydrogen bonds as far away from each other as possible. The highest-energy, least stable conformation is called **eclipsed** and has the carbon–hydrogen bonds as close as possible ("eclipsing" in a Newman projection). Any conformation

[4]Melvin S. Newman (1908–); b. New York; Ph.D. (1932), Yale University; professor, Ohio State University.

partway between staggered and eclipsed is referred to as a **skew** conformation. Since the magnitude of the barrier to rotation is 2.9 kcal/mol, and since the barrier is caused by three equal hydrogen–hydrogen eclipsing interactions, we can assign a value of approximately 1 kcal/mol for each single interaction.

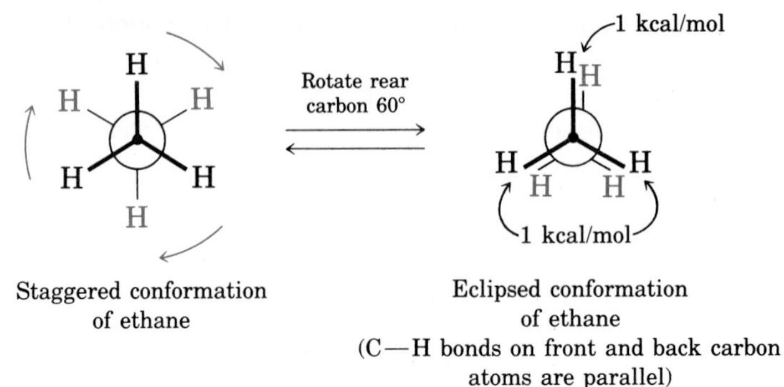

The 2.9 kcal/mol of **strain energy** present in the eclipsed ethane conformation is called **torsional strain,** and the barrier to rotation that

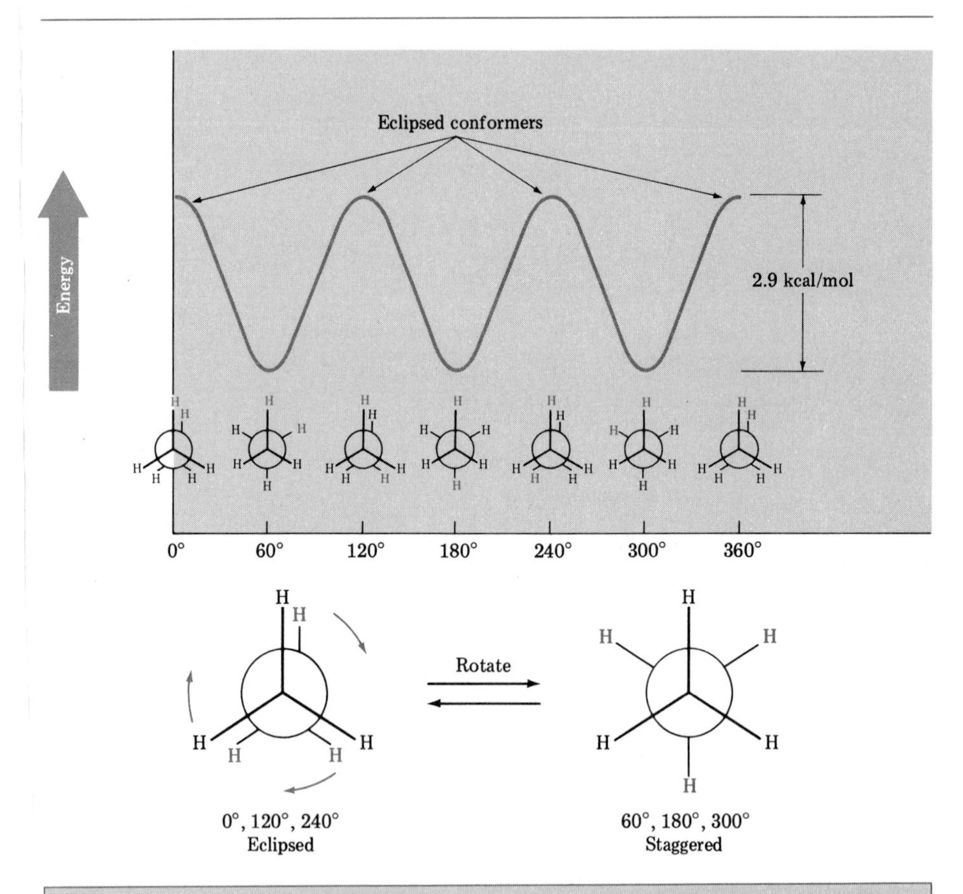

Figure 3.12. A graph of energy versus bond rotation in ethane. The staggered conformer is 2.9 kcal/mol lower in energy than the eclipsed conformer.

results from torsional strain can be represented on a graph of potential energy versus degree of rotation. Energy minima occur at staggered conformations; energy maxima occur at eclipsed conformations (Figure 3.12).

To what is torsional strain due? The reasons for its existence have been the subject of much controversy, but most theoretical chemists now hold that the strain is due to the slight repulsion between electron clouds in the carbon–hydrogen bonds as they pass by each other at close quarters in the eclipsed conformer. Calculations indicate that the internuclear hydrogen–hydrogen distance in the staggered conformer is 2.55 Å, but that this distance decreases to about 2.29 Å in the eclipsed conformer.

3.8 Conformations of propane

Propane is the next higher member in the alkane series, and we again find a torsional barrier that results in slightly hindered rotation about the carbon–carbon bonds. In the eclipsed conformer there are two ethane-type hydrogen–hydrogen interactions and one additional interaction between a carbon–hydrogen bond and a carbon–carbon bond. A slightly higher barrier to rotation is found in propane than is found in ethane—3.4 kcal/mol. Since our study of ethane showed that each eclipsing hydrogen–hydrogen interaction has an energy "cost" of 1 kcal/mol, we can assign to the interaction between the carbon–methyl bond and the carbon–hydrogen bond a value of 3.4 − (2 × 1 kcal/mol) = 1.4 kcal/mol (Figure 3.13).

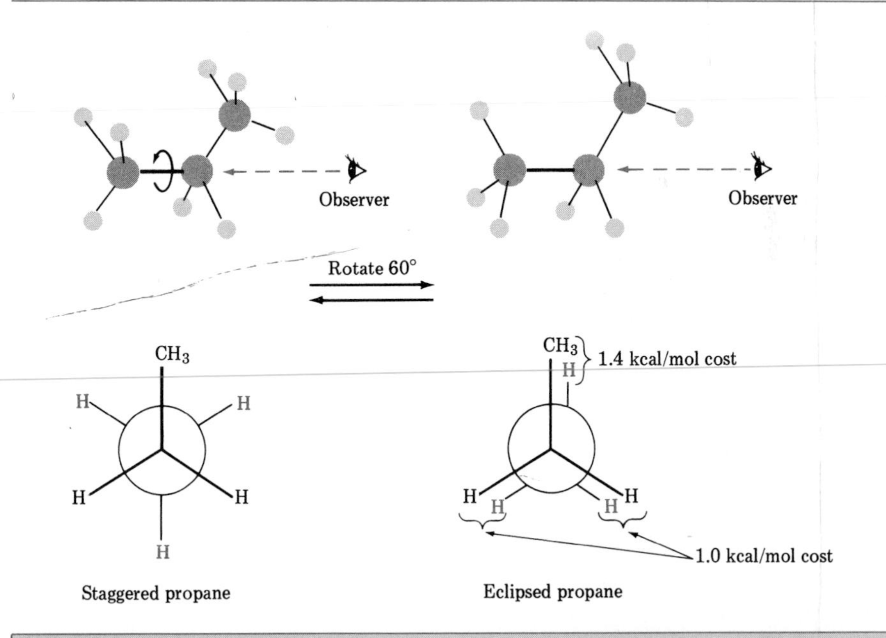

Figure 3.13. Newman projections of propane showing staggered and eclipsed conformations. The staggered conformation is lower in energy by 3.4 kcal/mol.

PROBLEM

3.10 Construct a graph plotting energy versus carbon–carbon bond rotation for propane. Assign quantitative values to the energy maxima.

3.9 Conformations of butane

The conformational situation becomes more complex for higher alkanes. A plot of potential energy versus rotation for rotation about the C2—C3 bond in butane is shown in Figure 3.14.

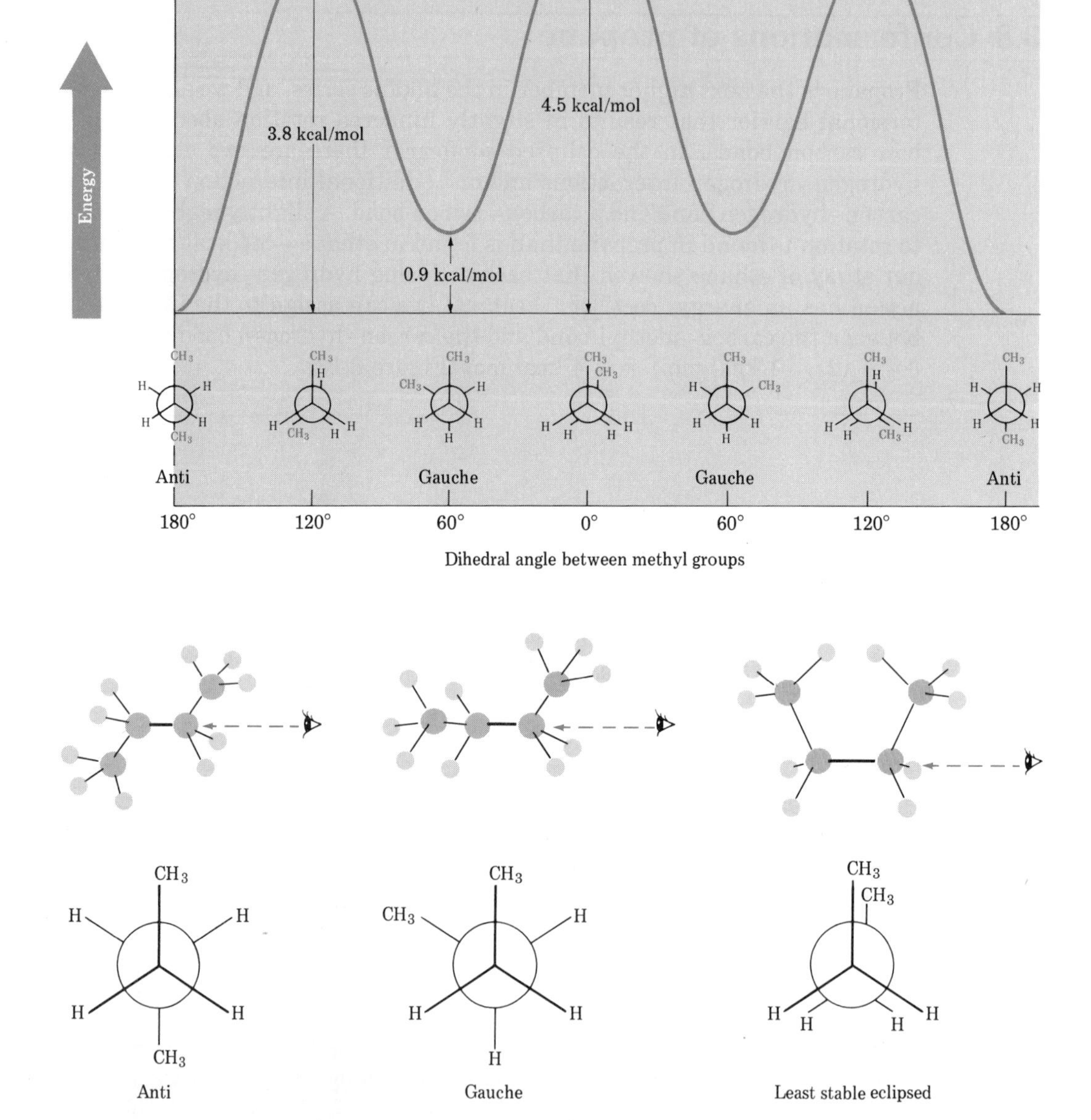

Figure 3.14. An energy versus rotation diagram for butane. The energy maximum occurs when the two methyl groups eclipse each other.

Not all the staggered conformations of butane are of equal energy, and not all eclipsed conformations are of equal energy. The lowest-energy conformation is the one in which the two large groups (methyls) are as far apart as possible—180°. This is called the **anti** conformation. As rotation around the C2—C3 bond occurs, an eclipsed conformation is reached in which there are two methyl–hydrogen interactions and one hydrogen–hydrogen interaction. If we assign the energy values for eclipsing interactions that we previously derived from ethane and propane, we can predict that this eclipsed conformation is more strained than the anti conformation by 2 × 1.4 kcal/mol (two methyl–hydrogen interactions) plus 1.0 kcal/mol (one hydrogen–hydrogen interaction) or a total of 3.8 kcal/mol. This is exactly what is observed.

When the rotation is continued, an energy minimum is reached at the staggered conformation where the methyl groups are 60° apart. This is called the **gauche** conformation, and it lies 0.9 kcal/mol higher in energy than the anti conformation *even though there are no eclipsing interactions.* This energy difference is due to the fact that the gauche conformation forces the large methyl groups to be near each other, resulting in **steric strain.** Steric strain is the repulsive interaction that occurs when two groups are forced to be closer to each other than their atomic radii allow; it is the result of trying to force two objects to occupy the same space (Figure 3.15).

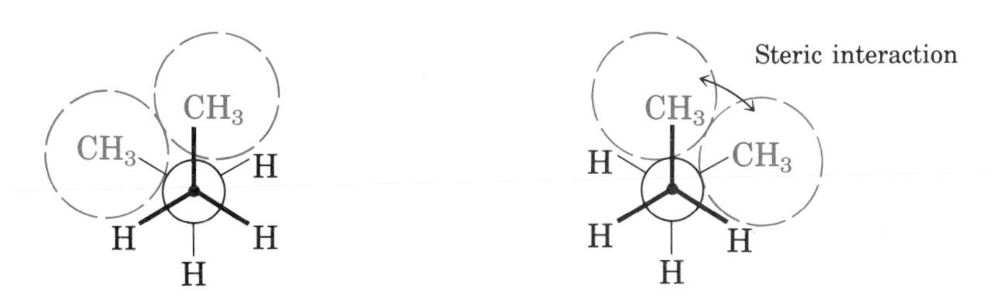

Figure 3.15. Gauche butane interactions. Steric strain results when the two methyl groups are forced too close together.

As the dihedral angle between the methyl groups approaches 0°, the energy maximum is reached. Since the methyl groups are forced even closer together than in the gauche conformation, a large amount of steric strain is present. A total strain energy of 4.5 kcal/mol has been estimated for this conformation, allowing us to calculate a value of 2.5 kcal/mol for the methyl–methyl eclipsing interaction [total strain (4.5 kcal/mol), less strain of two hydrogen–hydrogen eclipsing interactions (2 × 1.0 kcal/mol), results in 2.5 kcal/mol].

After 0°, the rotation becomes a **mirror image** of what we have already discussed. Thus, another gauche conformation is reached, another staggered conformation, and finally a return to the anti conformation occurs.

The concept of assigning definite energy values to specific interactions within a molecule is a very useful one, and we will return to it on occasion throughout the book. A summary of what we have found thus far is given in Table 3.6.

TABLE 3.6 Energy costs for interactions in alkane conformers

Interaction		*Cause*	*Energy cost (kcal/mol)*
H–H	eclipsed	Torsional strain	1.0
H–CH_3	eclipsed	Mostly torsional strain	1.4
CH_3–CH_3	eclipsed	Torsional plus steric strain	2.5
CH_3–CH_3 (gauche butane)	staggered	Steric strain	0.9

The same principles developed for butane apply to pentane and to all higher alkanes. The most favored conformation for any alkane is the one in which all carbon–carbon bonds have staggered arrangements, and in which large substituents are arranged anti to each other. A generalized alkane structure is shown in Figure 3.16.

One final point: It is important to remember that when we speak of a particular conformer as being "more stable" than another, we do not mean that the molecule in question adopts and maintains *only* the more stable conformation. At room temperature, sufficient thermal energy is present to ensure that rotation around sigma bonds occurs very rapidly and that all

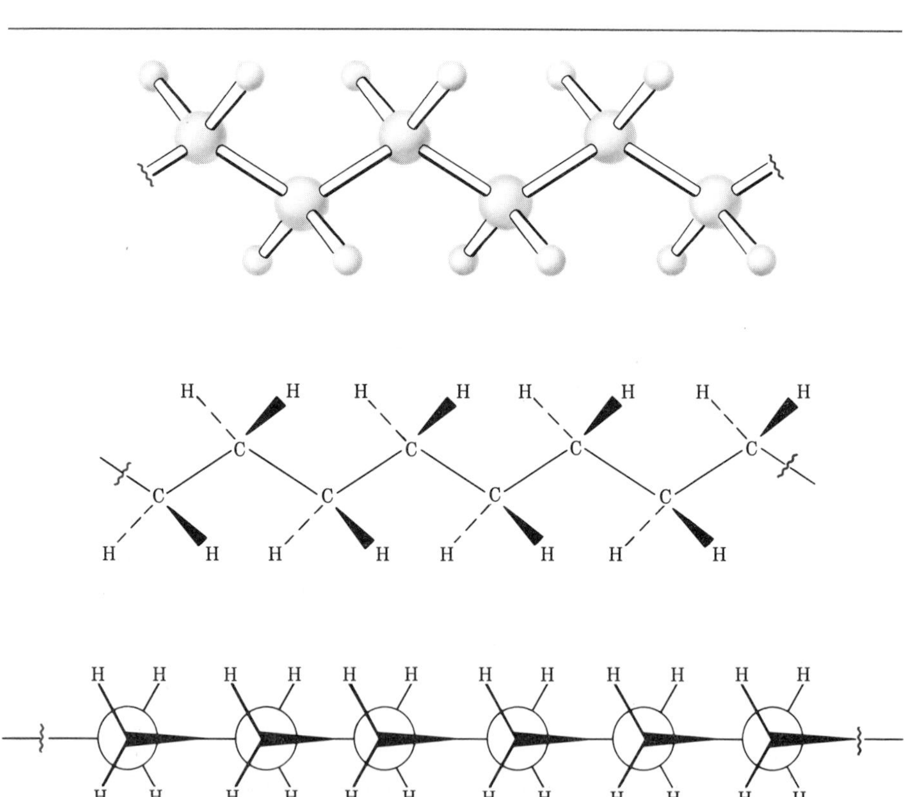

Figure 3.16. Most stable alkane conformation in sawhorse and Newman projections.

possible conformers are in a fluid equilibrium. At any given instant, however, a larger percentage of molecules will be found in a more stable conformation than in a less stable one. The exact relationship between the percentage of more and less stable conformers at room temperature is expressed in Table 3.7. For example, the energy difference between eclipsed and staggered conformations of ethane is 2.9 kcal/mol. Table 3.7 therefore says that at any given instant, 99% of all ethane molecules have a staggered conformation and only 1% of ethane molecules have an eclipsed conformation.

TABLE 3.7 **Relationship between isomer percentages at equilibrium and thermodynamic stability**

Stable isomer (%)	*Less stable isomer (%)*	*Energy difference between isomers (kcal/mol at 25°C)*
50	50	0
55	45	0.119
60	40	0.240
65	35	0.367
70	30	0.502
75	25	0.651
80	20	0.821
85	15	1.028
90	10	1.302
95	5	1.744
98	2	2.306
99	1	2.722
99.9	0.1	4.092

PROBLEM

3.11 Sight along the C2—C3 bond of 2,3-dimethylbutane and draw a Newman projection of the most stable conformation.

PROBLEM

3.12 Consider 2-methylpropane (isobutane). Sighting along the C1—C2 bond:

(a) Draw a Newman projection of the most stable conformation.

(b) Draw a Newman projection of the least stable conformation.

(c) Construct a qualitative graph of energy versus rotation about the C1—C2 bond.

(d) Since we know that a hydrogen–hydrogen eclipsing interaction "costs" 1.0 kcal/mol and a hydrogen–methyl eclipsing interaction "costs" 1.4 kcal/mol, assign quantitative values to your graph.

PROBLEM

3.13 In Problem 3.12, you constructed a quantitative diagram of potential energy versus rotation about the C1—C2 bond of 2-methylpropane. On the basis of that knowledge, answer the following:

(a) What is the energy difference between the most stable and least stable conformation? (Use Table 3.6.)

(b) At any given instant, what percentage of molecules would adopt the most stable conformation?

(c) What percentage adopt the least stable conformation?

3.10 Cycloalkanes: cis–trans isomerism

Up to this point, only open-chain alkanes have been discussed, but chemists have known for over 100 years that rings of carbon atoms are also capable of existence. Such compounds are called **cycloalkanes** or **alicyclic compounds** (**ali**phatic **cyclic**). Since cycloalkanes are composed of rings of —CH_2— units (methylene groups), they have the general formula $(CH_2)_n$ or C_nH_{2n}.

The smallest possible ring, *cyclopropane,* contains three carbon atoms, and is a well-known and much-studied compound that was first prepared by August Freund[5] in 1881 by reaction of sodium metal with 1,3-dibromopropane:

$$2\,Na + \underset{\text{1,3-Dibromopropane}}{BrCH_2CH_2CH_2Br} \longrightarrow \underset{\text{Cyclopropane}}{(CH_2)_3} + 2\,NaBr$$

Higher members of the cycloalkane series have also been prepared, and some physical data are given in Table 3.8.

TABLE 3.8 **Physical properties of some cycloalkanes**

Name	*Formula*	*Melting point (°C)*	*Boiling point (°C)*	*Density (g/mL)*
Cyclopropane	C_3H_6	−127.6	−32.7	
Cyclobutane	C_4H_8	−50.0	−12.0	0.720
Cyclopentane	C_5H_{10}	−93.9	49.3	0.7457
Cyclohexane	C_6H_{12}	6.6	80.7	0.7786
Cycloheptane	C_7H_{14}	−12.0	118.5	0.8098
Cyclooctane	C_8H_{16}	14.3	148.5	0.8349

In many respects, the chemistry of the cycloalkanes mimics that of open-chain (acyclic) alkanes. Both classes of compounds are relatively nonpolar and are chemically inert to most reagents. There are, however, some marked differences between cyclic and acyclic alkanes that deserve brief comment.

One difference is that cycloalkanes have less conformational mobility than their open-chain relatives. Although open-chain alkanes have nearly free rotation around carbon–carbon single bonds and are able to adopt any one of a large number of conformations, the same is not true of cycloalkanes. Small-ring cycloalkanes have much less freedom of rotation around bonds and are therefore constrained to fewer conformations. For example, cyclopropane *must* be a flat, planar molecule with a rigid structure (three points define a plane). No bond rotation about cyclopropane carbon–carbon bonds is possible without breaking the ring. Cyclopropane must be triangular in shape, and the carbon–hydrogen bonds on each carbon must eclipse the carbon–hydrogen bonds on its neighbors (Figure 3.17).

[5]August Freund (1835–1892); b. Kéty, Galicia; a pupil of Pebal; worked in Kéty and in Vienna.

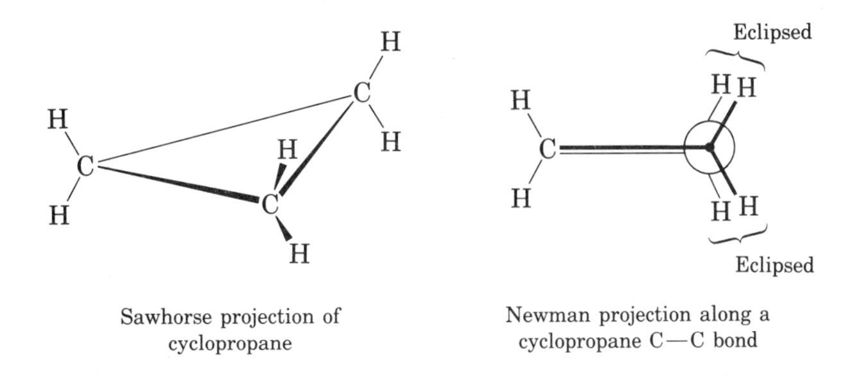

Figure 3.17. Cyclopropane.

Higher cycloalkanes have increasingly more freedom, and the very large rings (C_{25} and up) are so floppy that they are nearly indistinguishable from open-chain alkanes. The common ring sizes (C_3, C_4, C_5, C_6, C_7), however, are quite restricted in their molecular motions.

Cycloalkanes have two distinct sides, a "top" side and a "bottom" side, and isomerism is therefore possible in substituted cycloalkanes. For example, two different 1,2-dibromocyclopropane isomers exist. One isomer has the two bromines on the same side of the ring, and one isomer has them on opposite sides. The two isomers cannot be interconverted without breaking and re-forming chemical bonds; both are therefore stable, isolable compounds. Make molecular models to prove this to yourself.

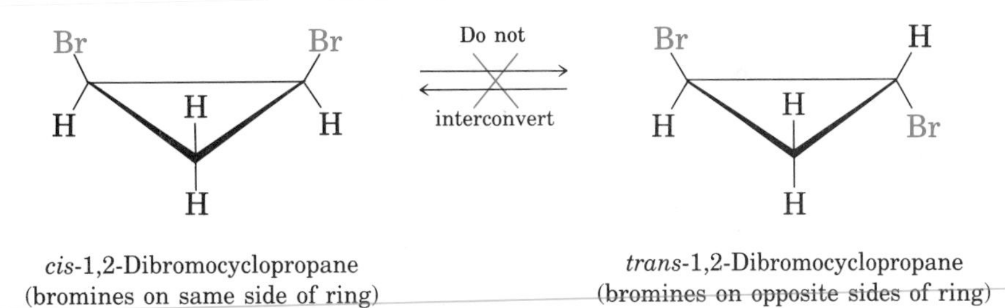

cis-1,2-Dibromocyclopropane (bromines on same side of ring)

trans-1,2-Dibromocyclopropane (bromines on opposite sides of ring)

We call the 1,2-dibromocyclopropanes **cis–trans isomers,** and we use the prefixes *cis*- (Latin, "on this side") and *trans*- (Latin, "across") to distinguish between them. Cis–trans isomerism is characteristic of all rings and can lead to important chemical consequences. More will be said about these consequences in Chapter 18.

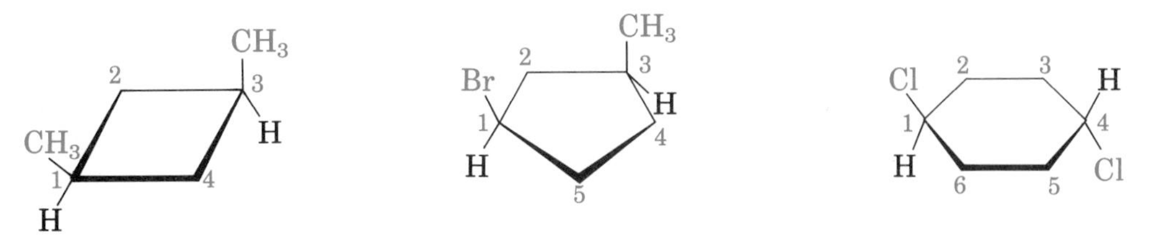

cis-1,3-Dimethylcyclobutane

cis-1-Bromo-3-methylcyclopentane

trans-1,4-Dichlorocyclohexane

3.11 Nomenclature of cycloalkanes

The systematic naming of cycloalkanes follows directly from the rules given for open-chain alkanes. For the majority of cases, there are only two rules:

1. Use the cycloalkane name as the base name. Compounds should be named as alkyl-substituted cycloalkanes, *not* as cycloalkyl-substituted alkanes.

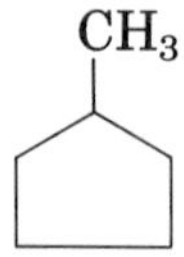

CH_3

Methylcyclopentane
(*not* cyclopentylmethane)

2. Number substituents on the ring so as to arrive at the lowest sum.

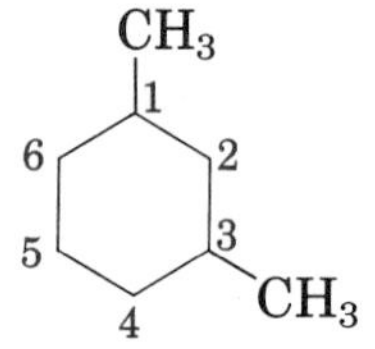

CH_3 1 2 3 CH_3 4 5 6 *NOT* CH_3 1 2 3 4 5 6 CH_3

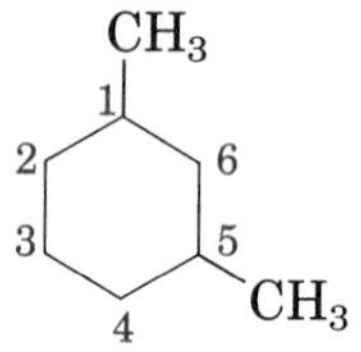

1,3-Dimethylcyclohexane 1,5-Dimethylcyclohexane

a. Functional groups usually take precedence over alkyl groups. Halogen substituents, however, are treated in the same manner as alkyl groups.

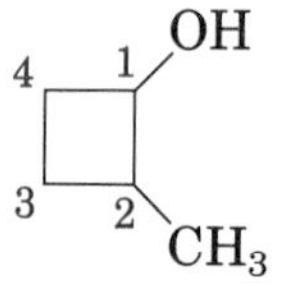

OH 1 2 CH_3 3 4 *NOT* OH 2 3 4 1 CH_3

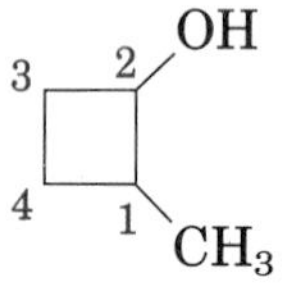

2-Methyl-1-cyclobutanol 1-Methyl-2-cyclobutanol

b. When two or more alkyl groups are present, they are numbered alphabetically.

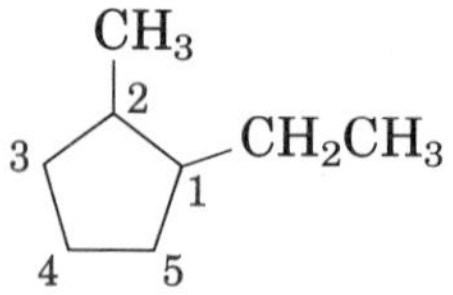

CH_3 2 CH_2CH_3 1 3 4 5 *NOT* CH_3 1 CH_2CH_3 2 3 4 5

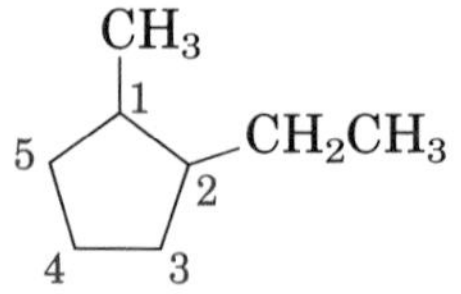

1-Ethyl-2-methylcyclopentane 2-Ethyl-1-methylcyclopentane

Some additional examples follow:

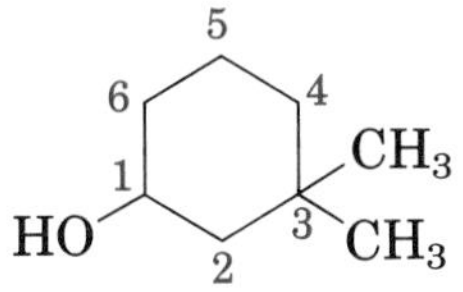

3,3-Dimethyl-1-cyclohexanol
(*not* 1,1-dimethyl-3-cyclohexanol)

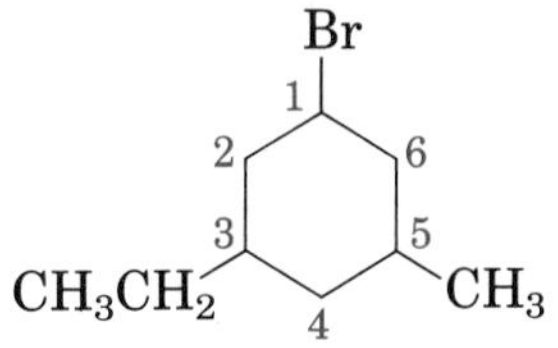

1-Bromo-3-ethyl-5-methylcyclohexane

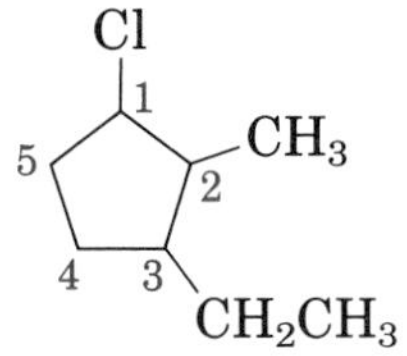

1-Chloro-2-ethyl-3-methylcyclopentane

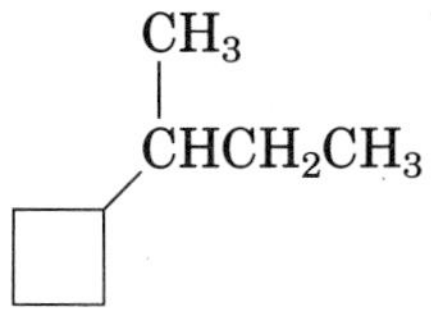

(1-Methylpropyl)cyclobutane
(or *sec*-butylcyclobutane)

PROBLEM

3.14 Give IUPAC names for the following cycloalkanes:

(a)

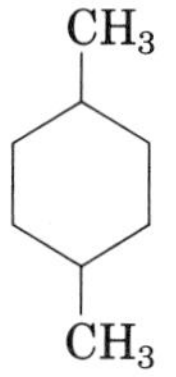

(b)

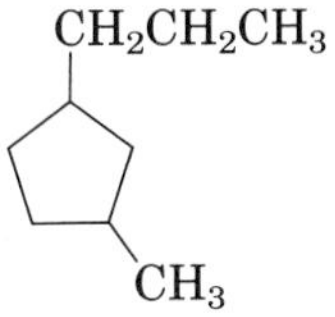

(c)

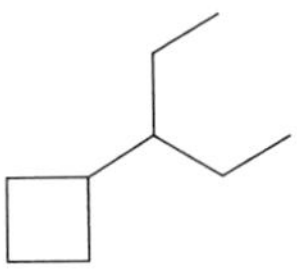

(d)

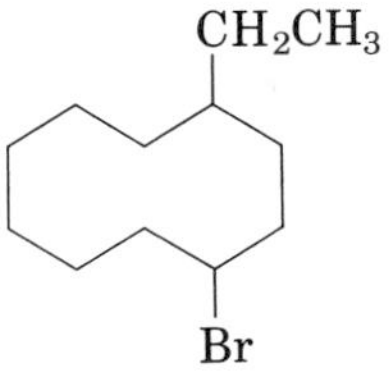

3.12 Summary

A functional group is an atom or group of atoms within a larger molecule. Functional groups have characteristic chemical reactivities that are approximately the same in all molecules of which they are a part. The chemical reactions an organic molecule undergoes are largely determined by the functional groups the molecule contains.

Alkanes are a class of hydrocarbons having the general formula C_nH_{2n+2}. They contain no functional groups and are chemically rather inert. Alkanes may be either straight-chain or branched, and they may be rationally and uniquely named by a series of rules.

The carbon–carbon bonds in alkanes are formed by sigma overlap of two carbon sp^3 hybrid orbitals. As a result of their symmetry, rotation is possible about sigma bonds, and alkanes therefore can adopt any of a large number of rapidly interconverting conformations. Newman projections allow us to visualize the spatial consequences of bond rotation by sighting directly along a carbon–carbon bond axis. The staggered conformation of ethane is 2.9 kcal/mol more stable than the eclipsed conformation as a result of torsional strain acting as a barrier to rotation.

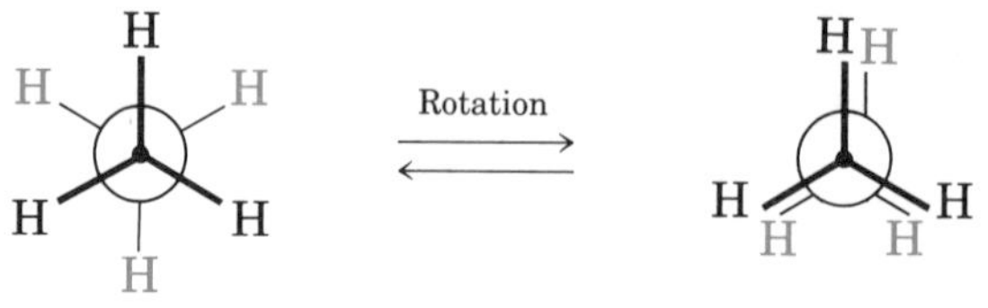

Rotation

Staggered ethane

Eclipsed ethane

Similar conformational descriptions can be made for propane, butane, and higher alkanes.

Cycloalkanes have the general formula C_nH_{2n} and contain rings of carbon atoms. One consequence of the ring structure is that conformational mobility is greatly reduced, and complete rotation around carbon–carbon bonds is not possible. Disubstituted cycloalkanes can therefore exist as cis–trans isomers. The cis isomer has both substituents on the same side of the ring; the trans isomer has substituents on opposite sides of the ring. For example:

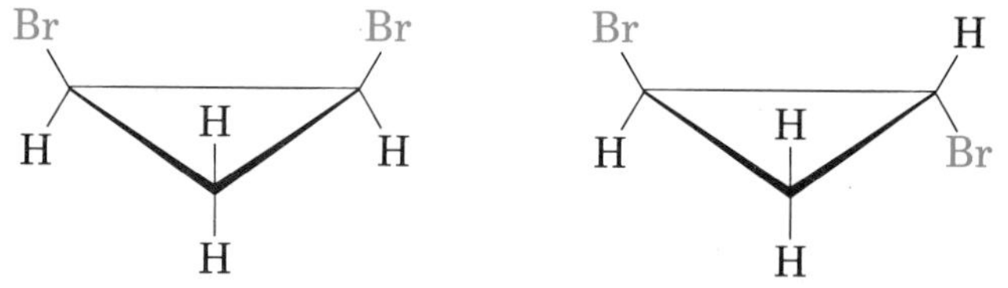

cis-1,2-Dibromocyclopropane

trans-1,2-Dibromocyclopropane

ADDITIONAL PROBLEMS

3.15 Locate and identify all of the functional groups present in these molecules:

(a) 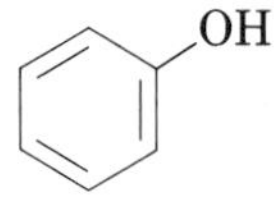 Phenol

(b) 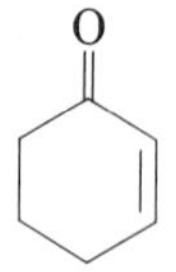 2-Cyclohexenone

(c) 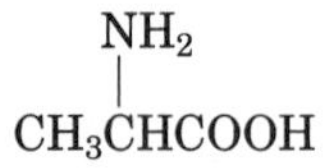 $CH_3CH(NH_2)COOH$

Alanine

(d) 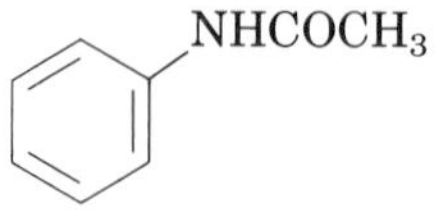 $C_6H_5NHCOCH_3$

Acetanilide

(e)

Nootkatone (from grapefruit)

(f)

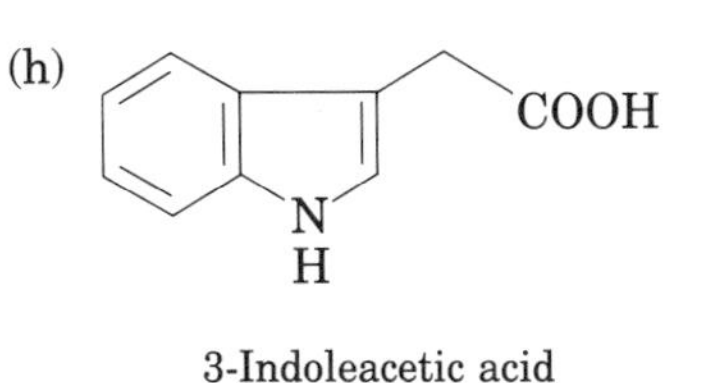

Estrone

(g)

Diethylstilbestrol

(h)

3-Indoleacetic acid

(i)

Morphine

3.16 Propose suitable structures for the following:

(a) A ketone with five carbons
(b) A four-carbon amide
(c) A five-carbon ester
(d) An aromatic aldehyde
(e) A keto ester
(f) An amino alcohol
(g) A five-carbon alkyne containing a nitrile function

3.17 Propose suitable structures for the following:

(a) A ketone with formula C_4H_8O
(b) A nitrile with formula C_5H_9N
(c) A dialdehyde with formula $C_4H_6O_2$
(d) A bromoalkene with formula $C_6H_{11}Br$
(e) An alkane with formula C_6H_{14}
(f) A cycloalkane with formula C_6H_{12}
(g) A cycloalkane with formula C_6H_{10}
(h) A diene (di-alkene) with formula C_5H_8
(i) A keto alkene (enone) with formula C_5H_6O

3.18 How many compounds can you write that fit these descriptions?

(a) Alcohols with formula $C_4H_{10}O$
(b) Amines with formula $C_5H_{13}N$
(c) Ketones with formula $C_5H_{10}O$
(d) Aldehydes with formula $C_5H_{10}O$
(e) Aromatic nitriles with formula C_8H_7N
(f) Esters with formula $C_4H_8O_2$
(g) Ethers with formula $C_4H_{10}O$

3.19 Draw compounds that contain the following:

(a) A primary alcohol
(b) A tertiary nitrile
(c) A secondary bromide
(d) Both primary and secondary alcohols (diol)
(e) An isopropyl group
(f) Four methyl groups
(g) A quaternary carbon

3.20 Draw all monobromo and dibromo derivatives of pentane.

3.21 What hybridization would you predict for the carbon atom in these functional groups?

(a) Ketone
(b) Nitrile
(c) Acid
(d) Ether
(e) Organometallic

3.22 Which of the following Kekulé structures represent the same compound, and which represent different compounds?

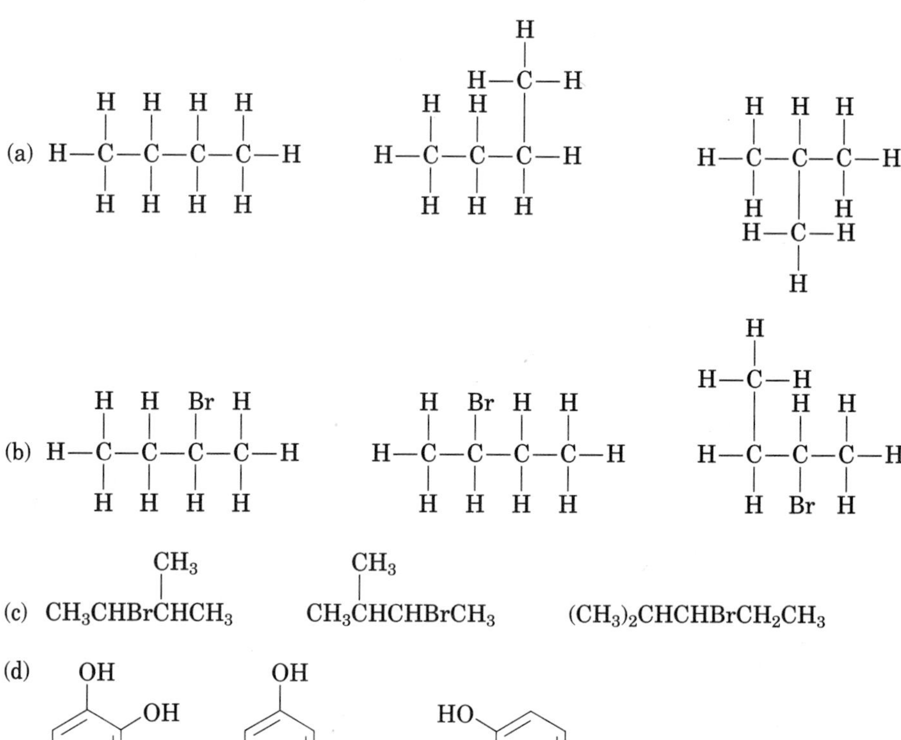

(d)

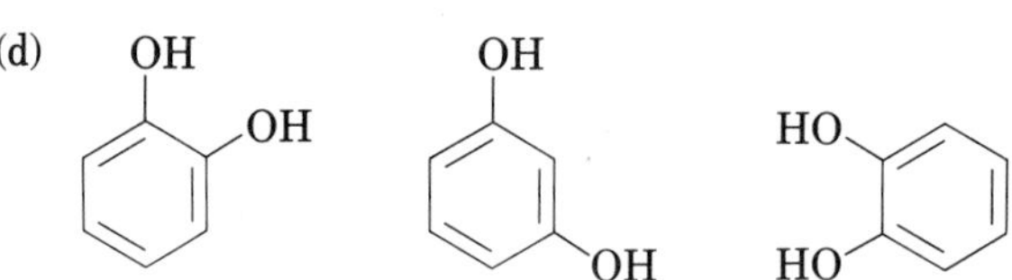

(e) $HOCH_2CH(CH_3)CH(CH_3)CH_3$ $\quad$ $CH_3CH_2CH(CH_3)CH_2CH(CH_3)CH_2OH$ $\quad$ $HOCH_2CH(CH_3)CH(CH_3)_2$

(f)

3.23 Draw structural formulas for the following:

(a) 2-Methylheptane
(b) 4-Ethyl-2,2-dimethylhexane
(c) 4-Ethyl-3,4-dimethyloctane
(d) 2,4,4-Trimethylheptane
(e) 3,3-Diethyl-2,5-dimethylnonane
(f) 4-Isopropyl-3-methylheptane
(g) 1,1-Dimethylcyclopentane

3.24 For each of the following compounds, draw an isomer having the same functional group.

(a) 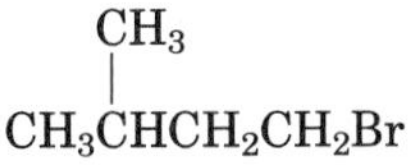 $CH_3\underset{}{\overset{CH_3}{\overset{|}{C}}}HCH_2CH_2Br$

(b) 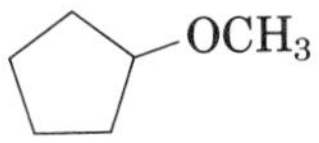 OCH_3

(c) $CH_3CH_2CH_2C{\equiv}N$

(d) 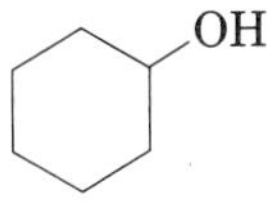 OH

(e) CH_3CH_2CHO

(f) 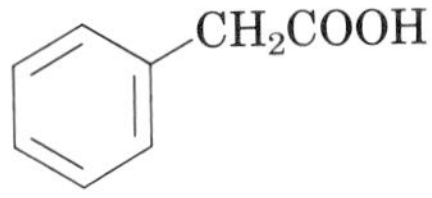 CH_2COOH

3.25 Identify the kinds of hydrogens (1°, 2°, or 3°) in these molecules:

(a) 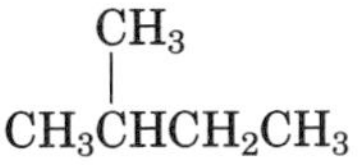 $CH_3\overset{CH_3}{\overset{|}{C}}HCH_2CH_3$

(b) $(CH_3)_2CHCH(CH_2CH_3)_2$

(c) 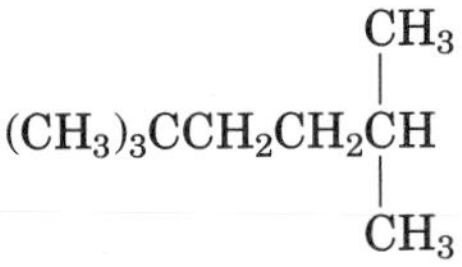 $(CH_3)_3CCH_2CH_2\underset{\underset{CH_3}{|}}{\overset{\overset{CH_3}{|}}{C}}H$

(d) 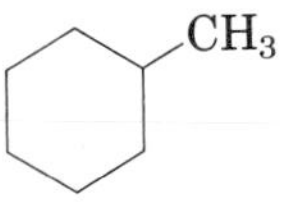 CH_3

(e)

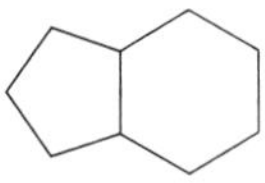

(f)  CH_3

3.26 Draw a compound that:

(a) Has only primary hydrogens
(b) Has both primary and tertiary (but no secondary) hydrogens
(c) Has no primary hydrogens
(d) Has four secondary hydrogens

3.27 Supply proper IUPAC names for compounds (a)–(g):

(a) 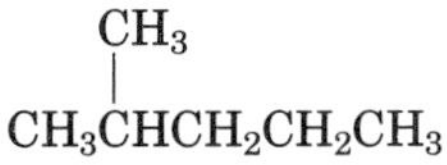 $CH_3\overset{CH_3}{\overset{|}{C}}HCH_2CH_2CH_3$

(b) $CH_3CH_2C(CH_3)_2CH_3$

(c) $(CH_3)_2CHC(CH_3)_2CH_2CH_2CH_3$

(d) 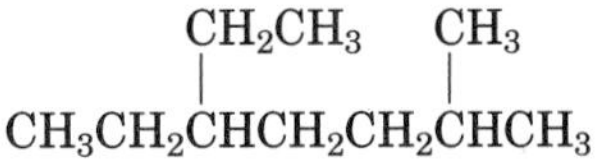 $CH_3CH_2\overset{CH_2CH_3}{\overset{|}{C}}HCH_2CH_2\overset{CH_3}{\overset{|}{C}}HCH_3$

(e) 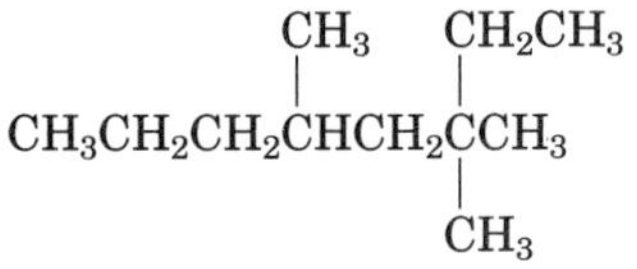 $CH_3CH_2CH_2\overset{CH_3}{\overset{|}{C}}HCH_2\underset{\underset{CH_3}{|}}{\overset{\overset{CH_2CH_3}{|}}{C}}CH_3$

(f) $(CH_3)_3CC(CH_3)_2CH_2CH_2CH_3$

```
                 CH2CH2CH3
                 |
(g) CH3CHCH2CCH2CH3
       |         |
    CH3CH2       CH3
```

3.28 Provide IUPAC names for the five isomers of C_6H_{14}.

3.29 Draw structures for the nine isomers of C_7H_{16}.

3.30 The following names are incorrect. Supply the proper IUPAC names.
(a) 2,2-Dimethyl-6-ethylheptane
(b) 4-Ethyl-5,5-dimethylpentane
(c) 3-Ethyl-4,4-dimethylhexane
(d) 5,5,6-Trimethyloctane
(e) 2-Isopropyl-4-methylheptane
(f) 1,5-Dimethylcyclohexane

3.31 Propose structures and give the correct IUPAC names for the following:
(a) A dimethyloctane
(b) A diethyldimethylhexane
(c) A cyclic alkane with three methyl groups
(d) A (3-methylbutyl)-substituted alkane

3.32 Consider 2-methylbutane (isopentane). Sighting along the C2—C3 bond:
(a) Draw a Newman projection of the most stable conformation.
(b) Draw a Newman projection of the least stable conformation.
(c) If a CH_3—CH_3 eclipsing interaction "costs" 2.5 kcal/mol and a CH_3—CH_3 gauche interaction "costs" 0.9 kcal/mol, construct a quantitative diagram of energy versus rotation about the C2—C3 bond.

3.33 What are the energy differences between the three possible staggered conformations around the C2—C3 bond in 2,3-dimethylbutane?

3.34 Construct a *qualitative* potential-energy diagram for rotation about the C—C bond of 1,2-dibromoethane. Which conformation would you expect to be more stable? Label the anti and the gauche conformations of 1,2-dibromoethane. Which would you expect to have the larger dipole moment? The observed dipole moment is $\mu = 1.0$ D. What does this tell you about the actual structure of the molecule?

3.35 The barrier to rotation about the C—C bond in bromoethane is 3.6 kcal/mol.
(a) What energy value can you assign to an H—Br eclipsing interaction?
(b) Construct a quantitative potential energy versus rotation diagram.

3.36 Malic acid, a compound of formula $C_4H_6O_5$, has been isolated from apples. Since this compound reacts with 2 equivalents (equiv) of base, it can be formulated as a dicarboxylic acid.
(a) Draw at least five possible structures.
(b) This compound can be shown to be a secondary alcohol. What is the structure of malic acid?

3.37 The compound α-methylenebutyrolactone is a skin irritant that has been isolated from the dogtooth violet. What functional groups does it contain?

O
O
CH_2

α-Methylenebutyrolactone

3.38 Formaldehyde, CH_2O, is a simple compound known to all biologists because of its usefulness as a tissue preservative. When pure, formaldehyde *trimerizes* to give trioxane, $C_3H_6O_3$. Trioxane, surprisingly enough, has no carbonyl groups. Only one monobromo derivative of trioxane is possible. Propose a structure that fits these data.

3.39 Locate and identify the functional groups in the following:

(a)

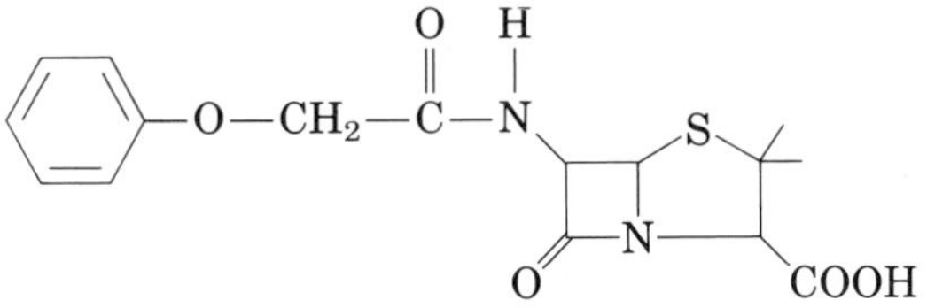

Penicillin V

(b)

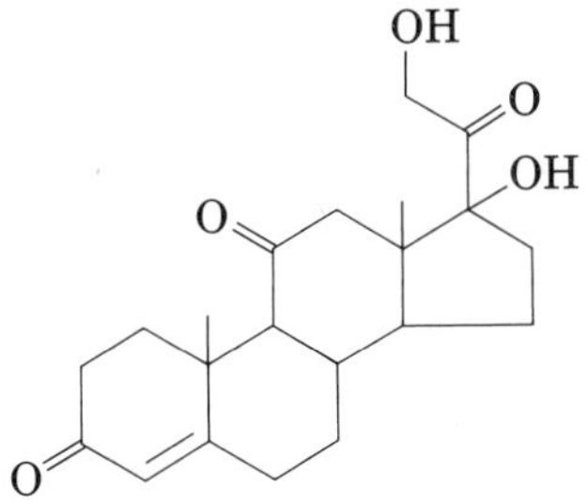

Cortisone

(c)

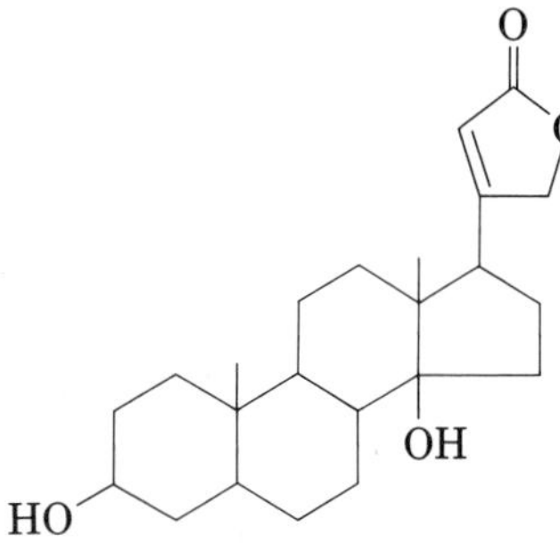

Digitoxigenin

(d)

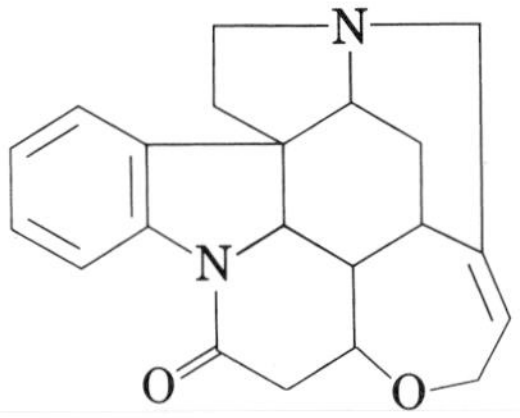

Strychnine

3.40 As mentioned earlier, cyclopropane was first prepared by reaction of 1,3-dibromopropane with sodium. What product might the following reaction give?

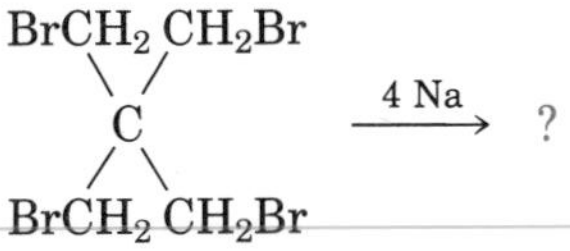

$$C(CH_2Br)_4 \xrightarrow{4\ Na} ?$$

What geometry would you predict for the product?

ORGANIC REACTIONS

When first approached, organic chemistry can seem like a bewildering collection of isolated facts—a collection of millions of compounds, dozens of functional groups, and a seemingly endless number of reactions. With study, however, it becomes evident that there are only a few fundamental concepts that underlie *all* organic reactions. Far from being a collection of isolated facts, organic chemistry is a beautifully logical subject unified by a few broad themes. When these themes are understood, learning organic chemistry becomes much easier; when the patterns are recognized, rote memorization can be avoided. It is the aim of this book to point out the themes that unify organic chemistry and to show how each new topic fits in with what has come before. We'll begin this process by seeing in this chapter what fundamental kinds of organic reactions take place and how organic reactions can be described.

Almost all organic chemistry can be explained in terms of three basic reaction types: *polar reactions, radical reactions,* and *pericyclic reactions*. It is important to realize, however, that we are categorizing reactions primarily for practical, pedagogical reasons; organic chemistry is much easier to learn when the material is suitably organized in a well-defined framework. In practice the basic categories are not always so sharply defined. Some reactions have characteristics of more than one category, and some reactions do not fit well in any category. Nevertheless, the categories of reactions do serve to group together most fundamental chemical processes.

Let's look at an example of each of the three fundamental reaction types. It's not important at this point to learn the specific details of each example, but it is very important to try to understand how the three examples differ.

4.1 Polar reactions

Polar reactions occur as the result of attractive forces between positive and negative charges on molecules. Polar processes are the most important reaction type in organic chemistry, and a large part of this book is devoted to their description.

In order to see how polar reactions occur, we need first to look deeper into the effects of bond polarity on organic molecules. Most organic molecules are electrically neutral; they have no net charge, either positive or negative. We saw in Section 2.4, however, that specific bonds within a molecule, particularly the bonds in functional groups, are often polar. Bond polarity is a consequence of unsymmetrical electron-density distribution in the bond, and is due to the intrinsic electronegativity of the atoms involved. A table of electronegativities (Table 4.1, which repeats Table 2.2 for convenience) shows that atoms such as oxygen, nitrogen, fluorine, chlorine, and bromine are more electronegative than carbon. A carbon atom bonded to one of these electronegative atoms has a partial positive charge (δ^+) and the electronegative atom has a slight negative charge (δ^-).

TABLE 4.1 Relative electronegativities of some common elements

Element	*Electronegativity*	*Element*	*Electronegativity*
H	2.2	Mg	1.3
Li	1.0	Al	1.6
B	2.0	Si	1.9
C	2.5	P	2.2
N	3.0	S	2.6
O	3.4	Cl	3.1
F	4.0	Br	3.0
Na	1.0	I	2.6

For example,

$$-\overset{\delta^+}{\underset{/}{\overset{\setminus}{C}}} \overset{\leftarrow\!\!+}{-} \overset{\delta^-}{X}$$

where X = O, N, F, Cl, or Br.

When carbon forms bonds to atoms that are *less* electronegative, polarity in the *opposite* sense results. For example, the reaction of bromomethane with magnesium metal yields methylmagnesium bromide—a so-called Grignard reagent. In Grignard reagents, and in most species that contain carbon–metal (**organometallic**) bonds, the carbon atom is *negatively* polarized with respect to the metal:

$$CH_3Br + Mg \longrightarrow \overset{\delta^-}{CH_3}-\overset{\delta^+}{MgBr}$$

A Grignard reagent

The polarity patterns of some common functional groups are shown in Table 4.2.

This discussion of bond polarity is oversimplified in that we have considered only bonds that are *inherently* polar due to electronegativity effects. Polar bonds can also result from the interaction of functional groups with solvents, with metal cations, and with acids. The polarity of the carbon–oxygen bond in methanol, for example, is greatly enhanced by protonation.

$$H_3C\overset{\leftarrow\!\!+}{-}\ddot{\underset{\cdot\cdot}{O}}-H \xrightarrow{H^+} H_3C\overset{\leftarrow\!\!+}{-}\underset{H}{\underset{|}{\ddot{O}^+}}-H$$

Methanol
(weakly polar C—O bond)

Protonated
methanol cation
(strongly polar C—O bond)

TABLE 4.2 Polarity patterns in some functional groups

Compound type	*Functional group structure*	*Compound type*	*Functional group structure*
Alcohol	$-\overset{\delta+}{C}-\overset{\delta-}{OH}$	Carbonyl	$\overset{\delta+}{C}=\overset{\delta-}{O}$
Alkene	$C=C$ Symmetrical, nonpolar	Carboxylic acid	$-\overset{\delta+}{C}(=\overset{\delta-}{O})-\overset{\delta-}{OH}$
Alkyl halide	$-\overset{\delta+}{C}-\overset{\delta-}{X}$	Carboxylic acid chloride	$-\overset{\delta+}{C}(=\overset{\delta-}{O})-\overset{\delta-}{Cl}$
Amine	$-\overset{\delta+}{C}-\overset{\delta-}{NH_2}$	Aldehyde	$-\overset{\delta+}{C}(=\overset{\delta-}{O})-H$
Ether	$-\overset{\delta+}{C}-\overset{\delta-}{O}-\overset{\delta+}{C}-$		
Nitrile	$-\overset{\delta+}{C}\equiv\overset{\delta-}{N}$	Ester	$-\overset{\delta+}{C}(=\overset{\delta-}{O})-\overset{\delta-}{O}-C$
Grignard reagent	$-\overset{\delta-}{C}-\overset{\delta+}{MgBr}$		
Alkyllithium	$-\overset{\delta-}{C}-\overset{\delta+}{Li}$	Ketone	$-\overset{\delta+}{C}(=\overset{\delta-}{O})-C$

In neutral methanol, the carbon atom is somewhat electron-poor because the electronegative oxygen attracts carbon–oxygen bond electrons. In the protonated methanol cation, a full positive charge on oxygen *strongly* attracts electrons in the carbon–oxygen bond and makes the carbon much more electron-poor.

Yet a further consideration is the **polarizability** (as opposed to polarity) of an atom. As the electric field around a given atom changes due to changing interactions with solvent or with other polar reagents, the electron distribution around that atom also changes in response. The measure of this response is the polarizability of the atom. Larger atoms with more loosely held electrons are more polarizable than smaller atoms with tightly held electrons; thus iodine is much more polarizable than fluorine. This means that the carbon–iodine bond, although electronically symmetrical according to the electronegativity table (Table 4.1), nevertheless can react as if it were polar.

What does functional-group polarity mean with respect to chemical reactivity? Since unlike charges attract, the fundamental characteristic of all polar organic reactions is that electron-rich sites in the functional groups of one molecule react with electron-poor sites in the functional groups of another molecule. Bonds are made when the electron-rich reagent donates a *pair* of electrons to the electron-poor reagent; conversely, bonds are broken when one of the two nuclei involved leaves with the electron *pair*. This electronically unsymmetrical bond breaking is referred to as a **heterolytic** process, and, in an electron-bookkeeping sense, we can follow the movement of the bonding electron pair by using curved arrows to indicate movement. By convention, a curved arrow means that an electron pair has moved from the tail to the head of the arrow. Thus, in a polar heterolytic bond-breaking reaction, the arrow indicates that the electron pair stays with one fragment. In a **heterogenic** polar bond-forming reaction, the arrow indicates that both electrons are donated by one reagent.

Polar (heterolytic) bond breakage $\quad A{:}\,B \longrightarrow \overset{-}{A}{:} + \overset{+}{B}$

Polar (heterogenic) bond formation $\quad \overset{-}{A}{:} + \overset{+}{B} \longrightarrow A{:}B$

In referring to this fundamental polar process and to the species involved, chemists have coined the words **nucleophile** and **electrophile.** A nucleophile is a reagent that is "nucleus-loving"; nucleophiles are reagents with electron-rich sites, which form a bond by *donating* a pair of electrons to an electron-poor reagent. Nucleophiles are often negatively charged, though this is not always the case. An electrophile, by contrast, is "electron-loving"; electrophiles are reagents with electron-poor sites that form a bond by *accepting* a pair of electrons from an electron-rich reagent.

$$\overset{-}{A}{:} + \overset{+}{B} \longrightarrow A{:}B$$

Nucleophile (electron-rich) Electrophile (electron-poor)

The definitions of nucleophiles and electrophiles are similar to those given in Section 2.7 for Lewis acids and Lewis bases, and there is indeed a correlation between electrophilicity/nucleophilicity and Lewis acidity/basicity. Thus, Lewis bases are electron donors and usually behave as nucleophiles, whereas Lewis acids are electron acceptors and usually behave as electrophiles. The major difference, however, is that the terms electrophile and nucleophile are used specifically when bonds to *carbon* are involved. We will explore these ideas in more detail in Chapter 10.

PROBLEM

4.1 Identify the functional groups present in these molecules and show the direction of polarity in each.

(a) Acetone, $CH_3\overset{O}{\overset{\|}{C}}CH_3$

(b) Chloroethylene, $H_2C{=}CHCl$

(c) Ethyl propenoate, $H_2C{=}CH\overset{O}{\overset{\|}{C}}OCH_2CH_3$

(d) Tetraethyllead, $(CH_3CH_2)_4Pb$ (the "lead" in gasoline)

4.2 An example of a polar reaction

Let's consider, as an example of a polar process, the reaction of sodium hydroxide with bromomethane. When the two compounds are warmed together in a suitable solvent, methanol is produced. Overall, the reaction can be formulated as follows:

$$\overset{\delta^-}{Br}-\overset{\delta^+}{CH_3} + \overset{+}{Na}\overset{-}{OH} \xrightarrow[\text{solvent}]{\text{Alcohol}} HO-CH_3 + \overset{+}{Na}\overset{-}{Br}$$

Bromomethane (electrophile) | Sodium hydroxide (nucleophile) | Methanol | Sodium bromide

This is an example of a general polar reaction type known as a **nucleophilic substitution,** and we can understand it in terms of the concepts of polar reactions just discussed. The carbon–bromine bond of bromomethane is polarized such that carbon is slightly positive and bromine is slightly negative. Bromomethane, therefore, has an electrophilic carbon atom. As organic chemists, we are primarily concerned with the reactivity of the carbon-containing portion, not the bromine, and we therefore use the term electrophile to refer to the *entire* bromomethane molecule rather than to just the carbon atom. Hydroxide ion, on the other hand, is a nucleophile since it has a negative charge and unshared pairs of electrons on oxygen. Nucleophilic hydroxide ion reacts with electrophilic bromomethane to give the observed products. The reaction is a substitution of one nucleophile (hydroxide ion) for another (bromide ion); hence the name, nucleophilic substitution reaction.

We will consider in Chapter 10 the exact details of how this sort of reaction occurs. For the present, however, the important point is that we can picture the reaction as occurring when nucleophilic hydroxide ion *donates an electron pair* from oxygen to form a new bond to the electrophilic carbon atom of bromomethane. At the same time, the bromine atom leaves and *takes with it the electron pair* from the carbon–bromine bond. We can use curved arrows to indicate the movement of electrons and can follow the movement of electron pairs by looking at Lewis electron-dot structures, as in Figure 4.1.

An electron pair from the nucleophilic hydroxide oxygen attacks the electrophilic carbon forming a new C—O bond. Bromine leaves, taking the C—Br bond electrons with it.

$$H:\ddot{O}:^- + H:\overset{H}{\underset{H}{C}}:\ddot{Br}: \longrightarrow H:\ddot{O}:\overset{H}{\underset{H}{C}}:H + :\ddot{Br}:^-$$

Figure 4.1. The nucleophilic substitution reaction of hydroxide ion with bromomethane.

We can also write the reaction in a simpler way in chemical shorthand:

$$HO^{-} + CH_3—Br \longrightarrow CH_3—OH + Br:^{-}$$

Note that all the reacting atoms—oxygen, carbon, and bromine—maintain their outer-shell electron octets throughout the reaction. Remember, also, that the curved arrows indicate an electron *pair* has moved from the tail to the head of the arrow.

This nucleophilic substitution reaction is only one example of a polar process; there are other types that we will study in detail later on, but the details of individual reactions are not important at the moment. The important point is that all polar reactions can be accounted for in the general terms just presented. Polar reactions take place between electron-poor sites and electron-rich sites and involve the donation of an electron pair from a nucleophile to an electrophile.

PROBLEM

4.2 Classify these reagents as either electrophiles or nucleophiles:

(a) H^+ (b) NH_3 (c) Mg^{2+}

(d) I^- (e) $CH_3\overset{\overset{O}{\|}}{C}Cl$ (f) CN^-

4.3 Radical reactions

All reagents in polar reactions contain an even number of electrons. Bonds are heterogenically made when a nucleophile donates an electron *pair* to an electrophile, and bonds are heterolytically broken when one fragment leaves with an electron *pair*. Chemical bonds can also be made **homogenically** by reaction between two odd-electron species (**radicals**) where each reactant donates *one* electron to the new bond. Conversely, bonds can be broken in such a way that each fragment leaves with *one* of the bonding electrons. This electronically symmetrical bond breaking is referred to as a **homolytic** breakage. Radical reactions constitute a second major class of organic transformations. They are not as common as polar reactions, but are nevertheless important and have been studied carefully (see Figure 4.2).

Polar (heterolytic) bond breakage	A:B	⟶	$\overset{-}{A}: + \overset{+}{B}$
Polar (heterogenic) bond formation	$\overset{-}{A}: + \overset{+}{B}$	⟶	A:B
Radical (homolytic) bond breakage	A:B	⟶	A· + B· Two radicals
Radical (homogenic) bond formation	A· + B· Two radicals	⟶	A:B

Figure 4.2. Radical versus polar reactions.

Although most radicals are electrically neutral, they are highly reactive since they contain only seven electrons in their outer shell rather than a stable octet. They can achieve the desired octet in several ways. For example, a radical can *abstract* an atom from another molecule, leaving behind a new radical.

$$R\cdot + X{:}Y \longrightarrow R{:}X + \cdot Y$$

Radical — Product radical

Let's look at an example of a radical reaction to see its characteristics. Once again, the specific details are not important at this point. What *is* important is to see how a radical reaction differs from a polar reaction.

4.4 An example of a radical reaction

The chlorination of methane is a typical **radical substitution reaction.** Although inert to most reagents, alkanes react readily with chlorine to give chlorinated alkane products:

$$CH_4 + Cl_2 \xrightarrow{\text{Light}} CH_3Cl + HCl$$

Methane — Chloromethane

A more detailed discussion of radical substitution reactions is given in Chapter 9. For the present, it is only important to know that careful studies have shown alkane chlorination to be a multistep process involving radicals, and we can view the reaction in the following way:

1. Initiation step:

$$:\ddot{Cl}:\ddot{Cl}: \xrightarrow{\text{Light}} 2\ :\ddot{Cl}\cdot$$

Chlorine *radicals*

Homolytic bond breaking

2. Propagation steps:

a. $Cl\cdot + H{:}CH_3 \longrightarrow H{:}Cl + \cdot CH_3$

b. $\cdot CH_3 + Cl{:}Cl \longrightarrow Cl{:}CH_3 + Cl\cdot$

c. Repeat steps a and b over and over.

3. Termination steps:

$$Cl\cdot + Cl\cdot \longrightarrow Cl{:}Cl$$

or $$Cl\cdot + \cdot CH_3 \longrightarrow Cl{:}CH_3$$

or $$\cdot CH_3 + \cdot CH_3 \longrightarrow CH_3{:}CH_3$$

Radical substitution reactions normally require these three kinds of steps—an **initiation** step, **propagation** steps, and **termination** steps. The initiation step starts the reaction by producing reactive radicals. In the present case, the relatively weak chlorine–chlorine bond is homolytically broken by irradiation with ultraviolet light. Two reactive chlorine radicals are produced, and further chemistry ensues.

Once chlorine radicals have been produced in small amounts, the reaction is self-propagating. Chlorine radicals are high-energy species and are therefore highly reactive. When a chlorine radical collides with a methane molecule, it abstracts a hydrogen atom to produce HCl and a methyl radical ($\cdot CH_3$). This methyl radical is highly energetic and reacts further with Cl_2 to give the new products, chloromethane and chlorine radical. This chlorine radical cycles back into the first propagation step, making the overall process a *chain reaction.* Once the sequence has been initiated, it becomes a self-sustaining cycle of repeating steps 1 and 2.

Occasionally, two radicals collide and combine to form stable products. When this happens, the reaction cycle is terminated and the chain is ended. Such termination steps occur infrequently, however, since the concentration of radicals in the reaction at any given moment is very small, and the likelihood that two radicals will collide is also small.

Alkane chlorination is not a generally useful reaction, since most alkanes (other than methane and ethane) have several different kinds of hydrogens, and *mixtures* of chlorinated products usually result in such cases. Nevertheless, radical chain reactions constitute a basic reaction type of considerable importance in numerous industrial processes.

The radical substitution reaction just discussed is only one of several different processes that radicals can undergo. The fundamental principle behind all radical reactions is the same, however: All bonds are broken and formed by reaction of odd-electron species.

PROBLEM

4.3 When a mixture of methane and chlorine is irradiated, reaction commences immediately. When irradiation is stopped, the reaction gradually slows down but does not stop immediately. How do you account for this behavior?

PROBLEM

4.4 Radical chlorination of pentane is a poor way to prepare 1-chloropentane, $CH_3CH_2CH_2CH_2CH_2Cl$, but radical chlorination of neopentane, $(CH_3)_4C$, is a good way to prepare neopentyl chloride, $(CH_3)_3CCH_2Cl$. How do you account for this difference?

4.5 Pericyclic reactions

Both polar and radical reactions have long been recognized as fundamental organic processes. By contrast, the underlying principles governing pericyclic reactions have been understood only since the mid-1960s. **Pericyclic reactions** are reactions that involve a redistribution of bonding electrons in a cyclic manner. It is not feasible at this point to discuss all of the kinds of pericyclic processes, but we should look carefully at one example to see what this definition means and to see how such reactions differ from polar and radical reactions. Let's consider a **cycloaddition reaction.**

Cycloaddition reactions involve the addition of one reactant to another to form a cyclic product. The reaction between a diene (a compound with two carbon–carbon double bonds) and an alkene is known as the Diels–Alder reaction. For example, 1,3-butadiene adds to methyl propenoate to give a cyclic product:

$$CH_2{=}CH{-}CH{=}CH_2 \;+\; CH_2{=}CH{-}CO_2CH_3 \xrightarrow[\text{benzene solvent}]{\text{Heat in}} \text{Methyl 3-cyclohexenecarboxylate}$$

1,3-Butadiene — Methyl propenoate (methyl acrylate) — Methyl 3-cyclohexenecarboxylate (a cyclic product)

The Diels–Alder reaction, like other pericyclic reactions, cannot be satisfactorily explained by the concepts used to explain polar and radical reactions. The two reagents are neither a nucleophile–electrophile combination nor are they radicals; they simply come together and form a new product in a single step. Such a reaction can best be understood by looking at what happens to the bonding electrons as the reaction commences. An orbital picture of the Diels–Alder reaction is shown in Figure 4.3.

H–C, C=CH₂, CH₂, CH₃OC(=O)–C(H)=C(H)–H → H–C, C–CH₂, CH₂, CH₂, CH, CH₃O₂C = (cyclohexene ring) CO₂CH₃

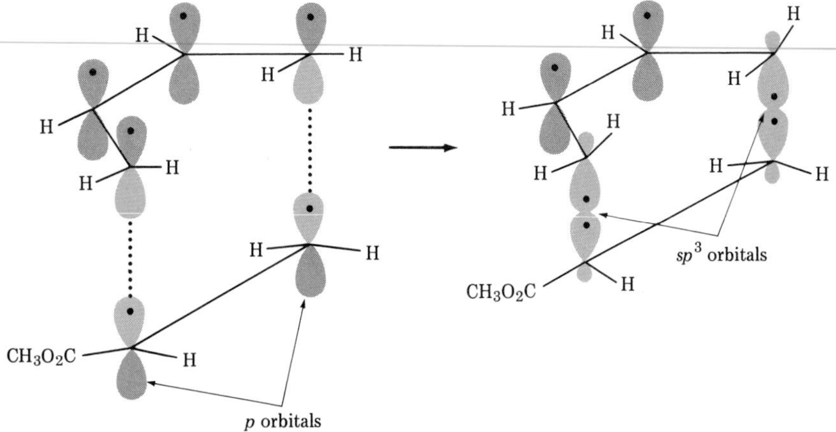

Figure 4.3. The Diels–Alder cycloaddition reaction. Overlap of diene *p* orbitals with alkene *p* orbitals occurs to give a cyclic product.

The diene component, 1,3-butadiene, enters the reaction with four *p* electrons in two double bonds, and methyl propenoate enters the reaction with two *p* electrons in one double bond. We can imagine that the two reactants might orient so that head-on orbital overlap occurs between the two terminal *p* orbitals of the diene and the two propenoate *p* orbitals. This is a *cyclic* orientation, and if the overlapping *p* orbitals rehybridize into sp^3 orbitals, the result is the formation of two new sigma bonds. The two remaining *p* orbitals of the diene are left to form a new pi bond. Taking into account all of the six *p* orbitals we started with, we find that three pi bonds have become one new pi bond and two new sigma bonds. A redistribution of bonding electrons has occurred in a cyclic manner and in a single step.

The details of the Diels–Alder reaction need not be learned now. The important point is to realize that there is a difference between the Diels–Alder cycloaddition reaction, polar reactions, and radical reactions. Pericyclic reactions, of which the Diels–Alder reaction exemplifies only one type, constitute a major class of reactions found in organic chemistry. Their study becomes much easier once the rules covering such reactions are known and the unifying mechanistic features are understood. We will study these rules in Chapter 30.

A summary of the three major types of organic reactions and their characteristics is shown in Table 4.3.

TABLE 4.3 Characteristics of the three major organic reaction types

Reaction type	*Characteristics*
Polar	Bonds are formed between electron-rich sites (nucleophiles) and electron-poor sites (electrophiles) when the nucleophile donates *both* electrons to the new bond. $\overset{-}{\text{A}}\!: + \overset{+}{\text{B}} \longrightarrow \text{A}\!:\!\text{B}$
Radical	Bonds are formed when each reactant donates *one* electron to the new bond. $\text{A}\cdot + \text{B}\cdot \longrightarrow \text{A}\!:\!\text{B}$
Pericyclic	Bonds are formed in a single step by a cyclic reorganization of electrons.

4.6 Rates and equilibria

An overall description of how a specific reaction occurs is called a **reaction mechanism.** A mechanism describes in detail exactly what takes place at each stage of a chemical transformation; it describes which bonds are broken and in what order, which bonds are formed and in what order, what the relative rates of each step are, and what the geometric position of each atom is at each moment. A complete mechanism must also account for all reactants used, all products formed, and the amounts of each. Let's begin a consideration of reaction mechanisms by seeing, in general terms, what must happen for a reaction to take place.

In principle, all chemical reactions can be written as equilibrium processes; starting materials can react to give products, and the products can react to give back starting materials. We can express a chemical equilibrium by an equation in which K_{eq}, the equilibrium constant (see Section 2.6), is equal to the concentration of products divided by the concentration of starting materials. For the reaction,

$$A + B \rightleftharpoons C + D$$

we have

$$K_{eq} = \frac{[\text{Products}]}{[\text{Reactants}]} = \frac{[C][D]}{[A][B]}$$

This equation tells us the position of the equilibrium—that is, which side of the reaction arrow is energetically more favored. If K_{eq} is large, then the product concentrations [C][D] are larger than the reactant concentrations [A][B], and the reaction proceeds as written from left to right. Conversely, if K_{eq} is small, the reaction does not take place as written. The equation does not, however, tell us the *rate* of reaction—how fast the equilibrium is established. For example, alkanes are generally considered to be chemically rather unreactive. Gasoline is stable indefinitely when stored because the rate of its reaction with air is slow under normal circumstances. Under other circumstances, however—ignition in an automobile engine, for example—gasoline reacts rapidly with oxygen and undergoes complete conversion to the equilibrium products water and carbon dioxide. Rates (*how fast* a reaction occurs) and equilibria (*how much* a reaction occurs) are entirely different.

In order for a reaction to take place, the energy level of the products must be lower than the energy level of the reactants; in other words, energy must be given off. (Reactions can sometimes be made to go the other way, but energy must be added for this to happen.) The reaction of bromomethane with hydroxide ion is now familiar (Section 4.2) and can serve as a good example. When hydroxide ion reacts with bromomethane in aqueous solution, methanol is produced. We can write this reaction as an equilibrium process and can determine experimentally that the equilibrium constant is approximately 10^{18}.

$$CH_3Br + {}^{-}\!:\!OH \rightleftharpoons Br\!:^{-} + CH_3\text{—}OH + \text{Energy}$$

$$\frac{[Br^-][CH_3OH]}{[CH_3Br][^-OH]} = K_{eq} \approx 10^{18}$$

Since K_{eq} is relatively large, the reaction proceeds as written and energy is given off. We speak of such reactions as "going to completion." This is imprecise terminology, since in practically no reaction does *every* molecule react, but equilibrium constants of greater than 10^3 can be considered to indicate "complete" reaction, since the amount of reactant left will be barely detectable (less than 1%). The total amount of energy change during a reaction is called the **standard Gibbs free-energy change, $\Delta G°$.**

(The Greek letter delta, Δ, is the mathematical symbol for the difference between two numbers—in this case the difference between the free energy of starting materials and the free energy of products.) For a reaction where $\Delta G°$ is negative, energy is evolved; for a reaction where $\Delta G°$ is positive, energy must be added. The equilibrium constant, K_{eq}, and the free-energy change, $\Delta G°$, both measure whether or not a reaction is favorable, and they are therefore mathematically related.

$$\Delta G° = \text{Free energy of reactants} - \text{Free energy of products}$$

$$\Delta G° = -RT \ln K_{eq} \quad \text{or} \quad K_{eq} = e^{-\Delta G°/RT}$$

where

$$R = 1.986 \text{ cal/degree mol (the gas constant)}$$

$$T = \text{Absolute temperature (in kelvins)}$$

$$e = 2.718 \text{ (the base of natural logarithms)}$$

$$\ln K_{eq} = \text{Natural logarithm of } K_{eq}$$

As an example of how we can use this mathematical relationship, the reaction of bromomethane with sodium hydroxide has $K_{eq} \approx 10^{18}$, and we can therefore calculate $\Delta G° \approx -24.7$ kcal/mol at room temperature (300 K). To what is this energy change due? It turns out that the Gibbs free-energy change is due to a combination of two factors, an **enthalpy** factor, **$\Delta H°$**, and an **entropy** factor, **$\Delta S°$**:

$$\Delta G° = \Delta H° - T\Delta S°$$

where T is the absolute temperature.

The enthalpy term, $\Delta H°$, is referred to as the **standard heat of reaction** and is a measure of the change in total bonding energy during a reaction. If $\Delta H°$ is negative, heat is evolved and the reaction is said to be **exothermic.** If $\Delta H°$ is positive, heat must be added and the reaction is said to be **endothermic.** For example, if a certain reaction breaks reactant bonds with a total strength of 100 kcal/mol and forms new product bonds with a total strength of 150 kcal/mol, then $\Delta H°$ for the reaction is −50 kcal/mol and the reaction is highly exothermic. (Remember that breaking bonds takes energy, and making bonds releases energy.)

$$\Delta H° = \text{Strength of bonds broken} - \text{Strength of bonds formed}$$

$$= 100 \text{ kcal/mol} - 150 \text{ kcal/mol} = -50 \text{ kcal/mol}$$

The entropy term, $\Delta S°$, is a measure of the amount of "disorder" caused by a reaction. To illustrate, in a reaction of the type

$$\text{A} \longrightarrow \text{B} + \text{C}$$

there is more freedom of movement (disorder) in the products than in the reactant because one molecule has split into two. Thus, there is a net gain in entropy during the reaction and $\Delta S°$ has a positive value. The catalytic cracking of long-chain alkanes to smaller hydrocarbon fragments during gasoline refining (Section 3.5) is an example of such a reaction.

On the other hand, for reactions of the type

$$A + B \longrightarrow C$$

exactly the opposite is true. Since such reactions restrict the freedom of movement of two molecules by joining them, the products have *less* disorder than the reactants and $\Delta S°$ has a negative value. The Diels–Alder reaction (Section 4.5) is one such example.

Of the two terms that make up $\Delta G°$, the enthalpy term ($\Delta H°$) is usually larger and more important than the entropy term ($\Delta S°$). Enthalpy changes during reactions are relatively easily measured, and large compilations of data are available. For these reasons, the entropy contribution, $T\Delta S°$, is sometimes ignored when making thermodynamic arguments, and we will do so at times in this book.

In summary, the standard Gibbs free-energy change, $\Delta G°$, is a measure of the overall amount of energy change during a reaction. It is the net result of two contributing terms; one term deals with changes in bond strengths between reactants and products ($\Delta H°$), and one term deals with the amount of disorder caused by the reaction ($\Delta S°$). Table 4.4 describes these terms more fully.

TABLE 4.4 Explanation of thermodynamic quantities: $\Delta G° = \Delta H° - T\Delta S°$

Term	*Name*	*Explanation*
$\Delta G°$	Gibbs free-energy change (kcal/mol)	Overall energy difference between reactants and products. When $\Delta G°$ is negative, a reaction can occur spontaneously. $\Delta G°$ is related to the equilibrium constant by the equation $\Delta G° = RT \ln K_{eq}$
$\Delta H°$	Enthalpy change (kcal/mol)	Heat of reaction; the energy difference between strengths of bonds broken in a reaction and bonds formed
$\Delta S°$	Entropy change (cal/degree mol)	Overall change in freedom of motion or "disorder" resulting from reaction; usually much smaller than $\Delta H°$

4.7 Bond dissociation energies

Energy is released (negative $\Delta H°$) when bonds are made, and energy is consumed (positive $\Delta H°$) when bonds are broken. The measure of this energy change is a quantity known as the **bond dissociation energy.** The bond dissociation energy is defined as the amount of energy required to *homolytically* break a given bond into two radical fragments when the molecule is in the gas phase at 25°C. Each specific bond has its own characteristic strength, and extensive compilations of bond data are available. For example, methane has a measured bond dissociation energy $\Delta H^\circ = +104$ kcal/mol, meaning that 104 kcal/mol is required to break a C—H bond of methane into the two radical fragments $\cdot CH_3$ and $H\cdot$. Conversely, 104 kcal/mol of energy is *released* when a methyl radical and a hydrogen atom combine to form methane (Figure 4.4). Table 4.5 lists some other bond strength data.

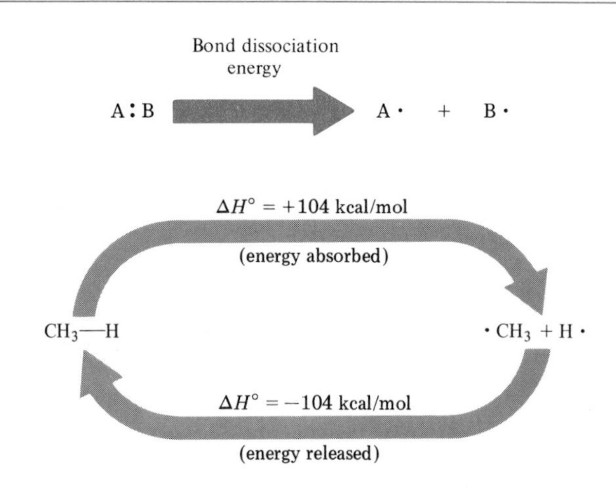

Figure 4.4. Formation of methane.

The data given in Table 4.5 are of great value, and, if enough bond strengths were known, it would seem possible to turn organic chemistry into a more quantitative science. Ideally, if we want to know whether a predicted reaction could occur, we could calculate ΔH° for the process and avoid the potential dangers, problems, and enjoyments of the laboratory. Unfortunately, there are two problems. The first is that the calculation says nothing about the probable rate of reaction; a reaction may have a favorable ΔH° and still not take place. The second problem is that bond dissociation energy data refer only to homolytic cleavages occurring in the gas phase, and are not directly relevant to solution chemistry.

In practice, the vast majority of organic reactions are carried out in solution, and solvent molecules can interact strongly with dissolved reagents (**solvation**). Solvation effects can weaken bonds and cause large changes in the value of ΔH° for a given reaction. The entropy term, ΔS°, can also be affected by solvent molecules, since the solvation of polar reagents by polar solvents causes a certain amount of orientation (reduces the amount of disorder) in the solvent. We can sometimes use bond dissociation energy data to get a rough idea of how thermodynamically favorable a given reaction might be, but we must be aware that the answer is only approximate.

To take a familiar example, what do the thermodynamics look like for the radical chlorination of methane? By totaling the energy required to

TABLE 4.5 Bond dissociation energy data for the reaction A—B $\longrightarrow$ A· + B·

Bond	$\Delta H°$ *(kcal/mol)*	*Bond*	$\Delta H°$ *(kcal/mol)*	*Bond*	$\Delta H°$ *(kcal/mol)*
H—H	104	$(CH_3)_3C$—I	50	CH_3—CH_3	88
H—F	136	H_2C=CH—H	108	C_2H_5—CH_3	85
H—Cl	103	H_2C=CH—Cl	88	$(CH_3)_2CH$—CH_3	84
H—Br	88	H_2C=$CHCH_2$—H	87	$(CH_3)_3C$—CH_3	81
H—I	71	H_2C=$CHCH_2$—Cl	69	H_2C=CH—CH_3	97
Cl—Cl	58			H_2C=$CHCH_2$—CH_3	74
Br—Br	46	C_6H_5—H	112		
I—I	36			C_6H_5—CH_3	102
CH_3—H	104				
CH_3—Cl	84	C_6H_5—Cl	97		
CH_3—Br	70			$C_6H_5CH_2$—CH_3	72
CH_3—I	56				
CH_3—OH	91	$C_6H_5CH_2$—H	85		
CH_3—NH_2	80			$CH_3C(=O)$—H	86
C_2H_5—H	98			HO—H	119
C_2H_5—Cl	81	$C_6H_5CH_2$—Cl	70	HO—OH	51
C_2H_5—Br	68			CH_3O—H	102
C_2H_5—I	53			CH_3S—H	88
C_2H_5—OH	91	C_6H_5—Br	82	C_2H_5O—H	103
$(CH_3)_2CH$—H	95				
$(CH_3)_2CH$—Cl	80			$CH_3C(=O)$—CH_3	77
$(CH_3)_2CH$—Br	68	C_6H_5—OH	112	CH_3CH_2O—CH_3	81
$(CH_3)_3C$—H	91	HC≡C—H	125	NH_2—H	103
$(CH_3)_3C$—Cl	79			H—CN	130
$(CH_3)_3C$—Br	65				

break old bonds and the energy released in making new bonds, we can calculate $\Delta H°$ for the overall reaction,

$$CH_4 + Cl_2 \xrightarrow{h\nu} CH_3Cl + HCl$$

The expression written over the arrow, $h\nu$, indicates irradiation with light (ν is the Greek letter nu).

Bonds broken		*Bonds formed*	
$H_3C{-}H$	$\Delta H° = 104$ kcal/mol	$H_3C{-}Cl$	$\Delta H° = 84$ kcal/mol
$Cl{-}Cl$	$\Delta H° = 58$ kcal/mol	$H{-}Cl$	$\Delta H° = 103$ kcal/mol
	$\Delta H° = 162$ kcal/mol		$\Delta H° = 187$ kcal/mol

$$\Delta H°_{\text{overall}} = \Delta H°_{\text{bonds broken}} - \Delta H°_{\text{bonds formed}}$$

$$\Delta H° = 162 - 187$$

$$= -25 \text{ kcal/mol}$$

We calculate that the gas-phase chlorination of methane is favorable by approximately -25 kcal/mol, and the reaction might therefore occur under suitable conditions. We cannot be *certain* that the reaction will actually occur, since our calculation says nothing about reaction rates or about how favorable the reaction might be in solution. In fact, however, the reaction does occur as written.

We have just seen an example of how bond dissociation energy data can be used to calculate the energetics of a homolytic (radical) substitution reaction. One further point that needs to be explored is the relationship of bond strength data to the *heterolytic* (polar) processes that account for the majority of organic reactions. In theory the overall thermodynamics of a reaction are not affected by the details of how the reaction occurs, and we should therefore be able to use bond strength data to calculate the overall changes in bonding energy for *any* reaction—radical, polar, or pericyclic. In practice, however, polar reactions are often far more susceptible to solvent influences than are radical reactions, and the accuracy of our calculations is correspondingly lower. We will see examples of solvent effects on reactions at numerous places in later chapters.

PROBLEM

4.5 Calculate $\Delta H°$ for each step in the radical chlorination of toluene:

(a) $Cl_2 \xrightarrow{h\nu} 2\,Cl\cdot$

(b) $C_6H_5CH_3 + Cl\cdot \longrightarrow C_6H_5CH_2\cdot + HCl$

(c) $C_6H_5CH_2\cdot + Cl_2 \longrightarrow C_6H_5CH_2Cl + Cl\cdot$

(d) What is the overall $\Delta H°$ for the reaction? [Consider only the propagation steps (b) and (c).]

PROBLEM

4.6 Calculate $\Delta H°$ for these reactions:

(a) $CH_3CH_2OCH_3 + HI \longrightarrow CH_3CH_2OH + CH_3I$

(b) $CH_3Cl + NH_3 \longrightarrow CH_3NH_2 + HCl$

(c) [phenol: benzene ring with OH] + HBr ⟶ [bromobenzene: benzene ring with Br] + H_2O

4.8 Reaction energy diagrams: transition states

In order for a reaction to take place, reactant molecules must collide, and reorganization of atoms and bonds must occur. Let's again consider the reaction of hydroxide ion with bromomethane as an example:

$$HO{:}^- + CH_3{-}Br \longrightarrow CH_3{-}OH + Br{:}^-$$

As the reaction proceeds, the two reagents must approach each other, the carbon–bromine bond must break, a new carbon–oxygen bond must form, and the products methanol and bromide ion must diffuse apart. Over the years, chemists have developed a method of graphically representing the energy changes that occur during a reaction. A **reaction energy diagram** for the reaction of hydroxide ion with bromomethane is shown in Figure 4.5.

The vertical axis of the diagram represents the total energy of all species involved in the reaction, and the horizontal axis represents the progress of the reaction from beginning (left) to end (right). Let's see what happens as the reaction occurs.

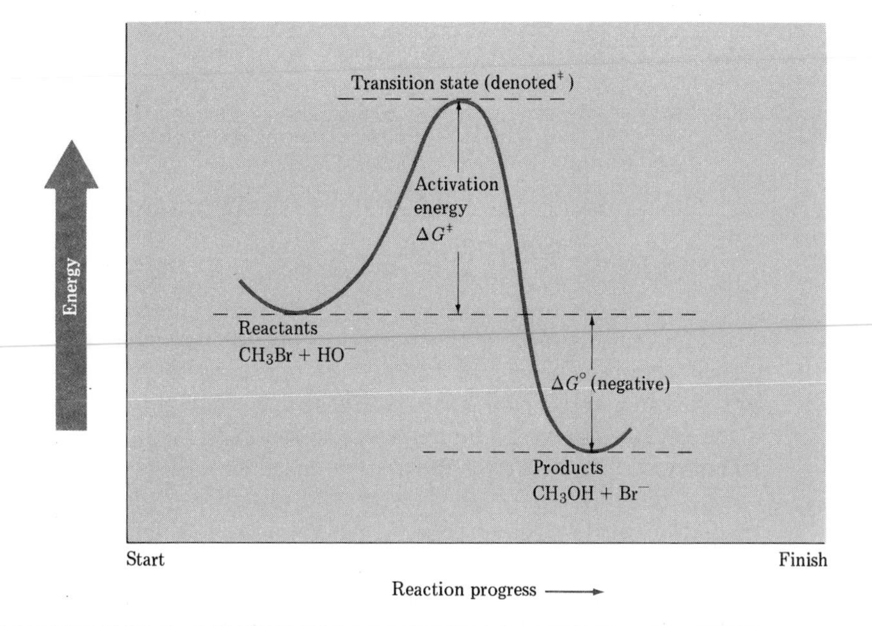

Figure 4.5. Reaction energy diagram for the reaction of bromomethane with hydroxide ion. The energy difference between reactants and transition state, $\Delta G^\ddagger$, controls the reaction rate. The energy difference between reactants and products, ΔG°, controls the position of the equilibrium.

At the start, hydroxide ion and bromomethane are both present in solution, and they have a certain total amount of energy indicated by the reactant level on the diagram. As the two reactants approach each other and reaction commences, a *repulsive* interaction occurs and the energy level therefore rises. This repulsive interaction is due to the spatial (or **steric**) problem of crowding the molecules too closely together. (Recall the similar steric strain in the eclipsed conformer of butane, discussed in Section 3.9.) In electronic terms, the electron clouds of the two reactants approach and repel each other. If the collision has occurred with sufficient force and proper orientation, however, the reactants continue to approach each other until the new carbon–oxygen bond starts to form and the carbon–bromine bond starts to break. At some point, a structure of maximum energy is reached, and we call this structure the **transition state** for the reaction.

Since the transition state represents the *highest*-energy structure involved in the reaction, *it is unstable and cannot be isolated.* We can get no direct information about the exact nature of the transition-state structure, but we can imagine it to be a kind of activated complex of the two reactants in which the new C—O bond is partially formed and the old C—Br bond is partially broken (Figure 4.6). The negative charge is divided between hydroxyl and bromine.

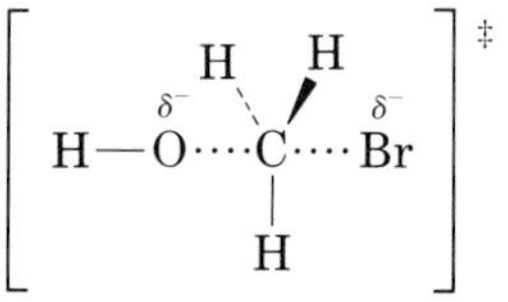

$$\left[\mathrm{H{-}\overset{\delta^-}{O}\cdots C(H)_3 \cdots \overset{\delta^-}{Br}} \right]^{\ddagger}$$

Figure 4.6. A hypothetical transition-state structure for the reaction of bromomethane with hydroxide ion. The C—O bond is just beginning to form, and the C—Br bond is just beginning to break.

The magnitude of the energy difference between reactants and transition state measures how rapidly the reaction takes place and is called the **activation energy, $\Delta G^{\ddagger}$** (the double dagger superscript, ‡, is used to refer to a transition state). If the activation energy is large, the reaction is slow, since very few reacting molecules come together with enough thermal energy to climb the high barrier and reach the high energy level of the transition state. Conversely, if $\Delta G^{\ddagger}$ is small, the reaction is fast, since almost all reacting molecules are energetic enough to climb to the transition state. The rate of a chemical reaction is dependent on the magnitude of the activation energy. Although it is difficult to generalize accurately, most organic reactions have activation energies in the range of 10–35 kcal/mol. Reactions with activation energies less than 20 kcal/mol take place spontaneously at room temperature or below, and reactions with higher activation energies normally require heating. Heat provides the necessary energy input for the reactants to climb the activation barrier.

Once the reactants have reached the transition state, they have an equal probability of going either way. Since both choices are "downhill" energetically once the high point has been reached, the transition-state complex can either come apart and revert to starting materials, or it can go on to give products. Energy is released as the new C—O bond forms fully and the C—Br bond completely breaks; the reaction is complete when the products methanol and bromide ion diffuse apart. The products have a total energy content that is lower than the reactant energy level, and the magnitude of the difference between the two is $\Delta G°$, the free-energy change for the reaction.

Not all reaction energy diagrams are like that for the reaction of bromomethane and hydroxide; each specific reaction has its own specific energy profile. Some reactions are very fast (low $\Delta G^{\ddagger}$) and some are very slow (high $\Delta G^{\ddagger}$); some have a negative value of $\Delta G°$, and some have a positive value of $\Delta G°$. Figure 4.7 illustrates some different possibilities for

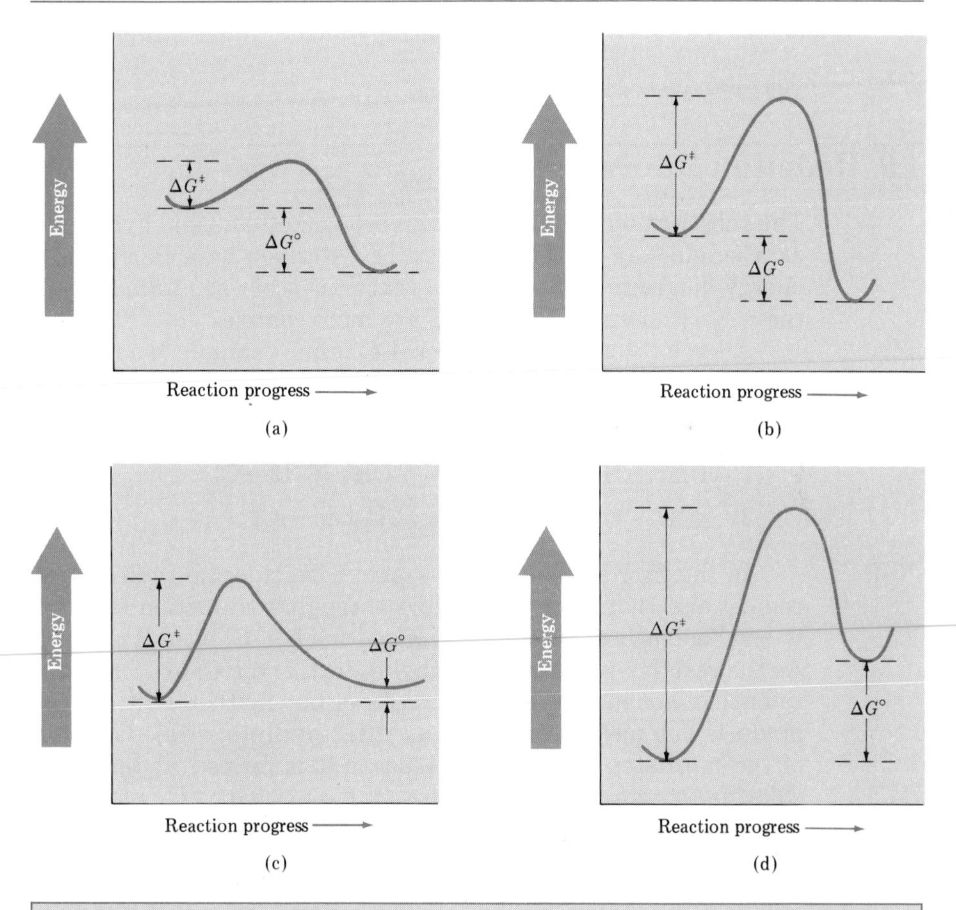

Figure 4.7. Hypothetical reaction energy diagrams for single-step processes. (a) A fast exothermic reaction (small $\Delta G^{\ddagger}$, negative $\Delta G°$); (b) a slow exothermic reaction (large $\Delta G^{\ddagger}$, negative $\Delta G°$); (c) an endothermic reaction (small $\Delta G^{\ddagger}$, positive $\Delta G°$); (d) a slow endothermic reaction (large $\Delta G^{\ddagger}$, positive $\Delta G°$). Note that our use of *exothermic* and *endothermic* to refer to negative and positive values, respectively, of $\Delta G°$ relies on the assumption that $\Delta G° \approx \Delta H°$.

energy profiles. Note in this figure the use of the words *exothermic* and *endothermic* to refer to reactions in which ΔG° is negative and positive, respectively. This usage is not strictly correct, since these words refer only to ΔH° and not to ΔG°. As stated earlier, however, chemists often make the simplifying assumption that ΔG° and ΔH° are approximately equal.

Two points in this discussion of reaction energy diagrams deserve special emphasis. The first concerns the difference between ΔG° and $\Delta G^\ddagger$: ΔG° measures the total amount of free-energy change during a reaction and indicates whether the reaction is endothermic or exothermic; ΔG° does *not* tell how fast the reaction will actually take place. The magnitude of the activation energy, $\Delta G^\ddagger$, *does* determine how rapidly the reaction will occur, but does not tell whether the process will be exothermic or endothermic.

The second point to be emphasized is that the overall energy change occurring during a reaction is measured by ΔG°, not ΔH°. Calculations of changes in bond dissociation energies (ΔH°) such as those illustrated in Section 4.7 are useful, but can only give an *indication* as to whether or not a given reaction will have a favorable equilibrium constant. Such calculations do not take solvent effects or entropy factors into account.

4.9 Reaction intermediates

The substitution reaction between bromomethane and hydroxide ion occurs as a single-step process and can be described by a fairly simple reaction energy diagram. Other kinds of reactions occur as *multistep* processes, and their reaction energy diagrams are more complex.

Alkane chlorination (Section 4.4) is one example of a multistep process. When methane gas reacts with chlorine under radical conditions, two propagation steps are involved:

1. $CH_3\text{—}H + Cl\cdot \longrightarrow \cdot CH_3 + H\text{—}Cl$
2. $\cdot CH_3 + Cl\text{—}Cl \longrightarrow CH_3Cl + Cl\cdot$

In the first step, methane reacts with chlorine radical to give methyl radical and HCl. This step occurs through a transition state in which the C—H bond is beginning to break and the H—Cl bond is beginning to form. As the reaction proceeds, a methyl radical and HCl are formed. The methyl radical is *not* itself a transition state, but neither is it a stable, isolable product. The methyl radical is an **intermediate,** formed during the course of the multistep reaction. As soon as it is formed, it reacts further with chlorine to give the final product, chloromethane. This second step occurs through a *second* transition state in which the new C—Cl bond is forming while the Cl—Cl bond is breaking.

We can construct a reaction energy diagram for the overall chlorination reaction, as shown in Figure 4.8. In essence, we draw diagrams for the two individual steps and then join them in the middle so that the *product* of step 1 serves as the *starting material* of step 2. As indicated, the intermediate methyl radical lies at the energy minimum between steps 1 and 2. Although intermediates in certain reactions may be stable enough to isolate, the intermediate in this case ($\cdot CH_3$) has only fleeting existence.

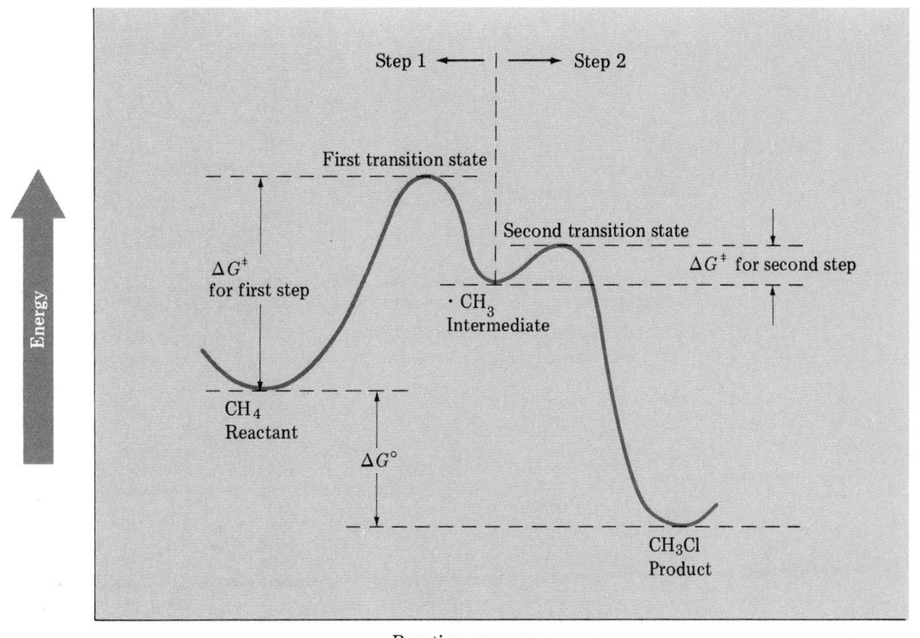

Figure 4.8. Overall reaction energy diagram for chlorination of methane. Two discrete steps are involved, each with its own transition state. The energy minimum between the two steps represents the intermediate methyl radical.

Each individual step in a multistep process can be considered separately. Each step has its own $\Delta G^{\ddagger}$ (rate) and its own ΔG° (equilibrium constant). The overall ΔG°, however, is the energy difference between initial reactants and final products. This is always true, regardless of the shape of the reaction energy curve, and Figure 4.9 (p. 116) illustrates some different possible cases. We will return for a more detailed look at a multistep reaction in Section 10.7.

PROBLEM

4.7 Draw a reaction energy diagram for the following Diels–Alder pericyclic reaction. Assume the reaction has negative ΔG°.

$$\text{(butadiene)} + \text{(methyl vinyl ketone)} \xrightarrow{\Delta} \text{(4-acetylcyclohexene)}$$

Make a rough drawing of what you imagine the transition state for this reaction might look like.

PROBLEM

4.8 Draw a reaction energy diagram for the light-induced cleavage of Cl_2 into two chlorine atoms. Is ΔG° for this reaction likely to be positive or negative? Label the parts of your diagram corresponding to $\Delta G^{\ddagger}$ and ΔG°.

$$Cl_2 \xrightarrow{\text{Light}} 2\ Cl\cdot$$

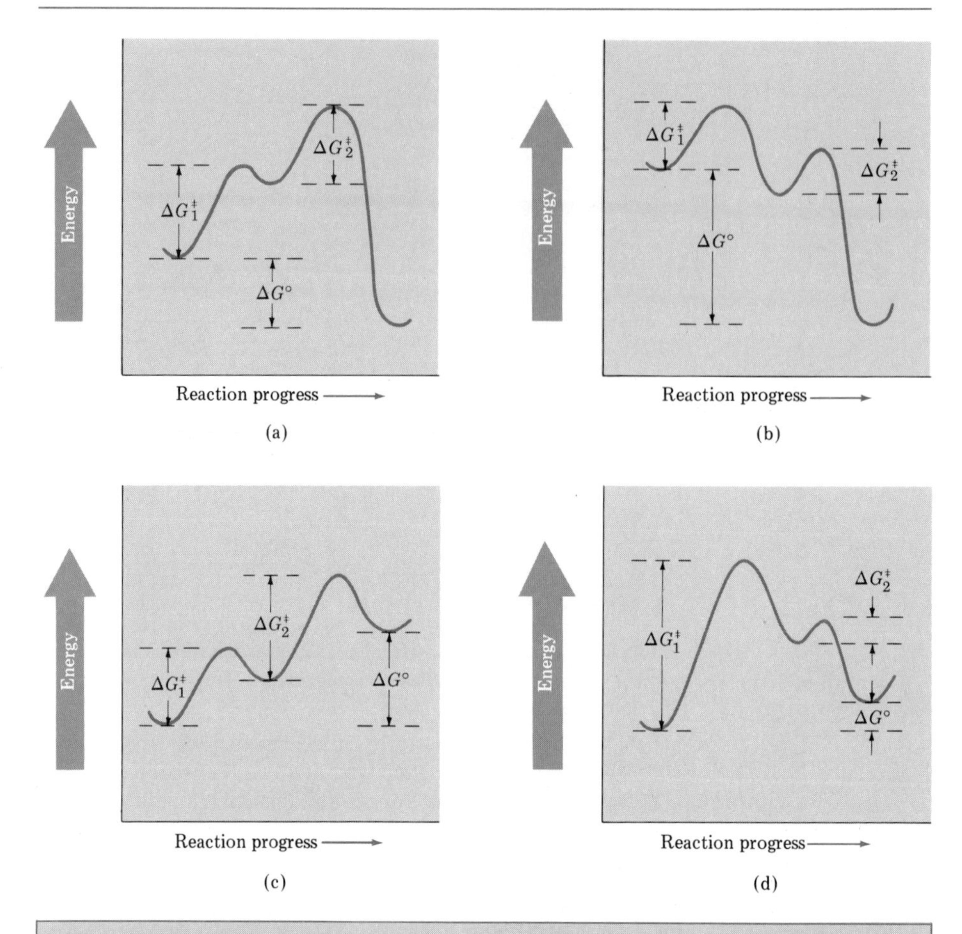

Figure 4.9. Hypothetical reaction energy diagrams for some two-step reactions. [Reactions (a) and (b) are exothermic, and reactions (c) and (d) are endothermic.] The overall ΔG° for any reaction, regardless of complexity, is the energy difference between starting material level and final product.

4.10 Summary

There are three fundamental kinds of reactions in organic chemistry: polar reactions, radical reactions, and pericyclic reactions. Polar reactions are the most common of the three and occur as the result of attractive interactions between an electron-rich site (nucleophile) in the functional group of one molecule and an electron-poor site (electrophile) in the functional group of another molecule. Bonds are formed in polar reactions when the nucleophile donates an electron pair to the electrophile. Radical reactions involve odd-electron species, and bonds are formed when each reactant donates one electron to the new bond. Pericyclic reactions involve neither nucleophile–electrophile interaction nor radicals, but instead occur in a single step by a redistribution of bonding electrons in a cyclic transition state.

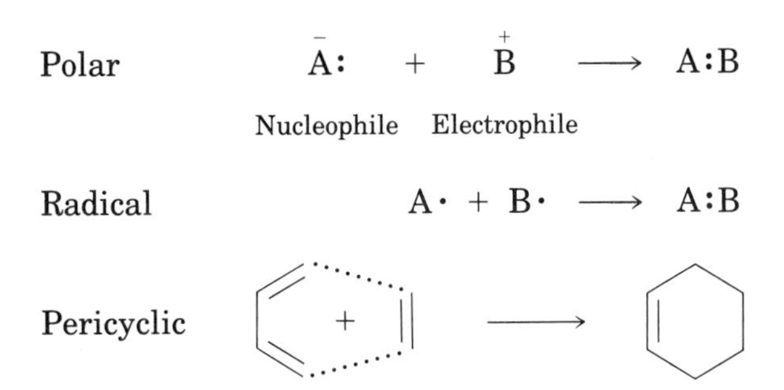

The energy changes taking place during reactions can be described by considering both rates (how fast reaction occurs) and equilibria (how much reaction occurs). All chemical reactions can be written as equilibrium processes, and the position of the equilibrium is determined by the Gibbs free-energy change (ΔG°) that takes place during reaction. The free-energy change is composed of two parts, $\Delta G^\circ = \Delta H^\circ - T\Delta S^\circ$. The enthalpy term ($\Delta H^\circ$) corresponds to the net change in strength of chemical bonds broken and formed during reaction. The entropy term (ΔS°) corresponds to the change in disorder during reaction. Since the enthalpy term is usually larger and more important than the entropy term, chemists often make the assumption that $\Delta G^\circ \approx \Delta H^\circ$.

Reaction energetics can be described pictorially using reaction energy diagrams that follow the reaction course from starting material through transition state to product (Figure 4.10).

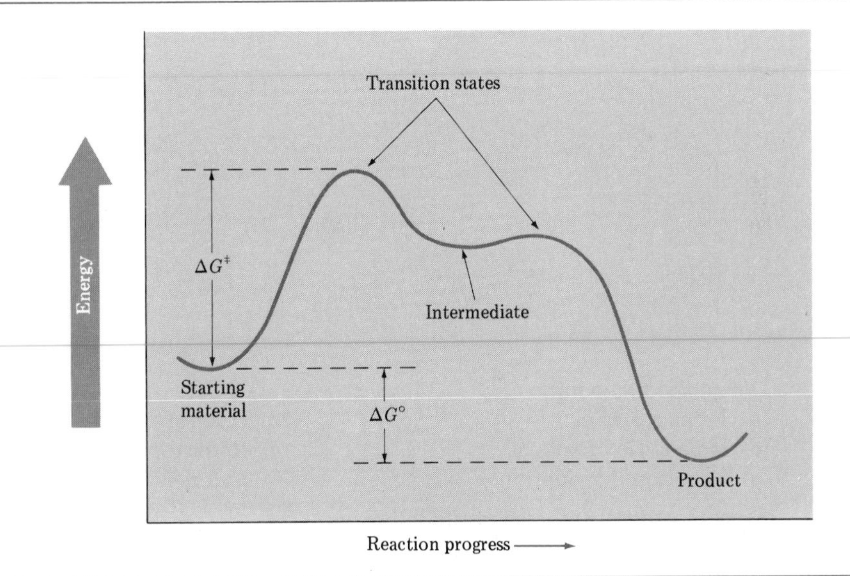

Figure 4.10. A reaction energy diagram.

The transition state is an activated complex occurring at the highest-energy point during reaction. The amount of energy needed by reactants to reach this high point is the activation energy, $\Delta G^\ddagger$, and it is the magnitude of $\Delta G^{\ddagger}$ that determines the rate of the reaction: The higher the activation energy, the slower the reaction. Most organic reactions have $\Delta G^{\ddagger} = 10\text{–}35$ kcal/mol. Many reactions take place in more than one step and involve the formation of intermediates. An intermediate is a species that lies at an energy minimum on the reaction curve and is formed briefly during the course of a reaction.

ADDITIONAL PROBLEMS

4.9 Identify the functional groups present in these molecules and predict the direction of polarity in each.

(a) CH_3CN

(b) 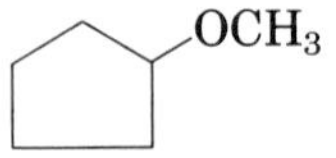 OCH_3

(c) 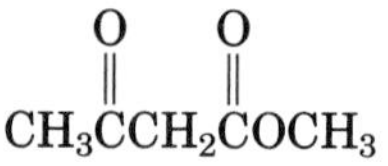 $CH_3\overset{O}{\overset{\|}{C}}CH_2\overset{O}{\overset{\|}{C}}OCH_3$

(d) 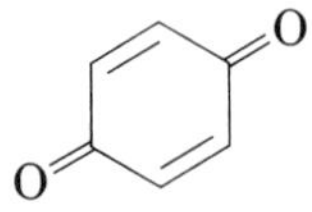 A quinone

4.10 Identify these reactions as being polar, radical, or pericyclic:

(a) 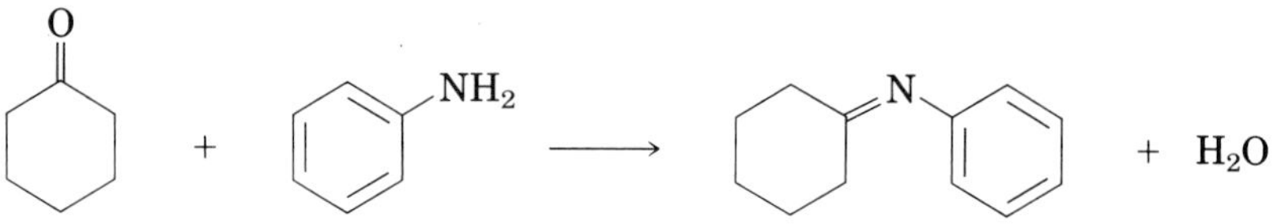 + NH_2 ⟶ N + H_2O

(b) 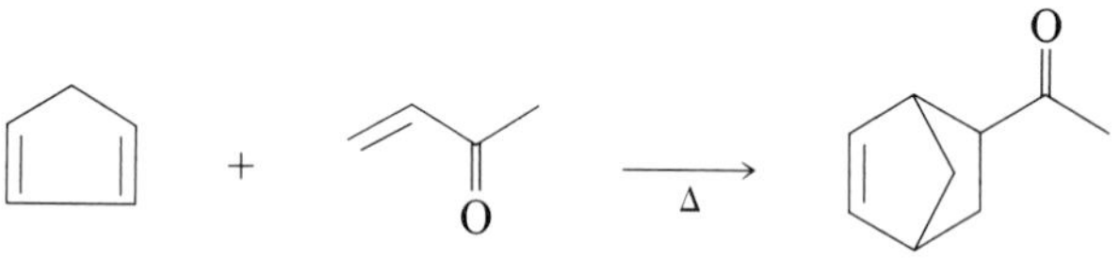 + ⟶ (Δ)

(c) 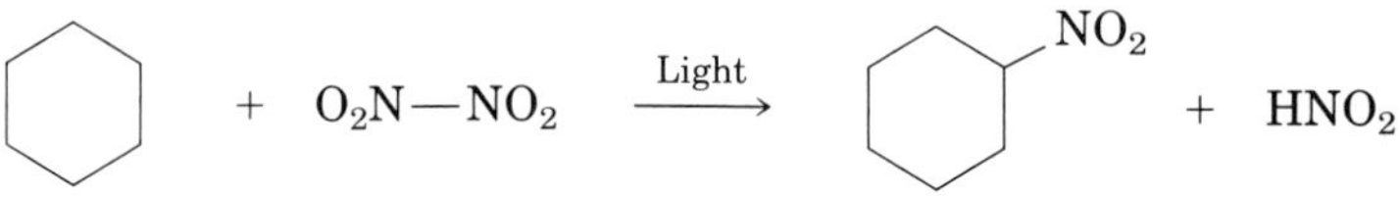 + $O_2N—NO_2$ ⟶ (Light) NO_2 + HNO_2

4.11 Define the following:

(a) Polar reaction
(b) Heterolytic breakage
(c) Homolytic breakage
(d) Radical reaction
(e) Functional group
(f) Polarization

4.12 Give an example of each of the following:

(a) A nucleophile
(b) An electrophile
(c) A polar reaction
(d) A substitution reaction
(e) A heterolytic bond breakage
(f) A homolytic bond breakage

4.13 Which of these would you classify as nucleophiles, and which as electrophiles?

(a) 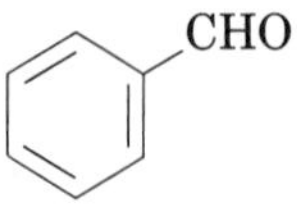 CHO

(b) Br^-

(c) Br^+

(d) $CH_3\overset{\overset{O}{\|}}{C}O^-$

(e) CH_3Br

(f) CH_3COCH_3

(g) $CH_3CH_2\ddot{N}H_2$

(h) BF_3

4.14 Draw a reaction energy diagram for a one-step endothermic reaction. Label the parts of the diagram corresponding to reactants, products, transition state, $\Delta G^\ddagger$, and ΔG°. Is ΔG° positive or negative?

4.15 Draw a reaction energy diagram for a two-step exothermic reaction. Label the overall $\Delta G^\ddagger$, ΔG°, transition states, and intermediate. Is ΔG° positive or negative?

4.16 State clearly the difference between a transition state and an intermediate.

4.17 Consider the reaction energy diagram shown here and answer the following questions:

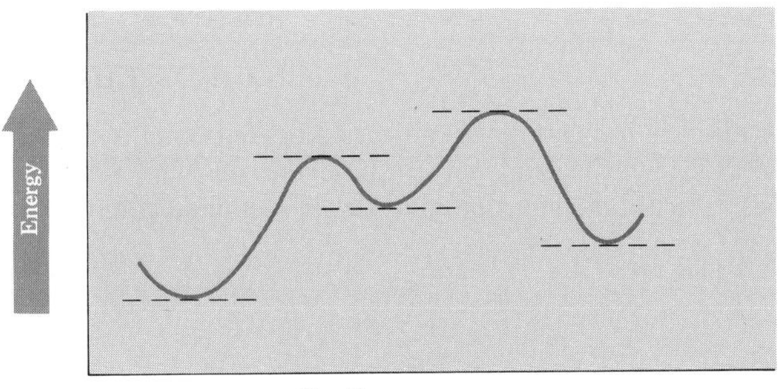

(a) Indicate ΔG° for the reaction. Is it positive or negative?
(b) How many steps are involved in the reaction?
(c) Which step is faster?
(d) How many transition states are there? Label them.

4.18 Calculate ΔH° for these reactions:

(a) $CH_3OH + HBr \longrightarrow CH_3Br + H_2O$

(b) $CH_3CH_2OH + CH_3Cl \longrightarrow CH_3CH_2OCH_3 + HCl$

(c) $t\text{-}C_4H_9Br + HI \longrightarrow t\text{-}C_4H_9I + HBr$

4.19 Calculate ΔH° for the following reactions:

(a) $CH_3CH_3 + Cl_2 \longrightarrow CH_3CH_2Cl + HCl$

(b) $CH_3CH_3 + Br_2 \longrightarrow CH_3CH_2Br + HBr$

(c) $CH_3CH_3 + I_2 \longrightarrow CH_3CH_2I + HI$

What can you conclude about the relative ease of chlorination, bromination, and iodination?

4.20 An alternative course for the reaction of bromine with ethane would result in the formation of bromomethane:

$$CH_3{-}CH_3 + Br_2 \longrightarrow 2\ CH_3Br$$

Calculate ΔH° for this reaction, and comment on how it compares with ΔH° for the formation of bromoethane (calculated in Problem 4.19).

4.21 On the basis of the information in Table 4.5, rank these radicals in order of stability:

$CH_3\cdot$ $\quad CH_3CH_2\cdot$ $\quad$ (phenyl radical) $\quad CH_2{=}CHCH_2\cdot$ $\quad t\text{-}C_4H_9\cdot$

4.22 Radical chlorination of alkanes is of interest mechanistically, but is of little general utility because mixtures of products usually result when more than one kind of C—H bond is present in the substrate. Calculate approximate $\Delta H°$ values for the possible monochlorination reactions of 2-methylbutane. You may use the bond dissociation energies measured for CH_3CH_2—H, H—$CH(CH_3)_2$, and H—$C(CH_3)_3$ as representative of typical primary, secondary, and tertiary C—H bonds, respectively.

$$CH_3CH_2\text{—}C(CH_3)_2\text{—}H \xrightarrow[\text{Light}]{Cl_2} CH_2ClCH_2CH(CH_3)_2 + CH_3CHClCH(CH_3)_2 + CH_3CH_2CCl(CH_3)_2 + CH_3CH_2CH(CH_3)CH_2Cl$$

4.23 Despite the limitations of radical halogenation of hydrocarbons, the reaction is still useful for synthesizing certain halogenated compounds. For which of the following does radical halogenation give single monohalogenation products?

(a) C_2H_6

(b) $(CH_3)_2CH_2$

(c) 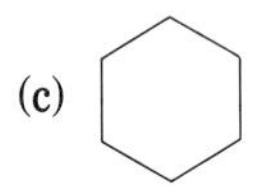

(d) $(CH_3)_3CCH_2CH_3$

(e) 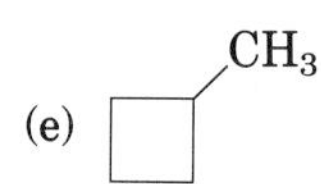

(f) $CH_3C{\equiv}CCH_3$

4.24 We have stated that the chlorination of methane proceeds by the following steps:

(a) $Cl_2 \xrightarrow{h\nu} 2\,Cl\cdot$

(b) $Cl\cdot + CH_4 \longrightarrow HCl + \cdot CH_3$

(c) $\cdot CH_3 + Cl_2 \longrightarrow CH_3Cl + Cl\cdot$

Alternatively, one might propose a different series of steps:

(d) $Cl_2 \xrightarrow{h\nu} 2\,Cl\cdot$

(e) $Cl\cdot + CH_4 \longrightarrow CH_3Cl + H\cdot$

(f) $H\cdot + Cl_2 \longrightarrow HCl + Cl\cdot$

Calculate $\Delta H°$ for each individual step in both possible routes. What insight does this provide into the relative merits of each route?

CHAPTER 5 ALKENES: STRUCTURE AND REACTIVITY

Alkenes are hydrocarbons that contain a carbon–carbon double bond functional group. The word **olefin** is often used as a synonym in the chemical literature, but alkene is the generally preferred term and the one we will use. Alkenes occur abundantly in nature, and many have important consequences in biology. For example, ethylene is a plant hormone that induces ripening in fruit, and α-pinene is the major component of turpentine.

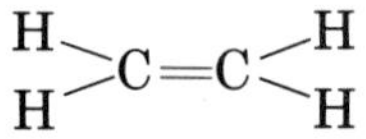

$$H_2C{=}CH_2$$

Ethylene

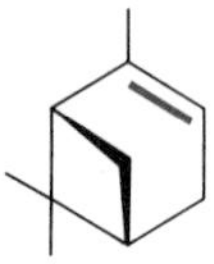

α-Pinene

Life itself would be impossible without such alkenes as β-carotene and squalene, compounds that contain *many* double bonds. β-Carotene, the orange pigment responsible for the color of carrots, has 11 double bonds, and serves as a valuable dietary source of vitamin A. Squalene, an alkene widely distributed throughout nature, has 6 double bonds and serves as the biological precursor from which all steroid hormones are made.

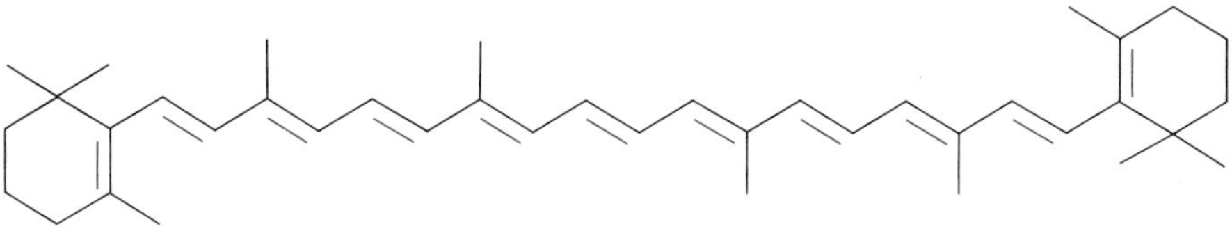

β-Carotene
(orange pigment and vitamin A precursor)

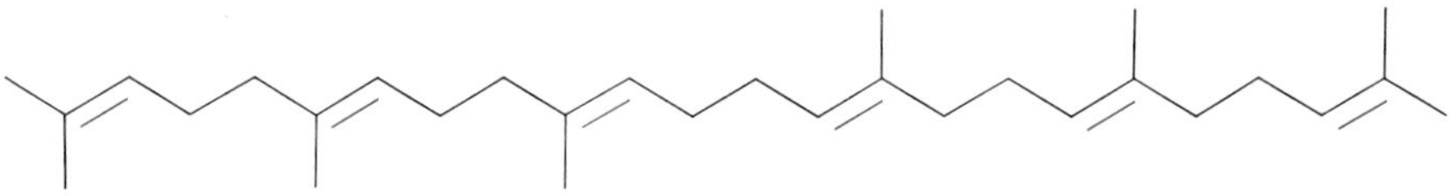

Squalene
(steroid hormone precursor)

5.1 Industrial preparation and use of alkenes

Ethylene and propylene, the two simplest alkenes, are the two most important organic chemicals produced industrially. More than 27 billion lb of ethylene and 13 billion lb of propylene are produced each year in the United States for use in the synthesis of polyethylene, polypropylene, ethylene glycol, acetic acid, acetaldehyde, and a host of other raw materials (Figure 5.1).

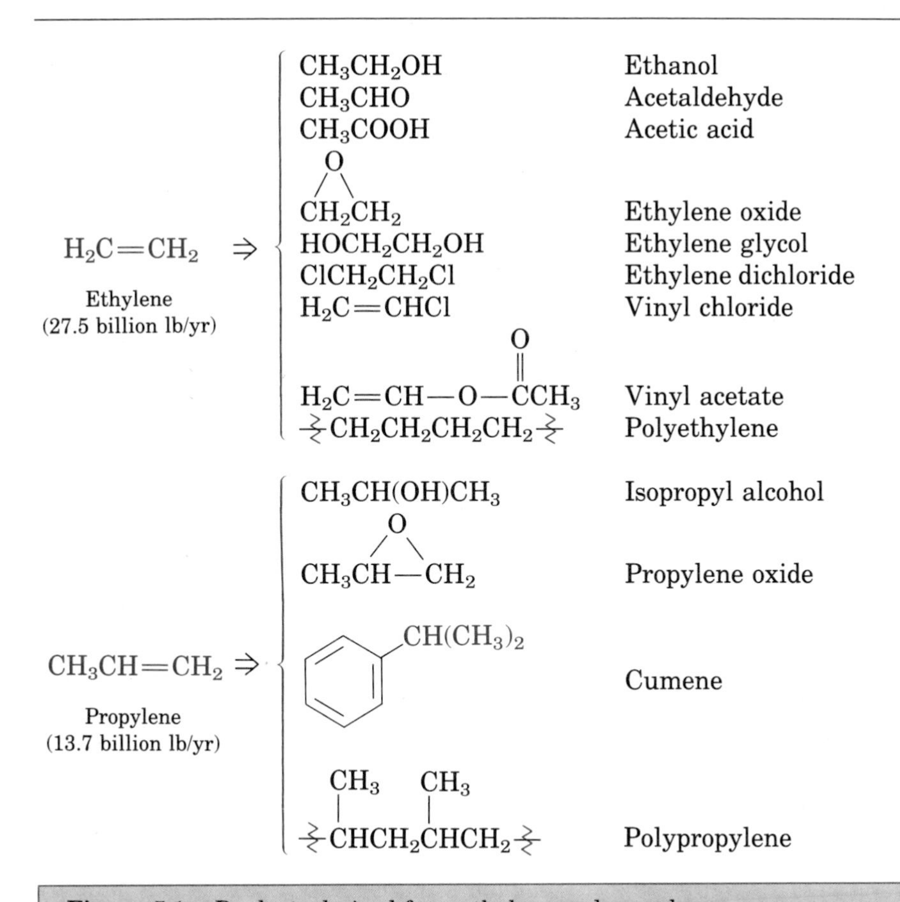

Figure 5.1. Products derived from ethylene and propylene.

Ethylene, propene, and butene are synthesized industrially by thermal cracking of both natural gas (C_1–C_4 alkanes) and straight-run gasoline (C_4–C_8 *n*-alkanes).

$$CH_3(CH_2)_nCH_3 \xrightarrow[\text{Steam}]{850\text{–}900^\circ C} H_2 + CH_4 + H_2C{=}CH_2 + CH_3CH{=}CH_2 + CH_3CH_2CH{=}CH_2 \qquad n = 0 - 6$$

Thermal cracking, introduced in 1912, takes place in the absence of catalysts at extremely high temperatures up to 900°C and undoubtedly involves radical reactions. Although the exact processes are complex, the severe conditions cause spontaneous homolysis of carbon–carbon and carbon–hydrogen bonds, with resultant formation of smaller fragments:

$$CH_3CH_2{-}CH_2CH_3 \xrightarrow{900^\circ C} 2\,\dot{C}H_2{-}CH_2\cdot \longrightarrow 2\,H_2C{=}CH_2 + H_2$$

As discussed earlier (Section 4.6), thermal cracking is an example of a reaction whose energetics are dominated by entropy ($T\Delta S^\circ$) rather than by enthalpy ($\Delta H°$) in the free-energy equation $\Delta G° = \Delta H° - T\Delta S°$. The large positive entropy change resulting from the fragmentation of one larger molecule into several smaller pieces, together with the extremely high temperature, T, makes the $T\Delta S°$ term larger than $\Delta H°$ and thus favors the cracking reaction.

5.2 Calculation of the degree of unsaturation

Alkanes have the general formula C_nH_{2n+2} and cycloalkanes have the formula C_nH_{2n}. Since a *pair* of hydrogen atoms must be removed from an alkane to generate an alkene, acyclic alkenes with one carbon–carbon double bond also have the formula C_nH_{2n}. Alkenes always have fewer hydrogens than do alkanes and are often referred to as being **unsaturated.** Alkanes, by contrast, have as many hydrogens as possible and are thus **saturated.** In general terms, each ring or double bond in a molecule causes a *pair* of hydrogens to be removed from the alkane formula C_nH_{2n+2}. This knowledge is quite useful since it allows us to work backwards from a molecular formula to calculate the **degree of unsaturation**—how many rings and/or double bonds are present in an unknown.

Let's assume, for example, that we wish to find the structure of an unknown hydrocarbon. A molecular weight determination on our unknown yields a value of 82, which corresponds to a molecular formula of C_6H_{10}. Since the fully saturated C_6 hydrocarbon, hexane, has the formula C_6H_{14}, we can calculate that our unknown compound has two fewer pairs of hydrogens ($H_{14} - H_{10} = H_4$). Our unknown therefore contains two double bonds, one ring and one double bond, or two rings (or one *triple* bond). We still have a long way to go to establish structure, but our simple calculation has greatly narrowed the field of choice.

Similar calculations can be carried out for compounds containing elements other than just carbon and hydrogen.

1. *Organohalogen compounds,* C, H, X, where X = Cl, Br, or I. Since a halogen substituent is simply a replacement for hydrogen (both are monovalent), we can *add* the number of halogens and the number of hydrogens to arrive at a base hydrocarbon formula from which the number of double bonds and/or rings can be found. For example,

$$C_4\underbrace{H_5Br_3}_{\text{Add}} = \text{“}C_4H_8\text{”} \qquad \text{One double bond and/or ring}$$

2. *Organooxygen compounds,* C, H, O. Since oxygen is divalent, it does not affect the formula of the parent hydrocarbon. The easiest way to convince yourself of this is to see what happens when an oxygen atom is inserted into an alkane C—C or C—H bond—there is *no* change in the number of hydrogen atoms. For example,

$$H_2C{=}CHCH{=}CHCH_2{-}O{-}H = H_2C{=}CHCH{=}CH{-}CH_2\overset{\text{O removed}}{\downarrow}{-}H$$

$$C_5H_8O = \text{“}C_5H_8\text{”} \qquad \text{Two double bonds or rings}$$

3. *Organonitrogen compounds,* C, H, N. Since nitrogen is trivalent, an organonitrogen compound has one more hydrogen than its base hydrocarbon. We therefore *subtract* the number of nitrogens from the number of hydrogens to arrive at a base hydrocarbon formula. Again, the best way to convince yourself of this is to see what happens when a nitrogen atom is inserted into an alkane bond—another hydrogen atom is required to fill the third valency on nitrogen, and we must therefore mentally subtract this extra hydrogen atom to arrive at the corresponding base hydrocarbon formula. For example,

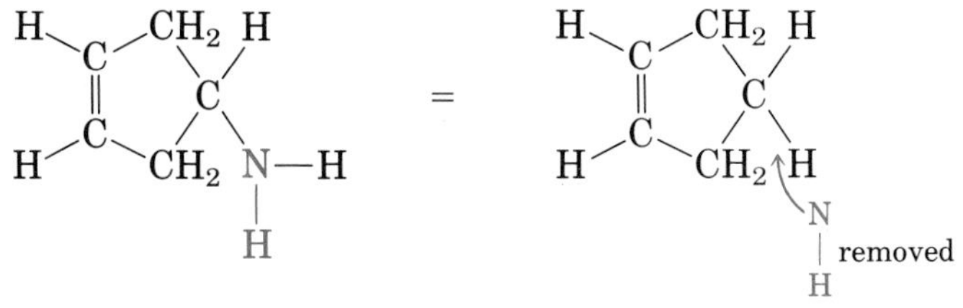

```
H\  /CH2  H              H\  /CH2  H
   C     \ /                C     \ /
   ||     C         =       ||     C
   C     / \                C     / \
H/  \CH2  N—H            H/  \CH2 ↑H
          |                        \N
          H                         | removed
                                    H
```

C_5H_9N = "C_5H_8" Two double bonds or rings

PROBLEM

5.1 Calculate the number of double bonds and/or rings in these formulas:

(a) C_8H_{14} (b) C_5H_6
(c) $C_{12}H_{20}$ (d) $C_{20}H_{32}$
(e) $C_{40}H_{56}$, β-carotene (f) $C_{30}H_{50}$, squalene

PROBLEM

5.2 Draw as many structures as you can that fit these formulas:

(a) C_4H_8 (b) C_4H_6
(c) C_3H_4 (d) C_5H_8

PROBLEM

5.3 Calculate the number of double bonds and/or rings in these formulas:

(a) C_6H_5N (b) $C_6H_5NO_2$
(c) $C_8H_9Cl_3$ (d) $C_9H_{16}Br_2$
(e) $C_{10}H_{12}N_2O_3$ (f) $C_{20}H_{32}O_2$, arachidonic acid

5.3 Electronic structure of alkenes

A carbon–carbon double bond consists of two parts, as we saw earlier—a sigma bond and a pi bond. The carbon atoms are sp^2 hybridized and have three equivalent orbitals directed to the corners of an equilateral triangle. The fourth carbon orbital is an unhybridized p orbital, which is perpendicular to the sp^2 plane. When two such carbon atoms approach each other, they form two kinds of bonds—a sigma bond formed by head-on overlap of sp^2 orbitals, and a pi bond formed by sideways overlap of p orbitals. The doubly bonded carbons and the four groups attached to them therefore lie in a plane, and the bond angles are approximately 120° (Figure 5.2, p. 126).

We might expect the carbon–carbon double bond to be both stronger and shorter than a carbon–carbon single bond, and this is the case. Table 5.1 compares the experimentally determined bond parameters of ethane and ethylene.

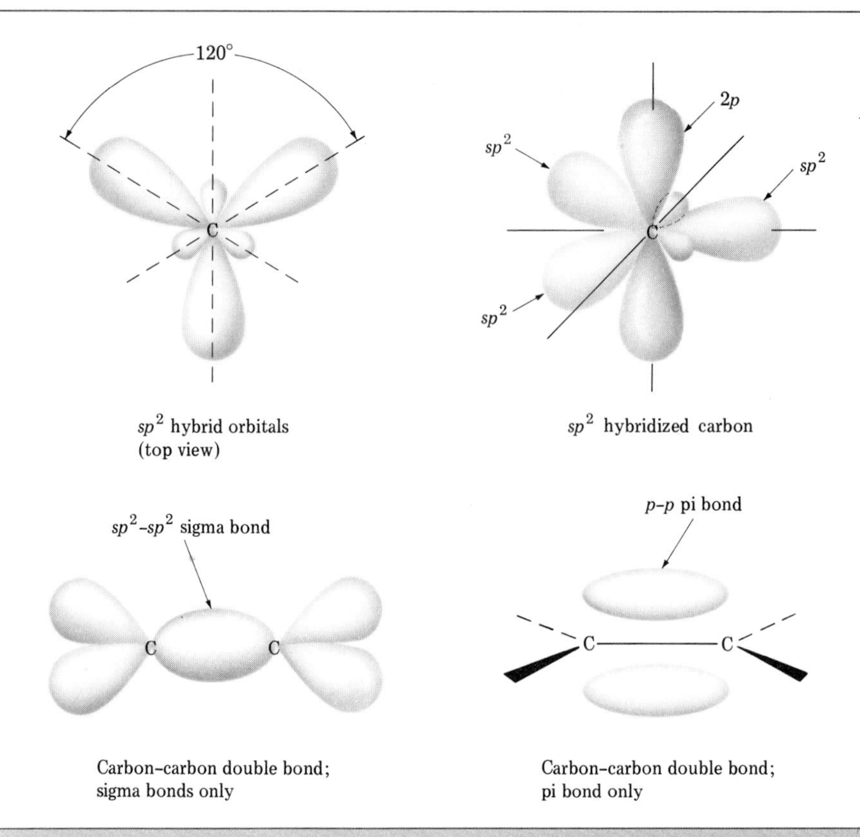

Figure 5.2. An orbital picture of the carbon–carbon double bond.

TABLE 5.1 Molecular parameters for ethylene and ethane[a]

	Ethylene	*Ethane*
H—C—H bond angle (degrees)	116.6	109.5
H—C—C bond angle (degrees)	121.7	109.5
C—C bond strength (kcal/mol)	152	88
C—C bond length (Å)	1.33	1.54
C—H bond strength (kcal/mol)	103	98
C—H bond length (Å)	1.076	1.10

[a]The double bond is both stronger and shorter than the single bond.

The presence of the double bond in alkenes has numerous consequences. One consequence is the phenomenon of **restricted rotation.** We know that relatively free rotation is possible around sigma bonds, and that open-chain alkanes such as butane therefore have an infinite number of rapidly interconverting conformations. The same is *not* true for double bonds. Carbon–carbon double bonds do not have circular cross-sections, and rotation cannot occur freely (Figure 5.3).

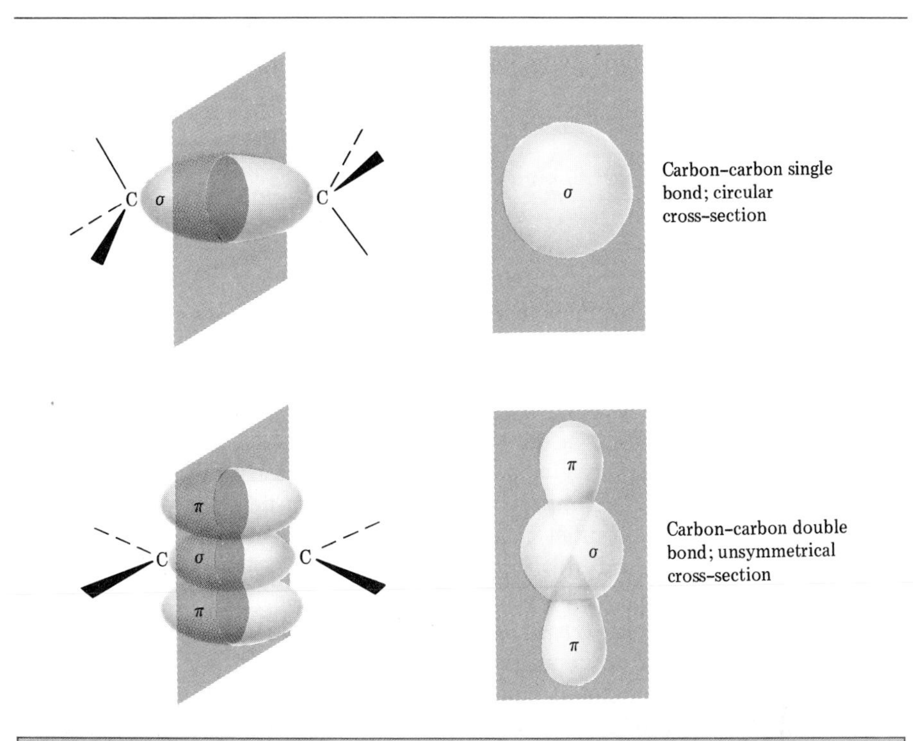

Figure 5.3. Cross sections of carbon–carbon single and double bonds.

If we were to *force* rotation to occur, we would need to break the pi bond temporarily (Figure 5.4, p. 128). Thus the barrier to rotation must be at least as great as the strength of the pi bond.

Breaking a chemical bond normally requires a large amount of energy. We can make a rough estimate of how much energy is required to break the pi bond of an alkene by subtracting the value for the strength of an average carbon–carbon sigma bond from the total bond strength value for ethylene.

Ethylene C=C bond strength (sigma + pi)	152 kcal/mol
Ethane C—C bond strength (sigma only)	− 88 kcal/mol
Difference (pi bond only)	64 kcal/mol

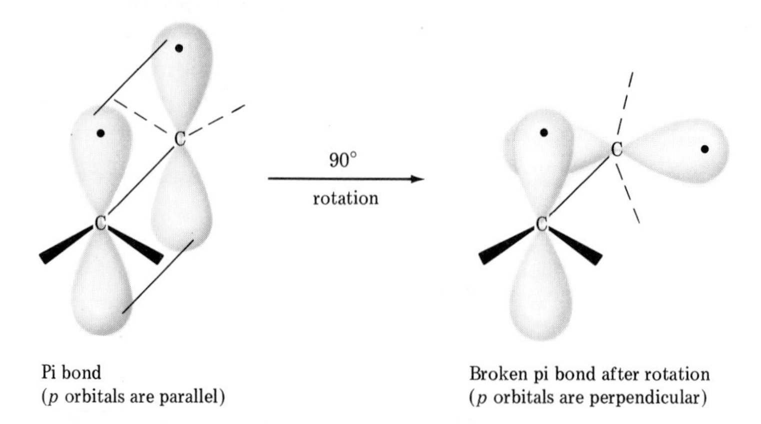

Figure 5.4. Breaking the pi bond to bring about rotation.

We predict an approximate bond strength of 64 kcal/mol for the ethylene pi bond, and it is therefore clear why rotation cannot occur (recall that the barrier to rotation for ethane is only 2.9 kcal/mol).

The lack of rotation around the carbon–carbon double bond is of more than just theoretical interest; it also has chemical consequences. Imagine the situation for a disubstituted alkene such as 2-butene. There are *two* possible structures for 2-butene: The two methyl groups can either be on the *same* side of the double bond, or they can be on *opposite* sides, a situation reminiscent of substituted cycloalkanes (Section 3.10; see Figure 5.5).

Since bond rotation cannot occur, the two 2-butenes do not spontaneously interconvert; they are distinct, isolable compounds. We call such compounds **cis–trans isomers,** as we saw in Chapter 3, because they have the same formula and overall skeleton, but they differ in the spatial arrangement of the atoms. The compound with substituents on the same side is referred to as *cis*-2-butene; the isomer with substituents on opposite sides is *trans*-2-butene. Cis–trans isomerism is a common feature of alkene

Side view: $H, H_3C > C{=}C < H, CH_3$ — Side view: $H, H_3C > C{=}C < CH_3, H$

Top view: $(H)(CH_3)C{=}C(H)(CH_3)$ — Top view: $(H)(CH_3)C{=}C(CH_3)(H)$

cis-2-Butene — *trans*-2-Butene

Figure 5.5. Geometric isomers of 2-butene.

chemistry and is not limited only to disubstituted alkenes. Isomerism can occur whenever each of the double-bond carbons is attached to two different groups. If one of the double-bond carbons is attached to two identical groups, however, then cis–trans isomerism is not possible (Figure 5.6).

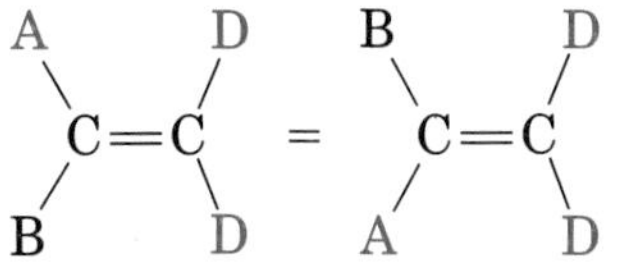

$$\begin{matrix} A & & D \\ & C{=}C & \\ B & & D \end{matrix} \quad = \quad \begin{matrix} B & & D \\ & C{=}C & \\ A & & D \end{matrix}$$

These two compounds are identical; they are not cis–trans isomers.

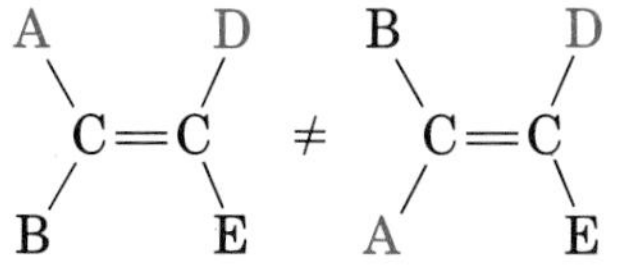

$$\begin{matrix} A & & D \\ & C{=}C & \\ B & & E \end{matrix} \quad \neq \quad \begin{matrix} B & & D \\ & C{=}C & \\ A & & E \end{matrix}$$

These two compounds are not identical; they are cis–trans isomers.

Figure 5.6. The requirement for cis–trans isomerism in alkenes.

PROBLEM

5.4 Which of the following compounds can exist as pairs of cis–trans isomers? Draw each cis–trans pair and indicate the geometry of each isomer.

(a) $CH_3CH{=}CH_2$
(b) $(CH_3)_2C{=}CHCH_3$
(c) $CH_3CH_2CH{=}CHCH_3$
(d) $(CH_3)_2C{=}C(CH_3)CH_2CH_3$
(e) $ClCH{=}CHCl$
(f) $BrCH{=}CHCl$

5.4 Alkene stability

Although the cis–trans interconversion of alkene isomers does not occur spontaneously, it can be made to happen under appropriate experimental conditions (for example, treatment with a strong acid catalyst). If we were to interconvert *cis*-2-butene with *trans*-2-butene and allow them to reach equilibrium, we would find that they are *not of equal stability.* At equilibrium, the ratio of isomers is 76% trans to 24% cis.

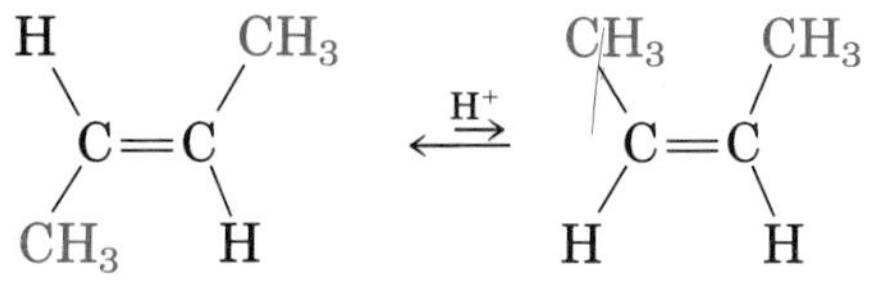

$$\begin{matrix} H & & CH_3 \\ & C{=}C & \\ CH_3 & & H \end{matrix} \quad \overset{H^+}{\rightleftarrows} \quad \begin{matrix} CH_3 & & CH_3 \\ & C{=}C & \\ H & & H \end{matrix}$$

Trans (76%) Cis (24%)

Using the relationship between equilibrium constants and free-energy differences shown in Table 5.2, we can calculate that *trans*-2-butene is more stable than *cis*-2-butene by 0.66 kcal/mol.

TABLE 5.2 Relationship between equilibrium constant and $\Delta G°$: $\Delta G° = -RT \ln K_{eq}$

K_{eq}	*Percentages of isomers*	$\Delta G°$ *(kcal/mol)*
1	50:50	0
2	67:33	0.41
3	75:25	0.65
4	80:20	0.82
5	83:17	0.96
10	91:9	1.37
25	96:4	1.91
50	98:2	2.32
100	99:1	2.74
1000	99.9:0.1	4.10

It turns out to be a general phenomenon that cis alkenes are less stable than their trans isomers because of steric (spatial) strain between the two bulky substituents on the same side of the double bond. This is the same kind of steric interference that we saw previously in the eclipsed conformations of butane (Section 3.9).

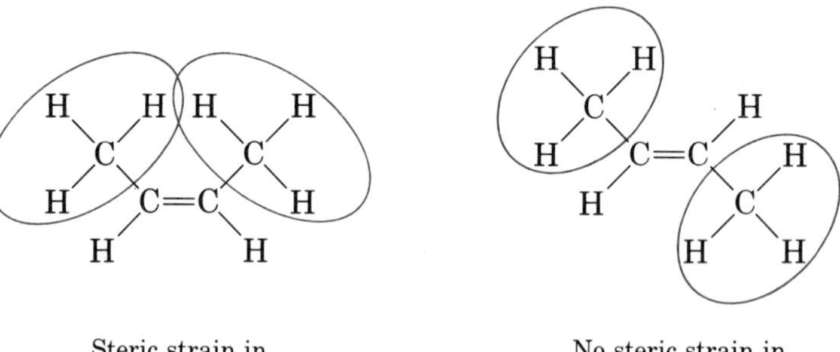

Steric strain in *cis*-2-butene

No steric strain in *trans*-2-butene

Although it is sometimes possible to interconvert double-bond isomers chemically to establish an equilibrium and thus obtain information about relative stabilities of alkenes, there is an easier way to gain nearly the same information. One of the more important reactions that alkenes undergo is **catalytic hydrogenation.** In the presence of catalysts such as palladium or platinum, hydrogen adds to carbon–carbon double bonds to yield the corresponding saturated alkanes.

$$\underset{trans\text{-2-Butene}}{(H)(CH_3)C{=}C(CH_3)(H)} \xrightarrow[Pd]{H_2} \underset{\text{Butane}}{CH_3CH_2CH_2CH_3} \xleftarrow[Pd]{H_2} \underset{cis\text{-2-Butene}}{(CH_3)(H)C{=}C(CH_3)(H)}$$

Consider the hydrogenation of *cis*- and *trans*-2-butene. Both alkenes react with hydrogen to give the *same* product, butane. Energy diagrams for the two reactions are shown in Figure 5.7. Since *cis*-2-butene is less stable than *trans*-2-butene by 0.66 kcal/mol, our energy diagram shows the cis alkene at a higher energy level. After reaction, however, both cis and trans isomers are at the same energy level (butane). It therefore follows that $\Delta G°$ for the cis isomer must be larger than $\Delta G°$ for the trans isomer. In other words, more energy is evolved in the hydrogenation of the cis isomer than of the trans isomer. This difference in isomer stability is due to differences in bond strengths ($\Delta H°$), and, if we were to measure the two heats of reaction and find their difference, we would have an independent means of determining the relative stabilities of cis and trans isomers without having to measure an equilibrium position. A large number of such **heats of hydrogenation ($\Delta H°_{hydrog}$)** have been measured, and the results bear out our expectation. For *cis*-2-butene, $\Delta H°_{hydrog}$ = 28.6 kcal/mol; for the trans isomer, $\Delta H°_{hydrog}$ = 27.6 kcal/mol. (Note that these $\Delta H°$ values are negative since heat is evolved, but that the minus sign is dropped to make comparisons easier.)

Trans isomer: H and CH_3 on opposite sides of C=C ($HC(CH_3)=C(CH_3)H$)

Trans isomer
$\Delta H°_{hydrog}$ = 27.6 kcal/mol

Cis isomer: CH_3 and CH_3 on the same side of C=C

Cis isomer
$\Delta H°_{hydrog}$ = 28.6 kcal/mol

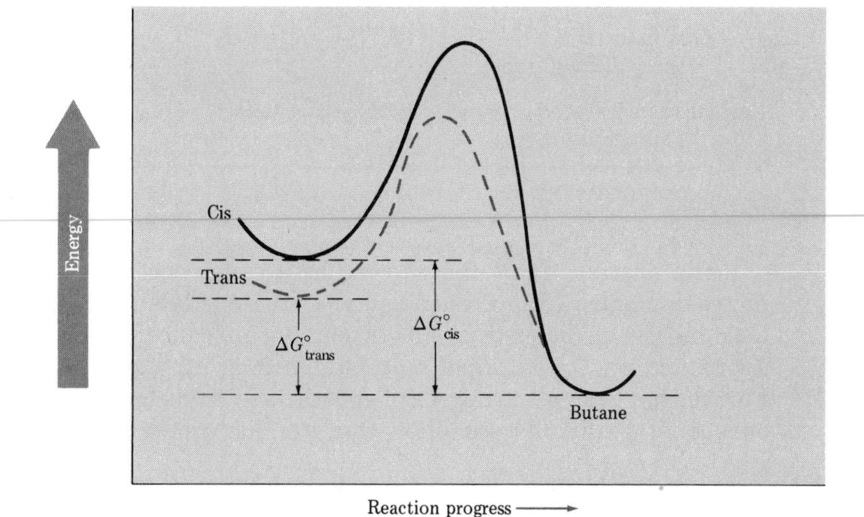

Figure 5.7. Reaction energy diagrams for hydrogenation of *cis*- and *trans*-2-butene. The cis isomer is higher in energy than the trans isomer and therefore gives off more energy in the reaction.

Although the energy difference in the heats of hydrogenation for the 2-butene isomers (1.0 kcal/mol) is in good agreement with the energy difference calculated from equilibrium data (0.66 kcal/mol), the two numbers are not exactly the same. There are two reasons for this. The first is simply experimental error; heats of hydrogenation require considerable expertise and specialized equipment to measure accurately, and we are looking at a small difference between two large numbers. The second reason is that heats of reaction and equilibrium constants do not measure exactly the same quantity. Heats of reaction measure enthalpy changes, $\Delta H°$, whereas equilibrium constants measure overall free-energy changes, $\Delta G°$ ($\Delta G° = \Delta H° - T\Delta S°$). We therefore *expect* a slight difference when comparing the two measurements.

Although heats of hydrogenation are not quite as accurate as we might like, we can nevertheless gain some useful and interesting information from them. Table 5.3 lists some representative data.

TABLE 5.3 Heats of hydrogenation of some alkenes

Substitution	*Alkene*	$\Delta H°_{hydrog}$ *(kcal/mol)*
	$H_2C{=}CH_2$	32.8
Monosubstituted (one alkyl group next to double bond)	$CH_3CH{=}CH_2$	30.1
	$CH_3CH_2CH{=}CH_2$	30.3
	$(CH_3)_2CHCH{=}CH_2$	30.3
	$(CH_3)_3CCH{=}CH_2$	30.3
Disubstituted (two alkyl groups)	Cis $CH_3CH{=}CHCH_3$	28.6
	Trans $CH_3CH{=}CHCH_3$	27.6
	$(CH_3)_2C{=}CH_2$	28.4
	$CH_3CH_2(CH_3)C{=}CH_2$	28.5
Trisubstituted (three alkyl groups)	$(CH_3)_2C{=}CHCH_3$	26.9
Tetrasubstituted (four alkyl groups)	$(CH_3)_2C{=}C(CH_3)_2$	26.6

The data in Table 5.3 show that alkenes become more stable with increasing substitution. For example, ethylene has $\Delta H°_{hydrog} = 32.8$ kcal/mol, but when one alkyl substituent is introduced, an alkene becomes approximately 2.5 kcal/mol more stable than ethylene ($\Delta H°_{hydrog} = 30.3$). Further increasing the degree of substitution leads to further stability, and as a general rule alkenes follow this stability order:

Tetrasubstituted > Trisubstituted > Disubstituted > Monosubstituted

$R_2C{=}CR_2$ > $R_2C{=}CHR$ > $RHC{=}CHR$, $R_2C{=}CH_2$ > $RHC{=}CH_2$

The reasons for this observed stability order for substituted alkenes are not well understood. Several different hypotheses have been advanced, but none have received universal acceptance. Some chemists feel that the order is due to a stabilizing interaction termed **hyperconjugation** (Figure 5.8). Hyperconjugation can be viewed as orbital overlap between the carbon–carbon pi bond and a properly oriented carbon–hydrogen sigma bond on a neighboring substituent. The more substituents that are present, the more opportunities exist for hyperconjugation and the more stable the alkene.

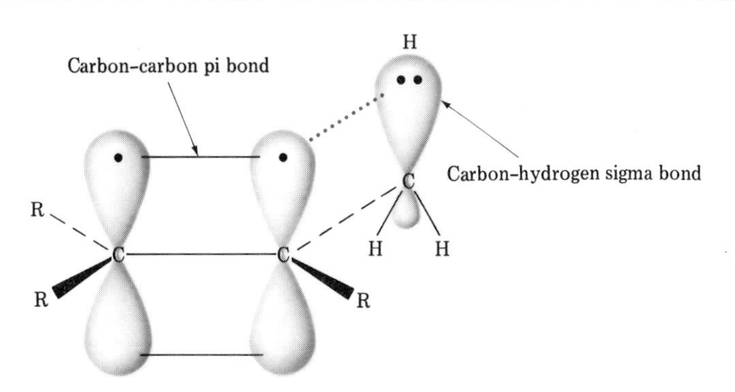

Figure 5.8. Hyperconjugation.

Others view hyperconjugation as unlikely and prefer a simple bond strength argument to account for the observed alkene stability order. Bonds between the sp^2 carbon and the sp^3 carbon are somewhat stronger than bonds between two sp^3 carbons. Thus, in comparing the two isomers, 1-butene and 2-butene, the monosubstituted isomer has one sp^3–sp^3 bond and one sp^3–sp^2 bond, and the disubstituted isomer has two sp^3–sp^2 bonds. Highly substituted alkenes always have a higher ratio of sp^3–sp^2 bonds to sp^3–sp^3 bonds and are therefore more stable than less substituted alkenes.

$$\underset{\substack{\text{2-Butene} \\ \text{(more stable)}}}{\overset{sp^3\text{–}sp^2 \downarrow \qquad\qquad sp^3\text{–}sp^2 \downarrow}{CH_3\text{—}CH\text{=}CH\text{—}CH_3}} \qquad \underset{\substack{\text{1-Butene} \\ \text{(less stable)}}}{\overset{sp^3\text{–}sp^3 \downarrow \quad sp^3\text{–}sp^2 \downarrow}{CH_3\text{—}CH_2\text{—}CH\text{=}CH_2}}$$

Regardless of which explanation ultimately gains acceptance, this controversy illustrates the point that chemistry is a living science. Chemists do not have explanations for everything, and some of what is now believed true might be proven false tomorrow.

PROBLEM

5.5 The double bonds in small-ring cycloalkenes must have cis geometry, because a stable trans double bond is impossible within the confines of a five- or six-membered ring. At some point, however, a ring becomes large enough to accommodate a trans double bond. The following heats of hydrogenation have been measured:

	$\Delta H^\circ_{\text{hydrog}}$ *(kcal/mol)*
cis-Cyclooctene	23.0
trans-Cyclooctene	32.2
cis-Cyclononene	23.6
trans-Cyclononene	26.5
cis-Cyclodecene	20.7
trans-Cyclodecene	24.0

How do you explain these data? Make molecular models of the trans cycloalkenes to see their conformations.

5.5 Nomenclature of alkenes

Alkenes are systematically named by following a series of rules similar to those developed for alkanes, with the suffix *-ene* used in place of *-ane*. Three basic steps suffice:

Step 1. Name the parent hydrocarbon. Find the longest carbon chain containing the double bond, and name the compound accordingly, using the suffix *-ene:*

$$(CH_3CH_2CH_2)(CH_3CH_2CH_2CH_2)C{=}CHCH_3$$

Named as a heptene

NOT

$$\boxed{CH_3CH_2CH_2}\quad C{=}CHCH_3 \quad \boxed{CH_3CH_2CH_2CH_2}$$

as an octene, since the double bond is not contained in the eight-carbon chain

Step 2. Number the carbon atoms in the chain. Beginning at the end nearer the double bond, assign numbers to the carbon atoms in the chain. If the double bond is equidistant from the two ends, begin at the end nearer the first branch point:

$$\overset{6}{C}H_3\overset{5}{C}H_2\overset{4}{C}H_2\overset{3}{C}H{=}\overset{2}{C}H\overset{1}{C}H_3$$

$$(\overset{1}{C}H_3)(CH_3)\overset{2}{C}H\overset{3}{C}H{=}\overset{4}{C}H\overset{5}{C}H_2\overset{6}{C}H_3$$

Correct numbering

Step 3. Write out the full name. Number the substituents according to their position in the chain and list them alphabetically. Indicate the position of the double bond by giving the number of the *first* alkene carbon. If more than one double bond is present, indicate the position of each and use the suffixes *-diene, -triene,* and so forth.

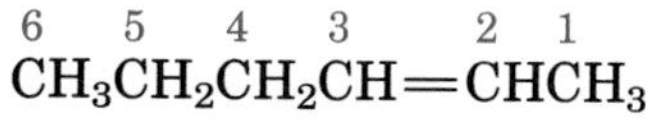

$\overset{6}{C}H_3\overset{5}{C}H_2\overset{4}{C}H_2\overset{3}{C}H{=}\overset{2}{C}H\overset{1}{C}H_3$

2-Hexene

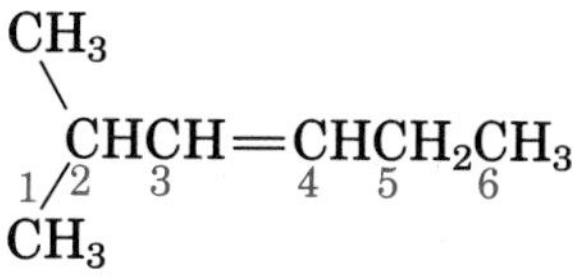

$(CH_3)_2\overset{2}{C}H\overset{3}{C}H{=}\overset{4}{C}H\overset{5}{C}H_2\overset{6}{C}H_3$ (the two CH_3 groups on C2 are numbered 1)

2-Methyl-3-hexene

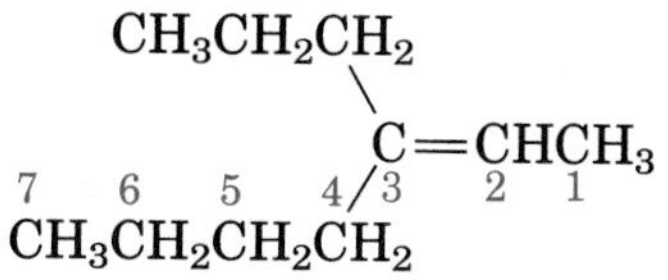

$(CH_3CH_2CH_2)(\overset{7}{C}H_3\overset{6}{C}H_2\overset{5}{C}H_2\overset{4}{C}H_2)\overset{3}{C}{=}\overset{2}{C}H\overset{1}{C}H_3$

3-Propyl-2-heptene

$H_2\overset{1}{C}{=}\overset{2}{C}(CH_3){-}\overset{3}{C}H{=}\overset{4}{C}H_2$

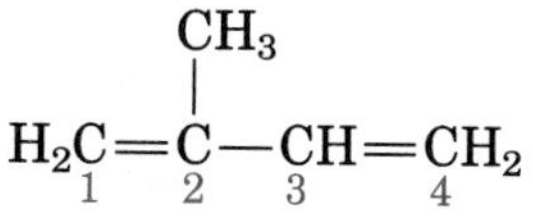

2-Methyl-1,3-butadiene

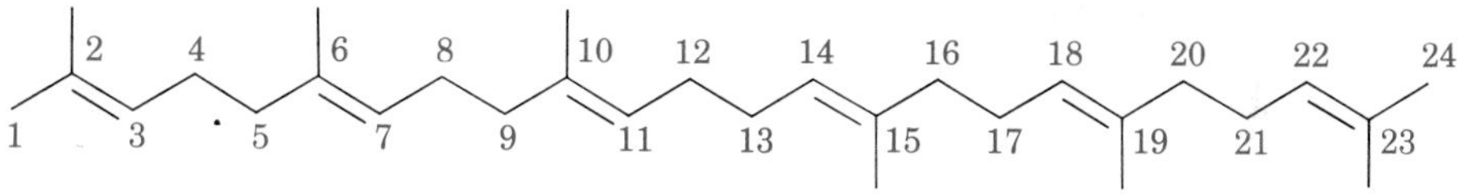

2,6,10,15,19,23-Hexamethyl-2,6,10,14,18,22-tetracosahexaene (squalene)

The polyene squalene that we saw earlier may be systematically named. It contains six double bonds and the longest chain has 24 carbons. It is therefore a tetracosahexaene ($C_{24}H_{50}$ = tetracosane; six double bonds = hexaene). The full name of squalene is 2,6,10,15,19,23-hexamethyl-2,6,10,14,18,22-tetracosahexaene. Simple!

Cycloalkenes are also named systematically. Since there is no chain end to begin from, we number the cycloalkene so that the double bond is between C1 and C2. Number so that the first branch point has as low a value as possible:

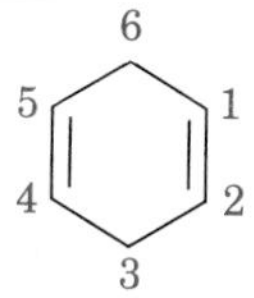

1,4-Cyclohexadiene

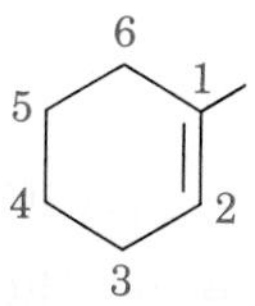

1-Methylcyclohexene

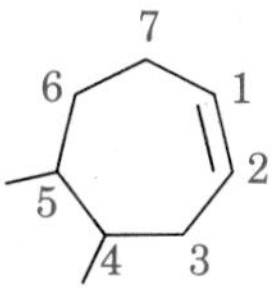

4,5-Dimethylcycloheptene *NOT* 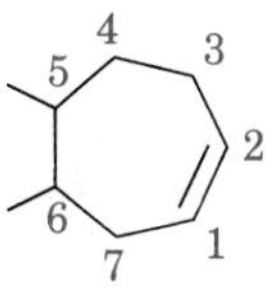 5,6-Dimethylcycloheptene

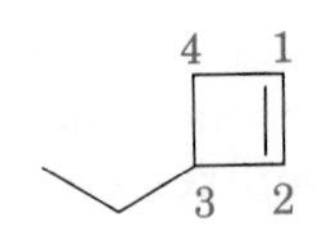

3-Ethylcyclobutene

There are a small number of alkenes whose names, though firmly entrenched in common usage, do not conform to strict rules of nomenclature. For example, the alkene derived from ethane should properly be called ethene, but the name ethylene has been used so long that it is accepted by IUPAC. Table 5.4 lists several other trivial names that are recognized by IUPAC.

TABLE 5.4 Trivial names of some common alkenes[a]

Compound	*Systematic name*	*Trivial name*
$H_2C{=}CH_2$	Ethene	Ethylene
$CH_3CH{=}CH_2$	Propene	Propylene
$(CH_3)_2C{=}CH_2$	2-Methylpropene	Isobutylene
$H_2C{=}C(CH_3){-}CH{=}CH_2$	2-Methyl-1,3-butadiene	Isoprene
$CH_3CH{=}CHCH{=}CH_2$	1,3-Pentadiene	Piperylene
$H_2C{=}CH{-}$	Ethenyl	Vinyl (an alkenyl group)
$H_2C{=}CH{-}CH_2{-}$	2-Propenyl	Allyl
$H_2C{=}$	Methylene	
$CH_3CH{=}$	Ethylidene	

[a]Both trivial and systematic names are recognized by IUPAC.

PROBLEM

5.6 Give the proper IUPAC names for these compounds:

(a) $H_2C{=}CHCH(CH_3)C(CH_3)_3$

(b) $CH_3CH_2CH{=}C(CH_3)CH_2CH_3$

(c) $CH_3CH{=}CHCH(CH_3)CH{=}CHCH(CH_3)_2$

PROBLEM

5.7 Draw structures corresponding to these IUPAC names:
(a) 2-Methyl-1,5-hexadiene
(b) 3-Ethyl-2,2-dimethyl-3-heptene
(c) 2,3,3-Trimethyl-1,4,6-octatriene
(d) 3,4-Diisopropyl-2,5-dimethyl-3-hexene

PROBLEM

5.8 Provide correct names for these cycloalkenes:

5.6 Sequence rules: the *E,Z* designation

In the discussion of isomerism in the 2-butenes, we used the terms *cis* and *trans* to denote alkenes whose substituents were on the same side and opposite side of a double bond, respectively. This cis–trans nomenclature is unambiguous and quite acceptable for all disubstituted alkenes. But how do we denote the geometry of trisubstituted and tetrasubstituted double bonds?

The answer is provided by the *E,Z* system of nomenclature, which uses a system of **sequence rules** to assign priorities to the substituent groups on the double-bond carbons. Considering each of the double-bond carbons separately, we use the sequence rules to decide which of the two substituent groups on each carbon is higher in priority. As Figure 5.9 shows, if the groups of higher priority on the two alkene carbons are on the same side of the double bond, the alkene is designated *Z* (for the German word *zusammen,* "together"). If the higher-priority groups are on opposite sides, the alkene is designated *E* (for the German word *entgegen,* "opposite"). (The easiest way to learn which is which is to remember *Z* = zame zide; *E* = the other guy.)

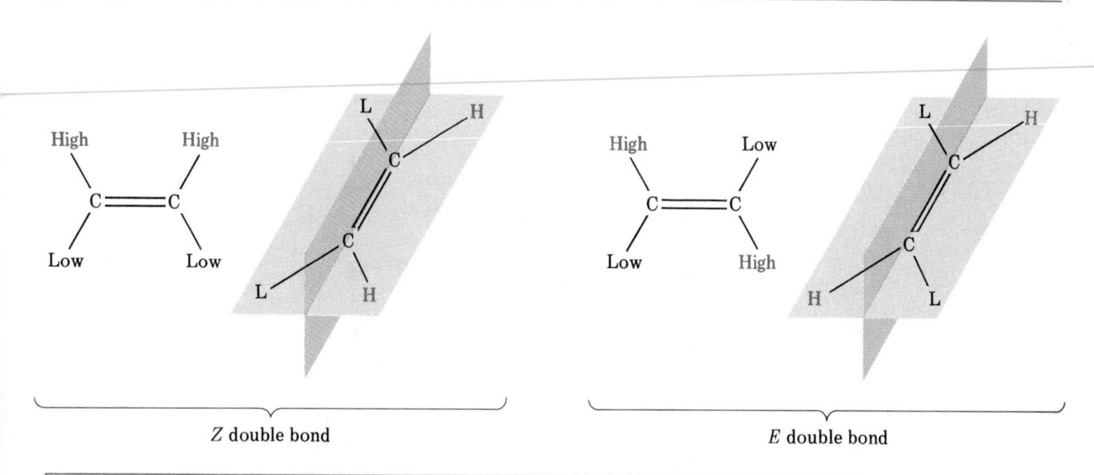

Figure 5.9. The *E,Z* system of nomenclature.

The sequence rules used in assigning priorities were introduced by Cahn, Ingold, and Prelog in 1964 and have proven extraordinarily useful. The rules are as follows:

Sequence rule 1. Look at the atoms directly attached to each carbon and rank them in order of *decreasing atomic number.* Thus, the common substituents that we might find on a double bond would be assigned the priority sequence Br > Cl > O > N > C > H. For example,

Low priority H, High priority CH_3	C=C	Cl High priority, CH_3 Low priority

(*E*)-2-Chloro-2-butene

Low priority H, High priority CH_3	C=C	CH_3 Low priority, Cl High priority

(*Z*)-2-Chloro-2-butene

Since chlorine has a higher atomic number than carbon, it receives higher priority than a methyl (CH_3) group. Methyl receives higher priority than hydrogen, however, and so the isomer on the left is *E* (high-priority groups on opposite sides of the double bond), whereas the one on the right is *Z* (high-priority groups on the same side of the double bond).

Sequence rule 2. If a decision cannot be reached by considering the first atoms in the substituent (rule 1), work outward to the first point of difference. Thus, an ethyl substituent, $—CH_2CH_3$, and a methyl substituent, $—CH_3$, are equivalent by rule 1 since both have carbon as the first atom. By rule 2, however, ethyl receives higher priority since the next atom is carbon rather than hydrogen. Look at each of the following pairs of substituents:

$—CH_3$ (Lower) vs. $—CH_2CH_3$ (Higher)

$—O—H$ (Lower) vs. $—O—CH_3$ (Higher)

$—CH(CH_3)CH_3$ (Higher) vs. $—CH_2CH_3$ (Lower)

H–C(H)(H)– : Lower	H–C(H)(H)–C(H)(H)– : Higher
–O–H : Lower	–O–C(H)(H)–H : Higher
–C(CH₃)(H)–CH₃ : Higher	–C(H)(H)–CH₃ : Lower

CH_3
$-C-NH_2$
H

Lower

H
$-C-Cl$
H

Higher

Sequence rule 3. Multiply bonded atoms are considered to be equivalent to the same number of singly bonded atoms. Thus, the following pairs are equivalent:

H H
$-C{=}C$
H

H H
$-C-C-H$
C C

$-C{\equiv}C-H$

C C
$-C-C-H$
C C

$-C{=}O$

$-C-O$
O C

According to the sequence rules, we can assign the configurations shown in the following examples:

H
H $C{=}CH_2$
$C{=}C$
CH_3 CH_3

(*E*)-3-Methyl-1,3-pentadiene

CH_3
H_3C-CH Br
$C{=}C$
$H_2C{=}C$ H
H

(*E*)-1-Bromo-2-isopropyl-1,3-butadiene

O
CH_3 $C-OH$
$C{=}C$
H CH_2OH

(*Z*)-2-Hydroxymethyl-2-butenoic acid

PROBLEM

5.9 Rank the sets of substituents in (a)–(d) in order of Cahn–Ingold–Prelog priorities:

(a) $-CH_3$, $-OH$, $-H$, $-Cl$

(b) $-CH_3$, $-CH_2CH_3$, $-CH{=}CH_2$, $-CH_2OH$

(c) $—COOH, —CH_2OH, —CN, —CH_2NH_2$

(d) $—CH_2CH_3, —C{\equiv}CH, —CN, —CH_2OCH_3$

PROBLEM

5.10 Assign *E* or *Z* configuration to these alkenes:

(a) CH_3 and CH_3CH_2 on one carbon, CH_2OH and Cl on the other: $(CH_3)(CH_3CH_2)C{=}C(CH_2OH)(Cl)$ — upper left CH_3, lower left CH_3CH_2, upper right CH_2OH, lower right Cl

(b) upper left Cl, lower left CH_3O, upper right CH_2CH_3, lower right $CH_2CH_2CH_3$: $(Cl)(CH_3O)C{=}C(CH_2CH_3)(CH_2CH_2CH_3)$

(c) 2-methylcyclopentylidene ring carbon $C{=}C$, upper right COOH, lower right CH_2OH (ring bears CH_3)

(d) upper left H, lower left CH_3, upper right CN, lower right CH_2NH_2: $(H)(CH_3)C{=}C(CN)(CH_2NH_2)$

5.7 Reactions of alkenes

Before beginning a detailed discussion of alkene reactions, let's see if we can make some predictions about the kind of chemistry we might expect. We know that *alkanes* are rather inert, since all outer-shell electrons are tied up in strong, relatively nonpolar, carbon–carbon and carbon–hydrogen bonds. Furthermore, the bonding electrons are sterically somewhat inaccessible, since they are localized in sigma orbitals between nuclei. The situation for *alkenes* is quite different, however. We have made a rough estimate of 64 kcal/mol for the strength of the carbon–carbon pi bond. This is 20–25 kcal/mol weaker than the average carbon–carbon sigma bond. Furthermore, the electrons in the pi bond are sterically accessible since they are located above and below the plane of the bond. Both low bond strength and electron accessibility might therefore lead us to predict high reactivity for carbon–carbon double bonds (Figure 5.10).

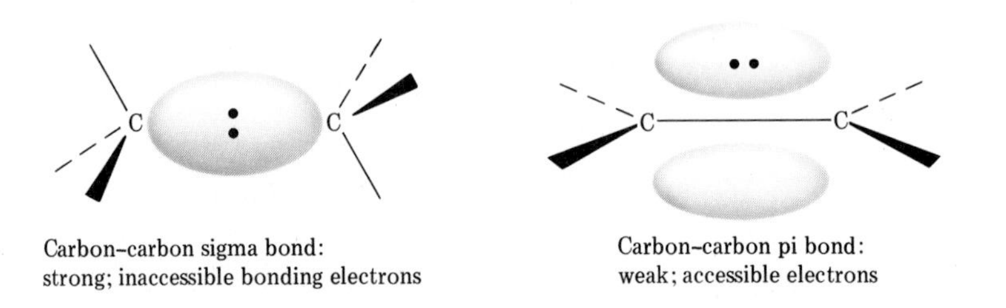

Figure 5.10. A comparison of carbon–carbon pi and sigma bonds.

Since there are four electrons in the carbon–carbon double bond, and since two of them are relatively available for reaction, the carbon–carbon double bond is a site of accessible electron density. In the terminology of polar reactions we have used (Section 4.1), alkenes should behave as nucleophiles. That is, alkene chemistry should be dominated by reaction of the electron-rich double bond with electron-poor species. This is exactly what we find. The most important reaction of alkenes is their reaction with electrophiles.

5.8 Addition of HX to alkenes: carbocations

Protic acids react with alkenes to give saturated addition products. Thus, 2-methylpropene reacts with gaseous HCl in ether solution to yield 2-chloro-2-methylpropane (*tert*-butyl chloride):

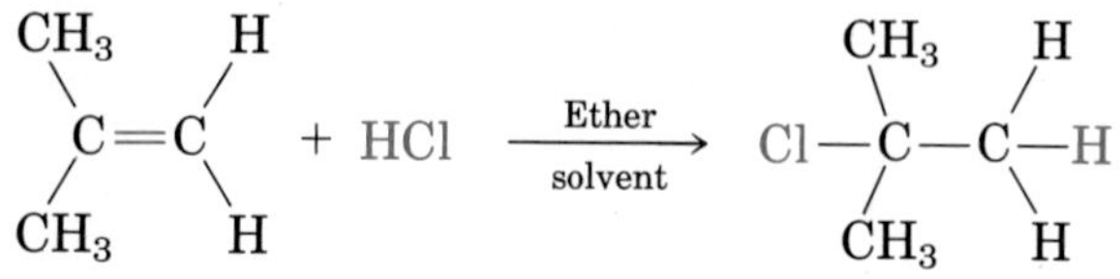

$$(CH_3)_2C{=}CH_2 + HCl \xrightarrow[\text{solvent}]{\text{Ether}} Cl{-}C(CH_3)_2{-}CH_3$$

2-Methylpropene

2-Chloro-2-methylpropane (94%)

Careful study of this and other **electrophilic addition reactions** by Sir Christopher Ingold[1] and others many years ago has led to the generally accepted mechanism shown in Figure 5.11.

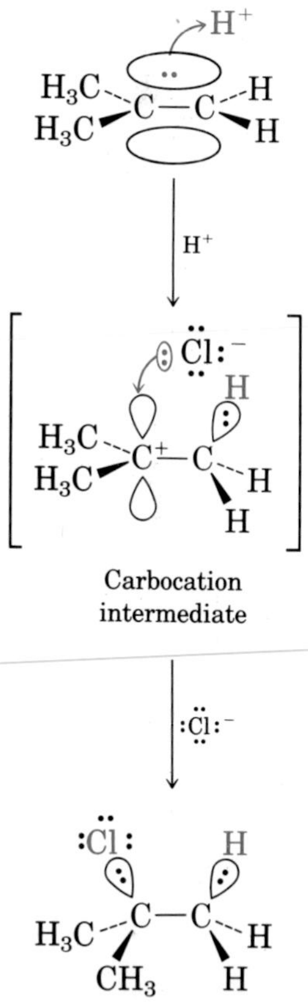

The electrophile, H^+, is attacked by the pi electrons of the alkene, and a new C—H sigma bond is formed. This leaves the other carbon atom with a ⊕ charge and a vacant *p* orbital.

Carbocation intermediate

$Cl:^-$ donates an electron pair to the positively charged carbon atom, forming a C—Cl bond and yielding a neutral addition product.

Figure 5.11. Mechanism of the electrophilic addition of HCl to an alkene. The reaction takes place in two steps and involves a carbocation intermediate.

[1]Sir Christopher Ingold (1893–1970); b. Ilford, England; D.Sc., London (Thorpe); professor, Leeds (1924–1930), University College, London (1930–1970).

The reaction begins with an attack on the electrophile, H^+, by the electrons of the nucleophilic pi bond. Two electrons from the pi bond form a new sigma bond between the entering hydrogen and an alkene carbon, as shown by the curved arrow in Figure 5.11, and an empty *p* orbital is left on the other alkene carbon. Since the double-bond pi electrons were used in formation of the new C—H bond, the trivalent carbon remaining has only six electrons in its outer shell and carries a formal positive charge. This positively charged intermediate is called a **carbocation** and is itself an electrophile that can accept an electron pair from nucleophilic chloride ion to form a C—Cl bond and yield a neutral product.

A reaction energy diagram for the overall electrophilic addition reaction is shown in Figure 5.12. Since two steps are involved, the diagram shows two peaks (transition states) separated by a valley (carbocation intermediate). The energy level of this intermediate is higher than that of the starting alkene, but the reaction as a whole is exothermic (negative $\Delta G°$). The first step, protonation of the alkene to yield the intermediate cation, is relatively slow, but, once formed, the cation intermediate rapidly reacts further to yield the final chloroalkane product. This is indicated in Figure 5.12 by the fact that $\Delta G_1^\ddagger$ is larger than $\Delta G_2^\ddagger$.

Electrophilic addition of HX to alkenes is a general reaction[2] that allows us to synthesize a variety of products. For example, addition of HCl and HBr is straightforward:

$$(CH_3)_2C{=}CH_2 \xrightarrow[\text{Ether}]{\text{HCl}} CH_3{-}C(CH_3)_2{-}Cl$$

2-Methylpropene → 2-Chloro-2-methylpropane (94%)

[2]Organic reaction equations may be written in different ways depending on the emphasis one wishes to achieve. For example, the reaction of 2-methylpropene with HCl might be written in the format A + B ⟶ C, emphasizing that both reaction partners are equally important for the purposes of the present discussion. The reaction solvent and any other reaction conditions such as temperature or concentration can then be noted above and below the reaction arrow:

$$(CH_3)_2C{=}CH_2 + HCl \xrightarrow[25°C]{\text{Ether}} (CH_3)_3CCl$$

Alternatively, an organic chemist might choose to write the same reaction in the format

$$A \xrightarrow{B} C$$

emphasizing that reagent A is the partner whose chemistry is of greater interest. Reagent B is then placed above the reaction arrow, together with notes about solvent and reaction conditions:

$$(CH_3)_2C{=}CH_2 \xrightarrow[\text{Ether, 25°C}]{\text{HCl}} (CH_3)_3CCl$$

(HCl — Reagent; Ether — Solvent)

Both reaction formats are used in this book, and it is important that the different roles of chemicals shown next to the reaction arrow be understood.

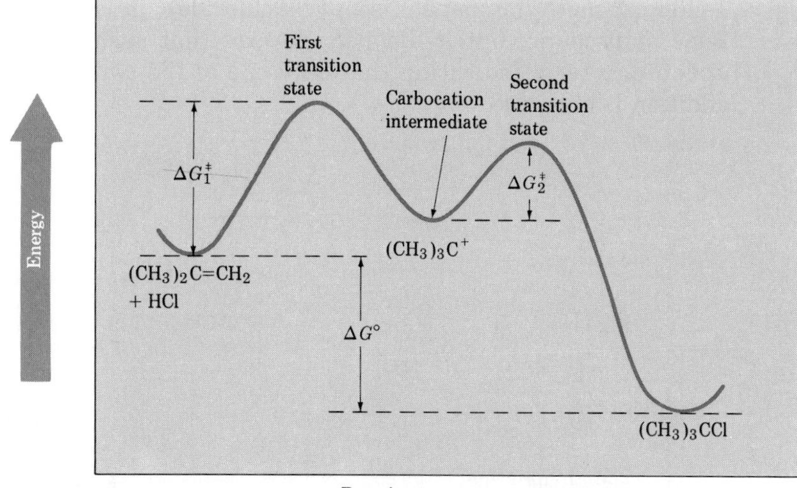

Figure 5.12. Reaction energy diagram for electrophilic addition of HCl to 2-methylpropene. The reaction proceeds in two steps and involves formation of a carbocation intermediate. The first step is slower than the second step (i.e., $\Delta G_1^\ddagger > \Delta G_2^\ddagger$.

$$\text{1-Methylcyclohexene} \xrightarrow[\text{Ether}]{\text{HBr}} \text{1-Bromo-1-methylcyclohexane (91\%)}$$

Addition of HI also occurs, but it is best to use slightly different experimental conditions. A mixture of phosphoric acid and potassium iodide has been found to be effective for generating HI, and the overall mechanism is the same as for the other additions.

$$\underset{\text{1-Pentene}}{CH_3CH_2CH_2CH{=}CH_2} \xrightarrow[H_3PO_4]{KI} \underset{\text{2-Iodopentane}}{CH_3CH_2CH_2\overset{\overset{\displaystyle I}{|}}{C}HCH_3}$$

(HI)

5.9 Orientation of electrophilic addition: Markovnikov's rule

In all of the examples just shown, an unsymmetrically substituted alkene has given a *single* addition product, rather than the mixture that might have been expected. For example, 2-methylpropene might have given 1-chloro-2-methylpropane (isobutyl chloride) in addition to 2-chloro-2-methylpropane, but it did not. We say that such reactions are **regiospecific,** a term indicating that only one of the two possible directions of addition is observed.

$$(CH_3)_2C{=}CH_2 + HCl \not\longrightarrow (CH_3)_2CHCH_2Cl$$

2-Methylpropene → 1-Chloro-2-methylpropane (*not formed*)

$$(CH_3)_2C{=}CH_2 + HCl \longrightarrow CH_3{-}C(CH_3)_2{-}Cl$$

2-Methylpropene → 2-Chloro-2-methylpropane (sole product; a regiospecific reaction)

From an empirical examination of many such reactions, the Russian chemist Vladimir Markovnikov[3] proposed in 1905 what has become known as **Markovnikov's rule:** In the addition of HX to an alkene, the acid hydrogen becomes attached to the carbon with fewer alkyl substituents. Conversely, the X group always bonds to the carbon with more alkyl substituents:

$$(CH_3)_2C{=}CHCH_2CH_3 + HCl \xrightarrow{\text{Ether}} (CH_3)_2CClCH_2CH_2CH_3$$

2-Methyl-2-pentene (2 alkyl groups on C-2; 1 alkyl group on C-3) → 2-Chloro-2-methylpentane

1-Methylcyclohexene (2 alkyl groups; 1 alkyl group) $\xrightarrow[H_3PO_4]{NaI}$ 1-Iodo-1-methylcyclohexane

When both ends of the double bond have the same degree of substitution, however, a mixture of products results:

[3] Vladimir Vassilyevich Markovnikov (or Morkovnikov or Markownikoff) (1833–1904); b. Nijni-Novgorod, Russia; pupil of Butlerov, Erlenmeyer, Baeyer, and Kolbe; professor in Odessa (1871) and Moscow (1873).

$$CH_3CH_2CH{=}CHCH_3 \xrightarrow[\text{Ether}]{\text{HBr}} CH_3CH_2CH_2\underset{}{\overset{\text{Br}}{|}}\!CHCH_3 + CH_3CH_2\overset{\text{Br}}{|}\!CHCH_2CH_3$$

2-Pentene → 2-Bromopentane + 3-Bromopentane

Since carbocations are involved as intermediates in these reactions, another way to express Markovnikov's rule is to say that, in the addition of HX to alkenes, the more substituted carbocation is formed as an intermediate in preference to the less substituted one. For example, addition of H^+ to 2-methylpropene yields the intermediate *tertiary* carbocation rather than the primary carbocation, and addition to 1-methylcyclohexene yields a tertiary rather than a secondary cation. Why should this be so?

2-Methylpropene $(CH_3)_2C{=}CH_2$ + H^+ →

- $[(CH_3)_2\overset{+}{C}{-}CH_2{-}H]$ *tert*-Butyl carbocation (tertiary; 3°) $\xrightarrow{Cl:^-}$ $Cl{-}C(CH_3)_3$ 2-Chloro-2-methylpropane
- ✕ $[H{-}C(CH_3)_2{-}\overset{+}{C}H_2]$ Isobutyl carbocation (primary; 1°) (*not formed*) $\xrightarrow{Cl:^-}$ $(CH_3)_2CHCH_2Cl$ 1-Chloro-2-methylpropane (*not formed*)

1-Methylcyclohexene + H^+ →

- Tertiary carbocation → 1-Iodo-1-methylcyclohexane
- ✕ Secondary carbocation (*not formed*) → 1-Iodo-2-methylcyclohexane (*not formed*)

PROBLEM

5.11 Predict the products of these reactions:

(a) [cyclohexene] + HCl ⟶

(b) $(CH_3)_2C{=}CHCH_2CH_3 \xrightarrow{H_2SO_4}$

(c) $CH_3CH_2CH_2CH{=}CH_2 \xrightarrow[KI]{H_3PO_4}$

(d) [methylenecyclohexane, $=CH_2$] + HBr ⟶

5.10 Carbocation structure and stability

To understand the reasons for the observed orientation of electrophilic addition reactions, we need to learn more about the structure and relative stabilities of various carbocations and about the general nature of reactions and transition states. The first point we need to explore involves structure.

A great deal of evidence points to the conclusion that carbocations are *planar;* the carbon is sp^2 hybridized and the three substituents are oriented to the corners of an equilateral triangle, as indicated in Figure 5.13. Since there are only six electrons in the carbon outer shell, and since all six are used in the three sigma bonds, the *p* orbital extending above and below the plane is unoccupied. [Note the similarity of carbocations to trivalent boron compounds such as BF_3 (Section 1.13).]

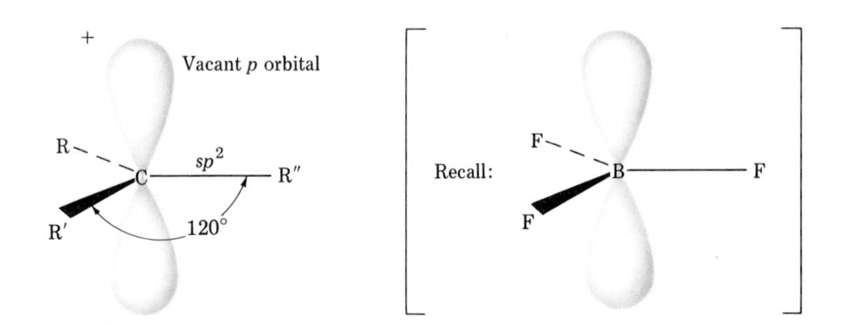

Figure 5.13. Carbocation structure. The carbon is sp^2 hybridized and has a vacant *p* orbital.

The second point we need to explore involves carbocation stability. 2-Methylpropene can react with HCl to form a carbocation having either three alkyl substituents (a tertiary ion, 3°) or one alkyl substituent (a primary ion, 1°). Since the tertiary chloride, 2-chloro-2-methylpropane, is the only product observed, formation of the tertiary cation is evidently favored over formation of the primary cation. Thermodynamic measurements show that, indeed, the stability of carbocations increases with increasing substitution: more highly substituted carbocations are more stable than less substituted ones. One way of determining carbocation stabilities is to measure the amount of energy required to ionize an alkyl halide into its corresponding cation, $R{-}X \rightarrow R^+ + {:}X^-$. Tertiary halides ionize to give carbocations much more readily than do primary ones (Table 5.5).

TABLE 5.5 Gas-phase ionization enthalpies:[a] **$R{-}Cl \rightarrow R^+ + {:}Cl^-$**

Type	*Reaction*	*ΔH° (kcal/mol)*
Methyl	$CH_3Cl \rightarrow CH_3^+ + {:}Cl^-$	227
Primary	$CH_3CH_2Cl \rightarrow CH_3CH_2^+ + {:}Cl^-$	195
Secondary	$(CH_3)_2CHCl \rightarrow (CH_3)_2CH^+ + {:}Cl^-$	173
Tertiary	$(CH_3)_3CCl \rightarrow (CH_3)_3C^+ + {:}Cl^-$	157

[a]Enthalpies are calculated in the following way:

$$CH_3{-}Cl \longrightarrow CH_3\cdot + Cl\cdot \quad \Delta H^\circ = 84 \text{ kcal/mol bond strength}$$
$$CH_3\cdot \longrightarrow CH_3^+ + e^- \quad \Delta H^\circ = 226.7 \text{ kcal/mol ionization energy}$$
$$Cl\cdot + e^- \longrightarrow Cl^- \quad \Delta H^\circ = -83.2 \text{ kcal/mol electron affinity}$$

$$\text{Net: } CH_3{-}Cl \longrightarrow CH_3^+ + Cl^- \quad \Delta H^\circ = 227 \text{ kcal/mol net ionization enthalpy}$$

As Table 5.5 shows, there are large differences in the gas-phase stabilities of substituted carbocations. The stability order is as follows:

$$\text{Tertiary} > \text{Secondary} > \text{Primary} > \text{Methyl}$$
$$(CH_3)_3C^+ > (CH_3)_2CH^+ > CH_3CH_2^+ > {}^+CH_3$$

Although the data quoted are taken from measurements made in the gas phase, a similar order is found in solution. The values for ionization are much lower in solution since polar solvents can stabilize the ions, but the order of carbocation stability remains the same.

Why should more heavily substituted carbocations be more stable than less substituted ones? Most chemists feel that the answer has to do with hyperconjugation. Hyperconjugation was discussed briefly in Section 5.4 in connection with the stability order of substituted alkenes, and it was described as the overlap between a *p* orbital and a neighboring C—H sigma bond. In the present situation, hyperconjugation can take place between the vacant carbocation *p* orbital and a neighboring C—H sigma bond (Figure 5.14).

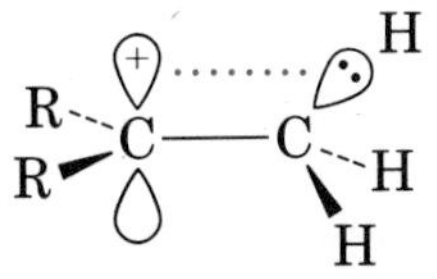

Figure 5.14. Hyperconjugative stabilization of a carbocation. The sigma electrons in the neighboring C—H bond help stabilize the positive charge.

The net effect of hyperconjugation in carbocations is to allow neighboring sigma bond electrons to stabilize the positive charge by spreading the charge out, or *delocalizing* it, over a greater volume of space. We will see repeatedly in later chapters that delocalizing charge over a greater volume invariably leads to greater stability. In the present instance, the more alkyl groups that are present on the carbocation, the more possibilities there are for hyperconjugation, and the more stable the ion (that is, tertiary > secondary > primary).

5.11 The Hammond postulate

To summarize our knowledge of electrophilic addition reactions up to this point, we know two facts:

1. We know that electrophilic addition reactions yield a product derived from the more substituted carbocation; a more substituted carbocation evidently forms faster than a less substituted one, and, once formed, rapidly goes on to give the final product.
2. We also know that the stability order of carbocations is tertiary > secondary > primary.

We must now see how these two pieces of information are related. We must see how the *stability* of the carbocation intermediate determines the *rate* at which carbocations form and thereby determines the structure of the final product.

The difficulty in relating these two pieces of information stems from the fact that stability and reactivity have no direct connection. Reactivity (point 1) is determined by the magnitude of the activation energy, $\Delta G^{\ddagger}$: A reaction leading to a more substituted carbocation has a lower $\Delta G^{\ddagger}$ than a reaction leading to a less substituted cation. Stability (point 2), on the other hand, is determined by ΔG°: A more stable carbocation has lower energy than a less stable carbocation. Although there is no precise thermodynamic relationship between product stability and reaction rate, there *is* an intuitive relationship between the two. It is a general observation that, when comparing two similar reactions, the more stable product usually forms faster than the less stable one. This situation is shown graphically in Figure 5.15; the reaction energy profile shown in part (a) represents the usual situation rather than the profile shown in part (b).

An explanation of this observation was advanced in 1955 and has come to be called the **Hammond postulate.**[4] The Hammond postulate is not a thermodynamic law; it is simply a reasonable explanation of observed facts that attempts to link reaction rate and product stability by looking at the energy level and structure of the transition state.

Transition states are high-energy activated complexes that occur fleetingly during the course of a reaction and immediately go on to some more

[4]George Simms Hammond (1921–); b. Auburn, Maine; Ph.D. (1947), Harvard University; professor, Iowa State University; California Institute of Technology; University of California, Santa Cruz; Allied Chemical Company.

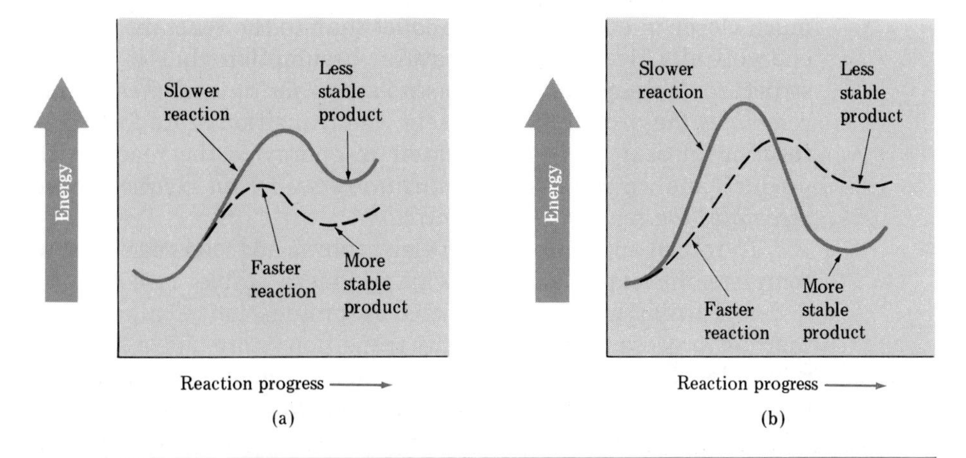

Figure 5.15. Reaction energy diagrams for two competing reactions. In part (a), the faster reaction yields the more stable product. In part (b), the faster reaction yields the less stable product. The case shown in part (a) is the usual situation.

stable species. We cannot actually *observe* transition states, because they have no finite lifetime. The Hammond postulate, however, states that we can get an idea of the structure of a particular transition state by looking at the structure of the nearest stable species. In terms of reaction energy diagrams, we can imagine the two cases shown in Figure 5.16. The reaction profile in part (a) shows the energy curve for an endothermic reaction step, and the profile in part (b) shows the curve for an exothermic step.

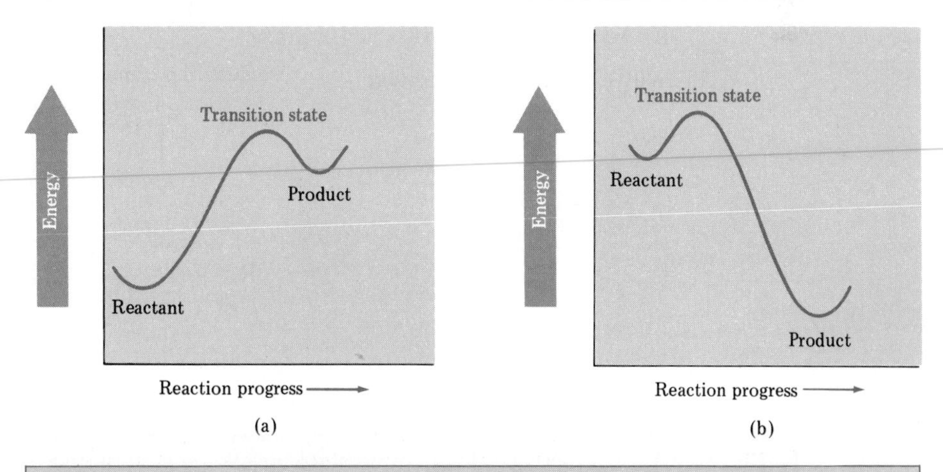

Figure 5.16. Reaction energy diagrams for endothermic and exothermic reactions: (a) An endothermic reaction; transition-state energy level and structure are close to product. (b) An exothermic reaction; transition-state energy level and structure are close to starting material.

In the endothermic reaction in Figure 5.16(a), the transition state is much closer in energy to the product than to the reactant. Since it is closer energetically, we make the natural assumption that it is also closer in structure. We say that *the transition state for an endothermic reaction step resembles the product.* Conversely, the transition state for the exothermic reaction in part (b) is much closer in energy to the reactant than to the product, and we say that *the transition state for an exothermic reaction step resembles the reactant in its structure.*

To return one more time to electrophilic addition reactions, we can now complete the explanation of why the more stable carbocation is formed faster. Formation of carbocations by protonation of alkenes is an endothermic process. Therefore, the transition state for alkene protonation resembles the carbocation product, and any factor that lowers the energy of the carbocation product should also lower the energy of the transition state.

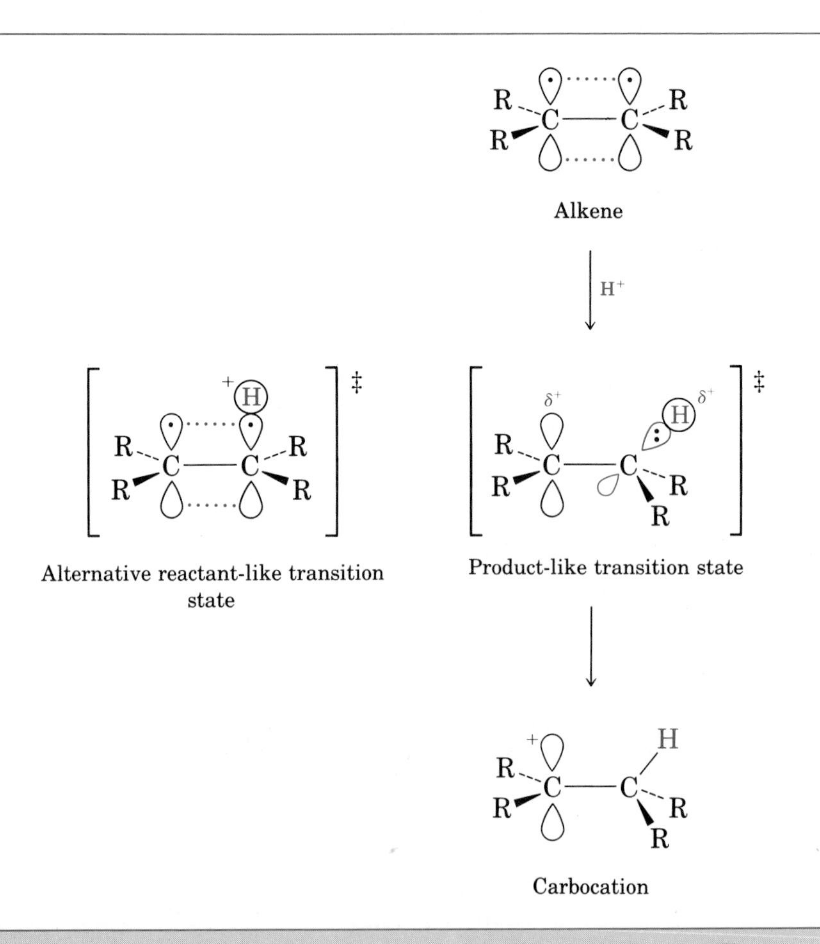

Figure 5.17. Hypothetical transition state for alkene protonation. The transition state for this endothermic protonation reaction is closer in both energy and structure to the carbocation product than to the reactant. Thus, an increase in cation stability (lower $\Delta G°$) also causes an increase in transition-state stability (lower $\Delta G^{\ddagger}$).

Since increasing alkyl substitution stabilizes carbocations, it also stabilizes the transition states leading to those ions; more stable carbocations form faster because their stability is reflected in the transition state leading to them. A hypothetical transition state for alkene protonation might be expected to look like that shown in Figure 5.17.

The transition state for alkene protonation shown in Figure 5.17 resembles the carbocation product. We can picture it to be a structure in which one of the alkene carbon atoms has almost completely rehybridized from sp^2 to sp^3 and in which the remaining alkene carbon bears a substantial portion of the positive charge. The positive charge in this transition state is delocalized and stabilized by hyperconjugation in the same way that the product carbocation is stabilized. The more alkyl groups that are present, the greater the extent of charge stabilization in the transition state and the faster it forms. Figure 5.18 summarizes the situation by showing competing reaction energy profiles in the reaction of 2-methylpropene with HCl.

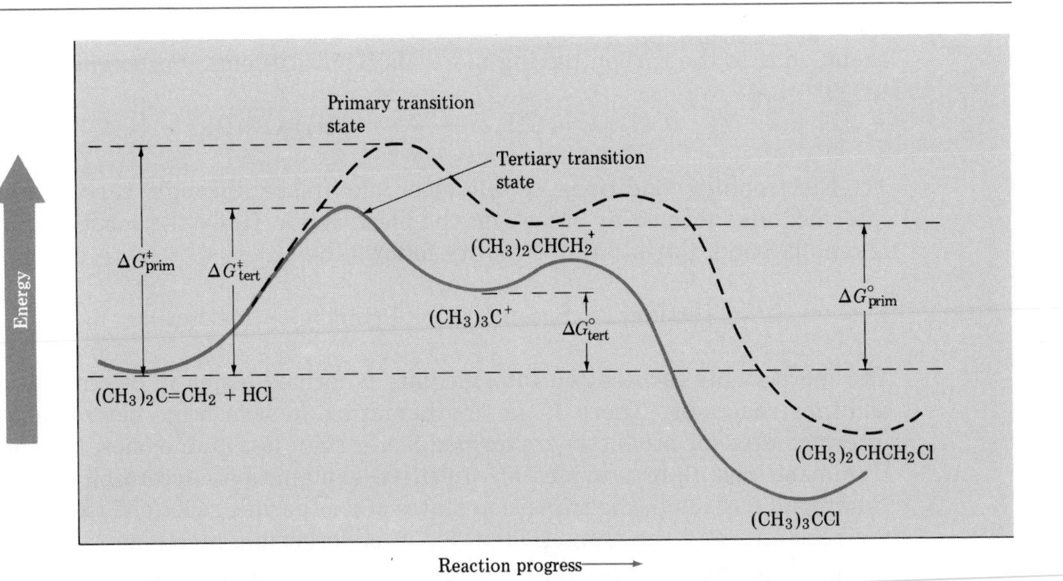

Figure 5.18. Reaction energy diagram for electrophilic addition of HCl to 2-methylpropene. The tertiary cation forms faster than the primary cation intermediate ($\Delta G^{\ddagger}_{tert} < \Delta G^{\ddagger}_{prim}$) because it is more stable ($\Delta G^{\circ}_{tert} < \Delta G^{\circ}_{prim}$). The same factors that make the tertiary cation more stable also make the transition state leading to it more stable.

PROBLEM

5.12 Consider the second step in the electrophilic addition of HCl to an alkene—the reaction of chloride ion with the carbocation intermediate. Is this step exothermic or endothermic? According to the Hammond postulate, should the transition state for this second step resemble starting material (carbocation) or product (chloroalkane)? Make a rough drawing of what you would expect the transition-state structure to look like.

5.12 Summary

Alkenes are hydrocarbons that contain one or more carbon–carbon double bonds. A double bond consists of two parts: a sigma bond formed by head-on overlap of two sp^2 orbitals, and a pi bond formed by sideways overlap of two *p* orbitals. The overall bond strength of an alkene double bond is greater than that of a carbon–carbon single bond, and the pi bond has an estimated strength of approximately 64 kcal/mol. Rotation around the double bond is restricted, and substituted alkenes can therefore exist as a pair of cis–trans isomers. The geometry of a double bond can be described as either *Z* (*zusammen*) or *E* (*entgegen*) by application of the Cahn–Ingold–Prelog sequence rules. Alkenes are named by IUPAC rules using the suffix *-ene*. The stability order of alkyl-substituted double bonds is

$$\text{Tetrasubstituted} > \text{Trisubstituted} > \text{Disubstituted} > \text{Monosubstituted}$$

The chemistry of alkenes is dominated by electrophilic addition reactions. When HX reacts with an alkene, Markovnikov's rule predicts that the hydrogen adds to the carbon having fewer alkyl substituents and the X group adds to the carbon having more alkyl substituents. For example,

$$(CH_3)_2C{=}CH_2 \xrightarrow{HCl} (CH_3)_3C{-}Cl$$

Electrophilic additions to alkenes take place through carbocation intermediates formed by attack on the electrophile $H^+$by the nucleophilic alkene pi bond. Carbocation stability follows the order

$$\text{Tertiary} > \text{Secondary} > \text{Primary} > \text{Methyl}$$

The more stable carbocation intermediate is formed faster in electrophilic addition reactions. There is no mathematical reason why more stable intermediates (or products) are formed faster than less stable ones, but the Hammond postulate provides an intuitive explanation. According to the Hammond postulate, the transition states of exothermic reactions resemble the reactant, and the transition states of endothermic reactions resemble the product. Alkene protonation is endothermic, and the stability of the more substituted carbocation is thus shared by the transition state leading to its formation.

ADDITIONAL PROBLEMS

5.13 Calculate the number of double bonds and/or rings in these formulas:
(a) Benzene, C_6H_6
(b) Cyclohexene, C_6H_{10}
(c) Myrcene (bay oil), $C_{10}H_{16}$
(d) Lindane, $C_6H_6Cl_6$
(e) Pyridine, C_5H_5N
(f) Safrole, $C_{10}H_{10}O_2$

5.14 Draw five possible structures for each of these formulas:
(a) $C_{10}H_{16}$
(b) C_8H_8O
(c) $C_7H_{10}Cl_2$
(d) $C_{10}H_{16}O_2$
(e) $C_5H_9NO_2$
(f) $C_8H_{10}ClNO$

5.15 A compound of formula $C_{10}H_{14}$ undergoes catalytic hydrogenation, but absorbs only two equivalents of hydrogen. How many rings does the compound have?

5.16 Provide IUPAC names for these alkenes:

(a) $CH_3CH{=}CH\overset{\overset{CH_3}{|}}{C}HCH_2CH_3$

(b) $CH_3CH{=}CH\overset{\overset{CH_2CH_2CH_2CH_3}{|}}{C}HCH_2CH_2\underset{\underset{CH_3}{|}}{C}HCH_3$

(c) $H_2C{=}C(CH_2CH_3)_2$

(d) $H_2C{=}CH\overset{\overset{CH_3}{|}}{C}H\underset{\underset{CH_3}{|}}{C}HCH{=}CHCH_3$

(e) $CH_3CH_2\overset{\overset{CH_3}{|}}{C}{=}CHCH{=}CH_2$

(f) $H_2C{=}C{=}CHCH_3$

(g) $H_2C{=}CHC(CH_3)_3$

(h) $(CH_3)_3CCH{=}CHC(CH_3)_3$

5.17 The triene ocimene is commonly found in the essential oil of many plants. What is its correct IUPAC name?

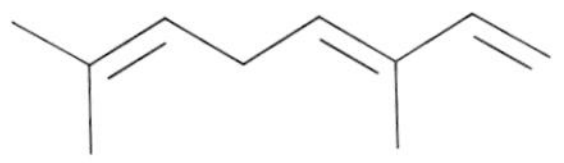

Ocimene

5.18 α-Farnesene is a constituent of the natural wax found on apples. What is its IUPAC name?

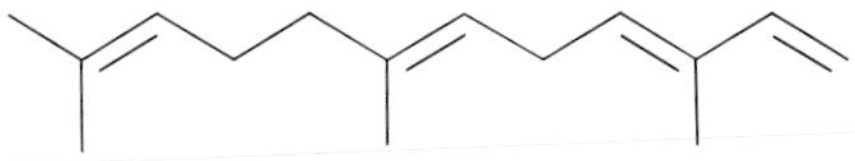

α-Farnesene

5.19 Draw structures corresponding to these systematic names:

(a) 2,4-Dimethyl-1,4-hexadiene
(b) 3,3-Dimethyl-4-propyl-1,5-octadiene
(c) 4-Methyl-1,2-pentadiene
(d) 3,7-Dimethyl-1,3,6-octatriene
(e) 3-Butyl-2-heptene
(f) 2,2,5,5-Tetramethyl-3-hexene

5.20 These names are incorrect. Draw structures and provide correct IUPAC names:

(a) 2-Methyl-2,4-pentadiene
(b) 3-Methylene-1-pentene
(c) 3,6-Octadiene
(d) 5-Ethyl-4-octene
(e) 3-Propyl-3-heptene
(f) 3-Vinyl-1-propene

5.21 Allene (1,2-propadiene), $H_2C{=}C{=}CH_2$, has two adjacent double bonds. What is the nature of the hybridization of the central carbon? Sketch the bonding orbitals in allene. What shape do you predict for allene?

5.22 According to heats of hydrogenation data, *trans*-2-butene is more stable than *cis*-2-butene by 1.0 kcal/mol. Hydrogenation measurements also show that *trans*-2,2,5,5-tetramethyl-3-hexene is more stable than its cis isomer by 9.3 kcal/mol. Explain this large difference.

		ΔH°_{hydrog} *(kcal/mol)*
Cis	$CH_3CH{=}CHCH_3$	28.6
Trans	$CH_3CH{=}CHCH_3$	27.6
Cis	$(CH_3)_3CCH{=}CHC(CH_3)_3$	36.2
Trans	$(CH_3)_3CCH{=}CHC(CH_3)_3$	26.9

5.23 1,4-Pentadiene, a compound with two nonadjacent double bonds, has ΔH°_{hydrog} = 60.8 kcal/mol, which is, as expected, approximately twice the value for propene (ΔH°_{hydrog} = 30.1 kcal/mol). However, 1,2-propadiene has ΔH°_{hydrog} = 71.3 kcal/mol. What does this tell you about the stability of 1,2-propadiene? Can you suggest an explanation?

5.24 Rank the following sets of substituents in order of priority according to the Cahn–Ingold–Prelog sequence rules:

(a) $-CH_3$, $-Br$, $-H$, $-I$

(b) $-OH$, $-OCH_3$, $-H$, $-COOH$

(c) $-COOH$, $-COOCH_3$, $-CH_2OH$, $-CH_3$

(d) $-CH_3$, $-CH_2CH_3$, $-CH_2CH_2OH$, $-\overset{O}{\overset{\|}{C}}CH_3$

(e) $-CH{=}CH_2$, $-CN$, $-CH_2NH_2$, $-CH_2Br$

(f) $-CH{=}CH_2$, $-CH_2CH_3$, $-CH_2OCH_3$, $-CH_2OH$

5.25 Assign *E* or *Z* configurations to the following alkenes:

(a) 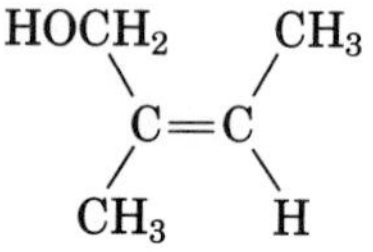 $(HOCH_2)(CH_3)C{=}C(CH_3)(H)$ — top: $HOCH_2$, CH_3; bottom: CH_3, H

(b) 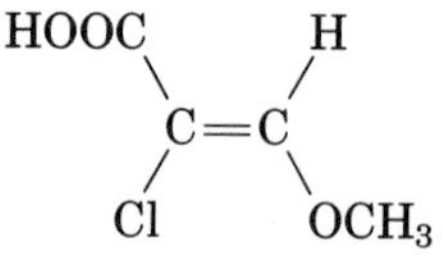 $(HOOC)(Cl)C{=}C(H)(OCH_3)$ — top: HOOC, H; bottom: Cl, OCH_3

(c) $(NC)(CH_3CH_2)C{=}C(CH_3)(CH_2OH)$ — top: NC, CH_3; bottom: CH_3CH_2, CH_2OH

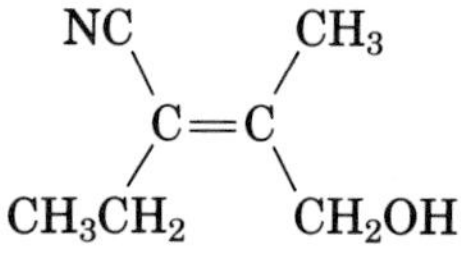

(d) 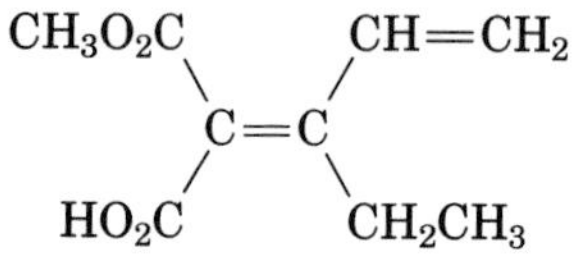 $(CH_3O_2C)(HO_2C)C{=}C(CH{=}CH_2)(CH_2CH_3)$ — top: CH_3O_2C, $CH{=}CH_2$; bottom: HO_2C, CH_2CH_3

5.26 Name these cycloalkenes according to IUPAC rules:

(a) 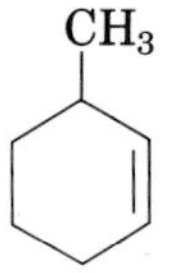 [cyclohexene with CH_3 substituent] (b) 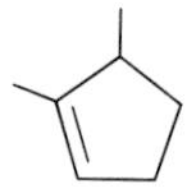 [methyl-substituted cyclopentene with second methyl] (c) 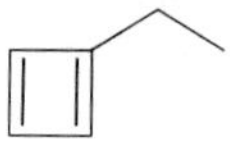 [ethylcyclobutene] (d) 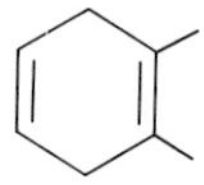 [dimethylcyclohexadiene] (e) 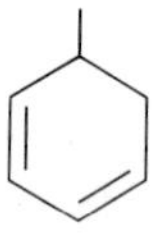 [methylcyclohexadiene] (f) 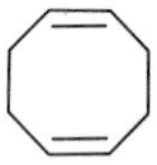 [cyclooctadiene]

5.27 Which of the given *E,Z* designations are correct, and which are incorrect?

(a) 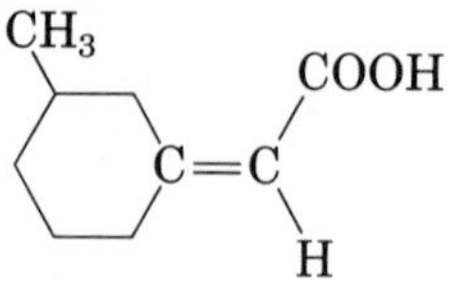 [2-methylcyclohexylidene]$C{=}C(COOH)(H)$ — CH_3 on ring; COOH, H

Z

(b) 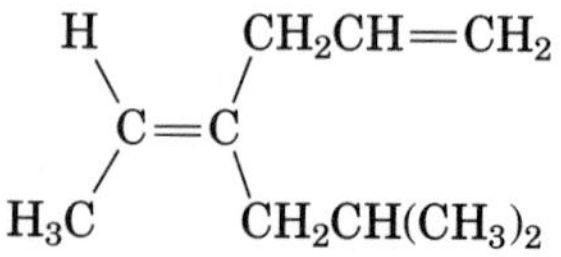 $(H)(H_3C)C{=}C(CH_2CH{=}CH_2)(CH_2CH(CH_3)_2)$ — top: H, $CH_2CH{=}CH_2$; bottom: H_3C, $CH_2CH(CH_3)_2$

E

(c) 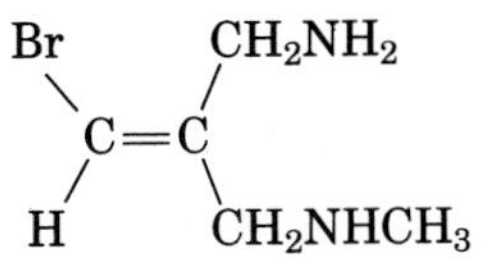
```
Br          CH2NH2
  \        /
   C == C
  /        \
H           CH2NHCH3
```
Z

(d) 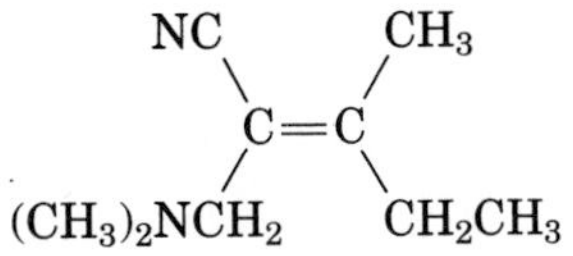
```
        NC          CH3
          \        /
           C == C
          /        \
(CH3)2NCH2          CH2CH3
```
E

(e) 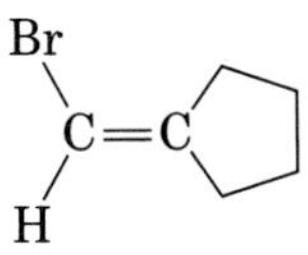
```
Br
  \
   C == C(cyclopentylidene)
  /
H
```
Z

(f) 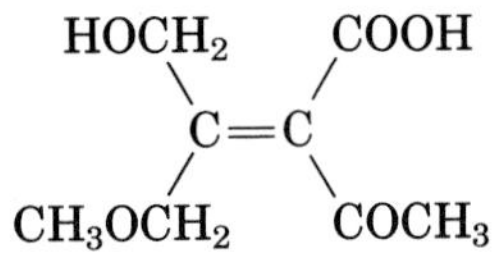
```
  HOCH2       COOH
       \     /
        C == C
       /     \
CH3OCH2       COCH3
```
E

5.28 Draw structures of the five possible pentene isomers, C_5H_{10}, and name them. Ignore *E,Z* isomers.

5.29 Draw structures of the 13 possible hexene isomers, C_6H_{12}, and name them. Ignore *E,Z* isomers.

5.30 Calculate the degree of unsaturation of these formulas:

(a) Cholesterol, $C_{27}H_{46}O$
(b) DDT, $C_{14}H_9Cl_5$
(c) Prostaglandin E, $C_{20}H_{34}O_5$
(d) Caffeine, $C_8H_{10}N_4O_2$
(e) Cortisone, $C_{21}H_{28}O_5$
(f) Atropine, $C_{17}H_{23}NO_3$
(g) Ascorbic acid, $C_6H_8O_6$

5.31 Draw a reaction energy diagram for the addition of HBr to 1-pentene. Let one curve on your diagram show the formation of 1-bromopentane product and another curve on the same diagram show the formation of 2-bromopentane product. Label the positions for all reactants, intermediates, and products.

5.32 Make sketches of the transition-state structures involved in the reaction of HBr with 1-pentene (Problem 5.31). Identify each structure as resembling either starting material or product.

CHAPTER 6
ALKENES: REACTIONS AND SYNTHESIS

We saw in the preceding chapter that the addition of electrophiles is one of the most important reactions of alkenes. We have studied only the addition of HX in detail thus far, but many other electrophiles also add to alkenes.

6.1 Addition of halogens

Halogens are quite reactive toward alkenes. Bromine and chlorine are particularly effective electrophilic addition reagents, and their reactions with alkenes provide a general method of synthesis of 1,2-dihaloalkanes. For example, each year more than 5 million tons of 1,2-dichloroethane (ethylene dichloride) are synthesized industrially by the addition of Cl_2 to ethylene. The product is used both as a solvent and as the starting material for the synthesis of the vinyl chloride needed in the manufacture of poly(vinyl chloride), PVC.

$$H_2C{=}CH_2 \xrightarrow{Cl_2} H_2C(Cl){-}CH_2(Cl)$$

Ethylene → 1,2-Dichloroethane (ethylene dichloride)

The addition of bromine to an alkene also serves as a simple and rapid laboratory test. A sample of unknown structure is dissolved in tetrachloromethane (carbon tetrachloride, CCl_4) and placed in a test tube. Several drops of bromine in CCl_4 are added. Immediate disappearance of the reddish bromine color signals a positive test and indicates that the sample is an alkene.

Cyclopentene $\xrightarrow{Br_2 \text{ in } CCl_4}$ 1,2-Dibromocyclopentane (95%)

Fluorine tends to be too reactive and difficult to control for most laboratory applications, and iodine does not react with alkenes.

Halogens react with alkenes by the electrophilic addition pathway shown in Figure 6.1 (p. 158). Although Br_2 is nonpolar, it is nevertheless highly polarizable, and, in the vicinity of a nucleophilic double bond, the bromine molecule becomes polarized. The pi electron pair of the alkene attacks the positive end of the polarized bromine molecule, displacing bromide ion. The net result is that electrophilic Br^+ adds to the alkene in the same way that H^+ adds. The intermediate carbocation then immediately reacts with bromide to yield the dibromo addition product.

This mechanistic description of halogen addition to alkenes is consistent with what we have learned thus far. Further examination, however, shows that it is not *fully* consistent with known data. In particular, the

The electron *pair* from the alkene bond attacks the positively polarized bromine, forming a C—Br bond and causing the Br—Br bond to break. Bromide ion departs with *both* Br—Br bond electrons.

Bromide ion then uses an electron pair to attack the carbocation intermediate, forming a C—Br bond and yielding a neutral addition product.

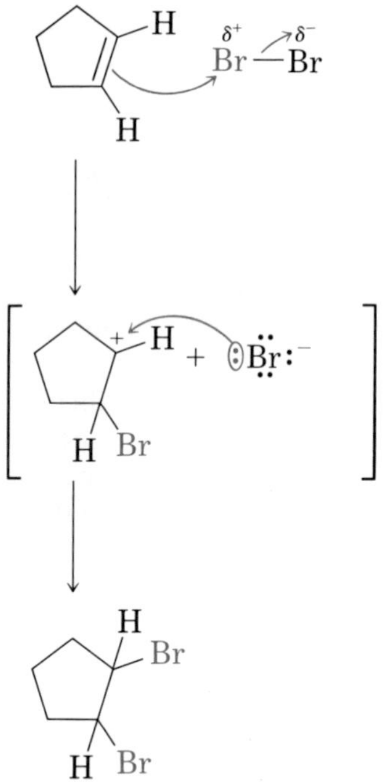

Figure 6.1. Mechanism of the electrophilic addition of bromine to cyclopentene.

mechanism proposed does not explain the stereochemistry of halogen addition. Let's look more closely at the addition of bromine to cyclopentene to see what is meant by "stereochemistry of addition."

Let's assume that Br^+ adds to cyclopentene from the bottom face to form the carbocation intermediate shown in Figure 6.2. Since this carbocation intermediate is planar and sp^2 hybridized, it could be attacked by bromide ion from either the top or the bottom to give a mixture of products, in which the two bromine atoms are either on the same side of the ring (cis) or on opposite sides (trans). However, only *trans*-1,2-dibromocyclopentane is produced. The two bromine atoms add to opposite faces of the double bond, and we say that the reaction occurs with **anti** stereochemistry. If the two bromines had added from the same face, the reaction would have had **syn** stereochemistry. Note that the word "anti" has a similar meaning in the present stereochemical context to the meaning it has in a butane conformational context (Section 3.9). In both cases, the two important substituents are 180° apart.

An explanation of the phenomenon of anti addition was suggested in 1937 by George Kimball and Irving Roberts, who postulated that the true reaction intermediate is not a carbocation, but a *bromonium ion*. A bromonium ion is a species containing a positively charged, divalent bromine, R_2Br^+. (A *chloronium ion,* similarly, is a species containing a positively charged, divalent chlorine, R_2Cl^+.) In the case of bromine addition to an alkene, the bromonium ion is in a three-membered ring, and is formed by

Figure 6.2. Stereochemistry of addition of bromine to cyclopentene. Only the trans product is formed.

the overlap of a bromine lone pair of electrons with the neighboring vacant carbocation p orbital (Figure 6.3). Although Figure 6.3 depicts three-membered-ring bromonium ion formation as stepwise, this is done only for

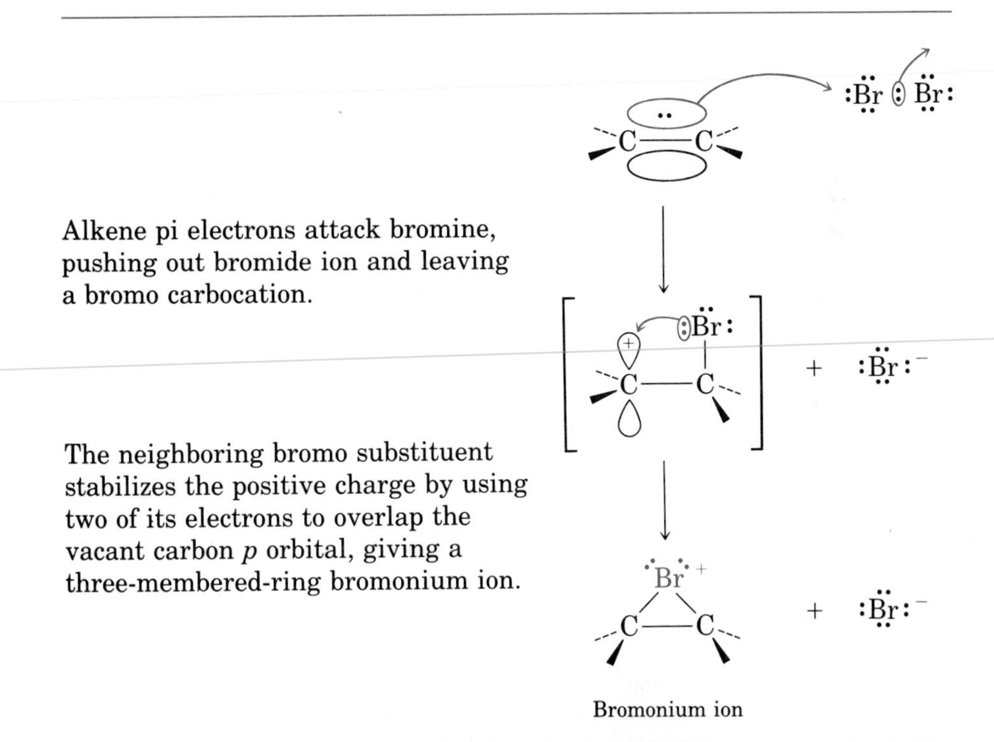

Alkene pi electrons attack bromine, pushing out bromide ion and leaving a bromo carbocation.
The neighboring bromo substituent stabilizes the positive charge by using two of its electrons to overlap the vacant carbon p orbital, giving a three-membered-ring bromonium ion.
:Br : Br:
C—C
Br:
C—C
+ :Br:⁻
Br +
C—C
+ :Br:⁻
Bromonium ion

Figure 6.3. Formation of a bromonium ion intermediate by electrophilic addition of Br^+ to an alkene.

the sake of clarity. It is likely that the bromonium ion is formed directly by the interaction of the alkene double-bond electrons with Br^+.

If a bromonium ion is formed as an intermediate in electrophilic addition reactions, the bromine ought to "shield" one face of the alkene. Further attack by bromide ion could then occur only from the opposite, sterically accessible, face, to give the anti product. The attack of the nucleophile, bromide, on the electrophilic bromonium ion can be considered a nucleophilic substitution reaction similar to the hydroxide-plus-bromomethane reaction discussed in Chapter 4:

$$HO^{-}: + CH_3{-}Br \longrightarrow HOCH_3 + :Br^{-}$$

This is just one example of how the same few basic reaction types keep showing up in organic chemistry.

$\xrightarrow[CCl_4]{Br_2}$ $\longrightarrow$

Cyclopentene

Bromonium ion intermediate (bottom side is blocked so reaction with bromide occurs from the top side)

trans-1,2-Dibromocyclopentane

The halonium ion postulate, made some 50 years ago to explain the stereochemistry of halogen addition to alkenes, is an excellent example of the use of deductive logic in chemistry. Arguing from known experimental results, chemists were able to make a hypothesis about the intimate mechanistic details of alkene electrophilic reactions. Much more recently, strong evidence supporting the postulate has come from the work of George Olah, who has observed and studied specially prepared *stable* solutions of cyclic bromonium ions. Thus, it appears that bromonium ions are indeed a reality.

$$(CH_3)_2C(F){-}CH(Br)CH_3 \xrightarrow[\text{Liquid } SO_2]{SbF_5} \text{(bromonium ion)} \quad SbF_6^-$$

A bromonium ion stable in SO_2 solution

PROBLEM

6.1 What product would you expect to obtain from addition of chlorine to 1,2-dimethylcyclohexene? Show the stereochemistry of the product.

PROBLEM

6.2 What product(s) would you expect to obtain from addition of deuterium chloride (DCl) to 1-methylcyclohexene? Indicate stereochemistry.

6.2 Halohydrin formation

Many other kinds of electrophilic additions to alkenes can take place, and among the more important is the addition of HOCl or HOBr to yield 1,2-halo alcohols, or **halohydrins.** When a solution of bromine in carbon tetrachloride reacts with an alkene, a cyclic bromonium ion is formed and then trapped by the only nucleophile present, bromide ion. If, however, we carry out the reaction in the presence of some other added nucleophile, the intermediate bromonium ion can be "intercepted" by the added nucleophile and diverted to a different product. For example, when an alkene reacts with bromine in the presence of water, water competes with bromide ion as nucleophile and reacts with the bromonium ion intermediate to yield a mixture of dibromide and bromo alcohol (a *bromohydrin*). The net effect is addition of hypobromous acid, HO—Br, to the alkene.

$$\text{trans-2-Butene} \xrightarrow[H_2O]{Br_2} [\text{bromonium ion}] \xrightarrow{:Br:^-} CH_3CHBrCHBrCH_3$$

$$[\text{bromonium ion}] \xrightarrow{H_2O:} CH_3CHBr{-}CH(OH)CH_3 + HBr$$

trans-2-Butene

A bromonium ion intermediate

A 1,2-dibromide (2,3-dibromobutane)

A bromohydrin (3-bromo-2-butanol)

In practice, few alkenes are soluble in water, and bromohydrin formation is often carried out in a solvent such as aqueous dimethyl sulfoxide (DMSO), with the less reactive *N*-bromosuccinimide (NBS) as a bromine source. Under these conditions, high yields of bromohydrins can be obtained, and the reaction is often used in the synthesis of complex organic molecules. For example,

$$C_6H_5CH{=}CH_2 \xrightarrow[H_2O/CH_3SOCH_3\ (DMSO)]{\text{N—Br (NBS)}} C_6H_5CH(OH){-}CH_2{-}Br$$

Styrene

2-Bromo-1-phenylethanol (76%)

Note that the aromatic ring in the example is inert to bromine under the conditions used, even though it contains three carbon–carbon double bonds. Aromatic rings are a good deal more stable than might be expected, and this stability will be examined in Chapter 14.

PROBLEM

6.3 When an unsymmetrical alkene such as propene is treated with *N*-bromosuccinimide in aqueous dimethyl sulfoxide, the major product is the one in which bromine is bonded to the less substituted carbon atom, as shown on p. 162.

$$CH_3CH{=}CH_2 \xrightarrow[H_2O/DMSO]{\text{N-bromosuccinimide (N—Br)}} CH_3CH(OH)CH_2Br$$

Propene — Major product, 1-bromo-2-propanol

Can you account for this result? Is this Markovnikov or non-Markovnikov orientation?

PROBLEM

6.4 It has been shown that iodine azide, IN_3, can add regiospecifically (only a single orientation of reaction is observed) to alkenes by an electrophilic mechanism:

$$C_6H_5CH{=}CH_2 \xrightarrow{I{-}N{=}\overset{+}{N}{=}\overset{-}{N}} C_6H_5CH(N_3){-}CH_2I$$

In light of this result, and assuming Markovnikov orientation, what is the polarity of the I—N_3 bond? Propose a mechanism for the addition reaction.

6.3 Oxymercuration

Thus far in the discussion of alkene reactions, the electrophilic addition of halogens and halogen acids have been considered, but the most important reaction of all has been neglected—the addition of water to a carbon–carbon double bond (**hydration**) to yield an alcohol.

Simple alkenes such as ethylene can be hydrated by reaction with hot sulfuric acid, followed by treatment of the intermediate sulfate with water. Ethanol was at one time produced commercially by this route, but a direct vapor-phase hydration at 300°C over an acid catalyst is now used.

$$H_2C{=}CH_2 \xrightarrow[250°C]{H_2SO_4} \left[H_2C(H){-}CH_2(OSO_3H) \right] \xrightarrow{H_2O} H_2C(H){-}CH_2(OH) + H_2SO_4$$

Ethylene — Sulfate intermediate — Ethanol

Although alkene hydration by the sulfuric acid route is suitable for large-scale industrial procedures, the reaction is of little value for most laboratory applications; the high temperatures and strongly acidic conditions required are simply too vigorous for many organic molecules to survive. In practice, most alkenes are best hydrated by the **oxymercuration** procedure.

When an alkene is treated with mercuric acetate [$Hg(O_2CCH_3)_2$, or commonly, $Hg(OAc)_2$] in aqueous tetrahydrofuran (THF) solvent,

Tetrahydrofuran (THF)

electrophilic addition of mercuric ion, Hg^{2+}, to the double bond rapidly occurs. The intermediate *organomercury* compound is then treated with sodium borohydride, $NaBH_4$, and an alcohol is produced:

1-Methylcyclopentene $\xrightarrow[H_2O\text{–}THF]{Hg(OAc)_2}$ [Organomercury intermediate] $\xrightarrow{NaBH_4}$ 1-Methylcyclopentanol (92%)

The oxymercuration reaction is initiated by electrophilic addition of mercuric ion to the alkene (Figure 6.4) to give an intermediate that is thought to be a cyclic acetoxymercurinium ion, much like the halonium ions we have seen. Like halonium ions, the cyclic acetoxymercurinium intermediate is opened by attack of water from the side opposite the mercury, and a stable

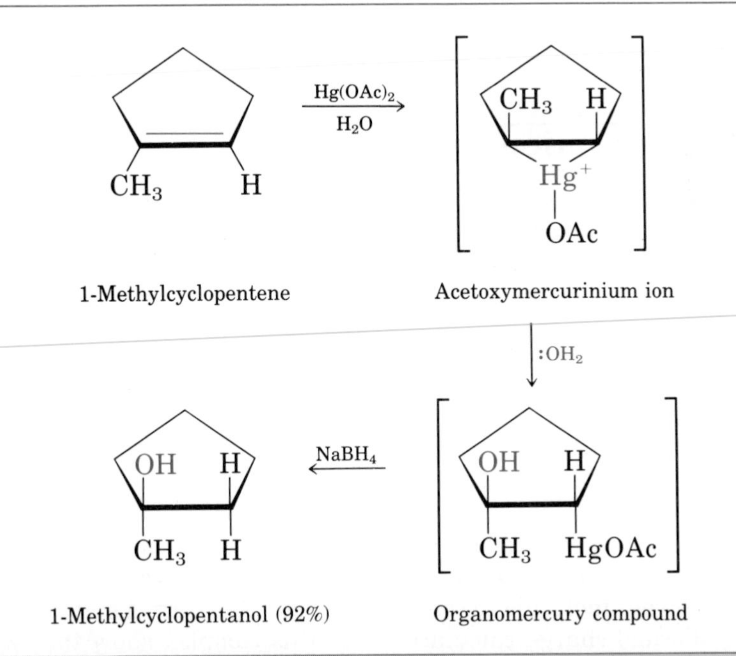

Figure 6.4. Mechanism of the oxymercuration reaction of alkenes. The mechanism of this electrophilic addition reaction is very similar to that of bromination.

organomercury product results. The final step, reaction of the organomercury compound with sodium borohydride, is not well understood but appears to involve radicals as intermediates. Note that the regiochemistry of the reaction corresponds to Markovnikov addition of water; that is, the hydroxyl group becomes attached to the more substituted carbon atom.

Just as cyclic bromonium ion intermediates can be intercepted by the addition of different nucleophiles to the reaction, mercurinium ions can also be intercepted. If the oxymercuration–demercuration sequence is carried out in methanol solution instead of water, a *methyl ether* is formed; if the reaction is carried out in acetic acid solvent, an *acetate ester* results. These reactions can be quite useful for converting the alkene double bond into a variety of other functional groups.

$$R_2C{=}CR_2 \xrightarrow[2.\ NaBH_4]{1.\ Hg(OAc)_2;\ CH_3OH} R_2C(H){-}C(OCH_3)R_2$$

An ether

$$R_2C{=}CR_2 \xrightarrow[2.\ NaBH_4]{1.\ Hg(OAc)_2;\ CH_3CO_2H} R_2C(H){-}C(O{-}CO{-}CH_3)R_2$$

An ester

6.4 Hydroboration

One of the most useful of all alkene addition reactions is **hydroboration.** First reported by H. C. Brown[1] in 1959, hydroboration involves addition of borane, BH_3, to an alkene to yield an *organoborane:*

$$H_2B{-}H + \;\rangle C{=}C\langle \;\longrightarrow\; H{-}C{-}C{-}BH_2$$

Borane

An organoborane

Borane is unstable by itself since boron has only six electrons in its outer shell. In THF solution, however, BH_3 can accept an electron pair from a solvent molecule to complete its octet and form a stable BH_3–THF complex. Formal-charge calculations for this complex show that we must assign a negative charge to boron and a positive charge to oxygen, but for all

[1]Herbert Charles Brown (1912–); b. London; Ph.D. (1938) University of Chicago (Schlessinger); professor, Purdue University (1947–); Nobel prize (1979).

practical purposes the BH_3–THF complex behaves chemically as if it were BH_3.

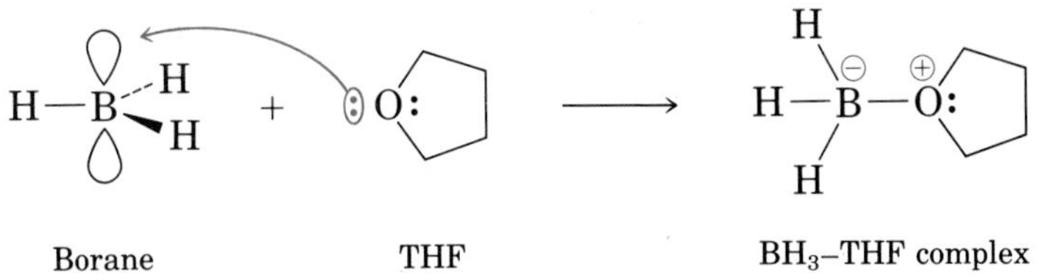

Borane THF BH_3–THF complex

When an alkene is treated with a THF solution of BH_3, rapid and quantitative addition of BH_3 to the double bond occurs. Since BH_3 has three hydrogens, addition occurs three times to produce a *trialkylborane* product, R_3B (Figure 6.5).

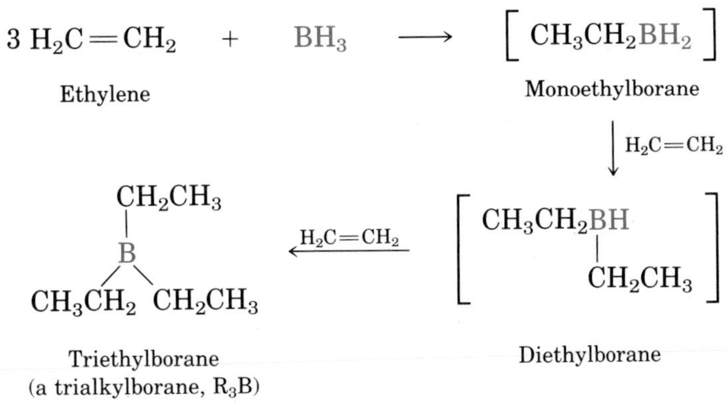

$$3\ H_2C{=}CH_2 + BH_3 \longrightarrow [CH_3CH_2BH_2]$$

Ethylene; Monoethylborane

$$[CH_3CH_2BH_2] \xrightarrow{H_2C=CH_2} [(CH_3CH_2)_2BH] \xrightarrow{H_2C=CH_2} (CH_3CH_2)_3B$$

Diethylborane; Triethylborane (a trialkylborane, R_3B)

Figure 6.5. Hydroboration of ethylene.

Alkylboranes are of great value in synthesis because of the further reactions they can be made to undergo. For example, when tricyclohexylborane is treated with aqueous hydrogen peroxide in basic solution, the carbon–boron bond is broken and cyclohexanol is produced. The net effect of this hydroboration–oxidation sequence is hydration of the alkene double bond.

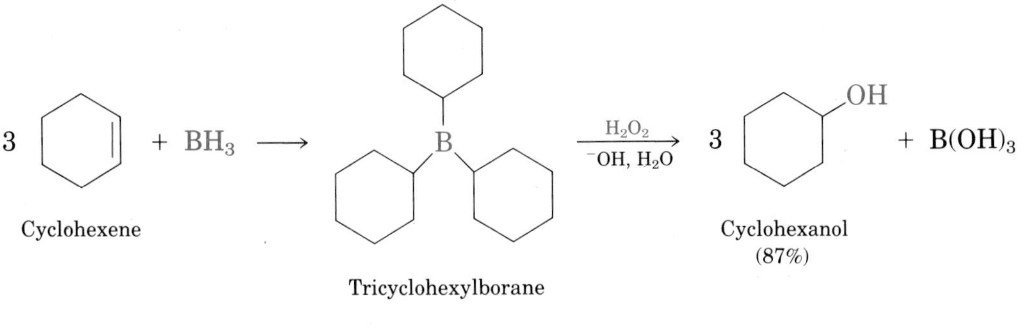

$$3\ \text{Cyclohexene} + BH_3 \longrightarrow \text{Tricyclohexylborane} \xrightarrow[{}^{-}OH,\ H_2O]{H_2O_2} 3\ \text{Cyclohexanol} + B(OH)_3$$

Cyclohexene; Tricyclohexylborane; Cyclohexanol (87%)

One of the features that makes the hydroboration reaction so useful is the regiochemistry that results when an unsymmetrical alkene is hydroborated. For example, *trans*-2-methylcyclopentanol results from the hydroboration–oxidation of 1-methylcyclopentene:

CH_3 $\xrightarrow[THF]{BH_3}$ [H B H CH_3] $\xrightarrow[H_2O_2]{^{-}OH}$ H OH H CH_3

1-Methylcyclopentene — Alkylborane intermediate — *trans*-2-Methylcyclopentanol (85%)

We find that the boron–hydrogen bond adds to the alkene in a syn manner, with boron attached to the less substituted carbon. During the oxidation step, the boron is replaced by a hydroxyl with the same stereochemistry, and the overall result is a non-Markovnikov syn addition of water, as opposed to the Markovnikov anti addition we have seen previously for other electrophilic additions. What makes hydroboration so apparently different?

The key word in this question is *apparently*. In fact, hydroboration is not really so different from other electrophilic addition reactions. Let's look at the mechanism of the reaction to see what this means (Figure 6.6). The first step is the formation of an acid–base complex between the nucleophilic alkene pi bond and the electrophilic BH_3. The interaction of the vacant boron orbital in BH_3 (electrophile) with the alkene pi electrons (nucleophile) is a polar process. Since boron has *accepted* electrons in forming the complex, it must gain a partial negative charge; since the alkene carbons have *donated* electrons in forming the complex, they gain a partial positive charge. This BH_3–alkene complex then reacts further to form carbon–hydrogen and carbon–boron bonds. We can view the reaction as taking place through a four-center cyclic transition state in which both carbon–hydrogen and carbon–boron bonds form at approximately the same time.

Since both carbon–hydrogen and carbon–boron bonds form simultaneously from the same face of the alkene, this mechanism satisfactorily explains why syn addition is observed. The regiochemistry of the reaction can also be accounted for by this mechanism, and is due to a combination of two factors. The first factor has to do with carbocation stabilities. In the addition of BH_3 to an unsymmetrically substituted alkene, there are two possible four-center transition states. From what we know about the nature of electrophilic additions, we would expect the transition state that places a partial positive charge on the more substituted carbon to be more favored. Thus boron bonds to the less substituted carbon (Figure 6.7, p. 168).

The second factor determining the regiochemistry of hydroboration is steric. The large, bulky boron attaches to the less sterically hindered carbon atom of the alkene. Both steric and electronic arguments predict the observed regiochemistry, and it is difficult to say which is more important. Evidence is accumulating, however, suggesting that steric factors probably have greater influence than electronic factors.

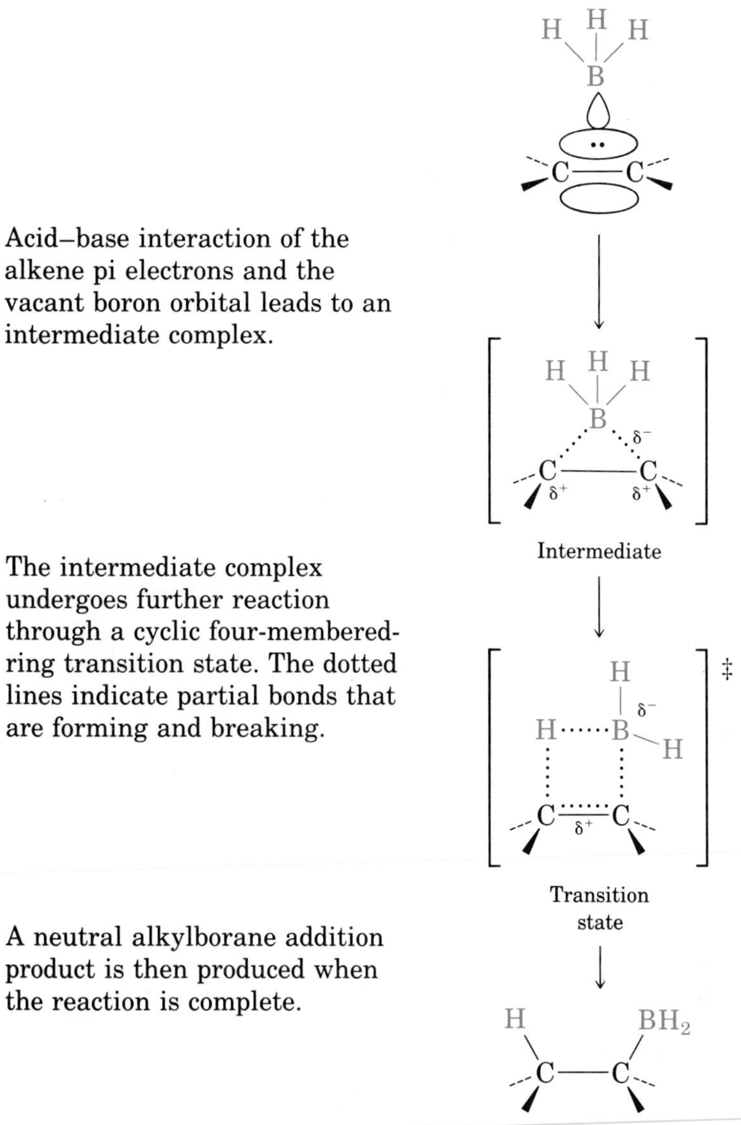

H H H
B
C C

Acid–base interaction of the alkene pi electrons and the vacant boron orbital leads to an intermediate complex.

H H H
B δ^-
C C
δ^+ δ^+

Intermediate

The intermediate complex undergoes further reaction through a cyclic four-membered-ring transition state. The dotted lines indicate partial bonds that are forming and breaking.

H
δ^-
H······B H
C C
δ^+
‡

Transition state

A neutral alkylborane addition product is then produced when the reaction is complete.

H BH_2
C—C

Figure 6.6. Mechanism of alkene hydroboration.

PROBLEM

6.5 Predict what product will be obtained when deuterated borane, BD_3, is used for the hydroboration of methylcyclopentene. Show both the stereochemistry (spatial arrangement) and regiochemistry (orientation) of the product.

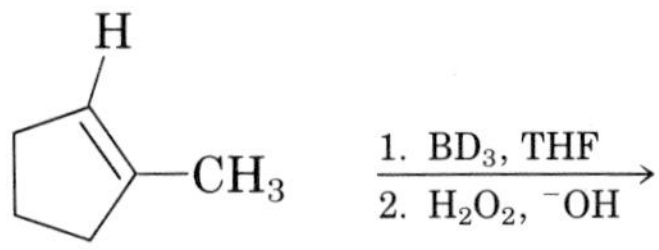

H
CH_3

$$\xrightarrow[\text{2. } H_2O_2,\ ^{-}OH]{\text{1. } BD_3,\ \text{THF}}$$

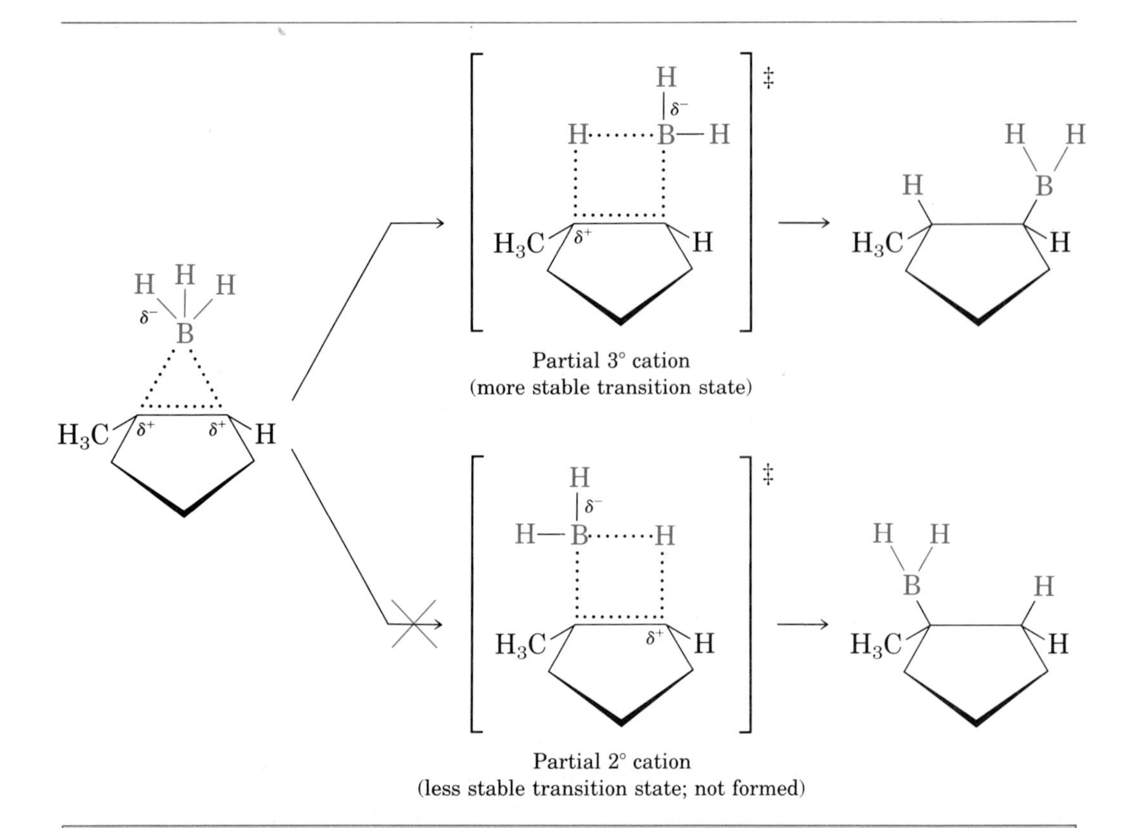

Figure 6.7. Mechanism of the hydroboration of 1-methylcyclopentene.

6.5 A radical addition to alkenes—HBr/peroxides

Chemists now consider the polar nature of electrophilic addition reactions to be well established; when HBr adds to an alkene under normal conditions, we know that an intermediate carbocation is involved and that Markovnikov orientation will be observed. This was not the case prior to 1933, however. In the 1920s the addition of HBr to 3-bromopropene had been studied several times by different workers, and conflicting results were reported in the chemical literature. The usual rule of Markovnikov addition did not appear to hold; *both* possible addition products were usually obtained, but in widely different ratios by different workers.

$$\underset{\text{3-Bromopropene}}{H_2C{=}CHCH_2Br} \xrightarrow{\text{Liquid HBr}} \underbrace{\underset{\text{1,3-Dibromopropane}}{\overset{Br\qquad Br}{H_2CCH_2CH_2}} + \underset{\text{1,2-Dibromopropane}}{\overset{Br\ Br}{H_3CCHCH_2}}}_{\text{Product mixture}}$$

A careful examination of the reaction ultimately resolved the problem. It was found that HBr (but not HCl or HI) can add to alkenes by two entirely different mechanisms. The compound 3-bromopropene is highly air sensitive and absorbs oxygen to form *peroxides* (compounds with oxygen–oxygen bonds). Since the oxygen–oxygen bond is weak and easily broken, peroxides are excellent sources of radicals and serve to catalyze a *radical addition* of HBr to the alkene, rather than the electrophilic addition. In subsequent work, it was found that non-Markovnikov radical addition of HBr occurs with many different alkenes. For example,

$$H_2C{=}CH(CH_2)_8COOH \xrightarrow[\text{Peroxides}]{\text{HBr}} BrCH_2CH_2(CH_2)_8COOH$$

10-Undecenoic acid → 11-Bromoundecanoic acid

This reaction proceeds by a radical chain process (recall the chlorination of methane in Section 4.4) involving addition of bromine radical, Br·, to the alkene. As with all radical chain reactions, both initiation steps and propagation steps are required:

Initiation steps. The reaction is initiated in two steps. The first step involves light-induced homolytic cleavage of the weak oxygen–oxygen peroxide bond to generate two alkoxy radicals. An alkoxy radical then abstracts a hydrogen atom from HBr in the second initiation step to give a bromine radical.

1. $R{-}O{-}O{-}R \xrightarrow{\text{Light}} 2\ RO\cdot$

   Peroxide → Alkoxy radicals

2. $R{-}O\cdot + H{:}Br \longrightarrow RO{:}H + Br\cdot$

Propagation steps. Once a bromine radical has formed in the initiation steps, a chain reaction of two repeating propagation steps commences. In the first propagation step, bromine radical adds to the alkene double bond, giving an alkyl radical. In the second propagation step, this alkyl radical reacts with HBr to yield addition product plus a bromine radical to carry on the chain reaction.

1. $H_2C{=}CHCH_2CH_3 + \boxed{Br\cdot} \longrightarrow BrCH_2{-}\dot{C}HCH_2CH_3$

2. $BrCH_2\dot{C}HCH_2CH_3 + HBr \longrightarrow BrCH_2CH_2CH_2CH_3 + \boxed{Br\cdot}$

According to this radical chain mechanism, the regiochemistry of addition is determined in the first propagation step when bromine radical adds to the alkene. In the case of 1-butene, this addition could conceivably take place at either of two carbons, to yield either a primary radical intermediate or a secondary radical. Experimental results, however, indicate that only the more substituted, secondary radical is formed (Figure 6.8). How can we explain these results?

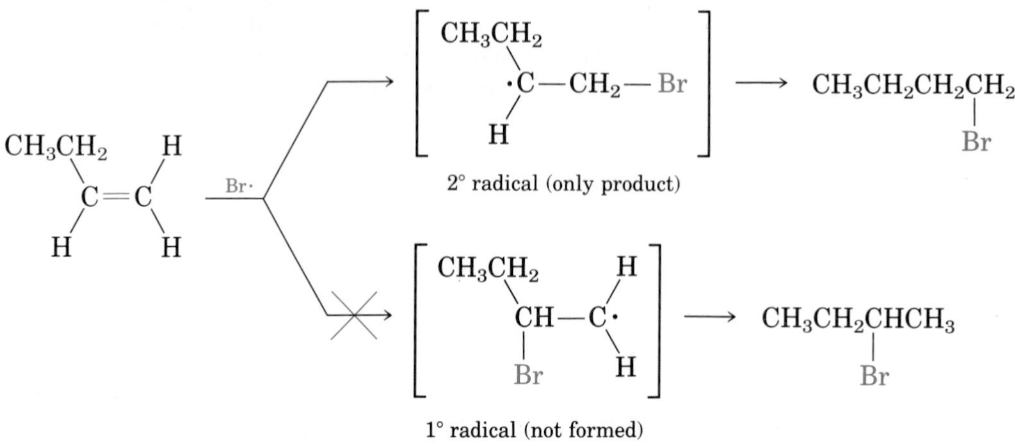

$CH_3CH_2(H)C{=}CH_2$ + Br· → $[CH_3CH_2\dot{C}H{-}CH_2{-}Br]$ → $CH_3CH_2CH_2CH_2Br$

2° radical (only product)

$CH_3CH_2(H)C{=}CH_2$ + Br· ⨯→ $[CH_3CH_2CH(Br){-}\dot{C}H_2]$ → $CH_3CH_2CH(Br)CH_3$

1° radical (not formed)

Figure 6.8. Addition of bromine radical to 1-butene. The reaction is regiospecific, leading to the secondary radical.

STABILITY OF RADICALS

In order to understand the reasons for the observed regiochemistry of the radical-catalyzed addition of HBr to alkenes, we need to examine the relative stabilities of substituted radicals. This is most easily done by comparing bond dissociation energies for different kinds of carbon–hydrogen bonds (Table 6.1).

TABLE 6.1 Bond dissociation energies of different kinds of C—H bonds in alkanes

Bond broken	*Radical products*	*Bond type*[a]	*Bond dissociation energy, $\Delta H°$(kcal/mol)*
$H{-}CH_3$	$H\cdot + CH_3\cdot$	Methyl	104
$H{-}CH_2CH_2CH_3$	$H\cdot + CH_3CH_2CH_2\cdot$	Primary	98
$CH_3CH(H)CH_3$	$H\cdot + CH_3\dot{C}HCH_3$	Secondary	95
$H{-}C(CH_3)_3$	$H\cdot + (CH_3)_3C\cdot$	Tertiary	92

[a]Tertiary C—H bonds are weaker than secondary ones, and secondary bonds are weaker than primary ones.

As we saw in Section 4.7, bond dissociation energy is the amount of energy that must be supplied to cleave a bond homolytically into two radical fragments. When $\Delta H°$ is high, the bond is strong and there is a large difference in stability between reactant and products. When $\Delta H°$ is low, the bond is weaker, and there is a smaller difference in stability between reactant and products. If we compare the energy required to break a primary C—H bond of propane (98 kcal/mol) with that required to break a secondary C—H bond of propane (95 kcal/mol), we find a difference of 3 kcal/mol. Since we are starting with the same reactant in both cases (propane), and since one product is the same in both cases (H·), the 3 kcal/mol energy difference is a direct measure of the difference in stability between the primary propyl radical and the secondary propyl radical. The secondary propyl radical is more stable than the primary propyl radical by 3 kcal/mol.

A similar comparison between a primary C—H bond of 2-methylpropane (98 kcal/mol) and the tertiary C—H bond in the same molecule (92 kcal/mol) leads us to conclude that the tertiary radical is more stable than the primary radical by 6 kcal/mol. We thus find that the stability order of radicals is as follows:

$$3° > 2° > 1° > CH_3\cdot$$

This is identical to the stability order of carbocations.

AN EXPLANATION OF THE REGIOCHEMISTRY OF RADICAL–CATALYZED ADDITION OF HBr TO ALKENES

Now that we know the stability order of radicals, we can complete an explanation for the observed regiochemistry of radical-catalyzed HBr additions to alkenes. Addition of bromine radical is an endothermic process and, according to the Hammond postulate (Section 5.11), the same factors that make a secondary radical more stable than a primary one also make the transition state leading to the secondary radical more stable. Thus, the more stable radical forms faster than the less stable one. We conclude that, although peroxide-catalyzed addition of HBr to an alkene gives an *apparently* non-Markovnikov-oriented product, the reaction in fact proceeds through the more stable intermediate. A reaction energy diagram illustrating this is shown in Figure 6.9 (p. 172).

PROBLEM

6.6 Draw a reaction energy diagram for the radical addition of HBr to 2-methyl-2-butene (consider only the propagation steps). Construct your diagram so that reactions leading to the two possible addition products are both shown. Which of the two curves has the lower $\Delta G^\ddagger$ for the first step? Which of the two curves has the lower $\Delta G°$ for the first step? Explain.

PROBLEM

6.7 Show how you would synthesize the following compounds (a)–(f). Identify the alkene starting material and indicate what reagents you would use.

(a) $CH_3CH_2C(CH_3)_2$ with Br on the C(CH₃)₂ carbon

(b) $CH_3CH_2CH_2CH_2Br$

(c) $CH_3CH_2CHCH_2CH_3$ with OH on the central CH carbon

(d) $CH_3CH_2CHCHCH_2CH_3$ with CH_3 on the third carbon and Br on the fourth carbon

(e) $CH_3CH_2C(CH_3)(OH)CH_2CH_2CH_3$

(f) CH_2OH (on a cyclohexane ring)

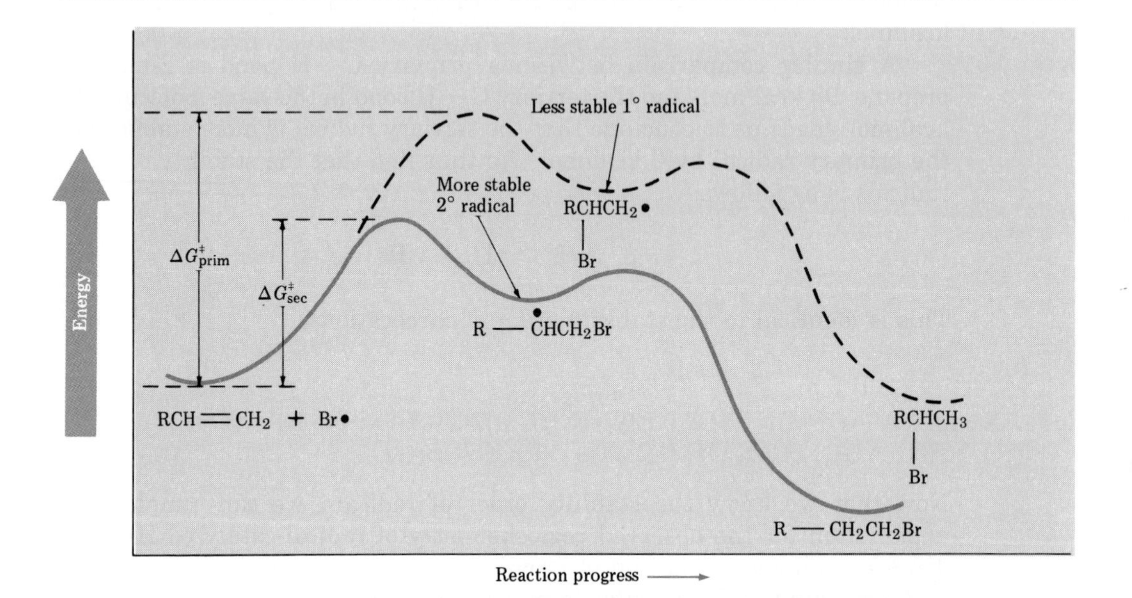

Figure 6.9. Reaction energy diagram for addition of bromine radical to an alkene. The more stable secondary radical forms faster than the less stable primary radical.

6.6 Hydrogenation

When alkenes are exposed to a hydrogen atmosphere in the presence of a suitable catalyst, addition of hydrogen to the double bond occurs, and a saturated product results. We have looked briefly at hydrogenation as a method of determining alkene stabilities, but it is also true that catalytic hydrogenation is a reaction of enormous practical value.

For most alkene reductions, either platinum or palladium is used as the catalyst. Palladium is normally employed in a very finely divided state "supported" on an inert material such as charcoal to maximize surface area (Pd/C). Platinum is normally used as PtO_2, a reagent known as *Adams catalyst* after its discoverer, Roger Adams.[2] Catalytic hydrogenation is unlike most other organic reactions in that it is a *heterogeneous* process rather than a homogeneous one. That is, the hydrogenation reaction occurs

[2]Roger Adams (1889–1971); b. Boston; Ph.D. Harvard (Torrey, Richards) (1912); professor, University of Illinois (1916–1971).

on the surface of solid catalyst particles rather than in solution. For this reason, catalytic hydrogenation has proven very difficult to study for mechanistic purposes. However, empirical observation has revealed that hydrogenation occurs with syn stereochemistry; both hydrogens add to the double bond from the same face.

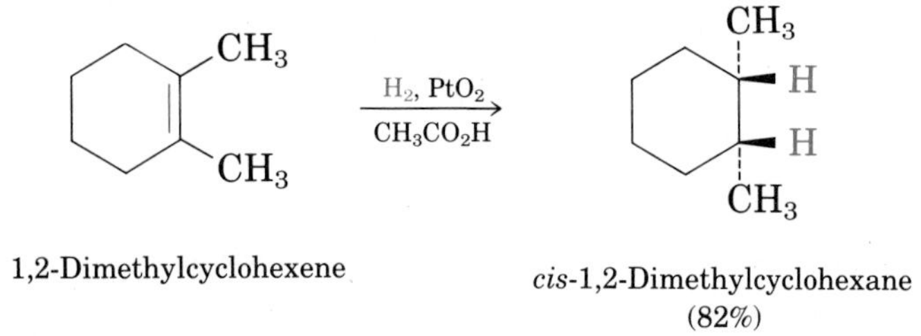

$$\text{1,2-Dimethylcyclohexene} \xrightarrow[CH_3CO_2H]{H_2,\ PtO_2} \textit{cis}\text{-1,2-Dimethylcyclohexane (82\%)}$$

The first step in the reaction is adsorption of hydrogen onto the catalyst surface. Complexation then occurs between catalyst and alkene by overlap of vacant metal orbitals with alkene pi electrons. In the final steps, hydrogen is inserted into the pi bond, and the saturated product diffuses away from the catalyst (Figure 6.10).

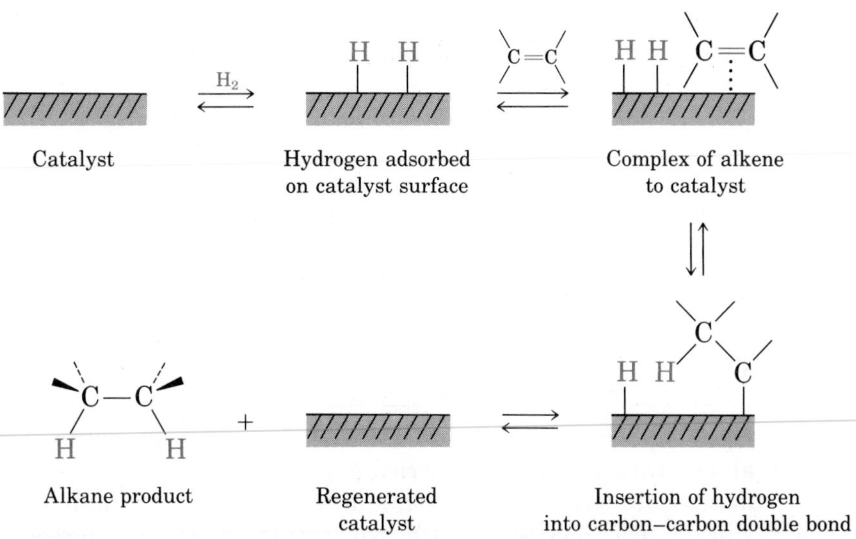

Catalyst $\xrightarrow{H_2}$ Hydrogen adsorbed on catalyst surface ⇌ Complex of alkene to catalyst ⇌ Insertion of hydrogen into carbon–carbon double bond ⇌ Regenerated catalyst + Alkane product

Figure 6.10. Mechanism of alkene hydrogenation. The reaction takes place on the surface of catalyst particles.

Alkenes are much more reactive than most other functional groups toward catalytic hydrogenation, and hydrogenation is therefore quite selective. Other functional groups such as ketones, esters, and nitriles survive normal alkene hydrogenation conditions unchanged. Reaction with these groups does occur under more vigorous conditions, however.

H_2
Pd/C in ethanol

2-Cyclohexenone → Cyclohexanone

Ketone not reduced

CO_2CH_3 → CO_2CH_3

H_2
Pd/C in ethanol

Methyl 3-phenylpropenoate → Methyl 3-phenylpropanoate

Benzene ring, ester not reduced

CN → CN

H_2
Pd/C in ethanol

Cyclohexylideneacetonitrile → Cyclohexylacetonitrile

Nitrile not reduced

Note that in the hydrogenation of methyl 3-phenylpropenoate the benzene ring functional group is not affected by hydrogen on palladium even though the ring contains three double bonds. This is yet another example of the remarkable stability of aromatic rings.

6.7 Hydroxylation

Hydroxylation of alkenes—the addition of a hydroxyl to each of the two alkene carbons—can be carried out with reagents such as potassium permanganate ($KMnO_4$) and osmium tetroxide (OsO_4). Both of these hydroxylation reactions occur with syn, rather than anti, stereochemistry and yield cis 1,2-dialcohols (diols).

Potassium permanganate is a common and inexpensive oxidant that reacts with alkenes under carefully controlled conditions in alkaline medium to yield 1,2-diols. The reaction occurs through an intermediate cyclic manganate species, which is then hydrolyzed to give the diol product. The oxidation of cyclohexene, for example, was first carried out in 1879 by Vladimir Markovnikov, and, as indicated, *cis*-1,2-cyclohexanediol was isolated in rather low yield. This result is typical; syn hydroxylation is always observed, but yields are sometimes poor because of the presence of side reactions.

Cyclohexene $\xrightarrow[\text{NaOH, } H_2O]{KMnO_4}$ [A cyclic manganate intermediate] $\xrightarrow{H_2O}$ *cis*-1,2-Cyclohexanediol (37%)

$$CH_3(CH_2)_7CH{=}CH(CH_2)_7CO_2H \xrightarrow[H_2O,\ NaOH]{KMnO_4} CH_3(CH_2)_7CH(OH){-}CH(OH)(CH_2)_7CO_2H$$

Oleic acid (from butter) → 9,10-Dihydroxyoctadecanoic acid (81%)

For small-scale laboratory preparation, the use of osmium tetroxide in place of potassium permanganate is much preferred. Although both toxic and expensive, OsO_4 reacts with alkenes to give good yields of cis 1,2-diols. As with permanganate hydroxylation, reaction with OsO_4 occurs through a cyclic intermediate, which is then cleaved to yield a cis diol. A cyclic *osmate* is more stable than a cyclic manganate, however, and must be cleaved in a separate step. Aqueous sodium bisulfite, $NaHSO_3$, is often used to accomplish this cleavage:

1,2-Dimethylcyclopentene $\xrightarrow[\text{Pyridine}]{OsO_4}$ [Cyclic osmate intermediate] $\xrightarrow[H_2O]{NaHSO_3}$ *cis*-1,2-Dimethyl-1,2-cyclopentanediol (87%)

PROBLEM

6.8 How would you prepare the following compounds? Show the starting alkene and the reagents you would use.

(a) [1-methyl-1,2-cyclohexanediol structure: CH_3, OH, OH, H]

(b) $CH_3CH_2CH(OH)C(OH)(CH_3)_2$

(c) $CH_2(OH)CH(OH)CH(OH)CH_2OH$

PROBLEM

6.9 Explain the observation that hydroxylation of *cis*-2-butene with OsO_4 yields a different product than hydroxylation of *trans*-2-butene. Draw the structure and show the stereochemistry of each product. (We will explore the stereochemistry of the products in more detail in Chapter 8.)

6.8 Oxidative cleavage of alkenes

In all the reactions of alkenes studied thus far, the carbon skeleton of the starting material has been left intact. We have seen the conversion of the carbon–carbon double bond into new functional groups (halides, alcohols, 1,2-diols) by adding different reagents, but no carbon structures have been broken or rearranged. There are, however, powerful oxidizing reagents that *cleave* carbon–carbon double bonds to produce two fragments.

Ozone (O_3) is the most useful cleavage reagent. Prepared conveniently in the laboratory by passing a stream of oxygen through a high-voltage electrical discharge, ozone adds rapidly to alkenes at low temperature. Studies have shown that an intermediate called a *molozonide* is first formed, and that the molozonide then rapidly rearranges to form an *ozonide:*

$$3\ O_2 \xrightarrow[\text{discharge}]{\text{Electric}} 2\ O_3$$

$$\text{C=C} \xrightarrow[\text{CH}_2\text{Cl}_2,\ -78^\circ\text{C}]{\text{O}_3} [\text{molozonide}] \longrightarrow \text{ozonide} \xrightarrow[\text{CH}_3\text{COOH/H}_2\text{O}]{\text{Zn}} \text{C=O} + \text{O=C}$$

A molozonide

An ozonide

Low-molecular-weight ozonides are highly explosive and are therefore not isolated. Instead, ozonides are usually further treated with a reducing agent such as zinc metal in acetic acid to convert them to carbonyl compounds. The net result of the *ozonolysis*–zinc reduction sequence is that the carbon–carbon double bond is cleaved, and oxygen becomes doubly bonded to each of the original alkene carbons. If a compound containing a tetrasubstituted double bond is ozonized, two ketone fragments result; if a compound with a trisubstituted double bond is ozonized, one ketone and one aldehyde result, and so on.

$$\text{C}_6\text{H}_{10}\text{=C(CH}_3)_2 \xrightarrow[\text{2. Zn, H}_3\text{O}^+]{\text{1. O}_3} \text{C}_6\text{H}_{10}\text{=O} + \text{CH}_3\overset{\text{O}}{\overset{\|}{\text{C}}}\text{CH}_3$$

Isopropylidenecyclohexane (tetrasubstituted)

Cyclohexanone Acetone

84%; two ketones

$$CH_3(CH_2)_7CH{=}CH(CH_2)_7COOCH_3 \xrightarrow[\text{2. Zn, } H_3O^+]{\text{1. } O_3} CH_3(CH_2)_7CH{=}O + O{=}CH(CH_2)_7COOCH_3$$

Methyl 9-octadecenoate (disubstituted) → Nonanal + Methyl 9-oxononanoate

78%; two aldehydes

β-Pinene (disubstituted) $\xrightarrow[\text{2. Zn, } H_3O^+]{\text{1. } O_3}$ Nopinone + Formaldehyde ($O{=}CH_2$)

75%; one ketone, one aldehyde

Another facet of the ozonolysis reaction is that *alcohols* are produced when ozonides are treated with sodium borohydride. For example, 1-butanol and 2-propanol are produced from 2-methyl-2-heptene. This flexibility, permitting the conversion of an alkene into either a carbonyl compound or an alcohol, adds greatly to the usefulness of ozone cleavages.

$$(CH_3)_2C{=}CHCH_2CH_2CH_2CH_3 \xrightarrow[\text{2. } NaBH_4,\ H_2O]{\text{1. } O_3\text{, ethyl acetate}} (CH_3)_2CH{-}OH + HO{-}CH_2CH_2CH_2CH_3$$

2-Methyl-2-heptene → 2-Propanol + 1-Butanol

Reagents other than ozone also cause double-bond cleavage reactions. For example, potassium permanganate in neutral or acidic solution cleaves alkenes, giving carbonyl fragments in low to moderate yield. If hydrogens are present on the double bond, carboxylic acids are produced; if two hydrogens are present on one carbon, CO_2 is formed:

$$(CH_3)_2CHCH_2CH_2CH_2CH(CH_3)CH{=}CH_2 \xrightarrow[H_2O]{KMnO_4} (CH_3)_2CHCH_2CH_2CH_2CH(CH_3)COOH + CO_2$$

3,7-Dimethyl-1-octene → 2,6-Dimethylheptanoic acid (45%)

PROBLEM

6.10 Predict the product of ozonolysis and zinc reduction of 1,2-dimethylcyclohexene.

6.9 Diol cleavage

1,2-Diols can be cleaved to carbonyl compounds by reaction with periodic acid (HIO_4). This reaction, combined with OsO_4 hydroxylation, is an excellent alternative to ozonolysis for the cleavage of alkenes. If the two hydroxyls are on an open chain, two carbonyl compounds result. If the two hydroxyls are on a ring, however, a *dicarbonyl* compound results. The reaction is believed to take place via cyclic periodate intermediates, as the following examples show:

1-Methylcyclohexene —(1. OsO_4; 2. $NaHSO_3$)→ A 1,2-diol —(HIO_4)→ [Cyclic periodate intermediate] → 6-Oxoheptanal (86%)

Cyclopentylidenecyclopentane —(1. OsO_4; 2. $NaHSO_3$)→ A 1,2-diol —(HIO_4)→ [Cyclic periodate intermediate] →

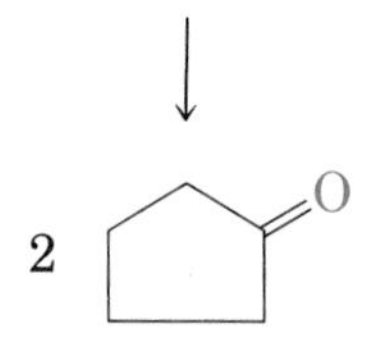

2 Cyclopentanone (81%)

PROBLEM

6.11 Evidence that cleavage of a 1,2-diol by HIO_4 occurs through a five-membered-ring intermediate is based on kinetic data—the measurement of reaction rates. Diols A and B were prepared, and their rates of reaction with $NaIO_4$ were measured. It was found that diol A cleaved approximately 1 million times faster than diol B. Make molecular models of A and B and of possible five-membered-ring periodate intermediates to see if you can explain these results.

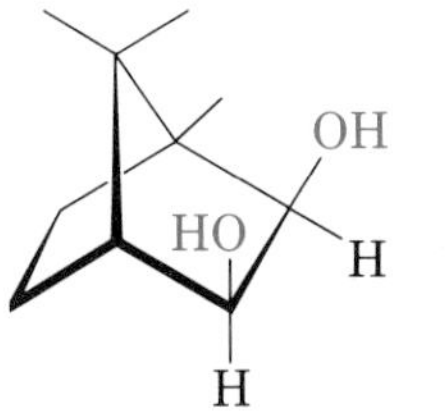

A
(cis diol)

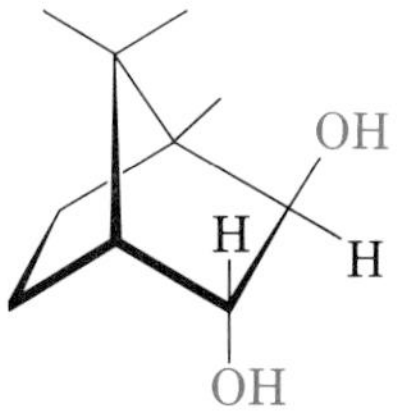

B
(trans diol)

6.10 Some biological addition reactions

The chemistry of living organisms is a fascinating field of study. Nature carries out complex chemistry on a grand scale with an ease that is humbling to laboratory chemists, and the simplest one-celled organism is capable of more sophisticated organic synthesis than any human. Yet as we learn more, it becomes clear that the same principles that apply to laboratory chemistry also apply to biochemistry. Biological organic chemistry takes place in aqueous media rather than in organic solvents, and involves complex catalysts called **enzymes,** but the *kinds* of reactions are remarkably similar. Thus, there are many cases of biological addition reactions to alkenes. For example, the enzyme fumarase catalyzes the addition of water to fumaric acid much as sulfuric acid catalyzes the addition of water to ethylene on an industrial scale:

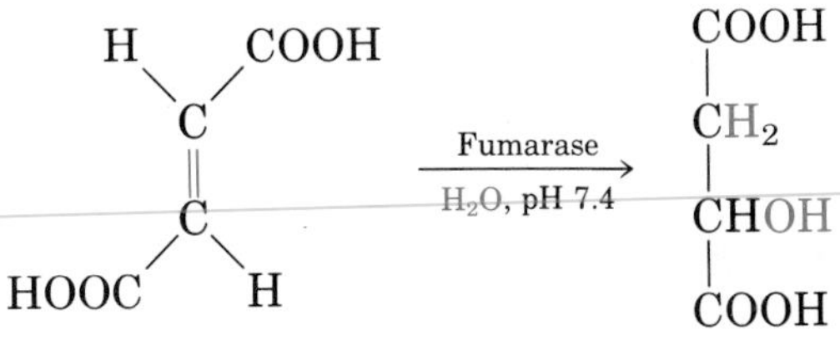

$$\text{HOOC-CH=CH-COOH (Fumaric acid)} \xrightarrow[\text{H}_2\text{O, pH 7.4}]{\text{Fumarase}} \text{HOOC-CH}_2\text{-CHOH-COOH (Malic acid)}$$

Fumaric acid

Malic acid

Enzymes can even catalyze the formation of bromohydrins from alkenes. The enzyme chloroperoxidase catalyzes the addition of HO—Br to certain double bonds such as that shown in Figure 6.11 (p. 180).

Enzyme-catalyzed reactions in organisms are usually much more chemically selective than are their laboratory counterparts. Fumarase, for example, is completely inert toward maleic acid, the cis isomer of fumaric acid. Nevertheless, the fundamental processes of organic chemistry are the same in the living cell and in the laboratory.

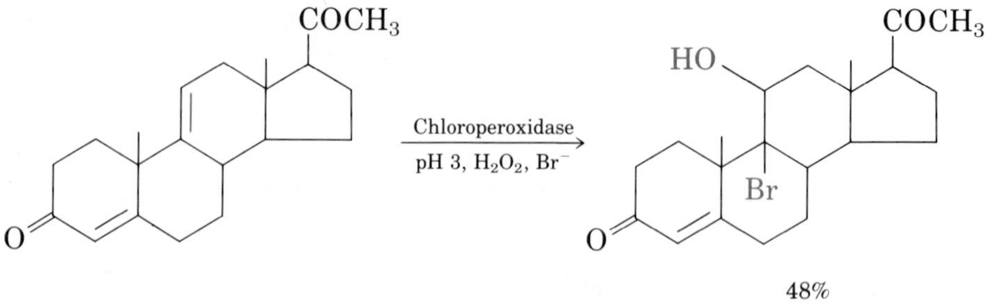

Figure 6.11. An enzyme-catalyzed addition reaction.

6.11 Preparation of alkenes

Just as the chemistry of alkenes is dominated by addition reactions, the preparation of alkenes is dominated by **elimination reactions.** Additions and eliminations are in many respects two sides of the same coin (Figure 6.12). Let's look briefly at two of the more common elimination reactions—the **dehydrohalogenation** (loss of HX) of an alkyl halide and the **dehydration** (loss of H_2O) of an alcohol. We'll take a closer look in Chapter 10.

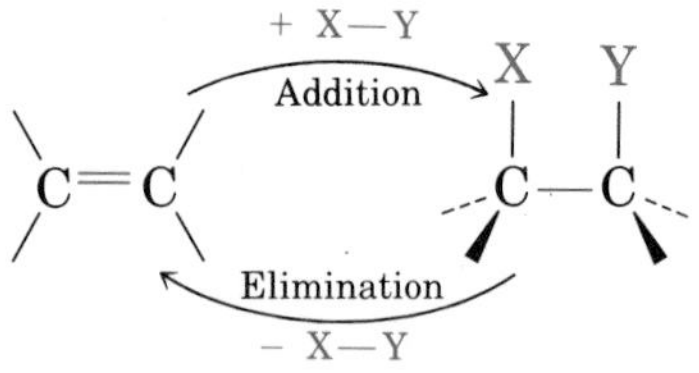

Figure 6.12. The relationship between addition and elimination reactions.

6.12 Elimination of HX: dehydrohalogenations

Alkyl halides can be synthesized by addition of HX to alkenes; conversely, alkenes can be synthesized by elimination of HX from alkyl halides. Elimination is normally effected by treating an alkyl halide with a strong base. Thus, bromocyclohexane yields cyclohexene when treated with potassium hydroxide in alcohol solution:

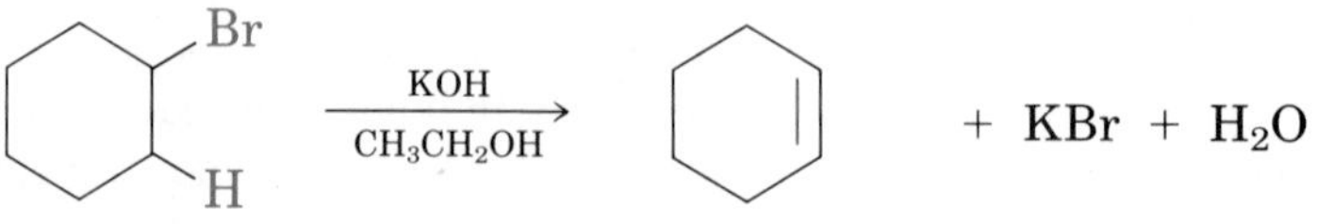

Bromocyclohexane Cyclohexene (81%)

Although the mechanism of elimination will not be discussed in detail until Chapter 10, we can for now view the reaction as occurring by means of an attack by a base on a hydrogen atom, as indicated in Figure 6.13. The electrons of the C—H bond form the new pi bond, and bromide ion departs with both of the C—Br bond electrons.

An electron pair from the hydroxide oxygen atom forms a bond to a hydrogen atom. The electron pair from the C—H bond then becomes the new alkene pi bond, and bromide ion departs with the two electrons from the C—Br bond.

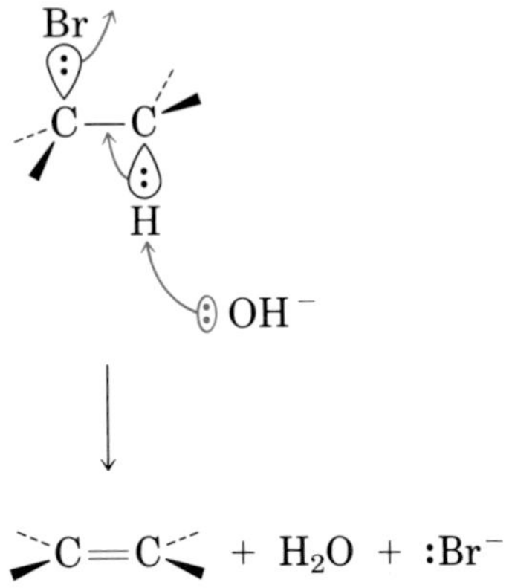

$$C{=}C + H_2O + :Br^-$$

Figure 6.13. Mechanism of a dehydrobromination reaction. The reaction occurs in a single step.

A transition state for this elimination might look like that shown in Figure 6.14. The C—H and C—Br bonds are both beginning to break at the same time that the carbon atoms are beginning to rehybridize from sp^3 to sp^2 and that the alkene pi bond is beginning to form.

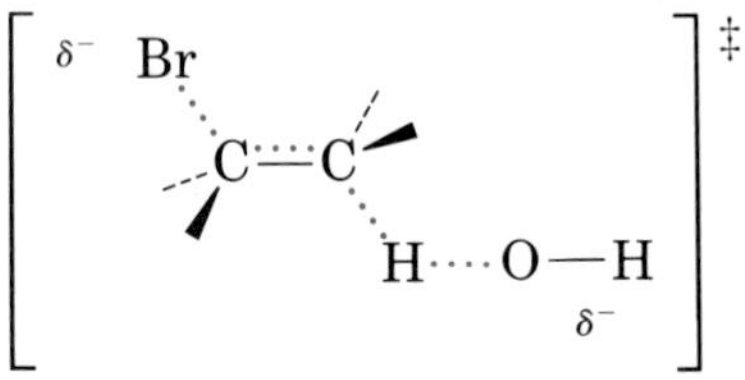

Figure 6.14. Hypothetical transition state for elimination of HBr to yield an alkene.

Elimination reactions are somewhat more complex than addition reactions for several reasons. There is, for example, the problem of regiochemistry of elimination—what products result from dehydrohalogenation of unsymmetrical halides? In fact, elimination reactions almost always give *mixtures* of alkene products, and the best we can usually do is to predict which the major product will be. According to a rule formulated by the Russian chemist Alexander Zaitsev,[3] base-induced elimination reactions generally give the more substituted alkene product, that is, the alkene

[3]Alexander M. Zaitsev (1841–1910); b. Kasan, Russia (name also spelled *Saytzeff*, according to the German pronunciation).

with more alkyl substituents on the double-bond carbons. In the following two examples, Zaitsev's rule is clearly applicable. The more substituted alkene product predominates in both cases when sodium ethoxide in ethanol is used as the base.

$$CH_3CH_2CH(Br)CH_3 \xrightarrow[CH_3CH_2OH]{CH_3CH_2O^-Na^+} CH_3CH{=}CHCH_3 + CH_3CH_2CH{=}CH_2$$

2-Bromobutane → 2-Butene (81%) + 1-Butene (19%)

$$CH_3CH_2C(Br)(CH_3)CH_3 \xrightarrow[CH_3CH_2OH]{CH_3CH_2O^-Na^+} CH_3CH{=}C(CH_3)_2 + CH_3CH_2C(CH_3){=}CH_2$$

2-Bromo-2-methylbutane → 2-Methyl-2-butene (70%) + 2-Methyl-1-butene (30%)

Many different bases effect dehydrohalogenation reactions, but strong oxygen bases such as hydroxide or alkoxide ($R—O^-$; salts of alcohols, R—OH) often lead to poor yields of products. In laboratory applications with complex molecules, chemists always try to use the mildest, most effective reagents possible. The amine base diazabicyclononene (DBN)

DBN

has been found to be an excellent choice for many dehydrohalogenation reactions.

$$CH_3CH_2CH_2CH_2CH_2CH(Br)CH_3 \xrightarrow[90°C]{DBN} CH_3(CH_2)_4CH{=}CH_2 + CH_3(CH_2)_3CH{=}CHCH_3$$

2-Bromoheptane → 1-Heptene (15%) + 2-Heptene (63%)

$$HC{\equiv}CCH_2CH(Cl)(CH_2)_8CO_2CH_3 \xrightarrow[90°C]{DBN} HC{\equiv}CCH{=}CH(CH_2)_8CO_2CH_3$$

Methyl 10-chloro-12-tridecynoate → Methyl 10-tridecen-12-ynoate (85%)

6.13 Elimination of H_2O: dehydrations

The **dehydration** of alcohols is perhaps the most useful of all elimination reactions, and chemists have devised many alternative ways of carrying out dehydrations. One of the more common methods, which works particularly well for tertiary alcohols, is **acid-catalyzed dehydration.** For example, when 1-methylcyclohexanol is warmed with aqueous sulfuric acid in a solvent such as tetrahydrofuran, loss of water occurs and 1-methylcyclohexene is formed:

$$\text{1-Methylcyclohexanol} \xrightarrow[50^\circ C]{H_3O^+,\ THF} \text{1-Methylcyclohexene (91\%)}$$

Acid-catalyzed dehydrations normally follow Zaitsev's rule and yield the more substituted alkene as the major product. Thus, 2-methyl-2-butanol gives primarily 2-methyl-2-butene (trisubstituted) rather than 2-methyl-1-butene (disubstituted):

$$H_3C-C(CH_3)(OH)-CH_2CH_3 \xrightarrow[25^\circ C]{H_3O^+,\ THF} (CH_3)_2C{=}CHCH_3 + CH_2{=}C(CH_3)-CH_2CH_3$$

2-Methyl-2-butanol → 2-Methyl-2-butene (trisubstituted), Major product + 2-Methyl-1-butene (disubstituted), Minor product

The reactivity order for acid-catalyzed dehydrations is

$$R_3COH > R_2CHOH > RCH_2OH$$

and, in normal laboratory practice, only tertiary alcohols are commonly dehydrated with acid. Secondary alcohols can be made to react, but the conditions are severe (75% H_2SO_4, 100°C) and sensitive molecules cannot survive.

$$CH_3CH_2CH(OH)CH_3 \xrightarrow[100^\circ C]{75\%\ H_2SO_4} CH_3CH{=}CHCH_3$$

2-Butanol → 2-Butene

Primary alcohols are even less reactive than secondary ones, and very harsh conditions are necessary to cause dehydration (95% H_2SO_4, 150°C).

The reasons for the observed reactivity order are best understood by looking at the mechanism of the dehydration reaction. These reactions are known to take place in three steps through carbocation intermediates. The reaction occurs by protonation of the alcohol oxygen, loss of water to generate a carbocation, and final loss of a neighboring hydrogen ion (Figure 6.15).

Two electrons from the oxygen atom bond to H^+, yielding a protonated alcohol intermediate.

The carbon–oxygen bond breaks, and the two electrons from the bond stay with oxygen, leaving a carbocation intermediate.

Two electrons from a neighboring carbon–hydrogen bond form the alkene pi bond, and H^+ (a proton) is eliminated.

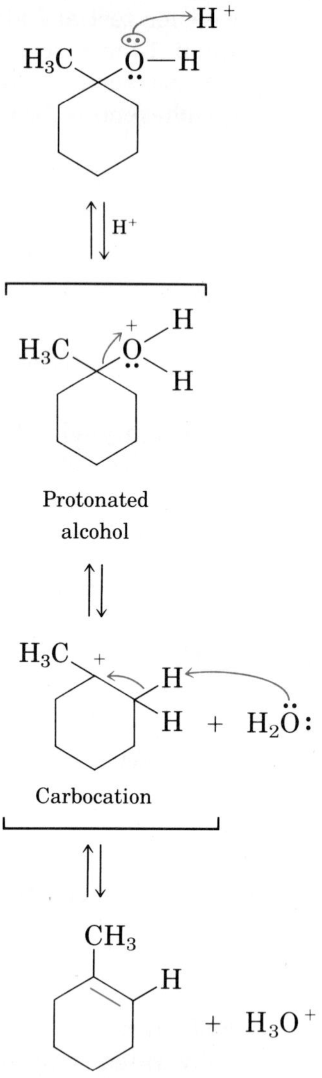

H_3C O—H H^+

H^+

H_3C $\overset{+}{O}$ H H

Protonated alcohol

H_3C + H H + $H_2\ddot{O}$:

Carbocation

CH_3 H + H_3O^+

Figure 6.15. Mechanism of acid-catalyzed dehydration of alcohols to yield alkenes.

The slowest of the three steps shown in Figure 6.15 is the actual loss of water from the protonated alcohol. It is this step that limits the rate at which reaction occurs. We have already seen (Section 5.10) that carbocations have the stability order 3° > 2° > 1°. Thus, according to the Hammond postulate, a more stable carbocation should form faster than a less stable one, and the order of alcohol reactivity therefore parallels the order of carbocation stability.

To circumvent strongly acidic dehydrating conditions, other reagents have been developed that are effective under mild, basic conditions. One such reagent, phosphorus oxychloride ($POCl_3$), is often able to effect the dehydration of secondary and tertiary alcohols at 0°C in the basic amine solvent, pyridine:

1-Methylcyclohexanol $\xrightarrow[\text{Pyridine, 0°C}]{POCl_3}$ 1-Methylcyclohexene (96%)

The mechanism of this dehydration reaction will be discussed in Chapter 10.

PROBLEM

6.12 Predict the product(s) you would expect to obtain from these elimination reactions. Indicate the major product in each case.

(a) $(CH_3)_2C(Br)CH_2CH_2CH_3 \xrightarrow[\Delta]{DBN}$

(b) $CH_3CH_2CH(OH)CH(CH_3)_2 \xrightarrow[\text{Pyridine}]{POCl_3}$

(c) Cyclohexyl$-C(Br)(CH_3)_2 \xrightarrow[\Delta]{DBN}$

PROBLEM

6.13 Compound A, $C_{10}H_{18}O$, undergoes reaction with dilute H_2SO_4 at 25°C to yield a mixture of two alkenes, $C_{10}H_{16}$. The major alkene product, B, gives only cyclopentanone after ozone treatment followed by reduction with zinc in acetic acid. Formulate the reactions involved and identify A and B.

Cyclopentanone (C_5H_8O)

6.14 Biological elimination reactions

Just as there are biological addition reactions, there are biological elimination reactions for preparing alkenes. Biological dehydrations are fairly common processes in living systems. To take just one example, the enzyme fumarate synthetase catalyzes the elimination of water from malic acid—a clear analogy with normal laboratory chemistry.

$$\text{HOOC--CH}_2\text{--CHOH--COOH} \xrightarrow{\text{Fumarate synthetase}} \text{(H)(COOH)C=C(HOOC)(H)} + H_2O$$

Malic acid (from apples) — Fumaric acid

6.15 Summary

Addition of electrophiles is the most important reaction of alkenes. For example, HCl, HBr, and HI add to carbon–carbon double bonds by a two-step mechanism involving initial reaction of the nucleophilic double bond with H^+ to form a carbocation intermediate, followed by attack of halide ion nucleophile on the cation intermediate. Many other electrophiles add to alkenes. Bromine and chlorine add via three-membered-ring halonium ion intermediates to give addition products having anti stereochemistry. If water is present during halogen addition reactions, the halonium ion intermediate can be intercepted to yield a halohydrin.

Hydration of (addition of water to) alkenes is best carried out either by the oxymercuration–demercuration procedure or by hydroboration–oxidation. Oxymercuration involves electrophilic addition of mercuric ion to an alkene, followed by trapping of the intermediate three-membered-ring mercurinium ion with water and reduction of mercury with $NaBH_4$. Hydroboration of alkenes involves electrophilic addition of boron and hydrogen, followed by oxidation of the intermediate organoborane. The two hydration methods are complementary, since oxymercuration gives the product of Markovnikov anti addition, whereas hydroboration–oxidation gives the product of non-Markovnikov syn addition.

HBr (but not HCl or HI) can also add to alkenes by a radical chain pathway to give the non-Markovnikov product. Radicals have been found to have the following stability order:

$$\text{Tertiary} > \text{Secondary} > \text{Primary} > \text{Methyl}$$

Hydrogen can be added to alkenes by reaction in the presence of a noble metal catalyst such as platinum or palladium. Catalytic hydrogenation is a heterogeneous process that occurs on the surface of catalyst particles, rather than in solution. The reaction occurs with syn stereochemistry.

Cis 1,2-diols can be made directly from alkenes by hydroxylation with OsO_4. 1,2-Diols can then be cleaved by reaction with HIO_4 to yield two carbonyl-containing compounds. Alkenes can also be cleaved directly to carbonyl-containing compounds by reaction with ozone, followed by reduction with zinc metal.

Methods for the preparation of alkenes generally involve elimination reactions. For example, treatment of an alkyl halide with a strong base effects elimination of HX (HCl, HBr, or HI) and yields the alkene. Alcohols can be similarly dehydrated (elimination of HOH) by treatment either with strong acid or with $POCl_3$ in pyridine. These elimination reactions usually give a mixture of alkene products in which the more substituted alkene predominates (Zaitsev's rule).

6.16 Summary of reactions

1. Synthesis of alkenes
   a. Dehydrohalogenation of alkyl halides (Section 6.12)

H X | | C—C $\xrightarrow{\text{Base}}$ C=C

The more substituted alkene predominates.

   b. Dehydration of alcohols (Section 6.13)

H OH | | C—C—R (R) $\xrightarrow[\text{Heat}]{H_3O^+}$ C=C(R)(R)

Best for 3° alcohols

$\xrightarrow[\text{Pyridine}]{POCl_3}$ C=C(R)(R)

Works for both 2° and 3° alcohols

2. Electrophilic addition reactions of alkenes
   a. Addition of HX, where X = Cl, Br, or I (Sections 5.8 and 5.9)

C=C $\xrightarrow[\text{Ether}]{HX}$ H—C—C—X

Markovnikov regiochemistry is observed: H adds to the less substituted carbon and X adds to the more substituted carbon.

   b. Addition of halogens (Section 6.1)

C=C $\xrightarrow[CCl_4]{X_2}$ X—C—C—X

where X_2 = Cl_2 or Br_2. Anti addition is observed.

c. Halohydrin formation (Section 6.2)

$$>C=C< \xrightarrow[H_2O]{Br_2} \text{Br}-\overset{|}{C}-\overset{|}{C}-\text{OH} + HBr$$

Markovnikov regiochemistry and anti stereochemistry are observed.

d. Addition of water by oxymercuration (Section 6.3)

$$>C=C< \xrightarrow[2.\ NaBH_4]{1.\ Hg(OAc)_2,\ H_2O} \text{HO}-\overset{|}{C}-\overset{|}{C}-\text{H}$$

The OH is at the more substituted carbon site.

e. Addition of methanol by oxymercuration (Section 6.3)

$$>C=C< \xrightarrow[2.\ NaBH_4]{1.\ Hg(OAc)_2,\ CH_3OH} CH_3O-\overset{|}{C}-\overset{|}{C}-\text{H}$$

f. Addition of water by hydroboration–oxidation (Section 6.4)

$$>C=C< \xrightarrow[2.\ H_2O_2,\ OH^-]{1.\ BH_3} \overset{H}{\overset{|}{C}}-\overset{OH}{\overset{|}{C}}$$

Non-Markovnikov syn addition is observed.

3. Radical addition of HBr to alkenes (Section 6.5)

$$>C=C< \xrightarrow[\text{Peroxides}]{HBr} \overset{H}{\overset{|}{C}}-\overset{Br}{\overset{|}{C}}$$

Non-Markovnikov addition is observed.

4. Hydrogenation of alkenes (Section 6.6)

$$>C=C< \xrightarrow{H_2/\text{catalyst}} \overset{H}{\overset{|}{C}}-\overset{H}{\overset{|}{C}}$$

Syn addition is observed.

5. Hydroxylation of alkenes (Section 6.7)

$$>C=C< \xrightarrow[\text{2. } NaHSO_3/H_2O]{\text{1. } OsO_4} -\underset{|}{\overset{OH}{\overset{|}{C}}}-\underset{|}{\overset{OH}{\overset{|}{C}}}-$$

Syn addition is observed.

6. Oxidative cleavage of alkenes (Section 6.8)

$$>C=C< \xrightarrow[\text{2. } Zn/H_3O^+]{\text{1. } O_3} >C=O + O=C<$$

7. 1,2-Diol cleavage (Section 6.9)

$$-\overset{OH}{C}-\overset{OH}{C}- \xrightarrow{HIO_4/H_2O} >C=O + O=C<$$

ADDITIONAL PROBLEMS

6.14 Predict the products of the following reactions. Indicate regiochemistry when relevant.

$C_6H_5CH=CH_2$

(a) $\xrightarrow{H_2/Pd}$

(b) $\xrightarrow{Br_2}$

(c) $\xrightarrow{HBr}$

(d) $\xrightarrow[\text{2. } NaHSO_3]{\text{1. } OsO_4}$

(e) $\xrightarrow{D_2/Pd}$

6.15 Suggest structures for alkenes that give the following reaction products:

(a) ? $\xrightarrow{H_2/Pd}$ 2-Methylhexane

(b) ? $\xrightarrow{H_2/Pd}$ 1,1-Dimethylcyclohexane

(c) ? $\xrightarrow{Br_2/CCl_4}$ 2,3-Dibromo-5-methylhexane

(d) ? $\xrightarrow[\text{2. } NaBH_4]{\text{1. } Hg(OAc)_2,\ H_2O}$ $CH_3CH_2CH_2CH(OH)CH_3$

(e) ? $\xrightarrow{\text{HBr/peroxides}}$ 2-Bromo-3-methylheptane

(f) ? $\xrightarrow{\text{HCl, ether}}$ 2-Chloro-3-methylheptane

6.16 Predict the products of the following reactions, indicating both regiochemistry and stereochemistry where appropriate.

(a) 1-methylcyclohexene (CH_3, H) $\xrightarrow[\text{2. Zn, } H_3O^+]{\text{1. } O_3}$

(b) cyclohexene $\xrightarrow[H^+]{KMnO_4}$

(c) 1-methylcyclohexene (CH_3) $\xrightarrow[\text{2. } H_2O, \ ^-OH]{\text{1. } BH_3}$

(d) 1-methylcyclohexene (CH_3) $\xrightarrow[\text{2. } NaBH_4]{\text{1. } Hg(OAc)_2, H_2O}$

(e) 3-methylcyclopentene (H_3C, H) $\xrightarrow[\text{2. } NaHSO_3]{\text{1. } OsO_4}$

6.17 How would you carry out the following transformations? Indicate the proper reagents.

(a) cyclopentene $\xrightarrow{?}$ trans-cyclopentane-1,2-diol (OH, H, OH, H)

(b) cyclopentene $\xrightarrow{?}$ cyclopentanol (H, OH)

(c) cyclopentene $\xrightarrow{?}$ methoxycyclopentane (H, OCH_3)

(d) $CH_3CH{=}CHCH(CH_3)_2 \xrightarrow{?} CH_3CH_2OH + (CH_3)_2CHCH_2OH$

(e) $H_2C{=}C(CH_3)_2 \xrightarrow{?} (CH_3)_2CHCH_2OH$

(f) 1-methylcyclohexanol (CH_3, OH) $\xrightarrow{?}$ 1-methylcyclohexene (CH_3)

6.18 Ethylidenecyclohexane, on treatment with strong acids, isomerizes to 1-ethylcyclohexene.

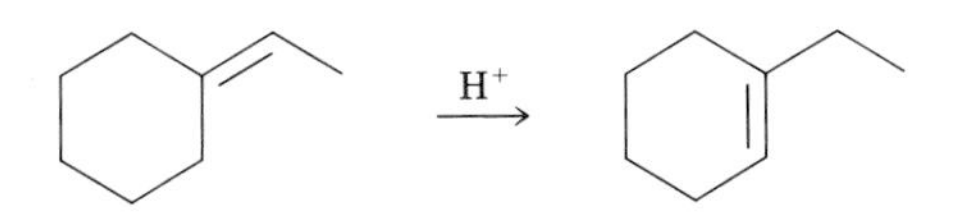

Ethylidenecyclohexane 1-Ethylcyclohexene

(a) Propose a mechanism for this reaction.
(b) Which is more stable, ethylidenecyclohexane or 1-ethylcyclohexene?

6.19 Give the structure of an alkene that provides only $(CH_3)_2C{=}O$ on ozonolysis.

6.20 An unknown hydrocarbon, A, formula C_6H_{12}, reacted with 1 equiv of hydrogen over a palladium catalyst. This hydrocarbon also reacted with OsO_4 to give a diol, B. When oxidized with $KMnO_4$, B gave two fragments. One fragment was identified as propanoic acid, CH_3CH_2COOH. The other fragment could be shown to be a ketone, C. What are the structures of A, B, and C? Write out all reactions, and show your reasoning.

6.21 Using an oxidative cleavage reaction, explain how you would distinguish between the following two isomeric dienes:

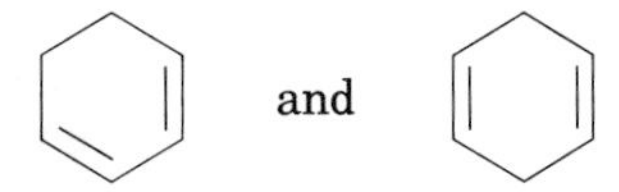

6.22 Draw the structure of a hydrocarbon that reacts with 1 mol equiv of hydrogen on catalytic hydrogenation and gives only pentanal, n-C_4H_9CHO, on ozonolysis. Write out the reactions involved.

6.23 In each case, decide which reaction you would expect to occur faster. Explain.
(a) Addition of HI to propene, or to 2-methylpropene.
(b) Addition of Br_2 to cyclohexene, or to 1-methylcyclohexene.
(c) Addition of HBr to ethylene, or to bromoethylene (vinyl bromide).
(d) Addition of HCl to propene, or to propenoic acid, $H_2C{=}CHCOOH$.

6.24 Predict the products of the following reactions, and indicate regiochemistry if relevant.

(a) $CH_3CH{=}CHCH_3 \xrightarrow{HBr}$

(b) $CH_3CH{=}CHCH_3 \xrightarrow{BH_3} A \xrightarrow[^{-}OH]{H_2O_2} B$

(c) $(CH_3)_2C{=}CH_2 \xrightarrow[\text{Peroxide}]{HBr}$

(d) $CH_3CH{=}C(CH_3)_2 \xrightarrow[\text{Peroxide}]{HI}$

6.25 Draw the structure of a hydrocarbon that absorbs 2 mol equiv of hydrogen on catalytic hydrogenation and gives only butanedial, $CHOCH_2CH_2CHO$, on ozonolysis.

6.26 In planning the synthesis of one compound from another, it is just as important to know what *not* to do as to know what to do. The following proposed reactions (a)–(f) all have serious drawbacks to them. Explain the potential problems of each reaction.

(a) $CH_3CHBrCH_2CH_3 \xrightarrow{\text{DBN}} H_2C{=}CHCH_2CH_3$

(b) $(CH_3)_2C{=}CHCH_3 \xrightarrow[\text{Peroxides}]{\text{HI}} (CH_3)_2CHCHICH_3$

(c) cyclopentene $\xrightarrow[\text{2. NaHSO}_3]{\text{1. OsO}_4}$ cyclopentane-1,2-diol (H, OH; HO, H)

(d) $CH_3CH_2CH_2CH_2OH \xrightarrow{H_2SO_4} CH_3CH_2CH{=}CH_2$

(e) 1,3-cyclohexadiene $\xrightarrow[\text{2. Zn}]{\text{1. O}_3}$ CHO CHO

(f) 1-methylcyclohexene (CH_3) $\xrightarrow[\text{2. H}_2\text{O}_2,\ ^{-}\text{OH}]{\text{1. BH}_3}$ (H, CH_3; H, OH)

6.27 Predict the products of the following reactions. Don't worry about the size of the molecule; concentrate on the functional groups.

HO

Cholesterol

$\xrightarrow{Br_2}$ A

$\xrightarrow{HBr}$ B

$\xrightarrow[\text{2. NaHSO}_3]{\text{1. OsO}_4}$ C

$\xrightarrow[\text{2. H}_2\text{O}_2,\ ^{-}\text{OH}]{\text{1. BH}_3\text{, THF}}$ D

6.28 Compound A has the formula $C_{10}H_{16}$. On catalytic hydrogenation over palladium, it reacts with only 1 equiv of hydrogen. Compound A undergoes reaction with ozone, followed by zinc treatment to yield a symmetrical diketone, B ($C_{10}H_{16}O_2$).
(a) How many rings does A have?
(b) What are the structures of A and B?
(c) Formulate the reactions.

6.29 Compound A has the formula C_8H_8. It reacts rapidly with $KMnO_4$, but reacts with only 1 equiv of H_2 on catalytic hydrogenation over a palladium catalyst. On hydrogenation under conditions that reduce aromatic rings, 4 equiv of H_2 are taken up, and hydrocarbon B (C_8H_{16}) is produced. The reaction of A with $KMnO_4$ gives CO_2 and a carboxylic acid, C ($C_7H_6O_2$). What are the structures of A, B, and C? Formulate the reactions.

6.30 When 4-penten-1-ol is treated with aqueous bromine, a cyclic bromo ether is formed, rather than the expected bromohydrin. Propose a mechanism for this transformation.

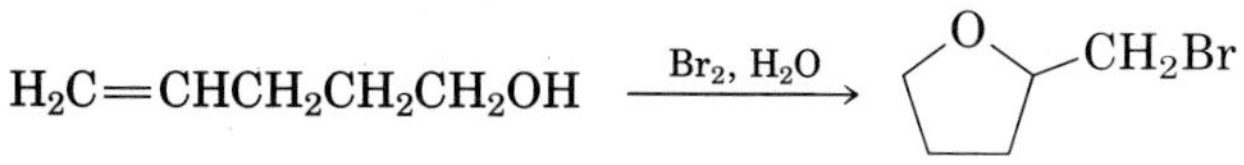

$$H_2C{=}CHCH_2CH_2CH_2OH \xrightarrow{Br_2,\ H_2O}$$ (tetrahydrofuran ring with CH_2Br substituent)

4-Penten-1-ol — 2-(Bromomethyl)tetrahydrofuran

6.31 How can you explain the observation that oxymercuration of 4-pentenoic acid yields a lactone (cyclic ester) product, rather than the expected alcohol? Propose a mechanism for this reaction.

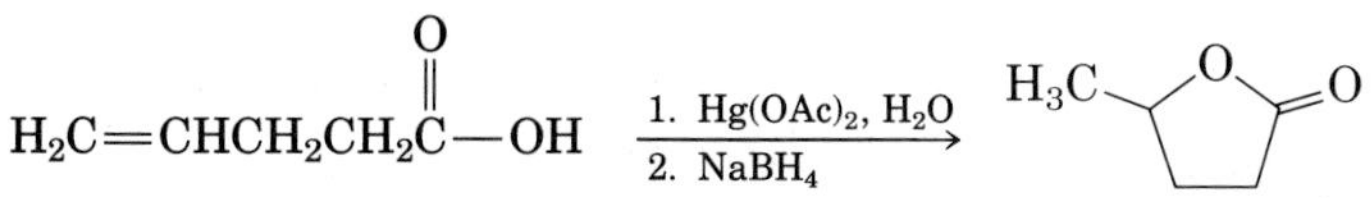

$$H_2C{=}CHCH_2CH_2\overset{\overset{\displaystyle O}{\|}}{C}{-}OH \xrightarrow[2.\ NaBH_4]{1.\ Hg(OAc)_2,\ H_2O}$$ (five-membered lactone ring with H_3C substituent and =O)

4-Pentenoic acid — 4-Methylbutanolide

6.32 α-Terpinene, $C_{10}H_{16}$, is a pleasant-smelling hydrocarbon that has been isolated from oil of marjoram. On hydrogenation over a palladium catalyst, α-terpinene reacts with 2 mol equiv of hydrogen to yield a new hydrocarbon, $C_{10}H_{20}$. On ozonolysis, followed by reduction with zinc and acetic acid, α-terpinene yields two products, glyoxal and 6-methylheptane-2,5-dione.

How many degrees of unsaturation does α-terpinene have? How many double bonds? How many rings? Propose a structure for α-terpinene that is consistent with the foregoing data.

$OHC{-}CHO$ — Glyoxal

$CH_3COCH_2CH_2COCH(CH_3)_2$ — 6-Methylheptane-2,5-dione

CHAPTER 7 ALKYNES

Alkynes (also called **acetylenes**) are hydrocarbons that contain carbon–carbon triple bonds. Acetylene (H—C≡C—H), the simplest alkyne, was once widely used in industry as the starting material for the preparation of acetaldehyde, acetic acid, vinyl chloride, and other high-volume chemicals, but more efficient routes using ethylene as the starting material are now more common. Acetylene is still used in the preparation of acrylic polymers, however, and is prepared industrially by high-temperature decomposition (*pyrolysis*) of methane. This method is not of general utility in the laboratory.

$$2\,CH_4 \xrightarrow[1200°C]{\text{Steam}} HC{\equiv}CH + 3\,H_2$$

Methane — Acetylene

A large number of naturally occurring acetylenic compounds have been isolated from the plant kingdom. For example, a *triyne* from the safflower, *Carthamus tinctorius* L., has considerable activity against nematodes and evidently forms part of the plant's chemical defenses against infestation:

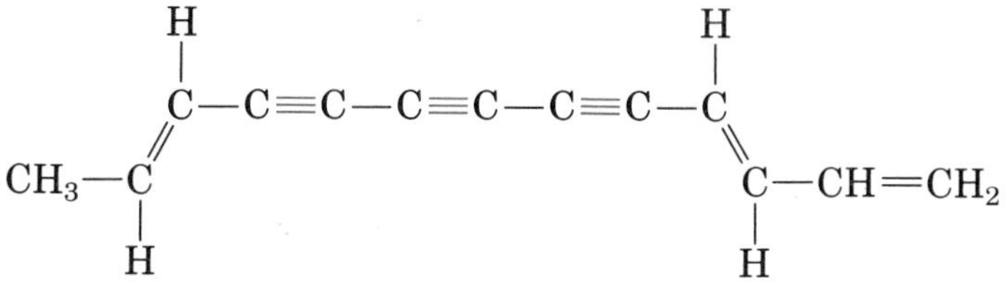

$$CH_3{-}CH{=}CH{-}C{\equiv}C{-}C{\equiv}C{-}C{\equiv}C{-}CH{=}CH{-}CH{=}CH_2$$

A triyne

7.1 Electronic structure of alkynes

A carbon–carbon triple bond results from the overlap of two *sp*-hybridized carbon atoms. Recall that the two *sp*-hybrid orbitals of carbon adopt a linear geometry. They lie at an angle of 180° to each other along an axis that is perpendicular to the axes of the two unhybridized $2p_y$ and $2p_z$ orbitals. When two *sp*-hybridized carbons approach each other for bonding, the geometry is perfect for the formation of one *sp–sp* sigma bond and two *p–p* pi bonds—a net *triple* bond (Figure 7.1, p. 196).

The two remaining *sp* orbitals form bonds to other atoms at an angle of 180° from the carbon–carbon bond. For example, acetylene, C_2H_2, has been shown by experimental measurement to be a linear molecule: H—C—C bond angles are 180° (Figure 7.2).

The bond strength for the carbon–carbon triple bond is approximately 200 kcal/mol, making it the strongest known carbon–carbon bond, and the bond length is 1.20 Å. In a purely bookkeeping sense, we can assign bond strengths to each of the three triple-bond "parts." We know that the carbon–carbon *single* bond in ethane has a strength of 88 kcal/mol and that the carbon–carbon *double* bond of ethylene has a strength of 152 kcal/mol.

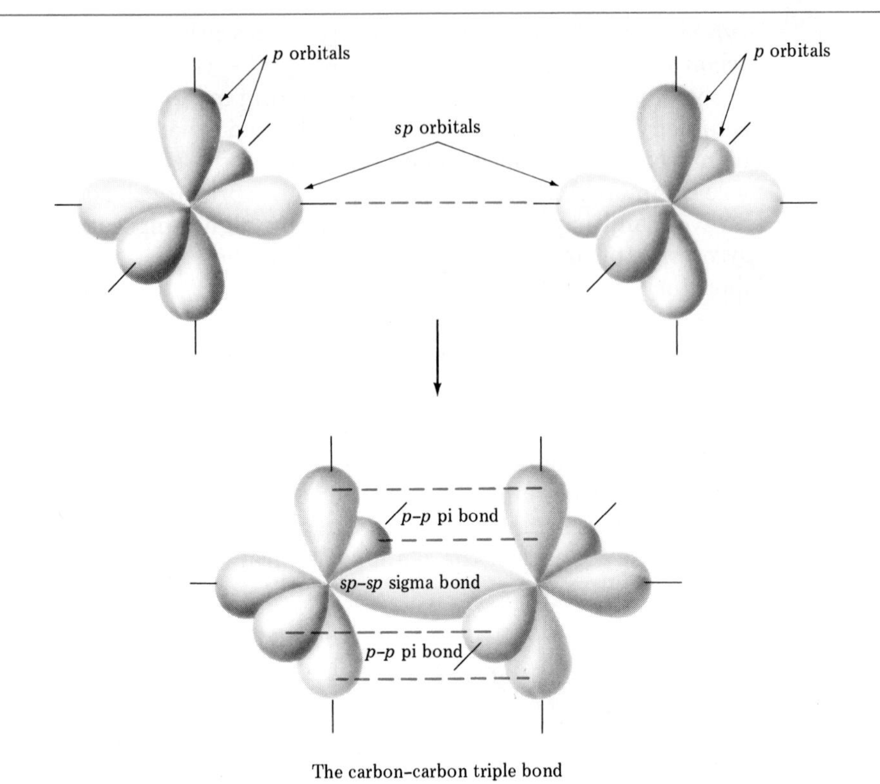

Figure 7.1. Formation of a carbon–carbon triple bond.

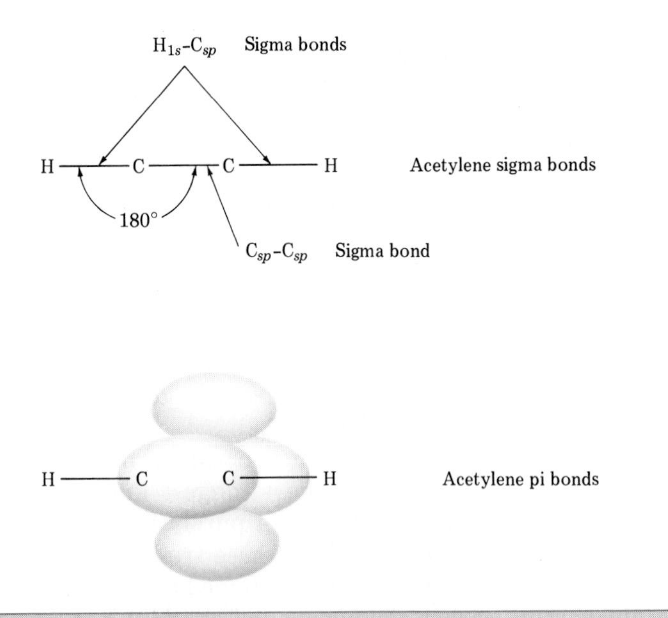

Figure 7.2. The structure of acetylene.

Using these figures, we can "dissect" the overall carbon–carbon triple bond:

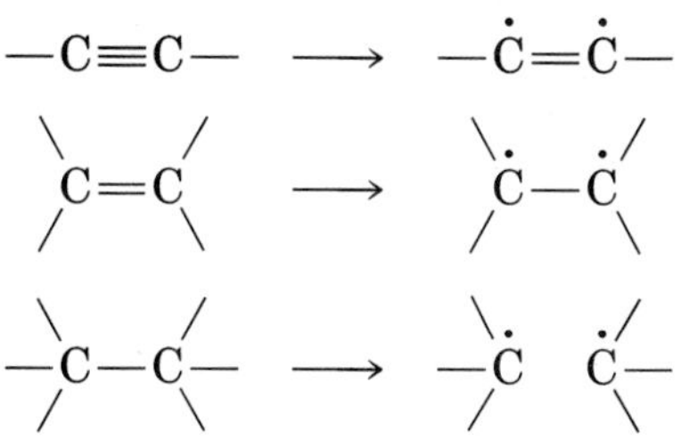

$$-C\equiv C- \longrightarrow -\dot{C}=\dot{C}- \qquad \Delta H° = 200 - 152 = 48 \text{ kcal/mol}$$

$$>C=C< \longrightarrow >\dot{C}-\dot{C}< \qquad \Delta H° = 152 - 88 = 64 \text{ kcal/mol}$$

$$-\overset{|}{\underset{|}{C}}-\overset{|}{\underset{|}{C}}- \longrightarrow -\overset{|}{\underset{|}{\dot{C}}} \quad \overset{|}{\underset{|}{\dot{C}}}- \qquad \Delta H° = 88 \text{ kcal/mol}$$

The two alkyne pi bonds are equivalent, but our crude calculation reveals that approximately 48 kcal/mol are needed to break the first of them. Since this is 64 − 48 = 16 kcal/mol *less* than the energy required to break an *alkene* pi bond, we might predict that alkynes should be highly reactive.

7.2 Nomenclature of alkynes

Alkynes closely follow the general rules of hydrocarbon nomenclature already discussed. The suffix *-yne* is substituted for *-ane* in the base hydrocarbon name to denote an alkyne, and the position of the triple bond is indicated by its number in the chain.

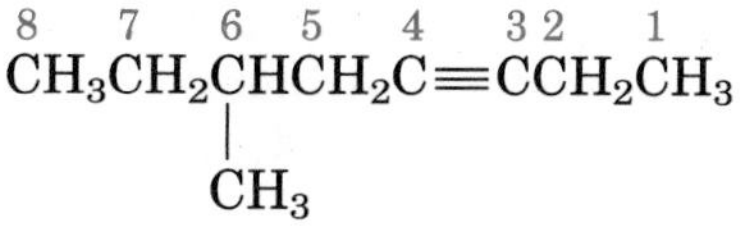

$$\overset{8}{C}H_3\overset{7}{C}H_2\overset{6}{C}H(CH_3)\overset{5}{C}H_2\overset{4}{C}\equiv\overset{3}{C}\overset{2}{C}H_2\overset{1}{C}H_3$$

Begin numbering at the end nearer the triple bond

6-Methyl-3-octyne

Compounds with more than one triple bond are called *diynes, triynes,* and so forth, and compounds containing both double and triple bonds are called *enynes* (not *ynenes*). Numbering of the hydrocarbon chain should always start from the end nearer the first multiple bond. When there is a choice in numbering, however, double bonds receive lower numbers than triple bonds. For example, the nematicide mentioned earlier is named as follows:

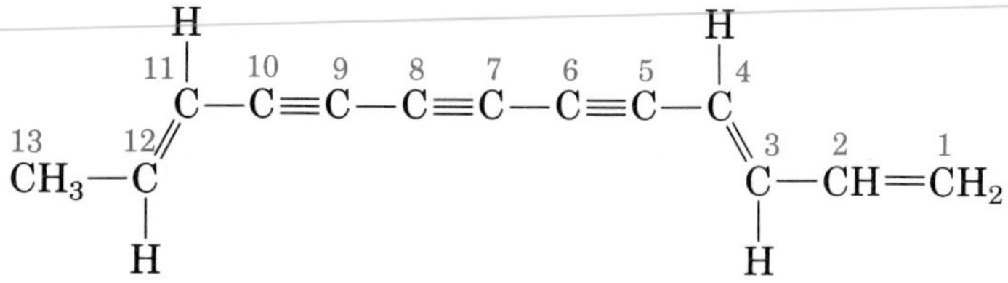

$$\overset{13}{C}H_3-\overset{12}{C}H=\overset{11}{C}H-\overset{10}{C}\equiv\overset{9}{C}-\overset{8}{C}\equiv\overset{7}{C}-\overset{6}{C}\equiv\overset{5}{C}-\overset{4}{C}H=\overset{3}{C}H-\overset{2}{C}H=\overset{1}{C}H_2$$

(1,3*E*,11*E*)-Tridecatrien-5,7,9-triyne

As was the case with hydrocarbon substituents derived from alkanes and alkenes, *alkynyl* groups are also possible:

$CH_3CH_2CH_2CH_2-$	$CH_3CH_2CH=CH-$	$CH_3CH_2C\equiv C-$
Butyl (an alkyl group)	1-Butenyl (a vinylic group)	1-Butynyl (an alkynyl group)

PROBLEM

7.1 Provide correct IUPAC names for the following compounds:

(a) $(CH_3)_2CHC{\equiv}CCH(CH_3)_2$

(b) $HC{\equiv}CC(CH_3)_3$

(c) $CH_3CH{=}CHCH{=}CHC{\equiv}CCH_3$

(d) $CH_3CH_2C(CH_3)_2C{\equiv}CCH_2CH_2CH_3$

(e) $CH_3CH_2C(CH_3)_2C{\equiv}CCH(CH_3)_2$

PROBLEM

7.2 There are seven isomeric alkynes with the formula C_6H_{10}. Draw them and name them according to IUPAC rules.

7.3 Reactions of alkynes: addition of HX and X_2

We have predicted, based on known thermodynamic values of carbon–carbon triple bond strengths, that the chemical reactivity of alkynes should be similar to that of alkenes. The pi part of the triple bond is weak (~48 kcal/mol) and the electrons are readily accessible to attacking reagents. Alkynes do indeed exhibit much chemistry similar to that of alkenes, but there are also differences.

As a general rule, alkynes react with electrophilic reagents in a manner similar to that of alkenes. With halogen acids, for example, alkynes give the expected addition products. Although the reactions can usually be stopped after addition of 1 equiv of HX, an excess of acid leads to the dihalide product. As the following examples indicate, the regiochemistry of addition follows Markovnikov's rule, with halogen adding to the more substituted side of the alkyne bond. Trans stereochemistry of H and X is normally (though not always) found in the product.

$$CH_3CH_2CH_2CH_2C{\equiv}CH \xrightarrow[CH_3COOH]{HBr} CH_3CH_2CH_2CH_2C(Br){=}CH_2$$

1-Hexyne → 2-Bromo-1-hexene

$$CH_3CH_2CH_2CH_2C(Br){=}CH_2 \xrightarrow{HBr} CH_3CH_2CH_2CH_2CBr_2{-}CH_3$$

2,2-Dibromohexane

$$CH_3CH_2C{\equiv}CCH_2CH_3 \xrightarrow[CH_3COOH]{HCl,\ NH_4Cl} (Cl)(CH_3CH_2)C{=}C(CH_2CH_3)(H)$$

3-Hexyne → (*Z*)-3-Chloro-3-hexene (95%)

The vinyl chloride needed as starting material for use in poly(vinyl chloride) polymers was once produced on an immense industrial scale by mercuric chloride–catalyzed addition of HCl to acetylene. Now, however, other synthetic methods are used.

$$H{-}C{\equiv}C{-}H + HCl \xrightarrow{HgCl_2} H_2C{=}CHCl$$

Acetylene — Vinyl chloride (chloroethylene)

Bromine and chlorine also add to alkynes to give addition products, and trans stereochemistry again results:

$$CH_3CH_2C{\equiv}CH \xrightarrow[CCl_4]{Br_2} (E)\text{-}CH_3CH_2(Br)C{=}C(Br)H \xrightarrow[CCl_4]{Br_2} CH_3CH_2CBr_2CHBr_2$$

1-Butyne — (*E*)-1,2-Dibromo-1-butene — 1,1,2,2-Tetrabromobutane

One of the more striking features of alkyne electrophilic addition reactions is the rate at which they occur. We might expect alkynes to be more reactive than alkenes in electrophilic additions, but as a general rule the reverse is true. Triple bonds are *less* subject to electrophilic addition than are double bonds. Thus, although 98% H_2SO_4 readily adds to alkenes, alkynes are inert to this reagent at room temperature. Why should this be?

The probable reason for the decreased reactivity of alkynes toward electrophilic reagents is the fact that *vinylic cations* are involved as intermediates:

$$R{-}CH{=}CH_2 \xrightarrow{H^+} \left[R{-}\overset{+}{C}HCH_3\right] \xrightarrow{:Br^-} R\overset{Br}{C}HCH_3$$

Alkene — An alkyl cation intermediate — An alkyl bromide

$$R{-}C{\equiv}CH \xrightarrow{H^+} \left[R{-}\overset{+}{C}{=}CH_2\right] \xrightarrow{:Br^-} R{-}\overset{Br}{C}{=}CH_2$$

Alkyne — A vinylic cation intermediate — A vinylic bromide

Secondary vinylic cations are known species that have been studied in detail and have been found to be less stable than secondary alkyl cations. The relative order of stability of carbocations is as follows:

$$R_3\overset{+}{C} > R_2\overset{+}{C}H > R\overset{+}{C}{=}CH_2 > R\overset{+}{C}H_2 > RCH{=}\overset{+}{C}H$$

Tertiary alkyl — Secondary alkyl — Secondary vinylic — Primary alkyl — Primary vinylic

The reasons for the relative instability of vinylic cations are not fully understood but involve, at least in part, the lack of any stabilizing interaction with neighboring groups. A priori, the vinylic cation could be either sp^2 hybridized or sp hybridized, but molecular orbital calculations indicate

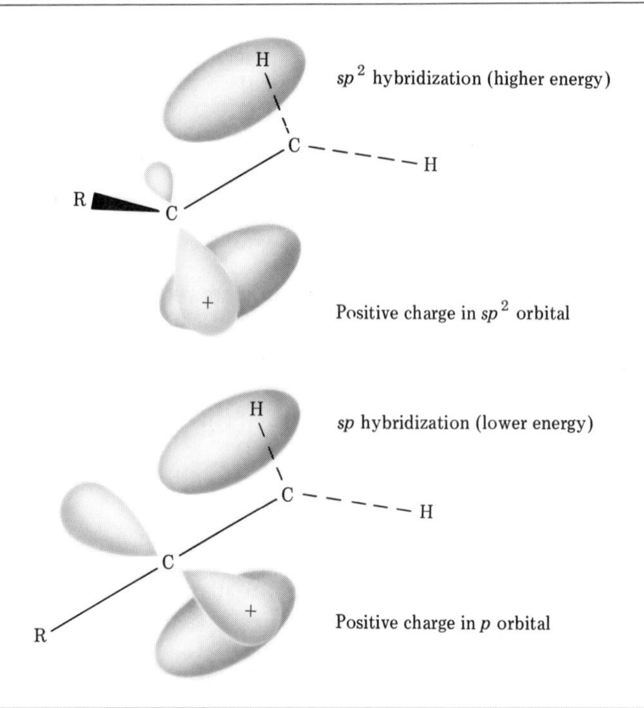

Figure 7.3. Electronic structure of a vinylic cation. The *sp* hybridization results in a lower energy state than sp^2 hybridization.

that linear *sp* hybridization is lower in energy. In this electronic configuration, the positive charge is in a vacant *p* orbital perpendicular to the double bond (Figure 7.3).

7.4 Hydration of alkynes

Alkynes cannot be hydrated as easily as simple alkenes, because of their lower reactivity toward electrophilic addition; aqueous sulfuric acid by itself has no effect on the carbon–carbon triple bond. In the presence of mercuric sulfate catalyst, however, hydration occurs readily:

$$\underset{\text{1-Hexyne}}{CH_3CH_2CH_2CH_2C{\equiv}CH} \xrightarrow[HgSO_4]{H_2O,\ H_2SO_4} \underset{\text{An enol}}{\left[CH_3CH_2CH_2CH_2\overset{\overset{OH}{|}}{C}{=}\underset{\underset{H}{|}}{C}H \right]}$$

$$\downarrow$$

$$\underset{\substack{\text{2-Hexanone} \\ (78\%)}}{CH_3CH_2CH_2CH_2\overset{\overset{O}{\|}}{C}{-}CH_3}$$

Markovnikov regiochemistry is observed, but the hydration product is not the expected vinylic alcohol or *enol* (*ene* + *ol*). The intermediate vinylic alcohol immediately rearranges to a more stable isomer, a ketone ($R_2C{=}O$). It turns out that enols and ketones rapidly achieve an equilibrium—a process called **tautomerism. Tautomers** are special kinds of structural isomers that are readily interconvertible through a rapid equilibration, and we will study them in more detail in Section 26.1. With few exceptions, the **keto–enol tautomeric equilibrium** lies heavily on the keto side. Vinylic alcohols are almost never isolated.

$$\underset{\text{Enol tautomer (less favored)}}{>C{=}C(OH)-} \;\underset{}{\overset{\text{Rapid}}{\rightleftarrows}}\; \underset{\text{Keto tautomer (more favored)}}{-C(=O)-CH<}$$

A mixture of both possible ketones results when an *internal* alkyne is hydrated, and the reaction is therefore most useful when applied to *terminal* alkynes ($R{-}C{\equiv}CH$), since only methyl ketones are formed.

An internal alkyne

$$R{-}C{\equiv}C{-}R' \xrightarrow[HgSO_4]{H_3O^+} \underbrace{R{-}\overset{O}{\overset{\|}{C}}{-}CH_2{-}R' + R{-}CH_2{-}\overset{O}{\overset{\|}{C}}{-}R'}_{\text{A mixture}}$$

A terminal alkyne

$$R{-}C{\equiv}C{-}H \xrightarrow[HgSO_4]{H_3O^+} \underset{\text{A methyl ketone}}{R{-}\overset{O}{\overset{\|}{C}}{-}CH_3}$$

Acetylene itself was formerly hydrated industrially on a large scale to produce acetaldehyde (ethanal), which can be further oxidized to acetic acid:

$$\underset{\text{Acetylene}}{H{-}C{\equiv}C{-}H} \xrightarrow[HgSO_4]{H_2O,\ H_2SO_4} \left[H{-}\overset{OH}{\overset{|}{C}}{=}CH_2\right] \longrightarrow \underset{\text{Acetaldehyde}}{H{-}\overset{O}{\overset{\|}{C}}{-}CH_3} \xrightarrow{\text{Oxidation}} \underset{\text{Acetic acid}}{HO{-}\overset{O}{\overset{\|}{C}}{-}CH_3}$$

7.5 Hydroboration of alkynes

We saw in Section 6.4 that hydroboration of alkenes occurs via a four-membered-ring transition state without the intermediacy of carbocations. Borane also adds to alkynes with great ease, and the resulting vinylic boranes are quite useful. Hydroboration of *symmetrical* internal alkynes gives vinylic boranes, which can be oxidized by basic hydrogen peroxide to ketones (via enols). Hydroboration of *unsymmetrical* internal alkynes gives a mixture of both possible ketones.

$$3\ CH_3CH_2C{\equiv}CCH_2CH_3 \xrightarrow[THF]{BH_3} \left[CH_3CH_2C(B{<})\!=\!C(H)CH_2CH_3 \right]$$

3-Hexyne — A vinylic borane

$$\xrightarrow{H_2O,\ HO^-,\ H_2O_2}\ 3\ CH_3CH_2C(HO)\!=\!C(H)CH_2CH_3 \longrightarrow 3\ CH_3CH_2\overset{O}{\overset{\|}{C}}CH_2CH_2CH_3$$

An enol — 3-Hexanone

Another reaction of organoboranes is **protonolysis.** That is, when an organoborane is heated with a carboxylic acid, the boron is replaced by a hydrogen atom. Thus, the net effect of hydroboration–protonolysis is a *reduction:*

$$R_3B \xrightarrow[100°C]{3\ CH_3COOH} 3\ R{-}H + B(OAc)_3$$

$$3\ CH_3CH_2C{\equiv}CCH_2CH_3 \xrightarrow[THF]{BH_3} \underset{CH_3CH_2\qquad CH_2CH_3}{\overset{{>}B\qquad H}{C{=}C}}$$

3-Hexyne

$$\xrightarrow[100°C]{CH_3COOH}\ 3\ \underset{CH_3CH_2\qquad CH_2CH_3}{\overset{H\qquad H}{C{=}C}}$$

cis-3-Hexene (68%)

As the preceding example indicates, borane adds to alkynes with syn stereochemistry. Protonolysis of the vinylic borane by acetic acid occurs with *retention* of double-bond stereochemistry and yields a cis alkene. The overall result, alkyne $\rightarrow$ cis alkene, provides an excellent synthetic method for preparing cis disubstituted alkenes.

Terminal alkynes can also be hydroborated, although in practice the reaction is difficult to stop at the vinylic borane stage. Under normal circumstances, a second addition of borane to the intermediate vinylic borane occurs:

$$R-C\equiv CH \xrightarrow[THF]{BH_3} [R-CH=CHBH_2] \xrightarrow[THF]{BH_3} R-CH_2-CH(B<)_2$$

Terminal alkyne — Vinylic borane

To prevent this double addition, a bulky, sterically hindered borane such as bis(1,2-dimethylpropyl)borane (known trivially as disiamylborane) can be used in place of borane. Thus, when a terminal alkyne reacts with disiamylborane, addition of B—H across the triple bond occurs with the expected non-Markovnikov regiochemistry. A second addition is rather slow, since the steric bulk of the large dialkylborane reagent makes approach to the double bond difficult. Oxidation of the intermediate then leads to an aldehyde.

$$CH_3(CH_2)_5C\equiv CH \xrightarrow[THF]{R_2BH} CH_3(CH_2)_5C(H)=CH(BR_2) \xrightarrow[H_2O]{H_2O_2,\ ^-OH} [CH_3(CH_2)_5C(H)=CH(OH)] \longrightarrow CH_3(CH_2)_5CH_2C(=O)-H$$

1-Octyne — A vinylic borane — An enol — Octanal (70%)

where R_2BH = $[(CH_3)_2CHCHCH_3]_2B-H$

Disiamylborane
[bis(1,2-dimethylpropyl)borane]

Note that the hydroboration–oxidation sequence is *complementary* to the direct hydration reaction of alkynes, since different products result. Direct hydration of a terminal alkyne leads to a methyl ketone, whereas hydroboration–oxidation of the same alkyne leads to an aldehyde:

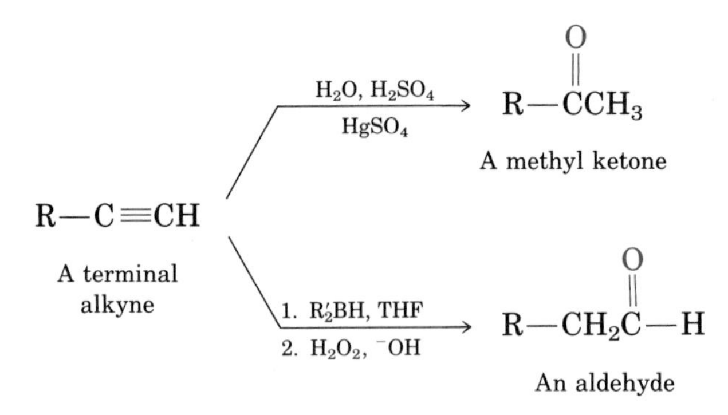

PROBLEM

7.3 Hydroboration, followed by deuterolysis with a deuterated carboxylic acid, is a convenient way to introduce deuterium into an organic molecule. Predict the product of the following reaction:

$$n\text{-}C_4H_9C{\equiv}C{-}C_4H_9\text{-}n \xrightarrow[\text{2. } CH_3COOD]{\text{1. } BH_3,\ THF}$$

PROBLEM

7.4 Disiamylborane is made by addition of BH_3 to 2 equiv of an alkene. What alkene is used, and why do you suppose the reaction can be stopped cleanly after 2 equiv have reacted?

$$[(CH_3)_2CHCH(CH_3)]_2BH$$

Disiamylborane

PROBLEM

7.5 What other organic product is formed after oxidation when disiamylborane is used to hydroborate terminal alkynes?

$$RC{\equiv}CH \xrightarrow[\text{2. } H_2O_2,\ ^-OH]{\text{1. } [(CH_3)_2CHCH(CH_3)]_2BH} RCH_2CHO + ?$$

7.6 Hydrogenation: reduction of alkynes

Alkynes are easily reduced to alkanes by catalytic hydrogenation over a metal catalyst. Heat of hydrogenation data indicate that the first step in the reaction has a larger $\Delta H°$ than the second step, and we might therefore expect alkynes to reduce more readily than alkenes. This turns out to be true.

$$HC{\equiv}CH \xrightarrow[\text{Catalyst}]{H_2} H_2C{=}CH_2 \qquad \Delta H°_{hydrog} = 42\text{ kcal/mol}$$

$$H_2C{=}CH_2 \xrightarrow[\text{Catalyst}]{H_2} CH_3{-}CH_3 \qquad \Delta H°_{hydrog} = 33\text{ kcal/mol}$$

Experimentally, alkynes reduce somewhat faster than alkenes. By using a suitable catalyst, triple-bond hydrogenation can be stopped at the alkene stage. The classic catalyst for this purpose is the Lindlar catalyst, a finely divided palladium metal that has been precipitated onto a calcium carbonate support and then deactivated by treatment with lead acetate and quinoline, an aromatic amine. Since hydrogenation occurs with syn stereochemistry, alkynes are catalytically reduced to give cis alkenes. This reaction has been used extensively in the retinoid (vitamin A) field by chemists at Hoffmann-LaRoche and Company in Basel, Switzerland. Figure 7.4 gives examples of the catalytic hydrogenation of alkynes.

$CH_3(CH_2)_3C{\equiv}C(CH_2)_3CH_3$

5-Decyne

$\xrightarrow[Pd/C]{2\ H_2}$ $CH_3(CH_2)_8CH_3$

Decane (96%)

$\xrightarrow[\text{Lindlar catalyst}]{H_2}$ H, H, C=C, $CH_3(CH_2)_3$, $(CH_2)_3CH_3$

cis-5-Decene (96%)

C≡C, CH_2OH

$\xrightarrow[\text{Lindlar catalyst}]{H_2}$

1, 2, 3, 4, 5, 6, 7, 8, 9, 10, 11, 12, 13, 14, CH_2OH 15

7-*cis*-Retinol
(7-*cis*-vitamin A; vitamin A has a trans double bond at C7)

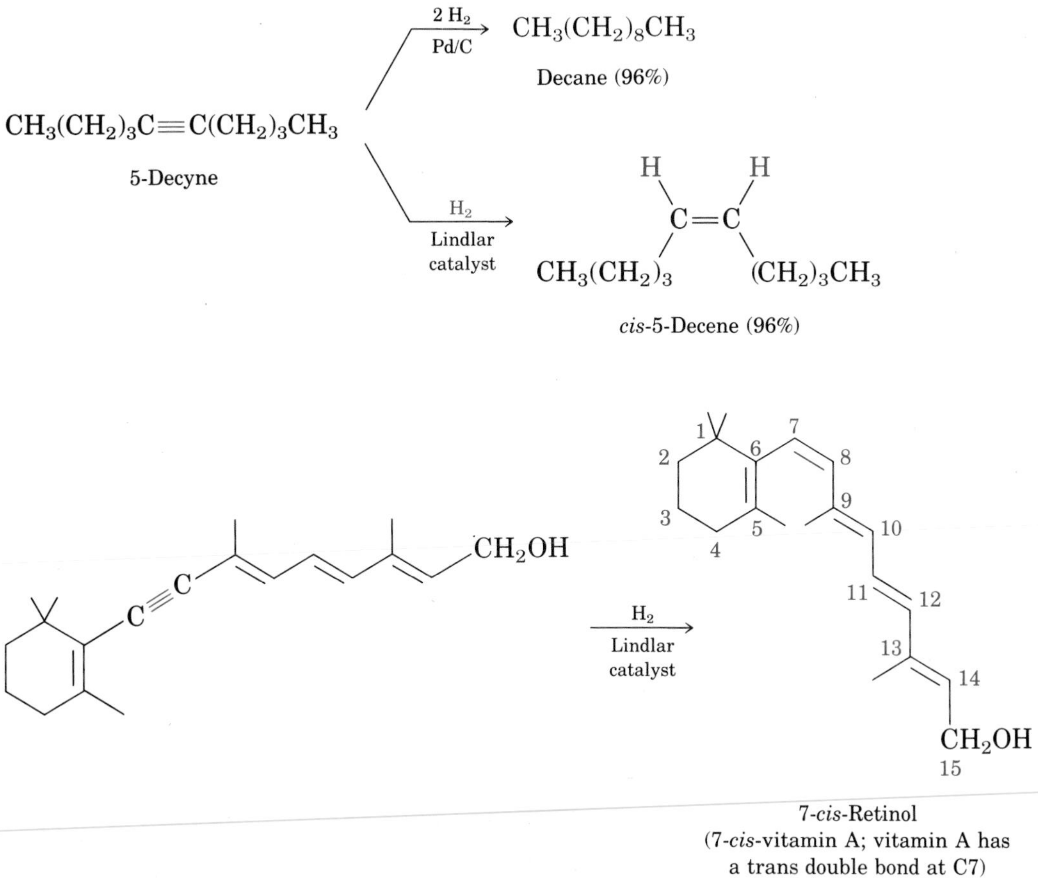

Figure 7.4. The catalytic hydrogenation of alkynes.

A second method for the partial reduction of alkynes to alkenes employs lithium metal in liquid ammonia as solvent. This method is complementary to the Lindlar reduction, since it produces trans rather than cis alkenes. Remarkably, alkali metals such as lithium and sodium dissolve in pure liquid ammonia at −33°C to produce a deep blue solution. When an alkyne is added to this blue solution, reduction of the triple bond occurs. The overall result constitutes an excellent method for the synthesis of trans alkenes. A typical reaction is shown at the top of the next page.

$$CH_3CH_2CH_2CH_2C{\equiv}CCH_2CH_2CH_2CH_3 \xrightarrow[2.\ H_2O]{1.\ Li/NH_3} (n\text{-}C_4H_9)HC{=}CH(C_4H_9\text{-}n)$$

5-Decyne

trans-5-Decene (78%)

7.7 Alkyne acidity: acetylides

The most striking difference between the chemistry of alkenes and that of alkynes is that *terminal alkynes are weakly acidic*. When a terminal alkyne is treated with a strong base such as sodium amide, $NaNH_2$, the terminal hydrogen is removed, and an **acetylide anion** is formed:

$$R{-}C{\equiv}C{-}H + :\bar{N}H_2Na^+ \longrightarrow R{-}C{\equiv}C{:}^-Na^+ + NH_3$$

Acetylide anion

According to the Brønsted–Lowry definition, an acid is a species that donates a proton, and we usually think of oxyacids (H_2SO_4, H_2O) or halogen acids (HCl, HBr) in this context. In fact, though, *any* compound containing a hydrogen atom can be considered an acid under the proper circumstances, and we can establish an **acidity order** by measuring dissociation constants and expressing the results as pK_a values (Table 7.1). Recall from Section 2.6 that a low pK_a corresponds to a strong acid, and a high pK_a corresponds to a weak acid.

At one end of the scale, the mineral acids such as sulfuric acid are very strong. Carboxylic acids such as acetic acid are moderately strong, water

TABLE 7.1 **Strengths of some common acids (pK_a values)**

Acid	*Formula*	pK_a	
Sulfuric acid	$H{-}O{-}S(=O)_2{-}O{-}H$	~−9	Strong acid
Acetic acid	$CH_3C(=O){-}O{-}H$	4.72	↑
Water	$H{-}O{-}H$	15.74	
Ethanol	$CH_3CH_2O{-}H$	16	
Acetylene	$H{-}C{\equiv}C{-}H$	25	
Ammonia	$H_2N{-}H$	35	Weak acid

and alcohols such as ethanol are intermediate, and ammonia is very weak. Since a stronger acid donates its proton to the anion of a weaker acid, a rank-ordered list allows us to know what bases are needed to deprotonate what acids. For example, since acetic acid ($pK_a = 4.72$) is a stronger acid than ethanol ($pK_a = 16$), we know that the ethanol anion (ethoxide) will remove a proton from acetic acid. Similarly, the anion of ammonia ($pK_a = 35$) will remove a proton from ethanol ($pK_a = 18$).

$$CH_3CH_2O{:}^- + H{-}OC(=O)CH_3 \rightleftharpoons CH_3CH_2O{-}H + {}^-{:}OC(=O)CH_3$$

$$H_2N{:}^- + H{-}OCH_2CH_3 \rightleftharpoons H_2N{-}H + {}^-{:}OCH_2CH_3$$

Stronger base | Stronger acid | Weaker acid | Weaker base

Where do hydrocarbons lie on the acidity scale? As the data in Table 7.2 indicate, both methane ($pK_a = 49$) and ethylene ($pK_a = 44$) are very weak acids and, for all practical purposes, cannot be deprotonated. Acetylene, however, has a pK_a of 25 and is thus quite susceptible to deprotonation by sufficiently strong bases.

TABLE 7.2 Acidity of simple hydrocarbons

Type	*Example*	K_a	pK_a	
Alkyne	$HC{\equiv}C{-}H \rightleftharpoons HC{\equiv}C{:}^- + H^+$	10^{-25}	25	Stronger acid
Alkene	$H_2C{=}CH{-}H \rightleftharpoons H_2C{=}\ddot{C}H^- + H^+$	10^{-44}	44	↑
Alkane	$CH_3{-}H \rightleftharpoons {:}CH_3^- + H^+$	10^{-49}	49	Weaker acid

The acidity data in Table 7.2 are approximate values, but they do indicate that alkynes are much more acidic than either alkanes or alkenes. Any base whose conjugate acid has a $pK_a > 25$ should be able to effect acetylide formation. Amide ion, NH_2^-, for example, is the conjugate base of ammonia (NH_3; $pK_a = 35$) and is therefore able to abstract a proton from terminal alkynes.

What makes terminal alkynes so acidic? The best intuitive explanation is simply that acetylide anions have more *s* character in the orbital containing the electron pair than do alkyl or vinylic anions. Alkane anions are sp^3 hybridized, so the negative charge resides in an orbital that has one-quarter *s* character and three-quarters *p* character. Vinylic anions are sp^2 hybridized and therefore have one-third *s* character, whereas acetylide anions (*sp*) have one-half *s* character. Since *s* orbitals are nearer the positively charged nucleus than are *p* orbitals, a negative charge can be stabilized to a greater extent in an orbital with high *s* character than in an orbital with low *s* character (Figure 7.5).

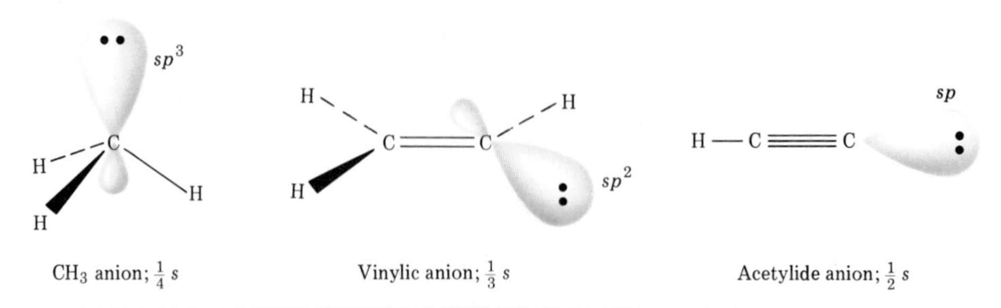

Figure 7.5. A comparison of alkyl, vinylic, and acetylide anions.

7.8 Alkylation of acetylide anions

Acetylide anions exhibit quite useful and interesting chemistry. We find, for example, that the presence of the unshared electron pair renders acetylides nucleophilic. We have previously discussed the reaction of hydroxide ion with bromomethane as an example of nucleophilic substitution. Acetylide nucleophiles carry out substitution reactions on alkyl halides in an analogous manner:

$$\underset{\text{Acetylide}}{HC{\equiv}\overset{-}{C}{:}\,Na^+} + \underset{\text{1-Iodobutane}}{I{-}CH_2CH_2CH_2CH_3} \longrightarrow \underset{\text{1-Hexyne (86\%)}}{HC{\equiv}C{-}CH_2CH_2CH_2CH_3} + NaI$$

Recall:

$$HO\overset{-}{:} + Br{-}CH_3 \longrightarrow HO{-}CH_3 + :Br^-$$

The acetylide ion attacks the carbon atom of 1-iodobutane and pushes out iodide ion, yielding the terminal alkyne, 1-hexyne, as product. We call such a reaction an **alkylation,** since a new alkyl group has become attached to the starting alkyne. The product 1-hexyne can itself be converted into an acetylide anion, and can be alkylated a second time to yield an internal alkyne product:

$$\underset{\text{1-Hexyne}}{CH_3CH_2CH_2CH_2C{\equiv}CH} \xrightarrow[\text{2. } CH_3CH_2CH_2CH_2Br,\text{ THF}]{\text{1. } NaNH_2,\ NH_3} \underset{\text{5-Decyne (76\%)}}{CH_3CH_2CH_2CH_2C{\equiv}CCH_2CH_2CH_2CH_3}$$

For reasons that are explained in greater detail in Section 10.5, acetylide alkylations are limited to primary alkyl bromides and iodides, $R{-}CH_2{-}X$. This is because, in addition to their reactivity as nucleophiles, acetylides are also strong bases and can cause dehydrohalogenation to occur in competition with substitution.

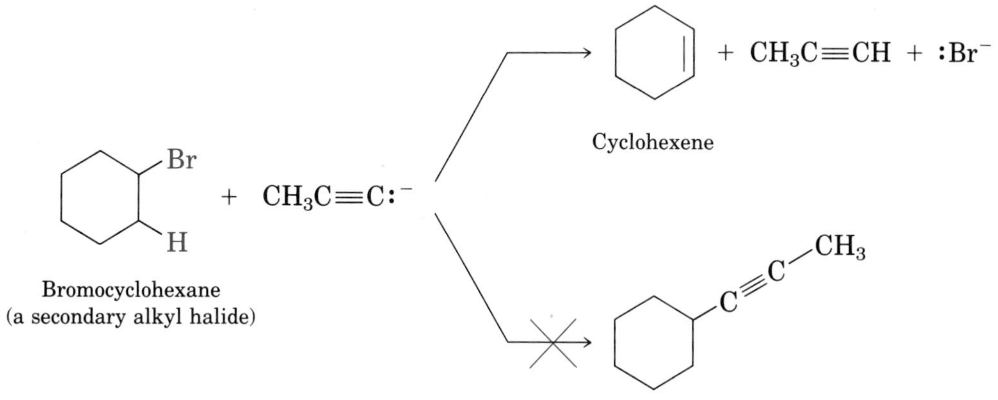

Bromocyclohexane (Br, H) + $CH_3C{\equiv}C:^-$ → Cyclohexene + $CH_3C{\equiv}CH$ + $:Br^-$

→ (crossed out) cyclohexyl–$C{\equiv}C{-}CH_3$

Bromocyclohexane
(a secondary alkyl halide)

Cyclohexene

Not formed

The competition between substitution and elimination is a complex matter, but, even with the restriction that only primary alkyl halides be used, acetylide alkylation is of considerable importance as a method of synthesis of internal alkynes.

PROBLEM

7.6 Show the terminal alkyne and alkyl halide starting materials from which the following products can be obtained. Where two routes are appropriate, list both choices.

(a) $CH_3CH_2CH_2C{\equiv}CCH_3$

(b) $(CH_3)_2CHC{\equiv}CCH_2CH_3$

(c) 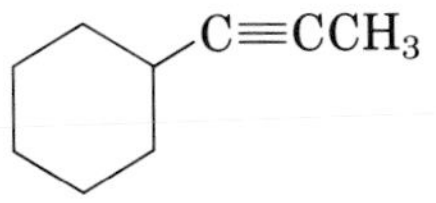 cyclohexyl–$C{\equiv}CCH_3$

(d) 5-Methyl-2-hexyne

(e) 2,2-Dimethyl-3-hexyne

PROBLEM

7.7 How would you prepare *cis*-2-butene, starting from 1-propyne, an alkyl halide, and any other reagents needed?

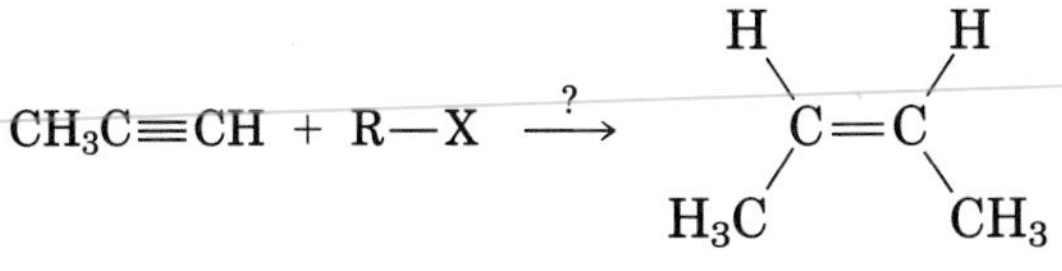

$$CH_3C{\equiv}CH + R{-}X \xrightarrow{?} \text{(H)(}H_3C\text{)C=C(H)(}CH_3\text{)}$$

cis-2-Butene

(This problem cannot be worked in a single step. You will have to carry out more than one chemical reaction.)

7.9 Alkylation of alkyne dianions

When an alkyne is treated with a very strong base, the protons *next* to the triple bond (*propargylic* protons, $—CH_2—C{\equiv}C—$) are found to be weakly acidic. For example, if a terminal alkyne such as 3-methyl-1-butyne is treated with 2 equiv of a very strong base such as butyllithium (BuLi; $CH_3CH_2CH_2CH_2^-Li^+$), a *dianion* is formed. Both the terminal alkyne proton and the propargylic proton are removed. If we then add 1 equiv of a primary alkyl halide, alkylation occurs exclusively at the highly reactive propargylic position, and not at the terminal alkyne position:

$$\underset{\text{3-Methyl-1-butyne}}{(CH_3)_2\overset{H}{\overset{|}{C}}-C{\equiv}C-H} \xrightarrow[\text{THF}]{\text{2 BuLi}} \underset{\text{A dianion}}{\left[(CH_3)_2\overset{-}{\ddot{C}}-C{\equiv}C:^-\right]} + 2\ C_4H_{10}$$

$$\downarrow \text{1 equiv } CH_3I$$

$$\underset{\text{3,3-Dimethyl-1-butyne}}{(CH_3)_2\overset{CH_3}{\overset{|}{C}}-C{\equiv}C-H} \xleftarrow{H_3O^+} \left[(CH_3)_2\overset{CH_3}{\overset{|}{C}}-C{\equiv}C:^-\right] + :I^-$$

The alkylation of terminal alkyne dianions in the propargylic position is a useful reaction that adds a new dimension to our ability to prepare substituted alkynes. It is particularly valuable since, as noted earlier, direct alkylation of acetylides does not occur when secondary and tertiary alkyl halides are used. Thus, alkynes that are substituted at the position next to the triple bond are best prepared by this dianion alkylation reaction.

PROBLEM

7.8 Show how you might prepare the following alkynes. Start with acetylene and use any alkyl halide needed.

(a) $CH_3CH_2\overset{CH_3}{\overset{|}{C}}HC{\equiv}CH$ (b) $CH_3\overset{CH_3}{\overset{|}{C}}HC{\equiv}C-CH_3$

PROBLEM

7.9 Butyllithium is a very powerful base that may be considered the lithium salt of butane, $Li^{+-}:CH_2CH_2CH_2CH_3$. It is prepared by reaction of lithium metal with 1-chlorobutane:

$$CH_3CH_2CH_2CH_2Cl + 2\ Li \xrightarrow[\text{Solvent}]{\text{THF}} CH_3CH_2CH_2CH_2Li + LiCl$$

Estimate the base strength of butyllithium by referring to Table 7.2.

7.10 Oxidative cleavage of alkynes

The reaction of alkynes with powerful oxidizing agents such as potassium permanganate or ozone affords cleavage products. A triple bond is generally less reactive than a double bond, however, and yields of cleavage products are sometimes low. In all cases, the products obtained are carboxylic acids. If a terminal alkyne is oxidized, CO_2 is formed as one product.

An internal alkyne $\quad R{-}C{\equiv}C{-}R' \xrightarrow{KMnO_4 \text{ or } O_3} RCOOH + R'COOH$

A terminal alkyne $\quad R{-}C{\equiv}C{-}H \xrightarrow{KMnO_4 \text{ or } O_3} RCOOH + CO_2$

These cleavage reactions are not often valuable for synthesis, since alkynes can be rather difficult to prepare in the first place, and it serves little purpose to then destroy them. The major application of these oxidation reactions is in structure determination. For example, alkyne cleavage occurred during the structure elucidation of tariric acid, a substance isolated from the seed fat of *Picramnia tariri* from Guatemala. Tariric acid was known to be an 18-carbon straight-chain acetylenic acid, but the position of the triple bond in the chain was unknown. Since oxidation of tariric acid with potassium permanganate gave a product identified as the known 6-carbon diacid, adipic acid, the position of the triple bond could be defined.

$$CH_3(CH_2)_{10}C{\equiv}C(CH_2)_4COOH \xrightarrow[H_2O]{KMnO_4} CH_3(CH_2)_{10}COOH + HOOC(CH_2)_4COOH$$

6-Octadecynoic acid (tariric acid) — Decanoic acid (lauric acid) — 1,6-Hexanedioic acid (adipic acid)

PROBLEM

7.10 Propose structures for hydrocarbons that give the following products on oxidative cleavage by $KMnO_4$:

(a) (benzene ring)–COOH $+ CO_2$

(b) $2\ CH_3(CH_2)_7COOH + HO_2C(CH_2)_7COOH$

(c) $CH_3(CH_2)_5COOH$ (sole product)

7.11 Preparation of alkynes: elimination reactions of dihalides

There are relatively few general methods for the preparation of alkynes; one of the best, as we have seen, is alkylation of acetylide anions. Remember that both terminal and internal alkynes can be prepared by suitable alkylation of primary alkyl halides:

$$HC{\equiv}C{:}^-Na^+ + RCH_2Br \longrightarrow RCH_2C{\equiv}CH \quad \text{A terminal alkyne}$$

$$RC{\equiv}C{:}^-Na^+ + R'CH_2Br \longrightarrow R'CH_2C{\equiv}CR \quad \text{An internal alkyne}$$

Alkynes can also be prepared in much the same manner as can alkenes—by elimination of the elements of HX from alkyl halides. Since an alkyne is doubly unsaturated, that is, has two fewer pairs of hydrogens than the corresponding alkane, we must eliminate *two* molecules of HX. Fortunately, this is easily done. When a 1,2-dihalide (also called a *vicinal* dihalide) is treated with excess strong base, twofold elimination occurs, and an alkyne is produced. The method is particularly useful because vicinal dihalides are readily available by addition of bromine or chlorine to alkenes. Thus, the overall sequence of halogenation–dehydrohalogenation provides an excellent method for going from an alkene to an alkyne. For example,

$$C_6H_5CH{=}CHC_6H_5 \xrightarrow{Br_2,\ CCl_4} C_6H_5CHBrCHBrC_6H_5 \xrightarrow[\text{Ethanol; }-2\ HBr]{2\ KOH} C_6H_5C{\equiv}CC_6H_5$$

1,2-Diphenylethylene (stilbene)

1,2-Dibromo-1,2-diphenylethane (A vicinal dibromide)

Diphenylacetylene (85%)

$$CH_3(CH_2)_7CH{=}CH(CH_2)_7COOH \xrightarrow[CCl_4]{Br_2} CH_3(CH_2)_7CHBr{-}CHBr(CH_2)_7COOH \xrightarrow[2.\ H_3O^+]{1.\ 2\ NaNH_2,\ NH_3} CH_3(CH_2)_7C{\equiv}C(CH_2)_7COOH$$

9-Octadecenoic acid (oleic acid from butter)

9,10-Dibromooctadecanoic acid

9-Octadecynoic acid (62%)

As the examples indicate, different bases can be used for dehydrohalogenation. Sodium amide is normally preferred, however, since it usually gives higher yields. The twofold dehydrohalogenation takes place in discrete steps, and vinylic halides are intermediates. This suggests that vinylic halides themselves should give alkynes when treated with strong base, and this is indeed the case. For example,

$$CH_3CCl{=}CHCH_2OH \xrightarrow[2.\ H_3O^+]{1.\ 2\ NaNH_2} CH_3C{\equiv}CCH_2OH$$

3-Chloro-2-buten-1-ol (a vinylic chloride)

2-Butyn-1-ol (85%)

7.12 Organic synthesis

The laboratory synthesis of organic molecules from simple precursors is carried out for many reasons. In the pharmaceutical industry, new organic molecules are designed and synthesized in the hope that useful new drugs will be found. In the chemical industry, many syntheses are undertaken to devise more economical routes to known compounds. In academic laboratories, the synthesis of highly complex molecules is often done purely for the intellectual challenge involved in mastering so difficult a subject; the successful synthesis route is a highly creative work that is sometimes described by such subjective terms as *elegant* or *beautiful*.

In this book, too, we will often devise syntheses of molecules from simpler precursors. The purpose, however, is purely pedagogical. The ability to plan workable synthetic sequences demands a thorough knowledge of a wide variety of organic reactions. Furthermore, it requires not just a theoretical knowledge, but also a practical grasp of the proper fitting together of steps in a sequence such that each reaction does only what is desired. Working synthesis problems is an excellent way to learn organic chemistry.

Some of the syntheses we plan may appear trivial. Here's an example:

Problem. Prepare octane starting from 1-pentyne.

Solution. A valid response would be to suggest the alkylation of the acetylide anion of 1-pentyne with 1-bromopropane, and then the reduction of the product using catalytic hydrogenation:

$$\underset{\text{1-Pentyne}}{CH_3CH_2CH_2C{\equiv}CH} \xrightarrow[\text{2. } BrCH_2CH_2CH_3\text{, THF}]{\text{1. } NaNH_2\text{, } NH_3} \underset{\text{4-Octyne}}{CH_3CH_2CH_2C{\equiv}CCH_2CH_2CH_3}$$

$$\xrightarrow{H_2/Pd \text{ in ethanol}} \underset{\text{Octane}}{CH_3CH_2CH_2CH_2{-}CH_2CH_2CH_2CH_3}$$

(Octane drawn with two H atoms shown above and below each of the two central carbons.)

The synthesis route presented should work perfectly well, but it has little practical value. A chemist can simply *buy* octane from several dozen chemical supply houses. The value of working the problem is that it makes us approach a chemical problem in a logical way, draw on our knowledge of chemical reactions, and organize that knowledge into a workable plan—it helps us *learn* organic chemistry.

There is no secret to planning organic syntheses. All it requires is a knowledge of the different reactions studied and lots of practice. However, here is a hint: *Work backwards*. Look at the end product and ask, "What was the *immediate* precursor of that product?" For example, if the end product is an alkyl halide, the immediate precursor might be an alkene (via HX addition). Having found an immediate precursor, proceed backwards again, one step at a time, until a suitable starting material is found.

Let's work some examples of increasing complexity.

Problem. Starting from 1-pentyne and any alkyl halide needed, synthesize *cis*-2-hexene. More than one step is required.

$$\underset{\text{1-Pentyne}}{CH_3CH_2CH_2C{\equiv}CH} + RX \overset{?}{\Rightarrow} \underset{cis\text{-2-Hexene}}{(CH_3CH_2CH_2)(H)C{=}C(CH_3)(H)}$$

Solution. First, we ask what an immediate precursor of a cis disubstituted alkene might be. We know that alkenes can be prepared from alkynes by partial reduction. The proper choice of experimental conditions will allow us to prepare either a trans disubstituted alkene (using lithium in liquid ammonia) or a cis disubstituted alkene (using catalytic hydrogenation over the Lindlar catalyst). Thus reduction of 2-hexyne by catalytic hydrogenation using the Lindlar catalyst should yield *cis*-2-hexene:

$$\underset{\text{2-Hexyne}}{CH_3CH_2CH_2C{\equiv}CCH_3} \xrightarrow[\text{Lindlar catalyst}]{H_2} \underset{cis\text{-2-Hexene}}{(CH_3CH_2CH_2)(H)C{=}C(CH_3)(H)}$$

Next, we ask, "What is an immediate precursor of 2-hexyne?" We have seen that internal alkynes can be prepared by alkylation of terminal alkyne anions (acetylides). In the present instance, we are told to start with 1-pentyne. Thus, alkylation of the anion of 1-pentyne with iodomethane should yield 2-hexyne:

$$\underset{\text{1-Pentyne}}{CH_3CH_2CH_2C{\equiv}C{-}H} + NaNH_2 \xrightarrow{\text{In } NH_3} CH_3CH_2CH_2C{\equiv}C{:}^-Na^+$$

$$CH_3CH_2CH_2C{\equiv}C{:}^-Na^+ + CH_3I \xrightarrow{\text{In THF}} \underset{\text{2-Hexyne}}{CH_3CH_2CH_2C{\equiv}C{-}CH_3}$$

In three steps we have synthesized *cis*-2-hexene from the given starting materials:

$$\underset{\text{1-Pentyne}}{CH_3CH_2CH_2C{\equiv}CH} \xrightarrow[\text{2. } CH_3I,\ THF]{\text{1. } NaNH_2,\ NH_3} \underset{\text{2-Hexyne}}{CH_3CH_2CH_2C{\equiv}CCH_3}$$

$$\downarrow H_2, \text{ Lindlar catalyst}$$

$$\underset{cis\text{-2-Hexene}}{(CH_3CH_2CH_2)HC{=}CH(CH_3)}$$

Problem. Starting from acetylene and any alkyl halide needed, synthesize 2-bromopentane. More than one step is required.

$$HC{\equiv}CH + RX \overset{?}{\Rightarrow} \underset{\text{2-Bromopentane}}{CH_3CH_2CH_2CHBrCH_3}$$

Solution. First, we ask, "What is an immediate precursor of an alkyl halide?" Perhaps an alkene:

$$\begin{matrix} CH_3CH_2CH_2CH{=}CH_2 \\ \text{or} \\ CH_3CH_2CH{=}CHCH_3 \end{matrix} \xrightarrow[\text{Ether}]{HBr} CH_3CH_2CH_2\underset{}{\overset{Br}{\overset{|}{C}}}HCH_3$$

Of the two alkene possibilities, addition of HBr to 1-pentene looks like a better choice than addition to 2-pentene, since the latter would give a mixture of isomers. "What is an immediate precursor of an alkene?" Perhaps an alkyne, which could be partially reduced:

$$CH_3CH_2CH_2C{\equiv}CH \xrightarrow[\text{Lindlar catalyst}]{H_2} CH_3CH_2CH_2CH{=}CH_2$$

"What is an immediate precursor of a terminal alkyne?" Probably sodium acetylide and an alkyl halide.

$$Na^+ :\bar{C}{\equiv}CH + BrCH_2CH_2CH_3 \longrightarrow CH_3CH_2CH_2C{\equiv}CH$$

In four steps we have synthesized the desired material from acetylene and 1-bromopropane, as shown at the top of p. 216.

$$\underset{\text{Acetylene}}{HC{\equiv}CH} \xrightarrow[\text{2. } CH_3CH_2CH_2Br, \text{ THF}]{\text{1. } NaNH_2, NH_3} \underset{\text{1-Pentyne}}{CH_3CH_2CH_2C{\equiv}CH} \xrightarrow[\text{Lindlar catalyst}]{H_2} \underset{\text{1-Pentene}}{CH_3CH_2CH_2CH{=}CH_2}$$

$$\downarrow \text{HBr, ether}$$

$$\underset{\text{2-Bromopentane}}{CH_3CH_2CH_2CH(Br)CH_3}$$

Problem. Synthesize *trans*-2-methyl-3-heptene from acetylene and any alkyl halide.

$$HC{\equiv}CH + RX \overset{?}{\Rightarrow} (CH_3)_2CH(H)C{=}C(H)CH_2CH_2CH_3$$

trans-2-Methyl-3-heptene

Solution. "What is an immediate precursor of a trans alkene?" Perhaps an alkyne, which could be reduced with Li/NH_3:

$$(CH_3)_2CHC{\equiv}CCH_2CH_2CH_3 \xrightarrow[NH_3]{Li} (CH_3)_2CH(H)C{=}C(H)CH_2CH_2CH_3$$

"What is an immediate precursor of an internal alkyne?" Perhaps a terminal alkyne, which could be alkylated:

$$(CH_3)_2CHC{\equiv}C{:}^- \, Na^+ + BrCH_2CH_2CH_3 \longrightarrow (CH_3)_2CHC{\equiv}CCH_2CH_2CH_3$$

$$(CH_3)_2CHBr + {}^-{:}C{\equiv}CCH_2CH_2CH_3 \not\longrightarrow (CH_3)_2CHC{\equiv}CCH_2CH_2CH_3$$

Of the two possibilities, only the top choice will work. Remember that acetylide anions will not alkylate secondary halides such as 2-bromopropane.

"What is an immediate precursor of 3-methyl-1-butyne?" Perhaps 1-butyne, which could undergo dianion alkylation:

$$CH_3CH_2C{\equiv}CH \xrightarrow[\text{2. } CH_3I \;\; \text{3. } H^+]{\text{1. 2 BuLi, THF}} (CH_3)_2CHC{\equiv}CH$$

"What is an immediate precursor of 1-butyne?" Perhaps sodium acetylide and bromoethane:

$$CH_3CH_2Br + Na^+ \,:\overset{-}{C}{\equiv}CH \longrightarrow CH_3CH_2C{\equiv}CH$$

We have completed the synthesis in four steps, by working backwards.

PROBLEM

7.11 Beginning with 4-octyne as your only source of carbon, and using any inorganic reagents necessary, how would you synthesize the following compounds?
(a) Butanoic acid (b) *cis*-4-Octene (c) 4-Bromooctane
(d) 4-Hydroxyoctane (e) 4,5-Dichlorooctane

PROBLEM

7.12 Beginning with acetylene and any alkyl halides needed, how would you synthesize the following compounds?
(a) Decane (b) 3,3-Dimethylhexane
(c) Hexanal (d) 2-Heptanone

7.13 Summary

Alkynes are hydrocarbons that contain one or more carbon–carbon triple bonds. Alkyne carbon atoms are *sp* hybridized, and the triple bond is formed when overlap occurs to give one *sp*–*sp* sigma bond and two *p*–*p* pi bonds.

Much of the chemistry of alkynes is similar to that of alkenes and is dominated by electrophilic addition reactions. For example, alkynes react with HBr and HCl to yield vinylic halides, and with Br_2 and Cl_2 to yield 1,2-dihalides (vicinal dihalides). Alkynes can be hydrated (addition of H_2O) by either of two procedures. Reaction of an alkyne with aqueous sulfuric acid in the presence of mercuric ion catalyst leads to an intermediate enol that immediately isomerizes (tautomerizes) to yield a ketone. The addition reaction occurs with Markovnikov regiochemistry, and a methyl ketone is therefore produced from a terminal alkyne. Alternatively, a net addition of water can also be effected by hydroboration of an alkyne, followed by oxidation with basic hydrogen peroxide. Disiamylborane, a sterically hindered dialkylborane, is often used in this reaction, and aldehydes can be prepared in good yield from terminal alkynes.

Alkynes can also be hydrogenated. Partial reduction of the triple bond is particularly useful, and catalytic hydrogenation of alkynes over a Lindlar catalyst yields cis alkenes. Reduction of the alkyne with lithium in ammonia yields trans alkenes.

One of the most striking differences in chemical reactivity between alkynes and alkenes is due to their different acidities; terminal alkynes contain an acidic hydrogen that can be removed by strong bases to yield an acetylide anion. Acetylide anions act as nucleophiles and can displace halide ion from primary halides (alkylation reaction). Acetylide anions are more stable than either alkyl anions or vinylic anions because the negative charge is in a hybrid orbital with much *s* character, allowing the charge to be closer to the nucleus.

There are relatively few general methods of alkyne synthesis. One of the most common is the twofold elimination of HX from a vicinal dihalide.

7.14 Summary of reactions

A. Reactions of alkynes

1. Addition of HX and X_2, where X = Br or Cl (Section 7.3)

$$R{-}C{\equiv}CH \xrightarrow[\text{Ether}]{HX} R{-}\overset{\displaystyle X}{\overset{|}{C}}{=}CH_2 \xrightarrow[\text{Ether}]{HX} R{-}\underset{\displaystyle X}{\underset{|}{\overset{\displaystyle X}{\overset{|}{C}}}}{-}CH_3$$

$$R{-}C{\equiv}C{-}R \xrightarrow[CCl_4]{X_2} R{-}\overset{\displaystyle X}{\overset{|}{C}}{=}\overset{\displaystyle X}{\overset{|}{C}}{-}R \xrightarrow[CCl_4]{X_2} R{-}CX_2CX_2{-}R$$

2. Mercuric sulfate–catalyzed hydration (Section 7.4)

$$R{-}C{\equiv}CH \xrightarrow[HgSO_4]{H_2SO_4,\ H_2O} \left[R{-}\overset{\displaystyle OH}{\overset{|}{C}}{=}CH_2\right] \longrightarrow R{-}\overset{\displaystyle O}{\overset{\|}{C}}{-}CH_3$$

A methyl ketone

3. Hydroboration

a. Hydroboration–oxidation (Section 7.5)

$$R{-}C{\equiv}C{-}R \xrightarrow[2.\ H_2O_2,\ ^-OH]{1.\ BH_3,\ THF} R{-}\overset{\displaystyle O}{\overset{\|}{C}}{-}CH_2R$$

$$R{-}C{\equiv}CH \xrightarrow[2.\ H_2O_2,\ ^-OH]{1.\ H{-}B[CH(CH_3)CH(CH_3)_2]_2} R{-}CH_2{-}CHO$$

An aldehyde

b. Hydroboration–protonolysis (Section 7.5)

$$R{-}C{\equiv}C{-}R \xrightarrow[2.\ CH_3COOH]{1.\ BH_3,\ THF} \begin{matrix} H & & H \\ & C{=}C & \\ R & & R \end{matrix}$$

A cis alkene

4. Reduction

a. Catalytic (Section 7.6)

$$R{-}C{\equiv}C{-}R \xrightarrow{H_2,\ Pd/C} R{-}CH_2CH_2R$$

$$R{-}C{\equiv}C{-}R \xrightarrow[\text{Lindlar catalyst}]{H_2} \begin{matrix} H & & H \\ & C{=}C & \\ R & & R \end{matrix}$$

A cis alkene

b. Lithium/ammonia (Section 7.6)

$$R{-}C{\equiv}C{-}R \xrightarrow{Li,\ NH_3} \underset{R}{\overset{H}{}}C{=}C\underset{H}{\overset{R}{}}$$

A trans alkene

5. Acetylide alkylation (Section 7.8)

a. $R{-}C{\equiv}C{-}H \xrightarrow[NH_3]{NaNH_2} R{-}C{\equiv}C{:}^-\ Na^+ + NH_3$

b. $H{-}C{\equiv}C{:}^- + RCH_2Br \xrightarrow{THF} R{-}CH_2{-}C{\equiv}CH + {:}Br^-$

c. $R{-}C{\equiv}C{:}^- + R'{-}CH_2Br \xrightarrow{THF} R{-}C{\equiv}C{-}CH_2R' + {:}Br^-$

6. Dianion alkylation of alkynes (Section 7.9)

$$R{-}CH_2{-}C{\equiv}C{-}H \xrightarrow[THF]{2\ BuLi} R{-}\ddot{C}H{-}C{\equiv}C{:}^-$$

$$\downarrow \text{1. } R'CH_2Br \quad \text{2. } H_3O^+$$

$$R{-}\underset{\displaystyle CH_2R'}{\underset{|}{C}}HC{\equiv}C{-}H + {:}Br^-$$

B. Preparation of alkynes

1. Acetylide alkylation (Section 7.8)

$$R{-}C{\equiv}C{:}^- + R'CH_2Br \xrightarrow{THF} R{-}C{\equiv}C{-}CH_2R' + {:}Br^-$$

2. Dehydrohalogenation (Section 7.11)

$$R{-}CHBrCHBr{-}R' \xrightarrow[\text{or } NaNH_2,\ NH_3]{2\ KOH,\ Ethanol} R{-}C{\equiv}C{-}R' + H_2O + KBr$$

$$R{-}\overset{\displaystyle Br}{\overset{|}{C}}{=}CHR' \xrightarrow[\text{or } NaNH_2,\ NH_3]{KOH,\ Ethanol} R{-}C{\equiv}C{-}R' + H_2O + KBr$$

ADDITIONAL PROBLEMS

7.13 Provide correct IUPAC names for compounds (a)–(f):

(a) $CH_3CH_2C{\equiv}CC(CH_3)_3$

(b) $CH_3C{\equiv}CCH_2C{\equiv}CCH_2CH_3$

(c) $CH_3CH{=}C(CH_3)C{\equiv}CCH(CH_3)_2$

(d) $HC{\equiv}CC(CH_3)_2CH_2C{\equiv}CH$

(e) $H_2C{=}CHCH{=}CHC{\equiv}CH$

(f) $CH_3CH_2CH(CH_2CH_3)C{\equiv}CCH(CH_2CH_3)CH(CH_3)_2$

7.14 Draw structures corresponding to these names:
(a) 3,3-Dimethyl-4-octyne
(b) 3-Ethyl-5-methyl-1,6,8-decatriyne
(c) 2,2,5,5-Tetramethyl-3-hexyne
(d) 3,4-Dimethylcyclodecyne
(e) 3,5-Heptadien-1-yne
(f) 3-Chloro-4,4-dimethyl-1-nonen-6-yne

7.15 The following names are incorrect. Draw the structures and give the correct names.
(a) 1-Ethyl-5,5-dimethyl-1-hexyne
(b) 2,5,5-Trimethyl-6-heptyne
(c) 3-Methylhept-5-en-1-yne
(d) 2-Isopropyl-5-methyl-7-octyne
(e) 3-Hexen-5-yne
(f) 5-Ethynyl-1-methylcyclohexane

7.16 These two hydrocarbons have been isolated from various plants in the Compositae family. Name them according to IUPAC rules.

(a) $CH_3CH{=}CHC{\equiv}CC{\equiv}CCH{=}CHCH{=}CHCH{=}CH_2$ (all trans)

(b) $CH_3C{\equiv}CC{\equiv}CC{\equiv}CC{\equiv}CC{\equiv}CCH{=}CH_2$

7.17 Predict the products of these reactions:

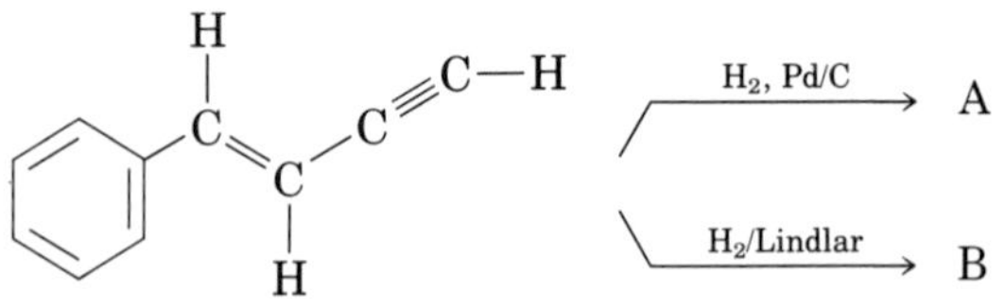

H, C, C, H, C, C—H

$\xrightarrow{H_2,\ Pd/C}$ A

$\xrightarrow{H_2/Lindlar}$ B

7.18 A hydrocarbon of unknown structure has the formula C_8H_{10}. On catalytic hydrogenation over the Lindlar catalyst, 1 equiv of H_2 is absorbed. On hydrogenation over a palladium catalyst, however, 3 equiv of H_2 are absorbed.
(a) How many rings/double bonds/triple bonds are present in the unknown?
(b) How many triple bonds are present?
(c) How many double bonds are present?
(d) How many rings are present?
Explain your answers and draw a possible structure that fits the data.

7.19 Predict the products of the following reactions of 1-hexyne:

$n\text{-}C_4H_9C{\equiv}CH$

(a) $\xrightarrow{\text{1 equiv HBr, } CCl_4}$

(b) $\xrightarrow{\text{1 equiv } Cl_2, CCl_4}$

(c) $\xrightarrow[\text{Lindlar catalyst}]{H_2}$

(d) $\xrightarrow[\text{2. } CH_3Br, \text{ THF}]{\text{1. } NaNH_2, NH_3}$

(e) $\xrightarrow[Hg^{2+}]{H_2O,\ H_2SO_4}$

7.20 Predict the products of the following reactions of 5-decyne:

$n\text{-}C_4H_9C{\equiv}CC_4H_9\text{-}n$

(a) $\xrightarrow[\text{Lindlar catalyst}]{H_2}$

(b) $\xrightarrow{Li/NH_3}$

(c) $\xrightarrow{\text{1 equiv } Br_2, CCl_4}$

$n\text{-}C_4H_9C{\equiv}CC_4H_9\text{-}n$

(d) $\xrightarrow[\text{2. } H_2O_2,\ ^-OH]{\text{1. } BH_3,\ THF}$

(e) $\xrightarrow[\text{2. } CH_3COOH]{\text{1. } BH_3,\ THF}$

(f) $\xrightarrow[\text{2. } CH_3COOD]{\text{1. } BH_3,\ THF}$

(g) $\xrightarrow[\text{2. } CH_3COOH]{\text{1. } BD_3,\ THF}$

(h) $\xrightarrow[Hg^{2+}]{H_2O,\ H_2SO_4}$

7.21 Predict the products of the following reactions of 2-hexyne:

$CH_3C{\equiv}CC_3H_7\text{-}n$

(a) $\xrightarrow{\text{2 equiv } Br_2,\ CCl_4}$

(b) $\xrightarrow[\text{2. } CH_3COOH]{\text{1. } BH_3,\ THF}$

(c) $\xrightarrow[\text{2. } H_2O_2,\ ^-OH]{\text{1. } BH_3,\ THF}$

(d) $\xrightarrow[\text{Ether}]{HBr}$

(e) $\xrightarrow{Li/NH_3}$

(f) $\xrightarrow[Hg^{2+}]{H_2O,\ H_2SO_4}$

7.22 Acetonitrile, CH_3CN, contains a carbon–nitrogen triple bond. Sketch the orbitals involved in the bonding in acetonitrile and indicate the hybridization of each atom.

7.23 How would you carry out these reactions?

(a) $CH_3CH_2C{\equiv}CH \xrightarrow{?} CH_3CH(CH_3)C{\equiv}CH$

(b) $CH_3CH_2C{\equiv}CH \xrightarrow{?} CH_3CH_2C(=O)CH_3$

(c) $CH_3CH_2C{\equiv}CH \xrightarrow{?} CH_3CH_2CH_2CHO$

(d) $C_6H_5C{\equiv}CH \xrightarrow{?} C_6H_5C{\equiv}C{-}CH_3$

(e) $C_6H_5C{\equiv}CCH_3 \xrightarrow{?}$ $C_6H_5CH{=}CHCH_3$ (H on the benzylic carbon; H and CH_3 on the other carbon)

(f) $CH_3CH_2C{\equiv}CH \xrightarrow{?} CH_3CH_2COOH$

(g) $CH_3CH_2CH_2CH_2CH{=}CH_2 \xrightarrow{\text{(2 steps)}} CH_3CH_2CH_2CH_2C{\equiv}CH$

7.24 Occasionally, chemists need to invert the stereochemistry of an alkene. That is, one might want to convert a cis alkene to a trans alkene, or vice versa. There is no one-step method for doing so, but by combining some of the reactions learned earlier into the proper sequence, alkenes can be inverted. How would you carry out these reactions?

(a) *trans*-5-Decene $\xrightarrow{?}$ *cis*-5-Decene

(b) *cis*-5-Decene $\xrightarrow{?}$ *trans*-5-Decene

7.25 Propose structures for hydrocarbons that give the following products on oxidative cleavage by $KMnO_4$ or O_3.

(a) $CO_2 + CH_3(CH_2)_5COOH$

(b) CH_3COOH + C_6H_5COOH (benzene ring bearing COOH)

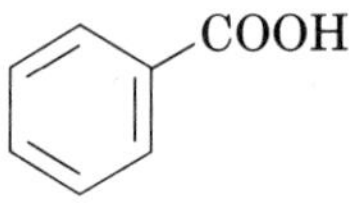

(c) $HOOC(CH_2)_8COOH$

(d) $CH_3CHO + CH_3\overset{\overset{O}{\|}}{C}CH_2CH_2COOH + CO_2$

(e) $OHCCH_2CH_2CH_2CH_2\overset{\overset{O}{\|}}{C}COOH + CO_2$

7.26 The syntheses that follow require more than one step. How would you carry them out?

(a) $HC{\equiv}CH \Rightarrow$ (cyclodecene ring with H and H on the double-bond carbons)

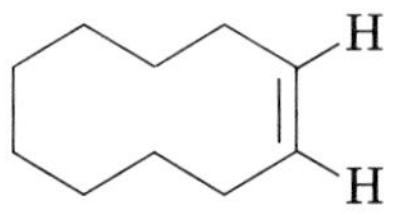

(b) $CH_3CH_2C{\equiv}CH \Rightarrow (CH_3)_3CC{\equiv}CCH_2CH_3$

(c) 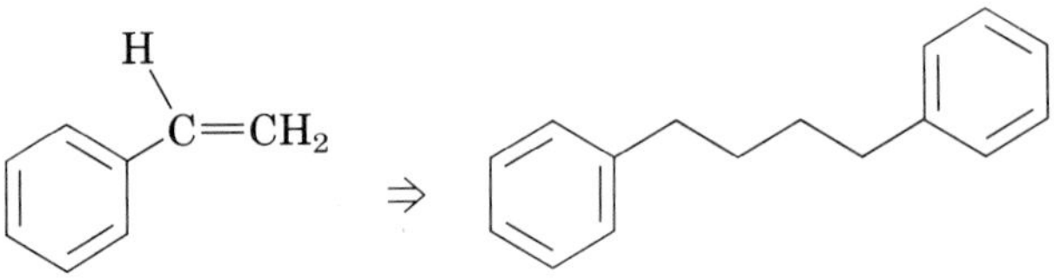 $C_6H_5CH{=}CH_2$ (benzene ring with $\overset{H}{\diagdown}C{=}CH_2$) $\Rightarrow$ $C_6H_5CH_2CH_2CH_2CH_2C_6H_5$ (two benzene rings joined by a four-carbon chain)

7.27 Organometallic reagents such as sodium acetylide undergo an addition reaction with ketones, giving alcohols (Section 20.7).

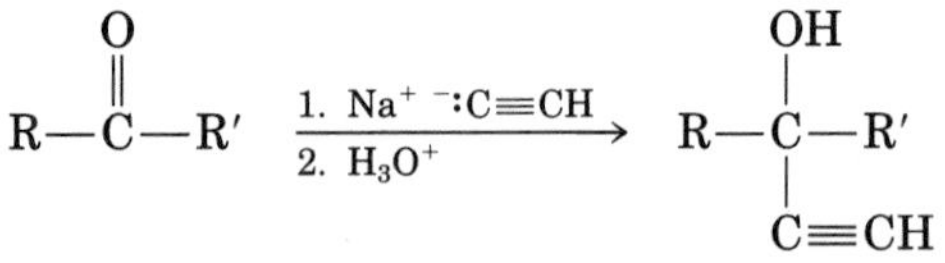

$$R{-}\overset{\overset{O}{\|}}{C}{-}R' \xrightarrow[2.\ H_3O^+]{1.\ Na^+\ ^-{:}C{\equiv}CH} R{-}\underset{\underset{C\equiv CH}{|}}{\overset{\overset{OH}{|}}{C}}{-}R'$$

How might you use this reaction to prepare 2-methyl-1,3-butadiene, the starting material used in the manufacture of synthetic rubber?

7.28 Using 1-butyne as the only source of carbon along with any inorganic reagents you need, synthesize the following compounds. More than one step may be needed.
(a) 1,1,2,2-Tetrachlorobutane
(b) Octane
(c) Butanal

7.29 Using acetylene and any alkyl halides that have four or fewer carbons as starting materials, how would you synthesize the following compounds? More than one step may be required.

(a) $CH_3CH_2CH_2C{\equiv}CH$

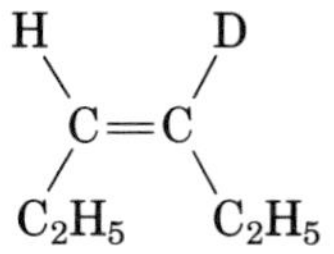

(b) $CH_3CH_2C{\equiv}CCH_2CH_3$

(c) $CH_3CH_2CH(CH_2CH_3)C{\equiv}CH$

(d) $CH_3CH_2CH_2C(=O)CH_3$

(e) $(CH_3)_2CHCH{=}CH_2$

(f) $CH_3CH_2CH_2C(=O)CH_2CH_2CH_2CH_3$

(g) $CH_3CH_2CH_2CH_2CH_2CHO$

7.30 How would you carry out these reactions to introduce deuterium into organic molecules?

(a) $CH_3CH_2C{\equiv}CCH_2CH_3 \longrightarrow$ (H)(C_2H_5)C=C(D)(C_2H_5)

(b) $CH_3CH_2C{\equiv}CCH_2CH_3 \longrightarrow$ (D)(C_2H_5)C=C(D)(C_2H_5)

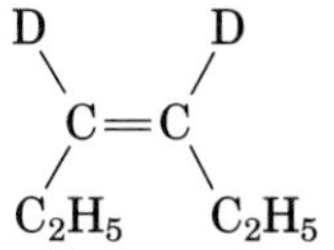

(c) $CH_3CH_2C{\equiv}CCH_2CH_3 \longrightarrow$ (D)(C_2H_5)C=C(C_2H_5)(D)

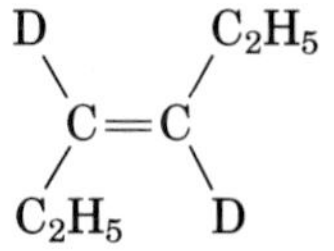

(d) $CH_3CH_2CH_2C{\equiv}CH \longrightarrow CH_3CH_2CH_2C{\equiv}CD$

(e) 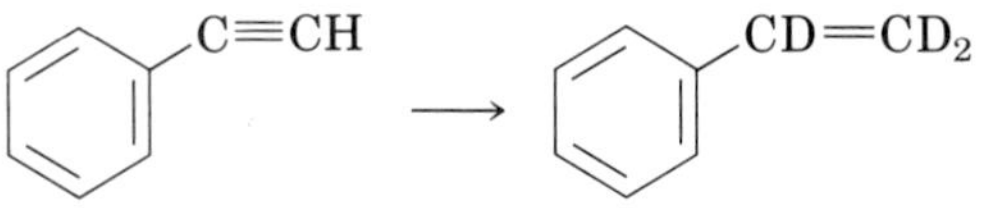 $C_6H_5C{\equiv}CH \longrightarrow C_6H_5CD{=}CD_2$

7.31 Predict the products of these reactions:

(a) $CH_3(CH_2)_4C{\equiv}CH \xrightarrow[\text{2. } H_2O_2,\ ^-OH]{\text{1. } H{-}B[CH(CH_3)CH(CH_3)_2]_2,\ THF}$

(b) 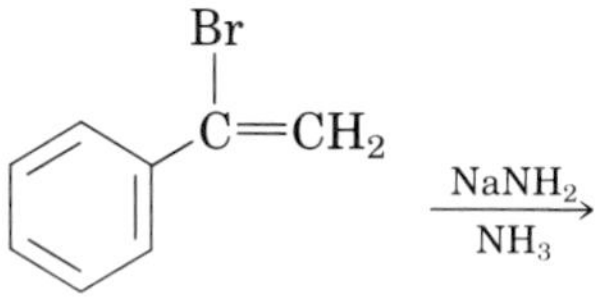 $C_6H_5C(Br){=}CH_2 \xrightarrow[NH_3]{NaNH_2}$

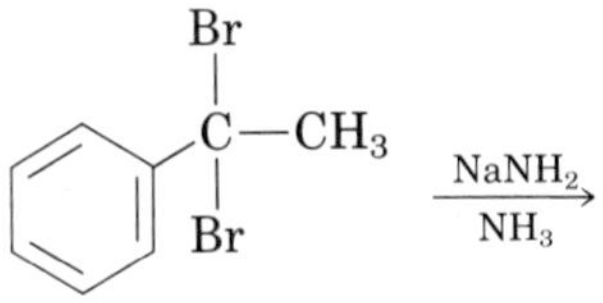

(c) $C_6H_5CBr_2{-}CH_3 \xrightarrow[NH_3]{NaNH_2}$

7.32 Cumulenes are compounds with three or more adjacent double bonds. Draw an orbital picture of a cumulene. What is the geometric relationship of the substituents on one end to the substituents on the other end? What kind of isomerism is possible? Make a model to help you see the answer.

$$R_2C{=}C{=}C{=}CR_2$$

A cumulene

7.33 The sex attractant given off by the common housefly is a simple alkene named *muscalure*. Propose a synthesis of muscalure starting from acetylene and any alkyl halides. What is the IUPAC name for muscalure?

$$\textit{cis}\text{-}CH_3(CH_2)_7CH{=}CH(CH_2)_{19}CH_3$$

Muscalure

7.34 Compound A (C_9H_{12}) absorbed 3 equiv of hydrogen on catalytic reduction over a palladium catalyst to give B (C_9H_{18}). On ozonolysis, A gave, among other things, a ketone that was identified as cyclohexanone. On treatment with $NaNH_2$ in NH_3, followed by addition of iodomethane, A gave a new hydrocarbon, C ($C_{10}H_{14}$). What are the structures of A, B, and C?

7.35 Hydrocarbon A has the formula $C_{12}H_8$. It absorbs 8 equiv of hydrogen on catalytic reduction over a palladium catalyst. On ozonolysis only two products are formed—oxalic acid, HOOC—COOH, and succinic acid, $HOOCCH_2CH_2COOH$. Formulate these reactions and propose a structure for A.

7.36 Erythrogenic acid, $C_{18}H_{26}O_2$, is an interesting acetylenic fatty acid. On exposure to light, it turns a vivid red. On catalytic hydrogenation over a palladium catalyst, 5 equiv of hydrogen are absorbed, and stearic acid, $CH_3(CH_2)_{16}COOH$, is produced. Ozonolysis of erythrogenic acid gives four products, which can be identified as follows: formaldehyde, CH_2O; oxalic acid, HOOC—COOH; azelaic acid, $HOOC(CH_2)_7COOH$; and the aldehyde acid $OHC(CH_2)_4COOH$. With this information, draw two possible structures for erythrogenic acid. Can you propose a way to tell them apart by carrying out some simple reactions?

CHAPTER 8
INTRODUCTION TO STEREOCHEMISTRY

Up to now, we have been concerned primarily with the general nature of chemical reactions and with the specific chemistry of some simple functional groups. Our concept of chemistry has been largely two-dimensional, and little thought has been given to the spatial arrangement of atoms in molecules. Many observations cannot be explained in such simple terms, however, and it is now time to add a third dimension to our considerations. **Stereochemistry** is the branch of chemistry concerned with the three-dimensional nature of molecules.

8.1 Optical activity

The study of stereochemistry can be traced to the work of the French scientist Jean Baptiste Biot[1] in the early nineteenth century. Biot, a physicist, was investigating the nature of **plane-polarized light.** A beam of ordinary light consists of electromagnetic waves that oscillate at right angles to the direction of light travel. Since ordinary light is unpolarized, this oscillation takes place in an infinite number of planes. When, however, a beam of ordinary light is passed through a device called a **polarizer,** only the light waves oscillating in a *single* plane pass through. Light waves in all other planes are blocked out. The light that passes through the polarizer has its electromagnetic waves vibrating in a well-defined plane, hence the name plane-polarized light. The polarization process is represented in Figure 8.1.

In 1815 Biot made the remarkable observation that when a beam of plane-polarized light is passed through solutions of certain organic molecules, such as sugar or camphor, the plane of polarization is *rotated.* He recognized that this phenomenon is due to some inherent molecular property. We call molecules that exhibit this property **optically active.**

The amount of rotation can be measured with an instrument known as a **polarimeter,** represented schematically in Figure 8.2. Optically active organic molecules are first placed in a sample tube, and plane-polarized light is passed through the tube, where rotation occurs. The light then goes through a second polarizer known as the **analyzer.** By rotating this analyzer until the light passes through it, we can find the new plane of polarization and can tell to what extent rotation has occurred. The observed rotation is denoted α (Greek alpha) and is expressed in degrees.

In addition to determining the extent of rotation, we can also find out what its direction is. Some optically active molecules rotate polarized light to the left (counterclockwise), and are said to be **levorotatory.** Others rotate polarized light to the right (clockwise), and are said to be **dextrorotatory.** By convention, rotation to the left is given a minus sign (−), and rotation to the right is given a plus sign (+). For example, (−)-morphine is levorotatory, and (+)-sucrose is dextrorotatory.

[1]Jean Baptiste Biot (1774–1862); b. Paris; physicist, Collège de France.

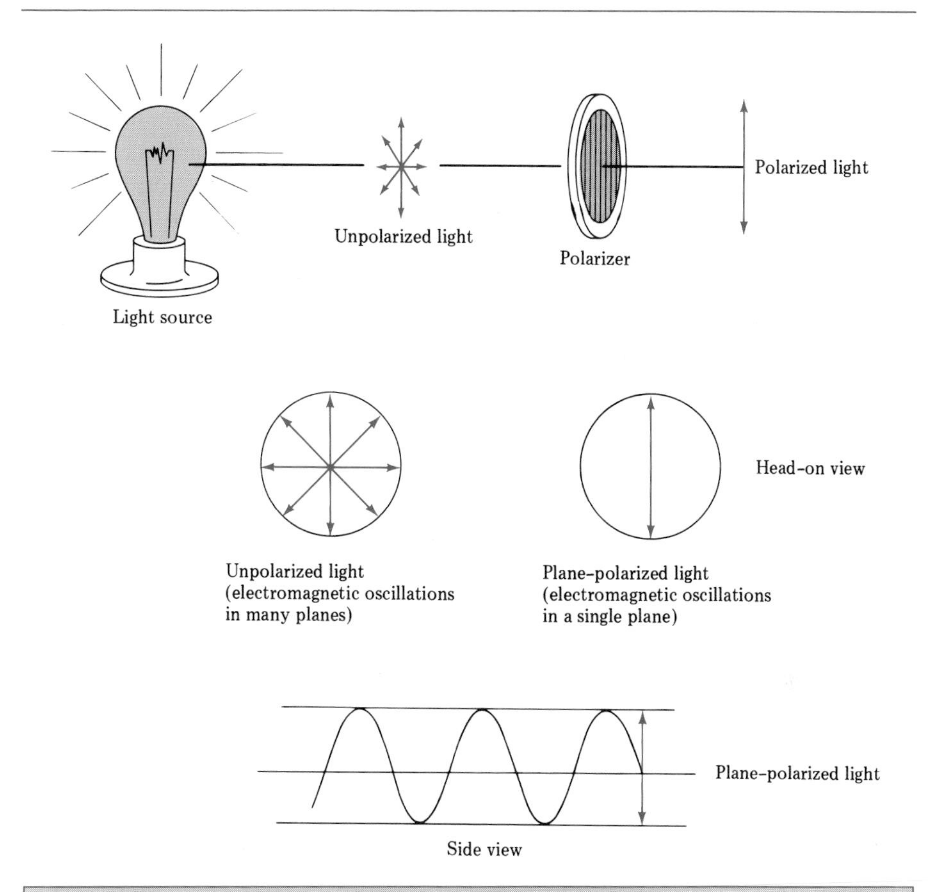

Figure 8.1. Plane-polarized light.

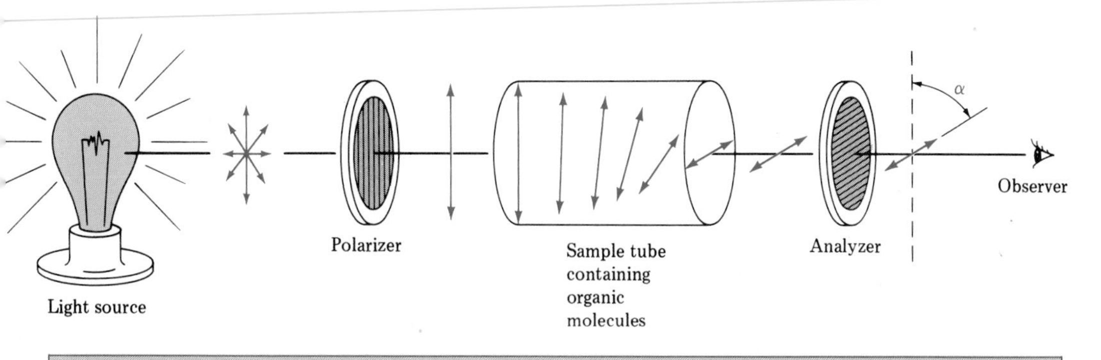

Figure 8.2. Schematic representation of a polarimeter.

8.2 Specific rotation

Since the rotation of plane-polarized light is an intrinsic property of optically active organic molecules, it follows that the amount of rotation depends on the number of molecules that the light beam encounters. Thus, the exact amount of rotation observed is dependent on both sample concentration and sample path length. A sample that is twice as concentrated as another shows twice the rotation. Similarly, a sample in a tube that is twice as long as another shows twice the rotation. The amount of rotation is also dependent on the wavelength of the light used. To express our data in a meaningful way so that comparisons can be made, we must choose standard conditions.

By convention, the **specific rotation, $[\alpha]_D$,** of a compound is defined as the observed rotation when light of 5896 Å wavelength (the yellow sodium D line) is used with a sample path length of 1 decimeter (1 dm; 10 cm) and a sample concentration of 1 g/mL:

Specific rotation,

$$[\alpha]_D = \frac{\text{Observed rotation, } \alpha}{\text{Path length, } l \text{ (dm)} \times \text{Concentration of sample, } C \text{ (g/mL)}}$$

$$[\alpha]_D = \frac{\alpha}{l \times C}$$

When optical rotation data are expressed in this standard way, the specific rotation, $[\alpha]_D$, is a physical constant that is characteristic of each optically active compound. Some examples are listed in Table 8.1.

TABLE 8.1 Specific rotations of some organic molecules

Compound	*$[\alpha]_D$ (degrees)*	*Compound*	*$[\alpha]_D$ (degrees)*
Camphor	+ 44.26	Penicillin V	+223
Morphine	−132	Monosodium glutamate	+ 25.5
Sucrose	+ 66.47	Benzene	0
Cholesterol	− 31.5	Acetic acid	0

PROBLEM

8.1 A 1.5 g sample of coniine, the toxic principle of poison hemlock, was dissolved in 10 mL ethanol and placed in a sample cell with a 5 cm path length. The observed rotation at the sodium D line was +1.2°. Calculate the specific rotation, $[\alpha]_D$, for coniine.

8.3 Pasteur's discovery of enantiomers

After Biot's discovery of optical activity in 1815, little was done until Louis Pasteur[2] entered the picture in 1848. Pasteur received his formal training in chemistry and became interested in the subject of crystallography. He began work on crystalline salts of tartaric acid derived from wine and was repeating some measurements published a few years earlier when he made a surprising observation. When he recrystallized a concentrated solution of sodium ammonium tartrate below 28°C, two distinct kinds of crystals precipitated. Furthermore, the two kinds of crystals were nonsuperimposable *mirror images* of each other. That is, the crystals were not symmetrical, but were related to each other in exactly the same way that a right hand is related to a left hand. (When a *right* hand is held up to a mirror, the image that you see looks like a *left* hand—try it.)

Working carefully with a pair of tweezers, Pasteur was able to separate the crystals into two piles, one of "right-handed" crystals and one of "left-handed" crystals like those shown in Figure 8.3. Although the original salt (a mixture of right and left) was optically inactive, the crystals in each of the individual piles were optically active, and their specific rotations were equal in degree but opposite in sign.

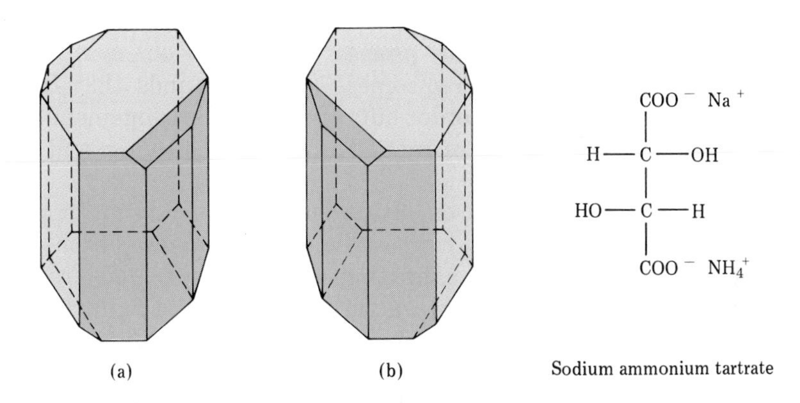

Figure 8.3. Crystals of sodium ammonium tartrate. The crystal in (a) is dextrorotatory, "right-handed," and the crystal in (b) is levorotatory, "left-handed." The drawings are taken from Pasteur's original sketches.

Pasteur was far ahead of his time. Although the structural theory of Kekulé had not yet been proposed, in explaining his results Pasteur spoke of the *molecules themselves,* saying: "It cannot be a subject of doubt that [in the *dextro* tartaric acid] there exists an asymmetric arrangement having a non-superposable image. It is not less certain that the atoms of the *levo* acid

[2]Louis Pasteur (1822–1895); b. Dôle, Jura, France; studied at Arbois, Besançon; professor, Dijon, Strasbourg (1849–1854), Lille (1854–1857), École Normale Supérieure (1857–1863).

possess precisely the inverse asymmetric arrangement." Pasteur's vision was extraordinary; it was not until 25 years later that the theories of van't Hoff and Le Bel confirmed his ideas regarding the asymmetric carbon atom.

Today, we would describe Pasteur's work by saying that he had discovered the phenomenon of **optical isomerism.** As noted earlier, isomers are different compounds having the same chemical formula; we have come across two types, constitutional isomers and cis–trans isomers. Constitutional isomers, such as butane and 2-methylpropane, are isomers because the atoms are connected to each other in a different order. Cis–trans isomers, such as *cis*-2-butene and *trans*-2-butene, have their atoms connected in the same order but have a different three-dimensional geometry.

```
                                   CH3
                                   |
CH3CH2CH2CH3        and        CH3CHCH3        Constitutional isomers

    Butane                   2-Methylpropane

  CH3      CH3                 CH3      H
     \    /                       \    /
      C=C           and            C=C          Cis–trans isomers
     /    \                       /    \
   H        H                   H        CH3

  cis-2-Butene                 trans-2-Butene
```

We must now be more precise. Cis–trans isomers are just special kinds of **stereoisomers.** Stereoisomers are compounds that have their atoms connected in the same order but differ in the arrangement of their atoms in space. Cis–trans isomers are one kind of stereoisomer, and Pasteur's optical isomers are another.

Optical isomers or **enantiomers** (Greek *enantio,* "opposite") are molecules that are *nonsuperimposable mirror images* of each other. The enantiomeric tartaric acid salts that Pasteur separated are identical in all respects except in their interaction with polarized light. They have identical physical properties such as melting points and boiling points, and identical spectroscopic properties. They are, however, related to each other as a right hand is to a left hand. Let's look further into the phenomenon of enantiomerism.

8.4 Enantiomers and the tetrahedral carbon

By 1874 a sufficient number of pieces were available to complete the puzzle of stereochemistry, but they had not yet been assembled into a coherent picture. Let's see what facts were known:

1. Kekulé's structural theory indicated that carbon was always tetravalent.

2. Only *one* isomer of the general formula CH_3X was known.
3. Only *one* isomer of the general formula CH_2XY was known.
4. *Two* isomers of the general formula CHXYZ were known. Pasteur's (+)- and (−)-tartaric acids were the first examples, but by 1874 twelve other pairs of enantiomers had been found, including (+)-lactic acid from muscle tissue and (−)-lactic acid from sour milk.

```
         H                 H
         |                 |
 CH3—C—COOH     (X)—C—(Z)
         |                 |
        OH                (Y)
```

Lactic acid; a molecule of general formula CHXYZ

Starting with these four facts, van't Hoff reasoned in the following way:

1. The fact that there is only one isomer for formula CH_3X indicates that all four valences on carbon are identical. There is only one known CH_3—(Cl), one known CH_3—(OH), one known CH_3—(COOH), and so on. If we imagine these molecules to be derived from methane by the replacement of one hydrogen atom with an X group, it does not matter which of the four methane hydrogens we replace. Replacement of any one of the four leads to the same CH_3—X product. All four hydrogens of methane are equivalent. What does this imply about the geometry of methane? Van't Hoff reasoned that there were only three possible geometries in which all methane hydrogens could be equivalent—planar geometry, pyramidal geometry, and tetrahedral geometry. (You can convince yourself of the correctness of van't Hoff's reasoning by building molecular models; Figure 8.4 will help.)

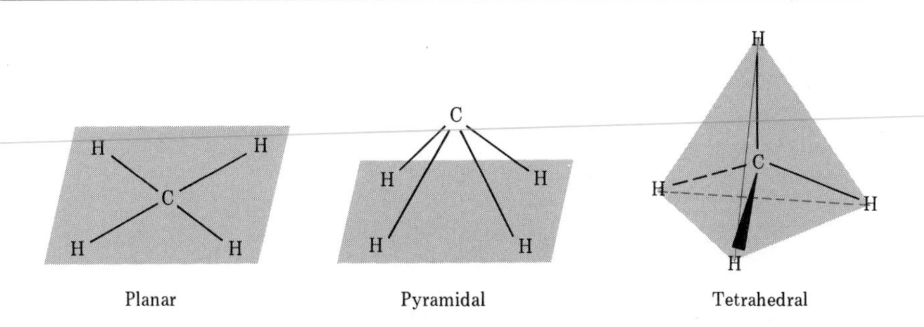

Figure 8.4. The three possible geometries of methane.

2. The fact that there is only one known isomer of formula CH_2XY indicates that the planar and pyramidal geometries are not correct. Therefore CH_4 must be tetrahedral. There is only one known

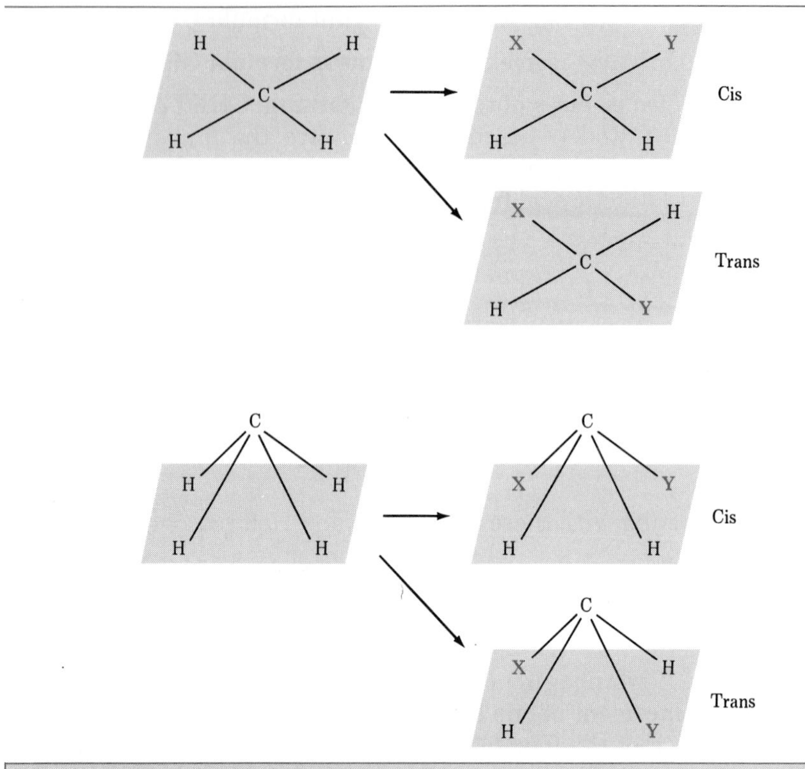

Figure 8.5. Hypothetical cis–trans isomers that would result from a planar or pyramidal methane molecule.

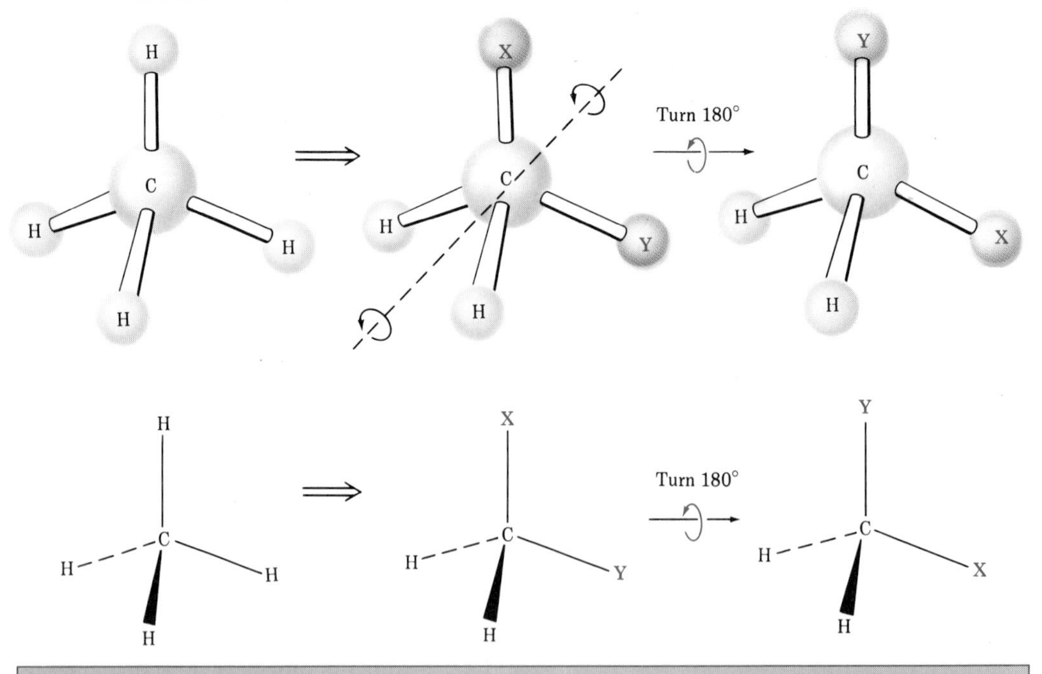

Figure 8.6. The tetrahedral methane molecule. A 180° rotation does not alter the geometry of the molecule.

(CH_3)—CH_2—(COOH), one known (CH_3)—CH_2—(Br), one known (CH_3CH_2)—CH_2—(OH), and so on. If these molecules were derived by replacement of two hydrogens from planar methane or pyramidal methane, two isomers of each would be possible, as Figure 8.5 shows.

Only tetrahedral methane allows for just one CH_2XY isomer. We therefore conclude with van't Hoff that carbon has tetrahedral geometry. Again, you can convince yourself that this reasoning is sound by studying molecular models (Figure 8.6).

The logic that led van't Hoff to postulate the tetrahedral carbon is an extraordinary piece of reasoning, but is still somewhat unsatisfying because its premises rest on *negative* evidence. Just because only one isomer with formula CH_2XY is known now doesn't mean that at some future time *two* isomers with that formula won't be found. Positive evidence would be much more convincing, and Pasteur's discovery of optical isomerism provides exactly this positive evidence. The tetrahedral geometry of carbon *predicts* the existence of two enantiomers of formula CHXYZ.

Let's look at the CH_3X, CH_2XY, and CHXYZ molecules to see why this should be so (Figure 8.7, p. 234).

On the left of Figure 8.7 are three molecules; on the right are their images reflected in a mirror. The CH_3X and CH_2XY molecules are identical with their mirror images. If we were to make molecular models of each molecule and of its mirror image, we would find that we could superimpose one on top of the other. The CHXYZ molecule is *not* identical with its mirror image. Try as we might, we cannot superimpose a model of the molecule on top of a model of its mirror image for the same reason that we cannot superimpose a left hand on a right hand. We might get two of the substituents superimposed, X and Y for example, but H and Z would be reversed. If the H and Z substituents were superimposed, X and Y would be reversed. Tetrahedral geometry predicts that a CHXYZ molecule can exist as a pair of enantiomers. Whenever a tetrahedral carbon is bonded to any four different substituents (one need not be H), optical activity can result.

Lactic acid, for example, exists as a pair of enantiomers because there are four different groups (H, OH, CH_3, COOH) attached to the central carbon atom:

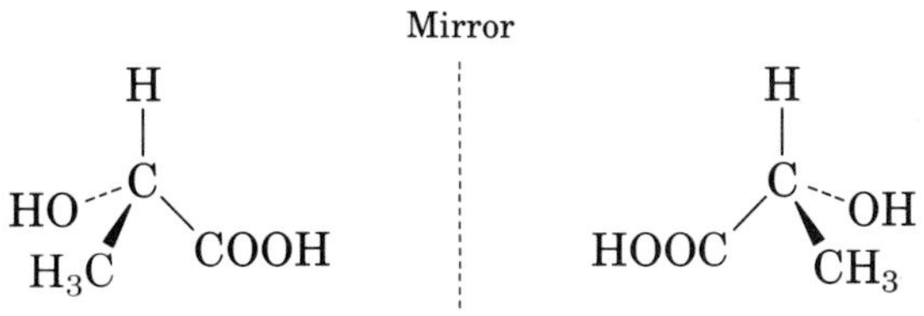

Mirror

H, HO, H_3C, C, COOH	H, C, OH, HOOC, CH_3
(+)-Lactic acid $[\alpha]_D = +3.82°$	(−)-Lactic acid $[\alpha]_D = -3.82°$

No matter how hard we try, we cannot superimpose a molecule of (+)-lactic acid on top of a molecule of (−)-lactic acid; the two molecules are simply not identical, as Figure 8.8 (p. 235) shows.

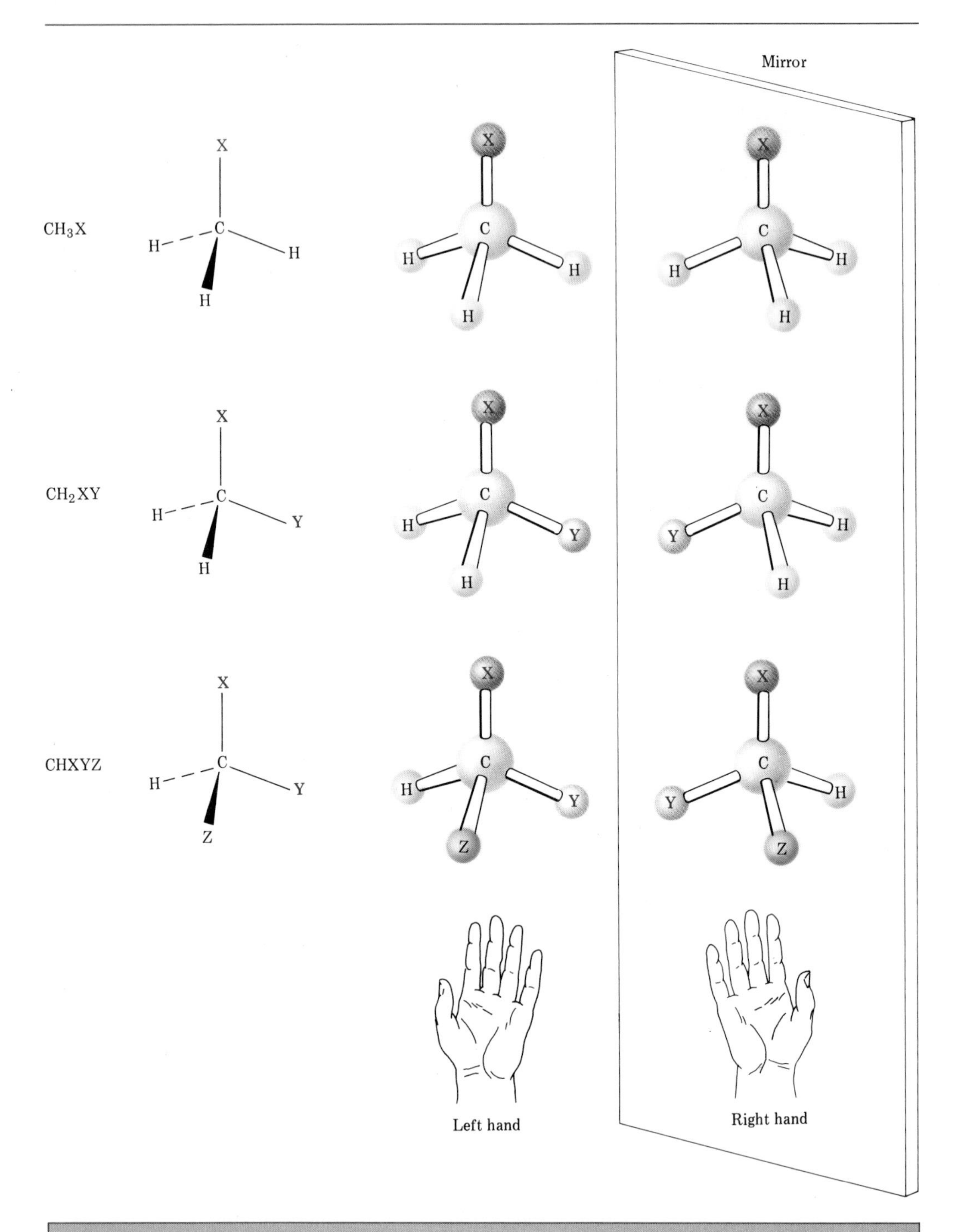

Figure 8.7. Tetrahedral carbons and their mirror images. A molecule is related to its mirror image in the same way that a right hand is related to a left hand.

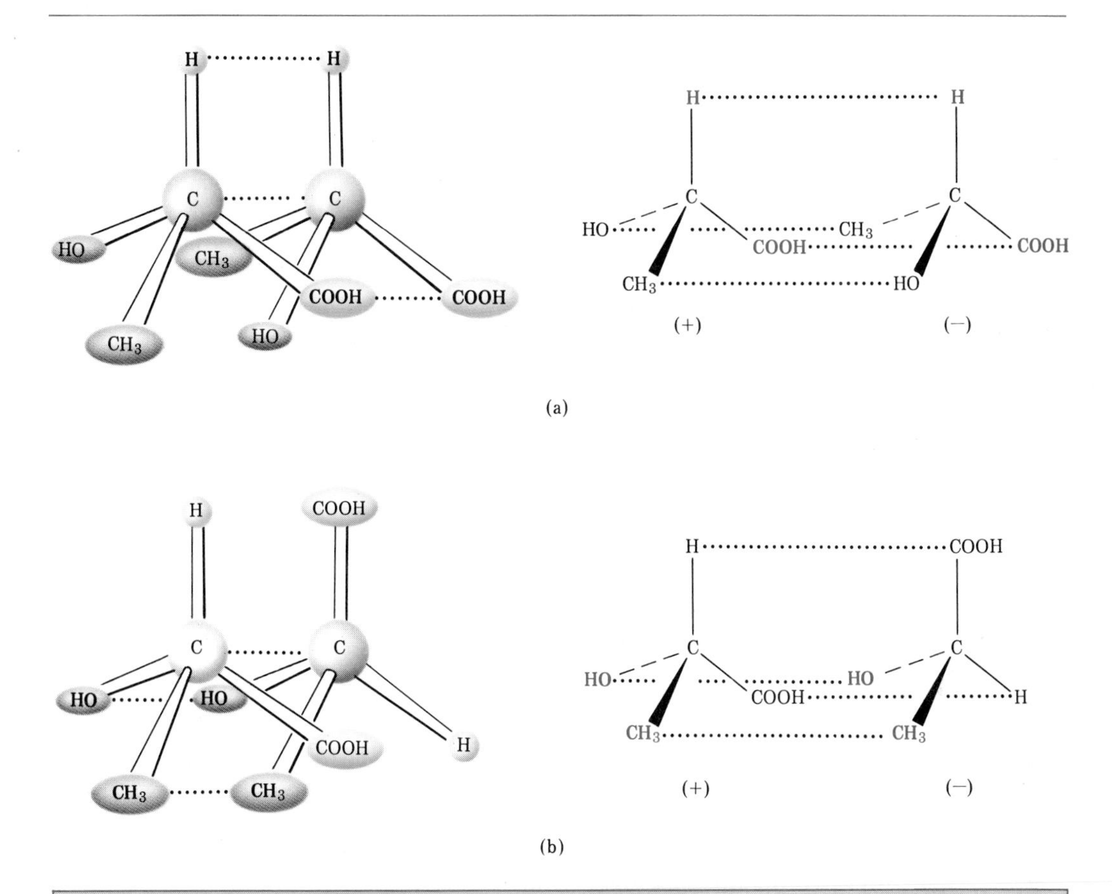

Figure 8.8. Attempts at superimposing (+)-lactic acid and (−)-lactic acid. (a) When —H and —COOH match up, —OH and —CH_3 do not. (b) When —OH and —CH_3 match up, —H and —COOH do not. Regardless of how the molecules are turned and oriented, they are not identical. (Note that, for clarity, the —OH, —CH_3, and —COOH substituent groups are shown as single ovals; the individual atoms are not shown.)

8.5 The nature of optical activity

Why do some compounds rotate light, whereas others do not? The answer is that *all* organic compounds do in fact rotate plane-polarized light, but we can only observe the rotation for certain compounds. To see what is meant by this, we need to look at the nature of optical activity at the molecular level. Constant electron movement causes all molecules to have tiny electric fields associated with them. When a beam of plane-polarized light encounters a molecule, its interaction with the electric field causes the light to rotate by a tiny amount. Since molecules in solution are randomly oriented, each successive encounter is slightly different and causes a different amount and direction of rotation. By the time the light leaves the sample tube it has been acted upon by a very large number of molecules, each of which has made its own small contribution to the net observed rotation. Now, for molecules that are symmetrical, the chances are high that, for every encounter of light with a given molecular orientation, there will also be an encounter with a mirror-image molecular orientation. The two encounters cancel each other out, leading to zero net rotation. For a single enantiomer of an optically active molecule, however, there can be no canceling effect because, *by definition,* no one molecule in solution is the mirror image of another.

In summary, when we observe optical rotation by a sample, we observe the net effect of millions of encounters between the light beam and individual molecules. Symmetrical molecules show no net rotation because the cumulative effects of all the individual rotations cancel to zero. The rotations caused by unsymmetrical molecules, however, cannot cancel to zero, and we therefore observe optical activity.

8.6 Chirality

The terms *symmetrical* and *unsymmetrical* have been used rather loosely so far; what exactly is meant by these terms? How can we predict whether a certain compound is or is not optically active? *A compound is optically active if it is not superimposable on its mirror image.* Such compounds are said to be **chiral** (pronounced ký-ral, from the Greek *cheir,* "hand"). We cannot take a chiral molecule and its mirror image (enantiomer) and place them on top of each other so that all atoms coincide.

A compound cannot be chiral if it contains a plane of symmetry. A **plane of symmetry** is an imaginary plane that bisects an object (or molecule) in such a way that one half of the object is an exact mirror image of the other half. For example, an object such as a flask has a plane of symmetry; if a plane were to cut the flask in half, one half would be an exact mirror image of the second half. An object such as a hand, however, does not have a plane of symmetry. One "half" of a hand is not a mirror image of the other half (Figure 8.9).

Molecules that have planes of symmetry *must* be superimposable on their mirror images and hence must be nonchiral or **achiral.** Thus, hydroxyacetic acid contains a plane of symmetry and is achiral, whereas lactic acid (2-hydroxypropanoic acid) has no plane of symmetry and is chiral (Figure 8.10).

The most common (although not the only) cause of chirality in organic molecules is the presence of a carbon that is bonded to four different groups. Such carbons are referred to as **chiral centers,** and we have seen an example of this kind of chirality in lactic acid. Detecting chiral centers in complex molecules can be difficult, because it is not always immediately apparent that four different groups are bonded to a given carbon. The differences do not necessarily appear right next to the chiral center. For

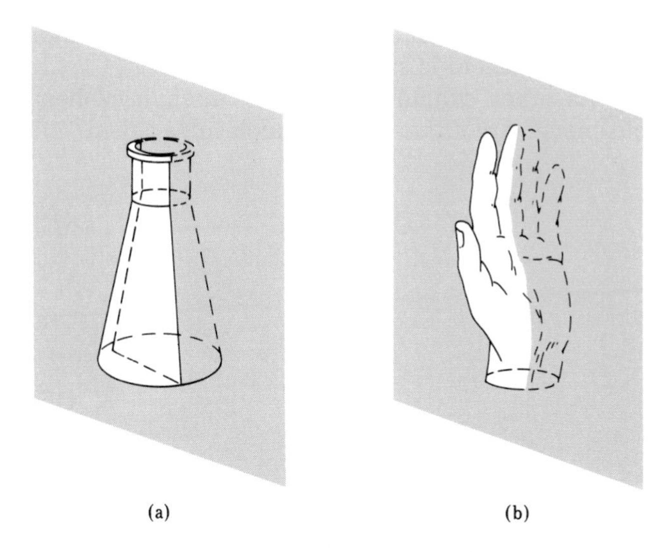

Figure 8.9. Symmetry planes. Objects like the flask (a) have planes of symmetry passing through them, making right and left halves mirror images. Objects like the hand (b) do not have symmetry planes; the right half is not a mirror image of the left half.

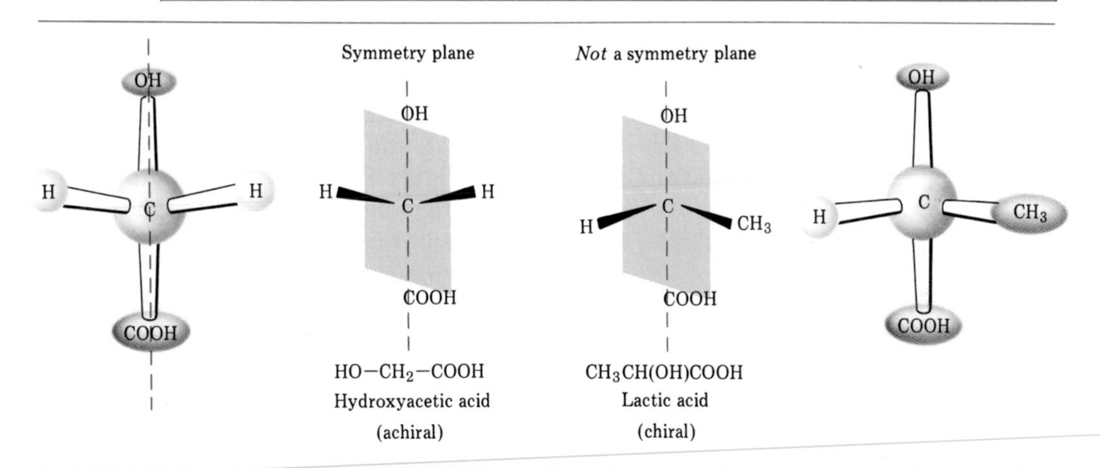

Figure 8.10. The achiral hydroxyacetic acid molecule versus the chiral lactic acid molecule. Note that hydroxyacetic acid has a plane of symmetry that makes one side of the molecule a mirror image of the other side; there is no plane of symmetry in lactic acid.

example, 5-bromodecane is a chiral molecule because four different groups are bonded to C5, the chiral center (marked by an asterisk):

$$CH_3CH_2CH_2CH_2CH_2\overset{\displaystyle Br}{\underset{\displaystyle H}{\overset{|}{\underset{|*}{C}}}}CH_2CH_2CH_2CH_3$$

5-Bromodecane (chiral)

Substituents on carbon 5
$—H$
$—Br$
$—CH_2CH_2CH_2CH_3$ (butyl)
$—CH_2CH_2CH_2CH_2CH_3$ (pentyl)

A butyl substituent is similar to a pentyl substituent, but it is not identical. The difference is not apparent until four carbon atoms away from the chiral center, but there is still a difference.

As other examples, consider methylcyclohexane and 2-methylcyclohexanone. Are either of these molecules chiral?

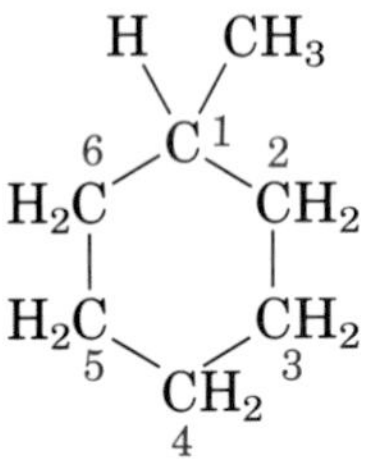

Methylcyclohexane (achiral) 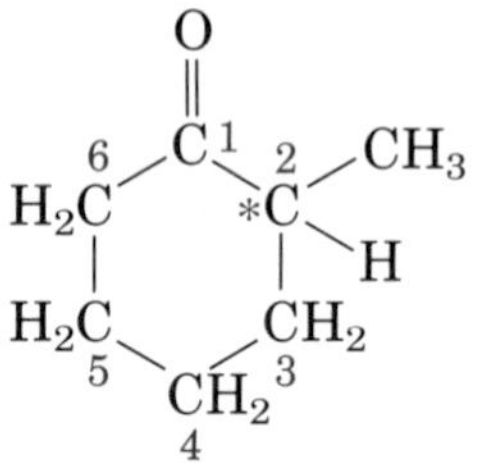   2-Methylcyclohexanone (chiral)

Methylcyclohexane is achiral because no carbon atom in the molecule is bonded to four different groups. We can eliminate all CH_2 groups and the CH_3 group from consideration, but what about C1 on the ring? It is bonded to a CH_3 group, an H atom, and to C2 and C6 of the ring. However, C2 and C6 are equivalent, as are C3 and C5. Thus the C6–C5–C4 "substituent" is equivalent to the C2–C3–C4 "substituent," and methylcyclohexane is therefore achiral.

The situation is not the same for 2-methylcyclohexanone, however. This molecule is chiral, since C2 is bonded to four different groups: a CH_3 group, an H atom, a —COCH— ring bond (C1), and a —CH_2CH_2— ring bond (C3). Several more examples appear in Figure 8.11. Check for yourself that the labeled centers are indeed chiral. (Remember: CH_2, CH_3, and C=C centers cannot be chiral, since they have at least two identical bonds.)

PROBLEM

8.2 Which of these compounds are chiral, and which are achiral? Build molecular models if you need help seeing spatial relationships.

(a) 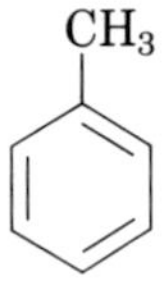 Toluene

(b) 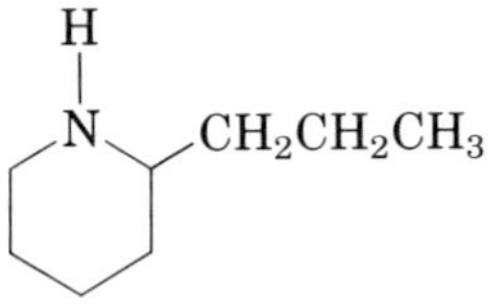 Coniine

(c) 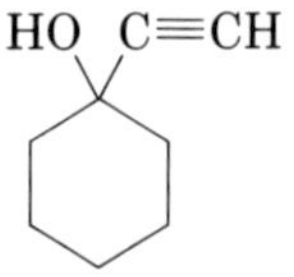 1-Ethynylcyclohexanol

(d) $HOCH_2CH(NH_2)COOH$

Serine

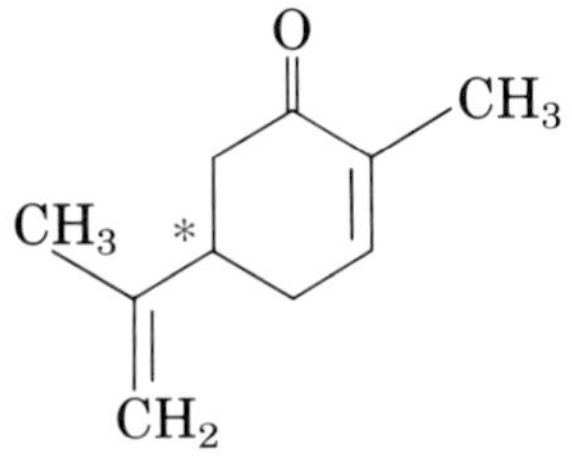

O
CH3
CH3 *
CH2

Carvone (spearmint oil)

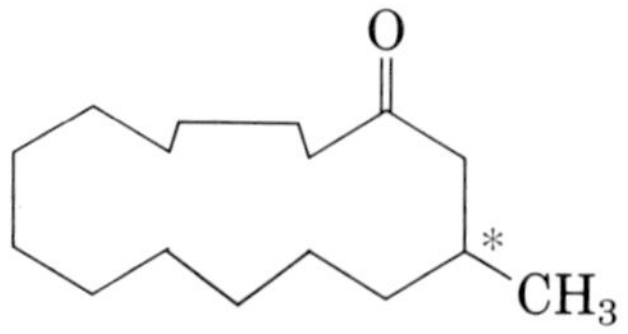

O
*
CH3

Muscone (musk oil)

H3C
CH3
CH2
*
*
*
CH3
O

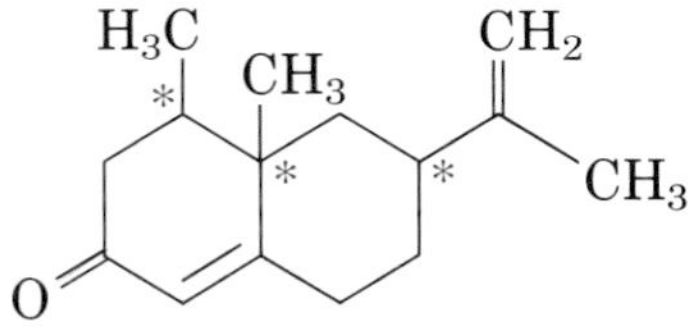

Nootkatone (grapefruit oil)

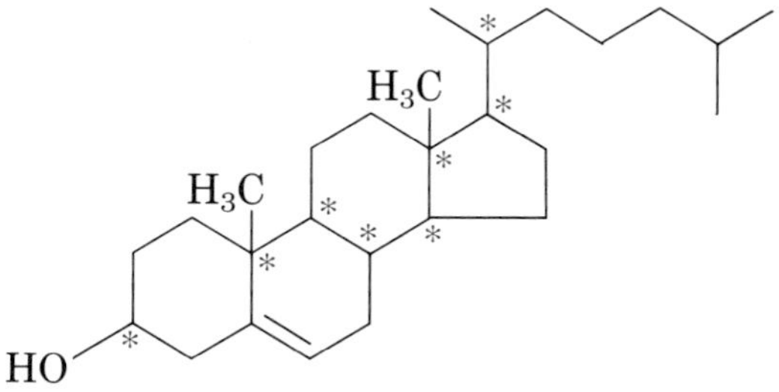

*
H3C
*
*
H3C
*
*
*
*
HO
*

Cholesterol

Figure 8.11. Some chiral molecules. Chiral carbon atoms are marked by asterisks. Both nootkatone and cholesterol have several chiral centers.

(e)
N
CH3
N

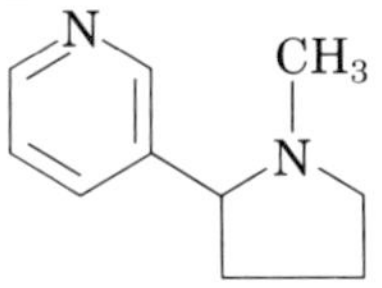

Nicotine

(f)
NH2

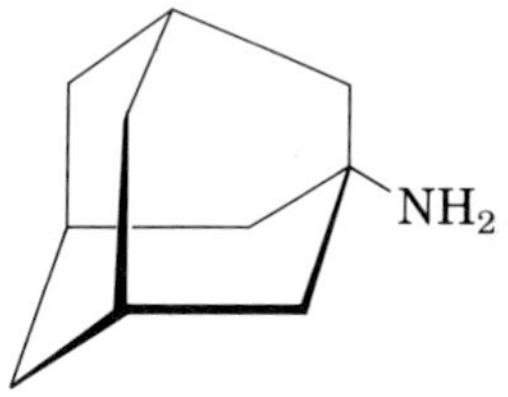

Aminoadamantane
(antiviral agent)

PROBLEM

8.3 Place asterisks at all of the chiral centers in molecules (a)–(d):

(a)
O
O
HN
NH
O

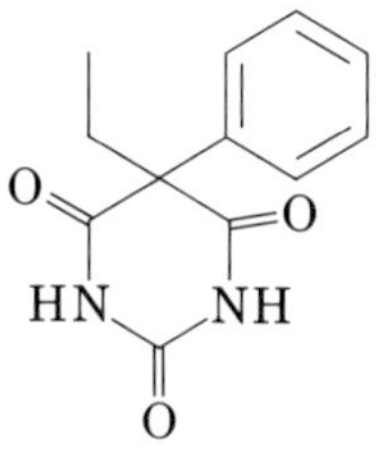

Phenobarbital

(b) H3C CH3
CH3
O

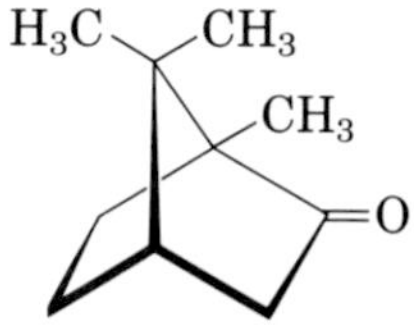

Camphor

(c) HO CH_3 H_3C CH_3

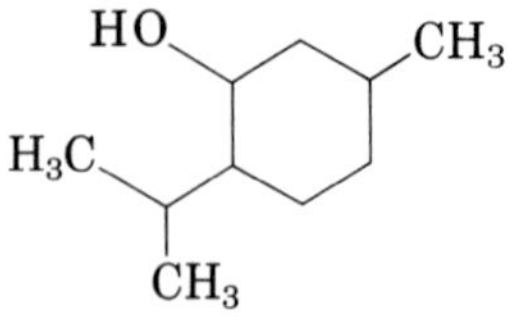

Menthol

(d) CH_3O NCH_3

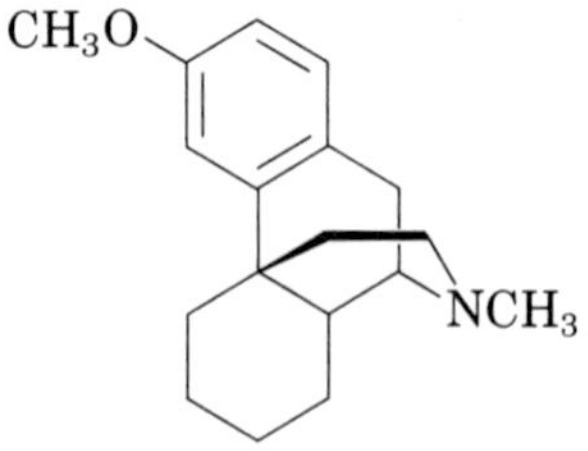

Dextromethorphan
(a cough suppressant)

8.7 Sequence rules for specification of configuration

Although drawings can provide a pictorial representation of stereochemistry, they are difficult to translate into words. Thus, a purely verbal method of specifying the stereochemical configuration at a chiral center is also necessary. The method used employs a system that is familiar to us—the Cahn–Ingold–Prelog sequence rules. We used these rules in connection with the specification of alkene geometry (Z versus E) in Section 5.6, but historically the Cahn–Ingold–Prelog sequence rules were proposed as a method of specifying the configuration of chiral carbon atoms. Let's briefly review the sequence rules presented in Section 5.6 and see how they can be used to specify the configuration of a chiral center:

1. Rank the atoms directly attached to the chiral center in order of decreasing atomic number. The group with highest atomic number is ranked first; the group with lowest atomic number is ranked fourth.
2. If a decision about priority cannot be reached by applying rule 1, work outward to the first point of difference.
3. Multiply bonded atoms are considered as if they were an equivalent number of singly bonded atoms. For example, a —CHO substituent is equivalent to $—CH(OC)_2$.

Following these sequence rules, we can assign priorities to the four substituent groups attached to a chiral carbon. To describe uniquely the stereochemical configuration around the chiral carbon, we mentally orient the molecule so that the group of lowest priority is pointing directly back, away from us. We then look at the three remaining substituents. If the direction of travel from highest to second-highest to third-highest substituent ($1 \rightarrow 2 \rightarrow 3$) is *clockwise,* we say that the chiral center has the R configuration (Latin *rectus,* "right"). If the direction of travel from highest to second-highest to third-highest substituent ($1 \rightarrow 2 \rightarrow 3$) is *counterclockwise,* the chiral center has the S configuration (Latin *sinister,* "left"). Think of a car's steering wheel when making a right (clockwise) or left (counterclockwise) turn.

Let's look, for example, at (+)-lactic acid:

$$\mathrm{H{-}C(OH)(CH_3)(COOH)}$$

(+)-Lactic acid

Priorities	
4 —H	(low)
3 —CH_3	
2 —C(=O)—OH	
1 —OH	(high)

Our first step is to assign priorities to the four substituents. Sequence rule 1 tells us that —OH is first priority (1) and —H is last priority (4), but does not allow us to distinguish between —CH_3 and —COOH, since both groups have carbon as their first atom. Sequence rule 2, however, tells us that —COOH is higher priority than —CH_3, since oxygen outranks hydrogen. We next orient the molecule so that the lowest-priority group (the H atom) is oriented toward the rear, away from the observer. When seen from this perspective, the three remaining substituents appear to radiate from the chiral center like the spokes on a steering wheel, as shown in Figure 8.12.

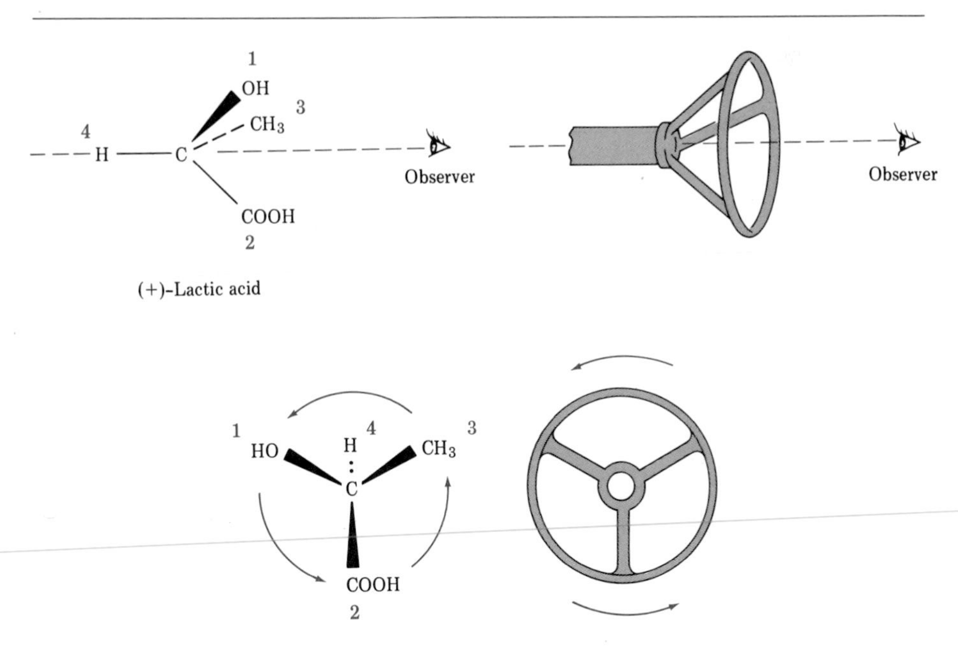

S configuration (left turn of steering wheel)

Figure 8.12. Assignment of configuration to *S* lactic acid.

Since the direction of travel from 1 (OH) to 2 (COOH) to 3 (CH_3) is counterclockwise, we assign the *S* configuration to (+)-lactic acid. Applying the same procedure to (−)-lactic acid should (and does) lead to the opposite assignment, as shown in Figure 8.13.

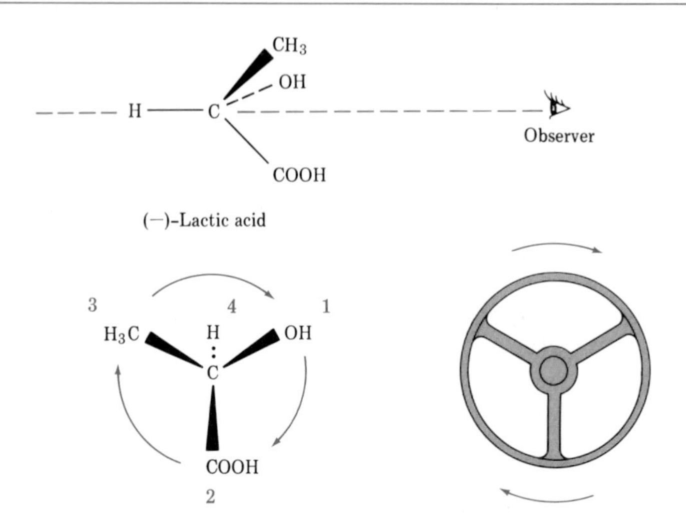

Figure 8.13. Assignment of configuration to *R* lactic acid.

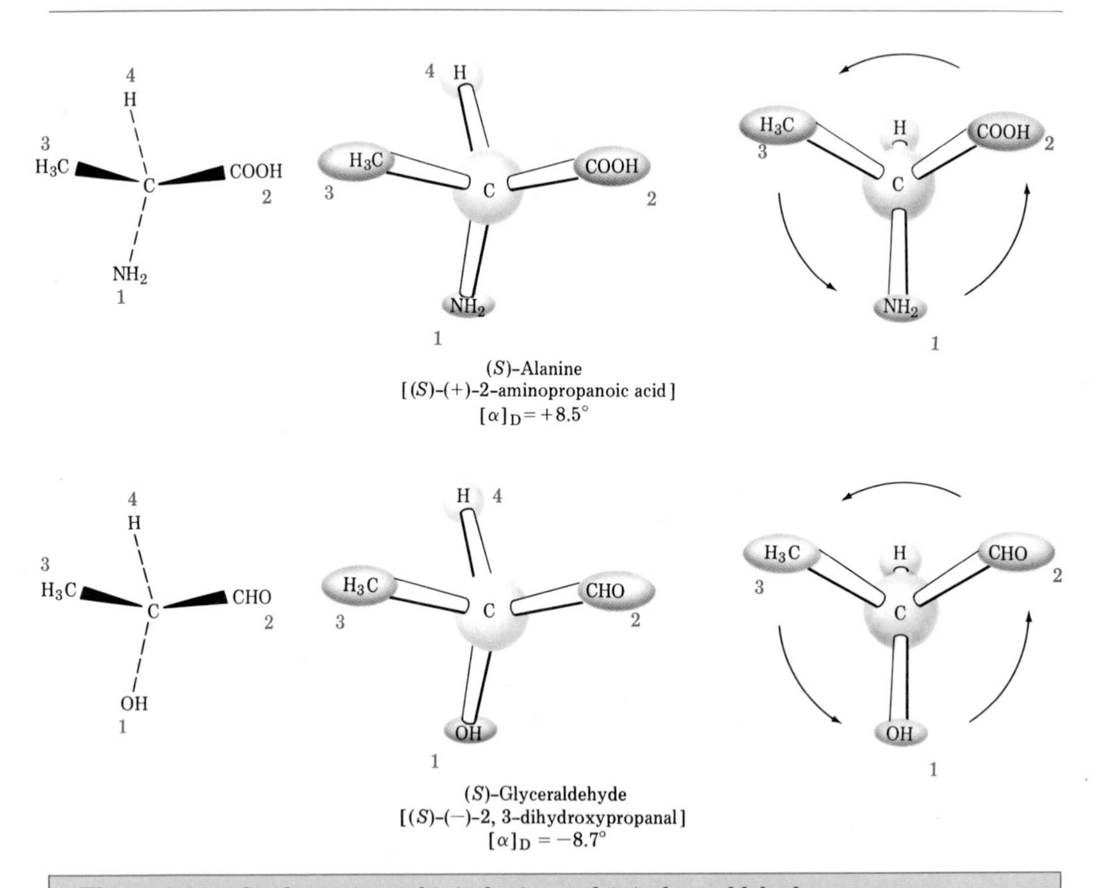

Figure 8.14. Configurations of (+)-alanine and (−)-glyceraldehyde.

Further examples are provided by naturally occurring (+)-alanine and (−)-glyceraldehyde, which have the *S* configurations shown in Figure 8.14.

Note that in the compounds shown in Figure 8.14 the sign of optical rotation is not related to the *R,S* designation. (*S*)-Alanine happens to be dextrorotatory (+), and (*S*)-glyceraldehyde happens to be levorotatory (−). There is no known method of predicting the direction or magnitude of rotation of a given enantiomer or of correlating the direction of rotation with the *R,S* configuration.

One further point bears mentioning—the matter of **absolute configuration.** We have satisfactory conventions for describing and differentiating the *relative* stereochemistry of pairs of enantiomers, but how do we know that our assignments of *R,S* configuration are correct in an *absolute* sense? How do we know for certain that it is the *R* enantiomer of lactic acid that is dextrorotatory, rather than the *S* enantiomer? This difficult question was not solved until 1951 when J. M. Bijvoet of the University of Utrecht reported an X-ray spectroscopic method for determining the absolute spatial arrangement of atoms in a molecule. Bijvoet carried out his work on the absolute stereochemistry of salts of tartaric acid, and, on the basis of his results we can say with certainty that our conventions are correct.

PROBLEM

8.4 Assign Cahn–Ingold–Prelog priorities to these sets of substituents:

(a) $—H$, $—Br$, $—CH_2CH_3$, $—CH_2CH_2OH$

(b) $—CO_2H$, $—CO_2CH_3$, $—CH_2OH$, $—OH$

(c) $—CN$, $—CH_2NH_2$, $—CH_2NHCH_3$, $—NH_2$

(d) $—Br$, $—CH_2Br$, $—Cl$, $—CH_2Cl$

PROBLEM

8.5 Assign *R,S* configurations to these molecules:

(a) CH_3 (top), Br (wedge, left), COOH (wedge, right), H (bottom) on C

(b) OH (top), H (wedge, left), COOH (wedge, right), CH_3 (bottom) on C

(c) NH_2 (top), H (wedge, left), CH_3 (wedge, right), CN (bottom) on C

(d) CN (top), H (wedge, left), CH_3 (wedge, right), NH_2 (bottom) on C

8.8 Diastereomers

Molecules such as lactic acid and alanine are relatively simple to deal with, since each has only one chiral center and can exist in only two enantiomeric configurations. The situation becomes more complex, however, for molecules having more than one chiral center.

Let's take the essential amino acid threonine (2-amino-3-hydroxybutanoic acid) as an example. Threonine has two chiral centers (C2 and C3) and there are four possible stereoisomers, which we can draw as shown in Figure 8.15. (Check for yourself that the *R*,*S* configurations are correct as shown.)

Inspection of the four threonine stereoisomers reveals that they can be classified into two pairs of enantiomers. The 2*R*,3*R* stereoisomer is the mirror image of 2*S*,3*S*, and the 2*R*,3*S* stereoisomer is the mirror image of 2*S*,3*R*. But what is the relationship between any two non-mirror-image configurations? For example, the 2*R*,3*R* compound and the 2*R*,3*S* compound are stereoisomers, yet they are not superimposable and they are not en-

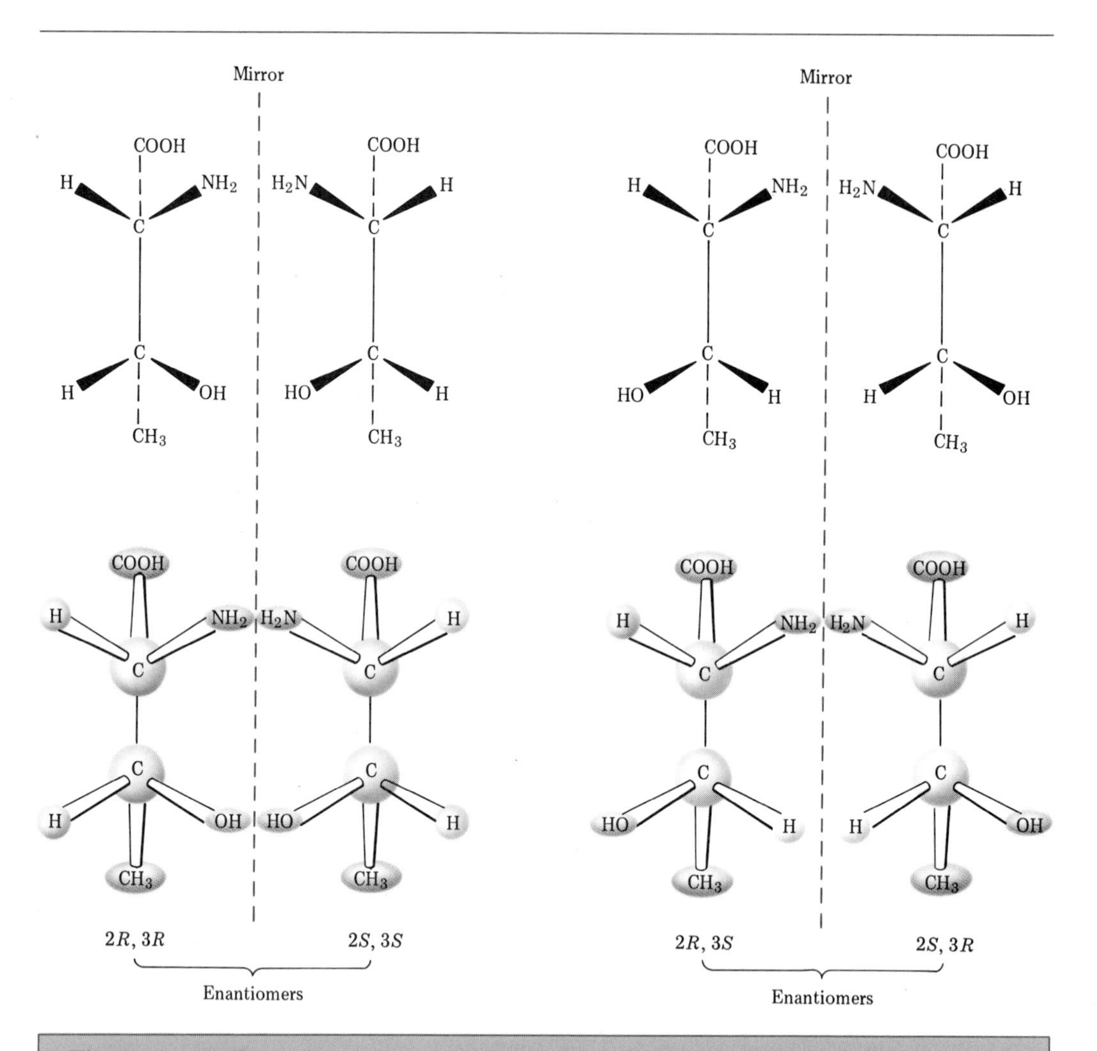

Figure 8.15. Four stereoisomers of threonine, 2-amino-3-hydroxybutanoic acid. Note that the —COOH, —CH_3, and —NH_2 substituent groups on the two chiral carbon atoms are depicted as single ovals for clarity; the individual atoms in these substituent groups are not shown.

antiomers. To describe such a relationship, we need a new term. **Diastereomers** are stereoisomers that are not mirror images of each other. Since we used the right-hand/left-hand analogy to describe the relationship between two enantiomers, we might extend the analogy further by saying that diastereomers have a hand–foot relationship. Hands and feet look very similar but they are not identical and are not mirror images. The same is true of diastereomers; they are similar but not identical. A full description of the four threonine stereoisomers is given in Table 8.2. Enantiomers must have opposite (mirror image) configurations around *all* chiral centers; diastereomers have the same configurations at one or more, but opposite configurations at the other, chiral centers.

TABLE 8.2 **Relationships between four stereoisomeric threonines**

Stereoisomer	*Enantiomeric with*	*Diastereomeric with*
2*R*,3*R*	2*S*,3*S*	2*R*,3*S*, 2*S*,3*R*
2*S*,3*S*	2*R*,3*R*	2*R*,3*S*, 2*S*,3*R*
2*R*,3*S*	2*S*,3*R*	2*R*,3*R*, 2*S*,3*S*
2*S*,3*R*	2*R*,3*S*	2*R*,3*R*, 2*S*,3*S*

Of the four possible stereoisomers of threonine, only one, the 2*S*,3*R* isomer, $[\alpha]_D = -28.3°$, occurs naturally in plants and animals and is an essential human nutrient. Most biologically important molecules are chiral, and usually only a single stereoisomer is found in nature.

PROBLEM

8.6 Assign *R*,*S* configurations to these molecules. Which are enantiomers and which are diastereomers?

(a) Br; H, C, CH_3; C; H, OH; CH_3

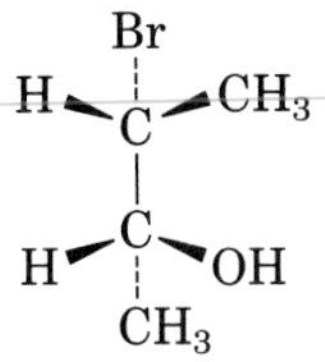

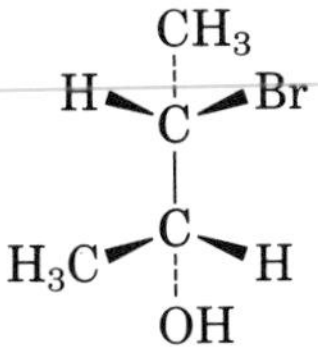

(b) CH_3; H, C, Br; C; H_3C, H; OH

(c) CH_3; Br, C, H; C; H, CH_3; OH

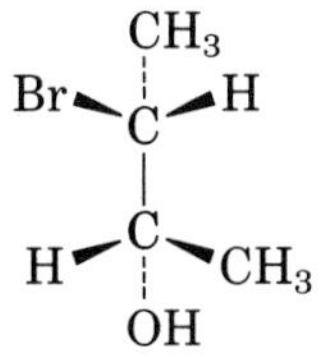

(d) H; Br, C, CH_3; C; H_3C, OH; H

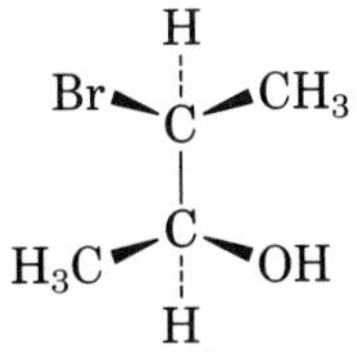

PROBLEM

8.7 Chloramphenicol is a powerful antibiotic isolated in 1949 from the *Streptomyces venezuelae* bacterium. It is active against a broad spectrum of bacterial infections and is particularly valuable against typhoid fever. Assign *R*,*S* configurations to the chiral centers in chloramphenicol.

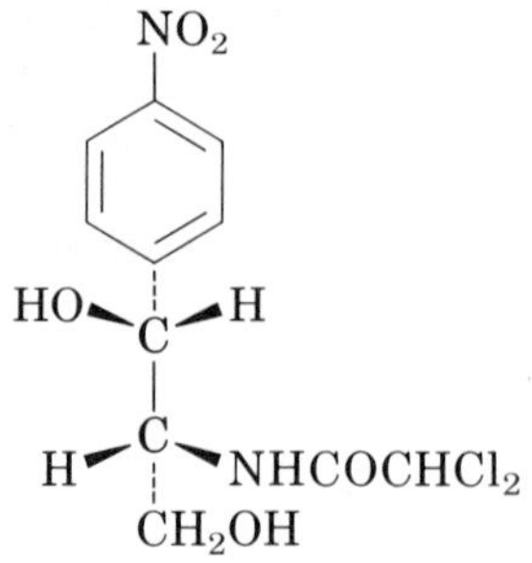

NO_2

HO C H

H C $NHCOCHCl_2$

CH_2OH

Chloramphenicol
$[\alpha]_D = +18.6°$

8.9 Meso compounds

Let's look at one more example of a compound with two chiral centers—tartaric acid. We are already acquainted with tartaric acid for its role in Pasteur's discovery of optical activity, and we can now draw the four possible stereoisomers:

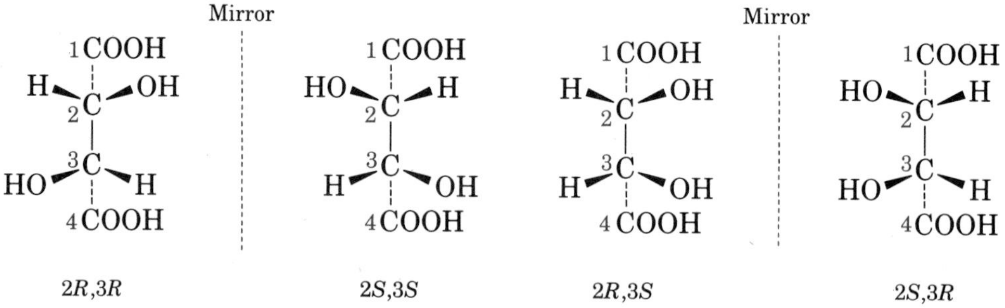

Mirror Mirror

1COOH, H 2C OH, HO 3C H, 4COOH — 2*R*,3*R*

1COOH, HO 2C H, H 3C OH, 4COOH — 2*S*,3*S*

1COOH, H 2C OH, H 3C OH, 4COOH — 2*R*,3*S*

1COOH, HO 2C H, HO 3C H, 4COOH — 2*S*,3*R*

The mirror-image 2*R*,3*R* and 2*S*,3*S* structures are not superimposable and therefore represent an enantiomeric pair. A close look, however, reveals that the 2*R*,3*S* and 2*S*,3*R* structures are *identical*. We can see this readily by rotating one structure by 180°:

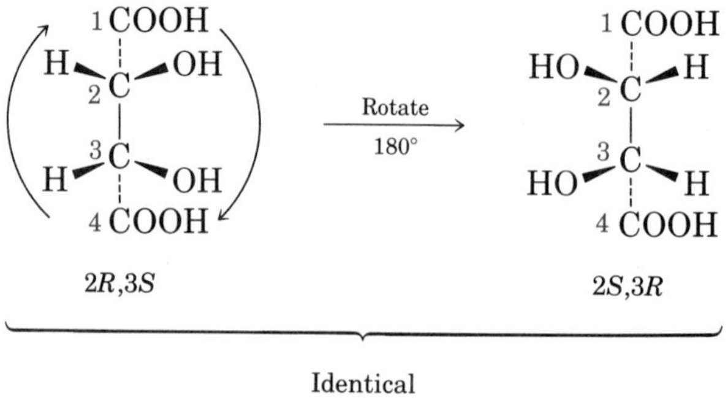

1COOH, H 2C OH, H 3C OH, 4COOH — 2*R*,3*S*

Rotate 180° →

1COOH, HO 2C H, HO 3C H, 4COOH — 2*S*,3*R*

Identical

The identity of the $2R,3S$ and $2S,3R$ structures is a consequence of the fact that there is a plane of symmetry through the molecule. The plane of symmetry cuts through the C2—C3 bond, making one half of the molecule a mirror image of the other half (Figure 8.16).

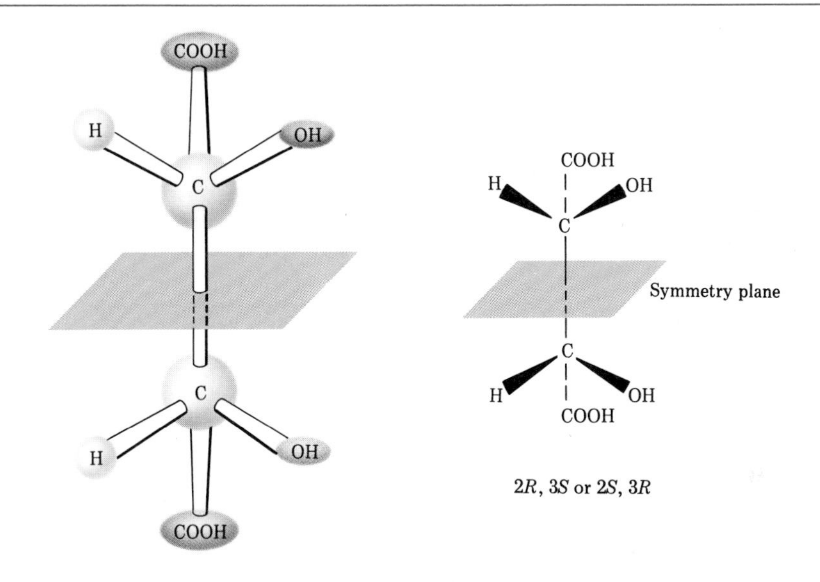

Figure 8.16. A symmetry plane through the C2—C3 bond of *meso* tartaric acid.

Because of the plane of symmetry, the molecule represented by the structures in Figure 8.16 must be achiral, despite the fact that it has two chiral centers. Compounds that are superimposable on their mirror images by virtue of a symmetry plane, yet contain chiral centers, are called **meso compounds.** Thus, tartaric acid exists in three stereoisomeric configurations: two enantiomers, and one meso form.

PROBLEM

8.8 Which, if any, of these structures represent meso compounds?

(a) H OH OH H

(b) H OH OH H

(c) CH_3 H

(d) H Br C CH_3 C H_3C H Br

8.10 Molecules with more than two chiral centers

We have seen that a single chiral center gives rise to two stereoisomers (one pair of enantiomers), and two chiral centers in a molecule give rise to a *maximum* of four stereoisomers (two pairs of enantiomers). Calculations show that a molecule with n chiral centers can give rise to a maximum of 2^n stereoisomers (2^{n-1} pairs of enantiomers). For example, cholesterol contains eight chiral centers. Thus $2^8 = 256$ stereoisomers (128 enantiomeric pairs) are possible. Only *one,* however, is produced in nature.

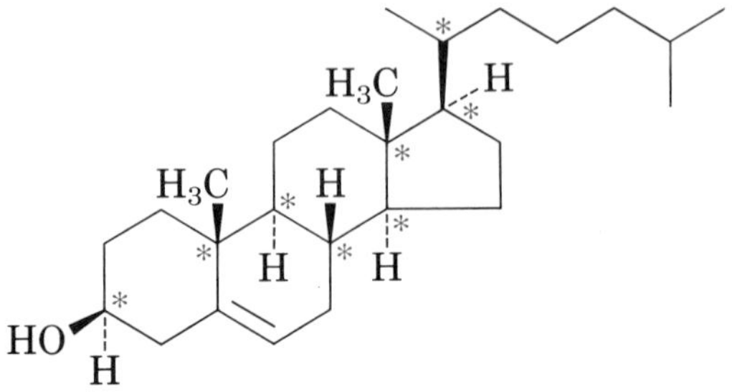

H_3C H H_3C H H H HO H

Cholesterol
(eight chiral centers)

8.11 Racemic mixtures

To conclude our discussion of stereoisomerism, let's return to Pasteur's pioneering work. Pasteur took an optically inactive form of a tartaric acid salt and found that he could crystallize two optically active forms from it. These two optically active forms were the 2*R*,3*R* and 2*S*,3*S* configurations just discussed—but what was the optically inactive form he started with? It could not have been *meso*-tartaric acid, since *meso*-tartaric acid is a different chemical compound from the two chiral enantiomers, and cannot interconvert with them unless chemical bonds are broken and re-formed—a most unlikely occurrence. The answer is that Pasteur started with a 50:50 mixture of the two chiral enantiomers of tartaric acid. Such a mixture is called a **racemic** (ray-seé-mic) **mixture** or **racemate,** and is often denoted by the symbol (±). Racemic mixtures *must* show zero optical rotation because equal amounts of both (+) and (−) forms are present, and the rotation from one enantiomer exactly cancels the rotation from the other enantiomer. Through good fortune, Pasteur was able to **resolve** (±)-tartaric acid into its (−) and (+) enantiomers. Unfortunately, the method he used (fractional crystallization) does not work for most racemic mixtures, and other techniques are required. The most often used method of resolution will be covered in Section 28.5.

8.12 Physical properties of stereoisomers

We have seen that seemingly simple compounds such as tartaric acid can exist in several different configurations. The question therefore arises whether the different stereoisomeric configurations of a compound give rise

TABLE 8.3 Some properties of the stereoisomers of tartaric acid

Stereoisomer	*Melting point (°C)*	*$[\alpha]_D$ (degrees)*	*Density (g/cm^3)*	*Solubility at 20°C (g/100 mL H_2O)*
(+)	168–170	+12	1.7598	139.0
(−)	168–170	−12	1.7598	139.0
meso	146–148	0	1.6660	125.0
(±)	206	0	1.7880	20.6

to different physical properties. The answer is yes, they can. Let's take the stereoisomeric tartaric acids as an example. We know that there are three different stereoisomers of tartaric acid: (+)-tartaric, (−)-tartaric, and *meso*-tartaric. To these we must add the racemic mixture, (±)-tartaric acid. Some physical properties of these four modifications are listed in Table 8.3.

As Table 8.3 illustrates, enantiomers are identical in their physical properties except for their interaction with polarized light. The (+)- and (−)-tartaric acids have identical melting points, solubilities, and densities, as they must since they are mirror images. They differ only in the sign of their rotation of plane-polarized light. The meso configuration, by contrast, is diastereomeric with the (+) and (−) forms. As such, it has no mirror-image relationship to (+)- and (−)-tartaric acids; it is a different compound altogether and therefore has different physical properties.

The racemic mixture is different still. Racemates comprise what are known as **eutectic mixtures,** and they act as though they were pure compounds. The theory of eutectic mixtures is complex and well beyond our present scope. As may be gathered from the table, however, the properties of racemic tartaric acid differ from those of the two pure enantiomers and from those of the meso form. Thus it is clear that different forms of tartaric acid differ significantly in their physical properties. As we shall see, they also differ in their *chemical* properties.

8.13 Fischer projections

When learning to visualize chiral molecules, it is best to begin by building molecular models. As more experience is gained it becomes easier to draw pictures and work with mental images. To do this successfully, however, we need a standard method for depicting the three-dimensional arrangement of atoms (the **configuration**) on a flat page, and **Fischer projections** are best used for this. By convention, a tetrahedral carbon atom is represented by two crossed lines. The horizontal lines represent bonds coming out of the page, and the vertical lines represent bonds going into the page.

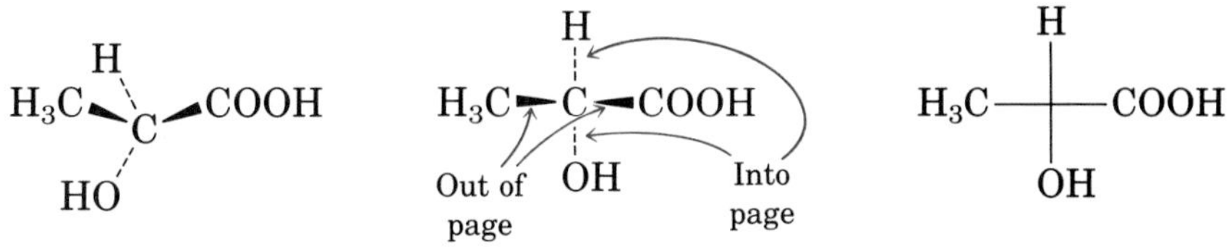

H
H_3C C COOH
HO

(+)-Lactic acid

H
H_3C C COOH
Out of page OH Into page

H
H_3C COOH
OH

Fischer projection

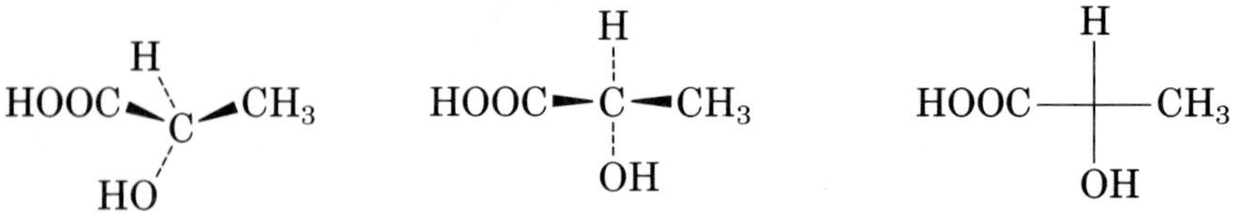

(−)-Lactic acid

Fischer projection

Many different Fischer projections can be used to draw a given molecule, and it is often necessary to compare two different projections to see if they represent the same or different enantiomers. To test for identity, Fischer projections may be moved around on the paper, but care must be taken not to change the meaning of the projection inadvertently. Only two kinds of motion are allowed:

1. We can rotate a Fischer projection on the page by 180°, *but not by 90° or 270°*. That is:

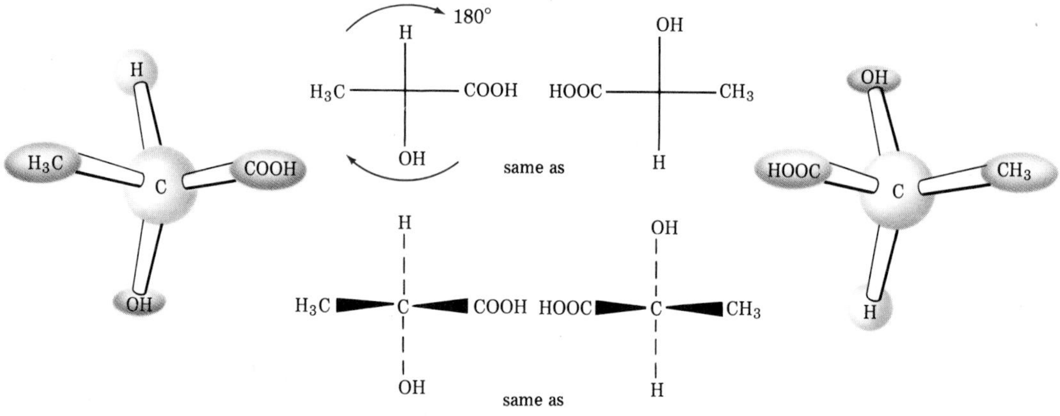

180°

same as

same as

but:

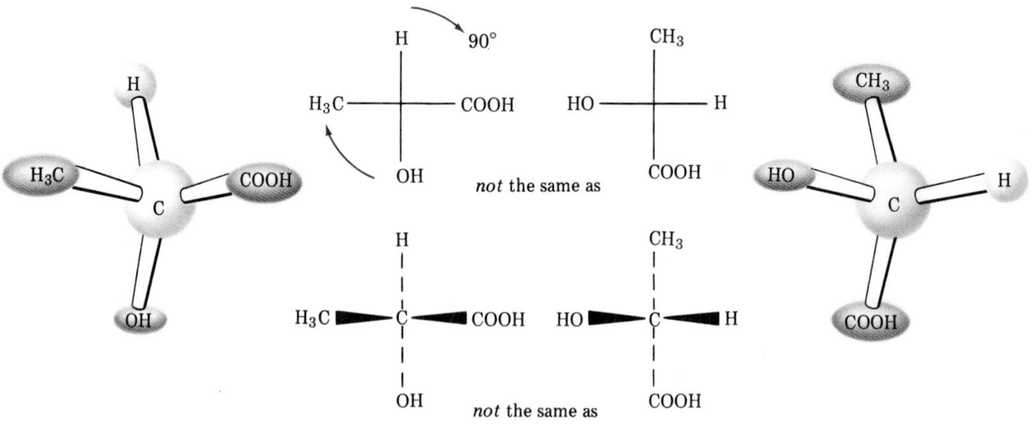

90°

not the same as

not the same as

2. We can hold any one group steady and rotate the other three in either a clockwise or a counterclockwise direction:

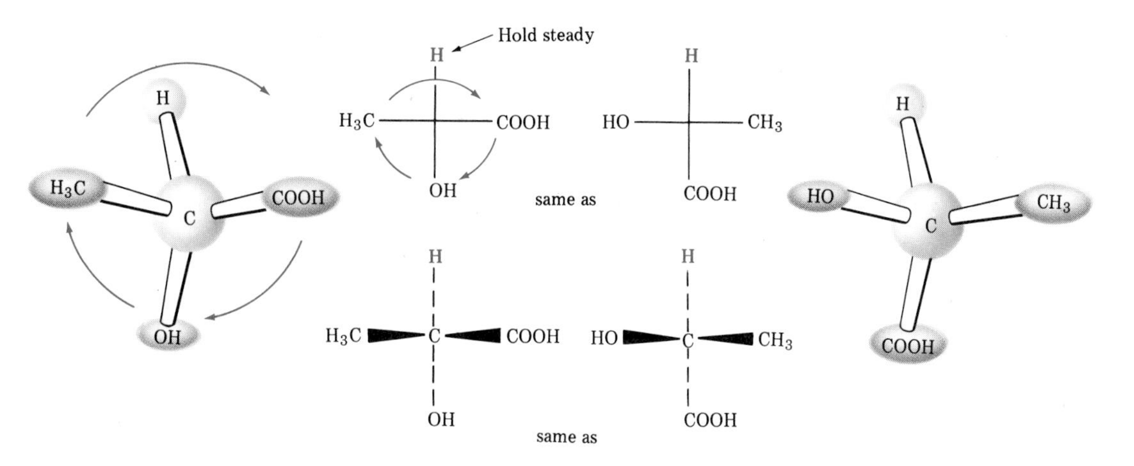

These are the only kinds of motion allowed. Moving a Fischer projection in any other way inverts its meaning. For example, if a Fischer projection of (+)-lactic acid were to be turned by 90°, a projection of (−)-lactic acid would result (Figure 8.17).

H
HO—CH$_3$ same as HO—C—CH$_3$
COOH

(+)-Lactic acid

↓ 90° rotation

OH
HOOC—H same as HOOC—C—H
CH$_3$

(−)-Lactic acid

Figure 8.17. Rotation of a Fischer projection by 90° inverts its meaning.

Knowing these rules provides us with an easy way to superimpose molecules mentally. For example, three different Fischer projections of 2-butanol follow. Do they all represent the same enantiomer, or is one different?

A	B	C
H (top)	CH$_2$CH$_3$ (top)	OH (top)
H$_3$C— —CH$_2$CH$_3$	HO— —H	H— —CH$_3$
OH (bottom)	CH$_3$ (bottom)	CH$_2$CH$_3$ (bottom)

The simplest way to determine if two Fischer projections represent the same enantiomer is to carry out allowed motions until *two* groups are superimposed. If the other two groups are also superimposed, then the Fischer projections are the same; if the other two groups are not superimposed, however, the Fischer projections are different.

For example, let's keep projection A unchanged and move B so that the $—CH_3$ and —H substituents match up with those in A:

B: top CH_2CH_3, left HO, right H, bottom CH_3 — Hold CH_3, Rotate other three groups clockwise → top OH, left H, right CH_2CH_3, bottom CH_3 — Hold CH_2CH_3, Rotate other three groups counterclockwise → A: top H, left H_3C, right CH_2CH_3, bottom OH

By performing two allowed movements on projection B, we find that it is *identical* to projection A. Now let's do the same thing to projection C:

C: top OH, left H, right CH_3, bottom CH_2CH_3 — Rotate 180° → top CH_2CH_3, left H_3C, right H, bottom OH — Hold CH_3, Rotate other three groups counterclockwise → Not A: top H, left H_3C, right OH, bottom CH_2CH_3

By performing two allowed movements on projection C, we can match up the —H and $—CH_3$ with A. The —OH and $—CH_2CH_3$ substituents do not match up, however, and projection C must therefore be enantiomeric with A and B.

PROBLEM

8.9 Build molecular models of (+)- and (−)-lactic acid and confirm the assertion that only two kinds of motion are allowed for Fischer projections without changing their meanings.

PROBLEM

8.10 Which of these Fischer projections of 2-bromopropanoic acid represent the same enantiomer?

(a) top CO_2H, left H, right Br, bottom CH_3

(b) top CH_3, left Br, right H, bottom CO_2H

(c) top CO_2H, left Br, right H, bottom CH_3

(d) top Br, left H, right CO_2H, bottom CH_3

PROBLEM

8.11 Are these pairs of Fischer projections the same or are they enantiomers?

(a) top CH_3, left Cl, right H, bottom CHO and top H, left OHC, right Cl, bottom CH_3

(b)

$$\begin{array}{c} CH_2OH \\ H—\!\!\!+\!\!\!—OH \\ CHO \end{array} \quad \text{and} \quad \begin{array}{c} CHO \\ HO—\!\!\!+\!\!\!—CH_2OH \\ H \end{array}$$

(c)

$$\begin{array}{c} CN \\ CH_3CH_2—\!\!\!+\!\!\!—OH \\ CH_3 \end{array} \quad \text{and} \quad \begin{array}{c} CH_3 \\ HO—\!\!\!+\!\!\!—CN \\ CH_2CH_3 \end{array}$$

8.14 Assignment of *R,S* configurations to Fischer projections

By following three steps, we can assign *R,S* designations to Fischer projections:

1. Assign priorities to the four substituents.
2. Perform one of the allowed motions to place the group of lowest (fourth) priority at the top of the Fischer projection.
3. Determine the direction of rotation in going from priority 1 to 2 to 3 and assign *R* or *S* configuration.

Here are two examples:

Hold CH_2OH, rotate counterclockwise

$$\begin{array}{c} {}^2COOH \\ {}^1H_2N—\!\!\!+\!\!\!—H^4 \\ {}_3CH_2OH \end{array} = \begin{array}{c} {}^4H \\ {}^2HOOC—\!\!\!+\!\!\!—NH_2^1 \\ {}_3CH_2OH \end{array} = \begin{array}{c} {}^4H \\ {}^2HOOC \blacktriangleright C \blacktriangleleft NH_2^1 \\ {}_3CH_2OH \end{array} = \begin{array}{c} {}^2HOOC \quad H \quad NH_2^1 \\ C \\ {}_3CH_2OH \end{array}$$

(*S*)-Serine (an amino acid)

Rotate 180°

$$\begin{array}{c} {}^1Br \\ {}^3H_3C—\!\!\!+\!\!\!—CH_2CH_3^2 \\ {}_4H \end{array} = \begin{array}{c} {}^4H \\ {}^2CH_3CH_2—\!\!\!+\!\!\!—CH_3^3 \\ {}_1Br \end{array} = \begin{array}{c} {}^4H \\ {}^2CH_3CH_2 \blacktriangleright C \blacktriangleleft CH_3^3 \\ {}_1Br \end{array}$$

(*R*)-2-Bromobutane

$$= \begin{array}{c} {}^2CH_3CH_2 \quad H \quad CH_3^3 \\ C \\ {}_1Br \end{array}$$

Fischer projections can also be used to specify more than one chiral center in a molecule simply by "stacking" the chiral centers on top of each other. The *R*,*S* designations can then be assigned in the usual way. For example, threose, a simple four-carbon sugar, has the 2*S*,3*R* configuration:

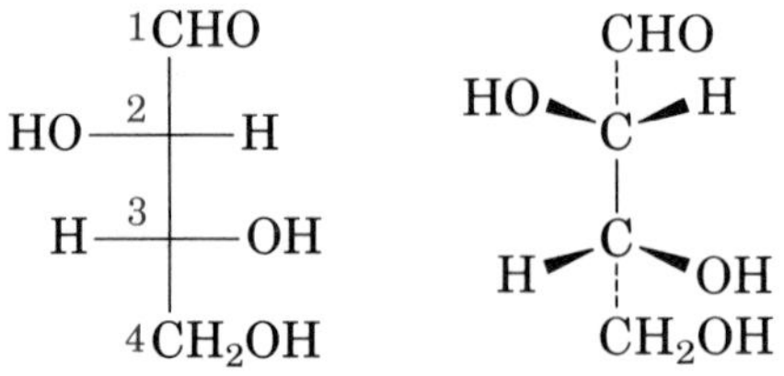

```
         1CHO                  CHO
        2 |                HO  |   H
   HO ----+---- H            \ C /
        3 |                    |
    H ----+---- OH           / C \
          |                 H  |  OH
        4CH2OH                CH2OH
```

Threose, (2*S*,3*R*)-2,3,4-trihydroxybutanal

PROBLEM

8.12 Assign the *R* or the *S* configuration to the chiral centers in these molecules:

(a) 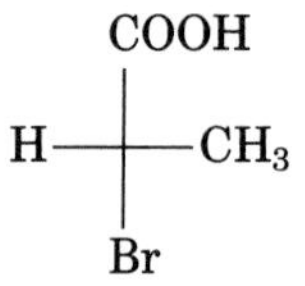 Fischer projection: top COOH, left H, right CH_3, bottom Br

(b) 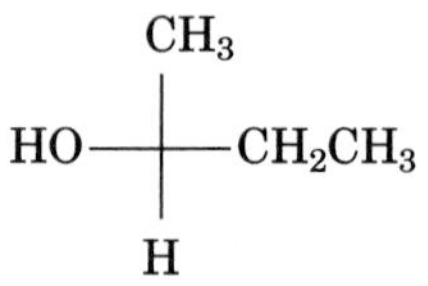 Fischer projection: top CH_3, left HO, right CH_2CH_3, bottom H

(c) 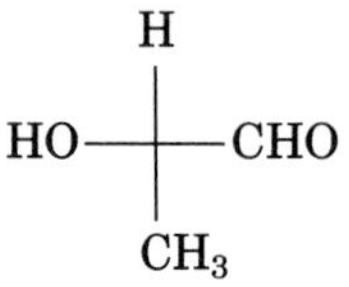 Fischer projection: top H, left HO, right CHO, bottom CH_3

(d) 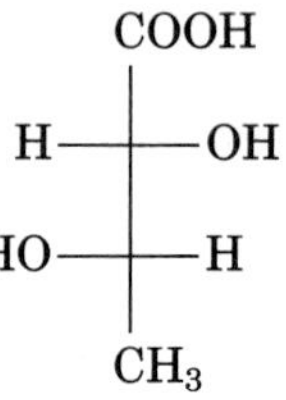 Fischer projection: top COOH; upper center H (left), OH (right); lower center HO (left), H (right); bottom CH_3

8.15 Stereochemistry of reactions: addition of HBr to alkenes

Many organic reactions, including some that we have studied, yield products with chiral centers. For example, HBr adds to 1-butene to yield 2-bromobutane, a chiral molecule. What predictions can we make about the stereochemistry of this chiral product? The answer is that the 2-bromobutane produced is a racemic mixture of *R* and *S* enantiomers.

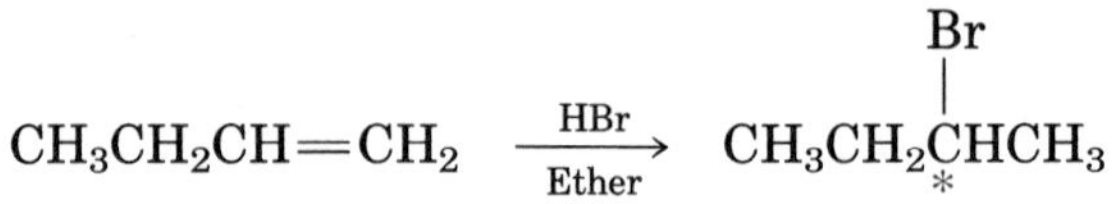

$$CH_3CH_2CH{=}CH_2 \xrightarrow[\text{Ether}]{\text{HBr}} CH_3CH_2\overset{*}{C}H(Br)CH_3$$

1-Butene (achiral) → (±)-2-Bromobutane (chiral)

To understand *why* a racemic product results, let's consider what happens during the reaction. 2-Butene is first protonated by H^+ to yield an intermediate 2° carbocation. This ion is sp^2 hybridized; it therefore has a plane of symmetry and is achiral. As a result, it can be attacked by bromide ion (also achiral) equally well from either the top or the bottom. Attack from the top leads to (*S*)-2-bromobutane, and attack from the bottom leads to (*R*)-2-bromobutane. Since both modes of attack occur with equal probability, a racemic mixture results (Figure 8.18).

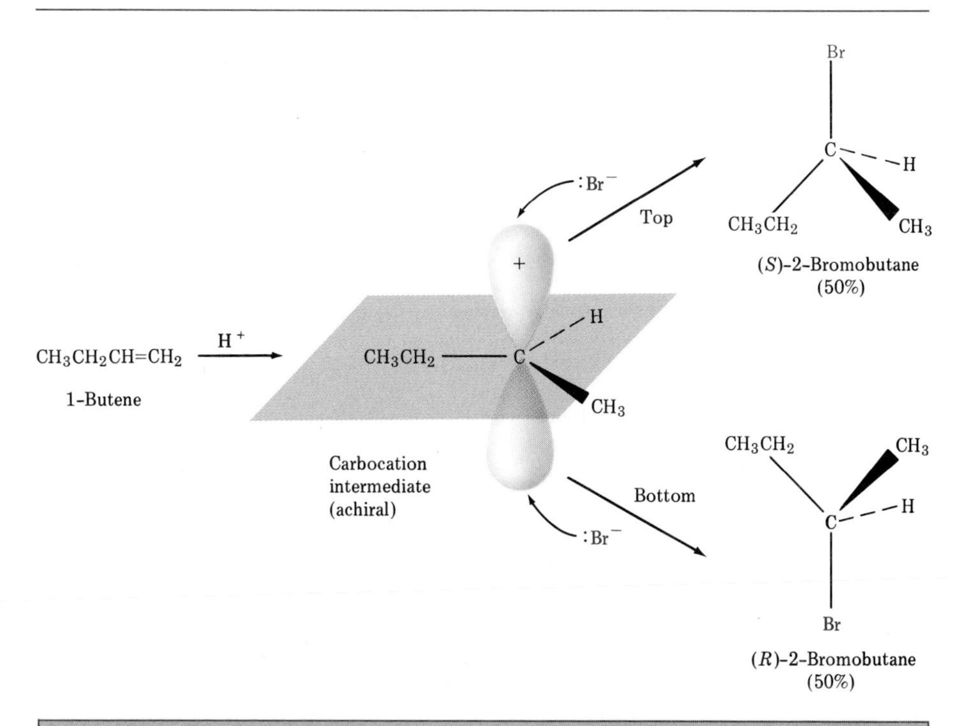

Figure 8.18. Stereochemistry of the addition of HBr to 1-butene. The intermediate carbocation is attacked equally well from both top and bottom sides.

It may help you to think about the two transition states leading to *R* and *S* products. If the intermediate 1-methylpropyl cation is attacked from the top, the attack leads to formation of *S* product through transition state 1 (TS 1) in Figure 8.19 (p. 256). Alternatively, if the cation is attacked from the bottom, the attack leads to *R* product through TS 2. The two transition states are mirror images; they therefore have identical activation energies and are equally likely to occur. Both transition states form at identical rates, leading to a 50:50 mixture of *R* and *S* products.

PROBLEM

8.13 Let's assume for a minute that carbocations are tetrahedral and sp^3 hybridized, rather than planar. Would you still expect bromide ion to attack the 1-methylpropyl cation equally well from both sides? What about the final product—is 2-bromobutane racemic or not?

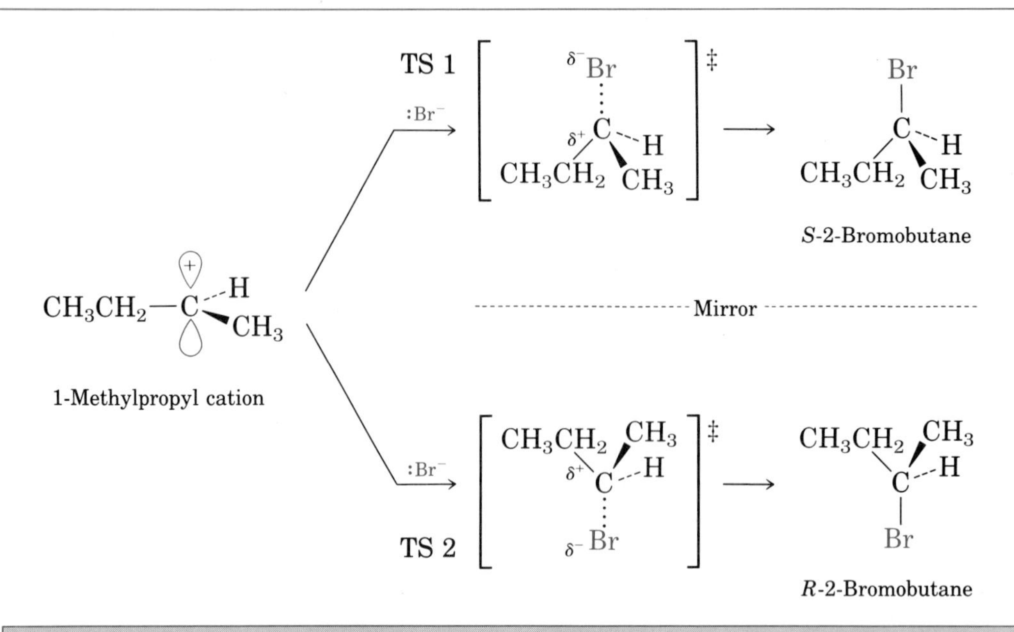

Figure 8.19. Bromide attack on the 1-methylpropyl cation. Attack from top is the mirror image of attack from the bottom. Both are equally likely, and racemic product is thus formed. The dotted C····Br bond in the transition state indicates that the bond is beginning to form.

8.16 Stereochemistry of reactions: addition of Br_2 to alkenes

The bromination of 2-butene leads to the generation of two chiral centers. What stereochemistry would we predict for such a reaction? If we begin with *cis*-2-butene (planar, achiral), we would expect bromine to add to the double bond equally well from either the top or the bottom face to generate two intermediate bromonium ions. For the sake of simplicity, let's consider only the result of attack from the top (Figure 8.20), keeping in mind that every structure we consider also has a mirror image.

The bromonium ion formed by top-side reaction of *cis*-2-butene can be attacked by bromide ion from either the right or the left side. Attack from the left (path a) leads to (2*S*,3*S*)-dibromobutane; attack from the right (path b) leads to (2*R*,3*R*)-dibromobutane. Since both modes of attack are equally likely, the two products are formed in equal amounts. Since the two products are enantiomers, the overall result is the formation of racemic (±)-2,3-dibromobutane.

PROBLEM

8.14 Formation of the enantiomeric bromonium ion (bottom-side attack on *cis*-2-butene) also leads to racemic product. Explain why this is so.

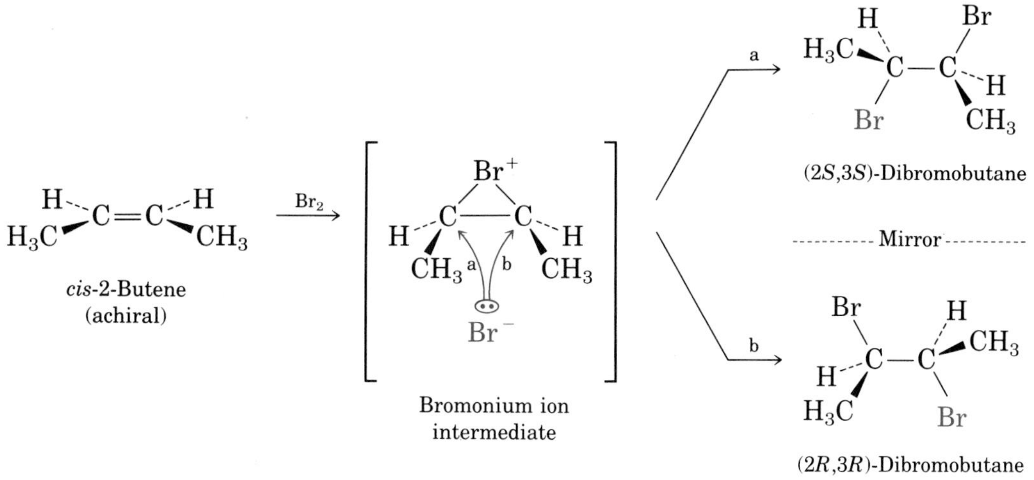

Figure 8.20. Addition of Br_2 to *cis*-2-butene. A racemic mixture of 2*S*,3*S* and 2*R*,3*R* products is formed.

PROBLEM

8.15 Bromination of an unsymmetrical alkene such as *cis*-2-hexene leads to racemic product, even though attack of bromide on the intermediate (unsymmetrical) bromonium ion is not equally likely at both ends. Why should this be so? Make drawings of the intermediate and the products.

What about the bromination of *trans*-2-butene? Is the same racemic product mixture formed? Perhaps surprisingly, the answer is no. *trans*-2-Butene attacks bromine to form a bromonium ion, and, once again, we will consider only top-side attack for simplicity (Figure 8.21). Attack of bromide

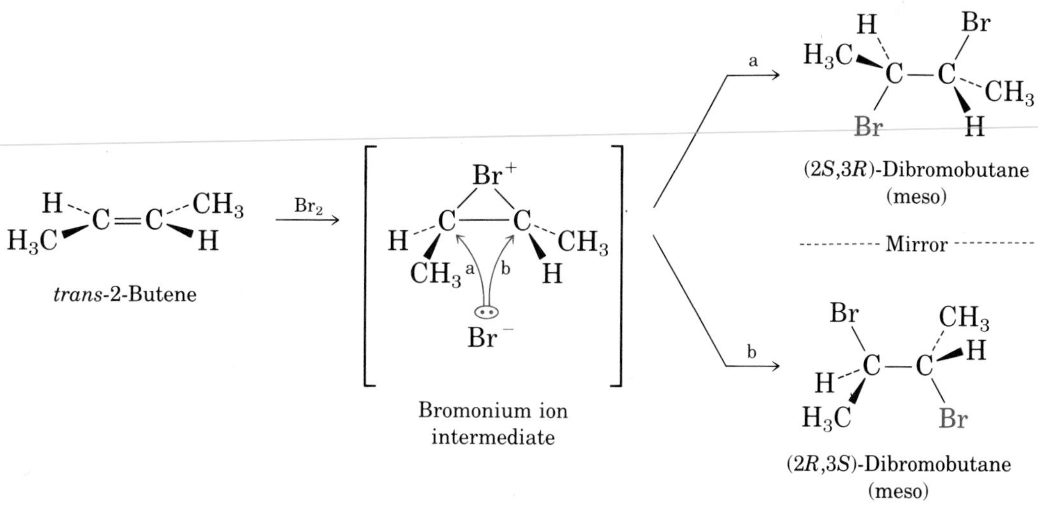

Figure 8.21. Addition of Br_2 to *trans*-2-butene. Meso product is formed.

on the bromonium ion intermediate leads to the formation of 2*R*,3*S* and 2*S*,3*R* products in equal amounts. Inspection of the two products, however, shows that they are *identical. meso*-2,3-Dibromobutane therefore results from bromination of *trans*-2-butene.

The key result in all three of the addition reactions just discussed is that optically inactive product has been formed. This is a general rule: Reaction between two achiral partners *always* leads to optically inactive products—either racemic or meso. Optically active products cannot be produced from optically inactive starting materials.

PROBLEM

8.16 Predict the stereochemical outcome of the reaction of Br_2 with *trans*-2-hexene. Show your reasoning.

8.17 Chirality at atoms other than carbon

If the most common cause of chirality is that four different substituents are bonded to a tetrahedral center, it follows that tetrahedral atoms other than carbon can also be chiral centers. Silicon, nitrogen, phosphorus, and sulfur are all commonly encountered in organic molecules, and under the proper circumstances, all can be chiral centers. We know, for example, that trivalent nitrogen is tetrahedral and contains a lone pair of electrons. Can trivalent nitrogen be chiral? Does a compound such as ethylmethylamine exist as a pair of enantiomers?

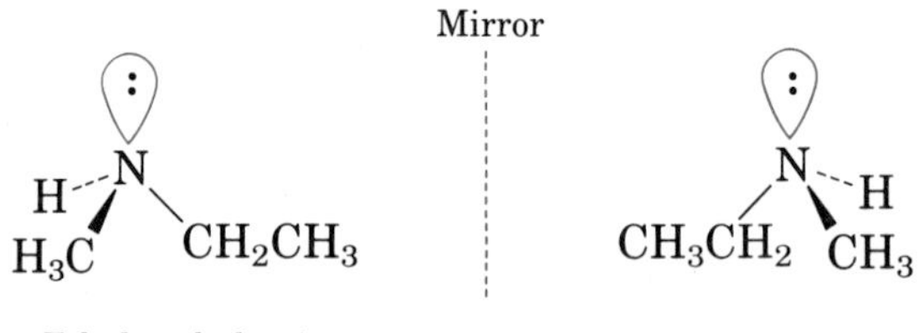

Mirror

H, H_3C, N, CH_2CH_3 | N, H, CH_3CH_2, CH_3

Ethylmethylamine

The answer is both yes and no. Yes in principle, but no in practice. Tetrahedral trivalent nitrogen undergoes a rapid "umbrella" inversion that interconverts enantiomers. We therefore cannot separate or isolate enantiomers except in special cases, and their existence is of little chemical consequence.

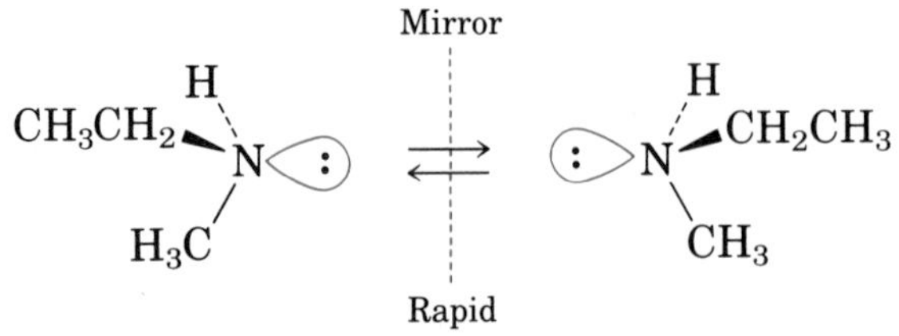

Mirror

H, CH_3CH_2, N, H_3C ⇌ H, N, CH_2CH_3, CH_3

Rapid

8.18 Summary

Stereochemistry is the study of the three-dimensional nature of molecules, and its origins can be traced to the early nineteenth century.

When a beam of plane-polarized light is passed through a solution of certain organic molecules, the plane of polarization is rotated. Compounds that exhibit this behavior are optically active. Optical activity is due to the asymmetric nature of the molecules themselves. The tetrahedral carbon atom proposed by van't Hoff predicts the existence of mirror-image isomers when four different groups are bonded to carbon. Compounds that are mirror images of each other, but are not identical, are called enantiomers and are said to be chiral. Enantiomers are identical in all physical properties except for the direction in which they rotate plane-polarized light.

The stereochemical configuration of chiral carbon centers can be depicted using Fischer projections, in which horizontal lines are understood to come out of the plane of the paper and vertical lines are understood to go back into the plane of the paper. For example,

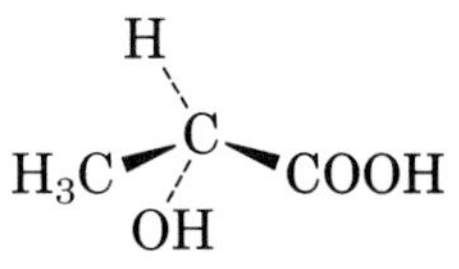

H
H_3C C COOH
OH

(+)-Lactic acid

H
H_3C — COOH
OH

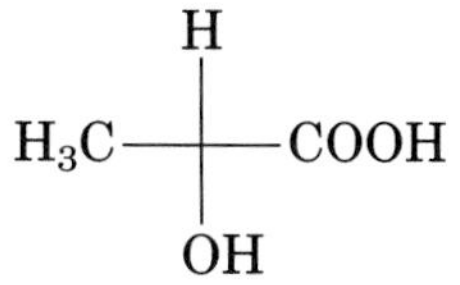

Fischer projection of (+)-lactic acid

The steric configuration of a chiral carbon atom can be specified as either *R* or *S* by using the Cahn–Ingold–Prelog sequence rules. This is done by first assigning priorities to the four substituents on the chiral carbon atom and then orienting the molecule so that the lowest-priority group points directly back away from the viewer. We then look at the remaining three substituents and let the eye travel from the group having the highest priority to the second highest and then to the third highest (i.e., from 1 to 2 to 3). If the direction of travel is clockwise, the configuration is labeled *R* (*rectus*); if the direction of travel is counterclockwise, the configuration is labeled *S* (*sinister*).

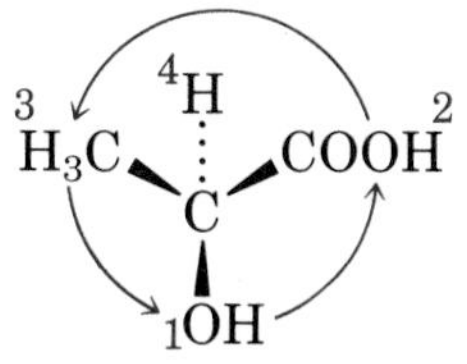

3 ^{4}H 2
H_3C C COOH
$_1$OH

(*S*)-(+)-Lactic acid

2 ^{4}H 3
HOOC C CH_3
$_1$OH

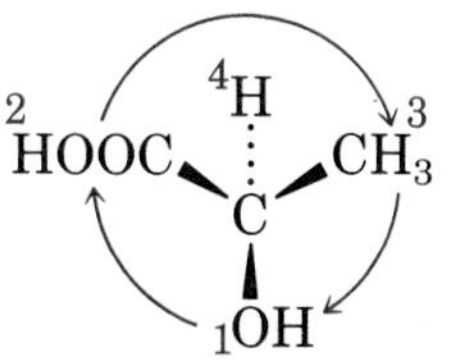

(*R*)-(−)-Lactic acid

Some molecules possess more than one chiral center. Enantiomers have opposite configuration at all chiral centers; diastereomers have the same configuration in at least one center but opposite configurations at the others.

A compound with n chiral centers can have 2^n stereoisomers. For example, threonine has two chiral centers, and the stereochemical relationships of the four stereoisomers are as follows:

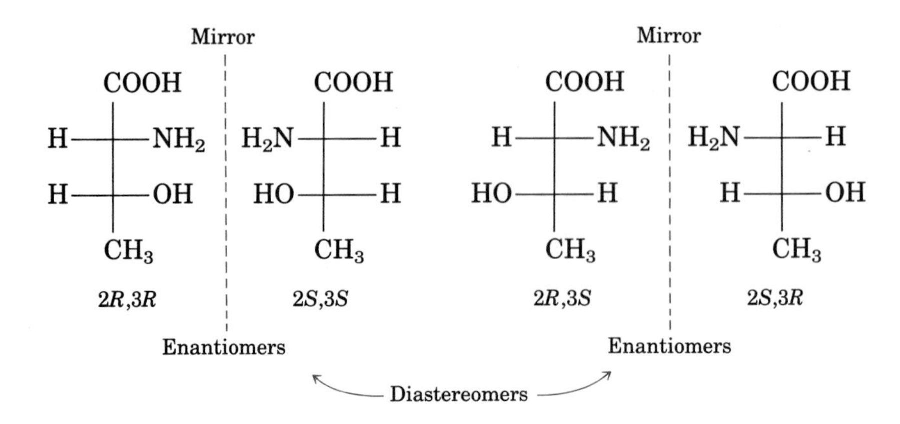

Meso compounds contain chiral centers, but are achiral overall because they contain a plane of symmetry. For example, *meso*-tartaric acid is achiral:

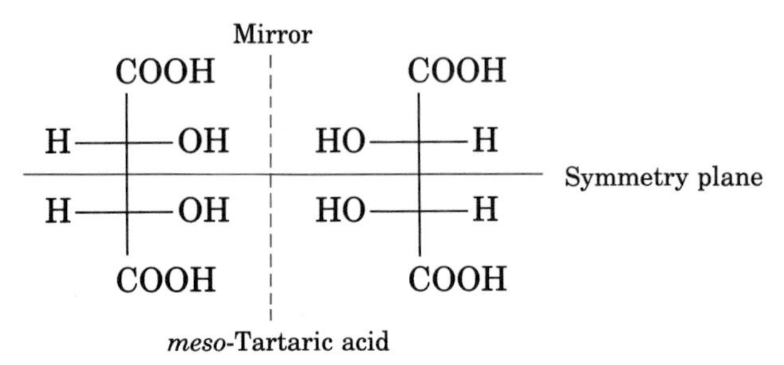

meso-Tartaric acid

Racemic mixtures are 50:50 mixtures of (+) and (−) enantiomers. Racemic mixtures and diastereomers all differ from each other in their physical properties such as solubility, melting point, and boiling point.

Most reactions give chiral products, but these products must be optically inactive (meso or racemic) if the starting materials are optically inactive.

ADDITIONAL PROBLEMS

8.17 Cholic acid, the major steroid found in bile, was observed to have a specific rotation of +2.22° when a 3 g sample was dissolved in 5 mL alcohol and the solution was placed in a sample tube with a 1 cm path length. Calculate $[\alpha]_D$ for cholic acid.

8.18 Polarimeters for measuring optical rotation are quite sensitive and can measure rotations to the thousandth of a degree. This can be important when only small amounts of sample are available. Ecdysone, for example, is an insect hormone that controls molting in the silkworm moth. When 7 mg ecdysone was dissolved in 1.0 mL chloroform and the solution was placed in a 2 cm path-length cell, an observed rotation of +0.087° was found. Calculate $[\alpha]_D$ for ecdysone.

8.19 Define these terms in your own words:

(a) Chirality (b) Chiral center

(c) Optical activity
(d) Diastereomer
(e) Enantiomer
(f) Racemate

8.20 Which of these compounds are chiral, and which are achiral? Label the chiral centers.

(a) $CH_3CH_2C(CH_3)_2CH(CH_3)CH_2CH_3$

(b) OCOCH$_3$ / COOH (benzene ring)

Aspirin

(c) CH$_3$ / H$_3$C (cyclohexane ring)

(d) $HOCH_2CH(OH)CHO$

Glyceraldehyde

(e) 1,2-Dibromo-2,3-diphenylbutane

(f) CH$_2$OH

Vitamin A

8.21 Which of these compounds are chiral? Draw them and label the chiral centers.
(a) 2,4-Dimethylheptane
(b) 3-Ethyl-5,5-dimethylheptane
(c) *cis*-1,4-Dichlorocyclohexane
(d) 4,5-Dimethyl-2,6-octadiyne

8.22 There are eight alcohols that have the formula $C_5H_{12}O$. Draw them. Which are chiral?

8.23 Draw the nine chiral molecules that have the formula $C_6H_{13}Br$.

8.24 Draw compounds that fit these descriptions:
(a) A chiral alcohol with four carbons
(b) A chiral carboxylic acid having the formula $C_5H_{10}O_2$
(c) A compound with two chiral centers
(d) A chiral aldehyde having the formula C_3H_5BrO

8.25 Which of these objects are chiral?
(a) A basketball
(b) A fork
(c) A wine glass
(d) A golf club
(e) A monkey wrench
(f) A screw
(g) A beanstalk
(h) An ear
(i) A coin
(j) A snowflake

8.26 Place asterisks by the chiral carbons in these molecules.

(a) OCH$_2$CONH, H, H, S, CH$_3$, CH$_3$, N, H, O, COOH

Penicillin V
(antibiotic)

(b)

Quinine
(antimalarial)

(c)

Aphidicolin
(antiviral)

(d)

Cortisone
(anti-inflammatory)

8.27 Draw examples of the following:
(a) A meso compound having the formula C_8H_{18}
(b) A meso compound having the formula C_9H_{20}
(c) A compound with two chiral centers, one *R* and the other *S*

8.28 Which of these pairs of Fischer projections represent the same molecule, and which represent different molecules?

(a) Top Br, left H_3C, right H, bottom CN **and** top CN, left H, right Br, bottom CH_3

(b) Top COOH, left H, right CN, bottom Br **and** top Br, left H, right COOH, bottom CN

(c) Top CH_3, left H, right CH_2CH_3, bottom OH **and** top OH, left H, right CH_3, bottom CH_2CH_3

(d) Top CH_3, left H, right NH_2, bottom COOH **and** top COOH, left H_3C, right NH_2, bottom H

(e)

```
       COOH                      CO2CH3
        |                          |
   H----+----CO2CH3    and   H-----+-----COOH
        |                          |
       CH3                        CH3
```

(f)

```
         H                    Cl
         |                    |
   H3C---+---Cl    and   H----+----CH3
         |                    |
         D                    D
```

8.29 Assign Cahn–Ingold–Prelog priorities to these sets of substituents:

(a) $-CH{=}CH_2$, $-CH(CH_3)_2$, $-C(CH_3)_3$, $-CH_2CH_3$

(b) $-C{\equiv}CH$, $-CH{=}CH_2$, $-C(CH_3)_3$, —(benzene ring)

(c) $-CO_2CH_3$, $-COCH_3$, $-CH_2OCH_3$, $-CH_2CH_3$

(d) $-CN$, $-CH_2Br$, $-CH_2CH_2Br$, $-Br$

8.30 Assign *R*,*S* configurations to these Fischer projections:

(a)

```
        CN
        |
   H----+----Br
        |
       CH3
```

(b)

```
        CH=CH2
          |
   H------+------CH2CH3
          |
         CO2H
```

(c)

```
        Br
        |
   H----+----(benzene ring)
        |
      CH2CH3
```

(d)

```
        CH3
         |
   H-----+-----CH=CH2
         |
   (benzene ring)
```

8.31 Assign *R*,*S* configurations to each chiral center in these molecules:

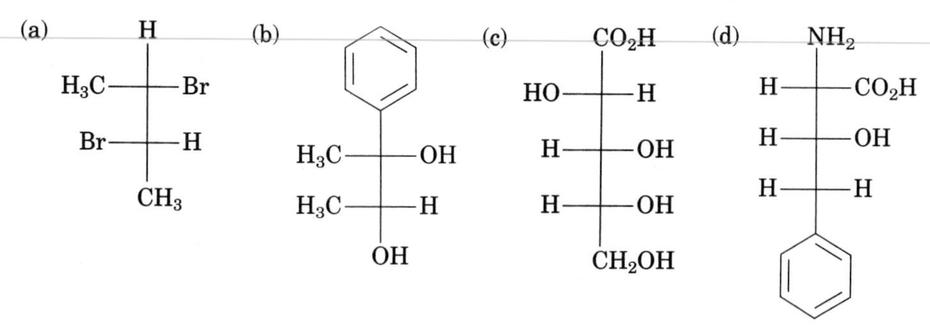

8.32 Draw Fischer projections that fit these descriptions:

(a) The *S* enantiomer of 2-bromobutane

(b) The *R* enantiomer of alanine, $CH_3CH(NH_2)CO_2H$

(c) The *R* enantiomer of 2-bromopropanoic acid

(d) The *S* enantiomer of 3-methylhexane

8.33 Xylose is a common sugar found in many woods (maple, cherry). Because it is much less prone to cause tooth decay than sucrose, xylose has been used in candy and chewing gum. Assign *R*,*S* configurations to the chiral centers in xylose:

```
       CHO
   H ──┼── OH
  HO ──┼── H
   H ──┼── OH
       CH2OH
```

(+)-Xylose, $[\alpha]_D = +92°$

8.34 What products would you expect to obtain from OsO_4 hydroxylation of *cis*-2-butene? Of *trans*-2-butene?

8.35 Alkenes undergo reaction with peroxycarboxylic acids (RCO_3H) to give three-membered-ring cyclic ethers called *epoxides*. For example, 4-octene reacts with peroxyacids to yield 4,5-epoxyoctane:

$$CH_3CH_2CH_2CH{=}CHCH_2CH_2CH_3 \xrightarrow{RCO_3H} CH_3CH_2CH_2\overset{O}{CH{-}CH}CH_2CH_2CH_3$$

4-Octene → 4,5-Epoxyoctane

Assuming that this epoxidation reaction occurs with syn stereochemistry, draw the structure obtained from epoxidation of *cis*-4-octene. Is this epoxide chiral? How many chiral centers does it have? How would you describe the product stereochemically?

8.36 Answer Problem 8.35, assuming that the epoxidation reaction was carried out on *trans*-4-octene.

8.37 Write the products of the following reactions and indicate the stereochemistry obtained in each instance.

Cyclopentene:

(a) $\xrightarrow[DMSO]{Br_2,\ H_2O}$

(b) $\xrightarrow{Br_2,\ CCl_4}$

(c) $\xrightarrow[2.\ NaHSO_3]{1.\ OsO_4}$

8.38 Draw Fischer projections that fit these descriptions:
(a) (2*R*,3*S*)-2,3-Dihydroxybutanoic acid
(b) *meso*-4,5-Dibromooctane
(c) (4*R*,5*R*)-4,5-Dichlorooctane

8.39 Glucose, a primary source of energy for living organisms, occurs naturally in most fruits and in other parts of plants. Stereochemically, glucose can be defined as (2*R*,3*S*,4*R*,5*R*)-6-pentahydroxyhexanal. Draw a Fischer projection of glucose showing the correct stereochemistry at the four chiral centers.

8.40 Draw all possible stereoisomers of cyclobutane-1,2-dicarboxylic acid and indicate the interrelationships. Which, if any, are optically active? Do the same for cyclobutane-1,3-dicarboxylic acid.

8.41 Compound A, C_7H_{12}, was found to be optically active. On catalytic reduction over a palladium catalyst, 2 equiv of hydrogen were absorbed, yielding compound B, C_7H_{16}. On ozonolysis, two fragments were obtained. One fragment was identified as acetic acid. The other fragment, compound C, was an optically active carboxylic acid, $C_5H_{10}O_2$. Formulate the reactions, and draw structures for A, B, and C.

8.42 Compound A, $C_{11}H_{16}O$, was found to be an optically active alcohol. Despite its apparent unsaturation, no hydrogen was absorbed on catalytic reduction over a palladium catalyst. On treatment of A with dilute sulfuric acid, dehydration occurred, and an optically inactive alkene B, $C_{11}H_{14}$, was produced as the major product. Alkene B, on ozonolysis, gave two products. One product was identified as propanal, CH_3CH_2CHO. Compound C, the other product, C_8H_8O, was shown to be a ketone. How many multiple bonds and/or rings does A have? Formulate the reactions and identify A, B, and C.

8.43 How many stereoisomers of 2,4-dibromo-3-chloropentane are there? Draw them and indicate which are optically active.

8.44 The so-called "tetrahedranes" are an interesting class of compounds. The first member of this class was synthesized in 1978. Construct a model (carefully!) of tetrahedrane. Consider a substituted tetrahedrane with four different substituents. Is it chiral? Explain your answer.

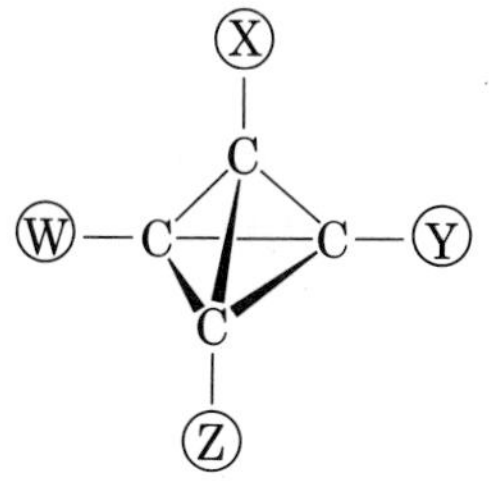

(X)
|
C
(W)—C—C—(Y)
C
|
(Z)

A tetrahedrane

8.45 *Allenes,* compounds with adjacent carbon–carbon double bonds, are well known. Many allenes are chiral, even though they do not contain chiral carbon atoms. Mycomycin, for example, is a naturally occurring antibiotic isolated from the bacterium *Nocardia acidophilus*. Mycomycin is chiral and has $[\alpha]_D = -130°$. Can you explain why mycomycin is chiral? Making a molecular model should be helpful.

$$HC{\equiv}C{-}C{\equiv}C{-}CH{=}C{=}CH{-}CH{=}CH{-}CH{=}CH{-}CH_2COOH$$

Mycomycin
(an allene)

8.46 Long before optically active allenes were known to exist (Problem 8.45), the resolution of 4-methylcyclohexylideneacetic acid into two enantiomers had been carried out. This was the first molecule to be resolved that was chiral yet did not contain a chiral center. Can you explain why this molecule is chiral? What geometric relation does this molecule have to allenes?

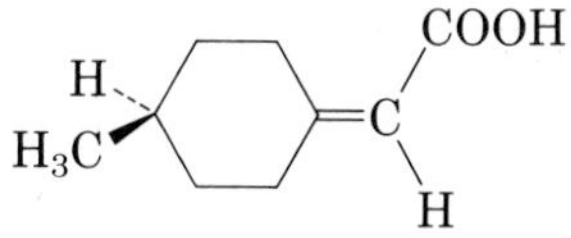

H, H_3C (ring) =C(COOH)(H)

4-Methylcyclohexylideneacetic acid

CHAPTER 9
ALKYL HALIDES

It would be difficult to study organic chemistry for long without becoming aware of the importance of halo-substituted alkanes; indeed, we have already touched on some of the chemistry of alkyl halides several times. Haloalkanes are common industrial chemicals. Tetrachloromethane, trichloromethane, and 1,1,1-trichloroethane, for instance, are widely used as solvents (although all are known to cause liver damage on chronic exposure). Freons such as dichlorodifluoromethane (Freon 12) are refrigerants and aerosol propellants, and halogenated polymers such as polytetrafluoroethylene (Teflon) and poly-1,1-dichloroethylene (Saran) are used as plastics.

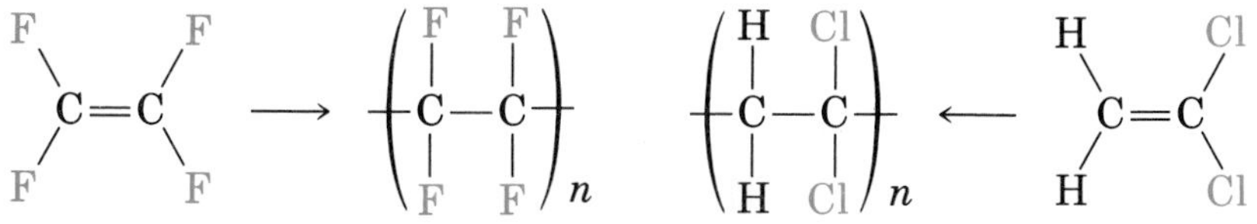

$$F_2C{=}CF_2 \longrightarrow \left(CF_2{-}CF_2 \right)_n \qquad \left(CH_2{-}CCl_2 \right)_n \longleftarrow H_2C{=}CCl_2$$

Tetrafluoroethylene — Teflon — Saran — 1,1-Dichloroethylene

Alkyl halides also occur widely in nature, mostly in marine rather than terrestrial sources. The 1970s saw an explosive growth in the chemical investigation of marine organisms, and the structures of many naturally occurring halogenated molecules were elucidated. For example, it has been shown that simple halomethanes such as $CHCl_3$, CCl_4, CBr_4, CH_3I, and CH_3Cl are constituents of the alga *Asparagopsis taxiformis,* which is found near Hawaii.

In addition, certain substances that have been isolated from marine organisms exhibit interesting biological activity. For example, plocamene B, a trichlorocyclohexene derivative isolated from the red alga *Plocamium violaceum,* is similar in potency to DDT in showing insecticidal activity against mosquito larvae.

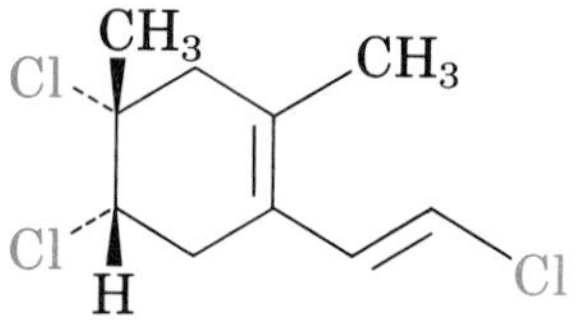

Plocamene B, a trichloride

Before discussing some chemistry of alkyl halides, we should point out that we will be referring only to compounds having halogen atoms bonded to saturated, sp^3-hybridized, carbon. Other classes, such as aromatic (*aryl*) and alkenyl (*vinylic*) halides also exist, but much of their chemistry is quite different.

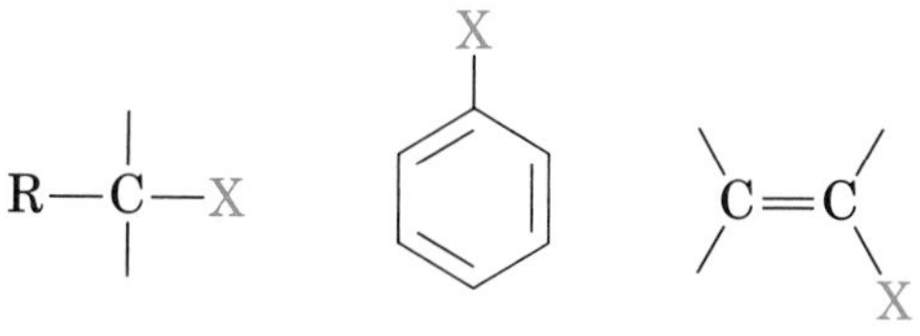

Alkyl halide — Aryl halide — Vinylic halide

9.1 Nomenclature of alkyl halides

Alkyl halides are named using a straightforward extension of the rules of alkane nomenclature (Section 3.3). According to IUPAC conventions, halogens are considered substituents on the parent chain, and three rules suffice:

1. Find and name the parent chain. As in naming alkanes, select the longest chain as the parent.
2. Number the parent chain. Always number the chain beginning at the end nearer the first substituent, regardless of whether it is alkyl or halo. For example,

```
                    Br                                  CH3
 1     3     5|    7                 1     3     5|    7
 CH3   CH2   CH    CH3               CH3   CH2   CH    CH3
   \  /   \  /  \  /                   \  /   \  /  \  /
    CH     CH    CH2                    CH     CH    CH2
   2|     4|     6                     2|     4|     6
    CH3    CH3                          Br     CH3
```

5-Bromo-2,4-dimethylheptane

2-Bromo-4,5-dimethylheptane

   a. If more than one of the same kind of halogen is present, number each and use one of the prefixes *di-*, *tri-*, *tetra-*, and so on. For example,

```
            Cl
           2|
  Cl        CH    4CH3
    \      /  \   /
     1CH2      CH
              3|
               CH3
```

1,2-Dichloro-3-methylbutane

   b. If different halogens are present, number each according to its position. List all substituents in alphabetical order. For example,

$$BrCH_2CH_2CH(Cl)CH(CH_3)CH_2CH_3$$

1-Bromo-3-chloro-4-methylhexane

3. If the parent chain can be properly numbered from either end, begin at the end nearer the substituent (either alkyl or halo) that has alphabetical precedence. For example,

```
                        Br
                       2|
  6CH3    4CH2          CH
      \   /    \       /  \
       CH       3CH2        1CH3
      5|
       CH3
```

2-Bromo-5-methylhexane
(*not* 5-bromo-2-methylhexane)

In addition to their systematic names, a small number of simple alkyl halides have alternative names that are well entrenched in the chemical literature and in common usage. Table 9.1 lists some of these, but they will not be used in this book.

TABLE 9.1 **Alternative names of some common alkyl halides**

Structure	*Systematic name*	*Alternative name*
CH_3I	Iodomethane	Methyl iodide
CH_3CH_2Br	Bromoethane	Ethyl bromide
$CH_3CHClCH_3$	2-Chloropropane	Isopropyl chloride
$CH_3CHBrCH_2CH_3$	2-Bromobutane	*sec*-Butyl bromide
(cyclohexane ring bearing H and Br on one carbon)	Bromocyclohexane	Cyclohexyl bromide

PROBLEM

9.1 Provide systematic IUPAC names for these alkyl halides:
(a) $(CH_3)_2CClCH_2CH_2Cl$
(b) $BrCH_2CH_2CH_2C(CH_3)_2CH_2Br$
(c) $CH_3CHICH(CH_2CH_2Cl)CH_2CH_3$

9.2 Structure of alkyl halides

The carbon–halogen bond in alkyl halides results from the overlap of a carbon sp^3 hybrid orbital with a halogen orbital. Thus, the carbon atoms have an approximately tetrahedral geometry with H—C—X bond angles near 109°. The halogens increase in size going down the periodic table, and this is reflected in the bond lengths of the halomethane series (Table 9.2). Table 9.2 also indicates that C—X bond dissociation energies decrease dramatically going down the periodic table.

TABLE 9.2 **Parameters for the C—X bond in halomethanes**[a]

Halomethane	*Bond length (Å)*	*Bond dissociation energy (kcal/mol)*
$CH_3—F$	1.39	109
$CH_3—Cl$	1.78	84
$CH_3—Br$	1.93	70
$CH_3—I$	2.14	56

[a]Bond lengths increase and bond strengths decrease going down the periodic table.

In an earlier discussion (Section 4.1) we noted that halogens are electronegative with respect to carbon, and that the C—X bond is polar, with the carbon atom bearing a slight positive charge and the halogen a slight negative charge:

$$\overset{\delta^+}{\text{C}}\text{—}\overset{\delta^-}{\text{X}}$$

Here X equals F, Cl, Br, or I. We can get a rough idea of the amount of bond polarity by measuring the dipole moments of the halomethanes, and Table 9.3 summarizes the results. As indicated in the table, the identity of the halogen does not have a large effect; all of the halomethanes have strong dipole moments.

TABLE 9.3 Dipole moments of halomethanes: —C ↔ X

Halomethane	*Dipole moment, μ (D)*
CH_3F	1.82
CH_3Cl	1.94
CH_3Br	1.79
CH_3I	1.64

Since the carbon atom of alkyl halides is positively polarized, alkyl halides are good electrophiles. As we will see in the next chapter, much of the chemistry alkyl halides undergo is dominated by their electrophilic behavior.

9.3 Preparation of alkyl halides

We have already looked at several methods of alkyl halide preparation and have seen that the electrophilic additions of HX and of X_2 to alkenes can provide excellent routes for the synthesis of haloalkanes (Sections 5.8 and 6.1). We know that HCl, HBr, and HI react with alkenes by a polar pathway to give the product of Markovnikov addition, and that HBr can also add by a radical pathway to give the non-Markovnikov product. Bromine and chlorine also add well to alkenes and give trans products. (See Figure 9.1.)

Another method of alkyl halide synthesis we have seen is the reaction of alkanes with chlorine or bromine (Section 4.4). To review briefly, alkane chlorination occurs by a radical chain reaction pathway (Figure 9.2). Once an *initiation* step has started the process by producing radicals, the reaction continues in a self-sustaining cycle. The cycle requires two repeating *propagation* steps, in which a radical, the halogen, and the alkane yield alkyl halide product plus more radical to carry on the chain. The chain is occasionally *terminated* by the combination of two radicals.

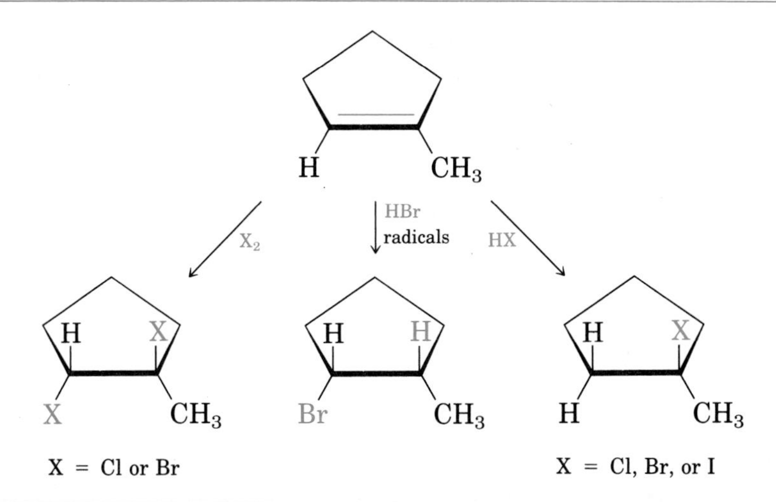

Figure 9.1. Synthesis of alkyl halides from alkenes.

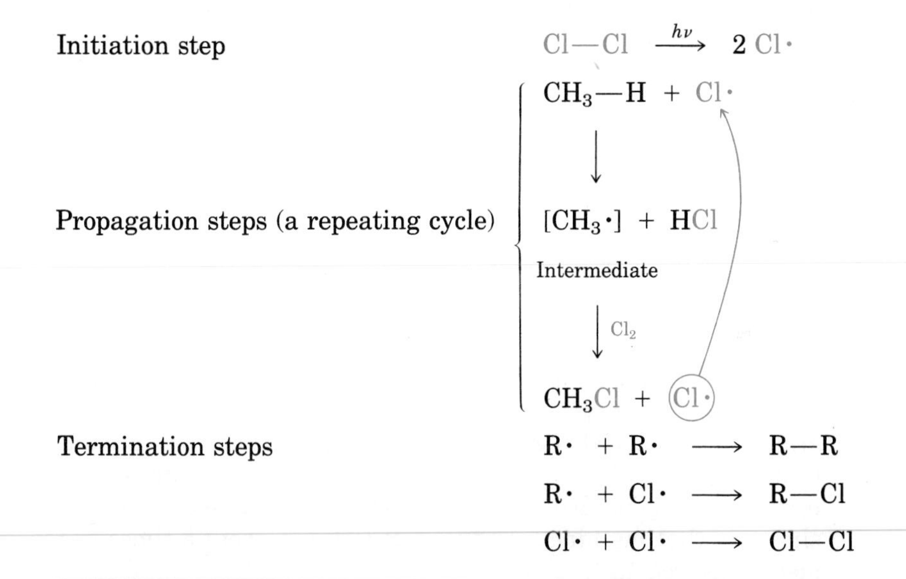

Figure 9.2. Mechanism of radical alkane chlorination.

Alkane chlorination is interesting from a mechanistic point of view because it is a radical chain process. Unfortunately, alkane chlorination is a poor synthetic method. Let's see why this is so.

9.4 Radical chlorination of alkanes

Alkane chlorination is a generally poor method of alkyl halide synthesis because mixtures of products invariably result. For example, chlorination of methane does not stop cleanly at the monochlorinated stage, but continues, giving a mixture of dichloro, trichloro, and even tetrachloro products:

$$CH_4 + Cl_2 \xrightarrow{h\nu} CH_4 + CH_3Cl + CH_2Cl_2 + CHCl_3 + CCl_4 \; (+ HCl)$$

The situation is even worse for chlorination of alkanes having more than one kind of hydrogen. For example, chlorination of butane gives two monochlorinated products in addition to dichlorobutane, trichlorobutane, and so on. Thirty percent of the monochloro product is 1-chlorobutane, and 70% is 2-chlorobutane:

$$\underset{\text{Butane}}{CH_3CH_2CH_2CH_3} + Cl_2 \xrightarrow{h\nu} \underbrace{\underset{\text{1-Chlorobutane}}{CH_3CH_2CH_2CH_2Cl} + \underset{\text{2-Chlorobutane}}{CH_3CH_2\overset{\overset{\displaystyle Cl}{|}}{C}HCH_3}}_{30:70} + \text{Dichloro-, trichloro-, tetrachloro-, and so on}$$

As a further example, 2-methylpropane yields 2-chloro-2-methylpropane and 1-chloro-2-methylpropane in a ratio of 35:65, along with more highly chlorinated products:

$$\underset{\text{2-Methylpropane}}{CH_3{-}\overset{\overset{\displaystyle CH_3}{|}}{\underset{\underset{\displaystyle CH_3}{|}}{C}}{-}H} + Cl_2 \xrightarrow{h\nu} \underbrace{\underset{\text{2-Chloro-2-methylpropane}}{CH_3{-}\overset{\overset{\displaystyle CH_3}{|}}{\underset{\underset{\displaystyle CH_3}{|}}{C}}{-}Cl} + \underset{\text{1-Chloro-2-methylpropane}}{ClCH_2{-}\overset{\overset{\displaystyle CH_3}{|}}{\underset{\underset{\displaystyle CH_3}{|}}{C}}H}}_{35:65} + \text{Dichloro-, trichloro-, tetrachloro-, and so on}$$

From these examples and other studies, we can calculate a reactivity order for different types of hydrogen atoms in a molecule. Thus, the chlorination of butane yields 2-chlorobutane as the major product and indicates that methylene hydrogens (R_2CH_2) are about 3.5 times as reactive as methyl hydrogens ($R{-}CH_3$). The 2-methylpropane chlorination indicates similarly that methine hydrogens (R_3CH) are about 5 times as reactive as methyl hydrogens.

These values are obtained by taking into account both the actual product ratios and the number of hydrogens of each type in the starting materials. For example, there are 1.5 times as many methyl hydrogens as methylene hydrogens in butane, and we therefore must divide the amount of methyl-group-chlorinated material by 1.5 to obtain an accurate reactivity value:

$$\frac{\text{Methyl}}{\text{Methylene}} = \frac{30 \div 1.5}{70} = \frac{20}{70} = \frac{1}{3.5}$$

Similarly, there are 9 times as many methyl hydrogens as methine hydrogens in 2-methylpropane:

$$\frac{\text{Methyl}}{\text{Methine}} = \frac{65 \div 9}{35} = \frac{7}{35} = \frac{1}{5}$$

Relative reactivity toward chlorination

$$\underset{5.0}{R_3CH} > \underset{3.5}{R_2CH_2} > \underset{1.0}{RCH_3}$$

The reactivity order of hydrogen atoms toward radical chlorination is exactly the same as the stability order of alkyl radicals.

As the bond dissociation energy data in Table 4.5 indicate, methine C—H bonds (91 kcal/mol) are weaker than methylene C—H bonds (95 kcal/mol), which are weaker than methyl C—H bonds (98 kcal/mol). Since less energy is required to break a tertiary C—H bond, the resultant tertiary radical is more stable than a primary or secondary radical.

Relative stability order

$$R_3C\cdot > R_2CH\cdot > RCH_2\cdot$$

An explanation of the correlation between reactivity and bond strength comes from the Hammond postulate and is similar to the argument used in Section 5.10 to account for the fact that addition of H^+ to an alkene yields the more stable carbocation. Abstraction of a hydrogen atom during alkane chlorination is an endothermic reaction whose reaction energy diagram should look like that of Figure 9.3. According to the Hammond postulate, the transition state for the hydrogen abstraction step resembles the radical

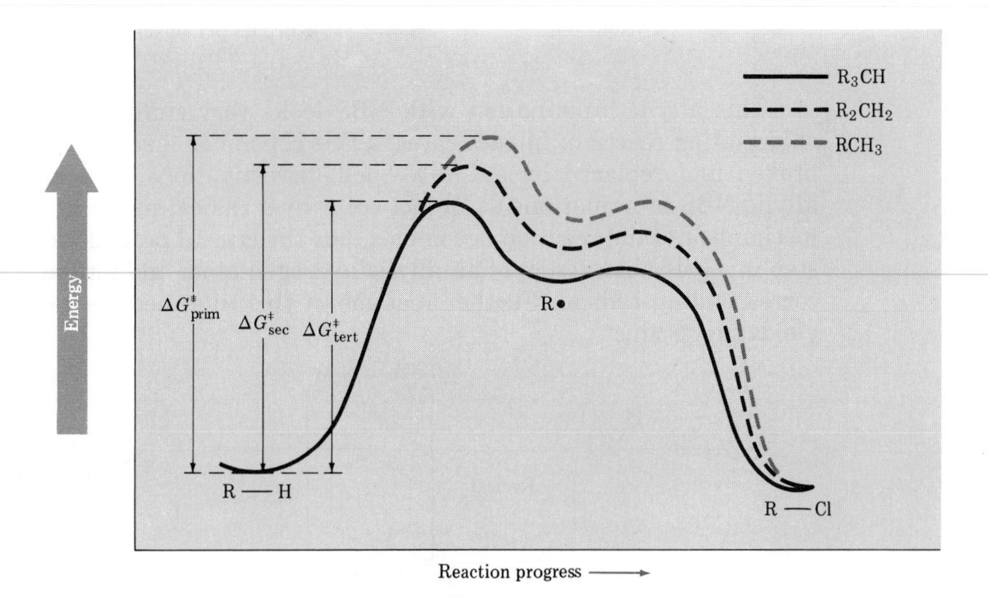

Figure 9.3. Reaction energy diagram for alkane chlorination. The stability order of tertiary, secondary, and primary radicals parallels their rates of formation.

intermediate, and the more stable radical should therefore form through the more stable (lower $\Delta G^{\ddagger}$) transition state. Once again, we see that the more stable product forms faster. Tertiary radicals are more stable, and methine hydrogen atoms are therefore more easily removed than other types.

PROBLEM

9.2 Taking the known relative reactivities of 1°, 2°, and 3° hydrogen atoms into account, what product(s) would you expect to obtain from monochlorination of 2-methylbutane? What would be the approximate percentages of each product? (Don't forget to take into account the number of each type of hydrogen.)

9.5 Allylic bromination of alkenes

In 1942, Karl Ziegler repeated some earlier work of Wohl[1] and found that alkenes react with *N*-bromosuccinimide (NBS) to give products resulting from substitution of hydrogen by bromine at the position *next to* the double bond (the *allylic* position). Cyclohexene, for example, gives 3-bromocyclohexene in 85% yield:

Cyclohexene

3-Bromocyclohexene (85%)

This allylic bromination with NBS looks very similar to the alkane chlorination reaction. In both cases, a C—H bond on a saturated carbon is broken and replaced by a C—X bond. Investigations have shown that allylic NBS brominations do in fact occur by a radical pathway. The exact mechanism of the reaction is complex, but the crucial product-determining step involves abstraction of an allylic hydrogen atom and formation of the corresponding radical. Further reaction of this intermediate radical then yields the product.

Intermediate allylic radical

[1]Alfred Wohl (1863–1933); b. Graudentz; Ph.D., Berlin (Hofmann); professor, University of Danzig.

The most important feature of this reaction concerns its regioselectivity. Why does bromination occur exclusively at the position next to the double bond, rather than at one of the other positions? The answer, once again, is provided by the Hammond postulate. In radical halogenation reactions, the product is determined by the stability of the intermediate radicals. The NBS brominations of alkenes occur in the allylic position because allylic radicals are more stable than other alkyl radicals. There are three kinds of C—H bonds in cyclohexene, and looking back at Table 4.5 gives us an idea of their relative bond dissociation energies. Although a typical alkyl C—H bond has a strength of about 95 kcal/mol, and a typical vinylic C—H bond has a strength of 108 kcal/mol, we find that an *allylic* C—H bond has a strength of only 87 kcal/mol. An allylic radical is therefore more stable than a typical alkyl radical by approximately 8 kcal/mol.

~87 kcal/mol (allylic)

~95 kcal/mol (alkyl)

108 kcal/mol (vinylic)

We can thus expand our stability ordering of radicals:

$$\text{Allylic} > R_3C\cdot > R_2CH\cdot > RCH_2\cdot > CH_3\cdot > \text{Vinylic}$$

9.6 Stability of the allyl radical: resonance

Why are allylic radicals so stable? To get an idea of the reason, look at the orbital picture of the allyl radical in Figure 9.4. The carbon atom next to the double bond can adopt sp^2 hybridization, placing the unpaired electron in a p orbital. This structure is electronically symmetrical. The p orbital on the

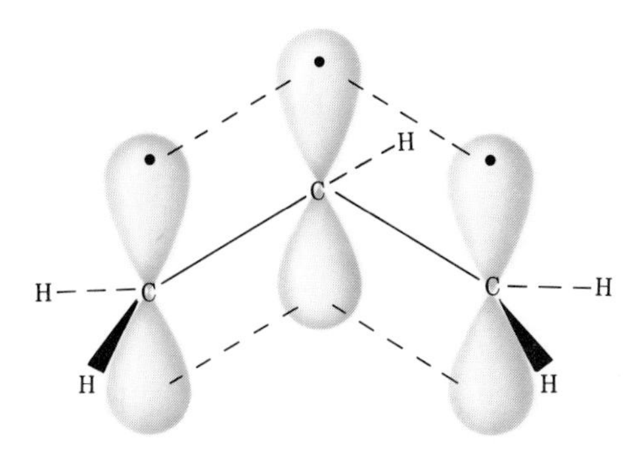

Figure 9.4. Electronic structure of the allyl radical. The structure is electronically symmetrical.

central carbon can overlap equally well with either of the two neighboring carbons.

Since the allyl radical is symmetrical, there are two ways in which we can draw it: with the unpaired electron on the left and the double bond on the right, or with the unpaired electron on the right and the double bond on the left. *Neither structure is correct by itself; the true structure of the allyl radical is somewhere in between.*

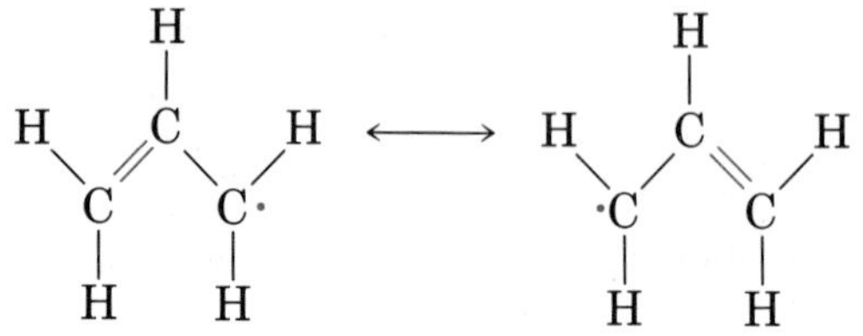

```
        H                          H
        |                          |
   H    C    H     <------>   H    C    H
    \ //  \ /                  \  / \\ /
     C     C·                  ·C     C
     |     |                    |     |
     H     H                    H     H
```

The two structures are called **resonance forms,** and their relationship is indicated by the double-headed arrow between them. The only difference between resonance forms is the placement of the bonding electrons. The nuclei themselves do not move, but occupy exactly the same places in both resonance forms.

Resonance is an extremely useful concept, and it is also a very simple concept if we keep in mind that a species like the allyl radical is no different from any other organic substance. An allyl radical does *not* jump back and forth between two resonance forms, spending part of its time looking like one and the rest of its time looking like the other; rather, an allyl radical has a single unchanging structure we call a **resonance hybrid.** The real difficulty in visualizing the resonance concept is that we cannot draw an accurate single picture of a resonance hybrid using familiar Kekulé structures. We might try to represent the allyl radical by using a dotted line to indicate that the two C—C bonds are equivalent and that each is approximately $1\frac{1}{2}$ bonds. Such a drawing is imprecise, however, and will be used infrequently in this book.

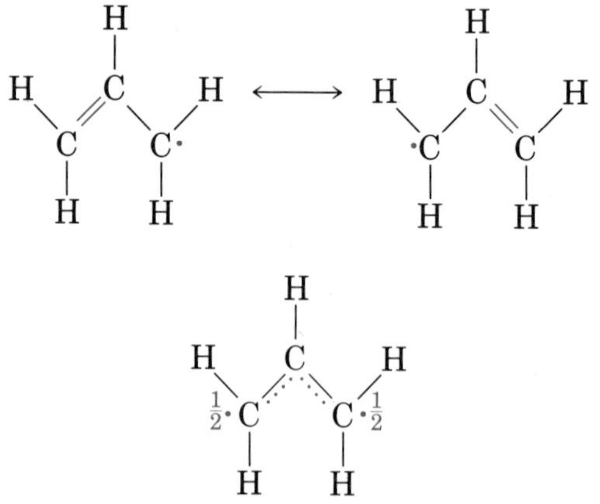

```
        H                          H
        |                          |
   H    C    H     <------>   H    C    H
    \ //  \ /                  \  / \\ /
     C     C·                  ·C     C
     |     |                    |     |
     H     H                    H     H

                   H
                   |
              H    C    H
               \ ·· ·· /
             ½·C       C·½
               |       |
               H       H
```

One of the most important postulates of resonance theory is that the greater the number of possible resonance forms, the greater the stability of the compound. For example, an allyl radical is a resonance hybrid of two equivalent Kekulé structures and is therefore more stable than a typical alkyl radical. In molecular orbital terms, the stability of the allyl radical is due to the fact that the unpaired electron can be *delocalized* over an extended pi orbital network rather than centered on only one site.

$$CH_2{=}CH{-}\dot{C}H_2 \longleftrightarrow \dot{C}H_2{-}CH{=}CH_2$$

Allyl radical (more stable)

$$CH_3CH_2\dot{C}H_2$$

Propyl radical (less stable)

In addition to affecting stability, the delocalization of the unpaired electron in the allyl radical has chemical consequences: Bromination of the allyl radical can occur *at either end* of the pi orbital system. This means that allylic bromination of an unsymmetrical alkene can lead to a *mixture* of products. For example, bromination of 1-octene leads to both 3-bromo-1-octene and 1-bromo-2-octene. We might expect the more stable disubstituted alkene to predominate over the monosubstituted alkene, and this is in fact observed, but the crucial point is that a mixture is formed.

$$CH_3(CH_2)_4CH_2CH{=}CH_2 \xrightarrow[CCl_4]{NBS} CH_3(CH_2)_4CH(Br)CH{=}CH_2 + CH_3(CH_2)_4CH{=}CHCH_2Br$$

1-Octene; 3-Bromo-1-octene (17%); 1-Bromo-2-octene (83%) (53:47, trans:cis)

Since product mixtures are formed by bromination of unsymmetrical alkenes, allylic bromination is best carried out on symmetrical alkenes such as cyclohexene. The products of these reactions are particularly useful for conversion into dienes by dehydrohalogenation with base:

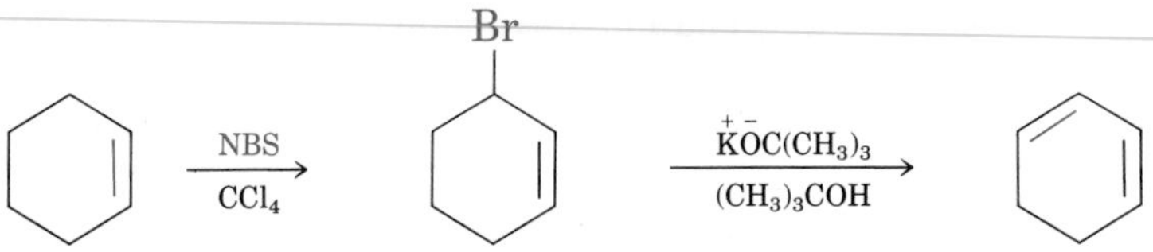

Cyclohexene $\xrightarrow[CCl_4]{NBS}$ 3-Bromocyclohexene $\xrightarrow[(CH_3)_3COH]{\overset{+}{K}\overset{-}{O}C(CH_3)_3}$ 1,3-Cyclohexadiene

PROBLEM

9.3 The major product of the reaction of methylenecyclohexane and *N*-bromosuccinimide is 1-(bromomethyl)cyclohexene. How do you explain this?

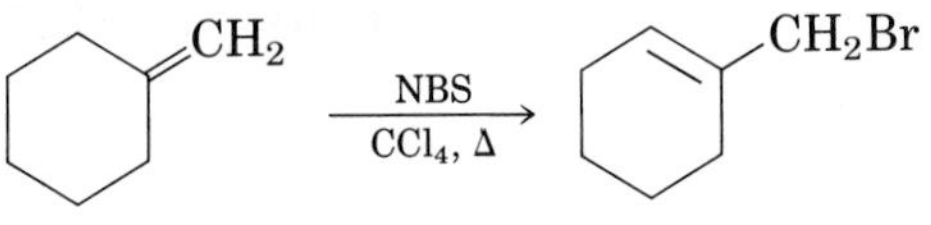

Methylenecyclohexane ($=CH_2$) $\xrightarrow[CCl_4,\ \Delta]{NBS}$ 1-(bromomethyl)cyclohexene (CH_2Br)

Major product

PROBLEM

9.4 How do you account for the fact that NBS is unreactive to alkanes, yet undergoes rapid reaction with alkylbenzenes? Refer to Table 4.5 for a hint.

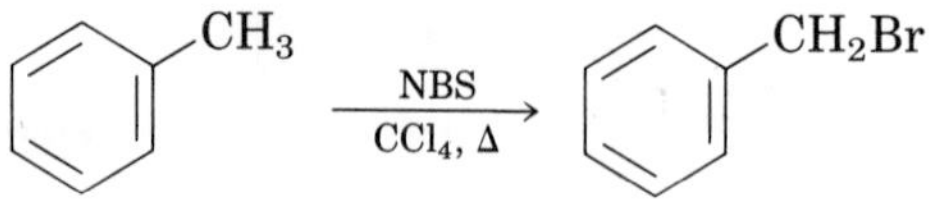

$$C_6H_5CH_3 \xrightarrow[CCl_4,\ \Delta]{NBS} C_6H_5CH_2Br$$

PROBLEM

9.5 Draw as many resonance structures as you can for these species:

(a) Carbonate ion, CO_3^{2-} (all oxygens are equivalent)

(b) Nitromethane, $CH_3-\overset{\oplus}{N}(=O)-O^{\ominus}$

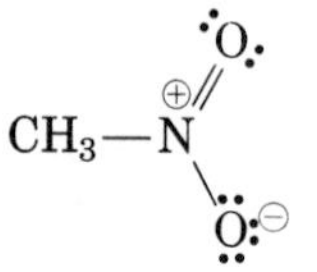

(c) Ozone, $O=\overset{\oplus}{O}-O^{\ominus}$

PROBLEM

9.6 Spectroscopic measurements indicate that the two oxygen atoms of sodium acetate are equivalent. Both C—O bonds have the same length (1.26 Å). Explain.

$$CH_3-\overset{O}{\overset{\|}{C}}-O^-\ Na^+$$

Sodium acetate

9.7 Alkyl halides from alcohols

The most valuable method for the preparation of alkyl halides is their synthesis from alcohols. A great variety of alcohols are commercially available, and we shall see later that many more can be obtained by reduction of carbonyl compounds. Because of the importance and generality of the reaction, many different reagents have been used for transforming alcohols into alkyl halides.

The simplest (but also least generally useful) method for carrying out the alcohol to alkyl halide conversion involves treating the alcohol with HX:

$$R-OH \xrightarrow{HX} R-X + H_2O \qquad (X = Cl, Br, \text{ or } I)$$

For reasons to be discussed in Section 10.14, the reaction works best when applied to tertiary alcohols. Primary and secondary alcohols also react, but at considerably slower rates. Although this may not be a problem in simple cases, more complicated molecules are sometimes acid sensitive and undergo unwanted side reactions.

Reactivity order:	Tertiary	>	Secondary	>	Primary
	R_3COH		R_2CHOH		RCH_2OH

The reaction of a hydrogen halide with a tertiary alcohol is so rapid that it is often carried out simply by bubbling the pure HX gas into a cold ether solution of the alcohol. Reaction is usually complete within minutes at 0°C.

$$\text{1-Methylcyclohexanol} \xrightarrow[\text{Ether, 0°C}]{\text{HCl (gas)}} \text{1-Chloro-1-methylcyclohexane (90\%)} + H_2O$$

Primary and secondary alcohols are best converted into their corresponding halides by treatment with such reagents as thionyl chloride ($SOCl_2$) or phosphorus tribromide (PBr_3). These reactions normally take place fairly readily under mild conditions; they are less acidic and therefore less likely than the HX method to cause acid-catalyzed rearrangements.

$$3\ CH_3CH_2CH(OH)CH_3 \xrightarrow[\text{Ether, 35°C}]{PBr_3} 3\ CH_3CH_2CH(Br)CH_3 + P(OH)_3$$

2-Butanol → 2-Bromobutane (86%)

2-(Hydroxymethyl)-tetrahydrofuran (CH_2OH):

$$\xrightarrow[\text{Ether, 35°C}]{PBr_3} CH_2Br + P(OH)_3$$

2-(Bromomethyl)-tetrahydrofuran (80%)

$$\xrightarrow[\text{Pyridine}]{SOCl_2} CH_2Cl + SO_2 + HCl$$

2-(Chloromethyl)-tetrahydrofuran (75%)

$$\text{Benzoin} \xrightarrow[\text{Pyridine}]{SOCl_2} \text{Desyl chloride (86\%)} + SO_2 + HCl$$

As the foregoing examples indicate, the yields of these PBr_3 and $SOCl_2$ reactions are generally high, and other functional groups such as ethers, carbonyls, and aromatic rings do not usually interfere.

9.8 Reactions of alkyl halides: Grignard formation

Earlier in this chapter we saw that alkyl halides are electrophiles. They therefore react with nucleophiles, as we have seen in the reaction of bromomethane with hydroxide ion (Section 4.2).

$$\overset{\delta+}{C}H_3\overset{\delta-}{Br} + :\overset{-}{O}H \longrightarrow CH_3OH + :Br^-$$

Electrophile Nucleophile

Reactions of alkyl halides with nucleophiles are extremely important, and the entire next chapter is devoted to their study. There are, however, a variety of other processes that alkyl halides undergo. One such process is their reaction with certain metals to form organometallic compounds—species that contain a carbon–metal bond. We'll take a brief look at the subject now, and then return for a more detailed examination in Section 20.7.

Organohalides of widely varying structure—alkyl, aryl, and vinylic—react with magnesium metal in ether or tetrahydrofuran (THF) solvent to yield organomagnesium halides:

$$R{-}X + Mg \xrightarrow[\text{or THF}]{\text{Ether}} R{-}Mg{-}X$$

where R = 1°, 2°, or 3° alkyl, aryl, or vinylic

X = Cl, Br, or I

For example,

$$C_6H_5Br \xrightarrow[\text{Ether}]{\text{Mg}} C_6H_5MgBr$$

Phenylmagnesium bromide

$$(CH_3)_3CCl \xrightarrow[\text{Ether}]{\text{Mg}} (CH_3)_3C{-}MgCl$$

t-Butylmagnesium chloride

These organomagnesium halides—named **Grignard reagents** after their discoverer, Victor Grignard[2]—are extraordinarily versatile and useful compounds. Steric hindrance in the halide does not appear to be a factor in Grignard formation, since alkyl halides of all description, 1°, 2°, and 3°, react with ease. Aryl and vinylic halides also react, although it is best to employ THF as solvent for these cases:

$$H_2C{=}C(CH_3){-}Br \xrightarrow[\text{THF}]{\text{Mg}} H_2C{=}C(CH_3){-}MgBr$$

2-Bromopropene — Isopropenylmagnesium bromide

[2]François Auguste Victor Grignard (1871–1935); b. Cherbourg, France; professor, University of Nancy, Lyons; Nobel prize, 1912.

The halogen may be Cl, Br, or I, although chlorides are somewhat less reactive than bromides and iodides. Organofluorides rarely react with magnesium.

As we might expect from our knowledge of electronegativities and bond polarities, the carbon–magnesium bond is highly polarized,

$$\overset{\delta^-}{C} \leftrightarrow \overset{\delta^+}{Mg}$$

making the organic residue both nucleophilic and strongly basic. In a formal sense, a Grignard reagent can be considered a carbon anion or **carbanion**—the magnesium salt of a hydrocarbon acid. This view is not fully accurate, however, and Grignard reagents are best viewed as containing a highly polar covalent C—Mg bond, rather than an ionic bond between C^- and Mg^+.

$$-\!\!\overset{\diagdown}{\underset{\diagup}{C}}\!\!-H \Rightarrow -\!\!\overset{\diagdown}{\underset{\diagup}{C}}\!:^{-\,+}MgBr$$

"Acid" Carbanion "salt"

Because of their nucleophilic/basic character, Grignard reagents react with a wide variety of electrophiles. For example, they react with proton donors such as H_2O, ROH, RCOOH, or RNH_2 to yield hydrocarbons. The overall sequence is a useful synthetic method for *reducing* organohalides (R—X → R—H):

$$R{-}X \xrightarrow{Mg} R{-}Mg{-}X \xrightarrow{H_2O} R{-}H + HOMgX$$

Alkyl halide — Grignard reagent — Alkane

For example,

$$CH_3(CH_2)_8CH_2Br \xrightarrow[2.\ H_2O]{1.\ Mg} CH_3(CH_2)_8CH_3$$

1-Bromodecane — Decane (85%)

Another method for converting primary and secondary halides into hydrocarbons is to displace the halide ion by hydride ion, H:⁻. A number of hydride-donor reagents are available; lithium aluminum hydride, $LiAlH_4$, is most often used. 1-Bromodecane, for example, reacts with lithium aluminum hydride in ether to give decane in 72% yield:

$$CH_3(CH_2)_8CH_2Br \xrightarrow[LiAlH_4,\ ether]{\text{"H:}^-\text{"}} CH_3(CH_2)_8CH_3 + :Br^-$$

1-Bromodecane — Decane (72%)

For the replacement of primary and secondary halogens, lithium triethylborohydride, $Li(CH_3CH_2)_3BH$, is preferred to $LiAlH_4$. The mechanisms of these displacement reactions will be discussed in the next chapter.

$$\text{Bromocyclohexane} \xrightarrow[\text{Li(CH}_3\text{CH}_2)_3\text{BH, ether}]{\text{"H:}^-\text{"}} \text{Cyclohexane (95\%)} + :Br^-$$

Bromocyclohexane — Cyclohexane (95%)

PROBLEM

9.7 How strong a base would you expect a Grignard reagent to be? Refer to Section 7.7 for a quantitative estimation.

PROBLEM

9.8 One major advantage of alkyl halide reduction via Grignard reagents is that the sequence can be used to introduce deuterium into a specific site in a molecule. How might you do this?

$$CH_3CH(Br)CH_2CH_3 \xrightarrow{?} CH_3CH(D)CH_2CH_3$$

PROBLEM

9.9 Why do you suppose one *cannot* prepare a Grignard reagent from a bromo alcohol such as 4-bromo-1-pentanol?

$$CH_3CH(Br)CH_2CH_2CH_2OH \overset{Mg}{\not\rightarrow} CH_3CH(MgBr)CH_2CH_2CH_2OH$$

Give another example of a molecule that will not form a Grignard reagent.

9.9 Organometallic coupling reactions

Alkyllithium reagents can be prepared by the reaction of lithium metal with alkyl halides in a manner analogous to the formation of Grignard reagents:

$$CH_3CH_2CH_2CH_2Br \xrightarrow[\text{Pentane}]{2\text{ Li}} CH_3CH_2CH_2\overset{\delta^-}{C}H_2\overset{\delta^+}{Li} + LiBr$$

1-Bromobutane — Butyllithium

These alkyllithiums, like alkylmagnesium halides, are both nucleophiles and bases, and their chemistry is quite similar in many respects to that of the alkylmagnesium halides. One valuable reaction of alkyllithiums is their use in preparing *lithium diorganocoppers,* or Gilman[3] reagents.

$$CH_3Br + 2\text{ Li} \xrightarrow{\text{Pentane}} CH_3Li + LiBr$$

Bromomethane — Methyllithium

$$2\text{ }CH_3Li + CuI \xrightarrow{\text{Ether}} (CH_3)_2Cu^-Li^+ + LiI$$

Methyllithium — Lithium dimethylcopper (a Gilman reagent)

[3]Henry Gilman (1893–); b. Boston; Ph.D. (1918) Harvard (Kohler); professor, Iowa State University (1923–).

Gilman reagents are easily prepared from alkyllithiums and cuprous iodide. Though rather unstable, they are soluble in ether solvents. They have the remarkable ability of undergoing organometallic *coupling* reactions with alkyl bromides and iodides (but not fluorides). One of the alkyl groups from the Gilman reagent replaces the halogen from alkyl halide, resulting in the formation of a hydrocarbon with a new carbon–carbon bond. Lithium dimethylcopper, for example, reacts with 1-iodododecane to give undecane in 90% yield:

$$(CH_3)_2CuLi + CH_3(CH_2)_8CH_2I \xrightarrow[0°C]{\text{Ether}} CH_3(CH_2)_8CH_2CH_3 + LiI + CH_3Cu$$

Lithium dimethylcopper 1-Iododecane Undecane (90%)

This coupling reaction is highly versatile and is of great use in organic synthesis. As the following examples indicate, the coupling reaction can be carried out on aryl and vinylic halides as well as on alkyl halides.

$$n\text{-}C_7H_{15}(H)C{=}C(H)I + (n\text{-}C_4H_9)_2CuLi \longrightarrow n\text{-}C_7H_{15}(H)C{=}C(H)C_4H_9\text{-}n + n\text{-}C_4H_9Cu + LiI$$

trans-1-Iodo-1-nonene *trans*-5-Tridecene (71%)

$$C_6H_5I + (CH_3)_2CuLi \longrightarrow C_6H_5CH_3 + CH_3Cu + LiI$$

Iodobenzene Toluene (91%)

An organocopper coupling reaction is carried out on a commercial scale to synthesize *muscalure,* (9*Z*)-tricosene, the sex attractant secreted by the common housefly. Minute amounts of this insect hormone, or **pheromone,** greatly increase the lure of insecticide-treated fly bait and provide an effective and species-specific means of insect control. Coupling of *cis*-1-bromo-9-octadecene with lithium dipentylcopper is used industrially to produce muscalure in 100 lb batches.

$$n\text{-}C_8H_7(H)C{=}C(H)(CH_2)_7CH_2Br + (n\text{-}C_5H_{11})_2CuLi \longrightarrow n\text{-}C_8H_{17}(H)C{=}C(H)C_{13}H_{27}\text{-}n$$

cis-1-Bromo-9-octadecene Lithium dipentylcopper Muscalure, (9*Z*)-tricosene (99%)

Although the details of the mechanism by which coupling occurs are not fully understood, it appears that radicals are involved. This coupling is *not* a typical nucleophilic substitution reaction of the type considered in the next chapter.

PROBLEM

9.10 By employing some of the reactions we have used, we can synthesize a wide variety of products from simple starting materials. How, for example, would you prepare these compounds?
(a) 3-Methylcyclohexene from cyclohexene
(b) Octane from 1-bromobutane
(c) Decane from 1-pentene

9.10 Summary

Alkyl halides are compounds containing halogen bonded to a saturated carbon atom. The C—X bond is polar, and alkyl halides can therefore behave as electrophiles.

Alkyl halides can be prepared by radical chlorination or bromination of alkanes, but this method is of little general value since mixtures of products always result. Studies have shown that the reactivity order of alkanes toward chlorination is identical to the stability order of radicals: tertiary $>$ secondary $>$ primary. According to the Hammond postulate, the more stable radical intermediate is formed faster because the transition state leading to it is more stabilized.

Alkyl halides can also be prepared from alkenes. Alkenes add HX, and they react with NBS to give the product of allylic bromination. The NBS bromination of alkenes is a complex radical process that takes place through an intermediate allyl radical. Allyl radicals are stabilized by resonance and can be drawn in two different ways:

$$CH_2{=}CH{-}\dot{C}H_2 \longleftrightarrow \dot{C}H_2{-}CH{=}CH_2$$

The only difference between the two resonance structures is in the location of bonding electrons—the nuclei remain in the same places. No single Kekulé drawing can depict the true structure of the allyl radical. The true structure is a resonance hybrid somewhere intermediate between the two Kekulé forms.

Alcohols react with HX to form alkyl halides, but this works well only for tertiary alcohols. Primary and secondary alkyl halides are normally prepared from alcohols using either $SOCl_2$ or PBr_3. Alkyl halides react with magnesium in ether solution to form organomagnesium halides—Grignard reagents. Grignard reagents are both nucleophilic and basic, and can react with proton sources to form hydrocarbons. The overall result of Grignard formation and protonation is reduction, that is,

$$-\overset{\diagdown}{\underset{\diagup}{C}}-X \longrightarrow -\overset{\diagdown}{\underset{\diagup}{C}}-H$$

Alkyl halides also react with lithium metal to form organolithiums. In the presence of CuI, these form diorganocoppers, or Gilman reagents. Gilman reagents react with alkyl halides by a radical process to yield coupled hydrocarbon products.

9.11 Summary of reactions

A. Preparation of alkyl halides

1. From alkenes by allylic bromination (Section 9.5)

$$>C{=}C{-}\underset{|}{\overset{H}{C}}< \xrightarrow[CCl_4]{NBS} >C{=}C{-}\overset{Br}{C}<$$

2. From alkenes by addition of HBr and HCl (Sections 5.8 and 5.9)

$$>C{=}C< + HBr \longrightarrow \overset{H}{C}{-}\overset{Br}{C}$$

3. From alcohols

a. Treatment with HX, where X = Cl, Br, or I (Section 9.7)

$$\overset{OH}{C} \xrightarrow[Ether]{HX} \overset{X}{C}$$

Reactivity order: 3° > 2° > 1°

b. Treatment with $SOCl_2$/pyridine, best for 1° and 2° alcohols (Section 9.7)

$$\overset{OH}{C}{-}H \xrightarrow[Pyridine]{SOCl_2} \overset{Cl}{C}{-}H$$

c. Treatment with PBr_3, best for 1° and 2° alcohols (Section 9.7)

$$\overset{OH}{C}{-}H \xrightarrow[Ether]{PBr_3} \overset{Br}{C}{-}H$$

B. Reactions of alkyl halides

1. Grignard formation (Section 9.8)

$$R{-}X \xrightarrow[Ether]{Mg} R{-}Mg{-}X$$

where X = Br, Cl, or I

R = 1°, 2°, 3° alkyl, aryl, or vinylic

2. Diorganocopper (Gilman reagent) formation (Section 9.9)

a. $R{-}X \xrightarrow[Pentane]{2\ Li} R{-}Li + LiX$

b. $2\ R{-}Li \xrightarrow[\text{Ether}]{\text{CuI}} R_2CuLi$

where R = 1°, 2°, 3° alkyl, aryl, or vinylic

3. Organometallic coupling (Section 9.9)

$$R_2CuLi + R'X \xrightarrow{\text{Ether}} R{-}R' + RCu + LiX$$

4. Halide reduction (Section 9.8)

a. $R{-}X \xrightarrow[\text{Ether}]{\text{Mg}} R{-}Mg{-}X \xrightarrow{H_2O} R{-}H$

b. $R{-}X \xrightarrow[\text{THF}]{\text{Li(CH}_3\text{CH}_2)_3\text{BH}} R{-}H$

where R = 1° or 2°

ADDITIONAL PROBLEMS

9.11 Provide correct IUPAC names for the following alkyl halides:

(a) $(CH_3)_2CHCHBrCHBrCH_2CH(CH_3)_2$

(b) $CH_3CH{=}CHCH_2CHICH_3$

(c) $(CH_3)_2CBrCH_2CHClCH(CH_3)_2$

(d) $CH_3CH_2CH(CH_2Br)CH_2CH_2CH_3$

(e) $ClCH_2CH_2CH_2C{\equiv}CCH_2Br$

9.12 A chemist requires a large amount of 1-bromo-2-pentene as starting material for a synthesis. She finds a supply of 2-pentene in the stockroom, and decides to carry out an NBS allylic bromination reaction:

$$CH_3CH_2CH{=}CHCH_3 \xrightarrow[CCl_4]{\text{NBS}} CH_3CH_2CH{=}CHCH_2Br$$

What is wrong with this synthesis plan? What side products would form in addition to the desired product?

9.13 What product(s) would you expect from the reaction of 1-methylcyclohexene with NBS? Would you use this reaction as part of a synthesis?

1-methylcyclohexene (ring with CH_3) $\xrightarrow[CCl_4]{\text{NBS}}$?

9.14 Table 4.5 shows that the methyl C—H bond of toluene is 13 kcal/mol weaker than the C—H bond of ethane. Another way of stating the same thing is to say that the $C_6H_5CH_2\cdot$ (benzyl) radical is 13 kcal/mol more stable than the $CH_3\cdot$ radical. Draw as many resonance structures as you can for the benzyl radical.

(benzene ring with CH_3)

Toluene

9.15 What product(s) would you expect from the reaction of 1-phenyl-2-butene with NBS? Explain.

(benzene ring–$CH_2CH{=}CH$–CH_3)

9.16 How would you prepare the following compounds, starting with cyclopentene and any other reagents needed?

(a) Chlorocyclopentane
(b) Methylcyclopentane
(c) 3-Bromocyclopentene
(d) Cyclopentanol
(e) Cyclopentylcyclopentane
(f) Monodeuterocyclopentane

9.17 Predict the product(s) of these reactions:

(a) 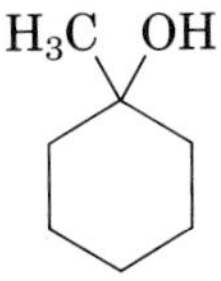 1-methylcyclohexanol (H_3C, OH) $\xrightarrow[\text{Ether}]{\text{HBr}}$?

(b) $CH_3CH_2CH_2CH_2OH \xrightarrow[\text{Pyridine}]{SOCl_2}$?

(c) 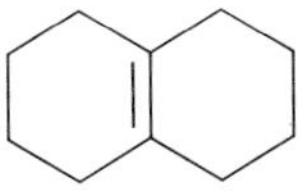 (bicyclic alkene) $\xrightarrow[CCl_4]{\text{NBS}}$?

(d) 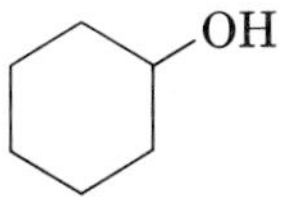 (cyclohexanol, OH) $\xrightarrow[\text{Ether}]{PBr_3}$?

(e) $CH_3CH_2CHBrCH_3 \xrightarrow[\text{Ether}]{\text{Mg}} A \xrightarrow{H_2O} B$

(f) $CH_3CH_2CH_2CH_2Br \xrightarrow[\text{Pentane}]{\text{Li}} A \xrightarrow{\text{CuI}} B$

(g) $CH_3CH_2CH_2CH_2Br + (CH_3)_2CuLi \xrightarrow{\text{Ether}}$?

9.18 Which of the following pairs of structures represent resonance forms?

(a) 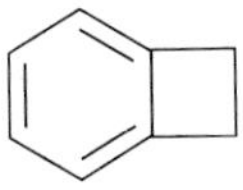 (structure) and (structure)

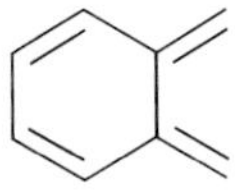

(b) 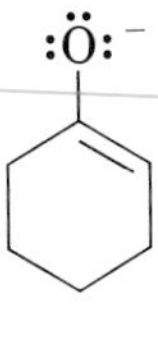 $:\ddot{O}:^-$ (structure) and $:O:$ (structure)

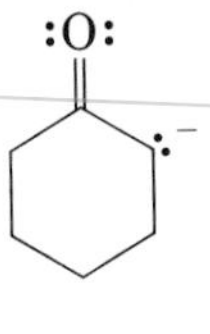

(c) 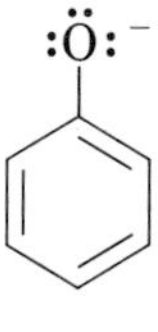 $:\ddot{O}:^-$ (structure) and $:O:$ (structure)

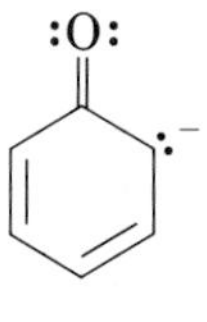

(d) 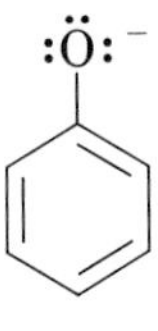 $:\ddot{O}:^-$ (structure) and $:O:$ (structure)

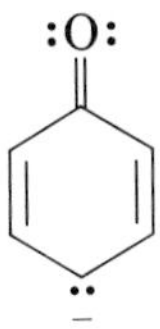

9.19 Draw as many resonance structures as you can for the following species:

(a) Acetone enolate, $CH_3{-}\overset{\overset{:O:}{\|}}{C}{-}\ddot{C}H_2^{-}$

(b) Diazomethane, $H_2C{=}\overset{\oplus}{N}{=}\ddot{\underset{..}{N}}^{\ominus}$

(c) Guanidinium ion, $H_2\ddot{N}{-}\overset{\overset{+NH_2}{\|}}{C}{-}\ddot{N}H_2$

9.20 How would you carry out these syntheses?
(a) Butylcyclohexane from cyclohexene
(b) Butylcyclohexane from cyclohexanol
(c) Butylcyclohexane from cyclohexane

9.21 The syntheses shown here are unlikely to occur as written. What is wrong with each?

(a) $(CH_3)_3CF \xrightarrow[\text{Ether}]{LiAlH_4} (CH_3)_3CH$

(b) 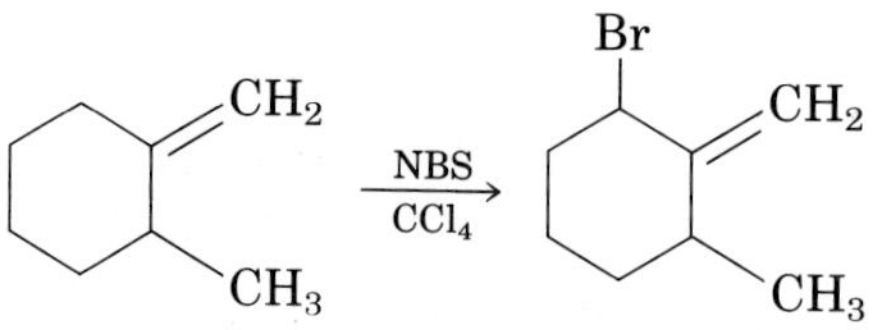 1-methylene-2-methylcyclohexane (ring with $=CH_2$ and CH_3) $\xrightarrow[CCl_4]{NBS}$ ring with Br, $=CH_2$ and CH_3

(c) 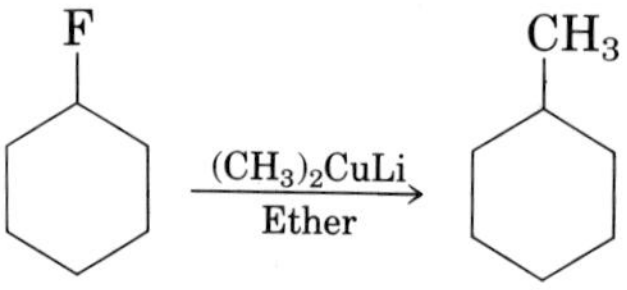 Fluorocyclohexane (F on ring) $\xrightarrow[\text{Ether}]{(CH_3)_2CuLi}$ methylcyclohexane (CH_3 on ring)

9.22 Which of the following pairs represent resonance structures?

(a) $CH_3C{\equiv}\overset{\oplus}{N}{-}\ddot{\underset{..}{O}}{:}^{\ominus}$ and $CH_3\overset{\oplus}{C}{=}\ddot{N}{-}\ddot{\underset{..}{O}}{:}^{\ominus}$

(b) $CH_3\overset{\overset{:O:}{\|}}{C}{-}\ddot{\underset{..}{O}}{:}^{-}$ and $:\overset{-}{C}H_2\overset{\overset{:O:}{\|}}{C}{-}\ddot{\underset{..}{O}}{-}H$

(c) 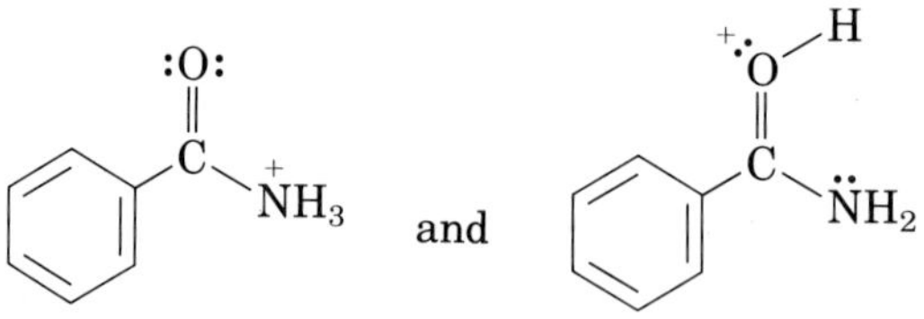 $C_6H_5{-}\overset{\overset{:O:}{\|}}{C}{-}\overset{+}{N}H_3$ and $C_6H_5{-}\overset{\overset{+\ddot{O}-H}{\|}}{C}{-}\ddot{N}H_2$

(d) 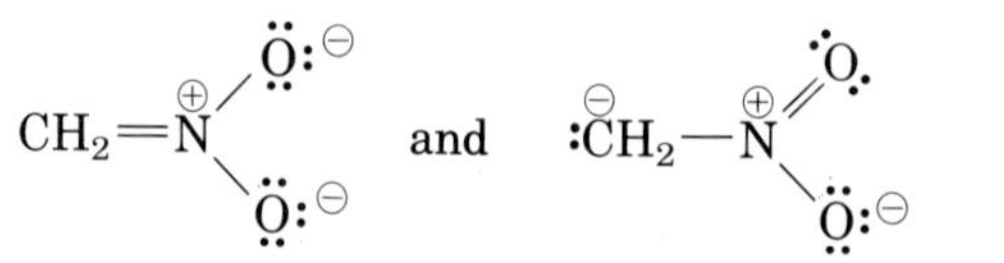 $CH_2{=}\overset{\oplus}{N}(\ddot{\underset{..}{O}}{:}^{\ominus})_2$ and $:\overset{\ominus}{C}H_2{-}\overset{\oplus}{N}({=}\ddot{O})(\ddot{\underset{..}{O}}{:}^{\ominus})$

CHAPTER 10 REACTIONS OF ALKYL HALIDES: NUCLEOPHILIC SUBSTITUTION REACTIONS

We saw in the last chapter that the carbon–halogen bond in alkyl halides is polar, and that the carbon atom bears a slight positive charge. Thus, alkyl halides are electrophiles, and much of their chemistry involves polar reactions with nucleophiles. We have already seen two examples of such reactions: the nucleophilic substitution of bromide ion by hydroxide ion, which occurs when bromomethane is treated with sodium hydroxide, and the elimination of HBr, which occurs when 2-bromo-2-methylpropane is treated with sodium ethoxide:

Substitution

$$CH_3{-}Br + {}^{-}{:}OH \longrightarrow CH_3{-}OH + Br{:}^{-}$$

Elimination

$$(CH_3)_2C(Br){-}CH_3 + {}^{-}{:}OCH_2CH_3 \longrightarrow CH_3{-}C(CH_3){=}CH_2 + {:}Br^{-} + HOCH_2CH_3$$

Nucleophilic substitution and elimination are two of the most important reactions in organic chemistry, and it is now time for a close look at them. We need to see how these reactions occur, what factors are involved, and how we can control the reactions for purposes of organic synthesis.

10.1 The discovery of the Walden inversion

In 1896, the German chemist Paul Walden[1] reported a remarkable discovery. He found that the pure enantiomeric (+)- and (−)-malic acids could be *interconverted* by a series of simple substitution reactions. When Walden treated (−)-malic acid with PCl_5, he isolated dextrorotatory (+)-chlorosuccinic acid. This, on treatment with wet silver oxide, gave (+)-malic acid. Similarly, (+)-malic acid gave levorotatory (−)-chlorosuccinic acid on reaction with PCl_5, and this was converted into (−)-malic acid when treated with wet silver oxide. The full cycle of reactions reported by Walden is shown in Figure 10.1.

At the time, the results were astonishing. The eminent chemist Emil Fischer called Walden's discovery "the most remarkable observation made in the field of optical activity since the fundamental observations of Pasteur." Since (−)-malic acid was being converted into (+)-malic acid, some reactions in the cycle must have occurred with a change in configuration at the chiral center. But which ones? (Recall that the direction of light rotation and absolute configuration are not related, and we cannot tell by looking at the sign of rotation whether or not a change in chirality has occurred during a reaction.)

[1] Paul Walden (1863–1957); b. Latvia; Ph.D., Leipzig; student and professor, Riga Polytechnic, Russia (1882–1919); professor, University of Rostock, University of Tübingen, Germany.

$$\underset{\substack{(-)\text{-Malic acid} \\ [\alpha]_D = -2.3^\circ}}{HO_2CCH_2\overset{\overset{OH}{|}}{C}HCO_2H} \xrightarrow[\text{Ether}]{PCl_5} \underset{(+)\text{-Chlorosuccinic acid}}{HO_2CCH_2\overset{\overset{Cl}{|}}{C}HCO_2H}$$

$$\uparrow Ag_2O,\ H_2O \qquad\qquad \downarrow Ag_2O,\ H_2O$$

$$\underset{(-)\text{-Chlorosuccinic acid}}{HO_2CCH_2\overset{\overset{Cl}{|}}{C}HCO_2H} \xleftarrow[\text{Ether}]{PCl_5} \underset{\substack{(+)\text{-Malic acid} \\ [\alpha]_D = +2.3^\circ}}{HO_2CCH_2\overset{\overset{OH}{|}}{C}HCO_2H}$$

Figure 10.1. Walden cycle interconverting (+)- and (−)-malic acids.

10.2 Stereochemistry of nucleophilic substitution

Although Walden realized that changes in configuration must have taken place during his reaction cycle, he did not know what steps were involved. In the 1920s, however, Joseph Kenyon and Henry Phillips began a series of investigations designed to elucidate the stereochemistry of substitution reactions. They recognized that the presence of the carboxylic acid group in Walden's work may have led to complications, and they therefore carried out their own work on simpler cases. Among reactions they studied was one that interconverted the two enantiomers of 1-phenyl-2-propanol (Figure 10.2, p. 292). Although this particular series of reactions involves nucleophilic substitution of an alkyl toluenesulfonate (a *tosylate*) rather than an alkyl halide, exactly the same type of reaction is involved as was studied by Walden:

$$R{-}X + Nu{:}^- \longrightarrow R{-}Nu + X{:}^-$$

where X = —Cl, —Br, —I, or *p*-toluenesulfonyl (Tos):

$$-O-\overset{\overset{O}{\|}}{\underset{\underset{O}{\|}}{S}}-C_6H_4-CH_3$$

$Nu{:}^-$ = A nucleophile

In this three-step reaction sequence, (+)-1-phenyl-2-propanol is converted into its levorotatory enantiomer. Therefore, at least one of the three steps must involve an *inversion* (change) of configuration at the chiral center. The first step, formation of a toluenesulfonate, occurs at the oxygen rather than at the chiral carbon, and the configuration around carbon is

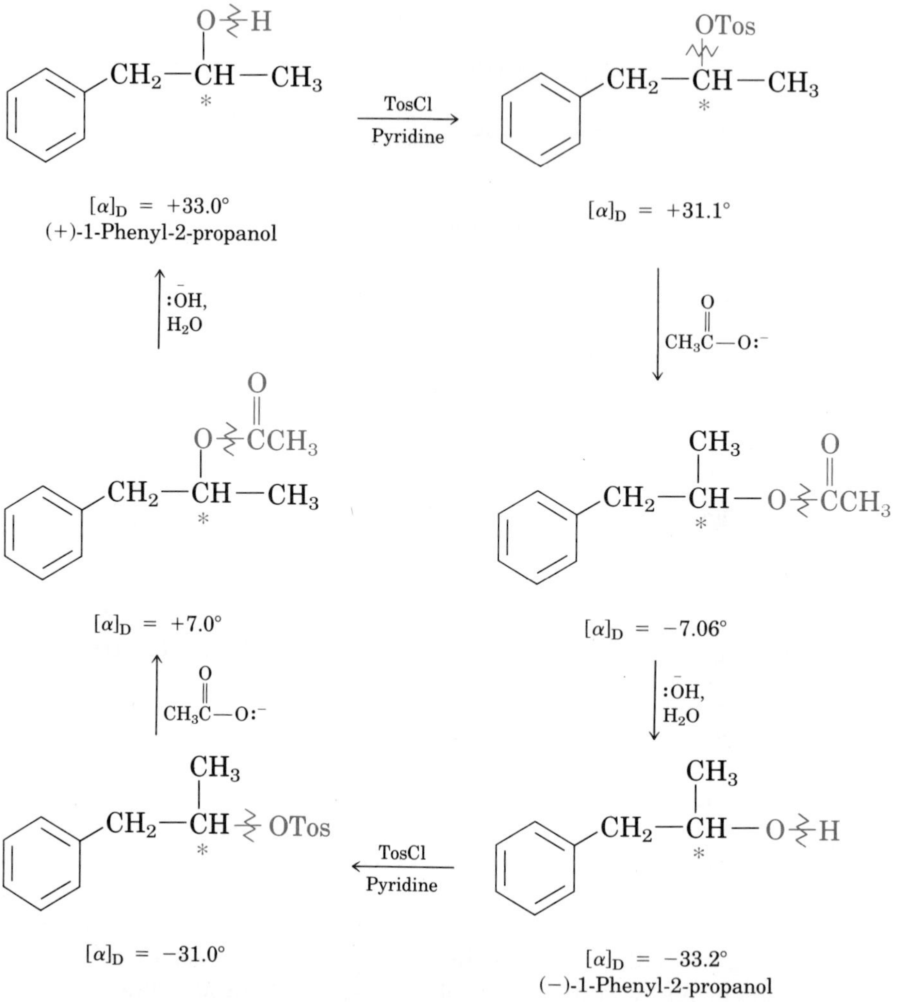

$C_6H_5CH_2—\overset{*}{C}H(O⫶H)—CH_3$ → (TosCl, Pyridine) → $C_6H_5CH_2—\overset{*}{C}H(⫶OTos)—CH_3$

$[\alpha]_D = +33.0°$
(+)-1-Phenyl-2-propanol

$[\alpha]_D = +31.1°$

↓ $CH_3\overset{O}{\overset{\|}{C}}—O:^-$

$C_6H_5CH_2—\overset{*}{C}H(CH_3)—O⫶\overset{O}{\overset{\|}{C}}CH_3$

$[\alpha]_D = -7.06°$

↓ $:\overset{-}{O}H$, H_2O

$C_6H_5CH_2—\overset{*}{C}H(CH_3)—O⫶H$

$[\alpha]_D = -33.2°$
(−)-1-Phenyl-2-propanol

← (TosCl, Pyridine) ←

$C_6H_5CH_2—\overset{*}{C}H(CH_3)⫶OTos$

$[\alpha]_D = -31.0°$

↑ $CH_3\overset{O}{\overset{\|}{C}}—O:^-$

$C_6H_5CH_2—\overset{*}{C}H(O⫶\overset{O}{\overset{\|}{C}}CH_3)—CH_3$

$[\alpha]_D = +7.0°$

↑ $:\overset{-}{O}H$, H_2O

Figure 10.2. Walden cycle on 1-phenyl-2-propanol. Chiral centers are marked by asterisks. The bonds broken in each reaction are indicated by wavy lines.

therefore unchanged. Similarly, it can be shown (Section 24.7) that the third step, hydroxide ion cleavage of the acetate, also takes place without breaking the C—O bond at the chiral center; thus, inversion cannot occur in this step. The inversion must therefore take place in the second step, the nucleophilic substitution of tosylate ion by acetate ion.

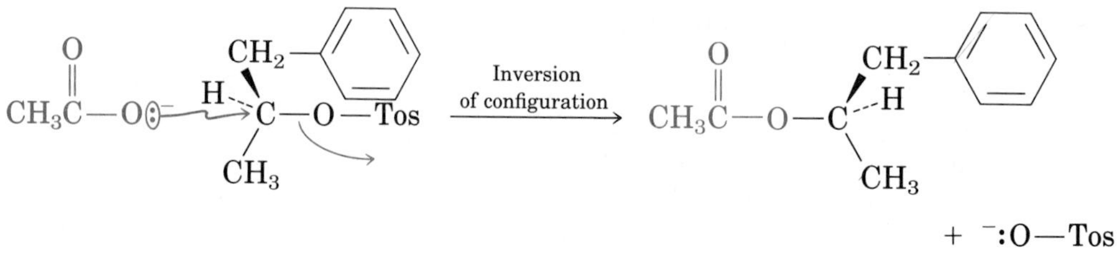

$CH_3\overset{O}{\overset{\|}{C}}—O:^-$ + H, $CH_2—C_6H_5$, CH_3 $>C—O—Tos$ $\xrightarrow{\text{Inversion of configuration}}$ $CH_3\overset{O}{\overset{\|}{C}}—O—C$ ($CH_2—C_6H_5$, H, CH_3)

$+ \ ^{-}:O—Tos$

From this and nearly a dozen other similar series of reactions, Kenyon and Phillips concluded that the nucleophilic substitution reaction of primary and secondary alkyl tosylates always proceeds with Walden inversion of configuration.

A second piece of stereochemical evidence concerning the mechanism of nucleophilic substitution reactions was provided by Hughes[2] and Ingold in 1935. It was well known at the time that optically active halides lose their optical activity and become racemic when treated with halide ions. For example, (+)-2-iodooctane racemizes to (±)-2-iodooctane when treated with lithium iodide. Hughes and Ingold reasoned that this racemization might be caused by a Walden inversion.

$$^{-}I: + (R)\text{-}(-)\text{-2-Iodooctane} \; [H, CH_3, n\text{-}C_6H_{13}, C\text{—}I] \xrightleftharpoons[\text{Walden inversion}]{\text{LiI}} (S)\text{-}(+)\text{-2-Iodooctane} \; [I\text{—}C, CH_3, H, n\text{-}C_6H_{13}] + :I^{-}$$

(R)-(−)-2-Iodooctane (S)-(+)-2-Iodooctane

If added iodide ion acts as a nucleophile and reacts with (R)-(+)-2-iodooctane via a Walden inversion, (S)-(−)-2-iodooctane will result. Hughes and Ingold reasoned that if *every* substitution of iodide by iodide results in an inversion then racemization will happen exactly twice as rapidly as iodide exchange. In other words, racemization will be complete when one-half of the starting (R)-2-iodooctane has been inverted by iodide exchange:

$$100\% \; (+) \text{ molecules} \xrightarrow{50\% \text{ inversion}} 50\% \; (+) \text{ and } 50\% \; (-) \text{ molecules}$$

Pure starting enantiomer Racemic mixture

This hypothesis was confirmed by an extremely clever experiment. The rate at which racemization occurs can be easily measured using a polarimeter by observing the loss of optical activity of 2-iodooctane. But how can one measure the rate of iodide exchange? Hughes and Ingold carried out the racemization reaction using radioactive iodide, ^{128}I, and normal 2-iodooctane, ^{127}I, and they were thus able to measure how rapidly the radioactive iodide became incorporated into the product. The result was exactly as predicted: The experimentally observed ratio of racemization to iodide exchange was 1.93 ± 0.16. Firm evidence had thus been obtained to show that such nucleophilic substitutions take place with *complete* inversion of configuration.

$$^{128}I:^{-} + [H, CH_3, n\text{-}C_6H_{13}]C\text{—}^{127}I \longrightarrow {}^{128}I\text{—}C[CH_3, H, n\text{-}C_6H_{13}] + {}^{127}:I^{-}$$

[2]Edward David Hughes (1906–); b. Caernarvonshire, North Wales; Ph.D., Wales (Watson); D.Sc., London (Ingold); professor, University College, London (1930–1974).

10.3 Kinetics of nucleophilic substitution

Chemists often speak in qualitative terms about a reaction being fast or slow. The speed at which a starting material reacts to give product is called the **reaction rate,** and is a quantity that can often be measured. The determination of reaction rates, and the dependence of reaction rate on reagent concentrations, can be enormously useful in determining mechanisms. Let's see what we can learn about the nucleophilic substitution reaction by a study of reaction rates.

For all chemical reactions, there is a mathematical relationship between the reaction rate and the concentrations of the reagents. When we measure this relationship, we measure the **kinetics** of the reaction. To return to a familiar reaction, let's look at some factors influencing the rate of the reaction of hydroxide ion with bromomethane. This is a simple reaction that involves the nucleophilic substitution of bromide ion by hydroxide ion:

$$H\bar{O}: + CH_3—Br \longrightarrow HO—CH_3 + :Br^-$$

At a given concentration of reagents, the reaction occurs at a certain rate. If we double the concentration of hydroxide ion, the frequency of encounter between the reagents is also doubled, and we might therefore predict that the reaction rate will double. Similarly, if we double the concentration of bromomethane, we might expect that the reaction rate will double. This behavior is exactly what is found, and we call such a reaction, in which the rate is linearly dependent on the concentrations of two reagents, a **second-order reaction.** Mathematically, we can express this second-order dependence of the nucleophilic substitution reaction by setting up a **rate equation:**

$$\begin{aligned}\text{Reaction rate} &= \text{Rate of disappearance of starting material}\\ &= k \times [\text{RX}] \times [^-\text{OH}]\end{aligned}$$

where $[\text{RX}] = CH_3Br$ concentration
$[^-\text{OH}] = HO^-$ concentration
k = A constant value

This equation says that the reaction rate is the same as the rate of disappearance of starting material, and is equal to a coefficient, k, times the alkyl halide concentration times the hydroxide ion concentration. The constant k is called the **rate coefficient** for the reaction and is measured in units of liters per mole second (L/mol sec). The rate equation simply tells us that as either [RX] or $[^-\text{OH}]$ changes, the rate of the reaction also changes. For example, if the alkyl halide concentration is doubled, the reaction rate also doubles; if the alkyl halide concentration is halved, the reaction rate is halved.

10.4 The S_N2 reaction

At this point, two important pieces of information have been obtained about the nature of nucleophilic substitution reactions on primary and secondary alkyl halides and tosylates, and we must now see how this information can be used to determine the mechanism of these reactions.

1. These reactions always occur with *complete* inversion of stereochemistry at the chiral carbon center.
2. These reactions show second-order kinetics and follow the rate law:

$$\text{Rate} = k\,[\text{RX}]\,[\text{Nu:}^-]$$

A satisfactory explanation of the observed results was first advanced in 1937 when Hughes and Ingold formulated the mechanism of what they called the **S_N2 reaction**—shorthand for substitution, nucleophilic, bimolecular. (*Bimolecular* means that two molecules—nucleophile and alkyl halide—are involved in the step whose kinetics are measured.)

The essential features of the S_N2 reaction formulated by Hughes and Ingold are that the entering nucleophile attacks the substrate from a position 180° away from the leaving group, and that the reaction takes place in a single step without intermediates (Figure 10.3).

The nucleophile Nu:^- uses its lone-pair electrons to attack the alkyl halide 180° away from the halogen. This leads to a transition state with a partially formed C—Nu bond and a partially broken C—X bond.

Transition state

The stereochemistry at carbon is inverted as the C—Nu bond forms fully and the halide departs with the electron pair from the original C—X bond.

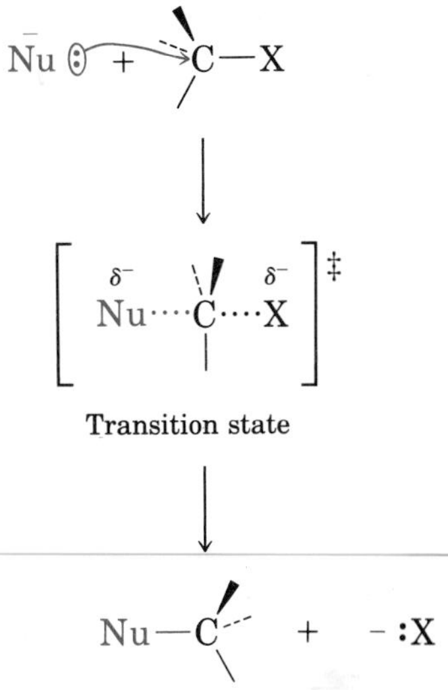

Figure 10.3. Mechanism of the S_N2 reaction.

We can picture the reaction as occurring when an electron pair on the nucleophile, Nu:, forces out the leaving group, X:, with its electron pair. This can occur through a transition state in which the new Nu—C bond is partially forming at the same time that the old C—X bond is partially breaking, and in which the negative charge is shared by the incoming nucleophile and the outgoing leaving group. The transition state for this inversion must have the remaining three bonds to carbon in a planar arrangement, as shown in Figure 10.4.

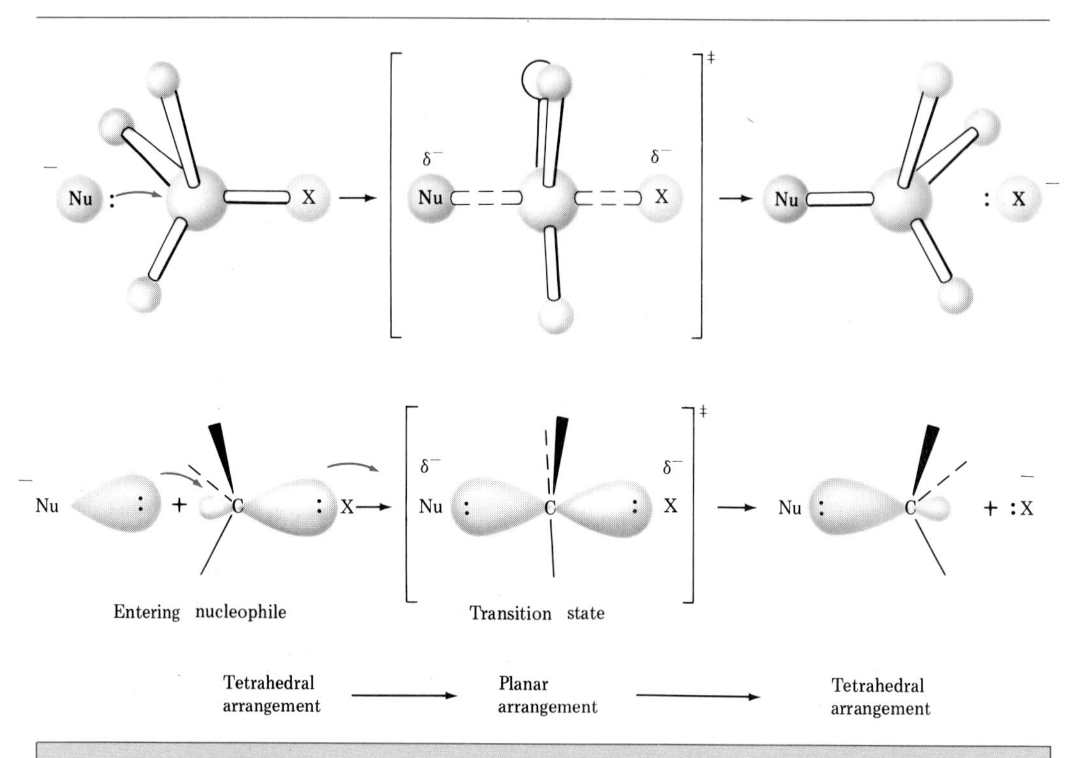

Figure 10.4. Hypothetical transition state for S_N2 reaction.

The mechanism proposed by Hughes and Ingold is consistent with experimental results, and explains both stereochemical and kinetic data. Thus, the requirement for back-side attack of the entering nucleophile from a direction 180° away from the departing X group causes the stereochemistry of the substrate to invert, much like an umbrella turning inside out in the wind (Figure 10.5).

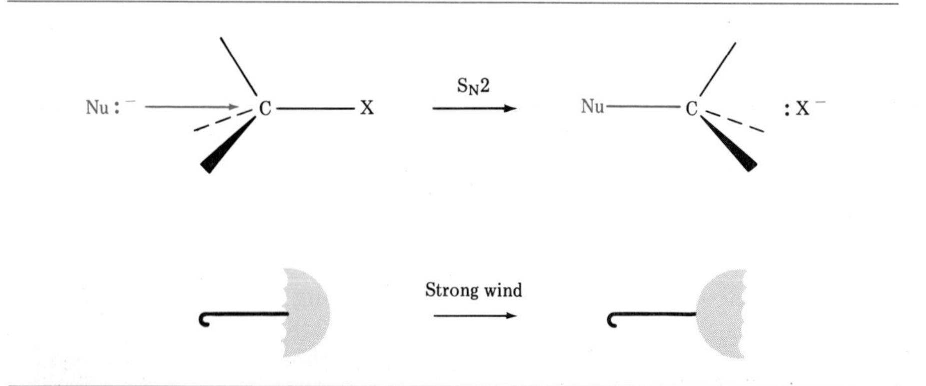

Figure 10.5. The inversion of configuration of a chiral center during an S_N2 reaction may be likened to the inversion of an umbrella in a strong wind.

The Hughes–Ingold mechanism also explains why second-order kinetics are found for S_N2 reactions. The reaction occurs in a single step, and that step is bimolecular. Two molecules (nucleophile and alkyl halide substrate) are involved in the step whose rate is being measured.

PROBLEM

10.1 A further piece of stereochemical evidence in support of the picture of back-side S_N2 displacement is the finding that the following alkyl bromide does not undergo reaction with ethoxide ion. Can you suggest a reason for the total lack of reactivity?

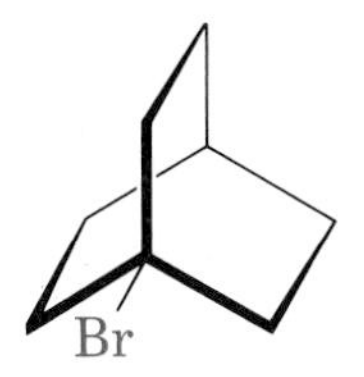

10.5 Characteristics of the S_N2 reaction

We now have a good picture of the mechanism of S_N2 reactions, but the chemical variables affecting potential uses of the S_N2 reaction in organic synthesis are also important. Some S_N2 reactions are fast and some are slow; some take place in high yield, and others in low yield. Understanding the different factors involved is of great value to chemists.

The rate of a chemical reaction is determined by $\Delta G^{\ddagger}$, the energy difference between reactant (ground state) and transition state. A change in reaction conditions can affect the magnitude of $\Delta G^{\ddagger}$ in two ways: either by changing the reactant energy level or by changing the transition-state energy level. If the reactant energy level is lowered, $\Delta G^{\ddagger}$ increases and reaction rate decreases; conversely, if the reactant energy level is raised, $\Delta G^{\ddagger}$ decreases and reaction rate increases, as can be seen in Figure 10.6(a).

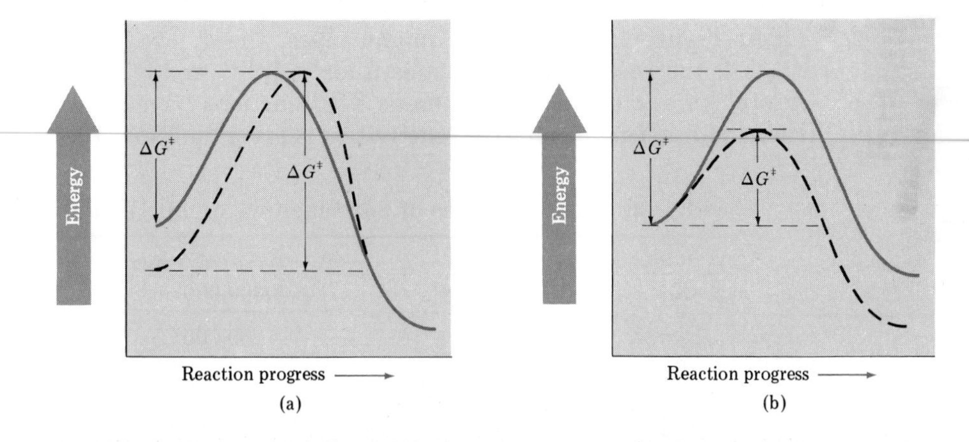

Figure 10.6. Effect of changes in reactant and transition-state energy levels on reaction rate. (a) Higher reactant energy level corresponds to faster reaction (lower $\Delta G^{\ddagger}$). (b) Higher transition-state energy level corresponds to slower reaction (higher $\Delta G^{\ddagger}$).

Similarly, stabilization of the transition state lowers $\Delta G^{\ddagger}$ and lowers reaction rate, whereas destabilization of the transition state raises $\Delta G^{\ddagger}$ and raises reaction rate [Figure 10.6(b)]. We will see examples of all of these effects as we look at S_N2 reaction variables.

STERIC EFFECTS IN THE S_N2 REACTION

The first S_N2 reaction variable we will look at is the steric bulk of the alkyl halide. We have said that an attacking nucleophile must approach the substrate closely to expel the leaving group; it therefore seems reasonable that the ease of approach should depend on the steric nature of the substrate. Sterically bulky substrates, in which the carbon atom is "shielded" from attack by the rest of the molecule, should react at a slower rate than substrates in which the carbon is more sterically accessible (Figure 10.7), since the transition state is of higher energy.

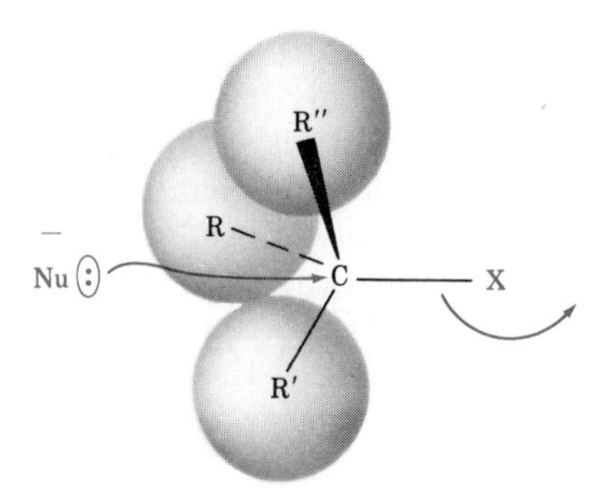

Figure 10.7. Steric hindrance to the S_N2 reaction. When R, R′, and R″ = H, a fast reaction occurs; when R, R′, and R″ = alkyl, a slow reaction occurs.

As Figure 10.7 shows, we might expect the difficulty of nucleophilic attack to increase as the three substituents, R, R′, and R″, increase in size. Careful kinetic studies of numerous S_N2 reactions have shown this to be true, and a listing of relative reactivities is given in Table 10.1.

TABLE 10.1 Relative rates of S_N2 reactions on alkyl halides

Alkyl halide	*Type*	*Relative rate of reaction*
CH_3—X	Methyl	3,000,000
CH_3CH_2—X	1°	100,000
$CH_3CH_2CH_2$—X	1°	40,000
$(CH_3)_2CH$—X	2°	2,500
$(CH_3)_3CCH_2$—X	1°—neopentyl	1
$(CH_3)_3C$—X	3°	~0

Table 10.1 shows that methyl halides are by far the most reactive in S_N2 reactions, followed by primary alkyl halides such as ethyl and propyl. Alkyl branching next to the leaving group, as in isopropyl halides (2°), slows the reaction greatly, and further branching, as in *tert*-butyl halides (3°), effectively halts the reaction. Even branching one carbon removed from the leaving group, as in 2,2-dimethylpropyl (neopentyl) halides, greatly slows nucleophilic displacement. With the information in Table 10.1, we can construct the following reactivity order for S_N2 reactions:

$$CH_3 > 1° > 2° >> \text{Neopentyl} > 3°$$

Clearly, S_N2 reactions can only occur at relatively unhindered sites; methyl and unsubstituted primary substrates are by far the most reactive.

Although not shown in Table 10.1, vinylic halides, $R_2C{=}CRX$, are completely unreactive toward attempted S_N2 displacements. This lack of reactivity is probably due to steric hindrance, since the incoming nucleophile would have to approach in the plane of the carbon–carbon double bond in order to be able to carry out a back-side displacement.

$R_2C{=}C(Cl)R$ + Nu: ⟶ (crossed arrow) No reaction

Vinylic halide
(back-side approach of attacking
nucleophile is hindered,
and S_N2 reaction cannot occur)

THE ATTACKING NUCLEOPHILE

The nature of the attacking nucleophile itself is a second variable whose effect on the S_N2 reaction we should consider. Any species, either neutral or negatively charged, can act as a nucleophile as long as it has an unshared pair of electrons (i.e., is a Lewis base). Thus, the S_N2 reaction is of great generality and use in organic synthesis. Table 10.2 (p. 300) lists some common nucleophiles and the products of their reactions with bromomethane.

When kinetic measurements are made, we find large differences in reactivity between various nucleophiles; some species are much more "nucleophilic" than others. What are the reasons for these reactivity differences? The answer to this question is not straightforward, and it is not yet fully understood by chemists. Part of the problem is that the term *nucleophilic* is imprecise. Most chemists use the term *nucleophilicity* to mean a measure of the affinity of a species for a carbon atom in the S_N2 reaction. However, the reactivity of a given nucleophile in a specific reaction depends on many factors, such as the nature of the substrate, the solvent, and even the concentration at which the reaction is run. In order to speak with any precision, therefore, we must study the relative reactivity of various nucleophiles on a single substrate in a single solvent system. Much work has been carried out on the S_N2 reactions of bromomethane in aqueous ethanol, and Table 10.3 lists some quantitative results.

TABLE 10.2 Some S_N2 reactions with bromomethane: $Nu:^- + CH_3Br \longrightarrow Nu{-}CH_3 + :Br^-$

Attacking nucleophile		*Product*	
Formula	*Name*	*Formula*	*Name*
$H:^-$	Hydride	CH_4	Methane
$CH_3S:^-$	Methanethiolate	$CH_3S{-}CH_3$	Dimethyl sulfide
$HS:^-$	Hydrosulfide	$HS{-}CH_3$	Methane thiol
$N{\equiv}C:^-$	Cyanide	$N{\equiv}C{-}CH_3$	Acetonitrile
$I:^-$	Iodide	$I{-}CH_3$	Iodomethane
$HO:^-$	Hydroxide	$HO{-}CH_3$	Methanol
$CH_3O:^-$	Methoxide	$CH_3O{-}CH_3$	Dimethyl ether
$N{=}N{=}N:^-$	Azide	$N_3{-}CH_3$	Azidomethane
$Cl:^-$	Chloride	$Cl{-}CH_3$	Chloromethane
$CH_3CO_2:^-$	Acetate	$CH_3CO_2{-}CH_3$	Methyl acetate
$H_3N:$	Ammonia	$H_3\overset{+}{N}{-}CH_3\ Br^-$	Methylammonium bromide
$(CH_3)_3N:$	Trimethylamine	$(CH_3)_3\overset{+}{N}{-}CH_3\ Br^-$	Tetramethylammonium bromide

TABLE 10.3 Relative nucleophilicities in S_N2 reactions on bromomethane: $Nu:^- + CH_3Br \longrightarrow Nu{-}CH_3 + :Br^-$

Nucleophile	*Relative reactivity*	*Nucleophile*	*Relative reactivity*
$HS:^-$	125,000	$C_6H_5O:^-$ (phenoxide)	8,000
$:CN^-$	125,000		
$:I^-$	100,000	$:Cl^-$	1,000
$CH_3CH_2O:^-$	25,000	$(CH_3)_3N:$	700
$HO:^-$	16,000	$CH_3CO_2:^-$	500
$:N_3^-$	10,000	$H_2O:$	1

The data in Table 10.3 disclose a large range of reactivity differences. Although we cannot offer precise explanations for the observed nucleophilicity order, we can detect some trends in the data:

1. Perhaps the most obvious trend is the observation that, in comparing nucleophiles that have the same attacking atom, nucleophilicity roughly parallels basicity. Since nucleophilicity is a measure of the affinity of a Lewis base for a carbon atom in the S_N2 reaction, and basicity is a measure of the affinity of a base for a proton, it is easy to see why there might be a rough correlation between the two kinds of behavior. Basicity can be quantitatively measured by obtaining the

acidity (pK_a) of the conjugate acid, and Table 10.4 provides a comparison between nucleophilicity and basicity for some oxygen nucleophiles. Recall that a strong base corresponds to a weak acid (high pK_a), and a weak base corresponds to a strong acid (low pK_a).

TABLE 10.4 Comparison of nucleophilicity and basicity for some oxygen nucleophiles. Stronger bases are better nucleophiles.

Decreasing nucleophilicity (↓) / Decreasing basicity (↓)

Nucleophile	*Relative nucleophilicity toward bromomethane*	*Conjugate acid*	pK_a
$CH_3CH_2O:^-$	25,000	CH_3CH_2OH	16
$HO:^-$	16,000	H_2O	15.7
$C_6H_5O:^-$ (phenoxide)	8,000	C_6H_5OH (phenol)	10
$CH_3CO_2:^-$	500	CH_3CO_2H	4.8
$H_2O:$	1	H_3O^+	−1.7

2. A second trend we can discern in the data of Table 10.3 is that nucleophilicity usually increases as we go down a column of the periodic table. Thus $HS:^-$ is more nucleophilic than $HO:^-$, and the halide reactivity order is $:I^- > :Br^- > :Cl^-$.

PROBLEM

10.2 The tertiary amine base quinuclidine reacts with CH_3I 50 times as fast as does triethylamine. Can you suggest a reason for this difference?

$$R_3N: + CH_3I \longrightarrow R_3\overset{+}{N}{-}CH_3\ I:^-$$

Quinuclidine $\qquad$ $(CH_3CH_2)_3N:$ Triethylamine

PROBLEM

10.3 Which reagent in each of the following pairs is more nucleophilic? Justify your choices.
(a) $(CH_3)_2\ddot{N}:^-$ and $(CH_3)_2\ddot{N}H$ (b) $(CH_3)_3B$ and $(CH_3)_3N:$ (c) $H_2\ddot{O}:$ and $H_2\ddot{S}:$

THE LEAVING GROUP

A further variable that can strongly affect the S_N2 reaction is the nature of the species expelled by the attacking nucleophile—the **leaving group.** In most S_N2 reactions, the leaving group is expelled with a negative charge, and we might therefore expect the best leaving groups to be those that best stabilize the negative charge. Furthermore, since anion stability is related to basicity, we can also say that the best leaving groups should be the weakest bases. The reason is that, in the transition state for S_N2 reactions, the charge is normally distributed over both the attacking and the leaving groups. The greater the extent of charge stabilization by the leaving group, the more stable the transition state and the more rapid the reaction.

$$\text{Nu:}^- + \text{C—X} \longrightarrow \left[\overset{\delta^-}{\text{Nu}} \cdots \text{C} \cdots \overset{\delta^-}{\text{X}} \right]^{\ddagger} \longrightarrow \text{Nu—C} + \text{:X}^-$$

Transition state

In fact, this prediction is borne out rather well. Table 10.5 lists a variety of potential leaving groups in order of reactivity and shows their correlation with basicity. The weakest bases (anions derived from the strongest acids) are the most reactive as leaving groups. The *p*-toluenesulfonate (tosylate) leaving group is the most reactive, although its basicity is slightly out of line with others in the table. Iodide and bromide ions are also excellent leaving groups, but chloride ion is much less effective. It is just as important to know which groups are *unreactive* as to know which are reactive, and Table 10.5 clearly indicates that alkyl fluorides, acetates, alcohols, ethers, and amines do not undergo displacement reactions under normal circumstances.

TABLE 10.5 Correlation of leaving-group ability with basicity. The anions of strong acids make good leaving groups in the S_N2 reaction.

Decreasing reactivity ↓

Leaving group	pK_a *of conjugate acid*	*Relative reactivity*
$CH_3-C_6H_4-SO_2-O:^-$ Tosylate	−6.5	60,000
$I:^-$	−9.5	30,000
$Br:^-$	−9	10,000
$Cl:^-$	−7	200
$F:^-$	3.2	1
$CH_3CO_2:^-$	4.8	~0
$HO:^-$	15.7	~0
$CH_3CH_2O:^-$	16	~0
$H_2N:^-$	35	~0

SOLVENT EFFECTS

Many different solvents may be used for S_N2 reactions, but methanol and ethanol are often preferred because they are inexpensive and easily removed after the reaction. However, the rates of many S_N2 reactions are strongly affected by the solvent used. Many S_N2 reactions are *slowest* in protic (hydroxylic) solvents, primarily because the protic solvents greatly decrease the reactivity of most nucleophiles by hydrogen bonding. A negatively charged nucleophile that is surrounded by a "cage" of solvent molecules is more stabilized and therefore less reactive than a bare nucleophile:

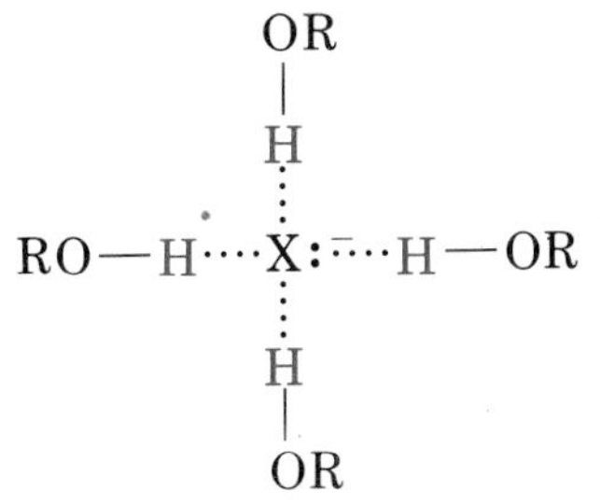

```
            OR
            |
            H
            :
RO—H····X:⁻····H—OR
            :
            H
            |
            OR
```

A solvated anion
(reduced nucleophilicity due to enhanced ground-state stability)

For this reason, polar *aprotic* (nonhydroxylic) solvents are often used to increase the rate of sluggish S_N2 reactions. Polar aprotic solvents are able to dissolve many salts because of their high dielectric constants, but they do *not* solvate nucleophiles strongly. Free anions therefore have a far greater effective nucleophilicity in these solvents, and S_N2 reactions take place at correspondingly faster rates. Particularly valuable polar aprotic solvents are acetonitrile, CH_3CN; dimethylformamide, $(CH_3)_2NCHO$; dimethyl sulfoxide, $(CH_3)_2SO$; and hexamethylphosphoramide, $[(CH_3)_2N]_3PO$. Rate increases of over a millionfold have been observed in going from methanol to hexamethylphosphoramide. Table 10.6 shows the results of a study of the reaction of azide ion with 1-bromobutane in different solvents. Hexamethylphosphoramide is clearly preferred.

TABLE 10.6 Relative rates for the reaction:
$$CH_3CH_2CH_2CH_2Br + :N_3^- \longrightarrow CH_3CH_2CH_2CH_2{-}N{=}\overset{+}{N}{=}\overset{-}{N} + :Br^-$$

Solvent	*Relative rate*	*Solvent*	*Relative rate*
CH_3OH	1	$(CH_3)_2NCHO$ (DMF)	2,800
H_2O	6.6	CH_3CN	5,000
$(CH_3)_2SO$ (DMSO)	1,300	$[(CH_3)_2N]_3PO$ (HMPA)	200,000

Explained in terms of reaction energy diagrams, the effect of polar aprotic solvents on S_N2 reactions is to lower $\Delta G^{\ddagger}$ by *destabilizing* (raising) the ground-state energy level of the nucleophile (Figure 10.8, p. 304).

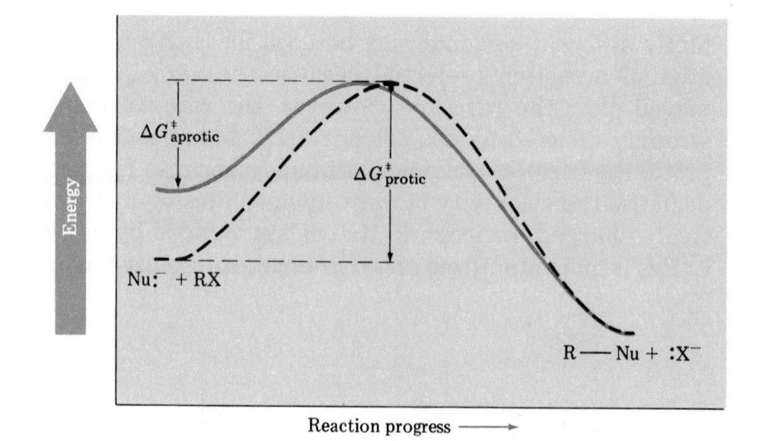

Figure 10.8. Effect of solvent on S_N2 reactions. The nucleophile is stabilized by solvation in protic solvents and is thus less reactive. (Solid curve, polar aprotic solvent; dashed curve, protic solvent.)

PROBLEM

10.4 Which of each of these pairs of reactions would you expect to progress faster?
(a) The S_N2 displacement by iodide on CH_3Cl, or on CH_3—OTos?
(b) The S_N2 displacement by acetate on bromoethane, or on bromocyclohexane?
(c) The S_N2 displacement on 2-bromopropane by ethoxide ion, or by cyanide ion?
(d) The S_N2 displacement by acetylide ion on bromomethane in ether, or in hexamethylphosphoramide?

10.6 The S_N1 reaction

We have seen that the S_N2 reaction occurs readily with primary halides but that it is sensitive to steric factors. Secondary halides often react adequately, but tertiary halides are essentially inert to back-side attack by nucleophiles. For example, bromomethane reacts with acetate ion in 80% aqueous ethanol approximately 40 times faster than does 2-bromopropane, and 2-bromo-2-methylpropane is unreactive toward S_N2 displacement by acetate ion. Remarkably, however, a completely different picture emerges when different reaction conditions are used. When the alkyl bromides are heated with acetate ion in acetic acid as solvent, displacement of bromide ion by acetate occurs, and rate measurements show that 2-bromo-2-methylpropane reacts several thousand times *faster* than does 2-bromopropane:

$$CH_3\overset{\overset{O}{\|}}{C}O:^- + (CH_3)_3CBr \xrightarrow[CH_3COOH]{\Delta} CH_3\overset{\overset{O}{\|}}{C}O—C(CH_3)_3 + :Br^-$$

This trend is noted in many displacement reactions in which halides are heated with *nonbasic nucleophiles in protic solvents:* Tertiary halides react considerably faster than do primary or secondary halides. For example, Table 10.7 gives the relative rates of reaction of some alkyl halides with

TABLE 10.7 **Relative rates of reaction of some alkyl halides with water: R—Br + H_2O: ⟶ R—OH + HBr**

Alkyl halide	*Type*	*Product*	*Relative rate of reaction*
CH_3Br	Methyl	CH_3OH	1.0
CH_3CH_2Br	1°	CH_3CH_2OH	1.0
$(CH_3)_2CHBr$	2°	$(CH_3)_2CHOH$	12
$(CH_3)_3CBr$	3°	$(CH_3)_3COH$	1,200,000

water. 2-Bromo-2-methylpropane is more than *1 million times* as reactive as bromoethane.

What are the reasons for this behavior? These reactions cannot be taking place by an S_N2 mechanism, and we must therefore conclude that *some alternative substitution process is operating.* This alternative mechanism is called the **S_N1 reaction** (for substitution, nucleophilic, unimolecular). Let's see what evidence is available concerning the S_N1 reaction, and what conclusions we can draw.

10.7 Kinetics of the S_N1 reaction

The reaction of acetate ion with 2-bromo-2-methylpropane is analogous to the reaction of hydroxide ion with bromomethane, and we might expect to observe second-order kinetics. In fact, we do not. We find instead that the reaction rate is dependent only on the alkyl halide concentration, and is independent of acetate ion concentration; the reaction is a **first-order process.** Only one molecule is involved in the step whose kinetics are measured, and we can write the rate expression as follows:

$$\text{Reaction rate} = \text{Rate of disappearance of alkyl halide}$$
$$= k \times [\text{RX}]$$

The rate of the reaction is equal to a rate coefficient k times the alkyl halide concentration. *Acetate ion concentration does not appear in the rate expression.*

By measuring the kinetics of the reaction of 2-bromo-2-methylpropane with acetate ion, we have found that the reaction appears to have a different mechanism than does the reaction of bromomethane with hydroxide ion. How can these results best be explained? To answer this question, we must first learn more about kinetics measurements.

Many organic reactions are rather complex and occur in successive steps. One of these steps is usually slower than the others, and we call this the **rate-limiting step.** No reaction can proceed faster than its rate-limiting step. The overall reaction rate that we actually measure in a kinetics experiment is determined by the height of the highest *cumulative* energy barrier ($\Delta G^{\ddagger}$) between a low point and a subsequent high point in the energy diagram of the reaction. The hypothetical reaction energy diagrams in Figure 10.9 illustrate the idea of the rate-limiting step.

In Figure 10.9(a), the rate-limiting step is simply the height of the barrier from the starting material to the first transition state, whereas in (b)

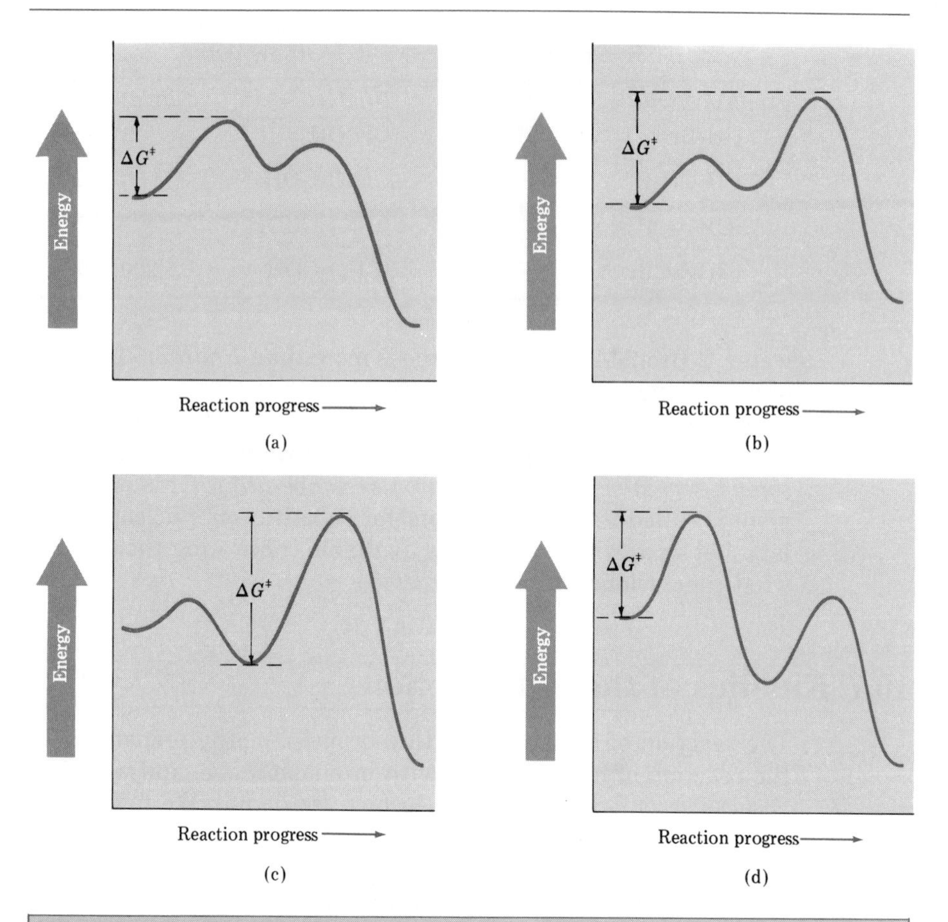

Figure 10.9. Hypothetical reaction energy diagrams showing rate-limiting steps. The rate-limiting step is determined by the highest energy difference between a low point and a subsequent high point.

the rate-limiting step represents the *cumulative* barrier from starting material to the highest-energy, *second* transition state. In (c) the step corresponding to the highest energy barrier is that from the stable intermediate to the second transition state, and in (d) the highest barrier is that from starting material to the first transition state. With this as background, let's look further into the kinetics of the S_N1 reaction.

The observation of first-order kinetics for the S_N1 reaction of 2-bromo-2-methylpropane with acetate indicates that the alkyl halide is involved in a unimolecular rate-limiting step. In other words, 2-bromo-2-methylpropane must undergo some manner of spontaneous unimolecular reaction *without assistance* from the nucleophile. The mechanism shown in Figure 10.10 can be postulated to account for the kinetic observations.

If 2-bromo-2-methylpropane spontaneously dissociates to the *tert*-butyl carbocation plus bromide ion in a slow, rate-limiting step, and if the intermediate ion is immediately trapped by nucleophilic acetate ion in a fast step, then first-order kinetics will be obtained; *acetate ion plays no role in the step that is measured by kinetics.* The reaction energy diagram is shown in Figure 10.11.

Spontaneous dissociation of the alkyl bromide occurs in a slow, rate-limiting step to generate a carbocation intermediate plus bromide ion.

$(CH_3)_3C{-}Br$

⇅ Slow (rate-limiting) step

[$(CH_3)_3C^+$ + $:Br^-$]

Intermediate

The carbocation intermediate reacts with added nucleophile (acetate ion) in a fast step to yield neutral product.

$:\overset{-}{O}\overset{O}{\overset{\|}{C}}CH_3$ ↓ Fast step

$(CH_3)_3C{-}O{-}\overset{O}{\overset{\|}{C}}CH_3$

Figure 10.10. Mechanism of the S_N1 reaction. Two steps are involved, and the first is rate limiting.

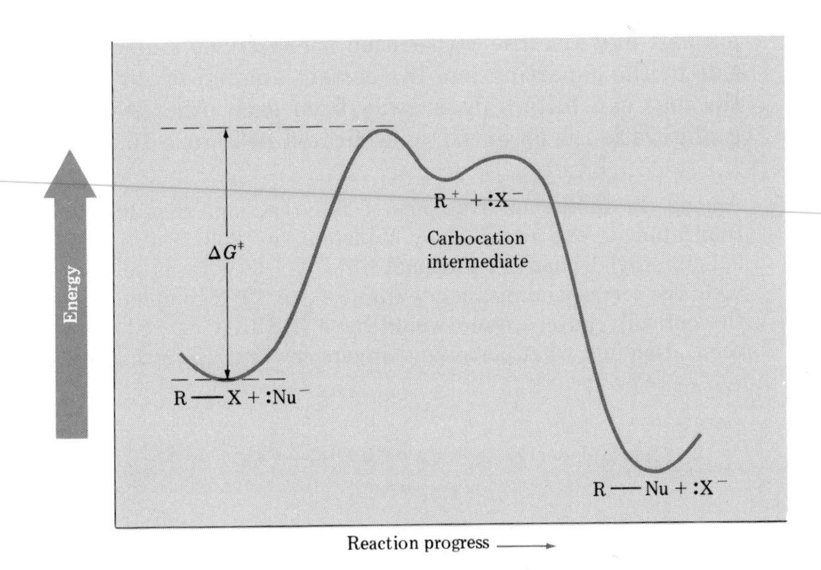

Figure 10.11. Reaction energy diagram for an S_N1 reaction. The rate-limiting step is the spontaneous dissociation of alkyl halide.

10.8 Stereochemistry of the S_N1 reaction

If our postulate is correct and S_N1 reactions occur through carbocation intermediates, there should be clear stereochemical consequences. Carbocations are planar, sp^2-hybridized species and are therefore *achiral.* If we carry out an S_N1 reaction on a *chiral* starting material and go through an *achiral* intermediate, then our product must be optically inactive. Another way of looking at this is to say that the symmetrical intermediate carbocation can be attacked by a nucleophile equally well from either the right or the left side, leading to a 50:50 mixture of enantiomers—a racemic mixture (Figure 10.12).

The prediction that S_N1 displacements on chiral substrates should lead to racemic products has been amply borne out by experiment. Surprisingly, however, few S_N1 displacements occur with complete racemization. Most give a minor (0–20%) amount of inversion. For example, the reaction of optically active (*R*)-6-chloro-2,6-dimethyloctane with water leads to an alcohol product that is approximately 80% racemized and 20% inverted (80% *R,S* + 20% *S* is the same as 40% *R* + 60% *S*):

$$\underset{\text{(R)-6-Chloro-2,6-dimethyloctane}}{(C_2H_5)(H_3C)C(Cl)(CH_2)_3CH(CH_3)_2} \xrightarrow[C_2H_5OH]{H_2O:} \underset{\substack{40\%\ R \\ \text{(retention)}}}{(C_2H_5)(H_3C)C(OH)(CH_2)_3CH(CH_3)_2} + \underset{\substack{60\%\ S \\ \text{(inversion)}}}{HO{-}C(C_2H_5)(CH_3)(CH_2)_3CH(CH_3)_2}$$

The situation is a complex one, and the reasons for the lack of complete racemization in most S_N1 reactions are not completely clear. An attractive suggestion, first proposed by Winstein,[3] is that **ion pairs** are involved. According to this suggestion, dissociation of the substrate occurs to give an ion pair in which the carbocation is effectively shielded from attack on one side by the departing ion. If a certain amount of substitution occurs before the ions can fully diffuse away from each other, then a net inversion of configuration is observed, as indicated in Figure 10.13.

PROBLEM

10.5 Among the numerous examples of S_N1 reactions that occur with incomplete racemization is one reported by Winstein in 1952. The optically pure tosylate of 2,2-dimethyl-1-phenyl-1-propanol ($[\alpha]_D = -30.3°$) was solvolyzed in acetic acid to yield the corresponding acetate ($[\alpha]_D = +5.3°$). If complete inversion had occurred, the optically pure acetate would have had $[\alpha]_D = +53.6°$. What percentage racemization and what percentage inversion occurred in this reaction?

$$\underset{[\alpha]_D = -30.3°}{(CH_3)_3C{-}CH(OTos){-}C_6H_5} \xrightarrow[AcO:^-]{HOAc} \underset{\substack{\text{Observed } [\alpha]_D = +5.3° \\ \text{(optically pure } [\alpha]_D = +53.6°)}}{(CH_3)_3C{-}CH(OAc){-}C_6H_5}$$

[3]Saul Winstein (1912–1973); b. Montreal; Ph.D., California Institute of Technology (Lucas); professor, University of California, Los Angeles (1942–1973).

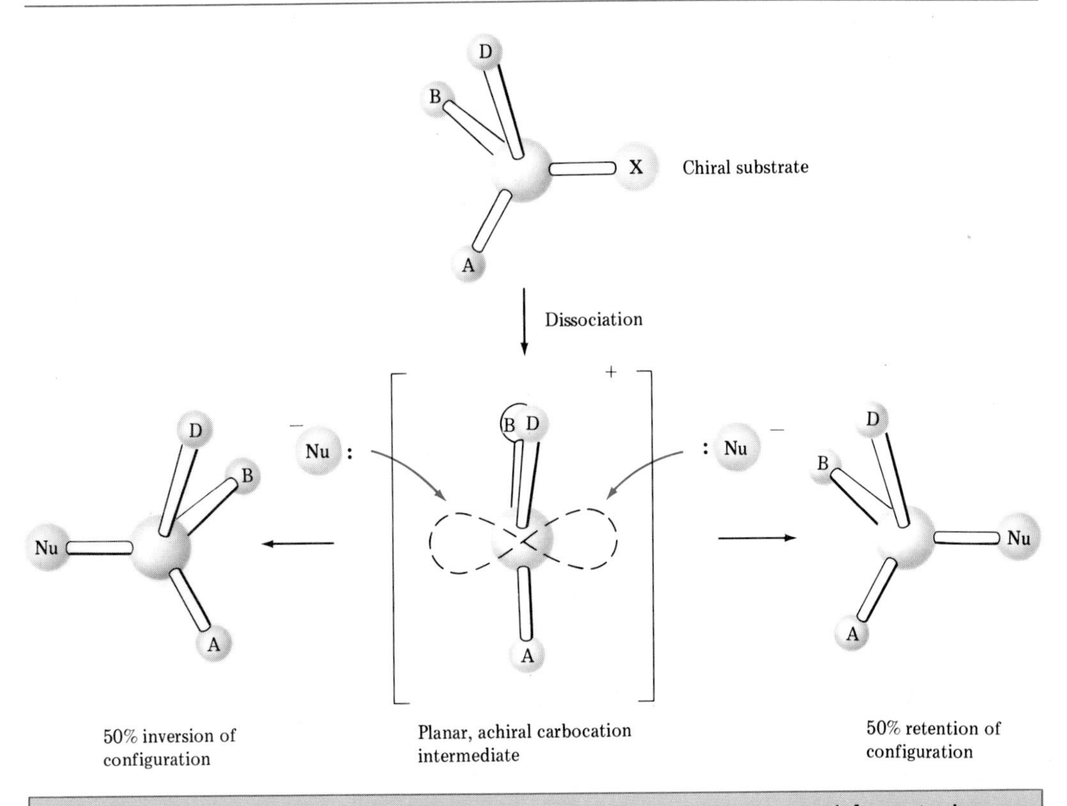

Figure 10.12. The S_N1 reaction. An optically active starting material must give a racemic product.

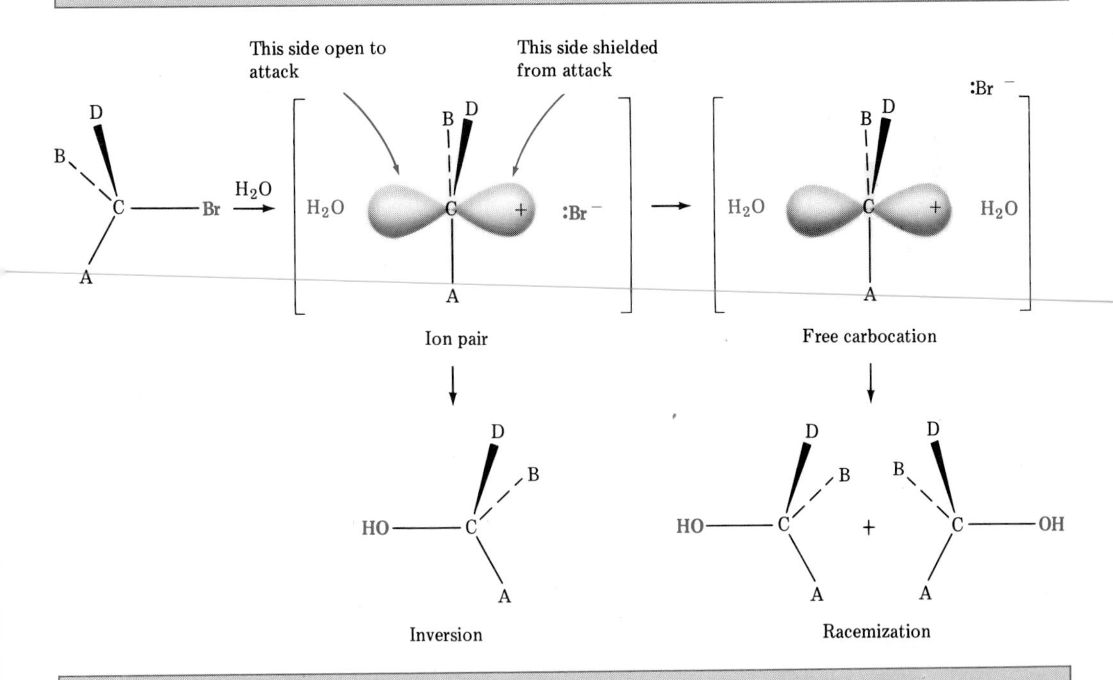

Figure 10.13. The ion-pair hypothesis. The leaving group effectively shields one side of the carbocation intermediate from attack by added nucleophile.

10.9 Characteristics of the S_N1 reaction

We saw in Section 10.5 that the S_N2 reaction is strongly influenced by such variables as solvent, leaving group, substrate structure, and nature of the attacking nucleophile. Similar observations have been made for the S_N1 reaction. Factors that either stabilize the transition state leading to carbocation formation or raise the reactant energy level lower $\Delta G^{\ddagger}$ and thus favor faster S_N1 reactions. Conversely, factors that either destabilize the transition state leading to a carbocation or lower reactant energy level raise $\Delta G^{\ddagger}$ and thus slow down the S_N1 reaction.

THE SUBSTRATE

According to the Hammond postulate, any factor that stabilizes a high-energy intermediate should also stabilize the transition state leading to that intermediate. Since the rate-limiting step in the S_N1 reaction is the spontaneous, unimolecular dissociation of the substrate, we would expect the reaction to be favored whenever stabilized carbocation intermediates are formed. This is exactly what is found—*the more stable the carbocation intermediate, the faster the S_N1 reaction.* We have already seen that the stability order of alkyl carbocations is 3° > 2° > 1° > CH_3, and to this list we now add the resonance-stabilized allyl and benzyl cations:

$$CH_2{=}CH{-}CH_2^+$$

Allyl carbocation

CH_2^+

Benzyl carbocation

We saw in Section 9.6 that an allylic radical has unusual stability because the unpaired electron can be delocalized over an extended pi orbital system. The same is true for the allyl and benzyl carbocations. Delocalization of the positive charge throughout the pi orbital system of these species results in unusual stability. As Figure 10.14 indicates, the allyl cation has two equivalent resonance forms. In one form the double bond is on the "left," and in the other form the double bond is on the "right." The benzyl cation, however, has *four* resonance forms that are *not* equivalent to each other. The resonance concept does not require that a molecule be symmetrical or that resonance forms be equivalent. The single requirement for resonance forms is that they differ only in the location of their electrons; the nuclei must remain in the same positions at all times. If, as is the case with the benzyl cation, the resonance forms are not equivalent, then the actual structure of the resonance hybrid resembles more closely the most stable resonance form, but is still a mix of all forms.

As a result of the resonance stabilization found for the allyl and benzyl carbocations, carbocations have the following stability order:

$$\text{Benzyl} \approx \text{Allyl} > 3° > 2° > 1° > CH_3$$

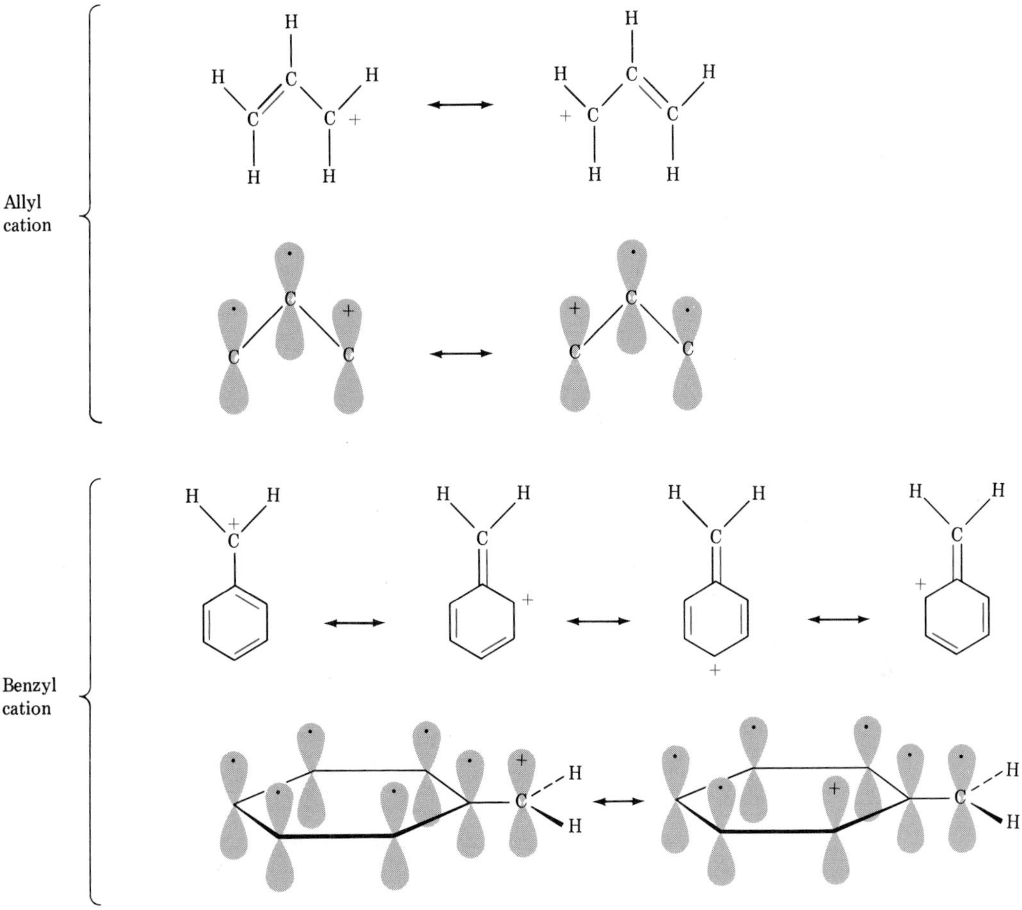

Allyl cation

Benzyl cation

Figure 10.14. Resonance forms of the allyl and benzyl carbocations.

This is precisely the order of S_N1 reactivity for alkyl halides. Table 10.8 (p. 312) shows the relative rates of reaction of some alkyl tosylates with ethanol, and indicates a strong correlation between the relative rates and carbocation stabilities. Note particularly the extraordinary rate at which triphenylmethyl tosylate reacts.

THE LEAVING GROUP

In the discussion of S_N2 reactivity, we reasoned that the best leaving groups should be those that are most stable. For the usual case of negatively charged leaving groups, the most stable anions are those corresponding to the strongest acids. An *identical* reactivity order is found for the S_N1 reaction, since in both cases the leaving group is intimately involved in the

TABLE 10.8 Relative rates of reaction of some alkyl tosylates with ethanol at 25°C

Alkyl tosylate		*Product*	*Relative rate*
CH_3CH_2OTos	$\longrightarrow$	$CH_3CH_2OCH_2CH_3$	1
$(CH_3)_2CHOTos$	$\longrightarrow$	$(CH_3)_2CHOCH_2CH_3$	3
$H_2C{=}CHCH_2OTos$	$\longrightarrow$	$H_2C{=}CHCH_2OCH_2CH_3$	35
$C_6H_5CH_2OTos$	$\longrightarrow$	$C_6H_5CH_2OCH_2CH_3$	400
$(C_6H_5)_2CHOTos$	$\longrightarrow$	$(C_6H_5)_2CHOCH_2CH_3$	10^5
$(C_6H_5)_3COTos$	$\longrightarrow$	$(C_6H_5)_3COCH_2CH_3$	10^{10}

rate-limiting step. Thus, we find an S_N1 reactivity order:

$$\text{Tosylate}^- > I\!:^- > Br\!:^- > Cl\!:^- > H_2O\!:$$

Note that in the S_N1 reaction, which is often carried out under acidic conditions, neutral water can act as a leaving group. This is the case, for example, when an alcohol is protonated and then loses water to generate a carbocation, as occurs when a tertiary halide is prepared from an alcohol by treatment with HX (Figure 10.15).

$$\underset{\text{2-Methyl-2-propanol}}{(CH_3)_3C\ddot{O}H} \xrightarrow[\text{Ether}]{HCl} \left[(CH_3)_3C\overset{+}{\ddot{O}}H_2\right] + :Cl^-$$

$$\downarrow S_N1 \text{ dissociation}$$

$$H_2O\!: + \left[(CH_3)_3C^+\right] \xrightarrow{:Cl^-} \underset{\text{2-Chloro-2-methylpropane}}{(CH_3)_3C{-}Cl}$$

Figure 10.15. The S_N1 reaction of a tertiary alcohol with HCl to yield an alkyl chloride.

THE ATTACKING NUCLEOPHILE: SOLVOLYSIS

The nature of the attacking nucleophile plays a major role in the S_N2 reaction. Should it play an equally major role in the S_N1 reaction? The answer is no. The S_N1 reaction, by its very nature, occurs through a rate-limiting step in which the added nucleophile plays no kinetic role. The nucleophile does not enter into the reaction until after rate-limiting dissociation has occurred. The reaction of 2-methyl-2-propanol with HX, for example, occurs at the same rate regardless of whether X is Cl, Br, or I:

$$(CH_3)_3COH \xrightarrow[\text{Ethanol}]{\text{HX}} (CH_3)_3C\text{—}X + H_2O$$

2-Methyl-2-propanol

Many S_N1 reactions occur simply upon heating the substrate in pure solvent *without* added nucleophile. In such **solvolysis** reactions, the solvent serves as both reaction medium and nucleophile and can greatly affect the rate of reaction. We can understand this solvent effect by considering the Hammond postulate.

The Hammond postulate says that any factor stabilizing the intermediate carbocation should increase the rate of reaction (i.e., will lower $\Delta G^{\ddagger}$). One factor that affects the stability of the cation is its structure, and this has already been discussed. Another factor is **solvation**—the interaction of the ion with solvent molecules. The exact nature of carbocation stabilization by solvent is not easily defined, but we can picture the solvent molecules in Figure 10.16 orienting themselves around the cation in such a manner that the electron-rich ends of the solvent dipoles are facing the positive charge.

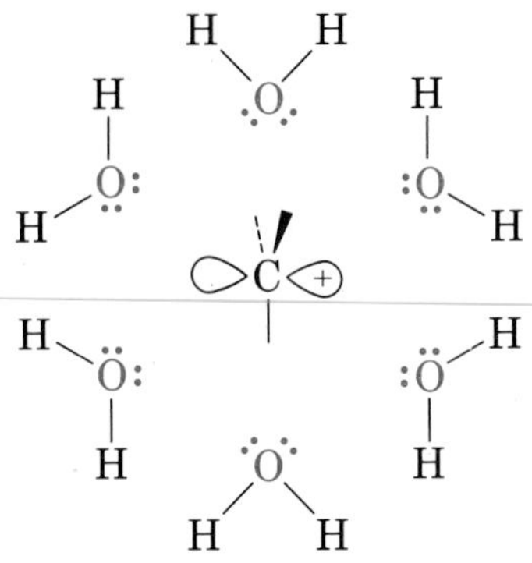

Figure 10.16. Solvation of a carbocation by water. The electron-rich oxygen atoms of solvent orient near the positively charged carbocation.

The properties of a solvent that contribute to its ability to stabilize ions by solvation are not fully understood, but are usually grouped under the term **polarity.** Polar solvents such as water and methanol are good at solvating ions, but nonpolar solvents such as hydrocarbons are poor at solvating ions.

Solvent polarity is usually defined in terms of **dielectric constants,** ε, which measure the ability of solvents to act as insulators of electric charges. In general, solvents of low dielectric constant such as hydrocarbons are nonpolar and are therefore good insulators, whereas solvents of high di-

TABLE 10.9 **Dielectric constants of some common solvents**

Name	*Structure*	*Dielectric constant,* ε
Aprotic (nonhydroxylic) solvents		
Hexane	$CH_3CH_2CH_2CH_2CH_2CH_3$	1.9
Benzene	(benzene ring)	2.3
Diethyl ether	$CH_3CH_2—O—CH_2CH_3$	4.3
Chloroform	$CHCl_3$	4.8
Ethyl acetate	$CH_3C(=O)—O—CH_2CH_3$	6.0
Acetone	$CH_3—C(=O)—CH_3$	20.7
Hexamethylphosphoramide (HMPA)	$(CH_3)_2N—P(=O)(N(CH_3)_2)—N(CH_3)_2$	30
Dimethylformamide (DMF)	$(CH_3)_2N—C(=O)—H$	38
Dimethyl sulfoxide (DMSO)	$CH_3—S(=O)—CH_3$	48
Protic (hydroxylic) solvents		
Acetic acid	$CH_3C(=O)—OH$	6.2
tert-Butyl alcohol	$(CH_3)_3COH$	10.9
Ethanol	CH_3CH_2OH	24.3
Methanol	CH_3OH	33.6
Formic acid	$H—C(=O)—OH$	58
Water	H_2O	80.4

electric constant such as water are polar and are therefore poor insulators. Table 10.9 lists the dielectric constants of some common solvents.

All S_N1 reactions are subject to large solvent effects. They generally take place much more rapidly in highly polar solvents than in nonpolar solvents. Table 10.10 lists some relative rates measured for the solvolysis of 2-chloro-2-methylpropane, and shows the magnitude of the rate differences due to solvent changes. A rate increase of 100,000 is observed on going from ethanol as solvent to the highly polar solvent water.

TABLE 10.10 **Relative rates for the solvolysis of 2-chloro-2-methylpropane**

Solvent	*Relative rate*
Ethanol	1
Acetic acid	2
Aqueous ethanol (40%)	100
Aqueous ethanol (80%)	14,000
Water	$\sim 10^5$

Note that although both S_N1 and S_N2 reactions show large solvent effects, they do so for different reasons. The S_N2 reactions (Section 10.5) are favored by polar aprotic solvents and disfavored by protic solvents, since the *ground-state energy level* of the attacking nucleophile is lowered by solvation. The S_N1 reactions, however, are favored by polar protic solvents, since the *transition-state energy level* leading to carbocation intermediate is lowered by solvation even more than the ground-state energy of the nucleophile (Figure 10.17). (Compare Figure 10.17 to Figure 10.8, where the effect of solvent on the S_N2 reaction is illustrated.)

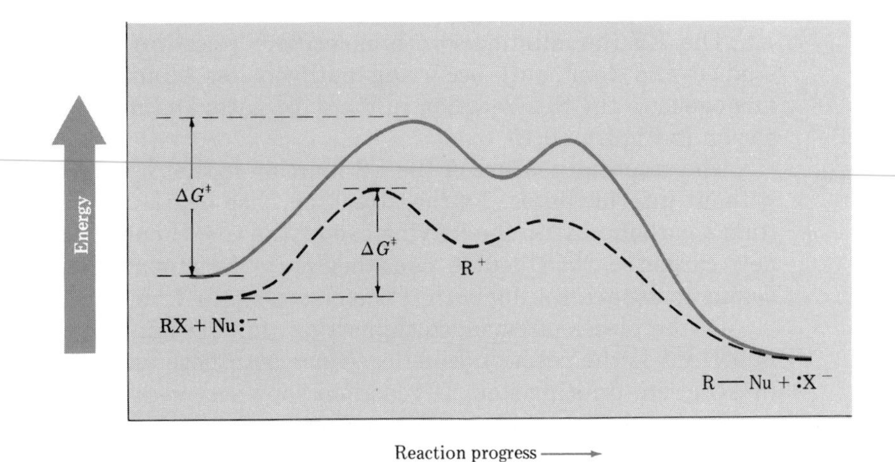

Figure 10.17. Solvent effect on S_N1 reactions. Both ground-state and transition-state energy levels are lowered by solvation in polar solvents, but the effect on the transition state is greater. (Solid curve, nonpolar solvent; dashed curve, polar solvent.)

PROBLEM

10.6 As indicated in Problem 10.1, halides such as the one shown here are inert to S_N2 displacement. Perhaps more surprisingly, they are also unreactive to S_N1 substitution, even though they are tertiary. Considering the electronic nature of carbocations, can you suggest a reason for this low reactivity? (Molecular models may be helpful.)

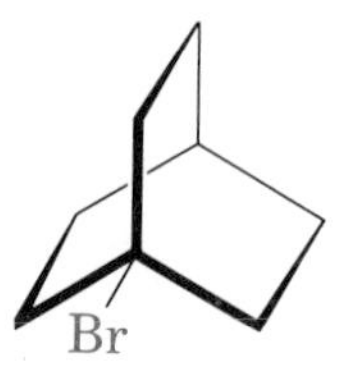

PROBLEM

10.7 How do you account for the fact that 1-chloro-1,2-diphenylethane reacts with the nucleophiles fluoride ion and triethylamine at exactly the same rate?

10.10 Elimination reactions of alkyl halides: the E2 reaction

We began this chapter by saying that two kinds of reactions are possible when a nucleophilic reagent (Lewis base) attacks an alkyl halide. Depending on the exact nature of the reaction and on the conditions used, the reagent may either attack at carbon and displace halide, or attack at hydrogen and eliminate HX to form an alkene (Section 6.12). The elimination of HX from alkyl halides to form alkenes is a general reaction of much importance, and we should now take a closer look at it. The subject is a complex one, and there is evidence that elimination reactions can take place through a variety of different mechanistic pathways. We will consider two of them—the E1 and E2 reactions.

The **E2** (for elimination, bimolecular) **reaction** is the most widely studied and commonly occurring pathway for elimination. It is closely analogous to the S_N2 reaction in many respects and may be formulated as shown in Figure 10.18.

The essential feature of the E2 reaction is that it is a one-step process without intermediates. As the attacking base begins to abstract a proton from a carbon next to the leaving group, the C—H bond begins to break, a new carbon–carbon double bond begins to form, and the leaving group begins to depart, taking with it the electron pair from the C—X bond.

Of the pieces of evidence supporting this mechanism, one of the most important is the reaction kinetics. Since both base and alkyl halide enter into the rate-limiting step, E2 reactions show second-order kinetics. In other words, E2 reactions follow the rate law:

$$\text{Rate} = k[\text{RX}][\text{Base}]$$

A second and more compelling piece of evidence involves the stereochemistry of E2 eliminations. As indicated by a large amount of experimental data, the E2 reaction is stereospecific. Elimination always occurs from a

Base (B:) attacks a neighboring C—H bond and begins to remove the H at the same time as the alkene double bond starts to form and the X group starts to leave.

Transition state

Neutral alkene is produced when the C—H bond is fully broken and the X group has departed with the C—X bond electron pair.

Figure 10.18. The E2 reaction of alkyl halides. (Dotted lines indicate partial bonding in the transition state.)

periplanar geometry, meaning that all five reacting atoms lie in the same plane. There are two such geometries possible: *syn periplanar* geometry, in which the H and the X leave from the *same* side; and *anti periplanar* geometry, in which the H and the X leave from *opposite* sides of the molecule:

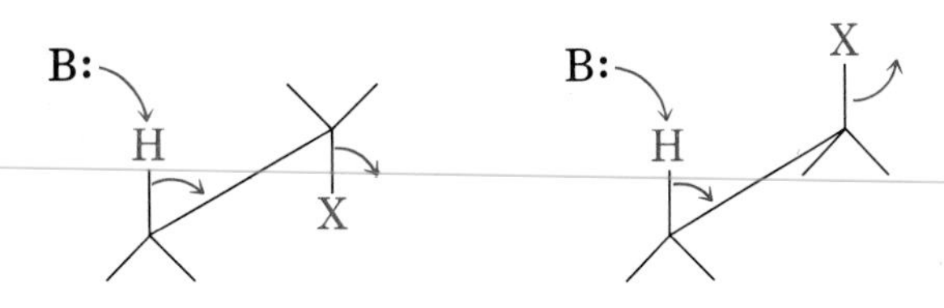

Anti periplanar geometry Syn periplanar geometry

A moment's reflection shows the reason for the requirement of periplanar geometry. Since the original C—H and C—X sp^3 sigma orbitals in the starting material must overlap and become p pi orbitals in the alkene product, there must also be partial overlap in the transition state. This can best occur if all of the orbitals are in the same plane to begin with (Figure 10.19, p. 318).

Of the two choices, anti periplanar and syn periplanar, anti periplanar geometry is much preferred. This is because anti geometry allows the two carbon centers to adopt a staggered relationship, whereas the higher-energy syn geometry requires the substituents on carbon to be eclipsed.

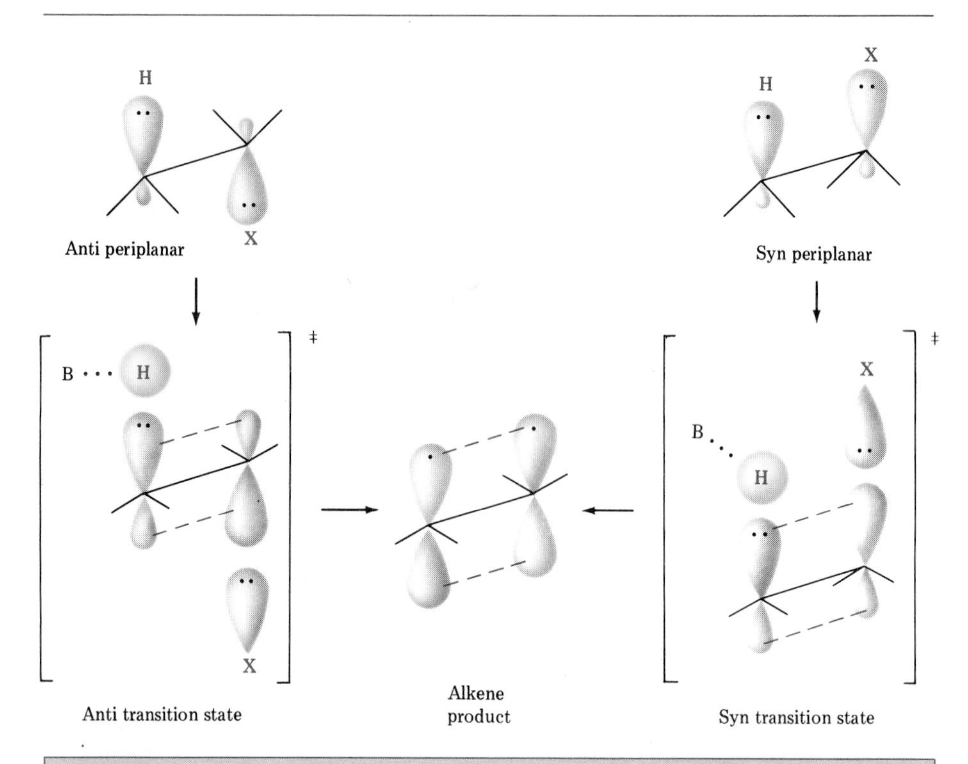

Figure 10.19. Transition state for the E2 reaction. Partial overlap of developing *p* orbitals in the transition state requires periplanar geometry.

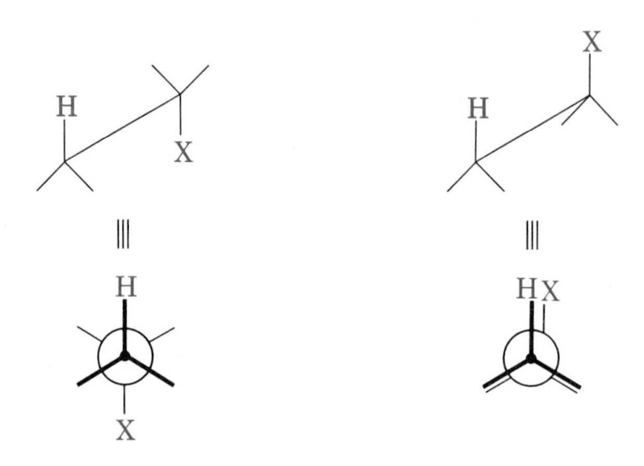

Anti periplanar geometry for E2 eliminations has definite stereochemical consequences that provide strong evidence for the proposed mechanism. To cite just one example, *meso*-1,2-dibromo-1,2-diphenylethane undergoes E2 elimination on treatment with base to give, *stereospecifically*, the pure *E* alkene (phenyl groups cis):

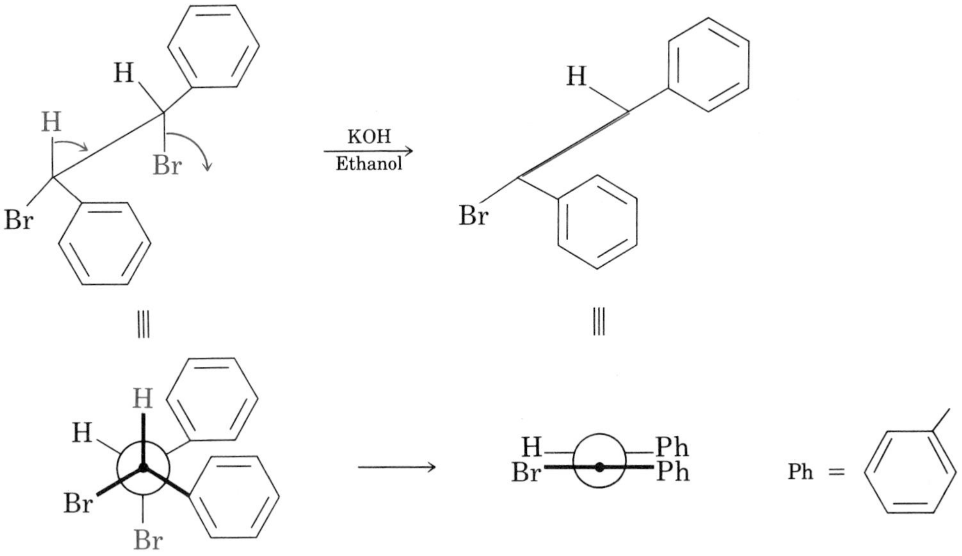

KOH / Ethanol

Ph =

meso-1,2-Dibromo-1,2-diphenylethane

(*E*)-1-Bromo-1,2-diphenylethylene

This preference for anti periplanar geometry is particularly noticeable in elimination reactions carried out on cyclic halides. Menthyl chloride, for example, undergoes E2 reaction to give a single alkene product, 2-menthene:

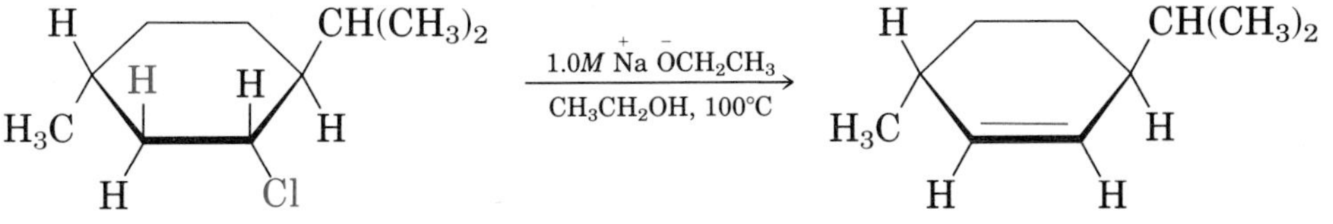

$1.0M\ \overset{+}{Na}\ \overset{-}{O}CH_2CH_3$ / CH_3CH_2OH, 100°C

Menthyl chloride

2-Menthene (100%)

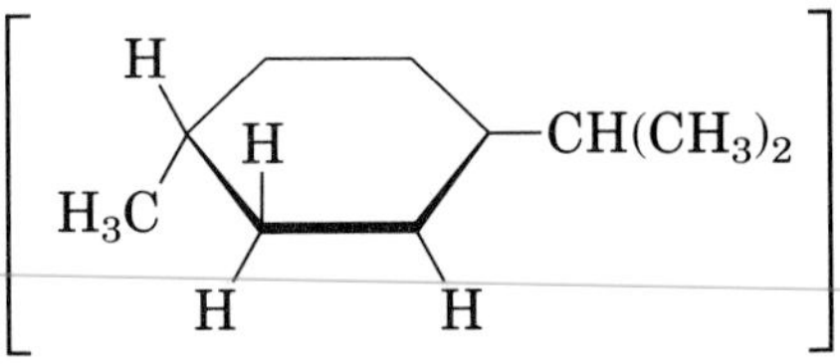

3-Menthene (*not formed*)

This result appears surprising at first glance if we remember that Zaitsev's rule predicts formation of the more substituted double-bond product. As we can see from looking at menthyl chloride, however, formation of the Zaitsev product (trisubstituted alkene) would require a syn elimination, whereas the observed formation of the non-Zaitsev product (disubstituted alkene) can occur through an energetically favored anti elimination. We will take a more detailed look at this reaction in Section 18.16.

PROBLEM

10.8 What alkene would you expect to obtain from E2 dehydrobromination of (1*R*,2*R*)-1,2-dibromo-1,2-diphenylethane? Draw a Newman projection of the reacting conformation.

PROBLEM

10.9 1-Chloro-1,2-diphenylethane can undergo E2 elimination to give either *cis*- or *trans*-1,2-diphenylethylene (stilbene). Draw Newman projections of the reactive conformations leading to both *cis*- and *trans*-1,2-diphenylethylene. From examination of these two conformations, can you suggest a reason why the trans alkene is the major product?

$$C_6H_5CH(Cl)CH_2C_6H_5 \xrightarrow{^-OCH_3} C_6H_5CH{=}CHC_6H_5$$

1-Chloro-1,2-diphenylethane → *trans*-1,2-Diphenylethylene

10.11 The deuterium isotope effect

One final piece of evidence in support of the E2 mechanism is provided by a phenomenon known as the **deuterium isotope effect.** For quantum mechanical reasons that we will not go into, a carbon–hydrogen bond is weaker by a small amount (about 1.2 kcal/mol) than a corresponding carbon–deuterium bond. Thus, a C—H bond is more easily broken than an equivalent C—D bond, and the rate of C—H bond cleavage is therefore faster. As an example of how this effect can be used to obtain mechanistic information, the base-induced elimination of HBr from 1-bromo-2-phenylethane proceeds 7.11 times as fast as the corresponding elimination of DBr from 1-bromo-2,2-dideuterio-2-phenylethane:

Faster reaction

$$C_6H_5CH_2{-}CH_2Br \xrightarrow{\text{Base}} C_6H_5CH{=}CH_2$$

1-Bromo-2-phenylethane

Slower reaction

$$C_6H_5CD_2{-}CH_2Br \xrightarrow{\text{Base}} C_6H_5CD{=}CH_2$$

1-Bromo-2,2-dideuterio-2-phenylethane

This result tells us that the C—H (or C—D) bond is broken *in the rate-limiting step.* Were it otherwise, we could not measure a rate difference. This result is fully consistent with our picture of the E2 reaction as a single-step process and illustrates the use of the deuterium isotope effect as a powerful tool in mechanistic investigations.

10.12 Elimination reactions of alkyl halides: the E1 reaction

Just as the E2 reaction is closely analogous to the S_N2 reaction, the S_N1 reaction also has a close analog—the **E1 reaction** (for elimination, unimolecular). The E1 reaction can be formulated as shown in Figure 10.20 for 2-chloro-2-methylpropane.

Spontaneous dissociation of the tertiary chloride yields an intermediate carbocation in a slow, rate-limiting step.

$$(CH_3)_3C{-}Cl \xrightleftharpoons{\text{Slow, rate-limiting}} \left[CH_3{-}\overset{CH_3}{\underset{CH_2-H}{C^+}} \right] + :Cl^-$$

Intermediate

Loss of a neighboring H^+ in a fast step yields neutral alkene product. The C—H bond electron pair goes to form the alkene pi bond.

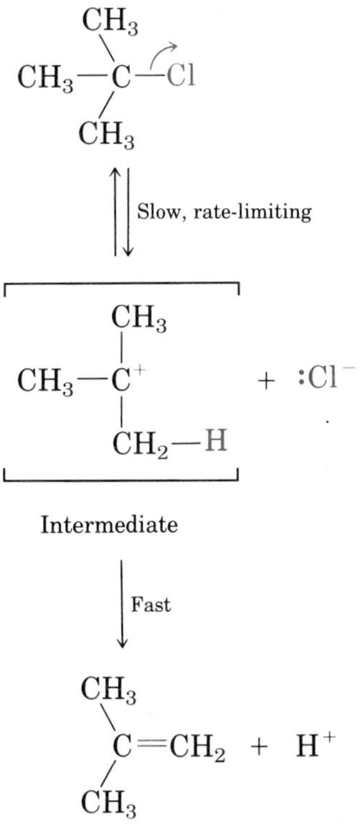

$$\xrightarrow{\text{Fast}} (CH_3)_2C{=}CH_2 + H^+$$

Figure 10.20. Mechanism of the E1 reaction. Two steps are involved, of which the dissociation step is rate limiting.

All E1 eliminations occur by spontaneous dissociation of a halide and loss of a proton from the intermediate carbocation. The E1 mechanism normally occurs under solvolysis conditions in the absence of an added base, and the best substrates are those that are also subject to S_N1 solvolysis. In practice, the E1 reaction usually occurs in competition with the S_N1 reaction when the substrate is simply heated in pure solvent, and mixtures of products are almost always obtained. For example, when 2-chloro-2-methylpropane is warmed to 65°C in 80% aqueous ethanol, a mixture of 2-methyl-2-propanol (S_N1) and 2-methylpropene (E1) is obtained:

$$(CH_3)_3CCl \xrightarrow[65°C]{H_2O,\ CH_3CH_2OH} (CH_3)_3COH + (CH_3)_2C{=}CH_2$$

2-Chloro-2-methylpropane → 2-Methyl-2-propanol (64%) + 2-Methylpropene (36%)

Much evidence has been obtained in support of the E1 mechanism. As expected, E1 reactions show first-order kinetics consistent with a spontaneous dissociation process:

$$\text{Rate} = k[\text{RX}]$$

A second piece of evidence involves the stereochemistry of elimination. Unlike the E2 reaction, where periplanar geometry is required, there is no geometric requirement on the E1 reaction. The intermediate carbocation can lose any available proton from a neighboring position, and we might therefore expect to obtain the more stable (Zaitsev) product. This is just what we find. To return to a familiar example, menthyl chloride loses HCl under E1 conditions to give a mixture of alkenes in which the Zaitsev product, 3-menthene, predominates (Figure 10.21).

One final piece of evidence is that *no* deuterium isotope effect is found for E1 reactions. This is because rupture of the C—H (or C—D) bond occurs

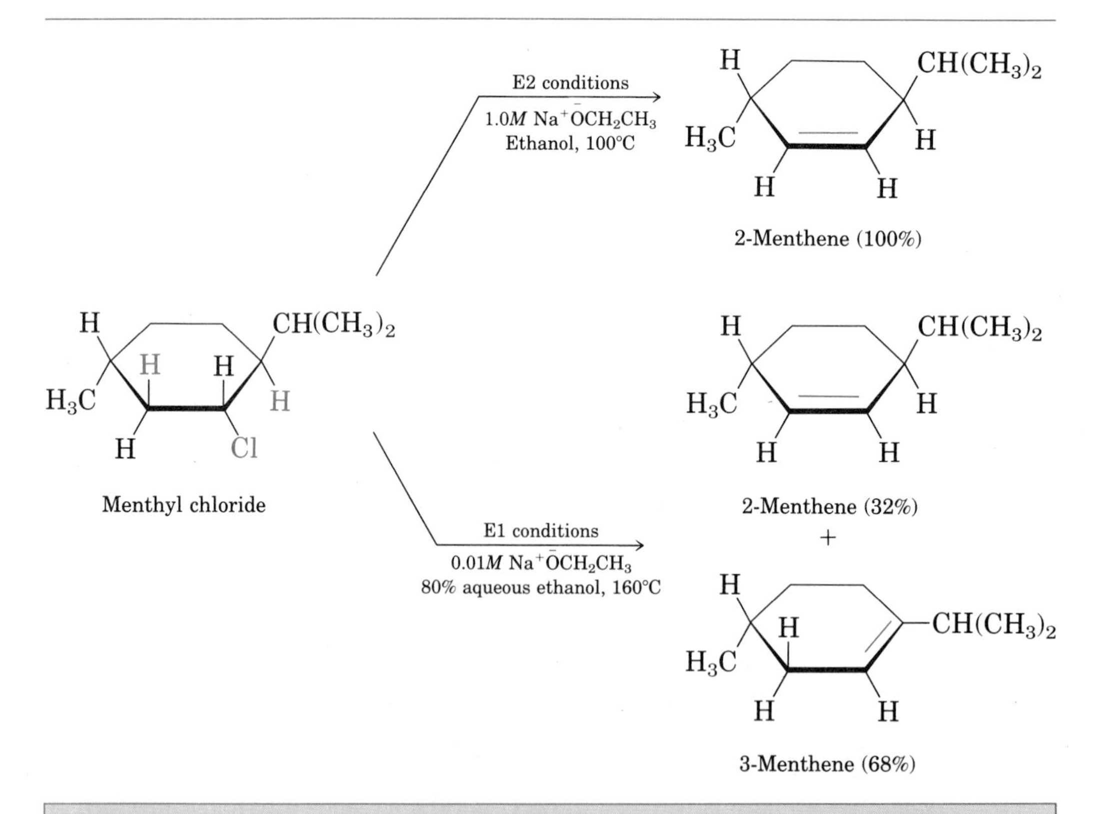

Figure 10.21. Elimination reactions of menthyl chloride. E2 conditions lead to 2-menthene, whereas E1 conditions lead to a mixture of 2-menthene and 3-menthene.

after the rate-limiting step, rather than during it, so we cannot measure a rate difference.

$$\text{(C}_6\text{H}_5)_2\text{C(Cl)CH}_3 \xrightarrow[-\text{Cl:}^-]{\text{Slow}} \left[(\text{C}_6\text{H}_5)_2\overset{+}{\text{C}}\text{—CH}_3\right] \xrightarrow[-\text{H}^+]{\text{Fast}} (\text{C}_6\text{H}_5)_2\text{C=CH}_2$$

$$\text{(C}_6\text{H}_5)_2\text{C(Cl)CD}_3 \xrightarrow[-\text{Cl:}^-]{\text{Slow}} \left[(\text{C}_6\text{H}_5)_2\overset{+}{\text{C}}\text{—CD}_3\right] \xrightarrow[-\text{D}^+]{\text{Fast}} (\text{C}_6\text{H}_5)_2\text{C=CD}_2$$

Both reactions take place at the same rate.

10.13 Summary of reactivity: S_N1, S_N2, E1, E2

We have examined four possible modes of reaction between an alkyl halide and a base/nucleophile, and you may well wonder how to predict what will happen in any given case. Will substitution or elimination occur? Will the reaction be bimolecular or unimolecular? Can a chemist choose the desired pathway by adjusting reaction conditions? There are no definitive answers to these questions, but it is possible to recognize some broad trends and make some generalizations about what to expect.

1. *Primary alkyl halides* react by either S_N2 or E2 mechanisms because they are relatively unhindered and their dissociation gives unstable carbocations. If a good nucleophile such as $RS:^-$, $I:^-$, $^-:CN$, or $Br:^-$ is used, nucleophilic substitution occurs to the virtual exclusion of elimination. Even strong bases such as hydroxide and ethoxide give largely the products of substitution. Only when strong, bulky bases such as *tert*-butoxide are used do E2 eliminations with primary halides occur. In such cases, nucleophilic substitution is prevented by the steric bulk of the base, but elimination can still occur:

$$CH_3CH_2CH_2CH_2Br \xrightarrow[\text{Ethanol}]{CH_3CH_2O:^-} CH_3CH_2CH{=}CH_2 + CH_3CH_2CH_2CH_2OCH_2CH_3$$

1-Bromobutane → 1-Butene (10%) + Butyl ethyl ether (90%)

$$CH_3CH_2CH_2CH_2Br \xrightarrow{(CH_3)_3CO:^-} CH_3CH_2CH{=}CH_2 + CH_3CH_2CH_2CH_2OC(CH_3)_3$$

1-Bromobutane → 1-Butene (85%) + Butyl *tert*-butyl ether (15%)

This generalization holds for all but the most hindered [i.e., neopentyl, $(CH_3)_3CCH_2X$] primary halides.

2. *Secondary alkyl halides* can react by any of the four mechanisms, and chemists can often make one or the other pathway predominate by choosing appropriate reaction conditions. When a secondary halide is treated with a strong base such as ethoxide, hydroxide, or amide ion, E2 elimination normally occurs. Conversely, when the same secondary halide is treated in a polar aprotic solvent such as hexamethylphosphoramide with a good nucleophile or weak base, S_N2 substitution usually occurs. For example, 2-bromopropane undergoes different reactions when treated with ethoxide ion and with acetate ion:

$$\underset{\text{2-Bromopropane}}{(CH_3)_2CHBr} \xrightarrow[\text{(weak base)}]{CH_3CO_2:^-} \underset{\text{Isopropyl acetate (100\%)}}{(CH_3)_2CHOCOCH_3} + \underset{\text{Propene (0\%)}}{CH_3CH{=}CH_2}$$

$$\underset{\text{2-Bromopropane}}{(CH_3)_2CHBr} \xrightarrow[\text{(strong base)}]{CH_3CH_2O:^-} \underset{\text{Ethyl isopropyl ether (20\%)}}{(CH_3)_2CHOCH_2CH_3} + \underset{\text{Propene (80\%)}}{CH_3CH{=}CH_2}$$

Secondary alkyl halides can be made to undergo S_N1 and E1 reactions under solvolysis conditions if weakly basic nucleophiles are used in polar protic solvents, but mixtures of products are usually obtained, and the reactions are of little use for synthesis.

3. *Tertiary halides* can react through three possible pathways—S_N1, E1, and E2—and one of the three can often be made to predominate by proper choice of reaction conditions. Steric hindrance at the tertiary carbon precludes S_N2 behavior. When a tertiary halide is treated with a strong base, bimolecular elimination predominates to the near exclusion of other possibilities. For example, 2-bromo-2-methylpropane gives only 3% substitution product when treated with ethoxide ion in ethanol:

$$\underset{\text{2-Bromo-2-methylpropane}}{(CH_3)_3CBr} \xrightarrow[CH_3CH_2OH]{CH_3CH_2O:^-} \underset{\text{Ethyl \textit{tert}-butyl ether (3\%)}}{(CH_3)_3COCH_2CH_3} + \underset{\text{2-Methylpropene (97\%)}}{CH_3\overset{\overset{\displaystyle CH_2}{\|}}{C}CH_3}$$

$$\underset{\text{2-Bromo-2-methylpropane}}{(CH_3)_3CBr} \xrightarrow[\Delta]{CH_3CH_2OH} \underset{\text{Ethyl \textit{tert}-butyl ether (80\%)}}{(CH_3)_3COCH_2CH_3} + \underset{\text{2-Methylpropene (20\%)}}{CH_3\overset{\overset{\displaystyle CH_2}{\|}}{C}CH_3}$$

By contrast, reaction under solvolytic conditions (heating in pure ethanol) leads to a mixture of products in which substitution predominates.

Table 10.11 summarizes these generalizations.

TABLE 10.11 Correlation of conditions and reactivity for substitution and elimination reactions

Halide type	*S_N1*	*S_N2*	*E1*	*E2*
Primary halide	Does not occur	Highly favored under most conditions	Does not occur	Favored when strong, hindered bases are used
Secondary halide	Can occur under solvolysis conditions in polar solvents	Favored by good nucleophiles in polar aprotic solvents	Can occur under solvolysis conditions in polar solvents	Favored when strong bases are used
Tertiary halide	Favored by nonbasic nucleophiles in polar solvents	Does not occur	Occurs under solvolysis conditions	Highly favored when bases are used

PROBLEM

10.10 Identify these reactions as to type:

(a) 1-Bromobutane + NaN_3 $\longrightarrow$ 1-Azidobutane

(b) Bromocyclohexane + NaOH $\longrightarrow$ Cyclohexene

(c) 2-Bromobutane + NaCN $\xrightarrow{\text{HMPA}}$ 2-Methylbutanenitrile

(d) Chlorodiphenylmethane $\xrightarrow{\text{HCOOH}}$ Diphenylmethyl formate

10.14 Substitution reactions in synthesis

We have discussed nucleophilic substitution reactions in much detail because they are of great importance in organic chemistry, and we have already seen a number of examples of S_N1 and S_N2 reactions used in organic synthesis, although they were not identified as such at the time.

One example is the alkylation of acetylide anions discussed in Section 7.8. As noted, acetylide anions react well with primary alkyl bromides, iodides, and tosylates, to provide the internal alkyne product:

$$R{-}CH_2{-}X + {}^{-}\!:\!C{\equiv}C{-}R' \longrightarrow R{-}CH_2C{\equiv}C{-}R' + :X^-$$

where X = Br, I, or OTos.

This is, of course, an S_N2 reaction, and it is quite understandable that only primary alkylating agents with good leaving groups react well. Since acetylide anion is a strong base as well as a good nucleophile, E2 elimination competes favorably with S_N2 alkylation when a secondary or tertiary substrate is used.

$$CH_3(CH_2)_3C{\equiv}C\!:^-Na^+ + CH_3\underset{}{\overset{\displaystyle Br}{\overset{|}{C}}}HCH_3 \longrightarrow CH_3(CH_2)_3C{\equiv}C{-}CH(CH_3)_2 \quad (7\%) \quad S_N2$$

$$+\ CH_3(CH_2)_3C{\equiv}CH + CH_3CH{=}CH_2 \quad (93\%) \quad E2$$

Other substitution reactions include some of the various methods used for preparing alkyl halides from alcohols. We saw in Section 9.7, for example, that halides can sometimes be prepared by treating alcohols with HX—reactions now recognizable as nucleophilic substitutions of halide ions on the protonated alcohols. It thus becomes clear why tertiary alcohols give by far the best reaction—a fast S_N1 process is strongly favored for tertiary alcohols, but primary alcohols must react by a slower S_N2 process (Figure 10.22).

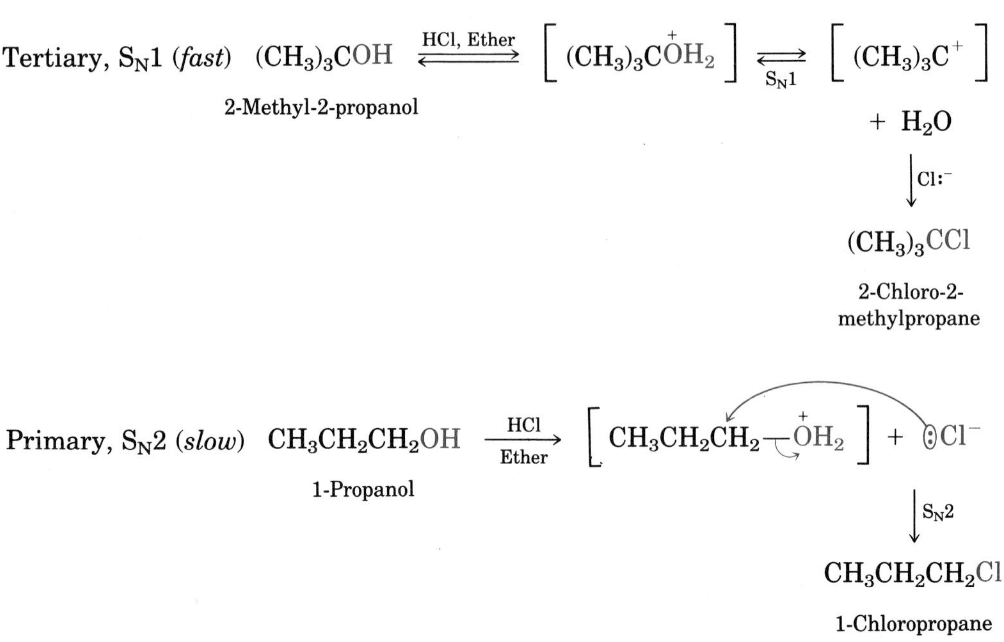

Tertiary, S_N1 (*fast*)

$$(CH_3)_3COH \underset{}{\overset{HCl,\ Ether}{\rightleftharpoons}} \left[(CH_3)_3C\overset{+}{O}H_2\right] \underset{S_N1}{\rightleftharpoons} \left[(CH_3)_3C^+\right] + H_2O \xrightarrow{Cl:^-} (CH_3)_3CCl$$

2-Methyl-2-propanol → 2-Chloro-2-methylpropane

Primary, S_N2 (*slow*)

$$CH_3CH_2CH_2OH \xrightarrow[Ether]{HCl} \left[CH_3CH_2CH_2-\overset{+}{O}H_2\right] + :Cl^- \xrightarrow{S_N2} CH_3CH_2CH_2Cl$$

1-Propanol → 1-Chloropropane

Figure 10.22. Mechanisms of reactions of HCl with tertiary and primary alcohols.

Primary alcohols can be *activated* toward displacement by transforming the hydroxyl into a better leaving group; this is exactly what reagents such as PBr_3 accomplish (Section 9.7):

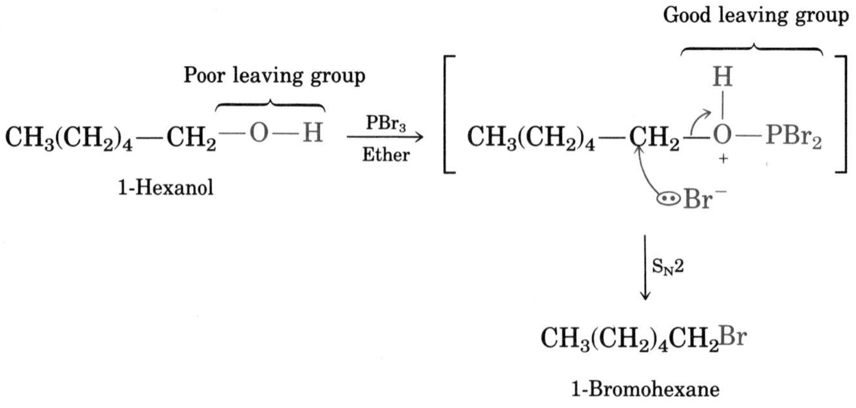

$$CH_3(CH_2)_4-CH_2-\underbrace{O-H}_{\text{Poor leaving group}} \xrightarrow[Ether]{PBr_3} \left[CH_3(CH_2)_4-CH_2-\underbrace{\overset{+}{O}(H)-PBr_2}_{\text{Good leaving group}}\right] + :Br^- \xrightarrow{S_N2} CH_3(CH_2)_4CH_2Br$$

1-Hexanol → 1-Bromohexane

Alcohols react with PBr_3 to give dibromophosphites ($R—O—PBr_2$). Since dibromophosphites are excellent leaving groups, S_N2 displacement by bromide occurs rapidly on the primary carbon, and alkyl bromides are produced in good yield.

We also saw in Section 10.2 that alcohols may be activated toward nucleophilic displacement by formation of their tosylates. Thus, another excellent method for preparing alkyl halides is the S_N2 reaction of a primary-alkyl tosylate with sodium bromide or sodium iodide (Figure 10.23).

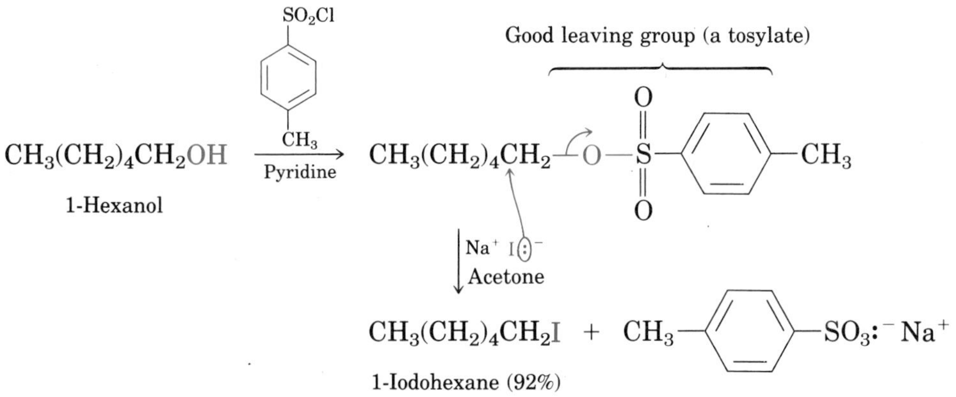

$$CH_3(CH_2)_4CH_2OH \xrightarrow[\text{Pyridine}]{p\text{-}CH_3C_6H_4SO_2Cl} CH_3(CH_2)_4CH_2—O—SO_2—C_6H_4—CH_3$$

1-Hexanol; Good leaving group (a tosylate)

$$\xrightarrow[\text{Acetone}]{Na^+\ I^-} CH_3(CH_2)_4CH_2I + CH_3—C_6H_4—SO_3^-\ Na^+$$

1-Iodohexane (92%)

Figure 10.23. Conversion of a primary alcohol into an alkyl iodide by S_N2 reaction of the tosylate.

One final example of an S_N2 reaction, alluded to in Section 9.8, is the displacement of halides by hydride ions. Now that we can recognize these reactions as S_N2 nucleophilic displacements, the reason why only primary and secondary halides react successfully is clear.

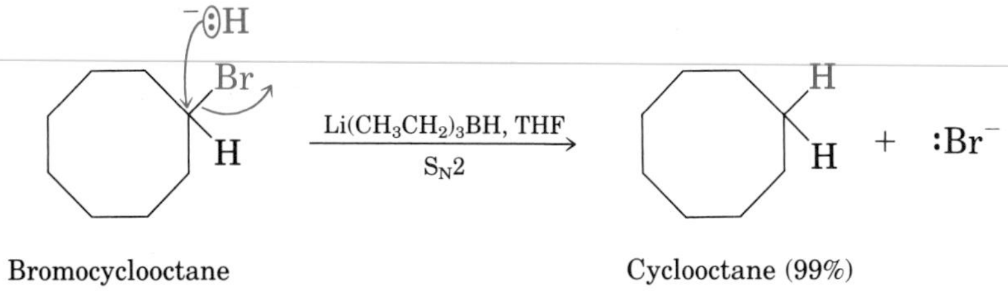

$$\text{Bromocyclooctane} \xrightarrow[S_N2]{Li(CH_3CH_2)_3BH,\ THF} \text{Cyclooctane (99\%)} + :Br^-$$

10.15 Substitution reactions in biological systems

Many biological processes occur by reaction pathways analogous to those carried out in the laboratory, and we have already seen some examples of biological additions and eliminations. The situation is similar with substitutions. A wide variety of reactions occurring in living organisms are known to take place by nucleophilic displacement mechanisms.

BIOLOGICAL METHYLATIONS

Perhaps the most ubiquitous of all biological substitutions is the **methylation** reaction—the transfer of a methyl group from an electrophilic donor to a nucleophile.

$$\sim Y{-}CH_3 + :Nu^- \longrightarrow CH_3{-}Nu + \sim Y:^-$$

Methyl donor — Methylated nucleophile

Although a laboratory chemist might choose iodomethane for such a reaction, living organisms operate in a more subtle way. The amino acid methionine is the biological methyl group donor. Methionine is first activated by reaction with adenosine triphosphate (S_N2 process) to form *S*-adenosylmethionine. Since the methionine sulfur atom now has a positive charge (a *sulfonium* ion), it is an excellent leaving group for S_N2 displacements on the methyl carbon. A biological nucleophile therefore attacks the methionine methyl by an S_N2 reaction (Figure 10.24).

One example of the action of *S*-adenosylmethionine in biological methylations takes place in the adrenal medulla during the formation of adrenaline from norepinephrine.

$$R(R')S^+{-}CH_3 + H_2\ddot{N}CH_2CH(OH){-}C_6H_3(OH)_2$$

S-Adenosylmethionine — Norepinephrine

$\downarrow$ S_N2 reaction

$$CH_3NHCH_2CH(OH){-}C_6H_3(OH)_2 + R{-}S{-}R' + H^+$$

Adrenaline

After becoming used to dealing with simple halides such as bromomethane used for laboratory alkylations, it can be something of a shock to encounter a molecule as complex as *S*-adenosylmethionine. The shock is only psychological, however. From the chemical standpoint, CH_3Br and *S*-adenosylmethionine are both doing exactly the same thing—transferring a methyl group by an S_N2 reaction. The same principles apply to both.

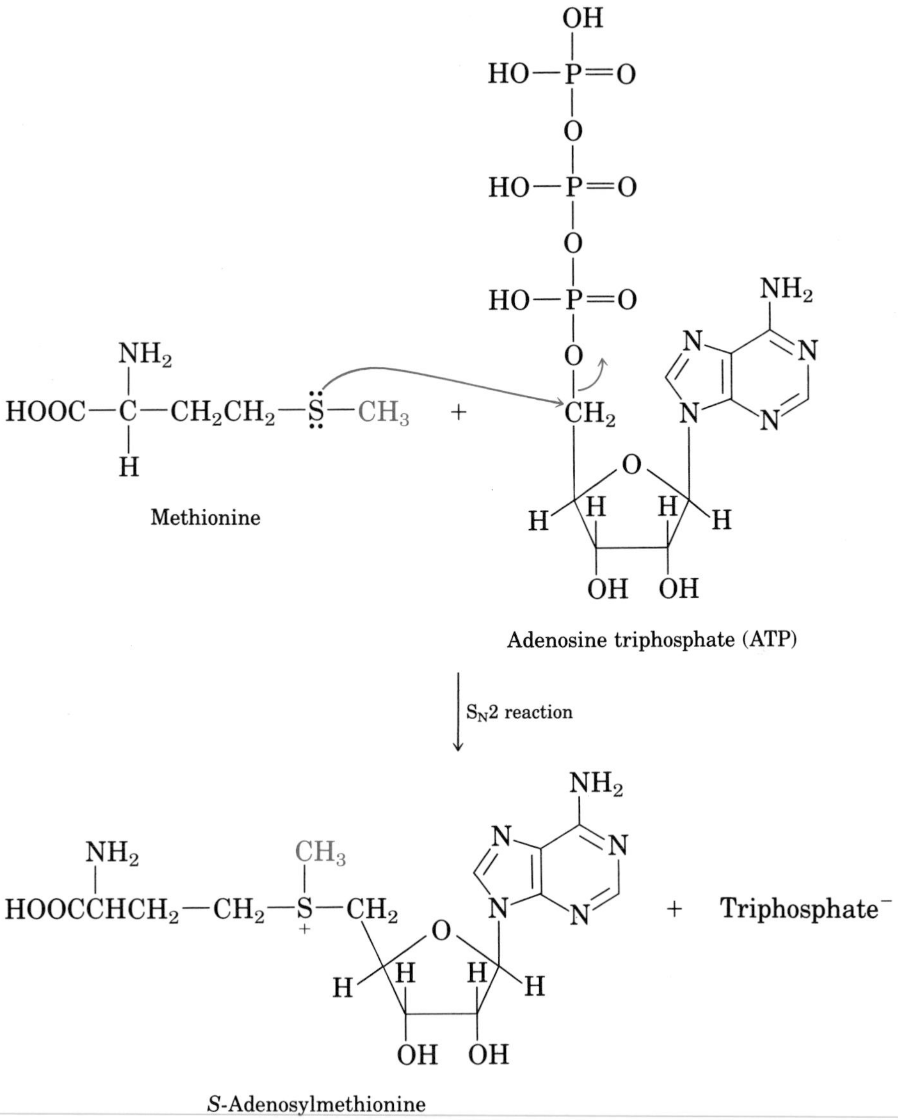

$HOOC{-}CH(NH_2){-}CH_2CH_2{-}\ddot{S}{-}CH_3$ + ATP

Methionine

Adenosine triphosphate (ATP)

S_N2 reaction

$HOOCCH(NH_2)CH_2{-}CH_2{-}\overset{+}{S}(CH_3){-}CH_2$–adenosine + Triphosphate^-

S-Adenosylmethionine

Figure 10.24. Formation of *S*-adenosylmethionine from methionine.

OTHER BIOLOGICAL ALKYLATIONS

As another example of some of the kinds of biological S_N2 reactions that can occur, let's examine some alkylations of biological molecules. Many reactive S_N2 substrates with deceptively simple structures are quite toxic to living organisms. For example, bromomethane is widely used as a fumigant to kill termites and chloromethyl methyl ether, $CH_3{-}O{-}CH_2Cl$, is thought to cause certain cancers. The toxicity of these chemicals is believed to derive from their ability to transfer alkyl groups to nucleophilic amino groups ($\sim\ddot{N}H_2$) and mercapto groups ($\sim\ddot{S}H$) on enzymes. With enzymes modified by alkylation, normal biological chemistry is altered, with dire consequences to the organism.

$$\text{Enzyme}\sim\ddot{N}H_2 \xrightarrow{R-X} \text{Enzyme}\sim\ddot{N}HR$$

$$\text{Enzyme}\sim\ddot{S}-H \xrightarrow[R-X]{} \text{Enzyme}\sim\ddot{S}-R$$

Alkylated enzymes

Historically, one of the best-known toxic alkylating agents is mustard gas, $ClCH_2CH_2SCH_2CH_2Cl$. Mustard gas gained notoriety because of its use as an agent of warfare during World War I, and it has been estimated that some 400,000 casualties resulted from its use. Mustard gas, as its structure might indicate (primary halide), is quite reactive toward S_N2 displacements by nucleophilic groups of proteins. It is thought to act through an intermediate sulfonium ion in much the same manner as does *S*-adenosylmethionine (Figure 10.25).

$$ClCH_2CH_2-\ddot{S}-CH_2CH_2-Cl \xrightleftharpoons{\text{Internal } S_N2 \text{ reaction}} \left[ClCH_2CH_2\overset{+}{S}(CH_2)_2 \quad :Cl^-\right]$$

Mustard gas

S_N2 reaction with $:NH_2\sim$Protein

$$ClCH_2CH_2S-CH_2CH_2NH\sim\text{Protein}$$

Alkylated protein

Figure 10.25. The alkylation of a protein by mustard gas.

10.16 Summary

Reaction of an alkyl halide with a nucleophile results either in substitution or in elimination. Both modes of reaction are of great importance in chemistry, and it is useful to be able to predict what will happen in specific cases.

Nucleophilic substitution reactions occur by two mechanisms—S_N2 and S_N1. In the S_N2 reaction, the entering nucleophile attacks the halide from a direction 180° away from the leaving group, resulting in an umbrella-like Walden inversion of configuration at the carbon atom. The reaction shows second-order kinetics, with rate = k[RX][Nu], and it is strongly inhibited by increasing steric bulk of the reagents. Therefore, S_N2

reactions are favored only for primary and secondary substrates. In the S_N1 reaction, the substrate spontaneously dissociates to a carbocation in a slow rate-limiting step, followed by a rapid attack of nucleophile. In consequence, S_N1 reactions show first-order kinetics, with rate $= k[RX]$, and they take place with racemization of configuration at the carbon atom. They are most favored for tertiary substrates.

Elimination reactions also can occur by two different mechanisms—E2 and E1. In the E2 reaction, a base abstracts a proton at the same time as the leaving group departs. The elimination takes place preferentially through an anti periplanar transition state. The reaction shows second-order kinetics and a deuterium isotope effect, and occurs when a secondary or tertiary substrate is treated with a strong base. In the E1 reaction, the substrate spontaneously dissociates to yield a carbocation in the slow rate-limiting step before losing a neighboring proton. The reaction shows first-order kinetics and no deuterium isotope effect, and occurs when a tertiary substrate reacts in polar, nonbasic solution.

All four reactions—S_N1, S_N2, E1, and E2—are strongly influenced by many factors, and Table 10.12 summarizes the effects of some important variables.

TABLE 10.12 **Effects of reaction variables on substitution and elimination reactions**

Reaction	*Solvent*	*Nucleophile/base*	*Leaving group*	*Substrate structure*
S_N1	Very strong effect; reaction favored by polar solvents	Weak effect; reaction favored by good nucleophile/weak base	Strong effect; reaction favored by good leaving group	Strong effect; reaction favored by 3°, allylic, and benzylic substrates
S_N2	Strong effect; reaction favored by polar aprotic solvents	Strong effect; reaction favored by good nucleophile/weak base	Strong effect; reaction favored by good leaving group	Strong effect; reaction favored by methyl and 1° substrates
E1	Very strong effect; reaction favored by polar solvents	Weak effect; reaction favored by weak base	Strong effect; reaction favored by good leaving group	Strong effect; reaction favored by 3°, allylic, and benzylic substrates
E2	Strong effect; reaction favored by polar aprotic solvents	Strong effect; reaction favored by poor nucleophile/strong base	Strong effect; reaction favored by good leaving group	Strong effect; reaction favored by 1° and 2° substrates

10.17 Summary of reactions

1. Nucleophilic substitutions
   a. S_N1 (Sections 10.6–10.9)

$$\text{C—X} \longrightarrow \left[\ \text{C}^+\ \right] \xrightarrow{Nu:^-} \text{C—Nu} + :X^-$$

Best for 3°, benzyl, and allyl halides

   b. S_N2 (Sections 10.4–10.5)

$$\text{C—X} \xrightarrow{Nu:^-} \text{Nu—C} + X:^-$$

Best for 1° and 2° halides

$Nu:^-$ = $H:^-$, $^-:CN$, $I:^-$, $Br:^-$, $Cl:^-$, $HO:^-$, $^-:NH_2$, $CH_3O:^-$, $CH_3CO_2:^-$, $HS:^-$, $H_2O:$, $:NH_3$, and so on.

2. Eliminations
   a. E1 (Section 10.12)

$$\text{H—C—C(X)(R)—R} \longrightarrow \left[\ \text{H—C—}\overset{+}{\text{C}}\text{(R)—R}\ \right] \longrightarrow \text{C=CR}_2 + \text{HX}$$

Best for 3° halides

   b. E2 (Section 10.10)

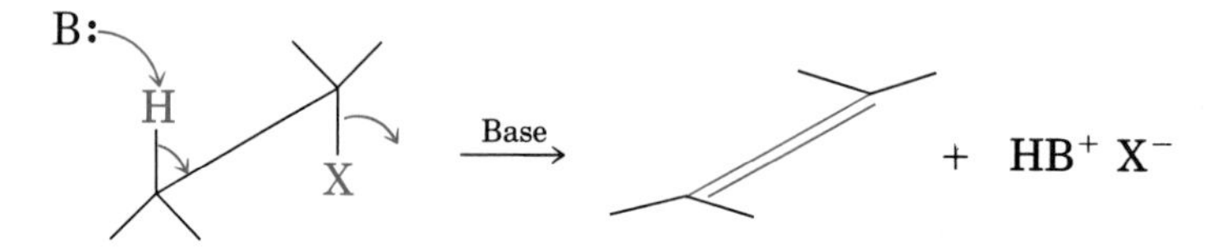

Best for 2° and 3° halides

ADDITIONAL PROBLEMS

10.11 Which reagent in each pair will react faster with hydroxide ion?
(a) CH_3Br or CH_3I
(b) CH_3CH_2I in ethanol or dimethyl sulfoxide
(c) $(CH_3)_3CCl$ or CH_3Cl
(d) $H_2C{=}CHBr$ or $H_2C{=}CHCH_2Br$

10.12 How might you prepare each of the following molecules using a nucleophilic substitution reaction at some step?

(a) $CH_3C{\equiv}CCH(CH_3)_2$
(b) $CH_3CH_2CH_2CH_2CN$
(c) $CH_3OC(CH_3)_3$

(d) $(C_6H_5)_3\overset{+}{P}{-}CH_3\ Br^-$

A phosphonium salt

(e) CH_3 Br (on cyclohexane ring)

(f) $CH_3CH_2CH(N_3)CH_3$

10.13 What products would you expect from the reaction of 1-bromopropane with:

(a) $NaNH_2$
(b) $K^{+\,-}O{-}C(CH_3)_3$
(c) NaI
(d) NaCN
(e) $NaC{\equiv}CH$
(f) $LiAlH_4$
(g) Mg, then H_2O
(h) $CH_3S{:}^-\ Na^+$

10.14 Which reagent in each of the following pairs is more nucleophilic? Explain.

(a) $N_3{:}^-$ or $H_3N{:}$
(b) $H_2O{:}$ or $CH_3C(=O){-}\ddot{O}H$
(c) BF_3 or $F{:}^-$
(d) $(CH_3)_3P{:}$ or $(CH_3)_3N{:}$
(e) $Cl{:}^-$ or $ClO_4{:}^-$
(f) $^-{:}CN$ or $^-{:}OCH_3$

10.15 Among the Walden cycles carried out by Phillips and Kenyon is the following series of reactions reported in 1923:

$CH_3CH(OH)CH_2C_6H_5$, $[\alpha]_D = +33.0°$ $\xrightarrow{TosCl}$ $CH_3CH(OTos)CH_2C_6H_5$, $[\alpha]_D = +31.1°$

$CH_3CH(OH)CH_2C_6H_5$ $\xrightarrow{K}$ $[CH_3CH(O^-K^+)CH_2C_6H_5]$ $\xrightarrow{CH_3CH_2Br}$ $CH_3CH(OCH_2CH_3)CH_2C_6H_5$, $[\alpha]_D = +23.5°$

$CH_3CH(OTos)CH_2C_6H_5$ $\xrightarrow{CH_3CH_2OH,\ \Delta}$ $CH_3CH(OCH_2CH_3)CH_2C_6H_5$, $[\alpha]_D = -19.9°$

Explain these results and indicate where Walden inversion is occurring.

10.16 The synthetic sequences shown here are unlikely to occur as written. What is wrong with each?

(a) $(CH_3)_3C{-}Cl \xrightarrow[THF]{Li(CH_3CH_2)_3BH} (CH_3)_3C{-}H$

(b) $CH_3CH(Br)CH_2CH_3 \xrightarrow[(CH_3)_3C{-}OH]{K^{+}\,{}^{-}O{-}C(CH_3)_3} CH_3CH(O\text{-}t\text{-}Bu)CH_2CH_3$

(c) 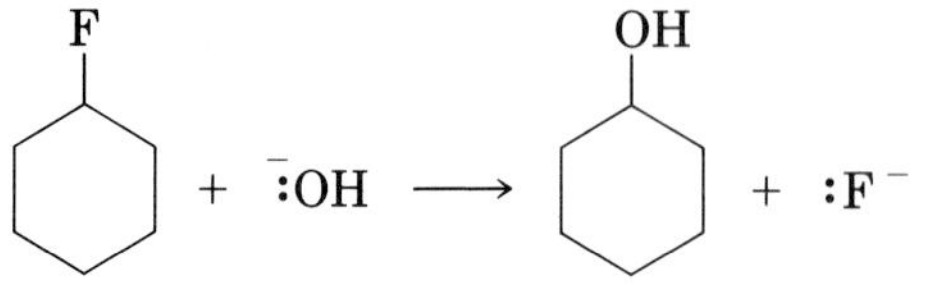 Fluorocyclohexane + $^{-}$:OH ⟶ Cyclohexanol + :F^{-}

(d) 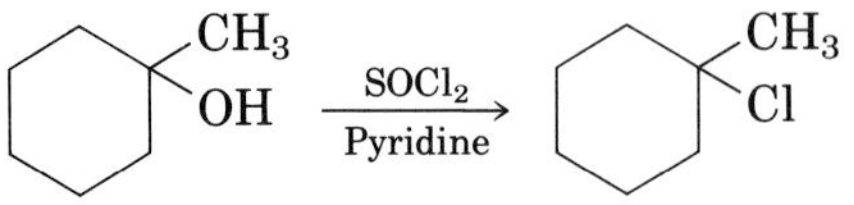 1-Methylcyclohexanol (CH$_3$, OH) $\xrightarrow[Pyridine]{SOCl_2}$ 1-Chloro-1-methylcyclohexane (CH$_3$, Cl)

10.17 Order each set of compounds with respect to S_N1 reactivity.

(a) $(CH_3)_3C{-}Cl$, 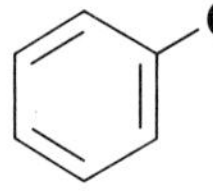 $C_6H_5C(CH_3)_2Cl$, $CH_3CH_2CH(NH_2)CH_3$

(b) $(CH_3)_3C{-}F$, $(CH_3)_3C{-}Br$, $(CH_3)_3C{-}OH$

(c) 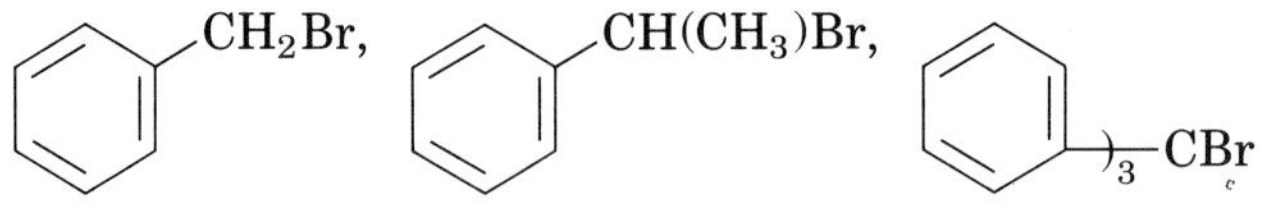 $C_6H_5CH_2Br$, $C_6H_5CH(CH_3)Br$, $(C_6H_5)_3CBr$

10.18 Order each set of compounds with respect to S_N2 reactivity.

(a) $H_2C{=}CHCH_2Cl$, $CH_3CH_2CH_2Cl$, $CH_3CH_2CH(Cl)CH_3$

(b) $(CH_3)_2CHCH(Br)CH_3$, $(CH_3)_2CHCH_2Br$, $(CH_3)_3CCH_2Br$

(c) $CH_3CH_2CH_2OCH_3$, $CH_3CH_2CH_2OTos$, $CH_3CH_2CH_2Br$

10.19 Predict the product and give the stereochemistry resulting from reaction of the following nucleophiles with (*R*)-2-bromooctane:

(a) $^{-}$:CN (b) CH_3CO_2:$^{-}$

(c) CH_3S:$^{-}$ (d) Br:$^{-}$

10.20 Describe in your own words the effects of the following variables on S_N2 and S_N1 reactions:

(a) Solvent (b) Leaving group

(c) Attacking nucleophile (d) Substrate structure

10.21 Ethers can often be prepared by S_N2 reaction of alkoxide ions with alkyl halides. Suppose you wanted to prepare cyclohexyl methyl ether. Which of the two possible routes shown here would you choose? Explain.

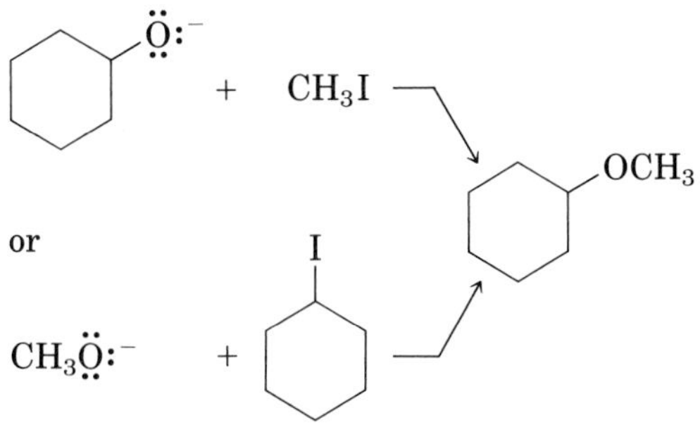

$C_6H_{11}\ddot{O}:^-$ + CH_3I → $C_6H_{11}OCH_3$

or

$CH_3\ddot{O}:^-$ + $C_6H_{11}I$ → $C_6H_{11}OCH_3$

10.22 The S_N2 reaction can also occur *intramolecularly*. What product would you expect from treatment of 4-bromo-1-butanol with base?

$$BrCH_2CH_2CH_2CH_2OH \xrightarrow[NaOH,\ CH_3CH_2OH]{} ?$$

10.23 In light of your answer to Problem 10.22, can you propose a synthesis of 1,4-dioxane starting only with a dihalide?

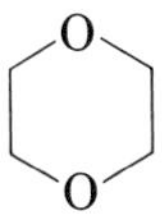

1,4-Dioxane

10.24 Propose structures for compounds that fit these descriptions:
(a) An alkyl halide that will give a mixture of three alkenes on E2 reaction
(b) An alkyl halide that will not undergo nucleophilic substitution
(c) An alkyl halide that gives the non-Zaitsev product on E2 reaction
(d) An alcohol that will react rapidly with HCl at 0°C

10.25 The tosylate of (2*R*,3*S*)-3-phenyl-2-butanol undergoes E2 elimination on treatment with ethoxide ion to yield (*Z*)-2-phenyl-2-butene.

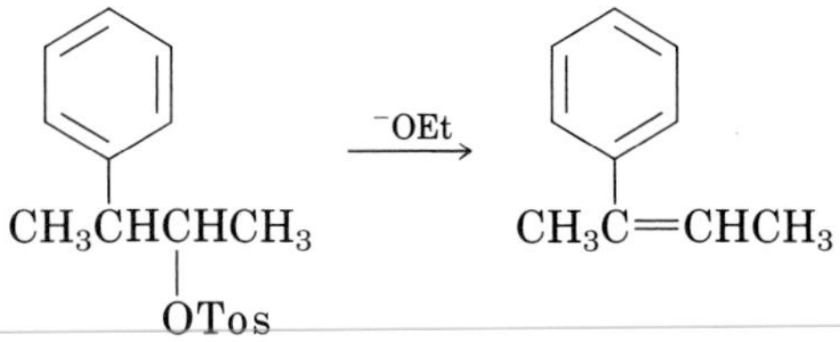

$$CH_3CH(C_6H_5)CH(OTos)CH_3 \xrightarrow{^-OEt} CH_3C(C_6H_5)=CHCH_3$$

Formulate the reaction, showing the proper stereochemistry. Explain the observed result using Newman projections.

10.26 In light of your answer to Problem 10.25, which alkene, *E* or *Z*, would you expect from the elimination reaction of the (2*R*,3*R*)-3-phenyl-2-butanol tosylate? Explain. Which alkene would result from the (2*S*,3*R*) and (2*S*,3*S*) tosylates?

10.27 We saw in Chapter 7 that alkynes can be made by dehydrohalogenation of vinylic halides in a reaction that is essentially an E2 process. In studying the stereochemistry of this elimination, it was found that (*Z*)-2-chloro-2-butene-1,4-dioic acid reacts 50 times as fast as the corresponding *E* isomer. What conclusion can you draw about the stereochemistry of eliminations? How does this result compare with eliminations of alkyl halides?

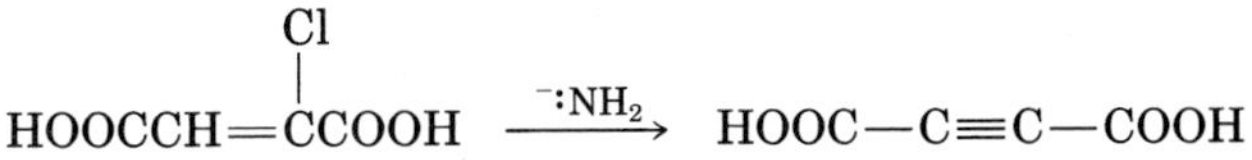

$$HOOCCH=C(Cl)COOH \xrightarrow{^-:NH_2} HOOC-C\equiv C-COOH$$

10.28 Predict the major alkene product from each of these elimination reactions.

(a) 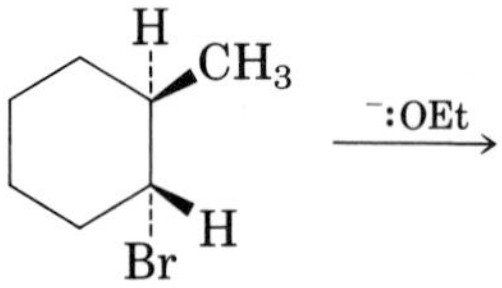 [structure: cyclohexane ring with H and CH_3 on one carbon, Br and H on adjacent carbon] $\xrightarrow{^-:OEt}$

(b) 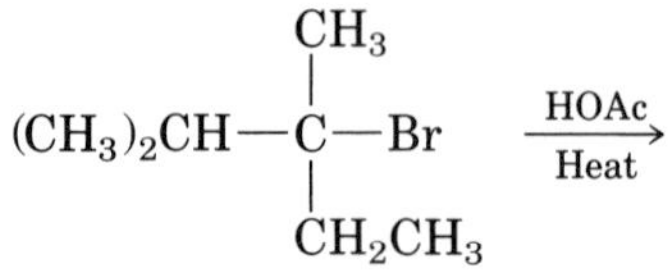 $(CH_3)_2CH—C(CH_3)(CH_2CH_3)—Br \xrightarrow[\text{Heat}]{\text{HOAc}}$

(c) [structure: cyclohexyl–$C(CH_3)_2Br$] $\xrightarrow{^-:OC(CH_3)_3}$

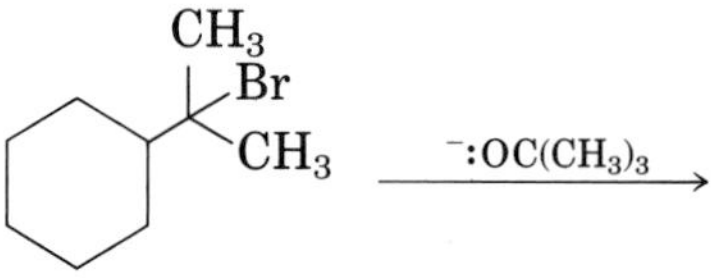

10.29 It has been observed that optically active 2-butanol slowly racemizes on standing in dilute sulfuric acid. Propose a mechanism to account for this observation.

10.30 Suggest an explanation for the observation that reaction of HBr with (*R*)-3-methyl-3-hexanol leads to (±)-3-bromo-3-methylhexane. [The formula for 3-methyl-3-hexanol is $CH_3CH_2CH_2C(OH)(CH_3)CH_2CH_3$.]

10.31 Account for the fact that treatment of 1-bromo-2-deuterio-2-phenylethane with strong base leads to a mixture of deuterated and nondeuterated phenylethylenes in which the deuterated product predominates by approximately 7:1.

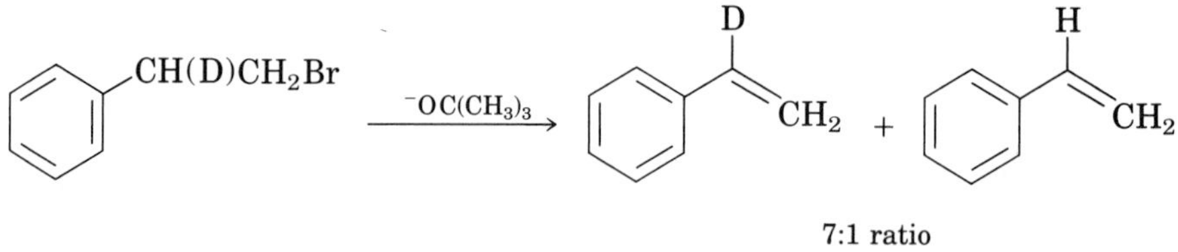

[Scheme: $C_6H_5CH(D)CH_2Br \xrightarrow{^-OC(CH_3)_3} C_6H_5C(D){=}CH_2 + C_6H_5C(H){=}CH_2$]

7:1 ratio

10.32 Although anti geometry is preferred for E2 reactions, it is not absolutely necessary. The deuterated bromo compound shown here reacts with strong base to yield an undeuterated alkene. Clearly, a syn elimination has occurred. Make a molecular model of the starting material and examine its geometry. Can you account for this result?

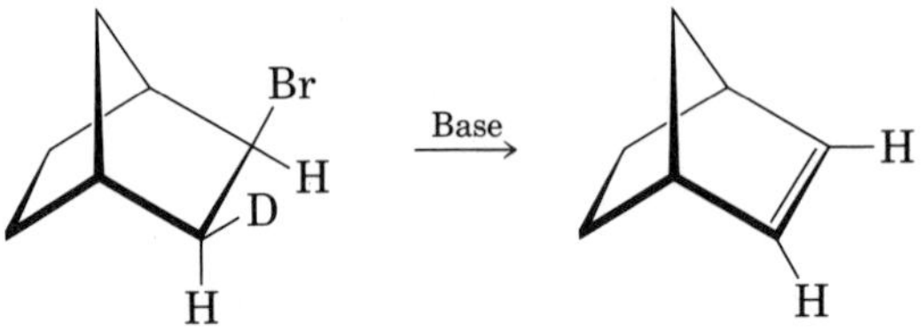

[Scheme: bicyclic bromo compound with Br, H, D, H labels $\xrightarrow{\text{Base}}$ bicyclic alkene with H, H labels]

10.33 In light of your answer to Problem 10.32, account for the observation that one of the following isomers undergoes E2 reaction approximately 100 times as fast as the other. Which isomer is the more reactive, and why?

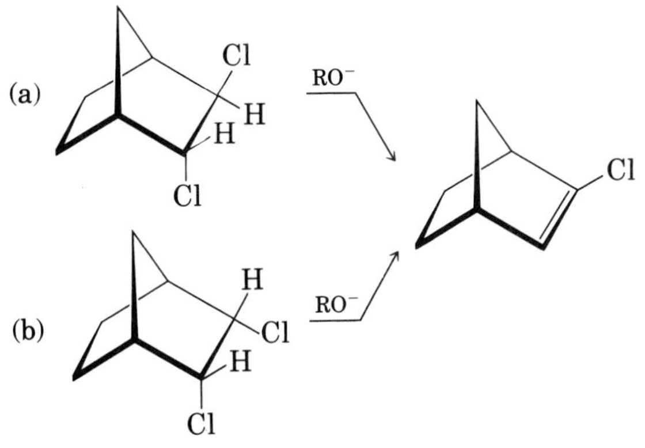

(a) Cl, H, H, Cl $\xrightarrow{RO^-}$ Cl

(b) H, Cl, H, Cl $\xrightarrow{RO^-}$

10.34 There are eight possible diastereomers of 1,2,3,4,5,6-hexachlorocyclohexane. Draw them. One isomer loses HCl in an E2 reaction nearly 1000 times more slowly than the others. Which isomer reacts so slowly, and why?

10.35 Consider the following methyl ester cleavage reaction:

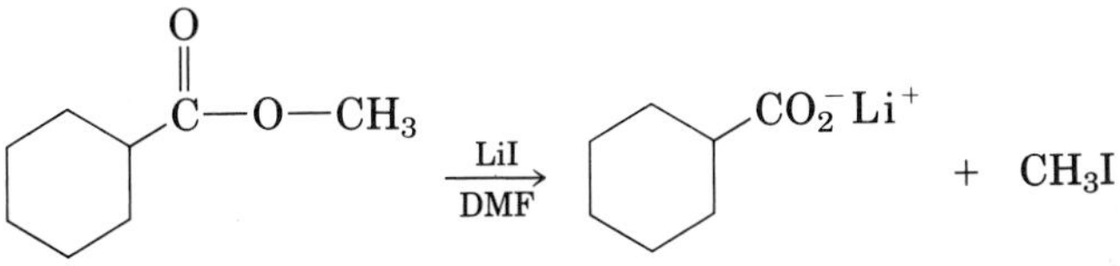

$$C_6H_{11}\text{—}\overset{O}{\overset{\|}{C}}\text{—O—}CH_3 \xrightarrow[DMF]{LiI} C_6H_{11}\text{—}CO_2^-\,Li^+ + CH_3I$$

The following pieces of evidence are available:

(a) The reaction occurs much faster in DMF than in ethanol.

(b) The corresponding ethyl ester cleaves approximately 10 times more slowly than the methyl ester.

Using this evidence, propose a mechanism for the reaction. What other kinds of experimental evidence could you gather to support your mechanistic hypothesis?

10.36 Account for the fact that the rate of reaction of 1-chlorooctane with acetate ion to give octyl acetate is greatly accelerated by the presence of a small quantity of iodide ion.

10.37 Compound X is optically inactive and has the formula $C_{16}H_{16}Br_2$. On treatment with strong base, X gives hydrocarbon Y, $C_{16}H_{14}$, which absorbs 2 equiv of hydrogen when reduced over a palladium catalyst, and reacts with ozone to give two fragments. One fragment, Z, is an aldehyde with formula C_7H_6O. The other fragment was identified as glyoxal, CHOCHO. Formulate the reactions involved and suggest structures for X, Y, and Z. What is the stereochemistry of X?

10.38 Propose a structure for an alkyl halide that gives (*E*)-3-methyl-2-phenyl-2-pentene and none of the *Z* isomer on E2 elimination. Make sure you indicate the stereochemistry.

CHAPTER 11 STRUCTURE DETERMINATION: MASS SPECTROSCOPY AND INFRARED SPECTROSCOPY

Structure determination is central to organic chemistry. Each time a reaction is run, the chemist must purify and identify the product. The desired product must be separated from solvents, from excess reagents, and from other products if more than one is formed, and its structure must be determined. In the nineteenth and early twentieth centuries, doing these things was a time-consuming process requiring great skill. In the past few decades, however, extraordinary advances have been made in chemical instrumentation. Sophisticated (and usually expensive) instruments are now available that greatly simplify the problems of purification and structure determination. Use of these instruments does not guarantee good results—skill and patience are still required—but it does ease the chemist's tasks.

11.1 Purification of organic compounds

Crystallization is a simple yet effective method for purifying a solid product. The crude reaction product is dissolved in a minimum amount of a suitable hot solvent, and the solution is allowed to cool slowly. Pure crystals form and precipitate; impurities remain in solution. The crystalline product is then isolated by filtration.

Distillation is an equally simple and effective method for purifying a volatile liquid product. The crude liquid reaction product is heated to a boil, and the vapors rise and are condensed into a receiver. Nonvolatile impurities remain in the still pot.

If the crude product is a mixture of two or more volatile compounds having different boiling points, **fractional distillation** can often effect a separation. The lower-boiling, more volatile component distills first, followed by the higher-boiling material. The theory of operation of distillation columns is complex, but as a practical matter good separation can usually be achieved in the laboratory if the components differ in boiling point by more than 5°C.

If neither crystallization nor distillation is effective, some form of **chromatography** is normally used to separate a mixture of organic compounds. The development of chromatography (literally, "color writing") as a separation technique dates back to the work of the Russian chemist Mikhail Tswett.[1] In 1903 Tswett described the separation of the pigments in green leaves by dissolving the leaf extract and allowing the solution to run down through a vertical glass tube packed with chalk powder. Different pigments passed down the column at different rates, leaving a series of colored bands on the white chalk column. A simple chromatographic column is shown in Figure 11.1 (p. 340).

The term *chromatography* is now used to refer to a variety of related separation techniques, but all work on a common principle: The mixture to be separated is dissolved in a *mobile phase* and passed through an adsorbent

[1]Mikhail Semenovich Tswett (or Tsvett) (1872–1919); b. Asti, Italy; Russian subject; University of Warsaw (1901–1907); Institute of Technology, Warsaw (1908); in Moscow (1915); professor and director, Botanical Garden, Estonia (1917).

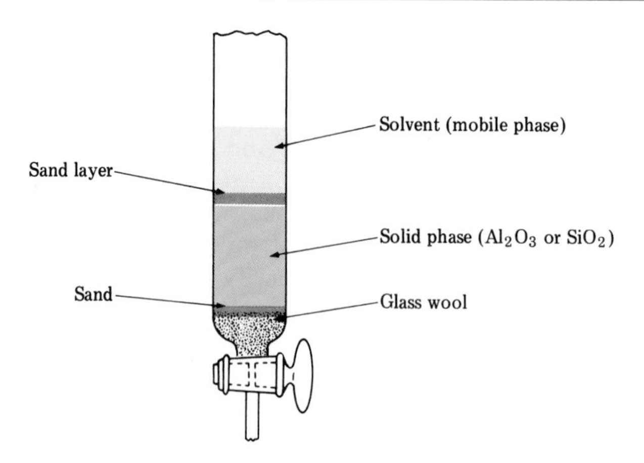

Figure 11.1. A simple chromatography column. A glass tube is filled with adsorbent material, and solvent is allowed to drip through the column.

stationary phase. Because different compounds adsorb to the stationary phase to differing extents, they migrate through the phase at different rates and are separated as they emerge. We will discuss briefly just three chromatographic techniques that are often used by organic chemists: liquid chromatography, high-pressure liquid chromatography, and gas chromatography.

11.2 Liquid chromatography

Liquid chromatography or **column chromatography** is one of the simplest and most often used chromatographic methods. As in Tswett's original experiments, a mixture of organic compounds is dissolved in a suitable solvent (the mobile phase) and is adsorbed onto a stationary phase such as alumina (Al_2O_3) or silica gel (hydrated SiO_2). More solvent is then passed down the column containing the stationary phase, and different compounds are removed **(eluted)** at different times. The time at which a compound is eluted is a function of many things, but has much to do with the polarity of the compound. As a general rule, molecules with polar functional groups (Section 4.1) are adsorbed more strongly and therefore migrate through the stationary phase more slowly than nonpolar molecules. For example, a mixture of an alcohol such as 1-heptanol and a related alkene such as 1-heptene can be easily separated by liquid chromatography. The relatively nonpolar alkene will pass through the column much faster than the more polar alcohol.

High-pressure liquid chromatography (HPLC) is a recent variant of the simple column technique. It has been found that the efficiency of column chromatography is vastly improved if the stationary phase is made up of very small and uniformly sized spherical particles. Small particle size ensures a large surface area for better adsorption, and a uniform spherical shape allows a tight, uniform packing. In practice, specially prepared silica microspheres of 10–25 micrometer (μm) size (1 μm = 10^{-6} m) are often used; 15 g of these microspheres have a surface area equivalent to the size of a football field. High-pressure pumps are required to force solvent through a tightly packed HPLC column, and sophisticated detectors monitor the appearance of material eluting from the column. These refinements are well worth it, however, for a good HPLC column can have up to several thousand times the separating power of a simple column. Figure 11.2 shows the results of HPLC analysis of a mixture of tetrachloromethane and five aromatic compounds, using silica microspheres as the stationary phase and hexane as the mobile phase. As each compound is eluted from the chromatography column, it passes through a detector, which registers its presence as a peak on a recorder chart.

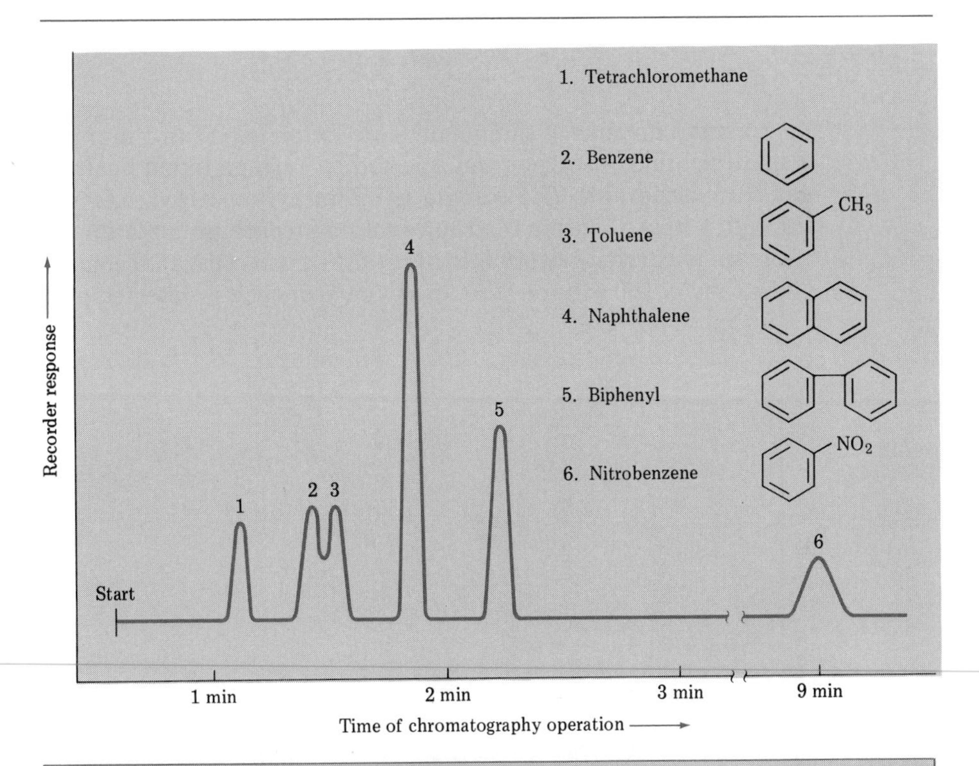

Figure 11.2. The HPLC analysis of a mixture of hydrocarbons in tetrachloromethane. The sample was dissolved in hexane and forced through a 4 mm × 60 cm column under a 2500 lb/in^2 pressure.

11.3 Gas chromatography

Gas chromatography differs from liquid chromatography in several respects. The most important difference is that a carrier *gas* such as nitrogen or helium, rather than a liquid solvent, is used as the mobile phase. Operationally, the technique employs an instrument known as a **gas chromatograph,** shown schematically in Figure 11.3.

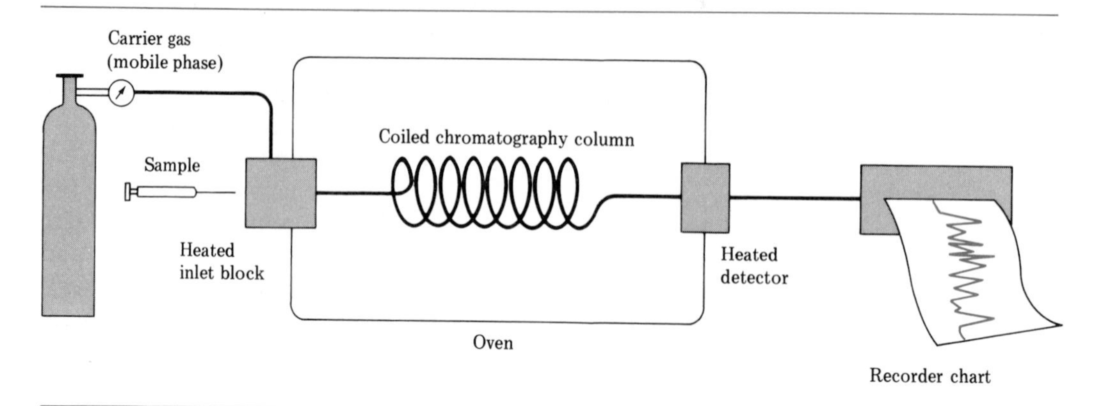

Figure 11.3. Schematic of a gas chromatograph.

A small amount of sample mixture (often less than 1 mg) is dissolved in a small volume of solvent and injected by syringe into a heated inlet of the gas chromatograph. The sample is instantly vaporized, and is then swept through a heated chromatography column (containing stationary phase) by a stream of carrier gas (mobile phase). As pure separated components come off the end of the column, the appearance of each is detected and registered

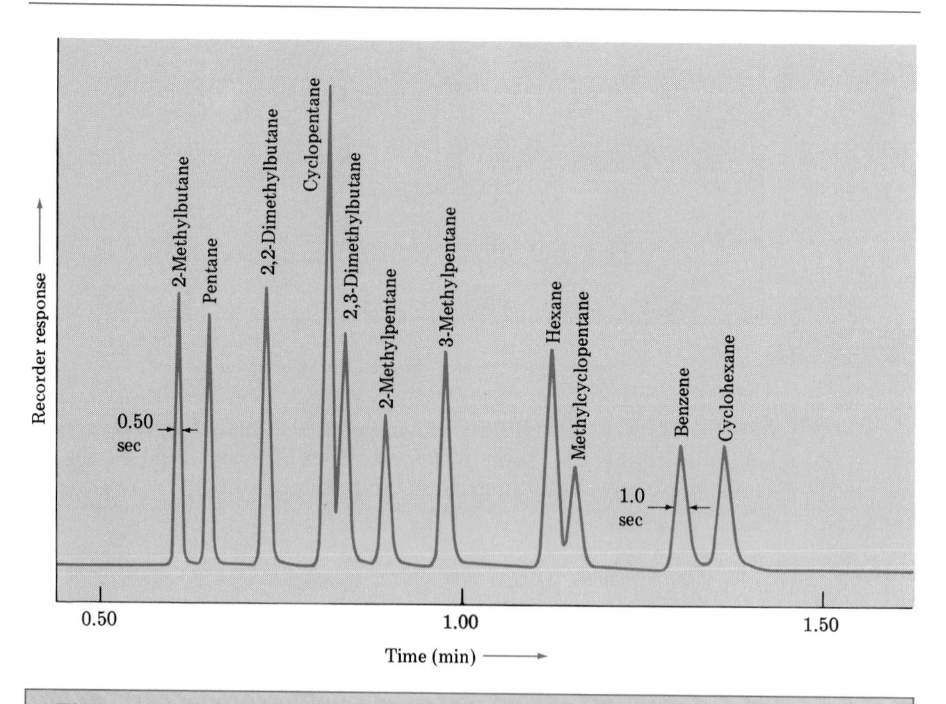

Figure 11.4. Gas chromatographic analysis of a hydrocarbon mixture.

as a peak on a recorder chart. Although only small amounts of material can be separated on a gas chromatograph, the separating power of modern instruments is phenomenal, and efficiencies up to 10 times that of HPLC and 100,000 times that of simple laboratory distillation can be achieved. Figure 11.4 shows the results of an analysis carried out on a mixture of C_5 and C_6 hydrocarbons, using a 10 m coiled-glass column.

11.4 Structure determination: spectroscopy

Once the products of a reaction have been separated and purified, the more difficult work begins. How do we identify what we have made? How do we *know,* for example, that the ionic addition of HBr to 1-butene gives 2-bromobutane and not 1-bromobutane?

$$CH_3CH_2CH{=}CH_2 \xrightarrow{HBr} CH_3CH_2\underset{}{\overset{\displaystyle Br}{\overset{|}{C}}}HCH_3 \quad \text{or} \quad CH_3CH_2CH_2CH_2Br\ ?$$

The answer is that we use **spectroscopy** to elucidate the structures of unknowns. Spectroscopy provides the tools that allow chemists to determine molecular structures. Organic chemists use several different kinds of spectroscopy to help determine structure. In this and the next two chapters we will look carefully at four of the most useful techniques, beginning with mass spectroscopy.

11.5 Mass spectroscopy

At its simplest, **mass spectroscopy** is a technique that allows us to measure the mass (molecular weight) of a molecule. In addition, we can often gain valuable structural information about unknowns by measuring the masses of the fragments produced when high-energy molecules fly apart. There are several different kinds of **mass spectrometers** available, but the most common is the electron-impact, magnetic-sector instrument, which operates as shown schematically in Figure 11.5 (p. 344).

A small amount of sample is introduced into the mass spectrometer and bombarded by a stream of high-energy electrons. The exact energy of the electron beam varies, but is commonly around 70 electron volts (eV) or 1600 kcal/mol. When a high-energy electron strikes an organic molecule, it dislodges a valence electron from the molecule, producing a **cation radical** (*cation* because the molecule has lost a negatively charged electron; *radical* because the molecule now has an odd number of electrons).

$$\underset{\text{Organic molecule}}{RH} \xrightarrow{e^-} \underset{\text{Cation radical}}{RH^{+\cdot}} + e^-$$

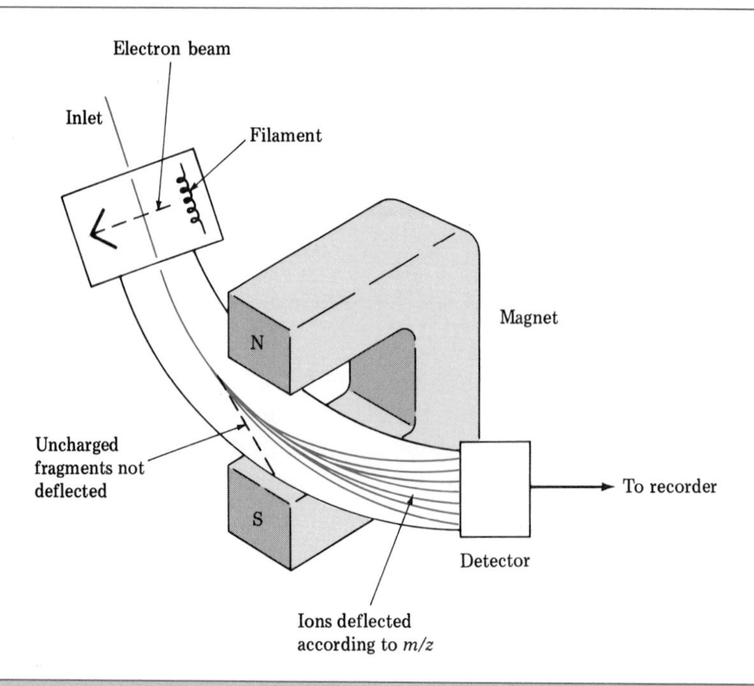

Figure 11.5. Schematic of an electron-impact, magnetic-sector mass spectrometer.

In addition to causing ionization, electron bombardment transfers such a large amount of energy to the sample molecules that the cation radicals *fragment;* they fly apart into numerous smaller pieces, some of which retain a positive charge, and some of which are neutral. The fragments then pass through a strong magnetic field, where they are deflected according to their mass-to-charge ratio (m/z). Neutral fragments are not deflected by the magnetic field and are lost on the walls of the instrument. Positively charged fragments, however, are sorted by the mass spectrometer onto a detector, which records them as peaks at the proper ratios. Since the number of charges, z, is usually 1, the peaks represent the masses of the ions. The **mass spectrum** of a compound is usually presented as a bar graph with **unit masses** (m/z values) on the x-axis, and **intensity** (number of ions of a given m/z striking the detector) on the y-axis.

The highest peak is called the **base peak** and is arbitrarily assigned an intensity of 100%. Figure 11.6 shows the mass spectra of methane and propane.

The mass spectrum of methane is relatively simple, since few fragmentations are possible. As Figure 11.6(a) shows, the base peak has $m/z = 16$, which corresponds to the unfragmented methane cation radical or the **molecular ion ($M^{+\cdot}$).** The mass spectrum also shows ions at $m/z = 15$ and 14, corresponding to cleavage of the molecular ion into CH_3^+ and $CH_2^{+\cdot}$ fragments.

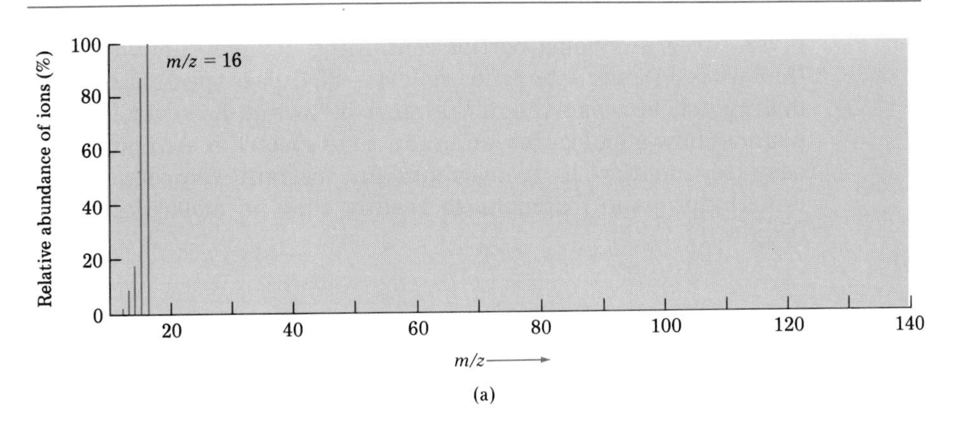

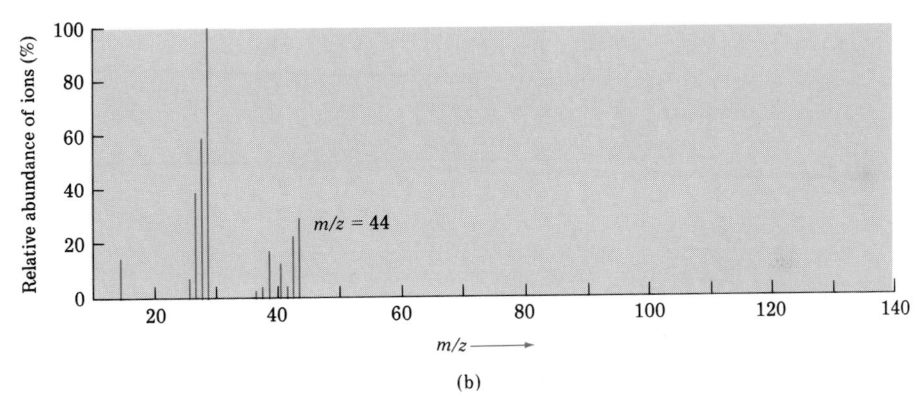

Figure 11.6. Mass spectra of (a) methane, CH_4 (mol wt = 16), and (b) propane, C_3H_8 (mol wt = 44).

$$CH_4 \xrightarrow{-e^-} \left[\begin{array}{c} H \\ H:\ddot{C}\cdot H \\ \ddot{H} \end{array} \right]^{+\cdot} \begin{cases} \longrightarrow [CH_3]^+ + H\cdot \quad m/z = 15 \\ \longrightarrow [CH_2]^{+\cdot} + 2\,H\cdot \quad m/z = 14 \end{cases}$$

$m/z = 16$ (molecular ion, $M^{+\cdot}$)

For larger molecules, the mass spectral fragmentation patterns are usually complex, and the molecular ion is often *not* the highest (base) peak. For example, the mass spectrum of propane, shown in Figure 11.6(b), has a molecular ion ($m/z = 44$) only about 30% as high as the base peak at m/z = 29. In addition, many other fragment ions are observed.

11.6 Interpreting mass spectra

What kinds of information can we get from studying the mass spectrum of a compound? Certainly the most obvious piece of information is the molecular weight (mol wt), and this in itself can be invaluable. For example, if we were given three unlabeled bottles containing hexane (mol wt = 86), 1-hexene (mol wt = 84), and 1-hexyne (mol wt = 82), mass spectroscopy would readily distinguish between them. We must be aware, however, that not all compounds show a molecular ion in the mass spectrum. Although $M^{+\cdot}$ is usually easy to identify if it is abundant, certain compounds such as 2,2-dimethylpropane fragment so readily that no molecular ion is observed (Figure 11.7).

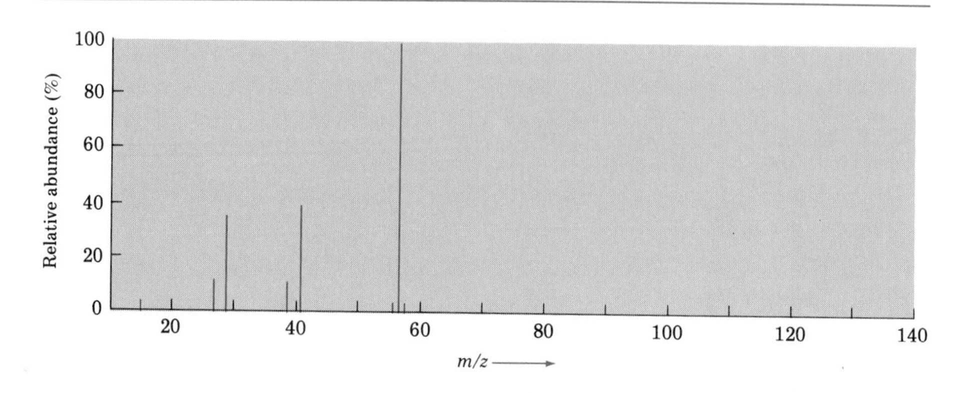

Figure 11.7. Mass spectrum of 2,2-dimethylpropane, C_5H_{12} (mol wt = 72). No molecular ion is observed.

Measurement of an unknown's molecular weight allows us to narrow the possibilities for molecular formula down to only a few choices. For example, if the mass spectrum of an unknown compound shows a molecular ion at *m/z* = 110, the molecular formula is likely to be C_8H_{14}, $C_7H_{10}O$, $C_6H_6O_2$, or $C_6H_{10}N_2$. There are always a number of possible molecular formulas for all but the lowest molecular weights, and tables are available that list all possible choices.

A further point about mass spectroscopy can be seen by looking carefully at the mass spectra of methane and propane in Figure 11.6. Perhaps surprisingly, $M^{+\cdot}$ is *not* the highest *m/z* peak in the two spectra; there is also a small peak in each spectrum at $(M + 1)^{+\cdot}$. This small peak is due to the presence in the samples of small amounts of isotopically substituted molecules. Although ^{12}C is the most abundant carbon isotope, a small amount (1.11% natural abundance) of ^{13}C is also present. Thus, a certain percentage of the molecules analyzed in the mass spectrometer are likely to contain a ^{13}C isotope, giving rise to the observed $(M + 1)^{+\cdot}$ peak.

A second piece of information provided by the mass spectrum is a kind of molecular fingerprint. Each organic molecule fragments in a unique pattern depending on its structure, and the chance that two compounds will have identical mass spectra is small. Thus, we can often identify unknowns by matching their mass spectra to reference spectra. The U.S. National Institutes of Health has established a reference library of more than 70,000 mass spectra, and the necessary computer-based pattern-recognition techniques have been developed to identify unknowns by matching their mass spectra with reference spectra.

PROBLEM

11.1 Write as many feasible molecular formulas as you can for compounds that have the following molecular ions in their mass spectra. (Assume that all of the compounds contain C and H, but that O may or may not be present.)

(a) $M^{+\cdot} = 86$ (b) $M^{+\cdot} = 128$

(c) $M^{+\cdot} = 156$ (d) $M^{+\cdot} = 180$

PROBLEM

11.2 Nootkatone, one of the chemicals responsible for the odor and taste of grapefruit, shows a molecular ion at $m/z = 218$ in the mass spectrum and is known to contain C, H, and O. Suggest several possible molecular formulas for nootkatone.

PROBLEM

11.3 By knowing the natural abundances of minor isotopes, it is possible to calculate the relative heights of $M^{+\cdot}$ and $(M + 1)^{+\cdot}$ peaks. If ^{13}C has a natural abundance of 1.11%, what relative heights would you expect for $M^{+\cdot}$ and $(M + 1)^{+\cdot}$ peaks in the mass spectrum of benzene, C_6H_6?

PROBLEM

11.4 The **nitrogen rule** of mass spectroscopy says that a compound containing an odd number of nitrogens has an odd-numbered molecular ion. Conversely, a compound containing an even number of nitrogens has an even-numbered $M^{+\cdot}$ peak. Can you explain why this should be so?

PROBLEM

11.5 In light of the nitrogen rule mentioned in Problem 11.4, what is the molecular formula of pyridine, $M^{+\cdot} = 79$?

11.7 Interpretation of fragmentation patterns

Mass spectroscopy would be of great value even if molecular weight and formula were the only information to be obtained, but, in fact, we can get much more. Proper interpretation of mass spectral fragmentation patterns allows us to derive structural information about the sample.

Fragmentation occurs when the high-energy cation radical flies apart by spontaneous cleavage of a chemical bond. One of the two fragments retains the positive charge (i.e., is a carbocation) and the other fragment is a neutral radical. In practice, the positive charge usually remains with the fragment that is better able to stabilize it. In other words, the more stable carbocation is usually formed during fragmentation. For example, propane tends to fragment in such a way that the positive charge remains with the ethyl group rather than with the methyl (the ethyl carbocation is more stable than the methyl carbocation; see Section 5.10). Propane therefore has a base peak at $m/z = 29$, and a barely detectable peak at $m/z = 15$ (Figure 11.6).

$$[CH_3CH_2CH_3]^{+\cdot} \longrightarrow CH_3CH_2^{+} + \cdot CH_3$$

$M^{+\cdot} = 44$ $m/z = 29$ Neutral; not observed

Since mass spectral fragmentation patterns are usually complex, it is often difficult to assign definite structures to fragment ions. Most hydrocarbons fragment in many ways, and the mass spectrum of hexane, a typical alkane, is shown in Figure 11.8.

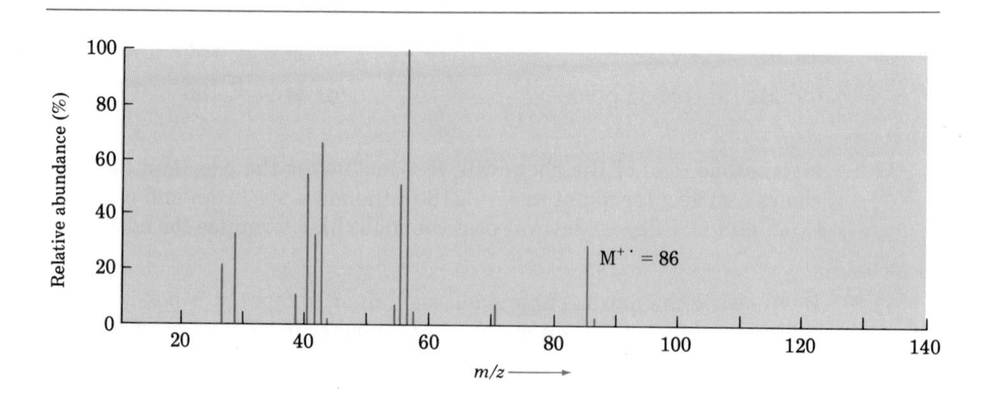

Figure 11.8. Mass spectrum of hexane, C_6H_{14} (mol wt = 86), $M^{+\cdot}$ = 86. The base peak is at *m/z* = 57.

The mass spectrum of hexane shows a moderately abundant molecular ion at *m/z* = 86, and substantial fragment ions at *m/z* = 71, 57, 43, and 29. Since all of the carbon–carbon bonds of hexane are electronically similar, all of them break to a similar extent, giving rise to the observed ions. Figure 11.9 shows how these fragments might arise.

The loss of a methyl radical from the hexane cation radical ($M^{+\cdot}$ = 86) gives rise to a fragment of mass 71; the loss of an ethyl radical accounts for a

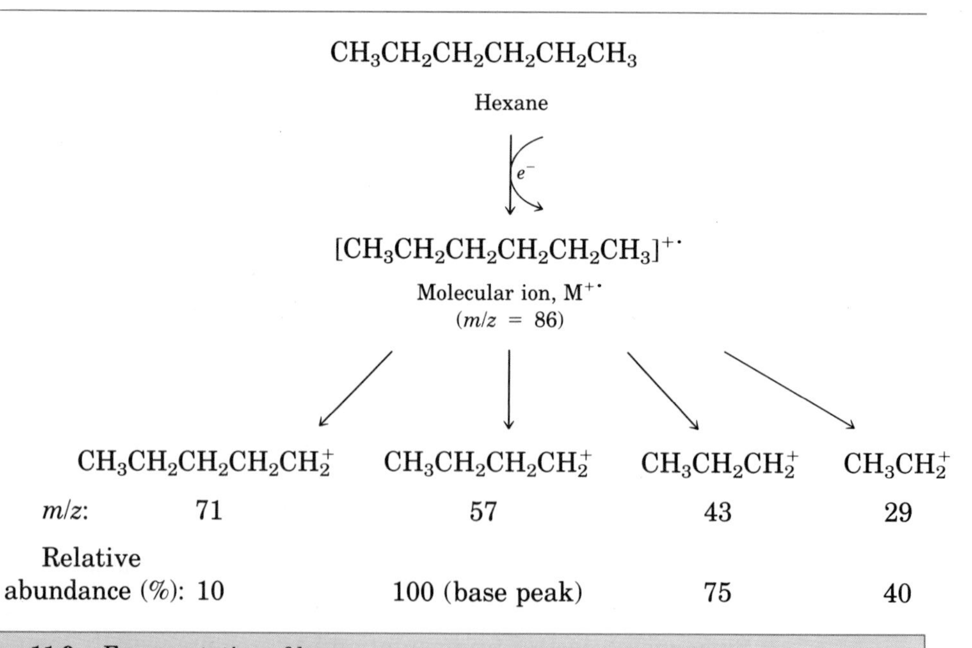

Figure 11.9. Fragmentation of hexane.

fragment of mass 57; the loss of a propyl radical accounts for a fragment of mass 43; and the loss of a butyl radical accounts for a fragment of mass 29. With skill and practice, chemists can learn to analyze the fragmentation patterns of unknown compounds and to work backwards to a structure that is compatible with the available data.

Figure 11.10 shows how we can use the information from fragmentation patterns. Assume that we have two unlabeled bottles, A and B. One contains methylcyclohexane; the other contains ethylcyclopentane, and we need to distinguish between them.

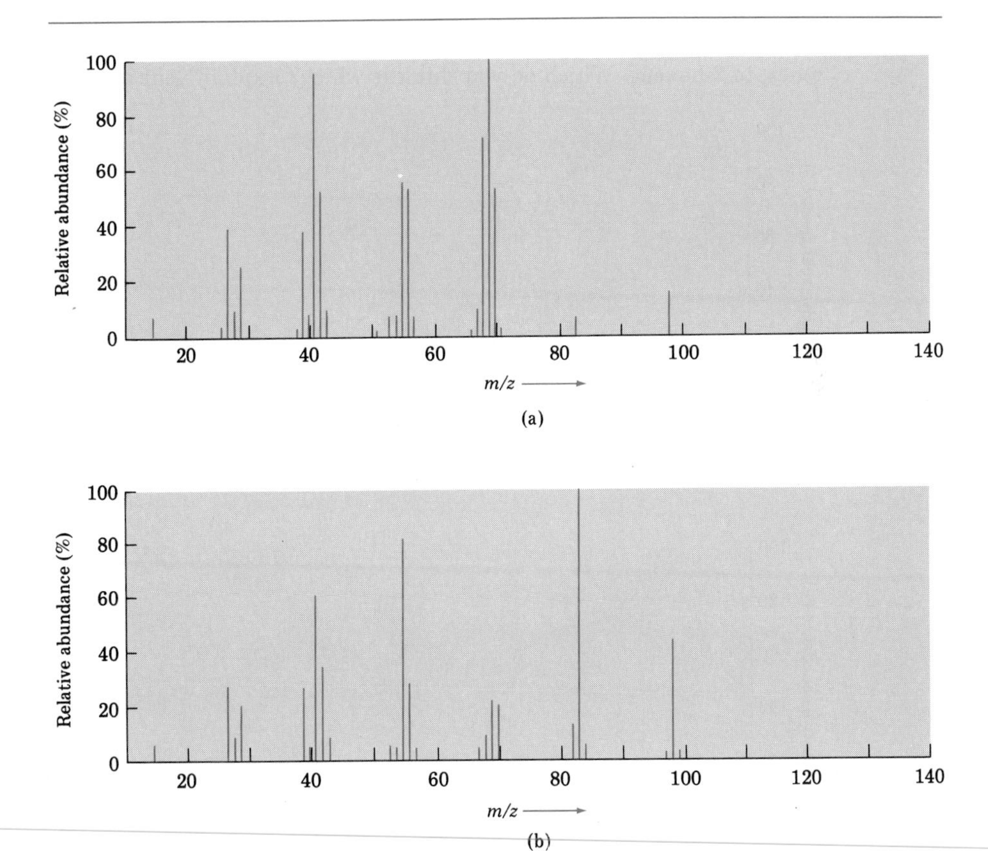

Figure 11.10. Mass spectra of (a) sample A, and (b) sample B.

The mass spectrum of sample A shows a molecular ion at $M^{+\cdot} = 98$, corresponding to C_7H_{14}, but this is of little help since the two samples are isomeric. Sample B also shows $M^{+\cdot} = 98$. The two mass spectra differ considerably in their fragmentation patterns, however. Sample B has its base peak at $m/z = 83$, corresponding to the loss of a $\cdot CH_3$ group (15 mass units) from the molecular ion, but sample A has only a small peak at $m/z = 83$. Conversely, B has a rather small peak at $m/z = 69$, but A has its base peak at 69, corresponding to the loss of a $\cdot CH_2CH_3$ group (29 mass units). We can therefore say with considerable confidence that B is methylcyclohexane and A is ethylcyclopentane. This example is, of course, a simple one, but the principles used are broadly applicable for organic structure determination by mass spectroscopy. As we will see in later chapters, specific functional group classes such as alcohols, ketones, aldehydes, and amines often show specific kinds of fragmentations that can be interpreted to provide structural information.

PROBLEM

11.6 The mass spectrum of 2,2-dimethylpropane (Figure 11.7) shows a base peak at m/z = 57. What molecular formula does this correspond to? Can you suggest a structure for the m/z = 57 fragment ion?

PROBLEM

11.7 Two mass spectra are shown. One spectrum corresponds to 2-methyl-2-pentene; the other, to 2-hexene. Which do you think is which? Explain your choices.

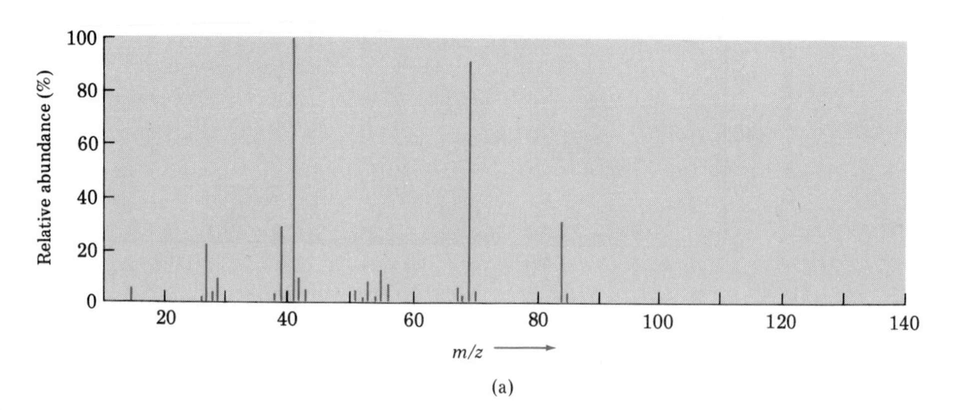

(a)

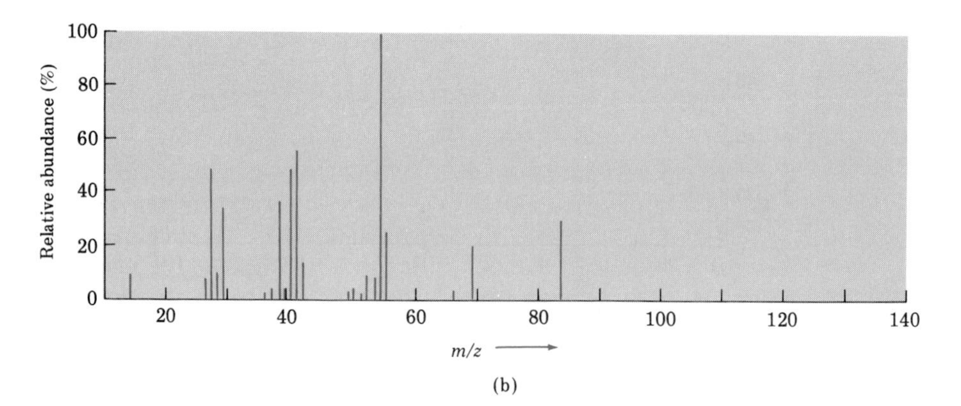

(b)

11.8 Infrared spectroscopy and the electromagnetic spectrum

Infrared spectroscopy differs from mass spectroscopy in that it concerns the interaction of molecules with infrared radiant energy rather than with a high-energy electron beam. Before beginning a study of infrared spectroscopy, however, we need to look into the nature of radiant energy and the electromagnetic spectrum.

Visible light, X rays, microwaves, radio waves, and so forth are all different kinds of **electromagnetic radiation.** Collectively, they make up the **electromagnetic spectrum,** shown in Figure 11.11.

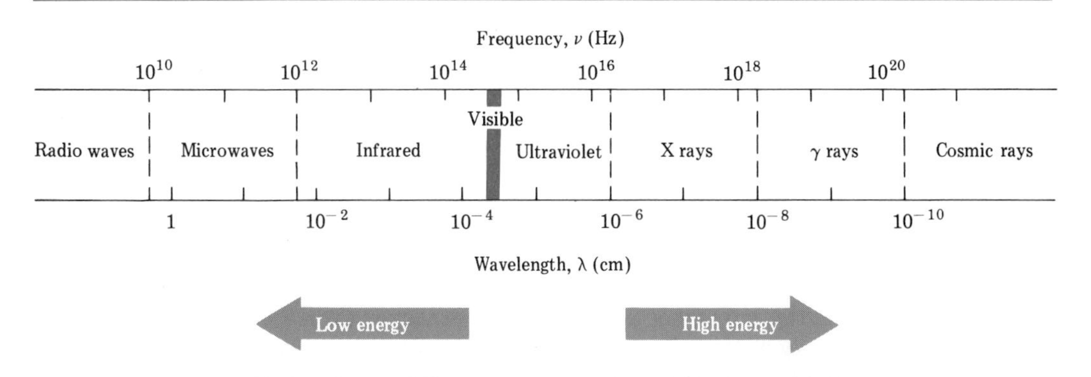

Figure 11.11. The electromagnetic spectrum.

Electromagnetic radiation can be considered to have dual behavior. In some respects it has the properties of a particle (called a **photon**), yet in other respects it behaves as a wave traveling at the speed of light. Electromagnetic waves can be described by their wavelength (λ) and frequency (ν). The wavelength is simply the length of one complete wave cycle from trough to trough, and the frequency is the number of wave cycles that travel past a fixed point in a certain unit of time (usually given in cycles per second, or **hertz, Hz**).

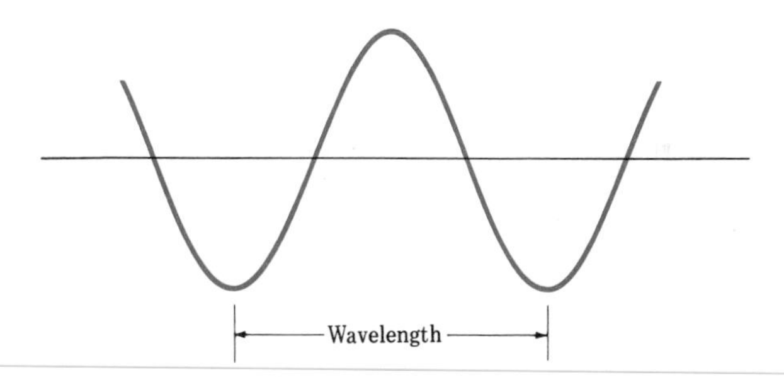

Wavelength and frequency are inversely related by the equation

$$\lambda = c/\nu$$

where λ = Wavelength in centimeters
c = Speed of light (3×10^{10} cm/sec)
ν = Frequency in hertz

As Figure 11.11 indicates, the electromagnetic spectrum is arbitrarily divided into various regions, with the familiar visible region accounting for only a small portion of the overall spectrum (from 3.8×10^{-5} cm to 7.8×10^{-5} cm in wavelength).

Electromagnetic energy is transmitted only in discrete energy packets, or **quanta,** and the amount of energy corresponding to 1 quantum of energy (or 1 photon) of a given frequency is expressed by the equation

$$\varepsilon = h\nu = hc/\lambda$$

where ε = The energy of 1 photon (1 quantum)
h = Planck's constant (6.62×10^{-27} erg sec)
ν = Frequency in hertz
λ = Wavelength in centimeters
c = Speed of light (3×10^{10} cm/sec)

Thus, the energy of a specific photon varies *directly* with its frequency, but *inversely* with its wavelength; high frequencies and short wavelengths correspond to high-energy radiation (gamma rays); low frequencies and long wavelengths correspond to low-energy radiation (radio waves). If we multiply ε by Avogadro's number, N, and convert to kilocalories per mole, we arrive at the same equation expressed in units familiar to organic chemists:

$$E = \frac{Nhc}{\lambda} = \frac{2.85 \times 10^{-3}\ \text{kcal/mol}}{\lambda\ (\text{cm})}$$

where E represents the energy of a "mole" of photons of wavelength λ.

When a sample of an organic compound is exposed to electromagnetic radiation, energy of certain wavelengths is absorbed by the sample, and energy of other wavelengths passes through. Whether the light energy is absorbed or not absorbed depends both on the structure of the sample compound and on the wavelength (energy level) of the radiation. If we irradiate the sample with energy of many different wavelengths and determine which wavelengths are absorbed and which pass through, we can determine the **absorption spectrum** of the compound. This is precisely what we do in a spectroscopic experiment, and the results are usually displayed on a graph that plots wavelength versus amount of radiation transmitted. For example, the infrared spectrum of ethyl alcohol is shown in Figure 11.12. The horizontal (x) axis records the wavelength, and the vertical (y) axis records the intensity of the various energy absorptions as they occur.

A molecule gains energy when it absorbs radiation, and this energy gain must be distributed over the molecule in some way. For example, energy absorption might result in increased molecular motions, causing bonds to stretch, bend, or rotate. Alternatively, energy absorption might cause electrons to be excited from a low-energy orbital to a higher one. Different radiation frequencies affect molecules in different ways, but each can provide structural information if we learn to interpret the results properly.

There are numerous kinds of spectroscopy, depending on which region of the electromagnetic spectrum is being examined. We will look closely at

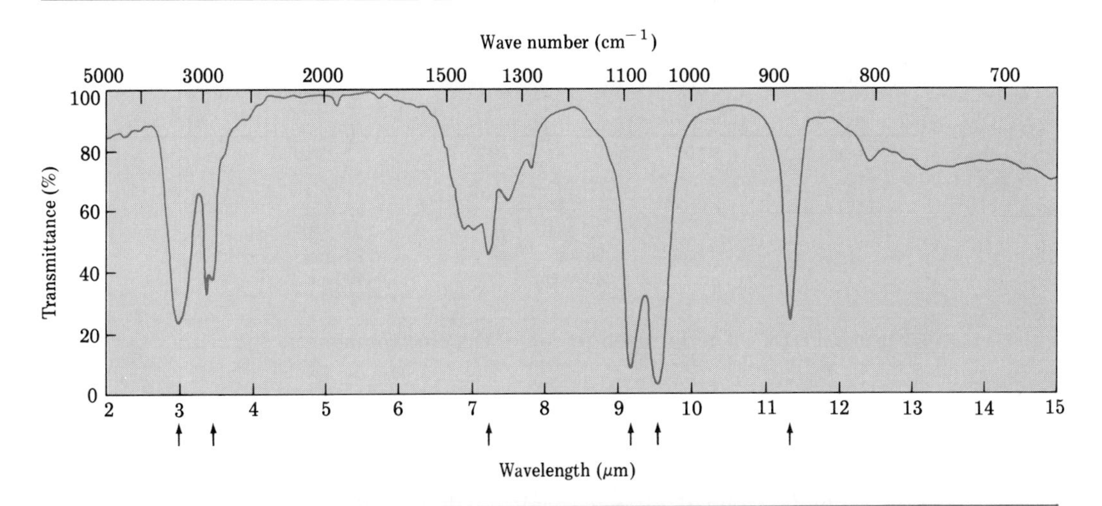

Figure 11.12. Infrared spectrum of ethyl alcohol. A transmittance of 100% on the infrared spectrum means that all the energy is passing through the sample. A lower transmittance means that some energy is being absorbed. Thus, each downward spike marked by an arrow corresponds to an energy absorption.

just two, infrared spectroscopy and nuclear magnetic resonance spectroscopy, and have a brief introduction to a third, ultraviolet spectroscopy.

Let's begin by seeing what happens when an organic sample absorbs infrared energy.

PROBLEM

11.8 It is useful to develop an intuitive feeling for the amounts of energy that correspond to different parts of the electromagnetic spectrum. Using the relationships

$$E = \frac{2.85 \times 10^{-3}\ \text{kcal/mol}}{\lambda\ (\text{cm})} \quad \text{and} \quad \nu = c/\lambda$$

calculate the energies of the following:
(a) A gamma ray with $\lambda = 5 \times 10^{-9}$ cm
(b) An X ray with $\lambda = 3 \times 10^{-7}$ cm
(c) Ultraviolet light with $\nu = 6 \times 10^{15}$ Hz
(d) Visible light with $\nu = 7 \times 10^{14}$ Hz
(e) Infrared radiation with $\lambda = 2 \times 10^{-3}$ cm
(f) Microwave radiation with $\nu = 10^{11}$ Hz
(g) Radio waves from your favorite local AM station

11.9 Infrared spectroscopy of organic molecules

The infrared (IR) region of the electromagnetic spectrum covers the range from just above the visible (7.8×10^{-5} cm) to approximately 10^{-2} cm, but only the middle of the region interests organic chemists (Figure 11.13). This

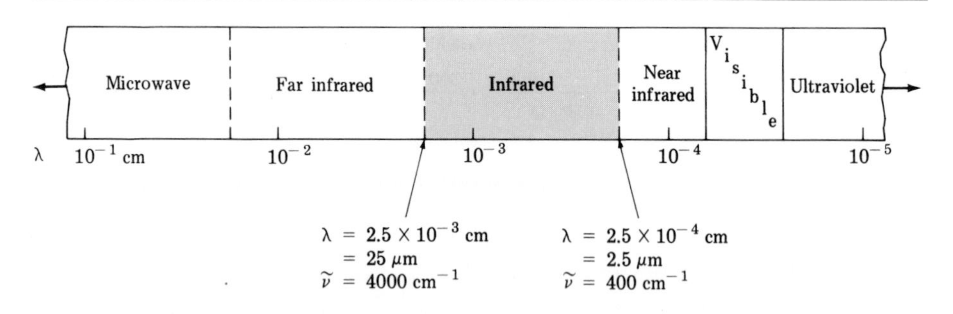

Figure 11.13. The infrared region of the electromagnetic spectrum.

midportion extends from 2.5×10^{-3} cm to 2.5×10^{-4} cm, and specific wavelengths are usually referred to in micrometers ($1\ \mu m = 10^{-4}$ cm). Specific frequencies are usually given in **wave numbers** or **reciprocal centimeters,** rather than in hertz. The wave number is the reciprocal of the wavelength in centimeters and is expressed in units of cm^{-1}.

$$\text{Wave number } (\tilde{\nu}) = 1/\lambda \text{ (cm)}$$

Using the equation $E = (2.85 \times 10^{-3}\ \text{kcal/mol})/\lambda$, we can calculate that the energy levels of infrared radiation range from 1.13 kcal/mol to 11.3 kcal/mol ($\lambda = 2.5 \times 10^{-3}$ cm to 2.5×10^{-4} cm). These energy levels correspond to the amount of radiation needed to increase molecular motions such as bond vibrations.

Every molecule is in constant motion—bonds vibrate back and forth, atoms rotate, bonds bend slightly. When a molecule absorbs infrared radiation, *these molecular motions increase in intensity.* Since each radiation frequency corresponds to a specific motion, we can see what kinds of molecular motions a sample has by measuring its **infrared spectrum.** By working backwards and interpreting the infrared spectrum, we can then find out what kinds of functional groups are present in the molecule.

PROBLEM

11.9 There is some disagreement among chemists about the best way to refer to infrared data. Some people prefer to think in terms of micrometers, whereas others prefer reciprocal centimeters. To converse with both groups, it is useful to be able to interconvert the two systems of measurement rapidly. Do the following conversions:

(a) 3.1 μm to cm^{-1}
(b) 5.85 μm to cm^{-1}
(c) 6.75 μm to cm^{-1}
(d) 2250 cm^{-1} to μm
(e) 970 cm^{-1} to μm
(f) 1560 cm^{-1} to μm

11.10 Interpretation of infrared spectra

The full interpretation of an infrared spectrum is not an easy task. Most organic molecules are so large that there are dozens or hundreds of different possible bond stretchings, rotations, and bending motions, and an infrared spectrum therefore contains dozens or hundreds of absorptions. In one sense this complexity is a valuable feature, since an infrared spectrum serves as a unique fingerprint of a specific compound. In fact, the complex region of the infrared spectrum below 1500 cm^{-1} is called the **fingerprint region;** if two samples have *identical* infrared spectra, the compounds are almost certainly identical. For structural purposes, however, the multitude of absorptions present in an infrared spectrum makes full interpretation difficult.

Fortunately, we do not need to interpret a spectrum fully to get useful structural information. Most *functional groups* give rise to characteristic infrared absorptions that change little in going from one compound to another. For example, the C=O absorption of ketones is almost always in the range 1690–1750 cm^{-1}, the O—H absorption of alcohols is almost always in the range 3200–3600 cm^{-1}, and the C=C bond of alkenes is almost always in the range 1640–1680 cm^{-1}. By learning to recognize where characteristic functional group absorptions occur, we can gain valuable structural information from infrared spectra. For example, Figure 11.14 (p. 356) shows the infrared spectra of hexane, 1-hexene, and 1-hexyne. Although all three spectra contain many absorptions, there are characteristic absorptions of the carbon–carbon double- and triple-bond functional groups that allow the compounds to be readily distinguished. Thus 1-hexene shows a characteristic carbon–carbon double-bond peak at 1660 cm^{-1} and a vinylic =C—H bond peak at 3100 cm^{-1}. 1-Hexyne exhibits a carbon–carbon triple bond absorption at 2100 cm^{-1} and a terminal alkyne ≡C—H bond absorption at 3300 cm^{-1}.

Table 11.1 (p. 357) lists the characteristic absorption frequencies of some common functional groups. Figure 11.15 (p. 358) shows how the infrared region from 4000 to 200 cm^{-1} can be roughly divided. The region from 4000 to 2500 cm^{-1} corresponds to absorptions caused by N—H, C—H, and O—H bond stretching and contracting motions. Of these, N—H and O—H bonds absorb in the 3300–3600 cm^{-1} range and C—H bond stretching occurs near 3000 cm^{-1}. The region from 2500 to 2000 cm^{-1} is where triple-bond stretching occurs, and both nitriles (R—C≡N) and alkynes show peaks here. The region from 2000 to 1500 cm^{-1} contains double-bond absorptions, and C=O, C=N, and C=C bonds show bands in this region due to stretching motions. Carbonyl groups generally absorb from 1670 to 1780 cm^{-1}, whereas alkene stretching normally occurs in a narrow range from 1640 to 1680 cm^{-1}. The region below 1500 cm^{-1} is called the fingerprint region and usually contains a large number of absorptions due to a variety of molecular bending, rocking, wagging, and scissoring motions. With the exception of hydrocarbon spectra, we will not go into further detail at the present time. As each new functional group is discussed, however, that group's infrared characteristics will be examined.

PROBLEM

11.10 Refer to Table 11.1, and make educated guesses as to what functional groups these molecules might contain:

(a) A compound with a strong absorption at 1710 cm^{-1}
(b) A compound with a strong absorption at 1540 cm^{-1}
(c) A compound with strong absorptions at 1720 cm^{-1} and 2500–3100 cm^{-1} (broad)
(d) A compound with strong absorptions at 3500 cm^{-1} and 1735 cm^{-1}

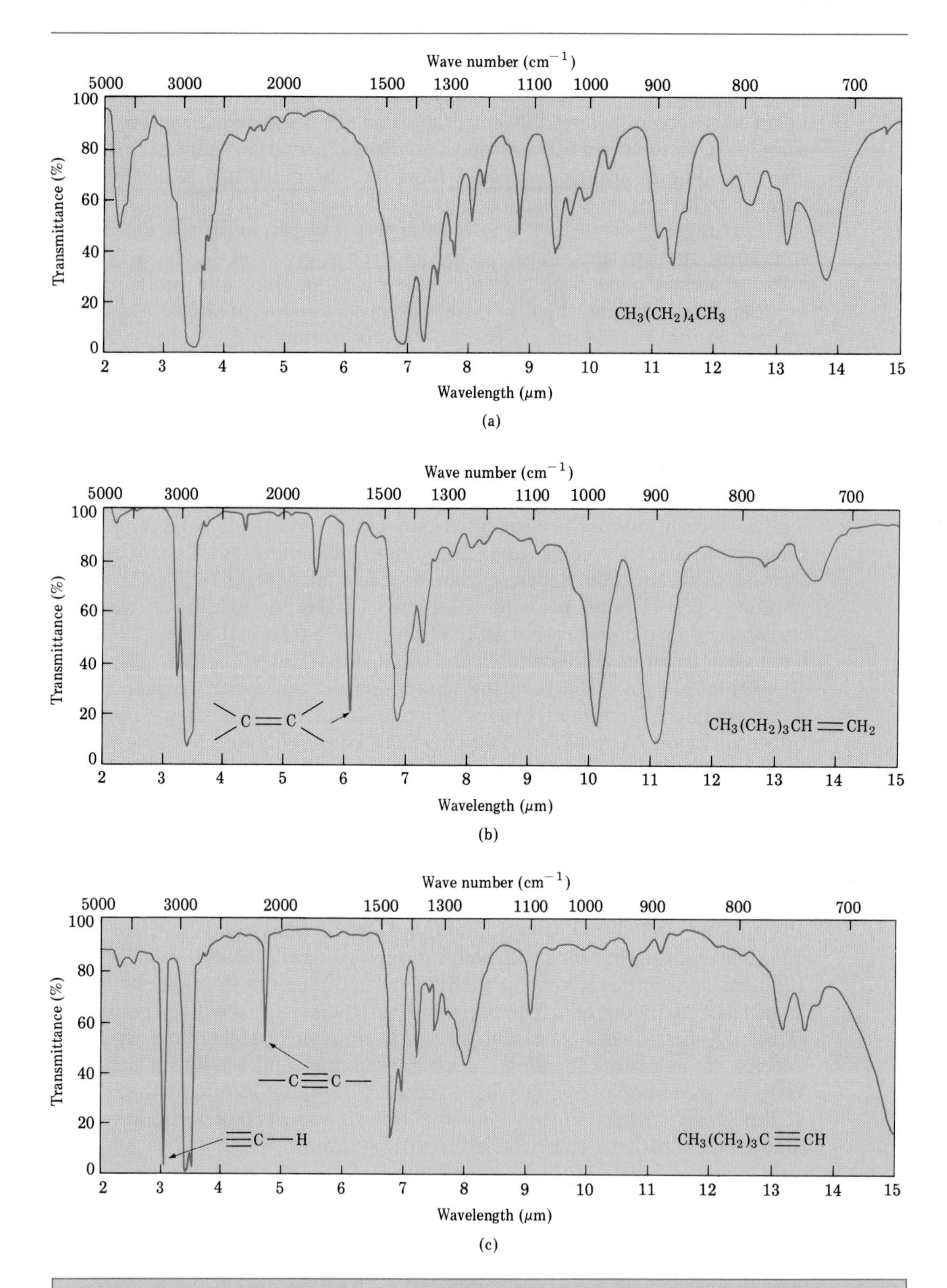

Figure 11.14. Infrared spectra of (a) hexane, (b) 1-hexene, and (c) 1-hexyne. Spectra such as these are easily obtained on 1 mg or 2 mg samples in a few minutes' time using routine, commercially available instruments.

TABLE 11.1 Characteristic infrared absorptions of some functional groups

Functional group class	*Band position* (cm^{-1})	*Intensity of absorption*
Alkanes, alkyl groups		
C—H	2850–2960	Medium to strong
Alkenes		
=C—H	3020–3100	Medium
C=C	1650–1670	Medium
Alkynes		
≡C—H	3300	Strong
—C≡C—	2100–2260	Medium
Alkyl halides		
C—Cl	600–800	Strong
C—Br	500–600	Strong
C—I	500	Strong
Alcohols		
O—H	3400–3640	Strong, broad
C—O	1050–1150	Strong
Aromatics		
C—H (aromatic ring)	3030	Medium
C=C (benzene ring)	1600, 1500	Strong
Amines		
N—H	3310–3500	Medium
C—N	1030, 1230	Medium
Carbonyl compounds[a]		
C=O	1670–1780	Strong
Carboxylic acids		
O—H	2500–3100	Strong, very broad
Nitriles		
C≡N	2210–2260	Medium
Nitro compounds		
NO_2	1540	Strong

[a]Acids, esters, aldehydes, and ketones.

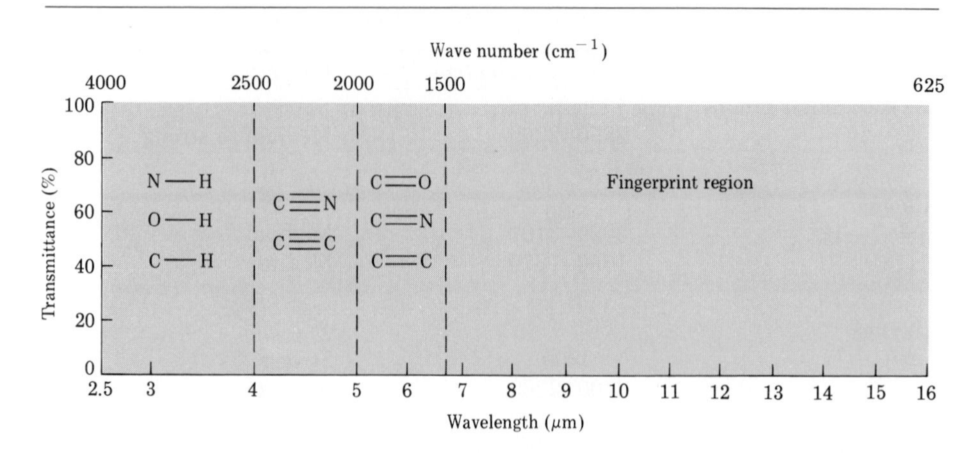

Figure 11.15. Regions in the infrared spectrum.

11.11 Infrared spectra of hydrocarbons

ALKANES

The infrared spectrum of a saturated hydrocarbon is rather uninformative, since no functional groups are present. All infrared absorptions are due to C—H and C—C bond stretching and bending. Many organic compounds contain saturated alkane-like portions, and therefore many organic compounds share the same characteristic peaks. For example, saturated C—H bonds always show a strong absorption from 2850 to 2960 cm^{-1}. This absorption is caused by a stretching and contracting motion of C—H bonds, and is clearly visible in all three of the spectra shown in Figure 11.14. Saturated C—C bonds show a number of absorptions in the 800–1300 cm^{-1} range, but it is difficult to assign them to specific motions.

Alkanes —C—H 2850–2960 cm^{-1}

—C—C— 800–1300 cm^{-1}

ALKENES

Alkenes show several characteristic stretching absorptions that can be used for structural identification purposes. For example, vinylic ═C—H bonds absorb from 3020 to 3100 cm^{-1}. Alkene C═C bonds usually show an absorption near 1650 cm^{-1}, although in some cases this can be rather weak and difficult to see clearly. Both ═C—H and C═C absorptions are diagnostic for alkenes, as the 1-hexene spectrum in Figure 11.14 shows. Mono-, di-, and trisubstituted alkenes have ═C—H bonds that give rise to out-of-plane bending absorptions in the 700–1000 cm^{-1} range. Monosubstituted alkenes such as 1-hexene show strong characteristic peaks at 910 and 990 cm^{-1}, and 2,2-disubstituted alkenes ($R_2C{=}CH_2$) have an intense band at 890 cm^{-1}.

Alkenes	$=C-H$	3020–3100 cm^{-1}
	$>C=C<$	1650–1670 cm^{-1}
	$RCH=CH_2$	910 and 990 cm^{-1}
	$R_2C=CH_2$	890 cm^{-1}

ALKYNES

Alkynes exhibit a $C\equiv C$ stretching absorption at 2100–2260 cm^{-1}, and this band is much more intense for terminal alkynes than for internal alkynes. Terminal alkynes such as 1-hexyne also have a characteristic $\equiv C-H$ stretch at 3300 cm^{-1}. This band is diagnostic for terminal alkynes, since it is fairly intense and quite sharp.

Alkynes	$-C\equiv C-$	2100–2260 cm^{-1}
	$\equiv C-H$	3300 cm^{-1}

One final point about infrared spectroscopy: We can derive much structural information from an infrared spectrum by noticing which characteristic absorptions are *not* present. For example, if the spectrum of an unknown does *not* contain absorptions at 3300 and 2150 cm^{-1}, it is not a terminal alkyne; if the spectrum has *no* absorption near 3400 cm^{-1}, the compound is not an alcohol.

PROBLEM

11.11 Shown here is the infrared spectrum of ethynylcyclohexane. Identify as many absorption bands as you can.

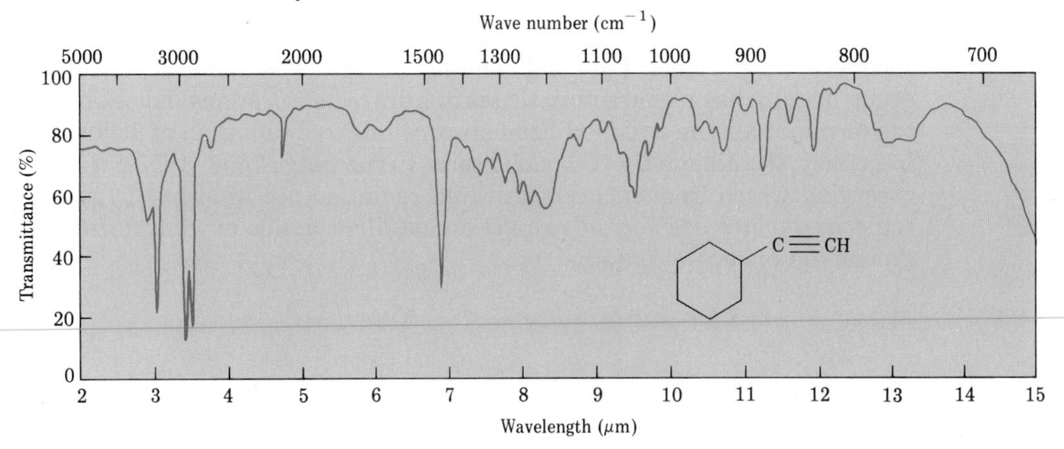

11.12 Summary

Product purification and structure elucidation are two of the most difficult tasks facing laboratory chemists. If possible, purification is done by a direct method such as crystallization or distillation. Often, however, some form of chromatography must be used. Although there are different kinds of chromatography, all operate on the same principle. The impure sample is applied to one end of a column containing an inert support material (the stationary phase), and the sample is carried along by a mobile phase (gas or liquid). Separation of mixtures is effected because different compounds adsorb to the stationary phase in differing degrees and therefore move along at different rates.

Once a reaction product has been purified, its structure must be determined. This is done using various spectroscopic methods such as mass spectroscopy and infrared spectroscopy. Mass spectroscopy gives us information about the molecular weight and formula of unknown samples, and infrared spectroscopy gives us information about the functional groups present.

Mass spectroscopy is a technique in which molecules are first ionized by collision with a high-energy electron beam. The ions fragment into smaller pieces, which are magnetically sorted according to their mass-to-charge ratio (*m/z*). The ionized sample molecule is called the molecular ion, $M^{+\cdot}$, and measurement of its mass gives us the molecular weight of the sample. Once the molecular weight is known, we can narrow the possibilities of molecular formula to a very few choices.

In addition to molecular weight information, we can obtain structural clues about unknown samples by interpreting the fragmentation pattern of the molecular ion. Mass spectral fragmentations can be complex, however, and skill is required for their correct interpretation. Alkanes, for example, fragment readily by breaking carbon–carbon bonds, and many ions are usually formed.

Infrared spectroscopy is another technique that is widely used for structure elucidation. When an organic molecule is irradiated with infrared light, certain frequencies of light are absorbed by the molecule and others are not absorbed. The energy absorbed by the sample molecule corresponds to the amount needed to increase the extent of certain molecular motions such as bond stretches and bond bendings. Each specific kind of functional group usually has a characteristic set of infrared absorptions. For example, the terminal alkyne $\equiv C—H$ bond absorbs infrared radiation of 3300 cm^{-1} frequency; the alkene $C{=}C$ bond absorbs in the range 1650–1670 cm^{-1}. By observing which frequencies of infrared radiation are absorbed by a molecule, *and which are not,* we can get an excellent notion of what functional groups the molecule contains.

ADDITIONAL PROBLEMS

11.12 Write as many possible molecular formulas as you can for hydrocarbons that show the following molecular ions in their mass spectra:

(a) $M^{+\cdot} = 64$ (b) $M^{+\cdot} = 186$
(c) $M^{+\cdot} = 158$ (d) $M^{+\cdot} = 220$

11.13 In Section 5.2, we calculated the degree of unsaturation of molecules according to their molecular formulas. Write the molecular formulas of all hydrocarbons corresponding to the following molecular ions. How many degrees of unsaturation (double bonds and/or rings) are indicated by each formula?

(a) $M^{+\cdot} = 86$ (b) $M^{+\cdot} = 110$
(c) $M^{+\cdot} = 146$ (d) $M^{+\cdot} = 190$

11.14 Draw the structure of a molecule that is consistent with the mass spectral data in each example:
(a) A hydrocarbon with $M^{+\cdot} = 132$ (b) A hydrocarbon with $M^{+\cdot} = 166$
(c) A hydrocarbon with $M^{+\cdot} = 84$

11.15 Write as many possible molecular formulas as you can for compounds that show the following molecular ions in their mass spectra. Assume that the elements C, H, N, and O might be present.
(a) $M^{+\cdot} = 74$ (b) $M^{+\cdot} = 131$

11.16 Camphor is an important compound isolated from the Asian camphor tree and used, among other things, as a moth repellent and as a constituent of embalming fluid. Camphor is a saturated monoketone that has $M^{+\cdot} = 152$ in its mass spectrum. What is a reasonable molecular formula for camphor? How many rings does camphor have?

11.17 Nicotine is a diamino compound that can be isolated from dried tobacco leaves. Nicotine has two rings and shows $M^{+\cdot} = 162$ in its mass spectrum. Propose a molecular formula for nicotine and calculate the number of double bonds present. (There are no oxygens present.)

11.18 Halogenated compounds are particularly easy to identify by their mass spectra because both chlorine and bromine occur naturally as mixtures of two abundant isotopes. Chlorine occurs as ^{35}Cl (75.5%) and as ^{37}Cl (24.5%), and bromine occurs as ^{79}Br (50.5%) and ^{81}Br (49.5%). At what masses do the molecular ion(s) occur for these formulas? What are the relative percentages of each molecular ion?
(a) Bromomethane, CH_3Br (b) Chlorohexane, $C_6H_{13}Cl$
(c) Vinyl chloride, C_2H_3Cl

11.19 The molecular ion(s) can be particularly complex for polyhalogenated compounds. Taking the natural abundances of Cl and Br into account (Problem 11.18), calculate the masses of the molecular ions of the following formulas. What are the relative percentages of each ion?
(a) Chloroform, $CHCl_3$
(b) Bromochloromethane, CH_2BrCl
(c) Freon 12, CF_2Cl_2 (Note that fluorine occurs only as ^{19}F.)

11.20 2-Methylpentane (C_6H_{14}, mol wt 86) has the mass spectrum shown here. What peak represents $M^{+\cdot}$? What peak is the base peak? Propose reasonable structures for fragment ions of m/z = 71, 57, 43, and 29. Suggest a reason to account for the base peak's having the mass it does.

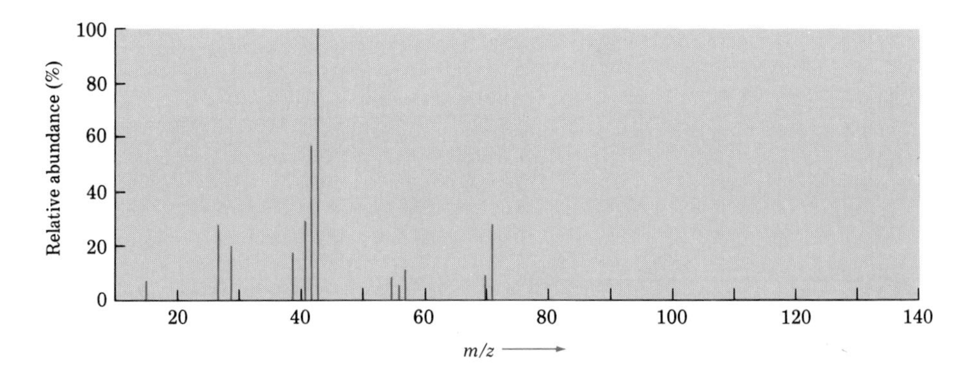

11.21 The combined gas chromatograph/mass spectrometer is a sophisticated instrument that uses a mass spectrometer to detect compounds as they are separated by a gas chromatograph. This technique allows chemists to inject a mixture of compounds onto a gas chromatography column and to determine automatically the mass spectrum of each compound present in the mixture. Assume that you are in the laboratory carrying out the catalytic hydrogenation of cyclohexene to cyclohexane. How could you use a gas chromatograph/mass spectrometer to determine when the reaction was complete?

11.22 Convert the following infrared absorption values from micrometers to reciprocal centimeters:

(a) An alcohol at 2.98 μm (b) An ester at 5.81 μm
(c) A nitrile at 4.93 μm

11.23 Convert the following infrared absorption values from reciprocal centimeters to micrometers:

(a) A cyclopentanone at 1755 cm^{-1} (b) An amine at 3250 cm^{-1}
(c) An aldehyde at 1725 cm^{-1}

11.24 Two infrared spectra are shown. One is the spectrum of cyclohexane and the other is the spectrum of cyclohexene. Identify them and explain your answer.

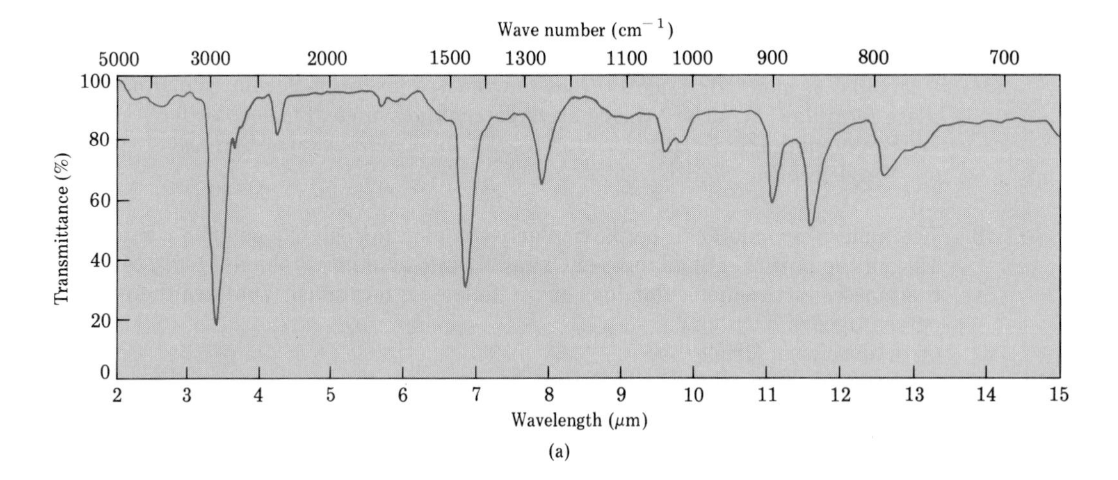

(a)

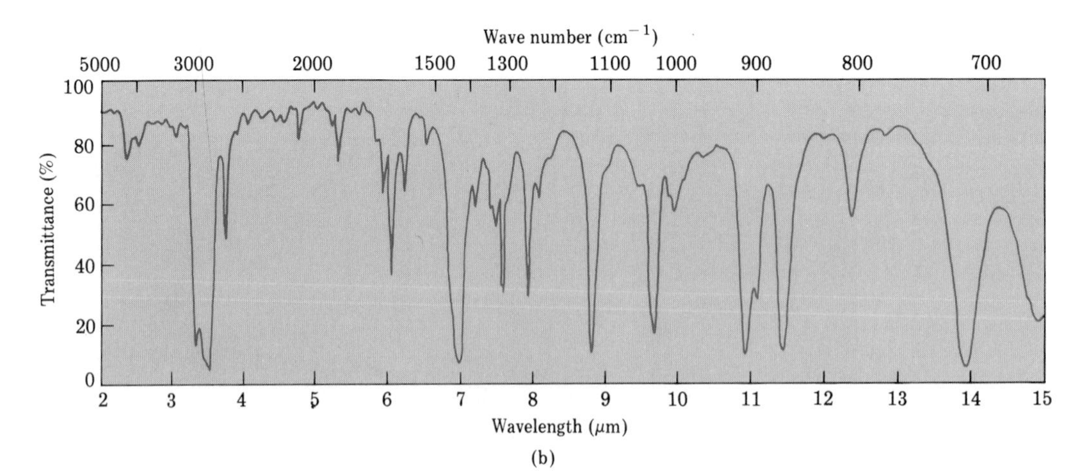

(b)

11.25 How would you use infrared spectroscopy to distinguish between these pairs of compounds?

(a) $(CH_3)_3N$ and $CH_3CH_2NHCH_3$

(b) $CH_3CH_2\overset{\overset{\displaystyle O}{\|}}{C}CH_3$ and $CH_3CH{=}CHCH_2OH$

(c) $H_2C{=}CHOCH_3$ and CH_3CH_2CHO

11.26 Assume you are carrying out the dehydration of 1-methylcyclohexanol to 1-methylcyclohexene. How could you use infrared spectroscopy to determine when the reaction was complete?

11.27 Shown are the mass spectrum (a) and the infrared spectrum (b) of an unknown hydrocarbon. Analyze and explain the data, and propose as many reasonable structures as you can.

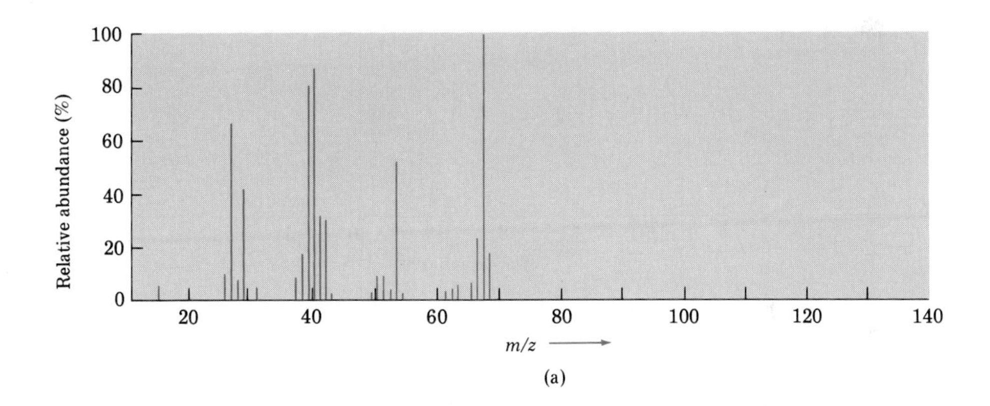

(a)

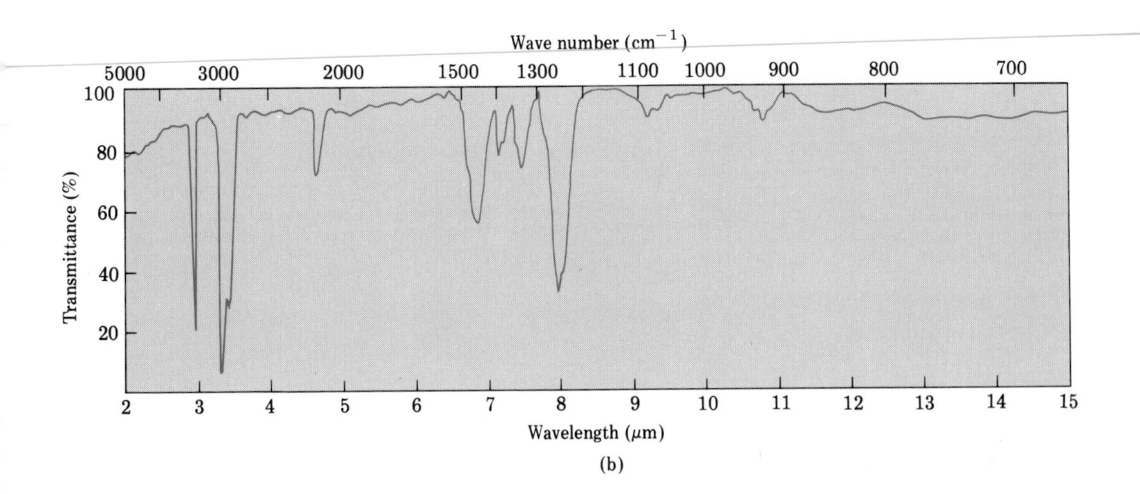

(b)

11.28 Shown are the mass spectrum (a) and infrared spectrum (b) of another unknown hydrocarbon. Analyze and explain the data, and propose as many reasonable structures as you can.

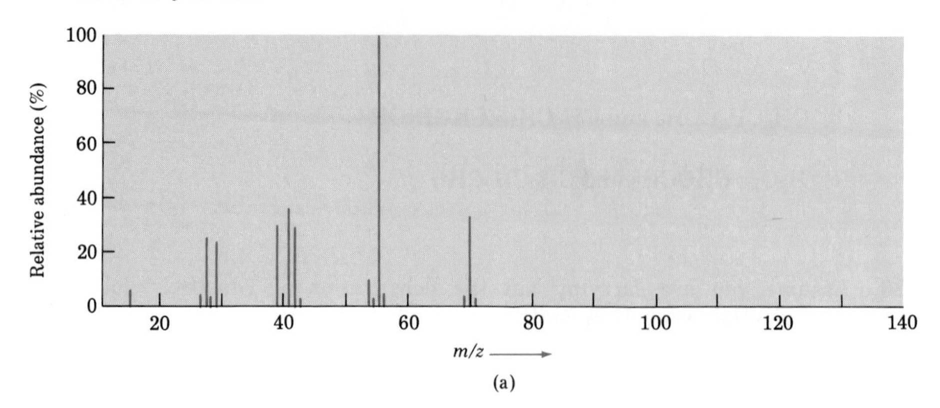

(a)

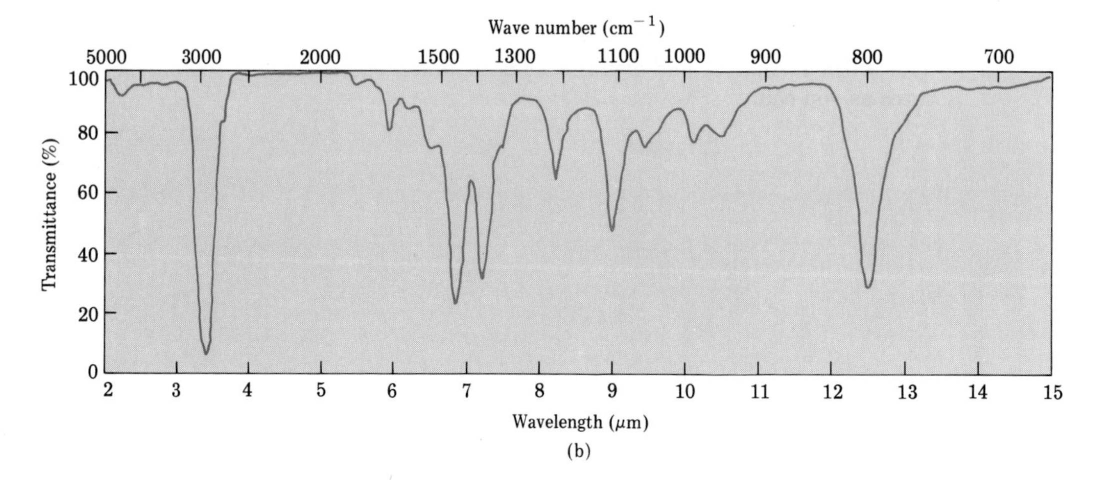

(b)

11.29 Carvone is an unsaturated ketone responsible for the odor of spearmint. If carvone has $M^{+\cdot} = 150$ in its mass spectrum, what molecular formulas are likely? If carvone has three double bonds and one ring, what molecular formula is correct?

CHAPTER 12 STRUCTURE DETERMINATION: NUCLEAR MAGNETIC RESONANCE SPECTROSCOPY

Nuclear magnetic resonance (NMR) spectroscopy is one of the most valuable spectroscopic techniques available to organic chemists. It is the method of structure determination to which chemists first turn for information.

We saw in Chapter 11 that mass spectroscopy provides information about the molecular weight and formula of a molecule of unknown structure, and infrared spectroscopy provides information about the kinds of functional groups in the unknown. Nuclear magnetic resonance spectroscopy does not replace or duplicate either of these techniques; rather, it complements them. NMR spectroscopy provides a "map" of the carbon–hydrogen framework of an organic molecule. Taken together, the three techniques often allow us to obtain complete solutions for the structures of even rather complex unknowns.

1. Mass spectroscopy	Molecular size and formula
2. Infrared spectroscopy	Functional groups present
3. NMR spectroscopy	Carbon–hydrogen framework

12.1 Nuclear magnetic resonance spectroscopy

Many kinds of nuclei behave as if they were spinning about an axis. Since they are positively charged, these spinning nuclei act like tiny magnets and can therefore interact with an externally applied magnetic field (H_0). Not all nuclei act this way but, fortunately for organic chemists, both the proton (^1H) and the ^{13}C nucleus do have spins. Let's see what the consequences of nuclear spin are, and how we can make use of the results.

In the absence of strong external magnetic fields, the nuclear spins of magnetic nuclei are oriented randomly. When these nuclei are placed between the poles of a strong magnet, however, they adopt specific orientations, much as a compass needle orients itself in the earth's magnetic field. A spinning ^1H or ^{13}C nucleus can orient so that its own tiny magnetic field is aligned either with (parallel to) the external field or against (antiparallel to) the external field. These two possible orientations do not have the same energy and therefore are not present in equal amounts. The parallel orientation is slightly lower in energy, and this spin state is slightly favored over the antiparallel orientation (Figure 12.1).

When the oriented nuclei are then irradiated with radio waves of the proper frequency, energy absorption occurs, and the lower energy state "spin-flips" to the higher energy state. When this spin-flip occurs, the nucleus is said to be in **resonance** with the applied radiation—hence the name, nuclear magnetic resonance.

The exact amount of radio-frequency (rf) energy necessary for resonance depends both on the strength of the external magnetic field and on the nucleus being irradiated. If a very strong magnetic field is applied, the energy separation between the two spin states is large, and higher-frequency (higher-energy) radiation is required. If a weaker magnetic field is applied, less energy is required to effect the transition between nuclear spin states (Figure 12.2). In practice, superconducting magnets producing

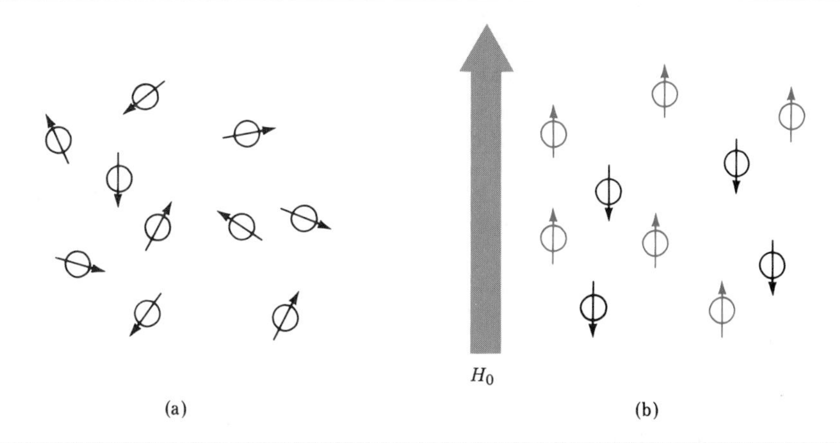

Figure 12.1. Orientation of nuclear spins: (a) random orientation, in the absence of an external magnetic field; (b) specific orientation, in the presence of an external magnetic field, H_0. Note that some of the spins (color) are aligned parallel to the external field and others are antiparallel. The parallel spin state is lower in energy.

enormously powerful fields up to 140,000 gauss are sometimes used, but a field strength of 14,100 gauss is more common. At this magnetic field strength, rf energy in the 60 MHz range (1 MHz = 1 megahertz = 1 million cycles per second) is required to bring a ^1H nucleus into resonance, and rf energy of 15 MHz is required to bring a ^{13}C nucleus into resonance.

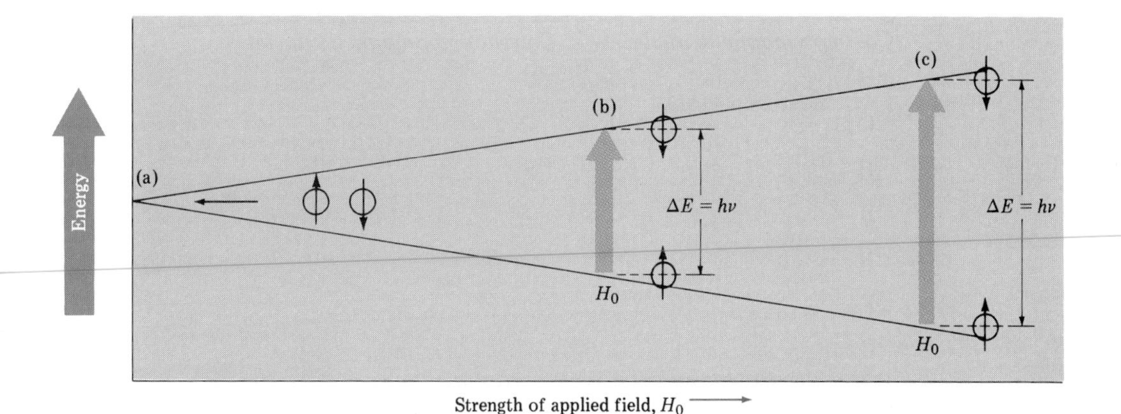

Figure 12.2. Variation of energy separation between nuclear spin states as a function of applied magnetic field strength. Absorption of rf energy of frequency ν converts a nucleus from a lower spin state to a higher spin state: (a) Spin states have equal energy in the absence of an applied magnetic field. (b) Spin states have unequal energy in the presence of an applied magnetic field. At ν = 60 MHz, $\Delta E = 5.7 \times 10^{-6}$ kcal/mol. (c) Energy separation between spin states is greater, the larger the applied field. At ν = 100 MHz, $\Delta E = 9.5 \times 10^{-6}$ kcal/mol.

These energy levels are much smaller than those required for infrared spectroscopy (1–10 kcal/mol). For example, 60 MHz rf energy corresponds to about 5.7×10^{-6} kcal/mol.

$$E = \frac{2.85 \times 10^{-3} \text{ kcal/mol}}{\lambda \text{ (cm)}}$$

Since

$$\lambda = \frac{c}{\nu}$$

$$= \frac{3 \times 10^{10} \text{ cm/sec}}{60 \text{ MHz}} = 5 \times 10^{2} \text{ cm}$$

Therefore,

$$E = \frac{2.85 \times 10^{-3}}{5 \times 10^{2}} = 5.7 \times 10^{-6} \text{ kcal/mol}$$

The ^{1}H and ^{13}C nuclei are not unique in their ability to exhibit the nuclear magnetic resonance phenomenon. All nuclei with odd-numbered masses, such as ^{1}H, ^{11}B, ^{13}C, ^{19}F, and ^{31}P, show magnetic properties. Similarly, all nuclei with even-numbered masses but odd atomic numbers show magnetic properties (^{2}H and ^{14}N, for example). Nuclei having both even masses and even atomic numbers (^{12}C, ^{16}O) do not give rise to magnetic phenomena (Table 12.1).

TABLE 12.1 The NMR behavior of some common nuclei

Common magnetic nuclei		*Common nonmagnetic nuclei*
^{1}H	Most important	^{12}C
^{13}C	Most important	^{16}O
^{2}H		^{32}S
^{11}B		
^{14}N		
^{19}F		
^{31}P		

PROBLEM

12.1 The exact amount of energy required to spin-flip a magnetic nucleus depends not only on the strength of the external magnetic field but also on the intrinsic properties of the specific isotope. We saw earlier that, at a field strength of 14,100 gauss, rf energy of 60 MHz is required to bring a ^{1}H nucleus into resonance. At the same field strength, rf energy of 56 MHz will bring a ^{19}F nucleus into resonance. Calculate the amount of energy required to spin-flip a ^{19}F nucleus. Is this amount greater or less than that required to spin-flip a ^{1}H nucleus?

PROBLEM

12.2 Calculate the amount of energy required to spin-flip a proton for a spectrometer operating at 100 MHz. Does increasing the spectrometer frequency from 60 MHz to 100 MHz increase or decrease the amount of energy necessary for resonance?

12.2 The nature of NMR absorptions

From the description given so far, one might expect all protons in a molecule to absorb rf energy at the same frequency and all ^{13}C nuclei to absorb at the same frequency. If this were true, we would observe a single NMR absorption band in the ^{1}H or ^{13}C spectrum of an unknown, and this would be of little use to us. In fact, the frequency is not the same for all nuclei.

All nuclei in molecules are surrounded by electron clouds. When we apply a uniform external magnetic field to a sample molecule, the circulating-electron clouds set up tiny local magnetic fields of their own. These local magnetic fields act in opposition to the applied field, so that the effective field actually felt by the nucleus is smaller than the applied field.

$$H_{\text{effective}} = H_{\text{applied}} - H_{\text{local}}$$

In describing this effect, we say that the nuclei are *shielded* by the circulating-electron clouds. Since each kind of nucleus in a molecule is in a slightly different electronic environment, it is shielded to a slightly different extent. This means that the effective magnetic field actually felt is not the same for each nucleus; at a given value of the applied field, every nonequivalent nucleus in a molecule feels a slightly different magnetic field. If our NMR instrument is sensitive enough, the tiny differences in the effective magnetic fields experienced by different nuclei can be observed, and we can see different NMR signals for each nucleus. Each unique kind of proton and each unique kind of ^{13}C in a molecule give rise to a unique NMR signal. Thus, the NMR spectrum of an organic compound provides us with a map of the carbon–hydrogen framework. With practice, we can learn how to read the map and thereby derive structural information about an unknown molecule.

Figure 12.3 (p. 370) shows both the ^{1}H and the ^{13}C NMR spectra of methyl acetate $CH_3CO_2CH_3$. The horizontal (x) axis shows the difference in effective field strength felt by the nuclei, and the vertical (y) axis indicates intensity of absorption of rf energy. Thus a peak in the NMR spectrum corresponds to a certain kind of nucleus in a molecule. Both ^{1}H and ^{13}C spectra cannot be observed at the same time on the same spectrometer, however, since different amounts of energy are required to spin-flip the different kinds of nuclei.

Note that in Figure 12.3 the ^{1}H NMR spectrum of methyl acetate shows only two distinct peaks even though the molecule contains six protons. One peak is due to the CH_3CO protons, and the other to the $COOCH_3$ protons. Since the three protons of each methyl group have the same chemical (and magnetic) environment, they are shielded to the same extent and show a single absorption. The two methyl groups themselves, however, are *not* equivalent; they therefore absorb at different positions. The ^{13}C spectrum of

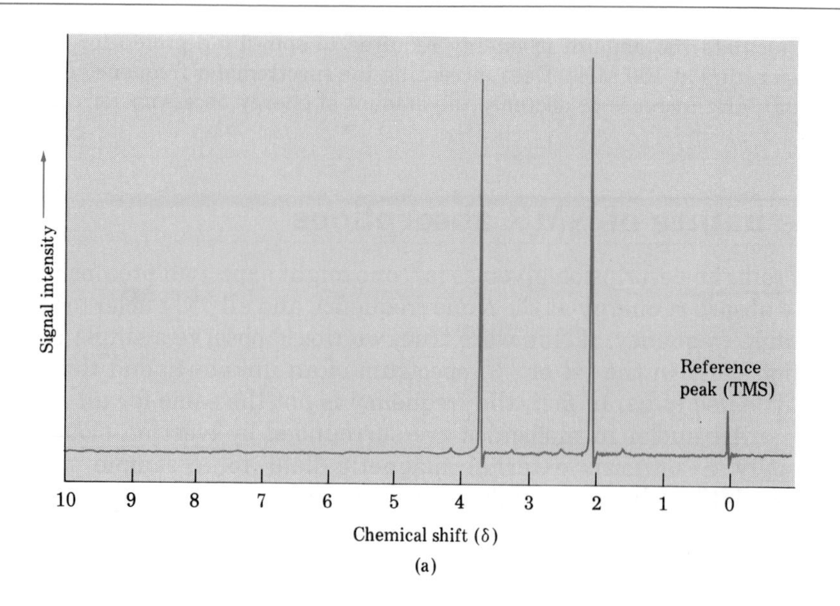

(a)

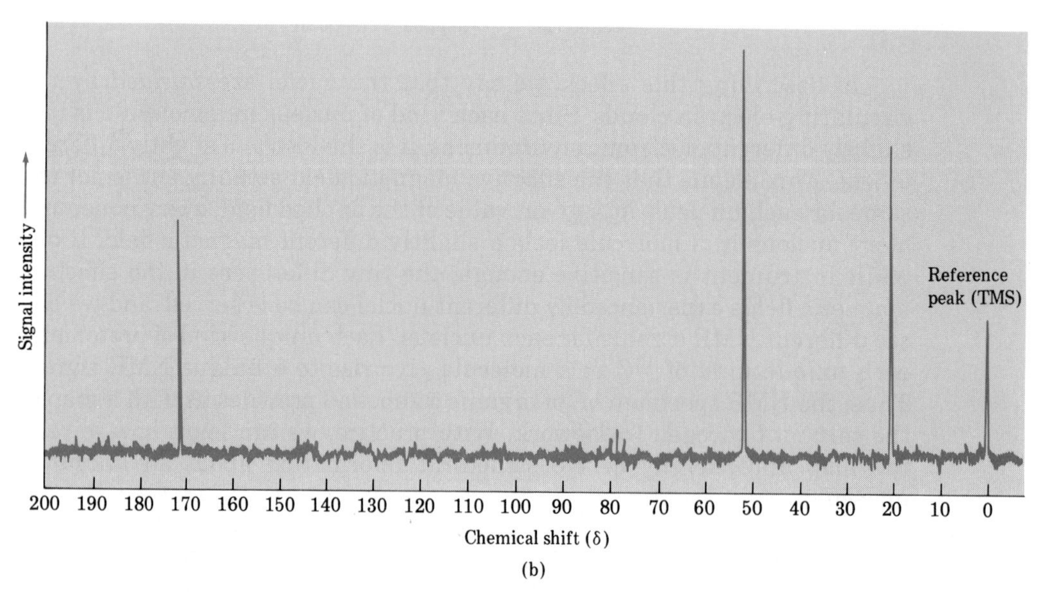

(b)

Figure 12.3. (a) The ^1H NMR spectrum, and (b) the ^{13}C NMR spectrum, of methyl acetate, $CH_3CO_2CH_3$.

methyl acetate shows three peaks, one for each of the three different carbon atoms present.

The operation of an NMR spectrometer is illustrated schematically in Figure 12.4. An organic sample is dissolved in a suitable solvent and placed in a thin glass tube between the poles of a magnet. The strong magnetic field causes the ^1H (and ^{13}C) nuclei in the molecule to align in one of the two possible orientations, and the sample is then irradiated with rf energy. The exact amount of energy required depends both on the strength of the magnetic field and on the kind of nucleus we intend to observe. As noted before, the two most commonly observed nuclei, ^{1}H and ^{13}C, absorb in quite different rf ranges, and we cannot observe both at the same time.

If the frequency of rf irradiation is held constant and the strength of the applied magnetic field is changed, each nucleus comes into resonance at a slightly different field strength. A sensitive detector monitors the absorption of rf energy, and the electronic signal is then amplified and displayed as a peak on a recorder chart.

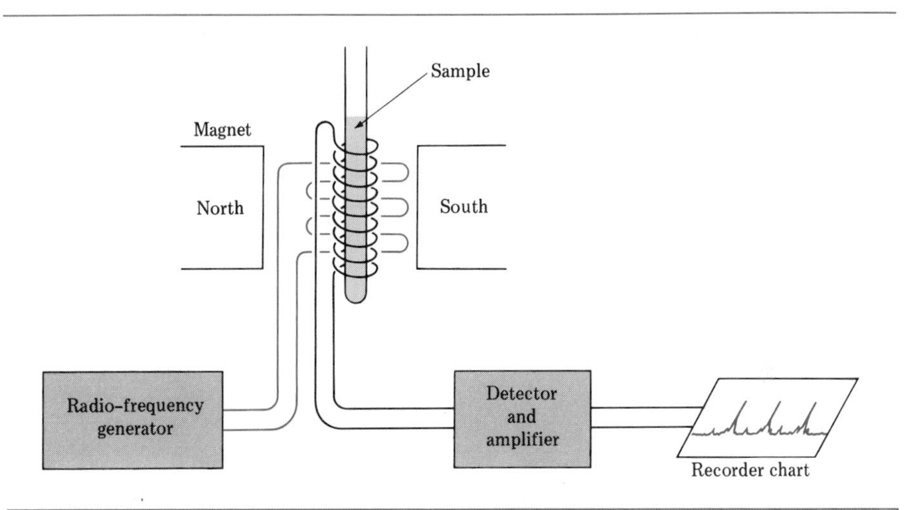

Figure 12.4. Schematic operation of an NMR spectrometer.

Nuclear magnetic resonance spectroscopy differs from infrared spectroscopy (Sections 11.8–11.10) in that the time scales of the two techniques are quite different. The absorption of infrared energy by a molecule, giving rise to a change in vibrational or rotational states, is an essentially instantaneous process (about 10^{-13} sec); the NMR process, however, requires far more time (about 10^{-3} sec). The difference in time scales between infrared spectroscopy and NMR spectroscopy can be compared to the difference between a camera operating at a very fast shutter speed and a camera operating at a very slow shutter speed. The fast camera (infrared) takes an instantaneous picture and "freezes" the action. If two interconverting species were present in a sample, infrared would record the spectrum of each. The slow camera (NMR), however, takes a blurred, "time-averaged" picture. If two species interconverting faster than 10^3 times per second were present in a sample, NMR would record only a single, averaged spectrum, rather than separate spectra of the two discrete species. Because of this "blurring" effect, NMR can sometimes be used to investigate the mechanisms of organic reactions.

PROBLEM

12.3 Predict how many signals compounds (a)–(f) show in their ^{1}H and ^{13}C spectra.
(a) $(CH_3)_2C{=}C(CH_3)_2$ (b) Cyclohexane

(c) $CH_3—O—CH_3$

(d) $(CH_3)_3C—\overset{O}{\overset{\|}{C}}—OCH_3$

(e) Benzene,

(f) 1,1,3,3-Tetramethylcyclobutane

12.3 Chemical shifts

All NMR spectra are displayed on charts that show the applied field strength increasing from left to right (Figure 12.5). Thus, the left part of the chart is the low-field side, and the right part is the high-field side. In discussing spectra, chemists often talk about a particular peak being either *upfield* (high) or *downfield* (low).

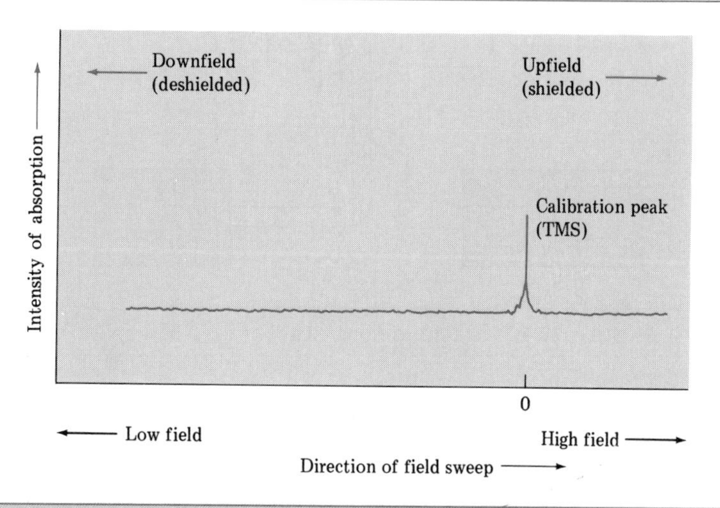

Figure 12.5. The NMR spectrum.

In order to define the position of absorptions, the NMR chart is calibrated, and a reference point is used. In practice, a small amount of TMS [tetramethylsilane, $(CH_3)_4Si$] is added to the sample so that an internal standard reference absorption line is produced when the spectrum is run. Tetramethylsilane is used as reference for both ^1H and ^{13}C spectra because it gives rise to a single peak that occurs upfield of all other absorptions normally found in organic compounds. The ^1H and ^{13}C spectra of methyl acetate (Figure 12.3) have the TMS reference peak indicated.

The exact place on the chart at which a nucleus absorbs is called its **chemical shift.** By convention, the chemical shift of TMS is arbitrarily set as the zero point, and all other absorptions normally occur downfield. For historical reasons, NMR charts are calibrated in units of frequency rather than in gauss (magnetic units). The arbitrary scale used is called the **delta scale,** and 1 delta unit (δ) is equal to 1 part per million (ppm) of the spectrometer frequency. For example, if we were measuring the ^1H NMR spectrum of a sample using an instrument operating at 60 MHz, 1 δ (or 1 ppm of 60×10^6 Hz) would equal 60 Hz. The following equation can be used for any absorption:

$$\delta = \text{Number of parts per million} = \frac{\text{Observed chemical shift (number of Hz away from TMS)}}{\text{Spectrometer frequency (MHz)}}$$

Similarly, if we were measuring the ^{13}C spectrum of a sample using an instrument operating at 15 MHz, then 1 ppm = 1 δ = 15 Hz. The spectra shown in Figure 12.3 are calibrated in these units.

Although this method of NMR calibration may seem needlessly complex, there is a good reason for it. In practice, many different kinds of spectrometers are available, operating at many different magnetic field strengths and rf frequencies. By employing a system of measurement in which NMR absorptions are expressed in terms relative to spectrometer frequency (ppm) rather than in absolute terms (Hz), we can avoid much confusion. The chemical shift of an NMR absorption given in ppm or δ units is constant, regardless of the operating frequency of the particular spectrometer. A given ^1H nucleus that absorbs at 2.0 δ on a 60 MHz instrument (2.0 ppm $\times$ 60 MHz = 120 Hz downfield from TMS) also absorbs at 2.0 δ on a 300 MHz instrument (2.0 ppm $\times$ 300 MHz = 600 Hz downfield from TMS). The difference between using a low-field-strength spectrometer (60 MHz) and a high-field-strength spectrometer (300 MHz), however, is that different NMR absorptions are more widely separated at high field strength. Thus, complexities due to accidental overlapping of signals are lessened, and interpretations of spectra become easier. For example, two signals that are only 6 Hz apart at 60 MHz (i.e., 0.1 ppm) are 30 Hz apart at 300 MHz (still 0.1 ppm).

The range in which most NMR absorptions occur is quite narrow. Almost all ^1H NMR absorptions occur 0–12 δ downfield from the proton absorption of TMS. Almost all ^{13}C absorptions occur 1–250 δ downfield from the carbon absorption of TMS. Let's look more closely at the interpretation of NMR spectra to see how to use this tool in organic structure determination. We will begin by looking at ^{13}C NMR. For technical reasons, ^{13}C spectra are more difficult to obtain than ^1H spectra, but they are also much easier to interpret. Although the complexities of spectrometer operation differ, the principles behind ^{13}C and ^1H NMR are the same. What we learn now about interpreting ^{13}C spectra will simplify the subsequent discussion of ^1H spectra.

PROBLEM

12.4 When the ^1H NMR spectrum of acetone, CH_3COCH_3, is recorded on an instrument operating at 60 MHz, a single sharp resonance line at 2.1 δ is observed.

(a) How many hertz downfield from TMS does the acetone resonance line correspond to?

(b) If the ^1H NMR spectrum of acetone were recorded at 100 MHz, what would be the position of the absorption in δ units?

(c) How many hertz downfield from TMS does this 100 MHz spectrum correspond to?

PROBLEM

12.5 The following ^1H NMR resonance lines were recorded on a spectrometer operating at 60 MHz. Convert each into δ units.

(a) $CHCl_3$ 436 Hz (b) CH_3Cl 183 Hz
(c) CH_3OH 208 Hz (d) CH_2Cl_2 318 Hz

12.4 ^{13}C NMR

At first glance, it appears surprising that carbon NMR is even possible. After all, ^{12}C, the most abundant carbon isotope, has no nuclear magnetic moment; ^{13}C is the only naturally occurring carbon isotope with a magnetic moment, but its natural abundance is only about 1.1%. Thus, only about 1 out of every 100 carbons in organic molecules is observable by NMR. The low abundance of ^{13}C means that ^{13}C instrumentation must be far more sensitive (and expensive) than that required for ^1H NMR, but these obstacles have been overcome through the use of improved electronics and computer techniques. Today ^{13}C NMR is a routine structural tool, and a ^{13}C NMR spectrum can often be obtained on 10 mg of sample in a few hours' time.

At its simplest, ^{13}C NMR allows us to count the number of carbons in an unknown structure. In addition, we can get information about the chemical (magnetic) environment of each carbon by observing its chemical shift. Let's see how we can do these things.

Several different modes of operation are possible with ^{13}C NMR instruments, but the most common is the **proton noise-decoupled mode.** The exact meaning of this phrase is not important; the consequence of operating in the proton noise-decoupled mode is that we observe a single sharp resonance line for each unique (nonequivalent) kind of carbon atom present in a molecule. The spectrum of methyl acetate (Figure 12.3) illustrates this fact. There are three carbon atoms in methyl acetate and three peaks in the ^{13}C NMR spectrum.

Most ^{13}C resonances are between 0 and 250 ppm downfield from the TMS reference line. The exact chemical shift of each ^{13}C resonance is dependent on that carbon's environment within the molecule. Table 12.2 and Figure 12.6 show how environment and chemical shift may be correlated.

Many factors affect chemical shifts, but as a general rule sp^3-hybridized carbons absorb in the 0–100 δ range and sp^2 carbons absorb in the range 100–210 δ. Carbonyl carbons are particularly distinct in ^{13}C NMR and are easily observed at the extreme low-field end of the spectrum in the range 170–210 δ. Figure 12.7 (p. 376) shows the proton noise-decoupled ^{13}C NMR spectra of 2-butanone and acetophenone and indicates the peak assignments. The carbonyl carbon of acetophenone appears at 197 δ and the methyl carbon absorbs at 26 δ. The six aromatic-ring carbon atoms occur as four peaks in the range 128–137 δ. One of the ring carbon atoms absorbs at 137 δ and one at 133 δ. The remaining four ring carbon atoms are accounted for by the two closely spaced peaks at 128.3 and 128.6 δ.

TABLE 12.2 **Carbon-13 NMR chemical shift correlations**

Type of carbon	*Chemical shift (δ)*	*Type of carbon*	*Chemical shift (δ)*
C—I	0–40	=C	100–150
C—Br	25–65	C—O	40–80
C—Cl	35–80	C=O	170–210
—CH_3	8–30	(benzene ring carbon)	110–160
—CH_2—	15–55	C—N	30–65
>CH—	20–60		
≡C	65–85		

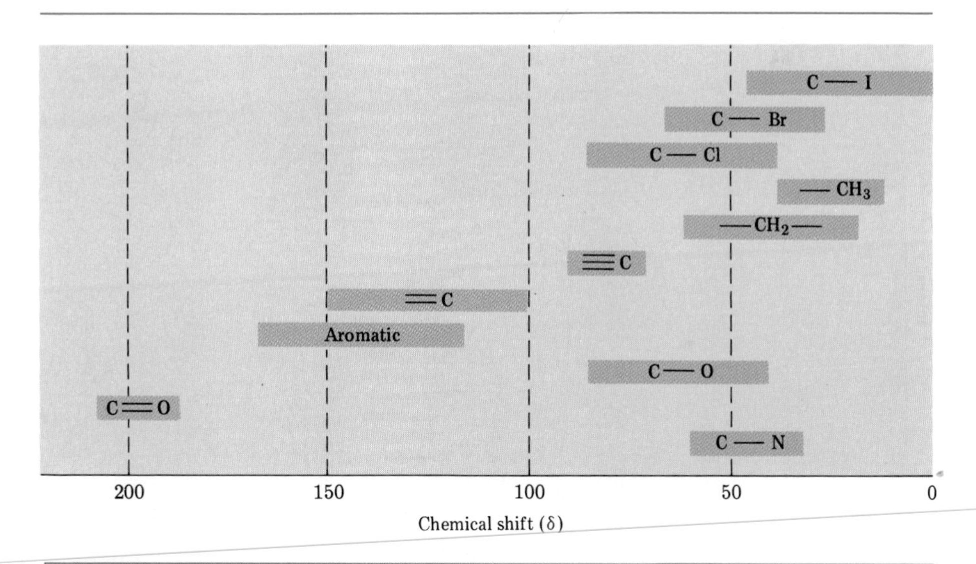

Figure 12.6. Chemical shift correlations for ^{13}C NMR.

The ^{13}C NMR spectrum of acetophenone is interesting in several ways. Note particularly that only six carbon absorptions are observed even though the molecule contains eight carbons. This is because the molecule has a symmetry plane that makes ring carbons 4 and 4′, and carbons 5 and 5′, chemically and magnetically equivalent. They therefore have identical resonance frequencies.

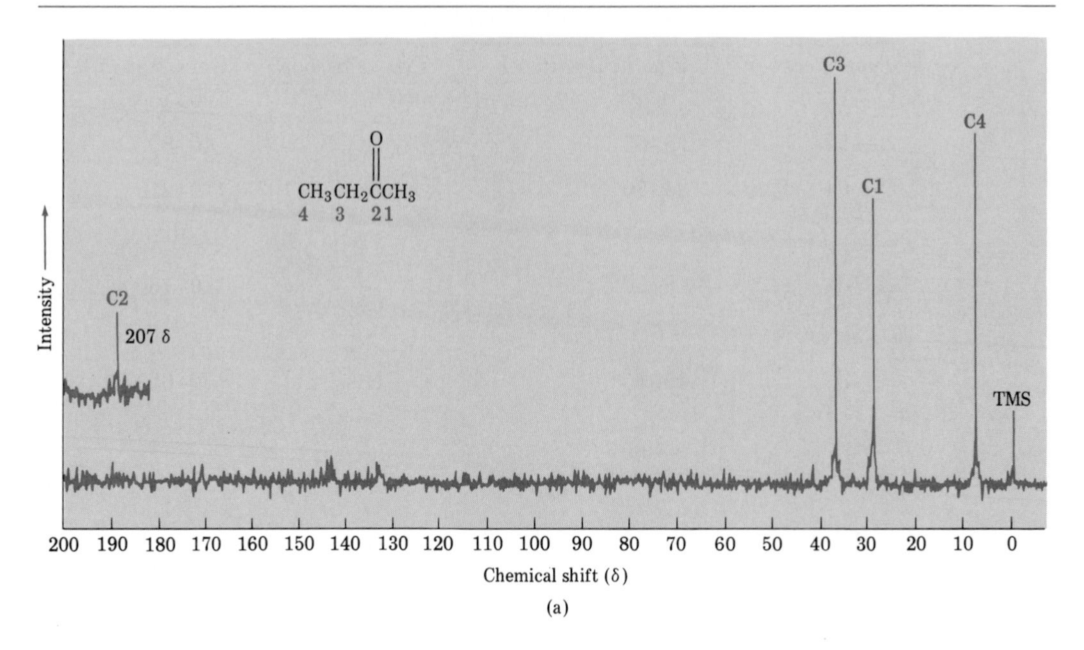

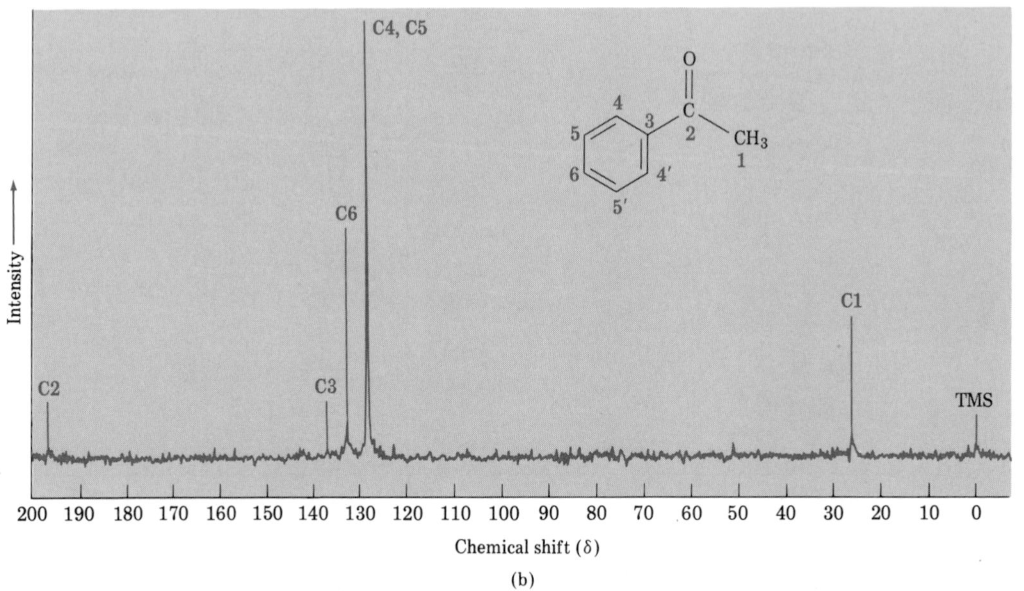

Figure 12.7. Proton noise-decoupled ^{13}C NMR spectra of (a) 2-butanone and (b) acetophenone.

Another interesting feature of both spectra is that the peaks are not uniform in height. Some peaks appear larger than others even though all are one-carbon resonances (except for the two-carbon peaks of acetophenone). This difference in peak size is caused by complex factors having to do with the electronics of the spectrometer, and is only indirectly related to the structure of the sample.

PROBLEM

12.6 The proton noise-decoupled ^{13}C NMR spectrum of methyl propanoate, $CH_3CH_2CO_2CH_3$, is shown. Assign all carbon resonances.

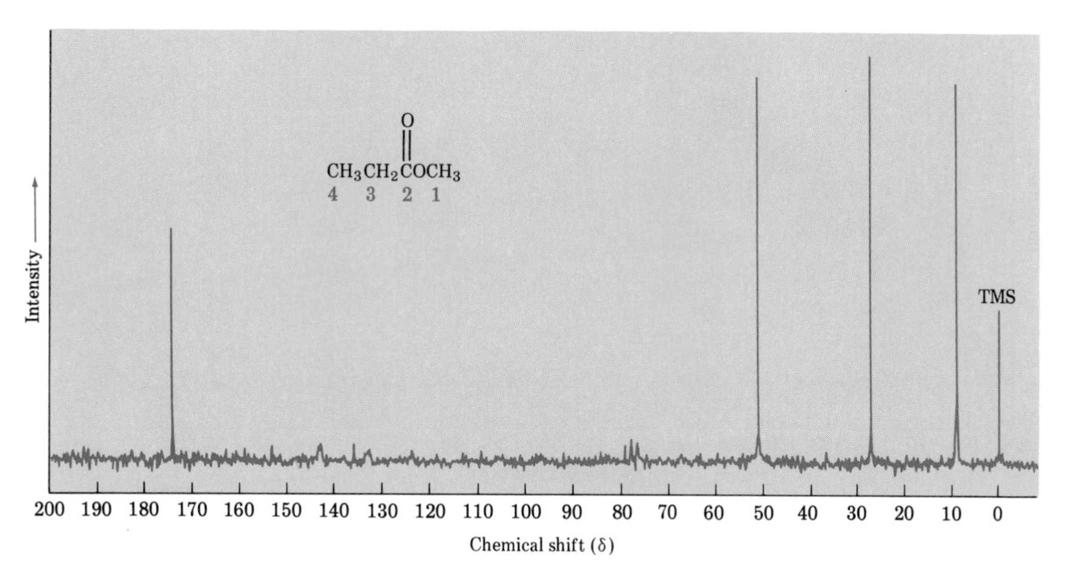

PROBLEM

12.7 Predict the number of carbon resonance lines you would expect to observe in the proton noise-decoupled ^{13}C NMR spectra of these compounds:

(a) Methylcyclopentane
(b) 1-Methylcyclohexene
(c) 1,2-Dimethylbenzene (*o*-xylene)
(d) 2-Methyl-2-butene

PROBLEM

12.8 Propose structures for compounds that fit these descriptions:

(a) A hydrocarbon with seven lines in its proton noise-decoupled ^{13}C NMR spectrum
(b) A six-carbon compound that shows only five resonance lines in its ^{13}C NMR spectrum
(c) A four-carbon compound that shows three resonance lines in its ^{13}C NMR spectrum

12.5 Measurements of NMR peak areas: integration

The relative sizes of the different peaks observed in a ^{13}C NMR spectrum depend to a considerable extent on the mode of spectrometer operation; not all single-carbon resonance lines are of equal peak area. For example, each of the four resonance lines in the spectrum of 2-butanone, Figure 12.7(a), is due to one carbon, but the peaks vary considerably in size. When the NMR spectrometer is operated in the **gated-decoupled** mode, however, all single-carbon resonances have the same areas. By electronically integrating (measuring) the area under each peak, we can determine the relative number of carbon atoms each peak represents. Integrated peak areas are presented on the chart in a "stair-step" fashion, with the height of each step proportional to the number of carbons causing that peak. Figure 12.8 shows the gated-decoupled ^{13}C NMR spectra of 2-butanone and acetophenone. Note that all four peaks in the 2-butanone spectrum have the same integrated peak area, but the acetophenone spectrum shows 4 equal one-carbon resonance lines and 2 two-carbon resonance lines.

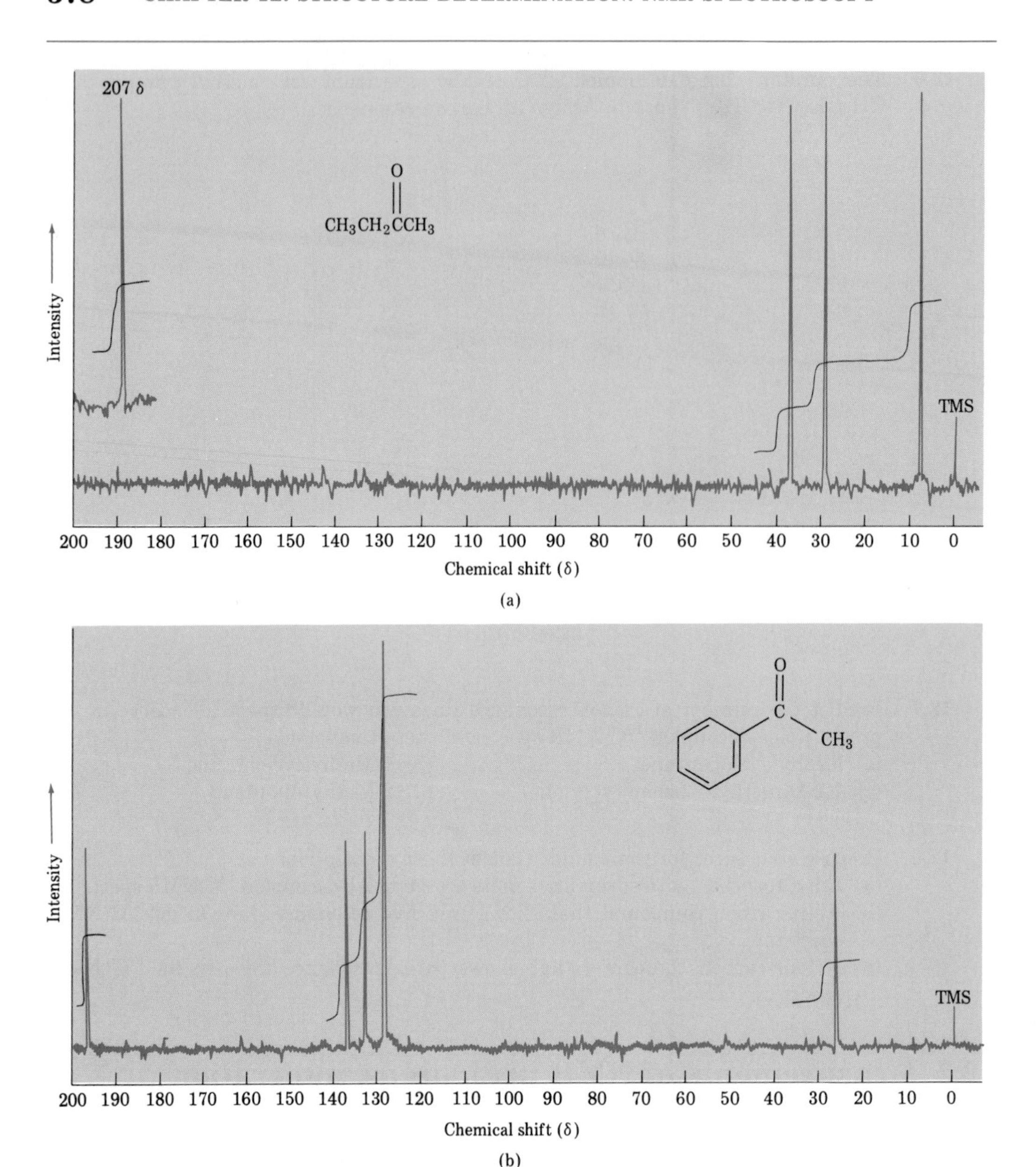

Figure 12.8. Gated-decoupled ^{13}C NMR spectra of (a) 2-butanone and (b) acetophenone.

Gated-decoupled spectra contain more information than normal proton noise-decoupled spectra since they can be integrated, but this information comes at a price. The NMR spectrometer is two or three times less sensitive in the gated-decoupled mode than in the proton noise-decoupled mode so that more sample and more time are required to obtain the spectrum. This price is not worth paying unless a particular ambiguity exists about the number of carbon atoms in a sample, and gated-decoupled spectra are rarely obtained in practice. The concept of NMR spectra integration is an important one, however, and we will soon see (Section 12.11) that peak integration is of great value in interpreting ^1H NMR spectra.

12.6 Spin–spin splitting of NMR signals

For most purposes, the proton noise-decoupled ^{13}C NMR spectrum is fully satisfactory. Obtaining the spectrum in this mode provides a carbon count of the sample molecule and gives information about the environment of each carbon. Occasionally, however, more detailed information is needed, and yet a third mode of spectrometer operation is employed. When the spectrometer is operated in the **off-resonance** mode, single-carbon resonance lines can split into multiple lines. What is this signal splitting due to, and what use can we make of this phenomenon?

Let's look carefully at both the proton noise-decoupled spectrum and the off-resonance spectrum of a simple molecule, dichloroacetic acid. When the two kinds of spectra are displayed on the same chart (Figure 12.9), it becomes clear that the carbonyl-carbon resonance at 170 δ remains a **singlet,** but the methine-carbon resonance (—C—H) at 64 δ splits into two peaks (a **doublet**).

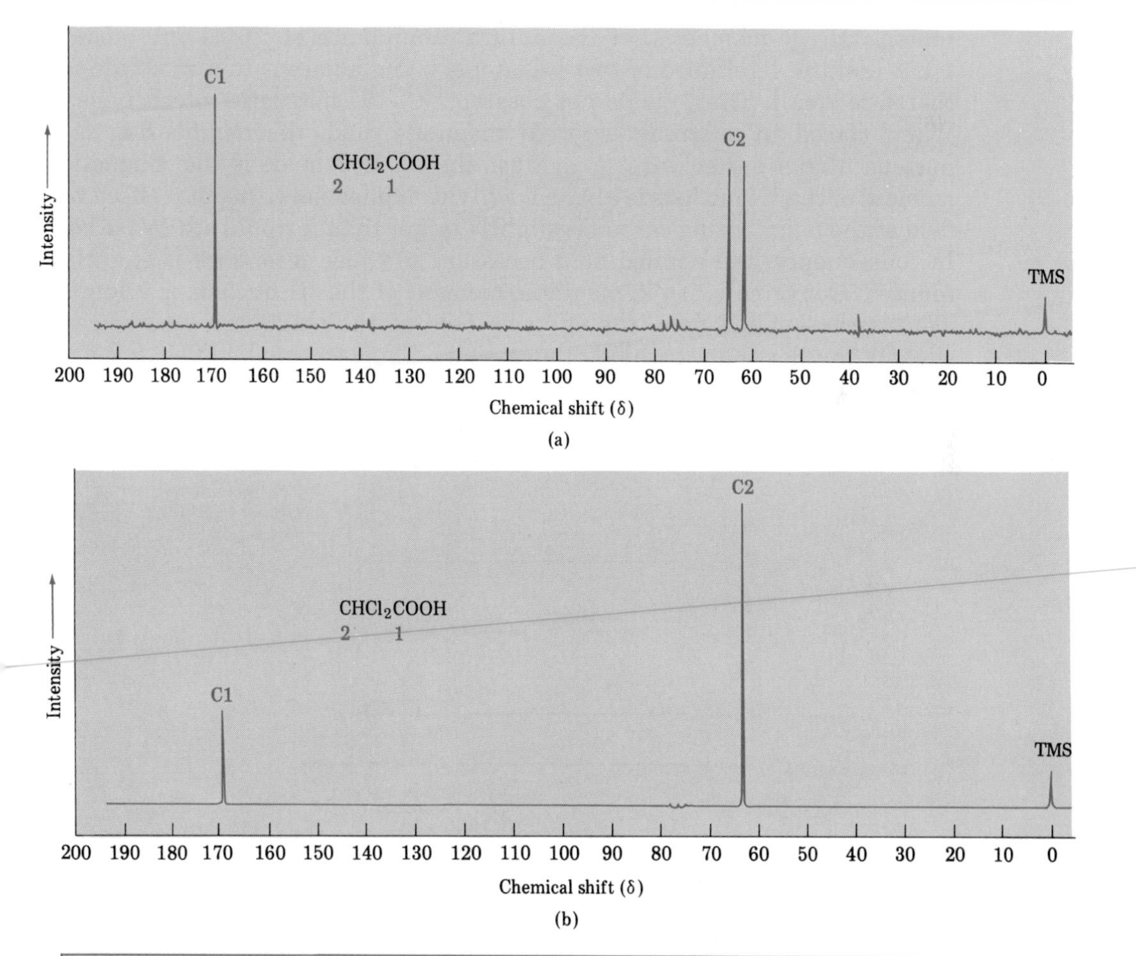

Figure 12.9. Proton noise-decoupled spectrum (lower scan) and off-resonance spectrum (upper scan) of dichloroacetic acid.

This phenomenon is known as **spin–spin splitting** and is due to the fact that the nuclear spin of one atom can interact with the nuclear spin of another nearby atom. The tiny magnetic field of one nucleus affects the magnetic field felt by a neighboring nucleus. We can understand spin–spin splitting by reviewing what we know and then looking more closely at what happens during nuclear magnetic resonance.

We have seen that when a magnetic nucleus such as ^{13}C is placed in a strong magnetic field the nucleus adopts one of two spin states. That is, the ^{13}C magnetic moment lines up either with or against the applied magnetic field. The applied magnetic field causes electrons in the molecule to *shield* the nucleus by setting up tiny local magnetic fields that act in opposition to the applied field. Because of this shielding, the effective field felt by the nucleus is slightly less than the applied field. Differences in the extent of shielding at each nucleus account for the differences in chemical shifts between nuclei that we observe in ^{13}C NMR spectra.

In addition to being affected by electron shielding, the field felt by a nucleus is affected by neighboring magnetic nuclei. For example, the carbonyl carbon of dichloroacetic acid is bonded only to nonmagnetic neighbor atoms, but the —$CHCl_2$ carbon is bonded to another magnetic nucleus—the proton, ^1H. (Remember that the natural abundance of ^{13}C is only about 1.1%, and the likelihood of two ^{13}C atoms being adjacent in a molecule is therefore small. Thus, we do not observe ^{13}C–^{13}C magnetic interactions.) When placed in a strong external magnetic field, the neighboring ^1H nucleus aligns either with or against the applied field. If the magnetic moment of the ^1H nucleus is aligned *with* the applied field, the total effective field at the neighboring carbon is slightly larger than it would otherwise be. In consequence, the applied field necessary to cause resonance is slightly reduced. Conversely, if the magnetic moment of the ^1H nucleus is aligned *against* the applied field, the effective field at the neighboring carbon is slightly smaller than it would otherwise be. Thus, the applied field needed to bring the carbon into resonance is slightly increased. Figure 12.10 shows schematically how spin–spin splitting arises.

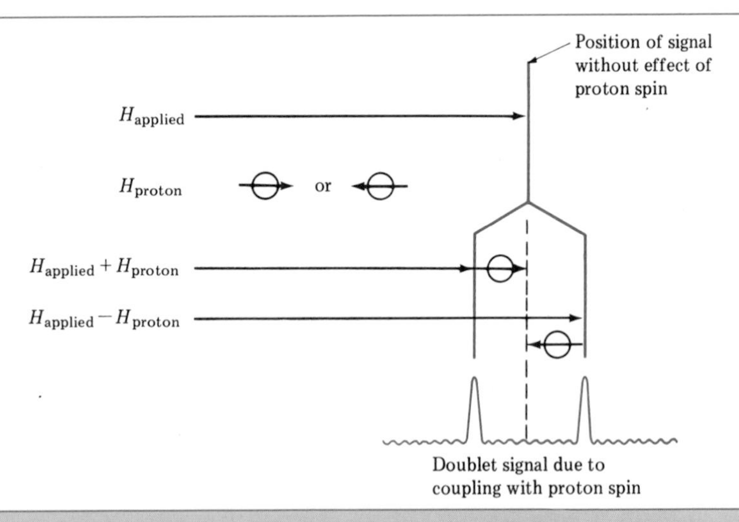

Figure 12.10. Origin of spin–spin splitting. The position of resonance of a nucleus is affected by neighboring magnetic nuclei.

The consequence of this interaction, or **coupling,** of adjacent nuclear spins is that the carbon comes into resonance at two slightly different values of the applied field. One resonance is a little above where it would be without coupling, and the other resonance is a little below where it would be without coupling. We therefore observe a doublet in the ^{13}C NMR spectrum of dichloroacetic acid.

If the carbon atom is bonded to more than one proton, more complex spin–spin splitting is observed. Figure 12.11 shows both the proton-noise decoupled spectrum and the off-resonance spectrum of 2-butanone.

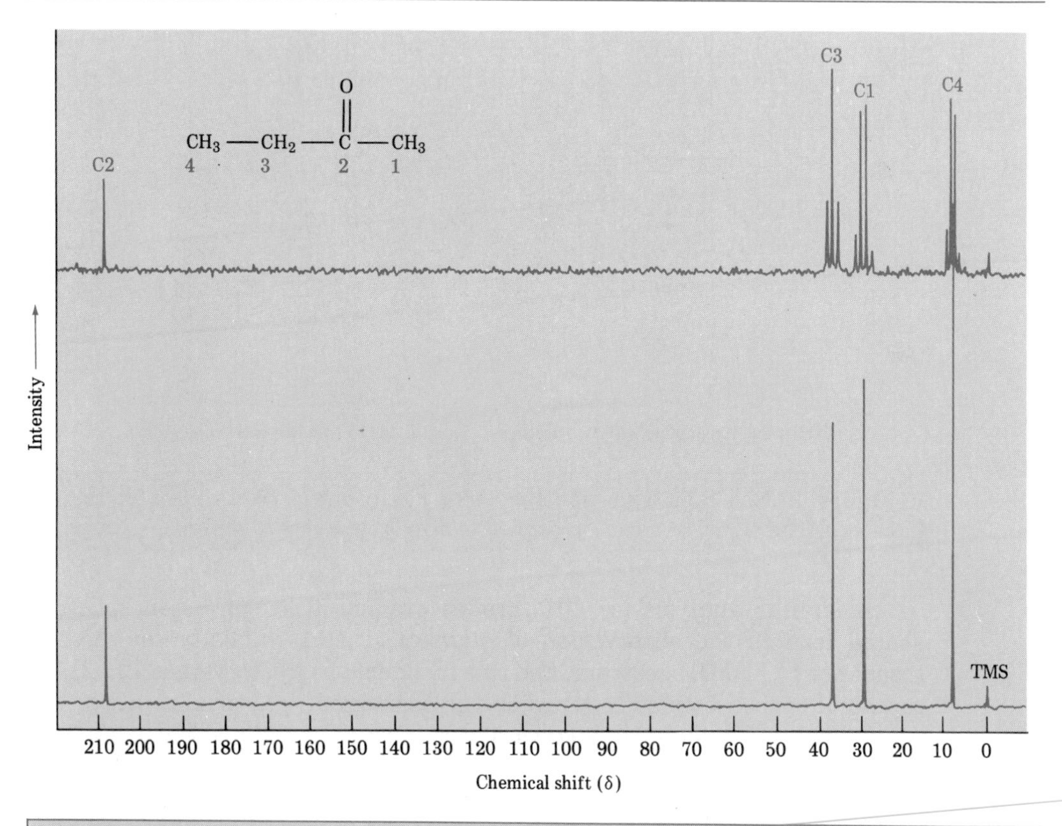

Figure 12.11. Proton noise-decoupled (lower) and off-resonance (upper) ^{13}C NMR spectra of 2-butanone.

The carbonyl-carbon (C2) resonance of 2-butanone remains a singlet in the off-resonance spectrum at 208 δ because it is not adjacent to any protons. However, the methylene-carbon (C3) resonance at 36.8 δ splits into a *triplet,* and the two methyl-carbon (C1 and C4) resonances at 6.6 and at 28.8 δ become *quartets*. As a general rule, a carbon bonded to n protons shows $n + 1$ peaks in the off-resonance ^{13}C spectrum. How does this rule arise?

A methylene carbon (—CH_2—) is bonded to two other magnetic nuclei (protons), and the magnetic moments of these two protons can orient in *three* different combinations. Both proton magnetic moments can orient against the applied field; one can orient against and one with the applied field (two possibilities); or both can orient with the applied field. This leads to three peaks of relative intensity (1:2:1) for the off-resonance spectrum of a methylene carbon (Figure 12.12).

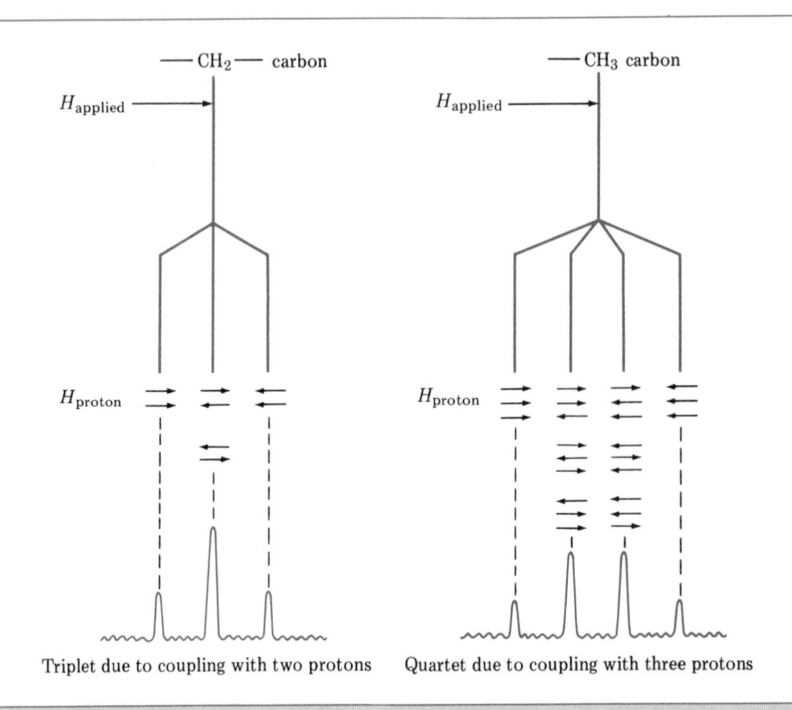

Figure 12.12. Spin–spin splitting of a methylene carbon and a methyl carbon. The signal of a carbon bonded to n hydrogens splits into $n + 1$ peaks.

A similar analysis for CH_3 groups predicts that spin–spin splitting should lead to the observation of *quartets* (1:2:2:1 intensity) in the off-resonance ^{13}C NMR spectrum, and this is in fact observed (Figure 12.12). To repeat: A carbon bonded to n protons gives a signal that is split into $n + 1$ peaks in the off-resonance spectrum.

PROBLEM

12.9 Sketch what the proton noise-decoupled spectrum of propene might look like.
(a) How many carbon resonances are there?
(b) Into how many peaks would each carbon resonance split if you ran the off-resonance spectrum?

PROBLEM

12.10 Sketch what the proton noise-decoupled spectrum of cyclopentene might look like.
(a) How many carbon resonances are there?
(b) What further information would the gated-decoupled spectrum give you, and how might this spectrum appear on a sketch? Make a sketch of what the off-resonance spectrum of cyclopentene might look like.

12.7 Summary of ^{13}C NMR

The information derived from ^{13}C NMR spectroscopy is of extraordinary value for structure determination. Not only are we able to count the number of carbon atoms in an unknown and deduce information about their chemical environment, we are also able to find how many protons are bonded to each carbon. This allows us to answer many structural questions that cannot be easily handled by infrared spectroscopy or mass spectroscopy.

As one example, how might we prove that E2 elimination of an alkyl halide gives the more substituted alkene? Does reaction of 1-chloro-1-methylcyclohexane with strong base lead predominantly to 1-methylcyclohexene or to methylenecyclohexane? These questions can easily be answered with ^{13}C NMR.

H_3C Cl — KOH / Ethanol → CH_3 or CH_2 ?

1-Chloro-1-methylcyclohexane → 1-Methylcyclohexene or Methylenecyclohexane?

1-Methylcyclohexene would be expected to show five sp^3-carbon resonances in the range 10–50 δ, and two sp^2-carbon resonances in the range 100–150 δ. Methylenecyclohexane, however, because of its symmetry, would be expected to have only three sp^3-carbon resonance peaks and two sp^2-carbon peaks. The proton noise-decoupled spectrum of the actual reaction product is shown in Figure 12.13, and clearly identifies 1-methylcyclohexene as the material formed in this E2 reaction.

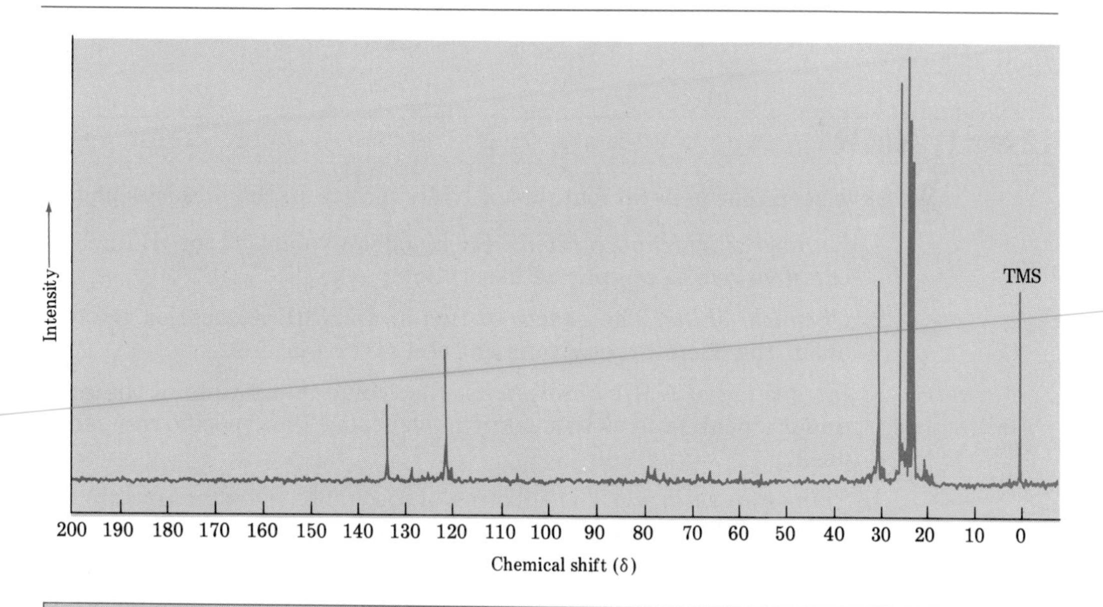

Figure 12.13. Proton noise-decoupled ^{13}C NMR spectrum of the elimination product from 1-chloro-1-methylcyclohexane.

A summary of the operating modes of ^{13}C NMR spectrometers and of the kind of information obtainable in each mode is shown in Table 12.3.

TABLE 12.3 **Summary of ^{13}C NMR and the modes of spectrometer operation**

Spectrometer operating mode	*Structural information obtained*
Proton noise-decoupled	Gives one resonance line for each kind of nonequivalent carbon atom present. This allows us to count number of carbons and deduce their environments from chemical shifts.
Gated-decoupled	All resonances are of equal peak area. This allows us to integrate the area under each peak to determine how many carbons each peak represents. This is rarely done in practice, however.
Off-resonance	Spin–spin splitting causes carbon resonances to split into multiplets. A carbon bonded to n protons gives a signal that is split into $n + 1$ peaks. This allows us to determine how many protons each carbon is bonded to.

PROBLEM

12.11 We saw in Section 6.5 that addition of HBr to alkenes under radical conditions leads to the non-Markovnikov product—the bromine bonds to the less substituted carbon. How do we know this is correct? How could you use ^{13}C NMR to identify the product of radical addition of HBr to 2-methylpropene?

$$(CH_3)_2C{=}CH_2 \xrightarrow[\text{Radicals}]{\text{HBr}} ?$$

12.8 ^1H NMR

We have seen four general features of NMR spectra in the previous pages:

1. *Number of NMR absorptions.* Each nonequivalent ^{13}C or ^1H nucleus can give rise to a separate absorption peak.
2. *Chemical shifts.* The exact position of an NMR absorption tells us about the chemical environment of a given nucleus.
3. *Integration of NMR absorptions.* Electronic integration of the area under a peak tells us how many nuclei cause that specific resonance peak.
4. *Spin–spin splitting.* The splitting of resonance lines into multiplets is due to the coupling of nearby nuclear spins and provides information about neighboring magnetic nuclei.

All four of these features are just as applicable to proton or ^1H NMR spectra as they are to ^{13}C NMR spectra. In fact, they are even easier to observe for ^1H spectra, since normal spectrometer operating conditions allow for signal integration and spin–spin splitting. Thus, we need not change the mode of spectrometer operation to observe these effects, and ^1H

NMR spectra can be obtained in a few minutes on 1–2 mg of sample. Let's look at each of the four features in more detail.

12.9 Number of NMR absorptions: proton equivalence

The great value of NMR spectroscopy is that it provides us with a carbon–hydrogen map of the sample molecule. In ^{13}C NMR, each unique kind of carbon atom normally gives rise to a distinct peak in the spectrum. Exactly the same is true of ^1H NMR. Each unique kind of proton gives rise to an absorption peak, and we can use this information to determine how many different kinds of protons are present. For example, Figure 12.14 shows the ^1H NMR spectrum of chloropropanone (a potent constituent of police tear gas). There are two kinds of protons present, $Cl—CH_2—$ protons and $—CH_3$ protons, and each gives rise to its own signal.

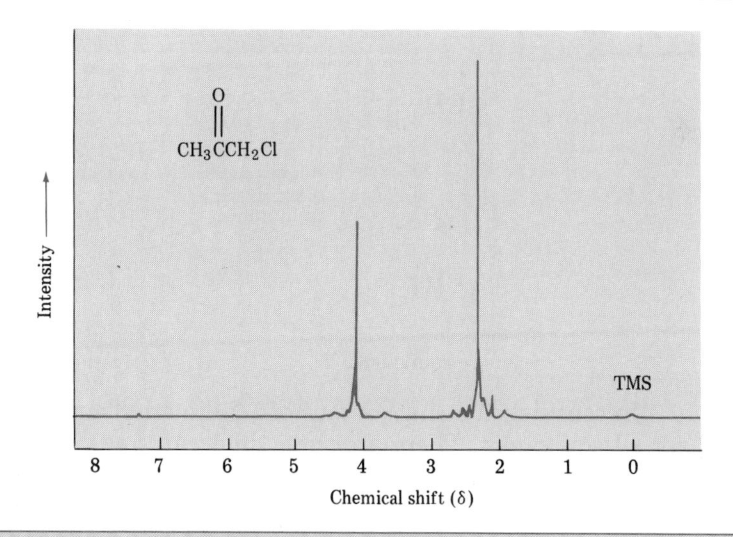

Figure 12.14. The ^1H NMR spectrum of chloropropanone, CH_3COCH_2Cl.

Simple visual inspection of a structure is usually enough to let us decide how many kinds of nonequivalent protons are present. If doubt exists, however, we can determine the equivalence or nonequivalence of protons by asking whether or not we would get the same structure or different structures if we mentally substituted an X group for one of the protons. If the protons are chemically equivalent, we would get the same product no matter which proton we substituted for. If they are not chemically equivalent, we would get different products on substitution.

For example, all four of the benzene-ring protons in 1,4-dimethylbenzene (*para*-xylene) are equivalent. No matter which ring proton we mentally replace by an X group, we get the same structure. The four protons thus give rise to a single, sharp ^1H NMR peak (Figure 12.15). Similarly, both methyl groups and all six methyl protons are equivalent.

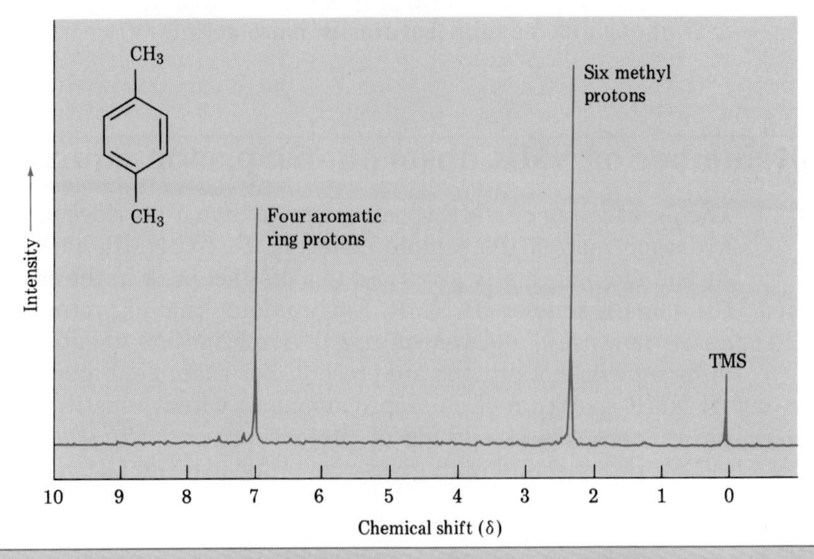

Figure 12.15. The ^{1}H NMR spectrum of 1,4-dimethylbenzene.

Mental substitution

All four ring C—H bonds are equivalent

Only one monosubstituted derivative possible

In contrast, the ring protons of 1,3-dimethylbenzene (*meta*-xylene) are *not* all equivalent. There are three different kinds of ring protons (C2; C4 and C6; and C5) leading to different signals. The six methyl protons, however, are all equivalent.

same

PROBLEM

12.12 How many nonequivalent kinds of protons are present in these compounds?
(a) CH_3CH_2Br
(b) $CH_3OCH_2CH(CH_3)_2$
(c) $CH_3CH_2CH_2NO_2$
(d) 2-Methyl-2-butene
(e) 2-Methyl-1-butene
(f) *cis*-3-Hexene

12.10 Chemical shifts in ^{1}H NMR

We have seen that the phenomenon of chemical shift is caused by small local magnetic fields due to electrons in the molecule. Nuclei that are strongly shielded by electrons require a higher applied field to bring them into resonance and therefore absorb at high field (to the *right* on the NMR chart). Conversely, nuclei that are not strongly shielded absorb in the low-field region of the spectrum (*left* on the NMR chart). Everything we have learned about ^{13}C chemical shifts is applicable to proton shifts. Proton chemical shifts tell us a great deal about the chemical (magnetic) environments within a molecule.

Proton chemical shifts are expressed in δ units (recall 1 δ = 1 ppm of spectrometer frequency), just as for ^{13}C shifts. In contrast to ^{13}C shifts, however, proton chemical shifts fall within the rather narrow range of 0–10 δ. The precise spot within this range is highly characteristic of environment. Table 12.4 (p. 388) and Figure 12.16 show the correlation of ^{1}H chemical shift with environment.

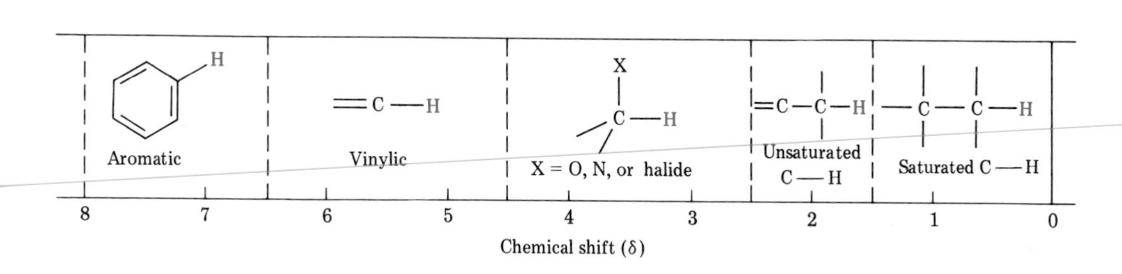

Figure 12.16. Chemical shifts of different kinds of protons.

The great majority of ^{1}H NMR absorptions occur from 0 to 8 δ, and this range can be conveniently divided into five regions, as shown in Table 12.5 (p. 389). It is very useful to memorize the positions of these five regions. With this knowledge, one can often tell at a glance what general kinds of protons a molecule contains.

TABLE 12.4 **Correlation of ^1H chemical shift with environment**

Type of proton	*Formula*	*Chemical shift (δ)*
Reference peak	$(CH_3)_4Si$	0
Saturated primary	$—CH_3$	0.7–1.3
Saturated secondary	$—CH_2—$	1.2–1.4
Saturated tertiary	$\geq C—H$	1.4–1.7
Allylic primary	$>C{=}C(—)—CH_3$	1.6–1.9
Methyl ketones	$—C(=O)—CH_3$	2.1–2.4
Aromatic methyl	$Ar—CH_3$	2.5–2.7
Alkyl chloride	$Cl—C—H$	3.0–4.0
Alkyl bromide	$Br—C—H$	2.5–4.0
Alkyl iodide	$I—C—H$	2.0–4.0
Alcohol, ether	$—O—C—H$	3.3–4.0
Alkynyl	$—C{\equiv}C—H$	2.5–2.7
Vinylic	$>C{=}C(—)—H$	5.0–6.5
Aromatic	$Ar—H$	6.5–8.0
Aldehyde	$—C(=O)—H$	9.7–10.0
Carboxylic acid	$—C(=O)—O—H$	11.0–12.0
Alcohol	$\geq C—O—H$	Extremely variable (2.5–5.0)

TABLE 12.5 Regions of the ^{1}H NMR spectrum

Region (δ)	*Proton type*	*Comments*
0–1.5	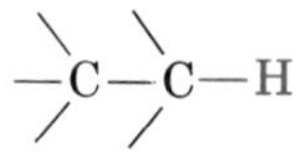 $-\overset{\backslash}{\underset{/}{C}}-\overset{\backslash}{\underset{/}{C}}-H$	Protons on carbon next to saturated centers absorb in this region. Thus the alkane portions of most organic molecules show complex absorption here.
1.5–2.5	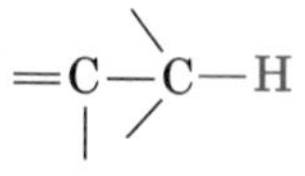 $=\underset{\vert}{C}-\overset{\backslash}{\underset{/}{C}}-H$	Protons on carbon next to unsaturated centers (allylic, benzylic, next to carbonyl) show characteristic absorptions in this region, just downfield from other alkane resonance.
2.5–4.5	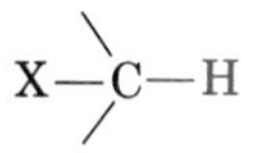 $X-\overset{\backslash}{\underset{/}{C}}-H$	Protons on carbon next to electronegative atoms (halogen, O, N) are deshielded because of the electron-withdrawing ability of these atoms. Thus the protons absorb in this midfield region.
4.5–6.5	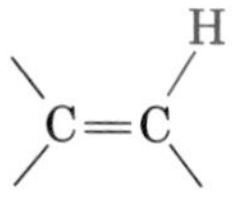 $>C=C<^{H}$	Protons on double-bond carbons (vinylic protons) are strongly deshielded by the neighboring pi bond and therefore absorb in this characteristic downfield region.
6.5–8.0	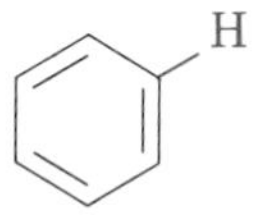 Ar–H (benzene ring–H)	Protons on aromatic rings (aryl protons) are strongly deshielded by the pi orbitals of the ring and absorb in this characteristic low-field range.

PROBLEM

12.13 Each of the following compounds exhibits a single ^{1}H NMR peak. Approximately where would you expect each compound to absorb?

(a) Cyclohexane

(b) Acetone, CH_3COCH_3

(c) Benzene

(d) Glyoxal, $H-\overset{O}{\overset{\|}{C}}-\overset{O}{\overset{\|}{C}}-H$

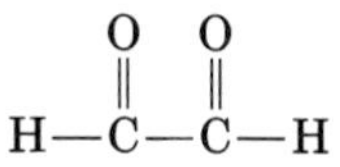

(e) CH_2Cl_2

(f) $(CH_3)_3N$

(g) Dioxane, (six-membered ring with O at positions 1 and 4)

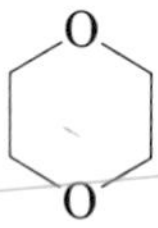

PROBLEM

12.14 Identify the different kinds of protons in the following molecule, and tell where you would expect each to absorb.

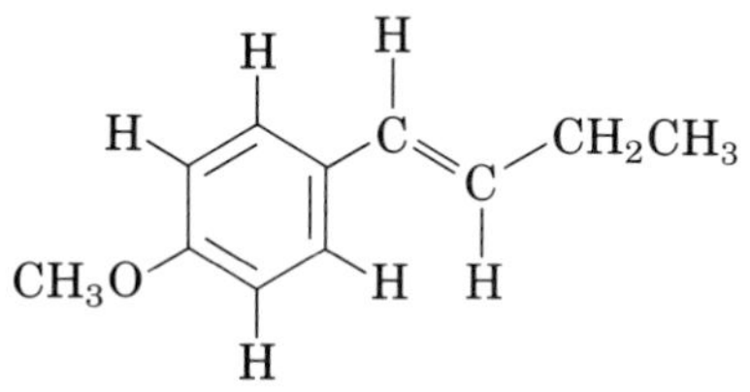

(structure: benzene ring bearing CH_3O and four ring H's, with a $-CH=CH-CH_2CH_3$ side chain)

12.11 Integration of NMR absorptions: proton counting

Electronic integration of the areas of ^{1}H NMR absorption peaks tells us the relative numbers of protons responsible for those peaks. The concept of peak integration was introduced during the discussion of ^{13}C NMR, but we remarked at that time that ^{13}C integration is not normally needed because each peak is usually due to a single carbon. By contrast, the integration of ^{1}H NMR spectra is of great value, since ^{1}H NMR spectra often show complicated patterns, and integration is necessary to help sort things out.

The integrated ^{1}H spectrum is presented in a stair-step manner, as shown in Figure 12.17. The height of each step is proportional to the number of protons represented by that peak. For example, the integrated spectrum of methyl 2,2-dimethylpropanoate given in Figure 12.17 shows that the two absorptions have an area ratio of 1:3. This is just what we would expect, since the three CH_3—O protons are equivalent and the nine $(CH_3)_3C$— protons are equivalent.

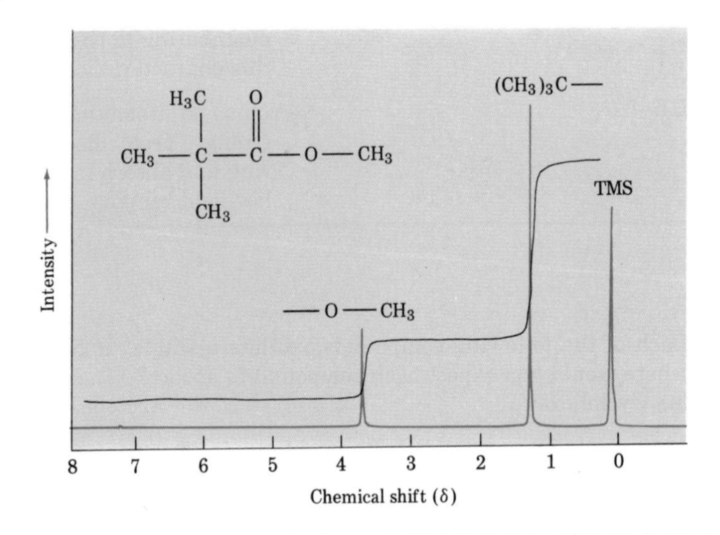

Figure 12.17. Integrated ^{1}H NMR spectrum of methyl 2,2-dimethylpropanoate. The two peaks have a ratio of 3:9 or 1:3.

12.12 Spin–spin splitting in ^{1}H NMR spectra

The spin–spin splitting of single absorption peaks into multiplets is due to the interaction or coupling of neighboring nuclear spins. For example, we have seen that a ^{13}C nucleus can couple with one or more nearby protons, leading to signal splitting when the ^{13}C off-resonance NMR spectrum is recorded. Spin–spin splitting is also observed in ^{1}H NMR.

Just as the spin of a ^{13}C nucleus can couple with the spins of neighboring protons, so the spin of one proton can couple with the spins of neighboring protons. (Because of the low natural abundance of ^{13}C, however, the coupling of a proton's spin to a ^{13}C nucleus is of too low an intensity to be observed.) The resultant splitting patterns can be complex,

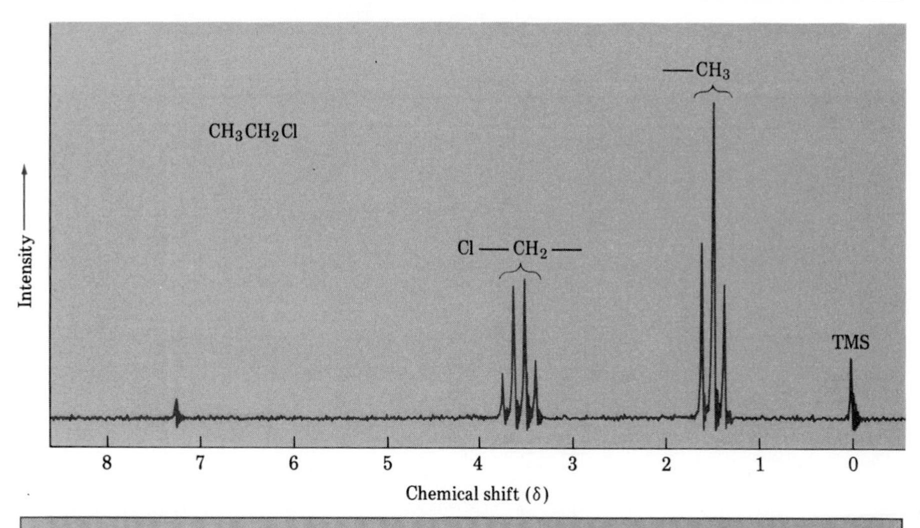

Figure 12.18. The ^{1}H NMR spectrum of chloroethane, CH_3CH_2Cl.

but they can also provide much information. For example, Figure 12.18 shows the ^{1}H NMR spectrum of chloroethane. There are two distinct groupings of peaks corresponding to the two different kinds of protons present, and we can account for the observed splitting pattern just as we accounted for splitting in ^{13}C spectra.

The two downfield $Cl—CH_2—$ protons are chemically equivalent, and their spins do not couple with each other. They do, however, couple with the spins of the three protons on the neighboring carbon. Since each of the three neighboring protons can have its own spin aligned either with or against the applied field, there are four possible coupling combinations.

Similarly, the $CH_3—$ protons of chloroethane are equivalent and their spins do not couple with each other. The methyl proton spins *do* couple with the neighboring $Cl—CH_2—$ proton spins in three possible combinations, leading to the observed triplet in the ^{1}H NMR spectrum (Figure 12.19, p. 392).

The distance between individual peaks in the multiplets is called the **coupling constant** and is denoted J. Coupling constants are measured in hertz and fall in the range 0–18 Hz. The exact value of the coupling constant between two groups of protons depends on several factors (such as geometric constraints on the molecule), but a typical value for an open-chain alkyl system is 6–8 Hz. Note that the same coupling constant is shared by both groups of nuclei and is independent of spectrometer field strength. In chloroethane, for example, the $ClCH_2—$ proton spins are coupled with the CH_3 proton spins and appear as a quartet with $J = 7$ Hz. The CH_3 protons appear as a triplet and must have the same coupling constant, $J = 7$ Hz. Coupling is a reciprocal interaction between the spins of two adjacent groups of protons, and we can sometimes use this fact to tell which multiplets in a complex spectrum are related to each other. It often happens that an NMR spectrum contains many multiplets and it is difficult to tell what is coupled with what. If two multiplets have exactly the same coupling constant, however, they are probably related, and the protons causing those multiplets are adjacent in the molecule.

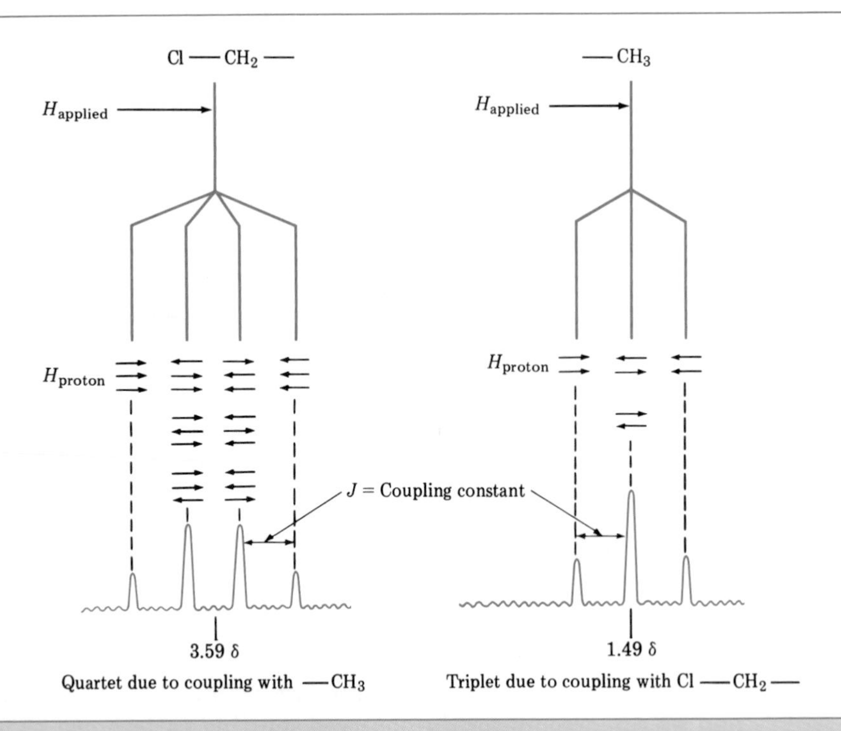

Figure 12.19. Spin–spin splitting in chloroethane.

The chloroethane spectrum in Figure 12.19 illustrates three important rules about spin–spin splitting in ^{1}H NMR:

1. Chemically equivalent protons do not exhibit spin–spin splitting. The equivalent protons may be on the same carbon or on different carbons, but their spins still do not couple; instead, the signal appears as a singlet.

Three C—H protons are chemically equivalent; no coupling occurs.

Four C—H protons are chemically equivalent; no coupling occurs.

2. A proton that has n equivalent neighboring protons gives a signal that is split into a multiplet of $n + 1$ peaks with coupling constant J. Protons that are farther than two carbon atoms apart do not usually couple. They sometimes show small coupling, however, when they are separated by pi bonds.

Coupling observed

Coupling not usually observed

The most commonly observed coupling patterns are shown in Table 12.6, as are the relative intensities of the multiplet signals.

TABLE 12.6 Some common spin multiplicities

Number of equivalent adjacent protons	*Type of multiplet observed*	*Ratio of intensities*
0	Singlet	1
1	Doublet	1:1
2	Triplet	1:2:1
3	Quartet	1:3:3:1
4	Quintet	1:4:6:4:1
5	Sextet	1:5:10:10:5:1
6	Septet	1:6:15:20:15:6:1

3. Two groups of protons coupled with each other must have the same coupling constant.

The spectra of 2-bromopropane and *para*-methoxypropiophenone in Figure 12.20 (p. 394) further illustrate these three rules.

The 2-bromopropane spectrum shows two proton signals split into a doublet at 1.71 δ and a septet at 4.32 δ. The downfield septet is due to splitting of the CHBr proton signal by six equivalent neighboring protons on the two methyl groups ($n = 6$ leads to $6 + 1 = 7$ peaks). The upfield doublet is due to signal splitting of the six equivalent methyl protons by the single CHBr proton ($n = 1$ leads to 2 peaks). Both multiplets have the same coupling constant, $J = 7$ Hz, and integration confirms the expected 6:1 ratio.

The *para*-methoxypropiophenone spectrum is more complex, but nevertheless can be interpreted in a straightforward way. The downfield absorptions at 6.98 and 8.0 δ are due to the aromatic ring protons. There are two kinds of protons, each of which gives a signal that is split into a doublet by its neighbor. Thus we see two doublets. The O—CH_3 signal is unsplit and appears as a sharp singlet at 3.90 δ. The

$$\mathrm{O{=}\overset{|}{C}{-}CH_2{-}}$$

proton signals at 2.95 δ are in the region expected for protons on carbon next to an unsaturated center, and appear as a quartet due to coupling with the neighboring methyl group. The methyl group appears as a triplet in the usual upfield region.

PROBLEM

12.15 Predict the splitting pattern you would expect for the protons indicated in these molecules.

(a) $CHBr_2CH_3$

(b) $CH_3OCH_2CH_2Br$

(c) $ClCH_2CH_2CH_2Cl$

(d) $CH_3CH_2O\overset{\overset{\large O}{\|}}{C}CH(CH_3)_2$

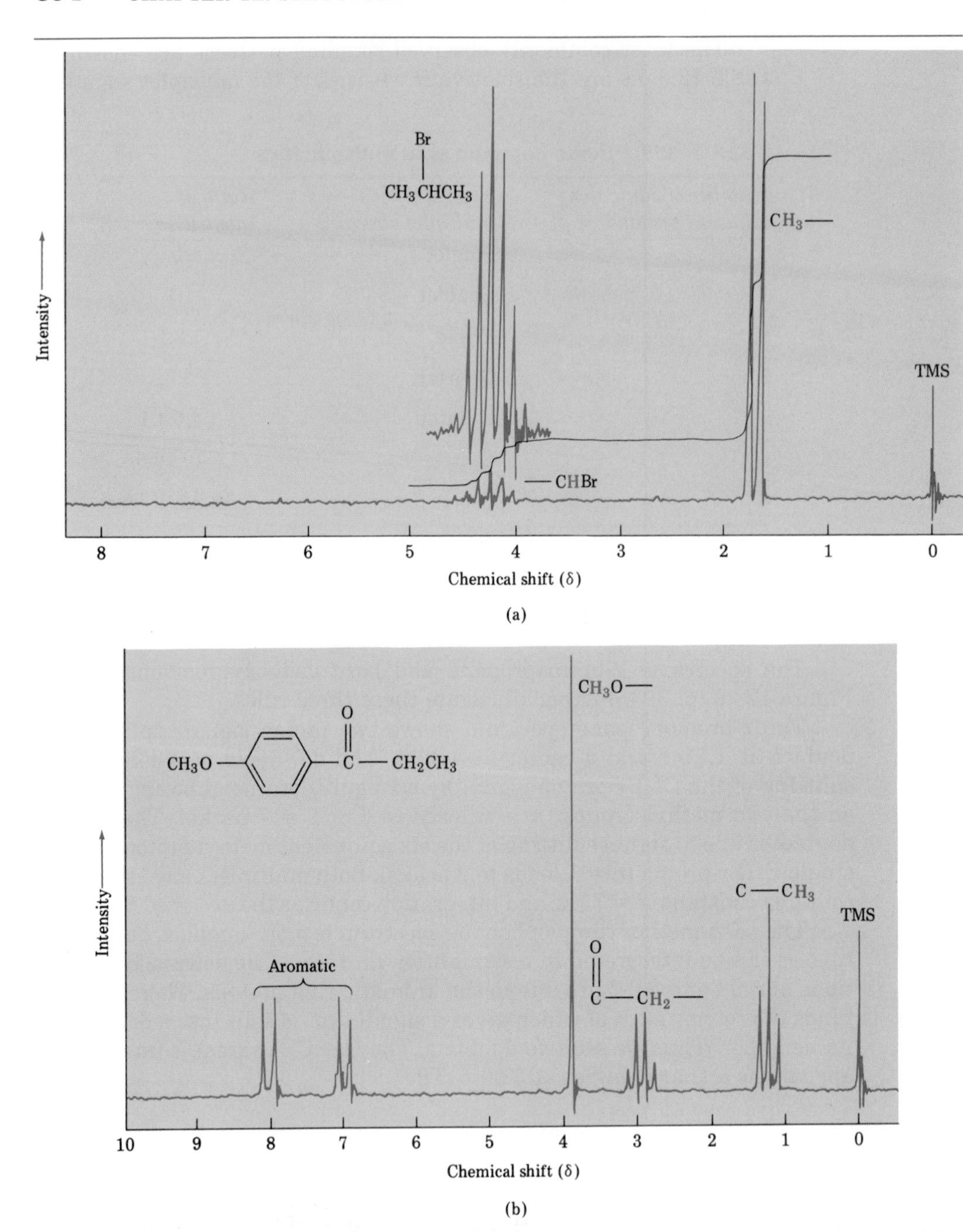

Figure 12.20. The ^1H NMR spectra of (a) 2-bromopropane and (b) *p*-methoxypropiophenone.

PROBLEM

12.16 Draw structures for compounds that meet these descriptions:
(a) C_2H_6O; one singlet
(b) C_3H_7Cl; one doublet and one septet
(c) $C_4H_8Cl_2O$; two triplets
(d) $C_4H_8O_2$; one singlet, one triplet, and one quartet

PROBLEM

12.17 The integrated ^1H NMR spectrum of a compound of formula $C_4H_{10}O$ is given here. Propose a structure that is consistent with the data.

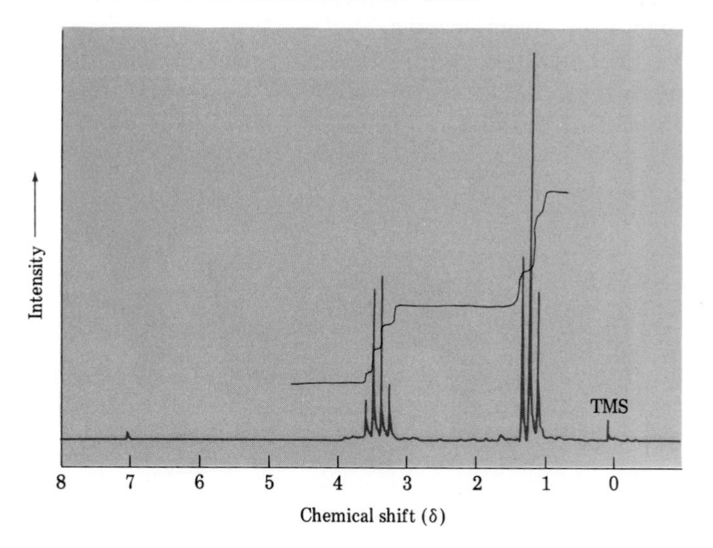

12.13 More complex spin–spin splitting patterns

In all the NMR spectra seen so far, the chemical shifts of different protons have been quite distinct, and the spin–spin splitting patterns have been relatively simple. Chemists are not always so lucky, and it often happens that different kinds of protons may have overlapping signals. Also, signals are often split because of coupling with more than one kind of neighboring proton.

The spectrum of toluene (methylbenzene) in Figure 12.21, for example,

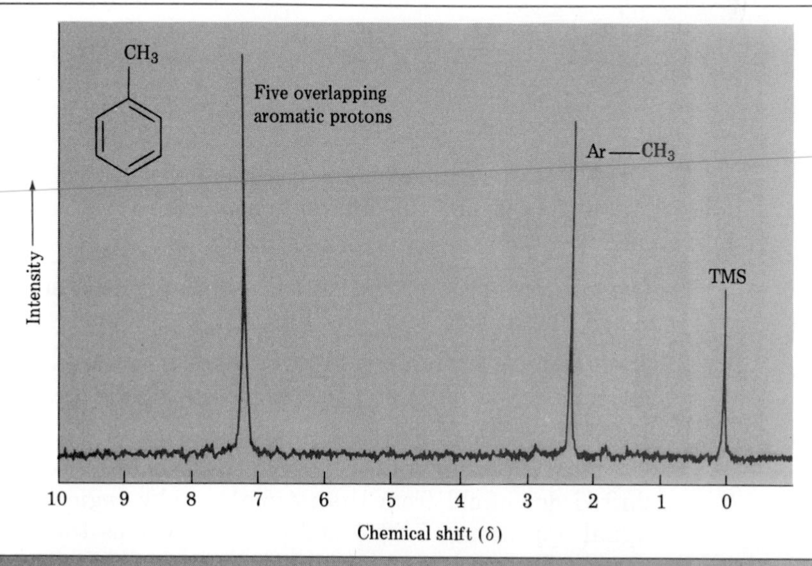

Figure 12.21. The ^1H NMR spectrum of toluene.

shows that all five aromatic ring protons give a single overlapping absorption, even though they are not all equivalent. This kind of accidental overlap of signals is something we must be aware of to avoid drawing false conclusions from spectra.

Yet another complication can arise when a signal is split by two or more *nonequivalent* kinds of protons, as is the case in the spectrum of *trans*-cinnamaldehyde, isolated from oil of cinnamon (Figure 12.22).

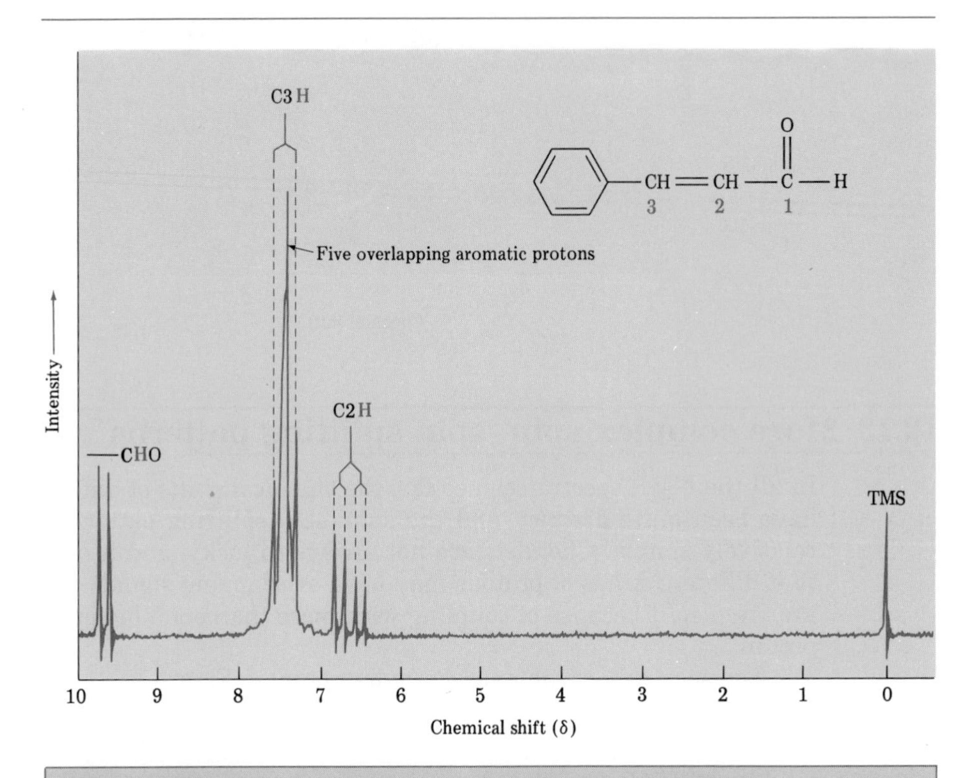

Figure 12.22. The ^{1}H NMR spectrum of *trans*-cinnamaldehyde.

The ^{1}H NMR spectrum of *trans*-cinnamaldehyde is complex, but we can understand it if we isolate the different parts and look at each kind of proton individually:

1. The five aromatic proton signals overlap into a single broad resonance line at 7.45 δ.
2. The aldehyde proton signal at C1 appears in the normal downfield position at $\delta = 9.67$ and is split into a doublet ($J = 7$ Hz) by the adjacent proton at C2.
3. The vinylic proton at C3 is next to the aromatic ring and is therefore shifted downfield from the normal vinylic region. This C3 proton signal appears at 7.42 δ and nearly overlaps the aromatic proton signals. Since it has one neighbor proton at C2, its signal is split into a doublet with $J = 15$ Hz.

4. The C2 vinylic proton signal appears at 6.66 δ and shows an interesting absorption pattern. It is coupled with two nonequivalent protons at C1 and C3, and, as we have seen, the two coupling constants are different—J_{1-2} = 7 Hz and J_{2-3} = 15 Hz.

The best way to see the effect of multiple coupling is to draw a **tree diagram** like that shown in Figure 12.23. Tree diagrams show the individual effects of each coupling constant on the overall pattern.

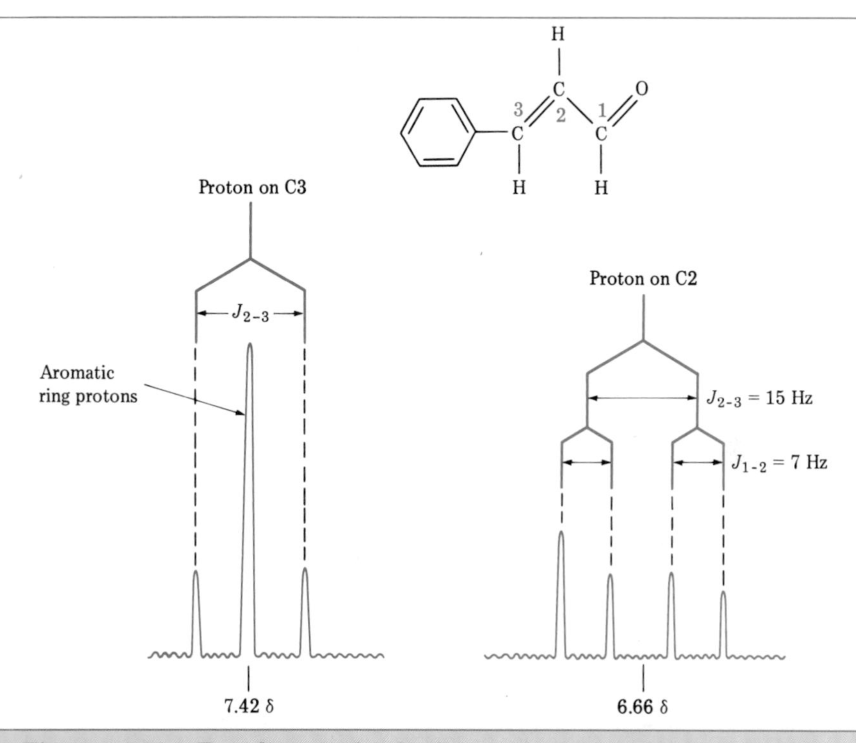

Figure 12.23. Tree diagram for the C2 proton of *trans*-cinnamaldehyde.

In the case of *trans*-cinnamaldehyde, the signal due to the C2 proton is split into a doublet by 15 Hz coupling with the C3 proton. The 7 Hz coupling with the aldehyde proton further splits each leg of the doublet into new doublets. Thus we observe a four-line spectrum for the C2 proton of *trans*-cinnamaldehyde. Multiple coupling can look quite complex, but is usually amenable to simplification using tree diagrams, which consider each coupling separately.

PROBLEM

12.18 3-Bromo-1-phenyl-1-propene shows a complex NMR spectrum in which the vinylic proton at C2 is coupled with both the C1 vinylic proton (J = 16 Hz) and the C3 methylene protons (J = 8 Hz). Draw a tree diagram for the C2 proton signal and account for the fact that a five-line multiplet is observed.

H
1 3
2 CH_2Br
H

3-Bromo-1-phenyl-1-propene

12.14 The NMR spectra of larger molecules

Digitoxigenin, the steroid responsible for the heart stimulant properties of digitalis preparations, has the formula $C_{22}H_{32}O_4$. Can a molecule this complex possibly give an interpretable NMR spectrum?

The answer is that we often can't *fully* interpret the spectrum of such a molecule, but NMR is useful nonetheless. The ^{1}H NMR spectrum of digitoxigenin taken on a 300 MHz instrument is shown in Figure 12.24. Even though we can't easily interpret the saturated alkane region between 1.5 and 2.0 δ, we can still recognize many structural features that provide a great deal of information about the compound. For example, we can readily spot two methyl-group singlets on saturated carbon at 0.84 and 0.92 δ. We can see one vinylic proton singlet at 5.85 δ, a pattern at 4.85 δ due to two protons on carbon next to oxygen, and a further broad absorption at 4.1 δ, due to one proton on carbon next to oxygen. Even in cases like this, NMR is an invaluable structural tool.

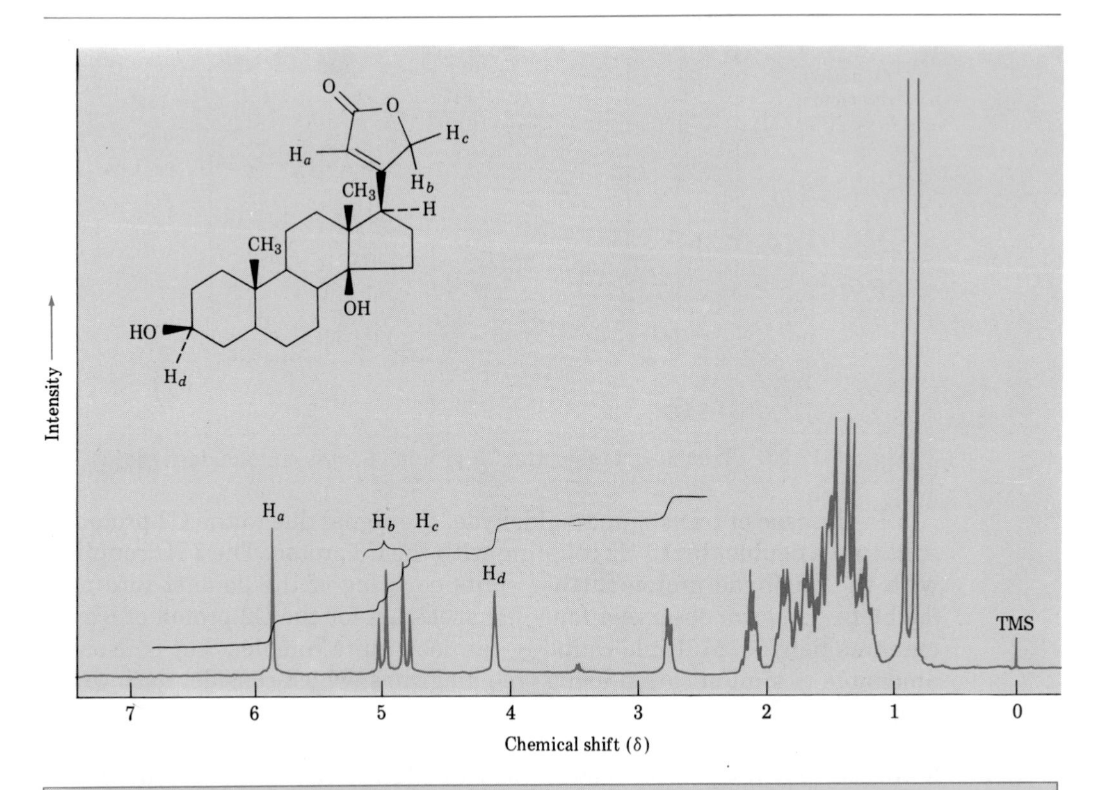

Figure 12.24. The ^{1}H NMR spectrum of digitoxigenin at 300 MHz.

12.15 Use of ^{1}H NMR spectra

We can use ^{1}H NMR to help identify the product of nearly every reaction we run in the laboratory. For example, in Chapter 6 the statement was made that addition of HCl to alkenes occurs with Markovnikov regiochemistry. That is, the more substituted alkyl chloride is formed. With the help of NMR, we are now able to prove this statement.

Does addition of HCl to 1-methylcyclohexene yield 1-chloro-1-methylcyclohexane or 1-chloro-2-methylcyclohexane?

CH_3 | $\xrightarrow[\text{Ether}]{\text{HCl}}$ | H_3C Cl | or | H CH_3 H Cl ?

1-Methylcyclohexene | 1-Chloro-1-methylcyclohexane (Markovnikov) | 1-Chloro-2-methylcyclohexane (non-Markovnikov)

The ^1H NMR spectrum of the reaction product is shown in Figure 12.25. The spectrum shows a large singlet absorption in the saturated methyl region at 1.5 δ, indicating that the product has a methyl group bonded to a quaternary carbon, $R_3C—CH_3$, rather than to a tertiary carbon, $R_2CH—CH_3$. Furthermore, the spectrum shows *no* absorptions in the range 4–5 δ, where we would expect the signal of a R_2CHCl proton to occur. Thus it is clear that the reaction gives 1-chloro-1-methylcyclohexane as product.

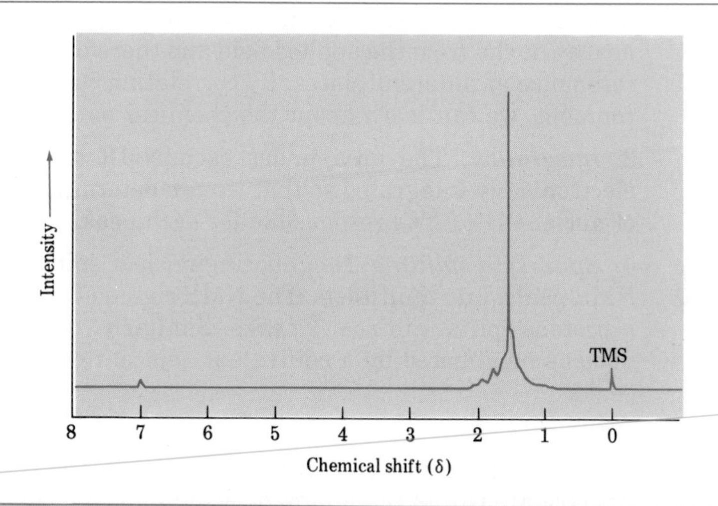

Figure 12.25. The ^1H NMR spectrum of the reaction product from 1-methylcyclohexene and HCl.

12.16 Summary

When certain nuclei such as ^1H and ^{13}C are placed in strong magnetic fields, their spins orient either with or against the applied field. On irradiation with radiofrequency (rf) waves, energy is absorbed and the nuclear spins "flip" from the lower energy state to the higher energy state. This absorption of rf energy is detected, amplified, and displayed as a nuclear magnetic resonance (NMR) spectrum.

The NMR spectrum is obtained by irradiating a sample with a constant-frequency rf energy and slowly changing the value of the applied magnetic field. Different kinds of ^{1}H and ^{13}C nuclei come into resonance at slightly different applied fields, and we therefore see a different absorption line for each different kind of ^{1}H and ^{13}C. The NMR chart is calibrated in delta (δ) units, where 1 δ = 1 part per million (ppm) of spectrometer frequency. For example, a ^{1}H spectrometer operating at 60 MHz frequency would have 1 δ = 60 Hz. Tetramethylsilane (TMS) shows both ^{1}H and ^{13}C absorptions at unusually high values of the applied magnetic field and is therefore used as a reference sample to which all other peaks are compared. The TMS absorption occurs at the right hand (upfield) edge of the chart and is arbitrarily assigned a value of 0 δ. All other NMR absorptions normally occur downfield (left) of the TMS reference line.

Both ^{1}H and ^{13}C NMR spectra display four general features:

1. *Number of resonance lines.* Each nonequivalent kind of ^{1}H or ^{13}C nucleus in a molecule can give rise to a different resonance line.

2. *Chemical shift.* The exact position of each peak is the chemical shift of the nucleus. Chemical shifts are due to the effects of electrons in the molecule. The electrons set up tiny local magnetic fields that shield the nearby nuclei from the applied field and therefore cause them to come to resonance at different places. By correlating chemical shifts with environment, we can learn about the chemical nature of each nucleus.

3. *Integration.* The area under each NMR absorption peak can be electronically integrated so that we can determine the relative number of nuclei (^{1}H or ^{13}C) responsible for each peak.

4. *Spin–spin splitting.* Neighboring nuclear spins can couple, splitting NMR peaks into multiplets. The NMR signal of a ^{13}C nucleus bonded to n protons splits into $n + 1$ peaks. Similarly, the NMR signal of a ^{1}H nucleus neighbored by n equivalent adjacent protons splits into $n + 1$ peaks.

Because of electronic constraints inherent in ^{13}C NMR spectrometers, most spectra are run in the proton noise-decoupled mode. Operating in this manner provides maximum sensitivity and gives a spectrum in which each nonequivalent carbon shows a single unsplit resonance line. Carbons that are sp^3 hybridized absorb in the upfield region from 0 to 100 δ; sp^2-hybridized carbons absorb from 100 to 200 δ. Operating in the gated-decoupled mode causes a loss of sensitivity, but provides a spectrum that can be electronically integrated to measure the number of carbon nuclei responsible for each peak. Operating in the off-resonance mode also causes a loss of sensitivity, but provides a spectrum in which spin–spin splitting is observed. Each carbon resonance is split into a multiplet depending on the number of protons to which it is bonded: Quaternary carbon resonances remain as singlets, tertiary carbons (R_3CH) appear as doublets, secondary carbons (R_2CH_2) appear as triplets, and primary carbons (RCH_3) appear as quartets.

Proton NMR spectra are even more useful than ^{13}C spectra. The sensitivity of ^1H instruments is high, and normal spectrometer operating conditions provide spectra that show spin–spin splitting and can be integrated. Proton resonances usually fall into the range 0–10 δ downfield from the TMS reference point, and the exact chemical shift of an absorption indicates the chemical environment of the nucleus responsible for that signal.

A specific resonance peak is often split into a multiplet due to spin–spin splitting with the spins of protons on adjacent carbons. The spins of equivalent protons do not couple with each other, but a proton with n equivalent neighboring protons gives a signal that is split into $n + 1$ peaks with coupling constant J.

ADDITIONAL PROBLEMS

12.19 The following ^1H NMR absorptions were determined on a spectrometer operating at 60 MHz and are given in hertz downfield from the TMS standard. Convert the absorptions to δ units.

(a) 131 Hz
(b) 287 Hz
(c) 451 Hz
(d) 543 Hz

12.20 The following ^1H NMR absorptions are given in δ units and were obtained on a spectrometer operating at 80 MHz. Convert the chemical shifts from δ units into hertz downfield from TMS.

(a) 2.1 δ
(b) 3.45 δ
(c) 6.30 δ
(d) 7.70 δ

12.21 When measured on a spectrometer operating at 60 MHz, chloroform ($CHCl_3$) shows a single sharp absorption at 7.3 δ.

(a) How many parts per million downfield from TMS does chloroform absorb?
(b) How many hertz downfield from TMS would chloroform absorb if the measurement were carried out on a spectrometer operating at 360 MHz?
(c) What would be the position of the chloroform absorption in δ units when measured on a 360 MHz spectrometer?

12.22 How many absorptions would you expect to observe in the proton noise-decoupled ^{13}C NMR spectra of compounds (a)–(d)?

(a) 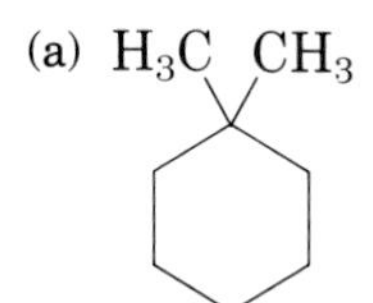

(b) $CH_3CH_2OCH_3$

(c) 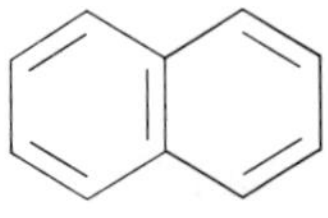 Naphthalene

(d) 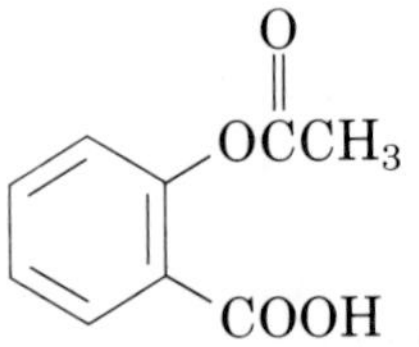 Aspirin

$OCCH_3$ (C=O)

$COOH$

12.23 Indicate the spin multiplicities you would expect to see for each carbon atom in the off-resonance ^{13}C NMR spectra of the molecules shown in Problem 12.22.

12.24 Why do you suppose accidental overlap of signals is much more common in ^1H NMR than in proton noise-decoupled ^{13}C NMR?

12.25 Tell what is meant by each of these terms:

(a) Chemical shift
(b) Spin–spin splitting
(c) Applied magnetic field
(d) Spectrometer operating frequency
(e) Coupling constant

12.26 How many types of nonequivalent protons are there in the following molecules?

(a) 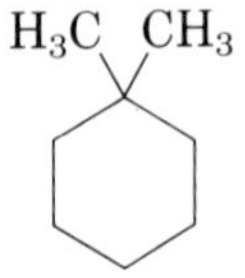 H_3C CH_3 (1,1-dimethylcyclohexane)

(b) $CH_3CH_2CH_2OCH_3$

(c) 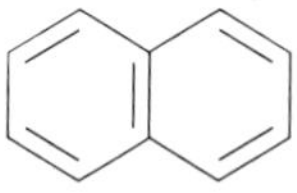 Naphthalene

(d) 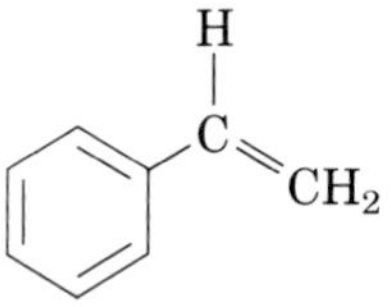 Styrene (C_6H_5–CH=CH_2)

(e) 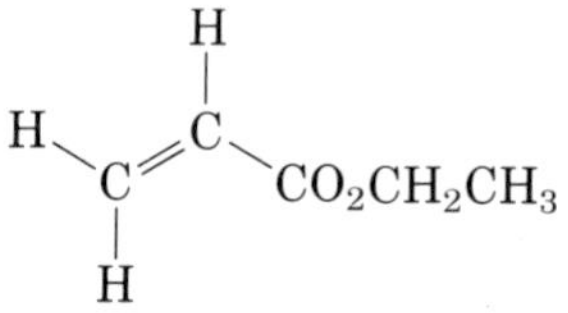 Ethyl acrylate (H(H)C=C(H)–$CO_2CH_2CH_3$)

12.27 How would you use NMR to help identify the product of the hydroboration/oxidation of 1-heptene?

$$\text{1-Heptene} \xrightarrow[\text{2. } H_2O_2,\ ^{-}OH]{\text{1. } BH_3} \text{2-Heptanol} \quad \text{or} \quad \text{1-Heptanol?}$$

12.28 The acid-catalyzed dehydration of 1-methylcyclohexanol yields a mixture of two alkenes as product. After you separated them by chromatography, how would you use ^1H NMR to help you decide which was which?

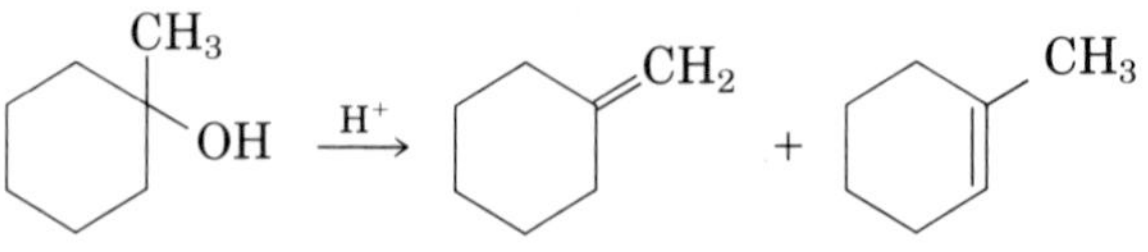

$$\text{1-methylcyclohexanol } (CH_3, OH) \xrightarrow{H^+} \text{methylenecyclohexane } (=CH_2) + \text{1-methylcyclohexene } (CH_3)$$

12.29 How would you use ^1H NMR to distinguish between these pairs of isomers?

(a) $CH_3CH{=}CHCH_2CH_3$ and $H_2C{-}CHCH_2CH_3$ (cyclopropane ring: the two ring carbons joined through a CH_2 above)

(b) $CH_3CH_2OCH_2CH_3$ and $CH_3OCH_2CH_2CH_3$

(c) $CH_3\overset{O}{\overset{\|}{C}}OCH_2CH_3$ and $CH_3CH_2\overset{O}{\overset{\|}{C}}OCH_3$

(d) $H_2C{=}C(CH_3)\overset{O}{\overset{\|}{C}}CH_3$ and $CH_3CH{=}CH\overset{O}{\overset{\|}{C}}CH_3$

12.30 Assume that you have a compound with formula C_3H_6O.
(a) How many double bonds and/or rings does your material contain?
(b) Propose as many structures as you can that fit the molecular formula.
(c) If your compound shows an infrared absorption peak at 1710 cm^{-1}, what inferences can you draw?
(d) If your compound shows a single ^1H NMR absorption peak at 2.1 δ, what is its structure?

12.31 How would you use NMR to help you distinguish between the following isomeric compounds of formula C_4H_8?

$$\begin{array}{l} CH_2{-}CH_2 \\ | \quad\quad\ \ | \\ CH_2{-}CH_2 \end{array} \qquad H_2C{=}CHCH_2CH_3 \qquad CH_3CH{=}CHCH_3 \qquad (CH_3)_2C{=}CH_2$$

12.32 The compound whose ^1H NMR spectrum is shown here has the molecular formula $C_3H_6Br_2$. Propose a plausible structure.

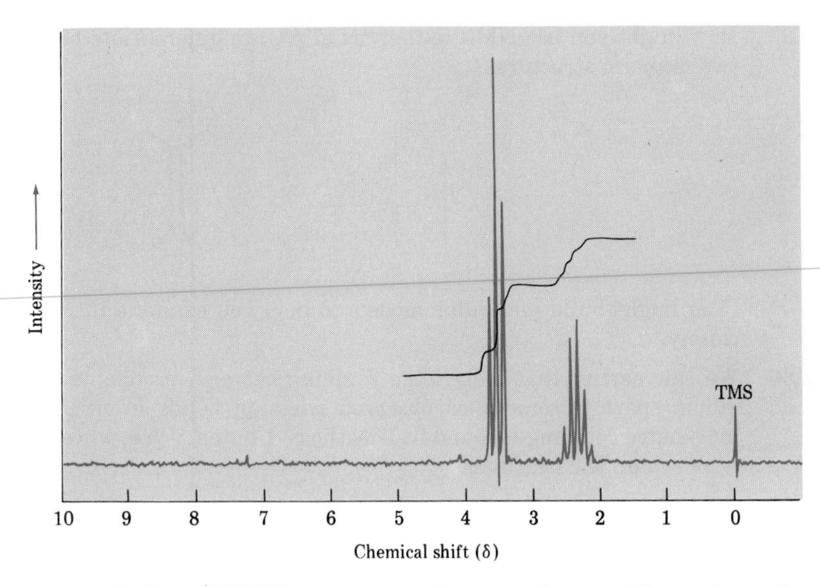

12.33 The compound whose ^1H NMR spectrum is shown at the top of the next page has the molecular formula $C_4H_7O_2Cl$ and shows an infrared absorption peak at 1740 cm^{-1}. Propose a plausible structure.

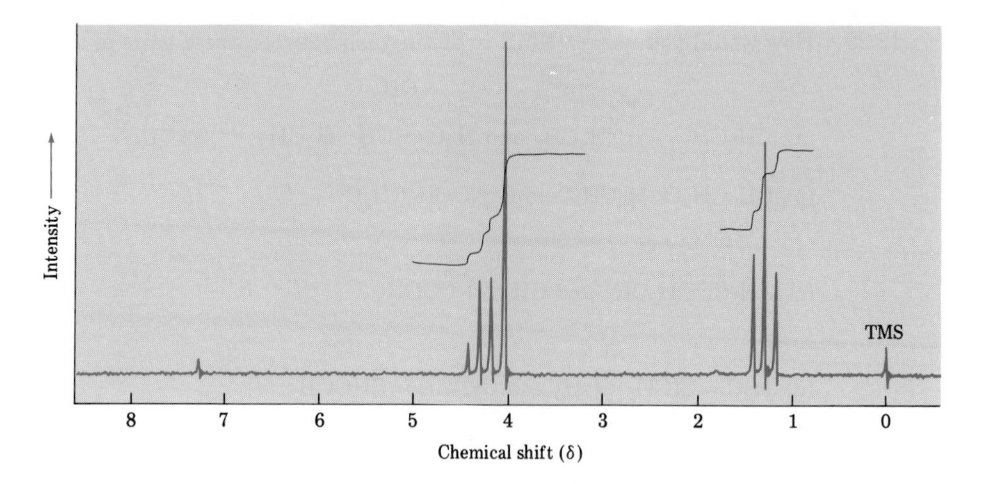

12.34 Propose structures for compounds that fit the following ^{1}H NMR data:

(a) $C_5H_{10}O$
- 6 H doublet at 0.95 δ, J = 7 Hz
- 3 H singlet at 2.10 δ
- 1 H multiplet at 2.43 δ

(b) C_3H_5Br
- 3 H singlet at 2.32 δ
- 1 H broad singlet at 5.35 δ
- 1 H broad singlet at 5.54 δ

(c) $C_4H_6Cl_2$
- 3 H singlet at 2.18 δ
- 2 H doublet at 4.16 δ, J = 7 Hz
- 1 H triplet at 5.71 δ, J = 7 Hz

(d) $C_{10}H_{14}$
- 9 H singlet at 1.30 δ
- 5 H singlet at 7.30 δ

(e) C_4H_7BrO
- 3 H singlet at 2.11 δ
- 2 H triplet at 3.52 δ, J = 6 Hz
- 2 H triplet at 4.40 δ, J = 6 Hz

(f) $C_9H_{11}Br$
- 2 H quintet at 2.15 δ, J = 7 Hz
- 2 H triplet at 2.75 δ, J = 7 Hz
- 2 H triplet at 3.38 δ, J = 7 Hz
- 5 H singlet at 7.22 δ

12.35 How might you use NMR (either ^{1}H or ^{13}C) to differentiate between the following two isomeric structures?

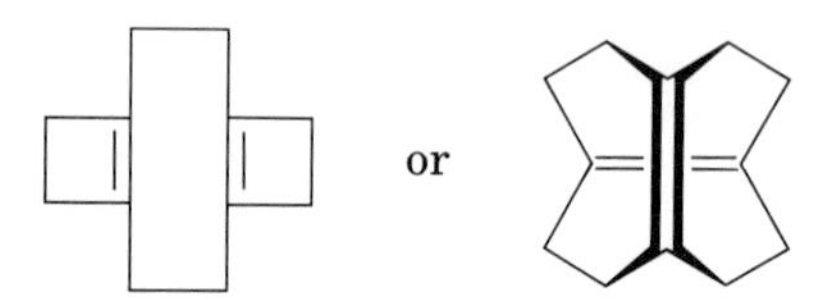

(You might build molecular models to help you examine the two structures more closely.)

12.36 We saw earlier that long-range coupling between protons more than two carbon atoms apart is sometimes observed when pi bonds intervene. One example of long-range coupling is found in 1-methoxy-1-buten-3-yne, whose ^{1}H NMR spectrum is shown on the next page.

CH_3O
$C—H_c$
$H_a—C\equiv C—C$
H_b

Not only does the acetylenic proton, H_a, couple with the vinylic proton H_b, but it also couples with the vinylic proton H_c (*four* carbon atoms away). The following coupling constants are observed:

$$J_{a-b} = 6 \text{ Hz}$$

$$J_{a-c} = 2 \text{ Hz}$$

$$J_{b-c} = 15 \text{ Hz}$$

Construct tree diagrams that account for the observed splitting patterns of H_a, H_b, and H_c.

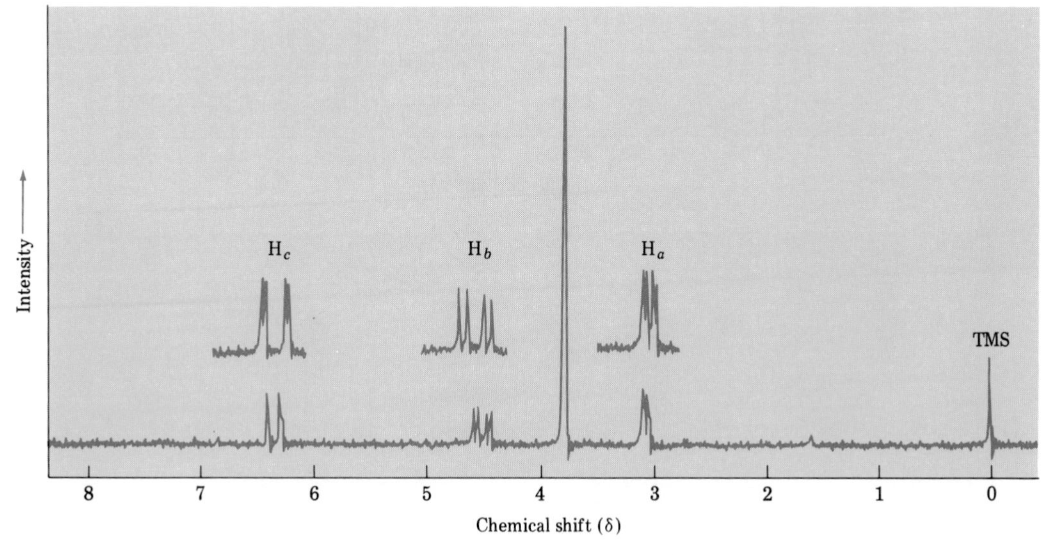

12.37 Assign as many of the resonances as you can to specific carbon atoms in the ^{13}C NMR spectrum of ethyl benzoate shown here.

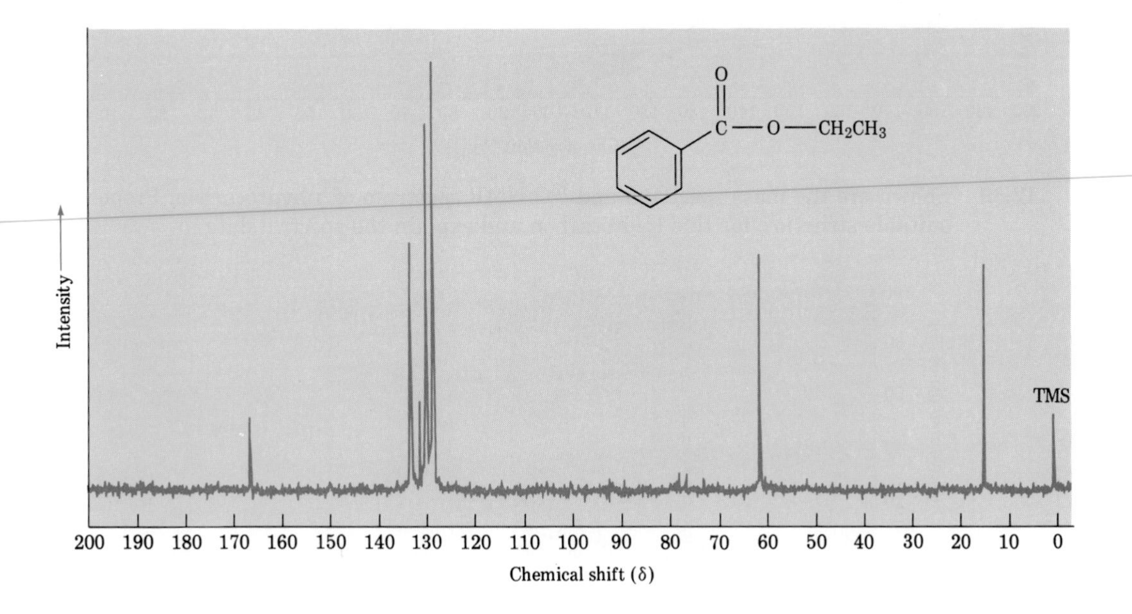

12.38 The ^{1}H and ^{13}C NMR spectra of compound A, C_8H_9Br, are shown. Propose a possible structure for A and assign peaks in the spectra to your structure.

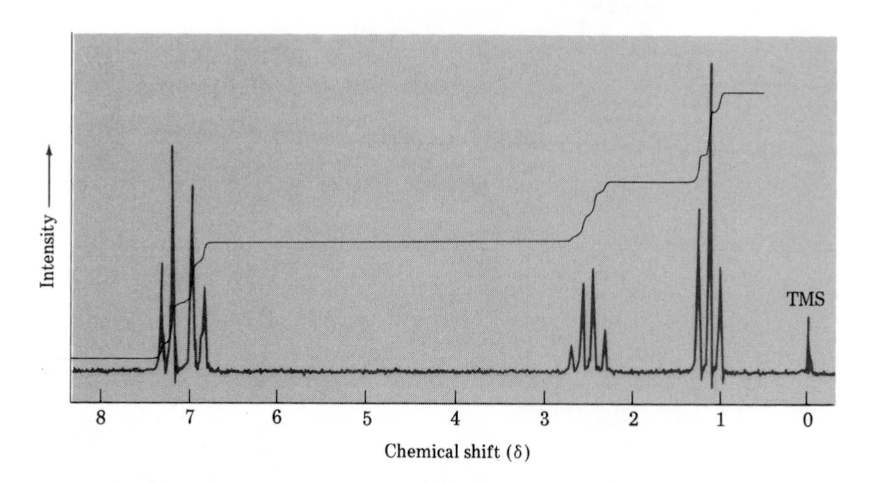

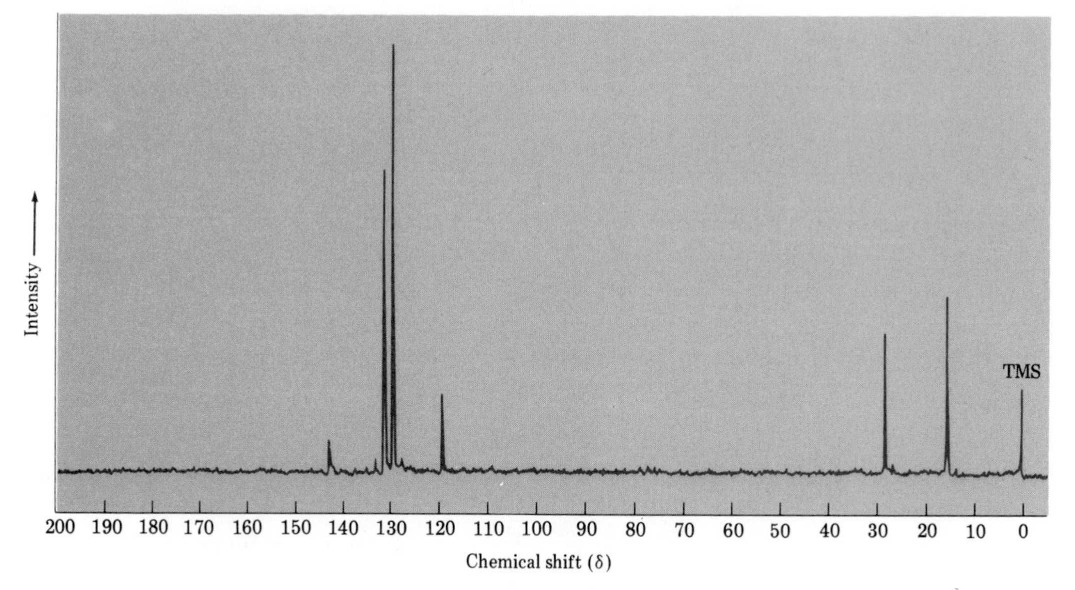

12.39 Shown are the mass spectrum and ^{13}C NMR spectrum of a hydrocarbon. Propose a suitable structure for this hydrocarbon and explain the spectral data.

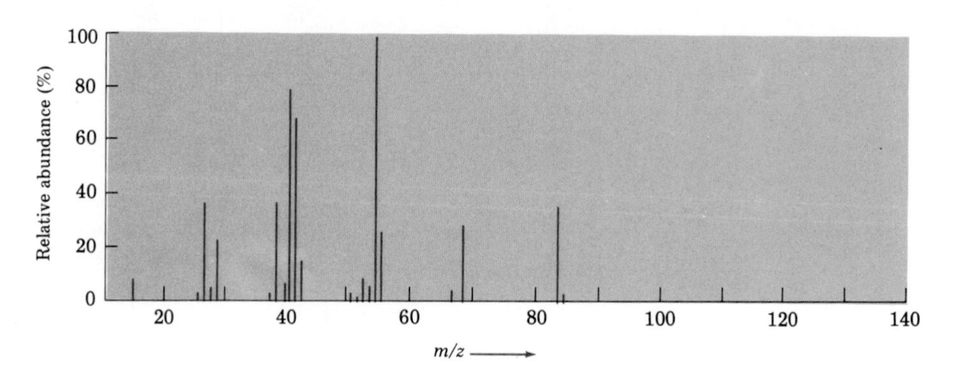

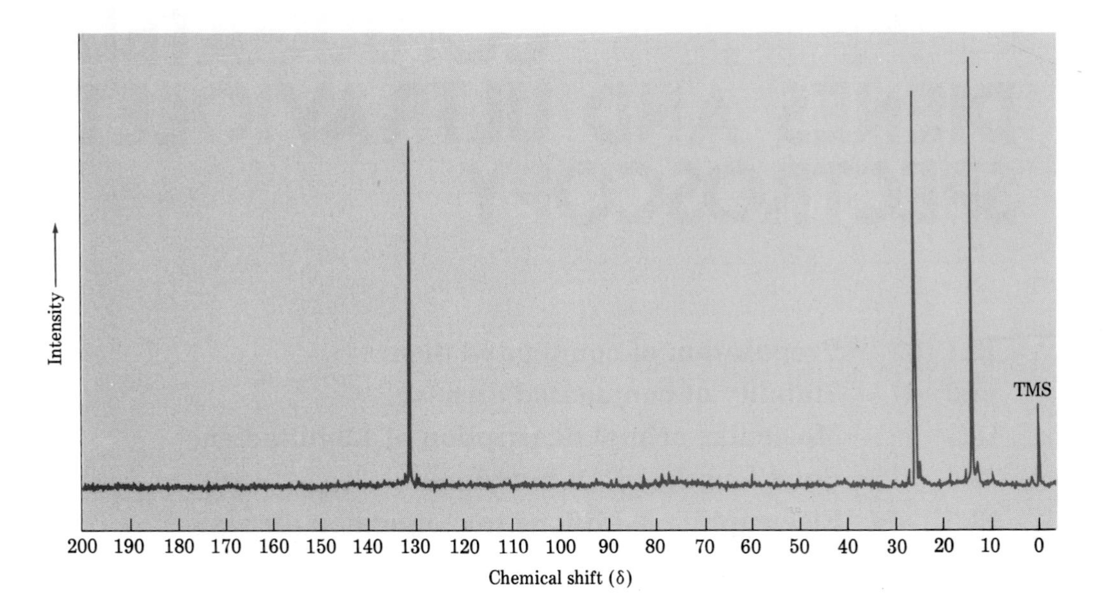
Intensity
TMS
200 190 180 170 160 150 140 130 120 110 100 90 80 70 60 50 40 30 20 10 0
Chemical shift (δ)

CHAPTER 13 CONJUGATED DIENES AND ULTRAVIOLET SPECTROSCOPY

Double bonds that alternate with single bonds are said to be **conjugated.** Thus, 1,3-butadiene is a **conjugated diene,** whereas 1,4-pentadiene is a nonconjugated diene with *isolated* double bonds.

$H_2C{=}CH{-}CH{=}CH_2$

1,3-Butadiene
(conjugated; alternating
double and single bonds)

$H_2C{=}CH{-}CH_2{-}CH{=}CH_2$

1,4-Pentadiene
(nonconjugated; nonalternating
double and single bonds)

There are other types of conjugated systems (Figure 13.1) and many play an important role in nature. For example, the pigments responsible for the brilliant reds and yellows of fruits and flowers are conjugated **polyenes;** lycopene is one such molecule. Conjugated **enones** (from alk**ene** + ket**one**) are common structural features of important molecules such as progesterone, the "pregnancy hormone." Conjugated cyclic molecules such as benzene are a major field of study in themselves and will be considered in detail in Chapter 14.

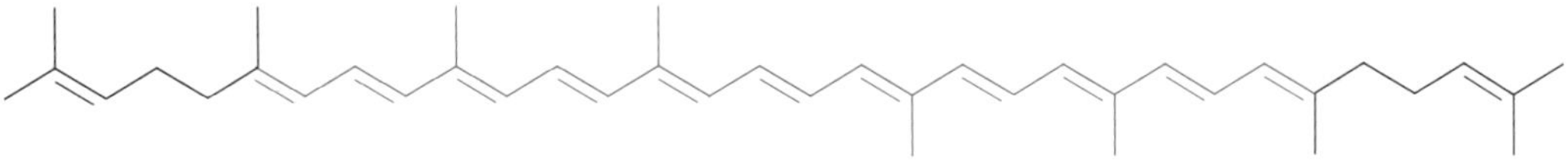

(a)

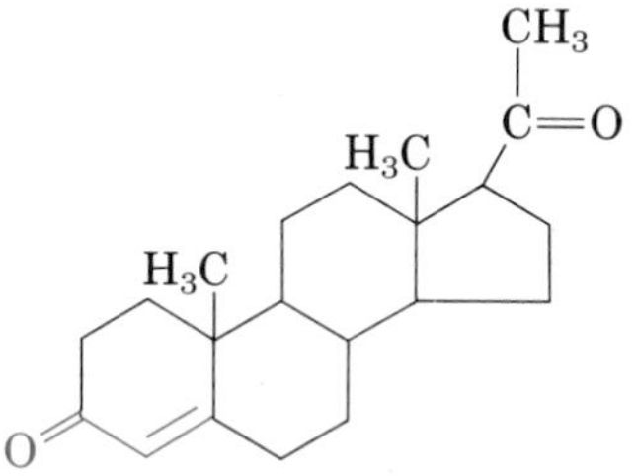

CH_3

C=O

H_3C

H_3C

O

(b)

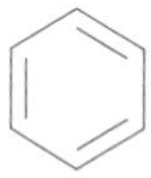

(c)

Figure 13.1. (a) Lycopene, a conjugated polyene from tomatoes; (b) progesterone, a conjugated enone; and (c) benzene, a conjugated cyclic molecule.

PROBLEM

13.1 Which of the following molecules (a)–(f) contain conjugated systems? Circle the conjugated portion.

(a)

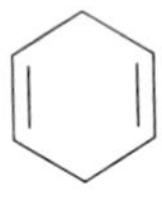

(b)

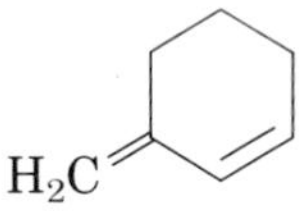

H_2C

(c) $H_2C{=}CH{-}CN$

(d) 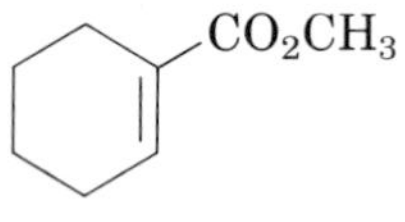 CO_2CH_3

(e) 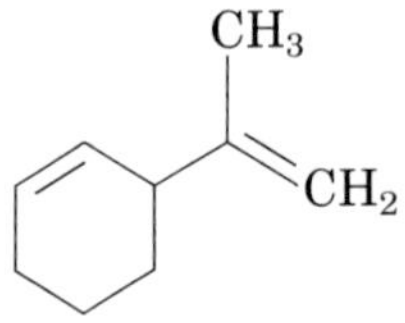 CH_3 CH_2

(f) 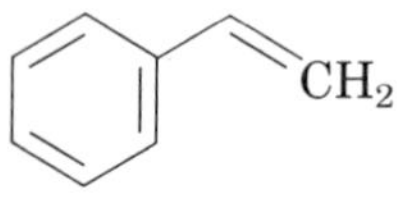 CH_2

Styrene

13.1 Preparation of conjugated dienes

Conjugated dienes are generally prepared by the methods already discussed for alkene synthesis. For example, the dehydration of allylic alcohols and the base-induced elimination of HX from allylic halides produce conjugated dienes.

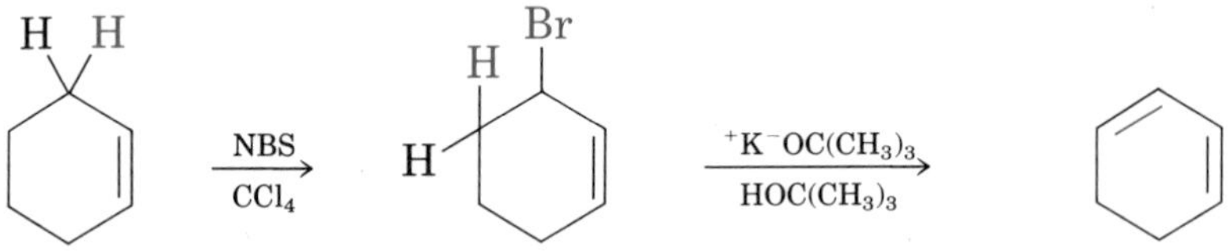

H H — $\xrightarrow[CCl_4]{NBS}$ — H, Br, H — $\xrightarrow[HOC(CH_3)_3]{^{+}K^{-}OC(CH_3)_3}$

Cyclohexene — 3-Bromocyclohexene — 1,3-Cyclohexadiene (76%)

1,3-Butadiene is prepared industrially on a vast scale for use in polymer synthesis. One industrial method involves thermal cracking of butane over a special chromium oxide–aluminum oxide catalyst, but this procedure is of no use in the laboratory.

$$CH_3CH_2CH_2CH_3 \xrightarrow[\text{Catalyst}]{600^\circ C} H_2C{=}CHCH{=}CH_2 + 2\,H_2$$

Butane — 1,3-Butadiene

Other simple conjugated dienes that have important uses in polymer synthesis include isoprene (2-methyl-1,3-butadiene) and chloroprene (2-chloro-1,3-butadiene). Isoprene has been prepared by a number of methods, including the dehydration of 2-methyl-3-buten-2-ol or the double dehydration of 3-methyl-1,2-butanediol over an alumina catalyst.

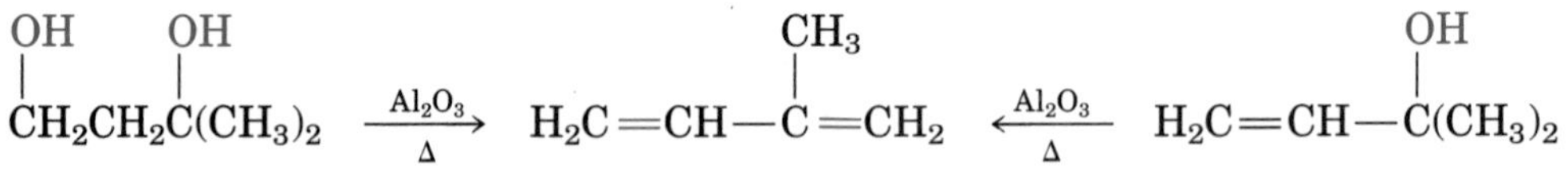

$$\underset{\text{OH}\quad\quad\text{OH}}{CH_2CH_2C(CH_3)_2} \xrightarrow[\Delta]{Al_2O_3} H_2C{=}CH{-}\overset{CH_3}{C}{=}CH_2 \xleftarrow[\Delta]{Al_2O_3} H_2C{=}CH{-}\overset{OH}{C}(CH_3)_2$$

3-Methyl-1,3-butanediol — Isoprene (2-methyl-1,3-butadiene) — 2-Methyl-3-buten-2-ol

The alumina-catalyzed dehydration of an alcohol to an alkene is sometimes used in the laboratory but is more commonly employed as an industrial procedure.

Chloroprene is prepared industrially by electrophilic addition of HCl to 1-buten-3-yne in the presence of cuprous chloride as catalyst.

$$H_2C{=}CHC{\equiv}CH \xrightarrow[CuCl]{HCl} H_2C{=}CH\overset{\overset{Cl}{|}}{C}{=}CH_2$$

1-Buten-3-yne → Chloroprene (2-chloro-1,3-butadiene)

13.2 Stability of conjugated dienes

Conjugated dienes are similar to other alkenes in their preparation and much of their chemistry. There are, however, a few important differences, one of which is stability. Conjugated dienes are somewhat more stable than nonconjugated dienes.

Evidence for the extra stability of conjugated dienes comes from measurements of heats of hydrogenation (Table 13.1). We saw in the earlier

TABLE 13.1 Heats of hydrogenation for some alkenes

Alkene		*Product*	ΔH°_{hydrog} *(kcal/mol)*
$CH_3CH_2CH{=}CH_2$ (1-Butene)	$\xrightarrow{H_2}$	$CH_3CH_2CH_2CH_3$	30.3
$CH_3CH_2\overset{\overset{CH_3}{\vert}}{C}{=}CH_2$ (2-Methyl-1-butene)	$\xrightarrow{H_2}$	$CH_3CH_2\overset{\overset{CH_3}{\vert}}{C}HCH_3$	26.9
$H_2C{=}CHCH{=}CH_2$ (1,3-Butadiene)	$\xrightarrow{H_2}$	$CH_3CH_2CH{=}CH_2$	26.7
$H_2C{=}CHCH{=}CH_2$ (1,3-Butadiene)	$\xrightarrow{2\,H_2}$	$CH_3CH_2CH_2CH_3$	57.1
$H_2C{=}CH\overset{\overset{CH_3}{\vert}}{C}{=}CH_2$ (2-Methyl-1,3-butadiene)	$\xrightarrow{2\,H_2}$	$CH_3CH_2\overset{\overset{CH_3}{\vert}}{C}HCH_3$	53.4
$H_2C{=}CHCH_2CH{=}CH_2$ (1,4-Pentadiene)	$\xrightarrow{2\,H_2}$	$CH_3CH_2CH_2CH_2CH_3$	60.8
$H_2C{=}CHCH_2CH_2CH{=}CH_2$ (1,5-Hexadiene)	$\xrightarrow{2\,H_2}$	$CH_3CH_2CH_2CH_2CH_2CH_3$	60.5

discussion of alkene stabilities (Section 5.4) that alkenes of similar structure have remarkably similar ΔH°_{hydrog} values. Monosubstituted alkenes such as 1-butene have values for ΔH°_{hydrog} near 30 kcal/mol, whereas disubstituted alkenes such as 2-methyl-1-butene show ΔH°_{hydrog} values approximately 3 kcal/mol lower. We concluded from these data that highly substituted alkenes are more stable than less substituted alkenes; that is, substituted alkenes release less heat on hydrogenation because they contain less energy to start with. A similar conclusion can be drawn for conjugated dienes.

Table 13.1 shows that 1-butene, a monosubstituted alkene, has ΔH°_{hydrog} = 30.3 kcal/mol. We might therefore predict that a compound with two monosubstituted double bonds should have a ΔH°_{hydrog} approximately twice this value, or 60.6 kcal/mol. This prediction is fully met by nonconjugated dienes: 1,4-pentadiene (ΔH°_{hydrog} = 60.8 kcal/mol) and 1,5-hexadiene (ΔH°_{hydrog} = 60.5 kcal/mol). The prediction is *not* met by the conjugated diene 1,3-butadiene, which has ΔH°_{hydrog} = 57.1 kcal/mol. 1,3-Butadiene is approximately 3.5 kcal/mol more stable than predicted.

Confirmation of this unexpected stability comes from data on the partial hydrogenation of 1,3-butadiene. If 1,3-butadiene is partially hydrogenated to yield 1-butene, 26.7 kcal/mol energy is released. This is 3.6 kcal/mol less than we would expect for a normal isolated monosubstituted double bond. The same is true of 2-methyl-1,3-butadiene (isoprene). Since an isolated monosubstituted alkene (1-butene) has a value of 30.3 kcal/mol and a disubstituted alkene (2-methyl-1-butene) has a value of 26.9 kcal/mol, we can calculate an expected value for isoprene of 26.9 + 30.3 = 57.2 kcal/mol. The measured value, however, is only 53.4 kcal/mol. Once again, the conjugated diene is more stable than expected (i.e., contains less energy), by 57.2 − 53.4 = 3.8 kcal/mol.

Compound	ΔH°_{hydrog} (kcal/mol)	
$H_2C{=}CHCH_2CH{=}CH_2$ (1,4-Pentadiene)	30.3 + 30.3 = 60.6	Expected
	60.8	Observed
	−0.2	Difference
$H_2C{=}CHCH{=}CH_2$ (1,3-Butadiene)	30.3 + 30.3 = 60.6	Expected
	57.1	Observed
	3.5	Difference
$H_2C{=}C(CH_3)CH{=}CH_2$ (2-Methyl-1,3-butadiene)	30.3 + 26.9 = 57.2	Expected
	53.4	Observed
	3.8	Difference

PROBLEM

13.2 From the data in Table 13.1, calculate an expected heat of hydrogenation value for allene, $H_2C{=}C{=}CH_2$. The measured value is 71.3 kcal/mol. How stable is allene? Rank a conjugated diene, a nonconjugated diene, and an allene in order of stability.

13.3 Molecular orbital description of 1,3-butadiene

The unexpected stability of conjugated dienes can be accounted for by a combination of two different explanations. One explanation says that 1,3-butadiene really does *not* have "unexpected" stability—that is, the problem is with our expectations. According to this view, our expectations are wrong because a nonconjugated diene such as 1,4-pentadiene is a poor choice for comparison with 1,3-butadiene. Nonconjugated dienes form carbon–carbon single bonds by sigma overlap of an sp^2 orbital from one carbon with an sp^3 orbital from the other carbon. The C2—C3 bond of 1,3-butadiene, however, results from sigma overlap of sp^2 orbitals on both carbons. Since sp^2 orbitals have more *s* character than sp^3 orbitals, they form somewhat shorter, stronger bonds. Thus the "extra" stability of a conjugated diene has more to do with the hybridization of the orbitals forming the central carbon–carbon single bond than with the double bonds themselves.

$H_2C{=}CH{-}CH_2{-}CH{=}CH_2$

Bonds formed by overlap of C_{sp^2} and C_{sp^3} orbitals

$H_2C{=}CH{-}CH{=}CH_2$

Bond formed by overlap of C_{sp^2} and C_{sp^2} orbitals

This hybridization hypothesis makes a valid point, but another factor is also important: the interaction between the pi orbitals of the conjugated diene system. Molecular orbital theory (Section 1.7) says that when a covalent bond is formed by overlap of two atomic orbitals the resultant new orbitals are the property of the molecule, not of the individual atoms. The electrons in those orbitals are shared between atoms, rather than localized on one atom. Thus, when two *p* atomic orbitals overlap to form a pi bond, the bonding electrons occupy a pi molecular orbital, rather than *p* atomic orbitals. Since two *p* atomic orbitals are involved, two pi molecular orbitals are formed. One is lower in energy than the starting *p* orbitals and is therefore a bonding molecular orbital; the other is higher in energy and is antibonding. When we fill the pi bond by assigning electrons to the orbitals, both electrons go into the low-energy bonding orbital, and a stable bond results.

We can represent the situation pictorially as in Figure 13.2. We saw earlier that a *p* orbital is dumbbell shaped, with the two lobes having different mathematical signs. When we allow two *p* orbitals to overlap, they can orient in either of two ways. Overlap of lobes with identical signs is additive and corresponds to the low-energy bonding molecular orbital, whereas overlap of lobes with different signs is subtractive and corresponds to the high-energy antibonding orbital. The change of sign between adjacent lobes in the antibonding orbital means that if there were electrons in this orbital they would not be shared between nuclei. Thus there is a region of **zero electron density** between the two nuclei, and we call this region a **node.** The bonding molecular orbital does not have a node between the nuclei.

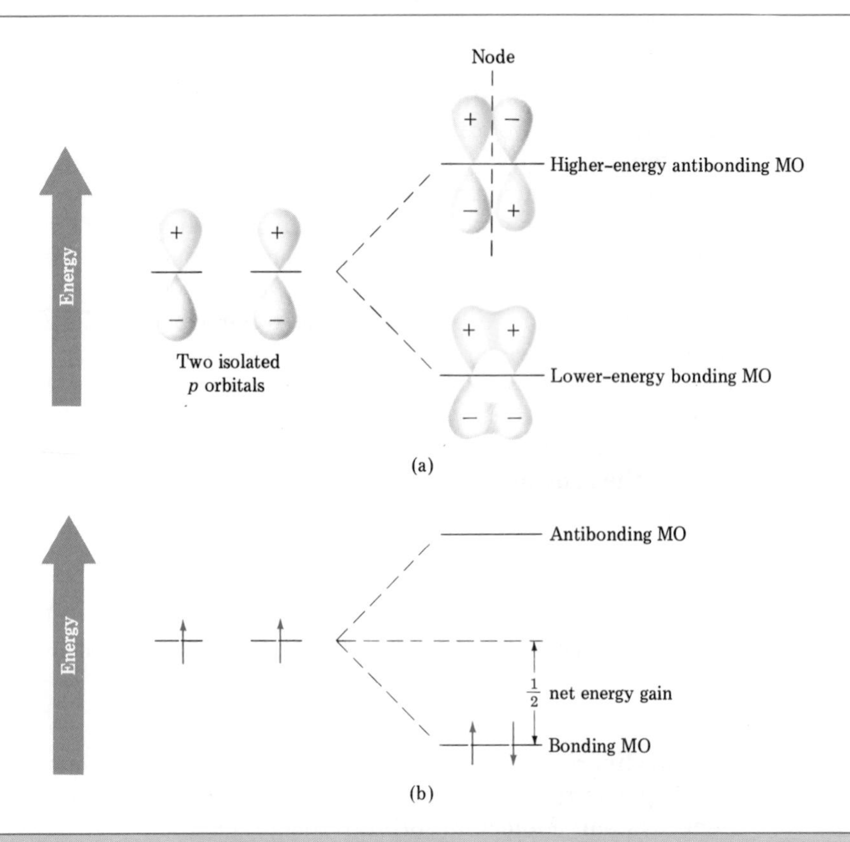

Figure 13.2. (a) Combination of isolated *p* orbitals into pi molecular orbitals of alkenes; (b) filling of molecular orbitals.

Now let's bring two pi bonds together and allow four *p* atomic orbitals to interact, as in a conjugated diene. In so doing, we generate a set of four molecular orbitals (Figure 13.3).

The lowest-energy molecular orbital (denoted ψ_1, Greek psi) is a fully additive combination that has no nodes between the nuclei. It is therefore a bonding orbital and holds two electrons. The next-lowest-energy molecular orbital, ψ_2, has one node, is also a bonding orbital, and holds the remaining two electrons. Above ψ_1 and ψ_2 in energy are the two antibonding molecular orbitals ψ_3^* and ψ_4^*. Of these, ψ_3^* has two nodes and ψ_4^*, the highest-energy molecular orbital, has three nodes. Note that the number of nodes increases as the energy level of the orbital increases.

Quantum mechanical calculations show that the sum of energy levels of the two bonding butadiene molecular orbitals is slightly lower than the sum of two isolated alkene molecular orbitals. In other words, placing the four electrons in the two lowest molecular orbitals results in a more stable arrangement than placing them in two isolated alkene orbitals (Figure 13.4). This "extra" stability is a consequence of a favorable bonding interaction across the C2—C3 bond of 1,3-butadiene.

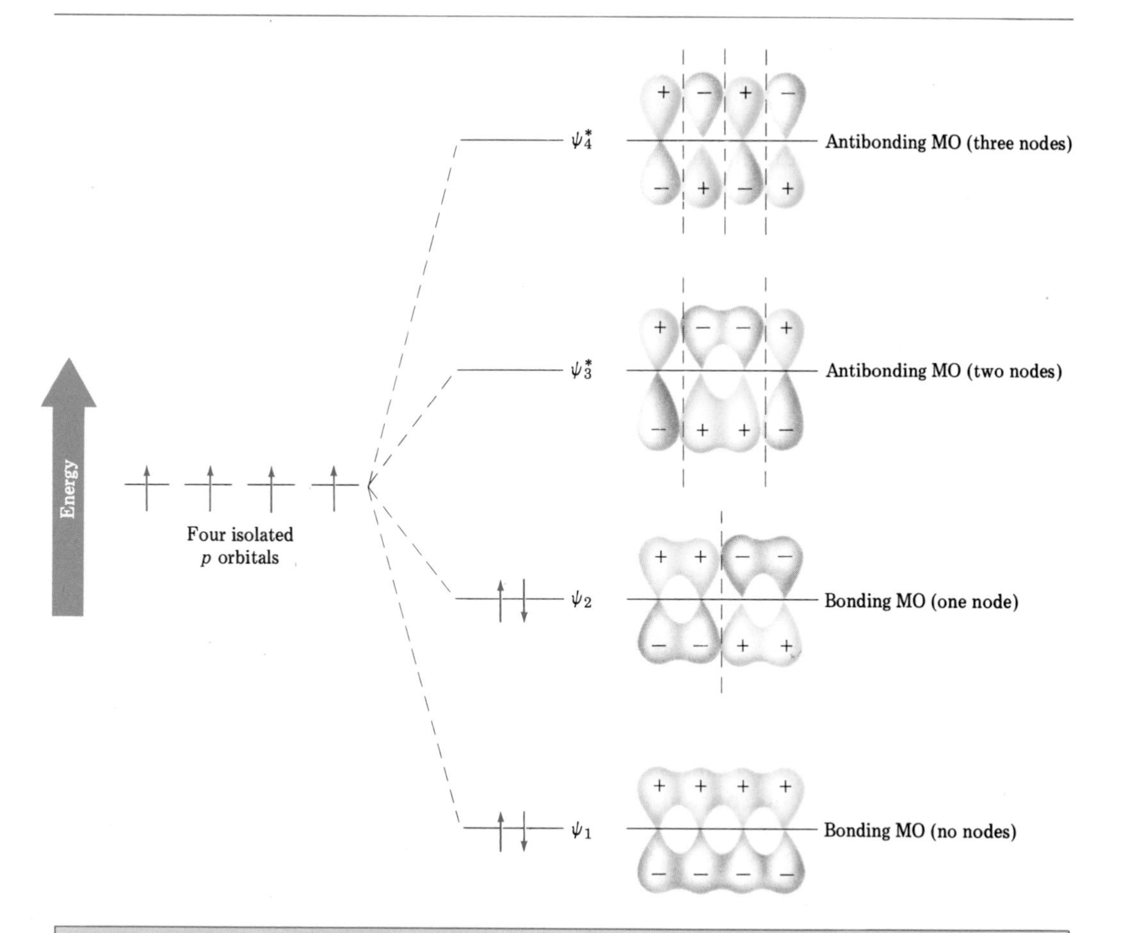

Figure 13.3. 1,3-Butadiene pi molecular orbitals. The asterisk on ψ_3^* and ψ_4^* indicates antibonding orbitals.

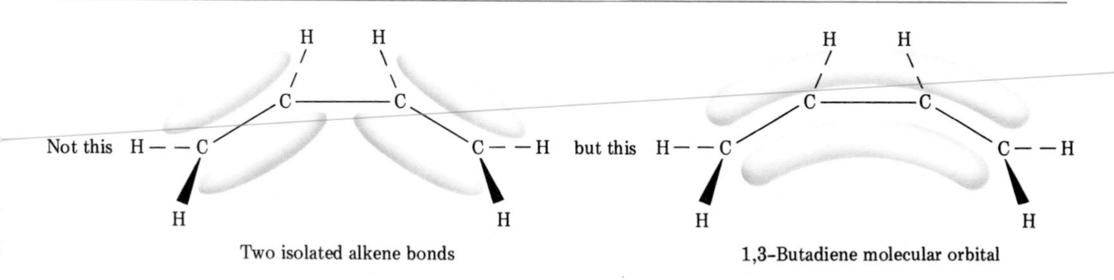

Figure 13.4. A molecular orbital view of 1,3-butadiene.

In describing the 1,3-butadiene molecular orbitals, we say that the pi electrons are **delocalized** over the entire pi framework rather than localized between two specific nuclei. Electron delocalization leads to lower-energy orbitals and greater stability.

13.4 Bond lengths in 1,3-butadiene

Further evidence for the special nature of conjugated dienes can be gathered from bond length data. Measurements have shown that the C2—C3 single bond in 1,3-butadiene has a length of 1.48 Å and that the two equivalent carbon–carbon double bonds have lengths of 1.34 Å (Table 13.2).

TABLE 13.2 **Some carbon–carbon bond lengths**

Bond	*Bond length (Å)*	*Bond hybridization*
$CH_3—CH_3$	1.54	C_{sp^3}–C_{sp^3}
$H_2C{=}CH_2$	1.33	C_{sp^2}–C_{sp^2}
$H_2C{=}CH—CH{=}CH_2$	1.48	C_{sp^2}–C_{sp^2}
$H_2C{=}CHCH{=}CH_2$	1.34	C_{sp^2}–C_{sp^2}

If we compare the length of the carbon–carbon single bond of 1,3-butadiene with that of ethane, we find that the 1,3-butadiene single bond is shorter by 0.06 Å. Two different explanations have been advanced to account for this bond shortening. One proposes that the shortening of the 1,3-butadiene single bond is due to the delocalization of pi electrons in the bonding molecular orbitals. According to this view, pi orbital overlap across the C2—C3 bond results in partial double-bond character (Figure 13.5) and consequent bond shortening to a value midway between a pure single bond (1.54 Å) and a pure double bond (1.33 Å). The partial double-bond character of the C2—C3 bond is sufficient to stabilize the molecule but not to prevent bond rotation from occurring.

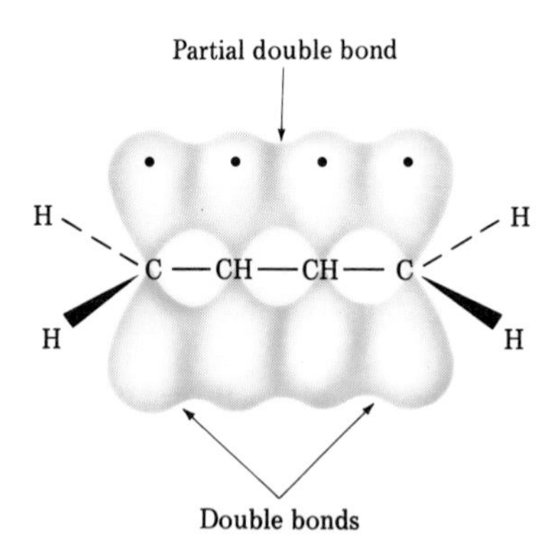

Figure 13.5. An orbital view of 1,3-butadiene. The C2—C3 bond has partial double-bond character.

Alternatively, it can be argued that the shortened 1,3-butadiene single bond is a natural consequence of the orbital hybridization involved. The C2—C3 bond results from sigma overlap of two carbon sp^2 orbitals, and a normal alkane bond results from sigma overlap of two carbon sp^3 orbitals.

Overlap of sp^2 orbitals results in a single bond that has more *s* character than usual; thus the 1,3-butadiene single bond is a bit shorter and stronger than usual. Both explanations are probably valid, and both contribute to the bond shortening observed for 1,3-butadiene.

13.5 Electrophilic additions to conjugated dienes: allylic cations

One of the striking differences between the chemistry of conjugated dienes and isolated alkenes is in their electrophilic addition reactions.

As we've seen, the addition of electrophilic reagents to isolated carbon–carbon double bonds is an important reaction in organic chemistry (Section 5.8). Markovnikov regiochemistry is observed for these reactions because the more highly substituted (more stable) carbocation is involved as an intermediate. Thus, addition of HCl to 2-methylpropene yields 2-chloro-2-methylpropane rather than 1-chloro-2-methylpropane, and addition of 2 mol equiv of HCl to the nonconjugated diene 1,4-pentadiene yields 2,4-dichloropentane.

$$(CH_3)_2C{=}CH_2 \xrightarrow[\text{Ether}]{\text{HCl}} \left[(CH_3)_3C^+\right] \longrightarrow (CH_3)_3C{-}Cl$$

2-Methylpropene → Tertiary carbocation intermediate → 2-Chloro-2-methylpropane

$$\left[(CH_3)_3C^+\right] \not\longrightarrow (CH_3)_2CHCH_2Cl$$

1-Chloro-2-methylpropane (*not formed*)

$$H_2C{=}CHCH_2CH{=}CH_2 \xrightarrow[\text{Ether}]{2\text{ HCl}} CH_3CH(Cl)CH_2CH(Cl)CH_3$$

1,4-Pentadiene (a nonconjugated diene) → 2,4-Dichloropentane

Conjugated dienes also undergo electrophilic addition reactions readily, but mixtures of products are invariably obtained. For example, addition of HBr to 1,3-butadiene yields a mixture of two products:

$$H_2C{=}CHCH{=}CH_2 \xrightarrow[0^\circ C]{\text{HBr}} H_2C{=}CHCH(Br)CH_2(H) + CH_2(H)CH{=}CHCH_2(Br)$$

1,3-Butadiene → 3-Bromo-1-butene (71%; 1,2 addition) + 1-Bromo-2-butene (29%; 1,4 addition)

3-Bromo-1-butene is the normal product of Markovnikov addition, but 1-bromo-2-butene appears unusual. The double bond in this product has moved to a position between carbons 2 and 3, and HBr has added to carbons 1 and 4. How can we account for the formation of this **1,4-addition** product?

The answer is that an *allylic carbocation* is involved as an intermediate in the reaction. When H^+ is attacked by an electron-rich pi bond of 1,3-butadiene, two carbocation intermediates are possible—a primary carbocation, and a secondary allylic cation (recall the meaning of *allylic:* "next to a double bond"). We saw in Section 10.9 that an allylic cation is highly stable, and it thus forms in preference to the less stable primary carbocation.

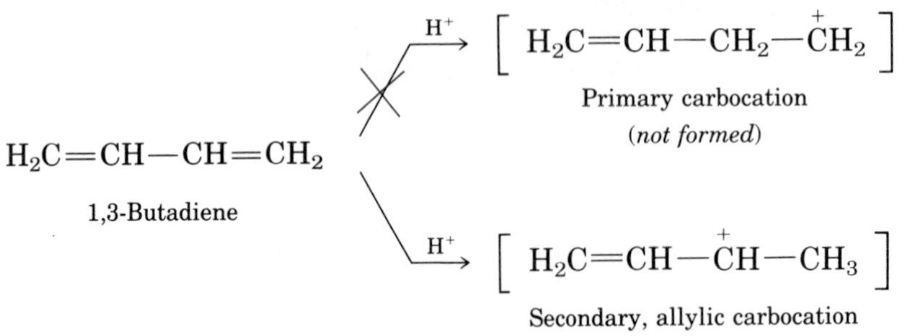

$$H_2C{=}CH{-}CH{=}CH_2 \xrightarrow{H^+} \left[H_2C{=}CH{-}CH_2{-}\overset{+}{C}H_2 \right]$$

1,3-Butadiene → Primary carbocation (*not formed*)

$$H_2C{=}CH{-}CH{=}CH_2 \xrightarrow{H^+} \left[H_2C{=}CH{-}\overset{+}{C}H{-}CH_3 \right]$$

Secondary, allylic carbocation

There are two ways to view an allylic carbocation and account for its stability. Resonance theory offers a pictorial representation of the situation through the use of different resonance forms. Recall from Section 9.6 that resonance forms differ only in the placement of bonding electrons—the nuclei remain in the same positions. The more resonance forms that are possible, the more stable the compound is. Thus, an allylic cation has two resonance forms and is more stable than a nonallylic cation. Neither of these two resonance forms is correct by itself; the true structure of the allylic cation is a combination, or resonance hybrid, of the two Kekulé structures. It is difficult to draw the true structure, but we might use a dotted line to indicate a partial bond, as shown in Figure 13.6, and use δ^+ to

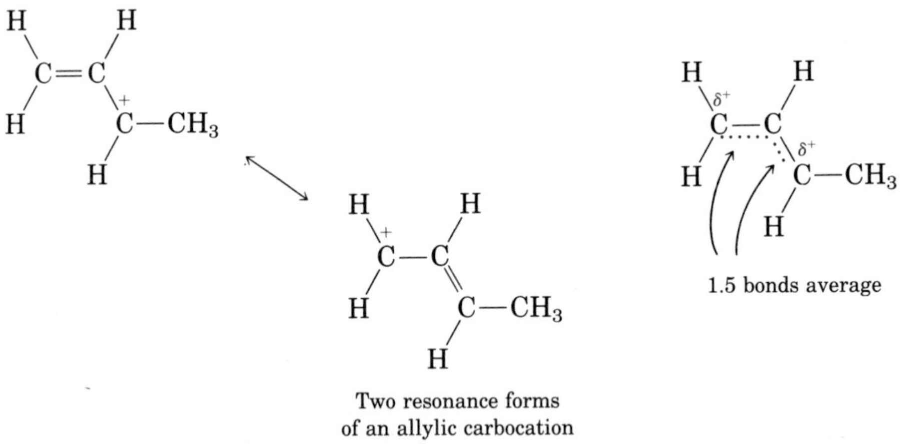

$$H_2C{=}CH{-}\overset{+}{C}H{-}CH_3 \longleftrightarrow H_2\overset{+}{C}{-}CH{=}CH{-}CH_3$$

Two resonance forms of an allylic carbocation

$$H_2\overset{\delta^+}{C}{\cdots}CH{\cdots}\overset{\delta^+}{C}H{-}CH_3$$

1.5 bonds average

Figure 13.6. Resonance forms for the allylic cation formed by protonation of 1,3-butadiene.

indicate that the positive charge is dispersed over both ends of the allyl system.

When the allylic cation is attacked by bromide ion to complete the electrophilic addition reaction, attack can occur at either carbon 1 or carbon 3, since both share the positive charge. The result is a mixture of 1,2- and 1,4-addition products.

$$\overset{+}{C}H_2{-}CH{=}CHCH_3 \longleftrightarrow CH_2{=}CH{-}\overset{+}{C}HCH_3$$

$$\downarrow :Br^-$$

$$\underset{\substack{\text{1,4 addition}\\(29\%)}}{BrCH_2{-}CH{=}CHCH_3} + \underset{\substack{\text{1,2 addition}\\(71\%)}}{H_2C{=}CH{-}\overset{Br}{\overset{|}{C}}HCH_3}$$

The alternative way of accounting for the stability of the allylic cation is to construct a molecular orbital description. When we allow three *p* orbitals to interact, three molecular orbitals are formed (Figure 13.7). Quantum mechanical calculations show that one low-energy bonding orbital, one nonbonding orbital, and one high-energy antibonding orbital result. The two available pi electrons occupy the low-energy bonding orbital, indicating that a partial bond exists between carbons 2 and 3. Thus,

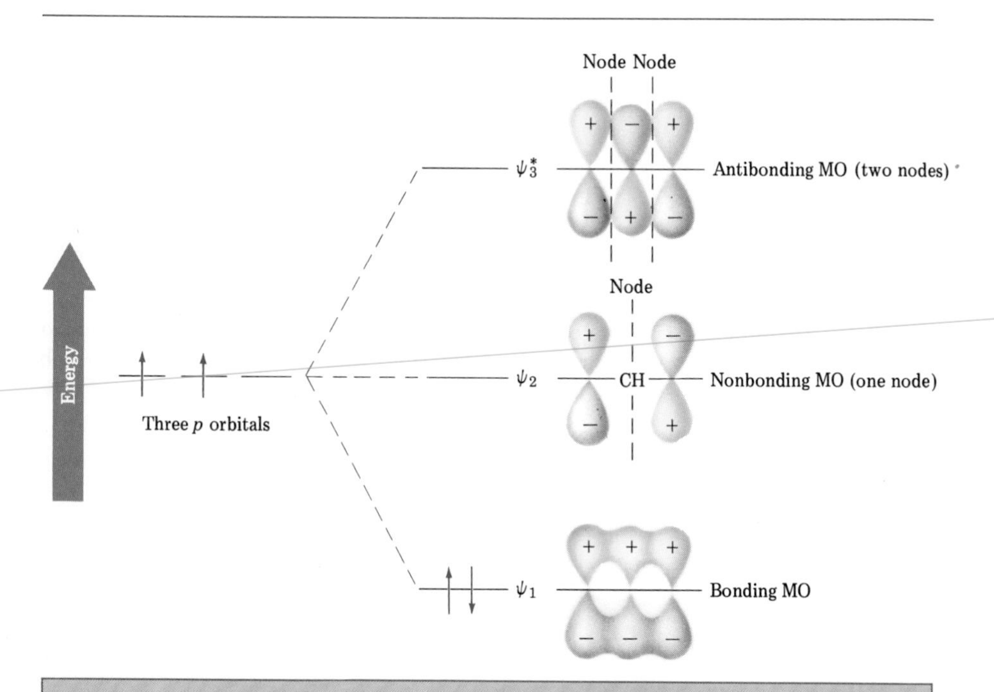

Figure 13.7. Molecular orbitals of an allylic carbocation.

the allylic cation is a conjugated system stabilized by electron delocalization in much the same way that 1,3-butadiene is stabilized by electron delocalization (Section 13.3).

Many other electrophiles besides HBr add to conjugated dienes. For example, Br_2 adds to 1,3-butadiene, and HCl adds to 2-methyl-1,3-butadiene, to give the products shown. Mixtures of products are formed in all cases, although the 1,2-addition product is usually favored at low temperature.

$$H_2C{=}CH{-}CH{=}CH_2 \xrightarrow[25°C]{Br_2} BrCH_2{-}CH{=}CH{-}CH_2Br \quad \text{1,4 addition}$$

1,3-Butadiene → 1,4-Dibromo-2-butene (45%)

+

$$BrCH_2{-}\underset{}{\overset{Br}{\overset{|}{C}}H}{-}CH{=}CH_2 \quad \text{1,2 addition}$$

3,4-Dibromo-1-butene (55%)

$$H_2C{=}\overset{CH_3}{\overset{|}{C}}{-}CH{=}CH_2 \xrightarrow[25°C]{HCl} (CH_3)C{=}CH{-}CH_2Cl \quad \text{1,4 addition}$$

2-Methyl-1,3-butadiene (isoprene) → 1-Chloro-3-methyl-2-butene (25%)

+

$$(CH_3)_2\overset{Cl}{\overset{|}{C}}{-}CH{=}CH_2 \quad \text{1,2 addition}$$

3-Chloro-3-methyl-1-butene (75%)

PROBLEM

13.3 Give the structures of all possible monoadducts of HCl and 1,3-pentadiene.

PROBLEM

13.4 Examine the possible carbocation intermediates produced during addition of HCl to 1,3-pentadiene, and predict which of the 1,2 adducts predominates. Which 1,4 adduct would be expected to predominate?

PROBLEM

13.5 Electrophilic addition of Br_2 to isoprene yields the following product mixture:

$$H_2C{=}\overset{CH_3}{\overset{|}{C}}CH{=}CH_2 \xrightarrow{Br_2} \underset{(3\%)}{H_2C{=}\overset{CH_3}{\overset{|}{C}}CHBrCH_2Br} + \underset{(21\%)}{BrCH_2\overset{CH_3}{\overset{|}{C}}BrCH{=}CH_2} + \underset{(76\%)}{BrCH_2\overset{CH_3}{\overset{|}{C}}{=}CHCH_2Br}$$

Of the 1,2-addition products, explain why 3,4-dibromo-3-methyl-1-butene (21%) predominates over 3,4-dibromo-2-methyl-1-butene (3%).

PROBLEM

13.6 Draw a molecular orbital diagram for the allyl radical $H_2C{=}CH{-}CH_2\cdot$. Indicate which orbitals the three pi electrons occupy.

13.6 Kinetic control versus thermodynamic control of reactions

Addition of electrophiles to conjugated dienes at or below room temperature normally leads to a mixture of products in which the 1,2 adduct predominates over the 1,4 adduct. When the same reaction is carried out at higher temperatures, however, the product ratio often changes and the 1,4 adduct may predominate. For example, addition of HBr to 1,3-butadiene at 0°C yields a 71:29 mixture of 1,2 and 1,4 adducts; the same reaction at 40°C yields a 15:85 mixture. Furthermore, when the product mixture formed at 0°C is heated to 40°C in the presence of more HBr, the ratio of adducts slowly changes from 71:29 to 15:85. How can we explain these observations?

$$H_2C{=}CHCH{=}CH_2 + HBr \xrightarrow[40^\circ C]{0^\circ C} \underset{\substack{\text{1,2 adduct}\\(71\%)\ \text{at } 0^\circ C;\ (15\%)\ \text{at } 40^\circ C}}{H_2C{=}CHCHBrCH_3} + \underset{\substack{\text{1,4 adduct}\\(29\%)\ \text{at } 0^\circ C;\ (85\%)\ \text{at } 40^\circ C}}{CH_3CH{=}CHCH_2Br}$$

(0°C mixture converted to 40°C mixture on heating at 40°C)

To understand the reasons for the effect of reaction temperature on electrophilic addition reactions of conjugated dienes, we need to review what we have already discussed about reactions and transition states. In principle, all reactions are reversible; an equilibrium distribution of products will result from any reaction if proper experimental conditions are found. In practice, however, it is often difficult or impossible to reach equilibrium, and a nonequilibrium product distribution results.

As an example, let's imagine a reaction that can give either of two products depending on the reaction conditions used:

$$A \begin{cases} \rightarrow B \\ \rightarrow C \end{cases}$$

Let's assume that B forms faster than C ($\Delta G^\ddagger_B < \Delta G^\ddagger_C$) but that C is a more stable product than B ($\Delta G^\circ_C > \Delta G^\circ_B$). A reaction energy diagram for the process might look like that shown in Figure 13.8 (p. 422). (This diagram starts in the middle to show more clearly that reactant A can follow either of two paths.)

Let's first carry out the reaction under conditions such that both processes are reversible and an equilibrium is reached. Since C is more stable

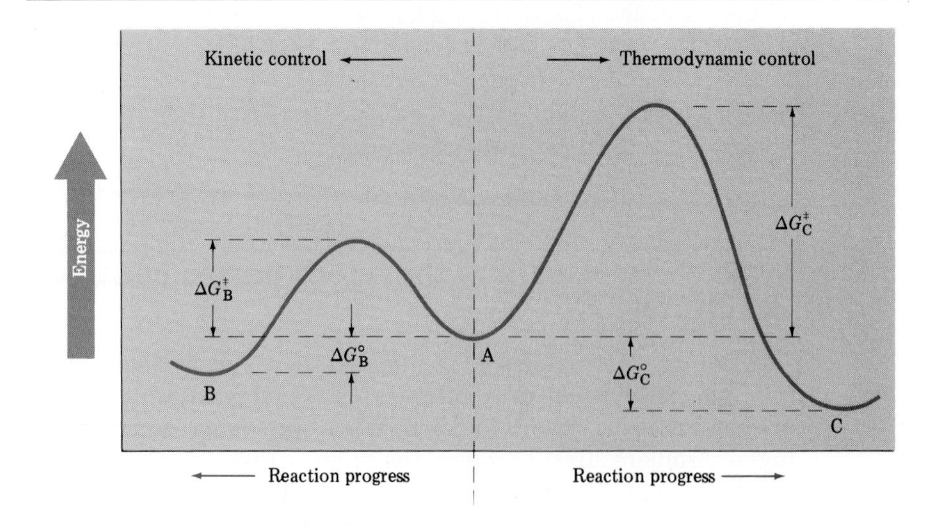

Figure 13.8. Reaction energy diagram showing two competing reactions. Reaction starts in the middle.

than B ($\Delta G^{\circ}_{C} > \Delta G^{\circ}_{B}$), C is the major product formed at equilibrium. It does not matter that C forms more slowly than B ($\Delta G^{\ddagger}_{B} < \Delta G^{\ddagger}_{C}$), because the reaction conditions have been chosen such that B and C are formed reversibly, and therefore interconvert rapidly. All that matters in a reversible reaction is the thermodynamic stability of the products at equilibrium. Such reactions are said to be under **thermodynamic control,** or equilibrium control.

$$B \rightleftharpoons A \rightleftharpoons C \qquad \text{Thermodynamic control (reversible)}$$

Let's now carry out the same reaction under milder conditions such that both processes are *irreversible,* and an equilibrium is *not* reached. Since B forms faster than C (lower $\Delta G^{\ddagger}$), B is the major product. It does not matter that C is more stable than B, because the reaction conditions have been chosen so that B and C are formed irreversibly and do not interconvert. All that matters in an irreversible process is the reaction rate. Such reactions are said to be under **kinetic control.**

$$B \longleftarrow A \longrightarrow C \qquad \text{Kinetic control (irreversible)}$$

We can now explain the effect of temperature on electrophilic addition reactions of conjugated dienes. At 0°C, HBr adds to 1,3-butadiene under kinetic control to give a 71:29 mixture of products with the 1,2 adduct predominating. Evidently these mild conditions do not permit reversal of the reaction and do not allow the products to reach equilibrium. At 40°C, however, the reaction occurs under thermodynamic control to give a 15:85 mixture of products, with the 1,4 adduct predominating. These more vigorous conditions permit a reversible reaction: The higher temperature provides more energy for product molecules to climb the high energy barrier

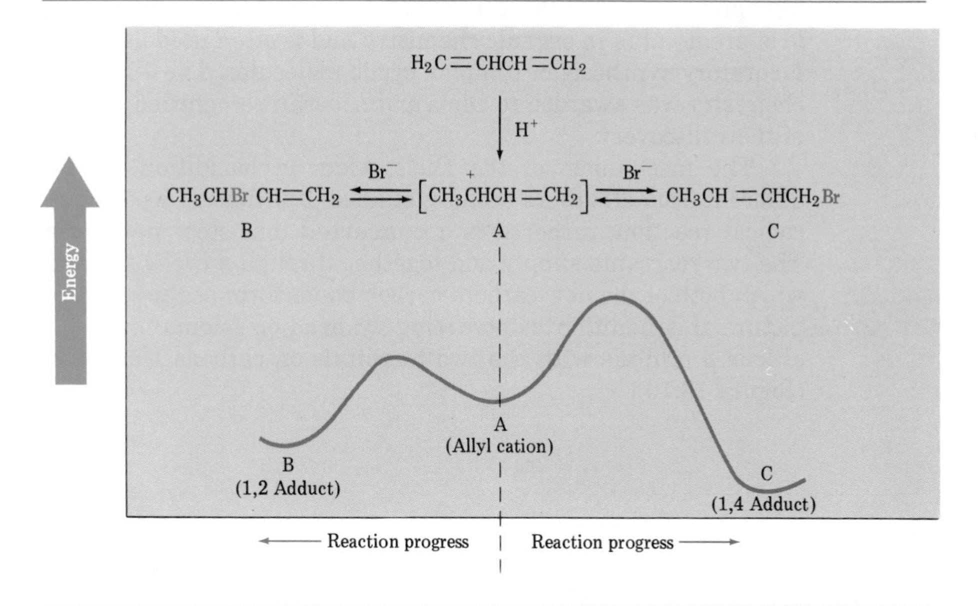

Figure 13.9. Reaction energy diagram for electrophilic addition of HBr to 1,3-butadiene. The 1,2 adduct is the kinetic product, and the 1,4 adduct is the thermodynamic product.

leading back to the allylic cation, and an equilibrium mixture of products therefore results. Figure 13.9 shows the situation on a reaction energy diagram.

The electrophilic addition of HBr to 1,3-butadiene is an excellent example of how experimental conditions can determine the product of a reaction. The concept of thermodynamic control versus kinetic control is a valuable one, which we can often use to advantage in the laboratory.

PROBLEM

13.7 We have seen that the 1,2 adduct and the 1,4 adduct of HBr with 1,3-butadiene are in equilibrium at 40°C, but we have not examined the mechanism by which equilibration takes place. Propose a pathway to account for the observed interconversion of 3-bromo-1-butene and 1-bromo-2-butene at 40°C. [*Hint:* See Section 10.6.]

13.7 The Diels–Alder cycloaddition reaction

Conjugated dienes react with isolated alkenes to yield substituted cyclohexene products. For example, 1,3-butadiene and 3-buten-2-one give 3-cyclohexenyl methyl ketone in nearly 100% yield:

$$\text{1,3-Butadiene} + \text{3-Buten-2-one} \xrightarrow[\Delta]{\text{Benzene}} \text{3-Cyclohexenyl methyl ketone (100\%)}$$

1,3-Butadiene: H(CH$_2$=)C–C(=CH$_2$)H

3-Buten-2-one: H(COCH$_3$)C=CH$_2$

3-Cyclohexenyl methyl ketone (100%): cyclohexene ring bearing C(=O)CH$_3$

This process, named the **Diels–Alder reaction** after its two discoverers,[1,2] is of great value in organic chemistry and is often used as a key step in the laboratory synthesis of complex cyclic molecules. The 1950 Nobel prize in chemistry was awarded to Diels and Alder in recognition of the importance of their discovery.

The mechanism of the Diels–Alder cycloaddition reaction is quite different from all others we have studied. It is neither a polar reaction nor a radical reaction; rather it is a **concerted** (one-step) **pericyclic process.** The two reactants simply add together through a cyclic transition state in which both of the new carbon–carbon bonds form at the same time. We can picture this addition as occurring by head-on (sigma) overlap of the two alkene *p* orbitals with the two *p* orbitals on carbons 1 and 4 of the diene (Figure 13.10).

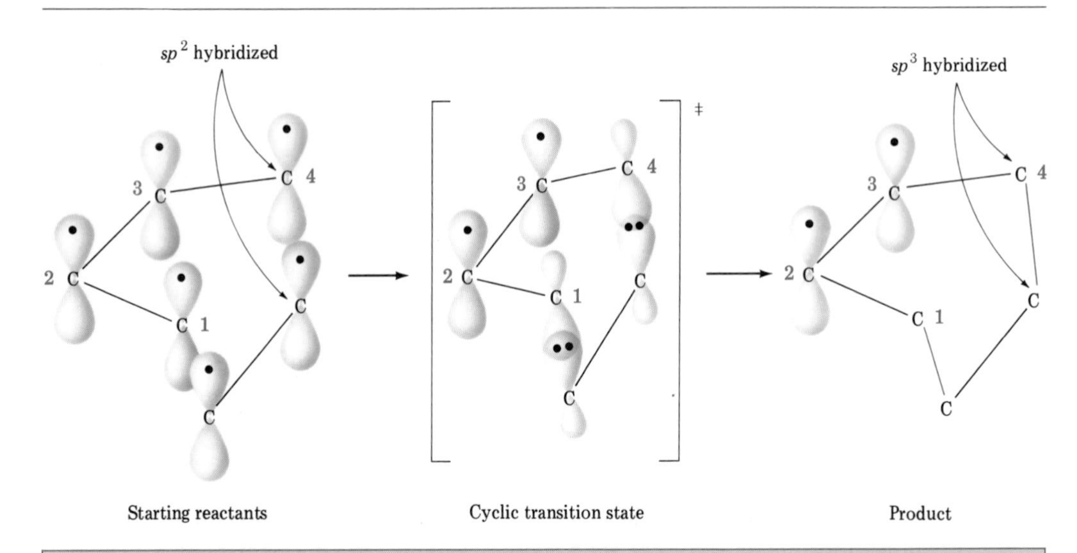

Figure 13.10. Mechanism of the Diels–Alder reaction. The reaction occurs in a single step through a cyclic transition state, in which the two new carbon–carbon bonds begin to form at the same time.

In the transition state, the alkene carbons and carbons 1 and 4 of the diene rehybridize from sp^2 to sp^3 to form the two new single bonds. Carbons 2 and 3 remain sp^2 hybridized and form the new double bond in the product. The Diels–Alder cycloaddition reaction is just one example of a large number of pericyclic reactions. We will study the mechanism of these reactions in Chapter 30 but will concentrate for the present on learning more about the chemistry of the process.

[1]Otto Diels (1876–1954); b. Hamburg; Ph.D. Berlin (E. Fischer); professor, University of Berlin (1906–1916), Kiel (1916–1948); Nobel prize (1950).

[2]Kurt Alder (1902–1958); b. Königshütte; Ph.D. Kiel (Diels); professor, University of Cologne (1940–1958); Nobel prize (1950).

13.8 Characteristics of the Diels–Alder reaction

The Diels–Alder cycloaddition reaction takes place most rapidly and in highest yield if the alkene component, or **dienophile** ("diene lover"), is substituted by electron-withdrawing groups. Thus, ethylene itself is unreactive in the Diels–Alder reaction, but propenal, ethyl propenoate, maleic anhydride, benzoquinone, propenenitrile, and others are highly reactive (Figure 13.11). Note also that alkynes such as methyl propynoate can act as Diels–Alder dienophiles. In all of these cases, the dienophile is bonded to a positively polarized carbon of a substituent that withdraws electron density from the double bond.

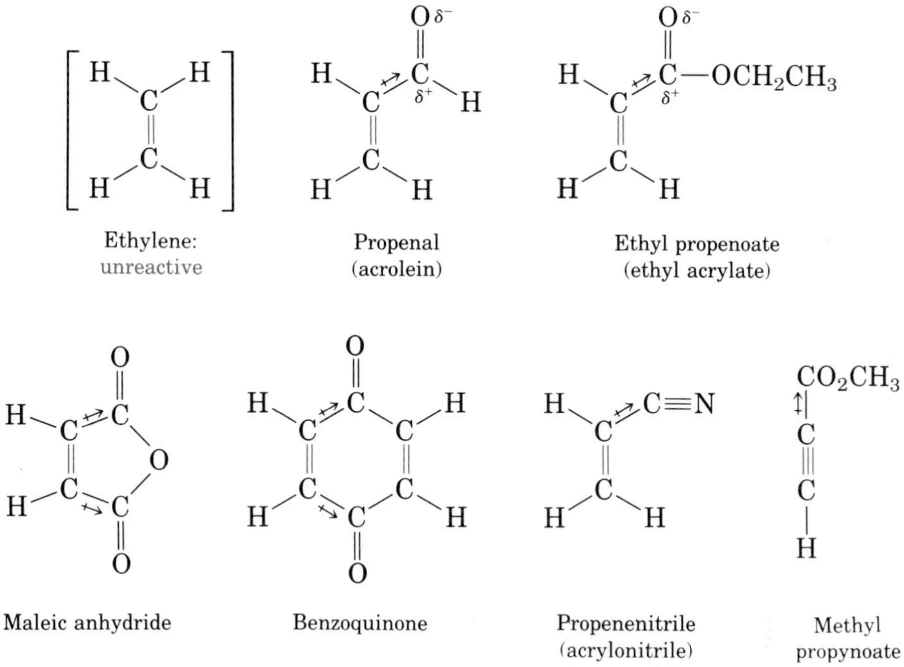

Ethylene: unreactive

Propenal (acrolein)

Ethyl propenoate (ethyl acrylate)

Maleic anhydride

Benzoquinone

Propenenitrile (acrylonitrile)

Methyl propynoate

Figure 13.11. Some Diels–Alder dienophiles. All contain electron-withdrawing groups.

An important feature of the Diels–Alder reaction is that it is *stereospecific:* The stereochemistry of the starting dienophile is maintained during the reaction. If we carry out the cycloaddition with a cis alkene such as methyl *cis*-2-butenoate, we produce only the cis-substituted cyclohexene. Conversely, Diels–Alder reaction with methyl *trans*-2-butenoate yields only the trans-substituted cyclohexene product.

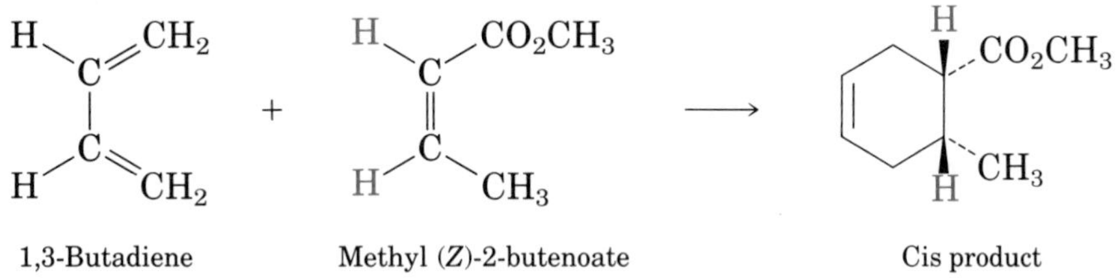

1,3-Butadiene + Methyl (*Z*)-2-butenoate ⟶ Cis product

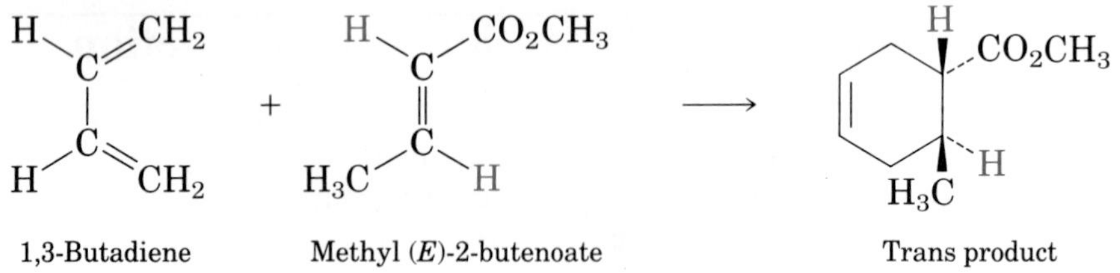

1,3-Butadiene + Methyl (*E*)-2-butenoate ⟶ Trans product

To undergo the Diels–Alder reaction, a diene must be able to adopt an *s*-cis geometry ("cis-like" about the single bond). Only in the *s*-cis conformation are carbons 1 and 4 of the diene close enough to react through a cyclic transition state to give a new ring. In the alternative *s*-trans geometry, the ends of the diene partner are too far apart to overlap the dienophile *p* orbitals successfully.

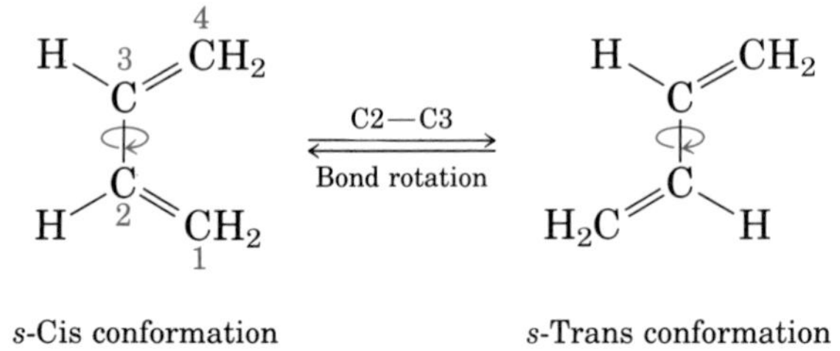

C2—C3 Bond rotation

s-Cis conformation

s-Trans conformation

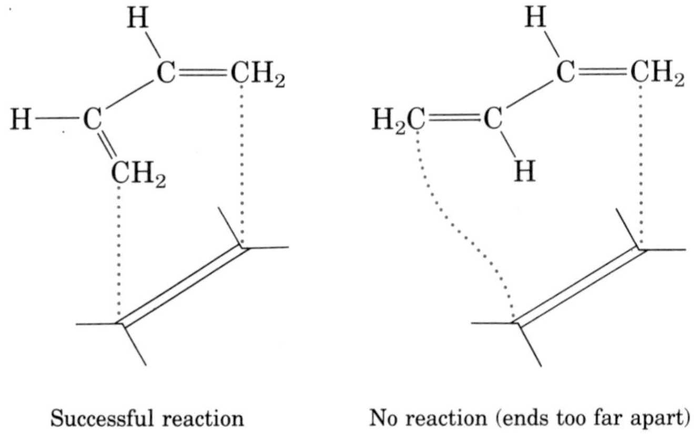

Successful reaction

No reaction (ends too far apart)

Examples of dienes that cannot adopt *s*-cis geometry and therefore do not undergo Diels–Alder reaction are shown in Figure 13.12. In the case of the bicyclic (two-ring) diene, the double bonds are rigidly fixed in the *s*-trans arrangement by geometric constraints of the rings. In the case of (2*Z*,4*Z*)-hexadiene, severe steric strain between the two methyl groups prevents the molecule from adopting *s*-cis geometry.

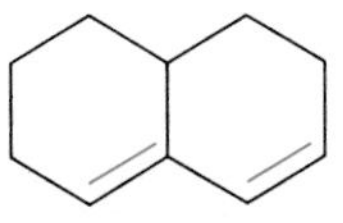

A bicyclic diene
(rigid *s*-trans diene)

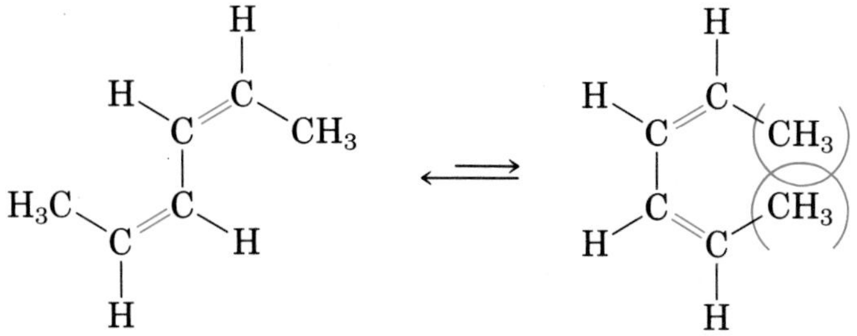

(2*Z*,4*Z*)-Hexadiene
(*s*-trans, more stable)

Severe steric strain
in *s*-cis form

Figure 13.12. Some *s*-trans dienes that cannot undergo Diels–Alder reactions.

PROBLEM

13.8 Indicate which of these alkenes you would expect to be good Diels–Alder dienophiles.

(a) $H_2C{=}CHNO_2$

(b)

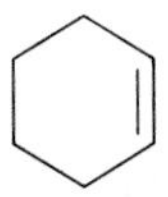

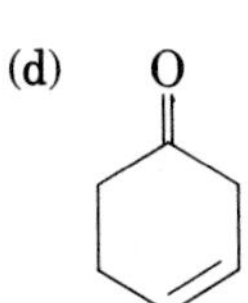

(c)

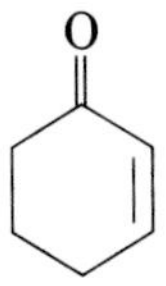

(d)

PROBLEM

13.9 Which of the following dienes have an *s*-cis conformation and which an *s*-trans conformation? Of the *s*-trans dienes, which can readily rotate to *s*-cis?

(a)

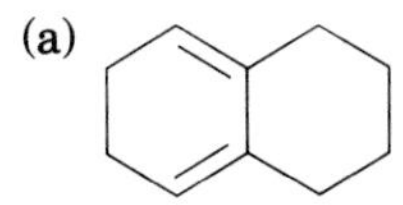

(b)

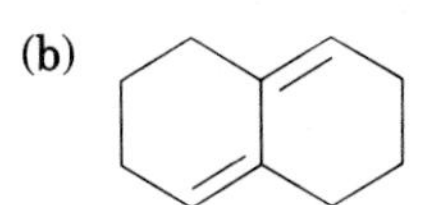

(c)

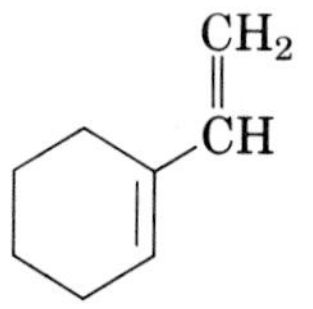

(d)

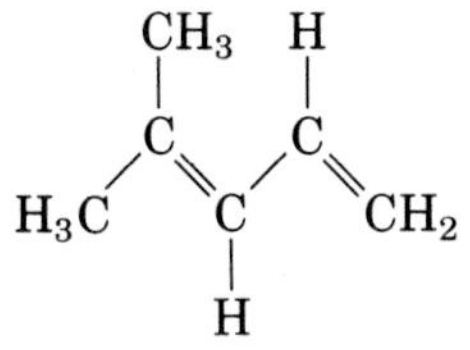

13.9 Cyclic dienes—chlorinated insecticides

In contrast to the unreactive *s*-trans dienes, other dienes are rigidly fixed in the correct *s*-cis geometry and are therefore highly reactive in the Diels–Alder cycloaddition reaction. Such, for example, is the case with cyclopentadiene. Cyclopentadiene is so reactive, in fact, that it reacts with itself! At room temperature (about 25°C), cyclopentadiene **dimerizes**—one molecule acts as diene and another acts as dienophile.

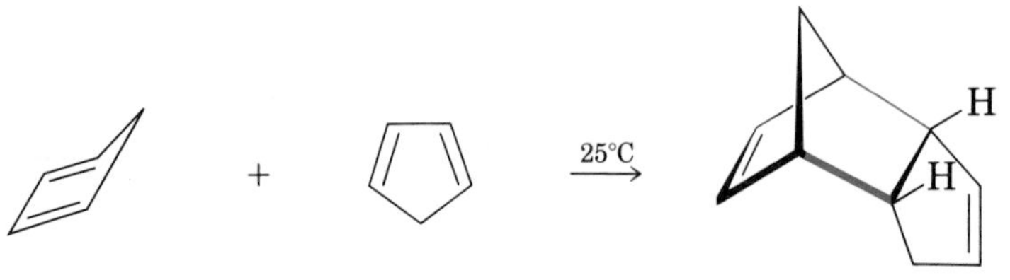

25°C

1,3-Cyclopentadiene (*s*-cis)

Bicyclopentadiene

This kind of Diels–Alder reaction is particularly important in the commercial production of a number of chlorinated insecticides. Thus, hexachlorocyclopentadiene, which is prepared by vapor-phase chlorination of cyclopentadiene, is allowed to undergo Diels–Alder reaction with cyclopentadiene. Further addition of chlorine to the Diels–Alder product yields the insecticide chlordane.

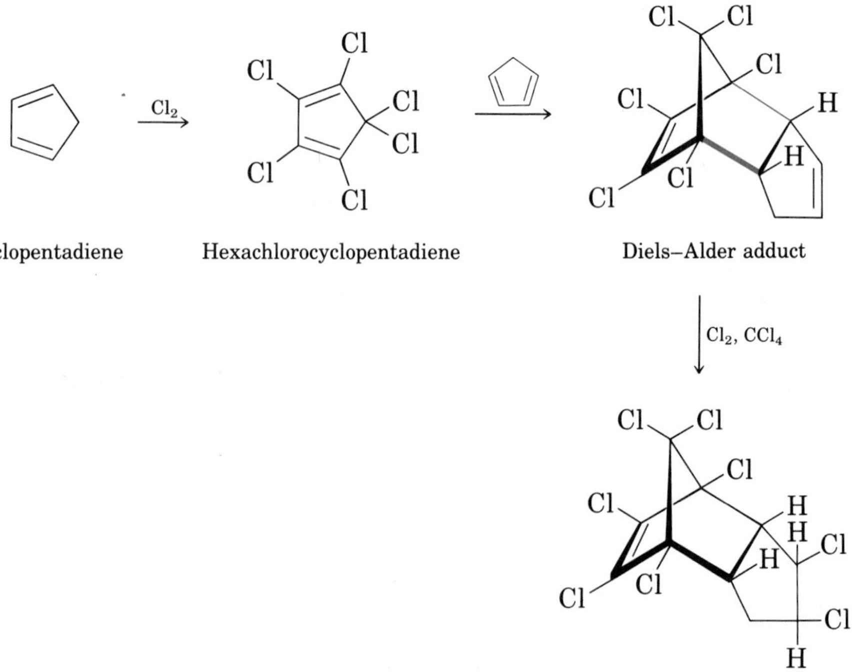

Cl_2

Cyclopentadiene

Hexachlorocyclopentadiene

Diels–Alder adduct

Cl_2, CCl_4

Chlordane—an insecticide

Similar reaction between hexachlorocyclopentadiene and the dienophile norbornadiene yields the insecticide aldrin. Treatment of aldrin with peroxyacetic acid yields another insecticide, dieldrin. The last reaction is an example of an *epoxidation,* which we will study in Section 19.7.

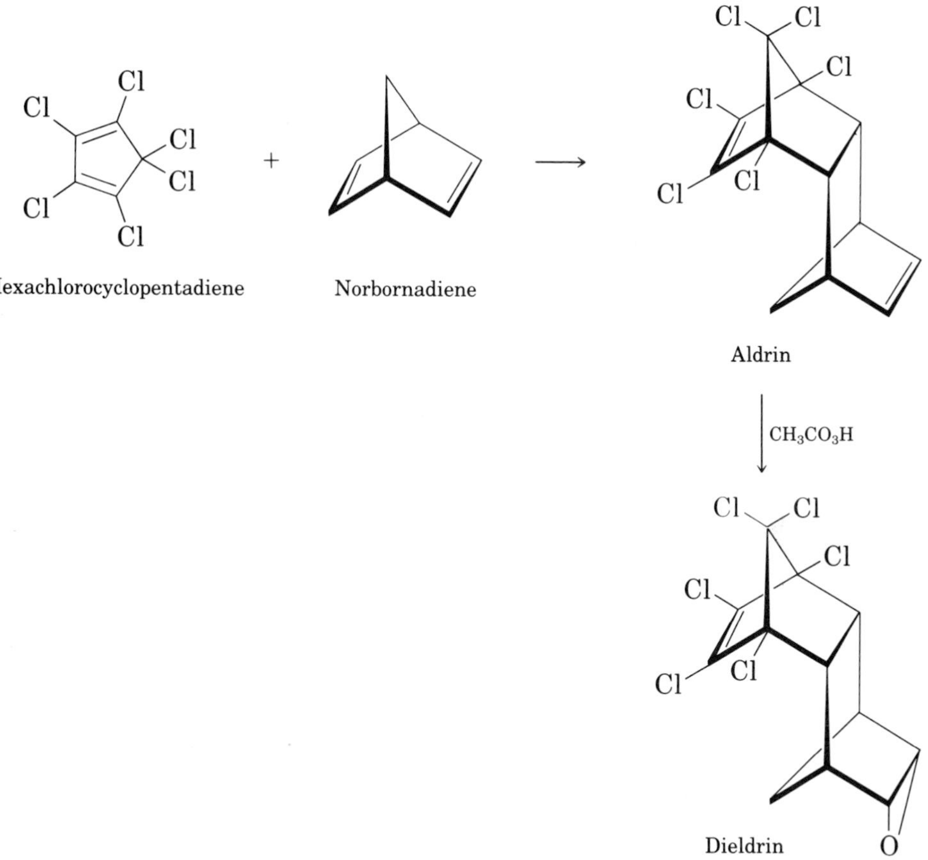

Hexachlorocyclopentadiene + Norbornadiene ⟶ Aldrin

Aldrin $\xrightarrow{CH_3CO_3H}$ Dieldrin

All three of these chlorinated Diels–Alder products are potent, nonspecific insecticides that are long-lasting and highly toxic to humans. The Environmental Protection Agency has banned most uses of chlordane, and usage of dieldrin and aldrin has been greatly curtailed.

PROBLEM

13.10 Although cyclopentadiene is highly reactive toward Diels–Alder cycloaddition reactions, 1,3-cyclohexadiene is less reactive, and 1,3-cycloheptadiene is nearly inert. Can you suggest a reason for this reactivity order? (Building molecular models should be helpful.)

13.10 Other conjugated systems

Early in this chapter a conjugated system was defined as one that consists of alternating single and double bonds. After considering a molecular orbital description of 1,3-butadiene, however, we might now more accurately describe a conjugated system as one that consists of *an extended series* of overlapping *p* orbitals (Figure 13.13). Thus a 1,3-diene and an allylic cation are both examples of conjugated systems, but there are other kinds as well.

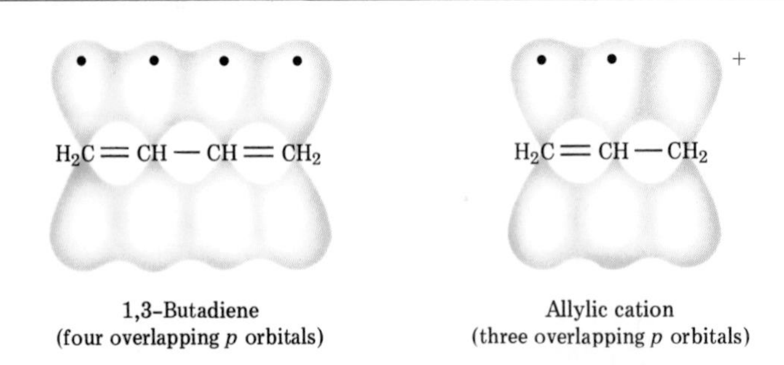

Figure 13.13. Overlapping *p* orbitals in 1,3-butadiene and the allylic cation.

One of the most important kinds of conjugated systems results from overlap of alkene *p* orbitals with a neighboring *p* orbital containing a lone pair of electrons. The neighboring *p* orbital is often on an atom such as oxygen or nitrogen, and the system can either be negatively charged or neutral (Figure 13.14).

An enamine

An enol ether

An enolate anion

Figure 13.14. Some conjugated systems resulting from overlap of alkene *p* orbitals with lone-pair electrons on neighboring atoms.

Each conjugated system will be examined in more detail at a later point. For the present, however, it is sufficient to note that the electronic nature of the alkene pi bond is greatly affected by conjugation with a neighboring lone pair. As the resonance forms shown in Figure 13.14 indicate, conjugation with a neighboring lone pair greatly increases the electron density of the carbon–carbon double bond. Thus enol ethers, enamines, and enolates are all strongly nucleophilic.

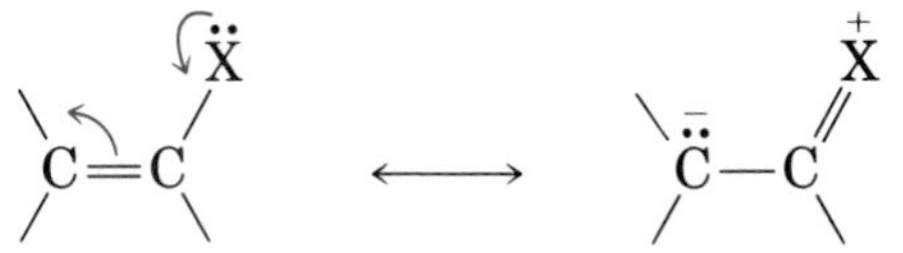

This resonance form puts extra electron density on *carbon,* making the carbon atom nucleophilic

The assertion that alkoxy groups (R—Ö—) and dialkylamino groups (R_2N̈—) donate electrons when conjugated with carbon–carbon double bonds may be surprising. After all, both oxygen and nitrogen are quite electronegative, and we have considered them as electron-withdrawing up to this point.

In fact, oxygen and nitrogen substituents behave in two different ways. They can *withdraw* electrons through sigma bonds as a result of their electronegativity, a phenomenon called the **inductive effect** (Section 2.4). These same substituents, however, *donate* electrons via pi bonds as a result of the **resonance effect.** The two effects operate in entirely different ways and pull in different directions.

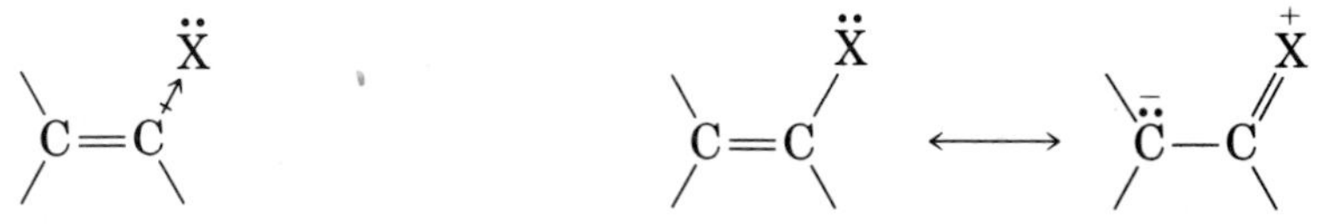

Inductive effect
(O and N *withdraw* sigma bond electrons)

Resonance effect
(O and N *donate* pi bond electrons)

13.11 Structure determination of conjugated systems: ultraviolet spectroscopy

Infrared, nuclear magnetic resonance, and mass spectroscopy all help in the structure determination of conjugated systems. In addition to these three generally useful spectroscopic techniques, there is a fourth—**ultraviolet (UV) spectroscopy**—that is applicable solely to conjugated systems.

1. Infrared spectroscopy	Functional groups present
2. Mass spectroscopy	Molecular size and formula
3. Nuclear magnetic resonance spectroscopy	Carbon–hydrogen framework
4. Ultraviolet spectroscopy	Nature of conjugated pi-electron system

Ultraviolet spectroscopy is less commonly used than the other spectroscopic techniques because of the rather specialized information it gives; we will therefore study it only briefly.

The ultraviolet region of the electromagnetic spectrum extends from the low wavelength end of the visible region (4×10^{-5} cm) down to 10^{-6} cm, but the portion of greatest interest to organic chemists is the narrow range from 2×10^{-4} cm to 4×10^{-4} cm. Absorptions in this region are usually measured in nanometers, nm (1 nm = 10^{-9} m = 10^{-7} cm). Thus, the ultraviolet range of interest is from 200 to 400 nm (Figure 13.15).

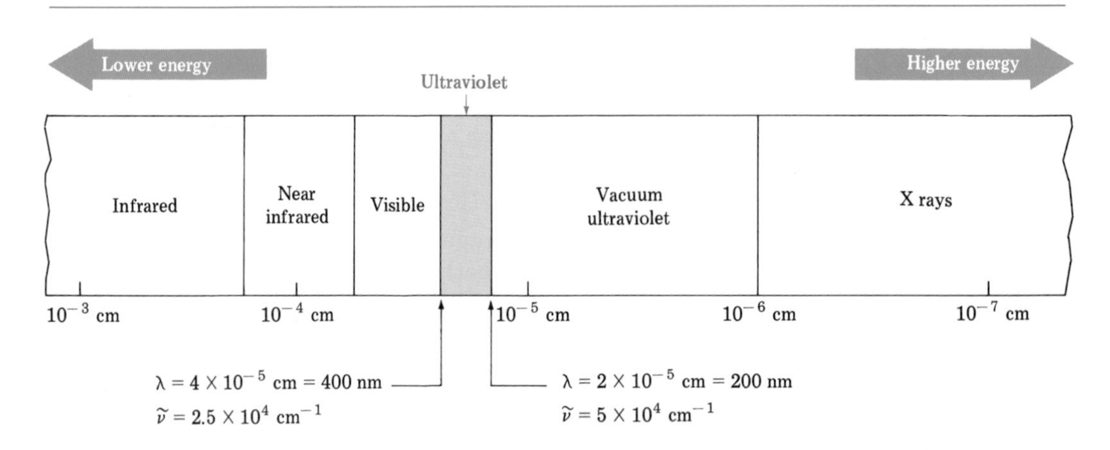

Figure 13.15. The ultraviolet region of the electromagnetic spectrum.

When an organic molecule is irradiated with electromagnetic waves, the radiation is either absorbed by the compound or passes through it, depending on the exact energy of the waves. We have already seen that when infrared radiation is used, the energy absorbed corresponds to the amount necessary to increase molecular motions—bendings, stretchings, twistings—of functional groups. On ultraviolet irradiation, the energy absorbed by a molecule corresponds to the amount necessary to excite electrons from one molecular orbital to another. Let's see what this means by looking first at 1,3-butadiene.

PROBLEM

13.11 Calculate the energy range of electromagnetic radiation in the ultraviolet region of the spectrum from 200 to 400 nm wavelength. Recall

$$E = \frac{Nhc}{\lambda} = \frac{2.85 \times 10^{-3}\ \text{kcal/mol}}{\lambda\ (\text{cm})}$$

PROBLEM

13.12 How does the energy you calculated (Problem 13.11) for ultraviolet radiation compare with the values calculated previously for infrared spectroscopy and nuclear magnetic resonance spectroscopy?

13.12 Ultraviolet spectrum of 1,3-butadiene

1,3-Butadiene has four pi molecular orbitals; the two lower-energy bonding molecular orbitals are fully occupied in the ground state, whereas the two higher-energy antibonding molecular orbitals are unoccupied, as illustrated in Figure 13.16 .

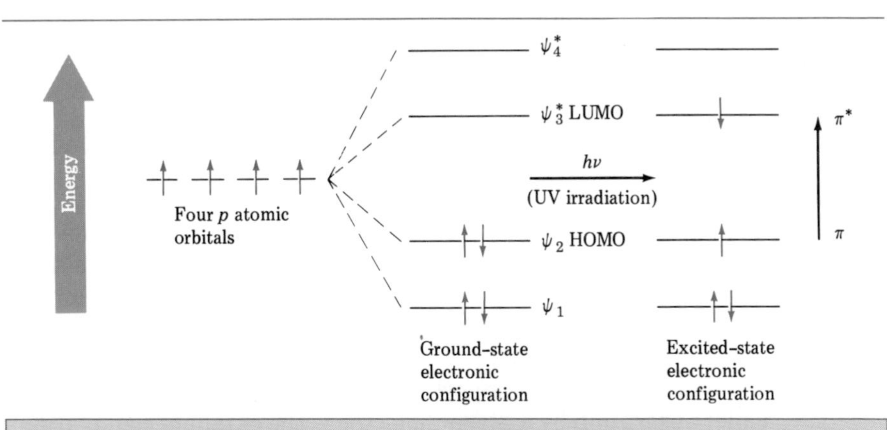

Figure 13.16. Ultraviolet excitation of 1,3-butadiene. An electron is promoted from ψ_2 to ψ_3^*.

On irradiation with ultraviolet light (denoted $h\nu$), 1,3-butadiene absorbs energy, and one pi electron is promoted from ψ_2, the highest occupied molecular orbital (HOMO), to ψ_3^*, the lowest unoccupied molecular orbital (LUMO). Since the electron is promoted from a bonding (π) molecular orbital to an antibonding (π^*) molecular orbital, we call this a $\pi \rightarrow \pi^*$ excitation (read as "pi to pi star"). The energy gap between the HOMO and the LUMO of 1,3-butadiene is such that ultraviolet light of 217 nm wavelength is required to accomplish the $\pi \rightarrow \pi^*$ electronic transition.

In practice, the ultraviolet spectrum of a conjugated molecule is recorded by irradiating a sample with ultraviolet light of continuously changing wavelength. When the wavelength of light corresponds to the energy level required to excite an electron to a higher level, energy is absorbed. This absorption is detected and displayed on a chart that plots wavelength versus percent radiation absorbed (Figure 13.17). Unlike infrared spectra

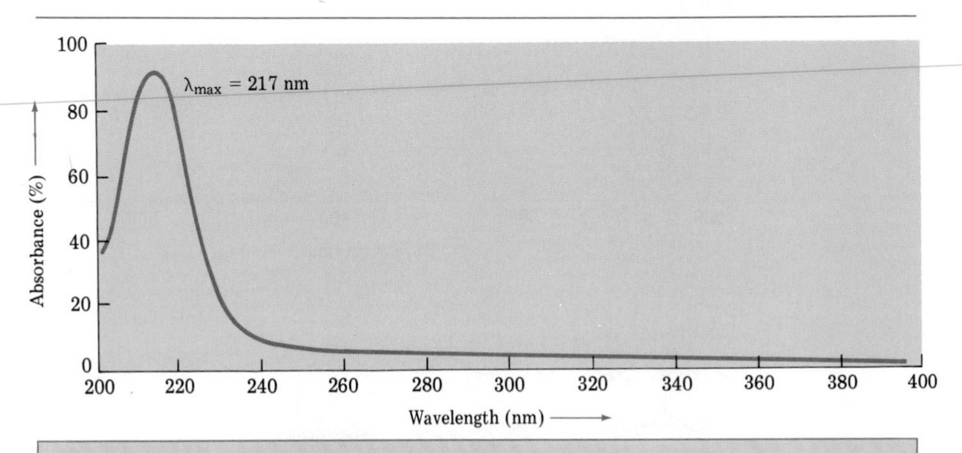

Figure 13.17. Ultraviolet spectrum of 1,3-butadiene.

and nuclear magnetic resonance spectra, which show many absorption lines for a given molecule, ultraviolet spectra are usually quite simple—often only one peak is produced. The peak is usually broad, and we identify its position by noting the wavelength (λ) at the very top of the peak (λ_{max}).

13.13 Interpreting ultraviolet spectra: the effect of conjugation

The exact wavelength of radiation necessary to effect the $\pi \rightarrow \pi^*$ transition in a conjugated molecule depends on the nature of the conjugated system. Thus, by measuring the ultraviolet spectrum of an unknown, we can derive structural information about the nature of any conjugated pi-electron system present.

One of the most important factors affecting the wavelength of ultraviolet absorption by a given molecule is the extent of conjugation. Molecular orbital calculations show that the energy difference between HOMO and LUMO decreases as the extent of conjugation increases; thus, 1,3-butadiene shows an absorption at $\lambda_{max} = 217$ nm, whereas 1,3,5-hexatriene absorbs at $\lambda_{max} = 258$ nm, and 1,3,5,7-octatetraene has $\lambda_{max} = 290$ nm. (Remember: Longer wavelength means lower energy.) β-Carotene, a conjugated polyene pigment responsible for the orange color of carrots, has such a long-

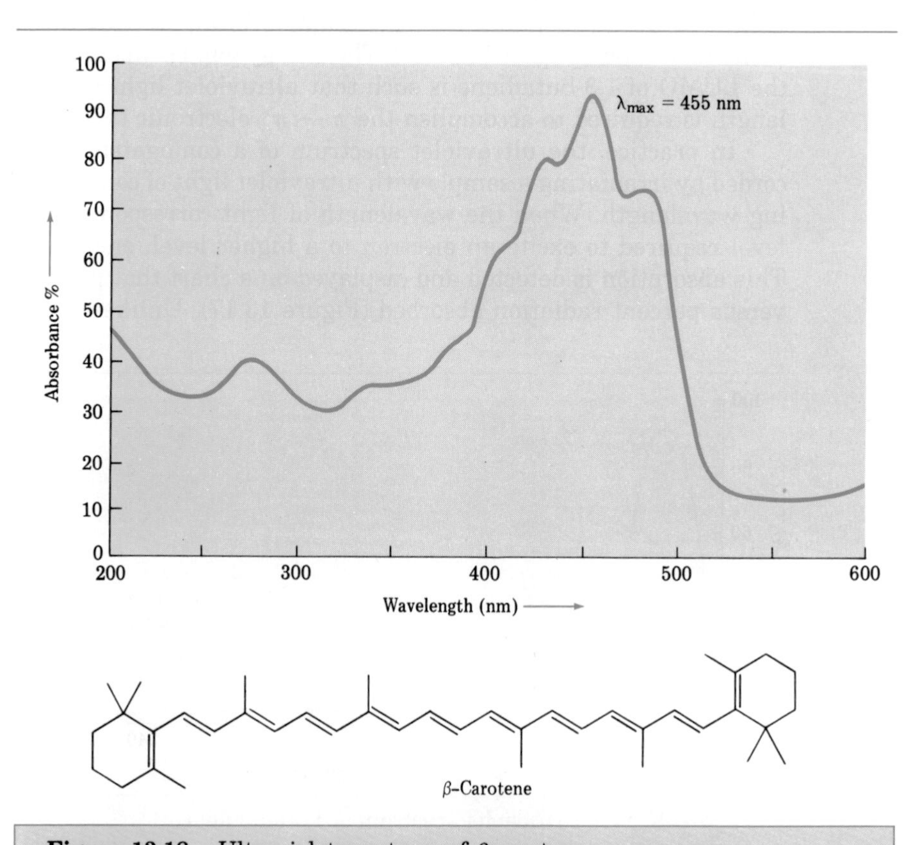

Figure 13.18. Ultraviolet spectrum of β-carotene.

wavelength absorption, $\lambda_{max} = 455$ nm, that the absorption actually occurs in the *visible* region of the electromagnetic spectrum (Figure 13.18). β-Carotene therefore is deep yellow in dilute solution and is used as a coloring agent in margarine.

Other kinds of conjugated systems besides dienes and polyenes also show ultraviolet absorptions. For example, conjugated enones and aromatic rings exhibit characteristic ultraviolet absorptions that aid in structure determination. More will be said about such compounds later, when the functional groups are discussed in more detail. The ultraviolet absorption maxima of some representative conjugated molecules are given in Table 13.3.

TABLE 13.3 Ultraviolet absorption maxima of some conjugated molecules

Name	*Structure*	λ_{max} *(nm)*
Ethylene	$H_2C=CH_2$	171
Cyclohexene	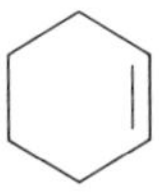	182
2-Methyl-1,3-butadiene	$H_2C=C(CH_3)-CH=CH_2$	220
1,3-Cyclohexadiene		256
1,3,5-Hexatriene	$H_2C=CH-CH=CH-CH=CH_2$	258
1,3,5,7-Octatetraene	$H_2C=CH-CH=CH-CH=CH-CH=CH_2$	290
2,4-Cholestadiene	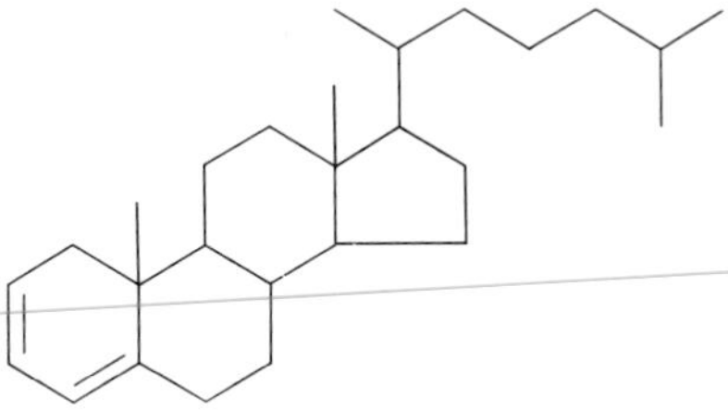	275
3-Buten-2-one	$H_2C=CH-C(CH_3)=O$	219
Benzene	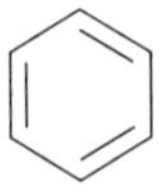	254
Naphthalene	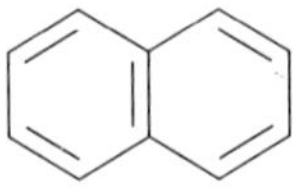	275

In addition to the $\pi \rightarrow \pi^*$ absorptions just discussed, other electronic transitions are observed in ultraviolet spectroscopy. Compounds with nonbonding electrons, such as the lone-pair electrons on oxygen, nitrogen, and halogen, show weak ultraviolet absorption. In these cases, a nonbonding electron (n) is promoted to an antibonding orbital (π^*). In acetone, for example, a nonbonding lone-pair electron on oxygen is excited by ultraviolet irradiation into the carbonyl antibonding π^* orbital. The resultant $n \rightarrow \pi^*$ transition shows an absorption peak at $\lambda_{max} = 272$ nm but is quite weak compared to the usual $\pi \rightarrow \pi^*$ absorption seen in conjugated systems. No further reference will be made to such absorptions, since they are not of great value for structure determination.

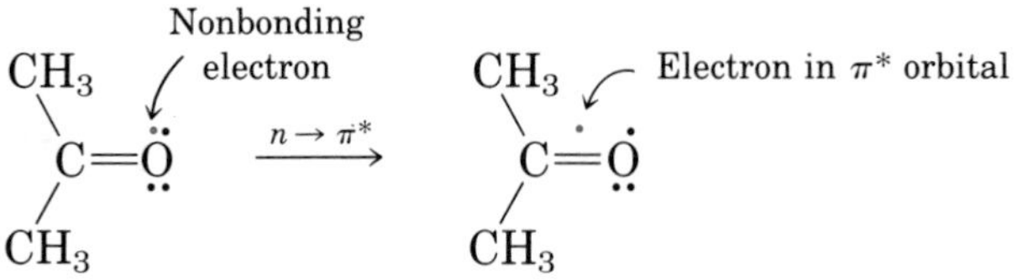

$$(CH_3)_2C{=}\ddot{O}: \xrightarrow{n \rightarrow \pi^*} (CH_3)_2C{=}\dot{O}:$$

Nonbonding electron

Electron in π^* orbital

PROBLEM

13.13 Which of the following compounds would you expect to show ultraviolet absorptions in the 200–400 nm range?

(a) 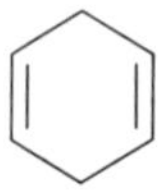 [1,3-cyclohexadiene]

(b) 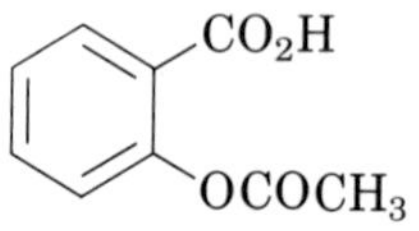 [benzene ring bearing CO_2H and $OCOCH_3$]

Aspirin

(c) $H_2C{=}CHNO_2$

(d) 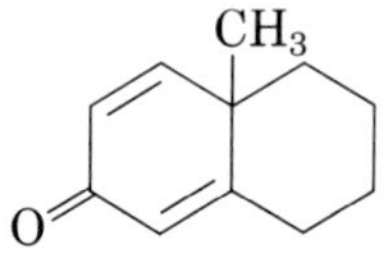 [bicyclic dienone with CH_3 and O]

(e) $H_2C{=}CHCN$

(f) 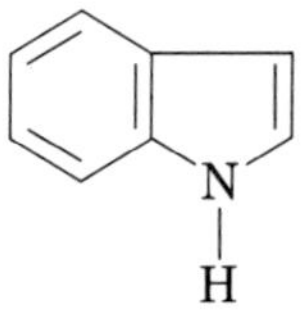 [indole structure, N–H]

Indole

13.14 Summary

A conjugated diene contains alternating single and double bonds. Thus, 1,3-butadiene is conjugated, whereas 1,4-pentadiene in nonconjugated.

$$H_2C{=}CH{-}CH{=}CH_2$$

A conjugated diene

For the most part, conjugated dienes are similar to other alkenes. One important difference between conjugated and nonconjugated dienes is that conjugated dienes are somewhat more stable than we might expect. Experiments show that 57.1 kcal/mol heat is released on hydrogenation of 1,3-butadiene. This value is about 3.5 kcal/mol less than we would expect for hydrogenation of two isolated double bonds, and we conclude that the conjugated diene has 3.5 kcal/mol extra stabilization.

The heat of hydrogenation data can be explained by a molecular orbital description of 1,3-butadiene. Four *p* atomic orbitals of the diene overlap to form four molecular orbitals, denoted ψ_1, ψ_2, ψ_3^*, and ψ_4^*. Two of the molecular orbitals are bonding and two are antibonding. When we assign four electrons to these orbitals, the two bonding orbitals, ψ_1 and ψ_2, are filled (Figure 13.19).

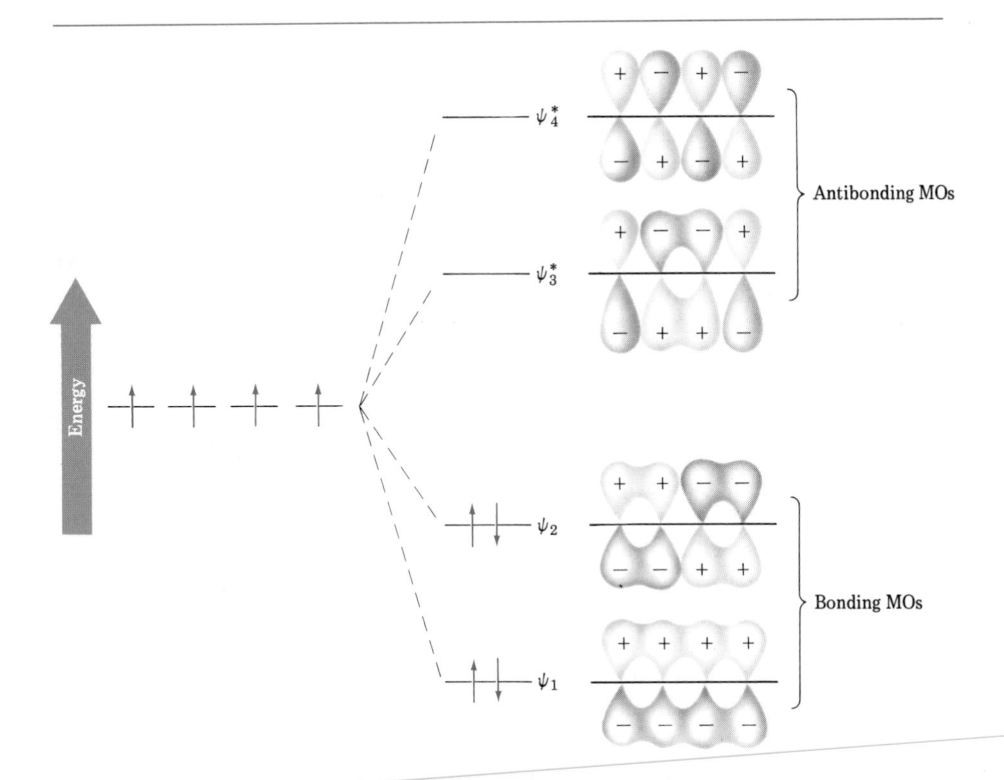

Figure 13.19. Molecular orbitals of 1,3-butadiene.

Calculations show that the sum of the energy levels of the two bonding 1,3-butadiene molecular orbitals is lower (i.e., the system is more stable) than the sum of the energy levels of two isolated double bonds. The reason for this extra stability is that in ψ_1 there is a bonding (stabilizing) interaction between carbons 2 and 3, which introduces some partial double-bond character (Figure 13.20, p. 438).

Conjugated dienes undergo two interesting reactions not observed for nonconjugated dienes. The first of these is 1,4 addition of electrophiles. When 1,3-butadiene is treated with HCl, both 1,2 and 1,4 adducts are

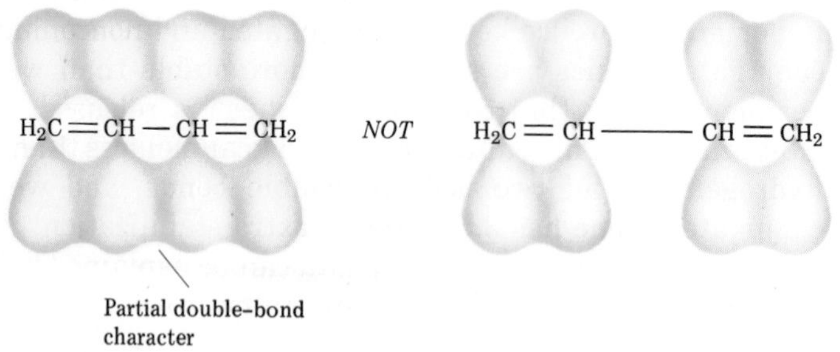

$H_2C{=}CH{-}CH{=}CH_2$ *NOT* $H_2C{=}CH{-}CH{=}CH_2$

Partial double-bond character

Figure 13.20. Partial double-bond character of the C2—C3 bond in 1,3-butadiene.

formed. Both products are formed from the same resonance-stabilized allylic carbocation intermediate, and are produced in varying ratios depending on the reaction conditions. The 1,2 adduct is usually formed faster and predominates at low temperature (kinetic control). The 1,4 adduct is usually more stable, however, and predominates when equilibrium is achieved at higher temperatures (thermodynamic control).

The second reaction unique to conjugated dienes is Diels–Alder cycloaddition. Conjugated dienes and electron-poor alkenes (dienophiles) react via a cyclic transition state to yield a cyclohexene product. This is an example of the general class of pericyclic reactions, which have neither polar nor radical mechanisms. Diels–Alder reactions can occur only if the diene is able to adopt an *s*-cis conformation. For this reason, cyclic dienes such as cyclopentadiene are highly reactive.

There are other kinds of conjugated systems besides polyenes. In general, a conjugated system is one that has an extended series of overlapping *p* orbitals. Thus enamines, enol ethers, and enolate anions are conjugated systems in which the double bond is made highly nucleophilic by resonance effects.

Ultraviolet spectroscopy is a method of structure determination uniquely applicable to conjugated systems. When a conjugated molecule is irradiated with ultraviolet light, energy absorption occurs and a pi electron is promoted from the highest occupied molecular orbital (HOMO) to the lowest unoccupied molecular orbital (LUMO). For 1,3-butadiene, radiation of λ_{max} = 217 nm is required. As a general rule, the greater the amount of conjugation, the less the energy needed (longer wavelength radiation).

ADDITIONAL PROBLEMS

13.14 Provide IUPAC names for the following alkenes:

(a) $CH_3CH{=}C(CH_3)CH{=}CHCH_3$

(b) $H_2C{=}CHCH{=}CHCH{=}CHCH_3$

(c) $CH_3CH{=}C{=}CHCH{=}CHCH_3$

(d) $CH_3CH{=}CHCH{=}CH_2$ with $CH_2CH_2CH_3$ on the third carbon

13.15 Circle any conjugated portions of these molecules:

(a) $CH_3CH{=}C{=}CHCH{=}CHCH_3$

(b) O, OCH_3

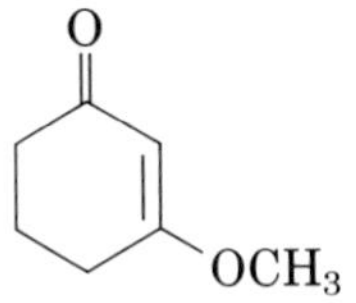

(c) OCH_3

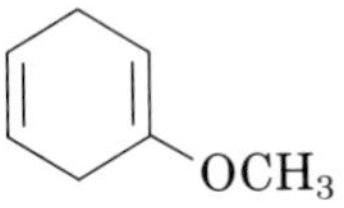

(d) O

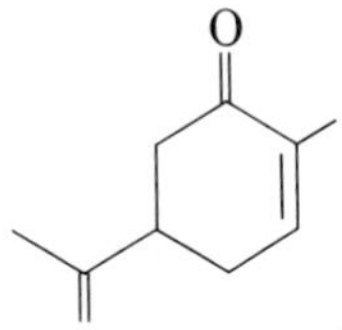

Carvone
(oil of spearmint)

(e) CH_3O

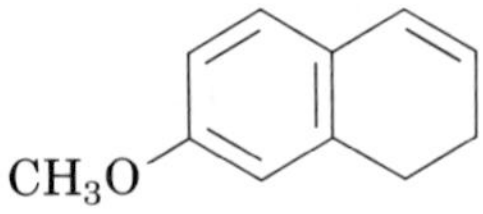

(f)

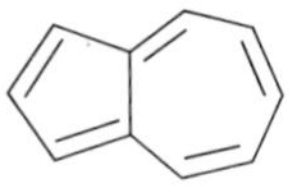

13.16 What product(s) would you expect to obtain from reaction of 1,3-cyclohexadiene with each of the following?
(a) 1 mol Br_2 in CCl_4
(b) O_3 followed by Zn
(c) 1 mol HCl in ether
(d) 1 mol DCl in ether
(e) 3-Buten-2-one ($H_2C{=}CHCOCH_3$)
(f) Excess OsO_4, followed by $NaHSO_3$

13.17 Draw and name the six possible diene isomers of formula C_5H_8. Which of the six are conjugated dienes?

13.18 A dihalide such as 3,4-dibromohexane can undergo base-induced double dehydrobromination to yield either 3-hexyne or 2,4-hexadiene. How would you use NMR spectroscopy to help you identify the product? How would you use ultraviolet spectroscopy?

13.19 Predict the products of these Diels–Alder reactions:

(a) + CHO ⟶

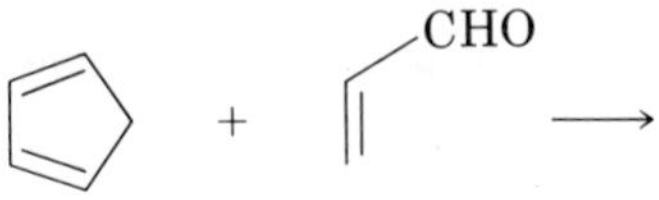

(b) + O, O, O ⟶

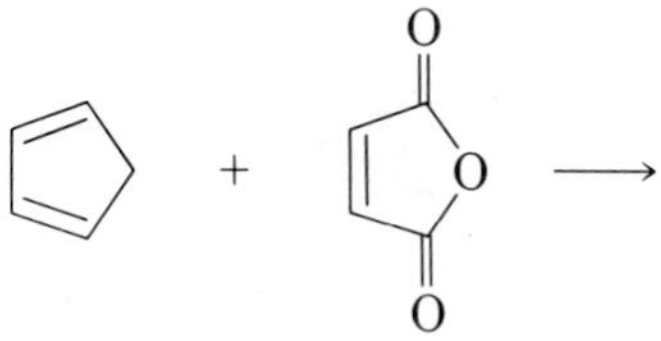

(c) + O O ⟶

(d) CH_3 CH_3 + $CO_2CH_3{-}C{\equiv}C{-}CO_2CH_3$ ⟶

13.20 How do you account for the fact that *cis*-1,3-pentadiene is much less reactive than *trans*-1,3-pentadiene in the Diels–Alder reaction?

13.21 Which of the following compounds would you expect to have $\pi \rightarrow \pi^*$ ultraviolet absorptions in the 200–400 nm range?

(a)

(b) $=CH_2$

(c) $H_2C{=}C{=}CHCH_3$
An allene

(d) $(CH_3)_2C{=}C{=}O$
A ketene

(e) N
Pyridine

13.22 Draw all possible products resulting from addition of 1 mol of HCl to 1-phenyl-1,3-butadiene. Which product or products would you expect to predominate, and why?

$CH{=}CH{-}CH{=}CH_2$

1-Phenyl-1,3-butadiene

13.23 How would you use Diels–Alder reactions to prepare these products? Show the starting dienes and dienophiles in each case.

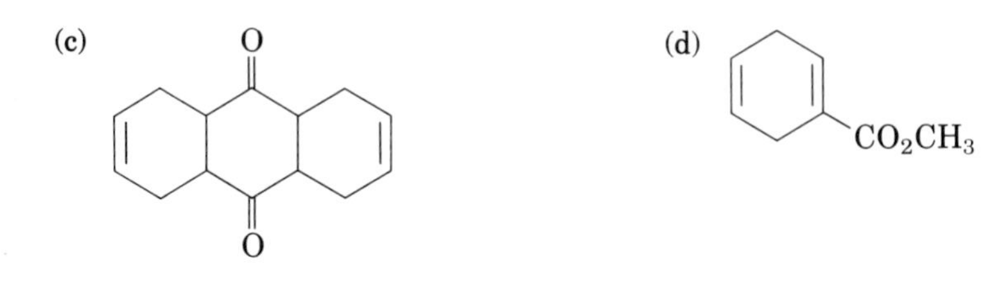

13.24 Would you expect 1,3-butadiyne to be a good Diels–Alder diene or a poor one? Explain.

13.25 We have seen that the Diels–Alder cycloaddition reaction is a pericyclic process that occurs in a concerted manner through a cyclic transition state. Depending on the exact energy levels of products and reactants, the Diels–Alder reaction can sometimes be made reversible. How can you account for the following reaction?

$\xrightarrow{\Delta}$ + $H_2C{=}CH_2$

13.26 Propose a mechanism to explain the following reaction. [*Hint:* Consider the reversibility of the Diels–Alder reaction.]

α-Pyrone + $CH_3O_2C{-}C{\equiv}C{-}CO_2CH_3$ $\xrightarrow{\Delta}$ (dimethyl phthalate: CO_2CH_3, CO_2CH_3) + CO_2

13.27 The following ultraviolet absorption maxima have been measured:

	λ_{max} (nm)
1,3-Butadiene	217
2-Methyl-1,3-butadiene	220
1,3-Pentadiene	223
2,3-Dimethyl-1,3-butadiene	226
2,4-Hexadiene	227
2,4-Dimethyl-1,3-pentadiene	232
2,5-Dimethyl-2,4-hexadiene	240

What conclusion can you draw from these data concerning the effect of alkyl substitution on ultraviolet absorption maxima? Approximately what effect does each added alkyl group have?

13.28 β-Ocimene is a pleasant-smelling hydrocarbon found in the leaves of certain herbs. It has the molecular formula $C_{10}H_{16}$ and exhibits an ultraviolet absorption maximum at 232 nm. On catalytic hydrogenation over palladium, a saturated hydrocarbon, $C_{10}H_{22}$, is obtained. This saturated product is identified as 2,6-dimethyloctane. Ozonolysis of β-ocimene, followed by treatment with hydrogen peroxide, produces four fragments: acetone, formic acid, pyruvic acid, and malonic acid.

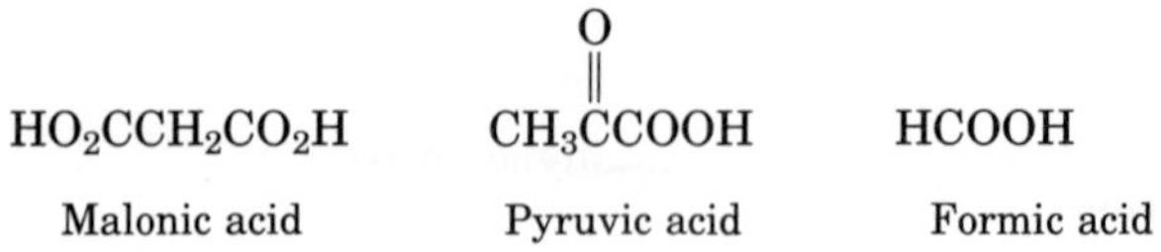

$HO_2CCH_2CO_2H$ — Malonic acid

$CH_3C(=O)COOH$ — Pyruvic acid

$HCOOH$ — Formic acid

(a) How many double bonds does β-ocimene have?
(b) Is β-ocimene conjugated or nonconjugated?
(c) Propose a structure for β-ocimene that is consistent with the observed data.
(d) Formulate all the reactions, showing starting material and products.

13.29 Myrcene, $C_{10}H_{16}$, is found in oil of bay leaves and is isomeric with β-ocimene (Problem 13.28). It shows an ultraviolet absorption at 226 nm and can be catalytically hydrogenated to yield 2,6-dimethyloctane. On ozonolysis followed by hydrogen peroxide treatment, myrcene yields formic acid, acetone, and 2-oxopentanedioic acid.

$HO_2CCH_2CH_2C(=O)CO_2H$

2-Oxopentanedioic acid

Propose a structure for myrcene, and formulate all the reactions, showing starting material and products.

13.30 Benzene has an ultraviolet absorption at λ_{max} = 204 nm and *p*-toluidine has λ_{max} = 235 nm. How do you account for this difference?

:NH_2

CH_3

Benzene (λ_{max} = 204 nm)

p-Toluidine (λ_{max} = 235 nm)

13.31 When the ultraviolet spectrum of *p*-toluidine (Problem 13.30) is measured in the presence of a small amount of HCl, the λ_{max} decreases to 207 nm, nearly the same value as for benzene. How do you account for the effect of acid?

13.32 Phenol is a weak acid with pK_a = 10.0. In ethanol solution, phenol has an ultraviolet absorption at λ_{max} = 210 nm. When dilute NaOH is added to this solution, the absorption increases to λ_{max} = 235 nm. How do you account for this shift?

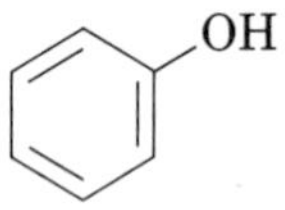

OH

Phenol

13.33 Hydrocarbon A, $C_{10}H_{14}$, has an ultraviolet absorption at $\lambda_{max} = 236$ nm and gives hydrocarbon B, $C_{10}H_{18}$, on catalytic hydrogenation. Ozonolysis of A followed by hydrogen peroxide treatment yields the following diketo diacid:

$$HO_2CCH_2CH_2CH_2\overset{O}{\overset{\|}{C}}-\overset{O}{\overset{\|}{C}}CH_2CH_2CH_2COOH$$

(a) Propose two possible structures for A.
(b) Hydrocarbon A reacts with maleic anhydride to yield a Diels–Alder adduct. Which of your structures for A is correct?
(c) Formulate all reactions showing starting material and products.

CHAPTER 14 BENZENE AND AROMATICITY

In the early days of organic chemistry, the word *aromatic* was used to describe fragrant substances such as benzaldehyde (from cherries, peaches, and almonds) and toluene (from Tolu balsam). It was soon realized, however, that substances grouped as aromatic behaved in a qualitatively different manner from most other organic compounds. Benzene, the simplest member of the class of aromatic compounds, was first isolated by Michael Faraday[1] in 1825 from the oily residue left by the illuminating gas used in London street lamps, and today we use the term **aromatic** to refer to the class of compounds composed of benzene and its structural relatives. We shall see in this and the next two chapters that aromatic substances show chemical behavior quite different from that of the aliphatic substances we have studied to this point. Thus, chemists of the early nineteenth century were correct when they realized that a chemical difference exists between aromatic compounds and other types, but the association of aromaticity with fragrance has long been lost.

Many compounds isolated from natural sources are aromatic in part. In addition to benzene, benzaldehyde, and toluene, complex compounds such as the female steroidal hormone, estrone, and the well-known analgesic, morphine, have aromatic rings.

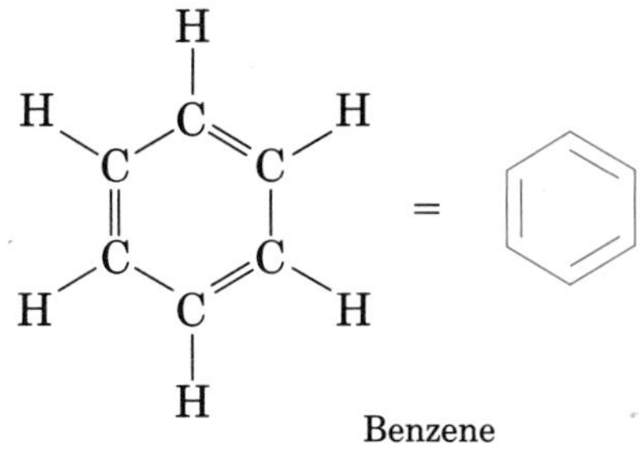

Benzene

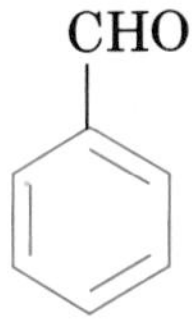

Benzaldehyde

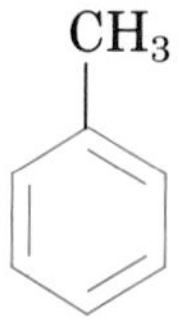

Toluene

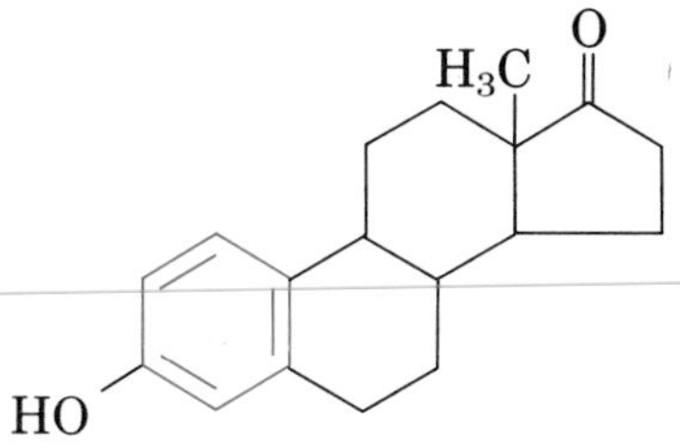

Estrone

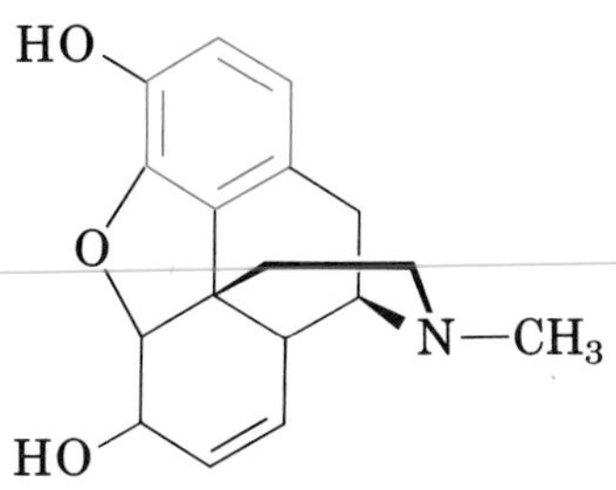

Morphine

Most synthetic drugs used medicinally are also aromatic in part. The local anesthetic procaine and the tranquilizer diazepam (Valium) are two of many examples.

[1]Michael Faraday (1791–1867); b. Newington Butts, Surrey, England; assistant to Sir Humphry Davy (1813); director, laboratory of the Royal Institution (1825); Fullerian Professor of Chemistry, Royal Institution (1833).

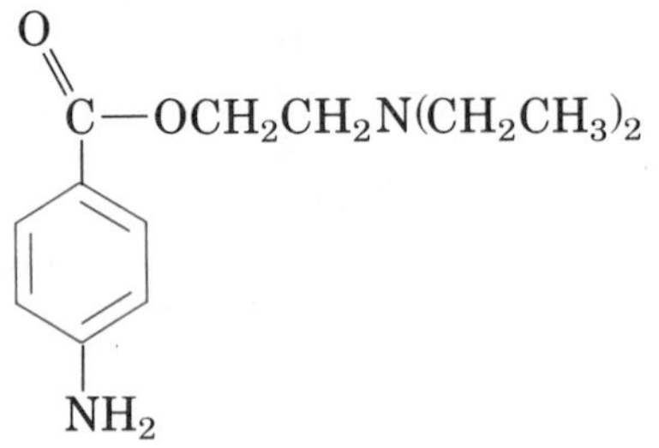

Procaine
(a local anesthetic)

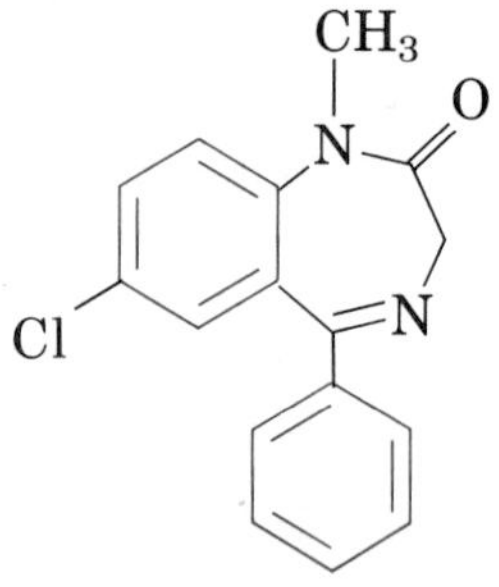

Diazepam
(Valium, a tranquilizer)

Benzene itself has been found to be toxic to humans. Prolonged exposure leads to bone-marrow depression and consequent leukopenia (depressed white-blood-cell count). Use of benzene as a solvent should therefore be avoided.

PROBLEM

14.1 Circle the aromatic portions of these molecules:

(a) Adrenaline (epinephrine)

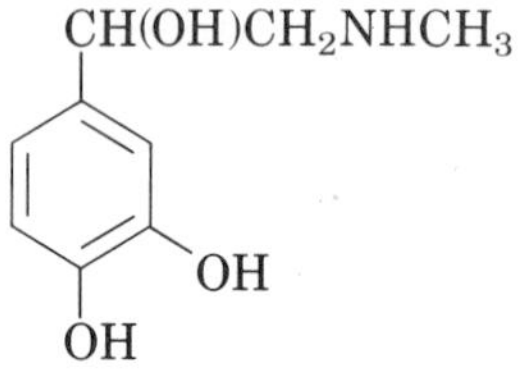

(b) Vitamin E

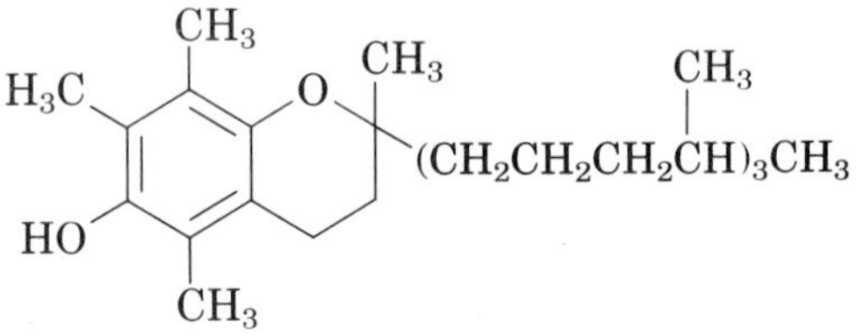

(c) Penicillin V

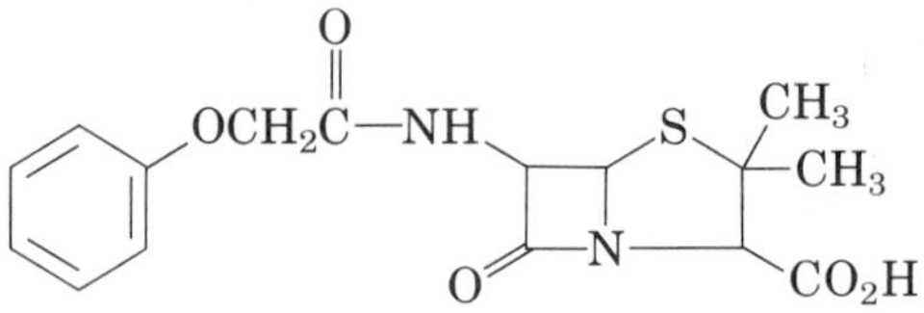

14.1 Nomenclature

Aromatic substances, more than any other class of organic compounds, have acquired a large number of trivial, nonsystematic names. Although the use of such names is discouraged, IUPAC rules allow for some of the more common ones to be retained. Some of these are shown in Table 14.1. Thus methylbenzene is known familiarly as toluene, hydroxybenzene as phenol, aminobenzene as aniline, and so on.

Monosubstituted benzene derivatives are systematically named in the same manner as other hydrocarbons, with *benzene* used as the parent name. Thus C_6H_5Br is bromobenzene, $C_6H_5NO_2$ is nitrobenzene, and $C_6H_5CH_2CH_2CH_3$ is propylbenzene.

TABLE 14.1 Trivial names of some common aromatic compounds

Formula	*Name*	*Formula*	*Name*
CH_3	Toluene (bp 110°C)	CHO	Benzaldehyde (bp 178°C)
OH	Phenol (mp 43°C)	COOH	Benzoic acid (mp 122°C)
NH_2	Aniline (bp 184°C)	CH_3, CH_3	*ortho*-Xylene (bp 144°C)
H_3C, CH_3, CH	Cumene (bp 152°C)	CH_3, CH_3	*meta*-Xylene (bp 139°C)
$CH{=}CH_2$	Styrene (bp 145°C)	CH_3, CH_3	*para*-Xylene (bp 138°C)

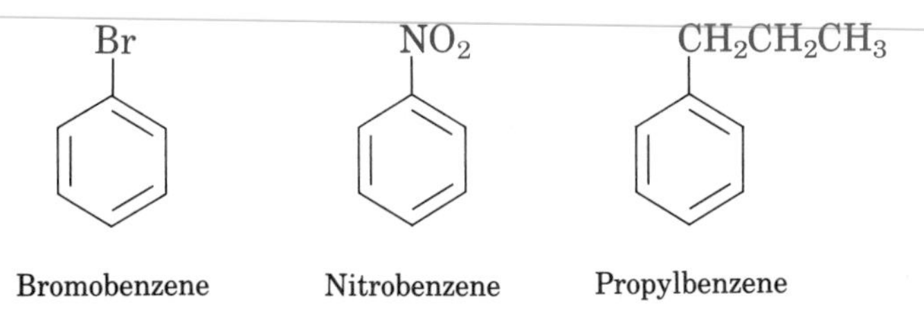

Bromobenzene Nitrobenzene Propylbenzene

Alkyl-substituted benzenes are sometimes referred to as **arenes,** and they are named in two different ways depending on the size of the alkyl group. If the alkyl substituent is small (six or fewer carbons), the arene is named as an alkyl-substituted benzene. If the alkyl substituent is larger than the ring (more than six carbons), it is also correct to name the compound as a phenyl-substituted alkane. The name **phenyl** is used for the C_6H_5 unit when the benzene ring is considered a substituent group. The word is derived from the Greek *pheno* ("I bear light"), in commemoration of Faraday's discovery of benzene from illuminating gas.

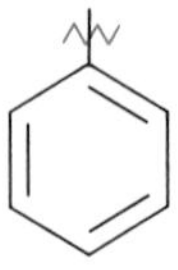

1 2 4 5 6 7 8
CH_3CH_2 $CH_2CH_2CH_2CH_2CH_3$
3
CH

A phenyl group 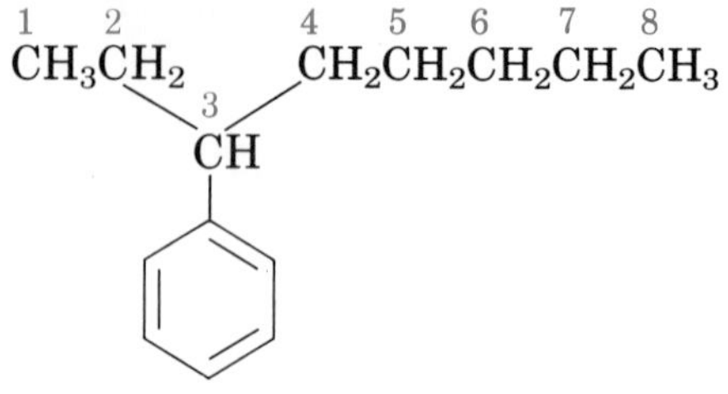 3-Phenyloctane

Disubstituted benzenes are named by using one of the prefixes *ortho-*, *meta-*, or *para-*. An *ortho-* or *o*-disubstituted benzene has the two substituents in a 1,2 relationship; a *meta* or *m*-disubstituted benzene has the substituents in a 1,3 relationship; and a *para-* or *p*-disubstituted benzene has the substituents in a 1,4 relationship.

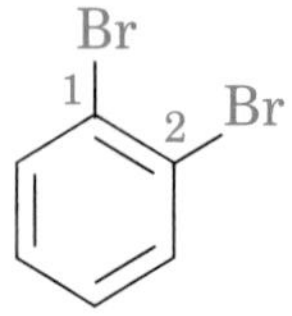

Br 1 2 Br

CH_3 1 2 3 CH_3

Br 1 2 3 4 Cl

ortho-Dibromobenzene, 1,2 disubstituted

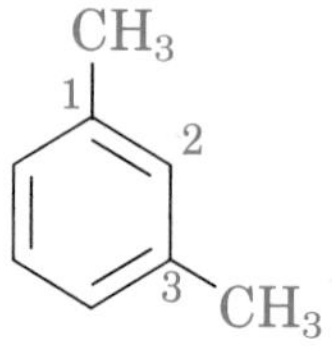

meta-Dimethylbenzene (*m*-xylene), 1,3 disubstituted

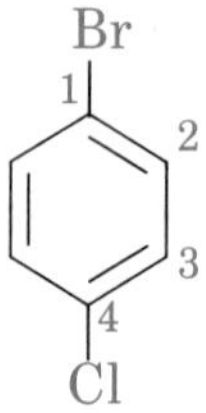

para-Bromochlorobenzene, 1,4 disubstituted

The ortho, meta, para system of nomenclature is valuable when discussing reactions, and chemists use it to refer to certain positions on the benzene ring. For example, we might describe the bromination of toluene by saying, "Reaction occurs in the para position"—in other words, para to the methyl group already present on the ring.

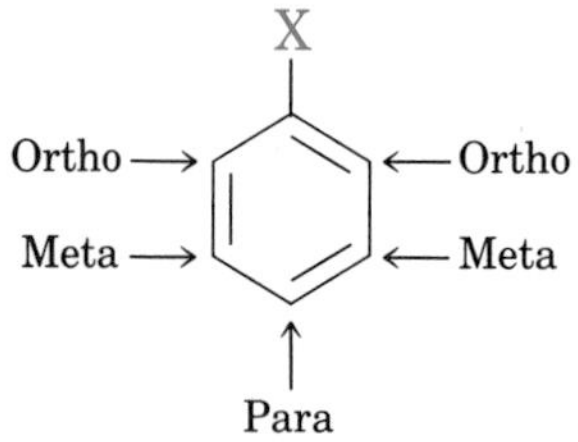

X

Ortho → ← Ortho

Meta → ← Meta

↑

Para

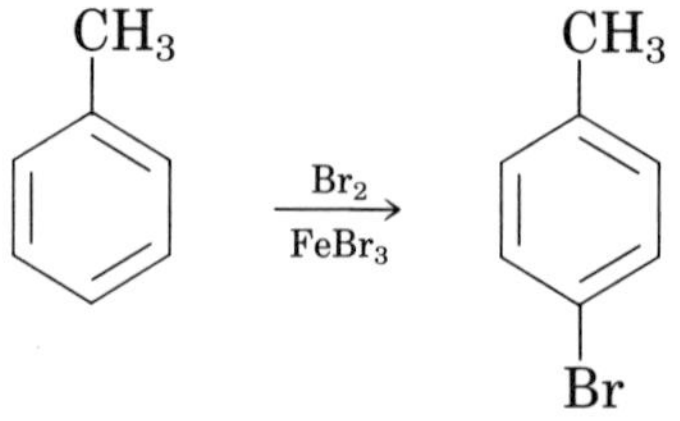

CH_3 $\xrightarrow[FeBr_3]{Br_2}$ CH_3

Br

Toluene *p*-Bromotoluene

Benzenes with *more than two* substituents must be named by numbering the position of each substituent on the ring. The numbering should be carried out in such a way that the substituents are listed alphabetically and the lowest possible numbers are used.

4-Bromo-1,2-dimethylbenzene 1-Chloro-2,4-dinitrobenzene 2,4,6-Trinitrotoluene (TNT)

In the third example shown, note that *toluene* is used as the base name rather than *benzene*. Any of the monosubstituted aromatic compounds shown in Table 14.1 can serve as a base name; in such cases the principal substituent (CH_3 in toluene) is assumed to be on carbon 1. The following two examples further illustrate this rule, and the base name is italicized.

2,6-Dibromo*phenol* *m*-Chloro*benzoic acid*

PROBLEM

14.2 Provide correct IUPAC names for these compounds:

(c) NH_2, Br (d) CH_3, Cl, Cl

(e) CH_2CH_3, O_2N, NO_2 (f) CH_3, CH_3, H_3C, CH_3

PROBLEM

14.3 The following names are incorrect. Draw the structure represented by each, and provide correct IUPAC names.
(a) 2-Bromo-3-chlorobenzene (b) 4,6-Dinitrotoluene
(c) 4-Bromo-1-methylbenzene (d) 2-Chloro-*p*-xylene

14.2 Structure of benzene: the Kekulé proposal

The story of the elucidation of the benzene structure is a fascinating chapter in the history of organic chemistry. Perhaps more than any other single substance, benzene has played a central role in the development of modern structural theory.

By the mid-1800s benzene was known to have the molecular formula C_6H_6, and its chemistry was being actively explored. For example, it was known that although benzene was relatively unreactive toward most reagents that attack alkenes, it would react with bromine in the presence of iron to give the *substitution* product C_6H_5Br, rather than the possible *addition* product $C_6H_6Br_2$. Furthermore, only one monobromo substitution product was known; no isomers had been prepared.

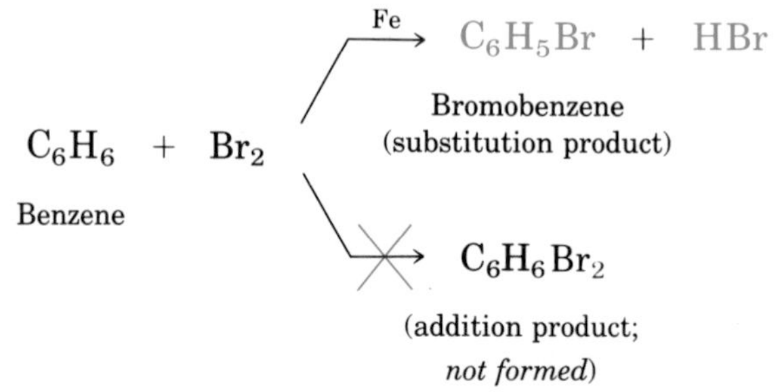

$$C_6H_6 + Br_2 \xrightarrow{Fe} C_6H_5Br + HBr$$

Benzene → Bromobenzene (substitution product)

$$C_6H_6 + Br_2 \not\rightarrow C_6H_6Br_2$$

(addition product; *not formed*)

On further reaction with bromine, disubstitution products were obtained, and three isomeric $C_6H_4Br_2$ compounds had been prepared. On the basis of these and similar results, Kekulé proposed in 1865 that benzene consists of a *ring* of carbon atoms and may be formulated as cyclohexatriene. Kekulé reasoned that this structure would readily account for the isolation of only a single monobromo substitution product, since all six carbon atoms and all six hydrogens in cyclohexatriene are equivalent.

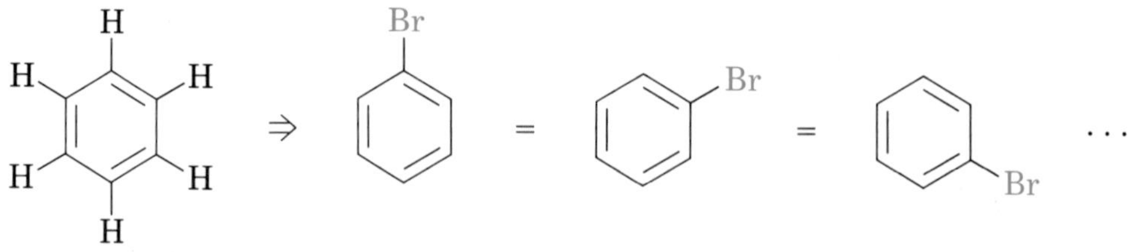

All six hydrogens of cyclohexatriene are equivalent

Only one possible monobromo product

The experimental observation that only three isomeric dibromo substitution products were known was more difficult to explain, since four structures can be written:

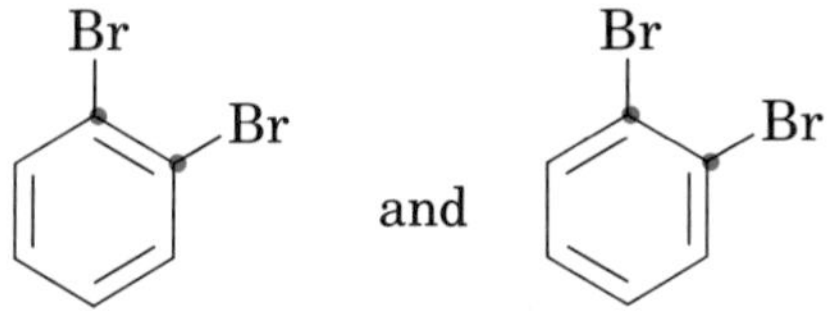

Two possible "1,2-dibromocyclohexatrienes"

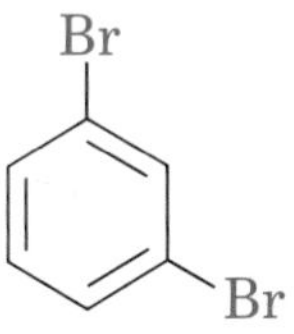

"1,3-Dibromocyclohexatriene" "1,4-Dibromocyclohexatriene"

Although there is only one possible 1,3 derivative and one possible 1,4 derivative, there appear to be two possible 1,2-dibromo substitution products, depending on the positions of the double bonds in the ring. Kekulé accounted for the formation of only three isomers by proposing that the double bonds in benzene rapidly "oscillate" between two positions. Thus, according to Kekulé, the two "1,2-dibromocyclohexatrienes" cannot be separated because they interconvert too rapidly (Figure 14.1).

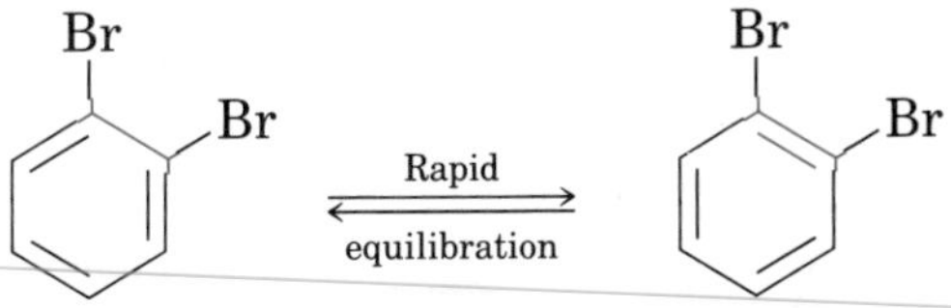

Rapid equilibration

Figure 14.1. The Kekulé proposal: Benzene double bonds rapidly oscillate back and forth between neighboring carbons. Distinct forms cannot be isolated.

Kekulé's proposed structure for benzene was widely criticized at the time. Although it satisfactorily accounts for the correct number of mono- and disubstituted benzene isomers, it does not answer the critical questions of why benzene is unreactive compared to other alkenes, and why benzene gives a substitution product rather than an addition product on reaction with bromine.

PROBLEM

14.4 According to Kekulé's theory, how many tribromo benzene derivatives are possible? Draw and name them.

PROBLEM

14.5 The following structures have the formula C_6H_6 and at one time were suggested for benzene. If we assume that bromine can be substituted for hydrogen in these structures, how many monobromo derivatives are possible for each? How many dibromo derivatives?

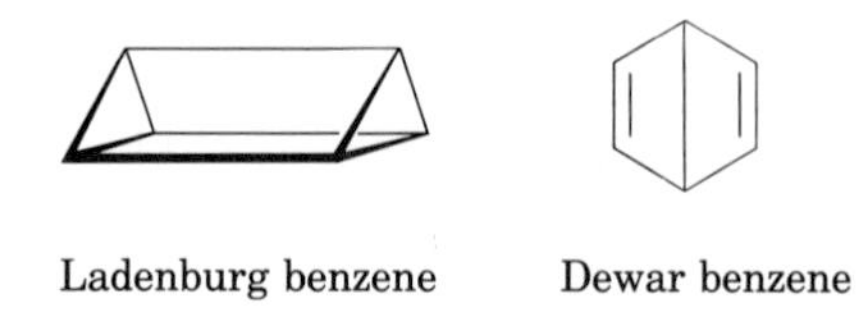

Ladenburg benzene Dewar benzene

14.3 Stability of benzene

The unusual chemical stability of benzene was a great puzzle to early chemists. Although its formula, C_6H_6, indicates the presence of multiple sites of unsaturation, and although the Kekulé structure proposes three carbon–carbon double bonds, benzene shows none of the behavior characteristic of alkenes. For example, alkenes react readily with potassium permanganate to give cleavage products; they react rapidly with osmium tetroxide to give 1,2 diols; and they react with gaseous HCl to give saturated alkyl chlorides. Benzene does none of these things (Figure 14.2). *Benzene does not undergo electrophilic addition reactions.*

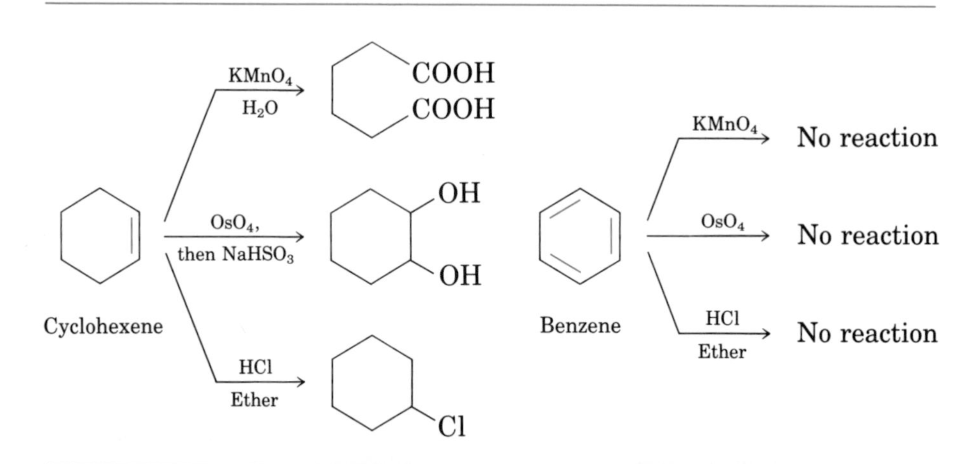

Figure 14.2. A comparison of the reactivity of cyclohexene and benzene.

We can get a quantitative idea of benzene's unusual stability by examining heats of hydrogenation (Table 14.2). We have seen that a compound containing an isolated double bond releases about 26–30 kcal/mol of heat upon catalytic hydrogenation, depending on its degree of substitution. Cyclohexene seems like a good model in the present case, and experimental data show a value for cyclohexene of $\Delta H^\circ_{\text{hydrog}} = 28.6$ kcal/mol. From our study of the effects of conjugation on alkene stability (Section 13.2), we would expect 1,3-cyclohexadiene to have a heat of hydrogenation somewhat

TABLE 14.2 **Heats of hydrogenation of cyclic alkenes**

Reactant	*Product*	ΔH°_{hydrog} *(kcal/mol)*
Cyclohexene	Cyclohexane	28.6
1,3-Cyclohexadiene	Cyclohexane	55.4
Benzene	Cyclohexane	49.8

less than twice the cyclohexene value, and this is exactly what is found:

$$\Delta H^\circ_{\text{hydrog}} \text{ for 1,3-cyclohexadiene} = 55.4 \text{ kcal/mol}$$

If we carry the analogy one step further, we might predict $\Delta H^\circ_{\text{hydrog}}$ for "cyclohexatriene" (benzene) to be a bit less than three times the cyclohexene value. The actual value is 49.8 kcal/mol, 36 kcal/mol *less* than expected. Since 36 kcal/mol less heat than expected is *released* during hydrogenation of benzene, benzene must *have* 36 kcal/mol less energy than expected. In other words, benzene has 36 kcal/mol "extra" stability (Figure 14.3).

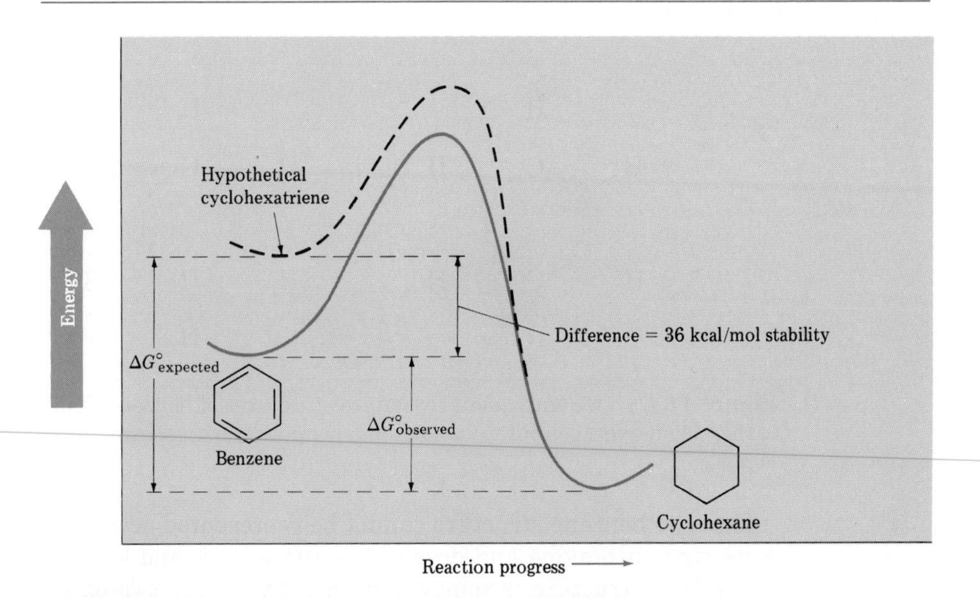

Figure 14.3. Reaction energy diagram for the hydrogenation of benzene and a hypothetical cyclohexatriene.

Further evidence for the unusual nature of benzene comes from spectroscopic studies. All carbon–carbon bonds in benzene have the same length, intermediate between a normal single and a normal double bond. Most carbon–carbon single bonds have lengths near 1.5 Å and most carbon–carbon double bonds are about 1.34 Å long. 1,3-Butadiene, for example, has a single-bond length of 1.48 Å and a double-bond length of

1.34 Å. All carbon–carbon bonds in benzene are 1.39 Å long—intermediate between single and double bonds.

All C—C—C bond angles = 120°
All C—C bond lengths = 1.39 Å

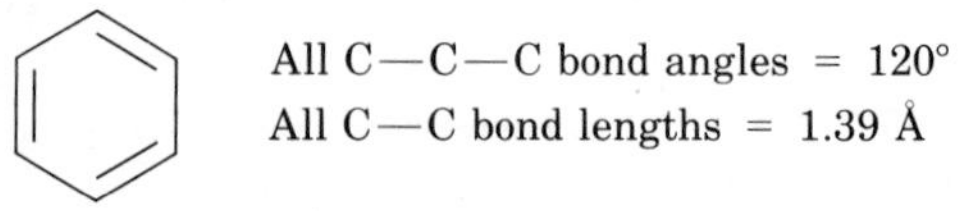

Benzene

14.4 Representations of benzene

How can we account for benzene's properties, and how can its structure best be represented? Resonance theory answers this question by saying that a *single* Kekulé structure is not satisfactory, and that benzene can best be described as a resonance hybrid of two equivalent Kekulé structures. Recall that resonance forms of a substance differ only in the placement of electrons; the nuclei themselves remain in the same position. This is exactly the situation in benzene. The six carbon nuclei form a regular hexagon while the pi electrons are shared equally between neighboring nuclei (Figure 14.4). Each carbon–carbon bond averages 1.5 electrons, and all bonds are equivalent.

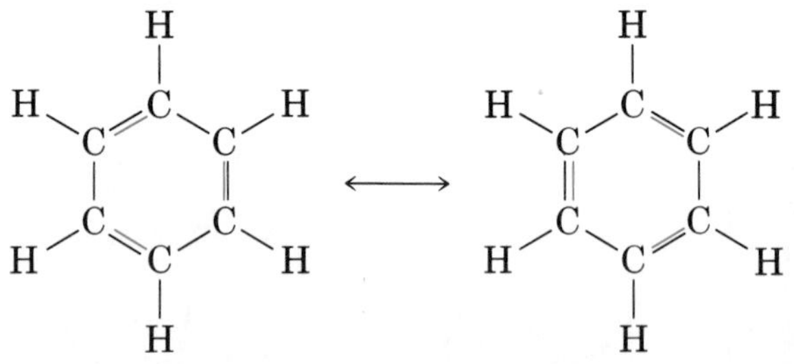

Figure 14.4. Two equivalent resonance structures of benzene. Each carbon–carbon bond averages out to be 1.5 bonds—midway between single and double bonds.

The true benzene structure cannot be represented accurately by either Kekulé structure alone and does *not* oscillate back and forth between the two; the true structure is somewhere in between the two extremes and is impossible to draw with our usual conventions. We might try to represent it by drawing benzene with either a full or dotted circle to indicate the equivalence of all carbon–carbon bonds, as in Figure 14.5, but these rep-

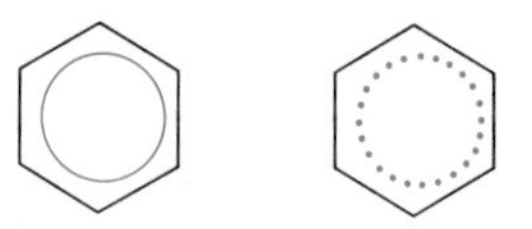

Figure 14.5. Alternative representations of benzene.

resentations can sometimes cause more problems than they solve. How many electrons does a circle represent? In this book, benzene and other arenes will be represented by a single Kekulé structure. In this way we will be able to keep track of all pi electrons during reactions, but we must be aware of the limitations of our drawings.

There is a subtle yet important difference between Kekulé's representation of benzene and our resonance representation. Kekulé considered benzene as "oscillating" rapidly back and forth between two cyclohexatriene structures, whereas resonance theory considers benzene a single "resonance hybrid" structure:

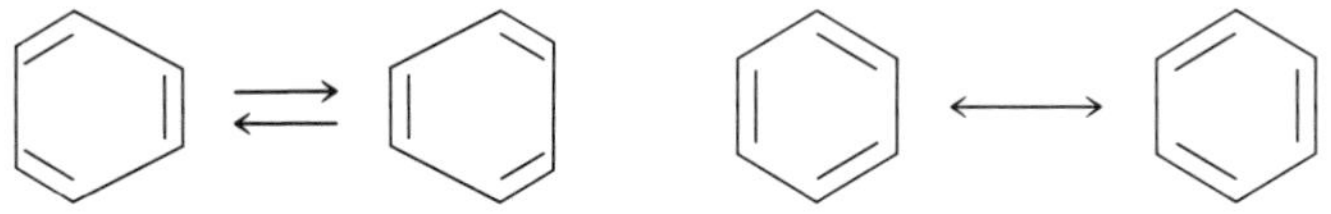

Kekulé benzene   Modern benzene

At any given instant, Kekulé's oscillating structures do not have all carbon–carbon bonds of the same length; three bonds are short and three are long. This bond length difference implies that the carbon atoms must change position in oscillating from one structure to another, and thus the two structures are not the same as resonance forms.

To complete this resonance description of benzene, we can now see why benzene is unusually stable and why all of its carbon–carbon bonds have the same 1.39 Å length. Resonance theory accounts for the stability of benzene by invoking the same postulate used to explain the stability of the allyl cation in Section 13.5: The more stable resonance forms a substance has, the more stable it is. Benzene, with two stable resonance forms of equal energy, is highly stabilized. All of the carbon–carbon bonds of benzene are equivalent and therefore have the same length.

The same reasoning explains why there is only one *ortho*-dibromobenzene, rather than two. The two possible Kekulé structures are different resonance forms of a single compound whose true structure is intermediate between the two (Figure 14.6).

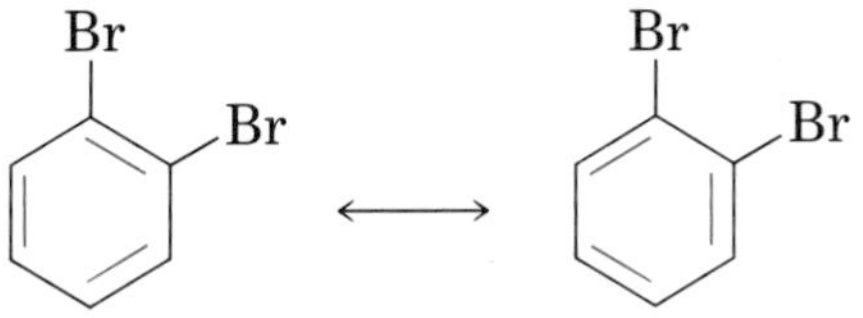

Figure 14.6. Two equivalent resonance forms of *o*-dibromobenzene.

PROBLEM

14.6 In 1932, A. A. Levine and A. G. Cole studied the ozonolysis of *o*-xylene. They isolated three products: glyoxal (OHC—CHO), butane-2,3-dione ($CH_3COCOCH_3$), and pyruvaldehyde (CH_3COCHO):

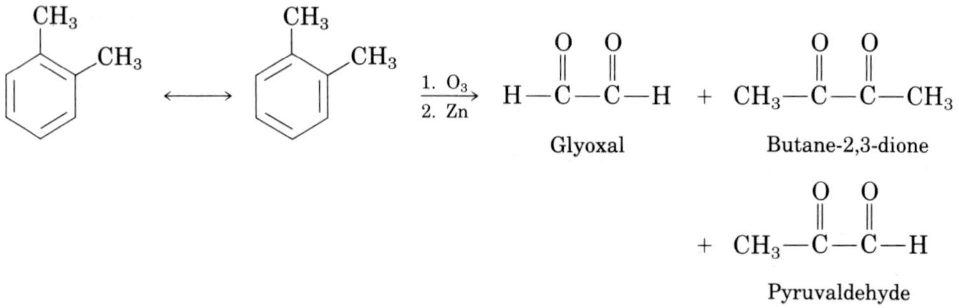

$$\text{o-xylene (Kekulé structure 1)} \longleftrightarrow \text{o-xylene (Kekulé structure 2)} \xrightarrow[\text{2. Zn}]{\text{1. } O_3} \underset{\text{Glyoxal}}{H-\overset{O}{\overset{\|}{C}}-\overset{O}{\overset{\|}{C}}-H} + \underset{\text{Butane-2,3-dione}}{CH_3-\overset{O}{\overset{\|}{C}}-\overset{O}{\overset{\|}{C}}-CH_3} + \underset{\text{Pyruvaldehyde}}{CH_3-\overset{O}{\overset{\|}{C}}-\overset{O}{\overset{\|}{C}}-H}$$

In what ratio would you expect these three products to be formed if *o*-xylene is a resonance hybrid of two Kekulé structures? The actual ratio found was 3 parts glyoxal, 2 parts pyruvaldehyde, and 1 part butane-2,3-dione. What conclusions can you draw about the structure of *o*-xylene?

14.5 Molecular orbital description of benzene

Molecular orbital theory provides a description of benzene that is in many respects superior to the simple resonance approach.

Benzene is a flat, symmetrical molecule having the shape of a regular hexagon, with all C—C—C bond angles having values of 120°. Each carbon atom is sp^2 hybridized, and an orbital picture looks like that in Figure 14.7. The six carbon–carbon sigma bonds are formed by C_{sp^2}–C_{sp^2} overlap, and each carbon has a *p* orbital perpendicular to the planar six-membered ring.

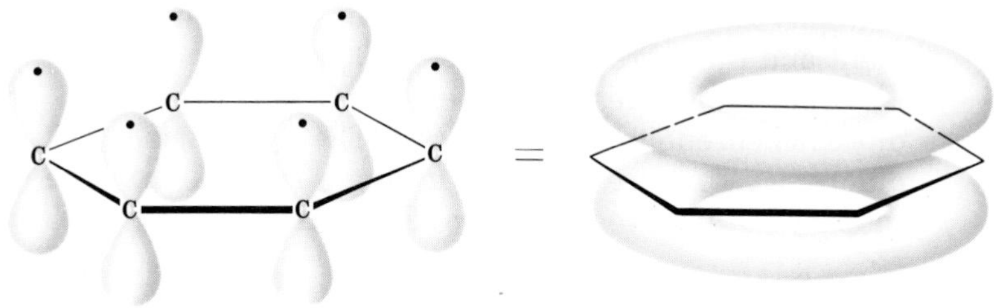

Figure 14.7. An orbital picture of benzene.

Since benzene is a cyclic conjugated system, and since all six *p* orbitals are equivalent, it is impossible to define three localized alkene pi bonds in which a given *p* orbital overlaps only one neighboring *p* orbital. Rather, each *p* orbital can overlap equally well with *both* neighboring *p* orbitals. This leads to a picture of benzene in which the pi electrons are completely delocalized around the ring. Benzene has two doughnut-shaped clouds of electrons, one above and one below the ring.

We can construct molecular orbitals for benzene just as we did for 1,3-butadiene in the preceding chapter. If we allow six *p* atomic orbitals to combine, six benzene molecular orbitals result, as illustrated in Figure 14.8. The three low-energy molecular orbitals, denoted ψ_1, ψ_2, and ψ_3, are of lower

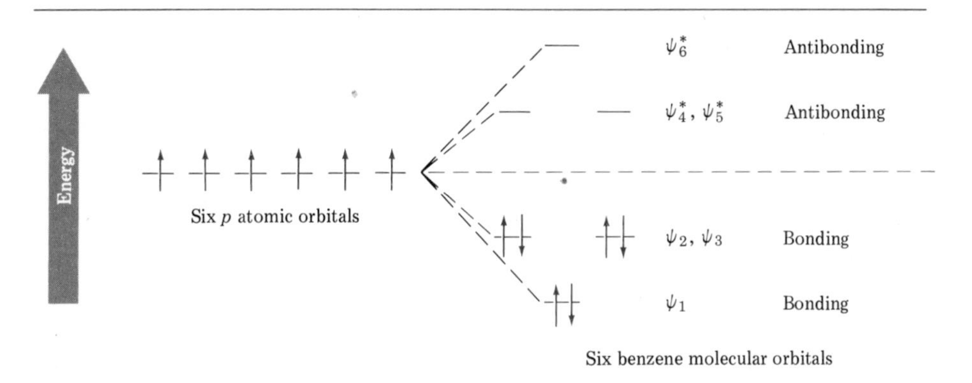

Figure 14.8. Molecular orbitals of benzene.

energy than the p atomic orbitals and are therefore bonding combinations. The three high-energy orbitals are antibonding. Note that two of the bonding orbitals, ψ_2 and ψ_3, have the same energy. Such orbitals are said to be **degenerate.** A similar situation holds for the degenerate antibonding orbitals, ψ_4^* and ψ_5^*.

The six p electrons occupy the three bonding molecular orbitals and are therefore delocalized over the entire conjugated system, leading to the observed 36 kcal/mol stabilization energy of benzene.

PROBLEM

14.7 Pyridine, C_5H_5N, is a flat, symmetrical molecule with bond angles of 120°. It undergoes electrophilic substitution rather than addition and generally behaves as an aromatic substance. Draw an orbital picture of pyridine to explain its properties. Check yourself by looking ahead to Section 14.8.

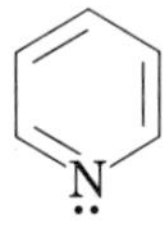

Pyridine

14.6 Aromaticity and the Hückel (4n + 2) rule

Let's review what we have learned thus far about benzene and, by extension, about other benzenoid aromatic molecules:

1. Benzene is a cyclic conjugated molecule with formula C_6H_6.
2. Benzene is unusually stable. Its heat of hydrogenation is 36 kcal/mol less than we might expect for a normal triene.
3. Benzene is a symmetrical, planar, hexagonal molecule. All C—C—C bond angles are 120°, and all C—C bonds are 1.39 Å in length.

4. Benzene undergoes substitution reactions that retain the cyclic conjugation, rather than electrophilic addition reactions that would destroy the conjugation.
5. Benzene can be described in terms of resonance theory as a hybrid whose structure is intermediate between two Kekulé structures:

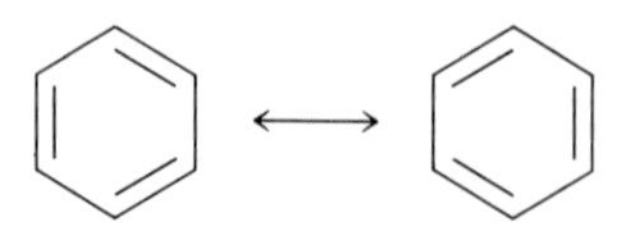

6. Benzene can be described more accurately by molecular orbital theory. The six atomic *p* orbitals combine to form six benzene molecular orbitals, with resultant electron delocalization throughout the conjugated system.

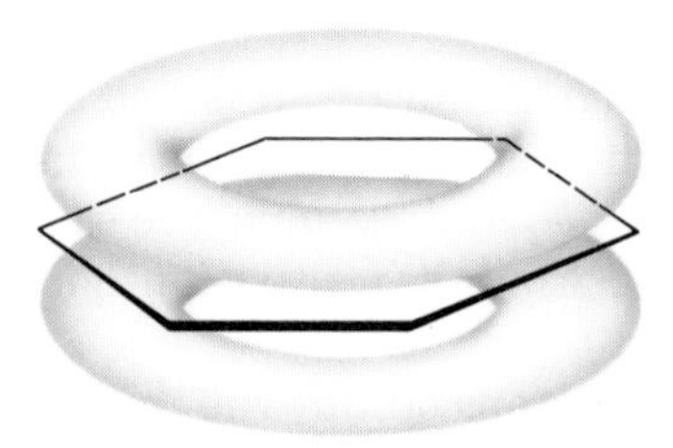

Taken together, these facts provide a description of benzene and of other aromatic molecules, but they are not sufficient. Something else is needed to complete our description of aromaticity.

According to theoretical calculations carried out by the German physicist Erich Hückel[2] in 1931, an aromatic molecule must be a cyclic conjugated species with $(4n + 2)$ pi electrons, where n is an integer ($n = 0, 1, 2, 3, \ldots$). In other words, only molecules with 2, 6, 10, 14, 18, . . . pi electrons can be aromatic. Molecules with $4n$ pi electrons (4, 8, 12, 14, . . .) *cannot* be aromatic.

Let's look at some examples to see how the $(4n + 2)$ rule works.

1. *Cyclobutadiene:*

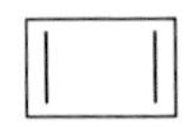

Four pi electrons

Cyclobutadiene is not aromatic. It is a $4n$ pi electron molecule ($n = 1$) and is highly unstable. A long history of attempts to synthesize the compound culminated in 1965 when cyclobutadiene was prepared, but not isolated, at low temperatures. Even at −78°C, cyclobutadiene dimerizes by Diels–Alder cycloaddition:

[2]Erich Hückel (1896–); b. Charlottenburg, Germany; Ph.D. Göttingen (Debye); professor at Stuttgart and Marburg.

Certainly, cyclobutadiene displays none of the properties we associate with aromaticity.

2. *Benzene:*

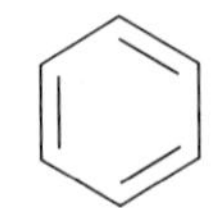

Six pi electrons

Benzene has already been discussed in detail. It is a ($4n + 2$) pi electron system, where $n = 1$, and is an excellent example of an aromatic compound.

3. *Cyclooctatetraene:*

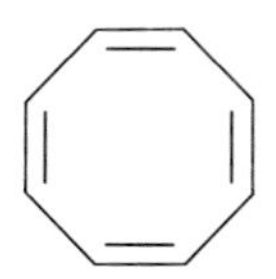

Eight pi electrons

Cyclooctatetraene is a $4n$-pi-electron molecule ($n = 2$), and it is not aromatic. Cyclooctatetraene was first prepared in 1911 by the great German chemist, Richard Willstätter.[3] Chemists at that time believed that the only requirement for aromaticity was the presence of a cyclic conjugated system, and it was therefore expected that cyclooctatetraene, as a close analog of benzene, would also prove to be unusually stable. The facts proved otherwise; cyclooctatetraene resembles open-chain polyenes in its reactivity. Cyclooctatetraene reacts readily with bromine, with potassium permanganate, and with hydrogen chloride, just as do other alkenes. We now know, in fact, that cyclooctatetraene is not even fully conjugated. It is a tub-shaped molecule in which there is no cyclic conjugation because neighboring orbitals do not have the proper geometry for overlap (Figure 14.9). The pi electrons are localized

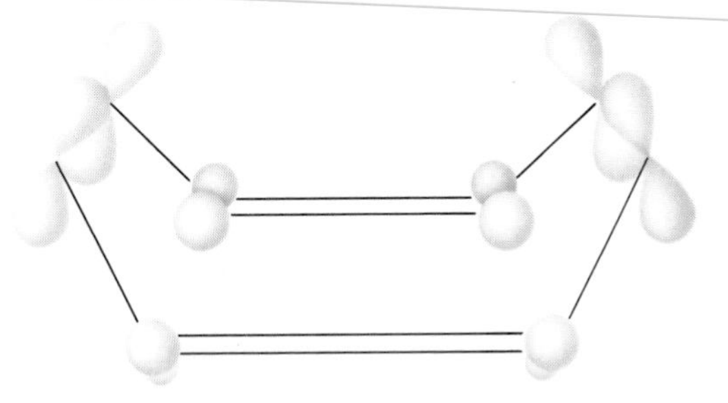

Figure 14.9. An orbital view of cyclooctatetraene. It is a tub-shaped molecule and has no cyclic conjugation.

[3]Richard Willstätter (1872–1942); b. Karlsruhe, Germany; Technische Hochschule, Munich (Einhorn) (1895); professor, Zurich, Dahlem, Munich; Nobel prize (1915)

in four distinct carbon–carbon double bonds rather than delocalized as in benzene. X-ray studies show that the carbon–carbon single bonds are 1.54 Å long; the double bonds, 1.34 Å long. In addition, the ^1H NMR spectrum shows a single sharp resonance line at 5.78 δ, characteristic of an alkene rather than of an aromatic molecule (Section 14.11).

PROBLEM

14.8 It is important to remember that in order to be aromatic a molecule must be flat, so that *p*-orbital overlap can occur, and must have $(4n + 2)$ pi electrons. Cyclodecapentaene fulfills one of these criteria but not the other. It is therefore not aromatic. Explain why this is so. (Molecular models may be useful.)

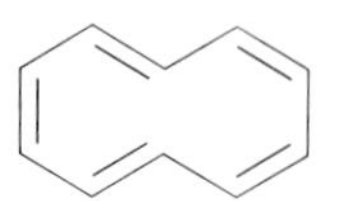

Cyclodecapentaene (not aromatic)

14.7 Aromatic ions

Hückel's rule is broadly applicable to many kinds of molecules; among the more interesting cases of aromatic species are those that are ionic. The cyclopentadienyl anion is one such example.

Cyclopentadiene itself is not aromatic because it is not a fully conjugated molecule; the CH_2 carbon in the ring is sp^3 hybridized. Imagine, though, that we *remove* one hydrogen from the saturated CH_2 group, and let that carbon become sp^2 hybridized. There are three ways we could do this:

1. We could remove the hydrogen and both electrons from the C—H bond. Since the hydrogen removed must carry a negative charge, the cyclopentadienyl group is left as a carbocation.
2. We could remove the hydrogen and one electron, leaving a cyclopentadienyl radical.
3. We could remove the hydrogen ion (H^+) with no electrons, leaving a cyclopentadienyl anion.

These three possibilities are shown in Figure 14.10.

Resonance theory predicts that all three species should be highly stabilized, since it is possible to draw five equivalent resonance structures for each. Hückel theory, however, predicts that *only* the six-pi-electron anion should be aromatic. The four-pi-electron cyclopentadienyl carbocation and the five-pi-electron cyclopentadienyl radical are predicted not to be aromatic, even though they are cyclic conjugated species.

In practice, both the cyclopentadienyl cation and the radical are highly reactive and difficult to prepare, and neither species shows any sign of the unusual stability expected of an aromatic system. The six-pi-electron cyclopentadienyl anion, by contrast, is a remarkably stable **carbanion** (carbon anion). Cyclopentadiene is one of the most acidic hydrocarbons known. Most hydrocarbons have a $pK_a > 45$, but cyclopentadiene has a pK_a of 16, a value comparable to that of water! Cyclopentadiene is acidic because the anion formed by ionization is so stable. It does not matter that the cyclopenta-

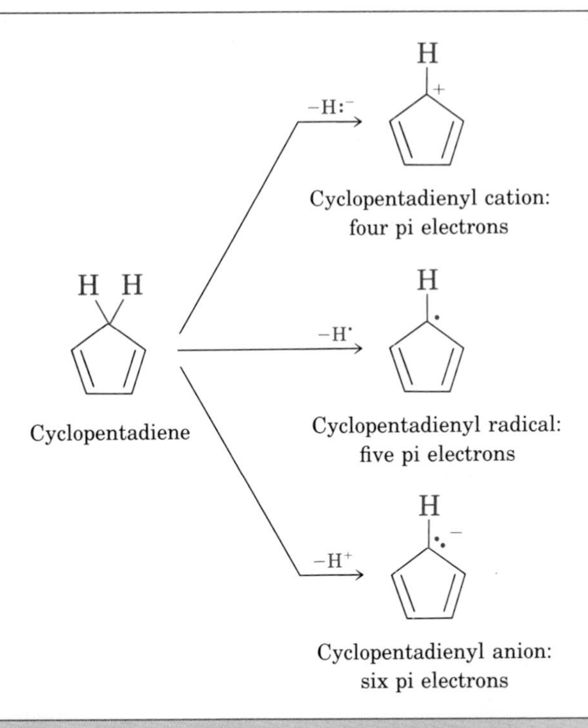

Figure 14.10. Cyclopentadienyl cation, radical, and anion.

Cyclopentadiene
($pK_a = 16$)

Aromatic six-pi-electron
cyclopentadienyl anion

Figure 14.11. An orbital view of the aromatic cyclopentadienyl anion.

dienyl anion has only five *p* orbitals; all that matters is that there are six pi electrons, a Hückel number (Figure 14.11).

Similar arguments can be used to predict the stability of the cycloheptatrienyl cation, radical, and anion. Removal of a hydrogen from cycloheptatriene can generate either the six-pi-electron cation, the seven-pi-electron radical, or the eight-pi-electron anion (Figure 14.12, p. 462).

Once again, resonance theory predicts a high level of stability for all three species, since seven equivalent resonance forms can be drawn for each.

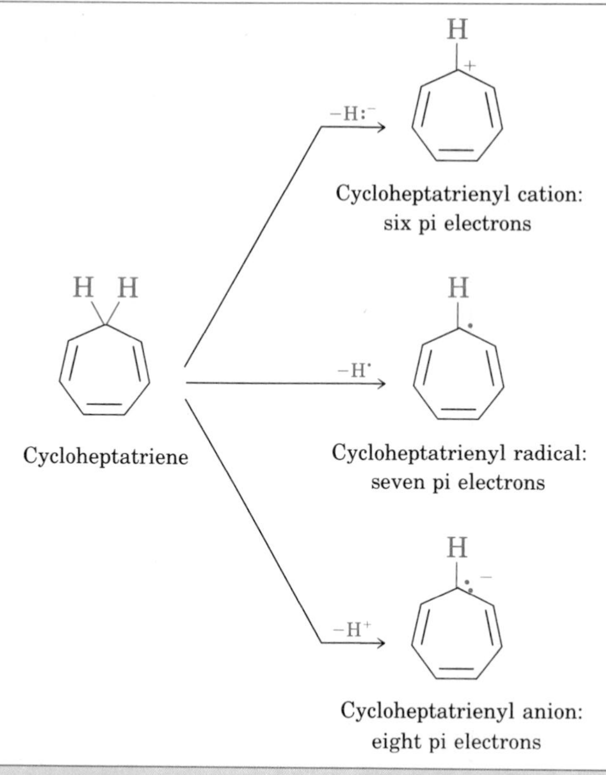

Figure 14.12. Cycloheptatrienyl cation, radical, and anion.

Hückel molecular orbital theory, however, predicts that only the six-pi-electron cycloheptatrienyl cation should have aromatic stability.

This is exactly what is found. Although the seven-pi-electron cycloheptatrienyl radical and the eight-pi-electron anion are highly reactive and difficult to prepare, the six-pi-electron cation is extraordinarily stable. In fact, the cycloheptatrienyl cation was first prepared in 1891, although its structure was not recognized at the time. Later investigation showed that the remarkably stable salt-like material prepared by action of bromine on cycloheptatriene is cycloheptatrienylium bromide (Figure 14.13).

H H

Cycloheptatriene $\xrightarrow{Br_2}$ HBr + [H, +] Br:⁻ = [ ]$^+$ Br:⁻

Cycloheptatrienylium bromide:
six pi electrons

Figure 14.13. Preparation of the aromatic cycloheptatrienyl cation.

PROBLEM

14.9 Draw all possible resonance structures of the cyclopentadienyl anion. Are all carbon–carbon bonds equivalent? How many absorption lines would you expect to see in the ^{1}H NMR and ^{13}C NMR spectra of the anion?

PROBLEM

14.10 Although cyclooctatetraene is not aromatic, it readily accepts two electrons from potassium to form the cyclooctatetraene dianion, $C_8H_8^{2-}$ Why do you suppose this is so? What geometry would you expect for the cyclooctatetraene dianion?

$$\xrightarrow{2\text{ K}} 2\text{ K}^+ \quad [\ \ ]^{2-}$$

14.8 Aromatic heterocycles

The cyclic compounds dealt with thus far have been **carbocycles**—all atoms in the rings have been carbon. **Heterocycles** are cyclic compounds that have one or more atoms other than carbon in the ring. The heteroatom is often nitrogen or oxygen, but sulfur, phosphorus, and other elements are also found. Many heterocycles are aromatic, and Figure 14.14 lists some common examples.

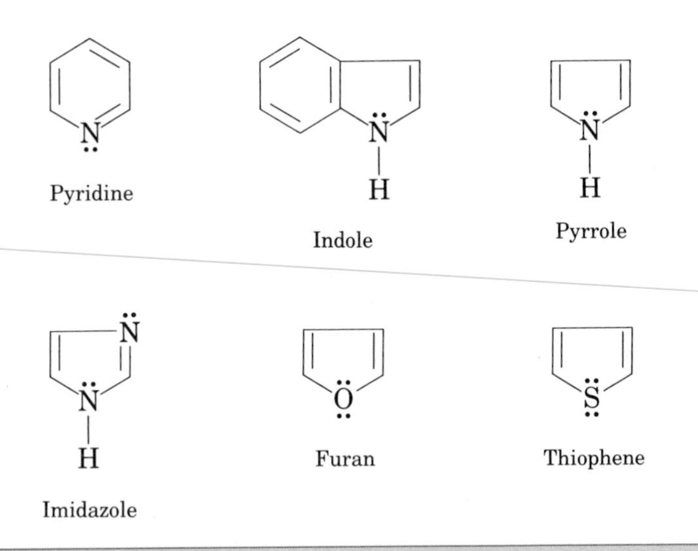

Figure 14.14. Some aromatic heterocycles.

Let's look at two aromatic heterocycles, pyridine and pyrrole, in more detail to see how they meet our definition of conjugated six-pi-electron systems. Pyridine is much like benzene in its pi electron structure. Each of the five sp^2-hybridized carbons has a p orbital perpendicular to the plane of the ring, and each contains one pi electron. The nitrogen atom also is sp^2

hybridized and has one electron in a *p* orbital, bringing the total to six pi electrons. The nitrogen lone-pair electrons are in an sp^2 orbital in the plane of the ring and are not involved with the aromatic pi system (Figure 14.15).

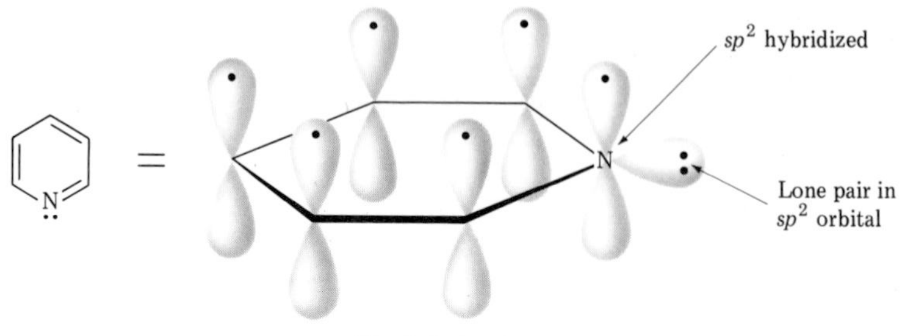

sp^2 hybridized
Lone pair in sp^2 orbital
Six pi electrons

Figure 14.15. Pyridine: an aromatic six-pi-electron heterocycle.

The five-membered heterocycle, pyrrole, is also an aromatic six-pi-electron substance, and its pi electron system is similar to that of the cyclopentadienyl anion. Each of the four sp^2-hybridized carbons has a *p* orbital perpendicular to the ring and each contributes one pi electron. The nitrogen atom is sp^2 hybridized, and the lone pair of electrons occupies the nitrogen *p* orbital. Thus there is a total of six pi electrons, making pyrrole an aromatic molecule. An orbital picture of pyrrole is shown in Figure 14.16.

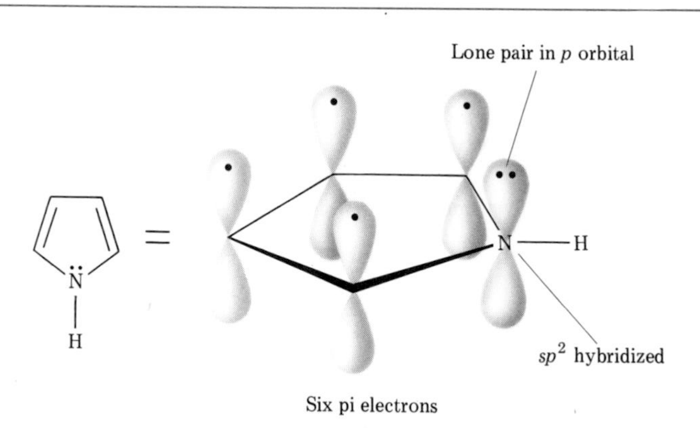

Figure 14.16. Pyrrole: an aromatic six-pi-electron heterocycle.

Note that although both pyridine and pyrrole are aromatic the nitrogen atom does not play the same role in both. The nitrogen atom in pyridine is part of a double bond and therefore contributes only *one* pi electron to the aromatic sextet. In pyrrole, however, the nitrogen atom is not part of a double bond and contributes *two* pi electrons (the lone pair) to the aromatic sextet. These and other aromatic heterocycles play an important role in many biochemical processes. Their chemistry will be discussed in more detail in Chapter 33.

PROBLEM

14.11 The aromatic five-membered heterocycle imidazole is of great importance in a number of biological processes. Draw an orbital picture of imidazole to account for its aromaticity. How many pi electrons does each nitrogen contribute? What is the hybridization of each nitrogen?

Imidazole

PROBLEM

14.12 Draw an orbital picture of furan, showing how this heterocycle can be a six-pi-electron aromatic substance. What is the hybridization of oxygen?

Furan

14.9 Why (4n + 2)?

What is so special about (4n + 2) pi electrons? Why is it that having 2, 6, 10, 14, and so on pi electrons leads to aromatic stability, but having other numbers of electrons does not? The answer to these questions has to do with the relative energy levels of the pi molecular orbitals.

When the energy levels of molecular orbitals for cyclic conjugated molecules are calculated, it turns out that there is always a *single* lowest-lying molecular orbital above which the molecular orbitals come in degenerate pairs (degenerate = same energy level). Thus, when we assign electrons to fill the various molecular orbitals, it takes two electrons (one pair) to fill the lowest-lying orbital and four electrons (two pairs) to fill each of n succeeding energy levels. The total is (4n + 2).

This is illustrated in Figure 14.17 for benzene. In benzene, six atomic p orbitals combine to give six molecular orbitals with the energy levels shown. Note that the lowest-energy molecular orbital, ψ_1, occurs singly and

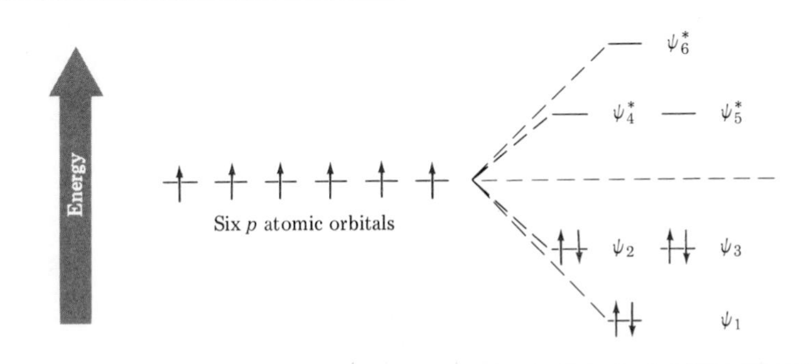

Figure 14.17. Energy levels of benzene pi molecular orbitals.

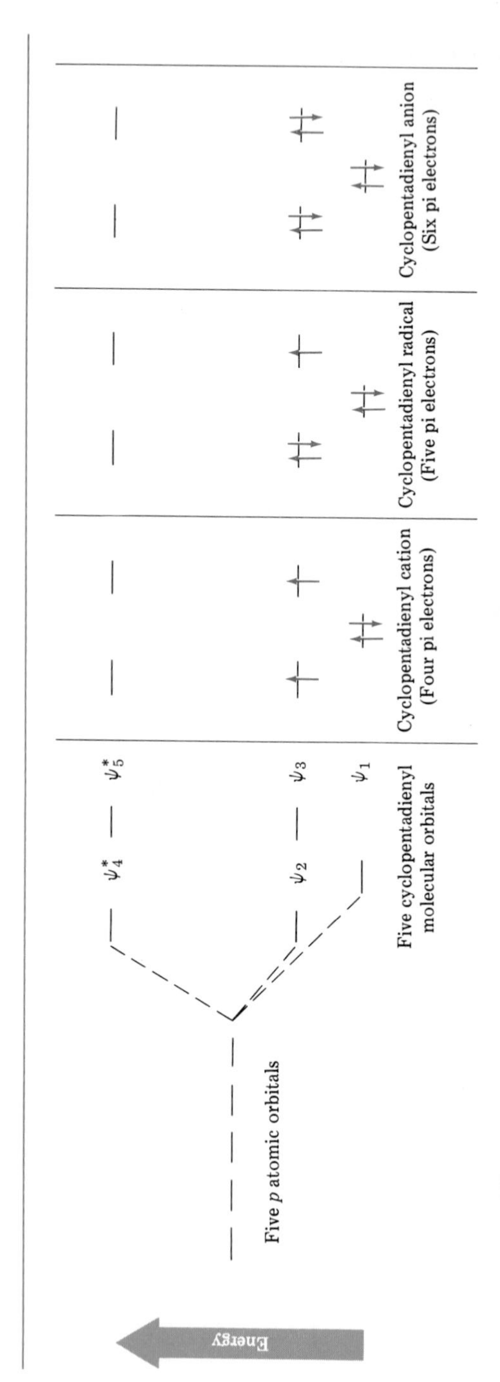

Figure 14.18. Energy levels of cyclopentadienyl molecular orbitals. Only the six-pi-electron cyclopentadienyl anion has a filled-shell configuration.

contains two electrons. The next two lowest-energy orbitals, ψ_2 and ψ_3, are degenerate, and it therefore takes four electrons to fill them. The result is a highly stabilized six-pi-electron aromatic molecule with filled bonding molecular orbitals.

A similar line of reasoning carried out for the cyclopentadienyl cation, radical, and anion leads to the conclusions illustrated in Figure 14.18. Once again, the atomic *p* orbitals (five) combine to give pi molecular orbitals (five) in which there is a *single* lowest-energy orbital, and higher-energy degenerate pairs of orbitals. In the four-pi-electron cation, there are two electrons in ψ_1 but only one electron each in ψ_2 and ψ_3. Thus the cation has two orbitals that are only partially filled, and it is therefore unstable. In the five-pi-electron radical, ψ_1 and ψ_2 are filled, but ψ_3 is still only half full. The radical is therefore unstable. Only in the six-pi-electron cyclopentadienyl anion are all of the bonding orbitals filled. This filled-shell configuration of the anion accounts for its aromatic behavior. Similar analyses can be carried out for all other aromatic species.

PROBLEM

14.13 Show the relative positions of the seven pi molecular orbitals of the cycloheptatrienyl system. Indicate which of the seven orbitals are filled in the cation, radical, and anion, and account for the aromaticity of the cycloheptatrienyl cation.

14.10 Naphthalene: a polycyclic aromatic compound

Benzene rings may be fused to form **polycyclic aromatic compounds.** Naphthalene, with two fused rings, is the simplest polycyclic aromatic molecule, but more complex substances such as anthracene, 1,2-benzpyrene, and coronene are known (Figure 14.19). 1,2-Benzpyrene is particularly

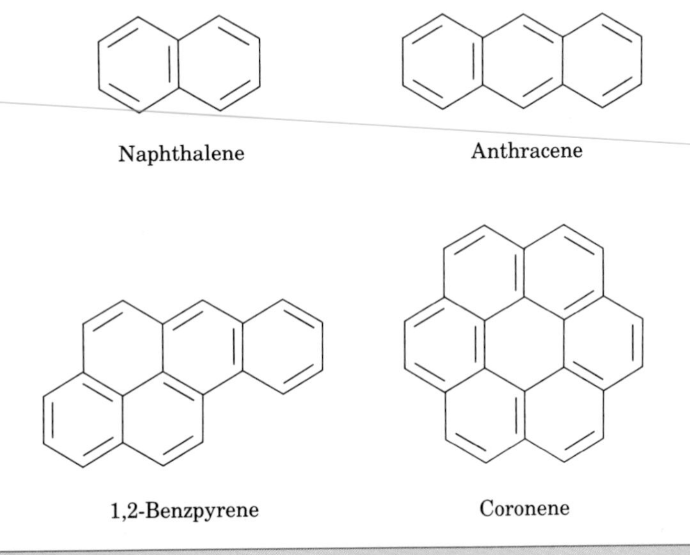

Figure 14.19. Some polycyclic aromatic hydrocarbons.

interesting because it is one of the cancer-causing substances that has been isolated from tobacco smoke.

All polycyclic aromatic hydrocarbons can be represented by a number of different Kekulé structures; naphthalene, for instance, has three resonance forms:

$\longleftrightarrow$ $\longleftrightarrow$

As was true for benzene with its two equivalent resonance forms, no single Kekulé structure is a true representation of naphthalene; the true structure of naphthalene lies somewhere in between the three resonance forms and is difficult to draw satisfactorily.

Naphthalene and many other polycyclic hydrocarbons show the chemical properties we associate with aromaticity. Thus, studies show an aromatic stabilization energy approximately 60 kcal/mol greater than might be expected if naphthalene had five isolated double bonds. Furthermore, naphthalene reacts slowly with electrophilic reagents such as bromine to give substitution products rather than double-bond addition products.

$$\text{Naphthalene} \xrightarrow[\Delta]{Br_2,\ Fe} \text{1-Bromonaphthalene (75\%)} + HBr$$

Br

Naphthalene 1-Bromonaphthalene (75%)

How can we explain the aromaticity of naphthalene? The orbital picture of naphthalene in Figure 14.20 shows a fully conjugated cyclic pi-electron system. There is *p*-orbital overlap both around the 10-carbon

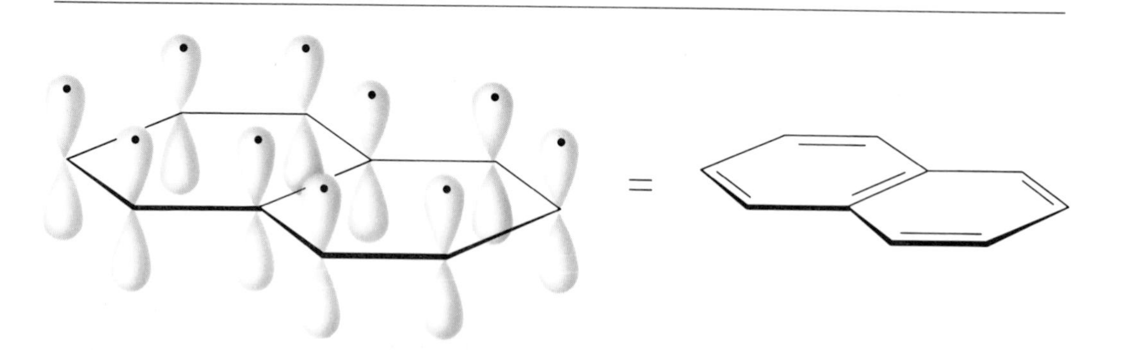

Figure 14.20. A pi orbital picture of naphthalene. The 10 pi electrons are fully delocalized throughout both rings.

periphery of the molecule and across the central bond. There are 10 pi electrons involved, and 10 is a Hückel number. Calculations show a high degree of pi-electron delocalization and clearly predict that naphthalene should be an aromatic molecule.

PROBLEM

14.14 Examine the three resonance structures of naphthalene and account for the fact that not all carbon–carbon bonds have the same length. The C1—C2 bond is 1.36 Å long, whereas the C2—C3 bond is 1.39 Å long.

PROBLEM

14.15 Naphthalene is sometimes represented with circles in each ring to represent aromaticity:

The difficulty with this representation is that one cannot tell immediately how many pi electrons are present. How many pi electrons are in each circle?

14.11 Spectroscopy of arenes

INFRARED SPECTROSCOPY

The presence of an aromatic ring functional group can be readily detected by infrared spectroscopy. Aromatic compounds show a characteristic C—H stretching absorption at 3030 cm^{-1} and a characteristic series of peaks in the 1450–1600 cm^{-1} range. The aryl C—H band at 3030 cm^{-1} is generally of rather low intensity and occurs just to the left of a normal saturated C—H band. As many as four aryl absorptions are sometimes observed in the 1450–1600 cm^{-1} region and are due to complex molecular motions of the ring itself. Two bands, one at 1500 cm^{-1} and one at 1600 cm^{-1}, are usually the most intense. In addition, aromatic compounds show strong absorptions in the 690–900 cm^{-1} range due to C—H out-of-plane bending. The exact position of these absorptions is diagnostic of the substitution pattern of the aromatic ring:

Monosubstituted:	690–710 cm^{-1} 730–770 cm^{-1}	*m*-Disubstituted:	690–710 cm^{-1} 810–850 cm^{-1}
o-Disubstituted:	735–770 cm^{-1}	*p*-Disubstituted:	810–840 cm^{-1}

The infrared spectrum of toluene, shown in Figure 14.21, shows these characteristic absorptions.

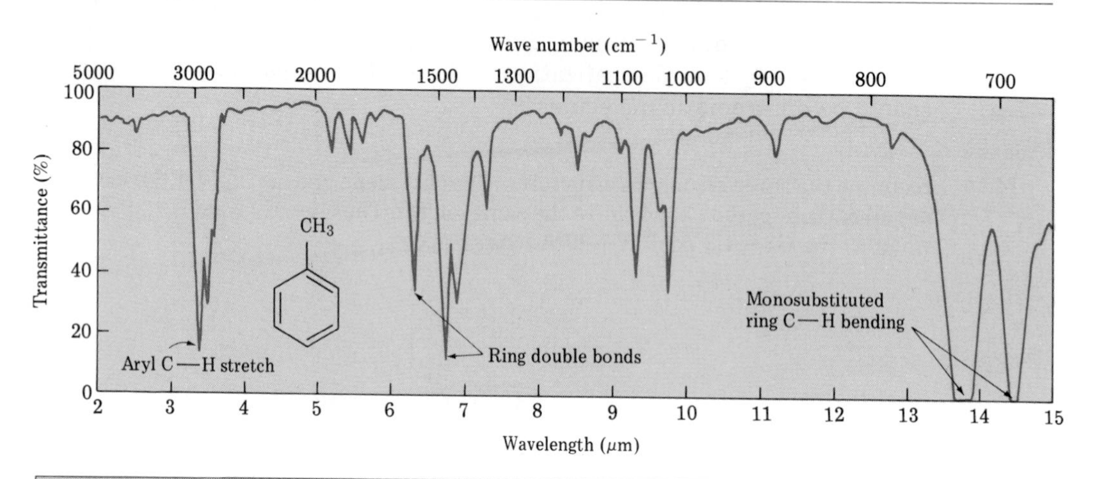

Figure 14.21. Infrared spectrum of toluene.

ULTRAVIOLET SPECTROSCOPY

Aromatic rings are also detectable by ultraviolet spectroscopy. The presence of a conjugated pi-electron system makes aromatic molecules suitable for study by ultraviolet techniques. In general, arenes show a series of bands, with a fairly intense absorption near 205 nm and a less intense absorption in the 255–275 nm range. The presence of these bands in the ultraviolet spectrum of a molecule of unknown structure is a sure indication that an aromatic ring is present.

NUCLEAR MAGNETIC RESONANCE SPECTROSCOPY

Hydrogens directly bonded to an aromatic ring are identifiable in the ^1H NMR spectrum. Aryl hydrogens are strongly deshielded by the aromatic ring and absorb between 6.5 and 8.0 ppm downfield from the TMS standard. The spins of nonequivalent aryl protons on substituted rings often couple with each other, giving rise to spin–spin splitting patterns that, when interpreted, can give information about the substitution pattern of the ring.

Much of the chemical shift difference between aryl protons (6.5–8.0 δ) and vinylic protons (4.5–6.5 δ) is due to a special property of aromatic rings called **ring current.** When an aromatic ring is oriented perpendicularly to a strong magnetic field, the pi electrons circulate around the ring in a direction such that they induce a tiny local magnetic field. This induced field opposes the applied field in the middle of the ring but *reinforces* the applied field outside the ring (Figure 14.22). Aryl protons are therefore slightly deshielded; they experience a magnetic field greater than the applied field and thus come into resonance at a slightly lower applied field (downfield).

Note that the existence of an aromatic ring current predicts different effects inside and outside the ring. If an aromatic ring were large enough to have both "inside" and "outside" protons, those protons on the outside should be deshielded and absorb at a lower-than-normal field, but those protons on the inside should be *shielded* and absorb at a higher-than-normal

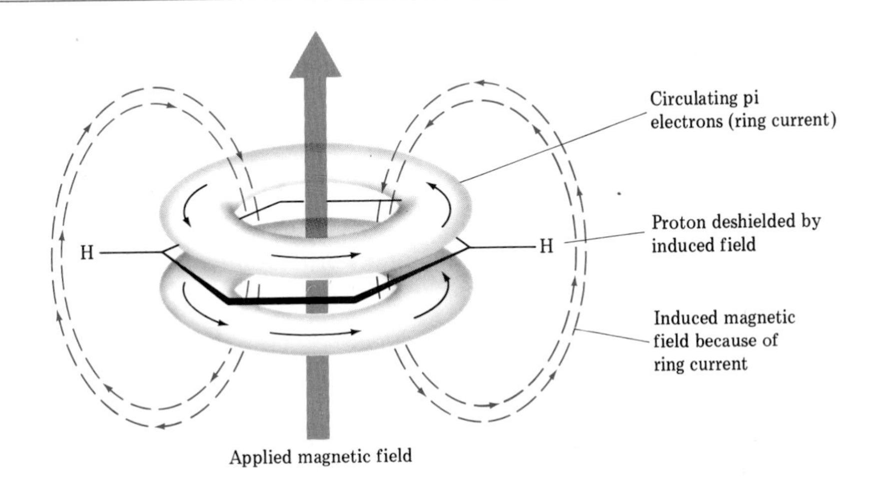

Figure 14.22. The origin of aromatic ring current. Aryl protons are deshielded because of the induced magnetic field caused by the circulating pi electrons.

field. This prediction has been strikingly borne out by studies on [18]annulene, an 18-pi-electron cyclic conjugated polyene that contains a Hückel number of electrons ($4n + 2$, where $n = 4$) and is large enough to have both inside and outside protons:

[18]Annulene

The ^1H NMR spectrum of [18]annulene is just as predicted—the 6 inside protons are strongly shielded by the aromatic ring current and absorb at 1.9 ppm upfield from TMS, whereas the 12 outside protons are strongly deshielded and absorb in the typical aryl region at 8.8 ppm downfield from TMS.

The presence of a ring current is characteristic of all Hückel aromatic molecules and serves as an excellent test of aromaticity. For example, benzene, a 6-pi-electron aromatic molecule, absorbs at 7.37 δ, but cyclooctatetraene, an 8-pi-electron nonaromatic molecule, absorbs at 5.78 δ.

Protons on carbon next to aromatic rings also show distinctive absorptions in the NMR spectrum. Benzylic protons normally absorb downfield from other alkane protons in the region from 2.3 to 3.0 δ.

Aryl protons, 6.5–8.0 δ

CH_2R

Benzylic protons, 2.3–3.0 δ

The ^{1}H NMR spectrum of *m*-bromotoluene shown in Figure 14.23 displays some of these features. The aryl protons absorb in a complex pattern from 6.9 to 7.6 δ, and the methyl protons absorb as a sharp singlet at 2.3 δ. Integration of the spectrum reveals the expected 4:3 ratio of peak areas.

Carbon atoms of an aromatic ring absorb in the range 110–160 δ in the ^{13}C NMR spectrum, as indicated by the examples in Figure 14.24. These

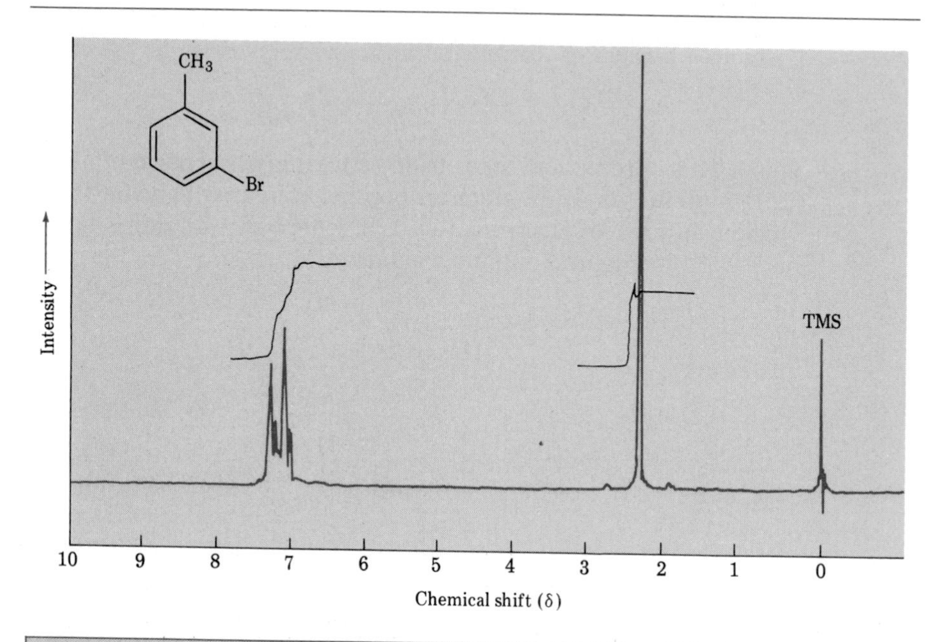

Figure 14.23. Proton NMR spectrum of *m*-bromotoluene.

21.3

CH_3 137.7 | Cl 133.8 | 133.7

128.4 | 129.3 | 127.6 | 128.1

128.5 | 128.4 | 126.0

125.6 | 125.4

Benzene | Toluene | Chlorobenzene | Naphthalene

Figure 14.24. Carbon-13 NMR absorptions of some aromatic compounds (δ units).

resonances are easily distinguished from those of alkane carbons, but occur in the same range as alkene carbons. Thus, the presence of ^{13}C absorptions at 110–160 δ does not *in itself* establish the presence of an aromatic ring in the sample. Confirming evidence from infrared, ultraviolet, or ^{1}H NMR is needed.

A summary of the kinds of information obtainable from different spectroscopic techniques is given in Table 14.3.

TABLE 14.3 **Summary of spectroscopic information on arenes**

Kind of spectroscopy	*Absorption position*	*Interpretation*
Infrared (cm^{-1})	3030	Aryl C—H stretch
	1500 and 1600	Two intense absorptions due to ring motions
	690–900	Intense C—H out-of-plane bending
Ultraviolet (nm)	205	Intense absorption
	260	Weak absorption
^{1}H NMR (δ)	2.3–3.0	Benzylic protons
	6.5–8.0	Aryl protons
^{13}C NMR (δ)	110–160	Aromatic ring carbons

14.12 Summary

The term *aromatic* is used for historical reasons to refer to the class of compounds related structurally to benzene. Many naturally occurring substances and most medicinally useful compounds contain aromatic (benzene-like) portions.

Benzene-like compounds may be systematically named according to IUPAC rules but many trivial names are used for common substances. Aromatic compounds are named as substituted benzenes, but it is also correct at times to consider the benzene ring itself as a substituent on a larger molecule. In these cases, the C_6H_5 unit is referred to as a phenyl group. Disubstituted benzenes are named as either ortho (1,2 disubstituted), meta (1,3 disubstituted), or para (1,4 disubstituted) derivatives.

Benzene is described by resonance theory as a resonance hybrid of two equivalent Kekulé structures. Neither structure is correct by itself—the true structure of benzene is intermediate between the two:

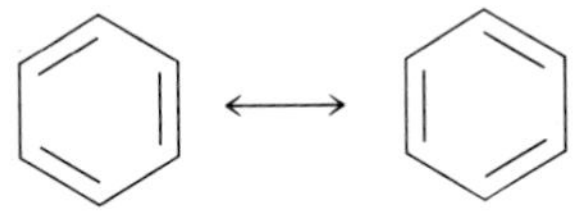

$$\longleftrightarrow$$

Benzene is described with more accuracy by molecular orbital theory—as a cyclic conjugated molecule with six pi electrons. According to the Hückel rule, a cyclic conjugated molecule must have $(4n + 2)$ pi electrons, where n = 0, 1, 2, 3, and so on, in order to be aromatic. This leads to complete delocalization of pi electrons throughout the ring.

Benzene and other arenes have five key characteristics:

1. Benzene is a cyclic conjugated molecule.
2. Benzene is unusually stable; its heat of hydrogenation is approximately 36 kcal/mol less than we might expect for a normal triene.
3. Benzene reacts slowly with electrophiles to give substitution products in which cyclic conjugation is retained, rather than alkene addition products in which conjugation is destroyed.
4. Benzene is symmetrical. All carbon–carbon bonds are equivalent and have a length of 1.39 Å, a value that is intermediate between normal single- and double-bond lengths.
5. Benzene has a Hückel number of pi electrons, $(4n + 2)$, where $n = 1$. Thus, the pi electrons are delocalized over all six carbons, and there is a doughnut-shaped ring of electron density above and below the plane of the ring.

Other kinds of molecules besides benzenoid compounds can be aromatic according to the Hückel $(4n + 2)$-pi-electron definition, since the number of pi electrons is the important factor, not the number of orbitals. For example, the cyclopentadienyl anion, with six pi electrons and five molecular orbitals (Figure 14.18), and the cycloheptatrienyl cation, with six pi electrons and seven molecular orbitals, are both $(4n + 2)$-pi-electron aromatic species.

Heterocyclic compounds, which have atoms other than carbon in the ring, can also be aromatic. Pyrrole and pyridine are examples of six-pi-electron heterocyclic aromatic compounds. Pyridine resembles benzene in its pi electron configuration, and pyrrole resembles the cyclopentadienyl anion.

The basis of the Hückel $(4n + 2)$-pi-electron rule is this: When molecular orbital calculations are carried out, there is always a single lowest-energy orbital, above which lie pairs of molecular orbitals. Thus, in order to obtain a filled-shell configuration leading to aromatic stability, it is necessary first to fill the single lowest orbital (two electrons), and then to fill successive pairs (four electrons at a time). The net result is $(4n + 2)$ pi electrons.

All of the spectroscopic techniques we have studied are applicable to the structure elucidation of aromatic compounds. Infrared, ultraviolet, and NMR spectroscopies all show characteristic aromatic absorption peaks.

ADDITIONAL PROBLEMS

14.16 Provide IUPAC names for these compounds:

(a) C_6H_5–$CH(CH_3)CH_2CH_2CH(CH_3)CH_3$

(b) CO_2H and Br on benzene ring (meta)

(c)

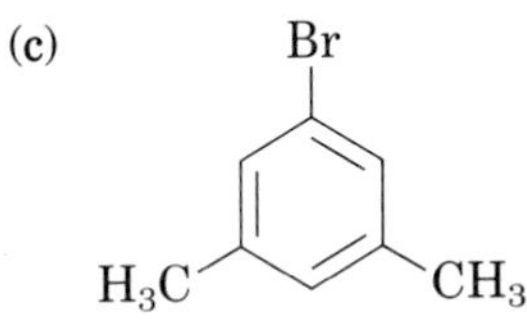

(d)

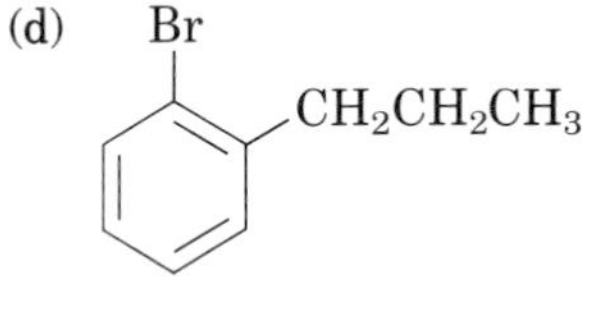

(e)

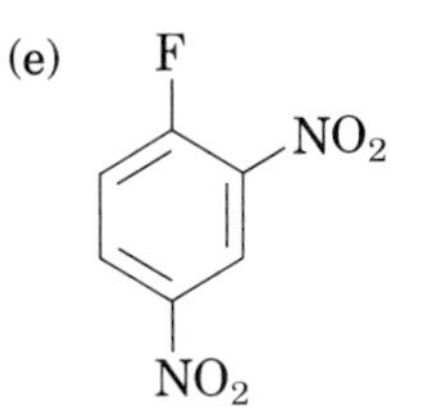

(f)

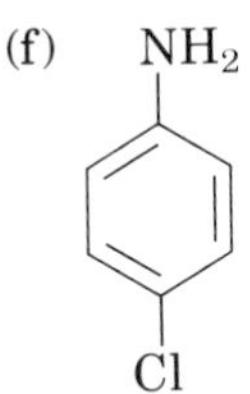

14.17 Draw the structures corresponding to these names:
(a) 2,3-Toluenediamine
(b) 1,3,5-Benzenetriol
(c) 3-Methyl-2-phenylhexane
(d) *o*-Aminobenzoic acid
(e) *m*-Bromophenol
(f) 2,4,6-Trinitrophenol (picric acid)
(g) *p*-Iodonitrobenzene

14.18 Draw and name all possible isomeric:
(a) Dinitrobenzenes
(b) Bromodimethylbenzenes
(c) Trinitrophenols

14.19 Draw and name all possible aromatic compounds with the formula C_7H_7Cl.

14.20 Draw and name all possible aromatic compounds with the formula C_8H_9Br. (There are 14.)

14.21 Propose structures for aromatic hydrocarbons that meet these descriptions:
(a) C_9H_{12}; gives only one product on aromatic substitution by bromine
(b) $C_{10}H_{14}$; gives only one product on aromatic substitution by chlorine
(c) C_8H_{10}; gives three products on aromatic substitution by bromine
(d) $C_{10}H_{14}$; gives two products on aromatic substitution by chlorine

14.22 Draw the five resonance structures of phenanthrene. On this basis, which of the carbon–carbon bonds should be shortest?

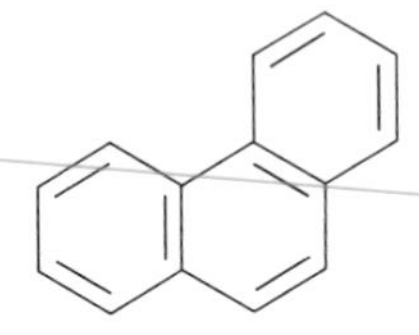

Phenanthrene

14.23 Define these terms in your own words:
(a) Aromaticity
(b) Conjugated
(c) Hückel ($4n + 2$) rule
(d) Resonance hybrid

14.24 Table 14.2 gives heat of hydrogenation data for benzene, 1,3-cyclohexadiene, and cyclohexene. On the basis of these data, calculate the heats of hydrogenation for the partial hydrogenation shown here. Estimate the chances of carrying out this reaction successfully in the laboratory.

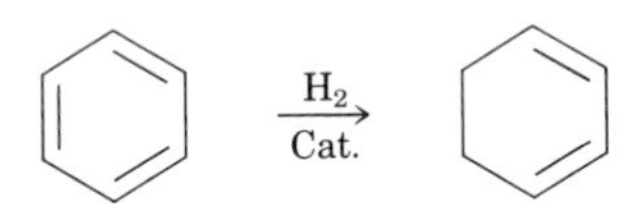

14.25 Cyclopropanone is an unstable and highly reactive molecule because of the large amount of angle strain it contains. Methylcyclopropenone, although more strained than cyclopropanone, is nevertheless quite stable. It was prepared by R. Breslow in 1966 and can even be distilled. Can you account for its stability? (Consider the polarity of the carbonyl group.)

Cyclopropanone Methylcyclopropenone

14.26 Cycloheptatrienone is a perfectly stable compound, but cyclopentadienone is so reactive that it cannot be isolated. What, do you suppose, accounts for the stability difference between the two?

Cycloheptatrienone Cyclopentadienone

14.27 Which member of the following sets of compounds would you expect to be most stable?

(a) Cyclopropenyl anion, cation, or radical?
(b) Cyclobutadiene dianion, radical, or cation?
(c) Cyclononatetraenyl radical, cation, or anion?

14.28 3-Chlorocyclopropene, on treatment with $AgBF_4$, gives a precipitate of AgCl and a stable solution of a species that shows only one absorption in the ^1H NMR spectrum at 11.04 δ. How do you explain this result? What is a likely structure for the product, and what is its relation to Hückel's rule?

3-Chlorocyclopropene

14.29 Draw an energy diagram for the three molecular orbitals of the cyclopropenyl system. How are these three molecular orbitals occupied in the cyclopropenyl anion, cation, and radical? Which of these three is aromatic according to Hückel's rule?

14.30 If we were to use the "circle" notation for aromaticity, we would draw the cyclopropenyl cation as shown here. How many pi electrons are represented by the circle in this instance?

14.31 Compound A, C_8H_{10}, yields three monobromo substitution products on treatment with $Br_2/FeBr_3$. Propose two possible structures. The ^1H NMR spectrum of A shows a complex four-proton multiplet at 7.0 δ and a six-proton singlet at 2.30 δ. What is the correct structure of A?

14.32 What is the structure of a hydrocarbon that shows a molecular ion at $m/z = 120$ in the mass spectrum and has the following ^1H NMR spectrum?

7.25 δ, broad singlet, five protons
2.90 δ, septet, $J = 7$ Hz, one proton
1.22 δ, doublet, $J = 7$ Hz, six protons

14.33 Azulene is a beautiful blue hydrocarbon that is isomeric with naphthalene. Unlike naphthalene, however, azulene has a large dipole moment ($\mu = 1.0$ D).
(a) Is azulene a Hückel aromatic compound?
(b) Draw an orbital picture of azulene.
(c) How can you account for the observed dipole moment of azulene?

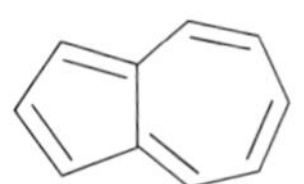

Azulene ($\mu = 1.0$ D)

14.34 Draw an orbital picture of indole.
(a) How many pi electrons are present?
(b) Is indole aromatic?
(c) What is the electronic relationship of indole to naphthalene?

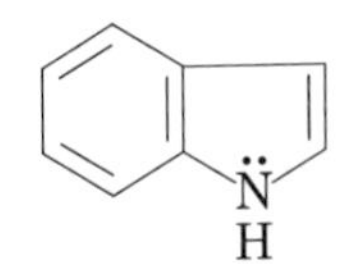

Indole

14.35 4-Pyrone is protonated by acid to give a stable cationic product. Propose a structure for this product and explain its stability.

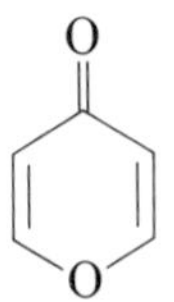

4-Pyrone

14.36 Pentalene is a most elusive molecule that has never been isolated. The pentalene dianion, however, is well known and quite stable. Explain.

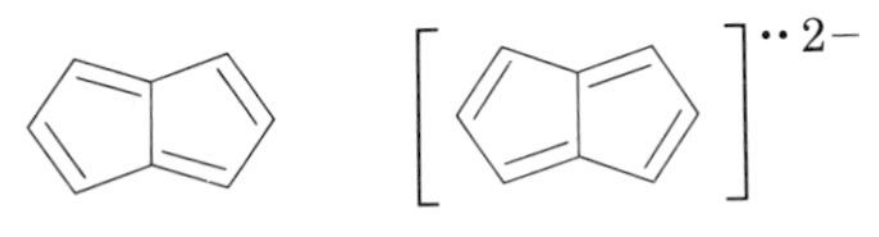

Pentalene Pentalene dianion

14.37 Purine is a heterocyclic aromatic compound that is a constituent of DNA and RNA. Why is purine considered to be aromatic? How many *p* electrons does each nitrogen donate to the aromatic pi system?

Purine

CHAPTER 15 CHEMISTRY OF BENZENE: ELECTROPHILIC AROMATIC SUBSTITUTION

The single most important reaction of aromatic compounds is **electrophilic substitution.** That is, an electrophile is attacked by an aromatic ring and substitutes for one of the hydrogens (Figure 15.1). In this chapter, we will discuss the details of how this polar process is thought to occur, and learn some of the important uses of the reaction.

$$C_6H_6 + E^+ \longrightarrow C_6H_5E + H^+$$

Figure 15.1. An electrophilic aromatic substitution (E^+ represents an electrophile).

Benzene, with six pi electrons in a cyclic conjugated system, is a site of electron density. Furthermore, the benzene pi electrons are sterically accessible to attacking reagents because of their location in circular clouds above and below the plane of the ring (Figure 15.2). Thus, benzene acts as an electron donor (a nucleophile) in most of its chemistry; most of the reactions of benzene take place with electron acceptors (electrophiles). The electrophilic substitution reaction is characteristic of all aromatic rings, not just of benzene and substituted benzenes. Indeed, the ability of a compound to undergo electrophilic substitution is an excellent test of aromaticity.

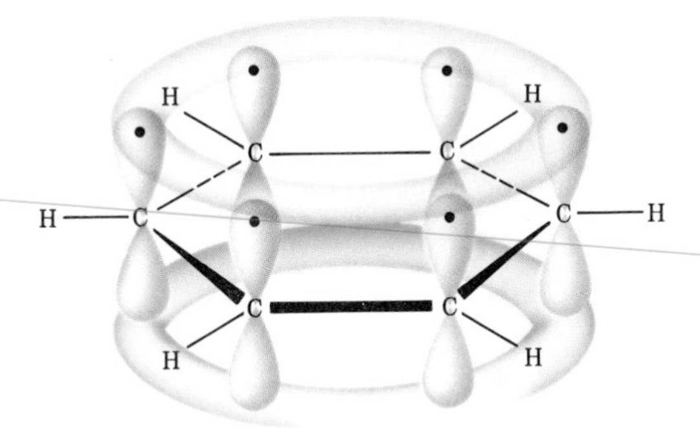

Figure 15.2. Benzene and its pi electrons, which are sterically unhindered and accessible for reaction with electrophiles.

There are many different kinds of electrophilic substitutions, and many different substituents may be introduced onto the aromatic ring by using this reaction. By choosing the proper conditions and reagents, we can **halogenate** (substitute a halogen: —F, —Cl, —Br, or —I), **nitrate** (substitute a nitro group: $—NO_2$), **sulfonate** (substitute a sulfonic acid group:

—SO_3H), **hydroxylate** (substitute a hydroxyl group: —OH), **alkylate** (substitute an alkyl group: —R), or **acylate** (substitute an acyl group: —COR) the aromatic ring. We can prepare many thousands of substituted aromatic compounds, starting from only a few simple materials. Table 15.1 lists some of these possibilities.

TABLE 15.1 Some electrophilic aromatic substitution reactions

Name	*Example*
Bromination	$Ar{-}H + Br_2 \xrightarrow{FeBr_3} Ar{-}Br$ (an aryl bromide)
Chlorination	$Ar{-}H + Cl_2 \xrightarrow{FeCl_3} Ar{-}Cl$ (an aryl chloride)
Nitration	$Ar{-}H + HNO_3 \xrightarrow{H_2SO_4} Ar{-}NO_2$ (a nitro aromatic compound)
Sulfonation	$Ar{-}H + SO_3 \xrightarrow{H_2SO_4} Ar{-}SO_3H$ (an aromatic sulfonic acid)
Hydroxylation	$Ar{-}H + H_2O_2 \xrightarrow{HOSO_2F} Ar{-}OH$ (a phenol)
Friedel–Crafts alkylation	$Ar{-}H + R{-}Cl \xrightarrow{AlCl_3} Ar{-}R$ (an arene)
Friedel–Crafts acylation	$Ar{-}H + R{-}\overset{O}{\overset{\Vert}{C}}Cl \xrightarrow{AlCl_3} Ar{-}\overset{O}{\overset{\Vert}{C}}{-}R$ (an aryl ketone)

All these reactions (and many more as well) take place by a similar mechanism, and it is therefore important to understand the principles behind this important polar reaction. Let's begin by studying one reaction in detail—the bromination of benzene.

15.1 Aromatic bromination

Benzene reacts with bromine in the presence of $FeBr_3$ as catalyst to yield the substitution product bromobenzene.

$$C_6H_6 + Br_2 \xrightarrow{FeBr_3} C_6H_5Br + HBr$$

Benzene — Bromobenzene (80%)

Before studying the mechanism of this electrophilic aromatic substitution reaction, let's briefly recall what we have learned about electrophilic additions to alkenes. When an electrophile such as H^+ adds to an alkene, it approaches perpendicular to the plane of the double bond and forms a bond to one carbon, leaving a positive charge at the other carbon. This carbocation intermediate is then attacked by a nucleophile such as $Cl{:}^-$ to yield the addition product (Figure 15.3).

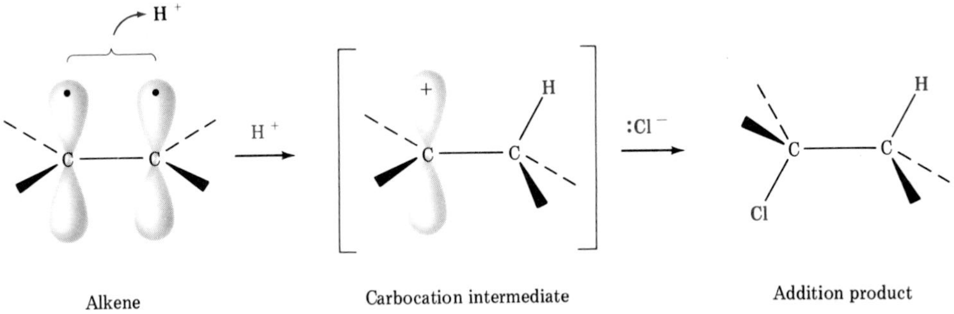

H⁺ → Carbocation intermediate → :Cl⁻ → Addition product

Alkene | Carbocation intermediate | Addition product

Figure 15.3. Mechanism of electrophilic addition reactions of alkenes.

An electrophilic aromatic substitution reaction begins in a similar way, but there are a number of differences. One difference is noticeable immediately—aromatic rings are much less reactive than alkenes toward electrophiles. For example, bromine in carbon tetrachloride solution reacts instantly with most alkenes but does not react with benzene. For bromination of benzene to take place, a catalyst is needed, and $FeBr_3$ is very effective. It exerts its catalytic effect by polarizing the Br_2 molecule, making it more electrophilic. The Lewis acid $FeBr_3$ complexes the bromine molecule, pulling off a $Br:^-$ and leaving a reactive Br^+.

$$\overset{\delta^-}{Br}-\overset{\delta^+}{Br} \xrightarrow{FeBr_3} Br_3\overset{\delta^-}{Fe}\cdots Br\cdots \overset{\delta^+}{Br} = FeBr_4^- \; Br^+$$

Bromine (a weak electrophile) | Polarized bromine (a strong electrophile)

The polarized electrophilic bromine is then attacked by the pi electron system of the nucleophilic benzene ring in a slow, rate-limiting step, yielding a nonaromatic carbocation intermediate. This carbocation is doubly allylic (recall the allyl cation, Section 10.9). It is stabilized by the two remaining double bonds, and can be written in three resonance forms.

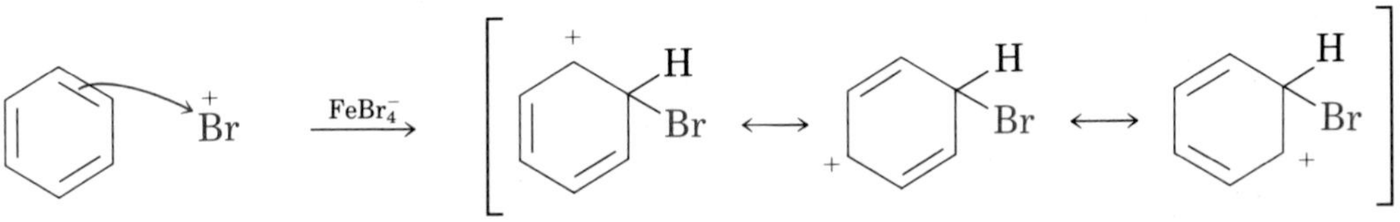

$$C_6H_6 + Br^+ \xrightarrow{FeBr_4^-} [\text{H, Br cation} \longleftrightarrow \text{H, Br cation} \longleftrightarrow \text{H, Br cation}]$$

Although stable by comparison with most other carbocations, the intermediate in electrophilic aromatic substitution is nevertheless much less stable than the starting benzene ring itself with its 36 kcal/mol of aromatic stability. Thus, electrophilic attack on a benzene ring is highly endothermic, has a high activation energy ($\Delta G^{\ddagger}$), and is therefore a rather

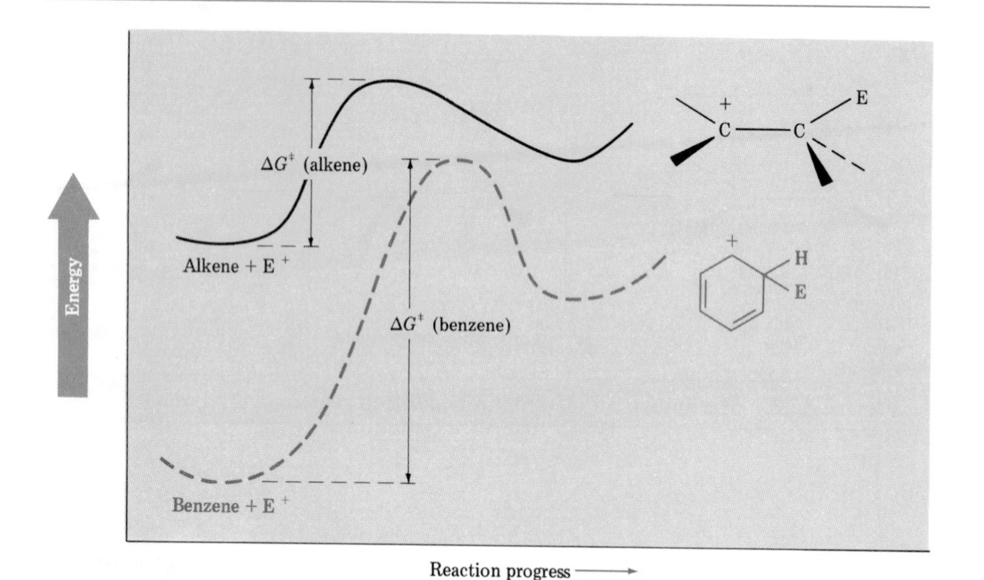

Figure 15.4. Comparison of electrophilic attack by an alkene and by benzene: $\Delta G^{\ddagger}$ (alkene) << $\Delta G^{\ddagger}$ (benzene).

slow reaction. Figure 15.4 gives reaction energy diagrams comparing attack on an electrophile, E^+, by an alkene and attack by benzene. The benzene reaction is slower (higher $\Delta G^{\ddagger}$) because the starting material is much more stable.

A second major difference between alkene addition and aromatic substitution occurs after an electrophile has been attacked by the benzene ring to give the carbocation intermediate. In theory, it is possible for a nucleophile such as bromide ion to attack the carbocation intermediate to yield a dibromocyclohexadiene, the product of electrophilic aromatic *addition*. In fact, this addition is not observed. Instead, the bromide ion (or some other base present in solution) abstracts a neighboring proton from the carbocation intermediate, yielding a neutral, aromatic product. The net effect of reaction of Br_2 with benzene is the electrophilic aromatic substitution of H^+ by Br^+. The overall mechanism is shown in Figure 15.5.

Why does the reaction of an electrophile with benzene and with other aromatic rings take a different course than reaction of the same electrophile with an alkene? The answer is quite simple: If addition occurred, the overall reaction would be endothermic, and the 36 kcal/mol aromatic stabilization would be lost. By losing a proton, however, the carbocation intermediate can revert to an aromatic ring structure and regain aromatic stabilization. A reaction energy diagram of the overall process is shown in Figure 15.6 (p. 484).

There are many other electrophilic aromatic substitutions besides bromination, and all are thought to occur by the same general mechanism. Let's see briefly what some of these other reactions are.

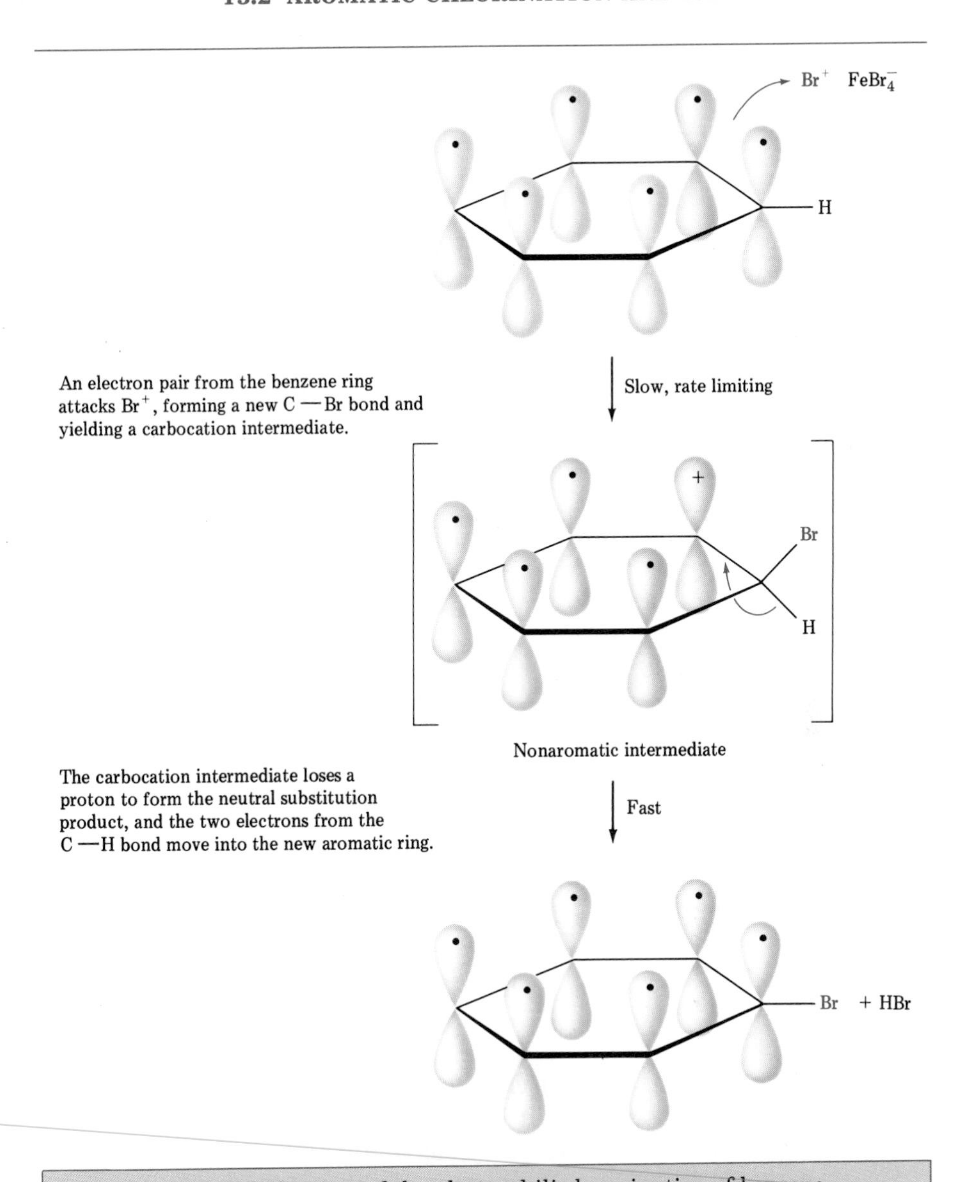

Br+ FeBr4−
H
An electron pair from the benzene ring attacks Br+, forming a new C—Br bond and yielding a carbocation intermediate.
Slow, rate limiting
+
Br
H
Nonaromatic intermediate
The carbocation intermediate loses a proton to form the neutral substitution product, and the two electrons from the C—H bond move into the new aromatic ring.
Fast
Br + HBr

Figure 15.5. Mechanism of the electrophilic bromination of benzene.

15.2 Aromatic chlorination and iodination

Chlorine and iodine can be introduced into aromatic rings by electrophilic substitution reactions if the proper conditions exist. As a general rule, fluorine is too reactive, and poor yields of monofluoroaromatic products are obtained. Chlorine, however, reacts smoothly and gives excellent yields of chloro-substituted aromatic derivatives. As with bromination, a catalyst is required to increase the rate of attack on chlorine by the aromatic ring, and $FeCl_3$ is usually chosen. $FeCl_3$ acts by polarizing the chlorine molecule, making it more electrophilic.

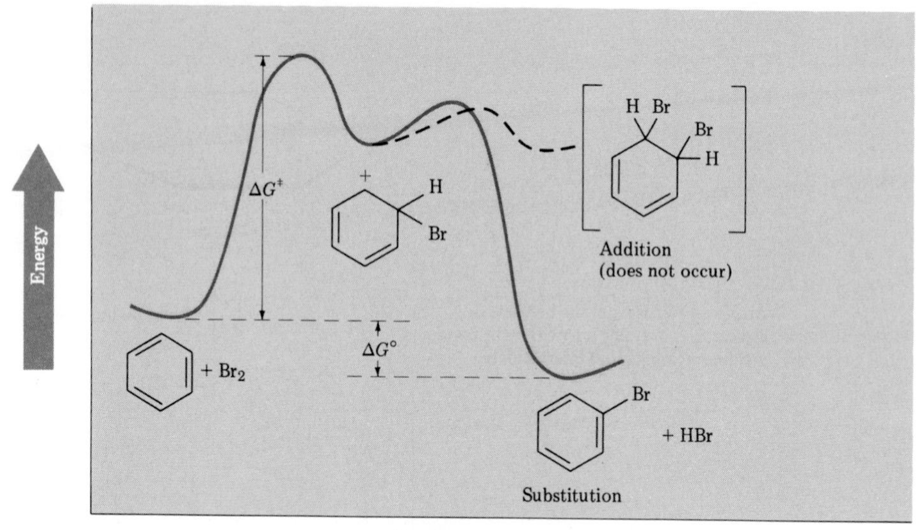

Figure 15.6. Reaction energy diagram for electrophilic bromination of benzene.

$$FeCl_3 + Cl{-}Cl \longrightarrow \overset{\delta^-}{Cl_3Fe}\cdots Cl \cdots \overset{\delta^+}{Cl} = FeCl_4^- \; Cl^+$$

$$\text{Benzene} + Cl{-}Cl \xrightarrow{FeCl_3} \left[\text{(cation, Cl, H)} \;\; :Cl^- \right] \longrightarrow \text{Chlorobenzene (86\%)} + HCl$$

Benzene

Chlorobenzene (86%)

Chlorine oxide, Cl_2O, in trifluoroacetic acid is a useful alternative to the $Cl_2/FeCl_3$ reagent. It is particularly valuable for chlorinating aromatic rings that are substituted by electron-withdrawing groups such as nitro or carbonyl. For example, chlorination of *p*-methylacetophenone takes place only on heating to 145°C when $Cl_2/FeCl_3$ is used, but occurs at room temperature and in higher yield when chlorine oxide in trifluoroacetic acid is used. Note that only one of the two possible chlorination products is formed. We will account for this regioselectivity in Section 15.10.

$$\textit{p}\text{-Methylacetophenone} \xrightarrow[\text{(65\% yield)}]{Cl_2,\ FeCl_3,\ 145^\circ C} \text{3-Chloro-4-methylacetophenone}$$

$$\textit{p}\text{-Methylacetophenone} \xrightarrow[\substack{25^\circ C \\ \text{(99\% yield)}}]{Cl_2O,\ CF_3COOH} \text{3-Chloro-4-methylacetophenone}$$

p-Methylacetophenone

3-Chloro-4-methylacetophenone

Iodine by itself is unreactive toward aromatic rings, and a promoter is required to obtain a suitable reaction. The best promoters for aromatic iodination are oxidizing agents such as hydrogen peroxide, H_2O_2, or copper salts such as $CuCl_2$. These promoters are thought to work by oxidizing molecular iodine to a more powerful electrophile—perhaps I^+. The aromatic ring then attacks I^+, yielding a normal substitution product. The details of this mechanism are still unclear, however.

$$I_2 + 2\,Cu^{2+} \longrightarrow 2\,\text{“}I^{+}\text{”} + 2\,Cu^{+}$$

$$\text{Benzene} \xrightarrow[(I_2\,+\,CuCl_2)]{\text{“}I^{+}\text{”}} [\text{intermediate}] \longrightarrow \text{Iodobenzene (65\%)} + H^{+}$$

Benzene

Iodobenzene (65%)

PROBLEM

15.1 Aromatic iodination can be carried out with a number of reagents, including ICl. What is the direction of polarization of ICl? Propose a mechanism to account for the iodination of an aromatic ring.

PROBLEM

15.2 Aryl fluorides are often prepared by a two-step process involving electrophilic **thallation** of the aromatic ring (substitution of thallium).

$$ArH + Tl(OCOCF_3)_3 \longrightarrow Ar{-}Tl(OCOCF_3)_2 \xrightarrow{KF,\ BF_3} ArF$$

Thallium tris(trifluoroacetate)

Propose a mechanism for the thallation of benzene.

15.3 Aromatic nitration

Aromatic rings can be nitrated by reaction with a mixture of concentrated nitric and sulfuric acids.

$$\text{Benzene} + HNO_3 \xrightarrow[50^{\circ}C]{H_2SO_4} \text{Nitrobenzene (85\%)} + H_2O$$

Benzene

Nitrobenzene (85%)

The electrophile in this reaction is thought to be the **nitronium ion, NO_2^+.** In the mixture of concentrated nitric and sulfuric acids, nitric acid is first protonated and then loses water to generate NO_2^+ as a highly reactive intermediate.

$$HO{-}NO_2 + H_2SO_4 \rightleftharpoons H{-}\overset{+}{O}(H){-}NO_2 + HSO_4^-$$

$$H{-}\overset{+}{O}(H){-}NO_2 \rightleftharpoons H_2O + NO_2^+$$

Nitronium ion

The nitronium ion is then attacked by benzene in much the same way that we discussed for Br^+, yielding a carbocation intermediate. Loss of a proton from this intermediate gives the neutral substitution product, nitrobenzene (Figure 15.7).

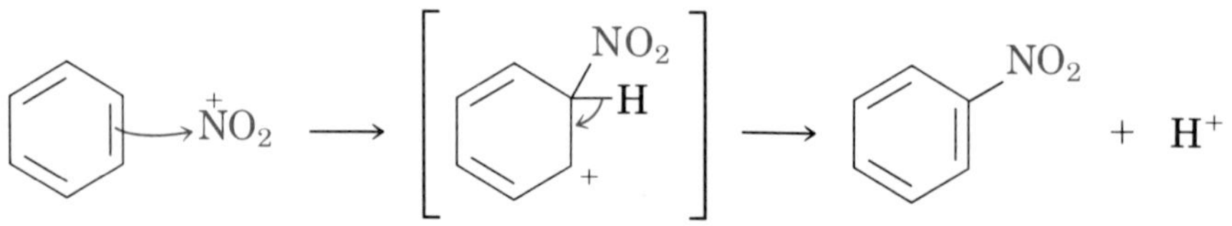

NO_2 H + H$^+$

Nitrobenzene

Figure 15.7. Mechanism of the nitration of benzene.

In the case of aromatic substances where, for some reason, nitration with HNO_3/H_2SO_4 is particularly sluggish, one can carry out rapid aromatic nitrations using a pure nitronium salt, $NO_2^+\ BF_4^-$. Nitronium tetrafluoroborate is a stable, white crystalline material that will smoothly nitrate many aromatic compounds at room temperature or below.

Nitration of aromatic rings is an important reaction for two reasons: nitroaromatics are valuable in themselves, and the nitro group can be converted into other functional groups. For example, reduction of the nitro group by reagents such as iron or stannous chloride yields the corresponding amine.

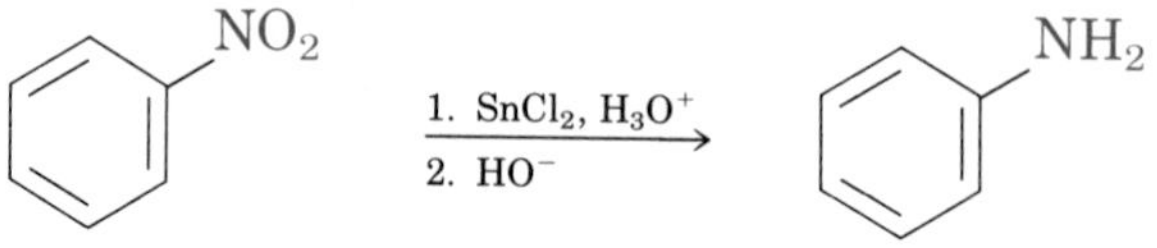

NO_2 1. $SnCl_2$, H_3O^+ 2. HO^- NH_2

Nitrobenzene Aniline (95%)

We will discuss this and other reactions of aromatic nitrogen compounds in Chapter 29.

15.4 Aromatic sulfonation

Aromatic rings can be sulfonated by reaction with fuming sulfuric acid (H_2SO_4 + SO_3):

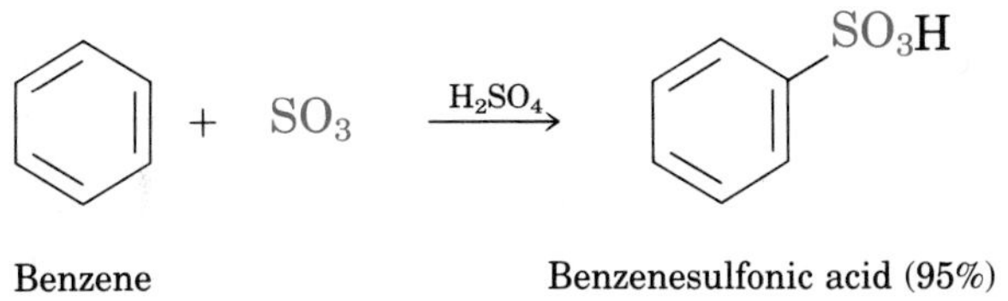

$$\text{Benzene} + SO_3 \xrightarrow{H_2SO_4} \text{Benzenesulfonic acid } (95\%)\ (C_6H_5SO_3H)$$

Benzene

Benzenesulfonic acid (95%)

The reactive electrophile is either HSO_3^+ or neutral SO_3, depending on reaction conditions, and substitution occurs by the same two-step mechanism we have seen previously for bromination and nitration (Figure 15.8). Note that the reaction is reversible and can occur either forwards or backwards, depending on the reaction conditions. Sulfonation is favored in strong acid, but desulfonation is favored in hot, dilute aqueous acid.

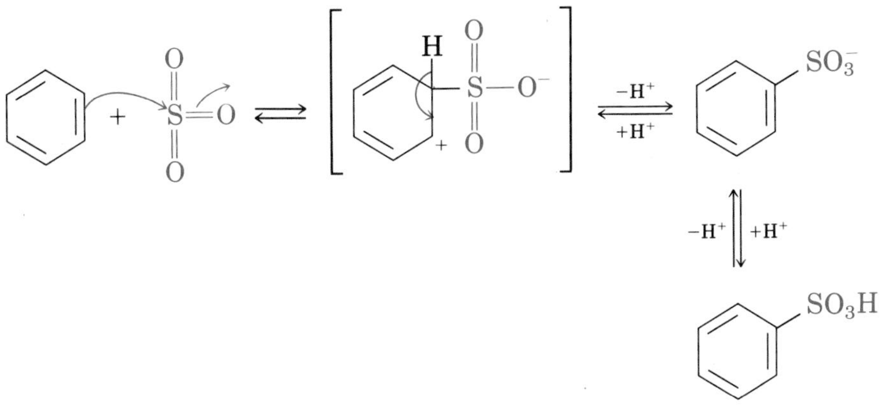

Figure 15.8. Mechanism of the sulfonation of benzene.

Aromatic sulfonic acids are valuable intermediates in the preparation of dyes and pharmaceuticals, and the sulfa drugs such as sulfanilamide were among the first useful antibiotics known. Although largely replaced today by more effective agents, sulfa drugs are still used in the treatment of meningitis. These drugs are prepared commercially by a process that involves aromatic sulfonation as the key step.

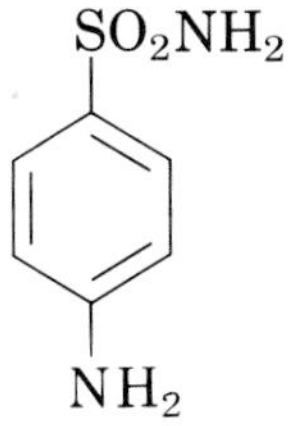

Sulfanilamide (an antibiotic)

Aromatic sulfonic acids are also valuable because of the further chemistry they undergo. Thus, **alkali fusion** of an arylsulfonic acid with NaOH at

300°C in the absence of solvent yields the corresponding phenol—a net replacement of the sulfonate group by hydroxyl. Yields in this process are generally good, but the conditions are so vigorous that the reaction is not compatible with the presence of substituents other than alkyl on the aromatic ring.

SO_3H / CH_3 — *p*-Toluenesulfonic acid $\xrightarrow[\text{2. } H^+]{\text{1. NaOH, 300°C}}$ OH / CH_3 — *p*-Cresol (72%) (a phenol)

PROBLEM

15.3 Show a detailed mechanism for the desulfonation reaction of benzenesulfonic acid to yield benzene. What is the electrophile in this reaction?

15.5 Aromatic hydroxylation

We have seen reactions for introducing halogens and nitrogen- and sulfur-containing functional groups onto an aromatic ring. It would be useful to be able to introduce an *oxygen*-containing functional group. By analogy with other electrophilic reagents, a positively polarized oxygen atom is required, and a mixture of hydrogen peroxide, H_2O_2, with a very strong acid such as fluorosulfonic acid, HSO_3F, is effective. The strong acid is thought to protonate H_2O_2, making it highly electrophilic.

$$H{-}O{-}O{-}H \xrightleftharpoons{HSO_3F} H{-}\overset{H}{\underset{+}{O}}{-}O{-}H \quad \overset{-}{S}O_3F \; = \; \text{“}^{+}OH\text{”} + H_2O$$

Benzene + $H\ddot{O}{-}\overset{+}{O}H_2$ ⟶ [(cation with OH, H) + H_2O] ⟶ Phenol (67%) + H^+

The hydroxylation reaction is not as general as the halogenation, nitration, and sulfonation reactions we have studied; hydroxylation is limited to reaction with benzene, halobenzenes, and alkylbenzenes, since other functional groups do not survive the reaction conditions.

15.6 Mechanism of electrophilic aromatic substitution

Let's review what we have learned about electrophilic aromatic substitution reactions:

1. Electrophilic substitutions proceed by a single common mechanism.

2. The reaction is a two-step process in which the first step is the attack on an electrophile, E^+, by the pi electrons of the aromatic ring, leading to a carbocation intermediate.

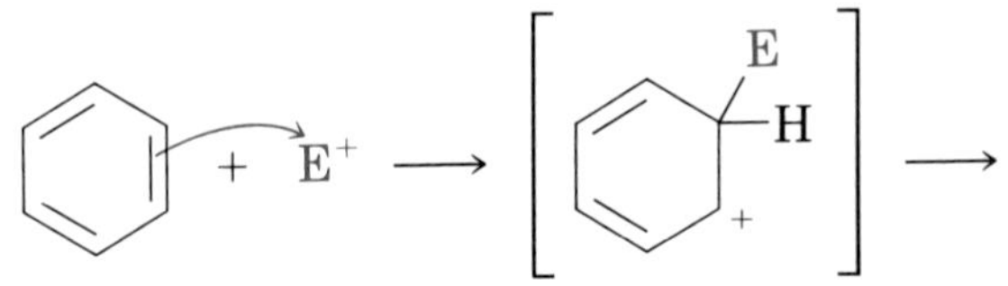

3. The second step is the loss of a proton to re-form the neutral aromatic ring. Loss of a proton occurs because of the stability of the aromatic product.

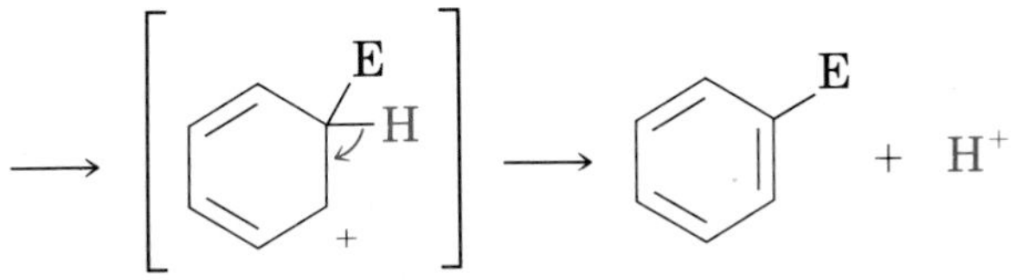

Let's see some supporting evidence for this mechanism.

1. All of these substitution reactions involve electrophiles. This is a very simple point, but it is nevertheless significant that a pure electrophilic reagent such as $NO_2^+ \ BF_4^-$ gives the same substitution products that HNO_3/H_2SO_4 does.

2. Electrophilic aromatic substitution is a bimolecular reaction—it shows second-order kinetics. In other words, these reactions follow the rate law:

$$\text{Rate of reaction} = k[\text{ArH}][\text{E}^+]$$

where k = A constant (the rate coefficient)

$[\text{ArH}]$ = Concentration of aromatic substrate

$[\text{E}^+]$ = Concentration of electrophile

The rates of electrophilic substitution reactions are dependent on the concentrations of both the aromatic substrate and the electrophilic reagent, which indicates that both reagents are involved in the rate-limiting step. These kinetic data are consistent with our two-step mechanism only if the first of the two steps (the bimolecular step) is rate limiting. Once the intermediate carbocation is formed, it rapidly decomposes to product.

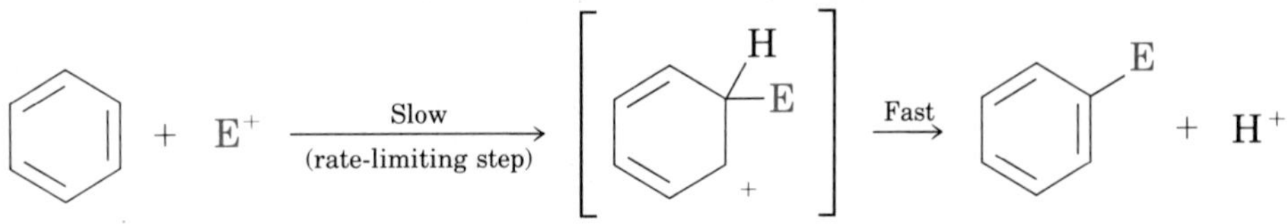

3. Electrophilic aromatic substitution reactions show no deuterium isotope effect. Recall that during our study of the elimination reactions of alkyl halides (Section 10.11) we used the deuterium isotope effect to distinguish between E1 and E2 reaction mechanisms. This effect is due to the fact that carbon–deuterium bonds are stronger than carbon–protium bonds and are therefore less easily broken in the course of a reaction.

To test for a deuterium isotope effect, we compare the rates of reaction of a deuterated molecule and a nondeuterated molecule. If the rate of reaction of the deuterated substrate is lower than the rate of reaction of the nondeuterated substrate, then an isotope effect is present, and a carbon–hydrogen bond is being broken in the rate-limiting step. Conversely, if the rates of the two reactions are identical, then no isotope effect is observed and a carbon–hydrogen bond is not being broken in the rate-limiting step.

It was shown in the early 1950s that benzene (C_6H_6) and deuterated benzene (C_6D_6) are both nitrated at exactly the same rate. Thus, no isotope effect is observed for this electrophilic aromatic substitution reaction. We conclude from this evidence that carbon–hydrogen bond breaking does not occur in the rate-limiting step, and that therefore another step must be rate limiting. This result fits perfectly with our two-step mechanism for electrophilic substitution.

$$C_6D_6 \xrightarrow[\Delta]{HNO_3,\ H_2SO_4} C_6D_5NO_2$$

$$C_6H_6 \xrightarrow[\Delta]{HNO_3,\ H_2SO_4} C_6H_5NO_2$$

Reactions occur at same rate

PROBLEM

15.4 How do you account for the fact that deuterium slowly replaces hydrogen in the aromatic ring when benzene is treated with D_2SO_4? Propose a mechanism for this reaction.

15.7 Reactivity and orientation

Only one monosubstitution product can result when electrophilic substitution occurs on benzene. But what happens when we carry out an electrophilic substitution reaction on an already-substituted benzene? Studies have shown that substituents already present on the benzene ring have two effects:

1. Substituents affect the *reactivity* of the aromatic ring; some substituents make the ring more reactive than benzene, and some make it less reactive.

2. Substituents affect the *orientation* of the reaction; three possible disubstituted products—ortho, meta, and para—can result. These three products are not formed at random; rather, a given substituent already present on the benzene ring usually directs the position of the second substitution.

Let's look at these two effects more closely.

15.8 Reactivity of aromatic rings

Substituents can be classified into two groups: those that *activate* the aromatic ring toward electrophilic substitution, and those that *deactivate* it. Rings that contain an activating substituent are more reactive than benzene, whereas rings with a deactivating substituent are less reactive than benzene. Table 15.2 lists some groups in these two categories.

TABLE 15.2 **Activating and deactivating substituents for electrophilic aromatic substitution**

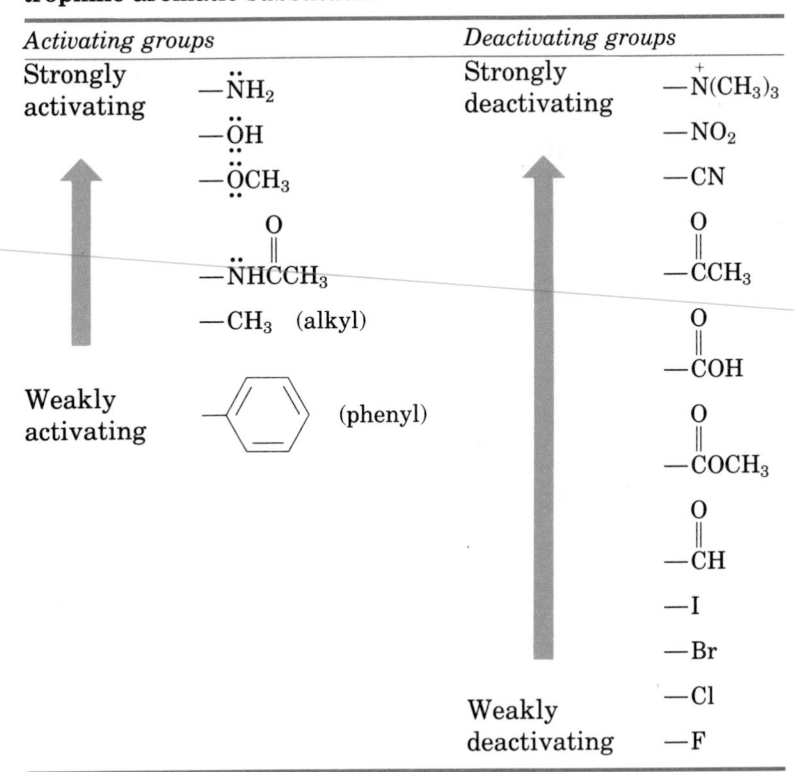

Activating groups		*Deactivating groups*	
Strongly activating	$-\ddot{N}H_2$	Strongly deactivating	$-\overset{+}{N}(CH_3)_3$
	$-\ddot{O}H$		$-NO_2$
	$-\ddot{O}CH_3$		$-CN$
	$-\ddot{N}HC(=O)CH_3$		$-C(=O)CH_3$
	$-CH_3$ (alkyl)		$-C(=O)OH$
Weakly activating	$-C_6H_5$ (phenyl)		$-C(=O)OCH_3$
			$-C(=O)H$
			$-I$
			$-Br$
			$-Cl$
		Weakly deactivating	$-F$

The common feature of all substituents within a category is that all activating groups are electron donors and all deactivating groups are electron acceptors. An aromatic ring with an electron-donating substituent is more electron-rich than benzene and more reactive toward electrophiles; an aromatic ring with an electron-withdrawing substituent is less electron-rich than benzene and less reactive.

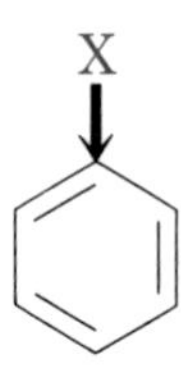

X is an electron donor; ring is electron-rich and very reactive

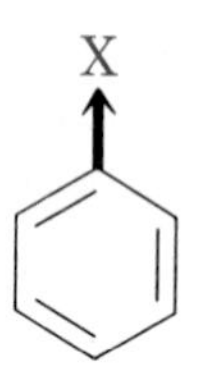

X is an electron acceptor; ring is electron-poor and less reactive

We can best understand the effects of activating and deactivating groups by focusing on the rate-limiting step of an electrophilic aromatic substitution reaction. We have seen that the rate-limiting step involves attack by the ring on an electrophile, E^+, to generate a resonance-stabilized cyclohexadienyl carbocation intermediate. If an electron-*donating* substituent (activating group) is present on the ring, then the carbocation intermediate is stabilized relative to a similar intermediate from benzene. If an electron-*withdrawing* substituent (deactivating group) is present on the ring, then the carbocation intermediate is destabilized relative to a similar intermediate from benzene.

According to the Hammond postulate (Section 5.11), the transition state for this endothermic step "looks like" the carbocation intermediate. The more stabilized carbocation intermediate forms faster, because the same factors that stabilize the carbocation also stabilize the transition state leading to it. Conversely, the less stabilized carbocation intermediate forms more slowly because the transition state leading to its formation is less stabilized. These effects are shown in Figure 15.9.

PROBLEM

15.5 Rank the compounds in each group in the order of their reactivity to electrophilic substitution.

(a) Nitrobenzene, phenol, toluene
(b) Phenol, benzene, chlorobenzene, benzoic acid
(c) Benzene, bromobenzene, benzaldehyde, aniline

15.9 Inductive and resonance effects

A substituent group can donate or withdraw electrons from the aromatic ring in two ways—by inductive effects and by resonance effects. We have

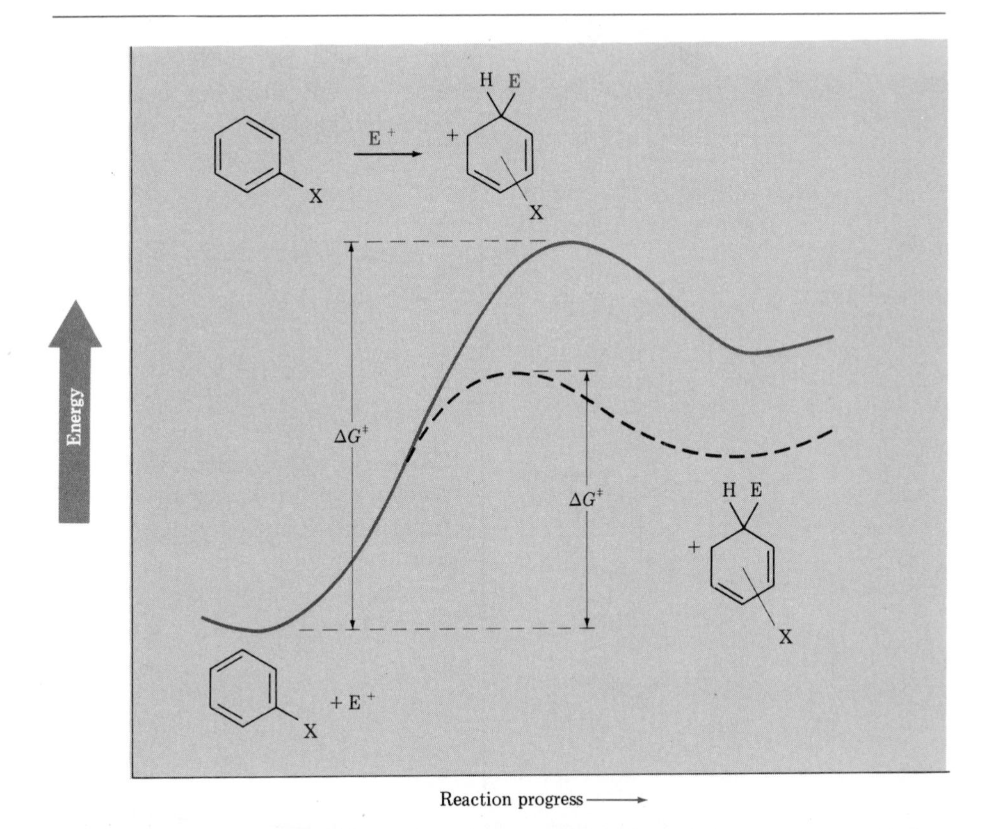

Figure 15.9. Reaction energy diagram for the rate-limiting step in electrophilic aromatic substitutions. Electron-donating substituents (dashed curve) stabilize the carbocation intermediate and thus activate the ring toward substitution; reaction is fast. Electron-withdrawing substituents (solid curve) destabilize the carbocation intermediate and thus deactivate the ring toward substitution; reaction is slow. Substituent X can be at any position on the ring.

seen both of these effects previously (Section 13.10), but let's see how they operate in the present instance.

Inductive effects are due to the intrinsic electronegativity of atoms and to the dipoles of functional groups; they operate by donating or withdrawing electrons either through sigma bonds or through space. For example, halogen substituents are deactivating because of their electronegativity; they withdraw electrons from an aromatic ring inductively through the carbon–halogen sigma bond. Substituents such as nitro, cyano, and carbonyl, however, are inductively electron withdrawing because of the dipoles of these functional groups. In these cases, the atom bonded directly to the aromatic ring bears a partial positive charge and therefore withdraws electrons from the ring.

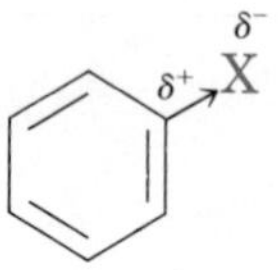

X = F, Cl, Br, or I; inductively electron withdrawing because of electronegativity

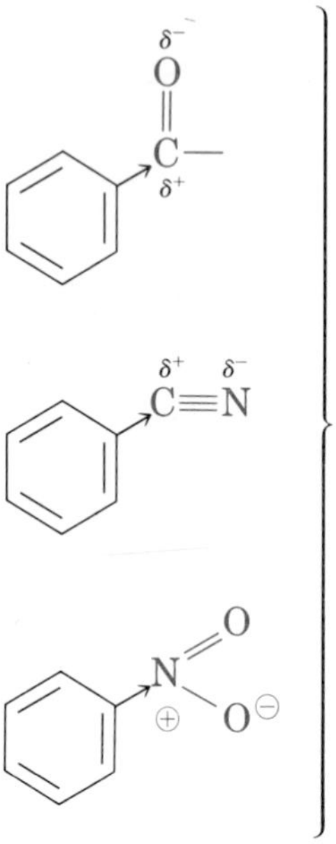

Carbonyl, cyano, nitro; inductively electron withdrawing because of functional-group polarity

Alkyl groups appear to be inductively electron donating, and they therefore activate the ring. The reasons for this are not fully understood but probably involve the same factors that cause alkyl substituents to stabilize alkenes. (Recall that more highly substituted alkenes are more stable.) Hyperconjugation may well be involved (Section 5.4).

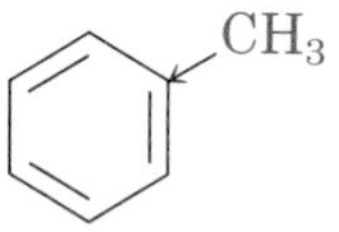

Alkyl group; inductively electron donating

Resonance effects operate by donating or withdrawing electrons by *p*-orbital overlap with the aromatic-ring pi electrons. Substituents such as nitro, carbonyl, and cyano are bonded to the aromatic ring through atoms that have *p* orbitals; aromatic-ring pi electrons can therefore be delocalized onto these substituents. For example, we can draw resonance structures of nitrobenzene and benzaldehyde in which aromatic-ring pi electrons move out onto the substituents, leaving a positive charge in the ring and deactivating the ring toward electrophilic attack.

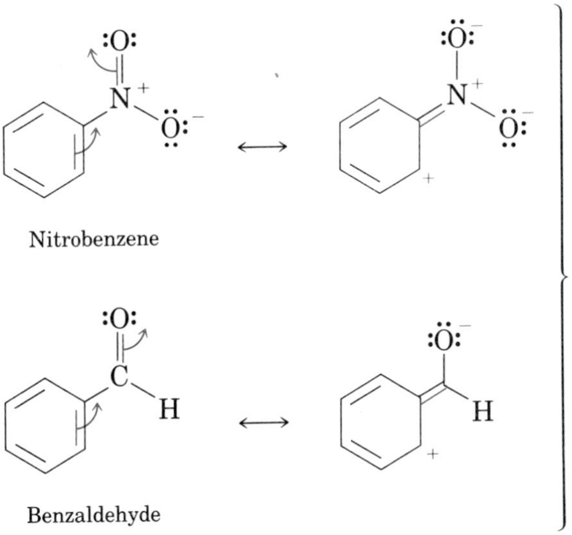

Nitrobenzene

Benzaldehyde

Carbonyl, nitro, and similar substituents deactivate the aromatic ring by resonance withdrawal of pi electrons from the ring

Conversely, substituents such as hydroxyl, methoxyl, and amino activate the aromatic ring by resonance effects that donate pi electrons from the substituents to the ring (Section 13.10).

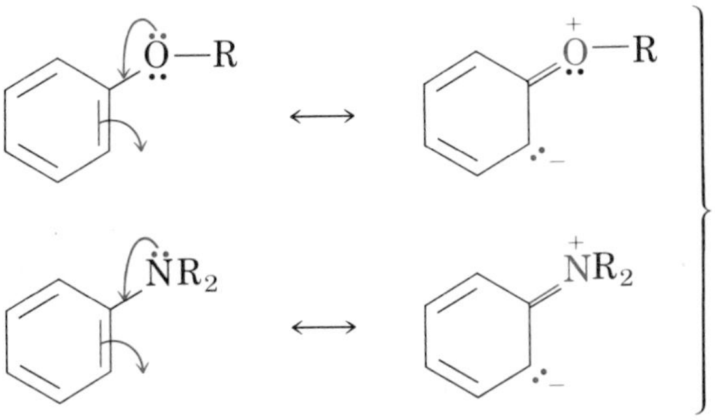

Hydroxyl, methoxyl, and amino substituents activate the ring by resonance donation of pi electrons

It may come as a surprise to learn that hydroxyl, methoxyl, and amino groups activate the ring. Both oxygen and nitrogen are electronegative, and we might therefore expect them to deactivate the ring inductively. However, the resonance electron-donation effect far outweighs the inductive electron-withdrawal effect for these substituents.

PROBLEM

15.6 Write as many resonance forms as you can for the carbocation intermediate resulting from the following reactions. Be sure to consider the resonance effect of the hydroxyl substituent.

(a) Bromination of phenol at the para position
(b) Bromination of phenol at the meta position
(c) Bromination of phenol at the ortho position

At which position(s) does reaction appear most favorable? Which position(s) are least favored?

PROBLEM

15.7 Write as many resonance forms as you can for nitration of the *N,N,N*-Trimethylanilinium ion at ortho, meta, and para positions. At which position(s) does reaction look more favorable?

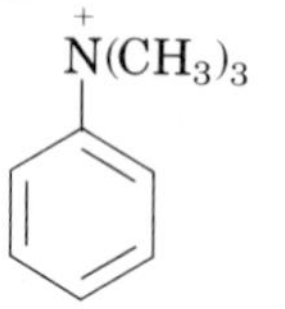

$\overset{+}{N}(CH_3)_3$

N,N,N-Trimethylanilinium ion

15.10 Orientation

The second effect a substituent can have is to direct the position of electrophilic substitution. For example, a methyl substituent shows a strong ortho- and para-directing effect. Nitration of toluene yields three products in the ratio 63% ortho, 34% para, and 3% meta.

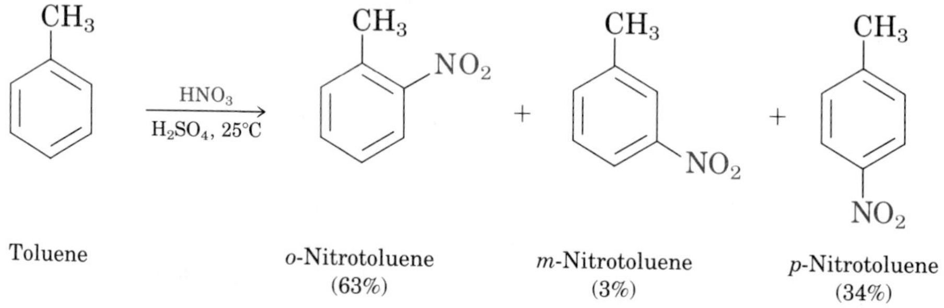

CH_3 $\xrightarrow[H_2SO_4,\ 25°C]{HNO_3}$ CH_3 NO_2 + CH_3 NO_2 + CH_3 NO_2

Toluene — *o*-Nitrotoluene (63%) — *m*-Nitrotoluene (3%) — *p*-Nitrotoluene (34%)

On the other hand, a cyano substituent shows a strong meta-directing effect. Nitration of benzonitrile yields 81% *meta*-, 17% *ortho*-, and 2% *para*-nitrobenzonitrile products.

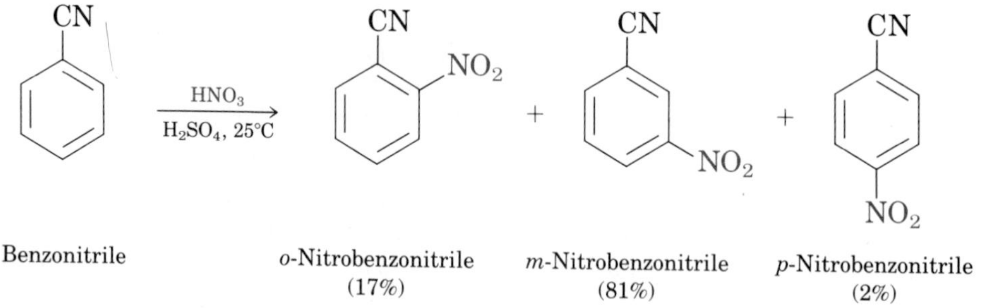

CN $\xrightarrow[H_2SO_4,\ 25°C]{HNO_3}$ CN NO_2 + CN NO_2 + CN NO_2

Benzonitrile — *o*-Nitrobenzonitrile (17%) — *m*-Nitrobenzonitrile (81%) — *p*-Nitrobenzonitrile (2%)

Substituents can be classified into three groups: ortho- and para-directing activators; ortho- and para-directing deactivators; and meta-directing deactivators. No meta-directing activators are known. Table 15.3 lists some of the groups in each category, and Table 15.4 shows experimental results of the orientation of nitration in substituted benzenes.

TABLE 15.3 Classification of directing effects for some common substituents

Ortho- and para-directing activators	*Ortho- and para-directing deactivators*	*Meta-directing deactivators*
$-\ddot{N}H_2$	$-\ddot{\underset{..}{I}}:$	$-\overset{+}{N}(CH_3)_3$
$-\ddot{\underset{..}{O}}H$	$-\ddot{\underset{..}{Br}}:$	$-NO_2$
$-\ddot{\underset{..}{O}}CH_3$	$-\ddot{\underset{..}{Cl}}:$	$-CN$
$-\ddot{N}HCOCH_3$	$-\ddot{\underset{..}{F}}:$	$-\overset{O}{\overset{\|}{C}}CH_3$
$-C_6H_5$		$-\overset{O}{\overset{\|}{C}}OCH_3$
$-CH_3$		$-\overset{O}{\overset{\|}{C}}OH$
		$-\overset{O}{\overset{\|}{C}}H$

TABLE 15.4 Orientation of nitration in substituted benzenes:

$$C_6H_5X \xrightarrow[H_2SO_4,\ 25°C]{HNO_3} X\text{-}C_6H_4\text{-}NO_2$$

X	Product (%) Ortho	Meta	Para
Meta-directing deactivators			
$-\overset{+}{N}(CH_3)_3$	2	89	11
$-NO_2$	7	91	2
$-COOH$	22	77	2
$-CN$	17	81	2
$-CO_2CH_2CH_3$	28	66	6
$-COCH_3$	26	72	2
$-CHO$	19	72	9
Ortho- and para-directing deactivators			
$-F$	13	1	86
$-Cl$	35	1	64
$-Br$	43	1	56
$-I$	45	1	54
Ortho- and para-directing activators			
$-CH_3$	63	3	34
$-\ddot{\underset{..}{O}}H$	50	0	50
$-\ddot{N}HCOCH_3$	19	2	79

We can understand how substituents exert their directing influence by looking further at the consequences of resonance effects and inductive effects. Before doing so, however, we should clarify our definitions of activating and deactivating. Ortho- and para-activating substituents activate *all* positions on the aromatic ring. They are ortho and para directors because they activate the ortho and para positions more than they activate the meta position. Substitution occurs where activation is felt most.

Similarly, deactivating substituents deactivate all positions on the ring. Meta directors deactivate the ortho and para positions more than they deactivate meta positions. Ortho and para directors deactivate the meta position more than they deactivate ortho and para. Substitution occurs where the deactivation is felt least.

PROBLEM

15.8 Predict the major products of these reactions:
(a) Nitration of bromobenzene
(b) Bromination of nitrobenzene
(c) Chlorination of phenol
(d) Hydroxylation of chlorobenzene
(e) Nitration of bromobenzene, followed by $SnCl_2$ reduction
(f) Bromination of aniline

ORTHO- AND PARA-ACTIVATING GROUPS: ALKYL

Electrophilic aromatic substitution on an alkyl-substituted benzene ring might occur at any of three positions. Let's consider the nitration of toluene as an example (Figure 15.10).

Nitration of toluene at any one of the three positions leads to resonance-stabilized carbocation intermediates, but the ortho and para intermediates are more stabilized. For both ortho and para (but not meta) attack, a resonance form places the positive charge directly on the methyl-substituted carbon, where it can best be stabilized by the methyl inductive effect. These intermediates are lower in energy and are therefore formed preferentially, as indicated by the reaction energy diagram in Figure 15.11.

PROBLEM

15.9 Which compound would you expect to be more reactive toward electrophilic substitution, toluene or (trifluoromethyl)benzene? Explain your answer.

CF_3

(Trifluoromethyl)benzene

ORTHO- AND PARA-ACTIVATING GROUPS: OH AND NH_2

Hydroxyl and amino groups (and their derivatives) are also ortho–para activators, but for a different reason than for alkyl groups. Hydroxyl and amino groups exert their activating influence through a strong resonance electron-donating effect, which is most pronounced at the ortho and para

Ortho 63%

Most stable

Meta 3%

Para 34%

Most stable

Figure 15.10. Carbocation intermediates in nitration of toluene. Ortho and para intermediates are more stable than meta intermediates.

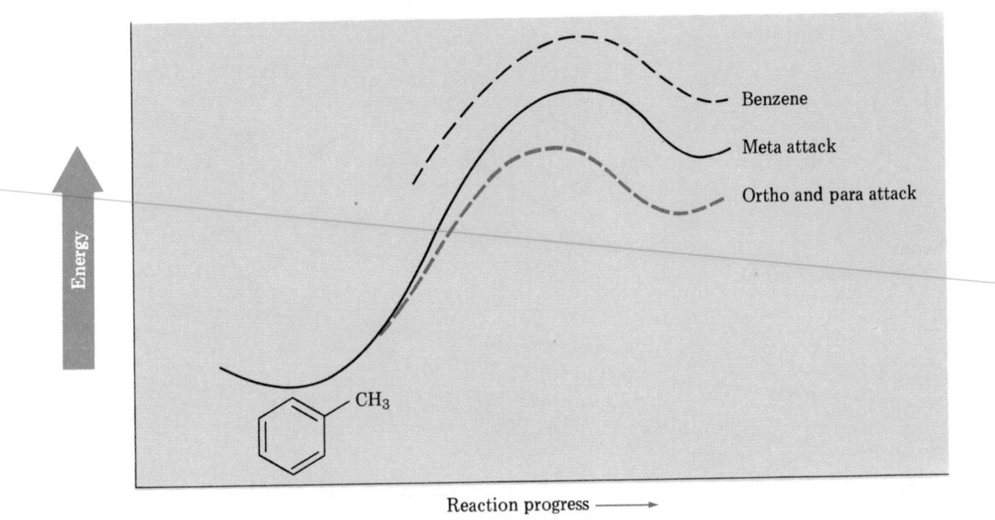

Figure 15.11. Reaction energy diagram for nitration of toluene. The carbocation intermediates resulting from attack at the ortho and para positions are more stable and are therefore formed more rapidly than the intermediate resulting from attack at the meta position. Note, however, that all three intermediates from attack on toluene are more stable than the intermediate from attack on benzene.

positions. The resonance forms of phenol in Figure 15.12 indicate why this is so; the oxygen lone-pair electrons can be shared only in the ortho and para positions, but not in the meta position.

When phenol is nitrated, the results shown in Figure 15.13 are obtained. Although three products are possible, only ortho and para attack is observed.

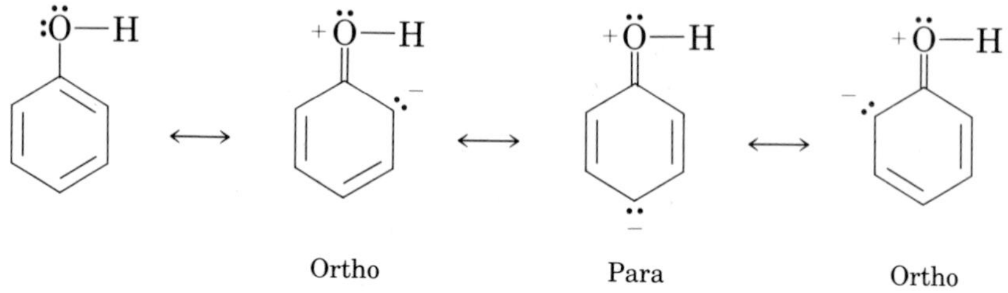

Ortho Para Ortho

Figure 15.12. Dipolar resonance forms of phenol. Oxygen electrons are shared in ortho and para positions only.

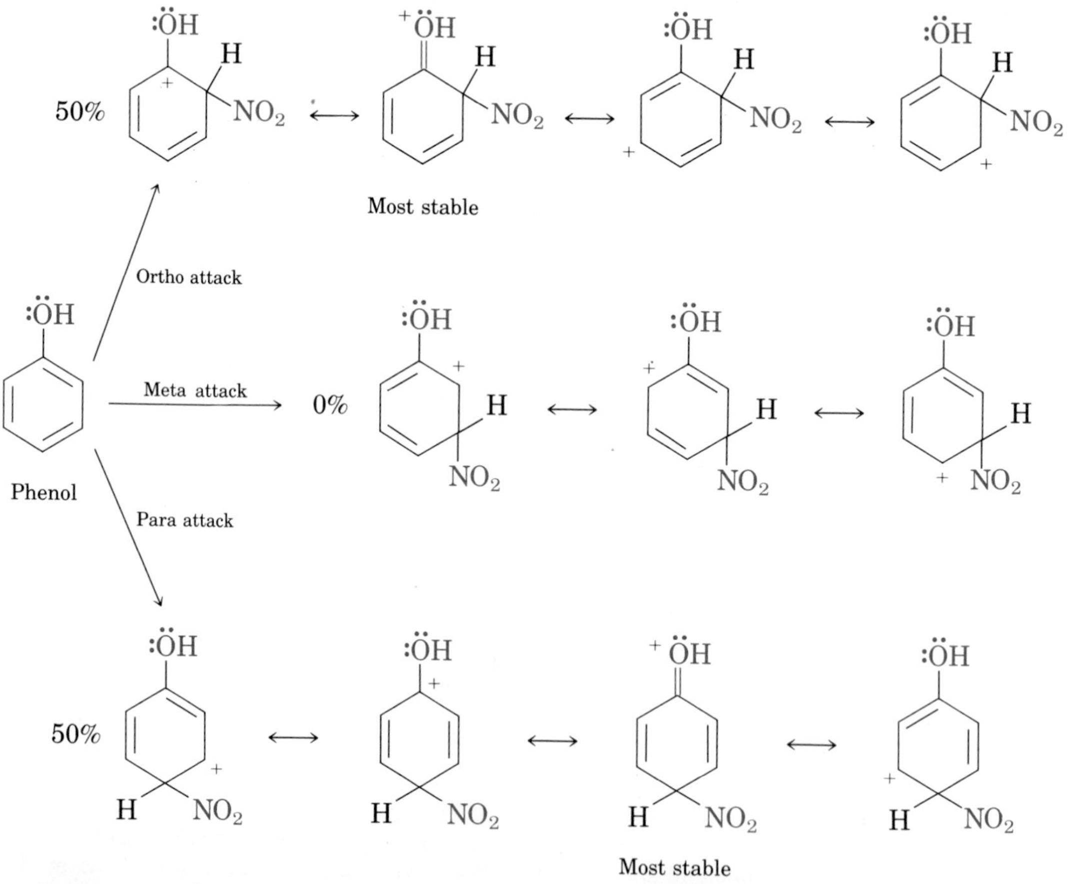

50% Most stable

Ortho attack

Meta attack 0%

Phenol

Para attack

50% Most stable

Figure 15.13. Intermediates in the nitration of phenol. Ortho and para intermediates are more stable than the meta intermediate.

All three of the possible carbocation intermediates are stabilized by resonance, but the intermediates from ortho and para attack are stabilized most. Only in ortho and para attack are there highly stabilized resonance forms in which the positive charge is stabilized by an electron pair from oxygen. The product of meta attack has no such stabilization. (The most stable resonance forms are indicated in Figure 15.13.)

PROBLEM

15.10 Acetanilide is much less reactive toward electrophilic substitution than is aniline. Can you account for this behavior?

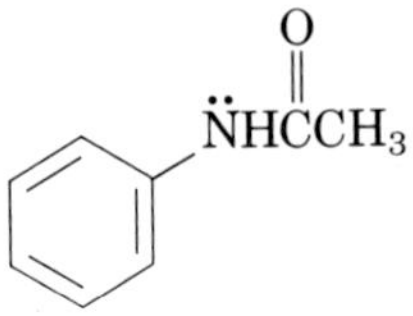

$C_6H_5\ddot{N}HCOCH_3$

Acetanilide

PROBLEM

15.11 What reactivity order would you expect for phenol, phenoxide ion, and phenyl acetate? Explain.

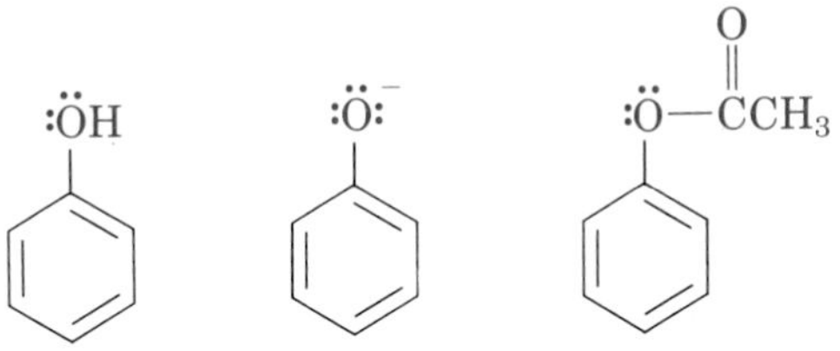

$C_6H_5\ddot{O}H$ $C_6H_5\ddot{O}:^-$ $C_6H_5\ddot{O}-COCH_3$

Phenol Phenoxide ion Phenyl acetate

META-DEACTIVATING GROUPS

We can explain the influence of meta deactivators by the same kinds of arguments used for ortho–para activators. As an example, let's look at the chlorination of benzaldehyde. Three possible modes of attack are shown in Figure 15.14 (p. 502).

Three resonance forms can be drawn for all three possible intermediates, but the intermediate from meta attack is most stable. In both ortho and para attack, the very unfavorable resonance forms indicated in Figure 15.14 place the positive charge directly on the carbon carrying the deactivating group. A severe electrostatic repulsive interaction between the positive charge and the positive end of the carbonyl dipole strongly disfavors these intermediates. All three intermediates are disfavored, but the meta intermediate is the least disfavored (hence, most favored). We therefore find a predominance (72%) of meta substitution.

Most meta deactivators also have an electron-withdrawing resonance effect that reinforces the inductive effect. As the dipolar resonance forms of benzaldehyde indicate (Figure 15.15), resonance withdrawal of electrons is felt only at the ortho and para positions. Reaction with an electrophile therefore occurs at the meta position.

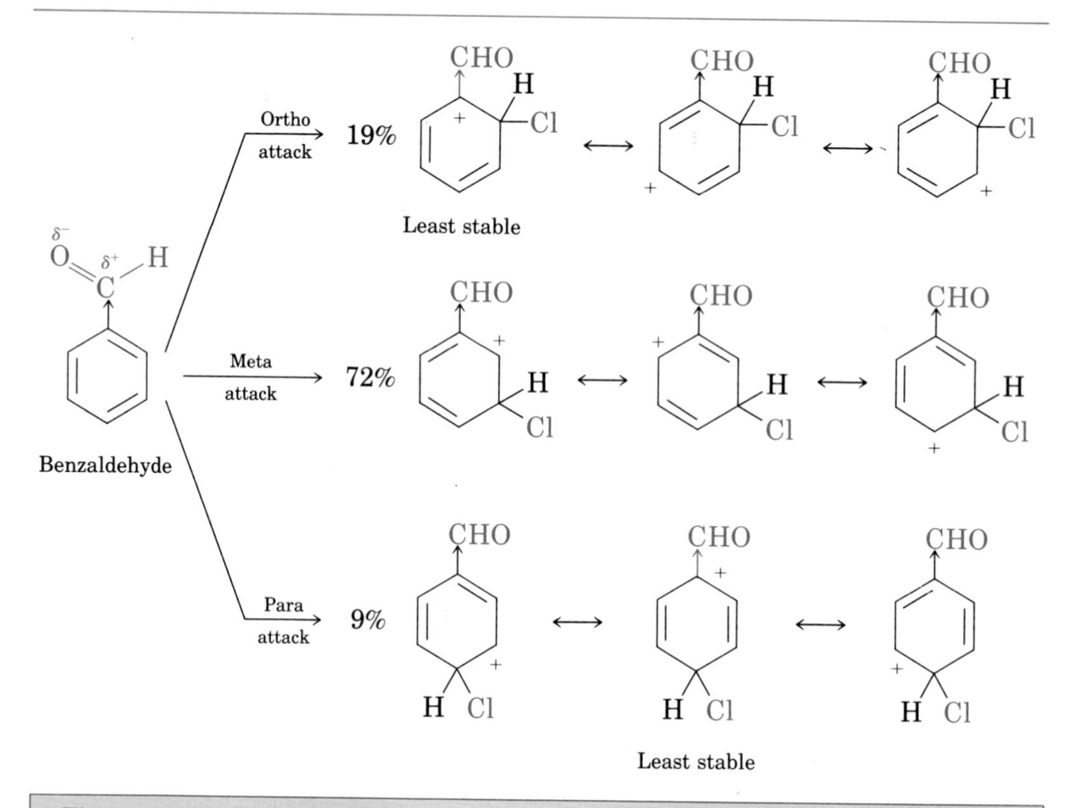

Figure 15.14. Intermediates in the chlorination of benzaldehyde. The meta intermediate is more stable than the ortho or para intermediate.

Ortho Para Ortho

Figure 15.15. Dipolar resonance forms of benzaldehyde. Electron density is lowered at ortho and para positions.

A reaction energy diagram for attack of benzaldehyde on an electrophile is shown in Figure 15.16. Note that reaction at the meta position is more favored than reaction at ortho or para positions, but that reaction at *any* of the three benzaldehyde positions is less favored than reaction on benzene.

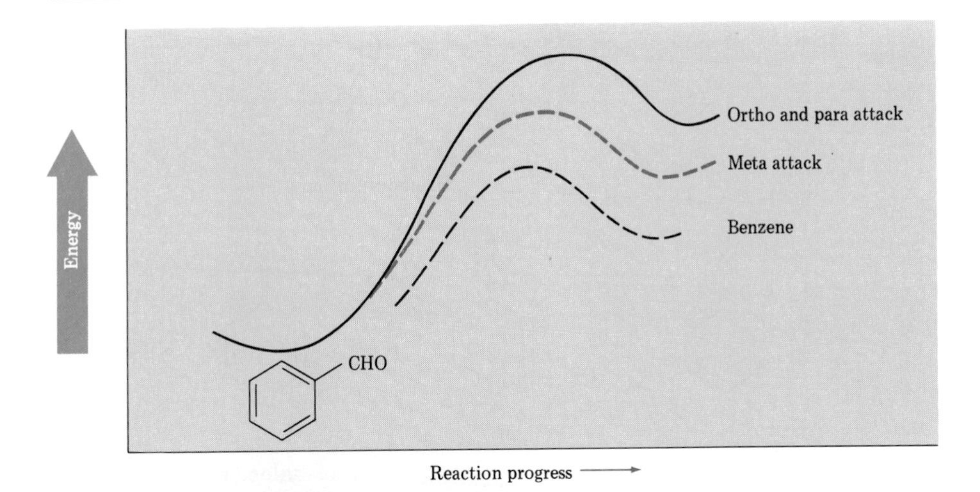

Figure 15.16. Reaction energy diagram for electrophilic aromatic substitution on benzaldehyde. Attack at the meta position is favored over attack at ortho and para positions, but reaction on benzaldehyde is slower than reaction on benzene.

ORTHO- AND PARA-DEACTIVATING GROUPS: HALOGENS

Halogen substituents occupy a unique position since they are deactivating yet have an ortho- and para-directing effect. Why should this be?

Halogen substituents are deactivating because of their strongly electron-withdrawing inductive effect. Halogens are electronegative atoms that remove electron density from the aromatic rings, thus making the ring less susceptible to electrophilic attack. Unlike other deactivating groups, however, the halogens deactivate the ortho and para positions *less* than they deactivate the meta position.

The reason for this behavior is that halogens exert both an electron-donating resonance effect and an electron-withdrawing inductive effect. The resonance effect is a consequence of the overlap of lone-pair electrons on halogen with the pi orbitals of the aromatic ring. As the resonance structures in Figure 15.17 (p. 504) show, this electron donation places a positive charge on the halogen (a halonium ion) and distributes the negative charge over the ortho and para positions. (We have seen halonium ions before as intermediates in electrophilic addition of halogens to alkenes, Section 6.1.)

Ortho- and para-activating substituents such as hydroxyl and amino also show the same two effects—inductive electron withdrawal and resonance electron donation. In these cases, the resonance effect far outweighs the inductive effect, so that hydroxyl and amino groups are activators. For the halogens, however, the inductive effect slightly outweighs the resonance effect. The inductive effect controls reactivity, but the resonance effect controls orientation.

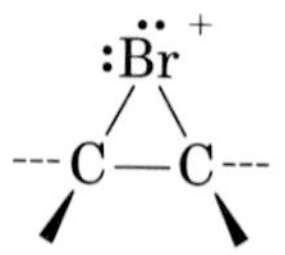

:Br:+

---C—C---

Bromonium ion

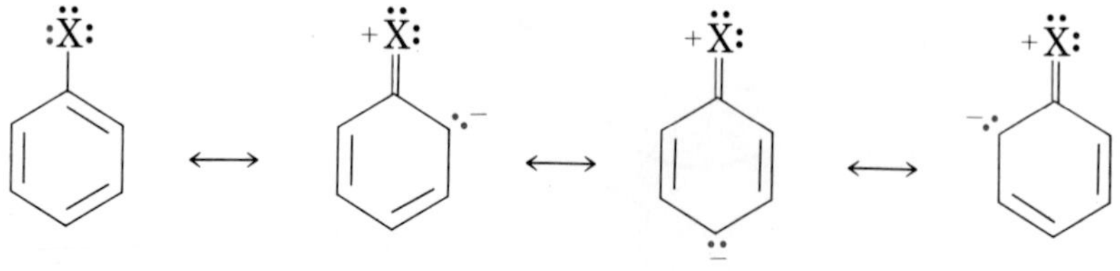

:X: ⟷ +X: ⟷ +X: ⟷ +X:

Figure 15.17. Dipolar resonance forms of halobenzenes. Electron density from halogen *p* electrons is shared by ortho and para ring positions.

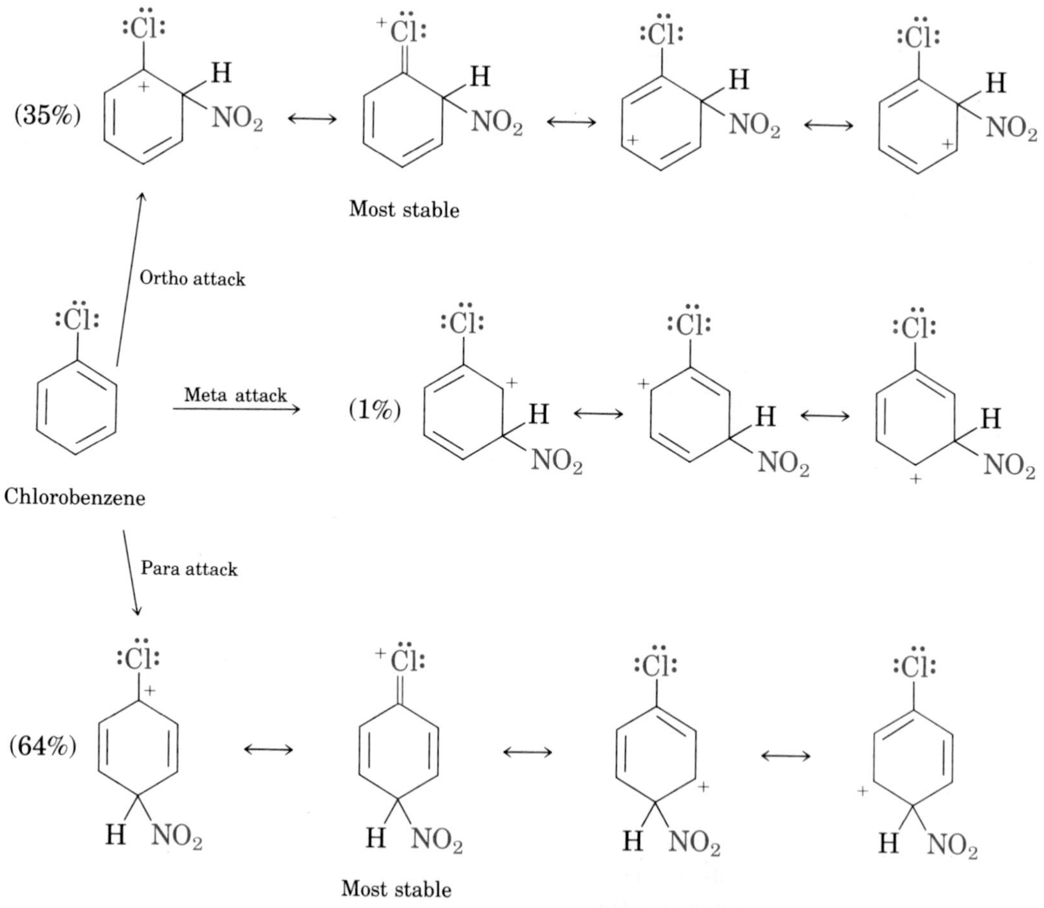

(35%) :Cl: H NO_2 ⟷ +:Cl: H NO_2 ⟷ :Cl: H NO_2 ⟷ :Cl: H NO_2

Most stable

Ortho attack

:Cl:

Meta attack

(1%) :Cl: H NO_2 ⟷ :Cl: H NO_2 ⟷ :Cl: H NO_2

Chlorobenzene

Para attack

(64%) :Cl: H NO_2 ⟷ +:Cl: H NO_2 ⟷ :Cl: H NO_2 ⟷ :Cl: H NO_2

Most stable

Figure 15.18. Carbocation intermediates from nitration of chlorobenzene. The ions resulting from ortho and para attack are most stable and therefore form faster.

If we consider the stabilization of carbocation intermediates from electrophilic substitution on halobenzenes, we find that the ortho- and para-substituted intermediates have their positive charge resonance-stabilized by the halogen substituent, whereas the meta-substituted intermediate has no such stabilization. For example, nitration of chlorobenzene takes the course illustrated in Figure 15.18. Ortho and para products strongly predominate over meta products by a ratio of 35% ortho, 1% meta, and 64% para.

A reaction energy diagram for the nitration of chlorobenzene is shown in Figure 15.19. The carbocation intermediates resulting from reaction at ortho and para positions are lower in energy than the intermediate from reaction at the meta position, but all three intermediates from the reaction of chlorobenzene are less stable than the intermediate from the reaction of benzene.

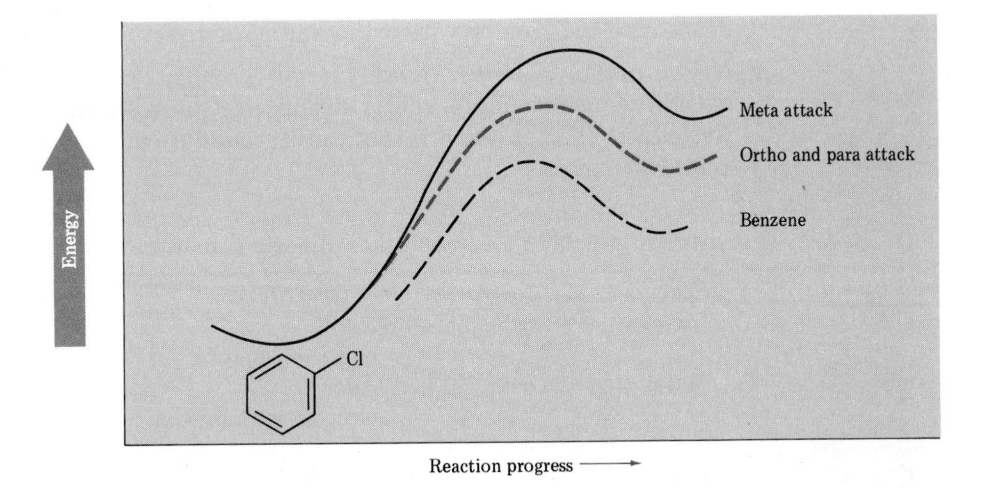

Figure 15.19. Reaction energy diagram for nitration of chlorobenzene. Reaction at ortho and para positions is favored, but chlorobenzene is less reactive than benzene.

PROBLEM

15.12 The nitroso group, —N̈=O, is one of the very few nonhalogens that is an ortho- and para-deactivating group. How can you explain this? Draw resonance structures of intermediates in ortho and para electrophilic attack on nitrosobenzene, and explain why they are favored over the intermediate from meta attack.

N̈=O

Nitrosobenzene

15.11 Substituent effects in aromatic substitution: a summary

Both orientation and reactivity effects in electrophilic aromatic substitution are controlled by the interplay of two factors—resonance effects and inductive effects (Table 15.5). Different substituents behave differently, depending on the direction and strength of the two effects. We can summarize the results as follows:

1. *Alkyl groups.* Electron-donating inductive effect; no resonance effect. The net result is that alkyl groups are activating and ortho and para directing.
2. *Hydroxyl and amino groups (and derivatives).* Powerful electron-donating resonance effect; weak electron-withdrawing inductive effect. The net result is that these groups are activating and ortho and para directing.
3. *Halogens.* Strong electron-withdrawing inductive effect; moderate electron-donating resonance effect. The net result is that halogens deactivate the ring but are ortho and para directing.
4. *Nitro, cyano, carbonyl, and similar groups.* Powerful electron-withdrawing resonance effect; strong electron-withdrawing inductive effect. The net result is that these groups are meta directing and deactivating.

TABLE 15.5 Substituent effects in electrophilic aromatic substitution

Substituent	*Reactivity*	*Orientation*	*Inductive effect*	*Resonance effect*
$-CH_3$	Activating	Ortho, para	Weak; electron donating	None
$-\ddot{\underset{..}{O}}H$, $-\ddot{N}H_2$	Activating	Ortho, para	Weak; electron withdrawing	Strong; electron donating
$-\ddot{\underset{..}{F}}:$, $-\ddot{\underset{..}{Cl}}:$, $-\ddot{\underset{..}{Br}}:$, $-\ddot{\underset{..}{I}}:$	Deactivating	Ortho, para	Strong; electron withdrawing	Weak; electron donating
$-\overset{+}{N}(CH_3)_3$	Deactivating	Meta	Strong; electron withdrawing	None
$-NO_2$, $-CN$, $-CHO$, $-CO_2CH_3$, $-COCH_3$	Deactivating	Meta	Strong; electron withdrawing	Strong; electron withdrawing

15.12 Trisubstituted benzenes: additivity of effects

Further electrophilic substitution of a disubstituted benzene is governed by the same resonance and inductive effects that we have just discussed. The only difference is that now we must consider the *additive* effects of two different directing groups. In practice, this is not as difficult as it sounds, and three rules are usually sufficient to predict the results of a reaction:

1. If both groups direct substitution toward the same position, there is no problem. Take, for example, the nitration of *p*-nitrotoluene. Both the methyl and the nitro group direct further substitution to the same position (ortho to the methyl = meta to the nitro), and a single product is formed during the reaction.

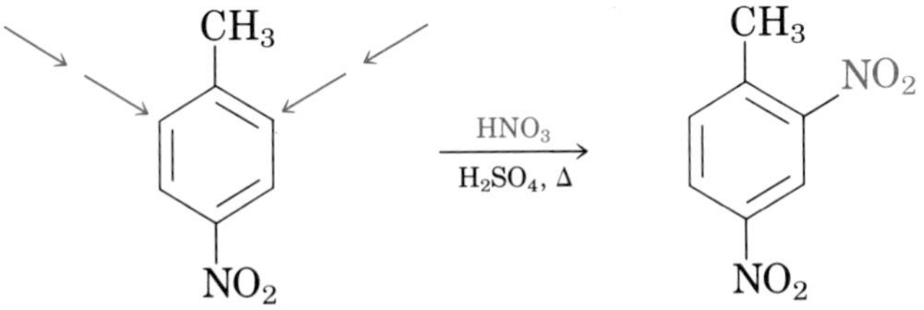

CH_3 NO_2 $\xrightarrow[H_2SO_4,\ \Delta]{HNO_3}$ CH_3 NO_2 NO_2

p-Nitrotoluene

2,4-Dinitrotoluene (sole product)

2. If the directing effects of the two groups oppose each other, the more powerful activating group usually has the dominant influence. Mixtures of products often result, however. For example, bromination of *p*-methylphenol yields largely 2-bromo-4-methylphenol, since hydroxyl is a more powerful activator than methyl.

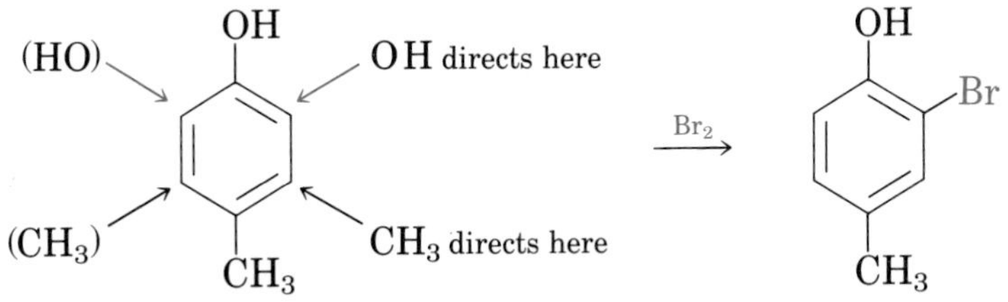

(HO) OH OH directs here (CH_3) CH_3 CH_3 directs here $\xrightarrow{Br_2}$ OH Br CH_3

p-Methylphenol (*p*-cresol)

2-Bromo-4-methylphenol (major product)

3. Further substitution rarely occurs between the two groups in a meta-disubstituted compound. The probable reason for this is steric—the position between the two groups is too hindered for reaction to occur easily.

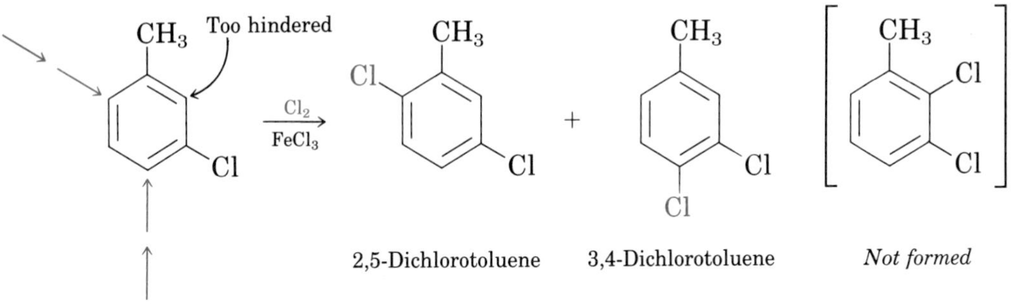

CH_3 Too hindered Cl $\xrightarrow[FeCl_3]{Cl_2}$ CH_3 Cl Cl + CH_3 Cl Cl [CH_3 Cl Cl]

m-Chlorotoluene

2,5-Dichlorotoluene

3,4-Dichlorotoluene

Not formed

PROBLEM

15.13 Where would you expect electrophilic substitution to occur in these substances?

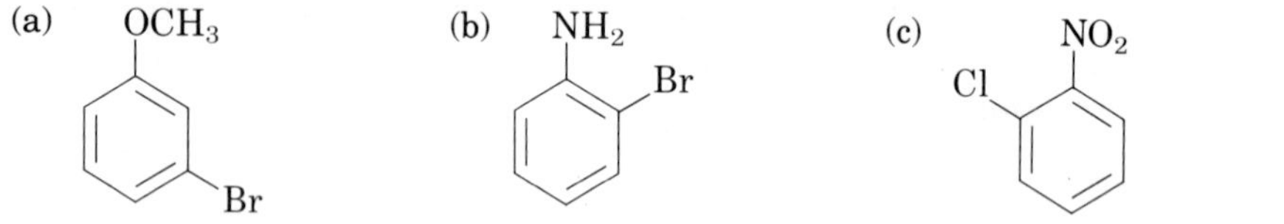

(a) OCH_3 / Br (b) NH_2 / Br (c) NO_2 / Cl

15.13 Nucleophilic aromatic substitution

The electrophilic substitutions just discussed are the most important and useful reactions of the aromatic ring. In certain cases, however, aromatic substitution can occur by a nucleophilic mechanism. Aryl halides that have electron-withdrawing substituents undergo **nucleophilic aromatic substitution.** For example, 2,4,6-trinitrochlorobenzene reacts with aqueous sodium hydroxide at room temperature to give 2,4,6-trinitrophenol in 100% yield. The nucleophile, hydroxide ion, has substituted for chloride ion:

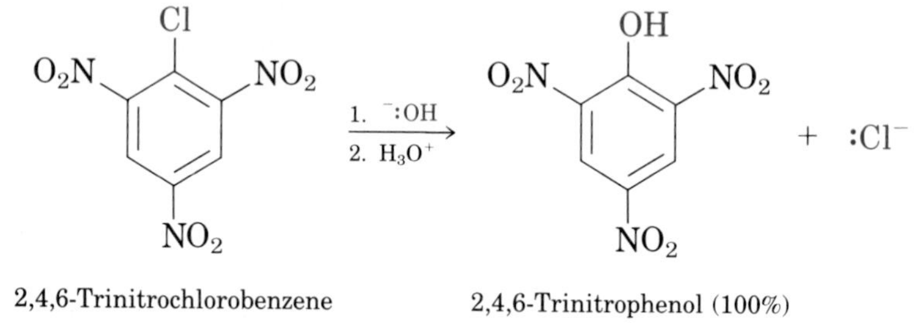

Cl, O_2N, NO_2, NO_2 — 1. $^-$:OH 2. H_3O^+ → OH, O_2N, NO_2, NO_2 + $:Cl^-$

2,4,6-Trinitrochlorobenzene 2,4,6-Trinitrophenol (100%)

How does this reaction take place? It appears similar to the S_N1 and S_N2 nucleophilic substitution reactions of alkyl halides (Chapter 10) but is in fact quite different, since aryl halides are inert to substitution by S_N1 and S_N2 mechanisms.

The S_N1 reactions of alkyl halides occur through a rate-limiting ionization of the alkyl halide to a stable carbocation. Aryl halides, however, do not ionize; the aryl cation is highly unstable because of the presence of the positive charge in an orbital with high *s* character (sp^2):

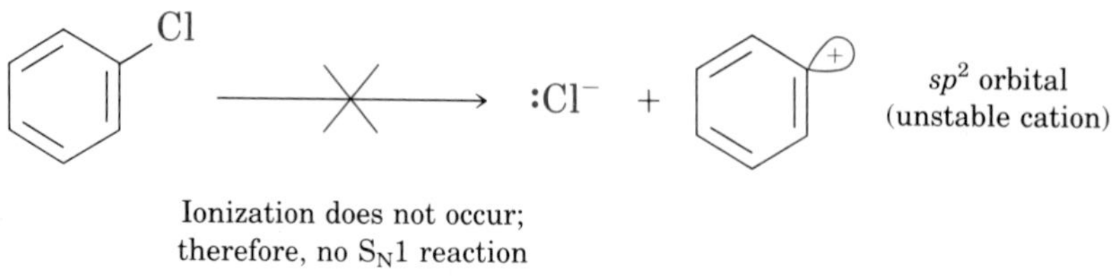

Cl —✕→ $:Cl^-$ + [aryl cation] sp^2 orbital (unstable cation)

Ionization does not occur; therefore, no S_N1 reaction

However,

$$(CH_3)_3C{-}Cl \longrightarrow (CH_3)_3C^+ + :Cl^-$$

Stable cation

The S_N2 reactions of alkyl halides occur through a rate-limiting back-side displacement of the leaving group by the attacking nucleophile. Aryl halides, however, are sterically shielded from back-side attack by the aromatic ring. In order for a nucleophile to attack an aryl halide, it would have to approach directly through the aromatic ring and invert the stereochemistry of the aromatic ring—a geometric impossibility. Nucleophilic aromatic substitution must therefore occur by a different mechanism.

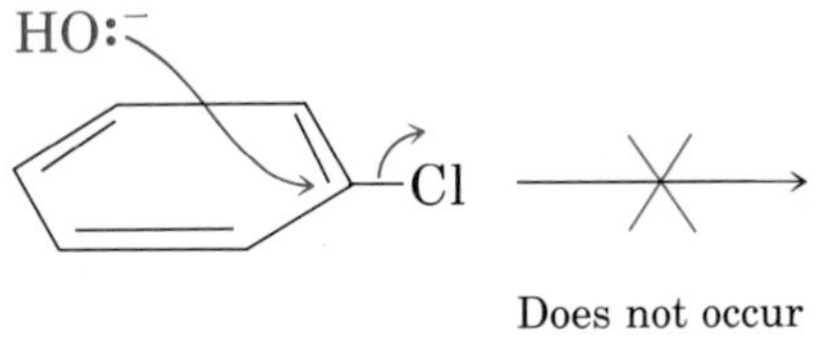

HO:⁻ + C₆H₅Cl ⟶ (crossed out)

Does not occur

However,

$$HO:^- + CH_3Cl \longrightarrow HO{-}CH_3 + :Cl^-$$

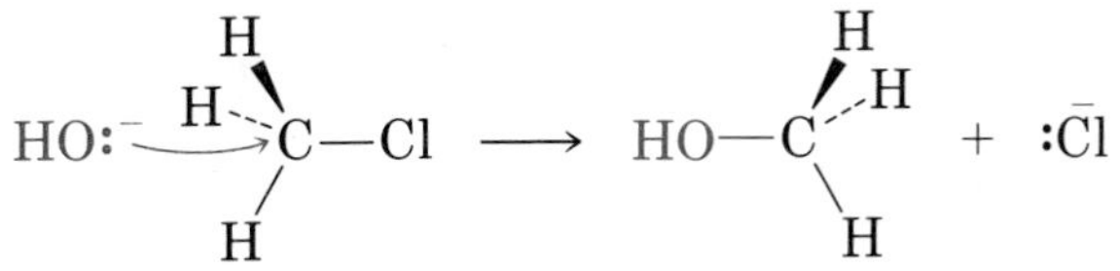

Studies have shown that nucleophilic aromatic substitutions proceed by the *addition–elimination* mechanism shown in Figure 15.20. The attacking nucleophile first adds to the electron-deficient aryl halide, forming a nega-

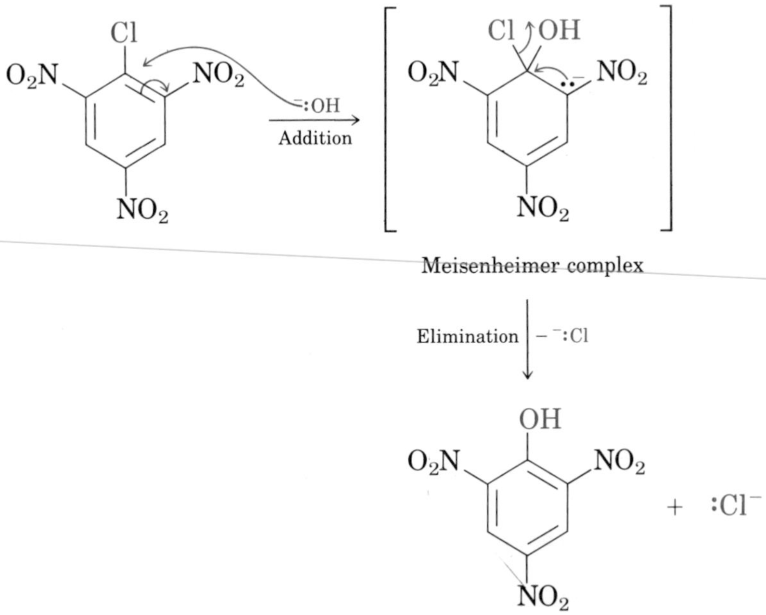

:OH⁻

Addition

Meisenheimer complex

Elimination

$- ^-:Cl$

$+ :Cl^-$

Figure 15.20. Mechanism of nucleophilic aromatic substitution on 2,4,6-trinitrochlorobenzene.

tively charged intermediate (**a Meisenheimer[1] complex**); halide ion is then eliminated in the second step.

Nucleophilic aromatic substitution occurs only if the halobenzene has electron-withdrawing substituents in the ortho and/or para positions; the more substituents there are, the faster the reaction goes. The reason for this requirement is that only ortho and para electron-withdrawing substituents can stabilize the anion intermediate through resonance. *p*-Chloronitrobenzene and *o*-chloronitrobenzene react with dilute hydroxide ion at 130°C to yield substitution products, but a meta substituent cannot offer resonance stabilization to the intermediate anion. *m*-Chloronitrobenzene is therefore inert to hydroxide ion (Figure 15.21).

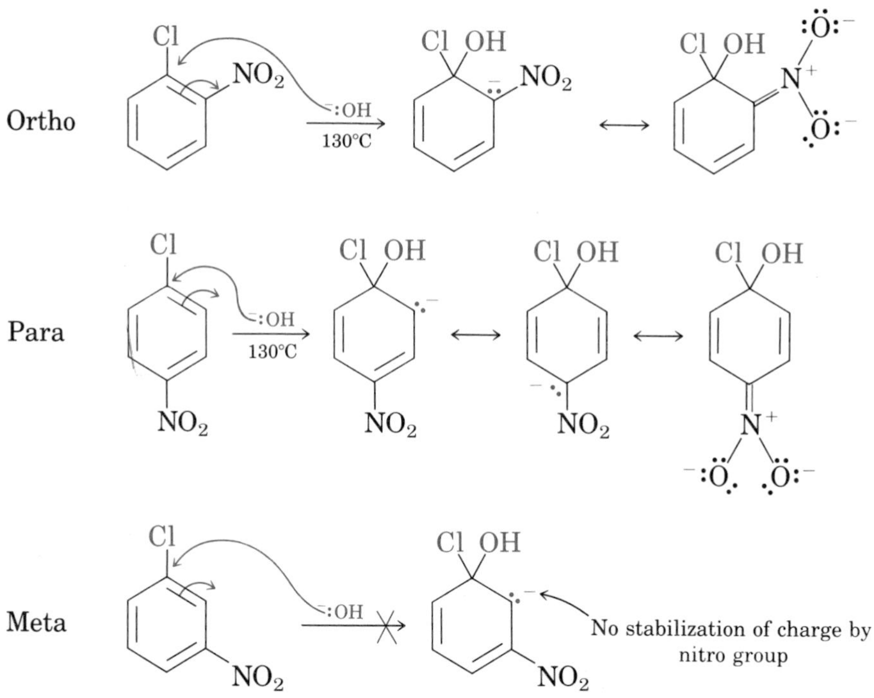

Figure 15.21. Nucleophilic attack on nitrochlorobenzenes. Only ortho and para isomers are attacked easily.

PROBLEM

15.14 Propose a mechanism to account for the observation that 1-chloroanthraquinone reacts with methoxide ion to give the substitution product, 1-methoxyanthraquinone.

[1]Jacob Meisenheimer (1876–1934); b. Greisheim; Ph.D. Munich; professor, universities of Berlin, Greifswald, Tübingen.

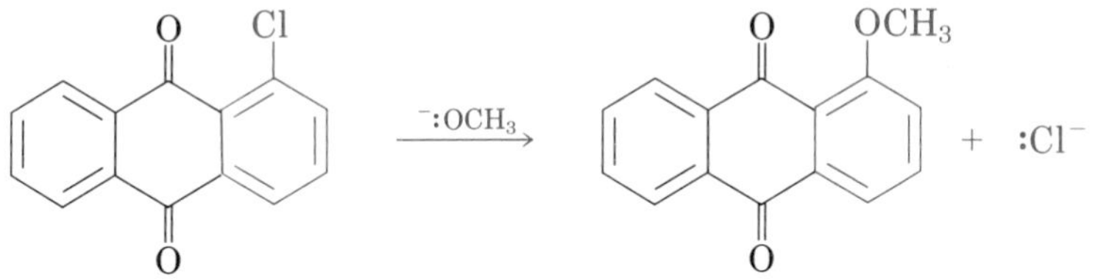

$$\text{1-Chloroanthraquinone} \xrightarrow{^{-}:OCH_3} \text{1-Methoxyanthraquinone} + :Cl^-$$

1-Chloroanthraquinone 1-Methoxyanthraquinone

PROBLEM

15.15 Draw a reaction energy diagram for the nucleophilic aromatic substitution reaction of *o*-nitrochlorobenzene with hydroxide ion. Assume that the first step is rate limiting.

15.14 Benzyne

Halobenzenes without electron-withdrawing substituents are inert to nucleophiles under normal conditions. Under conditions of high temperature and pressure, however, even chlorobenzene can be forced to react. Scientists at the Dow Chemical Company announced in 1928 that phenol could be prepared on a large industrial scale by treatment of chlorobenzene with dilute aqueous sodium hydroxide at 340°C under 2500 psi (pounds per square inch) pressure.

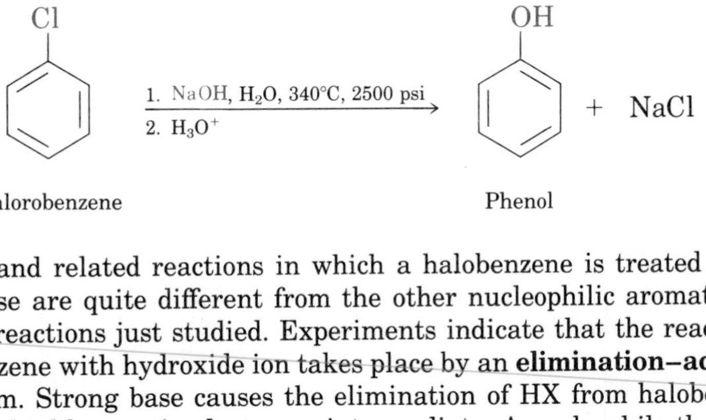

$$\text{Chlorobenzene} \xrightarrow[2.\ H_3O^+]{1.\ NaOH,\ H_2O,\ 340°C,\ 2500\ psi} \text{Phenol} + NaCl$$

Chlorobenzene Phenol

This and related reactions in which a halobenzene is treated with a strong base are quite different from the other nucleophilic aromatic substitution reactions just studied. Experiments indicate that the reaction of chlorobenzene with hydroxide ion takes place by an **elimination–addition** mechanism. Strong base causes the elimination of HX from halobenzene, yielding a highly reactive **benzyne** intermediate. A nucleophile then adds to benzyne to yield the product. The two steps are the same as in other nucleophilic aromatic substitutions, but the order of steps is reversed (addition before elimination for the usual reaction versus elimination before addition for the benzyne reaction).

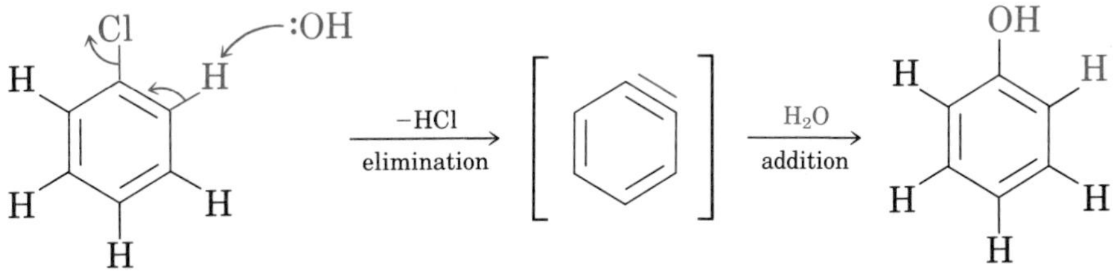

$$\text{Chlorobenzene} \xrightarrow[\text{elimination}]{-HCl} [\text{Benzyne}] \xrightarrow[\text{addition}]{H_2O} \text{Phenol}$$

Chlorobenzene Benzyne Phenol

Powerful evidence in support of the benzyne mechanism has been obtained by studying the reaction between bromobenzene and the strong base, potassium amide. When bromobenzene labeled with a radioactive ^{14}C carbon atom at the 1 position is used, the product has the label scrambled between positions 1 and 2. This result requires that the reaction proceed through a symmetrical intermediate in which positions 1 and 2 are equivalent. Only benzyne fits this requirement:

Bromobenzene $\xrightarrow[NH_3\ (-HBr)]{:\ddot{N}H_2^-}$ [Benzyne (symmetrical)] $\xrightarrow{NH_3}$ 50% Aniline + 50% Aniline

Further evidence for a benzyne intermediate comes from trapping experiments. Although benzyne is far too unstable and reactive to be isolated as a pure compound, it can be intercepted as a Diels–Alder adduct if furan is added to the reaction. This is just the kind of behavior we would expect for so strained and reactive a species as benzyne.

[Benzyne (a dienophile)] + Furan (a diene) ⟶ Diels–Alder adduct

The electronic structure of benzyne, shown in Figure 15.22, may be compared to that of a highly distorted alkyne. The normal alkyne triple

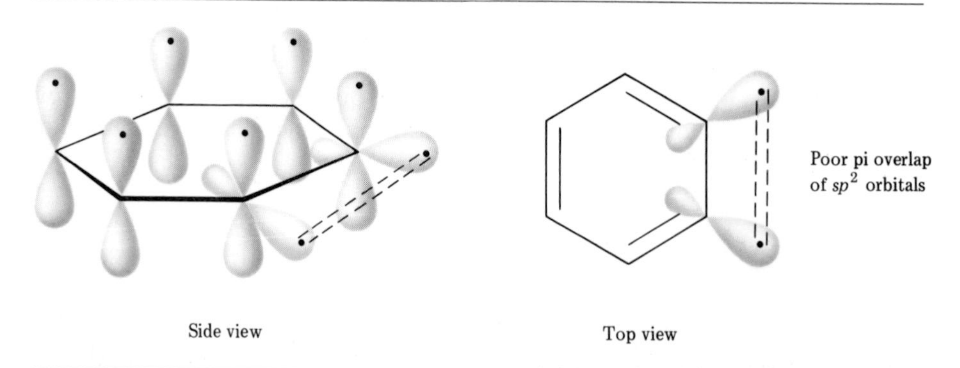

Figure 15.22. Orbital picture of benzyne. The benzyne carbons are sp^2 hybridized, and the "third" bond results from weak overlap of two adjacent sp^2 orbitals.

bond consists of a sigma bond formed by *sp–sp* overlap and two mutually perpendicular pi bonds formed by *p–p* overlap. The benzyne triple bond, however, consists of a sigma bond formed by sp^2–sp^2 overlap, one pi bond formed by *p–p* overlap, and one pi bond formed by sp^2–sp^2 overlap. The latter pi bond is in the plane of the ring and is very weak because of poor orbital overlap.

PROBLEM

15.16 Account for the fact that treatment of *p*-bromotoluene with NaOH at 300°C yields a mixture of two products, but treatment of *m*-bromotoluene with NaOH yields a mixture of three products.

15.15 Summary

Electrophilic aromatic substitution, a polar process that occurs by a two-step mechanism, is the single most important reaction of aromatic compounds. The pi electrons of the aromatic ring first attack the electrophile, E^+, in a slow, rate-limiting step. The resonance-stabilized intermediate carbocation loses a proton to regenerate an aromatic ring:

E H E E^+ + + H^+

Many different substituents can be introduced onto the ring by this process. Bromination, chlorination, iodination, nitration, sulfonation, and hydroxylation can all be carried out with the proper choice of reagent. The primary evidence for the mechanism of the reaction is of two sorts:

1. Electrophilic aromatic substitutions show second-order kinetics.
2. The reactions show no deuterium isotope effect.

We therefore conclude that the aryl C—H bond is not broken in the rate-limiting step and that at least two steps must be involved.

Substituents on the benzene ring affect both the reactivity of the ring toward further substitution, and the orientation of further substitution. We can classify substituents into three groups: ortho- and para-directing activators, ortho- and para-directing deactivators, and meta-directing deactivators.

Substituent effects are due to an interplay of both resonance and inductive effects. Resonance effects are transmitted by pi orbital overlap, whereas inductive effects are transmitted via sigma bonds.

When electrophilic substitution is carried out on a disubstituted benzene, both groups already present exert their orienting effects independently. If both groups direct substitution toward the same position, reaction occurs at that site. If the groups have conflicting directional effects, the more powerful activating substituent exerts a controlling influence.

In special cases, halobenzenes undergo nucleophilic aromatic substitution through either of two mechanisms. If the halobenzene has strongly electron-withdrawing substituents in the ortho and/or para position, substitution occurs by addition of a nucleophile to the ring, followed by elimination of halide from the intermediate anion (Meisenheimer complex).

If the halobenzene is not activated by electron-withdrawing substituents, nucleophilic substitution can occur by elimination of HX, followed by addition of a nucleophile to the intermediate benzyne. The Dow process for the industrial preparation of phenol is an example.

15.16 Summary of reactions

1. Electrophilic aromatic substitution
   a. Bromination (Section 15.1)

$$C_6H_6 + Br_2 \xrightarrow{FeBr_3} C_6H_5Br + HBr$$

   b. Chlorination (Section 15.2)

$$C_6H_6 \xrightarrow[\text{Or } Cl_2O,\ CF_3COOH]{Cl_2,\ FeCl_3} C_6H_5Cl + HCl$$

   c. Iodination (Section 15.2)

$$C_6H_6 + I_2 \xrightarrow{CuCl_2} C_6H_5I + HI$$

   d. Nitration (Section 15.3)

$$C_6H_6 + HNO_3 \xrightarrow[\text{(or } NO_2^+BF_4^-\text{)}]{H_2SO_4} C_6H_5NO_2 + H_2O$$

   e. Sulfonation (Section 15.4)

$$C_6H_6 + SO_3 \xrightarrow{H_2SO_4} C_6H_5SO_3H$$

f. Hydroxylation (Section 15.5)

$$C_6H_6 + H_2O_2 \xrightarrow{HOSO_2F} C_6H_5OH + H_2O$$

2. Reduction of aromatic nitro groups (Section 15.3)

$$C_6H_5NO_2 \xrightarrow[2.\ HO^-]{1.\ SnCl_2,\ H_3O^+} C_6H_5NH_2$$

3. Alkali fusion of aromatic sulfonates (Section 15.4)

$$C_6H_5SO_3H \xrightarrow[2.\ H_3O^+]{1.\ NaOH} C_6H_5OH$$

4. Nucleophilic aromatic substitution
   a. Via addition/elimination to activated aryl halides (Section 15.13)

$$\text{Cl-}C_6H_2(NO_2)_3 \xrightarrow[2.\ H_3O^+]{1.\ ^-\!:OH} \text{HO-}C_6H_2(NO_2)_3 + :Cl^-$$

   b. Via benzyne intermediate for unactivated aryl halides (Section 15.14)

$$C_6H_5Br \xrightarrow[NH_3]{^-\!:NH_2} C_6H_5NH_2 + :Br^-$$

ADDITIONAL PROBLEMS

15.17 Predict the major product(s) of mononitration of these substances. Which react faster, and which slower, than benzene?

(a) Bromobenzene (b) Benzonitrile
(c) Benzoic acid (d) Nitrobenzene
(e) Benzenesulfonic acid (f) Methoxybenzene

15.18 Rank the compounds in each group according to their reactivity toward electrophilic substitution.

(a) Chlorobenzene, *o*-dichlorobenzene, benzene
(b) *p*-Bromonitrobenzene, nitrobenzene, phenol
(c) Fluorobenzene, benzaldehyde, *o*-xylene
(d) Benzonitrile, *p*-methylbenzonitrile, *p*-methoxybenzonitrile

15.19 Suggest a reason for the observation that bromination of biphenyl occurs at ortho and para positions rather than at meta. Use resonance structures of the intermediates to explain your answers.

Biphenyl

15.20 At what position, and on what ring, would you expect nitration of 4-bromobiphenyl to occur?

Br

4-Bromobiphenyl

15.21 How do you explain the fact that electrophilic attack on 3-phenylpropanenitrile occurs at the ortho and para positions, whereas attack on 3-phenylpropenenitrile occurs at the meta position?

CH_2CH_2CN CN

3-Phenylpropanenitrile 3-Phenylpropenenitrile

15.22 At what position, and on what ring, would you expect these substances to undergo electrophilic substitution?

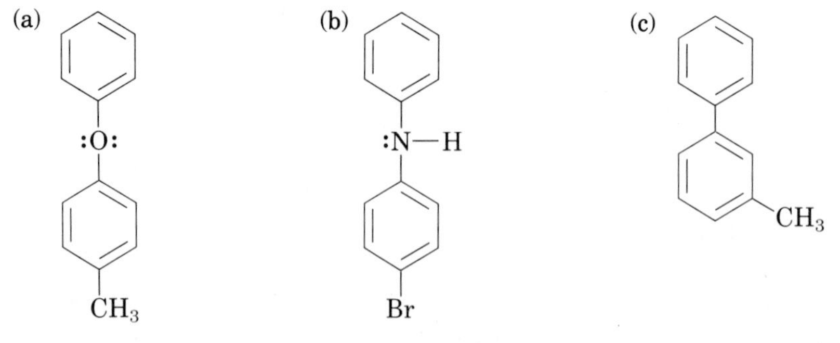

15.23 At what position, and on what ring, would you expect bromination of benzanilide to occur? Explain your answer by drawing resonance structures of the intermediates.

Benzanilide

15.24 In the next chapter, we will study the Friedel–Crafts reaction, by which aromatic rings can be alkylated. From your knowledge of aromatic substitution, can you propose a mechanism to account for the methylation of benzene by CH_3Cl and $AlCl_3$ catalyst?

$$C_6H_6 + CH_3Cl \xrightarrow{AlCl_3} C_6H_5CH_3 + HCl$$

15.25 In light of your answer to Problem 15.24 and your knowledge of carbocation stability, suggest a mechanism for the following reaction of *p*-methylphenol with 2-methyl-2-propanol in the presence of H_2SO_4.

$$p\text{-}CH_3C_6H_4OH + (CH_3)_3COH \xrightarrow{H_2SO_4} \text{2,6-}[(CH_3)_3C]_2\text{-4-}CH_3C_6H_2OH$$

BHT (butylated hydroxytoluene)

15.26 Give the name and draw the structure of the major product(s) of electrophilic chlorination of these materials:

(a) *m*-Nitrophenol
(b) *o*-Xylene
(c) *p*-Nitrobenzoic acid
(d) *m*-Xylene
(e) 2,4-Dibromophenol
(f) Salicylic acid (*o*-hydroxybenzoic acid)
(g) Sulfanilic acid (*p*-aminobenzenesulfonic acid)

15.27 At what position, and on what ring, would you expect these compounds to undergo electrophilic attack?

(a)

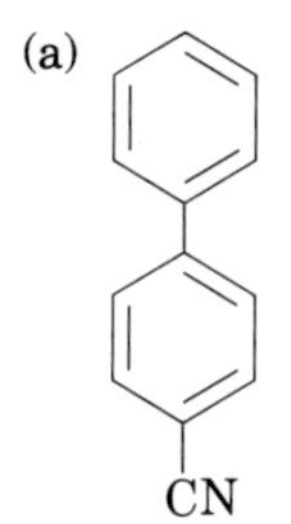

(b)

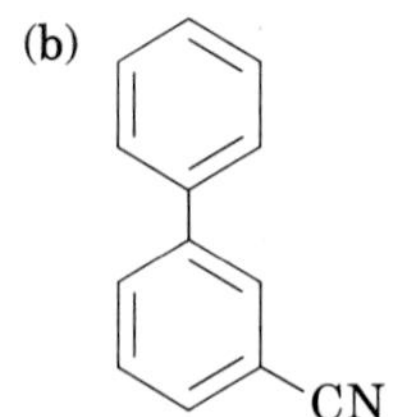

(c)

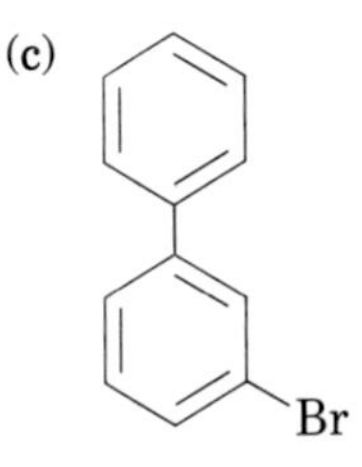

15.28 Starting with benzene as your only source of aromatic compounds, how would you synthesize these substances? Assume that you can separate ortho and para isomers if necessary.

(a) *p*-Chlorophenol
(b) *m*-Bromonitrobenzene
(c) *m*-Bromoaniline
(d) *p*-Bromoaniline
(e) *o*-Bromobenzenesulfonic acid
(f) *m*-Chlorobenzenesulfonic acid

15.29 Starting with either benzene or toluene, how would you synthesize these materials? Assume that ortho and para isomers can be separated.

(a) 2-Bromo-4-nitrotoluene
(b) 1,3,5-Trinitrobenzene
(c) 2-Chloro-4-methylaniline
(d) 2,4,6-Tribromoaniline
(e) 2-Chloro-4-methylphenol

15.30 When heated, benzenediazonium carboxylate decomposes to yield N_2, CO_2, and a reactive organic substance that cannot be isolated. When benzenediazonium carboxylate is heated in the presence of furan, the following reaction is observed:

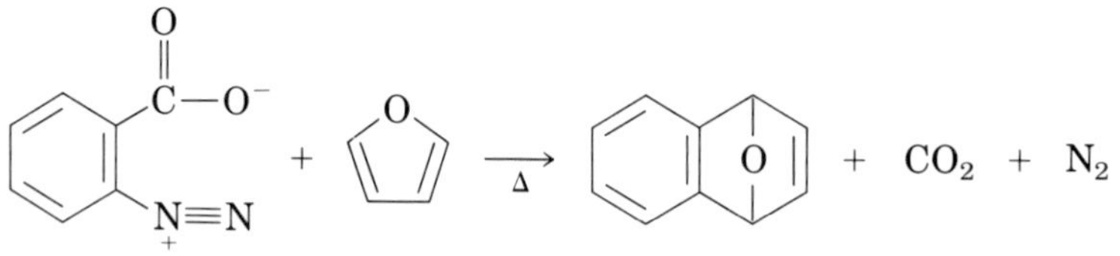

O
C—O⁻
N≡N
+
O
Δ
O
+ CO_2 + N_2

What intermediate is involved in this reaction? Propose a mechanism for formation of this intermediate; use arrows to show the movement of electron pairs during the reaction.

15.31 Phenylboronic acid is nitrated by HNO_3 to give 15% ortho substitution product and 85% meta. Account for the meta-directing effect of the —$B(OH)_2$ group.

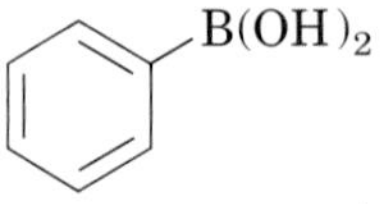

$B(OH)_2$

Phenylboronic acid

15.32 Draw resonance structures of the intermediate carbocations and account for the fact that naphthalene undergoes electrophilic attack at C1 rather than C2.

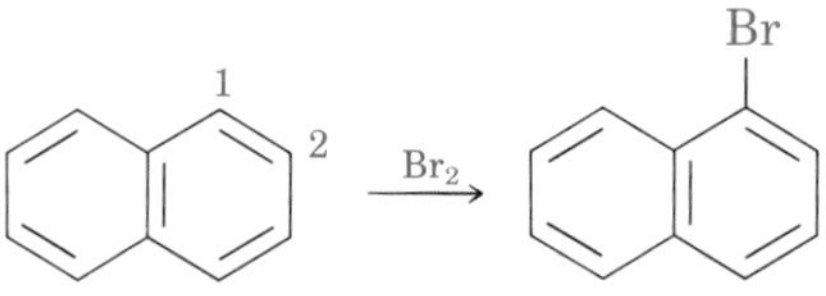

1
2
Br_2
Br

15.33 4-Chloropyridine undergoes reaction with dimethylamine to yield 4-dimethylaminopyridine. Propose a mechanism to account for this result.

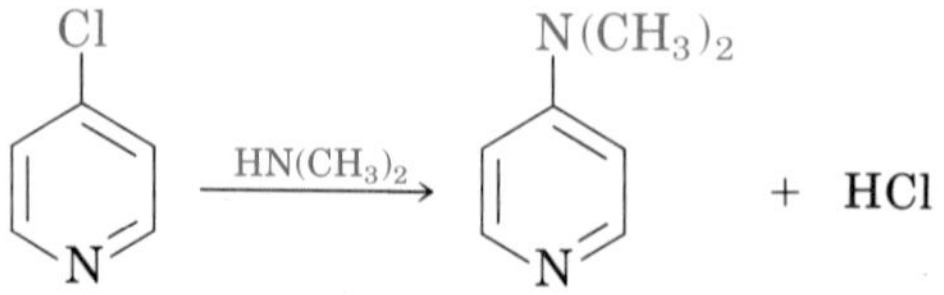

Cl
N
$HN(CH_3)_2$
$N(CH_3)_2$
N
+ HCl

15.34 How do you account for the fact that *p*-bromotoluene reacts with potassium amide to give a mixture of *m*- and *p*-methylaniline?

15.35 Dimethyl 3-chlorophthalate reacts with methoxide ion in methanol to yield dimethyl 3-methoxyphthalate. Propose a mechanism for this reaction.

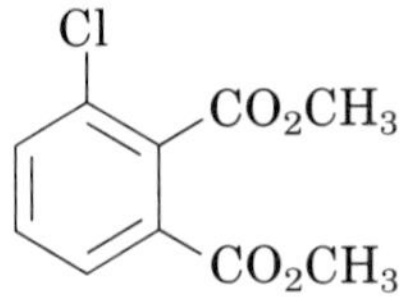

Cl
CO_2CH_3
CO_2CH_3

Dimethyl 3-chlorophthalate

15.36 How do you explain the fact that hydroxylation of toluene with $H_2O_2/HOSO_2F$ occurs readily, but hydroxylation of phenol under the same conditions does not occur? [*Hint:* Fluorosulfonic acid, $HOSO_2F$, is a very strong acid.]

15.37 Suggest an explanation of the observation that bromination of aniline gives 2,4,6-tribromoaniline, whereas nitration of aniline with HNO_3/H_2SO_4 gives *m*-nitroaniline. What is the nature of the directing substituent in the two reactions?

15.38 Triptycene is an unusual molecule that has been prepared by reaction of benzyne with anthracene. What kind of reaction is involved? Show the mechanism of the transformation.

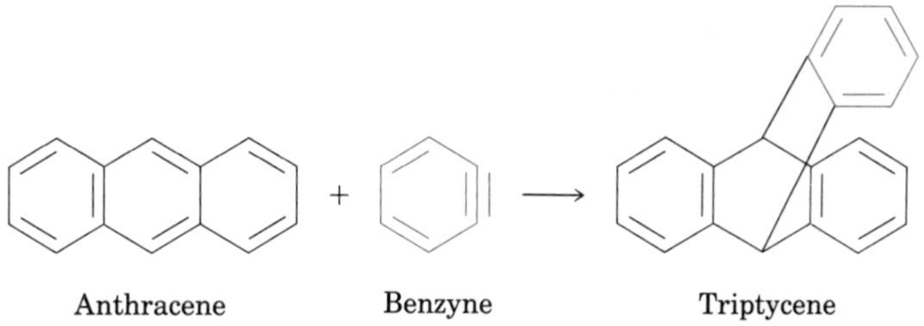

Anthracene + Benzyne ⟶ Triptycene

CHAPTER 16 ARENES: SYNTHESIS AND REACTIONS OF ALKYLBENZENES

A**renes** are hydrocarbons that contain one or more benzene rings. Such arenes as ethylbenzene, styrene, and cumene (Figure 16.1) are among the most important industrial chemicals, and we will study their preparation and chemistry in this chapter.

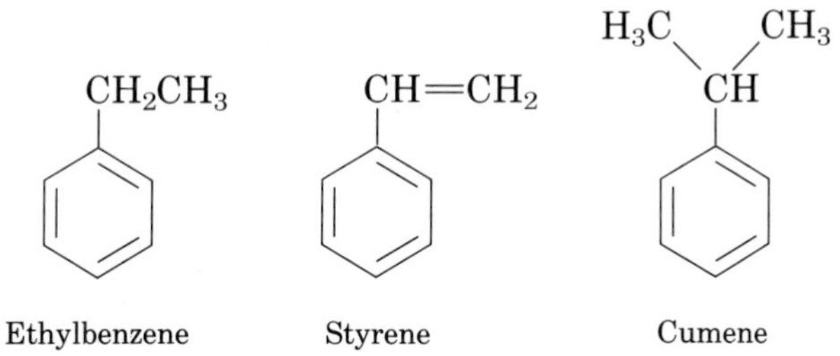

CH_2CH_3 $CH{=}CH_2$ H_3C CH_3 CH

Ethylbenzene Styrene Cumene

Figure 16.1. Three arenes widely used in industry. Yearly U.S. production of ethylbenzene is approximately 8.4 billion lb; of styrene, 6.7 billion lb; and of cumene, 3.3 billion lb.

16.1 Sources of arenes

The simple aromatic hydrocarbons used as starting materials for the preparation of more complex products come from two main sources, coal tar and petroleum. When bituminous coal is heated to 1000°C in the absence of air, volatile materials are driven off, and coke is produced. A hard material consisting primarily of carbon, coke is used to fuel the blast furnaces of the steel industry. Of more interest to chemists than coke are the volatile materials, consisting of coal gas (largely hydrogen and methane) and a residue called **coal tar.** Coal tar accounts for up to 6% of the weight of the coal and is rich in aromatic hydrocarbons. Distillation of coal tar yields benzene, toluene, xylene, naphthalene, and a host of other aromatic compounds (Figure 16.2, p. 522).

Petroleum is an additional abundant source of simple arenes. Arenes are formed during petroleum refining by a catalytic reforming process in which alkanes are passed over a catalyst at about 500°C under high pressure. Heptane, for example, can be dehydrogenated and cyclized to yield toluene. More than 11 billion lb of benzene and 9 billion lb of xylene are produced each year in the United States, and petroleum is the major source of both.

16.2 Friedel–Crafts alkylation

Complex substituted aromatic compounds are almost always synthesized from simpler, readily available aromatics. Since there are relatively few methods for synthesizing a benzene ring itself, the chemist uses an existing

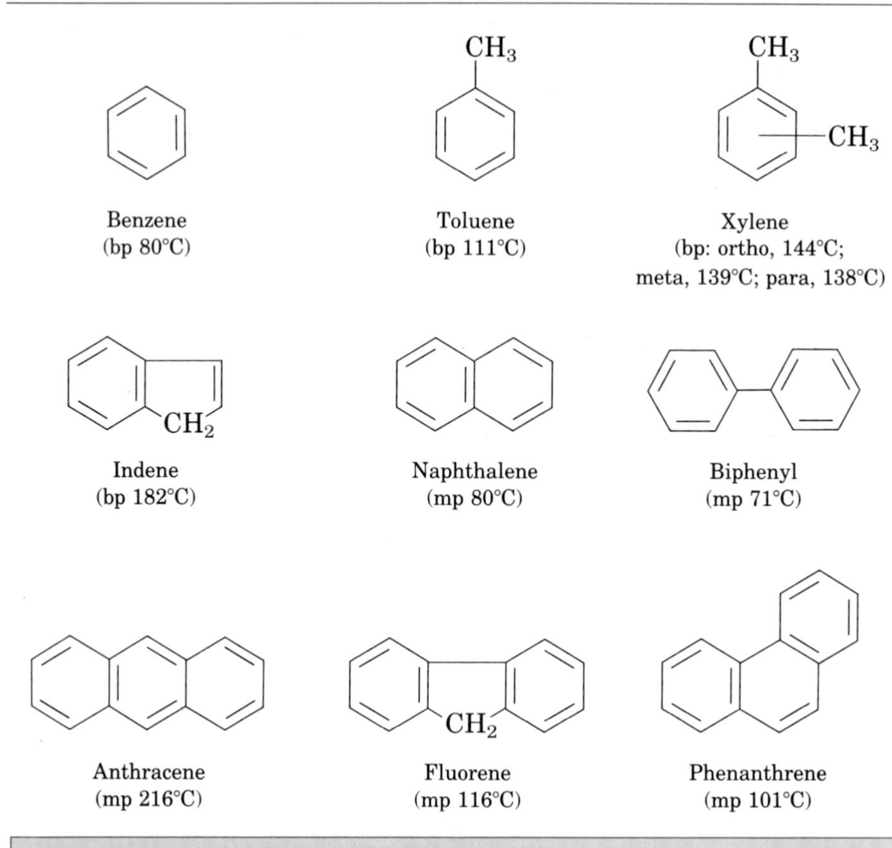

Figure 16.2. Some coal tar hydrocarbons.

ring as starting material and introduces the required substituents. We have already discussed reactions for attaching halogen-, oxygen-, nitrogen-, and sulfur-based functional groups to the aromatic ring, and we will now extend this list to include the most important reaction of all—the **alkylation of aromatic rings.**

Charles Friedel[1] and James Crafts[2] reported in 1877 that benzene rings can be alkylated by reaction with an alkyl chloride in the presence of aluminum chloride catalyst. For example, benzene reacts with 2-chloropropane in the presence of $AlCl_3$ to yield cumene (isopropylbenzene).

$$\text{Benzene} + (CH_3)_2CHCl \xrightarrow{AlCl_3} \text{Cumene } (C_6H_5CH(CH_3)_2)$$

Benzene 2-Chloropropane Cumene (85%)

[1]Charles Friedel (1832–1899); b. Strasbourg, France; studied at the Sorbonne; professor, Ecole des Mines (1876–1884) and at Paris (1884–1899).

[2]James M. Crafts (1839–1917); b. Boston; L.L.D., Harvard (1898); professor, Cornell University (1868–1871); Massachusetts Institute of Technology (1871–1900).

The Friedel–Crafts reaction in its most general form involves electrophilic attack by an aromatic ring on a carbocation. Many different substituted aromatic rings undergo the reaction, many different catalysts besides $AlCl_3$ can be used, and many different sources of carbocations besides alkyl chlorides can be used. Let's look first at the reaction mechanism.

16.3 Mechanism of the Friedel–Crafts reaction

The **Friedel–Crafts alkylation reaction** is an electrophilic aromatic substitution in which the aromatic ring attacks a carbocation. Loss of a proton from the intermediate then yields the alkylated aromatic ring. The carbocation is normally generated by reaction of an alkyl chloride with a Lewis acid catalyst such as aluminum chloride. It is thought that the $AlCl_3$ catalyst acts by helping the alkyl chloride to ionize, in much the same way that $FeCl_3$ catalyzes aromatic chlorinations by polarizing Cl_2 (Section 15.2). The overall Friedel–Crafts mechanism for the synthesis of cumene is shown in Figure 16.3.

$$(CH_3)_2CHCl + AlCl_3 \longrightarrow (CH_3)_2CH^+ \ AlCl_4^-$$

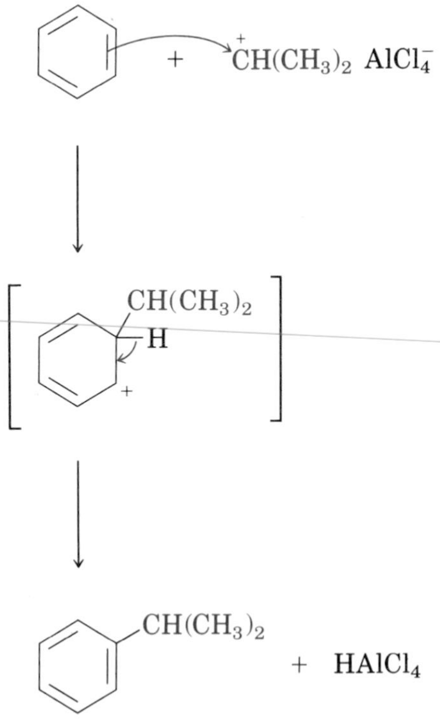

$$C_6H_6 + \overset{+}{C}H(CH_3)_2 \ AlCl_4^-$$

An electron pair from the aromatic ring attacks the carbocation, forming a C—C bond and yielding a new carbocation intermediate.

$$\left[\text{arenium ion: } CH(CH_3)_2, \ H, \ + \right]$$

Loss of a proton then gives the neutral arene product.

$$C_6H_5CH(CH_3)_2 + HAlCl_4$$

Figure 16.3. Mechanism of the Friedel–Crafts alkylation reaction in the synthesis of cumene.

Since the key step in the Friedel–Crafts alkylation reaction is the attack on an alkyl carbocation by an aromatic ring, we might expect that carbocations generated in other ways should also be attacked by aromatic rings. This is just what we find. We have seen, for example, that alkenes are protonated by strong acid to form carbocations (the first step in electrophilic addition reactions, Section 5.8). When an alkene and an aromatic ring react in the presence of a strong acid, alkylation occurs. Cumene is produced industrially as an intermediate in the synthesis of phenol, and in the United States more than 3 billion lb/yr are synthesized by the reaction of benzene with propene (Figure 16.4).

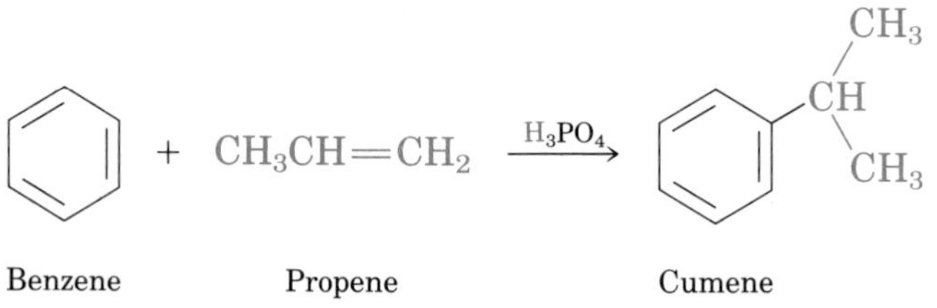

$$\text{Benzene} + CH_3CH{=}CH_2 \xrightarrow{H_3PO_4} \text{Cumene } (C_6H_5CH(CH_3)_2)$$

Benzene Propene Cumene

Figure 16.4. Synthesis of cumene from benzene and propene by an acid-catalyzed Friedel–Crafts reaction. Reaction occurs via:

$$CH_3CH{=}CH_2 \xrightarrow{H^+} CH_3\overset{+}{C}HCH_3$$

We have also seen that carbocations are generated as intermediates in the acid-catalyzed dehydration of alcohols (Section 6.13). Thus, we can readily account for the fact that aromatic rings are alkylated when they are treated with alcohols in the presence of strong acid. Tertiary alcohols such as 2-methyl-2-propanol work best, since the intermediate tertiary carbocations resulting from their dehydration are most stable, but primary and secondary alcohols can also be used. The food preservative BHT (butylated hydroxytoluene) can be prepared by treatment of a mixture of *p*-methylphenol and 2-methyl-2-propanol with strong acid (Figure 16.5).

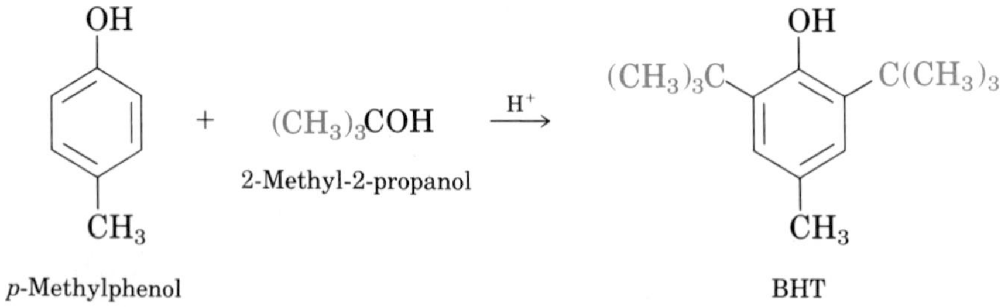

$$p\text{-}CH_3C_6H_4OH + (CH_3)_3COH \xrightarrow{H^+} \text{BHT}$$

p-Methylphenol 2-Methyl-2-propanol BHT

Figure 16.5. Synthesis of BHT via electrophilic aromatic substitution of *p*-methylphenol with 2-methyl-2-propanol. Reaction takes place via:

$$(CH_3)_3COH + H^+ \longrightarrow (CH_3)_3C^+ + H_2O$$

PROBLEM

16.1 Draw structures of the major monoalkylation product(s) of the reaction of these aromatic compounds with chloroethane and $AlCl_3$.

(a) Toluene (b) Phenol
(c) *m*-Xylene (d) Chlorobenzene

PROBLEM

16.2 If benzene is allowed to react with excess 2-methylpropene in the presence of H_3PO_4, *p*-di-*tert*-butylbenzene is formed as the major product. Formulate a mechanism for all steps in this reaction, and suggest a reason for the fact that no *o*-di-*tert*-butylbenzene is formed.

16.4 Limitations of the Friedel–Crafts reaction

THE ALKYL HALIDE

Friedel–Crafts alkylations are limited to alkyl halides. Alkyl fluorides, chlorides, bromides, and iodides all react well, but aryl halides and vinylic halides do not react. Aryl and vinylic carbocations are too unstable to form under Friedel–Crafts conditions.

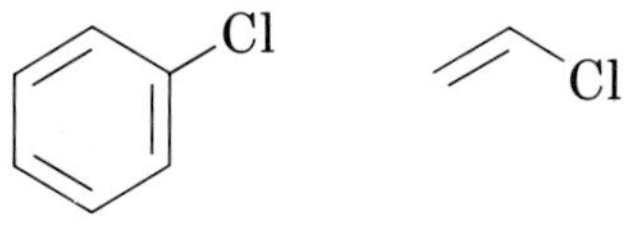

Cl Cl

An aryl halide A vinylic halide
Not reactive

THE AROMATIC REACTANT

Friedel–Crafts reactions do not succeed on aromatic rings substituted by strongly deactivating groups (Figure 16.6); deactivated aromatic rings are simply not reactive enough to attack carbocations.

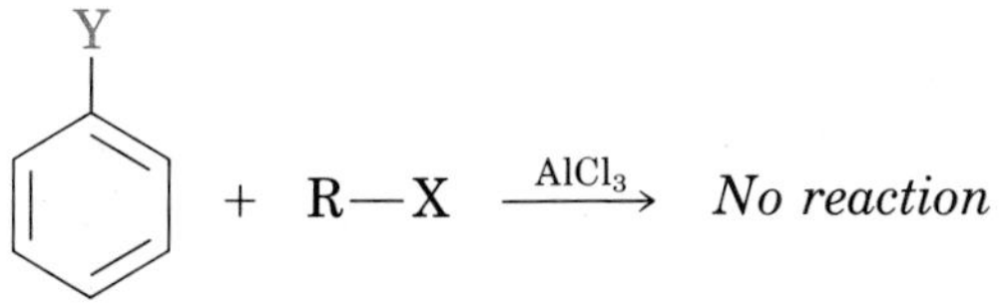

Y

$+ \text{ R—X} \xrightarrow{AlCl_3}$ *No reaction*

where $Y = —\overset{+}{N}R_3, —NO_2, —CN, —SO_3H, —CHO,$

$—COCH_3, —COOH, —COOCH_3$

$(—NH_2, —NHR, —NR_2)$

Figure 16.6. Limitations on the aromatic substrate in Friedel–Crafts reactions.

Substituents more strongly deactivating than halogen prevent Friedel–Crafts alkylations from occurring, as Figure 16.6 indicates. Note that the reaction also fails on amino-substituted aromatic rings. Basic amino groups form acid–base complexes with Lewis acid Friedel–Crafts catalysts; this complexation puts a positive charge on nitrogen and converts the amino group from an activator into a powerful deactivator.

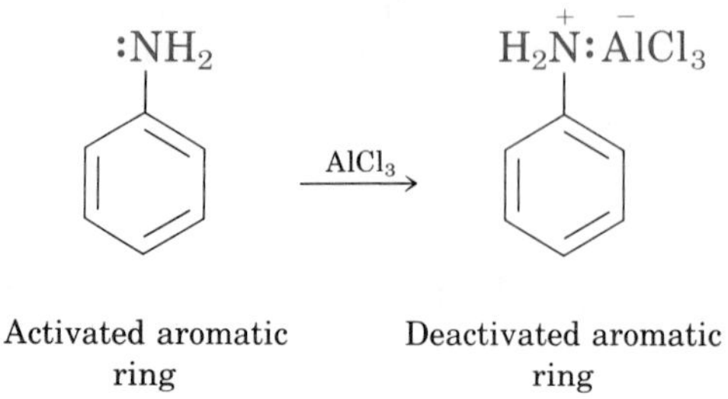

$:NH_2$ (benzene ring) $\xrightarrow{AlCl_3}$ $H_2\overset{+}{N}:\overset{-}{Al}Cl_3$ (benzene ring)

Activated aromatic ring — Deactivated aromatic ring

PROBLEM

16.3 Rank the following aromatic compounds in the expected order of their reactivity toward Friedel–Crafts alkylation. Which compounds are unreactive?

(a) Bromobenzene (b) Toluene (c) Phenol
(d) Aniline (e) Nitrobenzene (f) *p*-Bromotoluene

POLYALKYLATION

A fundamental difficulty of the Friedel–Crafts reaction is that the product is usually more reactive than the starting material. This is due to the fact that alkyl substituents are activating groups; once the first group is on the ring, a second substitution reaction is facilitated. Thus, a major drawback to the Friedel–Crafts reaction is that we often observe **polyalkylation.** For example, reaction of benzene with 1 mol equiv of 2-chloro-2-methylpropane yields *p*-di-*tert*-butylbenzene as the major product, along with a small amount of *tert*-butylbenzene and unreacted starting material. High yields of monoalkylation product are obtained only when a large excess of benzene is used.

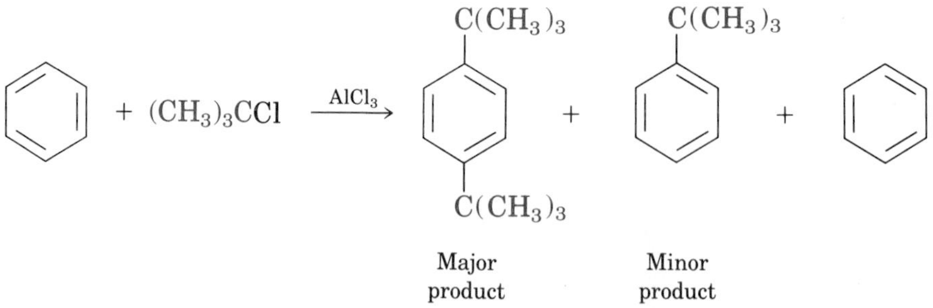

Benzene + $(CH_3)_3CCl$ $\xrightarrow{AlCl_3}$ *p*-$C(CH_3)_3$/$C(CH_3)_3$ benzene + $C(CH_3)_3$ benzene + benzene

Major product — Minor product

16.5 Carbocation rearrangements during Friedel–Crafts reactions

In 1878, less than one year after the discovery of the Friedel–Crafts reaction, G. Gustavson reported that skeletal *rearrangements* of the alkyl group can occur during reaction. When Gustavson attempted to prepare propylbenzene by reaction of benzene and 1-bromopropane in the presence of $AlCl_3$, he isolated isopropylbenzene (cumene) instead.

$$\text{Benzene} + CH_3CH_2CH_2Br \xrightarrow{AlCl_3} \text{Cumene } (C_6H_5CH(CH_3)_2)$$

Benzene 1-Bromopropane Cumene

Subsequent investigations have shown that isomerization of the alkyl group is a common occurrence in Friedel–Crafts reactions, particularly when primary halides are used. The amount of rearrangement is variable and depends on catalyst, reaction temperature, and even reaction solvent. Thus, less rearrangement is usually found at lower reaction temperatures, but mixtures of products are often obtained. For example, treatment of benzene with 1-chlorobutane gives an approximately 2:1 ratio of rearranged (*sec*-butyl) to unrearranged (*n*-butyl) products when the reaction is carried out at 0°C using $AlCl_3$ as catalyst.

$$\text{Benzene} + CH_3CH_2CH_2CH_2Cl \xrightarrow[0°C]{AlCl_3} C_6H_5CH(CH_3)CH_2CH_3 + C_6H_5CH_2CH_2CH_2CH_3$$

Benzene 1-Chlorobutane

sec-Butylbenzene (~65%)

Butylbenzene (~35%)

Pioneering studies by Whitmore[3] and others during the 1930s indicated that these isomerizations occur by carbocation rearrangements. Whitmore suggested that less stable carbocations can rearrange to their more stable isomers by **hydride shifts.** For example, the relatively unstable primary butyl carbocation can rearrange to a more stable secondary carbocation by the shift of a hydrogen atom and its electron pair (**a hydride ion, H:$^-$**) from C2 to C1.

[3]Frank C. Whitmore (1887–1947); b. North Attleboro, Mass.; Ph.D. Harvard (E. L. Jackson); professor, Pennsylvania State University.

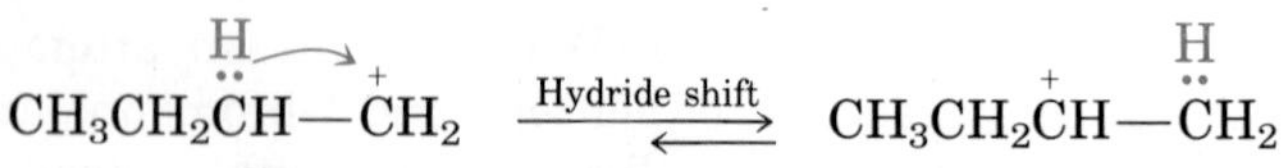

$$CH_3CH_2\overset{\text{H}}{C}H-\overset{+}{C}H_2 \xrightleftharpoons{\text{Hydride shift}} CH_3CH_2\overset{+}{C}H-\overset{\text{H}}{C}H_2$$

1° cation (less stable) — 2° cation (more stable)

Similarly, carbocation rearrangements can occur by *alkyl* shifts. For example, Friedel–Crafts alkylation of benzene with 1-chloro-2,2-dimethylpropane yields (1,1-dimethylpropyl)benzene as the sole product. The initially formed primary carbocation rearranges to a tertiary carbocation by shift of a methyl group and its electron pair from C2 to C1 (Figure 16.7).

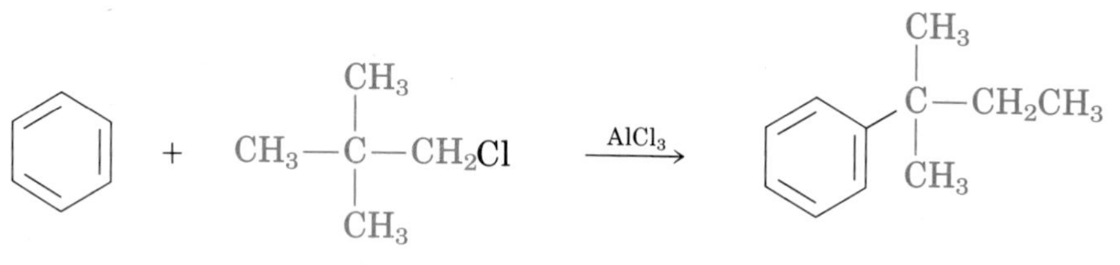

$$C_6H_6 + CH_3-C(CH_3)_2-CH_2Cl \xrightarrow{AlCl_3} C_6H_5-C(CH_3)_2-CH_2CH_3$$

Benzene — 1-Chloro-2,2-dimethylpropane — (1,1-Dimethylpropyl)benzene

$$\left[CH_3-C(CH_3)_2-\overset{+}{C}H_2\right] \rightleftharpoons CH_3-\overset{+}{C}(CH_3)-CH_2(CH_3)$$

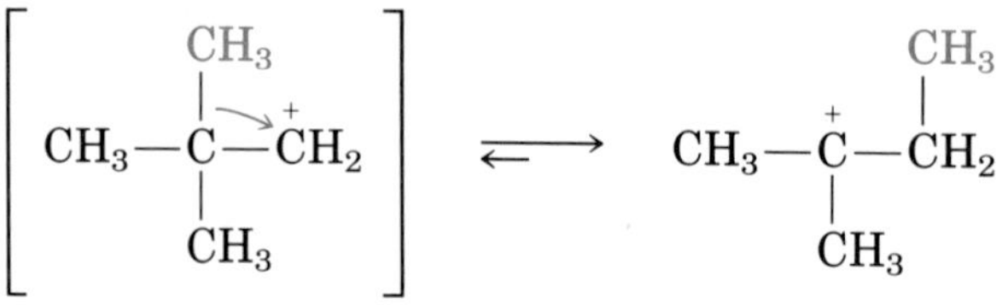

1° carbocation — 3° carbocation

Figure 16.7. Rearrangement of a primary to a tertiary carbocation.

Rearrangement from a less stable ion to a more stable ion is a general reaction of carbocations. It is not limited to Friedel–Crafts reactions but can occur in many other processes, such as the acid-catalyzed dehydration of certain alcohols. For example, treatment of 3,3-dimethyl-2-butanol with sulfuric acid yields 2,3-dimethyl-2-butene. The reaction occurs by protonation of the hydroxyl group, ionization to a secondary carbocation, and shift of a methyl group, followed by loss of a proton from the more stable tertiary cation (Figure 16.8).

Carbocation rearrangements significantly decrease the utility of the Friedel–Crafts alkylation. Fortunately, though, this difficulty can be avoided by using a variation known as the Friedel–Crafts acylation reaction, which will be discussed in the next section.

PROBLEM

16.4 What is the major monosubstitution product that you would expect to obtain from the Friedel–Crafts reaction of benzene and 1-chloro-2-methylpropane in the presence of $AlCl_3$?

$$\underset{\text{3,3-Dimethyl-2-butanol}}{CH_3C(CH_3)_2-CH(OH)CH_3} \underset{}{\overset{H^+}{\rightleftharpoons}} CH_3-C(CH_3)_2-CH(\overset{+}{O}H_2)CH_3 \xrightarrow{-H_2O} \underset{2^\circ \text{ carbocation}}{CH_3C(CH_3)_2-\overset{+}{C}HCH_3}$$

Methyl shift

$$\underset{\text{2,3-Dimethyl-2-butene}}{(H_3C)_2C=C(CH_3)_2} \xleftarrow{-H^+} \underset{3^\circ \text{ carbocation}}{CH_3-\overset{+}{C}(CH_3)-C(CH_3)_2-H}$$

Figure 16.8. Acid-catalyzed dehydration of 3,3-dimethyl-2-butanol.

PROBLEM

16.5 How can you account for the fact that electrophilic addition of HCl to 3,3-dimethyl-1-butene yields 2-chloro-2,3-dimethylbutane as the major product? Propose a mechanism for this reaction.

16.6 Friedel–Crafts acylation

An acyl group is introduced onto the ring when an aromatic compound is allowed to react with a carboxylic acid chloride in the presence of a Lewis acid catalyst. For example, reaction of benzene with acetyl chloride in the presence of $AlCl_3$ yields the ketone, acetophenone. Rings that bear strongly deactivating substituents, however, do not react.

$$\underset{\text{Benzene}}{C_6H_6} + \underset{\text{Acetyl chloride}}{CH_3\overset{O}{\overset{\|}{C}}-Cl} \xrightarrow[80^\circ C]{AlCl_3} \underset{\text{Acetophenone (95\%)}}{C_6H_5-\overset{O}{\overset{\|}{C}}-CH_3} + HCl$$

The mechanism of **Friedel–Crafts acylation** is similar to those of other electrophilic aromatic substitutions we have studied. The reactive electrophile is the resonance-stabilized **acylium ion,** generated by reaction between the acyl chloride and $AlCl_3$ (Figure 16.9, p. 530).

An acylium ion is stabilized by overlap of the vacant orbital on carbon with lone-pair electrons of the neighboring oxygen, as the resonance structures in Figure 16.9 indicate. Once formed, the acylium ion does not

$$R{-}\overset{\overset{\displaystyle :O:}{\|}}{C}{-}\ddot{\underset{\cdot\cdot}{Cl}}: + AlCl_3 \rightleftharpoons \left[R{-}\overset{\overset{\displaystyle :O:}{\|}}{C}{-}\overset{\delta^+}{Cl}\cdots\overset{\delta^-}{AlCl_3}\right]$$

$$\rightleftharpoons \left[R{-}\overset{+}{C}{=}\ddot{O}: \longleftrightarrow R{-}C{\equiv}\overset{+}{O}:\right] AlCl_4^-$$

An acylium ion

Figure 16.9. Generation of an acylium ion in Friedel–Crafts acylation.

$$C_6H_6 + R{-}C{\equiv}O^+ \longrightarrow \left[\text{arenium ion (H, COR, +)}\right] \xrightarrow{-H^+} C_6H_5{-}\overset{O}{\overset{\|}{C}}{-}R + HAlCl_4$$

$AlCl_4^-$ $^-{:}AlCl_4$

An aryl alkyl ketone

Figure 16.10. Mechanism of the Friedel–Crafts acylation reaction.

rearrange; rather, it is attacked by an aromatic ring to give unrearranged substitution product (Figure 16.10).

The acid chlorides necessary for acylation are easily prepared by treatment of carboxylic acids with either thionyl chloride, $SOCl_2$, or oxalyl chloride, $(COCl)_2$. The thionyl chloride route is normally used, since the reagent is inexpensive, but the oxalyl chloride method is much milder and takes place at lower temperatures. We will discuss both reactions in more detail in Section 24.4.

$$R{-}\overset{O}{\overset{\|}{C}}{-}OH \xrightarrow[80^\circ C]{SOCl_2} R{-}\overset{O}{\overset{\|}{C}}{-}Cl + SO_2 + HCl$$

A carboxylic acid   Carboxylic acid chloride

$$R{-}\overset{O}{\overset{\|}{C}}{-}OH \xrightarrow[25^\circ C]{Cl-\overset{O}{\overset{\|}{C}}-\overset{O}{\overset{\|}{C}}-Cl} R{-}\overset{O}{\overset{\|}{C}}{-}Cl + CO_2 + CO + HCl$$

Carboxylic acid chloride

Carboxylic acid anhydrides can be used as an alternative to carboxylic acid chlorides for Friedel–Crafts acylation reactions. The mechanism of the reaction is the same—an acylium ion is the attacking electrophile. Acyclic acid anhydrides such as acetic anhydride can be used, but the reaction is rather inefficient, since half of the acylating agent goes unused.

$$\text{Benzene} + CH_3\overset{O}{\overset{\|}{C}}-O-\overset{O}{\overset{\|}{C}}CH_3 \xrightarrow[80°C]{AlCl_3} C_6H_5\overset{O}{\overset{\|}{C}}-CH_3 + CH_3\overset{O}{\overset{\|}{C}}OH$$

Benzene — Acetic anhydride — Acetophenone (85%) — Unused by-product

Much more interesting and useful are the cases where *cyclic* anhydrides are used. Here the product is a **keto acid,** and we can often take advantage of these functional groups to do further chemistry. For example, phthalic anhydride reacts with benzene in the presence of $AlCl_3$ to yield *o*-benzoylbenzoic acid. Treatment of the Friedel–Crafts product with concentrated sulfuric acid causes a second, internal, Friedel–Crafts reaction to take place, producing anthraquinone, an important material used in the preparation of many dyes.

$$\text{Phthalic anhydride} + \text{Benzene} \xrightarrow[2.\ H_2O]{1.\ AlCl_3} \text{o-Benzoylbenzoic acid (92\%)}\ (CO_2H)$$

Phthalic anhydride — Benzene — *o*-Benzoylbenzoic acid (92%)

$$\xrightarrow{H_2SO_4\ \text{(internal Friedel–Crafts reaction)}}$$

Anthraquinone (100%)

PROBLEM

16.6 Propose a mechanism to account for the acid-catalyzed cyclization of *o*-benzoylbenzoic acid to anthraquinone. How might concentrated H_2SO_4 react with a carboxylic acid to generate an acylium ion?

PROBLEM

16.7 How do you account for the fact that polyalkylation often occurs during Friedel–Crafts reaction, but that polyacylation never occurs?

16.7 Reduction of aryl alkyl ketones

The reduction of a carbonyl group to a methylene (CH_2) group is one of the most important reactions of aryl alkyl ketones. This reaction can be carried out by a number of different methods, but catalytic hydrogenation over a palladium catalyst is one of the simplest and best. For example, propiophenone is reduced to propylbenzene in 100% yield by catalytic hydrogenation. The net effect of Friedel–Crafts acylation followed by reduction is the preparation of a primary alkylbenzene. This sequence of reactions allows us to circumvent the carbocation rearrangement problems associated with direct Friedel–Crafts alkylation using primary alkyl halides (Figure 16.11).

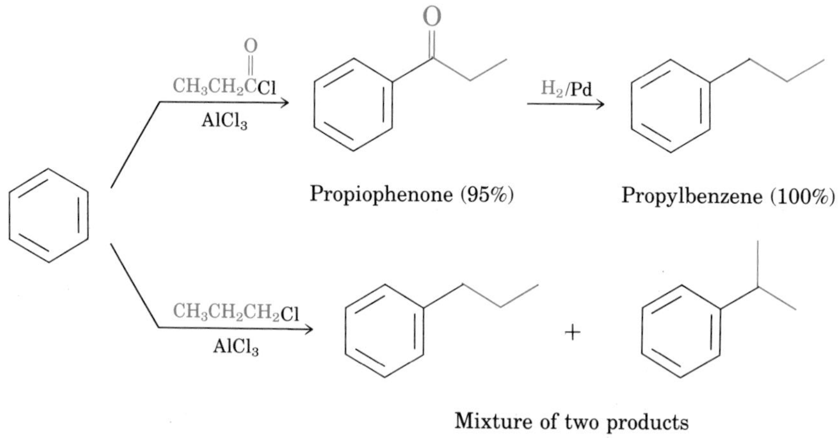

CH_3CH_2COCl, $AlCl_3$ → Propiophenone (95%) —H_2/Pd→ Propylbenzene (100%)

$CH_3CH_2CH_2Cl$, $AlCl_3$ → Mixture of two products

Figure 16.11. Use of the Friedel–Crafts acylation reaction to prepare propylbenzene.

Note, however, that the catalytic hydrogenation of acyl groups is limited to aryl alkyl ketones. The presence of the aromatic ring increases the reactivity of the neighboring carbonyl group toward hydrogenation, and dialkyl ketones are not reduced under these conditions. It should also be pointed out that the catalytic reduction of aryl alkyl ketones is not compatible with the presence of a nitro substituent on the aromatic ring, since nitro groups are reduced to amino groups under the reaction conditions. We will see a more general method for reducing ketone carbonyl groups to yield alkanes in Section 22.10.

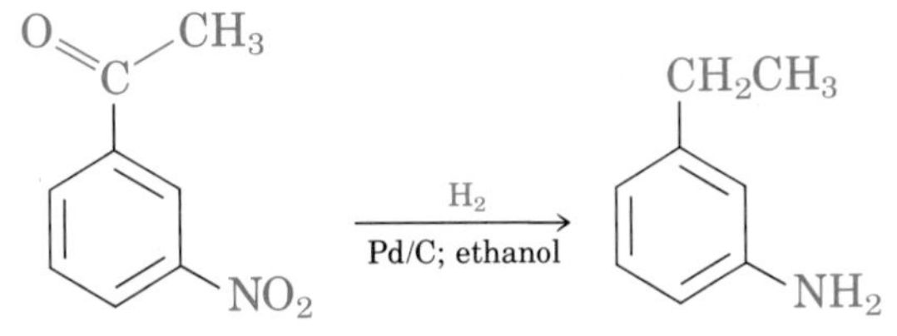

$$\text{O=C(CH}_3\text{)-C}_6\text{H}_4\text{-NO}_2 \xrightarrow[\text{Pd/C; ethanol}]{H_2} \text{CH}_2\text{CH}_3\text{-C}_6\text{H}_4\text{-NH}_2$$

m-Nitroacetophenone *m*-Ethylaniline

PROBLEM

16.8 How might you synthesize α-tetralone starting from benzene and succinic anhydride?

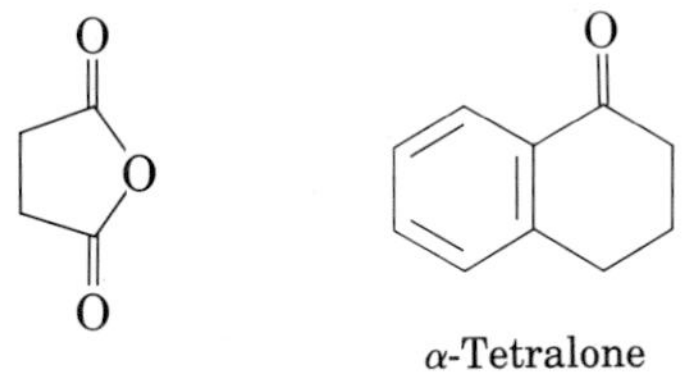

Succinic anhydride α-Tetralone

16.8 Bromination of alkylbenzene side chains

The catalytic hydrogenation of aryl alkyl ketones is just one example of how an aromatic ring can affect the reactivity of the neighboring side chain carbon atom. Another example is **benzylic oxidation.** The benzylic position of alkylbenzenes is readily attacked by a variety of oxidizing agents.

N-Bromosuccinimide (NBS) reacts with alkylbenzenes to brominate the benzylic position through a radical chain mechanism. For example, (3-bromopropyl)benzene gives (1,3-dibromopropyl)benzene in 99% yield on reaction with NBS in the presence of benzoyl peroxide, $(PhCO_2)_2$, as a radical initiator. Note that bromination occurs *exclusively* in the benzylic position and does not give a mixture of products.

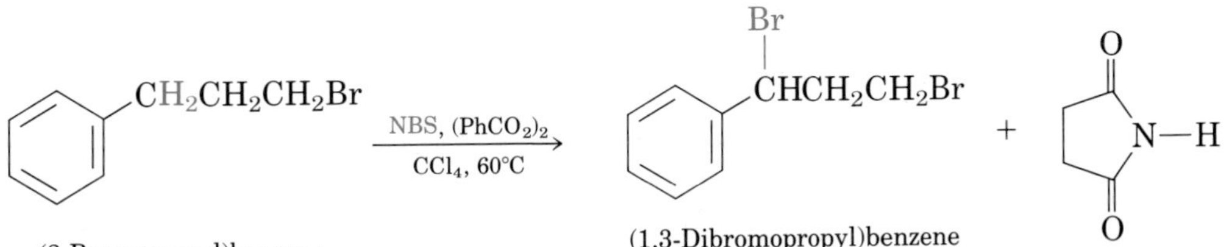

$$\text{C}_6\text{H}_5\text{CH}_2\text{CH}_2\text{CH}_2\text{Br} \xrightarrow[\text{CCl}_4,\ 60°\text{C}]{\text{NBS, (PhCO}_2)_2} \text{C}_6\text{H}_5\text{CHBrCH}_2\text{CH}_2\text{Br} + \text{succinimide (N—H)}$$

(3-Bromopropyl)benzene (1,3-Dibromopropyl)benzene (99%)

Recall the process of allylic bromination:

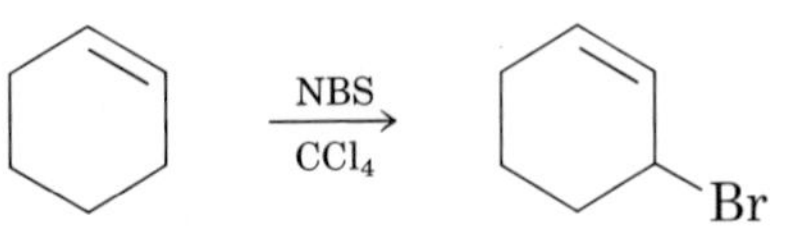

$$\text{cyclohexene} \xrightarrow[\text{CCl}_4]{\text{NBS}} \text{3-bromocyclohexene (Br)}$$

The mechanism of benzylic bromination is exactly the same as that seen previously for allylic bromination of alkenes (Section 9.5). Although the overall mechanism is somewhat complex, the critical step involves abstraction of a benzylic hydrogen atom of the alkylbenzene to generate an intermediate benzyl radical. The stabilized radical then reacts with Br_2 to yield product and a bromine radical, which cycles back into the reaction to carry on the chain. The Br_2 necessary for reaction with the benzyl radical is produced by a concurrent reaction of HBr with NBS, as shown in Figure 16.12.

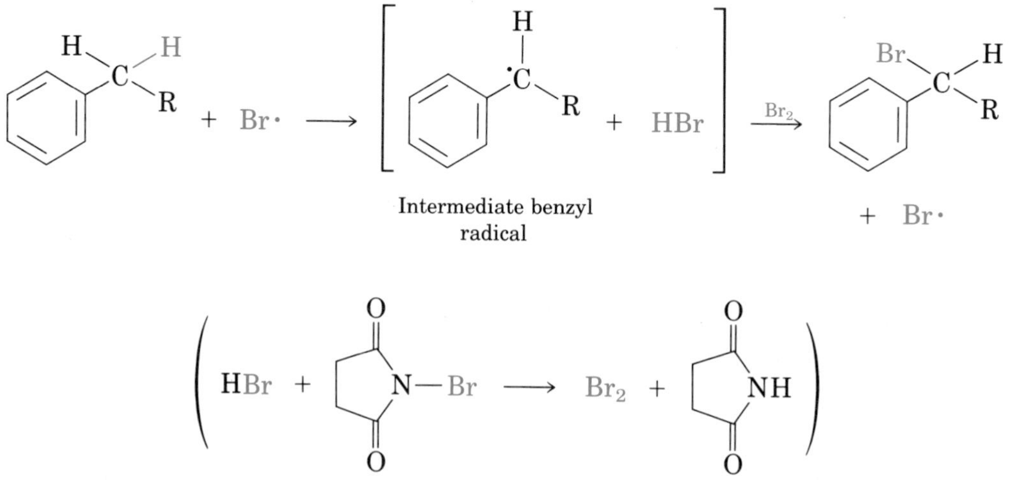

Intermediate benzyl radical

Figure 16.12. Mechanism of benzylic bromination by *N*-bromosuccinimide.

Reaction occurs at the benzylic position because the benzyl radical is highly stabilized by resonance. Figure 16.13 shows how this resonance stabilization arises, and Figure 16.14 shows an orbital view indicating how the radical can be stabilized by overlap of its *p* orbital with the ring pi electron system.

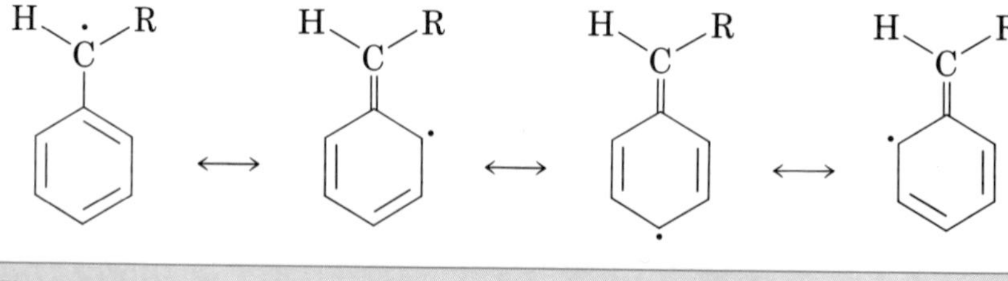

Figure 16.13. Resonance stabilization of a benzyl radical.

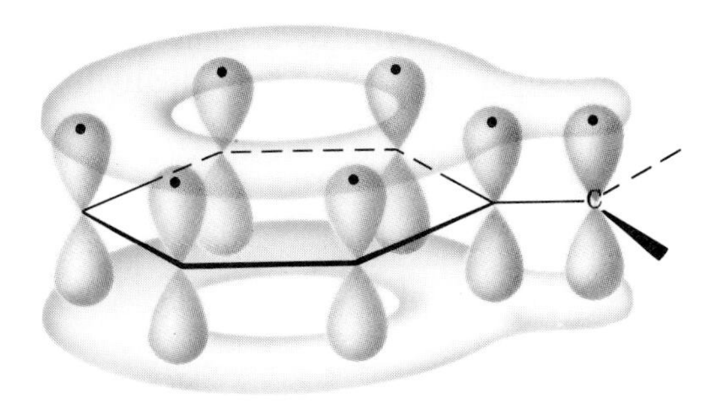

Figure 16.14. Orbital picture of a benzyl radical.

Bromoalkylbenzenes prepared by NBS reactions can be useful in synthesis and can serve as precursors for a great variety of molecules. All of the reactions of alkyl halides that we studied in Chapter 9, including nucleophilic substitution and base-catalyzed elimination, are applicable to benzylic bromides.

PROBLEM

16.9 Styrene, the simplest alkenylbenzene, is prepared commercially for use in plastics manufacture by dehydrogenation of ethylbenzene over a special catalyst. How might you prepare styrene on a laboratory scale using reactions you have studied?

$CH{=}CH_2$

Styrene

PROBLEM

16.10 Refer to Table 4.5 for a more quantitative idea of the stability of a benzyl radical. Approximately how much more stable (kcal/mol) is the benzyl radical than a primary alkyl radical? How does a benzyl radical compare in stability to an allyl radical?

16.9 Oxidation of alkylbenzene side chains

Another way to functionalize alkylbenzene side chains is to oxidize them to benzoic acids. This reaction is usually best carried out either with aqueous potassium permanganate or with acidic aqueous sodium dichromate. The benzene ring itself is normally unreactive to these strongly oxidizing conditions, despite its unsaturation. This provides yet another example of the remarkable inertness of aromatic rings. Side chains are readily attacked, however, and the reaction can generally be applied to the oxidation of all except tertiary alkyl substituents. Thus, *p*-nitrotoluene and butylbenzene are oxidized by $KMnO_4$ in high yield, but *tert*-butylbenzene is not affected.

The precise mechanism of the reaction is not fully understood, but probably involves attack on benzylic hydrogens to form intermediate benzylic radicals. *tert*-Butylbenzene has no benzylic hydrogens and is therefore inert.

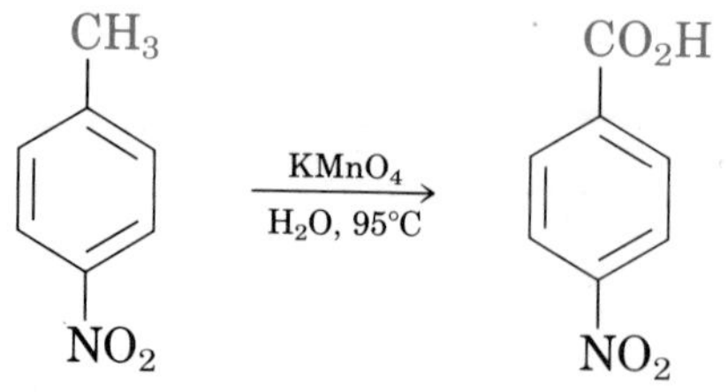

CH_3 / NO_2 $\xrightarrow[H_2O,\ 95°C]{KMnO_4}$ CO_2H / NO_2

p-Nitrotoluene → *p*-Nitrobenzoic acid (88%)

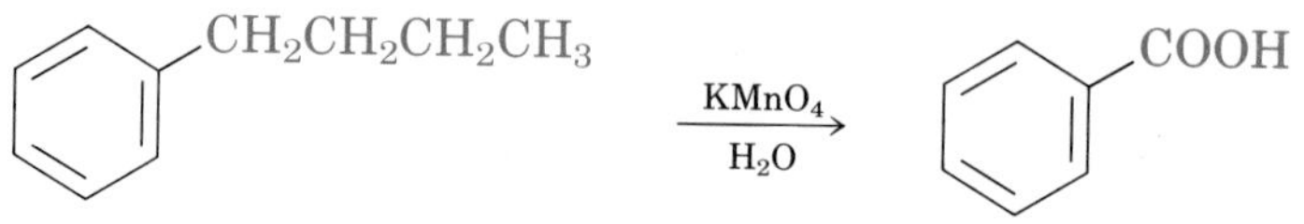

$CH_2CH_2CH_2CH_3$ $\xrightarrow[H_2O]{KMnO_4}$ COOH

Butylbenzene → Benzoic acid (85%)

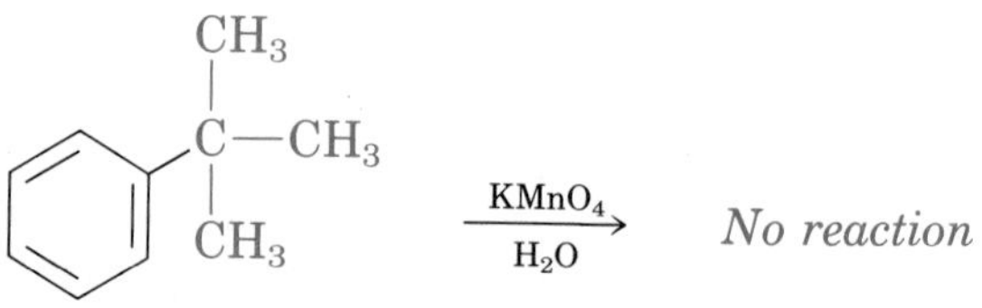

CH_3—C(CH_3)—CH_3 $\xrightarrow[H_2O]{KMnO_4}$ *No reaction*

t-Butylbenzene

A similar oxidation is employed industrially for the preparation of terephthalic acid, used in production of polyester fibers. Approximately 6 billion lb/yr of *p*-xylene are oxidized in this manner using air as the oxidant and Co(III) salts as catalyst.

Industrial procedure CH_3 / CH_3 $\xrightarrow[Co(III)]{O_2}$ CO_2H / CO_2H

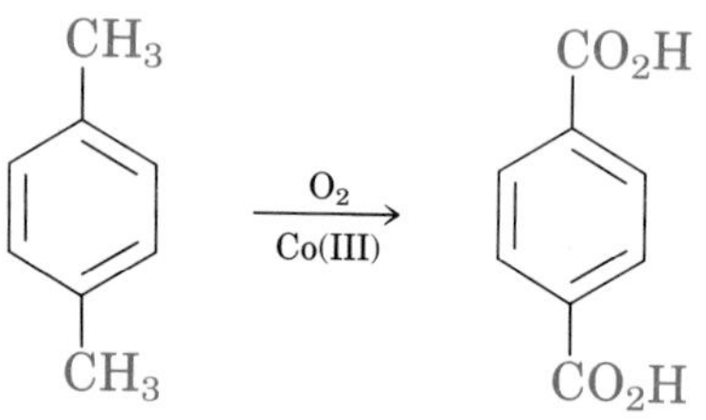

p-Xylene → Terephthalic acid

PROBLEM

16.11 What aromatic products would you expect to obtain from the $KMnO_4$ oxidation of these substances?

(a) Tetralin,

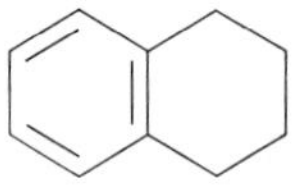

(b) *m*-Nitrocumene

(c) Dimestrol,

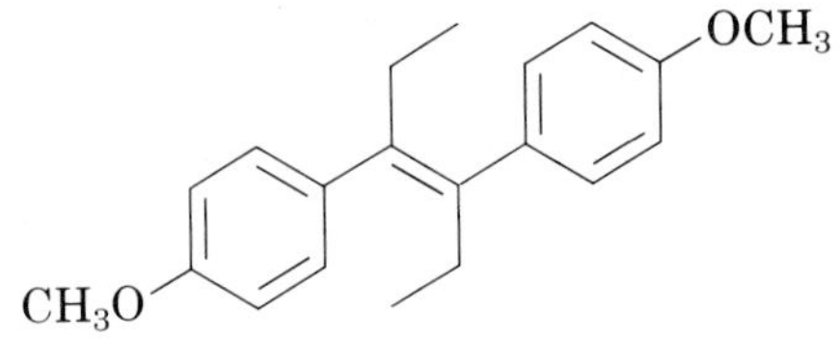

OCH_3

CH_3O

16.10 Catalytic hydrogenation of aromatic rings

Aromatic rings are reduced to cyclohexanes by hydrogenation over a rhodium-on-carbon catalyst. For example, *o*-xylene yields 1,2-dimethylcyclohexane, and 4-*tert*-butylphenol gives 4-*tert*-butylcyclohexanol in 100% yield.

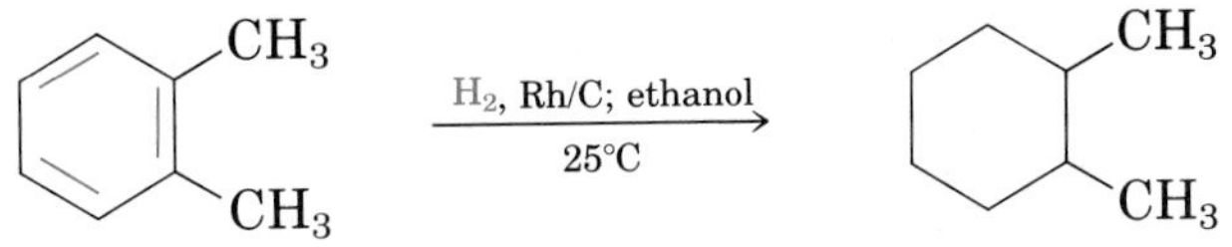

CH_3 CH_3 —H_2, Rh/C; ethanol, 25°C→ CH_3 CH_3

o-Xylene   1,2-Dimethylcyclohexane (100%)

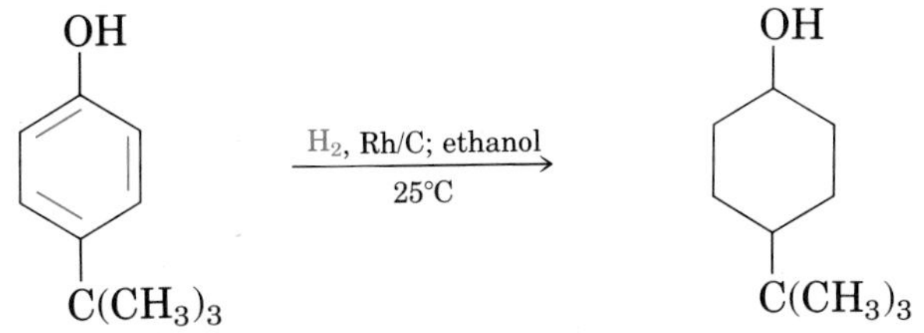

OH $C(CH_3)_3$ —H_2, Rh/C; ethanol, 25°C→ OH $C(CH_3)_3$

4-*tert*-Butylphenol   4-*tert*-Butylcyclohexanol (100%)

We have seen that the chemistry of benzene rings is normally ruled by a tendency to maintain aromatic stability, and thus aromatic rings are far more difficult to reduce than are isolated alkene double bonds. The usual platinum and palladium alkene hydrogenation catalysts do not affect aromatic rings, and it is therefore possible to selectively reduce alkenes in the presence of aromatic rings. For example, 4-phenyl-3-buten-2-one is selectively reduced to 4-phenyl-2-butanone over a palladium catalyst. Neither the benzene ring nor the ketone carbonyl group is affected.

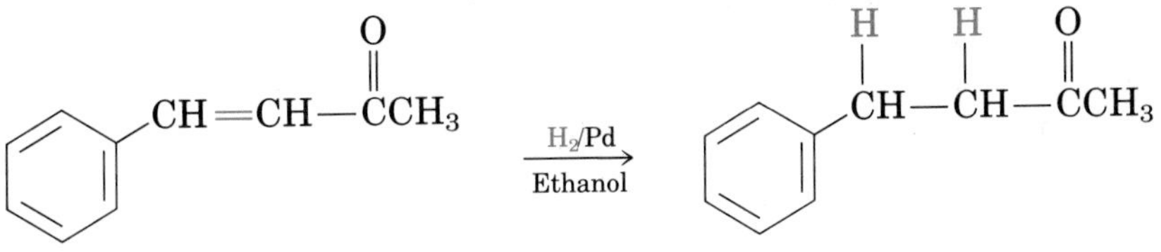

$CH{=}CH{-}\overset{O}{\overset{\|}{C}}CH_3$ —H_2/Pd, Ethanol→ $\overset{H}{\overset{|}{C}}H{-}\overset{H}{\overset{|}{C}}H{-}\overset{O}{\overset{\|}{C}}CH_3$

4-Phenyl-3-buten-2-one   4-Phenyl-2-butanone (100%)

Rhodium is a far more powerful hydrogenation catalyst than palladium, however, and is therefore used when aromatic ring reduction is desired.

16.11 Synthesis of substituted benzenes

One of the surest ways to acquire a command of organic chemistry is to work synthesis problems. The ability to plan a successful multistep synthesis of a complex molecule requires a working knowledge of the uses and limitations of many hundreds of organic reactions. Not only must we know *which* reactions to use, we must also know *when* to use them. The order in which reactions are carried out is often critical to the success of the overall scheme.

The ability to plan a sequence of reactions in the correct order is particularly valuable in the synthesis of substituted aromatic rings, where the introduction of one substituent is strongly affected by the directing effects of other substituents. Planning syntheses of substituted aromatic compounds is therefore an excellent way to gain facility with the many reactions learned in the past several chapters.

During our earlier discussion of the strategies that can be used in working synthesis problems (Section 7.12), we said that it is usually best to work problems backwards. Look at the target molecule and ask the question, "What is a possible immediate precursor of this compound?" Choose a likely answer and continue working backwards, one step at a time, until you arrive at a simple starting material. Let's try some examples.

Problem. Synthesize *p*-bromobenzoic acid starting from benzene.
Solution. Ask, "What is an immediate precursor of *p*-bromobenzoic acid?"

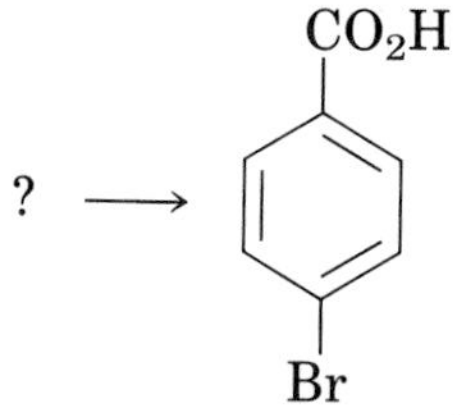

? $\longrightarrow$ [benzene ring with CO_2H and Br para]

p-Bromobenzoic acid

There are only two substituents on the ring, a carboxyl group (COOH), which is meta directing, and a bromine, which is ortho–para directing. We could not brominate benzoic acid, since the wrong isomer (*m*-bromobenzoic acid) would be produced. We know, however, that oxidation of alkylbenzene side chains yields benzoic acids. An immediate precursor of our target molecule might be *p*-bromotoluene.

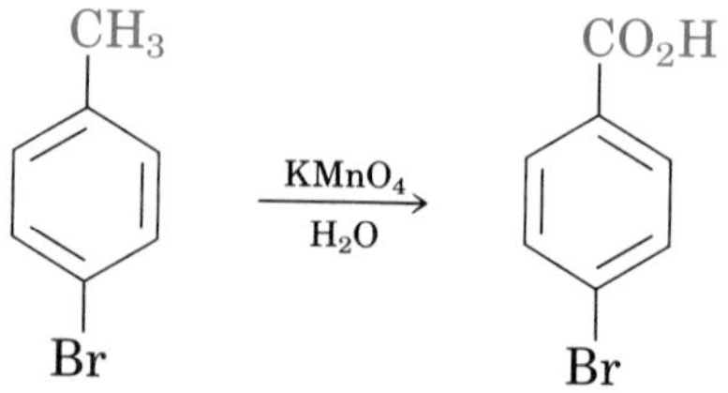

$$p\text{-Bromotoluene } (CH_3,\ Br) \xrightarrow[H_2O]{KMnO_4} p\text{-Bromobenzoic acid } (CO_2H,\ Br)$$

p-Bromotoluene *p*-Bromobenzoic acid

"What is an immediate precursor of *p*-bromotoluene?" Perhaps toluene is an immediate precursor, since the methyl group directs bromination to the ortho and para positions, and we could then separate isomers. Alternatively, bromobenzene might be an immediate precursor, since we could carry out a Friedel–Crafts methylation and obtain para product. Both answers are satisfactory. However, in view of the difficulties often observed with polyalkylation in Friedel–Crafts reactions, bromination of toluene may well be the more efficient route (Figure 16.15).

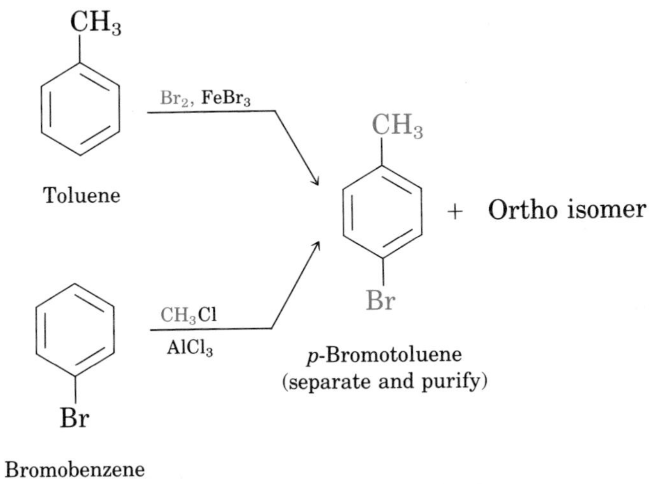

CH_3

Toluene

Br_2, $FeBr_3$

CH_3

\+ Ortho isomer

Br

CH_3Cl

$AlCl_3$

p-Bromotoluene
(separate and purify)

Br

Bromobenzene

Figure 16.15. Two routes for the synthesis of *p*-bromotoluene.

"What is an immediate precursor of toluene?" Benzene, which could be methylated in a Friedel–Crafts reaction.

CH_3

CH_3Cl / $AlCl_3$

Benzene Toluene

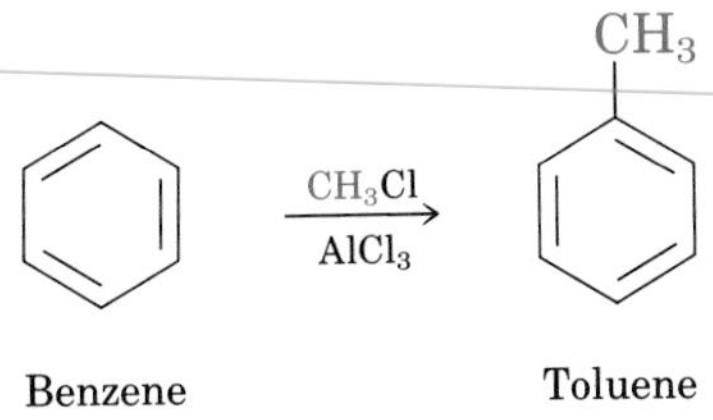

Alternatively, what is an immediate precursor of bromobenzene? Benzene, which could be brominated.

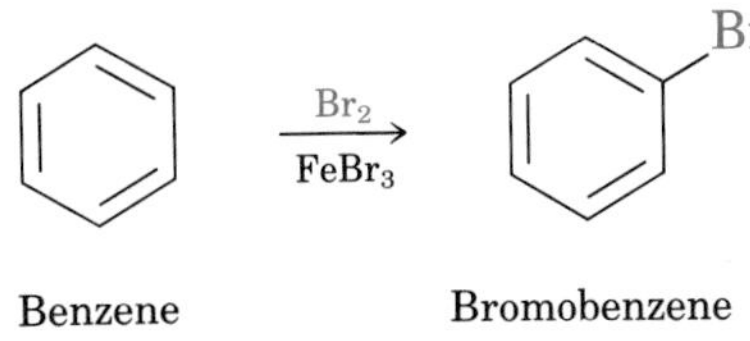

Br

Br_2 / $FeBr_3$

Benzene Bromobenzene

Our backwards synthetic (*retrosynthetic*) analysis has provided two valid routes from benzene to *p*-bromobenzoic acid (Figure 16.16).

Benzene

Br_2 / $FeBr_3$ → Br-benzene; CH_3Cl / $AlCl_3$ →

CH_3Cl / $AlCl_3$ → CH_3-benzene; Br_2 / $FeBr_3$ →

CH_3, Br (p-bromotoluene) → $KMnO_4$ → CO_2H, Br

p-Bromobenzoic acid

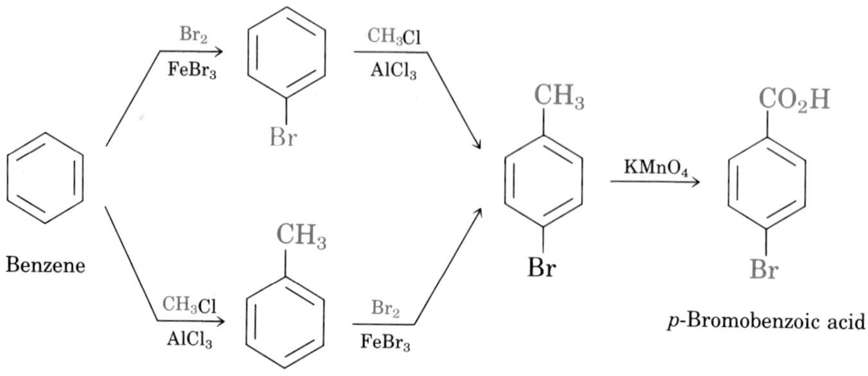

Figure 16.16. Synthesis of *p*-bromobenzoic acid from benzene.

Problem. Propose a synthesis of 4-chloro-1-nitro-2-propylbenzene starting from benzene.

Solution. "What is an immediate precursor of the target?" Since chlorination of *o*-nitropropylbenzene occurs at the wrong position, and the Friedel–Crafts reaction cannot be carried out on nitro-substituted (deactivated) benzenes, the immediate precursor of our desired product is probably *m*-chloropropylbenzene, which can then be nitrated. This nitration gives a mixture of product isomers, which can then be separated (Figure 16.17).

"What is an immediate precursor of *m*-chloropropylbenzene?" Since the two substituents have a meta relationship, one of the two must be derived from a meta-orienting substituent. Furthermore, since primary alkyl groups such as propyl cannot be introduced directly by Friedel–Crafts alkylation, the precursor of *m*-chloropropylbenzene is probably *m*-chloropropiophenone, which could undergo catalytic reduction of the acyl group.

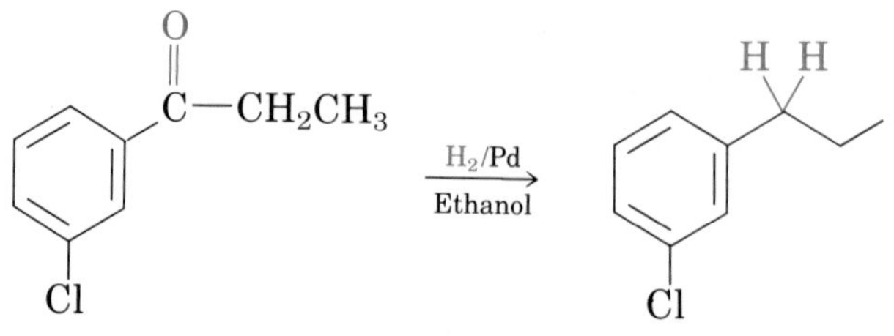

O ‖ C—CH_2CH_3, Cl → H_2/Pd, Ethanol → H H, Cl

m-Chloropropiophenone   *m*-Chloropropylbenzene

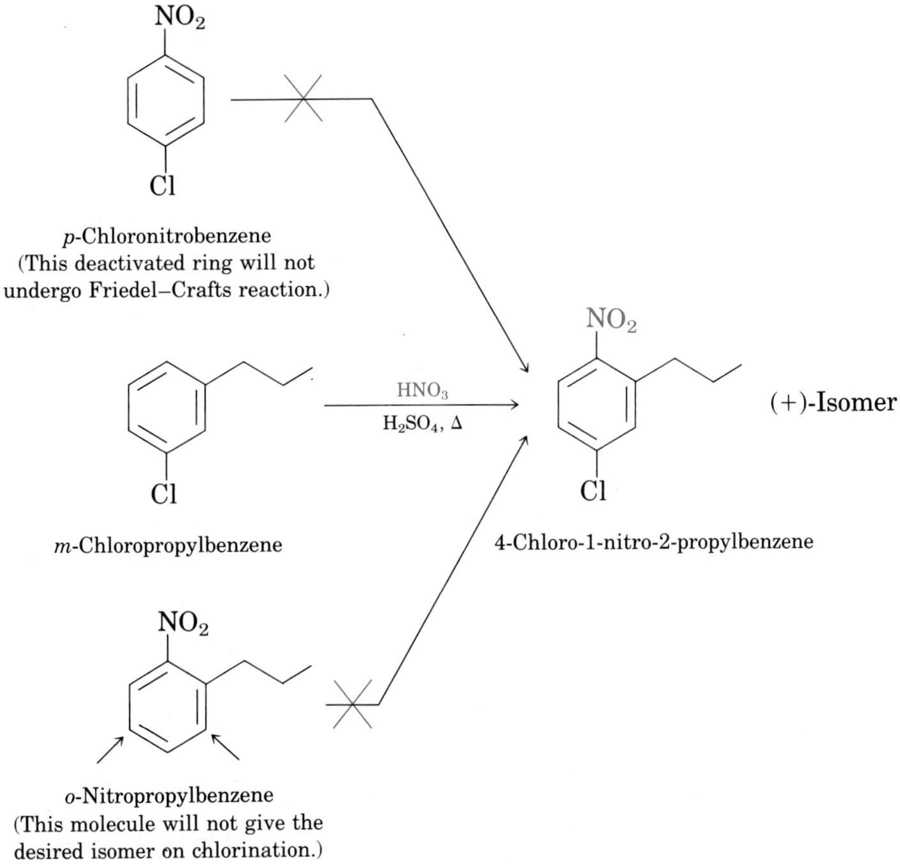

NO_2

Cl

p-Chloronitrobenzene
(This deactivated ring will not undergo Friedel–Crafts reaction.)

NO_2

$\xrightarrow[H_2SO_4,\ \Delta]{HNO_3}$

(+)-Isomer

Cl

Cl

m-Chloropropylbenzene

4-Chloro-1-nitro-2-propylbenzene

NO_2

o-Nitropropylbenzene
(This molecule will not give the desired isomer on chlorination.)

Figure 16.17. Possible routes for the synthesis of 4-chloro-1-nitro-2-propylbenzene.

"What is an immediate precursor of *m*-chloropropiophenone?" Perhaps propiophenone, which could be chlorinated.

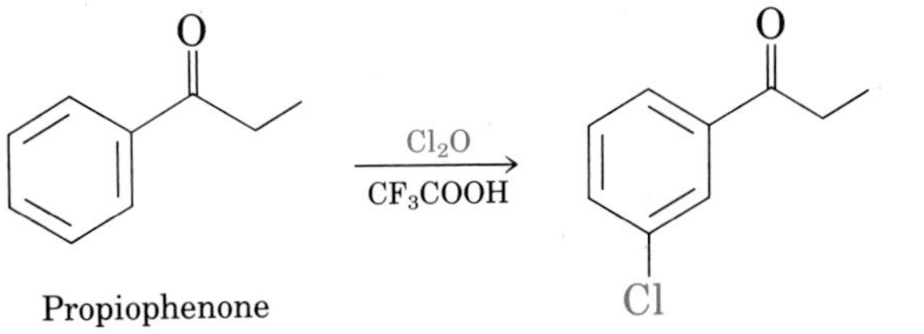

O

O

$\xrightarrow[CF_3COOH]{Cl_2O}$

Propiophenone

Cl

m-Chloropropiophenone

"What is an immediate precursor of propiophenone?" Benzene, which could undergo Friedel–Crafts acylation with propanoyl chloride and $AlCl_3$.

$$\text{Benzene} \xrightarrow[\text{AlCl}_3]{\text{CH}_3\text{CH}_2\overset{\text{O}}{\overset{\|}{\text{C}}}\text{Cl}} \text{Propiophenone}$$

Our final synthesis is a four-step route from benzene:

$$\text{Benzene} \xrightarrow[\text{AlCl}_3]{\text{CH}_3\text{CH}_2\overset{\text{O}}{\overset{\|}{\text{C}}}\text{Cl}} \text{Propiophenone} \xrightarrow[\text{CF}_3\text{COOH}]{\text{Cl}_2\text{O}} m\text{-Chloropropiophenone} \xrightarrow{\text{H}_2/\text{Pd}} m\text{-Chloropropylbenzene} \xrightarrow[\text{H}_2\text{SO}_4]{\text{HNO}_3} \text{(NO}_2\text{, Cl, propyl-substituted benzene)}$$

Planning organic syntheses has been compared to playing chess. There are no tricks involved; all that is required is a knowledge of the allowable moves (the organic reactions) and the discipline to work backwards and to evaluate carefully the consequences of each move. Practicing is not always easy, but there is no surer way to learn organic chemistry.

PROBLEM

16.12 Propose syntheses of the following substances, starting from benzene:
(a) *m*-Chloronitrobenzene
(b) *m*-Chloroethylbenzene
(c) *p*-Chloropropylbenzene

PROBLEM

16.13 Propose a synthesis of aspirin (acetylsalicylic acid) starting from benzene. You will need to use an acetylation reaction at some point in your scheme.

Aspirin (COOH, $\text{O}-\overset{\text{O}}{\overset{\|}{\text{C}}}-\text{CH}_3$)

$$\left[\ \text{Ar}-\text{OH} \xrightarrow{\text{CH}_3\overset{\text{O}}{\overset{\|}{\text{C}}}-\text{O}-\overset{\text{O}}{\overset{\|}{\text{C}}}\text{CH}_3} \text{Ar}-\text{OCOCH}_3\ \right]$$

An acetylation reaction

PROBLEM

16.14 In planning syntheses, it is as important to know what not to do as to know what to do. As written, the following reaction schemes have flaws that make their success unlikely. What is wrong with each one?

(a) Benzonitrile $\xrightarrow[\text{2. } HNO_3,\ H_2SO_4,\ \Delta]{\text{1. } CH_3CH_2\overset{O}{\overset{\|}{C}}Cl,\ AlCl_3}$ 3-nitro-5-propanoylbenzonitrile (NO_2, CN, $\overset{O}{\overset{\|}{C}}CH_2CH_3$)

(b) Chlorobenzene $\xrightarrow[\text{2. } Cl_2,\ FeCl_3]{\text{1. } CH_3CH_2CH_2Cl,\ AlCl_3}$ 1,2-dichloro-4-propylbenzene ($CH_3CH_2CH_2$, Cl, Cl)

16.12 Summary

The Friedel–Crafts reaction involves electrophilic attack of an aromatic ring on a carbocation. The carbocation is usually generated by reaction of an alkyl chloride with a Lewis acid catalyst such as aluminum chloride, but it can be generated in other ways as well. Thus, reaction of either an alcohol or an alkene with a strong acid leads to carbocations that can be attacked by aromatic rings. For example, cumene can be prepared by Friedel–Crafts alkylation of benzene using either 2-chloropropane/aluminum chloride, propene/strong acid, or 2-propanol/strong acid.

$$\text{Benzene} \xrightarrow{(CH_3)_2CHCl,\ AlCl_3 \ \text{or}\ (CH_3)_2CHOH,\ H^+ \ \text{or}\ CH_3CH{=}CH_2,\ H^+} \text{Cumene}\ (C_6H_5CH(CH_3)_2)$$

Benzene Cumene

Friedel–Crafts alkylation is highly useful but is nevertheless subject to certain limitations:

1. Only alkyl halides can be used. Vinylic and aryl halides do not react.
2. The aromatic ring must be at least as reactive as a halobenzene. Strongly deactivated rings do not react.
3. Polyalkylation often occurs, since the product alkylbenzene is often more reactive than the starting material.
4. Carbocation rearrangements can occur when primary alkylating agents are used.

Rearrangement of a carbocation from a less stable to a more stable structure is a general reaction process that occurs by migration of a hydrogen or alkyl group and its electron pair. For example, the primary propyl carbocation rapidly rearranges to the more stable secondary isopropyl cation by hydrogen migration.

$$\underset{1°}{CH_3CH_2\overset{+}{C}H_2} \xrightarrow[\text{shift}]{\text{Hydrogen}} \underset{2°}{CH_3\overset{+}{C}HCH_3}$$

Acyl cations, generated by reaction of either acid chlorides or acid anhydrides with $AlCl_3$, also undergo electrophilic substitution onto an aromatic ring. The ketones prepared in this way can be reduced to alkylbenzenes by hydrogenation over a palladium catalyst. Thus, Friedel–Crafts acylation, followed by catalytic hydrogenation, allows us to prepare primary alkylbenzenes that cannot be prepared directly because of carbocation rearrangement.

The side chain of alkylbenzenes has unique reactivity because of the neighboring aromatic ring. Thus, the benzylic position can be brominated by *N*-bromosuccinimide, and the entire side chain can be degraded to a carboxylic acid by oxidation with aqueous potassium permanganate. Although aromatic rings are unreactive to these strong oxidizing reagents, they can be reduced to cyclohexanes by hydrogenation over a rhodium catalyst. Alkenes react much more rapidly than benzene rings, however, and can be selectively reduced over a palladium catalyst.

16.13 Summary of reactions

1. Friedel–Crafts alkylation (Sections 16.2–16.5)

$$C_6H_6 + CH_3Cl \xrightarrow{AlCl_3} C_6H_5CH_3 + HCl$$

Aromatic ring: Must be at least as reactive as a halobenzene. Deactivated rings do not react.

Alkyl halide: Can be methyl, ethyl, 2°, or 3°; primary halides undergo carbocation rearrangement.

2. Friedel–Crafts acylation (Section 16.6)

$$C_6H_6 + CH_3\overset{O}{\overset{\|}{C}}Cl \xrightarrow{AlCl_3} C_6H_5\overset{O}{\overset{\|}{C}}CH_3 + HCl$$

Reaction also occurs when an acid anhydride is used in place of an acyl chloride:

$$C_6H_6 + CH_3\overset{O}{\overset{\|}{C}}{-}O{-}\overset{O}{\overset{\|}{C}}CH_3 \xrightarrow{AlCl_3} C_6H_5\overset{O}{\overset{\|}{C}}{-}CH_3 + CH_3COOH$$

3. Reduction of aryl alkyl ketones (Section 16.7)

$$C_6H_5COCH_3 \xrightarrow[\text{Ethanol}]{H_2/Pd} C_6H_5CH_2CH_3$$

Reaction is specific for alkyl aryl ketones; dialkyl ketones are not affected.

4. *N*-Bromosuccinimide bromination of alkylbenzenes (Section 16.8)

$$C_6H_5CH_3 \xrightarrow[CCl_4]{\text{NBS}} C_6H_5CH_2Br$$

5. Oxidation of alkylbenzene side chain (Section 16.9)

$$C_6H_5CH_3 \xrightarrow[H_2O]{KMnO_4} C_6H_5COOH$$

Reaction occurs with 1° and 2°, but not 3°, alkyl side chains.

6. Catalytic hydrogenation of aromatic ring (Section 16.10)

$$C_6H_6 \xrightarrow[\text{Ethanol}]{H_2/Rh/C} C_6H_{12}$$

7. Preparation of acyl chlorides (Section 16.6)

$$CH_3COOH + SOCl_2 \longrightarrow CH_3\overset{O}{\overset{\|}{C}}Cl + HCl + SO_2$$

Reaction also works using oxalyl chloride, $(COCl)_2$.

ADDITIONAL PROBLEMS

16.15 Predict the major monoalkylation products you would expect to obtain from reaction of the following substances with chloromethane and $AlCl_3$:

(a) Bromobenzene
(b) *m*-Bromophenol
(c) *p*-Chloroaniline
(d) 2,4-Dichloronitrobenzene
(e) 2,4-Dichlorophenol
(f) Benzoic acid
(g) *p*-Methylbenzenesulfonic acid
(h) 2,5-Dibromotoluene
(i) Biphenyl
(j) Naphthalene

16.16 What product(s) would you expect to obtain from these reactions?

(a) o-Nitroacetophenone (ring bearing CCH_3 with C=O, and NO_2) $\xrightarrow{H_2/Pd}$?

(b) p-Chlorophenol (ring bearing Cl and OH) $\xrightarrow[AlCl_3]{CH_3CH_2CH_2Cl}$?

(c) m-Bromotoluene (ring bearing CH_3 and Br) $\xrightarrow[AlCl_3]{CH_3COCCH_3 \text{ (O, O)}}$?

(d) CH_3O–benzene + phthalic anhydride (benzene fused with C(=O)–O–C(=O)) $\xrightarrow{AlCl_3}$?

(e) Product of (d) $\xrightarrow{H_2SO_4}$?

(f) 1,1,4,4-Tetramethyltetralin (H_3C, CH_3; H_3C, CH_3) $\xrightarrow[H_2O]{KMnO_4}$?

(g) 1,1,4,4-Tetramethyltetralin (H_3C, CH_3; H_3C, CH_3) $\xrightarrow[CCl_4]{NBS}$?

16.17 Triphenylmethane can be prepared by reaction of benzene and chloroform in the presence of $AlCl_3$. Propose a mechanism for this reaction:

$$\text{Benzene} + CHCl_3 \xrightarrow{AlCl_3} (C_6H_5)_3C\text{–}H$$

16.18 How would you synthesize these substances, starting from benzene? Assume that ortho and para substitution products can be separated.

(a) *o*-Methylaniline
(b) 2,4,6-Trinitrophenol
(c) 2,4,6-Trinitrobenzoic acid
(d) 3,5-Dibromoaniline
(e) *p-tert*-Butylbenzoic acid
(f) *m*-Butylaniline

16.19 When heated with aqueous acid, styrene undergoes the dimerization reaction shown. Propose a mechanism to account for the formation of this dimer.

$C_6H_5CH{=}CH_2 \xrightarrow[\Delta]{H^+}$ 1-methyl-3-phenylindane (CH_3 on the five-membered ring)

16.20 One method of formylating an aromatic ring is to allow the aromatic substance to react with formyl fluoride (F—C(=O)—H) in the presence of BF_3 as Lewis acid catalyst. Propose a mechanism for this reaction.

$C_6H_6 + HC(=O){-}F \xrightarrow[\text{Ether}]{BF_3} C_6H_5CHO$

16.21 As written, the following syntheses have certain flaws. What is wrong with each one?

(a) Toluene (CH_3) $\xrightarrow{\text{1. } Cl_2,\ FeCl_3;\ \text{2. } KMnO_4}$ *m*-chlorobenzoic acid (CO_2H, Cl)

(b) Chlorobenzene (Cl) $\xrightarrow{\text{1. } HNO_3,\ H_2SO_4,\ \Delta;\ \text{2. } CH_3Cl,\ AlCl_3;\ \text{3. } SnCl_2,\ H_3O^+;\ \text{4. } NaOH,\ H_2O}$ product with Cl, CH_3 (ortho to Cl), NH_2 (para to Cl)

(c) Toluene (CH_3) $\xrightarrow{\text{1. } CH_3C(=O)Cl,\ AlCl_3;\ \text{2. } HNO_3,\ H_2SO_4,\ \Delta;\ \text{3. } H_2/Pd;\ \text{ethanol}}$ product with CH_3, NO_2 (ortho to CH_3), CH_2CH_3 (para to CH_3)

(d) 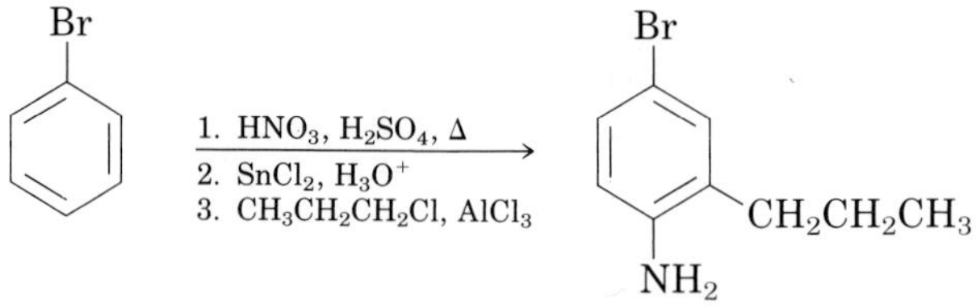 Br-benzene $\xrightarrow[\text{2. } SnCl_2,\ H_3O^+ \quad \text{3. } CH_3CH_2CH_2Cl,\ AlCl_3]{\text{1. } HNO_3,\ H_2SO_4,\ \Delta}$ (Br, $CH_2CH_2CH_3$, NH_2 substituted benzene)

16.22 Would you expect the reaction of benzene with optically active 2-chlorobutane to yield optically active or racemic product? Explain your answer.

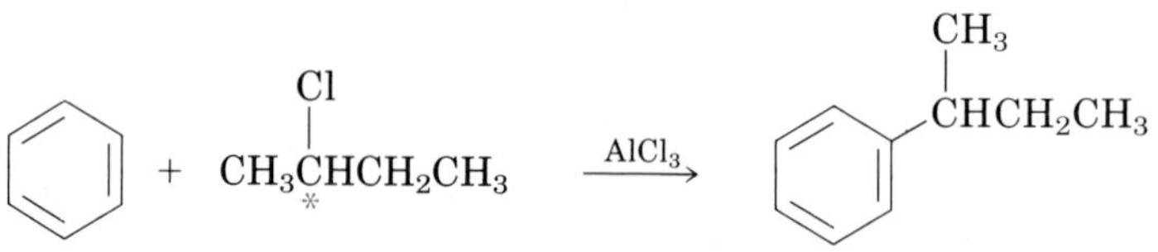

$$\text{benzene} + CH_3\overset{Cl}{\overset{|}{C}}HCH_2CH_3^{*} \xrightarrow{AlCl_3} C_6H_5\text{-}\overset{CH_3}{\overset{|}{C}}HCH_2CH_3$$

16.23 Suggest a mechanism to account for the fact that acid-catalyzed dehydration of the alcohol shown yields a rearranged alkene.

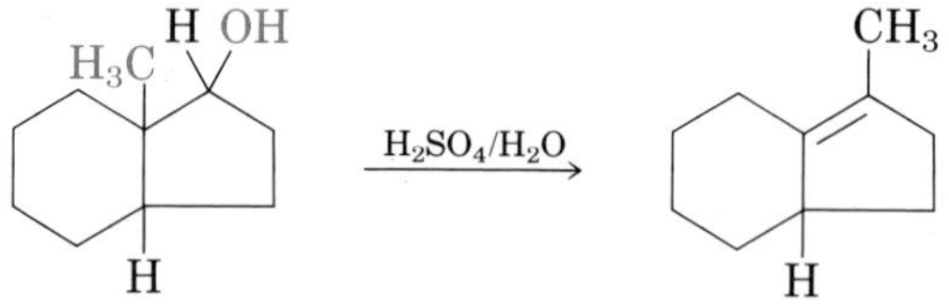

(bicyclic alcohol with H_3C, H, OH, H) $\xrightarrow{H_2SO_4/H_2O}$ (rearranged alkene with CH_3, H)

16.24 How would you synthesize these substances starting from benzene?

(a) 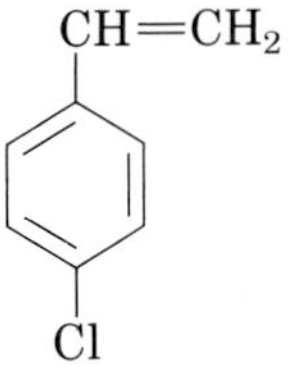 p-ClC$_6$H$_4$CH=CH$_2$ ($CH{=}CH_2$, Cl)

(b) 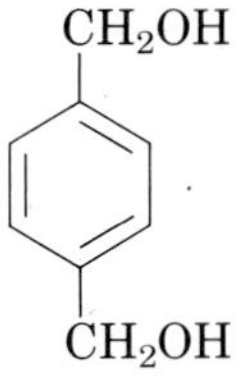 (CH_2OH, CH_2OH para-disubstituted benzene)

(c) 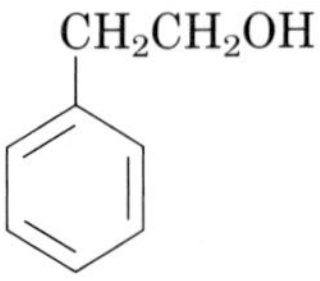 (CH_2CH_2OH on benzene)

(d) 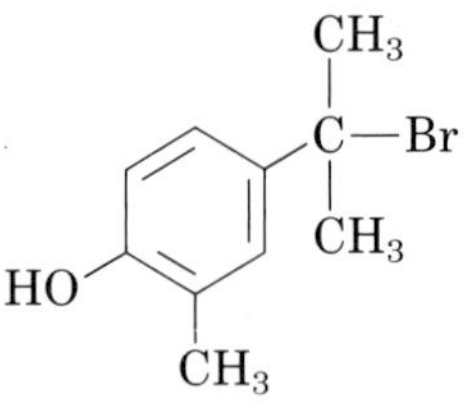 (benzene ring bearing $\overset{CH_3}{\underset{CH_3}{C}}$—Br, HO, CH_3)

16.25 The compound MON-0585 is a nontoxic, biodegradable larvicide that is highly selective against mosquito larvae. How could you synthesize MON-0585 using only benzene as a source of the aromatic rings?

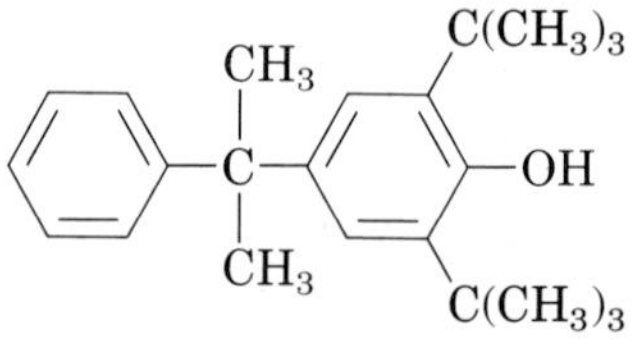

(phenyl–$C(CH_3)_2$–ring bearing OH and two $C(CH_3)_3$ groups)

MON-0585

16.26 Fenipentol is a pharmaceutical agent used in the treatment of mild chronic pancreatitis. How might you synthesize it from benzene?

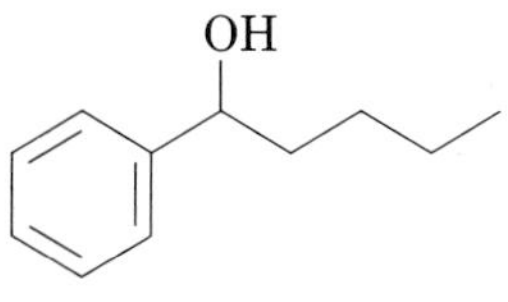

Fenipentol

16.27 How might you prepare the epoxy resin intermediate, bisphenol A, from benzene?

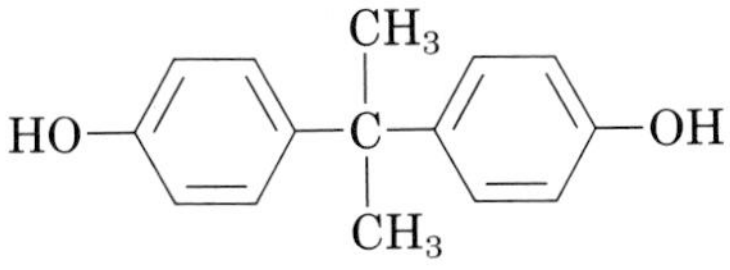

Bisphenol A

16.28 Hexachlorophene, a substance used in the manufacture of germicidal soaps, is prepared by reaction of 2,4,5-trichlorophenol with formaldehyde in the presence of concentrated sulfuric acid. Propose a mechanism to account for the reaction.

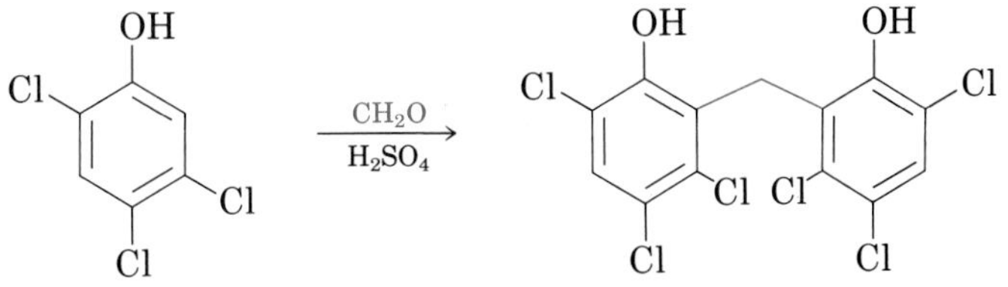

$\xrightarrow[H_2SO_4]{CH_2O}$

Hexachlorophene

16.29 Propose a mechanism to account for the following reaction of benzene with 2,2,5,5-tetramethyltetrahydrofuran.

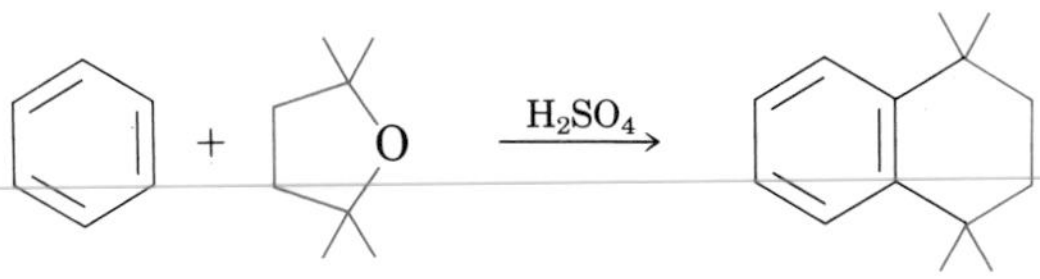

$\xrightarrow{H_2SO_4}$

16.30 Aromatic rings can be chloromethylated by treatment with formaldehyde and HCl gas in the presence of $ZnCl_2$ catalyst. Propose a mechanism for this reaction.

$$C_6H_6 + CH_2O + HCl \xrightarrow{ZnCl_2} C_6H_5CH_2Cl$$

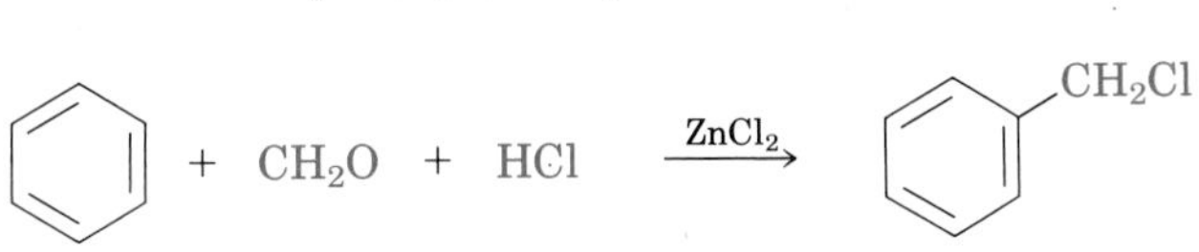

16.31 In the Gatterman–Koch reaction, a formyl group (—CHO) is introduced directly onto a benzene ring. For example, reaction of toluene with carbon monoxide and HCl in the presence of mixed $CuCl/AlCl_3$ gives *p*-methylbenzaldehyde in 55% yield. Propose a mechanism for this reaction (shown at the top of the next page).

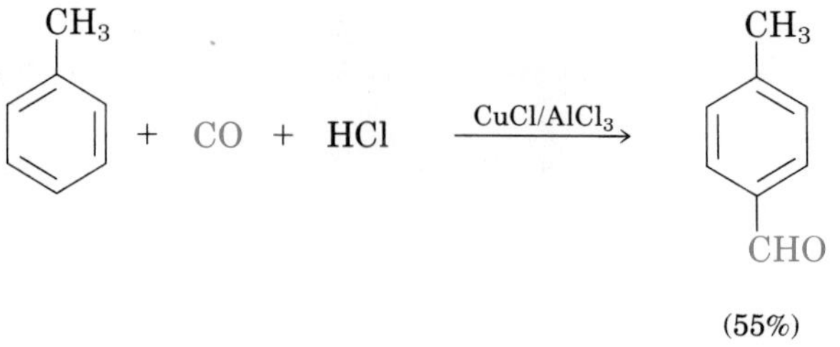

CH_3 + CO + HCl $\xrightarrow{CuCl/AlCl_3}$ CH_3 / CHO

(55%)

16.32 When 1-phenyl-1-pentanol is heated with phosphoric acid, dehydration and cyclization to α-methyltetralin occurs. Propose a mechanism to account for this reaction.

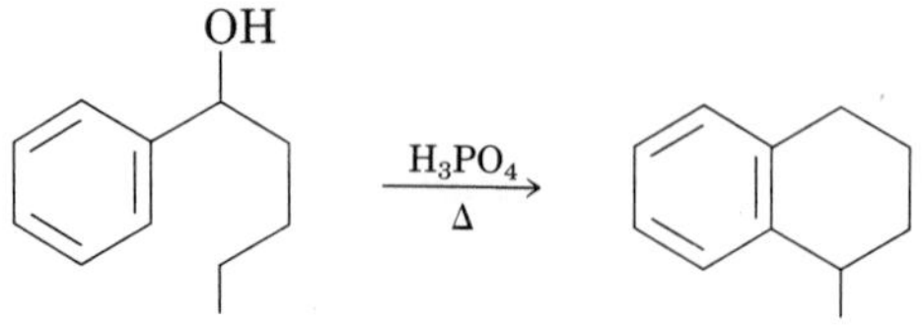

OH

$\xrightarrow[\Delta]{H_3PO_4}$

1-Phenyl-1-pentanol

α-Methyltetralin
(85%)

16.33 Propose a mechanism to account for the following reaction:

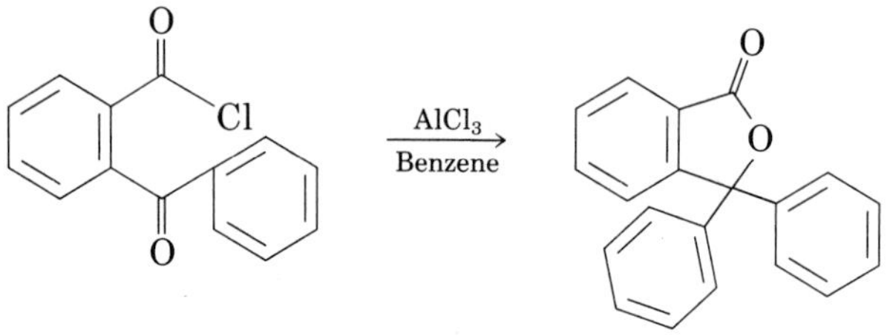

O Cl O $\xrightarrow[Benzene]{AlCl_3}$ O O

CHAPTER 17
ORGANIC REACTIONS: A BRIEF REVIEW

We began the study of organic chemistry with the assertion that there are three fundamental reaction types—polar reactions, radical reactions, and pericyclic reactions. Having seen many different reactions by now, let's see how this assertion stands up.

17.1 Polar reactions

Polar reactions are those between electron-rich reagents (nucleophiles/Lewis bases) and electron-poor reagents (electrophiles/Lewis acids). Polar reactions are heterolytic processes and involve even-numbered-electron species. Bonds are made when a nucleophile donates an electron pair to an electrophile. Conversely, bonds are broken when one fragment leaves with an electron pair (Figure 17.1).

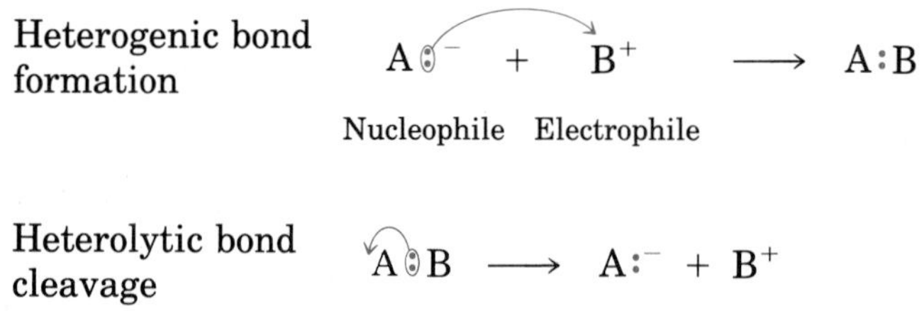

Heterogenic bond formation $A{:}^- + B^+ \longrightarrow A{:}B$

Nucleophile Electrophile

Heterolytic bond cleavage $A{:}B \longrightarrow A{:}^- + B^+$

Figure 17.1. Polar reactions.

The polar reactions studied thus far can be classified into five general categories:

1. Electrophilic addition reactions
2. Elimination reactions
3. Electrophilic aromatic substitution reactions
4. Nucleophilic substitution reactions
5. Nucleophilic aromatic substitution reactions

Let's review these five categories briefly to see where specific reactions fit in.

ELECTROPHILIC ADDITION REACTIONS

Table 17.1 lists some of the electrophilic addition reactions we have studied. The first five reactions in the table are mechanistically similar; all proceed by attack on an electrophile by an electron-rich alkene double bond to generate an intermediate carbocation. The electrophile may be H^+ (HX addition; reaction 1), X^+ (halogen addition and halohydrin formation; reactions 2 and 3), or Hg^{2+} (oxymercuration and alkyne hydration; reactions 4 and 5), but the basic process is the same.

TABLE 17.1 Some electrophilic addition reactions

1. Addition of HX (Sections 5.8 and 5.9)
   (X = Cl, Br, or I)

$$(CH_3)_2C{=}CH_2 \xrightarrow[\text{Ether}]{HCl} (CH_3)_2\overset{\overset{Cl}{|}}{C}{-}CH_3$$

2. Addition of X_2 (Section 6.1)
   (X = Cl or Br)

$$CH_3CH{=}CHCH_3 \xrightarrow[CCl_4]{Br_2} CH_3\overset{\overset{Br}{|}}{C}H{-}\overset{\overset{Br}{|}}{C}HCH_3$$

3. Halohydrin formation; addition of HOX (Section 6.2)
   (X = Cl, Br, or I)

$$(CH_3)_2C{=}CH_2 \xrightarrow[H_2O]{HOBr} (CH_3)_2\overset{\overset{OH}{|}}{C}{-}CH_2Br$$

4. Hydration by oxymercuration; addition of —Hg—OH (Section 6.3)

$$(CH_3)_2C{=}CH_2 \xrightarrow[H_2O]{Hg(OAc)_2} (CH_3)_2\overset{\overset{OH}{|}}{C}{-}CH_2HgOAc \xrightarrow{NaBH_4} (CH_3)_2\overset{\overset{OH}{|}}{C}{-}CH_3$$

5. Alkyne hydration (Section 7.4)

$$CH_3C{\equiv}CCH_3 \xrightarrow[H_3O^+]{HgSO_4} \left[CH_3\overset{\overset{OH}{|}}{C}{=}CHCH_3 \right] \longrightarrow CH_3\overset{\overset{O}{\|}}{C}CH_2CH_3$$

6. Hydroxylation (Section 6.7)

$$CH_3CH{=}CHCH_3 \xrightarrow[\text{2. } NaHSO_3]{\text{1. } OsO_4} CH_3\overset{\overset{OH}{|}}{C}H{-}\overset{\overset{OH}{|}}{C}HCH_3$$

7. Hydroboration; addition of $\rangle$B—H (Section 6.4)

$$(CH_3)_2C{=}CH_2 \xrightarrow[THF]{BH_3} (CH_3)_2\overset{\overset{H}{|}}{C}{-}CH_2BH_2 \xrightarrow{H_2O_2,\ ^-OH} (CH_3)_2\overset{\overset{H}{|}}{C}CH_2OH$$

$$\rangle C{=}C\langle + E^+ \longrightarrow \left[\rangle\overset{+}{C}{-}\overset{|}{\underset{|}{C}}{-}E \right] \xrightarrow{Nu:^-} {-}\overset{\overset{Nu}{|}}{C}{-}\overset{\overset{E}{|}}{C}{-}$$

Cation intermediate

where E^+ = Electrophile (H^+, X^+, or Hg^{2+})

$Nu:^-$ = Nucleophile ($HO:^-$ or $X:^-$)

The remaining two reactions in Table 17.1 differ from the first five in that they occur by one-step mechanisms without proceeding through cation intermediates. Nevertheless, it is still convenient to think of hydroxylation and hydroboration as electrophilic addition reactions, since they proceed by attack of the nucleophilic alkene double bond on electron-poor reagents.

ELIMINATION REACTIONS

The elimination reactions we have studied are illustrated in Table 17.2, and all occur in a similar way. This type of polar reaction involves attack of a basic reagent (nucleophile) on a hydrogen atom (electrophile), either next to a leaving group (E2 reaction) or next to a positive charge (E1 reaction). The

TABLE 17.2 **Some elimination reactions**

1. Dehydrohalogenation
   a. Elimination of HX from alkyl halides (Section 10.10)

$$(CH_3)_2C(Br)-CH_2-H \xrightarrow[CH_3CH_2OH]{^{-}:OCH_2CH_3} (CH_3)_2C=CH_2 + :Cl^- + HOCH_2CH_3$$

   b. Alkyne formation (Section 7.11)

$$CH_3C(H)=CH(Cl) \xrightarrow[NH_3]{KNH_2} CH_3C\equiv CH + KCl + NH_3$$

   c. Benzyne formation (Section 15.14)

$$C_6H_5Cl \xrightarrow[H_2O]{^{-}:OH} [\text{benzyne}] \xrightarrow{H_2O} C_6H_5OH$$

2. Dehydration (Section 6.13)

$$(CH_3)_2C(OH)-CH_2-H \xrightarrow[Pyridine]{POCl_3} [(CH_3)_2C(OPOCl_2)-CH_2-H] \xrightarrow{\ddot{}Pyridine} (CH_3)_2C=CH_2$$

3. Elimination of proton next to cationic carbon; acid-catalyzed dehydration (Section 6.13)

$$(CH_3)_2C(OH)CH_3 \xrightarrow[H_2O]{H^+} [(CH_3)_2C(^{+}OH_2)CH_3] \xrightarrow{-H_2O} [(CH_3)_2\overset{+}{C}-CH_2-H] \xrightarrow{H_2\ddot{O}} (CH_3)_2C=CH_2$$

usual E2 dehydrohalogenation reaction (reaction 1, Table 17.2) involves attack of a base on a hydrogen atom and simultaneous loss of halide ion. This may occur on an alkyl halide to yield an alkene, on a vinylic halide to yield an alkyne, or on an aryl halide to yield a benzyne. The mechanism is similar in all cases.

Dehydrations accomplished by reagents such as $POCl_3$ proceed similarly (reaction 2). In this case, normal E2 elimination of water is slow because hydroxide ion is a poor leaving group. The function of $POCl_3$ is to react with the hydroxyl and turn it into a better leaving group so that a normal E2 elimination can occur in the second step.

$POCl_3$, Pyridine; E2

Hydroxyl
(poor leaving group;
slow elimination)

Dichlorophosphate
(good leaving group;
rapid elimination)

Base attack and loss of a proton from the position next to a carbocation (reaction 3) is another typical elimination mechanism often encountered in organic chemistry.

ELECTROPHILIC AROMATIC SUBSTITUTION REACTIONS

All of the electrophilic aromatic substitution reactions shown in Table 17.3 (p. 556) occur by a common two-step mechanism. These two steps are not new to us; they are a combination of an electrophilic addition step and an elimination step:

E^+ Electrophilic addition; :Base Elimination

where E^+ is an electrophile.

The first step in electrophilic aromatic substitution is similar to the first step in electrophilic addition to alkenes—an electron-poor reagent is attacked by the electron-rich aromatic ring. The second step is identical to what happens during E1 elimination—a base present in solution attacks a hydrogen atom next to the positive charge, and elimination of the proton occurs. When viewed in this way, the mechanistic unity of electrophilic aromatic substitution reactions with other polar reactions is easier to see and understand.

NUCLEOPHILIC SUBSTITUTION REACTIONS

The nucleophilic substitution reaction is one of the most common reactions encountered in organic chemistry, and Table 17.4 illustrates some of the

TABLE 17.3 Some electrophilic aromatic substitution reactions

1. Halogenation (Sections 15.1 and 15.2) (Cl_2, Br_2, or I_2)

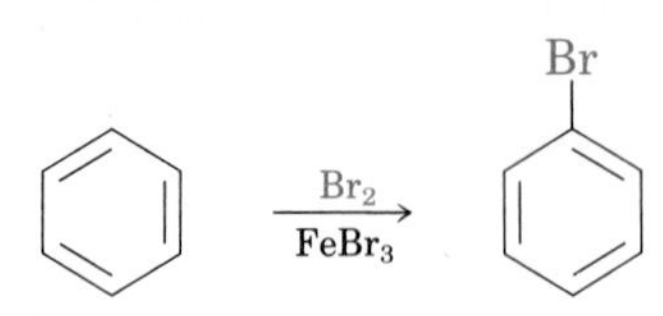

2. Hydroxylation (Section 15.5)

Benzene $\xrightarrow[HOSO_2F]{H_2O_2}$ phenol (OH)

3. Nitration (Section 15.3)

Benzene $\xrightarrow[H_2SO_4,\ \Delta]{HNO_3}$ nitrobenzene (NO_2)

4. Sulfonation (Section 15.4)

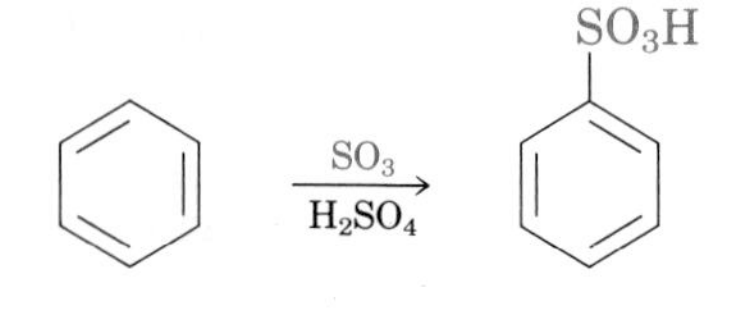

5. Friedel–Crafts alkylation (Sections 16.2–16.5)

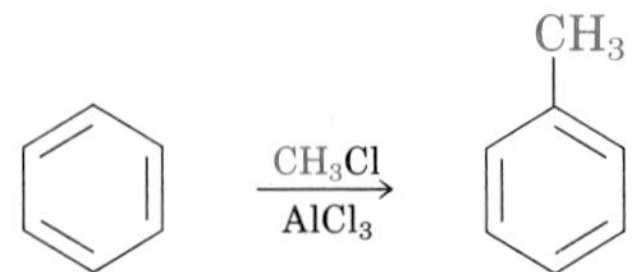

6. Friedel–Crafts acylation (Section 16.6)

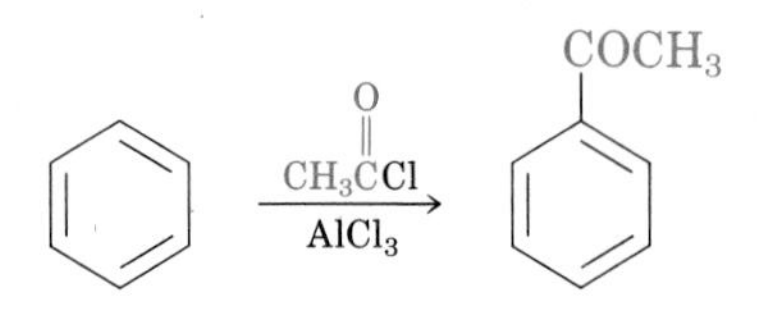

TABLE 17.4 Some nucleophilic substitution reactions

1. S_N2 halide displacement
   a. Hydroxide substitution (Sections 10.4 and 10.5)

$$HO:^- + CH_3I \longrightarrow HO{-}CH_3 + :I^-$$

   b. Hydride substitution (Section 9.8)

$$LiB(CH_2CH_3)_3^+H:^- + CH_3CH_2Br \longrightarrow CH_3CH_2{-}H + :Br^-$$

   c. Acetylide substitution (Sections 7.7–7.9)

$$CH_3C{\equiv}C:^- + CH_3CH_2Br \longrightarrow CH_3C{\equiv}CCH_2CH_3 + :Br^-$$

2. Alcohol $\longrightarrow$ Halide
   a. HX treatment; S_N1 reaction (Section 9.7)

$$(CH_3)_3C{-}OH + HCl \longrightarrow (CH_3)_3C{-}Cl + H_2O$$

   b. PBr_3 treatment; S_N2 reaction (Section 9.7)

$$\underset{CH_3CHCH_3}{\overset{OH}{|}} \xrightarrow[Ether]{PBr_3} \left[\underset{CH_3CHCH_3}{\overset{OPBr_2}{|}}\right] \xrightarrow{Br:^-} \underset{CH_3CHCH_3}{\overset{Br}{|}}$$

examples we have already studied. As reaction 1 indicates, most primary halides and tosylates, and many secondary ones, undergo substitution reactions with a variety of different nucleophiles. The nucleophile might be hydroxide ion (reaction 1a), hydride ion (reaction 1b), acetylide ion (reaction 1c), or many others. The electrophile might be an alkyl halide or an alkyl tosylate.

Many other processes involve nucleophilic substitution reactions, often as one step in a multistep sequence. For example, in reaction 2 halide ion does not effect nucleophilic replacement of a hydroxyl group, since hydroxyl is such a poor leaving group. Thus, in order to convert an alcohol into an alkyl halide, we must first transform the hydroxyl into a good leaving group. Treatment with HX accomplishes this transformation (reaction 2a) by protonating the oxygen atom; loss of water (a good leaving group) then gives an intermediate carbocation (S_N1 reaction). Treatment with PBr_3 accomplishes a similar transformation (reaction 2b) by converting the hydroxyl into a dibromophosphite ($-OPBr_2$) leaving group, which can be readily displaced in typical S_N2 fashion. Whatever the precise details of a given reaction, nucleophilic substitutions are common and all are mechanistically similar.

NUCLEOPHILIC AROMATIC SUBSTITUTION REACTIONS

Nucleophilic aromatic substitution is a polar reaction that must be placed in a unique category, since it is not related to the other general processes we have studied. In this reaction, a nucleophile attacks an electrophilic aromatic ring. The ring is made electrophilic, and hence reactive, only when substituted by strongly electron-withdrawing groups such as nitro, cyano, and carbonyl. Although this electrophilic behavior of the aromatic ring is a reversal of its normal reactivity, we can readily account for the observed behavior. This process still involves reaction between an electrophile and a nucleophile, as with all polar reactions (Table 17.5).

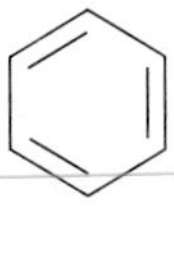

NO_2 Cl O_2N NO_2

"Normal" aromatic ring; electron-rich and nucleophilic

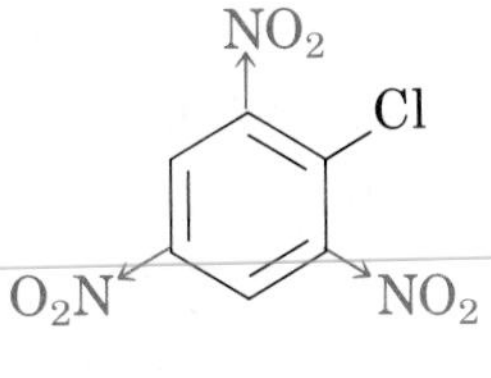

Electron-poor aromatic ring; electrophilic

TABLE 17.5 Nucleophilic aromatic substitution

1. Halide substitution (Section 15.13)

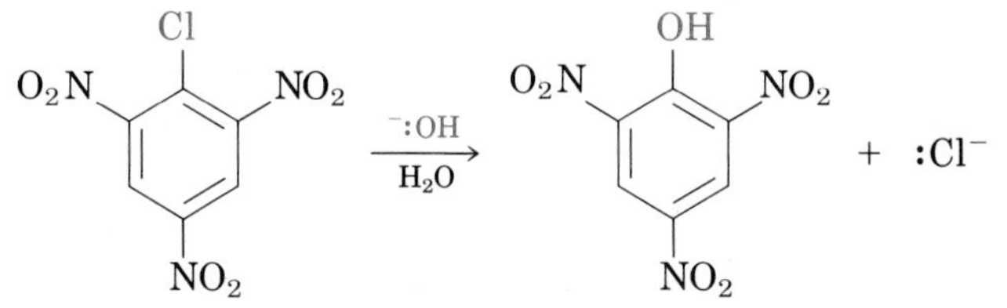

Cl O_2N NO_2 NO_2 $\xrightarrow[H_2O]{^-:OH}$ OH O_2N NO_2 NO_2 + $:Cl^-$

17.2 Radical reactions

Radical reactions are homolytic processes and involve odd-numbered-electron species. Bonds are made when each reactant donates one electron, and bonds are broken when each fragment leaves with one electron (Figure 17.2).

Homogenic bond formation	A· + B· ⟶ A:B
Homolytic bond cleavage	A:B ⟶ A· + B·

Figure 17.2. Radical reactions.

We have discussed rather few radical reactions. Those we have studied can be classified into three general categories: radical addition reactions, radical substitution reactions, and radical coupling reactions. These are shown in Table 17.6.

TABLE 17.6 Classification of some radical reactions

1. Radical addition reactions: $>C{=}C< + A\cdot \longrightarrow >\dot{C}{-}C{<}{-}A$
   a. Peroxide-catalyzed addition of HBr to alkenes (Section 6.5)

$$CH_3CH{=}CH_2 \xrightarrow[\text{Peroxide}]{\text{HBr}} CH_3CH_2{-}CH_2Br$$

2. Radical substitution reactions: A:B + C· ⟶ A:C + B·
   a. NBS bromination of allylic and benzylic positions (Sections 9.5 and 16.8)

$$CH_3CH{=}CH_2 \xrightarrow[CCl_4]{\text{NBS}} BrCH_2CH{=}CH_2$$

$$C_6H_5CH_3 \xrightarrow[CCl_4]{\text{NBS}} C_6H_5CH_2Br$$

3. Radical coupling reactions: A· + B· ⟶ A:B
   a. Lithium diorganocopper coupling (Section 9.9)

$$CH_3Cl + Li \longrightarrow CH_3Li$$

$$2\ CH_3Li + CuI \longrightarrow (CH_3)_2Cu^-Li^+$$

$$CH_3CH_2CH_2CH_2Br \xrightarrow[\text{Ether}]{(CH_3)_2CuLi} CH_3CH_2CH_2CH_2CH_3$$

Radical addition reactions involve the addition of a radical to an unsaturated substrate such as an alkene. These reactions are of great importance in the preparation of plastics such as polystyrene and polypropylene, and we will study them in Chapter 34. Up to this point, however, we have seen only one radical addition reaction—the peroxide-catalyzed non-Markovnikov addition of HBr to alkenes. This addition is a **chain reaction** and involves three distinct kinds of steps: (1) initiation, (2) propagation, and (3) termination.

1. Initiation steps:

a. $RO{-}OR \xrightarrow{\text{Heat}} 2\,RO\cdot$

A peroxide

b. $RO\cdot + H{-}Br \longrightarrow RO{-}H + Br\cdot$

2. Propagation steps:

a. $\text{>C=C<} + \boxed{Br\cdot} \longrightarrow \text{>}\dot{C}{-}C(Br)\text{<}$

b. $\text{>}\dot{C}{-}C(Br)\text{<} + H{-}Br \longrightarrow \text{–}C(H){-}C(Br)\text{–} + \boxed{Br\cdot}$

3. Termination steps:

a. $Br\cdot + Br\cdot \longrightarrow Br{-}Br$

b. $\text{–}C\cdot + Br\cdot \longrightarrow \text{–}C{-}Br$

c. $\text{–}C\cdot + \text{–}C\cdot \longrightarrow \text{–}C{-}C\text{–}$

The reaction is initiated by thermal homolytic cleavage of a peroxide, which forms two radicals. These radicals abstract H· from HBr, yielding a Br· radical. The Br· adds to the alkene, generating a new carbon radical and a carbon–bromine bond. Each reactant donates one electron to the new carbon–bromine bond. The reaction is then completed by reaction of the carbon radical with HBr to yield neutral product and a bromine radical to continue the chain. Note that the new carbon–hydrogen bond is also formed by donation of one electron from each reactant.

Radical substitution reactions are also common, and we have seen an example in the NBS bromination of allylic and benzylic positions. The key feature of these reactions is that one radical abstracts an atom from a neutral molecule, leaving a new radical. For example, NBS allylic brominations involve a step in which a bromine radical abstracts an allylic hydrogen atom from an alkene. The new carbon radical then reacts further to abstract a bromine atom from Br_2, leaving a new bromine radical to carry on the chain. The net effect of these two abstractions is radical substitution.

Propagation steps in NBS brominations

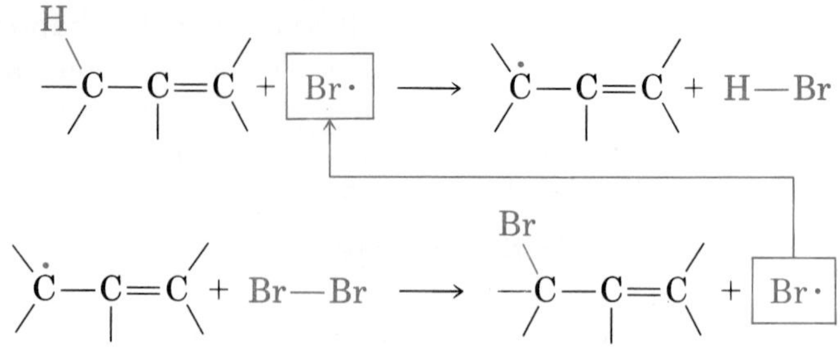

Radical couplings are the third category of radical reactions. The reaction of a diorganocopper reagent with an alkyl halide (reaction 3) is thought to involve the coupling of two radicals in the key step, with formation of a new bond, but the precise sequence of steps leading to radical coupling is not well understood.

The key point about all of these radical reactions is that they are *fundamentally different* from polar reactions. Radical reactions involve highly energetic odd-electron species, whereas polar reactions involve electrostatic attraction between electron-rich and electron-poor species.

17.3 Pericyclic reactions

Pericyclic reactions such as the Diels–Alder cycloaddition involve neither radicals nor nucleophile–electrophile interactions. Rather, they take place in a single step by a reorganization of bonding electrons through a cyclic transition state.

1,3-Butadiene

Methyl propenoate

Cyclic transition state

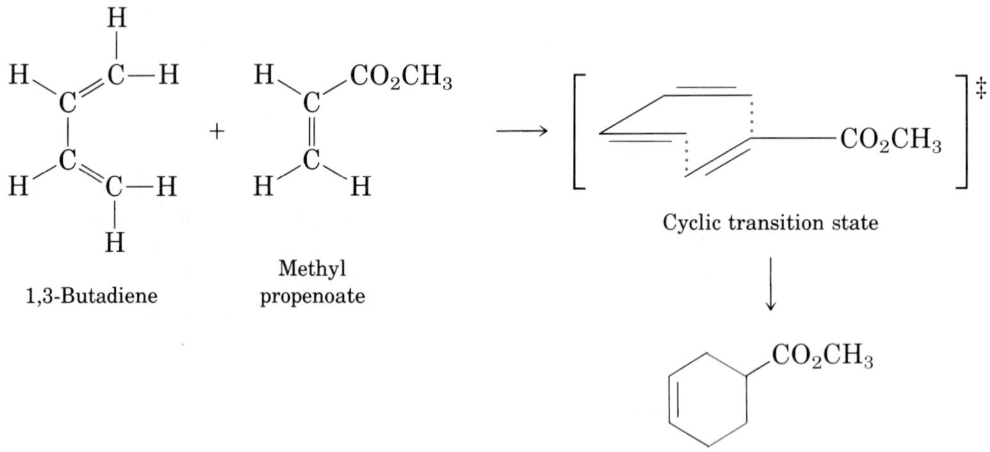

A fuller discussion of pericyclic reactions will be given in Chapter 30, and we will point out other examples as they occur.

PROBLEMS

17.1 Identify the electrophile and the nucleophile in these reactions:

(a) $:I^- + CH_3Cl \longrightarrow CH_3I + :Cl^-$

(b) [benzene] + Br_2 $\xrightarrow{AlBr_3}$ [bromobenzene, Br] + HBr

(c) [cyclohexene] + $CH_3\overset{O}{\overset{\|}{C}}OOH$ $\longrightarrow$ [cyclohexene oxide, O] + CH_3COOH

(d) [cyclohexene oxide, O] + HBr $\longrightarrow$ [2-bromocyclohexanol, OH, Br]

(e) $CH_3C{\equiv}C:^- + CO_2 \longrightarrow CH_3C{\equiv}C{-}CO_2^-$

(f) $CH_3S:^- + CH_3CH_2\overset{O}{\overset{\|}{C}}{-}O{-}CH_3 \longrightarrow CH_3{-}S{-}CH_3 + CH_3CH_2\overset{O}{\overset{\|}{C}}{-}O:^-$

(g) $:N_3^- + CH_3CH_2Br \longrightarrow CH_3CH_2N_3 + :Br^-$

(h) $3\ CH_3CH{=}CH_2 + BH_3 \longrightarrow (CH_3CH_2CH_2)_3B$

(i) [F, NO_2, NO_2 benzene] + $CH_3O:^-$ $\longrightarrow$ [OCH_3, NO_2, NO_2 benzene] + $:F^-$

17.2 Which of the following reagents would you expect to behave as nucleophiles and which as electrophiles? Might some reagents be either, depending on the reaction?

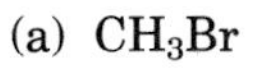

(a) CH_3Br (b) CH_3MgBr (c) $CH_3\overset{:O:}{\overset{\|}{C}}CH_3$

(d) [benzene] (e) $CH_3\ddot{N}H_2$ (f) [aniline, $\ddot{N}H_2$]

(g) $^-:CN$ (h) $CH_3C{\equiv}CH$ (i) $CH_3C{\equiv}C:^-$

(j) CO_2 (k) [benzonitrile, $C{\equiv}N:$] (l) $(CH_3)_3C^+$

17.3 Ketones are weak acids that can be converted into their anions **(enolate anions)** by treatment with strong base. Ketone anions undergo alkylation when treated with alkyl halides. Propose a mechanism for this alkylation and categorize it by type.

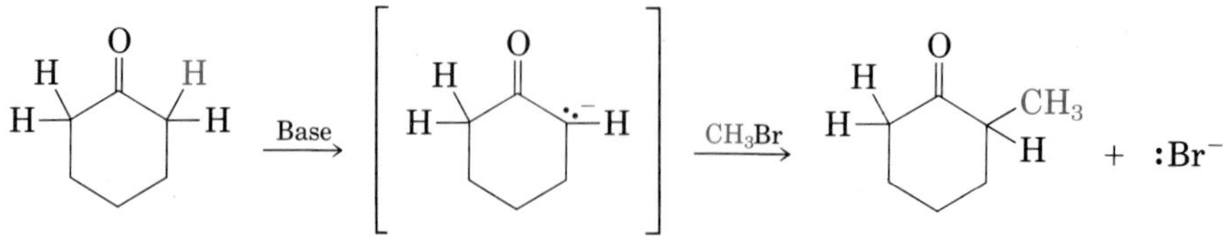

Ketone $\xrightarrow{\text{Base}}$ [Enolate anion] $\xrightarrow{CH_3Br}$ (2-methylcyclohexanone) + $:Br^-$

17.4 Ethylene and other alkenes polymerize when treated with a radical initiator. Propose a mechanism for this reaction.

$$H_2C{=}CH_2 \xrightarrow[\text{initiator}]{\text{Radical}} \text{Rad}{-}CH_2CH_2CH_2CH_2CH_2CH_2{-}$$

Polyethylene

17.5 The Cope rearrangement involves the thermal reaction of a 1,5-diene to yield a new 1,5-diene.

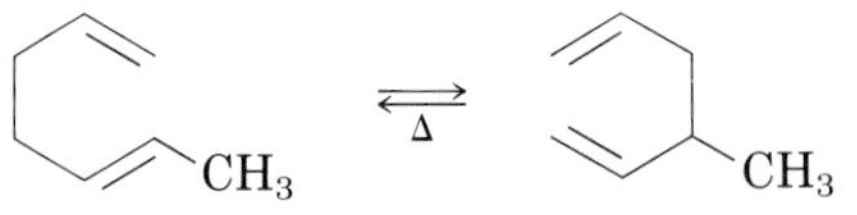

A 1,5-diene (CH_3) $\underset{\Delta}{\rightleftarrows}$ A new 1,5-diene (CH_3)

Propose a mechanism for this reaction, and identify it as polar, radical, or pericyclic.

17.6 We will see in Section 20.10 that alcohols can be easily oxidized to ketones by CrO_3 in pyridine. The reaction has been shown to proceed through formation of an intermediate chromate:

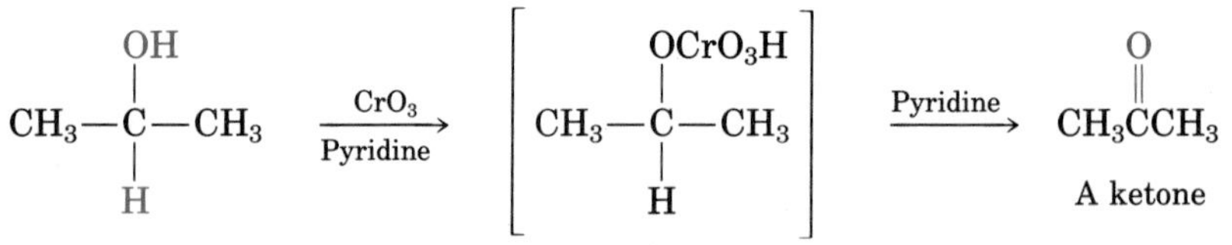

$$CH_3{-}\underset{H}{\overset{OH}{C}}{-}CH_3 \xrightarrow[\text{Pyridine}]{CrO_3} \left[CH_3{-}\underset{H}{\overset{OCrO_3H}{C}}{-}CH_3\right] \xrightarrow{\text{Pyridine}} CH_3\overset{O}{\overset{\|}{C}}CH_3$$

An alcohol — Chromate intermediate — A ketone

Propose a mechanism for the transformation of the chromate intermediate into ketone product. Into what category of reaction does this fall?

CHAPTER 18 ALICYCLIC MOLECULES

Up to this point we have made little distinction between open-chain and cyclic compounds. Much of the chemistry we have discussed is equally applicable to aliphatic and to **alicyclic** (aliphatic cyclic) molecules. There are, however, some special characteristics of alicyclic compounds that affect their chemistry, and we will now consider some of these features.

The majority of organic chemicals contain rings, and alicyclic compounds with many different ring sizes abound in nature. For example, chrysanthemic acid contains a three-membered ring (cyclopropane). Various esters of chrysanthemic acid occur naturally as the active insecticidal constituents of pyrethrum flowers.

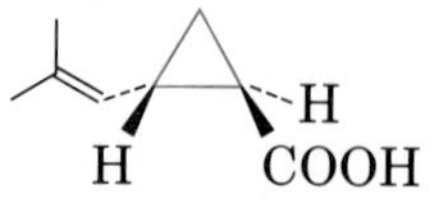

H COOH H

Chrysanthemic acid

The prostaglandins such as PGE_1 contain a five-membered ring (cyclopentane). Prostaglandins are potent hormones that control a wide variety of physiological functions in humans, including blood platelet aggregation, bronchial dilation, and inhibition of gastric secretions.

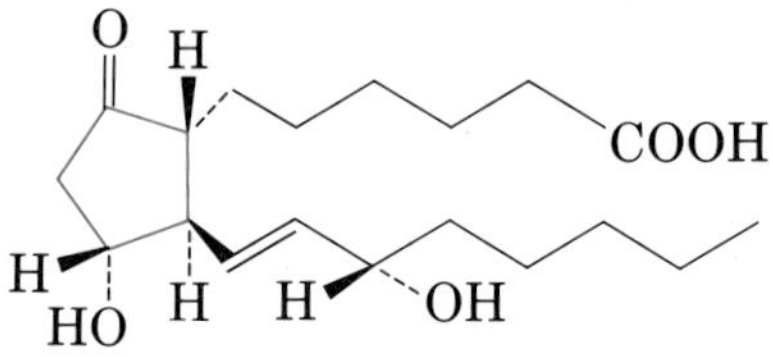

O H COOH H HO H H OH

Prostaglandin E_1 (PGE_1)

Muscone (3-methylcyclopentadecanone), highly valued for its use in perfumery, is isolated from a special gland of the male Tibetan musk deer and possesses a 15-membered ring.

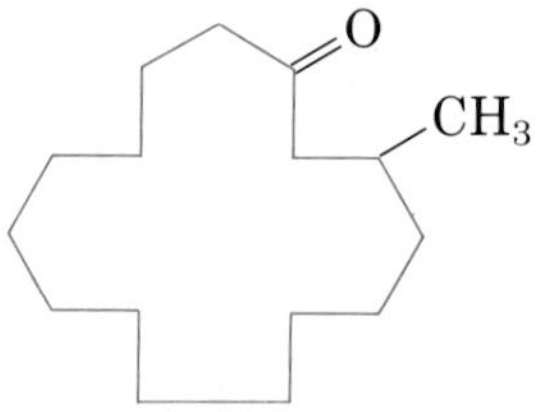

O CH_3

Muscone

The steroid hormones such as cortisone consist of four rings fused together into a compact **polycyclic** array.

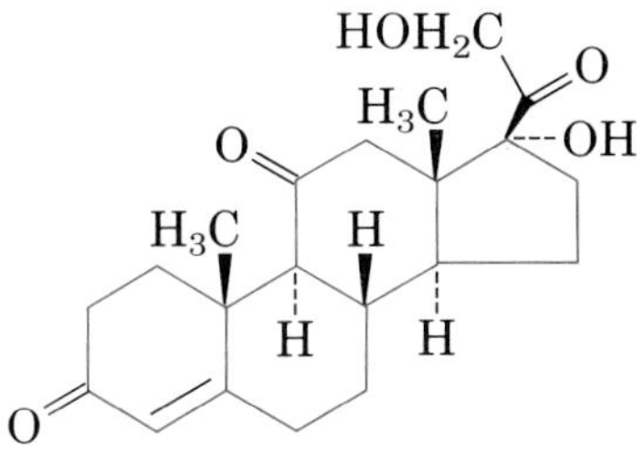

Cortisone

The chemistry of these alicyclic molecules can provide a fascinating field of study for the organic chemist.

18.1 Nomenclature of alicyclic molecules

We have already discussed the systematic nomenclature of cycloalkanes (Section 3.11) and will give only a brief review here:

1. Use the cycloalkane name as the base name. Compounds are normally named as alkyl-substituted cycloalkanes rather than as cycloalkyl-substituted alkanes. The only exception occurs when the alkyl side chain contains a greater number of carbon atoms than the ring. In such cases, the ring is considered a substituent on the parent acyclic alkane:

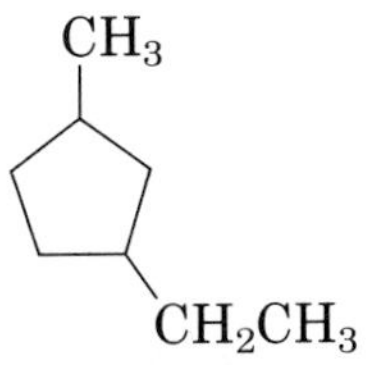

1-Ethyl-3-methylcyclopentane

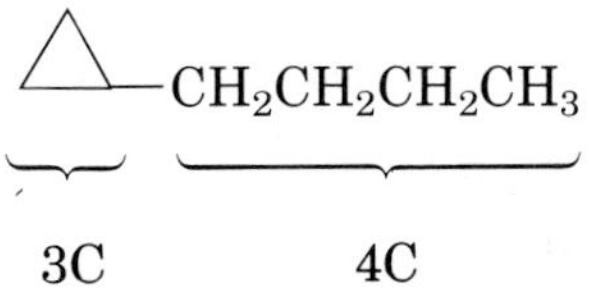

1-Cyclopropylbutane

2. Number the substituents on the ring so as to arrive at the lowest sum. Functional groups such as hydroxyl, amino, and cyano take precedence over alkyl groups and are given lower numbers.

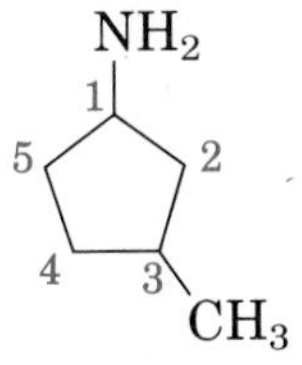

3-Methylcyclopentanamine
(*not* 1-methyl-3-cyclopentanamine)

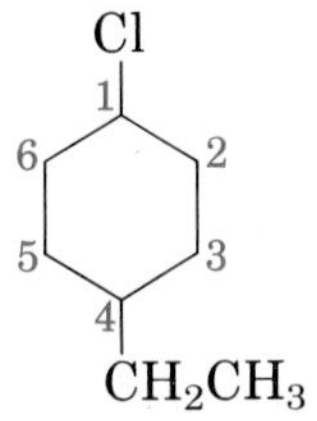

1-Chloro-4-ethylcyclohexane
(*not* 4-chloro-1-ethylcyclohexane)

PROBLEM

18.1 Name these compounds according to the IUPAC system:

(a) 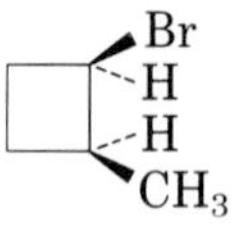 Br, H, H, CH_3

(b) 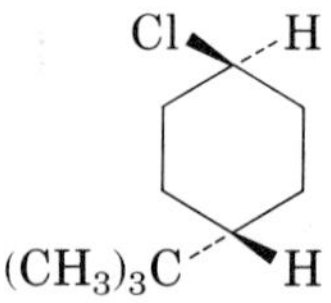 Cl, H, $(CH_3)_3C$, H

(c) 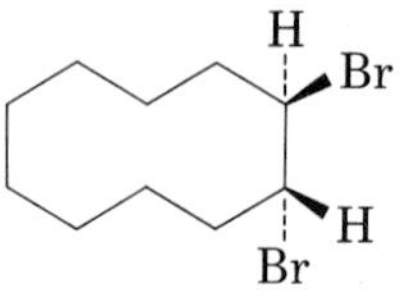 H, Br, H, Br

(d) 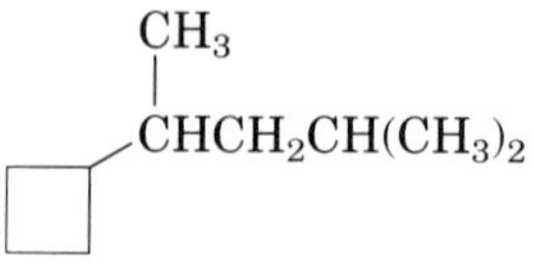 CH_3, $CHCH_2CH(CH_3)_2$

(e) 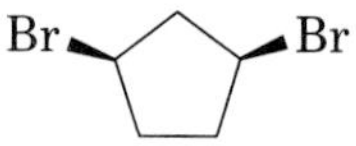 Br, Br

PROBLEM

18.2 Draw structures corresponding to these IUPAC names:
(a) 1,1-Dimethylcyclooctane
(b) 3-Cyclobutylhexane
(c) *trans*-1,2-Dichlorocyclopentane
(d) 1,3-Dibromo-5-methylcyclohexane (stereochemistry undefined)

18.2 Stability of cycloalkanes: the Baeyer strain theory

Chemists in the late 1800s had accepted the idea that cyclic molecules existed, but the limitations on feasible ring sizes were unclear. Numerous compounds containing five-membered and six-membered rings were known, but smaller and larger ring sizes had not been prepared. For example, no substituted cyclopropanes or cyclobutanes were known, despite numerous efforts to prepare them.

A theoretical interpretation of this observation was proposed in 1885 by Adolf von Baeyer.[1] Baeyer suggested that if carbon prefers to have tetrahedral geometry with bond angles of 109°, ring sizes other than five and six may be too *strained* to exist. Baeyer based his hypothesis on the simple geometric notion that an equilateral triangle (cyclopropane skeleton) must have bond angles of 60°, a square (cyclobutane skeleton) must have bond angles of 90°, a regular pentagon (cyclopentane skeleton) must have bond angles of 108°, and so on. According to this analysis, cyclopropane, with a bond-angle compression of $109° - 60° = 49°$, has a large amount of **angle strain** and would therefore be highly reactive. Cyclobutane ($109° - 90° = 19°$ angle strain) would be similarly reactive, but cyclopentane ($109° - 108° = 1°$ angle strain) would be nearly strain-free. Cyclohexane ($109° - 120° = -11°$ angle strain) would be somewhat strained, but cycloheptane

[1]Adolf von Baeyer (1835–1917); b. Berlin; Ph.D. Berlin (1858); professor, Berlin, Strasbourg (1872–1875), Munich (1875–1917); Nobel prize (1905).

($109° - 128° = -19°$ angle strain) and higher cycloalkanes would have bond angles that are forced to be too large. Carrying this line of reasoning further, Baeyer suggested that very large rings should be impossibly strained and incapable of existence.

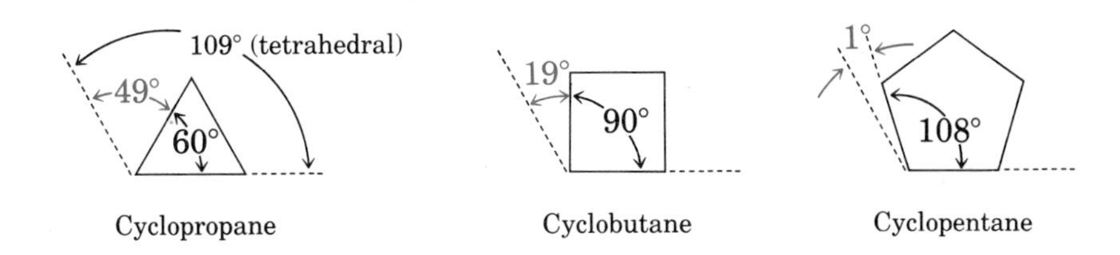

We know today that Baeyer's ideas were only partially correct. Rings of all sizes from 3 through 30 and beyond are known, but the concept of angle strain—the resistance of a bond angle to compression or expansion from the ideal tetrahedral angle—is a very useful one. Let's see what the facts are.

18.3 Heats of combustion of cycloalkanes

How can we measure the amount of strain in a cycloalkane? To do this, we must measure the total amount of energy in a compound and then subtract the amount of energy in a hypothetical strain-free reference compound. The difference between the two values should represent the amount of extra energy due to strain possessed by the molecule in question.

The simplest way to do this is to measure the **heats of combustion** of the cycloalkanes. The heat of combustion of a compound is the amount of heat (energy) released when the compound burns completely with oxygen:

$$—CH_2— + \tfrac{3}{2}O_2 \longrightarrow CO_2 + H_2O + \text{Heat}$$

The more energy (strain) the sample contains, the more energy (heat) is released on combustion. This relationship between strain and heat of combustion is shown schematically in Figure 18.1. If we compare the heats of combustion of two isomeric substances, more energy is released during combustion of the more strained substance because that compound has a higher energy level to begin with. Note that heats of combustion are a measure of the *enthalpy* change ($\Delta H°$) that occurs during oxidation, rather than of total free energy change ($\Delta G°$). Thus we are plotting only $\Delta H°$ in Figure 18.1 (p. 568).

Since the heat of combustion of a hydrocarbon depends on its molecular weight, it is more useful to look at heats of combustion per CH_2 unit. In this way, the size of the hydrocarbon is not a factor, and we can therefore compare cycloalkane rings of different sizes to a standard, strain-free, acyclic alkane. Table 18.1 shows the results of this comparison. Total strain energies are calculated by taking the difference between sample heat of combustion per CH_2 and reference heat of combustion per CH_2, and multiplying by the number of carbons, n, in the sample ring.

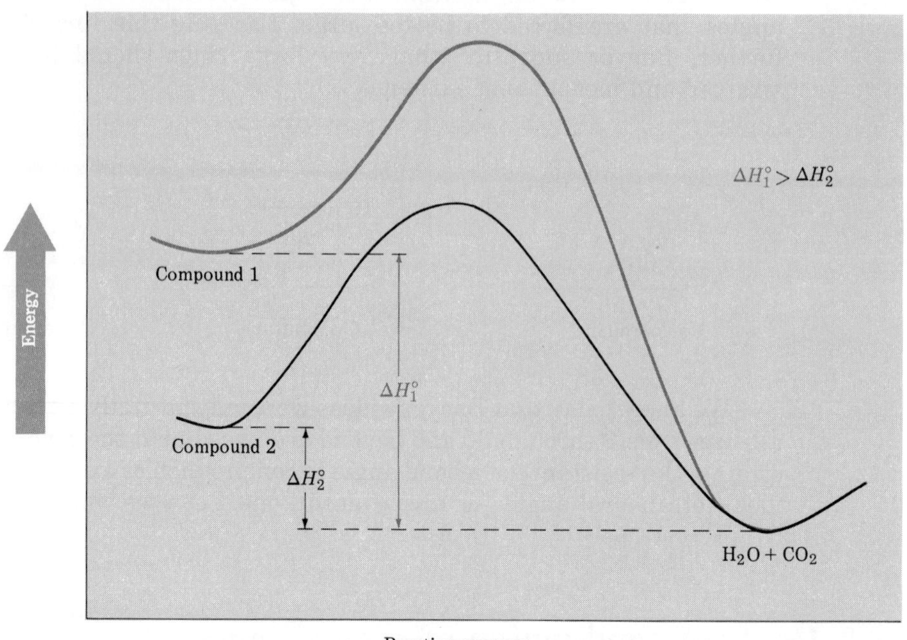

Figure 18.1. Comparison of heats of combustion of two substances. Compound 1 is more strained, and is higher in energy, than compound 2. The vertical axis represents the enthalpy change (ΔH°) during combustion.

TABLE 18.1 **Heats of combustion of cycloalkanes**

Cycloalkane $(CH_2)_n$	*Ring size, n*	*Heat of combustion (kcal/mol)*	*Heat of combustion per* CH_2 *(kcal/mol)*	*Total strain energy (kcal/mol)*
Cyclopropane	3	499.8	166.6	27.6
Cyclobutane	4	655.9	164.0	26.4
Cyclopentane	5	793.5	158.7	6.5
Cyclohexane	6	944.5	157.4	0
Cycloheptane	7	1108	158.3	6.3
Cyclooctane	8	1269	158.6	9.6
Cyclononane	9	1429	158.8	12.6
Cyclodecane	10	1586	158.6	12.0
Cycloundecane	11	1742	158.4	11.0
Cyclododecane	12	1891	157.6	2.4
Cyclotridecane	13	2051	157.8	5.2
Cyclotetradecane	14	2204	157.4	0
Alkane (reference)			157.4	0

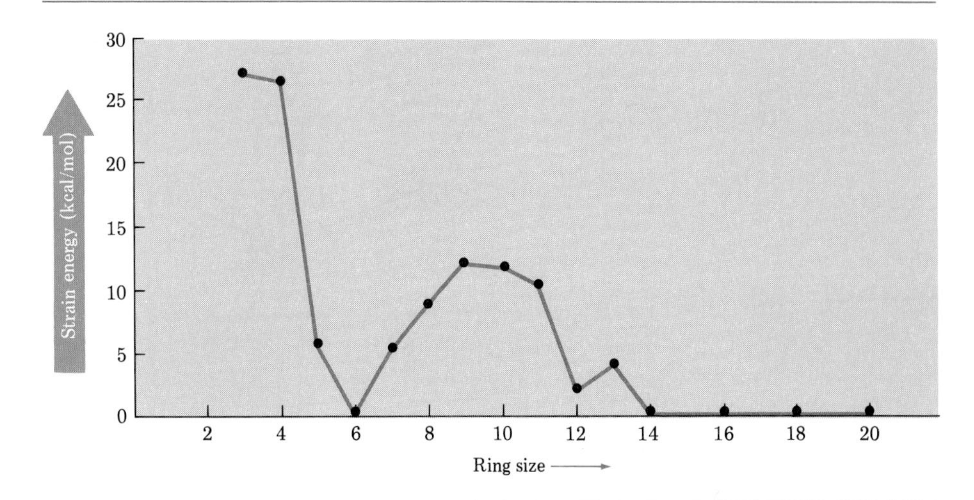

Figure 18.2. Cycloalkane strain energy as a function of ring size.

The data in Table 18.1 and the graph of Figure 18.2 show clearly that Baeyer's theory is not fully correct. Cyclopropane and cyclobutane are indeed quite strained, just as predicted. Cyclopentane, however, is more strained than predicted, and cyclohexane is perfectly strain-free. For rings of larger size, there is no regular increase in strain, and rings having more than 14 members are again strain-free. Why is Baeyer's theory wrong?

18.4 The nature of ring strain

Baeyer was wrong for a very simple reason—he assumed that all rings were flat. In fact, most cycloalkanes are not flat; they adopt puckered three-dimensional conformations, which allow bond angles to be nearly tetrahedral. Nevertheless, the concept of angle strain is a valuable one that goes far toward explaining the reactivity of three- and four-membered rings.

There are several other factors in addition to angle strain that are important in determining the shape and total strain energy of rings. One of these is **eclipsing strain** (also called **torsional strain**). Eclipsing strain was encountered earlier in our discussion of alkane conformations (Section 3.7). We said at that time that acyclic alkanes are most stable in the staggered conformation and least stable in the eclipsed conformation. A similar conclusion holds for cycloalkanes—eclipsing strain is present in a cycloalkane unless all the bonds have a staggered arrangement. For example, cyclopropane must have considerable eclipsing strain (in addition to angle strain), since C—H bonds on neighboring carbon atoms are eclipsed (Figure 18.3, p. 570). Larger cycloalkanes attempt to minimize this strain by adopting puckered, nonplanar conformations.

Steric strain (Section 3.9) is a third factor that contributes to the overall strain energy of a molecule. Two nonbonded groups repel each other if they approach too closely and attempt to occupy the same point in space.

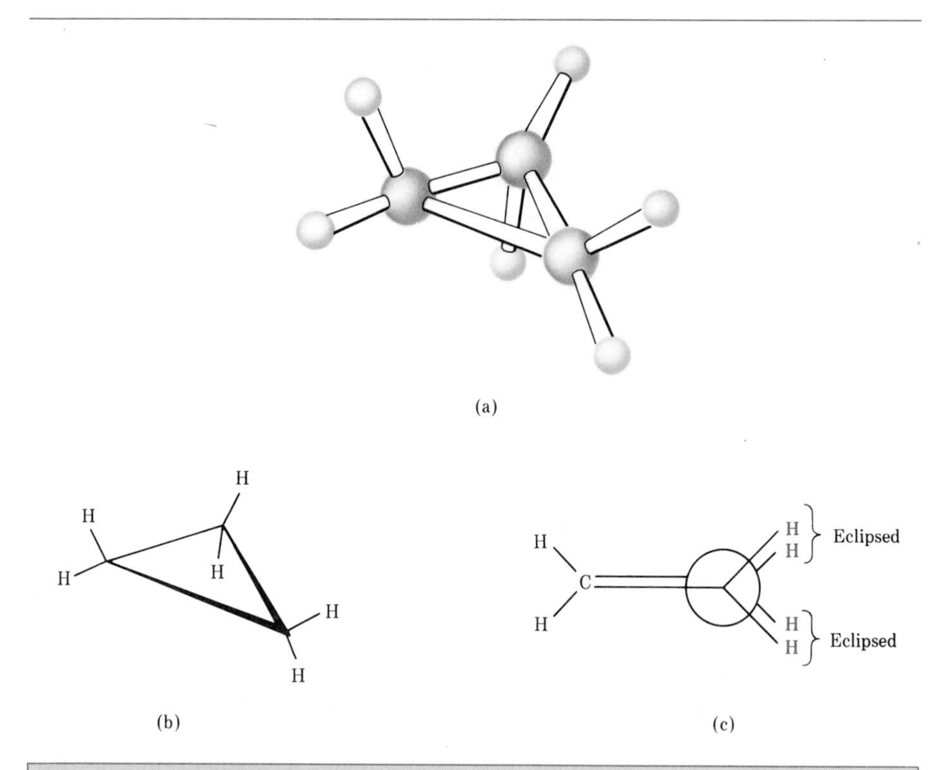

Figure 18.3. Conformation of cyclopropane. Part (c) is a Newman projection along a C—C bond.

Such nonbonded steric interactions are particularly important in determining the minimum-energy conformation of medium-ring (C_7–C_{11}) cycloalkanes.

In summary, the Baeyer theory is insufficient to explain the observed strain energies and geometries of cycloalkanes. Cycloalkanes adopt their minimum-energy conformations for a combination of reasons:

1. Angle strain, the strain due to expansion or compression of bond angles
2. Eclipsing strain, the strain due to eclipsing of neighboring bonds
3. Steric strain, the strain due to repulsive interaction of atoms approaching too closely

With this background, let's go on to examine the chemistry of the cycloalkanes.

PROBLEM

18.3 We saw in Section 3.7 that each hydrogen–hydrogen eclipsing interaction in the eclipsed conformation of ethane "costs" about 1.0 kcal/mol. How many such eclipsing interactions are present in cyclopropane? What fraction of the overall 27.6 kcal/mol strain energy of cyclopropane can be ascribed to eclipsing strain?

PROBLEM

18.4 *cis*-1,2-Dimethylcyclopropane has a higher heat of combustion than *trans*-1,2-dimethylcyclopropane. Can you account for this difference? Which of the two compounds is more stable?

PROBLEM

18.5 If both propane and cyclopropane were equally available and equally priced, which would be the more efficient fuel? Explain.

18.5 Cyclopropane: an orbital view

Cyclopropane is a colorless gas (bp = −33°C), that was first prepared by reaction of sodium with 1,3-dibromopropane:

$$BrH_2C-CH_2-CH_2Br \xrightarrow{2\ Na} \underset{\text{Cyclopropane}}{\text{cyclo-}(CH_2)_3} + 2\ NaBr$$

1,3-Dibromopropane → Cyclopropane

Since three points (the carbon atoms) define a plane, cyclopropane *must* be a flat, symmetrical molecule, and must have C—C—C bond angles of 60°. How, though, can molecular orbital theory account for this great distortion of the bonds from the normal 109° tetrahedral angle?

The answer is that cyclopropane may be viewed as having *bent bonds*. In an unstrained alkane, maximum bonding efficiency is achieved when two atoms are located so that the overlapping orbitals point directly toward each other. In cyclopropane, however, the orbitals cannot point directly toward each other; rather, they must overlap at a slight angle (Figure 18.4). The result of this poor overlap is that cyclopropane bonds are weaker and more reactive than normal alkane bonds.

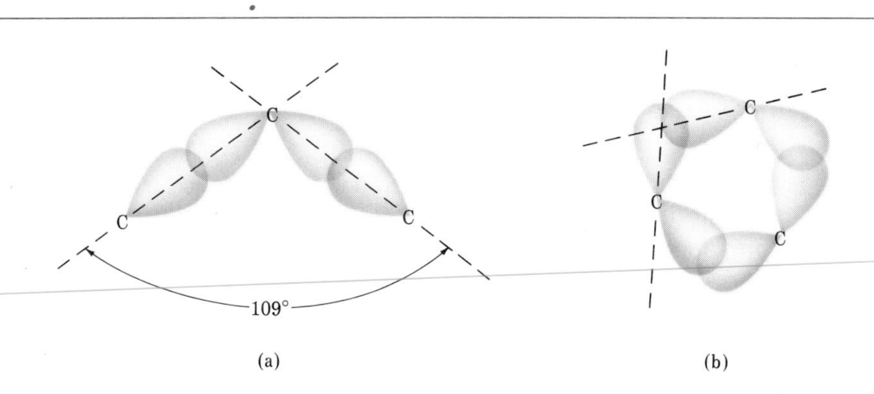

Figure 18.4. An orbital view of cyclopropane bent bonds: (a) normal alkane C—C—C bonds (good overlap); (b) cyclopropane bent bonds (poor overlap).

18.6 Preparation of cyclopropanes; carbenes

The most general method for preparing cyclopropanes is the addition of a **carbene** to an alkene. Carbenes are neutral, divalent carbon species that have only six electrons in their outer shell. They are therefore highly reactive and can be generated only as reaction intermediates, rather than as stable, isolable substances. Because they have only six valence electrons, carbenes are electron-deficient and behave as electrophiles.

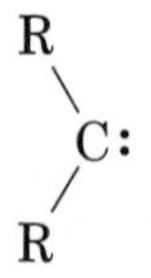

R_2C:

A carbene

One of the best methods for generating carbenes is the **alpha-elimination** route. For example, chloroform is weakly acidic ($pK_a = 25$) and, when treated with a strong base such as potassium *tert*-butoxide, is deprotonated, giving trichloromethanide anion, $^{-}:CCl_3$. This anion is unstable and expels a chloride ion, yielding dichlorocarbene, $:CCl_2$ (Figure 18.5). The overall loss of HCl from chloroform is referred to as an alpha elimination because both H and Cl are lost from the *same* carbon atom (as opposed to beta eliminations such as the E2 reaction, where H and X are lost from *neighboring* carbon atoms).

The dichlorocarbene carbon atom is sp^2 hybridized. A vacant p orbital extends above and below the plane of the three atoms, and the unshared pair of electrons occupies the third sp^2 lobe. Note that this electronic description of dichlorocarbene is similar to what we saw earlier for carbocations (Section 5.8), with respect to both hybridization of carbon and the presence of a vacant p orbital.

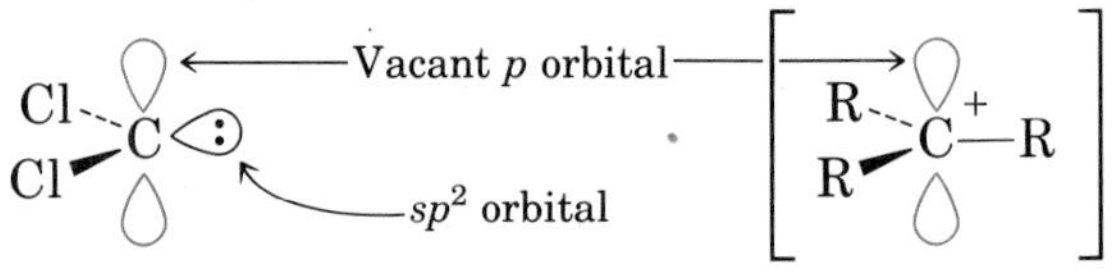

Vacant p orbital

sp^2 orbital

Dichlorocarbene

A carbocation (sp^2 hybridized)

If dichlorocarbene is generated in the presence of an alkene, insertion of the electrophilic carbene into the double bond occurs, and a dichlorocyclopropane is formed:

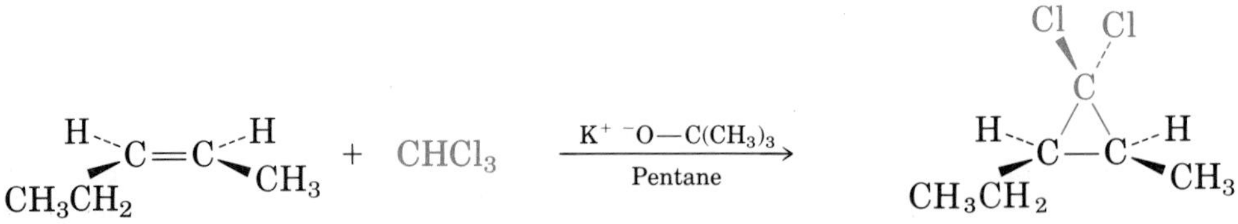

$$\textit{cis}\text{-2-Pentene} + CHCl_3 \xrightarrow[\text{Pentane}]{K^+\ ^-O{-}C(CH_3)_3} \text{1,1-dichloro-2-ethyl-3-methylcyclopropane (65\%)}$$

cis-2-Pentene

65%

$$\text{Cyclohexene} + CHCl_3 \xrightarrow[\text{Pentane}]{K^+\ ^-O{-}C(CH_3)_3} \text{7,7-dichlorobicyclo[4.1.0]heptane (60\%)}$$

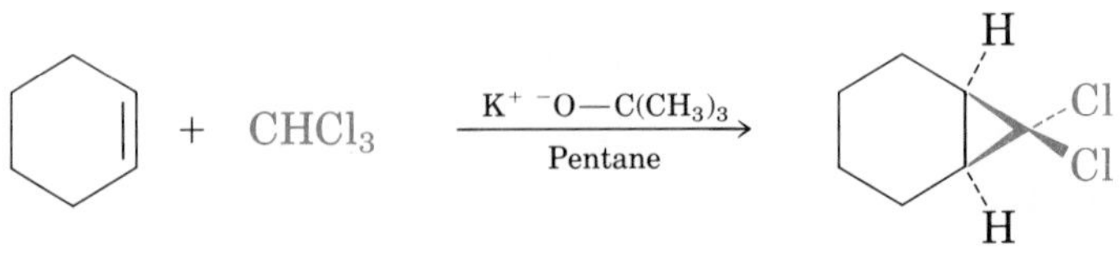

Cyclohexene

60%

Alkoxide base abstracts the chloroform proton, leaving behind the electron pair from the C—H bond and forming the trichloromethanide anion.

$$\mathrm{CHCl_3} \xrightarrow[\text{Pentane}]{\overset{+}{K}\ \overset{-}{:O}-C(CH_3)_3} [\mathrm{Cl_3C:^-}]$$

Trichloromethanide anion

Loss of a chloride ion and associated electrons from the C—Cl bond yields the neutral dichlorocarbene.

$$[\mathrm{Cl_3C:^-}] \longrightarrow \mathrm{Cl_2C:} + \mathrm{:Cl^-}$$

$$\text{H–C}_\alpha\text{–X} \xrightarrow[-HX]{\text{Base}} \text{C:} \qquad \text{Alpha elimination}$$

$$\text{H–C}_\beta\text{–C}_\alpha\text{–X} \xrightarrow[-HX]{\text{Base}} \text{C=C} \qquad \text{Beta elimination}$$

Figure 18.5. Mechanism of the formation of dichlorocarbene from chloroform.

The reaction of dichlorocarbene with *cis*-2-pentene is particularly interesting, since it demonstrates that the addition is stereospecific. Starting from the cis alkene, only cis-disubstituted cyclopropane is produced.

The best method for preparing nonhalogenated cyclopropanes is the Simmons–Smith reaction. This reaction, first investigated at Du Pont, does not involve a free carbene. Rather, it utilizes a **carbenoid**—a reagent with carbene-like reactivity. When diiodomethane is treated with a specially prepared zinc–copper couple, (iodomethyl)zinc iodide is formed. If an alkene is present, (iodomethyl)zinc iodide transfers a CH_2 group to the double bond and yields the cyclopropane. For example, both cyclohexene and *p*-methylstyrene react cleanly and in good yield to give the corresponding cyclopropanes. Aromatic rings, however, are not attacked by the reagent.

$$CH_2I_2 + Zn(Cu) \xrightarrow{\text{Ether}} I{-}CH_2{-}Zn{-}I = \text{“}{:}CH_2\text{”}$$

Diiodomethane — (Iodomethyl)zinc iodide (a carbenoid)

$$\text{Cyclohexene} + CH_2I_2 \xrightarrow[\text{Ether}]{\text{Zn(Cu)}} \text{Bicyclo[4.1.0]heptane (92\%)} + ZnI_2$$

$$p\text{-Methylstyrene} + CH_2I_2 \xrightarrow[\text{Ether}]{\text{Zn(Cu)}} p\text{-Cyclopropylmethylbenzene (67\%)} + ZnI_2$$

PROBLEM

18.6 Dichlorocarbene can also be generated by heating sodium trichloroacetate:

$$CCl_3CO_2^- \ Na^+ \xrightarrow{70°C} :CCl_2 + CO_2 + Na^+ \ Cl^-$$

Propose a mechanism to account for this reaction, and use curved arrows to indicate the movement of electrons. What relation does your mechanism bear to the alpha elimination of HCl from chloroform?

PROBLEM

18.7 Simmons–Smith reaction of cyclohexene with diiodomethane gives a single cyclopropane product. Reaction with 1,1-diiodoethane, however, gives (in low yield) a mixture of two isomeric methylcyclopropane products. Formulate the reactions, and account for the formation of the product mixture.

18.7 Reactions of cyclopropanes

What kind of reactivity would we expect cyclopropanes to have? The orbital picture of cyclopropane we discussed in Section 18.6 indicates that the bent cyclopropane bonds are weaker (more strained) than normal alkane sigma bonds, and that the sigma electrons are more sterically accessible to attacking reagents than normal. For both of these reasons, we might predict that cyclopropanes should be unusually reactive toward electrophiles. This is exactly what is found.

Although cyclopropanes are generally less reactive than alkenes, the reactivity of the cyclopropane ring toward electrophiles mimics the reactivity of the alkene pi bond in many respects. For example, cyclopropanes undergo ring-opening addition reactions with acids in much the same way that alkenes do. Thus, 1,1-dimethylcyclopropane reacts with HBr to give 2-bromo-2-methylbutane, the product of Markovnikov 1,3 addition. Presumably the reaction occurs through the intermediate tertiary carbocation, although the mechanism has not been studied in great detail (Figure 18.6). Bromine and chlorine also add to cyclopropanes, but complex mixtures of reaction products are usually obtained.

An electron pair from one of the bent cyclopropane sigma bonds attacks H^+, forming a tertiary carbocation intermediate.

This tertiary carbocation intermediate accepts an electron pair from bromide ion to yield neutral addition product.

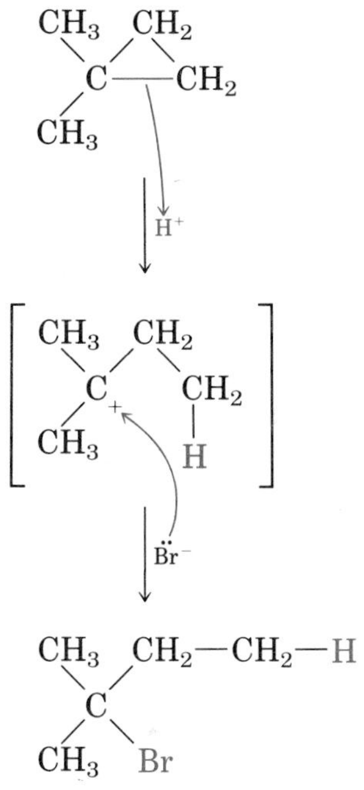

Recall:

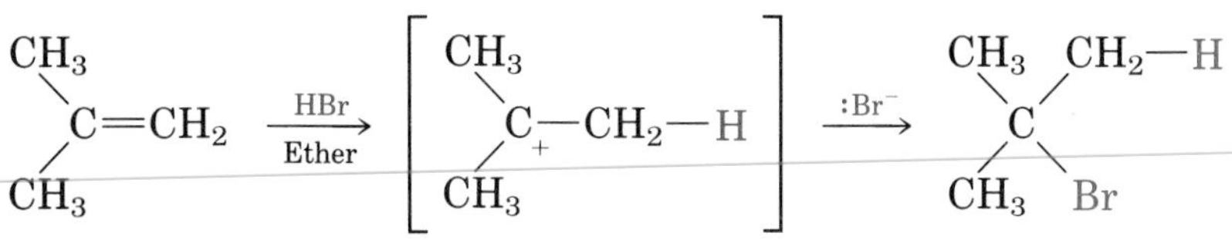

Figure 18.6. Mechanism of addition of HBr to cyclopropane.

Cyclopropanes undergo other alkene-like addition reactions, including catalytic addition of hydrogen. The bond between the least substituted carbon atoms is cleaved, as the following example illustrates.

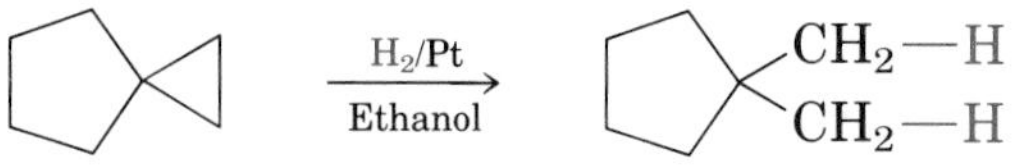

Spiro[4.2]hexane $\xrightarrow[\text{Ethanol}]{H_2/Pt}$ 1,1-Dimethylcyclopentane (91%)

18.8 Cyclobutane and cyclopentane

Cyclobutane has nearly the same total amount of strain as cyclopropane (26.4 kcal/mol for cyclobutane versus 27.6 kcal/mol for cyclopropane). The reason for this is that cyclobutane has more eclipsing strain than cyclopropane by virtue of its larger number of ring hydrogens, even though it has less angle strain. Spectroscopic measurements indicate that cyclobutane is not quite flat, but slightly bent so that one carbon atom lies about 25° above the plane of the other three (Figure 18.7). The effect of this slight bend is to *increase* angle strain but to *decrease* eclipsing strain, until a minimum-energy balance between the two opposing effects is achieved.

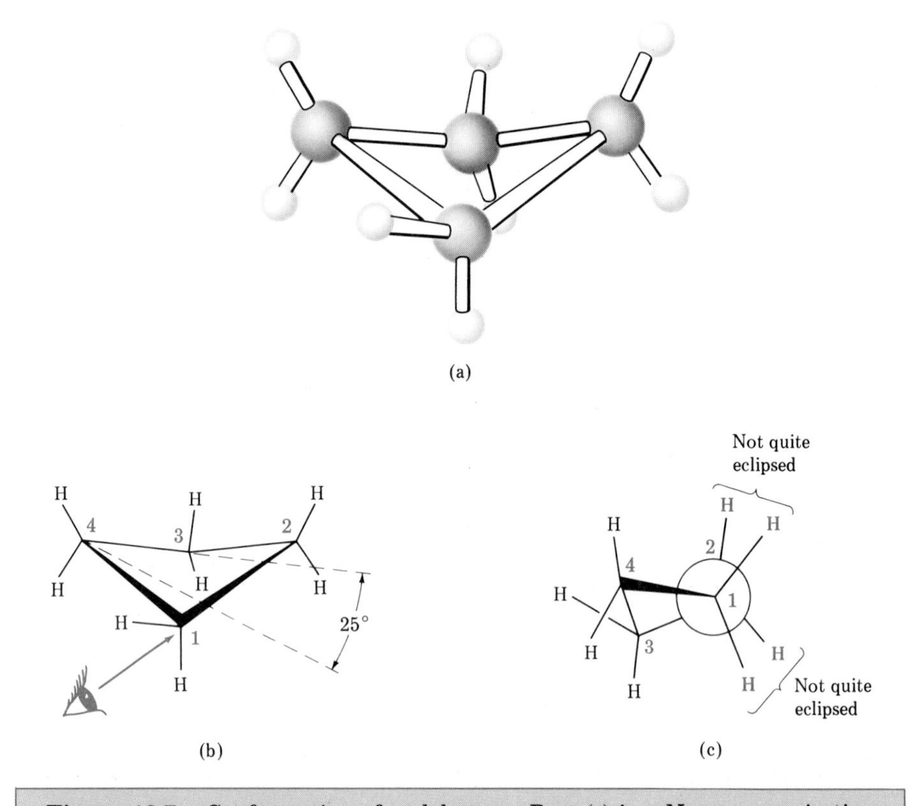

Figure 18.7. Conformation of cyclobutane. Part (c) is a Newman projection along the C1—C2 bond.

Cyclopentane was predicted by Baeyer to be nearly strain-free, but heat-of-combustion data indicate that this is not the case. Although cyclopentane has practically no angle strain, the 10 pairs of neighboring hydrogens introduce considerable eclipsing strain into a planar conformation. Cyclopentane therefore adopts a puckered, out-of-plane conformation that strikes a balance between increased angle strain and decreased eclipsing strain. Four of the cyclopentane carbon atoms are in approximately the same plane, and the hydrogens of the out-of-plane methylene group are nearly staggered with respect to their neighbors (Figure 18.8).

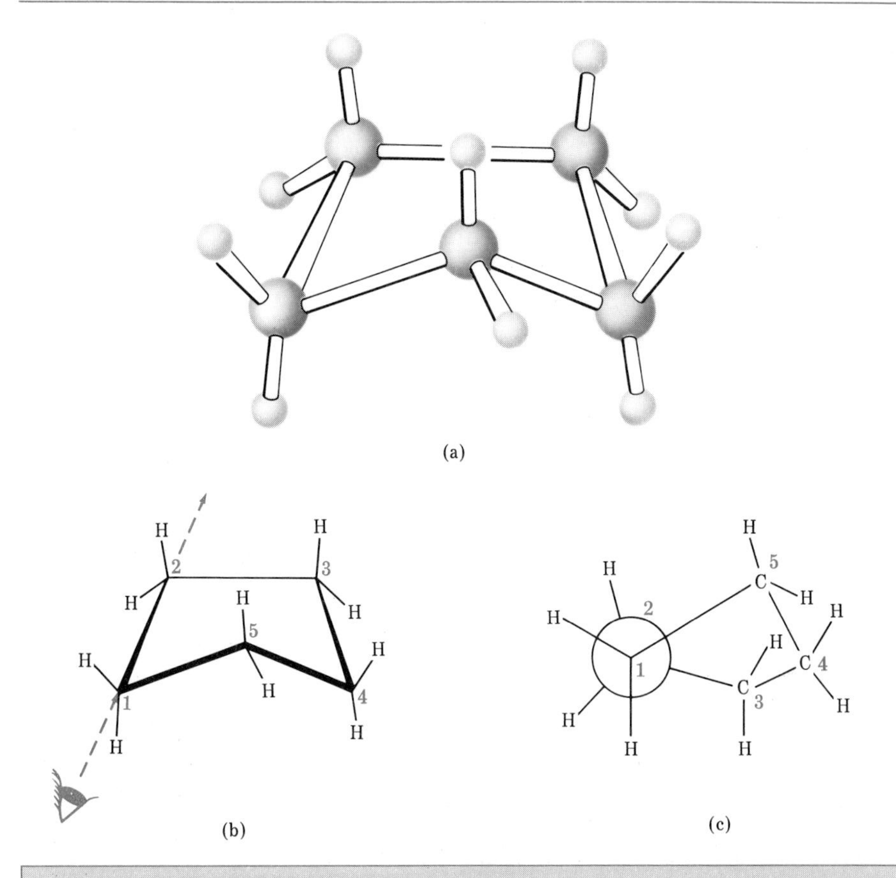

Figure 18.8. Conformation of cyclopentane: carbons 2, 3, 4, and 5 are nearly planar, but carbon 1 is out of the plane. Part (c) is a Newman projection along the C1—C2 bond.

PROBLEM

18.8 Account for the fact that *cis*-1,2-dimethylcyclobutane is less stable than its trans isomer, but *cis*-1,3-dimethylcyclobutane is more stable than its trans isomer. Draw your expectation of the favored conformation of *cis*-1,3-dimethylcyclobutane.

PROBLEM

18.9 How many hydrogen–hydrogen eclipsing interactions would be present if cyclopentane were planar? Assuming an energy cost of 1.0 kcal/mol for each eclipsing interaction, how much total strain would you expect cyclopentane to have? How much of this strain is relieved by puckering if the measured total strain is 6.5 kcal/mol?

18.9 Cyclohexane conformation

Cyclohexane rings are the most important of all cycloalkanes because of their wide occurrence in nature. A large number of compounds, including many important pharmaceutical agents, contain cyclohexane rings. Experimental data show that cyclohexane rings are strain-free; they have neither angle strain nor eclipsing strain. Why should this be?

The answer to this question was first suggested in 1890 by H. Sachse and later expanded on by Ernst Mohr:[2] Cyclohexane rings are not flat; they are puckered into a strain-free, three-dimensional conformation. The C—C—C angles of cyclohexane can reach the strain-free 109° tetrahedral angle if the ring adopts what is called the **chair conformation.** Furthermore, if we sight along any one of the carbon–carbon bonds in a Newman projection, we find that chair cyclohexane has no eclipsing strain; all neighboring C—H bonds are perfectly staggered (Figure 18.9).

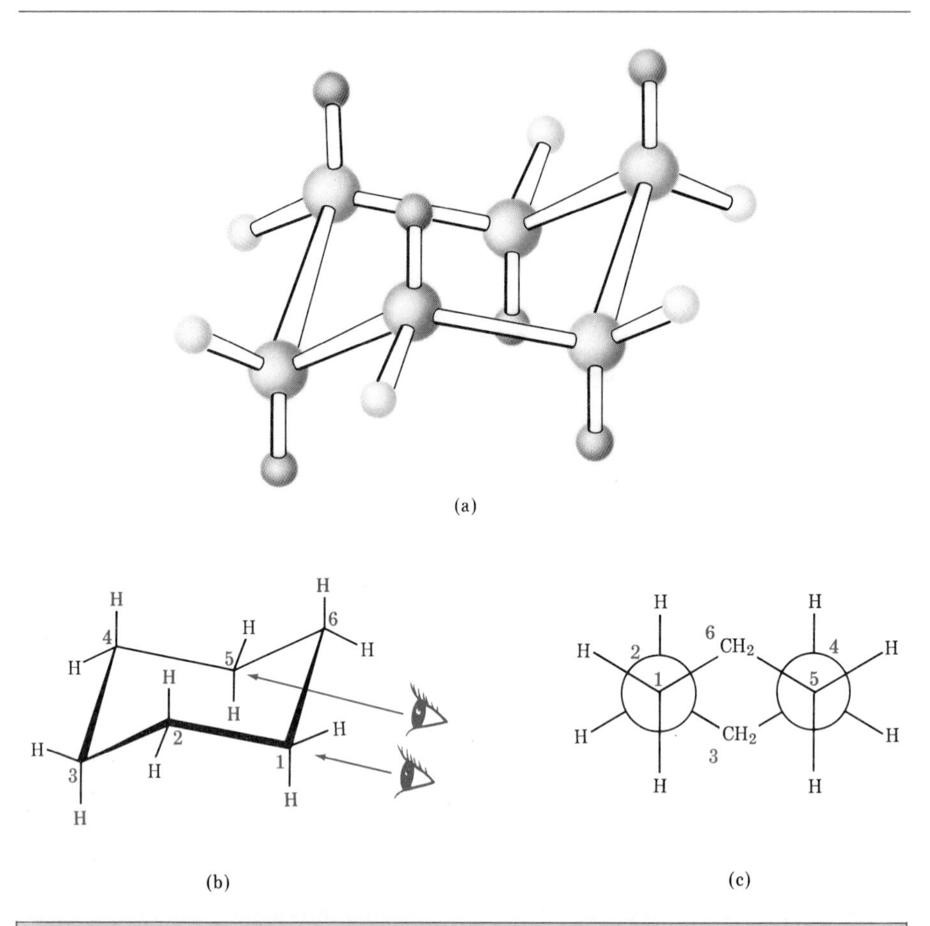

Figure 18.9. The chair conformation for cyclohexane. This conformation has no eclipsing strain and no angle strain. All C—C—C bond angles are 109°.

The simplest way to see this strain-free cyclohexane is to build and examine molecular models. Two-dimensional drawings such as Figure 18.9 are useful, but there is no substitute for holding, twisting, and turning a three-dimensional model in your hands.

The chair conformation of cyclohexane is of such great importance that all organic chemists must learn how to draw it properly. This is most simply done in three steps, as shown in Figure 18.10.

[2]Ernst Mohr (1873–1926); b. Dresden; Ph.D. Kiel (1897); professor, University of Heidelberg.

1. Draw two parallel lines, slanted downward and slightly offset from each other. This means that four of the cyclohexane carbon atoms lie in a plane.

2. Locate the topmost carbon atom above and to the right of the plane of the other four and connect the bonds.

3. Locate the bottommost carbon atom below and to the left of the plane of the middle four and connect the bonds. Note that the bonds to the bottommost carbon atom are parallel to the bonds to the topmost carbon.

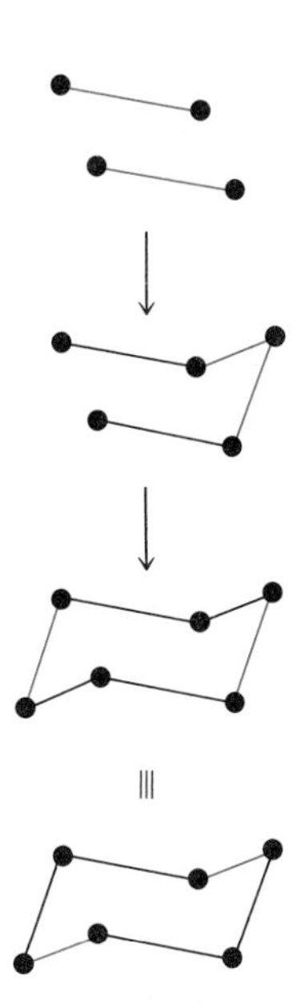

Figure 18.10. Drawing the cyclohexane chair conformation.

It is important to remember when viewing cyclohexane that the lower bond is considered to be in front; the upper bond is in back. If this convention is not defined, an optical illusion can make it appear that the reverse is true.

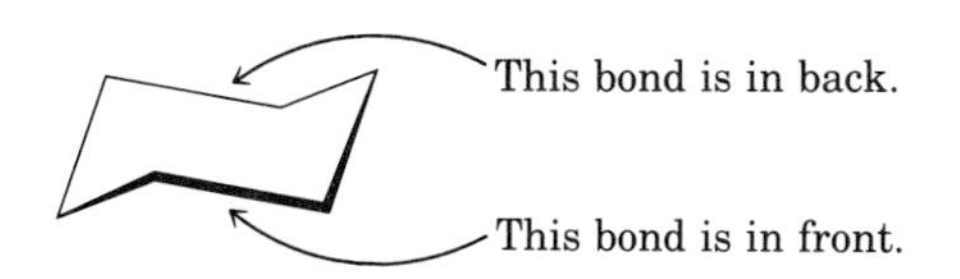
This bond is in back.
This bond is in front.

18.10 Axial and equatorial bonds

There are many consequences of the chair conformation of cyclohexane; we will see, for example, that the chemical behavior of substituted cyclohexanes is intimately involved with conformation. Another consequence of the chair cyclohexane conformation is that there are two kinds of hydrogen atoms on the ring—**axial hydrogens** and **equatorial hydrogens** (Figure 18.11, p. 580).

As Figure 18.11 indicates (and molecular models illustrate much better), cyclohexane has six axial hydrogens that are perpendicular to the ring (on the axis) and six equatorial hydrogens that are more or less in the rough plane of the ring (around the equator). Each carbon atom has one axial and one equatorial hydrogen. If we look only at the positions on the top face of the ring, as in Figure 18.12 (p. 581), we see an alternating axial–equatorial–axial–equatorial disposition of the groups; carbons 1, 3, and 5 have axial hydrogens on the top face, and carbons 2, 4, and 6 have equatorial hydrogens. Exactly the reverse is true for the bottom face; carbons 1, 3,

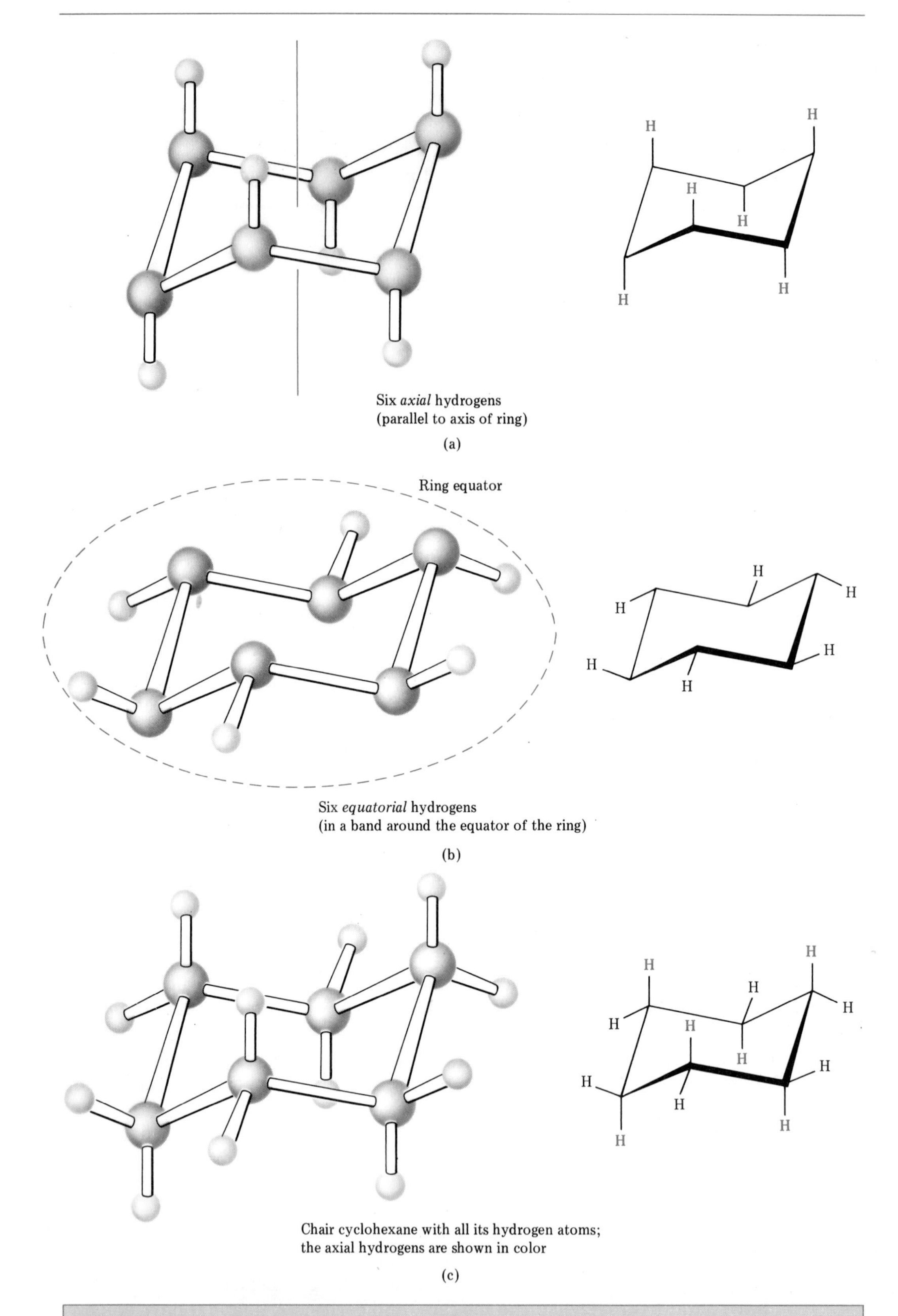

Six *axial* hydrogens
(parallel to axis of ring)
(a)
Ring equator
Six *equatorial* hydrogens
(in a band around the equator of the ring)
(b)
Chair cyclohexane with all its hydrogen atoms;
the axial hydrogens are shown in color
(c)

Figure 18.11. Axial and equatorial cyclohexane hydrogen atoms.

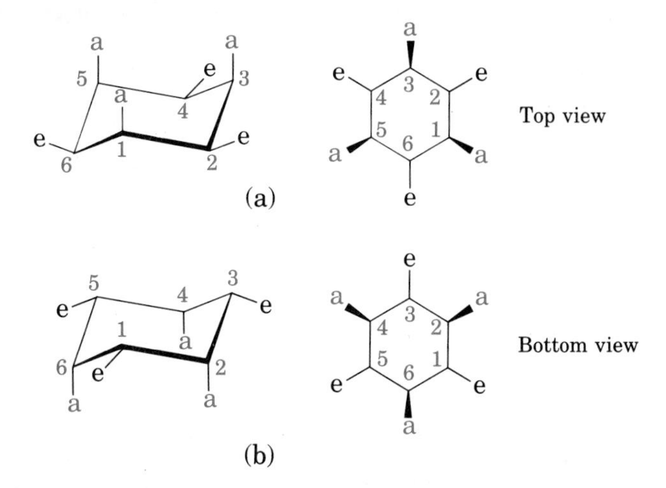

Figure 18.12. Alternating axial and equatorial positions on cyclohexane: (a) hydrogens on the top of the ring; (b) hydrogens on the bottom of the ring. In both views axial and equatorial positions alternate.

and 5 have equatorial hydrogens, and carbons 2, 4, and 6 have axial hydrogens.

It is important to practice drawing axial and equatorial bonds (Figure 18.13), and this is most easily done using the following guidelines (look at a molecular model as you practice):

1. *Axial bonds:* Keep in mind that all six axial bonds (one on each carbon) are parallel.
2. *Equatorial bonds:* Keep in mind that equatorial bonds come in three sets of two parallel lines, and that each set is also parallel to two ring bonds. (Parallel bonds are shown in color in Figure 18.13.)

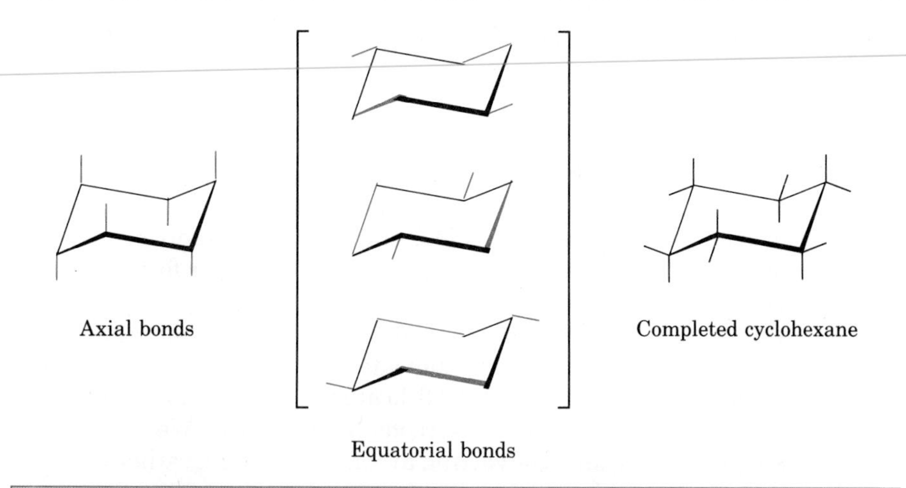

Figure 18.13. Drawing axial and equatorial chair cyclohexane bonds.

18.11 Conformational mobility of cyclohexane

How can we explain the fact that there is only one methylcyclohexane, one bromocyclohexane, one cyclohexanol, and so forth? In light of the fact that chair cyclohexane has two kinds of positions, axial and equatorial, we might expect a monosubstituted cyclohexane to exist in two isomeric forms, but this expectation is clearly wrong. The answer to this paradox is that cyclohexane is conformationally mobile. Different chair conformations can readily interconvert, with the result that axial and equatorial positions become interchanged.

This interconversion of chair conformations is usually referred to as a **ring-flip** and is shown in Figure 18.14. Molecular models show the process more clearly, and you should practice with models while studying this material.

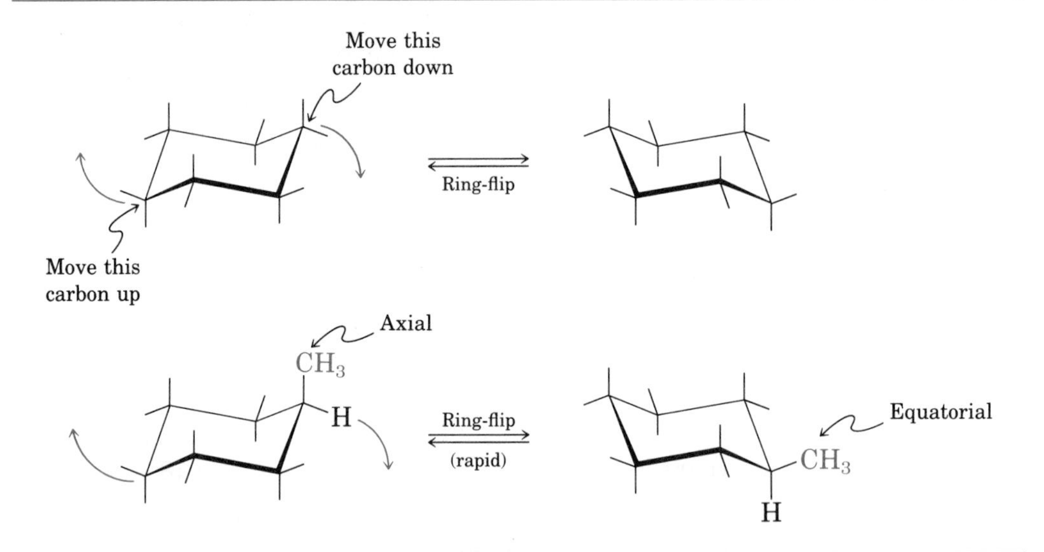

Figure 18.14. Cyclohexane ring-flip. Axial and equatorial positions interconvert.

We can mentally ring-flip a chair cyclohexane by holding the middle four carbon atoms in place, and folding the two ends in opposite directions. The net result of carrying out a ring-flip is the interconversion of axial and equatorial positions; an axial position in one chair form is equatorial in the ring-flipped chair form, and vice versa. For example, axial methylcyclohexane becomes equatorial methylcyclohexane after ring-flip. Spectroscopic measurements indicate that the barrier to chair–chair interconversions is about 10.8 kcal/mol, a value low enough to make the process extremely rapid at room temperature. We therefore see only a single, rapidly interconverting average structure, rather than distinct axial and equatorial isomers.

PROBLEM

18.10 Assume that you have a variety of cyclohexanes substituted in the positions indicated. Identify the substituents as either axial or equatorial. For example, a 1,2-cis relationship means that one substituent must be axial and one equatorial, whereas a 1,2-trans relationship means that both substituents are axial or both are equatorial.

(a) 1,3-Trans disubstituted
(b) 1,4-Cis disubstituted
(c) 1,3-Cis disubstituted
(d) 1,5-Trans disubstituted
(e) 1,5-Cis disubstituted
(f) 1,6-Trans disubstituted

PROBLEM

18.11 Draw two different chair conformations of *trans*-1,4-dimethylcyclohexane and label all positions as axial or equatorial.

18.12 Conformation of monosubstituted cyclohexanes

Let's look at the consequences of axial and equatorial cyclohexane bonds. We have said that cyclohexane rings rapidly flip conformations at room temperature. At any given instant, however, the substituent of a monosubstituted cyclohexane is either axial or equatorial. The two conformers are not equally stable; the equatorially substituted conformer is more stable than the axially substituted conformer. In methylcyclohexane, for example, the energy difference between axial and equatorial methyl is 1.8 kcal/mol. At any given instant, therefore, we know from Table 3.7 that about 95% of the methylcyclohexane molecules have the methyl group equatorial, and only 5% have the methyl axial (Figure 18.15).

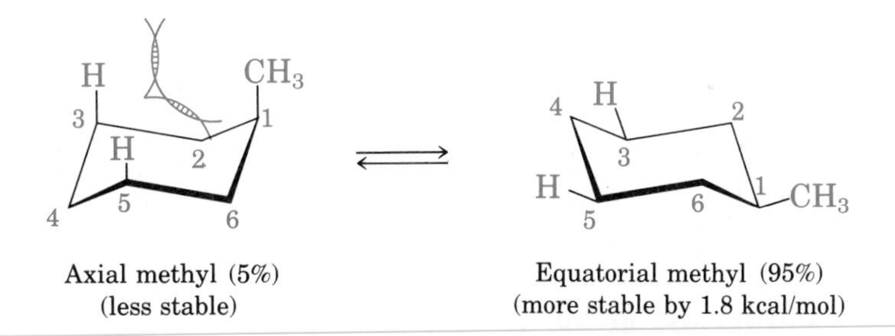

Figure 18.15. Axial and equatorial methylcyclohexane. The equatorial conformation is more stable by 1.8 kcal/mol.

The reason for this energy difference is steric strain due to **1,3-diaxial interactions.** The axial methyl group on carbon 1 suffers steric interference from the axial hydrogens three carbons away (carbons 3 and 5).

This kind of 1,3-diaxial steric strain is familiar to us, and we have seen it before in the guise of gauche butane steric strain (Section 3.9). Recall that during our discussion of alkane conformations we remarked that gauche butane is less stable than anti butane by 0.9 kcal/mol because of steric interference between hydrogen atoms on the two methyl groups. If we look at a four-carbon fragment of axial methylcyclohexane as a gauche butane, it is apparent that the steric interaction is the same. Since methylcyclohexane has two such interactions, however, there is $2 \times 0.9 = 1.8$ kcal/mol steric strain. The origin of 1,3-diaxial steric strain is shown in Figure 18.16.

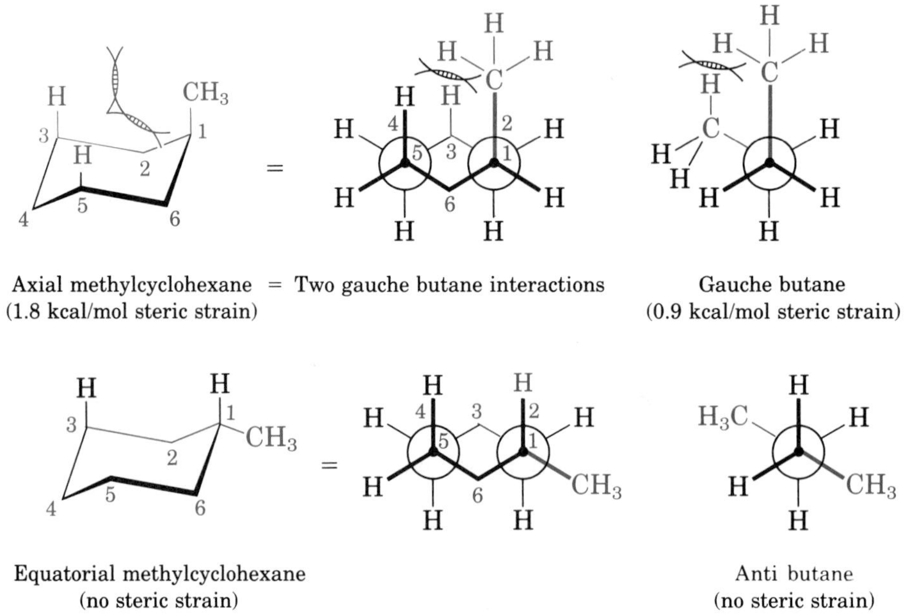

Axial methylcyclohexane = Two gauche butane interactions
(1.8 kcal/mol steric strain)

Gauche butane
(0.9 kcal/mol steric strain)

Equatorial methylcyclohexane
(no steric strain)

Anti butane
(no steric strain)

Figure 18.16. Origin of 1,3-diaxial cyclohexane interactions.

Sighting along the C1—C2 bond of axial methylcyclohexane shows that the axial hydrogen at C3 has a gauche butane interaction with the axial methyl group at C1. Sighting similarly along the C1—C6 bond shows that the axial hydrogen at C5 has a gauche butane interaction with the axial methyl group. Both of these interactions are absent in equatorial methylcyclohexane, and we therefore predict an energy difference of 1.8 kcal/mol between the two forms. Experiment agrees perfectly with this prediction.

What is true for methylcyclohexane is also true for all other monosubstituted cyclohexanes: A substituent is almost always more stable in an equatorial position than in an axial position. The precise amount of 1,3-diaxial steric strain depends on the size of the group, and Table 18.2 lists some values for common substituents. As would be expected, the amount of steric strain increases through the series $H_3C—$ < $CH_3CH_2—$ < $(CH_3)_2CH—$ << $(CH_3)_3C—$ in parallel with the bulk of the successively larger alkyl groups. Note that the values in Table 18.2 refer to 1,3-diaxial interactions with a single hydrogen atom. These values must therefore be doubled to arrive at the amount of strain in a monosubstituted cyclohexane.

TABLE 18.2 **Steric strain due to 1,3-diaxial interactions**

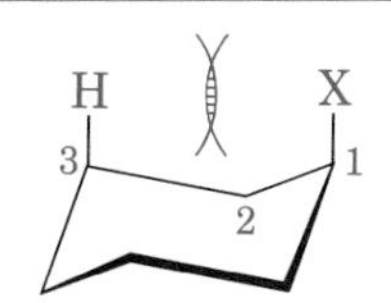

X	*Strain of one* H—X *1,3-diaxial interaction (kcal/mol)*
—F	0.12
—Cl	0.25
—Br	0.25
—OH	0.5
$—CH_3$	0.9
$—CH_2CH_3$	0.95
$—CH(CH_3)_2$	1.1
$—C(CH_3)_3$	2.7
$—C_6H_5$	1.5
—COOH	0.7
—CN	0.1

PROBLEM

18.12 How can you account for the fact (Table 18.2) that an axial *tert*-butyl substituent has much larger 1,3-diaxial interactions than isopropyl, but isopropyl is fairly similar to ethyl and methyl? Use molecular models to help with your answer.

PROBLEM

18.13 Why do you suppose an axial cyano substituent causes practically no 1,3-diaxial steric strain (0.1 kcal/mol)?

18.13 Conformational analysis

Monosubstituted cyclohexanes almost always prefer to have the substituent equatorial. In disubstituted cyclohexanes, however, the situation is more complex because the steric effects of both substituents must be taken into account; all of the steric interactions in the possible conformations must be analyzed before deciding which conformation is more favorable.

Let's look first at 1,2-dimethylcyclohexane. Recall from our earlier discussion (Section 3.10) that there are two isomers, *cis*-1,2-dimethylcyclohexane and *trans*-1,2-dimethylcyclohexane, and we have to consider them separately. In the cis isomer, both methyl groups are on the same side of the ring, and the compound can exist in either of the two chair con-

cis-1,2-Dimethylcyclohexane

One gauche interaction between CH_3 groups = 0.9 kcal/mol plus two CH_3—H 1,3-diaxial interactions; total strain = 0.9 + 1.8 = 2.7 kcal/mol

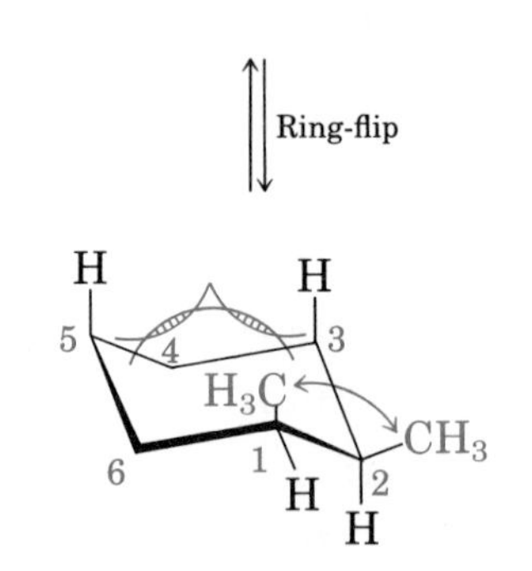

One gauche interaction between CH_3 groups = 0.9 kcal/mol plus two CH_3—H 1,3-diaxial interactions; total strain = 0.9 + 1.8 = 2.7 kcal/mol

Figure 18.17. Conformations of *cis*-1,2-dimethylcyclohexane. Both conformations are of equal energy.

formations shown in Figure 18.17. (Note that in Figure 18.17 it is often easier to see whether a compound is cis- or trans-disubstituted by first drawing a *flat* projection and then converting to chair conformations.)

Both conformations in Figure 18.17 have one methyl group axial and one methyl group equatorial. The conformation on the left has an axial methyl group at C2 that has 1,3-diaxial interactions with hydrogens on C4 and C6. The ring-flipped conformation on the right has an axial methyl group at C1 that has 1,3-diaxial interactions with hydrogens on C3 and C5. In addition, both conformations have gauche butane interactions between the two methyl groups. *The two conformations are exactly equal in energy.*

In the trans isomer, the two methyl groups are on opposite sides of the ring, and the compound can exist in either of the two chair conformations shown in Figure 18.18.

The situation here is quite different from that of the cis isomer. The trans conformation on the left in Figure 18.18 has both methyl groups equatorial and therefore has one gauche butane interaction, but has no 1,3-diaxial interactions. The conformation on the right, however, has *both methyl groups* axial. The axial methyl group at C1 interacts with axial

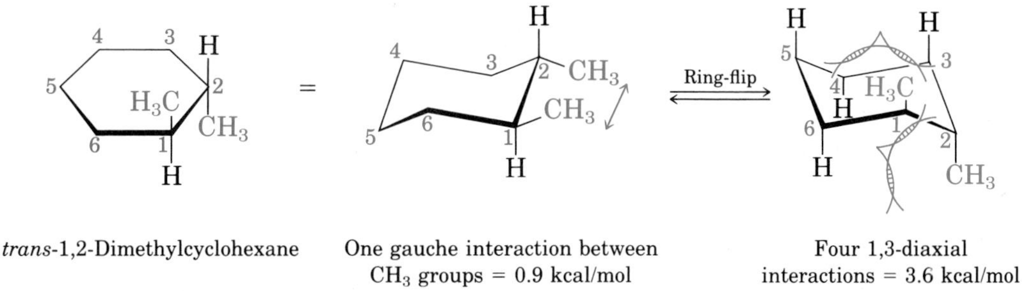

trans-1,2-Dimethylcyclohexane

One gauche interaction between CH_3 groups = 0.9 kcal/mol

Four 1,3-diaxial interactions = 3.6 kcal/mol

Figure 18.18. Conformations of *trans*-1,2-dimethylcyclohexane. The conformation in which both methyl groups are equatorial is favored by 3.6 − 0.9 = 2.7 kcal/mol.

hydrogens at C3 and C5, and the axial methyl group at C2 interacts with axial hydrogens at C4 and C6. These four 1,3-diaxial interactions make the diaxial conformation 3.6 − 0.9 = 2.7 kcal/mol less favorable than the diequatorial conformation. We can therefore predict with certainty that *trans*-1,2-dimethylcyclohexane exists almost exclusively (> 99%) in the diequatorial conformation.

The same kind of **conformational analysis** just carried out for 1,2-dimethylcyclohexane can be carried out for any substituted cyclohexane. For example, let's look at *cis*-1-chloro-4-*tert*-butylcyclohexane. This substance can exist in either of the two chair conformations shown in Figure 18.19.

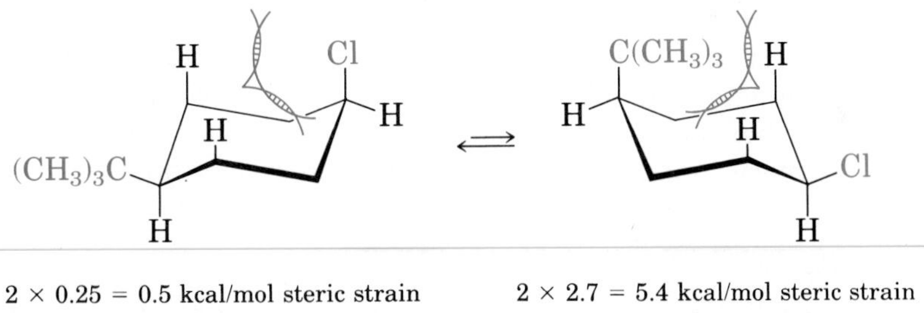

2 × 0.25 = 0.5 kcal/mol steric strain

2 × 2.7 = 5.4 kcal/mol steric strain

Figure 18.19. Conformations of 1-chloro-4-*tert*-butylcyclohexane. The conformation with an equatorial *tert*-butyl group is more stable.

In the left-hand conformation of Figure 18.19, the *tert*-butyl group is equatorial and the chlorine is axial. In the right-hand conformation, the *tert*-butyl group is axial and the chlorine is equatorial. These conformations are not of equal energy because an axial *tert*-butyl substituent and an axial chloro substituent produce different amounts of steric strain. Table 18.2 shows that a single *tert*-butyl–hydrogen 1,3-diaxial interaction "costs" 2.7 kcal/mol, whereas a single chlorine–hydrogen 1,3-diaxial interaction "costs" only 0.25 kcal/mol. An axial *tert*-butyl group therefore induces $(2 \times 2.7) - (2 \times 0.25) = 4.7$ kcal/mol more steric strain than an axial chlorine, and the compound adopts the left-hand conformation in Figure 18.19. The extremely large amount of steric strain caused by an axial *tert*-butyl group locks the cyclohexane ring into a specific conformation, and we can sometimes take advantage of this steric locking if we wish to study the chemical reactivity of an immobile cyclohexane ring.

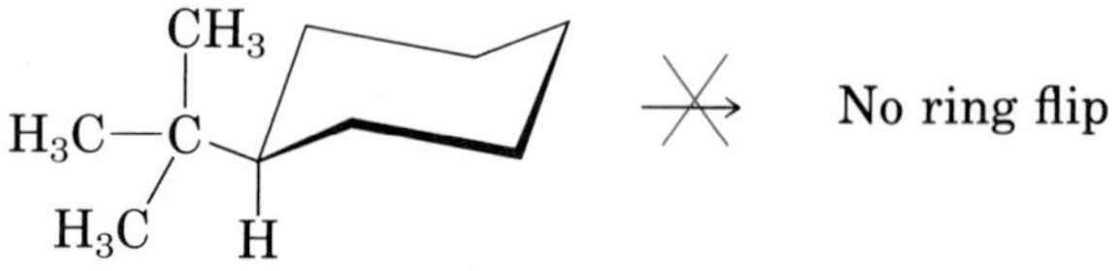

CH_3 H_3C—C H_3C H — No ring flip

Locked conformation

PROBLEM

18.14 Draw the most stable chair conformation of these molecules and estimate the amount of 1,3-diaxial strain in each.
(a) *trans*-1-Chloro-3-methylcyclohexane
(b) *cis*-1-Ethyl-2-methylcyclohexane
(c) *cis*-1-Bromo-4-ethylcyclohexane
(d) *trans*-4-Methylcyclohexanol

18.14 Stereoisomerism and chirality in substituted cyclohexanes

We have seen in the past few sections (Sections 18.10–18.13) that cyclohexane rings adopt a specific three-dimensional geometry, and that we can predict the stable conformation of a substituted ring by examining steric interactions in the molecule. What we must now do to complete our study of cyclohexane conformational analysis is to examine the questions of stereoisomerism and chirality in substituted cyclohexanes.

1,4-DISUBSTITUTED CYCLOHEXANES

1,4-Disubstituted cyclohexanes have no chiral centers by virtue of a symmetry plane passing through the middle of the ring. Thus, only cis and trans stereoisomers are possible, and these are diastereomers. (Recall that diastereomers are stereoisomers that are not related in a mirror-image fashion.) The symmetry plane is evident whether we use a flat view of the molecule or a three-dimensional view of the chair conformation (Figure 18.20).

1,3-DISUBSTITUTED CYCLOHEXANES

1,3-Disubstituted cyclohexanes have two chiral centers, and four stereoisomers are therefore possible. (Recall from Section 8.10 that n chiral centers in a molecule leads to 2^n stereoisomers.) *cis*-1,3-Dimethylcyclo-

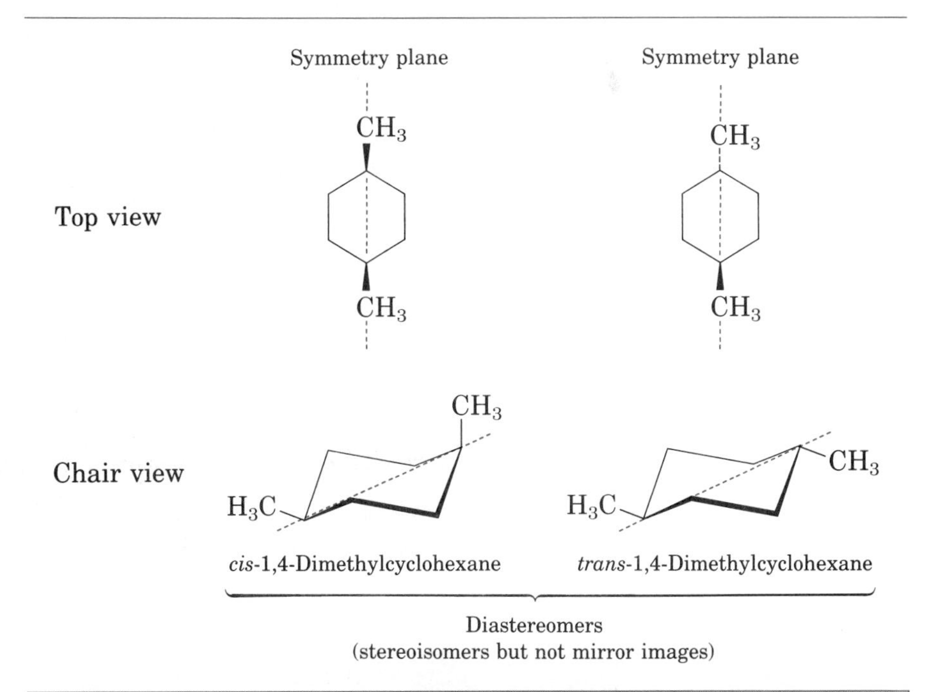

Figure 18.20. The relationships among 1,4-dimethylcyclohexane stereoisomers.

hexane, however, has a symmetry plane and is thus a meso compound, whereas *trans*-1,3-dimethylcyclohexane has no symmetry plane and must therefore exist as a pair of enantiomers. Again, these symmetry properties are evident both from flat views and from conformational views (Figure 18.21, p. 590).

1,2-DISUBSTITUTED CYCLOHEXANES

1,2-Disubstituted cyclohexanes have two chiral centers, and four stereoisomers are again possible. The situation here is more complex, however. When we considered the stereoisomeric relationships in the 1,3- and 1,4-disubstituted cyclohexanes, we were able to look at flat structures to obtain the correct answers about the optical activity (or lack thereof) of these compounds. With 1,2-disubstituted cyclohexanes, however, we must be more careful (Figure 18.22, p. 591). *trans*-1,2-Dimethylcyclohexane has no symmetry plane and exists as a pair of (+) and (−) enantiomers. A top view of *cis*-1,2-dimethylcyclohexane, on the other hand, reveals an apparent symmetry plane, and leads to the conclusion that the cis isomer is an optically inactive meso compound. When the cis isomer is viewed in chair conformation, however, the symmetry plane is no longer present because of the puckering of the ring, and we now predict that *cis*-1,2-dimethylcyclohexane exists as a pair of (+) and (−) enantiomers.

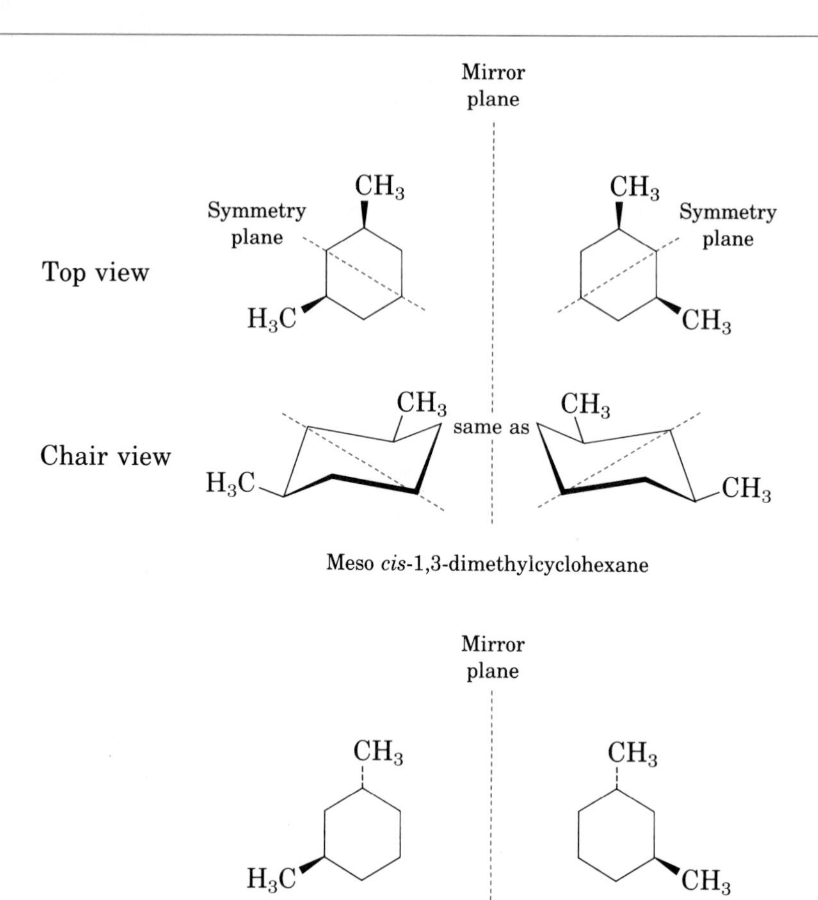

Figure 18.21. The relationships among 1,3-dimethylcyclohexane stereoisomers.

Although the prediction that *cis*-1,2-dimethylcyclohexane exists as a pair of optically active enantiomers is true in principle, we observe no optical activity in practice because the two enantiomers cannot be separated; they are interconverted by a ring-flip. (This interconversion is much easier to see with molecular models.)

In general, it is possible to predict the presence or absence of optical activity in any substituted cycloalkane merely by looking at flat structures,

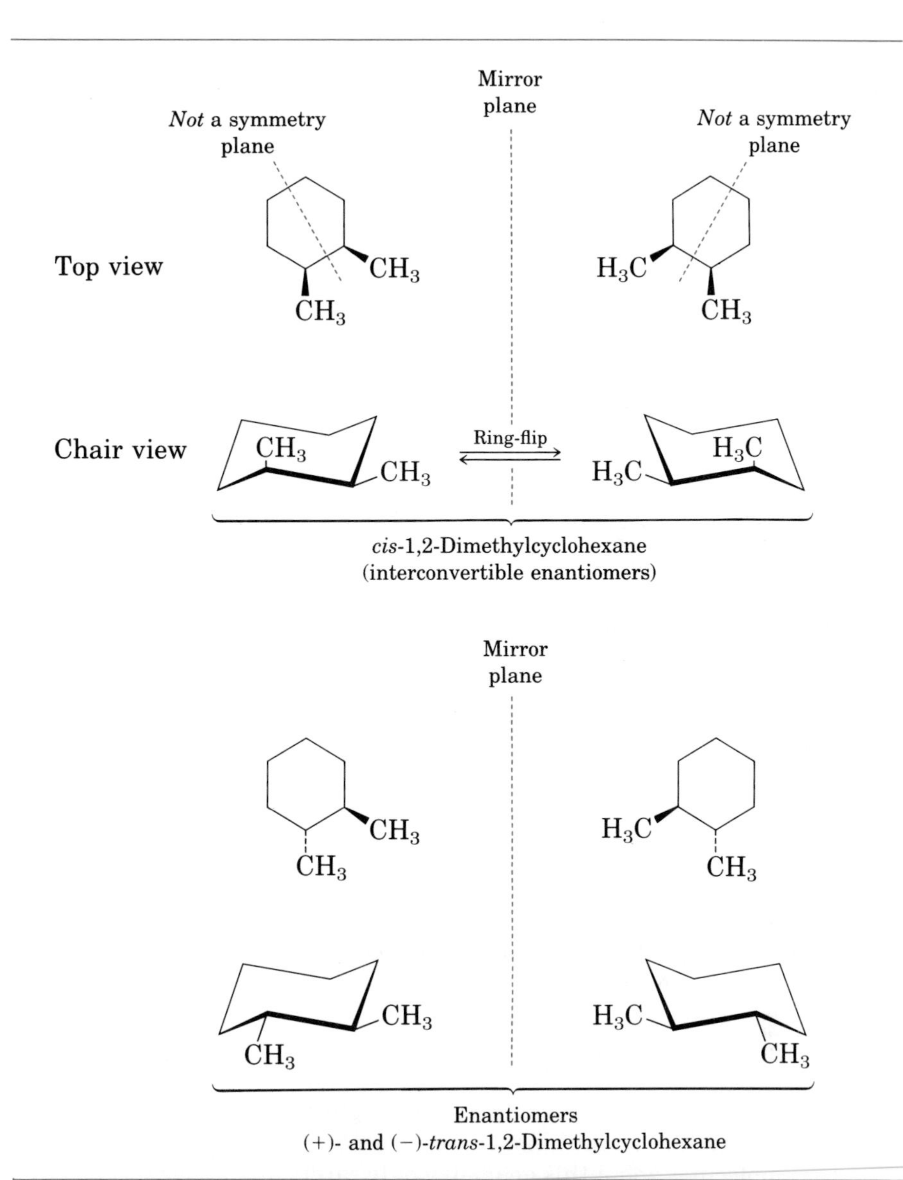

Figure 18.22. The relationships among 1,2-dimethylcyclohexane stereoisomers.

without considering the precise three-dimensional conformations. The exact reasons for the lack of optical activity in a given case may not be correct, however, as the situation present in *cis*-1,2-dimethylcyclohexane shows.

PROBLEM

18.15 How many stereoisomers of 1-chloro-3,5-dimethylcyclohexane are possible? Draw the most stable conformation.

18.15 Boat cyclohexane

Molecular models indicate that, in addition to chair cyclohexane, a second conformation known as **boat cyclohexane** is free of angle strain. We have not paid it any attention thus far, however, because boat cyclohexane is much less stable than chair cyclohexane. As Figure 18.23 shows, boat cyclohexane has a high degree of both steric strain and eclipsing strain.

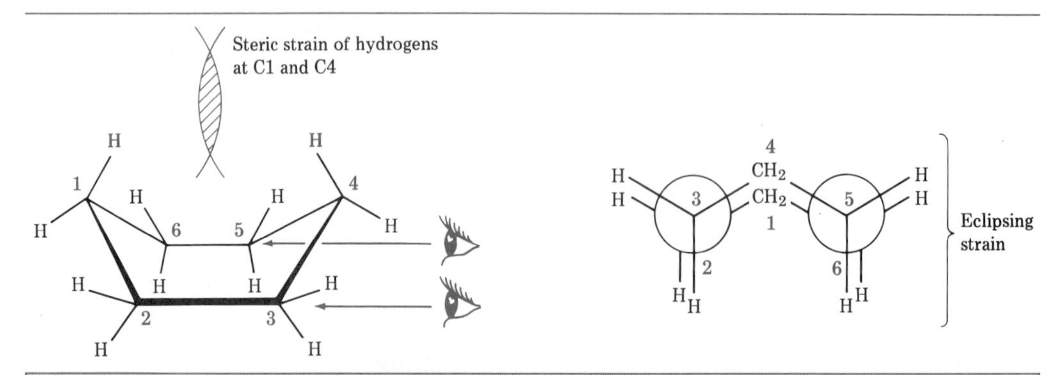

Figure 18.23. The boat conformation of cyclohexane. There is no angle strain, but there is severe steric and eclipsing strain.

There are two kinds of carbon atoms in boat cyclohexane. Carbons 2, 3, 5, and 6 lie in a plane, with carbons 1 and 4 above the plane. The inside hydrogen atoms on carbons 1 and 4 approach each other closely enough to result in considerable steric strain, and the four pairs of hydrogens on carbons 2, 3, 5, and 6 are eclipsed. The Newman projection in Figure 18.23, obtained by sighting along the C2—C3 and C5—C6 bonds, shows this eclipsing clearly.

Spectroscopic measurements indicate that boat cyclohexane is approximately 7.0 kcal/mol less stable than chair cyclohexane, although this value can be reduced to about 5.5 kcal/mol by twisting slightly and thereby relieving some eclipsing strain (Figure 18.24). Even the **twist-boat conformation** is still far more strained than the chair conformation, and molecules adopt this geometry only rarely and under special circumstances.

PROBLEM

18.16 There is good evidence for believing that *trans*-1,3-di-*tert*-butylcyclohexane exists largely in a twist-boat conformation. Explain why this might be so, and draw the likely twist-boat conformation.

18.16 Conformation and chemical reactivity

Cyclohexane conformational analysis is of much more than theoretical importance—we have discussed the subject in great detail for a very practical reason. As pointed out by Derek Barton[3] in a landmark 1950 paper,

[3]Derek H. R. Barton (1918–); b. Gravesend, England; Ph.D. and D.Sc. London (Heilbron, E. R. H. Jones); professor, Birkbeck College, Harvard, Glasgow, Imperial College, London, Institut de Chimie des Substances Naturelles, Gif-sur-Yvette, France; Nobel prize (1969).

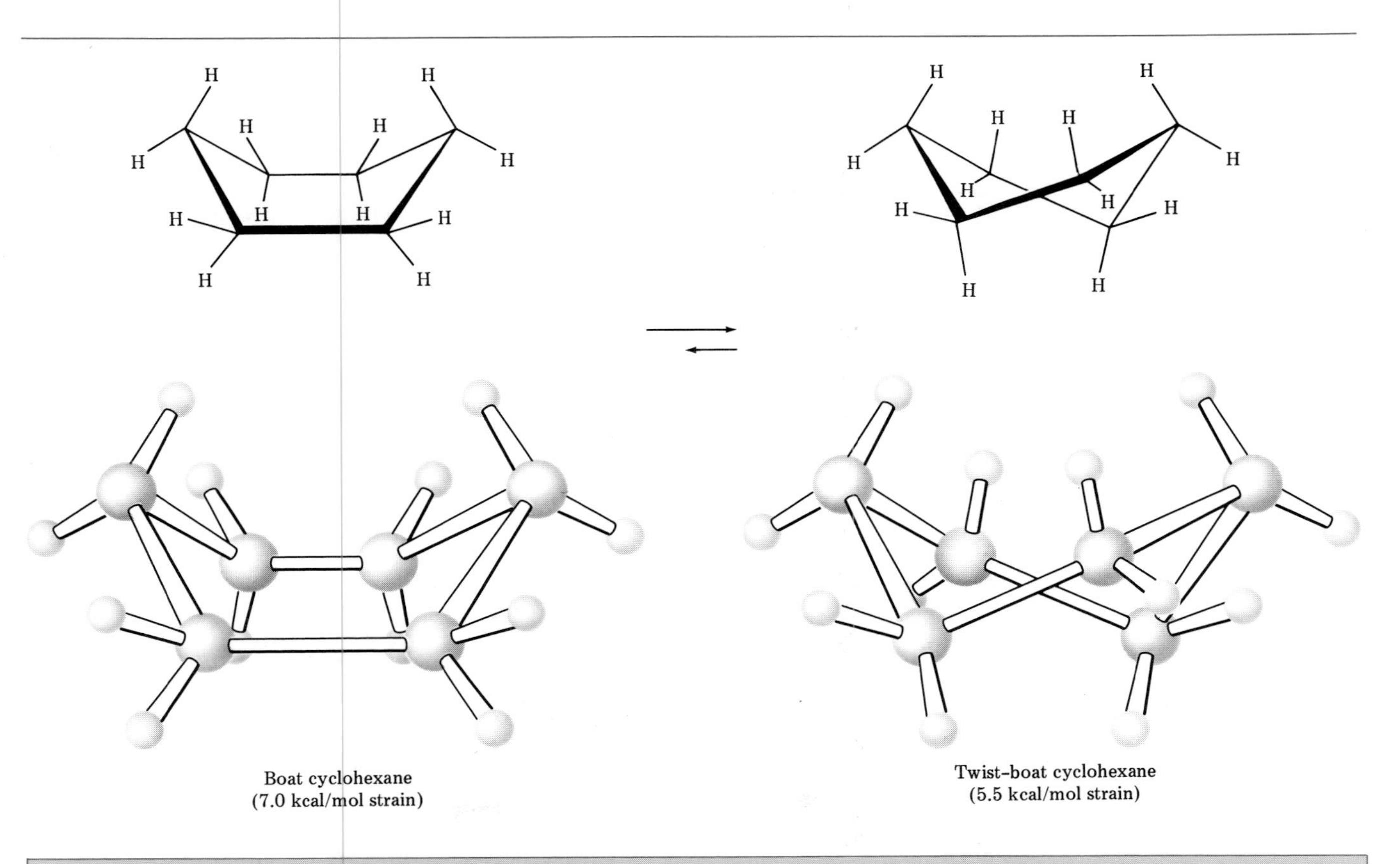

Figure 18.24. Boat and twist-boat conformations of cyclohexane. The twist-boat conformation is lower in energy than the boat conformation by 1.5 kcal/mol. Both conformations are more strained than chair cyclohexane.

much of the chemical reactivity of substituted cyclohexanes is controlled by conformational effects. Let's look at just one example—the E2 dehydrohalogenation of chlorocyclohexanes—to see the effect of conformation on reactivity.

We have already seen (Section 10.10) that E2 reaction requires anti periplanar geometry. In other words, the carbon–hydrogen and the carbon–halogen bonds that break must be 180° apart.

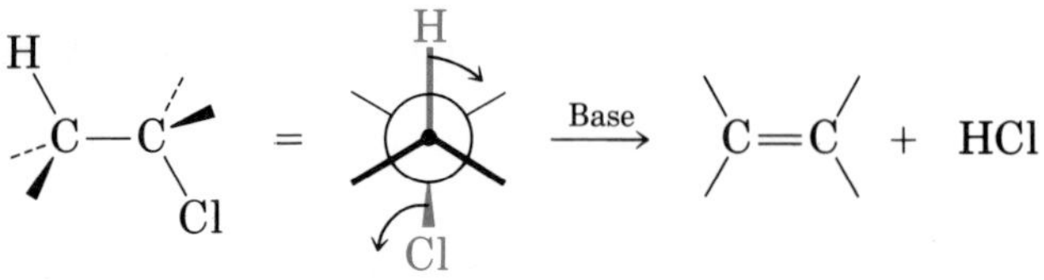

Anti periplanar geometry

The anti periplanar requirement can be met in cyclohexanes only if the hydrogen and the leaving group are trans diaxial (Figure 18.25). If the leaving group is equatorial, E2 elimination cannot occur.

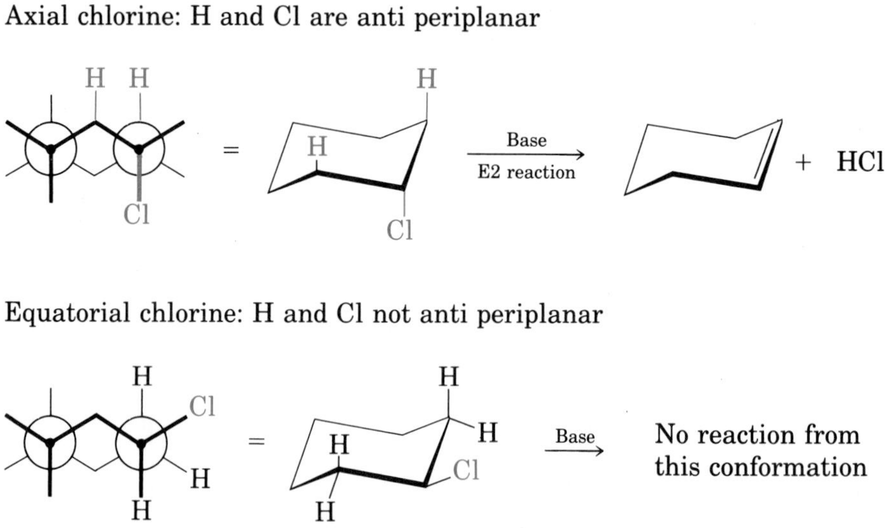

Axial chlorine: H and Cl are anti periplanar

Equatorial chlorine: H and Cl not anti periplanar

No reaction from this conformation

Figure 18.25. Geometry of E2 reaction in cyclohexanes. The leaving group must be axial for anti periplanar elimination to occur.

The elimination of HCl from the menthyl chlorides (Figure 18.26) provides a good illustration of this trans-diaxial requirement. Neomenthyl chloride reacts with ethoxide ion 200 times as fast as menthyl chloride. Furthermore, neomenthyl chloride yields 3-menthene as the sole alkene product, whereas menthyl chloride yields 2-menthene.

We can understand this difference in reactivity by looking at the most favorable chair conformations of the reactant molecules. Neomenthyl

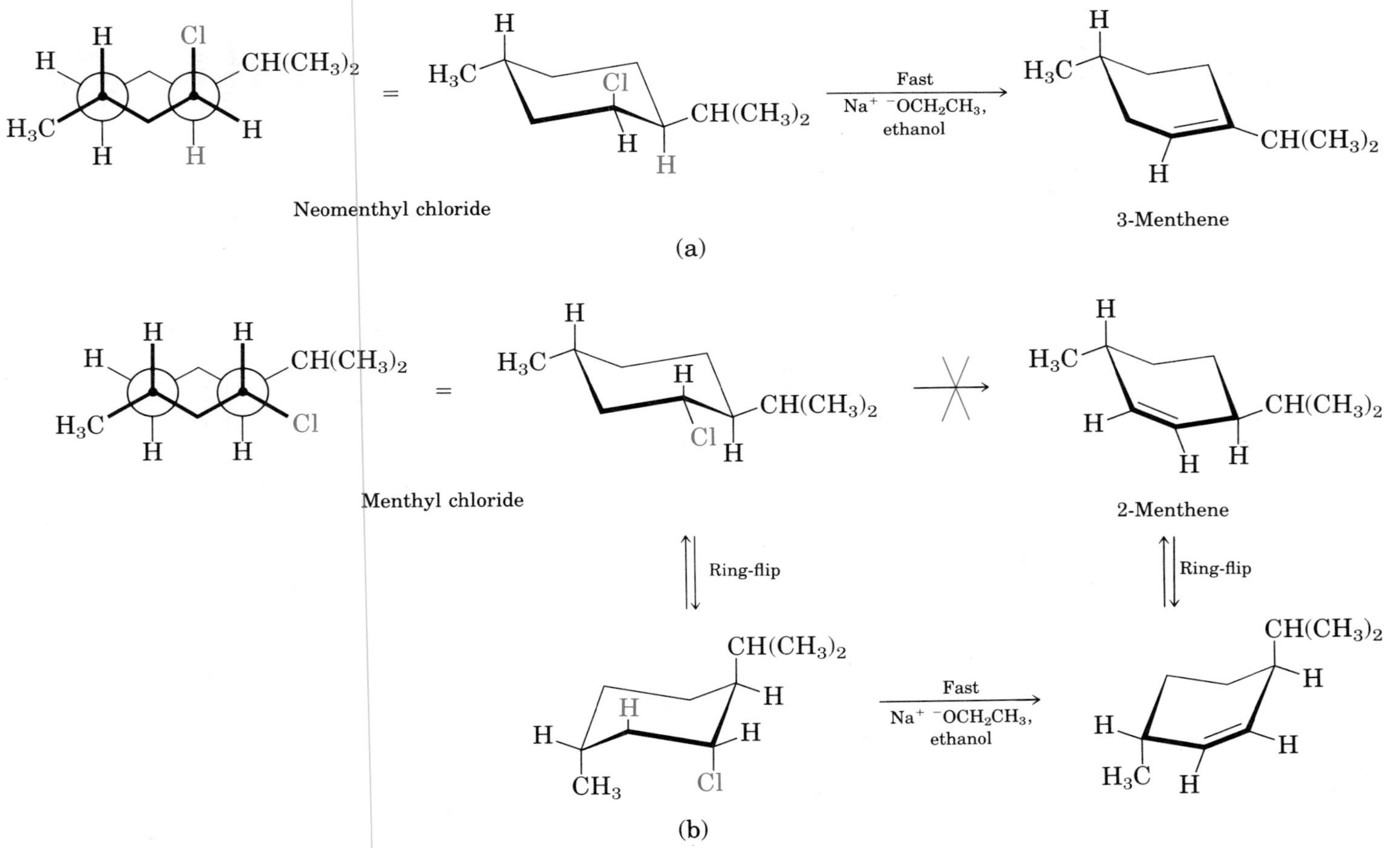

H H Cl $CH(CH_3)_2$
H_3C H H H
= H H_3C Cl $CH(CH_3)_2$ H H

Neomenthyl chloride

Fast
$Na^+\ ^-OCH_2CH_3$,
ethanol

H H_3C $CH(CH_3)_2$ H

3-Menthene

(a)

H H H $CH(CH_3)_2$
H_3C H H Cl
= H H_3C H $CH(CH_3)_2$ Cl H

Menthyl chloride

H H_3C H $CH(CH_3)_2$ H H

2-Menthene

Ring-flip

Ring-flip

$CH(CH_3)_2$ H H H H CH_3 Cl

Fast
$Na^+\ ^-OCH_2CH_3$,
ethanol

$CH(CH_3)_2$ H H H H_3C H

(b)

Figure 18.26. Dehydrochlorination of menthyl and neomenthyl chlorides. Neomenthyl chloride can lose HCl from its most stable conformation, but menthyl chloride must first ring-flip before HCl loss can occur.

chloride should have the conformation shown in Figure 18.26(a), since the two large substituents, methyl and isopropyl, are equatorial. In this conformation, chlorine is rigidly held in an axial orientation—the perfect geometry for E2 elimination with loss of the hydrogen atom at C4 to yield the more substituted alkene product, 3-menthene.

Menthyl chloride, on the other hand, has the geometry shown in Figure 18.26(b), in which all three substituents are equatorial. Since the chlorine is equatorial, it cannot undergo E2 elimination. In order to achieve the necessary geometry for elimination, menthyl chloride must ring-flip to a highly energetic chair conformation, in which all three substituents are axial. Then E2 elimination occurs with loss of the only possible trans-diaxial hydrogen, leading to 2-menthene. The net effect of the simple change in chlorine stereochemistry is a 200-fold decrease in reaction rate and a complete change of product. The chemistry of the molecule is controlled by conformational effects.

PROBLEM

18.17 Which isomer would you expect to react faster under E2 elimination conditions, *trans*-1-bromo-4-*tert*-butylcyclohexane or *cis*-1-bromo-4-*tert*-butylcyclohexane? Explain your answer.

18.17 Conformation of polycyclic molecules

The last point we will consider is what happens when two or more cycloalkane rings are fused together to construct polycyclic molecules such as decalin.

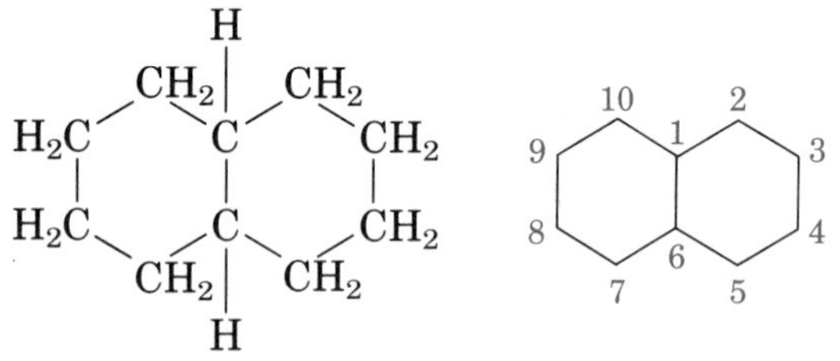

Decalin (two fused cyclohexane rings)

Decalin consists of two cyclohexane rings joined together to share two carbon atoms (the **bridgehead** carbons, C1 and C6) and a common bond. Decalin can exist in either of two isomeric forms, depending on whether the rings are trans fused or cis fused. In *trans*-decalin, the hydrogen atoms at the bridgehead carbons are on opposite sides of the rings; in *cis*-decalin, the bridgehead hydrogens are on the same side. Figure 18.27 shows how both compounds can be represented using chair cyclohexane conformations.

Note that *trans*- and *cis*-decalin are not interconvertible by ring-flips or other rotations. They are diastereomers and bear the same relationship to each other that *cis*- and *trans*-1,2-dimethylcyclohexane do.

Polycyclic compounds are of great importance, and many valuable substances have fused-ring structures. For example, the steroids such as

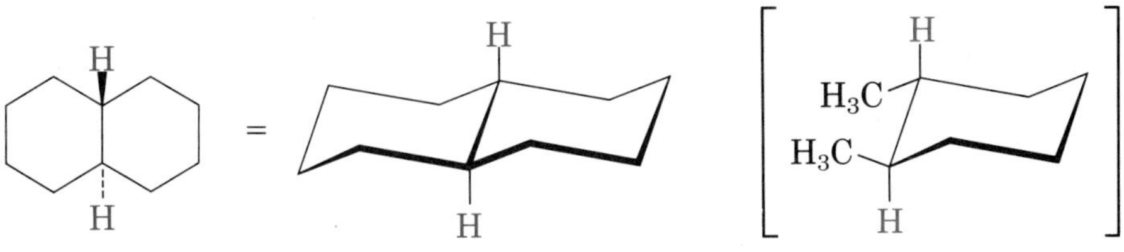

trans-Decalin

Bridgehead hydrogens on opposite sides of cyclohexane rings

trans-1,2-Dimethylcyclohexane

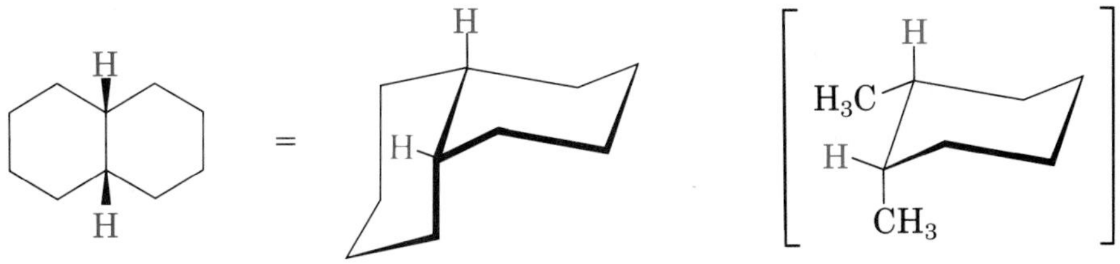

cis-Decalin

Bridgehead hydrogens on same side of cyclohexane rings

cis-1,2-Dimethylcyclohexane

Figure 18.27. Representations of *trans*- and *cis*-decalin, compared with *trans*- and *cis*-1,2-dimethylcyclohexane.

cholesterol have four rings—three 6-membered and one 5-membered—fused together. Though steroids look quite complicated in comparison to cyclohexane or decalin, they consist simply of chair cyclohexane rings locked together. The same principles that we applied to the conformational analysis of simple cyclohexane rings apply equally well (and often better) to steroids.

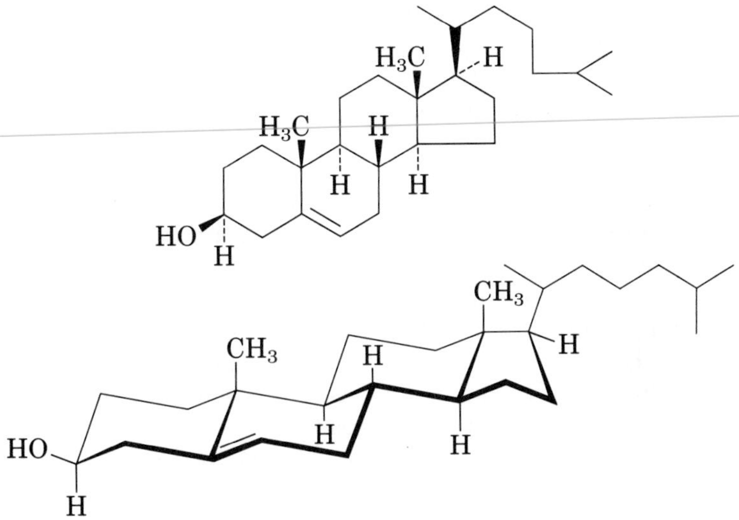

Cholesterol (a steroid)

Another common fused-ring system is the norbornane or bicyclo[2.2.1]heptane structure. Norbornane can be viewed as a conformationally locked boat cyclohexane in which carbons 1 and 6 are joined by an extra methylene group.

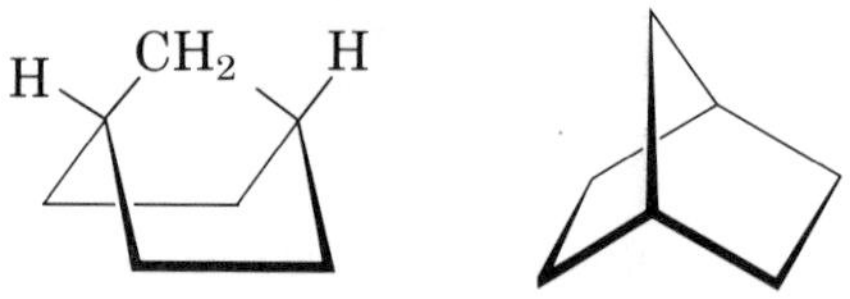

Norbornane, bicyclo[2.2.1]heptane

Substituted norbornanes such as camphor are found widely in nature, and many have played an important historical role in developing organic structural theories.

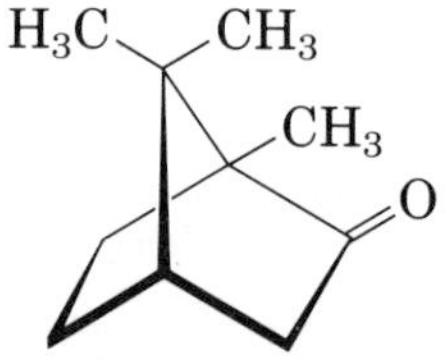

Camphor

PROBLEM

18.18 Which isomer do you think is the more stable, *trans*-decalin or *cis*-decalin? Explain your answer in terms of the steric interactions present in the two molecules.

18.18 Summary

Most organic molecules contain saturated rings. These alicyclic (aliphatic cyclic) molecules have special characteristics that affect the chemistry they undergo.

Data obtained from heat-of-combustion studies indicate that not all cycloalkanes are equally stable. Cyclopropane and cyclobutane are the least stable rings, whereas cyclopentane and the medium-sized rings (cycloheptane through cyclotridecane) all have varying degrees of strain. Cyclohexane and large rings (cyclotetradecane and above) are virtually strain-free. Three different kinds of strain contribute to the overall energy level of a cycloalkane: (1) angle strain, the resistance of a bond angle to compression or expansion from the normal 109° tetrahedral value; (2) eclipsing strain (torsional strain), the energy cost of having neighboring C—H bonds eclipsed rather than staggered; and (3) steric strain, the result of the repulsive van der Waals interaction that arises when two groups try to occupy the same spatial position.

Cyclopropanes are highly strained (27.6 kcal/mol) because of both angle strain and eclipsing strain. As a result, cyclopropanes are highly reactive and undergo some of the same electrophilic addition reactions that alkenes do. Thus, cyclopropanes add HBr and can be catalytically hydrogenated.

Cyclopropanes are best prepared by addition of a carbene to an alkene. Dichlorocarbene, which is usually prepared by alpha elimination of HCl from chloroform, adds to alkenes to give dichlorocyclopropanes. Non-halogenated cyclopropanes are usually prepared by the Simmons–Smith reaction, in which a zinc–copper couple reacts with diiodomethane to yield a carbenoid (carbene-like) reagent that attacks an alkene.

Cyclobutane is highly strained (26.4 kcal/mol) because of both angle strain and eclipsing strain. Cyclopentane is free of angle strain but suffers from a large number of eclipsing interactions. Both cyclobutane and cyclopentane pucker slightly away from planarity in order to relieve eclipsing strain.

Cyclohexane rings are the most important of all ring sizes because of their wide occurrence. Cyclohexane is strain-free by virtue of its puckered chair conformation, in which all bond angles are 109° and all neighboring C—H bonds are staggered. Chair cyclohexane has two kinds of hydrogens—axial and equatorial. Axial hydrogens are directed up and down, parallel to the ring axis, whereas equatorial hydrogens lie in a belt more or less along the equator of the ring. Each carbon atom has one axial and one equatorial hydrogen, and the positions alternate around the ring. Chair cyclohexanes are conformationally mobile, and can undergo a ring-flip that interconverts axial and equatorial positions.

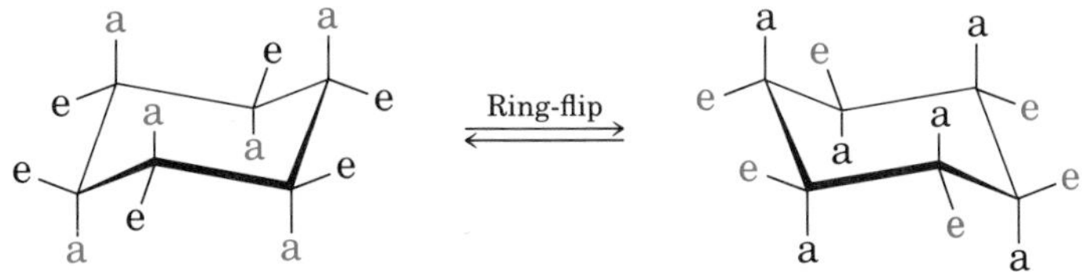

Ring-flip

Substituents on the ring are more stable in the equatorial position, since axial substituents suffer 1,3-diaxial steric strain. The amount of 1,3-diaxial strain caused by an axial substituent depends on the bulk of the substituent.

The stereochemistry of cyclohexane and its derivatives is best learned by using molecular models to examine conformational relationships. The great importance of a thorough understanding of cyclohexane conformational analysis lies in the fact that conformation has a strong effect on chemical reactivity.

18.19 Summary of reactions

1. Preparation of cyclopropanes
   a. Dichlorocarbene addition to alkenes (Section 18.6)

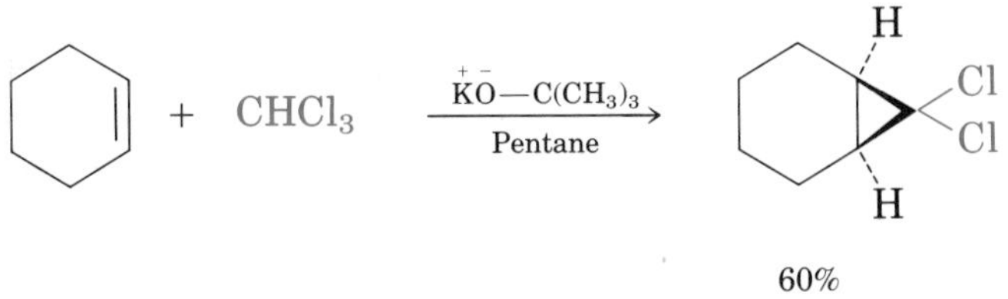

$$\text{Cyclohexene} + CHCl_3 \xrightarrow[\text{Pentane}]{\overset{+}{K}\overset{-}{O}\text{—}C(CH_3)_3} \text{7,7-dichlorobicyclo[4.1.0]heptane}$$

60%

b. Simmons–Smith reaction (Section 18.6)

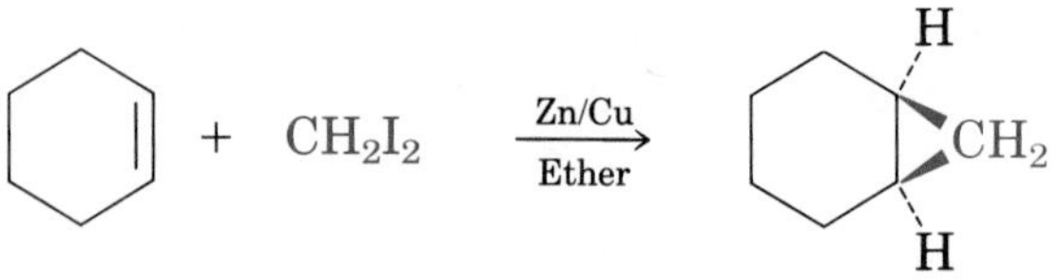

+ CH_2I_2 $\xrightarrow[\text{Ether}]{\text{Zn/Cu}}$ H, CH_2, H

92%

2. Reactions of cyclopropanes
   a. Electrophilic addition of HBr (Section 18.7)

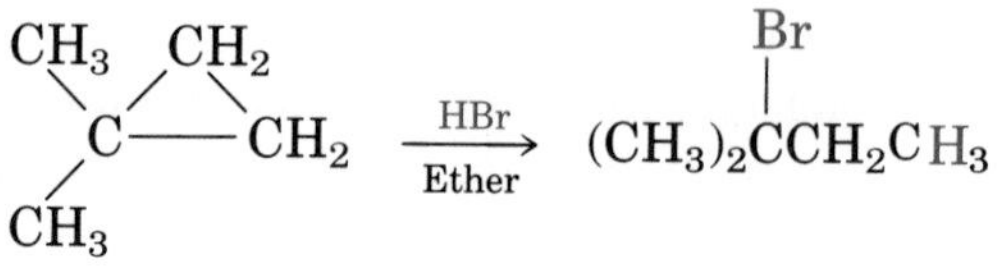

CH_3 CH_2 C CH_2 CH_3 $\xrightarrow[\text{Ether}]{\text{HBr}}$ $(CH_3)_2\overset{Br}{C}CH_2CH_3$

   b. Catalytic hydrogenation (Section 18.7)

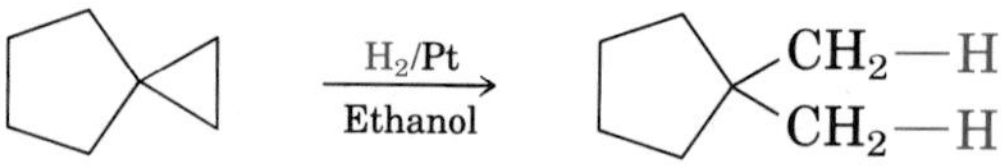

$\xrightarrow[\text{Ethanol}]{H_2/Pt}$ $CH_2—H$ $CH_2—H$

91%

ADDITIONAL PROBLEMS

18.19 Provide IUPAC names for these substances:

(a) CH_3

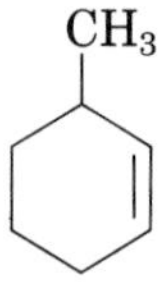

(b) HO H HO H

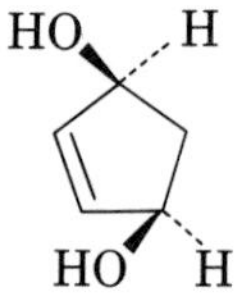

(c) H I CH_3 H

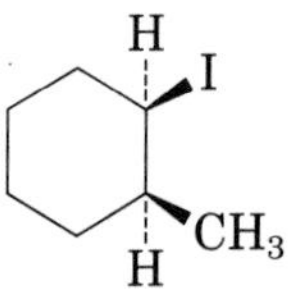

(d)

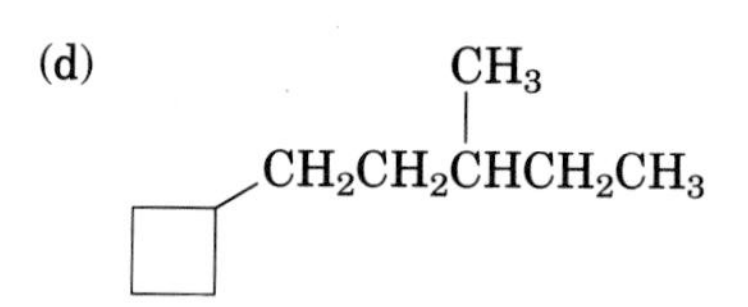

(e)

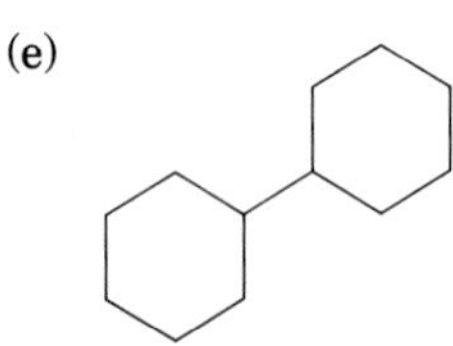

(f) $-CH_2CH_3$

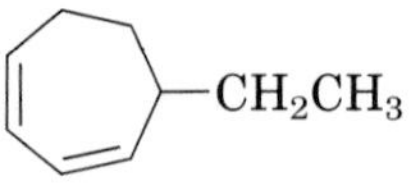

(g) CH_3 CH_3

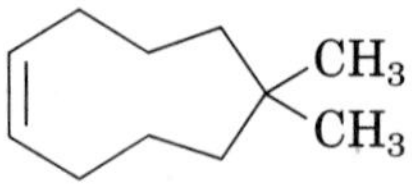

(h) $-CH_2CH_2-$

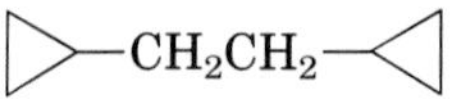

18.20 Draw structures corresponding to these names:
(a) 4-Methylcycloheptene
(b) 2-Cyclobutyl-1-pentene
(c) *trans*-1,3-Dibromocyclopentane
(d) *cis*-1,3-Cyclohexanediol
(e) 1,3,5-Cyclododecatriene
(f) *trans*-1-Isopropyl-4-methylcyclohexane
(g) Tricyclohexylmethane

18.21 Give the product(s) from the reaction of 1-methylcyclopentene with the following reagents. Indicate stereochemistry.
(a) OsO_4, then $NaHSO_3$
(b) BH_3, then $H_2O_2/^-OH$
(c) Cl_2 in CCl_4
(d) HBr in ether
(e) DBr in ether
(f) $KMnO_4$, H_3O^+
(g) $Hg(OAc)_2/H_2O$, then $NaBH_4$
(h) CH_2I_2, Zn/Cu in ether
(i) $CHCl_3$, $K^+ \ ^-OC(CH_3)_3$ in pentane

18.22 Define these terms in your own words:
(a) Angle strain
(b) Steric strain
(c) Heat of combustion
(d) Axial position
(e) Equatorial position

18.23 Draw a chair cyclohexane ring and label all positions as axial or equatorial.

18.24 *N*-Methylpiperidine is known to have the conformation shown. What does this tell you about the relative steric requirements of a methyl group versus an electron lone pair?

N—CH_3

18.25 There are four stereoisomers of 1-chloro-2-methylcyclohexane. Draw each isomer in its more stable chair conformation, and indicate the stereochemical relationships among isomers.

18.26 How can you account for the fact that *cis*-2-phenyl-1-cyclopentyl tosylate undergoes base-catalyzed E2 elimination approximately 15 times as fast as the trans isomer? Both isomers yield 1-phenylcyclopentene.

Ph, O—S(=O)(=O)—C₆H₄—CH_3 $\xrightarrow[HOC(CH_3)_3]{\overset{+}{K}\overset{-}{O}C(CH_3)_3}$ Ph

2-Phenyl-1-cyclopentyl tosylate

1-Phenylcyclopentene

18.27 Consider *cis*-1,2-cyclopentanediol. How many chiral centers are present? Is this substance chiral or meso? Explain.

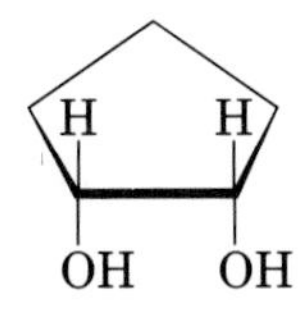

Answer these questions for *trans*-1,2-cyclopentanediol.

18.28 Can you propose an explanation for the observation that bromocyclohexane undergoes S_N2 reaction much faster than bromocyclopropane does?

18.29 Glucose contains a six-membered ring in which all of the substituents are equatorial. Draw glucose in its stable chair conformation.

HO O CH_2OH

HO OH

OH

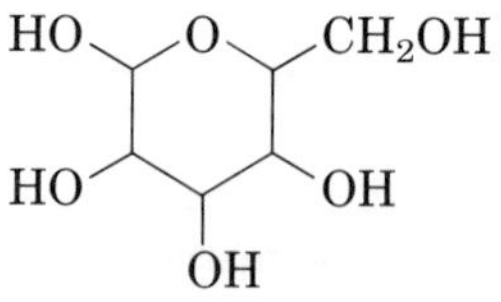

Glucose

18.30 From the data in Table 18.2 and in Table 3.7, calculate the percentages of molecules that have their substituents in an axial orientation for these compounds:
(a) *tert*-Butylcyclohexane (b) Bromocyclohexane
(c) Cyclohexanecarbonitrile (d) Cyclohexanol

18.31 Draw the two possible chair conformations of *cis*-1,3-dimethylcyclohexane. The diaxial conformation is approximately 5.4 kcal/mol less stable than the diequatorial conformation. Can you suggest a reason for this large energy difference? Approximately how much steric strain does the 1,3-diaxial interaction between the two methyl groups introduce into the diaxial conformation?

18.32 In light of your answer to Problem 18.31, draw the two chair conformations of 1,1,3-trimethylcyclohexane and estimate the amount of strain energy in each. Which conformation is favored?

18.33 We saw in Section 6.1 that addition of Br_2 to an alkene occurs with anti periplanar geometry; the two bromine atoms end up 180° apart in the product. In light of this knowledge, predict the stereochemistry of the product that results from addition of Br_2 to 4-*tert*-butylcyclohexene.

18.34 We saw in Problem 18.18 that *cis*-decalin is less stable than *trans*-decalin. Assume that the 1,3-diaxial interactions in *trans*-decalin are similar to those in axial methylcyclohexane (one CH_3—H interaction costs 0.9 kcal/mol) and calculate the magnitude of the energy difference between *cis*- and *trans*-decalin.

18.35 Using molecular models as well as structural drawings, explain why *trans*-decalin is rigid and cannot ring-flip, whereas *cis*-decalin can easily ring-flip.

18.36 Chlorocyclohexane exhibits a multiplet absorption at 3.95 δ when its ^1H NMR spectrum is recorded at room temperature. At −150°C, however, the 3.95 δ absorption disappears and is replaced by two absorptions at 4.50 δ and 3.80 δ. Explain. [*Hint:* See Section 12.2.]

18.37 Consider *cis*- and *trans*-1-methyl-4-*tert*-butylcyclohexane. How many chiral centers are present in each? Are these compounds meso or chiral? Draw each in its more stable chair conformation.

18.38 Answer Problem 18.37 for *cis*- and *trans*-1-methyl-3-*tert*-butylcyclohexane.

18.39 How many geometric isomers of 1,2,3,4,5,6-hexachlorocyclohexane are there? One of these isomers is inert to E2 elimination. Draw its structure and explain its lack of reactivity.

18.40 Propose an explanation of the observation that the all-cis isomer of 4-*tert*-butylcyclohexane-1,3-diol reacts readily with acetone to form an acetal, but that other stereoisomers do not react.

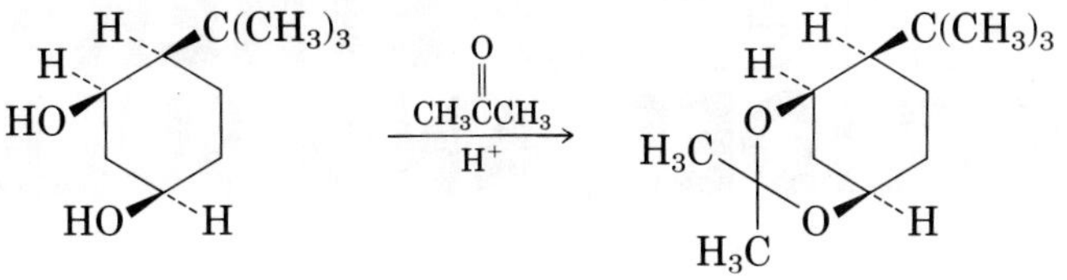

$$\xrightarrow[\text{H}^+]{\text{CH}_3\overset{\text{O}}{\overset{\|}{\text{C}}}\text{CH}_3}$$

An acetal

In formulating your answer, draw the stable chair conformations of all four stereoisomers and of the product acetal. Use molecular models for help.

CHAPTER 19 ETHERS, EPOXIDES, AND SULFIDES

An **ether** is a substance that has two organic residues bonded to the same oxygen atom, R—O—R′. The organic residues may be alkyl, aryl, or vinylic, and the oxygen atom can be part of either an open chain or a ring. Diethyl ether is a familiar substance that has been used medicinally as an anesthetic and is much used industrially as a solvent; anisole is a pleasant-smelling aromatic ether used in perfumery; tetrahydrofuran (THF) is a cyclic ether often used as a solvent.

$CH_3CH_2—O—CH_2CH_3$

Diethyl ether

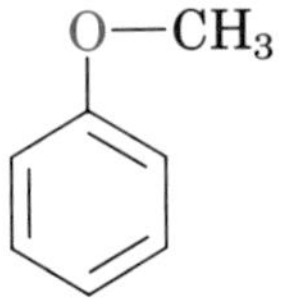

O—CH_3

Anisole
(methyl phenyl ether)

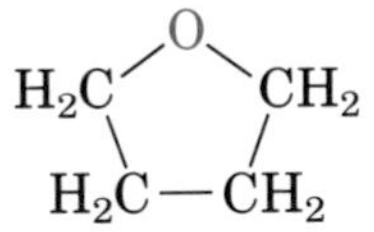

O
H_2C CH_2
$H_2C—CH_2$

Tetrahydrofuran
(a cyclic ether)

19.1 Nomenclature of ethers

The naming of ethers is complicated by the fact that two different systems are allowed by IUPAC rules. Relatively simple ethers are best named by identifying the two organic residues and adding the word *ether*. For example,

$CH_3OC(CH_3)_3$

tert-Butyl methyl ether

$CH_3CH_2—O—CH{=}CH_2$

Ethyl vinyl ether

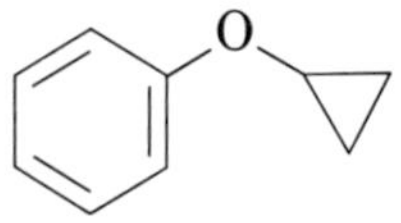

O

Cyclopropyl phenyl ether

If more than one ether linkage is present in the molecule, or if other functional groups are present, the ether is better named as an alkoxy-substituted parent compound. For example,

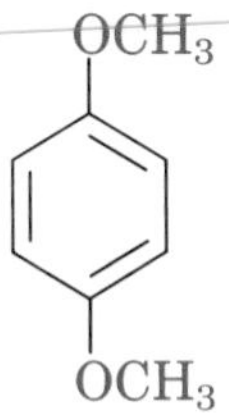

OCH_3
OCH_3

p-Dimethoxybenzene

1 2 3 4
O
$C(CH_3)_3$

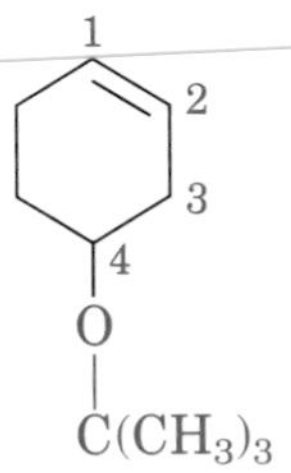

4-*tert*-Butoxy-1-cyclohexene

PROBLEM

19.1 Name ethers (a)–(f) according to IUPAC rules. For simple dialkyl ethers, name the compounds in two ways.

(a) $(CH_3)_2CH—O—CH(CH_3)_2$

(b) 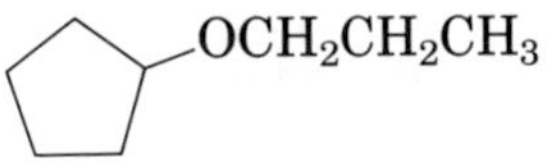 $OCH_2CH_2CH_3$ (on cyclopentane)

(c) OCH_3 / Br (para-bromoanisole structure)

(d) OCH_3 (1-methoxycyclohexene structure)

(e) $(CH_3)_2CHCH_2OCH_2CH_3$

(f) $H_2C{=}CHCH_2OCH{=}CH_2$

19.2 Properties of ethers

Ethers may be considered derivatives of water in which the hydrogen atoms have been replaced by organic residues, H—O—H ⇒ R—O—R. As such, ethers have nearly the same geometry as water. The R—O—R bonds have an approximately tetrahedral bond angle (112° in dimethyl ether), and the oxygen atom is sp^3 hybridized.

H_3C
112° O:
H_3C

The presence of the electronegative oxygen atom causes ethers to have a slight dipole moment, and the boiling points of ethers are therefore somewhat higher than the boiling points of comparable alkanes. Table 19.1 compares the boiling points of some common ethers with hydrocarbons of similar molecular weight.

TABLE 19.1 Comparison of boiling points of ethers and hydrocarbons

Compounds	*Boiling point (°C)*	
CH_3OCH_3 (vs. $CH_3CH_2CH_3$)	−25	(−45)
$CH_3CH_2OCH_2CH_3$ (vs. $CH_3CH_2CH_2CH_2CH_3$)	34.6	(36)
Tetrahydrofuran (vs. cyclopentane)	65	(49)
Anisole (vs. ethylbenzene)	158	(136)

Ethers are unusually stable toward most reagents, but certain ethers react slowly with air to give peroxides, compounds that contain oxygen–oxygen bonds. The peroxides from low-molecular-weight ethers such as diisopropyl ether and tetrahydrofuran are highly explosive and extremely dangerous even in tiny amounts. This sensitivity to air and their volatility and flammability make ether solvents potentially dangerous unless handled by skilled persons. Ether solvents are very useful in the laboratory, but they must always be treated with care.

19.3 Preparation of ethers

Diethyl ether and other simple symmetrical ethers are prepared industrially by the sulfuric acid catalyzed dehydration of alcohols:

$$2\ CH_3CH_2OH \xrightarrow{H_2SO_4} CH_3CH_2OCH_2CH_3 + H_2O$$

Ethanol — Diethyl ether

The reaction probably occurs by S_N2 displacement of water from a protonated ethanol molecule by the oxygen atom of a second ethanol:

$$CH_3CH_2\ddot{O}H + CH_2CH_3\ (\overset{+}{O}H_2) \xrightarrow{S_N2\ reaction} CH_3CH_2OCH_2CH_3 + H_2O + H^+$$

This method of ether preparation is limited to the industrial synthesis of symmetrical ethers from primary alcohols, since secondary and tertiary alcohols dehydrate readily to yield alkenes. The reaction conditions must be carefully controlled, and the method is of little practical value in the laboratory. Most laboratory syntheses of ethers are carried out by either of two methods—the Williamson synthesis and the oxymercuration reaction (Section 6.3). Let's see how these two reactions occur.

PROBLEM

19.2 Why do you suppose only symmetrical ethers can be prepared by the sulfuric acid catalyzed dehydration procedure? What product(s) would you expect to get if ethanol and 1-propanol were allowed to react together? In what ratio would the products be formed if the two alcohols were of equal reactivity?

19.4 The Williamson ether synthesis

Metal alkoxides react with alkyl halides and tosylates to yield ethers, a process known as the **Williamson ether synthesis.** The Williamson[1] synthesis was discovered in 1850 and is still the best method for the preparation of ethers. The reaction is widely used for the synthesis of both symmetrical and unsymmetrical ethers.

$$C_5H_9\ddot{O}^-K^+ + CH_3I \xrightarrow[\text{solvent}]{\text{Ether}} C_5H_9OCH_3 + KI$$

Potassium cyclopentoxide — Iodomethane — Cyclopentyl methyl ether (74%)

[1]Alexander W. Williamson (1824–1904); b. Wandsworth, England; Ph.D. Giessen (1846); professor, University College, London (1849–1904).

$$(CH_3)_3CCH_2\ddot{\underset{\cdot\cdot}{O}}{:}^-\ Na^+ + C_6H_5CH_2Cl \xrightarrow[\text{solvent}]{\text{THF}} C_6H_5CH_2{-}O{-}CH_2{-}C(CH_3)_3 + NaCl$$

Sodium 2,2-dimethylpropoxide

Benzyl chloride

Benzyl 2,2-dimethylpropyl ether (68%)

The alkoxide ion needed in the reaction is normally prepared by reaction of an alcohol with a strong base such as sodium hydride, NaH. An acid–base reaction occurs between the alcohol and sodium hydride to generate the sodium salt of the alcohol.

$$R{-}O{-}H + NaH \longrightarrow R{-}O^-\ Na^+ + H_2$$

An important variation of the Williamson synthesis involves the use of silver oxide, Ag_2O. Under these conditions, the free alcohol reacts directly with alkyl halide, and there is no need to preform the metal alkoxide salt. For example, glucose reacts with iodomethane in the presence of Ag_2O to generate a pentaether in 85% yield.

$$\alpha\text{-D-Glucose} \xrightarrow[Ag_2O]{CH_3I} \alpha\text{-D-Glucose pentamethyl ether} + AgI$$

α-D-Glucose

α-D-Glucose pentamethyl ether (85%)

Mechanistically, the Williamson synthesis occurs by S_N2 displacement of halide ion by the alkoxide ion nucleophile. The Williamson synthesis is thus subject to all of the normal constraints on S_N2 reactions that we have discussed (Section 10.5). Primary halides and tosylates work best, since competitive E2 elimination of HX can occur with more hindered substrates. For this reason, unsymmetrical ethers should be synthesized by reaction between the more hindered alkoxide partner and less hindered halide partner, rather than vice versa. For example, *tert*-butyl methyl ether is best prepared by reaction of *tert*-butoxide with iodomethane, rather than by reaction of methoxide with 2-chloro-2-methylpropane.

$$\text{S}_\text{N}2\ \text{reaction}\quad (CH_3)_3C\text{—}\ddot{\underset{\cdot\cdot}{O}}{:}^- + CH_3\text{—}I \longrightarrow (CH_3)_3C\text{—}O\text{—}CH_3 + {:}I^-$$

tert-Butoxide ion — Iodomethane — *tert*-Butyl methyl ether

$$\text{E2 reaction}\quad CH_3\text{—}\ddot{\underset{\cdot\cdot}{O}}{:}^- + H\text{—}CH_2\text{—}C(CH_3)_2\text{—}Cl \longrightarrow H_2C{=}C(CH_3)_2 + CH_3OH + {:}Cl^-$$

Methoxide ion — 2-Chloro-2-methylpropane — 2-Methylpropene

PROBLEM

19.3 How would you prepare the following compounds using a Williamson ether synthesis?
(a) Anisole (methyl phenyl ether)
(b) Benzyl isopropyl ether
(c) Ethyl 2,2-dimethylpropyl ether

PROBLEM

19.4 Rank the following compounds in order of their reactivity in the Williamson reaction.
(a) Bromoethane, 2-bromopropane, benzyl bromide
(b) Chloroethane, bromoethane, 3-bromopropene

19.5 Alkoxymercuration–demercuration of alkenes

We saw in Section 6.3 that alkenes react with water in the presence of mercuric acetate to yield the **oxymercuration** product. Subsequent treatment of this addition product with sodium borohydride breaks the carbon–mercury bond and yields the alcohol. A similar **alkoxymercuration** reaction occurs when an alkene is treated with an alcohol in the presence of mercuric trifluoroacetate, followed by sodium borohydride-induced demercuration. The net result is Markovnikov addition of alcohol to the alkene. For example, methanol adds to styrene in the alkoxymercuration–demercuration reaction to yield 1-methoxy-1-phenylethane.

$$C_6H_5CH{=}CH_2 \xrightarrow[CH_3OH]{Hg(O_2CCF_3)_2} \left[C_6H_5CH(OCH_3)CH_2HgO_2CCF_3\right] \xrightarrow{NaBH_4} C_6H_5CH(OCH_3)CH_2\text{—}H$$

Styrene — 1-Methoxy-1-phenylethane (97%)

$$\text{Cyclohexene} \xrightarrow[\text{2. NaBH}_4]{\text{1. Hg(O}_2\text{CCF}_3)_2,\ \text{CH}_3\text{CH}_2\text{OH}} \text{Cyclohexyl ethyl ether (100\%)}$$

Cyclohexene

Cyclohexyl ethyl ether (100%)

The mechanism of the alkoxymercuration–demercuration reaction is analogous to the mechanism described earlier (Section 6.3) for the oxymercuration procedure. The reaction is initiated by electrophilic addition of mercuric ion to the alkene, and the intermediate mercurinium ion is then attacked by alcohol. Displacement of mercury with sodium borohydride completes the process.

A wide variety of alcohols and alkenes can be used in the alkoxymercuration reaction. Primary, secondary, and even tertiary alcohols react smoothly, but di-tertiary ethers cannot be prepared because of steric hindrance to reaction.

PROBLEM

19.5 Show in detail the mechanism of the reaction between 1-methylcyclopentene, ethanol, and mercuric trifluoroacetate. Why is the Markovnikov adduct formed? What is the stereochemistry of the initial ethoxymercurated adduct prior to $NaBH_4$ reduction? Explain your answer. (A review of Section 6.3 may be helpful.)

PROBLEM

19.6 How would you prepare the following ethers? Use whichever method you think is most appropriate, the Williamson synthesis or the alkoxymercuration reaction.
(a) Butyl cyclohexyl ether
(b) Ethyl phenyl ether
(c) *tert*-Butyl *sec*-butyl ether
(d) Tetrahydrofuran

19.6 Reactions of ethers: acidic cleavage

Ethers are unusually stable to most reagents used in organic chemistry, and it is this property that accounts for their wide use as inert reaction solvents. Halogens, mild acids, bases, and nucleophiles have no effect on most ethers. In fact, ethers undergo only one reaction of general use—ethers are cleaved by strong acids.

The first example of ether cleavage by an acid was observed in 1861 by Alexander Butleroff,[2] who found that 2-ethoxypropanoic acid reacts with aqueous HI at 100°C to yield iodoethane and lactic acid:

$$CH_3CH(OCH_2CH_3)CO_2H + HI \xrightarrow[H_2O]{100°C} CH_3CH_2I + CH_3CH(OH)CO_2H$$

2-Ethoxypropanoic acid

Lactic acid

Aqueous HI is still the preferred reagent for cleaving simple ethers, although aqueous HBr can also be used. However, HCl does not cleave ethers readily.

[2]Alexander M. Butleroff (1828–1886); b. Tschistopol, Russia; Ph.D., University of Moscow (1854); professor, University of Kazan (1854–1867), University of St. Petersburg (1867–1880).

$$CH_3CH_2OCH(CH_3)_2 \xrightarrow[\text{Reflux}]{\text{HI, } H_2O} CH_3CH_2I + (CH_3)_2CHOH$$

Ethyl isopropyl ether → Iodoethane + Isopropyl alcohol

$$C_6H_5OCH_2CH_3 \xrightarrow[\text{Reflux}]{\text{HBr, } H_2O} C_6H_5OH + CH_3CH_2Br$$

Ethyl phenyl ether → Phenol + Bromoethane

$$C_6H_{11}OC(CH_3)_3 \xrightarrow[0°C]{\text{HBr, } H_2O} C_6H_{11}OH + (CH_3)_2C{=}CH_2$$

tert-Butyl cyclohexyl ether → Cyclohexanol + 2-Methylpropene

Many other acidic reagents besides HI and HBr cleave ethers; one of the most useful of these is the Lewis acid boron trichloride, BCl_3. Boron trichloride is a mild reagent that cleaves many ethers at 0°C. It is therefore often used for the cleavage of delicate ethers, which might be destroyed by HI at reflux. For example, the propyl ether of vitamin E reacts smoothly with BCl_3 at 0°C. Note, however, that the more hindered cyclic ether also present in vitamin E is not cleaved by these mild conditions.

Propyl ether of vitamin E: $CH_3CH_2CH_2O$–(trimethyl chroman ring, 2-CH_3)–$(CH_2)_3CH(CH_3)(CH_2)_3CH(CH_3)(CH_2)_3CH(CH_3)_2$

$$\xrightarrow[\text{2. } H_2O]{\text{1. } BCl_3\text{, 0°C}}$$

HO–(trimethyl chroman ring, 2-CH_3, 2-$C_{16}H_{33}$) + $CH_3CH_2CH_2Cl$

α-Tocopherol (vitamin E)

Acidic ether cleavage processes are typical nucleophilic substitution reactions; they take place by either an S_N1 pathway or an S_N2 pathway, depending on the structure of the ether. Primary and secondary alkyl ethers react by an S_N2 pathway in which iodide or bromide ion attacks the protonated ether at the less substituted site. This usually results in a selective cleavage into a single alcohol and a single alkyl halide, rather than a mixture of products. The ether oxygen atom stays with the more hindered alkyl group, and the halide attacks the less hindered group. For example, butyl isopropyl ether yields exclusively isopropyl alcohol and

1-iodobutane on cleavage by HI, since nucleophilic attack by iodide occurs at the less hindered primary site rather than at the more hindered secondary site. Similarly, anisole is cleaved to give phenol and iodomethane.

$$(CH_3)_2CH-\ddot{O}-CH_2CH_2CH_2CH_3 \xrightarrow[H_2O,\ 100°C]{HI} \left[(CH_3)_2CH-\overset{+}{\ddot{O}}(H)-CH_2CH_2CH_2CH_3\right] \xrightarrow{:I^-\ (S_N2\ reaction)} (CH_3)_2CHOH + CH_3CH_2CH_2CH_2I$$

Butyl isopropyl ether; Isopropyl alcohol; 1-Iodobutane

$$C_6H_5OCH_3 \xrightarrow[H_2O,\ 100°C]{HI} C_6H_5OH + CH_3I$$

Anisole; Phenol; Iodomethane

Tertiary, benzylic, and allylic ethers tend to cleave by an S_N1 mechanism, since the intermediate carbocations are so stable. These reactions are often fast and take place at moderate temperatures. *tert*-Butyl ethers, for example, can often be cleaved at room temperature or below. Trifluoroacetic acid, rather than HBr or HI, seems to be best for these reactions, as the following example shows:

$$C_6H_{11}-O-C(CH_3)_3 \xrightarrow[0°C]{CF_3COOH} C_6H_{11}OH + (CH_3)_2C{=}CH_2$$

tert-Butyl cyclohexyl ether; Cyclohexanol (90%); 2-Methylpropene

PROBLEM

19.7 Write a detailed mechanism for the trifluoroacetic acid catalyzed cleavage of a *tert*-butyl ether. Account for the fact that isobutylene is formed.

PROBLEM

19.8 Can you suggest an explanation of the observation that HI and HBr are much more effective than HCl in cleaving ethers? [*Hint:* See Section 10.5.]

19.7 Cyclic ethers; epoxides

For the most part, cyclic ethers behave like acyclic ethers. The chemistry of the ether functional group is the same, whether it is in an open chain or in a ring. For example, common cyclic ethers such as tetrahydrofuran and dioxane are often used as solvents because of their inertness, yet they can be cleaved by strong acids.

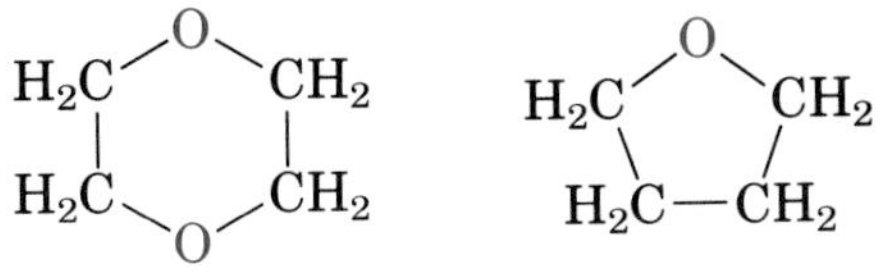

1,4-Dioxane

Tetrahydrofuran

The one group of cyclic ethers that behaves differently includes the three-membered, oxygen-containing rings called **epoxides,** or **oxiranes.** The strain of the three-membered ether ring makes epoxides highly reactive and confers unique chemical reactivity on them (recall the reactivity of cyclopropanes, Section 18.7).

Ethylene oxide, the simplest epoxide, is an intermediate in the manufacture of both ethylene glycol (automobile antifreeze) and polyester polymers. More than 2.5 billion lb of ethylene oxide are produced industrially each year by air oxidation of ethylene over a silver oxide catalyst at 300°C. This process is not of general utility for other epoxides, however, and is of little value in the laboratory.

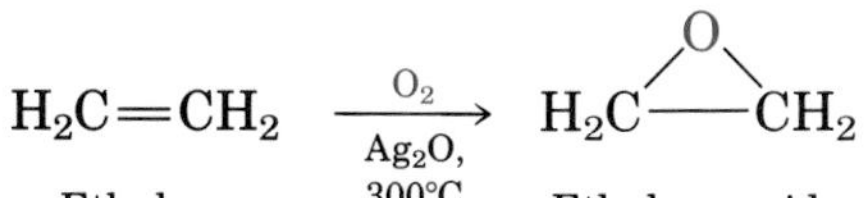

$$H_2C{=}CH_2 \xrightarrow[Ag_2O,\ 300°C]{O_2} H_2C\!-\!CH_2 \text{ (with bridging O)}$$

Ethylene → Ethylene oxide

In the laboratory, epoxidation is normally carried out by treatment of the alkene with a peroxyacid, RCO_3H:

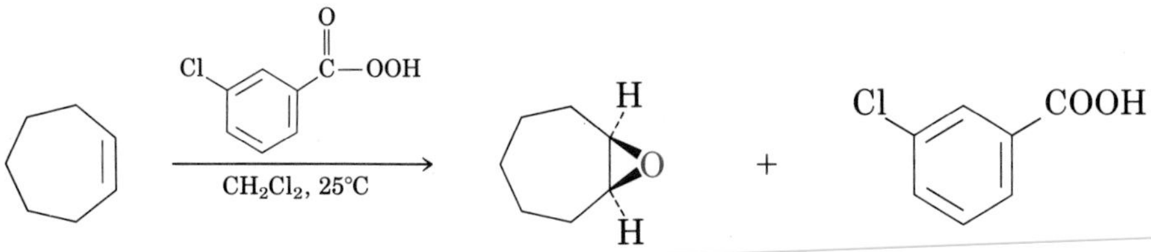

Cycloheptene → (m-ClC$_6$H$_4$CO$_3$H, CH_2Cl_2, 25°C) → 1,2-Epoxycycloheptane (78%) (cycloheptene epoxide) + m-ClC$_6$H$_4$COOH

Many different peroxyacids accomplish the epoxidation, but *m*-chloroperoxybenzoic acid is the preferred reagent on a laboratory scale. As opposed to most other peroxyacids, which are unstable and readily decompose, *m*-chloroperoxybenzoic acid is a stable, crystalline, easily handled material.

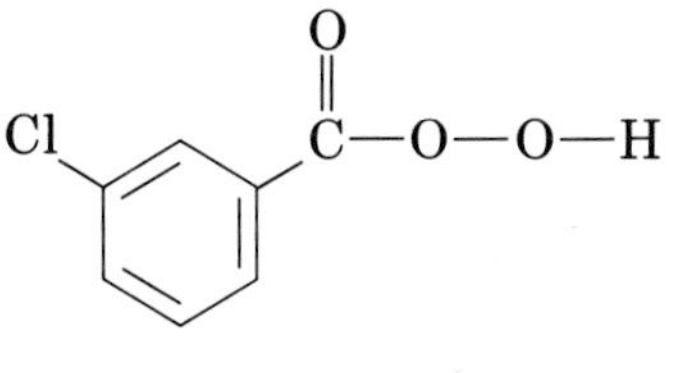

m-Chloroperoxybenzoic acid

Peroxyacids are thought to transfer oxygen to an alkene through a rather complex one-step mechanism without intermediates. The epoxidation reaction is different from other alkene addition reactions we have studied, since no carbocation intermediate is involved, and the details of the process are not fully understood. There is good evidence to show, however, that the oxygen farthest from the carbonyl group is the one that is transferred.

$$\text{Cycloheptene} + \text{Peroxyacid} \xrightarrow[25°C]{CH_2Cl_2} \text{1,2-Epoxycycloheptane} + \text{Acid}$$

Cycloheptene Peroxyacid 1,2-Epoxycycloheptane Acid

Another method for the synthesis of epoxides is through the use of halohydrins, prepared by electrophilic addition to alkenes. When halohydrins are treated with base, HX is eliminated, and an epoxide is produced.

$$\text{Cyclohexene} \xrightarrow[H_2O]{Cl_2} \text{2-Chlorocyclohexanol} \xrightarrow[H_2O]{NaOH} \text{1,2-Epoxycyclohexane} + H_2O + NaCl$$

Cyclohexene 2-Chlorocyclohexanol 1,2-Epoxycyclohexane (73%)

Note that the synthesis of epoxides by base treatment of halohydrins is actually an *intramolecular* Williamson ether synthesis. The nucleophilic oxygen atom and electrophilic carbon atom are in the same molecule, rather than in different molecules:

$$\text{Bromohydrin} \xrightarrow[H_2O]{NaOH} [\text{Intramolecular substitution}] \longrightarrow \text{Epoxide} + :Br^-$$

Bromohydrin Intramolecular substitution (within the same molecule) Epoxide

Recall the following:

$$CH_3O^- + CH_3{-}Br \longrightarrow CH_3O{-}CH_3 + :Br^-$$

Intermolecular substitution (between different molecules)

19.8 Ring-opening reactions of epoxides

Epoxide rings can be opened by treatment with acid in much the same way that other ethers are cleaved. The major difference is that epoxides react under much milder conditions because of ring strain. Dilute aqueous mineral acid at room temperature is sufficient to cause the hydrolysis of epoxides to 1,2-diols (also called *glycols*); 2 million tons of ethylene glycol are produced each year by acid-catalyzed hydration of ethylene oxide.

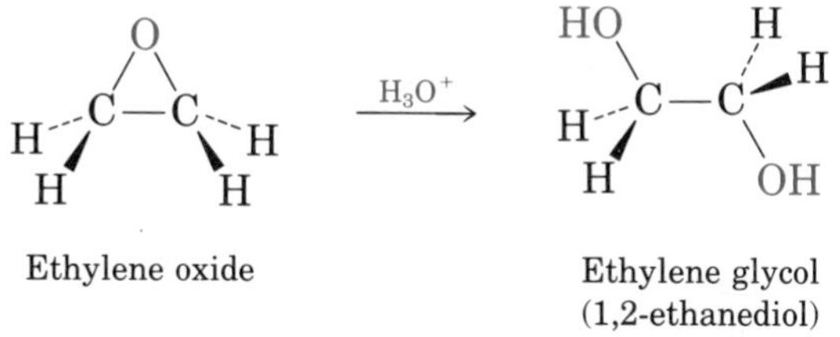

$\xrightarrow{H_3O^+}$

Ethylene oxide

Ethylene glycol (1,2-ethanediol)

Acid-induced epoxide ring-opening takes place by S_N2 attack of a nucleophile on the protonated epoxide, in a manner analogous to the final step of alkene bromination, where a three-membered ring bromonium ion is opened by nucleophilic attack (Section 6.1). When a cycloalkane epoxide is opened by aqueous acid, a *trans*-1,2-diol results (just as a *trans*-1,2-dibromide results from alkene bromination).

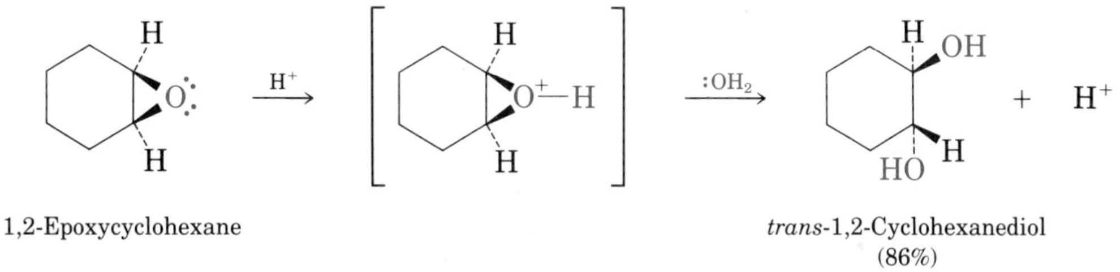

$\xrightarrow{H^+}$ $\xrightarrow{:OH_2}$ + H^+

1,2-Epoxycyclohexane

trans-1,2-Cyclohexanediol (86%)

Recall the following reaction:

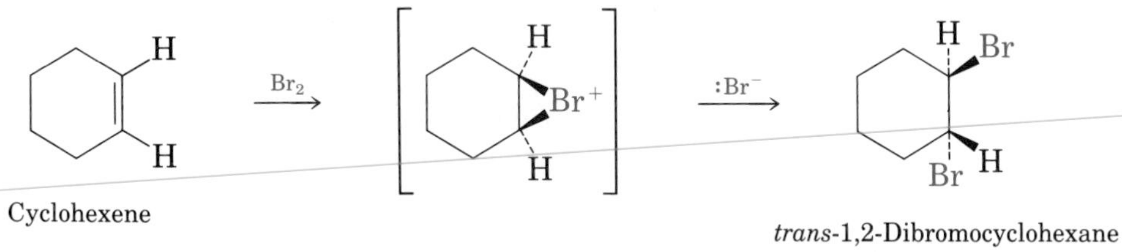

$\xrightarrow{Br_2}$ $\xrightarrow{:Br^-}$

Cyclohexene

trans-1,2-Dibromocyclohexane

Protonated epoxides can also be opened by nucleophiles other than water. For example, if anhydrous HX is used, epoxides can be converted into trans halohydrins:

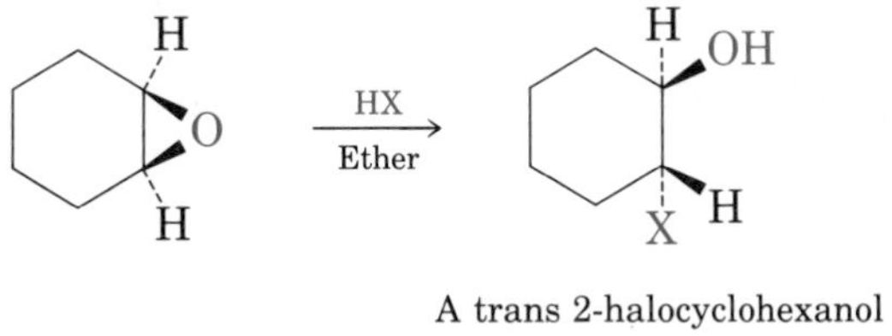

$\xrightarrow[\text{Ether}]{HX}$

A trans 2-halocyclohexanol

Here X = F, Br, Cl, or I.

Epoxides, unlike other ethers, can also be cleaved by base. An ether oxygen is normally a very poor leaving group in an S_N2 reaction (Section 10.5), but the reactivity of the three-membered ring is sufficient to cause epoxides to react with hydroxide ion at elevated temperatures.

S_N2 reaction

$$\text{Methylenecyclohexane oxide} \xrightarrow[100^\circ C]{:\ddot{O}H^-,\ H_2O} \text{1-Hydroxymethylcyclohexanol (70\%)}$$

Methylenecyclohexane oxide

1-Hydroxymethylcyclohexanol (70%)

PROBLEM

19.9 Show all the steps involved in the acidic hydrolysis of *cis*-5,6-epoxydecane. What is the stereochemistry of the product, assuming normal back-side S_N2 attack? Do the same for *trans*-5,6-epoxydecane and describe the product stereochemistry.

PROBLEM

19.10 Acid-induced hydrolysis of 1,2-epoxycyclohexanes is known to produce trans-diaxial 1,2-diols. Draw the product you would expect to obtain from acidic hydrolysis of *cis*-3-*tert*-butyl-1,2-epoxycyclohexane. (Recall that the bulky *tert*-butyl group locks the cyclohexane ring into a specific conformation.) Answer this question for *trans*-3-*tert*-butyl-1,2-epoxycyclohexane also.

REGIOCHEMISTRY OF EPOXIDE RING-OPENING

The direction in which unsymmetrical epoxide rings are opened depends on the conditions used. If base is used in a typical S_N2-type reaction, attack of the nucleophile takes place at the less hindered epoxide carbon. For example, 1,2-epoxypropane is attacked by ethoxide ion exclusively at the primary carbon to give 1-ethoxy-2-propanol.

$$CH_3-\overset{O}{CH-CH_2} \xrightarrow[CH_3CH_2OH]{^-:\ddot{O}CH_2CH_3} CH_3\overset{OH}{\overset{|}{C}}HCH_2OCH_2CH_3$$

No attack here (2°)

1-Ethoxy-2-propanol (83%)

If acid conditions are used, however, a different reaction course is followed, and attack of the nucleophile occurs primarily at the more substituted carbon atom.

$$CH_3\overset{O}{CH-CH_2} \xrightarrow[Ether]{HCl} CH_3\underset{Cl}{\underset{|}{C}}H\overset{OH}{\overset{|}{C}}H_2 + CH_3\overset{OH}{\overset{|}{C}}HCH_2Cl$$

1,2-Epoxypropane

2-Chloro-1-propanol (89%)

1-Chloro-2-propanol (11%)

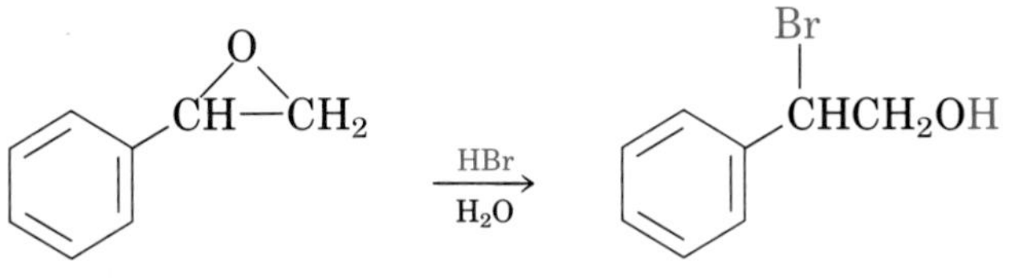

$$\text{C}_6\text{H}_5\text{-}\overset{\text{O}}{\text{CH}-\text{CH}_2} \xrightarrow[\text{H}_2\text{O}]{\text{HBr}} \text{C}_6\text{H}_5\text{-}\underset{|}{\overset{\text{Br}}{\text{C}}}\text{HCH}_2\text{OH}$$

Phenyl-1,2-epoxyethane 2-Bromo-2-phenylethanol

This difference in regiochemistry between the acidic and basic reactions can be explained by postulating that the transition state for acid-induced opening has a high degree of carbocationic character. Attack of the nucleophile therefore occurs at the more stabilized (more highly substituted) cationic site.

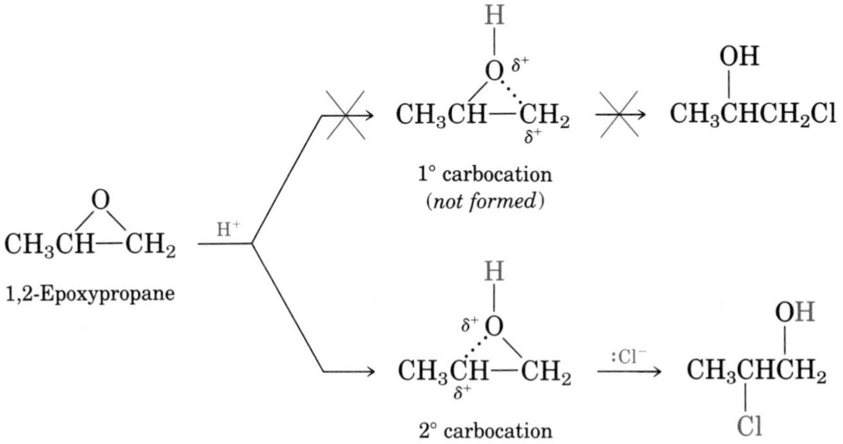

$$\text{CH}_3\overset{\text{O}}{\text{CH}-\text{CH}_2} \xrightarrow{\text{H}^+} \text{CH}_3\text{CH}-\overset{\text{O}^{\delta+}\text{H}}{\text{CH}_2^{\delta+}} \not\longrightarrow \text{CH}_3\overset{\text{OH}}{\text{C}}\text{HCH}_2\text{Cl}$$

1,2-Epoxypropane

1° carbocation
(not formed)

$$\text{CH}_3\text{CH}^{\delta+}-\overset{\delta^+\text{O}\text{H}}{\text{CH}_2} \xrightarrow{:\text{Cl}^-} \text{CH}_3\underset{\text{Cl}}{\text{C}}\text{H}\overset{\text{OH}}{\text{C}}\text{H}_2$$

2° carbocation
(more stable)

PROBLEM

19.11 Predict the major product of each of the following reactions.

(a) 1-methyl-1,2-epoxycyclohexane (CH$_3$, O) $\xrightarrow{\text{HBr}}$

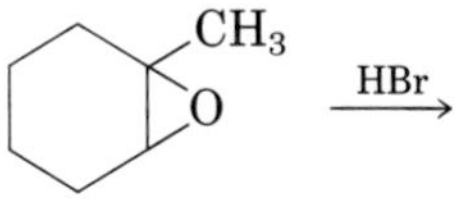

(b) $\text{CH}_3\text{CH}_2\overset{\text{O}}{\text{CH}-\text{CH}_2} \xrightarrow[\text{NH}_3]{^-:\text{NH}_2}$

(c) $\text{CH}_3\text{CH}_2\overset{\text{O}}{\text{CH}-\text{CH}_2} \xrightarrow{^+\text{H}_3\,^{18}\text{O}}$

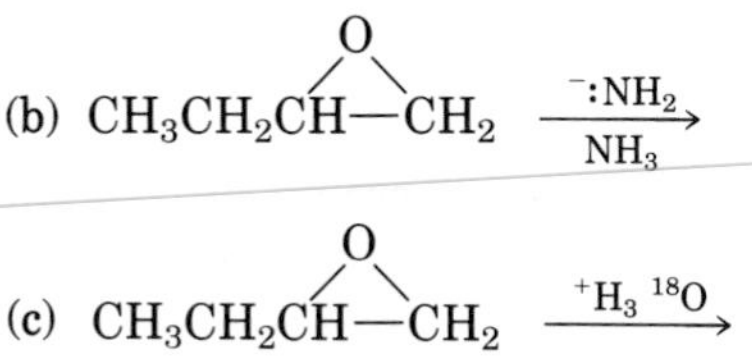

19.9 Crown ethers

Crown ethers are a relatively recent addition to the ether family. First discovered in the early 1960s at Du Pont, crown ethers have become important over the last decade. They can be looked at as cyclic polymers of ethylene glycol, and they are named according to the general format *x*-crown-*y*, where *x* is the total number of atoms in the ring and *y* is the number of oxygen atoms. Thus, 18-crown-6 ether is an 18-membered ring containing 6 ether oxygen atoms, as shown at the top of the next page.

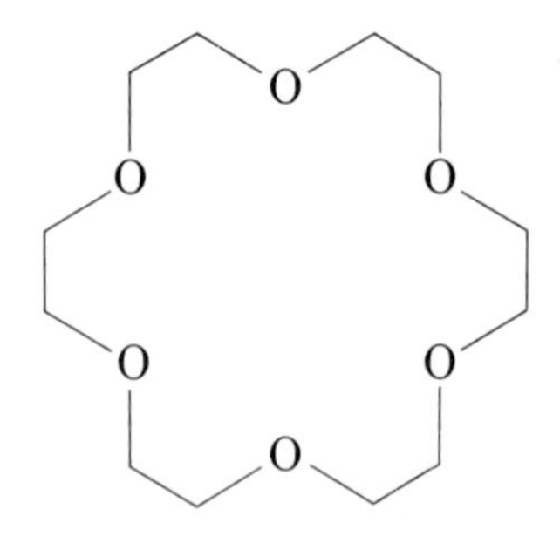

18-Crown-6 ether

The great importance of crown ethers derives from their extraordinary ability to solvate metal cations by sequestering the metal in the center of the polyether cavity. Different crown ethers solvate different metal cations, depending on the match between ion size and cavity size. For example, 18-crown-6 is able to complex strongly with potassium ion. Complexes between crown ethers and inorganic salts are soluble in nonpolar organic solvents, thus allowing many reactions to be carried out under aprotic conditions that would otherwise have to be carried out in aqueous solution. For example, potassium permanganate actually dissolves in benzene in the presence of 18-crown-6. The resulting solution of "purple benzene" is a valuable reagent for oxidizing alkenes.

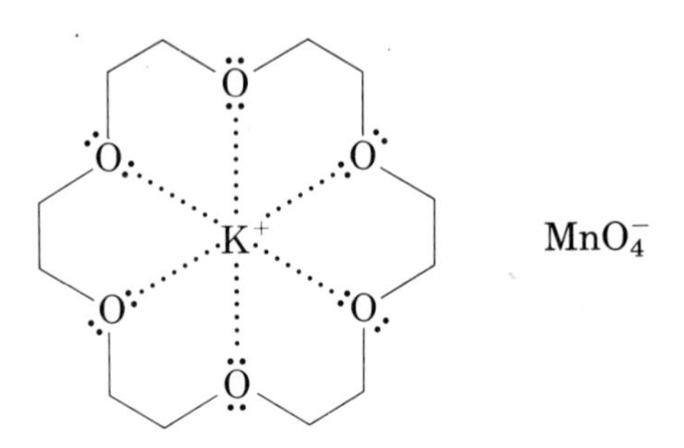

$KMnO_4$ solvated by 18-crown-6
(this solvate is soluble in benzene)

Many other inorganic salts, such as KF, KCN, and NaN_3, can be solubilized in organic solvents by the use of crown ethers, and the nucleophilicities of the anions are often dramatically increased.

PROBLEM

19.12 Ethers 15-crown-5 and 12-crown-4 complex Na^+ and Li^+, respectively. Make models of these crown ethers and compare the sizes of the cavities.

19.10 Spectroscopic analysis of ethers

INFRARED SPECTROSCOPY

Ethers are difficult to distinguish by infrared spectroscopy. Although they show a characteristic absorption due to carbon–oxygen single-bond stretching in the range 1050–1150 cm^{-1}, many other kinds of absorptions occur in the same range. Figure 19.1 shows the infrared spectrum of diethyl ether and identifies the C—O stretch.

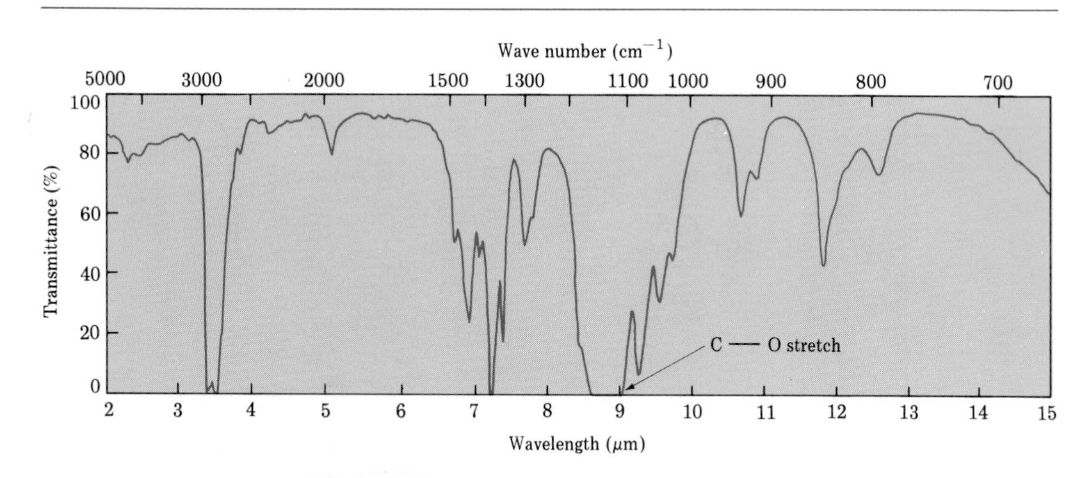

Figure 19.1. Infrared spectrum of diethyl ether, $CH_3CH_2—O—CH_2CH_3$.

NUCLEAR MAGNETIC RESONANCE SPECTROSCOPY

Protons on carbon next to the ether oxygen are shifted downfield from the normal alkane resonance and show characteristic ^{1}H NMR absorption in the region 3.5–4.5 δ. This downfield shift is clearly indicated in the spectrum of dipropyl ether shown in Figure 19.2.

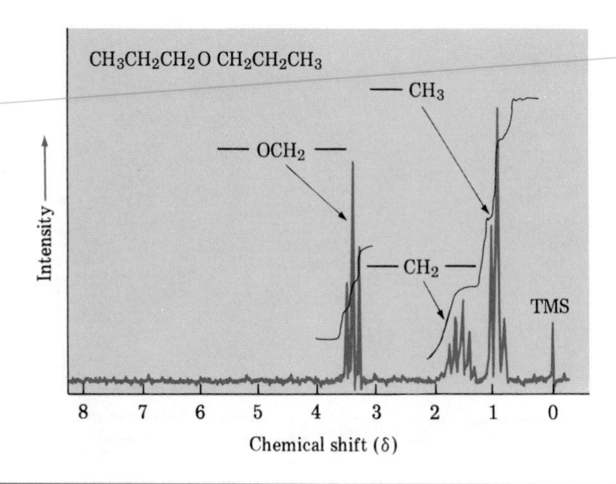

Figure 19.2. Proton NMR spectrum of dipropyl ether.

Epoxides absorb at a slightly higher field than other ethers and show characteristic resonances at 2.5–3.5 δ, as indicated for 1,2-epoxypropane in Figure 19.3.

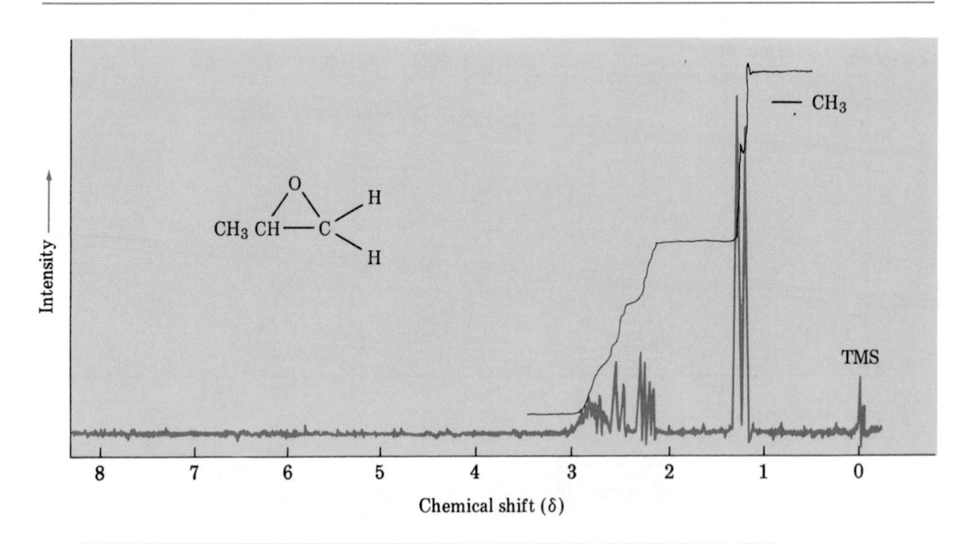

Figure 19.3. Proton NMR spectrum of 1,2-epoxypropane.

Ether carbon atoms also exhibit a downfield shift in the ^{13}C NMR spectrum, where they usually absorb in the range 40–80 δ.

PROBLEM

19.13 The ^1H NMR spectrum shown is for a substance having the formula C_4H_8O. Propose a structure compatible with the observed spectrum.

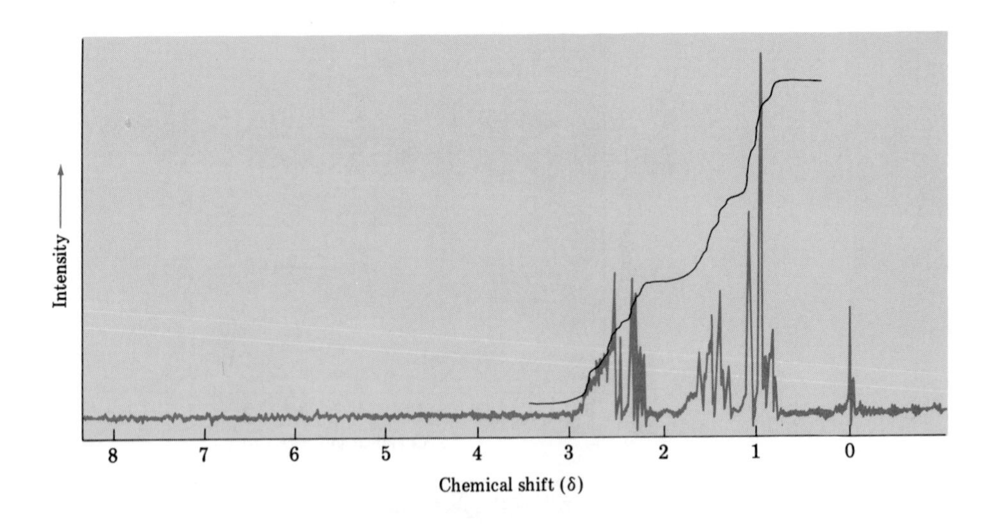

19.11 Sulfides

Sulfur is the element just below oxygen in the periodic table, and many oxygen-containing organic compounds have sulfur analogs. For example, **sulfides** are sulfur analogs of ethers, R—S—R′.

Sulfides are named by following the same rules of nomenclature used for ethers. *Sulfide* is used in place of *ether* for simple compounds, and *alkylthio* is used in place of *alkoxy* for more complex substances.

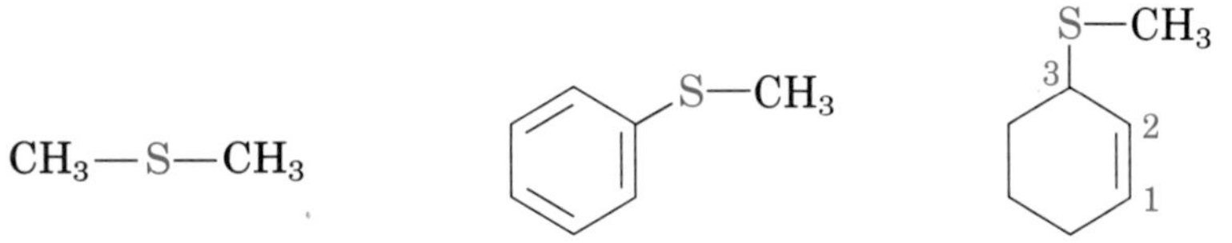

$CH_3—S—CH_3$ Dimethyl sulfide

Methyl phenyl sulfide

3-(Methylthio)cyclohexene

Sulfides are best prepared by treatment of a primary or secondary alkyl halide with a **thiolate** anion, R—S:⁻. Reaction occurs by an S_N2 mechanism, analogous to the Williamson synthesis of ethers (Section 19.4). Thiolate anions are among the best nucleophiles known (Section 10.5 and Table 10.3) and product yields are usually high in these substitution reactions.

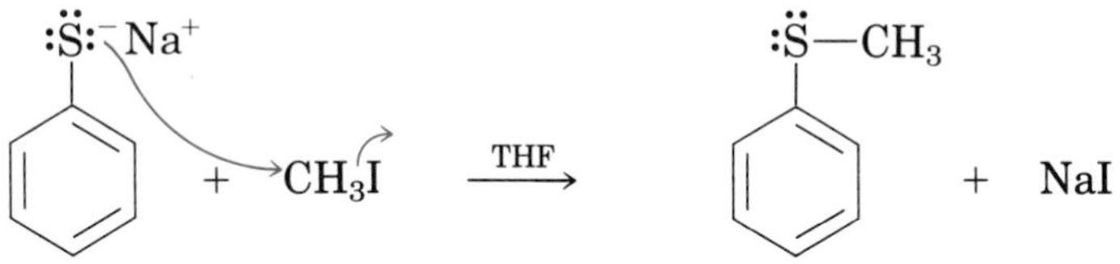

Sodium benzenethiolate + CH_3I $\xrightarrow{THF}$ Methyl phenyl sulfide (96%) + NaI

The outer-shell electrons on sulfur are farther from the nucleus and are less tightly held than those on oxygen (3*p* electrons versus 2*p* electrons). As a result, there are some important differences between the chemistries of ethers and sulfides. For example, sulfur is more polarizable than oxygen, and sulfur compounds are thus more nucleophilic than their oxygen analogs. Dialkyl sulfides are good nucleophiles that react rapidly with primary alkyl halides by an S_N2 mechanism. The products of such reactions are **trialkylsulfonium salts.**

$$CH_3—\ddot{S}—CH_3 + CH_3—I \xrightarrow{THF} CH_3—\overset{CH_3}{S^+}—CH_3 \ :I^-$$

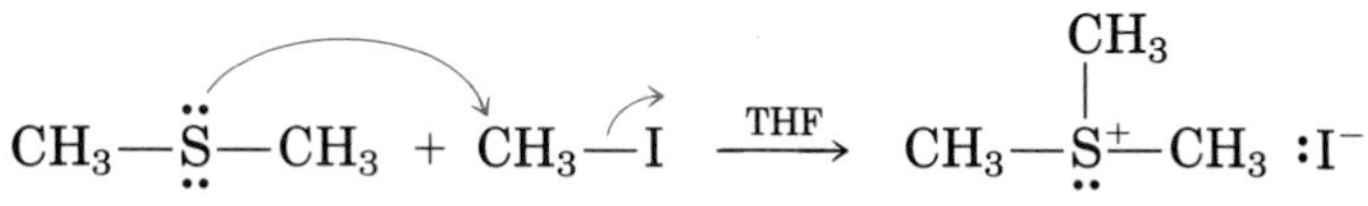

Dimethyl sulfide, Iodomethane, Trimethylsulfonium iodide

Nature makes extensive use of the trialkylsulfonium salt, *S*-adenosylmethionine, as a biological methylating agent (Section 10.15).

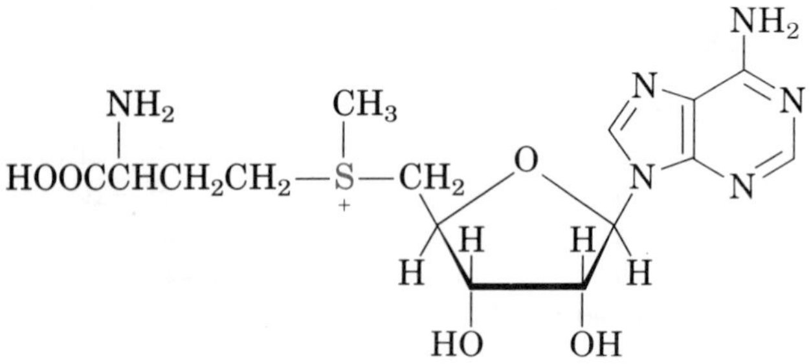

S-Adenosylmethionine (a sulfonium salt)

A second difference between sulfides and ethers is that sulfides are easily oxidized. Treatment of a sulfide with hydrogen peroxide, H_2O_2, at room temperature yields the corresponding **sulfoxide** (R_2SO), and further oxidation of the sulfoxide with a peroxyacid yields a **sulfone** (R_2SO_2).

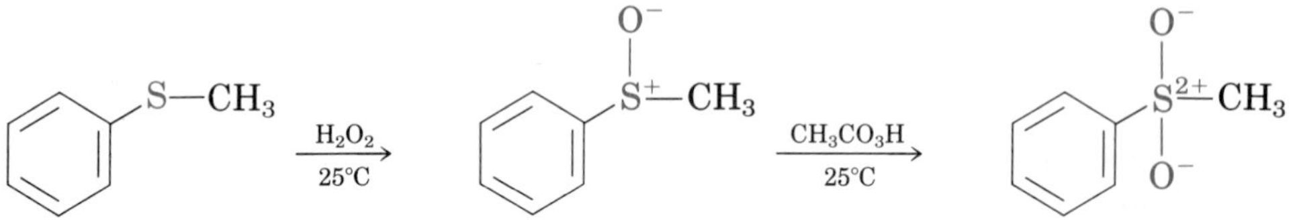

$$\xrightarrow[25^\circ C]{H_2O_2} \qquad \xrightarrow[25^\circ C]{CH_3CO_3H}$$

Methyl phenyl sulfide — Methyl phenyl sulfoxide — Methyl phenyl sulfone

Dimethyl sulfoxide (DMSO) is a particularly well-known sulfoxide that finds wide use as a polar aprotic solvent. It has a remarkable ability to penetrate the skin, carrying along with it whatever impurities are present; it must therefore be handled cautiously.

PROBLEM

19.14 Name the following compounds by IUPAC rules:

(a) $CH_3CH_2SCH_3$ (b) $(CH_3)_3CSCH_2CH_3$

(c) 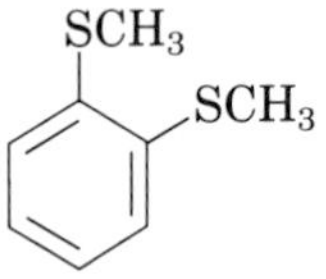 (d)

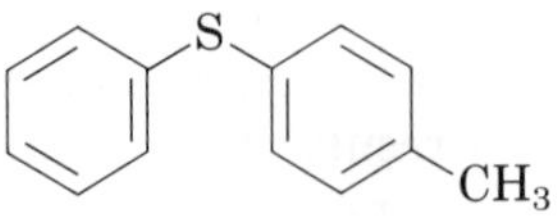

PROBLEM

19.15 How can you account for the fact that dimethyl sulfoxide has a boiling point of 189°C and is miscible with water, whereas dimethyl sulfide has a boiling point of 37°C and is immiscible with water?

19.12 Summary

Ethers are compounds that have two organic groups bonded to the same oxygen atom, R—O—R′. The organic groups may be alkyl, vinylic, or aryl, and the oxygen atom may be in a ring or in an open chain.

Diethyl ether is prepared industrially for use as a solvent by the sulfuric acid catalyzed dehydration of ethanol. In the laboratory, however, ethers are normally prepared either by a Williamson synthesis or by the alkoxymercuration–demercuration sequence. The Williamson synthesis involves S_N2 attack of an alkoxide ion on a primary alkyl halide, and the reaction is thus limited to the preparation of ethers in which one organic group is primary. The alkoxymercuration–demercuration sequence involves the formation of an intermediate alkoxy organomercurial compound, followed by sodium borohydride reduction of the carbon–mercury bond. The net result is Markovnikov addition of an alcohol to an alkene.

Ethers are inert to most common reagents but are attacked by strong acids to give cleavage products. Both HI and HBr are often used, but BCl_3 is the mildest reagent for ether cleavage and should be used in sensitive cases. Epoxides differ from other ethers in their ease of cleavage. The high reactivity of the three-membered ether ring because of ring strain allows epoxide rings to be opened by nucleophilic attack of strong bases as well as by acids. Base-catalyzed epoxide ring-opening occurs by S_N2 attack of a nucleophile at the less hindered epoxide carbon, whereas acid-induced epoxide ring-opening occurs by attack at the more substituted epoxide carbon.

Sulfides are sulfur analogs of ethers. They are prepared by a Williamson-type S_N2 reaction between a thiolate anion and a primary or secondary alkyl halide. Sulfides are much more nucleophilic than ethers and can be oxidized to sulfoxides (R_2SO) and sulfones (R_2SO_2).

Ethers are not readily identifiable by infrared spectroscopy, but show characteristic downfield ^1H NMR absorptions that are easily detected.

19.13 Summary of reactions

1. Preparation of ethers
   a. Williamson synthesis (Section 19.4)

$$CH_3CH_2O{:}^- + C_6H_5CH_2Cl \longrightarrow C_6H_5CH_2OCH_2CH_3 + {:}Cl^-$$

Alkyl halide should be primary.

   b. Alkoxymercuration–demercuration (Section 19.5)

$$C_6H_5CH{=}CH_2 + CH_3OH \xrightarrow[\text{2. NaBH}_4]{\text{1. Hg(O}_2\text{CCF}_3)_2} C_6H_5CH(OCH_3)CH_3$$

Markovnikov orientation is observed.

c. Epoxidation of alkenes with peroxyacids (Section 19.7)

$$\text{cyclohexene} \xrightarrow[CH_2Cl_2]{m\text{-}ClC_6H_4C(=O){-}OOH} \text{cyclohexene oxide (H, O, H)}$$

2. Reaction of ethers

a. Cleavage by HX (Section 19.6)

$$CH_2CH_2OCH(CH_3)_2 \xrightarrow[H_2O]{HBr} CH_3CH_2Br + (CH_3)_2CHOH$$

where HX = HBr or HI.

b. Cleavage by BCl_3 (Section 19.6)

$$C_6H_5OCH_3 \xrightarrow[2.\ H_2O]{1.\ BCl_3} C_6H_5OH + CH_3Cl$$

c. Base-catalyzed epoxide ring-opening (Section 19.8)

$$CH_3CH\overset{O}{-}CH_2 \xrightarrow[CH_3CH_2OH]{^{-}:OCH_2CH_3} CH_3\overset{OH}{C}HCH_2OCH_2CH_3$$

d. Acid-catalyzed hydrolysis of epoxides (Section 19.8)

$$\text{cyclohexene oxide (H, O, H)} \xrightarrow{H_3O^+} \text{cyclohexane-1,2-diol (H, OH, HO, H)}$$

trans-1,2-Diols are produced.

e. Acid-induced epoxide ring-opening (Section 19.8)

$$(CH_3)_2C\overset{O}{-}CH_2 \xrightarrow[H_2O]{HBr} (CH_3)_2\overset{Br}{C}CH_2OH$$

3. Preparation of sulfides (Section 19.11)

$$C_6H_5\ddot{S}:^- + CH_3I \xrightarrow{THF} C_6H_5S{-}CH_3 + :I^-$$

4. Oxidation of sulfides (Section 19.11)

a. $C_6H_5-S-CH_3 \xrightarrow[H_2O]{H_2O_2} C_6H_5-S^{+}(O^{-})-CH_3$

A sulfoxide

b. $C_6H_5-S-CH_3 \xrightarrow{2\ CH_3CO_3H} C_6H_5-S^{2+}(O^{-})_2-CH_3$

A sulfone

ADDITIONAL PROBLEMS

19.16 Draw structures corresponding to these IUPAC names:
(a) Ethyl 1-ethylpropyl ether
(b) Di(*p*-chlorophenyl) ether
(c) 3,4-Dimethoxybenzoic acid
(d) 4-Allyl-2-methoxyphenol (eugenol; oil of cloves)
(e) Cyclopentyloxycyclohexane

19.17 Provide correct IUPAC names for these structures:

(a) Benzene ring with two adjacent OCH_3 groups

(b) $(CH_3)_2CH-O-$cyclopropyl

(c) Five-membered ring numbered 1 (O), 2, 3, 4, 5 with CH_3 on C-2

(d) Cyclohexyl$-S-$cyclopropyl

(e) $(CH_3)_2C(OCH_3)_2$

(f) Cyclopentane ring fused with an epoxide O

(g) p-$NO_2-C_6H_4OCH_2CH_3$

(h) Cyclohexane with two SCH_3 groups on the same carbon

19.18 When 2-methylpentane-2,5-diol is treated with sulfuric acid, dehydration occurs and 2,2-dimethyltetrahydrofuran is formed. Suggest a mechanism for this reaction. Which of the two oxygen atoms is most likely to be eliminated, and why?

Tetrahydrofuran ring with two CH_3 groups on C-2

2,2-Dimethyltetrahydrofuran

19.19 Predict the likely products of these cleavage reactions:

(a) 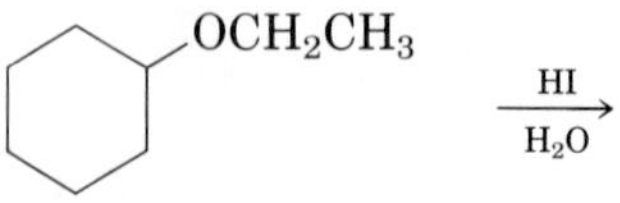 Cyclohexyl–OCH_2CH_3 $\xrightarrow[H_2O]{HI}$

(b) 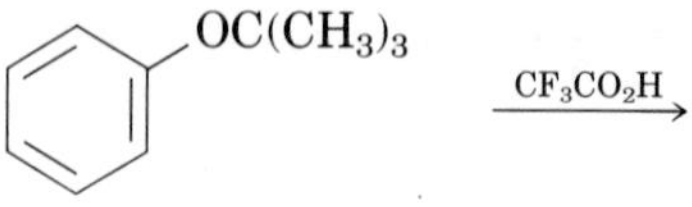 Phenyl–$OC(CH_3)_3$ $\xrightarrow{CF_3CO_2H}$

(c) $H_2C{=}CHOCH_2CH_3$ $\xrightarrow[H_2O]{HI}$

(d) $(CH_3)_3CCH_2{-}O{-}CH_2CH_3$ $\xrightarrow[H_2O]{HI}$

(e) 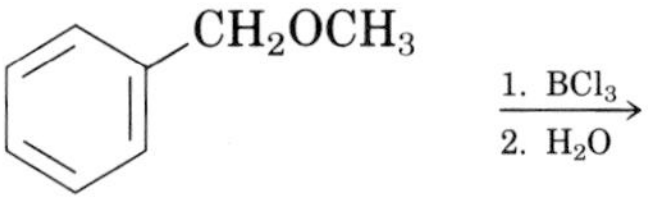 Phenyl–CH_2OCH_3 $\xrightarrow[2.\ H_2O]{1.\ BCl_3}$

19.20 The **Zeisel method** is a classic analytical procedure for determining the number of methoxyl groups in a compound. In this method, a weighed amount of the compound is heated with concentrated HI. Ether cleavage occurs, and the iodomethane formed is distilled off and passed into an alcohol solution of $AgNO_3$. The silver iodide that precipitates is then collected and weighed, and the percentage of methoxyl groups in the sample is thereby determined. For example, vanillin, the material responsible for the characteristic odor of vanilla, gives a positive Zeisel reaction; 1.06 g vanillin yields 1.60 g AgI. If vanillin has a molecular weight of 152, how many methoxyls does it contain?

19.21 How would you prepare these ethers?

(a) 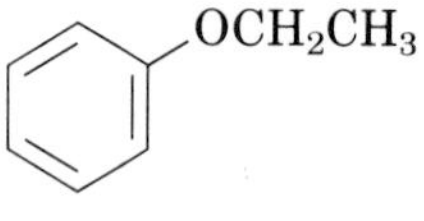 Phenyl–OCH_2CH_3

(b) 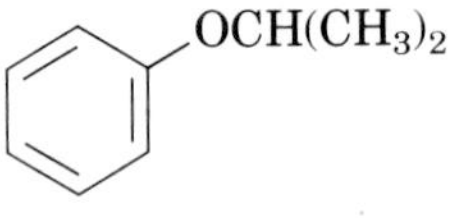 Phenyl–$OCH(CH_3)_2$

(c) 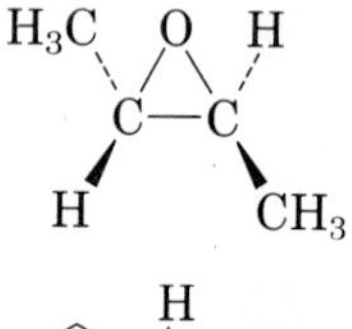 Epoxide: H_3C, O, H; C—C; H, CH_3

(d) 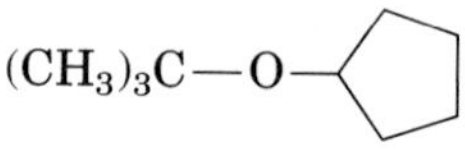 $(CH_3)_3C{-}O{-}$cyclopentyl

(e) 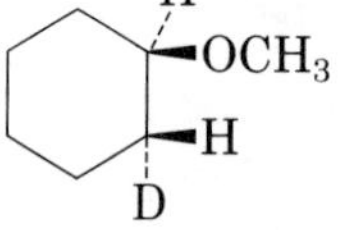 Cyclohexane with H, OCH_3, H, D

(f) 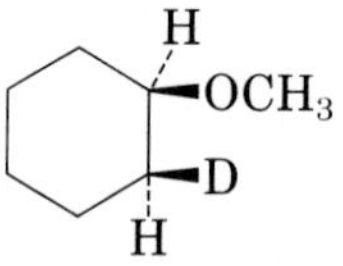 Cyclohexane with H, OCH_3, D, H

19.22 **Meerwein's reagent,** triethyloxonium tetrafluoroborate, is a powerful ethylating agent that converts alcohols into ethyl ethers at neutral pH. Formulate the reaction of Meerwein's reagent with cyclohexanol and account for the fact that trialkyloxonium salts are much more reactive alkylating agents than are alkyl iodides.

$$(CH_3CH_2)_3O^+\ BF_4^-$$

Meerwein's reagent

19.23 Propose a mechanism to account for the cleavage of an ether with BCl_3.

19.24 How would you carry out these transformations? More than one step may be required.

(a) 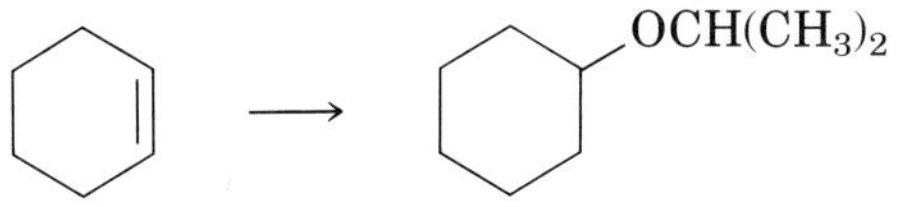 Cyclohexene ⟶ cyclohexyl–$OCH(CH_3)_2$

(b) 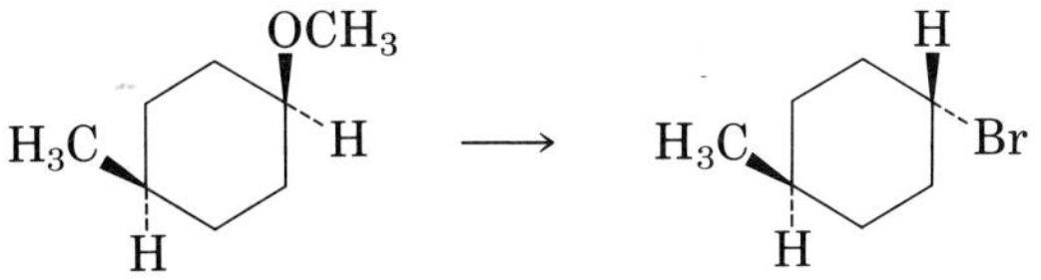 H_3C–cyclohexane–OCH_3 (with H's) ⟶ H_3C–cyclohexane–Br (with H's)

(c) 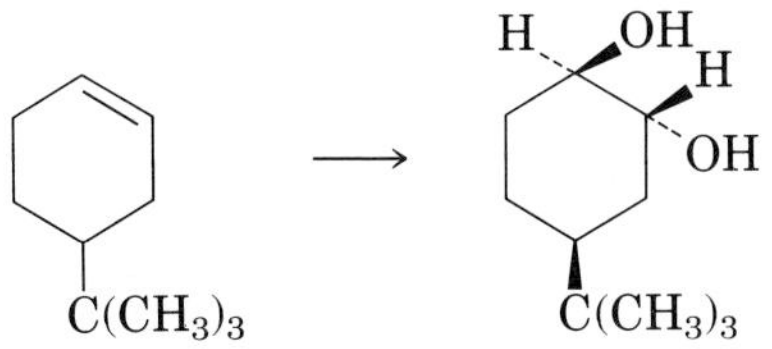 4-$C(CH_3)_3$-cyclohexene ⟶ diol (H, OH, H, OH) with $C(CH_3)_3$

(d) $CH_3CH_2CH_2CH_2C{\equiv}CH \longrightarrow CH_3CH_2CH_2CH_2CH_2CH_2OCH_3$

(e) $CH_3CH_2CH_2CH_2C{\equiv}CH \longrightarrow CH_3CH_2CH_2CH_2CH(OCH_3)CH_3$

19.25 What product(s) would you expect from these reactions?

(a) 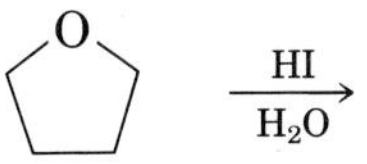 Tetrahydrofuran $\xrightarrow[H_2O]{HI}$

(b) 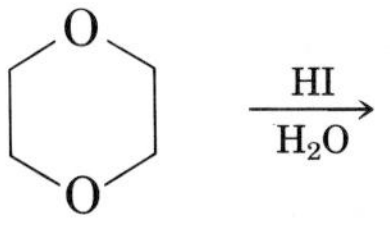 1,4-Dioxane $\xrightarrow[H_2O]{HI}$

19.26 Methyl aryl ethers can be cleaved to methyl iodide and the phenoxide anion by treatment with LiI in hot dimethylformamide solvent. Propose a mechanism for this reaction. What type of reaction is involved?

19.27 *tert*-Butyl ethers can be prepared by the reaction of an alcohol with 2-methylpropene in the presence of an acid catalyst. Propose a mechanism for this reaction.

19.28 Safrole, isolated from oil of sassafras, is used as a perfumery agent. Propose a synthesis of safrole, starting from catechol (1,2-dihydroxybenzene).

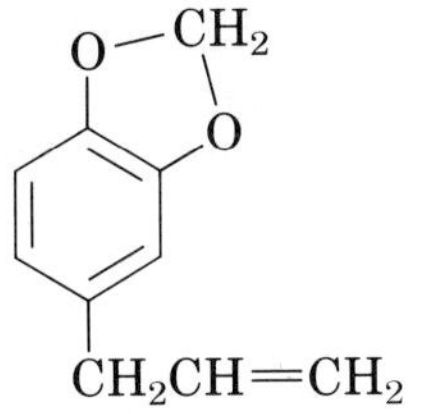

O—CH_2—O (methylenedioxy) benzene ring with $CH_2CH{=}CH_2$

Safrole

19.29 Ethyl vinyl ether reacts with ethanol in the presence of an acid catalyst to yield 1,1-diethoxyethane, rather than 1,2-diethoxyethane. How can you account for the observed regioselectivity of addition?

19.30 We will see in Section 22.11 that ketones and aldehydes react with alcohols to yield acetals, $R_2C(OR')_2$. For example, cyclohexanone reacts with ethylene glycol in the presence of an acid catalyst to yield an ethylene acetal. Suggest a mechanism for this reaction.

Cyclohexanone + $HOCH_2CH_2OH$ $\xrightarrow{H^+}$ (acetal) + H_2O

An acetal

19.31 In light of your answer to Problem 19.30, propose a mechanism to account for the reaction of 5-hydroxypentanal with methanol to yield 2-methoxytetrahydropyran.

HO–(CH₂)₄–CHO $\xrightarrow[H^+]{CH_3OH}$ 2-methoxytetrahydropyran (OCH_3)

19.32 Anethole, $C_{10}H_{12}O$, a major constituent of the oil of anise, has the NMR spectrum shown here. On oxidation with hot sodium dichromate, anethole yields a compound identified as *p*-methoxybenzoic acid. What is the structure of anethole? Assign all peaks in the NMR spectrum and account for the observed splitting patterns.

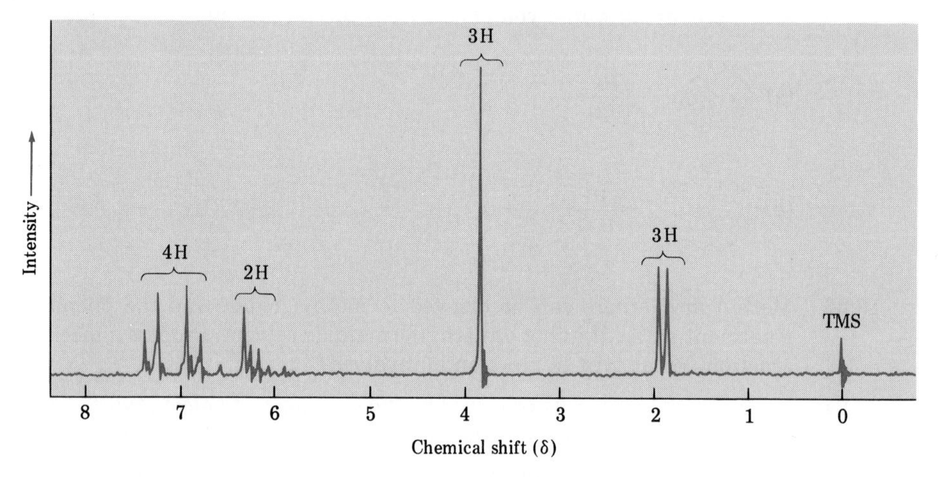

19.33 How would you synthesize anethole (Problem 19.32) from benzene?

19.34 The red fox (*Vulpes vulpes*) uses a chemical communication system based on scent marks in urine. Recent work has shown one component of fox urine to be a sulfide. Mass spectral analysis of the pure scent-mark component shows $M^{+\cdot} = 116$. Infrared spectroscopy shows an intense band at 890 cm^{-1} and ^1H NMR spectroscopy reveals the following peaks:

3-proton singlet at 1.74 δ
3-proton singlet at 2.11 δ
2-proton triplet at 2.27 δ, $J = 4.2$ Hz
2-proton triplet at 2.57 δ, $J = 4.2$ Hz
2-proton broad peak at 4.73 δ

Propose a structure consistent with these data. [*Note:* $(CH_3)_2S$ absorbs at 2.1 δ in ^1H NMR.]

19.35 How can you explain the fact that treatment of bornene with a peroxyacid yields a different epoxide than is obtained by reaction of bornene with aqueous bromine, followed by base treatment?

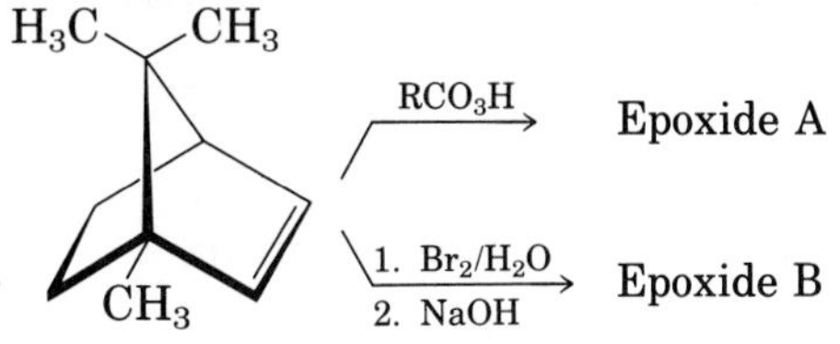

H_3C CH_3 CH_3 Bornene $\xrightarrow{RCO_3H}$ Epoxide A

$\xrightarrow[2.\ NaOH]{1.\ Br_2/H_2O}$ Epoxide B

Propose structures for epoxides A and B. [*Hint:* Build molecular models of bornene.]

19.36 Disparlure, $C_{19}H_{38}O$, is a sex attractant released by the female gypsy moth, *Lymantria dispar*. The ^1H NMR spectrum of disparlure shows a large absorption in the alkane region, 1–2 δ, and a triplet at 2.8 δ. Treatment of disparlure, first with aqueous acid and then with periodic acid, HIO_4, yields two carbonyl-containing compounds identified as undecanal and 6-methylheptanal. Neglecting stereochemistry, propose a structure for disparlure consistent with these data. The actual natural product is a chiral molecule with 7*R*,8*S* stereochemistry. Draw disparlure, showing the correct stereochemistry.

19.37 How would you synthesize racemic disparlure (Problem 19.36) from compounds having 10 or fewer carbons?

CHAPTER 20
ALCOHOLS AND THIOLS

Alcohols are compounds that have hydroxyl functional groups bonded to saturated, sp^3-hybridized carbon atoms. This definition purposely excludes phenols (hydroxyl groups bonded to an aromatic ring) and enols (hydroxyl groups bonded to a vinylic carbon), because these two types of compounds have characteristic chemistry quite different from that of saturated alcohols.

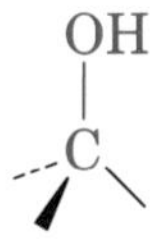

An alcohol 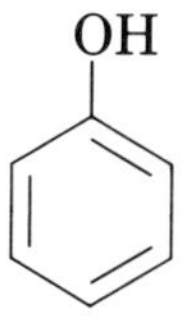 — A phenol 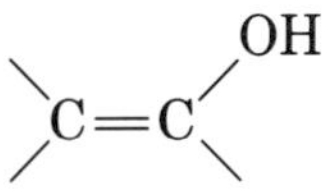 — An enol

Alcohols are an exceedingly important class of compounds in organic chemistry. They occur widely in nature and have a variety of valuable industrial and pharmaceutical applications. Ethanol, for example, is one of the simplest, yet best known, of all organic substances; menthol, a 10-carbon alcohol from peppermint oil, is widely used as a flavoring and perfumery agent; and cholesterol, a complicated-looking steroidal alcohol, has been implicated as a causative agent in heart disease. The chemistry of all these compounds is much the same, however, since the hydroxyl functional group behaves the same, regardless of the complexity of the remainder of the molecule.

CH_3CH_2OH

Ethanol — Menthol 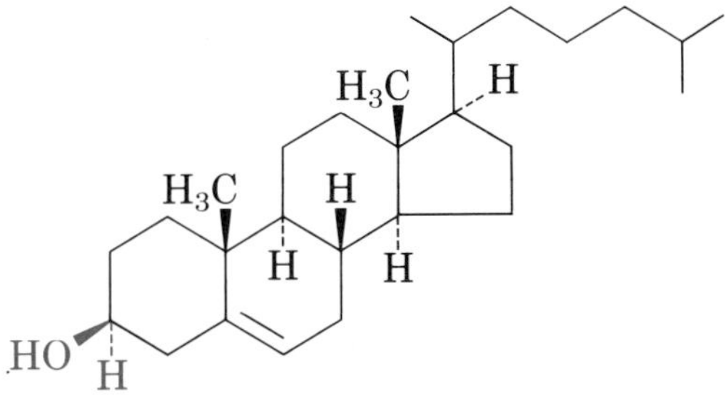 — Cholesterol (a steroid)

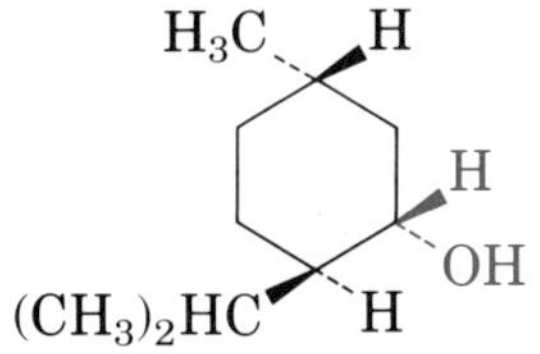

20.1 Nomenclature of alcohols

Alcohols are either primary (1°), secondary (2°), or tertiary (3°), depending on the number of carbon substituents bonded to the hydroxyl carbon.

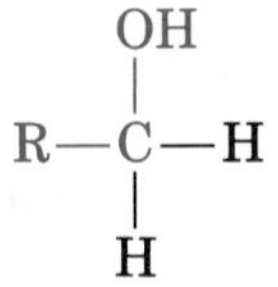

A primary alcohol 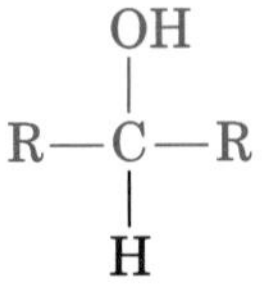 — A secondary alcohol 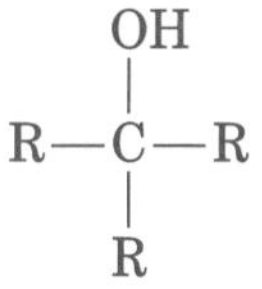 — A tertiary alcohol

Simple alcohols are named by the IUPAC system as derivatives of the parent alkane:

1. Select the longest carbon chain *containing the hydroxyl group,* and derive the parent name by replacing the *-e* ending of the corresponding alkane with *-ol.*
2. Number the alkane chain beginning at the end nearer the hydroxyl group.
3. Number all substituents according to their position on the chain and write the name listing the substituents in alphabetical order.

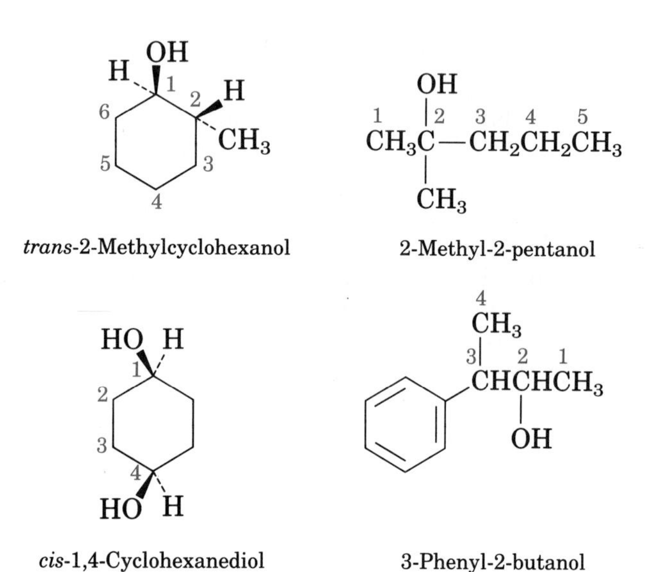

trans-2-Methylcyclohexanol

2-Methyl-2-pentanol

cis-1,4-Cyclohexanediol

3-Phenyl-2-butanol

Certain simple and commonly occurring alcohols have trivial names that are accepted by IUPAC. For example,

$C_6H_5CH_2OH$ — Benzyl alcohol (phenylmethanol)

$H_2C{=}CHCH_2OH$ — Allyl alcohol (2-propen-1-ol)

$(CH_3)_3C{-}OH$ — *tert*-Butyl alcohol (2-methyl-2-propanol)

$HO{-}CH_2CH_2{-}OH$ — Ethylene glycol (1,2-ethanediol)

$HO{-}CH_2CH(OH)CH_2{-}OH$ — Glycerol (1,2,3-propanetriol)

PROBLEM

20.1 Provide IUPAC names for these alcohols:

(a) 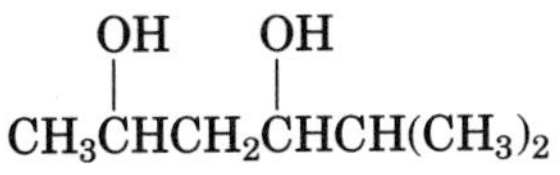 $CH_3CH(OH)CH_2CH(OH)CH(CH_3)_2$

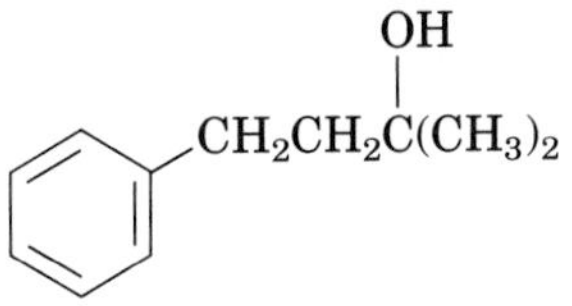

(b) $C_6H_5CH_2CH_2C(OH)(CH_3)_2$

(c) 4,4-dimethylcyclohexanol structure: OH on ring carbon; H_3C, CH_3 on opposite ring carbon

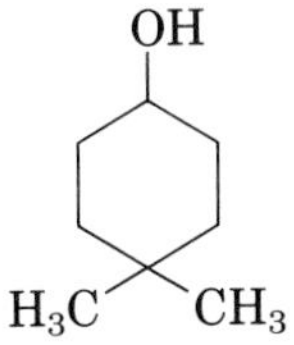

(d) 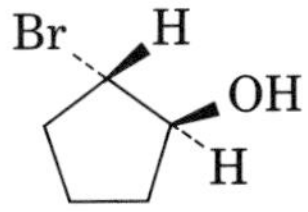 cyclopentane ring with Br, H on one carbon and OH, H on adjacent carbon

(e) 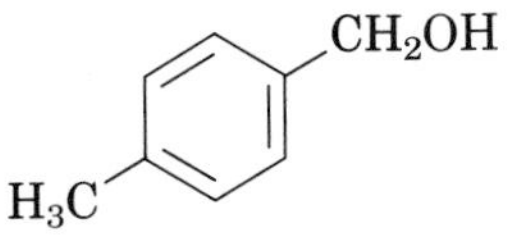 $H_3C-C_6H_4-CH_2OH$

PROBLEM

20.2 Draw structures corresponding to these IUPAC names:

(a) 2-Ethyl-2-buten-1-ol
(b) 3-Cyclohexen-1-ol
(c) *trans*-3-Chlorocycloheptanol
(d) 2-(Hydroxymethyl)-1,4-butanediol

20.2 Properties of alcohols

Alcohols are quite different from the hydrocarbons and alkyl halides we have studied thus far. Not only is their chemistry much richer, but their physical properties are also very different. Table 20.1 (p. 634) provides a comparison of the boiling points of some simple alcohols, alkanes, and chloroalkanes, and shows that alcohols have much higher boiling points. For example, 1-propanol (mol wt = 60), butane (mol wt = 58), and chloroethane (mol wt = 65) are close in weight, yet 1-propanol boils at 97°C, compared to −0.5°C for the alkane and 12.5°C for the chloroalkane.

The reason for their high boiling points is that alcohols, like water, are highly *associated* in solution because of the formation of hydrogen bonds. The positively polarized hydroxyl hydrogen atom from one molecule forms a weak hydrogen bond to the negatively polarized oxygen atom of another molecule (Figure 20.1). Hydrogen bonds have a strength of only about 5

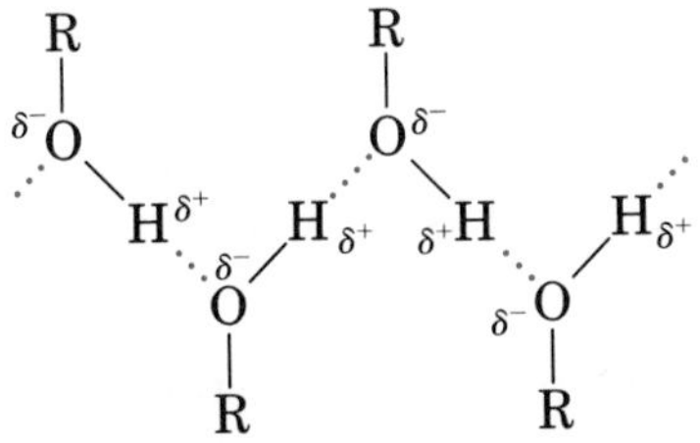

Figure 20.1. Hydrogen bonding in alcohols.

TABLE 20.1 **Boiling points of alkanes, chloroalkanes, and alcohols (°C)**

Alkyl group, R	*Alkane,* R—H	*Chloroalkane,* R—Cl	*Alcohol,*[a] R—OH
CH_3—	−162	−24	64.5 (−97)
CH_3CH_2—	−88.5	12.5	78.3 (−115)
$CH_3CH_2CH_2$—	−42	46.6	97 (−126)
$(CH_3)_2CH$—	−42	36.5	82.5 (−86)
$CH_3CH_2CH_2CH_2$—	−0.5	83.5	117 (−90)
$(CH_3)_3C$—	−12	51	83 (25.5)
$CH_3(CH_2)_8CH_2$—	174	223	228 (6)
Cyclohexyl—	81	142.5	161.5 (24)
$C_6H_5CH_2$—	111	179	205 (−15)
CH_2— | CH_2—	−88.5	83.5	197 (−16)

[a]Melting points are given in parentheses.

kcal/mol and are therefore much weaker than the 103 kcal/mol of the usual alcohol O—H covalent bond. Nevertheless, the presence of many hydrogen bonds means that extra energy must be added to break them during the boiling process, and boiling points are therefore raised.

PROBLEM

20.3 The following data for three isomeric four-carbon alcohols show that there is a decrease in boiling point with increasing substitution:

1-Butanol, bp 117.5°C
2-Butanol, bp 99.5°C
2-Methyl-2-propanol, bp 82.2°C

Propose an explanation to account for this trend. [*Hint:* Consider the effect of hydrogen bonding, among other things.]

20.3 Sources of simple alcohols

Methanol and ethanol are two of the most important of all industrial chemicals, and both are manufactured on a vast scale for a variety of uses. Prior to the development of the modern chemical industry, methanol was prepared by heating wood in the absence of air and thus came to be called *wood alcohol.* Today, approximately 1.2 billion gal of methanol are manufactured each year in the United States by catalytic reduction of carbon monoxide with hydrogen gas.

$$CO + 2\ H_2 \xrightarrow[\text{Zinc oxide/chromia}]{400°C} CH_3OH$$

Methanol is toxic to humans, causing blindness in low doses and death in larger amounts. Industrially, it is used both as a solvent and as a starting material for production of formaldehyde, CH_2O, and acetic acid, CH_3COOH.

Ethanol is one of the oldest known pure organic chemicals. Its production by fermentation of grains and sugars, and its subsequent purification by distillation, go back at least as far as the ancient Greeks. Fermentation is carried out by adding yeast to an aqueous sugar solution. Enzymes in the yeast break down carbohydrates into ethanol and CO_2.

$$\underset{\text{A carbohydrate}}{C_6H_{12}O_6} \xrightarrow{\text{Yeast}} 2\ CH_3CH_2OH + 2\ CO_2$$

Only about 5% of the ethanol produced industrially comes from fermentation, although that figure may well change drastically in the next decade as demand for use in automobile fuel increases. Most ethanol is currently obtained by acid-catalyzed hydration of ethylene. Nearly 300 million gal of ethanol a year are produced in the United States for use as a solvent or as a chemical intermediate in other industrial reactions.

$$\underset{\text{Ethylene}}{H_2C{=}CH_2} + H_2O \xrightarrow{H_2SO_4} CH_3CH_2OH$$

20.4 Preparation of alcohols

In many ways, alcohols occupy a central position in organic chemistry. They can be prepared from a variety of functional-group classes (alkenes, alkyl halides, ketones, esters, and aldehydes, among others) and they can be transformed into a wide assortment of compound types. Let's review briefly some of the methods of alcohol preparation that we have already seen.

Alcohols can be prepared by hydration of alkenes. The direct hydration of alkenes with aqueous acid is generally a poor reaction in the laboratory, and indirect methods are therefore preferred. Two methods are commonly used, and a choice between the two is made depending on the product desired. Hydroboration–oxidation (Section 6.4) yields the product of syn non-Markovnikov hydration, whereas oxymercuration–reduction (Section 6.3) yields the product of Markovnikov hydration. Both reactions are mild and are generally applicable to a number of alkenes (Figure 20.2, p. 636).

1,2-Diols can be prepared either by direct hydroxylation of an alkene with osmium tetroxide followed by sodium bisulfite reduction, or by acid-catalyzed epoxide ring-opening. Both reactions take place readily and are routinely used. Direct hydroxylation with osmium tetroxide yields a cis 1,2-diol, whereas acid-catalyzed epoxide ring-opening yields a trans 1,2-diol (Figure 20.3).

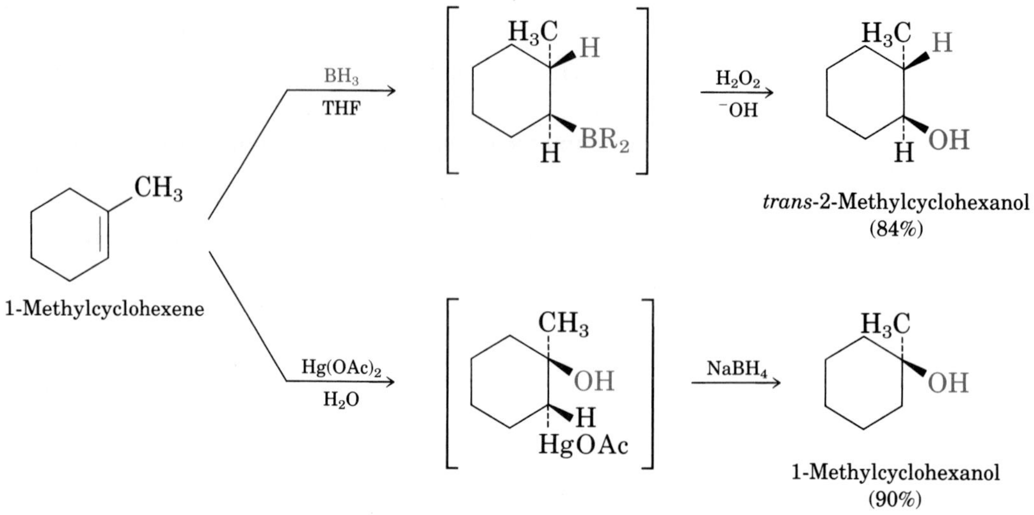

CH_3

1-Methylcyclohexene

BH_3 / THF

H_3C H H BR_2

H_2O_2 / ^-OH

H_3C H H OH

trans-2-Methylcyclohexanol (84%)

$Hg(OAc)_2$ / H_2O

CH_3 OH H HgOAc

$NaBH_4$

H_3C OH

1-Methylcyclohexanol (90%)

Figure 20.2. Two complementary methods for the hydration of an alkene.

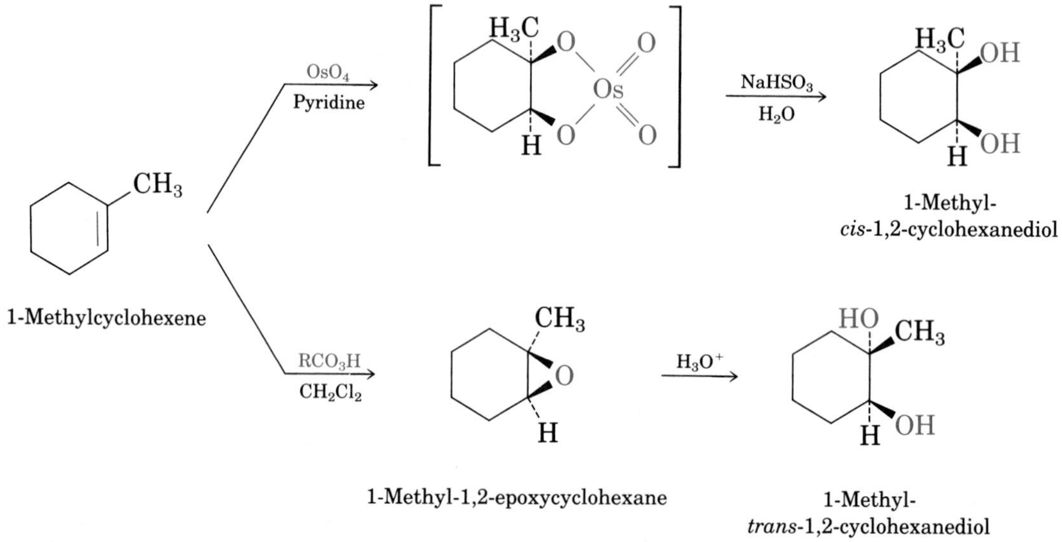

CH_3

1-Methylcyclohexene

OsO_4 / Pyridine

H_3C O O Os H O O

$NaHSO_3$ / H_2O

H_3C OH H OH

1-Methyl-*cis*-1,2-cyclohexanediol

RCO_3H / CH_2Cl_2

CH_3 O H

1-Methyl-1,2-epoxycyclohexane

H_3O^+

HO CH_3 H OH

1-Methyl-*trans*-1,2-cyclohexanediol

Figure 20.3. Two complementary methods for the preparation of 1,2-diols from an alkene.

PROBLEM

20.4 Predict the products of these reactions:

(a) Hydroboration of styrene, followed by treatment with basic hydrogen peroxide

(b) Oxymercuration of styrene, followed by sodium borohydride reduction

(c) Reaction of *cis*-5-decene with OsO_4, followed by $NaHSO_3$ reduction (indicate stereochemistry)

(d) Reaction of *trans*-5-decene with *m*-chloroperoxybenzoic acid, followed by treatment with aqueous acid (indicate stereochemistry)

20.5 Alcohols from reduction of carbonyl groups

The most valuable method for preparing alcohols is by **reduction** of carbonyl compounds:

$$\text{C=O} \xrightarrow{[H]} \text{H–C–OH}$$

where [H] is a reducing agent.

In inorganic chemistry, reduction is defined as the gain of electrons by an atom, and oxidation as the loss of electrons by an atom. In organic chemistry, it is often difficult to decide whether an atom gains or loses electrons during a reaction, and the terms oxidation and reduction have less precise meanings. For our purposes, an **organic reduction** is a reaction that either adds hydrogen or removes an electronegative element (oxygen, nitrogen, or halogen) from a molecule. Conversely, an **organic oxidation** is a reaction that either removes hydrogen or adds an electronegative element (oxygen, nitrogen, or halogen) to a molecule. For example, catalytic hydrogenation of an alkene is clearly a reduction, since two hydrogens are added to the starting material. Epoxidation and NBS brominations, however, are oxidations, since oxygen and bromine are added, respectively. Hydration of an alkene is neither an oxidation nor a reduction, since both hydrogen and oxygen are added to the alkene in the same reaction.

$$\text{C=C} \xrightarrow[\text{Ethanol}]{H_2/Pd} \text{H–C–C–H}$$

Reduction (addition of H_2)

$$\text{C=C} \xrightarrow[CH_2Cl_2]{RCO_3H} \text{C–C (epoxide, O)}$$

Oxidation (addition of O)

$$C_6H_5CH_3 \xrightarrow[CCl_4]{\text{NBS}} C_6H_5CH_2Br$$

Oxidation (removal of H and addition of Br)

$$\text{C=C} \xrightarrow{H_2O,\ H^+} \text{H–C–C–OH}$$

Neither oxidation nor reduction (addition of both H and —OH)

A list of functional-group classes of increasing oxidation state is shown in Figure 20.4. Any reaction that converts a functional group from a lower to a higher level is an oxidation; any reaction converting a functional group from a higher level to a lower level is a reduction; and any reaction that does not change the oxidation level is neither an oxidation nor a reduction.

	$H_2C{=}CH_2$	$HC{\equiv}CH$		
CH_3CH_3	CH_3CH_2OH	$CH_3CH{=}O$	CH_3CO_2H	CO_2
	$CH_3CH_2NH_2$	$CH_3CH{=}NH$	$CH_3C{\equiv}N$	
	CH_3CH_2Cl	CH_3CHCl_2	CH_3CCl_3	CCl_4

Low oxidation level → High oxidation level

Figure 20.4. Oxidation states for some functional-group classes.

PROBLEM

20.5 Rank the following series of compounds in order of increasing oxidation state:

(a) 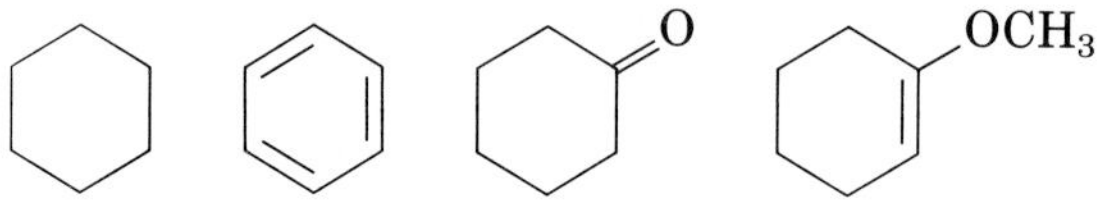 cyclohexane, benzene, cyclohexanone, 1-methoxycyclohexene (OCH$_3$)

(b) CH_3CN, $CH_3CH_2NH_2$, $NH_2CH_2CH_2NH_2$

PROBLEM

20.6 Are these reactions oxidations, reductions, or neither?

(a) Cyclohexene + NBS $\longrightarrow$

(b) 3-Bromocyclohexene + Base $\longrightarrow$

(c) Benzene + $Cl_2/FeCl_3$ $\longrightarrow$

(d) 1-Bromobutane + Mg, then H^+ $\longrightarrow$

(e) Benzene + $CH_3Cl/AlCl_3$ $\longrightarrow$

REDUCTION OF ALDEHYDES AND KETONES

Aldehydes and ketones are the most easily reduced carbonyl functional groups, and literally dozens of different reagents convert them into alcohols. In the laboratory, one always seeks the mildest, safest, and most convenient reagent available for a given task, and sodium borohydride, $NaBH_4$, is therefore usually chosen for aldehyde and ketone reductions. Sodium borohydride is a white, crystalline solid that can be safely handled and weighed. It can be used either in water or in alcohol solution. Aldehydes are reduced by $NaBH_4$ to give primary alcohols, and ketones are reduced to give secondary alcohols. High yields are usually obtained, as the following examples indicate:

$$CH_3CH_2CH_2\overset{\overset{\Large O}{\|}}{C}{-}H \xrightarrow[2.\ H_3O^+]{1.\ NaBH_4,\ ethanol} CH_3CH_2CH_2\overset{\overset{\Large OH}{|}}{\underset{\underset{\Large H}{|}}{C}}{-}H$$

Butanal → 1-Butanol (85%)

$$\text{m-Hydroxybenzaldehyde} \xrightarrow[2.\ H_3O^+]{1.\ NaBH_4,\ ethanol} \text{m-Hydroxybenzyl alcohol (93\%)}$$

m-Hydroxybenzaldehyde → *m*-Hydroxybenzyl alcohol (93%)

$$\text{Dicyclohexyl ketone} \xrightarrow[2.\ H_3O^+]{1.\ NaBH_4,\ ethanol} \text{Dicyclohexylmethanol (88\%)}$$

Dicyclohexyl ketone → Dicyclohexylmethanol (88%)

Lithium aluminum hydride, $LiAlH_4$, is another reducing agent that is often used. A white solid that is normally used in ether or tetrahydrofuran solution, $LiAlH_4$ is an *extremely dangerous* reagent. It reacts violently with water, decomposes explosively when heated above 120°C, and must therefore be handled only by skilled persons. Despite these drawbacks, $LiAlH_4$ is an extremely valuable reagent that is used daily in thousands of laboratories. It is particularly useful for reducing α,β-unsaturated ketones (ketones conjugated with carbon–carbon double bonds). α,β-Unsaturated ketones often undergo overreduction with $NaBH_4$ to give a mixture of both unsaturated alcohol and saturated alcohol. With $LiAlH_4$, however, clean reduction to the allylic alcohol occurs. Thus, 2-cyclohexenone gives a 59:41 mixture of two products when $NaBH_4$ is used as a reducing agent, but gives largely one product when $LiAlH_4$ is used (Figure 20.5, p. 640).

REDUCTION OF ESTERS AND CARBOXYLIC ACIDS

Esters and carboxylic acids can be reduced to give primary alcohols:

$$\underset{\text{Ester}}{R{-}CO_2CH_3} \quad \text{or} \quad \underset{\text{Carboxylic acid}}{R{-}CO_2H} \xrightarrow{[H]} R{-}\overset{\overset{\Large H}{|}}{\underset{\underset{\Large H}{|}}{C}}{-}OH$$

2-Cyclohexenone (an α,β-unsaturated ketone)

1. $NaBH_4$, ethanol 2. H_3O^+ → 2-Cyclohexenol (59%) + Cyclohexanol (41%)

1. $LiAlH_4$, ether 2. H_3O^+ → 2-Cyclohexenol (94%) + Cyclohexanol (2%)

Figure 20.5. The reduction of 2-cyclohexenone with $NaBH_4$ and with $LiAlH_4$.

These reactions are more difficult than the corresponding reductions of aldehydes and ketones; for example, sodium borohydride only slowly reduces esters and does not reduce acids at all. Ester and carboxylic acid reductions are therefore usually carried out with lithium aluminum hydride. All carbonyl groups, including esters, acids, ketones, and aldehydes, are reduced by $LiAlH_4$ in high yield, as some of the following examples indicate. Note that *one* hydrogen atom is delivered to the carbonyl carbon atom during reductions of ketones and aldehydes, but that *two* hydrogens become bonded to the carbonyl carbon during ester and carboxylic acid reductions.

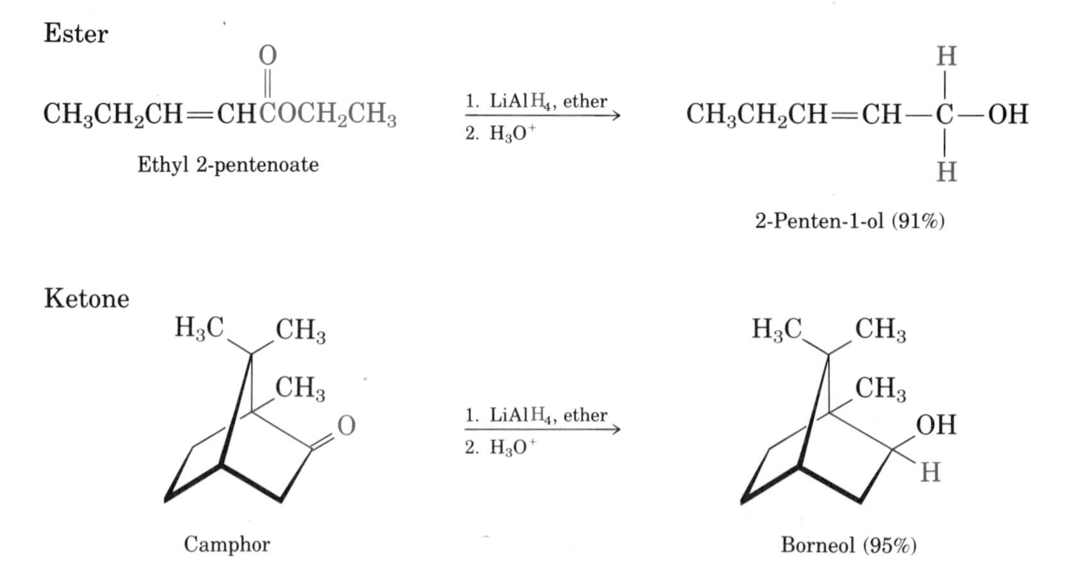

Aldehyde

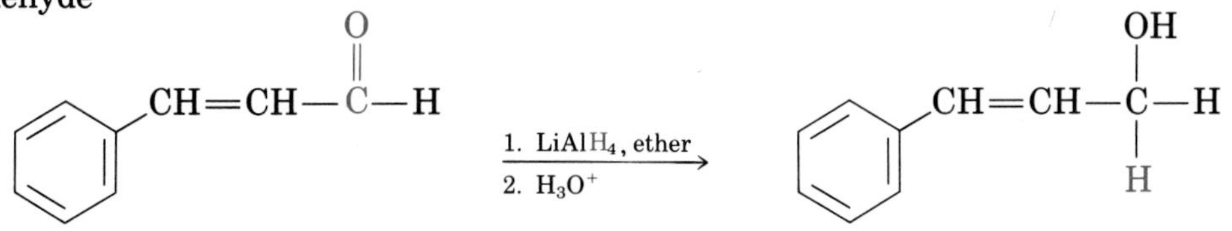

$$C_6H_5CH{=}CH{-}\overset{O}{\overset{\|}{C}}{-}H \xrightarrow[2.\ H_3O^+]{1.\ LiAlH_4,\ ether} C_6H_5CH{=}CH{-}CH_2OH$$

3-Phenylpropenal → 3-Phenyl-2-propen-1-ol (90%)

Acid

$$CH_3(CH_2)_7CH{=}CH(CH_2)_7CO_2H \xrightarrow[2.\ H_3O^+]{1.\ LiAlH_4,\ THF} CH_3(CH_2)_7CH{=}CH(CH_2)_7CH_2{-}OH$$

Oleic acid → 9-Octadecen-1-ol (87%)

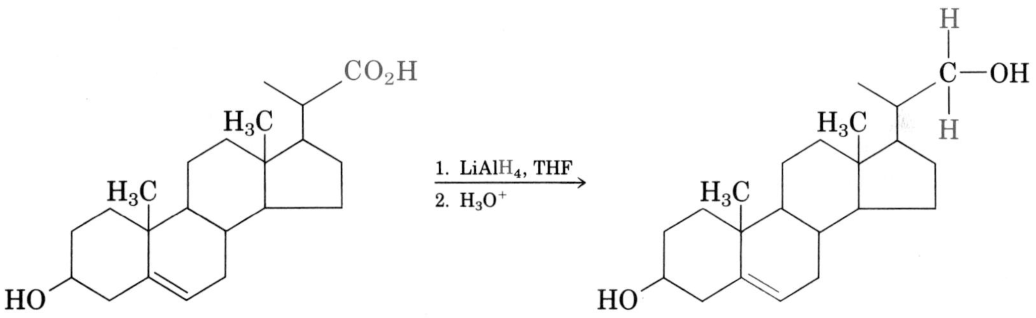

$$\text{(steroid carboxylic acid: } HO, H_3C, H_3C, CO_2H) \xrightarrow[2.\ H_3O^+]{1.\ LiAlH_4,\ THF} \text{(corresponding primary alcohol: } HO, H_3C, H_3C, CH_2{-}OH)$$

(87%)

PROBLEM

20.7 What reagent would you use to accomplish each of these reactions? Explain your choices.

(a) $CH_3\overset{O}{\overset{\|}{C}}CH_2CH_2CO_2CH_3 \xrightarrow{?} CH_3\overset{OH}{\overset{|}{C}}HCH_2CH_2CO_2CH_3$

(b) $CH_3\overset{O}{\overset{\|}{C}}CH_2CH_2CO_2CH_3 \xrightarrow{?} CH_3\overset{OH}{\overset{|}{C}}HCH_2CH_2CH_2OH$

(c) 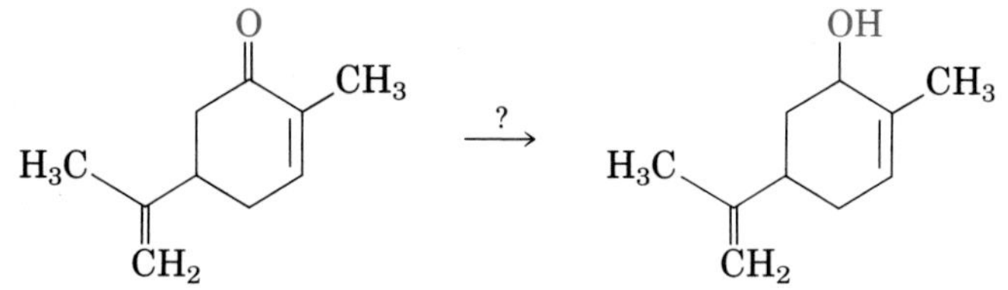 Carvone (from spearmint oil) $\xrightarrow{?}$ corresponding alcohol (C=O reduced to CH–OH; ring C=C, CH_3, and $H_3C(C{=}CH_2)$ groups retained)

Carvone
(from spearmint oil)

(d) 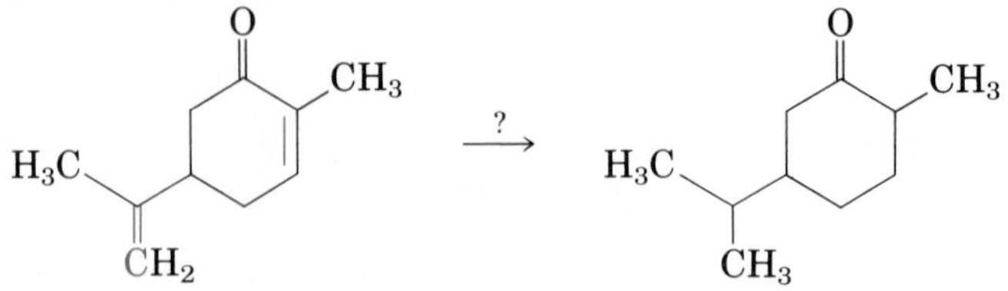 [structure: 2-methyl-5-isopropenylcyclohex-2-enone] $\xrightarrow{?}$ [structure: 2-methyl-5-isopropylcyclohexanone]

20.6 Mechanism of carbonyl reduction: nucleophilic addition reactions

The reduction of a carbonyl group by a complex metal hydride reagent such as $NaBH_4$ or $LiAlH_4$ is an example of a **nucleophilic addition reaction.** Nucleophilic addition reactions comprise one of the broadest and most general of all polar reaction types, and we shall see them repeatedly during our study of carbonyl chemistry in following chapters. As the name indicates, a nucleophilic addition reaction involves the addition of a nucleophile to an electrophile. The electrophile is a carbonyl carbon, and the nucleophile in the present instance is a hydride ion, $:H^-$. Figure 20.6 shows the general formulation of a nucleophilic addition to the positively polarized carbon of the carbonyl group. All carbonyl compounds—ketones, aldehydes, esters, and carboxylic acids—undergo nucleophilic addition reactions, and we will look into the details of specific examples in Chapter 22.

PROBLEM

20.8 What carbonyl compounds give the following alcohols on reduction with $LiAlH_4$? Show all possibilities.

(a) $C_6H_5CH_2OH$ (benzene ring with CH_2OH)

(b) $C_6H_5CH(OH)CH_3$ (benzene ring with $\overset{OH}{|}CHCH_3$)

(c) cyclohexane ring bearing OH and H

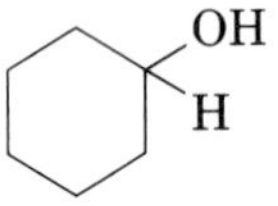

(d) $(CH_3)_2CHCH_2OH$

20.7 Grignard addition to carbonyl groups

Grignard reagents, RMgX, add to carbonyl compounds in much the same manner as hydride reagents. The result is a highly useful and general method for synthesizing alcohols.

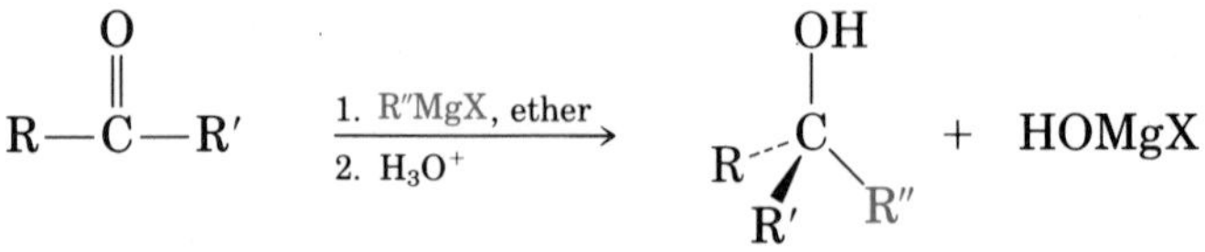

$$R-\overset{O}{\overset{\|}{C}}-R' \xrightarrow[2.\ H_3O^+]{1.\ R''MgX,\ ether} R(R')(R'')C-OH + HOMgX$$

An electron pair from the nucleophile attacks the electrophilic carbonyl carbon, pushing an electron pair from the C=O bond out onto oxygen. The carbonyl carbon rehybridizes from sp^2 to sp^3.

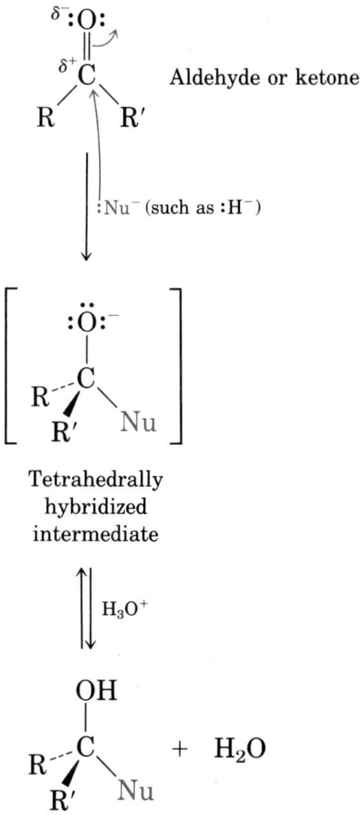

Aldehyde or ketone

:Nu⁻ (such as :H⁻)

Tetrahedrally hybridized intermediate

Protonation of the anion resulting from nucleophilic attack yields the neutral alcohol addition product.

H_3O^+

+ H_2O

Example:

1. :H⁻
2. H_3O^+

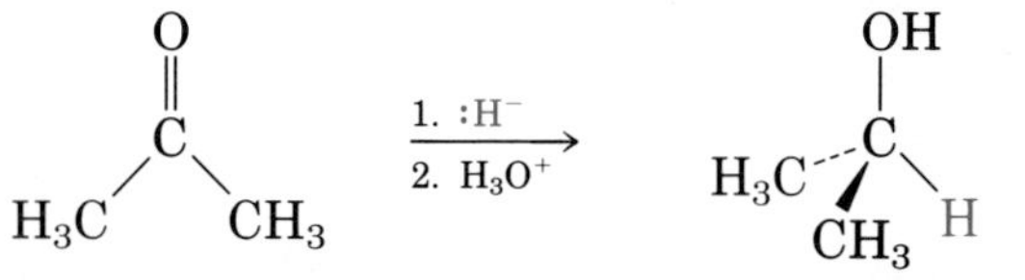

Figure 20.6. General mechanism of a nucleophilic addition reaction.

We saw in Section 9.8 that most alkyl, aryl, and vinylic halides react with magnesium in ether or tetrahydrofuran solution to generate Grignard reagents.

$$R{-}X + Mg \longrightarrow \overset{\delta^-}{R}{-}\overset{\delta^+}{Mg}X$$

A Grignard reagent

where R = 1°, 2°, or 3° alkyl, aryl, or vinylic

X = Cl, Br, or I

The carbon–magnesium bond of Grignard reagents is strongly polarized, making the carbon atom both nucleophilic and strongly basic.

Grignard reagents therefore react as if they were carbanions, R:$^-$, and they undergo nucleophilic addition reactions with carbonyl compounds just as hydride reagents do. Nucleophilic addition to the carbonyl group first produces a tetrahedrally hybridized magnesium alkoxide intermediate, which is then hydrolyzed to the alcohol by treatment with aqueous acid.

:R$^-$ MgX
Nucleophilic addition

H_3O^+

+ HO—Mg—X

Carbonyl

Tetrahedral intermediate

Alcohol

A large number of alcohol products can be obtained from Grignard reactions, depending on the reagents used. For example, Grignard reagents react with formaldehyde, CH_2O, to give primary alcohols.

1. CH_2O (formaldehyde)
2. H_3O^+

Cyclohexylmagnesium bromide

Cyclohexylmethanol (65%) (a primary alcohol)

Aldehydes react with Grignard reagents to give secondary alcohols. Ketones react similarly to yield tertiary alcohols:

$(CH_3)_2CHCH_2CHO$ + C_6H_5MgBr $\xrightarrow{\text{Ether, then } H_3O^+}$ $(CH_3)_2CHCH_2CH(OH)C_6H_5$

3-Methylbutanal

Phenylmagnesium bromide

3-Methyl-1-phenyl-1-butanol (73%) (a 2° alcohol)

1. CH_3CH_2—MgBr (ethylmagnesium bromide), ether
2. H_3O^+

Cyclohexanone

1-Ethylcyclohexanol (89%) (a 3° alcohol)

Esters are also attacked by Grignard reagents to yield a tertiary alcohol in which two of the substituents have come from the Grignard reagent (just as $LiAlH_4$ reduction of esters adds *two* hydrogens). For example,

$$CH_3CH_2CH_2CH_2\overset{O}{\overset{\|}{C}}OCH_2CH_3 \xrightarrow[2.\ H_3O^+]{1.\ 2\ CH_3MgBr} CH_3CH_2CH_2CH_2\underset{CH_3}{\underset{|}{\overset{OH}{\overset{|}{C}}}}-CH_3 + CH_3CH_2OH$$

Ethyl pentanoate

2-Methyl-2-hexanol (85%)

Carboxylic acids do not give addition products with Grignard reagents because the acidic carboxyl proton reacts with the Grignard reagent to produce a hydrocarbon and the magnesium salt of the acid. We have seen this reaction previously (Section 9.8) as a means of reducing organohalides to hydrocarbons.

$$R-Br + Mg \longrightarrow R-MgX$$

$$R-MgX + R'\overset{O}{\overset{\|}{C}}-O-H \longrightarrow R-H + R'\overset{O}{\overset{\|}{C}}-O-MgBr$$

Carboxylic acid — Hydrocarbon — Acid salt (unreactive)

PROBLEM

20.9 What carbonyl compound would react with what Grignard reagent to yield these compounds?

(a) 2-Methyl-2-propanol
(b) 1-Methylcyclohexanol
(c) 3-Methyl-3-pentanol
(d) 2-Phenyl-2-butanol
(e) Benzyl alcohol

20.8 Limitations on the Grignard reaction

The Grignard reaction is broad in scope, but it also has severe limitations. A Grignard reagent cannot be prepared from an organohalide if there are other reactive functional groups in the same molecule. For example, a compound that is both an alkyl halide and a ketone will not form a Grignard reagent—instead, it reacts with itself. Similarly, a compound that is both an alkyl halide and a carboxylic acid, alcohol, or amine cannot form a Grignard reagent because the acidic RCO_2H, ROH, or $R-NH_2$ protons present in the same molecule simply react with the basic Grignard reagent as it is formed. Generally speaking, Grignard reagents cannot be prepared from alkyl halides that also contain acidic hydrogens or multiple bonds to oxygen, nitrogen, or sulfur.

1. Grignard reagents *cannot* be prepared from compounds with these functional groups (FG):

$$Br-(\text{Molecule})-FG$$

where FG = $-OH$, $-NH$, $-SH$, $-CO_2H$, $-NO_2$, $-CHO$, $-COR$, $-CN$, $-CONH_2$, or $-SO_2R$

2. Grignard reagents *can* be prepared from compounds with these functional groups:

Br—(Molecule)—FG

where FG = Alkyl, aryl, vinylic, —OR (ether), —NR_2, —SR, or —F

PROBLEM

20.10 How can you explain the observation that treatment of 4-hydroxycyclohexanone with 1 equiv of methylmagnesium bromide yields none of the expected addition product, whereas treatment with an excess of Grignard reagent and then dilute acid leads to a good yield of 1-methylcyclohexane-1,4-diol?

20.9 Reactions of alcohols

Reactions of alcohols can be subdivided into two groups—reactions that occur at the C—O bond and reactions that occur at the O—H bond:

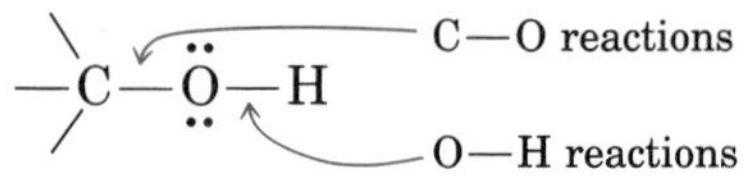

C—O reactions

—C—Ö—H

O—H reactions

Let's review some of the reactions of alcohols we have seen in previous chapters.

CARBON–OXYGEN BOND REACTIONS OF ALCOHOLS

Alcohols can be dehydrated to give alkenes (Section 6.13). Tertiary alcohols lose water when treated with mineral acid under fairly mild conditions, but primary and secondary alcohols require more severe acidic conditions. The most commonly used method of dehydration involves treatment of the secondary or tertiary alcohol with phosphorus oxychloride, $POCl_3$, in pyridine solvent. These conditions are mild and nonacidic, and most other functional groups are not affected.

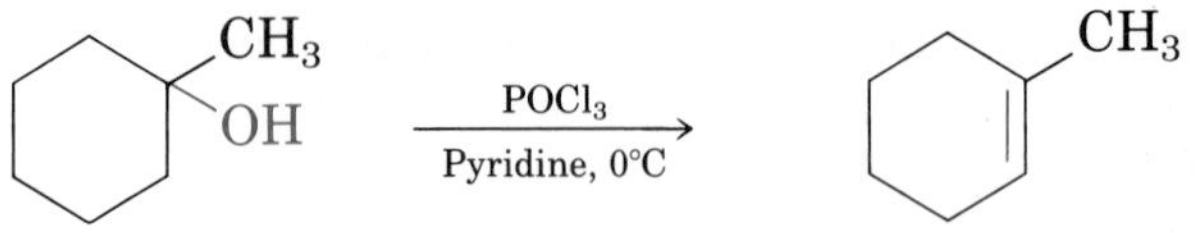

CH_3, OH —($POCl_3$ / Pyridine, 0°C)→ CH_3

1-Methylcyclohexanol — 1-Methylcyclohexene (89%)

Alcohols can be converted into alkyl halides (Section 9.7). Tertiary alcohols are readily converted into alkyl halides by treatment with either HCl or HBr at 0°C. Primary and secondary alcohols are much more resistant to acid, however, and are best converted into halides by treatment with either $SOCl_2$ or PBr_3.

OXYGEN–HYDROGEN BOND REACTIONS OF ALCOHOLS

Alcohols can be converted into ethers (Section 19.4). The Williamson ether synthesis discussed previously involves S_N2 reaction of an alkoxide anion with a primary halide.

$$CH_3CH_2\ddot{\underset{\cdot\cdot}{O}}{:}^{-}\,Na^{+} + C_6H_5CH_2Cl \longrightarrow C_6H_5CH_2OCH_2CH_3 + {:}Cl^{-}$$

Sodium ethoxide — Benzyl chloride — Benzyl ethyl ether (72%)

The Williamson reaction depends heavily on the fact that alcohols are weakly acidic, a subject we will now consider in more detail.

Acidity of alcohols. Alcohols, like water, are weakly acidic. In dilute aqueous solution, alcohols dissociate by donating a proton to the base, water.

$$R\ddot{\underset{\cdot\cdot}{O}}{-}H + H_2\ddot{\underset{\cdot\cdot}{O}} \rightleftharpoons R\ddot{\underset{\cdot\cdot}{O}}{:}^{-} + H_3\underset{\cdot\cdot}{O}^{+}$$

Recall from Section 2.6 that the acidity of a given alcohol in water can be defined by the following expression:

$$K_a = \frac{[RO^-][H_3O^+]}{[ROH]} \qquad pK_a = -\log K_a$$

where K_a is the acidity constant. Alcohols with a small K_a (or high pK_a) are weakly acidic, whereas alcohols with a larger K_a (smaller pK_a) are more strongly acidic. Table 20.2 gives the pK_a values of some common alcohols and compares them with water and with HCl.

TABLE 20.2 Acidity constants of some alcohols

Alcohol	pK_a	
$(CH_3)_3COH$	18.00	Weaker acid
CH_3CH_2OH	16.00	↑
HOH (water)[a]	(15.74)	
CH_3OH	15.54	
CF_3CH_2OH	12.43	
$(CF_3)_3COH$	5.4	
HCl (hydrochloric acid)[a]	(−7.00)	Stronger acid

[a]Values for water and hydrochloric acid are shown for reference.

Note that the K_a for water is given in the table as 15.74. This value is obtained by dividing the familiar dissociation constant of water, $[H^+][OH^-] = 10^{-14}$, by the concentration of water in a pure sample, 55.5*M*.

$$K_a = \frac{[H^+][^-OH]}{[H_2O]} = \frac{10^{-14}}{55.5} = 1.80 \times 10^{-16}$$

$$pK_a = -\log(1.8 \times 10^{-16}) = 15.74$$

Methanol and ethanol are about as acidic as water, whereas *tert*-butyl alcohol is slightly less acidic. The effect of alkyl substitution on acidity is thought to be due to solvation. Water is able to surround the sterically accessible oxygen atom of unhindered alcohols and to stabilize the alkoxide by solvation. Hindered alkoxides such as *tert*-butoxide, however, prevent solvation by their bulk and are therefore less stable. Inductive effects are also important in determining alcohol acidities. For example, electron-withdrawing halogen substituents stabilize the alkoxide anion by charge delocalization and thus raise the acidity of the alcohol. This inductive effect can be observed by comparing the acidities of ethanol ($pK_a = 16$) and 2,2,2-trifluoroethanol ($pK_a = 12.43$), or of *tert*-butyl alcohol ($pK_a = 18$) and nonafluoro-2-methyl-2-propanol ($pK_a = 5.4$).

Electron-withdrawing groups stabilize alkoxide and lower pK_a

$$(CF_3)_3C{-}O^- \quad (pK_a = 5.4) \qquad \text{versus} \qquad (CH_3)_3C{-}O^- \quad (pK_a = 18)$$

Alcohols are generally much weaker acids than carboxylic acids or mineral acids, and they do not react with weak bases such as amines or bicarbonate ion. Alcohols do, however, react with alkali metals and with strong bases like sodium amide ($NaNH_2$), sodium hydride (NaH), alkyllithium reagents (R—Li), or Grignard reagents (R—MgX). The metal salts of alcohols are much used as reagents in organic chemistry.

$$2\ CH_3OH + 2\ Na\ (\text{or } Na^+\ H^-) \longrightarrow 2\ CH_3O^-\ Na^+ + H_2$$

Sodium methoxide

$$2\ (CH_3)_3COH + 2\ K \longrightarrow 2\ (CH_3)_3CO^-\ K^+ + H_2$$

Potassium *tert*-butoxide

Tosylate formation. Although alcohols undergo many useful reactions, they do *not* undergo S_N2 displacement by nucleophiles. Hydroxide ion is the conjugate base of a weak acid (H_2O) and is simply too poor a leaving group to be displaced (Section 10.5):

$$R{-}OH + Nu{:}^- \not\longrightarrow R{-}Nu + {}^-{:}OH$$

One way to circumvent this lack of S_N2 reactivity is to convert the alcohol into a halide and then to carry out the desired displacement reaction. Alternatively, it is often better and more convenient to convert the alcohol into its *p*-toluenesulfonate or tosylate:

$$R{-}OH + H_3C{-}C_6H_4{-}SO_2Cl \xrightarrow[\text{Pyridine}]{} R{-}O{-}\overset{O}{\underset{O}{\overset{\|}{\underset{\|}{S}}}}{-}C_6H_4{-}CH_3$$

An alcohol — *p*-Toluenesulfonyl chloride — A tosylate

We saw in Section 10.2 that primary, secondary, and many tertiary alcohols react with *p*-toluenesulfonyl chloride (tosyl chloride, *p*-TosCl) in pyridine solution to yield tosylates, R—O—Tos. Alkyl tosylates behave much like alkyl halides in their chemistry, and they undergo S_N1 and S_N2 displacement reactions with ease.

One of the most important reasons for using tosylates instead of halides is stereochemical. In a sense, the two sequences of reactions,

$$\text{Alcohol} \longrightarrow \text{Halide} \longrightarrow S_N2 \text{ reaction}$$

and

$$\text{Alcohol} \longrightarrow \text{Tosylate} \longrightarrow S_N2 \text{ reaction}$$

are stereochemically complementary. The S_N2 reaction via the halide proceeds with two Walden inversions and yields a product with the same absolute stereochemistry as the starting material. The S_N2 reaction via the tosylate proceeds with only one Walden inversion and yields a product of opposite absolute stereochemistry from the starting material. Figure 20.7 (p. 650) gives a series of reactions on optically active 2-octanol that illustrates these stereochemical relationships.

20.10 Oxidation of alcohols

The most important reaction of alcohols is their oxidation to yield carbonyl compounds. Primary alcohols yield aldehydes or carboxylic acids; secondary alcohols yield ketones; and tertiary alcohols do not react with most oxidizing agents except under the most vigorous conditions.

$$1^\circ \quad R{-}CH_2OH \xrightarrow{[O]} R{-}CHO \xrightarrow{[O]} R{-}CO_2H$$

An aldehyde — A carboxylic acid

$$2^\circ \quad R{-}\overset{OH}{\overset{|}{C}H}{-}R' \xrightarrow{[O]} R{-}\overset{O}{\overset{\|}{C}}{-}R'$$

A ketone

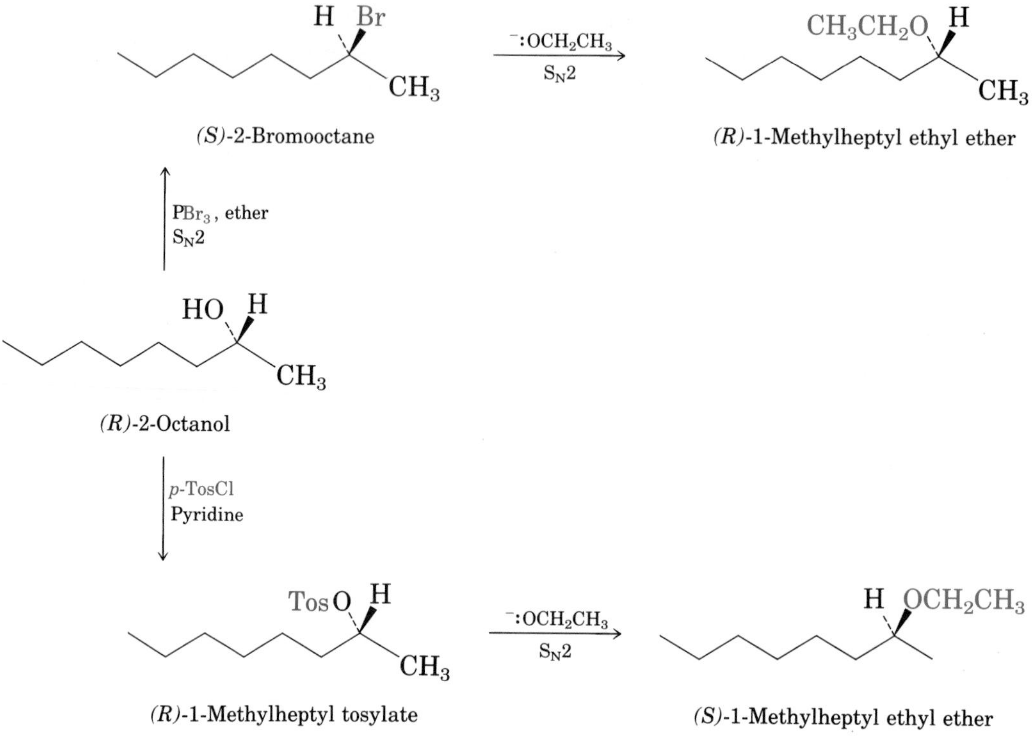

(S)-2-Bromooctane $\xrightarrow[S_N2]{^{-}:OCH_2CH_3}$ (R)-1-Methylheptyl ethyl ether

(R)-2-Octanol $\xrightarrow[S_N2]{PBr_3,\ ether}$ (S)-2-Bromooctane

(R)-2-Octanol $\xrightarrow[Pyridine]{p\text{-TosCl}}$ (R)-1-Methylheptyl tosylate

(R)-1-Methylheptyl tosylate $\xrightarrow[S_N2]{^{-}:OCH_2CH_3}$ (S)-1-Methylheptyl ethyl ether

Figure 20.7. S_N2 reactions on derivatives of *(R)*-2-octanol.

$$3^\circ \quad R{-}\overset{OH}{\underset{R'}{C}}{-}R'' \xrightarrow{[O]} \text{No reaction under most conditions}$$

Oxidation of primary and secondary alcohols can be accomplished by a large number of reagents, including $KMnO_4$, CrO_3, $Na_2Cr_2O_7$, and dilute HNO_3. Primary alcohols are oxidized either to aldehydes or to carboxylic acids, depending on the reagents chosen and on the conditions used. Probably the best method for preparing aldehydes from primary alcohols on a laboratory scale is by use of the reagent pyridinium chlorochromate (PCC), $C_5H_6NCrO_3Cl$. This reagent was introduced in 1975 and is now widely used.

$$CH_3(CH_2)_5CH_2OH \xrightarrow[CH_2Cl_2]{PCC} CH_3(CH_2)_5CHO$$

1-Heptanol → Heptanal (78%)

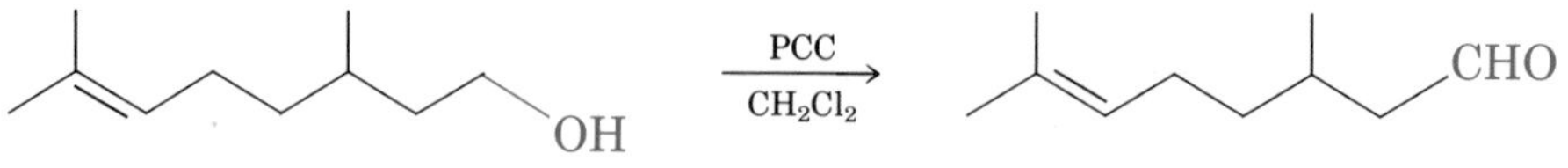

Citronellol (from rose oil) $\xrightarrow[CH_2Cl_2]{PCC}$ Citronellal (82%)

Most other oxidizing agents, such as chromium trioxide (CrO_3) in aqueous sulfuric acid (Jones' reagent), oxidize primary alcohols to carboxylic acids. Aldehydes are involved as intermediates in the Jones oxidation, but they cannot usually be isolated because they are further oxidized too readily.

$$CH_3(CH_2)_8CH_2OH \xrightarrow[\text{Acetone}]{\text{Jones' reagent } (CrO_3,\ H_2SO_4,\ H_2O)} CH_3(CH_2)_8CO_2H$$

1-Decanol → Decanoic acid (93%)

(1-Phenylcyclopentyl)methanol $\xrightarrow{\text{Jones' reagent}}$ 1-Phenylcyclopentanecarboxylic acid (85%)

Secondary alcohols are oxidized easily and in high yields to give ketones. For large-scale oxidations, an inexpensive reagent such as sodium dichromate in aqueous acetic acid is used.

4-*tert*-Butylcyclohexanol $\xrightarrow[H_2O,\ CH_3CO_2H,\ \Delta]{Na_2Cr_2O_7}$ 4-*tert*-Butylcyclohexanone (91%)

For more sensitive alcohols, however, pyridinium chlorochromate or Jones' reagent is often used, since the reactions are milder and occur at lower temperatures.

Cyclooctanol $\xrightarrow[\text{Acetone, 0°C}]{\text{Jones' reagent}}$ Cyclooctanone (96%)

Testosterone (steroid; male sex hormone) $\xrightarrow[CH_2Cl_2,\ 25°C]{PCC}$ 4-Androstene-3,17-dione (82%)

All these oxidations occur by an E2 reaction pathway. The first step involves reaction between the alcohol and a chromium(VI) reagent to form an intermediate chromate. Bimolecular elimination then yields the carbonyl product.

Alcohol $\xrightarrow{CrO_3}$ [A chromate, with :Base] $\xrightarrow[\text{reaction}]{\text{E2}}$ Carbonyl + $^{-}CrO_3H$

Alcohol

A chromate

Carbonyl

Although we usually think of the E2 reaction as a means of generating carbon–carbon double bonds by dehydrohalogenation of alkyl halides, it is also useful for preparing carbon–oxygen double bonds. This is just one more example of how the same few fundamental mechanistic types recur in different variations.

PROBLEM

20.11 One problem encountered in the oxidation of primary alcohols to acids is that esters are sometimes produced as by-products. For example, oxidation of ethanol yields acetic acid and ethyl acetate:

$$CH_3CH_2OH \xrightarrow{\text{Jones' reagent}} CH_3CO_2H + CH_3\overset{O}{\overset{\|}{C}}OCH_2CH_3$$

Propose a mechanism to account for the formation of ethyl acetate. Take into account the reversible reaction between aldehydes and alcohols:

$$R{-}CHO + RCH_2OH \rightleftarrows R{-}\overset{OH}{\overset{|}{C}H}{-}O{-}CH_2R$$

20.11 Protection of alcohols

It often happens, particularly during the synthesis of complex molecules, that one functional group in a molecule interferes with an intended reaction on a second functional group. For example, we have seen that one cannot prepare a Grignard reagent from a halo alcohol because the carbon–magnesium bond is not compatible with the presence of an acidic hydroxyl group in the same molecule.

HO—(Molecule)—Br $\xrightarrow{Mg}$ (not formed) HO—(Molecule)—MgBr

Not formed

When this kind of incompatibility arises, we can sometimes circumvent the problem by *protecting* the interfering functional group. Protection involves three steps: (1) formation of an inert derivative; (2) carrying out the desired reaction; and (3) removal of the protecting group.

One common method of alcohol protection involves ether formation by reaction with dihydropyran:

Cyclohexanol + Dihydropyran $\xrightarrow[\text{THF}]{H^+}$ Cyclohexyl 2-tetrahydropyranyl ether (95%)

Alcohols react readily with dihydropyran in the presence of anhydrous acid catalyst to yield tetrahydropyranyl (THP) ethers. Like other ethers, THP ethers are inert. They have no acidic protons and are therefore protected against reaction with oxidizing agents, reducing agents, nucleophiles, and Grignard reagents. Like other ethers, however, the THP group can be cleaved by reaction with acid to regenerate the alcohol. The cleavage is made easier by the presence of the second oxygen, and the reaction is normally carried out with aqueous mineral acid.

Cyclohexyl THP ether $\xrightarrow[25°C]{HCl,\ H_2O}$ Cyclohexanol + Tetrahydropyran-2-ol

To complete our earlier example, it is now possible to use halo alcohols in Grignard reactions by employing a three-step sequence: (1) protection; (2) Grignard formation and reaction; (3) deprotection.

PROBLEM

20.12 Propose a mechanism for the formation of a THP ether by acid-catalyzed addition of an alcohol to dihydropyran. How can you account for the regiochemistry of the reaction?

PROBLEM

20.13 Write out all the steps required for carrying out the following transformation:

4-Bromocyclohexanol (Br, OH) $\overset{?}{\Longrightarrow}$ 4-(1-hydroxy-1-methylethyl)cyclohexanol (H_3C, OH, CH_3; OH)

20.12 Spectroscopic analysis of alcohols

INFRARED SPECTROSCOPY

Alcohols show a characteristic O—H stretch absorption at 3300–3600 cm^{-1} in the infrared spectrum that simplifies their spectroscopic identification. The exact position of the absorption band depends on the extent of hydrogen bonding in the sample. Unassociated alcohols show a fairly sharp absorption near 3600 cm^{-1}, whereas hydrogen-bonded alcohols show a broader absorption in the 3300–3400 cm^{-1} range. The hydrogen-bonded hydroxyl absorption is easily seen at 3350 cm^{-1} in the spectrum of cyclohexanol (Figure 20.8). Alcohols also show a strong C—O stretching absorption near 1050 cm^{-1}.

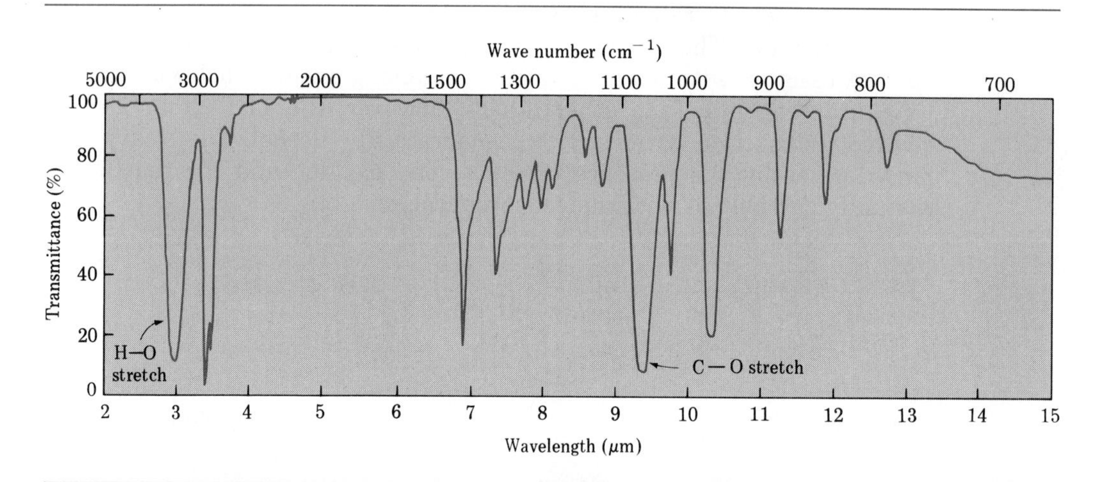

Figure 20.8. Infrared spectrum of cyclohexanol.

PROBLEM

20.14 Let's assume that you needed to prepare 5-cholestene-3-one from cholesterol by Jones oxidation. How could you use infrared spectroscopy to tell if the reaction was successful? What differences would you look for in the infrared spectra of starting material and product?

HO

Jones' reagent

O

Cholesterol

5-Cholestene-3-one

NUCLEAR MAGNETIC RESONANCE SPECTROSCOPY

Carbon atoms bearing hydroxyl substituents are somewhat deshielded and absorb at lower field than normal alkane carbon atoms in the ^{13}C NMR spectrum. Most alcohol carbon absorptions fall in the range 40–80 δ, as the spectral data in Figure 20.9 illustrate for cyclohexanol.

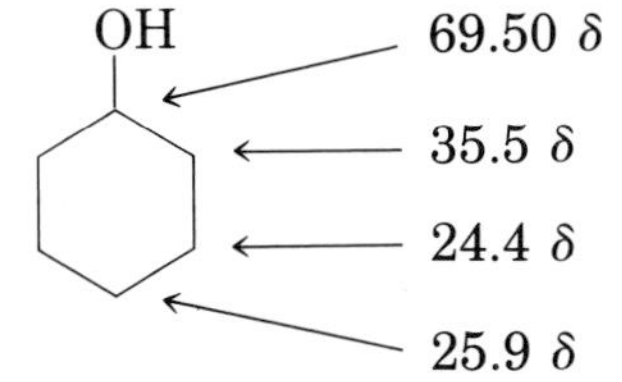

OH — 69.50 δ
35.5 δ
24.4 δ
25.9 δ

Figure 20.9. Carbon-13 NMR data for cyclohexanol.

Alcohols show characteristic absorptions in the ^{1}H NMR spectrum also. Protons on the oxygen-bearing carbon atom are deshielded, and their absorptions occur in the region from 3.5 to 4.5 δ. One of the interesting features of alcohol ^{1}H NMR spectra is that we do not usually observe coupling between the hydroxyl proton and the neighboring protons on carbon. This is because most samples contain small amounts of acidic impurities that catalyze a rapid exchange of the hydroxyl proton. The net result of this rapid exchange is that spin–spin splitting is removed and no coupling is observed.

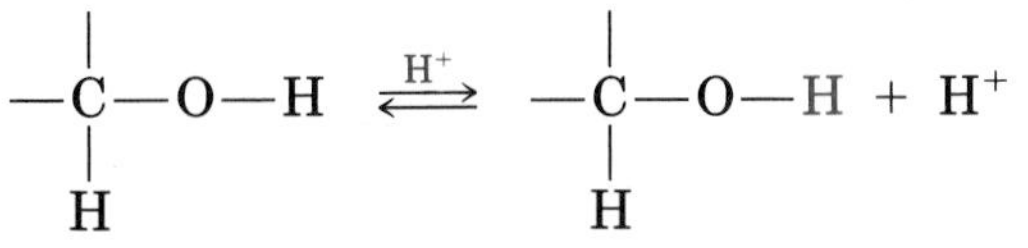

$$-\overset{|}{\underset{\text{H}}{\text{C}}}-\text{O}-\text{H} \overset{H^+}{\rightleftarrows} -\overset{|}{\underset{\text{H}}{\text{C}}}-\text{O}-\text{H} + H^+$$

No NMR coupling observed

We can also take advantage of this rapid exchange to identify the position of the O—H absorption. When a small amount of deuterated water, D_2O, is added to the NMR sample tube, the O—H proton is rapidly exchanged for deuterium, and the hydroxyl absorption disappears from the spectrum.

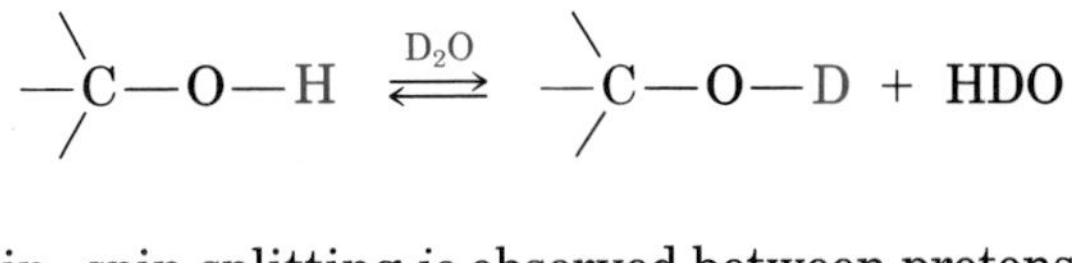

$$\gt\text{C}-\text{O}-\text{H} \overset{D_2O}{\rightleftarrows} \gt\text{C}-\text{O}-\text{D} + \text{HDO}$$

Normal spin–spin splitting *is* observed between protons on the oxygen-bearing carbon and other neighbors. This coupling leads to the broadening observed for the CHOH resonance centered at 3.6 δ in the spectrum of cyclohexanol (Figure 20.10).

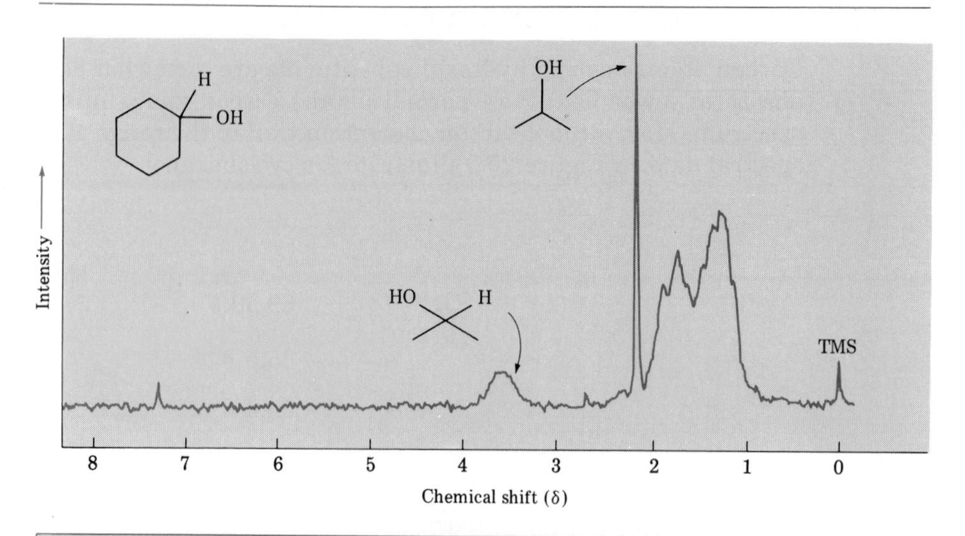

Figure 20.10. Proton NMR spectrum of cyclohexanol. The proton on the oxygen-bearing carbon absorbs at 3.6 δ.

PROBLEM

20.15 When the ^1H NMR spectra of alcohols are run in dimethyl sulfoxide (DMSO) solvent, exchange of the hydroxyl proton is slow, and spin–spin splitting is observed between the hydroxyl and neighboring protons, >CH—OH. What spin multiplicities would you expect for the hydroxyl protons in the following alcohols?

(a) 2-Methyl-2-propanol (b) Cyclohexanol (c) Ethanol
(d) 2-Propanol (e) Cholesterol (f) 1-Methylcyclohexanol

MASS SPECTROSCOPY

Alcohols undergo fragmentation in the mass spectrometer by two characteristic pathways, **alpha cleavage** and **dehydration.** In the alpha-cleavage pathway, a carbon–carbon bond nearest the hydroxyl group is broken, yielding a neutral radical plus a charged oxygen-containing fragment:

Alpha cleavage $$\left[R-\underset{R}{\overset{OH}{C}} \overset{\alpha}{\wr} CH_2R \right]^{+\cdot} \longrightarrow \left[R-\underset{R}{\overset{OH}{C}} \right]^{+} + \cdot CH_2R$$

In the dehydration pathway, water is eliminated, yielding an alkene radical cation:

Dehydration $$\left[\overset{H}{>C}-\overset{OH}{C<} \right]^{+\cdot} \longrightarrow H_2O + \left[>C=C< \right]^{+\cdot}$$

Both of these characteristic alcohol-fragmentation modes are apparent in the mass spectrum of 1-butanol (Figure 20.11). The peak at $m/z = 56$ is due to loss of water from the molecular ion, and the peak at $m/z = 43$ is due to an alpha cleavage.

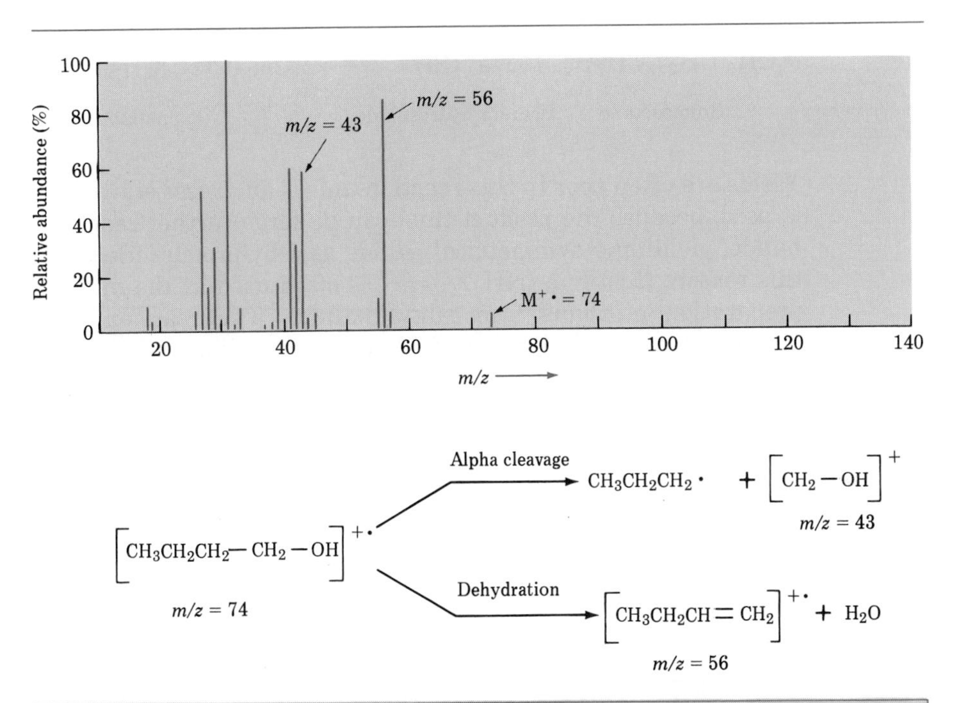

Figure 20.11. Mass spectrum of 1-butanol. The two common modes of alcohol fragmentation are visible. Dehydration of 1-butanol yields a peak at $m/z = 56$; alpha cleavage yields a peak at $m/z = 43$.

20.13 Thiols

Thiols, R—SH, are sulfur analogs of alcohols in the same sense that sulfides are sulfur analogs of ethers. Thiols are named by the same system used for alcohols, with the suffix *-thiol* used in place of *-ol*. The —SH group itself is referred to as a **mercapto** group.

CH_3CH_2SH
Ethanethiol

Cyclohexanethiol

m-Mercaptobenzoic acid

The outstanding physical characteristic of thiols is their truly appalling odor. For example, skunk scent is caused primarily by the simple thiols,

3-methyl-1-butanethiol and 2-butene-1-thiol. Small amounts of low-molecular-weight thiols are added to natural gas to serve as an easily detectable warning in case of leaks.

Thiols are usually prepared from the corresponding alkyl halides by S_N2 displacement with a sulfur nucleophile such as hydrosulfide anion, $^{-}$:SH.

$$CH_3(CH_2)_6CH_2Br + Na^{+}{:}SH^{-} \longrightarrow CH_3(CH_2)_6CH_2{-}SH + NaBr$$

1-Bromooctane — Sodium hydrosulfide — 1-Octanethiol

Yields are often poor in this reaction unless an excess of hydrosulfide anion is used, because the product thiol can undergo further reaction with alkyl halide, yielding a symmetrical sulfide as a by-product (Section 19.11). For this reason, thiourea, $(NH_2)_2C{=}S$, is often used as the nucleophile in the preparation of thiols from alkyl halides. The reaction occurs by displacement of the halide to yield an intermediate alkylisothiourea salt, followed by hydrolysis with aqueous base.

$$CH_3(CH_2)_6CH_2{-}Br + H_2N{-}\overset{\overset{:S:}{\|}}{C}{-}\ddot{N}H_2 \longrightarrow \left[CH_3(CH_2)_6CH_2{-}\overset{+}{S}{=}\overset{\overset{NH_2}{|}}{C}{-}NH_2 \quad Br^{-} \right]$$

1-Bromooctane — Thiourea — Alkylisothiourea salt

$$\xrightarrow{H_2O,\ NaOH} CH_3(CH_2)_6CH_2{-}SH + H_2N{-}\overset{\overset{O}{\|}}{C}{-}NH_2$$

1-Octanethiol (83%) — Urea

Thiolate anions, like alkoxide anions, are nucleophiles, and we have already studied their use in sulfide synthesis by reaction with alkyl halides (Section 19.10). Thiols can also be oxidized by mild reagents such as bromine or iodine to yield **disulfide** products. The reaction is easily reversed; disulfides can be reduced back to thiols by treatment with zinc and acid.

$$2\,R{-}SH \underset{Zn,\ H^{+}}{\overset{Br_2}{\rightleftharpoons}} R{-}S{-}S{-}R + 2\,HBr$$

A thiol — A disulfide

We will see in Section 31.7 that the thiol–disulfide interconversion is extremely important in biochemistry, where disulfide "bridges" form the cross-links between protein chains that help stabilize the three-dimensional conformations of proteins. For example, hair is particularly rich in mercapto groups; giving a permanent wave involves the oxidation of some of the —SH groups into disulfide linkages, thereby altering protein conformations and imparting a "wave" to the hair.

$$\text{(Protein)}{-}SH + HS{-}\text{(Protein)} \longrightarrow \text{(Protein)}{-}S{-}S{-}\text{(Protein)}$$

A cross-linked protein

PROBLEM

20.16 Name the following compounds by IUPAC rules:

(a) $CH_3CH_2CH(CH_3)SH$

(b) 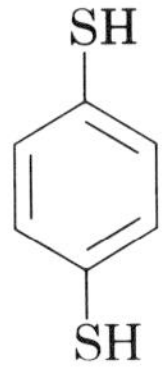 SH, SH

(c) $(CH_3)_3CCH_2CH(SH)CH_2CH(CH_3)_2$

(d) 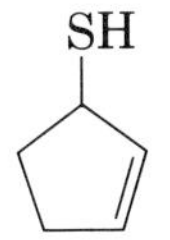 SH

PROBLEM

20.17 2-Butene-1-thiol is one component of skunk spray. How would you synthesize this substance from methyl 2-butenoate? From 1,3-butadiene?

$$CH_3CH{=}CHCO_2CH_3 \overset{?}{\Longrightarrow} CH_3CH{=}CHCH_2SH$$

Methyl 2-butenoate 2-Butene-1-thiol

20.14 Summary

Alcohols are among the most versatile of all organic compounds. They occur widely in nature; they are important industrially; and they have an unusually rich chemistry. Alcohols can be prepared in a number of different ways, including osmium tetroxide hydroxylation of alkenes, hydroboration–oxidation of alkenes, and oxymercuration–sodium borohydride reduction of alkenes.

The most important methods of alcohol synthesis involve nucleophilic addition reactions to carbonyl compounds. For example, aldehydes, ketones, esters, and carboxylic acids can all be reduced by nucleophilic addition of hydride ion:

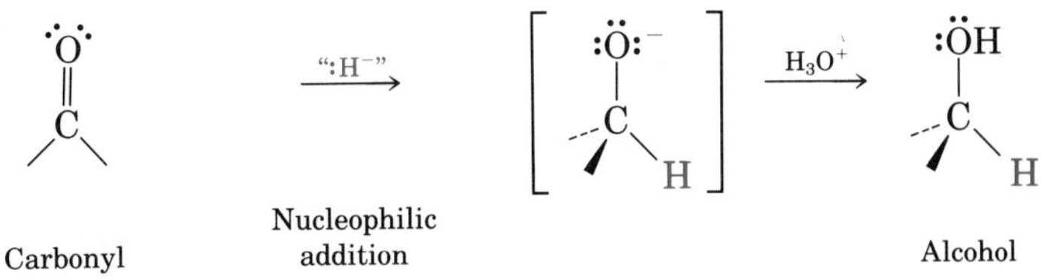

$$\text{C{=}O} \xrightarrow{\text{“:H}^-\text{”}} \left[\text{:}\ddot{O}\text{:}^- {-}C{-}H \right] \xrightarrow{H_3O^+} \text{:}\ddot{O}H{-}C{-}H$$

Carbonyl Nucleophilic addition Alcohol

Aldehydes, esters, and acids yield primary alcohols on reduction, and ketones yield secondary alcohols. A great many hydride reducing agents are available, but only a few are routinely used:

$NaBH_4$ Sodium borohydride is safe to use and relatively unreactive. It reduces aldehydes and ketones rapidly, reduces esters slowly, and does not reduce carboxylic acids.

$LiAlH_4$ Lithium aluminum hydride is a powerful and dangerous reagent that must be handled with great care. It rapidly reduces all types of carbonyl groups, including carboxylic acids.

The nucleophilic addition of Grignard reagents to carbonyl compounds is another important method for preparing alcohols. Grignard addition to formaldehyde yields a primary alcohol; addition to an aldehyde yields a secondary alcohol, and addition to a ketone or ester yields a tertiary alcohol. Carboxylic acids do not give Grignard addition products. The Grignard synthesis of alcohols is limited by the fact that Grignard reagents cannot be prepared from alkyl halides that contain reactive functional groups in the same molecule. This problem can sometimes be avoided, however, by protecting the interfering functional group. For example, alcohols are often protected by formation of tetrahydropyranyl ethers.

Alcohols undergo a wide variety of different reactions. They can be dehydrated by treatment with $POCl_3$ and can be transformed into alkyl halides by treatment with PBr_3 or $SOCl_2$. Furthermore, alcohols are weakly acidic ($pK_a \approx 16$–18). They react with strong bases and with alkali metals to form alkoxide anions, which are much used in organic synthesis. For example, the Williamson reaction between an alkoxide anion and an alkyl halide is a valuable method for preparing ethers.

The most important reaction of alcohols is their oxidation to carbonyl compounds. Primary alcohols yield either aldehydes or carboxylic acids, depending on the conditions used; secondary alcohols yield ketones; and tertiary alcohols do not react under the usual oxidizing conditions. Many oxidation methods are available, but only a few are commonly used:

$C_5H_6NCrO_3Cl$ Pyridinium chlorochromate (PCC) in dichloromethane is often used for oxidizing primary alcohols to aldehydes and secondary alcohols to ketones. The conditions are mild (room temperature) and reaction is rapid (5–10 min).

$CrO_3/H_2SO_4/H_2O$ (Jones' reagent) The Jones reagent is much used for oxidizing primary alcohols to carboxylic acids and secondary alcohols to ketones.

$Na_2Cr_2O_7/CH_3CO_2H/H_2O$ Sodium dichromate in aqueous acetic acid is an inexpensive reagent often used for simple oxidations on a large scale.

All of these oxidation reactions take place via bimolecular elimination (E2 reaction) of an intermediate chromate.

Alcohols show characteristic absorptions in both infrared and nuclear magnetic resonance spectra, and their spectroscopic analysis is relatively straightforward.

Thiols, which are sulfur analogs of alcohols, are usually prepared by S_N2 reaction of an alkyl halide with thiourea. Mild oxidation of a thiol yields the disulfide, R—S—S—R.

20.15 Summary of reactions

A. Synthesis of alcohols

1. Reduction of carbonyl groups (Section 20.5)

a. Aldehydes

$$CH_3CH_2CH_2CHO \xrightarrow[2.\ H_3O^+]{1.\ NaBH_4,\ ethanol\ (or\ LiAlH_4,\ ether)} CH_3CH_2CH_2CH_2OH$$

85%

b. Ketones

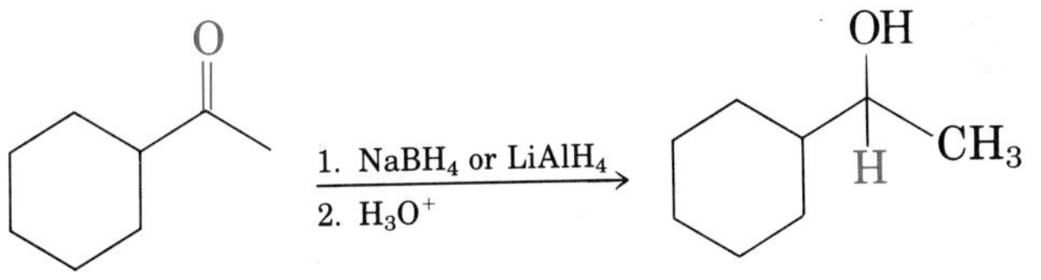

1. $NaBH_4$ or $LiAlH_4$

2. H_3O^+

OH

H

CH_3

(92%)

c. Esters

CO_2CH_3

1. $LiAlH_4$, THF

2. H_3O^+

CH_2OH

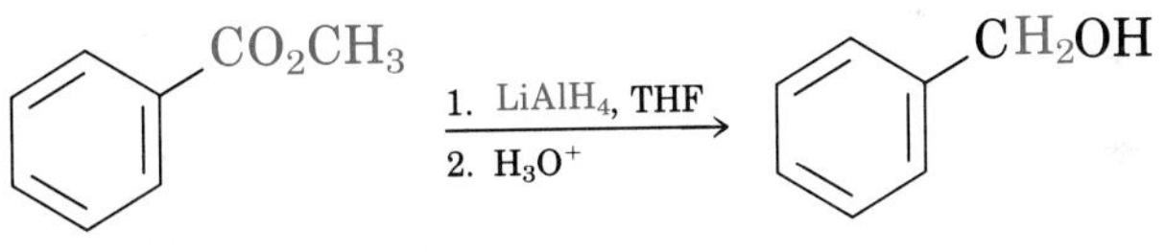

(89%)

d. Carboxylic acids

$$CH_3(CH_2)_7CH{=}CH(CH_2)_7CO_2H \xrightarrow[2.\ H_3O^+]{1.\ LiAlH_4,\ THF} CH_3(CH_2)_7CH{=}CH(CH_2)_7CH_2OH$$

87%

e. α,β-Unsaturated ketones

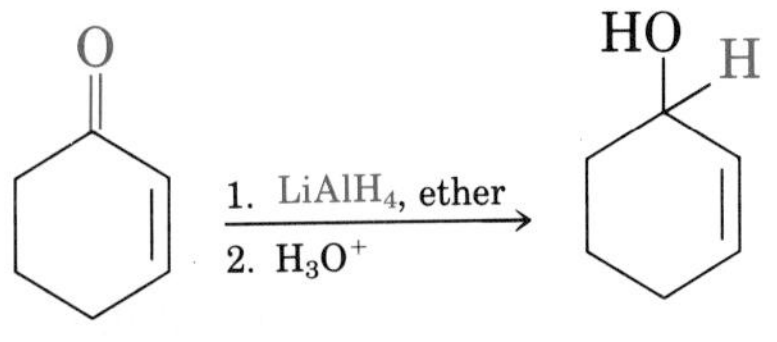

O

1. $LiAlH_4$, ether

2. H_3O^+

HO H

(97%)

2. Grignard addition (Section 20.7)
   a. Formaldehyde

$$CH_2O + \text{(cyclohexyl)}MgBr \xrightarrow[\text{then } H_3O^+]{\text{Ether,}} \text{(cyclohexyl)}CH_2OH$$

(65%)

   b. Aldehydes

$$(CH_3)_2CHCH_2CHO + \text{(phenyl)}MgBr \xrightarrow[\text{then } H_3O^+]{\text{Ether,}} (CH_3)_2CHCH_2CH(OH)C_6H_5$$

(73%)

   c. Ketones

$$\text{cyclohexanone} + CH_3CH_2MgBr \xrightarrow[\text{then } H_3O^+]{\text{Ether,}} \text{1-ethylcyclohexanol (HO, } CH_2CH_3\text{)}$$

(89%)

   d. Esters

$$CH_3CH_2CH_2CH_2CO_2CH_2CH_3 + CH_3MgBr \xrightarrow[\text{then } H_3O^+]{\text{Ether,}} CH_3CH_2CH_2CH_2C(OH)(CH_3)CH_3$$

85%

B. Reactions of alcohols
   1. Acidity (Section 20.9)

$$\text{(cyclohexyl)}OH + NaH \xrightarrow{\text{THF}} \text{(cyclohexyl)}O^- Na^+ + H_2$$

$$2\,(CH_3)_3COH + 2\,K \longrightarrow 2\,(CH_3)_3CO^-K^+ + H_2$$

2. Tosylate formation (Section 20.9)

$$\text{Cyclohexanol } (\ddot{O}H) + p\text{-}CH_3C_6H_4SO_2Cl \xrightarrow{\text{Pyridine}} C_6H_{11}\text{–O–}S(=O)_2\text{–}C_6H_4\text{–}CH_3$$

(86%)

3. Oxidation (Section 20.10)
   a. Primary alcohol

$$CH_3(CH_2)_5CH_2OH \xrightarrow[CH_2Cl_2]{PCC,} CH_3(CH_2)_5CHO$$

78%

$$CH_3(CH_2)_8CH_2OH \xrightarrow{\text{Jones' reagent}} CH_3(CH_2)_8CO_2H$$

93%

   b. Secondary alcohol

$$\text{Cyclooctanol (OH)} \xrightarrow[\text{(or PCC)}]{\text{Jones' reagent}} \text{Cyclooctanone (O)}$$

(96%)

C. Synthesis of thiols (Section 20.13)

$$CH_3CH_2Br \xrightarrow[2.\ ^{-}OH,\ H_2O]{1.\ (H_2N)_2C{=}S} CH_3CH_2SH$$

85%

D. Oxidation of thiols to disulfides (Section 20.13)

$$CH_3CH_2\text{—}SH \xrightarrow{Br_2} CH_3CH_2\text{—}S\text{—}S\text{—}CH_2CH_3$$

ADDITIONAL PROBLEMS

20.18 Name the following compounds according to the IUPAC system:

(a) $HOCH_2CH_2\overset{\overset{\large CH_3}{|}}{C}HCH_2OH$

(b) $CH_3CH(OH)\underset{\underset{\large CH_2CH_2CH_3}{|}}{C}HCH_2CH_3$

(c) 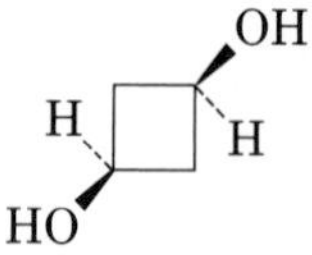 [structure: cyclobutane-1,3-diol with OH, H, H, HO substituents]

(d) 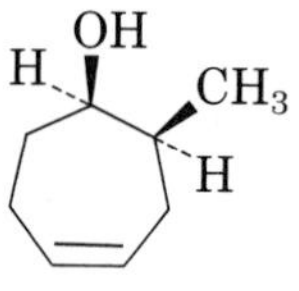 [structure: cycloheptene ring with OH, H, CH_3, H substituents]

(e) 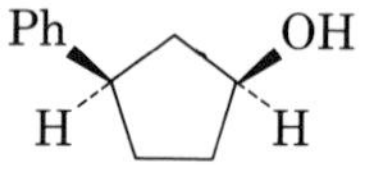 [structure: cyclopentane with Ph, H, OH, H substituents]

(f) 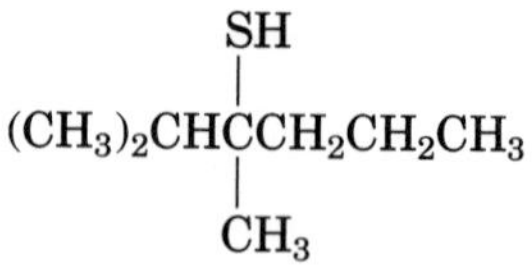 $(CH_3)_2CHC(SH)(CH_3)CH_2CH_2CH_3$

20.19 Draw and name the eight isomeric alcohols having the formula $C_5H_{12}O$.

20.20 Which of the eight alcohols that you identified in Problem 20.19 react with Jones' reagent (CrO_3, H_2O, H_2SO_4)? Show the products you would expect from each reaction.

20.21 How would you prepare the following compounds from 2-phenylethanol?
(a) Styrene
(b) Phenylacetaldehyde ($C_6H_5CH_2CHO$)
(c) Phenylacetic acid ($C_6H_5CH_2CO_2H$)
(d) Benzoic acid
(e) Ethylbenzene
(f) Benzaldehyde
(g) 1-Phenylethanol
(h) 2-Bromo-1-phenylethane

20.22 How would you carry out these transformations?

(a) 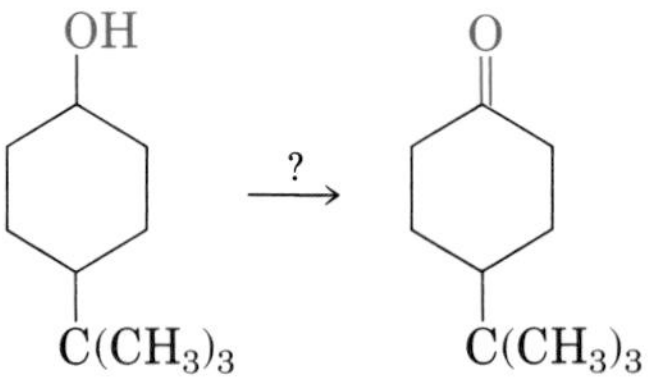 4-$C(CH_3)_3$-cyclohexanol (OH) $\xrightarrow{?}$ 4-$C(CH_3)_3$-cyclohexanone (O)

(b) 4-$C(CH_3)_3$-cyclohexanol (OH) $\xrightarrow{?}$ 4-$C(CH_3)_3$-cyclohexyl bromide (Br)

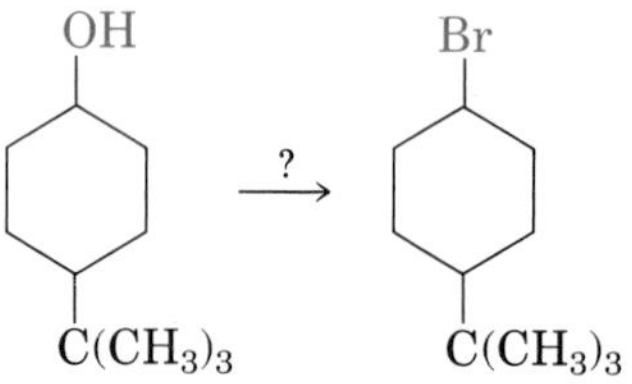

(c) 4-$C(CH_3)_3$-cyclohexanol (OH) $\xrightarrow{?}$ 4-$C(CH_3)_3$-cyclohexene

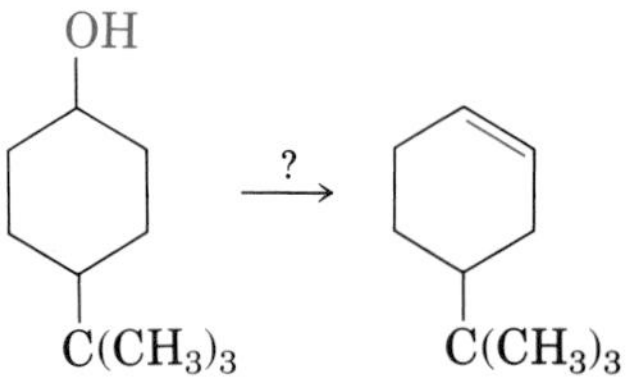

(d) $CH_3CH_2CH_2OH \xrightarrow{?} CH_3CH_2CHO$

(e) $CH_3CH_2CH_2OH \xrightarrow{?} CH_3CH_2CO_2H$

(f) $CH_3CH_2CH_2OH \xrightarrow{?} CH_3CH_2CH_2OTos$

(g) $CH_3CH_2CH_2OH \xrightarrow{?} CH_3CH_2CH_2Cl$

(h) $CH_3CH_2CH_2OH \xrightarrow{?} CH_3CH_2CH_2O^- Na^+$

(i) $CH_3CH_2CH_2OH \xrightarrow{?} CH_3CH_2CH_2OCH_2CH_2CH_3$

20.23 What Grignard reagent and what carbonyl compound might you start with to prepare the following alcohols?

(a) $CH_3CH(OH)CH_2CH_3$

(b) $C_6H_5C(OH)(CH_3)_2$

(c) $H_2C{=}C(CH_3){-}CH_2OH$

(d) $(Ph)_3COH$

(e) $CH_3CH(OH)CH_2CH_2CH_2Br$

20.24 Assume that you have been given a sample of *(S)*-2-octanol. How could you prepare *(R)*-2-chlorooctane? How could you prepare *(R)*-2-octanol?

20.25 When 4-chloro-1-butanol is treated with a strong base such as sodium hydride, NaH, tetrahydrofuran is produced. Suggest a mechanism for this reaction.

$$ClCH_2CH_2CH_2CH_2OH \xrightarrow[\text{Ether}]{\text{NaH}} \text{(tetrahydrofuran)} + H_2 + NaCl$$

20.26 How would you carry out these transformations?

(a) $C_6H_5CH{=}CHCO_2H \longrightarrow C_6H_{11}CH_2CH_2CO_2H$

(b) $C_6H_5CH{=}CHCO_2H \longrightarrow C_6H_5CH{=}CHCH_2OH$

(c) $C_6H_5CH{=}CHCO_2H \longrightarrow C_6H_5CH_2CH_2CO_2H$

20.27 Grignard reagents react with ethylene oxide to produce primary alcohols. Propose a mechanism for this reaction.

$$\underset{\text{Ethylene oxide}}{CH_2{-}CH_2 \text{ (O bridged)}} \xrightarrow[\text{2. } H_3O^+]{\text{1. RMgX}} R{-}CH_2CH_2OH$$

20.28 Grignard reagents also react with oxetane to produce primary alcohols, but the reaction is much slower than with ethylene oxide (Problem 20.27). Can you suggest a reason for the difference in reactivity between oxetane and ethylene oxide?

$$\text{Oxetane} \xrightarrow[\text{2. } H_3O^+]{\text{1. RMgX}} R{-}CH_2CH_2CH_2OH$$

20.29 Testosterone is one of the most important male steroid hormones. When testosterone is dehydrated by treatment with $POCl_3$ in pyridine, rearrangement occurs to yield the product indicated. Propose a mechanism to account for this reaction. [*Hint:* See Section 16.5.]

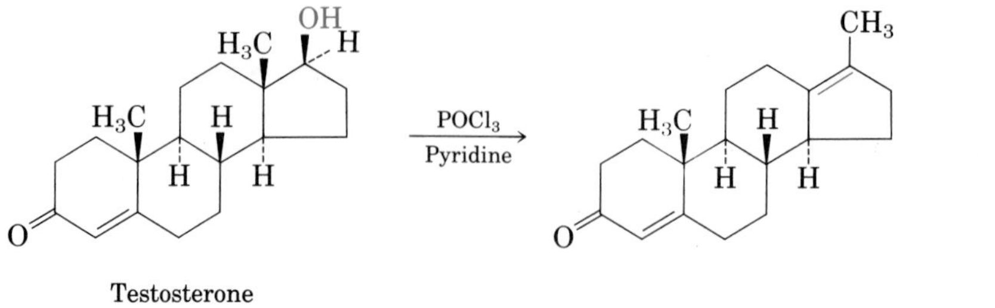

Testosterone

20.30 Starting from testosterone (Problem 20.29), how would you prepare the following substances?

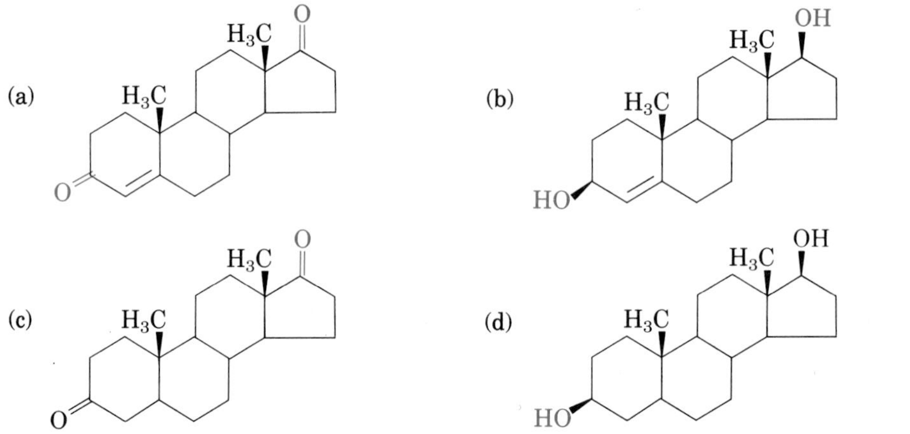

(a) (b)

(c) (d)

20.31 Dehydration of *trans*-2-methylcyclopentanol with $POCl_3$ in pyridine yields predominantly 3-methylcyclopentene. What is the stereochemistry of this dehydration? Can you suggest a reason for formation of the observed product? Inspection of molecular models should be quite helpful.

20.32 We have seen in this chapter that nucleophiles, such as hydride and Grignard reagents, add to carbonyl groups. In light of this knowledge, propose a mechanism for the observed exchange reaction between isotopically labeled H_2O^* and acetone.

$$(CH_3)_2C{=}O + H_2O^* \rightleftharpoons (CH_3)_2C{=}O^* + H_2O$$

20.33 The dehydration of testosterone (Problem 20.29) occurs slowly and in poor yield. If the isomeric alcohol is dehydrated, however, the reaction is much faster and higher yielding. Can you propose a reason for the difference in behavior between the two alcohols? [*Hint:* Build molecular models and look at the conformations of the two alcohols.]

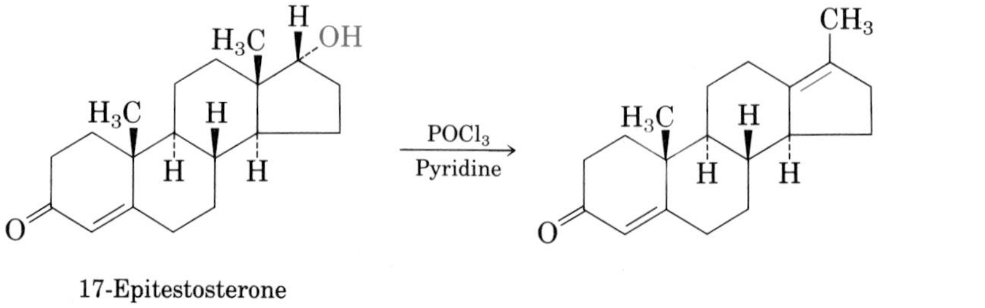

17-Epitestosterone

20.34 The ^1H NMR spectrum shown is that of 3-methyl-3-buten-1-ol. Assign all the observed resonance peaks to specific protons and account for the splitting patterns.

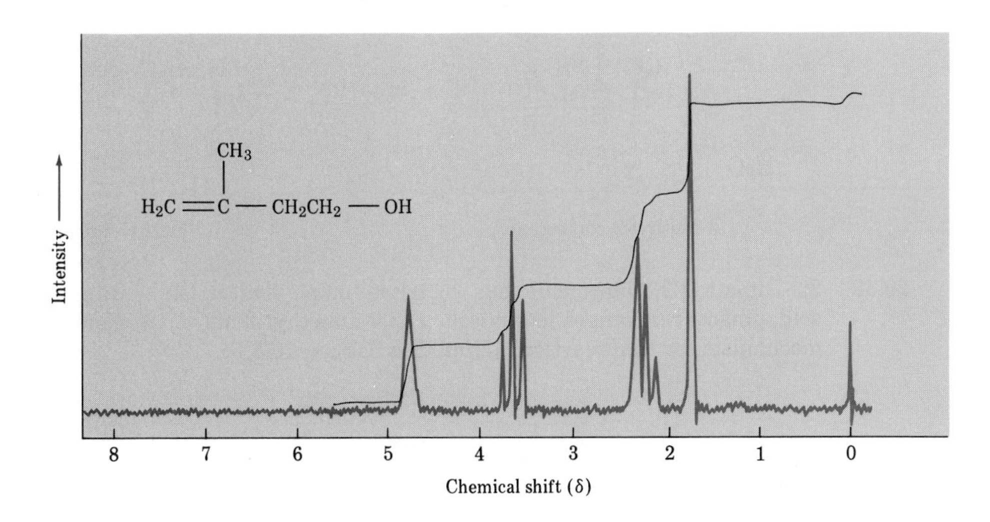

20.35 Propose a structure that is consistent with the following spectral data for a compound of formula $C_8H_{18}O_2$:

Infrared	3350 cm^{-1}
^1H NMR	1.24 δ (12-proton singlet)
	1.56 δ (4-proton singlet)
	1.95 δ (2-proton singlet)

20.36 The ^1H NMR spectrum shown is of an alcohol of formula $C_8H_{10}O$. Propose a structure that is consistent with the observed spectrum.

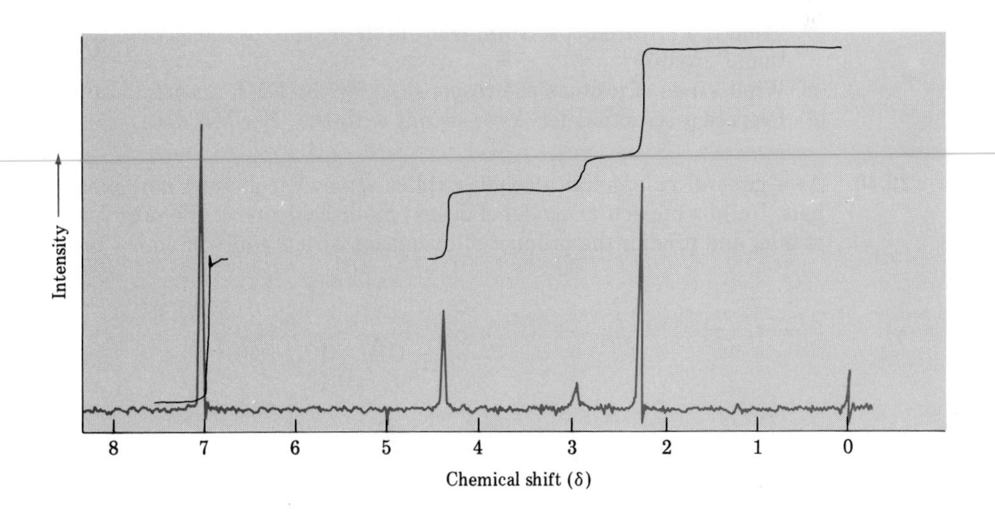

20.37 3-Methyl-2-cholestene can be hydroborated from either the top face or the bottom face to yield two possible products. In practice, the reaction occurs largely from the bottom face to yield, after treatment with basic hydrogen peroxide,

3β-methyl-2α-cholestanol. Can you suggest a reason for the observed stereoselectivity of the reaction? [*Hint:* Build a molecular model.]

1. BH_3, THF
2. H_2O_2/^-OH

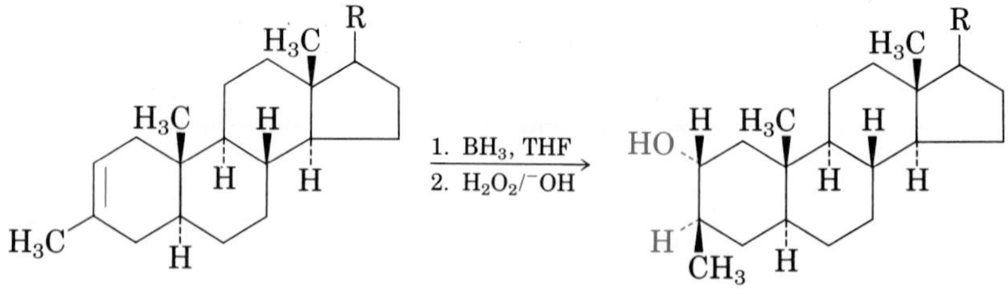

3-Methyl-2-cholestene

3β-Methyl-2α-cholestanol

20.38 2,3-Dimethyl-2,3-butanediol has the trivial name *pinacol.* On heating with aqueous acid, pinacol rearranges to *pinacolone,* 3,3-dimethyl-2-butanone. Can you suggest a mechanism for this reaction? [*Hint:* See Section 16.5.]

$$(CH_3)_2\underset{}{\overset{OH}{C}}-\overset{OH}{C}(CH_3)_2 \xrightarrow{H^+} CH_3\overset{O}{\overset{\|}{C}}-C(CH_3)_3 + H_2O$$

Pinacol — Pinacolone

20.39 Compound A, $C_5H_{10}O$, is one of the basic building blocks of nature. All steroids and many other naturally occurring compounds are built up from compound A. Spectroscopic analysis of A yields the following information:

Infrared	3400 cm^{-1} (strong); 1640 cm^{-1} (weak)
^1H NMR	1.63 δ (3-proton singlet)
	1.70 δ (3-proton singlet)
	3.83 δ (1-proton broad singlet)
	4.15 δ (2-proton doublet, J = 7 Hz)
	5.70 δ (1-proton triplet, J = 7 Hz)

(a) How many double bonds and/or rings does A have?
(b) From the infrared spectrum, what is the nature of the oxygen-containing functional group?
(c) What kinds of protons are responsible for the NMR absorption listed?
(d) Propose a structure for A consistent with the ^1H NMR data.

20.40 As a general rule, axial alcohols oxidize somewhat faster than do equatorial alcohols. Build a molecular model of desoxycholic acid, one of the principal constituents of bile, and predict the product of oxidation with 1 equiv of Jones' reagent.

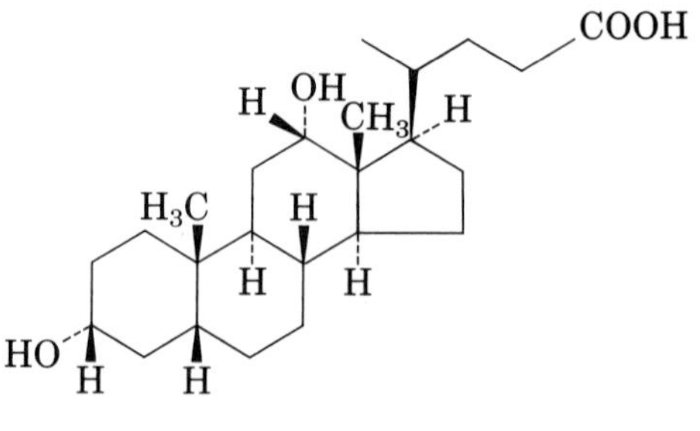

Desoxycholic acid

20.41 Propose a synthesis of bicyclohexylidene, starting from cyclohexanone as the only source of carbon.

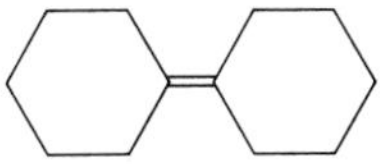

Bicyclohexylidene

20.42 Since all hamsters look pretty much alike, attraction between sexes is controlled by chemical secretions. The sex attractant exuded by the female hamster has been isolated and shown to have the following spectral properties:

Mass spectrum	$M^{+\cdot} = 94$
Infrared	Nothing higher than 1500 cm^{-1} except for saturated C—H absorptions at 2950 cm^{-1}
^1H NMR	Singlet at 2.1 δ

Propose a structure for the hamster sex attractant.

CHAPTER 21 CHEMISTRY OF CARBONYL COMPOUNDS: AN OVERVIEW

In this and the next five chapters, we will discuss the most important functional group in organic chemistry—the **carbonyl group,** $\gt C{=}O$.

Carbonyl compounds are ubiquitous. The majority of biologically important molecules contain carbonyl groups, as do many pharmaceutical agents and many of the synthetic chemicals that touch our everyday lives. Glycine, an amino acid constituent of many proteins, phenacetin, an over-the-counter headache remedy, and Dacron, the polyester material used in clothing, all contain different kinds of carbonyl groups.

$$H_2NCH_2C(=O){-}OH$$

Glycine
(an amino carboxylic acid)

$$CH_3C(=O){-}NH{-}C_6H_4{-}O{-}CH_2CH_3$$

Phenacetin
(an amide)

$$\left[{-}O{-}C(=O){-}C_6H_4{-}C(=O){-}O{-}CH_2CH_2{-}\right]_n$$

Dacron (a polyester)

There are many different kinds of carbonyl compounds, depending on what groups are bonded to the C=O unit. The chemistry of carbonyl groups is quite similar, however, regardless of their exact structure.

21.1 Kinds of carbonyl compounds

Table 21.1 shows some of the many different kinds of carbonyl compounds. All contain an **acyl fragment,** $R{-}C(=O){-}$, bonded to another residue. The R group of the acyl fragment may be alkyl, aryl, alkenyl, or alkynyl, and the other residue to which the acyl moiety is bonded may be a carbon, hydrogen, oxygen, halogen, nitrogen, or sulfur substituent.

TABLE 21.1 **Some common types of carbonyl compounds**

Name	*General formula*	*Name ending*
Aldehyde	$R-\overset{O}{\overset{\Vert}{C}}-H$	*-al*
Ketone	$R-\overset{O}{\overset{\Vert}{C}}-R'$	*-one*
Carboxylic acid	$R-\overset{O}{\overset{\Vert}{C}}-O-H$	*-oic acid*
Acid chloride	$R-\overset{O}{\overset{\Vert}{C}}-Cl$	*-yl* or *-oyl chloride*
Acid anhydride	$R-\overset{O}{\overset{\Vert}{C}}-O-\overset{O}{\overset{\Vert}{C}}-R'$	*-oic anhydride*
Ester	$R-\overset{O}{\overset{\Vert}{C}}-O-R'$	*-oate*
Lactone (cyclic ester)	$\rangle C-\overset{O}{\overset{\Vert}{C}}-O$ (ring)	None
Amide	$R-\overset{O}{\overset{\Vert}{C}}-N\langle$	*-amide*
Lactam (cyclic amide)	$\rangle C-\overset{O}{\overset{\Vert}{C}}-N/$ (ring)	None

It is very useful to classify carbonyl compounds into two general categories, based on the kinds of chemistry they undergo:

Aldehydes Ketones	The acyl units in these two functional groups are bonded to substituents (H and R, respectively) that *cannot serve as leaving groups.* Aldehydes and ketones therefore behave similarly and undergo many of the same reactions.
Carboxylic acids Esters Acid chlorides Acid Anhydrides Amides	The acyl units in carboxylic acids and their derivatives are bonded to substituents (oxygen, halogen, nitrogen) that *can serve as leaving groups* in substitution reactions. The chemistry of these compounds is therefore similar.

21.2 Nature of the carbonyl group

The carbon–oxygen double bond of carbonyl groups is similar in many respects to the carbon–carbon double bond of alkenes (Figure 21.1). The

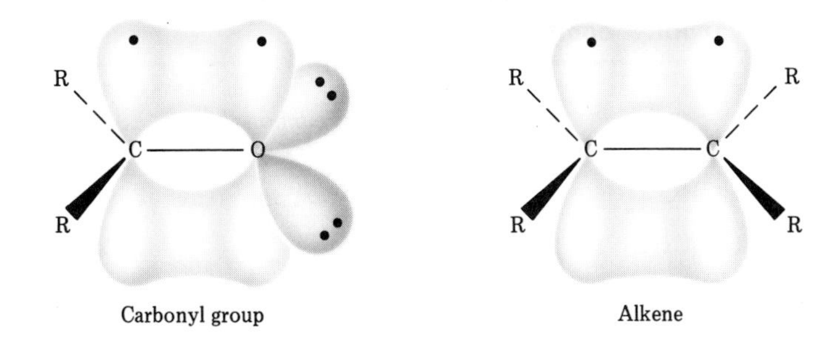

Figure 21.1. Electronic structure of the carbonyl group.

carbonyl carbon atom is sp^2 hybridized and forms three sigma bonds. The fourth valence electron remains in a carbon p orbital and forms a pi bond to oxygen by overlap with an oxygen p orbital. The oxygen atom also has two nonbonding pairs of electrons, which occupy its remaining two orbitals. (The oxygen atom is probably sp^2 hybridized, though there is some disagreement about this point.)

Like alkenes, carbonyl compounds are planar about the double bond and have bond angles of approximately 120°. Table 21.2 shows the structure of acetaldehyde and indicates the experimentally determined bond lengths and angles. As we would expect, the carbon–oxygen double bond is shorter (1.22 Å versus 1.43 Å) than the normal carbon–oxygen single bond; it is also stronger (175 kcal/mol versus 92 kcal/mol).

TABLE 21.2 **Structure of acetaldehyde**

$H_3C(H)C=\ddot{O}:$

Bond angle (degrees)		*Bond length (Å)*	
H—C—C	118	C=O	1.22
C—C—O	121	C—C	1.50
H—C—O	121	OC—H	1.09

Carbon–oxygen double bonds are polarized $C^{\delta+}—O^{\delta-}$ because of the high electronegativity of oxygen relative to carbon, and carbonyl compounds therefore have substantial dipole moments. Table 21.3 lists the observed dipole moments for a variety of different types of carbonyl compounds and indicates that all are strongly polarized.

The most important consequence of carbonyl-group polarization is the chemical reactivity of the carbon–oxygen double bond. Since the carbonyl carbon is positively polarized, it is an electrophilic site and is attacked by nucleophiles. Conversely, the carbonyl oxygen is negatively polarized and is a nucleophilic (basic) site.

Electrophilic carbon reacts with bases and nucleophiles → $\overset{\delta+}{C}=\overset{\delta-}{\ddot{O}}:$ ← Nucleophilic oxygen reacts with acids and electrophiles

TABLE 21.3 **Observed dipole moments of some carbonyl compounds, $R_2C \overset{\leftrightarrow}{=} O$**

Carbonyl compound	*Type of carbonyl group*	*Observed dipole moment (D)*
HCHO	Aldehyde	2.33
CH_3CHO	Aldehyde	2.72
$(CH_3)_2CO$	Ketone	2.88
$PhCOCH_3$	Ketone	3.02
Cyclobutanone	Ketone	2.99
CH_3COOH	Carboxylic acid	1.74
CH_3COCl	Acid chloride	2.72
$CH_3CO_2CH_3$	Ester	1.72
CH_3CONH_2	Amide	3.76
$CH_3CON(CH_3)_2$	Amide	3.81

We shall see in the next five chapters that the majority of carbonyl-group reactions can be rationalized in terms of simple bond-polarization arguments.

21.3 General reactions of carbonyl compounds

Most reactions of carbonyl groups occur by one of only four general reaction mechanisms—nucleophilic addition, nucleophilic acyl substitution, alpha substitution, and carbonyl condensation. These mechanisms have many variations, just as alkene electrophilic addition reactions and S_N2 reactions do, but the variations are easier to learn when the fundamental features of the mechanisms are understood. Let's see what these four reaction mechanisms are, and what kinds of chemistry carbonyl groups undergo.

21.4 Nucleophilic addition reactions

A **nucleophilic addition reaction** involves the addition of a nucleophile to the electrophilic carbon of the carbonyl group. Since the nucleophile uses its electron pair to attack carbon, two electrons from the carbon–oxygen double bond must move onto the electronegative oxygen atom, where they can be stabilized as an alkoxide anion. The carbonyl carbon rehybridizes from sp^2 to sp^3 during the reaction, and the initial product therefore has tetrahedral geometry.

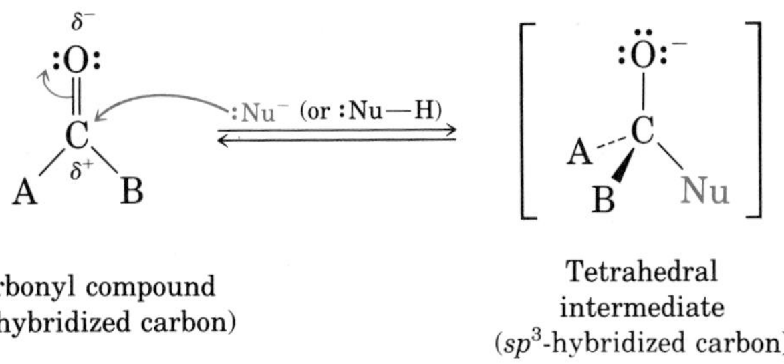

δ⁻
:O:
C
δ⁺
A
B
:Nu⁻ (or :Nu—H)
:Ö:⁻
A
C
B
Nu
Carbonyl compound
(sp²-hybridized carbon)
Tetrahedral
intermediate
(sp³-hybridized carbon)

Once formed, the tetrahedral intermediate can undergo either of the reactions indicated in Figure 21.2. The tetrahedral intermediate might be protonated to form a stable alcohol, or it might expel oxygen as water or hydroxide ion to form a new kind of double bond. We will see many examples of both possibilities in future chapters. (Note that this overview of carbonyl-group reactivity is being simplified to illustrate the fundamental similarities of many reactions. In practice, some carbonyl nucleophilic addition reactions are acid catalyzed and the initial tetrahedral intermediate is a neutral alcohol. The details of specific reactions will be dealt with in the next chapter.)

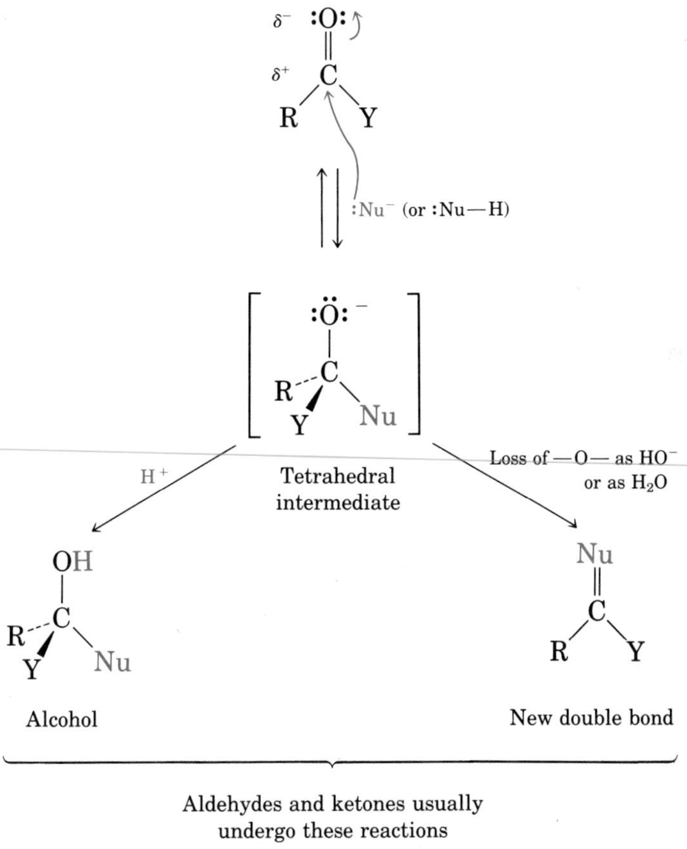

δ⁻ :O:
δ⁺ C
R
Y
:Nu⁻ (or :Nu—H)
:Ö:⁻
C
R
Y
Nu
H⁺
Tetrahedral
intermediate
Loss of —O— as HO⁻
or as H₂O
OH
C
R
Y
Nu
Nu
C
R
Y
Alcohol
New double bond
Aldehydes and ketones usually
undergo these reactions

Figure 21.2. General kinds of carbonyl-group nucleophilic addition reactions.

ALCOHOL FORMATION

The simplest reaction of a tetrahedral intermediate is protonation to yield a stable alcohol as final product. This happens only with aldehydes and ketones, and we have already seen examples of alcohol formation during reduction of these functional groups with hydride reagents such as $NaBH_4$ and $LiAlH_4$, and during Grignard reactions (Section 20.7).

Hydride reduction

:O:, C, H_3C, CH_3 — $:H^-$ ($NaBH_4$) → [$:\ddot{O}:^-$, C, H_3C, CH_3, H] Na^+ — H_3O^+ → OH, C, H_3C, CH_3, H

Carbonyl compound — Tetrahedral intermediate — Stable alcohol

Grignard addition

:O: (cyclohexanone) — $\bar{C}H_3$: $\overset{+}{M}gBr$ → [$:\ddot{O}:^-$, CH_3] $\overset{+}{M}gBr$ — H_3O^+ → HO, CH_3 + HOMgBr

Carbonyl compound — Tetrahedral intermediate — Stable alcohol

PROBLEM

21.1 Propose a mechanism to account for the fact that acetone cyanohydrin is isolated when acetone is allowed to react with aqueous acidic KCN.

$$\underset{\text{Acetone}}{CH_3\overset{\overset{O}{\|}}{C}CH_3} \xrightleftharpoons{KCN,\ H_2O} \underset{\text{Acetone cyanohydrin}}{CH_3\overset{OH}{\underset{CN}{C}}CH_3}$$

IMINE FORMATION

Alternatively, the tetrahedral intermediate might not be stable; further reaction often occurs when the attacking nucleophile has a lone pair of electrons. In such cases, oxygen can be expelled from the tetrahedral intermediate, leading to formation of a new carbon–nucleophile double bond. For example, ketones and aldehydes react with primary amines, $R—\ddot{N}H_2$, to form imines, $>C=\ddot{N}—R$. These reactions are usually acid catalyzed and occur in several steps, as shown in Figure 21.3.

Nucleophilic attack on the carbonyl group by an electron pair from nitrogen gives a tetrahedral intermediate.

Transfer of a proton from nitrogen to oxygen gives a different tetrahedral intermediate (a carbinolamine), which is not stable.

It undergoes further reaction when the oxygen atom is protonated by acid.

The lone-pair electrons from nitrogen expel water to yield an iminium ion.

Loss of a proton gives a neutral imine as the final product

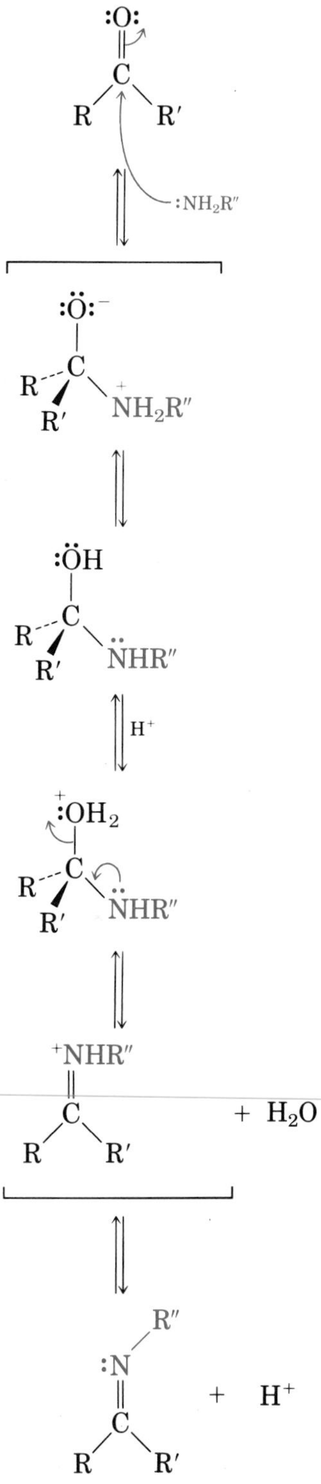

:O:
C
R R′
$:NH_2R''$
:Ö:⁻
C
R R′ $\overset{+}{N}H_2R''$
:ÖH
C
R R′ $\ddot{N}HR''$
H^+
$:\overset{+}{O}H_2$
C
R R′ $\ddot{N}HR''$
$^+NHR''$
C
R R′ $+ H_2O$
R″
:N
C
R R′ $+ H^+$

Figure 21.3. General mechanism of imine formation.

The net effect of reaction between the carbonyl compound and the amine is loss of water (dehydration) and formation of a carbon–nitrogen double bond in place of a carbon–oxygen double bond.

PROBLEM

21.2 Write all the steps involved in the synthesis of cyclohexanone oxime from cyclohexanone and hydroxylamine. The reaction is acid catalyzed.

$$\text{Cyclohexanone} + NH_2OH \overset{H^+}{\rightleftharpoons} \text{Cyclohexanone oxime} + H_2O$$

Cyclohexanone oxime

PROBLEM

21.3 Propose a mechanism to account for the formation of *N*-cyclopentylidenemethylamine oxide from cyclopentanone and *N*-methylhydroxylamine. The reaction is acid catalyzed.

$$\text{Cyclopentanone} + CH_3NHOH \overset{H^+}{\rightleftharpoons} \text{N-Cyclopentylidenemethylamine oxide} + H_2O$$

N-Cyclopentylidenemethylamine oxide

21.5 Nucleophilic acyl substitution reactions

The second fundamental reaction of carbonyl compounds, **nucleophilic acyl substitution,** is related to the nucleophilic addition reaction we have just seen but occurs with derivatives of carboxylic acids in which one of the carbonyl-group substituents can act as a leaving group. This is the case, for example, with esters, acid chlorides, acid anhydrides, and amides (but not with ketones or aldehydes). When such a carbonyl group is attacked by a nucleophile, addition occurs, but the initially formed tetrahedral intermediate is not stable. Instead, the tetrahedral intermediate reacts further by expelling the leaving group and forming a new carbonyl compound:

$$RC(=O)X + :Nu^- \overset{\text{(or :Nu—H)}}{\rightleftharpoons} [R(X)(Nu)C{-}O^-] \longrightarrow RC(=O)Nu + :X^-$$

Tetrahedral intermediate

New carbonyl compound

where X = OR (ester), Cl (acid chloride), NH_2 (amide), or OCOR′ (acid anhydride).

The net effect of such a process is the substitution of the leaving group by the attacking nucleophile. This result is similar to that which occurs during an S_N2 reaction, but the mechanism by which the substitution occurs is quite different. For example, we will see in Chapter 24 that acid chlorides are rapidly converted into carboxylic acid anions by treatment with hydroxide ion. This reaction is simply a nucleophilic acyl substitution in which hydroxide replaces chloride (Figure 21.4) and is then deprotonated.

Nucleophilic addition of hydroxide ion to an acid chloride yields a tetrahedral intermediate.

An electron pair from oxygen pushes out chloride ion and yields the substitution product—a carboxylic acid—that is subsequently deprotonated by hydroxide ion.

$$\mathrm{RCOCl} \xrightarrow{^-:\ddot{O}H} [\mathrm{R{-}C(O^-)(OH)Cl}] \longrightarrow \mathrm{Cl:^-} + \mathrm{RCOOH} \xrightarrow{HO^-} \mathrm{RCOO^-} + H_2O$$

Figure 21.4. Mechanism of the reaction of acid chlorides with hydroxide ion.

PROBLEM

21.4 We saw in Chapter 20 that esters react with Grignard reagents to form tertiary alcohols. Write all the steps in the reaction of methyl benzoate with methylmagnesium bromide to yield 2-phenyl-2-propanol. Identify the nucleophilic acyl substitution step.

PROBLEM

21.5 Propose a mechanism for the base-induced hydrolysis of methyl benzoate to benzoate anion.

$$\mathrm{C_6H_5CO_2CH_3} \xrightarrow[H_2O]{NaOH} \mathrm{C_6H_5CO_2^-\,Na^+} + CH_3OH$$

Methyl benzoate   Benzoate ion

21.6 Alpha-substitution reactions

The third major reaction of carbonyl compounds, **alpha substitution,** occurs at the position next to the carbonyl group, the alpha (α) position, and involves substitution of an alpha hydrogen by some other group. These reactions take place by formation of intermediate **enols** or **enolate ions:**

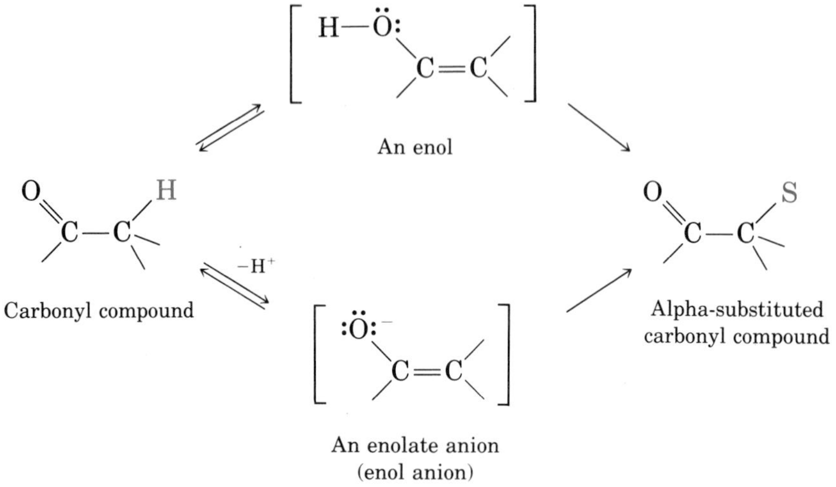

An enol

Carbonyl compound

$-H^+$

Alpha-substituted carbonyl compound

An enolate anion (enol anion)

We have already seen enols as the initial products of mercuric ion catalyzed hydration of alkynes in Section 7.4, where it was noted that enols rapidly equilibrate with the corresponding ketone:

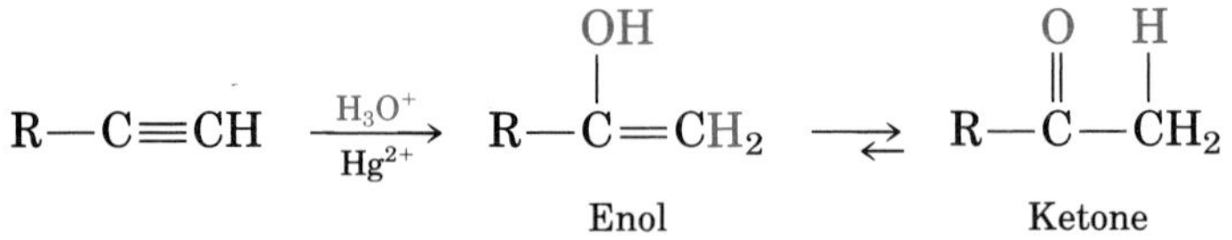

$$R{-}C{\equiv}CH \xrightarrow[Hg^{2+}]{H_3O^+} R{-}C(OH){=}CH_2 \rightleftharpoons R{-}C({=}O){-}CH_3$$

Enol Ketone

We now need to learn more about enol chemistry: The presence of a carbonyl group renders acidic the protons on the alpha carbon. Carbonyl compounds react with strong base to yield enolate ions.

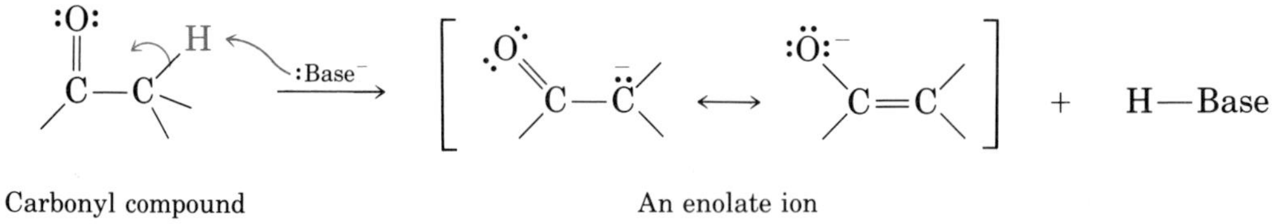

$:Base^-$

$+ \quad H{-}Base$

Carbonyl compound An enolate ion

Note that only the protons on the alpha position of carbonyl compounds are acidic. The protons at beta, gamma, delta, and so on are not acidic because the resulting anions cannot be stabilized by the carbonyl group.

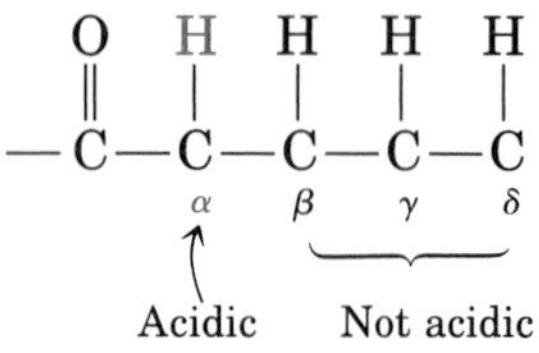

α β γ δ

Acidic Not acidic

What chemistry might we expect of enolates? Since they carry negative charges and have high electron density on the alpha carbon, enolates react as nucleophiles and take part in many of the reactions we have already studied. For example, enolates react with primary alkyl halides in the S_N2 reaction. The nucleophilic carbanion displaces halide ion and forms a new carbon–carbon bond:

Enolate + Alkyl halide ($R—CH_2—X$) $\xrightarrow[\text{reaction}]{S_N2}$ $C(=O)—C—CH_2—R$ + $:X^-$

For example,

Cyclohexanone $\overset{\text{Base}}{\rightleftharpoons}$ [Cyclohexanone enolate ion] $\xrightarrow{CH_3I}$ 2-Methylcyclohexanone (65%) + $:I^-$

The S_N2 reaction between an enolate and an alkyl halide is an example of an **alkylation** reaction, and is one of the most powerful methods available for making carbon–carbon bonds, thereby building up larger molecules from small precursors. We will study the alkylation of many kinds of carbonyl groups in detail in Chapter 25.

PROBLEM

21.6 Good evidence for the formation of enolates can be obtained from deuteration experiments. For example, when cyclohexanone is treated with a catalytic amount of sodium deuteroxide in D_2O, 2,2,6,6-tetradeuteriocyclohexanone is formed. Propose a mechanism to account for this reaction.

21.7 Carbonyl condensation reactions

The fourth and last fundamental reaction of carbonyl groups, **carbonyl condensation,** takes place when two carbonyl compounds come together *(condense)*. One very important such process is the **aldol reaction** between two aldehydes. For example, when acetaldehyde is treated with base, two molecules of aldehyde react with each other to yield the hydroxy aldehyde product known trivially as *aldol* (aldehyde + alcohol):

$$2\ CH_3CHO \xrightarrow{\text{Base}} CH_3\overset{\overset{\displaystyle OH}{|}}{C}HCH_2CHO$$

Acetaldehyde   Aldol

How does this dimerization occur? Although the aldol dimerization reaction appears different from the three general carbonyl-group processes already discussed, it is not. The aldol reaction is simply a combination of a nucleophilic addition step and an alpha-substitution step. The initially formed enolate ion of acetaldehyde acts as a nucleophile and adds to the carbonyl group of another acetaldehyde molecule. Reaction occurs by the pathway shown in Figure 21.5.

Hydroxide ion abstracts an acidic alpha proton from one molecule of acetaldehyde, yielding an enolate ion.

$$H{-}CH_2{-}\overset{\overset{\displaystyle :O:}{\|}}{C}{-}H \quad {}^{-}:OH/H_2O$$

$$\updownarrow$$

$$H_2C{=}\overset{\overset{\displaystyle :\ddot{O}:^-}{|}}{C}{-}H + H_2O$$

$$CH_3{-}\overset{\overset{\displaystyle :O:}{\|}}{C}{-}H$$

The enolate ion adds as a nucleophile to the carbonyl group of a second molecule of acetaldehyde, producing a tetrahedral intermediate.

$$\updownarrow$$

$$CH_3{-}\underset{\underset{\displaystyle H}{|}}{\overset{\overset{\displaystyle :\ddot{O}:^-}{|}}{C}}{-}CH_2{-}\overset{\overset{\displaystyle O}{\|}}{C}{-}H$$

The intermediate is protonated by water solvent to yield the neutral aldol product and regenerate hydroxide ion.

$$\updownarrow H_2O$$

$$CH_3{-}\underset{\underset{\displaystyle H}{|}}{\overset{\overset{\displaystyle OH}{|}}{C}}{-}CH_2\overset{\overset{\displaystyle O}{\|}}{C}{-}H + {}^{-}OH$$

Figure 21.5. Mechanism of the aldol reaction—a carbonyl condensation process.

There are many variations of the aldol reaction, including the dimerization of ketones and esters and the mixed aldol reaction between two different kinds of carbonyl compounds. The important point to keep in mind is that all the carbonyl reactions we will encounter are simply variations or combinations of four fundamental reaction types—nucleophilic addition, nucleophilic acyl substitution, alpha substitution, and carbonyl condensation.

21.8 Summary

Carbonyl compounds contain carbon–oxygen double bonds, $\rangle C{=}O$, and can be classified into two general categories:

R_2CO RCHO	Ketones and aldehydes are similar in their reactivity and are distinguished by the fact that the substituents on the acyl carbon cannot serve as leaving groups.
$R{-}CO_2H$ $R{-}CO_2R'$ $R{-}CONH_2$ $R{-}CO_2COR'$ $R{-}COCl$	Carboxylic acids and their derivatives—esters, amides, anhydrides, and acid chlorides—are distinguished by the fact that the heteroatom substituents on the acyl carbon can serve as leaving groups.

Structurally, the carbon–oxygen double bond is similar to the carbon–carbon double bond of alkenes. The carbonyl carbon atom is sp^2 hybridized and forms both an sp^2 sigma bond and a p pi bond to oxygen. Carbonyl groups are strongly polarized because of the electronegativity of oxygen, and the carbonyl carbon atom is therefore electrophilic.

Carbonyl compounds undergo four fundamental types of reactions—nucleophilic addition to the carbon–oxygen double bond, nucleophilic acyl substitution, alpha substitution of protons next to the carbon–oxygen double bond, and carbonyl condensation. There are many possible variations of these reaction types, but the fundamental mechanistic processes remain the same.

All nucleophilic addition reactions are initiated by attack of a nucleophile on the electrophilic carbonyl carbon atom, yielding a tetrahedral intermediate. This tetrahedral intermediate reacts further in either of two ways, depending on the substituents present. The tetrahedral intermediates from aldehydes and ketones may be protonated to give a stable alcohol product, or they may lose hydroxide ion to give carbon–nucleophile double-bond products.

Nucleophilic acyl substitution reactions result when nucleophiles add to carboxylic acid derivatives. The initially formed tetrahedral intermediate is not stable, but instead eliminates one of the substituents originally bound to the carbonyl carbon.

Alpha substitution reactions occur on the carbon atom next to the carbonyl group and are a consequence of the acidity of alpha protons. All carbonyl compounds with alpha protons can react with strong bases to form enolate anions. Enolates are nucleophilic, and they take part in S_N2 reactions with alkyl halides and in addition reactions with other carbonyl compounds (condensation).

ADDITIONAL PROBLEMS

21.7 Identify the different types of carbonyl groups in the following molecules:

(a) Aspirin

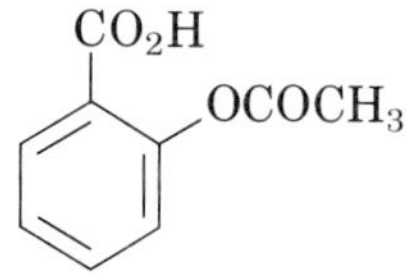

(b)

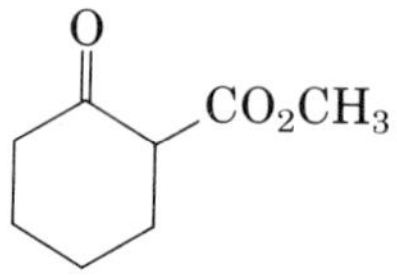

(c)

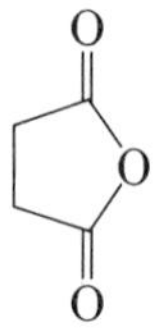

(d)

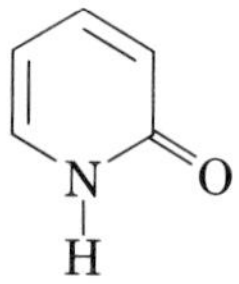

(e) 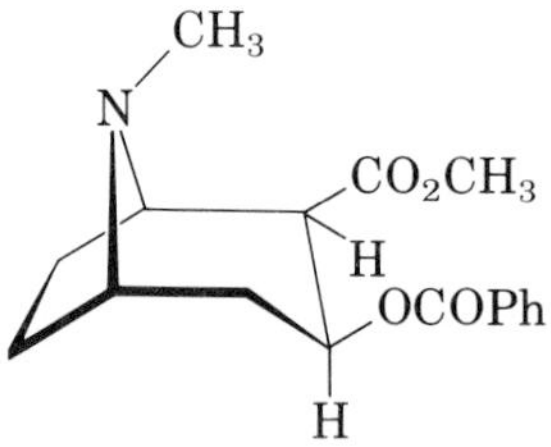 Cocaine

(f) Retinal

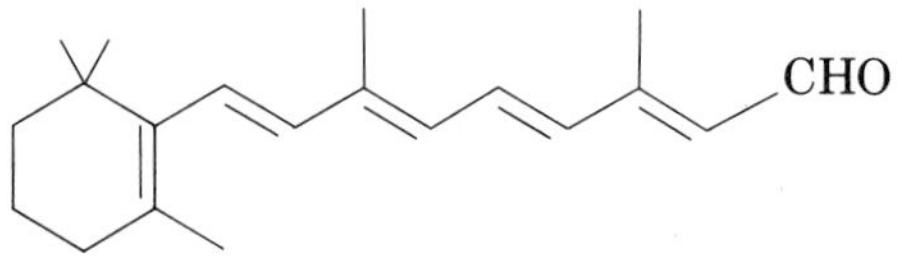

(g)

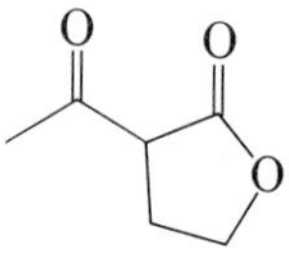

(h) 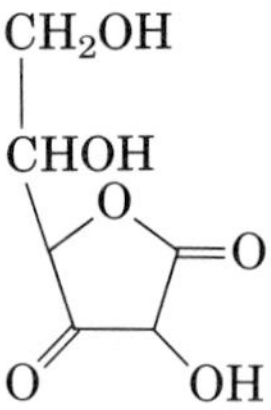 Ascorbic acid (vitamin C)

(i) 6-Aminopenicillanic acid (precursor of penicillins)

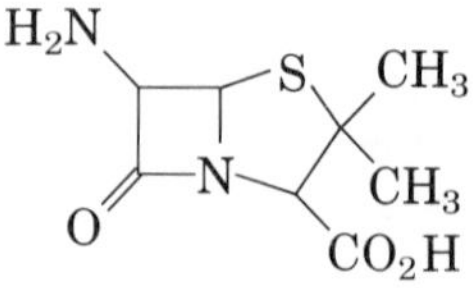

21.8 Propose a mechanism to explain the cyclization reaction that occurs when 2,5-hexanedione is treated with dilute base. What kind of carbonyl reaction is involved?

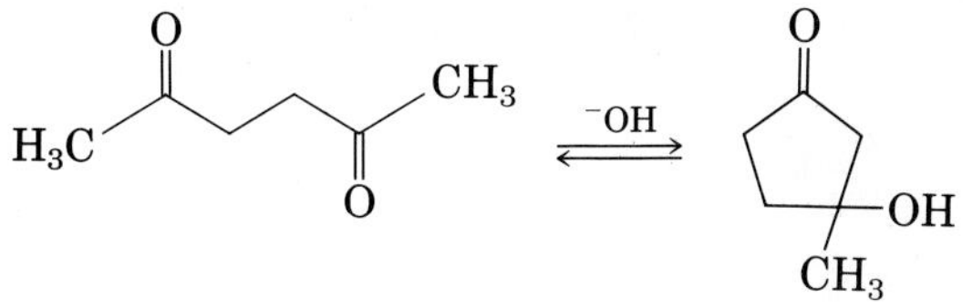

$$H_3C\text{–}CO\text{–}CH_2CH_2\text{–}CO\text{–}CH_3 \overset{^-OH}{\rightleftharpoons} \text{3-hydroxy-3-methylcyclopentanone}$$

21.9 When cyclohexanone is heated in the presence of 2-hydroxy-2-methylpropanenitrile (acetone cyanohydrin) and a catalytic amount of base, 1-hydroxycyclohexanecarbonitrile and acetone are produced. Suggest a mechanism to explain this reaction.

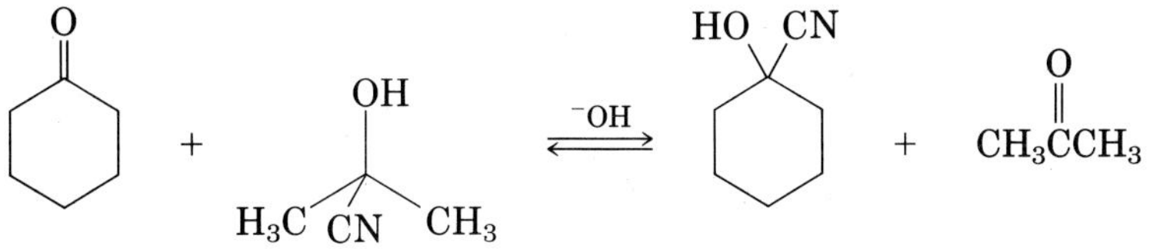

$$\text{cyclohexanone} + (H_3C)_2C(OH)CN \overset{^-OH}{\rightleftharpoons} \text{1-hydroxycyclohexanecarbonitrile (HO, CN)} + CH_3\overset{O}{\overset{\|}{C}}CH_3$$

21.10 Ketones react with dimethylsulfonium methylide to yield epoxides. Suggest a mechanism for this reaction.

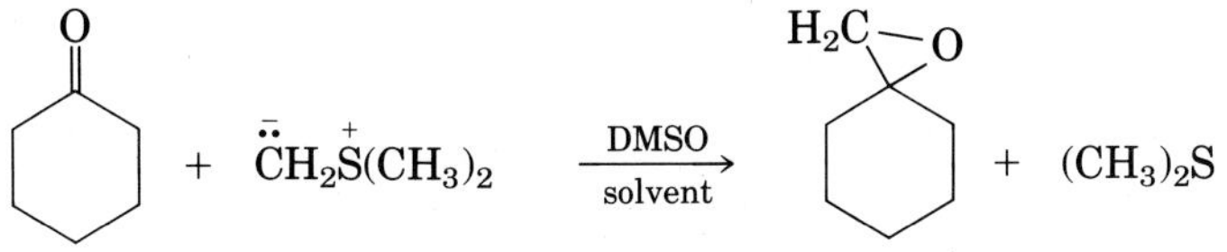

$$\text{cyclohexanone} + \ddot{C}H_2\overset{+}{S}(CH_3)_2 \xrightarrow[\text{solvent}]{\text{DMSO}} \text{spiro epoxide } (H_2C\text{—}O) + (CH_3)_2S$$

21.11 How can you account for the fact that 2,4-pentanedione is much more acidic than acetone, whereas 2,5-hexanedione has about the same pK_a as acetone? Which protons are most acidic in 2,4-pentanedione?

	pK_a
CH_3COCH_3, acetone	20
$CH_3COCH_2COCH_3$, 2,4-pentanedione	11
$CH_3COCH_2CH_2COCH_3$, 2,5-hexanedione	19

21.12 Most alcohols are inert to treatment with hydroxide ion, but β-hydroxy ketones dehydrate readily. How do you explain the difference in reactivity?

$$(CH_3)_2C(OH)CH_2CH_2CH_3 \xrightarrow[H_2O]{KOH} \text{No reaction}$$

$$(CH_3)_2\underset{\beta}{C}(OH)\underset{\alpha}{C}H_2\overset{O}{\overset{\|}{C}}CH_3 \xrightarrow[H_2O]{KOH} (CH_3)_2C{=}CHCOCH_3$$

CHAPTER 22 ALDEHYDES AND KETONES: NUCLEOPHILIC ADDITION REACTIONS

Aldehydes and ketones are among the most important of all compounds, both in nature and in the chemical industry. In nature, many substances required by living systems are aldehydes or ketones. In the chemical industry, simple aldehydes and ketones are synthesized in great quantity for use both as solvents and as starting materials for a host of other products. For example, approximately 6.5 billion lb/yr of formaldehyde, $H_2C{=}O$, are manufactured for use in building insulation materials and in the adhesive resins that bind particle board and plywood. Concern over the possible toxicity of such materials may sharply curtail their uses, however. Acetone, $(CH_3)_2C{=}O$, is also widely used in industry, and 2 billion lb/yr are prepared for use as a solvent.

Formaldehyde is synthesized industrially by catalytic dehydrogenation of methanol, and acetone was at one time prepared by dehydrogenation of 2-propanol. This method is of little use in the laboratory, however, because of the high temperatures required.

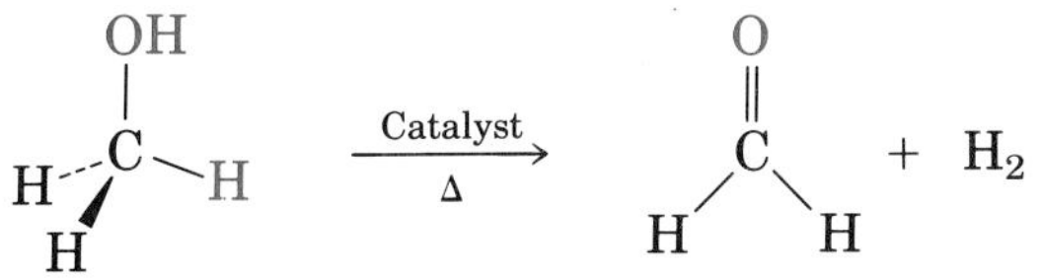

$$H_3C{-}OH \xrightarrow[\Delta]{\text{Catalyst}} H_2C{=}O + H_2$$

Methanol → Formaldehyde

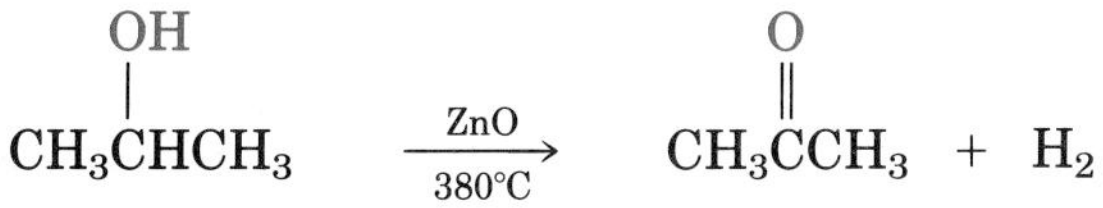

$$CH_3CH(OH)CH_3 \xrightarrow[380^\circ C]{ZnO} CH_3C(=O)CH_3 + H_2$$

2-Propanol → Acetone

22.1 Properties of aldehydes and ketones

We saw in the previous chapter that the carbonyl functional group is planar, and that the carbon–oxygen double bond is polar.

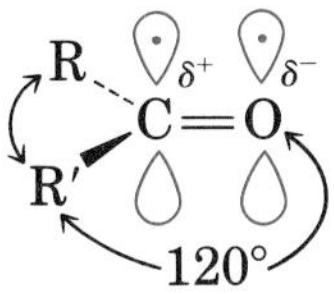

$$\overset{\delta+}{C}{=}\overset{\delta-}{O} \quad (120^\circ)$$

R = Alkyl, aryl, or alkenyl
R′ = Alkyl (ketone), or H (aldehyde)

One consequence of carbonyl bond polarity is that aldehydes and ketones are weakly associated and therefore have higher boiling points than do alkanes of similar molecular weight. Since they cannot form hydrogen bonds, however, their boiling points are lower than those of corresponding alcohols. Formaldehyde, the simplest aldehyde, is a gas at room temperature, but all other simple aldehydes and ketones are liquid (Table 22.1).

TABLE 22.1 **Physical properties of simple aldehydes and ketones**

Compound name	*Structure*	*Boiling point (°C)*	*Melting point (°C)*
Formaldehyde	HCHO	−21	−92
Acetaldehyde	CH_3CHO	21	−121
Propanal	CH_3CH_2CHO	49	−81
Butanal	$CH_3(CH_2)_2CHO$	76	−99
Pentanal	$CH_3(CH_2)_3CHO$	103	−92
Benzaldehyde	C_6H_5CHO	178	−26
Acetone	CH_3COCH_3	56	−95
2-Butanone	$CH_3CH_2COCH_3$	80	−86
2-Pentanone	$CH_3CH_2CH_2COCH_3$	102	−78
3-Pentanone	$CH_3CH_2COCH_2CH_3$	102	−40
Cyclohexanone	(cyclohexane ring)=O	156	−16

22.2 Nomenclature of aldehydes and ketones

Aldehydes may be named in several different ways. Systematic names for aldehydes are derived by replacing the terminal *-e* of the corresponding alkane name with *-al*. Note that the longest chain selected to be the base name must contain the CHO group, and that the CHO carbon is always numbered as carbon 1. For example,

$$CH_3\overset{\overset{\large O}{\|}}{C}{-}H \qquad CH_3CH_2\overset{\overset{\large O}{\|}}{C}{-}H$$

Ethanal (acetaldehyde) Propanal (propionaldehyde)

$$CH_3(CH_2)_4\overset{\overset{\large O}{\|}}{C}{-}H \qquad \underset{5}{CH_3}\underset{4}{\overset{\overset{\large CH_3}{|}}{C}H}\underset{3}{CH_2}\underset{2}{\overset{\overset{\overset{\large CH_3}{|}}{CH_2}}{\overset{|}{C}}H}\underset{1}{C}(=O)H$$

Hexanal 2-Ethyl-4-methylpentanal

Note that the longest chain in 2-ethyl-4-methylpentanal is a hexane, but that this chain does not include the CHO group.

For more complex aldehydes in which the CHO group is attached to a ring, the suffix *-carbaldehyde* is used:

Cyclohexanecarbaldehyde

2-Naphthalenecarbaldehyde

Certain simple and well-known aldehydes have trivial names that are recognized by IUPAC. These trivial names are given in Table 22.2.

TABLE 22.2 Trivial names of some simple aldehydes

Formula	*Trivial name*	*Systematic name*
HCHO	Formaldehyde	Methanal
CH_3CHO	Acetaldehyde	Ethanal
CH_3CH_2CHO	Propionaldehyde	Propanal
$CH_3CH_2CH_2CHO$	Butyraldehyde	Butanal
$CH_3CH_2CH_2CH_2CHO$	Valeraldehyde	Pentanal
$H_2C{=}CHCHO$	Acrolein	2-Propenal
C_6H_5CHO	Benzaldehyde	Benzenecarbaldehyde

Ketones are named by replacing the terminal *-e* of the corresponding alkane name with *-one*. The chain selected for the base name is the longest one that contains the ketone group, and the numbering begins at the end nearer the carbonyl carbon. For example,

$$CH_3{-}C({=}O){-}CH_3$$

Propanone (acetone)

$$\underset{1}{CH_3}\underset{2}{CH_2}\underset{3}{C}({=}O)\underset{4}{CH_2}\underset{5}{CH_2}\underset{6}{CH_3}$$

3-Hexanone

$$\underset{6}{CH_3}\underset{5}{CH}{=}\underset{4}{CH}\underset{3}{CH_2}\underset{2}{C}({=}O)\underset{1}{CH_3}$$

4-Penten-2-one

$$\underset{6}{CH_3}\underset{5}{CH_2}\underset{4}{C}({=}O)\underset{3}{CH_2}\underset{2}{C}({=}O)\underset{1}{CH_3}$$

2,4-Hexanedione

Certain ketones are allowed to retain their trivial names, though these are few:

$CH_3\overset{O}{\overset{\|}{C}}CH_3$

Acetone

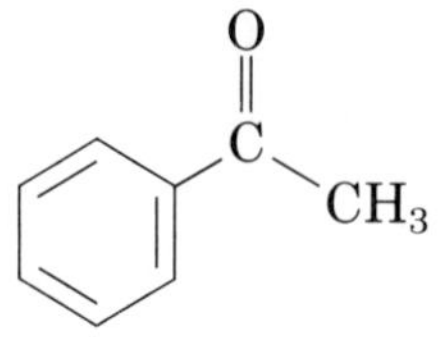

Acetophenone

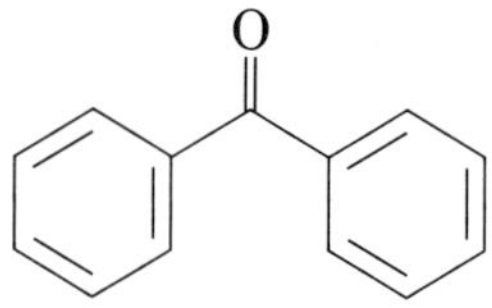

Benzophenone

When it becomes necessary to refer to the RCO— group as a substituent, the term *acyl* is used. Similarly, CHO— is called a *formyl* group, and ArCO— is referred to as an *aroyl* group.

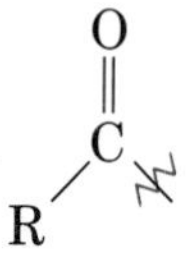

Acyl
(R = alkyl, alkenyl, or alkynyl)

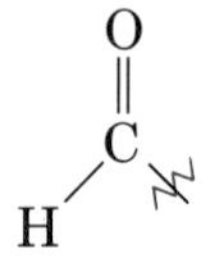

Formyl

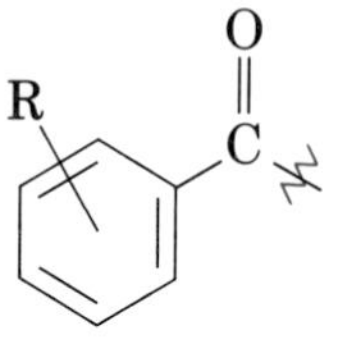

Aroyl

In some cases, the doubly bonded oxygen is considered a substituent, and the prefix *oxo-* is used. For example,

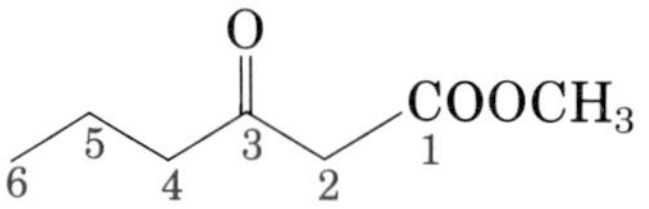

Methyl 3-oxohexanoate

PROBLEM

22.1 Name these aldehydes and ketones according to IUPAC rules:

(a) $CH_3CH_2\overset{O}{\overset{\|}{C}}CH(CH_3)_2$

(b) 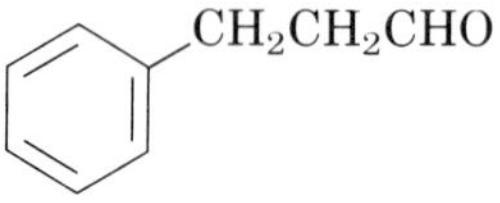 $C_6H_5CH_2CH_2CHO$

(c) $CH_3\overset{O}{\overset{\|}{C}}CH_2CH_2CH_2\overset{O}{\overset{\|}{C}}CH_2CH_3$

(d)

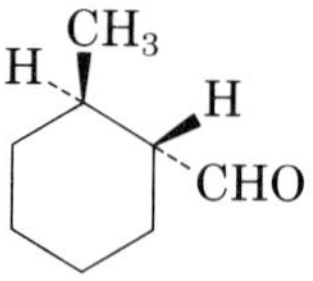

(e) $OHCCH_2CH_2CH_2CHO$

(f)

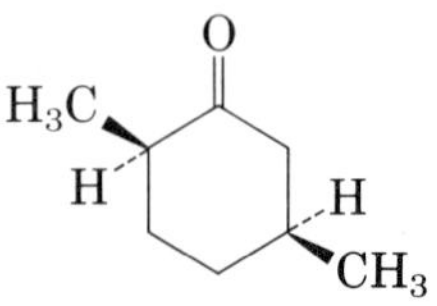

(g) $CH_3CH_2CH(CH_3)CH(CH_2CH_2CH_3)\overset{O}{\overset{\|}{C}}CH_3$

(h) $CH_3CH{=}CHCH_2CH_2CHO$

PROBLEM

22.2 Draw structures corresponding to these names:
(a) 3-Methyl-3-butenal
(b) 4-Chloro-2-pentanone
(c) Phenylacetaldehyde
(d) *cis*-3-*tert*-Butylcyclohexanecarbaldehyde
(e) 1-Hydroxy-2-methyl-3-hexanone
(f) 2-(1-Chloroethyl)-5-methylheptanal

22.3 Preparation of aldehydes

We have already discussed two of the best methods of aldehyde synthesis—oxidation of primary alcohols, and oxidative cleavage of alkenes. Let's review these methods briefly.

1. Primary alcohols can be oxidized to give aldehydes (Section 20.10). Many reagents accomplish the desired transformation, but pyridinium chlorochromate (PCC) is the most common choice. The reaction is carried out at room temperature in dichloromethane solution and is usually complete within a few minutes.

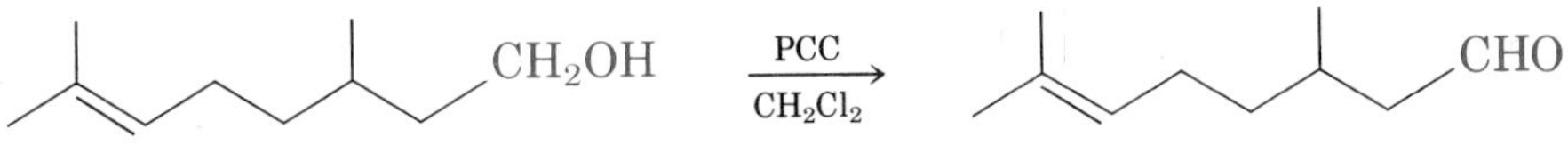

CH_2OH $\xrightarrow[CH_2Cl_2]{PCC}$ CHO

Citronellol   Citronellal (82%)

2. Alkenes that have at least one vinylic proton undergo oxidative cleavage when treated with ozone to yield aldehydes (Section 6.8). If the ozonolysis reaction is carried out on cyclic alkenes, dicarbonyl compounds result.

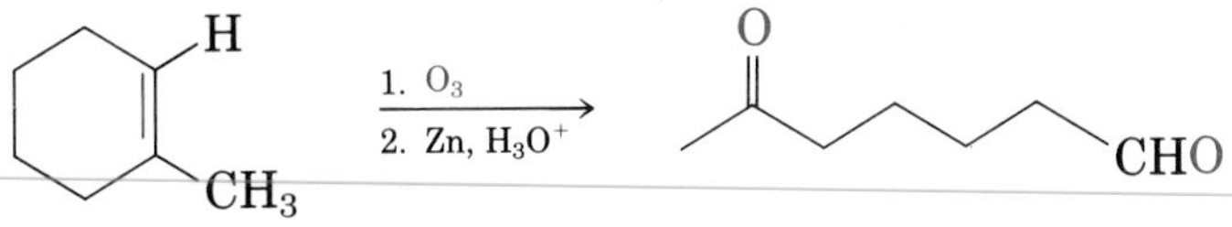

H, CH_3 $\xrightarrow[2.\ Zn,\ H_3O^+]{1.\ O_3}$ O, CHO

1-Methylcyclohexene   6-Oxoheptanal (86%)

Yet a third method of aldehyde synthesis is one that we will mention here just briefly and then return to for a more detailed explanation in Section 24.5. Certain carboxylic acid derivatives can be partially reduced to yield aldehydes.

$$R-\overset{\overset{O}{\|}}{C}-Y \xrightarrow{:H^-} R-\overset{\overset{O}{\|}}{C}-H + :Y^-$$

For example, acid chlorides can be catalytically hydrogenated over a catalyst of palladium on barium sulfate to yield aldehydes (**the Rosenmund reduction**):

$$\text{Cyclohexyl-}CH_2\overset{O}{\overset{\|}{C}}-Cl \xrightarrow[\text{Ethyl acetate}]{H_2;\ Pd/BaSO_4} \text{Cyclohexyl-}CH_2\overset{O}{\overset{\|}{C}}-H$$

Cyclohexylacetyl chloride — Cyclohexylacetaldehyde (70%)

The Rosenmund reduction is particularly suited for large-scale work because it is simple and relatively inexpensive to carry out. The main drawback, however, is that the acid chloride must first be prepared from the carboxylic acid in a separate step.

Alternatively, esters are usually readily available, and their partial reduction by diisobutylaluminum hydride (DIBAH) is an extremely important laboratory-scale method of aldehyde synthesis. The reaction is normally carried out at −78°C (dry ice temperature) in toluene solution, and yields are often excellent. The major disadvantages of the method, however, are that DIBAH is expensive and that considerable experimental skill must be exercised by the chemist to prevent overreduction to the primary alcohol.

$$CH_3(CH_2)_{10}\overset{O}{\overset{\|}{C}}-OCH_3 \xrightarrow[2.\ H_3O^+]{1.\ \text{DIBAH, toluene, } -78^\circ C} CH_3(CH_2)_{10}\overset{O}{\overset{\|}{C}}-H$$

Methyl dodecanoate — Dodecanal (88%)

where DIBAH = $[(CH_3)_2CHCH_2]_2Al-H$

22.4 Preparation of ketones

For the most part, methods of ketone synthesis are analogous to those for aldehydes:

1. We have already seen that secondary alcohols are oxidized by a variety of reagents to give ketones (Section 20.10). The Jones reagent (CrO_3 in aqueous sulfuric acid), pyridinium chlorochromate, and sodium dichromate in aqueous acetic acid are all effective, and the specific choice of reagent depends on factors such as reaction scale, cost, and acid or base sensitivity of the alcohol.

$$\text{4-}C(CH_3)_3\text{-cyclohexanol} \xrightarrow[CH_2Cl_2]{PCC} \text{4-}C(CH_3)_3\text{-cyclohexanone}$$

4-*tert*-Butylcyclohexanol — 4-*tert*-Butylcyclohexanone (90%)

2. We have also seen that alkene ozonolysis yields ketones if one of the unsaturated carbon atoms is disubstituted (Section 6.8):

$$\xrightarrow[\text{2. Zn/H}_3\text{O}^+]{\text{1. O}_3} \quad + \ CH_2O$$

Starting alkene: O, CH_2, CH_3; product: O, O, CH_3

70%

3. Aryl ketones can be prepared by Friedel–Crafts reaction between an acyl chloride and an aromatic ring (Section 16.6):

$$\text{Benzene} + CH_3\overset{\overset{\large O}{\|}}{C}Cl \xrightarrow[\Delta]{AlCl_3} C_6H_5\overset{\overset{\large O}{\|}}{C}CH_3$$

Benzene, Acetyl chloride, Acetophenone (95%)

This aromatic acylation reaction is subject to all of the directing and reactivity effects previously seen for Friedel–Crafts reactions but can nevertheless be quite useful.

4. Methyl ketones can be prepared by mercuric ion catalyzed hydration of terminal alkynes (Section 7.4). Internal alkynes can also be hydrated, but a mixture of ketone products usually results.

$$CH_3(CH_2)_3C\equiv CH \xrightarrow[Hg(OAc)_2]{H_3O^+} CH_3(CH_2)_3\overset{\overset{\large O}{\|}}{C}-CH_3$$

1-Hexyne, 2-Hexanone (78%)

5. Ketones can also be prepared from certain carboxylic acid derivatives, just as aldehydes can be prepared; we will discuss this subject in more detail in Section 24.5.

$$R-\overset{\overset{\large O}{\|}}{C}-Y \xrightarrow{:R^-} R-\overset{\overset{\large O}{\|}}{C}-R' + :Y^-$$

Among the most useful syntheses of this type is the reaction between an acid chloride and a diorganocopper reagent. This process is reminiscent of the coupling reaction between alkyl halides and diorganocoppers that we saw earlier (Section 9.9):

$$CH_3(CH_2)_4\overset{\overset{\large O}{\|}}{C}-Cl + (CH_3)_2\overset{-}{Cu}Li^+ \longrightarrow CH_3(CH_2)_4\overset{\overset{\large O}{\|}}{C}-CH_3$$

Hexanoyl chloride, Dimethylcopper lithium, 2-Heptanone (81%)

Recall the following reaction:

$$R-X + (R')_2CuLi \longrightarrow R-R'$$

PROBLEM

22.3 How would you carry out the following reactions? More than one step may be required.

(a) 1-Hexene ⟶ Hexanal
(b) 1-Hexene ⟶ 2-Hexanone
(c) 1-Hexene ⟶ Pentanal
(d) Benzene ⟶ Acetophenone
(e) Benzene ⟶ Benzaldehyde
(f) Bromobenzene ⟶ Benzaldehyde

22.5 Reactions of aldehydes: oxidation

Aldehydes are readily oxidized to yield carboxylic acids, but ketones are unreactive toward oxidation except under the most vigorous conditions. This difference in behavior is a consequence of the structural difference between the two functional groups; aldehydes have a CHO proton that can be readily abstracted during oxidation; ketones do not.

Many oxidizing agents such as hot nitric acid and potassium permanganate convert aldehydes into carboxylic acids, but the Jones reagent, CrO_3 in aqueous sulfuric acid, is a more common choice on a small laboratory scale. Jones oxidations occur rapidly at room temperature and give good yields of product.

$$CH_3(CH_2)_4\overset{\overset{\displaystyle O}{\|}}{C}—H \xrightarrow[\text{Acetone, 0°C}]{\text{Jones' reagent}} CH_3(CH_2)_4\overset{\overset{\displaystyle O}{\|}}{C}—OH$$

Hexanal — Hexanoic acid (85%)

One drawback to the Jones oxidation is that the conditions under which the reaction occurs are acidic, and sensitive molecules sometimes undergo acid-catalyzed decomposition. Where such decomposition is a possibility, aldehyde oxidations are often carried out using a very mild reagent, silver ion. A dilute ammonia solution of silver oxide, Ag_2O (the **Tollens[1] reagent**) oxidizes aldehydes in high yield without harming carbon–carbon double bonds or other functional groups.

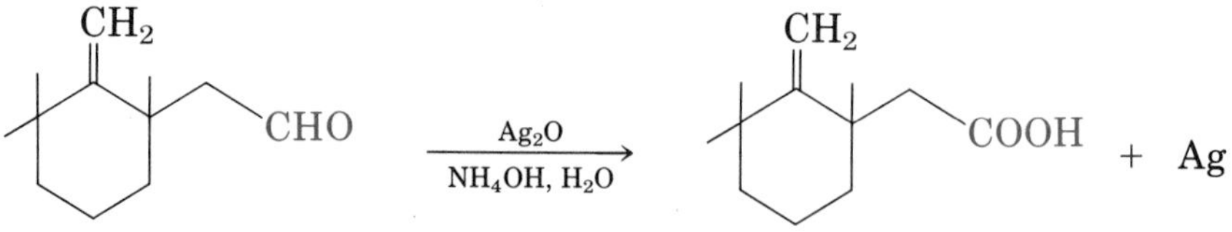

[Structure: CH_2= substituted cyclohexane bearing CH_2CHO] $\xrightarrow[\text{NH}_4\text{OH, H}_2\text{O}]{\text{Ag}_2\text{O}}$ [Structure: CH_2= substituted cyclohexane bearing CH_2COOH] + Ag

72%

[1]Bernhard Tollens (1841–1918); b. Hamburg, Germany; Ph.D. Göttingen; professor, University of Göttingen.

A shiny mirror of metallic silver is deposited on the walls of the flask during a Tollens oxidation; observation of such a mirror forms the basis of an old qualitative test for the presence of an aldehyde functional group in a molecule of unknown structure. A small sample of the unknown is dissolved in ethanol in a test tube, and a few drops of the Tollens reagent are added. If the test tube becomes silvery, the unknown is presumed to be an aldehyde.

Aldehyde oxidations are thought to occur through intermediate 1,1-diols or hydrates, which are formed by nucleophilic addition of water to the carbonyl group. Hydrate formation is reversible—the hydrate can eliminate water to re-form aldehyde—and the equilibrium position of the reaction usually favors the aldehyde:

$$\underset{\text{Aldehyde}}{R-\overset{O}{\overset{\|}{C}}-H} \xrightleftharpoons[]{:OH_2,\ H^+} \left[R-\overset{OH}{\overset{|}{C}}(H)-\overset{+}{O}H_2 \right] \xrightleftharpoons[]{-H^+} \underset{\text{Hydrate}}{R-\overset{OH}{\overset{|}{C}}(H)-OH}$$

Once formed, and even though present in small amounts, the hydrate reacts like any normal primary or secondary alcohol and is oxidized to a carbonyl compound (Section 20.10).

$$\underset{\text{Aldehyde}}{R-\overset{O}{\overset{\|}{C}}-H} \xrightleftharpoons[]{H_2O} \underset{\text{Hydrate}}{R-\overset{OH}{\overset{|}{C}}(HO)-H} \xrightarrow[H^+,\ H_2O]{CrO_3} \left[R-\overset{O-CrO_3H}{\overset{|}{C}}(HO)-H \right] \longrightarrow \underset{\text{Carboxylic acid}}{R-\overset{O}{\overset{\|}{C}}-OH}$$

Ketones are inert to most common oxidizing agents but undergo a slow cleavage reaction when treated with hot alkaline $KMnO_4$. The carbon–carbon bond next to the carbonyl group is broken and carboxylic acid fragments are produced. The reaction is only useful for symmetrical ketones such as cyclohexanone, however, since product mixtures are formed from unsymmetrical ketones.

$$\text{Cyclohexanone} \xrightarrow[^{-}OH,\ \Delta]{KMnO_4,\ H_2O} \text{Hexanedioic acid (79\%)}\ (HOOC-(CH_2)_4-COOH)$$

22.6 Nucleophilic addition reactions of aldehydes and ketones

We saw, in Section 20.6 and in the reaction overview of Chapter 21, that carbonyl compounds undergo nucleophilic addition reactions. A nucleophile attacks the electrophilic carbon atom of the polar carbonyl group from a direction approximately perpendicular to the plane of the carbonyl sp^2 orbitals. Rehybridization of the carbonyl carbon from sp^2 to sp^3 then occurs, and a tetrahedral alkoxide intermediate is produced (Figure 22.1).

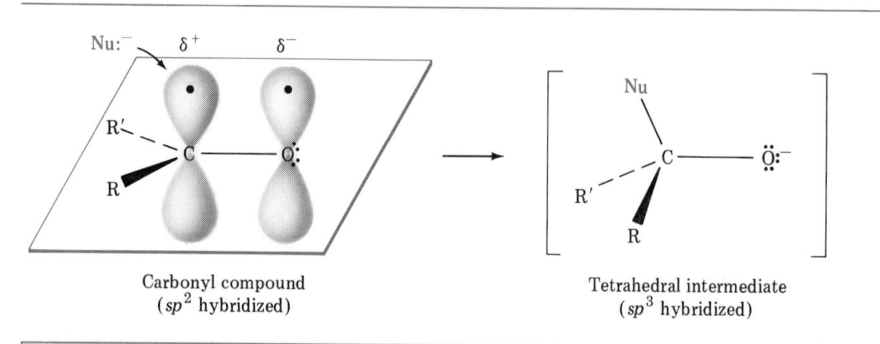

Figure 22.1. A nucleophilic addition reaction.

The attacking nucleophile can be either negatively charged ($^{-}$:Nu) or neutral (:Nu—H). If it is neutral, however, the nucleophile must be attached to a hydrogen atom that can subsequently be eliminated. Hydroxide ion (HO:$^{-}$), hydride ion (H:$^{-}$), and carbanions (R_3C:$^{-}$) are examples of negatively charged nucleophiles, whereas water (H—$\ddot{O}$—H), alcohols (R—$\ddot{O}$—H), ammonia (:NH_3), and primary amines (R—$\ddot{N}H_2$) are examples of neutral nucleophiles.

Nucleophilic addition to ketones and aldehydes usually has two variations: (1) The tetrahedral intermediate can be protonated to give a stable alcohol, or (2) the carbonyl oxygen atom can be expelled (as HO^- or as H_2O) to give a new carbon–nucleophile double bond. These two possibilities are shown in Figure 22.2.

In the remainder of this chapter, we will examine some specific examples of nucleophilic addition reactions. In looking at the details of each reaction, we will be concerned with two key points—the *reversibility* of a given reaction, and the acid or base *catalysis* of that reaction. Some nucleophilic addition reactions take place without catalysis, but many others require acid or base to proceed.

22.7 Relative reactivity of aldehydes and ketones

Aldehydes are generally more reactive than ketones in nucleophilic addition reactions, for both steric and electronic reasons. Sterically, the presence of two relatively large substituents in ketones versus only one

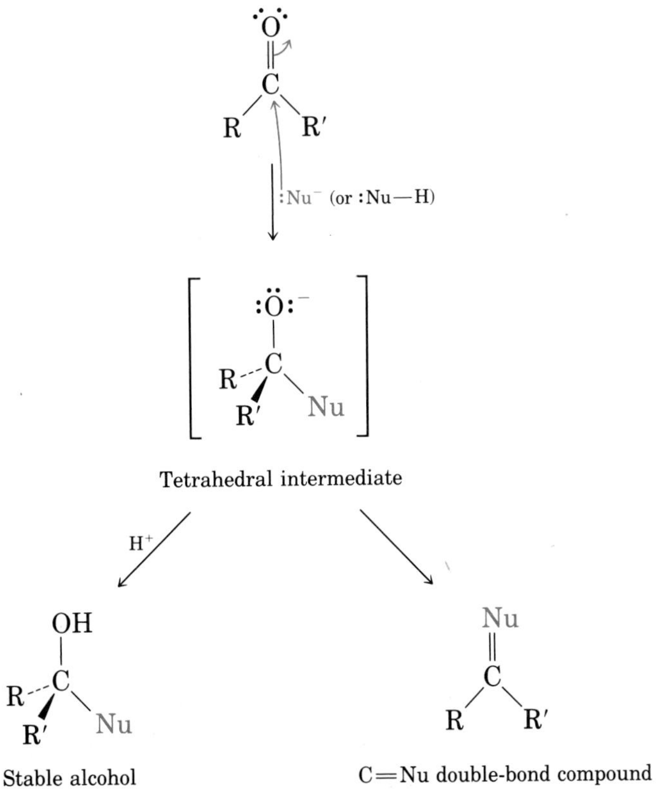

$$R(C{=}O)R' \xrightarrow{:Nu^- \text{ (or } :Nu{-}H)} \left[R(R')C(Nu){-}O:^- \right]$$

Tetrahedral intermediate

$\xrightarrow{H^+}$ R(R')C(OH)Nu — Stable alcohol

$\longrightarrow$ R(R')C=Nu — C=Nu double-bond compound

Figure 22.2. Possible reactions following nucleophilic carbonyl addition to ketones and aldehydes.

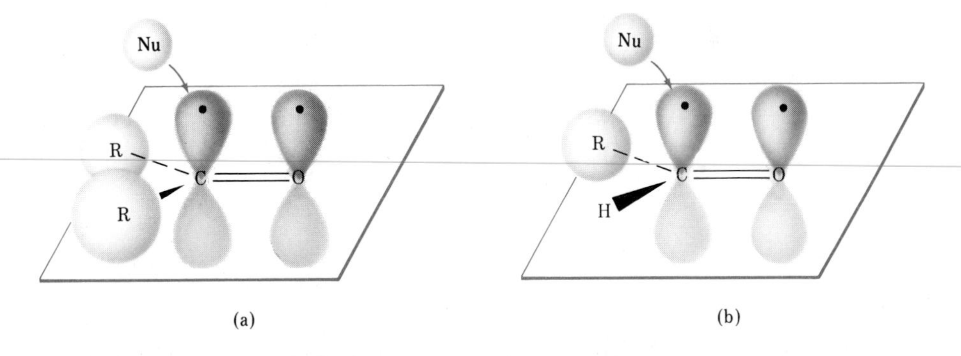

Figure 22.3. Nucleophilic attack on a ketone (a) is sterically hindered because of the two relatively large substituents; an aldehyde (b) has one substituent and is less hindered.

substituent in aldehydes means that attacking nucleophiles are able to approach aldehydes more readily (Figure 22.3), and that the transition state leading to the tetrahedral intermediate is less crowded for aldehydes than for ketones.

Electronically, aldehydes are more reactive than ketones because of the somewhat greater degree of polarity of aldehyde carbonyl groups. The easiest way to see this polarity difference is to recall the stability order of carbocations. Primary carbocations are less stable than secondary carbocations because there is only one alkyl group inductively stabilizing the positive charge, rather than two. For similar reasons, aldehydes are less stable than ketones because there is only one alkyl group inductively stabilizing the partial positive charge on the carbonyl carbon, rather than two.

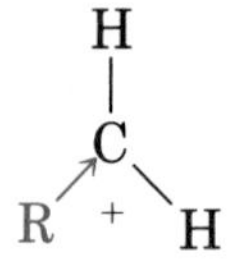

H
C
R + H

1° carbocation
(less stable, more reactive)

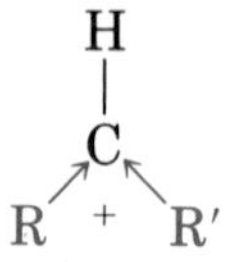

H
C
R + R′

2° carbocation
(more stable, less reactive)

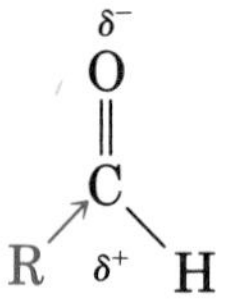

δ^-
O
C
R δ^+ H

Aldehyde
(less stabilization of δ^+, more reactive)

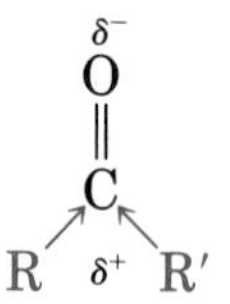

δ^-
O
C
R δ^+ R′

Ketone
(more stabilization of δ^+, less reactive)

PROBLEM

22.4 Would you expect an aromatic aldehyde such as benzaldehyde to be more reactive or less reactive in nucleophilic addition reactions than an aliphatic aldehyde? Explain your answer.

22.8 Addition of HCN: cyanohydrins

Aldehydes and unhindered ketones react with HCN to yield **cyanohydrins, R—CH(OH)C≡N.** For example, benzaldehyde gives the cyanohydrin mandelonitrile in 88% yield on treatment with HCN:

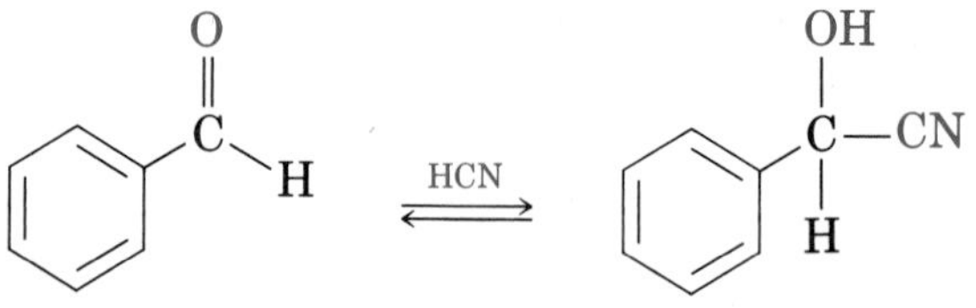

$$C_6H_5CHO \xrightleftharpoons{HCN} C_6H_5CH(OH)CN$$

Benzaldehyde

Mandelonitrile (88%)
(a cyanohydrin)

Detailed studies carried out by Arthur Lapworth[2] in the early 1900s showed that cyanohydrin formation is reversible and is base catalyzed.

[2]Arthur Lapworth (1872–1941); b. Galashiels, Scotland; Ph.D., Birmingham; professor, University of Manchester (1913–1941).

Reaction occurs very slowly when pure HCN is used, but rapidly when a trace amount of base or cyanide ion is added. We can understand this result by recalling that HCN is a weak acid ($pK_a = 9.1$) and therefore is neither dissociated nor nucleophilic. Cyanide ion, however, is strongly nucleophilic, and addition to benzaldehyde or other carbonyl compounds occurs by a typical nucleophilic addition pathway. Protonation of the anionic tetrahedral intermediate yields the stable tetrahedral cyanohydrin product plus regenerated cyanide ion:

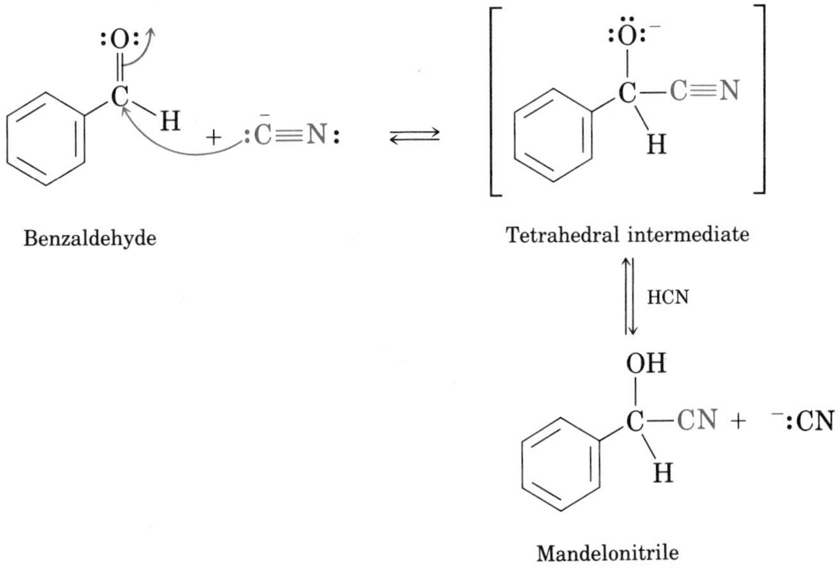

:O:
C
H
+ :C≡N:
:Ö:⁻
C—C≡N
H

Benzaldehyde

Tetrahedral intermediate

HCN

OH
C—CN + ⁻:CN
H

Mandelonitrile

In order to avoid the dangers inherent in handling pure toxic gases such as hydrogen cyanide, HCN is usually generated during the reaction by adding 1 equiv of mineral acid to a mixture of carbonyl compound and excess sodium cyanide.

Cyanohydrin formation is particularly interesting because it is one of the few examples of the addition of an acid to a carbonyl group. Acids such as HBr, HCl, H_2SO_4, and CH_3COOH do not form stable carbonyl adducts because the equilibrium constant for reaction is unfavorable. With HCN, however, the equilibrium lies in favor of the adduct.

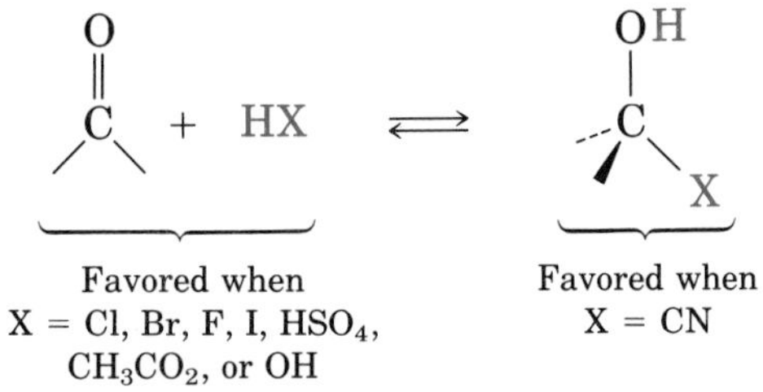

O
C + HX ⇌
OH
C
X

Favored when
X = Cl, Br, F, I, HSO_4,
CH_3CO_2, or OH

Favored when
X = CN

Cyanohydrin formation is useful because of the further chemistry that can be carried out. For example, nitriles (R—CN) can be reduced with

$LiAlH_4$ to yield primary amines, and they can be hydrolyzed to yield carboxylic acids. Thus, cyanohydrin formation provides a method for transforming a ketone or aldehyde into a different functional group while lengthening the carbon chain by one unit.

Benzaldehyde $\xrightarrow{HCN}$ Mandelonitrile $\xrightarrow[2.\ H_2O]{1.\ LiAlH_4,\ THF}$ 2-Amino-1-phenylethanol

Mandelonitrile $\xrightarrow{H_3O^+,\ \Delta}$ Mandelic acid (90%)

Cyanohydrins also play an interesting role in the chemical defense mechanisms used by certain insects against predators. It has been shown that when the millipede *Apheloria corrugata* is stimulated, it secretes a sticky substance that is strongly repellent to ants. Chemical analysis indicates that this defensive secretion contains mandelonitrile and an enzyme that catalyzes the decomposition of mandelonitrile into benzaldehyde and HCN. The millipede actually protects itself by discharging hydrogen cyanide at would-be attackers!

Mandelonitrile (from *Apheloria corrugata*) $\xrightarrow[H_2O]{Enzyme}$ C_6H_5CHO + HCN

PROBLEM

22.5 How can you account for the observation that cyclohexanone forms a cyanohydrin in good yield but that 2,2,6-trimethylcyclohexanone is unreactive to HCN/KCN?

22.9 Addition of amines: imine and enamine formation

Primary amines, $R{-}\ddot{N}H_2$, add to aldehydes and ketones to yield **imines**, $\mathbf{R{-}\ddot{N}{=}C\langle}$, as we saw in Chapter 21. Secondary amines add similarly to yield **enamines** (*-ene* + *amine*; unsaturated amine):

$$H_2O + \underset{\text{An imine}}{\overset{\ddot{N}-R}{\overset{\|}{\text{H}-\text{C}-\text{C}}}} \underset{}{\overset{R-\ddot{N}H_2}{\rightleftharpoons}} \underset{\text{Ketone/aldehyde}}{\text{H}-\text{C}-\text{C}=\text{O}} \overset{R_2\ddot{N}H}{\rightleftharpoons} \underset{\text{An enamine}}{\text{C}=\text{C}-\ddot{N}R_2} + H_2O$$

These two reactions may appear different, since one leads to a carbon–nitrogen double-bond product and the other leads to a carbon–carbon double-bond product. In fact, however, they are quite similar. Both are typical examples of nucleophilic addition reactions in which the initially formed tetrahedral intermediate is not stable. Instead, the carbonyl oxygen is eliminated, and a new carbon–nucleophile double bond is formed.

Imines are formed by a reversible, acid-catalyzed process involving nucleophilic attack on the carbonyl group by the primary amine, followed by transfer of a proton from nitrogen to oxygen to yield a neutral carbinolamine. Protonation of the carbinolamine oxygen by the acid catalyst present converts the hydroxyl into a better leaving group, and loss of water produces an iminium ion. Loss of a proton then gives the final product (Figure 22.4, p. 702) and regenerates acid catalyst.

Enamines are formed similarly from secondary amines and ketones or aldehydes. The process is identical up to the iminium ion stage, but at this point, there is no proton on nitrogen that can be lost to yield neutral product. Instead, a proton is lost from the alpha-carbon atom, yielding an enamine (Figure 22.5, p. 703).

As indicated in Figures 22.4 and 22.5, imine and enamine formation is normally an acid-catalyzed process. Studies of this reaction have revealed a pH versus reaction rate profile indicating that reaction is very slow at both high and low pH, but reaches a maximum rate at weakly acidic pH. Figure 22.6 (p. 703) shows the profile obtained for reaction between acetone and hydroxylamine, $H_2N—OH$, and indicates that maximum reaction rate is obtained at pH 4.5.

We can explain the observed maximum at pH 4.5 by looking at each of the individual steps in the overall mechanism. Acid is required to protonate the intermediate carbinolamine and thereby to convert the hydroxyl into a better leaving group. Thus, reaction cannot occur at low acid concentration (high pH). On the other hand, if too much acid is present (low pH), the attacking amine nucleophile is completely protonated and the initial nucleophilic addition step cannot occur.

$$\underset{\text{Base}}{H_2\ddot{N}—OH} + \underset{\text{Acid}}{H^+} \rightleftharpoons \underset{\text{Nonnucleophilic}}{H_3\overset{+}{N}—OH}$$

Evidently, pH 4.5 represents a compromise between the need for some acid to catalyze the rate-limiting dehydration step, and the need for not too much acid to avoid complete protonation of the amine. Each individual nucleophilic addition reaction has its own specific requirements, and reaction conditions must often be carefully controlled if maximum yields are to be obtained.

Ketone/aldehyde

Nucleophilic attack on the ketone or aldehyde by the lone-pair electrons of an amine leads to a dipolar tetrahedral intermediate.

$:NH_2R$

An H^+ ion is then transferred from nitrogen to oxygen, yielding a neutral carbinolamine.

Proton transfer

Carbinolamine

Acid catalyst protonates the hydroxyl oxygen.

H^+

The nitrogen lone-pair electrons expel water, giving an iminium ion.

$-H_2O$

Iminium ion

Loss of H^+ from nitrogen then gives the neutral imine product.

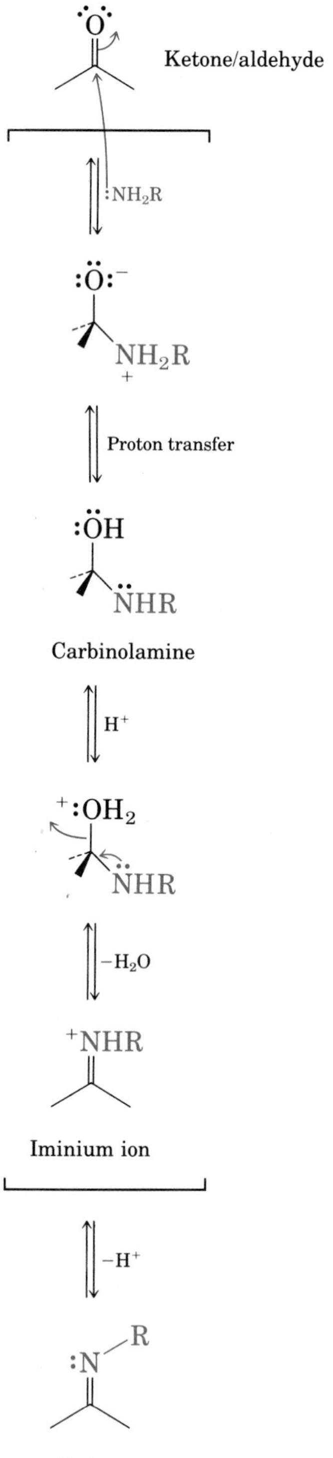

$-H^+$

Imine

Figure 22.4. Mechanism of imine formation from ketones and aldehydes.

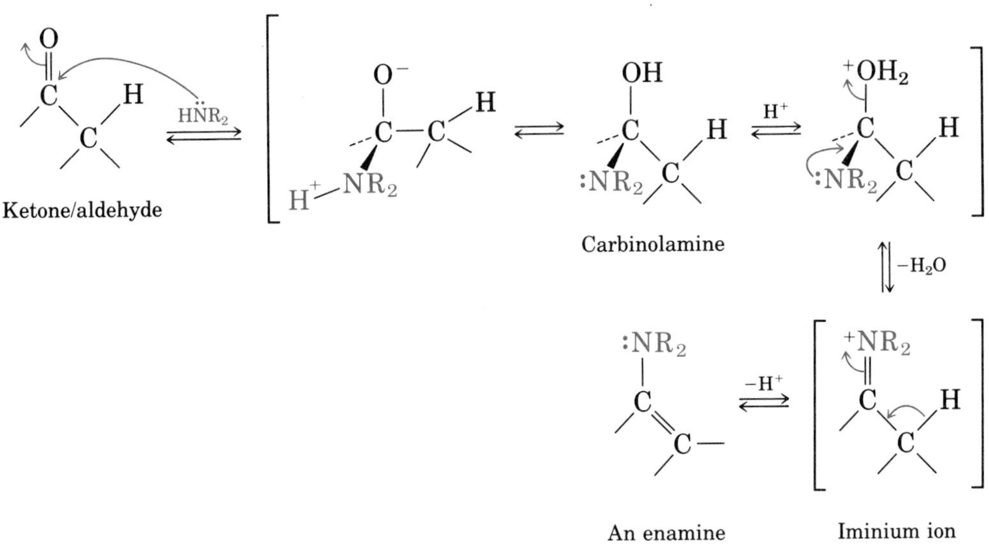

Ketone/aldehyde

HNR_2

Carbinolamine

$-H_2O$

An enamine

Iminium ion

Figure 22.5. Mechanism of enamine formation from ketones and aldehydes.

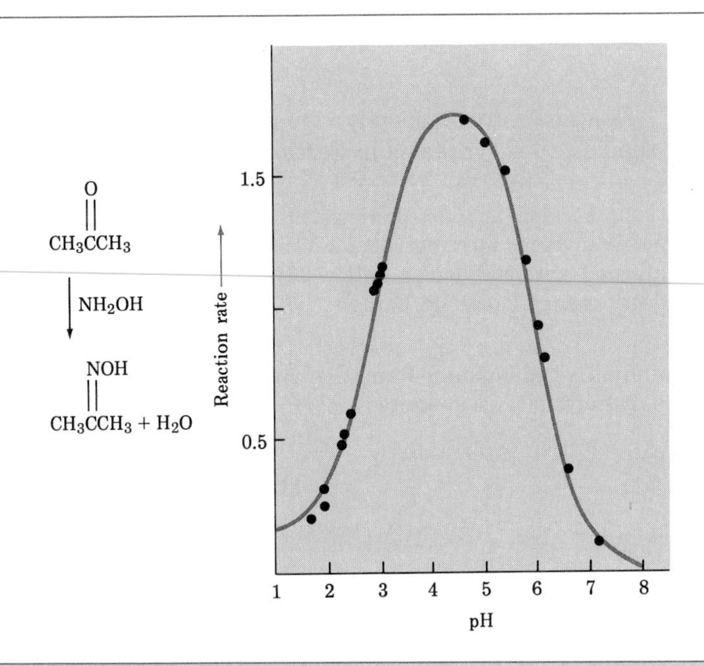

Figure 22.6. Dependence on pH of the rate of reaction between acetone and hydroxylamine.

Imine derivatives, such as oximes, semicarbazones, and 2,4-dinitrophenylhydrazones (2,4-DNPs), are usually crystalline and easy to handle, and are sometimes prepared as a means of identifying liquid ketones or aldehydes.

$$\text{Cyclohexanone} + \underset{\text{Hydroxylamine}}{H_2\ddot{N}{-}OH} \longrightarrow \underset{\text{(mp 90°C)}}{\text{Cyclohexanone oxime}}$$

$$C_6H_5CHO + \underset{\text{Semicarbazide}}{H_2\ddot{N}{-}NHCONH_2} \longrightarrow \underset{\text{(mp 222°C)}}{\underset{\text{Benzaldehyde semicarbazone}}{C_6H_5CH{=}N{-}NHCONH_2}}$$

$$CH_3COCH_3 + \underset{\text{2,4-Dinitrophenylhydrazine}}{H_2\ddot{N}{-}NH{-}C_6H_3(NO_2)_2} \longrightarrow \underset{\text{(mp 126°C)}}{\underset{\text{Acetone 2,4-dinitrophenylhydrazone}}{(CH_3)_2C{=}N{-}NH{-}C_6H_3(NO_2)_2}}$$

Enamines are extremely valuable compounds, and we will see examples of their use for synthesis in Section 26.13.

PROBLEM

22.6 A simple color test for distinguishing between saturated and α,β-unsaturated ketones involves forming the 2,4-dinitrophenylhydrazone of the unknown sample. Saturated ketones yield a yellow 2,4-DNP, but α,β-unsaturated ketones yield a reddish orange 2,4-DNP. Explain. [*Hint:* See Section 13.13.]

PROBLEM

22.7 Both imine and enamine formation are reversible. Show all of the steps involved in the hydrolysis of an enamine.

$$\text{(}{:}NR_2{-}C{=}C\text{)} \xrightarrow{H_3O^+} \text{(}O{=}C{-}C{-}H\text{)} + R_2\ddot{N}H$$

22.10 Addition of hydrazine: the Wolff–Kishner reaction

Aldehydes and ketones can be reduced to the corresponding alkanes on reaction with hydrazine, $H_2N{-}NH_2$, followed by treatment with strong base at high temperature. This reaction, discovered independently in 1911 by Ludwig Wolff[3] in Germany and N. M. Kishner[4] in Russia, is an extremely valuable synthetic method for converting ketones or aldehydes into alkanes.

$$\xrightarrow[\text{KOH, } H_2O,\ (HOCH_2CH_2)_2O,\ 240°C]{H_2N{-}NH_2}$$

90%

Cyclopropanecarbaldehyde —CHO $\xrightarrow[\text{reaction}]{\text{Wolff–Kishner}}$ —CH_3 Methylcyclopropane (72%)

The Wolff–Kishner reaction is often carried out at 240°C in refluxing diethylene glycol solvent, but a modification in which dimethyl sulfoxide is used as solvent allows the process to take place near room temperature.

The Wolff–Kishner reaction involves formation of a hydrazone intermediate, followed by base-catalyzed double-bond migration, loss of N_2 gas, and formation of alkane product.

Propiophenone ($C_6H_5COCH_2CH_3$) $\xrightarrow[\text{KOH, DMSO}]{H_2N{-}NH_2}$ [A hydrazone ($C_6H_5C(=N{-}NH_2)CH_2CH_3$) $\xrightleftharpoons{\text{KOH}}$ $C_6H_5CH(CH_2CH_3){-}N{=}N{-}H$] $\xrightarrow{-N{\equiv}N}$ Propylbenzene ($C_6H_5CH_2CH_2CH_3$) (82%)

Note that the Wolff–Kishner reduction accomplishes the same overall transformation that we saw earlier during catalytic hydrogenation of acyl

[3]Ludwig Wolff (1857–1919); b. Neustadt/Hardt; Ph.D. Strasbourg (Fittig); professor, University of Jena.
[4]N. M. Kishner (1867–1935); b. Moscow; Ph.D. Moscow (Markovnikov); professor, universities of Tomsk and Moscow.

benzenes to arenes (Section 16.7). The Wolff–Kishner reduction is more general and more useful than catalytic hydrogenation, however, since it works well with both alkyl and aryl ketones.

22.11 Addition of alcohols: acetal formation

Ketones and aldehydes react reversibly with alcohols in the presence of an acid catalyst to yield **acetals, $R_2C(OR')_2$,** also called *ketals* in older literature.

$$\text{R}_2\text{C{=}O} + 2\ \text{R'OH} \underset{}{\overset{H^+}{\rightleftharpoons}} \text{R}_2\text{C(OR')}_2 + H_2O$$

Ketone/aldehyde — An acetal

Acetal formation is fundamentally similar to the other nucleophilic addition reactions we have studied, but is distinguished by the fact that the *addition* step is acid catalyzed. Alcohols are relatively weak nucleophiles that add to ketones and aldehydes only slowly under neutral conditions. Under acidic conditions, however, the nucleophilic carbonyl oxygen is protonated, and the resultant protonated carbonyl compound is far more reactive than its neutral carbonyl precursor. Addition of alcohol therefore occurs rapidly.

$$\overset{\delta+}{C}{=}\overset{\delta-}{\ddot{O}}: \ \overset{H^+}{\rightleftharpoons}\ C{=}\overset{+}{\ddot{O}}{-}H \longleftrightarrow \overset{+}{C}{-}\ddot{O}{-}H$$

Neutral carbonyl group (moderately nucleophilic)

Protonated carbonyl group (strongly electrophilic and highly reactive toward nucleophiles)

As indicated in Figure 22.7, the initial nucleophilic addition of alcohol to the carbonyl group yields a hydroxy ether called a **hemiacetal.** Hemiacetals are formed reversibly, and equilibrium normally favors the carbonyl compound. In the presence of acid, however, a further reaction can occur. Protonation of the hydroxyl group, followed by loss of water, leads to a new carbon–oxygen double-bond compound. This process is analogous to imine formation (Section 22.9). Addition of a second equivalent of alcohol then yields the acetal.

All the steps of acetal formation are reversible: the reaction can be driven either forward (from carbonyl compound to acetal) or backward (from acetal to carbonyl compound), depending on the conditions chosen. The forward reaction is favored by choosing conditions that remove water from the medium. In practice, this is often done by distilling off water as it forms in the reaction. Conversely, acetals can be hydrolyzed back into carbonyl compound plus alcohol upon treatment with aqueous mineral acid.

Ketone/aldehyde

Protonation of the carbonyl oxygen strongly polarizes the carbonyl group and . . .

H^+

activates the carbonyl group for nucleophilic attack by oxygen lone-pair electrons from alcohol.

ROH

Loss of a proton yields a neutral hemiacetal tetrahedral intermediate.

$-H^+$

Hemiacetal

Protonation of the hemiacetal hydroxyl converts it into a good leaving group.

H^+

Dehydration yields an intermediate oxonium ion.

$-H_2O$

Addition of a second equivalent of alcohol gives protonated acetal.

R—O—H

Loss of a proton yields neutral acetal product.

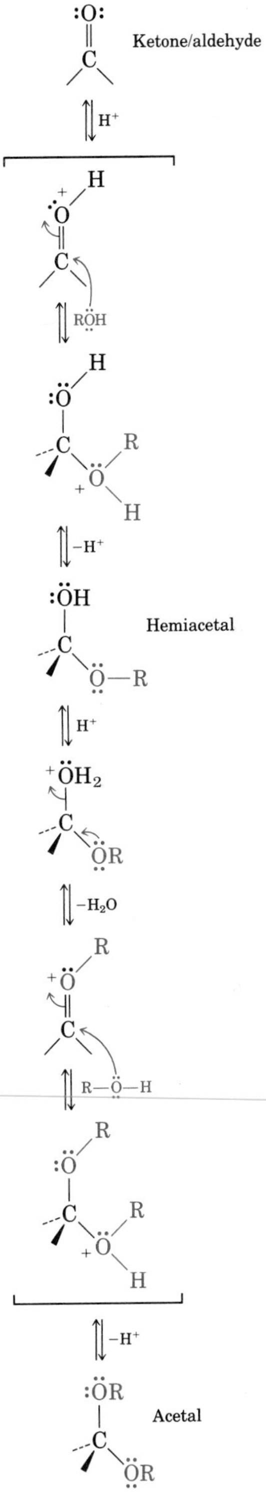

$-H^+$

Acetal

Figure 22.7. Mechanism of acid-catalyzed acetal formation.

Acetals are extremely valuable because they can serve as protecting groups for carbonyls. As we saw in Section 20.11, one functional group may interfere with intended chemistry elsewhere in a complex molecule. We can often circumvent the problem in such cases by first protecting the interfering functional group, carrying out the desired reaction, and then removing the protecting group. Like other ethers, acetals are stable to bases, hydride reducing agents, Grignard reagents, and catalytic reducing conditions, but they are acid-sensitive. For example, if we wished to reduce the ester group of ethyl 4-oxopentanoate by treatment with $LiAlH_4$, the ketone would interfere, since both carbonyl groups would be reduced. If we first form an acetal, however, subsequent ester reduction proceeds normally, and the acetal can be cleaved by hydrolysis.

$$CH_3\overset{O}{\overset{\|}{C}}CH_2CH_2\overset{O}{\overset{\|}{C}}OCH_2CH_3 \xrightarrow[\text{done directly}]{\text{Cannot be}} CH_3\overset{O}{\overset{\|}{C}}CH_2CH_2CH_2OH$$

Ethyl 4-oxopentanoate

$\downarrow$ H^+, $HOCH_2CH_2OH$ $\qquad\qquad$ $\uparrow$ H_3O^+

$$CH_3\overset{\text{O}-CH_2-CH_2-\text{O (ring)}}{C}CH_2CH_2CO_2C_2H_5 \xrightarrow[\text{Ether}]{LiAlH_4} CH_3\overset{\text{O}-CH_2-CH_2-\text{O (ring)}}{C}CH_2CH_2CH_2OH$$

In practice, most chemists find it convenient to use ethylene glycol as the alcohol and to form a cyclic acetal. The mechanism of cyclic acetal formation using 1 equiv of ethylene glycol is exactly the same as that using 2 equiv of methanol or other monoalcohol.

$$\text{4-}tert\text{-Butylcyclohexanone} \xrightarrow[H^+]{HOCH_2CH_2OH} \text{4-}tert\text{-Butylcyclohexanone ethylene acetal} + H_2O$$

$C(CH_3)_3$ $\qquad$ $C(CH_3)_3$

4-*tert*-Butylcyclohexanone

4-*tert*-Butylcyclohexanone ethylene acetal (88%)

PROBLEM

22.8 Draw all the steps in the acid-catalyzed formation of a cyclic acetal from ethylene glycol and a ketone or aldehyde.

PROBLEM

22.9 Use your knowledge of the reactivity difference between aldehydes and ketones to show how the following two selective reductions might be carried out. One of the two schemes requires a protection step.

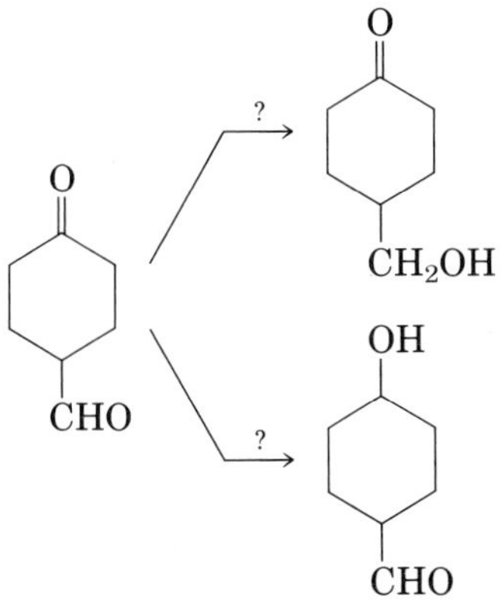

O, CHO → ? → O, CH_2OH

→ ? → OH, CHO

22.12 Addition of thiols: thioacetal formation and reduction

Thiols, RSH, add to ketones and aldehydes by a reversible, acid-catalyzed pathway to yield **thioacetals, $R'_2C(SR)_2$.** Ethanedithiol is often used, and the resultant cyclic thioacetals form rapidly in high yield.

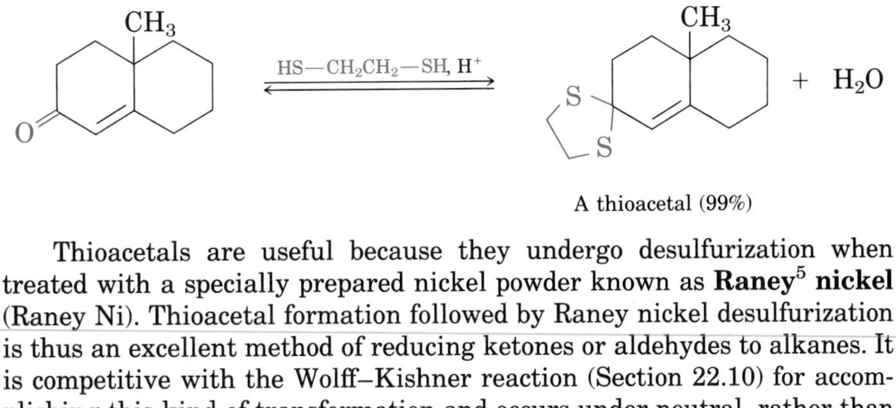

CH_3 ... $\xrightarrow{HS-CH_2CH_2-SH,\ H^+}$ ... CH_3, S, S $+ H_2O$

A thioacetal (99%)

Thioacetals are useful because they undergo desulfurization when treated with a specially prepared nickel powder known as **Raney[5] nickel** (Raney Ni). Thioacetal formation followed by Raney nickel desulfurization is thus an excellent method of reducing ketones or aldehydes to alkanes. It is competitive with the Wolff–Kishner reaction (Section 22.10) for accomplishing this kind of transformation and occurs under neutral, rather than basic, conditions.

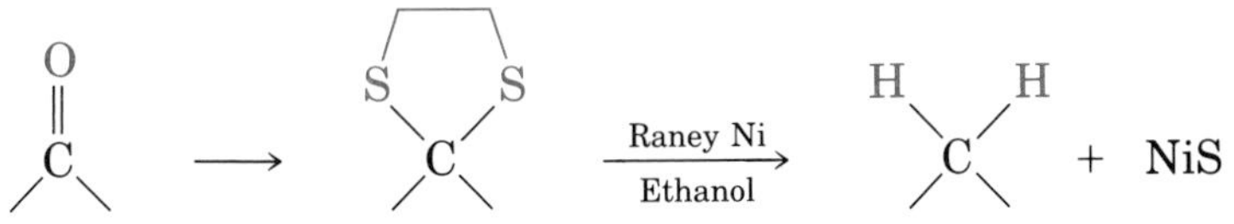

Aldehyde/ketone → Thioacetal $\xrightarrow[\text{Ethanol}]{\text{Raney Ni}}$ Alkane + NiS

[5]Murray Raney (1885–1966); b. Tennessee; B.A. University of Kentucky; D.Sc. (Hon.) University of Kentucky; Gilman Paint and Varnish Company (1925–1950); Raney Catalyst Company (1950–1966).

For example,

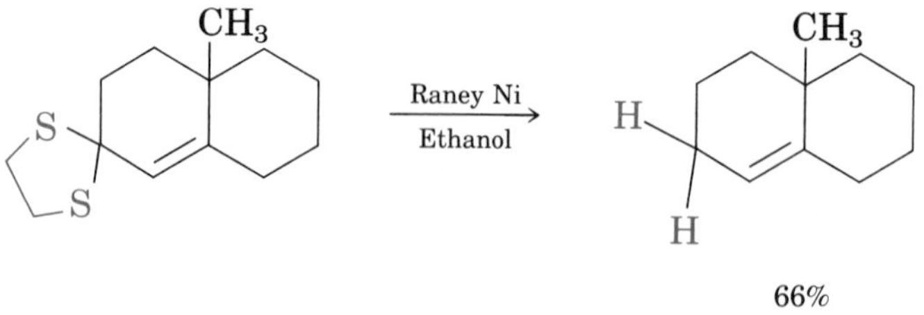

$$\xrightarrow[\text{Ethanol}]{\text{Raney Ni}}$$

66%

Raney nickel desulfurization is a general method for reducing *any* R—S bond to an R—H bond. The mechanism of the process is not fully understood, but it undoubtedly involves radical intermediates. The hydrogen atoms in the desulfurized products come from hydrogen gas, which is adsorbed onto the Raney nickel surface during preparation.

PROBLEM

22.10 Show three different methods by which you might prepare cyclohexane from cyclohexanone.

PROBLEM

22.11 How might you use Raney nickel desulfurization of a thioacetal to carry out the following reduction?

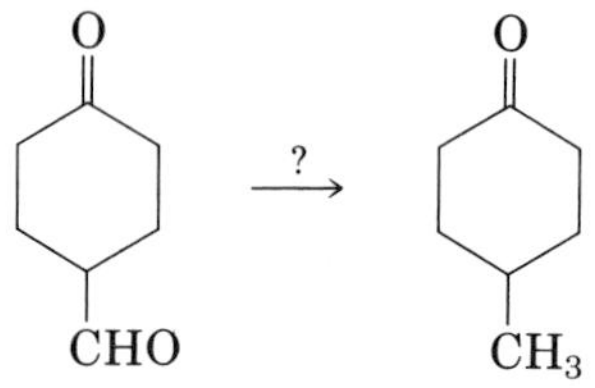

$$\xrightarrow{?}$$

22.13 Addition of phosphorus ylides: the Wittig reaction

Perhaps the most useful of all methods of alkene synthesis is the **Wittig reaction,** named for Georg Wittig.[6] In this process a phosphorus **ylide** (also called a **phosphorane**) adds to a ketone or aldehyde, yielding a dipolar **betaine.** (An ylide—pronounced *ill'-id*—is a dipolar compound with adjacent plus and minus charges; a betaine—pronounced *bay'-ta-een*—is a dipolar compound in which the charges are nonadjacent.) The betaine intermediate in the Wittig reaction is unstable and decomposes at temperatures above 0°C to yield alkene and triphenylphosphine oxide. The net result is replacement of carbonyl oxygen by the organic fragment originally bonded to phosphorus (Figure 22.8).

[6]Georg Wittig (1897–); b. Berlin; Ph.D. Marburg (von Auwers); professor, universities of Freiburg, Tübingen, and Heidelberg; Nobel prize (1979).

$$\underset{\text{Ketone/aldehyde}}{\mathrm{C{=}O}} + \underset{\text{An ylide}}{\mathrm{R_2\ddot{C}^{-}{-}\overset{+}{P}(C_6H_5)_3}} \xrightarrow{\text{THF}} \underset{\text{A betaine}}{\left[\begin{array}{cc} :\ddot{O}:^{-} & \overset{+}{P}(C_6H_5)_3 \\ | & | \\ C & -C-R_2 \end{array}\right]}$$

$$\downarrow$$

$$\underset{\text{Alkene}}{\mathrm{R_2C{=}C}} + \underset{\text{Triphenylphosphine oxide}}{\mathrm{(C_6H_5)_3P{=}O}} \longleftarrow \left[\begin{array}{cc} O- & P(C_6H_5)_3 \\ | & | \\ C & -C-R_2 \end{array}\right]$$

Figure 22.8. The mechanism of the Wittig reaction.

The phosphorus ylides necessary for Wittig reaction are easily prepared by S_N2 reaction of primary (but not secondary or tertiary) alkyl halides with triphenylphosphine, followed by treatment with base. Triorganophosphines are generally excellent nucleophiles in S_N2 reactions, and yields of stable, crystalline tetraorganophosphonium salts are high. The proton on the carbon next to the positively charged phosphorus is acidic and can be removed by a base such as sodium hydride or butyllithium (BuLi) to generate the neutral ylide. For example,

$$\underset{\text{Triphenylphosphine}}{\mathrm{(C_6H_5)_3P:}} + \underset{\text{Bromomethane}}{\mathrm{CH_3{-}Br}} \xrightarrow{S_N2\ \text{reaction}} \underset{\text{Methyltriphenylphosphonium bromide (99\%)}}{\mathrm{(C_6H_5)_3\overset{+}{P}{-}CH_3\ \overset{-}{Br}}} \xrightarrow[\text{THF}]{\text{BuLi}} \underset{\text{Methylenetriphenylphosphorane}}{\mathrm{(C_6H_5)_3\overset{+}{P}{-}\overset{-}{C}H_2}}$$

The Wittig reaction is broad in scope, and a great variety of mono-, di-, and trisubstituted alkenes can be prepared from the appropriate combination of phosphorane and ketone or aldehyde. Tetrasubstituted alkenes cannot be prepared, however, presumably for reasons having to do with steric hindrance during the reaction.

The great value of the Wittig reaction is that pure alkenes of known structure are prepared—the alkene double bond is *always* exactly where the carbonyl group was in the precursor, and no product mixtures (other than *E,Z* isomers) are formed. For example, addition of methylmagnesium bromide to cyclohexanone, followed by dehydration with $POCl_3$, yields a mixture of two alkenes. Wittig reaction of cyclohexanone with methylenetriphenylphosphorane, however, yields only the single pure alkene product, methylenecyclohexane.

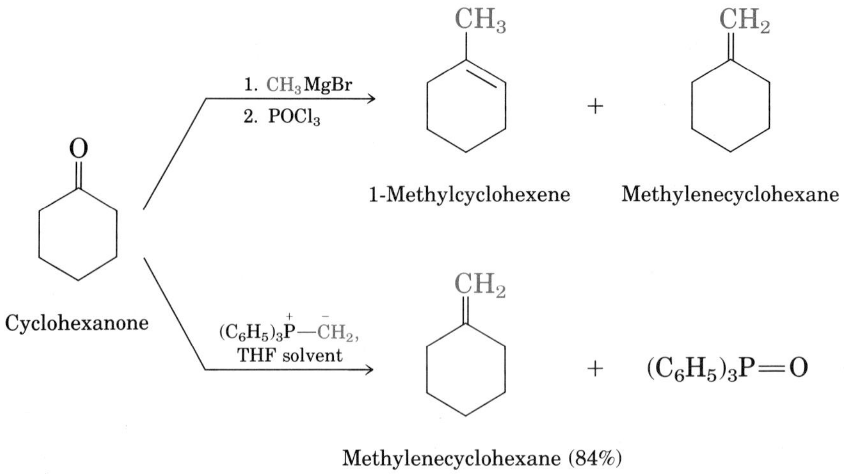

Cyclohexanone

1. CH_3MgBr
2. $POCl_3$

CH_3 1-Methylcyclohexene + CH_2 Methylenecyclohexane

$(C_6H_5)_3\overset{+}{P}—\overset{-}{C}H_2$, THF solvent

CH_2 Methylenecyclohexane (84%) + $(C_6H_5)_3P{=}O$

Wittig reactions are so important that they are used commercially. For example, the Swiss chemical firm of Hoffmann–La Roche prepares β-carotene, a yellow food-coloring agent, by Wittig reaction between retinal (vitamin A aldehyde) and retinylidenetriphenylphosphorane (Figure 22.9).

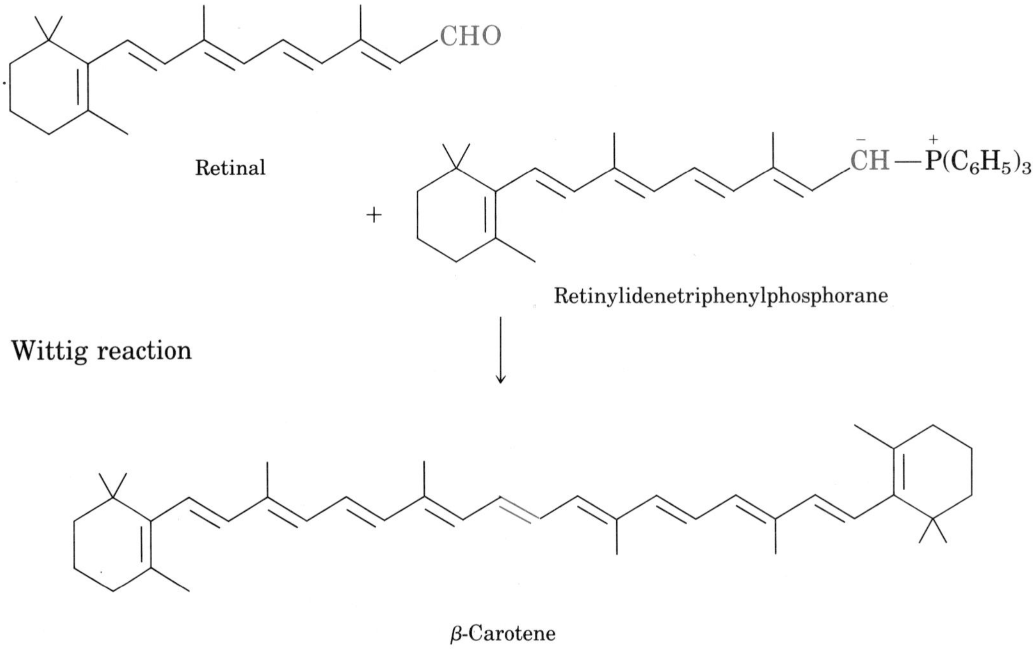

CHO

Retinal

+

$\overset{-}{C}H—\overset{+}{P}(C_6H_5)_3$

Retinylidenetriphenylphosphorane

Wittig reaction

β-Carotene

Figure 22.9. Preparation of β-carotene using the Wittig reaction.

PROBLEM

22.12 What carbonyl compounds and what phosphorus ylides might you use to prepare the following compounds?

(a) 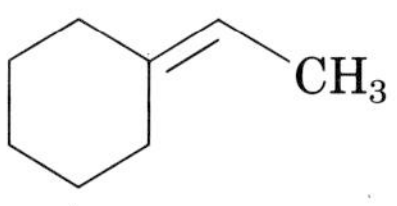 [cyclohexane ring with =CH–CH_3]

(b) 2-Methyl-2-hexene

(c) 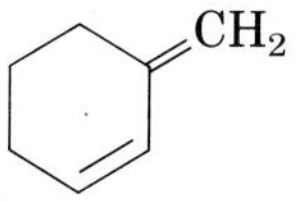 [cyclohexene ring with =CH_2]

(d) $C_6H_5CH{=}C(CH_3)_2$

(e) Stilbene (1,2-diphenylethylene)

PROBLEM

22.13 Why do you suppose triphenylphosphine is used to prepare Wittig reagents rather than, say, trimethylphosphine? What problems might you run into if trimethylphosphine were used?

PROBLEM

22.14 Another route to β-carotene involves a double Wittig reaction between 2 equiv of β-ionylideneacetaldehyde and a diylide. Formulate the reaction.

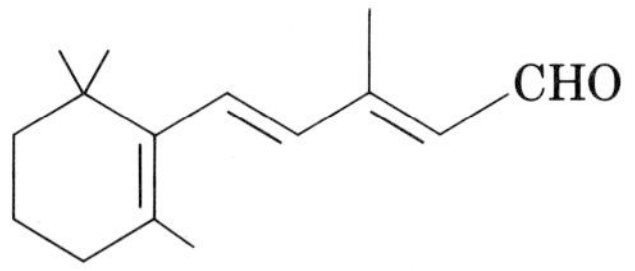

CHO

β-Ionylideneacetaldehyde

22.14 The Cannizzaro reaction

We said in the overview of carbonyl chemistry (Section 21.5) that nucleophilic acyl substitution reactions are characteristic of carboxylic acid derivatives, but not of ketones and aldehydes. The reason for this is that alkyl and hydrogen substituents (group Y in the general reaction scheme in Figure 22.10) cannot usually act as leaving groups. The **Cannizzaro reaction,** discovered in 1853, is one exception to this rule.[7]

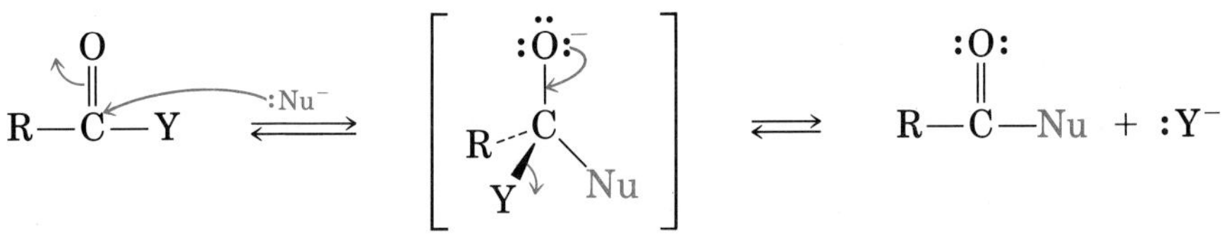

$$R{-}\overset{\overset{O}{\|}}{C}{-}Y \xrightleftharpoons{:Nu^-} \left[\, R{-}\overset{\overset{:\ddot{O}:^-}{|}}{C}{<}^{Nu}_{Y} \,\right] \rightleftharpoons R{-}\overset{\overset{:O:}{\|}}{C}{-}Nu + :Y^-$$

Figure 22.10. General reaction scheme for nucleophilic acyl substitution.

[7]Stanislao Cannizzaro (1826–1910); b. Palermo, Italy; studied Pisa (Piria); professor, universities of Genoa, Palermo, and Rome.

When an aldehyde having no alpha protons is heated with hydroxide ion, a disproportionation reaction occurs, yielding 1 equiv of carboxylic acid and 1 equiv of alcohol.

$$2\ C_6H_5CHO \xrightarrow[2.\ H_3O^+]{1.\ ^-OH,\ H_2O} C_6H_5CO_2H + C_6H_5CH_2OH$$

Benzaldehyde — Benzoic acid — Benzyl alcohol

The Cannizzaro reaction takes place by nucleophilic addition of hydroxide ion to the aldehyde to give a tetrahedral intermediate, which expels hydride ion. A second equivalent of aldehyde accepts the hydride ion. The net result is that one molecule of aldehyde has undergone an acyl substitution reaction of hydroxide for hydride and is thereby oxidized; a second molecule of aldehyde has accepted the hydride and is thereby reduced to alcohol.

$$C_6H_5CHO \xrightarrow{:OH^-} \left[C_6H_5CH(OH)O^-\right] \xrightarrow{C_6H_5CHO} C_6H_5C(=O)OH + C_6H_5CH_2O^-$$

Tetrahedral intermediate

Oxidized + Reduced

The Cannizzaro reaction is effectively limited to formaldehyde and substituted benzaldehydes, since aldehydes that have alpha protons undergo other processes involving enolization of their acidic alpha hydrogens.

PROBLEM

22.15 When *o*-phthalaldehyde is treated with base, *o*-(hydroxymethyl)benzoic acid is formed. Propose a mechanism to account for this reaction.

$$o\text{-}C_6H_4(CHO)_2 \xrightarrow[2.\ H_3O^+]{1.\ ^-OH} o\text{-}C_6H_4(CO_2H)(CH_2OH)$$

o-Phthalaldehyde — *o*-(Hydroxymethyl)benzoic acid

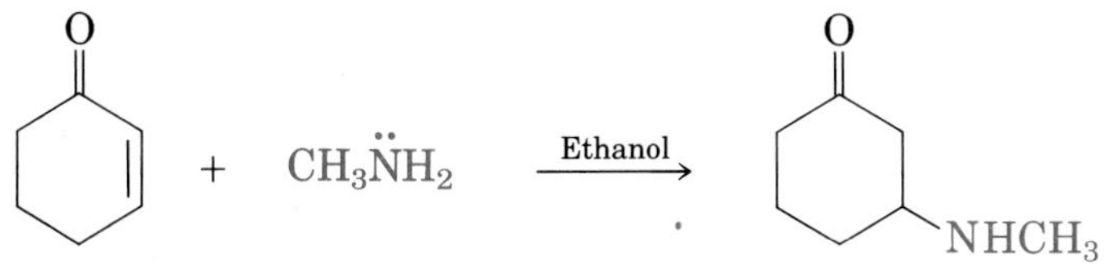

$$\text{2-Cyclohexenone} + CH_3\ddot{N}H_2 \xrightarrow{\text{Ethanol}} \text{3-(}N\text{-Methylamino)cyclohexanone}$$

2-Cyclohexenone Methylamine 3-(*N*-Methylamino)cyclohexanone

CONJUGATE ADDITION OF HCN

The elements of HCN can be added to α,β-unsaturated ketones and aldehydes, giving saturated keto nitriles:

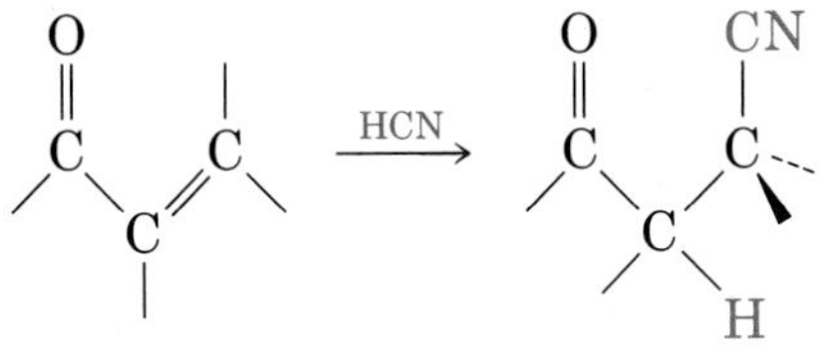

$$\text{Ketone/aldehyde} \xrightarrow{\text{HCN}} \text{keto nitrile (CN at } \beta\text{, H at } \alpha\text{)}$$

Ketone/aldehyde

Although this process can be carried out using sodium cyanide in aqueous alcohol, higher yields are obtained using a method introduced in 1966 by Wataru Nagata. The Nagata procedure for conjugate addition of HCN involves the use of diethylaluminum cyanide as the active reagent, and yields are generally excellent, as the following examples indicate. Note that only conjugate addition is observed; no cyanohydrin products of direct addition are formed.

$$CH_3\overset{O}{\overset{\|}{C}}-CH=C(CH_3)_2 \xrightarrow[2.\ H_3O^+]{1.\ (C_2H_5)_2Al-CN,\ \text{toluene}} CH_3\overset{O}{\overset{\|}{C}}-CH_2-\overset{CN}{\overset{|}{C}}(CH_3)_2$$

4-Methyl-3-penten-2-one 2,2-Dimethyl-4-oxopentanenitrile (88%)

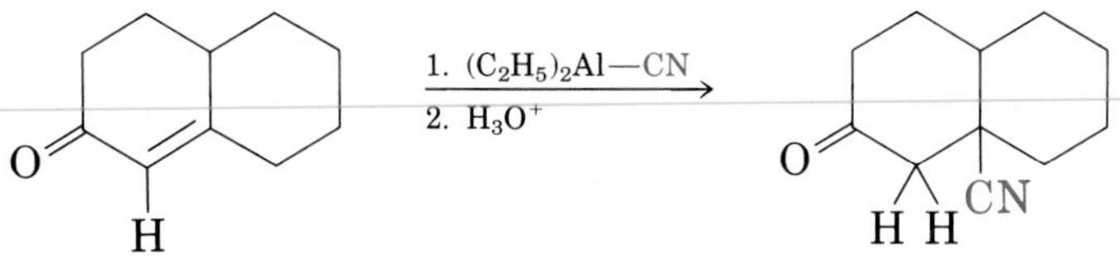

$$\xrightarrow[2.\ H_3O^+]{1.\ (C_2H_5)_2Al-CN}$$

89%

CONJUGATE ADDITION OF ALKYL GROUPS: ORGANOCOPPER REACTIONS

Conjugate addition of an alkyl group to an α,β-unsaturated ketone is one of the most important 1,4-addition reactions:

$$\alpha,\beta\text{-Unsaturated ketone/aldehyde} \xrightarrow[2.\ H^+]{1.\ \text{“}:R^-\text{”}} \text{product with R added at the } \beta \text{ carbon and H at the } \alpha \text{ carbon}$$

α,β-Unsaturated
ketone/aldehyde

This reaction is carried out by treating the α,β-unsaturated ketone with a lithium diorganocopper (Gilman reagent), and yields are excellent. A wide variety of diorganocopper reagents can be prepared by reaction between 1 equiv of cuprous iodide and 2 equiv of organolithium (Section 9.9):

$$R{-}X \xrightarrow[\text{Pentane}]{2\text{ Li}} R{-}Li + Li^+X^-$$

$$2\,R{-}Li \xrightarrow[\text{Ether}]{\text{CuI}} Li^+(R{-}\overset{-}{Cu}{-}R) + Li^+I^-$$

A lithium diorganocopper
(Gilman reagent)

Primary, secondary, and even tertiary alkyl groups undergo the addition reaction, as do aryl and alkenyl groups. Alkynyl groups, however, react poorly in the conjugate addition process.

$$CH_3\overset{O}{\overset{\|}{C}}CH{=}CH_2 \xrightarrow[2.\ H_3O^+]{1.\ Li(CH_3)_2Cu,\ \text{ether}} CH_3\overset{O}{\overset{\|}{C}}CH_2CH_2{-}CH_3$$

3-Buten-2-one → 2-Pentanone (97%)

$$\text{2-Cyclohexenone} \xrightarrow[2.\ H_3O^+]{1.\ Li(H_2C{=}CH)_2Cu,\ \text{ether}} \text{3-Vinylcyclohexanone}$$

2-Cyclohexenone

3-Vinylcyclohexanone
(65%)

$$\text{2-Cyclohexenone} \xrightarrow[2.\ H_3O^+]{1.\ Li(C_6H_5)_2Cu,\ \text{ether}} \text{3-Phenylcyclohexanone}$$

2-Cyclohexenone

3-Phenylcyclohexanone
(70%)

The mechanism of diorganocopper addition is still a matter of controversy, but it appears that radicals are involved; the reaction is not a typical nucleophilic addition process. One piece of evidence for this statement comes from the observation that diorganocoppers are unique in their ability to undergo conjugate addition. Other organometallic reagents such as organomagnesiums (Grignard reagents) and organolithiums normally give direct carbonyl addition on reaction with α,β-unsaturated enones.

2-Cyclohexenone $\xrightarrow[\text{2. } H_3O^+]{\text{1. } CH_3MgBr\text{, ether or } CH_3Li}$ 1-Methyl-2-cyclohexen-1-ol (95%)

2-Cyclohexenone $\xrightarrow[\text{2. } H_3O^+]{\text{1. } Li(CH_3)_2Cu\text{, ether}}$ 3-Methylcyclohexanone (97%)

PROBLEM

22.16 Show how conjugate addition reactions of lithium diorganocopper reagents might be used to synthesize these compounds:

(a) 2-Heptanone

(b) 3,3-Dimethylcyclohexanone

(c) 4-*tert*-Butyl-3-ethylcyclohexanone

(d) [bicyclic ketone bearing $CH{=}CH_2$ at the ring-fusion carbon]

22.16 Some biological nucleophilic addition reactions

Nature synthesizes the molecules of life using many of the same reactions that chemists use in the laboratory. This is particularly true of carbonyl-group reactions, and nucleophilic addition steps play an intimate role in the biosynthesis of many vital molecules.

For example, one of the pathways by which amino acids are made involves reductive amination of α-keto acids. To choose one specific case, alanine is synthesized from pyruvic acid and ammonia by bacterial enzymes from *Bacillus subtilis:*

$$CH_3\overset{O}{\overset{\|}{C}}{-}COOH + :NH_3 \xrightarrow{B.\ subtilis} CH_3\overset{NH_2}{\overset{|}{C}}HCOOH$$

Pyruvic acid (an α-keto acid) — Alanine (an amino acid)

The key step in this biological transformation is the nucleophilic addition of ammonia to the ketone carbonyl group of pyruvic acid. The tetrahedral intermediate loses water to yield an imine, which is further reduced by an enzymatic reaction to yield alanine.

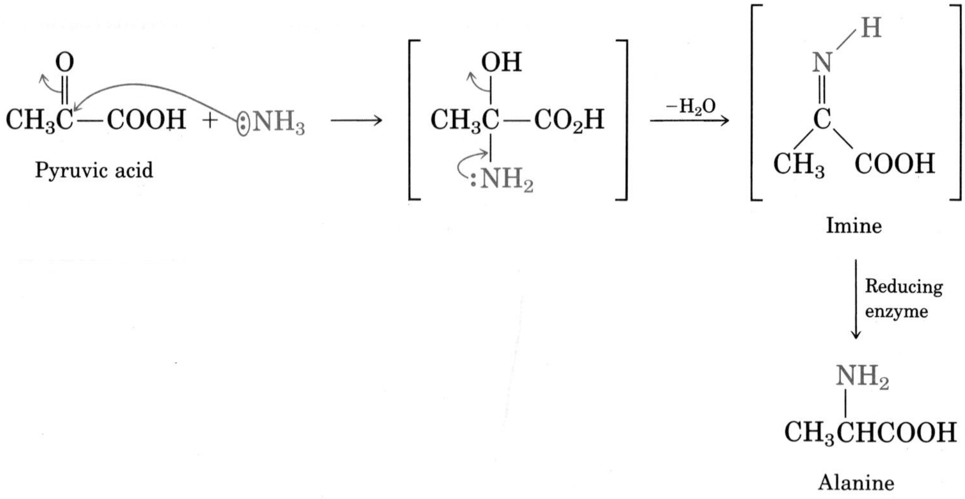

$$CH_3C(=O)-COOH + :NH_3 \longrightarrow \left[CH_3C(OH)(:NH_2)-CO_2H \right] \xrightarrow{-H_2O} \left[CH_3C(=N-H)COOH \right] \xrightarrow{\text{Reducing enzyme}} CH_3CH(NH_2)COOH$$

Pyruvic acid — Imine — Alanine

Another example of nucleophilic carbonyl addition occurs frequently in carbohydrate chemistry. The six-carbon sugar, glucose, acts in some respects as if it were an aldehyde. For instance, glucose can be oxidized to a carboxylic acid. Spectroscopic examination of glucose, however, indicates that no aldehyde group is present. In fact, glucose exists as a *cyclic hemiacetal.* The hydroxyl group at carbon 5 adds to the aldehyde at carbon 1.

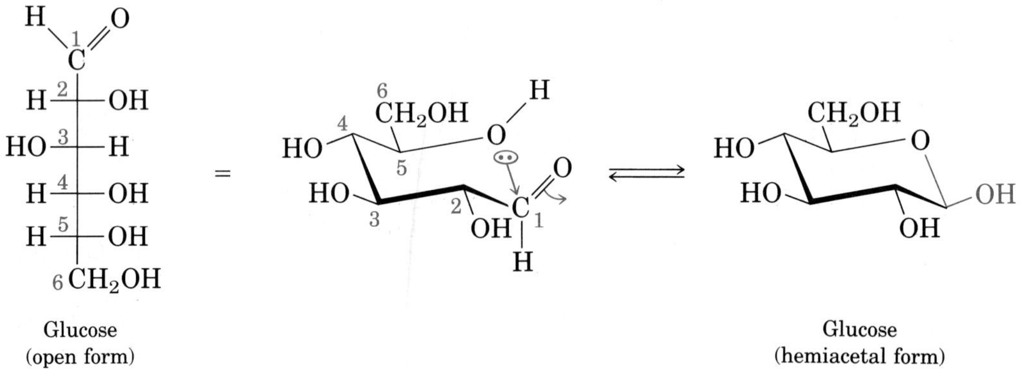

Glucose (open form) — Glucose (hemiacetal form)

Further reaction between molecules of glucose leads to the carbohydrate polymer cellulose. Cellulose, which constitutes the major building block of plant cell walls, consists simply of glucose units joined by acetal linkages between carbon 1 of one glucose with the hydroxyl group at carbon 4 of another glucose.

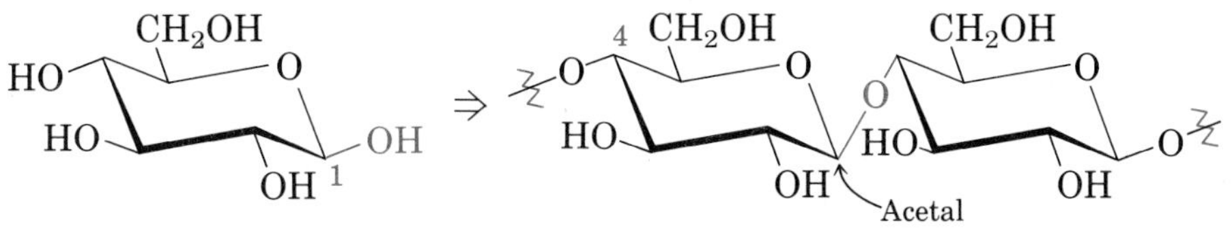

Glucose ⇒ Cellulose

We will study this and other reactions of carbohydrates in more detail in Chapter 27.

22.17 Spectroscopic analysis of ketones and aldehydes

INFRARED SPECTROSCOPY

Ketones and aldehydes show a strong C═O bond absorption in the infrared region 1660–1770 cm^{-1}, as the spectra of benzaldehyde and cyclohexanone (Figures 22.12 and 22.13, p. 722) demonstrate. In addition, aldehydes show two characteristic C—H absorptions in the range 2720–2820 cm^{-1}, due to stretching of the aldehyde —CO—H bond. The exact position of the C═O bond absorption varies slightly from compound to compound but is highly diagnostic of the precise nature of the carbonyl group. Table 22.3 shows the correlation between the infrared absorption maximum and carbonyl-group structure.

TABLE 22.3 Infrared absorptions of some ketones and aldehydes

Carbonyl type	*Example*	*Infrared absorption (cm^{-1})*
Aliphatic aldehyde	Acetaldehyde	1730
Aromatic aldehyde	Benzaldehyde	1705
α,β-Unsaturated aldehyde	$H_2C{=}CH{-}CHO$	1705
Aliphatic ketone	Acetone	1715
Six-membered ring ketone	Cyclohexanone	1715
Five-membered ring ketone	Cyclopentanone	1750
Four-membered ring ketone	Cyclobutanone	1785
Aromatic ketone	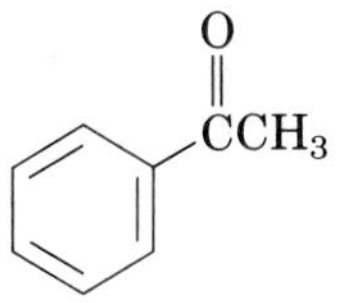 $C_6H_5CCH_3$ (O above C)	1690
α,β-Unsaturated ketone	$H_2C{=}CHCCH_3$ (O above C)	1685

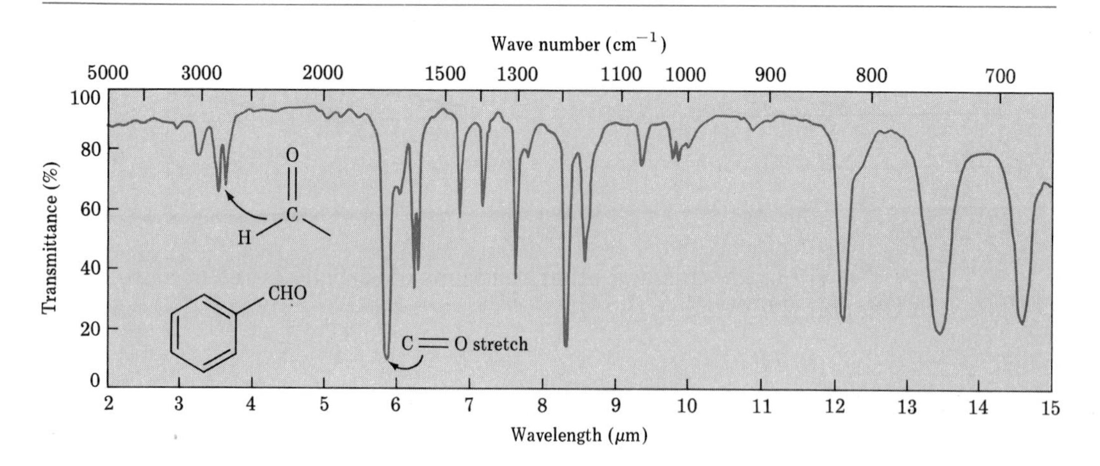

Figure 22.12. Infrared spectrum of benzaldehyde.

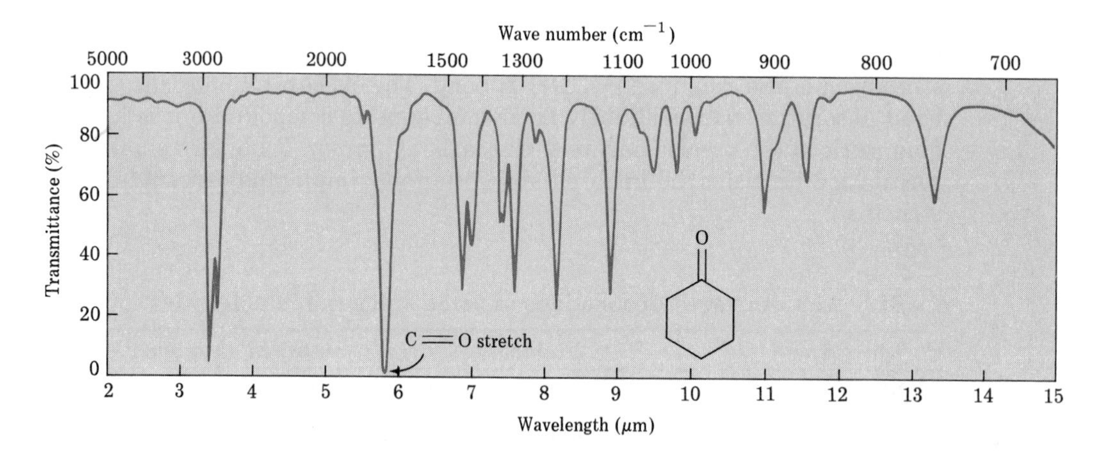

Figure 22.13. Infrared spectrum of cyclohexanone.

As the data in Table 22.3 indicate, saturated aldehydes usually show carbonyl absorptions near 1730 cm^{-1} in the infrared spectrum; conjugation of the aldehyde to an aromatic ring or a double bond lowers the absorption by 25 cm^{-1} to near 1705 cm^{-1}. Saturated aliphatic ketones and cyclohexanones both absorb near 1715–1720 cm^{-1}, and conjugation with a double bond or aromatic ring again lowers the absorption by 30 cm^{-1} to 1685–1690 cm^{-1}. Additional angle strain in the carbonyl group, caused by reducing the ring size of cyclic ketones to four or five, results in a marked raising of the absorption position.

The values given in Table 22.3 are remarkably constant from one ketone to another or from one aldehyde to another. As a result, infrared spectroscopy is an extraordinarily powerful tool for diagnosing the nature and chemical environment of a carbonyl group in a molecule of unknown structure. An unknown that shows an infrared absorption at 1730 cm^{-1} is almost certainly an aldehyde rather than a ketone; an unknown that shows an infrared absorption at 1750 cm^{-1} is almost certainly a cyclopentanone, and so on.

PROBLEM

22.17 How might you use infrared spectroscopy to determine whether reaction between 2-cyclohexenone and lithium dimethylcopper gives the direct addition product or the conjugate addition product?

PROBLEM

22.18 Tell where you would expect each of these compounds to absorb in the infrared spectrum:

(a) 4-Penten-2-one
(b) 3-Penten-2-one
(c) 2,2-Dimethylcyclopentanone
(d) *m*-Chlorobenzaldehyde
(e) 3-Cyclohexenone

PROBLEM

22.19 Dehydration of 3-hydroxy-3-phenylcyclohexanone by treatment with acid leads to a keto alkene. What possible structures are there for the product? At what position in the infrared spectrum would you expect each to absorb? If the actual product has an absorption at 1670 cm^{-1}, what is its structure?

NUCLEAR MAGNETIC RESONANCE SPECTROSCOPY

Carbonyl-group carbon atoms show readily identifiable and highly characteristic ^{13}C NMR resonance peaks in the range 190–210 δ. No other functional-group carbon absorbs in this range, and the presence of an NMR absorption near 200 δ is strong evidence for a carbonyl group. Isolated carbonyl-group carbons usually absorb in the region from 200 to 210 δ, whereas α,β-unsaturated carbonyl carbons absorb in the 190–200 δ region. Table 22.4 lists specific examples.

TABLE 22.4 **Carbon-13 NMR absorptions of some carbonyl carbons**

Carbonyl compound	*Carbon-13 NMR absorption of* $>C{=}O$ (δ)
Acetaldehyde	201
Benzaldehyde	192
2-Butanone	207
Cyclohexanone	211
Acetophenone	196

Proton NMR is also of considerable use for analysis of aldehydes, though less so for ketones. Aldehyde protons (R—CHO) absorb near 10 δ in the ^1H NMR spectrum and are highly distinctive, since no other kind of proton absorbs in this region. The aldehyde proton usually shows spin–spin coupling to neighbor protons, with coupling constant $J \approx 3$ Hz. Observation of the splitting pattern of the aldehyde proton enables us to tell the degree of substitution at the alpha position. Acetaldehyde, for example, shows a quartet at 9.8 δ for the aldehyde proton, indicating that there are three protons neighboring the CHO group (Figure 22.14).

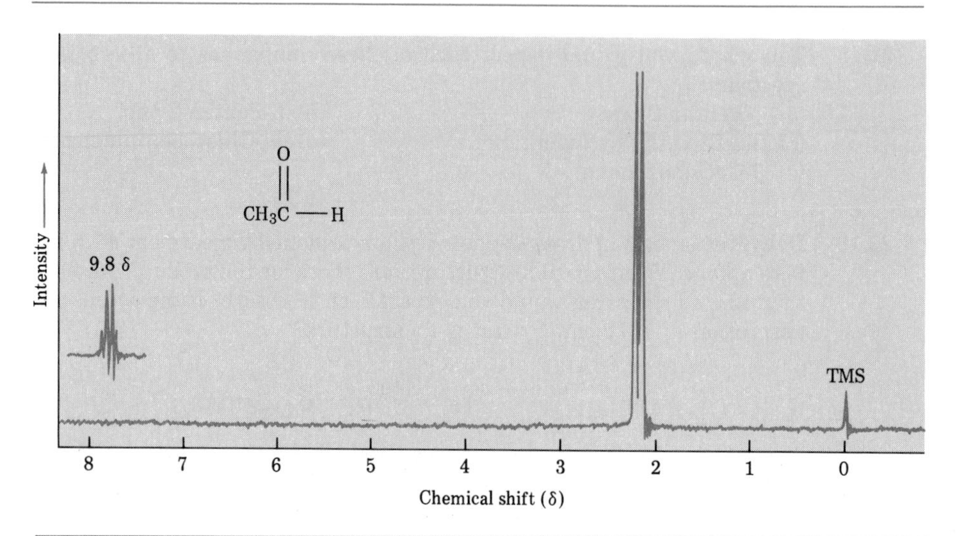

Figure 22.14. Proton NMR spectrum of acetaldehyde.

Protons on the carbon next to a carbonyl group are slightly deshielded and normally absorb near 2.0–2.3 δ (note that the acetaldehyde methyl group in Figure 22.14 absorbs at 2.20 δ). Methyl ketones are particularly distinctive, since they show a large, sharp, three-proton singlet near 2.1 δ. Complex spin–spin splittings often obscure the absorption patterns of other ketones, however, and reduce the diagnostic usefulness of ^1H NMR.

MASS SPECTROSCOPY

Aliphatic ketones and aldehydes having hydrogens on their gamma-carbon atoms undergo a characteristic mass spectral cleavage called the **McLafferty rearrangement.**[8] In this rearrangement, a hydrogen atom is transferred from the gamma carbon to the carbonyl-group oxygen, the bond between the alpha and beta carbons is broken, and a neutral alkene fragment is produced. The charge remains with the oxygen-containing fragment.

[8]Fred Warren McLafferty (1923–); b. Evanston, Ill.; Ph.D., Cornell University (1950); Dow Chemical (1950–1964), professor, Purdue University (1964–1968), Cornell University (1968–).

$$\left[\begin{array}{c} R' \quad H \\ \gamma CH \quad O \\ \beta CH_2 \quad C \\ \alpha CH_2 \quad R \end{array}\right]^{+\cdot} \xrightarrow[\text{rearrangement}]{\text{McLafferty}} \begin{array}{c} R' \\ \gamma CH \\ \| \\ \beta CH_2 \end{array} + \left[\begin{array}{c} H \\ O \\ | \\ C \\ \alpha CH_2 \quad R \end{array}\right]^{+\cdot}$$

In addition to fragmentation by the McLafferty rearrangement, ketones and aldehydes undergo cleavage of the bond between the alpha carbon and the carbonyl-group carbon (an alpha-cleavage reaction). Alpha cleavage yields a neutral radical; the charge remains with the oxygen-containing cation.

$$\left[R{-}CH_2{-}\overset{O}{\overset{\|}{C}}{-}R'\right]^{+\cdot} \xrightarrow{\text{Alpha cleavage}} R{-}CH_2\cdot + \left[\overset{O}{\overset{\|}{C}}{-}R\right]^{+}$$

Fragment ions resulting from both alpha cleavage and McLafferty rearrangement can be seen in the mass spectrum of 5-methyl-2-hexanone (Figure 22.15). Alpha cleavage occurs primarily at the more substituted side

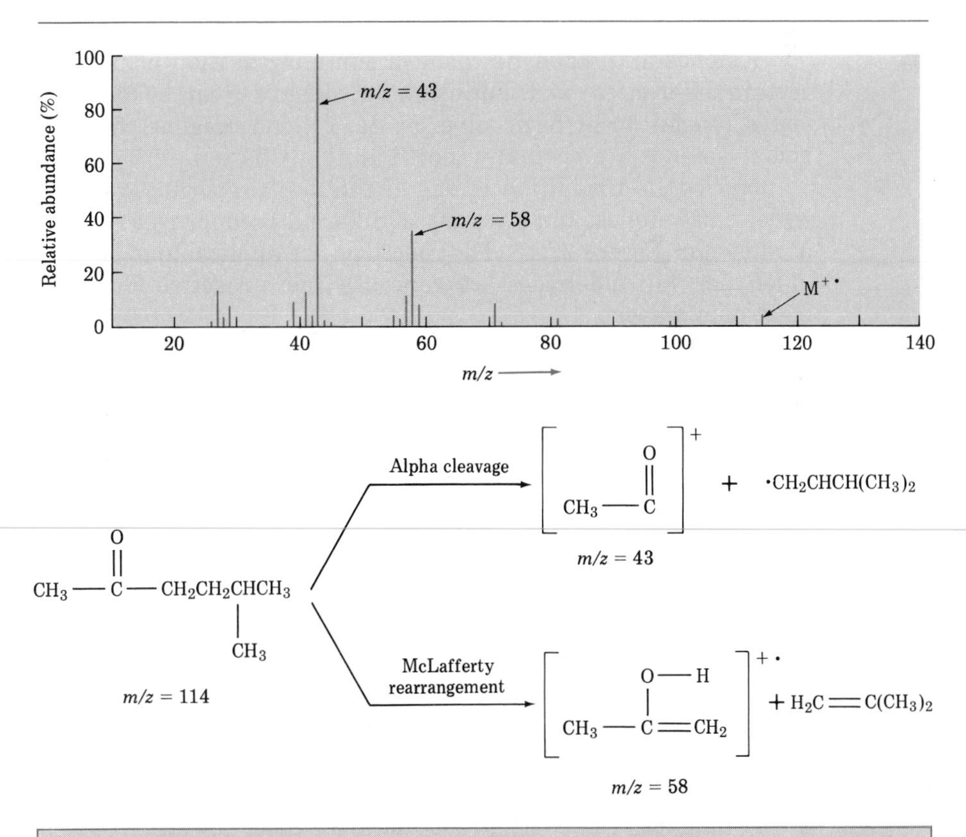

Figure 22.15. Mass spectrum of 5-methyl-2-hexanone. The abundant peak at $m/z = 43$ is due to alpha cleavage on the more highly substituted side of the carbonyl group. The peak at $m/z = 58$ is due to McLafferty rearrangement. Note that the peak due to the molecular ion is very small.

of the carbonyl group, leading to a $[CH_3CO]^+$ fragment with $m/z = 43$. McLafferty rearrangement and loss of 2-methylpropene yields a fragment with $m/z = 58$.

PROBLEM

22.20 How might you use mass spectroscopy to distinguish between these pairs of isomers?
(a) 3-Methyl-2-hexanone and 4-methyl-2-hexanone
(b) 3-Heptanone and 4-heptanone
(c) 2-Methylpentanal and 4-methylpentanal

22.18 Summary

Aldehydes and ketones are important, both in biochemistry and in the chemical industry. Aldehydes are normally prepared in the laboratory by oxidative cleavage of alkenes, by oxidation of primary alcohols with pyridinium chlorochromate (PCC), or by partial reduction of acid chlorides or esters. Ketones are similarly prepared by oxidative cleavage of alkenes, by oxidation of secondary alcohols, or by addition of diorganocopper reagents to acid chlorides. There are, however, certain specialized ketone syntheses such as Friedel–Crafts acylations for preparing aryl ketones, and mercuric ion catalyzed alkyne hydrations for preparing methyl ketones.

Ketones and aldehydes behave similarly in their chemistry, although certain differences exist. For example, aldehydes can be oxidized, either by Jones' reagent or by basic silver oxide (Tollens' reagent), to give carboxylic acids. Ketones are normally inert to mild oxidation.

Nucleophilic addition is one of the most important reactions of aldehydes and ketones, and a variety of different product types can be prepared, as shown in Figure 22.16. The reactions are applicable to both ketones and aldehydes, but aldehydes are generally more reactive for both steric and electronic reasons.

Ketones and aldehydes are reduced by $NaBH_4$ or $LiAlH_4$ to yield secondary and primary alcohols, respectively. Addition of Grignard reagents also lead to alcohols (tertiary and secondary, respectively), and addition of HCN yields cyanohydrins. Primary amines add to carbonyl compounds yielding imines, and secondary amines yield enamines. The latter two reactions are particularly valuable; oximes ($R_2C{=}NOH$), semicarbazones ($R_2C{=}N{-}NCONH_2$), and 2,4-dinitrophenylhydrazones [$R_2C{=}N{-}NH{-}C_6H_4{-}(NO_2)_2$] are often prepared as crystalline derivatives of liquid ketones or aldehydes. Reaction of a ketone or aldehyde with hydrazine and base yields an alkane. This Wolff–Kishner reduction proceeds through an intermediate imine, and is an important synthetic method. Alcohols and thiols (RSH) add to carbonyl groups to yield acetals and thioacetals, respectively. Acetals are valuable as protecting groups, and thioacetals are valuable because they can be desulfurized by Raney nickel treatment to produce alkanes. Phosphoranes add to ketones and aldehydes, giving alkenes. This Wittig reaction is the most powerful method available for preparing pure alkenes.

α,β-Unsaturated ketones and aldehydes often react with nucleophiles to give the product of conjugate addition, or 1,4 addition. The most impor-

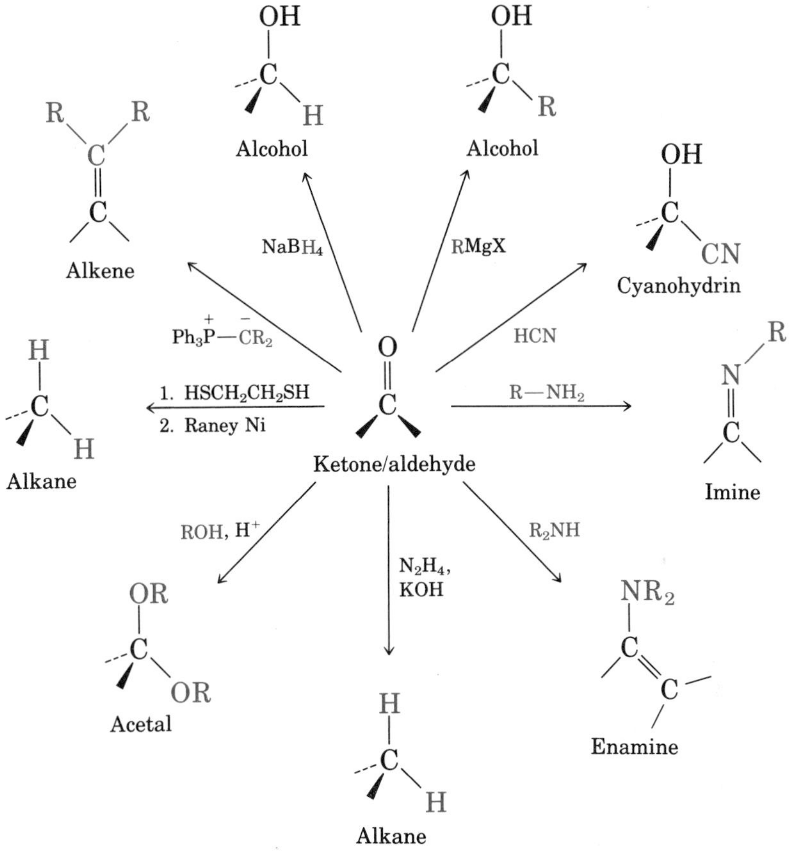

Alkene

$Ph_3\overset{+}{P}—\overset{-}{C}R_2$

Alcohol

$NaBH_4$

Alcohol

RMgX

Cyanohydrin

HCN

Alkane

1. $HSCH_2CH_2SH$
2. Raney Ni

Ketone/aldehyde

R—NH_2

Imine

ROH, H^+

N_2H_4, KOH

R_2NH

Acetal

Alkane

Enamine

Figure 22.16. Some nucleophilic addition reactions of ketones and aldehydes.

tant conjugate addition process is the reaction of diorganocopper reagents with enones. This process is broad in scope, and is successful for the addition of alkyl, aryl, and alkenyl (but not alkynyl) groups.

Ketones and aldehydes may be analyzed by infrared and nuclear magnetic resonance spectroscopy. Carbonyl groups absorb in the range 1660–1770 cm^{-1}, and the exact infrared absorption position is highly diagnostic of the precise kind of carbonyl group present in the molecule. Carbon-13 NMR spectroscopy is also highly diagnostic for aldehydes and ketones; carbonyl carbons show resonances in the 190–210 δ range. Proton NMR is useful largely for analysis of aldehydes. Aldehyde protons (R—CHO) absorb near 10 δ, and the splitting pattern observed indicates the degree of substitution on the alpha carbon. Ketones and aldehydes undergo two characteristic kinds of fragmentation in the mass spectrometer—alpha cleavage and McLafferty rearrangement.

22.19 Summary of reactions

1. Preparation of aldehydes (Section 22.3)
   a. Oxidation of primary alcohols (Section 20.10)

$$R{-}CH_2OH \xrightarrow[CH_2Cl_2]{PCC} R{-}CHO$$

   b. Ozonolysis of alkenes (Section 6.8)

$$R{-}CH{=}C\langle \xrightarrow[2.\ Zn,\ CH_3COOH]{1.\ O_3} R{-}CHO$$

   c. Rosenmund reduction of acid chloride (Section 24.5)

$$R{-}COCl \xrightarrow[\text{Ethyl acetate}]{H_2/Pd/BaSO_4} R{-}CHO$$

   d. Partial reduction of esters (Section 22.3)

$$R{-}CO_2R' \xrightarrow[2.\ H_3O^+]{1.\ \text{DIBAH, toluene}} R{-}CHO$$

2. Preparation of ketones (Section 22.4)
   a. Oxidation of secondary alcohols (Section 20.10)

$$R_2CHOH \xrightarrow{Cr(VI)} R_2C{=}O$$

   b. Ozonolysis of alkenes (Section 6.8)

$$R_2C{=}C\langle \xrightarrow[2.\ Zn,\ CH_3COOH]{1.\ O_3} R_2C{=}O$$

   c. Friedel–Crafts acylation (Section 16.6)

$$C_6H_6 + RCOCl \xrightarrow{AlCl_3} C_6H_5{-}\overset{O}{\overset{\|}{C}}{-}R$$

   d. Alkyne hydration (Section 7.4)

$$R{-}C{\equiv}CH \xrightarrow[H_3O^+]{Hg^{2+}} R{-}\overset{O}{\overset{\|}{C}}{-}CH_3$$

   e. Diorganocopper reaction with acid chlorides (Sections 22.4 and 24.5)

$$RCOCl + R'_2CuLi \xrightarrow{\text{Ether}} R{-}\overset{O}{\overset{\|}{C}}{-}R'$$

3. Reactions of aldehydes
   a. Oxidation (Section 22.5)

$$R{-}CHO \xrightarrow[\text{or } Ag^+,\ NH_4OH]{\text{Jones' reagent}} R{-}COOH$$

   b. Cannizzaro reaction (Section 22.14)

$$Ar{-}CHO \xrightarrow[2.\ H_3O^+]{1.\ HO^-,\ H_2O} ArCOOH + ArCH_2OH$$

4. Nucleophilic addition reactions of aldehydes and ketones
   a. Addition of hydride: reduction (Section 20.5)

$$R_2C{=}O \xrightarrow[2.\ H_3O^+]{1.\ NaBH_4,\ \text{ethanol}} R_2CH{-}OH$$

   b. Addition of Grignard reagents (Section 20.7)

$$R_2C{=}O \xrightarrow[2.\ H_3O^+]{1.\ R'MgX,\ \text{ether}} R_2C(OH)R'$$

   c. Addition of HCN: cyanohydrins (Section 22.8)

$$R_2C{=}O \underset{}{\overset{HCN}{\rightleftharpoons}} R_2C(OH)CN$$

   d. Addition of primary amines: imines (Section 22.9)

$$R_2C{=}O \overset{R'\ddot{N}H_2}{\rightleftharpoons} R_2C{=}N{-}R' + H_2O$$

For example: oximes, $R_2C{=}N{-}OH$;
semicarbazones, $R_2C{=}N{-}NHCONH_2$;
2,4-dinitrophenylhydrazones, $R_2C{=}N{-}NH{-}C_6H_4(NO_2)_2$

   e. Addition of secondary amines: enamines (Section 22.9)

$$R{-}\overset{O}{\overset{\|}{C}}{-}CH\langle \overset{R'_2\ddot{N}H}{\rightleftharpoons} R{-}\overset{NR'_2}{\overset{|}{C}}{=}C\langle + H_2O$$

   f. Wolff–Kishner reduction (hydrazine addition) (Section 22.10)

$$R_2C{=}O \xrightarrow[KOH]{H_2NNH_2} R_2CH_2 + N_2 + H_2O$$

g. Addition of alcohols: acetals (Section 22.11)

$$R_2C{=}O \xrightleftharpoons{2\ R'OH,\ H^+} R_2C(OR')_2 + H_2O$$

h. Addition of thiols: thioacetals (Section 22.12)

$$R_2C{=}O \xrightleftharpoons{R'SH,\ H^+} R_2C(SR')_2 + H_2O$$

i. Desulfurization of thioacetals with Raney nickel (Section 22.12)

$$R_2C(SR')_2 \xrightarrow[\text{Ethanol}]{\text{Raney Ni}} R_2CH_2 + NiS$$

j. Addition of phosphorus ylides: Wittig reaction (Section 22.13)

$$R_2C{=}O \xrightarrow[\text{THF}]{(C_6H_5)_3\overset{+}{P}{-}\overset{-}{C}R'_2} R_2C{=}CR'_2 + (C_6H_5)_3\overset{+}{P}{-}\overset{-}{O}$$

5. Conjugate additions to α,β-unsaturated ketones and aldehydes (Section 22.15)

a. Addition of HCN (Section 22.15)

$$R{-}\overset{O}{\overset{\|}{C}}{-}\overset{|}{C}{=}C\langle \xrightarrow[2.\ H_3O^+]{1.\ (C_2H_5)_2AlCN} R{-}\overset{O}{\overset{\|}{C}}{-}\overset{|}{C}H{-}\overset{CN}{\overset{|}{\underset{|}{C}}}{-}$$

b. Addition of amines (Section 22.15)

$$R{-}\overset{O}{\overset{\|}{C}}{-}\overset{|}{C}{=}C\langle \xrightleftharpoons{R'\ddot{N}H_2} R{-}\overset{O}{\overset{\|}{C}}{-}\overset{|}{C}H{-}\overset{NHR'}{\overset{|}{\underset{|}{C}}}{-}$$

c. Addition of alkyl groups: diorganocopper reaction (Section 22.15)

$$R{-}\overset{O}{\overset{\|}{C}}{-}\overset{|}{C}{=}C\langle \xrightarrow[2.\ H_3O^+]{1.\ R'_2CuLi,\ ether} R{-}\overset{O}{\overset{\|}{C}}{-}\overset{|}{C}H{-}\overset{R'}{\overset{|}{\underset{|}{C}}}{-}$$

ADDITIONAL PROBLEMS

22.21 Draw structures corresponding to these names:
(a) Bromoacetone
(b) 3,5-Dinitrobenzenecarbaldehyde
(c) 2-Methyl-3-heptanone
(d) 3,5-Dimethylcyclohexanone
(e) 2,2,4,4-Tetramethyl-3-pentanone
(f) 4-Methyl-3-penten-2-one
(g) Butanedial
(h) 3-Phenyl-2-propenal (cinnamaldehyde)
(i) 6,6-Dimethyl-2,4-cyclohexadienone
(j) *p*-Nitroacetophenone
(k) *(S)*-2-Hydroxypropanal
(l) 2,3,4,5-Tetraphenyl-2,4-cyclopentadienone
(m) (2*S*,3*R*)-2,3,4-Trihydroxybutanal (D-threose)

22.22 Draw and name the seven ketones and aldehydes having the formula $C_5H_{10}O$.

22.23 Provide IUPAC names for these structures:

(a) O

CH_3

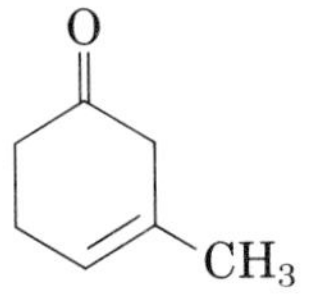

(b) CHO

H—OH

CH_2OH

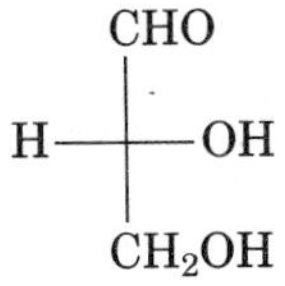

(c) O

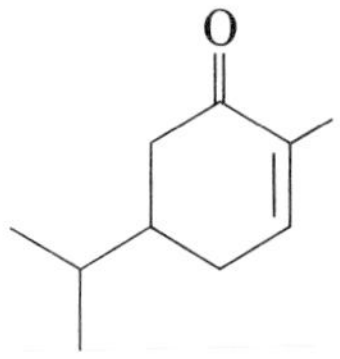

(d) $CH_3CH(CH_3)COCH_2CH_3$

(e) $CH_3CH(OH)CH_2CHO$

(f) CHO

CHO

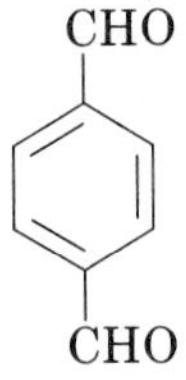

22.24 Give structures for:
(a) An α,β-unsaturated ketone, C_6H_8O
(b) An α-diketone
(c) An aromatic ketone, $C_9H_{10}O$

22.25 Predict the products of the reaction of phenylacetaldehyde with each of the following reagents:
(a) $NaBH_4$, then H_3O^+
(b) Tollens' reagent
(c) Hydroxylamine, H^+
(d) Methylmagnesium bromide, then H_3O^+
(e) Methanol plus acid catalyst
(f) H_2NNH_2/KOH
(g) Methylenetriphenylphosphorane
(h) HCN, KCN
(i) Sodium acetylide, then H_3O^+

22.26 Answer Problem 22.25 for acetophenone.

22.27 How would you carry out the following transformations on 2-cyclohexenone? More than one step may be required.

(a) O $\xrightarrow{?}$

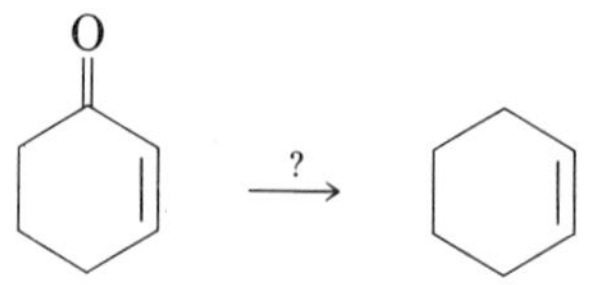

(b) O $\xrightarrow{?}$ O, C_6H_5

(c) O $\xrightarrow{?}$ O, COOH

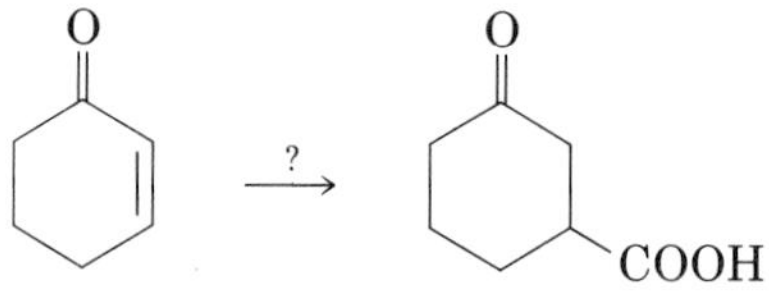

(d) O $\xrightarrow{?}$ CH_3 (two ways)

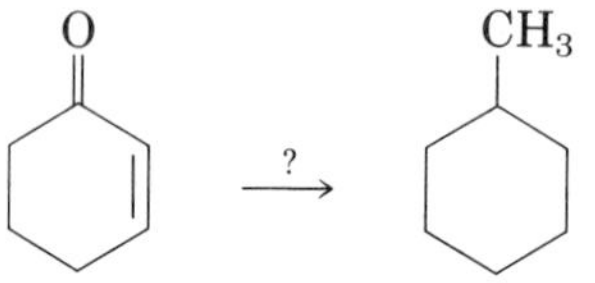

22.28 How can you account for the fact that glucose reacts with the Tollens reagent to give a silver mirror, but glucose α-methyl glycoside does not?

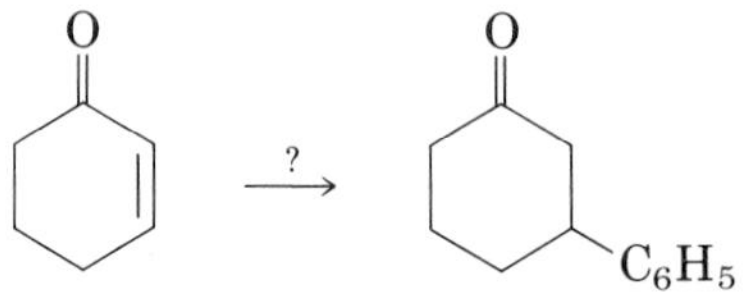

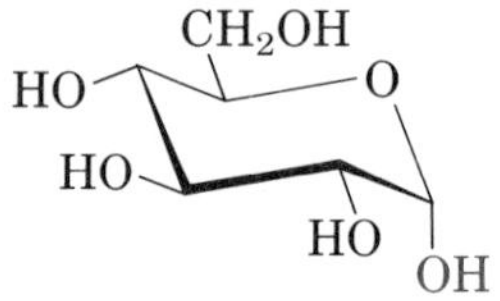

CH_2OH, HO, O, HO, HO, OH

Glucose

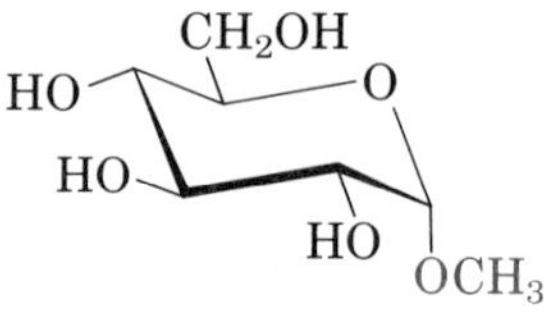

CH_2OH, HO, O, HO, HO, OCH_3

Glucose α-methyl glycoside

22.29 Can you propose a mechanism to account for the formation of 3,5-dimethylpyrazole from hydrazine and 2,4-pentanedione?

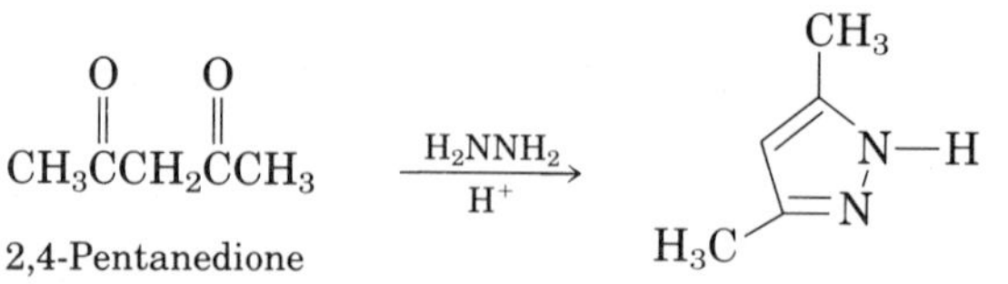

$$CH_3\overset{O}{\overset{\|}{C}}CH_2\overset{O}{\overset{\|}{C}}CH_3 \xrightarrow[H^+]{H_2NNH_2}$$

2,4-Pentanedione

CH_3, N—H, N, H_3C

3,5-Dimethylpyrazole

22.30 In light of your answer to Problem 22.29, can you account for the formation of 3,5-dimethylisoxazole from hydroxylamine and 2,4-pentanedione?

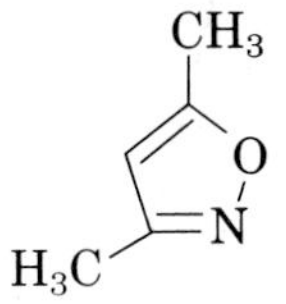

3,5-Dimethylisoxazole

22.31 Show how the Wittig reaction might be used to prepare these alkenes. Identify the alkyl halides and carbonyl components that would be used.

(a) $C_6H_5CH{=}CH{-}CH{=}CHC_6H_5$

(b)

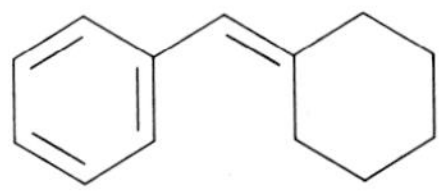

(c) 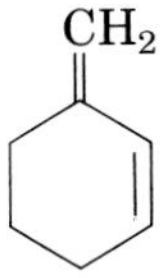 CH_2

(d) 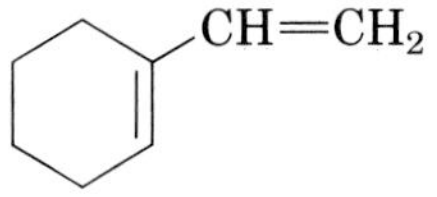 $CH{=}CH_2$

22.32 The Wittig reaction can be used to prepare aldehydes as well as alkenes. This is done by using (methoxymethylene)triphenylphosphorane as the Wittig reagent and hydrolyzing the product with acid. For example,

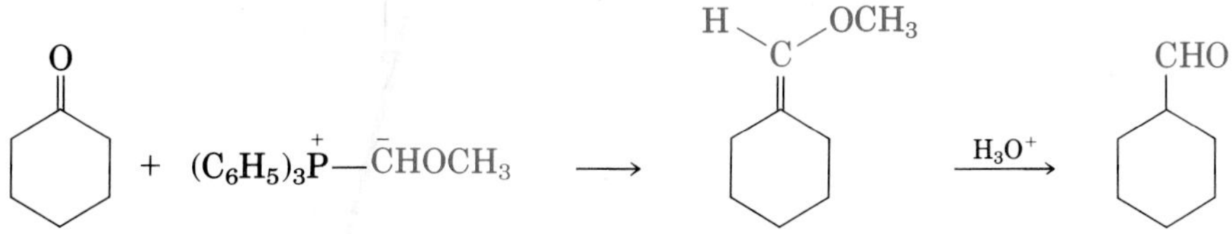

$$+ \ (C_6H_5)_3\overset{+}{P}{-}\overset{-}{C}HOCH_3 \longrightarrow \xrightarrow{H_3O^+}$$

(Methoxymethylene)triphenylphosphorane

(a) How would you prepare the required phosphorane?
(b) Propose a mechanism to account for the hydrolysis step.

22.33 When 4-hydroxybutanal is treated with methanol in the presence of an acid catalyst, 2-methoxytetrahydrofuran is formed. Explain.

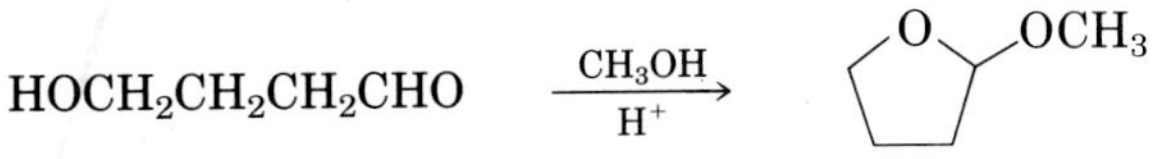

$$HOCH_2CH_2CH_2CHO \xrightarrow[H^+]{CH_3OH}$$

22.34 When crystals of pure α-glucose are dissolved in water, isomerization slowly occurs to produce β-glucose. Propose a mechanism to explain this isomerization.

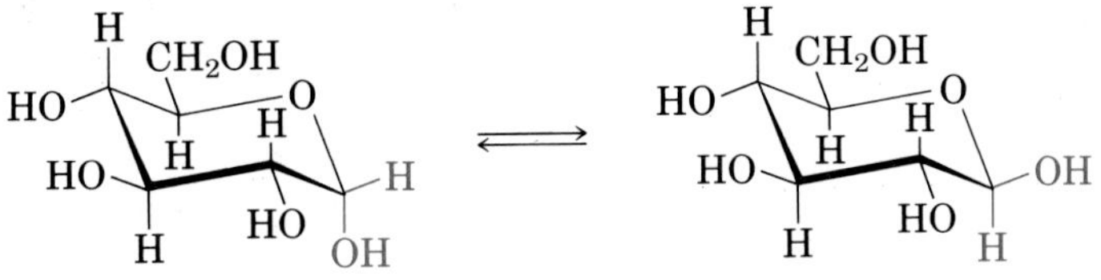

α-Glucose ⇌ β-Glucose

22.35 Give at least four methods for reducing a carbonyl group to a methylene group, $C{=}O \rightarrow CH_2$. What are the advantages and disadvantages of each?

22.36 Carvone is the major constituent of spearmint oil. What products would you expect from reaction of carvone with the following reagents?

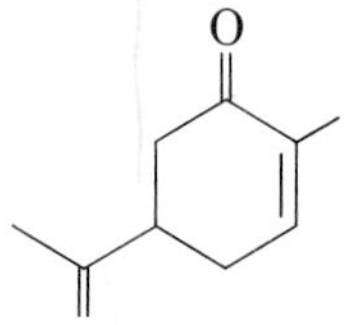

Carvone

(a) $(CH_3)_2Cu^+ Li^-$, then H_3O^+
(b) $LiAlH_4$, then H_3O^+
(c) $(C_2H_5)_2AlCN$, then H_3O^+
(d) CH_3NH_2
(e) C_6H_5MgBr, then H_3O^+
(f) H_2/Pd
(g) Jones' reagent (CrO_3 in H_2O/H_2SO_4)
(h) $(C_6H_5)_3\overset{+}{P}—\overset{-}{C}HCH_3$
(i) $HSCH_2CH_2SH$, then Raney nickel
(j) $HOCH_2CH_2OH$, H^+

22.37 Compound A, mol wt = 86, shows an infrared absorption at 1730 cm^{-1} and a very simple ^1H NMR spectrum with peaks at 9.7 δ (1 H, singlet) and 1.2 δ (9 H, singlet). Propose a structure for A.

22.38 Compound B is isomeric with A (Problem 22.37) and shows an infrared peak at 1720 cm^{-1}. The ^1H NMR spectrum of B has peaks at 2.4 δ (1 H, septet, J = 7 Hz), 2.1 δ (3 H, singlet), and 1.2 δ (6 H, doublet, J = 7 Hz). What is the structure of B?

22.39 At what position would you expect to observe infrared absorptions for the following molecules?

(a) 4-Androstene-3,17-dione

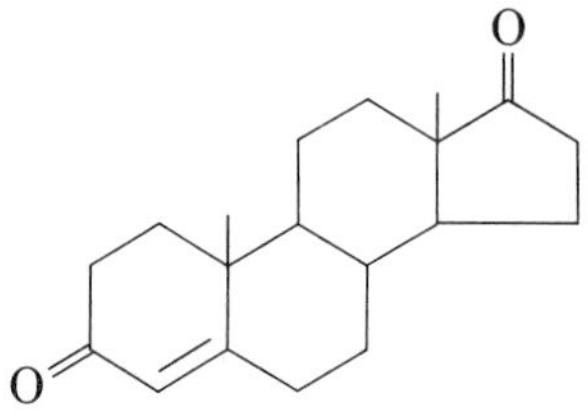

(b) 1-Indanone

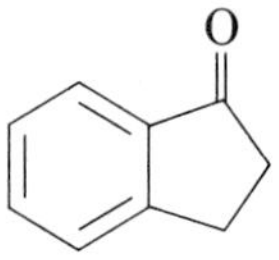

(c) 2-Indanone

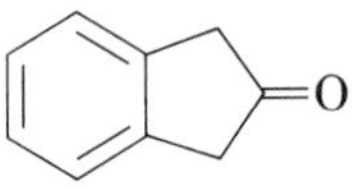

(d) OHC, CH_3, O

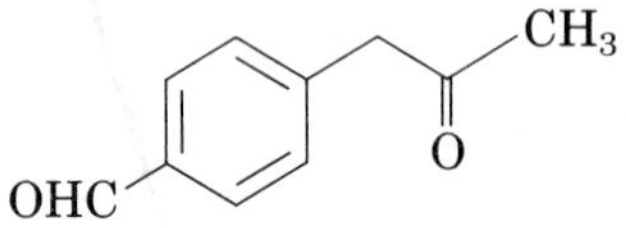

22.40 The Meerwein–Ponndorf–Verley reaction involves reduction of a ketone by treatment with an excess of aluminum triisopropoxide.

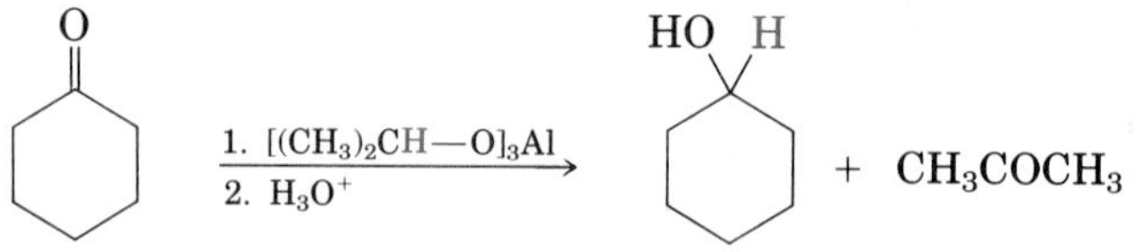

$$\text{cyclohexanone} \xrightarrow[\text{2. } H_3O^+]{\text{1. } [(CH_3)_2CH—O]_3Al} \text{cyclohexanol (HO H)} + CH_3COCH_3$$

If deuterated aluminum isopropoxide is used, deuterium is incorporated into the product. Propose a mechanism for this reaction. What relationship does your proposed mechanism for the Meerwein–Ponndorf–Verley reduction have to the Cannizzaro reaction?

22.41 As written, each of the following reaction schemes contains one or more flaws. What is wrong in each case? How would you correct each scheme?

(a) HO–(bicyclic enol) $\xrightarrow{\overset{+}{Ag},\ NH_4OH}$ O=(bicyclic enone) $\xrightarrow[2.\ H_3O^+]{1.\ CH_3MgBr}$ O=(bicyclic ketone with angular CH_3)

(b) $C_6H_5CH{=}CHCH_2OH \xrightarrow[\text{reagent}]{\text{Jones'}} C_6H_5CH{=}CHCHO$

$\xrightarrow{HOCH_2CH_2OH,\ H^+} C_6H_5CH{=}CHCH$(O–CH₂CH₂–O, cyclic acetal)

(c) $CH_3COCH_3 \xrightarrow[KCN]{HCN}$ $CH_3C(OH)(CN)CH_3 \xrightarrow{H_3O^+}$ $CH_3C(OH)(CH_2NH_2)CH_3$

(d) Cyclohexanone $\xrightarrow{H_2N{-}NH_2}$ cyclohexanone hydrazone ($C{=}N{-}NH_2$) $\xrightarrow{\text{Raney Ni}}$ cyclohexane (H H)

22.42 6-Methyl-5-hepten-2-one is a common constituent of many essential oils, particularly the lemongrass species. How could you synthesize this natural product from methyl 4-oxopentanoate?

$$CH_3\overset{O}{\overset{\|}{C}}CH_2CH_2CO_2CH_3$$

Methyl 4-oxopentanoate

22.43 The NMR spectrum shown is that of a compound with formula $C_9H_{10}O$. How many double bonds and/or rings does this compound contain? If the unknown has an infrared absorption at 1690 cm^{-1}, what is the likely structure?

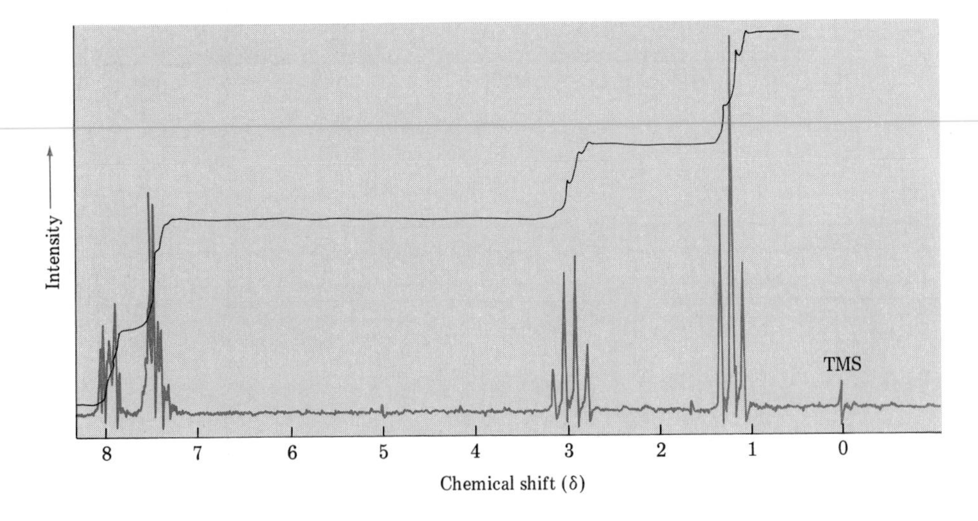

22.44 The NMR spectrum shown is that of a compound isomeric with that of Problem 22.43. This isomer has an infrared absorption at 1725 cm^{-1}. Propose a suitable structure.

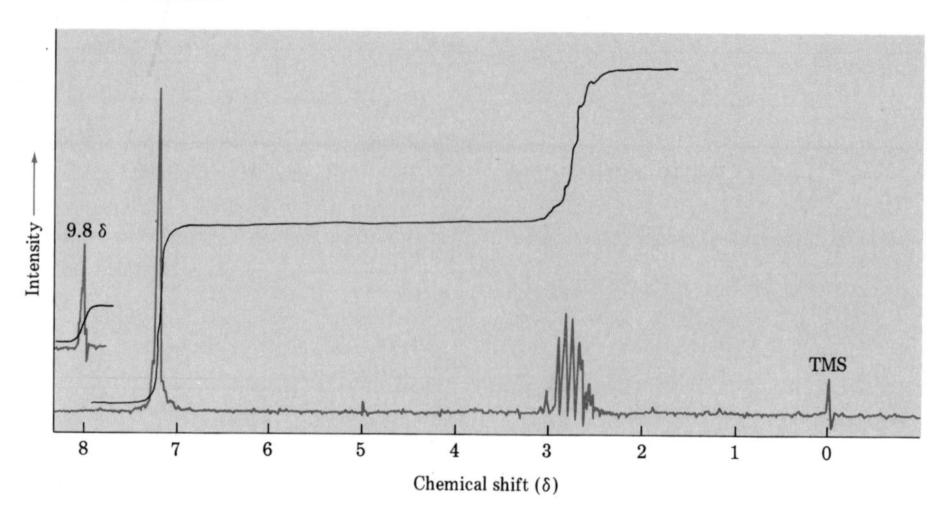

22.45 Compound A, $C_8H_{10}O_2$, has an intense infrared absorption at 1750 cm^{-1} and gives the ^{13}C NMR spectrum shown. Propose a suitable structure for A.

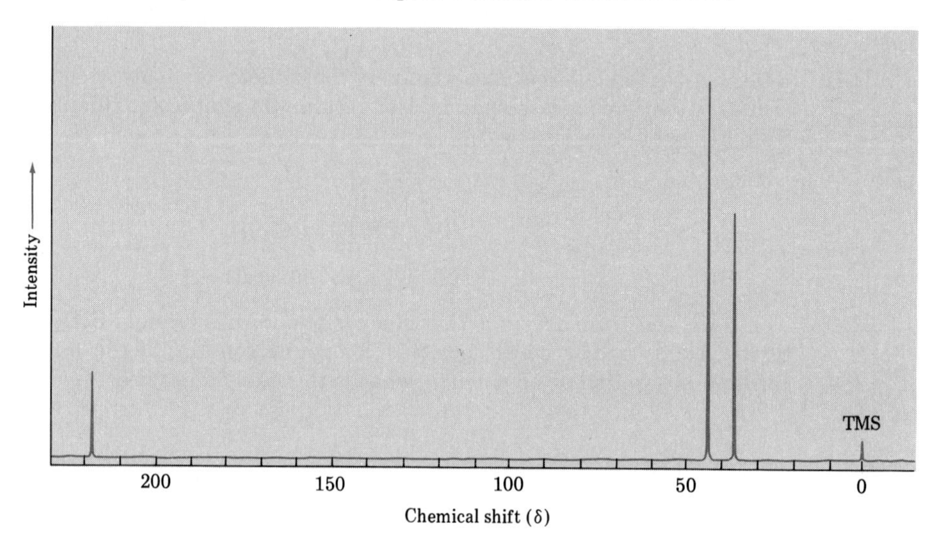

CHAPTER 23 CARBOXYLIC ACIDS

Carboxylic acids occupy a central place among acyl derivatives. Not only are they important compounds themselves, but they also serve as building blocks for preparing related acyl derivatives such as esters and amides. For example, cholic acid is a major component of human bile, and long-chain aliphatic acids such as oleic acid and linoleic acid are biological precursors of fats and other lipids (Figure 23.1).

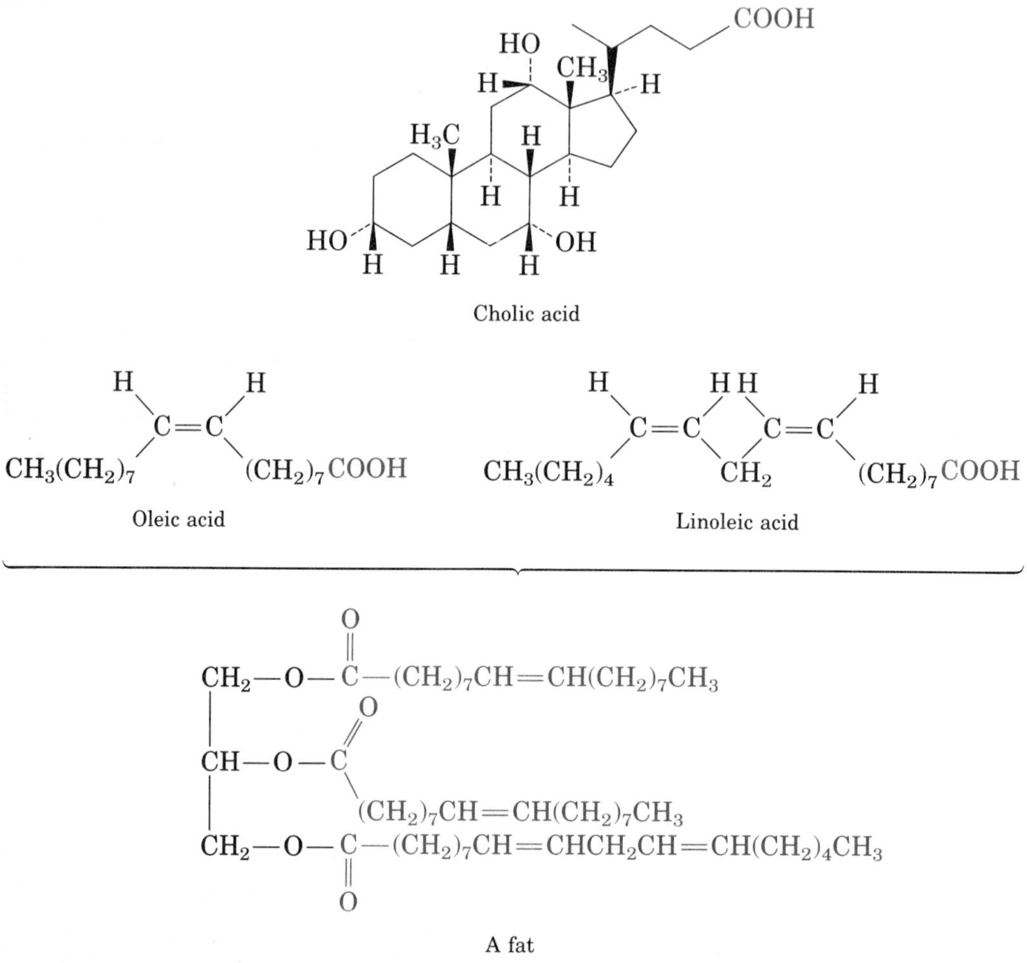

Cholic acid

Oleic acid

Linoleic acid

A fat

Figure 23.1. Some carboxylic acids. Shown are cholic acid, a bile acid, and oleic and linoleic acids, which are fatty acids.

Most of the simpler saturated carboxylic acids are also found in nature. For example, vinegar is a dilute aqueous solution of acetic acid, CH_3COOH; butanoic acid, $CH_3CH_2CH_2COOH$, is responsible for the rancid odor of sour butter; and hexanoic acid (caproic acid), $CH_3(CH_2)_4COOH$, is partially responsible for the unmistakable aroma of goats (Latin *caper,* "goat"). Some

3.3 billion lb/yr of acetic acid are produced industrially for a variety of purposes, including use as a raw material for preparing vinyl acetate polymers used in paints and adhesives. The industrial method of acetic acid synthesis involves a cobalt acetate catalyzed air oxidation of acetaldehyde, but this method is not used in the laboratory.

$$CH_3CHO + O_2 \xrightarrow[80°C]{\text{Cobalt acetate}} CH_3COOH + H_2O$$

The Monsanto Company has developed an even more efficient synthesis involving a direct rhodium catalyzed carbonylation of methanol.

$$CH_3OH + CO \xrightarrow{\text{Rh catalyst}} CH_3COOH$$

23.1 Nomenclature of carboxylic acids

The IUPAC rules allow for several different systems of nomenclature, depending on the complexity of the acid molecule. Carboxylic acids that are derived from alkanes by replacing a methyl group with a carboxyl (COOH) group are systematically named by replacing the terminal *-e* of the corresponding alkane name with *-oic acid*. Note that the carboxyl carbon atom is always numbered C1.

$$\underset{6}{CH_3}\underset{5}{CH_2}\underset{4}{CH_2}\underset{3}{CH_2}\underset{2}{CH_2}\underset{1}{COOH}$$

Hexanoic acid
(caproic acid)

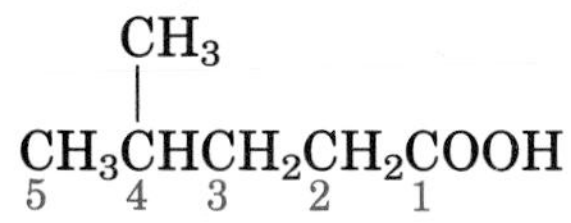

$$\underset{5}{CH_3}\underset{4}{\overset{\overset{CH_3}{|}}{CH}}\underset{3}{CH_2}\underset{2}{CH_2}\underset{1}{COOH}$$

4-Methylpentanoic acid

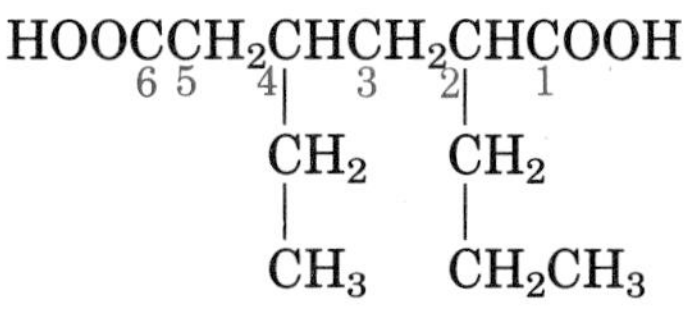

$$\underset{6}{HOOC}\underset{5}{CH_2}\underset{\underset{\underset{CH_3}{|}}{\underset{CH_2}{|}}}{\overset{4}{CH}}\underset{3}{CH_2}\underset{\underset{\underset{CH_2CH_3}{|}}{\underset{CH_2}{|}}}{\overset{2}{CH}}\underset{1}{COOH}$$

4-Ethyl-2-propylhexanedioic acid

Alternatively, compounds can be named by using the suffix *-carboxylic acid*. Any carboxylic acid can be named by using this method, but the alkanoic acid system just discussed is preferred for simple aliphatic acids. Note that in this alternative system, the carboxylic carbon is attached to C1 and is not itself numbered.

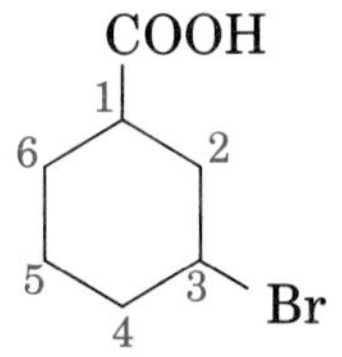

3-Bromocyclohexanecarboxylic acid

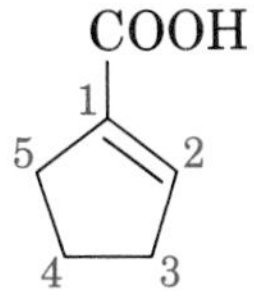

1-Cyclopentenecarboxylic acid

The IUPAC nomenclature rules also make allowance for a large number of well-entrenched trivial names, some of which are given in Table 23.1. Also listed in Table 23.1 are the trivial names used for acyl groups when the RCO— function is considered as a substituent.

TABLE 23.1 Some trivial names of common carboxylic acids and acyl radicals

Carboxylic acid		*Acyl group*	
Structure	*Name*	*Name*	*Structure*
HCOOH	Formic	Formyl	HCO—
CH_3COOH	Acetic	Acetyl	CH_3CO—
CH_3CH_2COOH	Propionic	Propionyl	CH_3CH_2CO—
$CH_3CH_2CH_2COOH$	Butyric	Butyryl	$CH_3(CH_2)_2CO$—
$CH_3CH_2CH_2CH_2COOH$	Valeric	Valeryl	$CH_3(CH_2)_3CO$—
$(CH_3)_3CCOOH$	Pivalic	Pivaloyl	$(CH_3)_3CCO$—
HOOCCOOH	Oxalic	Oxalyl	—OCCO—
$HOOCCH_2COOH$	Malonic	Malonyl	—$OCCH_2CO$—
$HOOCCH_2CH_2COOH$	Succinic	Succinyl	—$OC(CH_2)_2CO$—
$HOOCCH_2CH_2CH_2COOH$	Glutaric	Glutaryl	—$OC(CH_2)_3CO$—
$HOOCCH_2CH_2CH_2CH_2COOH$	Adipic	Adipoyl	—$OC(CH_2)_4CO$—
$H_2C{=}CHCOOH$	Acrylic	Acryloyl	$H_2C{=}CHCO$—
$CH_3CH{=}CHCOOH$	Crotonic	Crotonoyl	$CH_3CH{=}CHCO$—
$H_2C{=}C(CH_3)COOH$	Methacrylic	Methacryloyl	$H_2C{=}C(CH_3)CO$—
$HC{\equiv}CCOOH$	Propiolic	Propioloyl	$HC{\equiv}CCO$—
$HOOCCH{=}CHCOOH$	*cis*-Maleic	Maleoyl	—$OCCH{=}CHCO$—
	trans-Fumaric	Fumaroyl	
C_6H_5COOH (benzene ring with COOH)	Benzoic	Benzoyl	$C_6H_5C(=O)$— (benzene ring with C(=O)—)
$C_6H_4(COOH)_2$ (benzene ring with two adjacent COOH)	Phthalic	Phthaloyl	$C_6H_4(C(=O)\text{—})_2$ (benzene ring with two adjacent C(=O)—)

PROBLEM

23.1 Provide IUPAC names for these compounds:
(a) $(CH_3)_2CHCH_2COOH$ (b) $CH_3CHBrCH_2CH_2COOH$

(c) 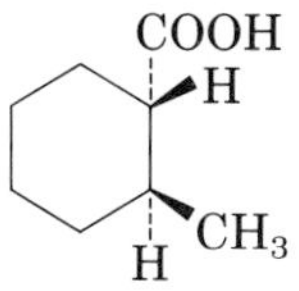 COOH, H, CH_3, H

(d) $CH_3CH{=}CHCH{=}CHCOOH$

(e) 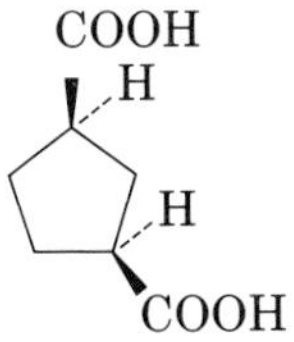 COOH, H, H, COOH

(f) 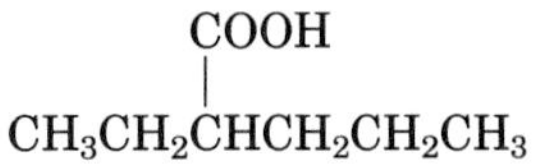 $CH_3CH_2CH(COOH)CH_2CH_2CH_3$

(g) 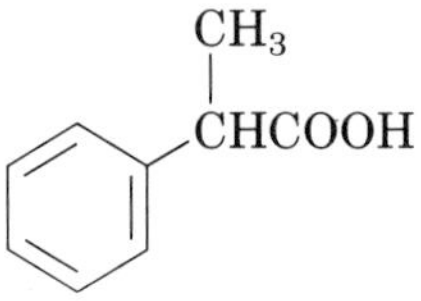 CH_3 CHCOOH

PROBLEM

23.2 Draw structures corresponding to these IUPAC names:
(a) 2,3-Dimethylhexanoic acid
(b) *trans*-1,2-Cyclobutanedicarboxylic acid
(c) (9*Z*,12*Z*)-Octadecadienoic acid (linoleic acid)
(d) *o*-Hydroxybenzoic acid (salicylic acid)
(e) 4-Methylpentanoic acid

23.2 Structure and physical properties of carboxylic acids

Carboxylic acid functional groups are structurally related to both ketones and alcohols, and we might therefore expect to see some familiar properties. In fact, carboxylic acids are indeed similar to both ketones and alcohols in some ways, but there are major differences. Like ketones, the carboxyl carbon is sp^2 hybridized, and carboxylic acid groups are therefore planar, with C—C—O and O—C—O bond angles of approximately 120°. The physical parameters of acetic acid are given in Table 23.2.

TABLE 23.2 **Structure of acetic acid**

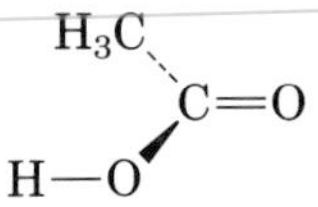

H_3C, C=O, H—O

Bond angle (degrees)		*Bond length (Å)*	
C—C=O	119	C—C	1.52
C—C—OH	119	C=O	1.25
O=C—OH	122	C—OH	1.31

Like alcohols, carboxylic acids are strongly associated because of intermolecular hydrogen bonding. Studies have shown that most carboxylic acids exist as dimers held together by two hydrogen bonds:

$$\mathrm{R{-}C(=O\cdots H{-}O){-}O{-}H\cdots O{=}C(O){-}R}$$

A carboxylic acid dimer

This strong hydrogen bonding has a noticeable effect on boiling points; carboxylic acids normally boil at much higher temperatures than the corresponding alcohols. Table 23.3 lists the observed properties of some common acids.

TABLE 23.3 Physical constants of some carboxylic acids

Structure	*Name*	*Melting point (°C)*	*Boiling point (°C)*
$HCOOH$	Formic	8.4	100.5
CH_3COOH	Acetic	16.6	118
CH_3CH_2COOH	Propanoic	−22	141
$CH_3CH_2CH_2COOH$	Butanoic	−4.2	163
$CH_3CH_2CH_2CH_2COOH$	Pentanoic	−34.5	187
$(CH_3)_3CCOOH$	2,2-Dimethylpropanoic	35.3	164
FCH_2COOH	Fluoroacetic	35.2	165
$BrCH_2COOH$	Bromoacetic	50	208
$HOCH_2COOH$	Glycolic	80	Decomposes
(±)-$CH_3CH(OH)COOH$	Lactic	18	122
$H_2C{=}CHCOOH$	Propenoic	13	141
$CH_3CH{=}CHCOOH$	*(E)*-2-Butenoic	71.5	185
C_6H_5COOH	Benzoic	122.4	249
$COOCCOOH$	Oxalic	189.5	Decomposes
$HOOCCH_2COOH$	Malonic	135	Decomposes
$HOOCCH_2CH_2COOH$	Succinic	188	Decomposes
(E)-$HOOCCH{=}CHCOOH$	Maleic	139	Decomposes

23.3 Dissociation of carboxylic acids

As the name implies, carboxylic acids are acidic; they therefore react with strong bases such as sodium hydroxide to give salts.

$$\mathrm{R\overset{O}{\overset{\|}{C}}{-}O{-}H + NaOH \xrightarrow{H_2O} R{-}\overset{O}{\overset{\|}{C}}{-}O^- \, Na^+ + H_2O}$$

Carboxylic acids with more than six carbon atoms are only slightly soluble in water; alkali metal salts of carboxylic acids, however, are generally quite water-soluble because of their ionic nature. We can often take advantage of this solubility to purify acids by extracting their salts into aqueous base, then reacidifying and extracting the pure acid back into an organic solvent.

Like the other Brønsted–Lowry acids that we discussed in Section 2.6, carboxylic acids dissociate slightly in dilute aqueous solution to give H_3O^+ and carboxylate anion, $RCOO^-$:

$$RCOOH + H_2O \rightleftharpoons RCOO^- + H_3O^+$$

For this reaction, we have

$$K_a = \frac{[RCOO^-][H_3O^+]}{[RCOOH]} \qquad pK_a = -\log K_a$$

Note that the concentration of water remains constant and does not appear in the equilibrium expression. The equilibrium constant for dissociation is the acidity constant, K_a, and for most carboxylic acids K_a is on the order of 10^{-5}. Acetic acid, for example, has $K_a = 1.8 \times 10^{-5}$, which corresponds to a pK_a of 4.72. In practical terms, K_a values near 10^{-5} mean that only about 1% of the molecules in a 0.1*M* solution are dissociated, as opposed to the 100% dissociation observed for strong mineral acids such as HCl and H_2SO_4.

Although much weaker than mineral acids, carboxylic acids are nevertheless much stronger acids than alcohols. For example, the K_a for ethanol is approximately 10^{-16}, making ethanol a weaker acid than acetic acid by a factor of 10^{11}.

H—Cl	$CH_3C(=O)$—O—H	CH_3CH_2O—H
$pK_a = -7$	$pK_a = 4.72$	$pK_a = 16$

Strong acid ← Weak acid

Why do carboxylic acids dissociate more readily than alcohols? The easiest way to answer this question is to look at the relative stability of carboxylate anions versus alkoxide anions.

Alkoxides are oxygen anions in which the negative charge is localized and is stabilized only by the electronegative oxygen atom:

$$CH_3CH_2\ddot{O}—H \rightleftharpoons CH_3CH_2—\ddot{O}:^- + H^+$$

Alcohol — Unstabilized alkoxide ion

Carboxylates are also oxygen anions, but the negative charge can be delocalized or spread out over both oxygen atoms, resulting in stabilization of the ion. In resonance terms (Section 9.6), a carboxylate ion is a stabilized resonance hybrid of two equivalent Kekulé structures:

$$CH_3C(=O)\ddot{O}H \rightleftharpoons CH_3{-}C(=O){-}\ddot{O}:^- \longleftrightarrow CH_3{-}C(-\ddot{O}:^-)(=O) + H^+$$

Carboxylic acid

Resonance-stabilized carboxylate ion
(Two equivalent resonance forms)

Carboxylic acids are more acidic than alcohols because resonance stabilization of carboxylate anions favors dissociation. Figure 23.2 compares reaction energy diagrams for the two dissociation reactions.

We saw in Section 4.6 that the equilibrium constant for a given reaction is related to the free-energy difference between reactant and product by the equation $\Delta G^\circ = -RT \ln K_{eq}$. Thus, in comparing two reactions, the reaction with the smaller equilibrium constant has the larger ΔG°. Conversely, the reaction with the larger equilibrium constant has the smaller ΔG°. Dis-

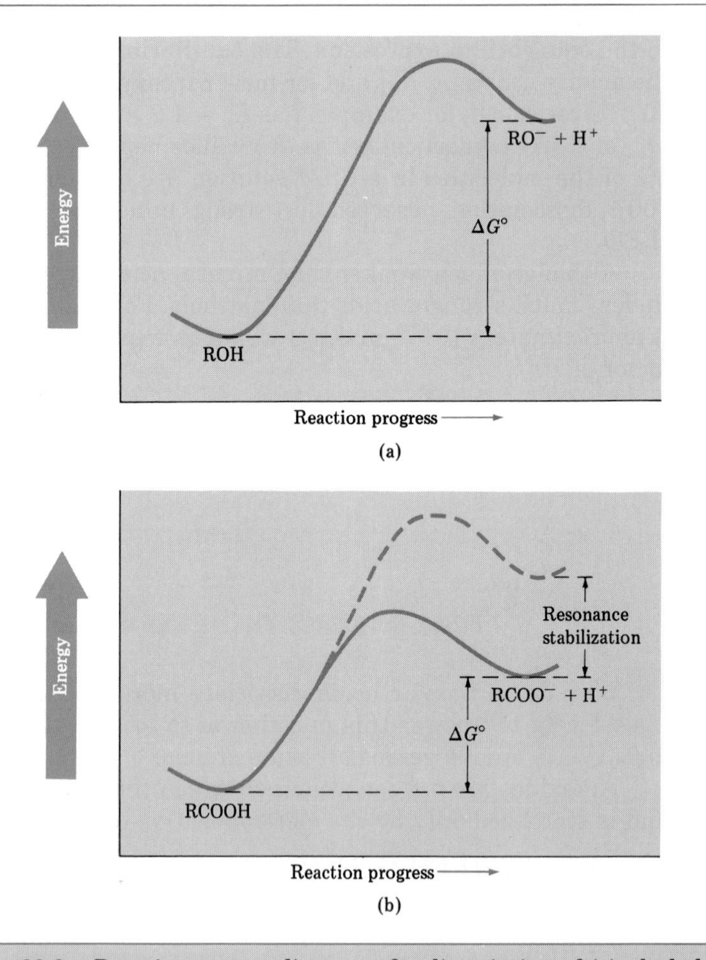

Figure 23.2. Reaction energy diagrams for dissociation of (a) alcohols and (b) carboxylic acids. Resonance stabilization of carboxylate anions lowers ΔG° for the dissociation of carboxylic acids. In the absence of resonance, ΔG° for the two reactions would be similar.

sociation of an alcohol ($K_a = 10^{-16}$) has a large $\Delta G°$ because alkoxide ion is much less stable than undissociated alcohol. Dissociation of a carboxylic acid ($K_a = 10^{-5}$), however, has a smaller $\Delta G°$ because resonance stabilization lowers the energy of the carboxylate anion relative to undissociated carboxylic acid.

We can't really draw an accurate representation of the carboxylate resonance hybrid using Kekulé structures, although we might show one-half negative charge on each oxygen and use dotted lines between carbon and oxygen to indicate one-and-one-half bonds:

$$CH_3-C\langle^{O^{\frac{1}{2}-}}_{O^{\frac{1}{2}-}}$$

Carboxylate resonance hybrid

A molecular orbital picture of acetate ion is more helpful in making it clear that the carbon–oxygen bonds are equivalent and that each is intermediate between single and double bonds (Figure 23.3). The *p* orbital on the carboxylate carbon atom overlaps equally well with *p* orbitals from *both* oxygens, and the four *p* electrons are delocalized throughout the three-center pi electron system.

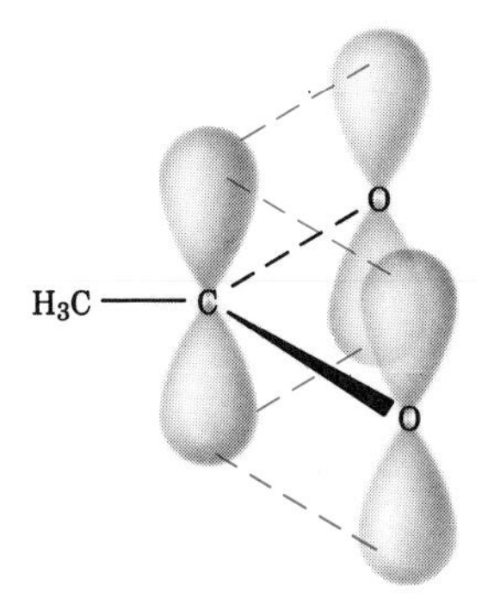

Figure 23.3. Orbital picture of acetate ion showing equivalence of oxygen atoms.

Physical evidence for the equivalence of the two carboxylate oxygens has been provided by X-ray studies on sodium formate. Both carbon–oxygen bonds are 1.27 Å in length, midway between the C=O bond (1.20 Å) and C—O bond (1.34 Å) of formic acid.

Sodium formate: H—C(O)(O)$^-$ Na^+ (both C–O bonds 1.27Å)

Formic acid: H—C(=O)—O—H (C=O 1.20Å; C—O 1.34Å)

Sodium formate　　　Formic acid

PROBLEM

23.3 Use the expression $\Delta G^\circ = -RT \ln K_a = -2.303\,RT \log K_a$ to calculate values of ΔG° for the dissociation of ethanol ($K_a \approx 10^{-16}$) and of acetic acid ($K_a \approx 10^{-5}$). Which reaction is favored at 25°C?

PROBLEM

23.4 Assume that you had a mixture of naphthalene and benzoic acid that you wished to separate. Tell how you might take advantage of the acidity of one component in the mixture to effect a simple separation.

PROBLEM

23.5 The K_a for dichloroacetic acid is 5.5×10^{-2}. Approximately what percentage of the acid is dissociated in a 0.1M aqueous solution?

23.4 Substituent effects on acidity

The pK_a values listed in Table 23.4 indicate that there is a considerable acidity range observed for different carboxylic acids. For example, trifluoroacetic acid ($K_a = 0.59$) is more than 32,000 times as strong as acetic acid ($K_a = 1.8 \times 10^{-5}$). How can we account for such differences?

TABLE 23.4 **Acidity of some carboxylic acids**

Structure	K_a	pK_a	
H—Cl (hydrochloric acid)[a]	(10^7)	(−7)	
F_3CCOOH	0.59	0.23	
Cl_3CCOOH	0.23	0.64	Strong acid ↑
$Cl_2CHCOOH$	5.5×10^{-2}	1.26	
FCH_2COOH	2.6×10^{-3}	2.59	
$ClCH_2COOH$	1.4×10^{-3}	2.85	
$BrCH_2COOH$	1.3×10^{-3}	2.89	
ICH_2COOH	7.5×10^{-4}	3.12	
HCOOH	1.77×10^{-4}	3.75	
$HOCH_2COOH$	1.5×10^{-4}	3.83	
$ClCH_2CH_2COOH$	1.04×10^{-4}	3.98	
C_6H_5COOH	6.46×10^{-5}	4.19	
$H_2C{=}CHCOOH$	5.6×10^{-5}	4.25	
$C_6H_5CH_2COOH$	4.9×10^{-5}	4.31	Weak acid
CH_3COOH	1.8×10^{-5}	4.72	
CH_3CH_2COOH	1.34×10^{-5}	4.87	
CH_3CH_2O—H (ethanol)[a]	(10^{-16})	(16)	

[a]Values for hydrochloric acid and ethanol are shown for reference.

Since the dissociation of a carboxylic acid is an equilibrium process, any factor that stabilizes carboxylate anion relative to undissociated carboxylic acid should drive the equilibrium toward increased dissociation and result in increased acidity. Conversely, any factor that destabilizes carboxylate relative to undissociated acid should result in decreased acidity. For example, an electron-withdrawing group attached to the carboxyl group should inductively withdraw electron density, thus stabilizing the carboxylate anion and increasing acidity. An electron-donating group, however, should have exactly the opposite effect by destabilizing the carboxylate and decreasing acidity. (This discussion assumes that the effect of the substituent on the stability of undissociated carboxylic acid is smaller than its effect on the stability of the carboxylate anion. Remember: It is the *difference* between carboxylate anion and carboxylic acid stabilities that is important.)

$$\text{(EWG)} \leftarrow \overset{\overset{\displaystyle O}{\|}}{C} - O^- \qquad \qquad \text{(EDG)} \rightarrow \overset{\overset{\displaystyle O}{\|}}{C} - O^-$$

Electron-withdrawing group stabilizes carboxylate and strengthens acid

Electron-donating group destabilizes carboxylate and weakens acid

The pK_a data of Table 23.4 show this predicted effect. Highly electronegative substituents such as the halogens tend to make the carboxylate anion more stable by inductively withdrawing electrons. Thus fluoroacetic, chloroacetic, bromoacetic, and iodoacetic acids are all stronger than acetic acid by factors of 50–150. Introduction of two electronegative substituents makes dichloroacetic acid some 3000-fold stronger than acetic acid, and trichloroacetic acid is more than 12,000 times stronger (Figure 23.4).

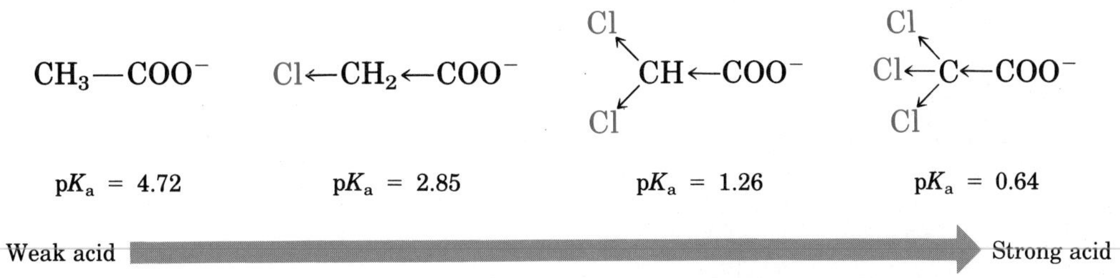

$CH_3—COO^-$	$Cl \leftarrow CH_2 \leftarrow COO^-$	$Cl_2CH \leftarrow COO^-$ (each Cl → CH)	$Cl_3C \leftarrow COO^-$ (each Cl → C)
$pK_a = 4.72$	$pK_a = 2.85$	$pK_a = 1.26$	$pK_a = 0.64$

Weak acid → Strong acid

Figure 23.4. Relative strengths (pK_a values) of chloro-substituted acetic acids.

Inductive effects are strongly dependent on distance, and the effect of halogen substitution therefore decreases as the substituent is moved farther from the carboxyl. The chlorobutanoic acids show clearly what happens as the electronegative substituent is moved successively farther from the carbonyl group (Table 23.5): 2-chlorobutanoic acid has a pK_a of 2.86, the 3-substituted acid has a pK_a of 4.05, and the 4-substituted acid, with a pK_a of 4.52, has an acidity similar to that of butanoic acid itself.

TABLE 23.5 Acidity of chloro-substituted butanoic acids

Structure	K_a	pK_a
$CH_3CH_2CH(Cl)COOH$	1.39×10^{-3}	2.86
$CH_3CH(Cl)CH_2COOH$	8.9×10^{-5}	4.05
$ClCH_2CH_2CH_2COOH$	3×10^{-5}	4.52
$CH_3CH_2CH_2COOH$	1.5×10^{-5}	4.82

PROBLEM

23.6 Rank the acids in the following groups in order of increasing acidity, without looking at a table of pK_a values:

(a) CH_3CH_2COOH, $BrCH_2COOH$, FCH_2COOH

(b) benzoic acid (COOH on benzene ring), p-cyanobenzoic acid (COOH, CN), p-hydroxybenzoic acid (COOH, OH)

(c) CH_3CH_2OH, $CH_3CH_2NH_2$, CH_3CH_2COOH

PROBLEM

23.7 How can you account for the fact that the first ionization constant of oxalic acid (HOOC—COOH) has a pK_1 of 1.2, and the second ionization constant has a pK_2 of 4.2? Why is the second carboxyl group so much less acidic than the first?

PROBLEM

23.8 Shown here are some pK_a data for simple dibasic acids. How do you account for the fact that the difference between the first and second ionization constants decreases with increasing distance between the carboxyl groups?

Name	*Structure*	pK_1	pK_2
Oxalic	HOOC—COOH	1.2	4.2
Succinic	$HOOC—CH_2CH_2—COOH$	4.2	5.6
Adipic	$HOOC—(CH_2)_4—COOH$	4.4	5.4

23.5 Substituent effects in substituted benzoic acids

We saw earlier during our discussion of electrophilic aromatic substitution (Section 15.8) that substituents on the aromatic ring play a large role in determining reactivity; aromatic rings with electron-donating groups are

TABLE 23.6 **Substituent effects on acidity of para-substituted benzoic acids**

X—C₆H₄—COOH

	X	K_a	pK_a	
Weak acid	—OH	2.8×10^{-5}	4.55	Activating groups
	—OCH_3	3.5×10^{-5}	4.46	
	—CH_3	4.3×10^{-5}	4.34	
↓	—H	6.46×10^{-5}	4.19	
	—Br	1.1×10^{-4}	3.96	Deactivating groups
	—Cl	1.1×10^{-4}	3.96	
Strong acid	—CHO	1.8×10^{-4}	3.75	
	—CN	2.8×10^{-4}	3.55	
	—NO_2	3.9×10^{-4}	3.41	

activated toward further electrophilic substitution, and aromatic rings with electron-withdrawing groups are deactivated. We notice exactly the same effect on the acidity of substituted benzoic acids. As Table 23.6 shows, electron-withdrawing (deactivating) groups increase acidity by stabilizing the carboxylate relative to undissociated carboxylic acid, and electron-donating (activating) groups decrease acidity by destabilizing the carboxylate anion. Thus, an activating group such as *p*-methoxy decreases the acidity of benzoic acid, but a deactivating group such as *p*-nitro increases the acidity.

COOH / OCH_3 — *p*-Methoxybenzoic acid (pK_a = 4.46)

COOH — Benzoic acid (pK_a = 4.19)

COOH / NO_2 — *p*-Nitrobenzoic acid (pK_a = 3.41)

It is much easier to measure the acidity of a substituted benzoic acid than it is to determine the relative electrophilic reactivity of a substituted benzene. For this reason, the correlation between the two effects can be valuable in predicting reactivity. If we want to know the effect of a certain substituent on electrophilic reactivity, we can simply find the acidity of the corresponding benzoic acid.

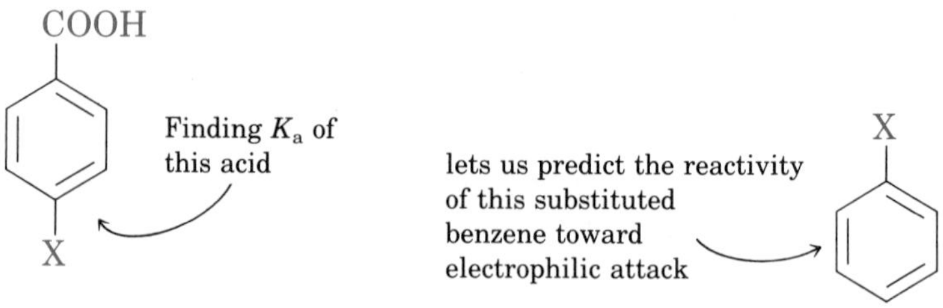

COOH

X

Finding K_a of this acid

lets us predict the reactivity of this substituted benzene toward electrophilic attack

X

PROBLEM

23.9 The K_a of *p*-(trifluoromethyl)benzoic acid has been measured as 2.2×10^{-4}. Would you expect the trifluoromethyl substituent to be an activating or deactivating group in the Friedel–Crafts reaction? Explain.

PROBLEM

23.10 Rank the following compounds in order of increasing acidity. Do not look at a table of pK_a data to help with your answer.
(a) Benzoic acid, *p*-methylbenzoic acid, *p*-chlorobenzoic acid
(b) *p*-Nitrobenzoic acid, acetic acid, benzoic acid

23.6 Preparation of carboxylic acids

We have already seen most of the common methods for preparing carboxylic acids, but let's review them briefly:

1. Oxidation of substituted alkylbenzenes with potassium permanganate or sodium dichromate gives substituted benzoic acids (Section 16.9). Primary and secondary alkyl groups may be oxidized in this manner, but tertiary groups are not affected.

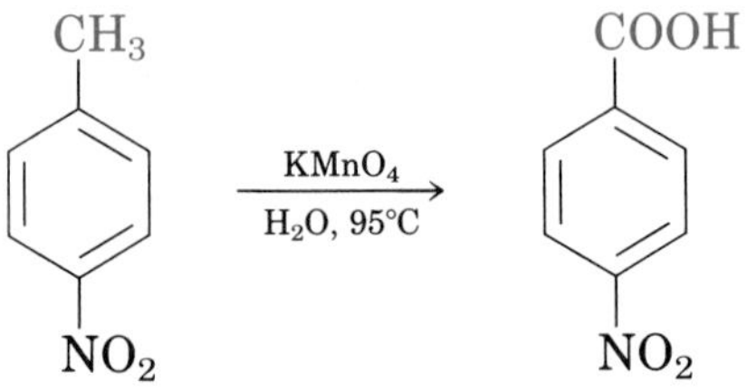

CH_3 / NO_2 $\xrightarrow[H_2O,\ 95^\circ C]{KMnO_4}$ COOH / NO_2

p-Nitrotoluene

p-Nitrobenzoic acid (88%)

2. Oxidative cleavage of alkenes gives carboxylic acids if the alkene has at least one vinylic hydrogen (Section 6.8). This oxidation can be carried out with sodium dichromate, with potassium permanganate, or with ozone, but one particularly useful method employs a mixture of sodium periodate and a small amount of potassium permanganate. The cleavage reaction takes place in good yields under mild conditions.

$$CH_3(CH_2)_7CH{=}CH(CH_2)_7COOH \xrightarrow[H_2O,\ K_2CO_3]{NaIO_4,\ KMnO_4} CH_3(CH_2)_7COOH + HOOC(CH_2)_7COOH$$

Oleic acid → Nonanoic acid + Nonanedioic acid

3. Oxidation of primary alcohols and aldehydes yields carboxylic acids (Sections 20.10 and 22.5). Primary alcohols are often oxidized with Jones' reagent (CrO_3, H_2O, H_2SO_4), and aldehydes are oxidized either with Jones' reagent or with basic silver oxide (Tollens' reagent). Both oxidations take place rapidly and in high yield.

$$CH_3(CH_2)_8CH_2OH \xrightarrow[\text{reagent}]{\text{Jones'}} CH_3(CH_2)_8COOH$$

1-Decanol → Decanoic acid (93%)

$$CH_3(CH_2)_4CHO \xrightarrow[\text{reagent}]{\text{Tollens'}} CH_3(CH_2)_4COOH$$

Hexanal → Hexanoic acid (85%)

HYDROLYSIS OF NITRILES

Nitriles, R—C≡N, can be hydrolyzed by strong aqueous acid or base to yield carboxylic acids. Since nitriles themselves are most often prepared by S_N2 reaction between an alkyl halide and cyanide ion, the two-step sequence of cyanide displacement followed by nitrile hydrolysis is an excellent method for preparing carboxylic acids from alkyl halides. The method works best with primary halides, since competitive E2 elimination reactions can occur when secondary and tertiary alkyl halides are used. Nevertheless, unhindered secondary halides react well, and the method is used commercially for synthesis of the antiarthritic drug, fenoprofen. Note that this method yields a carboxylic acid product having one more carbon than the starting alkyl halide.

$$C_6H_5\text{–O–}C_6H_4\text{–}CH(Br)CH_3 \xrightarrow[\text{2. }^{-}OH/H_2O\text{; 3. }H_3O^+]{\text{1. NaCN}} C_6H_5\text{–O–}C_6H_4\text{–}CH(CH_3)\text{–}COOH$$

Fenoprofen
(an antiarthritic agent)

CARBOXYLATION OF GRIGNARD REAGENTS

The last method of carboxylic acid synthesis we will discuss is the carboxylation of Grignard reagents. Both Grignard and organolithium reagents add to carbon dioxide to yield carboxylate salts, which can be protonated to give carboxylic acids. The reaction is carried out either by pouring the Grignard reagent over dry ice (solid CO_2), or by bubbling a stream of dry CO_2 through the Grignard reagent solution. Grignard carboxylation generally gives good yields of acids from alkyl halides but is clearly limited in use to those alkyl halides that can form Grignard reagents (Section 20.8).

1-Bromo-2,4,6-trimethylbenzene $\xrightarrow[\text{Ether}]{\text{Mg}}$ (2,4,6-trimethylphenyl)MgBr $\xrightarrow[\text{2. } H_3O^+]{\text{1. } CO_2\text{, ether}}$ 2,4,6-Trimethylbenzoic acid (87%)

$$CH_3CH_2CH_2CH_2Cl \xrightarrow[\text{Ether}]{\text{Mg}} CH_3CH_2CH_2CH_2MgCl \xrightarrow[\text{2. } H_3O^+]{\text{1. } CO_2\text{, ether}} CH_3CH_2CH_2CH_2COOH$$

1-Chlorobutane — Pentanoic acid (73%)

The mechanism of this carboxylation reaction is similar to that of other Grignard reactions; the nucleophilic organomagnesium halide adds to one of the C=O bonds of carbon dioxide, yielding the product:

$$\overset{+}{X}Mg\ \overset{-}{R}: + O{=}C{=}O \longrightarrow R{-}C(=O)O^- \ \overset{+}{Mg}X \xrightarrow{H_3O^+} RCOOH$$

Recall:

$$\overset{+}{X}Mg\ \overset{-}{R}: + {>}C{=}O: \longrightarrow R{-}C{-}O^-\ \overset{+}{Mg}X \xrightarrow{H_3O^+} R{-}C{-}OH$$

PROBLEM

23.11 We have now seen two methods of converting an alkyl halide into a carboxylic acid having one more carbon atom: (1) substitution with cyanide followed by hydrolysis, and (2) formation of a Grignard reagent followed by carboxylation. What are the strengths and weaknesses of the two methods? Under what circumstances might one method be better than the other?

PROBLEM

23.12 In light of your answer to Problem 23.11, what methods would you use to prepare the following carboxylic acids from organohalides?

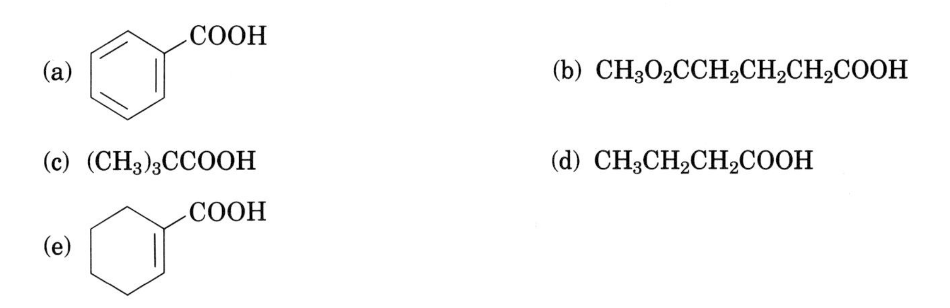

23.7 Reactions of carboxylic acids

We commented earlier in this chapter that carboxylic acids are structurally similar to both alcohols and ketones. There are chemical similarities as well. Like alcohols, carboxylic acids can be deprotonated to give anions; like ketones, carboxylic acids can undergo nucleophilic attack. In addition, carboxylic acids undergo other reactions characteristic neither of alcohols nor of ketones. Figure 23.5 shows some of the general types of reactions of carboxylic acids.

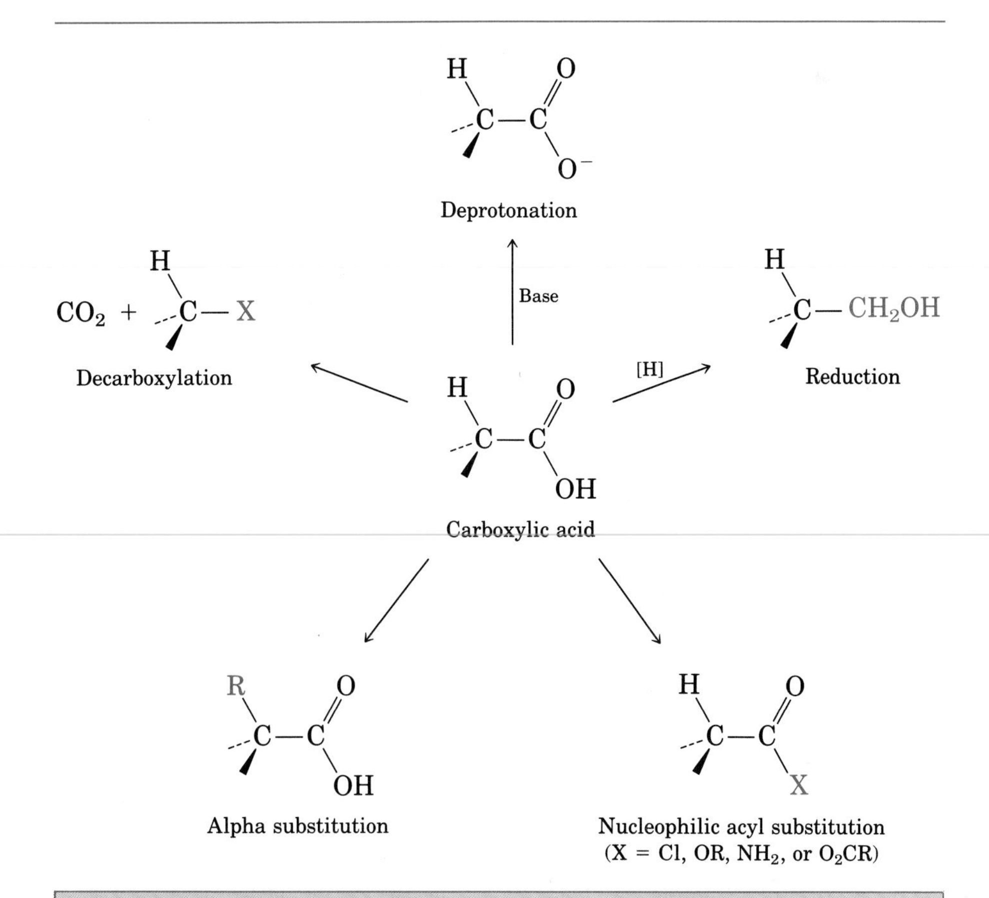

Figure 23.5. General reactions of carboxylic acids.

Reactions of carboxylic acids can be classified into five categories, as indicated in Figure 23.5. Of the five, we have already discussed the acidic behavior of carboxylic acids in Sections 23.3–23.5, and we will discuss the decarboxylation reaction and reduction reaction later in this chapter. The remaining two general reactions of carboxylic acids—nucleophilic acyl substitution and alpha substitution—are two of the four fundamental carbonyl-group reaction mechanisms and will be discussed in detail in Chapters 24 and 25.

23.8 Reduction of carboxylic acids

Carboxylic acids are reduced by hydride reagents such as lithium aluminum hydride (but not $NaBH_4$) to yield primary alcohols (Section 20.5). The reaction is difficult, however, and often requires refluxing in tetrahydrofuran solvent to go to completion.

$$CH_3(CH_2)_7CH{=}CH(CH_2)_7COOH \xrightarrow[2.\ H_3O^+]{1.\ LiAlH_4,\ THF,\ \Delta} CH_3(CH_2)_7CH{=}CH(CH_2)_7CH_2OH$$

Oleic acid — *cis*-9-Octadecen-1-ol (87%)

Borane, BH_3, is also used for converting carboxylic acids into primary alcohols. Reaction of an acid with borane occurs rapidly at room temperature, and this procedure is often preferred to reduction with $LiAlH_4$ because of the relative ease and safety of using BH_3.

$$p\text{-}O_2N\text{-}C_6H_4\text{-}CH_2COOH \xrightarrow[2.\ H_3O^+]{1.\ BH_3,\ THF} p\text{-}O_2N\text{-}C_6H_4\text{-}CH_2CH_2OH$$

p-Nitrophenylacetic acid — 2-(*p*-Nitrophenyl)ethanol (94%)

$$CH_3(CH_2)_4COOH \xrightarrow[2.\ H_3O^+]{1.\ BH_3,\ THF} CH_3(CH_2)_4CH_2OH$$

Hexanoic acid — 1-Hexanol (100%)

23.9 Decarboxylation of carboxylic acids

Carboxylic acids lose carbon dioxide under certain conditions to give a product having one less carbon atom than the starting acid. The **Hunsdiecker reaction**[1,2] is one example of such a decarboxylation process. The Hunsdiecker reaction involves heating the heavy metal salt of a carboxylic acid with bromine or iodine. Carbon dioxide is lost, and an alkyl halide

[1] Heinz Hunsdiecker (1904–); b. Cologne; Ph.D. Cologne (Wintgen); private laboratory, Cologne, Germany.

[2] Cläre Hunsdiecker (1903–); b. Kiel; Ph.D. Cologne (Wintgen); private laboratory, Cologne, Germany.

having one less carbon atom than the starting acid is formed. The metal ion may be either silver, mercuric, or lead(IV)—all work equally well.

$$CH_3(CH_2)_{15}CH_2COOH \xrightarrow[CCl_4]{HgO,\ Br_2} CH_3(CH_2)_{15}CH_2Br + CO_2$$

Octadecanoic acid → 1-Bromoheptadecane (93%)

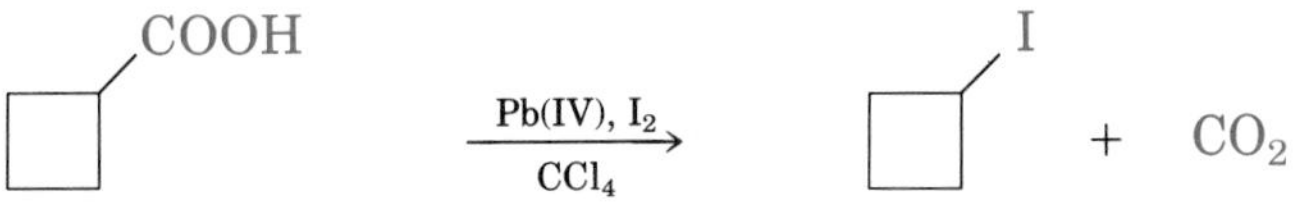

$$\text{Cyclobutanecarboxylic acid} \xrightarrow[CCl_4]{Pb(IV),\ I_2} \text{Iodocyclobutane} + CO_2$$

Cyclobutanecarboxylic acid → Iodocyclobutane (100%)

The Hunsdiecker reaction is thought to occur by a radical chain pathway. The initially formed acyl hypobromite undergoes homolytic cleavage of the weak O—Br bond to yield a carboxyl radical, which loses carbon dioxide to form an alkyl radical. The alkyl radical then propagates the chain by abstracting a bromine from acyl hypobromite (Figure 23.6).

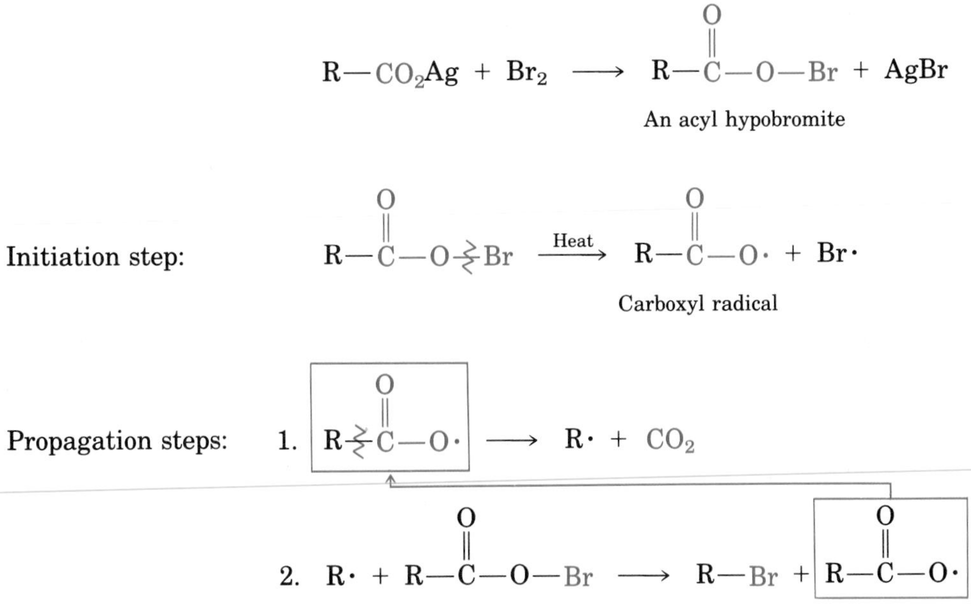

$$R{-}CO_2Ag + Br_2 \longrightarrow R{-}\overset{O}{\overset{\|}{C}}{-}O{-}Br + AgBr$$

An acyl hypobromite

Initiation step:

$$R{-}\overset{O}{\overset{\|}{C}}{-}O{\wr}Br \xrightarrow{Heat} R{-}\overset{O}{\overset{\|}{C}}{-}O\cdot + Br\cdot$$

Carboxyl radical

Propagation steps:

1. $$R{\wr}\overset{O}{\overset{\|}{C}}{-}O\cdot \longrightarrow R\cdot + CO_2$$

2. $$R\cdot + R{-}\overset{O}{\overset{\|}{C}}{-}O{-}Br \longrightarrow R{-}Br + R{-}\overset{O}{\overset{\|}{C}}{-}O\cdot$$

Figure 23.6. Mechanism of the Hunsdiecker reaction.

A second general mechanism of acid decarboxylation occurs when β-keto acids are heated above 100°C. Although most carboxylic acids are stable when heated, β-keto acids can decarboxylate by a cyclic mechanism. The initial product is an enol, which rapidly converts to the corresponding ketone, as indicated at the top of the next page.

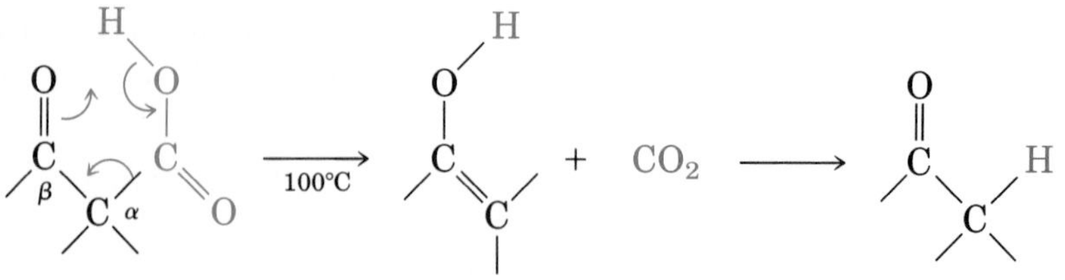

$$\text{A } \beta\text{-keto acid} \xrightarrow{100^\circ C} \text{An enol} + CO_2 \longrightarrow \text{A ketone}$$

The decarboxylation reaction of β-keto acids is important for organic synthesis, and we will see examples of its use in Section 25.9.

23.10 Spectroscopic analysis of carboxylic acids

INFRARED SPECTROSCOPY

Carboxylic acids show two highly characteristic absorptions in the infrared spectrum that make this functional group easily identifiable. The O—H bond of the carboxyl group gives rise to a very broad absorption over the range 2500–3300 cm^{-1}, and the C=O bond shows an absorption somewhere between 1710 cm^{-1} and 1760 cm^{-1}. The exact position of carbonyl absorption depends both on the structure of the molecule being studied and on whether the acid is free (monomeric) or associated (dimeric). Free carboxyl groups absorb at 1760 cm^{-1}, but the more commonly encountered associated carboxyl groups absorb at 1710 cm^{-1}:

Free carboxyl (uncommon), 1760 cm^{-1} — R—C(=O)—O—H

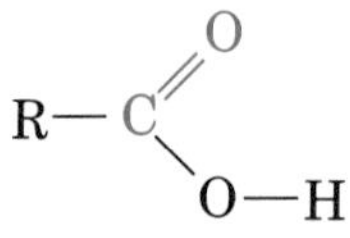

Associated carboxyl (usual case), 1710 cm^{-1} — R—C(=O····H—O)(—O—H····O=)C—R

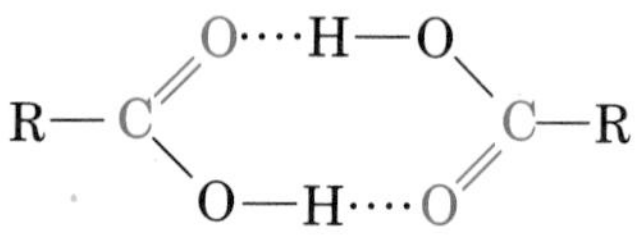

Figure 23.7 shows the infrared spectrum of butanoic acid and identifies both the broad O—H absorption and the C=O absorption at 1710 cm^{-1} (associated).

NUCLEAR MAGNETIC RESONANCE SPECTROSCOPY

Carboxylic acid groups can be detected by both ^{1}H and ^{13}C NMR spectroscopy. Carboxyl carbon atoms absorb in the range 165–185 δ in the ^{13}C NMR spectrum; the acidic proton normally absorbs near 12 δ in the ^{1}H NMR spectrum. Since the carboxylic acid proton has no neighbor protons, it is unsplit and occurs as a singlet. Figure 23.8 indicates the positions of the ^{13}C absorptions for several carboxylic acids, and Figure 23.9 shows the ^{1}H NMR

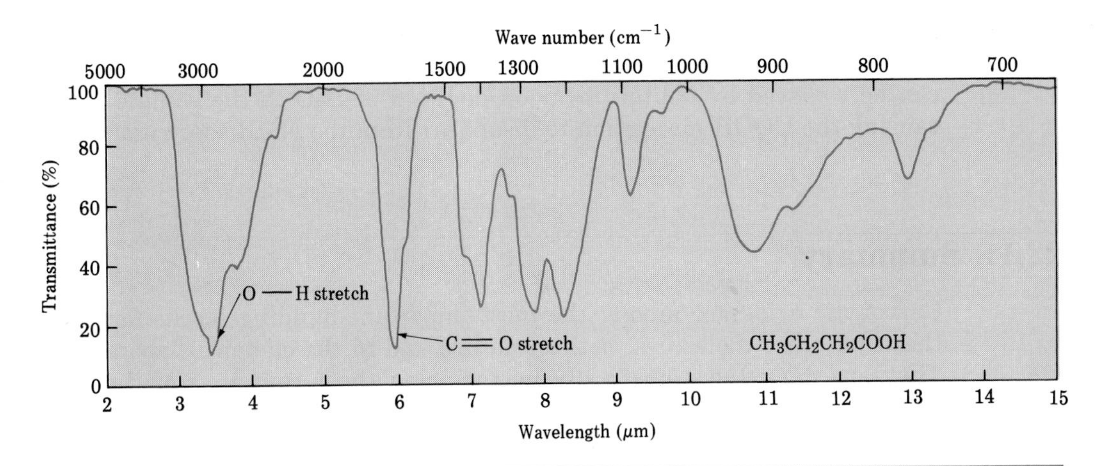

Figure 23.7. Infrared spectrum of butanoic acid, $CH_3CH_2CH_2COOH$.

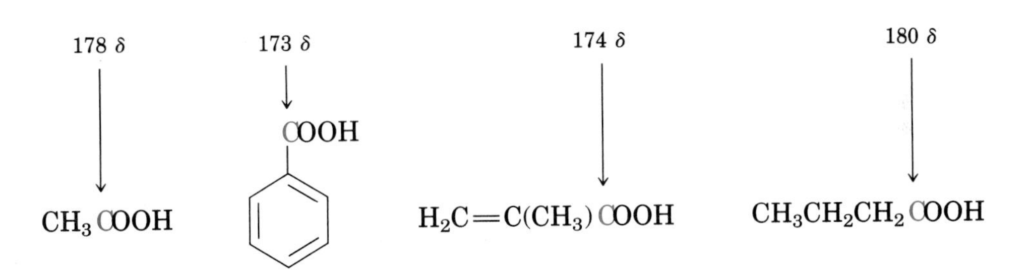

Figure 23.8. Carbon-13 NMR absorptions of some carboxylic acids.

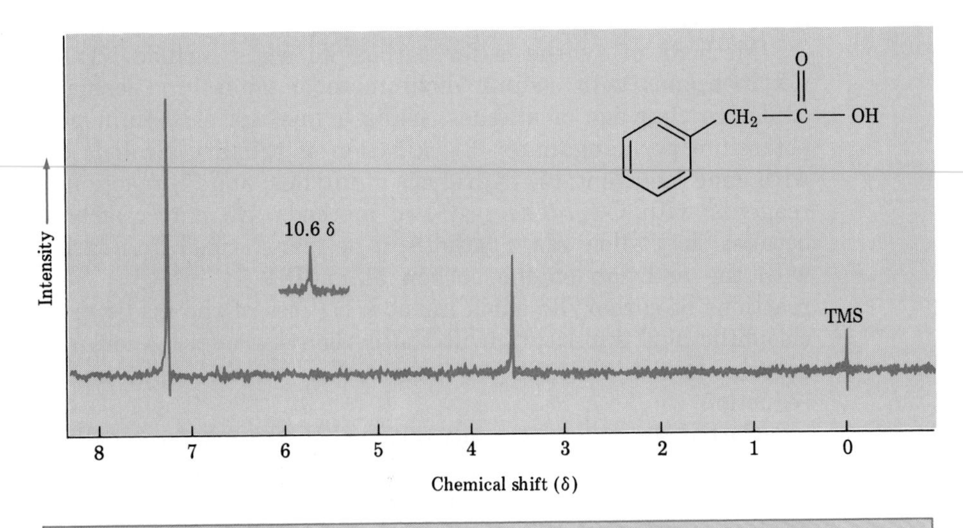

Figure 23.9. Proton NMR spectrum of phenylacetic acid.

spectrum of phenylacetic acid. Note that the carboxyl proton occurs at 10.6 δ and that all five aromatic ring protons show an accidental overlap as an apparent singlet at 7.3 δ. As with alcohols (Section 20.12), the COOH proton can be replaced by deuterium upon addition of D_2O to the sample tube, causing the COOH absorption to disappear from the NMR spectrum.

23.11 Summary

Carboxylic acids are among the most important building blocks for synthesizing other molecules, both in nature and in the chemical laboratory. They are named systematically by replacing the terminal *-e* of the corresponding alkane name with *-oic acid*. As in ketones and aldehydes, the carboxyl carbon atom is sp^2 hybridized; like alcohols, carboxylic acids are associated via hydrogen bonding and therefore have high boiling points.

The distinguishing characteristic of carboxylic acids is their acidity. Although weaker than mineral acids such as HCl, carboxylic acids nevertheless dissociate far more readily than do alcohols. The reason for this difference lies in the stability of carboxylate ions: Carboxylate ions are stabilized by resonance between two equivalent forms:

$$\mathrm{R{-}C({=}\ddot{O}{:})({-}\ddot{O}{:}^-)} \longleftrightarrow \mathrm{R{-}C({-}\ddot{O}{:}^-)({=}\ddot{O}{:})}$$

Most alkanoic acids have pK_a values near 5, but the exact acidity constant of a given acid is subject to considerable variation depending on structure. Carboxylic acids substituted by electron-withdrawing groups are more acidic (have a lower pK_a) because their carboxylate ions are stabilized. Carboxylic acids substituted by electron-donating groups are less acidic (have a higher pK_a) because their carboxylate ions are destabilized.

Methods of synthesis for carboxylic acids include: (1) oxidation of alkylbenzenes with sodium dichromate or potassium permanganate; (2) oxidative cleavage of alkenes, using a mixture of sodium periodate and potassium permanganate; (3) oxidation of primary alcohols or aldehydes with Jones' reagent; (4) hydrolysis of nitriles; and (5) reaction of Grignard reagents with CO_2. The last two methods are particularly noteworthy because they allow alkyl halides to be transformed into carboxylic acids with the addition of one carbon atom, R—Br → R—COOH. General reactions of carboxylic acids include: (1) loss of the acidic proton; (2) nucleophilic acyl substitution at the carbonyl group; (3) loss of CO_2 via a decarboxylation reaction; (4) substitution on the alpha carbon; and (5) reduction.

Carboxylic acids are easily distinguished spectroscopically. They exhibit characteristic infrared absorptions at 2500–3300 cm^{-1} (due to the O—H) and at 1710–1760 cm^{-1} (due to the C=O). Acids also show ^{13}C NMR absorptions at 165–185 δ due to the COOH carbon atom and ^{1}H NMR absorptions near 12 δ due to the acid proton.

23.12 Summary of reactions

1. Preparation of carboxylic acids (Section 23.6)
   a. Oxidation of alkylbenzenes (Section 16.9)

$$\text{Ar CHR}_2 \xrightarrow[\text{H}_2\text{O},\ \Delta]{\text{KMnO}_4} \text{Ar COOH}$$

   b. Oxidative cleavage of alkenes (Section 6.8)

$$\text{RCH}{=}\text{CHR}' \xrightarrow[\text{KIO}_4,\ \text{KMnO}_4]{\text{KMnO}_4\ \text{or}} \text{RCOOH} + \text{R}'\text{COOH}$$

   c. Oxidation of primary alcohols (Section 20.10)

$$\text{RCH}_2\text{OH} \xrightarrow[(\text{CrO}_3,\ \text{H}^+,\ \text{H}_2\text{O})]{\text{Jones' reagent}} \text{RCOOH}$$

   d. Oxidation of aldehydes (Section 22.5)

$$\text{RCHO} \xrightarrow[\text{Ag}^+,\ \text{NH}_4\text{OH}]{\text{Jones' reagent or}} \text{RCOOH}$$

   e. Hydrolysis of nitriles (Section 23.6)

$$\text{R—CN} \xrightarrow[\text{or}\ ^-\text{OH}]{\text{H}_3\text{O}^+} \text{RCOOH}$$

   f. Carboxylation of Grignard reagents (Section 23.6)

$$\text{R—MgX} \xrightarrow[2.\ \text{H}_3\text{O}^+]{1.\ \text{CO}_2} \text{RCOOH}$$

2. Reactions of carboxylic acids
   a. Deprotonation (Section 23.3)

$$\text{RCOOH} \xrightarrow{\text{Base}} \text{RCOO}^- + \text{H}^+$$

   b. Reduction to primary alcohols (Section 23.8)

$$\text{RCOOH} \xrightarrow[2.\ \text{H}_3\text{O}^+]{1.\ \text{LiAlH}_4\ \text{in THF or BH}_3\ \text{in THF}} \text{RCH}_2\text{OH}$$

   c. Decarboxylation: Hunsdiecker reaction (Section 23.9)

$$\text{RCOOH} \xrightarrow[\text{Br}_2,\ \text{CCl}_4]{\text{HgO}} \text{R—Br} + \text{CO}_2$$

ADDITIONAL PROBLEMS

23.13 Provide IUPAC names for compounds (a)–(f):

(a) $CH_3CH(COOH)CH_2CH_2CH(COOH)CH_3$

(b) $(CH_3)_3CCOOH$

(c) $CH_3CH_2CH_2CH(CH_2CH_2CH_3)CH_2COOH$

(d) COOH, NO_2 (para-substituted benzene ring)

(e) COOH (on cyclodecene ring)

(f) $BrCH_2CHBrCH_2CH_2COOH$

23.14 Draw structures corresponding to these IUPAC names:
(a) *cis*-1,2-Cyclohexanedicarboxylic acid
(b) Heptanedioic acid
(c) 2-Hexen-4-ynoic acid
(d) 4-Ethyl-2-propyloctanoic acid
(e) 3-Chlorophthalic acid
(f) Triphenylacetic acid
(g) 2,4-Dibromo-5-isopropyloctanoic acid

23.15 Acetic acid boils at 118°C, but its ethyl ester boils at 77°C. Why is the boiling point of the acid so much higher, even though it has the lower molecular weight?

23.16 Draw and name the eight carboxylic acid isomers having the formula $C_6H_{12}O_2$.

23.17 Order the compounds in each set with respect to increasing acidity:
(a) Acetic acid, oxalic acid, formic acid
(b) *p*-Bromobenzoic acid, *p*-nitrobenzoic acid, 2,4-dinitrobenzoic acid
(c) Phenylacetic acid, diphenylacetic acid, 3-phenylpropanoic acid
(d) Fluoroacetic acid, 3-fluoropropanoic acid, iodoacetic acid

23.18 Arrange the compounds in each set in order of increasing basicity:
(a) Sodium acetate, ammonia, sodium hydroxide
(b) Sodium benzoate, sodium *p*-nitrobenzoate, sodium acetylide
(c) Lithium hydroxide, lithium methoxide, lithium formate

23.19 Account for the fact that phthalic acid has $pK_2 = 5.4$ but terephthalic acid has $pK_2 = 4.8$.

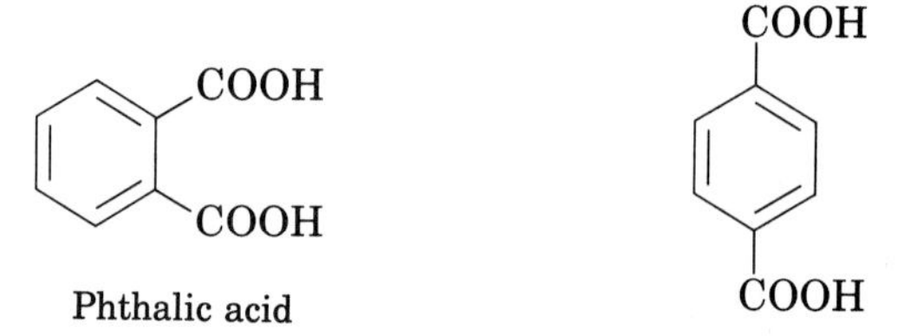

Phthalic acid

Terephthalic acid

23.20 How could you convert butanoic acid into the following compounds? Write out each step showing the reagents needed.
(a) 1-Butanol
(b) 1-Bromobutane
(c) Pentanoic acid
(d) 1-Butene
(e) 1-Bromopropane

23.21 Predict the product of the reaction of *p*-methylbenzoic acid with each of the following reagents:
(a) BH_3, then H_3O^+
(b) *N*-Bromosuccinimide in CCl_4
(c) HgO, I_2 in CCl_4
(d) CH_3MgBr in ether, then H_3O^+

23.22 Using $^{13}CO_2$ as your only source of labeled carbon, how would you synthesize these compounds?
(a) $CH_3CH_2{}^{13}COOH$
(b) $CH_3{}^{13}CH_2COOH$

23.23 Propose a satisfactory structure for an organic compound, $C_6H_{12}O_2$, that dissolves in dilute sodium hydroxide and that shows the following ^1H NMR spectrum: 1.08 δ (9 H, singlet), 2.2 δ (2 H, singlet), and 11.2 δ (1 H, singlet).

23.24 How would you carry out these transformations?

Methylenecyclohexane ($=CH_2$) $\xrightarrow{?}$ (a) cyclohexyl-CH_2COOH

Methylenecyclohexane ($=CH_2$) $\xrightarrow{?}$ (b) 1-methylcyclohexanecarboxylic acid (H_3C, $COOH$ on the same ring carbon)

23.25 What spectroscopic method might you use to distinguish among the following three isomeric acids? Tell exactly what characteristic features you would expect for each acid.

$CH_3(CH_2)_3COOH$	$(CH_3)_2CHCH_2COOH$	$(CH_3)_3CCOOH$
Pentanoic acid	3-Methylbutanoic acid	2,2-Dimethylpropanoic acid

23.26 Which method of acid synthesis, Grignard carboxylation or nitrile hydrolysis, would you use for each of the following reactions? Explain your choice.

(a) o-(CH_2Br)phenol (benzene ring with CH_2Br and OH) $\longrightarrow$ o-(CH_2COOH)phenol (benzene ring with CH_2COOH and OH)

(b) $CH_3CH_2CHBrCH_3 \longrightarrow CH_3CH_2CH(CH_3)COOH$

(c) $CH_3C(=O)CH_2CH_2CH_2I \longrightarrow CH_3C(=O)CH_2CH_2CH_2COOH$

(d) $HOCH_2CH_2CH_2Br \longrightarrow HOCH_2CH_2CH_2COOH$

23.27 As written, the following synthetic schemes (a)–(d) all have at least one flaw in them. What is wrong with each?

(a) Toluene (benzene ring with CH_3) $\xrightarrow[2.\ H_3O^+]{1.\ BH_3}$ benzoic acid (benzene ring with COOH)

(b) $CH_3CH_2CHBrCH_2CH_3 \xrightarrow[2.\ NaCN \quad 3.\ H_3O^+]{1.\ Mg} CH_3CH_2CH(COOH)CH_2CH_3$

(c) Phenylacetic acid (benzene ring with CH_2COOH) $\xrightarrow[2.\ H_3O^+]{1.\ LiAlH_4}$ ethylbenzene (benzene ring with CH_2CH_3)

(d) $(CH_3)_2C(OH)CH_2CH_2Cl \xrightarrow[2.\ H_3O^+]{1.\ NaCN} (CH_3)_2C(OH)CH_2CH_2COOH$

23.28 *p*-Aminobenzoic acid (PABA) is widely used as a sunscreen agent. Propose a synthesis of PABA starting from toluene.

23.29 Lithocholic acid is a steroid found in human bile:

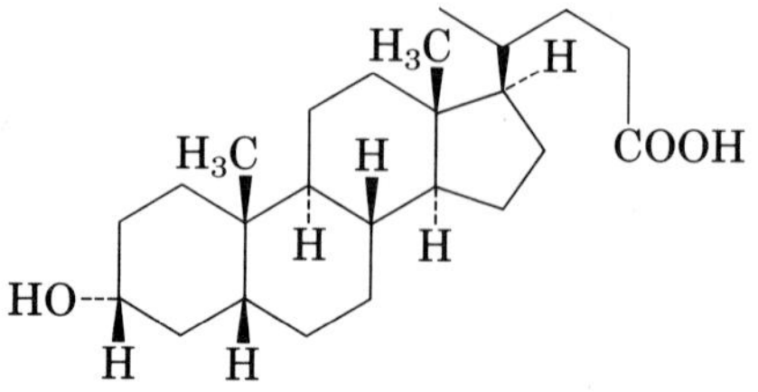

Lithocholic acid

Predict the product of reaction of lithocholic acid with each of the following reagents:

(a) Jones' reagent
(b) Tollens' reagent
(c) BH_3, then H_3O^+
(d) Dihydropyran, H^+
(e) CH_3MgBr, then H_3O^+
(f) $LiAlH_4$, then H_3O^+

23.30 Tranexamic acid, a drug useful for aiding blood clotting, is prepared commercially from *p*-methylbenzonitrile. Formulate the steps that are likely to be involved in the synthesis. [*Note:* The cis and trans isomers of tranexamic acid are thermally interconvertible at 300°C, and the trans isomer is more stable.]

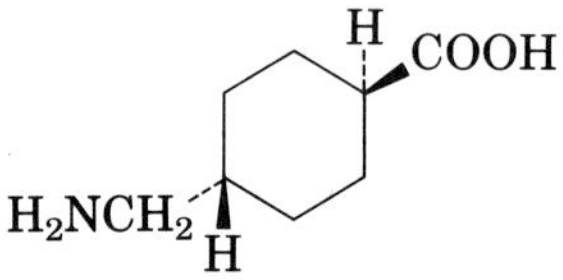

Tranexamic acid

23.31 Propose a synthesis of the anti-inflammatory drug, fenclorac, from phenylcyclohexane.

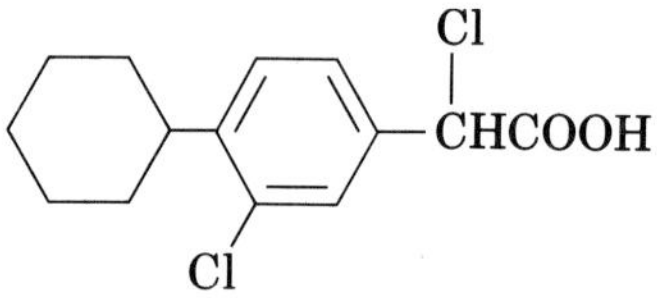

Fenclorac

23.32 How would you use NMR (either ^{13}C or ^{1}H) to distinguish between the following isomeric pairs?

(a)

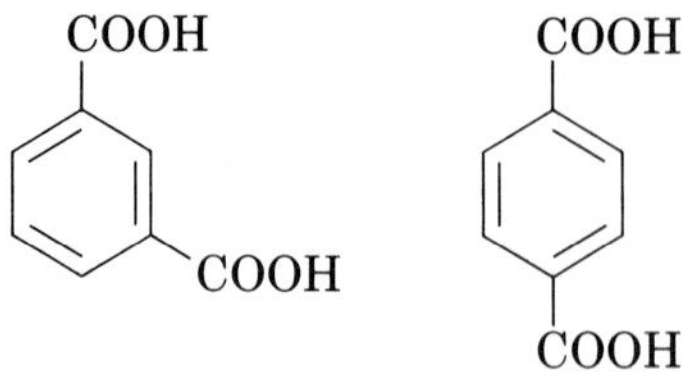

(b) $HOOCCH_2CH_2COOH$ $\quad$ $CH_3CH(COOH)_2$

(c) $CH_3CH_2CH_2COOH$ $HOCH_2CH_2CH_2CHO$

(d) $(CH_3)_2C{=}CHCH_2COOH$ [cyclopentanecarboxylic acid: cyclopentane ring–COOH]

23.33 Although most β-keto acids lose CO_2 when heated, the acid shown here does not. Use molecular models to help you arrive at an explanation of this observation.

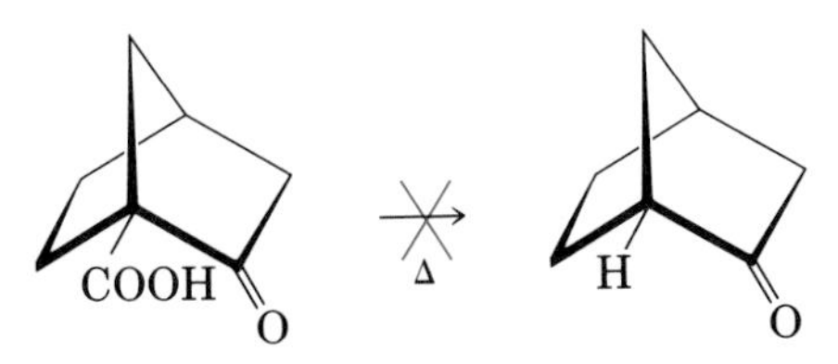

23.34 Compound A, $C_4H_8O_3$, has infrared absorptions at 1710 and 2500–3100 cm^{-1}, and exhibits the ^1H NMR spectrum shown. Propose a structure for A that is consistent with the data.

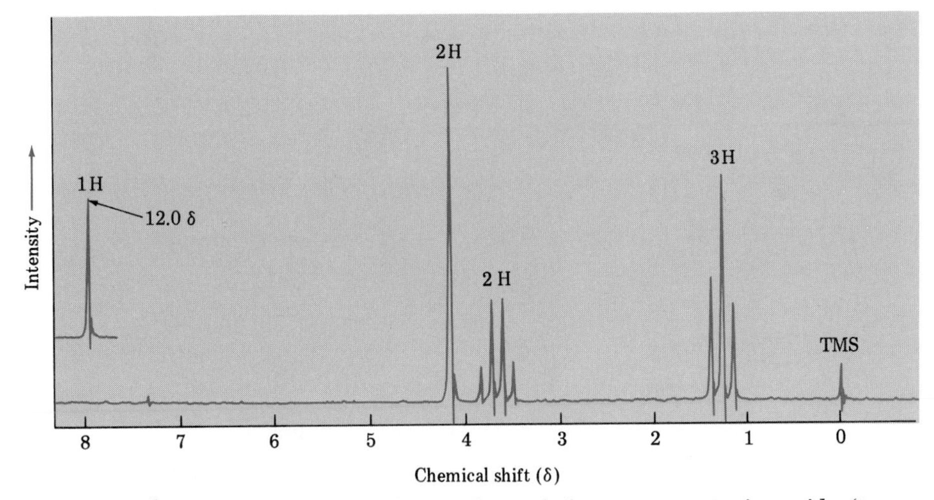

23.35 The two ^1H NMR spectra shown here belong to crotonic acid (*trans*-$CH_3CH{=}CHCOOH$) and to methacrylic acid [$H_2C{=}C(CH_3)COOH$]. Which spectrum corresponds to which acid? Explain your answer.

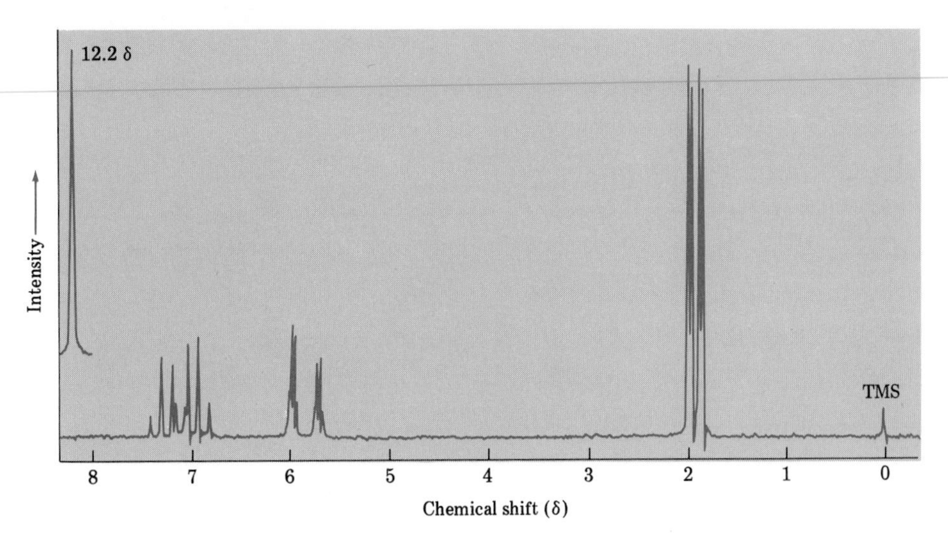

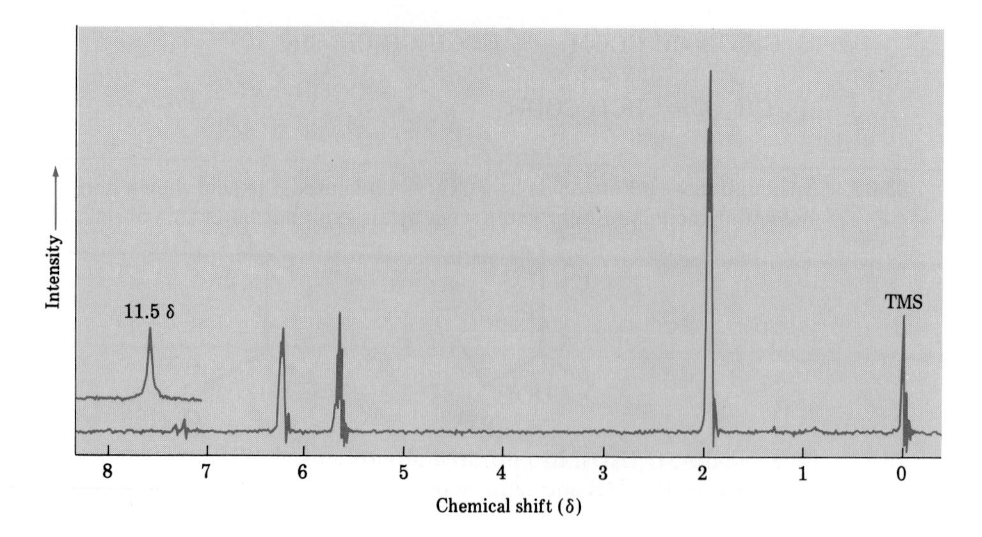

Intensity
11.5 δ
TMS
8
7
6
5
4
3
2
1
0
Chemical shift (δ)

CHAPTER 24 CARBOXYLIC ACID DERIVATIVES AND NUCLEOPHILIC ACYL SUBSTITUTION REACTIONS

Carboxylic acids are just one member of a class of **acyl derivatives, RCOX,** where the X substituent may be oxygen, halogen, nitrogen, or sulfur. Numerous acyl derivatives are possible, but we will be concerned only with four of the more common ones, in addition to carboxylic acids themselves: acid halides, acid anhydrides, esters, and amides. All these derivatives contain an acyl group, RCO—, bonded to an electronegative atom.

$$\mathrm{R{-}\overset{\overset{\displaystyle O}{\|}}{C}{-}OH}$$

Carboxylic acid

$$\mathrm{R{-}\overset{\overset{\displaystyle O}{\|}}{C}{-}X}$$

Acid halide (X = F, Cl, Br, I)

$$\mathrm{R{-}\overset{\overset{\displaystyle O}{\|}}{C}{-}O{-}\overset{\overset{\displaystyle O}{\|}}{C}{-}R'}$$

Acid anhydride

$$\mathrm{R{-}\overset{\overset{\displaystyle O}{\|}}{C}{-}OR'}$$

Ester

$$\mathrm{R{-}\overset{\overset{\displaystyle O}{\|}}{C}{-}NH_2}$$

Amide

The chemistry of these acyl derivatives is quite similar and is dominated by a single general reaction type—the **nucleophilic acyl substitution reaction:**

$$\mathrm{R{-}\overset{\overset{\displaystyle O}{\|}}{C}{-}X} + :\mathrm{Nu}^- \longrightarrow \mathrm{R{-}\overset{\overset{\displaystyle O}{\|}}{C}{-}Nu} + :\mathrm{X}^-$$

Let's first learn more about acyl derivatives and then explore the chemistry of acyl substitution reactions. In addition, we will explore the chemistry of nitriles, R—C≡N, in this chapter.

24.1 Nomenclature of carboxylic acid derivatives

ACID HALIDES

Acid halides are named by identifying the acyl group, and then citing the halide. The acyl group name is derived from the acid name by replacing the *-ic acid* ending with *-yl,* or the *-carboxylic acid* ending with *-carbonyl.* For example,

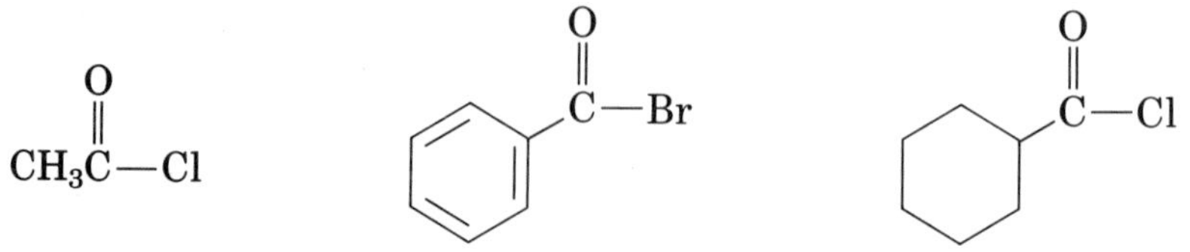

$$\mathrm{CH_3\overset{\overset{\displaystyle O}{\|}}{C}{-}Cl}$$

Acetyl chloride
(from acetic acid)

$$\mathrm{C_6H_5{-}\overset{\overset{\displaystyle O}{\|}}{C}{-}Br}$$

Benzoyl bromide
(from benzoic acid)

$$\mathrm{C_6H_{11}{-}\overset{\overset{\displaystyle O}{\|}}{C}{-}Cl}$$

Cyclohexanecarbonyl chloride
(from cyclohexanecarboxylic acid)

ACID ANHYDRIDES

Symmetrical anhydrides of straight-chain monocarboxylic acids, and cyclic anhydrides of dicarboxylic acids, are named by replacing the word *acid* with *anhydride:*

$$CH_3C(=O){-}O{-}C(=O)CH_3$$

Acetic anhydride

$$CH_3(CH_2)_5C(=O){-}O{-}C(=O)(CH_2)_5CH_3$$

Heptanoic anhydride

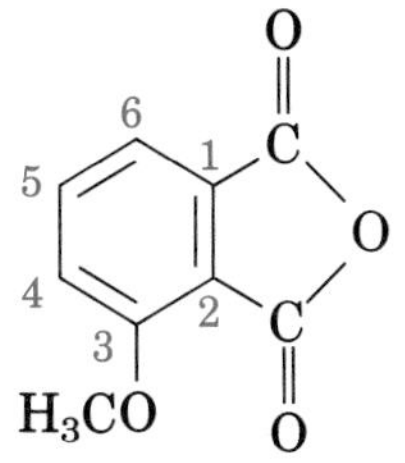

3-Methoxyphthalic anhydride

If the anhydride is derived from a substituted monocarboxylic acid, it is named by adding the prefix *bis-* to the acid name:

$$ClCH_2C(=O){-}O{-}C(=O)CH_2Cl$$

Bis(chloroacetic) anhydride

AMIDES

Amides with an unsubstituted $-NH_2$ group are named by replacing the *-oic acid* or *-ic acid* ending with *-amide,* or by replacing the *-carboxylic acid* ending with *-carboxamide:*

$$CH_3C(=O){-}NH_2$$

Acetamide
(from acetic acid)

$$CH_3(CH_2)_4C(=O){-}NH_2$$

Hexanamide
(from hexanoic acid)

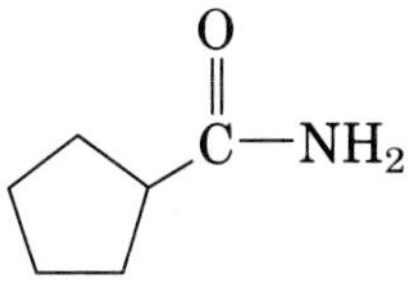

Cyclopentanecarboxamide
(from cyclopentanecarboxylic acid)

If the nitrogen atom is further substituted, the compound is named by first identifying the substituent groups and then citing the base name. Note that the substituents are preceded by the letter *N* to identify them as being directly attached to nitrogen.

$$CH_3CH_2C(=O){-}NHCH_3$$

N-Methylpropanamide

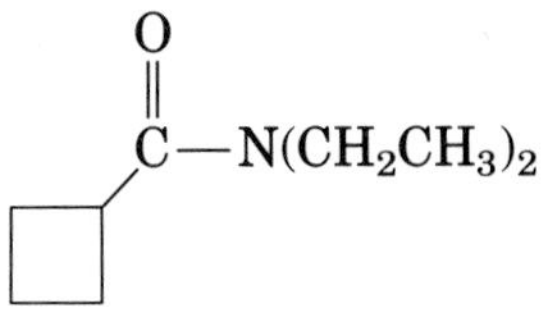

N,N-Diethylcyclobutanecarboxamide

ESTERS

Systematic names for esters are derived by first citing the alkyl-group name from the alcohol, and then identifying the carboxylic acid. In so doing, the *-ic acid* ending is replaced by *-ate:*

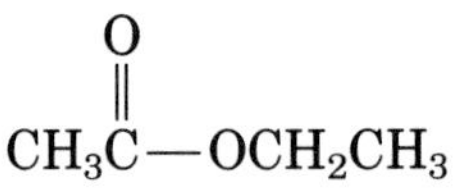

$$CH_3C(=O)—OCH_2CH_3$$

Ethyl acetate
(the ethyl ester of acetic acid)

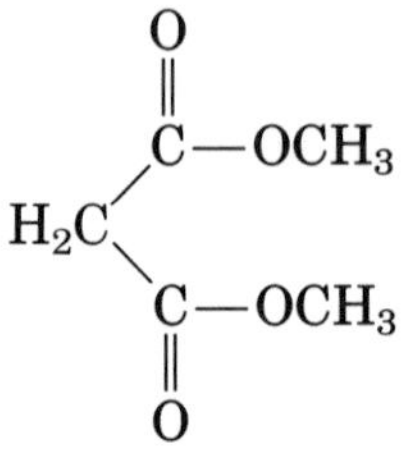

$$H_2C(C(=O)—OCH_3)_2$$

Dimethyl malonate
(the dimethyl ester of malonic acid)

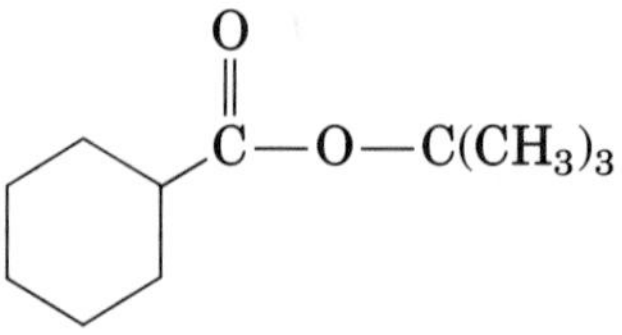

$$C_6H_{11}—C(=O)—O—C(CH_3)_3$$

tert-Butyl cyclohexanecarboxylate
(the *tert*-butyl ester of cyclohexanecarboxylic acid)

NITRILES

Compounds containing the —C≡N functional group are known as **nitriles** and are named by one of several systems. Simple acyclic alkane nitriles are named by adding *-nitrile* as a suffix to the alkane name. Note that the nitrile carbon itself is considered as C1.

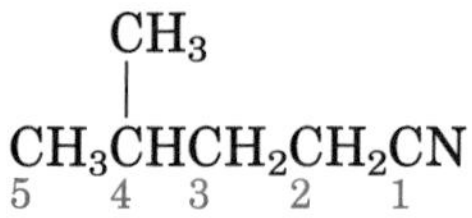

$$\underset{5}{CH_3}\underset{4}{CH}(CH_3)\underset{3}{CH_2}\underset{2}{CH_2}\underset{1}{CN}$$

4-Methylpentanenitrile

More complex nitriles are usually considered as derived from carboxylic acids, and are named either by replacing the *-ic acid* or *-oic acid* ending with *-onitrile,* or by replacing the *-carboxylic acid* ending with *-carbonitrile.* Note that in the latter system, the nitrile carbon atom is not numbered:

$$CH_3—C≡N$$

Acetonitrile
(from acetic acid)

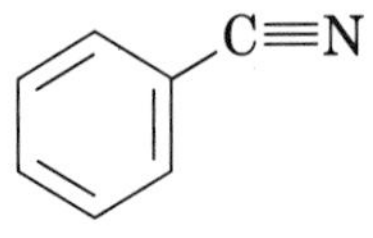

$$C_6H_5—C≡N$$

Benzonitrile
(from benzoic acid)

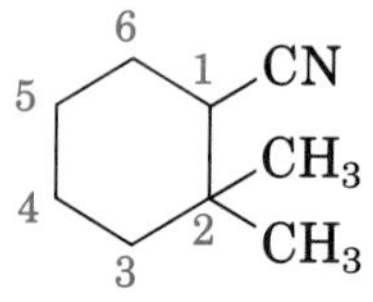

(cyclohexane ring numbered 1–6; C1 bears CN, C2 bears two CH_3)

2,2-Dimethylcyclohexanecarbonitrile
(from 2,2-dimethylcyclohexanecarboxylic acid)

A summary of the rules of nomenclature for carboxylic acid derivatives is given in Table 24.1.

PROBLEM

24.1 Provide IUPAC names for these structures:

(a) $(CH_3)_2CHCH_2CH_2COCl$

(b) 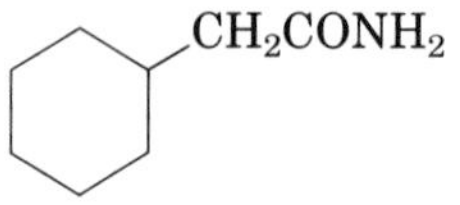 $C_6H_{11}—CH_2CONH_2$ (cyclohexane ring bearing CH_2CONH_2)

TABLE 24.1 **Nomenclature of carboxylic acid derivatives**

Functional group	*Structure*	*Name ending*
Carboxylic acid	$R{-}C({=}O){-}OH$	*-ic acid* (*-carboxylic acid*)
Acid halide	$R{-}C({=}O){-}X$	*-yl halide* (*-carbonyl halide*)
Acid anhydride	$R{-}C({=}O){-}O{-}C({=}O){-}R$	*anhydride*
Amide	$R{-}C({=}O){-}NH_2$	*-amide* (*-carboxamide*)
Ester	$R{-}C({=}O){-}OR'$	*-ate* (*-carboxylate*)
Nitrile	$R{-}C{\equiv}N$	*-onitrile* (*-carbonitrile*)

(c) $CH_3CH_2CH(CH_3)CN$

(d) 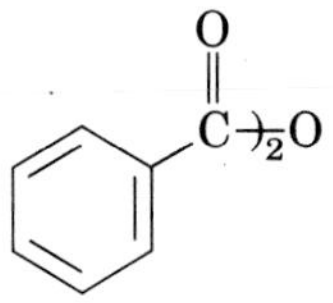 $(C_6H_5{-}C({=}O){\rangle}_2O$ (benzene ring bearing $C({=}O)$, $(C(=O))_2O$)

(e) 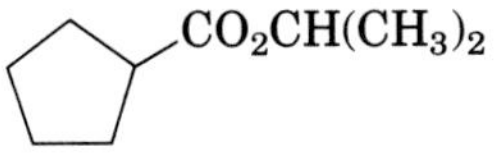 cyclopentyl$-CO_2CH(CH_3)_2$

(f) 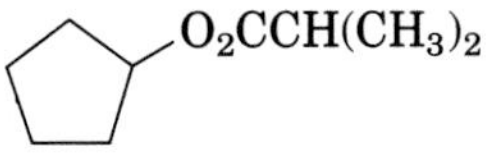 cyclopentyl$-O_2CCH(CH_3)_2$

(g) $H_2C{=}CHCH_2CH_2CONH_2$

(h) 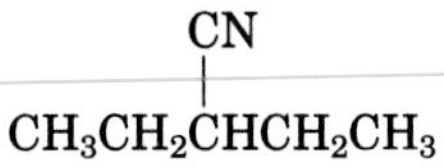 $CH_3CH_2CH(CN)CH_2CH_3$

(i) 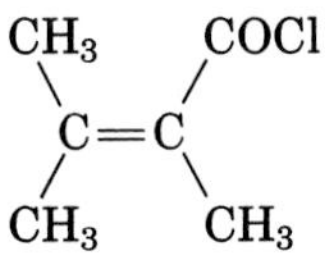 $(CH_3)_2C{=}C(CH_3)COCl$

(j) 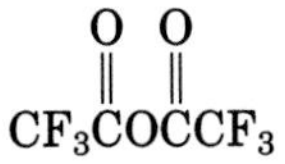 $CF_3C({=}O)OC({=}O)CF_3$

PROBLEM

24.2 The names for structures (a)–(d) are incorrect. Provide the correct names.

(a) $\overset{4}{C}H_3\overset{3}{C}H_2\overset{2}{C}H{=}\overset{1}{C}HCN$

1-Pentenenitrile

(b) $CH_3CH_2CH_2CONHCH_3$

Methylbutanamide

(c) $(CH_3)_2CHCH_2CH(CH_3)COCl$

2,4-Methylpentanoyl chloride

(d) [cyclohexane ring bearing CH_3 and CO_2CH_3 on the same carbon]

Methyl-2-methylcyclohexane carboxylate

24.2 Nucleophilic acyl substitution reactions

We said during our introductory look at carbonyl-group chemistry in Chapter 21 that the addition of a nucleophile to the polar C=O bond is a general feature of carbonyl-group reactions and is the first step in two of the four major carbonyl reaction types. When nucleophiles add to aldehydes and ketones, as we saw in Chapter 23, the initially formed tetrahedral intermediate can do one of two things: (1) It can be protonated to yield a stable alcohol, or (2) it can eliminate the carbonyl oxygen, leading to a new C=Nu bond (Figure 24.1). When nucleophiles add to carboxylic acid derivatives, however, a different reaction course, also shown in Figure 24.1, is followed. The initially formed tetrahedral intermediate expels one of the two sub-

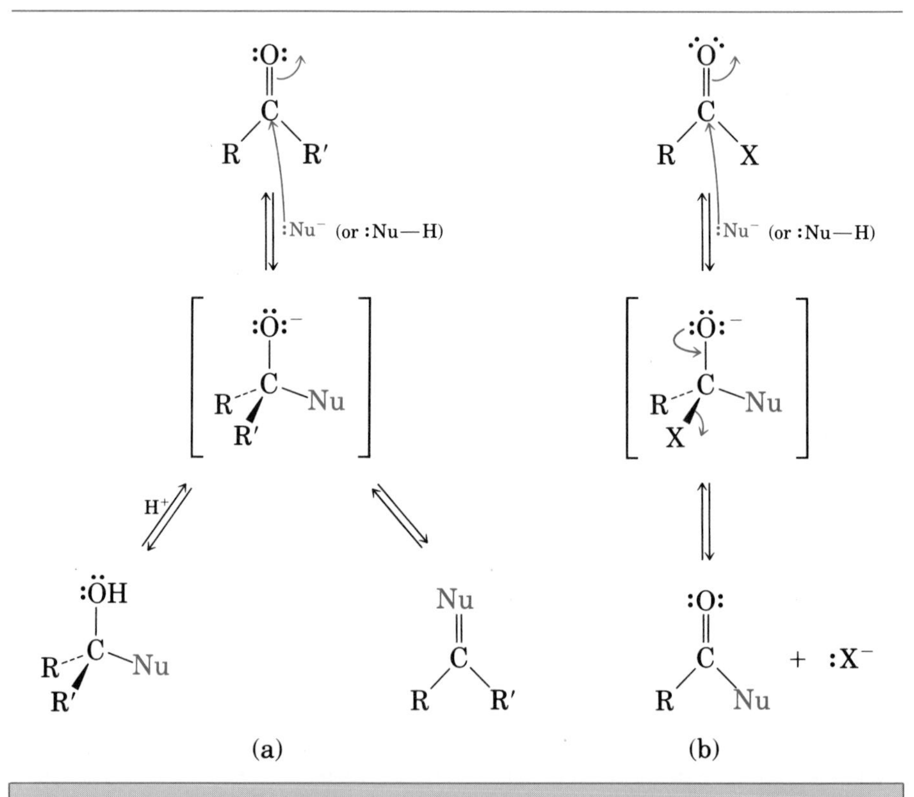

Figure 24.1. General reactions of carbonyl groups with nucleophiles: (a) nucleophilic addition reactions, undergone by ketones and aldehydes, and (b) nucleophilic acyl substitution reactions, undergone by carboxylic acids and acid derivatives.

stituents originally bonded to the carbonyl carbon, leading to a net nucleophilic acyl substitution.

What is the reason for the different behavior of ketones/aldehydes and carboxylic acid derivatives? The difference is simply a consequence of structure. Carboxylic acid derivatives have an acyl function bonded to a potential leaving group (a group that can leave as a stable anion), and the negative charge on oxygen in the tetrahedral intermediate can readily expel this leaving group. Ketones and aldehydes have no such leaving group, however, and therefore do not undergo elimination. The general mechanism for nucleophilic acyl substitution of carboxylic acid derivatives is shown in Figure 24.2.

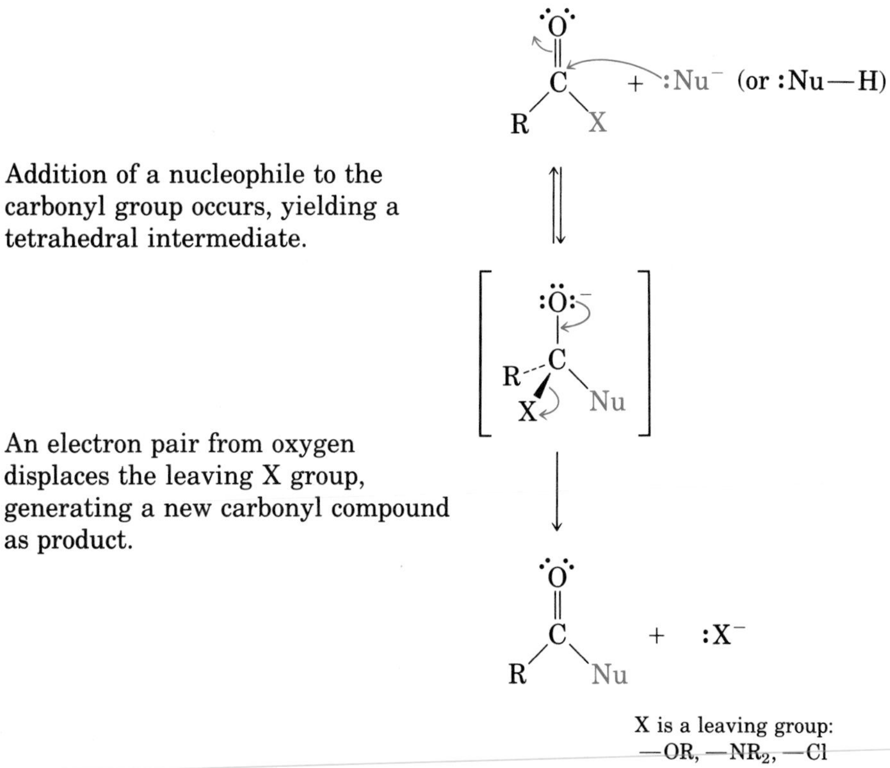

Addition of a nucleophile to the carbonyl group occurs, yielding a tetrahedral intermediate.

An electron pair from oxygen displaces the leaving X group, generating a new carbonyl compound as product.

$RC(=O)X + :Nu^-$ (or $:Nu{-}H$) $\rightleftharpoons$ $[RC(O^-)(X)(Nu)]$ $\rightarrow$ $RC(=O)Nu + :X^-$

X is a leaving group: $-OR$, $-NR_2$, $-Cl$

Figure 24.2. General mechanism of nucleophilic acyl substitution.

24.3 Relative stability of carboxylic acid derivatives

Nucleophilic acyl substitution reactions take place in two steps—addition of the nucleophile and elimination of a leaving group. Both steps can affect the overall rate of reaction. In general, though, it is the first step that is rate limiting, and both steric and electronic factors are important in determining reactivity. Sterically, we find within a series of the same acid derivatives that more hindered carbonyl groups are attacked less readily than sterically accessible groups:

$$\text{Less reactive}\quad R_3C\overset{O}{\overset{\|}{C}}\!- \;<\; R_2CH\overset{O}{\overset{\|}{C}}\!- \;<\; RCH_2\overset{O}{\overset{\|}{C}}\!- \;<\; CH_3\overset{O}{\overset{\|}{C}}\!- \quad\text{More reactive}$$

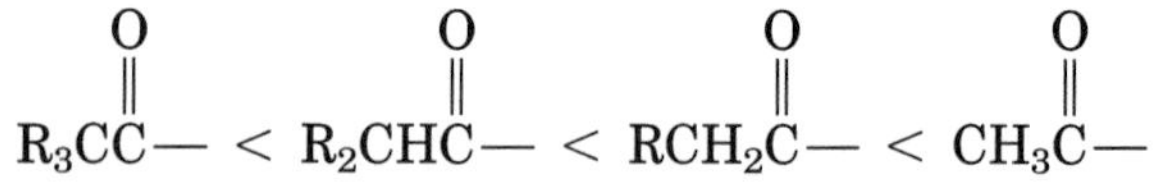

Electronically, we find that more strongly polar acid derivatives are attacked more readily than less polar groups. Thus the observed reactivity order is:

More stable, less reactive

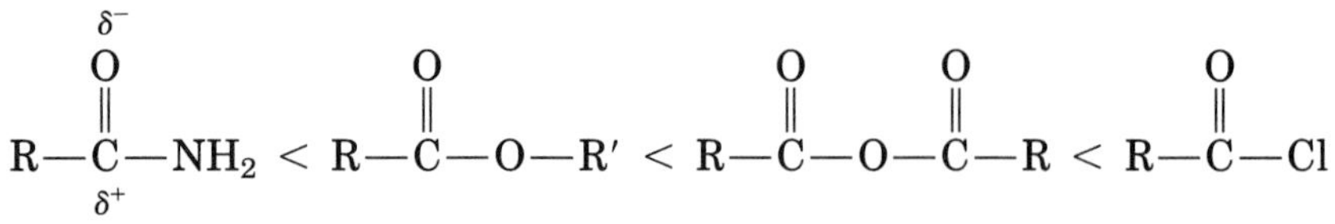

$$R{-}\underset{\delta^+}{\overset{\overset{\delta^-}{O}}{\overset{\|}{C}}}{-}NH_2 \;<\; R{-}\overset{O}{\overset{\|}{C}}{-}O{-}R' \;<\; R{-}\overset{O}{\overset{\|}{C}}{-}O{-}\overset{O}{\overset{\|}{C}}{-}R \;<\; R{-}\overset{O}{\overset{\|}{C}}{-}Cl$$

More reactive, less stable

These reactivity differences are best seen in terms of reaction energetics. Since reaction rate is a measure of $\Delta G^{\ddagger}$, the energy difference between ground state (reactant) and transition state, any factor that raises $\Delta G^{\ddagger}$ lowers reaction rate. Amides are the least reactive acid derivatives, largely because of resonance stabilization of their ground state. This resonance stabilization is lost in the transition state and $\Delta G^{\ddagger}$ is therefore high. Esters are less stabilized by resonance and therefore have a higher-energy ground state and lower $\Delta G^{\ddagger}$; whereas acid chlorides have little ground-state resonance stabilization and are least stable. Figure 24.3 shows these effects graphically using reaction energy diagrams.

One consequence of the observed reactivity differences is that, as a general rule, it is possible to transform a less stable acid derivative into a more stable acid derivative. As we will see in the next few sections, carboxylic acid chlorides can be converted into acids, esters, and amides, but amides cannot readily be converted into esters or acid chlorides.

A second consequence of the reactivity differences among acid derivatives is that only esters and amides are commonly found in nature; acid halides and acid anhydrides do not occur naturally. The reason is that acid halides and anhydrides are simply too reactive; they rapidly undergo nucleophilic attack by water and do not have sufficient stability to exist in living organisms. Esters and amides, however, have exactly the right balance between reactivity and stability to allow them to occur widely and to be vitally important in many life processes.

In studying the chemistry of acid derivatives, we will find that there are striking similarities among the various types of compounds. We will be concerned largely with the reactions of just a few nucleophiles, and will see that the same kinds of reactions keep occurring (Figure 24.4, p. 774).

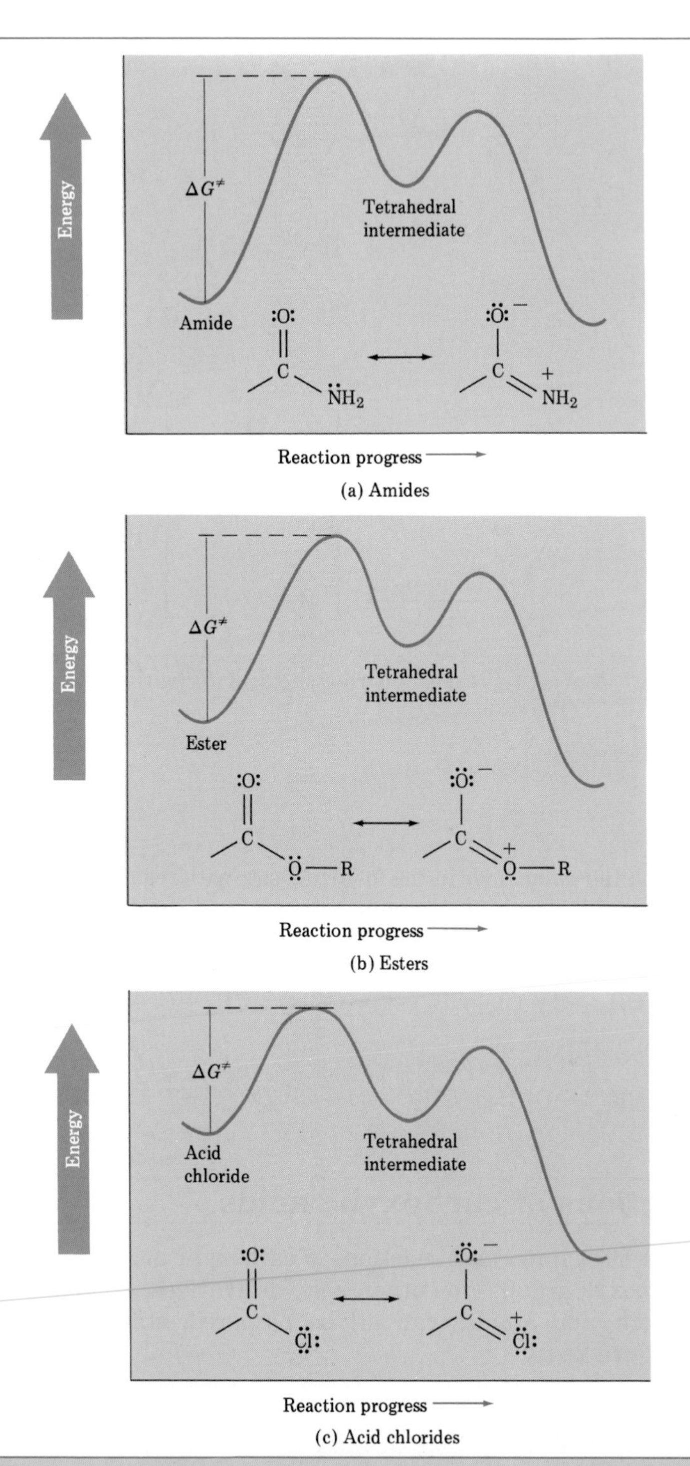

Figure 24.3. Energy diagrams for nucleophilic acyl substitution reactions of amides, esters, and acid chlorides. (a) For the amide, an important contribution from the dipolar resonance form stabilizes the ground state and raises $\Delta G^{\ddagger}$. (b) For the ester, a modest contribution from the dipolar resonance form results in a higher ground state and a lower $\Delta G^{\ddagger}$. (c) For the acid chloride, a minor contribution from the dipolar resonance form results in a low $\Delta G^{\ddagger}$.

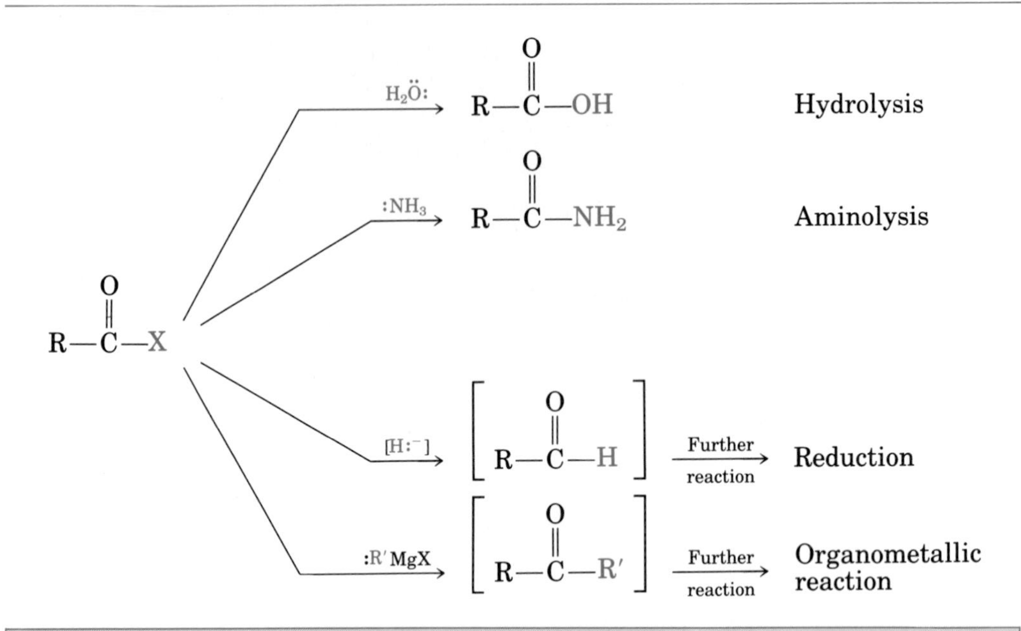

Figure 24.4. General reactions of carboxylic acid derivatives.

PROBLEM

24.3 Rank the compounds in the following sets with regard to expected reactivity toward nucleophilic acyl substitution:

(a) $CH_3—C(=O)—Cl$, $CH_3—C(=O)—OCH_3$, $CH_3—C(=O)—NH_2$

(b) $CH_3C(=O)—OCH_3$, $CH_3C(=O)—O—CH_2CCl_3$, $CH_3C(=O)—OCH(CF_3)_2$

24.4 Reactions of carboxylic acids

The most important reactions of carboxylic acids are those that convert the carboxyl group into other acid derivatives. Acid chlorides, anhydrides, esters, and amides can all be prepared, starting from carboxylic acids (Figure 24.5).

ACID CHLORIDES

Carboxylic acids are easily converted into carboxylic acid chlorides by treatment with reagents such as thionyl chloride ($SOCl_2$), phosphorus trichloride (PCl_3), or oxalyl chloride (ClCOCOCl). Thionyl chloride is both cheap and convenient to use but is strongly acidic; only acid-stable molecules can survive the reaction conditions. Oxalyl chloride, on the other

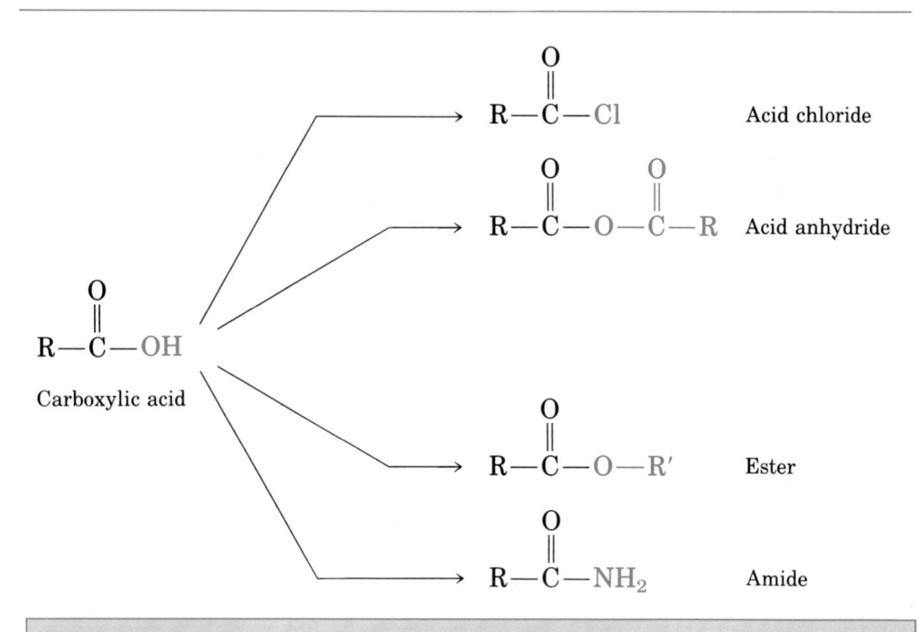

Figure 24.5. Derivatives of carboxylic acids.

hand, is much more expensive but gives higher yields and reacts under milder conditions.

$$\underset{\text{Oleic acid}}{CH_3(CH_2)_7CH{=}CH(CH_2)_7COOH} \xrightarrow[\text{Benzene}]{ClC(=O)C(=O)Cl} \underset{\text{Oleyl chloride (97\%)}}{CH_3(CH_2)_7CH{=}CH(CH_2)_7C(=O)Cl} + HCl + CO + CO_2$$

$$\underset{\text{2,4,6-Trimethylbenzoic acid}}{2,4,6\text{-}(CH_3)_3C_6H_2COOH} \xrightarrow[CHCl_3]{SOCl_2} \underset{\text{2,4,6-Trimethylbenzoyl chloride (90\%)}}{2,4,6\text{-}(CH_3)_3C_6H_2C(=O)Cl} + HCl + SO_2$$

These reactions occur by nucleophilic acyl substitution pathways in which the carboxylic acid is converted into a reactive derivative, which is then attacked. For example,

$$R{-}\overset{O}{\overset{\|}{C}}OH + SOCl_2 \xrightarrow{CHCl_3} R{-}\overset{O}{\overset{\|}{C}}{-}O{-}\overset{O}{\overset{\|}{S}}{-}Cl + HCl$$

Then,

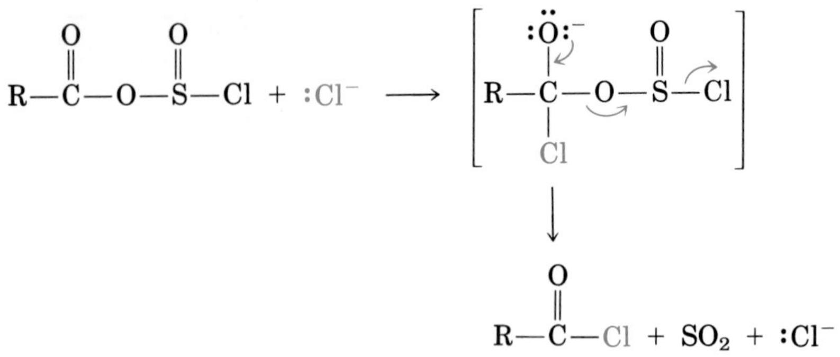

$$\mathrm{R{-}\overset{O}{\overset{\|}{C}}{-}O{-}\overset{O}{\overset{\|}{S}}{-}Cl} + :\mathrm{Cl}^- \longrightarrow \left[\mathrm{R{-}\overset{:\ddot{O}:^-}{\overset{|}{\underset{\underset{Cl}{|}}{C}}}{-}O{-}\overset{O}{\overset{\|}{S}}{-}Cl}\right] \longrightarrow \mathrm{R{-}\overset{O}{\overset{\|}{C}}{-}Cl} + SO_2 + :\mathrm{Cl}^-$$

ACID ANHYDRIDES

Acid anhydrides have the general structure $\mathrm{R\overset{O}{\overset{\|}{C}}{-}O{-}\overset{O}{\overset{\|}{C}}R'}$, and are formally derived from two molecules of carboxylic acid by removing 1 equiv of water. Acyclic anhydrides are difficult to prepare directly from the corresponding acids, and only acetic anhydride is commercially available.

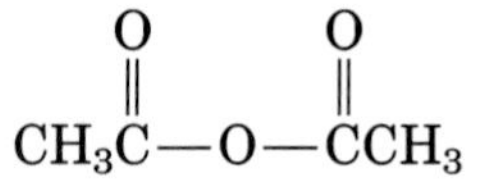

$$\mathrm{CH_3\overset{O}{\overset{\|}{C}}{-}O{-}\overset{O}{\overset{\|}{C}}CH_3}$$

Acetic anhydride

Cyclic anhydrides of ring size five or six are readily obtained by high-temperature dehydration of the diacids.

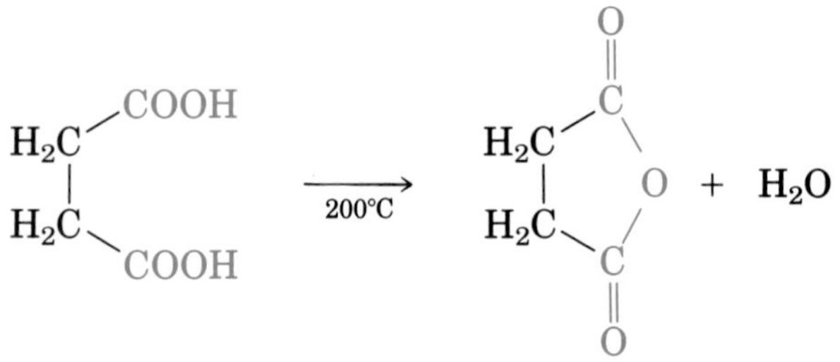

$$\mathrm{H_2C(COOH){-}H_2C(COOH)} \xrightarrow[200°C]{} \mathrm{H_2C{-}C(=O){-}O{-}C(=O){-}CH_2}\ (\text{ring}) + H_2O$$

Succinic acid — Succinic anhydride

ESTERS

One of the most important reactions of carboxylic acids is their conversion into esters, $R{-}CO_2R'$. There are many excellent methods for accomplishing this transformation, and we have already studied one such method—the S_N2 reaction between a carboxylate anion nucleophile and a primary alkyl halide (Section 10.5):

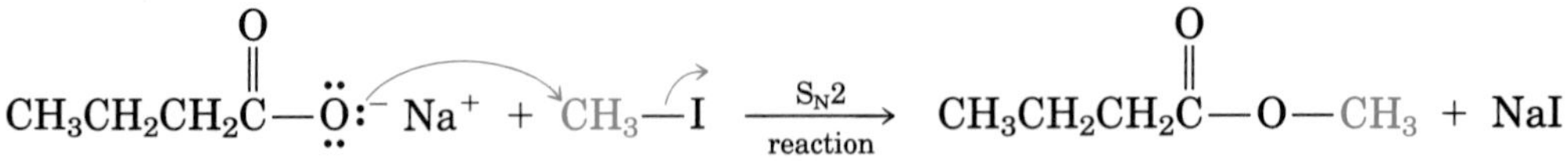

$$\mathrm{CH_3CH_2CH_2\overset{O}{\overset{\|}{C}}{-}\ddot{\underset{..}{O}}:^-\ Na^+} + \mathrm{CH_3{-}I} \xrightarrow[\text{reaction}]{S_N2} \mathrm{CH_3CH_2CH_2\overset{O}{\overset{\|}{C}}{-}O{-}CH_3} + \mathrm{NaI}$$

Sodium butanoate — Methyl butanoate, an ester (97%)

Alternatively, esters can be synthesized by direct reaction between a carboxylic acid and an alcohol. Fischer[1] and Speier discovered in 1895 that esters result from simply heating a carboxylic acid in methanol or ethanol solution containing a small amount of mineral acid catalyst. Yields are quite good in the Fischer esterification reaction, but the need to use excess alcohol as solvent effectively limits the method to the synthesis of methyl, ethyl, and propyl esters.

$$C_6H_5CH(OH)COOH + CH_3CH_2OH \xrightarrow[HCl]{Ethanol} C_6H_5CH(OH)COOCH_2CH_3 + H_2O$$

Mandelic acid — Ethyl mandelate (86%)

$$HOOC(CH_2)_4COOH + CH_3CH_2OH \xrightarrow[HCl]{Ethanol} CH_3CH_2OOC(CH_2)_4COOCH_2CH_3 + H_2O$$

Hexanedioic acid (adipic acid) — Diethyl hexanedioate (95%)

The Fischer esterification reaction, whose mechanism is shown in Figure 24.6 (p. 778), is perhaps the best known example of a nucleophilic acyl substitution reaction carried out under acidic conditions. Protonation of the basic carbonyl oxygen activates the carboxylic acid toward nucleophilic attack by alcohol, and subsequent loss of water yields the ester product. The net effect is substitution of —OH by —OR′. All steps are reversible, and the reaction can be driven in either direction by proper choice of reaction conditions. Ester formation is favored when alcohol is used as solvent, but carboxylic acid formation is favored when water is used as solvent.

One of the best pieces of evidence in support of the mechanism shown in Figure 24.6 comes from isotopic labeling experiments. When radiolabeled methanol, $CH_3\,^{18}OH$, reacts with benzoic acid under Fischer esterification conditions, the methyl benzoate produced is found to be ^{18}O labeled, but the water produced is unlabeled. This experiment shows unequivocally that it is the CO—OH bond of the carboxylic acid that is cleaved, rather than the COO—H bond, and that it is the RO—H bond of the alcohol that is cleaved, rather than the R—OH bond.

$$C_6H_5C(=O)-OH + CH_3\overset{*}{O}-H \overset{H^+}{\rightleftharpoons} C_6H_5C(=O)-\overset{*}{O}CH_3 + H_2O$$

A final method of ester synthesis that should be mentioned is the reaction between a carboxylic acid and diazomethane, CH_2N_2 (Hans von

[1]Emil Fischer (1852–1919); b. Euskirchen, Germany; Ph.D. Strasbourg (Baeyer); professor, universities of Erlangen, Würzburg, and Berlin; Nobel prize (1902).

Protonation of the carbonyl oxygen activates the carbonyl group . . .

. . . toward nucleophilic addition by alcohol, yielding a tetrahedral intermediate.

Transfer of a proton from one oxygen to another yields a second tetrahedral intermediate and converts the hydroxyl into a good leaving group.

Loss of water yields a protonated ester.

Loss of a proton regenerates acid catalyst, and gives the free ester product.

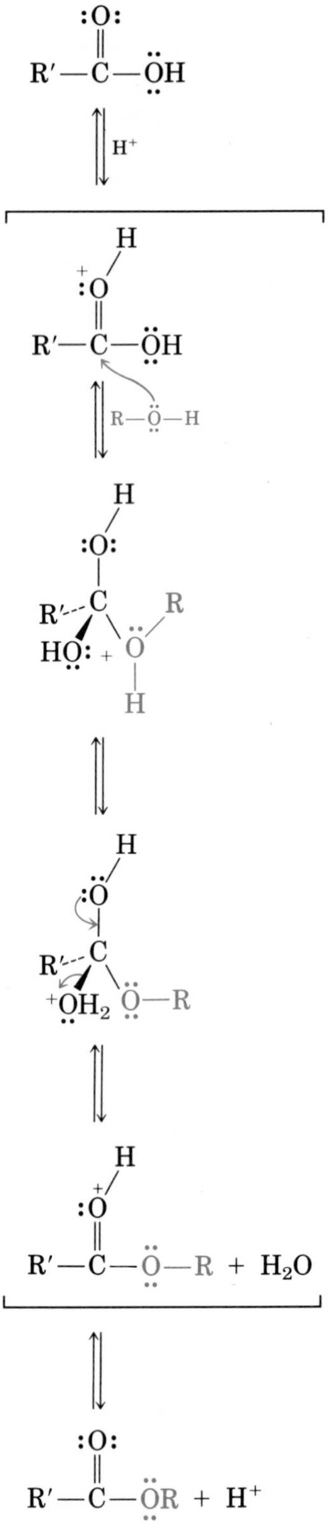

Figure 24.6. Mechanism of Fischer esterification.

Pechmann,[2] 1894). The reaction takes place instantly at room temperature to give a high yield of the methyl ester. Though quite useful, this process does not involve a nucleophilic acyl substitution reaction, since it is the —COO—H bond of the carboxylic acid that is broken.

$$C_6H_5COOH + CH_2N_2 \xrightarrow{\text{Ether}} C_6H_5C(=O)-O-CH_3 + N_2$$

Benzoic acid — Methyl benzoate (100%)

The diazomethane method of ester synthesis is ideal, since it occurs cleanly under mild, neutral conditions and gives nitrogen gas as the only by-product. Unfortunately, diazomethane is both toxic and explosive, and it should only be handled in small amounts by skilled persons.

PROBLEM

24.4 How can you account for the fact that when a carboxylic acid is dissolved in isotopically labeled water, the label rapidly becomes incorporated into *both* oxygen atoms of the carboxylic acid?

$$R-C(=O)-OH + H_2{*}O \rightleftharpoons R-C(={*}O)-{*}OH + H_2O$$

PROBLEM

24.5 If 5-hydroxypentanoic acid is treated with acid catalyst, an intramolecular esterification reaction occurs. What is the structure of the product? (*Intramolecular* means within the same molecule.)

AMIDES

Amides are carboxylic acid derivatives in which the acid hydroxyl group has been replaced by a nitrogen substituent, $-NH_2$ (or $-NR_2$). Amides are prepared directly from carboxylic acids with difficulty and at high temperatures. The reason for this lack of reactivity is simply that amines are basic reagents (Section 28.4) that convert acidic carboxyl groups into their unreactive carboxylate anions.

$$R-C(=O)-OH + :NH_3 \rightleftharpoons R-C(=O)-O^- + NH_4^+$$

PROBLEM

24.6 Why do you suppose carboxylate anions, $RCOO^-$, are much less reactive toward nucleophilic addition than are neutral carboxylic acids?

[2] Hans von Pechmann (1850–1902); b. Nuremberg; Ph.D. Greiswald; professor, universities of Munich and Tübingen.

24.5 Acid halides

PREPARATION OF ACID HALIDES

Acid chlorides are prepared from carboxylic acids by reaction with thionyl chloride ($SOCl_2$), oxalyl chloride (ClCOCOCl), or phosphorus trichloride (PCl_3) (Section 24.4). Reaction of the acid with phosphorus tribromide (PBr_3) yields the acid bromide.

$$R-\overset{O}{\overset{\|}{C}}-OH \xrightarrow[\text{ClCOCOCl, or } PCl_3]{SOCl_2,} R-\overset{O}{\overset{\|}{C}}-Cl$$

$$R-\overset{O}{\overset{\|}{C}}-OH \xrightarrow[\text{Ether}]{PBr_3} R-\overset{O}{\overset{\|}{C}}-Br + HOPBr_2$$

REACTIONS OF ACID HALIDES

Acid halides are among the most reactive of carboxylic acid derivatives and undergo many useful reactions. We have already seen the great value of acid chlorides in preparing aryl alkyl ketones via the Friedel–Crafts reaction (Section 16.6).

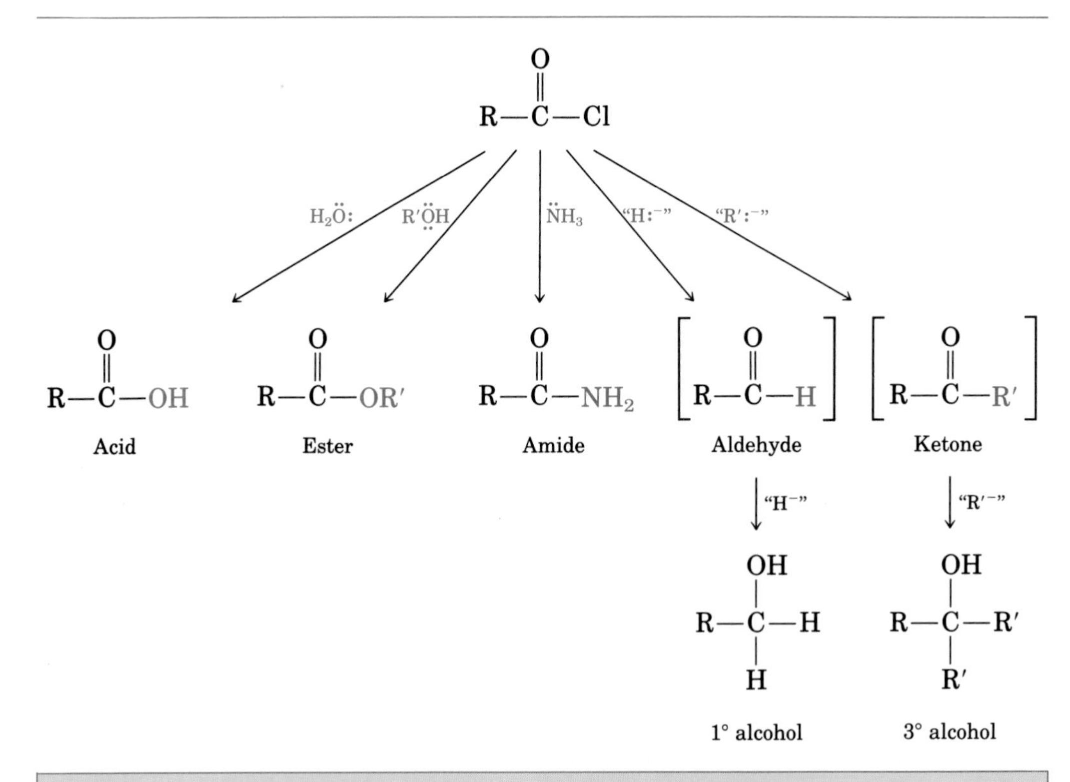

Figure 24.7. Nucleophilic acyl substitution reactions of acid chlorides.

$$\text{Ar—H} + \text{R}\overset{\overset{\displaystyle O}{\|}}{\text{C}}\text{Cl} \xrightarrow{AlCl_3} \text{Ar—}\overset{\overset{\displaystyle O}{\|}}{\text{C}}\text{—R} + \text{HCl}$$

The largest number of acid halide reactions occur by nucleophilic acyl substitution mechanisms; Figure 24.7 illustrates some possibilities. The reactions are illustrated only for acid chlorides since these are the most common, but they also take place with all other acid halides.

Hydrolysis. Acid chlorides react with water to generate carboxylic acids. This hydrolysis reaction is a typical nucleophilic acyl substitution process and is initiated by attack of water on the acid chloride carbonyl group. The initially formed tetrahedral intermediate undergoes elimination of chloride ion and loss of a proton to give the product carboxylic acid.

$$\underset{\text{An acid chloride}}{\text{R—}\overset{\overset{\displaystyle :\ddot{O}:}{\|}}{\text{C}}\text{—Cl}} + :OH_2 \longrightarrow \left[\text{R—}\underset{\underset{\displaystyle Cl}{|}}{\overset{\overset{\displaystyle :\ddot{O}:^-}{|}}{\text{C}}}\text{—}\overset{+}{O}H_2 \longrightarrow \text{R—}\overset{\overset{\displaystyle O}{\|}}{\text{C}}\text{—}\overset{+}{O}H_2\right] \xrightarrow{-H^+} \underset{\text{A carboxylic acid}}{\text{R—}\overset{\overset{\displaystyle O}{\|}}{\text{C}}\text{—OH}} + \text{HCl}$$

Since HCl is generated during the hydrolysis, the reaction is often carried out in the presence of pyridine as a base to scavenge the HCl.

Alcoholysis. The reaction of an acid chloride with an alcohol is analogous to the reaction with water, and alcoholysis reactions constitute an excellent method for preparing esters.

$$\underset{\text{Benzoyl chloride}}{C_6H_5COCl} + \underset{\text{Cyclohexanol}}{C_6H_{11}OH} \xrightarrow{\text{Pyridine}} \underset{\substack{\text{Cyclohexyl benzoate} \\ (97\%)}}{C_6H_5CO_2C_6H_{11}} + \text{HCl}$$

$$\underset{\text{Acetyl chloride}}{CH_3\overset{\overset{\displaystyle O}{\|}}{C}\text{—Cl}} + \underset{\text{1-Butanol}}{CH_3CH_2CH_2CH_2OH} \xrightarrow{\text{Pyridine}} \underset{\text{Butyl acetate (90\%)}}{CH_3\overset{\overset{\displaystyle O}{\|}}{C}\text{—O—}CH_2CH_2CH_2CH_3}$$

As with hydrolysis, alcoholysis reactions are usually carried out in the presence of pyridine to react with the HCl formed and prevent it from causing side reactions. If this were not done, the HCl might react with the alcohol to form an alkyl chloride, or it might add to a carbon–carbon double bond if one is present elsewhere in the molecule.

The esterification of alcohols with acid chlorides is subject to steric hindrance. Bulky groups on either reaction partner slow down the rate of reaction considerably, resulting in a reactivity order among alcohols of primary > secondary > tertiary. In consequence, it is often possible to esterify an unhindered alcohol selectively in the presence of a more hindered one. This can be important in complex synthesis where it is necessary to distinguish chemically between similar functional groups. For example,

Secondary (more hindered and less reactive) OH

Primary (less hindered and more reactive) CH_2OH

+ 1 equiv C_6H_5COCl $\xrightarrow{\text{Pyridine}}$ 4-(benzoyloxymethyl)cyclohexanol ($CH_2OC(=O)C_6H_5$, OH)

~80%

Aminolysis. Acid chlorides react rapidly with ammonia and with amines to give amides. The reaction is rapid and yields are usually excellent. Both mono- and disubstituted amines can be used, as well as ammonia.

$$(CH_3)_2CHC(=O){-}Cl + 2\,{:}NH_3 \xrightarrow{H_2O} (CH_3)_2CHC(=O){-}NH_2 + \overset{+}{N}H_4\overset{-}{Cl}$$

2-Methylpropanoyl chloride → 2-Methylpropanamide (83%)

$$C_6H_5C(=O){-}Cl + 2\,{:}NH(CH_3)_2 \longrightarrow C_6H_5C(=O){-}N(CH_3)_2 + \overset{+}{N}H_2(CH_3)_2\overset{-}{Cl}$$

Benzoyl chloride → *N,N*-Dimethylbenzamide (92%)

Since HCl is generated during the reaction, 2 equiv of the amine must be used; 1 equiv reacts with the acid chloride, and 1 equiv reacts with HCl to form an ammonium chloride salt. If, however, the amine component is valuable, amide synthesis is often carried out by using 1 equiv of the desired amine plus 1 equiv of an inexpensive base to scavenge HCl. Sodium hydroxide is often used for this purpose, and aminolysis reactions carried out in this manner are sometimes referred to as **Schotten–Baumann reactions** after their discoverers.[3] For example, the medically useful sedative, trimetozine, is prepared by reaction of 3,4,5-trimethoxybenzoyl chloride with the amine, morpholine, in the presence of 1 equiv NaOH.

3,4,5-Trimethoxybenzoyl chloride + Morpholine $\xrightarrow[H_2O]{NaOH}$ Trimetozine (an amide) + NaCl

CH_3O, CH_3O, CH_3O — C(=O)—Cl + H—N̈(morpholine ring, O) → CH_3O, CH_3O, CH_3O — C(=O)—N(morpholine ring, O)

3,4,5-Trimethoxybenzoyl chloride

Morpholine

Trimetozine
(an amide)

+ NaCl

PROBLEM

24.7 Write the steps in the mechanism of the reaction between 3,4,5-trimethoxybenzoyl chloride and morpholine to form trimetozine.

PROBLEM

24.8 Trisubstituted amines such as triethylamine can be used in place of NaOH to scavenge HCl during aminolysis reactions. Why does triethylamine not react with acid chlorides to yield amides?

Reduction of acid chlorides. Acid chlorides are reduced by lithium aluminum hydride to yield primary alcohols. The reaction is of little practical value, however, since the parent carboxylic acids are generally more readily available and are themselves reduced by $LiAlH_4$ to yield alcohols.

$$C_6H_5C(=O)Cl \xrightarrow[2.\ H_3O^+]{1.\ LiAlH_4,\ ether} C_6H_5CH_2OH$$

Benzoyl chloride

Benzyl alcohol
(96%)

Reduction occurs via a typical nucleophilic acyl substitution mechanism in which hydride ion ($H:^-$) first attacks the carbonyl group and chloride ion is subsequently eliminated. This substitution leads to an aldehyde intermediate, which is then rapidly reduced by $LiAlH_4$ to yield the primary alcohol, as shown at the top of the next page.

[3]Carl Schotten (1853–1910); b. Marburg, Germany; Ph.D. Berlin (Hofmann); professor, University of Berlin.

$$R{-}\overset{\overset{:O:}{\|}}{C}{-}Cl + \text{“}\ddot{H}^{-}\text{”} \xrightarrow[\text{Ether}]{LiAlH_4} \left[R{-}\overset{\overset{:\ddot{O}:^-}{|}}{\underset{\underset{Cl}{|}}{C}}{-}H \longrightarrow \underset{\substack{\text{An aldehyde} \\ \textit{(not isolated)}}}{R{-}\overset{\overset{\ddot{O}}{\|}}{C}{-}H} \xrightarrow{\text{“}\ddot{H}^{-}\text{”}} \right] R{-}\overset{\overset{:\ddot{O}:^-}{|}}{\underset{\underset{H}{|}}{C}}{-}H \xrightarrow{H^+} \underset{\text{A 1° alcohol}}{R{-}\overset{\overset{OH}{|}}{\underset{\underset{H}{|}}{C}}{-}H}$$

The aldehyde intermediate can be isolated if less powerful hydride reducing agents are used. Lithium tri-*tert*-butoxyaluminum hydride, obtained from reaction of $LiAlH_4$ with 3 equiv of *tert*-butyl alcohol, is particularly effective for carrying out the partial reduction of acid chlorides to aldehydes:

$$3\ (CH_3)_3COH + LiAlH_4 \longrightarrow \underset{\text{Lithium tri-}\textit{tert}\text{-butoxyaluminum hydride}}{Li^+\ {}^-{:}AlH[O{-}C(CH_3)_3]_3} + 3\ H_2$$

$$\underset{p\text{-Nitrobenzoyl chloride}}{O_2N{-}C_6H_4{-}\overset{\overset{O}{\|}}{C}{-}Cl} \xrightarrow[2.\ H_3O^+]{1.\ LiAlH[OC(CH_3)_3]_3,\ \text{ether}} \underset{\substack{p\text{-Nitrobenzaldehyde} \\ (81\%)}}{O_2N{-}C_6H_4{-}\overset{\overset{O}{\|}}{C}{-}H}$$

Alternatively, the Rosenmund reduction of acid chlorides under catalytic hydrogenation conditions can be used, but this procedure is mechanistically quite different (Section 22.3).

$$\underset{\text{Cyclohexylacetyl chloride}}{C_6H_{11}{-}CH_2\overset{\overset{O}{\|}}{C}{-}Cl} \xrightarrow[\text{Ethyl acetate}]{H_2,\ Pd/BaSO_4} \underset{\substack{\text{Cyclohexylacetaldehyde} \\ (70\%)}}{C_6H_{11}{-}CH_2\overset{\overset{O}{\|}}{C}{-}H} + HCl$$

Reaction with organometallic reagents. Grignard reagents react with acid chlorides to yield tertiary alcohols in which two of the substituents are identical:

$$\mathrm{RMgX} + \mathrm{R'C(=O)}\text{—}\mathrm{Cl} \longrightarrow \mathrm{R'}\text{—}\mathrm{C(OH)(R)}\text{—}\mathrm{R} + \mathrm{MgXCl}$$

The mechanism by which this Grignard reaction occurs is similar to that we have just seen for $LiAlH_4$ reduction. The first equivalent of Grignard reagent attacks the acid chloride. Loss of chloride ion then yields a ketone intermediate, which reacts with a second equivalent of the organometallic reagent:

$$\mathrm{C_6H_5C(=O)Cl} \xrightarrow[\text{Ether}]{\mathrm{CH_3MgBr}} \left[\mathrm{C_6H_5C(=O)CH_3}\right] \xrightarrow[\text{2. } \mathrm{H_3O^+}]{\text{1. } \mathrm{CH_3MgBr}} \mathrm{C_6H_5C(OH)(CH_3)}\text{—}\mathrm{CH_3}$$

Acid chloride — Ketone (not isolated) — 3° alcohol

The ketone intermediate cannot usually be isolated in Grignard reactions, since addition of the second equivalent of organomagnesium reagent occurs so rapidly. Ketones can, however, be isolated from the reaction of acid chlorides with diorganocopper reagents (Section 22.4):

$$\mathrm{R}\text{—}\mathrm{C(=O)}\text{—}\mathrm{Cl} + \mathrm{R'_2Cu\!:^-\ Li^+} \longrightarrow \mathrm{R}\text{—}\mathrm{C(=O)}\text{—}\mathrm{R'}$$

Reactions between acid chlorides and a wide variety of lithium diorganocoppers, including dialkyl-, diaryl-, and dialkenylcoppers, are possible. Despite their apparent similarity to Grignard reactions, however, these diorganocopper reactions are almost certainly not typical nucleophilic acyl substitution processes. Rather, it is believed that diorganocopper reactions occur via a radical pathway. The reactions are generally carried out at −78°C in ether solutions, and yields are often excellent. For example, manicone, a substance secreted by male ants to coordinate ant pairing and mating, has been synthesized by reaction of lithium diethylcopper with *(E)*-2,4-dimethyl-2-hexenoyl chloride:

$$\mathrm{CH_3CH_2CH(CH_3)CH{=}C(CH_3)C(=O)Cl} + \mathrm{(CH_3CH_2)_2CuLi} \xrightarrow[-78^\circ\mathrm{C}]{\text{Ether}} \mathrm{CH_3CH_2CH(CH_3)CH{=}C(CH_3)C(=O)CH_2CH_3}$$

2,4-Dimethyl-2-hexenoyl chloride — Manicone (92%)

Note that lithium diorganocoppers react only with acid chlorides. Esters and amides are inert to diorganocopper reagents.

PROBLEM

24.9 It has been reported in the chemical literature that ketones can be prepared by reaction of acid chlorides with 1 equiv of Grignard reagent. It appears, however, that the Grignard reagent must be slowly added to a solution of the acid chloride, rather than vice versa. Draw on your knowledge of the relative reactivity of ketones and acid chlorides to explain this observation.

24.6 Acid anhydrides

PREPARATION OF ACID ANHYDRIDES

The most general method of preparation of acid anhydrides is by nucleophilic acyl substitution reaction of an acid chloride with a carboxylate salt. Both symmetrical and unsymmetrical acid anhydrides can be prepared in this manner; yields are usually high.

$$C_6H_5CH_2\overset{O}{\overset{\|}{C}}{-}O^-\ Na^+ + C_6H_5CH_2\overset{O}{\overset{\|}{C}}{-}Cl$$

Sodium phenylacetate   Phenylacetyl chloride

$$\xrightarrow{\text{Ether}} C_6H_5CH_2\overset{O}{\overset{\|}{C}}{-}O{-}\overset{O}{\overset{\|}{C}}CH_2C_6H_5 + NaCl$$

Bis(phenylacetic) anhydride (87%)

$$H{-}\overset{O}{\overset{\|}{C}}{-}O^-\ Na^+ + CH_3\overset{O}{\overset{\|}{C}}{-}Cl \xrightarrow[25^\circ C]{\text{Ether}} H\overset{O}{\overset{\|}{C}}{-}O{-}\overset{O}{\overset{\|}{C}}CH_3 + NaCl$$

Sodium formate   Acetyl chloride   Acetic formic anhydride (64%)

REACTIONS OF ACID ANHYDRIDES

The chemistry of acid anhydrides is similar to that of acid chlorides; anhydrides react more slowly than acid chlorides, but the kinds of reactions the two groups undergo are the same. Thus, acid anhydrides are reduced by

$LiAlH_4$ to form primary alcohols; they react with water to form acids, with alcohols to form esters, and with amines to form amides (Figure 24.8).

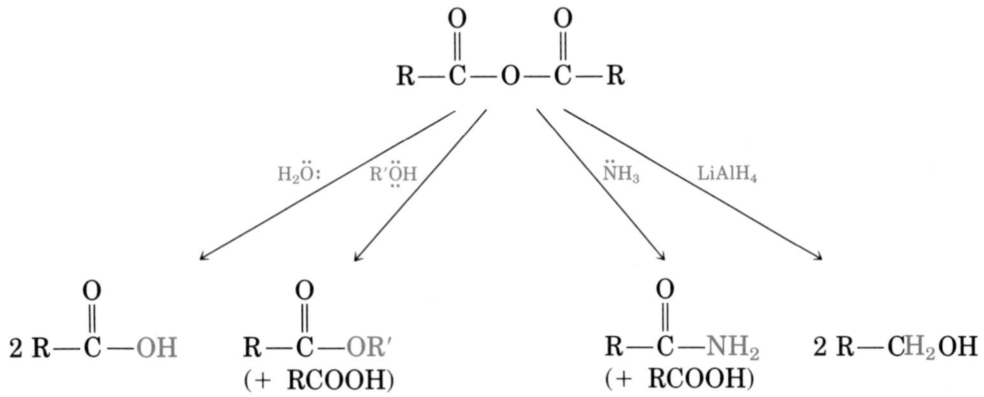

$$R-\overset{O}{\overset{\|}{C}}-O-\overset{O}{\overset{\|}{C}}-R$$

$H_2\ddot{O}$: → $2\,R-\overset{O}{\overset{\|}{C}}-OH$

$R'\ddot{O}H$ → $R-\overset{O}{\overset{\|}{C}}-OR'$ (+ RCOOH)

$\ddot{N}H_3$ → $R-\overset{O}{\overset{\|}{C}}-NH_2$ (+ RCOOH)

$LiAlH_4$ → $2\,R-CH_2OH$

Figure 24.8. Some reactions of acid anhydrides.

Acetic anhydride is often used to prepare acetate esters of complex alcohols and to prepare substituted acetamides from amines. For example, phenacetin, a drug found in headache remedies, is prepared commercially by reaction of *p*-ethoxyaniline with acetic anhydride. Aspirin (acetylsalicylic acid) is prepared similarly by the acetylation of *o*-hydroxybenzoic acid with acetic anhydride.

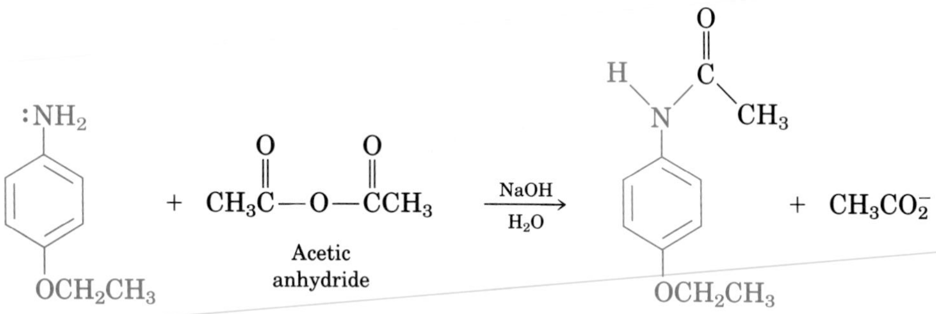

$$p\text{-}CH_3CH_2O{-}C_6H_4{-}\ddot{N}H_2 + CH_3\overset{O}{\overset{\|}{C}}-O-\overset{O}{\overset{\|}{C}}CH_3 \xrightarrow[H_2O]{NaOH} p\text{-}CH_3CH_2O{-}C_6H_4{-}N(H){-}\overset{O}{\overset{\|}{C}}{-}CH_3 + CH_3CO_2^-$$

p-Ethoxyaniline; Acetic anhydride; Phenacetin (an amide)

$$o\text{-}HO{-}C_6H_4{-}COOH + CH_3\overset{O}{\overset{\|}{C}}O\overset{O}{\overset{\|}{C}}CH_3 \xrightarrow[H_2O]{NaOH} o\text{-}CH_3\overset{O}{\overset{\|}{C}}{-}O{-}C_6H_4{-}COOH + CH_3CO_2^-$$

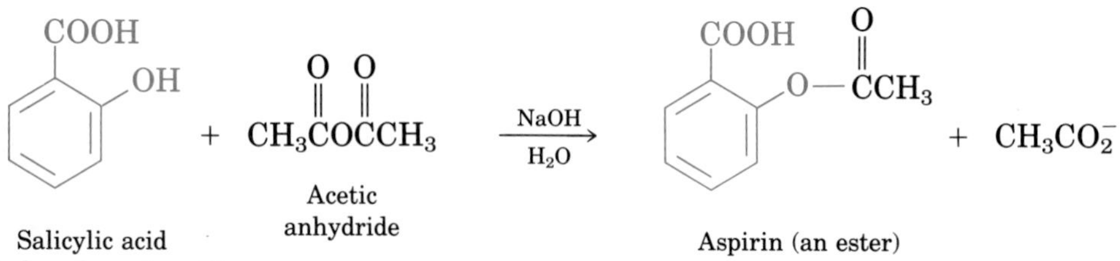

Salicylic acid (*o*-hydroxybenzoic acid); Acetic anhydride; Aspirin (an ester)

Note in the previous two examples that only "half" of the anhydride molecule is used; the other "half" acts as the leaving group during the nucleophilic acyl substitution step and produces carboxylate anion as a by-product. Thus, anhydrides are inefficient to use, and acid chlorides are normally preferred for introducing acyl substituents other than acetyl groups.

PROBLEM

24.10 What product would you expect to obtain from reaction of 1 equiv of methanol with a cyclic anhydride such as phthalic anhydride? What is the fate of the second "half" of the anhydride in such cases?

PROBLEM

24.11 Write the steps involved in the mechanism of reaction between *p*-ethoxyaniline and acetic anhydride to prepare phenacetin.

PROBLEM

24.12 Why is 1 equiv of a base such as sodium hydroxide required for the reaction between an amine and an anhydride to go to completion?

24.7 Esters

Esters are among the most important and most widespread of naturally occurring compounds. Many simple low-molecular-weight esters are pleasant-smelling liquids that are responsible for the fragrant odors of fruits and flowers. Methyl butanoate, for example, has been isolated from pineapple oil, and isopentyl acetate is a constituent of banana oil. The ester linkage is also present in many biologically important molecules such as animal fats.

The chemical industry uses esters for a variety of purposes. For example, ethyl acetate is a common solvent found in nail-polish remover and in rubber cement, and dialkyl phthalates are used to keep plastics from turning brittle.

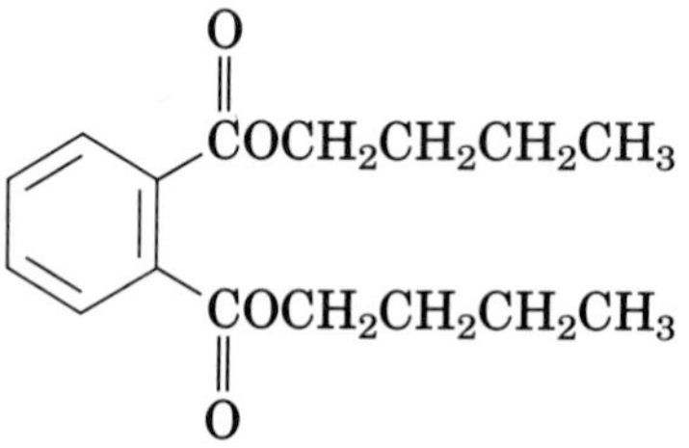

O
||
$COCH_2CH_2CH_2CH_3$

$COCH_2CH_2CH_2CH_3$
||
O

Dibutyl phthalate (a plasticizer)

PREPARATION OF ESTERS

Esters are prepared either from acids or from acid anhydrides; we have already discussed the best and most general of these methods in prior sections. Carboxylic acids are converted directly into esters by three methods: (1) S_N2 reaction of a carboxylate salt with a primary alkyl halide, (2) Fischer esterification of a carboxylic acid with a low-molecular-weight alcohol in the presence of a mineral acid catalyst, and (3) reaction of a carboxylic acid with diazomethane (Section 24.4). In addition, acid chlorides can be converted into esters by treatment with an alcohol in the presence of base (Section 24.5).

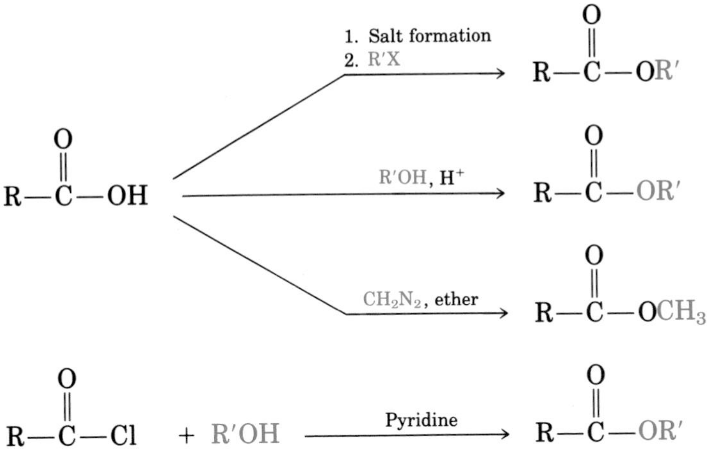

$$R{-}\overset{O}{\overset{\|}{C}}{-}OH \xrightarrow{\text{1. Salt formation; 2. } R'X} R{-}\overset{O}{\overset{\|}{C}}{-}OR'$$

$$R{-}\overset{O}{\overset{\|}{C}}{-}OH \xrightarrow{R'OH,\ H^+} R{-}\overset{O}{\overset{\|}{C}}{-}OR'$$

$$R{-}\overset{O}{\overset{\|}{C}}{-}OH \xrightarrow{CH_2N_2,\ \text{ether}} R{-}\overset{O}{\overset{\|}{C}}{-}OCH_3$$

$$R{-}\overset{O}{\overset{\|}{C}}{-}Cl + R'OH \xrightarrow{\text{Pyridine}} R{-}\overset{O}{\overset{\|}{C}}{-}OR'$$

REACTIONS OF ESTERS

Esters exhibit the same kinds of chemistry that we have seen for other acid derivatives, but they are less reactive toward nucleophiles than either acid chlorides or anhydrides. Figure 24.9 shows some general reactions of esters. All of these reactions are equally applicable to both acyclic and cyclic esters (**lactones**).

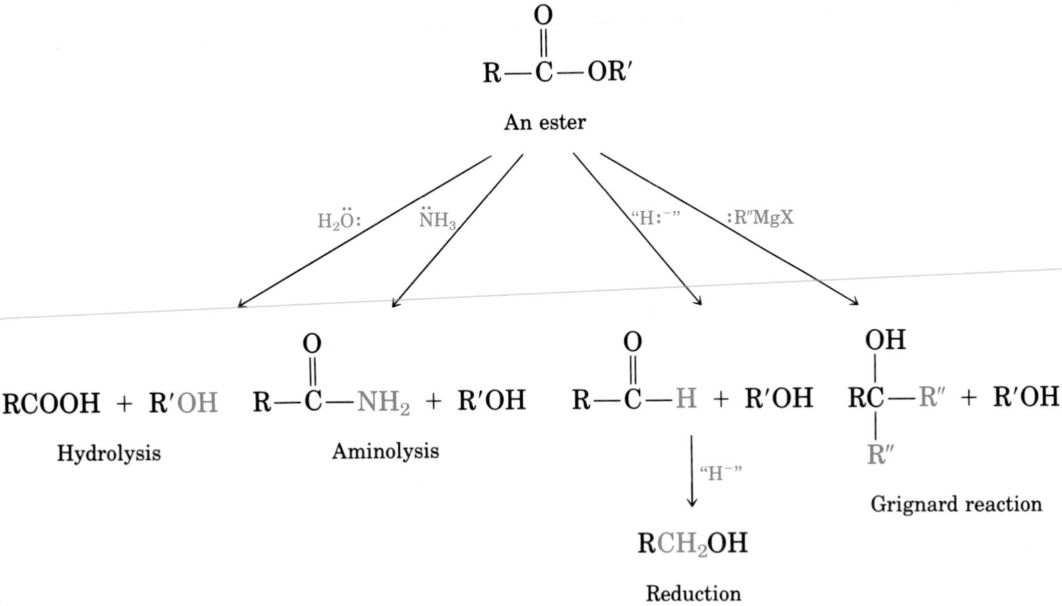

$$R{-}\overset{O}{\overset{\|}{C}}{-}OR'$$

An ester

$H_2\ddot{O}$: → $RCOOH + R'OH$ — Hydrolysis

$\ddot{N}H_3$ → $R{-}\overset{O}{\overset{\|}{C}}{-}NH_2 + R'OH$ — Aminolysis

"H:⁻" → $R{-}\overset{O}{\overset{\|}{C}}{-}H + R'OH \xrightarrow{\text{"}H^-\text{"}} RCH_2OH$ — Reduction

:R″MgX → $RC(OH)(R'')_2$, i.e. $R{-}\overset{OH}{\overset{|}{C}}(R''){-}R'' + R'OH$ — Grignard reaction

Figure 24.9. Some reactions of esters.

Hydrolysis. Esters are hydrolyzed either by aqueous base or by aqueous acid to yield the component acid and alcohol fragments.

$$R{-}\overset{\overset{\displaystyle O}{\|}}{C}{-}O{-}R' \xrightarrow[H^+ \text{ or } ^-OH]{H_2O} R{-}\overset{\overset{\displaystyle O}{\|}}{C}{-}OH + R'OH$$

Ester hydrolysis in alkaline solution is called **saponification,** after the Latin *sapo,* "soap." [The boiling of wood ash extract with animal fat to make soap (Section 32.2) is indeed a saponification, since wood ash contains hydroxide ion.] Because of its historical importance, ester hydrolysis has been extensively studied, and a large body of information has been amassed to show that alkaline ester hydrolysis occurs through a nucleophilic acyl substitution mechanism:

$$R{-}\overset{\overset{\displaystyle :O:}{\|}}{C}{-}O{-}R' + {}^-:\ddot{O}H \rightleftharpoons \left[R{-}\overset{\overset{\displaystyle :\ddot{O}:^-}{|}}{\underset{\underset{\displaystyle OH}{|}}{C}}{-}OR' \longrightarrow R{-}\overset{\overset{\displaystyle O}{\|}}{C}{-}OH + {}^-:\ddot{O}R' \right]$$

$$\downarrow$$

$$R{-}\overset{\overset{\displaystyle O}{\|}}{C}{-}\ddot{O}:^- + HOR'$$

One of the most elegant experiments in support of this mechanism involves isotope labeling. The Russian chemist D. N. Kursanov[4] showed that when ethyl propanoate labeled with ^{18}O in the ether-type oxygen was hydrolyzed in aqueous hydroxide, the ^{18}O label showed up exclusively in the ethanol fragment. None of the label remained with the propanoate fragment, indicating that saponification occurred by cleavage of the acyl–oxygen bond ($R\overset{\overset{O}{\|}}{C}{-}OR'$) rather than cleavage of the alkyl–oxygen bond ($R\overset{\overset{O}{\|}}{C}O{-}R'$). This result is just what we would expect, based on our knowledge of the nucleophilic acyl substitution mechanism.

$$CH_3CH_2\overset{\overset{\displaystyle O}{\|}}{C}{-}{}^{18}O{-}CH_2CH_3 + {}^-:OH \xrightarrow{H_2O} CH_3CH_2\overset{\overset{\displaystyle O}{\|}}{C}{-}O:^- + H{-}{}^{18}OCH_2CH_3$$

Acidic hydrolysis of esters can occur by more than one mechanism, depending on the structure of substrate. The usual pathway, however, is just the reverse of the Fischer esterification reaction (Section 24.4). The

[4]D. N. Kursanov (1899–); graduate of Moscow University (1924); professor, Moscow Textile Institute (1930–1947); Institute of Organic Chemistry (1947–1953); Institute of Scientific Information, USSR Academy of Sciences (1953–).

ester is first activated toward nucleophilic attack by protonation of the carboxyl oxygen atom. Nucleophilic attack by water, followed by transfer of a proton and elimination of alcohol, then yields the carboxylic acid (Figure 24.10).

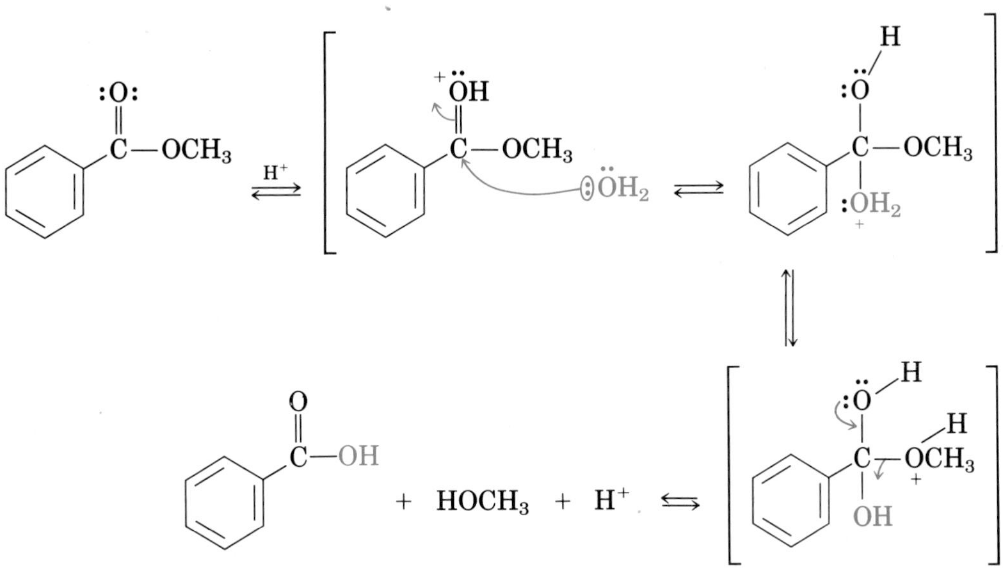

Figure 24.10. Mechanism of acidic ester hydrolysis (the forward reaction is a hydrolysis; the back-reaction is Fischer esterification).

PROBLEM

24.13 How would you synthesize the ^{18}O-labeled ethyl propanoate used by Kursanov in his mechanistic studies? Assume that ^{18}O-labeled acetic acid is your only source of isotopic oxygen.

PROBLEM

24.14 Explain the observation that alkaline hydrolysis of esters is irreversible, whereas acidic hydrolysis is reversible.

Aminolysis. Ammonia reacts with esters via a typical nucleophilic acyl substitution pathway to yield amides. The reaction is not often used, however, since higher yields are normally obtained by aminolysis of acid chlorides (Section 24.5).

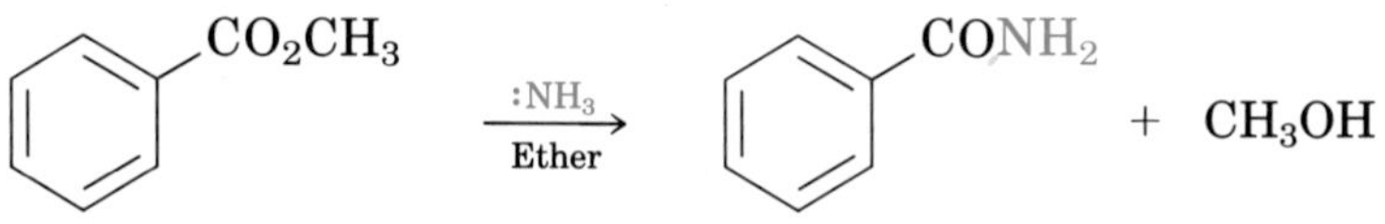

$$C_6H_5CO_2CH_3 \xrightarrow[\text{Ether}]{:NH_3} C_6H_5CONH_2 + CH_3OH$$

Methyl benzoate — Benzamide

Reduction. Treatment of esters with lithium aluminum hydride is an excellent method for preparing primary alcohols (Section 20.5).

$$\underset{\text{Ethyl 2-pentenoate}}{CH_3CH_2CH{=}CHC(=O)OCH_2CH_3} \xrightarrow[\text{2. } H_3O^+]{\text{1. LiAlH}_4\text{, ether}} \underset{\text{2-Penten-1-ol (91\%)}}{CH_3CH_2CH{=}CHCH_2OH} + CH_3CH_2OH$$

$$\text{(5-methyl-γ-butyrolactone)} \xrightarrow[\text{2. } H_3O^+]{\text{1. LiAlH}_4\text{, ether}} \underset{\text{1,4-Pentanediol (86\%)}}{HOCH_2CH_2CH_2CH(OH)CH_3}$$

The mechanism by which ester and lactone reductions occur is similar to that we saw earlier for acid chloride reduction: Hydride ion first adds to the carbonyl group, followed by elimination of alkoxide ion to yield an aldehyde intermediate. Further addition of hydride to this aldehyde produces the primary alcohol.

$$R{-}C(=O){-}OR' + H^- \xrightarrow{(\text{LiAlH}_4)} \left[R{-}C(O^-)(H){-}OR' \right]$$

$$\downarrow$$

$$R{-}C(=O){-}H + R'O:^- \xrightarrow{H_3O^+} R'OH$$

$$\downarrow\ H^-$$

$$\left[R{-}C(O^-)(H){-}H \right] \xrightarrow{H_3O^+} RCH_2OH$$

The aldehyde intermediate can often be isolated if DIBAH (diisobutylaluminum hydride) is used as the reducing agent in toluene solution. Great care must be taken—precisely 1 equiv of hydride reagent must be used, and the reaction must be carried out at −78°C. If these conditions are met, however, the DIBAH reduction of esters can be an excellent method of aldehyde synthesis.

$$\underset{\text{Ethyl dodecanoate}}{CH_3(CH_2)_{10}C(=O)OCH_2CH_3} \xrightarrow[\text{2. } H_3O^+]{\text{1. DIBAH in toluene}} \underset{\text{Dodecanal (88\%)}}{CH_3(CH_2)_{10}CHO} + CH_3CH_2OH$$

where DIBAH = $[(CH_3)_2CHCH_2]_2AlH$.

PROBLEM

24.15 What product would you expect from the reaction of butyrolactone with DIBAH?

Butyrolactone

Reaction with organometallic reagents. Esters and lactones react with 2 equiv of Grignard reagent or organolithium reagent to yield tertiary alcohols (Section 20.7). The reactions occur readily and give excellent yields of products by the usual nucleophilic addition mechanism.

$$C_6H_5COOCH_3 \xrightarrow[2.\ H_3O^+]{1.\ 2\ C_6H_5MgBr,\ ether} (C_6H_5)_3COH$$

Methyl benzoate → Triphenylmethanol (96%)

$$\text{Valerolactone} \xrightarrow[2.\ H_3O^+]{1.\ 2\ CH_3MgBr,\ ether} (CH_3)_2C(OH)CH_2CH_2CH_2CH_2OH$$

Valerolactone → 5-Methyl-1,5-hexanediol

24.8 Amides

PREPARATION OF AMIDES

Amides are usually prepared by reaction of an acid chloride with an amine. Ammonia, monosubstituted amines, and disubstituted amines all undergo this reaction (Section 24.5).

$$R{-}\overset{O}{\overset{\|}{C}}{-}Cl \xrightarrow{NH_3} R{-}\overset{O}{\overset{\|}{C}}{-}NH_2$$

$$R{-}\overset{O}{\overset{\|}{C}}{-}Cl \xrightarrow{R'NH_2} R{-}\overset{O}{\overset{\|}{C}}{-}NHR'$$

$$R{-}\overset{O}{\overset{\|}{C}}{-}Cl \xrightarrow{R'_2NH} R{-}\overset{O}{\overset{\|}{C}}{-}NR'_2$$

REACTIONS OF AMIDES

Amides are considerably less reactive than acid chlorides, acid anhydrides, or esters. Thus, the amide linkage is stable enough to serve as the basic unit from which all proteins are made (Chapter 31).

$$H_2N-\underset{}{\overset{R}{\overset{|}{C}}}H-COOH \Rightarrow \sim NH-\overset{R}{\overset{|}{C}}H-\underset{O}{\underset{\|}{C}}-NH-\overset{R'}{\overset{|}{C}}H-\underset{O}{\underset{\|}{C}}-NH-\overset{R''}{\overset{|}{C}}H-\underset{O}{\underset{\|}{C}}\sim$$

Amino acids

A protein
(a polyamide)

Amides can be hydrolyzed to carboxylic acids by refluxing in either aqueous acid or aqueous base. The conditions required for hydrolysis are more severe than those required for the hydrolysis of acid chlorides or esters, but the mechanisms are similar. The basic hydrolysis of amides yields an amine and a carboxylate ion as products, and occurs via nucleophilic addition of hydroxide to the amide carbonyl group, followed by elimination of amide ion ($^-:NH_2$). The acidic hydrolysis reaction occurs by nucleophilic addition of water to the protonated amide.

Acidic hydrolysis

$$R-\overset{O}{\overset{\|}{C}}-NH_2 \overset{H^+}{\rightleftharpoons} \left[R-\overset{\overset{+}{O}H}{\overset{\|}{C}}-NH_2 \overset{:\ddot{O}H_2}{\rightleftharpoons} R-\underset{\overset{+}{O}H_2}{\underset{|}{\overset{OH}{\overset{|}{C}}}}-NH_2 \rightleftharpoons R-\underset{OH}{\underset{|}{\overset{OH}{\overset{|}{C}}}}-\overset{+}{N}H_3 \right]$$

$$\downarrow$$

$$NH_4^+ \overset{H^+}{\longleftarrow} NH_3 + R-\overset{O}{\overset{\|}{C}}-OH$$

Basic hydrolysis

$$R-\overset{O}{\overset{\|}{C}}-NH_2 + {}^-:\ddot{O}H \rightleftharpoons \left[R-\underset{OH}{\underset{|}{\overset{O^-}{\overset{|}{C}}}}-NH_2 \longrightarrow R-\overset{O}{\overset{\|}{C}}-OH + {}^-:NH_2 \right]$$

$$\downarrow$$

$$R-\overset{O}{\overset{\|}{C}}-O^- + NH_3$$

Like other carboxylic acid derivatives, amides can be reduced by lithium aluminum hydride. The product of this reduction, however, is an amine rather than an alcohol.

$$CH_3(CH_2)_{10}\overset{O}{\overset{\|}{C}}-NHCH_3 \xrightarrow[2.\ H_2O]{1.\ LiAlH_4,\ ether} CH_3(CH_2)_{10}\underset{H}{\underset{|}{\overset{H}{\overset{|}{C}}}}-NHCH_3$$

N-Methyldodecanamide

Dodecylmethylamine (95%)

The net effect of this amide reduction reaction is to convert the amide carbonyl group into a methylene group ($C{=}O \rightarrow CH_2$). This kind of reaction is specific for amides and does not occur with other carboxylic acid derivatives. The reaction is thought to occur by initial nucleophilic addition of hydride ion to the amide carbonyl group, followed by expulsion of the oxygen atom as an aluminate anion. The intermediate imine is then further reduced by $LiAlH_4$ to yield the amine. Note that this reduction is mechanistically similar to the reaction of a ketone or aldehyde with an amine to yield an imine ($C{=}O \rightarrow C{=}NR$). In both cases, the carbonyl-group oxygen atom is removed and the C=O bond becomes a C=Nu bond:

$$R-\overset{O}{\overset{\|}{C}}-NH_2 \xrightarrow[Ether]{:H^-\ \overset{+}{Li}AlH_3} \left[R-\underset{H}{\underset{|}{\overset{O-\overset{-}{A}lH_3}{\overset{|}{C}}}}-\ddot{N}H_2 \longrightarrow R-\overset{^+NH_2}{\overset{\|}{C}}-H \underset{}{\overset{-H^+}{\rightleftharpoons}} R-\overset{\ddot{N}H}{\overset{\|}{C}}-H \right]$$

Iminium ion — Imine

$$\downarrow :H^-$$

$$R-\underset{H}{\underset{|}{\overset{:NH_2}{\overset{|}{C}}}}-H \xleftarrow{H^+} \left[R-\underset{H}{\underset{|}{\overset{^-:\ddot{N}H}{\overset{|}{C}}}}-H \right]$$

Recall the following:

$$R-\overset{O}{\overset{\|}{C}}-H \xrightarrow{:NH_3} R-\overset{NH}{\overset{\|}{C}}-H$$

Imine

Lithium aluminum hydride reduction is equally effective with both acyclic and cyclic amides **(lactams).** Lactam reductions provide cyclic amines in good yield, and constitute a valuable method of synthesis.

$$\text{A lactam} \xrightarrow[2.\ H_2O]{1.\ LiAlH_4,\ ether} \text{A cyclic amine (80\%)}$$

A lactam

A cyclic amine (80%)

PROBLEM

24.16 How would you use the reaction between an amide and $LiAlH_4$ as the key step in going from bromocyclohexane to (dimethylaminomethyl)cyclohexane? Formulate all steps involved in the reaction sequence.

$$\text{C}_6\text{H}_{11}\text{–Br} \overset{?}{\Longrightarrow} \text{C}_6\text{H}_{11}\text{–CH}_2\text{N(CH}_3)_2$$

(Dimethylaminomethyl)cyclohexane

24.9 Nitriles

Nitriles are not related to carboxylic acids in the same sense that acyl derivatives are; nevertheless, the chemistry of nitriles and carboxylic acids is so entwined that the two classes of compounds should be considered together.

PREPARATION OF NITRILES

The simplest method of nitrile preparation is S_N2 reaction of cyanide ion with a primary alkyl halide, a reaction we have already discussed (Section 23.6). This method is limited by the usual S_N2 steric constraints to the synthesis of α-unsubstituted nitriles.

$$\text{R—CH}_2\text{—Br} + \text{Na}^+\text{CN}^- \xrightarrow[\text{reaction}]{S_N2} \text{R—CH}_2\text{—CN} + \text{NaBr}$$

Another excellent method for preparing nitriles is dehydration of a primary amide. Thionyl chloride is often used to effect this reaction, although other dehydrating agents such as P_2O_5, $POCl_3$, or acetic anhydride can be used.

$$\underset{\text{2-Ethylhexanamide}}{\text{CH}_3\text{CH}_2\text{CH}_2\text{CH}_2\underset{\text{CH}_2\text{CH}_3}{\text{CH}}\text{—}\overset{\text{O}}{\overset{\|}{\text{C}}}\text{—NH}_2} \xrightarrow[80°\text{C}]{\text{SOCl}_2,\ \text{benzene}} \underset{\text{2-Ethylhexanenitrile (94\%)}}{\text{CH}_3\text{CH}_2\text{CH}_2\text{CH}_2\underset{\text{CH}_2\text{CH}_3}{\text{CH}}\text{—C}\equiv\text{N}} + \text{SO}_2 + 2\ \text{HCl}$$

This dehydration is thought to occur by initial reaction on the amide oxygen atom, followed by an elimination reaction.

$$\text{R—}\overset{\text{O}}{\overset{\|}{\text{C}}}\text{—}\ddot{\text{N}}\text{H}_2 + \text{O=SCl}_2 \longrightarrow \left[\text{R—}\overset{\text{O—}\overset{\text{O}}{\overset{\|}{\text{S}}}\text{—Cl}}{\overset{|}{\text{C}}}\text{=N—H} + \text{HCl}\right] \longrightarrow \text{R—C}\equiv\text{N} + \text{SO}_2 + \text{HCl}$$

Although both methods of nitrile synthesis—cyanide displacement on an alkyl halide and amide dehydration—are useful, the synthesis from amides is more general since it is not limited by steric hindrance.

REACTIONS OF NITRILES

The two most important reactions of nitriles are hydrolysis and reduction. In addition, nitriles can be partially reduced and hydrolyzed to yield aldehydes, and can be treated with Grignard reagents to yield ketones (Figure 24.11).

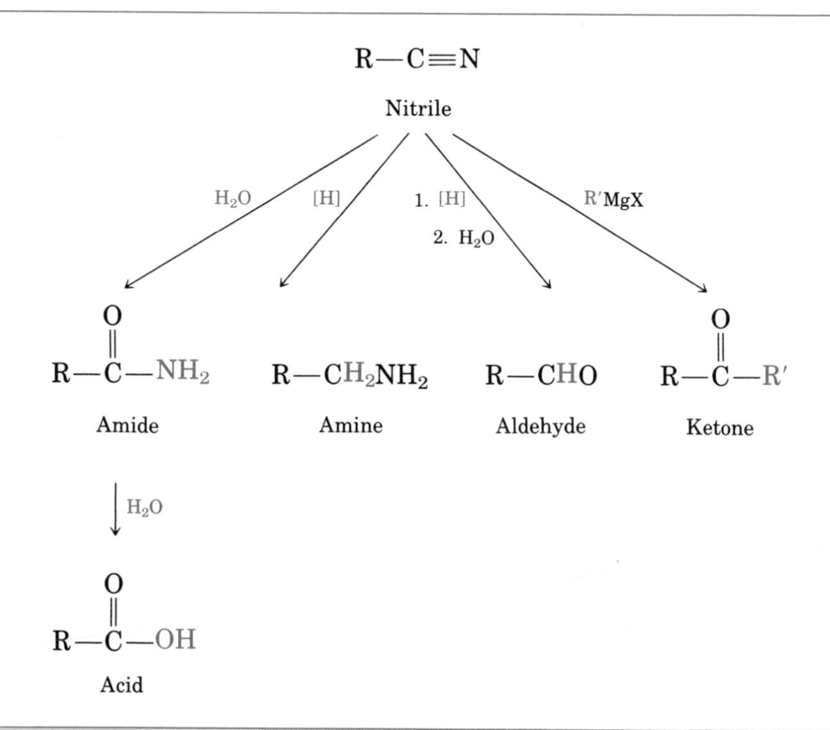

Figure 24.11. Some reactions of nitriles.

Hydrolysis. Nitriles are hydrolyzed in either acidic or basic aqueous solution to yield carboxylic acids and ammonia:

$$R{-}C{\equiv}N \xrightarrow{H_3O^+ \text{ or } {}^-{:}OH} RCOOH + NH_3$$

The mechanism of the alkaline hydrolysis involves nucleophilic addition of hydroxide ion to the polar $^{\delta+}C{\equiv}N^{\delta-}$ bond in a manner analogous to that of nucleophilic addition to a polar carbonyl C=O bond. The initial product is a hydroxy imine, which is rapidly converted to an amide in a step similar to the conversion of an enol to a ketone. Further hydrolysis of the amide then yields the carboxylic acid:

$$R{-}C{\equiv}N: + :\ddot{O}H^- \underset{}{\overset{H_2O}{\rightleftharpoons}} \left[R{-}\underset{}{\overset{O{-}H}{C}}{=}\ddot{N}^- \overset{H_2O}{\rightleftharpoons} R{-}\overset{O{-}H}{C}{=}\ddot{N}{-}H \right] + {}^-OH$$

Nitrile (δ+ on C, δ− on N) — Hydroxy imine

$$\text{Hydroxy imine} \rightleftharpoons R{-}\overset{O}{\overset{\|}{C}}{-}\ddot{N}H_2 \text{ (Amide)} \xrightarrow[2.\ H_3O^+]{1.\ HO^-,\ H_2O} NH_3 + R{-}\overset{O}{\overset{\|}{C}}OH \text{ (Carboxylic acid)}$$

Recall the following:

$$R{-}C{\equiv}CH \xrightarrow[Hg^{2+}]{H_3O^+} R{-}\overset{O{-}H}{C}{=}CH_2 \longrightarrow R{-}\overset{O}{\overset{\|}{C}}{-}CH_3$$

Enol — Ketone

The conditions required for nitrile alkaline hydrolysis are severe (KOH, 200°C), and the amide intermediate can sometimes be isolated if milder conditions are employed.

PROBLEM

24.17 Formulate the steps involved in the hydrolysis of a nitrile with aqueous acid to yield the carboxylic acid.

Reduction. Treatment of nitriles with lithium aluminum hydride gives primary amines in high yields:

$$\text{o-}CH_3C_6H_4C{\equiv}N \xrightarrow[2.\ H_2O]{1.\ LiAlH_4,\ ether} \text{o-}CH_3C_6H_4CH_2NH_2$$

o-Methylbenzonitrile — *o*-Methylbenzylamine (88%)

The reaction occurs by nucleophilic addition of hydride ion to the polar C≡N bond, yielding an imine anion. This intermediate undergoes further addition of a second equivalent of hydride, giving the final product. If, however, a reducing agent less powerful than $LiAlH_4$ is used, the second addition of hydride does not occur, and the imine can be hydrolyzed to yield an aldehyde. The reagent DIBAH is particularly useful for this partial reduction of nitriles and provides an excellent method of preparing aldehydes:

$$R{-}C{\equiv}N + \ddot{H}^- \longrightarrow \left[R{-}\overset{\overset{:\ddot{N}^-}{\|}}{C}{-}H \right] \begin{cases} \xrightarrow[2.\ H_2O]{1.\ LiAlH_4,\ ether} R{-}CH_2NH_2 \ \text{(Primary amine)} \\ \xrightarrow{H_2O} R{-}\overset{\overset{O}{\|}}{C}{-}H \ \text{(Aldehyde)} \end{cases}$$

Nitrile — Imine anion

For example,

$$\text{2-(1-methylethenyl)cyclohexyl}{-}\underset{CH_3}{CH}CH_2C{\equiv}N \xrightarrow[2.\ H_2O]{1.\ DIBAH,\ toluene,\ -78°C} \text{2-(1-methylethenyl)cyclohexyl}{-}\underset{CH_3}{CH}{-}CH_2{-}CHO$$

96%

Reaction with organometallic reagents. Grignard reagents add to nitriles, giving intermediate imine anions that can be hydrolyzed to ketones:

$$R{-}C{\equiv}N: + \ddot{R}'^- \overset{+}{Mg}X \longrightarrow \left[R{-}\overset{\overset{:\ddot{N}^-\ \overset{+}{Mg}X}{\|}}{C}{-}R' \right] \xrightarrow{H_3O^+} R{-}\overset{\overset{O}{\|}}{C}{-}R' + NH_3$$

Nitrile — Imine anion — Ketone

The reaction is similar to the DIBAH reduction of nitriles, except that the attacking nucleophile is an alkyl anion rather than a hydride ion. Yields are generally high. For example,

$$C_6H_5{-}C{\equiv}N \xrightarrow[2.\ H_3O^+]{1.\ CH_3CH_2MgBr,\ ether} C_6H_5{-}\overset{\overset{O}{\|}}{C}CH_2CH_3$$

Benzonitrile — Propiophenone (89%)

PROBLEM

24.18 Propose an explanation of the observation that only 1 equiv of Grignard reagent adds to nitriles. Why does a second Grignard reagent not add to the initially formed imine anion?

24.10 Thiol esters: biological carboxylic acid derivatives

Nucleophilic acyl substitution reactions take place in living organisms just as they take place in the chemical laboratory, and the same principles apply in both cases. Nature, however, uses a thiol ester as the reactive acylating agent, rather than an acid chloride or acid anhydride. The pK_a of a typical alkane thiol (R—SH) is about 10, placing thiols midway between carboxylic acids ($pK_a \approx 5$) and alcohols ($pK_a \approx 16$) in acid strength. As a result, thiol esters are intermediate in reactivity between acid anhydrides and esters. They are not so reactive that they hydrolyze rapidly like anhydrides, yet they are more reactive than normal esters.

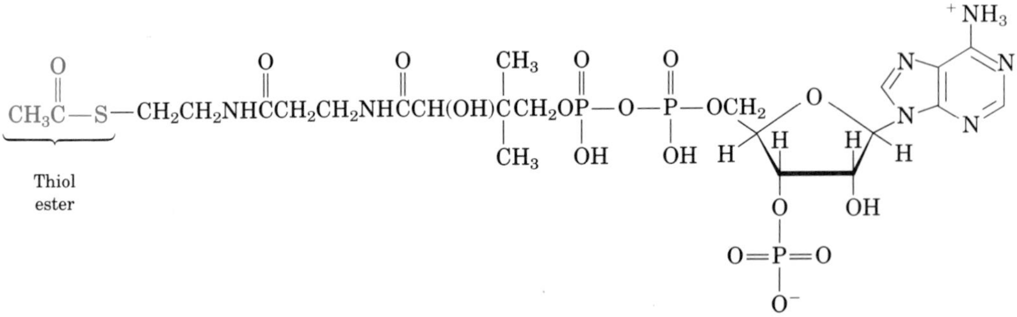

$$CH_3\overset{O}{\overset{\|}{C}}-S-CH_2CH_2NH\overset{O}{\overset{\|}{C}}CH_2CH_2NH\overset{O}{\overset{\|}{C}}CH(OH)C(CH_3)_2CH_2O\overset{O}{\overset{\|}{P}}(OH)-O-\overset{O}{\overset{\|}{P}}(OH)-OCH_2\text{-(ribose 3'-}OPO_3^{-}\text{)-adenine}$$

Thiol ester

Figure 24.12. Acetyl CoA. The structural formula is abbreviated as

$$CH_3\overset{O}{\overset{\|}{C}}-S-CoA$$

Acetyl coenzyme A (abbreviated acetyl CoA) is the most common thiol ester found in nature. Figure 24.12 shows its structure, which was determined in 1953. Acetyl CoA is an enormously complex molecule by comparison with acetyl chloride or acetic anhydride, yet it serves exactly the same purpose. Nature uses acetyl CoA as a reactive acylating agent in nucleophilic acyl substitution reactions:

$$CH_3\overset{O}{\overset{\|}{C}}-S-CoA + \ddot{Nu}^- \longrightarrow CH_3-\overset{O}{\overset{\|}{C}}-Nu + {}^-\!:S-CoA$$

For example, *N*-acetylglucosamine, an important constituent of surface membranes in mammalian cells, is synthesized in nature by an aminolysis reaction between glucosamine and acetyl CoA:

$$\text{Glucosamine} + CH_3C(=O)\text{—S—CoA} \longrightarrow N\text{-Acetylglucosamine} + \text{H—S—CoA}$$

Glucosamine
(an amine)

N-Acetylglucosamine
(an amide)

24.11 Spectroscopy of carboxylic acid derivatives

INFRARED SPECTROSCOPY

All carbonyl-containing compounds have intense infrared absorptions in the range 1650–1850 cm^{-1}, and the exact position of the absorption provides information about the nature of a given carbonyl group. Table 24.2 (p. 802) shows the typical absorption positions of some acid derivatives. For comparison, the absorptions of ketones, aldehydes, and acids are included in the table.

As the data in the table indicate, acid chlorides are readily detected in the infrared by their characteristic carbonyl-group absorption near 1800 cm^{-1}. Acid anhydrides can be identified by the fact that they show two absorptions in the carbonyl region, one at 1820 cm^{-1} and the second at 1760 cm^{-1}. Esters are detected by their absorption at 1735 cm^{-1}, a position somewhat higher than for either ketones or aldehydes. Amides, by contrast, absorb near the low end of the carbonyl region, and the degree of substitution on nitrogen affects the exact position of the infrared band. Nitriles are easily recognized by the presence of an intense absorption near 2250 cm^{-1}. Few other functional groups absorb in this region, and infrared spectroscopy is therefore highly diagnostic for nitriles.

PROBLEM

24.19 What kinds of functional groups might compounds have if they exhibit the following infrared spectral properties?
(a) Absorption at 1735 cm^{-1} (b) Absorption at 1810 cm^{-1}
(c) Two absorptions, at 2500–3300 cm^{-1} and at 1710 cm^{-1}
(d) Absorption at 2250 cm^{-1} (e) Absorption at 1715 cm^{-1}

PROBLEM

24.20 Propose structures for compounds having the following formulas and infrared absorptions:
(a) C_3H_5N, 2250 cm^{-1} (b) $C_6H_{12}O_2$, 1735 cm^{-1}
(c) C_4H_9NO, 1650 cm^{-1} (d) C_4H_5ClO, 1780 cm^{-1}

TABLE 24.2 Infrared absorptions of some carbonyl compounds

Carbonyl type	*Example*	*Infrared absorption (cm^{-1})*
Aliphatic acid chloride	Acetyl chloride	1810
Aromatic acid chloride	Benzoyl chloride	1770
Aliphatic acid anhydride	Acetic anhydride	1820, 1760
Aliphatic ester	Ethyl acetate	1735
Aromatic ester	Ethyl benzoate	1720
Six-membered ring lactone	(six-membered ring lactone structure: O, O)	1735
Aliphatic amide	Acetamide	1690
Aromatic amide	Benzamide	1675
N-Substituted amide	*N*-Methylacetamide	1680
N,N-Disubstituted amide	*N,N*-Dimethylacetamide	1650
Aliphatic nitrile	Acetonitrile	2250
Aromatic nitrile	Benzonitrile	2230
Aliphatic aldehyde	Acetaldehyde	1730
Aliphatic ketone	Acetone	1715
Aliphatic carboxylic acid	Acetic acid	1710

NUCLEAR MAGNETIC RESONANCE SPECTROSCOPY

Carbon-13 NMR is a valuable technique for determining the presence or absence of carbonyl groups in molecules of unknown structure, but precise information about the nature of the carbonyl group is difficult to obtain. Carbonyl carbon atoms show resonances in the range 160–210 δ, as Table 24.3 shows.

TABLE 24.3 Positions of ^{13}C NMR absorptions in some carbonyl compounds

Compound	*Absorption (δ)*	*Compound*	*Absorption (δ)*
Acetic acid	177.7	Acetic anhydride	166.9
Ethyl acetate	170.7	Acetonitrile	117.4
Acetyl chloride	170.3	Acetone	205.6
Acetamide	172.6	Acetaldehyde	201.0

Protons on the carbon next to a carbonyl group are slightly deshielded and absorb near 2 δ in the ^{1}H NMR spectrum. The exact nature of the carbonyl group cannot be distinguished by ^{1}H NMR, however, since all acyl derivatives absorb in the same range. Figure 24.13 shows the ^{1}H NMR spectrum of ethyl acetate and indicates that the acetate protons absorb at 2.1 δ.

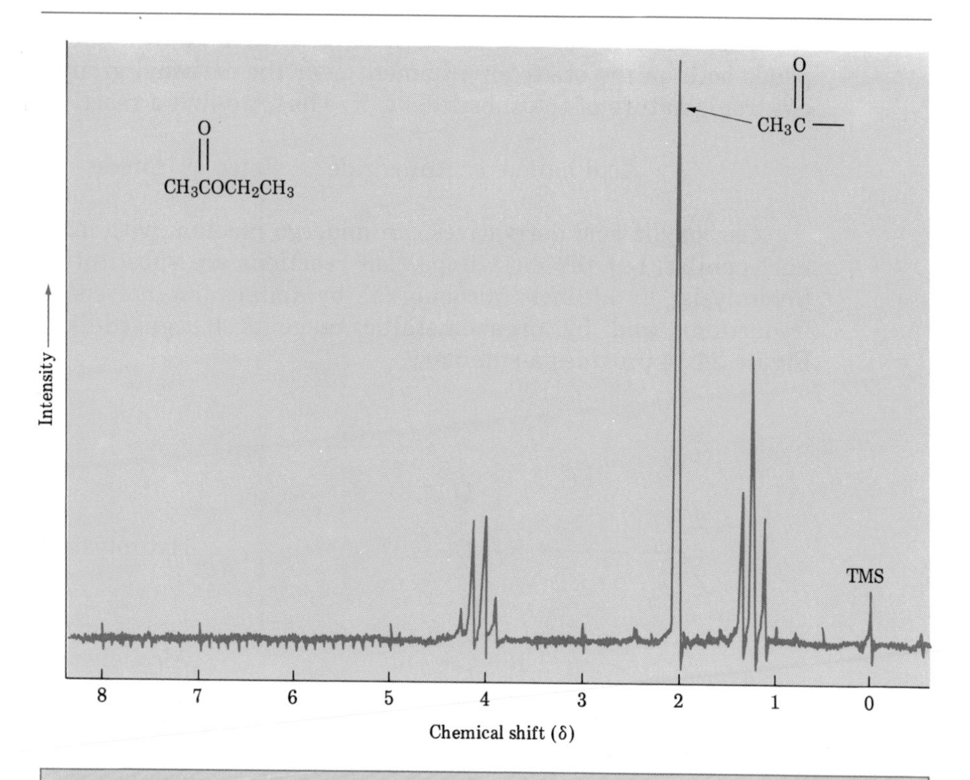

Figure 24.13. Proton NMR spectrum of ethyl acetate.

24.12 Summary

Carboxylic acids can be transformed into a variety of acid derivatives in which the acid —OH group has been replaced by other substituents. Acid chlorides, acid anhydrides, esters, and amides are the most important derivatives.

The chemistry of the different acid derivatives is similar and is dominated by a single general reaction type—the nucleophilic acyl substitution reaction. Mechanistically, these substitutions take place by addition of a nucleophile to the polar carbonyl group of the acid derivative, followed by expulsion of a leaving group from the tetrahedral intermediate, as shown at the top of the next page.

$$\mathrm{R{-}\overset{\overset{\Large\cdot\cdot}{O}}{\overset{\|}{C}}{-}X} + :\mathrm{Nu}^- \longrightarrow \left[\mathrm{R{-}\overset{:\ddot{O}:^-}{\overset{|}{C}}(X){-}Nu}\right] \longrightarrow \mathrm{R{-}\overset{O}{\overset{\|}{C}}{-}Nu} + :\mathrm{X}^-$$

Tetrahedral intermediate

where X = Cl, Br, I (acid halide); OR (ester); OCOR (anhydride); or NH_2 (amide).

The reactivity of an acid derivative, RCOX, toward substitution depends both on the steric environment near the carbonyl group and on the electronic nature of the substituent, X. Thus, we find a reactivity order:

$$\text{Acid halide} > \text{Anhydride} > \text{Ester} > \text{Amide}$$

Carboxylic acid derivatives can undergo reaction with many different nucleophiles, but the most important reactions are substitution by water (hydrolysis), by alcohols (alcoholysis), by amines (aminolysis), by hydride (reduction), and by organometallic reagents (Grignard-like reaction). Figure 24.14 provides a summary.

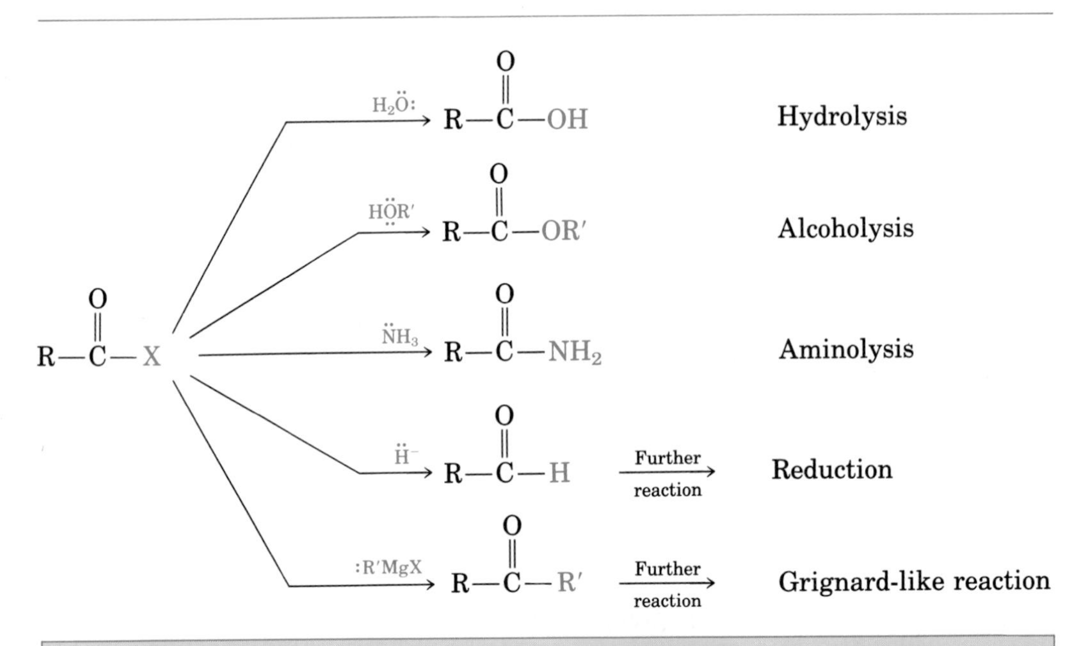

Figure 24.14. General reactions of carboxylic acid derivatives.

Nitriles can also be considered carboxylic acid derivatives, and they undergo nucleophilic addition to the polar C≡N bond in the same way that carbonyl compounds do. The most important reactions of nitriles are their hydrolysis to carboxylic acids, their reduction to primary amines, their partial reduction to aldehydes, and their reaction with organometallic reagents to yield ketones.

Nature employs nucleophilic acyl substitution reactions in the biosynthesis of many molecules and uses thiol esters for the purpose. Acetyl coenzyme A (acetyl CoA) is a complex thiol ester that is employed in living systems to acetylate amines and alcohols.

Infrared spectroscopy is an extremely valuable tool for the structure analysis of acid derivatives. Acid chlorides, anhydrides, esters, amides, and nitriles all show characteristic infrared absorptions that can be used to identify these functional groups in unknowns. Carbon-13 NMR is useful for establishing the presence or absence of a carbonyl carbon atom, but this spectroscopic technique does not usually allow exact identification of the different functional groups.

24.13 Summary of reactions

1. Reactions of carboxylic acids
   a. Conversion into acid chlorides (Section 24.4)

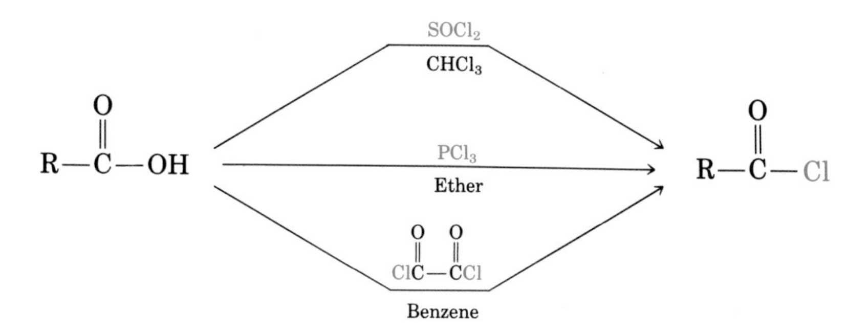

   b. Conversion into cyclic acid anhydrides (Section 24.4)

$$\mathrm{(CH_2)_n(COOH)_2 \xrightarrow{200^\circ C} (CH_2)_n(C{=}O)_2O + H_2O}$$

   where $n = 2$ or 3.

   c. Conversion into esters (Section 24.4)

$$\mathrm{RC(=O)O{:}^- + R'X \longrightarrow RC(=O)-O-R'}$$

Via S_N2 reaction

$$\mathrm{RC(=O)OH + R'OH \xrightarrow{H^+} RC(=O)-O-R' + H_2O}$$

Via Fischer esterification

$$\mathrm{RCOH} + \mathrm{CH_2N_2} \longrightarrow \mathrm{RCOCH_3} + \mathrm{N_2}$$

(In RCOH and $RCOCH_3$, the C bears =O.)

Via diazomethane

d. Conversion into amides (Section 24.4)

$$\mathrm{R\overset{O}{\overset{\|}{C}}OH} + \mathrm{NH_3} \xrightarrow[200°C]{} \mathrm{R\overset{O}{\overset{\|}{C}}{-}NH_2} + \mathrm{H_2O}$$

2. Reactions of acid chlorides
   a. Hydrolysis to yield acids (Section 24.5)

$$\mathrm{R{-}\overset{O}{\overset{\|}{C}}{-}Cl} + \mathrm{H_2O} \longrightarrow \mathrm{R{-}\overset{O}{\overset{\|}{C}}{-}OH} + \mathrm{HCl}$$

   b. Alcoholysis to yield esters (Section 24.5)

$$\mathrm{R{-}\overset{O}{\overset{\|}{C}}{-}Cl} + \mathrm{R'OH} \xrightarrow{\text{Pyridine}} \mathrm{R{-}\overset{O}{\overset{\|}{C}}{-}OR'} + \mathrm{HCl}$$

   c. Aminolysis to yield amides (Section 24.5)

$$\mathrm{R{-}\overset{O}{\overset{\|}{C}}{-}Cl} + 2\,\mathrm{NH_3} \longrightarrow \mathrm{R{-}\overset{O}{\overset{\|}{C}}{-}NH_2} + \mathrm{NH_4Cl}$$

   d. Reduction to yield primary alcohols (Section 24.5)

$$\mathrm{R{-}\overset{O}{\overset{\|}{C}}{-}Cl} \xrightarrow[\text{2. } \mathrm{H_3O^+}]{\text{1. } \mathrm{LiAlH_4}\text{, ether}} \mathrm{R{-}CH_2OH}$$

   e. Partial reduction to yield aldehydes (Section 24.5)

$$\mathrm{R{-}\overset{O}{\overset{\|}{C}}{-}Cl} \xrightarrow[\text{2. } \mathrm{H_3O^+}]{\text{1. } \mathrm{LiAlH(O\text{-}\textit{tert}\text{-}Bu)_3}\text{, ether}} \mathrm{R{-}CHO}$$

   f. Rosenmund reduction to yield aldehydes (Section 24.5)

$$\mathrm{R{-}\overset{O}{\overset{\|}{C}}{-}Cl} \xrightarrow[\text{Ethyl acetate}]{\mathrm{H_2}\text{, Pd, }\mathrm{BaSO_4}} \mathrm{R{-}CHO}$$

   g. Grignard reaction to yield tertiary alcohols (Section 24.5)

$$R{-}\overset{\overset{\displaystyle O}{\|}}{C}{-}Cl \xrightarrow[2.\ H_3O^+]{1.\ 2\ R'MgX,\ ether} R{-}\underset{\underset{\displaystyle R'}{|}}{\overset{\overset{\displaystyle OH}{|}}{C}}{-}R'$$

h. Diorganocopper reaction to yield ketones (Section 24.5)

$$R{-}\overset{\overset{\displaystyle O}{\|}}{C}{-}Cl \xrightarrow[Ether]{R'_2CuLi} R{-}\overset{\overset{\displaystyle O}{\|}}{C}{-}R'$$

3. Reactions of acid anhydrides
   a. Hydrolysis to yield acids (Section 24.6)

$$R{-}\overset{\overset{\displaystyle O}{\|}}{C}{-}O{-}\overset{\overset{\displaystyle O}{\|}}{C}{-}R + H_2O \longrightarrow 2\ R{-}\overset{\overset{\displaystyle O}{\|}}{C}{-}OH$$

   b. Alcoholysis to yield esters (Section 24.6)

$$R{-}\overset{\overset{\displaystyle O}{\|}}{C}{-}O{-}\overset{\overset{\displaystyle O}{\|}}{C}{-}R + R'OH \xrightarrow{Pyridine} R{-}\overset{\overset{\displaystyle O}{\|}}{C}{-}OR' + RCOOH$$

   c. Aminolysis to yield amides (Section 24.6)

$$R{-}\overset{\overset{\displaystyle O}{\|}}{C}{-}O{-}\overset{\overset{\displaystyle O}{\|}}{C}{-}R + 2\ NH_3 \longrightarrow R{-}\overset{\overset{\displaystyle O}{\|}}{C}{-}NH_2 + RCO_2^-NH_4^+$$

   d. Reduction to yield primary alcohols (Section 24.6)

$$R{-}\overset{\overset{\displaystyle O}{\|}}{C}{-}O{-}\overset{\overset{\displaystyle O}{\|}}{C}{-}R \xrightarrow[2.\ H_3O^+]{1.\ LiAlH_4,\ ether} 2\ RCH_2OH$$

4. Reactions of amides and lactams
   a. Hydrolysis to yield acids (Section 24.8)

$$R{-}\overset{\overset{\displaystyle O}{\|}}{C}{-}NH_2 \xrightarrow[or\ ^-OH]{H_3O^+} R{-}\overset{\overset{\displaystyle O}{\|}}{C}{-}OH + NH_3$$

   b. Reduction to yield amines (Section 24.8)

$$R{-}\overset{\overset{\displaystyle O}{\|}}{C}{-}NH_2 \xrightarrow[2.\ H_2O]{1.\ LiAlH_4,\ ether} R{-}CH_2NH_2$$

c. Dehydration of primary amides to yield nitriles (Section 24.9)

$$R{-}\overset{\overset{\displaystyle O}{\|}}{C}{-}NH_2 \xrightarrow{SOCl_2} R{-}C{\equiv}N + SO_2 + HCl$$

5. Reactions of esters and lactones

a. Hydrolysis to yield acids (Section 24.7)

$$R{-}\overset{\overset{\displaystyle O}{\|}}{C}{-}OR' + H_2O \xrightarrow{H^+ \text{ or } ^-OH} R{-}\overset{\overset{\displaystyle O}{\|}}{C}{-}OH + HOR'$$

b. Aminolysis to yield amides (Section 24.7)

$$R{-}\overset{\overset{\displaystyle O}{\|}}{C}{-}OR' + NH_3 \longrightarrow R{-}\overset{\overset{\displaystyle O}{\|}}{C}{-}NH_2 + HOR'$$

c. Reduction to yield primary alcohols (Section 24.7)

$$R{-}\overset{\overset{\displaystyle O}{\|}}{C}{-}OR' \xrightarrow[\text{2. } H_3O^+]{\text{1. } LiAlH_4\text{, ether}} R{-}CH_2OH + R'OH$$

d. Partial reduction to yield aldehydes (Section 24.7)

$$R{-}\overset{\overset{\displaystyle O}{\|}}{C}{-}OR' \xrightarrow[\text{2. } H_3O^+]{\text{1. DIBAH, toluene}} R{-}CHO + R'OH$$

e. Grignard reaction to yield tertiary alcohols (Section 24.7)

$$R{-}\overset{\overset{\displaystyle O}{\|}}{C}{-}OR' \xrightarrow[\text{2. } H_3O^+]{\text{1. 2 } R''MgX\text{, ether}} R{-}\underset{\underset{\displaystyle R''}{|}}{\overset{\overset{\displaystyle OH}{|}}{C}}{-}R'' + R'OH$$

6. Reactions of nitriles

a. Hydrolysis to yield carboxylic acids (Section 24.9)

$$R{-}C{\equiv}N + H_2O \xrightarrow{H^+ \text{ or } HO^-} R{-}\overset{\overset{\displaystyle O}{\|}}{C}{-}OH + NH_3$$

b. Partial hydrolysis to yield amides (Section 24.9)

$$R{-}C{\equiv}N + H_2O \xrightarrow{H^+ \text{ or } ^-OH} R{-}\overset{\overset{\displaystyle O}{\|}}{C}{-}NH_2$$

c. Reduction to yield primary amines (Section 24.9)

$$R{-}C{\equiv}N \xrightarrow[2.\ H_2O]{1.\ LiAlH_4,\ ether} R{-}CH_2NH_2$$

d. Partial reduction to yield aldehydes (Section 24.9)

$$R{-}C{\equiv}N \xrightarrow[2.\ H_3O^+]{1.\ DIBAH,\ toluene} R{-}CHO + NH_3$$

e. Reaction with Grignard reagents to yield ketones (Section 24.9)

$$R{-}C{\equiv}N \xrightarrow[2.\ H_3O^+]{1.\ R'MgX,\ ether} R{-}\overset{\overset{\displaystyle O}{\|}}{C}{-}R' + NH_3$$

ADDITIONAL PROBLEMS

24.21 Provide IUPAC names for these compounds:

(a) H_3C–(p-C_6H_4)–$CONH_2$

(b) $(CH_3CH_2)_2CHCH{=}CHCN$

(c) $CH_3O_2CCH_2CH_2CO_2CH_3$

(d) C_6H_5–$CH_2CH_2CO_2CH(CH_3)_2$

(e) C_6H_5–C(=O)–O–C_6H_5

(f) $CH_3CHBrCH_2CONHCH_3$

(g) 3,5-(Br)$_2C_6H_3$–C(=O)–Cl

(h) cyclopentenyl–CN

(i) cyclohexadienyl–CO_2CH_3

24.22 Draw structures corresponding to these names:

(a) *p*-Bromophenylacetamide
(b) *m*-Benzoylbenzonitrile
(c) 2,2-Dimethylhexanamide
(d) Cyclohexyl cyclohexanecarboxylate
(e) 2-Cyclobutenecarbonitrile
(f) 1,2-Pentanedicarbonyl dichloride

24.23 Draw and name compounds meeting these descriptions:
(a) Three different acid chlorides having the formula C_6H_9ClO
(b) Three different amides having the formula $C_7H_{11}NO$
(c) Three different nitriles having the formula C_5H_7N

24.24 The following reactivity order has been found for the saponification of alkyl acetates by aqueous hydroxide ion:

$$CH_3CO_2CH_3 > CH_3CO_2CH_2CH_3 > CH_3CO_2CH(CH_3)_2 > CH_3CO_2C(CH_3)_3$$

How can you explain this reactivity order?

24.25 In the basic hydrolysis of para-substituted methyl benzoates,

$$p\text{-}X\text{-}C_6H_4\text{-}CO_2CH_3 \xrightarrow{^-OH} p\text{-}X\text{-}C_6H_4\text{-}COO^- + CH_3OH$$

the following reactivity order has been found for X: $NO_2 > Br > H > CH_3 > OCH_3$. How can you explain this reactivity order? Where would you expect X = CN, X = CHO, and X = NH_2 to be in the reactivity list?

24.26 How can you explain the observation that attempted Fischer esterification of mesitoic acid (2,4,6-trimethylbenzoic acid) with methanol/HCl is unsuccessful? No ester is obtained and the acid is recovered unchanged. Can you suggest an alternative method of esterification that would be successful?

24.27 Outline methods for the preparation of acetophenone (phenyl methyl ketone) from the following:
(a) Benzene
(b) Bromobenzene
(c) Methyl benzoate
(d) Benzonitrile
(e) Styrene

24.28 What product would you expect to obtain from Grignard reaction of phenylmagnesium bromide with dimethyl carbonate, $CH_3OCOOCH_3$?

24.29 In the iodoform reaction, a triiodomethyl ketone reacts with aqueous base to yield a carboxylate ion and iodoform (triiodomethane). Propose a mechanism for this reaction.

$$R\text{—}\overset{\overset{\displaystyle O}{\|}}{C}\text{—}CI_3 \xrightarrow[H_2O]{^-OH} R\text{—}COO^- + CHI_3$$

24.30 Which of the two products shown would you expect from the reaction of methoxycarbonyl chloride with ammonia? Would the nucleophile replace the chlorine or the methoxy group? Explain your answer.

$$CH_3O\text{—}\overset{\overset{\displaystyle O}{\|}}{C}\text{—}Cl \xrightarrow{:NH_3} CH_3O\text{—}\overset{\overset{\displaystyle O}{\|}}{C}\text{—}NH_2 \quad \text{or} \quad H_2N\text{—}\overset{\overset{\displaystyle O}{\|}}{C}\text{—}Cl$$

24.31 Which compound would you expect to react faster with nucleophiles? Explain your answer.

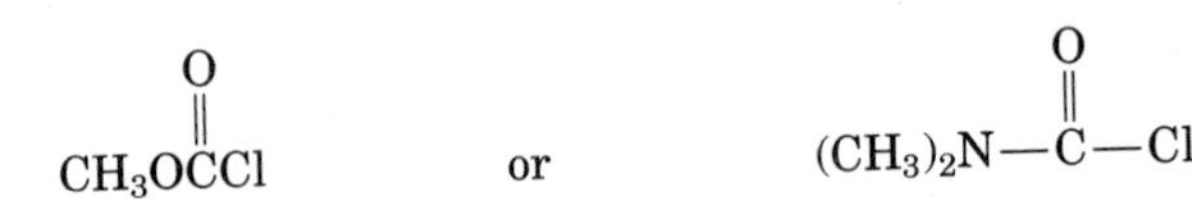

Methoxycarbonyl chloride   *N,N*-Dimethylaminocarbonyl chloride

24.32 *tert*-Butoxycarbonyl azide is an important reagent used in protein synthesis. It is prepared by treating *tert*-butoxycarbonyl chloride with sodium azide. Propose a mechanism for this reaction.

$$(CH_3)_3C{-}O{-}COCl + NaN_3 \longrightarrow (CH_3)_3C{-}O{-}CON_3 + NaCl$$

24.33 Predict the product, if any, of reaction between propanoyl chloride and the following reagents:

(a) $(Ph)_2CuLi$ in ether
(b) $LiAlH_4$, then H_3O^+
(c) CH_3MgBr, then H_3O^+
(d) H_2, Pd, $BaSO_4$
(e) H_3O^+
(f) Cyclohexanol
(g) Aniline
(h) Cholesterol

24.34 Answer Problem 24.33 for reaction between methyl propanoate and the listed reagents.

24.35 A particularly mild method of esterification involves the use of trifluoroacetic anhydride. Treatment of a carboxylic acid with trifluoroacetic anhydride leads to a mixed anhydride that rapidly reacts with alcohol:

$$R{-}COOH + (CF_3CO)_2O \longrightarrow R{-}\overset{O}{\overset{\|}{C}}{-}O{-}\overset{O}{\overset{\|}{C}}{-}CF_3 \xrightarrow{R'OH} R{-}\overset{O}{\overset{\|}{C}}{-}OR' + CF_3COOH$$

(a) Propose a mechanism for formation of the mixed anhydride.
(b) Why is the mixed anhydride unusually reactive?
(c) Why does the mixed anhydride react specifically as indicated, rather than giving trifluoroacetate esters plus carboxylic acid?

24.36 How would you accomplish the following transformations? More than one step may be required.

(a) $CH_3CH_2CH_2CH_2CN \longrightarrow CH_3CH_2CH_2CH_2CH_2NH_2$

(b) $CH_3CH_2CH_2CH_2CN \longrightarrow CH_3CH_2CH_2CH_2CH_2N(CH_3)_2$

(c) $CH_3CH_2CH_2CH_2CN \longrightarrow CH_3CH_2CH_2CH_2C(CH_3)_2OH$

(d) $CH_3CH_2CH_2CH_2CN \longrightarrow CH_3CH_2CH_2CH_2CH(OH)CH_3$

(e) $CH_3CH_2CH_2CH_2CN \longrightarrow CH_3CH_2CH_2CH_2CHO$

24.37 List as many ways as you can think of for transforming cyclohexanol into cyclohexanecarbaldehyde (try to get at least four).

24.38 Succinic anhydride yields succinimide when heated with ammonium chloride. Propose a mechanism for this reaction.

$$\begin{array}{l} CH_2{-}\overset{O}{\overset{\|}{C}} \\ | \qquad\quad \diagdown O \\ CH_2{-}\underset{O}{\underset{\|}{C}} \diagup \end{array} \xrightarrow[200^\circ C]{NH_4Cl} \begin{array}{l} CH_2{-}\overset{O}{\overset{\|}{C}} \\ | \qquad\quad \diagdown N{-}H \\ CH_2{-}\underset{O}{\underset{\|}{C}} \diagup \end{array} + H_2O + HCl$$

24.39 At least as far back as the sixteenth century, the Incas chewed the leaves of the coca bush, *Erythroxylon coca,* to combat fatigue. Chemical studies of *Erythroxylon coca* by Friedrich Wöhler in 1862 resulted in the discovery of cocaine as the active component. It was soon found that basic hydrolysis of cocaine led to methanol, benzoic acid, and another compound called *ecgonine.* Chromium trioxide oxidation of ecgonine led to a keto acid that readily lost CO_2 on heating, giving tropinone. These transformations are shown at the top of the next page.

$$\underset{\text{Cocaine}}{C_{17}H_{21}NO_4} \xrightarrow[H_2O]{^-OH} CH_3OH + C_6H_5COOH + \underset{\text{Ecgonine}}{C_9H_{15}NO_3}$$

$$\text{Ecgonine} \xrightarrow{CrO_3} \underset{\text{Keto acid}}{C_9H_{13}NO_3} \xrightarrow[\Delta]{} CO_2 + \text{Tropinone}$$

H₃C–N (tropinone structure)

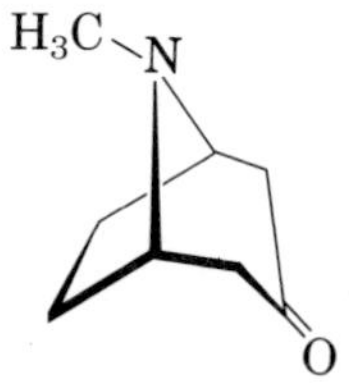

Tropinone

(a) What is a likely structure for the keto acid?
(b) What is a likely structure for ecgonine?
(c) What is a likely structure for cocaine?
(d) Formulate the reactions involved.

24.40 Butacetin is an analgesic (pain-killing) agent that is synthesized commercially from *p*-fluoronitrobenzene. Propose a likely synthesis route.

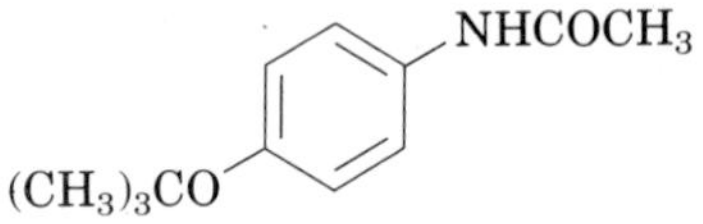

$NHCOCH_3$, $(CH_3)_3CO$

Butacetin

24.41 Phenyl 4-aminosalicylate is a drug used in the treatment of tuberculosis. Propose a synthesis of this compound starting from 4-nitrosalicylic acid.

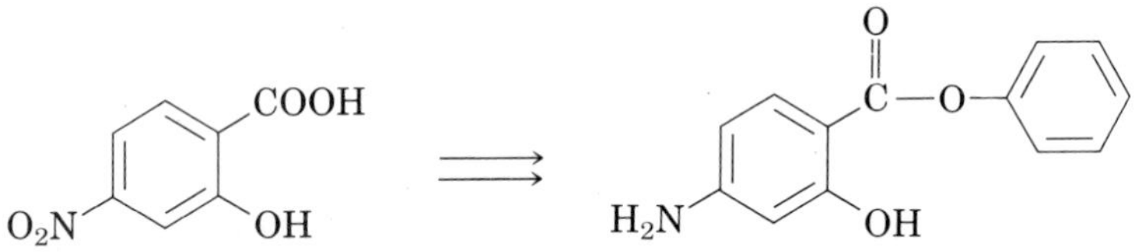

COOH, O_2N, OH $\Longrightarrow$ O, C—O—Ph, H_2N, OH

4-Nitrosalicylic acid — Phenyl 4-aminosalicylate

24.42 What spectroscopic technique would you use to distinguish between the following isomer pairs? Tell exactly what difference you would expect to see.
(a) *N*-Methylpropanamide and *N,N*-dimethylacetamide
(b) 5-Hydroxypentanenitrile and cyclobutanecarboxamide
(c) 4-Chlorobutanoic acid and 3-methoxypropanoyl chloride
(d) Ethyl propanoate and propyl acetate

24.43 *N,N*-Diethyl-*m*-toluamide is the active ingredient in many insect-repellent preparations. How might you synthesize this substance starting from *m*-bromotoluene?

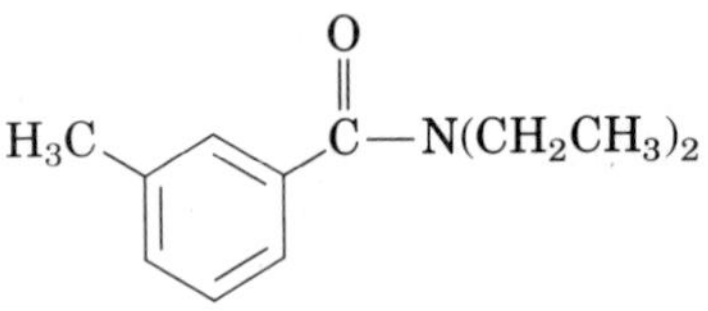

H_3C, O, C—$N(CH_2CH_3)_2$

N,N-Diethyl-*m*-toluamide

24.44 The pharmaceutical agent lifibrate is useful for controlling cholesterol levels in the body. Propose a synthesis of lifibrate, starting from any needed compound having six or fewer carbon atoms.

$$\left[Cl-C_6H_4-O\right]_2CH-\overset{O}{\overset{\|}{C}}-O-C_5H_9N-CH_3$$

Lifibrate

24.45 Account for the fact that treatment of an ethyl ester with an acid catalyst in methanol solution yields the methyl ester product.

$$R\overset{O}{\overset{\|}{C}}OCH_2CH_3 \xrightarrow[CH_3OH]{H^+} R-\overset{O}{\overset{\|}{C}}-OCH_3 + CH_3CH_2OH$$

24.46 Propose a structure for a compound, $C_4H_7ClO_2$, that has the infrared and NMR spectra shown.

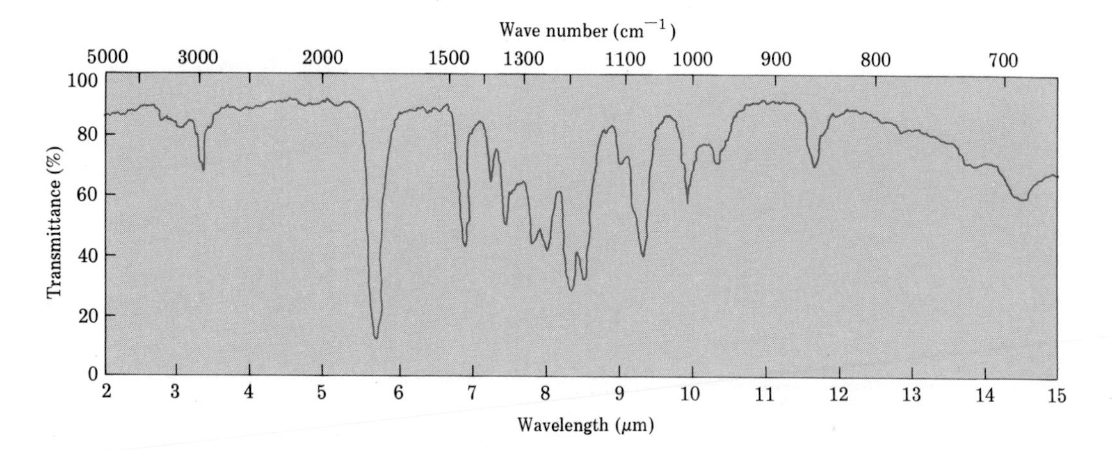

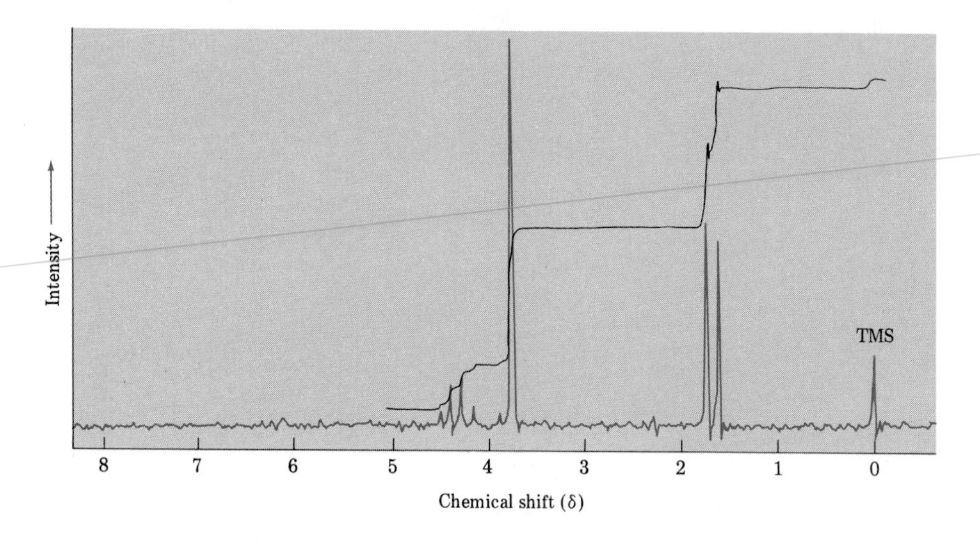

24.47 Propose a structure for a compound, C_4H_7N, that has the infrared and NMR spectra shown.

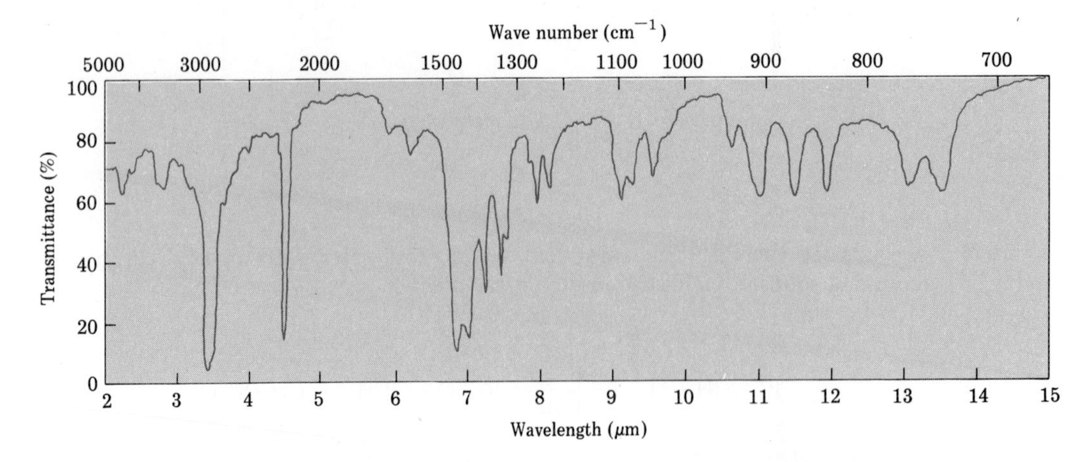

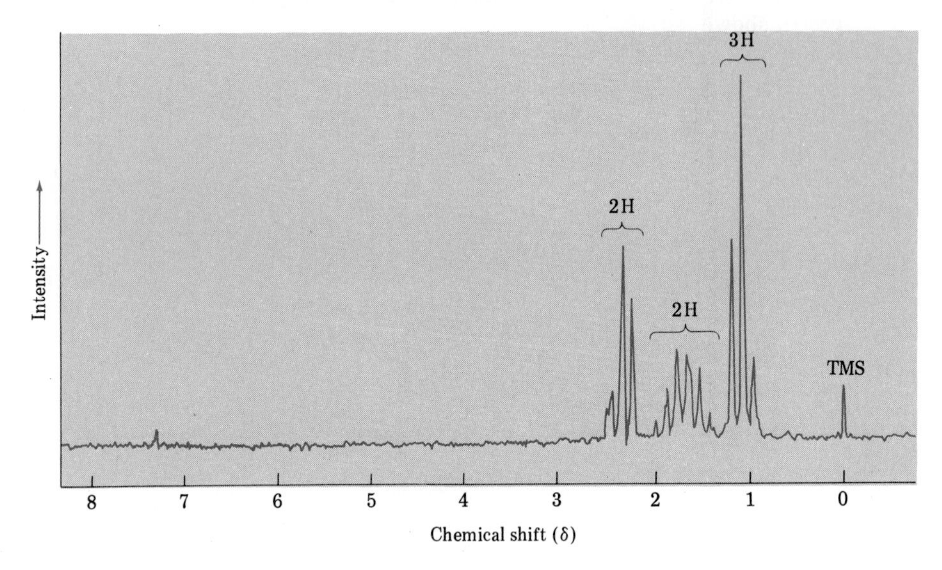

CHAPTER 25 CARBONYL ALPHA-SUBSTITUTION REACTIONS

We said, in the overview of carbonyl reactions in Chapter 21, that most of the chemistry of carbonyl compounds can be explained in terms of just four fundamental reactions—nucleophilic additions, nucleophilic acyl substitutions, alpha substitutions, and carbonyl condensations. We have already looked at the chemistry of nucleophilic addition reactions and nucleophilic acyl substitution reactions in detail and have seen numerous variations of these two processes. In this chapter we will look at the chemistry of the third major carbonyl-group process—the **alpha-substitution reaction.**

R—C(=O)—Y + Nu⁻ ⇌ [Tetrahedral intermediate] → R—C(=O)—Nu (Nucleophilic acyl substitution); → R—C(OH)(Y)(Nu) or R—C(=Nu)—Y (Nucleophilic addition)

Tetrahedral intermediate

Nucleophilic acyl substitution

Nucleophilic addition

An enol

or

An enolate anion

E^+ = an electrophile

Alpha substitution

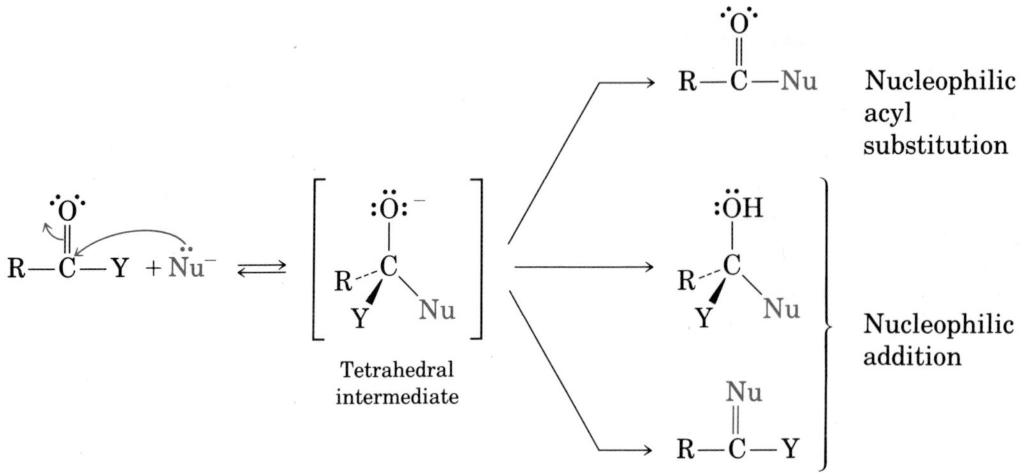

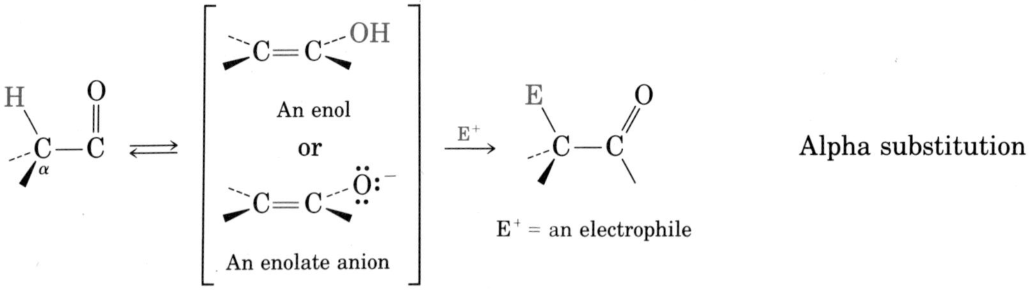

The key feature of alpha-substitution reactions is that they take place through the formation of either an enol or an enolate ion intermediate. Let's begin our study by learning more about these two species.

25.1 Keto–enol tautomerism

Carbonyl compounds that have hydrogen atoms on their alpha carbons are rapidly interconvertible with their corresponding **enols** (*ene* + *ol*, unsaturated alcohol). This rapid interconversion between two chemically distinct species is a special kind of isomerism known as **tautomerism** (from the Greek *tauto,* "the same," and *meros,* "part"); individual isomers are called **tautomers.** Note that tautomerism requires the two different isomeric forms to be *rapidly* interconvertible. Thus, keto and enol carbonyl isomers are tautomers, but two isomeric alkenes such as 1-butene and 2-butene are not, since they do not rapidly interconvert.

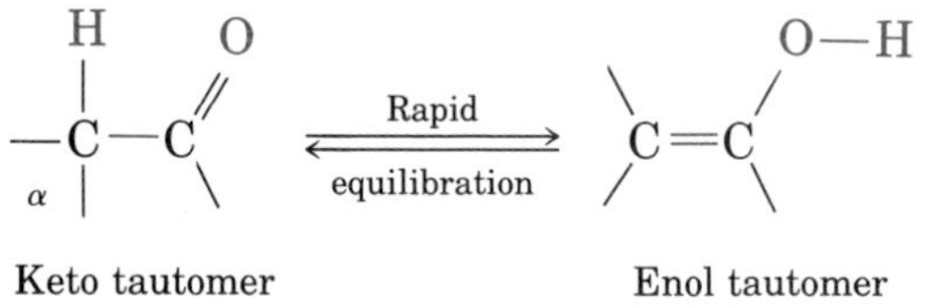

Rapid equilibration

Keto tautomer

Enol tautomer

However, these compounds are *not* tautomers:

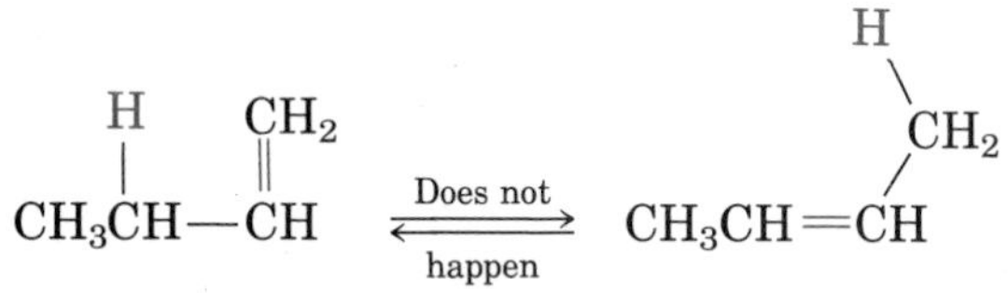

$$CH_3\overset{H}{C}H{-}\overset{CH_2}{CH} \xrightarrow[\text{happen}]{\text{Does not}} CH_3CH{=}CH{-}CH_2{-}H$$

At equilibrium, most carbonyl compounds exist almost exclusively in the keto form, and it is difficult to isolate the pure enol form. For example, cyclohexanone contains only about 0.001% of its enol tautomer at room temperature, and acetone contains about 0.0001% enol. The percentage of enol tautomer is even less for carboxylic acids and acyl derivatives such as esters and amides. Even though enols are difficult to isolate and are present to only a small extent at equilibrium, they are nevertheless extremely important and are involved in much of the chemistry of carbonyl compounds.

Cyclohexanone

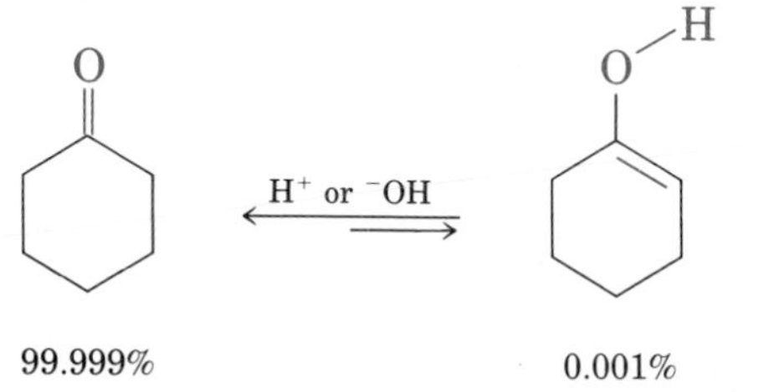

H^+ or ^-OH

99.999%

0.001%

Acetone

$$CH_3\overset{O}{\overset{\|}{C}}CH_3 \xrightleftharpoons{H^+ \text{ or } ^-OH} CH_3\overset{OH}{C}{=}CH_2$$

99.9999%

0.0001%

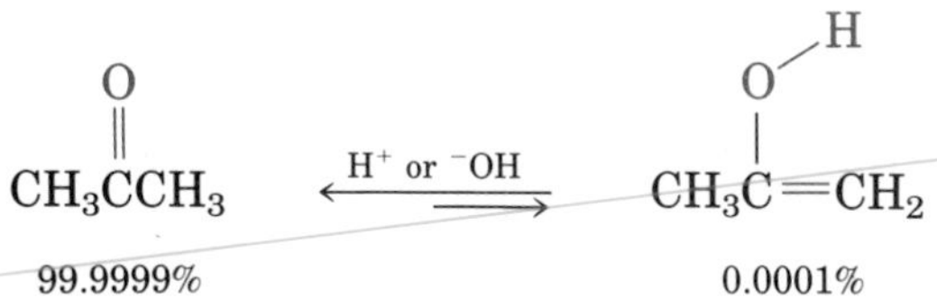

Keto–enol tautomerism of carbonyl compounds is catalyzed by both acids and bases. Acid catalysis involves protonation of the carbonyl oxygen atom (a Lewis base), followed by loss of a proton from the alpha carbon to yield neutral enol (Figure 25.1). Note that the proton loss from the positively charged intermediate is analogous to what occurs when a carbocation loses a proton to form an alkene in an E1 reaction (Section 10.12).

Base-catalyzed enol formation occurs by an acid–base reaction between catalyst and carbonyl compound. The carbonyl compound acts as a weak protic acid and donates one of its alpha hydrogens to the base. The resultant

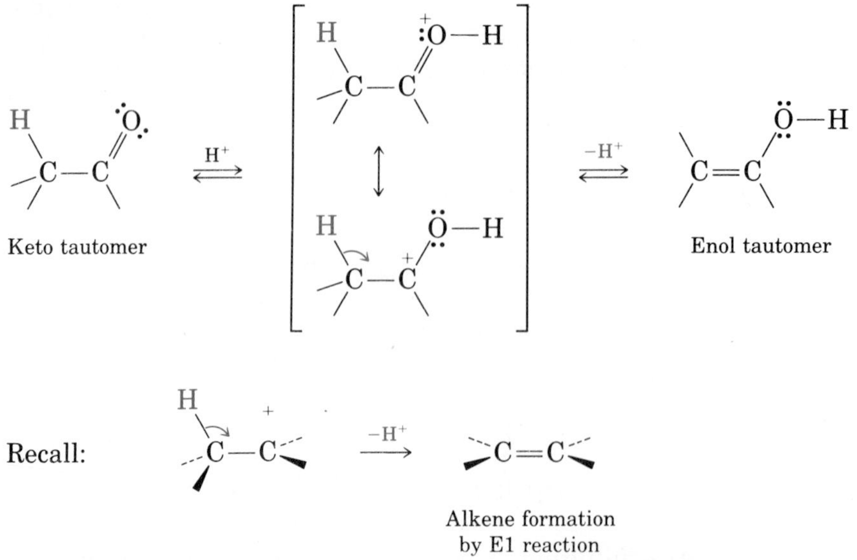

Keto tautomer $\underset{}{\overset{H^+}{\rightleftharpoons}}$ $\overset{-H^+}{\rightleftharpoons}$ Enol tautomer

Recall: $\overset{-H^+}{\longrightarrow}$

Alkene formation by E1 reaction

Figure 25.1. Mechanism of acid-catalyzed enol formation.

anion—an **enolate ion**—is then reprotonated to yield a neutral compound. If protonation of the enolate ion takes place on the alpha carbon, the keto tautomer is regenerated and no net change has occurred. If, however, protonation takes place on the oxygen atom, an enol tautomer is formed (Figure 25.2).

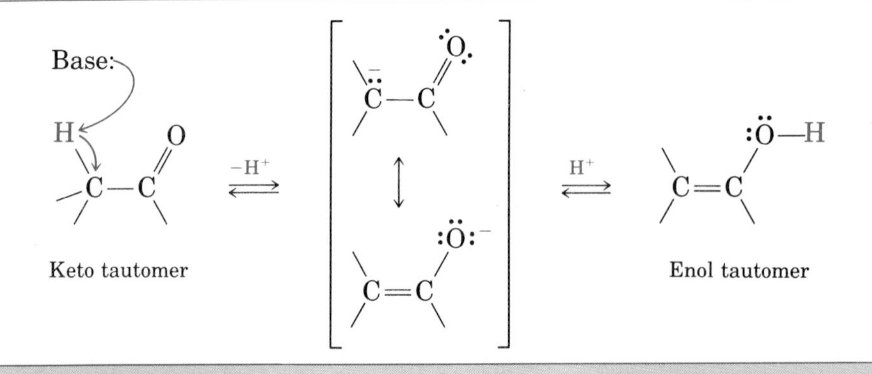

Figure 25.2. Base-catalyzed enol formation.

PROBLEM

25.1 Draw structures for the enol tautomers of these compounds:
(a) Cyclopentanone
(b) Acetyl chloride
(c) Ethyl acetate
(d) Propanal
(e) Acetic acid
(f) Phenylacetone
(g) Acetophenone (methyl phenyl ketone)

PROBLEM

25.2 Draw structures for the possible mono-enol forms of 1,3-cyclohexanedione. How many enol forms are possible? Which would you expect to be most stable? Explain your answer.

1,3-Cyclohexanedione

25.2 Reactivity of enols

What kind of chemistry should we expect enols to exhibit? Since their double bonds are electron-rich, enols behave as nucleophiles. They therefore react with electrophiles in much the same way that alkenes do, but they are more reactive than alkenes because of electron donation from the oxygen pi electrons. When an alkene reacts with an electrophile such as bromine, addition of Br^+ occurs to give an intermediate cation, which reacts with $Br{:}^-$. When an enol reacts with an electrophile, however, the intermediate cation loses the hydroxyl proton to regenerate a carbonyl compound. The net result of the reaction of an enol with an electrophile is alpha substitution. The general mechanism is shown in Figure 25.3 (p. 820).

25.3 Alpha halogenation of ketones and aldehydes

Alpha halogenation provides one of the best examples of enol reactivity. Ketones and aldehydes can be halogenated at their alpha positions by reaction with chlorine, bromine, or iodine in acidic solution. Bromine is most often used, and acetic acid is often employed as solvent.

$$C_6H_5C(=O)-CH_2-H \xrightarrow[CH_3COOH]{Br_2} C_6H_5C(=O)-CH_2Br + HBr$$

Acetophenone → Phenacyl bromide (72%)

$$\text{Cyclohexanone} \xrightarrow[H_2O,\ HCl]{Cl_2} \text{2-Chlorocyclohexanone} + HCl$$

Cyclohexanone → 2-Chlorocyclohexanone (66%)

Acid-catalyzed enolization occurs.

An electron pair from the enol attacks an electrophile, forming a new bond and leaving a positively charged intermediate that can be stabilized by two resonance forms.

Loss of a proton from oxygen yields the neutral alpha-substitution product, and the O—H bond electrons form a new C=O bond.

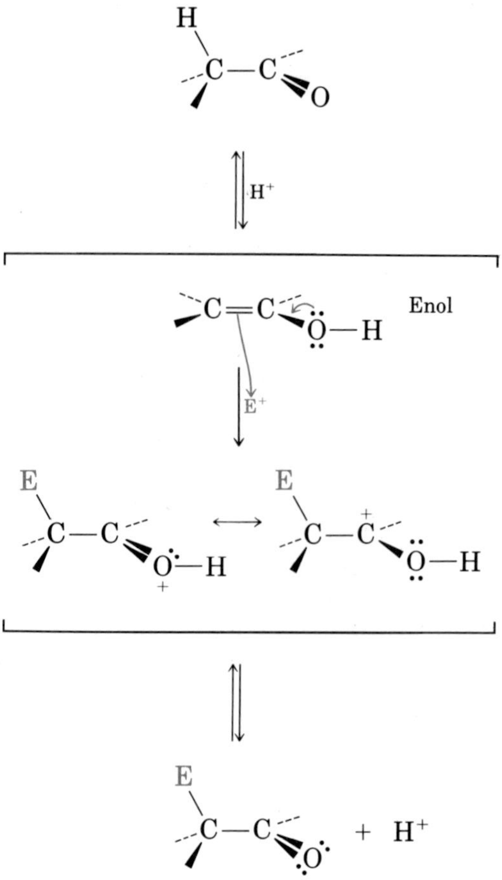

Recall:

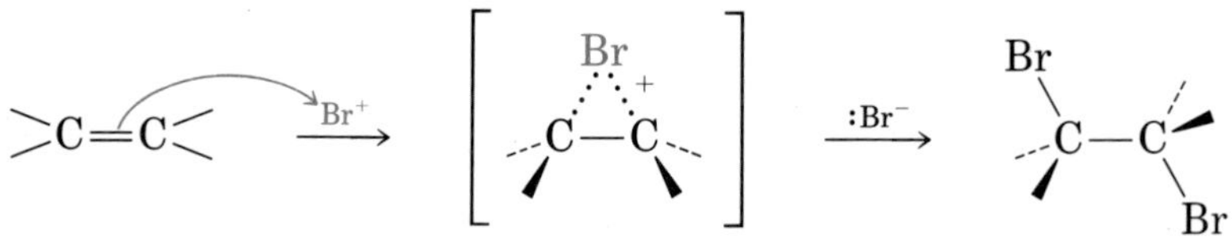

Figure 25.3. General mechanism of a carbonyl alpha-substitution reaction.

α-Bromo ketones are useful in organic synthesis because they can be dehydrobrominated by base treatment to yield α,β-unsaturated ketones. For example, 2-bromo-2-methylcyclohexanone gives 2-methyl-2-cyclohexenone in 62% yield when refluxed in pyridine. The reaction takes place by an E2 elimination pathway (Section 10.10) and is an excellent method of introducing carbon–carbon double bonds into molecules.

E2 reaction

2-Bromo-2-methylcyclohexanone $\xrightarrow[\Delta]{\text{pyridine}}$ 2-Methyl-2-cyclohexenone (62%) (an α,β-unsaturated ketone) + HBr

The alpha halogenation of ketones is a typical alpha-substitution reaction that proceeds by acid-catalyzed formation of enol intermediates, as shown in Figure 25.4.

Acid–base reaction between a catalyst and the carbonyl oxygen forms a protonated carbonyl compound.

$$CH_3-\overset{O}{\overset{\|}{C}}-CH_3 \underset{}{\overset{H^+}{\rightleftharpoons}} H-CH_2-\overset{^+OH}{\overset{\|}{C}}-CH_3$$

This compound loses an acidic proton from the alpha carbon to yield an enol intermediate.

$$\overset{-H^+}{\rightleftharpoons} H_2C{=}\overset{OH}{\overset{|}{C}}-CH_3 \quad \text{Enol}$$

An electron pair from the enol attacks a positively polarized bromine atom, giving an intermediate cation that can be drawn as a resonance hybrid of two forms.

$$\xrightarrow{\overset{\delta+}{Br}-\overset{\delta-}{Br}} BrCH_2-\overset{^+OH}{\overset{\|}{C}}-CH_3 \longleftrightarrow BrCH_2-\overset{OH}{\overset{|}{C^+}}-CH_3$$

Loss of a proton then gives the alpha-halogenated product.

$$\rightleftharpoons BrCH_2\overset{O}{\overset{\|}{C}}CH_3 + H^+$$

Figure 25.4. Mechanism of the acid-catalyzed halogenation of ketones.

There is much evidence to indicate that this mechanism is correct. For example, the rate of the halogenation reaction is independent of the nature of the halogen; chlorination, bromination, and iodination of a given ketone all occur at exactly the same rate. This information tells us that the same rate-limiting step is involved in chlorination, bromination, and iodination, and that this rate-limiting step must occur *before* interaction of the ketone with halogen. Corroborative evidence comes from measurements indicating that acid-catalyzed ketone halogenations exhibit second-order kinetics and follow the rate law:

$$\text{Reaction rate} = k[\text{Ketone}][\text{H}^+]$$

This information tells us that ketone halogenations depend only on ketone and acid concentrations and are independent of halogen concentration. Thus, again we see that halogen is not involved in the rate-limiting step.

A final piece of evidence comes from deuteration experiments. If a ketone is treated with D_3O^+ instead of H_3O^+, the acidic alpha-hydrogen atoms are replaced by deuterium atoms. For a given ketone, the rate of deuteration is identical to the rate of halogenation. This tells us that the same intermediate is involved in both deuteration and halogenation. The common intermediate that satisfies all the evidence just presented can only be an enol.

Ketone/aldehyde (H–C–C=O) $\xrightarrow[\text{Slow, rate limiting}]{H^+ \text{ (or } D^+)}$ [Enol (C=C–OH (D))] $\xrightarrow[\text{Fast}]{D_3O^+}$ D–C–C=O ; $\xrightarrow[\text{Fast}]{X_2}$ X–C–C=O

X = Cl, Br, or I

PROBLEM

25.3 Show in detail the mechanism of the deuteration of acetone on treatment with D_3O^+.

$$CH_3COCH_3 \xrightarrow{D_3O^+} CH_3COCH_2D$$

PROBLEM

25.4 How can you account for the fact that optically active 3-phenyl-2-butanone is racemized by treatment with acid?

PROBLEM

25.5 In light of your answer to Problem 25.4, would you expect optically active 3-methyl-3-phenyl-2-pentanone to be racemized by acid treatment? Explain.

25.4 Alpha bromination of carboxylic acids: the Hell–Volhard–Zelinskii reaction

Direct alpha bromination of carbonyl compounds by molecular bromine in acetic acid is limited to ketones and aldehydes, since acids, esters, and amides do not enolize rapidly enough for halogenation to take place. Carboxylic acids, however, can be α-brominated by a mixture of bromine and phosphorus tribromide in the Hell–Volhard–Zelinskii reaction:

$$CH_3CH_2CH_2CH_2CH_2\overset{\displaystyle H}{\overset{|}{C}}HCOOH \xrightarrow[2.\ H_2O]{1.\ Br_2,\ PBr_3} CH_3CH_2CH_2CH_2CH_2\overset{\displaystyle Br}{\overset{|}{C}}HCOOH + HBr$$

Heptanoic acid — 2-Bromoheptanoic acid (90%)

The Hell–Volhard–Zelinskii reaction involves several steps. The first step is reaction between PBr_3 and carboxylic acid to yield an intermediate acid bromide plus HBr (Section 24.5). The HBr catalyzes enolization of the acid bromide, and the enol reacts rapidly with bromine. Hydrolysis of the α-bromo acid bromide by addition of water then gives the α-bromo carboxylic acid product.

$$R{-}\overset{\displaystyle H}{\overset{|}{C}}HCOOH \xrightarrow{PBr_3} \left[R{-}\overset{\displaystyle H}{\overset{|}{C}}H\overset{\displaystyle O}{\overset{\|}{C}}{-}Br \underset{}{\overset{HBr}{\rightleftharpoons}} R{-}CH{=}\overset{\displaystyle OH}{\overset{|}{C}}{-}Br \right]$$

Acid bromide — Acid bromide enol

$$\downarrow Br_2$$

$$R{-}\overset{\displaystyle Br}{\overset{|}{C}}HCOOH \xleftarrow{H_2O} \left[R{-}\overset{\displaystyle Br}{\overset{|}{C}}H{-}\overset{\displaystyle O}{\overset{\|}{C}}{-}Br + HBr \right]$$

The overall result of this process is the transformation of an acid into an α-bromo acid. Note, however, that the key step involves bromination of an acid bromide enol in a manner analogous to that which occurs during ketone bromination.

PROBLEM

25.6 If an optically active carboxylic acid such as 2-phenylpropanoic acid were brominated under Hell–Volhard–Zelinskii conditions, would you expect the product to be optically active or racemic? Explain your answer.

25.5 Acidity of alpha-hydrogen atoms: enolate formation

During our preliminary discussion of base-catalyzed enol formation in Section 25.1, we said that carbonyl compounds act as weak protic acids. Strong bases can abstract acidic alpha protons from carbonyl compounds to yield enolate anions:

$$:Base \quad \overset{H}{\underset{/}{\overset{\backslash}{-C}}}{-}C(=\ddot{O}:) \underset{}{\overset{Base}{\rightleftharpoons}} \overset{\backslash\ddot{-}}{\underset{/}{C}}{-}C(=\ddot{O}:) \longleftrightarrow \overset{\backslash}{\underset{/}{C}}{=}C(-\ddot{O}:^-) + Base{-}H^+$$

An enolate anion

Why are carbonyl compounds slightly acidic? If we compare acetone, $pK_a \approx 20$, with ethane, $pK_a \geq 50$, we find that the presence of a neighboring carbonyl group increases the acidity of a ketone over an alkane by a factor of 10^{30}. The reason for this increased acidity is best seen by viewing an orbital picture of the enolate-forming reaction (Figure 25.5). Proton abstraction from a carbonyl compound occurs when the alpha C—H sigma bond is oriented parallel to the carbonyl-group *p* orbitals. The alpha-carbon atom of the enolate product is sp^2 hybridized and has a *p* orbital that overlaps the neighboring carbonyl-group *p* orbitals, allowing the negative charge to be shared by the electronegative oxygen atom. Thus, enolate anions are highly stabilized by resonance between two forms, and $\Delta G°$ for the deprotonation reaction is low.

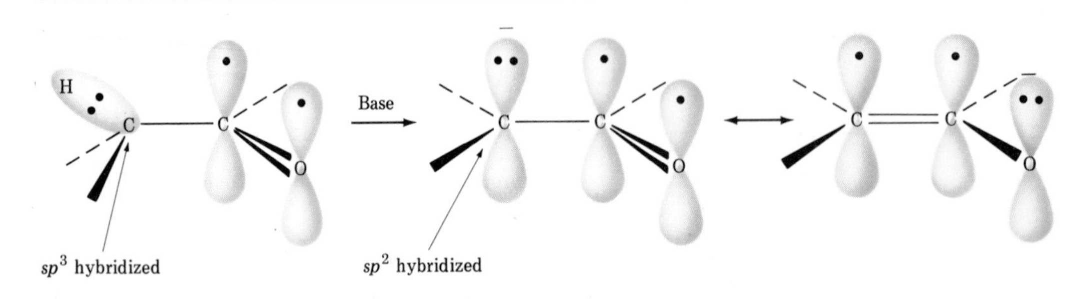

Figure 25.5. Mechanism of enolate ion formation.

Resonance stabilization of enolate anions is similar to the resonance stabilization we saw earlier for carboxylate anions (Section 23.3). Carbonyl compounds are more acidic than alkanes for the same reason that carboxylic acids are more acidic than alcohols. Enolate ions differ from carboxylate ions, however, in that their two resonance forms are not equivalent—the form in which the charge is on the oxygen atom is undoubtedly of lower energy—but the principle behind resonance stabilization is the same in both cases.

$$\underset{pK_a \geq 50}{CH_3CH_2\text{—}H} \quad \text{versus} \quad \underset{pK_a \approx 20}{CH_3\overset{\overset{\Large O}{\|}}{C}\text{—}CH_2\text{—}H} \rightleftharpoons CH_3\overset{\overset{:O:}{\|}}{C}\text{—}\ddot{C}H_2^{-} \longleftrightarrow CH_3\overset{\overset{:\ddot{O}:^-}{|}}{C}{=}CH_2 + H^+$$

Nonequivalent resonance forms of ketone enolate ion

Recall the following:

$$\underset{pK_a \approx 16}{CH_3O\text{—}H} \quad \text{versus} \quad \underset{pK_a \approx 5}{CH_3\overset{\overset{\Large O}{\|}}{C}O\text{—}H} \rightleftharpoons CH_3\overset{\overset{:O:}{\|}}{C}\text{—}\ddot{\underset{..}{O}}:^- \longleftrightarrow CH_3\overset{\overset{:\ddot{O}:^-}{|}}{C}{=}\ddot{\underset{..}{O}} + H^+$$

Equivalent resonance forms of carboxylate ion

Alpha-hydrogen atoms of carbonyl compounds are only weakly acidic when compared with carboxylic acids, and strong bases must therefore be used to effect enolate formation. If an alkoxide ion base such as sodium ethoxide is used, ionization takes place only to the extent of about 0.01%, since ethanol ($pK_a = 16$) is a stronger acid than acetone. If, however, a powerful base such as sodium hydride (NaH), sodium amide ($NaNH_2$), or lithium diisopropylamide, LDA [$LiN(i\text{-}C_3H_7)_2$], is used, a carbonyl compound can be completely converted into its enolate ion.

$$\text{Cyclohexanone} \xrightarrow[\text{THF}]{Li^+ \ ^-N(i\text{-}C_3H_7)_2} \text{Cyclohexanone enolate } (:\ddot{O}:^- Li^+) \ (100\%) + H{-}N(i\text{-}C_3H_7)_2$$

Cyclohexanone — Cyclohexanone enolate (100%)

Lithium diisopropylamide is commonly used as a base for preparing enolate ions from carbonyl compounds. It is easily prepared by reaction between butyllithium and diisopropylamine and has nearly ideal properties: It is an exceedingly powerful base since diisopropylamine has $pK_a \approx 40$; it is soluble in organic solvents such as THF; it is too hindered to add to carbonyl groups in nucleophilic addition reactions but is nevertheless extremely reactive, even at −78°C.

$$H{-}N[CH(CH_3)_2]_2 + n\text{-}C_4H_9Li \xrightarrow[\text{solvent}]{\text{THF}} Li^+ \ :N^-[CH(CH_3)_2]_2 + C_4H_{10}$$

Diisopropylamine — Lithium diisopropylamide (LDA)

A wide variety of carbonyl compounds can be converted into their enolates by reaction with LDA. Table 25.1 (p. 826) lists the approximate pK_a values of the alpha-hydrogen atoms of different kinds of carbonyl compounds and shows how these values compare with other acidic substances we have seen.

Table 25.1 indicates that *all* types of carbonyl compounds, including aldehydes, ketones, esters, acid chlorides, and dialkylamides, have greatly enhanced alpha-hydrogen acidity compared to alkanes. When a hydrogen atom is flanked by two carbonyl groups, the acidity is enhanced even more. Thus, compounds such as 1,3-diketones (known as β-diketones), 1,3-keto esters (β-keto esters), and 1,3-diesters are much more acidic than water. This seems reasonable, since the enolate ions derived from these β-dicarbonyl compounds are highly stabilized by delocalization of the charge onto both neighboring carbonyl groups.

TABLE 25.1 Acidity constants for some organic compounds

Compound type	*Compound*	pK_a
Carboxylic acid	$CH_3COO—H$	5
1,3-Diketone	$CH_2(COCH_3)_2$	9
1,3-Keto ester	$CH_3COCH_2CO_2C_2H_5$	11
1,3-Dinitrile	$CH_2(CN)_2$	11
1,3-Diester	$CH_2(CO_2C_2H_5)_2$	13
Water	HOH	16
Primary alcohol	CH_3CH_2OH	16
Acid chloride	CH_3COCl	16
Aldehyde	CH_3CHO	17
Ketone	CH_3COCH_3	20
Ester	$CH_3CO_2C_2H_5$	25
Nitrile	CH_3CN	25
Dialkylamide	$CH_3CON(CH_3)_2$	30
Ammonia	NH_3	35
Dialkylamine	$HN(i\text{-}C_3H_7)_2$	40
Alkyne	$HC{\equiv}CH$	25
Alkene	$CH_2{=}CH_2$	49
Alkane	CH_3CH_3	50

For example, the enolate from 2,4-pentanedione has three resonance forms:

$$CH_3\overset{\overset{:O:}{\|}}{C}—\underset{\alpha}{\overset{\overset{H}{|}}{C}H}—\overset{\overset{:O:}{\|}}{C}CH_3 \xrightleftharpoons{\text{Base}} CH_3\overset{\overset{:O:}{\|}}{C}—\ddot{C}H^{-}—\overset{\overset{:O:}{\|}}{C}CH_3 \longleftrightarrow CH_3\overset{\overset{:\ddot{O}:^-}{|}}{C}{=}CH—\overset{\overset{:O:}{\|}}{C}CH_3 \longleftrightarrow CH_3\overset{\overset{:O:}{\|}}{C}—CH{=}\overset{\overset{:\ddot{O}:^-}{|}}{C}CH_3$$

(β and α label the carbonyl carbon and central carbon, respectively.)

2,4-Pentanedione,
a β-diketone (pK_a = 9)

Similar resonance forms can be drawn for other doubly stabilized enolates.

PROBLEM

25.7 Indicate the acidic protons in the following molecules:
(a) CH_3CH_2CHO
(b) $(CH_3)_3CCOCH_3$

(c) CH_3COOH

(d) [benzene ring]$-CONH_2$

(e) $CH_3CH_2CH_2CN$

(f) $CH_3CON(CH_3)_2$

(g) 1,3-Cyclohexanedione

PROBLEM

25.8 Propose a mechanism to explain the observation that optically active 2-methylcyclohexanone is racemized by treatment with base. Would you expect optically active 3-methylcyclohexanone to be similarly racemized?

25.6 Reactivity of enolates

Enolate ions are much more useful than enols for two reasons. The first is that pure enols cannot normally be isolated; they are only generated as fleeting intermediates in small concentration. By contrast, solutions of pure enolate ions are easily prepared from most carbonyl compounds by treatment with a strong base. More important, though, is the fact that enolate ions are much more reactive than enols; they undergo many important reactions that enols do not undergo. The reason for this is simply that enolate ions carry a full negative charge, which makes them much more nucleophilic than neutral enols. Thus, the alpha position of enolate ions is highly reactive toward electrophiles.

$$\mathrm{C{=}C{-}\ddot{O}H} \qquad\qquad \mathrm{C{=}C{-}\ddot{O}{:}^-} \longleftrightarrow \mathrm{\bar{C}{-}C{=}O}$$

Enol: neutral, moderately reactive, very difficult to isolate

Enolate: negatively charged, very reactive, easily prepared

Enolate ions can be looked at either as vinylic alkoxides ($C{=}C{-}O^-$) or as α-keto carbanions ($\bar{C}{-}C{=}O$) and can react with electrophiles either on oxygen or on carbon. Reaction on oxygen would yield an enol derivative, whereas reaction on carbon would yield an α-substituted carbonyl compound (Figure 25.6, p. 828). In general, enolates react as carbanions to give α-substituted carbonyl products. Both kinds of reactivity are well known, however, and the factors that determine the course of a particular reaction are complex.

With this background knowledge, let's now see some specific examples of carbonyl alpha-substitution reactions via enolate intermediates.

25.7 Halogenation of enolates: the haloform reaction

Halogenation of ketones is promoted by base as well as by acid. The base-promoted reaction occurs through enolate ion intermediates. Even

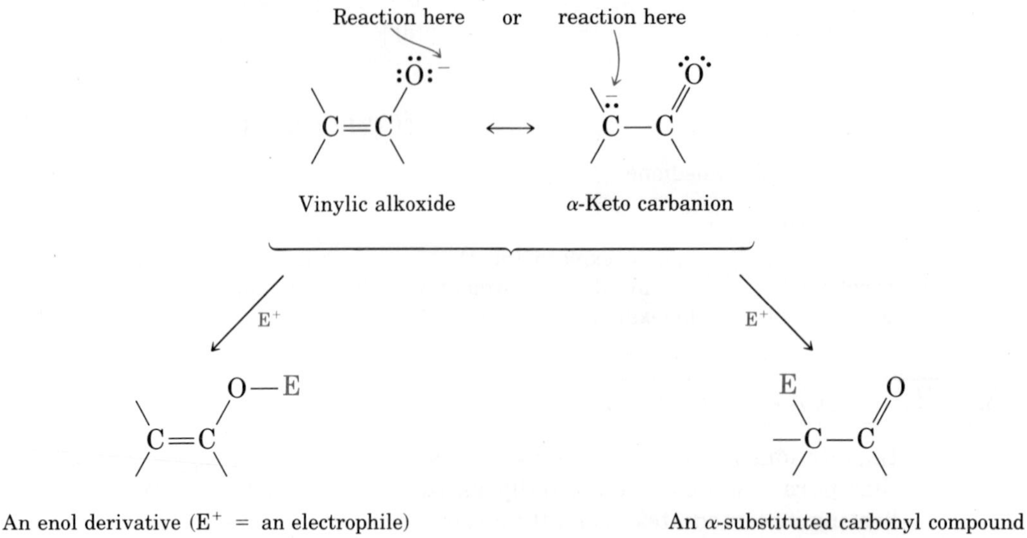

Reaction here or reaction here

Vinylic alkoxide $\longleftrightarrow$ α-Keto carbanion

E^+ E^+

An enol derivative (E^+ = an electrophile) An α-substituted carbonyl compound

Figure 25.6. Two modes of enolate reactivity.

relatively weak bases such as hydroxide ion are effective for halogenation, since it is not necessary to convert the ketone completely into its enolate ion; it is only necessary to generate a small amount of enolate at any one time because the reaction with halogen occurs as soon as the enolate is formed.

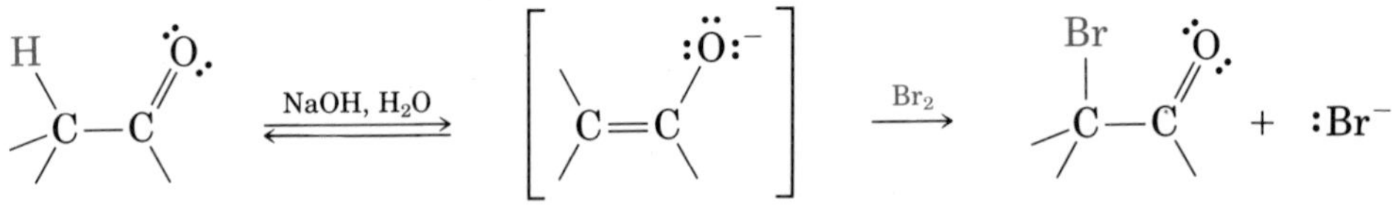

$$\text{NaOH, } H_2O \rightleftharpoons \qquad \xrightarrow{Br_2} \qquad + :Br^-$$

Base-promoted halogenation of ketones is little used in practice since it is difficult to stop the reaction at the monosubstituted product. An α-halogenated ketone product is generally more acidic than the starting unsubstituted ketone because of the electron-withdrawing inductive effect of the halogen atom. Thus, the monohalo products are themselves rapidly turned into enolates and further halogenated.

If excess base and halogen are used, methyl ketones are trihalogenated and then cleaved by base in the **haloform reaction:**

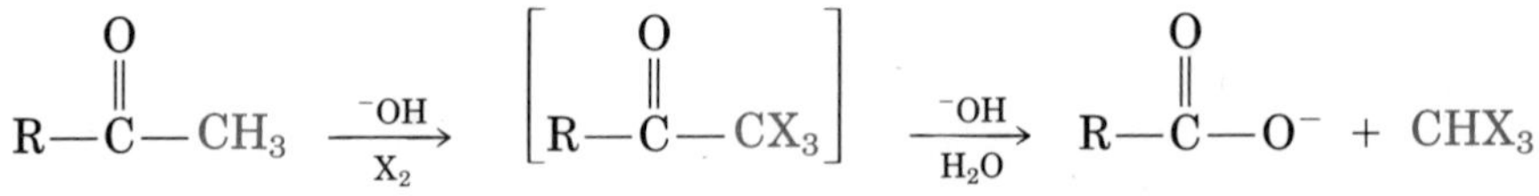

$$R{-}\overset{O}{\overset{\|}{C}}{-}CH_3 \xrightarrow[X_2]{^-OH} \left[R{-}\overset{O}{\overset{\|}{C}}{-}CX_3\right] \xrightarrow[H_2O]{^-OH} R{-}\overset{O}{\overset{\|}{C}}{-}O^- + CHX_3$$

where X = Cl, Br, or I.

The haloform reaction converts methyl ketones into carboxylic acids plus a **haloform** (chloroform, $CHCl_3$; bromoform, $CHBr_3$; or iodoform,

CHI_3) and is the basis for a qualitative test for methyl ketones. A sample of unknown structure is dissolved in THF and placed in a test tube. Dilute solutions of aqueous sodium hydroxide and iodine are then added, and the test tube is observed. Formation of a yellow precipitate of iodoform, CHI_3, signals a positive test and indicates that the sample is a methyl ketone.

PROBLEM

25.9 How can you explain the fact that base-promoted chlorination and bromination of a given ketone occur at the same rate? What kinetic rate law would you expect base-promoted halogenations of ketones to follow?

PROBLEM

25.10 Why do you suppose we have referred to ketone halogenations in acidic media as being acid *catalyzed,* whereas halogenations in basic media are base *promoted*? Why is the basic reaction not a true catalysis?

PROBLEM

25.11 Propose a mechanism to account for the second step of the haloform reaction—cleavage of the trihalomethyl ketone intermediate to yield a carboxylate ion and haloform. What general kind of carbonyl-group reaction is involved?

25.8 Selenenylation of enolates: enone synthesis

One of the most important advances in enolate chemistry occurred in the mid 1970s when it was found that carbonyl compounds can be **selenenylated.** Solutions of pure lithium enolates are prepared by reaction between a carbonyl compound and a strong base such as LDA in THF solvent. One equivalent of benzeneselenenyl bromide (C_6H_5SeBr) is added, and immediate reaction yields an α-phenylseleno-substituted product. Upon treatment with hydrogen peroxide, elimination occurs, and an α,β-unsaturated carbonyl compound is formed. The net result is introduction of a carbon–carbon double bond into the α,β position of a saturated carbonyl compound. Yields are generally excellent, and the method is often superior to the alternative alpha-bromination/dehydrobromination route.

$$\text{H–C–C(H)–C(=O)} \xrightarrow[\text{2. } C_6H_5SeBr]{\text{1. LDA, THF}} \text{H–C–C(SeC}_6\text{H}_5\text{)–C(=O)} \xrightarrow{H_2O_2} \text{C=C–C(=O)} + C_6H_5SeOH$$

An α-phenylseleno product

Mechanistically, the alpha-phenylselenenylation reaction is similar to halogenation except that the reactive enolate ion intermediate is prepared in a separate step prior to reaction, rather than during reaction, as in halogenation. The overall process is shown in Figure 25.7.

Note that the key step in the alpha phenylselenenylation of carbonyl compounds is similar to an S_N2 reaction (Section 10.5). The alpha-carbon

The strong base LDA abstracts an acidic alpha-hydrogen atom from the carbonyl compound, yielding an enolate ion.

The nucleophilic enolate ion then attacks electrophilic selenium and carries out a displacement of bromide, yielding the α-phenylseleno product.

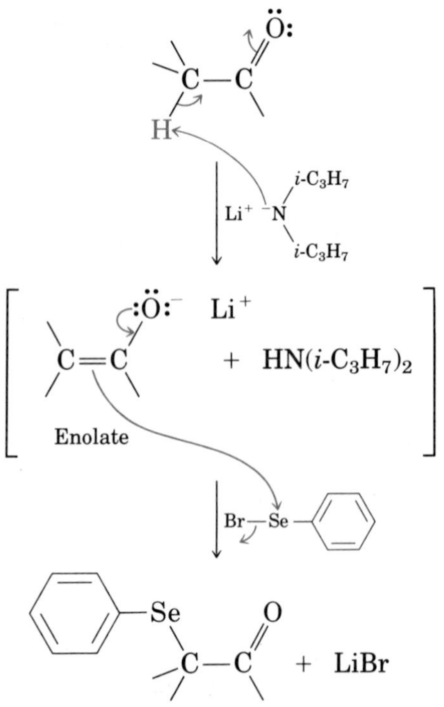

Figure 25.7. Mechanism of reaction of an enolate ion with benzeneselenenyl bromide.

atom of the enolate ion is the nucleophile that displaces bromide by attack on selenium:

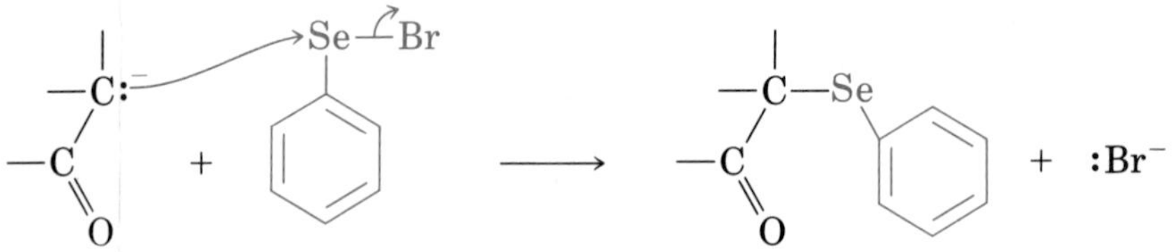

Recall the following:

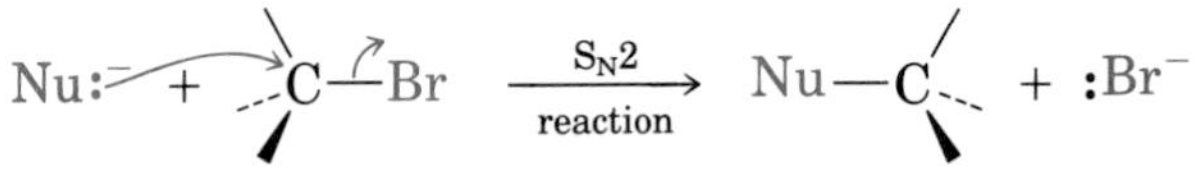

The elimination step involves oxidation of the phenylseleno intermediate to a phenyl selenoxide, which undergoes a spontaneous intramolecular elimination reaction. No added base is required (as it is in dehydrobromination), and the reaction occurs at room temperature under very mild conditions.

$\xrightarrow[25°C]{H_2O_2}$ $+ C_6H_5SeOH$

α-Phenylseleno ketone

A selenoxide

An enone

Ketones, esters, and nitriles all undergo the alpha-phenylselenenylation/elimination reaction in good yield. Aldehydes, however, often give product mixtures because their enolates are too reactive.

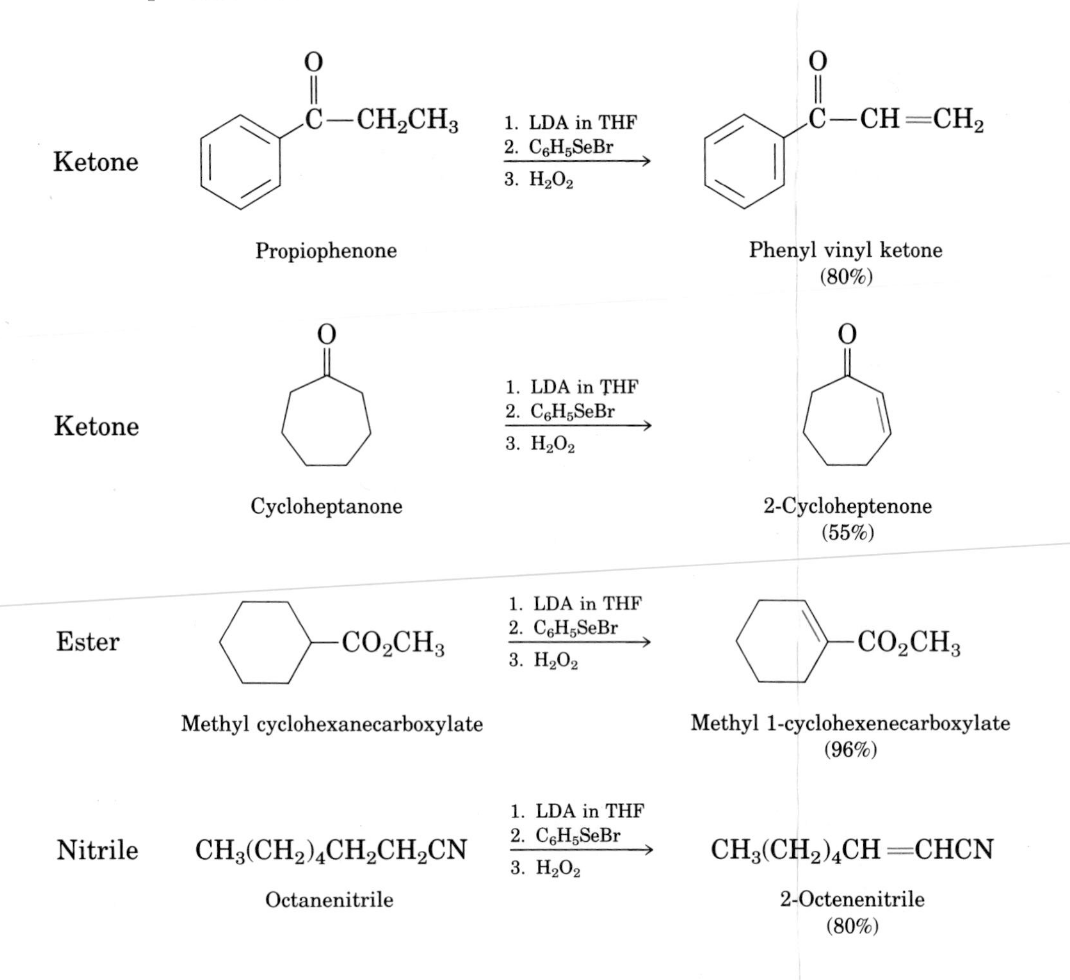

PROBLEM

25.12 Ketones react slowly with benzeneselenenyl chloride in the presence of HCl to yield α-phenylseleno ketones. Propose a mechanism for this reaction. To what other reaction we have studied is this process analogous?

25.9 Alkylation of enolates

One of the most important reactions of enolates is their **alkylation** by treatment with an alkyl halide. The reaction is extraordinarily useful for synthesis purposes because it allows the formation of a new carbon–carbon bond, thereby joining two smaller pieces into one larger molecule. Alkylation occurs when the nucleophilic enolate anion reacts with the electrophilic alkyl halide and displaces the leaving group by an S_N2 back-side attack. Reaction can occur either on the enolate oxygen atom or on the alpha carbon, but it normally takes place on carbon.

$$\text{—}\overset{\overset{\displaystyle O}{\|}}{C}\text{—}\ddot{C}^{-} + \text{C—X} \xrightarrow[\text{reaction}]{S_N2} \text{—}\overset{\overset{\displaystyle O}{\|}}{C}\text{—}\overset{|}{\underset{|}{C}}\text{—C} + :X^-$$

Enolate ion   Alkyl halide

Alkylation reactions are subject to the same constraints that affect all S_N2 reactions (Section 10.5). Thus, the leaving group, X, in the alkylating agent can be chloride, bromide, iodide, or tosylate. The alkyl group, R, must be primary or methyl and should preferably be allylic or benzylic (Figure 25.8). Secondary halides react poorly, and tertiary halides do not alkylate at all since competing E2 elimination of HX occurs instead. Vinylic and aryl halides are also unreactive, since back-side attack is sterically prevented.

$$\text{R—X}\begin{cases}\text{—X:} & \text{Tosylate} > \text{—I} > \text{—Br} > \text{—Cl}\\ \text{—R:} & \text{Allyl} \sim \text{Benzyl} > \text{—CH}_3 > \text{—CH}_2\text{R}'\end{cases}$$

Figure 25.8. Reactivity order of alkylating agents in carbonyl alpha-substitution reactions.

THE MALONIC ESTER SYNTHESIS

The **malonic ester synthesis** is one of the oldest and best known of the carbonyl alkylation reactions. It is an excellent method for preparing α-substituted acetic acids from alkyl halides:

$$\underset{\text{Alkyl halide}}{R{-}X} \xrightarrow[\text{ester synthesis}]{\text{Via malonic}} \underset{\alpha\text{-Substituted acetic acid}}{R{-}CH_2CO_2H}$$

Diethyl malonate, commonly called *malonic ester,* is more acidic than most other carbonyl compounds ($pK_a = 13$), since its alpha-hydrogen atoms are flanked by two carbonyl groups. The enolate anion of malonic ester can therefore be formed by reaction with sodium ethoxide in ethanol, and this anion is readily alkylated by treatment with an alkyl halide, yielding an α-substituted malonic ester. The product itself has one acidic alpha-hydrogen atom left, and the alkylation process can therefore be repeated to yield a dialkylated malonic ester:

$$\underset{\text{Malonic ester}}{CH_2(CO_2CH_2CH_3)_2} \xrightarrow[CH_3CH_2OH]{Na^+\ ^-OCH_2CH_3} \underset{\text{Sodio malonic ester}}{Na^+\ ^-\!:\!CH(CO_2CH_2CH_3)_2}$$

$$\xrightarrow{RX} \underset{\text{An alkylated malonic ester}}{R{-}CH(CO_2CH_2CH_3)_2} + NaX$$

$$R{-}CH(CO_2CH_2CH_3)_2 \xrightarrow[2.\ R'X]{1.\ Na^+\ ^-OCH_2CH_3} \underset{\text{A dialkylated malonic ester}}{RR'C(CO_2CH_2CH_3)_2}$$

What makes this reaction so useful is that malonic esters can be hydrolyzed and decarboxylated when heated with aqueous acid. The product is a substituted monoacid.

$$RR'C(CO_2CH_2CH_3)_2 \xrightarrow[\Delta]{H_3O^+} RR'CHCOOH + CO_2 + 2\ HOCH_2CH_3$$

The decarboxylation reaction occurs in two steps and involves initial acid-catalyzed hydrolysis of the diester to a diacid. The diacid then loses carbon dioxide by a cyclic mechanism in a step similar to that we have already seen for the decarboxylation of β-keto acids (Section 23.9).

$$R_2C(CO_2CH_2CH_3)_2 \xrightarrow[\Delta]{H_3O^+} \left[\underset{\text{Diacid}}{R_2C(COOH)_2} \xrightarrow{-CO_2} \underset{\text{Acid enol}}{R_2C{=}C(OH)OH} \longrightarrow \underset{\text{Acid}}{R_2CHCOH(=O)} \right]$$

Recall the following:

$$\underset{\beta\text{-Keto ester}}{\beta\text{-keto acid}} \xrightarrow[\Delta]{-CO_2} \underset{\text{Enol}}{C(OH){=}C} \longrightarrow \underset{\text{Ketone}}{C({=}O)CH}$$

The malonic ester synthesis is an excellent method for converting alkyl halides into carboxylic acids while lengthening the carbon chain by two atoms. Some specific examples follow:

$$\underset{\text{1-Bromobutane}}{CH_3(CH_2)_2CH_2Br} + Na^+\ ^-{:}CH(CO_2C_2H_5)_2 \longrightarrow \underset{84\%}{CH_3(CH_2)_3CH(CO_2C_2H_5)_2} \xrightarrow{H_3O^+,\ \Delta} \underset{\text{Hexanoic acid (75\%)}}{CH_3(CH_2)_4COOH}$$

$$CH_3CH_2CH_2CH_2CH(CO_2C_2H_5)_2 + CH_3I \xrightarrow[\text{Ethanol}]{Na^+\ ^-OCH_2CH_3} CH_3CH_2CH_2CH_2C(CH_3)(CO_2C_2H_5)_2 \xrightarrow{H_3O^+,\ \Delta} \underset{\text{2-Methylhexanoic acid (74\%)}}{CH_3CH_2CH_2CH_2CH(CH_3)COOH}$$

The malonic ester synthesis can be used to prepare cycloalkanecarboxylic acids by the proper choice of alkyl halide. For example, if 1,4-dibromobutane is treated with diethyl malonate in the presence of 2 equiv of sodium ethoxide base, the first alkylation occurs as expected, but the second alkylation step occurs internally to yield a five-membered-ring product. Hydrolysis and decarboxylation then lead to cyclopentanecarboxylic acid (Figure 25.9). Three-, four-, five-, and six-membered rings can be prepared in this manner, but yields decrease drastically for larger ring sizes.

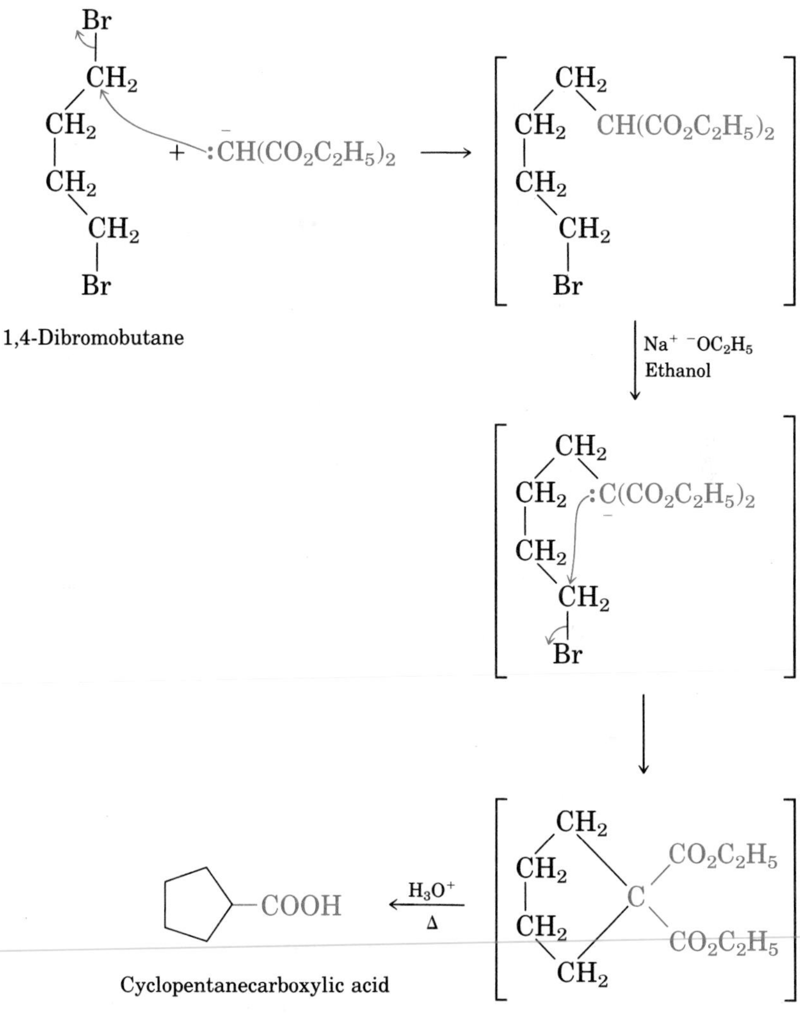

1,4-Dibromobutane + $:\bar{C}H(CO_2C_2H_5)_2$ ⟶ [$BrCH_2CH_2CH_2CH_2CH(CO_2C_2H_5)_2$]

$Na^+ \ ^-OC_2H_5$, Ethanol

[$BrCH_2CH_2CH_2CH_2\bar{C}(CO_2C_2H_5)_2$] ⟶ [cyclopentane-1,1-dicarboxylate, $C(CO_2C_2H_5)_2$ ring]

$\xleftarrow[\Delta]{H_3O^+}$ Cyclopentanecarboxylic acid (—COOH)

Figure 25.9. Malonic ester synthesis of cyclopentanecarboxylic acid.

PROBLEM

25.13 How could you use a malonic ester synthesis to prepare these compounds?
(a) 3-Phenylpropanoic acid
(b) 2-Methylpentanoic acid
(c) 4-Methylpentanoic acid
(d) Ethyl cyclobutanecarboxylate

PROBLEM

25.14 Monoalkylated and dialkylated acetic acids can be prepared by the malonic ester synthesis, but trialkylated acetic acids (R_3CCOOH) cannot be prepared. Explain.

THE ACETOACETIC ESTER SYNTHESIS

The **acetoacetic ester synthesis** is closely related to the malonic ester synthesis and provides a method for preparing α-substituted acetone derivatives from alkyl halides.

$$R{-}X \xrightarrow[\text{ester synthesis}]{\text{Via acetoacetic}} R{-}CH_2\overset{O}{\overset{\|}{C}}CH_3$$

α-Substituted acetone

Ethyl acetoacetate (acetoacetic ester) is much like malonic ester in that its alpha hydrogens are flanked by two carbonyl groups. Acetoacetic ester, $pK_a = 11$, is therefore readily converted into its enolate anion by treatment with sodium ethoxide, and this enolate can be alkylated by reaction with an alkyl halide. A second alkylation can be carried out, if desired, since acetoacetic ester has two acidic alpha protons that can be replaced.

$$CH_3\overset{O}{\overset{\|}{C}}{-}CH_2{-}\overset{O}{\overset{\|}{C}}OC_2H_5 + Na^+ \ {}^-OC_2H_5 \xrightarrow{\text{Ethanol}} \left[CH_3\overset{O}{\overset{\|}{C}}{-}\ddot{\bar{C}}HCO_2C_2H_5\right]$$

Acetoacetic ester

$$\downarrow R{-}X$$

$$CH_3{-}\overset{O}{\overset{\|}{C}}{-}\underset{R}{\underset{|}{\overset{R'}{\overset{|}{C}}}}{-}CO_2C_2H_5 \xleftarrow[\text{2. R'X}]{\text{1. Na}^+\ ^-\text{OC}_2\text{H}_5} CH_3\overset{O}{\overset{\|}{C}}{-}\underset{R}{\underset{|}{C}}HCO_2C_2H_5$$

Upon acid hydrolysis and heating, the alkylated acetoacetic ester loses carbon dioxide via a β-keto acid intermediate, and an α-substituted acetone product is formed. If a monoalkylated acetoacetic ester is hydrolyzed and decarboxylated, an α-monosubstituted acetone is formed; if a dialkylated acetoacetic ester is hydrolyzed and decarboxylated, an α,α-disubstituted acetone is formed.

$$\underset{CO_2C_2H_5}{\underset{|}{C}}H_2\overset{O}{\overset{\|}{C}}CH_3 \xrightarrow[\text{2. RX}]{\text{1. Na}^+\ ^-\text{OC}_2\text{H}_5} R{-}\underset{CO_2C_2H_5}{\underset{|}{C}}H\overset{O}{\overset{\|}{C}}CH_3 \xrightarrow[\Delta]{H_3O^+} \left[\text{cyclic transition state: } R{-}CH(C(=O)CH_3)(C(=O)O{-}H)\right]$$

Acetoacetic ester

$$\downarrow$$

$$R{-}CH_2\overset{O}{\overset{\|}{C}}CH_3$$

An α-monosubstituted acetone

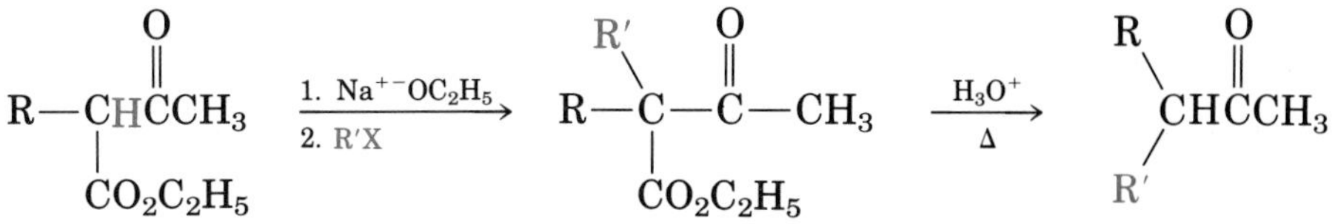

$$R{-}CH(CO_2C_2H_5)C(=O)CH_3 \xrightarrow[\text{2. } R'X]{\text{1. } Na^{+}\,{}^{-}OC_2H_5} R{-}C(R')(CO_2C_2H_5){-}C(=O){-}CH_3 \xrightarrow[\Delta]{H_3O^{+}} (R)(R')CHC(=O)CH_3$$

An α,α-disubstituted acetone

For example,

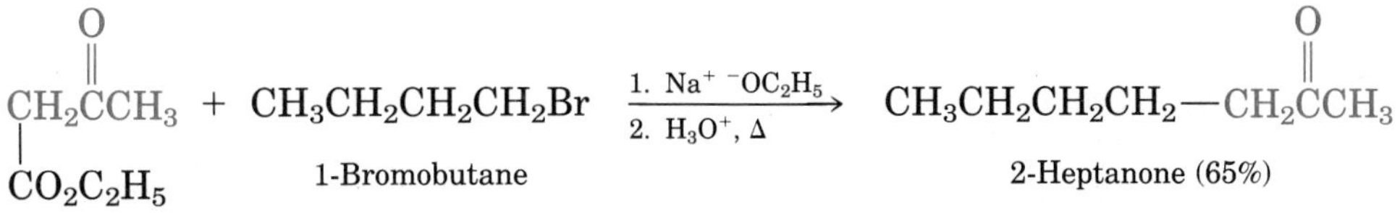

$$CH_2(CO_2C_2H_5)C(=O)CH_3 + CH_3CH_2CH_2CH_2Br \xrightarrow[\text{2. } H_3O^{+},\ \Delta]{\text{1. } Na^{+}\ {}^{-}OC_2H_5} CH_3CH_2CH_2CH_2{-}CH_2C(=O)CH_3$$

Acetoacetic ester; 1-Bromobutane; 2-Heptanone (65%)

The sequence shown—(1) enolate formation, (2) alkylation, and (3) hydrolysis/decarboxylation—is applicable to *all* β-keto esters with acidic alpha hydrogens and is not limited to just acetoacetic ester itself. Thus, cyclic β-keto esters such as ethyl 2-oxocyclohexanecarboxylate can be alkylated and decarboxylated to give 2-substituted cyclohexanones in high yield. For example,

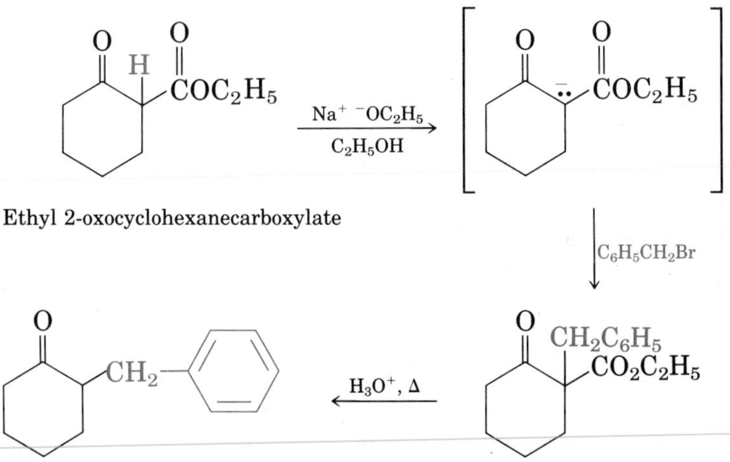

Ethyl 2-oxocyclohexanecarboxylate $\xrightarrow[C_2H_5OH]{Na^{+}\ {}^{-}OC_2H_5}$ [enolate anion, COC_2H_5] $\xrightarrow{C_6H_5CH_2Br}$ 2-($CH_2C_6H_5$)-2-($CO_2C_2H_5$)cyclohexanone $\xrightarrow{H_3O^{+},\ \Delta}$ 2-Benzylcyclohexanone (77%)

PROBLEM

25.15 How would you prepare methyl cyclopentyl ketone, using an acetoacetic ester synthesis?

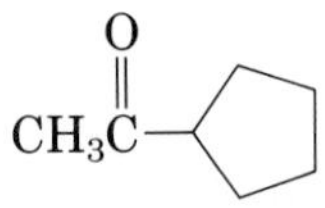

$CH_3C(=O){-}C_5H_9$

Methyl cyclopentyl ketone

PROBLEM

25.16 Which of the following compounds cannot be prepared by an acetoacetic ester synthesis? Explain.

(a) 2-Butanone
(b) Phenylacetone
(c) Acetophenone
(d) 3,3-Dimethyl-2-butanone
(e) 4-Phenyl-2-pentanone

ALKYLATION OF KETONES, ESTERS, AND NITRILES

The malonic ester synthesis and the acetoacetic ester synthesis are rather special, since both alkylation reactions take place at doubly carbonyl-activated centers. By contrast, it is also possible in certain cases to alkylate at the alpha position of monoketones, monoesters, and nitriles. The experimental conditions necessary to carry out such alkylations are precise, and much skill is required. Solvent and reaction temperature are both important, and the exact nature of the base used to generate the enolate ion is critical. The base must be sufficiently strong to quickly and completely convert a carbonyl compound ($pK_a \approx 20$–25) into its enolate anion, yet it must also be sufficiently bulky that it will not add to the carbonyl group in a nucleophilic addition reaction. Research carried out in the 1970s has shown that LDA in THF solvent is a highly effective base for promoting alkylation reactions of carbonyl compounds. It rapidly abstracts acidic alpha protons from carbonyl compounds to generate enolate anions, and addition of an alkyl halide then yields the desired α-substituted product in a typical S_N2 process.

Ketone/ester —(LDA, in THF)→ [enolate anion] —(R–X)→ α-Substituted ketone/ester + $:X^-$

In general, it is often possible to alkylate ketones, esters, and nitriles by using LDA or related dialkylamide bases in THF. Aldehydes, however, rarely give high yields of pure products, and their alkylations are not usually successful. Some specific examples of alkylation reactions are shown in Table 25.2.

Note in the examples of Table 25.2 that alkylation of unsymmetrically substituted ketones often leads to a mixture of products, depending on which of two possible enolate ions is formed. In general, though, the major product is derived from alkylation at the less hindered, more accessible, position. Thus, alkylation of 2-methylcyclohexanone occurs primarily at the secondary 6-position, rather than at the tertiary 2-position. It should also be reiterated that the alpha alkylation of carbonyl compounds is an experimentally delicate process that can give mixtures of by-products if care is not taken. Experimental conditions are constantly being improved, however, and carbonyl-compound alkylations are unquestionably among the most useful and important reactions in organic chemistry.

TABLE 25.2 Some alkylation reactions

Lactone

Butyrolactone $\xrightarrow[\text{THF}]{\text{LDA}}$ [enolate anion] $\xrightarrow{CH_3I}$ 2-Methylbutyrolactone (88%)

Ester

$$(CH_3)_2\overset{H}{C}-\overset{O}{\overset{\|}{C}}-OC_2H_5 \xrightarrow[\text{THF}]{\text{LDA}} \left[(CH_3)_2\ddot{\bar{C}}-\overset{O}{\overset{\|}{C}}-OC_2H_5\right] \xrightarrow{CH_3I} (CH_3)_2\overset{CH_3}{C}-\overset{O}{\overset{\|}{C}}OC_2H_5$$

Ethyl 2-methylpropanoate → Ethyl 2,2-dimethylpropanoate (87%)

Nitrile

$C_6H_5CH_2C\equiv N \xrightarrow[\text{THF}]{\text{LDA}} [C_6H_5\ddot{\bar{C}}HC\equiv N] \xrightarrow{CH_3I} C_6H_5CH(CH_3)C\equiv N$

Phenylacetonitrile → 2-Phenylpropanenitrile (71%)

Ketone

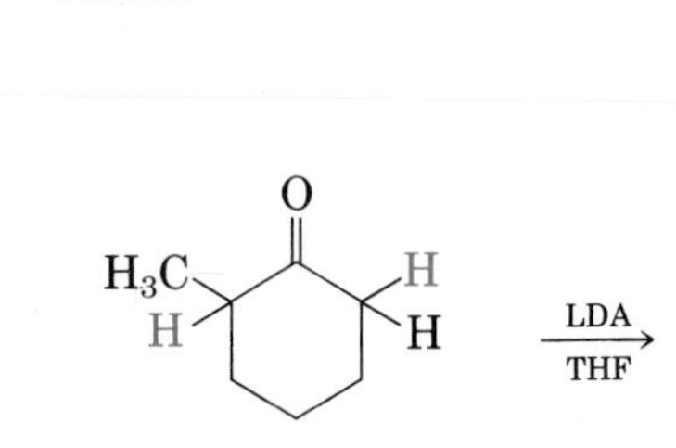

2-Methylcyclohexanone

$\xrightarrow[\text{THF}]{\text{LDA}}$

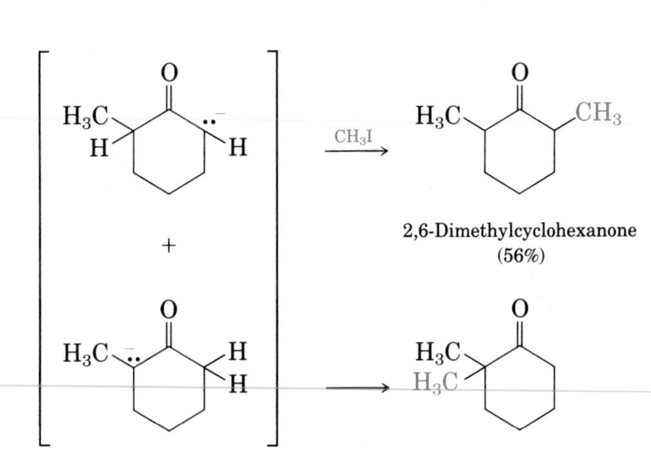

2,6-Dimethylcyclohexanone (56%)

2,2-Dimethylcyclohexanone (6%)

PROBLEM

25.17 How might you prepare the following compounds, using an alkylation reaction as the key step?

(a) 3-Phenyl-2-butanone (b) 2-Ethylpentanenitrile

(c) 2-Allylcyclohexanone (d) 2,2,6,6-Tetramethylcyclohexanone

25.10 Summary

The alpha substitution of carbonyl compounds via enol or enolate ion intermediates represents one of the four fundamental reaction types in carbonyl-group chemistry.

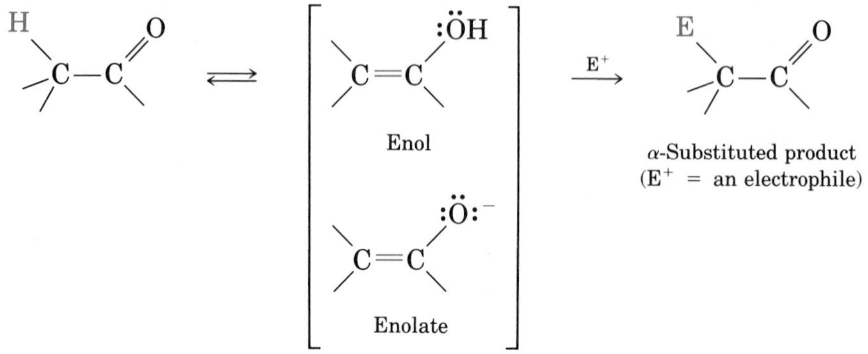

Enol

Enolate

α-Substituted product
(E^+ = an electrophile)

Carbonyl compounds are in an equilibrium with their enols in a process known as tautomerism. Enol tautomers are normally present to only a small extent and cannot usually be isolated in pure form. Nevertheless, enols contain a highly nucleophilic double bond and react rapidly with a variety of electrophiles. For example, ketones and aldehydes are rapidly halogenated at the alpha position by reaction with chlorine, bromine, or iodine in acetic acid solution. Much evidence indicates the participation of enol intermediates in these alpha-halogenation reactions. Thus, bromination, chlorination, iodination, and deuteration all occur at exactly the same rate and have a common rate-limiting step. Furthermore, kinetic measurements indicate that halogenation reactions are second order but are independent of halogen concentration.

Alpha bromination of carboxylic acids can be accomplished by the Hell–Volhard–Zelinskii reaction, in which an acid is treated with Br_2/PBr_3. α-Brominated ketone, aldehyde, and carboxylic acid products are valuable because they can be dehydrobrominated by amine bases in E2-type reactions to yield α,β-unsaturated carbonyl compounds.

Alpha-hydrogen atoms of carbonyl compounds are acidic and can be abstracted by strong bases to yield enolate ions. Ketones, aldehydes, esters, amides, and nitriles can all be deprotonated in this manner if a sufficiently powerful base such as lithium diisopropylamide (LDA) is used. Enolate ions can be generated quantitatively, in contrast to enols, and they are highly reactive as nucleophiles because of their negative charge. For example, enolates react with benzeneselenenyl bromide to yield α-phenylselenenylated products. These, in turn, yield α,β-unsaturated carbonyl compounds when treated with H_2O_2.

The most important reaction of enolates is their S_N2 alkylation by alkyl halides; the nucleophilic enolate ion attacks an alkyl halide from the back side and displaces the leaving halide group to yield an α-alkylated carbonyl product.

$$\text{C(=O)}\!-\!\ddot{\text{C}}^{-} + \text{C}\!-\!\text{X} \longrightarrow \text{C(=O)}\!-\!\text{C}\!-\!\text{C}$$

The malonic ester synthesis involves the alkylation of diethyl malonate with alkyl halides and provides a method for preparing monoalkylated or dialkylated acetic acids. The acetoacetic ester synthesis is similar to the malonic ester synthesis and provides a method for preparing monoalkylated or dialkylated acetone derivatives. Most important of all, however, is the fact that many carbonyl compounds, including ketones, esters, and nitriles, can be directly alkylated. Lithium diisopropylamide is normally used as the base, and the experimental conditions must be carefully controlled. These carbonyl alkylation reactions are extremely versatile and constitute what is perhaps the single most important method in organic chemistry for synthesizing complex molecules.

25.11 Summary of reactions

1. Ketone/aldehyde halogenation (Section 25.3)

$$\text{R}\!-\!\text{C(=O)}\!-\!\text{C(H)}\!- \xrightarrow[\text{CH}_3\text{COOH}]{\text{X}_2} \text{R}\!-\!\text{C(=O)}\!-\!\text{C(X)}\!- + \text{HX}$$

where X = Cl, Br, or I.

2. Hell–Volhard–Zelinskii bromination of acids (Section 25.4)

$$\text{H}\!-\!\text{C}\!-\!\text{COOH} \xrightarrow[\text{2. H}_2\text{O}]{\text{1. Br}_2 + \text{PBr}_3} \text{Br}\!-\!\text{C}\!-\!\text{COOH}$$

3. Dehydrobromination of α-bromo ketones (Section 25.3)

$$\text{R}\!-\!\text{C(=O)}\!-\!\text{C(Br)}\!-\!\text{C(H)} \xrightarrow[\Delta]{\text{Pyridine}} \text{R}\!-\!\text{C(=O)}\!-\!\text{C}\!=\!\text{C}$$

4. Haloform reaction (Section 25.7)

$$\text{R}\!-\!\text{C(=O)}\!-\!\text{CH}_3 \xrightarrow[^{-}\text{OH}]{\text{X}_2} \text{R}\!-\!\text{C(=O)}\!-\!\text{O}^{-} + \text{HCX}_3$$

where X = Cl, Br, or I.

5. Phenylselenenylation/elimination (Section 25.8)

$$R{-}\overset{O}{\overset{\|}{C}}{-}\overset{H}{\overset{|}{\underset{|}{C}}}{-}\overset{H}{\overset{|}{\underset{|}{C}}}{-} \xrightarrow[2.\ C_6H_5SeBr]{1.\ LDA\ in\ THF} R{-}\overset{O}{\overset{\|}{C}}{-}\overset{SeC_6H_5}{\overset{|}{\underset{|}{C}}}{-}\underset{|}{C}{\overset{/}{\underset{\backslash H}{}}} \xrightarrow{H_2O_2} R{-}\overset{O}{\overset{\|}{C}}{-}\underset{|}{C}{=}C\langle$$

6. Alkylation of enolates (Section 25.9)
 a. Malonic ester synthesis

$$CH_2(CO_2C_2H_5)_2 + RX \xrightarrow[Ethanol]{NaOC_2H_5} R{-}CH(CO_2C_2H_5)_2 \xrightarrow{H_3O^+} R{-}CH_2COOH$$

$$R{-}CH(CO_2C_2H_5)_2 + R'X \xrightarrow[Ethanol]{NaOC_2H_5} (R)(R')C(CO_2C_2H_5)_2 \xrightarrow{H_3O^+} (R)(R')CHCOOH$$

 b. Acetoacetic ester synthesis

$$CH_3\overset{O}{\overset{\|}{C}}CH_2\overset{O}{\overset{\|}{C}}OC_2H_5 + RX \xrightarrow[Ethanol]{NaOC_2H_5} CH_3\overset{O}{\overset{\|}{C}}\underset{R}{\underset{|}{C}H}\overset{O}{\overset{\|}{C}}OC_2H_5 \xrightarrow{H_3O^+} CH_3\overset{O}{\overset{\|}{C}}CH_2{-}R$$

$$CH_3\overset{O}{\overset{\|}{C}}\underset{R}{\underset{|}{C}H}\overset{O}{\overset{\|}{C}}OC_2H_5 + R'X \xrightarrow[Ethanol]{NaOC_2H_5} CH_3\overset{O}{\overset{\|}{C}}{-}\underset{R\ \ R'}{C}{-}\overset{O}{\overset{\|}{C}}OC_2H_5 \xrightarrow{H_3O^+} CH_3\overset{O}{\overset{\|}{C}}CH(R)(R')$$

 c. Alkylation of ketones

$$R{-}\overset{O}{\overset{\|}{C}}{-}C(H)\langle \xrightarrow[2.\ R'X]{1.\ LDA\ in\ THF} R{-}\overset{O}{\overset{\|}{C}}{-}C(R')\langle$$

 d. Alkylation of esters

$$\rangle C(H){-}\overset{O}{\overset{\|}{C}}OC_2H_5 \xrightarrow[2.\ RX]{1.\ LDA\ in\ THF} \rangle C(R){-}\overset{O}{\overset{\|}{C}}{-}OC_2H_5$$

 e. Alkylation of nitriles

$$\rangle C(H){-}C{\equiv}N \xrightarrow[2.\ RX]{1.\ LDA\ in\ THF} \rangle C(R){-}C{\equiv}N$$

ADDITIONAL PROBLEMS

25.18 All attempts to isolate primary and secondary nitroso compounds result only in the formation of oximes. Tertiary nitroso compounds, however, are quite stable. Explain these facts.

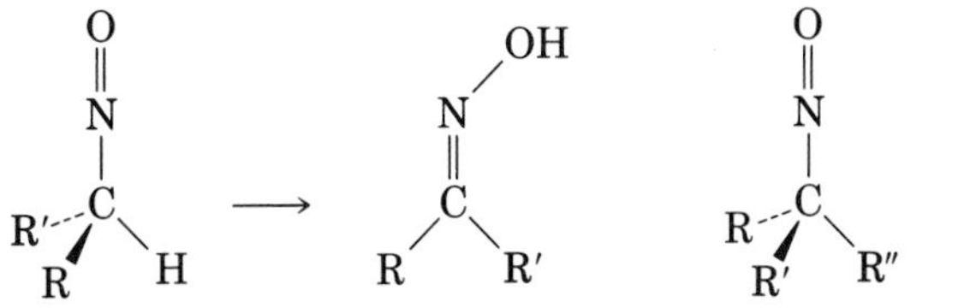

$R'RCH\text{—}N\text{=}O \longrightarrow RR'C\text{=}N\text{—}OH$ $\qquad RR'R''C\text{—}N\text{=}O$

1° or 2° nitroso compound (unstable) Oxime 3° nitroso compound (stable)

25.19 Suggest an explanation of the observation that acetone is enolized only to the extent of about 0.0001% at equilibrium, whereas 2,4-pentanedione is 76% enolized.

25.20 Write resonance structures for these anions:

(a) 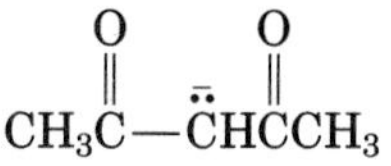 $CH_3C(=O)\text{—}\ddot{C}HC(=O)CH_3$ (b) $:\bar{C}H_2C\equiv N$

(c) 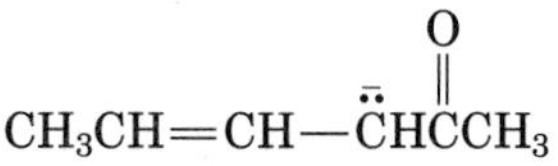 $CH_3CH\text{=}CH\text{—}\ddot{C}HC(=O)CH_3$ (d) $N\equiv C\text{—}\ddot{C}HCO_2C_2H_5$

(e) 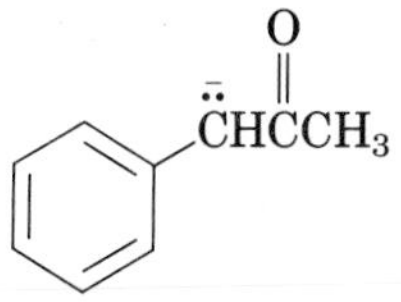 $C_6H_5\ddot{C}HC(=O)CH_3$

25.21 Indicate all the acidic hydrogen atoms in the following structures:

(a) 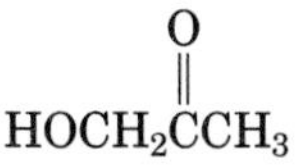 $HOCH_2C(=O)CH_3$ (b) 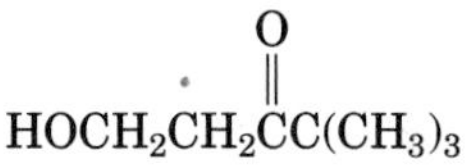 $HOCH_2CH_2C(=O)C(CH_3)_3$

(c) 1,3-Cyclopentanedione (d) $CH_3CH\text{=}CHCHO$

(e) $(CH_3)_2NH$ (f) 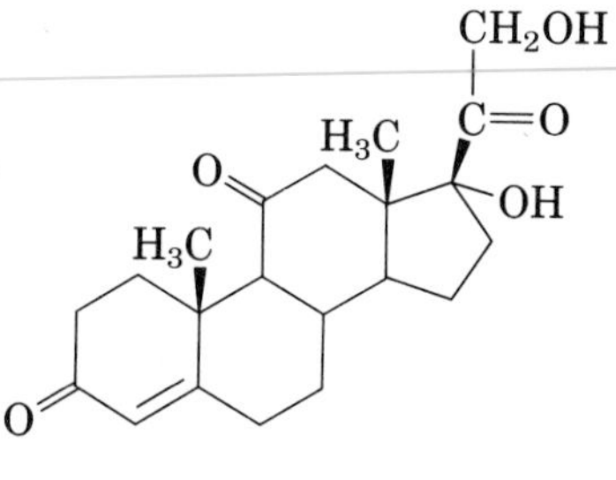 [structure of cortisone]

Cortisone

25.22 One method of determining the number of acidic hydrogen atoms in a molecule is to treat the compound with NaOD in D_2O, isolate the product, and determine its molecular weight by mass spectroscopy. For example, if cyclohexanone is treated with NaOD in D_2O, the product has a molecular weight of 102. Explain how this method works. What is the structure of the product of deuteration of cyclohexanone?

25.23 2-Methylcycloheptanone and 3-methylcycloheptanone are nearly indistinguishable by spectroscopic techniques. How could you differentiate them? [*Hint:* See Problem 25.22.]

25.24 Rank the following compounds in order of increasing acidity:

(a) CH_3CH_2COOH (b) CH_3CH_2OH

(c) $(CH_3CH_2)_2NH$ (d) CH_3COCH_3

(e) $CH_3C(=O)CH_2C(=O)CH_3$ (f) CCl_3COOH

25.25 Show how you would prepare the following esters. Which of the compounds cannot be prepared by a malonic ester synthesis?

(a) Ethyl pentanoate (b) Ethyl 3-methylbutanoate

(c) Ethyl 2-methylbutanoate (d) Ethyl 2,2-dimethylpropanoate

25.26 Nonconjugated β,γ-unsaturated ketones such as 3-cyclohexenone are in an acid-catalyzed equilibrium with their conjugated α,β-unsaturated isomers. Propose a mechanism for the acid-catalyzed interconversion of the two isomers.

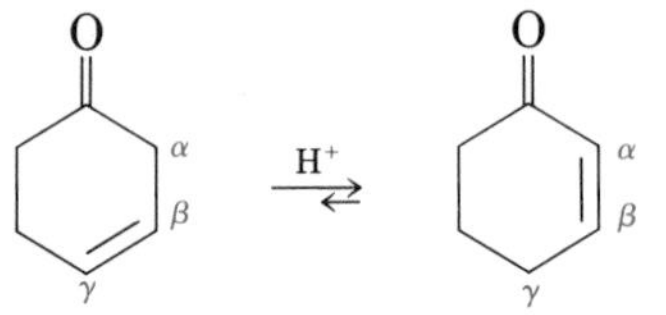

O, α, β, γ; H^+; O, α, β, γ

25.27 The $\alpha,\beta-\beta,\gamma$ interconversion of unsaturated ketones (Problem 25.26) can also be catalyzed by base. Explain.

25.28 One interesting consequence of the base-catalyzed $\alpha,\beta-\beta,\gamma$ isomerization of unsaturated ketones (Problem 25.27) is that 2-substituted 2-cyclopentenones can be interconverted with 5-substituted 2-cyclopentenones. Propose a mechanism to account for this isomerization.

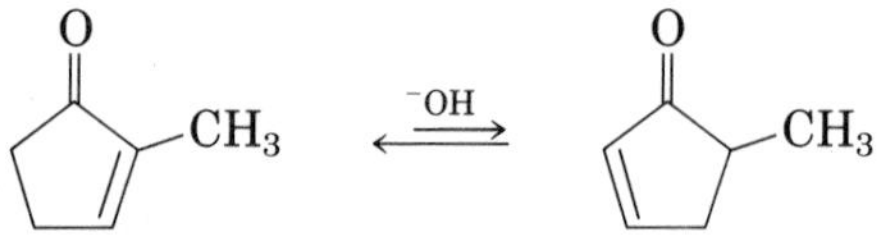

O, CH_3; ^-OH; O, CH_3

25.29 How would you prepare these compounds, using either an acetoacetic ester synthesis or a malonic ester synthesis?

(a) $(CH_3)_2C(CO_2C_2H_5)_2$ (b) 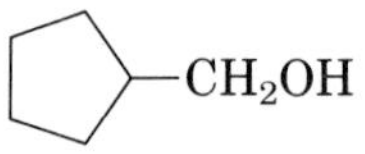 cyclopentyl—CH_2OH

(c) 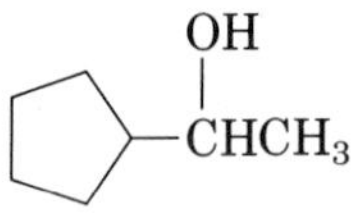 cyclopentyl—$CH(OH)CH_3$ (d) $HOOCCH_2CH_2COOH$

(e) $H_2C{=}CH{-}CH_2CH_2COCH_3$

25.30 Predict the product(s) of the following reactions:

(a) 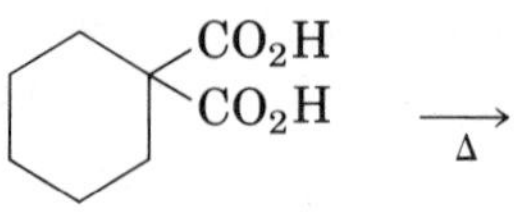 cyclohexane-1,1-$(CO_2H)_2$ $\xrightarrow{\Delta}$

(b) $(CH_3)_2CHCO_2C_2H_5 \xrightarrow[\text{2. } C_6H_5SeBr]{\text{1. LDA, THF}} A \xrightarrow{H_2O_2} B$

(c) Cyclopentane-1,3-dione $\xrightarrow[\text{2. } CH_3I]{\text{1. } NaOC_2H_5}$

(d) $CH_3CH_2CH_2COOH \xrightarrow{Br_2,\ PBr_3} A \xrightarrow{H_2O} B$

(e) $C_6H_5COCH_3 \xrightarrow[I_2]{^{-}OH,\ H_2O}$

25.31 Show how you might convert geraniol into either ethyl geranylacetate or geranylacetone.

$(CH_3)_2C{=}CHCH_2CH_2C(CH_3){=}CHCH_2OH$ (Geraniol) $\xrightarrow{?}$ $(CH_3)_2C{=}CHCH_2CH_2C(CH_3){=}CHCH_2CH_2CO_2C_2H_5$ (Ethyl geranylacetate)

$(CH_3)_2C{=}CHCH_2CH_2C(CH_3){=}CHCH_2OH$ (Geraniol) $\xrightarrow{?}$ $(CH_3)_2C{=}CHCH_2CH_2C(CH_3){=}CHCH_2CH_2COCH_3$ (Geranylacetone)

25.32 How would you synthesize the following compounds from cyclohexanone? More than one step may be required.

(a) Cyclohexane with exocyclic $=CH_2$

(b) Cyclohexane–CH_2Br

(c) 2-($CH_2C_6H_5$)cyclohexanone

(d) Cyclohexane–CH_2CH_2COOH

(e) Cyclohexene–$COOH$

(f) Cyclohexenone

(g) Cyclohexane–$CH{=}CHCO_2CH_3$

(h) Cyclohexane–$COOH$

25.33 How can you account for the fact that *cis*- and *trans*-4-*tert*-butyl-2-methylcyclohexanone are interconverted by base treatment?

cis-4-tert-Butyl-2-methylcyclohexanone [$(CH_3)_3C$, H; CH_3, H] $\overset{^{-}OH}{\rightleftharpoons}$ trans-4-tert-Butyl-2-methylcyclohexanone [$(CH_3)_3C$, H; CH_3, H]

Which of the two isomers do you think is more stable, and why? Use molecular models to help formulate your answer.

25.34 In Problem 25.28, we said that 2-substituted 2-cyclopentenones are in a base-catalyzed equilibrium with their 5-substituted 2-cyclopentenone isomers. Why do you suppose the analogous isomerization is not observed for 2-substituted 2-cyclohexenones?

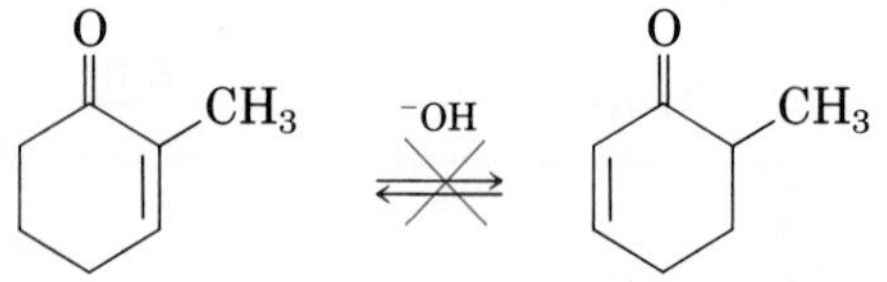

2-methyl-2-cyclohexenone $\underset{\times}{\overset{^{-}OH}{\rightleftharpoons}}$ 6-methyl-2-cyclohexenone

25.35 The following synthetic routes are incorrect as drawn. What is wrong with each?

(a) $CH_3CH_2CH_2CH_2CO_2CH_3 \xrightarrow[2.\ \Delta,\ pyridine]{1.\ Br_2,\ CH_3COOH} CH_3CH_2CH{=}CHCO_2CH_3$

(b) $CH_3CH(CO_2CH_2CH_3)_2 \xrightarrow[2.\ C_6H_5Br]{1.\ ^{-}OC_2H_5/C_2H_5OH} C_6H_5C(CH_3)(CO_2C_2H_5)_2$

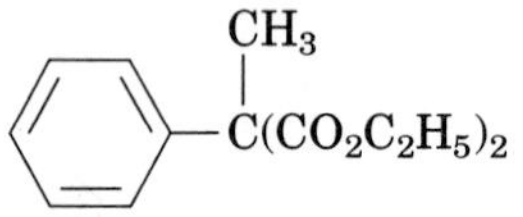

(c) $CH_3COCH_2CO_2C_2H_5 \xrightarrow[2.\ H_2C{=}CHCH_2Br;\ 3.\ H_3O^+]{1.\ ^{-}OC_2H_5/C_2H_5OH} H_2C{=}CHCH_2CH_2COOH$

(d) $CH_3CH_2CH_2CN \xrightarrow[2.\ C_6H_5SeBr;\ 3.\ \Delta,\ pyridine]{1.\ ^{-}OC_2H_5/C_2H_5OH} CH_3CH{=}CHCN$

(e) cyclohexanone $\xrightarrow[2.\ \Delta,\ pyridine]{1.\ Br_2,\ NaOH}$ 2-cyclohexenone

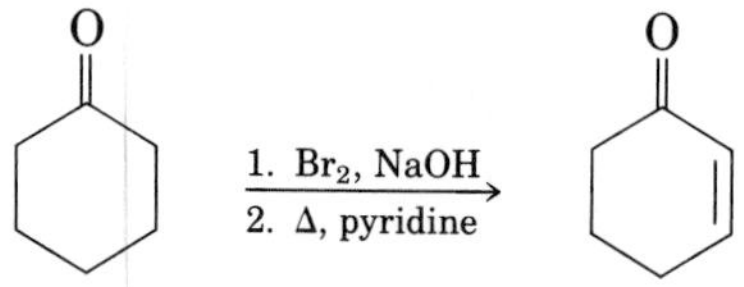

(f) $(CH_3)_2CHCOCH_3 \xrightarrow[2.\ C_6H_5CH_2Br]{1.\ LDA,\ THF} (CH_3)_2C(CH_2C_6H_5)COCH_3$

25.36 The β-diketone shown has no detectable enol content and is about as acidic as acetone. Explain this behavior. Molecular models should prove helpful.

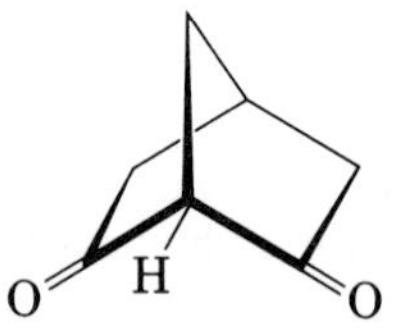

[bicyclic β-diketone structure; labels: O, H, O]

25.37 Methylmagnesium bromide adds to cyclohexanone to give the expected tertiary alcohol product in high yield. *tert*-Butylmagnesium bromide, however, gives only about a 1% yield of the addition product, along with 99% recovered starting material. Furthermore, if D_3O^+ is added to the reaction mixture after a suitable period, one deuterium atom is incorporated into the recovered cyclohexanone. Explain these results.

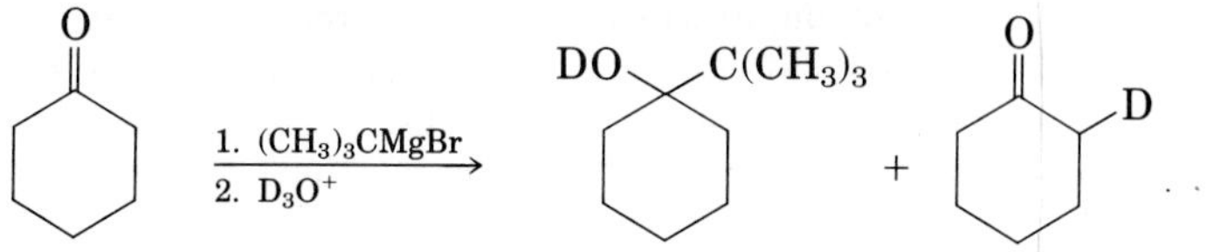

1. $(CH_3)_3CMgBr$
2. D_3O^+

DO C$(CH_3)_3$ + D

25.38 The final step in an attempted synthesis of laurene, a hydrocarbon isolated from the alga *Laurencia glandulifera,* involved the Wittig reaction shown. The product obtained, however, was not laurene, but an isomer of laurene. Propose a mechanism to account for these unexpected results.

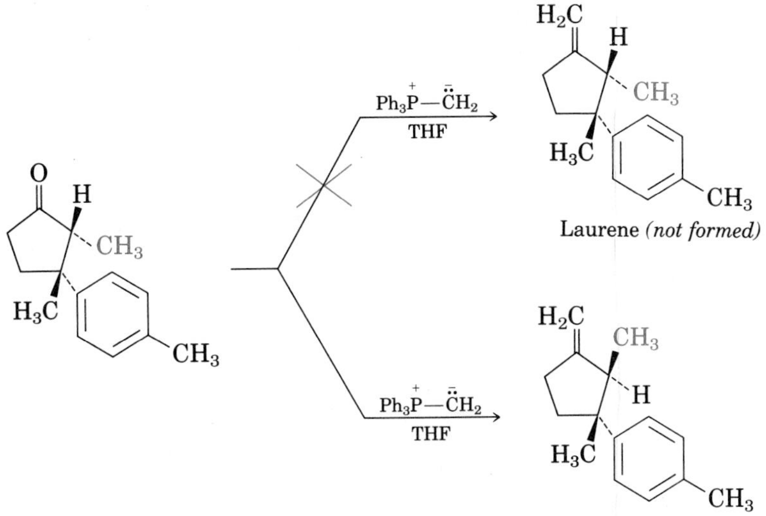

$Ph_3\overset{+}{P}—\ddot{C}H_2^-$ THF

Laurene *(not formed)*

$Ph_3\overset{+}{P}—\ddot{C}H_2^-$ THF

25.39 The key step in a reported synthesis of sativene, a hydrocarbon isolated from the mold *Helminthosporium sativum,* is as follows:

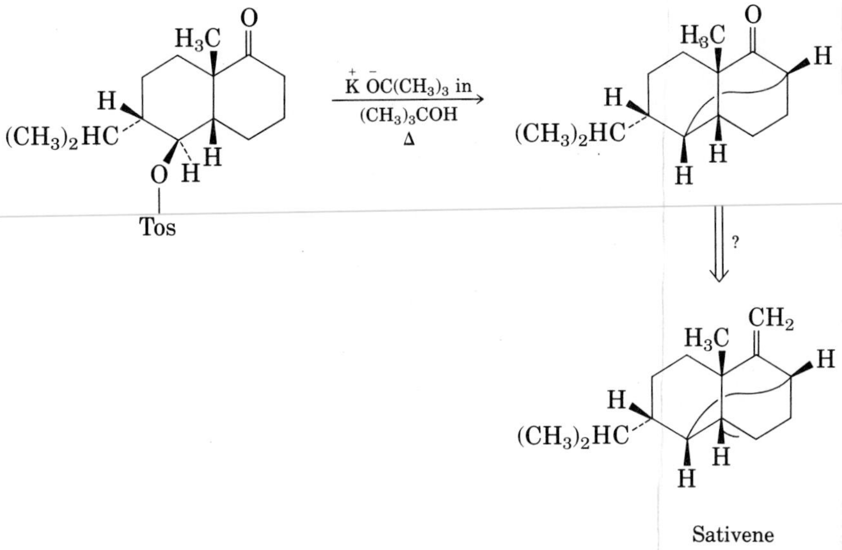

$\overset{+}{K}\ \overset{-}{O}C(CH_3)_3$ in $(CH_3)_3COH$ Δ

?

Sativene

What kind of reaction is occurring? How would you complete the synthesis?

25.40 The Favorskii reaction, in which an α-bromo ketone is treated with base, is a useful method for accomplishing a ring contraction. For example, treatment of 2-bromocyclohexanone with aqueous NaOH yields cyclopentanecarboxylic acid. Can you propose a mechanism to account for this reaction?

$$\text{2-bromocyclohexanone} \xrightarrow[\text{2. } H_3O^+]{\text{1. NaOH}} \text{cyclopentane-COOH}$$

25.41 Treatment of a cyclic ketone with diazomethane is a method for accomplishing a ring expansion and serves as a useful complement to the Favorskii reaction (Problem 25.40). For example, treatment of cyclohexanone with diazomethane yields cycloheptanone. Can you propose a mechanism to account for this reaction?

$$\text{cyclohexanone} \xrightarrow[\text{Ether}]{:\overset{-}{C}H_2-\overset{+}{N}\equiv N} \text{cycloheptanone} + N_2$$

CHAPTER 26 CARBONYL CONDENSATION REACTIONS

In the previous five chapters we have seen three general kinds of carbonyl-group reactions and have studied two general kinds of behavior:

1. Carbonyl groups can behave as electrophiles as a consequence of the polarity of the $^{\delta+}C{=}O^{\delta-}$ bond. This sort of reactivity is seen in nucleophilic addition reactions and in nucleophilic acyl substitution reactions.
2. Carbonyl compounds can behave as nucleophiles when they are converted into their enolate ions or enol tautomers. This sort of reactivity is seen during alpha-substitution reactions.

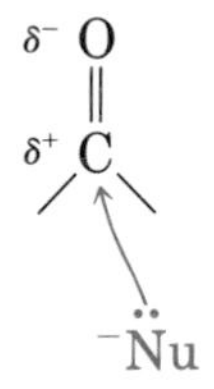

Electrophilic carbonyl is attacked by nucleophiles (:Nu$^-$)

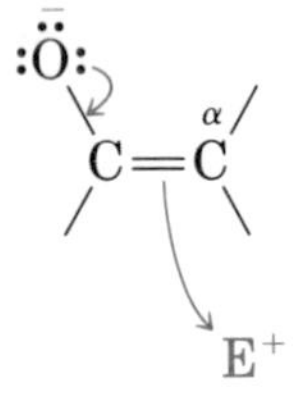

Nucleophilic alpha position of enolates attacks electrophiles (E$^+$)

In the present chapter we will study the fourth and last general polar reaction of carbonyl compounds—the **carbonyl condensation reaction.**

26.1 General mechanism of carbonyl condensation reactions

Carbonyl condensation reactions take place between two carbonyl components and involve a combination of nucleophilic addition and alpha-substitution steps. One component acts as a nucleophilic electron donor, and one component acts as an electrophilic electron acceptor. The general mechanism by which reaction occurs is shown in Figure 26.1.

From the point of view of the donor component, a carbonyl condensation reaction is simply an alpha-substitution process. From the point of view of the acceptor component, a carbonyl condensation reaction is a nucleophilic addition process. However one chooses to view the reaction, carbonyl condensations are among the most useful reactions in organic chemistry. All manner of carbonyl compounds, including aldehydes, ketones, esters, amides, acid anhydrides, thiol esters, and nitriles, enter into condensation reactions. Nature uses carbonyl condensation reactions as a key step in the biosynthesis of most naturally occurring compounds, and chemists use the same reactions in the laboratory.

There are numerous variations of carbonyl condensation reactions, depending on the exact structure of the two carbonyl components, but the general mechanism is similar in all cases. Let's see some examples.

26.2 Condensations of aldehydes and ketones: the aldol reaction

When acetaldehyde is treated in a protic solvent with a basic catalyst such as sodium ethoxide or sodium hydroxide, a rapid and reversible con-

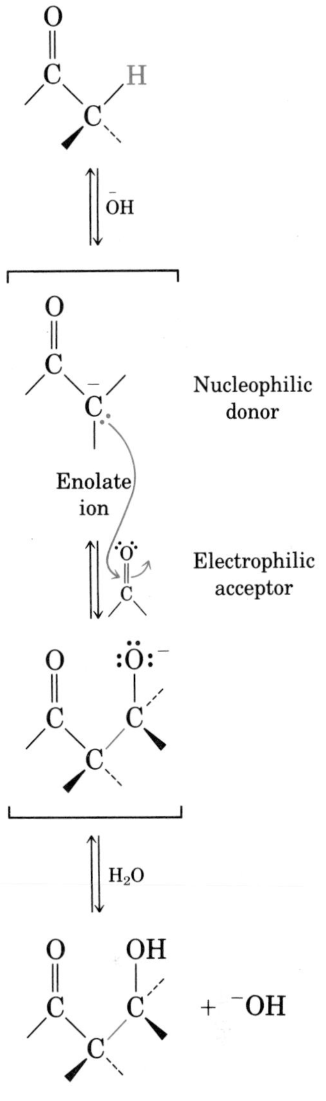

One carbonyl component with an alpha-hydrogen atom is converted by base into its enolate anion.

This enolate ion acts as a nucleophilic donor and adds to the electrophilic carbonyl group of the acceptor component.

Protonation of the tetrahedral alkoxide ion intermediate gives the neutral condensation product.

Figure 26.1. General mechanism of a carbonyl condensation reaction.

densation reaction occurs, yielding the β-hydroxy aldehyde product known trivially as **aldol** (aldehyde + alcohol).

$$2\ CH_3CHO \xrightleftharpoons[\text{Ethanol}]{Na^{+}\bar{O}C_2H_5} CH_3\overset{\overset{OH}{|}\beta}{\underset{\underset{H}{|}}{C}}-\overset{\alpha}{CH_2}CHO$$

Acetaldehyde

Aldol (a β-hydroxy aldehyde)

Base-catalyzed dimerization is general for all ketones and aldehydes having alpha-hydrogen atoms and is known as the **aldol reaction.** If the ketone or aldehyde does not have an alpha-hydrogen atom, however, aldol condensation cannot occur. As the following examples indicate, the aldol equilibrium generally favors condensation product in the case of monosubstituted acetaldehydes ($R—CH_2CHO$). The aldol equilibrium favors starting material, however, for disubstituted acetaldehydes (R_2CHCHO) and for most ketones. Steric factors are probably responsible for these trends, since increased substitution near the reaction site greatly increases steric congestion.

2 Cyclohexanone $\xrightleftharpoons{\text{NaOH, ethanol}}$ (product: 2-(1-hydroxycyclohexyl)cyclohexanone, OH) 22%

$$2\ CH_3\overset{O}{\overset{\|}{C}}CH_3 \xrightleftharpoons{\text{NaOH}} (CH_3)_2\overset{OH}{\overset{|}{C}}—CH_2\overset{O}{\overset{\|}{C}}CH_3$$

Acetone — 75%

2 $C_6H_5CH_2CHO$ (Phenylacetaldehyde) $\xrightleftharpoons{\text{NaOH, ethanol}}$ $C_6H_5CH_2\overset{OH}{\overset{|}{C}H}—\underset{C_6H_5}{\underset{|}{CH}}—CHO$ 90%

2 H_3C, H_3C (2,2-Dimethylcyclohexanone) $\xrightleftharpoons{\text{NaOH, ethanol}}$ (product with H_3C, CH_3, O, H_3C, CH_3, OH) Low yield

$$2\ H—\overset{CH_3}{\underset{CH_3}{C}}—CHO \xrightleftharpoons{\text{NaOH, ethanol}} (CH_3)_2CH\overset{OH}{\overset{|}{C}H}—\overset{CH_3}{\underset{CH_3}{C}}—CHO$$

2-Methylpropanal — Low yield

$$\left.\begin{array}{l} R_3C{-}CHO \\ ArCHO \\ ArCOCR_3 \\ ArCOAr \\ R_3CCOCR_3 \end{array}\right\} \xrightarrow[\times]{\text{NaOH, ethanol}} \text{No aldol condensation products, since the reactants have no alpha hydrogens}$$

Aldol reactions are typical carbonyl condensations. They occur by nucleophilic addition of the enolate ion of the donor molecule to the carbonyl group of the acceptor molecule, yielding a tetrahedral intermediate that is protonated to give the stable alcohol as final product. The reverse process occurs in exactly the opposite manner: Base abstracts the hydroxyl proton to yield an alkoxide, which fragments to give one molecule of enolate ion and one molecule of neutral carbonyl compound (Figure 26.2).

$$H{-}CH_2CHO \underset{\text{Enolization}}{\overset{:OR^-}{\rightleftharpoons}} \left[{}^-{:}CH_2CHO \underset{\text{Nucleophilic addition}}{\overset{CH_3{-}C(=O){-}H}{\rightleftharpoons}} H_3C{-}CH(O^-){-}CH_2CHO \right]$$

Tetrahedral intermediate

$$\Big\Updownarrow \text{HOR}$$

$$CH_3CH(OH)CH_2CHO + {}^-OR$$

Figure 26.2. Mechanism of the aldol reaction.

PROBLEM

26.1 Predict the product of aldol reaction of the following compounds:
(a) Butanal
(b) 2-Butanone
(c) Cyclopentanone
(d) Acetophenone

PROBLEM

26.2 The aldol reaction is catalyzed by acid as well as by base. What is the reactive nucleophilic species in the acid-catalyzed aldol reaction? Propose a possible mechanism for this acidic reaction.

PROBLEM

26.3 Show by means of curved arrows how the base-catalyzed reverse aldol reaction of 4-methyl-4-hydroxy-2-pentanone can take place. What is the product of this reaction?

26.3 Carbonyl condensation reactions versus alpha-substitution reactions

Two of the four general carbonyl-group reactions—carbonyl condensation and alpha substitution—take place under basic conditions and involve the formation of enolate ion intermediates. Since reaction conditions for both processes are similar, how can we be sure which of the two possible reaction courses will be followed in a given case? How, when we generate an enolate ion with the intention of carrying out an alpha alkylation, can we be sure that a carbonyl condensation reaction does not occur instead?

There are no simple answers to these questions. As a general rule, however, the reaction conditions chosen have much to do with the result. Alpha-substitution reactions require a full equivalent of base, and are normally carried out in such a manner that the carbonyl compound is rapidly and completely converted into its enolate ion at as low a temperature as possible. An electrophile is then added rapidly to ensure that the reactive enolate ion is quenched as quickly as possible. For example, in a malonic ester synthesis, we might use 1 equiv of sodium ethoxide in ethanol solution at room temperature. Quantitative generation of the malonic ester enolate ion would happen instantly, and no unreacted starting material would be left so that no condensation reaction could occur. We would then immediately add an alkyl halide to complete the alkylation reaction.

$$CH_2(CO_2CH_2CH_3)_2 \xrightarrow[\text{Ethanol, 25°C (very fast reaction)}]{\text{1 equiv } NaOCH_2CH_3} [Na^{+}\ {}^{-}{:}CH(CO_2CH_2CH_3)_2] \ (100\%)$$

$$\xrightarrow{\text{Add R—X}} R\text{—}CH(CO_2CH_2CH_3)_2$$

Similarly, if we wish to carry out a direct alkylation of a ketone such as 2-methylcyclohexanone, we might use 1 equiv of the strong base LDA in THF solution at very low temperature. Again, quantitative generation of the ketone enolate ion would occur rapidly, and no unreacted starting material would remain. Rapid addition of an alkyl halide would then allow for the desired alkylation (alpha-substitution) reaction to occur.

$$\text{2-methylcyclohexanone} \xrightarrow[\text{THF, }-78°C]{\text{1 equiv LDA}} \text{lithium enolate } (O^{-}Li^{+}) \xrightarrow{R\text{—}X} \text{2-methyl-6-R-cyclohexanone}$$

On the other hand, we might want to carry out a carbonyl condensation reaction, and this too we could accomplish by selecting the proper reaction conditions. The aldol reaction requires only a *catalytic* amount of base, rather than 1 full equiv, since we need to generate only small amounts of enolate ion in the presence of unreacted carbonyl compound. Once condensation has occurred, the basic catalyst is regenerated. For example, to carry out an aldol reaction on propanal we might dissolve propanal in methanol, add 5% of 1 equiv of sodium methoxide, and then warm the reaction. A high yield of aldol product would result.

$$CH_3CH_2CHO \xrightleftharpoons[CH_3OH,\ \Delta]{5\%\ NaOCH_3} [CH_3\ddot{C}H{-}CHO]$$

Present in tiny amount

$\updownarrow CH_3CH_2CHO$

$$^{-}OCH_3 + CH_3CH_2CH(OH)CH(CH_3){-}CHO \xrightleftharpoons{HOCH_3} [CH_3CH_2CH(O^-)CH(CH_3){-}CHO]$$

Regenerated catalyst

26.4 Dehydration of aldol products: synthesis of enones

The β-hydroxy ketones and β-hydroxy aldehydes formed in aldol reactions can be readily dehydrated to yield conjugated enones, and it is this loss of water that gives the aldol *condensation* its name.

$$-\overset{HO}{\underset{\beta}{C}}-\overset{H}{\underset{\alpha}{C}}-C(=O) \xrightarrow{H^+ \text{ or } ^-OH} \overset{\beta}{C}=\overset{\alpha}{C}-\overset{O}{\overset{\|}{C}}- + H_2O$$

A β-hydroxy aldehyde/ketone   A conjugated enone

Most alcohols are resistant to dehydration by dilute acid or base, and powerful reagents like $POCl_3$ must therefore be used (Section 6.13). Hydroxyl groups that are beta to a carbonyl group are special, however, because of the nearby carbonyl group. Under vigorous *basic* conditions, an acidic alpha hydrogen is abstracted, and an oxygen electron pair on the resultant enolate ion expels the hydroxide leaving group. Under *acidic* conditions, the hydroxyl group is protonated and then expelled by the neighboring enol. The conditions required to effect aldol dehydration reactions are often only a bit more vigorous (slightly higher temperature, for example) than the conditions required for the aldol dimerization itself. As a result, conjugated enones are often obtained directly from aldol reactions, and the intermediate β-hydroxy carbonyl compounds are not isolated.

Basic elimination

$$-\overset{HO}{C}-\overset{H}{C}-\overset{O}{\overset{\|}{C}}- \xrightleftharpoons{^-OH} \left[-\overset{HO}{C}-C=C-\ddot{O}:^-\right] \xrightleftharpoons{-OH^-} C=C-\overset{O}{\overset{\|}{C}}-$$

Enolate

Acidic elimination

$$-\overset{HO}{C}-\overset{H}{C}-\overset{O}{\overset{\|}{C}}- \xrightleftharpoons{H^+} \left[-\overset{H_2O^+}{C}-C=C-O-H\right] \xrightleftharpoons{-H_3O^+} C=C-\overset{O}{\overset{\|}{C}}-$$

Enol

Conjugated enones are more stable than nonconjugated enones for the same reasons that conjugated dienes are more stable than nonconjugated dienes (Section 13.3). Interaction between the pi electrons of the carbon–carbon double bond and the pi electrons of the carbonyl group leads to a molecular orbital description of conjugated enones, which shows a delocalization of the pi electrons over all four atomic centers, thus providing stability.

C=C—C=O

Conjugated enone
(more stable)

C=C—C—C=O

Nonconjugated enone
(less stable)

Recall the following:

C=C—C=C

Conjugated diene
(more stable)

C=C—C—C=C

Nonconjugated diene
(less stable)

PROBLEM

26.4 What enone products would you expect from aldol condensation of the following compounds?
(a) Cyclopentanone
(b) 2-Methylcyclohexanone
(c) Acetophenone
(d) 3-Methylbutanal
(e) 3-Methylcyclohexanone

26.5 Recognizing aldol products

The aldol condensation reaction yields either β-hydroxy ketones (aldehydes) or α,β-unsaturated ketones (aldehydes), depending on the specific case and on the reaction conditions. By learning how to think *backwards,* we can learn to predict when the aldol reaction might be useful in synthesis. Any time the target molecule contains either a β-hydroxy ketone or a conjugated enone functional group, it might come from an aldol reaction:

HO(β)C—C(α)(H)—C=O

or

(β)C=C(α)—C=O

Products

$\Longleftarrow$

C=O + H—C̈—C(=O)—

Starting materials

We can extend this kind of reasoning even further by considering that subsequent transformations might be carried out on the aldol products. For example, a saturated ketone might be prepared by catalytic hydrogenation of an enone product. A good example can be found in the industrial preparation of 2-ethyl-1-hexanol, an alcohol used in the synthesis of plasticizers. 2-Ethyl-1-hexanol bears little resemblance to an aldol product at first glance, yet it is in fact prepared commercially from butanal by an aldol reaction. Working backwards, we can reason that 2-ethyl-1-hexanol might come from 2-ethylhexanal by a reduction. 2-Ethylhexanal, in turn, might be prepared by catalytic reduction of 2-ethyl-2-hexenal, which is the aldol self-condensation product of butanal. The reactions that follow show the sequence in reverse order.

$$\underset{\text{Target: 2-ethyl-1-hexanol}}{CH_3CH_2CH_2CH_2\underset{\displaystyle CH_2CH_3}{\underset{|}{C}}HCH_2OH} \xleftarrow[\text{(industrially, by } H_2/Pt)]{[H]} \underset{\text{Starting material: 2-ethylhexanal}}{CH_3CH_2CH_2CH_2\underset{\displaystyle CH_2CH_3}{\underset{|}{C}}HCHO}$$

$$\underset{\text{2-Ethylhexanal}}{CH_3CH_2CH_2CH_2\underset{\displaystyle CH_2CH_3}{\underset{|}{C}}HCHO} \xleftarrow{H_2/Pt} \underset{\text{2-Ethyl-2-hexenal}}{CH_3CH_2CH_2CH{=}\underset{\displaystyle CH_2CH_3}{\underset{|}{C}}CHO}$$

$$\underset{\text{2-Ethyl-2-hexenal}}{CH_3CH_2CH_2CH{=}\underset{\displaystyle CH_2CH_3}{\underset{|}{C}}CHO} \xleftarrow{KOH} \underset{\text{Butanal}}{2\ CH_3CH_2CH_2CHO}$$

PROBLEM

26.5 Which of the following compounds are aldol self-condensation products? What is the ketone or aldehyde precursor to each?
(a) 2,2,3-Trimethyl-3-hydroxybutanal
(b) 2-Methyl-2-hydroxypentanal
(c) 5-Ethyl-4-methyl-4-hepten-3-one

26.6 Mixed aldol reactions

Until now, we have only considered symmetrical aldol reactions, in which the two carbonyl components have been the same. What would happen, though, if we attempted to carry out a mixed aldol reaction between two different carbonyl partners? There is no single answer to this question. Many mixed aldol reactions give mixtures of products, yet others give a single product in high yield. The results in any specific case depend on the nature of the two reactants.

In general, a mixed aldol reaction between two similar ketone or aldehyde components leads to a mixture of all possible products. For example, base treatment of a mixture of acetaldehyde and propanal gives a complex product mixture containing two "symmetrical" aldol products and two "mixed" aldol products. Clearly, such a reaction is of no practical value in the laboratory.

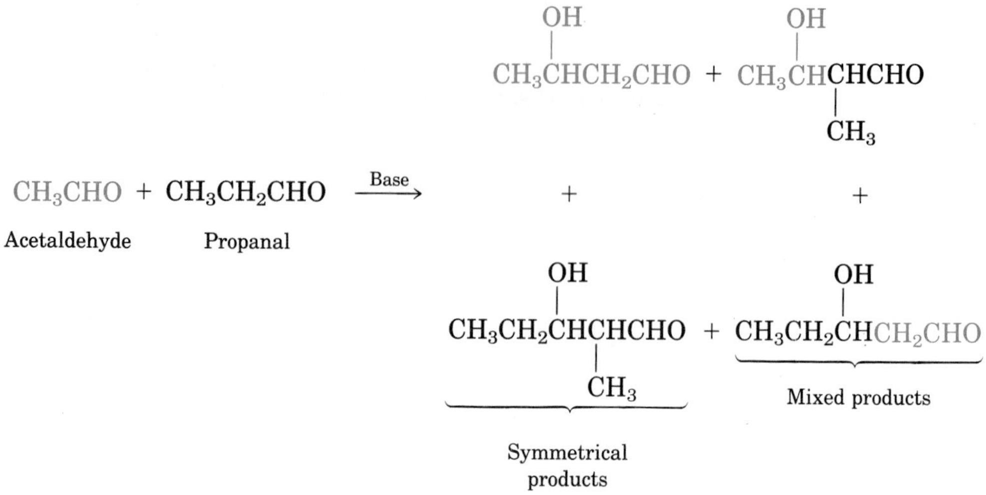

$$CH_3CHO + CH_3CH_2CHO \xrightarrow{\text{Base}} CH_3CH(OH)CH_2CHO + CH_3CH(OH)CH(CH_3)CHO + CH_3CH_2CH(OH)CH(CH_3)CHO + CH_3CH_2CH(OH)CH_2CHO$$

Acetaldehyde, Propanal; Symmetrical products; Mixed products

On the other hand, mixed aldol reactions *can* lead cleanly to a single product, if one of two conditions is met:

1. If one of the carbonyl components contains no alpha hydrogens (and thus cannot form an enolate ion to become a donor), but does contain a reactive carbonyl group that is a good acceptor of nucleophiles, then a mixed aldol reaction is likely to be successful. This is the case, for example, when benzaldehyde or formaldehyde is used as one of the carbonyl components:

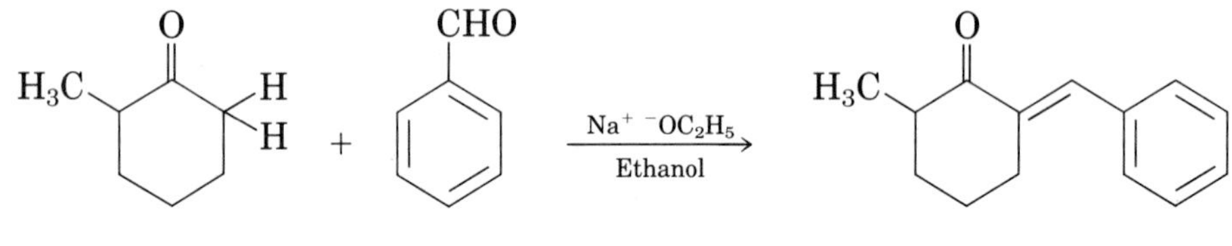

2-Methylcyclohexanone (donor) + Benzaldehyde (acceptor) $\xrightarrow[\text{Ethanol}]{Na^+ \ ^-OC_2H_5}$ 2-benzylidene-6-methylcyclohexanone, 78%

Benzaldehyde (or formaldehyde) cannot form an enolate ion to condense with itself or with another partner, yet its carbonyl group is unhindered and reactive. Thus, in the presence of a ketone such as 2-methylcyclohexanone, the ketone enolate adds preferentially to benzaldehyde giving the mixed aldol product.

2. One of the carbonyl components is unusually acidic and easily transformed into its enolate ion. This is the case, for example, when acetoacetic ester is used as one of the carbonyl components:

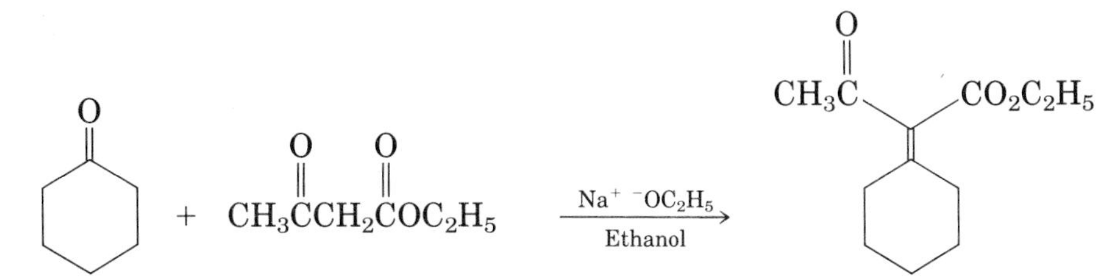

$$\text{Cyclohexanone} + CH_3\overset{O}{\overset{\|}{C}}CH_2\overset{O}{\overset{\|}{C}}OC_2H_5 \xrightarrow[\text{Ethanol}]{Na^+\ {}^-OC_2H_5} (CH_3CO)(CO_2C_2H_5)C{=}C_6H_{10}$$

Cyclohexanone (acceptor) | Ethyl acetoacetate (donor) | 80%

Acetoacetic ester is completely converted into its enolate ion (Section 25.9) in preference to enolate formation from other carbonyl partners such as cyclohexanone. Aldol condensation therefore occurs preferentially to give the mixed product.

We can summarize the situation by saying that a mixed aldol reaction between two different carbonyl partners leads to a mixture of products unless one of the partners is an unusually good nucleophilic donor (such as acetoacetic ester) or has no alpha protons and is a good electrophilic acceptor (such as benzaldehyde).

PROBLEM

26.6 Which of the following compounds can probably be prepared by a mixed aldol reaction?

(a) 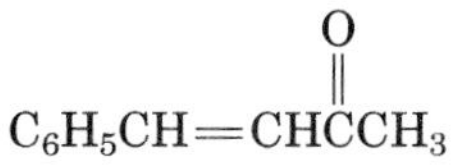 $C_6H_5CH{=}CH\overset{O}{\overset{\|}{C}}CH_3$

(b) $C_6H_5C(CH_3){=}CH\overset{O}{\overset{\|}{C}}CH_3$

(c) 2-(propylidene)cyclohexanone, $=CHCH_2CH_3$

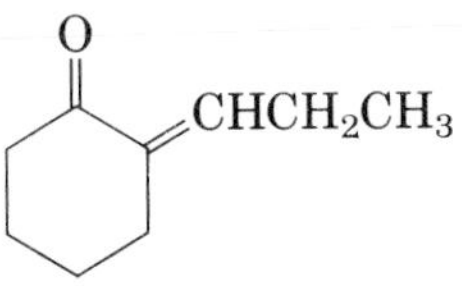

(d) 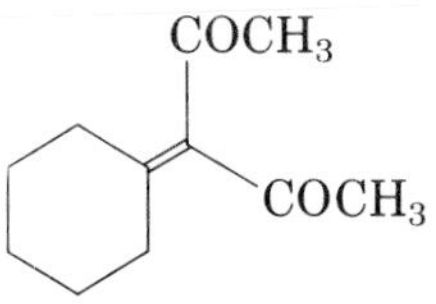 cyclohexylidene with $COCH_3$, $COCH_3$

26.7 Internal aldol reactions

Five- and six-membered-ring cyclic enones can be prepared by intramolecular aldol reactions of 1,4- or 1,5-diketones.

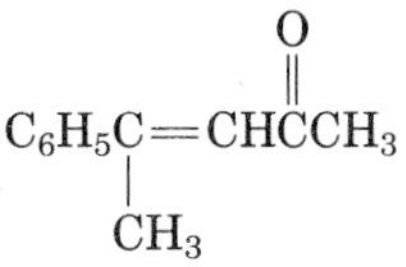

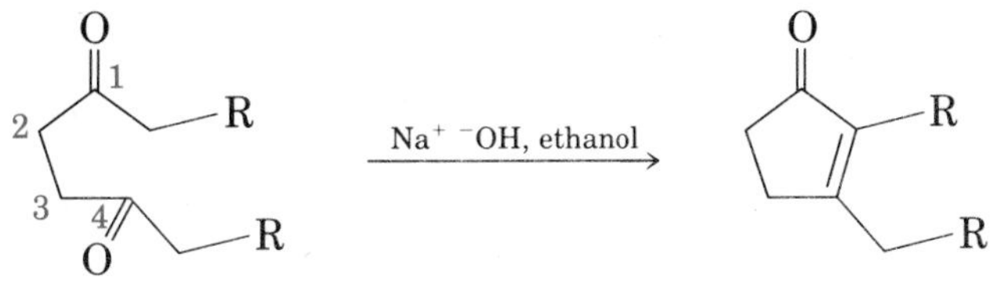

$$RCH_2\overset{O}{\overset{\|}{C}}CH_2CH_2\overset{O}{\overset{\|}{C}}CH_2R \xrightarrow{Na^+\ {}^-OH,\ \text{ethanol}} \text{cyclopentenone}$$

(carbons numbered 1, 2, 3, 4 on the diketone)

A 1,4-diketone | A cyclopentenone

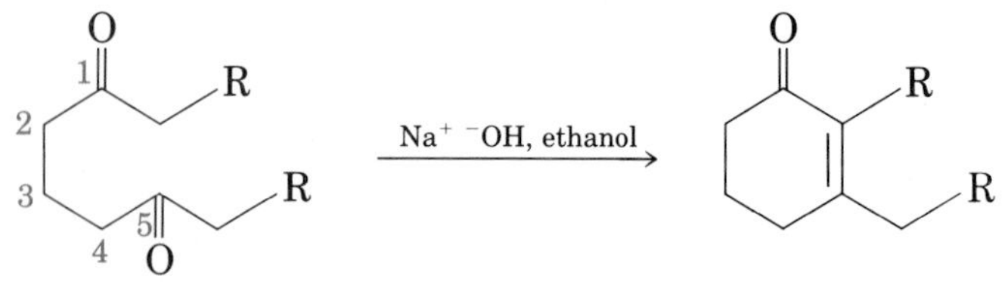

Na^+ ^-OH, ethanol

A 1,5-diketone → A cyclohexenone

Both the nucleophilic carbonyl anion donor and the electrophilic carbonyl acceptor are in the same molecule in these reactions, and internal aldol condensation effects a cyclization reaction. In a sense, these intramolecular condensations can be considered mixed aldol reactions, since the carbonyl groups need not be equivalent. For example, jasmone, the fragrant constituent of the jasmine flower used in perfumery, has been prepared by cyclization of the appropriate 1,4-diketone:

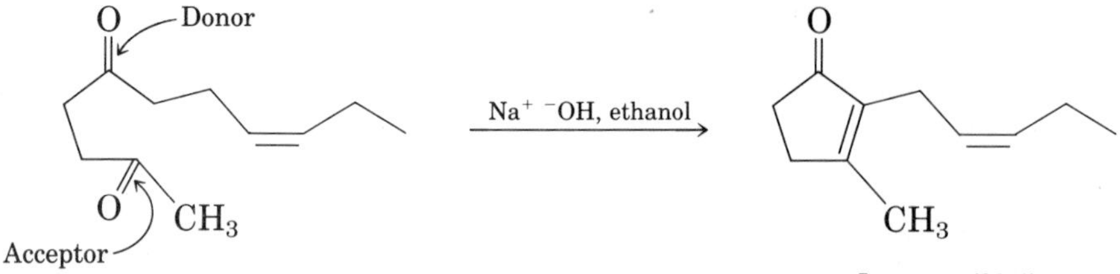

Donor

Acceptor

CH_3

Na^+ ^-OH, ethanol

CH_3

Jasmone (90%)
(a cyclopentenone)

PROBLEM

26.7 Can you suggest an explanation of the fact that 1,3-diketones do not undergo internal aldol condensation to yield cyclobutenones?

PROBLEM

26.8 What product would you expect to obtain from base treatment of 1,6-cyclodecanedione?

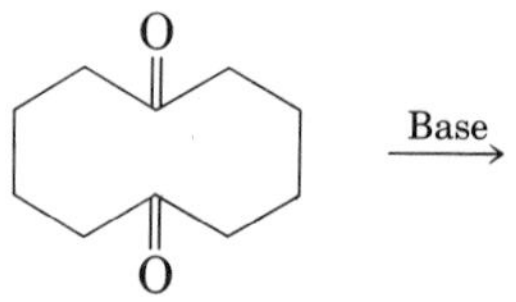

Base

26.8 Reactions similar to the aldol condensation

The aldol reaction is usually defined as the condensation of aldehydes and ketones. In a more general sense, however, there are many possible similar reactions in which a compound with acidic alpha hydrogens condenses with a carbonyl component. For example, diethyl malonate can condense with aldehydes and unhindered ketones to yield α,β-unsaturated diesters—the **Knoevenagel reaction.**[1] Like other malonates, these diesters can then be decarboxylated (Section 25.9).

[1]Emil Knoevenagel (1865–1921); b. Linden/Hannover, Germany; Ph.D. Göttingen; professor, University of Heidelberg.

$$\text{C}_6\text{H}_5\text{CHO} + \text{CH}_2(\text{CO}_2\text{C}_2\text{H}_5)_2 \xrightarrow[\text{Ethanol}]{\text{Na}^+\ ^-\text{OCH}_2\text{CH}_3} \text{C}_6\text{H}_5\text{CH}{=}\text{C}(\text{CO}_2\text{C}_2\text{H}_5)_2$$

Benzaldehyde; Diethyl malonate; 91%

$$\xrightarrow{\text{H}_3\text{O}^+} \text{C}_6\text{H}_5\text{CH}{=}\text{CHCOOH} + \text{CO}_2 + 2\ \text{C}_2\text{H}_5\text{OH}$$

Cinnamic acid

Many other possibilities exist, including condensations with nitriles, anhydrides, and nitro compounds as the acidic donor components. The many possibilities lead to an extremely rich and varied chemistry, but the important fact to remember is that all the different possibilities are simply variations of the generalized aldol mechanism in which a nucleophilic carbon donor adds to an electrophilic carbonyl acceptor.

PROBLEM

26.9 Show in detail the mechanism of the Knoevenagel reaction of diethyl malonate and benzaldehyde.

PROBLEM

26.10 In the Perkin reaction, an anhydride condenses with an aromatic aldehyde to yield a cinnamic acid. Propose a mechanism for this reaction.

$$\text{C}_6\text{H}_5\text{CHO} + \text{CH}_3\overset{\text{O}}{\overset{\|}{\text{C}}}\text{O}\overset{\text{O}}{\overset{\|}{\text{C}}}\text{CH}_3 \xrightarrow[\text{2. H}_2\text{O}]{\text{1. CH}_3\text{COONa, }\Delta} \text{C}_6\text{H}_5\text{CH}{=}\text{CHCOOH}$$

Benzaldehyde; Acetic anhydride; Cinnamic acid (64%)

26.9 The Claisen condensation reaction

Esters, like aldehydes and ketones, are weakly acidic. When an ester having an alpha hydrogen is treated with 1 equiv of a base such as sodium ethoxide, a reversible condensation reaction occurs to yield a β-keto ester product. This condensation between two ester components is known as the **Claisen condensation reaction.**[2] For example, ethyl acetate yields ethyl acetoacetate on base treatment, as shown at the top of the next page.

[2]Ludwig Claisen (1851–1930); b. Cologne; Ph.D. Bonn (Kekulé); professor, University of Bonn, Owens College (Manchester), universities of Munich, Aachen, Kiel, and Berlin; Goddesberg (private laboratory).

$$2\ CH_3\overset{O}{\overset{\|}{C}}OCH_2CH_3 \xrightarrow[2.\ H_3O^+]{1.\ Na^+\ ^-OCH_2CH_3,\ ethanol} CH_3\underset{\beta}{\overset{O}{\overset{\|}{C}}}-\underset{\alpha}{CH_2}-\overset{O}{\overset{\|}{C}}-OCH_2CH_3 + CH_3CH_2OH$$

Ethyl acetate

Ethyl acetoacetate,
a β-keto ester (75%)

The mechanism of the Claisen reaction is similar to that of the aldol reaction and involves the nucleophilic addition of an ester enolate ion donor to the carbonyl group of a second ester molecule. We can view the Claisen mechanism as shown in Figure 26.3. From the point of view of the donor component, the Claisen condensation is simply an alpha-substitution reaction; from the point of view of the acceptor component, the Claisen condensation is a nucleophilic acyl substitution reaction.

The only difference between an aldol condensation and a Claisen condensation involves the fate of the initially formed tetrahedral intermediate. The tetrahedral intermediate in the aldol reaction is protonated to give a stable alcohol product—exactly the behavior previously seen for ketones (Section 22.6). The tetrahedral intermediate in the Claisen reaction, however, expels an alkoxide leaving group to yield the acyl substitution product—exactly the behavior previously seen for esters (Section 24.7).

Note that if the starting ester has more than one acidic alpha hydrogen the product β-keto ester has a highly acidic, doubly activated hydrogen atom, which can be abstracted by base. This deprotonation of the product requires that a full equivalent of base, rather than a catalytic amount, be used in the reaction, and serves to drive the Claisen equilibrium completely to the product side so that high yields are often obtained.

PROBLEM

26.11 The Claisen reaction is reversible. That is, a β-keto ester can be cleaved by base into two fragments. Show in detail the mechanism by which this cleavage occurs.

$$C_6H_5\overset{O}{\overset{\|}{C}}-CH_2CO_2C_2H_5 \xrightarrow[Ethanol]{1\ equiv\ NaOH} C_6H_5COO^- + CH_3CO_2C_2H_5$$

PROBLEM

26.12 How can you account for the fact that ethyl α,α-dimethylacetoacetate cleaves instantly at room temperature when treated with ethoxide ion, but that ethyl acetoacetate itself requires temperatures of over 150°C to cleave?

$$CH_3\overset{O}{\overset{\|}{C}}C(CH_3)_2CO_2C_2H_5 \xrightarrow[25°C]{Na^+\ ^-OC_2H_5,\ ethanol} CH_3CO_2C_2H_5 + (CH_3)_2CHCO_2C_2H_5$$

$$CH_3\overset{O}{\overset{\|}{C}}CH_2CO_2C_2H_5 \xrightarrow[150°C]{Na^+\ ^-OC_2H_5,\ ethanol} 2\ CH_3CO_2C_2H_5$$

Ethoxide base abstracts an acidic alpha hydrogen atom from an ester molecule, yielding an ester enolate ion.

$$CH_3\overset{\overset{\displaystyle O}{\|}}{C}OC_2H_5 \xrightleftharpoons{^{-}OC_2H_5} {:}\overset{-}{C}H_2\overset{\overset{\displaystyle O}{\|}}{C}OC_2H_5 + C_2H_5OH$$

Nucleophilic donor

This ion does a nucleophilic addition to a second ester molecule, giving a tetrahedral intermediate.

$$\xrightleftharpoons{CH_3\overset{\overset{O}{\|}}{C}OC_2H_5} \quad \text{Electrophilic acceptor}$$

$$CH_3\overset{\overset{\displaystyle :\ddot{O}:^-}{|}}{\underset{\underset{\displaystyle OC_2H_5}{|}}{C}}-CH_2\overset{\overset{\displaystyle O}{\|}}{C}OC_2H_5$$

The tetrahedral intermediate is not stable. It expels ethoxide ion to yield the new carbonyl compound, ethyl acetoacetate.

$$\rightleftharpoons CH_3\overset{\overset{\displaystyle O}{\|}}{C}-CH_2\overset{\overset{\displaystyle O}{\|}}{C}OC_2H_5 + C_2H_5O{:}^-$$

But ethoxide ion is a base. It therefore converts the β-keto ester product into its enolate, thus shifting the equilibrium and driving the reaction to completion.

$$\xrightleftharpoons{^{-}OC_2H_5} CH_3\overset{\overset{\displaystyle O}{\|}}{C}-\overset{\overset{-}{\cdot\cdot}}{C}H-\overset{\overset{\displaystyle O}{\|}}{C}-OC_2H_5$$

Protonation by addition of acid yields the final product.

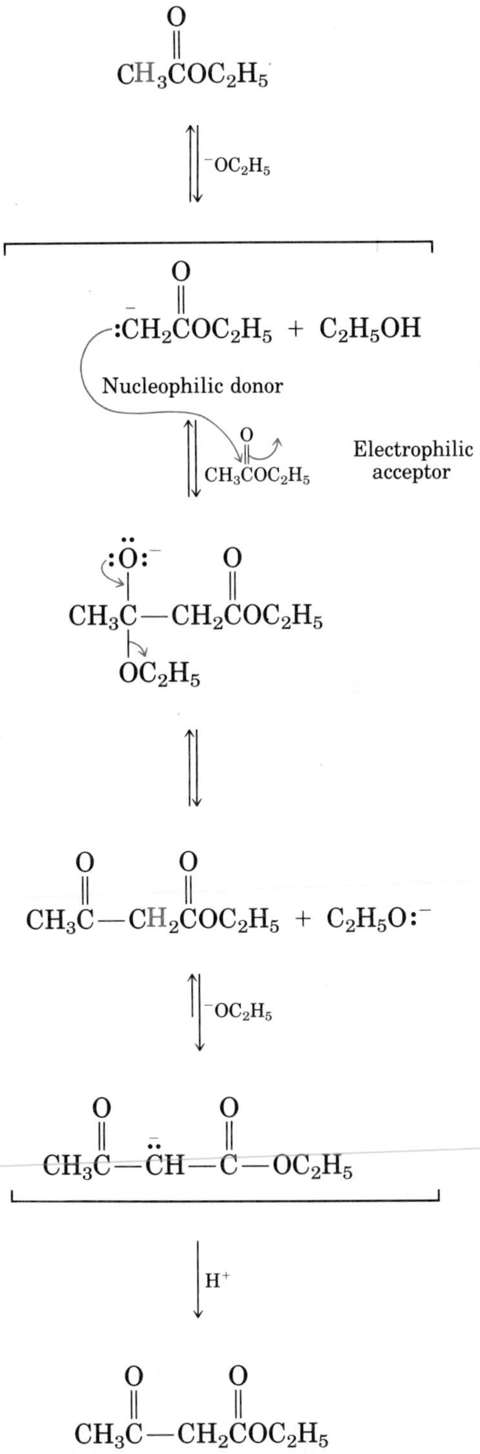

$$\xrightarrow{H^+} CH_3\overset{\overset{\displaystyle O}{\|}}{C}-CH_2\overset{\overset{\displaystyle O}{\|}}{C}OC_2H_5$$

Figure 26.3. Mechanism of the Claisen reaction.

26.10 Mixed Claisen condensations

The mixed Claisen condensation of two different esters is similar to the mixed aldol condensation (Section 26.6). Mixed Claisen reactions are generally successful only when one of the two ester components has no alpha hydrogens and thus cannot form an enolate ion. For example, ethyl benzoate and ethyl formate cannot form enolate ions and thus cannot serve as donors. They can, however, act as the electrophilic acceptor components in reactions with other ester anions to give good yields of mixed β-keto ester products.

$$C_6H_5COOC_2H_5 + CH_3COOC_2H_5 \xrightarrow[2.\ H_3O^+]{1.\ NaH/THF} C_6H_5CO{-}CH_2{-}COOC_2H_5$$

Ethyl benzoate (acceptor) + Ethyl acetate (donor) → Ethyl benzoylacetate

Mixed Claisen-type reactions can also be carried out between esters and ketones. The result is an excellent synthesis of β-diketones. The reaction works best when the ester component has no alpha hydrogens and thus cannot act as the nucleophilic donor. For example, ethyl formate and ethyl benzoate give particularly high yields in mixed Claisen condensations with ketones.

$$\text{2,2-Dimethylcyclohexanone} + HCOOC_2H_5 \xrightarrow[2.\ H^+]{1.\ Na^+\ {}^-OC_2H_5,\ \text{ethanol}} \text{A } \beta\text{-keto aldehyde}$$

2,2-Dimethylcyclohexanone (donor) + Ethyl formate (acceptor) → A β-keto aldehyde (91%)

$$CH_3COCH_3 + C_6H_5CO_2C_2H_5 \xrightarrow[2.\ H^+]{1.\ Na^+\ {}^-OC_2H_5,\ \text{ethanol}} C_6H_5CO{-}CH_2COCH_3$$

Acetone (donor) + Ethyl benzoate (acceptor) → A β-diketone

Often, however, even esters that do have acidic alpha hydrogens condense well with ketones, since ketones are more acidic than esters and are more likely to serve as the nucleophilic donor component. This kind of reaction works particularly well when it occurs internally, and cyclic β-diketones can thus be prepared by treatment of the appropriate keto esters with base.

$$\text{Ethyl 5-oxohexanoate} \xrightarrow[\text{2. } H_3O^+]{\text{1. } Na^+\ {}^-OC_2H_5,\ \text{ethanol}} \text{1,3-Cyclohexanedione (90\%)} + C_2H_5OH$$

Ethyl 5-oxohexanoate

1,3-Cyclohexanedione
(90%)

PROBLEM

26.13 Would you expect diethyl oxalate, $C_2H_5O_2CCO_2C_2H_5$, to give good yields in mixed Claisen reactions? Explain your answer. What product would you expect to obtain from mixed Claisen reaction of ethyl acetate with diethyl oxalate?

PROBLEM

26.14 Formulate the mechanism of the base-induced cyclization of ethyl 5-oxohexanoate to 1,3-cyclohexanedione. Show all of the steps involved.

26.11 Internal Claisen condensations: the Dieckmann cyclization

Internal Claisen condensations can be carried out with diesters, just as internal aldol condensations can be carried out with diketones (Section 26.7). These **Dieckmann cyclizations**[3] work best on 1,6-diesters and 1,7-diesters. Five-membered-ring cyclic β-keto esters result from Dieckmann cyclization of 1,6-diesters, and six-membered rings result from 1,7-diesters.

$$\text{Dimethyl hexanedioate} \xrightarrow[\text{2. } H_3O^+]{\text{1. } Na^+\ {}^-OCH_3,\ CH_3OH} \text{Methyl 2-oxocyclopentanecarboxylate (82\%)}$$

Dimethyl hexanedioate
(a 1,6-diester)

Methyl 2-oxocyclopentanecarboxylate
(82%)

$$\text{Dimethyl heptanedioate} \xrightarrow[\text{2. } H_3O^+]{\text{1. } Na^+\ {}^-OCH_3,\ CH_3OH} \text{Methyl 2-oxocyclohexanecarboxylate}$$

Dimethyl heptanedioate
(a 1,7-diester)

Methyl 2-oxocyclohexanecarboxylate

The Dieckmann cyclization is particularly valuable for preparing 2-substituted cyclopentanones and cyclohexanones, since the initially

[3]Walter Dieckmann (1869–1925); b. Hamburg, Germany; Ph.D. Munich (Bamberger); professor, University of Munich.

formed cyclic β-keto ester products can be further alkylated and then decarboxylated (Section 25.9).

$$\text{Methyl 2-oxocyclohexanecarboxylate} \xrightarrow[\text{2. } H_2C{=}CHCH_2Br]{\text{1. } Na^+ \ ^-OCH_3,\ CH_3OH} \text{2-(}CO_2CH_3\text{)-2-(}CH_2{-}CH{=}CH_2\text{)cyclohexanone} \xrightarrow[\Delta]{H_3O^+} \text{2-Allylcyclohexanone}$$

Methyl 2-oxocyclohexanecarboxylate

2-Allylcyclohexanone
(83%)

PROBLEM

26.15 How can you account for the fact that Dieckmann cyclization of diethyl 3-methylheptanedioate gives a mixture of two β-keto ester products? What are their structures, and how is each formed?

26.12 The Michael reaction

We saw earlier (Section 22.15) that certain nucleophiles such as amines can react with α,β-unsaturated ketones to give the conjugate addition product, rather than the direct addition product:

$$H{-}Nu{:} + {>}C{=}C{-}\overset{O}{\overset{\|}{C}}{-} \longrightarrow Nu{-}\overset{|}{C}{-}\overset{H}{\overset{|}{C}}{-}\overset{O}{\overset{\|}{C}}{-}$$

Conjugate addition product

Exactly the same kind of conjugate addition can occur when enolate ion nucleophiles react with α,β-unsaturated carbonyl electrophiles—a process known as the **Michael reaction.**[4]

The best and most useful Michael reactions take place when stabilized enolate anions derived from dialkyl malonates or from β-keto esters add to unhindered α,β-unsaturated ketones. For example, ethyl acetoacetate adds to 3-buten-2-one in 94% yield in the presence of sodium ethoxide catalyst.

[4]Arthur Michael (1853–1942); b. Buffalo, New York; studied Heidelberg, Berlin, École de Médecine, Paris; professor, Tufts University (1882–1889 and 1894–1907), Harvard University (1912–1936).

$$\underset{\text{Ethyl acetoacetate}}{CH_3\overset{O}{\overset{\|}{C}}-\underset{CO_2C_2H_5}{\underset{|}{CH_2}}} + \underset{\text{3-Buten-2-one}}{H_2C{=}CH\overset{O}{\overset{\|}{C}}CH_3} \xrightarrow[\text{2. } H_3O^+]{\text{1. } Na^+\ {}^-OC_2CH_3\text{, ethanol}} \underset{94\%}{CH_3\overset{O}{\overset{\|}{C}}-\underset{CO_2C_2H_5}{\underset{|}{C}}HCH_2CH_2\overset{O}{\overset{\|}{C}}CH_3}$$

Michael reactions take place by addition of a nucleophilic enolate ion donor to the beta carbon of the α,β-unsaturated carbonyl acceptor, as indicated in Figure 26.4.

The base catalyst removes an acidic alpha proton from the starting β-keto ester to generate a stabilized enolate nucleophile.

$$CH_3\overset{O}{\overset{\|}{C}}-CH_2-\overset{O}{\overset{\|}{C}}-OC_2H_5$$

$$\Big\Updownarrow\ {}^-OC_2H_5\text{, ethanol}$$

The nucleophile takes part in a conjugate (Michael) addition to the α,β-unsaturated ketone electrophile. The product anion is a ketone enolate.

$$CH_3\overset{O}{\overset{\|}{C}}-\overset{-}{\underset{\cdot\cdot}{C}}H-\overset{O}{\overset{\|}{C}}OC_2H_5$$

$$\overset{\delta^+}{H_2C}{=}\overset{\delta^-}{CH}-\underset{\delta^+}{\overset{O\delta^-}{\overset{\|}{C}}}-CH_3$$

$$CH_3\overset{O}{\overset{\|}{C}}-\underset{CH_2-\overset{\cdot\cdot}{\overset{-}{C}}H-\overset{O}{\overset{\|}{C}}-CH_3}{\underset{|}{CH}}-\overset{O}{\overset{\|}{C}}-OC_2H_5$$

The ketone enolate abstracts an available proton, either from solvent or from the starting β-keto ester, to yield the final product.

$$\Big\Updownarrow\ H^+$$

$$CH_3\overset{O}{\overset{\|}{C}}-\underset{CH_2CH_2\overset{O}{\overset{\|}{C}}CH_3}{\underset{|}{CH}}-\overset{O}{\overset{\|}{C}}OC_2H_5$$

Figure 26.4. Mechanism of the Michael reaction.

The Michael reaction is general and is not limited to conjugated enones. Conjugated aldehydes, esters, nitriles, amides, and nitro compounds can all act as the electrophilic acceptor component in the Michael reaction (Table 26.1).

TABLE 26.1 Some Michael acceptors and Michael donors

Michael acceptors		*Michael donors*	
$H_2C{=}CHCHO$	Propenal	$RCOCH_2COR'$	β-Diketone
$H_2C{=}CHCO_2CH_3$	Methyl propenoate	$RCOCH_2CO_2CH_3$	β-Keto ester
$H_2C{=}CHC{\equiv}N$	Propenenitrile	$CH_3O_2CCH_2CO_2CH_3$	Malonic ester
$H_2C{=}CHCOCH_3$	3-Buten-2-one	$RCOCH_2C{\equiv}N$	β-Keto nitrile
$H_2C{=}CH{-}NO_2$	Nitroethylene	RCH_2NO_2	Nitro compound
$H_2C{=}CHCONH_2$	Propenamide		

PROBLEM

26.16 How might the following compounds be prepared using Michael reactions?

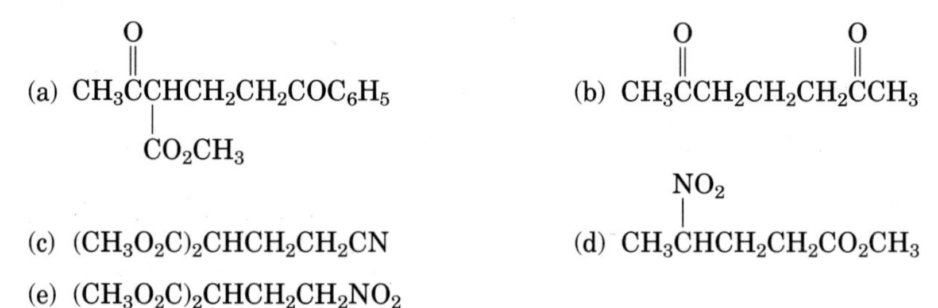

PROBLEM

26.17 The Michael reaction works best when highly acidic donor molecules (β-keto esters or β-diketones) are used. Monoketones can also be used but yields are generally much lower. Propose an explanation of this difference. [*Hint:* Consider the relative reactivities of starting materials versus products in both cases.]

26.13 The Stork enamine reaction

Other kinds of carbon nucleophiles besides enolate ions add to α,β-unsaturated ketones in the Michael reaction, and among the most important of these are enamines. Enamines are readily prepared by reaction between a ketone and a secondary amine (Section 22.9) and are electronically similar to enolate ions. As the following resonance structure indicates, delocalization of the nitrogen lone-pair electrons onto the alpha carbon renders the alpha carbon nucleophilic:

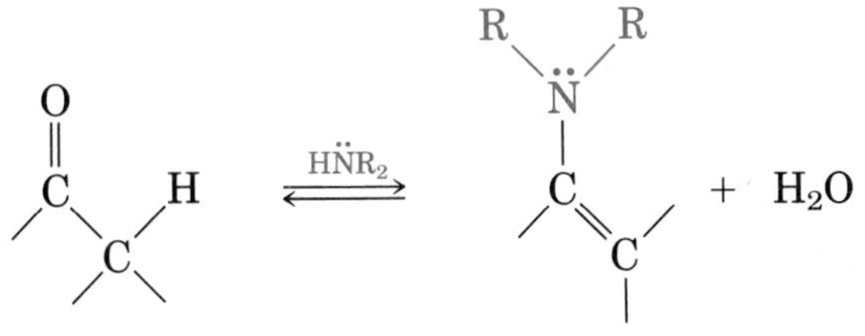

O, C, H, C, HNR_2, R, R, N, C, C, + H_2O

An enamine

R, R—N, C, C ⟷ R, R—N^+, C, C^-

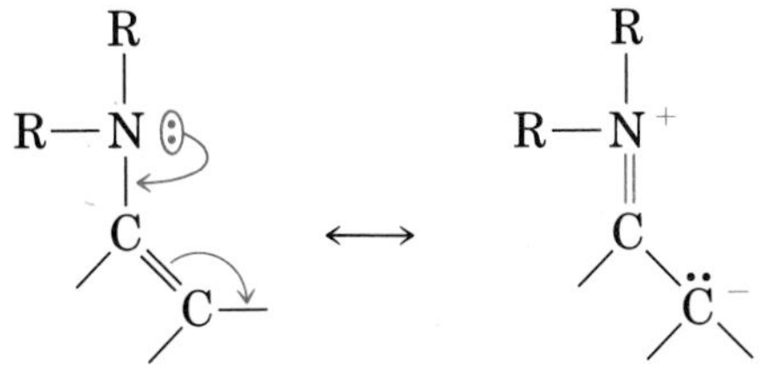

Nucleophilic alpha carbon

This is similar to

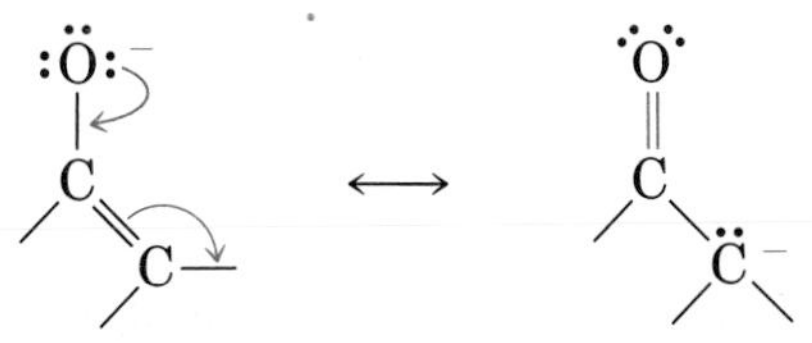

:O:$^-$, C, C ⟷ O, C, C^-

Enamines behave much as enolates and enter into many of the same kinds of reactions. In the **Stork enamine reaction,**[5] an enamine adds to an α,β-unsaturated carbonyl acceptor in a Michael-type process. Aqueous hydrolysis of the Michael adduct then yields a 1,5-dicarbonyl compound. The overall Stork enamine reaction is a three-step sequence:

1. Enamine formation from a ketone
2. Michael-type addition to an enone
3. Enamine hydrolysis back to a ketone

The net effect of the sequence is to carry out a Michael reaction of a monoketone. For example, cyclohexanone reacts with the cyclic amine pyrrolidine to yield an enamine; further reaction with an enone such as 3-buten-2-one yields the Michael-type adduct, and aqueous hydrolysis completes the sequence to provide a 1,5-diketone product, as shown on p. 870.

[5]Gilbert Stork (1921–); b. Brussels, Belgium; Ph.D. Wisconsin (McElvain); professor, Harvard University, Columbia University (1953–).

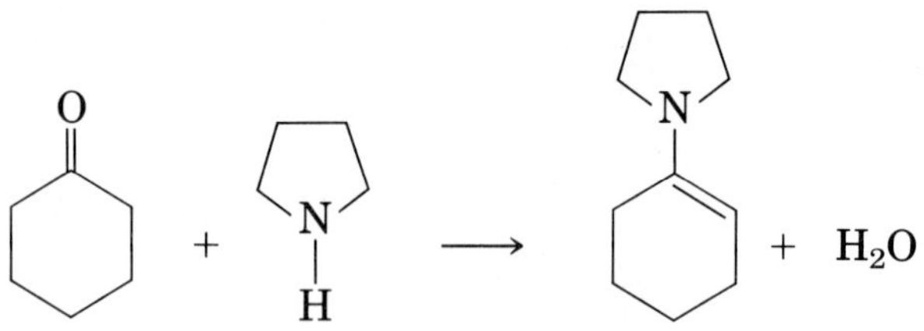

Cyclohexanone + Pyrrolidine ⟶ An enamine + H_2O

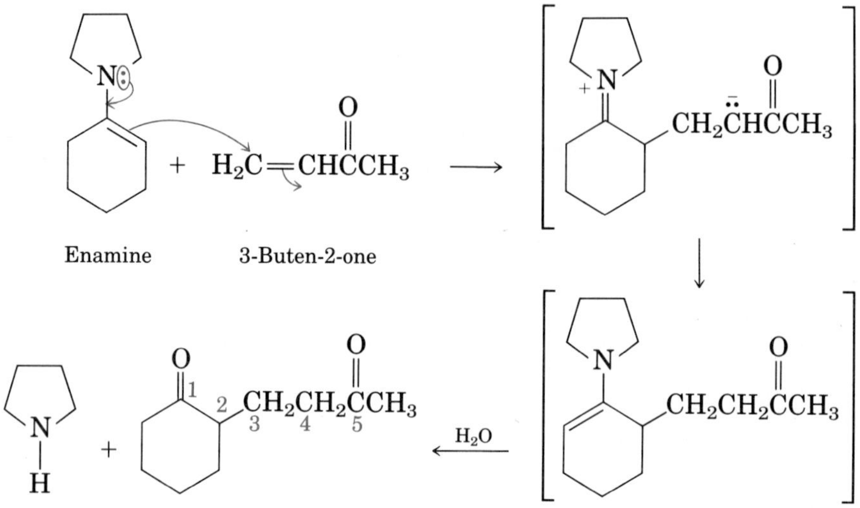

Enamine + 3-Buten-2-one ($H_2C{=}CHCCH_3$, O) ⟶ [iminium $CH_2\ddot{C}HCCH_3$] ⟶ [enamine $CH_2CH_2CCH_3$] $\xrightarrow{H_2O}$ pyrrolidine (N–H) + A 1,5-diketone (71%) ($CH_2CH_2CCH_3$; 1, 2, 3, 4, 5)

PROBLEM

26.18 When 2-methylcyclohexanone is converted into its pyrrolidine enamine, the only product formed has the double bond toward the unsubstituted side, rather than toward the methyl-substituted side. Build molecular models of the two possible products and explain the observed result.

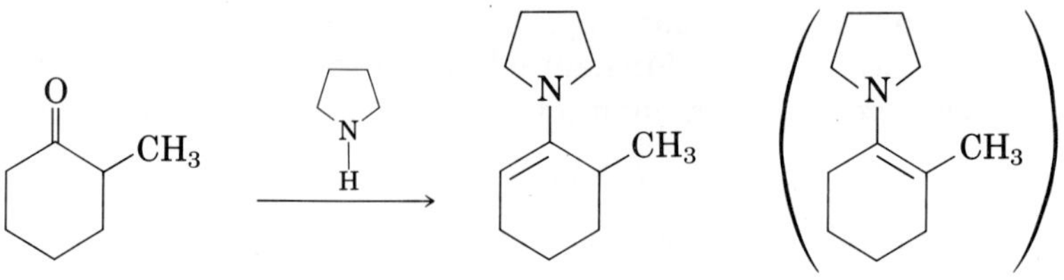

2-methylcyclohexanone (CH_3) ⟶ (pyrrolidine, N–H) enamine (CH_3) (enamine CH_3 — *Not formed*)

PROBLEM

26.19 Show how you might use an enamine reaction to prepare these compounds:

(a) 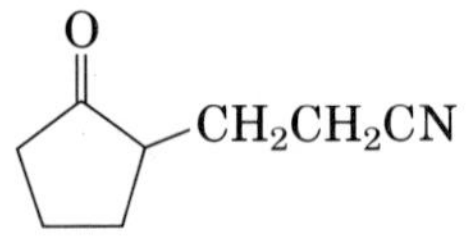 2-(CH_2CH_2CN)cyclopentanone

(b) 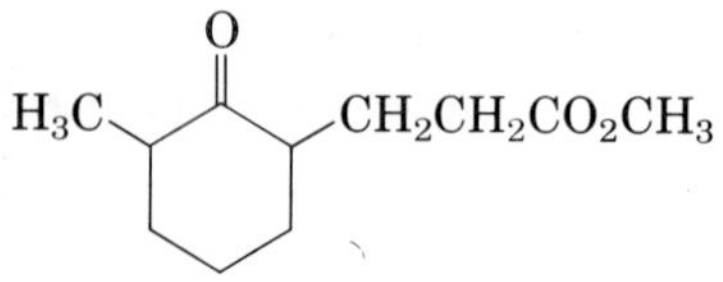 H_3C–cyclohexanone–$CH_2CH_2CO_2CH_3$

26.14 Carbonyl condensation reactions in synthesis: the Robinson annulation reaction

Carbonyl condensation reactions are among the most valuable methods available for the synthesis of complex molecules. One example of their use is the **Robinson annulation reaction.**[6] Many synthetic target molecules contain rings, and an annulation reaction (Latin *annulus*, "ring") is one that builds a new ring onto a molecule. The Robinson annulation reaction is a two-step process that takes place between a Michael-type nucleophilic donor such as a β-keto ester or β-diketone and an α,β-unsaturated ketone acceptor such as 3-buten-2-one:

$CO_2C_2H_5$ + $CO_2C_2H_5$

$Na^+ \ {}^-OC_2H_5$, ethanol

Michael reaction

3-Buten-2-one

Ethyl acetoacetate

$CO_2C_2H_5$

Michael product

Aldol reaction

$Na^+ \ {}^-OC_2H_5$, ethanol

$CO_2C_2H_5$

+ H_2O

Annulation product

The first step of the Robinson annulation is simply a Michael reaction—a stabilized anion effects a conjugate addition to an α,β-unsaturated ketone. The Michael product, however, is a 1,5-diketone, and as we have seen (Section 26.6), 1,5-diketones undergo internal aldol condensation when treated with base. Thus, the final product contains a six-membered ring, and an annulation has been accomplished.

An excellent example of the practical importance of the Robinson annulation reaction occurs as a key step during the synthesis of the female steroid hormone estrone, carried out by the French pharmaceutical company, Roussel-UCLAF (Figure 26.5, p. 872).

In this example, 2-methyl-1,3-cyclopentanedione (a β-diketone) serves as precursor for the stabilized enolate ion required for Michael reaction, and an aryl-substituted α,β-unsaturated ketone serves as the acceptor component. Michael reaction between the two yields an intermediate triketone, which immediately cyclizes in an internal aldol condensation. Several further transformations are then required to complete the synthesis of

[6]Sir Robert Robinson (1886–1975); b. Rufford/Chesterfield, England; Ph.D. Manchester (Perkin); professor, Liverpool, Manchester (1922–1928), University College, Oxford (1930–1955); Nobel prize (1947).

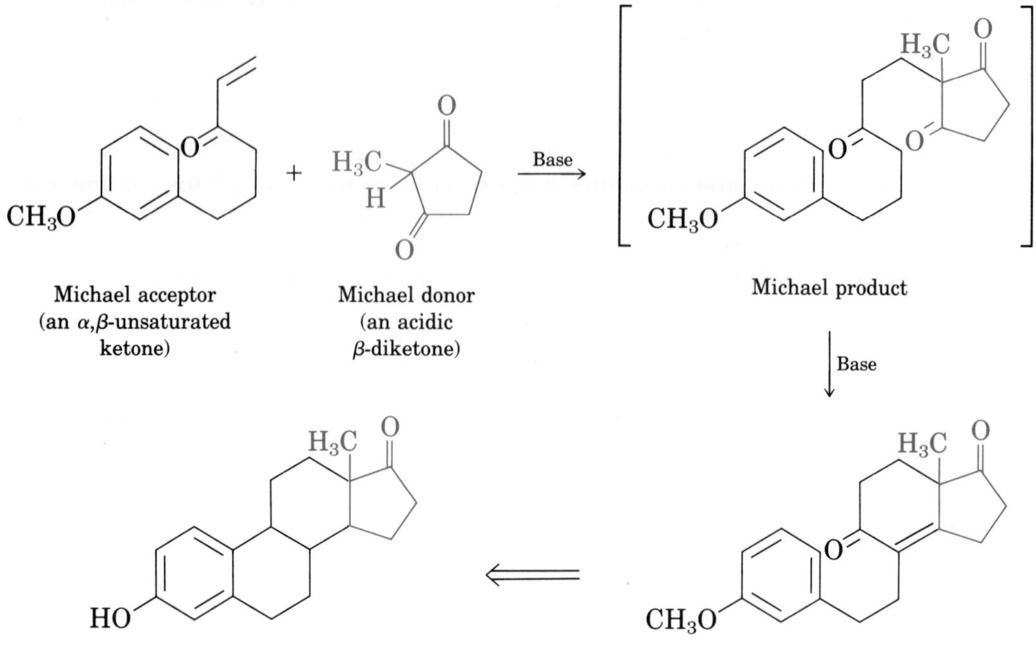

CH_3O ... + H_3C H ... O O —Base→ [CH_3O ... H_3C O]

Michael acceptor (an α,β-unsaturated ketone)

Michael donor (an acidic β-diketone)

Michael product

Base

Estrone

Robinson annulation product

Figure 26.5. Robinson annulation in the synthesis of estrone.

estrone, but the Robinson annulation serves as the key step for assembling much of the molecule.

PROBLEM

26.20 Show how you might combine a Stork enamine reaction and a Robinson annulation to prepare the following compounds:

(a) 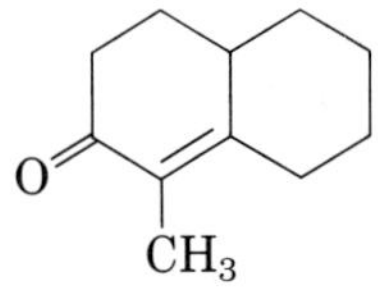 CH_3 (b) 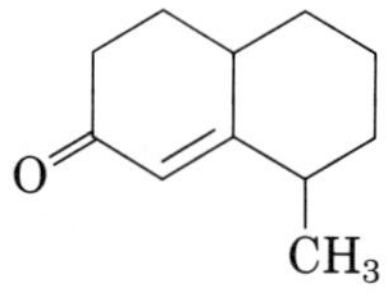 CH_3

26.15 Biological carbonyl condensation reactions

Carbonyl condensation reactions are used by nature for the biological synthesis of a wide variety of different molecules. Fats, amino acids, steroid hormones, and many other kinds of compounds are synthesized by plants and animals, using carbonyl condensation reactions as the key step.

Nature uses the two-carbon acetate fragment of acetyl CoA as the major building block for synthesis. Not only can acetyl CoA serve as an electrophilic acceptor for attack of nucleophiles at the acyl carbon (Section 24.10), but it can also serve as a nucleophilic donor by loss of its acidic alpha proton to generate an enolate anion. The enolate anion of acetyl CoA can then add to another carbonyl group in a condensation reaction. For example, citric acid is biosynthesized by addition of acetyl CoA to the ketone carbonyl group of oxaloacetic acid in a kind of mixed aldol reaction, followed by hydrolysis of the thiol ester group.

$$CH_3-\overset{O}{\overset{\|}{C}}-S-CoA \longrightarrow \text{“}:\overset{-}{C}H_2-\overset{O}{\overset{\|}{C}}-S-CoA\text{”}$$

Acetyl CoA,
a thiol ester

$$O{=}C(COOH)(CH_2COOH) + :\overset{-}{C}H_2-\overset{O}{\overset{\|}{C}}-S-CoA \longrightarrow \longrightarrow HO-C(COOH)(CH_2COOH)-CH_2COOH$$

Oxaloacetic acid — Citric acid

Acetyl CoA is involved as a primary building block in the biosynthesis of steroids, fats, and other lipids, and the key step is a Claisen-like condensation. We will go into more detail of this process in Section 32.4.

$$CH_3\overset{O}{\overset{\|}{C}}-S-CoA + :\overset{-}{C}H_2-\overset{O}{\overset{\|}{C}}-S-CoA \Longrightarrow CH_3\overset{O}{\overset{\|}{C}}-CH_2-\overset{O}{\overset{\|}{C}}-S-CoA$$

Acetyl CoA — Acetoacetyl CoA

Acetoacetyl CoA $\Rightarrow$ Fats, steroids, prostaglandins

26.16 Summary

A carbonyl condensation reaction is a polar process that takes place between two carbonyl components, one that involves a combination of nucleophilic addition and alpha-substitution steps. One carbonyl component (the donor) is converted into a nucleophilic enolate ion, which adds to the electrophilic carbonyl group of the second component (the acceptor). The donor molecule undergoes an alpha substitution, and the acceptor molecule undergoes a nucleophilic addition.

$$\underset{\text{Acceptor}}{>C{=}\ddot{O}} + \underset{\text{Donor}}{-\overset{-}{\ddot{C}}-C(=O)-} \rightleftarrows\ :\ddot{O}:^- -C-C-C(=O)-$$

The aldol reaction is a carbonyl condensation that occurs between two ketone or aldehyde components. Aldol reactions are reversible and lead first to β-hydroxy ketone (or aldehyde) products and then to α,β-unsaturated ketones.

$$\underset{\diagup\ \diagdown}{\overset{\overset{\displaystyle O}{\|}}{C}} + R{-}CH_2{-}\overset{\overset{\displaystyle O}{\|}}{C}{-} \rightleftharpoons {-}\overset{\overset{\displaystyle HO}{|}}{\underset{|}{\overset{\beta}{C}}}{-}\overset{\overset{\displaystyle H}{|}}{\underset{\underset{\displaystyle R}{|}}{\overset{\alpha}{C}}}{-}\overset{\overset{\displaystyle O}{\|}}{C}{-} \rightleftharpoons \diagdown\!\!\!\!\diagup\overset{\beta}{C}{=}\underset{\underset{\displaystyle R}{|}}{\overset{\alpha}{C}}{-}\overset{\overset{\displaystyle O}{\|}}{C}{-} + H_2O$$

A β-hydroxy ketone — An α,β-unsaturated ketone

Mixed aldol condensations between two different ketones generally give a mixture of all four possible products. A mixed reaction can be successful, however, if one of the two components is an unusually good donor (as with ethyl acetoacetate), or if it can act only as an acceptor (as with formaldehyde and benzaldehyde). Internal aldol condensations of 1,4- and 1,5-diketones are also successful and provide an excellent method for preparing five- and six-membered enone rings.

The Claisen reaction is a carbonyl condensation that occurs between two ester components. It leads to β-keto ester products:

$$R{-}\overset{\overset{\displaystyle O}{\|}}{C}{-}OC_2H_5 + R'{-}CH_2\overset{\overset{\displaystyle O}{\|}}{C}OC_2H_5 \rightleftharpoons R{-}\underset{\beta}{\overset{\overset{\displaystyle O}{\|}}{C}}{-}\underset{\underset{\displaystyle R'}{|\,\alpha}}{CH}\overset{\overset{\displaystyle O}{\|}}{C}OC_2H_5$$

A β-keto ester

Mixed Claisen condensations between two different esters are successful only when one of the two components has no acidic alpha hydrogens (as with ethyl benzoate and ethyl formate) and thus can function only as the acceptor component. Internal Claisen condensations (Dieckmann reactions) provide excellent syntheses of five- and six-membered cyclic β-keto esters, starting from 1,6- and 1,7-diesters.

The conjugate addition of carbon nucleophiles to α,β-unsaturated acceptors is known as the Michael reaction. The best Michael reactions take place between acidic donors (β-keto esters or β-diketones) and unhindered α,β-unsaturated ketone products. Enamines, prepared by reaction of a ketone with a disubstituted amine, are also excellent Michael donors.

Carbonyl condensation reactions enjoy widespread use in synthesis; one example of their versatility is the Robinson annulation reaction. Treatment of a β-diketone or β-keto ester with an α,β-unsaturated ketone leads to an initial Michael addition. This is followed by internal aldol cyclization of the product to yield a cyclohexenone ring. Condensation reactions are used widely in nature for the biosynthesis of many biologically important molecules, including fats and steroids.

26.17 Summary of reactions

1. Aldol reaction—condensation between two ketones, two aldehydes, or one ketone and one aldehyde
   a. Ketones (Section 26.2)

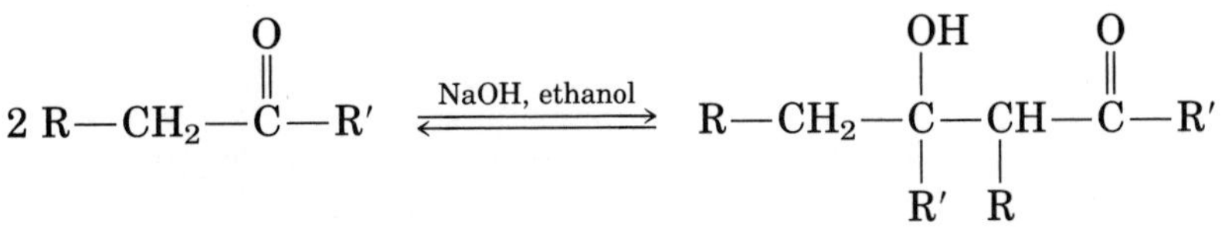

$$2\,R{-}CH_2{-}\overset{\overset{\displaystyle O}{\|}}{C}{-}R' \xrightleftharpoons{\text{NaOH, ethanol}} R{-}CH_2{-}\underset{\underset{\displaystyle R'}{|}}{\overset{\overset{\displaystyle OH}{|}}{C}}{-}\underset{\underset{\displaystyle R}{|}}{CH}{-}\overset{\overset{\displaystyle O}{\|}}{C}{-}R'$$

   b. Aldehydes (Section 26.2)

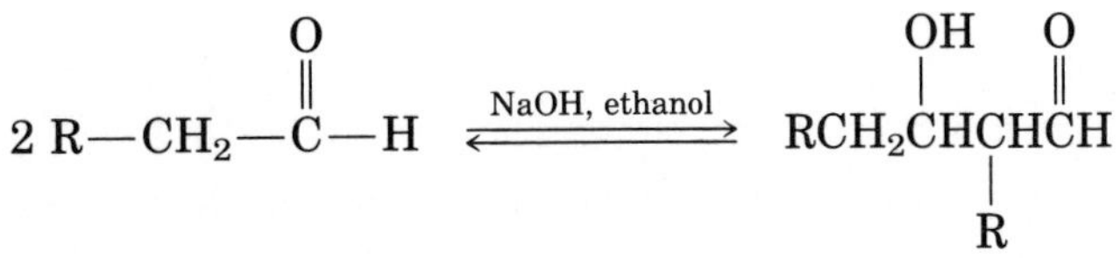

$$2\,R{-}CH_2{-}\overset{\overset{\displaystyle O}{\|}}{C}{-}H \xrightleftharpoons{\text{NaOH, ethanol}} RCH_2\overset{\overset{\displaystyle OH}{|}}{C}H\underset{\underset{\displaystyle R}{|}}{C}H\overset{\overset{\displaystyle O}{\|}}{C}H$$

   c. Mixed aldol reaction (Section 26.6)

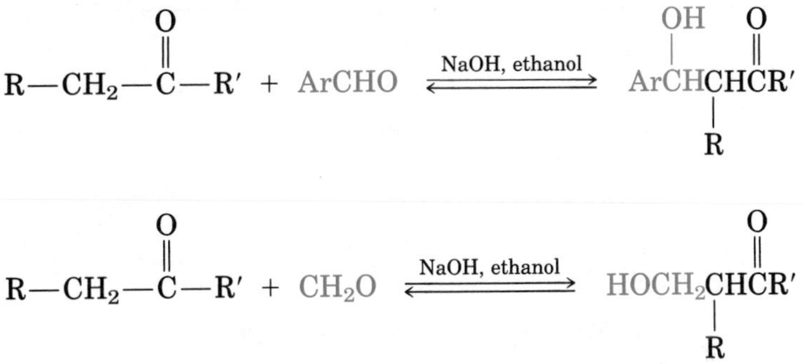

$$R{-}CH_2{-}\overset{\overset{\displaystyle O}{\|}}{C}{-}R' + ArCHO \xrightleftharpoons{\text{NaOH, ethanol}} Ar\overset{\overset{\displaystyle OH}{|}}{C}H\underset{\underset{\displaystyle R}{|}}{C}H\overset{\overset{\displaystyle O}{\|}}{C}R'$$

$$R{-}CH_2{-}\overset{\overset{\displaystyle O}{\|}}{C}{-}R' + CH_2O \xrightleftharpoons{\text{NaOH, ethanol}} HOCH_2\underset{\underset{\displaystyle R}{|}}{C}H\overset{\overset{\displaystyle O}{\|}}{C}R'$$

   d. Internal aldol reaction (Section 26.7)

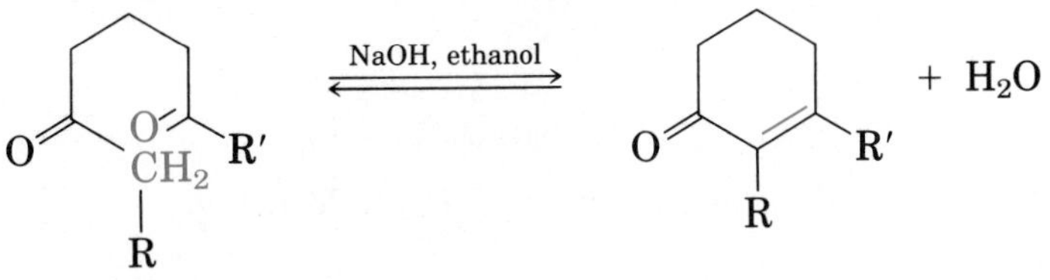

(Cyclic diketone with CH_2R and R' substituents) $\xrightleftharpoons{\text{NaOH, ethanol}}$ (cyclohexenone bearing R and R') $+ H_2O$

2. Dehydration of aldol products (Section 26.4)

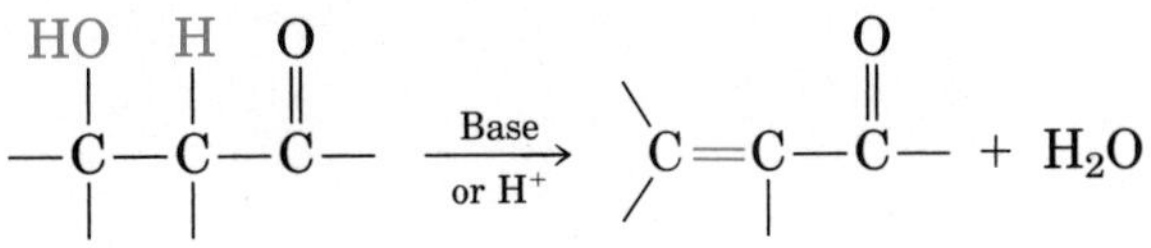

$$-\overset{\overset{\displaystyle HO}{|}}{\underset{|}{C}}-\overset{\overset{\displaystyle H}{|}}{\underset{|}{C}}-\overset{\overset{\displaystyle O}{\|}}{C}- \xrightarrow[\text{or } H^+]{\text{Base}} \;>C{=}\underset{|}{C}{-}\overset{\overset{\displaystyle O}{\|}}{C}{-} + H_2O$$

3. Claisen reaction—condensation between two esters, or one ester and one ketone (Section 26.9)

a. $$2\,R{-}CH_2CO_2R' \xrightleftharpoons[\text{ethanol}]{NaOCH_2CH_3} R{-}\overset{\overset{O}{\|}}{C}{-}\underset{\underset{R}{|}}{C}HCO_2R' + HOR'$$

b. Mixed Claisen reaction (Section 26.10)

$$R{-}CH_2CO_2R' + HCO_2R' \xrightleftharpoons{NaOCH_2CH_3,\ \text{ethanol}} H{-}\overset{\overset{O}{\|}}{C}{-}\underset{\underset{R}{|}}{C}HCO_2R' + HOR'$$

4. Dieckmann cyclization; internal Claisen condensation (Section 26.11)

$$RO_2C(CH_2)_4CO_2R \xrightleftharpoons{NaOCH_2CH_3,\ \text{ethanol}} \text{2-oxocyclopentane-}CO_2R + HOR$$

$$RO_2C(CH_2)_5CO_2R \xrightleftharpoons{NaOCH_2CH_3,\ \text{ethanol}} \text{2-oxocyclohexane-}CO_2R + HOR$$

5. Michael reaction (Section 26.12)

$$-\overset{\overset{O}{\|}}{C}CH_2\overset{\overset{O}{\|}}{C}- + \;>C{=}\underset{|}{C}{-}\overset{\overset{O}{\|}}{C}- \xrightarrow{NaOCH_2CH_3,\ \text{ethanol}} -\overset{\overset{O}{\|}}{C}\underset{\underset{\underset{|}{C=O}}{|}}{C}H{-}\underset{|}{\overset{|}{C}}{-}\overset{|}{C}H\overset{\overset{O}{\|}}{C}-$$

6. Enamine reaction (Section 26.13)

$$\overset{:NR_2}{>C}{=}\underset{|}{C}< \; + \; >C{=}\underset{|}{C}{-}\overset{\overset{O}{\|}}{C}< \xrightarrow{H_2O} >\overset{\overset{O}{\|}}{C}{-}C{-}C{-}C{-}\overset{\overset{O}{\|}}{C}<$$

7. Robinson annulation reaction (Section 26.14)

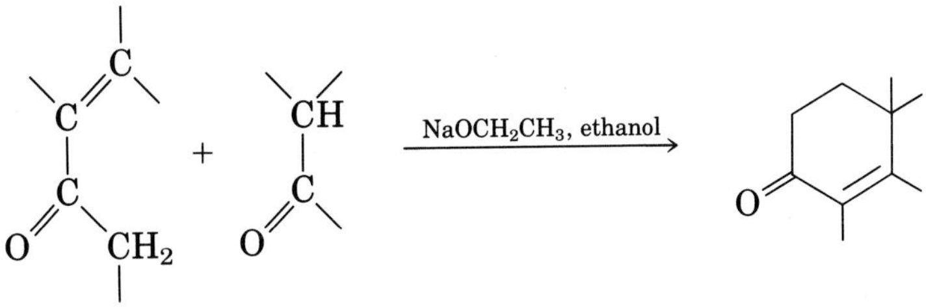

$$\text{C=C-C(=O)-CH}_2 + \text{CH-C(=O)} \xrightarrow{\text{NaOCH}_2\text{CH}_3\text{, ethanol}} \text{cyclohexenone}$$

ADDITIONAL PROBLEMS

26.21 Which of the following compounds would be expected to undergo aldol self-condensation?
(a) Trimethylacetaldehyde
(b) Cyclobutanone
(c) Benzophenone (diphenyl ketone)
(d) 2,5-Hexanedione
(e) Decanal
(f) Cinnamaldehyde (3-phenyl-2-propenal)

26.22 What product would you expect to obtain from aldol cyclization of hexanedial ($CHOCH_2CH_2CH_2CH_2CHO$)?

26.23 How might you synthesize the following compounds using aldol reactions?
(a) $C_6H_5CH{=}CHCOC_6H_5$
(b) 2-Cyclohexenone
(c) 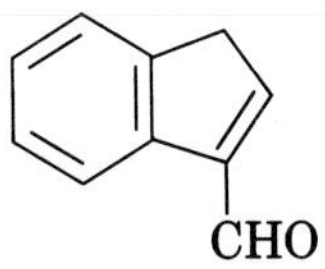 indene with CHO
(d) 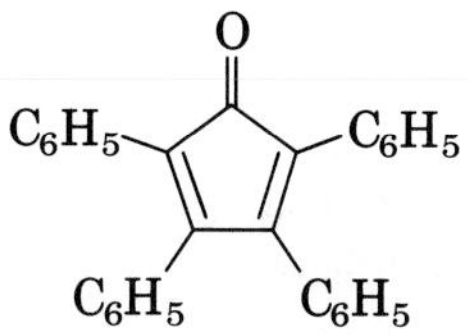 cyclopentadienone with four C_6H_5 groups

26.24 How would you prepare the following compounds from cyclohexanone?
(a) 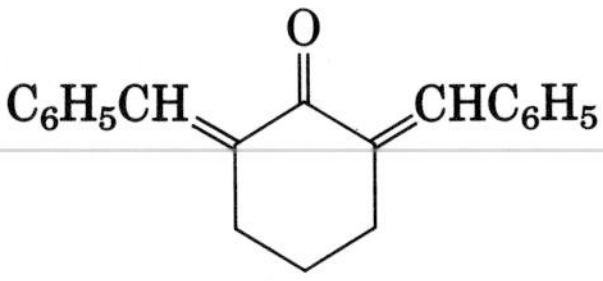 cyclohexanone with two $=CHC_6H_5$ groups (C_6H_5CH, CHC_6H_5)
(b) cyclohexanone with $CO_2C_2H_5$
(c) cyclohexanone with CH_3
(d) 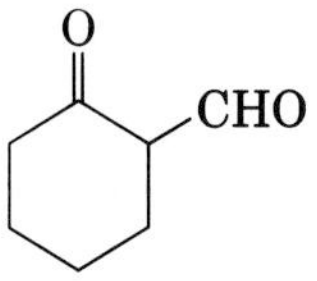 cyclohexanone with CHO
(e) 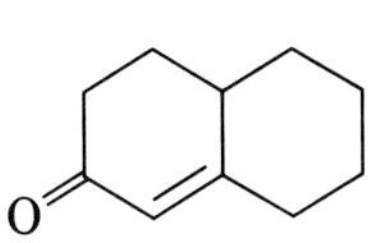 bicyclic enone (octalone) with O
(f) 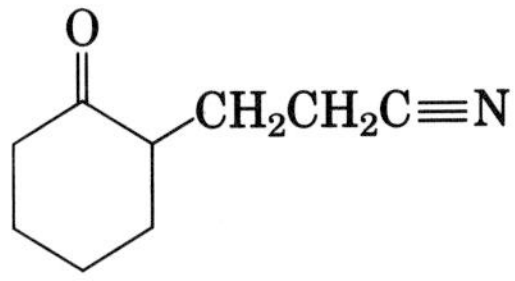 cyclohexanone with $CH_2CH_2C{\equiv}N$

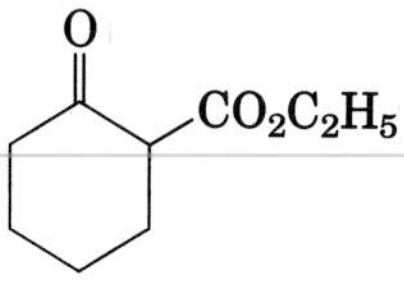

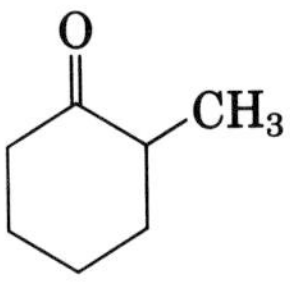

26.25 How might you carry out the following transformation? More than one step is required.

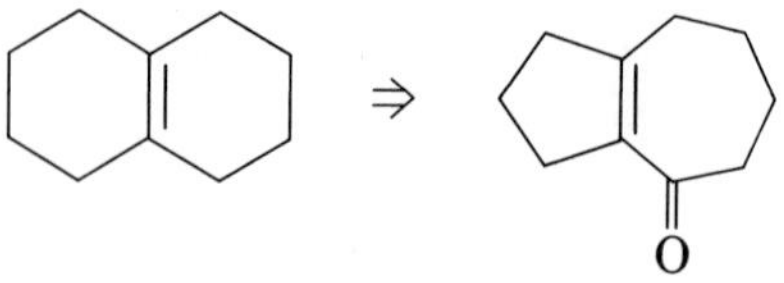

26.26 How can you account for the fact that 2,2,6-trimethylcyclohexanone does not yield an aldol product even though it has an acidic alpha hydrogen?

26.27 The Wieland–Miescher ketone shown is a valuable starting material used in steroid synthesis. How might you prepare it from 1,3-cyclohexanedione?

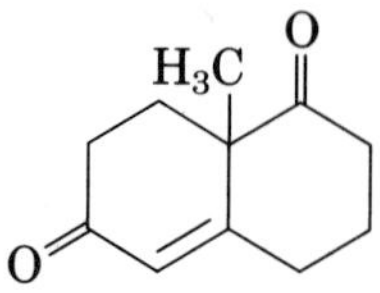

H_3C O

O

Wieland–Miescher ketone

26.28 1-Butanol is prepared commercially by a three-step process that involves an aldol condensation. Formulate the three steps.

26.29 The bicyclic ketone shown does not undergo aldol self-condensation even though it has two alpha-hydrogen atoms. Explain. [*Hint:* You may need to build molecular models.]

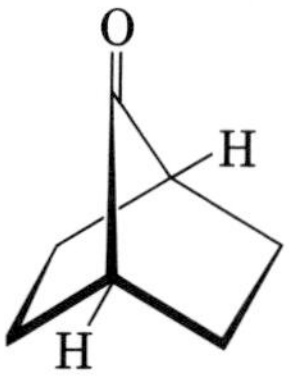

O H H

26.30 Give the structures of all the possible Claisen condensation products from the following reactions. Tell which, if any, you would expect to predominate in each case.

(a) $CH_3CO_2CH_3$ + $CH_3CH_2CO_2CH_3$ (b) $C_6H_5CO_2CH_3$ + $C_6H_5CH_2CO_2CH_3$

(c) $CH_3OCO_2CH_3$ + Cyclohexanone (d) C_6H_5CHO + $CH_3CO_2CH_3$

26.31 As written, the following reactions are unlikely to provide the desired product in high yield. What is wrong with each?

(a) $CH_3CHO + CH_3COCH_3 \xrightarrow[\text{Ethanol}]{Na^{+\,-}OC_2H_5} CH_3CH(OH)CH_2COCH_3$

(b) $CH_2(CO_2C_2H_5)_2 + H_2C{=}CHCOCH_3 \xrightarrow{\text{Base}} H_2C{=}CHC(OH)(CH_3){-}CH(CO_2C_2H_5)_2$

(c) 3-methylcyclohexanone (O, CH_3) + $H_2C{=}CHCOCH_3$ $\xrightarrow{\text{Base}}$ 2-($CH_2CH_2COCH_3$)-3-(CH_3)cyclohexanone (O, $CH_2CH_2COCH_3$, CH_3)

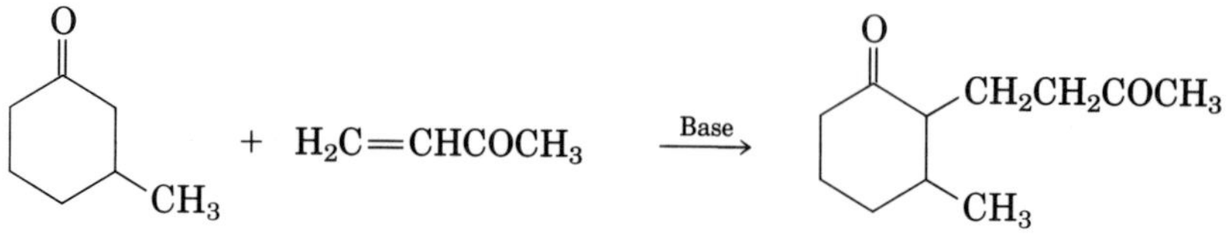

(d) 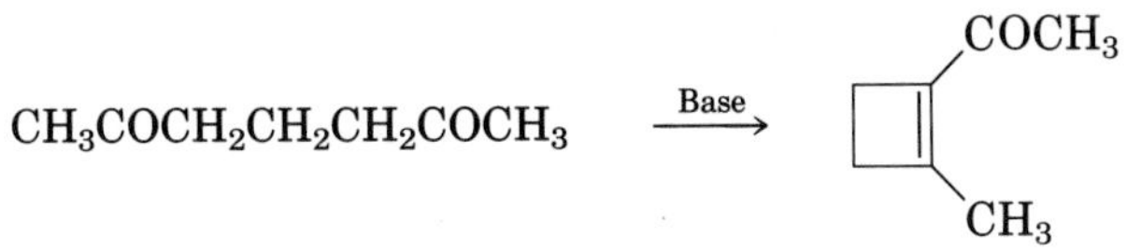 $CH_3COCH_2CH_2CH_2COCH_3 \xrightarrow{\text{Base}}$ 1-acetyl-2-methylcyclobutene ($COCH_3$, CH_3)

(e) 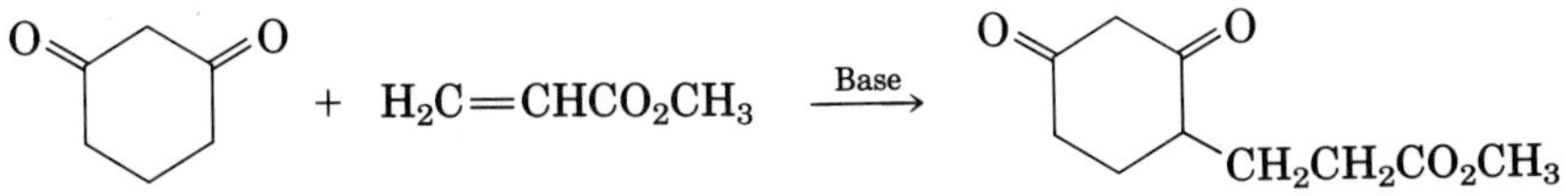 1,3-cyclohexanedione + $H_2C{=}CHCO_2CH_3 \xrightarrow{\text{Base}}$ 2-($CH_2CH_2CO_2CH_3$)-1,3-cyclohexanedione

26.32 In the mixed Claisen reaction of cyclopentanone with ethyl formate, a much higher yield of the desired product is obtained by first mixing the two carbonyl components and then adding base, rather than by first mixing base with cyclopentanone and then adding ethyl formate. Explain.

26.33 The Darzens[7] reaction involves a base-catalyzed condensation of ethyl chloroacetate with a ketone to yield an epoxy ester. Can you suggest a mechanism by which this reaction might occur?

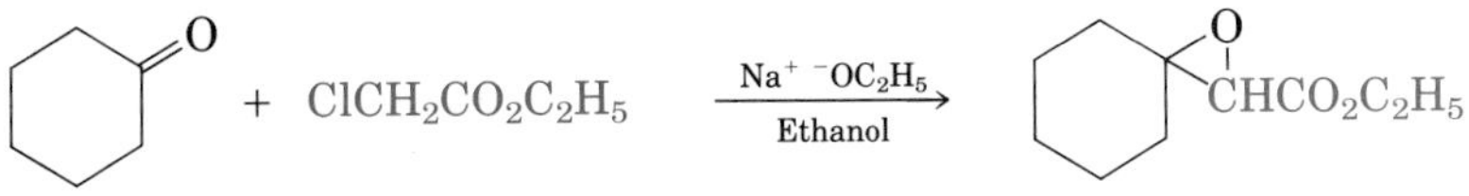

Cyclohexanone + $ClCH_2CO_2C_2H_5 \xrightarrow[\text{Ethanol}]{Na^+\ ^-OC_2H_5}$ spiro epoxide with $CHCO_2C_2H_5$

26.34 We saw earlier (Section 23.9) that β-keto acids decarboxylate by a cyclic mechanism when heated with aqueous acid. The anions of β-keto acids also lose CO_2 on warming. What is the mechanism of this process?

$$R{-}\overset{O}{\overset{\|}{C}}{-}CH_2{-}\overset{O}{\overset{\|}{C}}{-}\ddot{\underset{..}{O}}{:}^- \xrightarrow{\Delta} R{-}\overset{O}{\overset{\|}{C}}{-}\ddot{C}H_2 + CO_2$$

How is this decarboxylation related to the reverse aldol reaction?

26.35 How would you carry out the following preparations?

(a) 1,4-Cyclohexanedione from diethyl succinate

(b) 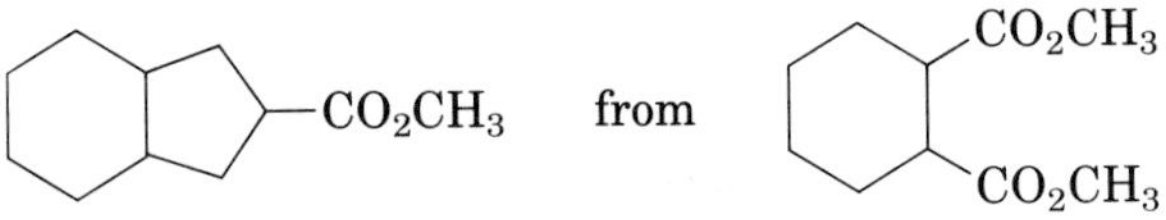 Bicyclic (cyclohexane-fused cyclopentane) $-CO_2CH_3$ from 1,2-cyclohexane with CO_2CH_3 and CO_2CH_3 substituents

26.36 The useful compound known as Hagemann's ester is prepared by treatment of a mixture of formaldehyde and ethyl acetoacetate with base, followed by acid-catalyzed decarboxylation. Propose a mechanism to account for the reactions involved.

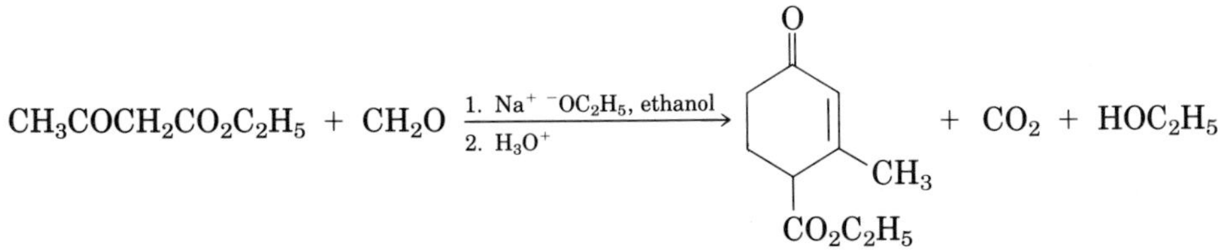

$$CH_3COCH_2CO_2C_2H_5 + CH_2O \xrightarrow[\text{2. } H_3O^+]{\text{1. } Na^+\ ^-OC_2H_5\text{, ethanol}} \text{(cyclohexenone with } CH_3\text{ and } CO_2C_2H_5\text{)} + CO_2 + HOC_2H_5$$

Hagemann's ester

[7] Georges Darzens (1886–1954); b. Moscow (French parentage); M.D. École Polytechnique, Paris; professor, École Polytechnique.

26.37 Propose a mechanism to account for the base-catalyzed isomerization of 3-ethyl-2-cyclopentenone to 2,3-dimethyl-2-cyclopentenone. [*Hint:* Remember that the aldol reaction is reversible.]

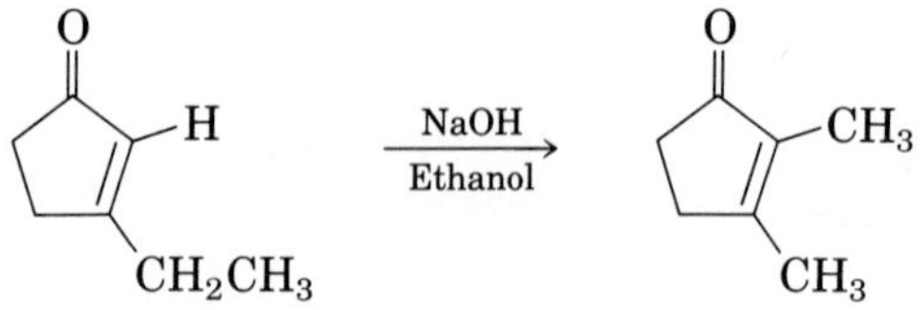

O, H, CH_2CH_3 $\xrightarrow[\text{Ethanol}]{\text{NaOH}}$ O, CH_3, CH_3

26.38 Propose a mechanism to account for the base-induced cleavage of β-diketones. For example:

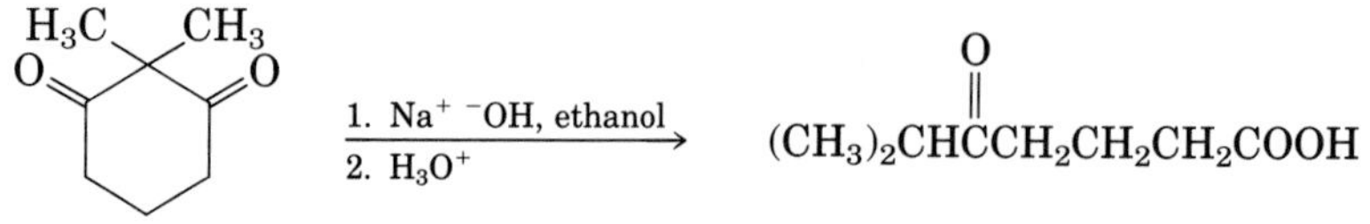

H_3C, CH_3, O, O $\xrightarrow[\text{2. } H_3O^+]{\text{1. } Na^+ \; ^-OH\text{, ethanol}}$ $(CH_3)_2CHCCH_2CH_2CH_2COOH$ (with C=O)

26.39 Can you suggest a mechanism by which the following isomerization reaction occurs? [*Hint:* Recall Problem 26.12.]

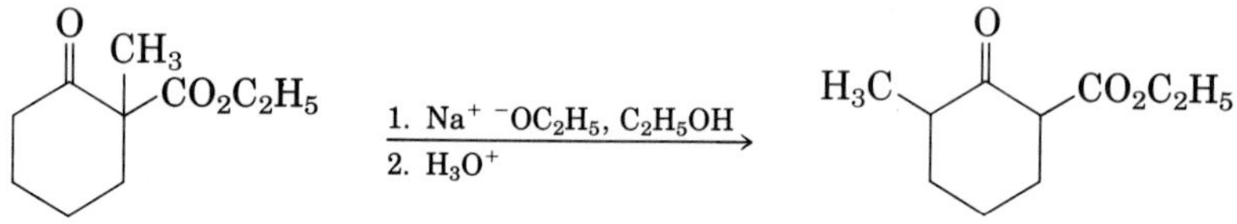

O, CH_3, $CO_2C_2H_5$ $\xrightarrow[\text{2. } H_3O^+]{\text{1. } Na^+ \; ^-OC_2H_5\text{, } C_2H_5OH}$ H_3C, O, $CO_2C_2H_5$

26.40 The base-catalyzed isomerization shown has been reported. Suggest a mechanism by which this reaction might occur. [*Hint:* The first step is a Claisen condensation.]

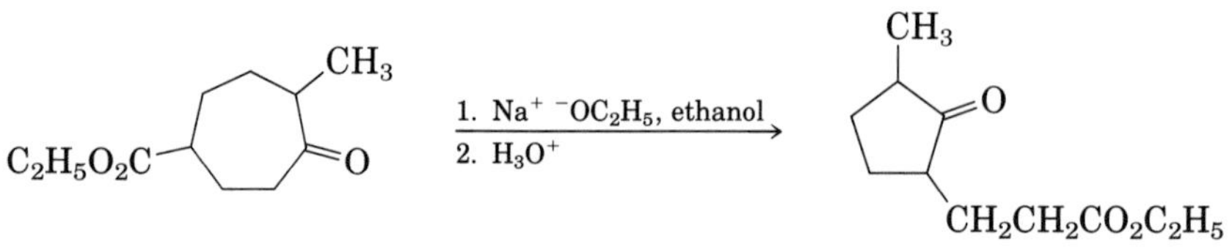

$C_2H_5O_2C$, CH_3, O $\xrightarrow[\text{2. } H_3O^+]{\text{1. } Na^+ \; ^-OC_2H_5\text{, ethanol}}$ CH_3, O, $CH_2CH_2CO_2C_2H_5$

26.41 How would you prepare 5,5-dimethyl-1,3-cyclohexanedione starting from 4-methyl-3-penten-2-one and diethyl malonate? Show all steps involved.

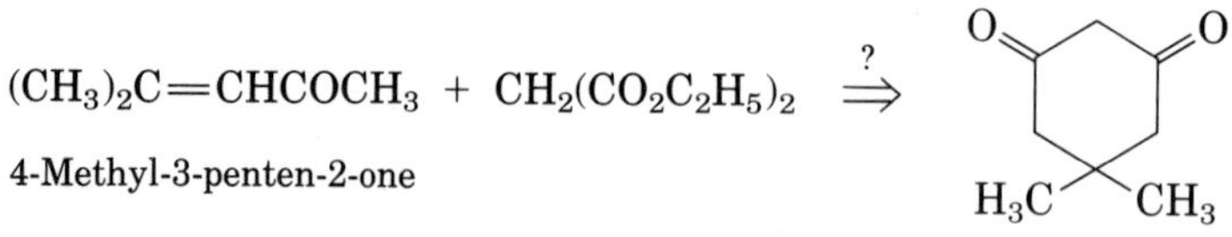

$$(CH_3)_2C{=}CHCOCH_3 + CH_2(CO_2C_2H_5)_2 \overset{?}{\Rightarrow}$$

4-Methyl-3-penten-2-one

O, O, H_3C, CH_3

26.42 Griseofulvin, an antibiotic produced by the mold *Penicillium griseofulvum* (Dierckx), has been synthesized by a route that employs the reaction shown as the key step. Propose a mechanism for this reaction.

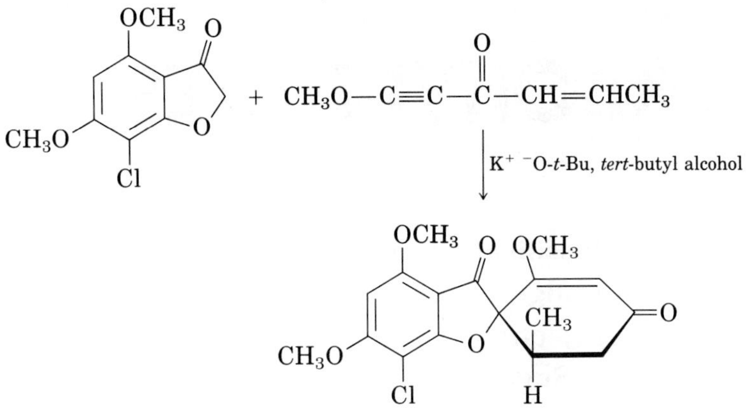

$$\text{(7-chloro-4,6-dimethoxybenzofuran-3(2H)-one)} + CH_3O{-}C{\equiv}C{-}\overset{O}{\overset{\|}{C}}{-}CH{=}CHCH_3 \xrightarrow{K^+\ {}^-O\text{-}t\text{-Bu},\ tert\text{-butyl alcohol}} \text{Griseofulvin}$$

Griseofulvin (10%)

CHAPTER 27
CARBOHYDRATES

Carbohydrates are ubiquitous; they occur in every living organism and are essential to life. The sugar and starch in food, and the cellulose in wood, paper, and cotton, are nearly pure carbohydrate. Some modified carbohydrates form part of the coating around living cells; others are found in the DNA that carries genetic information; and still others, such as gentamicin, are invaluable as medicines.

The word **carbohydrate** derives from the fact that glucose, the first simple carbohydrate to be purified, has the molecular formula $C_6H_{12}O_6$ and was originally thought to be a "hydrate of carbon," $C_6(H_2O)_6$. This view was soon abandoned, but the name persisted and is used to refer loosely to the broad class of polyhydroxylated aldehydes and ketones commonly called sugars.

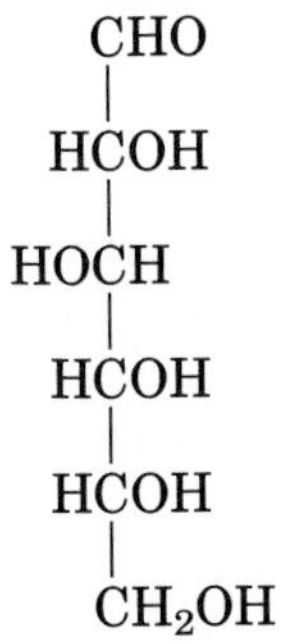

```
  CHO
   |
  HCOH
   |
 HOCH
   |
  HCOH
   |
  HCOH
   |
  CH2OH
```

Glucose (also called dextrose),
a pentahydroxyhexanal

Carbohydrates are the chemical intermediaries by which solar energy is stored and used to support life. Carbohydrates are synthesized by green plants during photosynthesis. When broken down in the cell, they provide the major source of energy required by living organisms.

The process of photosynthesis is a complex one in which carbon dioxide is converted into glucose. Once formed, many molecules of glucose can be chemically linked together for storage by the plant in the form of either cellulose or starch. It has been estimated that more than 50% of the dry weight of the earth's biomass—all plants and animals—consists of glucose polymers.

$$6\ CO_2 + 6\ H_2O \xrightarrow{\text{Sunlight}} 6\ O_2 + \underset{\text{Glucose}}{C_6H_{12}O_6} \longrightarrow \text{Cellulose, starch}$$

When ingested as food, glucose can be metabolized in the body to provide energy, or it can be stored by the body in the form of glycogen for use at a later time. Most mammals lack the enzymes needed for digestion of cellulose, and humans therefore require starches as their dietary source of carbohydrates. Grazing animals such as cows, however, contain in their rumen microorganisms that are able to digest cellulose. The energy stored in cellulose can thus be moved up the biological food chain when these animals are used for food.

27.1 Classification of carbohydrates

Simple sugars, or **monosaccharides,** such as glucose and fructose, are carbohydrates that cannot be hydrolyzed into smaller molecules. Monosaccharides can be linked together by acetal bonds, however, to form **di-, tri-,** and even **polysaccharides.** For example, sucrose (table sugar) is a disaccharide made up of one glucose molecule linked to one fructose molecule; cellulose is a polysaccharide made up of several thousand glucose molecules linked together. Hydrolysis of these polysaccharides breaks them down into their constituent monosaccharide units.

$$\text{1 Sucrose} \xrightarrow{H_3O^+} \text{1 Glucose} + \text{1 Fructose}$$

$$\text{Cellulose} \xrightarrow{H_3O^+} \sim\text{3000 Glucose}$$

Monosaccharides can be further classified into **aldoses** and **ketoses;** the *-ose* suffix is used to designate a carbohydrate and the *ald-* and *ket-* prefixes designate the nature of the carbonyl group (aldehyde or ketone). The number of carbon atoms in the monosaccharide is given by using *tri-, tetr-, pent-, hex-,* and so forth as the base name. When prefix, base, and suffix are combined, a monosaccharide is fully classified. For example, glucose is an aldohexose (a six-carbon aldehydic sugar); fructose is a ketohexose (a six-carbon ketonic sugar); and ribose is an aldopentose (a five-carbon aldehydic sugar). Most commonly occurring sugars are either aldopentoses or aldohexoses.

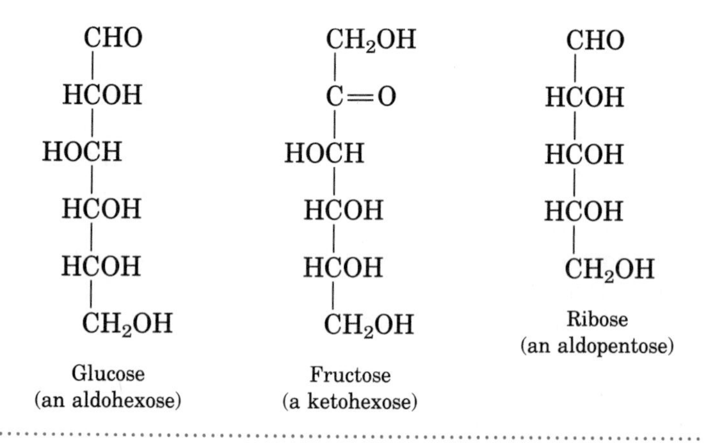

Glucose (an aldohexose) Fructose (a ketohexose) Ribose (an aldopentose)

PROBLEM

27.1 Classify each of the following monosaccharides:

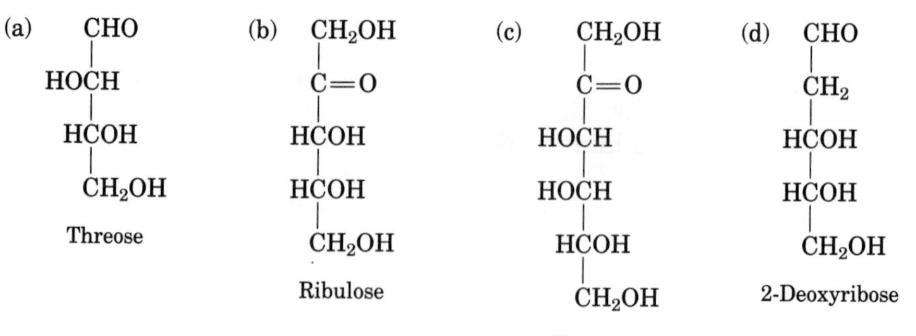

Threose Ribulose Tagatose 2-Deoxyribose

27.2 Configuration of monosaccharides: Fischer projections

Since all carbohydrates have chiral carbon atoms, it was recognized long ago that a standard method of representation was needed to designate carbohydrate stereochemistry. In 1891, Emil Fischer suggested a convention based on the projection of a tetrahedral carbon atom onto a flat surface. These **Fischer projections** were soon adopted and are now a standard means of depicting stereochemistry at chiral centers.

Recall from Section 8.13 that a tetrahedral carbon atom in a Fischer projection is represented by two crossed lines. The horizontal lines represent bonds coming out of the page, and the vertical lines represent bonds going into the page. By convention, the carbonyl carbon is placed at the top in Fischer projections. Thus, (*R*)-glyceraldehyde, the simplest monosaccharide, can be represented as follows:

$$\begin{array}{c} CHO \\ H-C-OH \\ CH_2OH \end{array} \equiv \begin{array}{c} {}^{1}CHO \\ H-{}^{2}C-OH \\ {}^{3}CH_2OH \end{array} \equiv \begin{array}{c} CHO \\ H-\!\!+\!\!-OH \\ CH_2OH \end{array}$$

Fischer projection

Recall also that Fischer projections can be rotated on the page by 180° (but not by 90° or 270°) without inverting their meaning.

$$180^\circ \left(\begin{array}{c} CHO \\ H-\!\!+\!\!-OH \\ CH_2OH \end{array} \right) \quad \text{same as} \quad \begin{array}{c} CH_2OH \\ HO-\!\!+\!\!-H \\ CHO \end{array}$$

(*R*)-Glyceraldehyde

Carbohydrates with more than one chiral center are depicted simply by "stacking" the atoms, one on top of the other. Again, however, we must obey the convention that the carbonyl carbon is at or near the top of the Fischer projection. Molecular models are particularly helpful in visualizing these structures.

$$\begin{array}{c} CHO \\ H-C-OH \\ HO-C-H \\ H-C-OH \\ H-C-OH \\ CH_2OH \end{array} \quad \begin{array}{c} CHO \\ H-\!\!+\!\!-OH \\ HO-\!\!+\!\!-H \\ H-\!\!+\!\!-OH \\ H-\!\!+\!\!-OH \\ CH_2OH \end{array}$$

Glucose (carbonyl group at top)

$$\begin{array}{c} CH_2OH \\ C{=}O \\ HO-C-H \\ H-C-OH \\ H-C-OH \\ CH_2OH \end{array} \quad \begin{array}{c} CH_2OH \\ {=}O \\ HO-\!\!+\!\!-H \\ H-\!\!+\!\!-OH \\ H-\!\!+\!\!-OH \\ CH_2OH \end{array}$$

Fructose (carbonyl group near top)

27.3 D,L Sugars

Glyceraldehyde has one chiral carbon atom and can therefore have two enantiomeric (mirror-image) forms. Only one of these enantiomers occurs naturally, however, and this natural enantiomer is dextrorotatory (Section 8.1). That is, a sample of naturally occurring glyceraldehyde placed in a polarimeter will rotate plane-polarized light in a clockwise direction, which we denote (+). (+)-Glyceraldehyde is known to have the *R* configuration at C2, and we can therefore represent it as shown in Figure 27.1. For historical reasons dating back long before the adoption of the *R*,*S* system, (*R*)-(+)-glyceraldehyde is also referred to as D-glyceraldehyde (D from dextrorotatory). The nonnatural enantiomer (*S*)-(−)-glyceraldehyde is known as L-glyceraldehyde (L from levorotatory). Glucose, fructose, and almost all other naturally occurring monosaccharides have the same configuration as D-glyceraldehyde at the highest-numbered chiral carbon atom (the one farthest from the carbonyl group). In Fischer projections, therefore, most naturally occurring sugars have the hydroxyl group at the lowest chiral carbon atom on the right (Figure 27.1), and they are referred to as **D sugars.** Although widely used by carbohydrate chemists, the D and L notations have no relation to the direction in which a given sugar rotates plane-polarized light. A D sugar may be either dextrorotatory or levorotatory.

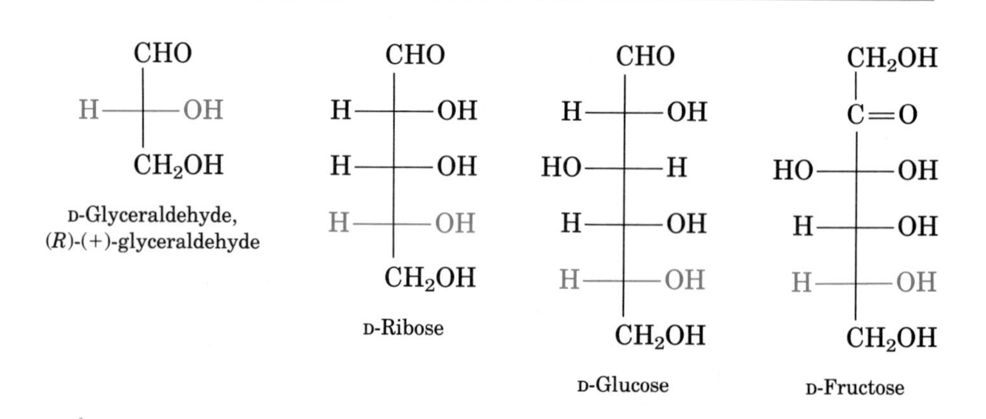

Figure 27.1. Some naturally occurring D sugars. The hydroxyl at the lowest chiral center is on the right in Fischer projections.

In contrast to the D sugars, all **L sugars** have the hydroxyl group at the lowest chiral carbon atom on the *left* and are mirror images (enantiomers) of D sugars.

CHO
HO—+—H
CH_2OH

L-Glyceraldehyde,
(*S*)-(−)-glyceraldehyde

CHO
HO—+—H
H—+—OH
HO—+—H
HO—+—H
CH_2OH

L-Glucose
(not naturally occurring)

CHO
H—+—OH
HO—+—H
H—+—OH
H—+—OH
CH_2OH

D-Glucose

The D,L system of carbohydrate nomenclature is of limited use, since it describes the configuration at only one chiral center and says nothing about other chiral centers that may be present. The advantage of the system, however, is that it allows us to relate one sugar to another rapidly and visually.

PROBLEM

27.2 Assign *R* or *S* configuration to each chiral carbon atom in the following sugars and tell if each is a D sugar or an L sugar.

(a)
CHO
HO—+—H
HO—+—H
CH_2OH

(b)
CHO
H—+—OH
HO—+—H
H—+—OH
CH_2OH

(c)
CH_2OH
C=O
HO—+—H
H—+—OH
CH_2OH

PROBLEM

27.3 (+)-Arabinose is an aldopentose that is widely distributed in plants. (+)-Arabinose can be systematically named as (2*R*,3*S*,4*S*)-5-tetrahydroxypentanal. Draw a Fischer projection of (+)-arabinose and identify it as a D or L sugar.

27.4 Configurations of aldoses

An aldotetrose is a four-carbon sugar that has two chiral centers. There are $2^2 = 4$ possible stereoisomers, or two D,L pairs of enantiomers called *erythrose* and *threose*. Aldopentoses have three chiral centers, leading to a total of $2^3 = 8$ possible stereoisomers, or four D,L pairs of enantiomers. These four pairs are called *ribose, arabinose, xylose,* and *lyxose,* and all except lyxose are widespread in nature. Ribose is an important constituent of RNA (ribonucleic acid); arabinose is found in many plants; and xylose is found in wood.

Aldohexoses have four chiral centers, for a total of $2^4 = 16$ possible stereoisomers, or eight D,L pairs of enantiomers. The names of these eight aldohexoses are *allose, altrose, glucose, mannose, gulose, idose, galactose,* and *talose.* Of the eight, only D-glucose, from starch and cellulose, and D-galactose, from gums and fruit pectins, are found widely in nature. Although D-mannose and D-talose also occur, the other four aldohexoses are not known to occur naturally.

Fischer projections of the four-, five-, and six-carbon aldoses can be constructed as shown in Figure 27.2 for the D series. Starting from D-glyceraldehyde, we can construct the two D-aldotetroses by inserting a new chiral carbon atom just below the aldehyde carbon. Each of the two D-aldotetroses can then lead to two D-aldopentoses (four total), and each of the four D-aldopentoses can lead to two D-aldohexoses (eight total).

Louis Fieser[1] of Harvard suggested this procedure for remembering the names and structures of the eight D-aldohexoses:

1. Set up eight Fischer projections with the aldehyde group on top and the CH_2OH group at the bottom.
2. Indicate stereochemistry at C5 by placing all eight hydroxyl groups to the right (D series).
3. Indicate stereochemistry at C4 by alternating four hydroxyl groups to the right and four to the left.
4. Indicate stereochemistry at C3 by alternating two hydroxyl groups to the right, two to the left, and so on.
5. Indicate stereochemistry at C2 by alternating hydroxyl groups right, left, right, left, and so on.
6. Name the eight isomers according to the mnemonic, "All altruists gladly make gum in gallon tanks."

The four D-aldopentose structures can be generated in a similar way and can be named by the mnemonic, "Ribs are extra lean."

PROBLEM

27.4 Only the D sugars are shown in Figure 27.2. Write Fischer projections for the following L sugars:

(a) L-Xylose (b) L-Galactose (c) L-Glucose

PROBLEM

27.5 How many aldoheptoses are possible? Draw Fischer projections for the two D-aldoheptoses whose stereochemistry at C3, C4, C5, and C6 corresponds to that of glucose at C2, C3, C4, and C5.

[1]Louis F. Fieser (1899–1977); b. Columbus, Ohio; Ph.D. Harvard (Conant); professor, Bryn Mawr College, Harvard University.

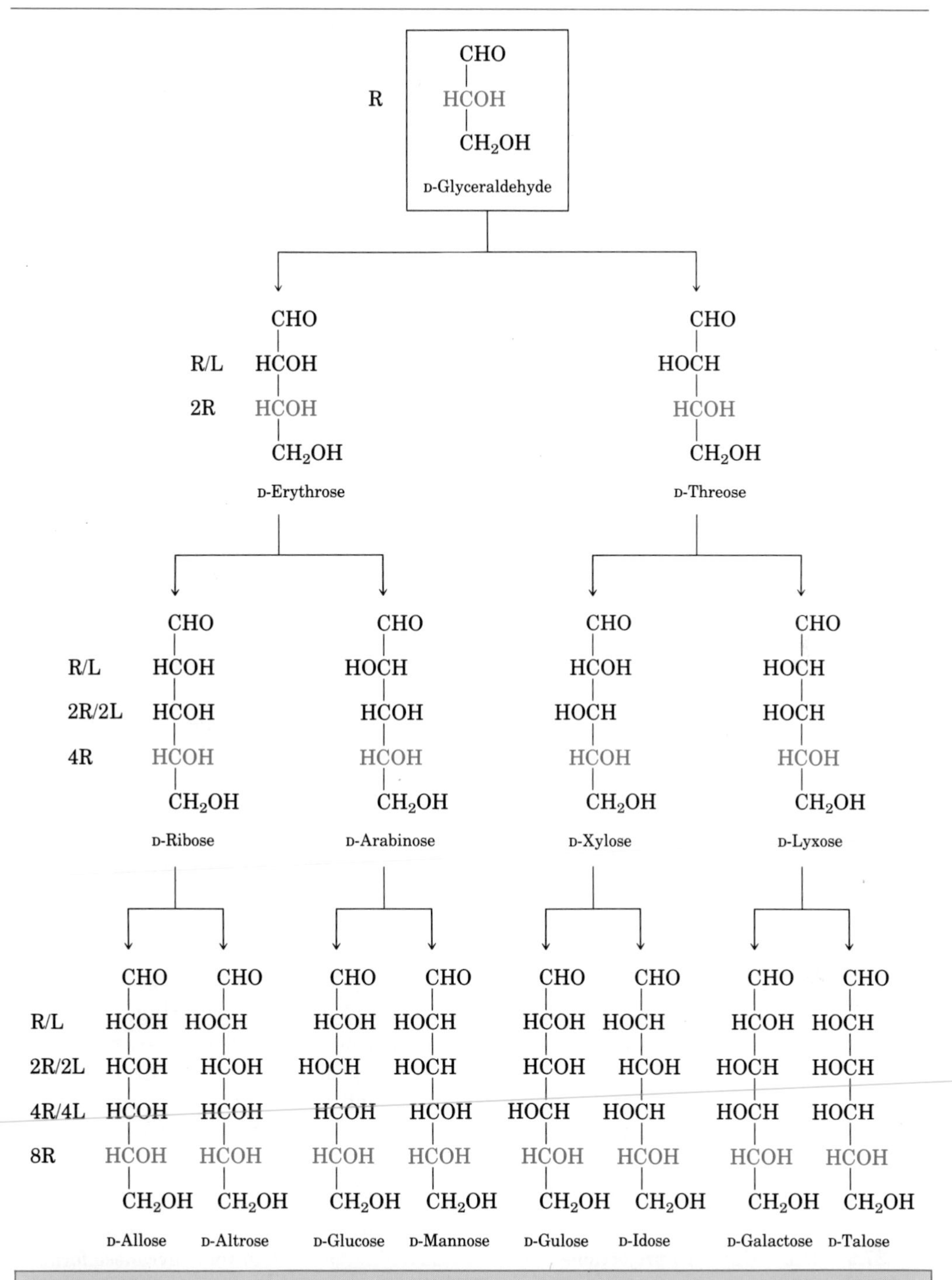

Figure 27.2. Configurations of D-aldoses. The structures are arranged in order from left to right so that the hydroxyl groups on C2 alternate right/left (R/L) in going across a series; the hydroxyl groups at C3 alternate two right/two left (2R/2L); the hydroxyl groups at C4 alternate four right/four left (4R/4L); and the hydroxyl groups at C5 are to the right in all eight (8R).

27.5 Cyclic structures of monosaccharides: hemiacetal formation

We said during our discussion of carbonyl-group chemistry (Section 22.11) that alcohols can undergo a rapid and reversible nucleophilic addition reaction with ketones and aldehydes to form hemiacetals:

$$R{-}O{-}H + R'{-}\overset{\overset{O}{\|}}{C}{-}H \underset{}{\overset{H^+}{\rightleftharpoons}} RO{-}C(OH)(R')(H)$$

A hemiacetal

If both the hydroxyl and the carbonyl group are in the same molecule, an intramolecular nucleophilic addition can take place, leading to formation of a cyclic hemiacetal. Five- and six-membered cyclic hemiacetals form particularly easily, and many carbohydrates exist in an equilibrium between open-chain and cyclic forms. For example, glucose exists in aqueous solution primarily as the six-membered ring (pyranose) formed by intramolecular nucleophilic addition of the hydroxyl group at C5 to the C1 aldehyde group. Fructose, on the other hand, exists to the extent of about 20% as the five-membered ring (furanose) formed by addition of the hydroxyl group at C5 to the C2 ketone. The words *pyranose* (six-membered ring) and *furanose* (five-membered ring) are derived from the names of the simple oxygen-containing cyclic compounds pyran and furan. Figure 27.3 shows both kinds of rings.

Carbohydrate chemists often represent pyranose and furanose forms by **Haworth projections,**[2] shown in Figure 27.3, rather than Fischer projections. In a Haworth projection, the hemiacetal ring is drawn as if it were flat, and is viewed edge-on with the oxygen atom at the upper right. This view is not completely accurate, since pyranose rings are actually chair shaped like cyclohexane (Section 18.9), rather than flat. Nevertheless, Haworth projections are widely used because they allow us to see at a glance the cis–trans relationships among hydroxyl groups on the ring.

When converting from one kind of projection to the other, remember that a hydroxyl on the right in a Fischer projection is down in a Haworth projection. Conversely, a hydroxyl on the left in a Fischer projection is up in a Haworth projection. For D sugars, the terminal CH_2OH group is always up, whereas L sugars have the group down. Figure 27.4 (p. 892) illustrates the conversion for glucose.

PROBLEM

27.6 Draw Haworth projections of D-galactose and D-mannose in their pyranose forms.

PROBLEM

27.7 Draw Haworth projections of L-glucose in its pyranose form and D-ribose in its furanose form.

[2]Sir Walter Norman Haworth (1883–1950); b. Chorley, Lancashire; Ph.D. Göttingen; D.Sc. Manchester; professor, University of Birmingham; Nobel prize (1937).

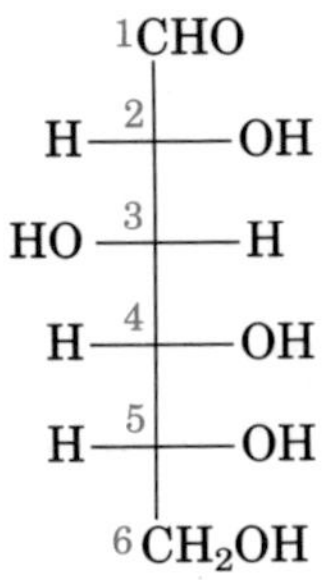

D-Glucose
(Fischer projection)

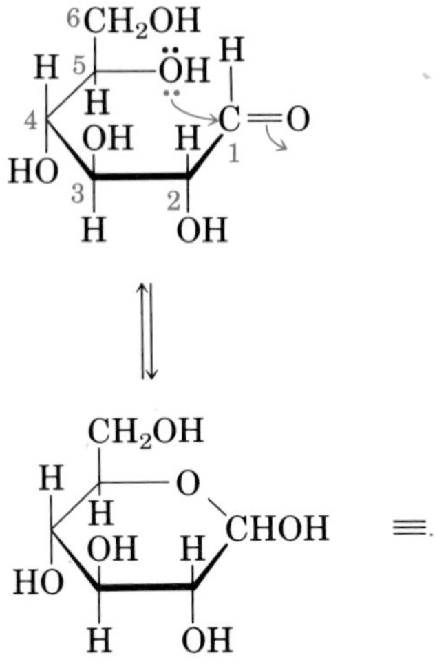

D-Glucose, pyranose form
(Haworth projection)

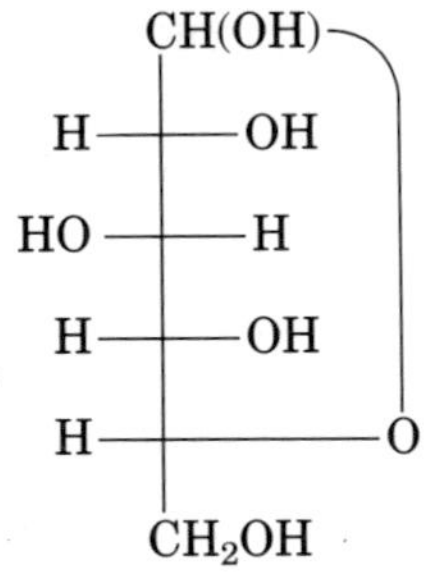

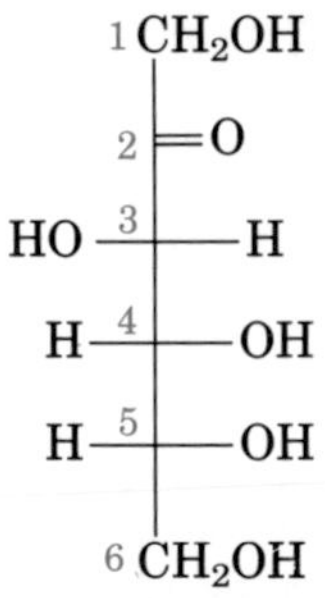

D-Fructose
(Fischer projection)

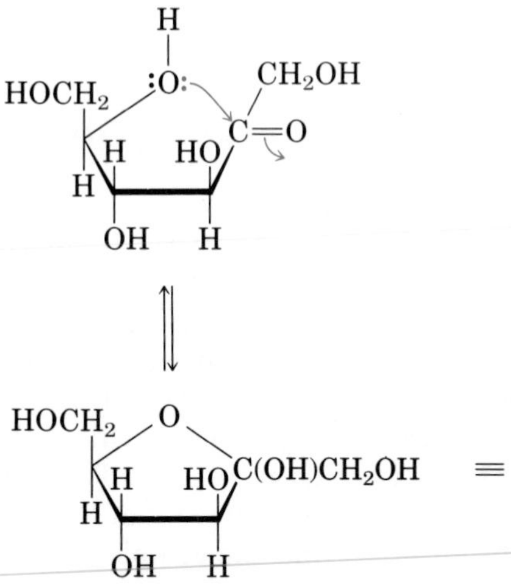

D-Fructose, furanose form
(Haworth projection)

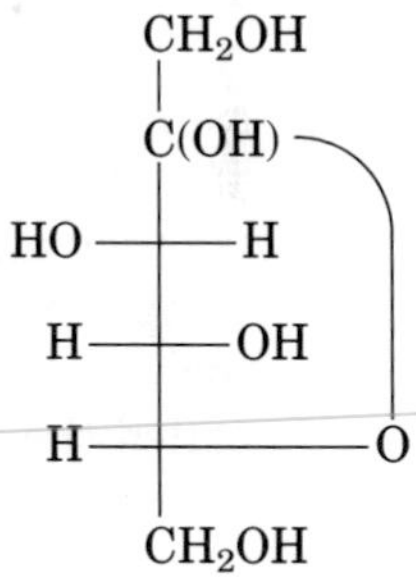

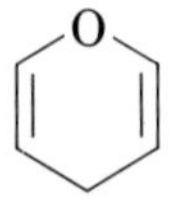

Pyran
(six-membered ring)

Furan
(five-membered ring)

Figure 27.3. Glucose and fructose; pyranose and furanose forms.

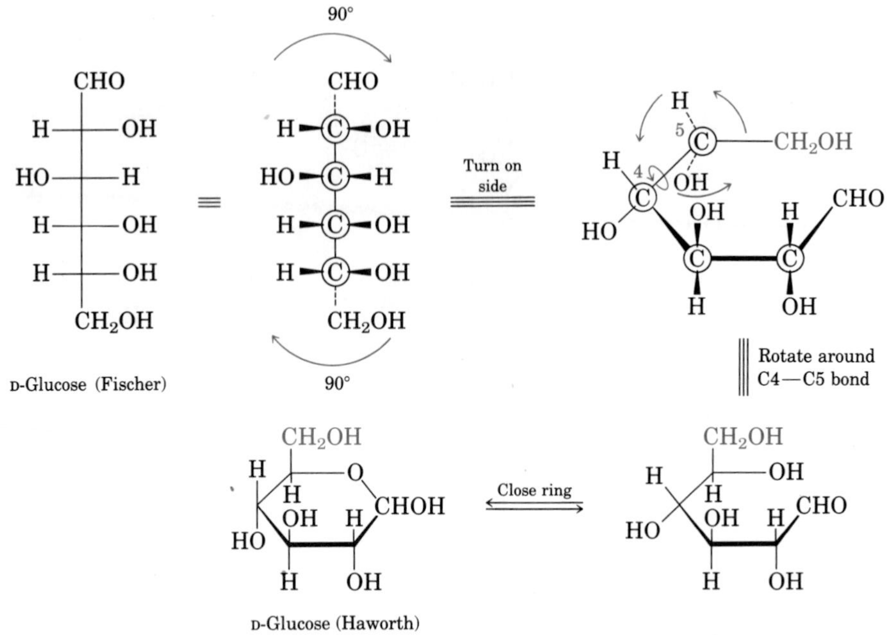

Figure 27.4. Interconversion of Fischer and Haworth projections of D-glucose.

27.6 Monosaccharide anomers: mutarotation

When an open-chain monosaccharide cyclizes to a furanose or pyranose form, a new chiral center is formed at what used to be the carbonyl carbon. The two diastereomers produced are called **anomers,** and the hemiacetal carbon atom is referred to as the **anomeric carbon.** For example, glucose cyclizes reversibly in aqueous solution to a 36:64 mixture of two anomers. The minor anomer with the C1 hydroxyl trans to the CH_2OH substituent at C5 (down in a Haworth projection) is called the **alpha anomer,** and the complete name is α-D-glucopyranose. The major anomer with the C1 hydroxyl cis to the CH_2OH substituent at C5 (up in a Haworth projection) is called the **beta anomer,** and the complete name is β-D-glucopyranose.

D-Glucose

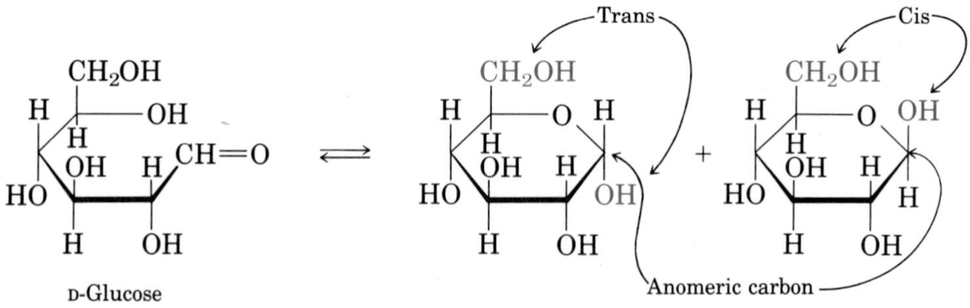

α-D-Glucopyranose (36%)
Alpha anomer: OH and CH_2OH are trans

β-D-Glucopyranose (64%)
Beta anomer: OH and CH_2OH are cis

Both anomers of D-glucose can be crystallized and purified. Pure α-D-glucose has a melting point of 146°C and a specific rotation, $[\alpha]_D$, of +112.2°; pure β-D-glucose has a melting point of 148–155°C and a specific rotation of +18.7°. When a sample of either pure α-D-glucose or pure β-D-glucose is dissolved in water, however, the optical rotations slowly change and ultimately converge to a constant value of +52.6°. The specific rotation of the alpha-anomer solution decreases from +112.2° to +52.6°, and the specific rotation of the beta-anomer solution increases from +18.7° to +52.6°. This phenomenon is known as **mutarotation,** and is due to the slow conversion of the *pure* anomers into the 36:64 equilibrium *mixture.* The equilibration occurs by a reversible ring-opening of each anomer to the open-chain aldehyde, followed by reclosure. Although equilibration is slow at neutral pH, it is catalyzed by either acid or base.

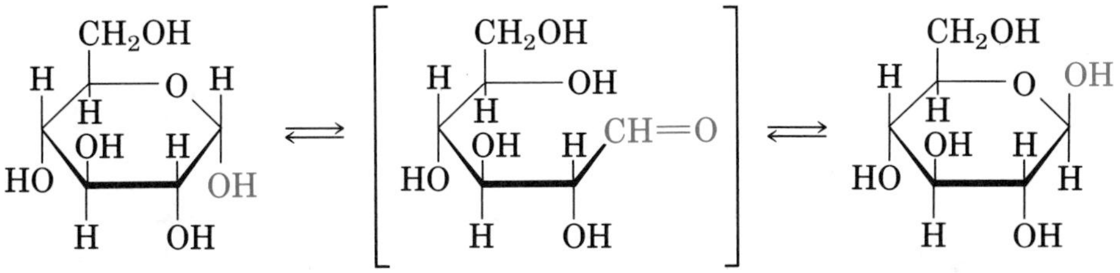

α-D-Glucose (36%)
$[\alpha]_D$ = +112.2°

β-D-Glucose (64%)
$[\alpha]_D$ = +18.7°

PROBLEM

27.8 Many other sugars besides glucose exhibit mutarotation. For example, α-D-galactose has $[\alpha]_D$ = +150.7°, and β-D-galactose has $[\alpha]_D$ = +52.8°. If either anomer is dissolved in water and allowed to reach equilibrium, the specific rotation of the solution is +80.2°. What is the percentage distribution between alpha and beta anomers at equilibrium? Draw the pyranose forms of both anomers using Haworth projections.

27.7 Conformation of monosaccharides

Haworth projections are relatively easy to draw, and they readily show cis–trans relationships between substituents on furanose and pyranose rings. They do not, however, give an accurate three-dimensional picture of molecular conformation. Pyranose rings, like cyclohexane rings (Section 18.9), can be viewed more accurately as having a chair-like geometry with axial and equatorial substituents. Any substituent that is up in a Haworth projection is also up in a chair conformational formula, and a substituent that is down in a Haworth projection is down in the chair formulation. Haworth projections can be converted into chair representations by following these steps:

1. Draw the Haworth projection with the ring oxygen atom at the upper right.
2. Raise the leftmost carbon atom (C4) *above* the ring plane.
3. Lower the anomeric carbon atom (C1) *below* the ring plane.

Figure 27.5 shows how this is done for α-D-glucopyranose and β-D-glucopyranose. (Make molecular models!)

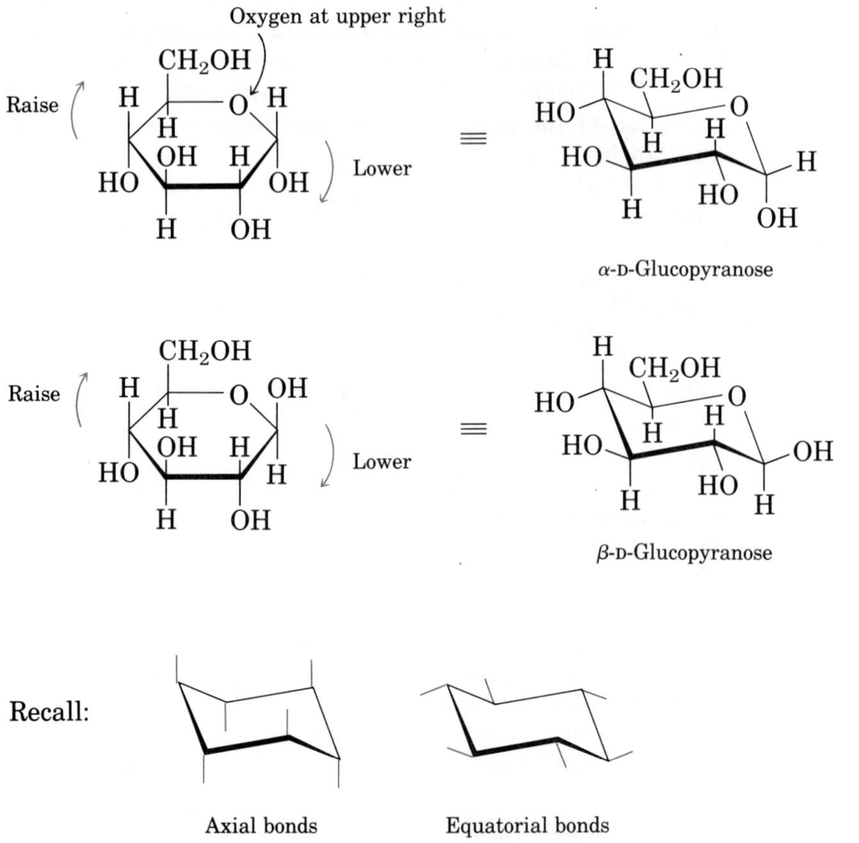

Oxygen at upper right

Raise

Lower

α-D-Glucopyranose

Raise

Lower

β-D-Glucopyranose

Recall:

Axial bonds

Equatorial bonds

Figure 27.5. Chair representations of α-D-glucopyranose and β-D-glucopyranose.

Note that in β-D-glucopyranose, all the substituents on the ring are equatorial. Thus, β-D-glucose is the least sterically crowded and most stable of the eight D-aldohexoses.

PROBLEM

27.9 Draw chair conformations of β-D-galactopyranose and of β-D-mannopyranose. Label the ring substituents as either axial or equatorial. Which would you expect to be more stable—galactose or mannose?

PROBLEM

27.10 Draw a chair conformation of β-L-glucopyranose and label the substituents as either axial or equatorial.

27.8 Reactions of monosaccharides

ESTER AND ETHER FORMATION

Monosaccharides behave as simple alcohols in much of their chemistry. For example, carbohydrate hydroxyl groups can be converted into esters and ethers.

Esterification is normally carried out by treating the carbohydrate with an acid chloride or acid anhydride in the presence of a base. *All* the hydroxyl groups react, including the anomeric one. For example, β-D-glucopyranose is converted into its pentaacetate by treatment with acetic anhydride in pyridine solution.

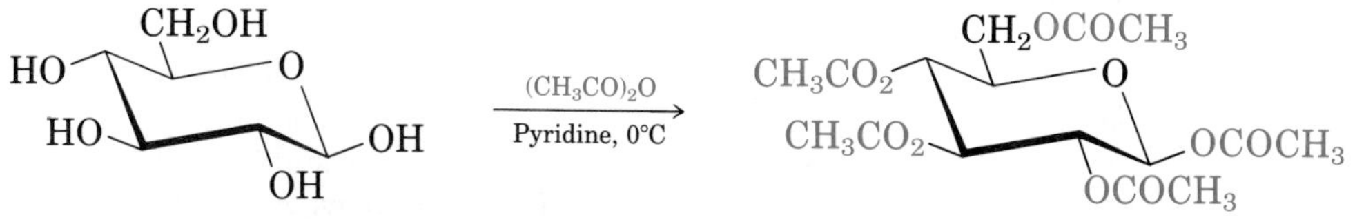

$$\xrightarrow[\text{Pyridine, 0°C}]{(CH_3CO)_2O}$$

β-D-Glucopyranose

Penta-*O*-acetyl-β-D-glucopyranose (91%)

Carbohydrates can be converted into ethers by treatment with an alkyl halide in the presence of base. Normal Williamson ether-synthesis conditions using a strong base are too harsh, but Purdie[3] showed in 1903 that silver oxide functions particularly well and that high yields of ethers are obtained. Under these conditions, α-D-glucopyranose is converted into its pentamethyl ether in 85% yield.

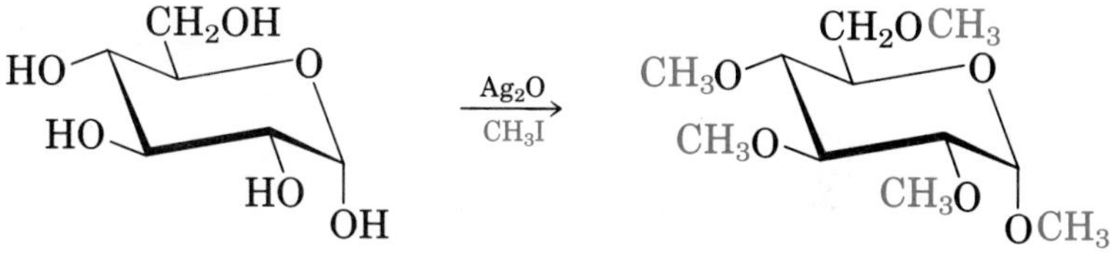

$$\xrightarrow[CH_3I]{Ag_2O}$$

α-D-Glucopyranose

α-D-Glucopyranose pentamethyl ether (85%)

Ester and ether derivatives of carbohydrates are often prepared because they are easier to work with than the free sugars. Because of their many hydroxyl groups, monosaccharides are usually quite soluble in water, but they are insoluble in organic solvents. They are also difficult to purify and have a tendency to form syrups rather than crystals when water is removed. Ester and ether derivatives, however, behave like most other organic compounds in that they tend to be soluble in organic solvents and to be readily purified and crystallized.

[3]Thomas Purdie (1843–1916); b. Biggar, Scotland; Ph.D. Wurzburg; professor, St. Andrews University.

GLYCOSIDE FORMATION

Treatment of a monosaccharide hemiacetal with an alcohol and an acid catalyst yields an acetal (Section 22.11) in which the anomeric hydroxyl has been replaced by an alkoxy group. For example, glucose reacts with methanol to give methyl β-D-glucopyranoside:

$$\beta\text{-D-Glucopyranose} \underset{H_3O^+}{\overset{CH_3OH,\ H^+}{\rightleftharpoons}} \text{Methyl } \beta\text{-D-Glucopyranoside} + H_2O$$

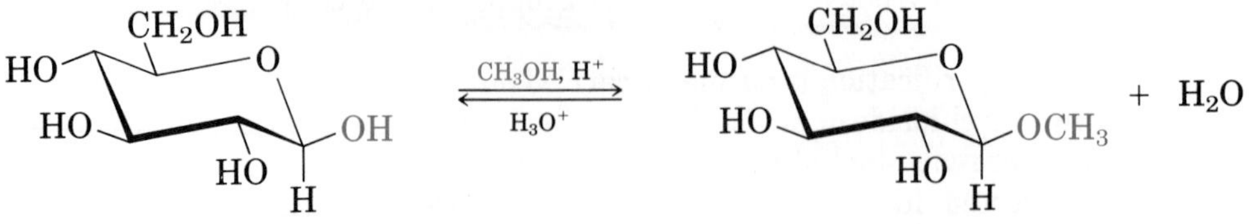

β-D-Glucopyranose (a hemiacetal)

Methyl β-D-Glucopyranoside (an acetal)

These carbohydrate acetals are called **glycosides.** They are named by citing the alkyl group and adding the *-oside* suffix to the name of the specific sugar. Note that glycosides, like all acetals, are stable and are *not* in equilibrium with an open-chain form. They can, however, be hydrolyzed to the original monosaccharide by treatment with aqueous acid.

Glycosides are widespread in nature, and a great many biologically important molecules contain glycosidic linkages. For example, digitoxin, the active component of the digitalis preparations used for treatment of heart disease, is a glycoside consisting of a complex steroid alcohol linked to a trisaccharide (Figure 27.6). The three sugars are linked to each other by glycosidic bonds.

Glycoside synthesis is a complex matter, and a successful reaction scheme is strongly dependent on the structures of both alcohol and sac-

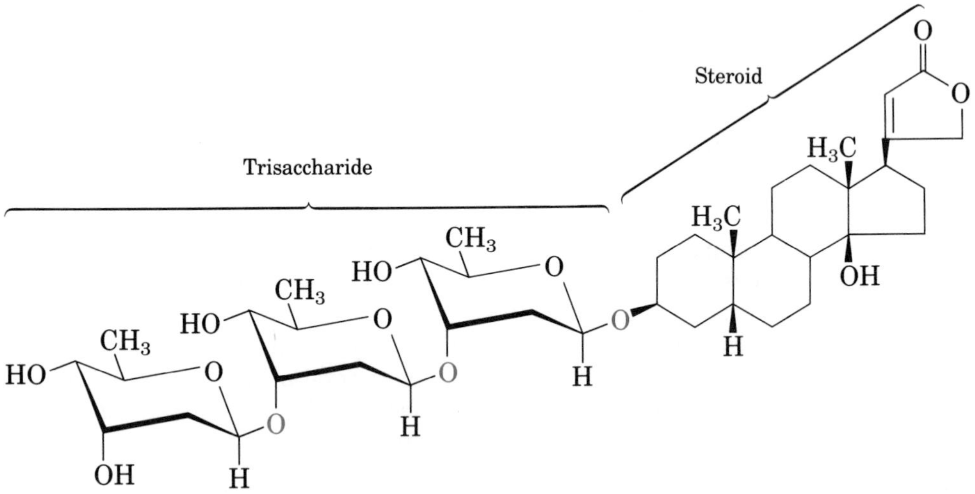

Trisaccharide

Steroid

Digitoxin, a complex glycoside

Figure 27.6. The structure of digitoxin.

charide components. One method (the **Koenigs–Knorr reaction**[4]) that is particularly suitable for preparation of glucose β-glycosides involves treatment of glucose pentaacetate with HBr, followed by addition of the appropriate alcohol in the presence of silver oxide. The reaction sequence involves formation of a pyranosyl bromide, followed by nucleophilic substitution, yielding a β-glycoside. For example, methylarbutin, a glycoside found in pear leaves, has been prepared by Koenigs–Knorr reaction between hydroquinone monomethyl ether and tetraacetyl-α-D-glucopyranosyl bromide (Figure 27.7).

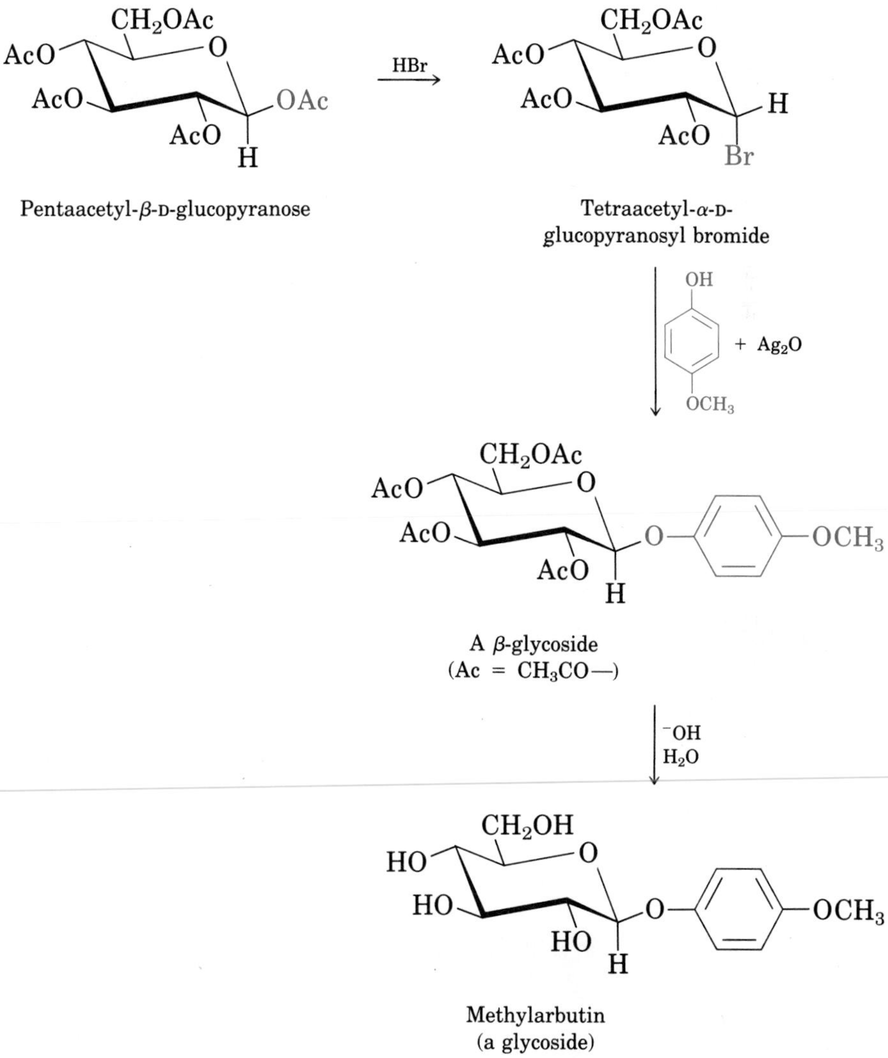

CH_2OAc, AcO, AcO, AcO, OAc, H, O

Pentaacetyl-β-D-glucopyranose

HBr →

CH_2OAc, AcO, AcO, AcO, H, Br, O

Tetraacetyl-α-D-glucopyranosyl bromide

OH, OCH_3 + Ag_2O

CH_2OAc, AcO, AcO, AcO, H, O, O, OCH_3

A β-glycoside
(Ac = CH_3CO—)

^{-}OH
H_2O

CH_2OH, HO, HO, HO, H, O, O, OCH_3

Methylarbutin
(a glycoside)

Figure 27.7. Synthesis of methylarbutin by Koenigs–Knorr reaction.

[4]Ludwig Knorr (1859–1921); D. Phil. Erlangen, 1882; professor, University of Jena.

Although the Koenigs–Knorr reaction appears to involve a simple back-side S_N2 displacement of bromide ion by alkoxide ion in a Williamson ether synthesis (Section 19.4), the actual situation is more complex, as can be gathered from the observation that both alpha and beta anomers of tetraacetyl-D-glucopyranosyl bromide give the *same* β-glycoside product. This observation is best explained by assuming that tetraacetyl-D-glucopyranosyl bromide (either alpha or beta anomer) undergoes a spontaneous S_N1 loss of bromide ion, followed by internal reaction of the cationic center at C1 with the ester group at C2 to form a stable oxonium ion intermediate. Since the ester group at C2 is on the bottom of the glucose ring (alpha orientation), the new carbon–oxygen bond also forms from the bottom. An S_N2 displacement of the oxonium ion by back-side attack at C1 then occurs with inversion of configuration, yielding a β-glycoside and regenerating the acetate ester group at C2 (Figure 27.8). This kind of participation in a reaction by a nearby group is referred to as a **neighboring-group effect** and is a common occurrence in organic chemistry.

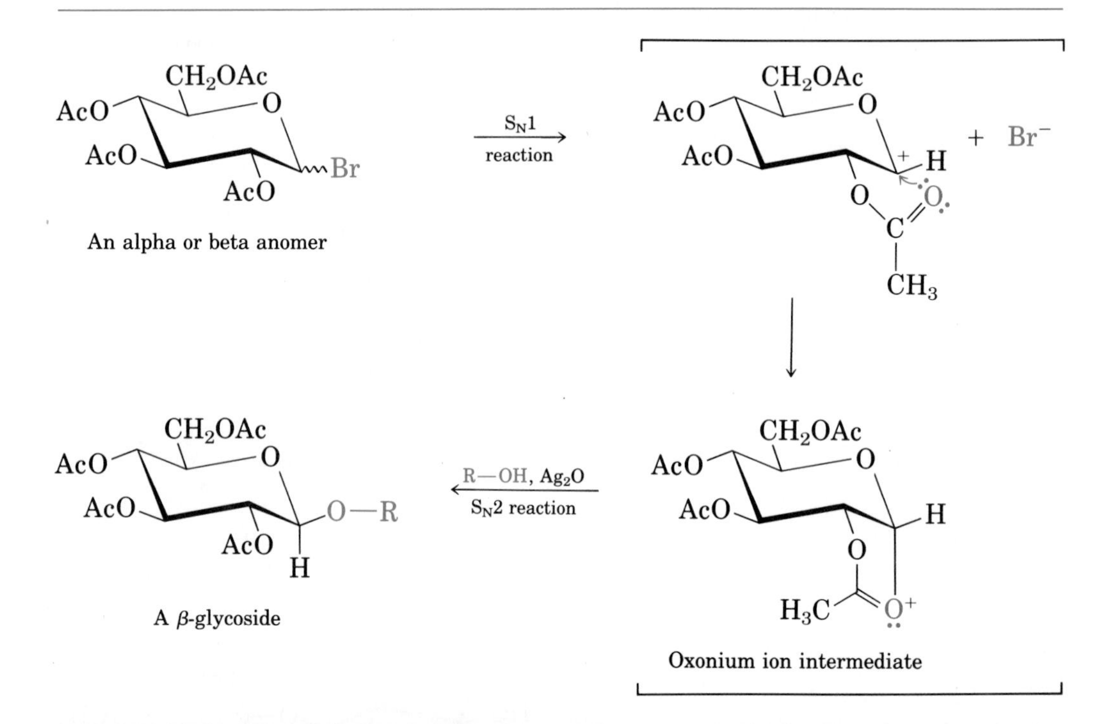

Figure 27.8. Mechanism of the Koenigs–Knorr reaction showing the neighboring-group effect of an acetoxyl.

REDUCTION OF MONOSACCHARIDES

Carbonyl groups of monosaccharides undergo many reactions characteristic of simple ketones and aldehydes (Chapter 22). For example, treatment of an aldose or ketose with sodium borohydride reduces it to a polyalcohol called an *alditol.* The reduction actually occurs by interception of the open-chain monosaccharide present in the aldehyde/ketone ⇄ hemiacetal equilibrium.

β-D-Glucopyranose ⇄ [D-Glucose] $\xrightarrow{\text{1. NaBH}_4 \text{ 2. H}_2\text{O}}$ D-Glucitol (D-sorbitol), an alditol

β-D-Glucopyranose		D-Glucose		D-Glucitol (D-sorbitol), an alditol
	⇄	CHO	1. $NaBH_4$ / 2. H_2O →	CH_2OH
		H—C—OH		H—C—OH
		HO—C—H		HO—C—H
		H—C—OH		H—C—OH
		H—C—OH		H—C—OH
		CH_2OH		CH_2OH

D-Glucitol, the alditol produced on reduction of D-glucose, is itself a natural product that has been isolated from many fruits and berries. It is used under an older name, D-sorbitol, as an artificial sweetener and sugar substitute in many foods.

PROBLEM

27.11 How can you account for the fact that reduction of D-galactose (Figure 27.2) with $NaBH_4$ leads to an alditol that is optically inactive?

PROBLEM

27.12 Reduction of one of the eight L-aldohexoses leads to the same alditol (D-glucitol) as reduction of D-glucose. Identify the L-aldohexose and explain this result.

OXIDATION OF MONOSACCHARIDES

Like other aldehydes, aldoses are easily oxidized to yield carboxylic acids. Aldoses reduce Tollens' reagent (silver ion in aqueous ammonia) and Fehling's reagent (an aqueous solution of cupric ion and tartrate salts). Both reactions serve as simple chemical tests for **reducing sugars.** Metallic silver is a product of Tollens' oxidation and is detected as a shiny mirror on the walls of the reaction flask or test tube. A reddish precipitate of cuprous oxide signals a positive reaction in the Fehling test. The diabetes self-test kits sold in drugstores for home use employ a simple variant of Fehling's test known as Benedict's test (aqueous cupric ion and citrate salts). As little as 0.1% glucose in urine gives a positive Benedict's test.

Any carbohydrate that reduces silver ion in the Tollens test or cupric ion in the Fehling/Benedict test is called a reducing sugar. All aldoses are reducing sugars, but some ketoses are reducing sugars as well. For example, fructose reduces Tollens' reagent even though it contains no aldehyde group. This occurs because fructose is readily isomerized to an aldose in basic solution by a series of keto ⇄ enol tautomeric shifts (Section 25.1). Once formed, the aldose is oxidized normally. Glycosides, however, are nonreducing; they do not react with the Tollens reagent because the acetal group cannot open to an aldehyde under basic conditions.

$$
\begin{array}{c}
\begin{array}{rcl} & CH_2OH & \\ & | & \\ & C{=}O & \\ HO & - & H \\ H & - & OH \\ H & - & OH \\ & CH_2OH & \end{array}
\underset{\text{Keto–enol tautomerism}}{\overset{^-OH/H_2O}{\rightleftharpoons}}
\begin{array}{rcl} & CHOH & \\ & \| & \\ & C{-}OH & \\ HO & - & H \\ H & - & OH \\ H & - & OH \\ & CH_2OH & \end{array}
\underset{\text{Keto–enol tautomerism}}{\overset{^-OH/H_2O}{\rightleftharpoons}}
\begin{array}{rcl} & CHO & \\ & | & \\ & CHOH & \\ HO & - & H \\ H & - & OH \\ H & - & OH \\ & CH_2OH & \end{array}
\end{array}
$$

D-Fructose — An enediol — An aldohexose

$$
\xrightarrow[NH_4OH]{Ag^+}
\begin{array}{rcl} & COOH & \\ & | & \\ & CHOH & \\ HO & - & H \\ H & - & OH \\ H & - & OH \\ & CH_2OH & \end{array} + Ag
$$

An aldonic acid

Although both the Tollens and Fehling reactions serve as useful tests for reducing sugars, they do not give good yields of carboxylic acid products because the alkaline conditions used cause decomposition of the carbohydrate skeleton. It has been found, however, that a buffered solution of aqueous bromine oxidizes aldoses to **aldonic acids** in high yield. The reaction is specific for aldoses; ketoses are not oxidized by bromine water.

$$
\alpha\text{-D-Galactose} \rightleftharpoons
\left[\begin{array}{rcl} & CHO & \\ H & - & OH \\ HO & - & H \\ HO & - & H \\ H & - & OH \\ & CH_2OH & \end{array}\right]
\xrightarrow[pH\ =\ 6]{Br_2,\ H_2O}
\begin{array}{rcl} & COOH & \\ H & - & OH \\ HO & - & H \\ HO & - & H \\ H & - & OH \\ & CH_2OH & \end{array}
$$

α-D-Galactose (an aldose)

D-Galactonic acid (an aldonic acid)

If a more powerful oxidizing agent such as warm dilute nitric acid is used, aldoses are oxidized to **aldaric acids.** Both the aldehyde group at C1 and the terminal —CH_2OH group are oxidized in this reaction, and a diacid results:

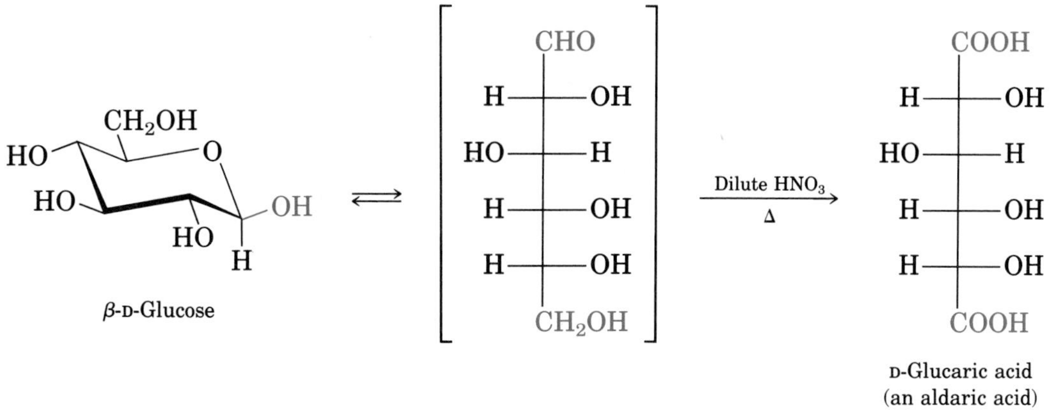

β-D-Glucose ⇌ [open-chain D-glucose: CHO; H—OH; HO—H; H—OH; H—OH; CH_2OH] $\xrightarrow[\Delta]{\text{Dilute } HNO_3}$ COOH; H—OH; HO—H; H—OH; H—OH; COOH

D-Glucaric acid (an aldaric acid)

The various kinds of carbohydrate derivatives are shown in Figure 27.9.

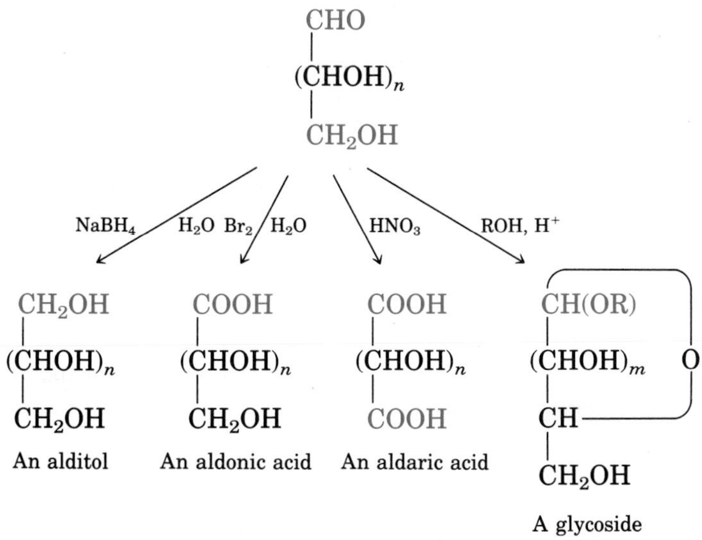

CHO—$(CHOH)_n$—CH_2OH

$NaBH_4$, H_2O → CH_2OH—$(CHOH)_n$—CH_2OH An alditol

Br_2/H_2O → COOH—$(CHOH)_n$—CH_2OH An aldonic acid

HNO_3 → COOH—$(CHOH)_n$—COOH An aldaric acid

ROH, H^+ → CH(OR)—$(CHOH)_m$—CH—CH_2OH (ring closed through O) A glycoside

Figure 27.9. Summary of carbohydrate derivatives.

PROBLEM

27.13 Which of the eight D-aldohexoses lead to optically active aldaric acids on oxidation, and which lead to meso aldaric acids?

CHAIN LENGTHENING: THE KILIANI–FISCHER SYNTHESIS

Much early activity in carbohydrate chemistry was devoted to unraveling the various stereochemical relationships among monosaccharides. One of the methods used was the **Kiliani–Fischer synthesis.** Heinrich Kiliani[5]

[5]Heinrich Kiliani (1855–1945); b. Wurzburg, Germany; Ph.D. Munich (Erlenmeyer); professor, University of Freiburg.

found in 1886 that aldoses react with HCN to form cyanohydrins (Section 22.8). Fischer immediately realized the importance of Kiliani's discovery, and in 1890 he published a method for converting the cyanohydrin nitrile group into an aldehyde group. The net effect of the Kiliani–Fischer synthesis is to lengthen the aldose chain by one carbon.

$$\begin{array}{ccccc} & & CN & & CHO \\ & & | & & | \\ CHO & \xrightarrow{HCN} & CHOH & \Longrightarrow & CHOH \\ | & & | & & | \\ CHOH & & CHOH & & CHOH \\ \text{An aldose} & & \text{A cyanohydrin} & & \text{A chain-lengthened aldose} \end{array}$$

Conversion of the nitrile into an aldehyde is accomplished by first hydrolyzing to a carboxylic acid and then forming a lactone ring by internal esterification with a hydroxyl group four carbon atoms away. Reduction of the lactone carbonyl group with sodium amalgam (an alloy of sodium and mercury) then yields the chain-lengthened aldose. Note that the cyanohydrin is formed as a mixture of stereoisomers at the new chiral center. Thus, *two* new aldoses result from Kiliani–Fischer chain extension, and these two aldoses differ only in their stereochemistry at C2. For example, chain extension of D-arabinose yields a mixture of D-glucose and D-mannose (Figure 27.10).

PROBLEM

27.14 What product(s) would you expect to obtain from Kiliani–Fischer reaction of D-ribose? Of L-xylose?

CHAIN SHORTENING: THE WOHL DEGRADATION

Just as the Kiliani–Fischer synthesis lengthens an aldose chain by one carbon, the **Wohl degradation**[6] shortens an aldose chain. The Wohl degradation process is almost exactly the opposite of the Kiliani–Fischer sequence: The aldose aldehyde carbonyl group is converted into a nitrile group, and the resulting cyanohydrin loses HCN under basic conditions (a retro nucleophilic addition reaction). Conversion of the aldehyde into a nitrile is accomplished by treatment of an aldose with hydroxylamine, followed by dehydration of the oxime product with acetic anhydride. The Wohl degradation does not give particularly high yields of chain-shortened aldoses, but the reaction is general for all aldopentoses and aldohexoses.

[6]Alfred Wohl (1863–1933); b. Graudentz; Ph.D. Berlin (Hofmann); professor, University of Danzig.

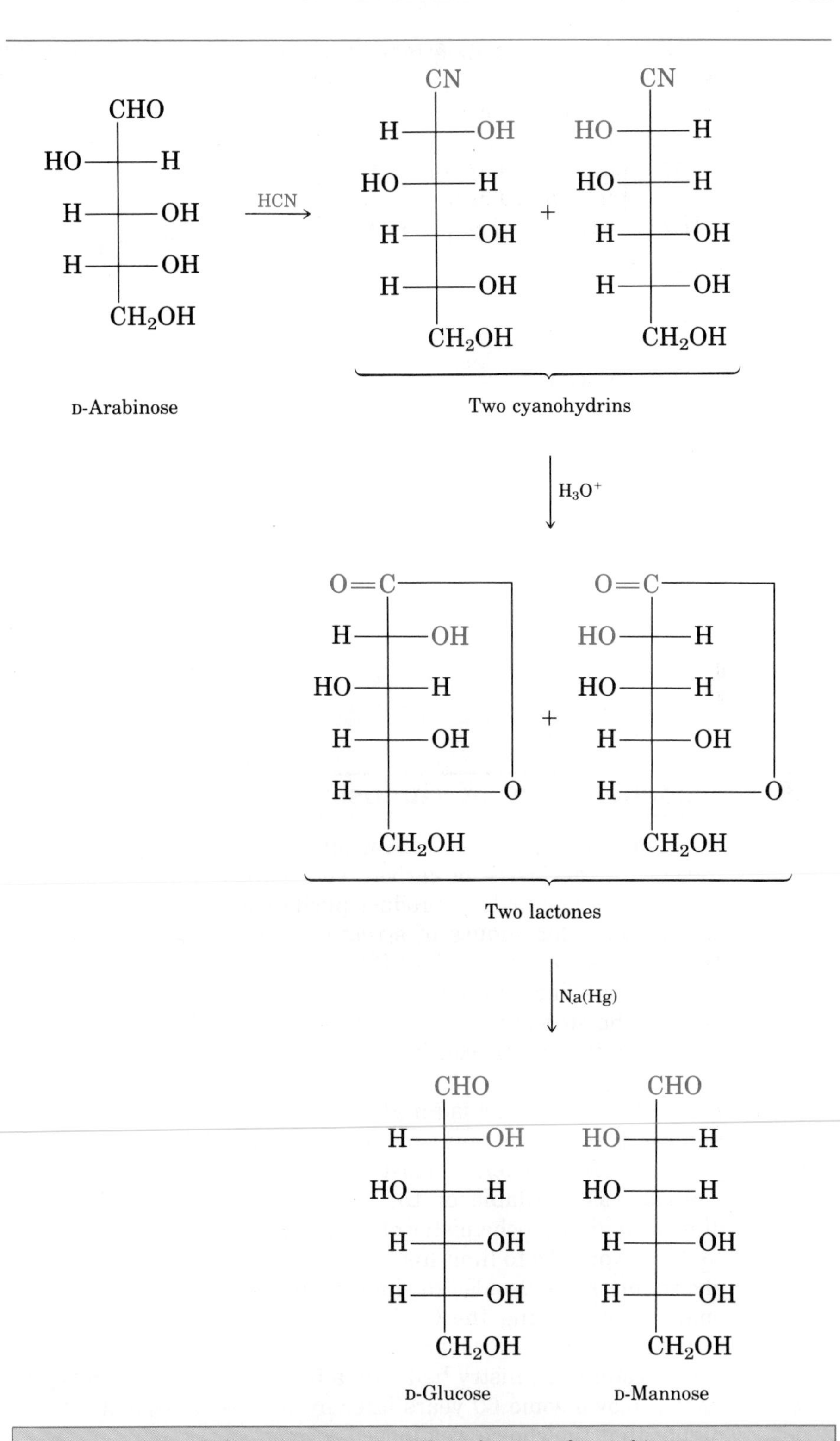

Figure 27.10. Kiliani–Fischer chain lengthening of D-arabinose.

For example, D-galactose is converted by Wohl degradation into D-lyxose:

$$\begin{array}{c} CHO \\ H-\!\!\!+\!\!\!-OH \\ HO-\!\!\!+\!\!\!-H \\ HO-\!\!\!+\!\!\!-H \\ H-\!\!\!+\!\!\!-OH \\ CH_2OH \\ \text{D-Galactose} \end{array} \xrightarrow{H_2NOH} \begin{array}{c} CH{=}NOH \\ H-\!\!\!+\!\!\!-OH \\ HO-\!\!\!+\!\!\!-H \\ HO-\!\!\!+\!\!\!-H \\ H-\!\!\!+\!\!\!-OH \\ CH_2OH \\ \text{D-Galactose oxime} \end{array} \xrightarrow[NaO_2CCH_3]{(CH_3CO)_2O} \begin{array}{c} C{\equiv}N \\ H-\!\!\!+\!\!\!-OH \\ HO-\!\!\!+\!\!\!-H \\ HO-\!\!\!+\!\!\!-H \\ H-\!\!\!+\!\!\!-OH \\ CH_2OH \\ \text{A cyanohydrin} \end{array}$$

$$\xrightarrow{NaOCH_3} \begin{array}{c} CHO \\ HO-\!\!\!+\!\!\!-H \\ HO-\!\!\!+\!\!\!-H \\ H-\!\!\!+\!\!\!-OH \\ CH_2OH \\ \text{D-Lyxose (37\%)} \end{array} + HCN$$

27.9 Stereochemistry of glucose: the Fischer proof

In the late 1800s, the stereochemical theories of van't Hoff and Le Bel on the tetrahedral geometry of carbon were barely a decade old. Modern chromatographic methods of product purification were unknown, and modern spectroscopic techniques of structure determination were undreamed of. Despite these obstacles, Emil Fischer published in 1891 what stands today as one of the finest examples of chemical logic ever recorded—a structural proof of the stereochemistry of glucose. Let's follow Fischer's logic and see how he arrived at his conclusions.

FACT 1. (+)-Glucose is an aldohexose. Glucose has four chiral centers and can therefore be any one of $2^4 = 16$ possible stereoisomers. These 16 possible stereoisomers consist of eight pairs of enantiomers. Since no method was available at the time for determining the *absolute* three-dimensional stereochemistry of a molecule, Fischer realized that the best he could do would be to limit his choices for the structure of glucose to a pair of enantiomers. He decided to simplify matters by considering only the eight enantiomers having the C5 hydroxyl group on the right in Fischer projections (D series). Fischer was well aware that this arbitrary choice of D-series stereochemistry had only a 50/50 chance of being right, but it was finally shown some 60 years later by the use of sophisticated X-ray techniques that the choice was indeed correct.

The four possible D-aldopentoses and the eight possible D-aldohexoses derived from them by Kiliani–Fischer synthesis are shown in Figure 27.11. One of the eight aldohexoses is glucose, but which one?

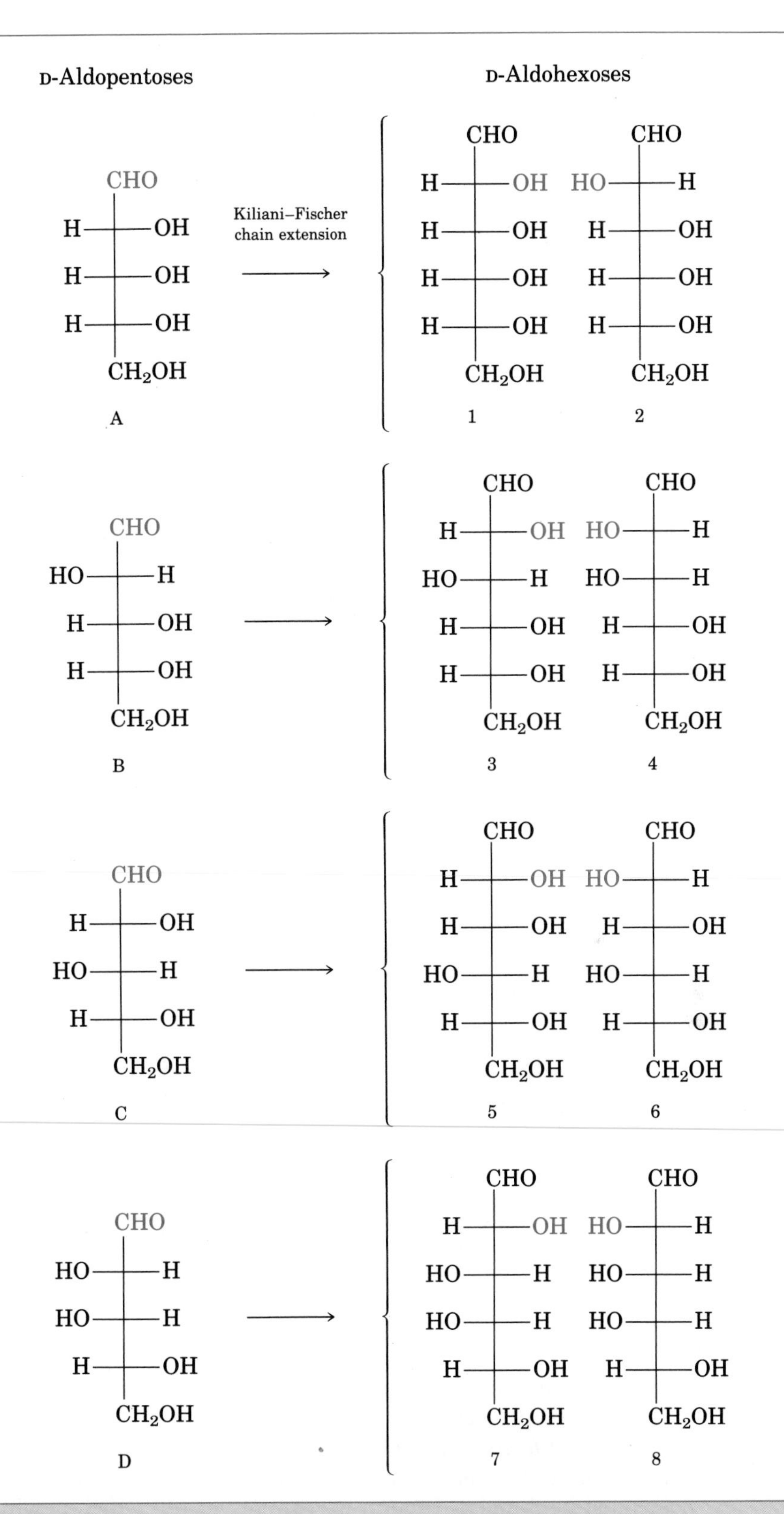

Figure 27.11. The four D-aldopentoses and eight D-aldohexoses.

FACT 2. Arabinose, an aldopentose, is converted by Kiliani–Fischer chain extension into a mixture of glucose and mannose. This means that glucose and mannose have the same stereochemistry at C3, C4, and C5, and differ only at C2. Glucose and mannose are therefore represented by one of the pairs of structures 1 and 2, 3 and 4, 5 and 6, or 7 and 8 in Figure 27.11.

FACT 3. Arabinose is converted by treatment with warm nitric acid into an optically active aldaric acid. Of the four possible aldopentose structures (A, B, C, and D in Figure 27.11), A and C give optically inactive meso aldaric acids, whereas B and D give optically active products. Thus, arabinose must be either B or D, and mannose and glucose must therefore be either 3 and 4 or 7 and 8 (Figure 27.12).

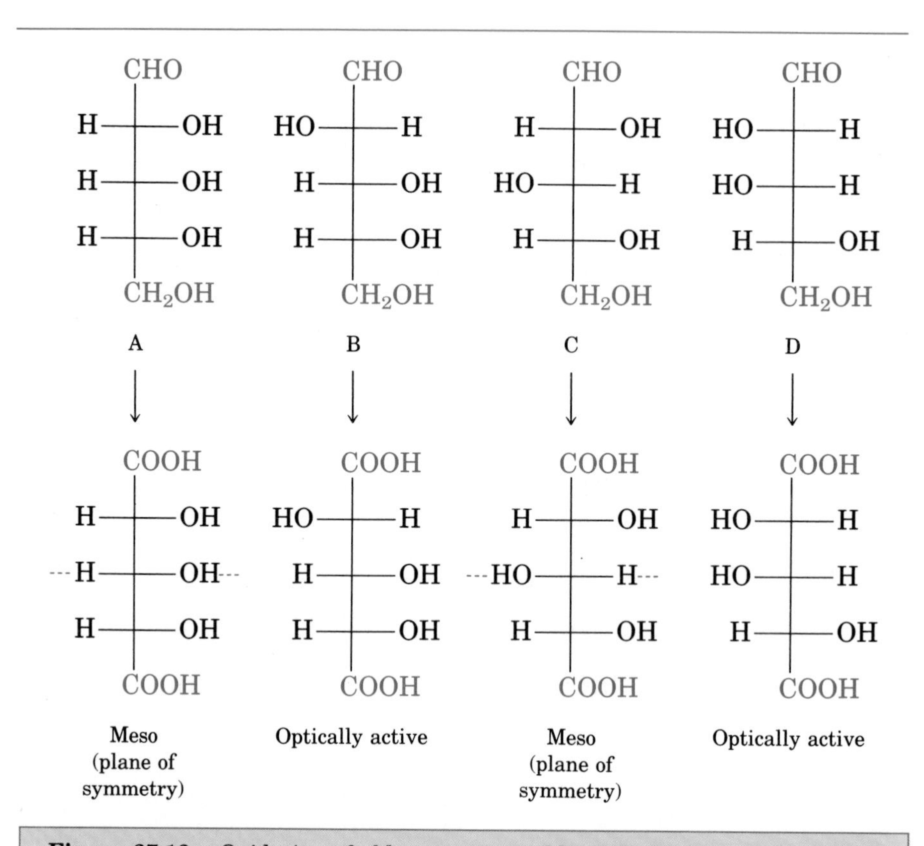

Figure 27.12. Oxidation of aldopentoses to aldaric acids.

FACT 4. Both glucose and mannose are oxidized by warm nitric acid to different optically active aldaric acids. Of the possibilities left at this point, the pair represented by structures 3 and 4 would *both* be oxidized to different optically active aldaric acids, but the pair represented by 7 and 8 would not *both* give optically active products. Compound 7 would give an optically inactive meso aldaric acid (Figure 27.13). Thus, glucose and mannose must be 3 and 4, though we cannot yet tell which is which.

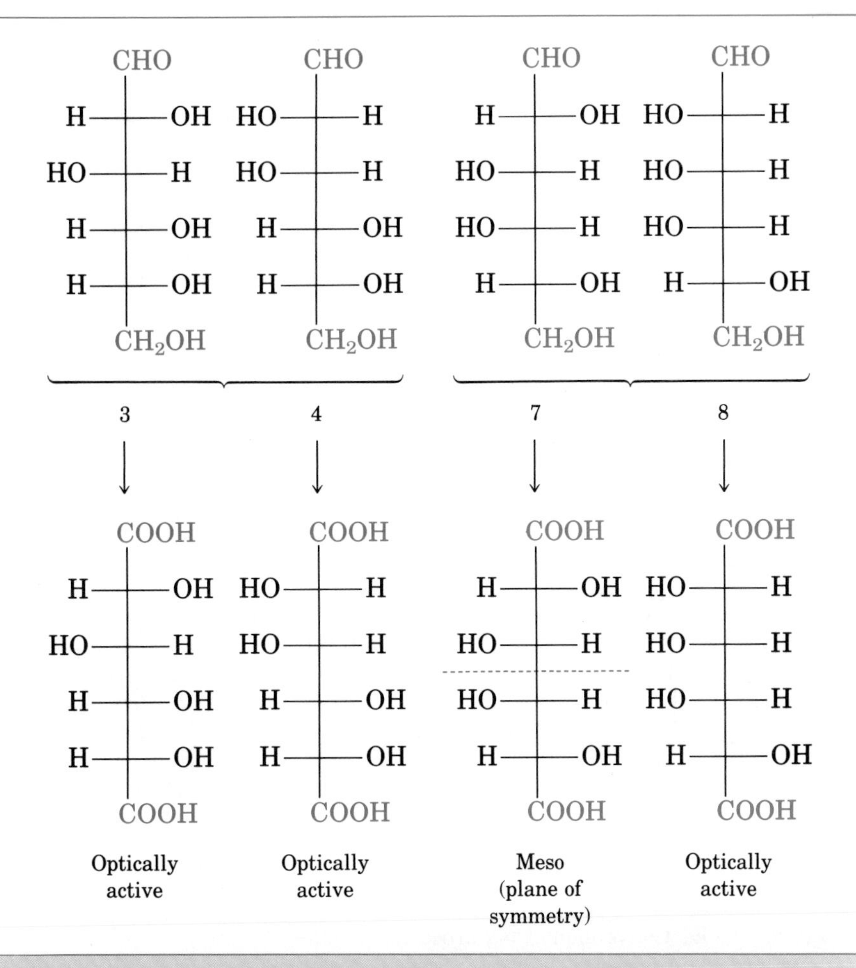

Figure 27.13. Oxidation of aldohexoses to aldaric acids.

FACT 5. (+)-Gulose, another aldohexose, is converted by nitric acid oxidation to the same aldaric acid as that derived from glucose. If we look at the aldaric acids derived from 3 and 4, we find that only the acid derived from 3 could also come from oxidation of another aldohexose. No other aldohexose could produce the same aldaric acid as the one from 4. Thus, glucose has structure 3 and mannose has structure 4 (Figure 27.14, p. 908).

Reasoning similar to that shown for D-glucose allowed Fischer to determine the stereochemistry of 12 of the 16 aldohexoses. For this remarkable achievement, he was awarded the 1902 Nobel prize in chemistry.

PROBLEM

27.15 The structures of the four aldopentoses, A, B, C, and D, are shown in Figure 27.11. In light of Fact 2 presented by Fischer, what is the structure of D-arabinose? In light of Fact 3, what is the structure of lyxose, another aldopentose that yields an optically active aldaric acid?

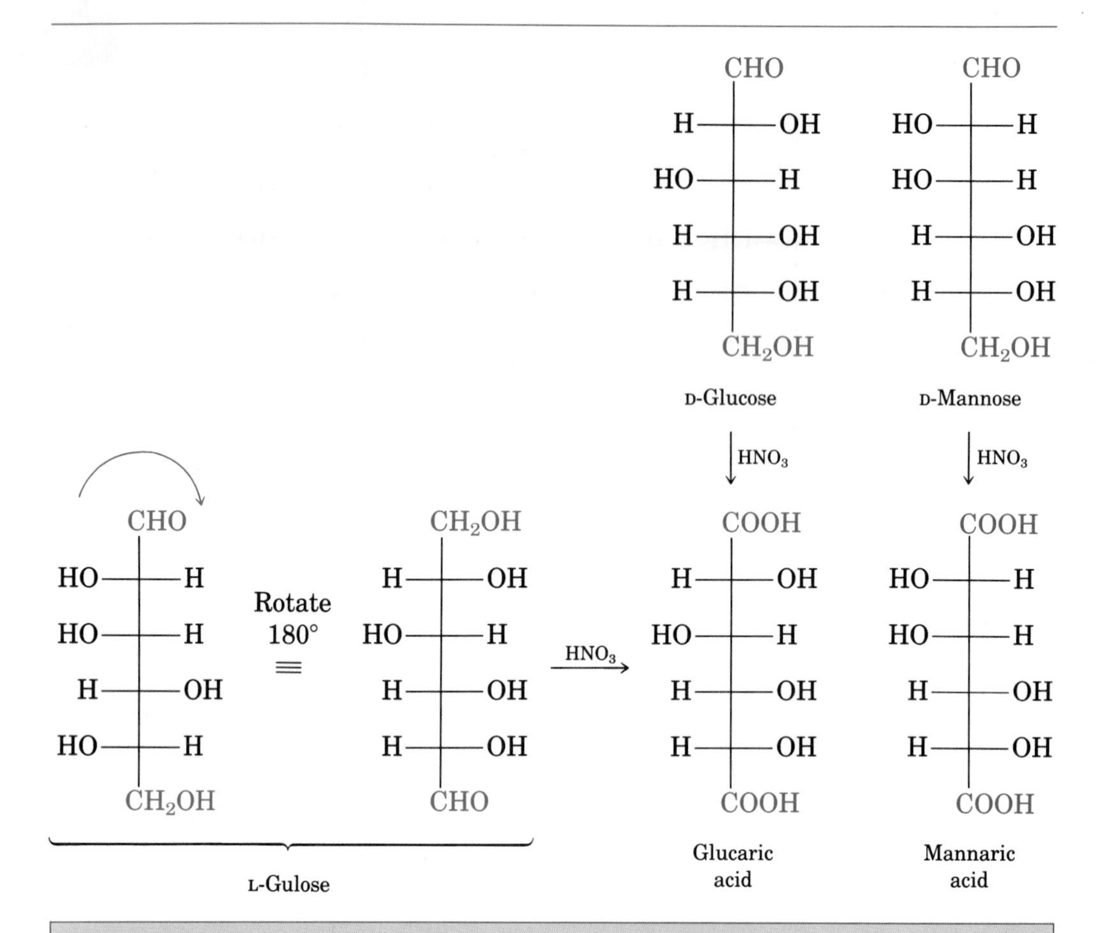

Figure 27.14. Conversion of D-glucose and L-gulose to glucaric acid.

PROBLEM

27.16 The aldotetrose D-erythrose yields a mixture of ribose and arabinose on Kiliani–Fischer chain extension.
(a) What is the structure of D-ribose?
(b) What is the structure of D-xylose, the fourth possible aldopentose?
(c) What is the structure of D-erythrose?
(d) What is the structure of D-threose, the other possible aldotetrose?

27.10 Determination of ring size

With the stereochemistry of glucose known, the only problem remaining is to determine the size of the cyclic hemiacetal ring. Is glucose a furanose or a pyranose? This problem was solved in 1926 by Haworth and Hirst by recourse to simple yet effective chemistry.

Methylation of glucose yields the pentamethyl ether derivative, and aqueous acid hydrolysis cleaves the methyl glycoside to a tetramethyl ether. In the ring-opened form of this tetramethyl ether, only the hydroxyl group that was part of the hemiacetal ring remains unmethylated.

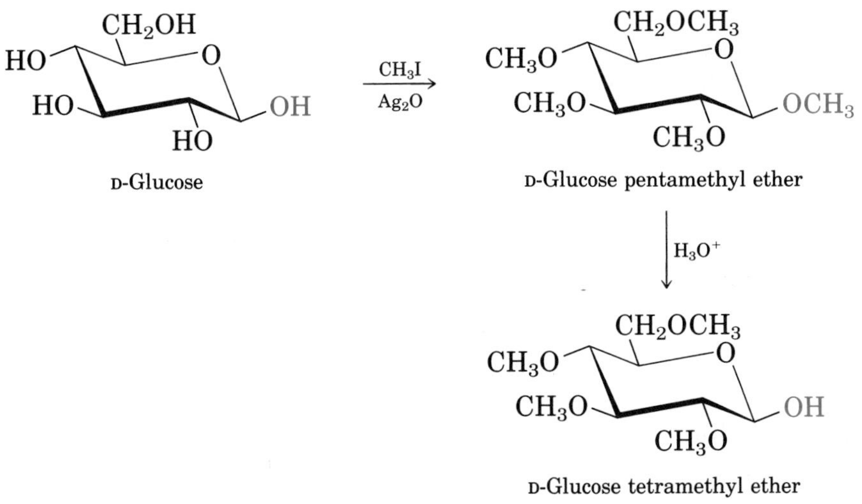

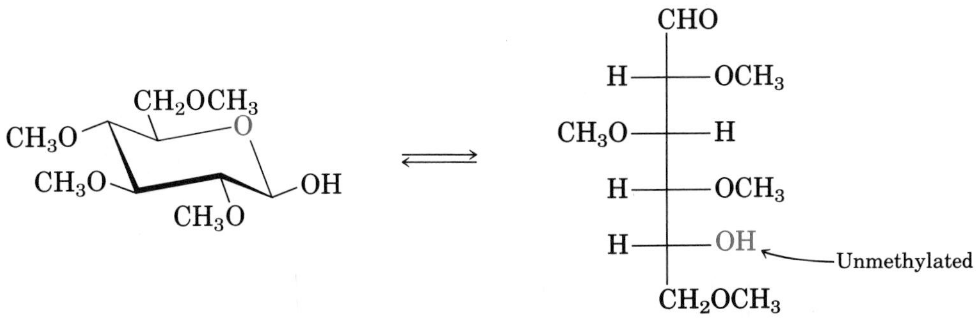

CH_2OH, HO, HO, O, OH, HO

D-Glucose

$\xrightarrow[Ag_2O]{CH_3I}$

CH_2OCH_3, CH_3O, CH_3O, O, OCH_3, CH_3O

D-Glucose pentamethyl ether

$\downarrow H_3O^+$

CH_2OCH_3, CH_3O, CH_3O, O, OH, CH_3O

D-Glucose tetramethyl ether

CH_2OCH_3, CH_3O, CH_3O, O, OH, CH_3O $\rightleftharpoons$ CHO; H—OCH_3; CH_3O—H; H—OCH_3; H—OH ← Unmethylated; CH_2OCH_3

The position of the free hydroxyl group was determined by oxidation of the tetramethyl ether. Under the conditions used, the aldehyde group at C1 and the free hydroxyl (at C5) were both oxidized, and cleavage occurred next to the ketone carbonyl group to yield a compound identified as dimethoxytartaric acid. This could happen only if the original hydroxyl group were at C5 and if glucose were therefore a pyranose. The structure of glucose was complete!

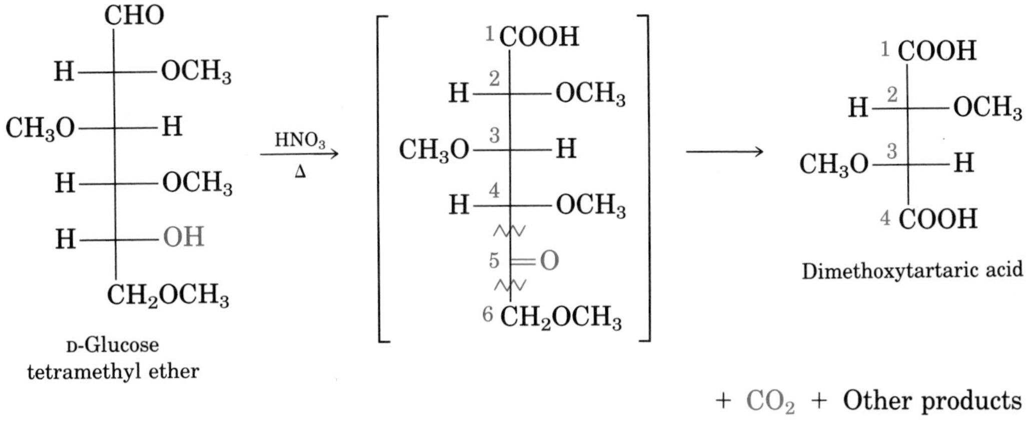

CHO; H—OCH_3; CH_3O—H; H—OCH_3; H—OH; CH_2OCH_3

D-Glucose tetramethyl ether

$\xrightarrow[\Delta]{HNO_3}$

[1 COOH; 2 H—OCH_3; 3 CH_3O—H; 4 H—OCH_3; 5 =O; 6 CH_2OCH_3]

$\longrightarrow$

1 COOH; 2 H—OCH_3; 3 CH_3O—H; 4 COOH

Dimethoxytartaric acid

+ CO_2 + Other products

PROBLEM

27.17 What product would you expect to obtain from oxidation of glucose tetramethyl ether if glucose were a furanose?

PROBLEM

27.18 Is the dimethoxytartaric acid obtained by degradation of glucose optically active or meso? Explain.

27.11 Disaccharides

We saw earlier that replacement of an anomeric hydroxyl group by an alkoxy substituent converts a monosaccharide into a glycoside. If the alkoxy substituent is itself a sugar, however, the glycosidic product is a **disaccharide.**

CELLOBIOSE AND MALTOSE

Disaccharides contain a glycosidic acetal bond between C1 of one sugar and a hydroxyl group at *any* position on the other sugar. A glycosidic bond between C1 of the first sugar and C4 of the second sugar is particularly common but is by no means required. Such a bond is called a **1,4′ link.** (The superscript, a *prime,* indicates that the 4 position is on a different sugar than the 1 position.)

A glycosidic bond can be either alpha or beta. For example, cellobiose, the disaccharide obtained by partial hydrolysis of cellulose, consists of two D-glucopyranoses joined by a 1,4′-β-glycoside bond; maltose, the disaccharide obtained by enzyme-catalyzed hydrolysis of starch, consists of two D-glucopyranoses joined by a 1,4′-α-glycoside bond.

Cellulose $\xrightarrow{H_3O^+}$

CH_2OH HO O HO 4′ CH_2OH O O HO 1 H HO OH HO H

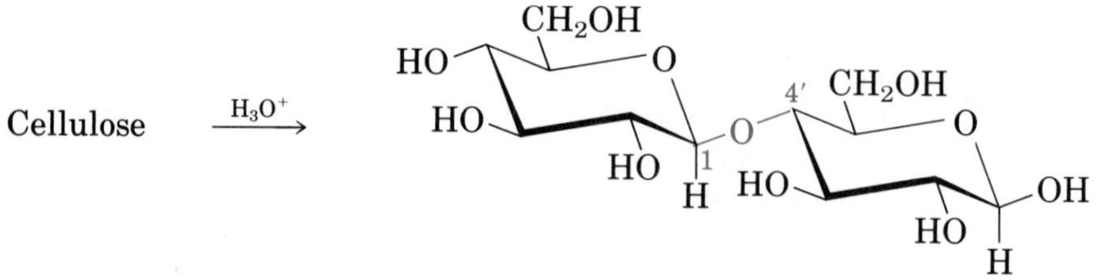

Cellobiose, a 1,4′-β-glycoside
[4-*O*-(β-D-glucopyranosyl)-β-D-glucopyranose]

Starch $\xrightarrow{\text{Enzyme}}$

CH_2OH HO O HO 1 H HO 4′ CH_2OH O O HO OH HO H

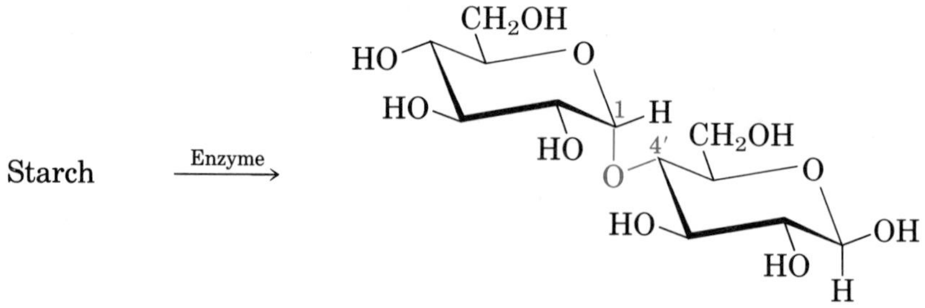

Maltose, a 1,4′-α-glycoside
[4-*O*-(α-D-glucopyranosyl)-β-D-glucopyranose]

Both maltose and cellobiose are reducing sugars because the anomeric carbons on the right-hand sugar are part of a hemiacetal. Both are therefore in equilibrium with aldehyde forms, which can reduce Tollens' reagent. For a similar reason, both maltose and cellobiose exhibit mutarotation of alpha and beta anomers of the glucopyranose unit on the right (Figure 27.15).

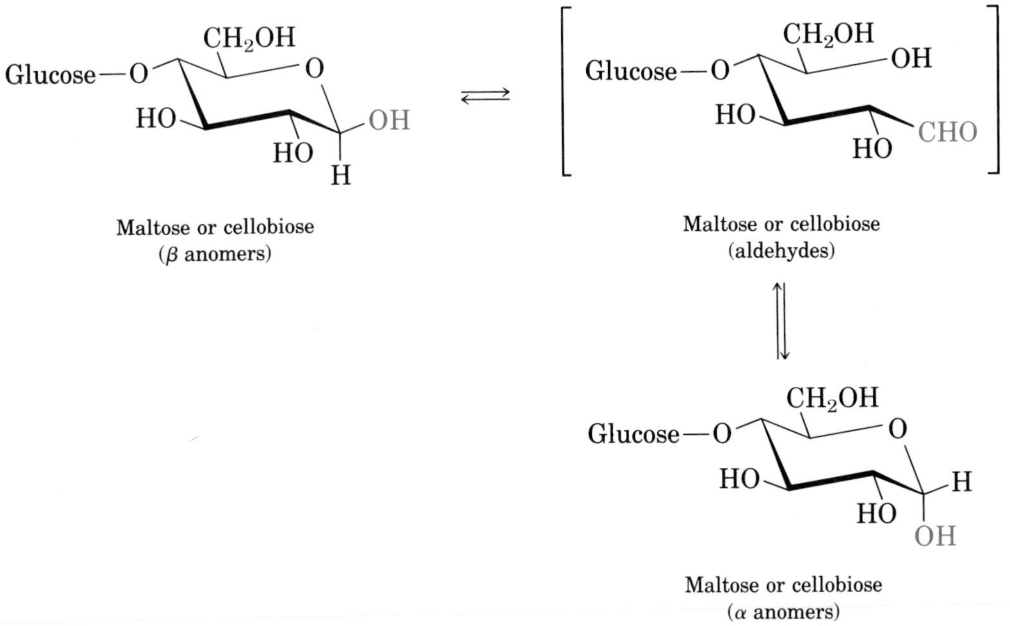

Glucose—O, CH_2OH, O, HO, HO, OH, H ⇌ [Glucose—O, CH_2OH, OH, HO, HO, CHO] ⇌ Glucose—O, CH_2OH, O, HO, HO, H, OH

Maltose or cellobiose (β anomers)

Maltose or cellobiose (aldehydes)

Maltose or cellobiose (α anomers)

Figure 27.15. Mutarotation of maltose and cellobiose.

Despite the similarities of their structures, cellobiose and maltose are dramatically different biologically. Cellobiose cannot be fermented by yeast and cannot be digested by humans. Maltose, however, is fermented readily by yeast and can be digested without difficulty.

PROBLEM

27.19 What product would you obtain from the reaction of cellobiose with aqueous bromine?

PROBLEM

27.20 The position of the glycosidic link can be determined by a modification of the method used by Haworth and Hirst to determine the ring size of glucose (Section 27.10). Reaction of cellobiose with methyl iodide and silver oxide yields an octamethyl ether derivative. Acid hydrolysis of this octamethyl ether yields a tri-*O*-methylglucopyranose and a tetra-*O*-methylglucopyranose. What are the structures of these octamethyl, trimethyl, and tetramethyl ethers? How can you use this information to determine the position of the glycoside link?

LACTOSE

Lactose is an important disaccharide that occurs naturally in both human and cow's milk. It is widely used in baking and in commercial infant-milk formulas. Like cellobiose and maltose, lactose is a reducing sugar; it exhibits mutarotation, and it is a 1,4′-β-linked glycoside. Unlike cellobiose and maltose, however, it consists of two *different* monosaccharide units. Acidic hydrolysis of lactose yields 1 equiv of D-glucose and 1 equiv of D-galactose; the two are joined by a β-glycosidic bond between C1 of galactose and C4 of glucose.

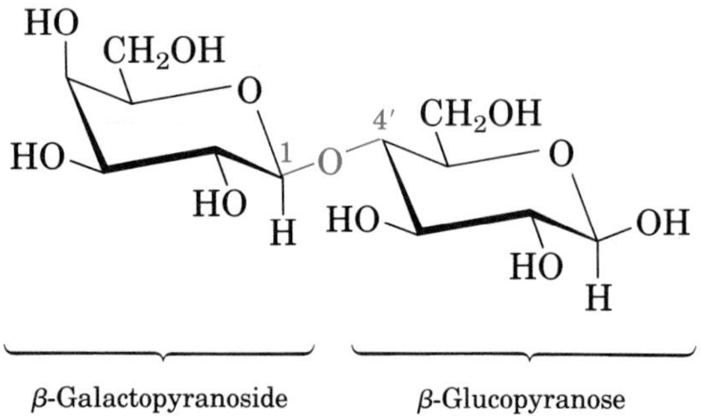

Lactose, a 1,4′-β-glycoside [4-*O*-(β-D-galactopyranosyl)-β-D-glucopyranose]

SUCROSE

Sucrose, ordinary table sugar, is the single most abundant pure organic chemical in the world and the one most widely known to nonchemists. Whether from sugarcane (20% by weight) or sugar beets (15% by weight), and whether raw or refined, common sugar is still sucrose.

Sucrose is a disaccharide that yields 1 equiv of glucose and 1 equiv of fructose on acidic hydrolysis. This 1:1 mixture of glucose and fructose is often referred to as **invert sugar,** since the sign of optical rotation changes (inverts) on going from sucrose ($[\alpha]_D = +66.5°$) to a glucose and fructose mixture ($[\alpha]_D \approx -22.0°$). Unlike most other disaccharides, however, sucrose is *not* a reducing sugar and does *not* exhibit mutarotation. This evidence implies that sucrose has no hemiacetal linkages and suggests that glucose and fructose must both be glycosides. This can happen only if the two sugars are joined by a glycoside link between C1 of glucose and C2 of fructose.

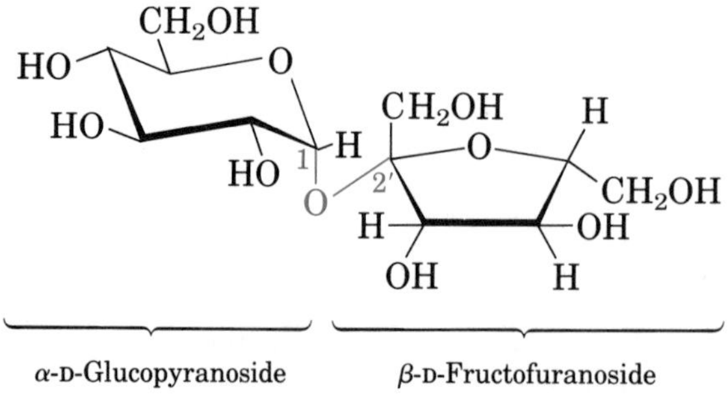

Sucrose, a 1,2′-glycoside [2-*O*-(α-D-glucopyranosyl)-β-D-fructofuranoside]

27.12 Polysaccharides

Polysaccharides are carbohydrates in which tens, hundreds, or even thousands of simple sugars are linked together through glycosidic bonds. Since they have no free anomeric hydroxyls (except for one at the end of the chain), polysaccharides are not reducing sugars and do not show mutarotation. Cellulose and starch are the two most widely occurring polysaccharides among plants.

Cellulose consists simply of D-glucose units linked by the 1,4′-β-glycoside bonds we saw in cellobiose. Several thousand glucose units can be linked to form one large molecule, and different molecules can then form a larger aggregate structure held together by hydrogen bonds.

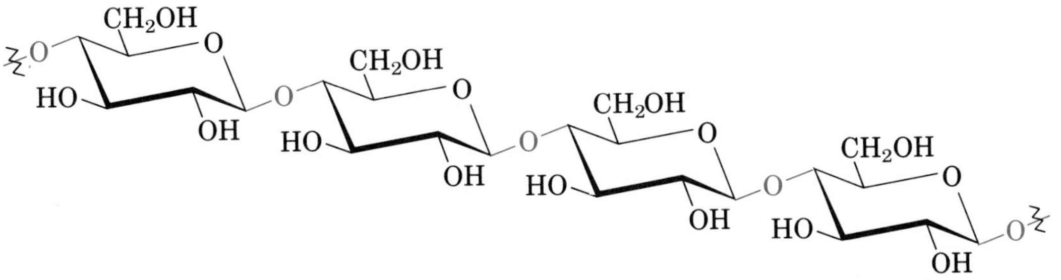

Cellulose, a 1,4′-*O*-(β-D-glucopyranoside) polymer

Starch is also a polymer of glucose, but the monosaccharide units are linked by the 1,4′-α-glycoside bonds we saw in maltose. Starch can be separated into a cold-water-soluble fraction called *amylopectin* and a cold-water-insoluble fraction called *amylose.* Amylose (20% of starch) consists of several hundred glucose molecules linked together by 1,4′-α-glycoside bonds. [Nutritionists remember the difference between cellulose and starch with the saying, "Up (β) with cellulose and down (α) with starch."]

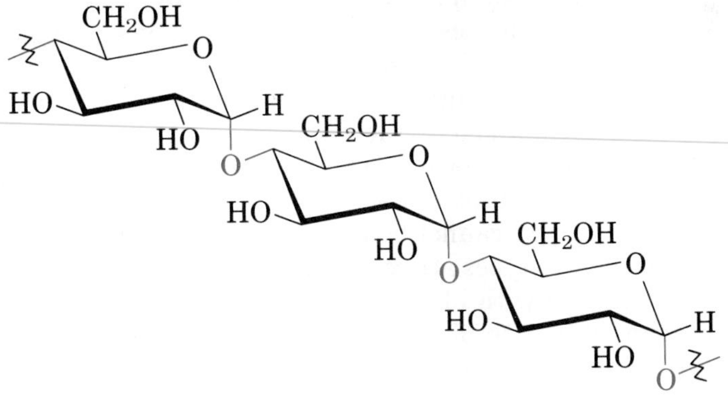

Amylose, a 1,4′-*O*-(α-D-glucopyranoside) polymer

Amylopectin accounts for the remaining 80% of starch and is more complex in structure than amylose. Unlike cellulose and amylose, which are linear or straight-chain polymers, amylopectin contains branches. These branches occur approximately every 25 glucose units and are formed by 1,6′-α-glycoside bonds (Figure 27.16). As a result, amylopectin has an exceedingly complex three-dimensional structure.

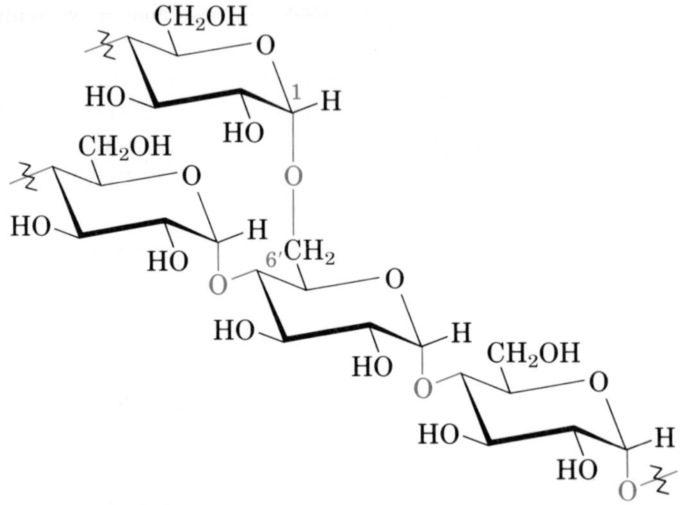

Figure 27.16. A 1,6′-α-glycoside branch in amylopectin.

27.13 Carbohydrates on cell surfaces

For many years, carbohydrates were thought to be rather dull compounds whose only biological purposes were to serve as structural materials and to be energy sources. Although carbohydrates do indeed fill these two roles, research has shown that they perform many other important biochemical functions. For example, polysaccharides are known to be centrally involved in the critical process by which one cell type recognizes another. Small polysaccharide chains, covalently bound by glycosidic links to hydroxyl groups on proteins **(glycoproteins),** act as biochemical labels on cell surfaces, as exemplified by the human blood-group antigens.

It has been known for over 80 years that human blood can be classified into four blood-group types—A, B, AB, and O—and that blood from a donor having one type cannot be transfused into a recipient having another type unless the two types are compatible. For example, blood from a type B donor is compatible with blood of either a type B or a type AB recipient, but is incompatible with blood of a type A or type O recipient. Should an incompatible mix be made, the red blood cells clump together, or *agglutinate*. This agglutination of incompatible types of red blood cells, which indicates that the body's immune system has recognized the presence of foreign cells in the body and has formed antibodies against them, results from the presence on the surface of foreign cells of polysaccharide markers. Type A, B, and O red blood cells each have characteristic markers **(antigenic determinants),** and type AB cells have both type A and type B markers. The structures of all three blood-group determinants have been elucidated and are shown in Figure 27.17.

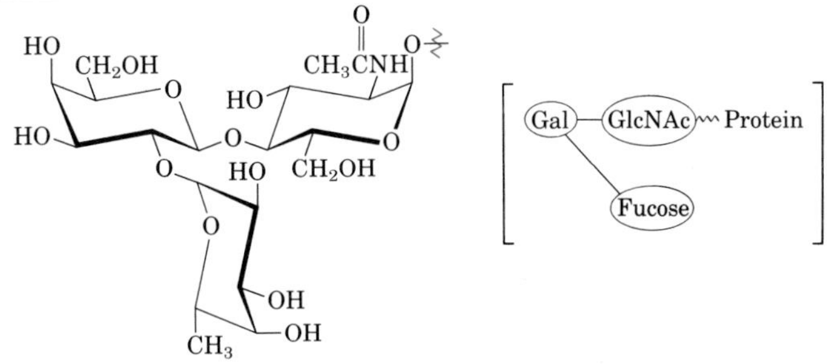

Blood group O

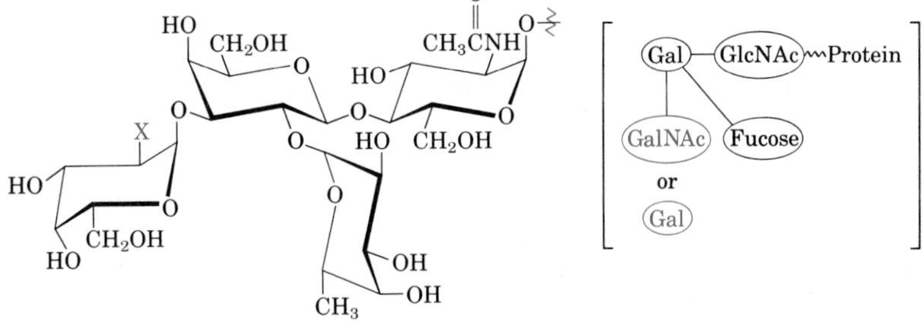

Blood group A, X = $NHCOCH_3$
Blood group B, X = OH

Figure 27.17. Structures of the A, B, and O blood-group antigenic determinants (Gal = D-galactose; GlcNAc = *N*-acetylglucosamine; GalNAc = *N*-acetylgalactosamine).

Note that some rather unusual carbohydrates are involved; thus, fucose is an L sugar, and *N*-acetylgalactosamine and *N*-acetylglucosamine are **amido sugars.**

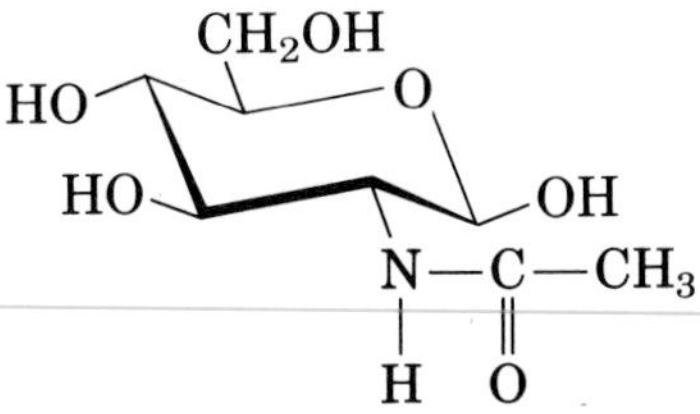

β-D-*N*-Acetylglucosamine
(D-2-acetamino-2-deoxyglucose)

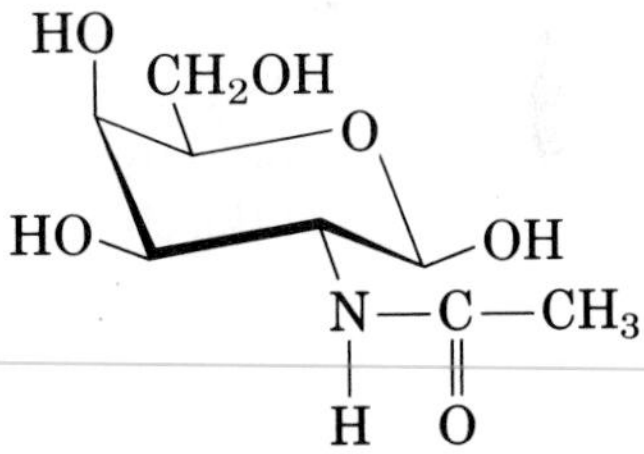

β-D-*N*-Acetylgalactosamine
(D-2-acetamino-2-deoxygalactose)

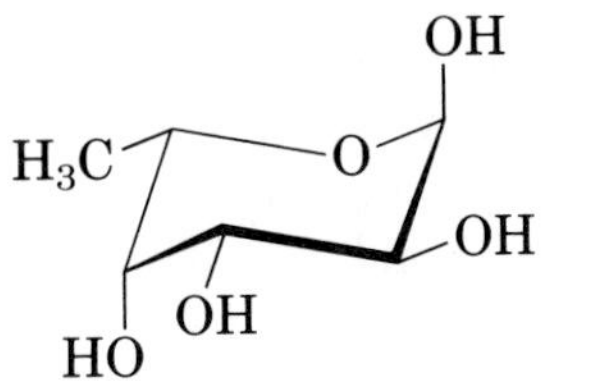

α-L-Fucose
(L-6-deoxygalactose)

The antigenic determinant of blood-group O is a trisaccharide, and the determinants of blood-groups A and B have an additional saccharide attached at C3 of the galactose unit. The type A and B determinants differ only in the substitution of an acetylamino group (—$NHCOCH_3$) for a hydroxyl in the terminal galactose residue.

Elucidation of the role of carbohydrates in cell recognition is an exciting area of current research that offers hope of breakthroughs in the understanding of a wide range of diseases from bacterial infections to cancer. All stages of this work—isolation, purification, structure determination, and chemical synthesis of the carbohydrate cell markers—are very difficult, but the ultimate rewards and benefits promise to be of great value in medicine.

27.14 Summary

Carbohydrates are polyhydroxy aldehydes and ketones. They can be classified according to the number of carbon atoms and the kind of carbonyl group they contain; thus, glucose is an aldohexose, a six-carbon aldehyde sugar. Monosaccharides are further classified as either D or L sugars, depending on the stereochemistry of the chiral carbon atom farthest from the carbonyl group. Monosaccharides normally exist as cyclic hemiacetals rather than as open-chain aldehydes or ketones. The hemiacetal linkage results from reaction of the carbonyl group with a hydroxyl group three or four carbon atoms away. A five-membered-ring hemiacetal is called a furanose and a six-membered-ring hemiacetal is called a pyranose. Cyclization leads to the formation of a new chiral center and production of two diastereomeric hemiacetals called alpha and beta anomers.

Stereochemical relationships among monosaccharides are portrayed in several ways. Fischer projections display chiral carbon atoms as a pair of crossed lines. These projections are useful in allowing us to quickly relate one sugar to another, but cyclic Haworth projections provide a more accurate view. Any group to the right in a Fischer projection is down in a Haworth projection.

Much of the chemistry of monosaccharides is familiar from our previous study of alcohols and aldehydes/ketones. Thus, the hydroxyl groups of carbohydrates form esters and ethers in the normal way. The carbonyl group of a monosaccharide can be reduced with sodium borohydride to form an alditol, oxidized with bromine water to form an aldonic acid, oxidized with warm nitric acid to form an aldaric acid, or treated with an alcohol in the presence of acid to form a glycoside. Monosaccharides can also be chain-lengthened by the multistep Kiliani–Fischer synthesis, and can be chain-shortened by the Wohl degradation.

Disaccharides are complex carbohydrates in which two simple sugars are linked by a glycoside bond between the anomeric carbon of one unit and a hydroxyl of the second unit. The two sugars can be the same, as in maltose and cellobiose, or different, as in lactose and sucrose. The glycosidic bond can be either α (maltose) or β (cellobiose, lactose), and can involve any hydroxyl of the second sugar. A 1,4′ link is most common (cellobiose, maltose), but others such as 1,6′ (lactose) and 1,2′ (sucrose) are also known.

27.15 Summary of reactions

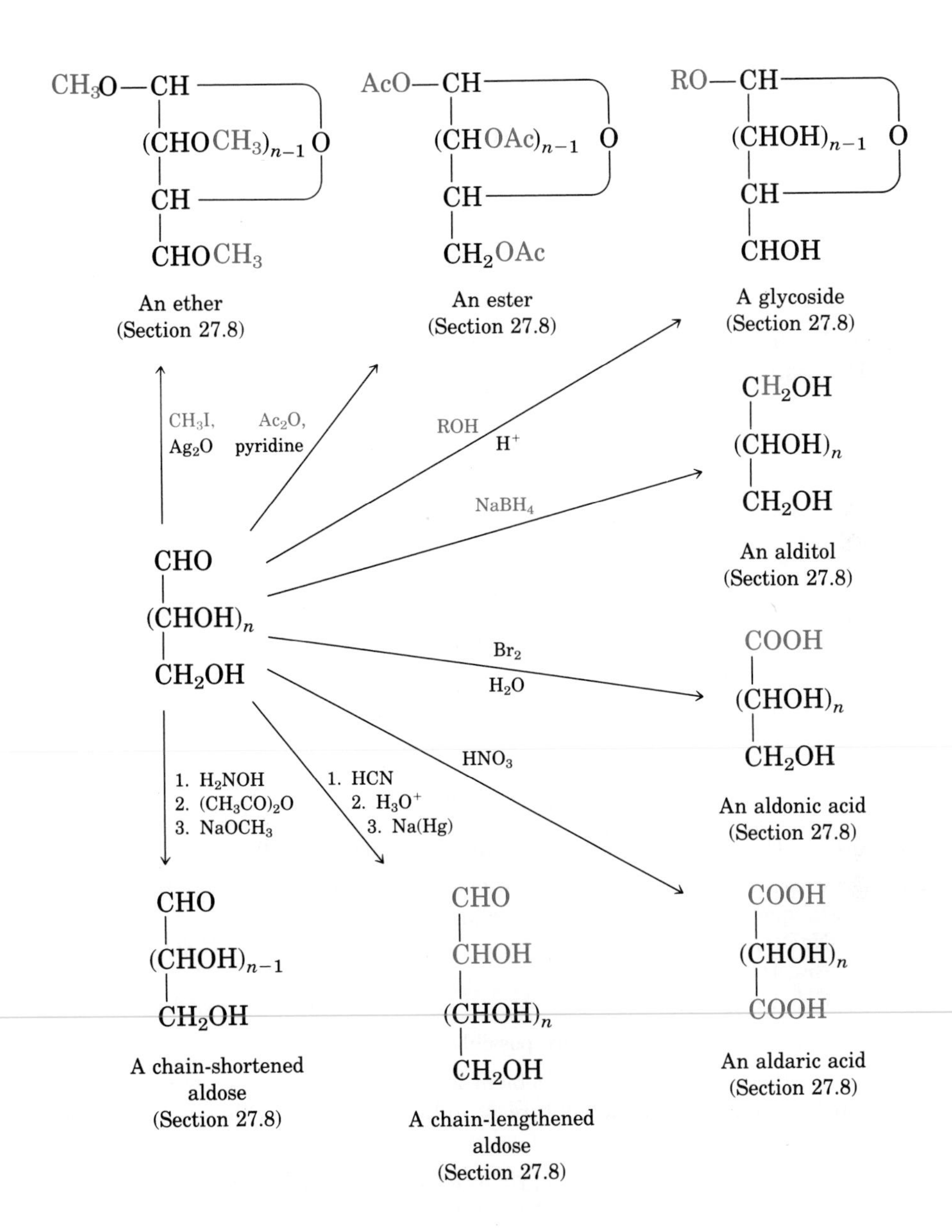

ADDITIONAL PROBLEMS

27.21 Classify the following sugars by type; for example, glucose is an aldohexose.

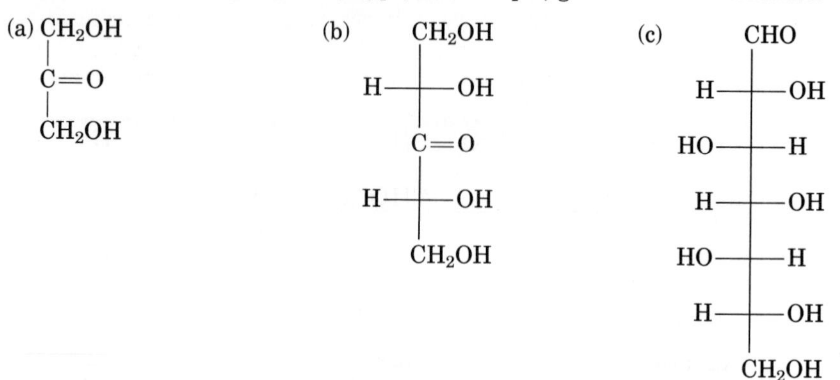

27.22 Assign R or S configuration to each chiral carbon atom in ascorbic acid (vitamin C). Does ascorbic acid have a D or an L configuration?

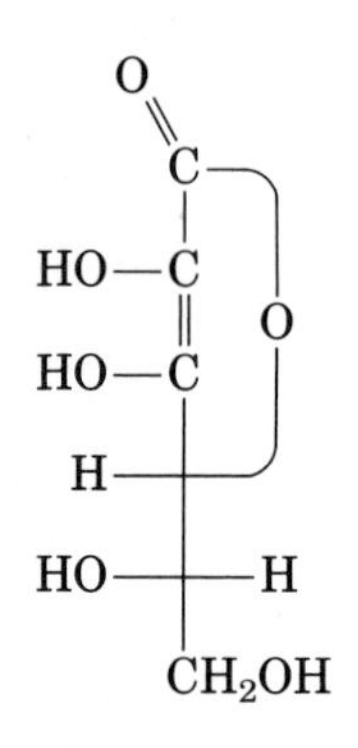

Ascorbic acid

27.23 Draw a Haworth projection of ascorbic acid (Problem 27.22).

27.24 Draw structures for the products you would expect to obtain from reaction of β-D-talopyranose with each of the following reagents:

(a) $NaBH_4$ in H_2O
(b) Warm dilute HNO_3
(c) Br_2, H_2O
(d) CH_3CH_2OH, H^+
(e) CH_3I, Ag_2O
(f) $(CH_3CO)_2O$, pyridine

27.25 How many D-2-ketohexoses are possible? Draw them.

27.26 One of the D-2-ketohexoses (Problem 27.25) is called *sorbose*. On treatment with $NaBH_4$, sorbose yields a mixture of gulitol and iditol. What is the structure of sorbose?

27.27 Another D-2-ketohexose, *psicose*, yields a mixture of allitol and altritol when reduced with $NaBH_4$. What is the structure of psicose?

27.28 Fischer prepared the gulose needed for his structure proof of glucose in the following way. D-Glucose was oxidized to D-glucaric acid, which can form two six-membered-ring lactones. These were separated and reduced with sodium amalgam to give D-glucose and L-gulose. What are the structures of the two lactones, and which one is reduced to L-gulose?

27.29 What other D-aldohexose would give the same alditol as D-talose?

27.30 Which of the eight D-aldohexoses give the same aldaric acids as their L enantiomers?

27.31 Which of the other three D-aldopentoses gives the same aldaric acid as D-lyxose?

27.32 Gentiobiose is a rare disaccharide found in saffron and gentian. It is a reducing sugar and forms only glucose on hydrolysis with aqueous acid. Reaction of gentiobiose with methyl iodide and silver iodide yields an octamethyl derivative, which can be hydrolyzed with aqueous acid to give 1 equiv of 2,3,4,6-tetra-*O*-methyl-D-glucopyranose and 1 equiv of 2,3,4-tri-*O*-methyl-D-glucopyranose. If gentiobiose contains a β-glycoside link, what is its structure?

27.33 Amygdalin, or Laetrile, is a glycoside isolated in 1830 from almond and apricot seeds. It is known as a cyanogenic glycoside since acidic hydrolysis liberates HCN, along with benzaldehyde and 2 equiv of glucose. Structural studies have shown amygdalin to be a β-glycoside of benzaldehyde cyanohydrin with gentiobiose (Problem 27.32). Draw the structure of amygdalin.

27.34 Trehalose is a nonreducing disaccharide that is hydrolyzed by aqueous acid to 2 equiv of D-glucose. Methylation followed by acidic hydrolysis yields 2 equiv of 2,3,4,6-tetra-*O*-methylglucose. How many possible structures are there for trehalose? Trehalose is cleaved by enzymes that hydrolyze α-glycosides but not by enzymes that cleave β-glycosides. What is the structure and systematic name of trehalose?

27.35 Isotrehalose and neotrehalose are chemically similar to trehalose (Problem 27.34) except for the fact that neotrehalose is hydrolyzed only by β-glycosidases, whereas isotrehalose is hydrolyzed by both α- and β-glycosidases. What are the structures of isotrehalose and neotrehalose?

27.36 Propose a scheme for the synthesis of gentiobiose methyl glycoside (Problem 27.32), starting from β-D-glucose and methyl 2,3,4-tri-*O*-acetyl-β-D-glucopyranoside.

27.37 D-Glucose reacts with acetone in the presence of acid to yield the nonreducing 1,2:5,6-diisopropylidene-D-glucofuranose. Propose a mechanism for this reaction.

$$\text{D-Glucose} + CH_3COCH_3 \xrightarrow[-H_2O]{H^+} \text{1,2:5,6-Diisopropylidene-D-glucofuranose}$$

1,2:5,6-Diisopropylidene-D-glucofuranose

27.38 D-Mannose reacts with acetone to give a diisopropylidene derivative that is still reducing toward Tollens' reagent. Propose a likely structure for this derivative.

27.39 Propose a mechanism to account for the fact that D-gluconic acid and D-mannonic acid are interconverted when either is heated in pyridine solvent.

27.40 The cyclitols are a group of carbocyclic sugar derivatives having the general formulation 1,2,3,4,5,6-cyclohexanehexol. How many stereoisomeric cyclitols are possible? Draw them in Haworth projection.

27.41 Compound A is a D-aldopentose that can be oxidized to an optically inactive aldaric acid, B. On Kiliani–Fischer chain extension, A is converted into C and D; C can be oxidized to an optically active aldaric acid E, but D is oxidized to an optically inactive aldaric acid, F. What are the structures of A–F?

CHAPTER 28
ALIPHATIC AMINES

Amines are organic derivatives of ammonia in the same way that alcohols and ethers are organic derivatives of water. Amines are classified as either **primary (RNH_2), secondary (R_2NH),** or **tertiary (R_3N),** depending on the degree of substitution at nitrogen. For example, methylamine (CH_3NH_2) is a primary amine and trimethylamine [$(CH_3)_3N$] is a tertiary amine. Note that this usage of the terms primary, secondary, and tertiary is different from our previous usage. When we speak of a tertiary alcohol or alkyl halide, we refer to the degree of substitution at the alkyl carbon atom; when we speak of a tertiary amine, however, we refer to the degree of substitution at the nitrogen atom.

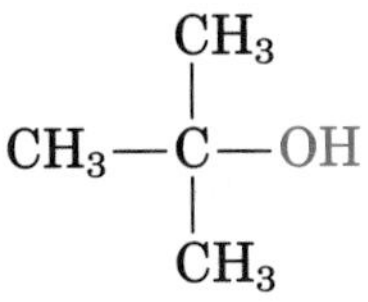

$$(CH_3)_3C{-}OH \qquad (CH_3)_3N \qquad (CH_3)_3C{-}NH_2$$

tert-Butyl alcohol (a tertiary alcohol) — Trimethylamine (a tertiary amine) — *tert*-Butylamine (a primary amine)

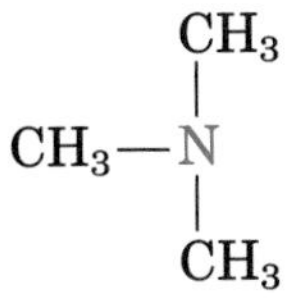

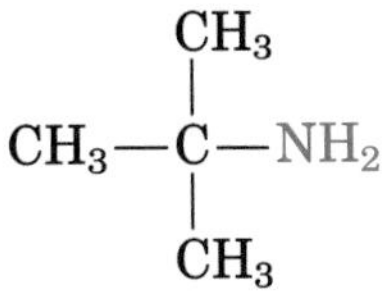

Compounds with four groups attached to nitrogen are also known, but the nitrogen atom must carry a positive charge. Such compounds are called **quaternary ammonium salts.**

$$R_4N^+ \; X^-$$

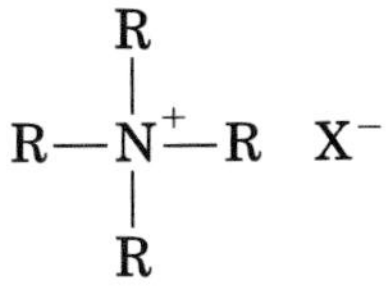

A quaternary ammonium salt

Amines can be either alkyl-substituted or aryl-substituted. Much of the chemistry of the two classes is similar, but there are sufficient differences that we will consider the classes separately. Arylamines will therefore be discussed in Chapter 29.

$$CH_3CH_2\ddot{N}H_2 \qquad C_6H_5{-}\ddot{N}H_2 \qquad C_6H_5{-}CH_2\ddot{N}H_2$$

Ethylamine (an aliphatic amine) — Aniline (an arylamine) — Benzylamine (an aliphatic amine)

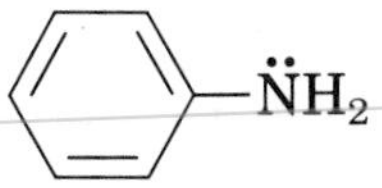

PROBLEM

28.1 Classify the following compounds (a)–(e) as either primary, secondary, or tertiary amines, or as quaternary ammonium salts.

(a) Cyclohexyl$-NH_2$

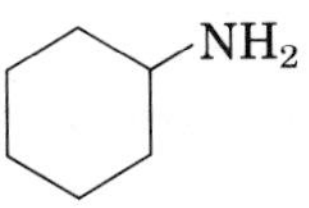

(b) Cyclohexyl$-N(CH_3){-}CH_3$

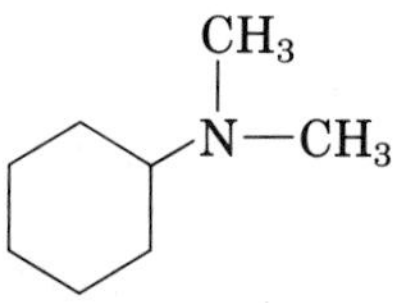

(c) $C_6H_5CH_2\overset{+}{N}(CH_3)_3\ {}^{-}I$

(d) $[(CH_3)_2CH]_2NH$

(e) Pyrrolidine ring with N—H

28.1 Nomenclature of amines

Primary amines, $R—NH_2$, are named in the IUPAC system in either of two ways. For simple amines, the suffix *-amine* is added to the name of the alkyl substituent.

$(CH_3)_3C—NH_2$ — *tert*-Butylamine

$C_6H_{11}—NH_2$ — Cyclohexylamine

$H_2N—CH_2CH_2CH_2CH_2—NH_2$ — 1,4-Butanediamine

Amines having more than one functional group are named by considering the $—NH_2$ as an amino substituent on a parent molecule.

$\overset{4}{C}H_3\overset{3}{C}H_2\overset{2}{C}H(NH_2)\overset{1}{C}OOH$ — 2-Aminobutanoic acid

$C_6H_3(COOH)(NH_2)_2$ — 2,4-Diaminobenzoic acid

$H_2N—\overset{4}{C}H_2\overset{3}{C}H_2\overset{2}{C}(=O)\overset{1}{C}H_3$ — 4-Amino-2-butanone

Symmetrical secondary and tertiary amines are named by adding the prefix *di-* or *tri-* to the alkyl group.

$(C_6H_5)_2N—H$ — Diphenylamine

$CH_3CH_2—N(CH_2CH_3)—CH_2CH_3$ — Triethylamine

Unsymmetrically substituted secondary and tertiary amines are named as *N*-substituted primary amines. The largest alkyl group is chosen as the parent name, and the other alkyl groups are considered *N*-substituents on the parent.

$CH_3-N(CH_3)-CH_2CH_2CH_3$

N,N-Dimethylpropylamine
(propylamine is the parent name; the two methyl groups are substituents on nitrogen)

$H_3C-N(CH_2CH_3)-C_6H_{11}$

N-Ethyl-*N*-methylcyclohexylamine
(cyclohexylamine is the parent name; methyl and ethyl are *N*-substituents)

There are relatively few trivial names for simple amines, but IUPAC rules do recognize the names *aniline* and *toluidine* for aminobenzene and aminotoluene, respectively.

Aniline

m-Toluidine

Heterocyclic amines—compounds in which the nitrogen atom occurs as part of a ring—are common, and each different heterocyclic ring system is given its own parent name. Some of the more common ones are shown. In all cases, the nitrogen atom is numbered as position 1.

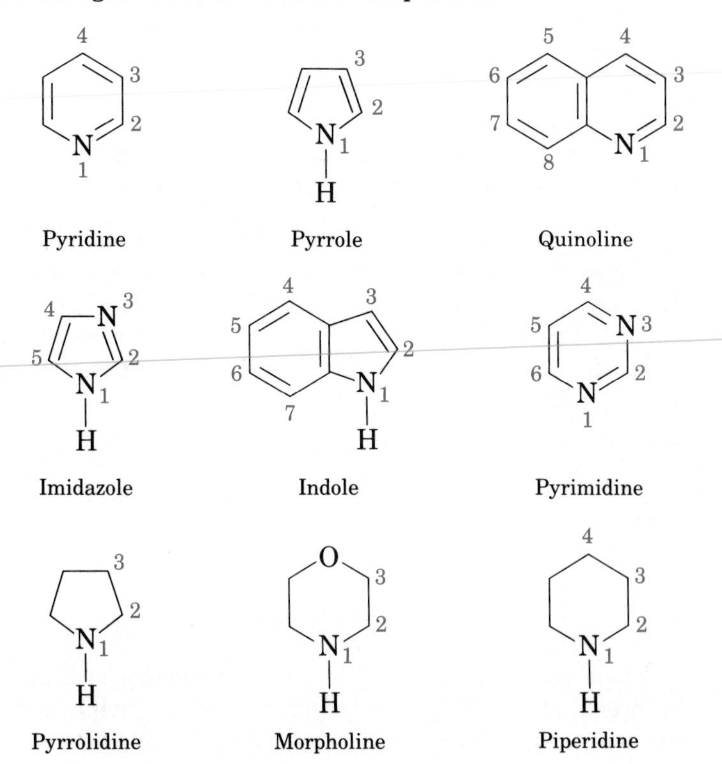

Pyridine

Pyrrole

Quinoline

Imidazole

Indole

Pyrimidine

Pyrrolidine

Morpholine

Piperidine

PROBLEM

28.2 Name the following compounds by IUPAC rules:

(a) $CH_3NHCH_2CH_3$

(b) 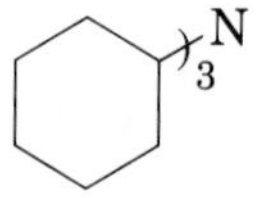 $(C_6H_{11})_3N$ (tricyclohexyl structure)

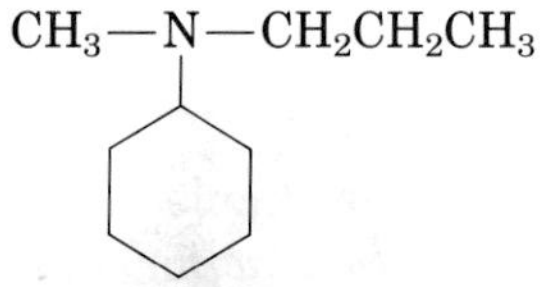

(c) $CH_3—N(C_6H_{11})—CH_2CH_2CH_3$

(d) 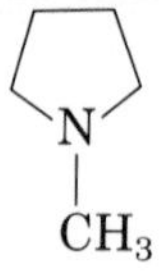 N-methylpyrrolidine ring structure with N—CH_3

(e) $[(CH_3)_2CH]_2NH$

(f) $H_2NCH_2CH_2CH(CH_3)NH_2$

(g) 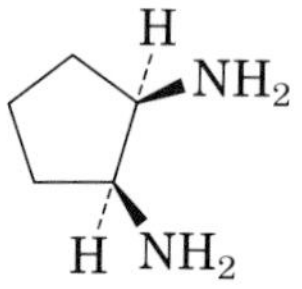 cyclopentane with H, NH_2 and H, NH_2 on adjacent carbons

PROBLEM

28.3 Draw structures corresponding to these IUPAC names:
(a) Triethylamine
(b) Triallylamine
(c) *N*-Methylaniline
(d) *N*-Ethyl-*N*-methylcyclopentylamine
(e) *N*-Isopropylcyclohexylamine
(f) *N*-Ethylpyrrole

28.2 Structure and bonding in amines

Bonding in amines is similar to bonding in ammonia—the nitrogen atom is sp^3 hybridized, and the three substituents occupy three corners of a tetrahedron; the nitrogen's nonbonding lone pair of electrons occupy the fourth corner. The bond angles are very close to the expected 109° tetrahedral value; for trimethylamine, the C—N—C angle is 108°, and the C—N bond is 1.47Å long:

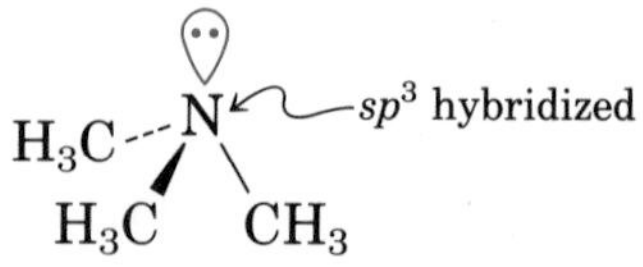

H_3C, H_3C, CH_3 on N — sp^3 hybridized

Trimethylamine

One consequence of tetrahedral geometry is that amines with three different substituents on nitrogen are chiral; such an amine has no plane of symmetry and therefore is not superimposable on its mirror image. If we consider the lone pair of electrons to be the fourth substituent on nitrogen, these chiral amines are analogous to chiral alkanes with four different substituents attached to carbon:

(W) W
X N Z Y X C Z Y

A chiral amine A chiral alkane

Unlike chiral alkanes, however, most chiral amines cannot be resolved into their two enantiomers, because the two enantiomeric forms rapidly interconvert by a **pyramidal inversion,** much as an umbrella inverts in a strong wind or an alkyl halide inverts in an S_N2 reaction. Pyramidal inversion presumably occurs by a momentary rehybridization of the nitrogen atom to a planar, sp^2 geometry, followed by rehybridization of the planar intermediate to a tetrahedral, sp^3 geometry (Figure 28.1).

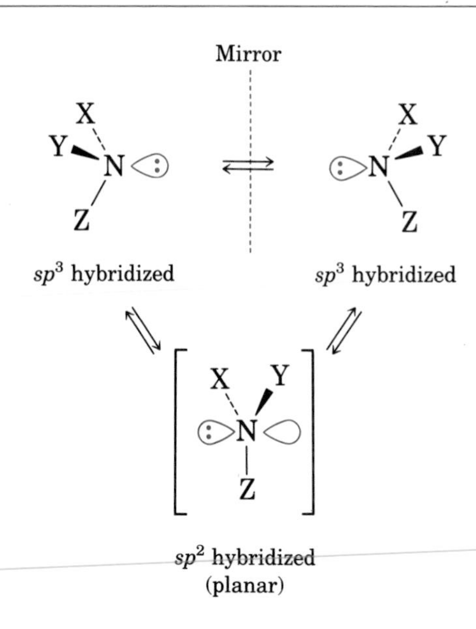

Figure 28.1. Pyramidal inversion of amines.

Spectroscopic studies have shown that the barrier to nitrogen inversion is about 6 kcal/mol, a figure that is only twice as large as the barrier to rotation about a carbon–carbon single bond. Pyramidal inversion is therefore rapid at room temperature, and the two optically active forms cannot normally be isolated.

PROBLEM

28.4 Although the barrier to nitrogen inversion is normally low enough to prevent synthesis of one pure enantiomeric form of a chiral amine, there are exceptions. One such exception is (+)-1-chloro-2,2-diphenylaziridine, which has been prepared optically pure and has been shown to be stable for several hours at 0°C. Propose an explanation of this unusual stability.

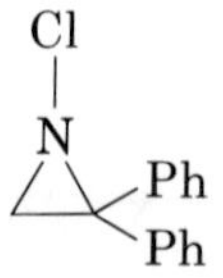

Cl N Ph Ph

1-Chloro-2,2-diphenylaziridine

28.3 Physical properties of amines

Amines are highly polar and therefore have higher boiling points than alkanes of equivalent molecular weight. Like alcohols, amines with fewer than five carbon atoms are generally water soluble; also like alcohols, primary and secondary amines form strong hydrogen bonds and are highly associated in the liquid state.

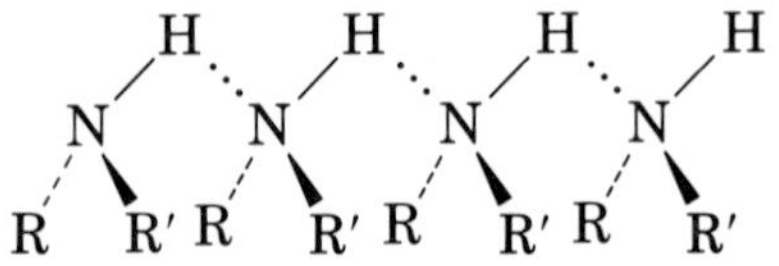

H N R R′ H N R R′ H N R R′ H N R R′

The physical properties of some simple amines are given in Table 28.1, but one property that does not show up in the table is *odor*. All low-molecular-weight amines have a characteristic and very distinctive fish-like aroma. Diamines such as putrescine (1,4-butanediamine) have names that are self-explanatory.

28.4 Amine basicity

The chemistry of amines is dominated by a single feature of their structure—the nitrogen lone pair of electrons. Because of the nitrogen lone pair, amines are both basic and nucleophilic. Amines react with Lewis acids to form salts, and they react with electrophiles in many of the polar reactions we have already studied.

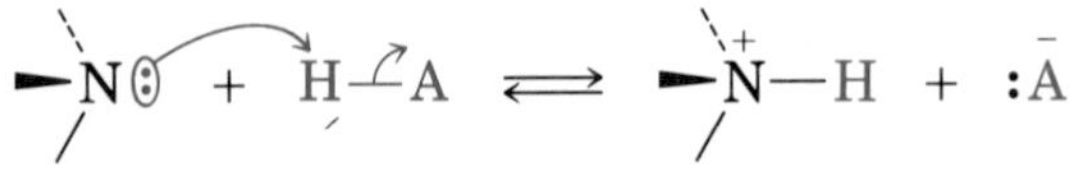

$$\text{—N:} + \text{H—A} \rightleftharpoons \text{—}\overset{+}{\text{N}}\text{—H} + :\bar{\text{A}}$$

An amine (a Lewis base) | An acid | A salt

Amines are much more basic than alcohols, ethers, or water. When an amine is dissolved in water, an equilibrium is established in which water acts as a protic acid and donates a proton to the amine.

TABLE 28.1 **Physical properties of some simple amines**

Name	*Structure*	*Melting point (°C)*	*Boiling point (°C)*
Ammonia	NH_3	−77.7	−33.3
Primary amines			
Methylamine	CH_3NH_2	−94	−6.3
Ethylamine	$CH_3CH_2NH_2$	−81	16.6
tert-Butylamine	$(CH_3)_3CNH_2$	−67.5	44.4
Aniline (an arylamine)	C_6H_5—NH_2	−6.3	184.1
Secondary amines			
Dimethylamine	$(CH_3)_2NH$	−93	7.4
Diethylamine	$(CH_3CH_2)_2NH$	−48	56.3
Diisopropylamine	$[(CH_3)_2CH]_2NH$	−61	84
Pyrrolidine	(five-membered ring) NH	2	89
Tertiary amines			
Trimethylamine	$(CH_3)_3N$	−117	3
Triethylamine	$(CH_3CH_2)_3N$	−114	89.3
N-Methylpyrrolidine	(five-membered ring) N—CH_3	−21	81

In discussing carboxylic acids (Section 23.3), we measured the ability of an acid to give up a proton to water, and we were able to establish a relative ordering of acid strengths expressed in terms of acidity constants, K_a. In the same way, we can measure the ability of an amine to accept a proton from water, and we can establish a relative ordering of base strengths expressed in terms of **basicity constants, K_b.**

$$R\text{—}\ddot{N}H_2 + H_2O \rightleftharpoons R\text{—}\overset{+}{N}H_3 + :\overset{-}{O}H$$

$$K_b = \frac{[R\text{—}\overset{+}{N}H_3][\overset{-}{O}H]}{[RNH_2]}$$

$$pK_b = -\log K_b$$

If the base is strong, the equilibrium is shifted toward the right; K_b is therefore larger and pK_b is smaller. Conversely, if the base is weak, K_b is smaller and pK_b is larger. Table 28.2 lists the measured pK_b's of some common amines and indicates that there is relatively little effect of substitution on alkylamine basicity. Most simple alkylamines have pK_b's in the narrow range 3–4, regardless of their substitution pattern.

TABLE 28.2 **Basicity of some common alkylamines**

Name	*Structure*	pK_b
Ammonia	$:NH_3$	4.74
Primary alkylamine		
Methylamine	$CH_3\ddot{N}H_2$	3.36
Ethylamine	$CH_3CH_2\ddot{N}H_2$	3.25
Secondary alkylamine		
Dimethylamine	$(CH_3)_2\ddot{N}H$	3.27
Diethylamine	$(CH_3CH_2)_2\ddot{N}H$	3.06
Pyrrolidine	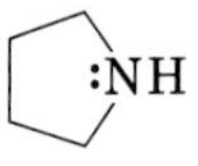 (five-membered ring) :NH	2.73
Tertiary alkylamine		
Trimethylamine	$(CH_3)_3N:$	4.21
Triethylamine	$(CH_3CH_2)_3N:$	3.25

In contrast to amines, however, **amides ($RCONH_2$)** are nonbasic. Amides do not form salts when treated with aqueous acids; their aqueous solutions are neutral; and they are very poor nucleophiles. There are two main reasons for the difference in basicity between amines and amides. First, the ground state of an amide is stabilized by delocalization of the nitrogen lone-pair electrons by the carbonyl group. In resonance terms, we can draw two contributing forms:

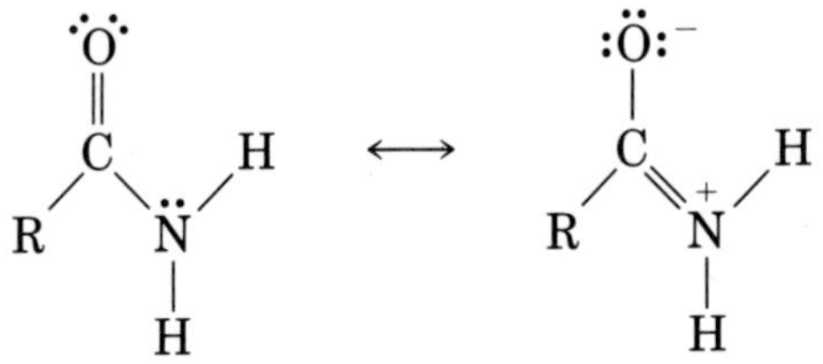

$$R-C(=\ddot{O}:)-\ddot{N}H_2 \longleftrightarrow R-C(-\ddot{O}:^-)=\overset{+}{N}H_2$$

This amide resonance stabilization is lost in the protonated product. Second, a protonated amide is higher in energy than a protonated amine because the electron-withdrawing carbonyl group inductively destabilizes the neighboring positive charge.

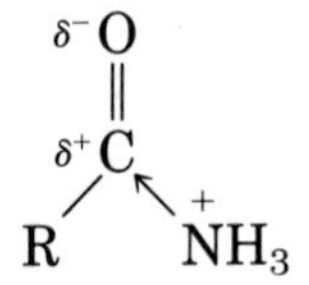

$$R-\overset{\delta^+}{C}(=\overset{\delta^-}{O})\leftarrow\overset{+}{N}H_3$$

Protonated amide (no resonance stabilization; inductive destabilization of positive charge)

Both factors—increased stability of an amide versus an amine, and decreased stability of a protonated amide versus a protonated amine—lead

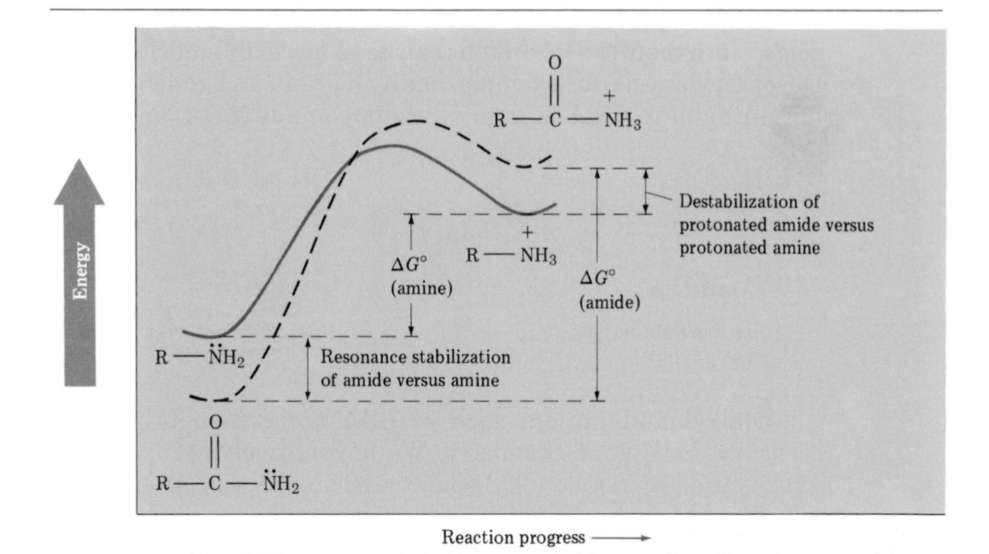

Figure 28.2. Reaction energy diagram for protonation reaction of amides and amines. The $\Delta G°$ for amide protonation is larger because of increased stability of amide reactant and decreased stability of protonated product.

to a large difference in $\Delta G°$ and consequently a large difference in basicity for amines and amides. Figure 28.2 shows these relationships on a reaction energy diagram.

We can often take advantage of the basicity of amines to purify them. For example, if we have a mixture of a basic amine and a neutral compound such as a ketone, alcohol, or ether, we can simply dissolve the mixture in an organic solvent such as ether and extract with aqueous acid. The basic amine dissolves in the water layer as its protonated salt, and the neutral compound remains in the organic solvent layer. Separation, basification, and extraction of the aqueous layer with organic solvent then allow us to recover pure amine (Figure 28.3).

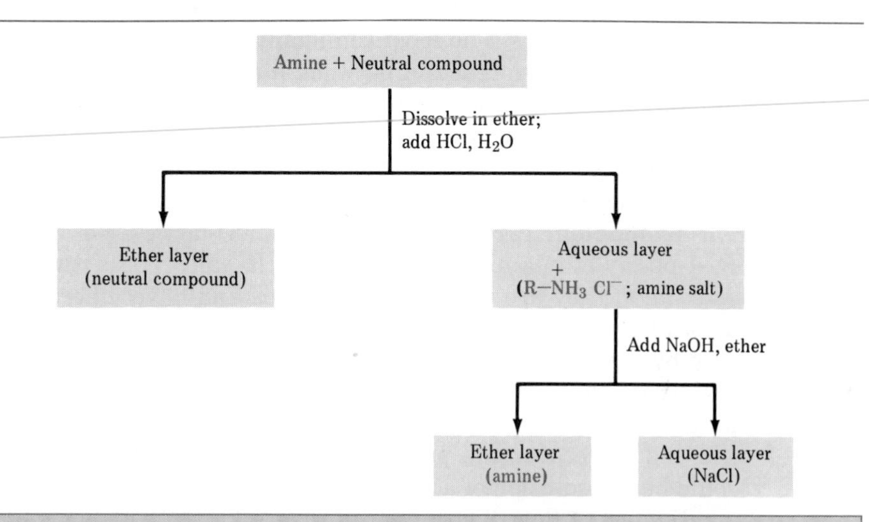

Figure 28.3. Purification of an amine component from a mixture.

Primary and secondary amines can also be considered extremely weak *acids*, since their N—H protons can be removed by sufficiently strong bases. Diisopropylamine, for example, has a $pK_a \approx 40$ and undergoes reaction with butyllithium to yield lithium diisopropylamide (LDA) and butane.

$$[(CH_3)_2CH]_2\ddot{N}{-}H + n\text{-}C_4H_9Li \longrightarrow [(CH_3)_2CH]_2\ddot{N}{:}^-\ Li^+ + n\text{-}C_4H_{10}$$

Diisopropylamine ($pK_a \approx 40$) — Lithium diisopropylamide (LDA) — Butane ($pK_a \approx 50$)

Dialkylamide anions such as LDA are extremely powerful bases of great use in organic chemistry. We have already seen, for example, how LDA is used in ketone alkylation reactions (Section 25.9).

PROBLEM

28.5 Protonation of an amide actually occurs on oxygen rather than on nitrogen. Can you suggest a reason for this behavior?

$$R{-}\overset{\overset{:O:}{\|}}{C}{-}\ddot{N}H_2 \xrightleftharpoons{H_2SO_4} R{-}\overset{\overset{+\ddot{O}-H}{\|}}{C}{-}\ddot{N}H_2 + HSO_4^-$$

28.5 Resolution of enantiomers via amine salts

We saw in the previous section that we can take advantage of the basic properties of an amine to carry out its purification. We can also take advantage of amine basicity in another important way, to carry out the **resolution** of a racemic carboxylic acid into its two pure enantiomers. Historically, Louis Pasteur was the first person to resolve a racemic mixture when he was able to crystallize a salt of (±)-tartaric acid and to separate two different kinds of crystals by hand (Section 8.3). Pasteur's method is not generally applicable, however, since few racemic compounds crystallize in mirror-image forms. The most commonly used method of resolution makes use of an acid–base reaction between a chiral carboxylic acid and a chiral amine. For example, let's see what would happen if a racemic mixture of (+)- and (−)-lactic acids were to react with a single enantiomer of a chiral amine base such as *(R)*-1-phenylethylamine (Figure 28.4).

(R)-Lactic acid would react with *(R)*-1-phenylethylamine, giving the *R,R* ammonium carboxylate salt, and *(S)*-lactic acid would react with *(R)*-1-phenylethylamine, giving the *S,R* salt. These two salts are *diastereomers;* they are different compounds and have different chemical and physical properties. It may therefore prove possible to separate them physically by fractional crystallization or by some other laboratory technique. Once they are separated, acidification of the two diastereomeric salts with mineral acid would allow us to isolate the two pure enantiomers of lactic acid and to recover the pure chiral amine for further use. The net effect would be a resolution of a racemic carboxylic acid mixture into its two pure enantiomers. A flow diagram for the overall process is shown in Figure 28.5.

(R) COOH, H—C—OH, CH_3

+

(S) COOH, HO—C—H, CH_3

Racemic lactic acid
(50% *R*, 50% *S*)

NH_2, H—C—CH_3, phenyl
(*R*)-1-phenylethylamine →

COO^-, H—C—OH, CH_3 ; $^+NH_3$, H—C—CH_3, phenyl

An *R*,*R* salt

+

COO^-, OH—C—H, CH_3 ; $^+NH_3$, H—C—CH_3, phenyl

An *S*,*R* salt

Figure 28.4. Reaction of racemic lactic acid with optically pure (*R*)-1-phenylethylamine. A mixture of diastereomers is produced.

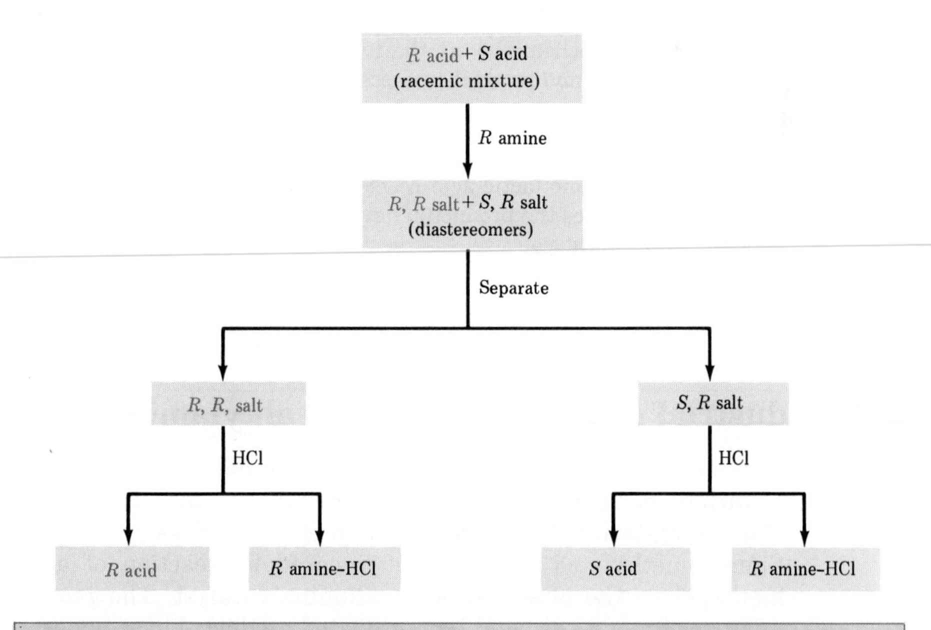

Figure 28.5. Flow diagram for resolution of a racemic carboxylic acid.

This reaction between both enantiomers of a chiral acid and one enantiomer of a chiral base to yield diastereomeric salts is one example of a general rule: A reaction between two chiral partners always yields chiral products; reaction between a racemic mixture and another chiral partner always yields a mixture of diastereomeric products.

A good way to visualize this general rule is to imagine what happens when a hand (chiral) reacts with a glove (also chiral). A right hand reacts with a right-handed glove, but a left hand reacts differently with the same right-handed glove. The results, right hand in right glove versus left hand in right glove, are not mirror images; they are totally different products (diastereomers).

$$\left.\begin{array}{l} \underset{\text{Right hand}}{R} + \underset{\text{Right glove}}{R} \longrightarrow R,R \\[2ex] \underset{\text{Left hand}}{S} + \underset{\text{Right glove}}{R} \longrightarrow S,R \end{array}\right\} \text{Diastereomers}$$

PROBLEM

28.6 When carboxylic acids react with alcohols under suitable conditions, esters are formed. Imagine that a chiral acid such as *(S)*-lactic acid reacts with a chiral alcohol such as *(R)*-2-butanol.

$$(S)\text{—}CH_3CH(OH)CO_2H + (R)\text{—}CH_3CH_2CH(OH)CH_3$$

$$\xrightarrow{H^+} CH_3CH(OH)\overset{\overset{\displaystyle O}{\|}}{C}\text{—}O\underset{\underset{\displaystyle CH_3}{|}}{C}HCH_2CH_3 + H_2O$$

What is the stereochemistry at the two chiral centers in the product? Draw the starting materials and product in Fischer projection.

PROBLEM

28.7 Imagine that racemic lactic acid reacts with *(S)*-2-butanol to form an ester. What stereochemistry does the product(s) have? What is the relationship of one product to another? How might you use this reaction to resolve (±)-lactic acid?

28.6 Industrial sources and uses of alkylamines

Alkylamines such as methylamine, ethylamine, and dimethylamine all find a variety of relatively minor applications in the chemical industry as starting materials for the preparation of insecticides and pharmaceuticals. Simple methylated amines are prepared by reaction of ammonia with methanol in the presence of an alumina catalyst. The reaction yields a mixture of mono-, di-, and trimethylated products but is nonetheless useful industrially since the separation of the three products by distillation is quite easy.

$$NH_3 + CH_3OH \xrightarrow[450°C]{Al_2O_3} CH_3NH_2 + (CH_3)_2NH + (CH_3)_3N + H_2O$$

28.7 Synthesis of amines

S_N2 REACTIONS OF ALKYL HALIDES

Ammonia and other alkylamines are excellent nucleophiles in S_N2 reactions. As a result, the simplest method of amine synthesis involves S_N2 alkylation of ammonia (or any alkylamine) with an alkyl halide. If ammonia is alkylated, a primary amine results; if a primary amine is alkylated, a secondary amine results, and so on. Even tertiary amines react rapidly with alkyl halides to yield quaternary ammonium salts, $R_4N^+ X^-$.

	$\ddot{N}H_3 + R{-}X \longrightarrow$	$RNH_3^+ \bar{X}$	$\xrightarrow{NaOH}$ RNH_2	Primary
Primary	$R\ddot{N}H_2 + R{-}X \longrightarrow$	$R_2NH_2^+ X^-$	$\xrightarrow{NaOH}$ R_2NH	Secondary
Secondary	$R_2\ddot{N}H + R{-}X \longrightarrow$	$R_3NH^+ X^-$	$\xrightarrow{NaOH}$ R_3N	Tertiary
Tertiary	$R_3\ddot{N} + R{-}X \longrightarrow$	$R_4N^+ X^-$	Quaternary ammonium salt	

S_N2 reaction

Unfortunately, the reaction does not stop cleanly after a single alkylation has occurred. Primary, secondary, and even tertiary amines are all of similar reactivity, and mixtures of products invariably result from amine alkylations. For example, treatment of 1-bromooctane with a twofold excess of ammonia leads to a mixture containing only 45% of the desired product, octylamine. A nearly equal amount of dioctylamine is produced by overalkylation, along with smaller amounts of trioctylamine and tetraoctylammonium bromide.

$$\underset{\text{1-Bromooctane}}{CH_3(CH_2)_6CH_2Br} + :NH_3 \longrightarrow \underset{\text{Octylamine (45\%)}}{CH_3(CH_2)_6CH_2\ddot{N}H_2} + \underset{\text{Dioctylamine (43\%)}}{[CH_3(CH_2)_6CH_2]_2\ddot{N}H}$$

$$+ \underset{\text{Trace}}{[CH_3(CH_2)_6CH_2]_3N:} + \underset{\text{Trace}}{[CH_3(CH_2)_6CH_2]_4\overset{+}{N}\overset{-}{Br}}$$

Higher yields can sometimes be obtained by using a large excess of the starting amine, but even so the reaction is a poor one, and better methods of synthesis are needed. Two such methods are often used—the **azide synthesis** and the **Gabriel synthesis.**

The azide synthesis is an excellent method for preparing primary amines from alkyl halides. Azide ion, N_3^-, is a nonbasic, highly reactive nucleophile in S_N2 reactions; it displaces halide ion from primary and even secondary halides in high yield to give alkyl azides, R—N_3. Since alkyl azides are not themselves nucleophilic, overalkylation cannot occur. Reduction of alkyl azides, either by catalytic hydrogenation over palladium or by reaction with $LiAlH_4$, leads to the desired primary amine. Yields are usually excellent, but the value of the process is tempered by the fact that low-molecular-weight alkyl azides are explosive and must be handled with great care.

$$C_6H_5CH_2CH_2Br \xrightarrow[S_N2\ reaction]{NaN_3,\ ethanol} C_6H_5CH_2CH_2N{=}\overset{+}{N}{=}\overset{-}{N} + NaBr$$

1-Bromo-2-phenylethane → 2-Phenylethyl azide

$$\xrightarrow[2.\ H_2O]{1.\ LiAlH_4,\ ether} C_6H_5CH_2CH_2NH_2$$

2-Phenylethylamine (89%)

$$C_6H_{11}Br \xrightarrow{NaN_3} C_6H_{11}N{=}\overset{+}{N}{=}\overset{-}{N} + NaBr \xrightarrow[2.\ H_2O]{1.\ LiAlH_4,\ ether} C_6H_{11}NH_2$$

Bromocyclohexane → Cyclohexyl azide → Cyclohexylamine (54%)

The Gabriel[1] amine synthesis provides an alternative method for preparing primary amines from alkyl halides. Though first introduced in 1887, the synthesis remains in common use. Potassium phthalimide is alkylated with a primary alkyl halide, yielding an *N*-alkyl phthalimide. **Imides** (**—CO—NH—CO—**) are structurally similar to acetoacetic ester in that the proton on nitrogen is flanked by two acidifying carbonyl groups. Thus, imides are readily deprotonated by bases such as KOH, and their anions are readily alkylated in a reaction similar to the acetoacetic ester synthesis (Section 25.9). Basic hydrolysis of an *N*-alkyl phthalimide yields a primary amine product. The reaction is best carried out in the highly polar solvent, dimethylformamide, DMF [$HCON(CH_3)_2$].

[1]Siegmund Gabriel (1851–1924); b. Berlin; Ph.D. University of Berlin (1874); assistant to A. W. von Hofmann; professor, University of Berlin.

$$\text{Phthalimide} \xrightarrow[\text{Ethanol}]{\text{KOH}} \text{Potassium phthalimide } (K^+) \xrightarrow[\text{DMF}]{R{-}X} \text{N-alkylphthalimide } (:N{-}R) + KX$$

$$\xrightarrow{^{-}OH/H_2O} R{-}NH_2 + \text{phthalate } (C_6H_4(CO_2^-)_2)$$

Phthalimide

Potassium phthalimide

Recall the acetoacetic ester synthesis:

$$C_2H_5O{-}\overset{O}{\overset{\|}{C}}{-}CH_2{-}\underset{O}{\underset{\|}{C}}{-}CH_3 \xrightarrow[HOC_2H_5]{NaOC_2H_5} C_2H_5O{-}\overset{O}{\overset{\|}{C}}{-}\ddot{C}H^{-}{-}\underset{O}{\underset{\|}{C}}{-}CH_3 \xrightarrow{RX} C_2H_5O{-}\overset{O}{\overset{\|}{C}}{-}CH(R){-}\underset{O}{\underset{\|}{C}}{-}CH_3$$

An example of the Gabriel synthesis follows:

$$C_6H_5CH_2Br \xrightarrow[\text{2. } ^{-}OH/H_2O]{\text{1. potassium phthalimide } (:N:^{-}\ K^{+}) + DMF} C_6H_5CH_2\ddot{N}H_2$$

Benzyl bromide

Benzylamine (81%)

PROBLEM

28.8 The last step in the Gabriel amine synthesis is the base-promoted hydrolysis of an imide to yield an amine plus phthalate ion. Propose a mechanism to account for this hydrolysis. [*Hint:* Review the mechanism of amide hydrolysis in Section 24.8.]

PROBLEM

28.9 The hydrolysis of phthalimides is often slow, and an alternative method is sometimes needed to liberate the primary amine in the last stage of a Gabriel synthesis. In the Ing–Manske modification, reaction of an *N*-alkylphthalimide with hydrazine is employed. Propose a mechanism for the hydrazinolysis shown on p. 936.

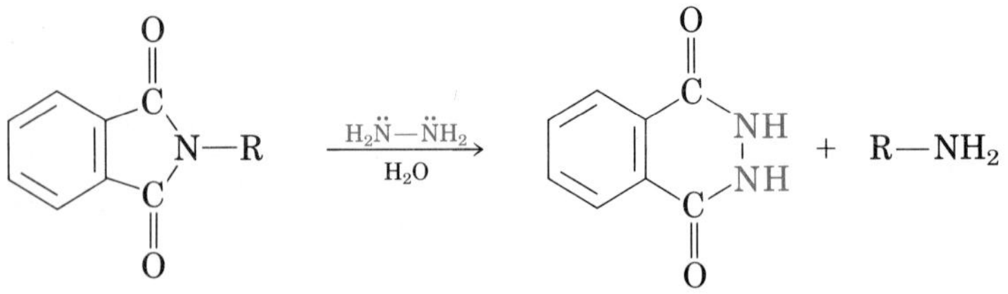

$$\text{N-alkylphthalimide (phthalimide-N-R)} \xrightarrow[H_2O]{H_2\ddot{N}-\ddot{N}H_2} \text{Phthalhydrazide} + R-NH_2$$

Phthalhydrazide

PROBLEM

28.10 Show two methods for the synthesis of dopamine, a neurotransmitter involved in regulation of the central nervous system. Use any alkyl halide needed.

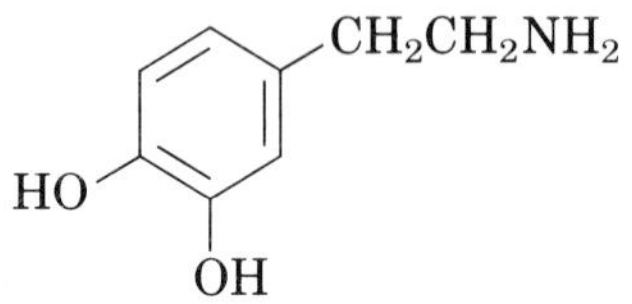

(3,4-dihydroxyphenyl)–$CH_2CH_2NH_2$ (HO, OH ring substituents)

Dopamine

REDUCTION OF NITRILES AND AMIDES

We have already studied the $LiAlH_4$ reduction of nitriles (Section 24.9) and of amides (Section 24.8) to give amines and have seen that both reactions usually result in high yields. The sequence of cyanide S_N2 displacement followed by reduction provides an excellent method for converting alkyl halides into primary amines having one more carbon atom; amide reduction provides an excellent method for converting carboxylic acids and their derivatives into amines:

$$\underset{\text{Alkyl halide}}{R-X} \xrightarrow{NaCN} R-CN \xrightarrow[2.\ H_2O]{1.\ LiAlH_4,\ ether} \underset{1^\circ\ \text{amine}}{R-CH_2NH_2}$$

$$\underset{\text{Carboxylic acid}}{R-COOH} \xrightarrow[2.\ NH_3]{1.\ SOCl_2} R-\overset{O}{\overset{\|}{C}}-NH_2 \xrightarrow[2.\ H_2O]{1.\ LiAlH_4,\ ether} \underset{1^\circ\ \text{amine}}{R-CH_2NH_2}$$

REDUCTIVE AMINATION OF KETONES AND ALDEHYDES

Amines can be synthesized from ketones and aldehydes in a single step by **reductive amination.** Reductive amination involves treating the ketone or aldehyde with ammonia or an amine in the presence of a reducing agent. For example, the central nervous system stimulant, amphetamine, is prepared commercially by reductive amination of phenyl-2-propanone with ammonia, using hydrogen gas over a Raney nickel catalyst as reducing agent. Reductive amination takes place by the pathway shown in Figure 28.6. As indicated, an imine intermediate is first formed by a nucleophilic addition reaction (Section 22.9), and the imine is then reduced.

Ammonia attacks the carbonyl group in a nucleophilic addition reaction to yield an intermediate carbinolamine.

$$C_6H_5CH_2C(=O)CH_3 \underset{}{\overset{:NH_3}{\rightleftharpoons}}$$

$$C_6H_5CH_2C(OH)(\ddot{N}H_2)CH_3$$

The intermediate loses water to give an imine.

$$\overset{-H_2O}{\rightleftharpoons} C_6H_5CH_2C(=NH)CH_3$$

The imine is reduced catalytically over Raney nickel to yield the amine product.

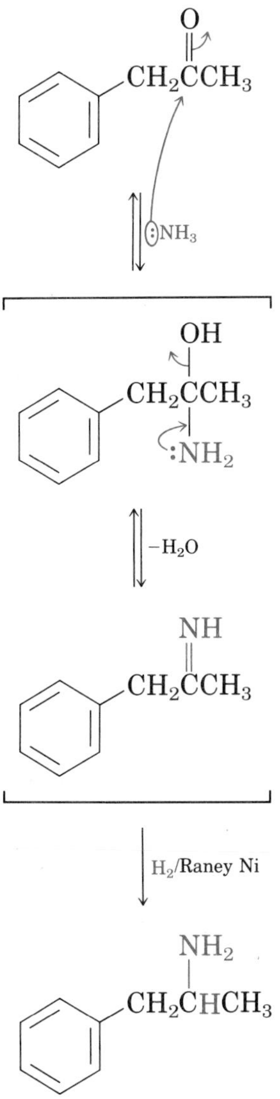

$$\xrightarrow{H_2/\text{Raney Ni}} C_6H_5CH_2CH(NH_2)CH_3$$

Figure 28.6. Mechanism of reductive amination.

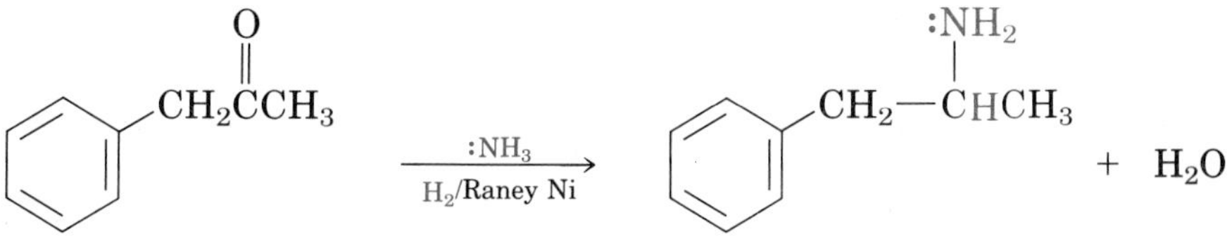

$$C_6H_5CH_2C(=O)CH_3 \xrightarrow[H_2/\text{Raney Ni}]{:NH_3} C_6H_5CH_2{-}CH(\ddot{N}H_2)CH_3 + H_2O$$

Phenyl-2-propanone → Amphetamine

Ammonia, primary amines, and secondary amines can all be used in the reductive amination reaction to yield primary, secondary, and tertiary amines, respectively.

$$R_2C{=}O \xrightarrow[H_2/cat.]{:NH_3} R_2CHNH_2 \quad \text{Primary amine}$$

$$R_2C{=}O \xrightarrow[H_2/cat.]{R'\ddot{N}H_2} R_2CHNHR' \quad \text{Secondary amine}$$

$$R_2C{=}O \xrightarrow[H_2/cat.]{R'_2\ddot{N}H} R_2CHNR'_2 \quad \text{Tertiary amine}$$

Many different reducing agents are effective, but the most common choice on a laboratory scale (as opposed to an industrial scale) is sodium cyanoborohydride, $NaBH_3CN$.

$$\text{Cyclohexanone} + H\ddot{N}(CH_3)_2 \xrightarrow[CH_3OH]{NaBH_3CN} N,N\text{-Dimethylcyclohexylamine (85\%)}$$

Cyclohexanone

N,N-Dimethylcyclohexylamine (85%)

PROBLEM

28.11 Show in detail the mechanism of reductive amination of cyclohexanone and dimethylamine with $NaBH_3CN$. What intermediates are involved?

PROBLEM

28.12 Ephedrine is an amino alcohol that is widely used for the treatment of bronchial asthma. Show how a reductive amination step might be used to synthesize ephedrine.

$C_6H_5CH(OH)CH(NHCH_3)CH_3$

Ephedrine

HOFMANN AND CURTIUS REARRANGEMENTS OF AMIDES

Both the **Hofmann rearrangement**[2] and the **Curtius rearrangement**[3] provide methods for degrading carboxylic acid derivatives to primary amines with the loss of one carbon atom, $RCOX \rightarrow R{-}NH_2$. The Hofmann rearrangement involves a primary amide, and the Curtius rearrangement involves an acyl azide. Both, however, proceed through similar mechanisms:

[2]August Wilhelm von Hofmann (1818–1892); b. Giessen, Germany; professor, Bonn, the Royal College of Chemistry, London (1845–1864), Berlin (1865–1892).
[3]Theodor Curtius (1857–1928); b. Duisberg; Ph.D. Leipzig; professor, universities of Kiel, Bonn, and Heidelberg (1898–1926).

$$\text{Hofmann rearrangement} \quad R{-}CONH_2 \xrightarrow[H_2O]{^-OH,\ Br_2} R{-}NH_2 + CO_2$$

$$\text{Curtius rearrangement} \quad R{-}CON_3 \xrightarrow[\Delta]{H_2O} R{-}NH_2 + CO_2 + N_2$$

Hofmann rearrangement occurs when a primary amide, $RCONH_2$, is treated with halogen and base. The reaction mechanism is shown in Figure 28.7 (p. 940). Although the overall mechanism is lengthy, most of the individual steps have been encountered before. Thus, the bromination of an amide in steps 1 and 2 is analogous to the base-promoted bromination of a ketone (Section 25.7), and the alpha elimination of HBr from the *N*-bromoamide intermediate in steps 3 and 4 is analogous to the alpha elimination of HCl from chloroform to form dichlorocarbene (Section 18.6). Once the acyl nitrene intermediate is formed, migration of the R group from carbon to electron-deficient nitrogen in step 5 is analogous to the kind of alkyl migrations that occur during carbocation rearrangements (Section 16.5). Nucleophilic addition of water to the isocyanate carbonyl group in step 6 is a typical carbonyl-group process (Section 22.6), as is the final decarboxylation step.

Despite its mechanistic complexity, the Hofmann rearrangement often gives high yields of both aryl- and alkylamines. For example, the appetite-suppressing drug phentermine is prepared commercially by Hofmann degradation of a primary amide.

$$C_6H_5{-}CH_2{-}C(CH_3)_2{-}CONH_2 \xrightarrow[H_2O]{^-OH,\ Cl_2} C_6H_5{-}CH_2{-}C(CH_3)_2{-}NH_2$$

2,2-Dimethyl-3-phenylpropanamide → Phentermine

The Curtius rearrangement, like the Hofmann rearrangement, involves an acyl nitrene as the key intermediate. The nitrene is produced thermally by loss of nitrogen from an acyl azide that is itself prepared by nucleophilic acyl substitution of an acid chloride.

$$R{-}\overset{O}{\overset{\|}{C}}{-}Cl \xrightarrow{NaN_3} R{-}\overset{O}{\overset{\|}{C}}{-}N{=}\overset{+}{N}{=}\overset{-}{N}$$

Acid chloride → Acyl azide

$$\downarrow \Delta$$

$$\left[N_2 + R{-}\overset{O}{\overset{\|}{C}}{-}\ddot{N}: \longrightarrow R{-}N{=}C{=}O \right] \xrightarrow{H_2O} R{-}NH_2 + CO_2$$

Acyl nitrene → Isocyanate

For example, Curtius rearrangement of 2-phenylcyclopropanecarbonyl

Base abstracts an acidic amide proton, yielding an anion.

The anion is brominated to give an *N*-bromoamide.

Base abstraction of the remaining amide proton gives a bromoamide anion.

The bromoamide anion spontaneously loses bromide ion (an alpha elimination) yielding a nitrene intermediate.

The R group then migrates from carbon to electron-deficient nitrogen.

The isocyanate formed adds water to give a carbamic acid.

The carbamic acid spontaneously loses CO_2, yielding the final product.

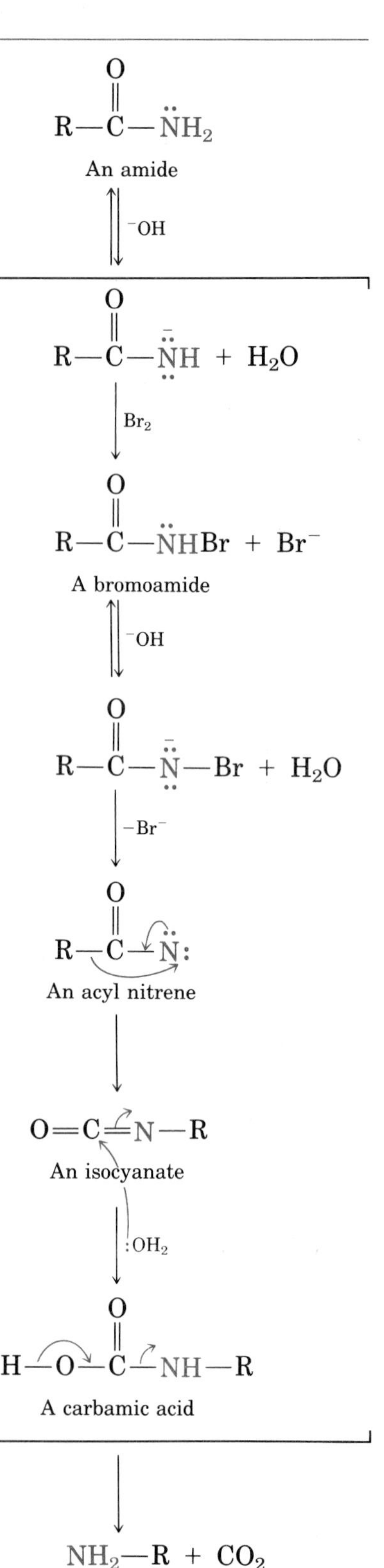

Figure 28.7. Mechanism of Hofmann rearrangement.

chloride leads to the antidepressant drug, tranylcypromine.

$$\text{trans-2-Phenylcyclopropanecarbonyl chloride} \xrightarrow[\text{3. } H_2O]{\text{1. } NaN_3 \quad \text{2. } \Delta} \text{Tranylcypromine}$$

trans-2-Phenylcyclopropanecarbonyl chloride

Tranylcypromine

PROBLEM

28.13 How many electrons does a nitrene nitrogen atom have in its valence shell? What relationship does a nitrene bear to a carbocation and to a carbene (Section 18.6)?

PROBLEM

28.14 Formulate the mechanism of the rearrangement of an acyl nitrene to an isocyanate, showing the origin and fate of all bonding electrons. Formulate also the mechanism of the addition of water to an isocyanate to yield a carbamic acid.

28.8 Reactions of amines

We have already studied the two most important reactions of alkylamines—alkylation and acylation. As we saw earlier in this chapter, primary, secondary, and tertiary amines can all be alkylated by reaction with primary alkyl halides. Alkylations of primary and secondary amines are rather difficult to control and often give mixtures of products, but tertiary amines are cleanly alkylated to give quaternary (tetrasubstituted) ammonium salts. These ammonium salts are ionic compounds and are therefore water soluble.

$$:NH_3 \xrightarrow{RX} RNH_3^+\ X^- \xrightarrow{NaOH} RNH_2 \quad 1°$$

$$1° \quad R\ddot{N}H_2 \xrightarrow{RX} R_2NH_2^+\ X^- \xrightarrow{NaOH} R_2NH \quad 2°$$

$$2° \quad R_2\ddot{N}H \xrightarrow{RX} R_3NH^+\ X^- \xrightarrow{NaOH} R_3N \quad 3°$$

$$3° \quad R_3\ddot{N} \xrightarrow{RX} R_4N^+\ X^- \quad \text{Quaternary salt}$$

Primary and secondary (but not tertiary) amines can also be acylated by reaction with acid chlorides or acid anhydrides (Section 24.5) to yield amides.

$$NH_3 \xrightarrow[\text{Pyridine}]{RCOCl} R\text{—}\overset{O}{\overset{\|}{C}}NH_2 + HCl$$

$$R'NH_2 \xrightarrow[\text{Pyridine}]{RCOCl} R\text{—}\overset{O}{\overset{\|}{C}}NHR' + HCl$$

$$R'_2NH \xrightarrow[\text{Pyridine}]{RCOCl} R\text{—}\overset{O}{\overset{\|}{C}}NR'_2 + HCl$$

If a sulfonyl chloride is used as the acylating agent, a sulfonamide ($R_2N—SO_2R'$) is produced. This reaction forms the basis of a classic laboratory test for distinguishing among primary, secondary, and tertiary amines (the **Hinsberg test**). Only primary and secondary amines react irreversibly with benzenesulfonyl chloride, and a tertiary amine is therefore identifiable by the fact that its reaction product is instantly hydrolyzed on addition of water, giving back the free amine. A primary amine yields a sulfonamide product that has one remaining acidic N—H proton and can therefore be identified by its base solubility. A secondary amine yields a sulfonamide that has no acidic protons and can be identified by its lack of base solubility.

Primary amine: $R\ddot{N}H_2 + C_6H_5SO_2Cl \xrightarrow{\text{Pyridine}} C_6H_5SO_2NHR$ — Base-soluble product

Benzenesulfonyl chloride; A sulfonamide

Secondary amine: $R_2\ddot{N}H + C_6H_5SO_2Cl \xrightarrow{\text{Pyridine}} C_6H_5SO_2NR_2$ — Base-insoluble product

Tertiary amine: $R_3N: + C_6H_5SO_2Cl \xrightarrow{\text{Pyridine}} [C_6H_5SO_2\overset{+}{N}R_3\ Cl^-] \xrightarrow{H_2O} R_3N: + C_6H_5SO_3H$

PROBLEM

28.15 Account for the fact that a quaternary ammonium salt such as allylbenzylmethylphenylammonium iodide can be resolved into two enantiomeric forms.

HOFMANN ELIMINATION

Amines can be made to undergo an elimination reaction under suitable conditions to yield alkenes. In the **Hofmann elimination,** an amine is first methylated with excess methyl iodide, yielding a quaternary ammonium iodide. This quaternary salt then produces an alkene when heated with silver oxide. For example, 1-hexene is formed from hexylamine in 60% yield.

$$\underset{\text{Hexylamine}}{CH_3CH_2CH_2CH_2CH_2CH_2\ddot{N}H_2} \xrightarrow[\text{(excess)}]{CH_3I} \underset{\text{Hexyltrimethylammonium iodide}}{CH_3(CH_2)_3CH_2CH_2\overset{+}{N}(CH_3)_3\ I^-}$$

$$\downarrow Ag_2O,\ H_2O$$

$$CH_3(CH_2)_3CH_2CH_2\overset{+}{N}(CH_3)_3\ \overset{-}{O}H + AgI$$

$$\downarrow \Delta$$

$$\underset{\text{1-Hexene (60\%)}}{CH_3CH_2CH_2CH_2CH{=}CH_2} + N(CH_3)_3$$

Silver oxide functions by exchanging hydroxide for iodide in the quaternary salt, thus providing the base necessary to effect elimination. The actual elimination step is probably an E2 reaction (Section 10.10) in which hydroxide ion removes a proton and the positively charged nitrogen acts as the leaving group (Figure 28.8).

$$HO{:}^- \quad H{-}\overset{|}{C}{-}CH_2{-}\overset{+}{N}(CH_3)_3 \xrightarrow{\text{E2 reaction}} \underset{\text{Alkene}}{{>}C{=}CH_2} + H_2O + N(CH_3)_3$$

Figure 28.8. Mechanism of Hofmann elimination.

An interesting feature of the Hofmann elimination is that its regiochemistry is different from that of most other E2 reactions. The less substituted alkene normally predominates in Hofmann eliminations, as opposed to the more substituted (Zaitsev rule) products formed from base-induced elimination reactions of alkyl halides (Section 10.10). The reasons for this selectivity are not well understood, but it may be due to steric factors that favor attack of base at the less hindered position. For example, (1-methylbutyl)trimethylammonium hydroxide yields 1-pentene and 2-pentene in a 94:6 ratio.

$$\underset{\text{(1-Methylbutyl)trimethylammonium hydroxide}}{CH_3CH_2CH_2\underset{\substack{|\\ {}^+N(CH_3)_3\ {}^-OH}}{C}HCH_3} \xrightarrow{\Delta} \underset{\text{1-Pentene}}{CH_3CH_2CH_2CH{=}CH_2} + \underset{\text{2-Pentene}}{CH_3CH_2CH{=}CHCH_3}$$

94:6 ratio

The Hofmann elimination reaction is important primarily because of its historical use as a degradative tool in the structure determination of many complex naturally occurring amines. The reaction is not often used today, however, since the product alkenes can usually be made more easily in other ways.

PROBLEM

28.16 What product would you expect from Hofmann elimination of a cyclic amine such as piperidine? Formulate all the steps involved.

$$\text{Piperidine (NH)} \xrightarrow[\text{3. } \Delta]{\text{1. } CH_3I \text{ (excess)} \quad \text{2. } Ag_2O,\ H_2O} $$

PROBLEM

28.17 Cyclooctatetraene was first synthesized by Willstätter in 1911 by a route that involved the following transformation:

$$H_3C{-}N \text{ (bicyclic amine)} \Longrightarrow \text{cyclooctatetraene}$$

How might you use the Hofmann elimination reaction to accomplish this? How would you finish the synthesis by converting cyclooctatriene into cyclooctatetraene?

28.9 Tetraalkylammonium salts as phase-transfer agents

Tetraalkylammonium salts, easily prepared by S_N2 reaction between a tertiary amine and an alkyl halide (Section 28.7), have come to be widely used in the past decade as catalysts for many different kinds of organic reactions. For example, we have seen that chloroform reacts with strong base to generate dichlorocarbene, which can then add to a carbon–carbon double bond to yield a dichlorocyclopropane (Section 18.6). Let's imagine an experiment in which we dissolve an alkene, cyclohexene, in chloroform and stir the organic solution with 50% aqueous sodium hydroxide. Since the organic layer and the water layer are immiscible, the strong base in the aqueous phase is unable to come into contact with chloroform in the organic phase; thus, there is no reaction. If, however, we add a small amount of a tetraalkylammonium salt such as benzyltriethylammonium chloride to the two-phase mixture, an immediate reaction occurs; the dichlorocyclopropane product is formed in 77% yield:

$$\text{Cyclohexene} \xrightarrow[C_6H_5CH_2\overset{+}{N}(CH_2CH_3)_3\ Cl^-\ \text{(trace)}]{CHCl_3,\ 50\%\ NaOH/H_2O} \text{7,7-dichlorobicyclo[4.1.0]heptane (Cl, Cl)}$$

77%

No reaction occurs without $C_6H_5\overset{+}{N}(CH_2CH_3)_3\ Cl^-$

How does the tetraalkylammonium salt exert its catalytic effect? Benzyltriethylammonium ion, even though positively charged, is nevertheless soluble in organic solvents because of the four hydrocarbon substituents on nitrogen. When the positively charged tetraalkylammonium ion goes into the organic layer, a negatively charged counter-ion must also move into the organic layer to preserve charge neutrality. Hydroxide ion, present in far greater amount than chloride ion, is thus transferred into the organic phase where reaction with chloroform immediately occurs (Figure 28.9).

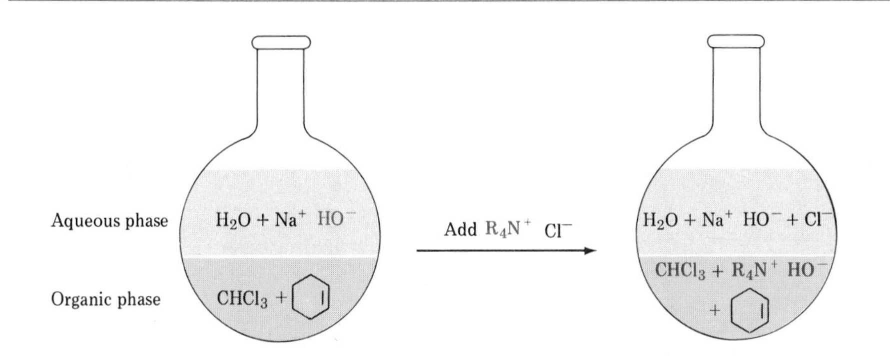

Figure 28.9. A phase-transfer catalysis. Addition of a small amount of a tetraalkylammonium salt to a two-phase mixture allows inorganic anion to be transferred from the aqueous phase to the organic phase, whereupon reaction occurs.

Transfer of an inorganic ion from one phase to another is called **phase transfer,** and the tetraalkylammonium salt is referred to as a **phase-transfer catalyst.** Many different kinds of organic reactions, including carbonyl-group alkylations, oxidations, reductions, and S_N2 reactions, are subject to phase-transfer catalysis, and product yields are often improved under these conditions. S_N2 reactions are particularly good candidates for phase-transfer catalysis, since inorganic nucleophiles can be transferred from an aqueous (protic) phase to an organic (aprotic) phase, where they are far more reactive (Section 10.5). For example,

$$CH_3(CH_2)_6CH_2Br + NaCN \xrightarrow[C_6H_5CH_2\overset{+}{N}(CH_2CH_3)_3\ Cl^-]{H_2O,\ benzene} CH_3(CH_2)_6CH_2CN + NaBr$$

1-Bromooctane — Nonanenitrile (92%)

28.10 Naturally occurring amines: alkaloids

Amines were among the first organic compounds to be isolated in pure form, and a great variety of amines are widely distributed among plants and animals. For example, trimethylamine occurs in animal tissues and is partially responsible for the distinctive odor of many fish; morphine is a powerful pain-killer isolated from the opium poppy; quinine is an important antimalarial drug isolated from the bark of the South American *Cinchona* tree; reserpine is a useful antihypertensive (blood-pressure-lowering) agent isolated from the Indian shrub *Rauwolfia serpentina.*

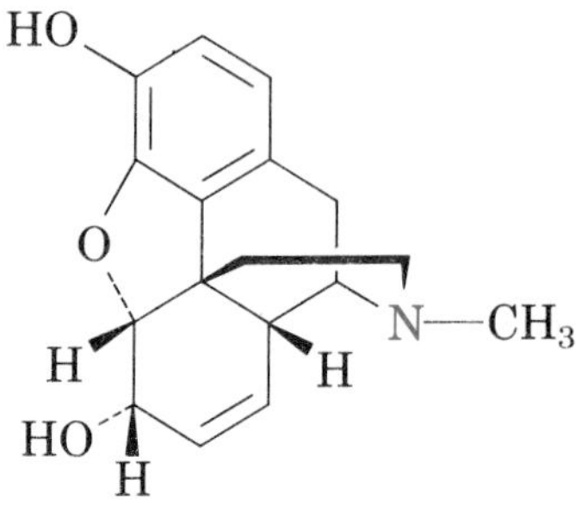

Morphine (analgesic)

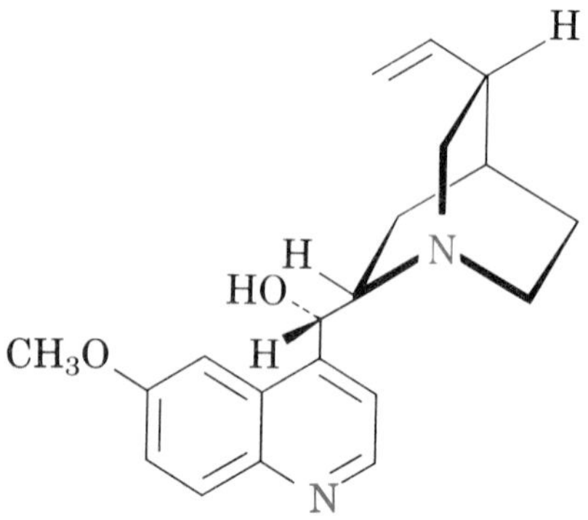

Quinine (antimalarial)

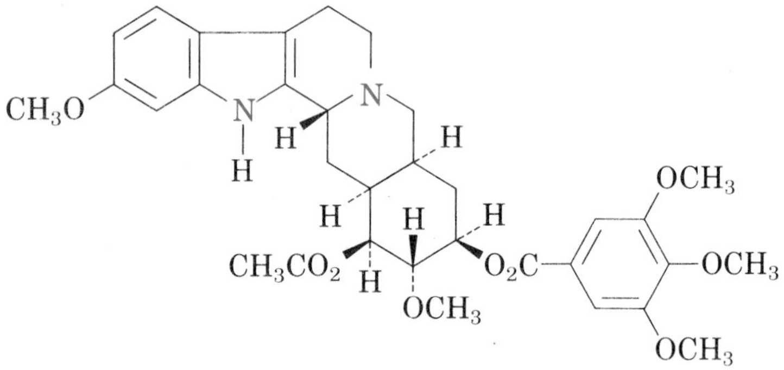

Reserpine (antihypertensive)

Naturally occurring amines derived from plant sources were once known as "vegetable alkali," since their aqueous solutions are basic, but they are now referred to as **alkaloids.** The study of alkaloids provided much of the impetus for the growth of organic chemistry in the nineteenth century and remains a fascinating area of research. Rather than attempt to classify the many kinds of alkaloids, let's look briefly at one particular group, the morphine alkaloids.

The medical uses of morphine have been known at least since the seventeenth century, when crude extracts of the opium poppy, *Papaver somniferum,* were employed for the relief of pain. Morphine was the first pure alkaloid to be isolated from the poppy, but its close relatives, codeine and thebaine, also occur naturally. Codeine is simply the methyl ether of morphine and is used in prescription cough medicines, and thebaine is a doubly unsaturated diether derivative. Heroin, another close relative of morphine, does not occur naturally but is synthesized by diacetylation of morphine.

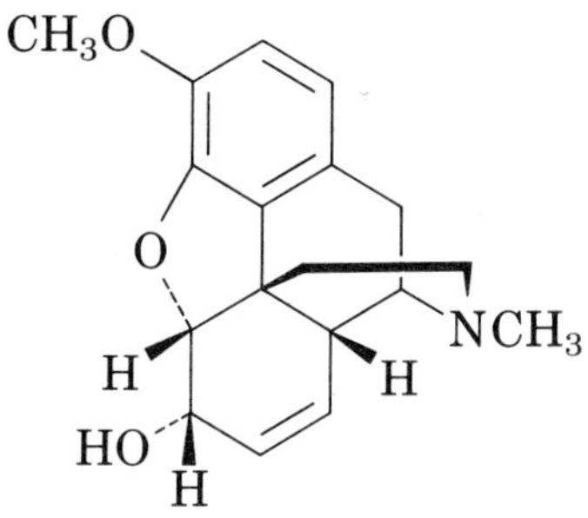

Codeine

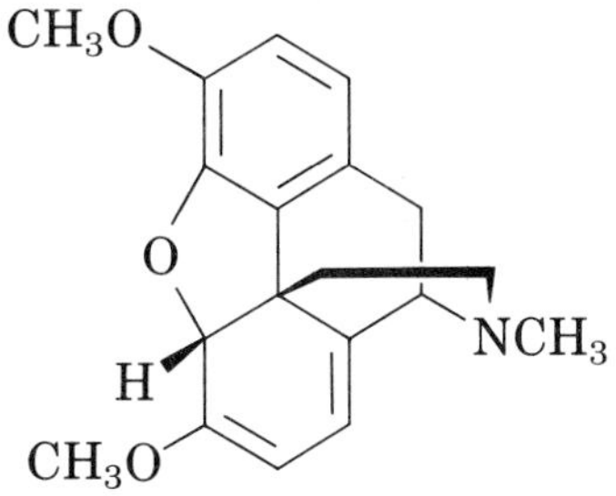

Thebaine

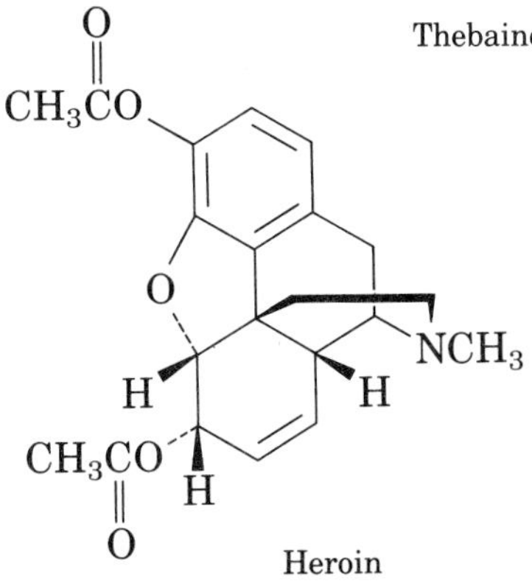

Heroin

Chemical investigations into the structure of morphine occupied some of the finest chemical minds of the nineteenth and early twentieth centuries, until the puzzle was finally solved by Robert Robinson in 1925. The key reaction used to establish structure was the Hofmann elimination.

Morphine and its relatives constitute a class of exceedingly useful pharmaceutical agents, yet they also pose a social problem of great proportion because of their addictive properties. Much effort has therefore gone into a search to understand the mode of action of morphine and to develop modified morphine analogs that retain the desired analgesic activity but do not cause physical dependence.

With respect to its biological mode of action, morphine appears to bind to opiate receptor sites in the brain; it does not interfere with or lessen the pain itself but rather changes the brain's perception of pain. With respect to the search for modified morphine-like agents, great progress has been made. For example, replacement of the morphine *N*-methyl group by an *N*-allyl substituent yields nalorphine, an analgesic agent that acts as a narcotic *antagonist* to reverse many of the undesirable side effects of morphine.

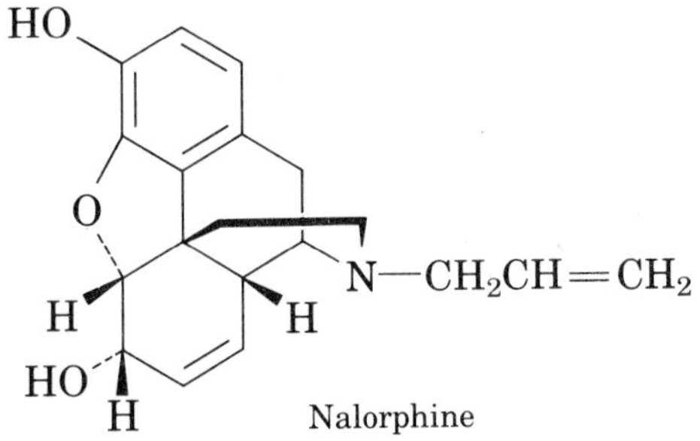

Nalorphine

Much work has shown that the entire tetracyclic framework of morphine is not necessary for biological activity. According to the "morphine rule," biological activity requires: (1) an aromatic ring attached to (2) a quaternary carbon atom, and (3) a tertiary amine (4) two carbon atoms farther away. Thus, the tricyclic amine cyclazocine has no narcotic properties but is still a powerful analgesic. Similarly, the bicyclic amine meperidine is widely used as a pain-killer.

C—C C
C N
C C C

The morphine rule: an aromatic ring, a quaternary carbon, two carbons, a tertiary amine

HO
CH_2
H_3C N—CH_2—CH
CH_3 CH_2

Cyclazocine

$C_2H_5O_2C$ N—CH_3

Meperidine

Although the morphine alkaloids are just one class of compounds, they are an excellent example of how chemists, following a lead provided by nature, have synthesized pharmaceutical agents of great benefit.

PROBLEM

28.18 Show how the morphine rule fits the structure of dextromethorphan, a common constituent of cough remedies.

CH_3O
N
H CH_3

Dextromethorphan

28.11 Spectroscopy of amines

MASS SPECTROSCOPY

The "nitrogen rule" of mass spectroscopy says that compounds with an odd number of nitrogen atoms have odd-numbered molecular weights. Thus we can be alerted to the presence of nitrogen in a molecule simply by observing its mass spectrum; an odd-numbered molecular ion usually means that the unknown compound has one or three nitrogen atoms, and an even-numbered molecular ion usually means that the compound has either zero or two nitrogen atoms. The logic behind this rule derives from the fact that nitrogen is trivalent, thus requiring an odd number of hydrogen atoms in the molecule. For example, methylamine has the formula CH_5N and a molecular weight of 31; morphine has the formula $C_{17}H_{19}NO_3$ and a molecular weight of 285.

Aliphatic amines undergo a characteristic alpha cleavage in the mass spectrometer, similar to the cleavage observed for aliphatic alcohols (Section 20.12). A carbon–carbon bond nearest the nitrogen atom is broken, yielding an alkyl radical and a nitrogen-containing cation:

$$\left[RCH_2 \text{-}\!\!\!\!\wr\, \underset{\alpha}{CH_2}\text{—}N\begin{matrix} R' \\ R' \end{matrix} \right]^{+\cdot} \xrightarrow{\text{Alpha cleavage}} RCH_2\cdot + \left[CH_2\text{—}N\begin{matrix} R' \\ R' \end{matrix} \right]^{+}$$

For example, the mass spectrum of *N*-ethylpropylamine shown in Figure 28.10 (p. 950) exhibits peaks at m/z = 58 and m/z = 72, corresponding to the two possible modes of alpha cleavage.

INFRARED SPECTROSCOPY

Primary and secondary amines can be identified by characteristic N—H bond stretching absorptions in the 3300–3500 cm^{-1} range of the infrared spectrum. Alcohols also absorb in this range (Section 20.12) but amine absorption bands are generally both sharper and less intense than hydroxyl bands. Primary amines show a pair of bands at about 3400 and 3500 cm^{-1}, and secondary amines show a single band at 3350 cm^{-1}. Tertiary amines show no absorption in this region, since they have no N—H protons. Representative infrared spectra of both primary and secondary amines are shown in Figure 28.11 (p. 951).

In addition to looking for characteristic N—H bands, we can also use a simple trick to tell whether or not a given unknown is an amine: addition of a small amount of mineral acid produces a broad and strong ammonium band in the 2200–3000 cm^{-1} range if the sample contains an amino group. All protonated amines give rise to this readily observable absorption caused by the ammonium $R_3\overset{+}{N}$—H bond. Figure 28.12 (p. 951) gives an example.

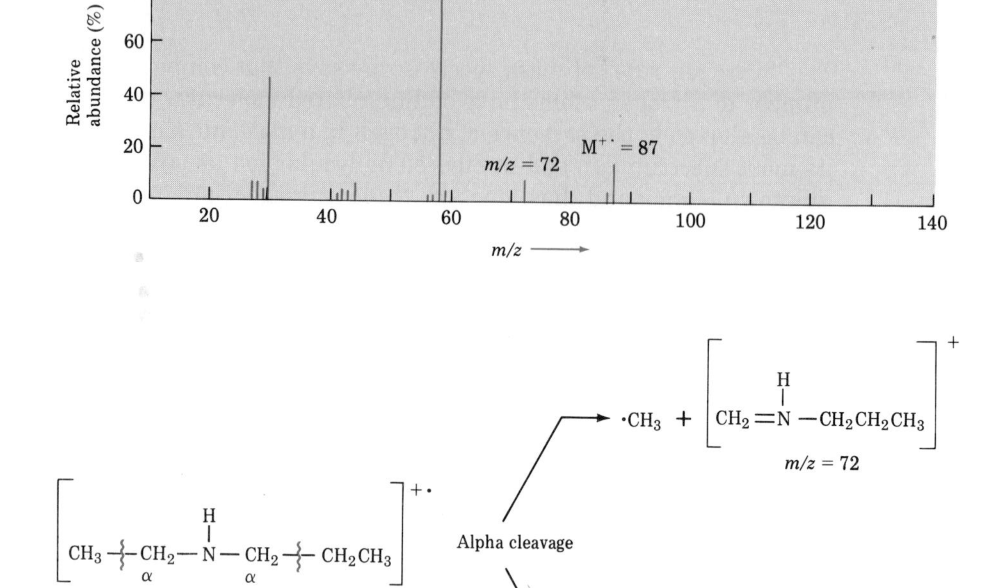

Figure 28.10. Mass spectrum of *N*-ethylpropylamine. The two possible modes of alpha cleavage occur, leading to the observed fragment ions at m/z = 58 and m/z = 72.

NUCLEAR MAGNETIC RESONANCE SPECTROSCOPY

Amines are often difficult to identify solely by ^{1}H NMR spectroscopy since N—H protons tend to appear as very broad resonances without clear-cut coupling to neighboring C—H protons. The situation is similar to that for hydroxyl protons (Section 20.12). As with hydroxyl O—H protons, amine N—H proton absorptions can appear over a wide range and are best identified by adding a small amount of D_2O to the sample tube. Exchange of N—D for N—H occurs, and the N—H signal disappears from the NMR spectrum.

$$\rangle N{-}H \xrightleftharpoons{D_2O} \rangle N{-}D + HDO$$

Protons on the carbon next to nitrogen are somewhat deshielded because of the electron-withdrawing nature of the nitrogen, and they therefore absorb at lower field than alkane protons. *N*-Methyl groups are particularly distinctive, since they absorb as a sharp three-proton singlet at 2.2–2.6 δ.

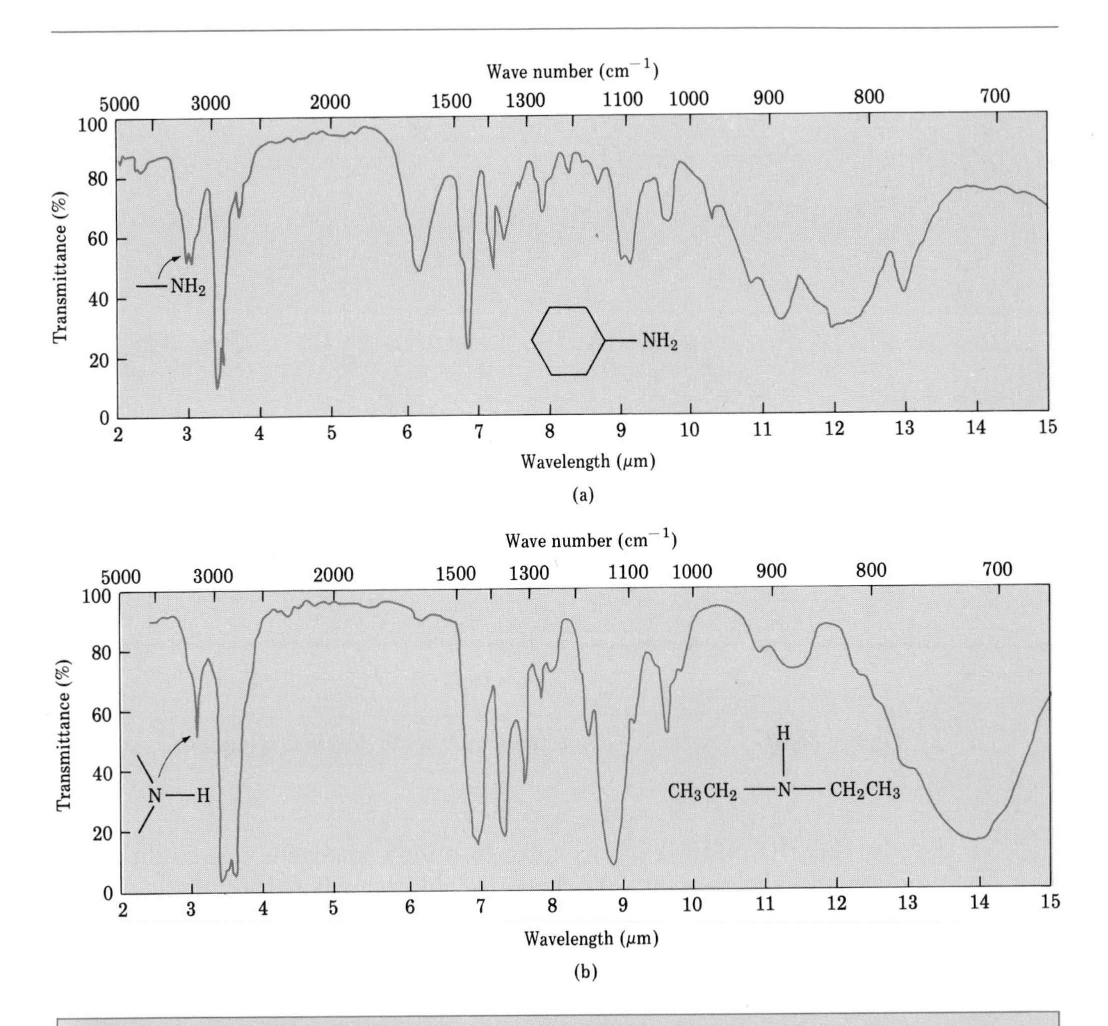

Figure 28.11. Infrared spectra of (a) cyclohexylamine and (b) diethylamine.

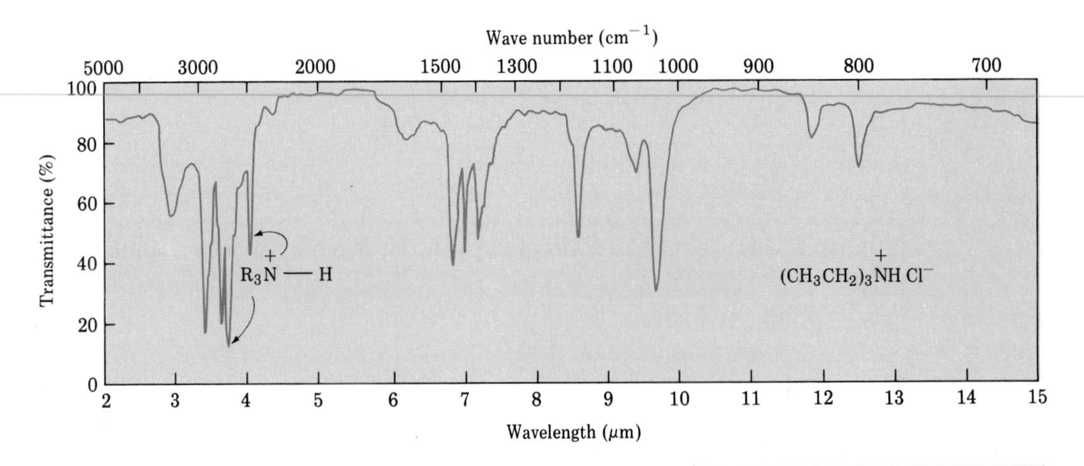

Figure 28.12. Infrared spectrum of triethylamine hydrochloride.

Figure 28.13 shows the ^{1}H NMR spectrum of *N*-methylcyclohexylamine; the *N*-methyl resonance at 2.42 δ is easily seen.

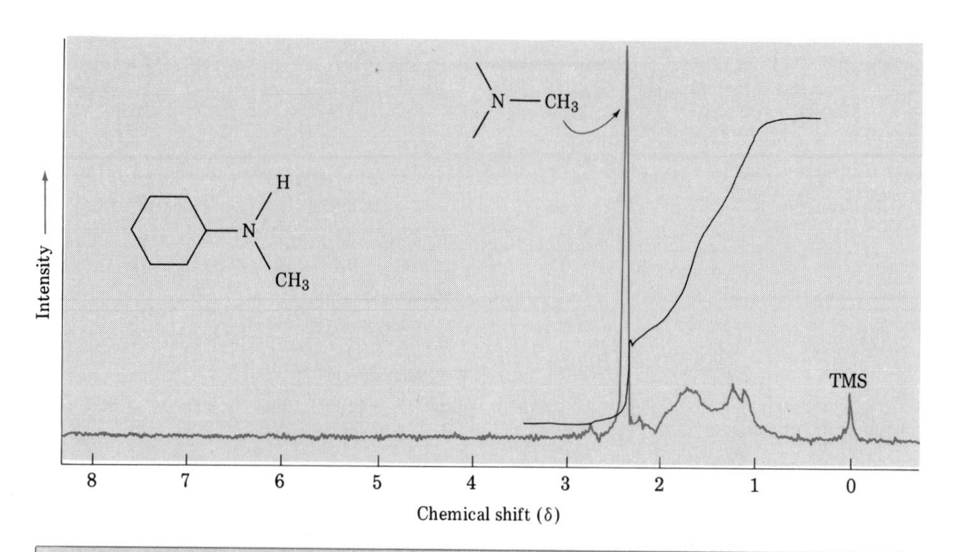

Figure 28.13. Proton NMR spectrum of *N*-methylcyclohexylamine.

In the ^{13}C NMR, carbons next to amine nitrogens are slightly deshielded and absorb about 20 ppm downfield from where they would absorb in an alkane of similar structure. For example, in *N*-methylcyclohexylamine, the ring carbon to which nitrogen is attached absorbs at a position 24 ppm lower than that of any other ring carbon (Figure 28.14).

H
33.3
N
33.4
25.2
CH$_3$
26.5
58.7

Figure 28.14. Carbon-13 NMR absorptions for *N*-methylcyclohexylamine (δ units).

PROBLEM

28.19 The infrared and ^{1}H NMR spectra shown on p. 953 are those of methamphetamine, which shows a molecular ion in the mass spectrum at $m/z = 149$. Propose a structure for methamphetamine and justify your answer. [*Note:* The signal at 1.2 δ disappears when D_2O is added to the sample.]

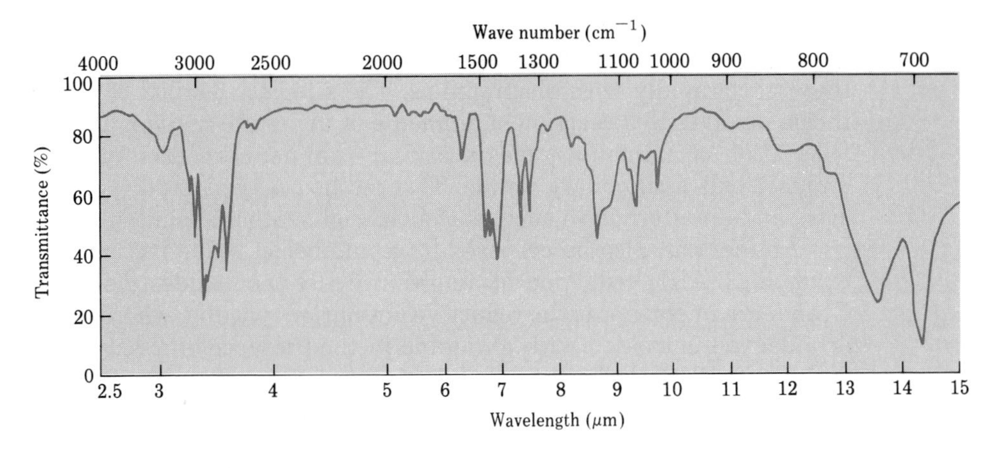

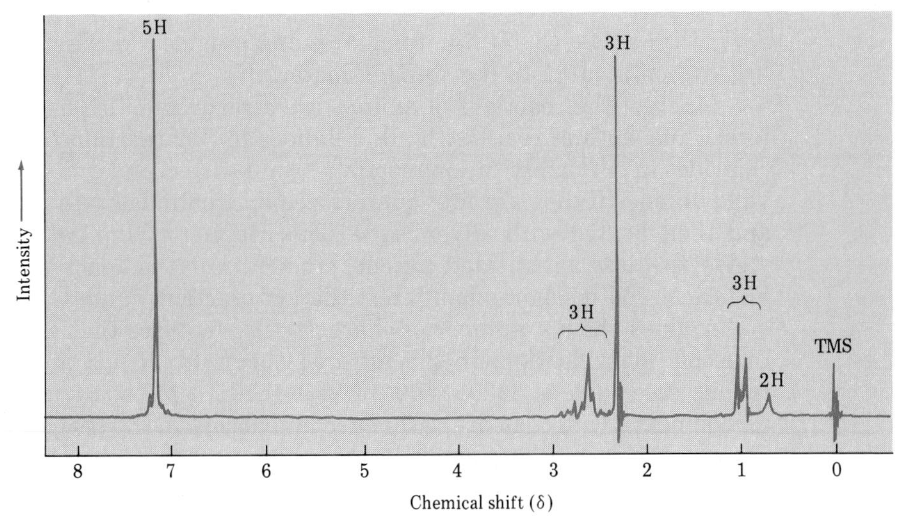

28.12 Summary

Amines are organo-substituted derivatives of ammonia. They are named in the IUPAC system either by adding the suffix *-amine* to the names of the alkyl substituents or by considering the amino group as a substituent on a more complex parent molecule. Bonding in amines is similar to that in ammonia. The nitrogen atom is sp^3 hybridized; the three substituents are directed to three corners of a tetrahedron; and the lone pair of nonbonding electrons occupies the fourth corner of the tetrahedron. An interesting feature of this tetrahedral structure is that amines undergo a rapid, umbrella-like pyramidal inversion, which interconverts mirror-image structures.

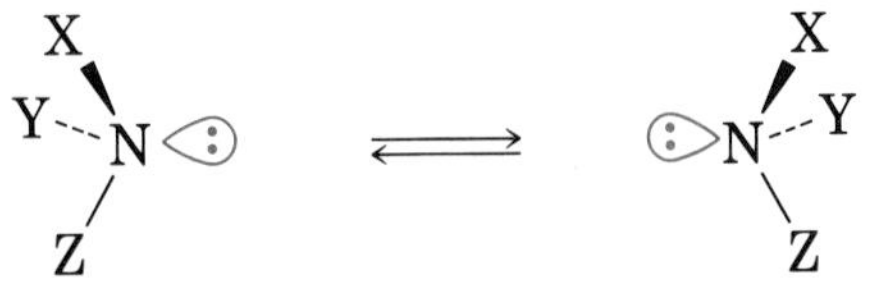

The chemistry of amines is dominated by the presence of the lone-pair electrons on nitrogen. Thus, amines are both basic and nucleophilic, and they react rapidly with electrophiles. The simplest method of amine synthesis involves S_N2 reaction of ammonia or an amine with an alkyl halide. Alkylation of ammonia yields a primary amine; alkylation of a primary amine yields a secondary amine. This method may give poor yields, however, and an alternative such as the Gabriel synthesis is often preferred.

Amines can also be prepared by a number of reductive methods. For example, $LiAlH_4$ reduction of amides, nitriles, and azides yields amines. Even more important is the reductive amination reaction, whereby a ketone or aldehyde is treated with an amine in the presence of a reducing agent such as $NaBH_3CN$. An intermediate imine is formed and is then immediately reduced by the hydride reagent present.

A final method of amine synthesis involves the Hofmann and Curtius rearrangements of carboxylic acid derivatives. Both methods proceed via acyl nitrene ($R{-}CO\ddot{N}{:}$) intermediates and provide a product that has one less carbon atom than the starting material.

Many of the reactions of amines are already familiar from past chapters. Thus, amines react with alkyl halides in S_N2 reactions and with acid chlorides in acyl substitution reactions. Amines also undergo elimination to yield alkenes if they are first quaternized by treatment with methyl iodide and then heated with silver oxide. This Hofmann elimination generally yields the less substituted alkene product, but the reaction is more of historical and mechanistic interest than of practical value.

Amines show a number of characteristic features that aid their spectroscopic identification. In the infrared spectrum, N—H absorptions are readily detectable at 3200–3500 cm^{-1}. In the ^1H NMR spectrum, protons on carbon next to nitrogen are deshielded and absorb in the range 2.2–2.6 δ. In the mass spectrum, amines undergo a characteristic alpha cleavage.

28.13 Summary of reactions

1. Preparation of amines (Section 28.7)
   a. The S_N2 alkylation of alkyl halides

$$:NH_3 + RX \longrightarrow RNH_3^+\,X^- \xrightarrow{NaOH} RNH_2 \quad \text{Primary}$$

$$\text{Primary} \quad :NH_2R + RX \longrightarrow R_2NH_2^+\,X^- \xrightarrow{NaOH} R_2NH \quad \text{Secondary}$$

$$\text{Secondary} \quad :NHR_2 + RX \longrightarrow R_3NH^+\,X^- \xrightarrow{NaOH} R_3N \quad \text{Tertiary}$$

$$\text{Tertiary} \quad :NR_3 + RX \longrightarrow R_4N^+\,X^- \quad \text{Quaternary ammonium salt}$$

   b. Gabriel phthalimide synthesis

$$\text{(phthalimide anion, } \ddot{N}{:}^-) + RX \xrightarrow[\text{solvent}]{DMF} \text{(N-alkylphthalimide, } {:}N{-}R) \xrightarrow[H_2O]{^-OH} R{-}NH_2$$

c. Reduction of azides

$$RX + NaN_3 \xrightarrow[\text{solvent}]{\text{Ethanol}} R{-}N_3 \xrightarrow[\text{2. } H_2O]{\text{1. } LiAlH_4\text{, ether}} R{-}NH_2$$

d. Reduction of nitriles

$$RX + NaCN \xrightarrow{\text{DMF}} R{-}CN \xrightarrow[\text{2. } H_2O]{\text{1. } LiAlH_4\text{, ether}} R{-}CH_2NH_2$$

e. Reduction of amides

$$R{-}CONH_2 \xrightarrow[\text{2. } H_2O]{\text{1. } LiAlH_4\text{, ether}} R{-}CH_2NH_2 \qquad \text{Primary}$$

$$R{-}CONHR \xrightarrow[\text{2. } H_2O]{\text{1. } LiAlH_4\text{, ether}} R{-}CH_2NHR \qquad \text{Secondary}$$

$$R{-}CONR_2 \xrightarrow[\text{2. } H_2O]{\text{1. } LiAlH_4\text{, ether}} R{-}CH_2NR_2 \qquad \text{Tertiary}$$

f. Reductive amination of ketones/aldehydes

$$NH_3 + R_2C{=}O \xrightarrow[\text{Ethanol}]{NaBH_3CN} R_2CH{-}NH_2 \qquad \text{Primary}$$

$$R'NH_2 + R_2C{=}O \xrightarrow[\text{Ethanol}]{NaBH_3CN} R_2CH{-}NHR' \qquad \text{Secondary}$$

$$R'_2NH + R_2C{=}O \xrightarrow[\text{Ethanol}]{NaBH_3CN} R_2CH{-}NR'_2 \qquad \text{Tertiary}$$

g. Hofmann rearrangement of amides

$$R{-}CONH_2 \xrightarrow[H_2O,\ \Delta]{^-OH,\ Br_2} R{-}NH_2 + CO_2$$

h. Curtius rearrangement of acyl azides

$$R{-}COCl + NaN_3 \xrightarrow{\text{Ethanol}} R{-}CON_3 \xrightarrow[\Delta]{H_2O} R{-}NH_2 + CO_2$$

2. Reactions of amines (Section 28.8)
   a. Alkylation of alkyl halides [see reaction 1(a)]
   b. Nucleophilic acyl substitution (see also Section 24.5)

Ammonia	$NH_3 + RCOCl \xrightarrow{\text{Ether}} RCONH_2 + HCl$	
Primary	$R'NH_2 + RCOCl \xrightarrow{\text{Ether}} RCONHR' + HCl$	
Secondary	$R'_2NH + RCOCl \xrightarrow{\text{Ether}} RCONR'_2 + HCl$	

   c. Sulfonamides (Hinsberg test)

Primary	$R'NH_2 + RSO_2Cl \longrightarrow$	$RSO_2NHR' + HCl$
Secondary	$R'_2NH + RSO_2Cl \longrightarrow$	$RSO_2NR'_2 + HCl$
Tertiary	$R'_3N + RSO_2Cl \longrightarrow$	Product hydrolyzes

d. Hofmann elimination

$$-\overset{H}{\underset{|}{\overset{|}{C}}}-\underset{{}^{+}NR_3}{\underset{|}{\overset{|}{C}}}- \xrightarrow[\Delta]{Ag_2O} \underset{\text{Alkene}}{C{=}C} + \underset{3^\circ\text{ amine}}{R_3N} + H_2O$$

ADDITIONAL PROBLEMS

28.20 Classify each of the amine nitrogen atoms in the following substances as either primary, secondary, or tertiary:

(a) $(C_2H_5)_2N-\overset{O}{\overset{\|}{C}}-$ [ring structure with $N-CH_3$ and $N-H$]

Lysergic acid diethylamide

(b) [ring structure with H_3C-N, $N-CH_3$, $N-CH_3$]

Caffeine

28.21 Draw structures corresponding to the following IUPAC names:
(a) *N,N*-Dimethylaniline
(b) (Cyclohexylmethyl)amine
(c) *N*-methylcyclohexylamine
(d) (2-Methylcyclohexyl)amine
(e) 3-(*N,N*-Dimethylamino)propanoic acid
(f) *N*-Isopropyl-*N*-methylcyclohexylamine

28.22 Name the following compounds by IUPAC rules:

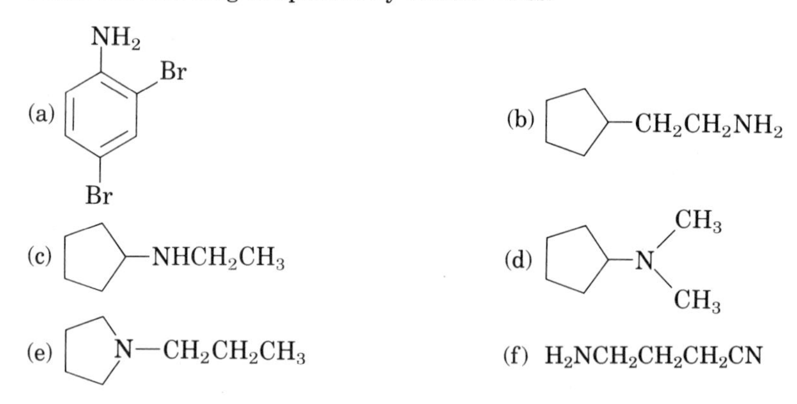

28.23 How can you explain the fact that trimethylamine (bp 3°C) boils lower than dimethylamine (bp 7°C)?

28.24 How would you prepare the following substances from 1-butanol?
(a) Butylamine
(b) Dibutylamine
(c) Propylamine
(d) Pentylamine
(e) *N,N*-Dimethylbutylamine
(f) Propene

28.25 How would you prepare the following substances from pentanoic acid?
(a) Pentanamide
(b) Butylamine
(c) Pentylamine
(d) 2-Bromopentanoic acid
(e) Hexanenitrile
(f) Hexylamine

28.26 Propose a mechanism to explain the observation that treatment of bromoacetone with ammonia yields a compound having the formula $C_6H_{10}N_2$, rather than the expected 1-amino-2-propanone. What is a likely structure for the product?

28.27 How might you prepare pentylamine from these starting materials?
(a) Pentanamide
(b) Pentanenitrile
(c) 1-Butene
(d) Hexanamide
(e) 1-Butanol
(f) 5-Decene
(g) Pentanoic acid

28.28 We said in Section 28.2 that most chiral trisubstituted amines cannot be resolved into enantiomers because nitrogen pyramidal inversion occurs too rapidly. One exception to this generalization is the substance known as Tröger's base. Make molecular models of Tröger's base and then explain why it is resolvable into enantiomers.

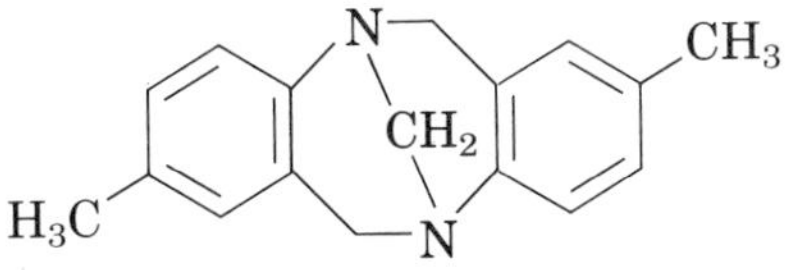

N, CH_3, CH_2, H_3C, N

Tröger's base

28.29 Predict the product(s) of these reactions. If more than one product is formed, tell which is the major one.

(a) 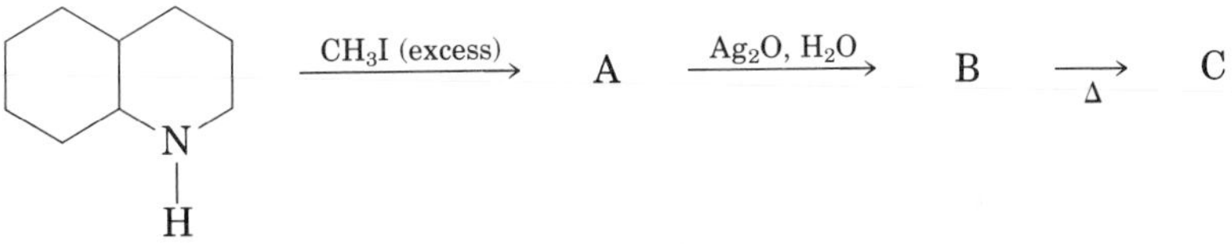 [decahydroquinoline, N–H] $\xrightarrow{CH_3I\ (excess)}$ A $\xrightarrow{Ag_2O,\ H_2O}$ B $\xrightarrow[\Delta]{}$ C

(b) 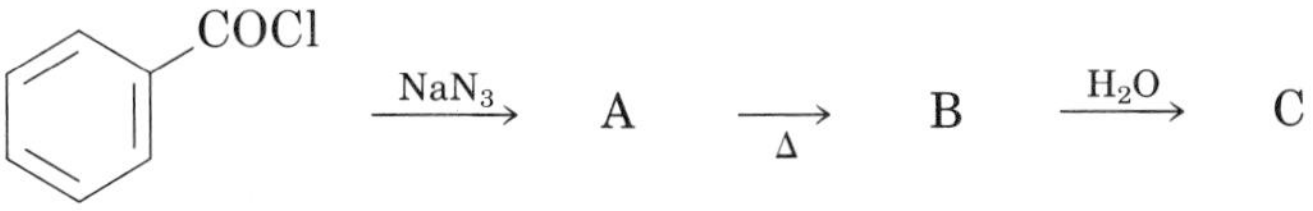 [benzene–COCl] $\xrightarrow{NaN_3}$ A $\xrightarrow[\Delta]{}$ B $\xrightarrow{H_2O}$ C

(c) [phthalimide, O, N–H, O] $\xrightarrow{KOH}$ A $\xrightarrow{C_6H_5CH_2Br}$ B $\xrightarrow[H_2O]{KOH}$ C

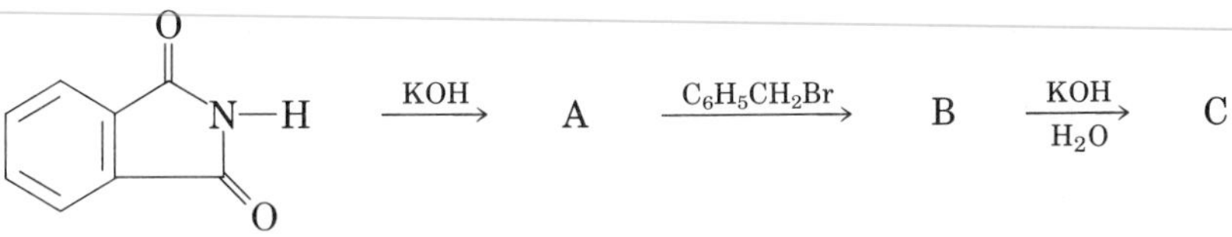

(d) $BrCH_2CH_2CH_2CH_2Br$ + 1 equiv CH_3NH_2 $\xrightarrow[H_2O]{NaOH}$

28.30 Phthalimide used in the Gabriel synthesis is prepared by reaction of ammonia with phthalic anhydride (1,2-benzenedicarboxylic anhydride). Propose a mechanism for the reaction.

28.31 The following syntheses are incorrect as written. What is wrong with each?

(a) $CH_3CH_2CONH_2 \xrightarrow[H_2O]{Br_2,\ ^-OH} CH_3CH_2CH_2NH_2$

(b) 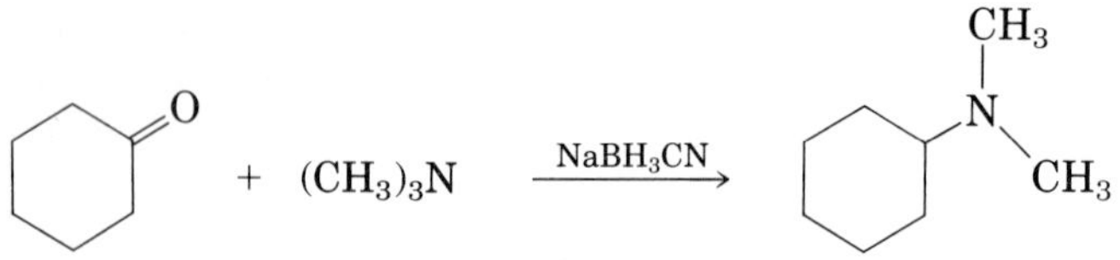 Cyclohexanone + $(CH_3)_3N$ $\xrightarrow{NaBH_3CN}$ N,N-dimethylcyclohexylamine (cyclohexyl–N(CH_3)CH_3)

(c) $(CH_3)_3C—Br + NH_3 \longrightarrow (CH_3)_3C—NH_2$

(d) 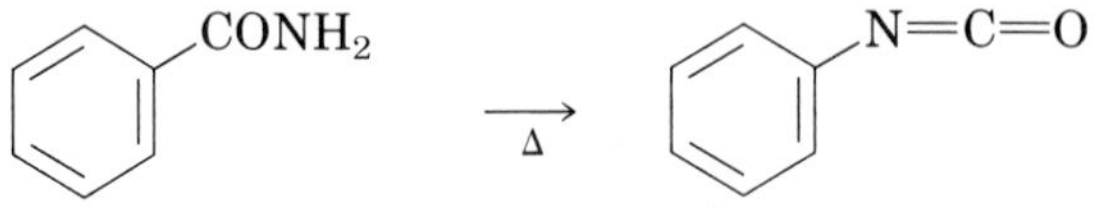 Benzene–$CONH_2$ $\xrightarrow{\Delta}$ Benzene–N=C=O

(e) $CH_3CH_2CH_2CH(NH_2)CH_3$ $\xrightarrow{\text{1. } CH_3I \text{ (excess)} \quad \text{2. } Ag_2O \quad \text{3. } \Delta}$ $CH_3CH_2CH{=}CHCH_3$

28.32 Coniine, $C_8H_{17}N$, is the toxic principle of poison hemlock. When subjected to Hofmann elimination conditions, coniine yields 5-(*N*,*N*-dimethylamino)-1-octene. When subjected to the Hinsberg test, coniine yields a benzenesulfonamide derivative that is insoluble in base. What is the structure of coniine?

28.33 Atropine, $C_{17}H_{23}NO_3$, is a poisonous alkaloid isolated from the leaves and roots of *Atropa belladonna,* the deadly nightshade. In low doses, atropine acts as a muscle relaxant; 0.5 ng (nanograms, 10^{-9} g) is sufficient to cause pupil dilation. On basic hydrolysis, atropine yields tropic acid, $C_6H_5CH(CH_2OH)COOH$, and tropine, $C_8H_{15}NO$. Tropine is optically inactive and on dehydration with H_2SO_4 yields tropidene. Propose a suitable structure for atropine.

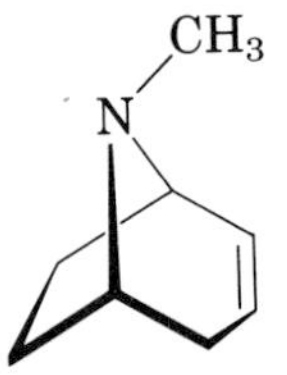

[Structure: bicyclic N–CH_3 bridged ring with one double bond]

Tropidene

28.34 Tropidene (Problem 28.33) has been converted by a series of steps into tropilidene (1,3,5-cycloheptatriene). How would you accomplish this conversion?

28.35 One problem with reductive amination as a method of amine synthesis is the fact that by-products are sometimes obtained. For example, reductive amination of benzaldehyde with methylamine leads to a mixture of methylbenzylamine and methyldibenzylamine. How do you suppose the tertiary amine by-product is formed? Propose a mechanism.

28.36 Cyclopentamine is an amphetamine-like central nervous system stimulant. Propose a synthesis of cyclopentamine from materials of five carbons or less.

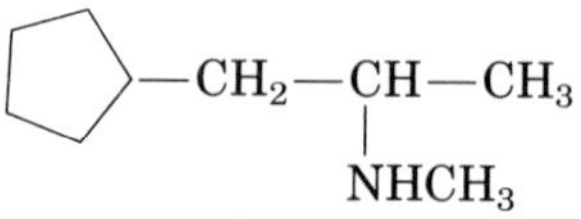

Cyclopentyl$—CH_2—CH(NHCH_3)—CH_3$

Cyclopentamine

28.37 Prolitane is an antidepressant drug that is prepared commercially by a route that involves a reductive amination. What amine and what carbonyl precursors are used?

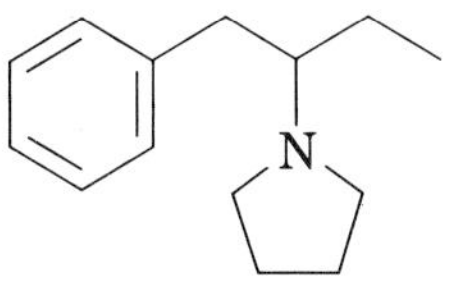

Prolitane

28.38 Tetracaine is a substance used medicinally as a spinal anesthetic during lumbar punctures (spinal taps).

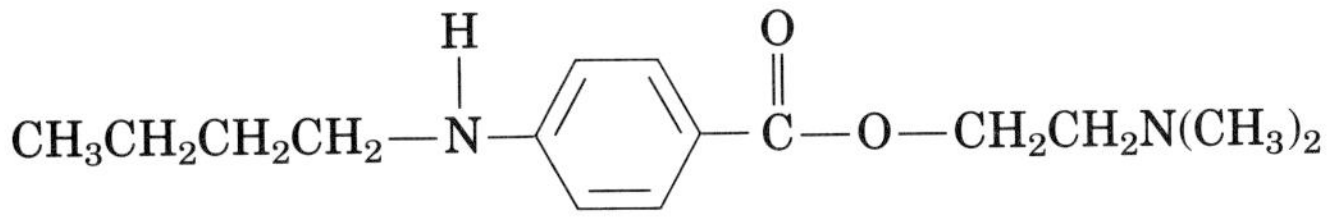

$CH_3CH_2CH_2CH_2{-}NH{-}C_6H_4{-}C(=O){-}O{-}CH_2CH_2N(CH_3)_2$

Tetracaine

(a) How would you prepare tetracaine from the corresponding aniline derivative, $ArNH_2$?
(b) How would you prepare tetracaine from *p*-nitrobenzoic acid?
(c) How would you prepare tetracaine from benzene?

28.39 Propose a structure for the product of formula $C_9H_{17}N$ that results when 2-(2-cyanoethyl)cyclohexanone is reduced catalytically.

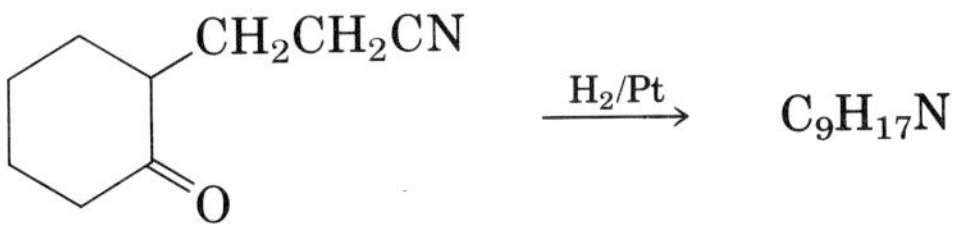

$$\text{2-(}CH_2CH_2CN\text{)cyclohexanone} \xrightarrow{H_2/Pt} C_9H_{17}N$$

28.40 How would you synthesize coniine (Problem 28.32) from acrylonitrile ($H_2C{=}CHCN$) and ethyl 3-oxohexanoate ($CH_3CH_2CH_2COCH_2CO_2C_2H_5$)? [*Hint:* See Problem 28.39.]

CHAPTER 29 ARYLAMINES AND PHENOLS

We have already studied the chemistry of aliphatic amines and aliphatic alcohols, and we have seen the kinds of reactions they undergo. Much of what we have learned about these compounds is applicable to aromatic amines **(arylamines)** and aromatic alcohols **(phenols),** but there are numerous differences, as we will see in this chapter.

29.1 Aniline and the discovery of synthetic dyes

The founding of the modern organic chemical industry can be traced to the need for a single organic compound—aniline—and to the activities of one person—Sir William Henry Perkin.[1] Perkin, a student of Hofmann's at the Royal College of Chemistry in London, worked during the day on problems assigned him by Hofmann but spent his free time working on his own ideas in an improvised home laboratory. He decided to examine the oxidation of aniline with potassium dichromate and carried out the reaction over Easter vacation in 1856. Although the reaction appeared unpromising at first and yielded a tarry black product, by careful extraction with methanol Perkin was able to isolate a few percent yield of a beautiful purple pigment with the properties of a dye.

The only dyes known at the time were the naturally occurring vegetable dyes such as indigo, and Perkin's synthetic purple dye, which he named *mauve,* created a sensation. Realizing the possibilities, Perkin resigned his post with Hofmann and, at the age of 18, formed a company to manufacture and exploit his remarkable discovery. Since there had never before been a need for synthetic chemicals, no chemical industry existed at the time. Large-scale chemical manufacture was unknown, and Perkin's first task was to devise a procedure for preparing the needed quantities of aniline. He therefore worked out the techniques of manufacture and soon learned to prepare aniline on a large scale by nitration of benzene, followed by reduction of nitrobenzene with iron and hydrochloric acid. A similar procedure is used today to prepare some 350,000 tons of aniline annually in the United States, although the reduction step is now carried out by catalytic hydrogenation.

$$\text{Benzene} \xrightarrow[H_2SO_4]{HNO_3} \text{Nitrobenzene } (C_6H_5NO_2) \xrightarrow[\text{or } H_2,\ Ni]{Fe,\ HCl} \text{Aniline } (C_6H_5NH_2)$$

Benzene Nitrobenzene Aniline

[1] Sir William Henry Perkin (1838–1907); b. London; studied at Royal College of Chemistry, London; industrial consultant, London.

Perkin's original mauve was in fact not derived from aniline but from a toluidine (methylaniline) impurity in his starting material. Pure aniline yields a similar dye, however, which came to be marketed under the name *pseudomauveine.*

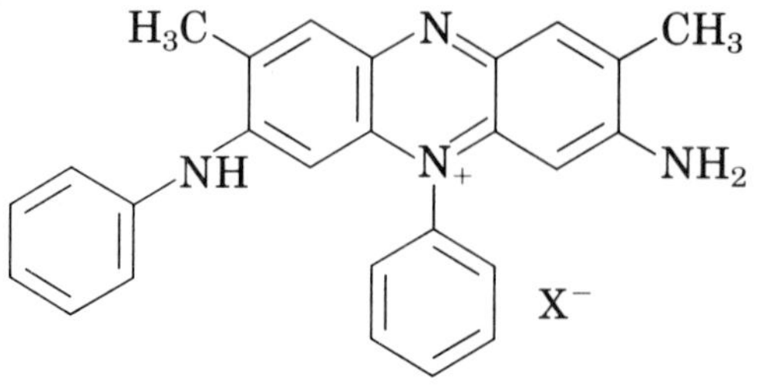

Perkin's mauve
(pseudomauveine has no methyl groups)

Today, dyestuff manufacture is a thriving and important part of the chemical industry, and many commonly used pigments are derived from aniline. Although aniline itself and substituted anilines are available naturally from coal tar, synthesis from benzene is the major source.

29.2 Basicity of arylamines

Arylamines, like their aliphatic counterparts, are basic; the lone pair of nonbonding electrons on nitrogen can form a bond with Lewis acids. The base strength of arylamines is generally lower than that of aliphatic amines, however, because the nitrogen lone-pair electrons are delocalized by orbital overlap with the aromatic ring pi electron system and are less available for bonding. Thus, methylamine has $pK_b = 3.36$, whereas aniline has $pK_b = 9.37$. In resonance terms, arylamines are stabilized relative to alkylamines because of the five contributing resonance structures that can be drawn:

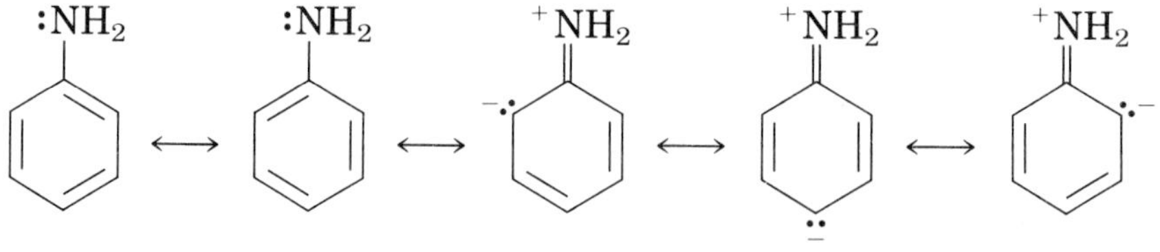

Resonance stabilization is lost on protonation, however, since only two resonance structures are possible for the arylammonium ion:

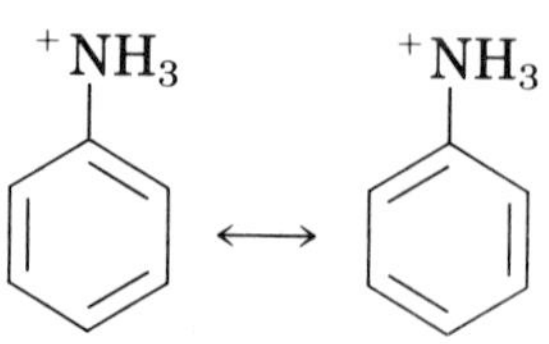

As a result, the energy difference ($\Delta G°$) between protonated and nonprotonated forms is higher for arylamines than it is for alkylamines, and arylamines are therefore less basic. Figure 29.1, which compares reaction energy diagrams for protonation of alkylamines and arylamines, illustrates the difference in $\Delta G°$ for the two reactions.

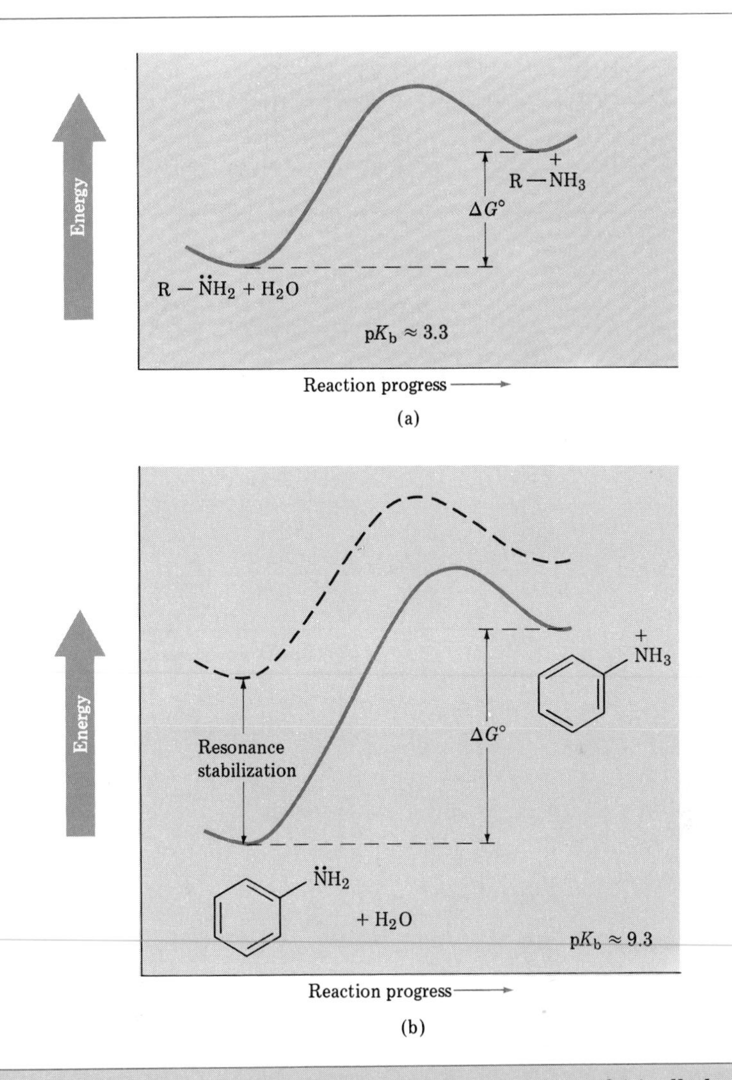

Figure 29.1. Reaction energy diagrams for protonation of (a) alkylamines and (b) arylamines. Arylamines are less basic (have a larger $\Delta G°$), primarily because of resonance stabilization of their ground state.

Substituted arylamines can be either more basic or less basic than aniline, depending on their structure. Table 29.1 presents data for a variety of para-substituted anilines. Recall that compounds with lower pK_b's are stronger bases than compounds with higher pK_b's.

TABLE 29.1 Base strength of para-substituted anilines

$$X-C_6H_4-\ddot{N}H_2 + H_2O \rightleftharpoons X-C_6H_4-\overset{+}{N}H_3 + {}^{-}OH$$

	Substituent (activating groups)	pK_b	*Substituent (deactivating groups)*	pK_b	
Stronger base ↑	—NH_2	7.85	—Cl	10.02	Stronger base ↑
	—OCH_3	8.66	—Br	10.14	
	—CH_3	8.92	—CN	12.26	
Weaker base	—H	9.37	—NO_2	13.0	Weaker base

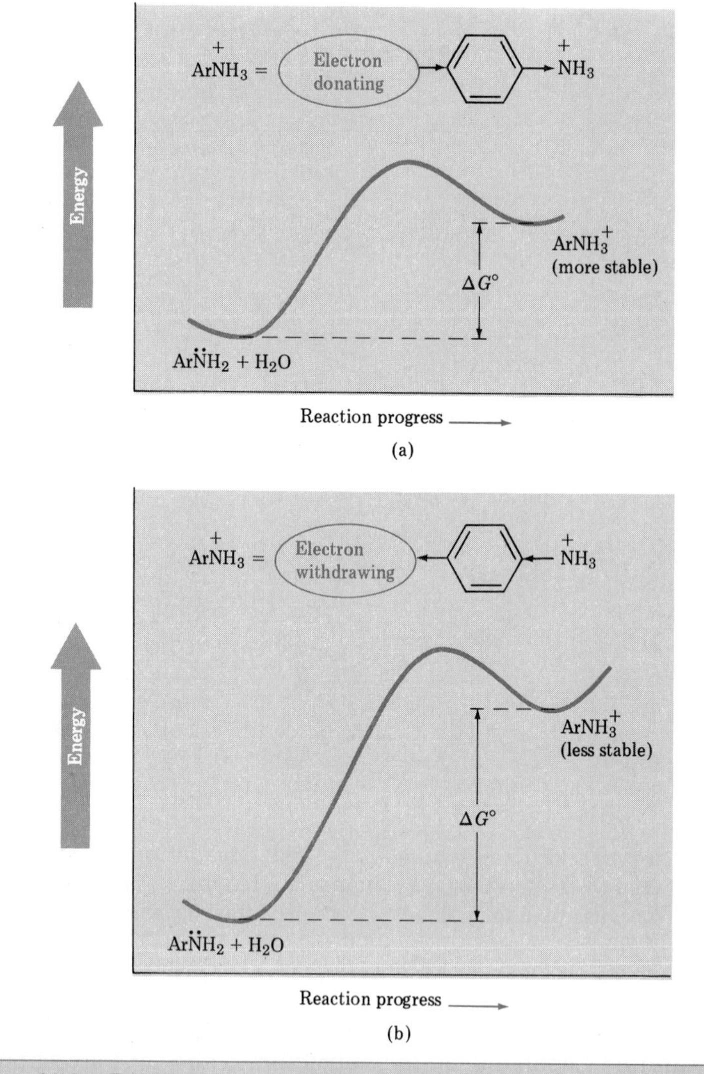

Figure 29.2. Reaction energy diagrams for substituted arylamines. The electron-donating substituent (a) stabilizes the ammonium salt much more than the electron-withdrawing substituent (b) does.

As a general rule, substituents that increase the reactivity of an aromatic ring toward electrophilic substitution (—CH_3, —NH_2, —OCH_3) also increase the basicity of the corresponding arylamine. Conversely, substituents that decrease ring reactivity (—Cl, —NO_2, —CN) also decrease basicity. Table 29.1 considers only para-substituted anilines, but the same general trends are observed for ortho and meta derivatives. The best way to see these effects is to look at reaction energy diagrams for the amine protonation step (Figure 29.2). Activating substituents make the aromatic ring electron-rich, and the nitrogen atom is therefore better able to sustain a positive charge. Deactivating substituents, however, make the aromatic ring electron-poor and decrease the stability of the positively charged nitrogen. We therefore find a lower $\Delta G°$ for protonation of an activated arylamine than for protonation of a deactivated arylamine.

PROBLEM

29.1 How do you account for the fact that *p*-nitroaniline (pK_b = 13.0) is less basic than *m*-nitroaniline (pK_b = 11.5) by a factor of 30? Draw resonance structures to support your answer.

PROBLEM

29.2 Rank the following compounds in order of ascending basicity. Do not look at Table 29.1 to decide your answers.
(a) *p*-Nitroaniline, *p*-aminobenzaldehyde, *p*-bromoaniline
(b) *p*-Chloroaniline, *p*-aminoacetophenone, *p*-methylaniline
(c) *p*-(Trifluoromethyl)aniline, *p*-methylaniline, *p*-(fluoromethyl)aniline

29.3 Preparation of arylamines

Arylamines are almost always prepared by nitration of an aromatic starting material, followed by reduction of the nitro group. No other method of synthesis approaches this nitration–reduction route with respect to versatility and generality. The reduction step can be carried out in many different ways, depending on the circumstances. Catalytic hydrogenation over platinum is clean and gives high yields, but is often not compatible with the presence of other reducible groups (such as carbon–carbon double bonds or carbonyl groups) elsewhere in the molecule. Iron, zinc, and tin metal are also effective, as is stannous chloride ($SnCl_2$). Stannous chloride is particularly mild and is often used when other reducible functional groups are present.

NO_2 / $C(CH_3)_3$ $\xrightarrow[\text{Ethanol}]{H_2/Pt}$ NH_2 / $C(CH_3)_3$

p-tert-Butylnitrobenzene — *p-tert*-Butylaniline (100%)

NO_2, NO_2, CH_3 $\xrightarrow[\text{2. NaOH, } H_2O]{\text{1. Fe, HCl}}$ NH_2, NH_2, CH_3

2,4-Dinitrotoluene → Toluene-2,4-diamine (74%)

NO_2, CHO $\xrightarrow[\text{2. NaOH, } H_2O]{\text{1. } SnCl_2\text{, } H_3O^+}$ NH_2, CHO

p-Nitrobenzaldehyde → *p*-Aminobenzaldehyde (90%)

29.4 Reactions of arylamines

ELECTROPHILIC AROMATIC SUBSTITUTION

An amino substituent is a strongly activating, ortho- and para-directing group in electrophilic aromatic substitution reactions (Section 15.10). The high reactivity of amino-substituted benzenes can be a drawback at times, since it is sometimes difficult to prevent polysubstitution. For example, bromination of aniline is an extremely rapid process that yields the 2,4,6-tribrominated product. The amino group is so strongly activating that it is not possible to stop cleanly at the monobromo stage.

NH_2 $\xrightarrow[H_2O]{Br_2}$ NH_2, Br, Br, Br

Aniline → 2,4,6-Tribromoaniline (100%)

Another drawback to the use of amino-substituted benzenes in electrophilic aromatic substitution reactions is that Friedel–Crafts reactions are not successful (Section 16.4). The amino group forms an acid–base complex with the aluminum chloride Friedel–Crafts catalyst, which prevents further reaction from occurring. Both of these drawbacks—high reactivity and amine basicity—can be overcome, however, by carrying out electrophilic aromatic substitution reactions on the amide, rather than on the free amine. Treatment of an arylamine with acetic anhydride yields an *N*-acetylated product. An amido substituent ($—\ddot{N}HCOR$) is still ortho- and para-directing and activating, since the lone-pair electrons on nitrogen can be shared by resonance with the ortho and para (but not meta) positions. However, it is much less strongly activating and much less basic than an amino group, since the nitrogen lone-pair electrons are delocalized by overlap with the neighboring carbonyl group (Section 28.4). As a result, bromination of an *N*-arylamide occurs cleanly to give a monobromo product, which can be hydrolyzed by aqueous base to give the free bromoamine. For example, *p*-toluidine (4-methylaniline) can be acetylated, brominated, and hydrolyzed to yield 2-bromo-4-methylaniline in 79% yield. None of the 2,6-dibrominated product is obtained.

:NH₂ / CH₃ (*p*-Toluidine) —$(CH_3CO)_2O$, Pyridine→ :NHCOCH₃ / CH₃ —Br_2→ :NHCOCH₃ / Br / CH₃ —^-OH, H_2O→ :NH₂ / Br / CH₃ (2-Bromo-4-methylaniline (79%)) + $CH_3CO_2^-$

Similarly, Friedel–Crafts alkylation and acylation of *N*-arylamides proceed normally. Benzoylation of acetanilide (*N*-acetylaniline) under Friedel–Crafts conditions gives *p*-aminobenzophenone in 80% yield after hydrolysis:

:NH₂ (Aniline) —$(CH_3CO)_2O$, Pyridine→ :NHCOCH₃ (Acetanilide) —C_6H_5COCl, $AlCl_3$→ :NHCOCH₃ / O=C —^-OH, H_2O→ $CH_3CO_2^-$ + :NH₂ / O=C (*p*-Aminobenzophenone (80%))

This simple chemical trick of modulating the reactivity of amino-substituted benzenes by forming an amide is extremely useful and allows many kinds of electrophilic aromatic substitutions to be carried out. A good example is the preparation of **sulfa drugs.** Sulfa drugs such as sulfanilamide were among the first antibiotics to be used clinically against infection. Although they have largely been replaced today by safer and more powerful antibiotics, sulfa drugs were widely used in the 1940s and were credited with saving the lives of thousands of wounded during World War II.

Sulfa drugs are prepared by chlorosulfonation of acetanilide, followed by reaction of *p*-(*N*-acetylamino)benzenesulfonyl chloride with ammonia or some other amine to give a sulfonamide (Section 28.8). Hydrolysis of the amide then yields the sulfa drug. (Note that this hydrolysis can be carried out in the presence of the sulfonamide group, because sulfonamides hydrolyze very slowly.)

:$NHCOCH_3$

Acetanilide

$\xrightarrow{HOSO_2Cl}$

:$NHCOCH_3$

O=S=O

Cl

p-(*N*-Acetylamino)-
benzenesulfonyl chloride

$\xrightarrow[H_2O]{NH_3}$

:$NHCOCH_3$

O=S=O

NH_2

A sulfonamide

$\downarrow$ ^-OH, H_2O

:NH_2

O=S=O

NH_2

Sulfanilamide
(a sulfa drug)

PROBLEM

29.3 Propose a synthesis of sulfathiazole from benzene and any necessary amine.

H_2N–C_6H_4–SO_2–NH–(thiazolyl)

Sulfathiazole

PROBLEM

29.4 Account for the fact that an amido substituent (—NHCOR) is ortho- and para-directing, by drawing resonance structures that share the nitrogen lone-pair electrons with the aromatic ring.

PROBLEM

29.5 Propose syntheses of these compounds from benzene:
(a) *N,N*-Dimethylaniline (b) *p*-Chloroaniline
(c) *m*-Chloroaniline (d) 2,4-Dimethylaniline

DIAZONIUM SALTS: THE SANDMEYER REACTION

Primary aromatic amines react with nitrous acid, HNO_2, to yield stable **arenediazonium salts, $Ar{-}\overset{+}{N}{\equiv}N\ X^-$.** The reaction is versatile, and is compatible with the presence of a wide variety of substituents on the aromatic ring.

$$ArNH_2 + HNO_2 + H_2SO_4 \longrightarrow Ar{-}\overset{+}{N}{\equiv}N\ HSO_4^- + 2\ H_2O$$

Alkylamines also react with nitrous acid, but the products of these reactions, alkanediazonium salts, are too unstable to isolate since they lose nitrogen instantly.

Arenediazonium salts are often highly reactive. They are extremely useful in synthesis, since the diazonio group (N_2^+) can be replaced by nucleophiles ($ArN_2^+ + \ddot{N}u^- \rightarrow ArNu + N_2$). A great many different nucleophiles react with arenediazonium salts, and a great many different substituted benzenes can be prepared. The mechanism by which these substitutions occur is probably a radical, rather than a polar, one. The overall sequence of (1) nitration, (2) reduction, (3) diazotization, and (4) nucleophile replacement is probably the single most versatile method of aromatic substitution (Figure 29.3).

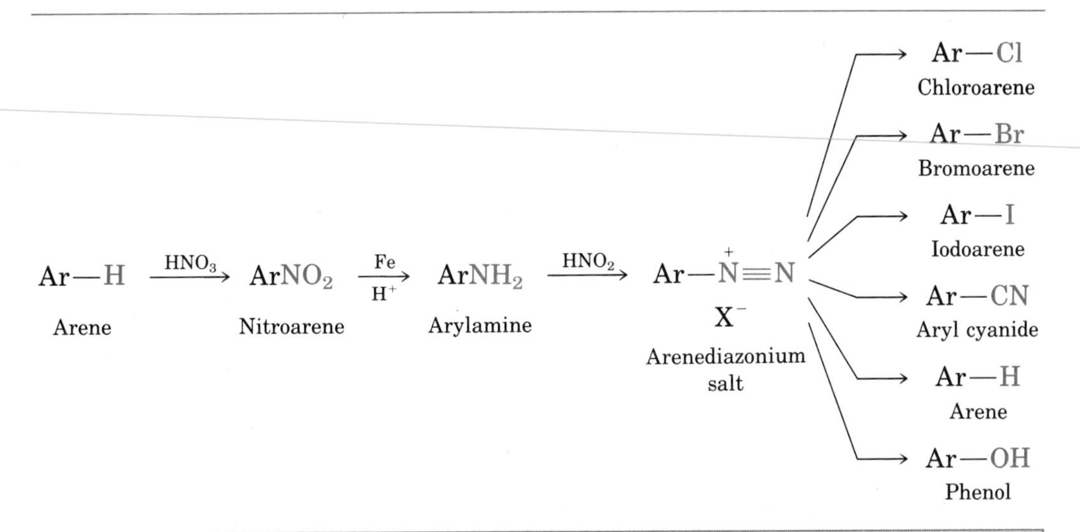

Figure 29.3. Preparation of substituted aromatic compounds by diazonio replacement reactions.

Aryl chlorides and bromides are prepared by reaction of an arenediazonium salt with the corresponding cuprous halide in a process called the **Sandmeyer reaction.**[2] Aryl iodides, however, can be prepared by direct reaction with sodium iodide without using a cuprous salt. Yields generally fall in the 60–80% range.

$$p\text{-}CH_3C_6H_4NH_2 \xrightarrow[H_2SO_4]{HNO_2} p\text{-}CH_3C_6H_4N_2^+\ HSO_4^- \xrightarrow{CuBr} p\text{-}CH_3C_6H_4Br$$

p-Toluidine → *p*-Bromotoluene (73%)

$$C_6H_5NH_2 \xrightarrow[H_2SO_4]{HNO_2} C_6H_5N_2^+\ HSO_4^- \xrightarrow{NaI} C_6H_5I$$

Aniline → Iodobenzene (67%)

Similar treatment of an arenediazonium salt with cuprous cyanide yields the aryl cyanide, ArCN. This reaction is particularly useful, since it allows the replacement of a nitrogen substituent by a carbon substituent, which can be further elaborated into other functional groups. For example, hydrolysis of *o*-methylbenzonitrile, produced by Sandmeyer reaction of *o*-methylbenzenediazonium bisulfate with cuprous cyanide, yields *o*-methylbenzoic acid. This product cannot be prepared from *o*-xylene by the usual arene oxidation route, since both methyl groups would be oxidized.

$$o\text{-}CH_3C_6H_4NH_2 \xrightarrow[H_2SO_4]{HNO_2} o\text{-}CH_3C_6H_4N_2^+\ HSO_4^- \xrightarrow{CuCN} o\text{-}CH_3C_6H_4CN \xrightarrow{H_3O^+} o\text{-}CH_3C_6H_4COOH$$

o-Toluidine → *o*-Methylbenzenediazonium bisulfate → *o*-Methylbenzonitrile (70%) → *o*-Methylbenzoic acid

The diazonio group can also be replaced by —OH to yield phenols and by —H to yield arenes. Phenols are usually prepared by addition of an arenediazonium salt to hot aqueous acid. This reaction is especially important, since few other general methods exist for introducing an —OH group onto an aromatic ring.

[2]Traugott Sandmeyer (1854–1922); b. Wettingen, Switzerland; Ph.D. Heidelberg; Geigy Company, Basel, Switzerland.

NH_2 $\xrightarrow[H_2SO_4]{HNO_2}$ $N_2^+\ HSO_4^-$ $\xrightarrow{H_3O^+}$ OH

NO_2 NO_2 NO_2

m-Nitroaniline *m*-Nitrophenol (86%)

Arenes are produced by reduction of the diazonium salt with hypophosphorous acid, H_3PO_2. This reaction is not of great use, however, unless one has a particular need for temporarily introducing an amino substituent onto a ring to take advantage of its activating effect. The preparation of 3,5-dibromotoluene from *p*-toluidine illustrates how this can be done. Dibromination of *p*-toluidine occurs ortho to the strongly directing amino substituent, and diazotization followed by treatment with hypophosphorous acid yields 3,5-dibromotoluene. This product cannot be prepared by direct bromination of toluene, however, since reaction would occur at positions 2 and 4.

NH_2 $\xrightarrow{2\ Br_2}$ Br NH_2 Br $\xrightarrow[H_2SO_4]{HNO_2}$ Br $N_2^+\ HSO_4^-$ Br $\xrightarrow{H_3PO_2}$ Br Br

CH_3 CH_3 CH_3 CH_3

p-Toluidine 3,5-Dibromotoluene

However,

Br

$\xrightarrow[FeBr_3]{2\ Br_2}$ Br

CH_3 CH_3

Toluene 2,4-Dibromotoluene

PROBLEM

29.6 Why do you suppose arenediazonium salts are more stable than alkanediazonium salts?

PROBLEM

29.7 How would you prepare these compounds from benzene, using a diazonium replacement reaction at some point?

(a) *p*-Bromobenzoic acid
(b) *m*-Bromobenzoic acid
(c) *m*-Bromochlorobenzene
(d) *p*-Methylbenzoic acid
(e) 1,2,4-Tribromobenzene

DIAZONIUM COUPLING REACTIONS

In addition to their reactivity in Sandmeyer-type substitution reactions, arenediazonium salts undergo a coupling reaction with activated aromatic rings to yield brightly colored **azo compounds, Ar—N=N—Ar′**:

$$C_6H_5-\overset{+}{N}\equiv N\ Cl^- + C_6H_5-\ddot{X} \longrightarrow C_6H_5-N=N-C_6H_4-\ddot{X}$$

An azo compound

where X = —OH or —NR_2.

Diazonium coupling reactions are typical electrophilic aromatic substitution processes in which the positively charged diazonium ion is attacked by the electron-rich ring of a phenol or an arylamine. Reaction almost always occurs at the para position, although ortho attack can take place if the para position is blocked.

$$C_6H_5-\overset{+}{N}\equiv\ddot{N}\ Cl^- \longleftrightarrow C_6H_5-\ddot{N}=\overset{+}{N}{:}\ Cl^- + C_6H_5-\ddot{O}H$$

Benzenediazonium chloride

Phenol

$$\downarrow$$

$$\left[C_6H_5-N=N-(H)C_6H_4=\overset{+}{O}-H\right]$$

$$\downarrow -H^+$$

$$C_6H_5-N=N-C_6H_4-\ddot{O}H$$

p-Hydroxyazobenzene (orange crystals, mp 152°C)

$$C_6H_5-\overset{+}{N}\equiv N\ Cl^- + C_6H_5-\ddot{N}(CH_3)_2 \longrightarrow C_6H_5-N=N-C_6H_4-\ddot{N}(CH_3)_2$$

Benzenediazonium chloride

N,N-Dimethylaniline

p-(Dimethylamino)azobenzene (yellow crystals, mp 127°C)

Azo-coupled products generally absorb in the visible region of the electromagnetic spectrum because of their extended conjugated pi electron system, and they are widely used as dyes. For example, *p*-(dimethylamino)azobenzene was at one time used as a yellow food-coloring agent in margarine and is now used commercially as a yellow dye added to gasoline; Alizarin Yellow R is used for dyeing wool.

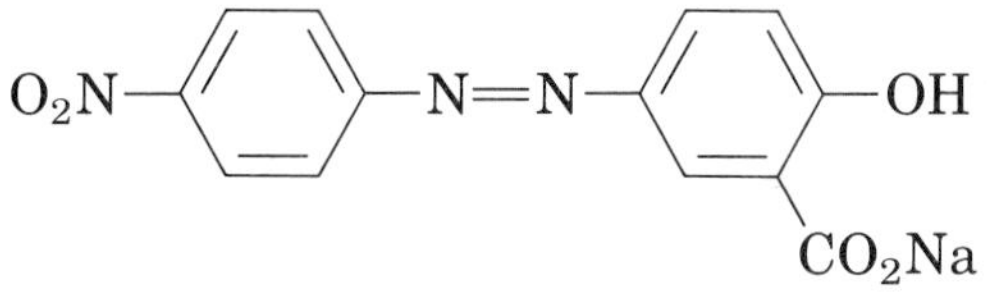

$O_2N-C_6H_4-N{=}N-C_6H_3(OH)(CO_2Na)$

Alizarin Yellow R

PROBLEM

29.8 Propose a synthesis of *p*-(dimethylamino)azobenzene from benzene.

PROBLEM

29.9 Methyl Orange is an azo dye that is widely used as a pH indicator. How would you synthesize Methyl Orange from benzene?

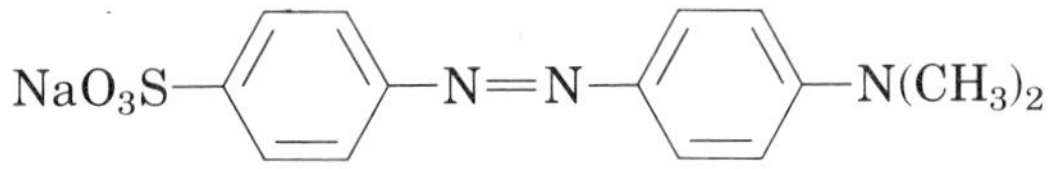

$NaO_3S-C_6H_4-N{=}N-C_6H_4-N(CH_3)_2$

Methyl Orange

29.5 Phenols

Phenols are aromatic alcohols. They occur widely throughout nature, and they serve as important intermediates in the industrial synthesis of products as diverse as adhesives and antiseptics. For example, phenol itself is found in coal tar and is used as a general disinfectant; methyl salicylate occurs in oil of wintergreen and is used both as a flavoring agent and as a liniment; the urushiols are the main allergenic constituents of poison oak and poison ivy.

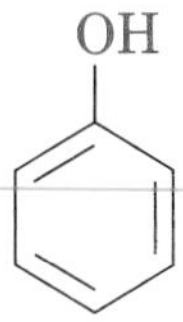

Phenol (also known as carbolic acid

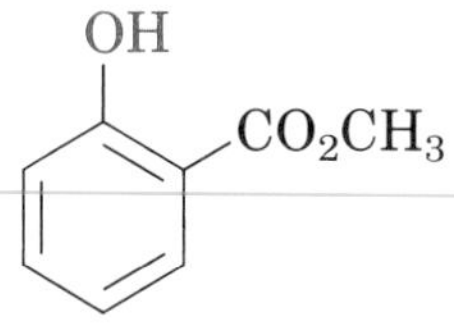

Methyl salicylate

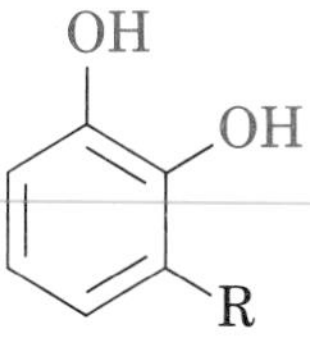

Urushiols (R = different C_{15} alkyl and alkenyl chains)

29.6 Industrial uses of phenols

Although phenol can be isolated in modest quantity from coal tar, the outbreak of World War I provided the stimulus for industrial preparation of large amounts of synthetic phenol as a raw material for manufacture of picric acid (2,4,6-trinitrophenol), which was needed for explosives. Today, approximately 1.5 million tons of phenol per year are manufactured for use in a variety of products, including Bakelite resin and adhesives for binding plywood.

For many years, phenol was manufactured by the Dow process (Section 15.14), in which chlorobenzene reacts with sodium hydroxide at high temperature and pressure. An alternative synthesis from cumene is now in use. Cumene reacts with air at high temperature by a radical mechanism to form cumene hydroperoxide; acid treatment then gives phenol and acetone. This is a particularly efficient process, since two valuable chemicals are prepared at the same time.

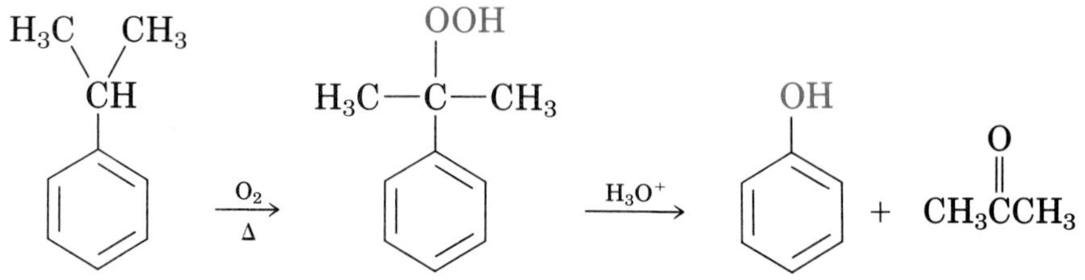

H_3C CH_3 CH; $\xrightarrow[\Delta]{O_2}$ OOH $H_3C—C—CH_3$; $\xrightarrow{H_3O^+}$ OH + O CH_3CCH_3

Cumene Cumene hydroperoxide Phenol Acetone

The reaction occurs by protonation of oxygen, followed by migration of the phenyl group and concurrent loss of water. Readdition of water then yields a hemiacetal, which breaks down to phenol and acetone (Figure 29.4).

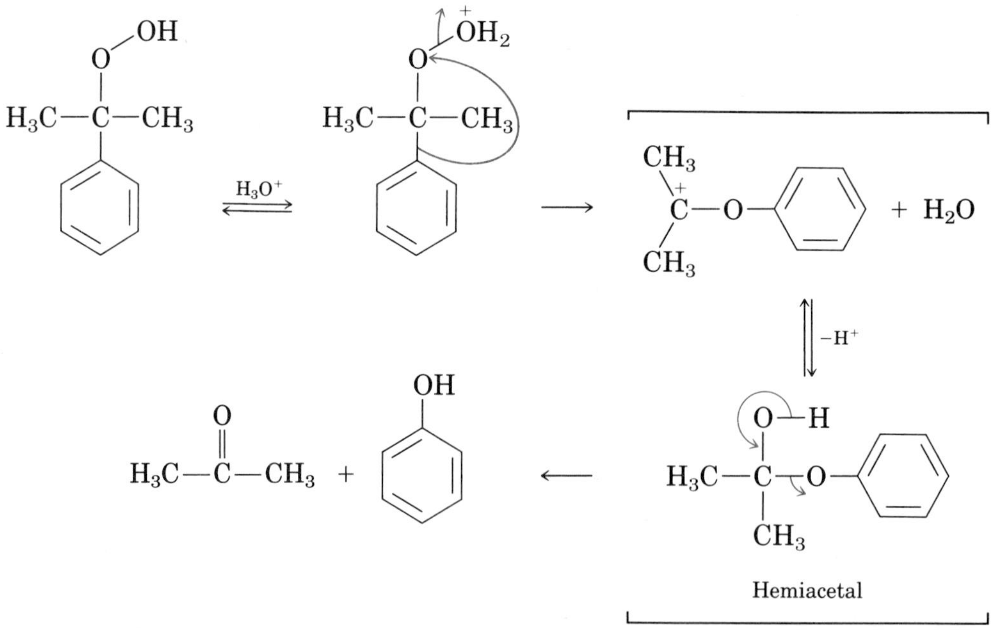

OH O $H_3C—C—CH_3$ $\xrightleftharpoons{H_3O^+}$ $\overset{+}{O}H_2$ O $H_3C—C—CH_3$ $\longrightarrow$ CH_3 $\overset{+}{C}$—O— CH_3 + H_2O

$-H^+$

O $H_3C—C—CH_3$ + OH $\longleftarrow$ O—H $H_3C—C—O—$ CH_3

Hemiacetal

Figure 29.4. Mechanism of the formation of phenol from cumene hydroperoxide.

In addition to its use in resins and adhesives, phenol serves as the starting material for the synthesis of chlorinated phenols and of the food preservatives BHT (butylated hydroxytoluene) and BHA (butylated hydroxyanisole). Thus, pentachlorophenol is prepared by reaction of phenol with excess chlorine and is widely used as a wood preservative. 2,4-Dichlorophenol is used to prepare the herbicide 2,4-D (2,4-dichlorophenoxyacetic acid), and 2,4,5-trichlorophenol is used to prepare the antiseptic agent hexachlorophene.

OH; Cl, Cl, Cl, Cl, Cl

Pentachlorophenol
(wood preservative)

OCH_2COOH; Cl; Cl

2,4-Dichlorophenoxyacetic acid,
2,4-D (herbicide)

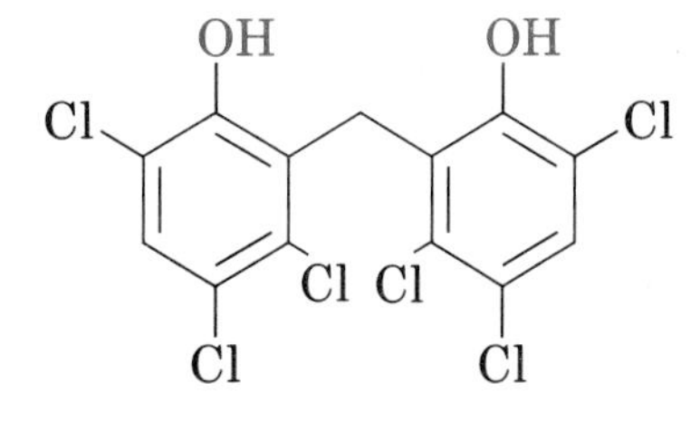

Hexachlorophene
(antiseptic)

The food preservative BHT is prepared by Friedel–Crafts alkylation of *p*-methylphenol (*p*-cresol) with 2-methylpropene in the presence of acid; BHA is prepared similarly by alkylation of *p*-methoxyphenol.

$$p\text{-Methylphenol} + (CH_3)_2C{=}CH_2 \xrightarrow{H^+} \text{BHT}$$

p-Methylphenol (OH, CH_3) — BHT (OH, $(CH_3)_3C$, $C(CH_3)_3$, CH_3)

$$p\text{-Methoxyphenol} + (CH_3)_2C{=}CH_2 \xrightarrow{H^+} \text{BHA}$$

p-Methoxyphenol (OH, OCH_3) — OH, $C(CH_3)_3$, OCH_3 + OH, $C(CH_3)_3$, OCH_3

BHA

PROBLEM

29.10 Propose a synthesis of 2,4-D from phenol.

29.7 Properties of phenols; acidity

The properties of phenols are similar in some respects to those of alcohols. For example, low-molecular-weight phenols are generally somewhat water soluble and are high boiling because of intermolecular hydrogen bonding. The most important property of phenols, however, is their acidity. Phenols are weak acids that can dissociate in aqueous solution to give a proton plus a phenoxide anion, ArO^-. Acidity values for some common phenols are given in Table 29.2.

TABLE 29.2 **Physical properties of some phenols**

	Phenol	*Melting point (°C)*	*Boiling point (°C)*	pK_a
	Acetic acid[a]			4.75
	2,4,6-Trinitrophenol	122	—	0.38
	p-Nitrophenol	115	—	7.15
Stronger acid	*o*-Nitrophenol	97	—	7.17
↑	*m*-Nitrophenol	45	216	8.28
	p-Chlorophenol	43	220	9.20
	p-Iodophenol	94	—	9.20
	p-Bromophenol	66	238	9.25
	p-Hydroxyphenol (hydroquinone)	173	286	9.70
	o-Aminophenol	174	—	9.70
Weaker acid	Phenol	43	182	9.89
	p-Fluorophenol	48	185	9.96
	p-Methylphenol	35	202	10.17
	Ethanol[a]			16.00

[a]Values for acetic acid and ethanol are given for reference.

The data in Table 29.2 show that phenols are much more acidic than alcohols. Indeed, some phenols, such as the nitro-substituted ones, even approach or surpass the acidity of carboxylic acids. One practical consequence of this acidity is that phenols are soluble in dilute aqueous sodium hydroxide. Thus, a phenolic component can often be separated from a mixture of compounds simply by basic extraction into aqueous solution, followed by reacidification.

Phenols are more acidic than alcohols for the same reason that carboxylic acids are more acidic—the phenoxide anion is resonance-stabilized by the aromatic ring. Sharing of the negative charge on oxygen with the ortho and para positions of the aromatic ring results in increased stability of the phenoxide anion relative to undissociated phenol, and in low $\Delta G°$ for the dissociation reaction.

Recall the following:

$$\text{R—C(=O)—O:}^- \longleftrightarrow \text{R—C(—O:}^-\text{)=O}$$

Figure 29.5 compares the acidity of phenols and alcohols through reaction energy diagrams.

The arguments we have just used to account for phenol acidity are similar to those used in Section 29.2 to account for arylamine basicity, but the two situations are opposite. Arylamines are less basic than alkylamines

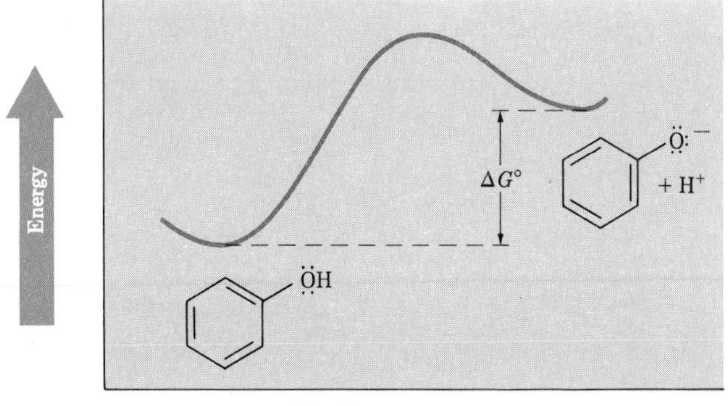

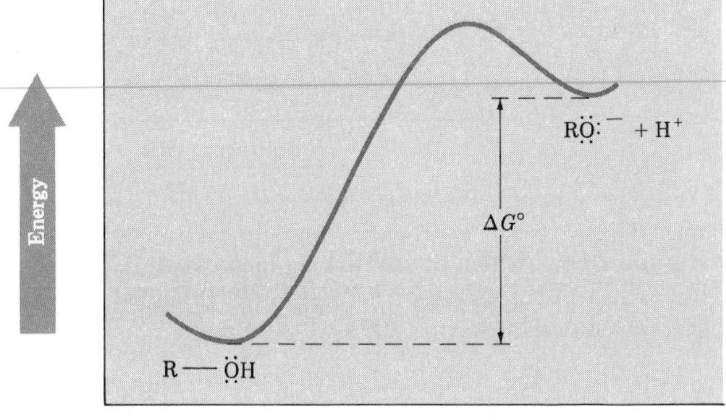

Figure 29.5. Phenol and alcohol acidities. The phenol is more acidic (has a lower $\Delta G°$) because the phenoxide anion is stabilized relative to free phenol more than the alkoxide is stabilized relative to alcohol.

because resonance stabilization of the free arylamine is greater than the stabilization of the arylammonium ion. Phenols, however, are more acidic than alcohols because resonance stabilization of the phenoxide ion is greater than that of the free phenol. In general, any effect that stabilizes the starting material more than it stabilizes the product raises $\Delta G°$. Conversely, any effect that stabilizes the product more than it stabilizes the starting material will lower $\Delta G°$. These effects are shown in Figure 29.6.

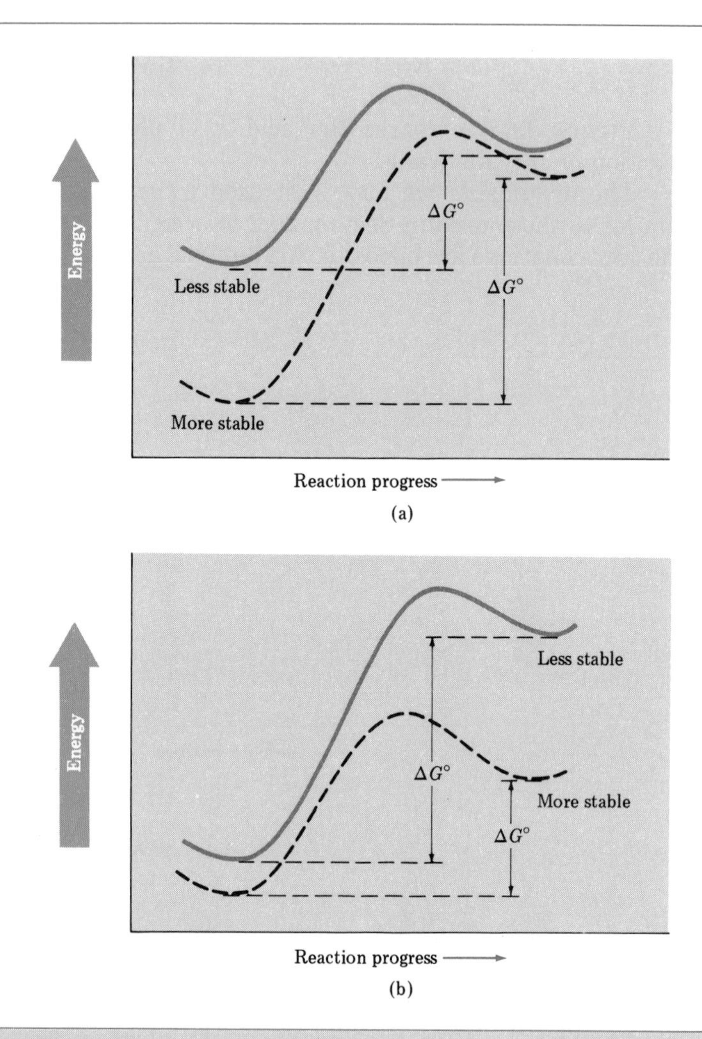

Figure 29.6. Effect of stabilizing factors on $\Delta G°$: (a) stabilization of starting material relative to product raises $\Delta G°$; (b) stabilization of product relative to starting material lowers $\Delta G°$.

Substituted phenols can be either more acidic or less acidic than phenol itself, depending on their structure. As a general rule, phenols that are substituted by an electron-withdrawing substituent are more acidic, and phenols with an electron-donating substituent are somewhat less acidic.

This is particularly true for phenols having a nitro group at the ortho or para positions because the phenoxide anion is strongly stabilized by delocalization of charge into the nitro group.

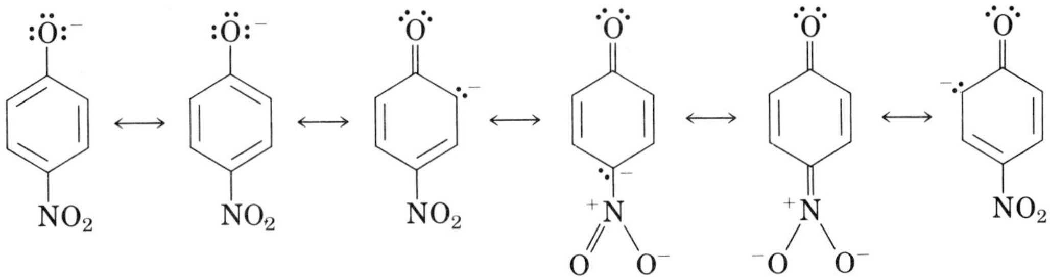

NO_2 ⟷ NO_2 ⟷ NO_2 ⟷ ^+N (O, O^-) ⟷ ^+N (^-O, O^-) ⟷ NO_2

Note that the effect of substituents on phenol acidity is the same as their effect on benzoic acid acidity (Section 23.4), but *opposite* to their effect on aniline basicity (Table 29.3). The same resonance factors operate in all three cases.

TABLE 29.3 Substituent effect on acidity of benzoic acids and phenols, and on basicity of arylamines

Type of compound	X
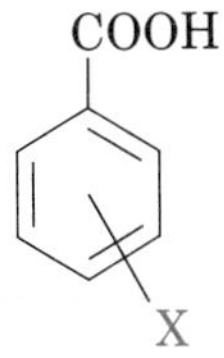 COOH (benzene ring with X)	Electron-withdrawing; increased acidity Electron-donating; decreased acidity
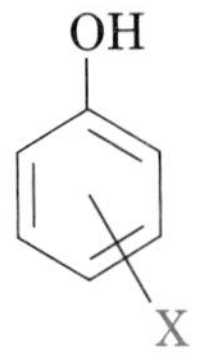 OH (benzene ring with X)	Electron-withdrawing; increased acidity Electron-donating; decreased acidity
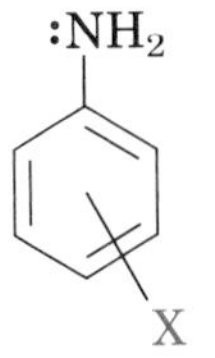 :NH_2 (benzene ring with X)	Electron-withdrawing; decreased basicity Electron-donating; increased basicity

PROBLEM

29.11 Rank the compounds in each group in order of increasing acidity:
(a) Phenol, *p*-methylphenol, *p*-(trifluoromethyl)phenol
(b) Benzyl alcohol, phenol, *p*-hydroxybenzoic acid
(c) *p*-Bromophenol, 2,4-dibromophenol, 2,4,6-tribromophenol

29.8 Preparation of phenols

We have already seen four methods of phenol synthesis. To review briefly:

1. Alkali fusion of aromatic sulfonates occurs when a sulfonic acid is melted with sodium hydroxide at high temperatures (Section 15.4). Few functional groups can survive such harsh conditions, and the reaction is therefore limited to the preparation of alkyl-substituted phenols.

SO_3H / OH / CH_3 / CH_3 / CH_3

$\xrightarrow[H_2SO_4]{SO_3}$ $\xrightarrow[200°C]{NaOH}$

Toluene — *p*-Toluenesulfonic acid — *p*-Methylphenol (72%)

2. Electrophilic aromatic substitution occurs when an arene is treated with H_2O_2 and fluorosulfonic acid (Section 15.5). The reaction is limited to arenes and haloarenes.

OH

+ H_2O_2 $\xrightarrow{HSO_3F}$

Benzene — Phenol (67%)

3. Nucleophilic aromatic substitution by hydroxide ion on halobenzenes can be used to prepare certain kinds of phenols (Section 15.13). Reaction on unactivated halobenzenes normally occurs through a benzyne mechanism and often produces a mixture of products. Reactions on nitro-substituted halobenzenes, however, occur through an addition–elimination mechanism and usually give a single product.

Cl / OH / OH

$\xrightarrow{NaOH}$ $\xrightarrow{H_2O}$ +

CH_3 / CH_3 / CH_3 / CH_3

p-Chlorotoluene — A benzyne — *m*-Methylphenol — *p*-Methylphenol

O_2N / Cl / NO_2 / NO_2 $\xrightarrow[H_2O]{NaOH}$ [HO Cl intermediate] $\xrightarrow{-Cl:^-}$ OH / O_2N / NO_2 / NO_2

2,4,6-Trinitrochlorobenzene

2,4,6-Trinitrophenol (picric acid)

4. Hydrolysis of arenediazonium salts in a Sandmeyer-type replacement reaction is the most versatile and widely used laboratory method of phenol synthesis (Section 29.4). Most functional groups are compatible with the reaction conditions needed, and yields are generally good.

NH_2 / Br / CH_3 $\xrightarrow[H_2SO_4]{HNO_2}$ $N_2^+HSO_4^-$ / Br / CH_3 $\xrightarrow{H_3O^+}$ OH / Br / CH_3

2-Bromo-4-methylaniline

2-Bromo-4-methylphenol (92%)

PROBLEM

29.12 Carvacrol, 5-isopropyl-2-methylphenol, is a natural product isolated from the herbs oregano, thyme, and marjoram. Propose two different syntheses of carvacrol from benzene.

29.9 Reactions of phenols

REACTIONS SIMILAR TO THOSE OF ALCOHOLS

The chemistry of phenols is similar in part to that of alcohols, but considerable differences exist. Thus, phenols can be converted into esters by reaction with acid chlorides or acid anhydrides, and into ethers by reaction with alkyl halides in the presence of base. Both reactions occur under relatively mild conditions since phenols are so much more acidic than alcohols, and since the reactive phenoxide ion intermediates are formed so much more readily than alkoxide ions. Direct esterification by acid-catalyzed reaction between a phenol and a carboxylic acid, however, is not usually successful.

Ester formation

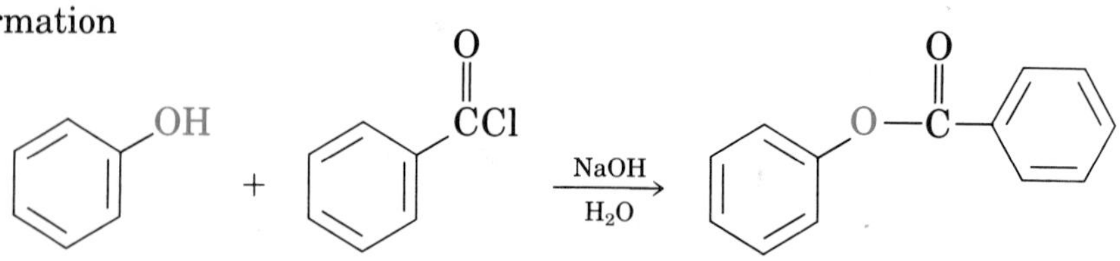

Phenol + Benzoyl chloride $\xrightarrow[H_2O]{NaOH}$ Phenyl benzoate (96%)

Ether formation

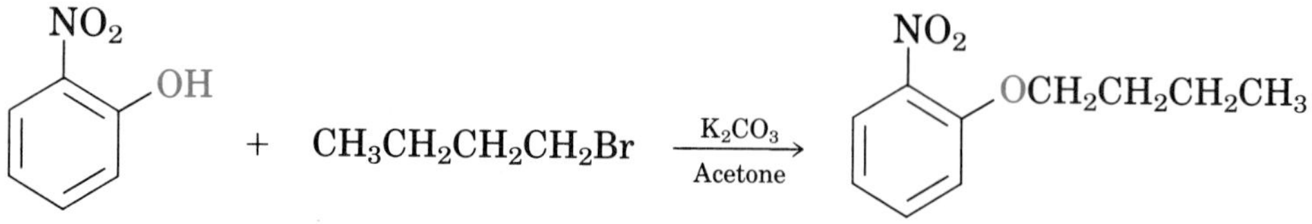

o-Nitrophenol + $CH_3CH_2CH_2CH_2Br$ (1-Bromobutane) $\xrightarrow[\text{Acetone}]{K_2CO_3}$ Butyl *o*-nitrophenyl ether (80%) ($OCH_2CH_2CH_2CH_3$)

ELECTROPHILIC AROMATIC SUBSTITUTION REACTIONS

The hydroxyl group is a strongly activating ortho- and para-directing substituent in electrophilic aromatic substitution reactions (Section 15.10). As a result, phenols are highly reactive substrates for electrophilic halogenation, nitration, and sulfonation, as well as for coupling with diazonium salts to produce azo dyes.

Not surprisingly, phenoxide anions are even more reactive toward electrophilic aromatic substitution than are neutral phenols because a full negative charge is present:

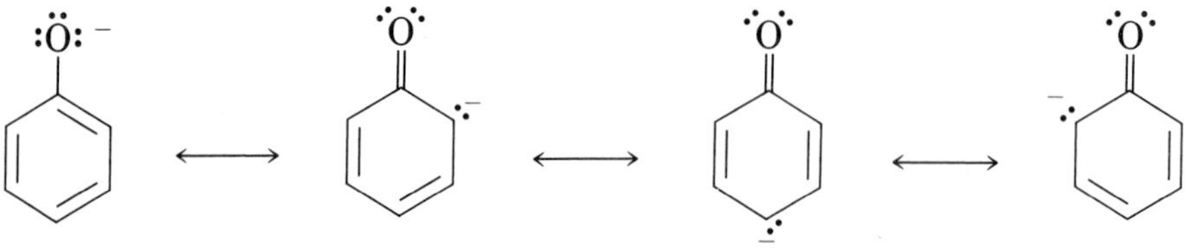

Phenoxide ion

Recall the following:

Ketone enolate ion

Resonance structures of phenoxide anion show a similarity to the resonance structures of a ketone enolate anion and suggest the possibility that phenoxides might undergo alpha-substitution reactions similar to those we saw for ketones (Section 25.6). In practice, phenoxides are less reactive than enolate ions because of the great stability of the benzene ring.

Nevertheless, there are examples of enolate-like reactivity of phenoxide anions. For example, in the **Kolbe–Schmitt carboxylation reaction,**[3,4] phenoxide ion adds to carbon dioxide under pressure to yield an intermediate keto acid anion that enolizes to give *o*-hydroxybenzoic acid (salicylic acid). This reaction is a key step in the industrial synthesis of aspirin (acetylsalicylic acid).

OH

NaOH

:Ö: Na^+ O=C=O

O H COO^-

Phenol

H^+

OH COOH

Salicylic acid

OH COOH

$(CH_3CO)_2O$

$OCOCH_3$ COOH

Acetylsalicylic acid (aspirin)

PROBLEM

29.13 When sodium phenoxide is alkylated with allyl bromide, a mixture of phenyl allyl ether and *o*-allylphenol is formed. How can you account for the formation of *o*-allylphenol?

OXIDATION OF PHENOLS: QUINONES

The susceptibility of phenols to electrophilic aromatic substitution is one consequence of the electron-rich nature of the phenol ring. Another consequence is the susceptibility of phenols to oxidation. Treatment of a phenol with any of a number of strong oxidizing agents yields a **quinone.** Older procedures employed sodium dichromate as oxidant, but Fremy's salt, potassium nitrosodisulfonate $[(KSO_3)_2NO]$, is now preferred. The reaction takes

[3]Herman Kolbe (1818–1884); b. Germany; Ph.D. Göttingen; professor, universities of Marburg and Leipzig.

[4]Rudolf Schmitt (1830–1898); b. Wippershain, Germany; Ph.D. Marburg; professor, University of Dresden.

place under mild conditions through a radical mechanism, and good yields are normally obtained. Arylamines are similarly oxidized to quinones.

$$\text{Phenol} \xrightarrow[H_2O]{(KSO_3)_2NO} \text{Benzoquinone (79\%)}$$

$$\text{2-Methyl-6-methoxyaniline} \xrightarrow[H_2O]{(KSO_3)_2NO} \text{2-Methyl-6-methoxybenzoquinone (96\%)}$$

Quinones are an interesting and valuable class of compounds because of their oxidation–reduction properties. They can be easily reduced to **hydroquinones** (*p*-dihydroxybenzenes) by reagents such as $NaBH_4$ or $SnCl_2$, and hydroquinones can be easily reoxidized back to quinones by Fremy's salt.

$$\text{Benzoquinone} \underset{\text{Fremy's salt}}{\overset{SnCl_2}{\rightleftharpoons}} \text{Hydroquinone}$$

These redox properties of quinones are important to the functioning of living cells, where compounds called **ubiquinones** mediate the electron-transfer processes involved in energy production. Ubiquinones are components of the cells of most aerobic organisms, from the simplest bacteria to man, and are so named because of their ubiquitous occurrence in nature.

CH_3O, CH_3O, CH_3, $(CH_2CH{=}C(CH_3)CH_2)_nH$

Ubiquinones
(n = 1–10)

PROBLEM

29.14 Early work on the structural elucidation of ubiquinones was complicated by the fact that extraction of the compounds from cells was carried out using basic ethanol solution. Under these conditions, the ubiquinone methoxyl groups became exchanged for ethoxyls. Propose a mechanism to account for this exchange.

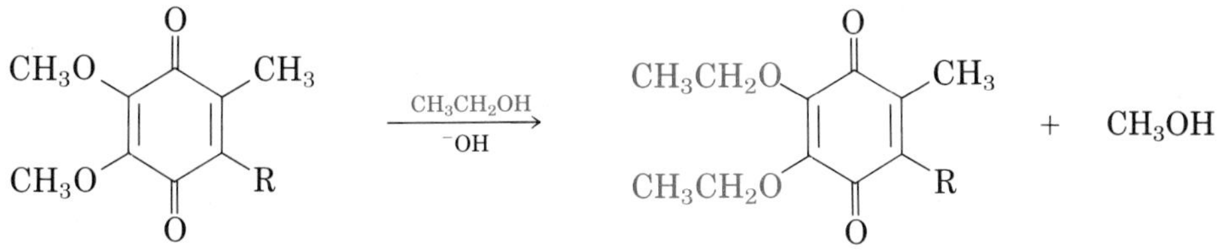

CH_3O, CH_3O, CH_3, R, O, O $\xrightarrow[^{-}OH]{CH_3CH_2OH}$ CH_3CH_2O, CH_3CH_2O, CH_3, R, O, O + CH_3OH

CLAISEN REARRANGEMENT

When a phenoxide ion is alkylated by treatment with an allyl bromide, Williamson ether synthesis occurs and an allyl phenyl ether is produced. Heating the allyl phenyl ether to 200–250°C then effects rearrangement leading to an *o*-allylphenol. The net effect of a **Claisen rearrangement** reaction is alkylation of the ortho position of the phenol.

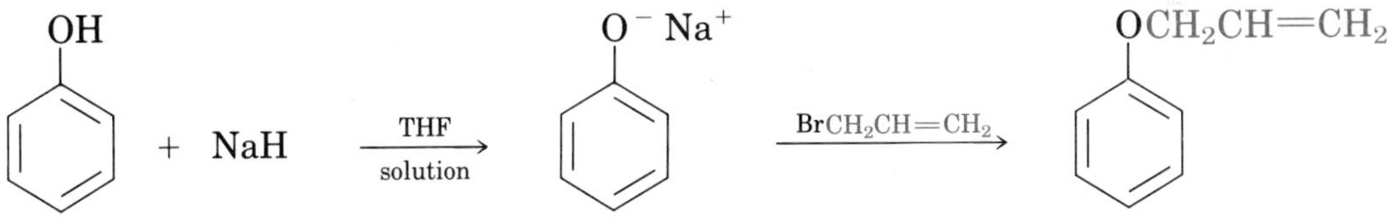

OH + NaH $\xrightarrow[\text{solution}]{\text{THF}}$ $O^- Na^+$ $\xrightarrow{BrCH_2CH{=}CH_2}$ $OCH_2CH{=}CH_2$

Phenol | Sodium phenoxide | Allyl phenyl ether

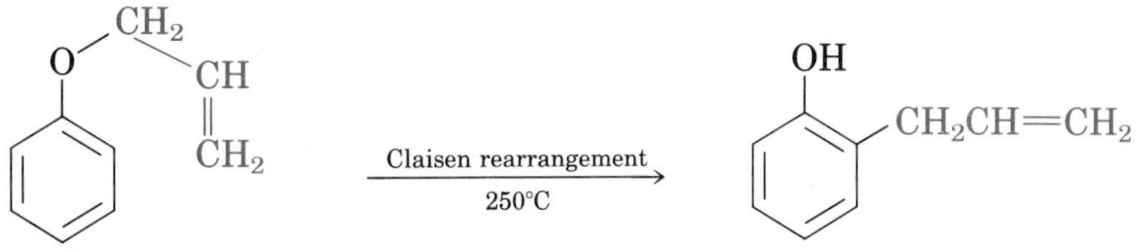

O, CH_2, CH, CH_2 $\xrightarrow[250°C]{\text{Claisen rearrangement}}$ OH, $CH_2CH{=}CH_2$

Allyl phenyl ether | *o*-Allylphenol

Claisen rearrangement of allyl phenyl ethers is a general reaction that is compatible with the presence of many other substituents on the benzene ring. The reaction proceeds through a pericyclic mechanism in which a concerted reorganization of bonding electrons occurs via a cyclic six-membered-ring transition state. The α-allylcyclohexadienone intermediate then tautomerizes to *o*-allylphenol. Good evidence for this mechanism comes from the observation that the rearrangement takes place with an *inversion* of the allyl unit. For example, phenyl allyl ether containing a ^{14}C label on the allyl ether carbon atom yields *o*-allylphenol in which the label is on the terminal carbon. It would be very difficult to explain this result by any mechanism other than a pericyclic one, and we will take a look at this reaction in more detail in Section 30.10.

Allyl phenyl ether → [Transition state]‡ → [Intermediate (6-allyl-2,4-cyclohexadienone)] → *o*-Allylphenol

PROBLEM

29.15 What product would you expect to obtain from Claisen rearrangement of phenyl 2-butenyl ether?

$$C_6H_5\text{—}O\text{—}CH_2CH{=}CHCH_3 \xrightarrow{250°C}$$

Phenyl 2-butenyl ether

29.10 Spectroscopy of arylamines and phenols

INFRARED SPECTROSCOPY

The infrared spectra of arylamines and phenols are little different from those of aliphatic amines and alcohols. For example, aniline shows infrared absorptions at 3400 and 3500 cm^{-1} that are characteristic of a primary amine, and it shows a pair of bands at 1500 and 1600 cm^{-1} that is characteristic of aromatic rings (Figure 29.7). Note that the infrared spectrum of aniline also shows the typical monosubstituted aromatic-ring peaks at 690 and 760 cm^{-1}.

Phenol shows a characteristic broad absorption at 3500 cm^{-1} due to the hydroxyl group, and the usual 1500 cm^{-1} and 1600 cm^{-1} aromatic bands (Figure 29.8). Here, too, the monosubstituted aromatic-ring peaks at 690 and 760 cm^{-1} are visible.

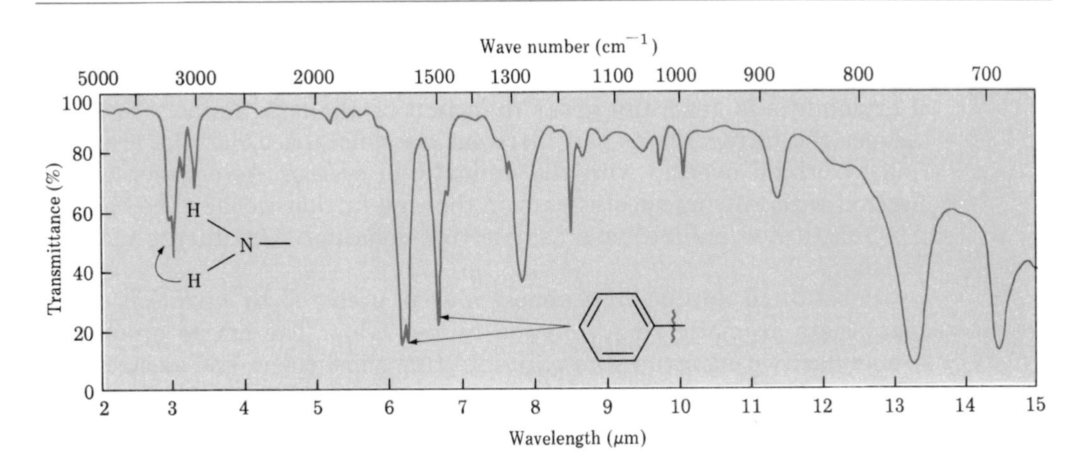

Figure 29.7. Infrared spectrum of aniline.

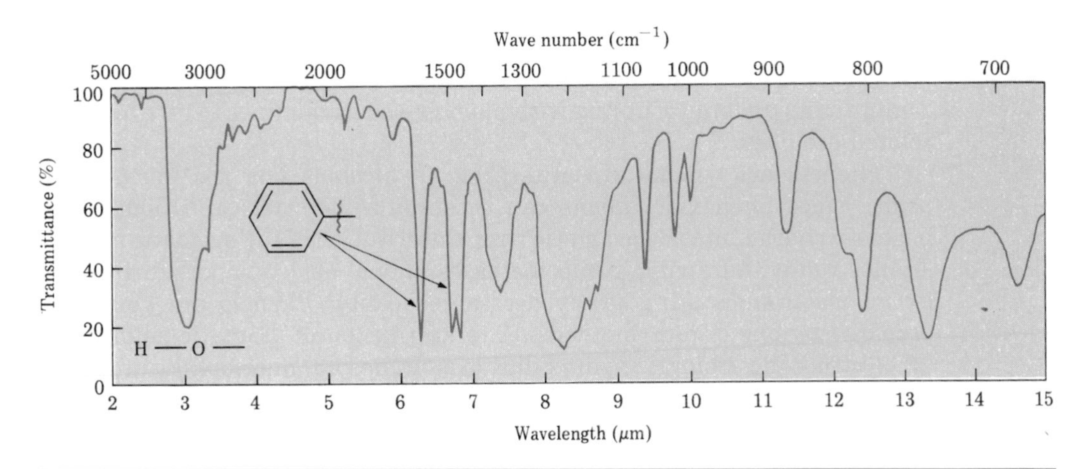

Figure 29.8. Infrared spectrum of phenol.

NUCLEAR MAGNETIC RESONANCE SPECTROSCOPY

Arylamines and phenols, like all aromatic compounds, show NMR absorptions near 7–8 δ, the expected position for aromatic-ring protons. In addition, amine N—H protons usually absorb in the 2–3 δ range, and phenol O—H protons absorb at 2.5–6 δ. In neither case are these absorptions uniquely diagnostic for arylamines or phenols, since other kinds of protons absorb in the same range. As a result, a combination of both NMR and infrared evidence is usually needed to assign structure.

As was true for alcohols (Section 20.12), the identity of the NMR peak due to NH and OH protons can be easily determined by adding a small amount of D_2O to the NMR sample tube. The OH and NH protons are rapidly exchanged with added D_2O, and their peaks disappear from the spectrum.

29.11 Summary

Arylamines, like their aliphatic counterparts, are basic. The base strength of arylamines is generally lower than that of aliphatic amines, however, because the nitrogen lone-pair electrons are delocalized into the aromatic ring by orbital overlap with the aromatic pi system. As a general rule, electron-withdrawing substituents on the ring further weaken the basicity of a substituted aniline, whereas electron-donating substituents increase basicity.

Substituted anilines are almost always prepared by nitration of the appropriate aromatic ring, followed by reduction. The amine group is a strongly activating ortho- and para-directing substituent, and electrophilic aromatic substitution is an important reaction of arylamines. If the amine group makes the ring too reactive, however, its reactivity can be modulated by converting it into a nonbasic amide.

The most important reaction of arylamines is conversion by nitrous acid into arenediazonium salts, $ArN_2^+X^-$. The diazonio group can then be replaced by many other substituents to give a wide variety of substituted aromatic compounds. For example, aryl chlorides, bromides, iodides, and nitriles can be prepared from arenediazonium salts, as can arenes and phenols. In addition to their reactivity toward substitution reactions, diazonium salts undergo coupling with phenols and arylamines to give brightly colored azo dyes.

Phenols are aromatic counterparts of alcohols but are much more acidic, since phenoxide anions can be stabilized by delocalization of the negative charge into the aromatic ring. Substitution of the aromatic ring by an electron-withdrawing group increases phenol acidity, and substitution by an electron-donating group decreases acidity. Phenols are generally prepared by one of four methods: (1) alkali fusion of aromatic sulfonates; (2) electrophilic aromatic hydroxylation of arenes; (3) nucleophilic aromatic substitution; or (4) hydrolysis of an arenediazonium salt.

Reactions of phenols can occur either at the hydroxyl group or on the aromatic ring. The phenol hydroxyl can be converted into an ester or an ether group. Phenyl allyl ethers are particularly interesting since they undergo Claisen rearrangement to give *o*-allylphenols when heated to 250°C. The hydroxyl group strongly activates the aromatic ring toward ortho and para electrophilic substitution reactions. In addition, phenols can be oxidized to quinones by reaction with Fremy's salt, potassium nitrosodisulfonate.

29.12 Summary of reactions

A. Preparation of arylamines

1. Reduction of nitrobenzenes (Section 29.3)

$$ArNO_2 + H_2 \xrightarrow[\text{Ethanol}]{\text{Pt}} ArNH_2$$

$$ArNO_2 + Fe \xrightarrow[\text{2. HO}^-]{\text{1. H}_3\text{O}^+} ArNH_2$$

$$ArNO_2 + SnCl_2 \xrightarrow[\text{2. HO}^-]{\text{1. H}_3\text{O}^+} ArNH_2$$

B. Reactions of arylamines

1. Electrophilic aromatic substitution (Sections 15.10 and 29.4)

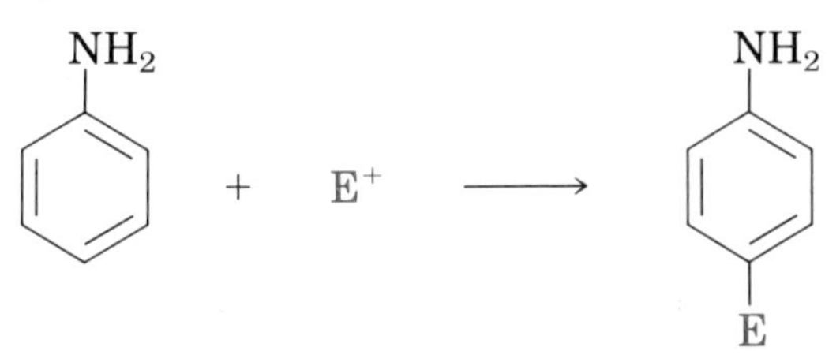

Ortho and para directing

2. Formation of arenediazonium salts (Section 29.4)

$$Ar{-}NH_2 + HNO_2 \xrightarrow{HX} ArN_2^+X^-$$

3. Reaction of arenediazonium salts (Section 29.4)

a. Aryl chlorides

$$ArN_2^+X^- + CuCl \longrightarrow ArCl + CuX + N_2$$

b. Aryl bromides

$$ArN_2^+X^- + CuBr \longrightarrow ArBr + CuX + N_2$$

c. Aryl iodides

$$ArN_2^+X^- + NaI \longrightarrow ArI + NaX + N_2$$

d. Aryl cyanides

$$ArN_2^+X^- + CuCN \longrightarrow ArCN + CuX + N_2$$

e. Phenols

$$ArN_2^+X^- + H_3O^+ \longrightarrow ArOH + HX + N_2$$

f. Arenes

$$ArN_2^+X^- + H_3PO_2 \longrightarrow ArH + HX + N_2$$

g. Diazonium coupling

$$C_6H_5{-}\overset{+}{N}{\equiv}NX^- + C_6H_5{-}OH \longrightarrow C_6H_5{-}N{=}N{-}C_6H_4{-}OH$$

$$C_6H_5{-}\overset{+}{N}{\equiv}NX^- + C_6H_5{-}NR_2 \longrightarrow C_6H_5{-}N{=}N{-}C_6H_4{-}NR_2$$

4. Oxidation to quinones (Section 29.9)

$$C_6H_5{-}NH_2 \xrightarrow[H_2O]{(KSO_3)_2NO} O{=}C_6H_4{=}O$$

C. Preparation of phenols

1. Alkali fusion of aryl sulfonates (Sections 15.4 and 29.8)

$$ArSO_3H + NaOH \xrightarrow[\Delta]{} ArOH$$

2. Hydrolysis of arenediazonium salts (Section 29.4)

$$ArN_2^+X^- + H_3O^+ \longrightarrow ArOH + N_2$$

3. Hydroxylation of arenes (Sections 15.5 and 29.4)

$$ArH + H_2O_2 \xrightarrow{HSO_3F} ArOH + H_2O$$

4. Nucleophilic aromatic substitution (Sections 15.13 and 29.8)

$$ArX + NaOH \longrightarrow ArOH$$

where X = Cl, Br, or I.

D. Reactions of phenols

1. Ester formation (Section 29.9)

$$ArOH + RCOCl \xrightarrow[H_2O]{NaOH} ArOCOR$$

2. Williamson ether synthesis (Sections 19.4 and 29.9)

$$ArOH \xrightarrow[Ethanol]{NaOH} ArO{:}^- \xrightarrow{RX} ArOR$$

3. Kolbe–Schmitt carboxylation (Section 29.9)

$$C_6H_5OH \xrightarrow[\substack{2.\ CO_2,\ \Delta \\ 3.\ H^+}]{1.\ NaOH} o\text{-}HOC_6H_4COOH$$

4. Oxidation to quinones (Section 29.9)

$$C_6H_5OH \xrightarrow[H_2O]{(KSO_3)_2NO} O{=}C_6H_4{=}O$$

5. Claisen rearrangement (Sections 29.9 and 30.12)

$$C_6H_5OCH_2CH{=}CH_2 \xrightarrow{250°C} o\text{-}(CH_2CH{=}CH_2)C_6H_4OH$$

ADDITIONAL PROBLEMS

29.16 Provide IUPAC names for these compounds:

(a)  OH, Br

(b) 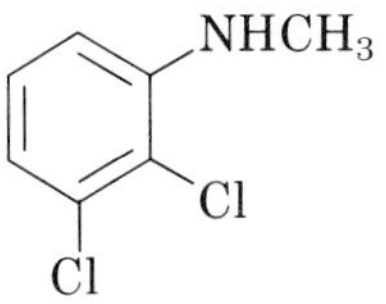 NHCH$_3$, Cl, Cl

(c) 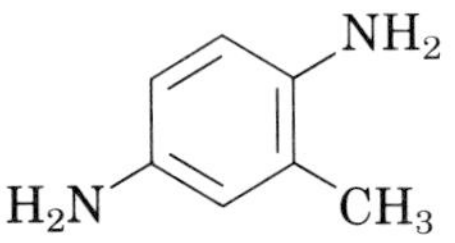 NH$_2$, H$_2$N, CH$_3$

(d) 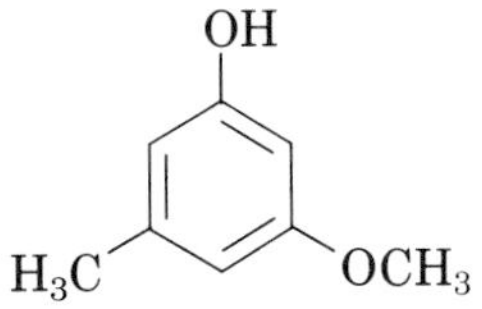 OH, H$_3$C, OCH$_3$

(e) 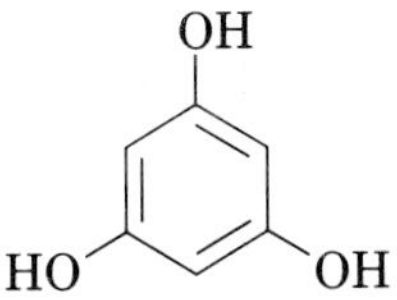 OH, HO, OH

(f) 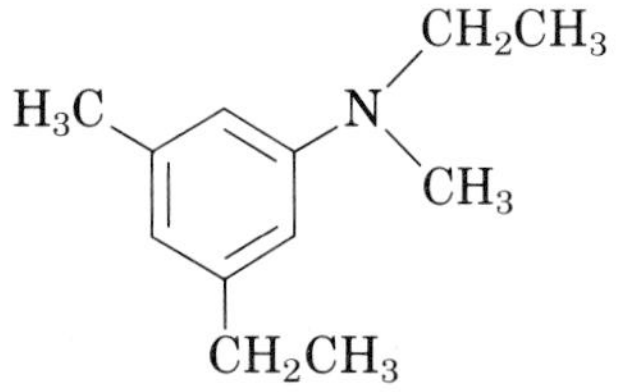 CH$_2$CH$_3$, H$_3$C, N, CH$_3$, CH$_2$CH$_3$

29.17 Tyramine is an alkaloid found, among other places, in mistletoe and ripe cheese. How would you synthesize tyramine from benzene? How would you synthesize it from toluene?

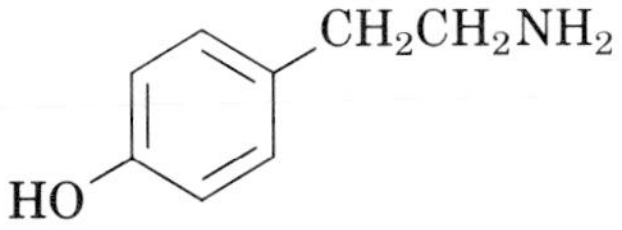

CH$_2$CH$_2$NH$_2$, HO

Tyramine

29.18 How would you prepare aniline from each of these starting materials?
(a) Benzene
(b) Chlorobenzene
(c) Benzoic acid
(d) Toluene

29.19 Suppose that you were given a mixture of toluene, aniline, and phenol and were asked to separate the mixture into its three pure components. Describe in detail how you would do this.

29.20 Give the structures of the major organic products you would expect to obtain from reaction of *m*-toluidine (*m*-methylaniline) with these reagents:
(a) Br_2 (1 mol)
(b) $(KSO_3)_2NO$
(c) CH_3I (excess)
(d) $CH_3Cl + AlCl_3$
(e) CH_3COCl in pyridine
(f) The product of (e), then HSO_3Cl

29.21 Benzoquinone is an excellent dienophile in the Diels–Alder reaction. What product would you expect to obtain from reaction of benzoquinone with 1 mol of butadiene? From reaction with 2 mol of butadiene?

29.22 When the product, A, of the Diels–Alder reaction of benzoquinone and 1 mol of butadiene (Problem 29.21) is treated with dilute acid or base, an isomerization occurs, and a new product, B, is formed. This new product shows a two-proton singlet in the ^{1}H NMR spectrum at 6.7 δ and an infrared absorption at 3500 cm^{-1}. What is the structure of the isomer B?

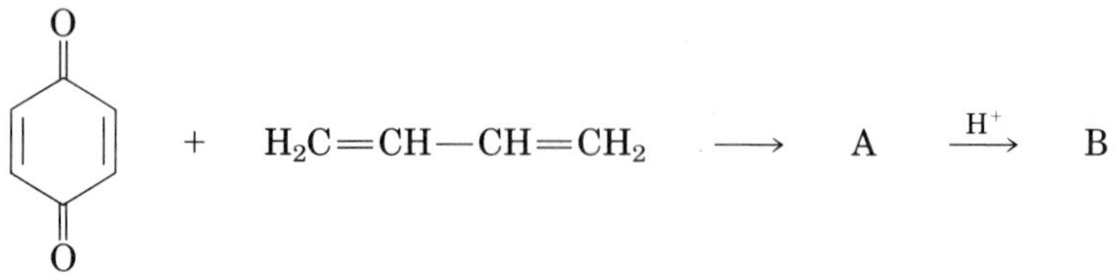

$$\text{(p-benzoquinone)} + H_2C{=}CH{-}CH{=}CH_2 \longrightarrow A \xrightarrow{H^+} B$$

29.23 In the Hoesch reaction, resorcinol (*m*-dihydroxybenzene) is treated with a nitrile in the presence of a Lewis acid catalyst. After hydrolysis, an acyl resorcinol is isolated. Propose a mechanism for the Hoesch reaction. To what other well-known reaction is this similar?

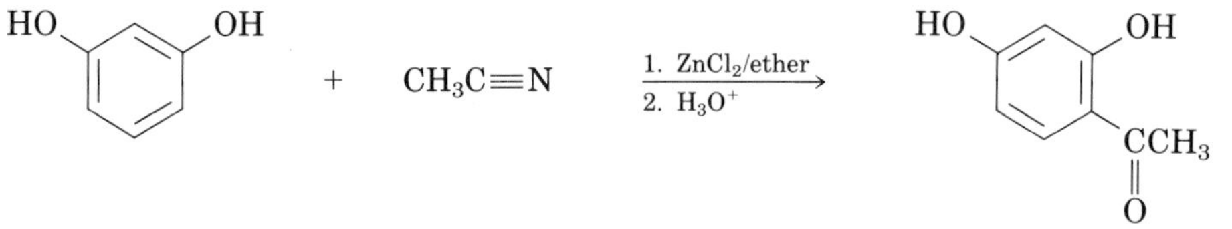

$$\text{(HO, OH resorcinol)} + CH_3C{\equiv}N \xrightarrow[\text{2. } H_3O^+]{\text{1. } ZnCl_2\text{/ether}} \text{(HO, OH, } \underset{\underset{O}{\|}}{C}CH_3 \text{ acyl resorcinol)}$$

29.24 How can you account for the fact that diphenylamine does not dissolve in dilute aqueous HCl and appears to be nonbasic?

29.25 Mephenesin is a drug used as a muscle relaxant and sedative. Propose a synthesis of mephenesin from benzene and any other reagents needed.

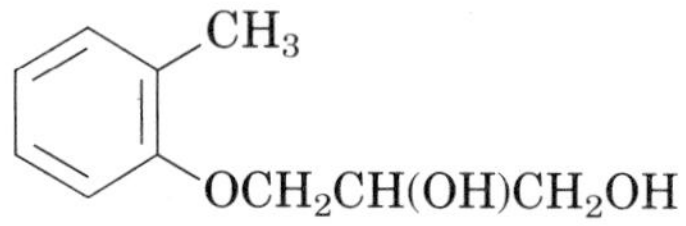

CH_3 / $OCH_2CH(OH)CH_2OH$

Mephenesin

29.26 Gentisic acid is a naturally occurring hydroquinone found in gentian, and its sodium salt is used medicinally as an antirheumatic agent. How would you prepare gentisic acid from benzene?

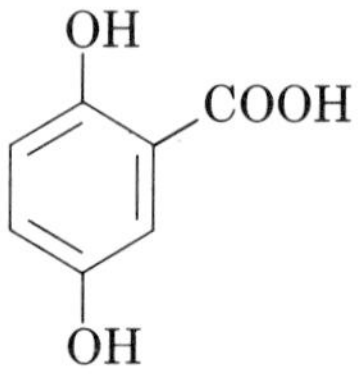

OH / COOH / OH

Gentisic acid

29.27 Prontosil is an antibacterial azo dye that was once used for urinary tract infections. How would you prepare prontosil from benzene?

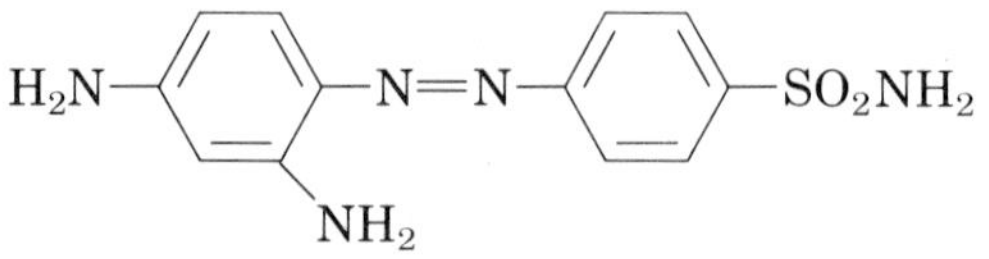

H_2N—(ring, NH_2)—N=N—(ring)—SO_2NH_2

Prontosil

29.28 How would you synthesize the dye Orange II from benzene and β-naphthol?

$\Longrightarrow$

β-Naphthol

Orange II

29.29 2-Nitro-3,4,6-trichlorophenol is used as a lampricide—a compound toxic to lampreys—to combat the intrusion of sea lampreys into the Great Lakes. How would you synthesize this material from benzene?

29.30 The germicidal agent hexachlorophene is prepared by condensation of two molecules of 2,4,5-trichlorophenol with one molecule of formaldehyde in the presence of sulfuric acid. Propose a mechanism to account for this reaction.

Hexachlorophene

29.31 Propose a route from benzene for the synthesis of the antiseptic agent trichlorosalicylanilide.

Trichlorosalicylanilide

29.32 Compound A, $C_8H_{10}O$, has the infrared and ^1H NMR spectra shown. Propose a structure for A consistent with the observed spectral properties. Assign each peak in the NMR spectrum. Note that the absorption at 5.4 δ disappears when D_2O is added.

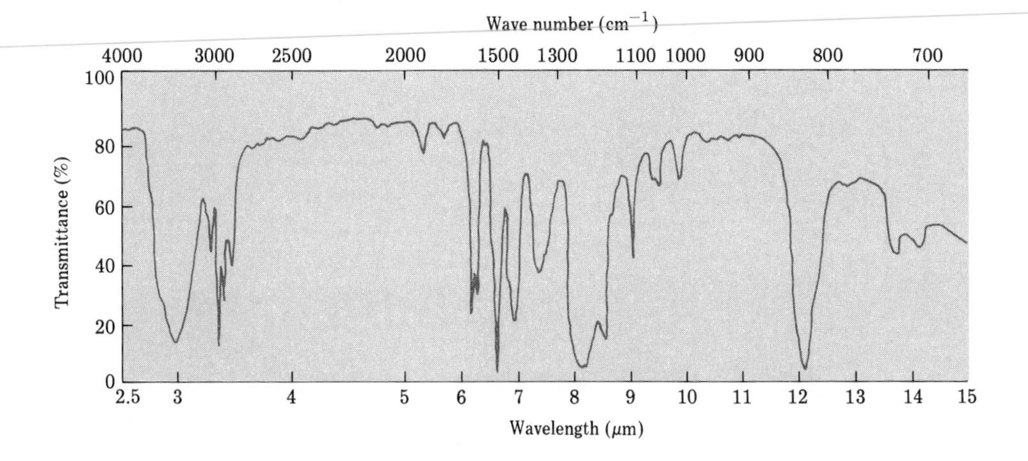

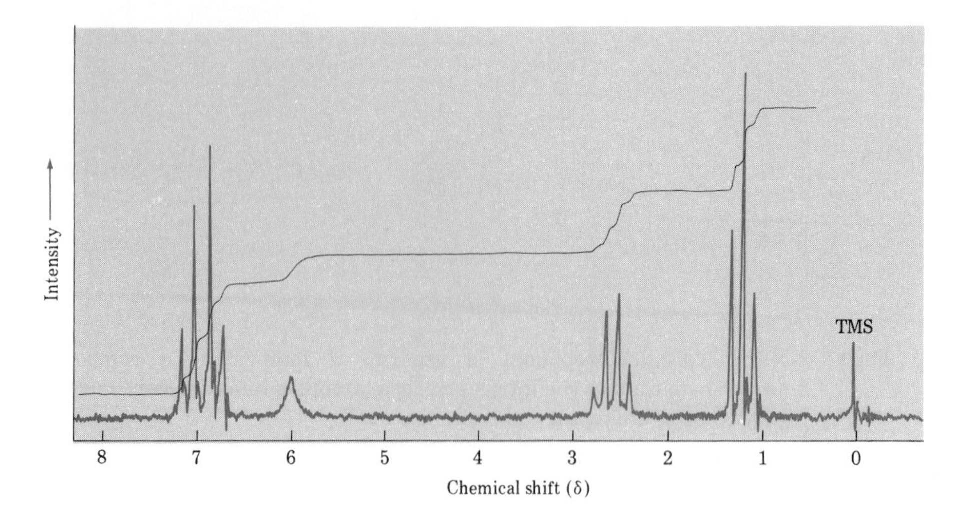

29.33 Phenacetin (the "P" in APC tablets for headaches) has the formula $C_{10}H_{13}NO_2$, and the infrared and NMR spectra shown. Phenacetin itself is neutral and does not dissolve in either acid or base. When warmed with aqueous hydroxide, phenacetin yields an amine, $C_8H_{11}NO$. When heated with HI, the amine is cleaved to an aminophenol C_6H_7NO, which, on treatment with Fremy's salt, yields benzoquinone. What is the structure of phenacetin, and what are the structures of the amine and the aminophenol?

$$\text{Phenacetin} \xrightarrow{^{-}\text{OH}} \text{Amine} \xrightarrow{\text{HI}} \text{Aminophenol} \xrightarrow[\text{salt}]{\text{Fremy's}} \text{Benzoquinone}$$

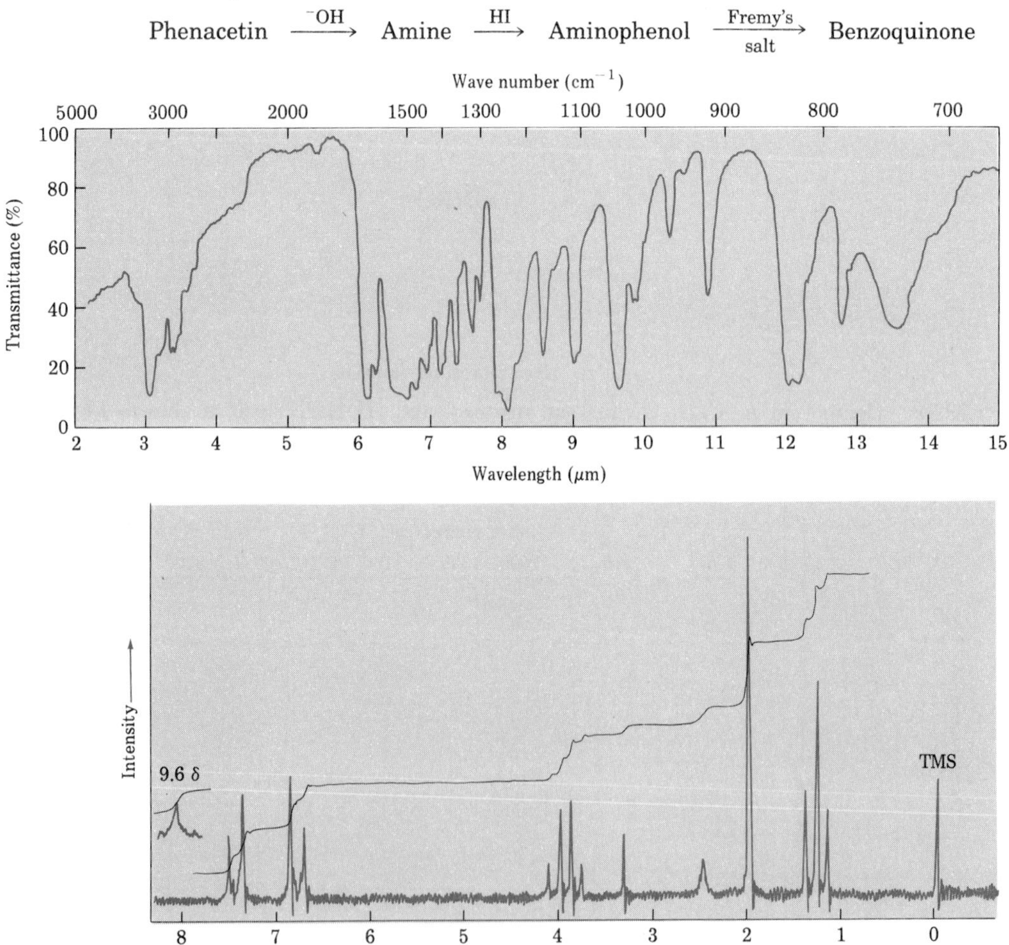

CHAPTER 30 ORBITALS AND ORGANIC CHEMISTRY: PERICYCLIC REACTIONS

Pericyclic reactions constitute the third major class of organic processes, along with polar and radical reactions.

Most organic reactions take place by polar mechanisms, in which an electron-rich nucleophile donates two electrons to form a bond to an electron-poor electrophile. Many other organic reactions take place by radical mechanisms, in which each of two reactants donates one electron to form a new bond. Although much remains to be learned about polar and radical reactions, the broad outlines of both classes have been studied for many years and seem relatively well understood.

The fundamental principles of pericyclic reactions, by contrast, have been understood only recently, beginning in the mid-1960s. Numerous individuals have made major contributions, but it was the work of Robert Woodward[1] and Roald Hoffmann[2] that made most chemists aware of the principles of pericyclic reactions.

30.1 Some examples of pericyclic reactions

We previously defined a pericyclic reaction as one that occurs by a concerted process through a cyclic transition state. The word *concerted* means that all bonding changes occur *at the same time* and *in a single step;* no intermediates are involved. Rather than try to expand this definition now, let's look at some examples of the three major classes of pericyclic reactions.

The Diels–Alder reaction (Section 13.7) is an intermolecular pericyclic reaction between a diene and a dienophile to yield a cyclohexene. Three new bonds are formed (two sigma bonds and one pi bond) and three bonds are broken (two diene pi bonds and one alkene pi bond) at the same time via a cyclic transition state. This is an example of a **cycloaddition reaction.**

Diels-Alder reaction, a cycloaddition

A diene + A dienophile ⟶ [Cyclic transition state]‡ ⟶ A cyclohexene

The Claisen rearrangement (Section 29.9) is an intramolecular pericyclic reaction of an allyl aryl ether. Three new bonds are formed (one sigma bond and two pi bonds) and three bonds are broken (one sigma bond and two pi bonds) at the same time via a cyclic transition state. This is an example of a **sigmatropic rearrangement.**

[1]Robert Burns Woodward (1917–1979); b. Boston, Mass.; Ph.D. Massachusetts Institute of Technology (1937); professor, Harvard University (1941–1979); Nobel prize (1965).

[2]Roald Hoffmann (1937–); b. Zloczow, Poland (now in Soviet Union); Ph.D. Harvard University (1962); professor, Cornell University; Nobel prize (1981).

Claisen rearrangement,
a sigmatropic rearrangement

An allyl aryl ether

Cyclic transition state

Enolize

An *o*-allylphenol

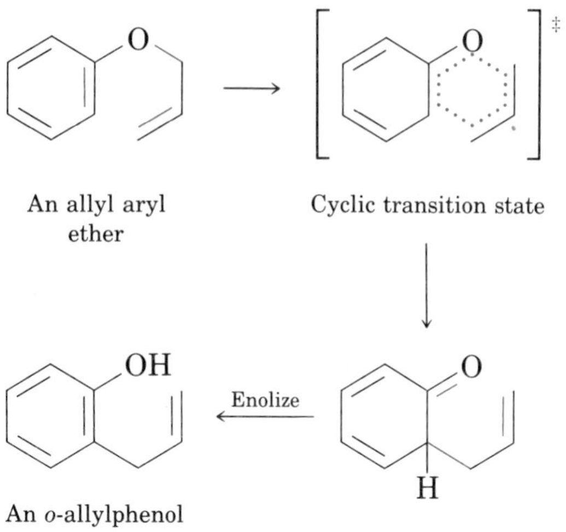

In addition to sigmatropic rearrangements and cycloaddition reactions, there are many other pericyclic processes, some of which have unusual stereochemical features. For example, the thermal cyclization of (2*E*,4*Z*,6*E*)-octatriene leads exclusively to formation of *cis*-5,6-dimethyl-1,3-cyclohexadiene. Cyclization of the (2*E*,4*Z*,6*Z*)-octatriene isomer, however, leads exclusively to formation of *trans*-5,6-dimethyl-1,3-cyclohexadiene. These are examples of **electrocyclic reactions.**

Electrocyclic reactions

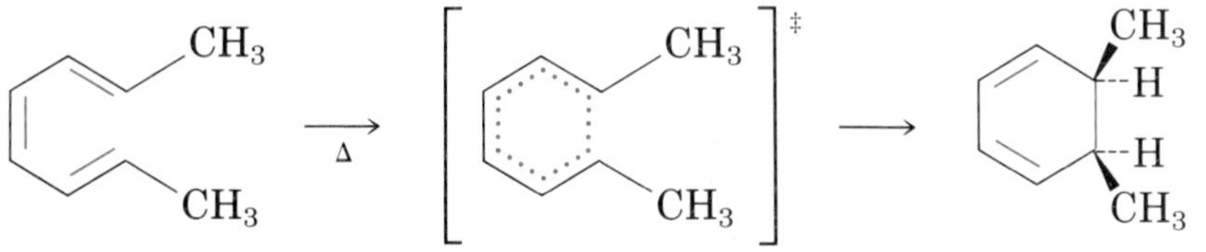

(2*E*,4*Z*,6*E*)-Octatriene

Cyclic transition state

cis-5,6-Dimethyl-1,3-cyclohexadiene

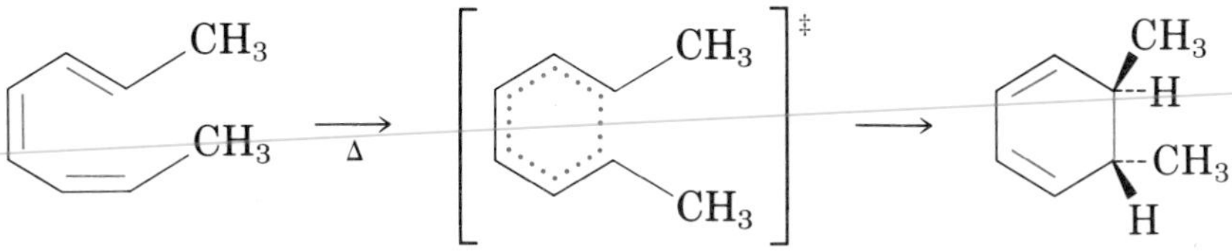

(2*E*,4*Z*,6*Z*)-Octatriene

Cyclic transition state

trans-5,6-Dimethyl-1,3-cyclohexadiene

Both triene cyclizations, and many other pericyclic reactions as well, are **stereospecific;** they yield only a *single* product stereoisomer rather than a mixture. Why should this be so? The answer to this question is fundamental to the nature of pericyclic reactions and has to do with symmetry properties of the reactant and product orbitals. To understand how orbital symmetry affects reactivity, we need to look more deeply into the mathematical description of orbitals.

30.2 Atomic and molecular orbitals

Let's begin by briefly reviewing some of the concepts introduced in Chapter 1. We have defined an orbital as the region of space where a given electron is most likely to be found. In mathematical terms, orbitals are derived from quantum mechanical calculations involving electron wave functions. The exact nature of these calculations is beyond the scope of this text and is not important for our purposes. What is important is this: When the wave functions are solved, the different lobes of an orbital turn out to have algebraic signs, + and −. For example, the two equivalent lobes of a *p* atomic orbital have plus and minus signs, and the two nonequivalent lobes of an sp^3 hybrid orbital also have plus and minus signs (Figure 30.1).

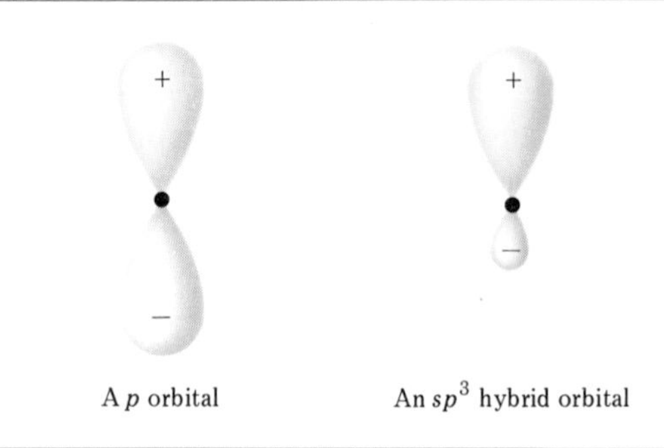

Figure 30.1. Signs of lobes in *p* and sp^3 orbitals.

When a bond is formed between two atoms by overlap of atomic orbitals, the two electrons in the bond no longer occupy atomic orbitals but instead occupy molecular orbitals (MOs). Overlap of two atomic orbitals leads to two molecular orbitals, one of which is lower in energy than either of the original two atomic orbitals, and one of which is higher in energy. The lower-energy molecular orbital, denoted ψ_1 (Greek psi), results from additive overlap of two lobes with the *same* algebraic sign and is a **bonding molecular orbital.** The higher-energy molecular orbital, denoted ψ_2, results from subtractive overlap, however, since the two lobes have *opposite* algebraic signs: ψ_2 is therefore an **antibonding molecular orbital.**

Figure 30.2 shows ψ_1 and ψ_2 for both sigma and pi bonds. Note that the antibonding orbitals have nodes between nuclei—planes of zero electron density between lobes of opposite sign. Higher-energy orbitals have more nodes than lower-energy orbitals and thus have fewer favorable bonding interactions.

Having constructed molecular orbitals, we can arrive at a complete description of sigma and pi bonds by assigning electrons to the orbitals in exactly the same way that we assigned electrons to atomic orbitals in describing the electronic configuration of atoms (the *aufbau* principle, Section 1.3). Each molecular orbital can hold two electrons of opposite spin.

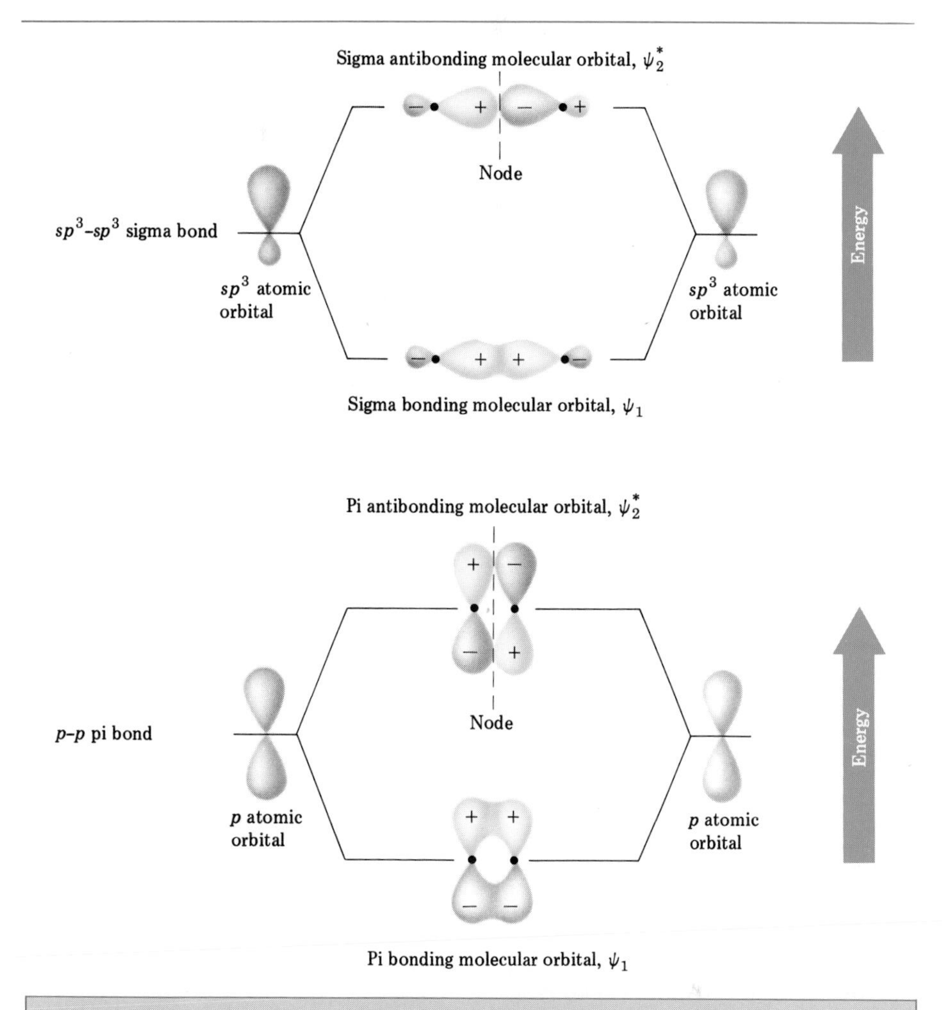

Figure 30.2. Sigma and pi molecular orbitals. Bonding orbitals result from overlap of like lobes, and antibonding orbitals result from overlap of unlike lobes.

The assignments are made by filling the lowest-energy molecular orbitals first and by filling each molecular orbital with two electrons before going on to the next higher orbital. For example, the ground-state electronic configuration of an alkene pi bond has ψ_1 filled and ψ_2 vacant. Irradiation with ultraviolet light, however, excites an electron from ψ_1 to ψ_2, leaving each orbital with one electron (Section 13.12). The electronic configurations of both ground and excited states of ethylene are shown in Figure 30.3, p. 1000.

30.3 Molecular orbitals of conjugated pi systems

The molecular orbital description of a conjugated pi system is more complex than that of a simple alkene because the pi electrons are delocalized over more than two atoms. In 1,3-butadiene, for example, four $2p$ atomic orbitals

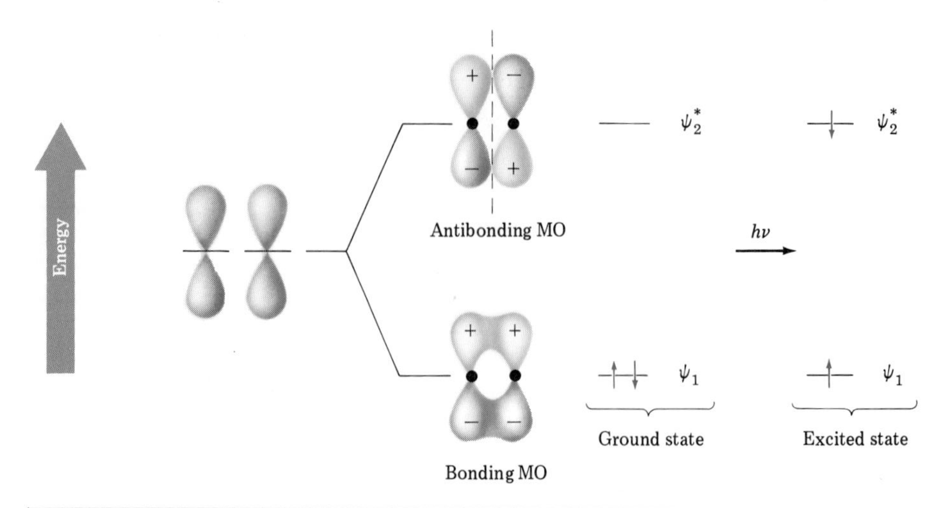

Figure 30.3. Ground-state and excited-state electronic configurations of the ethylene pi bond.

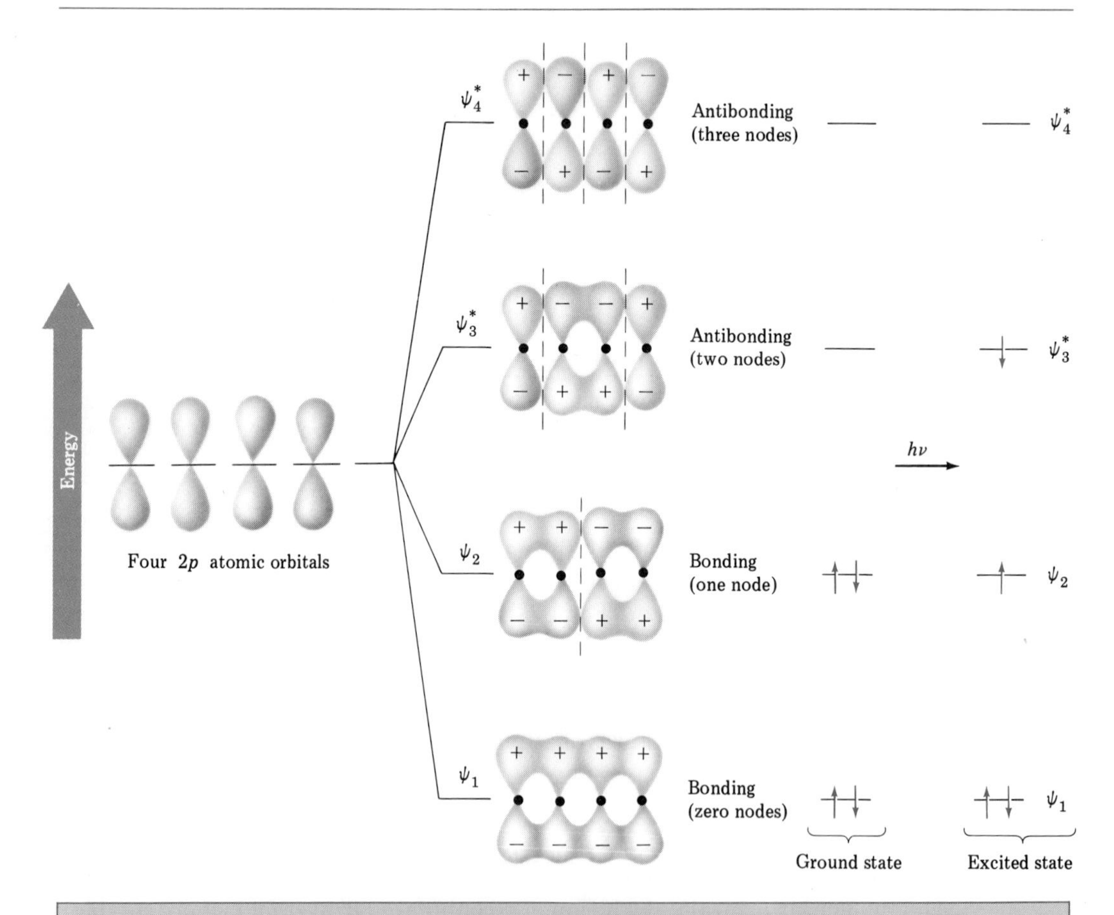

Figure 30.4. Pi molecular orbitals of 1,3-butadiene. The two bonding orbitals are occupied and the two antibonding orbitals are vacant in the ground state.

combine into four diene molecular orbitals spanning the entire pi system. Two of these molecular orbitals are bonding (lower energy, ψ_1 and ψ_2), and two are antibonding (higher energy, ψ_3 and ψ_4), as shown in Figure 30.4. The two bonding orbitals in ground-state butadiene are occupied by four electrons, whereas the two antibonding orbitals are unoccupied. Note that, once again, the higher the energy of a molecular orbital, the more nodes it has.

A similar sort of molecular orbital description can be derived for a conjugated triene, or for *any* conjugated pi electron system. The six molecular orbitals of 1,3,5-hexatriene are shown in Figure 30.5. Only the three bonding orbitals, ψ_1, ψ_2, and ψ_3, are occupied in the ground state, whereas ψ_3 and ψ_4 have one electron each in the excited state.

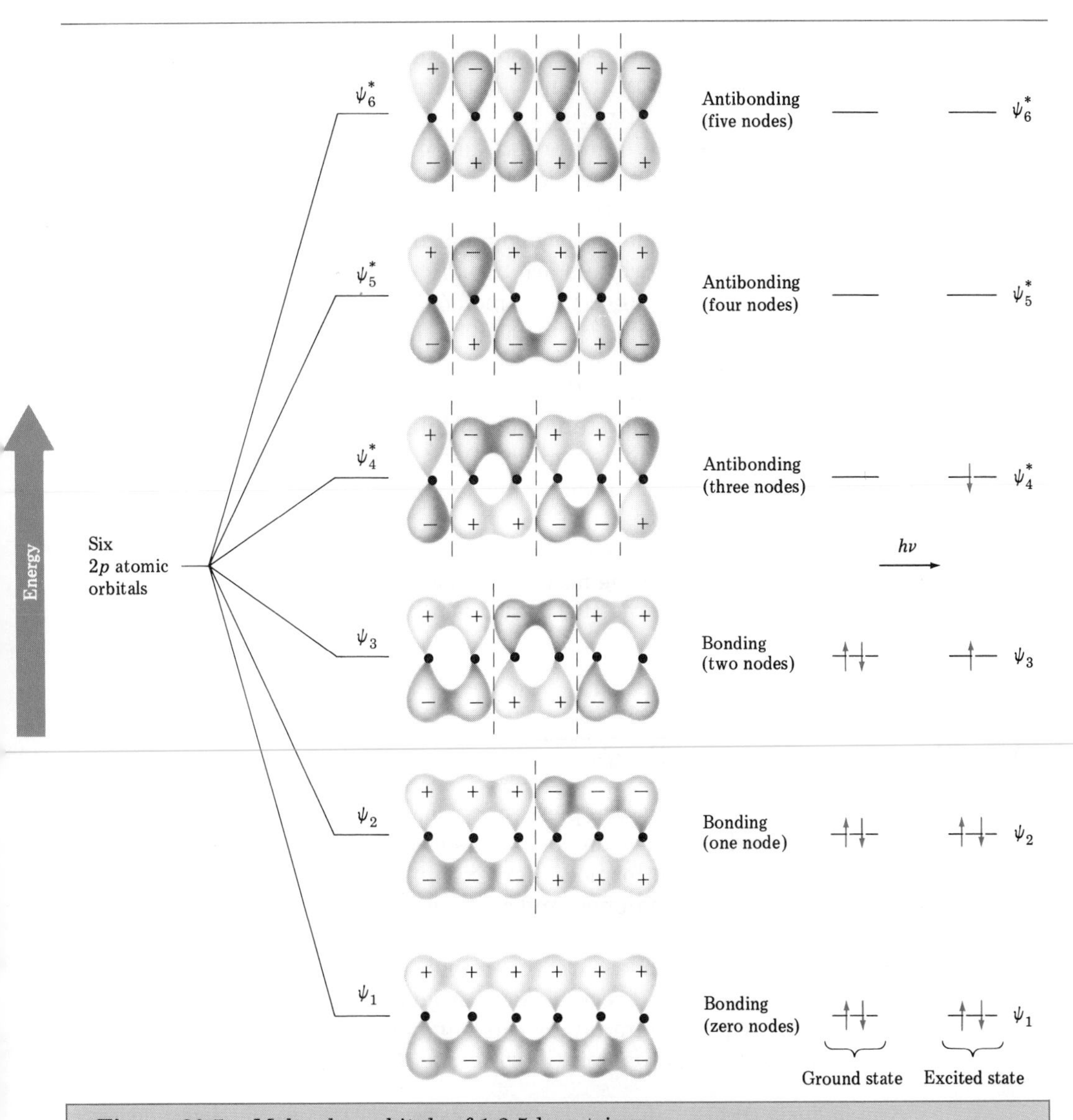

Figure 30.5. Molecular orbitals of 1,3,5-hexatriene.

30.4 Molecular orbitals and pericyclic reactions

What do molecular orbitals and the signs of their lobes have to do with pericyclic reactions?

The answer is, *everything!* According to a series of rules formulated by Woodward and Hoffmann, a pericyclic reaction can take place only if the symmetry of all reactant molecular orbitals is the same as the symmetry of the product molecular orbitals. In other words, the lobes of reactant molecular orbitals must be of the correct algebraic sign for bonding overlap to occur in the transition state leading to the product. If the orbital symmetries of both reactant and product match up, or "correlate," the reaction is said to be a **symmetry-allowed process.** If the orbital symmetries of reactant and product do not correlate, however, the reaction is a **symmetry-disallowed process.** Symmetry-allowed reactions often occur under relatively mild conditions, but symmetry-disallowed reactions cannot occur by concerted paths. They take place either by nonconcerted pathways under forcing, high-energy conditions, or not at all.

The Woodward–Hoffmann rules for pericyclic reactions require an analysis of *all* reactant and product molecular orbitals, but Kenichi Fukui[3] at Kyoto University in Japan has introduced a simplified version. According to Fukui, we need consider only *two* molecular orbitals called the *frontier orbitals.* These frontier orbitals are the highest occupied molecular orbital (HOMO) and the lowest unoccupied molecular orbital (LUMO). In ground-state ethylene, for example, ψ_1 is the HOMO since it has two electrons, and ψ_2 is the LUMO since it is vacant. In ground-state 1,3-butadiene, ψ_2 is the HOMO and ψ_3 is the LUMO (Figure 30.6).

The best way to understand how orbital symmetry affects pericyclic reactions is to look at some examples. Let's look first at some electrocyclic reactions of polyenes and then go on to cycloaddition reactions and sigmatropic rearrangements.

PROBLEM

30.1 Refer to Figure 30.5 to find the molecular orbitals of a conjugated triene, and tell which molecular orbital is the HOMO and which the LUMO for both ground and excited states.

30.5 Electrocyclic reactions

Electrocyclic reactions are pericyclic processes that involve the cyclization of conjugated polyenes; one pi bond is broken, the other pi bonds change position, and a ring is formed. For example, conjugated trienes can be converted into cyclohexadienes, and conjugated dienes can be converted into cyclobutenes.

Both of these reactions are reversible, and the position of the equilibrium depends on the specific case. In general, however, the triene $\rightleftarrows$ cyclohexadiene equilibrium favors the ring-closed product and the diene $\rightleftarrows$ cyclobutene equilibrium favors the ring-opened product.

[3]Kenichi Fukui (1918–); b. Nara Prefecture, Japan; Ph.D. Kyoto University; professor, Kyoto University; Nobel prize (1981).

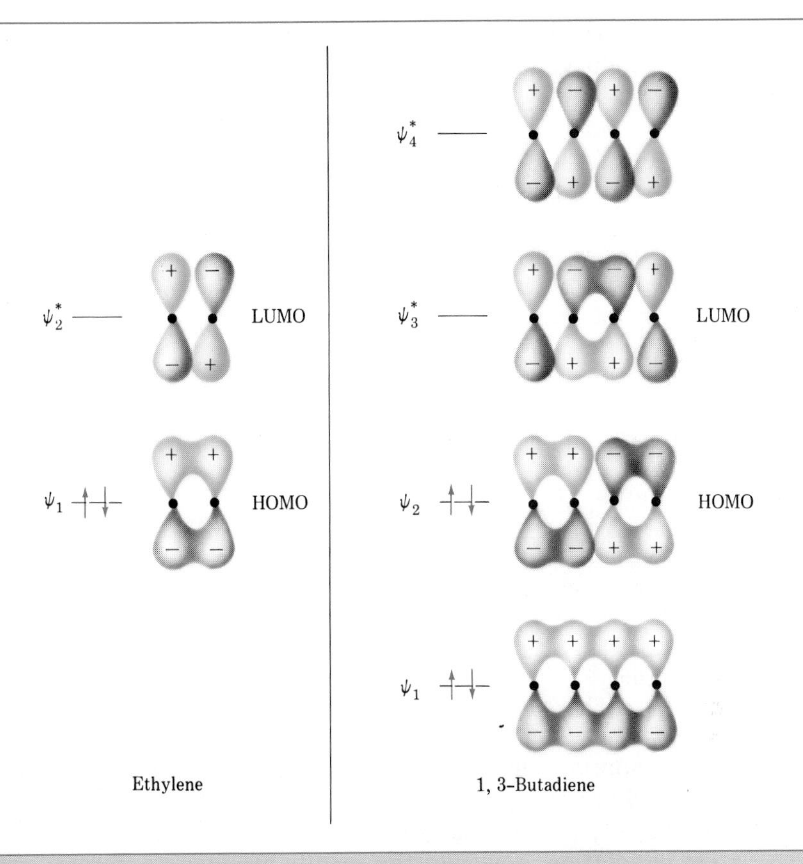

Figure 30.6. The HOMO and LUMO orbitals of ground-state ethylene and 1,3-butadiene.

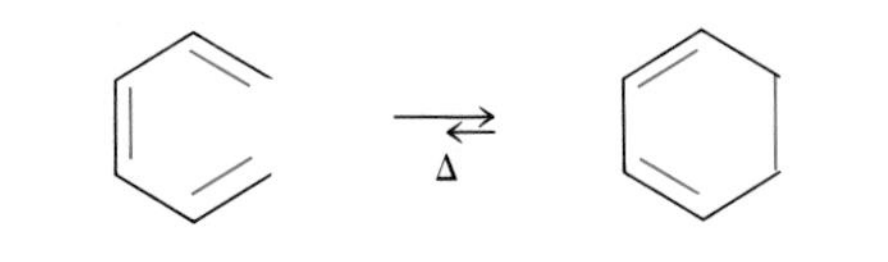

A conjugated triene A cyclohexadiene

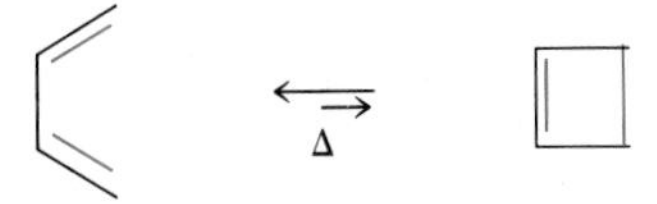

A conjugated diene A cyclobutene

The key feature of electrocyclic reactions is their stereochemistry. For example, (2*E*,4*Z*,6*E*)-octatriene yields only *cis*-5,6-dimethyl-1,3-cyclohexadiene, and (2*E*,4*Z*,6*Z*)-octatriene yields only *trans*-5,6-dimethyl-1,3-cyclohexadiene when heated. Remarkably, however, the stereochemical results change when the reactions are carried out under photochemical, rather than thermal, conditions. Thus, irradiation of (2*E*,4*Z*,6*E*)-octatriene with ultraviolet light yields *trans*-5,6-dimethyl-1,3-cyclohexadiene (Figure 30.7).

(2*E*,4*Z*,6*E*)-Octatriene ⇄ (Δ) *cis*-5,6-Dimethyl-1,3-cyclohexadiene

(2*E*,4*Z*,6*E*)-Octatriene ⇄ ($h\nu$) *trans*-5,6-Dimethyl-1,3-cyclohexadiene

(2*E*,4*Z*,6*Z*)-Octatriene ⇄ (Δ) *trans*-5,6-Dimethyl-1,3-cyclohexadiene

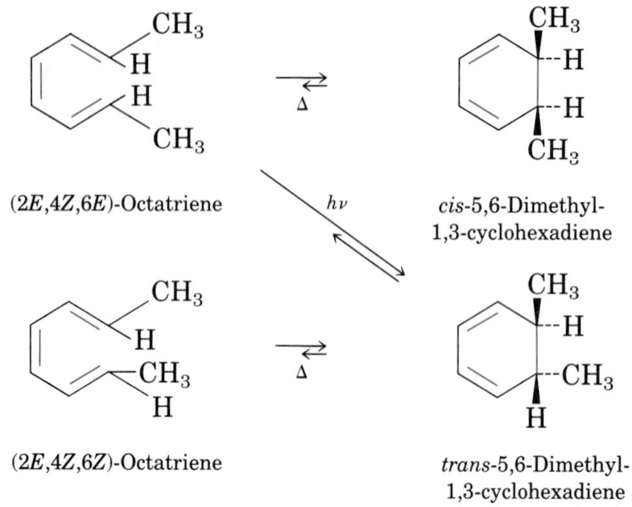

Figure 30.7. Electrocyclic reactions of 2,4,6-octatrienes and 5,6-dimethyl-1,3-cyclohexadienes.

A similar result is obtained for the thermal electrocyclic ring opening of the 3,4-dimethylcyclobutenes. The trans isomer yields only (2*E*,4*E*)-hexadiene, and the cis isomer yields only (2*E*,4*Z*)-hexadiene when heated. On irradiation, however, the results are again different; cyclization of the 2*E*,4*E* isomer under photochemical conditions yields cis product (Figure 30.8).

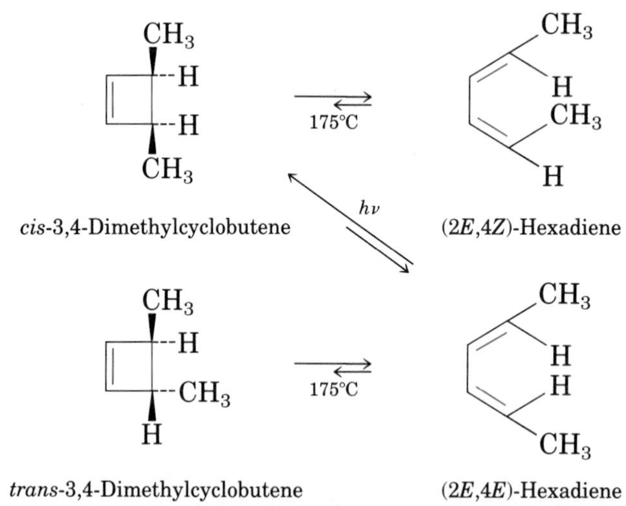

cis-3,4-Dimethylcyclobutene ⇄ (175°C) (2*E*,4*Z*)-Hexadiene

cis-3,4-Dimethylcyclobutene ⇄ ($h\nu$) (2*E*,4*E*)-Hexadiene

trans-3,4-Dimethylcyclobutene ⇄ (175°C) (2*E*,4*E*)-Hexadiene

Figure 30.8. Electrocyclic interconversions of 2,4-hexadienes and 3,4-dimethylcyclobutenes.

All these stereospecific electrocyclic reactions can be accounted for by orbital-symmetry arguments. To do so, we need to look only at the symmetries of the two outermost lobes of the polyene. There are two possibilities; the lobes of like sign can be either on the same side or on opposite sides of the molecule.

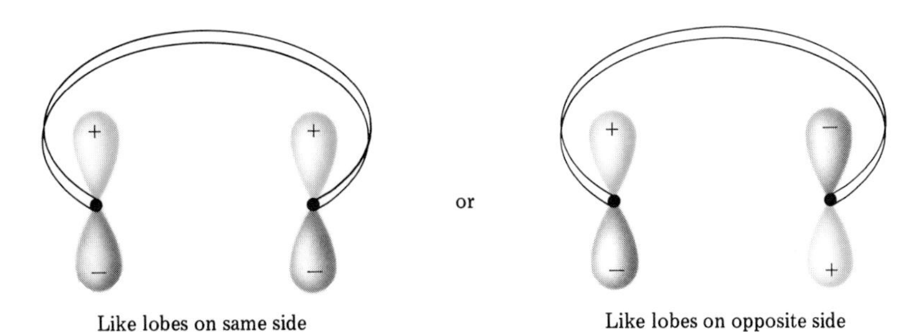

For a bond to form, the outermost pi lobes must rotate so that favorable bonding overlap is achieved—a positive lobe overlapping a positive lobe or a negative lobe overlapping a negative lobe. If two lobes of like sign are on the same side of the molecule, the two orbitals must rotate in *different* directions; one orbital must rotate clockwise and one must rotate counterclockwise. This kind of motion is referred to as **disrotatory:**

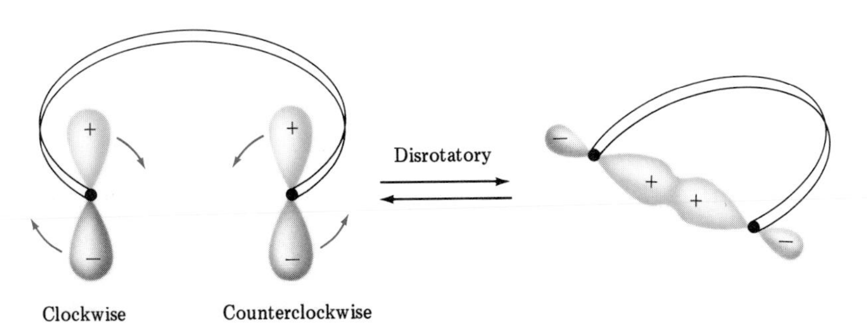

Conversely, if lobes of like sign are on opposite sides of the molecules, both orbitals must rotate in the *same* direction, either both clockwise or both counterclockwise. This kind of motion is called **conrotatory:**

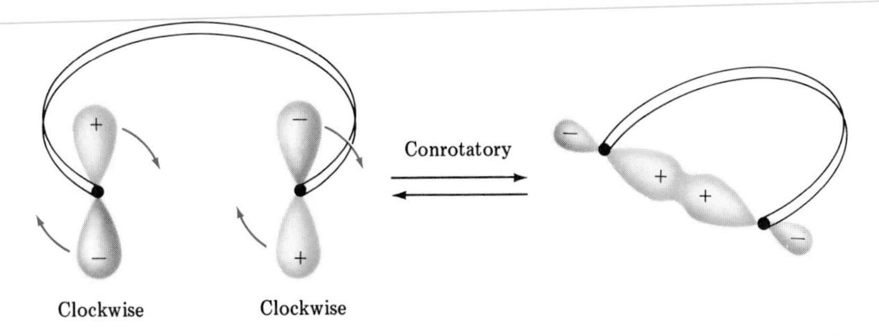

Now let's see what happens to substituents on the polyene carbon atoms when cyclization occurs. To choose a specific example, let's look at the thermal cyclization of (2*E*,4*Z*,6*E*)-octatriene. If a disrotatory cyclization were to occur, *cis*-5,6-dimethyl-1,3-cyclohexadiene would result. If a conrotatory cyclization were to occur, *trans*-5,6-dimethyl-1,3-cyclohexadiene would result (Figure 30.9).

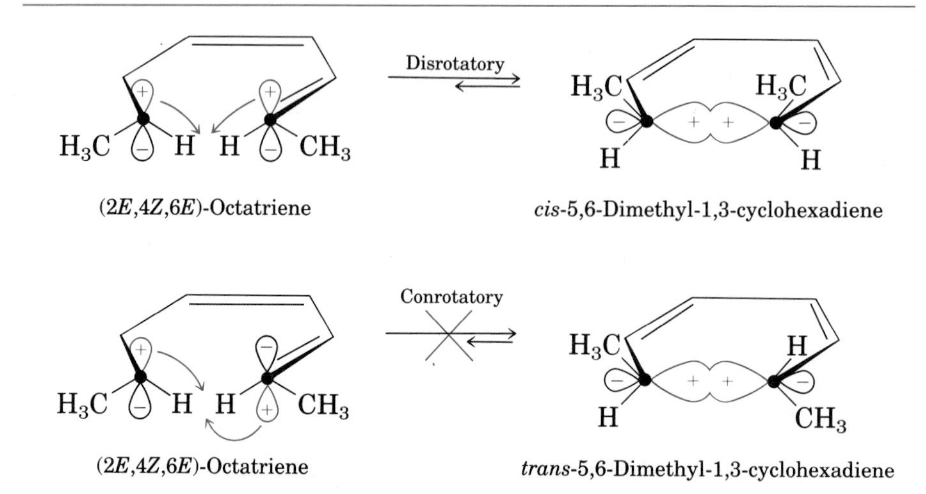

Figure 30.9. Stereochemistry of disrotatory and conrotatory cyclization of (2*E*,4*Z*,6*E*)-octatriene.

In fact, only the disrotatory cyclization of (2*E*,4*Z*,6*E*)-octatriene is observed to occur. We therefore conclude that the stereochemistry of an electrocyclic reaction is determined by the mode of ring closure. The mode of ring closure, in turn, is determined by the symmetry of reactant molecular orbitals.

30.6 Stereochemistry of thermal electrocyclic reactions

How can we predict which mode of ring closure, conrotatory or disrotatory, will occur in a given case? How can we tell whether the terminal lobes of like sign will be on the same side or on opposite sides of the molecule?

According to frontier orbital theory, the stereochemistry of an electrocyclic reaction is determined by the symmetry of the polyene's HOMO. The electrons in the HOMO are the highest-energy, most loosely held electrons, and are therefore most easily moved during reaction. For thermal ring openings and closings, the ground-state electronic configuration is used to identify the HOMO; for photochemical ring openings and closings, the excited-state electronic configuration is used.

Let's look again at the thermal ring closure of conjugated trienes. According to Figure 30.5, the HOMO of a conjugated triene in its ground state has a symmetry that predicts a disrotatory ring closure:

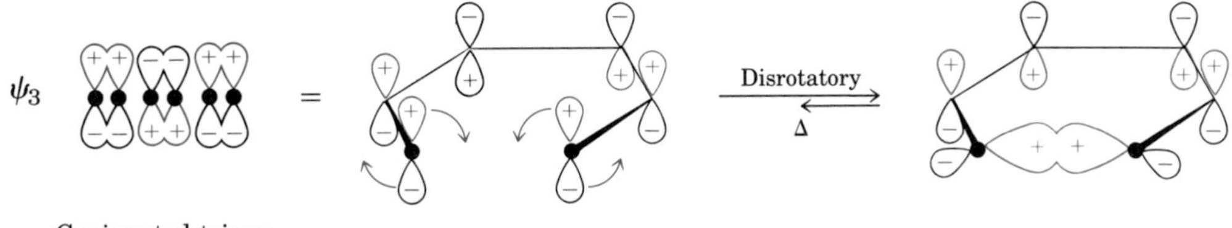

Conjugated triene
ground-state HOMO

This disrotatory cyclization is exactly what is observed in the thermal cyclization of 2,4,6-octatriene. The 2*E*,4*Z*,6*E* isomer yields cis product; the 2*E*,4*Z*,6*Z* isomer yields trans product (Figure 30.10).

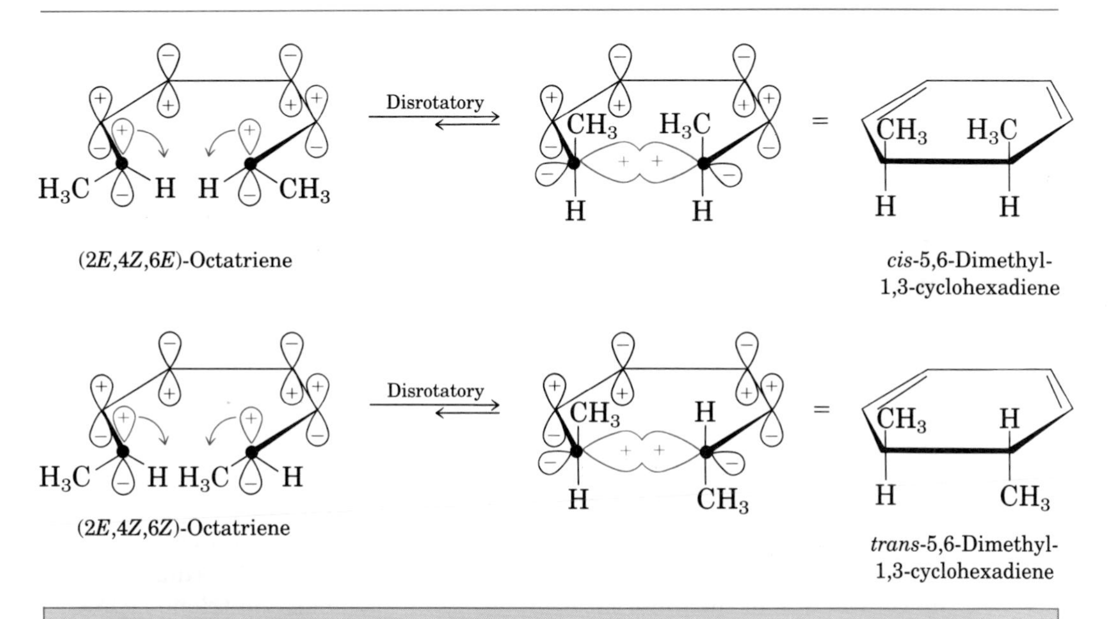

Figure 30.10. Thermal disrotatory ring closure of 2,4,6-octatrienes.

In a similar manner, the ground-state HOMO of a conjugated diene (Figure 30.4) has a symmetry that predicts conrotatory ring closure:

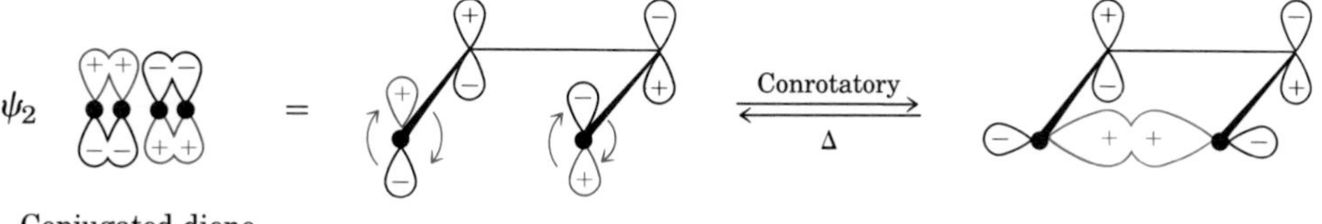

Conjugated diene
ground-state HOMO

In practice, of course, the conjugated diene reaction can only be observed in the reverse direction (cyclobutene → butadiene) because of the position of the equilibrium. We therefore predict that 3,4-dimethylcyclobutene will *ring-open* in a conrotatory fashion, and this is exactly what is observed.

cis-3,4-Dimethylcyclobutene yields (2*E*,4*Z*)-hexadiene, and *trans*-3,4-dimethylcyclobutene yields (2*E*,4*E*)-hexadiene by conrotatory opening (Figure 30.11).

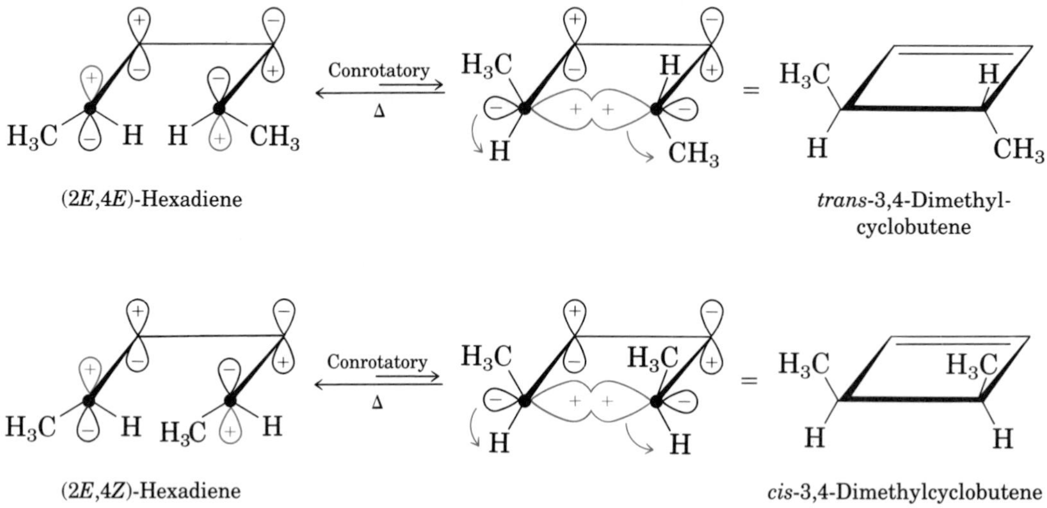

(2*E*,4*E*)-Hexadiene ⇌ (Conrotatory, Δ) *trans*-3,4-Dimethylcyclobutene

(2*E*,4*Z*)-Hexadiene ⇌ (Conrotatory, Δ) *cis*-3,4-Dimethylcyclobutene

Figure 30.11. Conrotatory ring opening of dimethylcyclobutene.

Note that a conjugated diene and a conjugated triene react in opposite stereochemical senses. The diene opens and closes by a conrotatory path, whereas the triene opens and closes by a disrotatory path. This difference is, of course, due to the different symmetries of the diene and triene HOMOs:

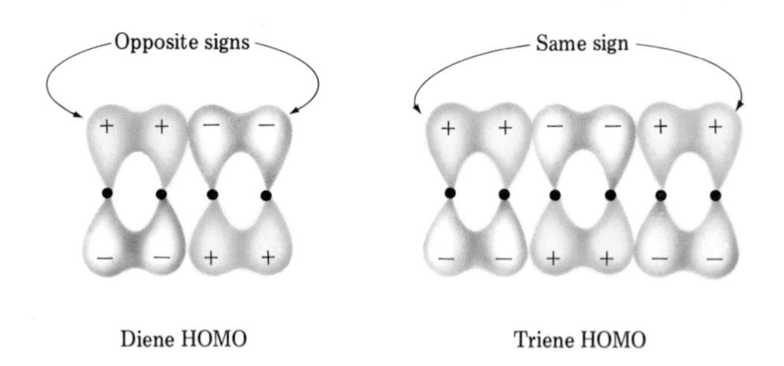

There is an alternating relationship between the number of electron pairs (double bonds) undergoing bond reorganization and the mode of ring closure (or opening). Polyenes with an even number of electron pairs undergo thermal electrocyclic reactions in a conrotatory sense, whereas polyenes with an odd number of electron pairs undergo the same reactions in a disrotatory sense.

PROBLEM

30.2 Draw the products you would expect to obtain from conrotatory and disrotatory cyclizations of (2*Z*,4*Z*,6*Z*)-octatriene. Which of the two paths would you expect the reaction to follow?

PROBLEM

30.3 In theory, *trans*-3,4-dimethylcyclobutene can open by two conrotatory paths to give either (2*E*,4*E*)-hexadiene or (2*Z*,4*Z*)-hexadiene. Explain why both products are symmetry allowed, and then account for the fact that only the 2*E*,4*E* isomer is obtained in practice.

30.7 Photochemical electrocyclic reactions

We noted previously that photochemical electrocyclic reactions take a different stereochemical course than their thermal counterparts. We are now in a position to explain this difference. Ultraviolet irradiation of a polyene causes an excitation of one electron from the ground-state HOMO to the ground-state LUMO. Thus, irradiation of a conjugated diene excites an electron from ψ_2 to ψ_3, and irradiation of a conjugated triene excites an electron from ψ_3 to ψ_4 (Figure 30.12).

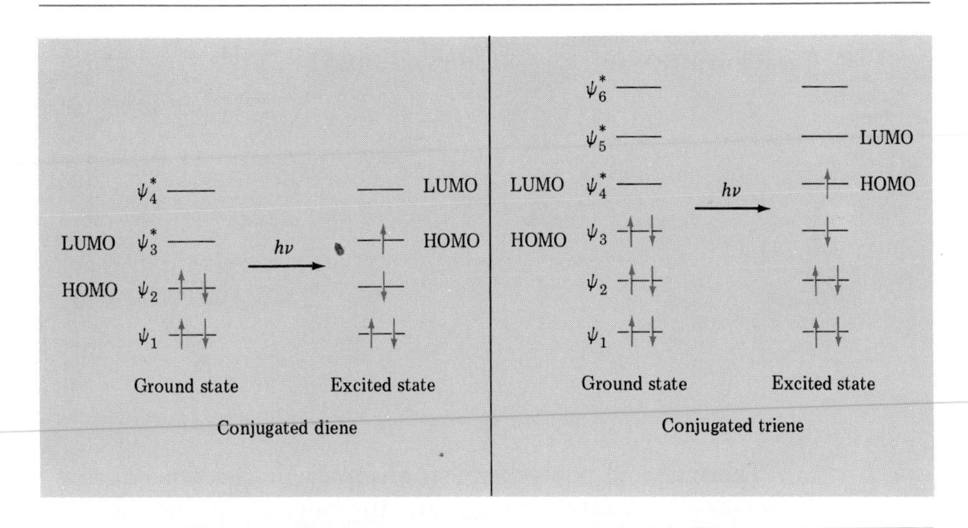

Figure 30.12. Ground- and excited-state electronic configurations of conjugated dienes and trienes.

Electronic excitation changes the symmetries of HOMO and LUMO and hence changes the reaction stereochemistry. For example, (2*E*,4*E*)-hexadiene undergoes photochemical cyclization by a disrotatory path (recall that the thermal reaction is conrotatory):

Different sign

Same sign

$h\nu$

H_3C H H CH_3

Ground-state HOMO

Excited-state HOMO

(2*E*,4*E*)-Hexadiene

Disrotatory

H_3C H_3C H H

cis-3,4-Dimethylcyclobutene

Similarly, (2*E*,4*Z*,6*E*)-octatriene undergoes photochemical cyclization by a conrotatory path (the thermal reaction is disrotatory):

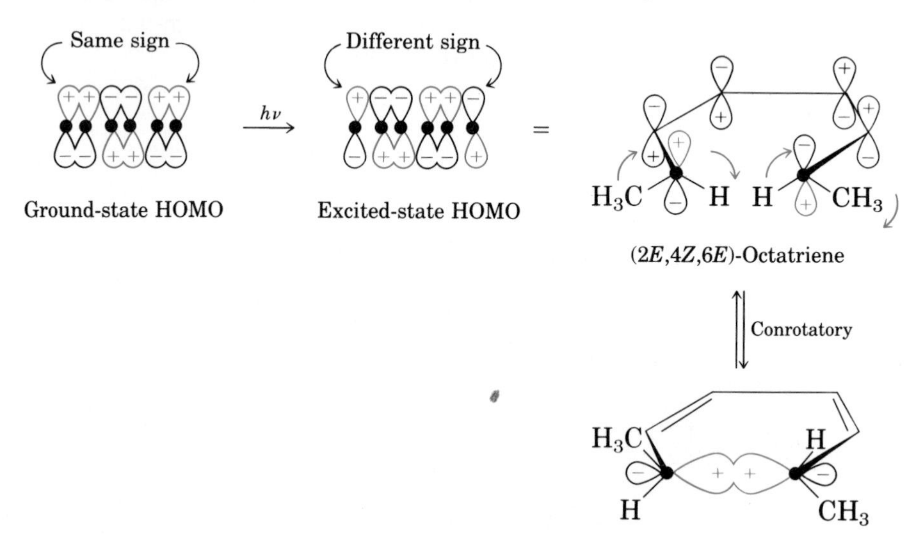

Thermal and photochemical electrocyclic reactions *always* take place with opposite stereochemistry, and the rules governing these processes are given in Table 30.1. Learning these simple rules allows us to predict the stereochemistry of a large number of organic reactions.

TABLE 30.1 Stereochemical rules for electrocyclic reactions

Electron pairs (double bonds)	*Thermal reaction*	*Photochemical reaction*
Even number	Conrotatory	Disrotatory
Odd number	Disrotatory	Conrotatory

PROBLEM

30.4 What products would you expect to obtain from the photochemical cyclizations of (2*E*,4*Z*,6*E*)-octatriene and of (2*E*,4*Z*,6*Z*)-octatriene?

PROBLEM

30.5 The following thermal isomerization has been reported to occur under relatively mild conditions. Identify the pericyclic reactions involved and show exactly how the rearrangement occurs.

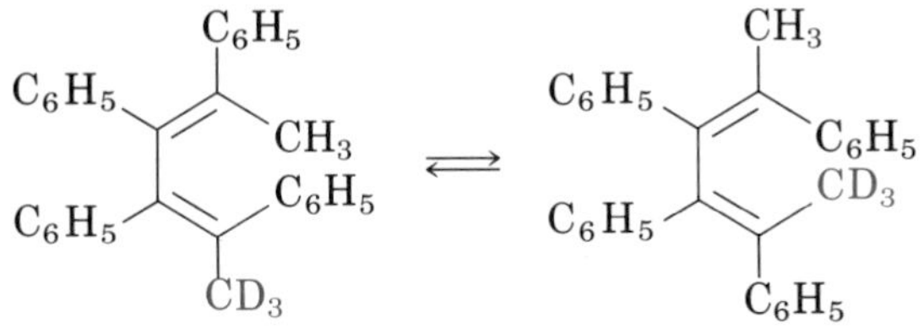

PROBLEM

30.6 Would you expect the following reaction to proceed in a conrotatory or disrotatory manner? Show the stereochemistry of the cyclobutene product and explain your answer.

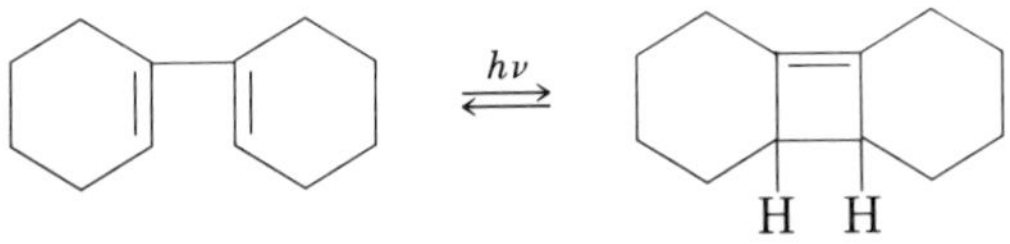

30.8 Cycloaddition reactions

Cycloaddition reactions are intermolecular pericyclic processes in which two molecules add to each other, yielding a cyclic product. As with electrocyclic reactions, cycloadditions are controlled by the orbital symmetry of the reactants. Symmetry-allowed processes often take place readily, but symmetry-disallowed processes take place with great difficulty, if at all, and then only by nonconcerted pathways. Let's look at two possible reactions to see how they differ.

The Diels–Alder cycloaddition reaction is a [4 + 2] pi electron process that takes place between a diene (4 pi electrons) and a dienophile (2 pi electrons) and yields a cyclohexene product. Hundreds of examples of Diels–Alder reactions are known; they often take place under mild conditions (room temperature or slightly above), and they are stereospecific with respect to substituents. For example, room-temperature reaction between 1,3-butadiene and diethyl maleate (cis) yields exclusively cis-disubstituted cyclohexene product; reaction between 1,3-butadiene and diethyl fumarate (trans) yields exclusively trans-disubstituted product (Figure 30.13).

In contrast to the [4 + 2] pi electron Diels–Alder reaction, thermal cycloaddition between two alkenes (2 pi + 2 pi) does not occur. Photochemical [2 + 2] pi electron cycloadditions often take place readily, however, and yield cyclobutane products. How can we use orbital symmetry arguments to explain these results?

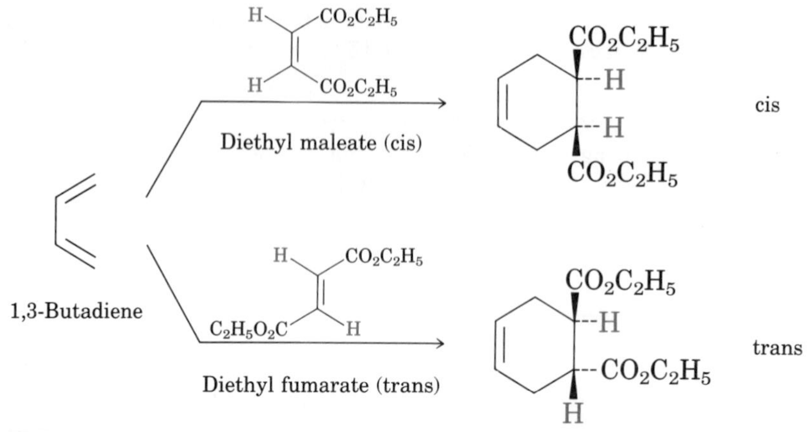

1,3-Butadiene

Diethyl maleate (cis)

cis

Diethyl fumarate (trans)

trans

Figure 30.13. Diels–Alder cycloaddition reactions of diethyl maleate and diethyl fumarate. The reactions are stereospecific.

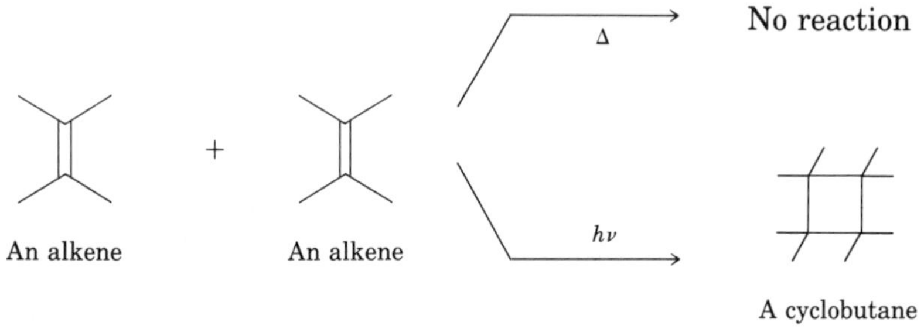

An alkene + An alkene

Δ → No reaction

hν → A cyclobutane

For a successful cycloaddition to take place, the terminal lobes of the two unsaturated reactants must have the correct symmetry for bonding overlap to occur. This can happen in either of two ways, designated **suprafacial** and **antarafacial.** Suprafacial cycloadditions occur when the orbital symmetries of the reactants are such that reactions occur between lobes on the *same* face of one component and lobes on the *same* face of the other component (Figure 30.14).

Antarafacial cycloadditions occur when the orbital symmetries of the reactants are such that reactions occur between lobes on the same face of one component and lobes on *opposite* faces of the other component, as Figure 30.15 shows.

Note that both suprafacial and antarafacial cycloadditions are allowed on orbital-symmetry grounds. Geometric constraints often make antarafacial reactions difficult, however, since there must be twisting of the *p* orbital system, and only suprafacial cycloadditions are possible for small pi systems.

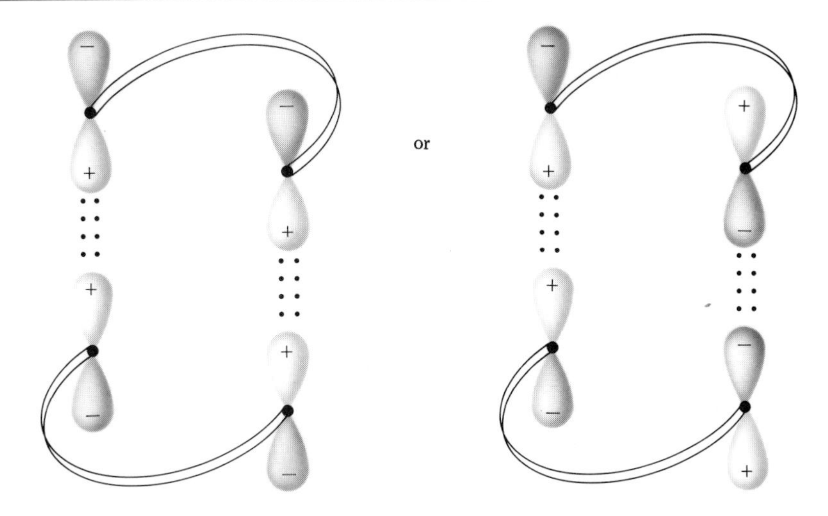

Figure 30.14. Suprafacial cycloaddition occurs when there is bonding overlap between lobes on the same face of one reactant and lobes on the same face of a second reactant.

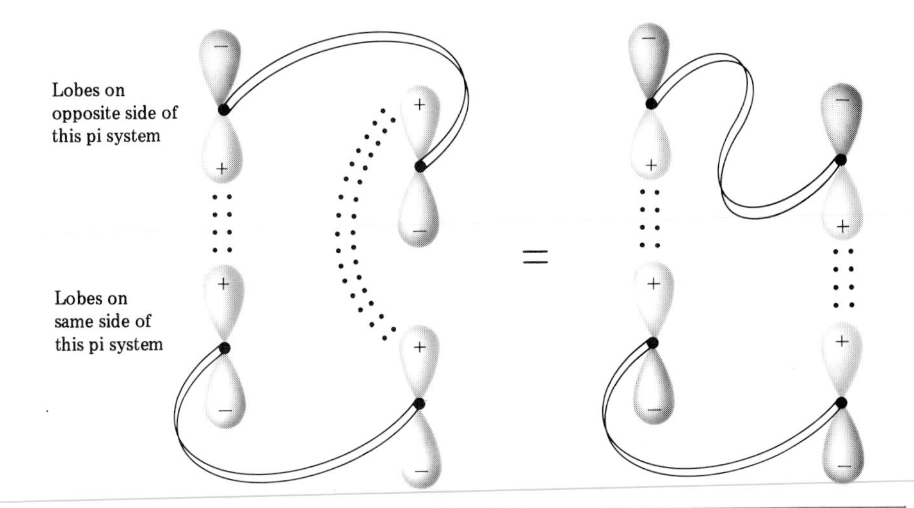

Figure 30.15. Antarafacial cycloaddition occurs when there is bonding overlap between lobes on the same face of one reactant and lobes on opposite faces of the second reactant.

30.9 Stereochemistry of cycloadditions

How can we predict whether a given reaction will occur with suprafacial or with antarafacial geometry? According to frontier orbital theory, cycloaddition reactions take place when the HOMO of one reaction partner overlaps the LUMO of the other reaction partner in a bonding manner. A good intuitive explanation of this rule is to imagine that one partner reacts by donating two electrons to the second partner. As with electrocyclic reactions, it is the electrons in the HOMO of the first partner that are least tightly held and most likely to be donated. Since only two electrons can be in any one orbital, these electrons must go into a vacant orbital of the second partner, and it is the LUMO that is lowest in energy and most likely to accept the electrons. Let's see how this rule applies to specific cases.

For the [4 + 2] pi-electron cycloaddition (Diels–Alder reaction), let's arbitrarily select the diene LUMO and the alkene HOMO. (We could equally well use the diene HOMO and the alkene LUMO.) The symmetries of these two orbitals are such that bonding overlap of the terminal lobes can occur with suprafacial geometry (Figure 30.16). The Diels–Alder reaction therefore takes place readily under thermal conditions. Note that, as with electrocyclic reactions, we need be concerned only with the symmetries of the *terminal* lobes. For purposes of prediction, it does not matter whether or not the interior lobes have bonding or antibonding overlap in the product.

In contrast to the [4 + 2] Diels–Alder reaction, the [2 + 2] cycloaddition of two alkenes to yield a cyclobutane does not occur thermally but can only be observed photochemically. The explanation follows from orbital-symmetry arguments. Looking at the HOMO of one alkene and the LUMO of the second alkene, it is apparent that a thermal [2 + 2] cycloaddition must take place by an antarafacial pathway (Figure 30.17). Geometric

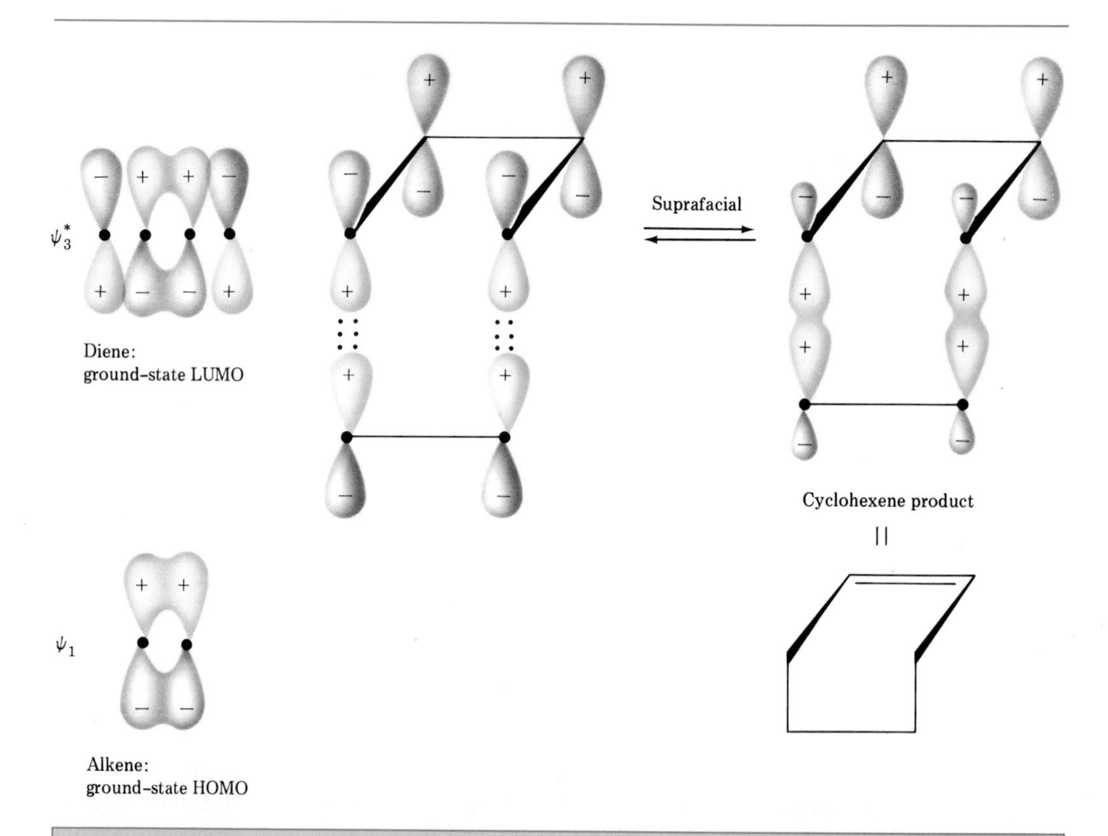

Figure 30.16. Correlation of diene LUMO and alkene HOMO in a suprafacial [4 + 2] cycloaddition reaction (Diels–Alder reaction).

constraints make the antarafacial transition state impossible, however, and concerted thermal [2 + 2] cycloadditions are therefore not observed.

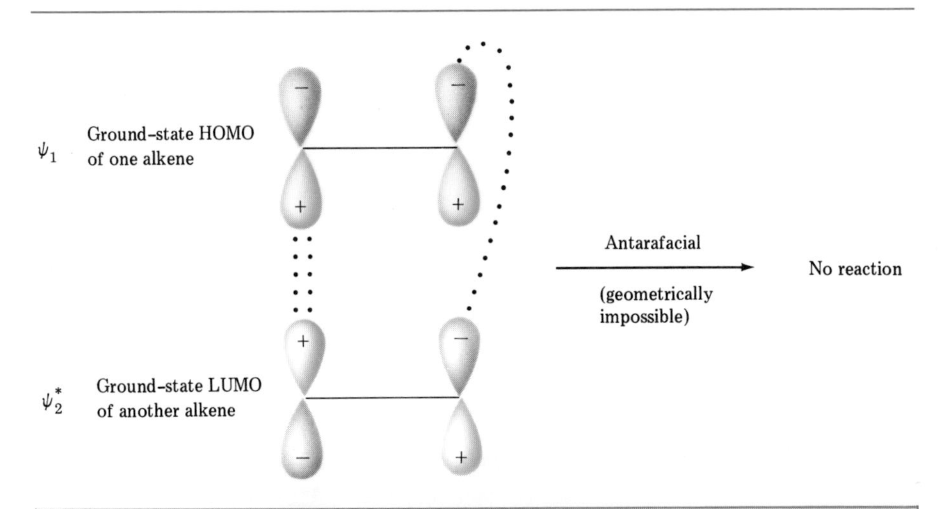

Figure 30.17. Correlation of HOMO and LUMO in thermal [2 + 2] cycloaddition. The reaction does not occur because antarafacial geometry is too strained.

Photochemical [2 + 2] cycloadditions, however, *are* observed. Irradiation of an alkene with ultraviolet light excites an electron from ψ_1, the ground-state HOMO, to ψ_2, the excited-state HOMO. Correlation between the excited-state HOMO of one alkene and the LUMO of the second alkene indicates that a photochemical [2 + 2] cycloaddition reaction can occur by a suprafacial pathway (Figure 30.18).

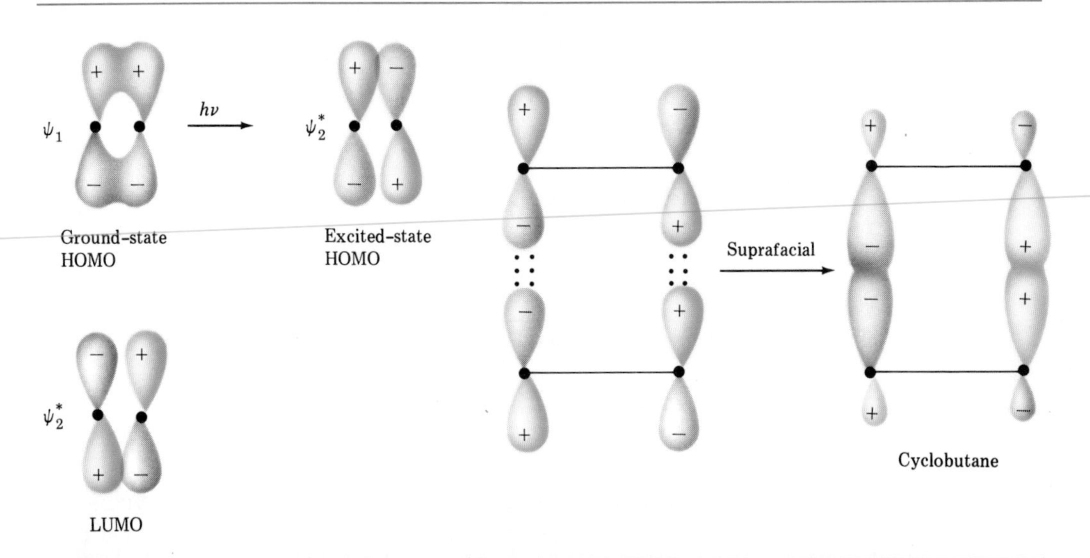

Figure 30.18. Correlation of HOMO and LUMO in photochemical [2 + 2] cycloaddition. The reaction occurs with suprafacial geometry.

The photochemical [2 + 2] cycloaddition reaction occurs smoothly and represents one of the best methods known for synthesizing cyclobutane rings. The reaction can take place either inter- or intramolecularly, as the following examples show:

2-Cyclohexenone + 2-Methylpropene $\xrightleftharpoons{h\nu}$ 40%

$\xrightarrow{h\nu}$ 55%

Thermal and photochemical cycloaddition reactions always take place by opposite stereochemical pathways, and the rules governing these processes are given in Table 30.2. As with electrocyclic reactions, we can categorize cycloadditions according to the total number of electron pairs (double bonds) involved in the rearrangement. Thus a Diels–Alder [4 + 2] reaction between a diene and a dienophile involves an odd number (three) of electron pairs and takes place by a ground-state suprafacial pathway, whereas the [2 + 2] reaction between two alkenes involves an even number (two) of electron pairs and must occur with ground-state antarafacial geometry.

TABLE 30.2 Stereochemical rules for cycloaddition reactions

Electron pairs (double bonds)	*Thermal reaction*	*Photochemical reaction*
Even number	Antarafacial	Suprafacial
Odd number	Suprafacial	Antarafacial

It should be reiterated that both suprafacial and antarafacial cycloaddition pathways are symmetry-allowed processes. Only the geometric constraints inherent in twisting a conjugated pi electron system out of planarity make antarafacial reaction geometry difficult in many cases.

PROBLEM

30.7 What stereochemistry would you expect for the product of the Diels–Alder reaction between (2*E*,4*E*)-hexadiene and ethylene? What stereochemistry would you expect if (2*E*,4*Z*)-hexadiene were used instead?

PROBLEM

30.8 Cyclopentadiene reacts with cycloheptatrienone to give the product shown. Tell what kind of reaction is involved and explain the observed result. Is the reaction suprafacial or antarafacial?

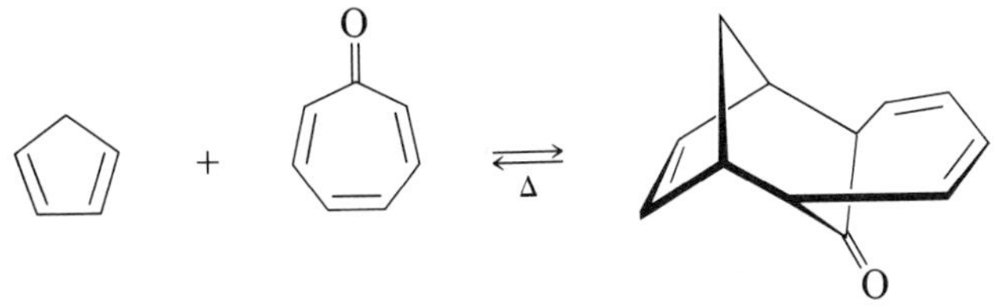

Δ

PROBLEM

30.9 The following reaction has been observed. Identify the two pericyclic reactions involved and show exactly how they occur.

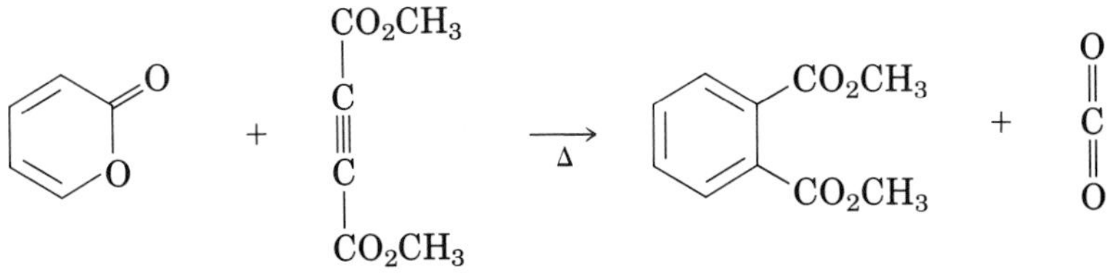

CO_2CH_3 C≡C CO_2CH_3 → Δ CO_2CH_3 CO_2CH_3 + O=C=O

30.10 Sigmatropic rearrangements

Sigmatropic rearrangements are pericyclic reactions in which a sigma-bonded substituent group (denoted here by a circled S) migrates across a pi electron system. The sigma-bonded group can be either at the end or in the middle of the pi system, as the following [1,3] and [3,3] rearrangements illustrate:

A [1,3] rearrangement

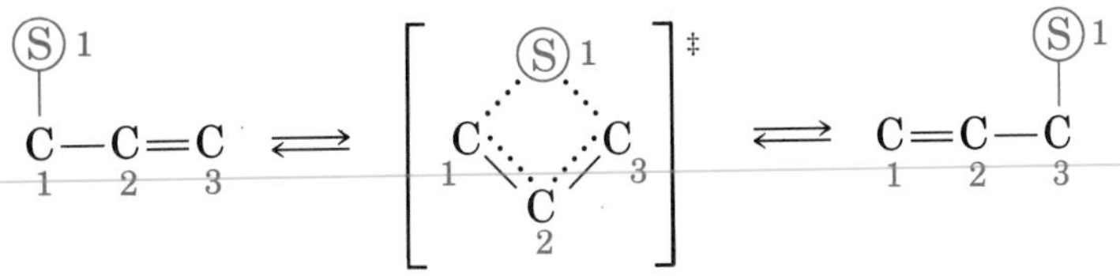

Cyclic transition state

A [3,3] rearrangement

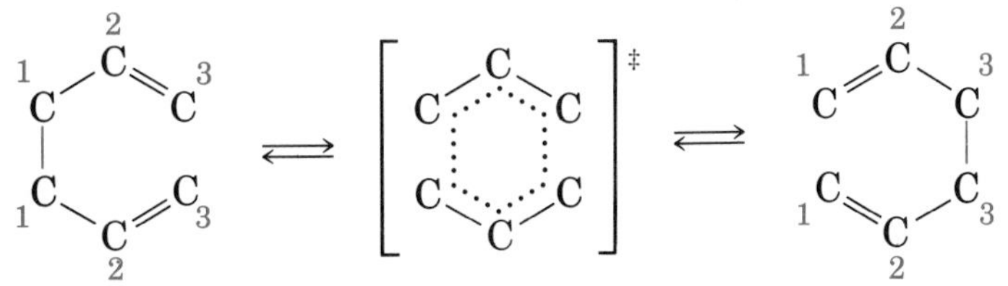

Cyclic transition state

The designations [1,3] and [3,3] describe the kind of rearrangement that has occurred. The two numbers in brackets refer to the two groups connected by the sigma bond and designate the positions in those groups *to which migration has occurred.* For example, in the [1,5] sigmatropic rearrangement of a diene, the two groups connected by the sigma bond are a hydrogen atom and a pentadienyl fragment. Migration occurs to position 1 of the H group and to position 5 of the pentadienyl group.

A [1,5] rearrangement

$$\underset{1}{\overset{H_1}{CH_2}}-\underset{2}{CH}=\underset{3}{CH}-\underset{4}{CH}=\underset{5}{CH_2} \rightleftharpoons \underset{1}{CH_2}=\underset{2}{CH}-\underset{3}{CH}=\underset{4}{CH}-\underset{5}{\overset{H_1}{CH_2}}$$

In the [3,3] Claisen rearrangement, the two groups connected by the sigma bond are an allyl group and a vinylic ether group. Migration occurs to position 3 of the allyl group and also to position 3 of the vinylic ether.

Claisen rearrangement, a [3,3] rearrangement

$$\overset{1}{O}-\overset{2}{CH}=\overset{3}{CH_2},\ \overset{1}{H_2C}-\overset{2}{CH}=\overset{3}{CH_2}\ (\sigma\ \text{bond } O-H_2C) \longrightarrow \overset{1}{O}=\overset{2}{CH}-\overset{3}{CH_2},\ \overset{1}{H_2C}=\overset{2}{CH}-\overset{3}{CH_2}\ (\sigma\ \text{bond } CH_2-CH_2)$$

30.11 Stereochemistry of sigmatropic rearrangements

Sigmatropic rearrangements are more complex than either electrocyclic or cycloaddition reactions but are nonetheless controlled by orbital-symmetry considerations. There are two possible modes of reaction: Migration of a group across the same face of the pi system is called a suprafacial rearrangement, and migration of a group from one face of the pi system to the other face is an antarafacial rearrangement (Figure 30.19).

The rules for sigmatropic rearrangements are identical to those for cycloaddition reactions and are summarized in Table 30.3. Both suprafacial and antarafacial sigmatropic rearrangements are symmetry-allowed processes, but suprafacial rearrangements are often easier, for geometric reasons.

TABLE 30.3 Stereochemical rules for sigmatropic rearrangements

Electron pairs	*Thermal reaction*	*Photochemical reaction*
Even number	Antarafacial	Suprafacial
Odd number	Suprafacial	Antarafacial

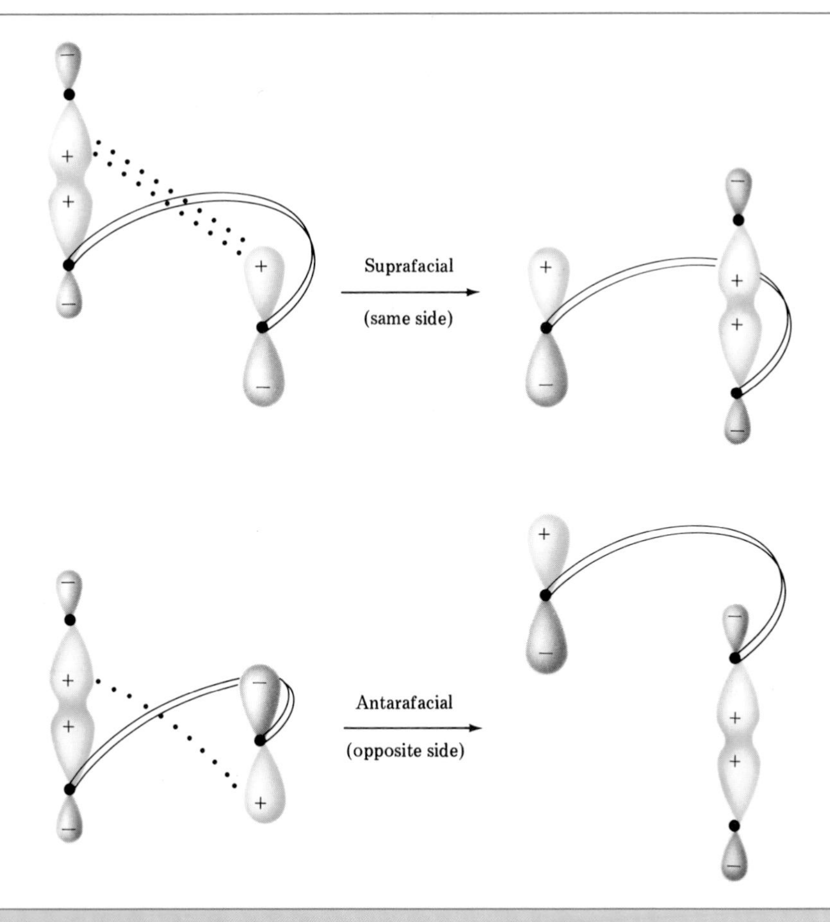

Figure 30.19. Suprafacial and antarafacial sigmatropic rearrangements.

30.12 Some examples of sigmatropic rearrangements

A [1,5] sigmatropic rearrangement involves three electron pairs (two pi bonds and one sigma bond) and the orbital-symmetry rules in Table 30.3 therefore predict suprafacial reaction. In fact, the [1,5] shift of a hydrogen atom across two double bonds of a pi system is one of the most commonly observed of all sigmatropic rearrangements. For example, 5-methylcyclopentadiene rapidly scrambles at room temperature to yield a mixture of 1-methyl-, 2-methyl-, and 5-methyl-substituted products.

5-Methylcyclopentadiene $\underset{25°C}{\overset{[1,5]\ Shift}{\rightleftharpoons}}$ 1-Methylcyclopentadiene $\underset{25°C}{\overset{[1,5]\ Shift}{\rightleftharpoons}}$ 2-Methylcyclopentadiene

As another example, heating 5,5,5-trideuterio-(1,3*Z*)-pentadiene causes scrambling of deuterium between positions 1 and 5.

$$\text{CD}_3,\ \text{CH}_2 \underset{\Delta}{\xrightleftharpoons{[1,5]\ \text{D Shift}}} \text{CD}_2,\ \text{CH}_2\text{D}$$

Both of these [1,5] hydrogen shifts occur by a symmetry-allowed suprafacial rearrangement, as illustrated in Figure 30.20.

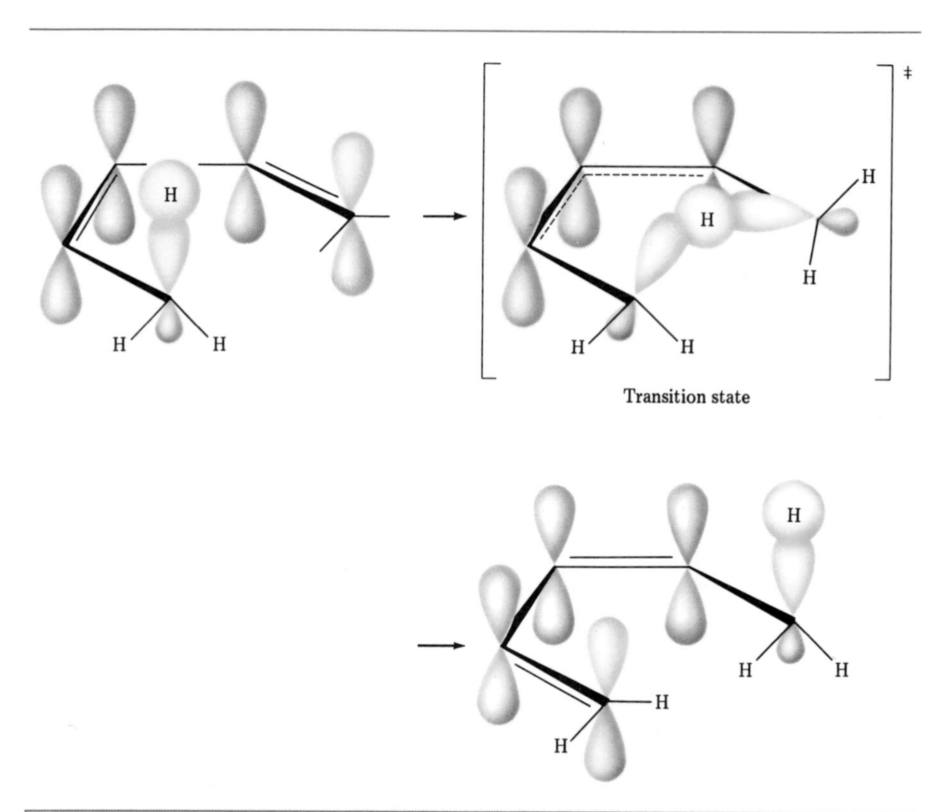

Figure 30.20. Orbital view of [1,5] hydrogen shift.

In contrast to the preceding two examples of [1,5] sigmatropic hydrogen shifts, thermal [1,3] hydrogen shifts are unknown. Were they to occur, they would have to proceed via an impossibly strained antarafacial reaction pathway.

Two other important sigmatropic reactions are the **Cope rearrangement** and the **Claisen rearrangement** (Section 29.9). These two, along with the Diels–Alder reaction, are the most useful pericyclic reactions for organic synthesis, and many hundreds of examples of all three are known. Note that the Claisen rearrangement works well with both allyl aryl ethers and with allyl vinylic ethers.

Claisen rearrangement

[3,3]

An allyl aryl ether

An *o*-allylphenol

[3,3]

An allyl vinylic ether

A γ,δ-unsaturated carbonyl compound

Cope rearrangement

[3,3]

A 1,5 diene

A new 1,5 diene

Both Cope and Claisen rearrangements involve reorganization of an odd number of electron pairs (two pi bonds and one sigma bond) and are therefore predicted to react by suprafacial pathways (Figure 30.21, p. 1022).

The ease with which these two rearrangements occur provides evidence for the correctness of orbital-symmetry predictions. For example, it has been estimated, based on spectroscopic measurements, that homotropilidene undergoes [3,3] rearrangement *several hundred times each second* at room temperature.

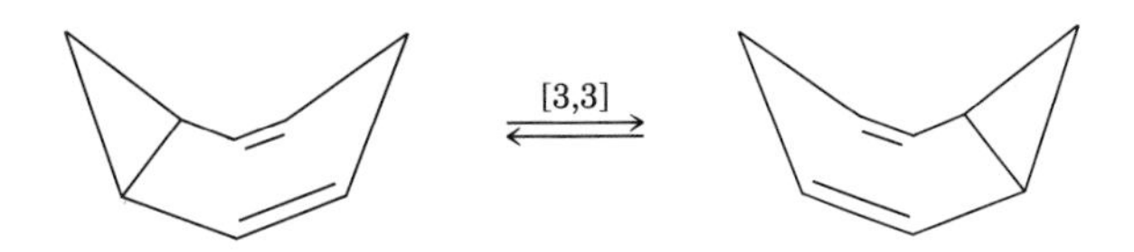

Homotropilidene

PROBLEM

30.10 The ^{13}C NMR spectrum of homotropilidene taken at room temperature shows only three peaks. Explain.

PROBLEM

30.11 How can you account for the fact that heating 1-deuterioindene scrambles the isotope label to all three positions on the five-membered ring?

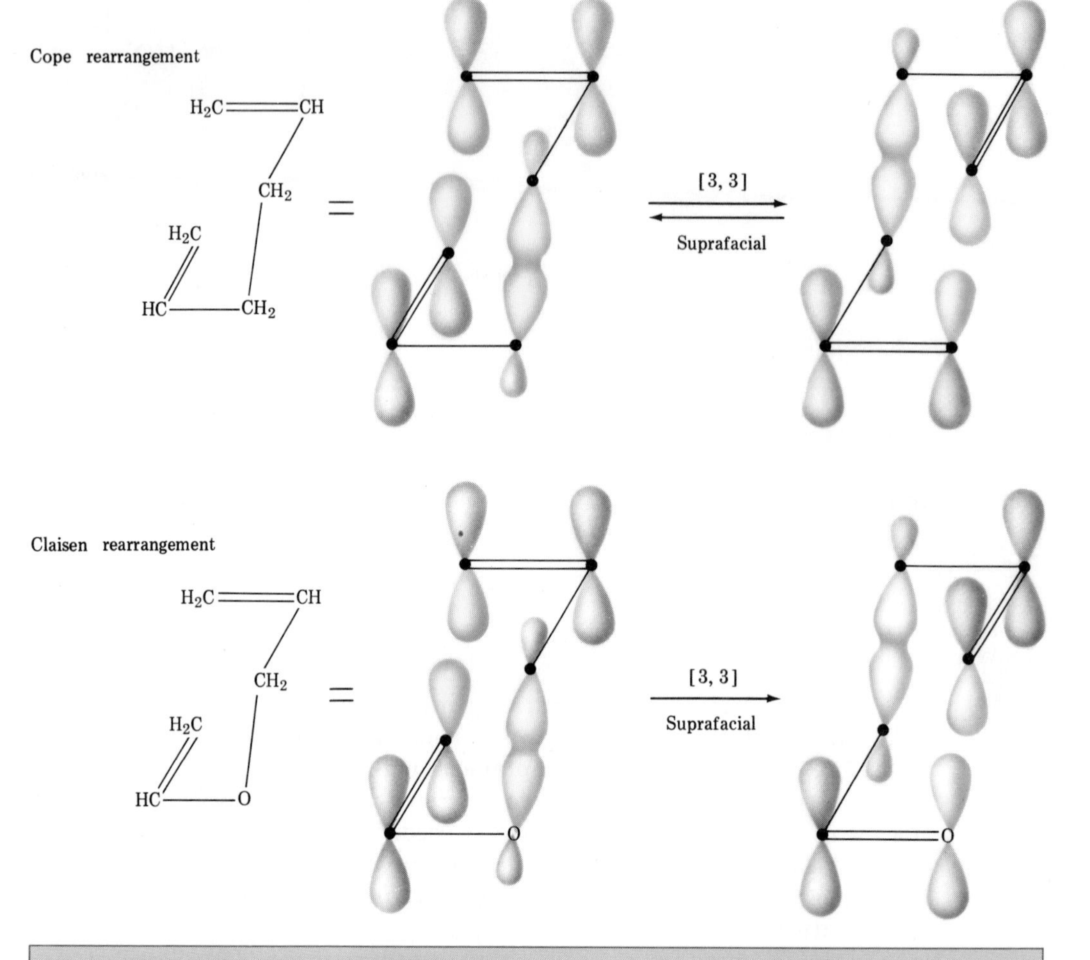

Figure 30.21. Suprafacial [3,3] Cope and Claisen rearrangements.

PROBLEM

30.12 Classify the following sigmatropic reaction by order [x,y] and indicate whether you would expect it to proceed with suprafacial or antarafacial stereochemistry.

CH_2—H ⇌ CH_2—H

PROBLEM

30.13 When a 2,6-disubstituted allyl phenyl ether is heated in an attempted Claisen rearrangement, migration occurs to give *p*-allyl product. Explain how this occurs.

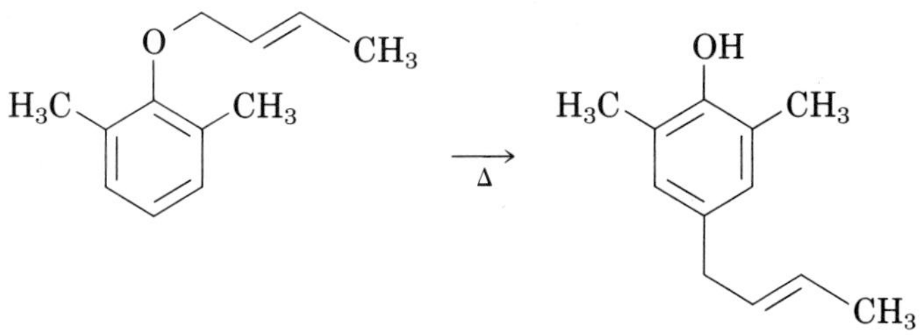

30.13 A summary of rules for pericyclic reactions

Pericyclic, electrocyclic, cycloaddition, sigmatropic, conrotatory, disrotatory, suprafacial, antarafacial . . . how can we keep it all straight?

Tables 30.1, 30.2, and 30.3 summarize the selection rules for electrocyclic, cycloaddition, and sigmatropic reactions, and the information provided leads us to the conclusion that pericyclic processes can be grouped according to whether they involve the reorganization of an even or an odd number of electron pairs (bonds). All this information can be distilled into one simple phrase that, when memorized, provides an easy and accurate way to predict the stereochemical outcome of any pericyclic reaction:

> For a ground-state (thermal) pericyclic reaction, the groupings are *odd–supra–dis* and *even–antara–con*.

Cycloaddition and sigmatropic reactions involving an odd number of electron pairs (bonds) occur with suprafacial geometry; electrocyclic reactions involving an odd number of electron pairs occur with disrotatory stereochemistry. Conversely, pericyclic reactions involving an even number of electron pairs occur with either antarafacial geometry or conrotatory stereochemistry.

Once the selection rules for thermal reactions have been memorized, the rules for photochemical reactions are easily derived by simply remembering that they are the opposite of the thermal rules:

> For an excited-state (photochemical) pericyclic reaction, the groupings are *odd–antara–con* and *even–supra–dis*.

Both rules are summarized in Table 30.4. Memorizing this table will give you the ability to predict the stereochemistry of literally thousands of pericyclic reactions.

TABLE 30.4 **Generalized selection rules for pericyclic reactions**

Electron state	*Electron pairs*	*Stereochemistry*
Ground state (thermal)	Even number	Antara–con
	Odd number	Supra–dis
Excited state (photochemical)	Even number	Supra–dis
	Odd number	Antara–con

PROBLEM

30.14 Predict the stereochemistry of the following pericyclic reactions:
(a) A thermal cyclization of a conjugated tetraene
(b) A photochemical cyclization of a conjugated tetraene
(c) A photochemical [4 + 4] cycloaddition
(d) A thermal [2 + 6] cycloaddition
(e) A photochemical [3,5] sigmatropic rearrangement

30.14 Summary

A pericyclic reaction is one that takes place by a concerted pathway involving a cyclic transition state. All bonding changes take place in a single step and no intermediates are involved. There are three major classes of pericyclic processes: electrocyclic reactions, cycloaddition reactions, and sigmatropic rearrangements. The stereochemistry of these reactions is controlled by the symmetry of the orbitals involved in bond reorganization.

Electrocyclic reactions involve the cyclization of conjugated polyenes. For example, 1,3,5-hexatriene cyclizes to 1,3-cyclohexadiene on heating. Electrocyclic reactions can occur by either conrotatory or disrotatory paths, depending on the symmetry of the terminal lobes of the pi system. Conrotatory cyclization requires that both lobes rotate in the same direction, and disrotatory cyclization requires that the lobes rotate in opposite directions (Figure 30.22). The reaction course for any given case can be found by looking at the symmetry of the highest occupied molecular orbital (HOMO).

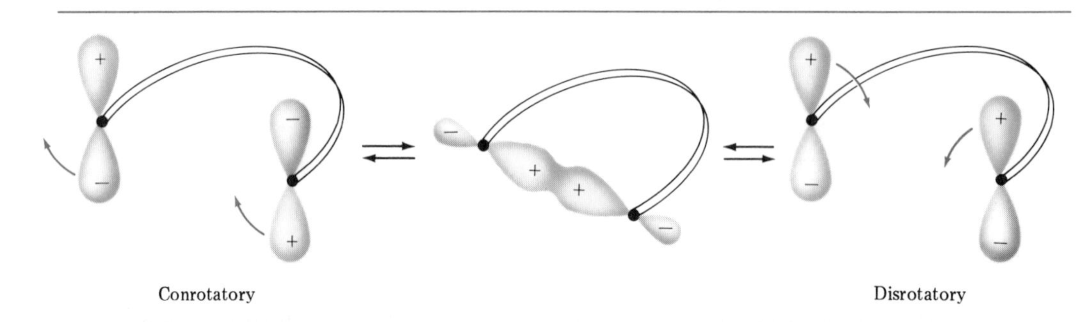

Figure 30.22. Conrotatory and disrotatory motions during electrocyclic reactions.

Cycloaddition reactions are those in which two molecules add together to yield a cyclic product. For example, Diels–Alder reaction between a diene (4 pi electrons) and a dienophile (2 pi electrons) yields a cyclohexene. Cycloadditions can take place either by suprafacial or antarafacial pathways. Suprafacial cycloaddition involves reaction between lobes on the same face of one component and on the same face of the second component. Antarafacial cycloaddition involves reaction between lobes on the same face of one component and on opposite faces of the other component

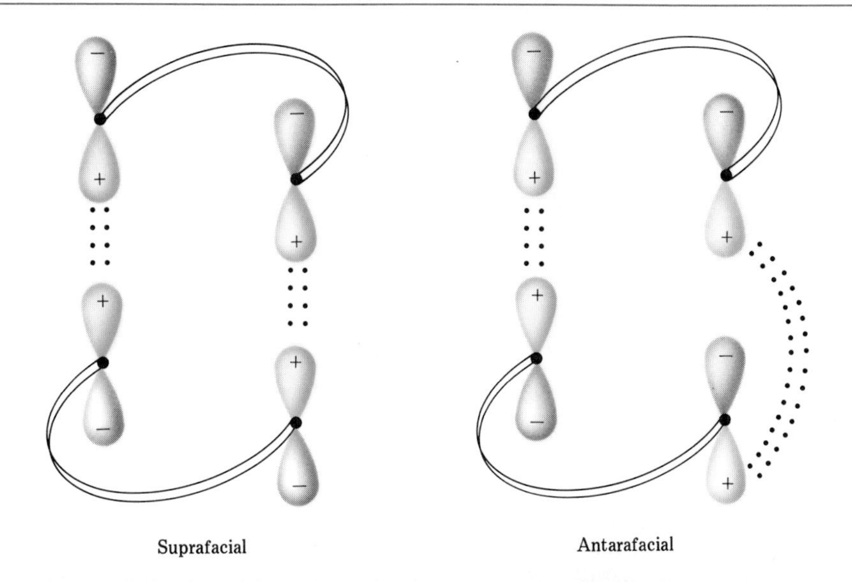

Figure 30.23. Suprafacial and antarafacial cycloadditions.

(Figure 30.23). The reaction course in a given case can be found by looking at the symmetry of the HOMO of one component and the LUMO of the other component.

Sigmatropic rearrangements involve the migration of a sigma-bonded group across a pi electron system. For example, Claisen rearrangement of an allyl vinylic ether yields an unsaturated carbonyl compound. Sigmatropic rearrangements can occur with either suprafacial or antarafacial stereochemistry, and the selection rules for a given case are the same as those for cycloaddition reactions.

The stereochemistry of any pericyclic reaction can be predicted by counting the total number of electron pairs (bonds) involved in bond reorganization and then applying some simple rules. Thermal (ground-state) reactions involving an even number of electron pairs occur with either antarafacial or conrotatory stereochemistry (even–antara–con); thermal reactions involving an odd number of electron pairs occur with suprafacial or disrotatory stereochemistry (odd–supra–dis). Exactly the opposite rules apply to photochemical (excited-state) reactions.

ADDITIONAL PROBLEMS

30.15 Define the following terms in your own words:

(a) Electrocyclic reaction (b) Conrotatory motion
(c) Suprafacial (d) Antarafacial
(e) Disrotatory motion (f) Sigmatropic rearrangement

30.16 Have the following reactions taken place in a conrotatory or in a disrotatory manner? Under what conditions, thermal or photochemical, would you carry out each reaction?

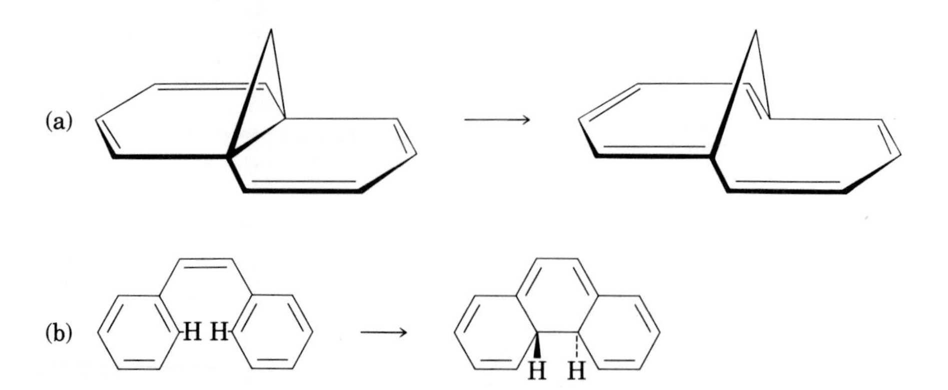

30.17 (2*E*,4*Z*,6*Z*,8*E*)-Decatetraene has been cyclized to 7,8-dimethyl-1,3,5-cyclooctatriene. Predict the manner of ring closure—conrotatory or disrotatory—for both thermal and photochemical reactions, and predict the stereochemistry of the product in each case.

30.18 Answer Problem 30.17 for the thermal and photochemical cyclizations of (2*E*,4*Z*,6*Z*,8*Z*)-decatetraene.

30.19 The cyclohexadecaoctaene shown isomerizes to two different isomers, depending on reaction conditions. Explain the observed results and indicate the conrotatory or disrotatory nature of each reaction.

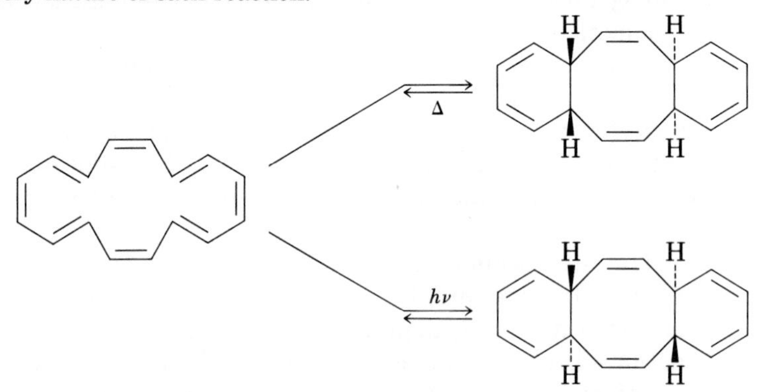

30.20 What stereochemistry would you expect to observe in these reactions?
(a) A photochemical [1,5] sigmatropic rearrangement
(b) A thermal [4 + 6] cycloaddition
(c) A thermal [1,7] sigmatropic rearrangement
(d) A photochemical [2 + 6] cycloaddition

30.21 Which of the following two reactions is more likely to occur? Explain.

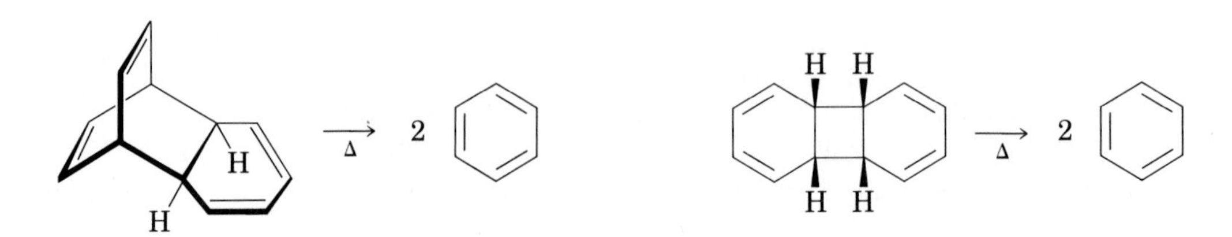

30.22 The following thermal rearrangement involves two pericyclic reactions in sequence. Identify them and propose a mechanism to account for the observed result.

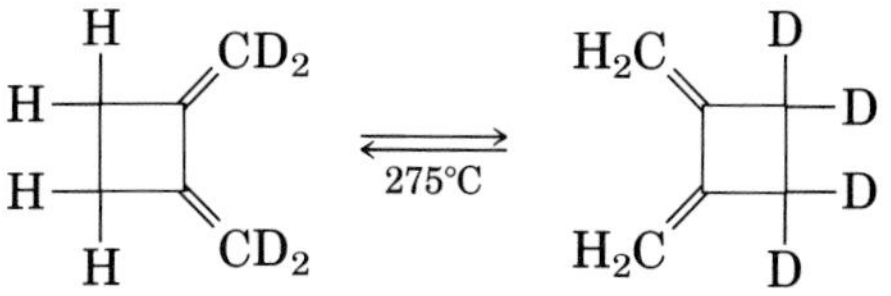

CD_2 H_2C D   H   275°C CD_2 H_2C

30.23 Predict the product you would expect to obtain from the following reaction. Should this [5,5] shift be a suprafacial or an antarafacial process?

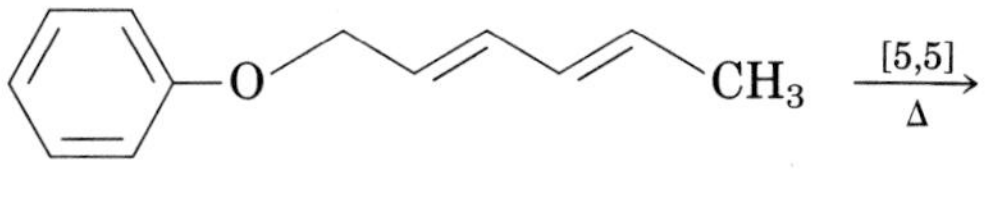

O CH_3 [5,5]   Δ

30.24 How can you account for the fact that ring opening of the *trans*-cyclobutene isomer shown takes place at much lower temperatures than a similar ring opening of the *cis*-cyclobutene isomer? Identify the stereochemistry of each reaction.

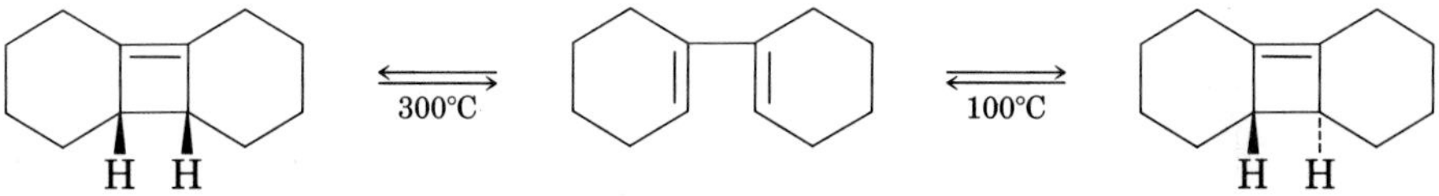

H H   300°C   100°C   H H

30.25 Photolysis of the *cis*-cyclobutene isomer in Problem 30.24 yields *cis*-cyclododecaen-7-yne, but photolysis of the trans isomer yields *trans*-cyclododecaen-7-yne. Explain these results and identify the type and stereochemistry of the pericyclic reaction.

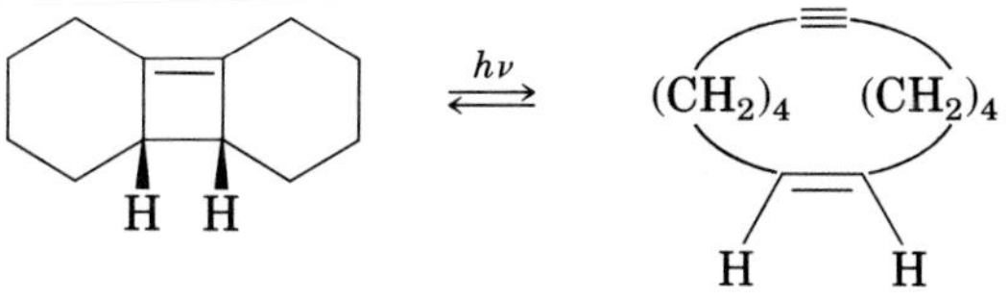

H H $h\nu$ $(CH_2)_4$ $(CH_2)_4$ H   H

Cis

H H $h\nu$ $(CH_2)_4$ H $(CH_2)_4$ H

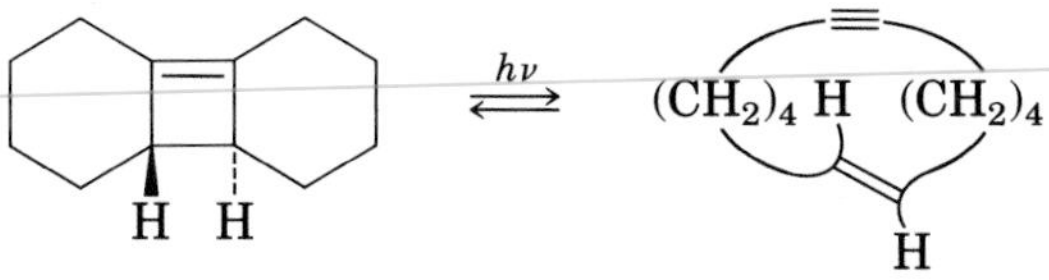

Trans

30.26 Two pericyclic reactions are involved in the furan synthesis shown. Identify them and propose a mechanism for the transformation.

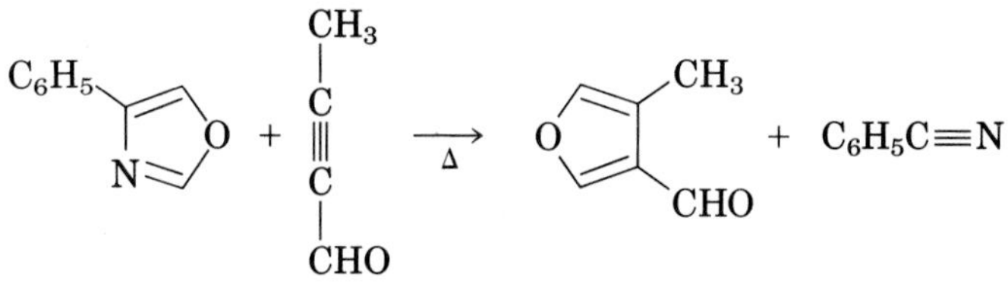

C_6H_5 O   N   + $CH_3-C{\equiv}C-CHO$ $\xrightarrow{\Delta}$ O CH_3 CHO   + $C_6H_5C{\equiv}N$

30.27 The following synthesis of dienones has been observed to occur readily. Propose a mechanism to account for the results, and identify the kind of pericyclic reaction involved.

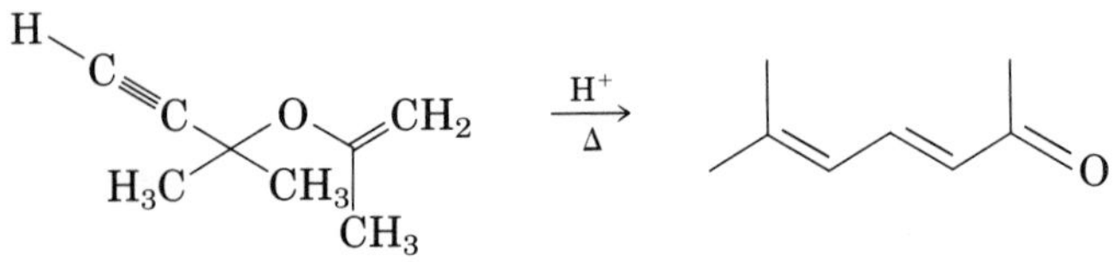

30.28 Karahanaenone, a compound isolated from oil of hops, has been synthesized by the thermal reaction shown. Identify the kind of pericyclic reaction and explain how karahanaenone is formed.

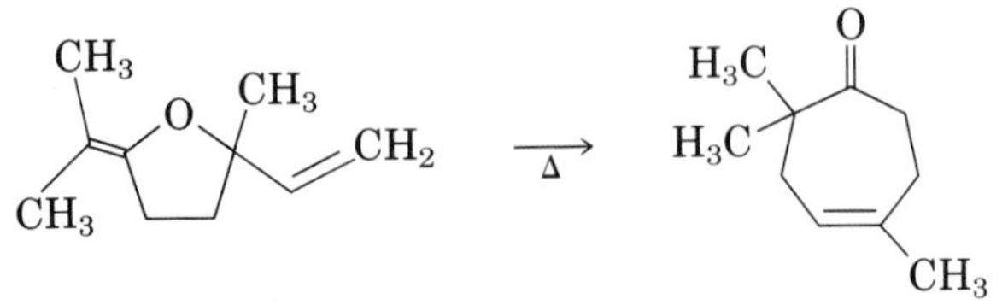

Karahanaenone

30.29 The ^{1}H NMR spectrum of bullvalene at 100°C consists only of a single peak at 4.22 δ. What conclusion can you draw from this? What explanation can you suggest to account for this result?

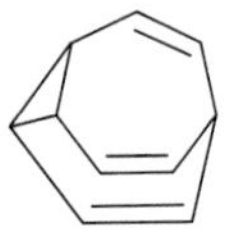

Bullvalene

30.30 The rearrangement shown was devised and carried out to prove the stereochemistry of [1,5] sigmatropic hydrogen shifts. Explain how the observed result confirms the predictions of orbital symmetry.

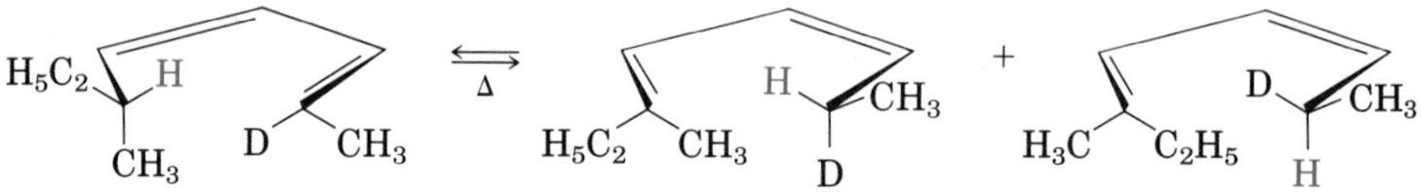

30.31 The reaction shown is an example of a [2,3] sigmatropic rearrangement. Would you expect the reaction to be suprafacial or antarafacial? Explain.

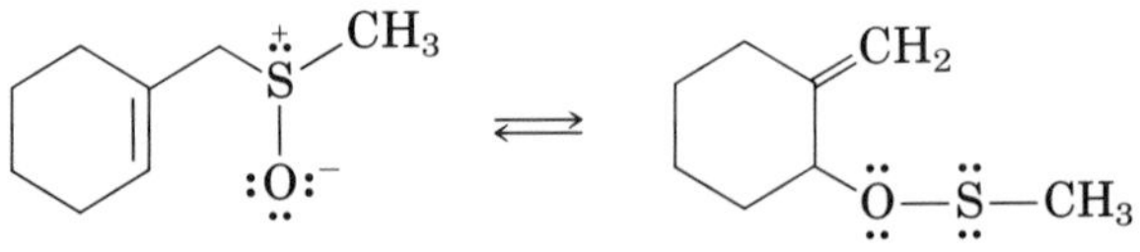

30.32 When the compound having a cyclobutene fused to a five-membered ring is heated, (1*Z*,3*Z*)-cycloheptadiene is formed. When the related compound having a cyclobutene fused to an eight-membered ring is heated, however, (1*E*,3*Z*)-cyclodecadiene is formed. Explain these results and suggest a reason why the eight-membered-ring opening occurs at lower reaction temperatures.

H H 270°C

H H 190°C

30.33 In light of your answer to Problem 30.32 explain why a mixture of products occurs in the following reactions:

H CH_3 H 190°C CH_3 + CH_3

30.34 Estrone, a major female sex hormone, has been synthesized by a route that involves the following step. Identify the pericyclic reactions involved and propose a mechanism.

CH_3O H_3C O H Δ CH_3O H_3C O H H H

Estrone methyl ether

30.35 Coronafacic acid, a bacterial toxin, was synthesized using a key step that involves three sequential pericyclic reactions. Identify them and propose a mechanism for the overall transformation. How would you complete the synthesis?

H C_2H_5 O O H O CH_3 CH_3

185°C

H_3C CH_3 + H_5C_2 H H O O O H ? HOOC H_5C_2 H H O H

92%

Coronafacic acid

30.36 The following rearrangement of *N*-allyl-*N*,*N*-dimethylanilinium ion has been observed to occur. Propose a mechanism to account for the reaction.

$$C_6H_5\overset{+}{N}(CH_3)_2CH_2CH{=}CH_2 \xrightarrow{\Delta} o\text{-}(CH_2CH{=}CH_2)C_6H_4\overset{+}{N}H(CH_3)_2$$

N-Allyl-*N*,*N*-dimethylanilinium ion → *o*-Allyl-*N*,*N*-dimethylanilinium ion

CHAPTER 31 AMINO ACIDS, PEPTIDES, AND PROTEINS

Proteins are large biomolecules that occur in every living organism; they are of many different types, and they serve many different biological roles. The keratin of skin and fingernails, the fibroin of silk and spider webs, and the collagen of tendons and cartilage are all different **structural proteins;** insulin is a **hormonal protein** that regulates glucose metabolism in the body; snake venoms and botulinum toxins are powerful poisonous proteins; still other proteins such as DNA polymerase and reverse transcriptase serve as **enzymes,** the biological catalysts needed to carry out all chemical reactions in the cell. Regardless of their appearance or their function, however, all proteins are chemically similar and are made up of many amino acid units linked together.

Amino acids are the building blocks from which all proteins are made. As their name implies, amino acids are difunctional; they contain both a basic amino group and an acidic carboxyl group.

$$H_2N{-}CH_2{-}COOH$$

Glycine, an amino acid

Their great value as biological building blocks derives from the fact that amino acids can link together by forming amide, or **peptide,** bonds. A **dipeptide** results when an amide bond is formed between the $-NH_2$ of one amino acid and the $-COOH$ of a second amino acid; a **tripeptide** results from linkage of three amino acids via two amide bonds, and so on. Any number of amino acids may link together to form large chains. For classification purposes, chains with fewer than 50 amino acids are called **polypeptides;** the term *protein* is usually reserved for larger chains.

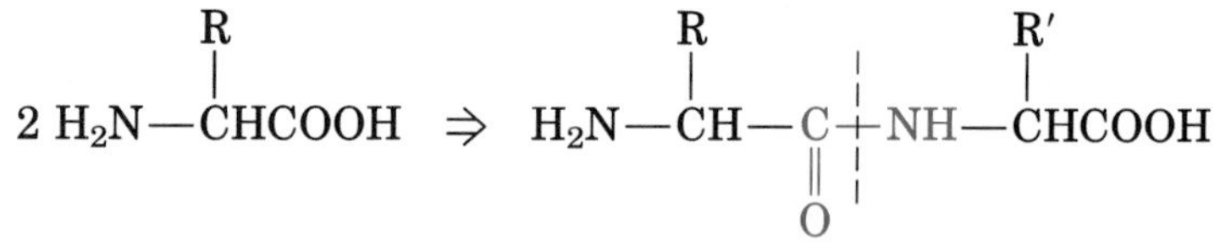

$$2\ H_2N{-}\underset{}{\overset{R}{\overset{|}{C}}}HCOOH \Rightarrow H_2N{-}\overset{R}{\overset{|}{C}}H{-}\underset{\underset{O}{\|}}{C}{-}NH{-}\overset{R'}{\overset{|}{C}}HCOOH$$

A dipeptide (one amide bond)

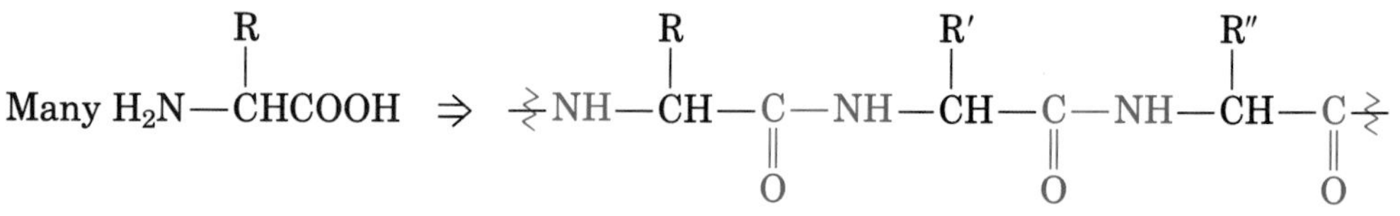

$$\text{Many } H_2N{-}\overset{R}{\overset{|}{C}}HCOOH \Rightarrow {-}NH{-}\overset{R}{\overset{|}{C}}H{-}\underset{\underset{O}{\|}}{C}{-}NH{-}\overset{R'}{\overset{|}{C}}H{-}\underset{\underset{O}{\|}}{C}{-}NH{-}\overset{R''}{\overset{|}{C}}H{-}\underset{\underset{O}{\|}}{C}{-}$$

A polypeptide (many amide bonds)

31.1 Structures of amino acids

There are 20 amino acids commonly found in proteins, and their structures are shown in Table 31.1 (pp. 1034–1035). All 20 are α-amino acids; the amino group in each is a substituent on the carbon atom alpha (α) to, or next to, the carbonyl. The amino acid structures differ only in the nature of the side chain. Note that 19 of the 20 are primary amines, $R—NH_2$, but that proline is a secondary amine whose nitrogen and α-carbon atoms are part of a pyrrolidine ring. Proline can still form amide bonds like the other 19 α-amino acids, however.

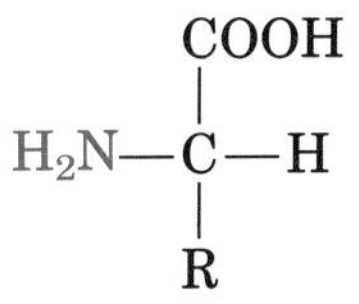

COOH
$H_2N—C—H$
R

Primary α-amino acids
(R = a side chain)

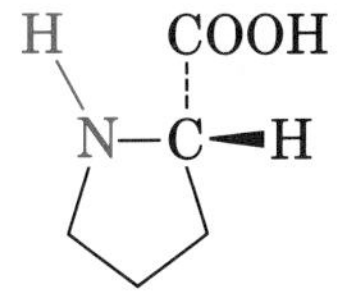

H COOH
N—C—H

Proline, a secondary
α-amino acid

Note also that each of the amino acids in Table 31.1 can be referred to by a mnemonic three-letter shorthand code—Ala for alanine, Gly for glycine, and so on. In addition, a new one-letter code is gaining popularity, and usage of the new system will probably increase.

With the exception of glycine, $H_2N—CH_2COOH$, the α carbons of amino acids are chiral, and two different enantiomeric forms are possible. Nature, however, uses only the *S* enantiomers to construct proteins. In Fischer projections, *S* amino acids are represented by placing the carboxyl group at the top as if drawing a carbohydrate and then placing the amino group on the left. Because of their stereochemical similarity to L sugars (Section 27.3), the naturally occurring α-amino acids are often referred to as L-amino acids.

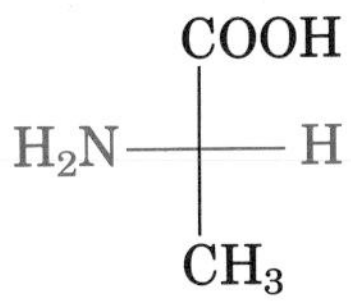

COOH
H_2N — H
CH_3

(S)-Alanine
(L-alanine)

COOH
H_2N — H
CH_2

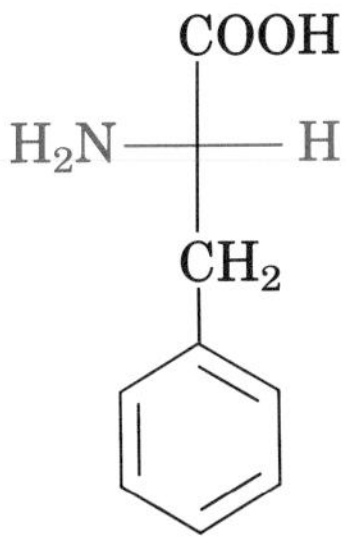

(S)-Phenylalanine
(L-phenylalanine)

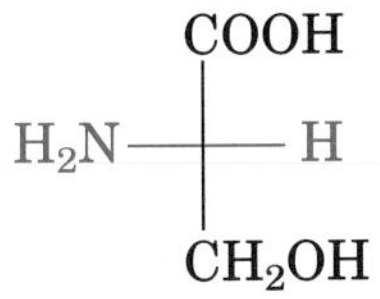

COOH
H_2N — H
CH_2OH

(S)-Serine
(L-serine)

CHO
HO — H
CH_2OH

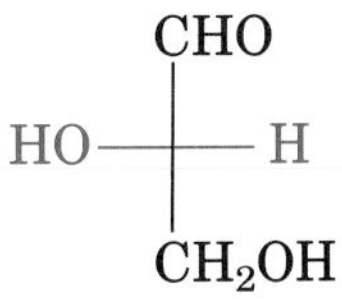

Stereochemically
similar to
L-glyceraldehyde

The 20 common amino acids can be further categorized as either neutral, acidic, or basic, depending on the nature of their specific side chains. Most have neutral side chains, but two (aspartic acid and glutamic acid) have an extra carboxylic acid function in their side chains, and three (lysine, arginine, and histidine) have basic amino groups in their side chains.

All 20 of the amino acids are important for protein synthesis, but humans are thought to be able to synthesize only 10 of the 20. The remaining 10 are called **essential amino acids** since they *must* be obtained from dietary sources. Failure to include an adequate dietary supply of these essential amino acids can lead to severe deficiency diseases.

TABLE 31.1 **The twenty common amino acids found in proteins; essential amino acids are shown in color**

Name	*Abbreviations*	*Molecular weight*	*Structure*	*Isoelectric point*
Neutral amino acids				
Alanine	Ala (A)	89	$CH_3CH(NH_2)COOH$	6.0
Asparagine	Asn (N)	132	$H_2N{-}C(=O)CH_2CH(NH_2)COOH$	5.4
Cysteine	Cys (C)	121	$HSCH_2CH(NH_2)COOH$	5.0
Glutamine	Gln (Q)	146	$H_2N{-}C(=O){-}CH_2CH_2CH(NH_2)COOH$	5.7
Glycine	Gly (G)	75	$CH_2(NH_2)COOH$	6.0
Isoleucine	Ile (I)	131	$CH_3CH_2CH(CH_3)CH(NH_2)COOH$	6.0
Leucine	Leu (L)	131	$(CH_3)_2CHCH_2CH(NH_2)COOH$	6.0
Methionine	Met (M)	149	$CH_3SCH_2CH_2CH(NH_2)COOH$	5.7
Phenylalanine	Phe (F)	165	$C_6H_5{-}CH_2CH(NH_2)COOH$	5.5
Proline	Pro (P)	115	$H_2C(CH_2)(CH_2)$ ring with $CH{-}COOH$ and $N{-}H$ (pyrrolidine-2-carboxylic acid)	6.3
Serine	Ser (S)	105	$HOCH_2CH(NH_2)COOH$	5.7
Threonine	Thr (T)	119	$CH_3CH(OH)CH(NH_2)COOH$	5.6

Name	Abbreviations	Molecular weight	Structure	Isoelectric point
Tryptophan	Trp (W)	204	(indol-3-yl)$CH_2CH(NH_2)COOH$	5.9
Tyrosine	Tyr (Y)	181	HO–C_6H_4–$CH_2CH(NH_2)COOH$	5.7
Valine	Val (V)	117	$(CH_3)_2CHCH(NH_2)COOH$	6.0
Acidic amino acids				
Aspartic acid	Asp (D)	133	$HOOCCH_2CH(NH_2)COOH$	3.0
Glutamic acid	Glu (E)	147	$HOOCCH_2CH_2CH(NH_2)COOH$	3.2
Basic amino acids				
Arginine	Arg (R)	174	$H_2N{-}C({=}NH){-}NHCH_2CH_2CH_2CH(NH_2)COOH$	10.8
Histidine	His (H)	155	(imidazol-4-yl)$CH_2CH(NH_2)COOH$	7.6
Lysine	Lys (K)	146	$H_2NCH_2CH_2CH_2CH_2CH(NH_2)COOH$	9.7

PROBLEM

31.1 The amino acid threonine, (2*S*,3*R*)-2-amino-3-hydroxybutanoic acid, has two chiral centers. Draw a Fischer projection of threonine. Draw the Fischer projection of a threonine diastereomer, and label the chiral centers as *R* or *S*.

PROBLEM

31.2 Draw a Fischer projection of L-proline.

31.2 Dipolar structure of amino acids

Amino acids contain both acidic and basic groups in the same molecule. For this reason, they undergo an intramolecular acid–base reaction and exist primarily as the dipolar ion or **zwitterion** (German *zwitter,* "hybrid"):

$$H_2\ddot{N}-\underset{}{\overset{R}{\overset{|}{C}}}H-COOH \rightleftharpoons H_3\overset{+}{N}-\overset{R}{\overset{|}{C}}H-COO^-$$

A zwitterion

Amino acid zwitterions are a kind of internal salt, and have many of the properties we associate with salts. Thus, amino acids have large dipole moments. They are soluble in water but insoluble in hydrocarbons, and are crystalline substances with high melting points. Furthermore, amino acids are **amphoteric;** they can react either as acids or as bases, depending on the circumstances. In aqueous acid solution, an amino acid zwitterion can accept a proton to yield a cation; in aqueous basic solution, the zwitterion can lose a proton to form an anion.

In acid solution

$$H_3\overset{+}{N}-\overset{R}{\overset{|}{C}}H-CO_2^- + H^+ \rightleftharpoons H_3\overset{+}{N}-\overset{R}{\overset{|}{C}}H-COOH$$

In base solution

$$H_3\overset{+}{N}-\overset{R}{\overset{|}{C}}H-CO_2^- + {}^-OH \rightleftharpoons H_2N-\overset{R}{\overset{|}{C}}H-CO_2^- + H_2O$$

Note that it is the carboxylate anion, $-COO^-$, rather than the amino group that acts as the basic site and accepts the proton in acid solution. Similarly, it is the ammonium cation rather than the carboxyl group that acts as the acidic site and donates a proton in base solution. This reversal in behavior is simply a consequence of the zwitterionic structure of amino acids.

31.3 Isoelectric point

At low pH an amino acid is protonated and exists as a cation; at high pH, an amino acid is deprotonated and exists as an anion. Thus, at some intermediate point, the amino acid must be exactly balanced between anionic and cationic forms and exist primarily as the neutral, dipolar zwitterion. This point is called the **isoelectric point.**

$$H_3\overset{+}{N}-\overset{R}{\overset{|}{C}}HCOOH \underset{}{\overset{H^+}{\rightleftharpoons}} H_3\overset{+}{N}-\overset{R}{\overset{|}{C}}HCOO^- \overset{-H^+}{\rightleftharpoons} H_2N-\overset{R}{\overset{|}{C}}HCOO^-$$

Low pH (protonated) → Isoelectric point (neutral zwitterion) → High pH (deprotonated)

The isoelectric point of a given amino acid depends on its structure; values for the 20 most common amino acids are given in Table 31.1. The 15 amino acids with neutral side chains have isoelectric points near neutrality, in the pH range 5.0–6.5. (These values are not exactly at neutral pH = 7, because carboxyl groups are stronger acids in aqueous solution than amino groups are bases.) The two amino acids with acidic side chains have isoelectric points at lower pH (3.2–3.5), and the three amino acids with basic side chains have isoelectric points at higher pH (7.6–10.8).

We can take advantage of the differences in isoelectric points to separate a mixture of amino acids. In the technique known as **paper electrophoresis,** a solution of different amino acids is placed near the center of a strip of paper. The paper is moistened with an aqueous buffer of a given pH, and electrodes are connected to the ends of the strip. When an electric field is applied, those amino acids with negative charges (those with isoelectric points below the pH of the buffer) migrate slowly toward the positive electrode, and those amino acids with positive charges (those with isoelectric points above the pH of the buffer) migrate toward the negative electrode. Different amino acids migrate at different rates, depending both on their isoelectric point and on the pH of the aqueous medium, and they can thus be separated. Figure 31.1 illustrates this separation for a mixture of lysine (basic), glycine (neutral), and aspartic acid (acidic).

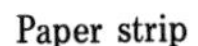

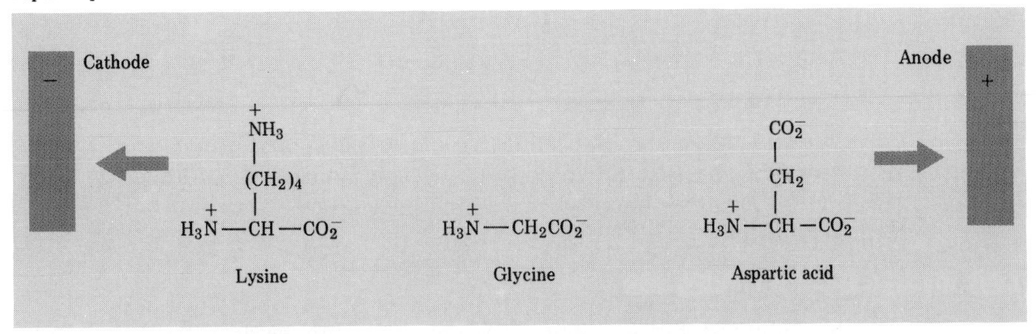

Figure 31.1. Separation of an amino acid mixture by electrophoresis. At pH 6.0, glycine is neutral and does not migrate. Lysine has a positive charge at pH 6.0 and migrates toward the cathode, whereas aspartic acid has a negative charge and migrates toward the anode. Lysine has its isoelectric point at 9.7; glycine, at 6.0; and aspartic acid, at 3.0.

PROBLEM

31.3 For the mixtures of amino acids indicated, predict the direction of migration of each component (toward anode or cathode) and relative rate of migration.

(a) Valine, glutamic acid, and histidine at pH = 7.6
(b) Glycine, phenylalanine, and serine at pH = 5.7
(c) Glycine, phenylalanine, and serine at pH = 5.5
(d) Glycine, phenylalanine, and serine at pH = 6.0

PROBLEM

31.4 How can you account for the fact that tryptophan has a lower isoelectric point than histidine, even though both have five-membered-ring nitrogen atoms? Which nitrogen in the five-membered ring of histidine is more basic?

31.4 Synthesis of α-amino acids

α-Amino acids can be synthesized using some of the standard chemical methods already discussed, but all methods starting from achiral reagents yield a racemic mixture of *R* and *S* products. For example, α bromination of carboxylic acids can be carried out by treatment of the acid with bromine and phosphorus (the Hell–Volhard–Zelinskii reaction, Section 25.4). Nucleophilic displacement of the α bromine by ammonia yields an α-amino acid.

$$(CH_3)_2CHCH_2CH_2COOH \xrightarrow[\text{2. } H_2O]{\text{1. } Br_2,\ P} (CH_3)_2CHCH_2CH(Br)COOH$$

4-Methylpentanoic acid → 2-Bromo-4-methylpentanoic acid

$$\downarrow NH_3 \text{ (excess)}$$

$$(CH_3)_2CHCH_2CH(NH_2)COOH$$

(R,S)-Leucine (45%)

Alternatively, higher product yields are obtained when the bromide displacement reaction is carried out by the Gabriel phthalimide method (Section 28.7) rather than by the ammonia method.

THE STRECKER SYNTHESIS

Another method for preparing racemic α-amino acids is the **Strecker synthesis.**[1] Developed in 1850, this versatile two-step process involves treatment of an aldehyde with KCN and aqueous ammonia to yield an intermediate α-amino nitrile. Hydrolysis of the nitrile then gives an α-amino acid:

$$C_6H_5CH_2CHO \xrightarrow[H_2O]{NH_4Cl/KCN} C_6H_5CH_2CH(NH_2)CN \xrightarrow{H_3O^+} C_6H_5CH_2CH(NH_2)COOH$$

Phenylacetaldehyde — An α-amino nitrile — *(R,S)*-Phenylalanine (53%)

[1] Adolph Friedrich Ludwig Strecker (1822–1871); Ph.D. Giessen (1842); assistant to Liebig at Tübingen.

PROBLEM

31.5 Propose a mechanism to account for the first step in the Strecker synthesis of α-amino acids. What kinds of general carbonyl-group reactions are involved?

REDUCTIVE AMINATION OF α-KETO ACIDS: BIOSYNTHESIS

Yet a third method for the synthesis of α-amino acids is reductive amination of an α-keto acid (Section 28.7):

$$CH_3\overset{\overset{O}{\|}}{C}COOH \xrightarrow[NaBH_4]{NH_3} CH_3\underset{\underset{NH_2}{|}}{C}HCOOH$$

Pyruvic acid (an α-keto acid) → (*R,S*)-Alanine

This reductive amination method is not widely used, but it is interesting because it is a close laboratory analogy of a pathway by which amino acids can be biosynthesized in nature. For example, the major route for glutamic acid synthesis in most organisms is reductive amination of α-ketoglutaric acid. The biological reducing agent is the rather complex molecule, nicotinamide adenine dinucleotide (NADH), and the reductive amination step is catalyzed by an enzyme, L-glutamate dehydrogenase. Nevertheless, the fundamental chemical principles of this biosynthetic reaction are identical to those of the laboratory reaction.

$$HOOCCH_2CH_2\overset{\overset{O}{\|}}{C}COOH + NH_3 \xrightarrow[\text{L-Glutamate dehydrogenase}]{NADH} HOOCCH_2CH_2\underset{\underset{NH_2}{|}}{C}HCOOH$$

α-Ketoglutaric acid → (*S*)-Glutamic acid

PROBLEM

31.6 The rare amino acid L-dopa (3,4-dihydroxyphenylalanine) is useful as a drug against Parkinson's disease. Show how (±)-dopa might be synthesized from 3,4-dihydroxyphenylacetaldehyde.

$$(HO)_2C_6H_3-CH_2\underset{\underset{NH_2}{|}}{C}HCOOH$$

Dopa

PROBLEM

31.7 Show how a Gabriel phthalimide synthesis (Section 28.7) might be used to prepare isoleucine.

THE AMIDOMALONATE SYNTHESIS

The amidomalonate synthesis is the most general method of preparation for α-amino acids. This route is a straightforward extension of the malonic ester synthesis (Section 25.9), and involves initial conversion of diethyl acetamidomalonate into its enolate anion by treatment with base, followed by reaction with a primary alkyl halide. Hydrolysis and decarboxylation occur when the alkylated product is warmed with aqueous acid, and a racemic α-amino acid results. For example, aspartic acid is prepared in good yield when diethyl acetamidomalonate is alkylated with ethyl bromoacetate, followed by hydrolysis and decarboxylation:

$$CH_3\overset{O}{\overset{\|}{C}}-NH-C(CO_2C_2H_5)_2-H \xrightarrow[\text{Ethanol}]{Na^+\ ^-OC_2H_5} \left[CH_3\overset{O}{\overset{\|}{C}}-NH-\ddot{C}^-(CO_2C_2H_5)_2\right]$$

Diethyl acetamidomalonate

$$\xrightarrow{Br-CH_2CO_2C_2H_5} CH_3\overset{O}{\overset{\|}{C}}-NH-C(CO_2C_2H_5)_2-CH_2CO_2C_2H_5$$

$$\xrightarrow{H_3O^+} CO_2 + 2\ C_2H_5OH + CH_3COOH + H_2N-CH(COOH)CH_2COOH$$

(R,S)-Aspartic acid (55%)

Recall the malonic ester synthesis:

$$CH_2(CO_2C_2H_5)_2 \xrightarrow[\text{2. RX} \quad \text{3. } H_3O^+]{\text{1. } Na^+\ ^-OC_2H_5} R-CH_2COOH$$

PROBLEM

31.8 What alkyl halides would you use to prepare these α-amino acids by the amidomalonate method?

(a) Leucine
(b) Histidine
(c) Tryptophan
(d) Methionine

PROBLEM

31.9 Serine can be synthesized by a simple variation of the amidomalonate method. Can you suggest how this might be done? [*Hint:* See Section 26.6.]

31.5 Resolution of *R,S* amino acids

The synthesis of chiral amino acids from achiral precursors by any one of the methods just described must yield a racemic mixture—an equal mixture of *S* and *R* products. In order to use these synthetic amino acids for the laboratory synthesis of naturally occurring peptides, however, we must first resolve the racemic mixture into pure enantiomers. Often, this resolution can be done by the general method discussed earlier (Section 28.5) whereby the racemic mixture is converted into a mixture of diastereomeric salts by reaction with a chiral acid or base, and the different diastereomers are then separated by fractional crystallization.

Alternatively, we can make use of biological methods of resolution. Enzymes are chiral catalysts that often show an astounding selectivity toward one enantiomer of an *R*,*S* mixture. For example, the enzyme carboxypeptidase selectively catalyzes the hydrolysis of *S* amido acids but not *R* amido acids. We can therefore resolve an *R*,*S* mixture of amino acids by first allowing the mixture to react with acetic anhydride to form the *N*-acetyl derivatives. Selective hydrolysis of the *R*,*S* amido acid mixture with carboxypeptidase then yields a mixture of the desired *S* amino acid and the unchanged *N*-acetyl *R* amido acid, which can be separated by usual chemical techniques.

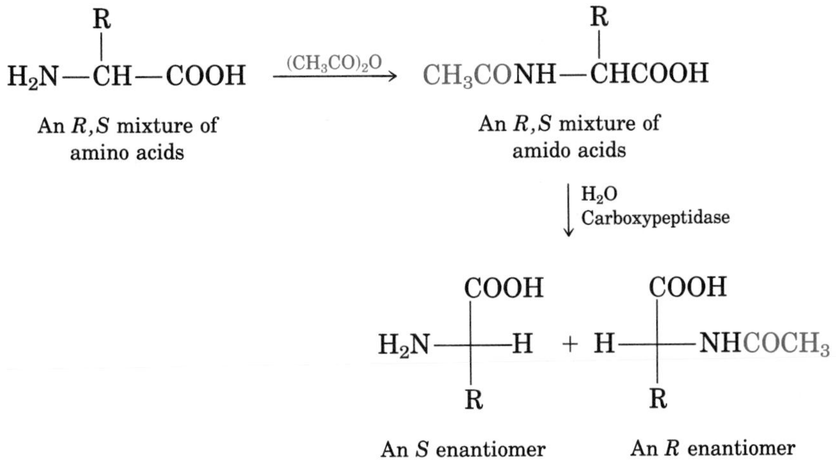

$$H_2N-\underset{\displaystyle R}{\overset{|}{C}H}-COOH \xrightarrow{(CH_3CO)_2O} CH_3CONH-\underset{\displaystyle R}{\overset{|}{C}H}COOH$$

An *R*,*S* mixture of amino acids

An *R*,*S* mixture of amido acids

$$\downarrow H_2O, \text{ Carboxypeptidase}$$

(Fischer projections: COOH top, R bottom; left: H_2N—C—H; right: H—C—$NHCOCH_3$)

$$H_2N-C(COOH)(R)-H \quad + \quad H-C(COOH)(R)-NHCOCH_3$$

An *S* enantiomer

An *R* enantiomer

31.6 Peptides

Peptides are amino acid polymers in which the individual amino acid units, called **residues,** are linked together by amide, or peptide, bonds. An amino group from one residue forms an amide bond with the carboxyl of a second residue; the amino group of the second forms an amide bond with the carboxyl of a third, and so on. For example, alanylserine is the dipeptide formed when an amide bond is made between the alanine carboxyl and the serine amino group:

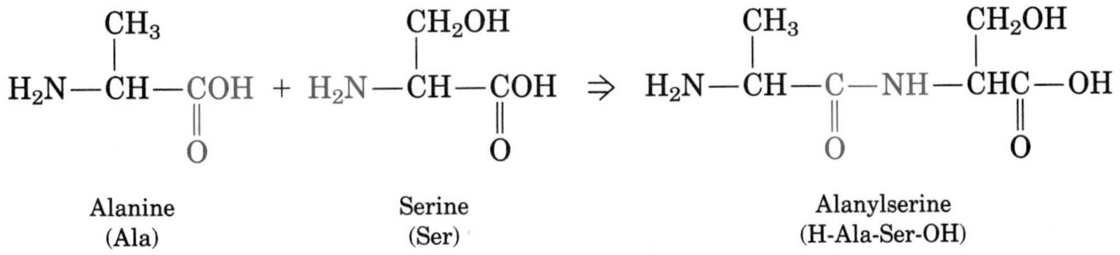

$$H_2N-\overset{CH_3}{\overset{|}{C}H}-\underset{\|}{C}OH + H_2N-\overset{CH_2OH}{\overset{|}{C}H}-\underset{\|}{C}OH \Rightarrow H_2N-\overset{CH_3}{\overset{|}{C}H}-\underset{\|}{C}-NH-\overset{CH_2OH}{\overset{|}{C}H}\underset{\|}{C}-OH$$

(each C=O carbon bears a doubly bonded O)

Alanine (Ala)

Serine (Ser)

Alanylserine (H-Ala-Ser-OH)

Note that two dipeptides can result from reaction between alanine and serine, depending on which carboxyl group reacts with which amino group. If the alanine amino group reacts with the serine carboxyl, serylalanine results:

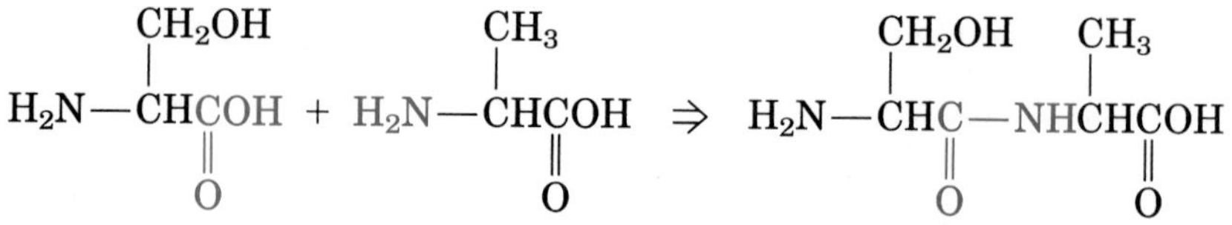

$$\underset{\substack{\text{Serine}\\\text{(Ser)}}}{H_2N{-}\overset{\overset{\displaystyle CH_2OH}{|}}{C}H\underset{\underset{\displaystyle O}{\|}}{C}OH} + \underset{\substack{\text{Alanine}\\\text{(Ala)}}}{H_2N{-}\overset{\overset{\displaystyle CH_3}{|}}{C}H\underset{\underset{\displaystyle O}{\|}}{C}OH} \Rightarrow \underset{\substack{\text{Serylalanine}\\\text{(H-Ser-Ala-OH)}}}{H_2N{-}\overset{\overset{\displaystyle CH_2OH}{|}}{C}H\underset{\underset{\displaystyle O}{\|}}{C}{-}NH\overset{\overset{\displaystyle CH_3}{|}}{C}H\underset{\underset{\displaystyle O}{\|}}{C}OH}$$

By convention, peptides are always written with the N-terminal amino acid (that with the free $-NH_2$ group) on the left, and the C-terminal amino acid (that with the free $-COOH$ group) on the right. The name of the peptide is usually indicated by using the three-letter abbreviations listed in Table 31.1 for each amino acid. The N terminus (left) and the C terminus

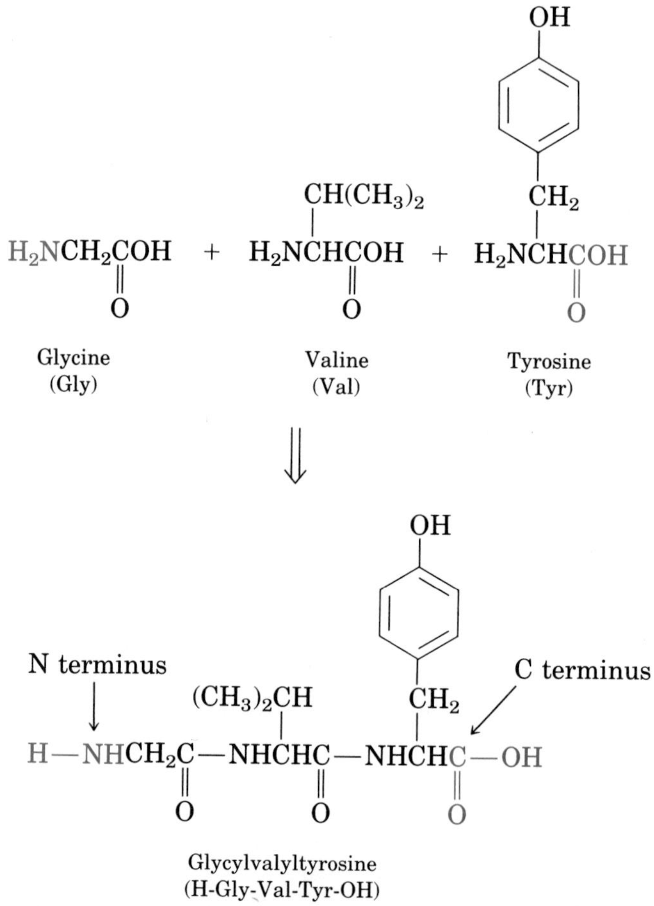

$$\underset{\substack{\text{Glycine}\\\text{(Gly)}}}{H_2NCH_2\underset{\underset{\displaystyle O}{\|}}{C}OH} + \underset{\substack{\text{Valine}\\\text{(Val)}}}{H_2N\overset{\overset{\displaystyle CH(CH_3)_2}{|}}{C}H\underset{\underset{\displaystyle O}{\|}}{C}OH} + \underset{\substack{\text{Tyrosine}\\\text{(Tyr)}}}{H_2N\overset{\overset{\displaystyle CH_2C_6H_4OH}{|}}{C}H\underset{\underset{\displaystyle O}{\|}}{C}OH}$$

$$\Downarrow$$

$$\underset{\substack{\text{Glycylvalyltyrosine}\\\text{(H-Gly-Val-Tyr-OH)}}}{\overset{\text{N terminus}}{H}{-}NHCH_2\underset{\underset{\displaystyle O}{\|}}{C}{-}NH\overset{\overset{\displaystyle (CH_3)_2CH}{|}}{C}H\underset{\underset{\displaystyle O}{\|}}{C}{-}NH\overset{\overset{\displaystyle CH_2C_6H_4OH}{|}}{C}H\underset{\underset{\displaystyle O}{\|}}{C}{-}\overset{\text{C terminus}}{OH}}$$

Figure 31.2. Structure of glycylvalyltyrosine.

(right) are indicated by H- and -OH, respectively. For example, H-Gly-Val-Tyr-OH is the tripeptide glycylvalyltyrosine, whose structure is shown in Figure 31.2.

The number of possible isomeric peptides increases rapidly as the number of amino acid units increases. Thus, there are six different ways in which three amino acids can be joined and more than 40,000 ways in which the eight amino acids present in the blood-pressure-regulating hormone angiotensin II can be joined (Figure 31.3).

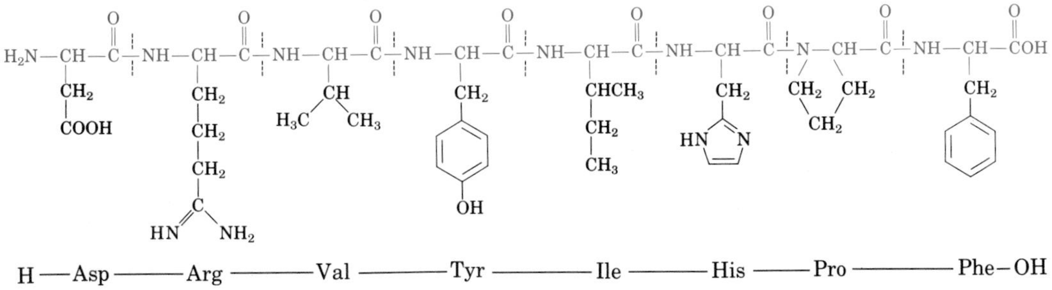

H—Asp——Arg——Val——Tyr——Ile——His——Pro——Phe—OH

Figure 31.3. Angiotensin II (present in blood plasma).

PROBLEM

31.10 Name the six possible isomeric tripeptides that contain valine, tyrosine, and glycine. Use the three-letter shorthand notation for each amino acid.

PROBLEM

31.11 Draw the full structure of H-Met-Pro-Val-Gly-OH.

31.7 Covalent bonding in peptides

Amide bonds in peptides are similar to the simple amide bonds we have already discussed (Section 28.4). Amide nitrogen atoms are nonbasic because their unshared electron pair is delocalized by orbital overlap with the carbonyl group. This overlap imparts a certain amount of double-bond character to the amide C—N bond and restricts its rotation (Figure 31.4).

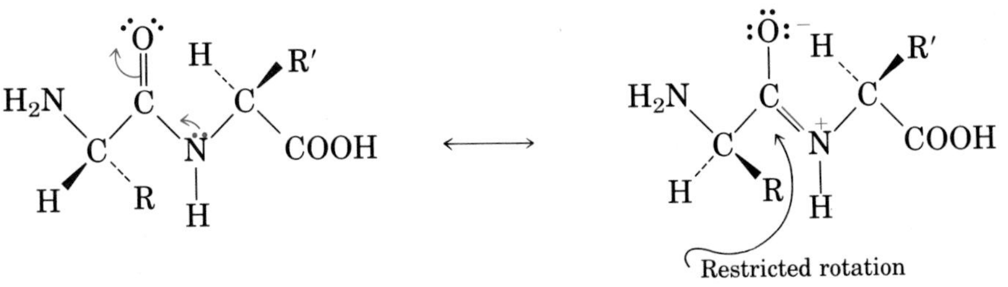

Restricted rotation

Figure 31.4. Amide resonance causes restricted rotation around C—N bond.

A second kind of covalent bonding in peptides occurs when a disulfide linkage, R—S—S—R, is formed between two cysteine residues. Disulfide bonds are easily formed by mild oxidation of thiols, R—SH, and are easily cleaved back to thiols by mild reduction (Section 20.13).

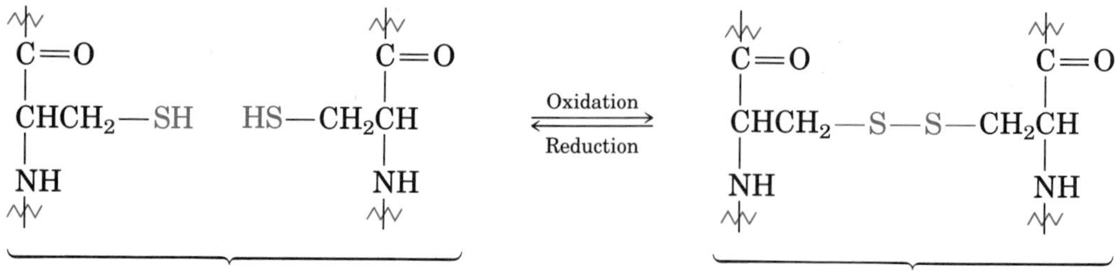

C=O | CHCH$_2$—SH HS—CH$_2$CH | C=O, NH ⇌ (Oxidation / Reduction) C=O | CHCH$_2$—S—S—CH$_2$CH | C=O, NH

Two cysteines (thiols) — Cystine (disulfide)

The new amino acid formed by coupling of two cysteines is called *cystine* (CyS). Cystine disulfide bonds between cysteine residues in two different peptide chains can link the otherwise separate chains together. Alternatively, a disulfide bond between two cysteine residues within the same chain can cause a loop in the chain. Such is the case with the nonapeptide vasopressin, an antidiuretic hormone involved in controlling water balance in the body. Note also that the C-terminal end of vasopressin occurs as the primary amide, —CONH$_2$, rather than as the free acid.

Disulfide bridge

H-CyS-Tyr-Phe-Glu-Asn-CyS-Pro-Arg-Gly-NH$_2$

Vasopressin

31.8 Peptide structure determination: amino acid analysis

Peptide structure determination is a challenging task that requires asking and answering three questions: What amino acids are present? How much of each is present? Where does each occur in the peptide chain? The answers to the first two questions are provided by a remarkable device, the **amino acid analyzer.**

The amino acid analyzer is an automated instrument based on analytical techniques worked out in the 1950s at the Rockefeller Institute by William Stein[2] and Stanford Moore.[3] One begins by reducing all cystine disulfide bonds and then hydrolyzing the polypeptide with 6*N* HCl to break it down into its constituent amino acids. Chromatography (Section 11.2) of the resultant amino acid mixture, using a series of aqueous buffers as the mobile phase, then effects a separation into component amino acids.

[2]William H. Stein (1911–1980); b. New York; Ph.D. Columbia; professor, Rockefeller Institute; Nobel prize (1972).

[3]Stanford Moore (1913–1982); b. Chicago; Ph.D. Wisconsin; professor, Rockefeller Institute; Nobel prize (1972).

As each different amino acid is eluted from the end of the chromatography column, it is allowed to mix with ninhydrin, a reagent that forms an intense purple color on reaction with α-amino acids. The purple color is detected by a spectrometer, and a plot of elution time versus spectrometer absorbance is obtained.

2 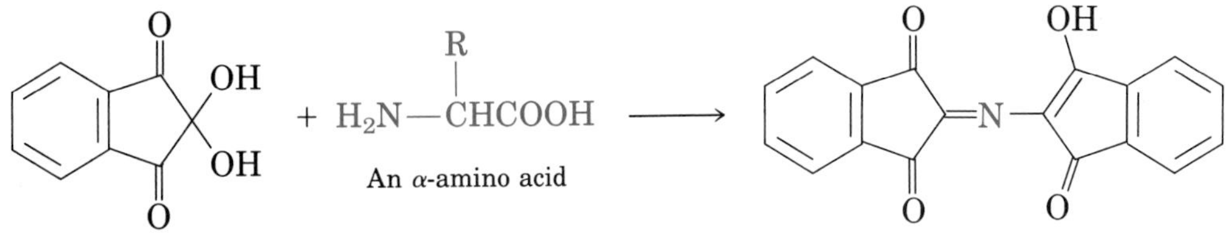 Ninhydrin + $H_2N-CH(R)COOH$ (An α-amino acid) $\longrightarrow$ Purple + $RCHO + CO_2$

The amount of time required for a given amino acid to elute from the chromatography column is reproducible from sample to sample, and the identity of all amino acids in a peptide of unknown composition can be determined by noting the various elution times. The amount of a given amino acid in a sample is proportional to the size of the peaks. Figure 31.5 (p. 1046) shows the results of amino acid analysis of a standard equimolar mixture of 17 α-amino acids and compares them to results obtained from analysis of methionine enkephalin, a pentapeptide with morphine-like analgesic activity.

PROBLEM

31.12 The amino acid analysis data in Figure 31.5(a) indicate that proline is not easily detected by reaction with ninhydrin; only a very small peak is seen on the chromatogram. Explain this observation.

31.9 Peptide sequencing: the Edman degradation

With the identity and amount of each amino acid known, the final task of structure determination is to **sequence** the peptide—to find out in what order the amino acids are linked together. The general idea of peptide sequencing is to selectively cleave one amino acid residue at a time from the end (either N terminus or C terminus) of the peptide chain. That terminal amino acid is then separated and identified, and the cleavage reactions are repeated on the chain-shortened peptide until the sequencing is finished.

Almost all peptide sequencing is now done by **Edman degradation,**[4] an efficient method of N-terminal analysis. Automated Edman *protein sequenators* are available that allow a series of 20 or more repetitive sequencing steps to be carried out before a buildup of unwanted by-products begins to interfere with the results.

[4]Pehr Edman (1916–); b. Stockholm; M.D. Karolinska Institute (E. Jorpes); professor, University of Lund.

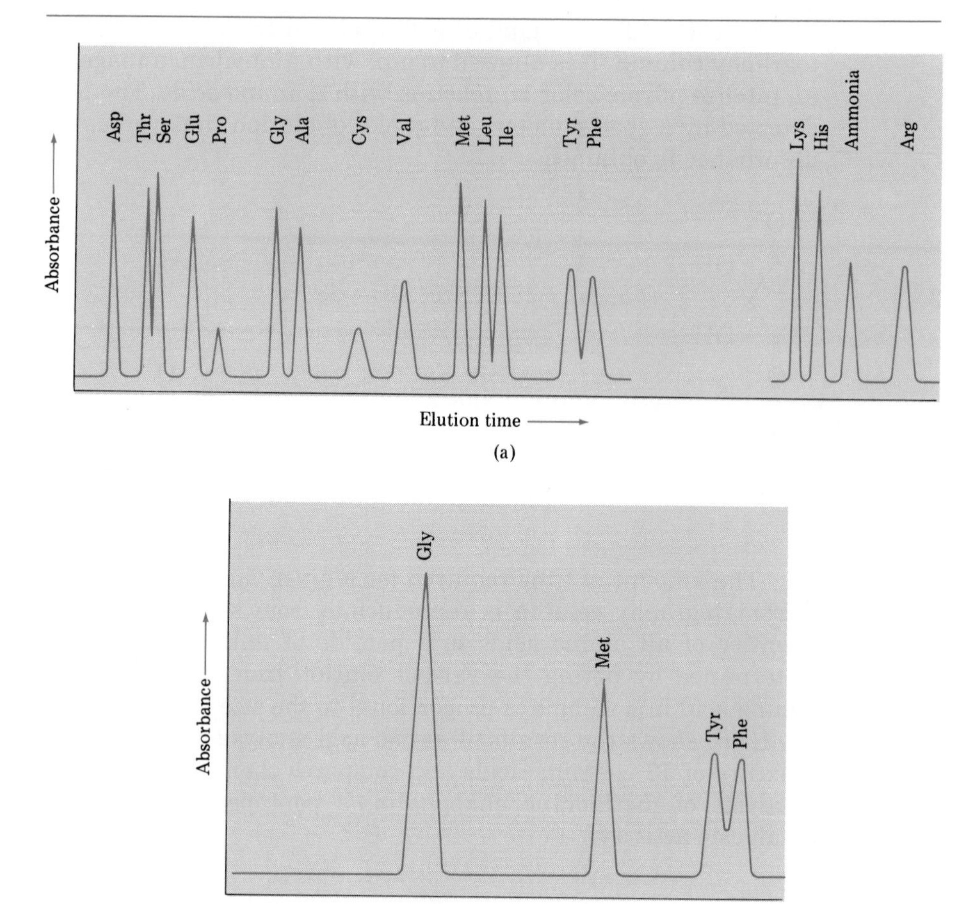

Figure 31.5. Amino acid analysis of (a) an equimolar amino acid mixture, and (b) methionine enkephalin (H-Tyr-Gly-Gly-Phe-Met-OH).

Edman degradation involves treatment of a peptide with phenyl isothiocyanate, followed by mild acid hydrolysis. These steps yield a phenylthiohydantoin derivative of the N-terminal amino acid plus the chain-shortened peptide (Figure 31.6). The phenylthiohydantoin is then identified chromatographically by comparison with known derivatives of the common amino acids.

Complete sequencing of large peptides and proteins by Edman degradation is impractical since the method is limited to about 20 cycles. Instead, the large peptide chain is first cleaved by partial hydrolysis into a number of smaller fragments. The sequence of each fragment is then determined, and the individual pieces are fitted together like a jigsaw puzzle.

The partial hydrolysis can be carried out either chemically with aqueous acid, or enzymatically with enzymes such as trypsin and chymotrypsin. Acidic hydrolysis is unselective and leads to a more or less random mixture of small fragments; enzyme hydrolysis, however, is quite specific. Thus,

HO—H

$C_6H_5—N=C=S$

+

R

$H_2N—CH—C(=O)—NH—$Peptide

Nucleophilic addition of the peptide terminal amino group to the isothiocyanate yields an *N*-phenylthiourea derivative.

H O H^+

$C_6H_5—N$ C—NH—Peptide

C C---R

S N H

H

Acid-catalyzed cyclization then yields a tetrahedral intermediate.

HCl/H_2O

H

O H^+

C_6H_5

N—C—NH—Peptide

H C C---R

S N H

H

The intermediate expels the chain-shortened peptide and forms the *N*-phenylthiohydantoin.

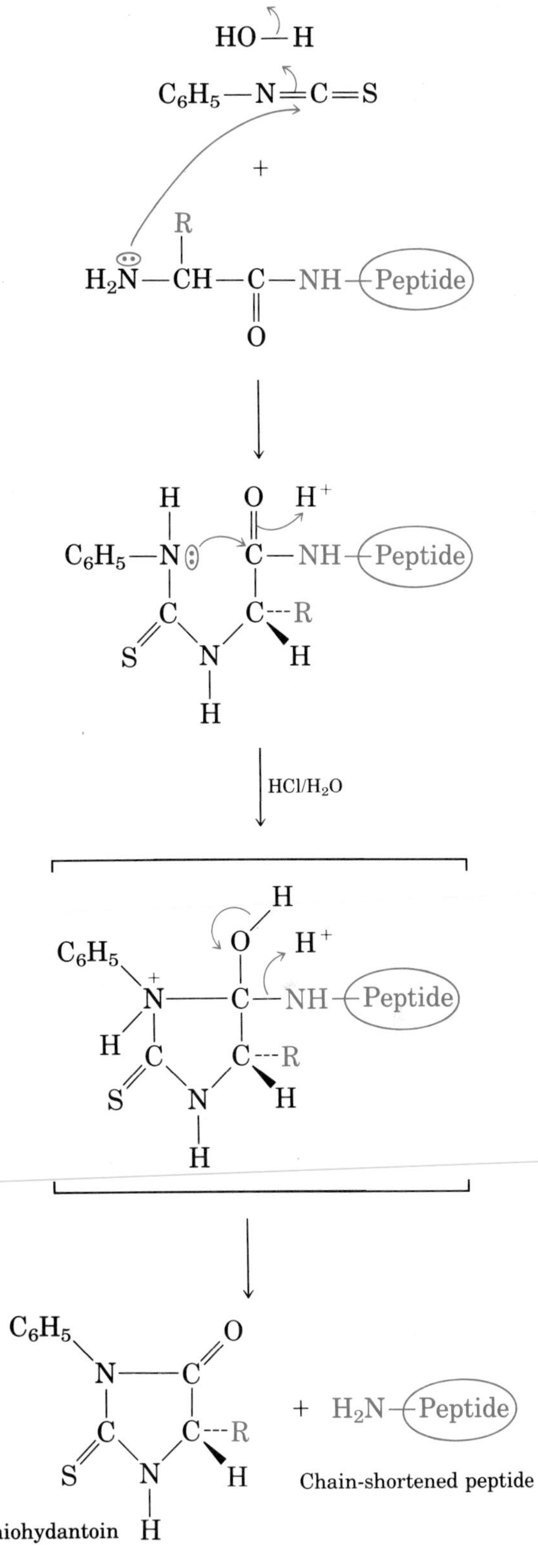

C_6H_5 O

N—C

C C---R + $H_2N—$Peptide

S N H

N-Phenylthiohydantoin H

Chain-shortened peptide

Figure 31.6. Mechanism of the Edman degradation.

trypsin catalyzes hydrolysis only at the carboxyl side of the basic amino acids arginine and lysine, and chymotrypsin cleaves only at the carboxyl side of the aryl-substituted amino acids phenylalanine, tyrosine, and tryptophan.

To take an example, let's look at a hypothetical structure determination of angiotensin II, a hormonal octapeptide involved in controlling hypertension by regulating the sodium–potassium salt balance in the body.

1. Amino acid analysis of angiotensin II would show the presence of eight different amino acids: Arg, Asp, His, Ile, Phe, Pro, Tyr, and Val in equimolar amounts.
2. An N-terminal analysis by the Edman method would show that angiotensin II has an aspartic acid residue at the N terminus.
3. Partial hydrolysis of angiotensin II with dilute hydrochloric acid might yield the following fragments, whose sequences could be determined by Edman degradation:
   a. H-Asp-Arg-Val-OH
   b. H-Ile-His-Pro-OH
   c. H-Arg-Val-Tyr-OH
   d. H-Pro-Phe-OH
   e. H-Val-Tyr-Ile-OH
4. Matching of overlapping fragment regions provides the full sequence of angiotensin II:

```
a.  H-Asp-Arg-Val-OH
c.      H-Arg-Val-Tyr-OH
e.          H-Val-Tyr-Ile-OH
b.                H-Ile-His-Pro-OH
d.                       H-Pro-Phe-OH
    H-Asp-Arg-Val-Tyr-Ile-His-Pro-Phe-OH
```

Angiotensin II

The structure of angiotensin II is relatively simple—the entire sequence could easily be done by a protein sequenator instrument—but the methods and the logic we used to solve this simple structure are the same as those used to solve more complex structures. Indeed, single protein chains with more than 400 amino acids have been sequenced by these methods.

PROBLEM

31.13 Give the amino acid sequence of hexapeptides that produce the following fragments on partial acid hydrolysis:

(a) Arg, Gly, Ile, Leu, Pro, Val gives H-Pro-Leu-Gly-OH, H-Arg-Pro-OH, H-Gly-Ile-Val-OH

(b) Asp, Leu, Met, Trp, Val_2 gives H-Val-Leu-OH, H-Val-Met-Trp-OH, H-Trp-Asp-Val-OH

PROBLEM

31.14 What fragments would result if angiotensin II were cleaved with trypsin? With chymotrypsin?

31.10 Peptide sequencing: C-terminal residue determination

The Edman degradation is an excellent method of analysis for the N-terminal residue, but a complementary method of C-terminal residue determination is also valuable. The best method currently available makes use of the enzyme carboxypeptidase to specifically cleave the C-terminal amide bond in a peptide chain.

$$\text{(Peptide)}-NH\overset{R'}{\overset{|}{C}}H\underset{\underset{O}{\|}}{C}-NH\overset{R}{\overset{|}{C}}HCOOH \xrightarrow[H_2O]{\text{Carboxypeptidase}} \text{(Peptide)}-NH\overset{R'}{\overset{|}{C}}HCOOH + H_2N\overset{R}{\overset{|}{C}}HCOOH$$

The analysis is carried out by incubating the polypeptide with carboxypeptidase and watching for the appearance of the first free amino acid produced. Of course, further degradation also occurs, since a new C terminus is produced when the first amino acid residue is cleaved off; ultimately the entire peptide is hydrolyzed.

PROBLEM

31.15 A hexapeptide with the composition Arg, Gly, Leu, Pro_3 is found to have proline at both C-terminal and N-terminal positions. Partial hydrolysis gives the following fragments:

H-Gly-Pro-Arg-OH H-Arg-Pro-OH H-Pro-Leu-Gly-OH

What is the structure of the hexapeptide?

PROBLEM

31.16 How can you explain the observation that a tripeptide that gives Leu, Ala, and Phe on hydrolysis does not react with carboxypeptidase, and does not react with phenyl isothiocyanate? Propose two structures for this tripeptide.

31.11 Peptide synthesis

Once the structure of a peptide has been determined, synthesis is often the next goal. This might be done either as a final proof of structure or as a means of obtaining larger amounts of a valuable peptide for biological evaluation.

Ordinary amide bonds are usually formed by reaction between amines and acylating agents (Section 24.4).

$$R'-\overset{O}{\overset{\|}{C}}-X + H_2N-R \longrightarrow \underset{\text{An amide}}{R'-\overset{O}{\overset{\|}{C}}-NH-R} + HX$$

Peptide synthesis is much more complex than simple amide synthesis, however, because of the specificity requirement. Many different amide links must be formed, and they must be formed in a specific order, rather than at random. The solution to the specificity problem is *protection*. We can force a reaction to take only the desired course by protecting all of the amine and acid functional groups except for those we wish to have react. For example, if we wished to couple alanine with leucine to synthesize H-Leu-Ala-OH, we could protect the amino group of leucine and the carboxyl group of alanine to render them unreactive, then form the desired amide bond, and then deprotect.

$$\underset{\text{Leucine}}{H_2N{-}\underset{\substack{|\\ CH_2\\ |\\ CH(CH_3)_2}}{C}HCOOH} \qquad \underset{\text{Alanine}}{H_2N{-}\underset{\substack{|\\ CH_3}}{CH}{-}COOH}$$

$$\downarrow \text{Protect} \qquad\qquad \downarrow \text{Protect}$$

$$\underset{\text{N-protected Leu}}{\boxed{H_2N}{-}CH(CH_2CH(CH_3)_2)COOH} \;+\; \underset{\text{O-protected Ala}}{H_2N{-}CH(CH_3){-}\boxed{COOH}}$$

$$\downarrow \text{1. Form amide; 2. Deprotect}$$

$$\underset{\text{H-Leu-Ala-OH}}{H_2N{-}CH(CH_2CH(CH_3)_2){-}\underset{\substack{\|\\ O}}{C}{-}NH{-}CH(CH_3){-}COOH}$$

Many different amino- and carboxyl-protecting groups have been devised but only a few are widely used. Carboxyl groups are often protected as their methyl or benzyl esters, since these groups are easily introduced by standard methods of ester formation and are easily removed by mild hydrolysis with aqueous hydroxide. As Figure 31.7 shows, benzyl esters can also be cleaved by catalytic hydrogenolysis of the weak benzylic C—O bond ($ArCH_2{-}OCOR + H_2 \rightarrow ArCH_3 + RCOOH$).

Amino groups are often protected as their *tert*-butoxycarbonyl amide (BOC) derivatives. The BOC protecting group is easily introduced by reac-

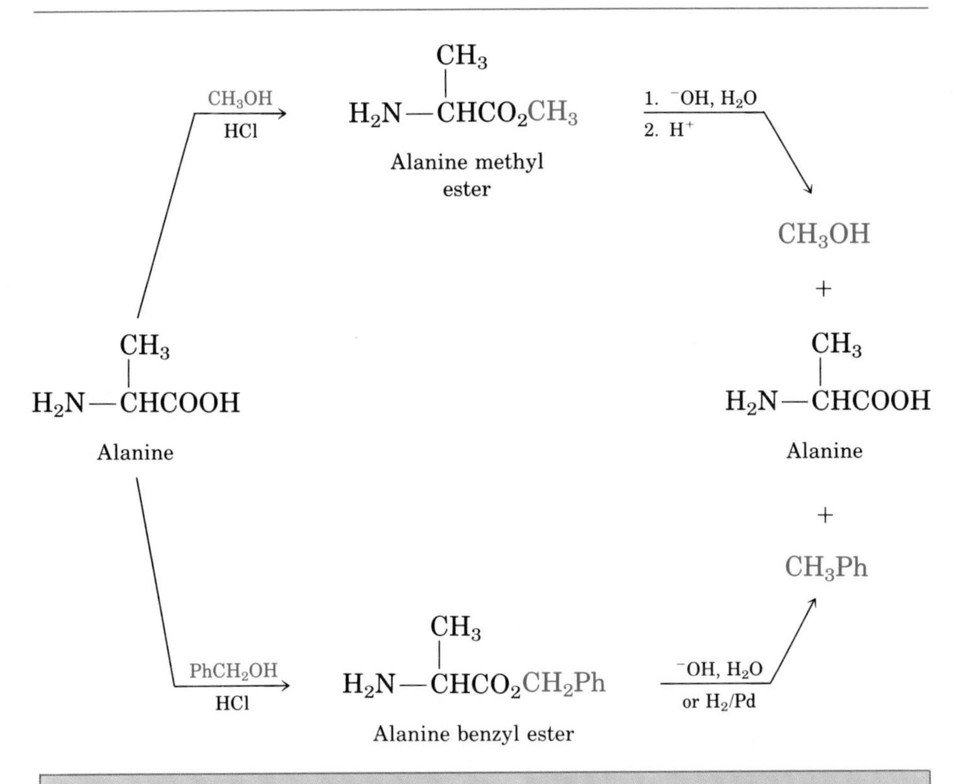

Figure 31.7. Protection of an amino acid carboxyl group by ester formation.

tion of the amino acid with di-*tert*-butyl dicarbonate, and is removed by brief treatment with a strong acid such as trifluoroacetic acid, CF_3COOH.

$$H_2N{-}\underset{\underset{\displaystyle CH(CH_3)_2}{|}}{\overset{\overset{\displaystyle CH(CH_3)_2}{|}}{CH_2}}\text{—}\ldots$$

Leucine: $H_2N{-}CH(CH_2CH(CH_3)_2){-}COOH$ + Di-*tert*-butyl dicarbonate: $((CH_3)_3C{-}O{-}C(=O))_2O$

$$\xrightarrow{(CH_3CH_2)_3N}$$

BOC-Leu: $(CH_3)_3C{-}O{-}C(=O){-}NH{-}CH(CH_2CH(CH_3)_2){-}COOH$

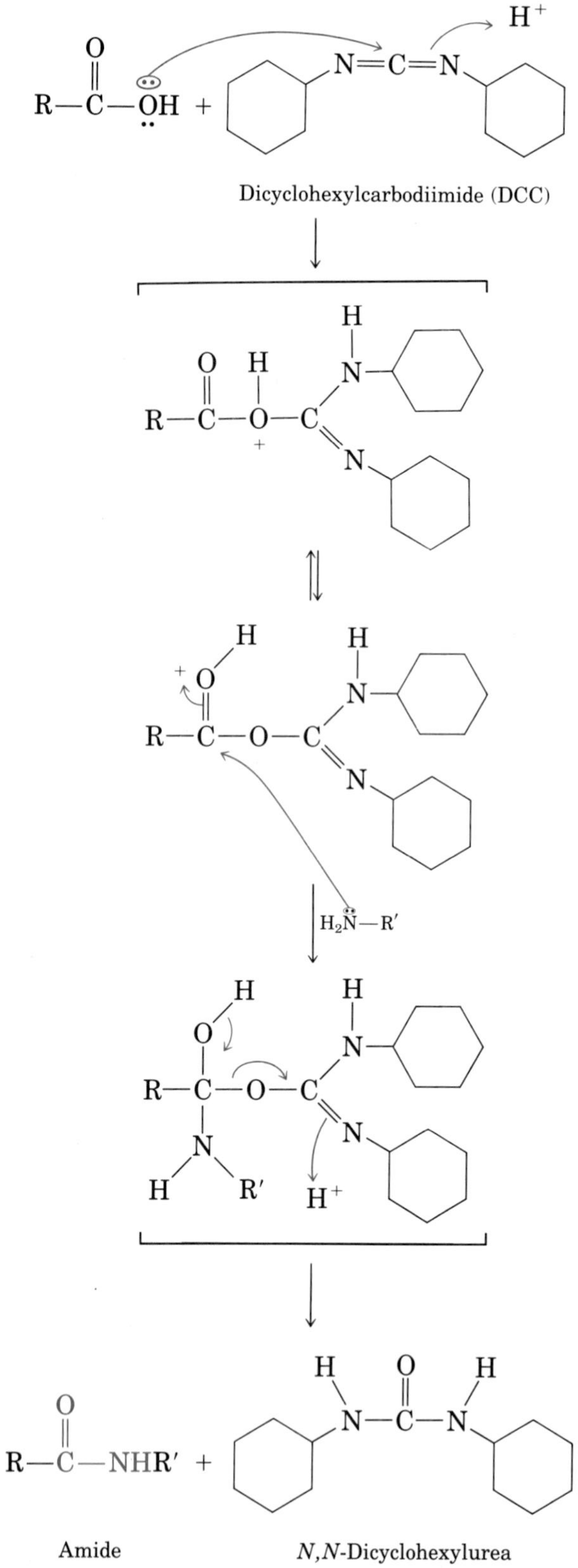

The carboxylic acid first adds to the carbodiimide reagent.

A reactive acylating derivative is produced.

Nucleophilic attack of the amine on the acylating agent gives a tetrahedral intermediate.

The intermediate loses dicyclohexylurea and yields the desired amide.

$$R{-}\overset{\overset{\displaystyle O}{\|}}{C}{-}\ddot{O}H + C_6H_{11}{-}N{=}C{=}N{-}C_6H_{11}$$

Dicyclohexylcarbodiimide (DCC)

H^+

$H_2\ddot{N}{-}R'$

H^+

$$R{-}\overset{\overset{\displaystyle O}{\|}}{C}{-}NHR' + C_6H_{11}{-}NH{-}\overset{\overset{\displaystyle O}{\|}}{C}{-}NH{-}C_6H_{11}$$

Amide

N,N-Dicyclohexylurea

Figure 31.8. Mechanism of amide formation with DCC.

The peptide-bond-forming step is best done by treating a mixture of protected acid and amine components with dicyclohexylcarbodiimide (DCC). Dicyclohexylcarbodiimide functions by converting the carboxylic acid group into a reactive acylating agent needed for amide bond formation. The mechanism of this coupling step is shown in Figure 31.8.

We now have the knowledge needed to complete a synthesis of H-Leu-Ala-OH. Five separate steps are required:

1. Protect the amino group of leucine as the BOC derivative:

$$\text{H-Leu-OH} + (t\text{-BuO}-\overset{\overset{\displaystyle O}{\|}}{C})_2\text{O} \longrightarrow \text{BOC-Leu-OH}$$

2. Protect the carboxyl group of alanine as the methyl ester:

$$\text{H-Ala-OH} + CH_3OH \xrightarrow{H^+} \text{H-Ala-OCH}_3$$

3. Couple the two protected amino acids using DCC:

$$\text{BOC-Leu-OH} + \text{H-Ala-OCH}_3 \xrightarrow{\text{DCC}} \text{BOC-Leu-Ala-OCH}_3$$

4. Remove the BOC protecting group by acid treatment:

$$\text{BOC-Leu-Ala-OCH}_3 \xrightarrow{CF_3COOH} \text{H-Leu-Ala-OCH}_3$$

5. Remove the methyl ester by basic hydrolysis:

$$\text{H-Leu-Ala-OCH}_3 \xrightarrow[H_2O]{^-OH} \text{H-Leu-Ala-OH}$$

These steps can be repeated to add one amino acid at a time to the growing chain or to link two peptide chains together. Many remarkable achievements in peptide synthesis have been reported, including a complete synthesis of human insulin. Insulin is composed of two chains totaling 51 amino acids linked by cystine disulfide bridges (Figure 31.9, p. 1054). Its structure was determined by Frederick Sanger,[5] who received the 1958 Nobel prize for his work.

PROBLEM

31.17 Write all five steps required for the synthesis of H-Leu-Ala-OH from alanine and leucine.

PROBLEM

31.18 How would you prepare these tripeptides?
(a) H-Leu-Ala-Gly-OH
(b) H-Gly-Leu-Ala-OH

[5]Frederick Sanger (1918–); b. Gloucestershire, England; Ph.D. Cambridge; professor, Cambridge University; Nobel prize (1958, 1980).

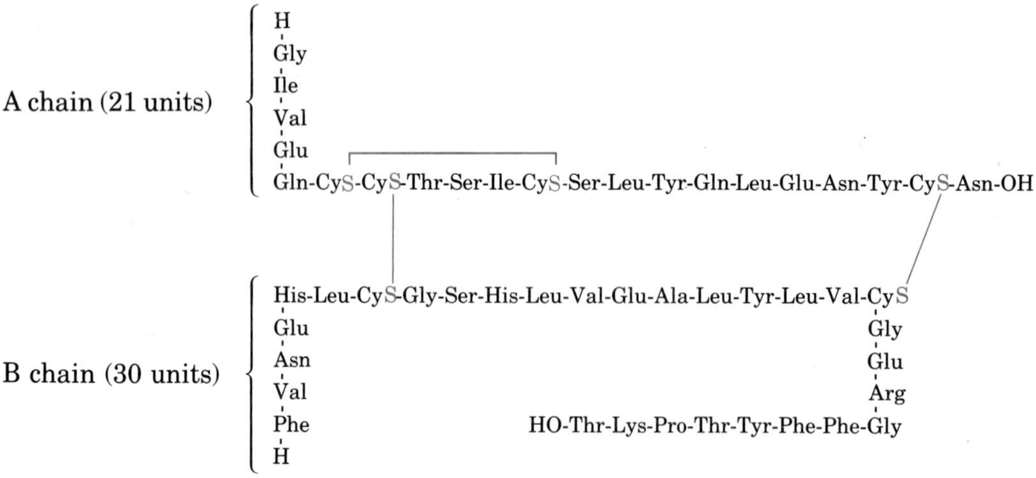

A chain (21 units)

H
Gly
Ile
Val
Glu
Gln-CyS-CyS-Thr-Ser-Ile-CyS-Ser-Leu-Tyr-Gln-Leu-Glu-Asn-Tyr-CyS-Asn-OH

B chain (30 units)

H
Phe
Val
Asn
Glu
His-Leu-CyS-Gly-Ser-His-Leu-Val-Glu-Ala-Leu-Tyr-Leu-Val-CyS
Gly
Glu
Arg
HO-Thr-Lys-Pro-Thr-Tyr-Phe-Phe-Gly

Figure 31.9. Structure of human insulin.

31.12 Automated peptide synthesis: the Merrifield solid-phase technique

The synthesis of large peptide chains by sequential addition of one amino acid at a time is a long and arduous task. An immense simplification is possible, however, using the **solid-phase technique** introduced by Bruce Merrifield.[6] In the Merrifield method, peptide synthesis is carried out on solid polymer beads of chloromethylated polystyrene, prepared so that one of every 100 or so benzene rings bears a chloromethyl group:

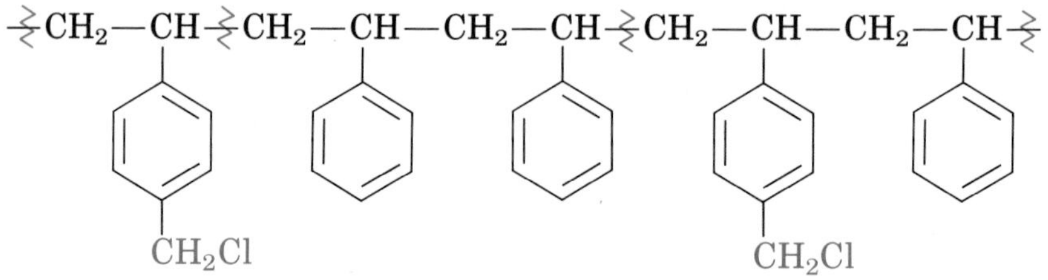

$$\sim CH_2—CH \sim CH_2—CH—CH_2—CH \sim CH_2—CH—CH_2—CH \sim$$

CH_2Cl CH_2Cl

Chloromethylated polystyrene

A BOC-protected amino acid first reacts with the polymer and becomes covalently bound by an ester linkage. In effect, the solid polymer serves as the ester-protecting group for the amino acid (step 1).

[6]Robert Bruce Merrifield (1921–); Ph.D. University of California, Los Angeles (1949); professor, Rockefeller Institute.

Step 1. $(CH_3)_3COC(=O)NH-CH(R)-COOH + ClCH_2-\text{Polymer}$

BOC-protected amino acid

$\xrightarrow{S_N2 \text{ reaction}}$

$\text{BOC}-NH-CH(R)-C(=O)-O-CH_2-\text{Polymer}$

Ester link to polymer

The insoluble, polymer-bonded amino acid is then washed free of excess reagents and treated with acid to remove the BOC group (step 2).

Step 2. $\text{BOC}-NH-CH(R)C(=O)-O-CH_2-\text{Polymer}$

$\xrightarrow{\text{1. Wash; 2. } CF_3COOH}$

$H_2N-CH(R)C(=O)-O-CH_2-\text{Polymer}$

Polymer-bonded amino acid

A second BOC-protected amino acid is added along with the coupling reagent, DCC, and peptide-bond formation occurs (step 3).

Step 3. $\text{BOC}-NH-CH(R')C(=O)-OH + H_2N-CH(R)C(=O)-O-CH_2-\text{Polymer}$

$\xrightarrow{\text{1. DCC; 2. Wash}}$

$\text{BOC}-NH-CH(R')-C(=O)-NHCH(R)C(=O)-OCH_2-\text{Polymer}$

Polymer-bonded dipeptide

Excess reagents are then removed by washing the insoluble polymer; step 2 is repeated to again remove a BOC group, and step 3 is repeated to add a third amino acid unit to the chain. In this way, dozens or even 100 amino acid units can be efficiently and specifically linked to synthesize the desired polymer-bonded peptide. At the end of the synthesis, treatment with anhydrous hydrogen fluoride cleaves the peptide from the polymer (step 4).

Step 4.

$$H_2N{-}\underset{\displaystyle R''}{CH}{-}\overset{}{\underset{\displaystyle \overset{\|}{O}}{C}}{-}(NH{-}\underset{\displaystyle R'}{CH}\underset{\displaystyle \overset{\|}{O}}{C})_n{-}NH{-}\underset{\displaystyle R}{CH}{-}\underset{\displaystyle \overset{\|}{O}}{C}{-}O{-}CH_2{-}\text{(Polymer)}$$

$$\Big\downarrow \text{HF}$$

$$H_2N{-}\underset{\displaystyle R''}{CH}{-}\underset{\displaystyle \overset{\|}{O}}{C}{-}(NH{-}\underset{\displaystyle R'}{CH}\underset{\displaystyle \overset{\|}{O}}{C})_n{-}NH\underset{\displaystyle R}{CH}\underset{\displaystyle \overset{\|}{O}}{C}OH + HO{-}CH_2{-}\text{(Polymer)}$$

Polypeptide

This solid-phase technique has now been automated, and peptide-growing machines are available for *automatically* repeating the coupling and deprotection steps with different amino acids as many times as desired. Each step occurs in extremely high yield, and mechanical losses are minimized since the peptide intermediates are never removed from the insoluble polymer until the final step. Among the many remarkable achievements recorded by Merrifield is the synthesis of bovine pancreatic ribonuclease, a protein containing 124 amino acid units. The entire synthesis required only six weeks and took place in 17% overall yield.

31.13 Classification of proteins

Proteins may be classified into two major types according to their composition. **Simple proteins** such as blood serum albumin are those that yield only amino acids and no other organic compounds on hydrolysis. **Conjugated proteins,** such as are found in cell membranes, yield other compounds in addition to amino acids on hydrolysis. Conjugated proteins are far more common than simple proteins and may be further classified with respect to the chemical nature of the non–amino acid portion. Thus there are **glycoproteins, lipoproteins, nucleoproteins,** and others (Table 31.2). Glycoproteins—those in which the polypeptide chain is covalently bound to a carbohydrate portion—are particularly widespread in nature and make up a large part of the membrane coating around living cells.

Proteins may also be classified according to their three-dimensional conformation as either **fibrous** or **globular.** Fibrous proteins such as collagen and α-keratin consist of polypeptide chains arranged side by side in long threads. These proteins are tough, and insoluble in water; they are used in nature for structural materials like tendons, hoofs, horns, and fingernails.

Globular proteins, by contrast, are usually coiled into compact, nearly spherical shapes. These proteins are generally soluble in water and are

TABLE 31.2 **Classification of some conjugated proteins**

Class	*Non–amino acid group*	*Weight of non–amino acid (%)*
Glycoproteins		
γ-Globulin	Carbohydrate	10
Carboxypeptidase Y	Carbohydrate	17
Interferon	Carbohydrate	20
Lipoproteins		
Plasma β lipoprotein	Fats, cholesterol	80
Nucleoproteins		
Ribosomal proteins	Ribonucleic acid	60
Tobacco mosaic virus	Ribonucleic acid	5
Phosphoproteins		
Casein	Phosphate esters	4
Metalloproteins		
Ferritin	Iron oxide	23
Hemoglobin	Iron	0.3

mobile within the cell. Most of the 2000 or so known enzymes, as well as hormonal and transport proteins, are globular. Table 31.3 lists some common examples of fibrous and globular proteins.

TABLE 31.3 **Conformational classes of proteins**

Protein	*Description*
Fibrous proteins (insoluble)	
Collagen	Connective tissue, tendons
α-Keratin	Hair, horn, skin, nails
Elastin	Elastic connective tissue
Globular proteins (soluble)	
Insulin	Hormone controlling glucose metabolism
Lysozyme	Hydrolytic enzyme
Ribonuclease	Enzyme controlling RNA synthesis
Albumins	Proteins coagulated by heat
Immunoglobulins	Proteins involved in immune response
Myoglobin	Protein involved in oxygen transport

31.14 Protein structure

Proteins are so large in comparison to simple organic molecules that the word *structure* takes on a broader meaning when applied to these immense macromolecules. At its simplest, protein structure is the sequence in which amino acid residues are bound together. This sequence is called the **primary structure** and is the most fundamental structural level. There is, however, much more to protein structure than just amino acid sequence. The chemical properties of a protein are also dependent on higher levels of structure, on exactly how the peptide backbone is folded to give the molecule a precise three-dimensional shape. Thus, the term **secondary structure** refers to the way in which segments of the peptide backbone are oriented into a regular pattern; **tertiary structure** refers to the way in which the entire protein molecule is coiled into an overall three-dimensional shape; and **quaternary structure** refers to the way in which several protein molecules come together to yield large aggregate structures.

We will look at three examples—α-keratin (fibrous), fibroin (fibrous), and myoglobin (globular)—to see how higher structure affects protein properties.

α-KERATIN

α-Keratin is the fibrous structural protein found in wool, hair, nails, and feathers. Much evidence has been amassed to show that α-keratin is coiled into a right-handed α-helical secondary structure (Figure 31.10). This **α helix** is held together by hydrogen bonding between amide N—H groups and other amide carbonyl groups several residues away. The strength of a single hydrogen bond (5 kcal/mol) is only about 5% of the strength of a C—C or C—H covalent bond, but the large number of hydrogen bonds made possible by helical winding imparts a large amount of stability to this structure. Each coil of the helix (the **repeat distance**) contains 3.6 amino acid residues, and the distance between coils is 5.4 Å.

Further evidence suggests that the α-keratins of wool and hair also have a definite quaternary structure. The individual helices are themselves coiled about one another to form a superhelix.

FIBROIN

Fibroin, the fibrous protein found in silk, has a secondary structure called a **β pleated sheet.** In the pleated-sheet structure, polypeptide chains line up in a parallel arrangement held together by hydrogen bonds (Figure 31.11). Although this arrangement is not as common as the α helix, small β-pleated-sheet regions are often found in proteins where sections of peptide chains double back on themselves.

MYOGLOBIN

The enzyme myoglobin is a rather small globular protein containing 153 amino acid residues in a single chain. A relative of hemoglobin, myoglobin

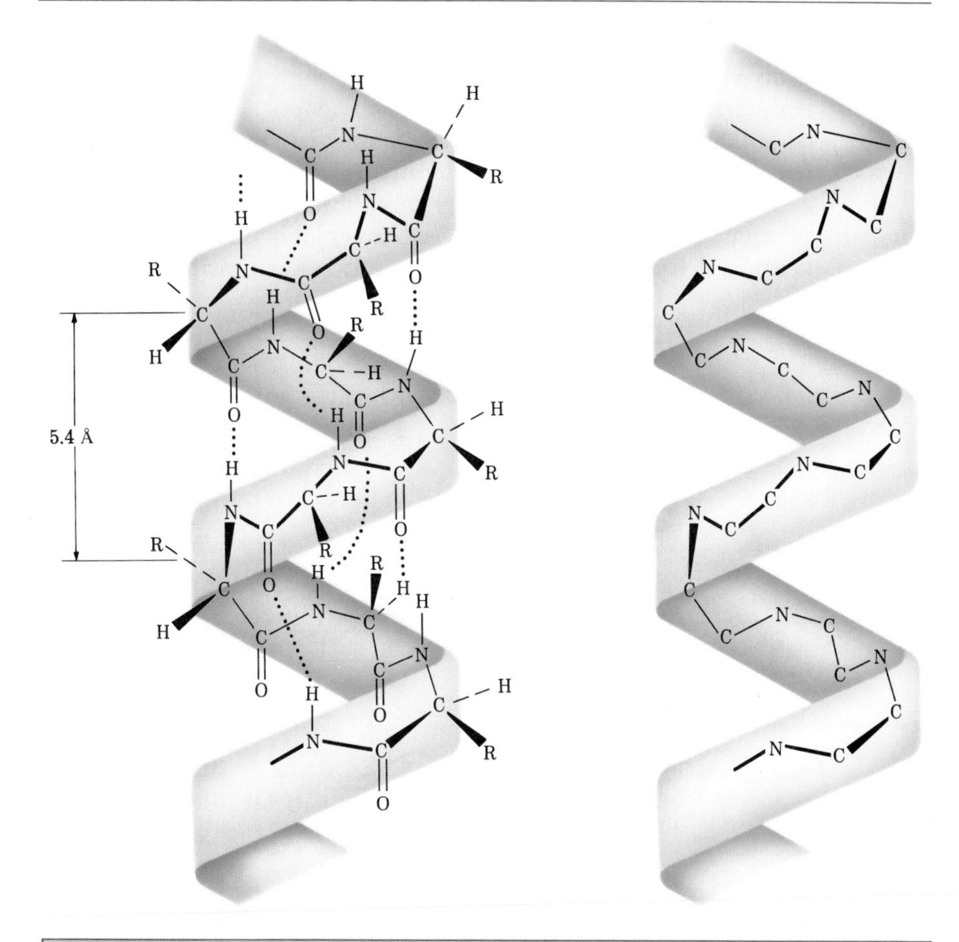

Figure 31.10. The helical secondary structure of α-keratin.

Chain 1

Chain 2

Figure 31.11. The β-pleated-sheet structure in silk fibroin.

is found in the skeletal muscles of sea mammals, where it stores oxygen needed to sustain the animals during dives. X-ray evidence obtained by Sir John Kendrew[7] and Max Perutz[8] has shown that myoglobin consists of eight straight segments, each of which adopts an α-helical secondary structure. These helical sections are connected by bends to form a compact, nearly spherical, tertiary structure (Figure 31.12). Although the bends appear to be irregular, and the three-dimensional structure appears to be random, this is not the case. All myoglobin molecules adopt this shape because it has a lower energy than any other possible shape.

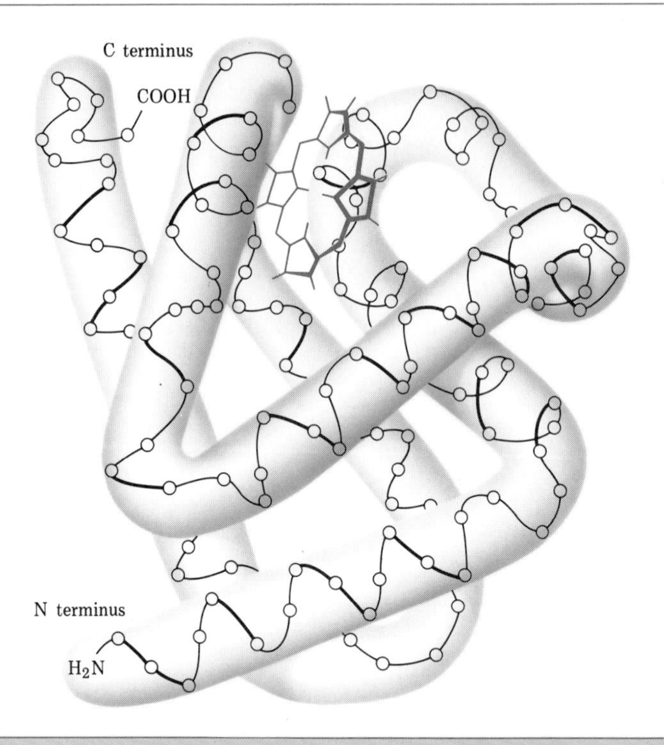

Figure 31.12. Secondary and tertiary structure of myoglobin.

Note that myoglobin is a conjugated protein that contains the covalently bound organic group, heme (Figure 31.13). Such a protein-bound group is called a **prosthetic group** and is crucial to the mechanism of enzyme action.

PROBLEM

31.19 How can you account for the fact that proline is never present in a protein α helix? The α-helix parts of myoglobin and other proteins stop when a proline residue is encountered in the chain.

[7]Sir John C. Kendrew (1917–); b. Oxford, England; Ph.D. Cambridge; professor, Cambridge University; Nobel prize (1962).

[8]Max Ferdinand Perutz (1914–); b. Vienna; universities of Vienna and Cambridge; professor, Cambridge University; Nobel prize (1962).

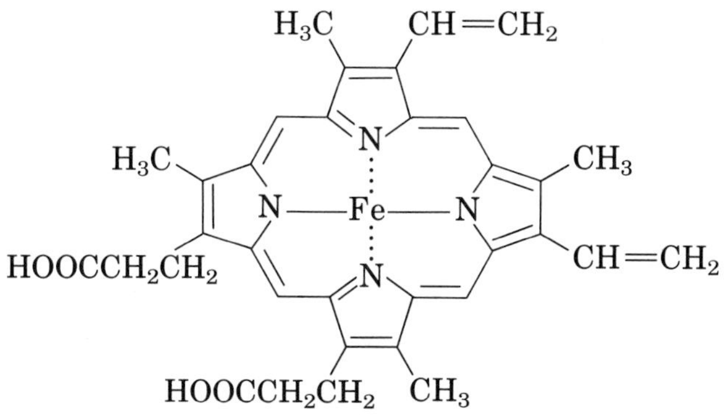

Figure 31.13. The structure of heme.

31.15 Protein denaturation

The tertiary structure of globular proteins is delicately held together by weak intermolecular attractions. Often, a modest change in temperature or pH will disrupt the tertiary structure and cause the protein to become **denatured.** Denaturation occurs under such mild conditions that covalent bonds are not affected; the polypeptide primary structure remains intact, but the tertiary structure unfolds from a well-defined spherical shape to a randomly looped chain.

Denaturation is accompanied by changes in both physical and biological properties. Solubility is drastically decreased, as occurs when egg white is cooked and the albumins unfold and coagulate to an insoluble white mass. Most enzymes also lose all catalytic activity when denatured, since a precisely defined tertiary structure is required for their action.

Most, but not all, denaturation is irreversible. Eggs do not become uncooked when their temperature is lowered, and curdled milk does not become homogeneous. Many cases have now been found, however, where spontaneous **renaturation** of an unfolded protein occurs. Renaturation is accompanied by a full recovery of biological activity in the case of enzymes, indicating that the protein has completely returned to its stable tertiary structure.

31.16 Summary

Proteins are large biomolecules that consist of α-amino acid residues linked together by amide, or peptide, bonds. Twenty amino acids are commonly found in proteins; all are α-amino acids and all except glycine have *S* stereochemistry similar to that of L sugars.

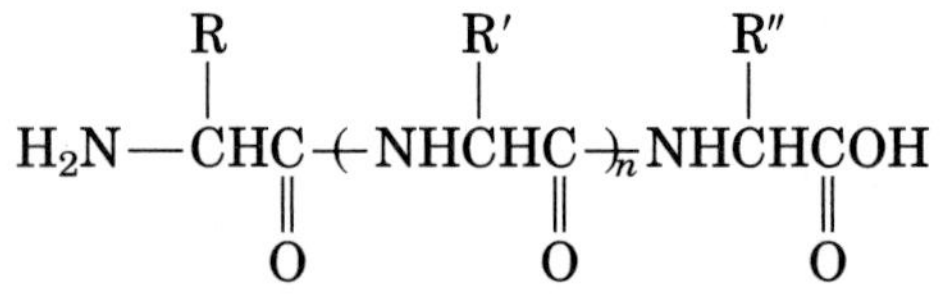

$$\mathrm{H_2N{-}\overset{R}{\overset{|}{C}}H\underset{\underset{O}{\|}}{C}{-}\!\left(NH\overset{R'}{\overset{|}{C}}H\underset{\underset{O}{\|}}{C}\right)_{\!n}\!{-}NH\overset{R''}{\overset{|}{C}}H\underset{\underset{O}{\|}}{C}OH}$$

A polypeptide or protein

Amino acids can be synthesized by several methods, including ammonolysis of α-bromo acids, reductive amination of α-keto acids, reaction of aldehydes with KCN/NH_4Cl followed by hydrolysis, and alkylation of diethyl acetamidomalonate. Resolution of the synthetic racemate then provides the optically active amino acid.

Determining the structure of a large polypeptide or protein is a challenging task that must be carried out in several steps. The identity and amount of each amino acid present in a peptide can be determined by amino acid analysis. The peptide is first hydrolyzed to its constituent α-amino acids, which are then chromatographically separated and identified. Next, the peptide is sequenced. Edman degradation by treatment with phenyl isothiocyanate cleaves off one residue from the N terminus of the peptide and forms an easily identifiable derivative of the N-terminal amino acid. A series of sequential Edman degradations allows us to sequence peptide chains up to 20 residues in length.

Peptide synthesis is an equally challenging task, and one that must be solved by the use of selective protecting groups. An N-protected amino acid having a free carboxyl group is treated with an O-protected amino acid having a free amino group in the presence of dicyclohexylcarbodiimide (DCC). Amide formation occurs, the protecting groups are removed, and the sequence is repeated. Amines are usually protected as their *tert*-butoxy carbonyl derivatives, and acids are protected as esters.

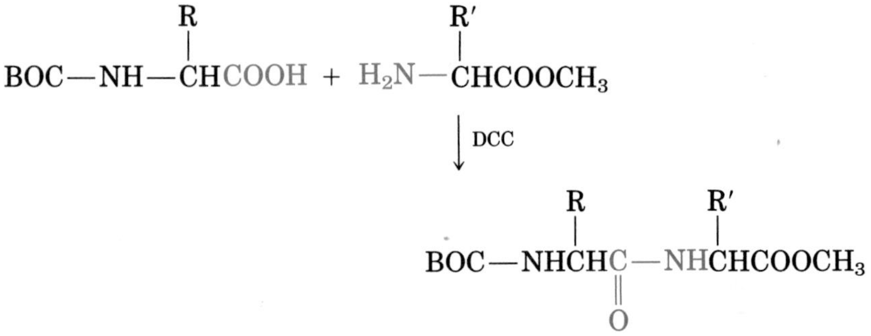

$$\mathrm{BOC{-}NH{-}\overset{R}{\overset{|}{C}}HCOOH + H_2N{-}\overset{R'}{\overset{|}{C}}HCOOCH_3 \xrightarrow{DCC} BOC{-}NH\overset{R}{\overset{|}{C}}H\underset{\underset{O}{\|}}{C}{-}NH\overset{R'}{\overset{|}{C}}HCOOCH_3}$$

This synthetic sequence is often carried out by the Merrifield solid-phase modification, in which the peptide is esterified to an insoluble polymeric support.

Proteins are classified as either globular or fibrous, depending on their secondary and tertiary structures. Fibrous proteins such as α-keratin are tough, rigid, and water insoluble, and are used in nature for forming structures such as hair and nails. Globular proteins such as myoglobin are water soluble, roughly spherical in shape, and are mobile within cells. Most of the 2000 or so known enzymes are globular proteins.

31.17 Summary of reactions

1. Amino acid synthesis
   a. From α-bromo acids (Section 31.4)

$$R\text{—}CH_2COOH \xrightarrow[P]{Br_2} R\text{—}\overset{\overset{\displaystyle Br}{|}}{C}HCOOH \xrightarrow{NH_3} R\text{—}\overset{\overset{\displaystyle NH_2}{|}}{C}HCOOH$$

   b. Strecker synthesis (Section 31.4)

$$R\text{—}CHO \xrightarrow[NH_4Cl]{KCN} R\text{—}\overset{\overset{\displaystyle NH_2}{|}}{C}HCN \xrightarrow{H_3O^+} R\text{—}\overset{\overset{\displaystyle NH_2}{|}}{C}HCOOH$$

   c. Reductive amination (Sections 28.7 and 31.4)

$$R\text{—}\overset{\overset{\displaystyle O}{\|}}{C}COOH \xrightarrow[NaBH_4]{NH_3} R\text{—}\overset{\overset{\displaystyle NH_2}{|}}{C}HCOOH$$

   d. Diethyl acetamidomalonate synthesis (Sections 25.9 and 31.4)

$$CH_3\overset{\overset{\displaystyle O}{\|}}{C}\text{—}NHCH(CO_2C_2H_5)_2 \xrightarrow[\text{2. RX} \quad \text{3. } H_3O^+]{\text{1. } Na^{+}\,{}^{-}OC_2H_5} H_2N\text{—}\overset{\overset{\displaystyle R}{|}}{C}HCOOH$$

2. Peptide synthesis
   a. Nitrogen protection (Section 31.11)

$$H_2N\text{—}\overset{\overset{\displaystyle R}{|}}{C}HCOOH + [(CH_3)_3C\text{—}O\text{—}\overset{\overset{\displaystyle O}{\|}}{C}\text{]}_2O$$

$$\downarrow$$

$$(CH_3)_3C\text{—}O\text{—}\overset{\overset{\displaystyle O}{\|}}{C}\text{—}NH\overset{\overset{\displaystyle R}{|}}{C}HCOOH$$

BOC-protected amino acid

The BOC protecting group can be removed by acid treatment:

$$(CH_3)_3C\text{—}O\text{—}\overset{\overset{\displaystyle O}{\|}}{C}\text{—}NH\overset{\overset{\displaystyle R}{|}}{C}HCOOH$$

$$\downarrow CF_3COOH$$

$$H_2N\text{—}\overset{\overset{\displaystyle R}{|}}{C}H\text{—}COOH + CO_2 + (CH_3)_2C\text{=}CH_2$$

b. Oxygen protection (Section 31.11)

$$H_2N-\underset{\displaystyle |}{\overset{\displaystyle R}{C}}HCOOH + CH_3OH \xrightarrow{H^+} H_2N-\overset{R}{\overset{|}{C}}HCOOCH_3$$

$$H_2N-\overset{R}{\overset{|}{C}}HCOOH + C_6H_5CH_2OH \xrightarrow{H^+} H_2N-\overset{R}{\overset{|}{C}}H-COOCH_2C_6H_5$$

The ester-protecting group can be removed by base hydrolysis:

$$H_2N-\overset{R}{\overset{|}{C}}HCOOCH_3 \xrightarrow[H_2O]{^-OH} H_2N-\overset{R}{\overset{|}{C}}H-COO^- + CH_3OH$$

c. Amide-bond formation (Section 31.11)

$$BOC-NH-\overset{R}{\overset{|}{C}}HCOOH + H_2N\overset{R'}{\overset{|}{C}}HCOOCH_3 \xrightarrow{DCC} BOC-NH\overset{R}{\overset{|}{C}}H-\underset{\underset{O}{\|}}{C}-NH\overset{R'}{\overset{|}{C}}HCOOCH_3$$

where DCC = $C_6H_{11}-N=C=N-C_6H_{11}$ (dicyclohexyl)

3. Peptide sequencing: Edman degradation (Section 31.9)

$$H_2N-\overset{R}{\overset{|}{C}}H\underset{\underset{O}{\|}}{C}-NH\overset{R'}{\overset{|}{C}}H\underset{\underset{O}{\|}}{C}\text{-}\!\!\sim + C_6H_5N=C=S \longrightarrow \text{(phenylthiohydantoin: } C_6H_5\text{–N in ring with C=O, C(R)(H), N–H, C=S)} + H_2N-\overset{R'}{\overset{|}{C}}H\underset{\underset{O}{\|}}{C}\text{-}\!\!\sim$$

ADDITIONAL PROBLEMS

31.20 Only *S* amino acids occur in proteins, but several *R* amino acids are also found in nature. Thus, (*R*)-serine is found in earthworms and (*R*)-alanine is found in insect larvae. Draw Fischer projections of (*R*)-serine and (*R*)-alanine.

31.21 Propose an explanation for the observation that amino acids exist as dipolar zwitterions in aqueous solution, but exist largely as true amino carboxylic acids in chloroform solution.

$$H_3\overset{+}{N}CH(R)CO_2^- \rightleftharpoons H_2N{-}CH(R){-}COOH$$

In H_2O In $CHCl_3$

31.22 At what pH would you carry out an electrophoresis experiment if you wished to separate a mixture of histidine, serine, and glutamic acid? Explain.

31.23 Cytochrome *c* is an enzyme that is found in the cells of all aerobic organisms. It is involved in respiration. Elemental analysis of cytochrome *c* reveals it to contain 0.43% iron. What is the minimum molecular weight of this enzyme?

31.24 Predict the product of the reaction of valine with these reagents:
(a) CH_3CH_2OH, H^+
(b) Di-*tert*-butyl dicarbonate
(c) KOH, H_2O

31.25 Write out full structures for these peptides:
(a) H-Val-Phe-Cys-Ala-OH
(b) H-Glu-Pro-Ile-Leu-OH

31.26 Arginine contains a guanidine functional group in its side chain and is by far the most basic of the 20 common amino acids. How can you account for this basicity?

$$H_2N{-}C({=}NH){-}NHCH_2CH_2CH_2CH(NH_2)COOH$$

Guanidino group

Arginine

31.27 The chloromethylated polystyrene resin used for Merrifield solid-phase peptide synthesis is prepared by treatment of polystyrene with chloromethyl methyl ether and a Lewis acid catalyst. Propose a mechanism for the reaction.

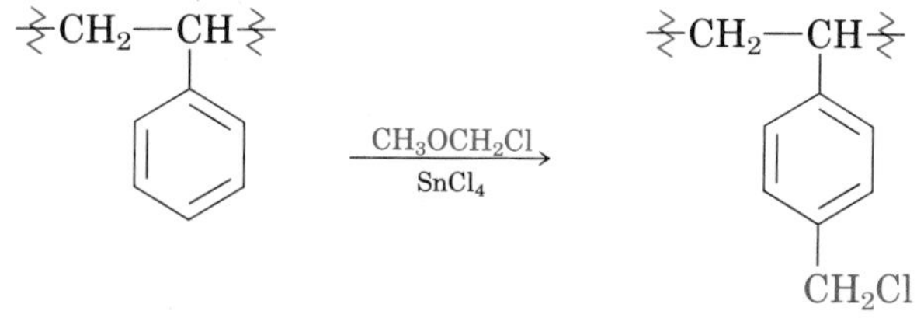

$$\text{-}CH_2{-}CH(C_6H_5)\text{-} \xrightarrow[SnCl_4]{CH_3OCH_2Cl} \text{-}CH_2{-}CH(C_6H_4CH_2Cl)\text{-}$$

31.28 Show the steps involved in a synthesis of H-Phe-Ala-Val-OH using the Merrifield procedure.

31.29 The synthesis of large peptides is much more efficient when done in a convergent manner. That is, higher overall yields of final products are obtained if several small chains are constructed and coupled, as opposed to constructing one long chain by stepwise addition of one amino acid residue at a time. For example, consider a synthesis of methionine enkephalin, H-Tyr-Gly-Gly-Phe-Met-OH, by two routes:

(a) Tyr + Gly $\longrightarrow$ Tyr-Gly $\xrightarrow{\text{Gly}}$ Tyr-Gly-Gly $\xrightarrow{\text{Phe}}$ Tyr-Gly-Gly-Phe $\xrightarrow{\text{Met}}$ Tyr-Gly-Gly-Phe-Met

(b) Tyr + Gly $\longrightarrow$ Tyr-Gly $\xrightarrow{\text{Gly}}$ Tyr-Gly-Gly

Phe + Met $\longrightarrow$ Phe-Met

Tyr-Gly-Gly + Phe-Met $\longrightarrow$ Tyr-Gly-Gly-Phe-Met

Assume a yield of 90% for each coupling step and calculate overall yields for the two routes.

31.30 The Sanger end-group determination is sometimes used as an alternative to the Edman degradation. In the Sanger method, a peptide is allowed to react with 2,4-dinitrofluorobenzene, the peptide is hydrolyzed, and the N-terminal amino acid is identified by separation as its *N*-2,4-dinitrophenyl derivative:

2,4-dinitrofluorobenzene (F, NO_2, NO_2) + $H_2N—CHR—C(=O)—NH—$Peptide

$\downarrow$

$O_2N—C_6H_3(NO_2)—NH—CHR—C(=O)—NH—$Peptide

$\downarrow H_3O^+$

$O_2N—C_6H_3(NO_2)—NHCHRCOOH$ + Amino acids

Propose a mechanism to account for the initial reaction between peptide and dinitrofluorobenzene.

31.31 Would you foresee any problems in using the Sanger end-group determination method (Problem 31.30) on a peptide such as H-Gly-Pro-Lys-Ile-OH? Explain.

31.32 When α-amino acids are treated with dicyclohexylcarbodiimide, DCC, 2,5-diketopiperazines result. Propose a mechanism for this reaction.

$$H_2N{-}\underset{}{\overset{R}{\overset{|}{C}}}HCOOH \xrightarrow{DCC}$$

A 2,5-diketopiperazine

31.33 Good evidence for restricted rotation around amide CO—N bonds comes from NMR studies. At room temperature, the ^{1}H NMR spectrum of *N,N*-dimethylformamide shows three peaks: 2.9 δ (singlet, 3 H), 3.0 δ (singlet, 3 H), 8.0 δ (singlet, 1 H). As the temperature is raised, however, the two singlets at 2.9 δ and 3.0 δ slowly merge. At 180°C, the ^{1}H NMR spectrum shows only two peaks: 2.95 δ (singlet, 6 H) and 8.0 δ (singlet, 1 H). Explain this temperature-dependent behavior. [*Hint:* See Section 12.2.]

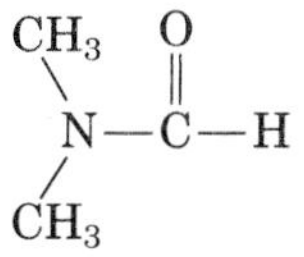

$(CH_3)_2N{-}C(=O){-}H$

N,N-Dimethylformamide

31.34 An octapeptide shows the composition Asp, Gly_2, Leu, Phe, Pro_2, Val on amino acid analysis. Edman analysis shows a glycine N-terminal group, and carboxypeptidase cleavage produces leucine as the first amino acid to appear. Acidic hydrolysis gives the following fragments:

1. H-Val-Pro-Leu-OH
2. H-Gly-OH
3. H-Gly-Asp-Phe-Pro-OH
4. H-Phe-Pro-Val-OH

Propose a suitable structure for the starting octapeptide.

31.35 Propose a mechanism to account for the reaction of ninhydrin with an α-amino acid:

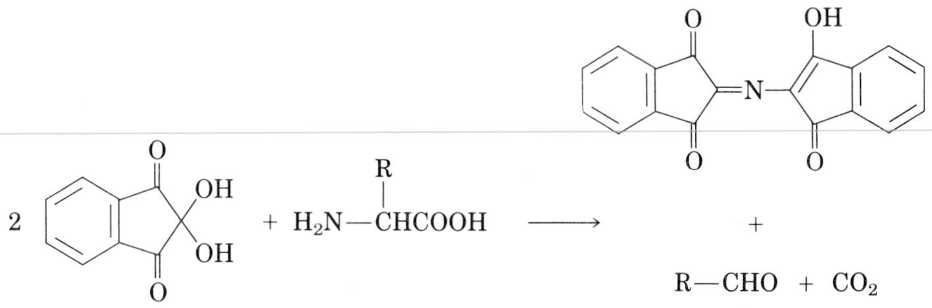

$$2\ \text{(ninhydrin)} + H_2N{-}CH(R)COOH \longrightarrow \text{(product)} + R{-}CHO + CO_2$$

31.36 Draw as many resonance forms as you can for the anion of the ninhydrin product (Problem 31.35).

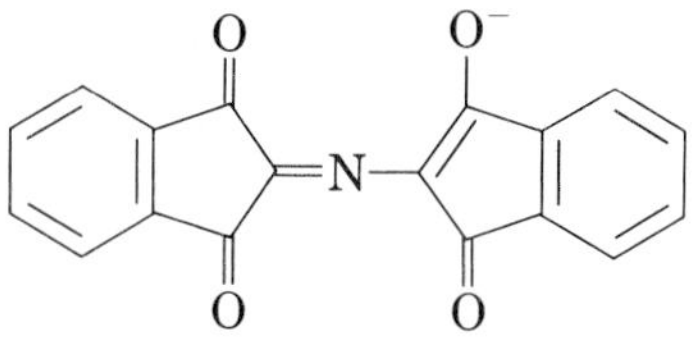

31.37 Oxytocin is a nonapeptide hormone secreted by the pituitary gland. Its function is to stimulate uterine contraction and lactation during childbirth, and its sequence was determined from the following evidence:

1. Oxytocin is a cyclic compound containing a disulfide bridge between two cysteine residues.
2. When the disulfide bridge is reduced, oxytocin has the constitution Asn, Cys_2, Gln, Gly, Ile, Leu, Pro, Tyr.
3. Partial hydrolysis of reduced oxytocin yields seven fragments:

H-Asp-Cys-OH	H-Ile-Glu-OH
H-Cys-Tyr-OH	H-Leu-Gly-OH
H-Tyr-Ile-Glu-OH	H-Glu-Asp-Cys-OH
H-Cys-Pro-Leu-OH	

4. Gly can be shown to be the C-terminal group.
5. Both Glu and Asp are present as their side-chain amides (Gln and Asn) rather than as free side-chain acids.

On the basis of this evidence, what is the amino acid sequence of reduced oxytocin? What is the structure of oxytocin itself?

CHAPTER 32 LIPIDS

Lipids are naturally occurring organic molecules isolated from cells and tissues by extraction with nonpolar organic solvents. They usually have large hydrocarbon portions in their structures and are thus insoluble in water but soluble in organic solvents. Note that this definition differs in kind from those we have used for other classes of biomolecules. Lipids are defined by *physical property* (solubility) rather than by structure, as with carbohydrates and proteins.

Lipids can be further classified into two general types. **Complex lipids** such as fats and waxes contain ester linkages and can be hydrolyzed to yield smaller molecules. **Simple lipids** such as cholesterol and other steroids cannot be hydrolyzed.

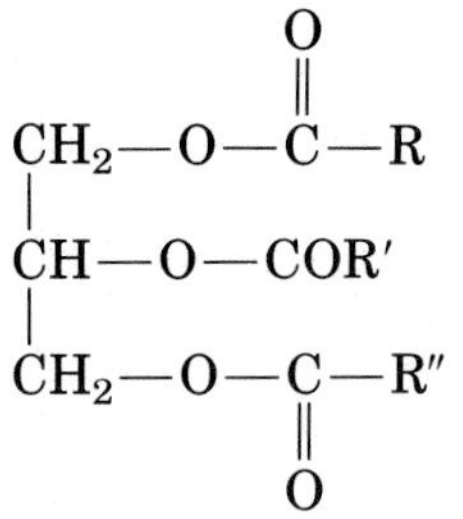

$$\begin{array}{l} CH_2-O-\overset{\displaystyle O}{\overset{\|}{C}}-R \\ | \\ CH-O-COR' \\ | \\ CH_2-O-\underset{\displaystyle O}{\underset{\|}{C}}-R'' \end{array}$$

Fat, a complex lipid
(R, R′, R″ = C_{11}–C_{19} chains)

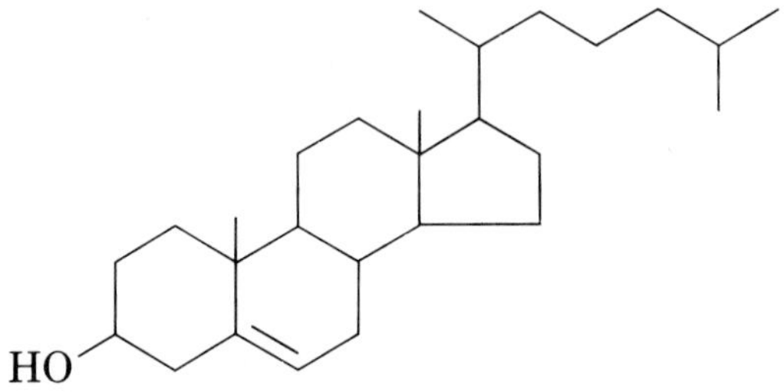

HO

Cholesterol, a simple lipid

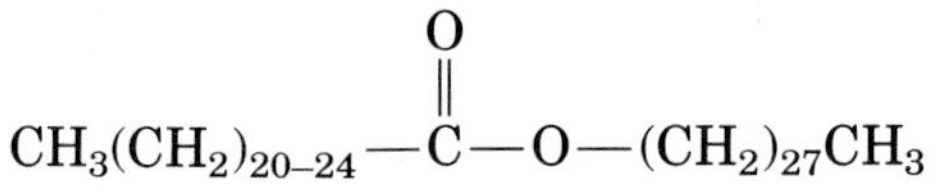

$$CH_3(CH_2)_{20-24}-\overset{\displaystyle O}{\overset{\|}{C}}-O-(CH_2)_{27}CH_3$$

Beeswax, a complex lipid

32.1 Fats and oils

Animal fats and vegetable oils are the most widely occurring lipids. Chemically, fats and oils are **triacylglycerols,** triesters between glycerol and long-chain carboxylic acids. Saponification of a fat with aqueous sodium hydroxide yields glycerol and three **fatty acids:**

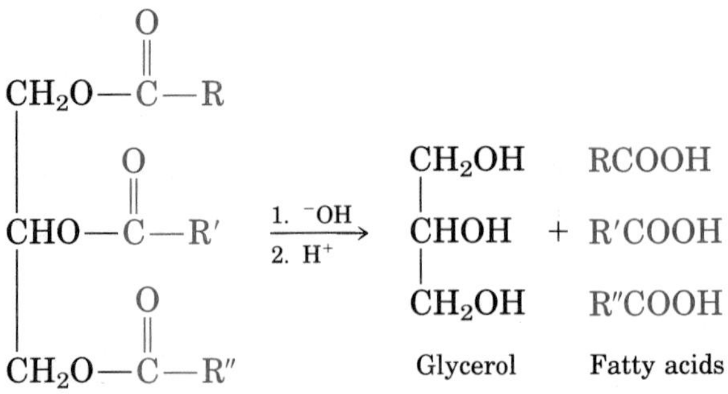

$$\begin{array}{l} CH_2O-\overset{\displaystyle O}{\overset{\|}{C}}-R \\ | \\ CHO-\overset{\displaystyle O}{\overset{\|}{C}}-R' \\ | \\ CH_2O-\overset{\displaystyle O}{\overset{\|}{C}}-R'' \end{array} \xrightarrow[2.\ H^+]{1.\ ^-OH} \begin{array}{l} CH_2OH \\ | \\ CHOH \\ | \\ CH_2OH \end{array} + \begin{array}{l} RCOOH \\ R'COOH \\ R''COOH \end{array}$$

A fat — Glycerol — Fatty acids

The fatty acids obtained by hydrolysis of triacylglycerols are generally unbranched, have between 12 and 20 carbon atoms, and may be either saturated or unsaturated. The three fatty acids of a given triacylglycerol need not be the same. Table 32.1 (p. 1072) lists some of the commonly occurring fatty acids, and Table 32.2 (p. 1073) lists the approximate composition of fats and oils from different sources.

The melting-point data in Table 32.1 show that unsaturated fatty acids generally have lower melting points than their saturated counterparts. The same trend holds true for triacylglycerols; vegetable oils generally have a higher proportion of unsaturated to saturated fatty acids than do animal fats (Table 32.2) and are therefore lower melting. This behavior is due to the fact that saturated fats have a uniform shape that allows them to pack together easily in a crystal lattice. Carbon–carbon double bonds in unsaturated vegetable oils, however, introduce bends and kinks into the hydrocarbon chains and make crystal formation difficult. Figure 32.1 illustrates this effect with space-filling molecular models.

The carbon–carbon double bonds present in vegetable oils can be reduced by catalytic hydrogenation to produce saturated solid or semisolid fats. Margarine and solid cooking fats such as Crisco are produced commercially in this way.

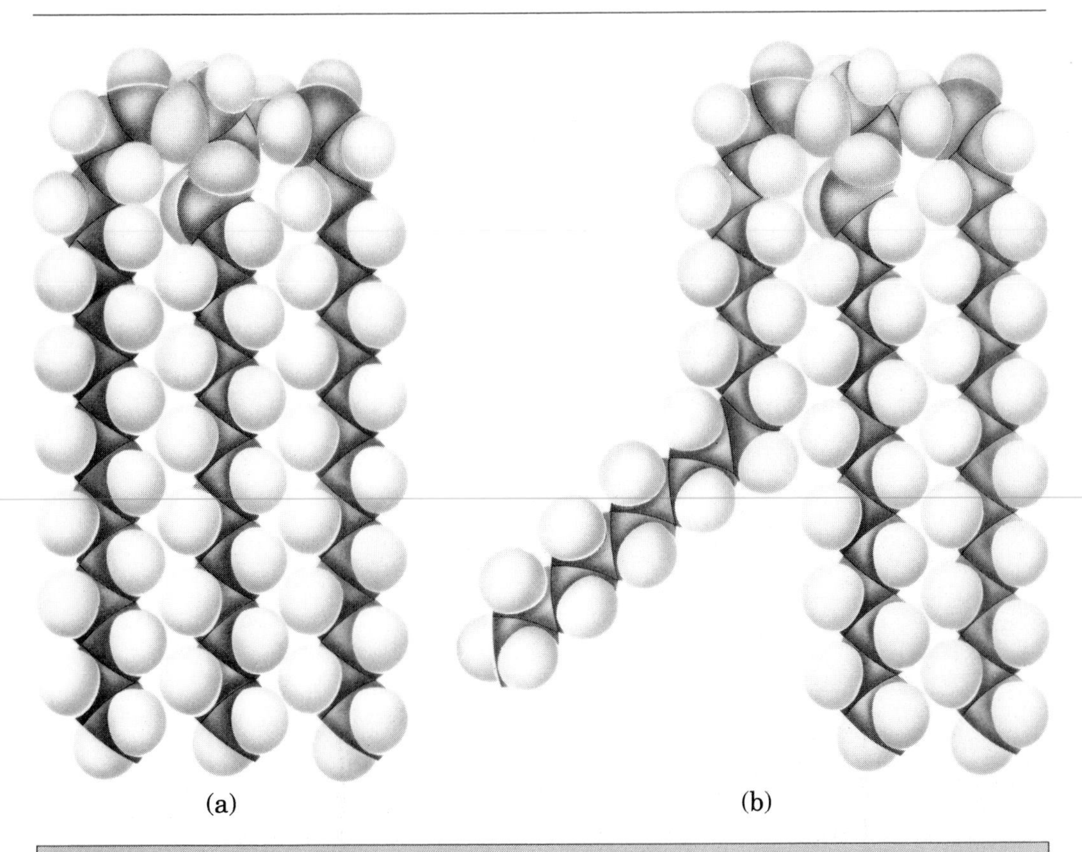

Figure 32.1. Space-filling molecular models of (a) saturated and (b) unsaturated triacylglycerols. There is one unsaturated acyl group in (b).

TABLE 32.1 Some common fatty acids

Name	*Carbons*	*Structure*	*Melting point* (°)
Saturated			
Lauric	12	$CH_3(CH_2)_{10}COOH$	44
Myristic	14	$CH_3(CH_2)_{12}COOH$	58
Palmitic	16	$CH_3(CH_2)_{14}COOH$	63
Stearic	18	$CH_3(CH_2)_{16}COOH$	70
Arachidic	20	$CH_3(CH_2)_{18}COOH$	75
Unsaturated			
Palmitoleic	16	$CH_3(CH_2)_5CH{=}CH(CH_2)_7COOH$ (cis)	32
Oleic	18	$CH_3(CH_2)_7CH{=}CH(CH_2)_7COOH$ (cis)	4
Ricinoleic	18	$CH_3(CH_2)_5CH(OH)CH_2CH{=}CH(CH_2)_7COOH$ (cis)	5
Linoleic	18	$CH_3(CH_2)_4CH{=}CHCH_2CH{=}CH(CH_2)_7COOH$ (cis,cis)	−5
Linolenic	18	$CH_3CH_2CH{=}CHCH_2CH{=}CHCH_2CH{=}CH(CH_2)_7COOH$ (cis,cis,cis)	−11
Arachidonic	20	$CH_3(CH_2)_4(CH{=}CHCH_2)_4CH_2CH_2COOH$ (all cis)	−50

TABLE 32.2 **Approximate fatty acid composition of some common fats and oils**

	Saturated fatty acids (%)				*Unsaturated fatty acids (%)*			
Source	C_{12} *Lauric*	C_{14} *Myristic*	C_{16} *Palmitic*	C_{18} *Stearic*	C_{18} *Oleic*	C_{18} *Ricinoleic*	C_{18} *Linoleic*	C_{18} *Linolenic*
Animal fat								
Lard	—	1	25	15	50	—	6	1
Butter	2	10	25	10	25	—	5	—
Human fat	1	3	25	8	46	—	10	—
Whale blubber	—	8	12	3	35	—	10	—
Vegetable oil								
Coconut	50	18	8	2	6	—	1	—
Corn	—	1	10	4	35	—	45	—
Olive	—	1	5	5	80	—	7	—
Peanut	—	—	7	5	60	—	20	—
Linseed	—	—	5	3	20	—	20	50
Castor bean	—	—	—	1	8	85	4	—

PROBLEM

32.1 Eleostearic acid, $C_{18}H_{30}O_2$, is a rare fatty acid found in the tung oil used for furniture finishing. On ozonolysis followed by treatment with zinc, eleostearic acid furnishes one part pentanal, two parts glyoxal (OHC—CHO), and one part 9-oxononanoic acid [$OHC(CH_2)_7COOH$]. What is the structure of eleostearic acid?

PROBLEM

32.2 Stearolic acid, $C_{18}H_{32}O_2$, yields stearic acid on catalytic hydrogenation and undergoes oxidative cleavage with ozone to yield nonanoic acid and nonanedioic acid. What is the structure of stearolic acid?

32.2 Soaps

Soap has been known since at least 600 BC, when the Phoenicians reportedly prepared a curdy material by boiling goat fat with extracts of wood ash. The cleansing properties of soap were not generally recognized for a long time, however, and the use of soap was not widespread until the eighteenth century.

Chemically, soap is a mixture of the sodium or potassium salts of long-chain carboxylic acids produced by saponification of animal fat with alkali. Wood ash was used as a source of alkali until the mid-1800s, when the Leblanc process for producing Na_2CO_3 was invented, and NaOH thus became commercially available:

$$\underset{\text{A fat}}{\begin{array}{l} CH_2-O-\overset{O}{\overset{\|}{C}}-R \\ | \\ CH-O-\overset{O}{\overset{\|}{C}}-R \\ | \\ CH_2-O-\overset{O}{\overset{\|}{C}}-R \end{array}} \xrightarrow[H_2O]{NaOH} \underset{\text{Soap}}{3\,R-\overset{O}{\overset{\|}{C}}O^-\,Na^+} + \underset{\text{Glycerol}}{\begin{array}{c} CH_2OH \\ | \\ CHOH \\ | \\ CH_2OH \end{array}}$$

where R = C_{15}–C_{17} aliphatic chains.

The crude soap curds contain glycerol and excess alkali as well as soap, but purification can be effected by boiling with a large amount of water, followed by precipitation of the pure sodium carboxylate salts on addition of sodium chloride. The smooth soap that precipitates is dried, perfumed, and pressed into bars for household use. Dyes can be added if a colored soap is desired, antiseptics can be added for medicated soaps, pumice can be added for scouring soaps, and air can be blown in for a soap that floats. Regardless of these extra treatments, though, all soaps are basically the same.

The mechanism by which soaps exert their cleansing action derives from the fact that the two ends of a soap molecule are so different. The sodium salt end of the long-chain molecule is ionic and therefore **hydrophilic** (water loving); it tries to dissolve in water. The long aliphatic chain portion of the molecule, however, is a nonpolar hydrocarbon and is therefore **lipophilic** (fat loving); it tries to dissolve in grease. The net effect of these two opposing tendencies is that soaps are attracted to both greasy dirt and water, and are therefore valuable as cleansers.

When soaps are dispersed in water, the long hydrocarbon tails cluster together in a lipophilic ball, while the ionic heads on the surface of the cluster stick out into the water layer. These spherical clusters, called **micelles,** are shown schematically in Figure 32.2. Soap micelle solutions are good cleansers because grease and oil droplets are coated by the non-polar tails of soap molecules and are then dispersed in water where they can be washed away.

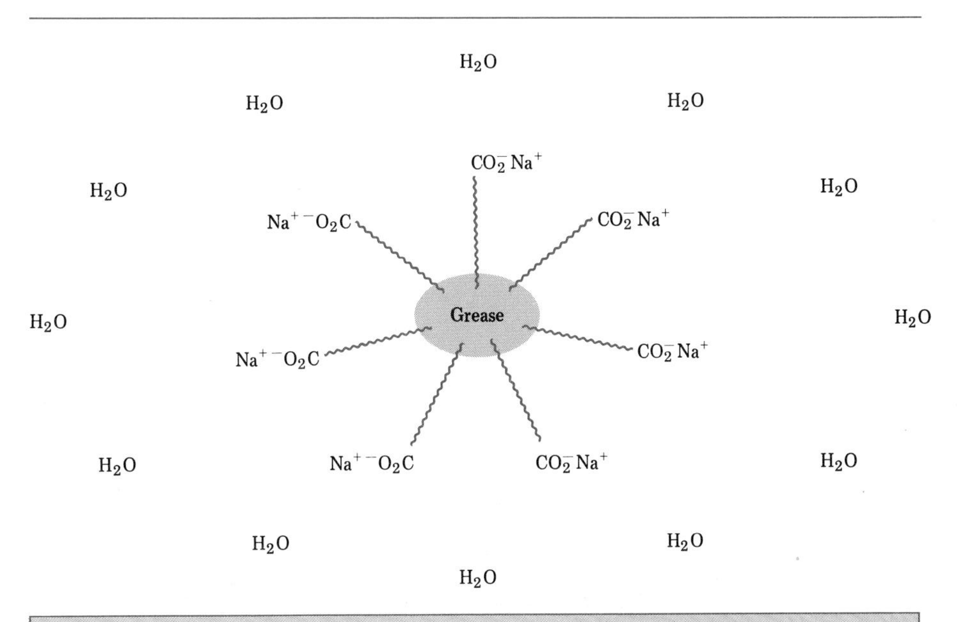

Figure 32.2. A soap micelle solubilizing a grease particle in water.

Soaps make life much more pleasant than it would otherwise be, but they have certain drawbacks. In hard water, which contains metal ions, soluble sodium carboxylates are converted into insoluble magnesium and calcium salts, leaving the familiar ring of scum around bathtubs and the "tattletale gray" on white clothes. Chemists have circumvented these problems by synthesizing a class of synthetic detergents based on salts of long-chain alkylbenzenesulfonic acids. The principle by which synthetic detergents operate is identical to the principle of soaps—the alkylbenzene end of the molecule is lipophilic and attracts grease, but the sulfonate salt end is ionic and is attracted to water. Unlike soaps, however, sulfonate detergents do not form insoluble metal salts in hard water.

$$\text{R}-\text{C}_6\text{H}_4-\text{S}(=\text{O})_2-\text{O}^-\ \text{Na}^+$$

A synthetic detergent

where R = a mixture of C_{12} aliphatic chains.

32.3 Phospholipids

Phospholipids are esters of phosphoric acid, H_3PO_4. Most phospholipids are closely related to fats and contain a glycerol backbone linked by ester bonds to two fatty acids and one phosphoric acid. The fatty acid residues in these so-called **phosphoglycerides** may be any of the C_{12}–C_{22} units normally present in fats, but the acyl group at carbon 1 is usually saturated, and that at carbon 2 is unsaturated. The phosphate group at carbon 3 is also bound by a separate ester link to an amino alcohol such as choline, $HOCH_2CH_2\overset{+}{N}(CH_3)_3$, or ethanolamine, $HOCH_2CH_2NH_2$. The most important phosphoglycerides are the lecithins and the cephalins. Note that these compounds are chiral and that they have the L configuration at carbon 2.

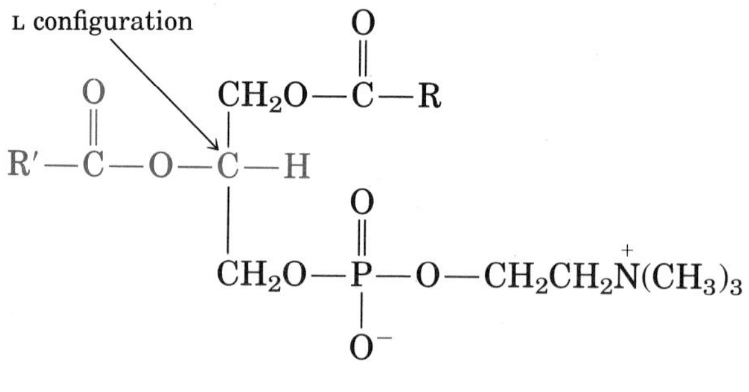

L configuration

$$\begin{array}{l} \qquad\qquad\qquad\quad CH_2O{-}\overset{O}{\overset{\|}{C}}{-}R \\ R'{-}\overset{O}{\overset{\|}{C}}{-}O{-}\underset{|}{\overset{|}{C}}{-}H \\ \qquad\qquad\qquad\quad CH_2O{-}\underset{O^-}{\underset{|}{\overset{O}{\overset{\|}{P}}}}{-}O{-}CH_2CH_2\overset{+}{N}(CH_3)_3 \end{array}$$

Phosphatidylcholine,
a lecithin

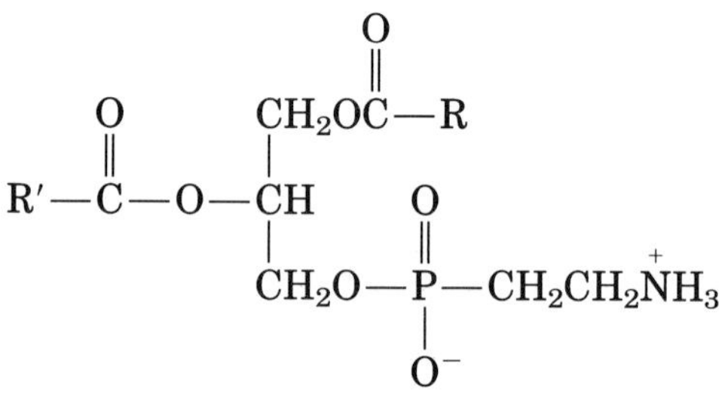

$$\begin{array}{l} \qquad\qquad\qquad\quad CH_2O\overset{O}{\overset{\|}{C}}{-}R \\ R'{-}\overset{O}{\overset{\|}{C}}{-}O{-}\underset{|}{\overset{|}{C}}H \\ \qquad\qquad\qquad\quad CH_2O{-}\underset{O^-}{\underset{|}{\overset{O}{\overset{\|}{P}}}}{-}CH_2CH_2\overset{+}{N}H_3 \end{array}$$

Phosphatidylethanolamine,
a cephalin

where R is saturated and R′ is unsaturated.

Phosphoglycerides, found widely in both plant and animal tissues, are the major lipid component of cell membranes. Like soaps, phosphoglycerides have a long, nonpolar hydrocarbon tail bound to a polar ionic head (the phosphate group). There is evidence to suggest that cell membranes are composed in large part of phosphoglycerides oriented in a lipid bilayer about 50 Å thick. The lipophilic tails aggregate in the center of the bilayer in much the same way that soaps aggregate in water (Section 32.2), and the hydrophilic heads lie together on the outside of the bilayer (Figure 32.3). This bilayer serves as an effective barrier to the passage of water and other components into and out of the cell.

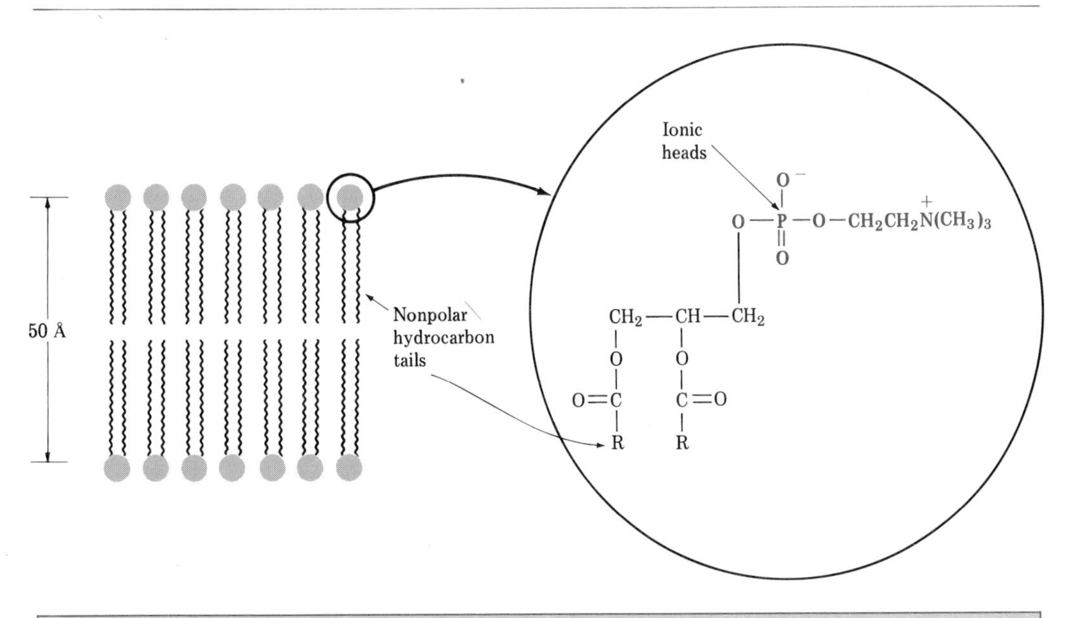

Figure 32.3. Aggregation of phosphoglycerides into a lipid bilayer.

The second major group of phospholipids are the **sphingolipids.** These complex lipids have sphingosine or a related dihydroxyamine as their backbones and are important constituents of plant and animal cell membranes. They are particularly abundant in brain and nerve tissue, where sphingomyelins are a major constituent of the coating around nerve fibers.

CH_2OH
|
$CHNH_2$
|
$CHOH$
|
$CH{=}CH(CH_2)_{12}CH_3$

Sphingosine

$CH_2—O—\overset{\overset{O}{\|}}{\underset{\underset{O^-}{|}}{P}}—CH_2CH_2\overset{+}{N}(CH_3)_3$
|
$CH_2—NHCO(CH_2)_{16–24}—CH_3$
|
$CHOH$
|
$CH{=}CH(CH_2)_{12}CH_3$

Sphingomyelin, a sphingolipid

32.4 Biosynthesis of fatty acids

A striking feature of the fatty acids in Table 32.1 is that all have an even number of carbon atoms. The reason for this is that all are derived biosynthetically from the simple two-carbon precursor, acetic acid. The pathway by which this is accomplished is shown in Figure 32.4.

The key starting material for fatty acid synthesis is the thiol ester, acetyl CoA (Section 26.15). The synthetic pathway begins with several priming reactions that convert acetyl CoA into more reactive species. In step 1, reaction with an enzyme, acyl carrier protein (ACP) transferase, exchanges the thiol ester linkage of acetyl CoA for a different thiol ester bond to ACP. Step 2 involves a further exchange of thiol ester linkages, resulting in the formation of acetyl synthase, a highly reactive acylating agent. In step 3, acetyl CoA is carboxylated by CO_2 and the enzyme acetyl CoA carboxylase to yield malonyl CoA. Step 4 is also a thiol ester exchange reaction and converts malonyl CoA into the more reactive malonyl ACP.

The key carbon–carbon bond-forming reaction to build the fatty acid chain occurs in step 5. This key step is simply a Claisen condensation (Section 26.9). An enolate ion derived from the doubly activated methylene group of malonyl ACP attacks the carbonyl group of acetyl synthase, yielding an intermediate β-keto acid, which decarboxylates to give the four-carbon product, acetoacetyl ACP.

$$CH_3C(=O)\text{—S synthase} \;+\; {}^{-}\!:\!CH(COOH)\text{—}C(=O)\text{—S—ACP} \longrightarrow \left[CH_3\text{—}C(=O)\text{—}CH(COOH)\text{—}C(=O)\text{—S—ACP} \right]$$

$$\downarrow$$

$$CH_3C(=O)\text{—}CH_2C(=O)\text{—S—ACP} + CO_2$$

Acetoacetyl ACP

Step 6 involves the NADPH (nicotinamide adenine dinucleotide phosphate) enzyme-catalyzed reduction of the β-keto carbonyl group to a hydroxyl group, followed by dehydration in step 7. The carbon–carbon double bond of crotonyl ACP is further reduced by NADPH in step 8 to yield butyryl ACP. The cycle is then repeated by condensation of butyryl synthase with another malonyl ACP, yielding a six-carbon unit, and further repetitions add two carbon atoms each cycle.

Fatty acid biosynthesis is a lengthy process—some 64 separate steps are required for a cell to produce stearic acid—but all the individual steps

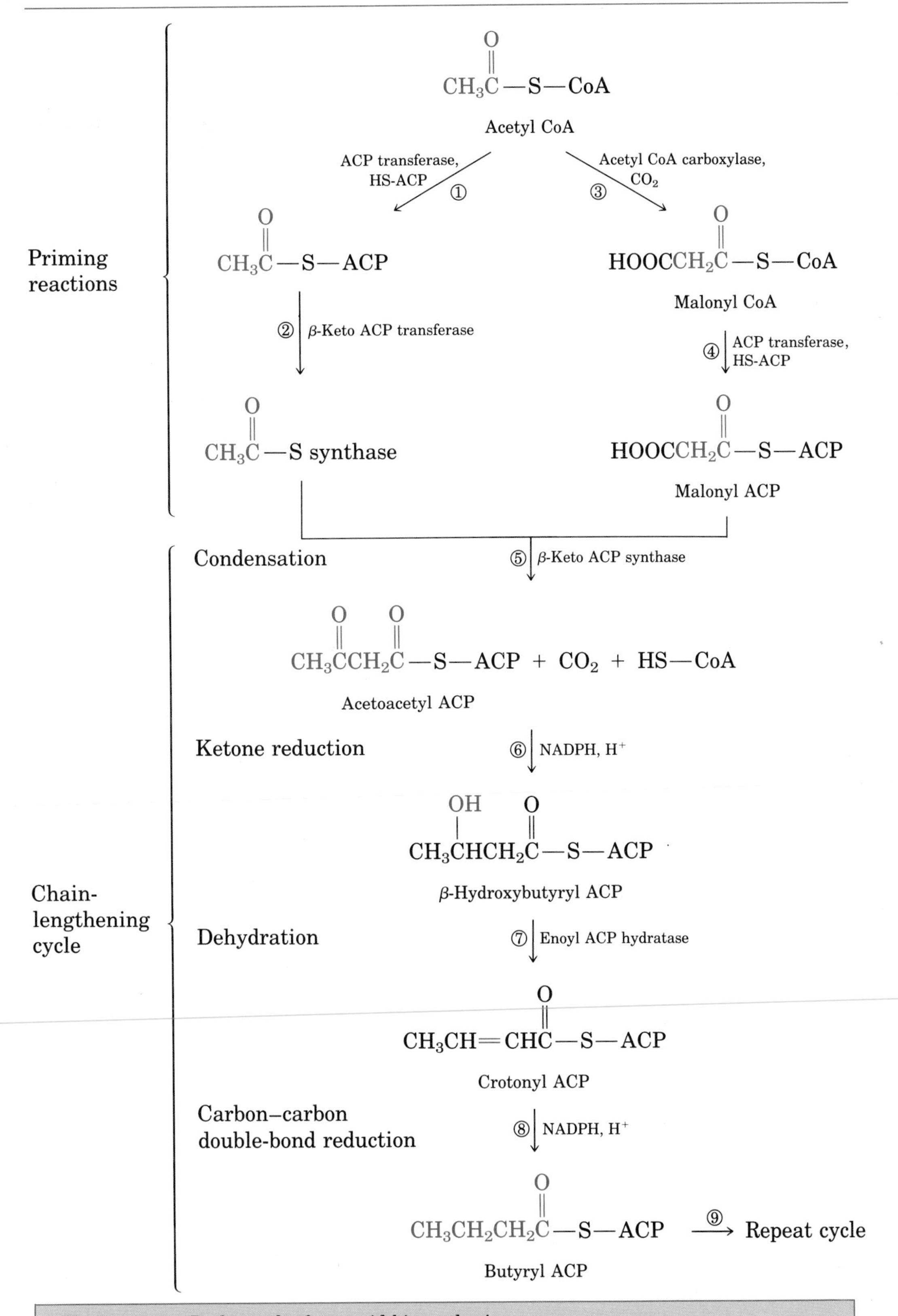

Figure 32.4. Pathway for fatty acid biosynthesis.

are precedented by close laboratory analogy. Biological chemistry and laboratory organic chemistry follow the same rules.

PROBLEM

32.3 Show a likely reaction mechanism for the transformation of acetyl CoA into acetyl ACP.

$$CH_3\overset{O}{\overset{\|}{C}}-S-CoA + HS-ACP \rightleftharpoons CH_3\overset{O}{\overset{\|}{C}}-S-ACP + HS-CoA$$

PROBLEM

32.4 Show a likely mechanism for the ready decarboxylation of the acylmalonyl ACP intermediate formed in step 5 of Figure 32.4.

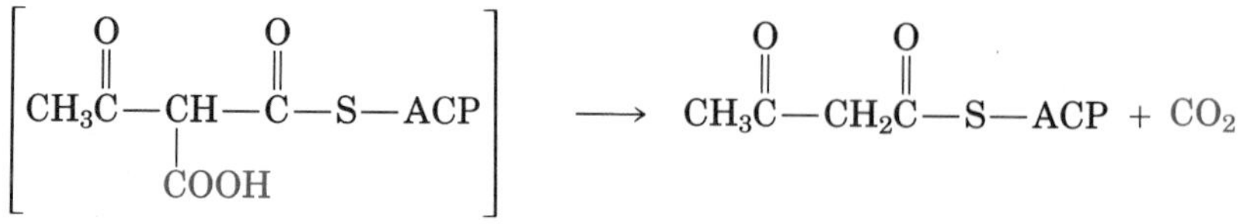

$$\left[CH_3\overset{O}{\overset{\|}{C}}-\underset{COOH}{\underset{|}{CH}}-\overset{O}{\overset{\|}{C}}-S-ACP\right] \longrightarrow CH_3\overset{O}{\overset{\|}{C}}-CH_2\overset{O}{\overset{\|}{C}}-S-ACP + CO_2$$

PROBLEM

32.5 Evidence for the proposed role of acetate in fatty acid biosynthesis comes from isotope-labeling experiments. If acetate labeled with ^{14}C in the methyl group ($^{14}CH_3COOH$) were incorporated into fatty acids, at what positions in the fatty acid chain would you expect the ^{14}C label to appear?

32.5 Prostaglandins

Few compounds have caused as much excitement among medical researchers in the past decade as have the **prostaglandins.** First isolated by Sune Bergstrom,[1] Bengt Samuelsson,[2] and their collaborators at the Karolinska Institute in Sweden, these simple lipids are synthesized in nature from the C_{20} fatty acid, arachidonic acid. The name prostaglandin derives from the fact that these compounds were first thought to be produced by the prostate gland, but they have subsequently been shown to be present in small amounts in all body tissues and fluids.

The prostaglandins are simple in structure; all have a cyclopentane ring with two long side chains, and they differ only in the number of oxygen atoms and number of double bonds present. Prostaglandin E_1 (PGE_1) and prostaglandin $F_{2\alpha}$ ($PGF_{2\alpha}$) are representative structures:

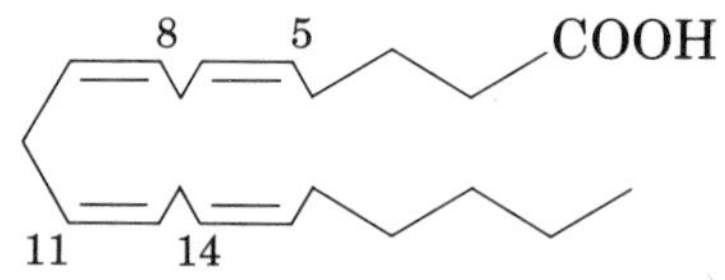

8 5 COOH
11 14

Arachidonic acid,
(5Z,8Z,11Z,14Z)-eicosatetraenoic acid

[1]Sune Bergstrom (1916–); M.D. Karolinska Institute; professor, Karolinska Institute; Nobel prize (1982).

[2]Bengt Samuelsson (1934–); M.D. Lund; professor, Karolinska Institute; Nobel prize (1982).

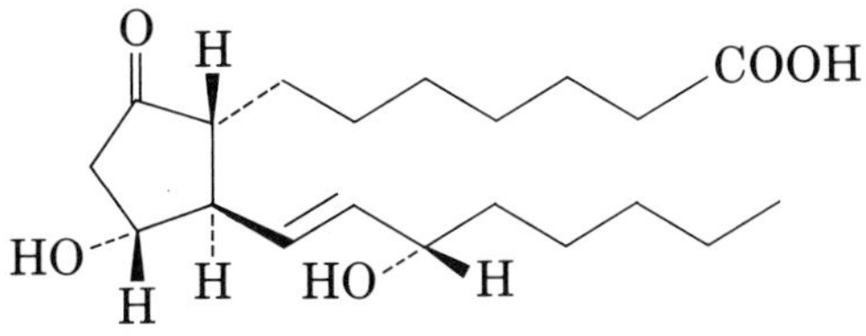

Prostaglandin E_1

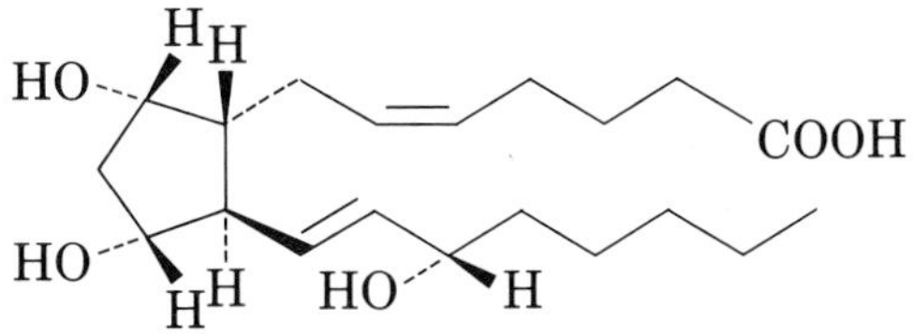

Prostaglandin $F_{2\alpha}$

The several dozen known prostaglandins have an extraordinarily wide range of biological activities. Among their known actions are their abilities to affect blood pressure, to affect blood-platelet aggregation during clotting, to affect gastric secretions, to control inflammation, to affect kidney function, to affect reproductive systems, and to stimulate uterine contractions. In addition, compounds that are closely related to the prostaglandins have still other effects. Interest has centered particularly on the thromboxanes and prostacyclin, as well as on the leukotrienes, whose release in the body appears to trigger the asthmatic response.

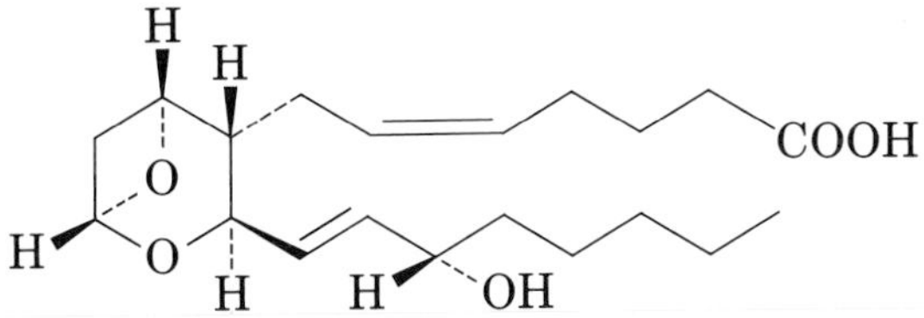

Thromboxane A_2

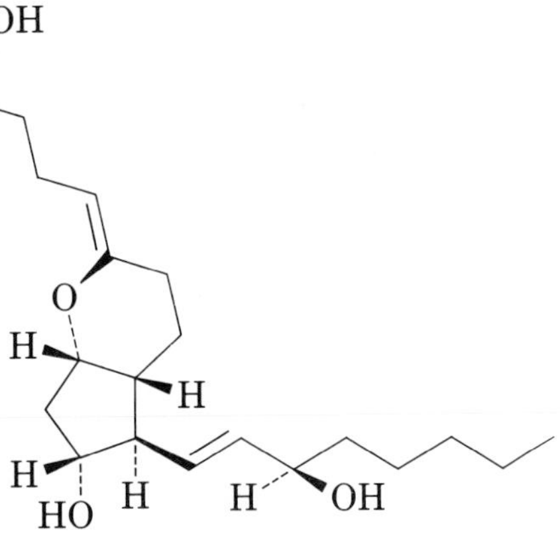

Prostacyclin

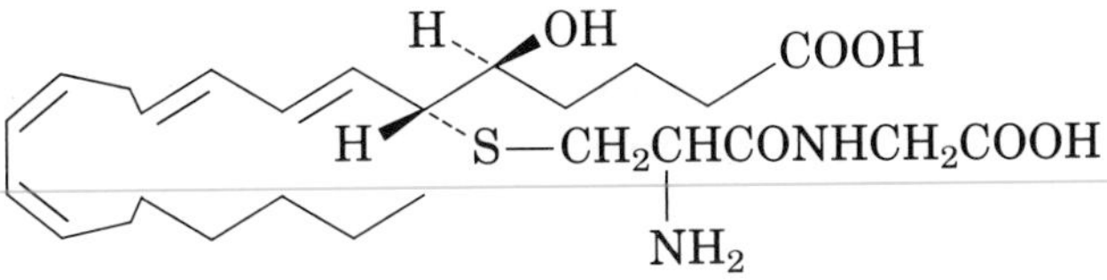

Leukotriene D_4

Prostaglandins have not yet been fully exploited in medicine, but their remarkable biological properties will surely lead to valuable new drugs in the future.

32.6 Terpenes

It has long been known that steam distillation of many plant materials produces a fragrant mixture of **essential oils.** Plant extracts have been used for thousands of years as medicines, spices, and perfumes, and the investigation of essential oils played a major role in the emergence of organic chemistry as a science during the nineteenth century.

Chemically, the essential oils of plants consist largely of mixtures of simple lipids called **terpenes.** Terpenes are relatively small organic molecules that have an immense diversity of structure. Thousands of different terpenes are known; some are hydrocarbons, and others contain oxygen; some are open-chain molecules, and others contain rings. Figure 32.5 gives some examples.

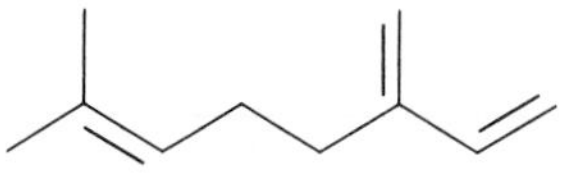

Myrcene (oil of bay)

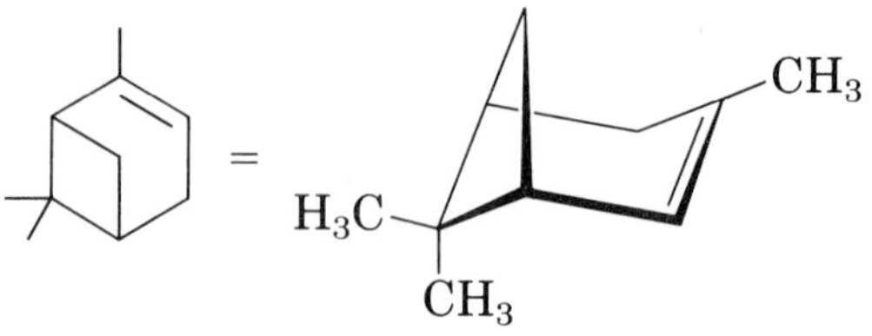

H_3C, CH_3, CH_3

α-Pinene (oil of turpentine)

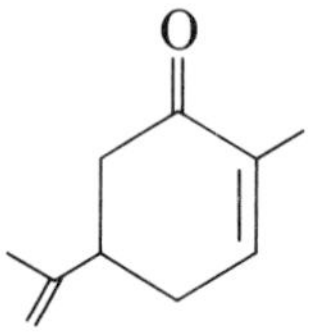

O

Carvone (oil of spearmint)

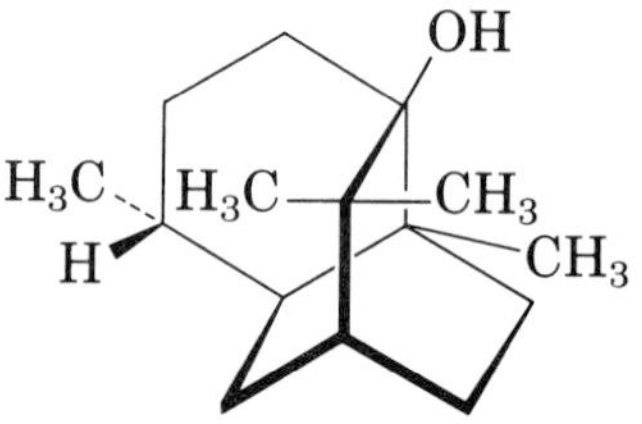

OH, H_3C, H, H_3C, CH_3, CH_3

Patchouli alcohol (patchouli oil)

Figure 32.5. Some terpenes from essential oils.

All terpenes are related, regardless of their apparent structural differences. According to the **isoprene rule,** terpenes can be considered to arise from head-to-tail joining of simple five-carbon isoprene units; in 2-methyl-1,3-butadiene, the head of the isoprene unit is carbon 1 and the tail is carbon 4. For example, myrcene contains two isoprene units joined head to tail, forming an eight-carbon chain with 2 one-carbon branches. α-Pinene similarly contains two isoprene units assembled into a more complex cyclic structure.

$$\overset{1}{H_2C}{=}\overset{CH_3}{\overset{|2}{C}}{-}\overset{3}{CH}{=}\overset{4}{CH_2}$$

Head ↗ (C1) ↖ Tail (C4)

Isoprene (2-methyl-1,3-butadiene)

Tail

Head

⇒

Myrcene

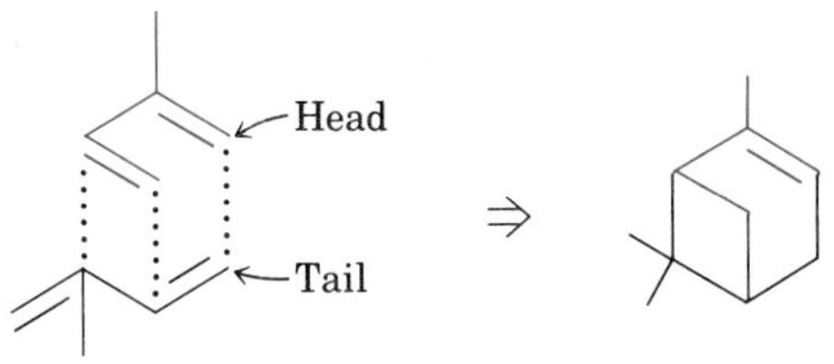

Head

Tail

Two isoprenes

α-Pinene

Terpenes are classified by the number of 5-carbon isoprene units they contain. Thus, monoterpenes are 10-carbon substances biosynthesized from two isoprene units, sesquiterpenes are 15-carbon molecules from three isoprene units, and so on (Table 32.3).

TABLE 32.3 Classification of terpenes

Carbon atoms	*Isoprene units*	*Classification*
10	2	Monoterpenes
15	3	Sesquiterpenes
20	4	Diterpenes
25	5	Sesterterpenes
30	6	Triterpenes
40	8	Tetraterpenes

Although mono- and sesquiterpenes are found primarily in plants, the higher terpenes occur in both plants and animals. Many of the higher terpenes have important biological activity. For example, the triterpene lanosterol is the precursor for biosynthesis of steroid hormones, and the tetraterpene β-carotene is a dietary source of vitamin A.

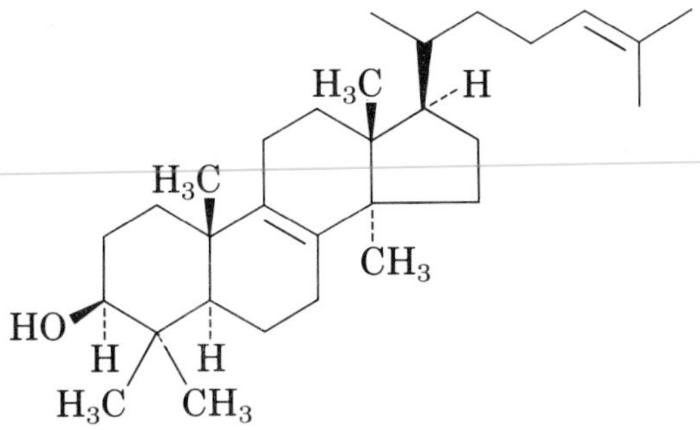

H_3C H H_3C CH_3 HO H H H_3C CH_3

Lanosterol, a triterpene (C_{30})

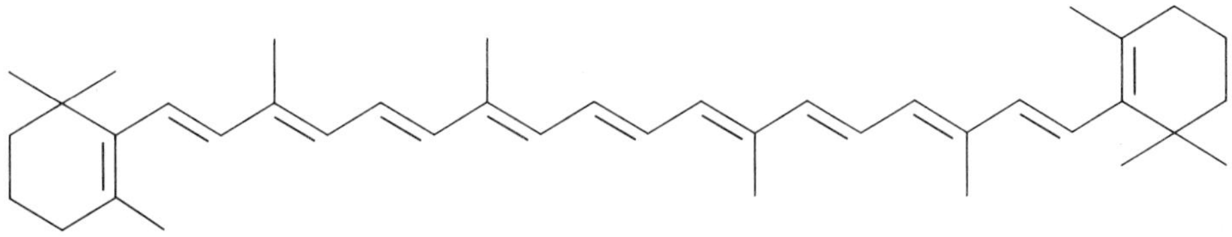

β-Carotene, a tetraterpene (C_{40})

PROBLEM

32.6 Show the positions of the isoprene units in the following terpenes:

(a) 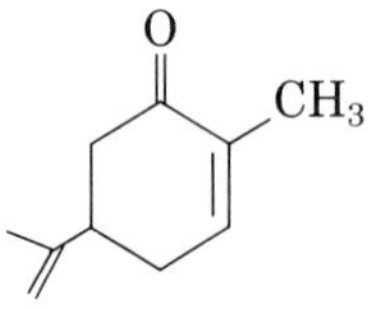 Carvone (spearmint oil)

(b) 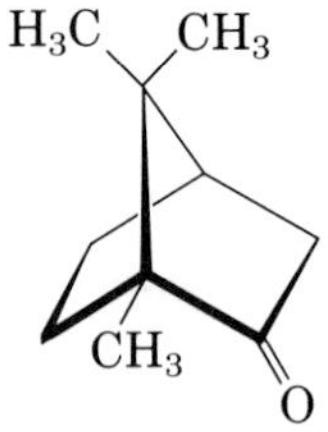 Camphor

(c) 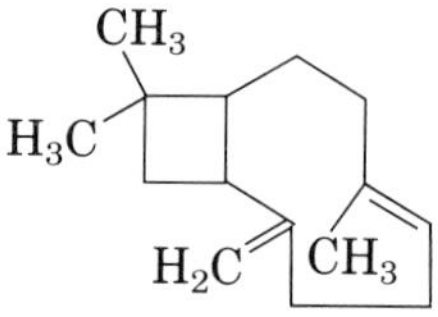 Caryophyllene (cloves)

32.7 Biosynthesis of terpenes

The isoprene rule is a convenient formalism for helping to determine new structures, but isoprene itself is not the biological precursor of terpenes. Nature instead uses two isoprene equivalents, isopentenyl pyrophosphate and dimethylallyl pyrophosphate, for the biosynthesis of terpenes. These five-carbon molecules are in turn made from condensation of three acetyl CoA units.

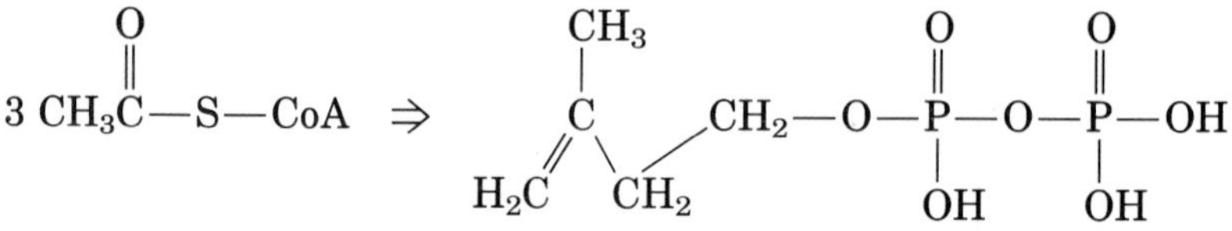

$$3\ CH_3C(=O){-}S{-}CoA \Rightarrow H_2C{=}C(CH_3){-}CH_2{-}CH_2{-}O{-}P(=O)(OH){-}O{-}P(=O)(OH){-}OH$$

Isopentenyl pyrophosphate

and

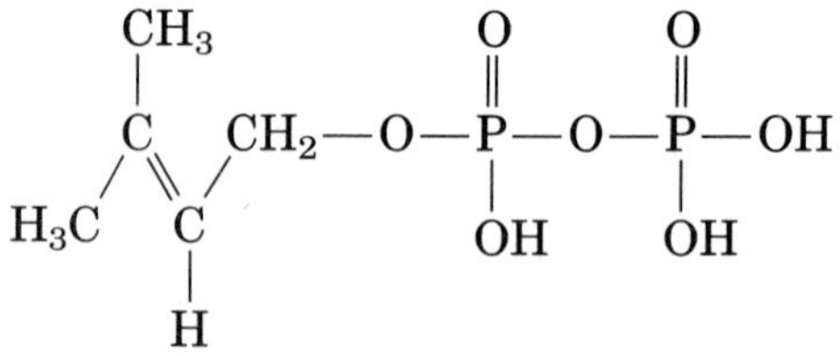

$$(H_3C)(CH_3)C{=}C(H){-}CH_2{-}O{-}P(=O)(OH){-}O{-}P(=O)(OH){-}OH$$

Dimethylallyl pyrophosphate

Dimethylallyl pyrophosphate is an excellent alkylating agent since the primary, allylic pyrophosphate group (OPP) is a good leaving group

(Section 10.5). We can therefore imagine a substitution (either S_N1 or S_N2) of this leaving group by the nucleophilic double bond of isopentenyl pyrophosphate. Loss of a proton from the reaction intermediate leads to the head-to-tail coupled 10-carbon unit, geraniol pyrophosphate. Geraniol itself is a terpene alcohol that occurs in rose oil.

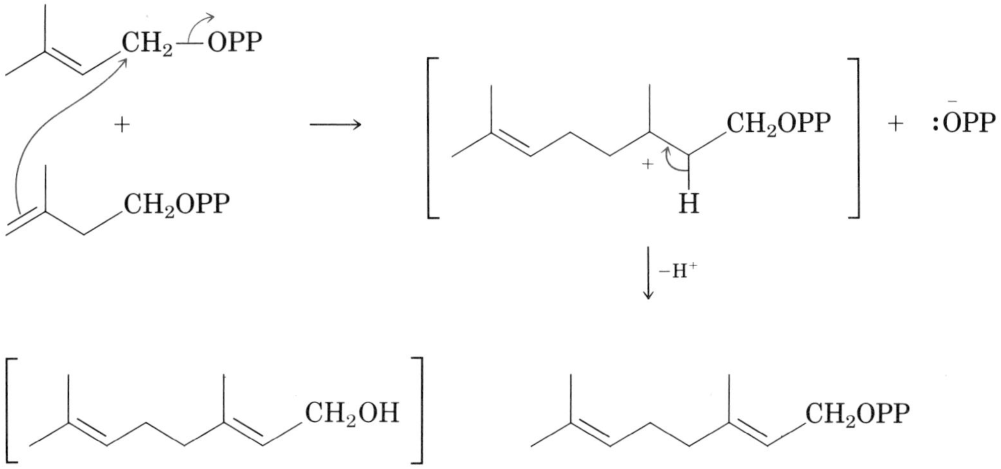

CH_2—OPP + CH_2OPP ⟶ [CH_2OPP H] + :$\bar{O}PP$

$-H^+$

[CH_2OH] CH_2OPP

Geraniol Geraniol pyrophosphate

Geraniol pyrophosphate is the precursor of all other monoterpenes, and the rough outlines of how the multitude of monoterpenes might arise can be rationalized by fundamental organic processes. For example, limonene, a monoterpene found in many citrus oils, can arise by a double-bond isomerization followed by internal nucleophilic substitution of geraniol pyrophosphate and subsequent loss of a proton (Figure 32.6).

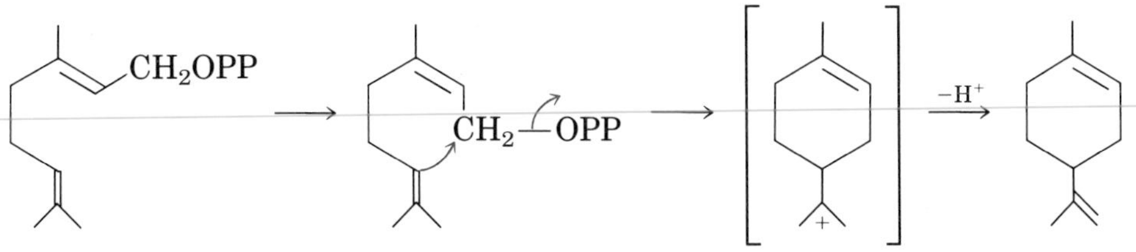

CH_2OPP ⟶ CH_2—OPP ⟶ [ ] $\xrightarrow{-H^+}$

Geraniol pyrophosphate

Nerol pyrophosphate

Limonene (from citrus oils)

Figure 32.6. Biosynthesis of limonene from geraniol pyrophosphate.

Geraniol pyrophosphate can react an additional time with isopentenyl pyrophosphate to yield the 15-carbon farnesol pyrophosphate. Farnesol pyrophosphate is the precursor of all sesquiterpenes.

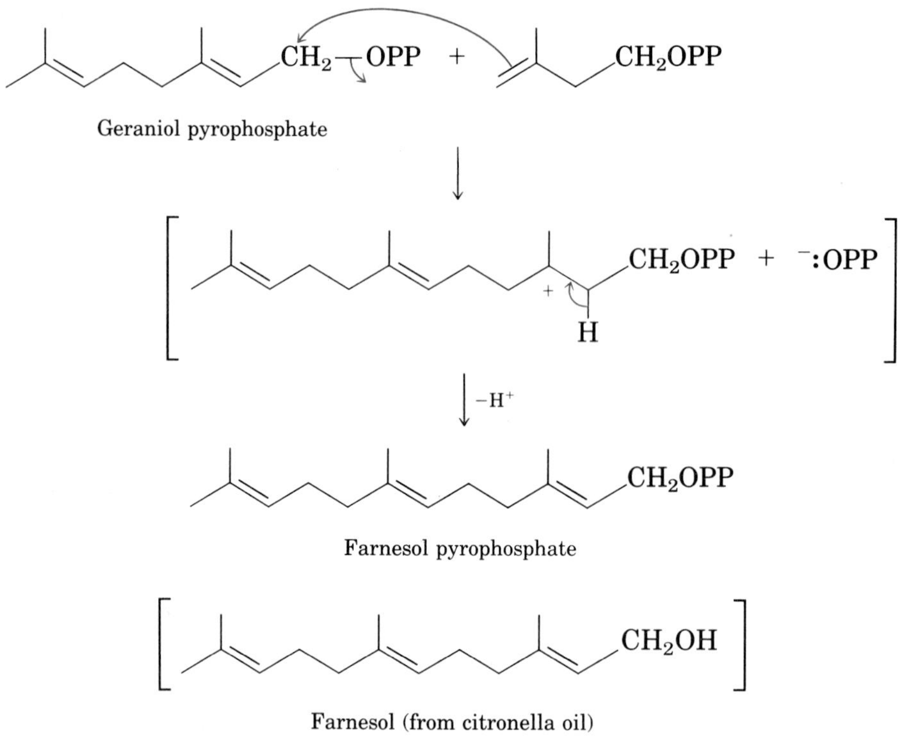

CH_2—OPP + CH_2OPP

Geraniol pyrophosphate

[CH_2OPP + $^-$:OPP]

H

$-H^+$

CH_2OPP

Farnesol pyrophosphate

[CH_2OH]

Farnesol (from citronella oil)

Addition to farnesol of further isoprene units to give 20-carbon (diterpene) and 25-carbon (sesterterpene) precursors can also take place. Triterpenes, however, arise biosynthetically by tail-to-tail coupling of two farnesol pyrophosphates to give squalene, a 30-carbon hexaene (Figure

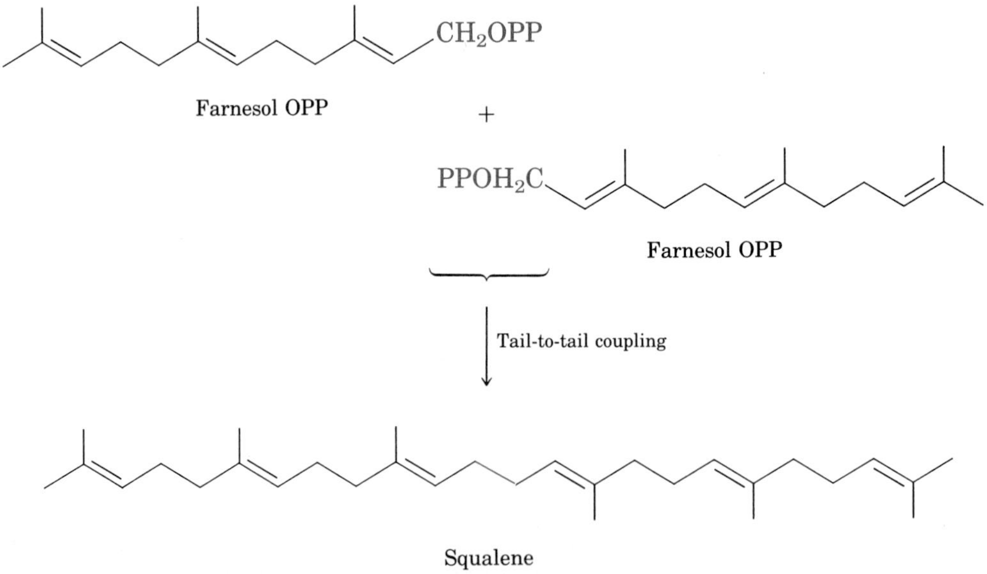

CH_2OPP

Farnesol OPP

+

$PPOH_2C$

Farnesol OPP

Tail-to-tail coupling

Squalene

Figure 32.7. Synthesis of squalene, a C_{30} precursor of triterpenes, by tail-to-tail coupling of two farnesol pyrophosphates.

32.7). Squalene, in turn, is the precursor from which all other triterpenes and steroids arise.

PROBLEM

32.7 Propose a plausible pathway to account for the biosynthetic formation of γ-bisabolene from farnesol pyrophosphate.

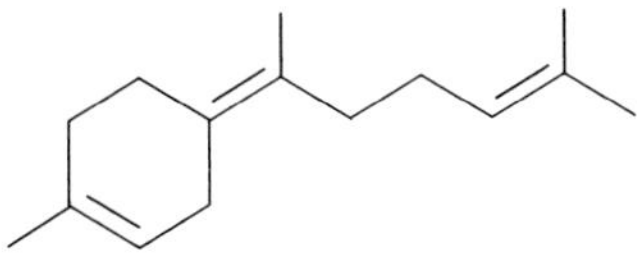

γ-Bisabolene

32.8 Steroids

In addition to fats, phospholipids, and terpenes, the lipid extracts of plants and animals contain **steroids.** A steroid is an organic molecule whose structure is based on the tetracyclic ring system shown in Figure 32.8. The four rings are designated A, B, C, and D, beginning at the lower left; the carbon atoms are numbered beginning in the A ring. Common examples are cholesterol, an animal steroid (and principal component of gallstones), and β-sitosterol, a ubiquitous plant steroid.

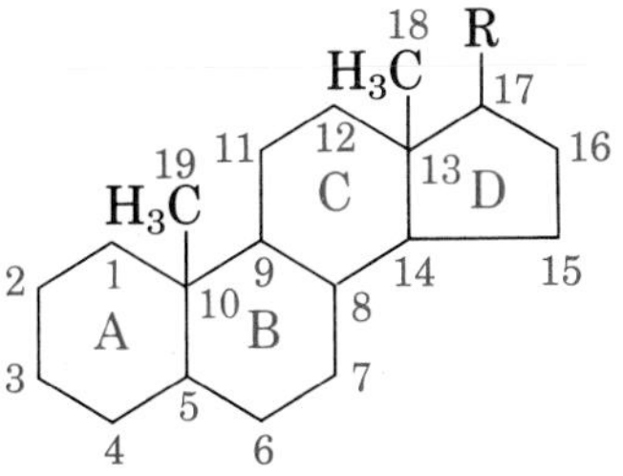

Steroid skeleton
(R = different side chains)

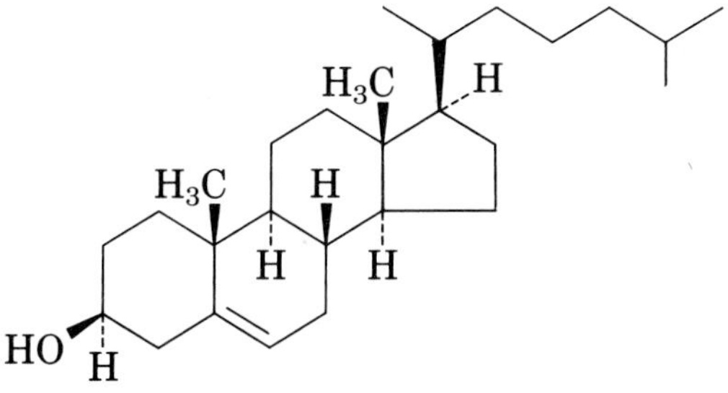

Cholesterol (animal sources)

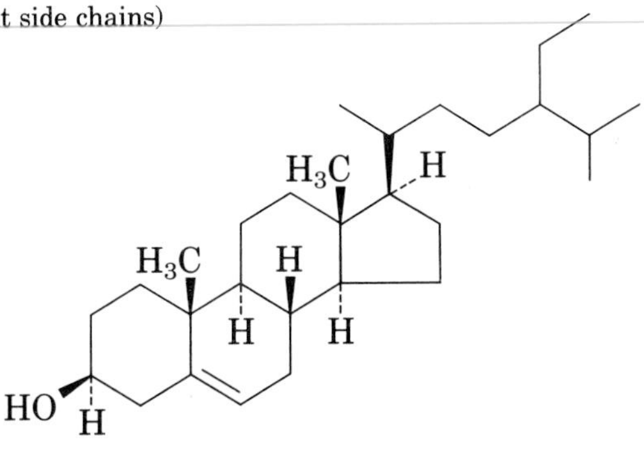

β-Sitosterol (plant sources)

Figure 32.8. Some representative steroids.

Steroids are widespread in both plant and animal kingdoms, and many steroids have interesting biological activity. For example, digitoxigenin, a plant steroid found in *Digitalis purpurea* (purple foxglove), is widely used medicinally as a heart stimulant; androsterone and estradiol are steroid sex hormones; and cortisone is a steroid hormone with anti-inflammatory properties.

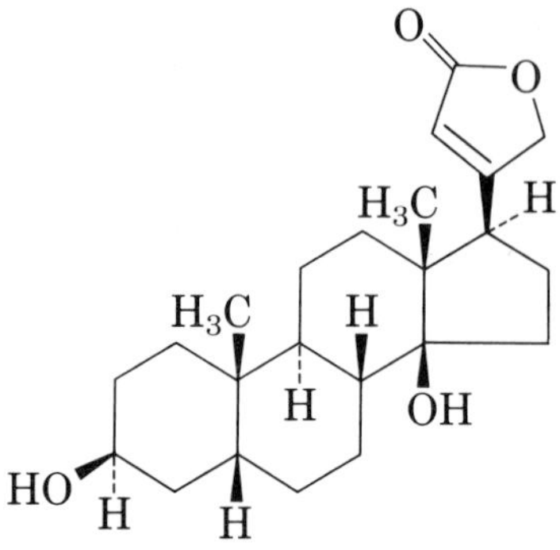

Digitoxigenin (heart stimulant)

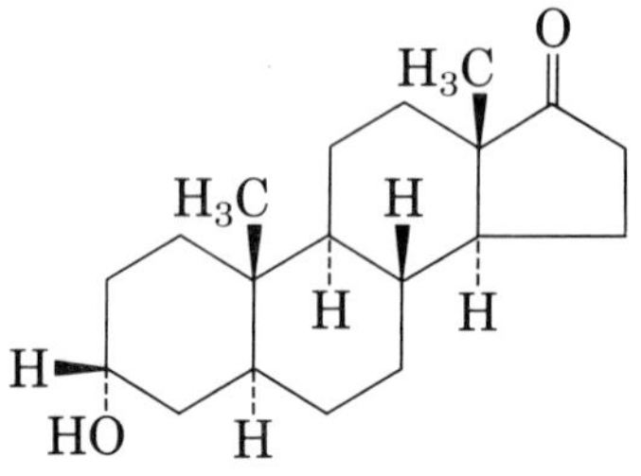

Androsterone (male sex hormone)

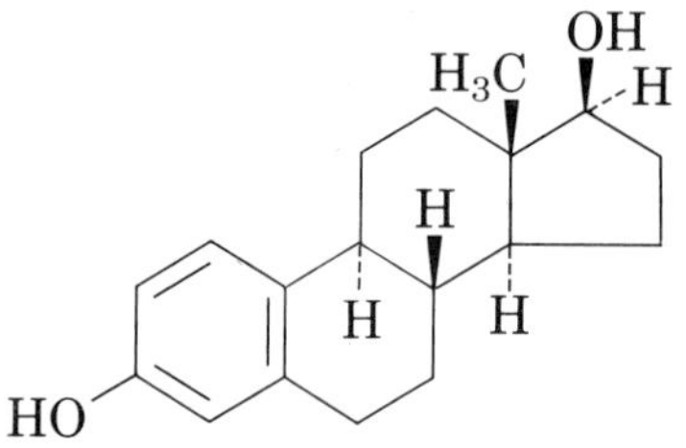

Estradiol (female sex hormone)

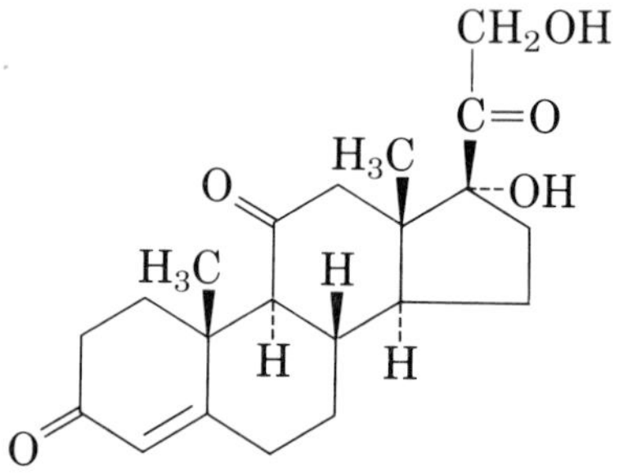

Cortisone (anti-inflammatory drug)

Many other steroids are produced synthetically by pharmaceutical companies. Even such nonnaturally occurring steroids as methandrostenolone (Dianabol, an anabolic or tissue-building steroid) and norethindrone (Norlutin, an oral contraceptive agent) have potent physiological effects.

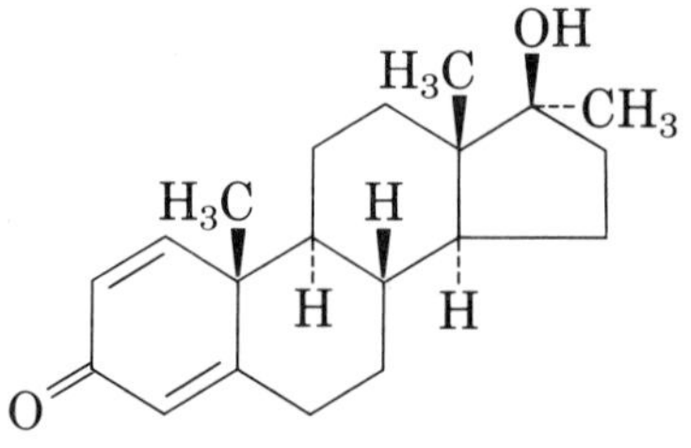

Methandrostenolone (anabolic)

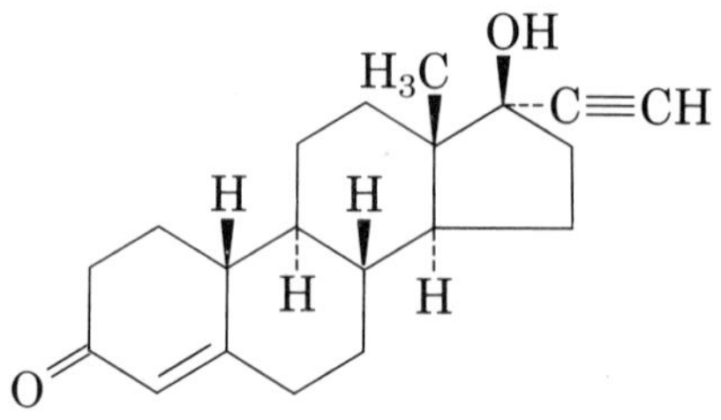

Norethindrone (oral contraceptive)

32.9 Stereochemistry of steroids

The steroid skeleton is composed of four rings fused together with a specific stereochemistry. All three of the six-membered rings (rings A, B, and C) can adopt strain-free chair conformations, as indicated in Figure 32.9. Unlike simple cyclohexane rings, which can undergo chair–chair interconversions (Section 18.11), however, steroids are constrained by a rigid conformation and cannot undergo ring-flips.

An A,B trans steroid

An A,B cis steroid

Figure 32.9. Steroid conformations.

Two cyclohexane rings can be joined in either a cis or a trans manner. In *cis*-decalin, both groups at the ring-junction positions (the *angular* groups) are on the same side. In *trans*-decalin, however, the groups at the ring junctions are on opposite sides. These spatial relationships are best grasped by building molecular models of *cis*- and *trans*-decalin.

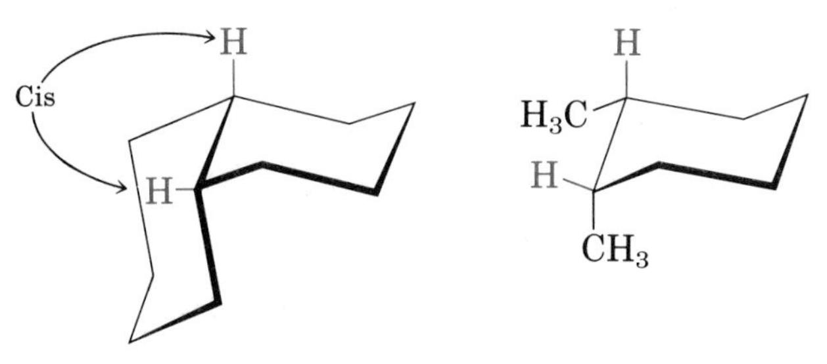

cis-Decalin *cis*-1,2-Dimethylcyclohexane

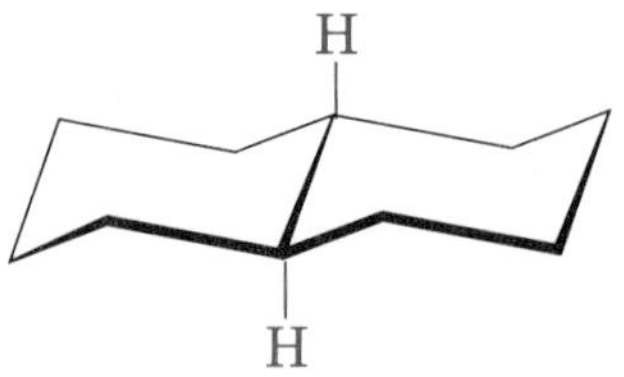

trans-Decalin 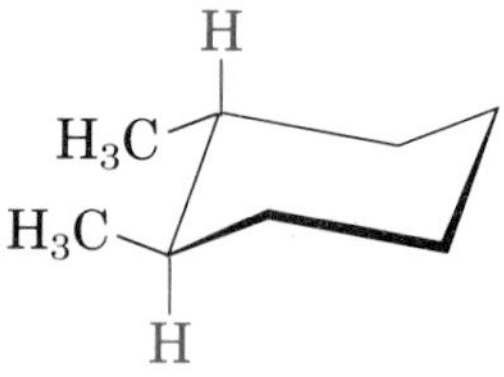 *trans*-1,2-Dimethylcyclohexane

Steroids can have either a cis or a trans fusion of the A and B rings, but the other ring fusions (B–C and C–D) are usually trans (Figure 32.9).

Trans-fused steroids have the C19 angular methyl group "up" (denoted β), and the hydrogen atom at C5 "down" (denoted α) on the opposite side of the molecule. Cis-fused steroids, by contrast, have both the C19 angular methyl group and the C5 hydrogen atom β on the same side of the molecule. Both kinds of steroids are relatively long, flat molecules that have their two methyl groups (C18 and C19) protruding axially above the ring system. The A,B trans-fused steroids are by far the more common, though cis-fused steroids are found in bile.

Substituent groups on the steroid ring system may be either axial or equatorial. As was true for simple cyclohexanes (Section 18.12), equatorial substitution is generally more favorable than axial substitution for steric reasons. Thus, the hydroxyl group at carbon 3 of cholesterol has the more stable equatorial orientation.

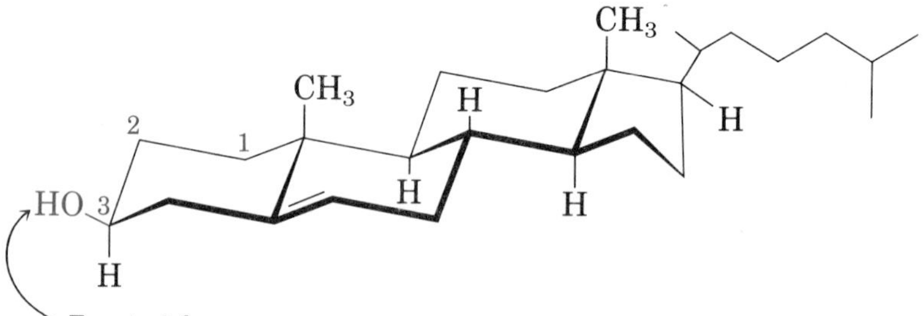

Equatorial

Cholesterol

PROBLEM

32.8 Draw the following molecules in chair conformations, and indicate whether the ring substituents are axial or equatorial.

(a)

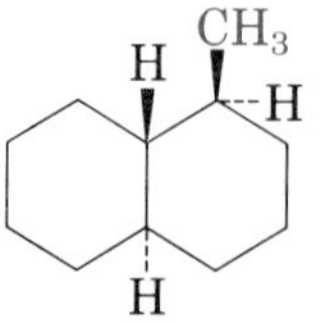

(b)

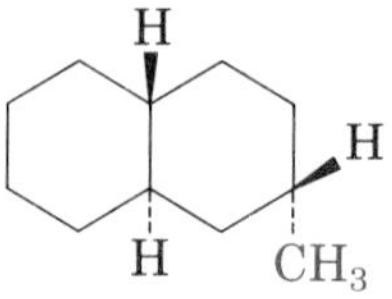

PROBLEM

32.9 Lithocholic acid is a steroid found in human bile. Draw lithocholic acid showing chair conformations, and tell whether the hydroxyl group at C3 is axial or equatorial.

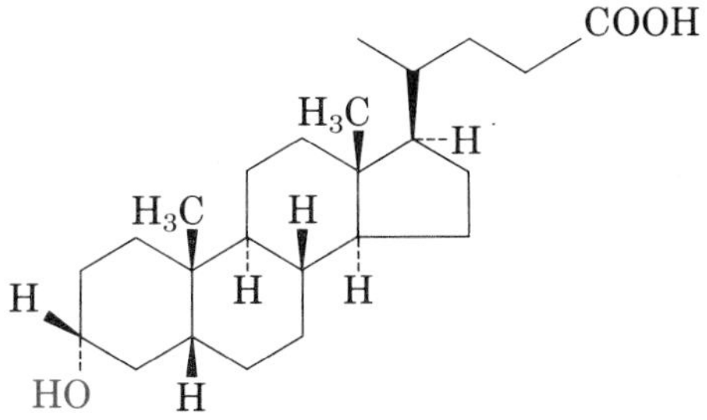

COOH

H_3C H

H_3C H

H H H

HO H

Lithocholic acid

32.10 Steroid biosynthesis

Steroids are closely related to triterpenes and are biosynthesized in living organisms from squalene (Section 32.7). The exact pathway by which this remarkable transformation is accomplished is lengthy and complex, but most of the key steps have now been worked out, with notable contributions made by Konrad Bloch[3] and John Cornforth,[4] who received Nobel prizes for their accomplishments.

In essence, steroid biosynthesis occurs by enzyme-catalyzed epoxidation of squalene followed by acid-catalyzed cyclization and carbocation rearrangement yielding lanosterol. Lanosterol is then degraded by other enzymes to produce cholesterol, which is itself converted by other enzymes to produce a host of different steroids (Figure 32.10, p. 1092).

PROBLEM

32.10 Rearrangement of the tetracyclic cation formed initially by cyclization of squalene oxide ultimately yields lanosterol. Four groups undergo 1,2 migrations, and a proton from C9 is eliminated. Assume that the steps occur in sequence and draw the structure of each intermediate involved.

PROBLEM

32.11 Examine the structures of lanosterol and cholesterol and catalog the changes that have occurred in the transformation.

[3]Konrad E. Bloch (1912–); b. Neisse, Germany; Ph.D. Columbia; professor, University of Chicago, Harvard University; Nobel prize in medicine (1964).

[4]John Warcup Cornforth (1917–); b. Australia; Ph.D. Oxford (Robinson); National Institute of Medical Research (Great Britain); Nobel prize (1975).

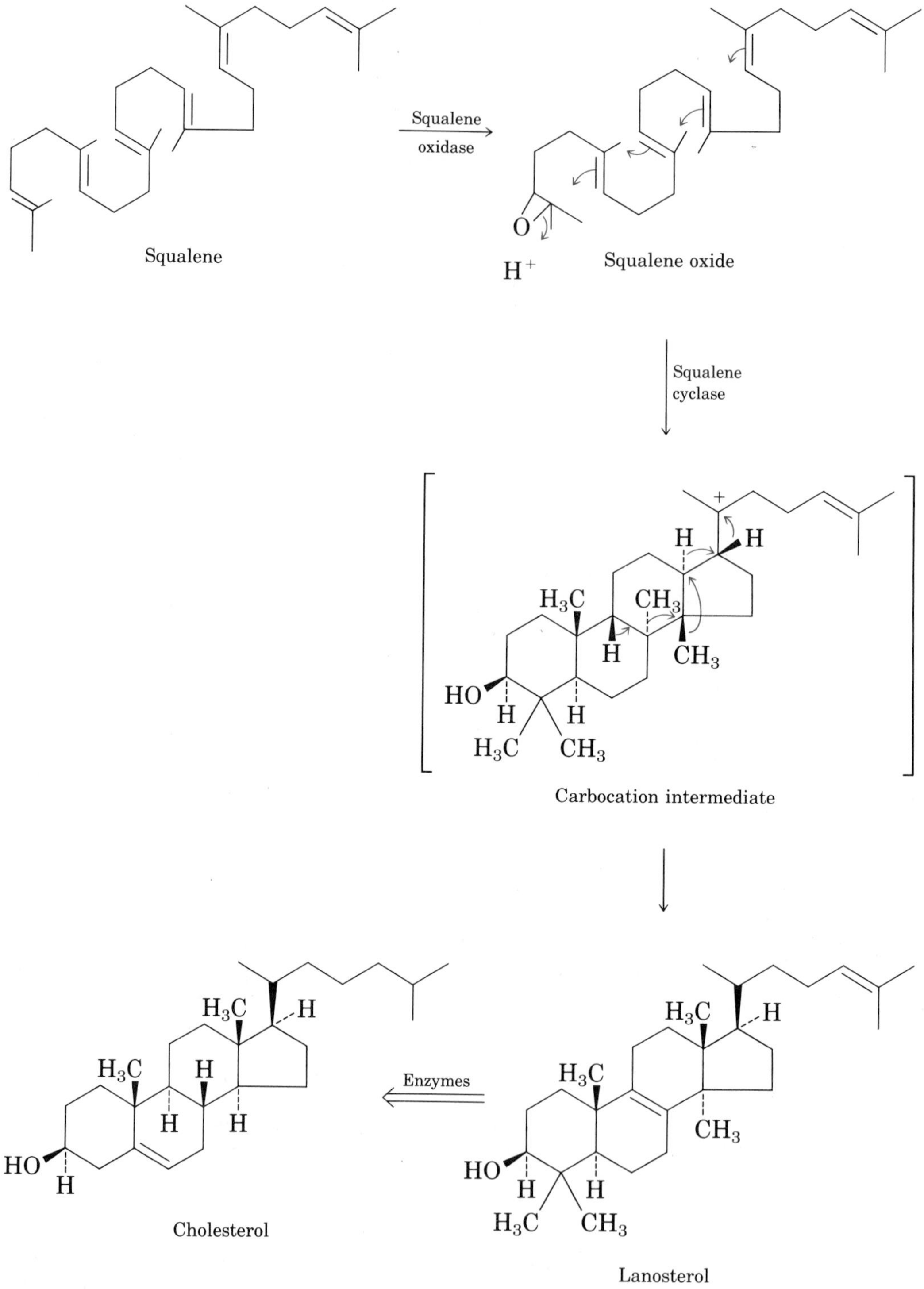

Squalene

Squalene oxidase

H^+

Squalene oxide

Squalene cyclase

+

H H

H_3C CH_3

H CH_3

HO

H H

H_3C CH_3

Carbocation intermediate

H_3C H

H_3C H

H H

HO

H

Cholesterol

Enzymes

H_3C H

H_3C

CH_3

HO

H H

H_3C CH_3

Lanosterol

Figure 32.10. Biosynthesis of cholesterol from squalene.

32.11 Summary

Lipids are the naturally occurring materials isolated from plant and animal cells by extraction with organic solvents. They usually have large hydrocarbon portions in their structure. Animal fats and vegetable oils are the most widely occurring lipids. Fats and oils are chemically similar; both are triesters of glycerol with long-chain fatty acids, but animal fats usually are saturated, whereas vegetable oils usually have unsaturated fatty acid residues.

Phosphoglycerides such as lecithin and cephalin are closely related to fats; the glycerol backbone in these molecules is esterified to two fatty acids (one saturated and one unsaturated) and to one phosphate ester. Sphingolipids, another major class of phospholipids, have an amino alcohol such as sphingosine for their backbone; they are important constituents of cell membranes.

Fatty acids are biosynthesized in nature by condensation of enzyme-bound two-carbon acetate units. The overall scheme is lengthy, but the individual steps are the fundamental organic reactions expected of the functional groups involved.

Prostaglandins and terpenes are still other classes of lipids. Prostaglandins, simple lipids found in all body tissues, have a wide range of physiological actions. Terpenes are often isolated from the essential oils of plants. They have an immense diversity of structure and are produced biosynthetically by head-to-tail coupling of the five-carbon "isoprene equivalents," isopentenyl pyrophosphate and dimethylallyl pyrophosphate.

Steroids are plant and animal lipids with a characteristic tetracyclic carbon skeleton. Like the prostaglandins, steroids occur widely in body tissue and have a large variety of physiological activities. Steroids are closely related to terpenes and arise biosynthetically from the triterpene precursor, lanosterol. Lanosterol, in turn, arises from cyclization of the acyclic hydrocarbon, squalene.

ADDITIONAL PROBLEMS

32.12 Fats can be either optically active or optically inactive, depending on their structure. Draw the structure of an optically active fat that yields 2 equiv of stearic acid and 1 equiv of oleic acid on hydrolysis. Draw the structure of an optically inactive fat that yields the same products.

32.13 Show the products you would expect to obtain from reaction of glyceryl trioleate with the following:

(a) Excess Br_2 in CCl_4
(b) H_2/Pd
(c) $NaOH/H_2O$
(d) O_3, then Zn/CH_3COOH
(e) $LiAlH_4$, then H_3O^+
(f) CH_3MgBr, then H_3O^+
(g) $NaOH/H_2O$, then CH_2N_2 (diazomethane)

32.14 Vaccenic acid, $C_{18}H_{34}O_2$, is a rare fatty acid that gives heptanal and 11-oxoundecanoic acid [$OHC(CH_2)_9COOH$] on ozonolysis followed by zinc treatment. When allowed to react with CH_2I_2/Zn–Cu, vaccenic acid is converted into lactobacillic acid. What are the structures of vaccenic and lactobacillic acids?

32.15 How would you convert oleic acid into these substances?

(a) Methyl oleate
(b) Methyl stearate
(c) Nonanal
(d) Nonanedioic acid
(e) 9-Octadecynoic acid (stearolic acid)
(f) 2-Bromostearic acid
(g) 18-Pentatriacontanone $[CH_3(CH_2)_{16}C(=O)(CH_2)_{16}CH_3]$

32.16 How would you synthesize stearolic acid [Problem 32.15(e)] from 1-decyne and 1-chloro-7-iodoheptane?

32.17 Show the location of the isoprene units in the following terpenes:

(a) 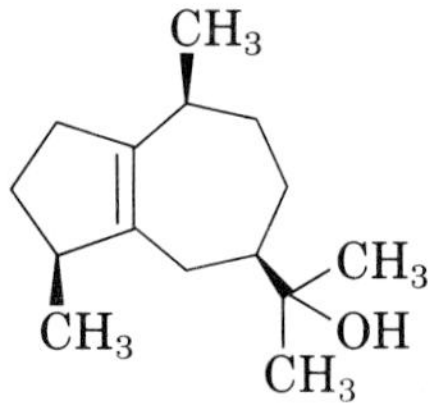 Guaiol

(b) 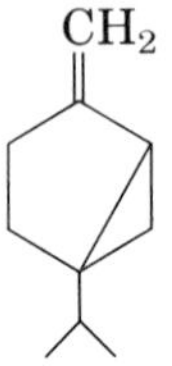 Sabinene

(c) 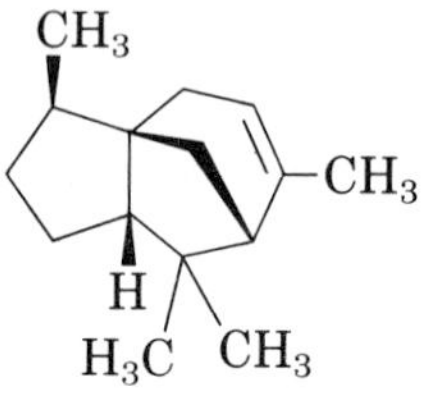 Cedrene

32.18 Indicate by asterisks the chiral centers present in each of the three terpenes shown in Problem 32.17. How many stereoisomers of each are theoretically possible?

32.19 Assume that the three terpenes in Problem 32.17 were derived biosynthetically from isopentenyl pyrophosphate and dimethylallyl pyrophosphate, each of which was isotopically labeled at the pyrophosphate-bearing carbon atom (carbon 1). At what positions would the terpenes be isotopically labeled?

32.20 Suggest a mechanistic pathway by which α-pinene might arise biosynthetically from geraniol pyrophosphate.

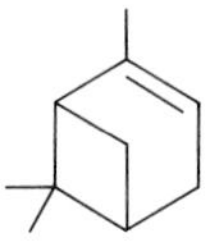

α-Pinene

32.21 Suggest a mechanism by which ψ-ionone is transformed into β-ionone on treatment with acid.

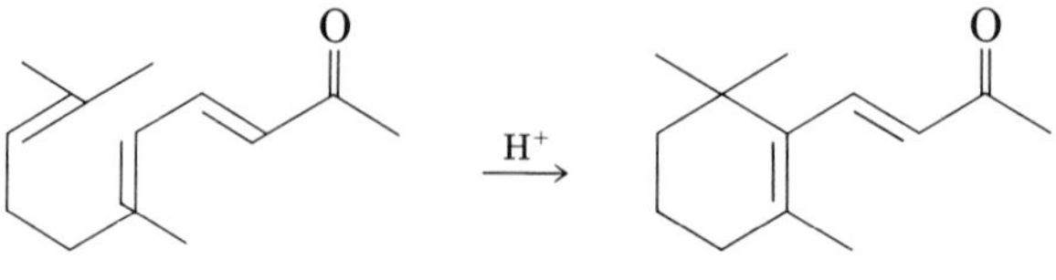

ψ-Ionone $\xrightarrow{H^+}$ β-Ionone

32.22 Which isomer would you expect to be more stable, *cis*-decalin or *trans*-decalin? Explain your answer.

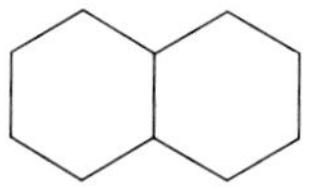

Decalin

32.23 Draw the most stable conformation of dihydrocarvone.

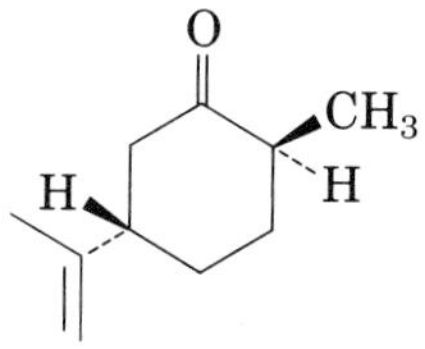

O
CH_3
H
H

Dihydrocarvone

32.24 Draw the most stable conformation of menthol and label each substituent as axial or equatorial.

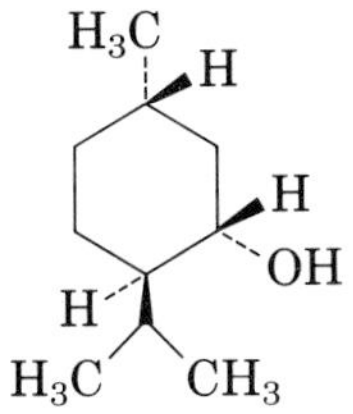

H_3C
H
H
OH
H
H_3C CH_3

Menthol (from peppermint oil)

32.25 Cholic acid, a major steroidal constituent of human bile, has the structure shown. Draw a conformational structure of cholic acid and label the three hydroxyl groups as axial or equatorial.

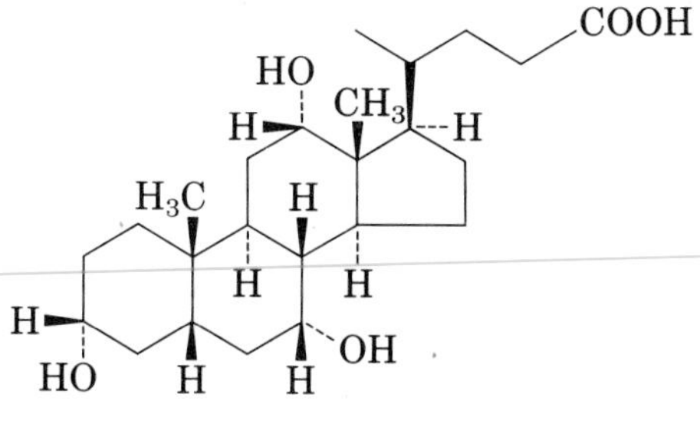

COOH
HO
CH_3
H H
H_3C H
H H
H
OH
HO H H

Cholic acid

32.26 How many chiral centers does cholic acid have? How many stereoisomers are possible?

32.27 Show the products you would expect to obtain from reaction of cholic acid with these reagents:
(a) C_2H_5OH, H^+
(b) Excess pyridinium chlorochromate in CH_2Cl_2
(c) BH_3 in THF, then H_3O^+

32.28 As a general rule, equatorial alcohols are esterified more readily than axial alcohols. What product would you expect to obtain from reaction of these two compounds with 1 equiv of acetic anhydride?

(a) HO H CH$_3$ OH H

H

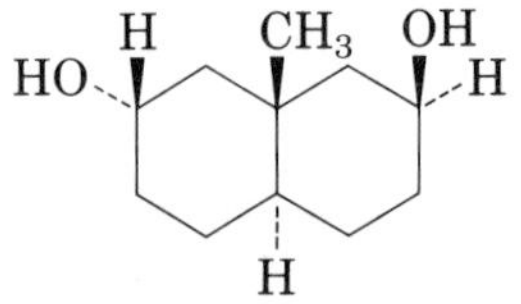

(b) Methyl cholate (see Problem 32.25)

32.29 Diethylstilbestrol (DES) exhibits estradiol-like activity even though it is structurally unrelated to steroids. Once used widely as an additive in animal feed, DES has been implicated as a causative agent in several types of cancer. Look up the structure of estradiol (Section 32.8) and show how DES can be drawn so that it is sterically similar to estradiol.

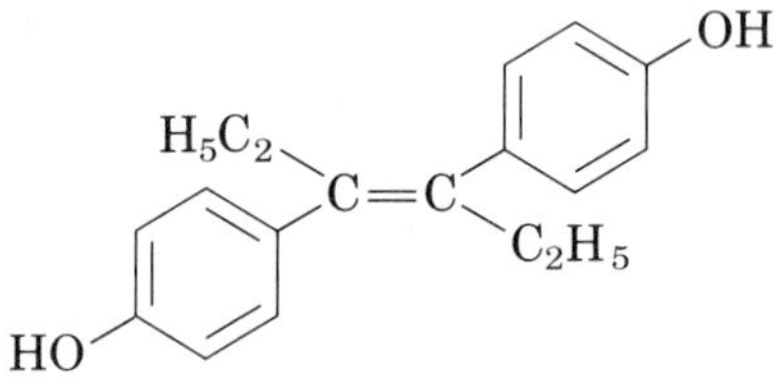

OH H$_5$C$_2$ C=C C$_2$H$_5$ HO

Diethylstilbestrol

32.30 Propose a synthesis of diethylstilbestrol (Problem 32.29) from phenol and any other organic compound required.

32.31 Cembrene, $C_{20}H_{32}$, is a diterpene hydrocarbon isolated from pine resin. Cembrene has an ultraviolet absorption at 245 nm, but dihydrocembrene ($C_{20}H_{34}$), the product of hydrogenation with 1 equiv of hydrogen, has no ultraviolet absorption. On exhaustive hydrogenation, 4 equiv of hydrogen react, and octahydrocembrene, $C_{20}H_{40}$, is produced. On ozonolysis of cembrene, followed by treatment of the ozonide with zinc, four carbonyl-containing products are obtained:

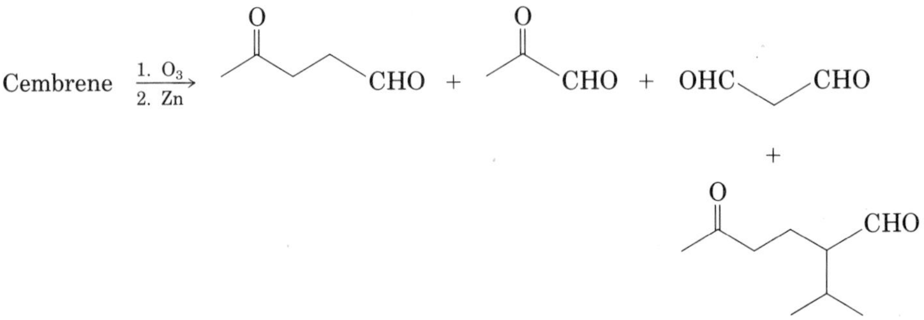

Cembrene $\xrightarrow[2.\ Zn]{1.\ O_3}$ O CHO + O CHO + OHC CHO + O CHO

Propose a suitable structure for cembrene that is consistent with the isoprene rule.

CHAPTER 33
HETEROCYCLES AND NUCLEIC ACIDS

Cyclic organic compounds are classified either as **carbocycles** or as **heterocycles.** Carbocyclic rings contain only carbon atoms, but heterocyclic rings contain one or more different atoms in addition to carbon. Nitrogen, oxygen, and sulfur are the most common heteroatoms, but many others are also found.

Heterocyclic compounds are common in organic chemistry, and many heterocycles have important biological properties. For example, the antibiotic penicillin, the antiulcer agent cimetidine, the sedative phenobarbital, and the nonnutritive sweetener saccharin are all heterocycles.

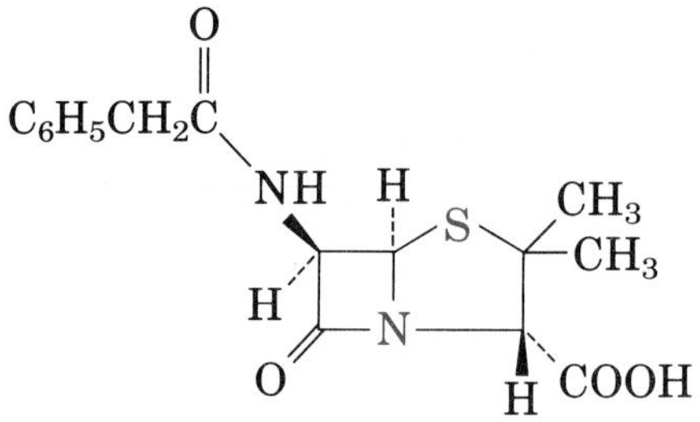

Penicillin G
(an antibiotic)

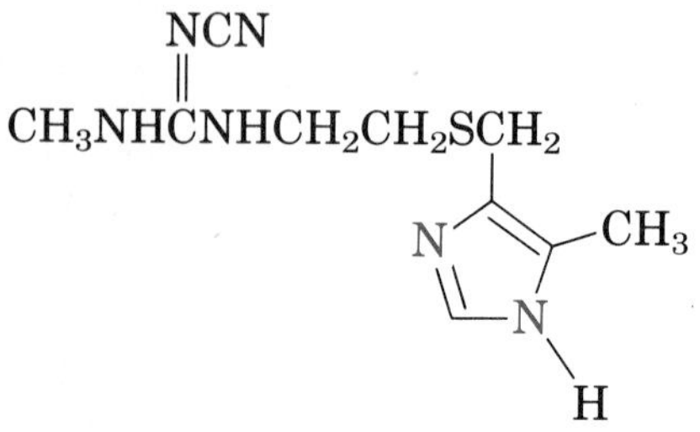

Cimetidine
(an antiulcer agent)

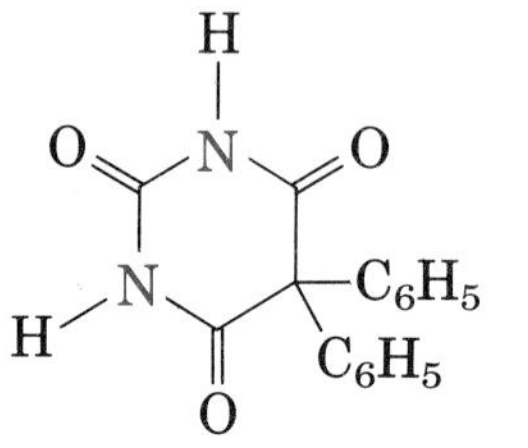

Phenobarbital
(a sedative)

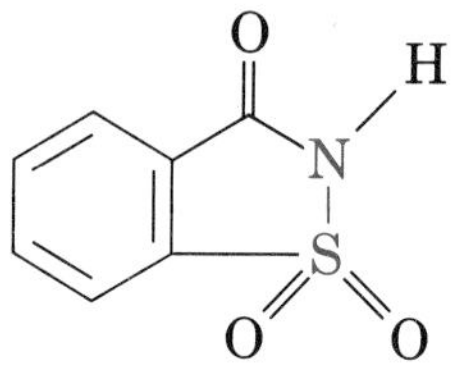

Saccharin
(an artificial sweetener)

Heterocycles are not new to us; we have encountered them many times in previous chapters, usually without comment. Thus, epoxides (three-membered-ring ethers, Section 19.7), lactones (cyclic esters), and lactams (cyclic amides) are heterocycles, as are the solvents tetrahydrofuran (a cyclic ether) and pyridine (a cyclic amine). In addition, most carbohydrates exist as heterocyclic hemiacetals (Section 27.5).

Most heterocycles have the same chemistry as their open-chain counterparts; lactones and acyclic esters behave similarly, lactams and acyclic amides behave similarly, and cyclic and acyclic ethers behave similarly. In certain cases, however, particularly when the ring is unsaturated, heterocycles have unique and interesting properties. Let's look first at the five-membered unsaturated heterocycles.

33.1 Five-membered unsaturated heterocycles

Pyrrole, furan, and thiophene are the simplest five-membered unsaturated heterocycles. Each of the three has two double bonds and one heteroatom (N, O, or S).

Pyrrole   Furan   Thiophene

Pyrrole is obtained commercially either by distillation of coal tar or by treatment of furan with ammonia over an alumina catalyst at 400°C.

$$\text{Furan} \xrightarrow[Al_2O_3,\ 400°C]{NH_3,\ H_2O} \text{Pyrrole}$$

Furan is synthesized by catalytic loss of carbon monoxide (decarbonylation) from furfural, which is itself prepared by acidic dehydration of the pentoses found in oat hulls and corncobs.

$$\underset{\text{Pentose mixture}}{C_5H_{10}O_5} \xrightarrow{H_3O^+} \underset{\text{Furfural}}{\text{CHO}} \xrightarrow[280°C]{\text{Ni catalyst}} \underset{\text{Furan}}{} + CO$$

Thiophene is found in small amounts in coal-tar distillates and is synthesized industrially by cyclization of butane or butadiene with sulfur at 600°C.

$$\underset{\text{1,3-Butadiene}}{CH_2{=}CH{-}CH{=}CH_2} \xrightarrow[600°C]{S} \underset{\text{Thiophene}}{} + H_2S$$

All three heterocycles are liquid at room temperature, and their physical properties are given in Table 33.1. The chemistry of these three heterocyclic ring systems contains some surprises. For example, pyrrole is both an amine and a conjugated diene, yet its chemical properties are not

TABLE 33.1 Physical properties of five-membered heterocycles

Name	*Molecular weight*	*Melting point (°C)*	*Boiling point (°C)*
Furan	68	−85	31
Pyrrole	67	−23	130
Thiophene	84	−38	

consistent with either of these structural features. Unlike most other amines, pyrrole is not basic; unlike most other conjugated dienes, pyrrole undergoes electrophilic substitution rather than addition reactions. The same is true of furan and thiophene; both tend to react with electrophiles to give substitution products.

How can we explain these observations?

33.2 Structures of pyrrole, furan, and thiophene

In fact, pyrrole, furan, and thiophene are *aromatic*. Each has six pi electrons ($4n + 2$, where $n = 1$) in a cyclic conjugated system. To choose pyrrole as an example, each of the four carbon atoms of pyrrole contributes one pi electron, and the sp^2-hybridized nitrogen atom contributes two (its lone pair). The six pi electrons occupy *p* orbitals, with lobes above and below the plane of the ring, as shown in Figure 33.1. Overlap of the five *p* orbitals forms aromatic molecular orbitals just as in benzene (Section 14.8).

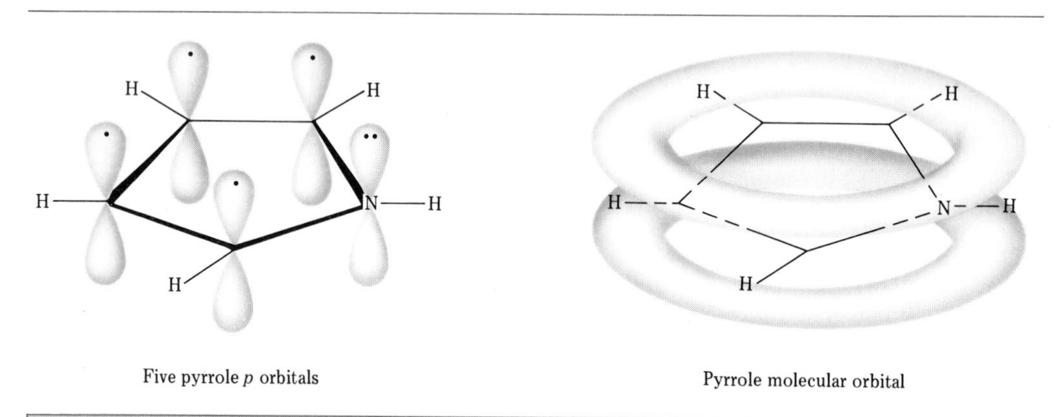

Figure 33.1. Pi bonds in pyrrole.

Note that the pyrrole nitrogen atom uses all five of its valence electrons in bonding. Three electrons are used in forming three sigma bonds (two to carbon and one to hydrogen), and the two lone-pair electrons are involved in aromatic pi bonding, as the following resonance structures indicate:

Because the nitrogen lone pair is a part of the aromatic sextet, it is not available for bonding to electrophiles. Pyrrole is therefore much less basic ($pK_b \approx 14$) and much less nucleophilic than aliphatic amines. By the same token, however, the carbon atoms of pyrrole are much more electron-rich and much more nucleophilic than are typical double-bond carbon atoms. The pyrrole ring is therefore highly reactive toward electrophiles in the same way that activated benzene rings are reactive.

PROBLEM

33.1 Draw an orbital picture of furan. Assume that the oxygen atom is sp^2 hybridized and show the orbitals that the two oxygen lone pairs are occupying.

PROBLEM

33.2 Pyrrole has a substantial dipole moment of 1.8 D. Which is the positive end and which is the negative end of the dipole? Explain your answer.

33.3 Electrophilic substitution reactions of pyrrole, furan, and thiophene

The chemistry of pyrrole, furan, and thiophene is similar to that of activated benzenoid aromatic rings. Like benzene, the five-membered aromatic heterocycles undergo electrophilic substitution rather than addition reactions. In general, these heterocycles are much more reactive toward electrophiles than are benzene rings, and mild, low-temperature reaction conditions are often used to prevent destruction of the starting material. Halogenation, nitration, sulfonation, and Friedel–Crafts acylation can all be accomplished if the proper reaction conditions are chosen; a reactivity order of furan > pyrrole > thiophene is normally found.

Bromination: Furan + Br_2 $\xrightarrow[0°C]{Dioxane}$ 2-Bromofuran (90%) + HBr

Nitration: Pyrrole + HNO_3 $\xrightarrow{Acetic\ anhydride}$ 2-Nitropyrrole (83%) + H_2O

Friedel–Crafts acylation: Thiophene + CH_3COCl $\xrightarrow[SnCl_4]{Benzene}$ 2-Acetylthiophene (83%) + HCl

Electrophilic substitution of these aromatic heterocycles normally occurs at C2, the position next to the heteroatom, because C2 is the most electron-rich (most nucleophilic) position on the ring. Another way of saying the same thing is to note that electrophilic attack at C2 leads to a more stable intermediate cation, with three resonance forms, than does attack at C3, which results in a cation with two resonance forms (Figure 33.2).

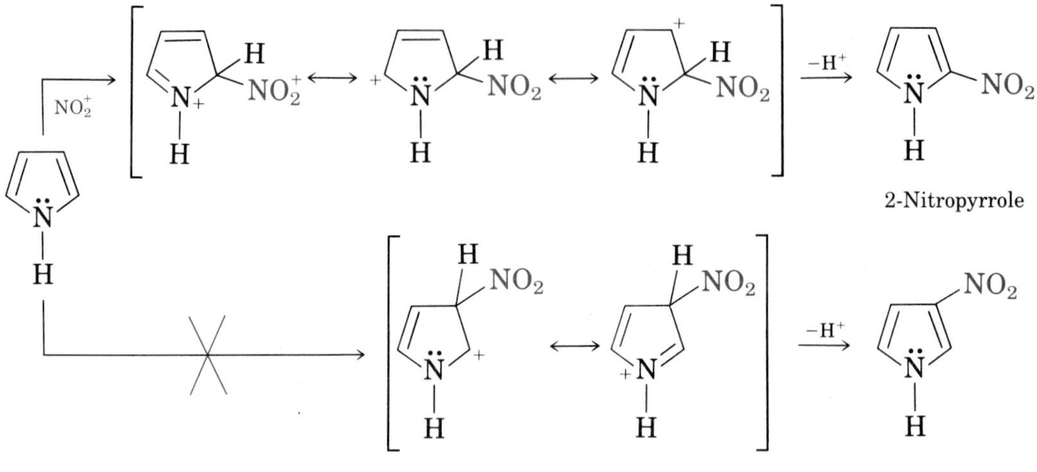

NO_2^+

2-Nitropyrrole

3-Nitropyrrole
(not formed)

Figure 33.2. Electrophilic nitration of pyrrole. The reaction takes place at C2, not at C3.

PROBLEM

33.3 Propose a mechanism to account for the fact that treatment of pyrrole with deuteriosulfuric acid, D_2SO_4, leads to formation of 2-deuteriopyrrole.

33.4 Pyridine, a six-membered heterocycle

Pyridine is the nitrogen-containing heterocyclic analog of benzene. It is obtained commercially by distillation of coal tar. Like benzene, pyridine is aromatic. It is a flat molecule with bond angles of 120° and with carbon–carbon bond lengths of 1.39 Å, intermediate between normal single and double bonds. Each of the five carbon atoms contributes one pi electron, and the sp^2-hybridized nitrogen atom also contributes one pi electron to complete the aromatic sextet. Unlike the situation in pyrrole, however, the lone pair of electrons on the pyridine nitrogen atom is not involved in bonding but occupies an sp^2 orbital in the plane of the ring (Figure 33.3).

Pyridine is a stronger base than pyrrole (pK_b = 8.75 versus 14 for pyrrole) because of its electronic structure. The pyridine nitrogen's lone-pair

Figure 33.3. Electronic structure of pyridine.

electrons are not involved in aromatic pi bonding but are instead available for donation to a Lewis acid. Pyridine is a weaker base than aliphatic amines, however, (pK_b = 8.75 versus 3) because of its hybridization. The lone-pair electrons in the sp^3-hybridized nitrogen orbital ($\frac{1}{4}$ *s* character) of an aliphatic amine are held less closely to the nucleus than are lone-pair electrons in an sp^2 hybrid orbital ($\frac{1}{3}$ *s* character), with the result that sp^3-hybridized nitrogen is more basic.

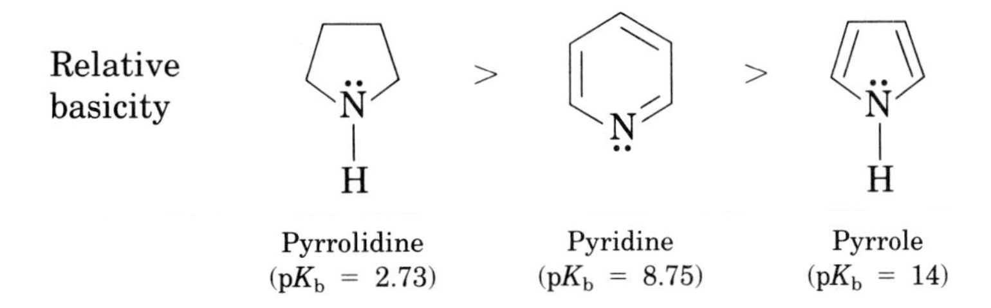

PROBLEM

33.4 Imidazole has $pK_b = 7.05$. Draw an orbital structure of imidazole and indicate which nitrogen is more basic.

Imidazole

33.5 Electrophilic substitution of pyridine

The pyridine ring undergoes electrophilic aromatic substitution reactions only with great difficulty. Halogenation and sulfonation can be carried out under drastic conditions, but nitration occurs in very low yield, and Friedel–Crafts reactions are not successful. Reactions usually give the 3-substituted product, as indicated at the top of the next page.

Br_2, 300°C → 3-Bromopyridine (30%)

SO_3, $HgSO_4$, H_2SO_4, 220°C → 3-Pyridinesulfonic acid (70%)

HNO_3, $NaNO_3$, 370°C → 3-Nitropyridine (5%)

The low reactivity of pyridine toward electrophilic aromatic substitution is due to a combination of factors. Most important is that the electron density of the ring is decreased by the electron-withdrawing inductive effect of the electronegative nitrogen atom. Pyridine has a dipole moment of 2.26 D, with the ring acting as the positive end of the dipole. Electrophilic attack on the positively polarized ring is therefore difficult.

μ = 2.26 D

A second factor is that acid–base complexation between the basic ring nitrogen atom and the attacking electrophile further serves to deactivate the ring.

PROBLEM

33.5 Electrophilic aromatic substitution reactions of pyridine normally occur at C3. Draw the intermediate cations resulting from electrophilic attack at all possible positions, and explain the observed result.

33.6 Nucleophilic substitution of pyridine

In contrast to a lack of reactivity toward electrophilic substitution, certain pyridines undergo nucleophilic aromatic substitution with relative ease. Both 2- and 4-halo-substituted (but not 3-substituted) pyridines react particularly well.

4-Chloropyridine $\xrightarrow[C_2H_5OH]{Na^+ \ ^-:OC_2H_5}$ 4-Ethoxypyridine (75%) + $:Cl^-$

2-Bromopyridine $\xrightarrow[NH_3]{NaNH_2}$ 2-Aminopyridine (67%) + $:Br^-$

These reactions are typical nucleophilic aromatic substitutions, analogous to those we saw earlier (Section 15.13) for halo-substituted benzenes. A benzene ring needs to be further activated by the presence of electron-withdrawing substituents for nucleophilic substitution to occur, but pyridines are already sufficiently activated. Reaction occurs by addition of the nucleophile to the C=N bond, followed by loss of halide ion from the anion intermediate.

2-Chloropyridine + $\ddot{N}H_2$ $\longrightarrow$ [anion intermediate] $\longrightarrow$ 2-Aminopyridine + $:Cl^-$

This nucleophilic aromatic substitution is in some ways analogous to the nucleophilic acyl substitution of an acid chloride (Section 24.5). In both cases, the initial addition step is favored by the ability of the electronegative atom (nitrogen or oxygen) to stabilize the anion intermediate:

2-Chloropyridine $\xrightarrow{^-:NH_2}$ Stabilized anion $\longrightarrow$ 2-Aminopyridine

Acid chloride $\xrightarrow{^-:NH_2}$ Stabilized anion $\longrightarrow$ Amide

PROBLEM

33.6 Draw the anion intermediates expected from nucleophilic attack at C4 of a 4-halopyridine and at C3 of a 3-halopyridine. How can you account for the fact that substitution of the 4-halopyridine occurs readily, but the 3-halopyridine does not react?

PROBLEM

33.7 If 3-bromopyridine is heated with $NaNH_2$ under forcing conditions, a mixture of 3- and 4-aminopyridine is obtained. Explain. [*Hint:* See Section 15.14.]

33.7 Fused-ring heterocycles

Quinoline, isoquinoline, and indole are **fused-ring heterocycles** that contain both a benzene ring and a heterocyclic aromatic ring. All three ring systems occur commonly among natural products, and many members of the class have pronounced biological activity. Thus, the quinoline alkaloid quinine is widely used as an antimalarial drug and the indole alkaloid *N,N*-dimethyltryptamine is a powerful hallucinogen.

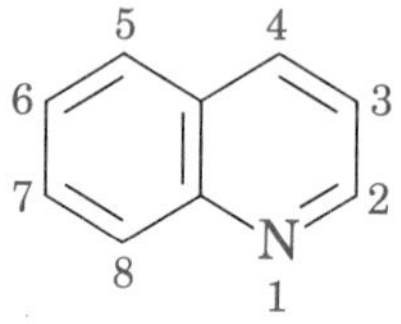

Quinoline 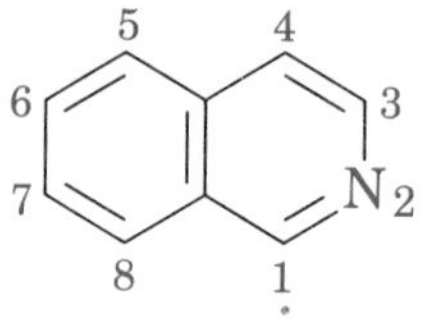   Isoquinoline 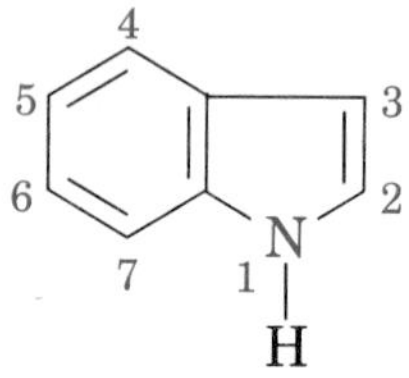   Indole

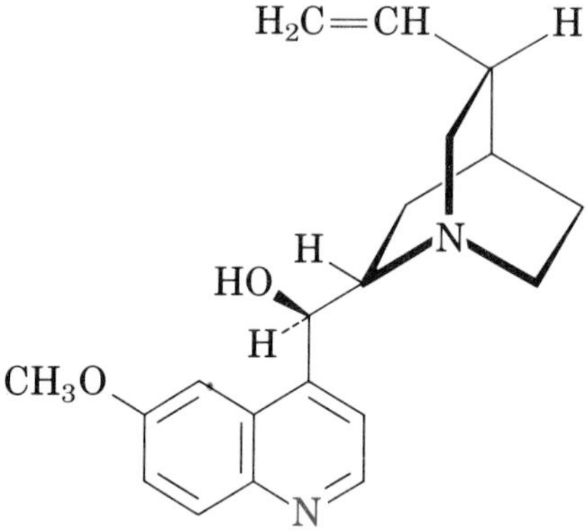

Quinine, an antimalarial drug (a quinoline alkaloid)

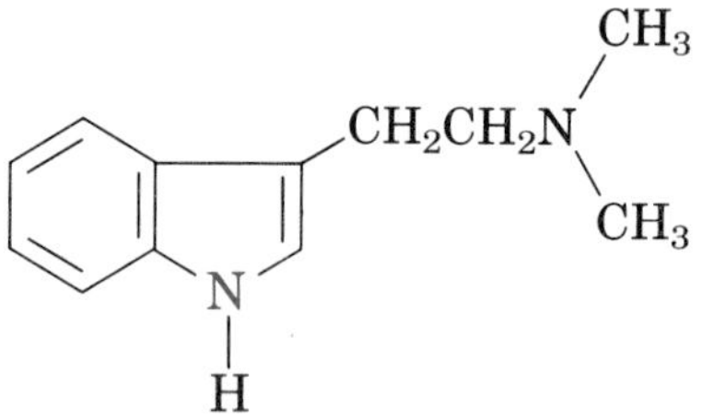

N,N-Dimethyltryptamine, a hallucinogen (an indole alkaloid)

The chemistry of these three classes of fused-ring heterocycles is just what we might expect from our knowledge of the simpler heterocycles, pyridine and pyrrole. All three undergo electrophilic aromatic substitution reactions. Quinoline and isoquinoline both undergo electrophilic substitution more easily than pyridine but less easily than benzene, consistent with our previous observation (Section 33.5) that pyridine rings are deactivated compared to benzene. Note that reaction occurs on the *benzene* ring (not the pyridine ring) and that a mixture of C5 and C8 substitution products are obtained.

Quinoline $\xrightarrow[H_2SO_4]{Br_2}$ 5-Bromoquinoline + 8-Bromoquinoline

A 51:49 ratio

Isoquinoline $\xrightarrow[H_2SO_4,\ 0°C]{HNO_3}$ 5-Nitroisoquinoline + 8-Nitroisoquinoline

A 90:10 ratio

Indole undergoes electrophilic substitution more easily than benzene but less easily than pyrrole. Again, this is consistent with our previous statement that pyrrole rings are more strongly activated than benzene rings. Substitution occurs at C3 of the electron-rich pyrrole ring, rather than on the benzene ring; reaction conditions must be chosen carefully to avoid destructive side reactions.

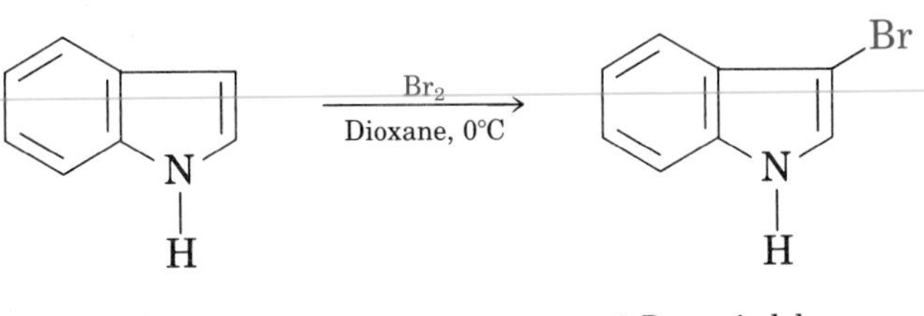

Indole 3-Bromoindole

PROBLEM

33.8 Which nitrogen atom in *N,N*-dimethyltryptamine is more basic? Explain.

PROBLEM

33.9 Indole reacts with electrophiles at C3 rather than at C2. Draw resonance forms of the intermediate cations resulting from attack at C2 and C3 and explain the observed results.

33.8 Pyrimidine and purine

The most important heterocyclic ring systems from a biological viewpoint are **pyrimidine** and **purine.** Pyrimidine contains two nitrogens in a six-membered aromatic ring, and purine has four nitrogens in a fused-ring structure.

Pyrimidine

Purine

Both heterocycles are essential components of the last major class of biomolecules we will consider—the nucleic acids.

33.9 Nucleic acids and nucleotides

The nucleic acids—**deoxyribonucleic acid (DNA)** and **ribonucleic acid (RNA)**—are the chemical carriers of a cell's genetic information. Coded in a cell's DNA is all the information that determines the nature of the cell, controls cell growth and division, and directs biosynthesis of the enzymes and other proteins required for all cellular functions.

Like proteins, nucleic acids are polymers. Mild enzyme-catalyzed hydrolysis cleaves a nucleic acid into its monomeric building blocks called **nucleotides.** Each nucleotide can be further cleaved by enzyme-catalyzed hydrolysis to give a **nucleoside** plus phosphoric acid, H_3PO_4, and each nucleoside can be hydrolyzed to yield a simple pentose sugar plus a heterocyclic purine or pyrimidine base.

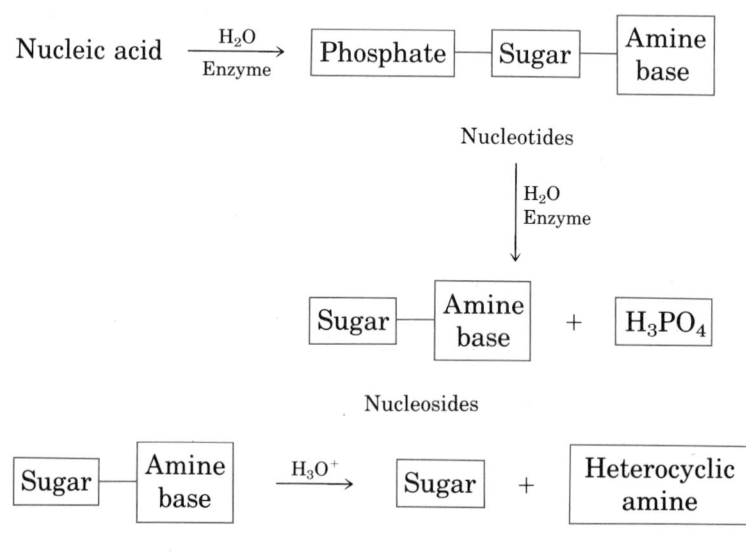

The sugar component in RNA is **ribose,** and the sugar in DNA is 2′-deoxyribose. (The expression "2′-deoxy" indicates that oxygen is missing from the 2′ position; numbers with a prime superscript refer to positions on the sugar component of a nucleotide, and numbers without a prime refer to positions on the heterocyclic amine base.)

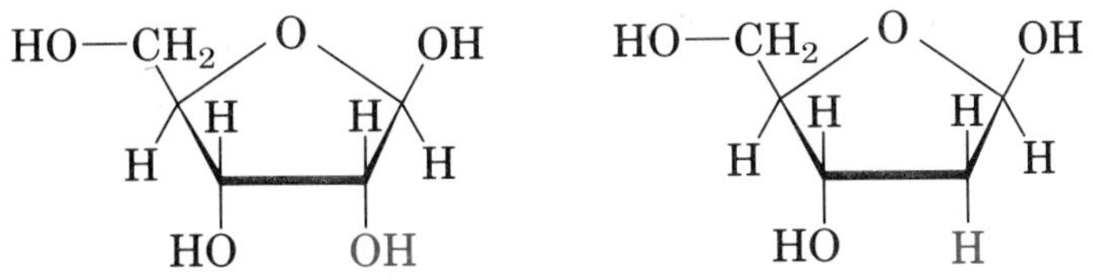

Ribose

2′-Deoxyribose

Four different heterocyclic amine bases are found in deoxyribonucleotides; two are purines (**adenine** and **guanine**) and two are pyrimidines (**cytosine** and **thymine**). Adenine, guanine, and cytosine also occur in ribonucleotides, but thymine is replaced by a different pyrimidine base called **uracil.**

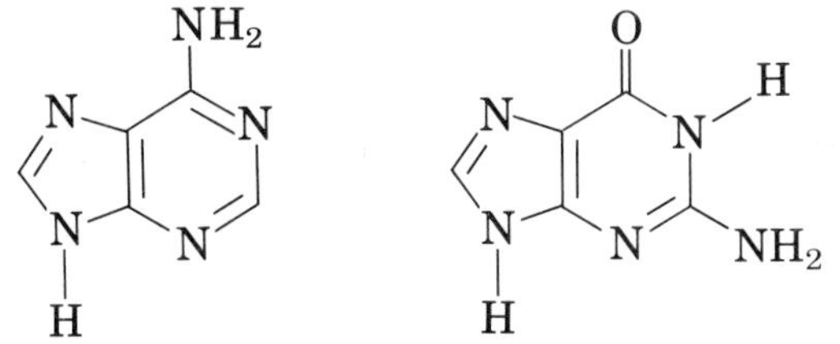

Adenine

Guanine

Purines

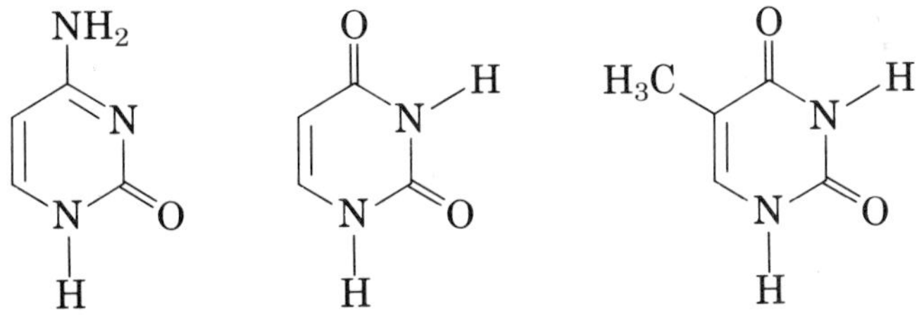

Cytosine

Uracil (RNA)

Thymine (DNA)

Pyrimidines

In both DNA and RNA, the heterocyclic amine base is bound at C1′ of the sugar, and phosphoric acid is bound by a phosphate ester linkage to the C5′ sugar position. Thus, nucleosides and nucleotides have the structure shown in Figure 33.4 (p. 1110).

The names of the bases, nucleotides, and corresponding nucleosides are given in Table 33.2 (p. 1110), and the complete structures of all four deoxyribonucleotides and all four ribonucleotides are shown in Figure 33.5 (p. 1111).

DNA and RNA, though chemically similar, are different in size and have different roles within the cell. Molecules of DNA are enormous; they have molecular weights of up to 1 trillion and are found mostly in the nucleus of the cell. Molecules of RNA, by contrast, are much smaller (as low

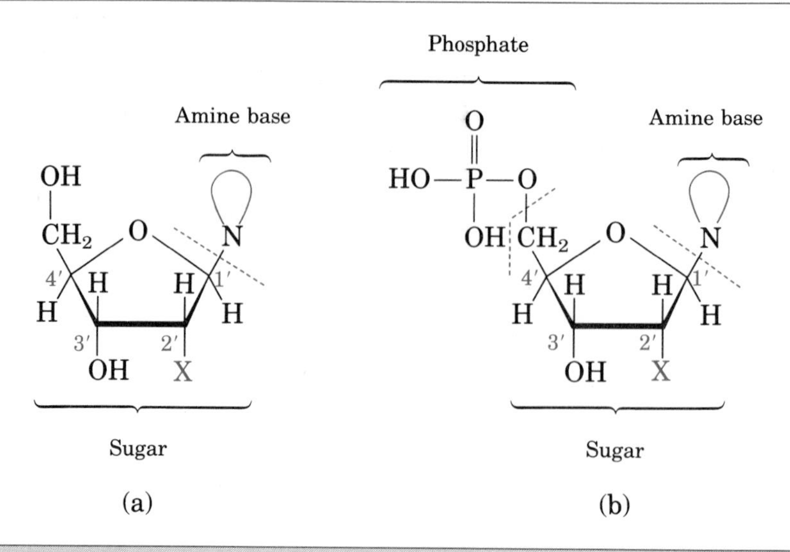

Figure 33.4. (a) A nucleoside, and (b) a nucleotide. When X = H, the sugar is deoxyribose; when X = OH, the sugar is ribose.

TABLE 33.2 Names of bases, nucleosides, and nucleotides

Heterocyclic base	*Source*	*Nucleoside*	*Nucleotide*
Adenine	RNA	Adenosine	Adenosine 5′-phosphate
	DNA	2′-Deoxyadenosine	2′-Deoxyadenosine 5′-phosphate
Guanine	RNA	Guanosine	Guanosine 5′-phosphate
	DNA	2′-Deoxyguanosine	2′-Deoxyguanosine 5′-phosphate
Cytosine	RNA	Cytidine	Cytidine 5′-phosphate
	DNA	2′-Deoxycytidine	2′-Deoxycytidine 5′-phosphate
Uracil	RNA	Uridine	Uridine 5′-phosphate
Thymine	DNA	2′-Deoxythymidine	2′-Deoxythymidine 5′-phosphate

as 35,000 mol wt) and are found mostly outside the cell nucleus. We will consider the two kinds of nucleic acids separately, beginning with DNA.

PROBLEM

33.10 2′-Deoxythymidine exists largely in the lactam form rather than in the tautomeric lactim form. Can you suggest a reason for this?

Lactam form ⇄ Lactim form

Deoxyribonucleotides

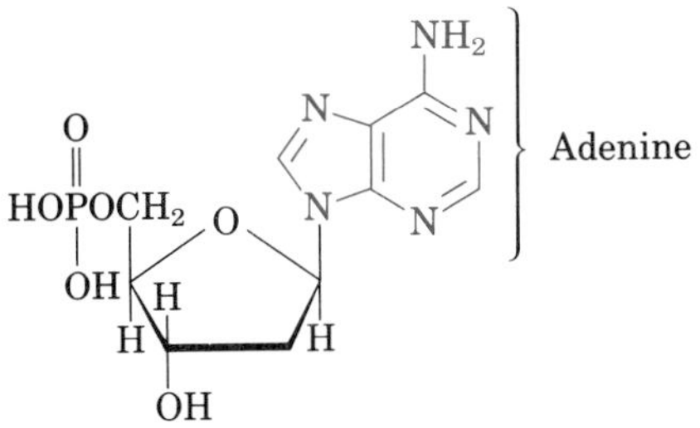

2′-Deoxyadenosine 5′-phosphate (Adenine)

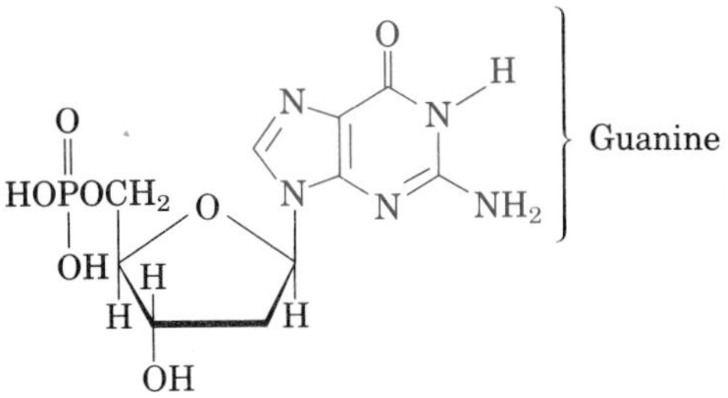

2′-Deoxyguanosine 5′-phosphate (Guanine)

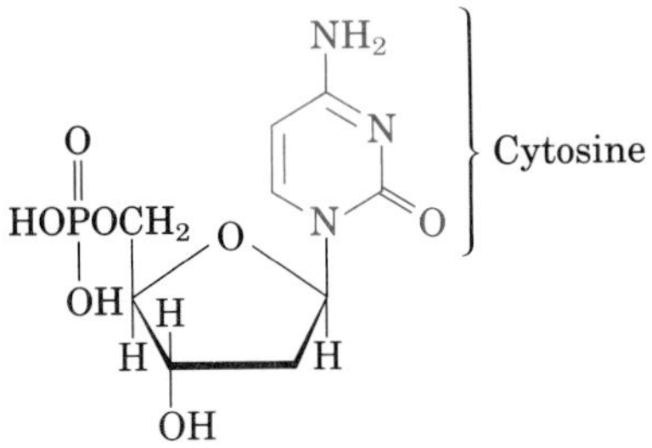

2′-Deoxycytidine 5′-phosphate (Cytosine)

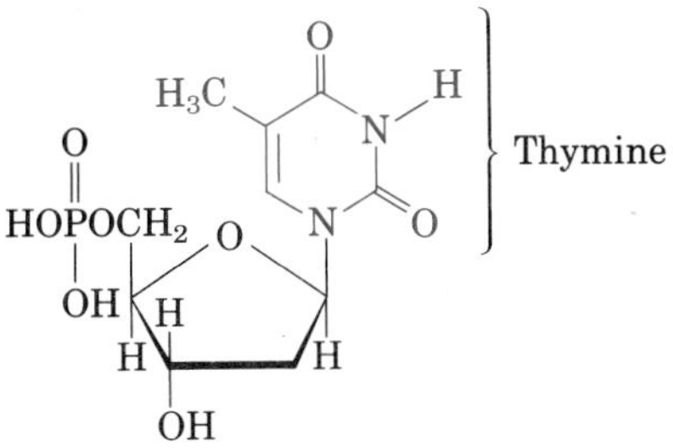

2′-Deoxythymidine 5′-phosphate (Thymine)

Ribonucleotides

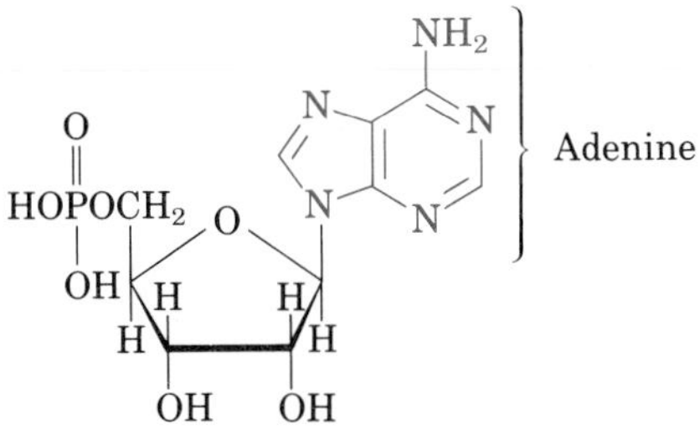

Adenosine 5′-phosphate (Adenine)

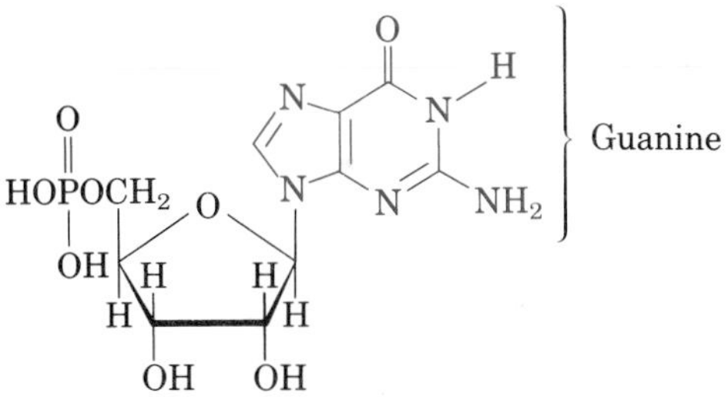

Guanosine 5′-phosphate (Guanine)

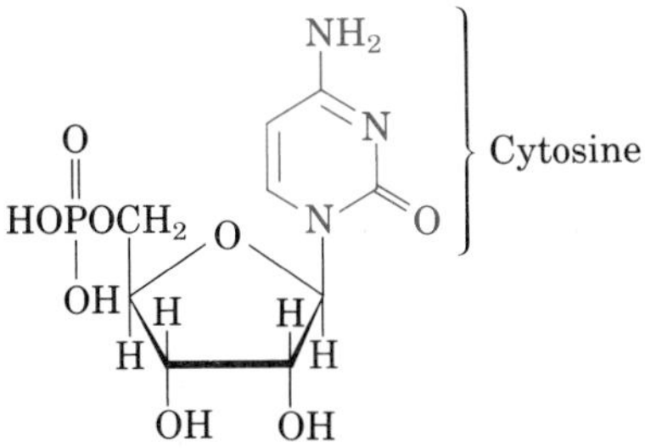

Cytidine 5′-phosphate (Cytosine)

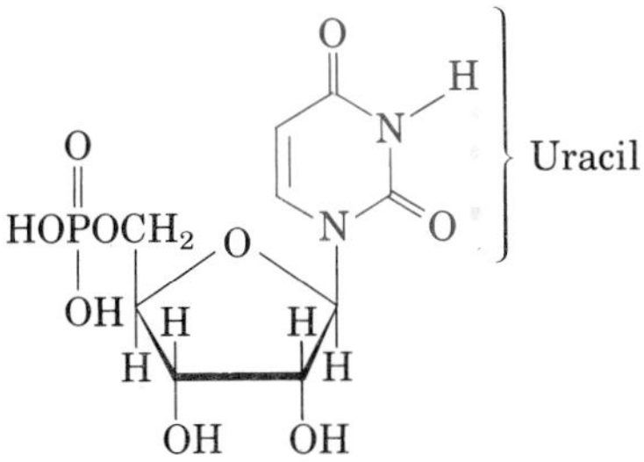

Uridine 5′-phosphate (Uracil)

Figure 33.5. Structures of the four deoxyribonucleotides and four ribonucleotides.

33.10 Structure of DNA

Nucleic acids consist of nucleotide units joined by a bond between the 5′-phosphate component of one nucleotide and the 3′-hydroxyl on the sugar component of another nucleotide (Figure 33.6). One end of the nucleic acid polymer has a free hydroxyl at C3′ (the 3′ end) and the other end has a phosphoric acid residue at C5′ (the 5′ end).

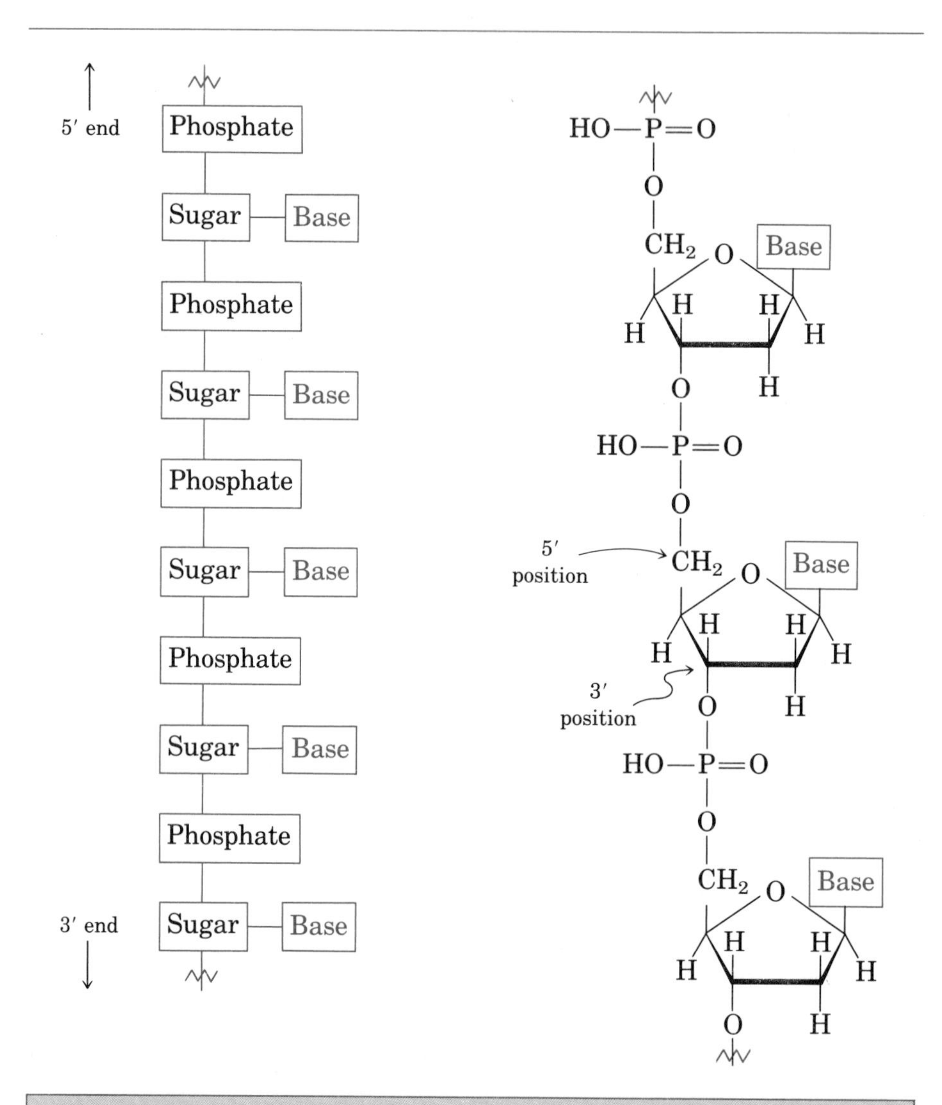

Figure 33.6. Generalized structure of DNA.

Just as the precise structure of a protein depends on the sequence in which individual amino acid residues are arranged, the precise structure of a nucleic acid depends on the sequence in which the individual nucleotides are arranged. Samples of DNA isolated from different tissues of the same species have the same proportions of heterocyclic bases, but samples from different species can have greatly different proportions of bases. Human DNA contains about 30% each of adenine and thymine and about 20% each of guanine and cytosine. The bacterium *Clostridium perfringens,* however, contains about 37% each of adenine and thymine and only 13% each of guanine and cytosine. Note that in both of these examples the bases occur in pairs; adenine and thymine are usually present in equal amounts, as are cytosine and guanine. The reason for this pairing of bases has much to do with the secondary structure of DNA.

In 1953, James Watson[1] and Francis Crick[2] made their now classic proposal for the secondary structure of nucleic acids. According to the Watson–Crick model, DNA consists of two polynucleotide strands coiled into a double helix. The two strands run in opposite directions and are held together by hydrogen bonds between bases. The hydrogen bonding occurs only between specific pairs of bases. Adenine (A) and thymine (T) form strong hydrogen bonds to each other but not to other bases; guanine (G) and cytosine (C) form strong hydrogen bonds to each other but not to other bases.

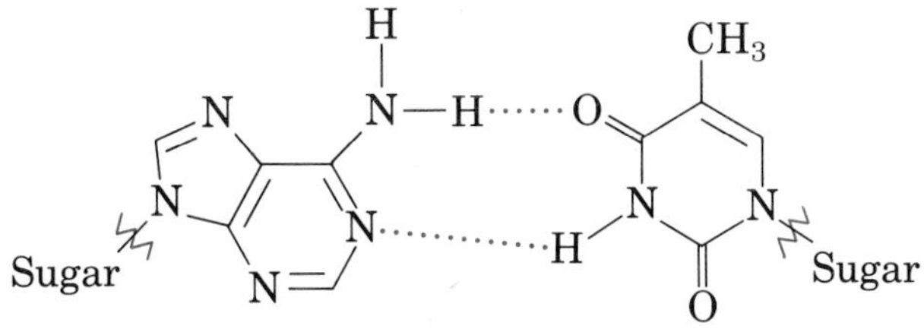

(Adenine) A : : : : : T (Thymine)

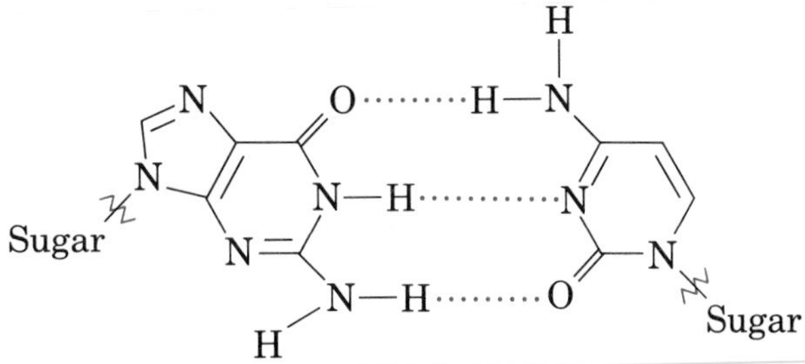

(Guanine) G ⋮ ⋮ ⋮ ⋮ ⋮ C (Cytosine)

The two strands of the DNA double helix are not identical; rather, they are complementary. Whenever a C base occurs in one strand, a G base occurs opposite it in the other strand; when an A base occurs in one strand, a T appears opposite it in the other strand. This complementary pairing of bases explains why A and T, and C and G, are always found in equal

[1]James Dewey Watson (1928–); b. Chicago, Ill.; Ph.D. Indiana; professor, Harvard University; Nobel prize in medicine (1960).

[2]Francis H. C. Crick (1916–); b. England; Ph.D. Cambridge; professor, Cambridge University; Nobel prize in medicine (1960).

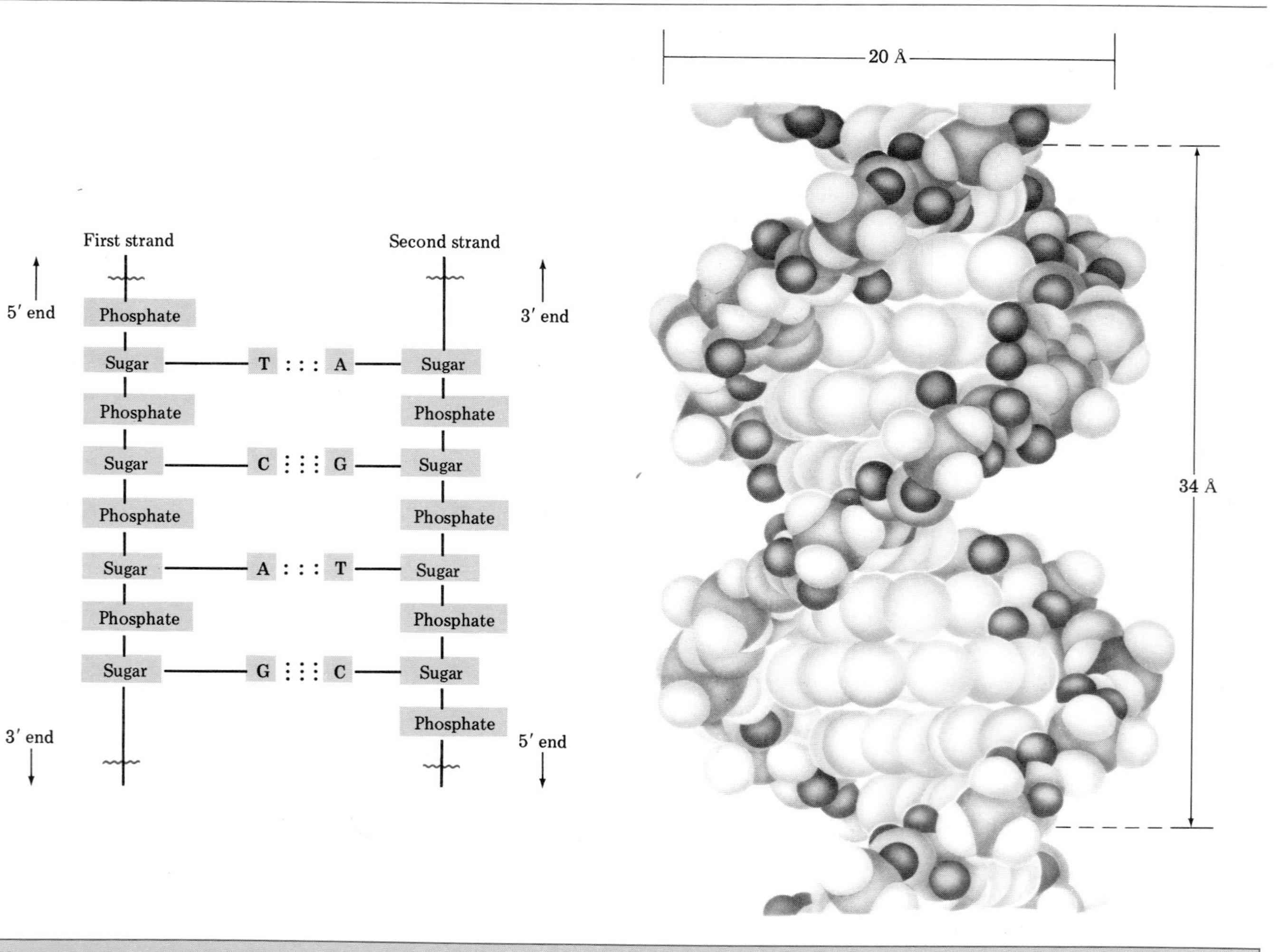

Figure 33.7. Complementarity of base pairing in the two DNA strands and the DNA double helix. The sugar–phosphate backbone of DNA is shown in gray; the atoms of the amine bases are shown in color, and lie inside the helix; the small black atoms are hydrogen.

amounts. Figure 33.7 illustrates this base pairing and shows how the two complementary strands are coiled into the double helix. X-ray measurements show that the DNA double helix is 20 Å wide, that there are exactly 10 base pairs in each full turn, and that each turn is 34 Å in height.

A helpful mnemonic device to remember the nature of the hydrogen bonding between the four DNA bases is the simple phrase "Pure silver taxi."

Pure	Silver	Taxi
Pur	Ag	TC
The purine bases,	A and G,	hydrogen bond to T and C.

PROBLEM

33.11 What sequence of bases on one strand of DNA would be complementary to the following sequence on another strand?

GGCTAATCCGT

33.11 Nucleic acids and heredity

The DNA molecule is the chemical repository of an organism's genetic information, which is stored as a sequence of deoxyribonucleotides strung together in the DNA chain. For this information to be *preserved,* mechanisms must exist for the DNA molecule to be copied and passed on to succeeding generations. For this information to be *used,* mechanisms must exist for "reading" the DNA, for decoding the instructions contained therein, and for using those instructions to carry out the myriad biochemical processes necessary for life.

What Crick has termed the *central dogma* of molecular genetics says that three major processes take place:

1. Replication is the process by which identical copies of DNA are made, forming daughter molecules and preserving genetic information.
2. Transcription is the process by which information in the DNA is read by RNA and carried from the nucleus to the ribosomes.
3. Translation is the process by which RNA decodes the genetic message and uses the information to build proteins.

Thus, information is stored in DNA but is read and used by RNA to make proteins:

$$\text{DNA} \longrightarrow \text{RNA} \longrightarrow \text{Protein}$$

33.12 Replication of DNA

Replication of DNA is an enzyme-catalyzed process that begins by a partial unwinding of the double helix. As the strands separate and the bases are exposed, new nucleotides line up on each strand in an exactly complementary manner, A to T and C to G, and two new strands begin to grow. Each new strand is complementary to its old template strand, and two new identical DNA double helices are produced (Figure 33.8).

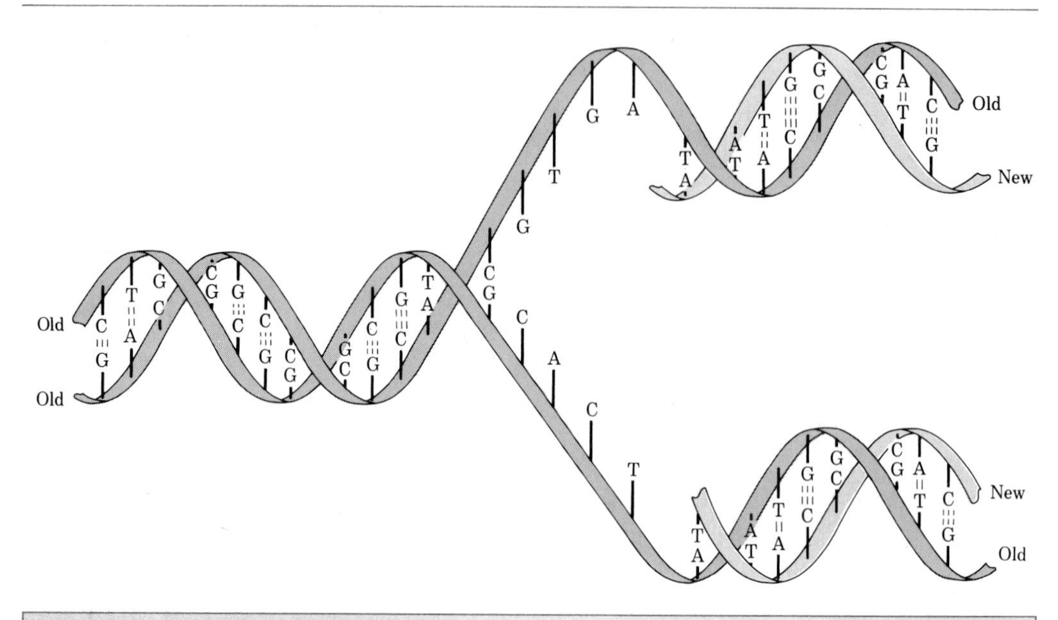

Figure 33.8. Schematic representation of DNA replication.

Crick probably described the process best when he used the analogy of the two DNA strands fitting together like a hand in a glove. The two separate, a new hand forms inside the glove, and a new glove forms around the hand. Two identical copies now exist where only one existed before.

The process by which the individual nucleotides are joined to create new DNA strands is complex, involving many steps and many different enzymes. Addition of new nucleotide units to the growing chain is catalyzed by the enzyme DNA polymerase, and has been shown to occur by addition of a 5′-mononucleotide triphosphate to the free 3′-hydroxyl group of the growing chain, as indicated in Figure 33.9. Both of the new DNA strands are synthesized in the same 5′-to-3′ direction, but this implies that the two strands cannot be synthesized in exactly the same way.

Since the two complementary DNA strands are lined up in opposite directions, one strand must have its 3′ end near the point of unraveling (the **replicating fork**) while the other strand has its 5′ end near the replicating fork. What evidently happens is that the complement of the original 3′ → 5′ strand is synthesized smoothly and in a single piece, but the complement of the original 5′ → 3′ strand is synthesized discontinuously in small pieces that are then linked at a later point by DNA ligase enzymes (Figure 33.10, p. 1118).

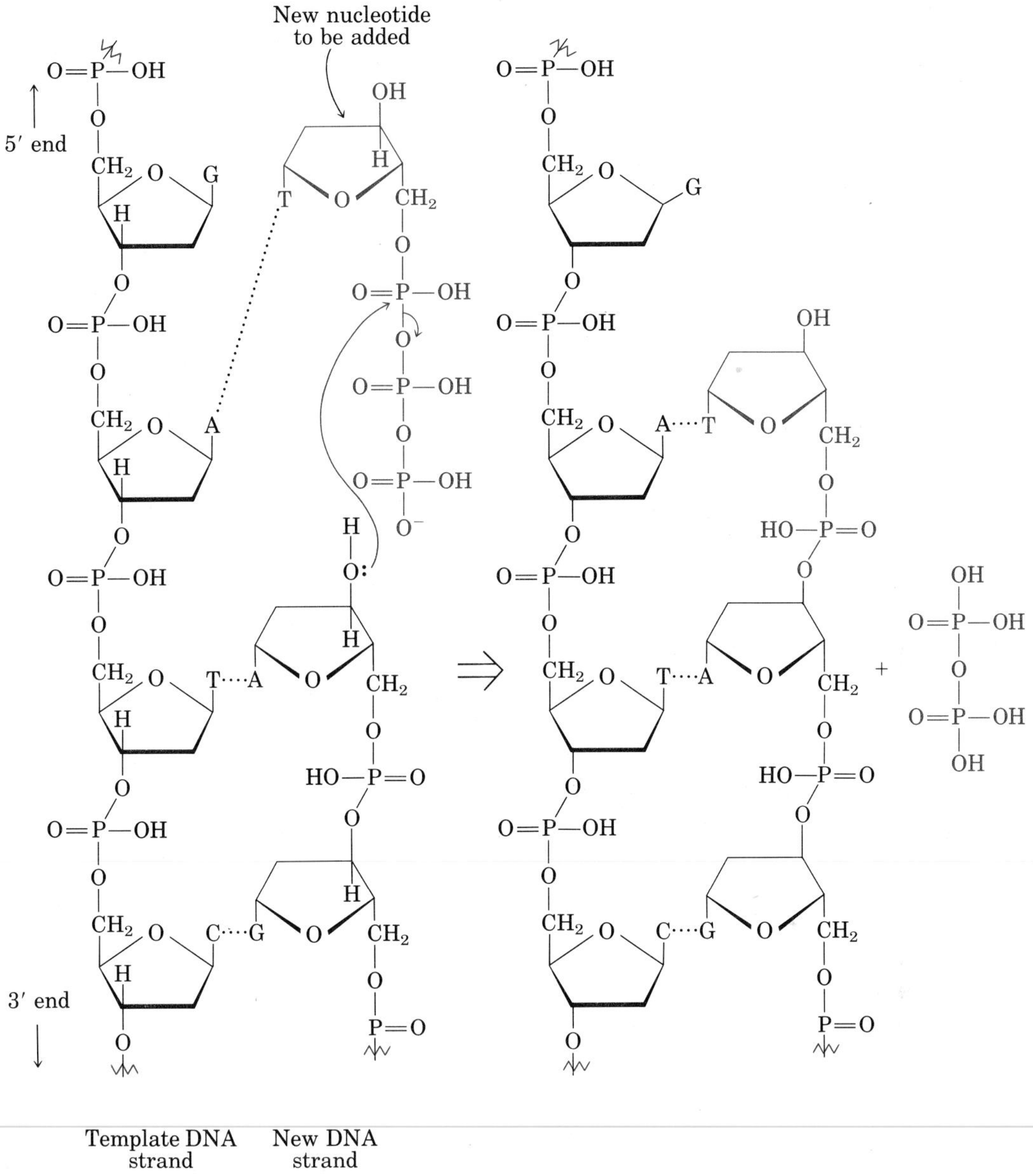

New nucleotide to be added

5′ end

3′ end

Template DNA strand

New DNA strand

Figure 33.9. Addition of a new nucleotide to a growing DNA strand.

It is difficult to conceive of the magnitude of the replication process. The nucleus of a human cell contains 46 chromosomes, each of which consists of one very large DNA molecule. The best current estimate of size places the molecular weight of a human DNA molecule as high as 1 trillion. A single molecule of human DNA is calculated to be about 1 m long and to contain 3 billion individual nucleotides. Regardless of the size of these massive molecules, it is the base sequence that is important, and this sequence is faithfully copied during replication.

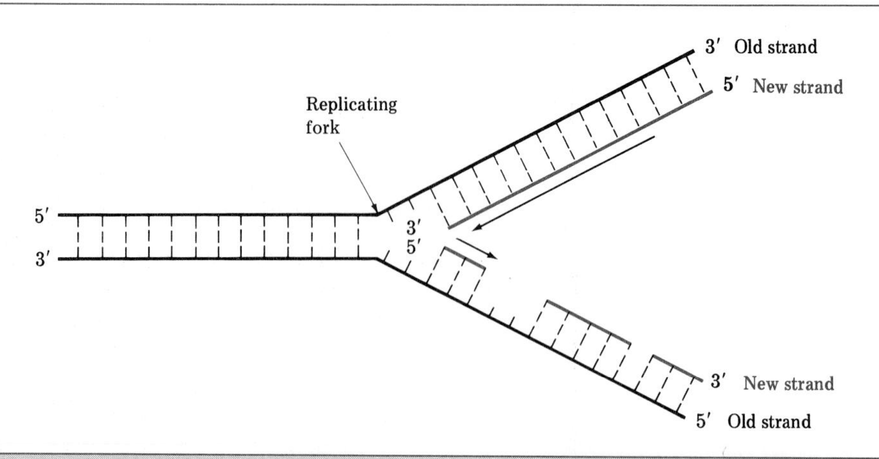

Figure 33.10. Replication of DNA. Both new DNA strands are synthesized in a 5′ → 3′ direction. The original strand whose 3′ end is near the point of unraveling is replicated smoothly, but the original strand whose 5′ end is near the unraveling point is replicated in small pieces that are later joined. The arrowheads are at the growing ends of the chains.

33.13 Structure and synthesis of RNA: transcription

Ribonucleic acid is structurally similar to DNA; both are sugar–phosphate polymers and both have heterocyclic bases attached. The only differences are that RNA contains ribose rather than deoxyribose, and uracil rather than thymine. Uracil in RNA forms strong hydrogen bonds to its complementary base, adenine, just as thymine does in DNA.

Uracil (in RNA) Thymine (in DNA)

There are three major kinds of ribonucleic acid: **messenger RNA (mRNA), transfer RNA (tRNA),** and **ribosomal RNA (rRNA).** All three kinds are much smaller molecules than DNA, and all occur as single polyribonucleotide strands, rather than as double helices.

Molecules of RNA are synthesized in the nucleus of the cell by **transcription** of DNA. A small portion of the DNA double helix unwinds, and one of the two DNA strands serves as a template for complementary ribonucleotides to line up. Bond formation then occurs in the 5′ → 3′ sense, as with DNA replication, and it is catalyzed by enzymes called **RNA polymerases.** Unlike DNA replication, however, the completed RNA molecule does not remain in a double helix with DNA but separates and migrates from the nucleus. The DNA then returns to its stable double-helix conformation (Figure 33.11).

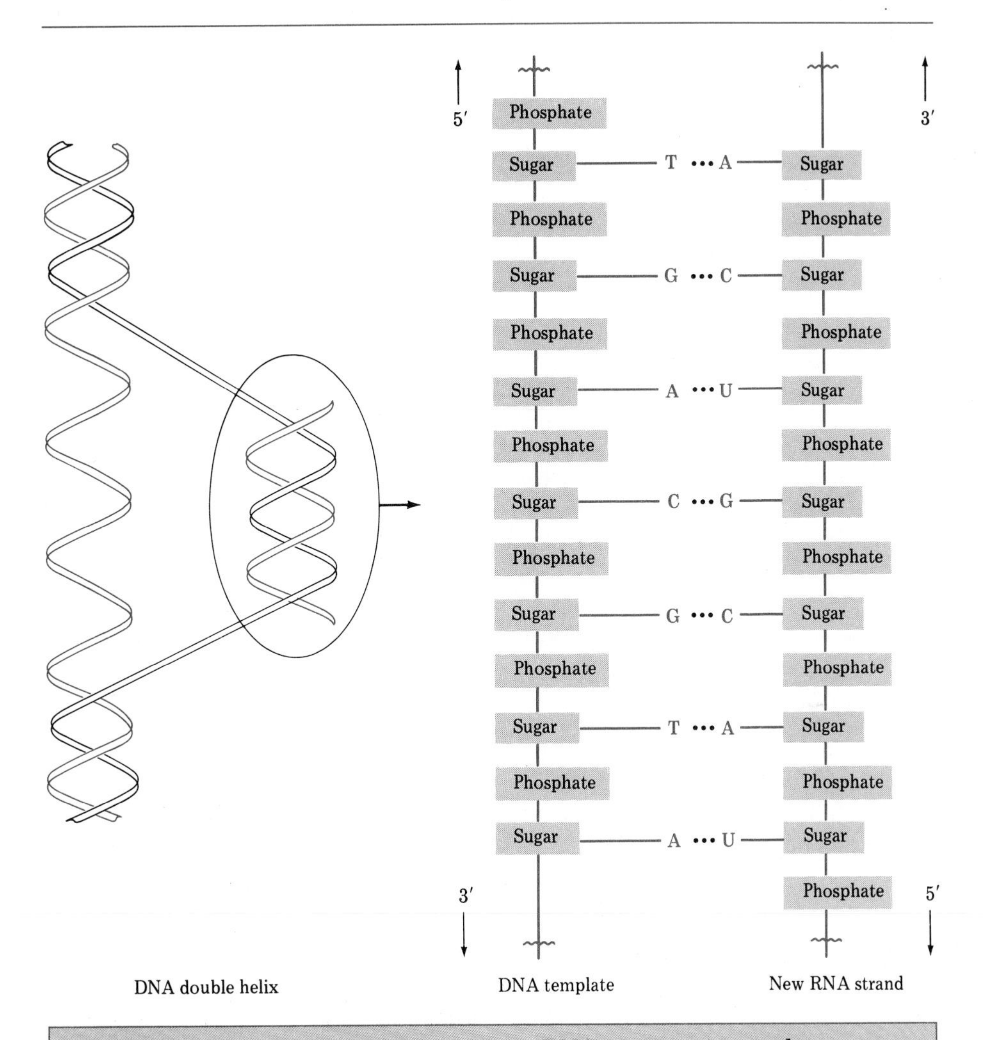

Figure 33.11. Synthesis of RNA using a DNA segment as template.

PROBLEM

33.12 Show how uracil can form strong hydrogen bonds to adenine.

PROBLEM

33.13 What RNA base sequence would be complementary to the following DNA base sequence?

GATTACCGTA

33.14 RNA and biosynthesis of proteins: translation

The primary cellular function of RNA is to direct biosynthesis of the thousands of diverse peptides and proteins required by an organism. These proteins in turn regulate all other biological processes. The mechanics of protein biosynthesis are directed by messenger RNA (mRNA) and take place on **ribosomes,** small granular particles in the cell's cytoplasm, consisting of about 60% ribosomal RNA and 40% protein. On the ribosome, mRNA serves as a template to pass on the genetic information it has transcribed from DNA.

The specific ribonucleotide sequence in mRNA forms a "code" that determines the order in which different amino acid residues are to be joined. Thus, each of the estimated 100,000 proteins in the human body is synthesized from a different mRNA that has been transcribed from a specific gene segment on DNA. Each "word" or **codon** along the mRNA chain consists of a series of three ribonucleotides that is specific for a given amino acid. For example, the series cytosine-uracil-guanine (C-U-G) on mRNA is a codon directing incorporation of the amino acid leucine into the growing protein, and guanine-adenine-uracil (G-A-U) codes for aspartic acid. Of the $4^3 = 64$ possible triads of the four bases in RNA, 61 code for specific amino acids (certain amino acids are specified by more than one codon). In addition, 3 of the 64 codons are known to code for chain termination. Table 33.3 shows the meaning of each codon.

TABLE 33.3 Codon assignments of base triads

First base (5′ end)	*Second base*	*Third base (3′ end)*			
		U	*C*	*A*	*G*
U	U	Phe	Phe	Leu	Leu
	C	Ser	Ser	Ser	Ser
	A	Tyr	Tyr	Stop	Stop
	G	Cys	Cys	Stop	Trp
C	U	Leu	Leu	Leu	Leu
	C	Pro	Pro	Pro	Pro
	A	His	His	Gln	Gln
	G	Arg	Arg	Arg	Arg
A	U	Ile	Ile	Ile	Met
	C	Thr	Thr	Thr	Thr
	A	Asn	Asn	Lys	Lys
	G	Ser	Ser	Arg	Arg
G	U	Val	Val	Val	Val
	C	Ala	Ala	Ala	Ala
	A	Asp	Asp	Glu	Glu
	G	Gly	Gly	Gly	Gly

The code expressed in mRNA is read by transfer RNA (tRNA) in a process called **translation.** There are at least 60 different transfer RNAs, one for each of the codons in Table 33.3. Each specific tRNA acts as a carrier to bring a specific amino acid into place so that it may be transferred to the growing protein chain. A typical tRNA is roughly the shape of a cloverleaf. It consists of about 70–100 ribonucleotides and is bound to a specific amino acid by an ester linkage through the free 3′-hydroxyl on ribose at the 3′ end of the tRNA. Each tRNA also contains in its structure a segment called an **anticodon,** a sequence of three ribonucleotides complementary to the codon sequence. For example, the codon sequence C-U-G present on mRNA would be "read" by a leucine-bearing tRNA having the complementary anticodon base sequence G-A-C. As each successive codon on mRNA is read, different tRNAs bring the correct amino acids into position for enzyme-mediated transfer to the growing peptide. When synthesis of the proper protein is completed, a "stop" codon signals the end, and the protein is released from the ribosome. The structure of a tRNA molecule is shown in Figure 33.12, and the entire process of protein biosynthesis is illustrated schematically in Figure 33.13.

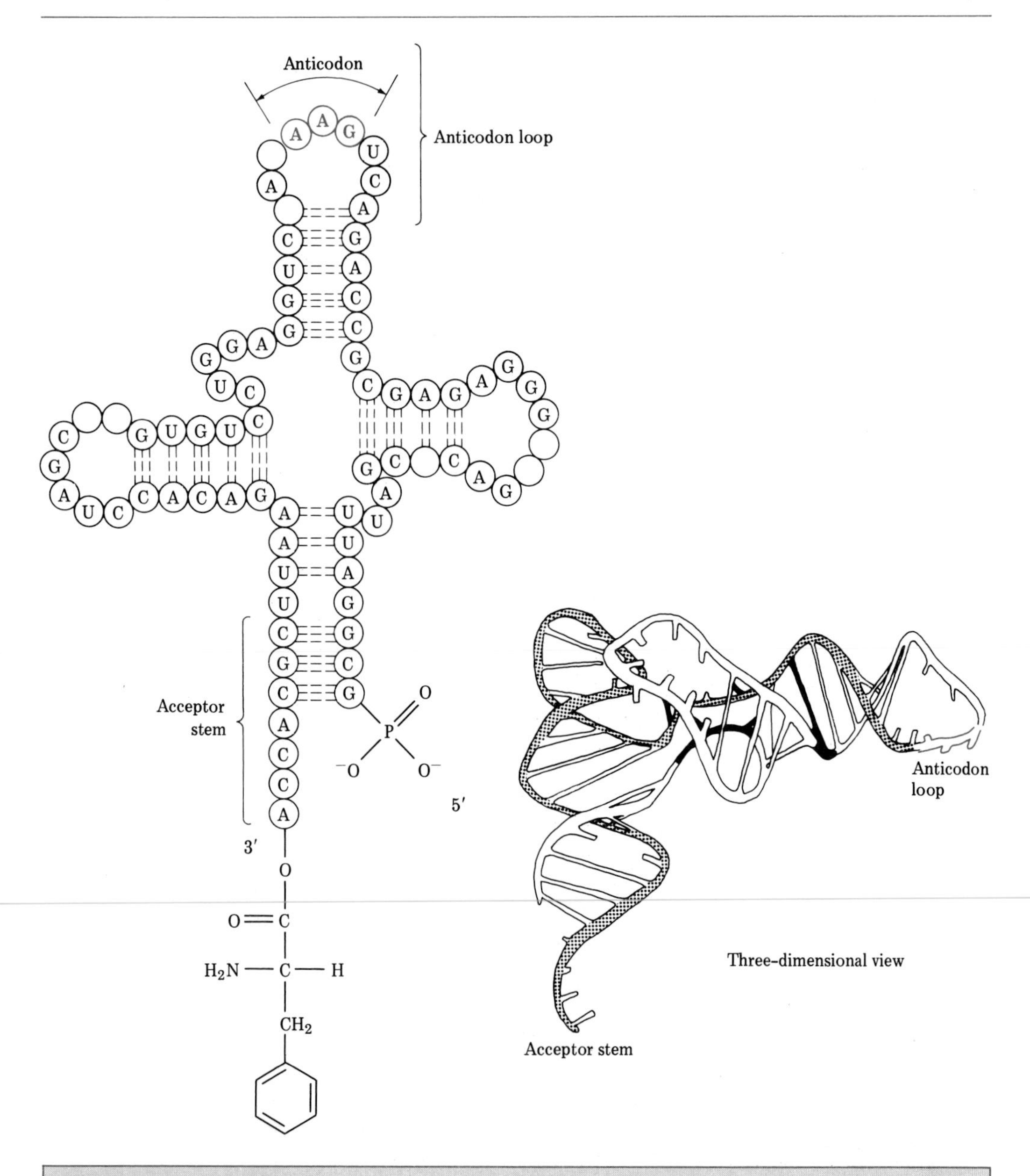

Figure 33.12. Structure of a tRNA molecule. The tRNA is a roughly cloverleaf-shaped molecule containing an anticodon triplet on one "leaf," and a covalently attached amino acid residue at its 3′ end. The tRNA shown is a yeast tRNA that codes for phenylalanine. (The nucleotides that are not specifically identified are chemically modified analogs of the four normal nucleotides.)

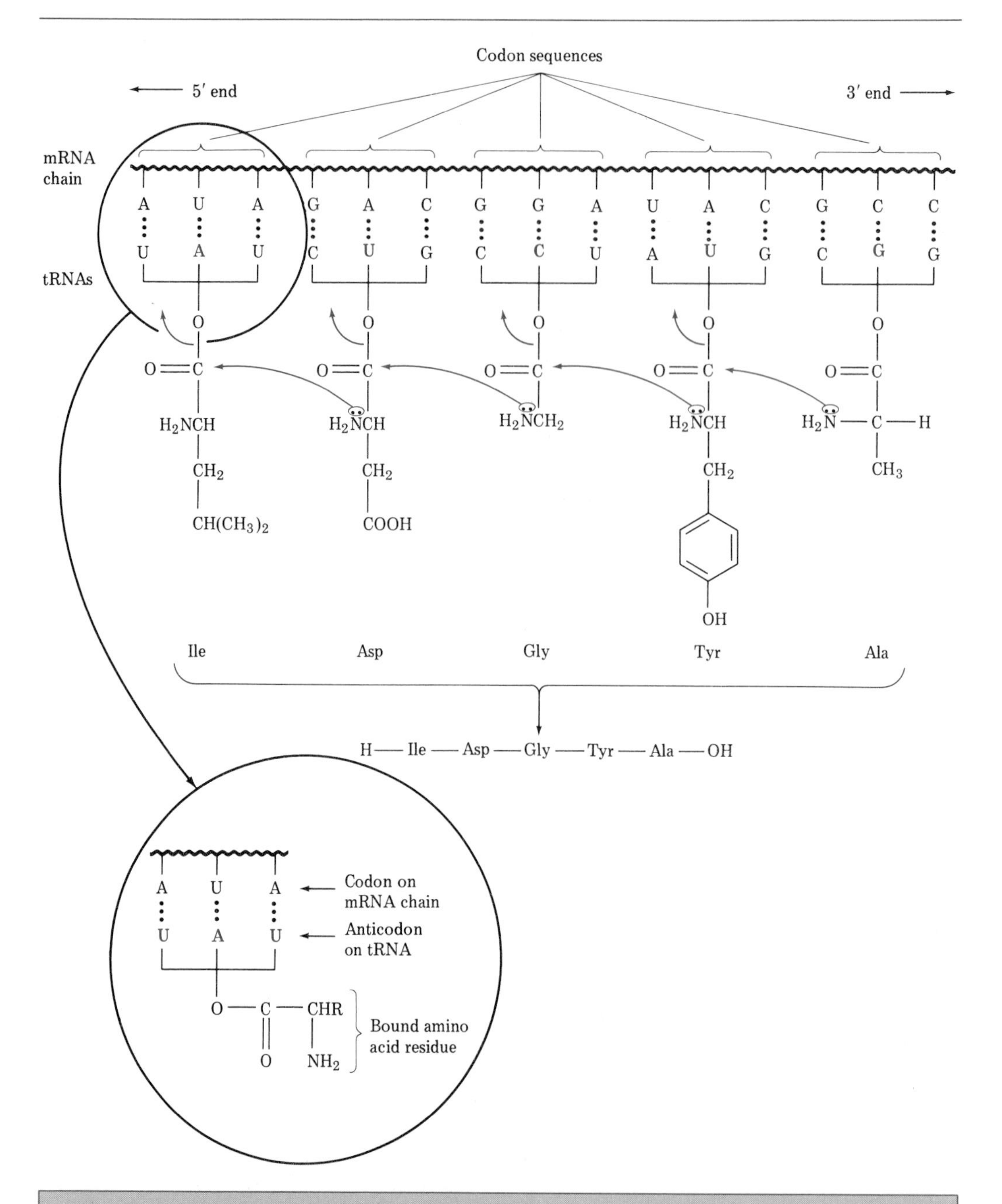

Figure 33.13. Schematic representation of protein biosynthesis. The mRNA, containing codon base sequences, is read by tRNA, containing complementary anticodon base sequences. Then tRNA assembles proper amino acids into position for incorporation into the peptide.

In summary, protein biosynthesis takes place in five discrete steps:

1. Messenger RNA, containing the genetic information transcribed from DNA, is synthesized in the nucleus and moves to the ribosomes.

2. Individual amino acids are activated by binding to specific tRNAs.
3. Transfer RNAs containing the correct anticodon sequences line up at the proper complementary codon sequences on one end of the mRNA chain and move the bound amino acid groups into position.
4. The polypeptide is produced as enzymes catalyze the addition of tRNA-bound amino acids to the growing peptide chain.
5. When the peptide is completed, a "stop" codon on the mRNA halts the biosynthesis, and the peptide is released from the ribosome.

PROBLEM

33.14 What amino acid sequence is coded for by the following mRNA base sequence?

CUU-AUG-GCU-UGG-CCC-UAA

PROBLEM

33.15 What anticodon sequences of tRNAs are coded for by the mRNA in Problem 33.14? What was the base sequence in the original DNA strand on which this mRNA was made?

33.15 Sequencing of DNA

When we work out the structure of DNA molecules, we examine the fundamental level that underlies all processes in living cells. DNA is the information store that ultimately dictates the structure of every gene product, delineates every part of the organism. The order of the bases along DNA contains the complete set of instructions that make up the genetic inheritance. (Walter Gilbert, Nobel Prize Lecture, 1980)

DNA sequencing is now carried out by a remarkably efficient and powerful method developed in 1977 by Allan Maxam and Walter Gilbert.[3]

Since molecules of DNA are so enormous—some molecules of human DNA contain up to 1 trillion base pairs—the first problem in DNA sequencing is to find a method for reproducibly and selectively cleaving the DNA chain at specific points to produce smaller, more manageable pieces. This problem has been solved by the use of enzymes called **restriction endonucleases.** Each different restriction enzyme, of which more than 200 are currently available, cleaves a DNA molecule between two nucleotides at well-defined points along the chain where specific base sequences occur. (The sequence required is usually four or more nucleotides long and thus is unlikely to occur very often in the overall DNA sequence.)

By incubation of large DNA molecules with a given restriction enzyme, many different and well-defined segments of manageable length (100–150 nucleotides) are produced. If the original DNA molecule is incubated with another restriction enzyme having a different specificity for cleavage, still other segments are produced, whose sequences partially overlap those produced by the first enzyme. Sequencing of all the segments, followed by identification of the overlapping sequences, then allows complete DNA

[3] Walter Gilbert (1932–); b. Boston; Ph.D. Cambridge University (1957); professor, Harvard University (1958–); Nobel prize (1980).

sequencing in a manner similar to that used for protein sequencing (Section 31.9). For example, the restriction enzyme Alu I cleaves the linkage between G and C in the four-base sequence AG-CT; the enzyme Hpa II cleaves the C-C linkage in the four-base sequence C-CGG (Figure 33.14).

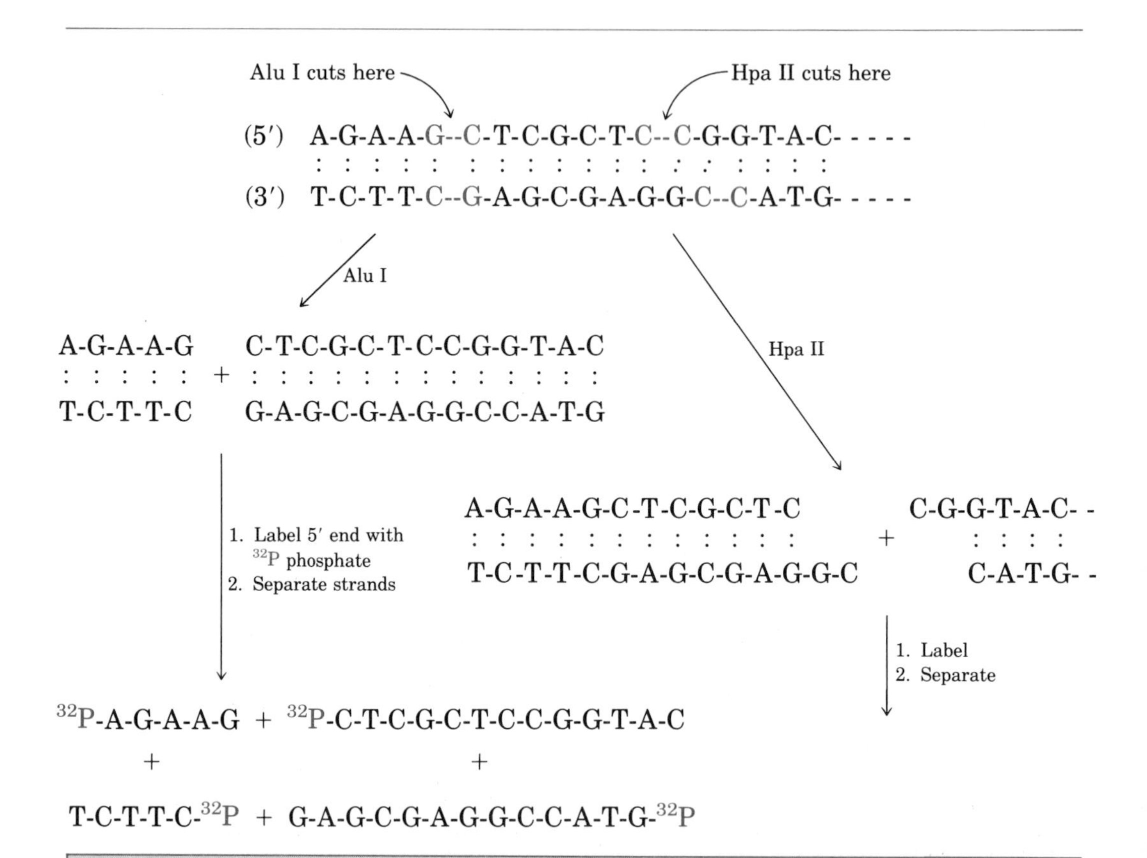

Figure 33.14. Cleavage of a DNA molecule with restriction enzymes. The enzyme Alu I cleaves at the sequence AG-CT, and the enzyme Hpa II cleaves at C-CGG. After cleavage of the double-stranded DNA, the fragments are isolated, and each fragment is labeled at its 5′ end by enzyme-catalyzed formation of a radioactive ^{32}P-containing phosphate ester. The strands are then separated.

After restriction enzymes have cleaved DNA into smaller, more manageable pieces (restriction fragments) the various double-stranded fragments are isolated, and each is radioactively tagged by enzymatically incorporating a labeled ^{32}P phosphate group onto the 5′-hydroxyl group of the terminal nucleotide. The fragments are then separated into two strands by heating, and the strands are isolated.

The heart of the sequencing problem is finding reaction conditions for obtaining specific DNA chain breakage next to each of the four nucleotide bases so that the restriction fragments can be further degraded and ultimately sequenced. Maxam and Gilbert solved this problem using two reagents, dimethyl sulfate and hydrazine.

Treatment of DNA with dimethyl sulfate results in methylation (S_N2 reaction) of the purine bases A and G, but does not affect the pyrimidine bases C and T. The adenine of deoxyadenosine is methylated at N3, and the guanine of deoxyguanosine is methylated at N7.

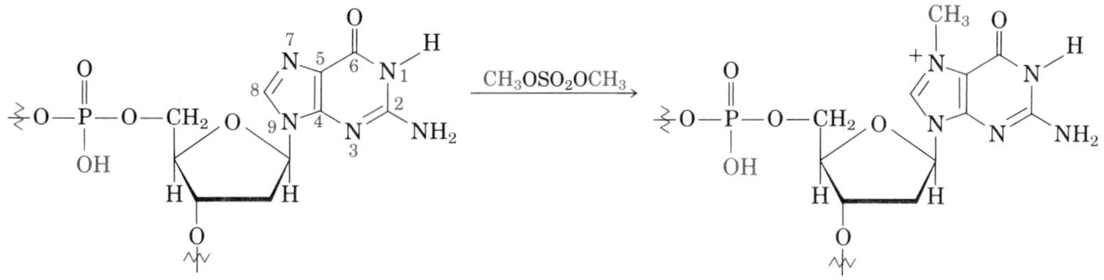

$CH_3OSO_2OCH_3$ →

Deoxyguanosine

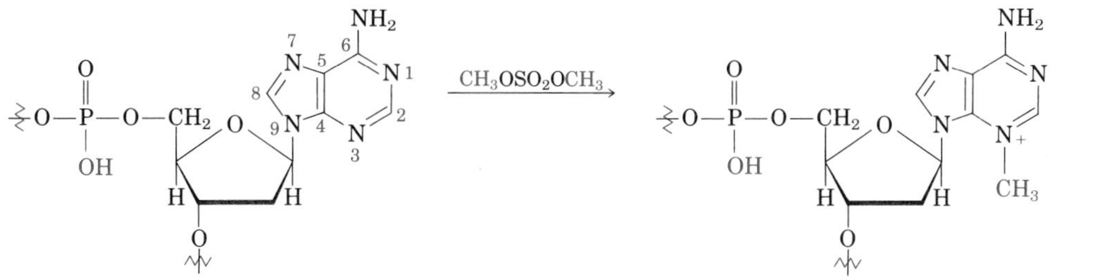

$CH_3OSO_2OCH_3$ →

Deoxyadenosine

Treatment of methylated DNA with an aqueous solution of the secondary amine piperidine then brings about destruction of the methylated nucleotides and specific opening of the DNA chain at both the 3′ and the 5′ positions next to the methylated bases. The mechanism of the cleavage process is complex and involves a number of different steps, as shown in Figure 33.15 for deoxyguanosine breakage: (1) Hydrolysis occurs, opening the five-membered heterocycle; (2) hydrolysis of the aminoglycoside sugar linkage then yields an open-chain 2-deoxyribose; (3) formation of an enamine between piperidine and the 2-deoxyribose aldehyde group occurs; and (4) two β eliminations of the 2-deoxyribose oxygen substituents at C3 and C5 take place. It is these two elimination reactions that break open the DNA chain.

By working carefully, Maxam and Gilbert were able to find reaction conditions that were selective for cleavage either at A or at G. Thus, deoxyguanosine methylates five times as rapidly as deoxyadenosine, but the hydrolytic breakdown of deoxyadenosine occurs more rapidly than breakdown of deoxyguanosine if the methylated product is first heated with dilute acid prior to base treatment.

Breaking the DNA chain next to both pyrimidine nucleotides C and T can be accomplished by treatment of DNA with hydrazine, followed by heating with aqueous piperidine. Although conditions that are selective for

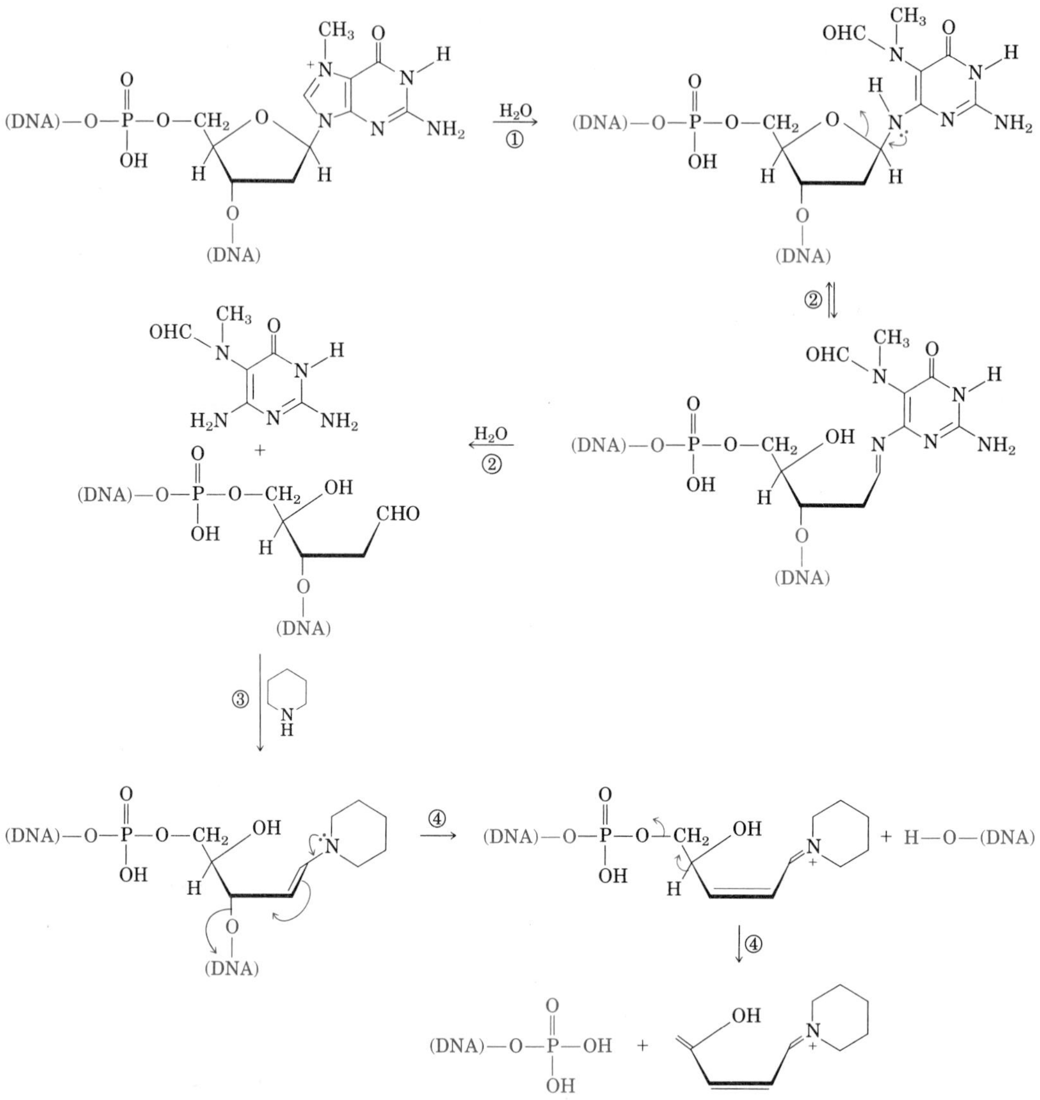

Figure 33.15. Mechanism of DNA cleavage at deoxyguanosine.

cleavage next to deoxythymidine have not been found, a selective cleavage next to deoxycytidine can be accomplished by carrying out the hydrazine reaction in 2*M* NaCl. The mechanism of deoxythymidine cleavage by hydrazine is shown in Figure 33.16. Once again, the breakdown reaction involves numerous steps: (1) Hydrazine first undergoes a Michael-type addition to thymine, followed by (2) an intramolecular nucleophilic acyl substitution reaction that opens the heterocyclic thymine ring; (3) breakage of the aminoglycoside linkage and (4) hydrolysis yields open-chain deoxyribose,

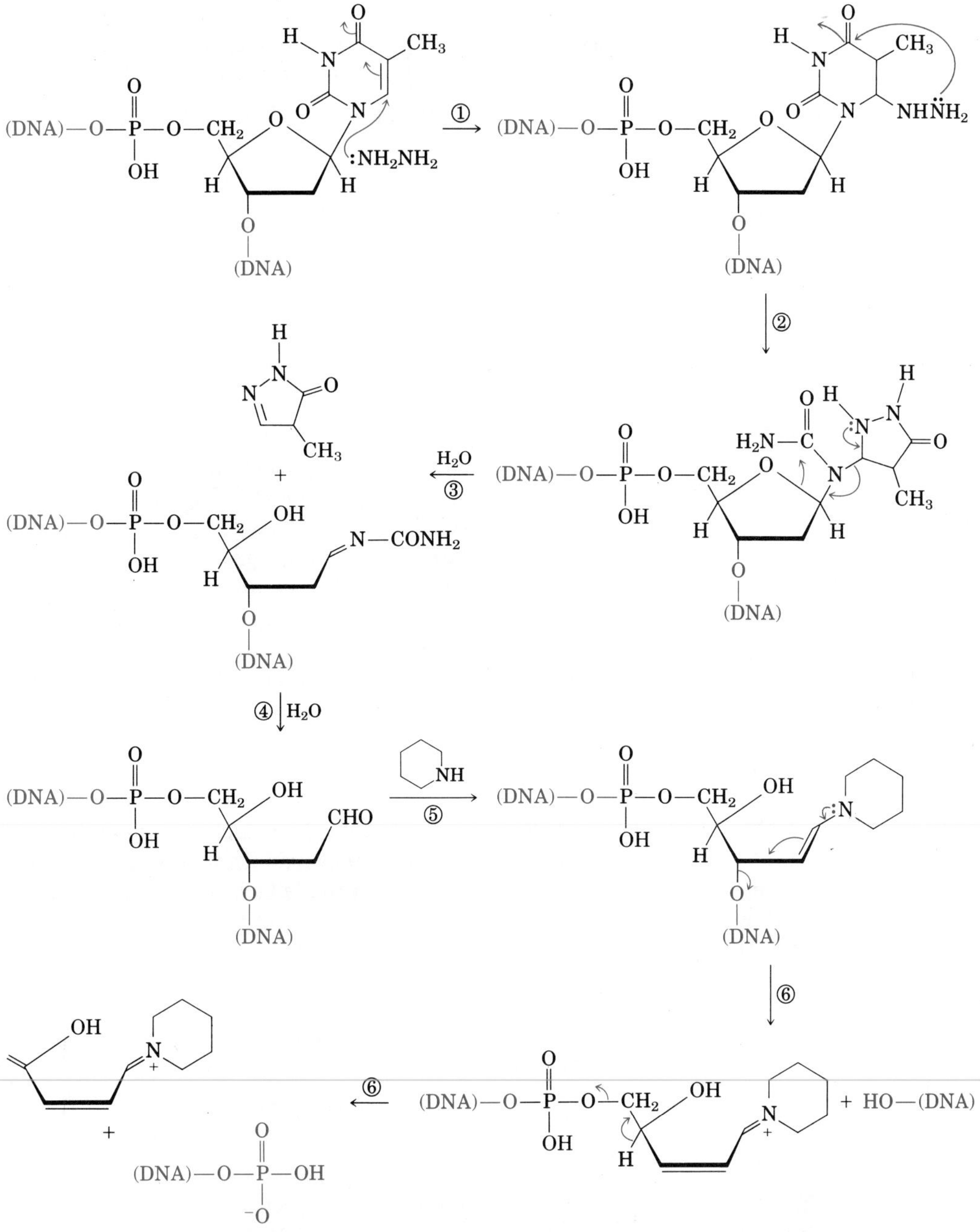

Figure 33.16. Mechanism of breakdown of deoxythymidine by reaction with hydrazine.

which (5) forms an enamine by reaction with piperidine; (6) two β-elimination reactions then break open the DNA chain.

In summary, four sets of reaction conditions have been devised for breaking a DNA chain at specific points:

1. At A > G. Methylation, followed first by treatment with dilute acid and then by heating with aqueous piperidine, preferentially breaks the chain on both sides of A. (Some breakage also occurs next to G.)
2. At G > A. Methylation, followed by heating with aqueous piperidine, preferentially breaks the chain on both sides of G. (Some breakage also occurs next to A.)
3. At C. Treatment with hydrazine in 2*M* NaCl, followed by heating with aqueous piperidine, breaks the chain on both sides of C.
4. At C + T. Treatment with hydrazine in the absence of NaCl, followed by heating with aqueous piperidine, breaks the chain next to *both* C and T.

After the restriction fragment has been broken down by selective cleavage reactions into a mixture of smaller pieces, the mixture is separated by electrophoresis (Section 31.3). When the mixture of DNA pieces is placed at one end of a strip of buffered gelatinous polyacrylamide and a voltage difference is applied across the ends of the strip, electrically charged pieces move along the gel. Each piece moves at a rate that depends both on its size and on the number of negatively charged phosphate groups (the number of nucleotides) it contains; smaller pieces move rapidly, and larger pieces move more slowly. The technique is so sensitive that up to 250 DNA pieces, differing in size by only one nucleotide, can be separated.

Once separation of the pieces has been accomplished, the position on the gel of each radioactive ^{32}P-containing piece is determined by exposing the gel to a photographic plate. Only the pieces containing the radioactively labeled 5′-end phosphate group are visualized. Unlabeled pieces from the middle of the chain do not appear.

At this point, we have discussed methods for selectively degrading a DNA restriction fragment, for separating the degradation products, and for visualizing those products. How can these methods be used for sequencing? Let's follow a DNA fragment through the series of steps just discussed to see how the steps lead ultimately to a sequence for the DNA fragment.

1. A DNA molecule is incubated with a restriction enzyme, which cuts the chain at specific places and yields DNA fragments containing 100–150 nucleotide pairs.

2. The double-stranded DNA restriction fragments are radioactively labeled by incorporation of a ^{32}P-containing phosphate group at the 5′-hydroxyl of the terminal nucleotide.

3. The labeled restriction fragments are isolated, and each is separated into its two complementary strands. For example, imagine that we have now isolated a single-stranded DNA segment of approximately 100 nucleotides with the following partial structure:

$$(5') \quad {}^{32}\text{P-C-T-C-A-G-T-A-C-C-G-}\,\text{- - - - - - - -} \quad (3')$$

4. The radioactively labeled single-stranded DNA segment is subjected to four parallel sets of cleavage experiments under conditions that lead to (a) preferential splitting next to A, (b) preferential splitting next to

G, (c) exclusive splitting next to C, and (d) splitting next to both T and C. Mild reaction conditions are chosen so that only a few of the many possible splittings occur in each reaction. In our example, the pieces shown in Table 33.4 would be produced.

TABLE 33.4 **Splitting of a DNA restriction fragment under four sets of conditions**[a]

Cleavage conditions	*Pieces produced*		
Original DNA segment	^{32}P-C-T-C-A-G-T-A-C-C-G- - - -		
A > G	^{32}P-C-T-C		
	^{32}P-C-T-C-A-G-T	+	Larger pieces
G > A	^{32}P-C-T-C-A		
	^{32}P-C-T-C-A-G-T-A-C-C	+	Larger pieces
C	^{32}P-C-T		
	^{32}P-C-T-C-A-G-T-A		
	^{32}P-C-T-C-A-G-T-A-C	+	Larger pieces
C + T	^{32}P-C		
	^{32}P-C-T		
	^{32}P-C-T-C-A-G		
	^{32}P-C-T-C-A-G-T-A		
	^{32}P-C-T-C-A-G-T-A-C	+	Larger pieces

[a]Only the pieces containing the radioactive end-label are considered. Other pieces are also produced but are not visualized.

5. Product mixtures from the four cleavage reactions are separated by gel electrophoresis, and the spots on the gel are visualized by exposing the gel to a photographic plate. The location of each radioactive piece appears as a dark band on the photographic plate; nonradioactive pieces are not visualized. The gel electrophoresis pattern shown in Figure 33.17 (p. 1130) would be obtained in our hypothetical example.

6. The DNA sequence is then read directly from the gel. The band that appears farthest from the origin is the terminal mononucleotide (the smallest piece) and cannot be identified. Since the terminal mononucleotide appears only in the T + C column, however, it must have been produced by splitting next to a T or a C. Thus, the second nucleotide in the sequence is a T or a C. Since this smallest piece does not appear in the C column, however, the second nucleotide is not a C and must therefore be a T.

The second farthest band from the origin is a dinucleotide that appears in both C and T + C columns. This dinucleotide is produced by splitting next to the third nucleotide, which must therefore be a C. The

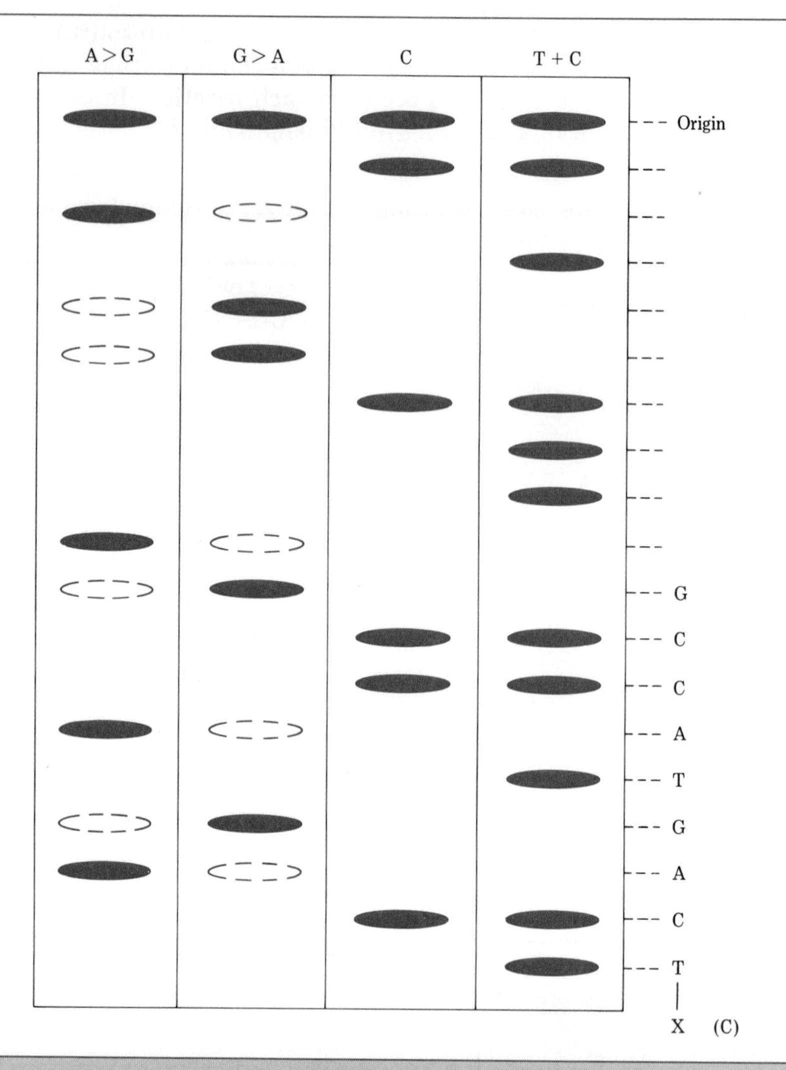

Figure 33.17. Representation of a gel electrophoresis pattern. The products of the four parallel cleavage experiments are placed at the top of the gel and a voltage is applied between top and bottom. Smaller products migrate along the gel at a faster rate and thus appear at the bottom. The DNA sequence can be "read" by visualizing the radioactive spots.

third farthest band appears mainly in the A > G column, which means that the fourth nucleotide is an A. (There is also a faint band in the G > A column, since the specificity of these splittings is not complete.)

Continuing in this manner, the entire sequence of the DNA can be read from the gel simply by noting in what columns the successively larger labeled polynucleotide pieces appear. Once read, the entire sequence can be checked by determining the sequence of the complementary strand. The identity of the 5′-terminal nucleotide can be determined by sequencing an overlapping segment produced by cleavage with another restriction enzyme.

The Maxam–Gilbert method of DNA sequencing is so efficient that sequencing rates of up to 150 nucleotides per day can be achieved. So powerful is the method, in fact, that some *protein* sequencing is now done by sequencing the DNA from which the protein's mRNA was transcribed (Section 31.13). Impressive achievements are being recorded at an ever-increasing rate, with the current (1983) record being the sequencing of the Epstein–Barr virus having 170,000 base pairs.

PROBLEM

33.16 Show all the labeled cleavage products you would expect to obtain if the following DNA segment were subjected to each of the four different cleavage reactions:

-^{32}P-A-A-C-A-T-G-G-C-G-C-T-T-A-T-G-A-C-G-A

PROBLEM

33.17 Finish assigning the sequence to the gel electrophoresis pattern shown in Figure 33.17.

33.16 Laboratory synthesis of DNA

The development of genetic engineering techniques in the 1970s brought with it an increased demand for efficient chemical methods for the synthesis of short DNA segments. Ideally, whole genes—specific DNA segments—might be synthesized in the laboratory and inserted into the DNA of microorganisms. This would direct the microorganisms to produce the specific protein coded for by that gene—perhaps insulin or some other valuable material.

The problems of DNA synthesis are similar to those of protein synthesis (Section 31.11) but are considerably more difficult because of the structural complexity of the deoxyribonucleotide monomers. Each nucleotide has several reactive sites that must be selectively protected and deprotected at the proper times, and coupling of the four nucleotides must be carried out in the proper sequence. Despite these difficulties, some extremely impressive achievements have been recorded, including the synthesis by Khorana[4] in 1979 of the tyrosine suppressor tRNA gene from the bacterium *Escherichia coli*. Some 207 base pairs were assembled in an effort that required 10 years of work.

More recently, automated "gene machines" have become available, which allow the fast and reliable synthesis of DNA sequences 10–20 nucleotides long. These DNA synthesizers, based on chemistry developed in the mid-1960s, operate on a principle similar to that of the Merrifield solid-phase peptide synthesizer (Section 31.12). In essence, a protected nucleotide is covalently bound to a solid support, and one nucleotide at a time is added to the chain by means of a coupling reagent. When the last nucleotide has been added, the protecting groups are removed, and the synthetic DNA is cleaved from the solid support.

The first step in DNA synthesis involves attachment of a protected nucleoside fragment to the polymer support by an ester linkage to the 3′

[4]Har Gobind Khorana (1922–); b. Raipur, India; Ph.D. University of Liverpool; professor, Massachusetts Institute of Technology; Nobel prize in medicine (1968).

position of a deoxyribonucleoside. Both the 5′-hydroxyl and free amino groups on the heterocyclic bases must be protected to accomplish this attachment. Adenine and cytosine bases are protected by benzoyl groups, guanine is protected by an isobutyryl group, and thymine requires no protection. The deoxyribose 5′-hydroxyl is protected as its DMT ether (*para*-dimethoxytrityl). Other protecting groups may also be used, but the ones mentioned here are common choices.

Step 1.

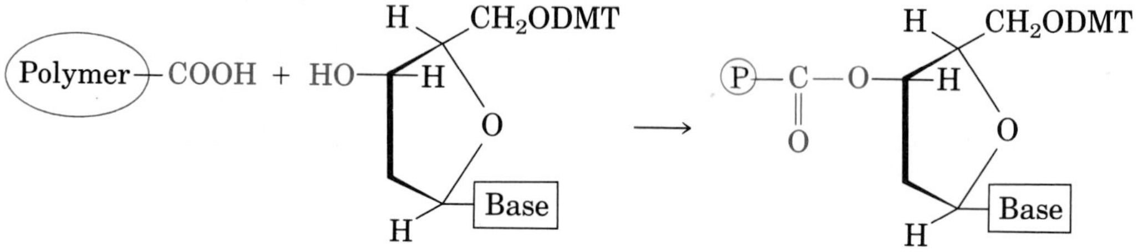

Polymer—COOH + HO–(deoxyribose with CH_2ODMT and Base) ⟶ Ⓟ—C(=O)—O–(deoxyribose with CH_2ODMT and Base)

where DMT = $-C(C_6H_5)(C_6H_4OCH_3)_2$

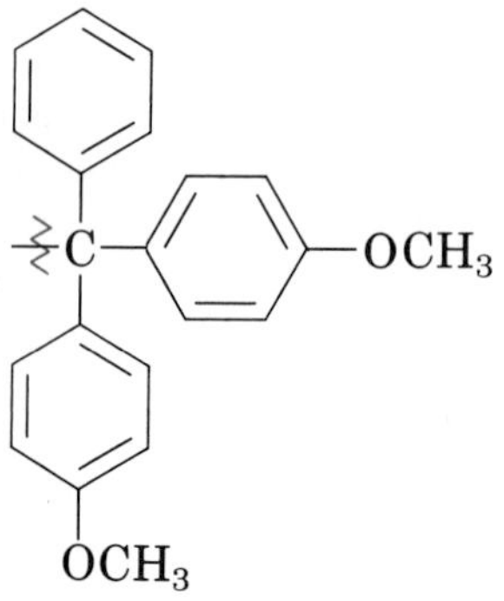

Base =

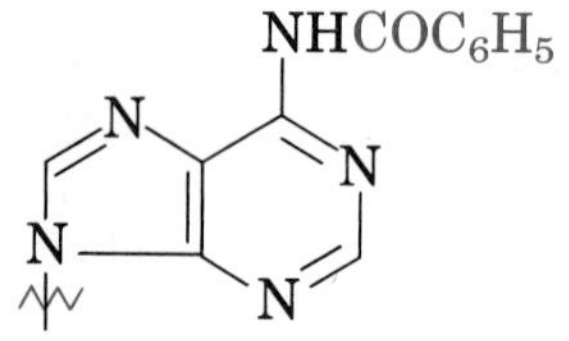

N-protected adenine ($NHCOC_6H_5$)

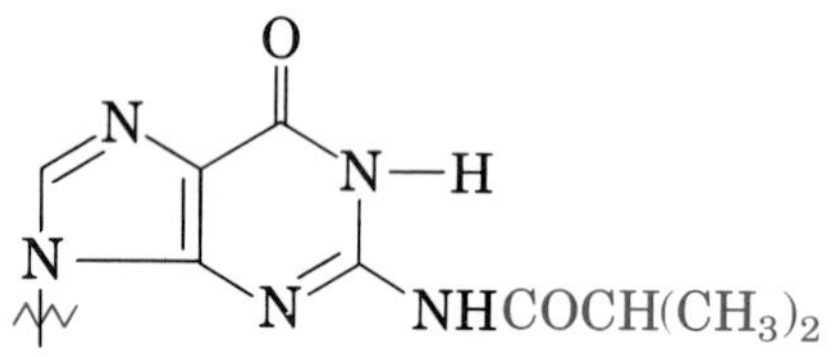

N-protected guanine ($NHCOCH(CH_3)_2$)

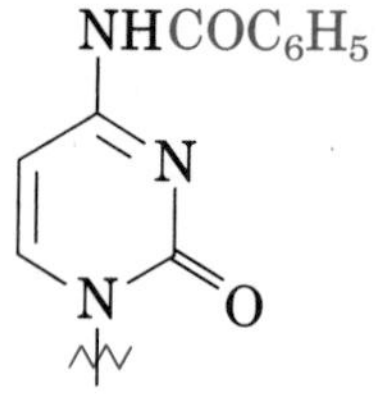

N-protected cytosine ($NHCOC_6H_5$)

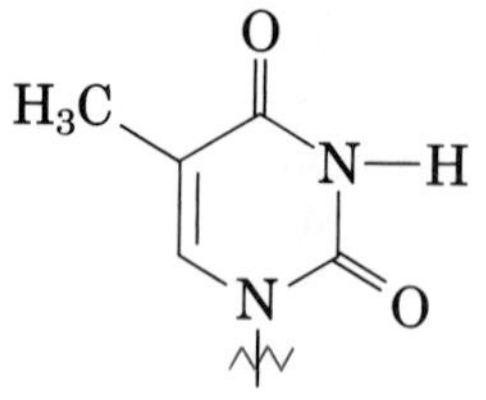

Thymine

Step 2 involves deprotection of the 5′-deoxyribose hydroxyl by Lewis acid catalyzed cleavage to remove the DMT group.

Step 2.

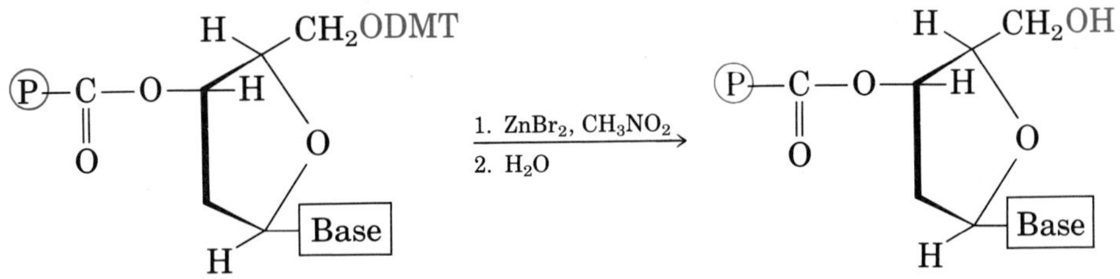

Step 3 involves phosphate ester formation between the free 5′ position of the polymer-bound nucleoside and the 3′-phosphate of a protected nucleotide. This ester-forming step involves use of a coupling agent such as 2,4,6-triisopropylbenzenesulfonyltetrazole (TPST), a reagent that serves the same purpose in nucleotide synthesis that DCC serves in protein synthesis (Section 31.11). Note that one of the phosphate oxygens is protected as the *o*-chlorophenyl ester in this coupling step.

Step 3.

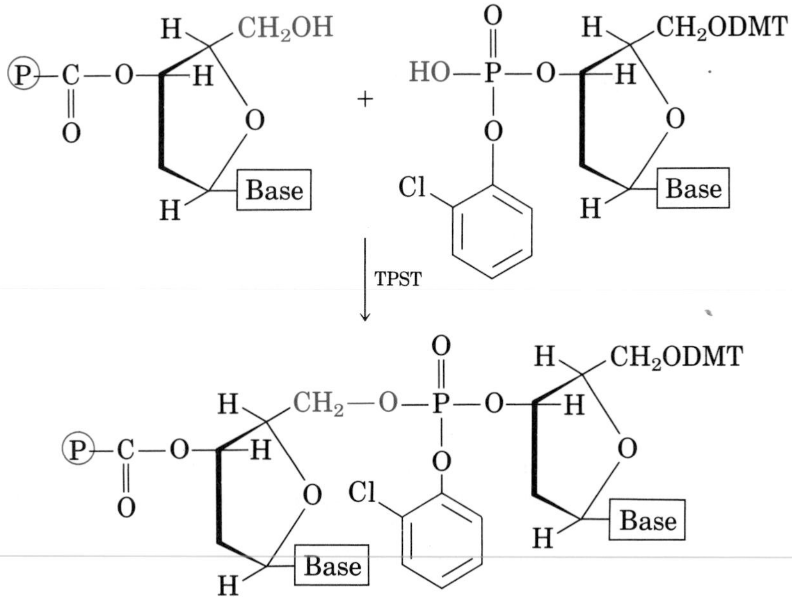

where TPST =

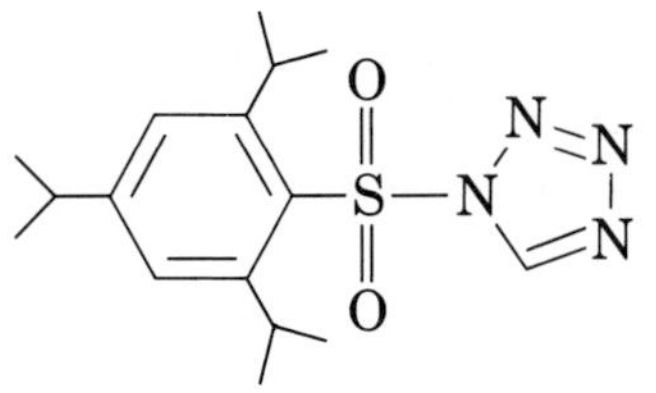

The cycle of steps 2 and 3 is then repeated until a polydeoxyribonucleotide chain of the desired length and sequence has been built. The final step is to cleave all protecting groups from the heterocyclic bases and from the phosphates and to cleave the ester bond holding the polynucleotide to the polymer. All these reactions can be effected by treatment with aqueous ammonia.

Final step.

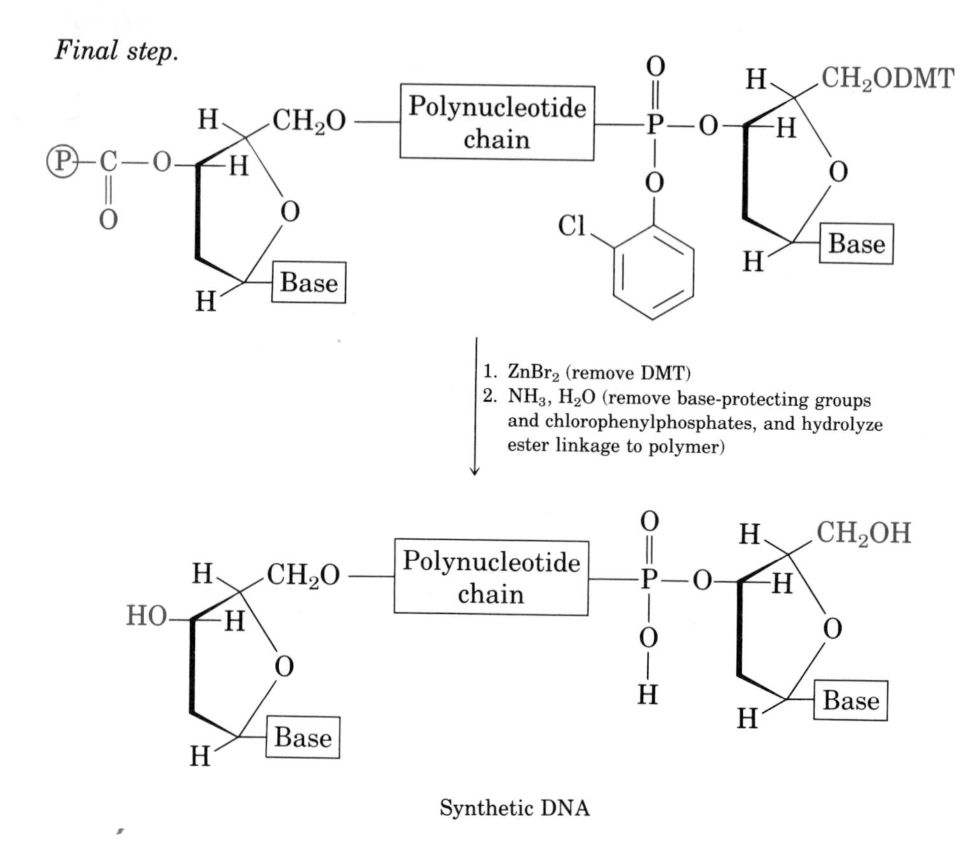

Synthetic DNA

PROBLEM

33.18 The DMT ethers are easily cleaved by mild Lewis acid, a step of great value in polydeoxyribonucleotide synthesis. Can you suggest a reason why this ether cleavage is unusually easy?

33.17 Summary

A heterocycle is a compound having a ring that contains more than one kind of atom. Nitrogen, oxygen, and sulfur are often found along with carbon in heterocyclic rings. Saturated heterocyclic amines, ethers, and sulfides usually display the same chemistry as their open-chain analogs, but unsaturated heterocycles often display aromaticity. Pyrrole, furan, and thiophene are the simplest five-membered aromatic heterocycles. All three are unusually stable, and all three undergo aromatic substitution when reacted with electrophiles. Reaction usually occurs at the highly activated position next to the heteroatom.

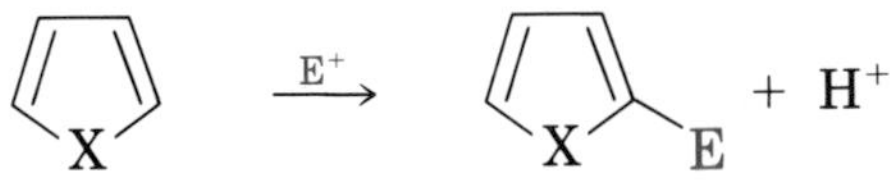

X (five-membered ring) $\xrightarrow{E^+}$ X (ring)—E + H^+

where X = N (pyrrole), O (furan), or S (thiophene).

Pyridine is the six-membered-ring, nitrogen-containing heterocyclic analog of benzene. The pyridine ring is electron-poor and undergoes electrophilic aromatic substitution reactions with great difficulty. Nucleophilic aromatic substitutions of 2- or 4-halopyridines take place readily, however.

The nucleic acids—DNA (deoxyribonucleic acid) and RNA (ribonucleic acid)—are biological polymers that act as chemical carriers of an organism's genetic information. Enzyme-catalyzed hydrolysis of nucleic acids yields nucleotides, the monomer units from which RNA and DNA are constructed. Each nucleotide consists of a purine or pyrimidine base linked to C1′ of a simple pentose sugar (ribose in RNA and 2′-deoxyribose in DNA), with the sugar in turn linked through its C5′ hydroxyl to a phosphate group:

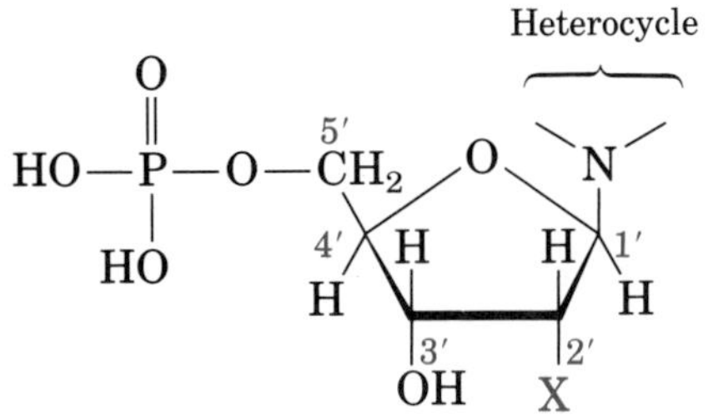

Heterocycle

HO—P(=O)(HO)—O—CH_2 (5′) — sugar ring (O, C4′ H H, C1′ H, N of heterocycle), 3′ OH, 2′ X

where X = OH (a ribonucleotide)

= H (a deoxyribonucleotide)

The nucleotides are joined by phosphate links between the phosphate of one nucleotide and the 3′-hydroxyl on the sugar of another nucleotide.

Molecules of DNA consist of two polynucleotide strands held together by hydrogen bonds between heterocyclic bases on the different strands and coiled into a double-helix conformation. Adenine and thymine form hydrogen bonds to each other, as do cytosine and guanine. The two strands of DNA are not identical but are complementary.

Three main processes take place in deciphering the genetic information of DNA:

1. Replication of DNA is the process by which identical DNA copies are made and genetic information is preserved. This occurs when the DNA double helix unwinds, complementary deoxyribonucleotides line up in order, and two new DNA molecules are produced.
2. Transcription is the process by which RNA is produced to carry genetic information from the nucleus to the ribosomes. This occurs when a segment of the DNA double helix unwinds, and complementary ribonucleotides line up to produce messenger RNA (mRNA).

3. Translation is the process by which mRNA directs protein synthesis. Each mRNA has segments called codons along its chain. These codons are ribonucleotide triads that are recognized by small amino acid carrying molecules of transfer RNA (tRNA), which then deliver the appropriate amino acids needed for protein synthesis.

Small DNA segments can be synthesized in the laboratory, and commercial "gene machines" are available for automating the work. Sequencing of DNA can be carried out rapidly and efficiently by the Maxam–Gilbert method.

33.18 Summary of reactions

1. Electrophilic aromatic substitution in five-membered heterocycles (Section 33.3)
   a. Bromination

   Furan $\xrightarrow[\text{Dioxane}]{Br_2}$ 2-bromofuran (Br)

   b. Nitration

   Pyrrole $\xrightarrow[\text{Acetic anhydride}]{HNO_3}$ 2-nitropyrrole (NO_2)

   c. Friedel–Crafts acylation

   Thiophene $\xrightarrow[SnCl_4]{CH_3COCl}$ 2-acetylthiophene ($\overset{O}{\overset{\|}{C}}CH_3$)

2. Nucleophilic aromatic substitution of halopyridines (Section 33.6)

   2-chloropyridine (Cl) $\xrightarrow[C_2H_5OH]{^{-}:OC_2H_5}$ 2-ethoxypyridine (OC_2H_5) $+ \ :Cl^-$

ADDITIONAL PROBLEMS

33.19 Although pyrrole is a much weaker base than most other amines, it is a much stronger acid ($pK_a \approx 15$ for pyrrole versus 35 for diethylamine). The N—H proton is readily abstracted by base to yield the pyrrole anion, $C_4H_4N^-$. Explain.

33.20 Substituted pyrroles are often prepared by treatment of the appropriate 1,4-diketone with ammonia. Suggest a mechanism by which this reaction occurs.

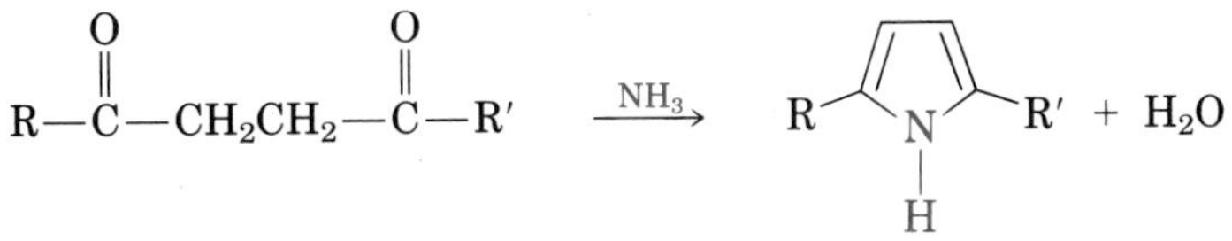

$$R-\overset{O}{\overset{\|}{C}}-CH_2CH_2-\overset{O}{\overset{\|}{C}}-R' \xrightarrow{NH_3} \text{2-R-5-R'-pyrrole (N–H)} + H_2O$$

33.21 Nitrofuroxime is a pharmaceutical agent used in the treatment of urinary tract infections. Propose a synthesis of nitrofuroxime from furfural.

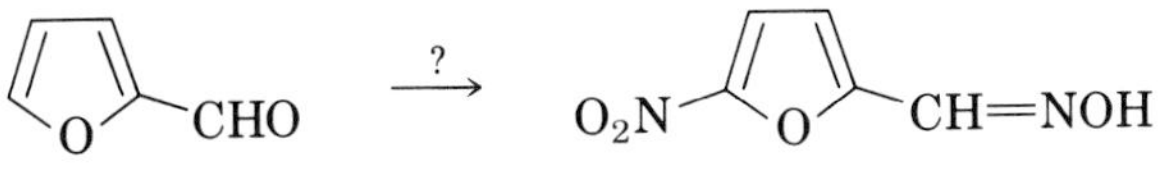

Furfural (2-furyl–CHO) $\xrightarrow{?}$ Nitrofuroxime (O_2N–furan–CH=NOH)

Furfural | Nitrofuroxime

33.22 Oxazole is a five-membered aromatic heterocycle. Draw an orbital picture of oxazole, showing all *p* orbitals and all lone-pair orbitals. Would you expect oxazole to be more basic or less basic than pyrrole? Explain.

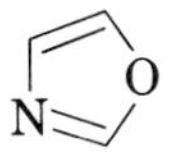

Oxazole

33.23 3,5-Dimethylisoxazole is prepared by reaction of 2,4-pentanedione with hydroxylamine. Propose a mechanism for this reaction.

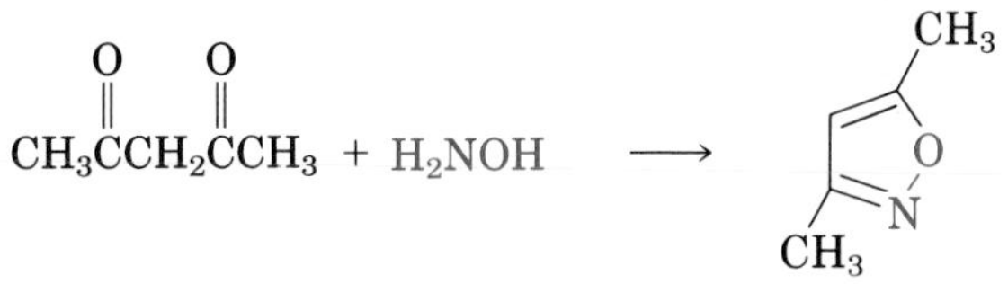

$$CH_3\overset{O}{\overset{\|}{C}}CH_2\overset{O}{\overset{\|}{C}}CH_3 + H_2NOH \longrightarrow \text{3,5-dimethylisoxazole}$$

3,5-Dimethylisoxazole

33.24 Write the products of the reaction of furan with each of the following reagents:

(a) Br_2, dioxane, 0°C
(b) HNO_3, acetic anhydride
(c) CH_3COCl, $SnCl_4$
(d) H_2/Pd
(e) SO_3, pyridine

33.25 Define the following terms:

(a) Heterocycle
(b) DNA
(c) Base pair
(d) Transcription
(e) Translation
(f) Replication (of DNA)
(g) Codon
(h) Anticodon

33.26 Draw the complete structure of the ribonucleotide codon UAC. For what amino acid does this sequence code?

33.27 Draw the complete structure of the deoxyribonucleotide sequence from which the mRNA codon in Problem 33.26 was transcribed.

33.28 Give an mRNA sequence that would code for synthesis of methionine enkephalin:

H-Tyr-Gly-Gly-Phe-Met-OH

33.29 Isoquinolines are often synthesized by the Bischler—Napieralski cyclization of an *N*-acyl-2-phenylethyl amine with strong acid and P_2O_5, followed by oxidation of the initially formed dihydroisoquinoline. Suggest a mechanism by which this cyclization occurs.

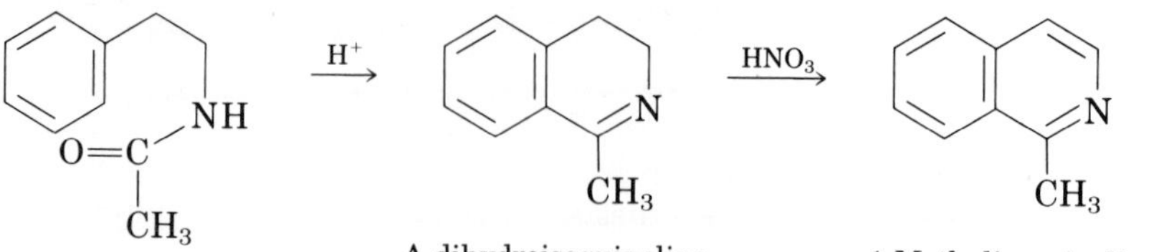

NH, O=C, CH_3 $\xrightarrow{H^+}$ N, CH_3 $\xrightarrow{HNO_3}$ N, CH_3

A dihydroisoquinoline — 1-Methylisoquinoline

33.30 Quinolines are often prepared by the Skraup synthesis,[5] in which an aniline reacts with an α,β-unsaturated aldehyde and the dihydroquinoline product is oxidized. Suggest a mechanistic pathway for the Skraup reaction.

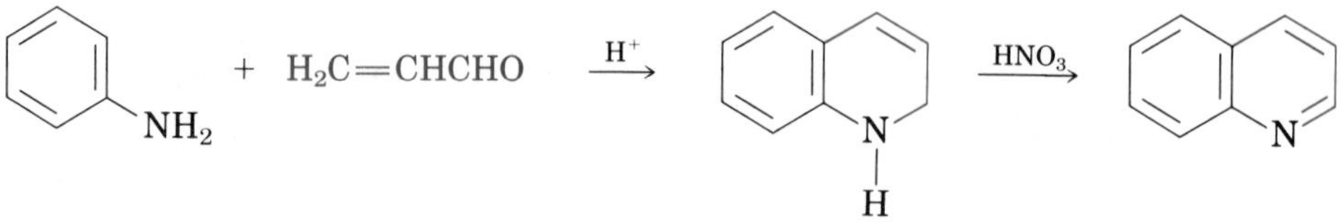

NH_2 + $H_2C{=}CHCHO$ $\xrightarrow{H^+}$ N–H $\xrightarrow{HNO_3}$ N

1,2-Dihydroquinoline — Quinoline

33.31 Show the steps involved in a laboratory synthesis of the DNA fragment having the sequence CTAG.

33.32 Review the mechanism shown in Figure 33.15 for the cleavage of deoxyguanosine residues, and propose a mechanism to account for the similar cleavage of deoxyadenosine residues in a DNA chain. Recall that deoxyadenosine is first methylated at N3 prior to hydrolysis.

[5] Hans Zdenko Skraup (1850–1910); b. Austria; professor, Graz, Vienna, Austria.

CHAPTER 34
SYNTHETIC POLYMERS

No other group of synthetic organic compounds has had as great an impact on our day-to-day living as the synthetic polymers. As plastics, adhesives, and paints, synthetic polymers have a multitude of uses, from the trivial foam coffee cup to the life-saving artificial heart valve. Polymer synthesis is a major part of the chemical industry, and annual production figures for some of the more important polymers are shown in Table 34.1.

TABLE 34.1 Production figures for some major polymers (1980)

Polymer	*U.S. production (lb/yr, millions)*	*Polymer*	*U.S. production (lb/yr, millions)*
Polyethylene		Acrylonitrile/butadiene/styrene	920
Low density	4400		
High density	7300	Acrylic fibers	780
Polypropylene	3600	Epoxy adhesives	320
Polystyrene	3500	Phenolic resins	1500
Poly(vinyl chloride)	5500	Urea/formaldehyde resins	1200
Nylon	3000		
Polyesters	4000		

Polymers are not new to us. A **polymer** is simply a large molecule built up by repetitive bonding together of many smaller units (**monomers**); we have already studied the major classes of biopolymers. Thus, cellulose is a large carbohydrate polymer built of repeating glucose units; proteins are large polyamides built of repeating amino acid units; and nucleic acids are large molecules built of repeating nucleotide units. Synthetic polymers are chemically much simpler than most biopolymers, since the repeating units tend to be small, simple, and inexpensive molecules. There is, however, an immense diversity in the structure and properties of synthetic polymers, depending on the exact nature of the monomer and on the reaction conditions used for polymerization.

Polymers can be classified in many ways—by method of synthesis, by structure, by physical properties, or by end use, to name a few. Let's begin by looking at the organic chemistry of polymer synthesis and then see how polymer structure can be correlated with uses and physical properties.

34.1 General classes of polymers

Synthetic polymers can be classified by their method of synthesis as either chain-growth polymers or step-growth polymers. These categories are necessarily broad and imprecise but nevertheless provide a useful distinction. **Chain-growth polymers,** called *addition polymers* in the older terminology, are produced by chain-reaction polymerization in which an initiator adds to a carbon–carbon double bond to yield a reactive intermediate. The polymer is built as more monomers add successively to the reactive end of the growing chain. The initiator may be either an anion, a cation, or a radical, and the polymers produced in this way usually have only carbon atoms in their backbones. Polyethylene, produced by radical-initiated polymerization of ethylene, is by far the most common example of a chain-growth polymer.

$$\text{In}\cdot + CH_2{=}CH_2 \longrightarrow [\text{In}{-}CH_2CH_2\cdot] \xrightarrow{CH_2=CH_2} [\text{In}{-}CH_2CH_2CH_2CH_2\cdot]$$

A radical initiator

Repeat many times

A section of a polyethylene chain

Step-growth polymers, called *condensation polymers* in the older terminology, are produced by polymerization processes in which the bond-forming step is one of the fundamental polar reactions we have studied. Reaction occurs between two difunctional molecules, and each bond in the polymer is formed independently of the others. The polymer produced normally has the two monomers in an alternating order and usually has other atoms in addition to carbon in the main chain. Nylon, a polyamide formed by reaction of a diacid and a diamine, is the most common example of a step-growth polymer.

$$H_2N{-}(CH_2)_n{-}NH_2 + HOOC{-}(CH_2)_m{-}COOH \xrightarrow{\Delta} {-}NH{-}(CH_2)_n{-}NH{-}\overset{O}{\overset{\|}{C}}{-}(CH_2)_m{-}\overset{O}{\overset{\|}{C}}{-} + H_2O$$

Nylon, a step-growth polymer (a polyamide)

34.2 Radical polymerization of alkenes

Many low-molecular-weight alkenes undergo a rapid polymerization reaction when treated with catalytic amounts of a radical initiator. Polyethylene, one of the first alkene chain-growth polymers to be manufactured commercially, and the simplest, has been produced since 1943. It has a current annual U.S. production volume of nearly 12 billion lb.

Polymerization of ethylene is usually carried out at high pressure (1000–3000 atm) and high temperature (100–250°C) with a radical catalyst such as benzoyl peroxide; the resultant polymer may have anywhere from a few hundred to a few thousand monomer units in the chain. As with all radical chain reactions, three kinds of steps are required: initiation steps, propagation steps, and termination steps.

Initiation occurs when trace amounts of radicals are generated by the catalyst (step 1). One of these radicals adds to ethylene to generate a new carbon radical (step 2), and the polymerization is off and running:

Step 1.

$$C_6H_5{-}\overset{\displaystyle O}{\overset{\|}{C}}{-}O{-}O{-}\overset{\displaystyle O}{\overset{\|}{C}}{-}C_6H_5 \xrightarrow[\Delta]{} 2\ C_6H_5\overset{\displaystyle O}{\overset{\|}{C}}{-}O\cdot$$

Benzoyl peroxide — In· Benzoyloxy radical

Step 2.

$$In\cdot + CH_2{=}CH_2 \longrightarrow In{-}CH_2{-}CH_2\cdot$$

Propagation occurs when the carbon radical adds to another ethylene molecule (step 3). Repetition of step 3 builds the polymer chain:

Step 3.

$$In{-}CH_2{-}CH_2\cdot + CH_2{=}CH_2 \longrightarrow In{-}CH_2{-}CH_2{-}CH_2{-}CH_2\cdot \xrightarrow[\text{many times}]{\text{Repeat}} In{-}(CH_2{-}CH_2)_nCH_2{-}CH_2\cdot$$

Eventually, the polymer chain is terminated by reactions that consume the radical. Combination (step 4) or disproportionation (step 5) of two radicals are possible chain-terminating reactions:

Step 4.

$$2\ In{-}(CH_2{-}CH_2)_n{-}CH_2CH_2\cdot \longrightarrow In(CH_2CH_2)_nCH_2CH_2{-}CH_2CH_2(CH_2CH_2)_nIn$$

Step 5.

$$2\ In(CH_2CH_2)_nCH_2CH_2\cdot \longrightarrow In(CH_2CH_2)_nCH{=}CH_2 + In(CH_2CH_2)_nCH_2CH_3$$

Many substituted ethylene monomers undergo radical reaction to yield polymers with substituent groups (denoted by a circled S) regularly spaced along the polymer backbone.

$$\underset{\text{Monomer}}{CH_2{=}\overset{Ⓢ}{CH}} \xrightarrow[\text{polymerization}]{\text{Radical}} \underset{\text{Polymer}}{\left(CH_2\overset{Ⓢ}{CH}CH_2\overset{Ⓢ}{CH}CH_2\overset{Ⓢ}{CH}\right)}$$

Table 34.2 shows some of the more important of these **vinyl monomers** and lists the industrial uses of the different polymers.

TABLE 34.2 **Some chain-growth polymers and their uses**

Monomer name	*Formula*	*Trade or common names of polymer*	*Uses*
Ethylene	$H_2C{=}CH_2$	Polyethylene	Packaging, bottles, cable insulation, films and sheets
Propene (propylene)	$H_2C{=}CHCH_3$	Polypropylene	Automotive moldings, rope, carpet fibers
Chloroethylene (vinyl chloride)	$H_2C{=}CHCl$	Poly(vinyl chloride), Tedlar	Insulation, films, pipes
Styrene	$H_2C{=}CHC_6H_5$	Polystyrene, Styron	Foam and molded articles
Tetrafluoroethylene	$F_2C{=}CF_2$	Teflon	Valves and gaskets, coatings
Acrylonitrile	$H_2C{=}CHCN$	Orlon, Acrilan	Fibers
Methyl methacrylate	$H_2C{=}C(CH_3)CO_2CH_3$	Plexiglas, Lucite	Molded articles, paints
Vinyl acetate	$H_2C{=}CHOCOCH_3$	Poly(vinyl acetate)	Paints, adhesives
Vinyl alcohol	"$H_2C{=}CHOH$"	Poly(vinyl alcohol)	Fibers, adhesives

Note that vinyl alcohol, the monomer corresponding to poly(vinyl alcohol), is an unstable enol isomer that, if prepared, would tautomerize rapidly to acetaldehyde. Poly(vinyl alcohol) is therefore made by hydrolysis of poly(vinyl acetate):

$$\left(\!-CH_2-\underset{\displaystyle OCOCH_3}{CH}-CH_2-\underset{\displaystyle OCOCH_3}{CH}-\right) \xrightarrow[H_2O]{^-OH} \left(\!-CH_2\underset{\displaystyle OH}{CH}CH_2\underset{\displaystyle OH}{CH}-\right)$$

Poly(vinyl acetate) Poly(vinyl alcohol)

PROBLEM

34.1 How might the polymers (a)–(c) be prepared? Show the monomer units you would use.

(a) $$\left(\!-CH_2-\underset{\displaystyle OCH_3}{CH}-CH_2-\underset{\displaystyle OCH_3}{CH}-CH_2-\underset{\displaystyle OCH_3}{CH}-\right)$$

(b) 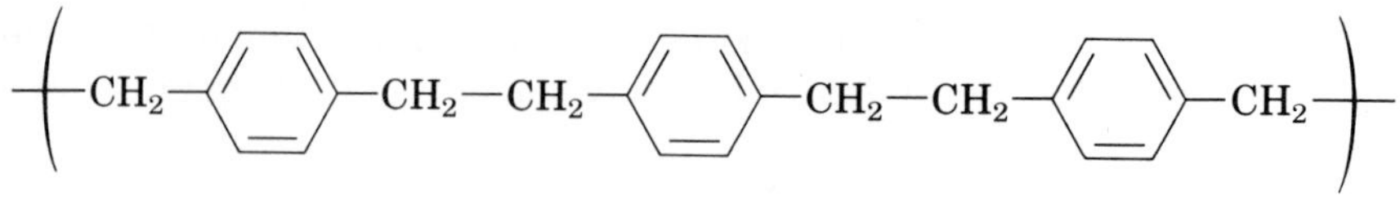 $\left(CH_2-C_6H_4-CH_2-CH_2-C_6H_4-CH_2-CH_2-C_6H_4-CH_2 \right)$

(c) $\left(\underset{Cl}{CH}-\underset{Cl}{CH}-\underset{Cl}{CH}-\underset{Cl}{CH}-\underset{Cl}{CH}-\underset{Cl}{CH} \right)$

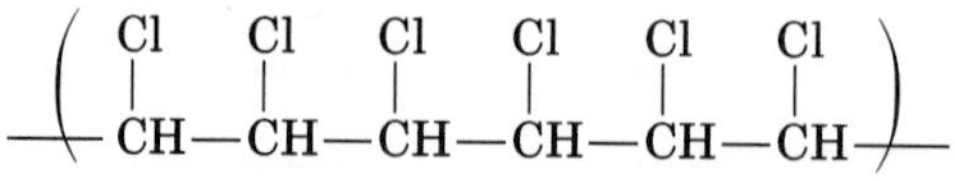

PROBLEM

34.2 How can you account for the fact that radical polymerization of styrene yields a product in which the phenyl substituents are on alternate carbon atoms rather than on neighboring carbons?

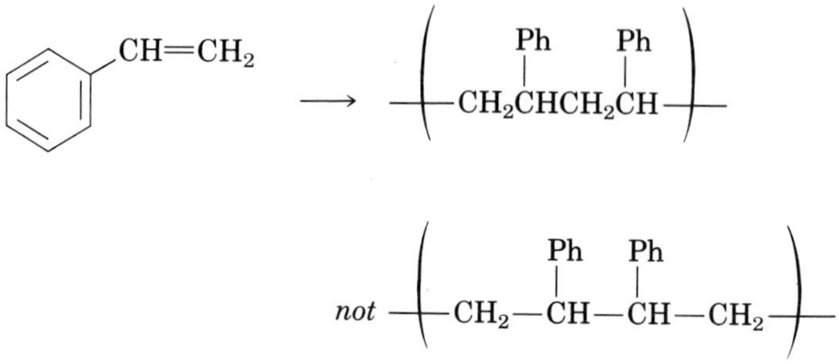

$$C_6H_5CH{=}CH_2 \longrightarrow \left(CH_2\overset{Ph}{C}HCH_2\overset{Ph}{C}H \right)$$

$$\textit{not} \left(CH_2-\overset{Ph}{CH}-\overset{Ph}{CH}-CH_2 \right)$$

34.3 Cationic polymerization

Certain alkene monomers can be polymerized by a cationic mechanism, as well as by a radical mechanism. Cationic polymerizations occur by a chain-reaction pathway and require the use of strong protic or Lewis acids as initiators. The key chain-carrying step is the electrophilic addition of a carbocation intermediate to the carbon–carbon double bond of another monomer unit.

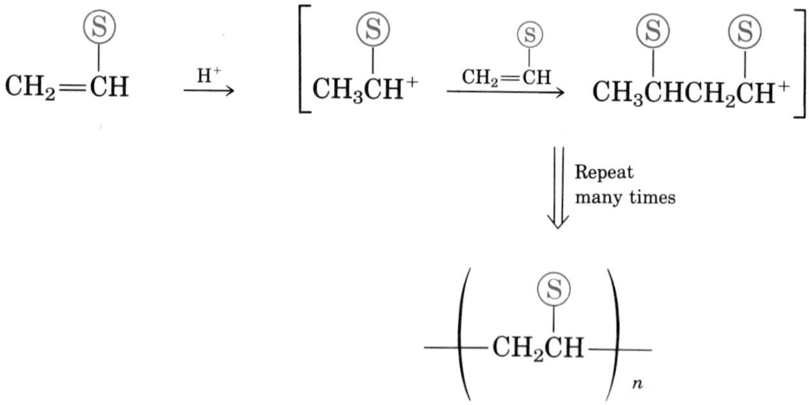

$$CH_2{=}\overset{(S)}{CH} \xrightarrow{H^+} \left[CH_3\overset{(S)}{C}H^+ \xrightarrow{CH_2{=}\overset{(S)}{C}H} CH_3\overset{(S)}{C}HCH_2\overset{(S)}{C}H^+ \right]$$

$$\Big\Downarrow \text{Repeat many times}$$

$$\left(CH_2\overset{(S)}{C}H \right)_n$$

As we might expect, vinyl monomers with electron-donating substituents polymerize much more readily than do monomers with electron-withdrawing substituents. Thus, ethylene, vinyl chloride, and acrylonitrile do not polymerize easily under cationic conditions, but 2-methylpropene polymerizes nicely. This difference in behavior simply reflects the electron densities of the carbon–carbon double bonds and the difference in stability of the potential chain-carrying intermediate cations.

$$CH_2{=}\overset{\overset{\text{(S)}}{\downarrow}}{C}H \xrightarrow{H^+} \left[H{-}CH_2{-}\overset{\overset{\text{(S)}}{\downarrow}}{C}H^+\right]$$

Electron-rich alkene; stabilized cation intermediate when substituent is electron-donating; good reaction

$$CH_2{=}\overset{\overset{\text{(S)}}{\uparrow}}{C}H \overset{H^+}{\not\longrightarrow} \left[H{-}CH_2{-}\overset{\overset{\text{(S)}}{\uparrow}}{C}H^+\right]$$

Electron-poor alkene; destabilized cation intermediate when substituent is electron-withdrawing; poor reaction

The polyisobutylene used for the manufacture of truck tire inner tubes is prepared commercially by treatment of isobutylene with BF_3 catalyst at $-80°C$.

$$\underset{\text{2-Methylpropene (isobutylene)}}{CH_2{=}C(CH_3)_2} \xrightarrow{H^+} \left[CH_3{-}\overset{+}{C}(CH_3)_2 \xrightarrow{CH_2{=}C(CH_3)_2} CH_3{-}\underset{CH_3}{\overset{CH_3}{C}}{-}CH_2{-}\underset{CH_3}{\overset{CH_3}{C^+}}\right]$$

$$\Downarrow$$

$$\left(CH_2{-}\underset{CH_3}{\overset{CH_3}{C}}\right)_n$$

Polyisobutylene

PROBLEM

34.3 List the expected reactivity order of the following monomers to cationic polymerization. Explain your ordering.

$H_2C{=}CHCH_3$, $H_2C{=}CHCl$, $H_2C{=}CH{-}C_6H_5$, $H_2C{=}CHCO_2CH_3$

34.4 Anionic polymerization

Alkene monomers with electron-withdrawing (anion-stabilizing) substituents can be polymerized by anionic catalysts. A chain reaction occurs in which the key step is nucleophilic addition of an anion to the unsaturated monomer (Michael reaction, Section 26.12).

$$CH_2{=}CH(Ⓢ) + Nu{:}^- \longrightarrow \left[Nu{-}CH_2{-}CH(Ⓢ){:}^- \longrightarrow Nu{-}CH_2CH(Ⓢ)CH_2CH(Ⓢ){:}^- \right] \Longrightarrow \left(CH_2{-}CH(Ⓢ) \right)_n$$

where Nu = A nucleophilic initiator
Ⓢ = An electron-withdrawing substituent

The monomers acrylonitrile ($H_2C{=}CHCN$), methyl methacrylate [$H_2C{=}C(CH_3)COOCH_3$], and styrene ($H_2C{=}CHC_6H_5$) can all be anionically polymerized, although radical-initiated polymerization is preferred commercially.

One particularly interesting example of anionic polymerization accounts for the remarkable properties of "super glue," one drop of which is claimed to support 2000 lb. Super glue is simply a solution of highly pure methyl α-cyanoacrylate. Since the carbon–carbon double bond has two electron-withdrawing groups, anionic addition is particularly easy and particularly rapid. Trace amounts of water or bases on the surface of an object are sufficient to initiate polymerization of the cyanoacrylate and bind articles together. Skin is a good source of the necessary basic initiators, and many people have found their fingers stuck together after inadvertently touching super glue.

$$CH_2{=}C(CN)(COOCH_3) + HO^- \longrightarrow \left[HO{-}CH_2{-}C(CN)(COOCH_3){:}^- \right] \Longrightarrow \left(CH_2{-}C(CN)(COOCH_3) \right)_n$$

Methyl α-cyanoacrylate

Super glue

PROBLEM

34.4 Place the following monomers in order with respect to their expected reactivity toward anionic polymerization. Explain your ordering.

$$H_2C{=}CHCH_3, \quad H_2C{=}CF_2, \quad H_2C{=}CHCN, \quad H_2C{=}CHC_6H_5$$

PROBLEM

34.5 Poly(ethylene glycol), or Carbowax, can be made by base-induced polymerization of ethylene oxide. Show the mechanism by which this reaction occurs.

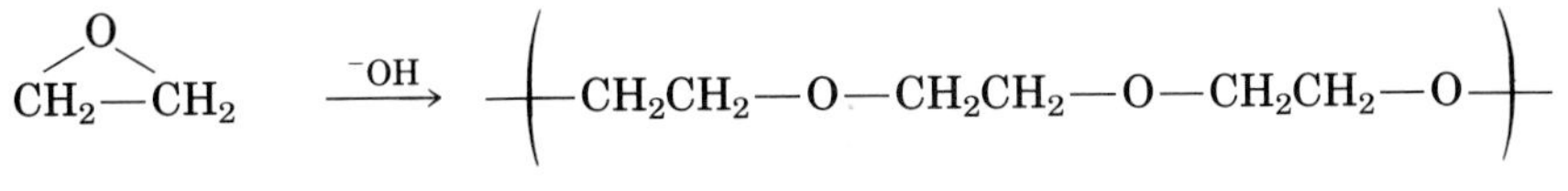

$$\underset{\text{Ethylene oxide}}{\overset{O}{CH_2{-}CH_2}} \xrightarrow{^{-}OH} \underset{\text{Carbowax}}{\left(CH_2CH_2{-}O{-}CH_2CH_2{-}O{-}CH_2CH_2{-}O \right)}$$

34.5 Chain branching during polymerization

Polymerization of unsaturated monomers is complicated in practice by several factors that greatly affect the properties of the product. One such problem is that radical polymerization yields a product that is not linear, but has numerous branches in it. Branches arise when the radical end of a growing chain abstracts a hydrogen atom from the middle of a chain to yield an internal radical site that continues the polymerization. The most common kind of branching, termed **short-chain branching,** arises from intramolecular hydrogen atom abstraction from a position four carbon atoms away from the chain end (Figure 34.1).

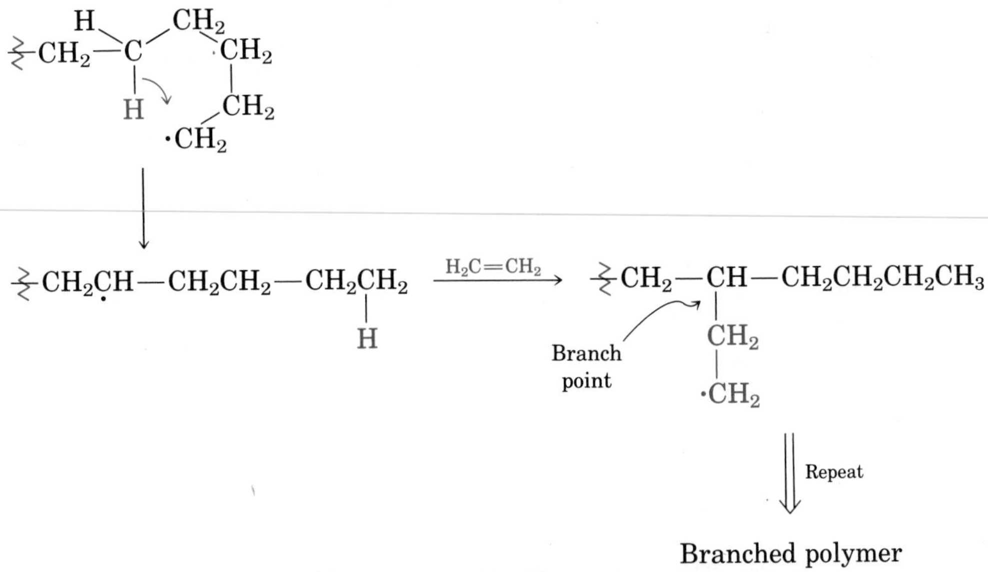

$$\sim CH_2{-}C(H)(H){-}CH_2{-}CH_2{-}CH_2{-}\cdot CH_2$$

$$\downarrow$$

$$\sim CH_2\dot{C}H{-}CH_2CH_2{-}CH_2CH_2{-}H \xrightarrow{H_2C=CH_2} \sim CH_2{-}CH(CH_2{-}\cdot CH_2){-}CH_2CH_2CH_2CH_3$$

Branch point

Repeat

Branched polymer

Figure 34.1. Short-chain branching during the polymerization of ethylene.

Alternatively, intermolecular hydrogen atom abstraction can take place by reaction of the radical end of one chain with the middle of another chain. **Long-chain branching** results from this kind of reaction (Figure 34.2). Studies have shown that short-chain branching occurs about 50 times more often than long-chain branching.

$$\sim CH_2CH_2CH(H)CH_2CH_2CH_2 \sim \; + \; \cdot CH_2CH_2 \sim$$

$$\downarrow$$

$$\sim CH_2CH_2\dot{C}HCH_2CH_2CH_2 \sim \; + \; H{-}CH_2CH_2 \sim$$

$$\downarrow H_2C{=}CH_2$$

$$\sim CH_2CH_2CH(CH_2CH_2\cdot)CH_2CH_2CH_2 \sim$$

Branch point

$\xrightarrow{\text{Repeat}}$ Branched polymer

Figure 34.2. Long-chain branching during the polymerization of ethylene.

Chain branching is a common occurrence during radical polymerizations and is not restricted to polyethylene. Polypropylene, polystyrene, and poly(methyl methacrylate) all contain branched chains.

34.6 Stereochemistry of polymerization: Ziegler–Natta catalysts

Yet another complication that arises during alkene polymerization has to do with stereochemistry. Although we did not point it out earlier, polymerization of substituted alkenes leads to polymers with numerous chiral centers on the backbone. For example, we can imagine that propylene might polymerize with any of the three stereochemical outcomes shown in Figure 34.3. The conformation in which all methyl groups are on the same side of the zigzag backbone is called **isotactic;** that in which the methyl groups regularly alternate on opposite sides of the backbone is called **syndiotactic;** and the conformation in which the methyl groups are randomly oriented is called **atactic.**

The three different stereochemical forms of polypropylene all have somewhat different properties, and all three can be made by the proper choice of polymerization conditions. Branched atactic polymers arise from normal radical chain polymerizations, but the use of special Ziegler–Natta catalysts[1,2] allows preparation of isotactic and syndiotactic forms.

[1] Karl Ziegler (1889–1976); b. Helsa, near Kassel, Germany; Ph.D. Marburg University; director, Max Planck Institute for Coal Research, Mülheim-Ruhr, Germany; Nobel prize (1963).

[2] Giulio Natta (1903–1979); b. Imperia, Italy; D.C.E. Milan Polytechnic Institute; professor, Milan Polytechnic Institute; Nobel prize (1963).

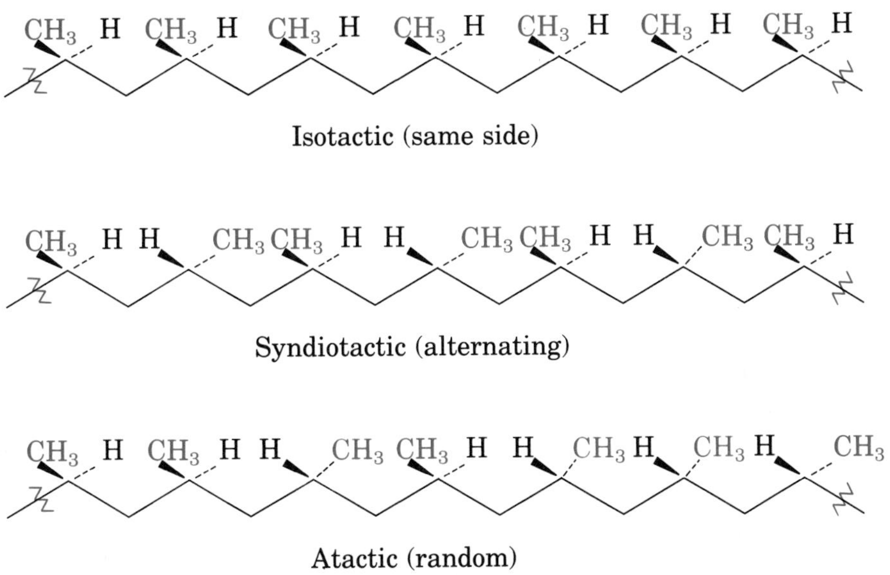

Isotactic (same side)

Syndiotactic (alternating)

Atactic (random)

Figure 34.3. Isotactic, syndiotactic, and atactic forms of polypropylene.

Ziegler–Natta catalysts are organometallic transition-metal complexes prepared by treating a trialkylaluminum with a titanium compound. Triethylaluminum and titanium trichloride form a typical preparation, although the precise structure of the active catalyst is still unknown.

$$(CH_3CH_2)_3Al + TiCl_3 \longrightarrow \text{Ziegler–Natta catalyst}$$

Ziegler–Natta catalysts were introduced in 1953 and immediately revolutionized the field of polymer chemistry, largely because of two advantages:

1. Ziegler–Natta polymers are linear and have practically no chain branching.
2. Ziegler–Natta polymers are stereochemically regular. Either isotactic or syndiotactic forms can be produced, depending on the exact catalyst system used. All commercial polypropylene is now produced by the Ziegler–Natta process, since the product has greater strength, stiffness, and resistance to cracking than the branched atactic polypropylene prepared by radical polymerization.

Linear polyethylene produced by the Ziegler–Natta process (called *high-density polyethylene*) is a highly crystalline polymer with 500–1000 ethylene units per chain. High-density polyethylene has greater strength and heat resistance than the product of radical-induced polymerization (*low-density polyethylene*) and is used to produce plastic squeeze bottles and molded housewares.

The precise mechanism by which Ziegler–Natta catalysts operate is not clear, but the key chain-lengthening steps undoubtedly involve formation of alkyltitanium species, followed by coordination of alkene monomer to the titanium and insertion of coordinated alkene into the carbon–titanium bond.

$$\text{-}CH_2CH_2\text{—}Ti \xrightarrow{CH_2=CH_2} \text{-}CH_2CH_2\text{—}Ti(\cdots H_2C{=}CH_2) \longrightarrow \text{-}CH_2CH_2CH_2CH_2\text{—}Ti$$

An alkyltitanium intermediate

Chain-extended alkyltitanium intermediate

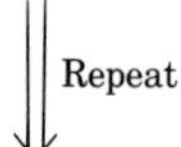

PROBLEM

34.6 Account for the fact that vinylidene chloride, $H_2C{=}CCl_2$, does not polymerize in isotactic, syndiotactic, and atactic forms.

PROBLEM

34.7 Polymers such as polypropylene contain a large number of chiral carbon atoms. Would you therefore expect samples of either isotactic, syndiotactic, or atactic polypropylene to rotate plane-polarized light? Explain.

34.7 Diene polymers: natural and synthetic rubbers

We have discussed only the polymerization of simple alkene monomers up to this point, but the same principles apply to the polymerization of conjugated dienes. Diene polymers are structurally more complex, however, since double bonds remain every four carbon atoms along the chain. These double bonds may be either cis or trans, and the proper choice of Ziegler–Natta catalyst allows preparation of either geometry. Note that the polymerization reaction corresponds to 1,4 addition of the growing chain to each conjugated diene monomer (recall 1,4 ionic addition to dienes, Section 13.5).

$$CH_2{=}CH{-}CH{=}CH_2 \quad CH_2{=}CH{-}CH{=}CH_2$$

1,3-Butadiene

$$\longrightarrow \left(CH_2{-}CH{=}CH{-}CH_2{-}CH_2{-}CH{=}CH{-}CH_2 \right)$$

cis-Poly(1,3-butadiene)

$$\longrightarrow \left(CH_2{-}CH{=}CH{-}CH_2{-}CH_2{-}CH{=}CH{-}CH_2 \right)$$

trans-Poly(1,3-butadiene)

Natural rubber is a polymer of isoprene in which the double bonds have cis stereochemistry. Gutta-percha, the all-trans isomer of natural rubber, is also known and occurs as the exudate of certain trees. Gutta-percha is harder and more brittle than rubber but finds a variety of minor applications, including use as the covering on golf balls.

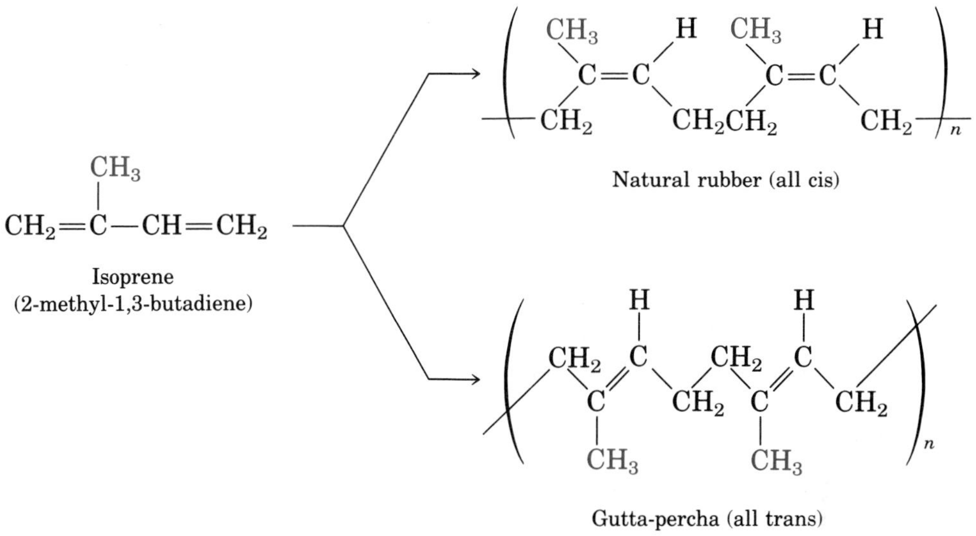

$$CH_2{=}C(CH_3){-}CH{=}CH_2$$

Isoprene
(2-methyl-1,3-butadiene)

$$\longrightarrow \left(-CH_2-C(CH_3){=}CH-CH_2CH_2-C(CH_3){=}CH-CH_2- \right)_n$$

Natural rubber (all cis)

$$\longrightarrow \left(-CH_2-C(CH_3){=}CH-CH_2-CH_2-C(CH_3){=}CH-CH_2- \right)_n$$

Gutta-percha (all trans)

A number of different synthetic rubbers are produced commercially by diene polymerization. Both *cis-* and *trans*-polyisoprene can be produced under Ziegler–Natta conditions, and synthetic rubber is quite similar to the natural material. Chloroprene (2-chloro-1,3-butadiene) is polymerized commercially to yield neoprene, an excellent, though expensive, synthetic rubber with good weather resistance. Neoprene is used in the production of industrial hoses and gloves, among other things.

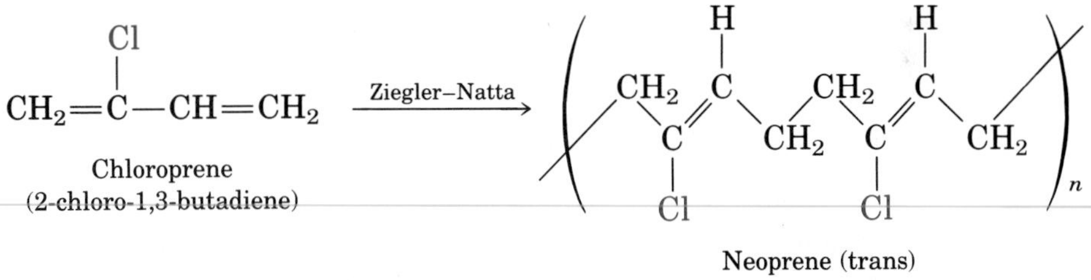

$$CH_2{=}C(Cl){-}CH{=}CH_2 \xrightarrow{\text{Ziegler–Natta}} \left(-CH_2-C(Cl){=}CH-CH_2-CH_2-C(Cl){=}CH-CH_2- \right)_n$$

Chloroprene
(2-chloro-1,3-butadiene)

Neoprene (trans)

Both natural and synthetic rubbers are soft and tacky unless hardened by the process of **vulcanization.** Vulcanization was discovered in 1839 by Charles Goodyear (of subsequent tire fame) and involves heating the polymer with a few percent by weight of sulfur. The result is a much harder rubber with greatly improved resistance to wear and abrasion. The chemistry of vulcanization is complex but involves formation of sulfur bridges or **cross-links** between polymer chains. Cross-linked polymers tend to be rigid, because the individual chains can no longer slip over each other but are instead locked together into immense single molecules (Figure 34.4). Note the similarity in structure between a vulcanized rubber and a peptide that has cysteine cross-links (Section 31.7).

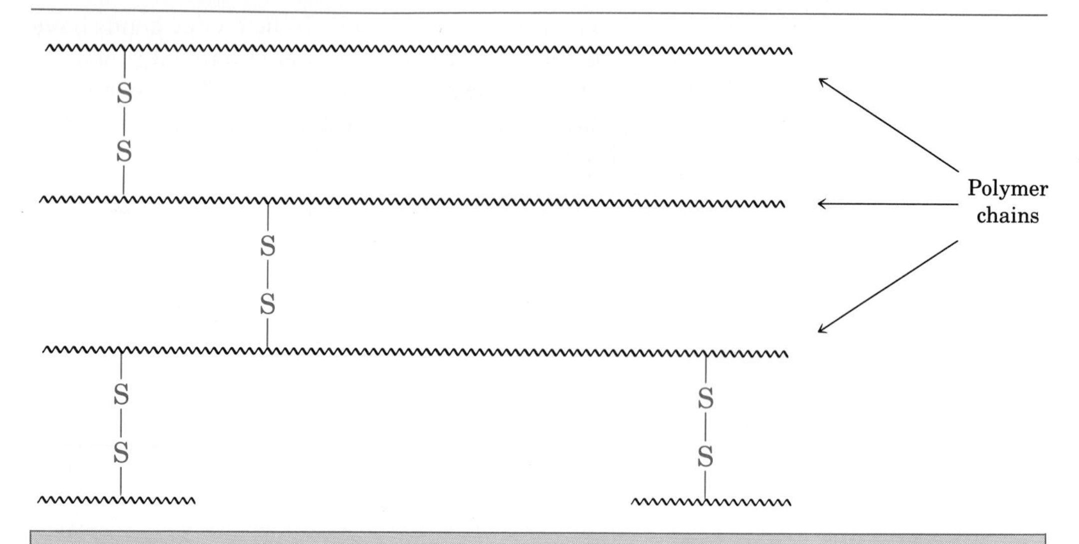

Figure 34.4. Sulfur-cross-linked chains resulting from the vulcanization of rubber.

PROBLEM

34.8 Diene polymers contain occasional vinyl branches along the chain. Show how these branches might arise.

$$CH_2{=}CH{-}CH{=}CH_2 \longrightarrow \sim CH_2CH{=}CHCH_2CH_2\underset{\underset{CH=CH_2}{|}}{C}HCH_2CH{=}CHCH_2 \sim$$

A vinyl branch

PROBLEM

34.9 Tires made of natural rubber tend to crack and weather rapidly in areas around major cities where high levels of ozone and other industrial pollutants are found. Explain.

34.8 Copolymers

Up to now we have discussed only **homopolymers,** polymers that are made up of identical repeating units. In practice, however, **copolymers** are more common and more important commercially. Copolymers are obtained when two or more different monomers are allowed to polymerize together. For example, copolymerization of vinyl chloride with vinylidene chloride (1,1-dichloroethylene) leads to the well-known polymer Saran.

$$\underset{\text{Vinyl chloride}}{CH_2{=}\overset{\overset{Cl}{|}}{C}H} + \underset{\text{Vinylidene chloride}}{CH_2{=}CCl_2} \longrightarrow \underset{\text{Saran}}{\left(CH_2\overset{\overset{Cl}{|}}{C}HCH_2\overset{\overset{Cl}{|}}{\underset{\underset{Cl}{|}}{C}}{-}CH_2\overset{\overset{Cl}{|}}{C}H \right)_n}$$

Copolymerization of monomer mixtures often leads to materials with properties quite different from those of either corresponding homopolymer and gives the polymer chemist a vast amount of flexibility for devising new materials. Table 34.3 lists some common copolymers and indicates their commercial applications.

TABLE 34.3 Some common copolymers and their uses

Monomer name	*Formula*	*Trade or common name of polymer*	*Uses*
Vinyl chloride Vinylidene chloride	$H_2C{=}CHCl$ $H_2C{=}CCl_2$	Saran	Food wrapping, fibers
Styrene (25%) Butadiene (75%)	$H_2C{=}CHC_6H_5$ $H_2C{=}CHCH{=}CH_2$	SBR (styrene–butadiene rubber)	Tires
Hexafluoropropene Vinylidene fluoride	$F_2C{=}CFCF_3$ $H_2C{=}CF_2$	Viton	Gaskets, rubber articles
Acrylonitrile Butadiene	$H_2C{=}CHCN$ $H_2C{=}CH{-}CH{=}CH_2$	Nitrile rubber	Latex, adhesives, gasoline hoses
Isobutylene Isoprene	$H_2C{=}C(CH_3)_2$ $H_2C{=}C(CH_3)CH{=}CH_2$	Butyl rubber	Inner tubes
Acrylonitrile Butadiene Styrene	$H_2C{=}CHCN$ $H_2C{=}CHCH{=}CH_2$ $H_2C{=}CHC_6H_5$	ABS (initials of three monomers)	Pipes, high-impact applications

Several different structural types of copolymers can be defined, depending on the distribution of monomer units in the chain. If we imagine, for example, that monomer A and monomer B are being copolymerized, the resultant product might have a random distribution of the two units throughout the chain, or it might have an alternating distribution.

$$\text{A + B} \longrightarrow \begin{cases} \sim\text{A—A—B—A—B—B—A—B—A—A—A—B}\sim & \text{Random copolymer} \\ \sim\text{A—B—A—B—A—B—A—B—A—B—A—B}\sim & \text{Alternating copolymer} \end{cases}$$

The exact distribution depends on such factors as the proportion of the two reactant monomers used and their relative reactivities. In practice, neither perfectly random nor perfectly alternating copolymers are usually found. Most copolymers tend more toward the alternating form but have many random imperfections.

Two other special forms of copolymers can be prepared under certain conditions—**block copolymers** and **graft copolymers.** Block copolymers are those in which different blocks of identical monomer units alternate with each other, and graft copolymers are those in which homopolymer branches of one monomer unit are "grafted" onto a homopolymer chain of another monomer unit.

~A—A—A—A—A—A—A—A—B—B—B—B—B—B—B—B~

Segment of a block copolymer

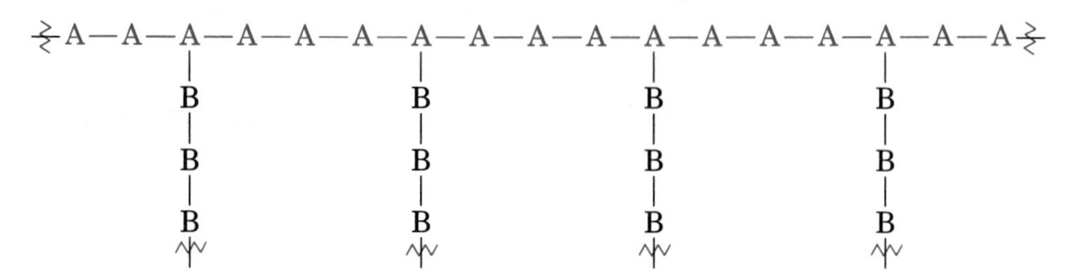

Segment of a graft copolymer

Block copolymers can be prepared by initiating the radical polymerization of one monomer to grow homopolymer chains, followed by addition of an excess of the second monomer. Graft copolymers can be made by gamma irradiation of a homopolymer chain in the presence of a second monomer. The high-energy irradiation knocks hydrogen atoms off the homopolymer chain at random points, thus generating radical sites that can initiate polymerization of the added monomer.

PROBLEM

34.10 Draw the structure of an alternating segment of butyl rubber, a copolymer of 2-methyl-1,3-butadiene and 2-methylpropene prepared under cationic conditions.

PROBLEM

34.11 One of the most important commercial applications of graft polymerization involves irradiation of polybutadiene, followed by addition of styrene. The product is used to make rubber soles for shoes. Draw the structure of a representative segment of this styrene–butadiene graft copolymer.

34.9 Step-growth polymers: nylon

Step-growth polymers are produced by polymerization reactions between two difunctional molecules, as we saw earlier. Each bond is formed in a discrete step, independent of all other bonds in the polymer; chain reactions are not involved. The key bond-forming step is usually one of the fundamental polar reactions that we studied earlier, as opposed to a radical reaction:

A~A + B~B ⟶ ~A~A—B~B—A~A—B~B~

where A and B are reactive functional groups.

A large number of different step-growth polymers have been made, and some of the more important are shown in Table 34.4.

TABLE 34.4 Some common step-growth polymers and their uses

Monomer name	*Formula*	*Trade or common name of polymer*	*Uses*
Adipic acid Hexamethylene diamine	$HOOC(CH_2)_4COOH$ $H_2N(CH_2)_6NH_2$	Nylon 66	Fibers, clothing, tire cord, bearings
Ethylene glycol Dimethyl terephthalate	$HOCH_2CH_2OH$ p-$C_6H_4(COOCH_3)_2$	Dacron, Terylene, Mylar	Fibers, clothing, tire cord, film
Caprolactam	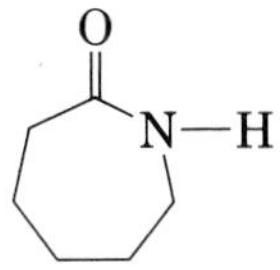 cyclic amide (C=O, N—H)	Nylon 6, Perlon	Fibers, large cast articles
Diphenyl carbonate Bisphenol A	$C_6H_5OCOOC_6H_5$ $HO-C_6H_4-C(CH_3)_2-C_6H_4-OH$	Lexan, polycarbonate	Molded articles, machine housings
Poly(2-butene-1,4-diol) Tolylene diisocyanate	$HO\mathord{+}CH_2CH{=}CHCH_2\mathord{+}_nOH$ $CH_3C_6H_3(N{=}C{=}O)_2$	Polyurethane, Spandex	Foams, fibers, coatings

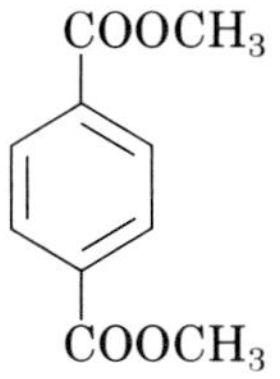

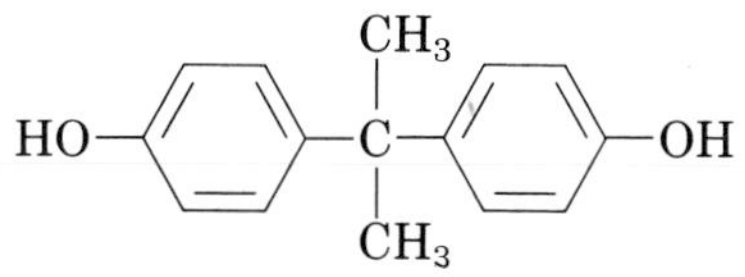

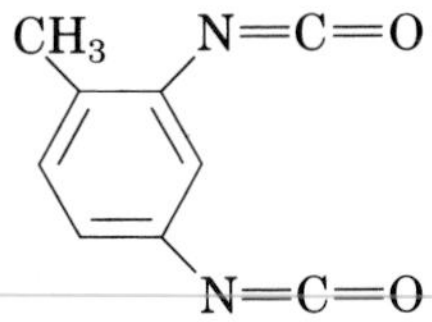

First synthesized by Wallace Carothers[3] at Du Pont, the best known step-growth polymers are the polyamides (nylons), prepared by heating diamines with diacids. For example, nylon 66 is prepared by reaction of the six-carbon adipic acid with the six-carbon hexamethylenediamine at 280°C as shown at the top of the next page.

[3]Wallace H. Carothers (1896–1937); b. Burlington, Iowa; Ph.D. Illinois (Adams); Du Pont Company.

$$\underbrace{\underset{\text{Adipic acid}}{HOOC-(CH_2)_4-COOH} + \underset{\text{Hexamethylenediamine}}{H_2N-(CH_2)_6-NH_2}}$$

$$\downarrow$$

$$\left(\overset{O}{\overset{\|}{C}}-(CH_2)_4-\overset{O}{\overset{\|}{C}}-NH-(CH_2)_6-NH \right)_n + n\,H_2O$$

Nylon 66

Nylon 6 is closely related in structure to nylon 66 and is prepared by polymerization of caprolactam. Water is first added to hydrolyze caprolactam to 6-aminohexanoic acid, and strong heating then brings about dehydration and polymerization.

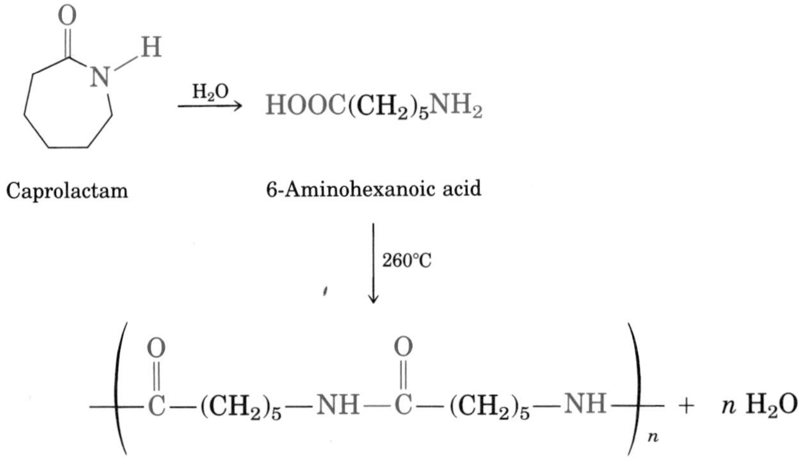

$$\underset{\text{Caprolactam}}{\text{(seven-membered ring: C=O, N–H)}} \xrightarrow{H_2O} \underset{\text{6-Aminohexanoic acid}}{HOOC(CH_2)_5NH_2}$$

$$\downarrow 260°C$$

$$\left(\overset{O}{\overset{\|}{C}}-(CH_2)_5-NH-\overset{O}{\overset{\|}{C}}-(CH_2)_5-NH \right)_n + n\,H_2O$$

Nylon 6

Nylons are used both in engineering applications and in making fibers. A combination of high impact strength and abrasion resistance makes nylon an excellent metal-substitute for bearings and gears. As fiber, nylon is used in a wide variety of applications, from clothing to tire cord to Perlon mountaineering ropes.

PROBLEM

34.12 Nylon is far more easily damaged by accidental spillage of acid or base than are chain-growth polymers such as Orlon or polyethylene. In other words, wearing nylon stockings in a chemistry laboratory can be very expensive. Explain.

PROBLEM

34.13 Draw structures of the step-growth polymers you would expect to obtain from the following reactions:

(a) $BrCH_2CH_2CH_2Br + HOCH_2CH_2CH_2OH \xrightarrow{\text{Base}}$

(b) $HOCH_2CH_2OH + HOOC(CH_2)_6COOH \xrightarrow{H^+}$

(c) $H_2N(CH_2)_6NH_2 + ClOC(CH_2)_4COCl \longrightarrow$

PROBLEM

34.14 Fibers of Kevlar, a nylon polymer prepared by reaction of 1,4-benzenedicarboxylic acid (terephthalic acid) with 1,4-diaminobenzene (*p*-phenylenediamine), are so strong that they are used to make bulletproof vests. Draw the structure of a segment of Kevlar.

34.10 Polyesters

Just as polyamides can be made by reaction between diacids and diamines, **polyesters** can be made by reaction between diacids and dialcohols. The most generally useful polyester is made by ester exchange reaction between dimethyl terephthalate and ethylene glycol. The product is widely used under the trade name Dacron to make clothing fiber and tire cord, and is used under the name Mylar to make plastic film. The tensile strength of poly(ethylene terephthalate) film is nearly equal to that of steel, and the film is unusually flex- and tear-resistant. These remarkable properties account for the major use of polyester in magnetic recording tape.

$$CH_3O_2C-C_6H_4-CO_2CH_3 + HOCH_2CH_2OH$$

Dimethyl terephthalate Ethylene glycol

$$\xrightarrow{200°C}$$

$$\left(-O-CH_2CH_2-O-\overset{O}{\overset{\|}{C}}-C_6H_4-\overset{O}{\overset{\|}{C}}-\right)_n + 2n\ CH_3OH$$

Polyester, Dacron, Mylar

Lexan, a polycarbonate prepared from diphenyl carbonate and bisphenol A, is another commercially valuable polyester. Lexan has an unusually high impact strength, making it valuable for use in machinery housings, telephones, and bicycle safety helmets.

$$C_6H_5-O-\overset{O}{\overset{\|}{C}}-O-C_6H_5 + HO-C_6H_4-C(CH_3)_2-C_6H_4-OH$$

Diphenyl carbonate Bisphenol A

$$\xrightarrow{300°C}$$

$$\left(-O-C_6H_4-C(CH_3)_2-C_6H_4-O-\overset{O}{\overset{\|}{C}}-O-\right)_n + 2n\ C_6H_5OH$$

Lexan

PROBLEM

34.15 Draw the structure of the polymer you would expect to obtain from reaction of dimethyl terephthalate with a triol such as glycerol. What structural feature would this new polymer have that was not present in Dacron? How do you think this new feature would affect the properties of the polymer?

34.11 Polyurethanes

A **urethane** is a carbonyl-containing functional group in which the carbonyl carbon is bound both to an ether oxygen and to an amine nitrogen. As such, a urethane can be considered intermediate between a carbonate and a urea:

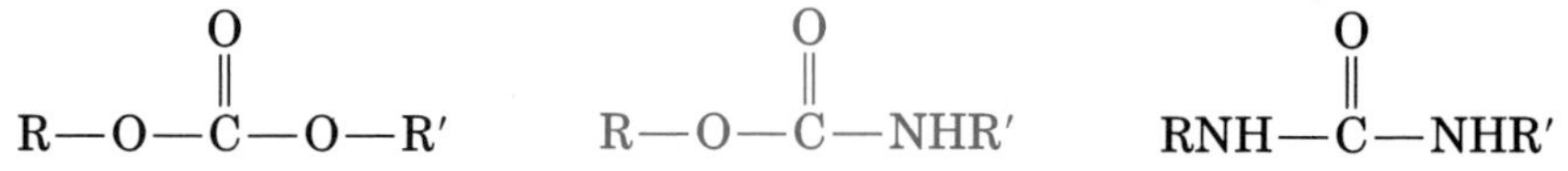

$$R{-}O{-}\overset{O}{\overset{\|}{C}}{-}O{-}R' \qquad R{-}O{-}\overset{O}{\overset{\|}{C}}{-}NHR' \qquad RNH{-}\overset{O}{\overset{\|}{C}}{-}NHR'$$

A carbonate — A urethane — A urea

Urethanes are prepared by nucleophilic addition of alcohols to isocyanates:

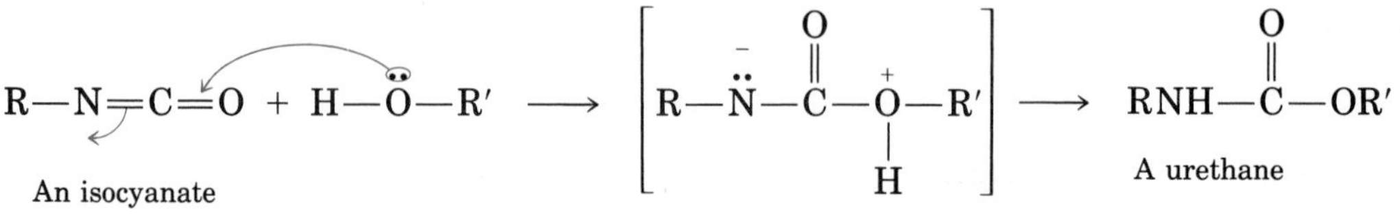

$$R{-}N{=}C{=}O + H{-}\ddot{O}{-}R' \longrightarrow \left[R{-}\ddot{N}^{-}{-}\overset{O}{\overset{\|}{C}}{-}\overset{+}{O}(H){-}R'\right] \longrightarrow RNH{-}\overset{O}{\overset{\|}{C}}{-}OR'$$

An isocyanate — A urethane

Polyurethanes, polymers containing urethane linkages, are prepared by reaction between a diol and a diisocyanate. The diol is itself a low-molecular-weight (mol wt ≈ 1000) polymer with hydroxyl end-groups, and the diisocyanate is often tolylene diisocyanate.

HO—(Polymer)—OH + [H_3C, N=C=O, N=C=O on benzene ring]

Tolylene diisocyanate

↓

$$\left(\overset{O}{\overset{\|}{C}}{-}NH{-}C_6H_3(CH_3){-}NH{-}\overset{O}{\overset{\|}{C}}{-}O{-}(\text{Polymer}){-}O\right)_n$$

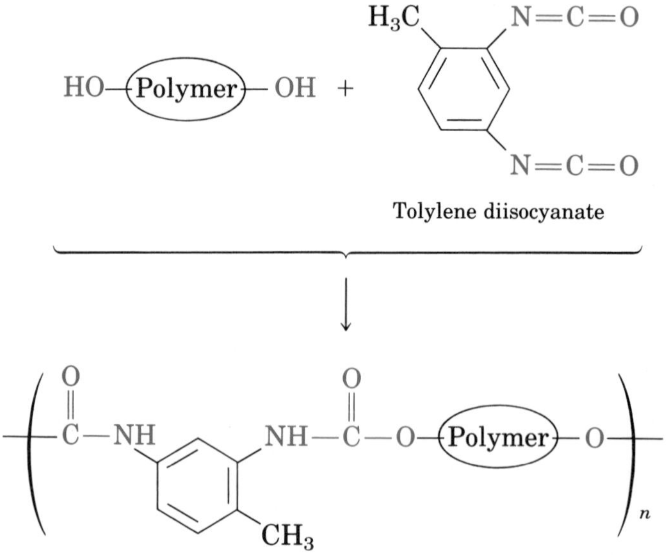

A polyurethane

A number of different kinds of polyurethanes are produced, depending on the nature of the polymeric alcohol used and on the degree of cross-linking achieved. One major use of polyurethane is in the stretchable Spandex and Lycra fibers used for bathing suits and leotards. These polyurethanes have a rather low degree of cross-linking so that the resultant polymer is soft and elastic.

A second major use of polyurethanes is in foams. Foaming occurs when a small amount of water is added during polymerization. Water adds to isocyanate groups giving carbamic acids, which spontaneously lose CO_2, thus generating the foam bubbles.

$$R{-}N{=}C{=}O + H_2O \longrightarrow \left[R{-}NH{-}\overset{\overset{\displaystyle O}{\|}}{C}{-}OH\right] \longrightarrow R{-}NH_2 + CO_2 \uparrow$$

A carbamic acid

Polyurethane foams generally have a higher degree of cross-linking than do polyurethane fibers, and the amount of cross-linking can be varied by using a polyalcohol (rather than a diol) as one of the reactive components. The result is a rigid but very light foam suitable for use as thermal insulation in building construction and in portable ice chests.

34.12 Polymer structure and chemistry

Polymers are not really so different from other organic molecules. They are much bigger, to be sure, but their chemistry is the same as that of analogous small molecules. The chemistry of polymers is the familiar chemistry of functional groups; molecular size plays little role. Thus, the ester linkages of a polyester such as Dacron are hydrolyzed by base; the aromatic rings of polystyrene undergo typical electrophilic aromatic substitution reactions; and the alkane chains of polyethylene undergo radical-initiated halogenation.

The major differences between small and very large organic molecules are in structure and in physical properties. Here too, though, the bulk structures and properties of polymers are the result of the same intermolecular forces that operate in small molecules.

The most important intermolecular forces between non-cross-linked polymer chains are **van der Waals forces.** These are the same forces that act between small molecules in solution or in the solid state, and they are due to weak attractive interactions between transient dipoles in nearby molecules (Section 3.6). Since van der Waals forces operate only at close distances, they are stronger in those polymers such as linear polyethylene in which chains can line up in a regular and close-packed way. Many polymers, in fact, have regions that are essentially crystalline. These regions, called **crystallites,** consist of highly ordered portions in which the zigzag polymer chains are bound together by van der Waals forces (Figure 34.5, p. 1160).

As we might expect, polymer crystallinity is strongly affected by the steric requirements of substituent groups on the chains. Thus poly(methyl methacrylate) is noncrystalline because the chains cannot pack closely together in a regular way, but linear polyethylene is highly crystalline.

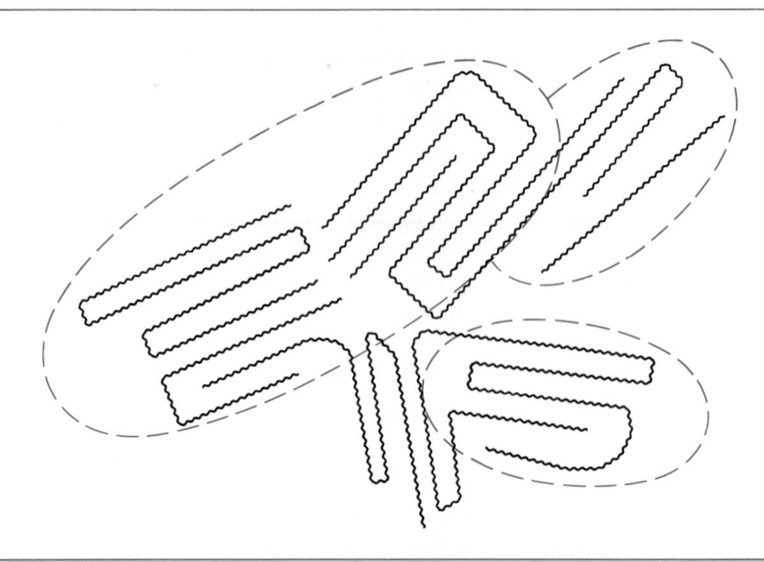

Figure 34.5. Crystallites in linear polyethylene. The long polymer chains are arranged in parallel lines in the crystallite regions.

PROBLEM

34.16 An improved polymeric resin used for Merrifield solid-phase peptide synthesis (Section 31.12) is prepared by treating polystyrene with *N*-(hydroxymethyl)-phthalimide and trifluoromethanesulfonic acid, followed by reaction with hydrazine. Show how these steps occur.

$(-CH_2-CH(C_6H_5)-)_n$ $\xrightarrow[CF_3SO_3H]{\text{phthalimide}-N-CH_2OH}$ $[(-CH_2CH(C_6H_4-CH_2-N(\text{phthalimide}))-)_n]$ $\xrightarrow{H_2NNH_2}$ $(-CH_2-CH(C_6H_4-CH_2NH_2)-)_n$

PROBLEM

34.17 What product would you expect to obtain from catalytic hydrogenation of natural rubber? How else might you obtain a similar polymer? Would the product be syndiotactic, atactic, or isotactic?

34.13 Polymer structure and physical properties

Classification of synthetic polymers according to their physical properties is a useful exercise because it allows us to make a rough correlation between structure and property. In general, we can divide polymers into four major categories: thermoplastics, fibers, elastomers, and thermosetting resins.

Thermoplastics are the polymers most people think of when the word *plastic* is mentioned. These polymers are hard at room temperature but become soft and viscous when heated. As a result, they can be molded into toys, beads, telephone housings, or into any one of thousands of other items. Because thermoplastics have little or no cross-linking, the individual chains can slip past one another on heating. Some thermoplastic polymers such as poly(methyl methacrylate), used in Plexiglas, are amorphous (i.e., non-crystalline); others, such as polyethylene and nylon, are partially crystalline. **Plasticizers**—small organic molecules that act as lubricants between chains—are usually added to plastics to keep them from becoming brittle at room temperature. Dialkyl phthalates are commonly used for this purpose and, in the past few decades, have become among the most widely dispersed of all environmental pollutants. Phthalate plasticizers have even been detected in the fat of antarctic penguins.

$COOCH_2CH_2CH_2CH_3$

$COOCH_2CH_2CH_2CH_3$

Dibutyl phthalate
(a plasticizer)

Fibers are thin threads produced by extruding a molten polymer through small holes in a die or *spinneret*. The fibers are then cooled and drawn out. Drawing has the effect of orienting the crystallite regions along the axis of the fiber, a process that adds considerable tensile strength (Figure 34.6, p. 1162). Nylon, Dacron, and polyethylene all have the semi-crystalline structure necessary for drawing into oriented fibers.

Elastomers are amorphous polymers that have the ability to stretch out and spring back to their original shapes. These polymers must have a modest amount of cross-linking to prevent the chains from slipping over one another, and the chains must have an irregular shape to prevent crystallite formation. When stretched, the randomly coiled chains straighten out and orient along the direction of the pull. Van der Waals forces are too weak and too few to maintain this orientation, however, and the elastomer therefore reverts to its random-coiled state when the stretching force is released (Figure 34.7).

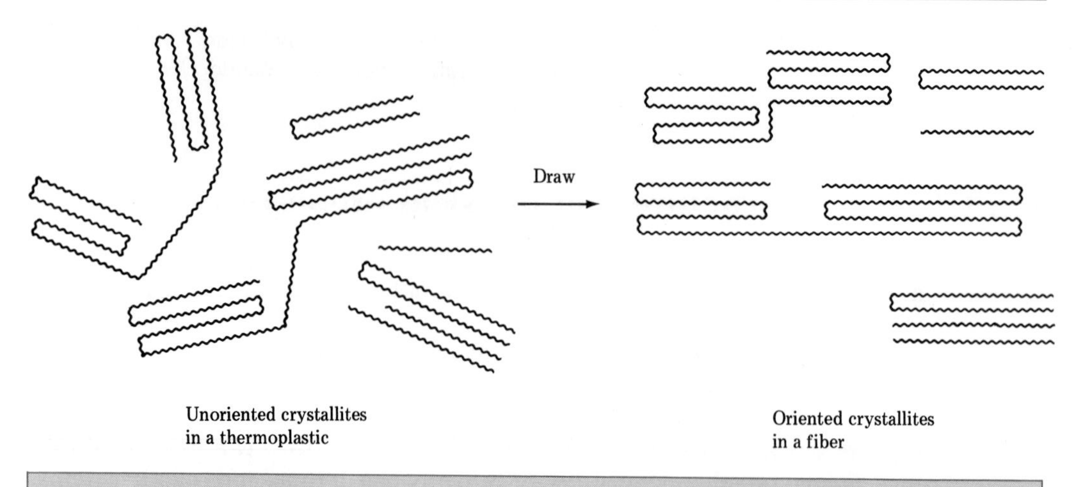

Figure 34.6. Oriented crystallites in a polymer fiber.

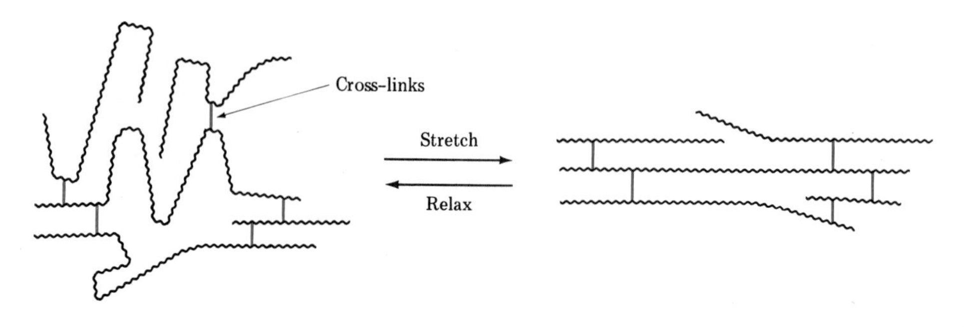

Figure 34.7. Unstretched and stretched forms of an elastomer.

Natural rubber is one example of an elastomer. Rubber has the long chains and occasional cross-links needed for elasticity, but its irregular geometry prevents close packing of the chains into crystallites. Gutta-percha, by contrast, is highly crystalline and is not an elastomer (Figure 34.8).

Thermosetting resins are polymers that become highly cross-linked and solidify into a hard, insoluble mass when heated. We have not paid particular attention to such polymers up to this point, but one example should suffice. Bakelite, a thermosetting resin that was first produced in 1907 by Leo Baekeland,[4] has been in commercial use longer than any other synthetic polymer. It is widely used for molded parts, for adhesives, for coatings, and even for high-temperature applications such as missile nose cones. Chemically, Bakelite is a phenolic resin produced by reaction of phenol and formaldehyde. On heating, water is eliminated, many cross-

[4]Leo Hendrik Baekeland (1863–1944); b. Ghent, Belgium; founder and president, Bakelite Corp., U.S. (1910–1939).

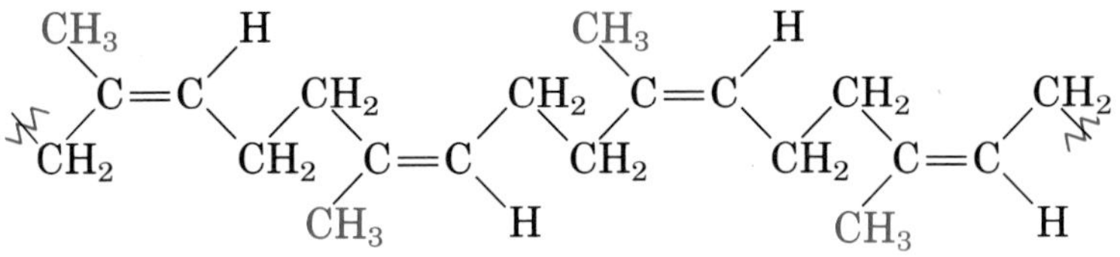

(a)

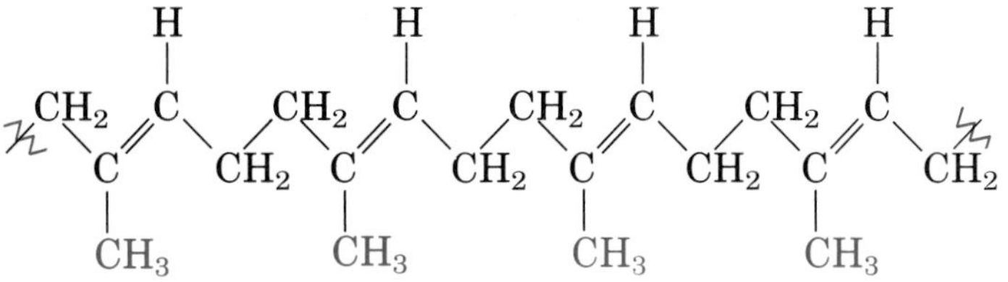

(b)

Figure 34.8. Geometry of (a) natural rubber (all cis) and (b) gutta-percha (all trans). Gutta-percha is crystalline and nonelastic, but natural rubber is elastic and noncrystalline.

links form, and the polymer sets into a stable, rock-like mass. The cross-linking in Bakelite and other thermosetting resins is three-dimensional and is so extensive that we cannot speak of polymer "chains." Bakelite is essentially one large molecule.

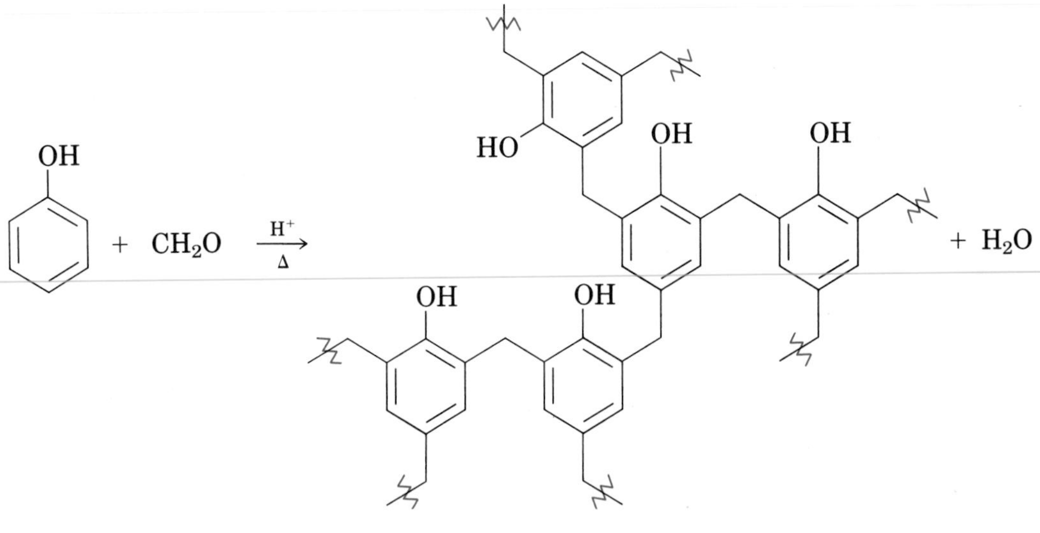

$$\text{C}_6\text{H}_5\text{OH} + CH_2O \xrightarrow[\Delta]{H^+} \text{Bakelite} + H_2O$$

Bakelite

PROBLEM

34.18 Propose a mechanism to account for the formation of Bakelite from acid-catalyzed polymerization of phenol and formaldehyde.

34.14 Summary

Synthetic polymers can be classified as either chain-growth polymers or step-growth polymers. Chain-growth polymers are prepared by chain reaction polymerization of unsaturated monomers, in the presence of a radical, an anion, or a cation initiator. Radical polymerization is the most commonly used method, but alkenes such as 2-methylpropene that have electron-donating substituents on the double bond polymerize easily by a cationic route. Similarly, monomers such as methyl α-cyanoacrylate that have electron-withdrawing substituents on the double bond polymerize by an anionic (Michael-type reaction) pathway. Conjugated dienes such as 1,4-butadiene, isoprene, and chloroprene also undergo polymerization, and the products have double bonds in their chains. Still another possibility is the copolymerization of two monomers to give a product that has properties different from those of either homopolymer.

Alkene polymerization can be carried out in a much more controlled fashion if Ziegler–Natta catalysts are used. Ziegler–Natta polymerization minimizes the amount of chain branching in the polymer and leads to stereoregular chains—either isotactic (substituents on the same side of the chain) or syndiotactic (substituents on alternate sides of the chain), rather than atactic (substituents randomly disposed).

Step-growth polymers, the second major class of polymers, are prepared by reactions between two different difunctional molecules; the individual bonds in the polymer are formed independently of one another. Polyamides (nylons) are formed by step-growth polymerization between a diacid and a diamine; polyesters are formed from a diester and a diol; and polyurethanes are formed from a diisocyanate and a diol.

The chemistry of synthetic polymers is similar to the chemistry of small molecules with the same functional groups, but the physical properties of polymers are greatly affected by size. Polymers can be classified by physical property into four groups: thermoplastics, fibers, elastomers, and thermosetting resins. The properties of each group can be accounted for by the structure, the degree of crystallinity, the geometry, and the amount of cross-linking in the specific polymers.

ADDITIONAL PROBLEMS

34.19 Identify the monomer units in each of the following polymers, and categorize each as either a step-growth or a chain-growth polymer.

(a) $\left(CH_2\overset{NO_2}{\overset{|}{C}H}—CH_2\overset{NO_2}{\overset{|}{C}H} \right)$

(b) $\left(CFCl—CF_2—CFCl—CF_2 \right)$, Kel-F

(c) $\left(CH_2—O—CH_2—O—CH_2—O \right)$, Delrin

(d) $\left(O—C_6H_4—\overset{O}{\overset{\|}{C}}—O—C_6H_4—\overset{O}{\overset{\|}{C}} \right)$

(e) $\text{+CH(CH}_2)_4\text{CH=N—(CH}_2)_6\text{—N+}$

34.20 Cyclopentadiene undergoes thermal polymerization to yield a polymer that has no double bonds in the chain. On strong heating, this polymer breaks down to regenerate cyclopentadiene. Propose a structure for the product. [*Hint:* See Section 13.7.]

34.21 Draw a three-dimensional representation of segments of these polymers:
(a) Syndiotactic polyacrylonitrile (b) Atactic poly(methyl methacrylate)
(c) Isotactic poly(vinyl chloride)

34.22 When styrene is copolymerized in the presence of a few percent *p*-divinylbenzene, a hard, insoluble, cross-linked polymer is obtained. Show how this cross-linking of polystyrene chains occurs.

34.23 One method for preparing the 1,6-hexanediamine needed in nylon production starts with 1,3-butadiene. How would you accomplish this synthesis?

$$H_2C{=}CH{-}CH{=}CH_2 \xrightarrow{?} H_2N(CH_2)_6NH_2$$

34.24 Nitroethylene, $H_2C{=}CHNO_2$, is a sensitive compound that must be prepared with great care. Attempted purification of nitroethylene by distillation often results in low recovery of product and a white coating on the inner walls of the distillation apparatus. Explain.

34.25 Poly(vinyl butyral) is used as the plastic laminate in the preparation of automobile windshield safety glass. How would you synthesize this polymer?

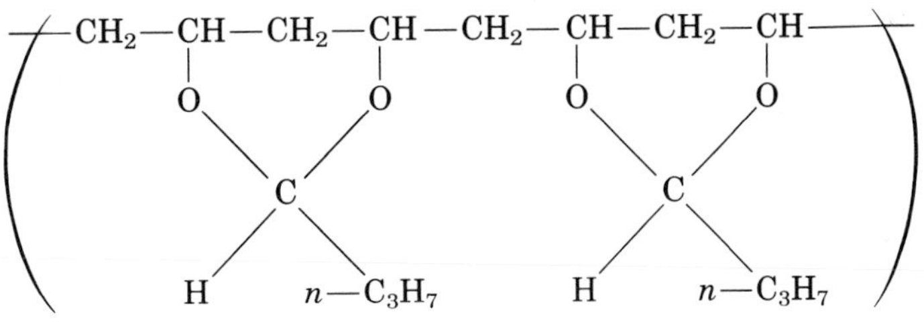

Poly(vinyl butyral)

34.26 Polyimides having the structure shown are used as coatings on glass and plastics to improve scratch resistance. How would you synthesize a polyimide?

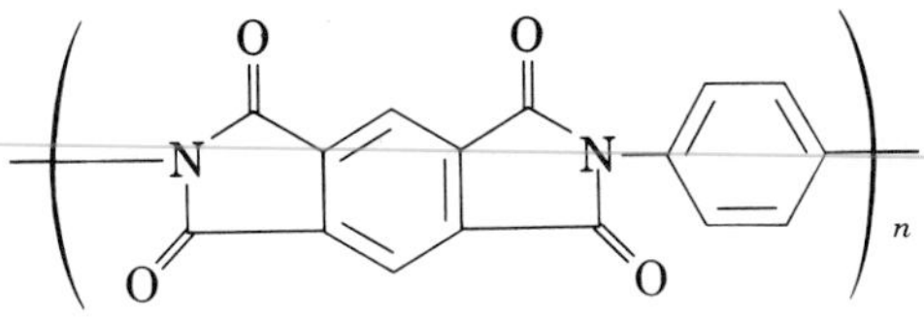

A polyimide

34.27 Qiana, a polyamide fiber with a silk-like feel, has the structure indicated. What are the monomer units used in the synthesis of Qiana? Can you suggest a synthesis of the necessary diamine from aniline and formaldehyde?

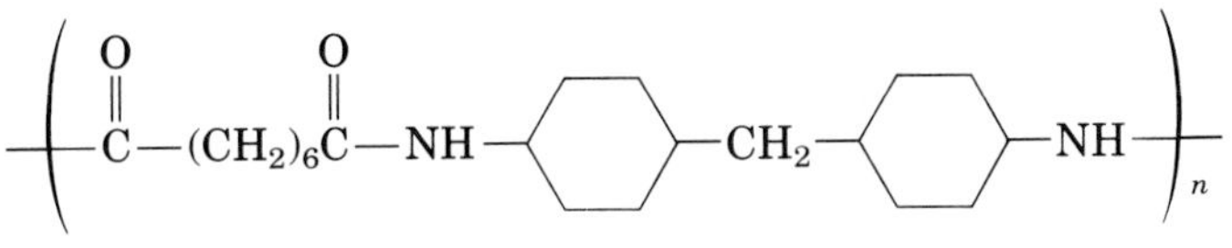

Qiana

34.28 What is the structure of the polymer produced by treatment of β-propiolactone with a catalytic amount of hydroxide ion?

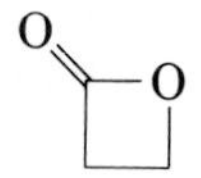

β-Propiolactone

34.29 Glyptal is a highly cross-linked thermosetting resin produced by heating glycerol and phthalic anhydride (1,2-benzenedicarboxylic acid anhydride). Show the structure of a representative segment of glyptal.

34.30 Melmac, a thermosetting resin often used to make plastic dishes, is prepared by heating melamine with formaldehyde. Propose a structure for Melmac. [*Hint:* See the structure of Bakelite, Section 34.13.]

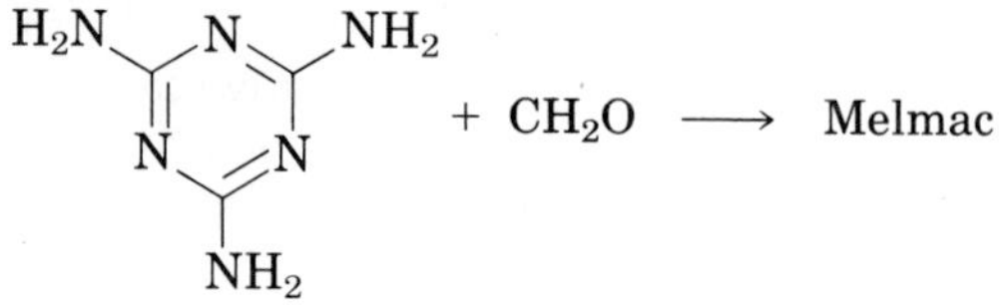

$$\text{Melamine} + CH_2O \longrightarrow \text{Melmac}$$

Melamine

34.31 Epoxy adhesives are cross-linked resins prepared in two steps. The first step involves S_N2 reaction of the disodium salt of bisphenol A with epichlorohydrin to form a low-molecular-weight prepolymer. This prepolymer is then "cured" into a cross-linked resin by treatment with a triamine such as $H_2NCH_2CH_2NHCH_2CH_2NH_2$. What is the structure of the prepolymer? How does addition of the triamine result in cross-linking?

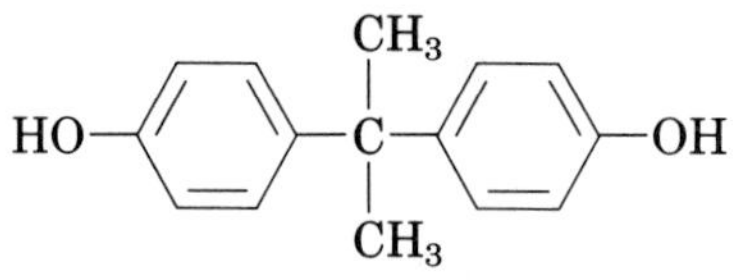

Bisphenol A

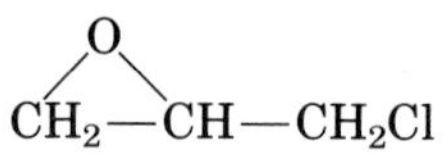

Epichlorohydrin

34.32 The polyurethane foam used for home insulation uses methanediphenyldiisocyanate (MDI) as monomer. The MDI is prepared by reaction of aniline with formaldehyde, followed by treatment with phosgene, $COCl_2$. Propose mechanisms for the two steps.

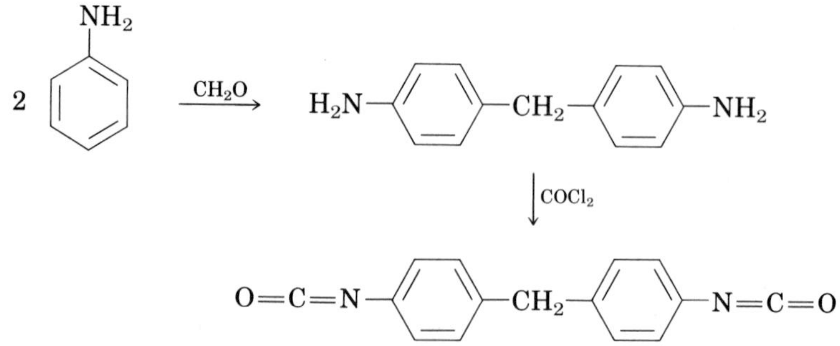

$$2\ C_6H_5NH_2 \xrightarrow{CH_2O} H_2N{-}C_6H_4{-}CH_2{-}C_6H_4{-}NH_2 \xrightarrow{COCl_2} O{=}C{=}N{-}C_6H_4{-}CH_2{-}C_6H_4{-}N{=}C{=}O$$

MDI

34.33 Urea–formaldehyde resins have been prepared commercially for over 60 years and have served a variety of purposes, including use as adhesives and insulating foams. For example, the smoking salons of the great hydrogen-filled dirigibles of the 1930s were insulated with urea–formaldehyde foams, and urea–formaldehyde adhesives were used extensively in boats and aircraft during World War II. The structure of the urea–formaldehyde polymer is highly cross-linked, like that of Bakelite (Section 34.13). Propose a structure for this polymer.

$$\underset{\text{Urea}}{H_2N{-}\overset{\overset{\displaystyle O}{\|}}{C}{-}NH_2} \quad + \quad \underset{\text{Formaldehyde}}{CH_2O} \quad \xrightarrow[\Delta]{}$$

APPENDIX ANSWERS TO SELECTED IN-TEXT PROBLEMS

Answers to all problems can be found in the *Solutions Manual and Study Guide* that accompanies this text.

CHAPTER 1

1.1 (a) Boron: $1s^2\ 2s^2\ 2p$ (b) Phosphorus: $1s^2\ 2s^2\ 2p^6\ 3s^2\ 3p^3$
(c) Iron: $1s^2\ 2s^2\ 2p^6\ 3s^2\ 3p^6\ 4s^2\ 3d^6$ (d) Selenium: $1s^2\ 2s^2\ 2p^6\ 3s^2\ 3p^6\ 4s^2\ 3d^{10}\ 4p^4$
1.3 Ionic: $CaCl_2$, KF; covalent: CH_3Cl, CH_3NH_2, BF_3, H_2NNH_2
1.4 Propene: C—C—C bond angle ≈ 120° **1.5** Propyne: C—C—C bond angle ≈ 180°
1.6 1,3-Butadiene: all carbons sp^2, all bond angles ≈ 120°
1.7 The N_2 molecule has a triple bond similar to that in alkynes.
1.8 (a) PH_3, tetrahedral (b) AlH_3, planar (c) CO_2, linear (d) CO_3^{2-}, planar
(e) CH_3SCH_3, both C and S are tetrahedrally hybridized
(f) CH_3OH, both C and O are tetrahedrally hybridized

CHAPTER 2

2.1 (a) Pyridine, C_5H_5N (b) Cyclohexanone, $C_6H_{10}O$ (c) Indole, C_8H_7N (d) BHT, $C_{15}H_{24}O$
2.2 There are many possibilities for each part. (a) C_5H_{12}, $CH_3CH_2CH_2CH_2CH_3$
(b) C_2H_7N, CH_3NHCH_3 (c) C_3H_6O, $H_2C{=}CHCH_2OH$ (d) C_4H_9Cl, $CH_3CH_2CH_2CH_2Cl$

2.3 (a) Diazomethane, $H_2C{=}\overset{\oplus}{N}{=}\overset{\ominus}{\ddot{N}}{:}$ (b) Nitromethane, $H_3C{-}\overset{\oplus}{N}(={\ddot{O}}{:})({-}{:}\ddot{O}{:}^{\ominus})$
(c) Methyl isocyanide, $H_3C{-}\overset{\oplus}{N}{\equiv}\overset{\ominus}{C}{:}$

2.5 (a) Boron formal charge $= 3 - \frac{8}{2} - 0 = -1$ (b) Aluminum formal charge $= 3 - \frac{8}{2} - 0 = -1$
Oxygen formal charge $= 6 - \frac{6}{2} - 2 = +1$ Nitrogen formal charge $= 5 - \frac{8}{2} - 0 = +1$

2.6 Lewis acids: $MgBr_2$, $B(CH_3)_3$; Lewis bases: $(CH_3)_3P{:}$, $(CH_3)_2\ddot{S}{:}$, $CH_3C{\equiv}N{:}$, $C_5H_5N{:}$
2.7 (a) Benzene: C = 92.25%, H = 7.75%
(b) Laetrile: C = 54.37%, H = 4.89%, N = 4.53%, O = 36.21%
(c) Quinine: C = 74.05%, H = 7.46%, N = 8.63%, O = 9.86%
(d) Diethylstilbestrol: C = 80.56%, H = 7.51%, O = 11.92%
2.8 Molecular formula of citral: $C_{10}H_{16}O$ **2.9** Molecular formula of squalene: $C_{30}H_{50}$

CHAPTER 3

3.2 There are many possible answers. (a) C_8H_{18}: $CH_3(CH_2)_6CH_3$
(b) $C_5H_{10}O_2$: $CH_3CH_2CH_2COOCH_3$, $CH_3CH_2COOCH_2CH_3$
(c) C_4H_7N: $CH_3CH_2CH_2CN$, $(CH_3)_2CHCN$
3.3 (a) C_3H_8O: two isomeric alcohols (b) C_4H_9Br: four isomeric bromoalkanes
(c) $C_4H_8Cl_2$: nine isomeric dichloroalkanes
3.4 (a) Pentane, 2-methylbutane, 2,2-dimethylpropane (b) 3,4-Dimethylhexane
(c) 2,4-Dimethylpentane (d) 2,2,5-Trimethylheptane
3.5 (a) $CH_3CH_2CH_2CH_2CH_2CH(CH_3)CH(CH_3)CH_2CH_3$ (b) $CH_3CH_2CH_2C(CH_3)_2CH(CH_2CH_3)_2$
(c) $CH_3CH_2CH_2CH_2CH(CH_2CH_2CH_3)CH_2C(CH_3)_3$ (d) $CH_3CH(CH_3)CH_2C(CH_3)_3$

3.6 (a) 2-Methylhexane (b) 4,5-Dimethyloctane (c) 3-Ethyl-2,2-dimethylpentane (d) 3,4-Dimethylheptane (e) 2,3-Dimethylhexane (f) 3,3-Dimethylpentane
3.7 Pentyl; 1-methylbutyl; 1-ethylpropyl; 2-methylbutyl; 3-methylbutyl; 1,1-dimethylpropyl; 1,2-dimethylpropyl; 2,2-dimethylpropyl
3.8 There are many possible answers. (a) 2,3,4-Trimethylpentane (b) 2,4-Dimethylpentane (c) 2,2-Dimethylbutane (d) 2-Chloropropane
3.11 **3.12** (a) (b)

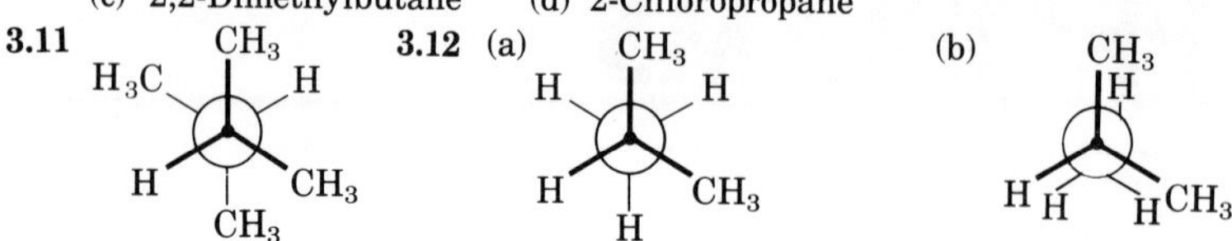

3.13 (a) 3.8 kcal/mol (b) 99.8% (c) 0.2%
3.14 (a) 1,4-Dimethylcyclohexane (b) 1-Methyl-3-propylcyclopentane (c) 3-Cyclobutylpentane (d) 1-Bromo-4-ethylcyclodecane

CHAPTER 4

4.2 Nucleophiles: NH_3, I^-, CN^-; electrophiles: H^+, Mg^{2+}, CH_3COCl
4.3 Alkane chlorination is a radical chain reaction that, once started, sustains itself. As time goes on, however, the radical intermediates combine in termination steps that remove them from the cycle and stop the reaction.
4.4 Pentane has three different kinds of hydrogen positions and thus yields a mixture of products on chlorination. Neopentane, however, has only a single type of hydrogen and thus yields a single monochlorination product.
4.5 (a) $\Delta H° = +58$ kcal/mol (b) $\Delta H° = -18$ kcal/mol (c) $\Delta H° = -12$ kcal/mol (d) $\Delta H° = -30$ kcal/mol
4.6 (a) $\Delta H° = -7$ kcal/mol (b) $\Delta H° = +4$ kcal/mol (c) $\Delta H° = -1$ kcal/mol

CHAPTER 5

5.1 (a) C_8H_{14}, two (b) C_5H_6, three (c) $C_{12}H_{20}$, three (d) $C_{20}H_{32}$, five (e) $C_{40}H_{56}$, thirteen (f) $C_{30}H_{50}$, six
5.2 There are many possible answers. (a) C_4H_8: $CH_3CH{=}CHCH_3$ (b) C_4H_6: $CH_3CH_2C{\equiv}CH$ (c) C_3H_4: $H_2C{=}C{=}CH_2$ (d) C_5H_8: $CH_3CH{=}CHCH{=}CH_2$
5.3 (a) C_6H_5N, five (b) $C_6H_5NO_2$, five (c) $C_8H_9Cl_3$, three (d) $C_9H_{16}Br_2$, one (e) $C_{10}H_{12}N_2O_3$, six (f) $C_{20}H_{32}O_2$, five
5.5 Trans double bonds are highly strained in rings of size eight and lower, but can be accommodated without strain by larger rings.
5.6 (a) 3,4,4-Trimethyl-1-pentene (b) 3-Methyl-3-hexene (c) 4,7-Dimethyl-2,5-octadiene
5.7 (a) $H_2C{=}CHCH_2CH_2C(CH_3){=}CH_2$ (b) $CH_3CH_2CH_2CH{=}C(CH_2CH_3)C(CH_3)_3$ (c) $CH_3CH{=}CHCH{=}CHC(CH_3)_2C(CH_3){=}CH_2$ (d) $[(CH_3)_2CH]_2C{=}C[CH(CH_3)_2]_2$
5.8 (a) 3,3-Dimethylcyclohexene (b) 2,3-Dimethyl-1,3-cyclohexadiene (c) 6,6-Dimethyl-1,3-cycloheptadiene (d) 1,2,5,5-Tetramethyl-1,3-cyclopentadiene
5.9 (a) (High) —Cl, —OH, —CH_3, —H (Low)
(b) (High) —CH_2OH, —$CH{=}CH_2$, —CH_2CH_3, —CH_3 (Low)
(c) (High) —COOH, —CH_2OH, —CN, —CH_2NH_2 (Low)
(d) (High) —CH_2OCH_3, —CN, —$C{\equiv}CH$, —CH_2CH_3 (Low)
5.10 (a) *Z* (b) *E* (c) *Z* (d) *E*
5.12 The second step is exothermic; the transition state should resemble the carbocation.

CHAPTER 6

6.1 Trans addition

6.3 $$CH_3CH{=}CH_2 \xrightarrow{Br^+} [CH_3\overset{+}{C}HCH_2Br] \xrightarrow{H_2O} CH_3\overset{OH}{\overset{|}{C}}HCH_2Br + H^+$$

Markovnikov addition by Br^+

6.4 $\overset{\delta^-}{I}{-}\overset{\delta^+}{N_3}$

6.5 Syn addition of D—OH to double bond

6.7 (a) $H_2C{=}C(CH_3)CH_2CH_3$ or $CH_3CH{=}C(CH_3)_2$ $\xrightarrow{HBr}$ $CH_3CH_2CBr(CH_3)_2$

(b) $CH_3CH_2CH{=}CH_2 \xrightarrow[\text{Peroxides}]{HBr} CH_3CH_2CH_2CH_2Br$

(c) $CH_3CH{=}CHCH_2CH_3 \xrightarrow[\text{2. } H_2O_2,\ ^-OH]{\text{1. } BH_3} CH_3CH(OH)CH_2CH_2CH_3 + CH_3CH_2CH(OH)CH_2CH_3$

(d) $CH_3CH_2CH{=}C(CH_3)CH_2CH_3 \xrightarrow[\text{Peroxides}]{HBr} CH_3CH_2CHBrCH(CH_3)CH_2CH_3$

(e) $CH_3CH{=}C(CH_3)CH_2CH_2CH_3 \xrightarrow[\text{2. } NaBH_4]{\text{1. } Hg(OAc)_2,\ H_2O} CH_3CH_2C(OH)(CH_3)CH_2CH_2CH_3$

6.8 (b) $CH_3CH_2CH{=}C(CH_3)_2 \xrightarrow[\text{2. } NaHSO_3,\ H_2O]{\text{1. } OsO_4} CH_3CH_2CH(OH)C(OH)(CH_3)_2$

(c) $H_2C{=}CHCH{=}CH_2 \xrightarrow[\text{2. } NaHSO_3,\ H_2O]{\text{1. 2 } OsO_4} CH_2(OH)CH(OH)CH(OH)CH_2OH$

6.9 The two diols are not identical. Make models to prove it.

6.10 $CH_3COCH_2CH_2CH_2CH_2COCH_3$

6.11 The cis diol (A) can easily form a five-membered-ring intermediate, but the oxygens of the trans diol (B) are too far apart to be in a five-membered ring.

6.12 (a) $(CH_3)_2C{=}CHCH_2CH_3$ (major); $H_2C{=}C(CH_3)CH_2CH_2CH_3$ (minor)
(b) $CH_3CH_2CH{=}C(CH_3)_2$ (major); $CH_3CH{=}CHCH(CH_3)_2$ (minor)

CHAPTER 7

7.1 (a) 2,5-Dimethyl-3-hexyne (b) 3,3-Dimethyl-1-butyne (c) 2,4-Octadien-6-yne
(d) 3,3-Dimethyl-4-octyne (e) 2,5,5-Trimethyl-3-heptyne

7.2 1-Hexyne; 2-hexyne; 3-hexyne; 3-methyl-1-pentyne; 4-methyl-1-pentyne; 4-methyl-2-pentyne; 3,3-dimethyl-1-butyne

7.3 (*Z*)-5-Deuterio-5-decene

7.4 Disiamylborane is so sterically hindered by the bulk of the two alkyl groups that addition to a third equivalent of alkene is very slow. The borane is prepared from 2-methyl-2-butene.

7.5 $(CH_3)_2CHCH(OH)CH_3$ from the alkyl groups

7.8 (a) $HC{\equiv}CH \xrightarrow[\text{2. } CH_3CH_2CH_2Br]{\text{1. } NaNH_2} CH_3CH_2CH_2C{\equiv}CH \xrightarrow[\text{2. } CH_3I \text{ 3. } H_3O^+]{\text{1. 2 BuLi}} CH_3CH_2CH(CH_3)C{\equiv}CH$

(b) $HC{\equiv}CH \xrightarrow[\text{2. } CH_3CH_2Br]{\text{1. } NaNH_2} CH_3CH_2C{\equiv}CH \xrightarrow[\text{2. 2 } CH_3Br]{\text{1. 2 BuLi}} CH_2CH(CH_3)C{\equiv}CCH_3$

7.9 Since methane, a typical alkane, has $pK_a \approx 49$, butane is probably similar. Thus, butyllithium is an extremely strong base.

7.10 (b) $CH_3(CH_2)_7C{\equiv}C(CH_2)_7C{\equiv}C(CH_2)_7COOH$ (c) $CH_3(CH_2)_5C{\equiv}C(CH_2)_5CH_3$

7.12 (a) $HC{\equiv}CH \xrightarrow[\text{2. } CH_3(CH_2)_6CH_2Br]{\text{1. } NaNH_2} CH_3(CH_2)_6CH_2C{\equiv}CH \xrightarrow{\text{2 } H_2/Pd} CH_3(CH_2)_8CH_3$

CHAPTER 8

8.1 $[\alpha]_D = +16.0°$ **8.2** Chiral: (b), (d), and (e)

8.3 (a) Achiral (b) Two (c) Three (d) Three

8.4 (a) (High) —Br, $—CH_2CH_2OH$, $—CH_2CH_3$, —H (low)
(b) (High) —OH, $—CO_2CH_3$, $—CO_2H$, $—CH_2OH$ (low)
(c) (High) $—NH_2$, —CN, $—CH_2NHCH_3$, $—CH_2NH_2$ (low)
(d) (High) —Br, —Cl, $—CH_2Br$, $—CH_2Cl$ (low)

8.5 (a) *S* (b) *S* (c) *R* (d) *S*

8.6 Starting from the top of each structure: (a) *R,R* (b) *R,S* (c) *S,R* (d) *S,S*
Structures (a) and (d) are enantiomers and are diastereomeric with (b) and (c). Structures (b) and (c) are enantiomers and are diastereomeric with (a) and (d).

8.7 Starting from the top, *R,R* **8.8** (a) and (d)
8.10 Structures (a) and (b) are identical and are enantiomeric with (c) and (d).
8.11 (a) Different (b) Different (c) Different **8.12** (a) *S* (b) *S* (c) *R* (d) *R,S*
8.13 Attack on a nonplanar carbocation would not occur equally well from both sides. The final product would still be racemic, however, since the carbocation itself would be racemic.
8.14 Bottom-side attack on *cis*-2-butene to yield a bromonium ion is a mirror image of top-side attack and gives the same (racemic) stereochemical result.
8.15 Bottom-side attack on an unsymmetrically substituted alkene leads to an unequal mixture of diastereomeric products. Top-side attack leads to the exactly opposite mixture. The sum of the two mixtures is 50:50 (racemic).
8.16 Bromination of *trans*-2-hexene yields a racemic product for the reason given in Problem 8.15.

CHAPTER 9

9.1 (a) 1,3-Dichloro-3-methylbutane (b) 1,5-Dibromo-2,2-dimethylpentane
(c) 1-Chloro-3-ethyl-4-iodopentane
9.2 1-Chloro-2-methylbutane (29%), 1-chloro-3-methylbutane (14%), 2-chloro-3-methylbutane (33%), and 2-chloro-2-methylbutane (24%)
9.3 The intermediate resonance-stabilized allylic radical yields the more stable trisubstituted double-bond product.
9.4 A C—H bond next to an aromatic ring is much weaker than an ordinary alkane C—H bond (85 kcal/mol versus 98 kcal/mol).
9.6 Acetate ion is a hybrid of two equivalent resonance forms. The two C—O bonds are therefore equivalent.
9.7 Alkanes have pK_a's ≈ 49. Grignard reagents are therefore very strong bases.
9.8 $R—Br \xrightarrow{Mg} [R—Mg—Br] \xrightarrow{D_2O} R—D + MgBrOD$

9.9 The hydroxyl proton is weakly acidic and reacts with Grignard reagent as soon as it is formed.
9.10 (b) $n\text{-}C_4H_9Br \xrightarrow{Li} n\text{-}C_4H_9Li \xrightarrow{CuI} (n\text{-}C_4H_9)_2CuLi$

$n\text{-}C_4H_9Br + (n\text{-}C_4H_9)_2CuLi \longrightarrow n\text{-}C_8H_{18}$

(c) $CH_3CH_2CH_2CH{=}CH_2 \xrightarrow[\text{Peroxides}]{HBr} CH_3CH_2CH_2CH_2CH_2Br$; then similar to (b)

CHAPTER 10

10.1 Back-side approach of the incoming nucleophile is blocked by the rigid carbon skeleton of the molecule. (Make a model.)
10.2 The nitrogen atom in quinuclidine is sterically more accessible for bonding than that in triethylamine because the alkyl "groups" are tied back in a rigid framework.
10.3 (a) $(CH_3)_2\ddot{N}\!:^-$ (negatively charged) (b) $(CH_3)_3N\!:$ (c) $H_2\ddot{S}\!:$
10.4 (a) CH_3—OTos (b) Bromoethane (c) Cyanide ion (d) Hexamethylphosphoramide
10.5 Inversion was 10% and racemization was 90%.
10.6 The carbocation formed by dissociation would be unstable, since steric constraints of the ring system would prevent it from becoming planar.
10.7 Reaction proceeds by an S_N1 mechanism. **10.8** (*Z*)-1-Bromo-1,2-diphenylethylene
10.9 The conformation leading to the trans product is less hindered.
10.10 (a) S_N2 (b) E2 (c) S_N2 (d) S_N1

CHAPTER 11

11.1 (a) C_6H_{14}, $C_5H_{10}O$, $C_4H_6O_2$, $C_3H_2O_3$ (b) C_9H_{20}, $C_8H_{16}O$, $C_7H_{12}O_2$, $C_6H_8O_3$, $C_5H_4O_4$
(c) $C_{11}H_{24}$, $C_{10}H_{20}O$, $C_9H_{16}O_2$, $C_8H_{12}O_3$, $C_7H_8O_4$, $C_6H_4O_5$
(d) $C_{13}H_{24}$, $C_{12}H_{20}O$, $C_{11}H_{16}O_2$, $C_{10}H_{12}O_3$, $C_9H_8O_4$, $C_8H_4O_5$
11.2 $C_{15}H_{22}O$ (correct), $C_{14}H_{18}O_2$, $C_{13}H_{14}O_3$, $C_{12}H_{10}O_4$
11.3 $(M + 1)^{+\cdot}$ is 6.7% of $M^{+\cdot}$.

11.4 Molecules with an odd number of nitrogens must have an odd number of hydrogens since nitrogen is trivalent. Hence, these molecules have odd-numbered molecular weights.
11.5 C_5H_5N **11.6** $(CH_3)_3C^+$, C_4H_9
11.7 (a) 2-Methyl-2-pentene (identified by loss of $—CH_3$)
(b) 2-Hexene (identified by loss of $—CH_2CH_3$)
11.8 (a) 5.7×10^5 kcal/mol (b) 9.5×10^3 kcal/mol (c) 5.7×10^2 kcal/mol (d) 67.8 kcal/mol
(e) 1.42 kcal/mol (f) 9.5×10^{-3} kcal/mol (g) 9.5×10^{-9} kcal/mol
11.9 (a) 3225 cm^{-1} (b) 1710 cm^{-1} (c) 1480 cm^{-1} (d) 4.44 μm (e) 10.3 μm (f) 6.41 μm
11.10 (a) Carbonyl (b) Nitro (c) Carboxylic acid (d) Carbonyl + alcohol, or amide
11.11 2100 cm^{-1}, $C{\equiv}C$ stretch; 2900 cm^{-1}, alkyl C—H; 3300 cm^{-1}, alkynyl $\equiv$C—H

CHAPTER 12

12.1 5.3×10^{-6} kcal/mol
12.2 9.53×10^{-6} kcal/mol. Increasing spectrometer frequency increases the amount of energy needed for spin flip.
12.3 (a) One ^1H and two ^{13}C signals (b) One ^1H and one ^{13}C (c) One ^1H and one ^{13}C
(d) Two ^1H and four ^{13}C (e) One ^1H and one ^{13}C (f) Two ^1H and three ^{13}C
12.4 (a) 126 Hz (b) 2.1 δ (c) 210 Hz
12.5 (a) 7.27 δ (b) 3.05 δ (c) 3.47 δ (d) 5.30 δ
12.6 C1 (51.4 δ); C2 (174.6 δ); C3 (27.6 δ); C4 (9.3 δ)
12.7 (a) Four (b) Seven (c) Four (d) Five
12.8 There are several possible answers. (a) 1-Methylcyclohexene (b) 2-Methylpentane
(c) 1-Chloro-2-methylpropane
12.9 (a) Three
(b) A signal near 20 δ would split into four peaks; a signal near 120 δ would split into two peaks; a second signal near 120 δ would split into three peaks.
12.10 (a) Three peaks at approximately 20, 30, and 120 δ
(b) The intensities of the three would be 1:2:2. The peaks at 20 δ and 30 δ would become triplets and the peak near 120 δ would become a doublet in the off-resonance spectrum.
12.11 2-Bromo-2-methylpropane: two ^{13}C peaks 1-Bromo-2-methylpropane: three ^{13}C peaks
12.12 (a) Two (b) Four (c) Three (d) Four (e) Five (f) Three
12.13 (a) 1.4 δ (b) 2.1 δ (c) 7.3 δ (d) 9.7 δ (e) 5.3 δ (f) 2.1 δ (g) 3.7 δ
12.15 (a) Quartet and doublet (b) Singlet, triplet, and triplet (c) Triplet and quintet
(d) Triplet, quartet, septet, and doublet
12.16 There are several possible answers. (a) CH_3OCH_3 (b) $CH_3CHClCH_3$
(c) $ClCH_2CH_2OCH_2CH_2Cl$ (d) $CH_3CH_2CO_2CH_3$
12.17 $CH_3CH_2OCH_2CH_3$

CHAPTER 13

13.1 Molecules (b), (c), (d), and (f) are conjugated.
13.2 Expected ΔH_{hydrog} = 60.6 kcal/mol; thus, allene is 10.7 kcal/mol higher in energy than expected.
13.3 $CH_3CH{=}CHCHClCH_3$; $CH_3CH_2CHClCH{=}CH_2$; $CH_3CHClCH_2CH{=}CH_2$;
$CH_3CH_2CH{=}CHCH_2Cl$; $CH_3CH{=}CHCH_2CH_2Cl$
13.4 $CH_3CH{=}CHCHClCH_3$ is the predominant 1,2 *and* 1,4 adduct.
13.5 The observed product predominates because it is formed from the more stable tertiary allylic carbocation intermediate.

13.7 $CH_3CHBrCH{=}CH_2 \xrightleftharpoons{S_N1} [CH_3\overset{+}{C}HCH{=}CH_2 + :Br^-] \xrightleftharpoons{S_N1} CH_3CH{=}CHCH_2Br$

13.8 Alkenes (a) and (c) are good dienophiles.
13.9 Diene (a) is *s*-cis; (b) is *s*-trans; (c) and (d) can rotate to *s*-cis.
13.10 The reactivity difference is due to steric factors; the terminal carbons of the diene are pushed too far apart as the ring size expands.
13.11 71–142 kcal/mol
13.12 Ultraviolet: 70–142 kcal/mol; infrared: 1.13–11.3 kcal/mol; ^1H NMR at 60 MHz: 5.7×10^{-6} kcal/mol
13.13 (b), (c), (d), (e), (f)

CHAPTER 14

14.2 (a) *m*-Bromochlorobenzene (b) (3-Methylbutyl)benzene (c) *p*-Bromoaniline (d) 2,4-Dichlorotoluene (e) 1-Ethyl-2,4-dinitrobenzene (f) 1,2,3,5-Tetramethylbenzene
14.3 (a) *o*-Bromochlorobenzene (b) 2,4-Dinitrotoluene (c) *p*-Bromotoluene (d) 2-Chloro-1,4-dimethylbenzene
14.4 There are four.
14.5 Ladenburg benzene: one monobromo and three dibromo derivatives; Dewar benzene: two monobromo and six dibromo derivatives
14.6 This experiment shows that both Kekulé resonance forms contribute to the structure of *o*-xylene.
14.7 Pyridine is a six-pi-electron aromatic molecule.
14.8 Cyclodecapentaene is not flat because of a severe cross-ring steric interaction.
14.9 All carbon–carbon bonds are equivalent; there is only one absorption in ^1H and ^{13}C NMR spectra.
14.10 Cyclooctatetraene dianion is a planar, regular octagon.
14.11 Imidazole has six pi electrons; both nitrogens are sp^2 hybridized.
14.12 Furan has six pi electrons; oxygen is sp^2 hybridized.
14.14 The C1—C2 bond is a double bond in two of the three resonance structures, but the C2—C3 bond is double in only one resonance structure. Thus, the C1—C2 bond is shorter than the C2—C3 bond.
14.15 Five pi electrons.

CHAPTER 15

15.5 (a) (Most) Phenol > Toluene > Nitrobenzene (Least)
(b) (Most) Phenol > Benzene > Chlorobenzene > Benzoic acid (Least)
(c) (Most) Aniline > Benzene > Bromobenzene > Benzaldehyde (Least)
15.6 Ortho and para are favored. **15.7** Meta attack is most favored.
15.8 (a) *o*-Bromonitrobenzene and *p*-bromonitrobenzene (b) *m*-Bromonitrobenzene (c) and (d) *o*-Chlorophenol and *p*-chlorophenol (e) and (f) *o*-Bromoaniline and *p*-bromoaniline
15.9 Toluene is more reactive than (trifluoromethyl)benzene. Fluorine causes the trifluoromethyl group to be deactivating.
15.10 The amide carbonyl group withdraws electron density from nitrogen, making the nitrogen a weaker activator.
15.11 (Most) Phenoxide ion > Phenol > Phenyl acetate (Least)
15.12 A nitroso group is inductively electron withdrawing and deactivating because of the electronegativity of nitrogen and oxygen. Resonance donation of the nitrogen lone-pair electrons is felt at ortho and para positions, however.

CHAPTER 16

16.1 (a) *o*-Ethyltoluene and *p*-ethyltoluene (b) *o*-Ethylphenol and *p*-ethylphenol (c) 1-Ethyl-2,4-dimethylbenzene (d) *o*-Chloroethylbenzene and *p*-chloroethylbenzene
16.2 Benzene is attacked by *tert*-butyl carbocation.
16.3 (Most) Phenol > Toluene > *p*-Bromotoluene > Bromobenzene (Least). Aniline and nitrobenzene are unreactive.
16.4 *tert*-Butylbenzene
16.7 Alkylbenzenes are more reactive than benzene, but acylbenzenes are less reactive than benzene.
16.10 Benzyl, 85 kcal/mol; allyl, 87 kcal/mol; alkyl, 98 kcal/mol
16.12 (a) 1. HNO_3; 2. Cl_2O (b) 1. CH_3COCl, $AlCl_3$; 2. Cl_2O; 3. H_2/Pd
(c) 1. Cl_2; 2. CH_3COCl, $AlCl_3$; 3. H_2/Pd
16.14 (a) Benzonitrile is too deactivated to undergo Friedel–Crafts reaction.
(b) Rearrangement occurs when primary alkyl halides are used in Friedel–Crafts reactions; chlorination would occur ortho to the alkyl group.

CHAPTER 18

18.1 (a) *cis*-1-Bromo-2-methylcyclobutane (b) *trans*-1-Chloro-4-*t*-butylcyclohexane (c) *cis*-1,2-Dibromocyclodecane (d) 2-Cyclobutyl-4-methylpentane (e) *cis*-1,3-Dibromocyclopentane

18.3 Six eclipsing interactions = 6.0 kcal/mol.
18.4 A methyl–methyl eclipsing interaction increases the energy of *cis*-1,2-dimethylcyclopropane relative to the trans isomer.
18.5 Cyclopropane has more energy than propane.
18.8 *cis*-1,2-Dimethylcyclobutane has an eclipsing interaction; *trans*-1,3-dimethylcyclobutane has a cross-ring steric interaction.
18.9 Planar cyclopentane has 10 eclipsing interactions, or 10 kcal/mol strain; 3.5 kcal is relieved by puckering.
18.10 (a), (b), (d): One axial and one equatorial (c), (e), (f): Both axial or both equatorial
18.13 Since —C≡N is linear, there are no interactions with axial hydrogens three carbons away.
18.14 (a) 0.5 kcal/mol (b) 1.8 kcal/mol (c) 0.5 kcal/mol (d) 0 kcal/mol
18.15 Four stereoisomers are possible. The all-equatorial isomer is most stable.
18.17 The cis isomer reacts faster than the trans isomer, because Br is axial.
18.18 Trans: 0 kcal/mol; Cis: 2.7 kcal/mol

CHAPTER 19

19.1 (a) Diisopropyl ether (b) Cyclopentyl propyl ether (c) *p*-Bromoanisole (d) 1-Methoxycyclohexene (e) Ethyl 2-methylpropyl ether (f) Allyl vinyl ether
19.2 A 1:2:1 mixture of ethers would be formed from dehydration of two different alcohols.
19.4 (a) (Most) Benzyl bromide > Bromoethane > 2-Bromopropane (Least)
(b) (Most) 3-Bromopropene > Bromoethane > Chloroethane (Least)
19.6 (a) Williamson synthesis of cyclohexanol with 1-bromobutane, or ethoxymercuration of cyclohexene
(b) Williamson synthesis of phenol with bromoethane
(c) Alkoxymercuration of 2-methylpropene with *sec*-butyl alcohol
(d) Cyclization of 4-bromo-1-butanol
19.8 Bromide ion and iodide ion are better nucleophiles than chloride ion.
19.9 *trans*-5,6-Epoxydecane gives a meso product; *cis*-5,6-epoxydecane gives a racemic product.

19.13 CH_3CH_2CH—CH_2 (epoxide: O bridging CH and CH_2)

19.14 (a) Ethyl methyl sulfide (b) Ethyl *tert*-butyl sulfide (c) *o*-(Dimethylthio)benzene (d) Phenyl *p*-tolyl sulfide
19.15 Dimethyl sulfoxide is a dipolar compound and can form hydrogen bonds to water.

CHAPTER 20

20.1 (a) 5-Methyl-2,4-hexanediol (b) 2-Methyl-4-phenyl-2-butanol (c) 4,4-Dimethylcyclohexanol (d) *trans*-2-Bromocyclopentanol (e) *p*-Methylbenzyl alcohol
20.3 More hindered alcohols are less highly associated.
20.5 (a) (High) Benzene > Cyclohexanone > Cyclohexyl methyl ether (Low)
(b) (High) CH_3CN > $H_2NCH_2CH_2NH_2$ > $CH_3CH_2NH_2$ (Low)
20.6 oxidations, (a), (c); reductions, (d); neither, (b), (e)
20.7 (a) $NaBH_4$ (b) $LiAlH_4$ (c) $LiAlH_4$ (d) H_2/Pd
20.9 (a) $CH_3MgBr + CH_3COCH_3$ (b) Cyclohexanone + CH_3MgBr
(c) $CH_3CH_2COCH_2CH_3 + CH_3MgBr$, or $CH_3CH_2COCH_3 + CH_3CH_2MgBr$, or $CH_3CO_2R + 2\ CH_3CH_2MgBr$
(d) $CH_3CH_2COCH_3 + PhMgBr$, or $PhCOCH_2CH_3 + CH_3MgBr$, or $PhCOCH_3 + CH_3CH_2MgBr$
(e) $PhMgBr + CH_2O$
20.10 One equivalent of Grignard reagent removes the acidic —OH proton; further reagent adds to the carbonyl group.

20.11 $$\text{R—CHO} + \text{RCH}_2\text{OH} \rightleftharpoons \text{R—CH(OH)OCH}_2\text{R} \xrightarrow[\text{reagent}]{\text{Jones'}} \text{R—}\overset{\text{O}}{\overset{\|}{\text{C}}}\text{—O—CH}_2\text{R}$$

20.14 Hydroxyl band at 3300 cm^{-1} would disappear; carbonyl at 1710 cm^{-1} would appear.
20.15 (a) Singlet (b) Doublet (c) Triplet (d) Doublet (e) Doublet (f) Singlet

20.16 (a) 2-Butanethiol (b) *p*-Benzenedithiol (c) 2,2,6-Trimethyl-4-heptanethiol
(d) 2-Cyclopentene-1-thiol

20.17 $CH_3CH{=}CHCO_2CH_3 \xrightarrow[\text{2. } PBr_3]{\text{1. } LiAlH_4} CH_3CH{=}CHCH_2Br \xrightarrow{NaSH} CH_3CH{=}CHCH_2SH$

CHAPTER 22

22.1 (a) 2-Methyl-3-pentanone (b) 3-Phenylpropanal (c) 2,6-Octanedione
(d) *trans*-2-Methylcyclohexanecarbaldehyde (e) Pentanedial
(f) *cis*-2,5-Dimethylcyclohexanone (g) 4-Methyl-3-propyl-2-hexanone (h) 4-Hexenal

22.2 (a) $H_2C{=}C(CH_3)CH_2CHO$ (b) $CH_3CHClCH_2COCH_3$ (c) $PhCH_2CHO$

(d) 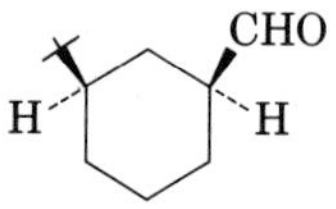 CHO, H, H

(e) $HOCH_2CH(CH_3)COCH_2CH_2CH_3$
(f) $CH_3CH_2CH(CH_3)CH_2CH_2CH(CHClCH_3)CHO$

22.3 (a) $C_4H_9CH{=}CH_2 \xrightarrow[\text{2. } H_2O_2,\ ^-OH]{\text{1. } BH_3} C_4H_9CH_2CH_2OH \xrightarrow{PCC} C_4H_9CH_2CHO$

(b) $C_4H_9CH{=}CH_2 \xrightarrow[\text{2. } NaBH_4]{\text{1. } Hg(OAc)_2,\ H_2O} C_4H_9CH(OH)CH_3 \xrightarrow{PCC} C_4H_9COCH_3$

(c) $C_4H_9CH{=}CH_2 \xrightarrow[\text{2. } Zn]{\text{1. } O_3} C_4H_9CHO$

22.4 Benzaldehyde is less reactive because the carbonyl carbon is less strongly polarized.
22.5 Steric hindrance prevents nucleophilic attack on 2,2,6-trimethylcyclohexanone.
22.6 The wavelength of an ultraviolet absorption moves to higher wavelengths as conjugation increases.
22.12 (a) Cyclohexanone + $CH_3CH{=}PPh_3$ (b) $(CH_3)_2C{=}O + CH_3CH_2CH{=}PPh_3$
(c) 2-Cyclohexenone + $H_2C{=}PPh_3$ (d) $(CH_3)_2C{=}O + PhCH{=}PPh_3$
(e) $PhCHO + PhCH{=}PPh_3$
22.13 Triphenylphosphine has no acidic hydrogens.
22.15 Internal Cannizzaro reaction
22.17 Starting material is a conjugated ketone; product is a saturated ketone.
22.18 (a) 1720 cm^{-1} (b) 1685 cm^{-1} (c) 1750 cm^{-1} (d) 1705, 2820 cm^{-1} (e) 1715 cm^{-1}
22.20 (a) Different peaks for McLafferty rearrangement (b) Different peaks for alpha cleavage
(c) Different peaks for McLafferty rearrangement

CHAPTER 23

23.1 (a) 3-Methylbutanoic acid (b) 4-Bromopentanoic acid
(c) *trans*-2-Methylcyclohexanecarboxylic acid (d) 2,4-Hexadienoic acid
(e) *cis*-1,3-Cyclopentanedicarboxylic acid (f) 2-Ethylpentanoic acid
(g) 2-Phenylpropanoic acid
23.3 Ethanol, $\Delta G^\circ = 21.8$ kcal/mol; acetic acid, $\Delta G^\circ = 6.8$ kcal/mol
23.4 Benzoic acid is soluble, and naphthalene insoluble, in aqueous base. **23.5** 52%
23.6 (a) (Strongest) $FCH_2COOH > BrCH_2COOH > CH_3CH_2COOH$ (Weakest)
(b) (Strongest) *p*-Cyanobenzoic > Benzoic > *p*-Hydroxybenzoic (Weakest)
(c) (Strongest) $CH_3CH_2COOH > CH_3CH_2OH > CH_3CH_2NH_2$ (Weakest)
23.7 Further ionization of the monoanion is unfavorable because the product of ionization is a dianion.
23.8 The dianions become more stable as the distance between the negative charges increases. **23.9** Deactivating
23.10 (a) (Strongest) *p*-Chlorobenzoic > Benzoic > *p*-Methylbenzoic (Weakest)
(b) (Strongest) *p*-Nitrobenzoic > Benzoic > Acetic (Weakest)
23.12 (a), (c), (e) Grignard carboxylation (b) Cyanide displacement (d) Either method

CHAPTER 24

24.1 (a) 4-Methylpentanoyl chloride (b) Cyclohexylacetamide (c) 2-Methylbutanenitrile (d) Benzoic anhydride (e) Isopropyl cyclopentanecarboxylate (f) Cyclopentyl 2-methylpropanoate (g) 4-Pentenamide (h) 2-Ethylbutanenitrile (i) 2,3-Dimethyl-2-butenoyl chloride (j) Trifluoroacetic anhydride

24.2 (a) 2-Pentenenitrile (b) *N*-Methylbutanamide (c) 2,4-Dimethylpentanoyl chloride (d) Methyl 1-methylcyclohexanecarboxylate

24.3 (a) (Most reactive) $CH_3COCl > CH_3CO_2CH_3 > CH_3CONH_2$ (Least reactive) (b) $CH_3CO_2CH(CF_3)_2 > CH_3CO_2CH_2CCl_3 > CH_3CO_2CH_3$

24.6 Carboxylate anions are inert to nucleophilic attack because of their negative charge.

24.8 Triethylamine has no N—H proton that can be lost from the tetrahedral intermediate, and thus cannot yield a neutral amide.

24.9 Acid chlorides are more reactive than ketones and react preferentially with Grignard reagent. If, however, an excess of Grignard reagent is present, the ketone will also react.

24.12 The carboxylic acid by-product must be scavenged by an added base to prevent protonation of the amine reactant.

24.13 $CH_3\overset{O^*}{\overset{\|}{C}}—OH \xrightarrow[2.\ H^+]{1.\ BH_3} CH_3CH_2\overset{*}{O}H \xrightarrow[\text{Pyridine}]{CH_3CH_2COCl} CH_3CH_2\overset{*}{O}—\overset{O}{\overset{\|}{C}}CH_2CH_3$

24.14 Alkaline hydrolysis yields a carboxylate anion that is inert to nucleophilic attack; acidic hydrolysis yields a free carboxylic acid that is susceptible to reverse reaction.

24.18 The initially formed imine anion has a negative charge and is thus inert to nucleophilic attack.

24.19 (a) Ester or six-membered-ring lactone (b) Acid chloride (c) Carboxylic acid (d) Nitrile (e) Aliphatic ketone or cyclohexanone

24.20 (a) $CH_3CH_2C{\equiv}N$ (b) $CH_3CH_2CH_2CH_2CO_2CH_3$ (c) $CH_3CON(CH_3)_2$ (d) $CH_3CH{=}CHCOCl$

CHAPTER 25

25.4 $CH_3\overset{*}{C}H(Ph)\overset{O}{\overset{\|}{C}}CH_3 \xrightarrow{H^+} \left[CH_3C(Ph){=}C(OH)Ph\right] \rightleftarrows CH_3CH(Ph)\overset{O}{\overset{\|}{C}}CH_3$

Optically active — Achiral — Racemic

25.5 No. The chiral center does not have an enolizable proton.

25.6 Racemic. The reaction proceeds through an achiral enol intermediate.

25.7 (a) CH_3CH_2CHO (b) $(CH_3)_3CCOCH_3$ (c) CH_3COOH (d) $PhCONH_2$ (e) $CH_3CH_2CH_2CN$ (f) $CH_3CON(CH_3)_2$

25.8 No; 3-Methylcyclohexanone would not racemize.

25.9 Rate = k[ketone][base]; rate is independent of halogen.

25.10 H^+ is generated during acidic ketone halogenation, but a molar equivalent of base is used up in the basic reaction.

25.11 Nucleophilic acyl substitution is involved in the reaction.

25.12 The reaction is similar to acid-catalyzed halogenation.

25.13 (a) $PhCH_2Br + CH_2(CO_2CH_2CH_3)_2 \xrightarrow[2.\ H_3O^+]{1.\ Base} PhCH_2CH_2COOH$

(b) $CH_3CH_2CH_2Br + CH_2(CO_2CH_2CH_3)_2 \xrightarrow{Base} CH_3CH_2CH_2CH(CO_2CH_2CH_3)_2$

$\xrightarrow[2.\ CH_3I;\ 3.\ H_3O^+]{1.\ Base} CH_3CH_2CH_2CH(CH_3)COOH$

(c) $(CH_3)_2CHCH_2Br + CH_2(CO_2CH_2CH_3)_2 \xrightarrow[2.\ H_3O^+]{1.\ Base} (CH_3)_2CHCH_2CH_2COOH$

(d) $BrCH_2CH_2CH_2Br + CH_2(CO_2CH_2CH_3)_2 \xrightarrow[2.\ H_3O^+]{1.\ Base}$ (cyclobutane)–COOH

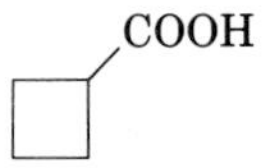

25.14 Malonic ester synthesis products must have a hydrogen on the alpha carbon. **25.16** None

25.17 (a) $PhCH_2COCH_3 \xrightarrow[2.\ CH_3I]{1.\ LDA} PhCH(CH_3)COCH_3$

(b) $CH_3CH_2CH_2CH_2CN \xrightarrow[2.\ CH_3CH_2Br]{1.\ LDA} CH_3CH_2CH_2CH(CH_2CH_3)CN$

CHAPTER 26

26.1 (a) $CH_3CH_2CH_2CH(OH)CH(CHO)CH_2CH_3$
(b) $CH_3CH_2C(OH)(CH_3)CH(CH_3)COCH_3$ and $CH_3CH_2C(OH)(CH_3)CH_2COCH_2CH_3$
(d) $PhC(OH)(CH_3)CH_2COPh$

26.3 $(CH_3)_2C(O{-}H)-CH_2-C(=O)-CH_3$ (with $:\bar{O}H$) $\rightleftharpoons$ $2\ (CH_3)_2C{=}O$

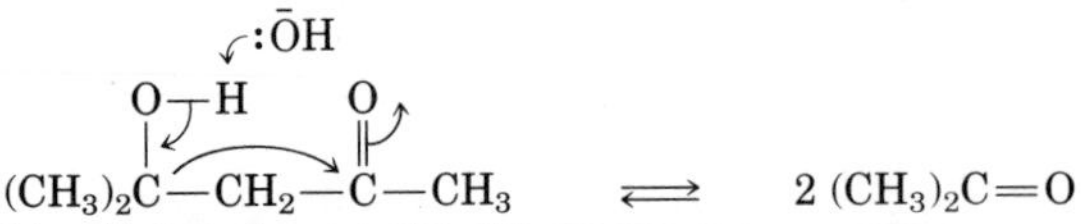

26.5 (c), aldol product of $CH_3CH_2COCH_2CH_3$ **26.6** (a) and (d)

26.7 When treated with base, 1,3-diketones form a stabilized anion that cannot cyclize.

26.12 Ethyl acetoacetate forms a stable anion on treatment with base, but ethyl α,α-dimethylacetoacetate cannot form an anion.

26.13 Yes; $CH_3CH_2O-C(=O)-C(=O)-CH_2C(=O)CH_2CH_3$

26.16 (a) $CH_3COCH_2COCH_2CH_3 + H_2C{=}CHCOPh$
(b) $CH_3COCH_2CO_2CH_2CH_3 + H_2C{=}CHCOCH_3$; then H_3O^+
(c) $CH_2(CO_2CH_3)_2 + H_2C{=}CHCN$
(d) $CH_3CH_2NO_2 + H_2C{=}CHCO_2CH_3$ (e) $CH_2(CO_2CH_3)_2 + H_2C{=}CHNO_2$

26.17 When monoketones are used, the product is similar in reactivity to the starting material.

26.18 H_2C N CH_2 CH_3 Steric hindrance

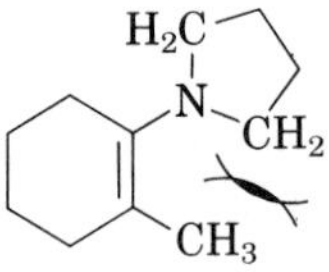

CHAPTER 27

27.1 (a) Aldotetrose (b) Ketopentose (c) Ketohexose (d) Aldopentose

27.2 (a) L; $2S,3S$ (b) D; $2R,3S,4R$ (c) D; $3S,4R$

27.3 CHO; H—OH; HO—H; HO—H; CH_2OH
L-Arabinose

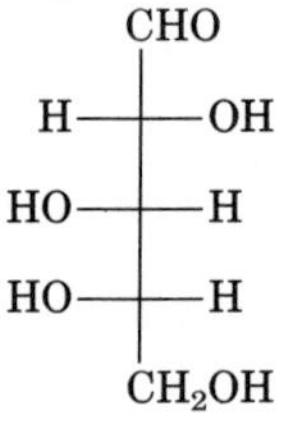

27.5 16 D and 16 L aldoheptoses are possible.

27.8 28% alpha; 72% beta

27.9 Similar in stability

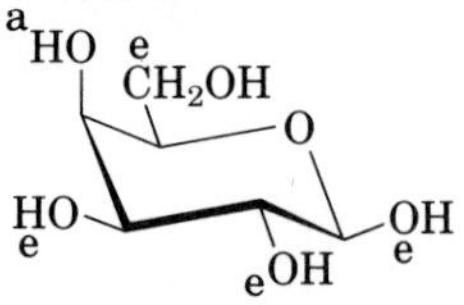

β-D-Galactopyranose (HO a; CH_2OH e; HO e; OH e; OH e)

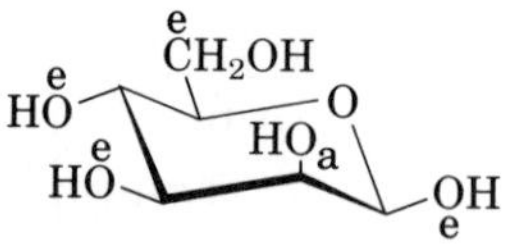

β-D-Mannopyranose (CH_2OH e; HO e; HO e; HO a; OH e)

27.10 $HOCH_2$ e; OH e; HO e; OH e; OH e

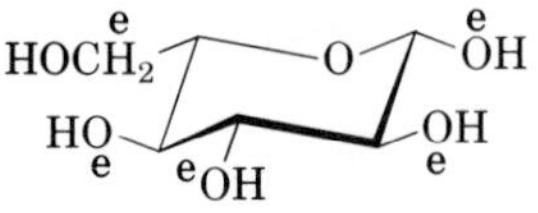

β-L-Glucopyranose

27.11 Galactitol is a meso compound. **27.12** L-Gulose and D-glucose yield the same alditol.
27.13 D-Allose and D-galactose yield meso aldaric acids. All others yield optically active aldaric acids.
27.14 D-Ribose → D-Allose + D-Altrose **27.15** See Figure 27.2. **27.16** See Figure 27.2.
L-Xylose → L-Idose + L-Gulose

27.17 H—C(COOH)(COOH)—OCH_3 (Fischer projection: COOH top, COOH bottom, H left, OCH_3 right) **27.18** Optically active

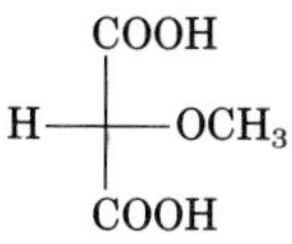

CHAPTER 28

28.1 (a) 1° (b) 3° (c) 4° (d) 2° (e) 2°
28.2 (a) *N*-Methylethylamine (b) Tricyclohexylamine (c) *N*-Methyl-*N*-propylcyclohexylamine
(d) *N*-Methylpyrrolidine (e) Diisopropylamine (f) 1,3-Butanediamine
(g) *cis*-1,2-Cyclopentanediamine
28.4 Pyramidal inversion involves transient rehybridization of nitrogen from sp^3 to sp^2, which would increase the strain in the three-membered ring.

28.5 Protonation on oxygen still allows for amide resonance: $R{-}C({=}\overset{+}{O}H){-}\ddot{N}H_2 \longleftrightarrow R{-}C(\ddot{O}H){=}\overset{+}{N}H_2$

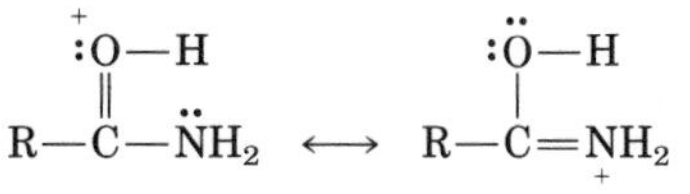

28.6 (Fischer projection, top to bottom) H; HO—C—CH_3; C=O; H—C—CH_3; CH_2CH_3

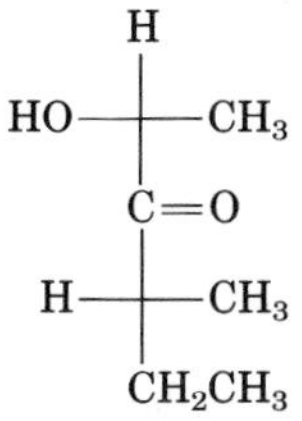

28.7 Products are diastereomeric. **28.13** $R{-}\ddot{N}{:}$ nitrene; six electrons
28.15 Quaternary ammonium salts, R_4N^+, are chiral.

28.16 Piperidine (N—H) ⟶ [N,N-dimethylpiperidinium, $N^+(CH_3)_2$] ⟶ $CH_2{=}CHCH_2CH_2CH_2N(CH_3)_2$

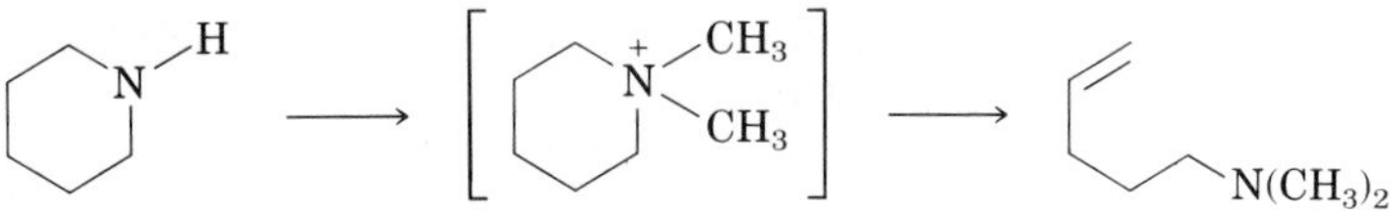

28.19 $C_6H_5CH_2CH(NHCH_3)CH_3$

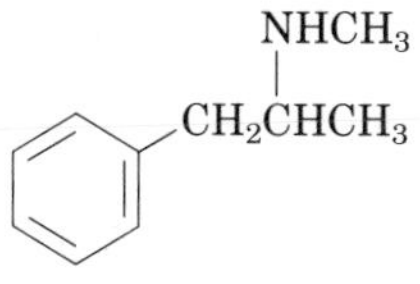

CHAPTER 29

29.1 The nitrogen lone-pair electrons in *p*-nitroaniline are delocalized onto the nitro group and are less available for bonding to an acid.
29.2 (a) (Most basic) *p*-Bromoaniline > *p*-Aminobenzaldehyde > *p*-Nitroaniline (Least basic)
(b) (Most basic) *p*-Methylaniline > *p*-Chloroaniline > *p*-Aminoacetophenone (Least basic)
(c) (Most basic) *p*-Methylaniline > *p*-(Fluoromethyl)aniline > *p*-(Trifluoromethyl)aniline (Least basic)

29.4 $H{-}\ddot{N}{-}COCH_3$ (attached to benzene ring) ⇌ $H{-}\overset{+}{N}{-}COCH_3$ (C=N, ortho carbanion) ⇌ $H{-}\overset{+}{N}{-}COCH_3$ (C=N, para carbanion)

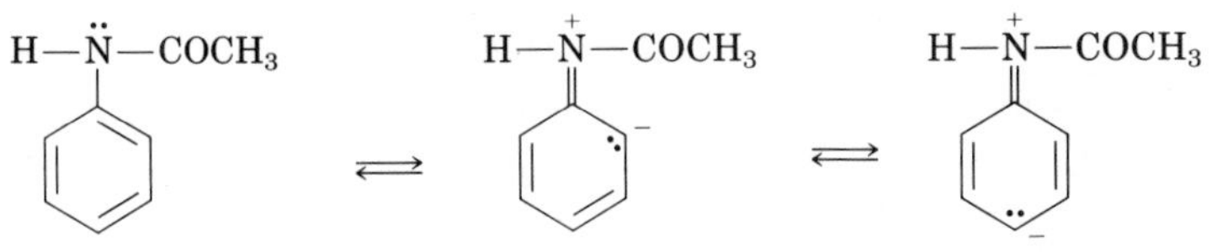

29.6 Aryl cations are less stable than alkyl cations.
29.11 (a) (Most acidic) *p*-Trifluoromethylphenol > Phenol > *p*-Methylphenol (Least acidic)
(b) (Most acidic) *p*-Hydroxybenzoic acid > Phenol > Benzyl alcohol (Least acidic)
(c) (Most acidic) 2,4,6-Tribromophenol > 2,4-Dibromophenol > *p*-Bromophenol (Least acidic)

29.12 Toluene $\xrightarrow[AlCl_3]{(CH_3)_2CHCl}$ p-$CH_3C_6H_4CH(CH_3)_2$ $\xrightarrow[H_2SO_4]{SO_3}$ (CH3, SO_3H, $CH(CH_3)_2$ ring) $\xrightarrow[200°C]{KOH}$ (CH3, OH, $CH(CH_3)_2$ ring)

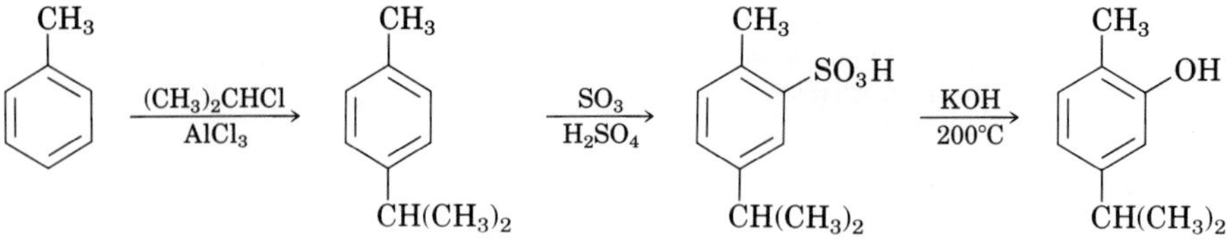

29.15 *o*-(1-Methyl-2-propenyl)phenol

CHAPTER 30

30.1 Ground state: ψ_3 HOMO, ψ_4^* LUMO; excited state: ψ_4^* HOMO, ψ_5^* LUMO

30.2 Disrotatory: 5,6-dimethyl-1,3-cyclohexadiene (CH_3, H / H, CH_3). Conrotatory *(not formed)*: (CH_3, H / CH_3, H).

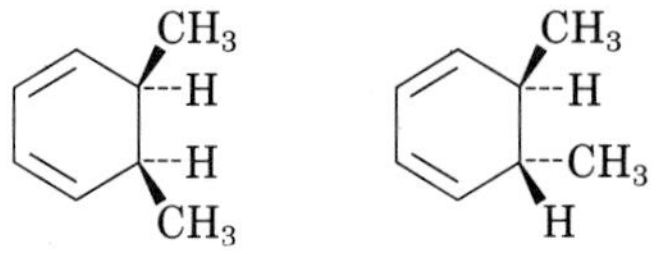

Disrotatory Conrotatory *(not formed)*

30.3 Both are products of conrotatory opening but the *E,E* product is much less strained.

30.4 (2*E*,4*Z*,6*E*)-Octatriene $\xrightarrow{h\nu}$ *trans*-5,6-Dimethyl-1,3-cyclohexadiene

(2*E*,4*Z*,6*Z*)-Octatriene $\xrightarrow{h\nu}$ *cis*-5,6-Dimethyl-1,3-cyclohexadiene

30.5 The dienes are equilibrated via conrotatory closure to a cyclobutene, followed by conrotatory opening in two different ways.
30.6 Disrotatory **30.8** Suprafacial; [6 + 4] cycloaddition
30.9 Diels–Alder reaction, followed by retro Diels–Alder reaction
30.10 Homotropilidene undergoes a rapid [3,3] rearrangement. Thus, one sees a time-averaged NMR spectrum.
30.11 A series of [1,5] sigmatropic hydrogen shifts takes place.
30.12 A [1,7] antarafacial sigmatropic rearrangement
30.14 (a) Conrotatory (b) Disrotatory (c) Suprafacial (d) Antarafacial (e) Suprafacial

CHAPTER 31

31.4 The lone-pair electrons of the tryptophan-ring nitrogen are less basic because they are part of an aromatic ring.

31.5 R—CHO + NH_3 $\rightleftharpoons$ [R—CH=NH] $\xrightarrow{:\bar{C}N}$ R—CH(CN)NH$_2$

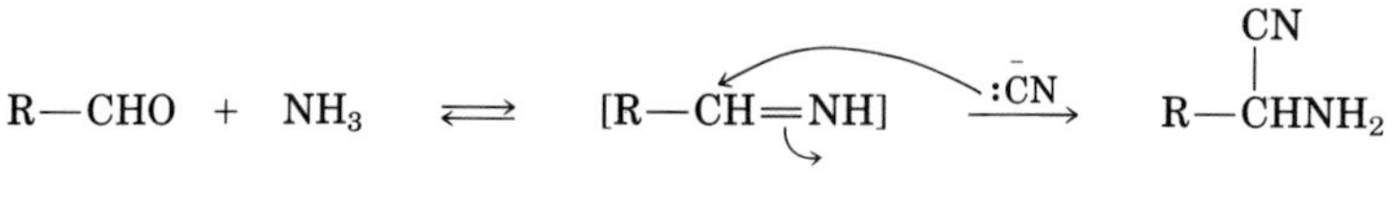

31.8 (a) $(CH_3)_2CHCH_2Br$ (b) 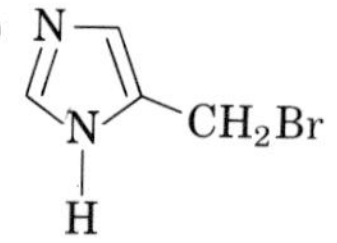 4-(bromomethyl)imidazole (CH_2Br, N—H) (c) 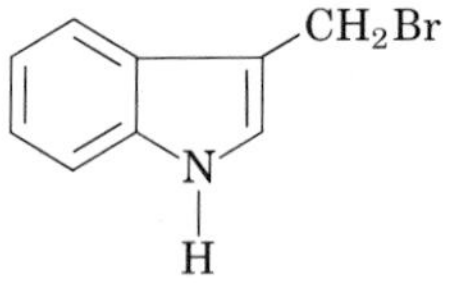 3-(bromomethyl)indole (CH_2Br, N—H)
(d) $CH_3SCH_2CH_2Br$

31.9 Add diethyl acetamidomalonate to formaldehyde.
31.10 H-Val-Tyr-Gly-OH; H-Val-Gly-Tyr-OH; H-Tyr-Gly-Val-OH; H-Tyr-Val-Gly-OH; H-Gly-Val-Tyr-OH; H-Gly-Tyr-Val-OH
31.12 Only primary amines yield the conjugated product.
31.13 (a) H-Arg-Pro-Leu-Gly-Ile-Val-OH (b) H-Val-Met-Trp-Asp-Val-Leu-OH
31.14 Trypsin: H-Asp-Arg-OH; H-Val-Tyr-Ile-His-Pro-Phe-OH
Chymotrypsin: H-Asp-Arg-Val-Tyr-OH; H-Ile-His-Pro-Phe-OH
31.15 H-Pro-Leu-Gly-Pro-Arg-Pro-OH

31.16 The peptide is cyclic: Leu-Ala \ / Phe or Ala-Leu \ / Phe

31.17 H-Leu-OH $\xrightarrow{\text{BOC}}$ BOC-Leu-OH; H-Ala-OH $\xrightarrow[\text{H}^+]{\text{CH}_3\text{OH}}$ H-Ala-OCH_3 $\xrightarrow{\text{DCC}}$ BOC-Leu-Ala-OCH_3 $\xrightarrow[2.\ ^-\text{OH}]{1.\ \text{H}^+}$ H-Leu-Ala-OH

31.18 BOC-Ala-OH + H-Gly-OCH_3 $\xrightarrow{\text{DCC}}$ BOC-Ala-Gly-OCH_3 $\xrightarrow{\text{H}^+}$ H-Ala-Gly-OCH_3 $\xrightarrow[\text{DCC}]{\text{BOC-Leu-OH}}$ BOC-Leu-Ala-Gly-OCH_3 $\xrightarrow[2.\ ^-\text{OH}]{1.\ \text{H}^+}$ H-Leu-Ala-Gly-OH

31.19 The amide nitrogen of proline has no hydrogen and thus interrupts the H-bonded helix.

CHAPTER 32

32.1 (9*Z*,11*E*,13*E*)-Octadecatrienoic acid **32.2** 9-Octadecynoic acid
32.3 Nucleophilic acyl substitution **32.4** Retro aldol reaction
32.5 Every other carbon: C2, C4, C6, C8, . . .

32.8 (a) H, e CH_3, H, H (b) H, H, H, CH_3 a

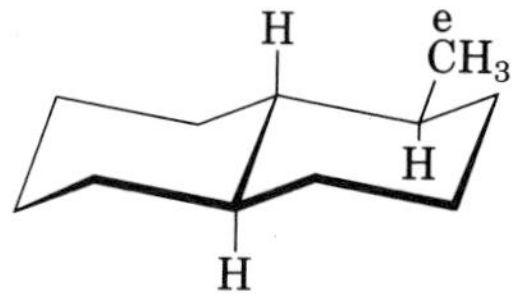

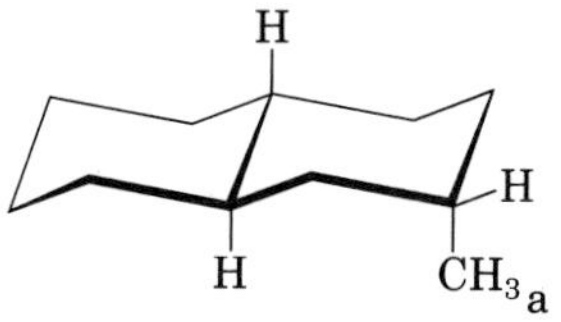

32.9 The hydroxyl group is equatorial.

CHAPTER 33

33.1 O

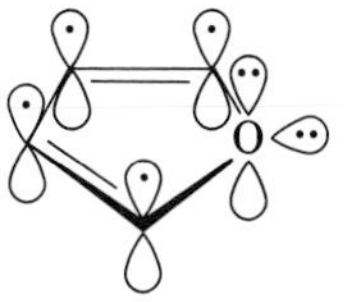

33.2 H, N

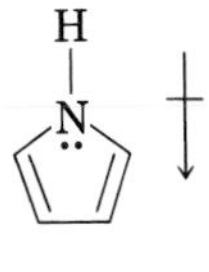

33.4 H—N, N, More basic

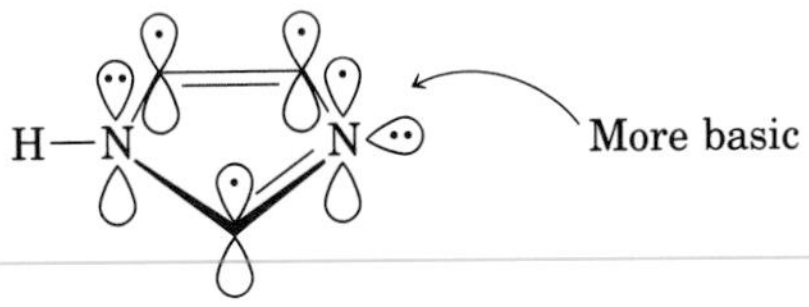

33.5 $\xrightarrow{\text{E}^+}$ H, E, N, +

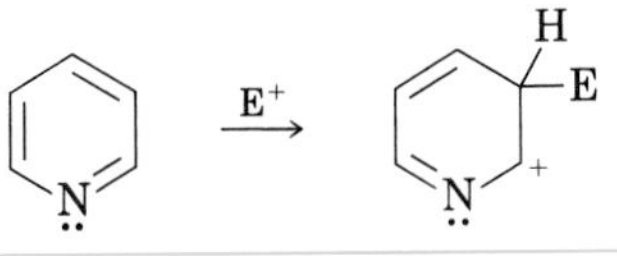

Charge cannot be delocalized on nitrogen.

33.10 Lactam form is stabilized by more amide resonance. **33.11** CCGATTAGGCA

33.12 O·······H, H, N, N, N, H, H, N, N, H···············N, N, O

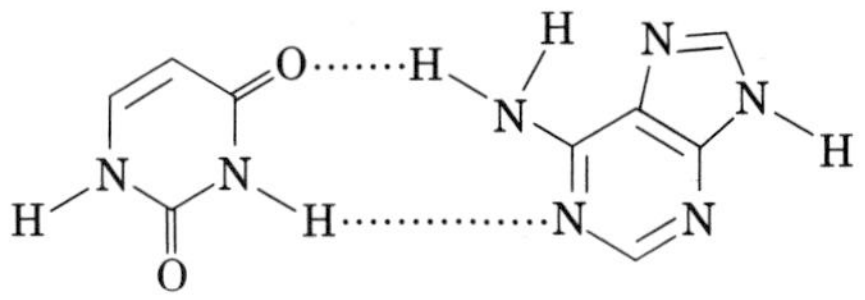

33.13 CUAAUGGCAU **33.14** Leu-Met-Ala-Trp-Pro (Stop)
33.15 Anticodon: GAA-UAC-CGA-ACC-GGG-AUU; DNA: GAA-TAC-CGA-ACC-GGG-ATT

33.16 A > G
A-C-A-T-G-G-C-G-C-T-T-A-T-G-A-C-G-A
A-A-C T-G-G . . . . .
A-A-C-A-T-G-G-C-G-C-T-T T-G-A-C-G-A, etc.

G > A
A-A-C-A-T G-C-G-C-T . . . . .
A-A-C-A-T-G C-G-C-T-T . . . . .
A-A-C-A-T-G-G-C C-T-T-A . . . . . , etc.

C
A-A A-T-G-G-C-G-C-T . . . . .
A-A-C-A-T-G-G G-C-T-T . . . . .
A-A-C-A-T-G-G-C-G T-T-A . . . . . , etc.

T + C
A-A A-T-G-G-C-G . . . . .
A-A-C-A G-G-C-G-C . . . . .
A-A-C-A-T-G-G G-C-T-T . . . . . , etc.

33.17 X-T-C-A-G-T-A-C-C-G-A-T-T-C-G-G-T-A-C

33.18 An extremely stable carbocation is formed in this S_N1 cleavage reaction.

CHAPTER 34

34.1 (a) $H_2C{=}CHOCH_3$ (b) $H_2C{=}C_6H_4{=}CH_2$ (c) $CHCl{=}CHCl$

34.2 The polymerization step is a Markovnikov addition that yields the most stable, benzylic radical.

34.3 (Most reactive) $H_2C{=}CHPh > H_2C{=}CHCH_3 > H_2C{=}CHCl > H_2C{=}CHCO_2CH_3$

34.4 (Most reactive) $H_2C{=}CHCN > H_2C{=}CF_2 > H_2C{=}CHPh > H_2C{=}CHCH_3$

34.6 Poly(vinylidene chloride) has no chiral centers.

34.7 No. Net rotation is zero since equal amounts of (+) and (−) forms are present.

34.9 Ozone attacks the double bonds in the polymer.

34.12 The amide bonds in nylon are hydrolyzed by acid or base.

34.13 (a) $\sim CH_2CH_2CH_2{-}O{-}CH_2CH_2CH_2{-}O\sim$

(b) $\sim C(=O)(CH_2)_6C(=O){-}O{-}CH_2CH_2{-}O\sim$ (c) $\sim C(=O)(CH_2)_4C(=O){-}NH(CH_2)_6NH\sim$

34.14 $\sim NH{-}C_6H_4{-}NH{-}C(=O){-}C_6H_4{-}C(=O)\sim$

34.15 This polymer is cross-linked and very rigid. **34.17** Atactic

INDEX

The references given in color refer either to boldface entries in the text (where terms are defined) or to mini-biographies.

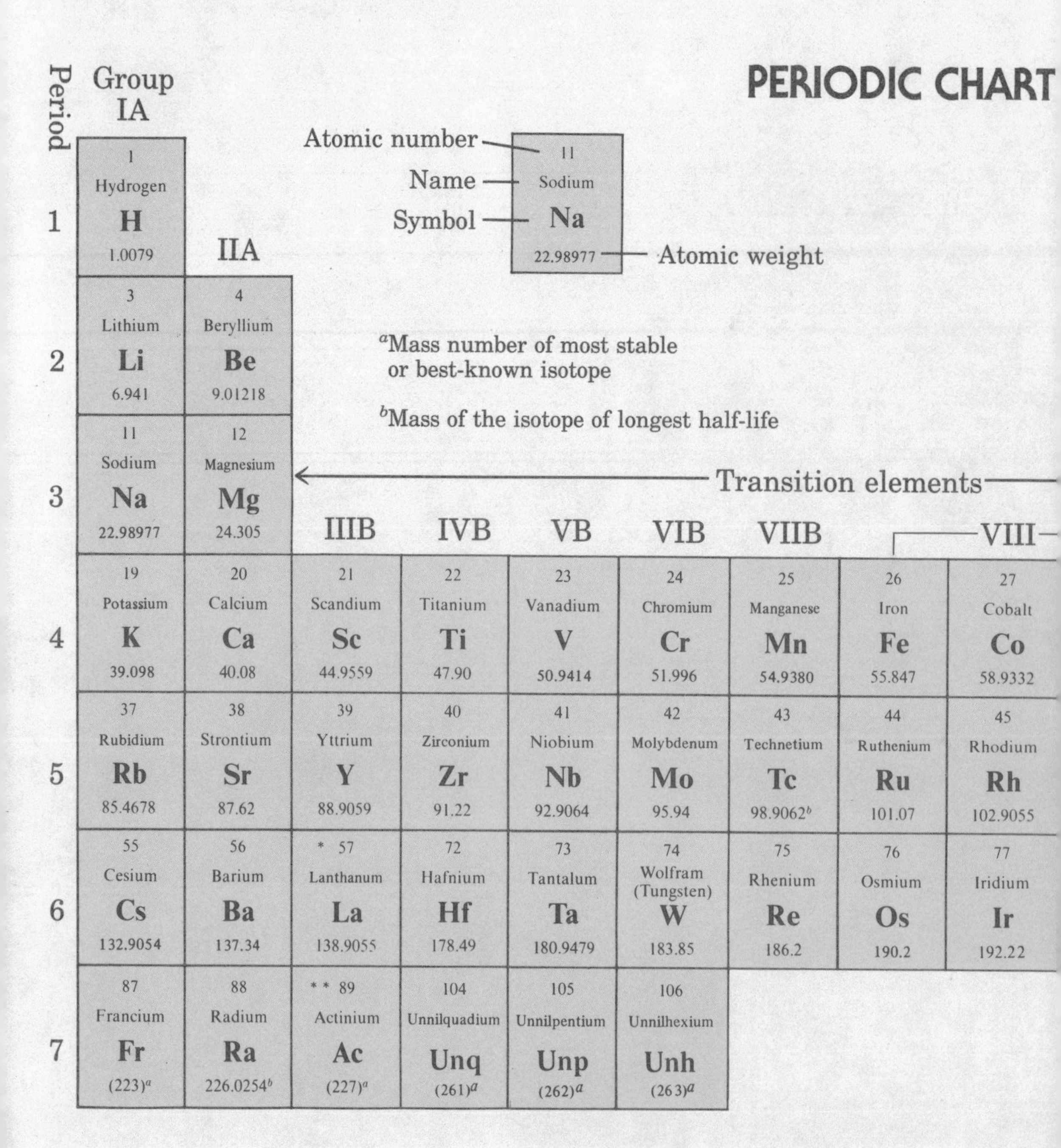

PERIODIC CHART

Atomic number — 11
Name — Sodium
Symbol — Na
22.98977 — Atomic weight

[a]Mass number of most stable or best-known isotope

[b]Mass of the isotope of longest half-life

← Transition elements —

Period	Group IA	IIA	IIIB	IVB	VB	VIB	VIIB	VIII	
1	1 Hydrogen **H** 1.0079								
2	3 Lithium **Li** 6.941	4 Beryllium **Be** 9.01218							
3	11 Sodium **Na** 22.98977	12 Magnesium **Mg** 24.305							
4	19 Potassium **K** 39.098	20 Calcium **Ca** 40.08	21 Scandium **Sc** 44.9559	22 Titanium **Ti** 47.90	23 Vanadium **V** 50.9414	24 Chromium **Cr** 51.996	25 Manganese **Mn** 54.9380	26 Iron **Fe** 55.847	27 Cobalt **Co** 58.9332
5	37 Rubidium **Rb** 85.4678	38 Strontium **Sr** 87.62	39 Yttrium **Y** 88.9059	40 Zirconium **Zr** 91.22	41 Niobium **Nb** 92.9064	42 Molybdenum **Mo** 95.94	43 Technetium **Tc** 98.9062[b]	44 Ruthenium **Ru** 101.07	45 Rhodium **Rh** 102.9055
6	55 Cesium **Cs** 132.9054	56 Barium **Ba** 137.34	* 57 Lanthanum **La** 138.9055	72 Hafnium **Hf** 178.49	73 Tantalum **Ta** 180.9479	74 Wolfram (Tungsten) **W** 183.85	75 Rhenium **Re** 186.2	76 Osmium **Os** 190.2	77 Iridium **Ir** 192.22
7	87 Francium **Fr** (223)[a]	88 Radium **Ra** 226.0254[b]	** 89 Actinium **Ac** (227)[a]	104 Unnilquadium **Unq** (261)[a]	105 Unnilpentium **Unp** (262)[a]	106 Unnilhexium **Unh** (263)[a]			

* Lanthanide series 6	58 Cerium **Ce** 140.12	59 Praseo-dymium **Pr** 140.9077	60 Neodymium **Nd** 144.24	61 Promethium **Pm** (145)[a]	62 Samarium **Sm** 150.4
** Actinide series 7	90 Thorium **Th** 232.0381[b]	91 Protactinium **Pa** 231.0359[b]	92 Uranium **U** 238.029	93 Neptunium **Np** 237.0482	94 Plutonium **Pu** (242)[a]

Identified information requirements may be expressed in various ways. The simplest approach is to list the various information items. A more useful approach, though, is to group the needed information into reports and other outputs. For example, information for the inventory management project may be stated as inventory reorder reports, inventory stock status reports, and so on.

SUBMISSION OF A SYSTEMS ANALYSIS REPORT

The requirements, relating both to the physical system and the needed information, are incorporated in a **systems analysis report.** Also included in the report would be a statement of the objectives and scope of the project, the role of the project within the strategic systems plan, a summary of the problems in the present system, a list of constraints and assumptions, and a revised time schedule and cost budget. Appendices would contain all relevant documentation concerning the present system, including detailed system flowcharts, copies of source documents, record layouts, and so on.

This report provides the basis for review by higher management, such as members of a steering committee. If the report is approved, the project team has the "green light" to move into the systems design phase. The report should be a main source of guidance during the design process.

TECHNIQUES FOR GATHERING FACTS

Techniques for gathering facts include reviews of documents and records, observations, interviews, and questionnaires. All of these techniques may be employed during systems surveys and analyses, as well as during the design phase. Usually the fact gatherers are systems analysts or accountants, although auditors employ most of the techniques in the course of audits.

REVIEWS OF DOCUMENTS AND RECORDS

Documents and records, such as journal vouchers and charts of accounts, provide considerable insights to systems analysts. For instance, a chart of accounts shows how the present accounting information system (AIS) is structured.

Reviews of documents and records should normally precede most of the other data-gathering techniques. In addition to affording an overview of the firm and its system, they provide a basis for comparison. Procedures manuals, for example, show how the system is designed to work. The procedures described therein can be compared with facts gathered concerning how the system actually is working at present. Any differences that are found point to possible problems.

Where possible, the reviews can extend to reports from previous studies. Examples are audit reports prepared by internal auditors and management letters prepared by independent external auditors. These reports and letters usually identify weaknesses and recommend improvements to controls and other aspects of the information system.

OBSERVATIONS

Through observations a systems analyst can become familiar with the setting, relationships, and constraints of a system. Often an experienced analyst can spot weaknesses, especially if the visit is unannounced. For instance, a walk through the production area may reveal that employees tend to be unproductive and careless. The act of observing also provides verification. For example, assertions made through documents or questionnaires may be checked firsthand.

Observations can be combined with data-organizing techniques to good effect. Work sampling is an example of a technique that can be used effectively with the observation technique. **Work sampling** consists of taking observations at randomly selected times, to determine the extent to which resources are being used productively. For instance, an analyst may observe the activity in the shipping department at random times for a week. During each observation the analyst counts the number of employees who are performing productive tasks. At the end of the week, he or she can extrapolate the sampled observations to estimate the productivity level of the department. Thus, the analyst might estimate that the department has an average level of 75 percent productivity. If expected productivity is at a level of 90 percent, the actual level indicates a productivity problem.

INTERVIEWS

Interviewing is perhaps the single most popular and important data-gathering technique. Interviews with managers enable information needs to be discovered. Interviews with employees as well as managers can clarify problems found through document reviews and observations. Equally important, interviews provide a natural means of involving managers and employees in systems development.

Good interviewing is an art that can be learned. It is based on sound psychological principles and interviewing theories. The guidelines listed in Figure 19-8 provide an introduction to this important technique; however, its mastery requires further study and practice.

1. Ascertain the interests, background, and responsibilities of the person to be interviewed *before* the interview.
2. Gather facts concerning the matters to be discussed *before*hand.
3. Prepare a list of the questions to be asked during the interview.
4. Obtain approval from the interviewee's superior for the interview.
5. Make an appointment that is convenient with the interviewee and be on time.
6. Notify the interviewee beforehand of the purpose of the interview and of the matters that it will cover.
7. Open the interview by explaining the interviewer's role in the study, then draw out information by pertinent questions, especially concerning the interviewee's knowledge of the situation, needs for information, and ideas for improvements.
8. Listen carefully to the interviewee's answers without interruption.
9. When conversing do not resort to jargon, broad generalizations, personal opinions, or irrelevant comments.
10. Be natural, but businesslike, so that the interview flows easily.
11. Maintain a courteous, respectful, tactful, and friendly manner throughout the interview.
12. Ask permission to take notes or use a tape recorder during the interview.
13. End the interview with a summary of the discussion, a thank you, and a prompt exit.
14. Shortly after the interview, review and complete the notes taken or conversation recorded; send a copy of the notes to the interviewee for review and correction.

FIGURE 19-8 A list of interviewing guidelines.

Although critical to systems development, interviews do have two potential shortcomings. On the one hand, interviewees may be biased or too anxious to please. Thus, they may provide false or misleading information. On the other hand, interviewees may be antagonistic or too busy. In such cases they are likely not to cooperate.

In addition, interviews can be quite time consuming for the analyst. This drawback especially applies to the unstructured or open-ended type of interview, in which the interviewee is encouraged to talk freely and to ramble from one topic to another. Furthermore, numerous interviews are often necessary in order to draw together all the "pieces of the puzzle" when problems are complex. If time is short, the questionnaire (discussed next) may be preferable.

QUESTIONNAIRES

A **questionnaire** is a standardized list of questions. It is an efficient means of surveying when brief answers are desired. For instance, an internal control questionnaire is used to ascertain which specific controls are absent in an accounting information system. That is, it determines the *facts* concerning a situation or question. A personnel questionnaire, by contrast, can be used to ascertain the *opinions* of employees and managers concerning a policy, procedure, or other matter relating to the firm.

The processes of administering and evaluating questionnaires are capable of automation. Respondents can answer questions that are displayed on terminal or microcomputer screens, using keyboards or light pens. The accumulated responses are evaluated by means of appropriate software, which in some cases can incorporate expert sytems.

Questionnaires have one major failing. They do not enable respondents to provide in-depth answers in an easy manner. Thus, it is often necessary to follow up questionnaires with interviews that explore the reasons for certain answers.

TECHNIQUES FOR ORGANIZING FACTS

Analyzing and designing information systems entails coping with numerous facts and complex relationships. During the past 50 years a variety of diagrammatic and graphic techniques have been devised to organize and model these facts and relationships. Most of the fact-organizing techniques are useful for both (1) analyzing a current system and (2) documenting the features of a new or improved design. Experienced systems analysts should be capable of applying the great majority of the techniques. Accountants who are involved in systems analysis and design should be familiar with most and able to apply several.

Categories of Techniques Fact-organizing techniques may be categorized in various ways. Certain techniques present *conceptual* and/or *logical* views, which focus on what processes are performed or what data are handled by an information system. Other techniques focus on *physical* aspects, which emphasize how an information system functions and what facilities it employs. Certain techniques present broad views that may span entire firms or major functions, other techniques provide very narrow and detailed views, while still other techniques range from very broad to quite detailed. Finally, some techniques are members of integrated frameworks known as structured and object-oriented systems development, while other techniques have been employed in traditional or classical systems development efforts.

In this section our major attention is devoted to structured techniques, especially those that are relatively broad and conceptual. Traditional techniques are briefly covered. Object-oriented techniques are discussed later in the chapter, while detailed design techniques are discussed in Chapter 20.

STRUCTURED SYSTEMS DEVELOPMENT TECHNIQUES

Structured systems development represents a disciplined application of the top-down approach, using a variety of structured techniques. Three major benefits are provided by structured systems development: (1) greater efficiency in the systems development process, (2) greater consistency throughout the designed system, and (3) better communication between systems analysts and the prospective users of the newly designed systems. Due to these benefits, structured techniques have become more widely used than traditional techniques. It should be mentioned, however, that some analysts question whether the resulting system designs are of significantly higher quality than when traditional techniques are employed; thus, both sets of techniques need be considered.

Characteristics Structured systems development begins with an integrated view of the overall information system, or of the portion of the system within the scope of a systems project (i.e., the module). Through systematic decomposition, the module is broken into smaller, more detailed submodules. Each submodule is more manageable and understandable for purposes of analysis and design, while still tying back to the integrated view. The focus of structured techniques is on the conceptual and logical, rather than the physical, data and information flows and relationships. As a result, the project team can focus on what needs to be done and produced, rather than on exactly how to process and generate information.

Specific Structured Techniques Most of the available structured techniques are listed in Figure 19-9, together with examples of traditional techniques. Since these techniques are used as documentation of an AIS as well as tools of analysis, several appeared in Chapter 4. Structured techniques introduced and illustrated included the entity-relationship diagram, data-flow diagram (with context diagram), structure chart, and data dictionary.* These diagrams developed through key structured techniques have close relationships to each other, as Figure 4-12 shows. Before reading further, it might be useful to review the structured techniques described in Chapter 4.

Example An example involving data-flow diagrams and structure charts should reinforce your understanding of the nature of structured techniques. Figure 19-10 presents a high-level structure chart having a hierarchy of three levels and pertaining to the sales transaction processing system.† Each successive level provides a more detailed (decomposed) view of activities specified at the next higher level. Submodules at the second level represent four activities involved in the sales processing system. In turn, submodules at the third level represent more detailed activities of each of the activities at the second level. Furthermore, each submodule at the third level can be decomposed into even more detailed activities at a fourth level. For instance,

---

*In addition, Chapter 6 expanded the discussion of the entity-relationship diagram. It also added the data structure diagram, another logical diagram. The data structure diagram is not listed as a structured technique, since it is directly related to a specific type of data structure.

†Structure charts that focus on the higher levels of activities are also known as *functional decomposition charts*. Those structure charts that focus on more detailed levels are discussed in Chapter 20.

Structured Techniques
Entity-relationship diagrams
Data-flow diagrams
Structure charts
Decision trees
Decision tables
Data dictionaries
HIPO diagrams
State-transition diagrams
Structured English
Pseudocode
Traditional Techniques
System flowcharts
Work distribution charts
Work measurement analyses
Input-output matrices
Space layout diagrams
Form analysis sheets
Record layouts
Hardware configuration diagrams

FIGURE 19-9 Systems analysis and design techniques.

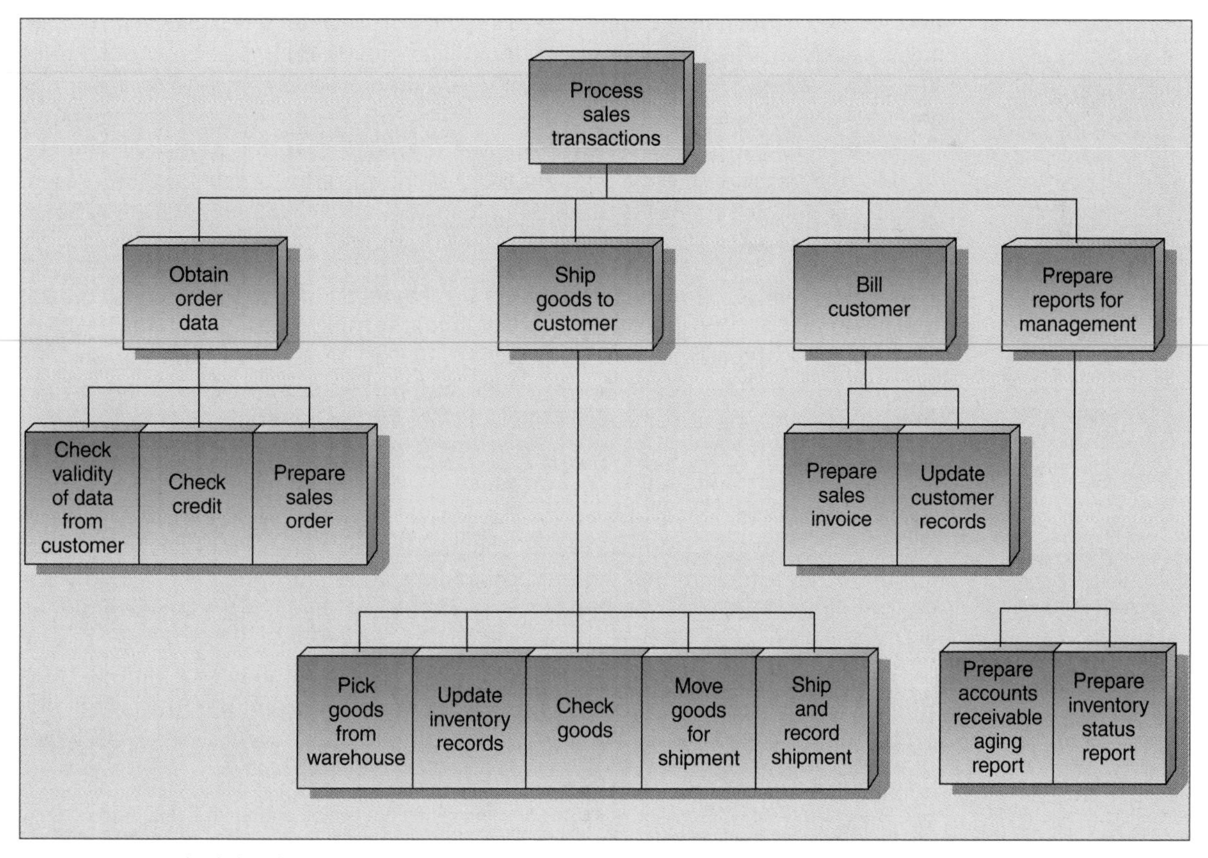

FIGURE 19-10 A high-level structure chart of the sales transaction processing system.

"prepare sales invoice" can be broken into several detailed steps or activities, such as "access sales order and shipping data" and "enter heading of sales invoice."

The activities portrayed by a high-level structure chart can alternatively be shown by means of data-flow diagrams. Each level of the structure chart can be paralleled by one or more separate data-flow diagrams. The second level of the structure chart, for instance, translates into a data-flow diagram similar to the one shown in Figure 12-20. Similarly, the third level of the structure chart can be represented by four data-flow diagrams. These data-flow diagrams articulate (interlock) with other data-flow diagrams at higher and lower levels, just as the submodules of a structure chart link to each other.

Figure 19-11 shows two data-flow diagrams that articulate with each other. The data-flow diagram at the left in the figure shows the details of the process in the structure chart submodule labeled "Ship goods to customer." As we can see, the processes (circles) in the data-flow diagram correspond to the submodules at the next lower level in the structure chart. The data-flow diagram at the right in the figure shows an even more detailed view of the process labeled "Ship and record shipment."

TRADITIONAL SYSTEMS DEVELOPMENT TECHNIQUES

Well before the introduction of structured techniques, a variety of techniques were devised and employed to meet the needs of specific systems development situations. These techniques have been loosely associated with what has been called traditional systems development.

Benefits of Traditional Techniques Techniques in this category generally involve physical aspects of an information system. Thus, they continue to have useful roles to play in systems analysis and design. First, they provide a means of organizing physical facts concerning systems components, such as inputs and storage. They therefore enable a project team to analyze the facts in a more efficient manner. Second, they provide clear documentation of the physical aspects of the present system, as well as of a new system being considered. This documentation is beneficial in gaining acceptance of systems analysis reports.

Specific Traditional Techniques A list of traditional techniques that are widely used appears in Figure 19-9. Perhaps the most familiar technique is the system flowchart. As noted in Chapter 4, the system flowchart appears in a variety of forms, such as the process flowchart, document flowchart, and computer system flowchart. Sometimes accompanying a computer system flowchart are program flowcharts, which provide the details of the processing steps or runs.*

Other useful traditional techniques include work distribution charts, work measurement analyses, input-output matrices, space layout diagrams, form analysis sheets, record layouts, and hardware configuration diagrams. Form analysis sheets and record layouts were illustrated in Chapter 4, while hardware configuration diagrams were described in Chapter 3. The remainder of the listed techniques will be briefly described.

Work distribution concerns the distribution of times spent on various tasks within an activity or organizational unit. A **work distribution chart** portrays in matrix format the times worked by each employee on each task. Figure 19-12, on pages

*Since program flowcharts can be viewed as decompositions of computer system flowchart processes, flowcharts are sometimes considered structured techniques. However, since flowcharts show physical aspects of a processing system, they clearly belong in the traditional category.

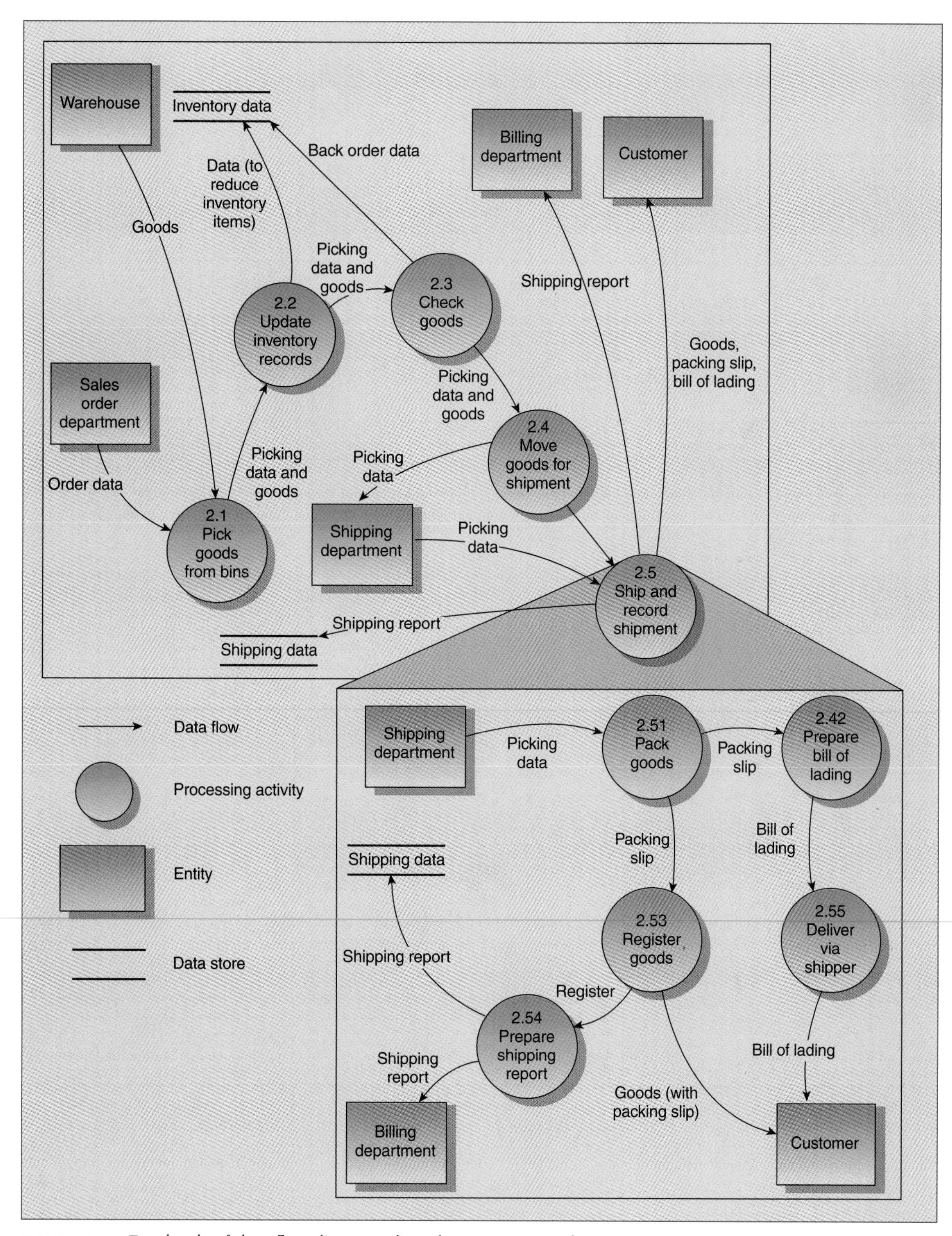

FIGURE 19-11 Two levels of data-flow diagrams, based on a structure chart.

Organizational Unit *Shipping Dept.* Date Prepared: June 1997			Employee(s) *M. T. Sullivan* Job title: supervisor		Employee(s) *Terry Frank &* *Jim Williams* Job title: shipping clerk		Employee(s) *Ralph Johnson* Job title: traffic clerk		Employee(s) *Paul Emerson* Job title: expeditor		Employee(s) *Linda Dent &* *Sandra Pyle* Job title: typist	
No.	Activity	Hours per week	Task	Hours per week	Task	Hours per week	Task	Hours per week	Task	Hours per week	Task	Hours per week
1	Controlling sales orders	35	Spot-checking shipping order file against outstanding order report from sales order department	2	Date-stamping and checking packing slip copies from warehouse to shipping copies; noting unshipped items on shipping copies	33						
2	Preparing shipping forms and reports	102	Reviewing summary report of shipments	1	Proofreading typed bills of lading for accuracy and completeness; preparing summary report of shipments from bills of lading copies	30 3					Typing bills of lading from packing slip copies; typing summary report of shipments	60 8
3	Scheduling, routing, and dispatching shipments	60	Preparing work assignments for shipping handlers	2	Collecting packing slips and spot-checking weights of packed shipments against weights noted on packing slip copies by handlers	8	Maintaining file of shipping rates; entering shipper, schedule, route, and charges for shipments on packing slip copies	1 37	Delivering bills of lading to shippers' representatives for signature	12		

4	Supervising shipping activities	25	Training employees and observing their performances; checking employees' attendance; answering employees' questions	12 2 2					Helping speed shipments and eliminating bottlenecks.	9		
5	Following status of sales orders and shipments	25	Handling customers' inquiries regarding shipments	8	Handling customers' inquiries regarding shipments	2	Answering inquiries of shippers regarding routes and charges	1	Following progress of special orders through production and into shipping area	14		
6	Performing miscellaneous activities	33	Preparing letters and memos; conferring with vice-president of marketing and other department heads	3 8	Collating bills of lading and shipping orders to be sent to billing department	4	Preparing special instructions for shippers	1	Picking up and delivering papers dealing with rush orders; filing	2 3	Typing correspondence and packing labels; filing	4 8
	Totals	280		40		80		40		40		80

FIGURE 19-12 A work distribution chart for a shipping department.

856–857, shows a work distribution chart for a shipping department. Upon analyzing the data in such a chart, a systems analyst can more easily spot inequities and inefficiencies. After determining improvements, the analyst can then reflect the revised allocation of tasks in a new chart.

Numerous measurements are necessary to trace and analyze the flows related to an information system. A typical data flow is the number of sales orders received in one week; an analysis of this flow may be in terms of the numbers of sales orders by various size ranges. Work measurement consists of measuring work flows, e.g., rates of output or levels of productivity. For example, if a clerical employee spends 20 hours processing 400 purchase orders, then his or her rate of output is 20 purchase orders per hour. **Work measurement analysis** provides a means of evaluating work performances. First, work performance standards are established, usually by means of time and motion studies. Then, the actual work measurements are compared against the established standards, and variances are computed.

An **input-output matrix** shows the relationships between data elements (inputs) and the reports (outputs) in which they appear. Figure 19-13 shows an input-output matrix pertaining to the purchases transaction system. It highlights reports that are redundant or that can be combined with other reports. For instance, the receiving register in effect duplicates the receiving report; it may therefore be replaced by a copy of the receiving report.

Other useful matrices show relationships between (1) users and reports they receive, (2) data items and files in which they are stored, and (3) organizational units and related responsibilities.

Floor-plan-like diagrams, called **space layout diagrams,** can aid in arranging the physical and paperwork flows. For purposes of analysis, the current flows can be superimposed onto the space layout. Then proposed flows can be drawn onto another copy of the space layout. Scaled cutouts of desks, machines, and other fixtures are helpful in trying different arrangements.

Output / Data Element and Source	Sales order–invoice	Bill of lading	Back order	Shipment record	Delayed order report	Inventory flash report	Sales flash report	Sales analysis by salesperson	Sales analysis by product
Date of event or output	X	X	X	X	X	X	X	X	X
Customer order number (sales branch)	X								
Name of customer (customer's order)	X	X							
Address of customer (customer's order)	X								
Place to be shipped (customer's order)	X	X							
Quantities ordered (customer's order)	X				X		X		
Product numbers (customer's order)	X	X	X	X		X	X	X	X
Product descriptions (product file)	X	X	X						

FIGURE 19-13 Portion of an input-output matrix.

ACTIVITIES DURING SYSTEMS DESIGN

Systems design is the creative phase of systems development. It consists of synthesizing the requirements into a cohesive and focused information framework. To be practical, the systems design must consider such constraints as available resources and technology.

The systems design process involves two levels: conceptual design and detailed design. A **conceptual design** provides the overall system structure or architecture, plus a relatively broad view of the combined system components. It is user oriented and logical in nature. A **detailed design** provides the physical details of each system component, such as reports and data and controls. Usually the detailed design includes such software as the application programs.

The conceptual design is developed during the systems design phase, while the detailed design is compiled in a succeeding phase. Steps in the conceptual design phase—illustrated in Figure 19-2—consist of evaluating design alternatives, preparing the conceptual design specifications, and obtaining approval for the design.

EVALUATION OF DESIGN ALTERNATIVES

Range of Alternatives Design alternatives may range from slight modifications of the present system to radically new structures. Assume that the present system involves manual processing. One design alternative could simply add a new control or revised source document without affecting the manual processing. Another design alternative could employ a real-time computer-based system with a network of computers. Still another alternative might include the capability for decision support as well as transaction processing. Because a system design incorporates a combination of features, such as input-output devices and data bases, the number of possible alternatives is large.

Example of Design Alternatives Two illustrative design alternatives are broadly portrayed in Figures 19-14 and 19-15. The first alternative employs a centralized mainframe computer at the home office. Each sales order is received by a salesperson from a customer and delivered to the appropriate sales branch. The branch mails the order to the home office, where clerks process the order and deliver a copy to the warehouse. After the goods are picked, they are shipped to the customer. Then the shipping copy is batched with other notices of shipped orders and processed daily on the centralized computer. The resulting sales invoices are mailed to the affected customers.

The second alternative employs a communications network with client/server architecture. Upon receiving a sales order, the salesperson transmits the order to the sales branch via his or her laptop computer. At the sales branch the order is checked for completeness and accuracy. Then it is transmitted to the home office on a microcomputer. The computer system automatically prints a copy of the sales order on a printer at the warehouse. After the goods have been picked and shipped, the shipping department clerk transmits shipping data back to the home office via a terminal. By reference to on-line customer accounts receivable files, the sales workstation in the home office prepares a sales invoice and stores it on the disk. The stored invoices are printed hourly via a print server and then mailed to customers.

Narrowing of Alternatives Selecting the single best design consists essentially of eliminating all of the second-best design alternatives. To minimize duplicate design

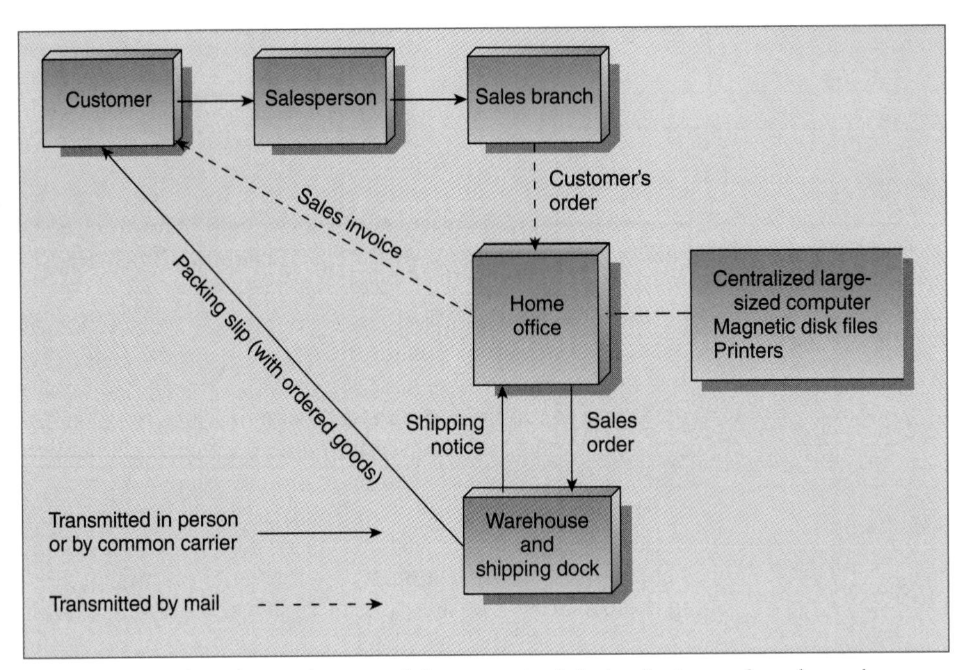

FIGURE 19-14 A broad-view diagram of the conceptual design features of a sales order processing system: centralized computer alternative.

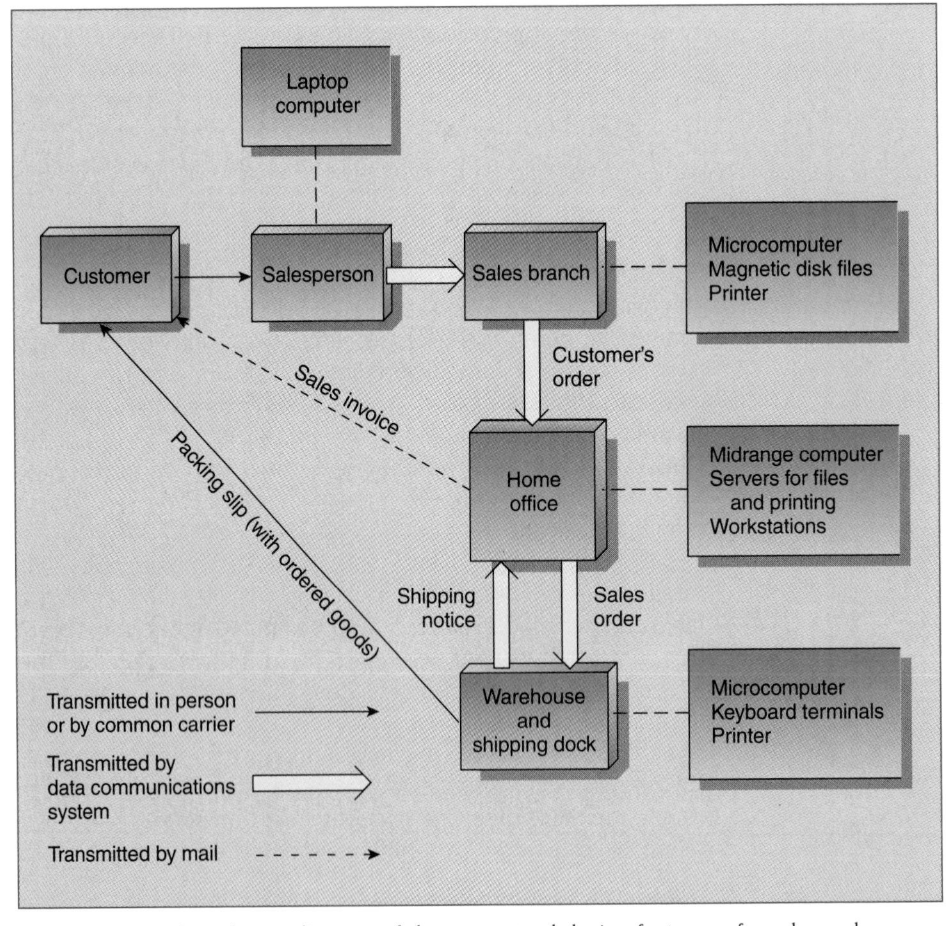

FIGURE 19-15 A broad-view diagram of the conceptual design features of a sales order processing system: distributed communications network alternative.

SPOTLIGHTING

SYSTEMS DESIGN *at Chase Manhattan Bank**

Chase Manhattan Bank is a large financial services institution whose major activities involve corporate lending, consumer lending, investment banking, and demand deposit banking. Of these activities, consumer lending in the early 1970s was the least profitable. Upon investigation it was determined that several critical problems existed:

1. The consumer lending department was small and buried in the organizational structure.
2. The management accounting system was weak, in that the managerial control reports were inadequate.
3. Data processing for all banking functions involved batch processing on a pooled basis, using large mainframe computers and service bureaus.

A systems project to develop new management accounting systems for consumer banking was approved. Two organizational changes were quickly made to enhance the success of the project: (1) A new consumer group was established at a high level in the organizational structure, with all consumer-related activities (e.g., ATMs, Visa bankcards, consumer loans) assigned to it, and (2) all branch banks and consumer services were set up as profit centers. Then the project was assigned to the MIS controllership group within the firm.

After surveying the situation, the project team first determined that one primary information need concerns profitability—analyzed by branch, service, and consumer. With respect to systems requirements, the team determined that the systems should provide interactive capacity, allow downloading from mainframe computers to workstations, process all requests from managers promptly, summarize detailed production data and upload, and incorporate adequate security measures.

Various alternative system designs were considered, including the continuation of then-current centralized networks. The design alternative that was selected involves a three-tier architecture. The top tier, called the backbone, links the information system services on a corporationwide basis. It includes mainframe computers, minicomputers, and a number of terminal emulators. The middle tier, called community-of-interest services, consists of linking minicomputers and high-speed communications lines and shared resources such as local-area network servers. These connector networks tie the backbone tier to the bottom tier. This bottom tier, called personal services, consists of the workstations employed by users in automated offices and branch banks. These workstations allow end users to do their own computing and to develop applications as needed.

After selecting the system architecture, the team then developed the design specifications and selected the specific hardware and software to be installed. Finally, the designed system was implemented.

*Michael Robin, "The Evolution of Management Accounting Systems at Chase Manhattan Bank," *Journal of Accounting and* EDP (Winter 1988), pp. 15–29.

efforts, the project team should narrow the alternatives as early as possible. Certain alternatives can be quickly discarded because they do not meet the systems requirements specified during the analysis phase. Others may be discarded because they would not provide all needed information. Still others may be unsatisfactory because they would consume disproportionate quantities of resources. The remaining design alternatives must be evaluated on the basis of feasibility.

As described in Chapter 18, questions of feasibility first arise during a feasibility study. These questions, listed as follows, are again considered when selecting the most suitable system design alternative:

1. Can and will the new system be used by the personnel for whom it is designed, i.e., is it **operationally feasible?** This type of feasibility is enhanced through the appropriate use of behavioral approaches, as discussed early in this chapter.
2. Are the expected economic benefits likely to exceed the economic costs of the new system, i.e., is it **economically feasible?** This type of feasibility, and the difficulties it involves, will be discussed in Chapter 20.
3. Can the new system be designed and implemented in the period scheduled, i.e., is it **time feasible?** This type of feasibility, also called schedule feasibility, is critical in keeping complex and user-developed projects from meandering onward interminably.

Other types of feasibility may also be judged, although they are of concern in fewer cases. A system is *technically feasible* if proven technology is available to implement the design. A system is *legally feasible* if there is no conflict between its design and the firm's ability to meet its legal obligations.

Because the details of the new system are clearer in the design phase than at the time of the feasibility study, these feasibilities can now be applied with more precision. Most of the design alternatives should therefore be readily eliminated by means of this screening process. However, two or more design alternatives may very likely remain. The final selection is then in the hands of higher-level management.

PREPARATION OF DESIGN SPECIFICATIONS

As mentioned earlier, the conceptual design for the selected design alternative is expressed in terms of user-oriented **conceptual design specifications.** These specifications are generally grouped around the components of the information system, (i.e., the data inputs, the data processing procedures, the data base, the data controls and security measures, and the information outputs).

Presumably the final selection of the best design alternative will have been made prior to the preparation of these specifications. If two or more alternatives survive all screening efforts throughout this phase, however, it will be necessary to prepare two or more sets of specifications.

Design specifications will typically include such features and capabilities as listed in Figure 19-16. Other specifications, such as those pertaining to coding systems and decision models, may be added as needed. These specifications should be as complete as possible. If the design is approved by management, in the final step, the specifications will be used when requesting bids from suppliers of computer hardware and software.

SUBMISSION OF SYSTEM DESIGN SPECIFICATIONS

System design specifications should be incorporated in a formal **systems design report.** Higher-level management should review this report thoroughly. If it receives final approval, the relatively expensive systems implementation activity will begin.

A systems design report should begin with a cover letter. This letter summarizes the findings of the design phase. Figure 19-17, on page 864, presents contents of the cover letter prepared by Infoage in connection with its inventory management project. The body of the report (not shown) contains the design specifications, together with system flowcharts and other documentation. In essence, the systems design report should contain all the information needed by management to make a sound decision concerning the proposed design.

System Components	Features
Output	Name Purpose Distribution to users Contents (information items) General format Frequency (or trigger) Timeliness (response time or delay after an event occurs) Output medium
Data base	File or table name File or table type File size (number of records or rows) Content of record or table Record or table layout File organization method Storage medium Data characteristics Updating frequency Data structures
Data processing	Sequence of steps or runs Processing modes, cycles, volumes Modes of data communication Processing capabilities at each physical location
Data input	Name Purpose Source Method of collecting data Volume (peak and average) Contents (data elements) General format Data entry method
Control and security	Type Purpose Specific system component affected Method of correcting error or establishing security

FIGURE 19-16 Typical conceptual design specifications.

SYSTEM DESIGN CONSIDERATIONS

Because the design phase is so crucial, we should pause to search for clues to a successful design. Even though much depends upon creativity, there are sound guidelines with respect to the sequence for designing a system, the principles underlying an effective system, and appropriate design methodologies.

DESIGN SEQUENCE

As the final products of an information system, the outputs are the key determinant of the remaining system components. Thus, as shown in Figure 19-18, on page 865,

May 1, 1998

Mr. George Freeman, President
Infoage, Inc.
Seattle, WA

Dear Mr. Freeman:
Enclosed is a systems design report concerning the inventory management systems project. The contents of the report are as follows:

Summary of the recommended conceptual design

Objectives to be achieved by the design

Expected benefits to be attained, including serious problems to be overcome

Key constraints and assumptions

Summarized system requirements and information needs

Cost-benefits analysis of design

Specifications of system components:
- Reports and other outputs
- Control procedures and security measures
- Data base
- Data and information processing
- Data inputs

Expected impact of design on information system, including audit procedures

Expected impact of design on formal and informal organization

Plan and schedule for implementation of design

Appendices:
- Entity-relationship diagrams
- Data-flow diagrams
- System flowcharts
- High-level structure charts
- Other diagrams and documentation

Sincerely,

(Signed)
Ralph Cannon
Office and Information Systems Manager

FIGURE 19-17 The cover letter containing the contents of a systems design report for Infoage's inventory management project.

the output design specifications should be considered first. The remaining specifications could be prepared in the sequence shown. Alternatively, they could be prepared concurrently, together with the control and security specifications.

DESIGN PRINCIPLES

A conceptual system design should reflect certain principles. Many of the relevant principles have been mentioned earlier, often in relation to such components of an information system as the data base. Nevertheless, the following list provides a useful summary of those principles already introduced and includes several new principles of design.

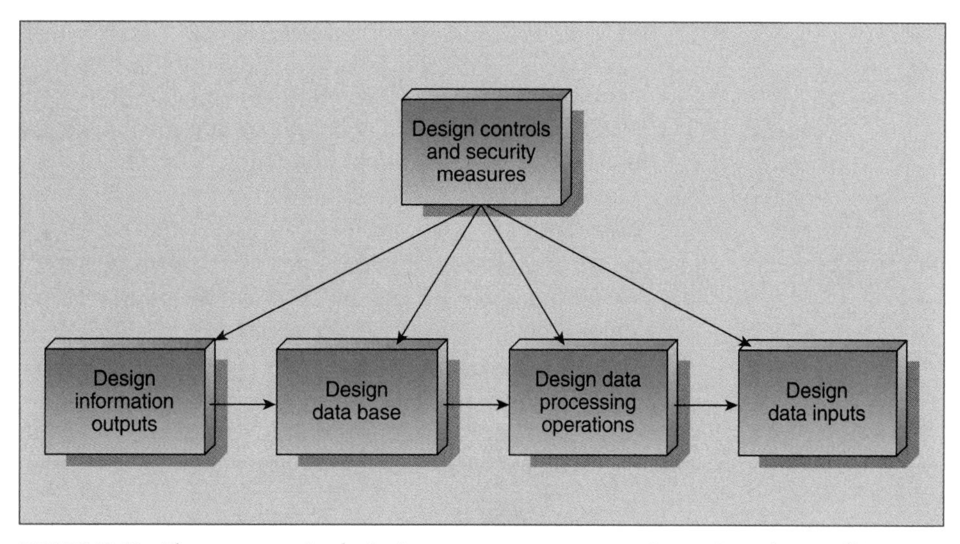

FIGURE 19-18 The sequence in designing system components. *Source*: Joseph W. Wilkinson, "Guidelines for Designing Systems." *Journal of Systems Management* (December 1974), p. 38. Reprinted with permission from the *Journal of Systems Management*.

1. ***Foster system objectives.*** The objectives that appear in the systems plan and project proposal should be achieved to the greatest feasible extent. These objectives should essentially describe the qualities that the newly designed information systems should possess.
2. ***Incorporate reasonable trade-offs.*** Since certain objectives conflict with each other, trade-offs are necessary. For instance, it may be desirable to gain greater timeliness at the expense of economy and reliability. If the added benefits realized from enhancing one objective exceed the losses from sacrificing another objective, the trade-off is viewed as reasonable.
3. ***Focus on functional requirements.*** Design specifications should be expressed in terms of capabilities and needs. That is, they should be functional rather than specific to hardware and software. A particular requirement might state, for instance, that the primary storage should have a designated level of capacity, rather than that a MAXI 9000 computer is needed.
4. ***Serve multiple purposes.*** Most designs should be multidimensional, in that they enable the system to serve more than a single purpose or type of user. Generally a system module should aid in processing transactions as well as providing decision or control information; it should serve the needs of employees, managers, and relevant outside parties.
5. ***Relate to users' concerns.*** A system design should be as simple as possible, so that the resulting system is easily usable by employees and managers. It should reduce the load on human users through prompts, menus, on-line help, preformatted screens, and so on. The design should also provide all information that the employees and managers need to fulfill their responsibilities.
6. ***Provide a tailored product.*** A system design should fit the particular circumstances of a firm. Thus, a general ledger system for a steel manufacturer should be quite different from a general ledger system for a university. A logistics system will differ significantly between a retail merchandiser and a manufacturer.

7. ***Integrate system modules and components.*** A system design for a single module should link the module to other modules within the overall information system. Also, the design should be relatively standardized, so that the system will perform consistently throughout and so that maintenance is simplified. If the system is to employ multiple computers, they should be compatible with each other, so that data and information can be easily interchanged.
8. ***Avoid design excesses.*** Although a system design should allow for anticipated growth and change, it should not be too high powered. For instance, a design should not specify complete automation if a mix of machines and people can achieve the same result at lower cost.
9. ***Apply sound methodology.*** Although the methodology used in creating a system design are not a part of the specifications, it is important in achieving desired results. The most suitable methodology should be employed when designing an information system. Design methodologies are discussed in the next section.

STATE-OF-THE-ART SYSTEMS DESIGN METHODOLOGIES

A **systems design methodology** is the collection of approaches, techniques, tools, and procedures by which a systems design is developed. A well-defined methodology provides a discipline that aids systems analysts in preparing a design. Thus, it increases the productivity of systems analysts. Moreover, it aids systems analysts in spotting inconsistencies and omissions, thereby reducing the likelihood of serious deficiencies in the design.

Numerous systems design methodologies have been devised during recent decades. Prominent examples are Information Engineering (developed by James Martin Associates and Ernst & Young), Business Systems Planning (developed by IBM), and Structured Analysis and Design Technique (developed by Sof Tech, Inc.). Each represents a combination of approaches, techniques, tools, and procedures.

Although certain methodologies draw on traditional techniques applied manually, the more recently devised methodologies employ structured or object-oriented techniques that are often applied by means of computerized tools. For instance, Information Engineering aids systems analysts in developing design specifications through the computerized application of a variety of structured techniques and documentation forms. Figure 19-19 portrays the steps and techniques and forms embodied in the methodology. Through the set of specifications, the developers receive aid in writing structured application programs.

State-of-the-art developments are properly the subject of an advanced AIS course. However, we can illustrate their usefulness through brief surveys of the object-oriented methodology, application generators, and CASE tools.

OBJECT-ORIENTED SYSTEMS DEVELOPMENT METHODOLOGY

Object-oriented systems development represents an alternative to structured systems development. Rather than emphasizing processes, the object-oriented methodology focuses primarily on the data to be handled by an information system. Moreover, the data are viewed in terms of *objects*—tangible or intangible things that can be uniquely classified. We encountered objects in Chapter 6, when we briefly surveyed the nature of object-oriented data-base structures. You may want to review the features of an object-oriented data base before continuing.

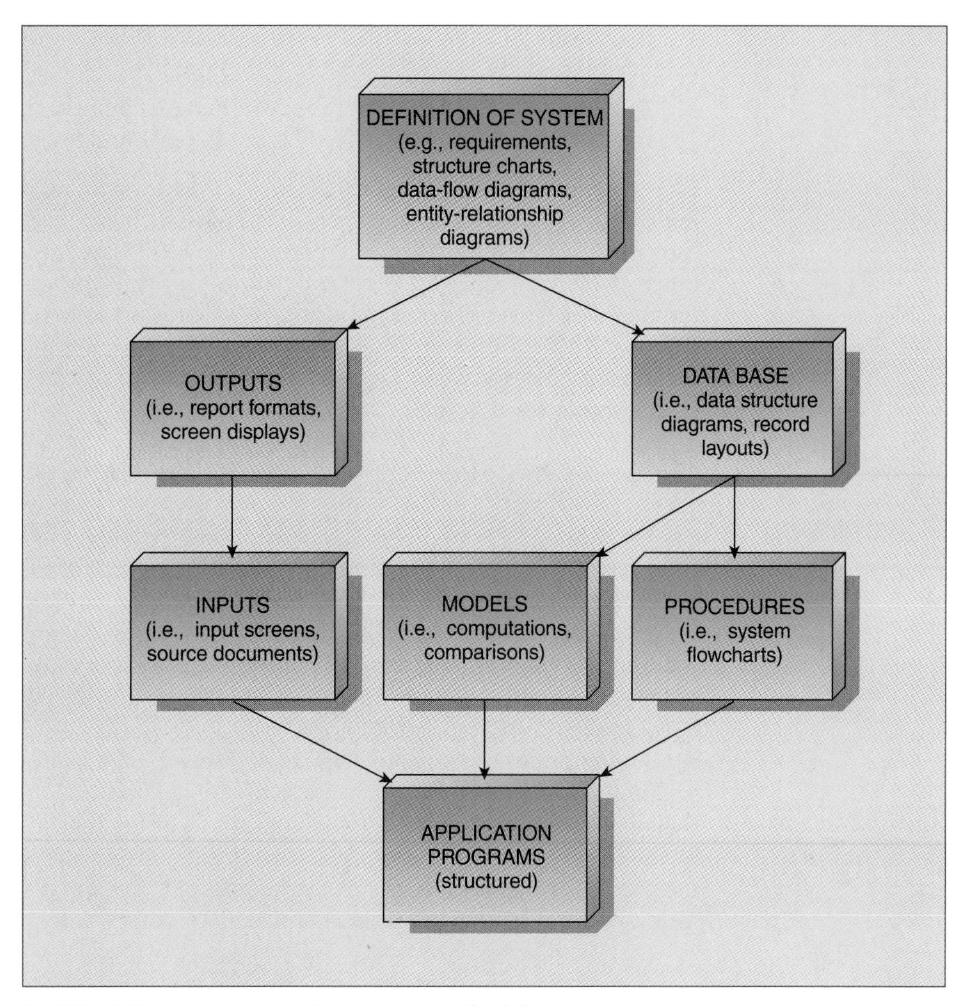

FIGURE 19-19 Components of a systems methodology.

Advantages of Object-Oriented Development The object-oriented methodology allows systems analysts to build systems easily and effectively. Many analysts believe that this new methodology is an improvement over structured systems development for these reasons:*

1. Once an object (e.g., customer, fixed asset, employee, general ledger account) has been identified, it does not change. This trait gives stability to a system design.
2. Both the data definition and the programming logic are "encapsulated" within an object. Building blocks are created that can be reused. Thus, the time required to write the program code for an application can be reduced, since an already-created building block can be pulled from "inventory" and "plugged in" to the system. Consequently, systems analysts can be more productive.
3. Objects can form hierarchies of classes, such as fixed assets, vehicles, trucks, and so on. Each member of the hierarchy inherits the attributes of its superior

*Linda Garceau, Elise Jancura, and John Kneiss, "Object-Oriented Analysis and Design: A New Approach to Systems Development," *Journal of Systems Management* (January 1993), pp. 29–30.

(e.g., a truck has the attributes of a vehicle). This inheritance trait increases the flexibility of the system development process.

4. Due to the modularity of objects and the reusability of their encapsulated logic, the system design is likely to be more reliable and free of coding errors.

Shortcomings of Object-Oriented Systems Development Object-oriented development employs special programming languages, such as Smalltalk and C++. Because they must incorporate such constructs as object classes, encapsulation, and inheritance, these languages are more difficult to use. Thus, programmers need extensive training. The languages also impose greater hardware demands, such as greater random-access memory and secondary storage. These shortcomings are expected to be temporary, however, and object-oriented systems development is likely to displace structured systems development in time.

APPLICATION GENERATORS

Current systems development generation methodologies usually incorporate powerful software tools. One broad category of tools is known as **fourth-generation languages (4GLs).** These languages are sufficiently user-friendly and powerful that users can design their own applications. For instance, an accountant can develop a reporting program for her use in a cost control situation. The 4GLs include a variety of languages, such as Interactive Financial Planning System, which enables users to develop their own application programs.

An important user-friendly type of software tool, usually incorporating a 4GL, is an **application generator.** An example is Application-By-Forms, a type of generator that allows users to define desired applications by filling in forms provided on screen by the software.* (Application-By-Forms is a feature of Ingres, a data-base management product.) Other application generators, which can be employed by users, are RAMIS II (from Mathematica, Inc.) and FOCUS (from Information Builders).

COMPUTER-AIDED SOFTWARE ENGINEERING (CASE) TOOLS

A broad range of automated systems software development tools are labeled **Computer-Aided Software Engineering (CASE).** CASE tools are generally special-purpose data-base management systems or dictionary systems that aid in making methodologies operational.† As described in Chapter 4, CASE tools can enhance the productivity of systems analysts and produce high-quality documentation for computer-based systems being developed. Because of the diversity of CASE tools available commercially, they are often grouped into several categories.

Basic CASE Tools The most limited CASE tools primarily generate code and thus are similar to application generators. More useful CASE tools support more than one phase of the systems development life cycle. Certain CASE tools emphasize the earlier phases, others emphasize the later phases, and still others span early and later phases.

Front-end CASE tools focus primarily on the systems analysis and design phases. These tools, also called *upper* CASE tools, provide conceptual design specifications by means of such techniques as entity-relationship diagrams, data-flow

*Chang-Yang Lin, "Systems Development with Application Generators: An End-User Perspective," *Journal of Systems Management* (April 1990), pp. 32–36.

†Michael L. Gibson, Charles A. Snyder, and Houston H. Carr, "CASE's Place in Financial Systems Management," *Financial and Accounting Systems* (Spring 1991), p. 15.

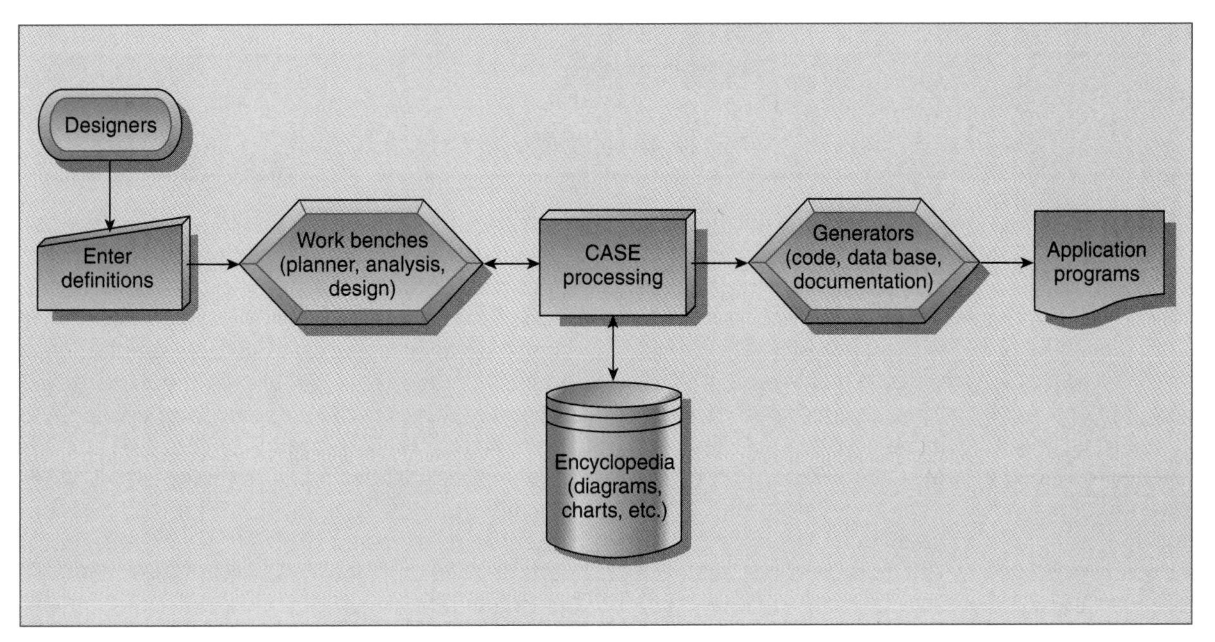

FIGURE 19-20 Components of an integrated CASE tool.

diagrams, and structure charts. Examples of front-end CASE tools are DESIGN/1 (developed by Andersen Consulting) and Excelerator (developed by Intersolv).

Back-end CASE tools focus on the detailed design and implementation phases of the systems development life cycle. These tools, also called *lower* CASE tools, generate structured program code from detailed design specifications such as tables and screen formats. They also perform testing and maintenance of the code. Examples are CA-Telon (developed by Computer Associates, International) and DesignAid II (developed by Transform Logic).

Integrated* CASE *Tools A truly integrated CASE tool package supports all phases of systems development. It incorporates various software modules called workbenches and generators. A **CASE workbench** accepts definitions from human systems analysts; then it converts the definitions into various structured diagrams and charts. These diagrammatic descriptions are stored in a centralized repository called a **CASE encyclopedia.** A **CASE generator** next draws particular system descriptions from the encyclopedia and prepares suitable coded application programs. The generator also develops the specifications for outputs, data base, and other components of the newly designed system. Thus, it automatically provides the documentation for a new product. Figure 19-20 shows the components of an integrated CASE package.

Examples of integrated CASE tools include Information Engineering Workbench (developed by KnowledgeWare) and Visible Analyst Workbench (developed by Visible Systems Corporation). Most such commercial CASE tools support a structured design methodology, although a package known as SALSA can also support the semantic object approach.*

Limitations of* CASE *Tools Although they improve productivity and the quality of developed systems, currently available CASE tools have definite limitations. Perhaps the most serious limitation is that they do not allow for the creativity needed for

*William S. Remington, Jong-Sung Lee, and Rajesh Aggarwal, "Semantic Objects with SALSA: Some Like It Hot," *Journal of Systems Management* (December 1994).

SPOTLIGHTING

SYSTEMS DEVELOPMENT WITH CASE TOOLS *at Bank of Montreal**

The Bank of Montreal (BOM) found in the early 1980s that its systems development process could not keep pace with the dynamic banking marketplace and that developments often cost too much and were inflexible. Its manager of the Development Technical Services department heard of CASE and its benefits to systems development. Through his efforts, CASE became a major catalyst for change within the bank's 400-person application development group. The hope was that CASE would cut in half the time and cost of developing a typical application. This hope was not achieved because of shortcomings in the available CASE tools, although productivity did increase by 10 to 15 percent per annum.

BOM encountered the limitations of CASE tools during the first year they were employed. Most vendors emphasized lower CASE tools, which automatically generated computer program code from design specifications. However, much of the time devoted to systems development elapsed in defining the requirements of the system—a systems analysis activity. In order to define requirements, systems analysts had to interview users over extended time periods. In addition to misunderstandings between the parties, the requirements tended to change frequently due to changes in circumstances. Since lower CASE tools simply did not address the matter of system requirements, it was necessary to await the emergence of upper CASE tools. Finally, BOM located a CASE tool that could focus on the systems analysis phase—ADW Workbench from Knowledgeware. This product develops an entity-relationship diagram, which reflects the enduring view of data and information needs. It does generate at least some of the requirements, although certain information must still be determined through manual means.

Recently BOM began assembling a suite of CASE tools that can work with BOM's mix of hardwares and technologies from several vendors. Part of the key in this effort has been to develop a repository—a central data bank or encyclopedia that stores the diagrams and other outputs from CASE tools, so that they can be reused and imported to other CASE tools. BOM is still looking for a fully integrated CASE tool, however. One promising prospect BOM is testing is High Productivity System from Seer Systems.

*Gerry Blackwell, "The Quest for CASE." I.T. *Magazine* (November 1993), pp. 22–26.

superior systems design. Integrated CASE tools are quite expensive, and the training costs can be high. Furthermore, CASE tools have been found most useful in designing totally new systems, which represent only a small percentage of all systems development work.

SUMMARY

Systems analysis is necessary to familiarize the systems analysts with the system module to be developed, to define fully the current problems, to gather detailed cost data, to generate ideas for improvement, and to gain the acceptance of the users. The systems analysis phase begins with a survey of the present system, leads to an analysis of the survey findings, moves to an identification of system requirements and information needs, and concludes with the submission of a systems analysis report. If the report is approved, the systems design phase begins. It consists of identifying design alternatives, narrowing these alternatives, preparing specifications for a conceptual design, and submitting the specifications for approval. The conceptual design specifications should be based on sound design principles and involve appropriate design methodologies. State-of-the-art systems development methodologies tend to be computer-based, to be grounded on structured or object-oriented techniques, and to employ such tools as application generators and CASE.

Systems analysis and design involve the use of fact-gathering and fact-organizing techniques. Fact-gathering techniques include reviews of documents and records, observations, interviews, and questionnaires. Fact-organizing techniques may be categorized as logical or physical, broad or detailed, structured or traditional. Structured techniques are logical in nature and include entity-relationship diagrams, structure charts, data-flow diagrams, data dictionaries, and structured English. Traditional techniques include system flowcharts, work distribution charts, work measurement analysis, input-output matrices, space layout diagrams, record layouts, and screen displays.

Review Problem with Solution

Precise Manufacturing Company, Second Installment

Statement

The sales order project team undertakes the analysis and design of the sales order processing system. After evaluating several alternative designs, the team proposes a design involving on-line computer input and processing. Accordingly, it prepares a set of conceptual design specifications for such a system. The following cover letter, which the information systems manager signs, describes the proposed design and accompanies the specifications.

(Date)

Mr. John Curtis, President
Precise Manufacturing Company
Chicago, Illinois

Dear Mr. Curtis:

Enclosed is a proposal pertaining to a redesigned sales order and management system. This design proposal is based on a thorough survey and analysis of the present system, together with a careful consideration of the objectives to be achieved.

Summary of Recommended Design. We propose that the sales order and management system be redesigned to employ on-line input and processing, and reconfigured to be a computer network. A mainframe computer would be installed in the home office, with powerful microcomputers in the warehouse and in the sales branches. These distributed processors would be linked by a data communications network consisting of leased lines. In addition, microcomputers would be located in the key departments (i.e., sales order, credit, billing, shipping, warehouse) and linked to the network. Each salesperson would transmit orders to the appropriate sales branch via a laptop computer. Sales and inventory files would be stored on magnetic disks and backed up on magnetic tapes. Sales orders, sales invoices, plus managerial reports and other documents, would be generated on printers located in the home office, warehouse, and sales branches. In addition, key managers would be provided with laptop computers for use in accessing the data base and retrieving desired sales information.

Objectives to Be Achieved. This redesigned system should achieve the following broad objectives stated in the initial project proposal.

1. To improve the processing of sales orders so that needed information is available to managers and interfacing systems.
2. To enhance sales processing efficiency, especially with respect to times, and the benefit-cost ratio.
3. To reduce sales processing and billing errors.

Summarized Resource Needs. *Hardware* needs consist of one midrange current-model computer with 200 megabyte primary storage capacity, and six microcomputers with eight megabyte primary storage capacity, 20 microcomputers, approximately 30 laptops, 12 laser and inkjet printers, 26 modems, and one front-end processor. *Software* needs consist of an operating system, a data communications software system, a network data-base management system, a sales application software package (including a credit-checking model), an inventory control package, a report generator, a sales forecast package, and a financial modeling package. Additional *personnel* needs include a data-base administrator, a systems programmer, and an applications programmer. *Costs* are expected to be approximately $2 million for systems development, hardware, software, plus the following for annual recurring operations.

(*Continued*)

Computer operations	$ 7,500
Maintenance contract	120,000
Communications lines	160,000
Information system maintenance	30,000
Data and information control	20,000
Information system administration	20,000
Total recurring costs	$357,500

These costs, however, should be offset by roughly $900,000 in annual benefits as a result of clerical savings, inventory carrying cost savings, savings in stockout costs, savings in interest charges on working capital and other funds, savings from reduced errors, and increased contribution margins from greater sales arising from better customer service and managerial decisions. Also, certain costs (including all one-time costs) are to be allocated to the inventory and financial management projects, since the hardware and part of the software should benefit those areas.

Expected Impacts. With respect to the *information system*, the mode of data input and processing will change from batch to on-line. More editing and processing and routine decision-making steps will be performed by the computer system. With respect to the *organization*, fewer clerks will be needed. On the other hand, the remaining clerks and managers will need intensive training in the use of the newly implemented system. This training should help to offset expected resistance from those affected.

We will be pleased to discuss this recommendation with you at your convenience.

Sincerely,

Tod Stuart
Information Systems Manager

Required

The benefits to be expected from the new system have been omitted from the cover letter. Prepare a list of benefits that should have been inserted after "Objectives to Be Achieved."

Solution

Expected Benefits to Be Attained. The recommended system design can easily meet all system requirements. Since the requirements have been derived by reference to observed weaknesses in the present system, the redesigned system is expected to overcome all such weaknesses. In specific terms, the redesigned system is expected to

a. Provide adequate capacity to meet sales volumes during the next five years.

b. Reduce the time needed to process sales order transactions so that almost all promised delivery dates are met and backlogs are minimized.

c. Provide greater efficiency and effectiveness in processing and maintaining finished-goods inventories so that sufficient goods are available to fill most orders and back orders are minimized.

d. Establish adequate controls over sales order transactions so that transaction data are not lost, and input and processing errors are almost completely eliminated.

e. Maintain up-to-date records pertaining to the status of open sales orders and finished-goods inventory, as well as sufficient backup records and audit trails.

f. Render greater accessibility to stored data so that inquiries by clerks, managers, and customers can be quickly answered.

g. Provide more timely, accurate, and relevant information (e.g., sales forecasts, sales analyses, profitability reports) for managerial decision making. (Although all information needed by the vice-president of marketing cannot be satisfied by this system, a large proportion can be.)

h. Provide tangible and intangible benefit values that exceed the relatively high level of one-time and recurring costs. (Although savings in operating costs—as listed in the project proposal—are not likely to be realized, the larger-than-expected benefits should more than offset the added costs.)

KEY TERMS

application generator (868)
back-end CASE (869)
CASE encyclopedia (869)
CASE generator (869)
CASE workbench (869)
Computer-Aided Software Engineering (CASE) (868)
conceptual design (859)
conceptual design specifications (862)
detailed design (859)
economic feasibility (862)
fourth-generation language (4GL) (868)
front-end CASE (868)
information needs analysis (847)
input-output matrix (858)
interviewing (850)
object-oriented systems development (866)
operational feasibility (862)
questionnaire (851)
space layout diagram (858)
structured systems development (852)
system requirements (846)
systems analysis report (849)
systems design methodology (866)
systems design report (862)
system survey (841)
time feasibility (862)
work distribution chart (854)
work measurement analysis (858)
work sampling (850)

REVIEW QUESTIONS

19-1. What are the deliverables from the systems analysis and design phases of the systems development life cycle?

19-2. In what roles can accountants serve during the systems analysis and design phases?

19-3. What are the steps in the systems analysis phase?

19-4. Why is a systems analysis phase necessary?

19-5. What are the steps in the systems design phase?

19-6. Identify several aspects of the scope of a system project.

19-7. Identify several internal and external sources of data.

19-8. What are several forms of resistance that employees and managers may employ with respect to a proposed system project?

19-9. Describe several behavioral actions that should be applied to minimize dysfunctional behavior in the face of a system project.

19-10. What questions does an analysis of the findings from a survey attempt to answer?

19-11. List several possible objectives of (a) an information system, and (b) a system project pertaining to a module within the information system.

19-12. List several typical requirements for the physical aspects of an information system.

19-13. Identify the steps of an information needs analysis.

19-14. Describe key aspects of information needed by managers for making decisions.

19-15. What is likely to be contained in a systems analysis report?

19-16. Describe the four basic techniques for gathering facts.

19-17. What are several guidelines to good interviewing?

19-18. What are the key features of structured systems analysis techniques?

19-19. What are the benefits of structured systems development?

19-20. Describe several structured systems analysis techniques.

19-21. Describe several traditional fact-organizing techniques used in systems development.

19-22. What is the purpose of a work distribution chart?

19-23. For what purposes may work measurement techniques be used?

19-24. Identify several systems design alternatives.

19-25. Contrast technical, operational, and economic feasibilities.

19-26. On what bases is a systems design alternative evaluated?

19-27. What types of specifications are included in a conceptual design of an information system?

19-28. What are the likely contents of a systems design report?

19-29. Identify several principles of conceptual system design.

19-30. Which of the system components should be specified first in designing a system?

19-31. What are the advantages of object-oriented systems development?

19-32. Describe the characteristics of computerized system design methodologies, including such tools as fourth-generation languages and application generators.

19-33. Identify the types and capabilities of CASE tools.

19-34. Describe the components and steps in applying an integrated CASE tool.

19-35. What are the limitations of currently available CASE tools?

DISCUSSION QUESTIONS

19-36. Discuss the reasons *against* analyzing the present information system.

19-37. Discuss the advantages and disadvantages of (a) stopping all reports and (b) requiring that each manager explicitly request the continuation of any report that he or she feels is really needed.

19-38. If designing an information system is essentially a creative process, why attempt to teach systems design in a formal course?

19-39. Discuss the following: "The physical operations should drive the accounting information system."

19-40. What added difficulties, if any, are likely to be encountered when designing an information system for a not-for-profit organization instead of a profit-oriented firm?

19-41. When managers leave a firm and are replaced by other managers, the information needs often change even though the decisions to be made remain unchanged. What difficulties does this situation pose for a systems analyst or team during the analysis phase?

19-42. What are several design alternatives that Infoage could consider with respect to the inventory management project? Which one appears to be the best?

PROBLEMS

19-1. Five members of the information systems department of a large city—Jane, Bill, Jack, Mary, and Sarah—are discussing a proposed new information systems project. The new information system is intended to replace the present manual system, perhaps with a system that uses information technology. Respond to each of their comments below, indicating whether it is sound or suspect. If suspect, explain why.

Jane: "We should focus on designing the new system. It is a waste of time to analyze the present system, since the new system will likely be very different because of the technology that we can incorporate."

Bill: "I believe we must analyze the present system and develop new system specifications before we can begin to develop a sound design for a new system."

Jack: "We should focus on the requirements for system hardware capabilities and the information that can be provided at a reasonable cost; users should not be involved since they cannot be expected to know all of their needs, and they do not appreciate the costs of providing timely information in any case."

Mary: "Our concern should be to define the problems in the present system, so that we can determine the system and information required to overcome the problems; then we can develop specifications for a new system."

Sarah: "First we should perform a feasibility study to determine if we can afford a new system. Then we can design a new system that uses as much of the present system as possible, in order to keep costs at the lowest feasible level."

19-2. Rainbow Manufacturing Co. has been experiencing declining sales for several years. It is essentially a single-product firm with 90 percent of its sales revenue from one product. The product is well protected by patents. However, substitute products have improved in quality, price, and performance, and foreign competition has increased.

The president called top management together to make a comprehensive effort to solve the firm's problems. The president indicated that they should carefully and thoughtfully define the problem, identify alternative solutions, select a solution, then implement and evaluate the solution. Problem definition entails describing the current situation, the desired situation, the difference, and the causes for the difference.

In the opening discussion, the following comments were noted:

1. The problem is foreign competition.
2. We're spending too much on health insurance.
3. We need another product.
4. Our prices are too high.
5. We need to spend more on advertising.
6. Our market share has fallen 10 points in the last 3 years.
7. Cash flow problems are becoming more frequent.
8. Our return on equity must be increased.
9. As a minimum, we need to get back to our profit levels of three years ago.
10. Return on assets this past year was only 1.2 percent.
11. We haven't introduced any product innovation for 3 years.
12. Within one year we need to increase net profit by 25 percent.

Required

a. The term "problem definition" is used sometimes, as in the statement above, to mean systems analysis.

What else does the systems analysis phase of a systems development project entail in addition to problem definition, and why is the phase critical to the systems development process?

b. The listed comments relate to various aspects, including current problems. Identify each of the following, based on the comments:

(1) A well-defined objective.

(2) The most critical problem.

(3) Two measures that reflect the effects of the critical problem.

(4) Two lesser problems, which do not appear to be directly related to the critical problem.

(5) Three possible strategies that may point to ways of solving the critical problem.

c. In what ways can an improved information system that incorporates state-of-the-art information technology aid in solving the critical problem?

(CIA *adapted*)

19-3. The Houseman Farms Company has decided to computerize its payroll, accounts payable, and general ledger operations. Houseman Farms consists of four separate farming firms owned and managed by the parent firm. All administrative and accounting functions are centralized at headquarters.

Two people work full-time in the accounting area. Payroll requires the most effort since labor constitutes 35 percent of total operating expense. The size of the labor force varies, reaching a peak of several hundred workers from August through October. Weekly paychecks are prepared and records are kept for each worker. Payroll deductions include group hospitalization, social security taxes, and, if the employee desires, state and federal income taxes.

Accounts payable are coded by firm and crop as well as the account to which they apply. Some bills are allocated across several firms. About 300 checks are written each month. Some are delayed by heavy payroll workload.

The general ledger consists mostly of payroll and accounts payable entries. A separate set of books is kept for each firm. These books provide the basis for financial statements and are used to prepare special reports, by crop, to respond to lender inquiries.

Because Houseman Farms lacks the in-house capability to perform the conversion to a computerized system, they plan to hire a consultant to analyze their needs and recommend equipment and software.

Required

a. Prepare a brief proposal for the described project, assuming that the consultant will require from now (September 1) until November 15 to complete the engagement and that the estimated costs will be $70,000.

Hint: Use Figure 19-3 as a guide and list the expected benefits in qualitative terms.

b. Describe the steps in the analysis to be performed by the consultant, including the data to be collected and the techniques to be employed.

(CIA *adapted*)

19-4. Go-Go, a firm that manufactures specialized lift-and-delivery vehicles, currently processes its cash disbursements manually. Because of added models and higher sales, the average volume of checks written each month has grown significantly. As a consequence, the treasurer proposes that the cash disbursements system be computerized.

A consultant is engaged to develop a new system. He discovers the following facts during an analysis of the present system:

(1) 16,000 suppliers have been paid this year.

(2) Of these 16,000 suppliers, 12,000 have received one order from the firm.

(3) A few suppliers receive in excess of 100 orders per month from the firm.

(4) Each order is paid by a separate check.

Required

a. What fact-gathering techniques were likely employed in obtaining these facts?

b. Describe the key problem in general terms, and indicate what its impact is likely to be on the current system.

c. What are the basic system requirements for a sound redesigned system?

d. What alternatives other than a computerized system should be considered in redesigning the cash disbursements procedure?

e. The president recognizes that other alternatives exist and may be preferable to a computerized system. However, she believes that the computerization option should be explored further. What steps and techniques may be employed in analyzing the current system and developing the specifications for a new computerized cash disbursements system?

19-5. For several years the Tasty Restaurant of Baltimore, Maryland, has served customers in a choice downtown location. Although everyone says that the food is excellent, the manager, Alvin Scott, has become uncomfortably aware that serious problems exist. Numerous complaints from customers have reached his ears. In essence, the customers are dissatisfied with the service, particularly with the lengthy waits during peak meal periods. In addition, there is antagonism between the chefs and the employees who wait on tables.*

*Chefs rank higher in a restaurant hierarchy than waiters.

Finally, the profits and cash receipts appear to be unsatisfyingly small, in spite of the fact that prices are comparable with those of other "fancy" restaurants in town.

Briefly, the restaurant operates as shown in the accompanying layout. Customers enter and seat themselves. A waiter writes their orders on a blank check form, departs and enters the kitchen, and calls out (loudly, because of the noise) the orders to the chefs. When the food has been prepared, the waiter brings the food from the kitchen and serves the customers. When the meal is completed, the waiter takes the payment to the cashier and returns with any change.

The chefs (two in number) handle the food purchasing tasks, although on a rather unorganized basis. For instance, because of separate preferences, the two chefs may purchase the same items from two different food vendors. In fact, both of the chefs—although good, experienced cooks—are opinionated and take orders only from Alvin. Unfortunately, he must frequently be away from the kitchen because of his many duties, including being gracious to regular customers.

Alvin learns that you, a close relative, are taking a course in accounting information systems at a nearby university. He calls and asks you for help in resolving his problems.

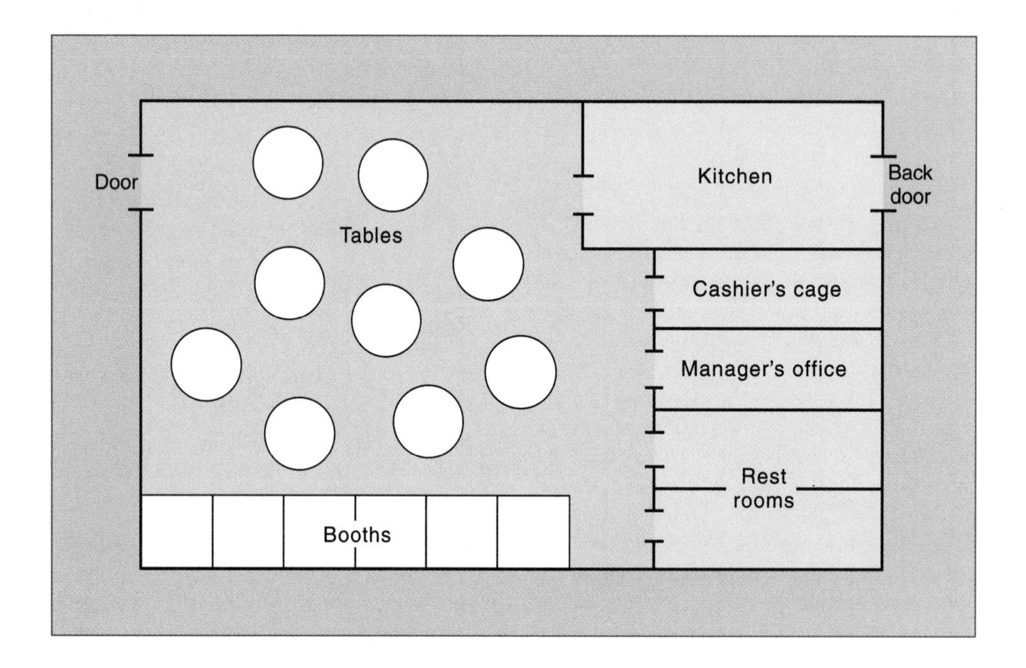

Required

a. Analyze the current weaknesses at the Tasty Restaurant, based on the given facts.

b. Identify other facts that you might want to gather to determine the degree of seriousness of the problems and to aid you in developing a suggested solution.

c. Describe the sources of the facts, as well as the techniques you might use in gathering the facts.

d. Suggest improvements to these aspects of the business:

(1) Organizational arrangements.

(2) Physical arrangements.

e. Suggest improvements to these aspects of the accounting information system:

(1) Documents.

(2) Controls.

(3) Data base.

(4) Procedures.

(5) Reports for the manager.

With respect to controls and procedures, indicate the linkages to the organizational and physical arrangements.

f. Discuss whether a restaurant of this size should consider installing a computerized system.

19-6. List the questions that should be asked in analyzing each of the following problem situations, which have occurred within a manufacturing firm.

a. Production costs have risen significantly because of increased overtime, idle time waiting for materials, and reworking of products.

b. Credit losses have increased because of a heightened delinquency rate.

c. Late shipment of orders has increased because of bottlenecks in production and the warehouse.
d. Customer complaints have increased because of the poor quality of products.
e. Finished-goods inventory levels have fluctuated widely because of large errors in forecasting sales.

19-7. Which fact-gathering technique(s) appear to be most appropriate in each of the following cases? Why?

a. Determining the procedure that should be followed by the cashier in preparing the bank deposit.
b. Determining the information needs of the personnel manager.
c. Determining the average number of work orders processed during a typical day.
d. Determining how the sales order processing procedure is actually performed.
e. Investigating the general level of productivity in the finished-goods warehouse.
f. Surveying the general adequacy of internal accounting controls in the cash disbursements procedure.
g. Surveying the opinions of all employees concerning the possible change to flextime.
h. Investigating the complaint that ordered products are frequently out of stock.
i. Determining the extent to which transaction documents are miscoded.

19-8. Rasher's Grocery Stores is a local chain of 10 stores in Tampa, Florida. Each store has an average of four cash registers and serves 600 customers daily. However, growth trends indicate that 800 to 900 customers likely will be served daily at each store within three years.

For several years the management of Rasher's has been watching developments in computerized POS systems. It has become convinced that the benefits of such systems—increased productivity, more effective use of checkout stations, improved control over inventories, and automatic verification of customer credit—will have a healthy effect on the firm's slim profits.

Recently, therefore, management requested the information systems manager to undertake an analysis and feasibility study. After several weeks of investigation, the information systems manager concludes, on the basis of a rough comparison of benefits and costs, that a centralized computer with disk files and POS terminals in each store is the most likely configuration. However, this determination depends on the ability of the POS system to accomplish certain feats.

Thus, the system should reduce the average checkout time from 4 minutes to 2.5 minutes. It should enable a credit check to be performed in 10 seconds or less, and the response to each keyed-in grocery sale should not exceed 2 seconds. No more than one error in 1000 transactions should be caused by the system, and the system should reduce human checkout errors by 25 percent.

In addition to aiding the checkout procedure, the POS system should provide inventory reorder reports and other daily management information within one hour after the stores close at 9:00 P.M. It should also have the capacity of storing all the price data, plus inventory records and sales history data; all these data might require as much as 100 million bytes of storage.

Required

a. List several objectives of the POS systems project.

b. Describe the fact-gathering techniques that were likely employed during the analysis and feasibility study, and associate at least one fact with each technique.

c. List questions that should have been asked during the analysis of the survey findings.

d. List the system requirements and information needs for the proposed POS system. Be as specific as possible.

19-9. Discuss key decisions and information needs pertaining to each of the following functions within the specified types of firms:

a. Loan function within a savings and loan institution.
b. Sales function within a discount department store.
c. Production function within a farm equipment manufacturer.
d. Patient care function within a hospital.
e. Construction function within a building contractor.

19-10. Thomas Kingman is the purchasing manager for Sparky Electronics, a manufacturer of high-quality electronics products. Before becoming purchasing manager, he was an electronics technician and quality control supervisor.

Thomas will be making decisions with regard to

a. Selecting the suppliers from whom to buy parts and subassemblies (tactical planning decisions).
b. Establishing purchasing policies (strategic planning decisions) to be made together with higher-level managers within the firm.
c. Hiring and evaluating supervisors, buyers, and other purchasing department personnel (management control decisions).
d. Negotiating purchase contracts (tactical planning decisions).

Required

a. Describe the analysis that should be performed to determine Thomas's information needs.

b. List several information items needed for making each of these decisions.

c. List the qualities of information needed for the first and second decisions.

d. Describe two key reports that Thomas should receive.

19-11. Adart Co., a privately held firm and wholesale distributor of specialty auto parts, began operations 10 years ago with a network of 20 suppliers. Adart's bookkeeper, Joan Halpert, has been with the firm since inception and has responsibility for recording the firm's transactions, paying all expenses, and reconciling the bank statements. In the last several years, Adart expanded beyond the specialty area and now handles a wider range of auto parts, which has increased its network to over 100 suppliers. As the firm grew, Halpert added staff to handle the increase in clerical activities. Rob Walters, president, signs the checks that Halpert prepares. Walters recently decided to hire a senior accountant, Jim Carlon, who has extensive computer experience, in order to begin automating Adart's accounting functions. Carlon reports directly to Walters for the duration of the automation project.

In reviewing and flowcharting the current operations, Carlon could not locate invoices for a number of checks indicated on the manual check register. The amount of checks totaled $37,000. Halpert advised Carlon that the invoices were probably misplaced when the files were moved to a new file cabinet room. When the canceled checks were reviewed, it was determined that these checks were issued to a supplier that was not listed on Adart's authorized supplier list. Carlon also determined that the supplier did not exist and that Halpert controlled this nonexistent supplier's bank account; Halpert confessed to the embezzlement and her employment was terminated.

As a result of these discoveries, Walters realized that automation of the accounting functions is even more critical now, and that a risk assessment and reevaluation of the accounting information system are necessary. Accordingly, Walters has decided to establish a steering committee for the overall automation project.

Required

a. Identify the weaknesses in internal control and explain the risk exposure of each weakness.

b. Identify the scope of the project that should be undertaken to correct the weaknesses and improve the system.

c. Describe the composition of a steering committee and the composition of a project team. What should be the relationship of Carlon to each of these groups?

d. Describe the nature of the analysis that should be undertaken if the project is to result in an automated system, including the techniques to be employed, types of data to be gathered, and so on.

e. Describe the steps in designing a new system, and list possible alternatives to an in-house automated system.

f. Describe improvements in internal control that must be included in any new design.

(CMA *adapted*)

19-12. Jack Ladd has recently joined the Wexler Home Products Company of Long Beach, California. His title is Systems Analyst I, a position for which he is qualified by virtue of receiving a degree in business systems analysis from the local university. He has performed exceedingly well in his first several assignments. Thus his supervisor feels confident in assigning Jack to perform an interview. He gives Jack the assignment late Thursday afternoon.

The interview is to take place with the manager of inventory, Colin Blunt. Jack's task is two-fold: (1) to uncover facts that can aid in correcting problems that have been traced to the inventory management function, and (2) to gain Blunt's cooperation in an analysis and redesign of the inventory management system.

Jack resolves to deal with these tasks in an expeditious manner. Therefore, he appears bright and early Friday morning at Blunt's office. He explains that his business is urgent, so Blunt cancels his scheduled staff meeting to talk with Jack.

Jack opens the interview by telling Blunt that top management is "concerned about serious problems in the inventory management function." He continues by explaining that he is there to help correct these problems. "In fact," he says, "our systems group can completely redesign your inventory procedures without bothering you at all. All we need is your story concerning the problems, and we can get started. We'll even keep mum during our study so that the employees won't know what is going on. That way, they can keep on with their work and won't be worried about possible layoffs when the new system is installed."

Jack continues by offering suggestions concerning several small problems pertaining to inventories that he had noted while walking through the area that morning. He also explains the workings of scientific inventory order models, including their ability to "minimize the array of inventory expenditures and optimize return on investment." Jack feels that this will show Blunt that he "knows his stuff."

After Jack has completed this exposition, he asks Blunt if he is willing to cooperate. "After all," as Jack added, "it's to your benefit to clear up these problems before more are uncovered." Blunt replies (as Jack clicks on his tape recorder) that he will have to discuss the matter with his superior, the production manager.

Jack feels that this reply is a rebuff to his efforts. Therefore he states that he has another meeting to attend, gets up, and leaves. When he sees his supervisor later that morning, Jack repeats what Blunt had said and ends by exclaiming, "I don't think that Mr. Blunt wants to cooperate. He certainly didn't offer any facts, and he seemed to be stalling when I put the question to him."

Required

Critique the interviewing approach of Jack Ladd.

19-13. Refer to Figures 19-10 and 19-11. Figure 19-10 shows a high-level structure chart for the function "process sales transactions." In Figure 19-11 the activity labeled "ship goods to customer" and one of its subactivities are portrayed.

Required

a. Prepare a level 1 logical data-flow diagram that details the activity "obtain order data" shown in Figure 19-10. Note in Figure 19-11 that this activity provides data to (and thus links with) the "ship goods to customer" activity.

b. Prepare a level 1 logical data-flow diagram that details the activity "bill customer" shown in Figure 19-10. Note that this activity receives data based on the shipment shown in Figure 19-11.

19-14. Masters Merchandising, Inc., of Worcester, Massachusetts, acquires and sells a wide variety of housewares. Inventory management is therefore a critical function. Recently the firm studied the inventory management function for the purpose of redesigning the key inputs, processing steps, and outputs. It began the study by defining the various activities and subactivities that are encompassed within the inventory management function. The results follow:

Activities (at the second level)

Store and record additions to merchandise.

Replenish inventory.

Prepare reports for management.

Monitor inventory usage.

Receive ordered merchandise.

Subactivities (at the third level)

Prepare supplier evaluation report.

Prepare shipping record and ship merchandise.

Prepare purchase requisition.

Determine quantity of merchandise received.

Determine quantity of merchandise moved to storage.

Prepare inventory status report.

Prepare receiving report.

Calculate quantity to reorder.

Prepare and transmit purchase order.

Update inventory records to show reduction of inventory on hand.

Prepare inventory aging report.

Determine supplier with whom to place order.

Match receipts with orders or returns with credit memos.

Update inventory records to show additional quantity on hand.

Update inventory records to show quantity ordered.

The study also found that six external parties are involved: inventory control clerks, suppliers, customers, the purchasing manager, shipping clerks, and the inventory manager. Three data stores are used to contain data concerning inventory, suppliers, and purchase orders.

Required

a. Prepare a high-level structure chart for inventory management.

b. Prepare a context diagram.

c. Prepare a level-zero data-flow diagram that portrays the logical flows through the described activities. Use your knowledge of sound purchasing procedures to label the flows correctly.

d. Prepare a level 1 data-flow diagram that details the activity "replenish inventory." Five processes are involved.

e. Prepare a level 2 data-flow diagram that details the subactivity "prepare and transmit purchase order."

19-15. Six employees compose the accounts payable department of the Hubbard Sales Company, a Stillwater, Oklahoma, firm. They perform the following sets of tasks for the designated number of hours daily:

Supervisor (Alice Whitespan)	
Supervise employees	4 hours
Aid in verifying invoices	2 hours
Approve disbursement vouchers	½ hour
Aid in preparing checks	½ hour
Other activities, such as correspondence	1 hour

Accounts Payable Documentation Clerks (Jackie Culver and Susan Lynch)	
Handle purchase order copies	2 hours
Handle receiving report copies	2 hours
Process suppliers' invoices	10½ hours
Assemble disbursement voucher packets	1½ hours
Other activities, such as filing	1 hour

Freight Bill Clerk (Joyce Itel)	
Prepare, verify, and post freight bills	4½ hours
Assemble and review disbursement vouchers	1½ hours
Other activities	1 hour

Disbursement Control Clerk (Pat Chase)	
Assemble, review, and approve disbursement vouchers	4 hours
Prepare and review checks	3 hours
Other activities	1 hour

Typist-Clerk (Betty Bush)	
Type checks (and accompanying voucher stubs)	5 hours
Other activities, such as typing correspondence and filing	2 hours

Required

a. Prepare a work distribution chart, using such tasks as supervision, purchase order processing, and voucher processing.

b. Identify weaknesses in the distribution of tasks, such as illogical assignments of tasks and assignments that appear to violate internal control concepts.

19-16. For each of the following described systems, propose an alternative system design and sketch overview diagrams of both the present and proposed systems.

a. A department store employs a point-of-sale system to process its sales transactions. The cash registers on each floor are in effect stand-alone intelligent terminals. Transactions are processed by each register and then captured on magnetic tape. At the end of each day the tapes are carried to the accounting department, read into the store's computer, and processed to update magnetic disk files and print sales and cash summaries.

b. A retail chain with ten outlets employs a centralized order-filling system. Each store prepares replenishment orders when stocks run low. These orders are mailed to the warehouse, which fills the orders and ships the merchandise to the ordering store. When warehouse stocks need to be replenished, the warehouse sends a purchase requisition to the purchasing department. Purchase orders are then prepared and mailed to suppliers. Copies of the orders filled by the warehouse and the purchase orders are sent to the computer data processing department, where they are keyed to magnetic tape and processed against files stored on magnetic disks.

c. A manufacturer with a factory and local warehouses employs a decentralized order-processing system. Minicomputers are located at each warehouse, whereas larger computers are located at the factory and the home office. All these computers utilize magnetic disk files. Orders are received from customers at the home office. Formal sales orders are prepared and mailed to the warehouse, which maintains an inventory file on magnetic disk. Next, the home office prepares sales invoices, which are mailed to customers. Periodically the home office initiates production orders to the factory. The factory maintains production records via its computer. It ships the finished goods directly to the warehouses according to preestablished proportions and provides information to the home office.

19-17. Tiger Manufacturing Co. intends to develop a new system to implement activity-based cost accounting for its three plants that produce earth-moving equipment to order. Two of the plants make parts that are shipped to the third plant for assembly and distribution to customers. The largest equipment is shipped in pieces and assembled at the customer's location. Parts and raw materials from suppliers go directly to the plants. The manufacturer provides maintenance service on its equipment anywhere in the world. There are 11 regional sales offices worldwide.

The existing chart of accounts contains four-digit account numbers, which are inadequate for coding the departmental and activity-related information required for activity-based cost accounting. The existing cost accounting system is based on allocation of plantwide overhead rates. Sales expenses are currently treated as administrative expense, but are to be treated as an activity in the new system. In the absence of compelling reasons not to do so, the manufacturer wants to relate all costs to the activities that drive them.

Required

a. What feasibilities must be ascertained before a new system design is approved?

b. Identify the specific features of the conceptual design specifications for this new system with respect to

(1) Account number coding—coding structure, examples of codes, plus outlines of the operator and user documentation.

(2) Outputs—

(3) Inputs—

(4) Procedures—

(5) Controls—

(6) Data base—

c. Provide examples of system design specifications that are sufficiently detailed that they do not appear in the systems design report, but are delayed until the implementation phase. (An example is the set of specifications relating to the data communications protocols between the plants and sales offices.)

(CIA *adapted*)

19-18. The Auto Rite Corp. of St. Paul replaces mufflers, transmissions, brakes, and other key automobile parts. Its dozen shops throughout the city service a total of about 200 car owners on an average day and 300 car owners on a busy day. Each shop maintains a standard inventory of 1000 different types and sizes of mufflers and other parts. A manager and an average of four mechanics staff each shop. All shop managers report to a shop operations superintendent.

After a thorough analysis by the firm's controller, the president decides that a microcomputer network is needed to handle the various transactions at the shops and to prepare such outputs as shop orders for the mechanics, itemized receipts of work done for customers, and daily analyses of jobs performed for managers. The microcomputer network will have workstations at the respective shops, with a midrange computer and disk files and printer located at the main office. The selected system should provide features that aid data entry, foster data control and security, and enable users to access specific shop orders. It should be capable of operating 12 hours per day, processing 60 transactions per hour, and responding to 90 percent of all data requests within 10 seconds. It should also accommodate a data-base management system that the president plans to acquire next year.

Required

a. List the system requirements that are identified in the statement, plus two others that are sufficiently important to the success of a new system.

b. Describe the feasibilities that should be ascertained before a new system design is approved.

c. Prepare specific conceptual design specifications for a new shop order-processing system, based on the above description. Be sufficiently specific to include needed files, inputs, and outputs.

19-19. Browning Companies of El Paso, Texas, is a construction-oriented firm that provides a variety of products and services. The products include sand, gravel, aggregates, regular concrete, and ready-mix concrete. The services consist of constructing pavements, earthworks such as parks, concrete structures, and other projects.

Raw materials are dredged from the Rio Grande River bottom and conveyed to the nearby crushing and sand plant, where they are crushed, screened, mixed, and otherwise processed. Finished products are checked for quality and then stored. Ordered products are dispatched by trucks to the points where needed, which may be the locations of customers' facilities or the sites of construction projects. Most of the trucks are ready-mix concrete trucks, which pour at the dispatched locations.

Dispatching is done by three dispatchers who prepare delivery tickets based on telephone calls from customers, on sales orders taken by salespersons, or on requisitions received from construction supervisors. The next available truck is assigned to fill an order. If more than one truck is available, the one that has been waiting the longest is assigned. On receiving his or her assignment, the driver is given two copies of the delivery ticket, which show the person to whom sold, the date, the truck driver, and the quantity sold. The driver then loads and gives the copies to a warehouse worker, who records the weight loaded on the tickets and returns a copy to the driver. After the load has been delivered, the driver initials the delivery ticket and returns the ticket to the dispatcher. Then the dispatcher stamps the time of return on the ticket. At the end of each week the tickets are delivered to the accounting department for processing.

Certain problems have been noted in this procedure. Customers on occasion dispute bills received for concrete or other delivered products. Truck drivers tend to have excessive idle time on some days and to work overtime hours (at time-and-a-half pay) on other days. The times required to deliver loads to the same locations vary considerably among truck drivers.

Required

a. Prepare a level-zero data-flow diagram of the current dispatching system.

b. Propose improvements to the dispatching system and state the expected benefits that they offer.

c. Describe the main features of a computerized dispatching system, including the suitable hardware components.

d. Specify two reports that would aid managers in evaluating and controlling the dispatching operations.

19-20. The B&B Company of Boone, North Carolina, manufactures and sells chemicals for agricultural and industrial use. The firm has grown significantly over the last 10 years but has made few changes in its information gathering and reporting system. Some of the managers have expressed concern that the system is essentially the same as it was when the firm was only half its present size. Others believe that much of the information from the system is not relevant and that more appropriate and timely information should be available.

Dora Hepple, chief accountant, has observed that the actual monthly cost data for most production processes are compared with the actual costs of the same processes for the previous year. Any variance not explained by price changes requires an explanation by the individual in charge of the cost center. She believes that this information is inadequate for good cost control.

George Vector, one of the production supervisors, contends that the system is adequate because it allows for explanation of discrepancies. The current year's costs seldom vary from the previous year's costs (as adjusted for price changes). This indicates that costs are under control.

Vern Hopp, general manager of the Fine Chemical Division, is upset with the current system. He has to request the same information each month regarding recurring operations. This is a problem that he believes should be addressed.

Walter Metts, president, has appointed a committee to review the system. The charge to this System Review Task Force is to determine whether the information needs of the internal management of the firm are being met by the existing system. Specific modifications in the existing system or implementation of a new system will be considered only if management's needs are not being met. William Afton, assistant to the president, has been put in charge of the task force.

Shortly after the committee was appointed, Afton overheard one of the cost accountants say, "I've been doing it this way for 15 years, and now Afton and his committee will try to eliminate my job." Another person replied, "That's the way it looks. John and Brownie in general accounting also think their positions are going to be eliminated or at least changed significantly." Over the next few days, Afton overheard a middle manager talking about the task force, saying, "That's all this firm thinks about, maximizing its profits—not the employees." He also overheard a production manager in the mixing department say that he believed the system was in need of revision because the most meaningful information he received came from Brad Cummings, a salesperson. He stated, "After they have the monthly sales meeting, Brad stops by the office and indicates what the sales plans and targets are for the next few months. This sure helps me in planning my mixing schedules."

Afton is aware that two problems of paramount importance to be addressed by his System Review Task Force are (1) to determine management's information needs for cost control and decision-making purposes and (2) to meet the behavioral needs of the firm and its employees.

Required

a. Discuss the behavioral implications of having an accounting information system that does not appear to meet the needs of management.

b. Identify and explain the specific problems B&B Company appears to have with regard to the perception of B&B's employees concerning:

(1) The accounting information system.

(2) The firm.

c. Assume that the initial review of the System Review Task Force indicates that a new accounting information system should be designed and implemented.

(1) Identify specific behavioral factors that B&B's management should address during the development of a new system.

(2) For each behavioral factor identified, discuss how B&B's management can address the behavioral factor.

(CMA *adapted*)

19-21. Hawkeye Hobby Shops is a chain of four stores located in Des Moines, Iowa. Each store carries a full line of hobby products; in addition, two of the stores specialize in radio control modeling. In 1995 the combined sales volume for the stores exceeded $2 million.

Each of the stores specializing in radio control modeling has a staff of four persons, and each of the remaining stores has a staff of three. In addition, Adele Tush, the owner, and three bookkeeping clerks reside at the flagship store.

Because the hobby market is very competitive, Ms. Tush places the highest priority on customer service. One key aspect of customer service, she feels, is the maintenance of adequate inventory on hand at all times. When the firm consisted of only one store and did not specialize, she was able to achieve this objective. With multiple stores and a much larger number of items sold, however, her control over inventory has slipped. As a result, stockouts and lost sales frequently occur. Reordering of merchandise is consuming much of her time and the time of the clerks.

Other related problems are also being experienced. Processing of transactions such as accounts payable and cash disbursements is becoming burdensome. Suppliers often complain that payments are received late, and many purchase discounts are lost. Moreover, Ms. Tush is becoming buried in masses of data and, consequently, has had increasing difficulty in finding or preparing information she needs for managing.

Another problem area is internal control. For instance, each cashier—the only nonsales employee in each outlying store—reconciles the cash receipts and the cash register tape at the end of each day and delivers them to the main store. The processing duties in the main store are inadequately divided, and Ms. Tush does not have time to provide close supervision. For instance, the one clerk handles both the general ledger and cash receipts.

Required

a. Based on the description of problems in the current system, prepare a list of requirements for a new computerized information system.

b. Identify the objectives of the new information system.

c. List several items, other than objectives, that should appear in a systems design report pertaining to this system. Include conceptual specifications with respect to the procedures, controls, and data base.

d. Describe several reports that would be useful to Ms. Tush in managing the stores.

19-22. Federated Department Stores decided in recent years to migrate from a mainframe environment to a client/server environment.* This decision led to a reengineering of its business processes, so that it could fit a suitable technology architecture to these processes and its hundreds of retail stores and warehouses. To aid its development, the firm first selected a well-known application development tool or generator known as PowerBuilder (marketed by Powersoft). This tool is oriented to the client/server environment and can accommodate the object-oriented design approach.

As the systems development activity progressed, however, Federated found that PowerBuilder was inadequate to handle the massive application. Thus, the firm looked at various CASE tools, narrowing its choice to Information Engineering (from Texas Instruments) and High Productivity System (from Seer Systems). It finally settled on High Productivity System, an integrated CASE tool that has an object-oriented repository (encyclopedia). It generates code and allows changes to be made to the code automatically when a designer makes changes to an entity-relationship diagram or other stored diagram. Other CASE tools that the firm investigated were less integrated; most required the addition of add-ons from other vendors in order to provide an integrated package.

Required

a. Summarize the benefits that an integrated CASE tool can offer to a systems development effort such as Federated, which is moving to a client/server object-oriented environment.

b. In addition to high costs (often in excess of $1 million for a large development project), what are the drawbacks of every CASE tool?

19-23. Sparks Manufacturing Corp. has established an object-oriented manufacturing data base within a client/server environment. The clients are primarily employees and managers in the various production departments. Although the system operates smoothly and provides recurring useful information relative to production activities, certain managers and employees desire to make individual ad hoc inquiries. After investigation the IS department discovers an applications development tool called GeODE (from Servio Corp.). This tool enables users to develop object-oriented applications very quickly without having to know and use a programming language. After GeODE (which stands for Gemstone Object Database Development Environment) is acquired, Nan, a cost accountant, develops a query application in a matter of hours. As a result, she can obtain customized reports concerning costs of specific production orders in progress and completed, analyzed in a variety of ways.

Required

a. What traits enable an object-oriented tool to facilitate end-user development of applications?

b. Why would a package such as GeODE not be suitable for most of the systems and reporting applications needed by Sparks?

c. Would a user such as Nan generally be helped by being provided a CASE tool for developing her individual applications?

19-24. *Datacruncher Office Equipment, Inc.* (*Continuing Case*)

Required

For the systems project that you have selected or been assigned to undertake, perform the following steps:

a. Prepare a plan of analysis, specifying (i) the techniques to be employed in gathering and organizing facts concerning the current system, (ii) the areas in which each technique is to be used, and (iii) the nature of results that each technique can be expected to provide.

b. Prepare an outline of questions to ask one key manager (who is within the project scope) during an interview pertaining to his or her information needs.

c. Prepare an analysis of the current system and its significant problems, based in part on the data-flow diagram and system flowchart prepared earlier (see requirements at the end of Chapters 12, 13, and 14).

---

*J. William Semich, "Big Development Jobs Need CASE," *Datamation* (September 1, 1994), pp. 72–80.

d. Prepare a list of requirements for an improved system. Assign reasonable, though arbitrary, values to system requirements involving capacities, timeliness, security, and so on. Also, include a set of information needed by the manager interviewed in requirement **b** above.

e. Describe two alternative designs for an improved system (within the scope of the systems project); include an overview diagram with each description; identify the relative benefits and drawbacks of each design alternative and the likely impacts on operations and the organization.

f. Prepare a menu that contains a variety of data-entry and retrieval choices, including several key transactions. Then draft a preformatted screen display, which is to aid in the entry of data pertaining to one key transaction.

g. Prepare a list of documents and reports to be provided by the improved system, including the purpose, frequency, distribution, and medium of each output.

h. Sketch the formats for (i) two key reports and (ii) one key document.

i. Prepare a list of needed files, showing for each file its organization method, updating frequency, and primary key.

j. Draw a computer system flowchart that portrays the processing steps in the preferable design alternative.

k. Prepare a list of suitable application controls, including programmed checks. Employ a matrix to show the data elements to which each programmed check is related.

l. Assume that a relational data base is to be installed. Design the set of normalized tables that are suitable for the project area. Also, specify several inquiries that the data base could answer and design several ad hoc reports that the data base could readily accommodate.

m. Prepare the cover letter of the design proposal that pertains to the preferable system design alternative.

CHAPTER 20

SYSTEMS JUSTIFICATION, SELECTION, AND IMPLEMENTATION

THE LEARNING OBJECTIVES FOR THIS CHAPTER ARE TO ENABLE YOU TO:

1. Discuss the factors affecting the determination of a system's economic feasibility.
2. Describe the options and steps in the selection of computer hardware and software for a new system.
3. Identify the steps in the implementation of a new system.
4. Identify various techniques employed in a detailed system design.
5. Contrast widely used control techniques pertaining to a system project.

INTRODUCTION

Most systems designs require firms to acquire resources before the designs can be transformed into reality. These resources often include new personnel to operate and maintain the newly designed systems, which are generally computer-based. In some cases firms are acquiring their first computer systems. In other cases they are replacing currently owned computer systems with more up-to-date computer systems. In still other cases they are adding new hardware or software to existing computer systems.

SURVEY OF THE POSTDESIGN PHASES

After management approves the specifications pertaining to a system design, the next step would seem to be the installation of a new physical system. However, before committing to the costs—which are likely to be considerable—it is necessary and prudent to take several key steps. Figure 20-1 shows the steps in relation to the systems design and systems implementation phases.

First, management should be convinced that the expenditures for the new system are justified, that is, that the project is economically feasible. Both the costs and benefits should have been considered at the outset of the project; however, at this point in the systems development the values can and should be sharpened and carefully evaluated. Also, two or more design alternatives may be under consideration; if so, they need to be compared on an economic basis.

Second, the specific computer-based resources must be selected. Requests for proposals should be prepared and sent to likely suppliers or vendors of hardware and software. When proposals are received back from the vendors, the soliciting firm should evaluate each proposal and select the most suitable one(s). In the process the evaluators must determine that the selected hardware and software are technically feasible.

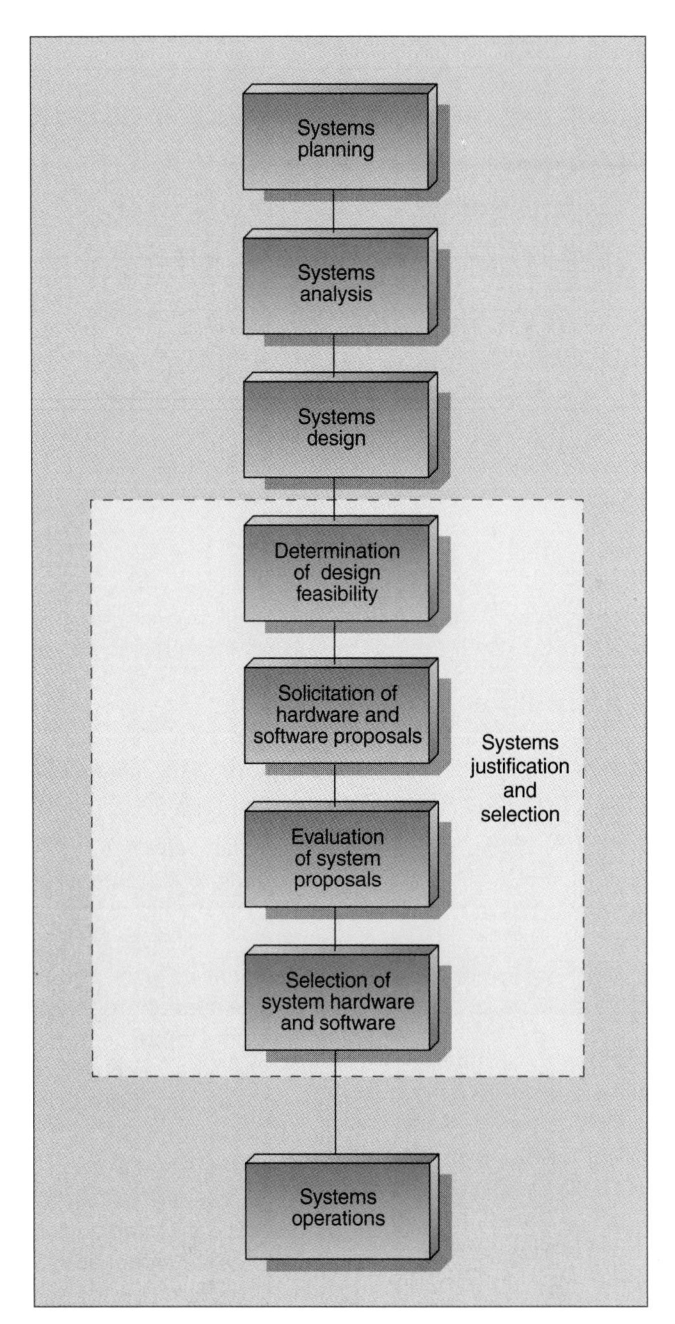

FIGURE 20-1 Systems justification and selection in the systems development life cycle.

Systems implementation takes place concurrently with the selection of hardware and software. This lengthy phase consists of numerous and varied activities, usually including the development of the detailed design features of the new system. Because of the lengthy and costly nature of a project, controls become quite important during the systems implementation phase. After all implementation activities have been completed, the system begins operations and other postimplementation activities.

ROLES OF ACCOUNTANTS IN THE POSTDESIGN PHASES

Accountants are likely to be less involved in the postdesign phases than they are in the earlier (front-end) phases. They have less interest in the details of hardware and most software. Also, accountants are not familiar with many of the activities involved in implementing a new system.

However, accountants can be very useful in four areas. First, they have a clear understanding of the concepts underlying economic feasibility, since it is based on techniques of capital investment analysis. Second, accountants can provide constructive views when accounting applications software packages are being evaluated. Third, accountants can aid in establishing and maintaining effective project controls. Fourth, accountants can evaluate the results of implemented projects and determine whether the decisions to undertake them were wise or not.

JUSTIFYING A DESIGNED INFORMATION SYSTEM

In this section we begin with the concepts related to economic feasibility, perhaps the most important of the feasibilities. We next identify the various types of costs and benefits related to information system resources. Then we illustrate sound computational methods used to ascertain economic feasibility.

ECONOMIC FEASIBILITY CONCEPTS

Both values and costs are attached to items of information. A typical value is the "surprise value" of new relevant information that pertains to an impending decision. Typical costs are those involved in collecting and processing needed information.

Information economics is the area of study concerning values and costs of information and their trade-offs. A principle of information economics is that added information should be gathered and provided for any purpose, such as making a decision, as long as the value of the added information exceeds the added costs of obtaining the information.

Consider the information pertaining to units of product scrapped during a production process. Each scrapped unit results in a lost sale. To combat these lost sales, we could prepare a daily report showing the scrap rate and causes of scrappage. This report will be useful if it provides the production manager with information that helps him or her to make decisions leading to reduced scrappage. If the added value exceeds the costs of providing the information, the report is economically feasible. However, added information concerning the exact time each unit is scrapped will likely not be worth its added cost of collection.

Since information systems collect and process information, they are affected by the principles of information economics. Consequently, design decisions concerning any aspect of an information system should be likewise governed. Assume, for example, that we are concerned about the number of internal accounting controls to design into a particular system. Our design decision should be based on a trade-off between (1) the benefit (value) of the added accuracy gained from each additional control and (2) the direct cost of each control plus the indirect cost resulting from a loss in processing efficiency.

At the level of a system project, the information economics principle can be restated as follows: A proposed investment in information system resources is economically feasible if the benefits (values) derived from the use of the acquired resources exceed the added costs related to the investment. Thus, the costs and benefits pertaining to investments in information systems should be explored.

COSTS RELATED TO INFORMATION SYSTEM RESOURCES

Relevant costs for determining economic feasibility consist of those involved in (1) the acquisition of system resources, (2) the development of a new or improved information system, and (3) the operation and maintenance of the new or improved information system during the lives of the resources. Most of these costs will be future cash outflows and hence must be estimated.

One-Time Costs Costs of acquiring system resources and developing an information system, taken together, represent the amount of the initial investment in a new or improved system. Since they normally are incurred only once during a life cycle of a system module, they are called **one-time costs.**

Figure 20-2 lists typical one-time costs. Many of the costs are for the salaries of the project team and a variety of other personnel, such as training specialists and construction workers. Other costs pertain to computer hardware and software, although all such items listed in the figure would not likely be included in a single project. Most of the one-time costs, however, are incurred during the activities that are necessary to implement the newly designed system. These activities are discussed in a later section of this chapter.

Recurring Costs Figure 20-3, on page 890, lists typical **recurring costs,** which can be broadly defined as all system-related costs other than one-time costs. Recurring costs are categorized in the figure according to system functions. Each system function requires costs grouped by object classes such as labor, supplies, equipment, and overhead. Those listed in the figure are illustrative only and not intended to be complete.

To illustrate recurring costs, assume that the current (though outdated) information system of the Precise Manufacturing Company had the following costs during the average month over the past year:

Salaries for three information systems employees	$2,000 each
Rental charge on the mainframe computer	600
Rental charge on the printers	500
Rental charge on the disk drives	700
Rental charge on the backup tape drives	550
Utilities, including electrical power	300
Rental charge for space needed for operations	$9 per square foot
Supplies, including paper and magnetic tapes	350
Purchase price of software packages and upgrades	800

If the information system facilities occupy 400 square feet of space, the total recurring costs per month would be as follows:

Salaries (3 × $2,000)	$6,000
Hardware ($600 + $500 + $700 + $550)	2,350
Software	800
Supplies	350
Utilities	300
Overhead (space occupancy cost) (400 × $9)	3,600
Total monthly recurring costs	$13,400

System Design Costs
Detailed design
Programming

System Installation and Conversion Costs
System and program testing
File conversion
Retraining of displaced employees
Training of newly hired analysts, programmers, and operators
Inefficiencies caused by learning new equipment and procedures

System Site Preparation Costs
Construction of wiring and piping systems
Construction of electrical power supply
Construction of air conditioning system
Construction of sprinkler system
Construction of other miscellaneous facilities, such as false flooring, file storage vault, and special lighting

System Hardware Costs
Central processing unit
Additional processors
Secondary storage devices
Input-output devices
Data communications equipment
Terminals
Peripheral equipment, such as key-to-disk devices
Transportation of equipment

System Software Costs
Operating system, utility routines, compilers
Data communications software
Application program packages
Data management software packages
Decision model software packages
Outside computer time-sharing rentals

FIGURE 20-2 One-time costs for a new or improved computer-based information system.

Recurring cost levels vary widely among firms and projects, being affected by such factors as the type of industry, degree of computerization, and system requirements. For instance, firms in the grocery industry tend to spend about 1 percent of their revenues for information system operations and maintenance, whereas firms in the banking industry spend about 5 percent of their revenues for the same purpose. Firms that have high transaction volumes and require short response times can have significantly higher recurring costs than firms with less stringent demands.

BENEFITS RELATED TO INFORMATION SYSTEM RESOURCES

Information systems can provide a variety of benefits or added values. These benefits may consist of savings in collecting or processing information. For instance, a computer-based system may replace a number of clerks and hence reduce labor costs.

Computer Operations Costs
Salaries for computer supervisors, operators, technicians, data-entry clerks, librarians, security guards, and others
Supplies, including forms, paper, ribbons, and tapes
Utilities, including power, water, and telephone
Rentals of computer hardware
Software purchases and upgrades
Communications equipment and services
Backup equipment and services

Information System Maintenance Costs
Salaries for systems analysts, programmers, repair technicians, and others
Replacement parts and upgrades
Printing costs for documentation

Information System Administration Costs
Salaries of systems management, data-base administrator, internal auditors, secretaries, and others
Insurance
Taxes
Space and building occupancy costs

FIGURE 20-3 Recurring costs related to a computer-based information system.

Benefits may consist of added revenues for the firm. They may be achieved through added processing capacity or more efficient use of resources. For example, a new system may process and fill customers' orders more speedily, thereby creating satisfied customers. Greater numbers of such customers may then give the firm their repeat business, hence leading to higher sales. Other benefits may include better information, hopefully leading to better decision making.

Benefits can be classified in various ways. For instance, though most benefits are likely to be recurring in nature, certain benefits provide one-time values. The most typical classification of benefits, however, is according to their tangible or intangible natures.

Tangible Benefits A **tangible benefit** is a cost saving or revenue increase that can be estimated in dollars. Displaced personnel cost savings, also called avoidable costs, can be measured by the reduced salaries. Reduced inventory levels result in cost savings that can be measured by expected savings in carrying costs. Revenue increases are usually measured by the higher contribution to margins provided through added sales. They may be measured in some cases by the increased number of customers, perhaps due to greater market share or the penetration of new markets.

Intangible Benefits An **intangible benefit** is one whose value cannot be measured with reasonable accuracy. Examples of intangible benefits that information systems may provide are

- Better and more selective information for decision making.
- More timely reporting to managers, leading to better control over operations.
- Higher morale of employees, leading to greater productivity.
- Improved control over operations, leading to fewer errors, higher product quality, and so on.

- Better utilization of equipment.
- Fewer stockouts of product, and hence fewer lost sales.
- Higher levels of service to customers.
- Greater flexibility in adapting to changes, leading to shorter lead times in developing new products, shipping goods to customers, and so on.
- Greater awareness of and responsiveness to competitive opportunities.

As you might suspect, the line between tangible and intangible benefits is rather blurred. The identification of the type will in many instances be subjective. For instance, a higher level of customer service might be either type of benefit. If better customer service arises from faster deliveries, and the consequent impact of added repeat sales can be measured, it is a tangible benefit. If better customer service is based on faster responses to customer inquiries, the impact is probably not measurable, and thus the benefit is intangible.

Intangible benefits are sometimes converted into intangible costs. For instance, a firm may hope to achieve an improvement in employee morale from a systems change to a user-friendly computer-based system. If the behavioral aspects of a systems project are not appropriately considered, however, employees may suffer a loss in morale. As a consequence, they may spend as much time in passing rumors and harboring resentments as in performing their duties.

APPROACHES TO THE ESTIMATION OF VALUES

Estimating costs and benefits can pose significant difficulties. In fact, in the case of intangible benefits, they cannot, by definition, be measured with any degree of reliability. Even tangible benefits are quite often difficult to estimate. Moreover, certain costs, such as the joint costs of sharing an integrated data base, cannot easily be determined.

Because of such difficulties and uncertainties, many firms may decide to forego formal analyses. Small firms in particular may feel that such analyses are beyond their capabilities or even a waste of time. After all, they are likely to view the acquisition of microcomputers and applications software to be vital necessities. Otherwise, they may not be able to compete effectively and to continue to grow efficiently. Nevertheless, most if not all firms should attempt to estimate the relevant costs and benefits. If they do not, their managements will be forced to substitute intuition and uninformed judgment for sound analytical procedures. Consequently, managements may either (1) acquire more information technology than they need or can afford or (2) reject investments in information technology that could be vital to their firms' long-term welfare.

Various approaches may be employed to estimate benefits and costs. One approach is to obtain the estimates of experts. Consultants may be employed who have experience with similar systems projects. Focus groups made up of accountants and managers may be formed, who then study the likely cost and benefit effects and place values on such effects. (In essence, the system project team is a focus group.) Another approach is to apply analytical techniques. For instance, managerial accountants can be assigned to perform incremental analyses or to develop benefit profile charts and pay-off analysis graphs. They may also perform risk analyses, in which costs and benefits are estimated in terms of probability distributions rather than single point values.

ECONOMIC FEASIBILITY COMPUTATIONS

Several methods are available by which to evaluate the economic feasibility of a proposed system design. Perhaps the most suitable method in most cases employs the net present value of an investment as the decision criterion. Other methods use the payback period and the benefit-cost ratio as the criteria. An investment in information systems resources is no different, economically speaking, from investments in a delivery truck or a headquarters building. That is, it is based on principles embodied in *capital budgeting*, an area in which accountants should be well versed.

Net Present Value A discounted cash-flow model is used to find the **net present value** of an investment. Future cash inflows and outflows are discounted to the present time and compared. If the total present value of cash inflows exceeds the total present value of cash outflows, the net present value is positive. The investment is then evaluated as being economically feasible. If the net present value is negative, however, the investment is considered to be economically infeasible.

Estimates or measures of the following factors are needed in order to compute the net present value of an investment:

1. The cash outflows, such as the acquisition cost of the system (e.g., computer hardware and software), plus the operating and maintenance costs of the system during its economic life.
2. The cash inflows, such as the cost savings or other benefits to be derived from the acquired system during its economic life.
3. The economic (as opposed to the physical) life of the system.
4. The salvage value of the system at the end of its economic life.
5. The salvage value (if any) of the system being replaced.
6. The tax considerations, such as the tax rate and depreciation.
7. The required rate of return on invested capital, also called the opportunity cost of capital.

To illustrate the net present value method, let us assume that a small firm has designed a new system that requires the acquisition of a midrange computer system. The relevant data are as follows:

Purchase price of the hardware and software, as well as the costs of systems development	$45,000
Annual recurring operating and maintenance costs of the present accounting information system	240,000
Expected annual recurring operating and maintenance costs of the proposed computer system	220,000
Salvage value of the present data processing equipment (equal to its book value)	5,000
Expected salvage value of the computer system at the end of its economic life (4 years)	10,000
Depreciation method	Straight-line
Required after-tax rate of return	14 percent

Figure 20-4 displays the computation of the net present value, assuming that the tax rate is zero. The present value (PV) factors have been taken from the table that appears on page 920. The net present value is a positive $24,200. However, this amount is suspect, since the tax rate has in effect been ignored.

One-time costs	($45,000)	
Less: salvage value of present equipment	5,000	
Net investment (cash outflow)*		($40,000)
Annual operating costs, present system	$240,000	
Annual operating costs, proposed system	220,000	
Annual savings in operating costs	$ 20,000	
Total cost savings, at present value (PV): (PV of $20,000 for 4 years at 14% = $20,000 × 2.914)		58,280
Salvage value of proposed computer system at PV: (PV of $10,000 to be received in 4 years = $10,000 × 0.592)		5,920
Total cash inflows, at PV		$64,200
Excess of returns at PV		$24,200

Note: Outflows are shown in parentheses.
*Assuming no taxable gain or loss.

FIGURE 20-4 Computations using the net present value method; the effects of income taxes are ignored.

Net investment (cash outflow)*		($40,000)
Before-tax annual costs savings = $20,000		
Aftertax annual cost savings: $20,000 × (1 − 0.34) =	$13,200	
Depreciation tax shield[a] =	2,975	
Aftertax annual cash inflows	$16,175	
Total cost savings at present value (PV): (PV of $16,175 for 4 years at 14% = $16,175 × 2.914)		47,134
Salvage value of proposed computer system at PV: (PV of $10,000 to be received in 4 years = $10,000 × 0.592)		5,920
Total cash inflows, at PV		$53,054
Excess of returns at PV		$13,054

[a]Depreciation expense per year $= \frac{\$45{,}000 - \$10{,}000}{4} = \$8{,}750$

Depreciation tax shield per year = depreciation expense × tax rate = $8,750 × 0.34 = $2,975

Note: Outflows are shown in parentheses.
*Assuming no taxable gain or loss.

FIGURE 20-5 Computations using the net present value method; the effects of income taxes are included.

Figure 20-5 displays the computation of the net present value if the tax rate is assumed to be 34 percent. The effect of taxes has been to reduce the degree of positive feasibility from $24,200 to $13,054. Taxes must be considered, however, if the computations are to be viewed as being closer to realism.

Even with the inclusion of taxes, it is clear that the example has been simplified. Its overriding purpose has been to illustrate what accountants view to be the most sound approach to determining economic feasibility. In a real-world situation, the following alterations can render the computations more realistic and provide more useful information to management:

1. The hardware, software, and developmental costs composing the one-time costs would be separately enumerated.

2. Instead of limiting the benefits to annual savings in operating costs, the benefits would be expanded to include intangible items. Although the intangible benefits may not be quantifiable, at least they can be listed.
3. The depreciation tax shield would be based on the Accelerated Cost Recovery System (ACRS), rather than the straight-line method.
4. The economic life of the investment may be expressed as a range rather than a single estimate of four years. Economic lives can vary considerably, depending on the type of computer hardware being selected. For instance, a mainframe may have a life that ranges from four to six years, while a microcomputer network may range from two years to eight or 10 years.

Payback Period The **payback period** method reflects the number of years required to recover the net investment. Although it does not measure the return on investment, the payback period can be useful as a screening device.

In the above example, the payback period is computed as

$$\frac{\text{Net investment}}{\text{After-tax annual cash inflow}} = \frac{\$40{,}000}{\$16{,}175} = 2.47 \text{ years}$$

Benefit-Cost Ratio The **benefit-cost ratio** measures the effectiveness gained from each invested dollar. A benefit-cost ratio is a better measure than net present value when competing investments of differing sizes are involved. In the above example, the benefit-cost ratio is

$$\frac{\text{Total present value of cash inflows}}{\text{Present value of net investment}} = \frac{\$53{,}054}{\$40{,}000} = 1.33$$

Since the ratio exceeds 1.00, the investment is economically feasible.

CONCLUSION

It should be very apparent that economic feasibility is not easy to ascertain. Because of their expertise in this area, accountants clearly have an important role to play. Many of the information systems disasters of the past might have been averted if accountants had been involved when projects were first being considered.

SELECTING SYSTEM HARDWARE AND SOFTWARE

During the process of selecting needed hardware and software for a newly designed system, the acquiring firm must consider the various acquisition options, solicit proposals concerning hardware and software, and evaluate the proposals. Both higher-level management and the project team will likely be involved in these steps. Accountants can also be involved, especially in the selection of software that pertains to financial processing applications.

In the following discussion we might keep in mind that a firm may be acquiring new computer hardware and software for the first time, may be adding to the hardware and software that it already has, or may simply be upgrading current computer systems. In the latter cases, the selection process will be affected by the hardware and software that are already in-house.

ACQUISITION OPTIONS

Several key decisions demand attention very early in the acquisition process. Choices must be made from such options as purchasing or leasing needed resources, dealing with a single vendor or with multiple vendors, installing an in-house system or outsourcing information services, and developing software in-house or purchasing commercial software packages.

Purchasing Versus Leasing Computer hardware and software can be purchased, as was tacitly assumed in the economic feasibility example. Alternatively, however, the hardware and software may be leased on a long-term financing contract. Each option has advantages. Purchasing generally requires a smaller cash outlay in the long run. On the other hand, leasing involves a smaller initial cash outlay, provides greater flexibility, lessens the risks of obsolescence, and in some cases yields greater tax benefits.

Single Vendors Versus Multiple Vendors A computer system includes hardware, software, input-output devices, on-line storage devices, communications lines and equipment, and business forms and supplies. A variety of vendors (suppliers), listed in Figure 20-6, are available as sources of these resources. An acquiring firm must decide which vendor or vendors can best satisfy its needs. Before determining the particular vendor or vendors, however, the firm should evaluate the advantages of dealing with a single vendor versus dealing with multiple vendors.

Using a single vendor simplifies the acquisition process and generally ensures that the various items will be compatible. Also, by selecting a single vendor for all of its needs, a firm is likely to receive better and more reliable service. However, few vendors are capable of providing a wide array of computer-related resources. Thus, in the case of complex acquisitions, a firm is often forced to deal with multiple vendors.

Buying from multiple vendors generally results in lower acquisition costs. Because the computer industry is so competitive, vendors are constantly vying for new business by reducing prices. For instance, discount dealers announce special prices on all types of hardware and software through publications such as *PC Week* and *PC Shopper* and on-line services such as Ziffnet. However, this approach may lead to inconveniences, incompatibilities, and less reliable service. For example, Ralph

Manufacturers of mainframe computers and servers, such as IBM and NCR

Manufacturers of minicomputers and midrange computers, such as IBM and DEC

Manufacturers of microcomputers and workstations, such as Compaq, Apple, and Sun Microsystems

Manufacturers of networks, such as Novell and Lantastic

Peripheral equipment manufacturers, including input-output devices and storage devices, such as Hewlett Packard and NEC

Commercial software development vendors, such as Computer Associates and Microsoft

Computer leasing firms, such as BRS Leasing Computer Service and Computers For Rent

Resellers and dealers, such as Computer Supermarket and Microage

Used-equipment brokers and dealers, such as Alltech Computers and Computer Multi-Systems

Business forms and supplies outlets, such as OfficeMax and CompuAdd

FIGURE 20-6 Sources of computer information system resources.

Cannon of Infoage may buy several bargain-priced microcomputers from a discount dealer. After the purchase he may discover that the microcomputers are not strictly compatible with the firm's mainframe computer because the operating systems are different. Hence, they cannot be tied together to form a computer network. When he attempts to obtain help in overcoming this problem, neither the dealer nor the microcomputer manufacturer offers any assistance.

In-house System Versus Outsourcing Computing Services Most firms with computerized information systems operate and maintain their own systems. By handling their systems in-house, such firms exert complete control over the information system activities. However, they also reap all the headaches and costs that information systems can sow—and often do.

For firms that do not have the experience or will to cope with their own information systems, several outsourcing options are available.* They may be broadly categorized as outside commercial computer services and facilities management firms.

Outside computing services are provided by resources that physically reside off a firm's premises. These services are available from two types of firms: service bureaus and time-sharing utilities. A **service bureau** is a firm that provides batch data processing services at a remote location. A subscribing firm prepares the source data records, such as time cards and sales orders. It then transports these records to the site of the service bureau. When the desired outputs are prepared, they are returned to the subscriber. A **time-sharing service center** is a firm that provides batch or on-line data processing service through one or more on-line terminals located on a subscriber's premises. The subscribing firm prepares the source data records and enters them via the terminals. After processing takes place within the center's computer system, the outputs are produced on the subscriber's terminals or connected printers. Certain of the subscriber's files are usually stored on-line within the center's computer system. Most of the applications suited for outside computer services involve transaction processing. Also, certain audit functions can be performed off the premises, for example, at the offices of an audit firm.

Perhaps the most important benefit gained from using an outside computing service is economy. Since the fees are related to usage, a subscriber pays only for what is used. Another benefit is that outside computer services generally make available professional assistance, specialized software, and specialized data bases. For instance, a small insurance firm may subscribe to a service that provides assistance in devising needed reports, that employs insurance-oriented processing programs, and that enables the firm to receive current economic and industry statistics from an on-line commercial data base. A third benefit is that outside computing services provide added capacity or backup. This benefit is particularly useful to a large firm that has fluctuating data processing loads.

Using outside computing services can create problems, however. Data security and accessibility may be weakened, since data records are turned over to an outside party. Outputs can be delayed, since outside services tend to require longer turnaround times. Also, processing costs can become excessive if transaction volumes continue to grow. Subscribing firms should therefore carefully monitor the use of outside services.

If a firm prefers to have all of its information system facilities in-house, but wants to avoid the problems of operating and managing the system, it can turn to a facilities

*As defined in Chapter 18, outsourcing consists of employing an outside firm to handle the information system activities.

SPOTLIGHTING

IMPLEMENTATION OF APPLICATION TEMPLATES *at Western Resources**

Application templates are existing systems that have been built with the aid of CASE tools. Thus, they are customized for a particular organization's or industry's features and needs. Templates provide an alternative approach to the systems development life cycle approach and the commercial software package approach. Because a firm that employs an application template starts with a customized design for an existing system, this approach offers two advantages. First, the costs are much lower than with the "build from scratch" approach. Second, the modifications are applied to a design that is already similar to the firm's needs and is thus more likely to lead to a satisfactory redesign.

Western Resources, a gas and electric utility located in Topeka, Kansas, decided a few years ago that it needed a new flexible customer processing system. After reviewing the alternatives, it selected a customer processing system product that Andersen Consulting had designed for another utility. This "system product" was called Customer/1 DesignWare and in effect was an application template or system design guide. The product was in electronic form with accompanying hard-copy documentation. Although customized for a utility, the system design guide required further customization for Western Resources. A system design team from Western Resources completed the customization over a three-year period, with assistance from Andersen Consulting. It was developed by the prototyping approach, with key functional segments of the working prototype being presented to the business units. Screens being presented were modified as necessary after comments from users.

Western Resources estimates that it saved over one year and $20 million by using this DesignWare application template, rather than starting from scratch to develop a customized customer processing system. In addition, the systems personnel at Western Resources learned new data-base technology (DB/2) and CASE tools during the customization process. The value of this approach was enhanced when Western Resources merged with another utility a few months after the new system was implemented. The customer processing system of the merged entity was adapted in about three months, since the templates from the customization process were readily available in stored electronic form.

*J. Debra Hofman and John F. Rockart, "Application Templates: Faster, Better, and Cheaper Systems." *Sloan Management Review* (Fall 1994), pp. 49–60.

management type of outsourcer. A **facilities management firm** is a professional information management organization that manages the information systems of subscribing firms on a fee basis. This service is especially useful to firms that do not have the time or expertise to operate the information system efficiently. Electronic Data Systems (EDS) Corporation, the largest information processing service firm in the United States, serves more than 4000 client firms throughout the world.

In-house Software Development Versus Commercial Software Packages Traditionally firms have employed their own programmers to develop the application programs for their computer systems. They have thus controlled the software development process. Also, the consequent programs have been customized to the procedures that have been designed by the systems analysts. As a result, the completed programs have usually been acceptable to the firm and to the users.

Recent trends, however, have led to a thriving and growing commercial software market. The salaries of in-house systems analysts and programmers have risen quite rapidly during the past decades, so that in-house software development has become very expensive. Many small firms simply cannot afford to employ full-time systems professionals. Many medium- and large-sized firms have installed distributed and client/server networks, with an emphasis on localized and workstation computing.

As commercial software development has bloomed, various types of software applications have emerged, with their attendant advantages and limitations relative to in-house software development:

1. ***Types of commercial software.*** The broadest types of commercial software packages are general accounting systems and turnkey software systems. General accounting systems handle the transaction processing for firms of all sizes. Packages usually contain modules pertaining to the general ledger, accounts receivable, accounts payable, payroll, fixed assets, and job order cost accounting. Turnkey software systems, mentioned in Chapter 18 under the Vendor-Developed Approach, are usually general-purpose software packages that are customized for particular firms and even tested and implemented.

   Other types of commercial software packages are more specialized in nature. Certain packages focus on particular economic segments and industries that face nonstandard accounting principles and practices as well as differing business rules and conditions. Another category of specialized packages focuses on ranges of business-related functions, such as document and text preparation and financial analysis.
2. ***Advantages of commercial software.*** Commercial packages are widely employed for several reasons. First, they represent products that are available to firms without lengthy developmental periods. Most packages can be implemented almost immediately, especially if they are not modified. Second, the software products are generally soundly designed and well tested, since software development firms tend to employ very competent programmers and systems analysts. Hence, they are likely to be efficient and reliable. Third, the prices of such packages are usually very reasonable. The sales of certain general accounting packages and spreadsheet packages have ranged into the hundreds of thousands and even millions of dollars; hence the development costs per copy are very low, and the prices can be set accordingly.
3. ***Limitations of commercial software packages.*** A main drawback to software packages is that they must of necessity be generalized. Thus, they cannot meet the precise needs of any particular firm as well as in-house-developed programs. Another limitation is that the acquiring firm is dependent on the software vendor for support and maintenance and upgrades. If adversity affects a vendor, the firm may also be adversely affected.

   Most systems applications today involve the use of commercial software packages. Often, however, this option becomes a hybrid choice. If a particular package is of a general-purpose type, or is not sufficiently customized, an acquiring firm usually is forced to modify the package extensively to render it usable. The resulting costs and time required can severely shrink the advantages offered by a commercial package.

SOLICITATION OF PROPOSALS

Most of the above-mentioned options involve vendors of computer resources. To acquire the needed computer resources, a firm must inform likely vendors of its needs and obtain responses from them. Since the needs can be rather involved and sometimes complex, a careful solicitation procedure should be employed.

The most important step in this procedure is to prepare a request for proposal (RFP). The purpose of an RFP is to portray, to the vendors, the system that has been conceptually designed. A key portion of an RFP is a set of resource (e.g., hardware, software) specifications. These specifications are evolved from the design specifications, which in turn are based on the system requirements and information needs. Figure 20-7 lists the resource specifications that might be used by a firm like Infoage. In addition to the specifications for hardware and software, the list includes those pertaining to such needed support activities as training for users and available facilities for testing the system.

A cover letter accompanying the RFP should summarize the contents of the RFP, specify critical constraints, and clarify the data to be provided by each supplier. Constraints might include the deadline date for the response and the ceiling on planned system expenditures. The data requested from each supplier should focus on how the supplier's hardware and/or software will satisfy the design specifications. Other data required of the supplier might be a list of existing clients and technical

System Design Specifications
Output specifications
Data-base specifications
Processing specifications
Input specifications
Control and security specifications

Hardware Specifications
Processor speeds and capabilities
Secondary storage capacities and access capabilities
Input-output speeds and capabilities
Compatibility features
Modularity (expandability) features
Error detection and correction techniques
Data communications capabilities
Special features, such as multiprogramming and virtual storage
Maximum allowable downtime (as a percentage of total time)

Software Specifications
Programming languages and compilers
Utility packages
Application packages
Operating system capabilities
Data-management packages

System Support Specifications
Programming assistance
Training programs
Test facilities and time available
Backup facilities
Maintenance assistance

FIGURE 20-7 A list of resource specifications for a firm such as Infoage.

requirements of the supplier's hardware. To aid the supplier in preparing a proposal, the soliciting firm will likely list the criteria by which each proposal will be evaluated.

Since possible vendors are numerous, some means of screening is necessary. The objective is to locate those reputable vendors whose products or services appear most capable of satisfying all of the specifications in the RFP. After narrowing the available vendors to a manageable number—perhaps three to six—the requesting firm then mails the RFP. When the deadline date has passed, the firm begins the evaluation process.

EVALUATION OF PROPOSALS

Proposals received from vendors essentially recommend specific hardware and/or software marketed by the vendors. The value of each proposal is in its detailed description of the manner in which the recommended hardware and/or software can best serve the specific needs of the soliciting firm.

The evaluation process must therefore consider the extent to which each proposal fulfills the firm's needs. A first step might be to review each proposal, to ensure that it responds directly to all specifications and requests. If a proposal is seriously deficient, it can and should be eliminated from further consideration. Another early step is to judge broadly the suitability of the hardware and/or software to the firm's situation. One useful approach is to talk with other firms who currently use the recommended products of the suppliers whose proposals are under consideration. Another approach is to compare reports concerning the recommended products of the various suppliers. For instance, *Datapro Reports* and *Auerbach* evaluate various software packages for microcomputers on a comparative basis.* Still another approach is to have the competing vendors make formal presentations on the firm's premises, including demonstrations of the capabilities of their products.

In most cases, however, each product and its supplier should be evaluated through techniques applied by the soliciting firm. Three proven techniques that may be applied are the benchmark problem, simulation model, and weighted-rating analysis.

Benchmark Problem Technique The **benchmark problem technique** is useful in showing how well proposed hardware or software can be expected to perform in typical circumstances. Assume that Infoage is evaluating computer processors to handle its various applications, including those involving inventory. Its inventory-updating program is chosen as the benchmark problem. This program is run on the hardware of each vendor whose proposal is being evaluated. Times required to process a given test data set are compared. The specific hardware having the lowest time "wins" the benchmark test.

Assume now that competing commercial software packages involving inventory processing are also to be evaluated. In this case their performance would be compared by running both on the same computer processor.

Simulation Model Technique Although the benchmark problem technique deals with an important aspect of performance, it ignores many equally important criteria. A broader-based technique involves the use of a mathematical model that simulates the computer system. When "stepped through" simulated processing activities, the

*Occasionally evaluation articles appear in accounting as well as microcomputer journals. A good example is the article entitled "A Shopper's Guide to Accounting Software," by Harley M. Courtney and Cheryl L. Flippen, which appeared in the *Journal of Accountancy* (February 1995), pp. 37–59.

simulation model technique generates data concerning access times, response times, run times, throughputs, cost levels, and equipment utilization. It is particularly useful in evaluating on-line real-time computer systems.

Weighted-Rating Analysis Technique Because the simulation model technique is expensive and overlooks key qualitative factors, a third technique has become dominant. The **weighted-rating analysis technique,** also known as weighted point scoring, provides an inexpensive means of evaluating all factors relevant to computer resource selection.

Figure 20-8 shows a matrix that contains a number of weighted factors. Included are factors pertaining to hardware, software, and support. Proposals from suppliers A and B are being compared. Weights have been assigned by the evaluators, based on their judgment concerning the relative importance of each factor. Raw scores are based on information gathered from the proposals and other sources. Weighted scores are computed by multiplying weights and raw scores. The weighted scores are totaled and compared. In the figure, A's proposal earns 368 points, while B's proposal earns 385 points. B's proposal thus appears to be better than A's, although the difference is relatively slight.

The weighted-rating analysis technique can be modified to evaluate hardware alone or software alone. It may be used to compare as many proposals as desired.

		A's Proposal		*B*'s Proposal	
Factor	Weight	Raw Score	Weighted Score	Raw Score	Weighted Score
Hardware					
Performance, e.g., response time	10	5	50	3	30
Compatibility with existing hardware and software	10	4	40	3	30
Modularity	5	3	15	4	20
Reliability	5	4	20	5	25
Ability to deliver on schedule	5	2	10	5	25
Special features, e.g., control features, security packages	5	5	25	3	15
Hardware subtotals	40		160		145
Software					
Range of capabilities	10	3	30	4	40
Efficiency in use	8	5	40	3	24
Ease in making changes	7	4	28	3	21
Advanced features, e.g., firmware	5	5	25	2	10
Software subtotals	30		123		95
Support					
Assistance, training, and documentation	5	2	10	5	25
Test arrangements	5	3	15	4	20
Backup facilities	5	3	15	5	25
Maintenance and service	7	3	21	5	35
Reputation, experience, and financial condition	8	3	24	5	40
Support subtotals	30		85		145
Totals	100		368		385

FIGURE 20-8 A weighted-rating table for hardware and software selection.

Its main drawback is that the evaluations are essentially subjective, since the weights and raw scores are assigned on the basis of judgment. Another drawback can be the laborious nature of the calculations. Fortunately, software packages are available to perform the calculations and even prepare graphs that rank the alternatives.*

The cost of the products is an important consideration. It can be included in the evaluation process by relating cost to the factors listed in the weighted-rating table. For instance, if the total costs for A's products equal $1 million, and those for B's products are $1.1 million, we can compute requirements-costs indexes as follows:

$$\text{Index for supplier A} = \frac{368}{\$1{,}000{,}000} = 0.368.$$

$$\text{Index for supplier B} = \frac{385}{\$1{,}100{,}000} = 0.350.$$

Thus, the proposal from supplier A appears to provide greater value for the required cost than does the proposal from supplier B.

SYSTEMS IMPLEMENTATION

After the needed system resources are selected by the evaluation team, a report is prepared and presented to management. If approval is given to acquire the selected resources, the project then enters the implementation phase. Implementing a new information system design consists of three major steps: (1) performing preliminary actions, (2) executing activities leading to an operational system, and (3) conducting follow-up activities and evaluations.

PRELIMINARY ACTIONS

The implementation phase generally engages the efforts of many persons over a period that is much longer than the preceding phases combined. Frequently it entails considerable costs. Therefore, several important actions should be taken before beginning the actual implementation activities.

Establish Implementation Plans and Controls Plans concerning the implementation activities should include cost budgets, time schedules, and work plans. Project controls, such as progress review meetings and periodic exception reports, can be based on the details established in these plans.

Two key techniques often used for controlling systems projects are Gantt charts and network diagrams. A **Gantt chart** is a bar chart with a calendar scale. Figure 20-9 presents a Gantt chart that schedules seven major implementation activities. Each planned activity appears as a bar that marks the scheduled starting date, ending date, and duration. Below each planned activity bar is a bar representing actual progress on the project.† The current status of the project can therefore be seen at a glance. For instance, if today is June 1 we can see that the activity labeled "select and train personnel" is ahead of schedule. However, the activity labeled "test programs" is behind schedule. The major drawback of the chart is that it does not show relationships among the various activities. Because of this drawback many firms

*Two useful software packages are Expert Choice and Criterion.

†Ideally, the Gantt chart should include the analysis, design, and earlier phases, as well as the time devoted to the preliminary actions currently being described.

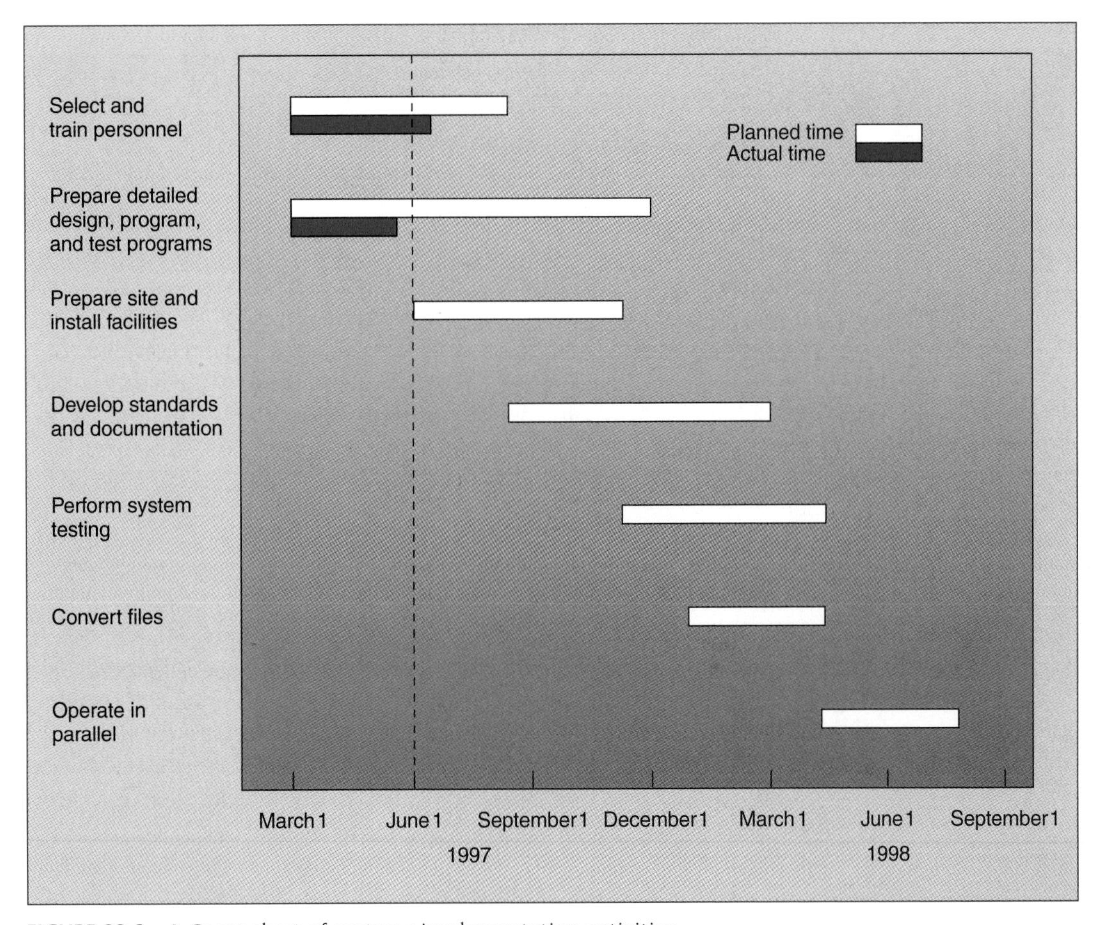

FIGURE 20-9 A Gantt chart of systems implementation activities.

employ **network diagrams,** which do reflect the relationships among the various project activities. Network diagrams are discussed in the appendix to this chapter.

Recognize Behavioral Concerns Since people will use and operate a new information system, their concerns and fears must be kept in mind. Chapter 19 discussed these concerns and fears; it also listed several desirable actions that should address these concerns and induce positive reactions. With respect to the implementation phase, the affected managers and employees should be informed beforehand about the implementation project. The expected impact on work units within the organization should be clarified. As the implementation phase progresses, several of the key activities—such as user-oriented training—are focused on the people who must make the system work.

Review the Organization of the Project Team In many cases the project team might of necessity be reorganized. Since the implementation phase requires managerial rather than technical skills, a new project leader may be appointed. A likely source from which to draw a new leader is the area being redesigned.

Complete Arrangements for Selected System Resources Proposals from suppliers have no legal status. Thus, contracts are normally necessary to affirm the terms between the parties. As soon as feasible, contracts should therefore be signed with the one or more suppliers whose hardware and software are to be acquired. If the soliciting

SPOTLIGHTING

SYSTEMS DEVELOPMENT AND IMPLEMENTATION *at Newton Construction Components, Inc.**

Newton Construction Components, Inc., manufactures metal components for contractors and subcontractors in the construction industry. Headquartered in San Jose, California, the firm acquired similar firms and designated them as the Los Angeles, Reno, Portland, and Seattle divisions. The divisions operate as decentralized profit centers.

Several problems were encountered with respect to the divisions. The headquarters did not receive information concerning important events on a timely basis. Each division maintained its own data base, which caused credit difficulties if a customer dealt with more than one division. Excessive time was spent in costing and analyzing inventory; also, inventory control was poor. On analysis, it was found that the causes were lack of standardization in operating procedures and lack of computerization in record keeping and report generation.

Top management decided to correct these problems. It determined that the new accounting information system should be both centralized and computerized. The key objectives were to provide immediate access by top management to the data of all the divisions and to standardize the operating rules and procedures among divisions. A system development committee consisting of the executive vice-president and controller and data processing manager was appointed to undertake the development. One of the committee's first decisions was to acquire software packages, rather than to develop software in house. After gathering internal data by reference to the various records, the committee established requirements by which to evaluate the needed software. In essence, the requirements or criteria specified compatibility with operating procedures, programming support from the supplier, adequate security and documentation, and a reasonable price. After considerable evaluation, the committee decided on a software package from Landmark, Inc. It then adopted IBM System 36 (a minicomputer system) as the hardware. The new software consists of five modules: order entry and sales, accounts receivable, purchasing, accounts payable, and inventory, which are tied together by a compatible general ledger package from IBM.

Implementation was scheduled over an eight-month period. The headquarters was chosen as the initial module for conversion. Implementation activities included creation and conversion of the inventory, supplier, and customer master files. Then employees were trained to use the new system. Training helped to overcome the fear generated among the employees. Another means of instilling confidence in the employees was to encourage their participation in the conversion activities. Most of the implementation proceeded smoothly. However, the hardware arrived late from IBM. Also, communications problems occurred in trying to tie together several different hardware components. Furthermore, the software required more modifications than expected.

After several years of operation, management believes that most of the problems in the system have been successfully overcome. The postimplementation evaluation has revealed the following benefits:

1. Top management can access data more quickly and has a better understanding of the business operations.
2. Operating procedures are completely standardized among divisions.
3. Control over accounts receivables and inventory has been greatly increased.

On the other hand, the evaluation pointed out two mistakes: (1) no written proposals from suppliers were obtained, and (2) employees were not encouraged to participate during the planning phase of the project.

*Bor-Yi Tsay and Robert J. Steverson, "Post Merger: Integrating the Accounting System." *Management Accounting* (January 1991), pp. 20–23.

firm objects to certain terms in the proposals, such as proposed prices or delivery dates, it may be necessary to enter into contract negotiations. In order to avoid undesirable terms and possible later disputes, an attorney should review the final negotiated contracts before they are signed.

IMPLEMENTATION ACTIVITIES

The activities that take place during a systems implementation vary widely from project to project. Also, the sequences in which the activities are executed cannot be standardized. Figure 20-10 presents a diagram of typical activities and their sequences. It is not a true network diagram, in that it lacks certain features as described in this chapter's appendix. However, it does suggest the overlapping and parallel manner in which activities are conducted during the implementation phase. Thus, it can aid in visualizing the respective activities, including the preliminary activity of establishing project controls.

Personnel Selection and Training Often newly installed systems require added personnel. Preferably new personnel should be drawn from the present employee force. Current employees already understand the firm's objectives and operations, and the morale of all employees is enhanced by a fill-from-within policy. Also, the costs are

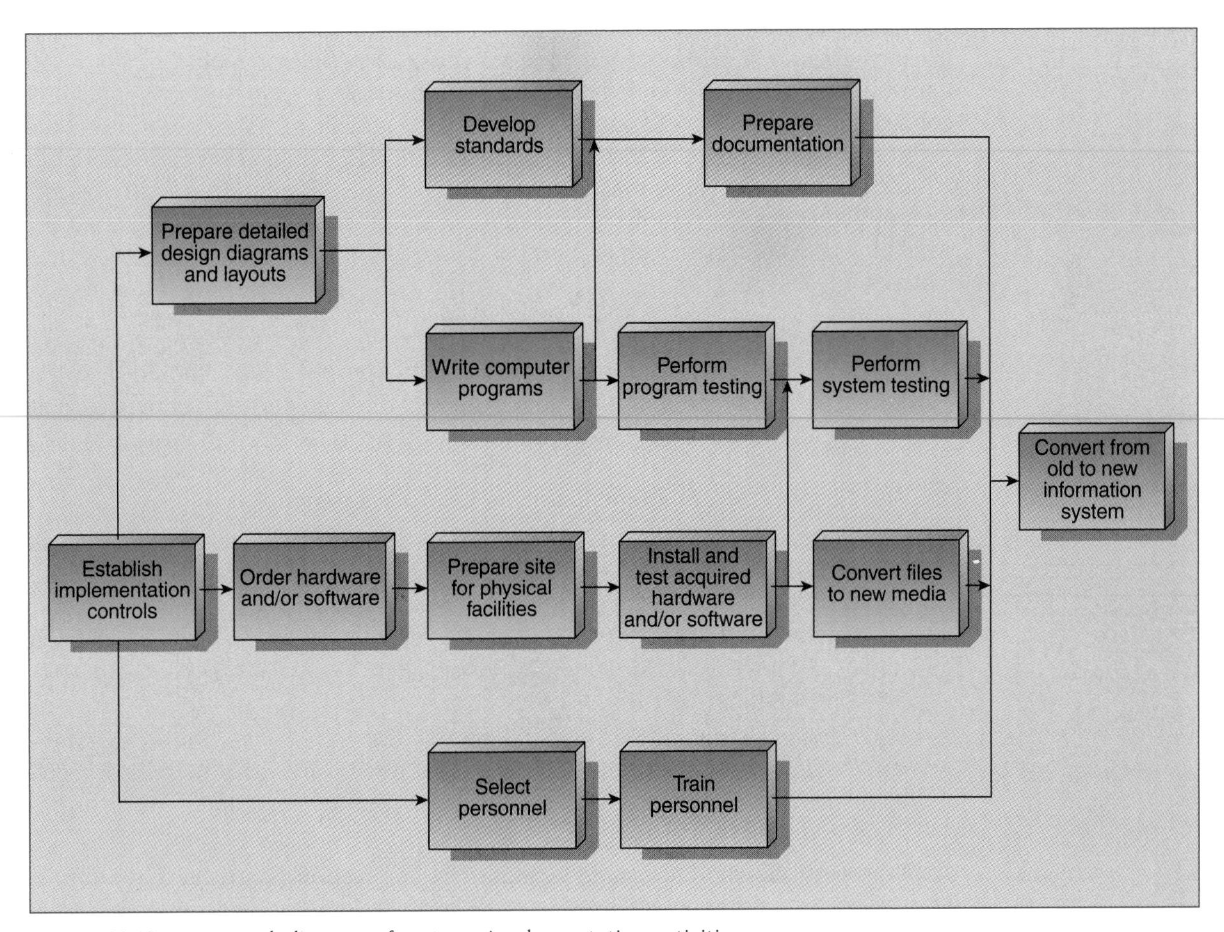

FIGURE 20-10 A network diagram of systems implementation activities.

less than hiring a person from outside the firm. However, in some cases no current employees have the needed skills or expertise. For instance, needed systems analysts are generally acquired from outside the firm for this reason.

Employees in new positions or faced with new systems always require training. Although costly, training is necessary if a new system is to function effectively. All affected employees need to be oriented with respect to the purposes and general operations of the new system. New employees should receive orientation concerning the business activities of the firm and its objectives and policies. Those who will directly interact with the new system need in addition to receive intensive training in its specific operations and rules.

Training may be provided through a variety of approaches. Proven approaches include video presentations, case studies, and hands-on simulated sessions with the system.

Physical Site Preparation Often the sites of the new system must be constructed or modified. Added facilities and accommodations—ranging from local-area network outlets to humidity controls and workstations—may need to be added. Space also may need to be allotted for new facilities, as well as for expected future expansions. If possible, these physical preparations should be completed before the delivery dates of ordered computer hardware and other facilities.

Detailed System Design The conceptual system design needs to be converted from logical and requirements-oriented specifications to the detailed features that physically compose a workable system. The activity of developing the design details, on paper or electronically, is called **detailed systems design.** As in the development of conceptual specifications, various techniques are available to aid in developing these detailed design features. These detailed system design techniques are discussed in a later section. Like conceptual system design, detailed system design may be categorized according to system outputs, data base, processing, inputs, and controls. Assuming that the new design is to be computer-based, we can briefly review typical features in each category.

1. ***Output design.*** Outputs include reports, documents, screen displays, voice responses, and archived information. For each output the design should show the format (e.g., tabular, narrative, graphical) and the medium (e.g., hard copy, soft copy, voice, microfilm). Figure 20-11 shows the format of a report as designed on a *spacing chart*. If hard copy is involved, the number of copies and stock (e.g., blank paper stock, preprinted forms) should be designated.
2. ***Data-base design.*** A data base includes files and other data structures, such as tables. (Also, a particular firm may maintain a number of separate data bases.) For each file and table the design should show the format (e.g., record layout, table layout) and medium (e.g., magnetic disk, magnetic tape, hard copy). Each format should include the names, modes, placements, and sizes of the respective fields or columns.
3. ***Input design.*** Inputs consist of data entering the system via documents, terminals, screens, optical scanning devices, voice inputs, and other means. For each input the design should show the format (e.g., source document layouts, preformatted screens, bar codes to be scanned) as well as the medium.
4. ***Processing design.*** Processing includes the application programs that are processed by the computer. Because of its relative importance, program development is treated as a separate activity. Every computer-based information system

SHARE PRINT CHART   PROG ID ______________   PAGE ______

DATE ______

PROGRAM TITLE ______________

PROGRAMMER OR DOCUMENTALIST ______________

CHART TITLE ______________

```
                                                                                                      11111111111111111111111111111111111
            11111111112222222222333333333344444444445555555555666666666677777777778888888888999999999900000000001111111111222222222233333
   12345678901234567890123456789012345678901234567890123456789012345678901234567890123456789012345678901234567890123456789012345678901234
 1
 2
 3
 4
 5
 6                                     JOURNAL VOUCHER LISTING
 7
 8                                         FOR  XX/XX/XX
 9
10
11                         ACCOUNT
12    J-V  NO.            NUMBER   DR/CR        DESCRIPTION                          DATE              AMOUNT
13
14     XXXX               XXX.X     XX     XXXXXXXXXXXXXXXXXXXXXXXXXXXXXX           XX/XX/XX       XXX,XXX,XXX.XX
15
16     XXXX               XXX.X     XX     XXXXXXXXXXXXXXXXXXXXXXXXXXXXXX           XX/XX/XX       XXX,XXX,XXX.XX
17
18     XXXX               XXX.X     XX     XXXXXXXXXXXXXXXXXXXXXXXXXXXXXX           XX/XX/XX       XXX,XXX,XXX.XX
19
20
21
22
23
24
25
26
27
28
29
30
31
32
33
34
35
36
37
38
39
40
41
42
```

FIGURE 20-11 A layout of a report format on a spacing chart.

also requires interactions by humans—ranging from sales order clerks to computer operators. These human interactions in essence are controlled procedures, consisting of a series of steps to be performed with respect to the various applications of the new system.

5. ***Controls design.*** Controls consist of general controls, application controls, and security measures. For each control being implemented, the details should specify its purpose (e.g., checking input data elements, verifying authorized access) and the aspect of the system being controlled (e.g., checking accuracy of entered customer numbers, verifying passwords pertaining to the accounts receivable table). All of the implemented controls should be fully documented.

Application Software Development Unless commercial software packages are to be acquired, the application programs must be developed. Program development consists essentially of writing the instructions, called **coding,** for the application programs of the designed system. Coding may be done automatically by means of a CASE tool, as described in Chapter 19. In the future this approach will increasingly be selected.

Currently, however, most firms that develop computer programs in-house employ programmers to do the coding. They are guided by the results of detailed system design techniques as decision tables, structure charts, HIPO (hierarchical input process output) diagrams, Warnier-Orr diagrams, and structured English. Before beginning the actual coding, however, two decisions must be made: (1) the method of programming to adopt, and (2) the particular programming language to use.

Consistent with the structured analysis approach, most firms choose **structured programming** as the method by which to develop application programs. This method consists of beginning with an overall view and then decomposing that view into successively more detailed segments or modules. Each module performs a separate well-defined logical function. For instance, in a payroll application program, one module may compute the net pay for an employee and another module may print the paycheck. Structured programming with modules allows programs to be coded easier and faster, while errors can be detected more quickly. Programs can be maintained in a simpler manner, since each module is semi-independent of the others.

Software Testing All developed (or acquired) software must be thoroughly tested, so that errors and malfunctions can be spotted and eliminated. Computer programs that the firm writes require a series of tests, ranging from desk checking to full string testing.

Desk checking consists of visually tracing through the instructions in each program, searching for errors in the logic or in keying. A related test involves compiling the program in order to detect syntax errors—errors that violate the rules of the programming language.

Further testing should employ data to simulate the functioning of the programs and their segments. The data to be used is called test data, and the method is similar to the test data technique applied by auditors. In the case of applications, the test data will include transactions that contain errors. The initial testing, called *module testing*, consists of having the various modules of the programs process the test data, in order to observe how both error-free and erroneous transactions are handled. If either type of transaction does not provide the results expected, the program module apparently has one or more "bugs." The next level of testing, called *program testing*, involves an entire application program. Test data are again entered and processed. Finally, several application programs are linked together in order to perform **string testing.** Testing at this level is intended to see that the programs interact properly as parts of a substantial application. The variety of test data should be expanded in order to provide a full test.

System Testing Newly acquired computer hardware must also be tested thoroughly. After the hardware has been checked, usually in conjunction with personnel from the hardware vendor, more complete tests are needed. **System testing** consists of testing the hardware together with the software, sample files and data bases, inputs, and outputs. Purposes of system testing are (1) to detect all errors and problems and make needed adjustments and corrections, and (2) to demonstrate to all concerned that all components function together smoothly as a viable system.

The final testing, called **acceptance testing,** fully involves the users of the new system. Users provide the data for acceptance testing. They also set the standards for acceptance of the system, which usually parallel the system requirements from the analysis phase. Actual data are preferably used in this final testing, so that the users can compare the results to those provided by the current system. When the users are satisfied that the system meets their standards of acceptance, the testing is complete. Users may be asked to sign a formal acceptance document to denote their approval and acceptance.

Standards Development Major system changes generally call for new standards. These system-related standards may pertain to

1. System components, such as standardized data elements and codes.
2. Performance, such as standardized employee productivity rates.
3. Documentation, such as standardized flowcharting techniques.

Documentation Although sometimes slighted, adequate documentation is an important implementation activity. It provides the basis for later system changes and aids new employees in performing their duties and responsibilities. Needed types of systems documentation were described in Chapters 4 and 8.

File Conversion Most system changes affect the files and data bases. Either files are to be converted from manual forms to computer-based forms, or they are to be converted from one computer medium to another. **File conversions** often involve such steps as (1) "cleaning up" the data elements in the present files, (2) writing special programs to perform the actual data transfer, (3) physically transferring the data from the present files to the new storage medium, (4) reconciling the new files with the control totals of the old files, and (5) storing the old files as backup.

System Conversion The activity in changing or converting from the current system to a new system is called **system conversion.** The point at which the new system becomes operational is known as **cutover.** Four conversion approaches leading to cutover are the direct conversion, parallel operation, modular, and phased approaches.

1. The **direct conversion approach** consists of cutting over to a new system as soon as all prior implementation activities are completed. This "cold turkey" approach is quite risky, being akin to using diskette files without backups. It is suitable only when a system change is very simple or the need for a new system is urgent.
2. The **parallel operation approach** consists of operating the new and present systems side by side, and comparing the resulting outputs (e.g., reports, account balances). If the results agree over a reasonable period (e.g., two or more operating cycles), the present system is abandoned. This approach is suitable when

the outputs are not drastically changed in the new system, e.g., in a basic transaction processing system. It provides a high degree of control over the conversion process, and hence greater assurance that the new system will function as designed. Without doubt the parallel operation approach is costly and burdensome, since dual staffs are often needed to operate the two systems. However, it is likely to be the best bargain in the long run.

3. The **modular conversion approach,** also called the pilot approach, consists of testing and converting the new system at one location initially. For instance, Infoage could convert a new point-of-sale system at one of its retail sales outlets. When the new system is operating satisfactorily there, with all problems resolved, conversion can take place at each of the other sales outlets in succession.

   The modular approach allows time for "debugging" and training, although it causes the conversion period to stretch over a longer time period.

4. The **phased conversion approach** consists of cutting over segments of the new system until the entire system is operational. For example, Infoage might phase in a new data base by cutting over one data area or transaction system at a time. Thus, the inventory tables may be converted first, followed by the purchasing tables and then the accounts payable tables. This approach is necessary when the system components (e.g., data base, outputs) of the new system are significantly different from those of the present system. As in the case of the modular approach, more time is allowed for adjusting to the new system and correcting problems. On the other hand, the conversion period can be awkward as well as lengthy. In the example, Infoage would be functioning with related data that are in different structures (files versus tables) and that are accessed by different means. Temporary interfaces must be established until conversion is completed.

Regardless of the conversion approach taken, the last step is to obtain the formal acceptance of the new system by the key user. This step is called **user signoff.**

POSTIMPLEMENTATION ACTIVITIES

The major activities that take place after an implementation is completed include fine-tuning, postimplementation evaluations, and operational activities.

Fine-Tuning Even though a new system is ready for operations, it should not be viewed as being in its final form. Typically it must undergo a "shakedown" period. Hidden quirks must be uncovered and ironed out. System components must be fine-tuned. Users must learn how to operate the system, even when unusual conditions occur. To ensure that these results are achieved, certain members of the project team should be assigned as troubleshooters. They can observe operations, provide assistance, and make necessary adjustments.

Postimplementation Evaluation After the new system has stabilized, the time has arrived for "second thoughts." A **postimplementation review and evaluation** consists of examining the newly operational information system and forming judgments. Among its purposes are the following:

1. To assess the degree to which the objectives of the system project have been met.
2. To spot any additional modifications that might be needed in the newly designed system.

3. To evaluate the project team's performance, both in terms of a quality product and adherence to the project schedule and work plan.
4. To serve as the basis for improving future systems developments and the accuracy of cost and benefit estimates.

Accountants can aid in determining the success of the newly operational systems. They can design measures for evaluating the information and reports provided to managers, the quality of service, the improvement in productivity due to the systems, and the financial payoff. Measures of productivity and utilization were described in Chapter 18, while financial payoffs were discussed early in this chapter. Evaluating the usefulness of information and the quality of service is more difficult. One approach is to have the recipients, such as managers, rank these factors on a normative scale ranging from excellent to poor.

Operational Activities Apart from fine-tuning and evaluations, the operational phase may seem to consist mainly of routine operations. However, it is necessary to manage and control the costs of the system through such means as the chargeback procedures described in Chapter 18. Also, the information system and its resources need to be continually maintained and modified as circumstances require. Many of the modifications consist of changing and updating the applications software. In addition, audits of the internal control structure and information outputs should be performed periodically, as discussed in Chapter 10.

DETAILED SYSTEMS DESIGN TECHNIQUES

Developing a detailed systems design is a vital step in the systems development life cycle. However, it can be quite time consuming and tedious. Like systems analysis, detailed systems design can be aided by a variety of techniques. Most currently employed techniques are structured. Included in this category are decision tables, Warnier-Orr diagrams, detailed structure charts, Jackson diagrams, decision trees, HIPO (hierarchical input process output) diagrams, state-transition diagrams, and structured English. Extensive coverage of these techniques can be found in most systems analysis and design textbooks. We will briefly introduce selected techniques to illustrate their usefulness.

DECISION TABLES

A **decision table** focuses on the "decision choices" inherent in many applications. It shows, within a matrix format, all of the rules pertaining to a transaction processing or decision situation. Decision tables can be viewed as replacements of *program flowcharts*, sequential flow representations of the steps performed by the instructions in computer programs.

The decision table in Figure 20-12 pertains to the processing of transactions relating to customers. Each decision point (which is represented by a diamond symbol in a program flowchart) appears as a separate condition. Each processing and input-output step (which is represented by a rectangle or parallelogram symbol in a program flowchart) appears as an action. Each logical construct (which is represented by a branching in a program flowchart) appears as a rule.

A decision table differs in two significant ways from a program flowchart. First, it does not specify the sequence in which the actions are to be performed. Thus, a decision table presents the unconstrained logic of the situation being portrayed.

Condition	Rule 1	2	3	4	5	6
C1. Has end-of-file indicator been reached?	N	N	N	N		Y
C2. Is customer number on transaction record larger than number on master record?	Y	N	N	N		
C3. Is customer number on transaction record equal to number on master record?		N	Y	Y		
C4. Is transaction a sale?			Y			
C5. Is transaction a payment?				Y		
C6. Is customer number the dummy 999?		N	N	N	Y	
Action						
A1. Read transaction record.		X	X	X		
A2. Read master record.	X				X	
A3. Write new master record.	X				X	
A4. Write error message.		X				
A5. Add amount of transaction to customer balance.			X			
A6. Deduct amount of transaction from customer balance.				X		
A7. Stop processing.						X

Note: Y means yes, N means no.

FIGURE 20-12 A decision table.

Second, its construction is based on mathematical principles. Thus, a "full" decision table consists of two rules, where R is the number of independent conditions. A "collapsed" table (such as shown in Figure 20-12) is then developed by eliminating all redundant rules.

In summary, a decision table can be described as a logic diagram of a decision-oriented situation. It reflects the variety of possibilities that a computer program might encounter with respect to transaction data. A decision table is particularly helpful when the situation being portrayed has numerous conditional branches.

If a situation to be portrayed in a decision table is extremely complex, however, a decision table could become too large to handle easily. For this reason, decision tables are often subdivided to reflect portions of a complex situation. The fractional decision tables are then linked together. Alternatively, an overall decision table may be prepared at a broad level and then partitioned into a set of detailed decision tables. Because decision tables are logical diagrams capable of being modularized, they qualify as a structured technique.

WARNIER-ORR DIAGRAMS

Among the most widely used of the detailed techniques are **Warnier-Orr diagrams,** devised by Jean-Dominique Warnier and Ken Orr. These diagrams are used to represent in graphical format a processing sequence, a file structure, or the structure of a document or report. Figure 20-13 displays the structure of a sales invoice. Each bracket indicates a grouping of data; as the brackets move to the right in the diagram, the data items become more detailed (i.e., at lower levels in the hierarchy).

Warnier-Orr diagrams provide easy-to-read data and processing documentation. They can be decomposed to very detailed levels. Hence, they are widely used to aid

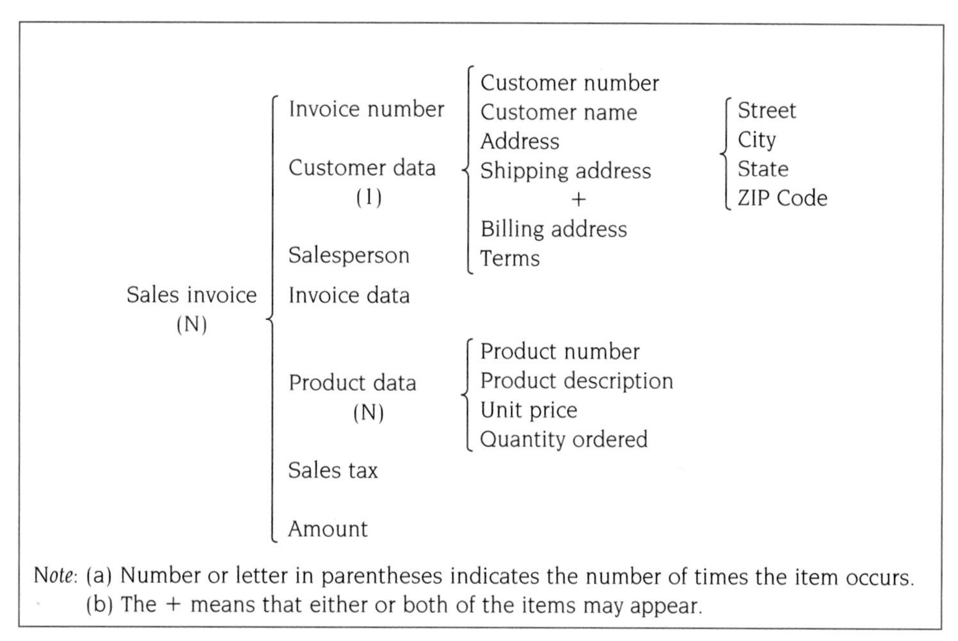

FIGURE 20-13 A Warnier-Orr diagram of the data hierarchy within a sales invoice.

in writing structured computer programs. Their main drawback is that they are not well suited for large-scale and data-base oriented systems.

DETAILED STRUCTURE CHARTS

As noted in Chapter 4, a structure chart portrays the hierarchy of levels and relationships within a system. A **detailed structure chart** depicts the program modules that will comprise the subroutines of a computer program. It can be used to guide the preparation of program instructions or English-like pseudoinstructions.

Figure 20-14 shows a detailed structure chart that graphically describes the subroutines of a program for preparing a sales invoice. The box at the top, called the boss procedure, represents the control module for the overall program. This module calls on the subroutines or individual instructions from left to right. The EOF (end-of-file) symbol specifies that the program is completed. Data and control flags are used to annotate the flows between the control module and the subroutines.

STRUCTURED ENGLISH

Structured English is an informal language that uses English-like statements to describe procedural logic. Its advantage is that the statements do not require the syntax of a specific programming language, such as COBOL. However, structured English employs statements involving such key terms as IF, THEN, ELSE, and CÔMPUTE, which are similar to the programming statements of commonly used languages. These statements can be easily converted into a formal programming language. Figure 20-15 lists a series of instructions written in structured English. The instructions are based on the detailed structure chart in Figure 20-14, which describes the preparation of a sales invoice.

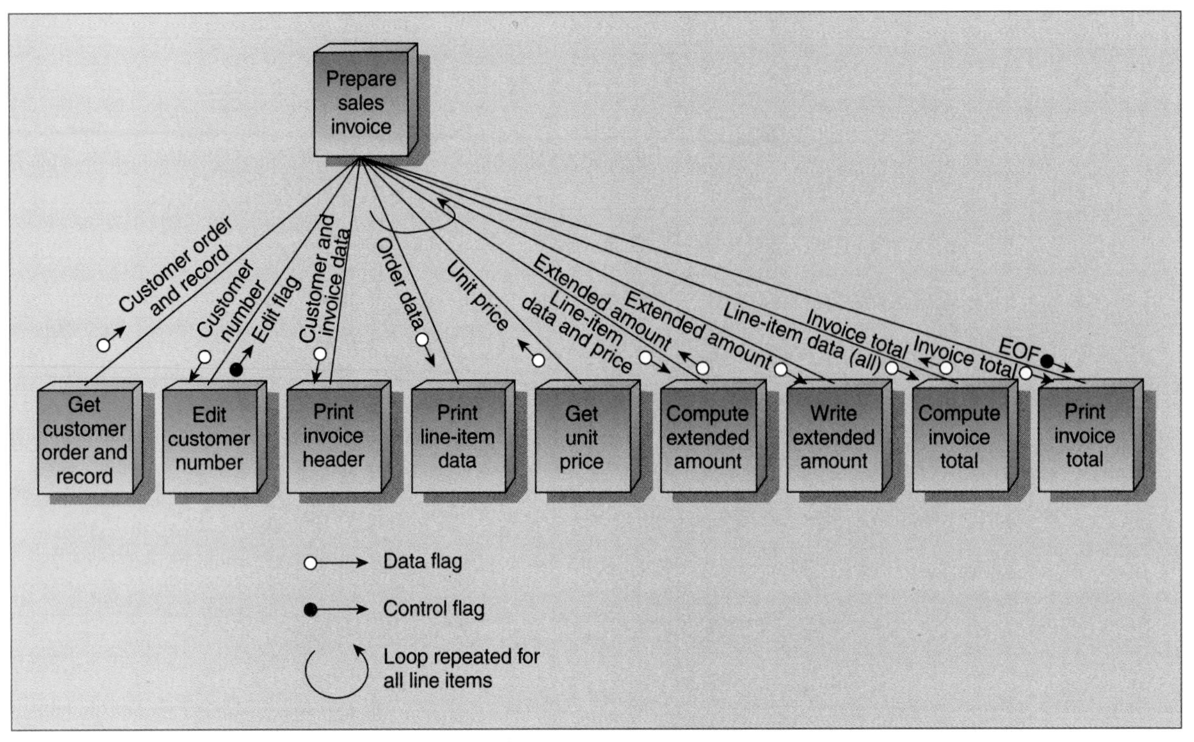

FIGURE 20-14 A detailed structure chart pertaining to the process of preparing a sales invoice.

```
GET CUSTOMER-ORDER
OPEN CUSTOMER file
GET CUSTOMER record
IF CUST-NO is valid
     THEN SET INVOICE-HEADER
     THEN ENTER CUST-NO, CUST-ADDRESS, ORDER-DATE,
     ORDER-NO
     THEN WRITE INVOICE-HEADER
ELSE
     DISPLAY ''CUSTOMER NUMBER NOT VALID''
ENDIF
FOR each line item
     WRITE ITEM-NO and ORD-QTY on INVOICE-LINE
     GET UNIT-PRICE-ITEM from pricing file
     COMPUTE ITEM-EXTENSION as ORD-QTY times
     UNIT-PRICE-ITEM
     WRITE ITEM-EXTENSION on INVOICE-LINE
ENDFOR
COMPUTE INVOICE-AMT as sum of ITEM-EXTENSION
WRITE INVOICE-AMT on INVOICE-FOOTING
EOF
```

FIGURE 20-15 A segment of structured English pertaining to a sales invoice.